SAUNDERS COMPLETE PACKAGE FOR TEACHING ORGANIC CHEMISTRY

Ternay: **Contemporary Organic Chemistry**

Francis: **Student Guide and Solutions Manual to Ternay's Contemporary Organic Chemistry**

Moore and Dalrymple: **Experimental Methods in Organic Chemistry—** *Second Edition*

Pavia, Lampman and Kriz: **Introduction to Organic Laboratory Techniques: A Contemporary Approach**

Banks: **Naming Organic Compounds: A Programed Introduction to Organic Chemistry—***Second Edition*

Weeks: **Electron Movement: A Guide for Students of Organic Chemistry**

 SAUNDERS GOLDEN SUNBURST SERIES

1976

W. B. SAUNDERS COMPANY

Philadelphia • London • Toronto

contemporary organic chemistry

ANDREW L. TERNAY, JR.

University of Texas, Arlington
University of Texas, Health Sciences Center, Dallas

W. B. Saunders Company: West Washington Square
Philadelphia, PA 19105

12 Dyott Street
London, WC1A 1DB

833 Oxford Street
Toronto, Ontario M8Z 5T9, Canada

Library of Congress Cataloging in Publication Data

Ternay, Andrew L

Contemporary organic chemistry.

(Saunders golden sunburst series)

Bibliography: p.

Includes index.

1. Chemistry, Organic. I. Title.

QD251.2.T47 1976 547 75-8187

ISBN 0-7216-8794-6

Cover artwork, a photomicrograph of sugar crystals, courtesy of Dr. Lewis R. Wolberg and ICN Pharmaceuticals, Inc., Life Sciences Group.

Contemporary Organic Chemistry ISBN 0-7216-8794-6

Last digit is the print number: 9 8 7 6 5 4 3 2 1

contemporary organic chemistry

ANDREW L. TERNAY, JR.

University of Texas, Arlington
University of Texas, Health Sciences Center, Dallas

W. B. Saunders Company: West Washington Square
Philadelphia, PA 19105

12 Dyott Street
London, WC1A 1DB

833 Oxford Street
Toronto, Ontario M8Z 5T9, Canada

Library of Congress Cataloging in Publication Data

Ternay, Andrew L

Contemporary organic chemistry.

(Saunders golden sunburst series)

Bibliography: p.

Includes index.

1. Chemistry, Organic. I. Title.

QD251.2.T47 1976 547 75-8187

ISBN 0-7216-8794-6

Cover artwork, a photomicrograph of sugar crystals, courtesy of Dr. Lewis R. Wolberg and ICN Pharmaceuticals, Inc., Life Sciences Group.

Contemporary Organic Chemistry ISBN 0-7216-8794-6

Last digit is the print number: 9 8 7 6 5 4 3 2 1

To Marilyn and Janice and Andrew

PREFACE

I suspect that every organic chemist has his own view of *what* should be included in a modern, one year course in organic chemistry and *how* it should be presented. Given the possible permutations and combinations, what could prompt anyone to enter another organic chemistry text into an arena that contains several good texts?

The book that I wanted for my organic chemistry classes (composed of about 30% chemistry majors and 65% biology majors, pre-dental and pre-medical students) had to read well. It had to be written at a reasonable level. It had to present topics in a suitable sequence. It had to have a good index and a good set of problems. Finally, it had to answer the student's question: "Why should I have to learn organic chemistry?" I wanted a text that made organic chemistry interesting as well as easy to understand. This book is the end result of my search for that text.

Is this text organized along mechanistic lines or does it follow the functional group approach? It seems to me that *structure* and *mechanism* are threads that should run through the course, but that they should not dominate. In order to permit the introduction of mechanism when it is needed, I have collected S_N and E reactions and discussed them, along with some synthetic implications, in Chapters 5 and 6. These chapters are surrounded, and permeated, with a generous quantity of stereochemistry. The material beyond Chapter 7 is couched in the functional group approach.

For what audience is the book intended? This text is written for any student who needs a sound, one year organic chemistry course. While some of the examples, discussions, and problems involve biological topics, this is not a "watered-down" presentation. I simply see no harm in exposing the chemistry major to ways in which his discipline interacts with other areas. The answer to the question raised by the non-chemistry major ("Why study organic?") can be a source of motivation (and pride) to the chemistry major.

What about the glossary? Each chapter contains a glossary of important terms introduced in that chapter. Since the terms are collected at the end of each chapter, they also may serve as one focus of review.

Where is spectroscopy discussed? Chapter 28 (ir, uv, and ms) and Chapter 29 (nmr) are "free-floating," and can be introduced almost anywhere in the course. Those problems which require a knowledge of spectroscopy are marked with an asterisk. I have chosen to present an extended discussion of nmr, since it can be used to re-introduce conformational analysis, symmetry, chemical exchange, and isotopes.

I would welcome suggestions for improvement from the readers of this text.

I am grateful to all those who reviewed this manuscript in its various stages of completion; their collective efforts are my co-authors. This group includes: J. Banks, E. Bellion, R. Dougherty, R. Fessenden, A. Krubsack, K. Mislow, J. A. Moore, R. Murray, D. Pavia, M. Pomerantz, D. Sargent, R. Scanio, E. T. Strom, J. Swenton, D. Tavares, and D. Weeks. Robert Francis, in working through each problem, proved himself to be both a help and a source of stimulation. Of over four hundred students who have been exposed to various stages of this book, Ms. C. Mischer, Ms. J. Blumentritt, Ms. S. Skeie, Mr. P. Rast, Mr. M. Campbell, Mr. M. Reardon, and Mr. G. Johnston have made particularly valuable contributions. This text could not have been published without the encouragement and assistance of the competent staff of W. B. Saunders. I must single out one individual from this team: Mr. J. J. Freedman, whose multi-faceted abilities have earned for him the title of "Maestro."

Finally, I must thank my family for their patience and understanding during those many evenings, weekends and holidays which were consumed with writing.

ANDREW L. TERNAY

ACKNOWLEDGMENT

Infrared spectra in this text have been provided, in the main, by the Sadtler Research Laboratories and the Aldrich Chemical Co. Mr. C. J. Pouchert, of Aldrich, was particularly helpful. Nuclear magnetic resonance spectra have been provided by the Sadtler Research Laboratories and Varian Associates. Mass spectra are taken from the literature of Varian Associates.

CONTENTS

ATOMS AND THINGS

It has been suggested, and not totally in jest, that the most important theory to develop from organic chemistry is that carbon always possesses four bonds in its compounds. If there is any truth to this, then before we can hope to understand the reactions of carbon compounds we must know (a) what a carbon atom is and (b) why it does tend to be quadrivalent (if, indeed, it does). Moreover, since organic chemistry has begun to encroach upon elements which were once the sole property of inorganic chemists (*e.g.*, boron, tin and magnesium), we should include a discussion of what *any* atom is and why it forms the number of bonds that it does. Although the nature of the atom is discussed in all elementary chemistry courses, this first chapter will *review* aspects of atomic structure. Such a review must, of necessity, be superficial when dealing with certain subjects; greater detail may be gotten from most elementary chemistry texts (see the list at the end of the text). Perhaps you will feel such a review unnecessary, in which case you can proceed immediately to Chapter 2. However, bear in mind that we are generally poor judges of what we remember; the problems at the end of this chapter will serve as an impartial referee.

1.1 THE BASIC STRUCTURE OF THE ATOM

The earliest views of the atom did not consider it to be divisible. Indeed, until about a century ago the atom was considered *the* fundamental particle of nature, and most scientists thought it a waste of time to speculate upon the existence of subatomic particles. However, the classic experiments of Lord Rutherford, on the scattering of α-particles by gold foil, firmly established that atoms possess some type of internal structure; they consist of a rather dense nucleus surrounded by a sparsely populated mantle of electrons. The major particles, or **nucleons,** which make up the nucleus are the **neutron** (electrically neutral) and the **proton** (carrying a unit positive charge). Since the atom is electrically neutral, the number of electrons (carrying unit negative charge) about the nucleus must equal the number of protons.

The mass[a] of all of these particles is rather small, and an arbitrary scale of mass has been established using u to represent the "unified atomic mass unit." The unit mass, u, is now *defined* as $\frac{1}{12}$ the mass of the ^{12}C atom (see below). Similarly, charges of subatomic particles are usually expressed in multiples of the charge on the electron (electronic charge units). Some pertinent data are summarized in Table 1–1.

The **mass number** of an atom is the total number of neutrons and protons in the nucleus. Since the mass number is an integral (whole) number, it can only approximate the true nuclear

[a] One should draw a clear distinction between *mass* and *weight.* The former is the amount of material present, while the latter is a measure of the pull of gravity on the former. On the moon your weight would decrease but your mass would remain the same.

TABLE 1-1 **Properties of Subatomic Particles**

charge on the electron	4.8×10^{-10} esu[a] (1 electronic charge unit)
rest mass of the electron[b]	9.1×10^{-28} g
wavelength of the electron	2.4×10^{-10} cm
atomic mass of the electron	5.5×10^{-4} *u*
atomic mass of the proton	1.0 *u*
mass of the proton	1.6725×10^{-24} g
wavelength of the proton	1.3×10^{-13} cm
charge on the proton	1 electronic charge unit (positive)
atomic mass of the neutron	1.0 *u*
charge on the neutron	0
mass of the neutron	1.6757×10^{-24} g

[a] esu represents *e*lectro*s*tatic *u*nit.
[b] Why should it be necessary to specify the "rest mass"?

mass because the masses of subatomic particles are not integral. In fact, if one were to calculate the total mass of any atom by adding the masses of the electrons, protons and neutrons, the result would be greater than the observed mass of the atom. The difference reflects the energy which is evolved when these particles are brought together from infinite separation to form an atom. This energy is called the **binding energy** and is related to the "lost" mass by the famous Einstein equation, $E = mc^2$. Let us note that one gram of mass, if completely converted to energy, is equivalent to 22 billion kcal.

The **atomic weight** of an element (it is generally improper to speak of the atomic weight of an atom, for reasons that should become clear shortly) tells us the "average" weight of an atom of an element relative to that of carbon-12. Considering that the masses of the neutron and the proton are almost 1 *u*, and that the mass of the electron is rather small, the deviation from whole numbers for the atomic weights of a number of elements is rather surprising. Illustrative of this are the following atomic weights: antimony (Sb), 121.75; boron (B), 10.811; chlorine (Cl), 35.45; and rubidium (Rb), 85.47. These non-integral values arise because most elements exist as mixtures of atoms of different masses, called **isotopes.** The number of protons (hence, electrons) present in all atoms of a given element must be the same. *Isotopes differ in the number of neutrons in the nucleus. This means that isotopes differ in their mass numbers* (number of protons plus neutrons) *but not in their atomic number* (number of protons). It is the existence and distribution of isotopes which dictates the precise value of the atomic weight. The relative amount of a given isotope is termed the "natural abundance" of that isotope.[a]

To illustrate how non-integral atomic weights arise, consider chlorine (A.W. = 35.5). Chlorine is a mixture of 75.5% chlorine-35 (^{35}Cl) and 24.5% chlorine-37 (^{37}Cl) and hence has an atomic weight of: $0.755(35.0) + 0.245(37.0) = 35.5$. Carbon has an atomic weight of 12.01. This might surprise you since the unified atomic mass unit is defined as $\frac{1}{12}$ of the mass of the carbon-12 atom. The apparent paradox is resolved when we realize that one isotope of carbon is *defined* as having a mass of 12.0000. The natural abundance of ^{12}C is 98.89%, the remainder being ^{13}C and ^{14}C. While there are six isotopes of carbon, only these three occur naturally. Carbon-14 is radioactive (half-life 5770 years), and has received much attention since it is used in radiocarbon dating procedures. It is the presence of these heavier isotopes which raises the average atomic weight of carbon slightly above that of the ^{12}C atom.

Aside from carbon, hydrogen is the element found most often in organic compounds. Many

---

[a] The exact distribution of isotopes throughout the universe is unknown. However, since various regions of our universe have different ages, it seems reasonable that the isotopic distribution will vary from one region of the universe to another.

TABLE 1-2 Common Isotopes*

Isotope	% Natural Abundance	Atomic Mass (u)	Nuclear Spin[a]
^{1}H	99.98	1.008	$\frac{1}{2}$
^{2}H	0.02	2.014	1
^{3}H	0.00	3.016	$\frac{1}{2}$
^{10}B	19.6	10.013	3
^{11}B	80.4	11.009	$\frac{3}{2}$
^{12}C	98.89	12.0000	0
^{13}C	1.11	13.003	$\frac{1}{2}$
^{14}N	99.63	14.003	1
^{15}N	0.37	15.000	$\frac{1}{2}$
^{16}O	99.76	15.995	0
^{17}O	0.037	16.999	$\frac{5}{2}$
^{18}O	0.20	17.999	0
^{19}F	100.00	18.998	$\frac{1}{2}$
^{31}P	100.00	30.974	$\frac{1}{2}$

[a] This value, given in multiples of $h/2\pi$, will be of use in our discussions of nuclear magnetic resonance spectroscopy (Chapter 29).

* A complete list of isotopes can be found in the current edition of "Handbook of Chemistry and Physics" published by The Chemical Rubber Co., Cleveland, Ohio. (This is sometimes called "The Rubber Handbook.")

reactions of organic compounds involve hydrogen, and the rates of these reactions sometimes are dependent upon the particular isotope of hydrogen involved. The three isotopes of hydrogen are: *protium* (hydrogen-1, ^{1}H or "just plain hydrogen"), *deuterium* (hydrogen-2, ^{2}H or D) and *tritium* (hydrogen-3, ^{3}H or T). While protium and deuterium occur in nature, radioactive tritium (half-life 12 years) must be prepared as required. Tritium can be traced (*i.e.*, detected) with a counter; tritiated compounds (compounds in which tritium replaces some or all of the hydrogen-1) are sometimes fed to animals in order to determine the distribution of organic compounds within the organism. The use of isotopes to help understand *how* organic reactions occur will be discussed in Chapters 5 and 6.

What do chemists do when they want to indicate a specific atomic species? Around the atomic symbol are placed the following values: *upper right,* charge of the species; *lower right,* number of atoms in a given species; *upper left,* mass number; *lower left,* atomic number. The number in the lower left indicates the element under study, while the number in the upper left defines the specific isotope.

$$^{\text{mass number}}_{\text{atomic number}}\mathrm{X}^{\text{charge (if other than zero)}}_{\text{number of atoms (if other than one)}}$$

Some examples of this type of designation include:

$^{12}_{6}C$	$^{35}_{17}Cl^{\ominus}$	$^{2}_{1}HO^{\ominus}$	$^{3}_{1}H_2$
carbon-12 atom	chlorine-35 anion	deuteroxide anion	tritium molecule

It should be recognized that the organic chemist usually takes shortcuts in representing various atoms. For example, isotopes which have common names (*e.g.*, deuterium and tritium) usually are designated by the appropriate abbreviations (*e.g.*, D and T). Also, atomic numbers normally are deleted from the structure. Thus the species shown above are normally represented by:

^{12}C	$^{35}Cl^{\ominus}$	$DO^{\ominus}$	T_2
carbon-12 atom	chlorine-35 anion	deuteroxide anion	tritium molecule

1.2 THE ELECTRONIC STRUCTURE OF THE ATOM

Our attention will now focus upon developing a picture of the arrangement of the electrons in an atom, passing from the older quantum theory (based upon the work of Niels Bohr) to the newer quantum theory (based upon the works of Heisenberg, de Broglie and Schrödinger). We shall begin, however, with the work of Lord Rutherford, since in his work we can see quite clearly the conflict between classical physics and an ever-maturing picture of the atom.

Death of a Concept—The Need for a New Idea. One of the earliest descriptions of the atom was that of J.J. Thomson, who suggested (1904) that it consisted of a homogeneous sphere of positive charge in which were imbedded negatively charged electrons. This has been called the "plum pudding" picture of the atom. Based upon his α-particle scattering experiment, Lord Rutherford characterized (1911) the atom as having a dense positive core surrounded by a sea of electrons. The nucleus was estimated to have a density approximating that now known to exist for certain types of stars—about 10^8 *tons* per milliliter!

By 1915 it was believed, in an expansion of the Rutherford model, that electrons traveled around the nucleus. We should note that classical physics dictates that if a charge is accelerated it should emit energy. These two ideas, taken together, suggest that electrons should constantly emit radiation since motion in a circle is tantamount to constant acceleration in a new direction. If an electron were to continuously emit energy, its orbit would continuously shrink ("decay") and bring the electron ever closer to the nucleus. In the end it should collapse into the nucleus! During the decay of the orbit the electron should also emit a constantly changing, continuous spectrum of radiation. These predictions were in conflict with observation. The electron is not to be dictated to by classical physics!

Atomic Emission Spectra. When an element is heated to a sufficiently high temperature it will emit light. If the emitted light is passed through a prism, the emergent light usually does not produce a continuous spectrum (typical, for example, of a rainbow). Instead, very discrete colored lines (a "bright line spectrum"), corresponding to characteristic wavelengths of light, are observed. To explain this observation, Niels Bohr (a student of Lord Rutherford) constructed a model of the atom in which the electron travels in circular orbits around the nucleus. According to Bohr, these orbits are limited in number and correspond to specific energy levels ("quantum levels"). In other words, electrons are not permitted to exist between these orbits, and the energy of the electrons is said to be *quantized.* Transition of an electron from a lower energy orbit to a higher energy orbit requires the absorption of a definite amount ("quantum") of energy. When an electron goes from a higher to a lower energy orbit, a clearly defined quantum of energy is emitted. This latter feature accounts for the bright line spectrum.

The Bohr picture of the atom explains the appearance of the bright line spectrum but requires the introduction of a number to describe each of the various orbits. This number is a constant for each orbit and reflects the energy of the orbit. It is termed the "main" or "principal" quantum number and arises because Bohr required the angular momentum of the electron (mvr) to be an integral multiple of $h/2\pi$:

$$mvr = n(h/2\pi)$$

where

m = mass of the electron
v = velocity of the electron
r = radius of the orbit ("Bohr radius")
h = Planck's constant, 6.626×10^{-27} erg sec
n = an integer (the main quantum number) = 1, 2, 3. . . .

Using the equation $\Delta E = hc/\lambda$, it is possible to calculate the difference in energy (ΔE) between two energy levels from the wavelength (λ) of light emitted when the electron moves from a higher to a lower level.

The Bohr model remained relatively intact until Sommerfeld suggested that some of the orbits are elliptical and that two quantum numbers should be used to describe the orbit. The first (the main quantum number, n) represents the major axis of the ellipse and the second, l, represents the minor axis.

Peter Zeeman observed that if the excited atoms used to generate bright line spectra were placed in a strong magnetic field, then the number of observed lines increased. The phenomenon, the Zeeman effect, suggested that a description of the energy states of an electron must also include another quantum number, m, the magnetic quantum number. Finally, the results of Stern and Gerlach demanded the introduction of a fourth quantum number, s, the spin quantum number. These workers found that in a nonhomogeneous magnetic field a stream of gaseous silver atoms was split into two beams, deviating from the original path in equal and opposite directions.

The exclusiveness of the four quantum numbers in an atom was proposed in 1925 by Wolfgang Pauli in what is now known as the **Pauli Exclusion Principle:** *no two electrons within the same atom may have the same four quantum numbers.*

The Wave Picture of an Electron. In 1924, de Broglie suggested a quantitative relationship between the mass of a particle and its wavelength. Put differently, he placed on firm ground the dualistic view of the nature of the electron, *i.e.*, that an electron can be viewed as exhibiting both wave-like and particle-like character.

$$\lambda = h/mv,$$

where λ = wavelength of the particle

Heisenberg noted (1927) the impossibility of exactly determining both the position and the momentum (energy) of the electron. The **Heisenberg Uncertainty Principle,** which is significant only for small particles, notes that an increasingly accurate determination of the position of a particle requires an increasingly erroneous determination of the momentum of the particle if these determinations are carried out simultaneously.

With the efforts of many individuals as background knowledge, Schrödinger was able to apply (1926) the mathematics of wave motion to the problem of the nature of the electron, thereby creating the foundation of modern quantum mechanics, the Schrödinger wave equation. The wave equation describes the subatomic particles in terms of their wave nature and, thus, moves away from the Bohr picture of an electron existing in a well-defined orbit. The wave equation allows one to discuss electron properties as probabilities (the "most probable velocity," "most probable distance from the nucleus," etc.). The equation generates a physical picture of characteristic energy values (eigenvalues) and the corresponding wave function (eigenfunction, ψ) for an electron. The probability of finding an electron in a region of space between x and $x + dx$ from the nucleus is directly related to ψ^2. (Max Born is credited with the interpretation of ψ^2 as a "probability.")

The Quantum Numbers. We have seen how there developed a need for four integers to describe any electron even before the advent of wave mechanics. It is gratifying, then, to note that an attempt to solve the wave equation for hydrogen also requires the introduction of four "quantum numbers": n (principal or main quantum number), l (azimuthal quantum number), m (magnetic quantum number) and s (spin quantum number).

The main quantum number tells us the general distance (and energy) from the nucleus to the electron. The azimuthal quantum number defines the angular momentum of the electron. Of greater significance to us, the value of l defines the "shape" of the region in which the electron is most likely to be found. The magnetic quantum number explains the orientations of the various orbitals (defined below) relative to one another. The spin quantum number describes the "spin character" (not strictly analogous to the ordinary meaning of *spin*) of the electron.

The four quantum numbers are related to one another by the following rules:

1. The main quantum number may take any integral value from + 1 to ∞, *i.e.*, $n = 1, 2, 3, 4 \ldots$
2. For any given value of n, the azimuthal quantum number (l) may take on values ranging from 0 to $n - 1$, *i.e.*, $l = 0, 1, 2, 3 \ldots (n - 1)$.
3. For any value of l, the magnetic quantum number (m) may take on any integral value (including zero) from $-l$ to $+l$, *i.e.*, $m = -l, -l + 1, -l + 2, \ldots 0, 1, 2 \ldots + l$.
4. For any value of m, the spin quantum number (s) may have only two values, either $-\frac{1}{2}$ or $+\frac{1}{2}$.

The first three quantum numbers, when taken together, describe a certain specific energy level called an **orbital.** Each orbital may hold a maximum of two electrons. Two electrons in the same orbital will have identical values of n, l and m; but one will have $s = -\frac{1}{2}$, while the other will have $+\frac{1}{2}$. If this were not so, the Pauli exclusion principle would be violated. By way of illustration, a breakdown of the orbitals which can be generated when $n = 3$ is presented in Figure 1–1.

On the basis of the periodic behavior of the elements, it has been suggested that the total number of electrons which could exist in a main energy level (Bohr orbit or "shell") is $2n^2$, where n is the Bohr quantum number. This observation can now be better understood when it is realized that the "shells" are made up of orbitals. Since a given value of n generates n^2 orbitals (see Figure 1–1) it follows that the *total* number of electrons is $2n^2$.

Organic chemists rarely refer to the azimuthal quantum numbers by their numerical values. Instead, they are MUCH more commonly identified by letter designation; an electron for which $l = 0$ is called an *s* electron, $l = 1$ is *p*, $l = 2$ is *d*, and $l = 3$ is *f*. Wave mechanics marked the end of the "orbit" as a significant term used to describe the electron; it also marked the birth of the "orbital." We shall now go on to describe, in a non-mathematical fashion, some properties of orbitals.

How Far from the Nucleus Is the Electron? In the old picture of atoms, one could draw a line from the nucleus to the circumference of an orbit and take that distance to be *the* distance to the nucleus. Having rejected the idea of an orbiting electron, how do we now speak of the distance to the nucleus?

Consider a plot of r versus $r^2\psi^2(r)$. Such a plot will tell us the probability of finding an electron in a thin shell of thickness dr at a given distance r from the nucleus. Such graphs, called "radial probability distribution" curves, indicate where one is most likely to find a given electron. The probability distributions for the orbitals of greatest interest to us are presented in Figure 1–2.

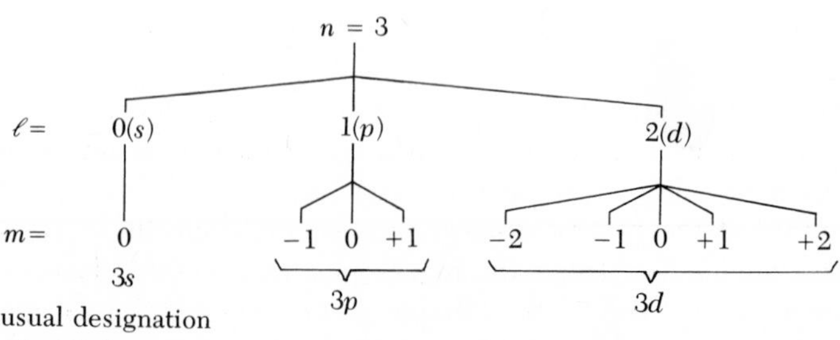

Figure 1–1 Orbitals within the third main quantum level.

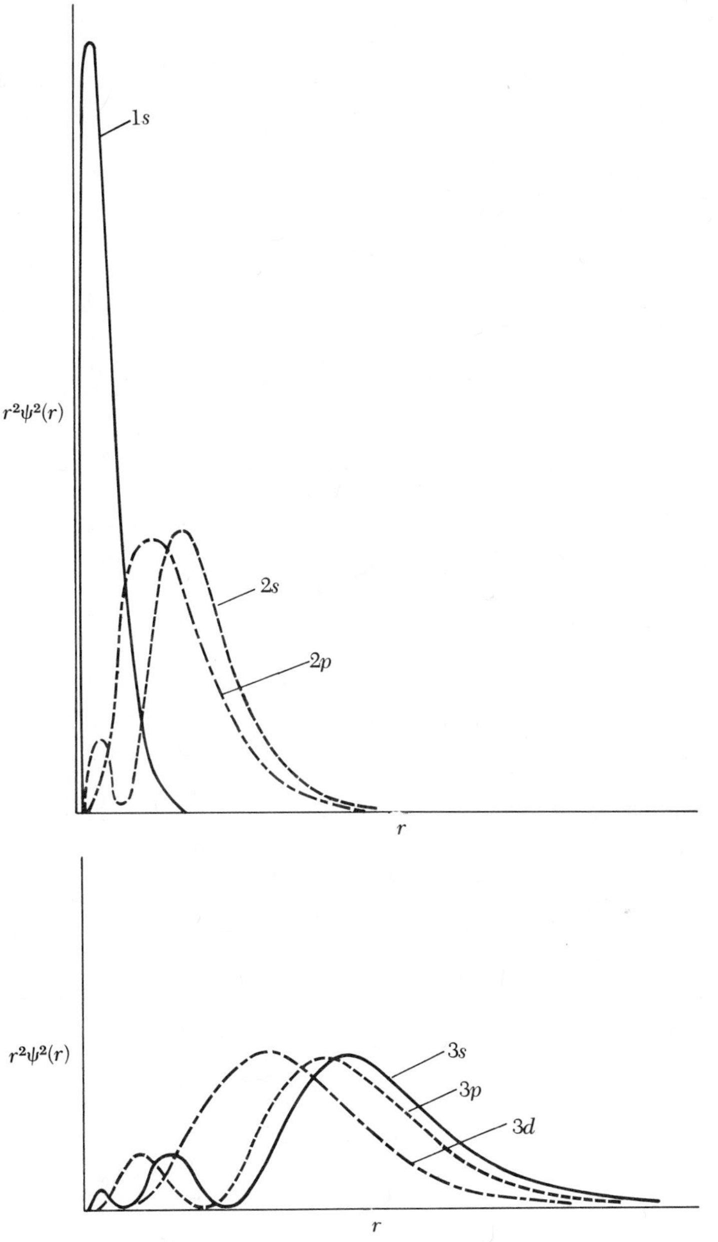

Figure 1-2 Probability distribution curves for 1*s* through 3*d* orbitals. These curves can be imagined to result from (a) rotating the orbit in all directions in space about the nucleus in order to remove any effects dependent upon a specific direction away from the nucleus, and (b) plotting the smeared electron density *vs.* distance from the nucleus. All curves are drawn to the same scale.

We note from these curves that for the 1*s* orbital there is a maximum probability of finding the electron within a shell approximately 0.5 Å from the nucleus. This distance corresponds to the Bohr radius of the hydrogen atom, so the Bohr picture and the wave mechanical picture have similarities—at least for hydrogen. Although there is a finite probability of finding the electron infinitely far from the nucleus, near 5 Å the probability is down to one in one million.

If we now examine the radial distribution plot of the 2*s* orbital, we observe a broad maximum at approximately 2.6 Å. However, there is another maximum at approximately 0.4 Å!

The existence of this inner probability maximum reflects the phenomenon known as *penetration*. Penetration requires that the electron spend more time near the nucleus than might be imagined simply from an examination of the main quantum number. Penetration is particularly important for *s* and *p* orbitals, lowering their energy by bringing them closer to the nucleus. Notice that in the 2*s* orbital there are two places where the probability of finding the electron within a volume element of thickness *dr* is zero. The first is the nucleus (as in the 1*s* orbital); the second place, between the two maxima, is termed a *node*. The three-dimensional equivalent is a *nodal surface* and, specifically for 2*s*, is a nodal sphere. In general, the number of nodes for *s* orbitals equals $n - 1$, where n is the main quantum number.

The 2*p* orbital shows one probability maximum and one place where the probability of finding the electron is zero (the nucleus). The 3*p* orbital shows two maxima, separated by a node. In general, the number of such nodes for *p* orbitals is $n - 2$. The 3*d* orbital shows one probability maximum.

For $n = 3$ the penetration of the *s* orbital is greater than is the penetration of the *p* or *d* orbitals; however, it is the *d* orbital which has its major probability maximum closest to the nucleus. This reflects the universal observation that the distance at which the maximum probability is found decreases in going from *s* to *p* to *d* within a given main quantum level.

WHAT ARE THE SHAPES OF ATOMIC ORBITALS? When discussing the electron in terms of probabilities, it is helpful (though not entirely accurate) to describe the electron as a "cloud" smeared out through space. We have just finished examining the positions of the densest parts of those clouds relative to the nucleus, and we will now consider the shapes of these clouds.

The *s* orbital is spherically symmetric about the nucleus; the 1*s*, 2*s* and 3*s* orbitals are drawn (as clouds) in Figure 1–3. These clouds enclose the regions of space in which the electron spends 95% of its time.

For the 2*p* energy level, m may take on values of -1, 0 and $+1$. Each of the resultant 2*p* orbitals has the same energy in the isolated atom (they are said to be *degenerate*), but they differ in their orientations in space. The shapes and orientations of the 2*p* orbitals are presented in Figure 1–4. The complexity of orbitals increases with increasing values of n and l, and the shapes of 3*p* and 3*d* orbitals will not be discussed until Chapter 23.

Although it is convenient to think in terms of these shapes of orbitals, let us emphasize

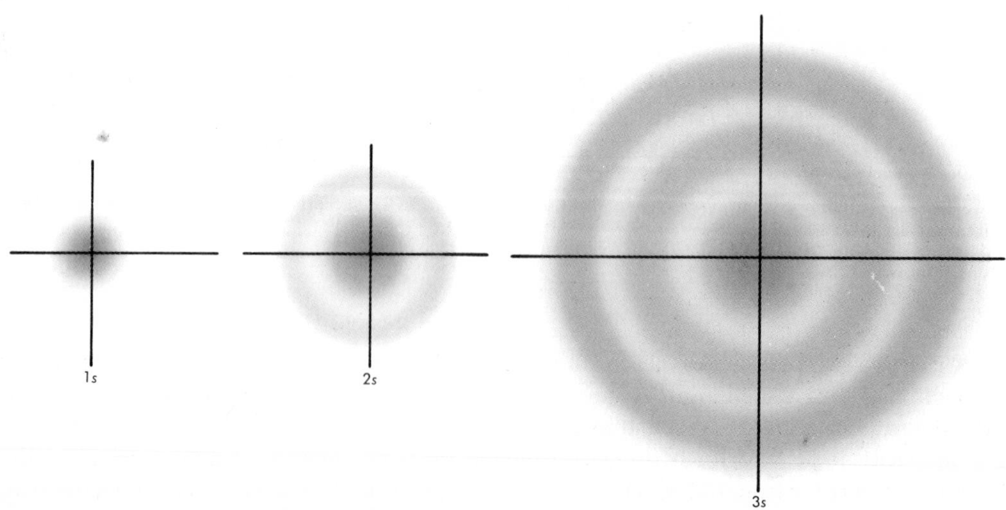

Figure 1–3 A comparison of 1*s*, 2*s*, and 3*s* orbitals. (From W.L. Masterton and E.J. Slowinski, *Chemical Principles*, 2nd edition, W.B. Saunders Co., Philadelphia, 1969.)

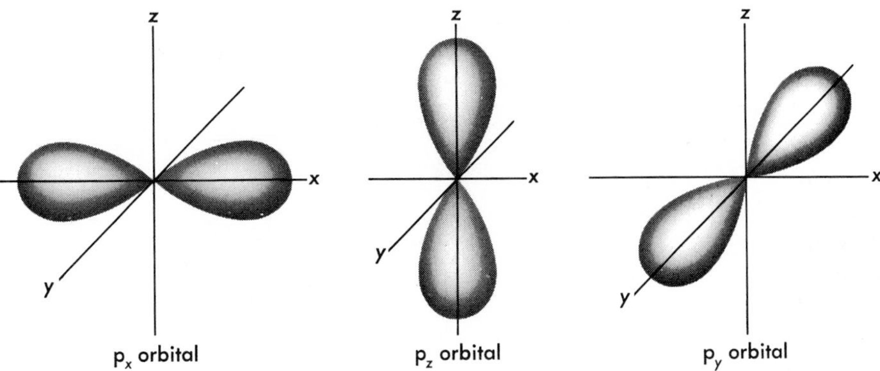

Figure 1–4 The three orthogonal $2p$ orbitals. (From W.L. Masterton and E.J. Slowinski, *Chemical Principles*, 2nd edition, W.B. Saunders Co., Philadelphia, 1969.)

that the volume shown is the space within which the probability of finding the electron is 95%. If we were to raise the probability to 100%, the orbital would extend to infinity.

How Many Orbitals Exist in the Hydrogen Atom? We can use Figure 1–1 to predict the total number of orbitals corresponding to a particular value of the principal quantum number. For example, when $n = 3$, l can take on values of 0, 1 and 2; these sets are labeled $3s$ ($n = 3$, $l = 0$), $3p$ ($n = 3$, $l = 1$) and $3d$ ($n = 3$, $l = 2$). Within the set $n = 3$, $l = 0$, m may only be zero, so there is only one $3s$ orbital. Within the set $n = 3$, $l = 1$, m may acquire values of -1, 0 and $+1$, giving three $3p$ orbitals. Within the set $n = 3$, $l = 2$, m may be -2, -1, 0, $+1$ and $+2$, giving a total of five $3d$ orbitals. We see, then, that in the principal level $n = 3$ there are a total of $1 + 3 + 5 = 9$ orbitals.

But how many of these orbitals actually exist in a hydrogen atom? All of them! Orbitals, like closets, do not go out of existence because they are empty. Unlike a closet, an orbital is limited in what it can hold, having a maximum capacity of two electrons. Thus, hydrogen has one half-filled orbital ($1s$) and a myriad of vacant ones. All of the orbitals discussed so far are called "hydrogen-like orbitals," representing calculations based upon one electron interacting with one proton. Because we are not yet capable of calculating orbital characteristics for most elements, we assign hydrogen-like orbitals to them.

Construction of the Electronic Ground State of the Elements. The rules for assigning quantum numbers which we have considered allow us to summarize the types of electrons which can be found in a particular atom, provided we know which orbitals are filled and which are empty. In constructing the ground state (lowest energy state) electronic configuration of the elements, we employ the **aufbau principle**—filling the available orbitals according to their potential energy, the lowest energy orbitals being filled first. In filling degenerate orbitals (orbitals having the same potential energy as each other), we start by placing one electron in each of the degenerate orbitals (with the same spin!); a degenerate orbital does not receive two electrons until each member of the degenerate set has received at least one. This order of filling follows what is called Hund's rule. Finally, but quite importantly, we assume that the electronic structure of an atom of atomic number $x + 1$ is like that of atom x with the addition of one more electron.

There are exceptions to the aufbau principle, but these are not of much interest to organic chemistry. Let us only bear in mind that we eventually predict the ground state electronic configuration of an ISOLATED atom. Although a detailed account of periodicity, aufbau filling, and so forth is beyond our needs, Table 1–3 presents the aufbau filling order while Table 1–4 contains ground state electronic configurations for atoms of interest to us.

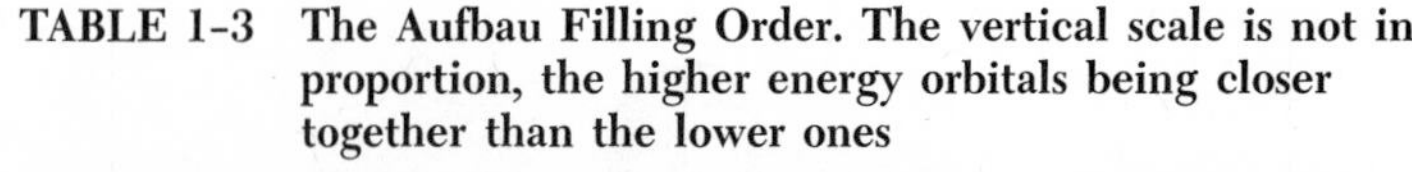

TABLE 1-3 The Aufbau Filling Order. The vertical scale is not in proportion, the higher energy orbitals being closer together than the lower ones

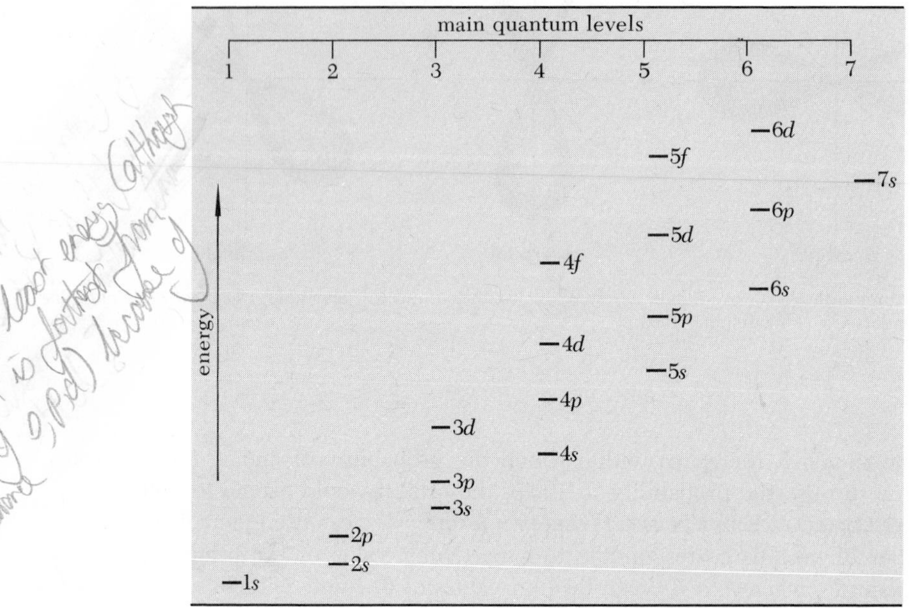

TABLE 1-4 Ground State Electronic Configuration For Selected Atoms

		$n = 1$			$n = 2$			$n = 3^a$		
ELEMENT	CONFIGURATION	l	m	s	l	m	s	l	m	s
H	$1s^1$	0	0	$-\frac{1}{2}$						
C	$1s^22s^22p^2$	0	0	$-\frac{1}{2}$	0	0	$-\frac{1}{2}$			
		0	0	$+\frac{1}{2}$	0	0	$+\frac{1}{2}$			
					1	−1	$-\frac{1}{2}$			
					1	0	$-\frac{1}{2}$			
N	$1s^22s^22p^3$	0	0	$-\frac{1}{2}$	0	0	$-\frac{1}{2}$			
		0	0	$+\frac{1}{2}$	0	0	$+\frac{1}{2}$			
					1	−1	$-\frac{1}{2}$			
					1	0	$-\frac{1}{2}$			
					1	+1	$-\frac{1}{2}$			
O	$1s^22s^22p^4$	0	0	$-\frac{1}{2}$	0	0	$-\frac{1}{2}$			
		0	0	$+\frac{1}{2}$	0	0	$+\frac{1}{2}$			
					1	−1	$-\frac{1}{2}$			
					1	0	$-\frac{1}{2}$			
					1	+1	$-\frac{1}{2}$			
					1	−1	$+\frac{1}{2}$			
S	$1s^22s^22p^63s^23p^4$	0	0	$-\frac{1}{2}$	0	0	$-\frac{1}{2}$	0	0	$-\frac{1}{2}$
		0	0	$+\frac{1}{2}$	0	0	$+\frac{1}{2}$	0	0	$+\frac{1}{2}$
					1	−1	$-\frac{1}{2}$	1	−1	$-\frac{1}{2}$
					1	0	$-\frac{1}{2}$	1	0	$-\frac{1}{2}$
					1	+1	$-\frac{1}{2}$	1	+1	$-\frac{1}{2}$
					1	−1	$+\frac{1}{2}$	1	−1	$+\frac{1}{2}$
					1	0	$+\frac{1}{2}$	1	0	$+\frac{1}{2}$*
					1	+1	$+\frac{1}{2}$	1	+1	$+\frac{1}{2}$*

[a] The two "starred" electrons are not a part of sulfur. They are included because they demonstrate the conversion of the sulfur electronic configuration to the noble gas (argon) configuration.

Spectroscopic Nomenclature. It becomes rather awkward to say, for example, "The hydrogen atom has one electron in the 1*s* orbital," so this statement is abbreviated by saying "Hydrogen is $1s^1$" (read *one s one*). Similarly, the electronic configuration of nitrogen is abbreviated $1s^22s^22p^3$, indicating two electrons in the 1*s* orbital, two in the 2*s* orbital, and three in 2*p* orbitals. This abbreviated nomenclature is sometimes called "spectroscopic nomenclature." When describing the electronic configuration of an element in this way, it is common to list the orbitals in order of increasing quantum numbers (n and l), rather than in the order of aufbau filling.

1.3 ATOMIC PROPERTIES

There are several atomic properties which are of importance to us including size, ionization potential, electron affinity, and electronegativity.

Size. The size of an atom cannot be specified precisely. This is inherent in the fact that the electron is not at a fixed, immutable distance from the nucleus but, rather, occurs at varying distances. As noted earlier, however, the probability of finding an electron far from the nucleus is remote.

The size of an atom can be estimated by several procedures, the most common being X-ray crystallography. Results of such studies lead to the general observation that atomic size decreases in going from left to right across a given period of the periodic table and increases in going down a family of elements within the periodic table.

Ionization Potential and Electron Affinity. The minimum amount of energy necessary to remove an electron from an atom in the gas phase is called the **ionization potential.** (The term "remove" implies taking the electron to extremely high values of n.) The ionization potential is, then, one means of stating the ease of producing a *cation* (positively charged ion) from a neutral atom. This suggests that ionization potential is an important parameter in determining the course of electron-transfer (ionic) reactions. Table 1–5 contains the ionization potentials for representative elements.

While it is relatively easy to measure ionization potential (a mass spectrometer will do), it is much more difficult to measure the energy released when an atom gains an electron, the **electron affinity.** Thus electron affinities, important in measuring the ease of *anion* (negatively charged ion) formation from an atom, are known for a limited number of elements.

Electronegativity. The chemical properties of an element are dependent upon the behavior of its electrons, especially its outer electrons. Therefore, anything which influences

TABLE 1–5 Ionization Potentials Of Selected Atoms[a]

Element	Ionization Potential (Kcal/Mole)
H	314
Li	124
C	260
N	335
O	314
F	402
Na	119
Cl	300
Br	273
I	241

[a] These values represent the loss of the first electron. Higher ionization processes will require more energy. (Why ?)

TABLE 1-6 Element Electronegativity (Pauling Scale)

ELEMENT	ELECTRONEGATIVITY
hydrogen (H)	2.1
helium (He)	0.00 (estimated)
lithium (Li)	1.0
boron (B)	2.0
carbon (C)	2.5
nitrogen (N)	3.0
oxygen (O)	3.5
fluorine (F)	4.0[a]
silicon (Si)	1.8
phosphorus (P)	2.1
sulfur (S)	2.5
chlorine (Cl)	3.0
bromine (Br)	2.8
iodine (I)	2.5

[a] The largest electronegativity in the Pauling scale.

the behavior of these electrons is important to the chemist. Atoms attract electrons to varying degrees, and the ability of an atom to attract its own outer electrons (sometimes limited to electrons used in bond formation) is a measure of that atom's **electronegativity.** The greater the electronegativity of an element, the stronger the attraction between that atom and its outer electrons.

There have been several attempts at quantitating the electronegativity of the elements, beginning with the efforts of Mulliken, who defined an element's electronegativity as the average of its ionization potential and electron affinity. Of these attempts, the Pauling scale of electronegativity, derived from bond energy calculations, is used most often. On this scale fluorine, the most electronegative element, has an electronegativity of 4. In general, one associates a rather high electronegativity with the classic non-metals (Table 1-6). The elements which are typical, reactive metals have rather small electronegativities (*e.g.*, Na, 0.9; Mg, 1.2). Because of this, the metals are often said to be "electropositive" in comparison to the non-metals, which are said to be "electronegative." Examining the periodic table, one notes that electronegativity decreases in going down a family of elements but increases in moving from left to right within a period.

You may recall from previous chemistry courses that attractive forces (bonds) between atoms are of two major types, "ionic" and "covalent," and that these actually are at the opposite ends of an entire range of bond types. In the simplest cases, the position of a given bond on this scale is a function of the difference in the electronegativities of the elements involved in the bond. If this difference is greater than 1.7, then the bond is more than 50% ionic; if it is less than 1.7, the bond is more than 50% covalent. For example, the electronegativity difference between cesium (Cs) and fluorine (F) is 3.3 ($|4.0 - 0.7|$) and the attractive force between cesium and fluorine is described as *ionic.* On the other hand, the attractive force between carbon and chlorine is *covalent* since the electronegativity difference between these elements is only 0.5 ($|2.5 - 3.0|$).

Do not be discouraged if you have forgotten about ionic and covalent bonds; they are at the heart of the following chapter!

IMPORTANT TERMS

α-Particle (alpha particle): A particle containing two neutrons and two protons, and bearing a +2 charge. These particles are identical with the nuclei of ordinary helium atoms (atomic number = 2; mass number = 4). A stream of such particles is called an α-ray. α-Rays are emitted during radioactive decay.

Atomic number: The number of protons in the nucleus of an atom. All isotopes of a given element must have the same atomic number. Atoms contain the same number of electrons outside the nucleus as they do protons within the nucleus. However, since electrons may be gained or lost during chemical reactions, while protons are *not* lost during chemical reactions, atomic number is defined in terms of protons within the nucleus rather than electrons outside the nucleus. Elements are ordered in the periodic table according to their atomic numbers.

Atomic orbital: An energy level around an atom which is described by three quantum numbers (main, azimuthal, and magnetic). Each atomic orbital can accommodate two electrons. These two electrons have common main, azimuthal, and magnetic quantum numbers, but different spin quantum numbers.

Atomic orbitals have different shapes. The most important atomic orbitals to the organic chemist are the 1*s*, 2*s* and 2*p* orbitals; their shapes are shown in Figures 1–3 and 1–4. In the following chapter we will learn how atoms may interact with one another using their atomic orbitals.

Atomic weight: The "average" weight of an atom of an element. It is necessary to speak of an average weight since almost all elements exist as a mixture of isotopes, and each isotope has a different weight. If an element contains a% of an isotope of weight A and b% of an isotope of weight B, its atomic weight equals $0.a \times A + 0.b \times B$. Atomic weights are rarely integers (whole numbers). The atomic weights of the elements are included in the periodic table on the inside back cover.

Binding energy: The energy which holds an atom together. It is energy which is lost when the atom's constituents (neutrons, protons, electrons) come together from infinite separation.

Bright line spectrum: A pattern of discrete lines of color which may extend from one end of the visible spectrum (red) to the other (violet). You can picture what a bright line spectrum looks like by imagining that you are looking at a rainbow through a picket fence with *very* little space between the pickets, and with pickets of varying width. What you will see might contain some lines of varying shades of red, orange, yellow, green, blue, indigo, and violet.

A bright line spectrum is produced when an electron comes down in energy from a higher energy orbital to a lower energy orbital. Since these orbitals differ in their energy by discrete amounts (one speaks of "energy levels"), the result is a series of discrete lines of color rather than a continuum of color such as is found in the rainbow. [There is a direct correlation between the energy separating two orbitals and the color (wavelength) of light emitted when an electron goes from the higher to the lower energy orbital.]

Degenerate: The term is used to describe two or more species or phenomena which are of equal energy under normal circumstances. For example, the three 2*p* orbitals are of equal energy and are said to be degenerate. However, the 1*s* and 2*s* orbitals are of unequal energy and are *not* degenerate.

Electronegativity: In its most general sense, electronegativity means "attraction for electrons." The chemist most commonly uses this term to imply attraction between an atom and electrons in its outer orbital(s). The most electronegative elements are found in the upper right hand corner of the usual form of the periodic table and include fluorine, oxygen and nitrogen. The least electronegative elements, described as *electropositive,* include the alkali (Group I) and alkaline earth (Group II) metals. Electronegativity plays a major role in determining the nature of interaction between atoms when they react, an aspect of chemistry which will be described in Chapter 2. When we begin to look at how ions react with organic compounds (Chapter

5), we will see a very important application of the concept of electronegativity to organic chemistry.

Half-life: The time required for 50% of a starting material to undergo a certain reaction. It can be used to describe conventional chemical reactions as well as the decay of radioactive isotopes.

Hund's rule: In the filling of orbitals of equal energy, each orbital receives one electron before any of these degenerate orbitals receives two electrons. The six illustrations below show how this is applied to the consecutive addition of six electrons to the $2p$ energy level.

	$2p$	$2p$	$2p$		$2p$	$2p$	$2p$
1.	↑			4.	↑↓	↑	↑
2.	↑	↑		5.	↑↓	↑↓	↑
3.	↑	↑	↑	6.	↑↓	↑↓	↑↓

Hund's rule is followed in developing the electronic structure of the elements by sequentially adding one electron at a time around a nucleus. The build-up follows what is called the aufbau principle.

Ionization potential. The work required to remove an electron from an atom. The energy required to remove the first electron (usually the outermost electron) is the first ionization potential; the energy required to remove a second electron is the second ionization potential, and so on. Metals have low ionization potentials, while non-metals have high ionization potentials.

Isotopes: Isotopes are atoms of the same element which differ in the number of neutrons in the nucleus and, therefore, in their mass number. Since they are the same element, isotopes have the same number of protons. While most elements possess several isotopes, there are exceptions. Perhaps the most famous example of an element which has only one naturally occurring isotope is fluorine (^{19}F).

Some isotopes of elements are radioactive and have been used in diagnosis and treatment of disease. Isotopes which are radioactive and which have a comparatively short lifetime must be created. (Of course, such isotopes normally are not expected to occur in nature.) Tritium has already been mentioned (p. 3) as one such synthetic isotope. Another is radioactive iodine (^{131}I) which, with an eight day half-life, is used to treat diseased thyroid glands. Strontium has sixteen isotopes, the most famous of which is strontium-90 (^{90}Sr). This isotope, with a 28 year half-life, is produced in nuclear explosions and is a serious health hazard when produced.

Neutron: A neutral particle, mass number 1, found in all nuclei with a mass greater than 1.

Proton: A positively charged particle bearing a charge of the same magnitude (but of opposite sign) as the electron. The proton occurs in all nuclei. The proton *is* the nucleus of the hydrogen atom and, because of this, a hydrogen atom which has lost its lone outer electron is aptly called a proton.

$$\underset{\text{hydrogen atom}}{H\cdot} \rightarrow \text{electron}^{\ominus} + \underset{\text{proton}}{H^{\oplus}}$$

ON PROBLEMS

There are two general types of questions in this text, those which are meant to elicit clear, crisp answers (the so-called "drill problems") and those requiring some discussion (clarity is welcomed here also). The purpose of the former is largely to test and improve your retention of material (even going back to your previous chemistry course), while the latter are designed to probe and improve your ability to think through material and to draw some conclusions related to that material. While it is difficult to establish a clear boundary, the more thought-provoking problems are found toward the end of problem sets located at the end of each chapter. Do not become frustrated if you cannot "answer" some of these so-called "thought questions." Many of the purposes of asking such questions are served if you think, contemplate, discuss and even argue about the problems yet fail to get the "right answer."

A very fine organic textbook of the early 1960's, known to many simply as "Gould,"[a] supports this point of view. It contains a series of problems which are not answered and, it seemed to many people using the book, were unanswerable. The stimulation provided by these questions far outweighed any disadvantage that one might associate with the absence of answers in the textbook.

Problem solving is essential to learning organic chemistry and, if done properly, can save you time in studying the subject material. I shall, therefore, suggest several "dos and don'ts." There is no associated guarantee of success because students are individuals and each combination of student, teacher, and textbook is unique. However, a number of my students have been helped by the following suggestions.

1. **DO NOT** attend a lecture without having at least scanned the appropriate sections in the text.

2. **DO NOT** attempt to solve the problems immediately after you read the text. By allowing time to pass, you will help yourself identify subject material which you did not retain.

3. When there are problems which you cannot solve, **DO NOT** scurry back to the text and try to discover *that* portion which may house the answer. Instead, re-read the entire chapter, scanning inappropriate sections and slowing down at those places which are pertinent to answering your problems. After all, you do not just want to answer the question, you also want to assimilate the appropriate information into your growing fund of chemical knowledge.

4. **DO NOT** run to the answer section or answer book immediately after deciding that you cannot solve a given problem. For many, the answers are best used as a last resort and not as a means to avoid re-examining the text.

5. **DO** try to identify any new information in a problem and make it a part of the chemistry which you have already learned. There are important ideas which have been incorporated into problems rather than into the body of the text. This has been done because of spatial limitations and, equally importantly, to permit you to practice your deductive powers on significant situations.

---

[a] E.S. Gould, *Mechanism and Structure in Organic Chemistry*, Holt, New York, 1959.

PROBLEMS

1. Distinguish between:
   (a) mass number and atomic number
   (b) atomic weight and atomic number
   (c) main quantum number and principal quantum number
   (d) electronegativity and electron affinity
   (e) cation and anion
   (f) deuterium and tritium
   (g) 2*s* orbital and 2*p* orbital
   (h) orbit and orbital
2. With the aid of the Handbook of Chemistry and Physics, identify ten elements below atomic

number 92 which have isotopes that
(a) are radioactive
(b) are not naturally occurring

3. What contributions to atomic theory were made by:

(a) Pauli	(b) de Broglie
(c) Bohr	(d) Rutherford
(e) Stern and Gerlach	(f) Zeeman
(g) Schrödinger	(h) Pauling

4. Following the aufbau process, provide the quantum numbers of the last electron to be added to each of the following ^{16}O, ^{14}N, ^{32}S, ^{31}P, and all of the isotopes of uranium.

5. What are the quantum numbers of all of the electrons in the silicon atom? Compare and contrast the electronic structures of carbon and silicon.

6. Using the data in Table 1–2, calculate the atomic weights of oxygen, boron, and nitrogen.

7. Suggest an equation to relate the number of nodal surfaces in an orbital to its main and azimuthal quantum numbers.

8. Which of the following pairs of atoms should form bonds that are more than 50% ionic? (a) HBr; (b) LiBr; (c) BrCl; (d) CO.

9. While the aufbau process provides the order of orbital filling, the ionization potential data suggest that electrons are lost from an atom in order of their quantum numbers. Offer an explanation for this difference.

10. It has been suggested that, within a given main quantum level, an *s* orbital is more "electronegative" than the corresponding *p* orbitals. Suggest a rationale for this difference in *orbital electronegativity*.

11. Copper, silver, and gold are all excellent electrical conductors. They have a common, atypical electronic configuration, being $d^{10}s^1$ rather than d^9s^2. What is the significance of this observation with regard to (a) electrical conductivity and (b) the stabilities of partially *vs.* completely filled orbitals?

12. The ionization potential for the halogens decreases in going from F to I, even though the number of protons in the nucleus increases in going from F to I. This might be expected to cause the electrons to be held more strongly. Explain.

13. What correlation exists between the manner in which formulas are written and the electronegativity of the elements constituting those formulas? Base your answer on the following formulas: NaCl; $CaBr_2$; HCl; BrCl; H_2SO_4. Can the same thing be said of complex anions? Base your answer on the following examples: CrO_4^{-2} (chromate ion); SO_4^{-2} (sulfate ion); NO_3^{-1} (nitrate ion); MnO_4^{-1} (permanganate ion).

FROM BONDS TO SMALL MOLECULES

2.1 WHERE ARE WE GOING?

We shall begin by examining the various ways in which atoms can "stick" together. Once these different types of bonds have been presented, we shall study the application of molecular orbital theory to the bonding in several small molecules. Our preliminary efforts will culminate in a magnificently wrong prediction. *Facts* will force us to discuss the concept of "hybridization" and the influence which it has upon covalent bonding schemes. This chapter ends with a presentation of the bonding in some **functional groups** (collections of atoms which often behave as a unit) and in some polyatomic molecules.

2.2 HOW MAY WE DEFINE THE TERM "BOND"?

To be able to say that a bond exists between two species, be they individual atoms or very large molecules, the adduct (*i.e.*, the new, combined species) formed from the components must have a "reasonable" lifetime. In the infancy of organic chemistry, a "reasonable lifetime" implied something which existed long enough to be bottled and put on a shelf. The term later took on a broader meaning, encompassing anything which had a long enough lifetime to permit detection. As our methods improved, the time that an adduct had to stay around in order to be detected decreased. It quickly becomes apparent that our experimental definition of a bond, like most definitions based upon experiments, is a function of the procedures currently available and, therefore, subject to change. Ultimately, the bond between two species is best defined in terms of the energy that the system loses by virtue of adduct formation. The greater the amount of energy liberated, the stronger is the bond.

2.3 WHAT TYPES OF BONDS ARE THERE?

The vast majority of chemical interactions and chemical reactions can be discussed in terms of three bond types: *electrostatic*, *covalent*, and *metallic*. While the metallic bond is currently not of sufficient interest to most organic chemists to be discussed further, electrostatic and covalent bonds are of great interest and, therefore, some of their properties will be reviewed in this chapter.

ELECTROSTATIC BONDS

We begin by considering the attraction between oppositely charged species and the accompanying formation of an electrostatic bond. This type of bond can be divided into three sub-categories: ion-ion bonds (ionic bonds), ion-dipole bonds, and dipole-dipole bonds.

Ionic Bonds. The reaction of an atom of sodium (an electropositive element) with chlorine (an electronegative element) is an electron-transfer process, the sodium transferring an electron to the chlorine. The sodium atom becomes a positively charged sodium ion, while the chlorine becomes a negatively charged chloride ion (Figure 2–1). These oppositely charged ions attract one another, and the potential energy of this ionic system decreases as these ions approach one another. The potential energy of the system begins to rise again when both ions begin to claim the same space. The energy decrease when the ions approach one another indicates the strength of the bond between $Na^{\oplus}$ and $Cl^{\ominus}$.

$$Na^0 + Cl^0 \rightarrow NaCl \; (\textit{or } Na^{\oplus}Cl^{\ominus})$$

or

$$(1s^22s^22p^63s^1)^0 + (1s^22s^22p^63s^23p^5)^0 \rightarrow (1s^22s^22p^6)^{\oplus} + (1s^22s^22p^63s^23p^6)^{\ominus}$$

or

$$Na\cdot + \cdot\ddot{\underset{\cdot\cdot}{Cl}}: \rightarrow Na^{\oplus} + :\ddot{\underset{\cdot\cdot}{Cl}}:^{\ominus}$$

Figure 2–1 Alternative representations of the reaction of a sodium atom with a chlorine atom. In the last representation the symbols for the elements also imply the inner electrons; only outer, or "valence," electrons are shown explicitly.

Ionic compounds are generally solids having high melting points, with ions occupying well-defined lattice points in the crystal lattice. The stability of a solid such as sodium chloride is dependent upon three factors: the ionization potential of sodium, the electron affinity of chlorine, and the energy released when these ions are brought *from* infinite separation *to* the lattice points (crystal lattice energy). Examination of Figure 2–2 reveals that in the ionic crystal no single cation belongs to any specific anion. Indeed, each ion has six equivalent neighbors and *there is no molecule!* Furthermore, the bond between oppositely charged ions is non-directional.

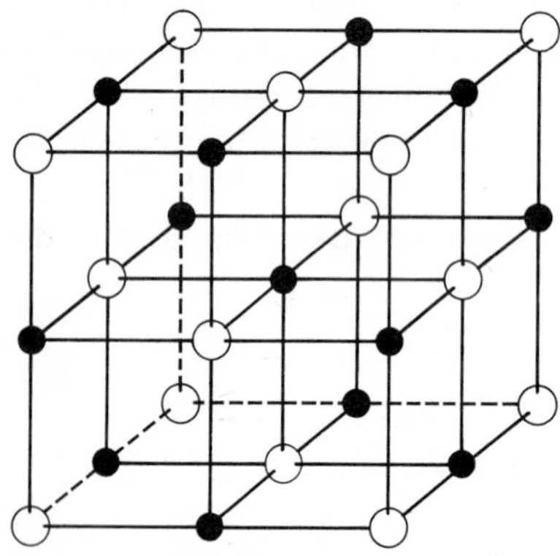

Figure 2–2 The crystal lattice of sodium chloride. The lines are not bonds, but are included to facilitate depth perception.

The representation in Figure 2–2 shows the distribution of $Na^{\oplus}$ and $Cl^{\ominus}$ ions in solid sodium chloride but, for clarity, does not attempt to compare their relative sizes. Actually, these ions are "touching" in the solid, with the chloride ions larger than the sodium ions. The term used to express the volume requirement of an ion is "ionic radius" (Figure 2–3). A cation should be smaller than its parent atom, and an anion should be larger than its parent atom. (Why?) The actual values of ionic radii (Table 2–1) must be determined experimentally—most commonly by X-ray crystallography.

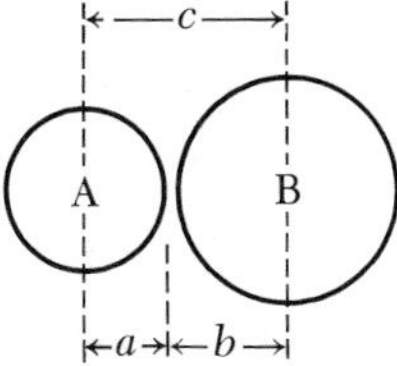

Figure 2–3 Establishment of the ionic radius. a = ionic radius of A; b = ionic radius of B; c = separation of nuclei at closest approach. When discussing one ion, the other is called the "counter" or "gegen" ion.

TABLE 2–1 Ionic Radii (Å)

Element	Charge	Ionic Radius
Li	+1	0.60
Na	+1	0.95
K	+1	1.33
Rb	+1	1.48
Mg	+2	0.65
Ca	+2	0.99
F	−1	1.36
Cl	−1	1.81
Br	−1	1.95
I	−1	2.16

Ion-Dipole Bonds. The water molecule is "polar," the oxygen end of the molecule being electron-rich relative to the hydrogen end (Figure 2–4). We can substitute such a polar molecule for one of the ions in the ion-ion bond, producing an electrostatic attraction between an ion and an oppositely charged end of a polar molecule. The polarity of the water molecule is a consequence of the polar nature of the O—H bond (discussed later), hence the name *ion-dipole bond*. The most common type of ion-dipole bond is formed when an inorganic compound (*e.g.*, NaCl) is dissolved in a highly polar solvent (*e.g.*, H_2O). Under these circumstances the ions become associated with the solvent (*i.e.*, *solvated*); the process is termed *hydration* when the solvent is water (Figure 2–5). The decrease in energy when an ion becomes solvated is the *solvation energy* of the process.

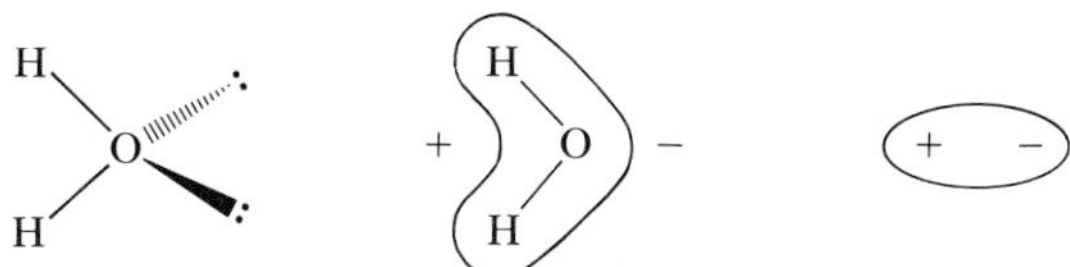

Figure 2–4 The polarity of water: various representations. The solid wedge (▬) represents a bond projecting toward the viewer, while the dashed wedge (IIIII) represents a bond projecting away from the viewer. Solid lines represent bonds in the plane of the paper. This convention, used throughout this book, is useful for showing the dimensional character of covalent bonds (defined in the following section).

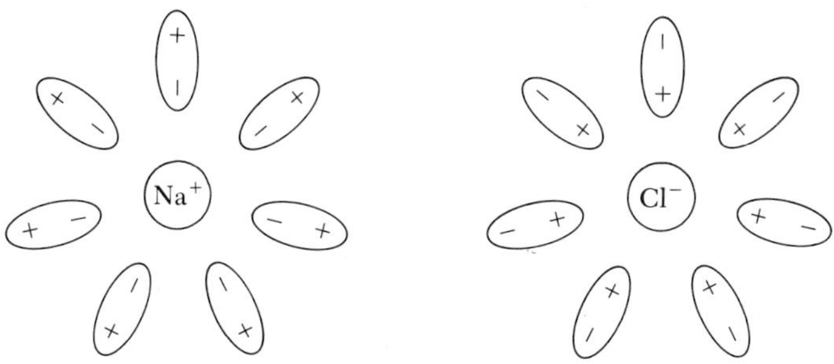

Figure 2–5 The solvation of sodium and chloride ions.

Dipole-Dipole Bonds. If the ion participating in an ion-dipole bond is replaced by a second polar molecule, the stage is set for the formation of a *dipole-dipole bond.* Notice that the water molecules in Figure 2–6 are aligned so as to bring together the oppositely charged ends of the polar molecules. This picture of a highly associated network of molecules accounts for the relatively high boiling point, high viscosity (resistance to flow), and low vapor pressure of water. This specific type of attraction, the *hydrogen bond,* is among the strongest types of dipole-dipole attraction, but is still much weaker than the ionic bond. (The factors which must be present in order for a hydrogen bond to exist are presented on p. 30.)

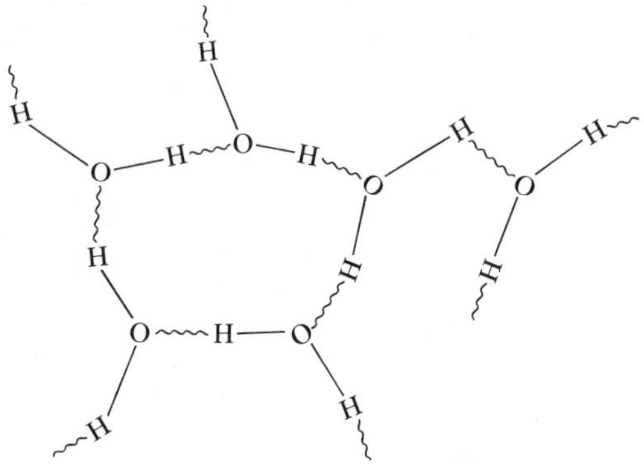

(a)

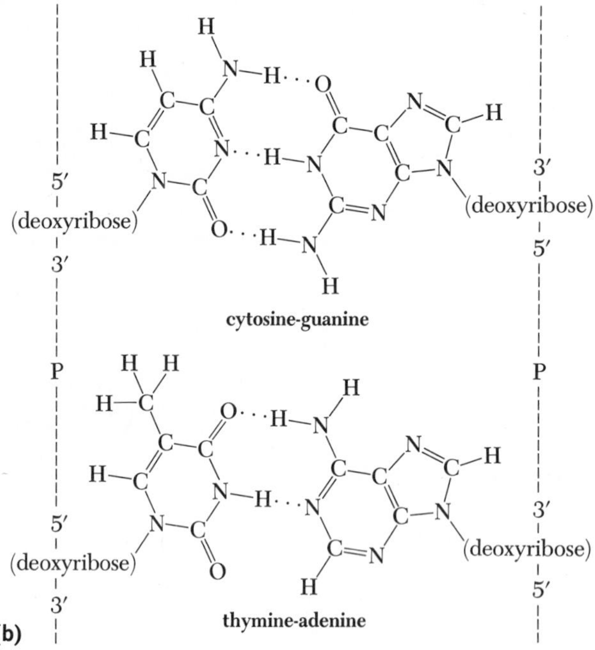

(b)

cytosine-guanine

thymine-adenine

Figure 2–6 (a) Intermolecular hydrogen bonding in water. (b) Hydrogen bonding in a fragment of a gene. The two very long chains are held in their proper geometry, relative to one another, by several forces, including hydrogen bonding. (From McGilvery, R.W.: Biochemical Concepts. Philadelphia, W.B. Saunders Company, 1975.)

Just to show that hydrogen bonding also is important in molecules more complex than water, and to show how it can keep complex molecules at a fixed distance from one another, Figure 2–6 also shows hydrogen bonding between two very small fragments of a gene. The gene (discussed in detail in Chapter 26) is a very complex molecule which must maintain a carefully specified geometry in order to transmit the proper genetic information. The hydrogen bonds such as those shown in Figure 2–6b play a critical role in assuring this geometry.

1. What are the characteristics which ionic, ion-dipole, and dipole-dipole bonds have in common? How do they differ?

COVALENT BONDS

The bonds which have been discussed thus far result from *obvious* attractions between oppositely charged species. One might wonder whether two atoms could be bonded together if one of them were unable to transfer one or more electrons to the other. (This situation could occur because of either the high ionization potential of the former or the low electron affinity of the latter—or perhaps both.) The answer is YES.

Two or more atoms can be bound together if they *share* electrons. The most common form of this type of bond involves the sharing of two electrons by two atoms. The resultant two-electron bond arises because of the increased electron density between the two nuclei.[a] One may imagine that each nucleus is strongly attracted to the two electrons, and the net result is that the two nuclei are maintained in close proximity. This bond, which requires the interaction (or "overlap") of two atomic orbitals (a.o.'s) of the atoms involved in the bond, is termed a **covalent bond.** In this bond an electron pair is shared equally by two atoms. In the final analysis, ionic bonds and covalent bonds represent two ends of a spectrum of bond types. The middle ground, where the shared electron pair is closer to one atom (the more electronegative one), is well represented by the *polar covalent bond.* We shall follow the common practice of calling both covalent and polar covalent bonds "covalent bonds." In the following paragraphs we shall see how we can produce a picture of covalent bonding with the aid of molecular orbitals.

Molecular Orbitals. The molecular orbital (m.o.) picture of a molecule is a logical extension of the atomic orbital picture of atoms (Chapter 1). In discussing a.o.'s, electrons around an atom were assigned to certain energy levels, the orbitals of the atom. This method, applied here but limited to an atom's outer or valence electrons, generates a molecular orbital for each atomic orbital that is combined with another atomic orbital. Thus, the overlap of two a.o.'s, one on each of two atoms, produces two m.o.'s; six overlapping a.o.'s on six atoms produce six m.o.'s, and so forth.

We apply the same kind of rules to the filling of these m.o.'s that we applied to a.o. filling:

1. Each m.o. can accommodate a maximum of two electrons; the m.o. associated with a covalent bond usually contains two electrons.
2. When filling degenerate m.o.'s, each m.o. receives one electron before any of them gets two.
3. The sequence of filling non-degenerate m.o.'s is such as to fill the lowest energy orbital first, and so on.

Using the molecular orbital approach to covalent bond description, we will now explain why hydrogen is a diatomic gas while helium is not, *and* why oxygen is attracted into a magnetic field (that is, why oxygen is paramagnetic).

---

[a] While this is an oversimplification, it is sufficient for our needs.

2.4 THE CONSTITUTION OF SOME SMALL MOLECULES

After looking at the structures of hydrogen, helium, and oxygen, we shall examine hydrogen sulfide and hydrogen oxide (water). Unfortunately, questions raised at the end of this section over a molecule as simple as water will lead to some delay in completing our discussion of the structure of H_2O.

The Hydrogen Molecule. The hydrogen atom has one electron which exists in the 1*s* a.o. When two hydrogen atoms, with electrons of opposite spin (*i.e.*, *spin-paired*), approach one another, the a.o.'s overlap to form a molecular orbital which encompasses both protons (Figure 2–7). This m.o. does not "belong" to either atom but, rather, to the molecule.

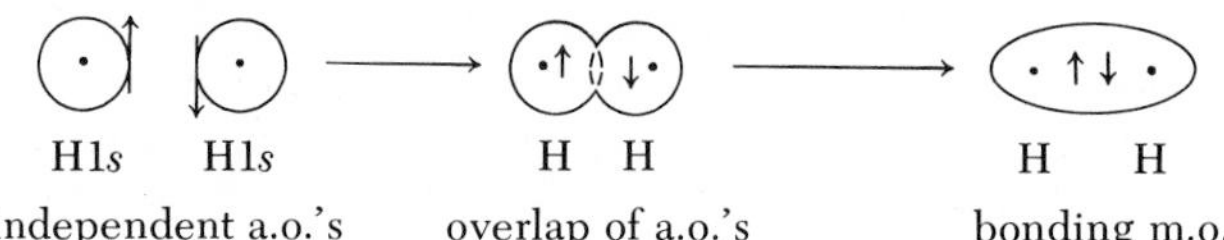

Figure 2–7 Formation of the H—H covalent bond. Arrows pointing up and down represent electrons of opposite spin.

The electron density contour map (Figure 2–8) of the bonding m.o. represented in Figure 2–7 clearly shows that the pair of electrons can be found between individual hydrogen nuclei—the m.o. completely enshrouds both protons. Although it is impossible to locate the two electrons within the m.o., the two protons occupy fairly well defined positions (a) within the m.o. *and* (b) relative to one another. This last statement demands that there be some rather constant distance between the two bonded nuclei. This distance (0.76 Å), called the *bond length* or *bond distance*, actually represents the distance corresponding to minimum potential energy between the nuclei. The last contour line does not indicate a zero probability of finding the electron, because the probability of finding an electron in an m.o., as in an a.o., extends to infinity. Rather, the last line is the 95% probability contour (*i.e.*, 95% of the time the probability of finding the electrons inside the last contour line is extremely high).

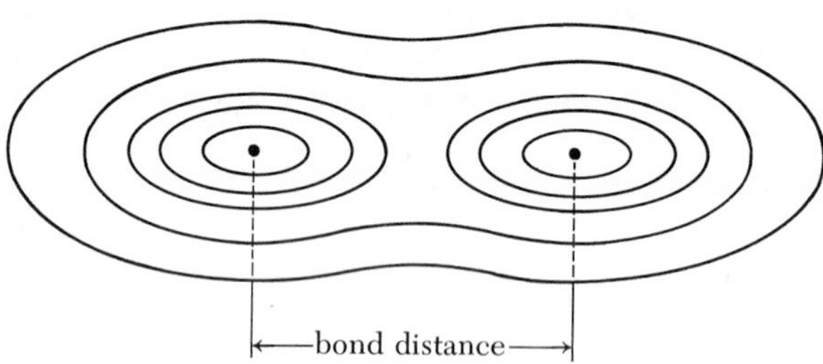

Figure 2–8 Electron density contour map of the bonding m.o. of H_2. The protons (*i.e.*, nuclei) reside at the points shown. What is significant is that electron density is permitted between the two nuclei.

Remembering that whenever we combine *n* atomic orbitals we must generate *n* molecular orbitals, our picture of the hydrogen molecule must be expanded to include a second m.o., one of higher energy than that one already described. The energies of this high energy m.o., of the low energy m.o., and of the atomic orbitals in atomic hydrogen are interrelated in Figure 2–9.

The low energy m.o. of hydrogen (H_2) accommodates both electrons used to form the bond and is the *bonding* molecular orbital. The high energy m.o., which possesses a discontinuity

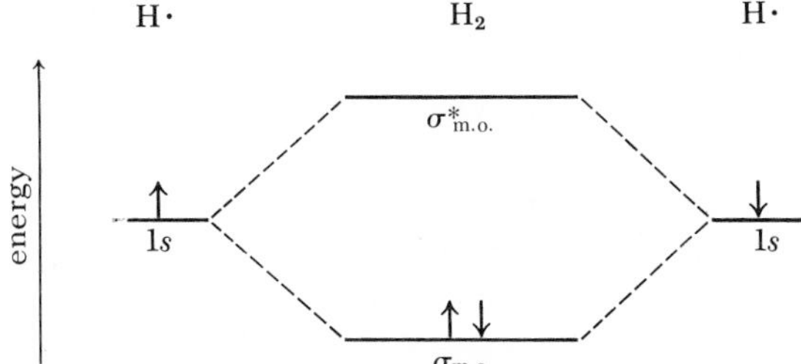

Figure 2-9 Relative energies of orbitals in H· and H_2.

of electron density probability between the hydrogen nuclei, is the *anti-bonding* molecular orbital (Figure 2-10). This orbital is vacant since the two electrons from the two hydrogen atoms are in the bonding molecular orbital. Both bonding and anti-bonding m.o.'s of hydrogen are designated as σ (Greek: sigma) orbitals since they possess cylindrical symmetry (axis of symmetry corresponds to the internuclear axis). In order to distinguish between them, the anti-bonding orbital is "starred", *i.e.*, σ^*. This terminology also is applied to the corresponding bonds; that is, the bond in hydrogen is a σ bond.

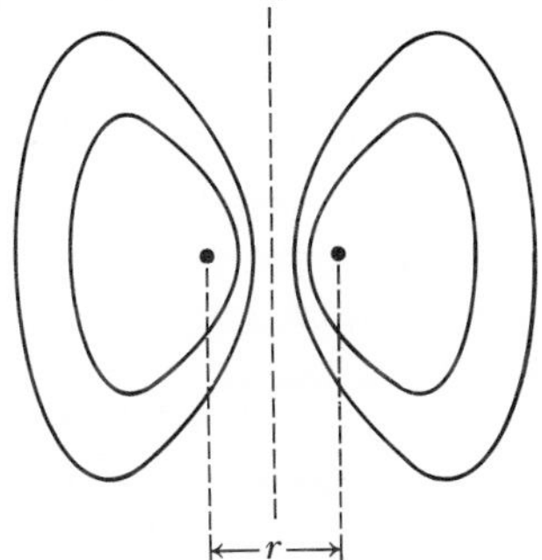

Figure 2-10 The anti-bonding molecular orbital ($1\sigma^*_{1s}$) for H_2. Here r = internuclear distance; a = end-on view of nodal plane. Compare this to Figure 2-8.

A straight line often is used to represent two electrons in a m.o. or in an a.o. Molecular hydrogen can, therefore, be represented by any of the following: H_2, H:H, H· ·H, and H—H. The organic chemist usually represents a covalent bond between two atoms by a straight line.

WHY ISN'T HELIUM A DIATOMIC GAS? We can answer this question by constructing the molecular orbital energy level diagram for He_2 (Figure 2-11). Each helium atom contributes two 1*s* electrons which must be accommodated in any attempt to construct He_2. Since the bonding molecular orbital can accept only two of these four electrons, the remaining two must go into the anti-bonding orbital. Any stabilization gained by filling the bonding molecular orbital is offset by the necessary filling of the anti-bonding molecular orbital. Therefore, there is no net decrease in energy when two helium atoms produce He_2 and so the latter is unstable.

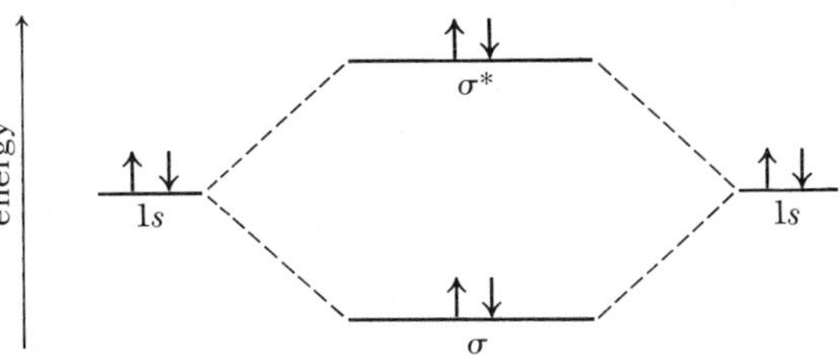

Figure 2-11 Molecular orbitals of the hypothetical He_2 molecule.

2. Would you expect $He_2^{\oplus}$ to be more stable than He_2? than He?

Orbital Overlap Involving *p* Orbitals. While the overlap involved in forming H_2 and the hypothetical He_2 employs only *s* orbitals, covalent bonds can be formed by the overlap of other orbitals. When *p* orbitals are involved, we may produce several different types of bonding and anti-bonding molecular orbitals, as shown in Figure 2–12.

In both of the σ bonding arrays (Figure 2–12), the two nuclei lie along the major axis of the *p* orbital(s) involved in the bond, ensuring a high degree of a.o. overlap. The third type of bond in Figure 2–12 is a π type bond and results from π type overlap of a.o.'s. A π bond is produced when two non-spherically symmetric orbitals overlap with their major axes parallel rather than collinear. One π bond consists of two sausage-shaped clouds of electron density, one above and one below a **nodal plane** (plane in which the probability of finding π electron density is always zero); both nuclei occur in this nodal plane. Because of diminished interaction, π bonds normally are weaker than σ bonds.

Why Is Oxygen Gas (O_2) Paramagnetic? Let us describe the various m.o.'s which result when two oxygen atoms are brought together from infinite separation. Neglecting the filled 1*s* orbitals, the two 2*s* orbitals interact to generate two molecular orbitals—one bonding and one anti-bonding. Assume that the oxygen atoms approach one another so that the major axis of one of the three 2*p* orbitals of each oxygen atom is coincidental with the internuclear axis between the atoms. These two 2*p* orbitals overlap to produce two σ m.o.'s. We shall arbitrarily designate these 2*p* a.o.'s as the $2p_x$ orbitals and the resulting m.o.'s as $\sigma 2p_x$ and $\sigma^* 2p_x$. The two $2p_y$ orbitals overlap with one another in a π fashion, producing a bonding ($\pi 2p_y$) and an anti-bonding ($\pi^* 2p_y$) orbital. The two $2p_z$ orbitals behave similarly. The energies of the various molecular orbitals produced by this overlap are presented in Figure 2–13 (p. 26).

In order to accommodate all of the eight electrons which were originally *p* electrons, two must be placed in a π^* level. Furthermore, since there are two degenerate π^* orbitals, each receives one electron. The resulting set of two unpaired electrons is the cause of oxygen's paramagnetism. In general, the demonstration of paramagnetism in a molecule is an indication of the presence of unpaired electrons in that molecule. The chemical significance of this will become apparent in Chapter 3.

One normally writes the formula for oxygen as O_2. If you were to draw the **Lewis structure** (structures in which the attainment of an outer octet of electrons is emphasized) for O_2, as shown below, it would not account for the paramagnetism of O_2. The explanation of the paramagnetism of oxygen is one of the triumphs of m.o. theory.

O_2 ⟨O=O⟩ :Ö=Ö:

3. The orbital energy diagrams below are those of nitric oxide and carbon monoxide. With the aid of these diagrams, explain why NO is paramagnetic while CO is not.

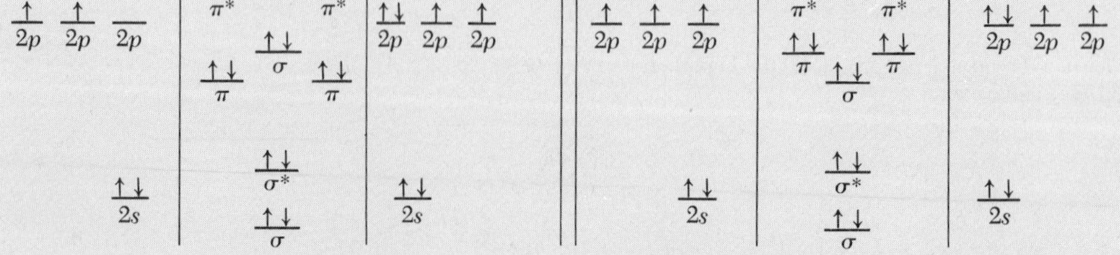

·Ċ·	CO	:Ȯ:	:Ṅ·	NO	:Ȯ:
	σ* (empty)			σ* (empty)	
	π* (empty), π* (empty)			π* ↑, π* (empty)	
2p ↑, 2p ↑, 2p (empty)	σ ↑↓	2p ↑↓, 2p ↑, 2p ↑	2p ↑, 2p ↑, 2p ↑	π ↑↓, π ↑↓	2p ↑↓, 2p ↑, 2p ↑
	π ↑↓, π ↑↓			σ ↑↓	
	σ* ↑↓			σ* ↑↓	
2s ↑↓		2s ↑↓	2s ↑↓		2s ↑↓
	σ ↑↓			σ ↑↓	

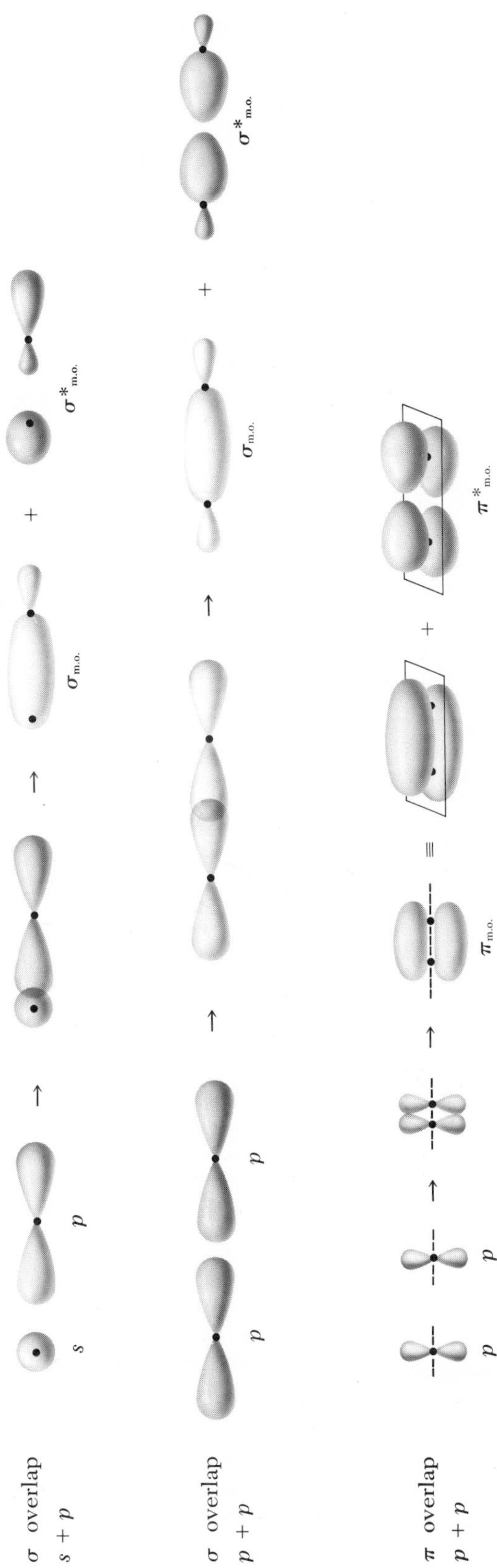

Figure 2-12 Bonding and anti-bonding orbitals formed with $2p$ orbitals. Remember that under most situations the bonding orbital is filled and the anti-bonding orbital is vacant.

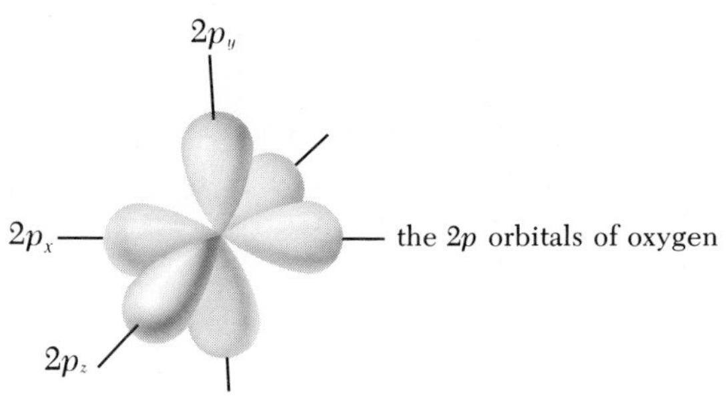

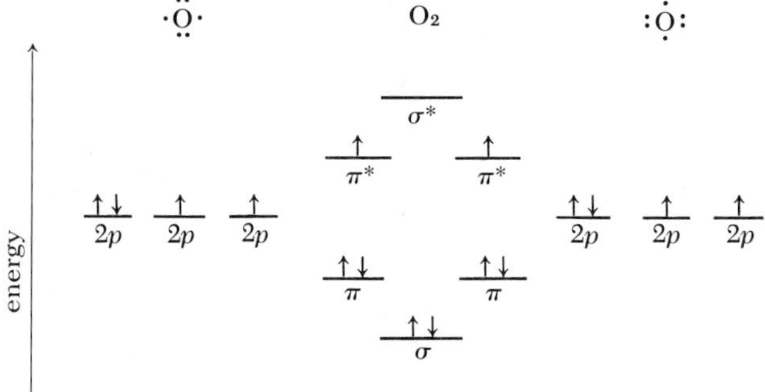

Figure 2-13 The molecular orbital energy diagram for O_2. Only valence electrons are shown. Examine the result of interaction of the two $2s$ filled orbitals. Is the suggested result inconsistent with the idea that "completely filled orbitals cannot overlap to form bonds"?

4. Draw Lewis structures, *i.e.*, structures which provide each atom (except hydrogen) with an outer octet of electrons, for
   (a) HCl (b) NH_3
   (c) Br_2 (d) H_2O
   (e) CH_3Cl (f) C_2Cl_4
   (g) C_2H_2

The Localized Bond. In discussing many π bonds and almost all σ bonds, we assume that a bond between two atoms in a polyatomic molecule can be treated as being made up of a.o.'s originating only from the two atoms under consideration. This type of bond is termed a *localized bond* because the electron pair which constitutes the bond is fixed between these two atoms. In this case the "molecular" orbital is really an orbital holding only two atoms together. Only later will we consider molecular orbitals which encompass a larger portion of a molecule, and these almost always will be π orbitals. Molecular orbitals which encompass more than two atoms are described as *delocalized,* since they permit an electron pair to associate with more than two atoms.

What Is the Structure of Hydrogen Sulfide? In making the transition from di- to tri-atomic molecules we will cease to construct energy level diagrams and, instead, will concentrate upon molecular geometry.

Hydrogen sulfide (H_2S), a poisonous gas, provides a good illustration of the necessity for discrete interatomic angles brought about as a consequence of the **principle of maximum overlap.** (This principle states that the strongest bond is achieved by maximum orbital overlap without generating severe interatomic repulsion.[a]) The electronic configuration of the ground state of sulfur is $1s^22s^22p^63s^23p^4$. If we add two electrons, each associated with a proton (*i.e.*, two hydrogen atoms), we complete the outer octet for sulfur, now $1s^22s^22p^63s^23p^6$. This covalent bonding of hydrogen to sulfur is depicted in Figure 2–14.

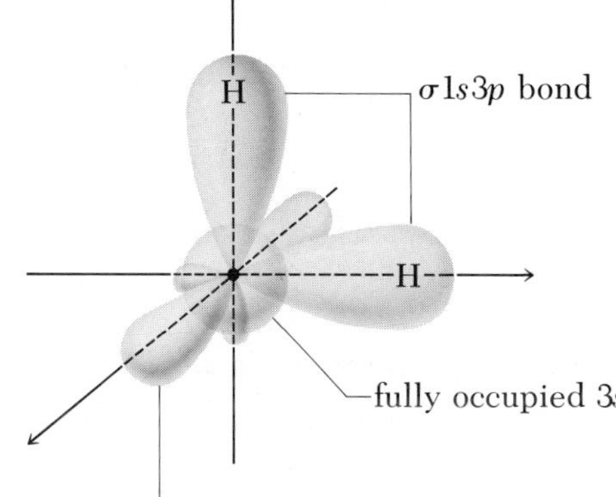

Figure 2–14 The bonding in hydrogen sulfide. Although we have depicted bonding with $3p_y$ and $3p_z$ orbitals, this distinction is unnecessary because all three p orbitals are equivalent.

The principle of maximum overlap demands that the protons reside on the major axis of each of two $3p$ orbitals. Since these orbitals have their major axes at right angles to one another, we conclude that the interatomic bond angle (H—S—H) should be 90°. This is quite close to the observed value, 92°.

Water. One would imagine, based upon the ground state electronic configuration for oxygen ($1s^22s^22p^4$), that water would have a bonding scheme (and geometry) identical to that of H_2S. However, the bond angle in water is 104.5°.

Because the angle between p orbitals is 90° and *if* (the weakest part of the argument) the hydrogens in water are bonded to p orbitals, the bond angle should be 90°. Since the bond angle in water is not 90°, to what kind of orbitals are those hydrogens bonded? The question is answered on p. 39; but before we can appreciate the answer, we must discuss covalent bonds in greater depth than before.

2.5 SOME PROPERTIES OF COVALENT BONDS

We shall limit this initial discussion to *bond distance, bond polarity,* and *bond strength.*

Bond Distance. It is tempting to define "bond distance" as "The distance between two covalently bonded nuclei," the implication being that there is some microscopic stick which keeps the atoms at their "proper" distance. Actually, the bond acts like a spring, allowing atoms to oscillate about some optimum distance (equilibrium bond distance) corresponding to an energy minimum for the system containing the two nuclei. An idealized plot of the energy

[a] There are more sophisticated, and perhaps more accurate, pictures of the factors responsible for the varying strengths of chemical bonds. Such representations are inappropriate at this level and will not be discussed here.

of a system composed of two nuclei capable of bonding together, as a function of nuclear separation, is presented in Figure 2–15.

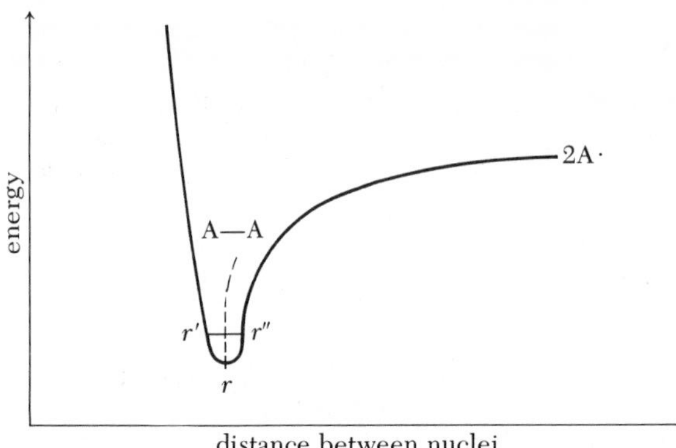

Figure 2-15 Energy relationship between two atoms as a function of internuclear distance. This type of plot is often termed a Morse curve. Point r represents an energy minimum, while the distance from r to the origin represents the equilibrium bond distance. The molecule A_2 is expected to exist in the vibrational ground state, v_0, unless it receives precisely enough energy ($v_1 - v_0$) to excite it to the next higher level, v_1. The dissociation energy is the amount of energy required to cleave the A—A bond when the molecule is in its vibrational ground state.

A small increase in the amount of energy in the molecule (thermal energy from the environment is sufficient) will cause the bond to exist in a vibrational excited state. Since the base of the curve is nearly parabolic, the bond length will still be given by r (Figure 2–15); however, deviation from a parabola at higher excitation levels does cause the equilibrium bond distance to increase.

Covalent bond distances are commonly determined by single crystal X-ray crystallography. Half of the covalent bond distance in a symmetric molecule, *e.g.*, Cl—Cl, is the **covalent radius** of that element. Once a covalent radius is measured, one can determine the bond distance between *that* atom and another and, by difference, determine the covalent radius of the second atom. With covalent compounds, as with ionic compounds, the volume requirement of an atom will change somewhat with a change in the atom to which it is bonded. The deviations from ideal behavior normally increase as the electronegativity difference between the bonded atoms increases (WHY?—see the following section). Therefore, if an accurate bond distance is required, it must be obtained by experiment and not by prediction.

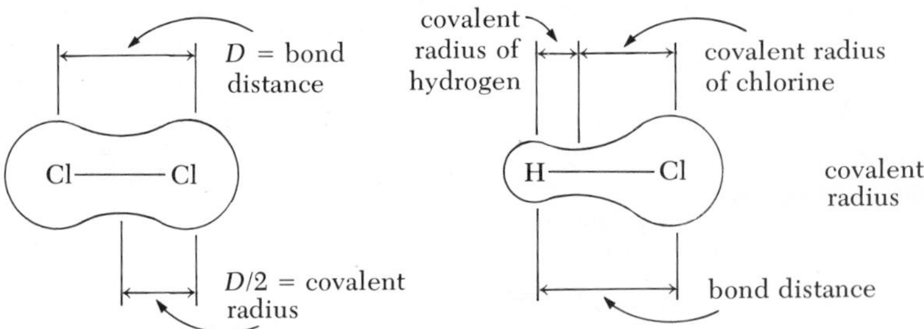

Before leaving the subject of bond distance, let us ask, "How closely can two atoms approach one another if they are not covalently bonded together?" The answer to this is, of course, a function of the size of the atoms involved. In turn, remember that most of the volume of an atom is derived from its electrons; thus, the original question might read, "How closely may the electrons in non-bonded atoms approach one another?" This question will be answered by beginning with two isolated helium atoms and moving them closer together, following the energy change with an **energy profile** (a plot of energy *vs.* a parameter to indicate the progress of the reaction), shown in Figure 2–16.

When the helium atoms are far apart they don't influence one another (**1** in Figure 2–16). As the helium atoms come closer, the electrons around one atom repel the electrons of the other, causing the second helium atom to develop some small, short-lived charge separation.

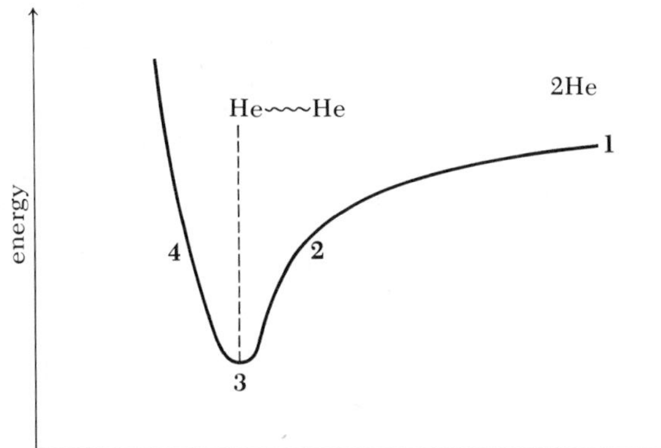

Figure 2-16 The van der Waals radius. The depth of the minimum is greatly exaggerated for legibility. The He_2 species is stable only below about 4° K.

The second atom has become *polarized.* The first atom, "feeling" this polarization, itself becomes polarized; the resultant dipoles attract one another. This transient attraction, due to *induced polarization,* is known as a London dispersion force and is responsible for the dip at **2** and the minimum at **3** in Figure 2–16. Finally, as the atoms come still closer (region **4**), the electrons begin to require the same space, and the energy of the system rises sharply. The distance corresponding to **3** is as close as these atoms can come without experiencing severe repulsion. One half of the distance of separation at this point approximates helium's **van der Waals radius,** our measure of the non-bonded size of an atom.

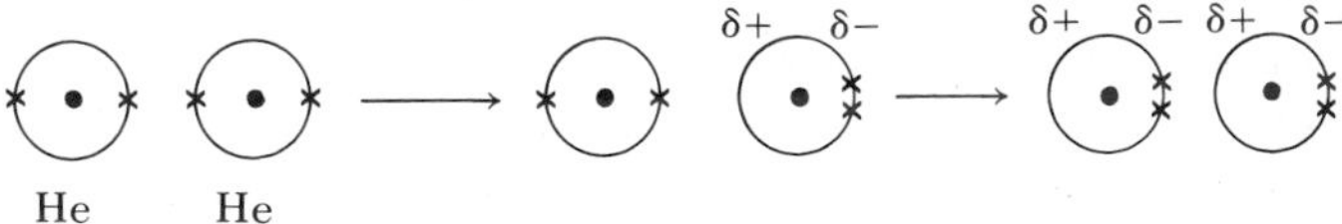

Bond Polarity. If two identical atoms are covalently bonded (*e.g.*, I_2), the electron density will be equally distributed between both atoms. However, if the atoms are of unequal electronegativity (see p. 12), the bonding pair will be closer to the more electronegative atom (giving rise to a polar covalent bond). This can be expressed in an m.o. diagram such as Figure 2–17.

Figure 2-17 The m.o.'s generated by overlap involving atoms of different electronegativities. B is more electronegative than A. While only the *s* orbitals are shown, it is generally true that as B becomes more electronegative (relative to A), the bonding molecular orbitals decrease in energy and resemble the atomic orbitals of B more closely than those of A.

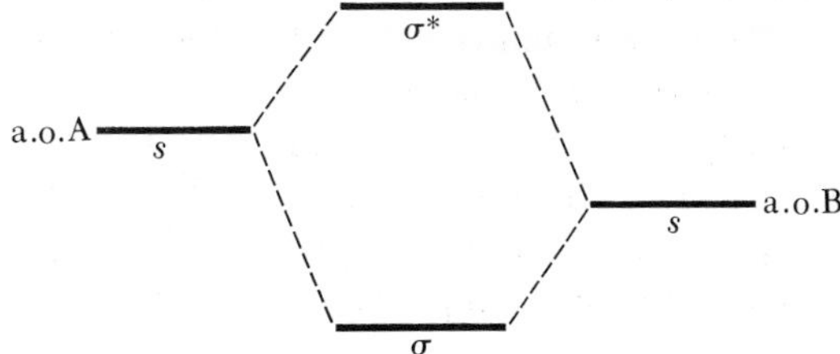

This polarity of the bond can be indicated by marking the more electron rich atom "δ−" and the more electron poor atom "δ+". Such charge separation can also be indicated by placing an arrowhead near the more electronegative atom and crossing the bond near the less electronegative atom. This is done below for iodine monochloride, I—Cl.

$$\text{ICl} \quad \text{or} \quad \overset{\delta+}{\text{I}}\text{—}\overset{\delta-}{\text{Cl}} \quad \text{or} \quad \text{I} \;{+\!\!\!\rightarrow}\; \text{Cl} \quad \text{or} \quad \overset{+\!\!\!\rightarrow}{\text{I—Cl}}$$

TABLE 2–2 Average Bond and Group Dipole Moments

Structural Unit[a]	Dipole Moment (D)
H—N	1.3
H—O	1.5
H—S	0.7
C—Cl	1.9
C—Br	1.8
C—N	1.0
C—O	1.2
C=O	2.7
C—OCH_3	1.3
C—NH_2	1.3
C—OH	1.5
C—CO_2H	1.7
C—CHO	2.7
C—NO_2	4.0
C—CN	4.2

[a] The more electronegative portion is to the right of the bond.

The bond in iodine monochloride is polar, the extent of charge separation being measured by the *bond dipole moment*. This equals the amount of charge which is separated multiplied by the distance of separation.[a] For a diatomic molecule like iodine monochloride, the bond moment must also be the *molecular dipole moment*. When a molecule contains more than one polar bond, the molecular dipole moment is given by the vector sum of the individual bond moments. Clearly, if a molecule has a point of symmetry (if you are unclear about what this is, see p. 87), it must have a net dipole moment (μ) of zero since all of the bond moments will cancel one another. Thus, carbon dioxide, which is a linear molecule, has a zero dipole moment but sulfur dioxide, a non-linear molecule, has a dipole moment of 1.6 D. A number of representative bond moments are collected in Table 2–2 along with the *group moments* of some functional groups.

In Chapter 5 we will learn that very many organic reactions are initiated by the attack of a pair of electrons of one atom upon another atom which is relatively electron poor. Since the ease with which such reactions occur is at least partially controlled by the magnitude of this electron paucity, and since this is mimicked in the polarity of bonds, it will help you in many future discussions if you learn to identify polar bonds early in this course.

When a proton is bonded to a highly electronegative element (F, O, N, or S) it becomes quite positive in character and can act as a hydrogen bond donor. All that is required to form a hydrogen bond, then, is a rather electronegative element to serve as a hydrogen acceptor. A molecule such as hydrogen fluoride, H—F, can act as both a proton donor and a proton acceptor; therefore, it hydrogen bonds strongly to itself. The unusually high boiling point of hydrogen fluoride compared to those of other hydrogen halides (hydrogen fluoride is the only one boiling near room temperature) confirms the existence of this strong intermolecular association. The hydrogen bond is so strong that the unusual anion $HF_2^{\ominus}$ exists in pure hydrogen fluoride. As already noted (see Figure 2–6), hydrogen bonding also explains some of the physical properties of water and part of the structure of the gene.

$$\text{H—F} \sim\!\text{H—F} \sim\!\text{H—F} \sim\!\text{H—F}$$

$$2\text{HF} \rightleftarrows \text{H}^{\oplus} + \text{HF}_2^{\ominus}$$

[a] The electronic charge is about 10^{-10} esu, and the distances are of the order of 10^{-8} cm. The product is of the order of 10^{-18} esu cm. For convenience, 10^{-18} esu cm is defined as 1 Debye unit (1 D) and dipole moments are given in Debyes (D).

The ability of compounds like water, sulfur dioxide, and ammonia to solvate ionic and polar materials is due, in large part, to the polarity of these solvents and to their hydrogen bonding potential (Figure 2–5).

5. Neither of the following have molecular dipole moments, although they contain polar bonds: BF_3, SF_6. Account for this.
6. Why does hydrogen fluoride undergo intermolecular self-association while the other hydrogen halides do not?

Bond Energy. The process whereby a two-electron bond is broken and each fragment receives one electron is called "homolysis" or "homolytic cleavage." The energy required to break a specific bond homolytically and form two neutral atoms is the **bond dissociation energy** of that bond. In organic chemistry this energy commonly is expressed in kcal/mole of bond. It often is impossible to determine the energy required to break a specific bond and, instead, one must settle for an average value, the **bond energy.** The distinction between these is nicely made by examining the bond cleavage steps for methane, CH_4:

a. $H{-}CH_3 + 102\text{ kcal} \rightarrow \cdot CH_3 + H\cdot$

b. $H{-}\dot{C}H_2 + 105\text{ kcal} \rightarrow \cdot\dot{C}H_2 + H\cdot$

c. $H{-}\underset{\cdot}{\dot{C}}H + 108\text{ kcal} \rightarrow \cdot\underset{\cdot}{\dot{C}}H + H\cdot$

d. $H{-}\underset{\cdot}{\dot{C}}\cdot + 83\text{ kcal} \rightarrow \cdot\underset{\cdot}{\dot{C}}\cdot + H\cdot$

Steps *a* through *d* contain energy terms which are bond dissociation energies. The average value $((a + b + c + d)/4)$, the bond energy of the C—H bond in methane, is not equal to any of the bond dissociation energies. Bond dissociation energies and bond energies are presented in Table 2–3.

TABLE 2–3 Bond Energies and Bond Dissociation Energies for Single Bonds (kcal/mole)[a]

	Bond Energies	
H—C (99)	C—C (80)	C—O (81)
H—N (84)	C—F (102)	C—S (65)
H—O (110)	C—Cl (77)	S—S (49)
H—S (81)	C—Br (64)	N—N (35)
C—N (62)	C—I (56)	O—O (35)

Bond Dissociation Energies	
H—H (104)	F—F (37)
H—F (135)	Cl—Cl (58)
H—Cl (103)	Br—Br (46)
H—Br (87)	I—I (36)
H—I (71)	
CH_3—H (102)	CH_3—CH_3 (88)
CH_3CH_2—H (98)	CH_3CH_2—CH_3 (85)
$(CH_3)_2CH$—H (95)	$(CH_3)_2CH$—CH_3 (84)
$(CH_3)_3C$—H (91)	$(CH_3)_3C$—CH_3 (80)
=CH—H (104)	=CH—CH_3 (92)
=$CHCH_2$—H (88)	=$CHCH_2$—CH_3 (72)
C_6H_5—H[b] (112)	C_6H_5—CH_3[b] (93)
$C_6H_5CH_2$—H[b] (85)	$C_6H_5CH_2$—CH_3[b] (70)

[a] Where more than one bond is shown, the bond of interest is in boldface (**—**).

[b] C_6H_5 represents the phenyl group, H—C_6H_4· (a ring of six carbons, C=C, C—C alternating, with H on each of four ring carbons shown).

A Graphic Representation of Bond Dissociation Energy. The energy change accompanying the formation of the symmetric A_2 molecule from two atoms of A is depicted in Figure 2–18. Note that the energy released in the process is the negative of the energy required to cleave the bond. Since very few reactions are encountered in which the absolute energy values are known for all pertinent species, such energy profiles are commonly plotted with an arbitrary ordinate scale. Compare Figures 2–15 and 2–18.

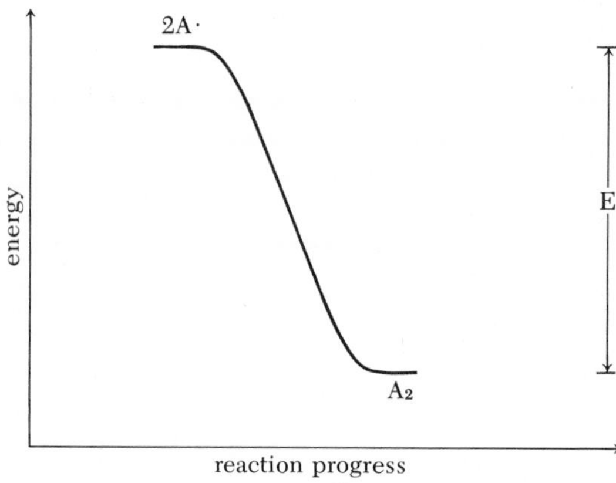

Figure 2–18 Energy profile for the formation of A_2. Here, E is the energy released when two atoms of A combine to form A_2.

2.6 HYBRIDIZATION

Let us return to a study of the structures of some small molecules.

What Is the Simplest Compound Containing Only Carbon and Hydrogen? The ground state configuration of carbon is $1s^2 2s^2 2p^2$. One might predict, then, that the addition of two electrons (associated with two protons as hydrogen atoms) to the partially occupied $2p$ orbitals should produce the compound CH_2 as the $1s$ orbital of each hydrogen overlaps with a $2p$ orbital of carbon.

1s	2s	2p	2p	2p	
↓↑	↓↑	↑	↑	—	the electronic ground state of an isolated carbon atom
↑↓	↑↓	H ↑↓	H ↑↓	—	suggested structure for the simplest stable compound containing only carbon and hydrogen

After what appears to be a rather logical answer, you might be surprised to find that:

a. CH_2 was well characterized less than a decade ago,
b. there are two compounds with the formula CH_2,
c. both CH_2's are *extremely* reactive, and,
d. neither CH_2 has a bond angle of 90°.

Given the facts, we can see that our prediction may not have been so sound. After all, our CH_2 is an *electron deficient species,* having only six outer electrons rather than the eight normally associated with a stable electronic array (the so-called "noble gas configuration"). We clearly require some new concept to explain the molecular formula for the simplest *stable* hydrocarbon (compound composed solely of C and H), CH_4. This concept also must explain why the four C—H bonds in CH_4 are equivalent.

Methane (CH_4) and Other Hydrocarbons. In order to generate a carbon atom which can accommodate four groups around it, we must deal with an electronic excited state of carbon. This excited state involves the construction of four new outer orbitals by "hybridizing" the $2s$ orbital and all three $2p$ orbitals. (Quantum mechanics dictates that we must create the same number of new orbitals as components that went into them.) These four hybrid orbitals are of equal energy, each being called $2sp^3$ (2 indicating the main quantum number and sp^3 indicating that the orbital is a composite of characteristics which are one quarter s and three quarters p.)

$$\underset{1s}{\underline{\uparrow\downarrow}}\quad \underset{2s}{\underline{\uparrow\downarrow}}\quad \underset{2p}{\underline{\uparrow}}\quad \underset{2p}{\underline{\uparrow}}\quad \underset{2p}{\underline{}}\qquad\qquad \underset{1s}{\underline{\uparrow\downarrow}}\quad \underset{2sp^3}{\underline{\uparrow}}\quad \underset{2sp^3}{\underline{\uparrow}}\quad \underset{2sp^3}{\underline{\uparrow}}\quad \underset{2sp^3}{\underline{\uparrow}}$$

ground state carbon sp^3 excited state

These four hybrid orbitals have their major axes directed along lines which, if connected at a constant distance from the nucleus, would form a tetrahedron. (The angle between any two of these intersecting lines is 109.5°.) For this reason, sp^3 hybridized carbon is also called "tetrahedral carbon."

What advantage is there to hybridization? *Firstly*, hybrid orbitals exhibit better overlap characteristics by concentrating the wave function in the direction of the bond. Compared to an unhybridized s or p orbital, the sp^3 orbital possesses one large lobe (the "front" lobe) which is well suited for bonding, and a small lobe (the "back" lobe) which is generally not even drawn, although it does play a role in some reactions. One sp^3 orbital is shown in Figure 2–19.

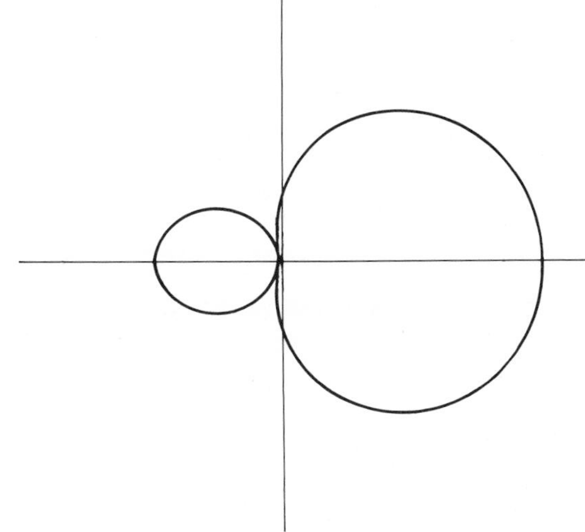

Figure 2–19 The $2sp^3$ orbital. The large lobe (on the right in the diagram) is termed the "front" lobe; this is the lobe that is important in bond formation. The small lobe (on the left) is termed the "back" lobe, and its role in controlling some reactions will become apparent during the reading of Chapter 5. (Remember that the p orbital has two lobes of equal size; see Figure 1–4.) The front and back lobes of sp^3 orbitals are often represented as two ellipses of unequal size, even though the lobes are not exactly elliptical.

Secondly, repulsion between valence electrons is known to follow the order: lone-pair–lone-pair > lone-pair–bonding-pair > bonding-pair–bonding-pair. Tetrahedral hybridization provides the maximum separation (and minimum repulsion) between four identical bonding electron pairs. Indeed, the geometries of all the hybrid states to be discussed provide the minimum possible amount of electron-pair–electron-pair repulsion. The establishment of a geometry which minimizes the repulsion between the various pairs of outer electrons is at the heart of what is called the *valence shell electron pair repulsion* (VSEPR) *theory*. (The student is referred to the glossary for further elaboration.)

Having four sp^3 orbitals of carbon, each containing one electron (one from each of the original orbitals), we now overlap these with the hydrogen 1*s* orbitals (each containing one electron) and completely fill the bonding m.o.'s that we generated.

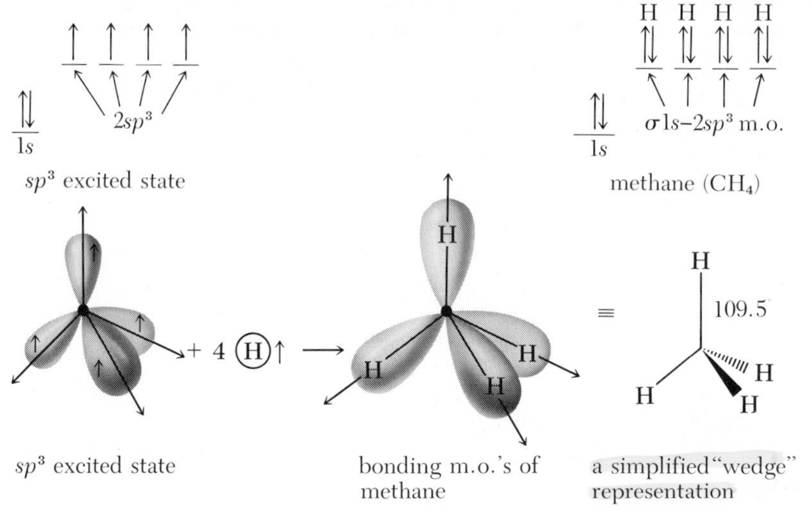

1s

$2sp^3$

sp^3 excited state

H H H H

1s

$\sigma 1s$–$2sp^3$ m.o.

methane (CH_4)

+ 4 (H)↑ →

H

H

H

H

≡

H

109.5°

H

H

H

sp^3 excited state

bonding m.o.'s of methane

a simplified "wedge" representation

Let us now build up a picture of the next higher homolog of methane, ethane (C_2H_6). In our hypothetical construction of methane, assume that we had stopped adding hydrogen atoms after only three had been bonded to the sp^3 hybrid. We could then combine two of the resulting units (methyl radical, $CH_3\cdot$), by overlapping their partially filled $2sp^3$ orbitals, to form a new σ bond between the two carbon atoms (Figure 2–20). Both the H—C—C and the H—C—H bond angles in ethane are approximately 109.5°.

≡ H_3C——CH_3

two overlapping methyl fragments

bonding m.o. of the CC fragment of ethane ($\sigma sp^3, sp^3$)

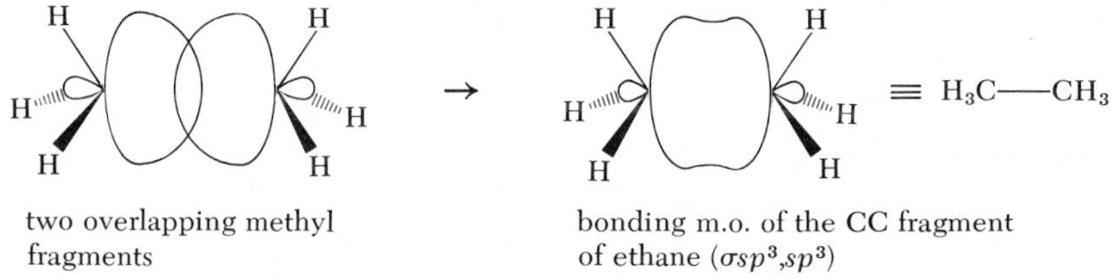

Figure 2–20 Formation of ethane from two $CH_3\cdot$ fragments.

Propane, C_3H_8, can be constructed in a related fashion, using two $CH_3\cdot$ fragments and one $\cdot CH_2\cdot$ fragment (a **methylene group**). This latter unit is made up of an sp^3 carbon and two hydrogen atoms. All interatomic angles in propane are *approximately* 109.5°.

or H_3C—CH_2—CH_3

~109.5° ~109.5°

a: $\sigma 1s$-$2sp^3$
b: $\sigma 2sp^3$-$2sp^3$

propane

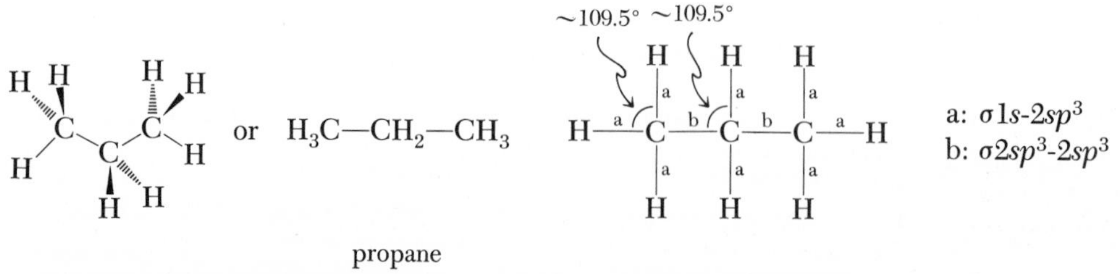

Methane, ethane, and propane, hydrocarbons containing only sp^3 hybridized carbons, are **alkanes** and have the general formula C_nH_{2n+2}. They represent a portion of a *homologous series,* since adjacent members differ from one another by one methylene group. Two or more members of such a series are "homologs" of one another, a term which was introduced to compare ethane to methane and could as correctly apply to ethane and propane.

Is Another Type of Hybridized Carbon Atom Possible? Ethylene, used in the artificial ripening of fruit and in the preparation of polyethylene, has the formula C_2H_4. It is the simplest member of the **alkene** family, compounds with the general formula C_nH_{2n}. Both of its carbons are described as sp^2 hybridized. How do we picture such orbital hybridization?

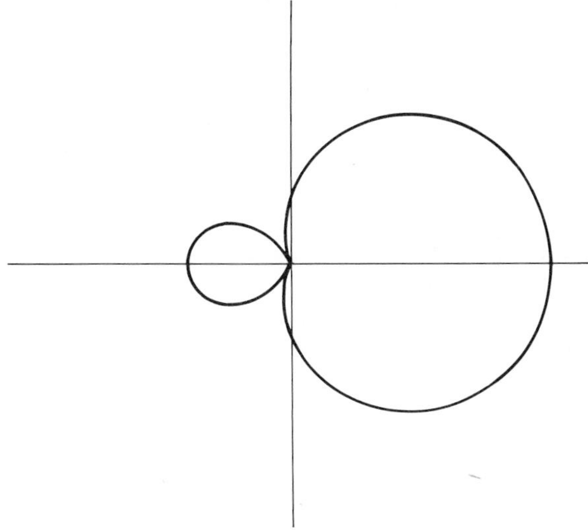

Figure 2-21 The sp^2 orbital. As with the sp^3 orbital, the large lobe is termed the front lobe and the small one is termed the back lobe. The most obvious difference between the sp^3 orbital (Figure 2-19) and the sp^2 orbital is the smaller back lobe in the latter. As with the sp^3 orbital, the front and back lobes are often shown as ellipses of unequal size.

Hybridization of the 2*s* orbital and two of the three 2*p* orbitals of carbon generates three sp^2 orbitals and leaves one unhybridized 2*p* orbital. The sp^2 orbitals are similar in shape to the sp^3 orbitals (see Figure 2-21) but are arranged differently in space. The major axes of the three sp^2 orbitals reside in one plane, separated by 120°. This provides maximum separation of the bonding electron pairs. The major axis of the remaining 2*p* orbital is perpendicular to that plane. This hybridization is called "trigonal" or "planar."

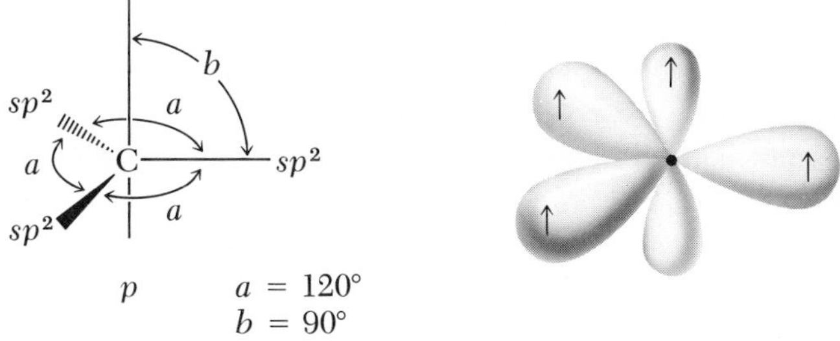

We can begin to construct C_2H_4 by adding one hydrogen atom to each of two sp^2 orbitals, leaving a partially filled sp^2 orbital and a partially filled *p* orbital. If we take two such units and bring them together, we construct two carbon-carbon bonds: one σ (formed by overlap of two $2sp^2$ orbitals) and one π (formed by overlap of two *p* orbitals—review p. 24 and Figure 2-12 if necessary). This is depicted below:

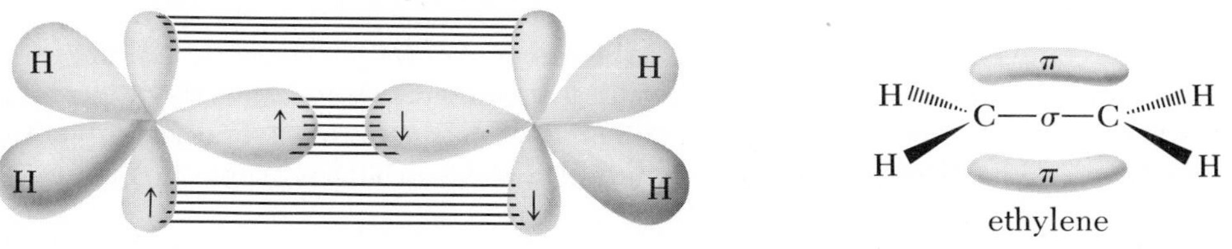

The two carbon atoms are bound by a σ bond and a π bond, a combination called a *double bond.* Although the double bond consists of non-identical bonds, it is usually written and printed in a manner which does not attempt to make this clear. All six atoms of ethylene reside in one plane, a result of the need for the p orbital major axes to remain parallel in order to maximize their overlap (to see this, build a model).

MUST AN sp^2 HYBRIDIZED CARBON ATOM ALWAYS BE DOUBLY BONDED TO ANOTHER sp^2 HYBRIDIZED CARBON ATOM? The essential characteristics of a double bond require a p orbital and an sp^2 orbital, both partially filled and originating from a common atom. This suggests that carbon could form a double bond to any element which could exist with partially filled p and sp^2 orbitals. Oxygen is a particularly important example, making the necessary orbitals available through the following hybridization:

↑↓	↑↓	↑↓	↑	↑		↑↓	↑↓	↑↓	↑	↑
$1s$	$2s$	$2p$	$2p$	$2p$		$1s$	$2sp^2$	$2sp^2$	$2sp^2$	$2p$

ground state — oxygen — sp^2 excited state

This state of oxygen is found in the many compounds known to contain a carbon-oxygen double bond. The simplest example is formaldehyde, CH_2O, the oxygen analog of ethylene and a common preservative for biological specimens (*e.g.*, frogs). (*Analogous* compounds possess the same basic bonding pattern but have one atom or group of atoms replaced by a different atom or group of atoms. For example, CO_2 and CS_2 are analogs.)

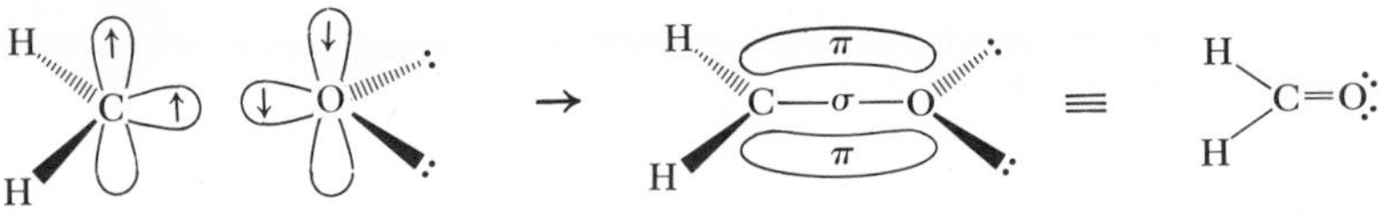

formaldehyde

While double bonds can be formed between elements in the second and third rows of the periodic table (*e.g.*, C=S), double bonds between two elements in the same row are generally stronger than double bonds involving elements in different rows. The double bonds of importance in organic chemistry are: C=C, C=O, C=N, C=S, and N=N.

HOW DOES AN sp^2 CARBON ATOM BOND TO AN sp^3 ATOM? Propylene (C_3H_6), used in the preparation of polypropylene, is the next higher homolog of ethylene. Its structure, shown below, reveals how it is possible to bond sp^2 and sp^3 carbons together. While the p orbital and one sp^2 orbital of each of two sp^2 hybridized carbons are committed to double bond formation, we can construct a σ bond (single bond) between Csp^2 and Csp^3 by overlapping an sp^2 and an sp^3 orbital.

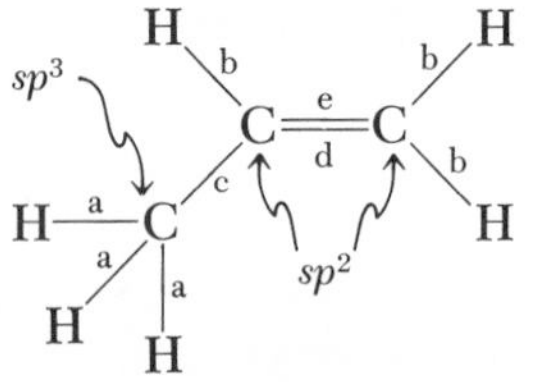

a: $\sigma 1s2sp^3$
b: $\sigma 1s2sp^2$
c: $\sigma 2sp^3 2sp^2$
d: $\sigma 2sp^2 2sp^2$
e: $\pi 2p2p$

propylene

sp HYBRIDIZATION. Acetylene, used in the oxyacetylene torch for high temperature welding, has the formula C_2H_2. It represents the simplest member of the **alkyne** family of hydrocarbons, all of the formula C_nH_{2n-2}. Both carbons of acetylene are sp hybridized.

To construct a picture of acetylene, we must first convert carbon to its sp hybridized state by hybridizing the $2s$ orbital with one $2p$ orbital. This produces two hybrid $2sp$ orbitals and leaves two $2p$ orbitals unhybridized. A picture of the $2sp$ orbital is shown in Figure 2–22.

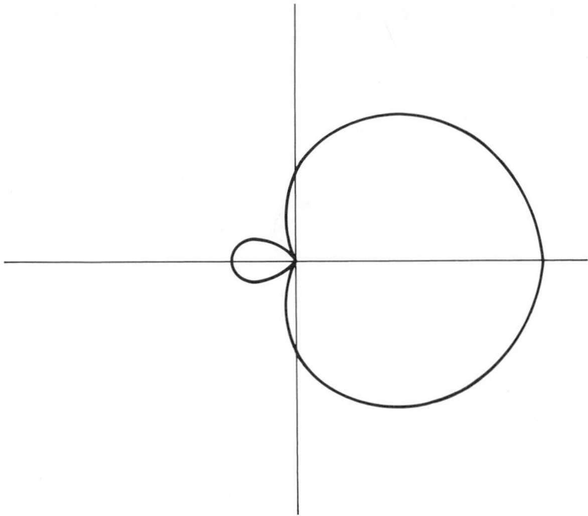

Figure 2-22 The $2sp$ orbital. As with the sp^3 and sp^2 orbitals, the large and small lobes are termed the front and back lobes, respectively. Note how small the back lobe of the sp orbital is, compared to those of the sp^3 and sp^2 orbitals (Figures 2–19 and 2–21). Once again, these lobes are often represented as ellipses of unequal size.

	$\uparrow\downarrow$	$\uparrow\downarrow$	$\uparrow$	$\uparrow$			$\uparrow\downarrow$	$\uparrow$	$\uparrow$	$\uparrow$	$\uparrow$
	$1s$	$2s$	$2p$	$2p$	$2p$		$1s$	$2sp$	$2sp$	$2p$	$2p$
		ground state				carbon		sp excited state			

The two sp orbitals are collinear (Why?), while the p orbitals are at right angles to one another *and* to the major axes of the sp orbitals.

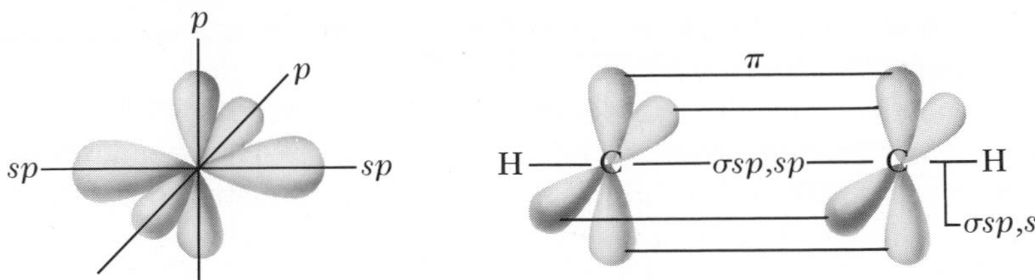

The hypothetical construction of acetylene is begun by adding a hydrogen atom to one of the sp orbitals. Two of the fragments so produced are then put together to yield one σ bond and two π bonds between the two carbons. The planes containing the two π bonds are mutually perpendicular and their intersection defines the direction of the σ bonding m.o. The digonal (linear) sp hybridized carbon imposes a linear geometry upon acetylene. As with the double bond, normal representations of the *triple* bond (one σ and two π) do not show two discrete bond types.

Can Carbon Form Triple Bonds to Other Elements? Nitrogen is an element which can form triple bonds to itself and to carbon. The ground state of nitrogen ($1s^22s^22p^3$) can be hybridized to the sp state.

	$\uparrow\downarrow$	$\uparrow\downarrow$	$\uparrow$	$\uparrow$	$\uparrow$		$\uparrow\downarrow$	$\uparrow\downarrow$	$\uparrow$	$\uparrow$	$\uparrow$
	$1s$	$2s$	$2p$	$2p$	$2p$		$1s$	$2sp$	$2sp$	$2p$	$2p$
		ground state				nitrogen		sp excited state			

The overlap necessary to produce the simplest compound containing the C≡N functional group, hydrogen cyanide, is shown below. Hydrogen cyanide is most famous because of its use as a poisonous gas.

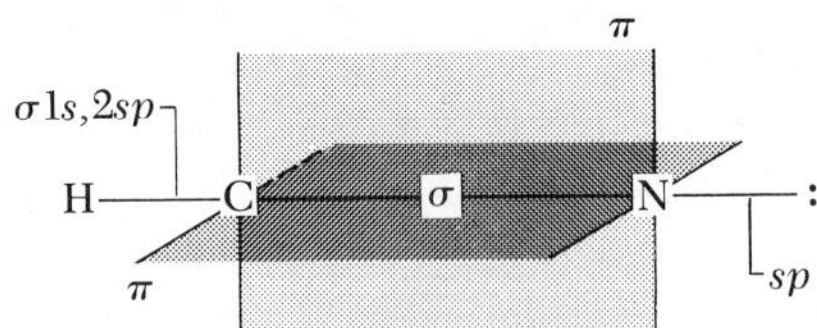

7. Complete the table below for the various hybrid states of carbon.

	sp^3	sp^2	sp
designation	tetrahedral	*a*	*b*
# of bonds to carbon	*c*	*d*	4
# of σ bonds	*e*	3	2
# of π bonds	0	*f*	*g*
# of substituents	*h*	3	*i*
bond angle	*j*	120°	*k*

2.7 BOND PROPERTIES—A SECOND LOOK

Hybridization requires that we expand our initial view of bond properties so as to distinguish, when necessary, between bonds of different hybridization states of the same atom.

Bond Distance. In general, as the multiplicity of the bond between two atoms increases, the bond length decreases. However, the decrease in bond length is not constant as we go from single to double to triple bond. The shortening of the bond in going from single to double is greater than the shortening in going from double to triple. Multiple bond distances involving C, N and O are found in Table 2–4. Single bond distances between different hybridized states of carbon and a given element (*e.g.*, C_{sp^3}—H, C_{sp^2}—H, and C_{sp}—H) are almost equal, although increasing *s* character in the hybrid orbital used in σ bond formation does lead to a *slightly* shorter bond (*i.e.*, the C—H bond in acetylene is slightly shorter than the C—H bond in methane).

TABLE 2-4 Average Multiple Bond Distances (Å)[a]

	C	N	O	Bond Type
	1.54	1.47	1.43	σ
C	1.34	1.28	1.23	σ + π
	1.21	1.16	—	σ + π + π
	1.47	1.46	1.41	σ
N	1.28	1.25	1.14	σ + π
	1.16	1.10	—	σ + π + π
	1.43	1.41	1.49	σ
O	1.23	1.14	1.21	σ + π
	—	—	—	σ + π + π

[a] Bond distances for isolated single, double, and triple bonds (read top-to-bottom). For example, the central entry in the central set of distances, 1.25 Å, is the nitrogen-nitrogen double bond (—N=N—) distance.

Bond Polarity. Not all bonds between identical elements have identical bond dipole moments. (Table 2–2 was generated before we considered hybridization.) Within a given main quantum level, *s* orbitals are more electronegative than *p* orbitals, a consequence of the relative extents of penetration of these orbitals (see p. 8). Therefore, the greater the *s* character of a hybrid orbital, the more electronegative will the orbital be. For carbon the order of orbital electronegativity is: $s > sp > sp^2 > sp^3 > p$. This means, for example, that the C—H bond in acetylene is more polar than the C—H bond in methane.

This last statement cannot be tested by determining the molecular dipole moments of acetylene and methane, since both must be zero (Why?). However, let us make the suggestion that the more polar a covalent bond is, the more readily it will undergo **heterolysis** or **heterolytic cleavage.** (In heterolysis, one atom involved in a bond receives both bonding electrons when the bond is broken.) The negative end of a polar bond broken in this way should become the anionic fragment. We then predict that it should be easier to remove a proton from acetylene than from methane—and this is observed! In the equation below, amide ion, $NH_2^{\ominus}$, is functioning as the proton acceptor (or base) while acetylene is depicted as a proton donor (or acid).

$$\underset{\text{acetylene}}{H{-}C{\equiv}C{-}H} + \underset{\text{amide ion}}{H_2\ddot{N}{:}^{\ominus}} \rightarrow \underset{\text{acetylide ion}}{H{-}C{\equiv}C{:}^{\ominus}} + \underset{\text{ammonia}}{{:}NH_3}$$

$$\underset{\text{methane}}{CH_4} + H_2\ddot{N}{:}^{\ominus} \rightarrow \text{no reaction}$$

Bond Strength. The strength of the bond between two atoms increases as the multiplicity of the bond between them increases. After all, we are speaking of the energy associated with *all* of the individual bonds when we talk of the "bond strength of a multiple bond." Some representative, average values are given in Table 2–5. One particularly important value is that of the nitrogen-nitrogen triple bond. This is the largest value in the table and reflects the great stability of N_2.

TABLE 2–5 Average Multiple Bond Energies (kcal/mole)[a]

	C	N	O	Bond Type
	80	70	85	σ
C	150	150	175	$\sigma + \pi$
	200	215	—	$\sigma + \pi + \pi$
	70	35	55	σ
N	150	100	145	$\sigma + \pi$
	215	230	—	$\sigma + \pi + \pi$
	85	55	35	σ
O	175	145	119	$\sigma + \pi$
	—	—	—	$\sigma + \pi + \pi$

[a] Bond energies for isolated single, double, and triple bonds (read top-to-bottom). For example, the central entry in the central set of energies, 100 kcal/mole, is the energy required to cleave *both* the σ and π bonds of the nitrogen-nitrogen double bond (—N=N—).

Bond Angle, or What Ever Happened to the Structure of Water? All of the properties discussed thus far are properties of a bond. The bond angle, however, actually is a property of two bonds sharing a common atom and should be viewed as a property of the common atom. Indeed, the bond angle at a given atom is a good indication of the state of hybridization of that atom.

While internuclear bond angles, Y—C—X, in organic compounds are basically 109.5°, 120°, or 180° (reflecting sp^3, sp^2, and sp hybridization, respectively), these angles are observed only in rare cases. Repulsions between pairs of valence electrons are sufficiently great to alter the geometries from the ideal pictures (see p. 34). Consider the structure of water . . . (do you remember WATER, p. 27?)

The oxygen in water has four "groups" about it (two hydrogens and two lone pairs of electrons). Since four groups are most effectively removed from one another in the tetrahedral geometry, one can suggest that the oxygen in water should be in the center of a deformed tetrahedron (*i.e.*, sp^3 hybridized). Distortion from an ideal tetrahedral geometry is expected because the oxygen's non-bonding electron pairs (sp^3) will repel one another more than will the bonding pairs (the O—H bonds) (review p. 34 if necessary). This will cause an increase in the angle between the orbitals bearing these non-bonded pairs and a concomitant decrease in the H—O—H bond angle. Thus, the question raised on p. 27 about the structure of water *vs.* the structure of hydrogen sulfide may now be answered by stating that the oxygen in water is sp^3 hybridized with some distortion present in order to minimize repulsion between valence shell electron pairs.

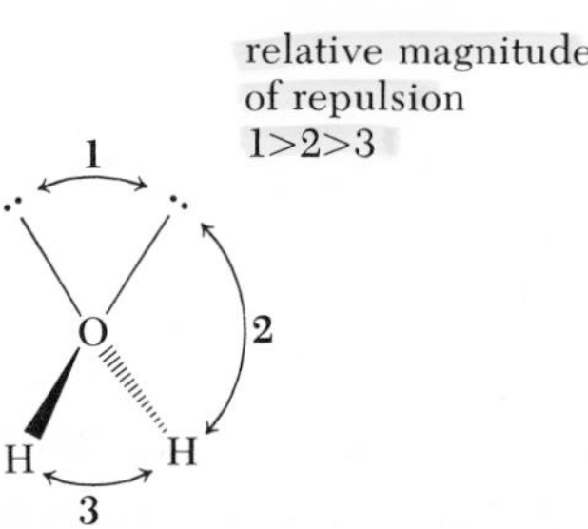

But why isn't the sulfur in hydrogen sulfide tetrahedral? The valence (outer) electrons of sulfur occupy the third main quantum level and are farther away from one another than are the outer electrons of oxygen. Presumably, they are sufficiently far apart that in sulfur, unlike oxygen, there is little repulsion between outer electron pairs.

8. Construct a picture of the structure of hydrogen sulfide, assuming that the sulfur is sp hybridized. In what orbitals are the non-bonding electron pairs on sulfur located?

Ammonia is also predicted to be approximately tetrahedral. However, the greatest repulsions are between the lone pair on nitrogen and the bonding pairs (N—H bonds). Since these bonds do not interact with a lone pair as strongly as does another lone pair, one predicts a geometry for ammonia that is less distorted than is that of water. This prediction is borne out by the observed bond angles: water, 104.5°; H—N—H in ammonia, 107.3°; ideal tetrahedron, 109.45°.

Methyl cation ($CH_3^{\oplus}$) is a highly reactive electron-deficient species. The central sp^2 hybridized carbon is attached to three identical atoms. A planar arrangement with a bond angle of 120° affords maximum separation of these groups. The p orbital on this carbon is vacant. This picture is consistent with the planar geometry assumed for this and most cations of this type.

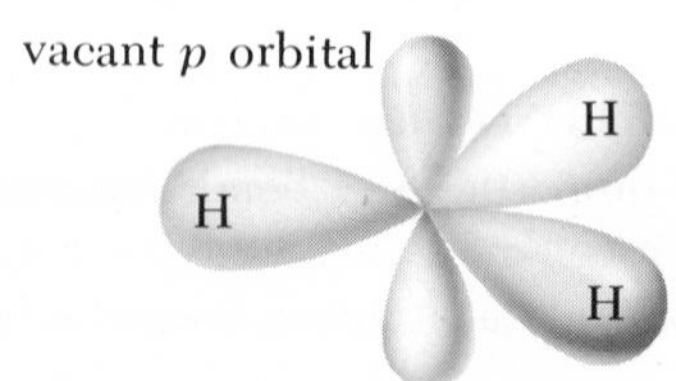

2.8 THE DATIVE OR COORDINATE COVALENT BOND

Thus far we have constructed covalent bonds by overlapping two half-filled a.o.'s. We can, however, envision a fully occupied m.o. resulting from the overlap of one fully occupied a.o. and one vacant a.o. One example of such a process is the reaction of ammonia with a proton to form the ammonium ion. The reaction of boron trifluoride, BF_3, with ammonia also illustrates this reaction type.

$$H_3N\!: \quad + \quad H^{\oplus} \quad \rightarrow NH_4^{\oplus}$$

$$\underset{\text{Lewis base}}{H_3N\!:} \quad + \quad \underset{\text{Lewis acid}}{BF_3} \quad \rightarrow H_3N^{\oplus}\text{—}BF_3^{\ominus}$$

These reactions are acid-base reactions in the Lewis sense. According to Lewis, an *acid* is any species capable of accepting a pair of electrons into an unfilled orbital, thereby forming a covalent bond. A *base*, by this theory, is any species which possesses a non-bonding or weakly bonding pair of electrons (*e.g.*, π bonds) and which is capable of donating those electrons to a Lewis acid. Clearly, the Lewis definition encompasses the most common (Brønsted) acid and base — $H^{\oplus}$ and $OH^{\ominus}$.

$$\underset{\text{base}}{H\text{—}\ddot{\underset{..}{O}}\!:^{\ominus}} + \underset{\text{acid}}{H^{\oplus}} \rightarrow H\text{—}\ddot{\underset{..}{O}}\text{—}H \qquad \text{an acid-base reaction}$$

9. When hydrogen chloride dissolves in water, the product is "hydrochloric acid." Such a solution contains water, $H_3O^{\oplus}$, and $Cl^{\ominus}$. Which of these species are Lewis acids? Lewis bases? If you have more than one compound per category, arrange them in order of decreasing strength.
10. Do elements of Groups VA, VIA, and VIIA normally act as Lewis acids or bases in the compounds in which they appear? Explain.
11. Magnesium reacts with fluorine to form the salt magnesium fluoride, MgF_2. Even though magnesium formed its bond with its two outer electrons, it is not functioning as a Lewis base. Why?

A third illustration of the formation of this type of bond is the oxidation of a trivalent nitrogen compound. The reaction is shown below in two ways: (a) conceptually, where the nitrogen (a Lewis base) reacts with an atom of oxygen (a Lewis acid) and (b) preparatively, where the nitrogen atom is oxidized with hydrogen peroxide. While it may not be clear from *b*, equation *a* suggests that this *is* an acid-base reaction.

$$(a)\quad \underset{\text{Lewis base}}{H_3C\text{—}\overset{CH_3}{\underset{CH_3}{N}}\!:} \; + \; \underset{\text{Lewis acid}}{\ddot{\underset{..}{O}}\!:} \; \rightarrow H_3C\text{—}\overset{CH_3}{\underset{CH_3}{N^{\oplus}}}\text{—}\ddot{\underset{..}{O}}\!:^{\ominus}$$

$$(b)\quad H_3C\text{—}\overset{CH_3}{\underset{CH_3}{N}}\!: \; + H\text{—}O\text{—}O\text{—}H \rightarrow H_3C\text{—}\overset{CH_3}{\underset{CH_3}{N^{\oplus}}}\text{—}\ddot{\underset{..}{O}}\!:^{\ominus} + H_2O$$

The σ bond produced in these three reactions is just like any other σ bond. In the latter two illustrations (NH_3/BF_3 and $N(CH_3)_3/H_2O_2$) the formation of the σ bond is accompanied by the generation of opposite charges on adjacent atoms. The resulting combination of a σ bond and electrostatic attraction is known as a **coordinate covalent bond.** By this definition the N—H bond produced by protonation of ammonia is not a coordinate covalent bond. The recognition of a coordinate covalent bond (also called a "dative bond") does not require an understanding of the reaction producing the bond. All that is required is the σ bond accompanied by the adjacent charges. Put differently, we don't care if the compound was made by the overlapping of a filled and a vacant orbital—*it might have been!*

Coordinate covalent bonds are extremely polar. Associated with compounds containing dative bonds are:

a. high melting point and boiling point,
b. low vapor pressure,
c. high molecular dipole moments (assuming non-cancellation of bond moments),
d. potent hydrogen bond accepting capability,
e. good solvent properties for polar compounds (assuming the compound is a liquid), and
f. relatively high water solubility.

2.9 RESONANCE

In all of our discussions thus far, we have tacitly assumed that there was only one structure possible for a molecule or ion in which all of the atoms had an outer octet of electrons. But what happens when two or more electron distributions provide satisfactory Lewis structures? A classic example of this problem is the nitrite ion, $NO_2^{\ominus}$, which can be drawn with two completely acceptable electron distributions:

⊖:Ö—N̈=Ö   Ö=N̈—Ö:⊖   alternate Lewis structures for the nitrite ion

Which of these is the correct representation of the electron distribution in the nitrite ion? NEITHER! *The real nitrite ion does not have either distribution but, instead, has some intermediate one.* The resulting intermediate structure has half of the negative charge on one oxygen and half on the other. The resulting intermediate structure contains two identical nitrogen-oxygen bonds. These bonds are neither single nor double but something in between. Adherence to the use of dots to represent electrons, coupled with the desire to associate an octet of electrons with as many atoms as possible, makes it difficult to show a single picture of what the nitrite ion really looks like. An electron or a charge may be "here" or "there," but how does one represent them as "half here" and "half there"?

To solve such problems, the chemist has invented the concept of **resonance.** In its simplest version it suggests that if we can draw two or more acceptable structures for a species, then the actual electron distribution does not correspond to any single one of them but, instead, to something in between. The real molecule is termed a **hybrid** of the structures which can be drawn but which have no real existence themselves. For this reason these hypothetical structures are sometimes called *contributing structures.* The idea that the real molecule is not adequately represented by a single contributing structure, but that it is a blend of the contributing structures, is conveyed by relating the contributing structures to one another with a so-called resonance arrow, ↔ . The energy of the real molecule is less than the energy of any of the individual contributing structures.

⊖:Ö—N̈=Ö ↔ Ö=N̈—Ö:⊖

What happens if you draw contributing structures with different charge characteristics, *i.e.*, in which one has no internal charge separation and one does? While both are contributing structures, it usually is true that the real molecule resembles the uncharged form more closely. This is because a contributor with internal charge separation is (usually) of higher energy than one without charge separation, and the hybrid is more accurately represented by the lower energy (more stable) of two non-equivalent contributing structures.

Furthermore, generation of charge separation often is accompanied by loss of an outer octet for some atoms, and this also is considered to make that contributing structure more energetic and, thus, less important. Of the five contributing structures shown below for carbon dioxide, **1** is more important in describing the real molecule than is either **2** or **3.** Structures **4** and **5** are even less important because, in addition to other shortcomings, they also place the negative charge on the less electronegative atom and the positive charge on the more electronegative element.

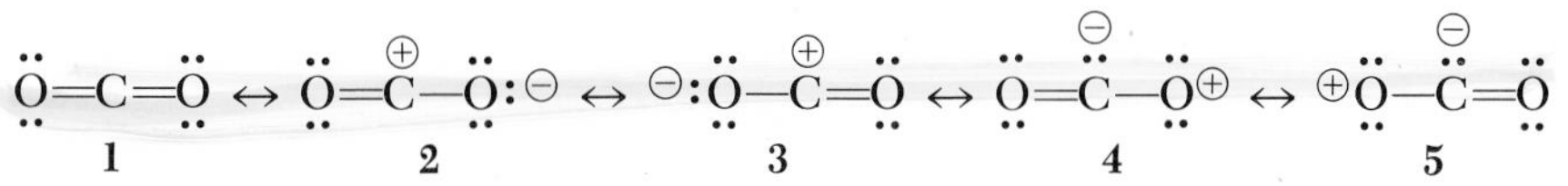

$$\underset{1}{O=C=O} \leftrightarrow \underset{2}{O=\overset{\oplus}{C}-O{:}^{\ominus}} \leftrightarrow \underset{3}{{}^{\ominus}{:}O-\overset{\oplus}{C}=O} \leftrightarrow \underset{4}{O=\overset{\ominus}{C}-O^{\oplus}} \leftrightarrow \underset{5}{{}^{\oplus}O-\overset{\ominus}{C}=O}$$

We will return to a more detailed discussion of resonance in future chapters.

12. Indicate which, if either, of the resonance contributing structures in each of the following pairs is a better approximation of the hybrid (real molecule). Explain your choice.

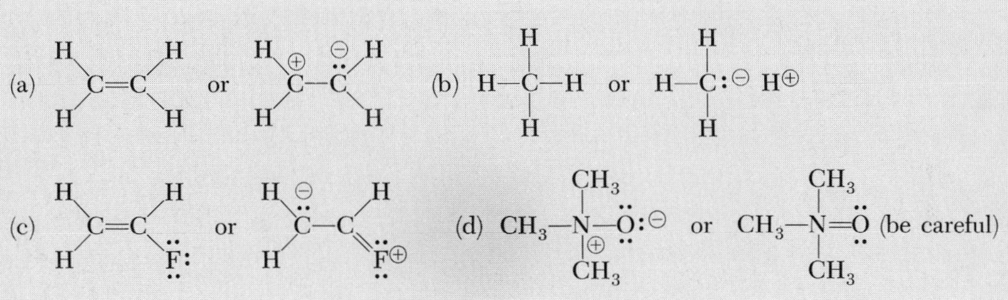

(a) $H_2C=CH_2$ or $H_2\overset{\oplus}{C}-\overset{\ominus}{C}H_2$

(b) $H-\underset{H}{\overset{H}{C}}-H$ or $H-\underset{H}{\overset{H}{C}}{:}^{\ominus}\ H^{\oplus}$

(c) $H_2C=CH-F{:}$ or $H_2\overset{\ominus}{C}-CH=F^{\oplus}$

(d) $CH_3-\underset{CH_3}{\overset{CH_3}{N^{\oplus}}}-O{:}^{\ominus}$ or $CH_3-\underset{CH_3}{\overset{CH_3}{N}}=O$ (be careful)

2.10 WHAT ARE SOME COMMON FUNCTIONAL GROUPS?

Most organic compounds consist of various assemblages of atoms bonded, *via* covalent bonds, to a framework of carbon atoms (called the "backbone" or "skeleton"). These assemblages are "functional groups" and provide the sites for most chemical reactions in organic compounds. A number of functional groups are presented in Table 2–6. For some of these the hybridization of important atoms has been provided. You should convince yourself of the correctness of the structures by applying the discussions of this chapter. The names of the functional groups are given and should be committed to memory. As with most foreign languages that we learn, we have to begin by simply memorizing the vocabulary—the names of these functional groups are a basic part of the vocabulary of organic chemistry.

TABLE 2-6 Common Organic Functional Groups

Group	Structure[a]	Group	Structure
bromo	—Br:	ethynyl [an acetylene]	—C≡C—H (C: sp)
chloro	—Cl:	phenyl	C_6 ring with four H: —C(ring)—H (—C_6H_5; —Ph; —ϕ) (ring C: sp^2)
deuterio	—D	azo	—N=N—
hydroxy	—O—H	nitroso	—N=O (—NO)
ether	—O—	azoxy[d]	—N=N$^{\oplus}$(—:O:$^{\ominus}$)
peroxy	—O—O—	nitro	—N$^{\oplus}$(=O)(—O:$^{\ominus}$) (—NO_2)
mercapto [thiol]	—S—H	formyl[e] [an aldehyde]	(:O:=)C—H (—CHO)
amino	—N:(H)H (—NH_2)	carboxy [an acid]	(:O:=)C—O—H (—CO_2H) (C: sp^2)
methyl[b]	—C(H)(H)—H (—CH_3; —Me)	cyano	—C≡N: (—CN)
ethyl[c]	—C(H)(H)—C(H)(H)—H (—C_2H_5; —Et)	carboxamido [an amide]	(:O:=)C—N(—H)— (—CONH—)
methoxy	—O—CH_3	alkoxycarbonyl [an ester]	(:O:=)C—O—R (—CO_2R)
carbonyl	>C=O (C: sp^2; O: sp^2)		
imino	>C=N—H		
vinyl	(H)C=C(H)H (—$CHCH_2$)		
allyl	—C(H)(H)—C(H)=C(H)H		

[a] Non-bonding electron pairs are shown in the expanded structures, although they normally are deleted when drawing organic structures. Often-used compressed formulas or abbreviations for selected functional groups are shown in parentheses. Names in brackets indicate the class of compounds in which these groups occur.

[b] While the methyl group can be viewed as one end of a carbon chain, thus part of the carbon skeleton, it will help to learn to recognize it.

[c] The ethyl group is another important carbon fragment that should be committed to memory. Both methyl and ethyl groups will be discussed further in Chapter 3.

[d] A more proper name is oxidodiazenediyl; however, azoxy is more frequently encountered.

[e] Sometimes called, less properly, carboxaldehyde.

Below are displayed the structures of several small molecules, with notations about utility. Why not build models of them and attempt to understand their hybridization, geometry, and other properties?

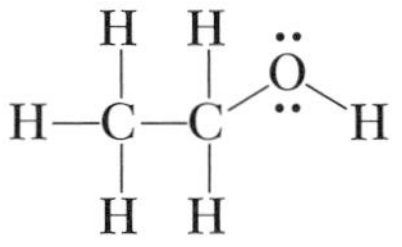

ethyl alcohol
(solvent, medicine)

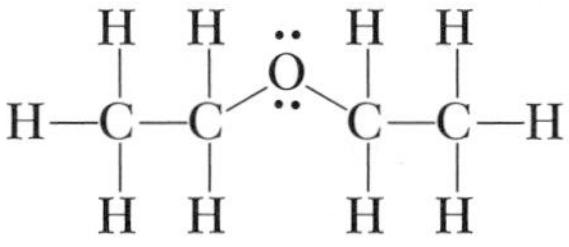

ethyl ether
(solvent, anesthetic)

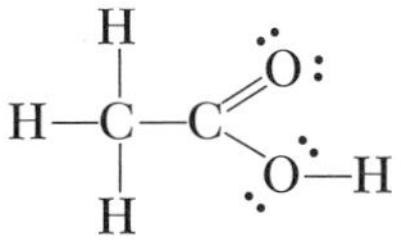

acetic acid
(cooking)

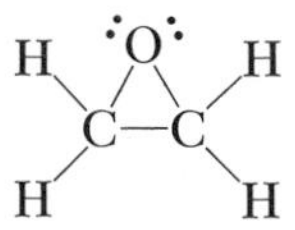

ethylene oxide
(sterilization of surgical instruments)

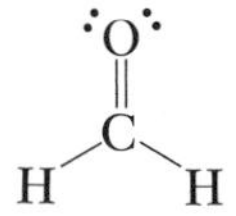

formaldehyde
(preservative)

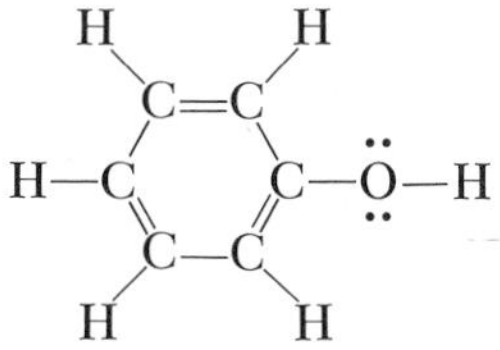

phenol
(disinfectant)

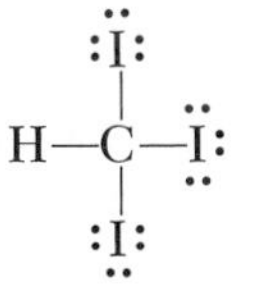

iodoform
(antiseptic)

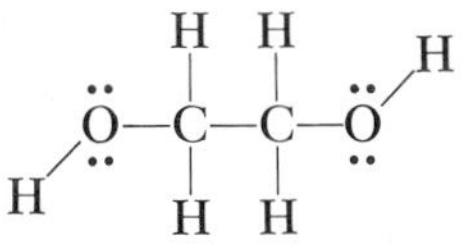

ethylene glycol
(antifreeze)

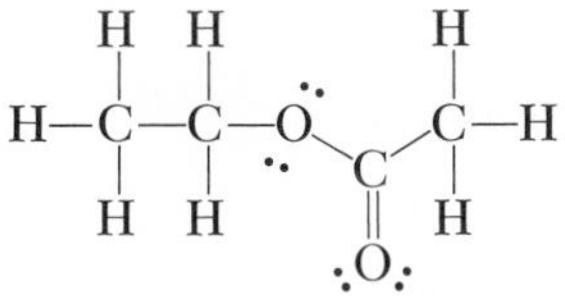

ethyl acetate
(nail polish remover)

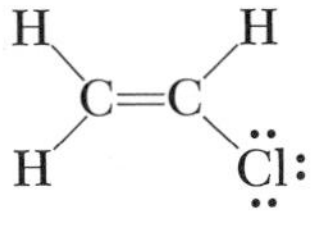

vinyl chloride
(starting material for polyvinyl chloride, PVC, a plastic; suspected carcinogen)

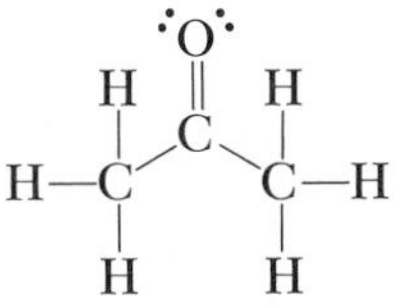

acetone
(solvent)

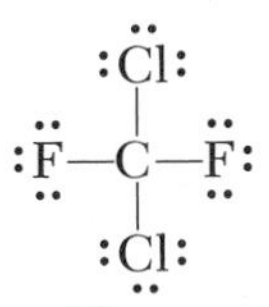

dichlorodifluoromethane
(Freon-12, a refrigerant)

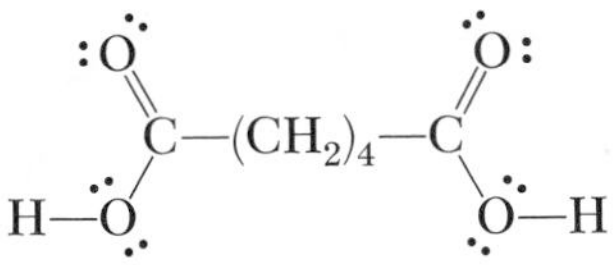

adipic acid
(preparation of nylon 66)

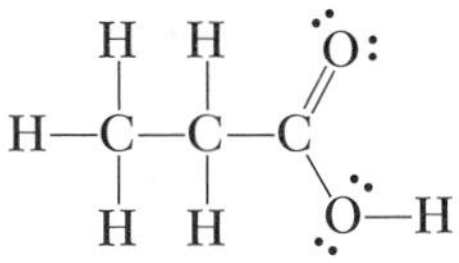

propionic acid
(preparation of calcium propionate, the common preservative in breads and related foods)

cortisone
(anti-inflammatory agent)

13. Identify the functional groups in the following compounds:

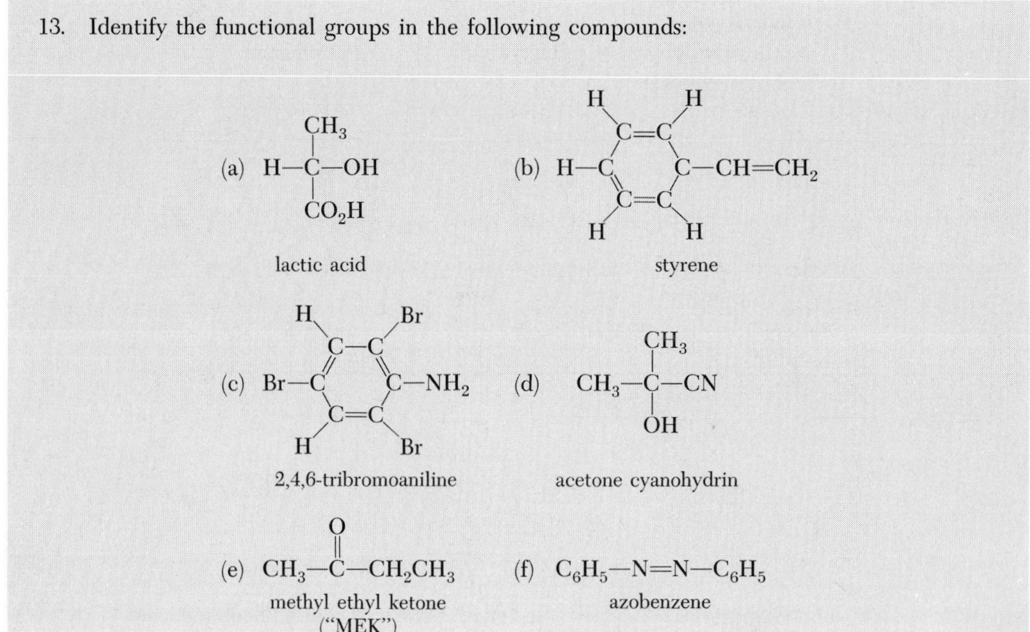

AN APPENDIX: THE CALCULATION OF FORMAL CHARGE

On occasion you may be confronted with a structure whose origin is unknown to you. One of the first questions you must ask (and answer) is, "Is that an ion or a molecule?" One way to answer this is to calculate the "formal charge" on each atom in the structure. The sum of the formal charges in a molecule will be zero.

To calculate formal charge, one determines the number of outer electrons in the atom of interest by locating it in the periodic table. (Remember that the number of outer electrons of an element equals its group number in the periodic table.) Next, we examine the number of covalent bonds to that atom and *assume* that the formation of each covalent bond provides the atom of interest with one additional electron. The sum of the number of outer electrons in the isolated atom and the number of covalent bonds to that atom should equal the number of electrons to be found around that atom. We count the number of outer electrons actually present around the atom. If the number of electrons actually present is less than the calculated number, the atom has a net positive charge, the magnitude equalling the difference between these two numbers. If the number of electrons actually present is more than that calculated, the atom has a net negative charge.

Let us illustrate this with methane, CH_4:

carbon:	(A)	number of outer electrons in isolated atom	4
	(B)	number of covalent bonds to atom	4
	(C)	number of electrons calculated (A + B)	8
	(D)	number of electrons present	8
	(E)	formal charge on atom (C − D)	0

Repeating this for hydrogen gives A = 1, B = 1, C = 2, D = 2, E = 0. Methane is, therefore, a neutral species.

The ammonium ion ($NH_4^{\oplus}$) is a valuable illustration:

nitrogen:	(A)	5
	(B)	4
	(C)	9
	(D)	8
	(E)	+1

As with methane, the hydrogens carry no formal charge.

14. Using the concept of formal charge, show why the hydrogen cation carries a unit positive charge. How might the concept of formal charge be used to calculate the charges on the ions in sodium chloride, NaCl?
15. The oxygen of a carbonyl group bears a double bond while the oxygen of an ether bears two single bonds. Show, by calculation, that both of these oxygens have the same formal charge. What is it?

IMPORTANT TERMS

Anti-bonding orbital: A molecular orbital whose energy is greater than that of its precursor atomic orbitals.

Bond: Any attractive interaction between two or more species. Bond formation is expected to be an exothermic (heat releasing) process.

Bond angle: Most commonly considered as the interatomic angle in the fragment X—Y—Z.

Bond dissociation energy: The exact energy (strength) required to cleave a specific bond homolytically.

Bond distance: The average distance between two covalently bonded nuclei. The exact distance between two nuclei is variable, over a rather narrow range. Also called the *bond length.*

Bond energy: An average energy (strength) required to cleave a certain type of bond homolytically.

Bond moment: A measure of the polarity of a bond, it equals the extent of charge separation in a bond multiplied by the distance separated.

Bonding orbital: A molecular orbital whose energy is lower than that of its precursor atomic orbitals.

Coordinate covalent bond: An attractive interaction involving both a covalent bond and the attraction between opposite charges. It may be imagined to be made by donation of an electron pair (from an uncharged donor) into a vacant orbital of an (uncharged) acceptor.

Covalent bond: The bond between two nuclei which share an electron pair between them. It may be viewed as resulting from the electrostatic attraction of two atoms to the electron pair between the atoms. Covalent bonds usually are of either the σ-type or the π-type.

Delocalized bond: A covalent bond in which the molecular orbital enshrouds more than two atoms. These are almost always π bonds.

Dipole-dipole bond: The bond between oppositely charged ends of two polar bonds or two polar molecules.

Functional group: Any collection of atoms which tends to survive for a reasonable period of time, exclusive of a carbon skeleton. Functional groups are responsible for providing the variation in chemical behavior among organic compounds. They occur bonded to the carbon skeleton. A molecule containing the chloro and nitro functional groups is:

$$\begin{array}{ccccccc} & & H & & H & & H \\ & & | & & | & & | \\ Cl & - & C & - & C & - & C - NO_2. \\ & & | & & | & & | \\ & & H & & H & & H \end{array}$$

Heterolysis: Any process which splits a covalent bond and leaves one participating atom with both electrons. Also called a *heterolytic cleavage.* The ionization of an acid is typical.

$$H{-}A \rightarrow H^{\oplus} + A^{\ominus}$$

Homologous series: A set of compounds whose nearest neighbors differ by one repeating unit, most often a methylene group.

$CH_3{-}H$ $CH_3CH_2{-}H$ $CH_3CH_2CH_2{-}H$ $CH_3CH_2CH_2CH_2{-}H$ a homologous series

Homolysis: Any process which splits a covalent bond and leaves each participating atom with one electron. Also called a *homolytic cleavage.* The homolysis of chlorine is typical.

$$Cl_2 \xrightarrow{\text{energy}} 2Cl\cdot$$

Hybrid orbital: An atomic orbital with characteristics other than those associated with only s, p, d, or f orbitals. All hybrid orbitals may be viewed as blends of varying proportions of two or more types of simple, unhybridized orbitals. The most common types of hybrid orbitals are sp, sp^2, and sp^3. Hybrid orbitals are associated with atoms in molecules but not with isolated atoms.

Hydrogen bond: A weak (about 5 kcal/mole) attractive interaction between a hydrogen atom that is covalently bonded to an electronegative element, and a second electronegative element.

The most commonly employed electronegative elements are nitrogen, oxygen, and fluorine. A hydrogen bond between two sites in one molecule is an *intramolecular* hydrogen bond. A hydrogen bond between two different molecules is an *intermolecular* hydrogen bond.

Ion-dipole bond: The bond between an ion and the oppositely charged end of a polar bond or polar molecule.

Ionic bond: The bond between oppositely charged ions. It is considered to be electrostatic in nature.

Lewis acid: Any species which possesses a vacant orbital and can accept an electron pair into that orbital. Typical Lewis acids include $H^{\oplus}$, BF_3, and $AlCl_3$.

Lewis base: Any species capable of donating a pair of electrons into a vacant orbital in a Lewis acid. Typical Lewis bases include $Cl^{\ominus}$, NH_3, H_2O, and $OH^{\ominus}$.

Lewis structure: A representation of a molecule or ion which provides as many atoms as possible (except hydrogen) with eight outer electrons. Lewis structures are considered to be excellent representations of the electron distribution in molecules unless resonance is present. Several Lewis structures are shown below.

F_2 $:\ddot{\underset{\cdot\cdot}{F}}\text{—}\ddot{\underset{\cdot\cdot}{F}}:$

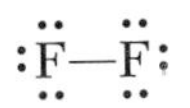

C_2N_2 $:N{\equiv}C\text{—}C{\equiv}N:$

CH_2O $(H)_2C{=}\ddot{O}:$

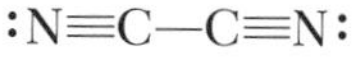

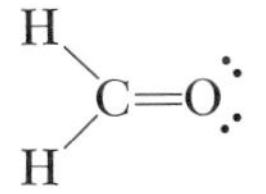

HNO_3 $H\text{—}\ddot{\underset{\cdot\cdot}{O}}\text{—}\overset{\oplus}{N}({=}\ddot{O}:)\text{—}\ddot{\underset{\cdot\cdot}{O}}:^{\ominus}$

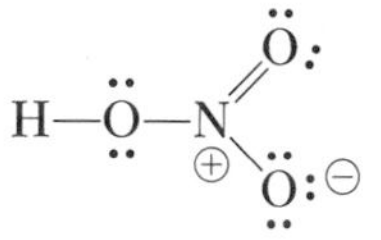

Localized bond: A covalent bond in which the molecular orbital enshrouds only two atoms. Most σ bonds are localized bonds.

London dispersion force: An attractive interaction caused by the distortion of electron clouds around atoms and by the ensuing electrostatic attraction. While the individual interactions are quite weak, the total effect of these forces can be quite strong. They explain, for example, why methane is a gas while hexane is a liquid.

Molecular dipole moment: The vector sum of all bond and group moments within a molecule.

Molecular orbital: An orbital formed by the overlap (interaction) of two or more atomic orbitals. There are two types, bonding and anti-bonding molecular orbitals. Molecules have even numbers of molecular orbitals.

Multiple bond: An attraction resulting from the sharing of more than one pair of electrons between two atoms. When two pairs are shared (one σ and one π) the result is a double bond. When three pairs are shared (one σ and two π) the result is a triple bond.

Paramagnetism: The tendency to be attracted into a magnetic field. It usually indicates the presence of one or more unpaired electrons.

π Bond: A covalent bond formed by overlapping two p orbitals with their major axes parallel. It consists of an upper and a lower lobe, both of which are shaped like frankfurters.

Polar covalent bond: A covalent bond in which the electron pair is, on the average, closer to one nucleus. Most covalent bonds are of this type.

Principle of maximum overlap: The view that with all other things equal, the greater the degree of overlap of atomic orbitals, the stronger will be the resultant bond.

Resonance: The representation of a real molecule through hypothetical "contributing structures." This phenomenon is employed when no single electronic arrangement can be used to adequately represent a given compound. The phenomenon is discussed in greater detail in Chapter 5.

σ Bond: A covalent bond possessing cylindrical symmetry, the type of symmetry found in a frankfurter.

Single bond: An attractive interaction resulting from the sharing of one pair of electrons between two atoms. Single bonds usually are σ bonds.

Van der Waals radius: A measure of how closely two non-bonded atoms may come to one another.

Valence shell electron pair repulsion theory: A theory which suggests that molecular geometry is controlled by the repulsion of pairs of outer electrons around one common atom. It is assumed that a molecule will adopt that geometry which minimizes the electron pair repulsion. This suggests that two electron pairs, three electron pairs, and four electron pairs will be separated by 180°, 120°, and 109.5°, respectively. Distortions from idealized geometries may arise because of variations in the kinds of electron pairs under consideration.

What is the relationship between valence shell electron pair repulsion theory and hybridization? VSEPR theory permits us to examine a structure and, based upon an attempt to minimize electron pair repulsions, to suggest a likely geometry for it. Once having done this, we then construct the appropriately hybridized central atom which will give the necessary geometry.

PROBLEMS

16. Draw Lewis structures for the following, remembering that all atoms except hydrogen should have an outer octet of electrons. Which of these structures contains a coordinate covalent bond? Which contain an ionic bond?

(a) CH_3OH (methanol, wood alcohol)
(b) CH_3CH_2OH (ethanol, grain alcohol)
(c) CH_3NO_2 (nitromethane, a racing fuel)
(d) HCOOH (formic acid)
(e) PCl_3
(f) $(CH_3)_3PO$
(g) $CaCO_3$
(h) H_2C_2O (ketene)

17. Label covalent bonds in the structures below as σ or π and also by the orbitals involved in bond formation. The first one is done by way of illustration.

σ 1s-2sp^3

(a) 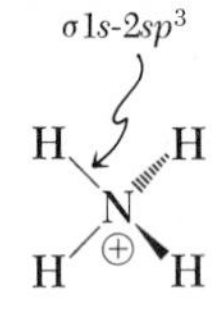 $H_2N^{\oplus}H_2$ (tetrahedral NH_4^+)

(b) :N≡N:

(c) 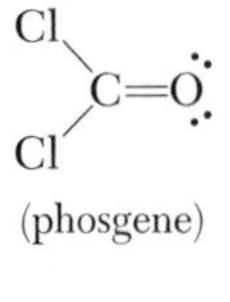 $Cl_2C{=}\ddot{O}$: (phosgene)

(d) 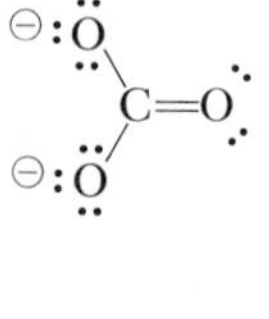 $(^{\ominus}:\ddot{O}:)_2C{=}\ddot{O}$:

(e) 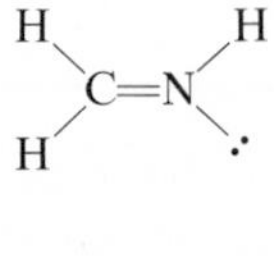 $H_2C{=}\ddot{N}H$

(f) 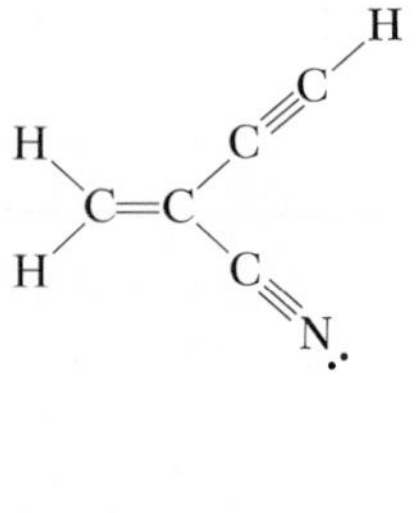 $H_2C{=}C(C{\equiv}CH)(C{\equiv}N:)$

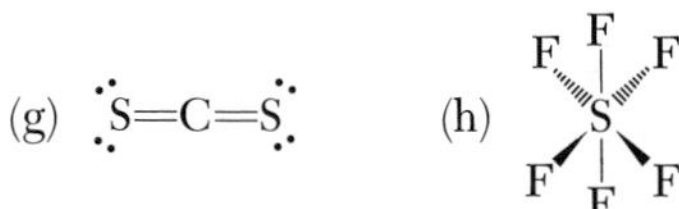

18. Draw three-dimensional representations for each of the following, and indicate which will have a net molecular dipole moment and in what direction it is aligned.
(a) $CHCl_3$ (b) CH_2Cl_2
(c) CH_2ClBr (d) $CFCl_3$
(e) CCl_4 (f) $CH_2{=}CCl_2$
(g) $CH_2{=}CHCl$ (h) $HC{\equiv}C{-}C{\equiv}CH$
(i) BrCN (j) BrCH=CHBr (two answers)
(k) $(CH_3)_2C{=}O$

19. Which of the following species can act as Lewis acids? Lewis bases?
(a) H_2O (b) NH_3
(c) $CH_3^{\ominus}$ (d) $H^{\ominus}$
(e) $H^{\oplus}$ (f) $Cl^{\ominus}$
(g) $AlCl_3$

20. Suggest a reason for the fact that solvent behavior of water usually becomes poorer as the temperature is lowered from 90° to 35°.

21. Dimethyl sulfoxide [DMSO, $(CH_3)_2SO$] is much more soluble in water than is dimethyl sulfide [$(CH_3)_2S$]. Explain.

22. Draw one important and one unimportant resonance contributing structure for each of the following.

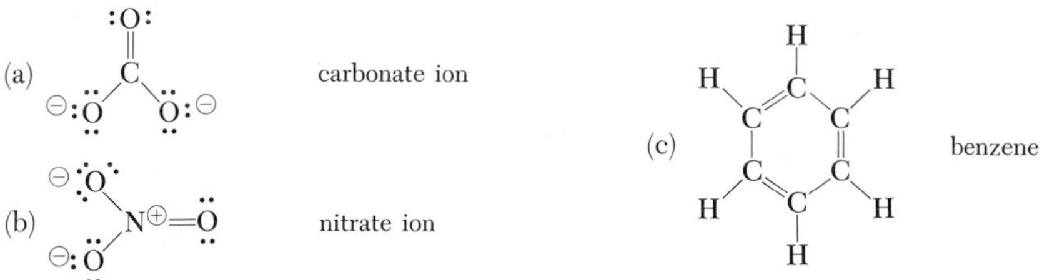

23. Why can helium be liquefied? Why does it have a lower boiling point than does neon?

24. Assign formal charges to each of the fragments below. Each line represents a covalent bond (σ or π) to some group.

(a) $-\overset{|}{\underset{|}{S}}-$ (b) $-\overset{|}{\underset{\cdot\cdot}{S}}-$ (c) $-\overset{\cdot\cdot}{\underset{\cdot\cdot}{S}}-$

(d) $-\overset{\cdot\cdot}{\underset{\cdot\cdot}{S}}:$ (e) $-\overset{|}{\underset{|}{N}}-$ (f) $-\overset{|}{\underset{\cdot\cdot}{N}}-$

(g) $-\overset{\cdot\cdot}{\underset{\cdot\cdot}{N}}-$ (h) $-\overset{\cdot\cdot}{\underset{\cdot\cdot}{N}}$ (i) $-\overset{\cdot\cdot}{\underset{|}{O}}-$

(j) $-\overset{\cdot\cdot}{\underset{\cdot\cdot}{O}}-$ (k) $-\overset{\cdot\cdot}{\underset{\cdot\cdot}{O}}:$ (l) $-\overset{|}{\underset{|}{C}}-$

(m) $-\overset{|}{\underset{\cdot\cdot}{C}}-$ (n) $-\overset{|}{C}-$ (o) $-\overset{\cdot\cdot}{C}-$

(p) $-\overset{|}{\underset{|}{Sn}}-$ (q) $-\overset{\cdot\cdot}{\underset{\cdot\cdot}{Br}}:$ (r) $-\overset{\cdot\cdot}{\underset{\cdot\cdot}{Br}}-$

(s) $-\overset{|}{\underset{\cdot\cdot}{Br}}-$ (t) $:\overset{\cdot\cdot}{\underset{\cdot\cdot}{Cl}}\cdot$ (u) H:

25. The equation below gives the "percent ionic character" in a polar covalent bond. Using this equation, calculate the percent ionic character in bonds in which $|\chi_A - \chi_B|$ is 0.2, 1.0, and 3.0, respectively.

$$\%\ \text{ionic character} = 16|\chi_A - \chi_B| + 3.5|\chi_A - \chi_B|^2$$

where χ_A and χ_B are the electronegativities of the atoms involved in the bond.

26. Some organic compounds are slightly soluble in water. In order to remove them, the aqueous phase can be saturated with salt (hence the name "salting out") and then washed with an appropriate solvent (*e.g.*, ethyl ether). Why does salting out assist in the extraction of the organic impurity from water? Why would methanol be a poor choice for the extraction solvent?
27. Ethyl ether, $H_5C_2—O—C_2H_5$, and *n*-butyl alcohol, $CH_3CH_2CH_2CH_2—OH$, are both equally soluble in water (about 80 g/liter), yet the alcohol has a much higher boiling point than does the ether (118° *vs.* 35°). How can hydrogen bonding be used to rationalize these data?
28. Methyl ether, CH_3OCH_3, dissolves in liquid hydrogen fluoride to form a solution from which methyl ether can be recovered by addition of water. What species are present in the original hydrogen fluoride-methyl ether solution? What is the function of water in the regeneration of methyl ether?
29. The following are van der Waals radii. Is there any relationship between ionic radii and van der Waals radii? H, 1.2; F, 1.4; O, 1.4; N, 1.5; Cl, 1.8; S, 1.9; Br, 2.0; I, 2.2.
30. Mercuric chloride, $HgCl_2$, is a covalent molecule with a linear geometry. Suggest the electronic configuration of Hg in $HgCl_2$. Dimethyl mercury, CH_3HgCH_3, is a compound implicated in the pollution of waters with mercury. Draw the structure for this, labeling all orbitals as in Problem 17.
31. 2-Methylpyrrolidine has a higher boiling point than does *N*-methylpyrrolidine. Explain.

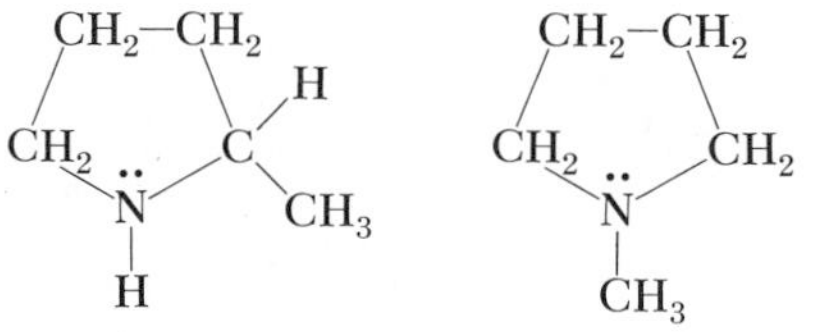

2-methylpyrrolidine   *N*-methylpyrrolidine

32. The following represent the order of increasing acidity among various sets of compounds. How can you explain these trends?
   (a) $H—CH_3 < H—NH_2 < H—OH < H—F$
   (b) $H—SH < H—Cl$
   (c) $H—OH > H—SH > H—SeH$
33. The reaction below can be described in terms of an acid-base reaction. Identify the Lewis acid and Lewis base. Identify the state of hybridization of each carbon in the reaction sequence.

$$CH_3^{\oplus} + CH_2{=}CH_2 \rightarrow CH_3—CH_2—CH_2^{\oplus}$$

34. Why is the species CH_2**:** (carbene) better considered as a Lewis acid than as a Lewis base?
35. The methyl radical, $CH_3\cdot$, has a carbon which does not have an outer octet of electrons, yet it is not a Lewis acid. Explain. What is the state of hybridization of the carbon in this nearly planar radical? What is the formal charge on this carbon?
36. The bond dissociation energy of the hydrohalic acids decreases in the order shown below. Explain this trend.

$$HF > HCl > HBr > HI$$

37. Since contributing structures have no independent existence, they cannot be separated from one another and cannot be in equilibrium with one another. Suggest a method and appropriate observations which would support the existence of the resonance shown.

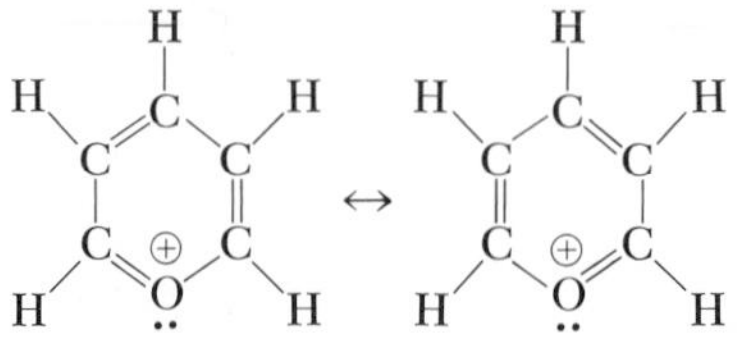

ALKANES

3.1 WHERE ARE WE GOING?

Alkanes are hydrocarbons having the formula C_nH_{2n+2} and containing only sp^3 hybridized carbons. While alkanes are rather important to all of us, representing our most abundant portable fuel, their direct importance to most chemists is rather limited. Alkanes do not undergo a variety of conveniently studied reactions, and those which have been studied are of limited utility in the laboratory. However, you can benefit from an early exposure to the chemistry of alkanes since some of their reactions can be used to introduce important physical-chemical concepts without excessive discomfort to the reader. We also will use alkanes to develop our first glimpse of the methods chemists employ to discover how starting material is converted to product in chemical reactions. Since the nomenclature of alkanes is at the root of the names of many organic compounds, this chapter also contains our first systematic treatment of the nomenclature of organic compounds. You will find this chapter divided into two main portions. The first deals with the structure of hydrocarbons, while the second deals with their chemical reactivity. The discussion of alkane nomenclature separates these two portions.

3.2 THE OCCURRENCE OF ALKANES IN NATURE

Several alkanes are quite abundant in nature. For example, natural gas, although varied in composition from well to well, is composed of approximately 75% methane, 15% ethane, and 5% propane; the remainder represents larger ("higher") alkanes and many other compounds. In spite of this, *petroleum* is the most important source of alkanes, it being likely that all crude oil contains some alkanes.

Crude oil is distilled into a variety of fractions (gasoline, kerosene, gas oils and light lubricating oils) and several types of non-volatile residues (lubricating oils, petrolatum, petroleum pitch and paraffin wax). These are listed in Table 3–1. While man's specific needs may change (kerosene was once the most important component of crude oil because of its use in lanterns), it seems clear that crude oil will continue to be the backbone of the chemical industry. Need for it is sufficiently great, and it can be such a potent weapon of international politics, that research is underway to discover methods to convert both coal and animal manure into artificial petroleum.

3.3 THE STRUCTURE OF SOME SIMPLE ALKANES

Methane, the simplest alkane, is the most abundant constituent of natural gas, marsh gas, and fire damp (found in coal mines). It arises from the anaerobic oxidation of plant and animal material. It can also be prepared by the reaction of aluminum carbide (considered to be an

TABLE 3-1 Petroleum Fractions[a]

FRACTION[b]	COMPOSITION[c]	DISTILLATION RANGE
natural gas	C_1–C_4	below 20°
liquefied petroleum	C_5–C_6	20–60°
gasoline[d]	C_4–C_8	40–200°
kerosene	C_{10}–C_{16}	175–275°
fuel oil, diesel oil	C_{15}–C_{20}	250–400°
lubricating oils	C_{18}–C_{22}	above 300°
asphalt	complex mixture (C_{20} and above)	non-volatile

[a] From *Chemistry, Man And Society*, by M.M. Jones, J.T. Netterville, D.O. Johnston, and J.L. Wood, W.B. Saunders Co., Philadelphia, 1972.

[b] Other descriptions and boiling ranges have been used, but all are quite similar. Crystallization of solids from less-volatile fractions produces *paraffin* and similar solid mixtures.

[c] The exact percentage composition varies from well to well and may vary considerably from one region or country to another.

[d] Sometimes called *straight-run gasoline.*

"inorganic" compound) and water, reminding us that the division between organic and inorganic chemistry often is arbitrary.

$$\underset{\text{aluminum carbide}}{Al_4C_3} + 12H_2O \rightarrow \underset{\text{methane}}{3CH_4} + 4Al(OH)_3$$

As noted in Chapter 2, the carbon of methane is sp^3 hybridized, the four hydrogens[a] being equivalent with regard to both geometry and reactivity. The three-dimensional representation of methane shown below illustrates this hydrogen equivalence. Since it is convenient to represent structures in two dimensions, we can make the moves shown below and produce a planar representation of CH_4 from this three dimensional representation.

3-D —rotation in space→ 3-D —projection onto a plane parallel to page→ structural formula Fischer projection

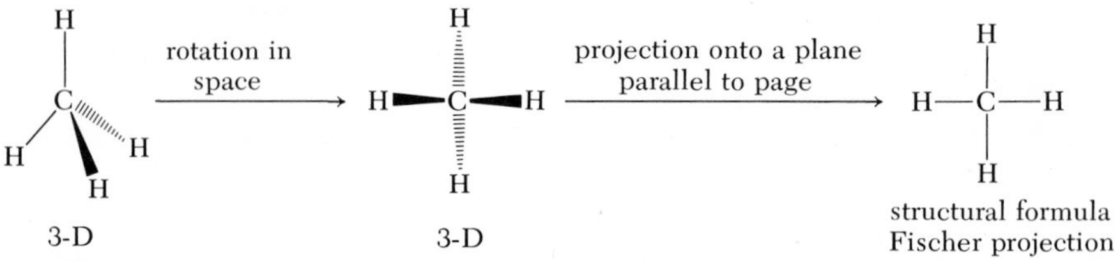

Such planar representations are almost always used to describe the bonding sequence in a molecule—*even when a three-dimensional structure is not implied.* Normally one should assume that such a *structural formula* does NOT represent a projection from a three-dimensional structure. When used to convey a three-dimensional picture, these structural formulas are specifically termed *Fischer projections* (discussed in more detail in Chapter 4). Because it does not explicitly show any bonds, "CH_4" is the *condensed* structural formula of methane.

The **molecular formula** of a compound conveys the actual number of atoms of different elements in the molecule, while the **empirical formula** conveys the lowest whole number ratio of elements in the molecule. For example, the molecular formula of hydrogen peroxide is H_2O_2,

[a] The terms "hydrogen," "hydrogen atom," and "proton" all are used to describe the first element in the periodic table covalently bonded to another element. The term *proton* also retains its classic meaning—$H^{\oplus}$.

while the empirical formula is HO ($H_2O_2/2$). Methane has identical empirical and molecular formulas, CH_4.

Ethane, C_2H_6, is the next most simple alkane. The conversion of a three-dimensional representation to the corresponding structural formula is shown below. The projection suggested above the arrow does not lead to a value of 90° for the angle α. However, chemists traditionally draw the four bonds to a central sp^3 hybrid atom with angles of 90°. You may recall that this practice was followed in drawing some structures in the previous chapter.

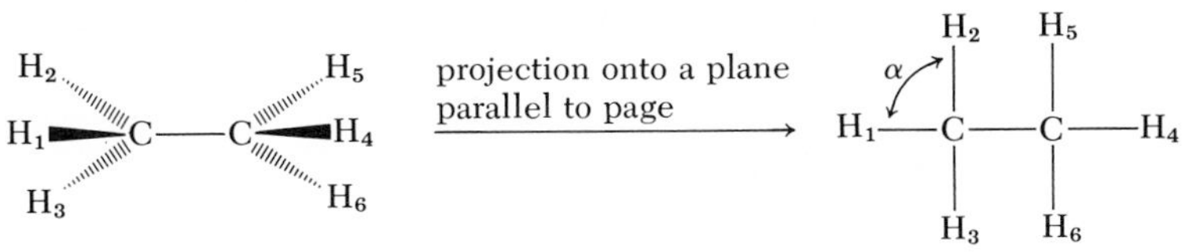

The condensed structural formula of ethane is not C_2H_6; *that* is its molecular formula. The condensed structural formula is CH_3CH_3, reflecting the custom of reporting substituents *after* the atom to which they are bonded. However, this practice may be abandoned when, for example, emphasis is placed on a specific bond, *e.g.*, $H_3C—CH_3$. Finally, the empirical formula of ethane is CH_3.

In the three-dimensional representation of ethane shown above, it appears as if some *numbered* hydrogens are different from other *numbered* hydrogens with respect to a third (reference) *numbered* hydrogen. For example, while H_1 and H_4 are "face to face" (*eclipsed*), H_1 and H_5 are not; therefore, one might argue that H_4 and H_5 are not equivalent with respect to H_1. However, notice that H_5 bears the same relationship to H_2 that H_4 bears to H_1. In reality there is no difference between H_1 and H_2 (except for the numbers given to them in order to facilitate our discussion) and H_4 and H_5 are, consequently, equivalent when one considers *both* H_1 and H_2. By reasoning such as this we eventually can conclude that all six hydrogens are equivalent to one another. (Simple symmetry arguments lead to the same result.)

One outgrowth of this is that there are a number of equivalent ways of drawing the structural formula of a compound possessing a central sp^3 hybridized carbon. By way of illustration, the six possible equivalent representations of the compound CH_2R_2, where R represents some anonymous organic residue, are presented below.

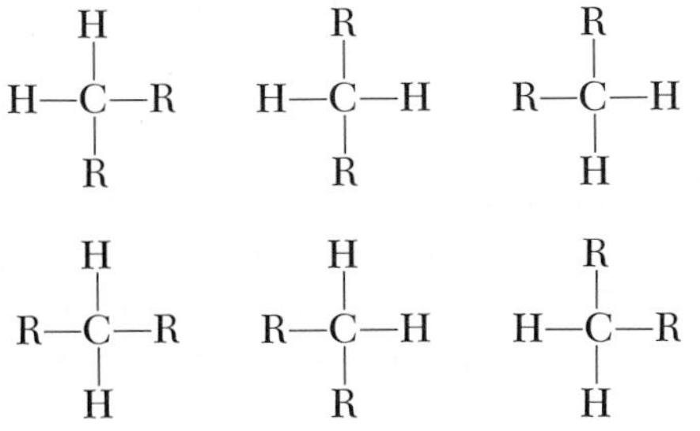

equivalent representations of CH_2R_2

Melvin Newman has suggested a two-dimensional representation of compounds which preserves some of their 3-D characteristics. Such *Newman projections* are generated by viewing along a convenient σ bond and drawing what you see. This is done below for one geometry of ethane. Since atoms "in the back" are eclipsed, they cannot be seen. It is, therefore, common to displace back atoms a bit when dealing with eclipsed geometries so that they can be seen. The front carbon atom is represented by the intersection of bonds from it, while the rear carbon atom appears as a circle.

line of sight

wedge projection
eclipsed form

Newman projection
eclipsed form

There is a rather low energy barrier to rotation around most single bonds, including the bond between the carbons in ethane. The rotational barrier in ethane can be described in terms of the change in the potential energy of the molecule as a function of the change in the *torsional angle*, τ, defined in the diagram below.

or

As you can see from Figure 3–1, a rotational barrier of only 3 kcal/mole separates one low energy form of ethane from another. These low energy forms, whose torsional angles are multiples of 60°, are called *staggered forms*. The staggered forms are separated by high energy forms recognizable as the eclipsed form of ethane. Since some collisions at room temperature may provide as much as 20 kcal, this 3 kcal barrier is easily surmounted and rotation in ethane is considered to be free. The periodicity of the curve results from the identity of molecular structures as we rotate through every 120°. All of the structures which represent points along the curve are *conformations* of ethane, since they can be readily interconverted by rotation around a single bond without breaking any bonds. When speaking of a **conformation** of a molecule, we mean a specific geometry of that molecule, particularly in terms of bond distances, bond angles, and torsional (or dihedral) angles.

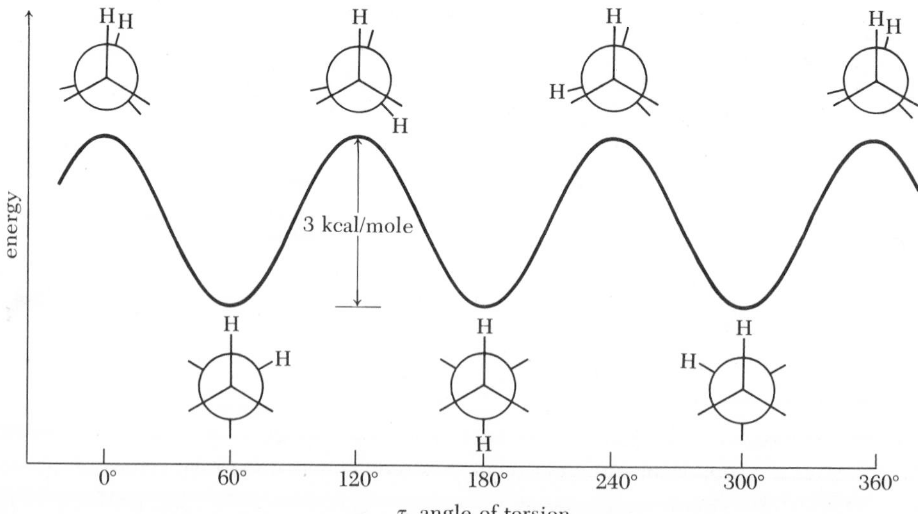

Figure 3-1 Energy profile for C—C bond rotation in ethane. Rotation of the back carbon is described by the changing torsional angle between the two hydrogens that are shown. The remaining hydrogens have been deleted for clarity. The upper figures are eclipsed conformations, while the lower figures are staggered conformations.

While the origin of the barrier to rotation in ethane is still a topic of discussion, it appears likely that it does not arise because hydrogens which are *vicinal* (on adjacent carbons) "bump" into one another in the high energy conformation. Such "eclipsing interactions" are small because vicinal hydrogen atoms are far apart (relative to their van der Waals radius) even in the eclipsed, high energy conformation. It currently is suggested that the origin of the barrier can be found in the electronic repulsions between C—H bonds as they pass by one another.

1. How many conformations are implied in Figure 3–1?

Propane, C_3H_8, is sold to campers, non-urban dwellers and "do-it-yourselfers" as "bottle gas" or "LP (liquefied petroleum) gas." It is the simplest of the alkanes which can be readily liquefied and stored as a liquid.

```
     H H                H  H  H                 H        H
H    |/    H            |  |  |                 |        |
 \   C    /          H—C——C——C—H    or     H—C————————C—H
H···C  \ C···H          |  |  |                 |        |
    |    |              H  H  H                 H  H—C—H
    H    H                                             |
                                                       H
```

wedge projection of a staggered conformation | structural formula | $CH_3CH_2CH_3$ condensed structural formula | propane

Rotation around a C—C single bond in propane (does it matter which?) is represented in Figure 3–2. The similarity of the barriers in ethane and propane suggests that H,H eclipsing interactions have about the same strength as H,CH_3 interactions in this system.

While every point on the curves in Figures 3–1 and 3–2 represents a specific conformation, those corresponding to energy minima are *conformational isomers* or *conformers*. Conformers are the most abundant species in a mixture of conformations. The height of the curve from a minimum to a maximum represents the least amount of energy required to pass from one

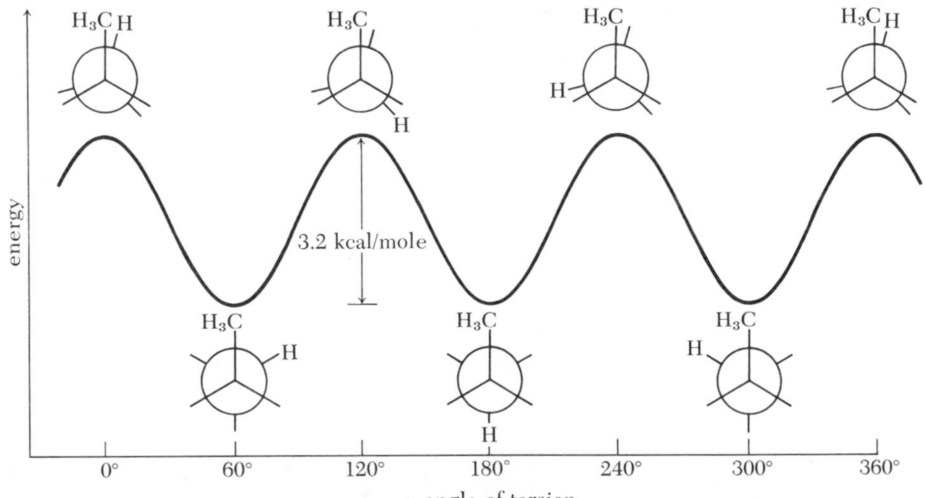

Figure 3-2 Energy profile for C—C bond rotation in propane. Note the similarity to Figure 3–1. Four hydrogens have been deleted for clarity.

conformer to an adjacent one, the maximum being the **transition state** between adjacent conformers. The species corresponding to the transition states are the least abundant species in a mixture of conformations. The high energy conformation (the transition state) of propane is an eclipsed conformation, while the low energy conformation is a staggered conformation. This has been noted for ethane and is generally true for other compounds. (Remember that eclipsed conformations can be identified by the 0° angle between vicinal groups, while a staggered conformation has a 60° angle between vicinal groups in a Newman projection.)

2. Which structures in Figure 3–1 represent transition states? Are they of equal energy? Why?

Butane, C_4H_{10}, like propane, serves as a portable energy source. It can be represented in two dimensions in any of the ways shown below. A detailed look at the conformations of butane is presented in Figure 3–3.

molecular formula: C_4H_{10}
empirical formula: C_2H_5
condensed structural formula: $CH_3CH_2CH_2CH_3$ *or* $CH_3(CH_2)_2CH_3$
structural formulas:

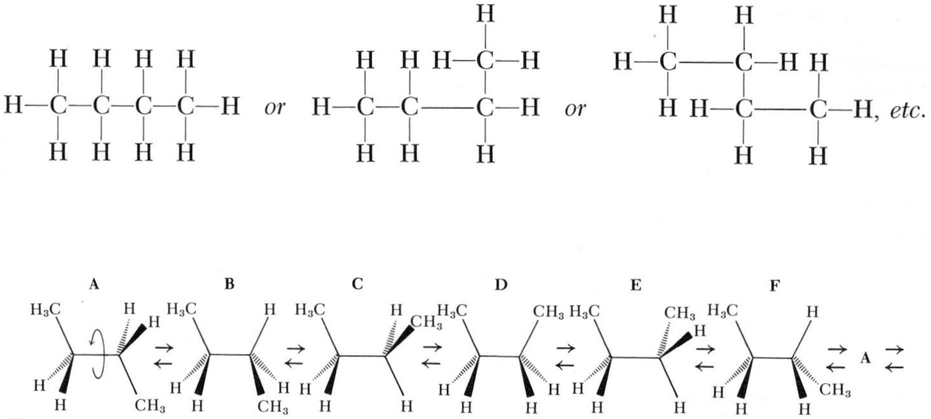

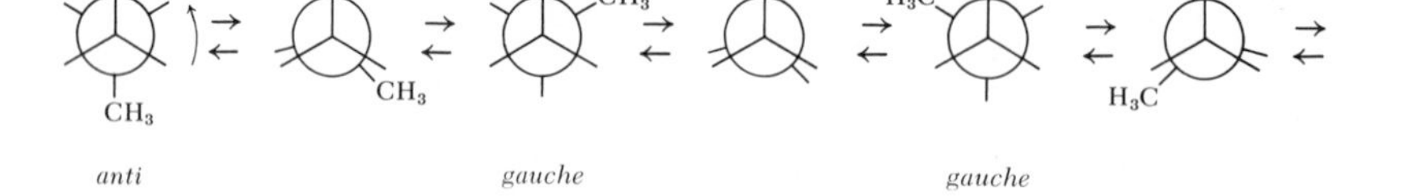

```
                                        H                H     H
                                        |                |     |
  H  H  H  H               H  H  H—C—H          H—C———C—H H
  |  |  |  |               |  |     |                |     |  |
H—C—C—C—C—H  or  H—C—C———C—H  or       H H—C———C—H, etc.
  |  |  |  |               |  |     |                   |     |
  H  H  H  H               H  H     H                   H     H
```

A B C D E F

anti gauche gauche

Figure 3-3 Conformational distribution in *n*-butane. Interconversion is accomplished by rotation in the direction of the curved arrow. Conformations **A, C,** and **E** are staggered conformers, while **B, D,** and **F** are eclipsed conformations.

Unlike ethane and propane, not all staggered conformers of butane are equivalent nor are all eclipsed conformations alike. **A, C,** and **E** (see Figure 3–3) are all staggered conformers; however, they are not alike. To distinguish among these staggered conformers, **A** (with its methyl groups separated by 180°) is dubbed the *anti* conformer, while **C** and **E** are called *gauche* conformers. The eclipsed conformations are distinguished by the presence of either "methyl, methyl eclipsing," as in **D**, or "methyl, hydrogen eclipsing," as in **B** and **F.** The energies of these conformations are interrelated in Figure 3–4.

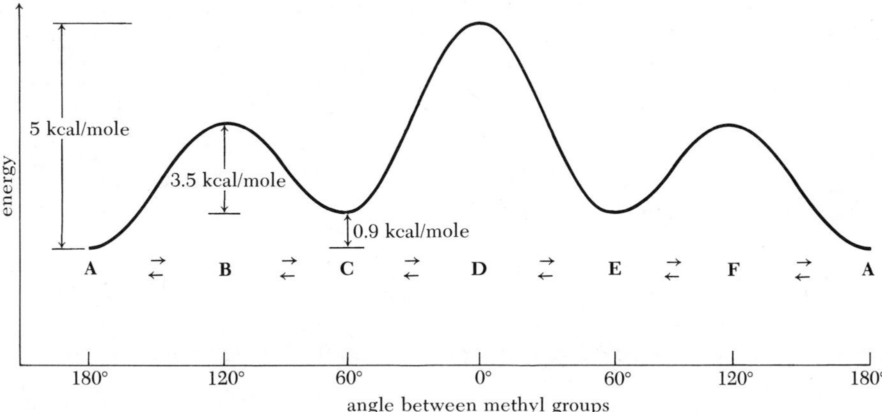

Figure 3-4 The rotational barrier around the central C—C bond in butane. The letters correspond to the structures in Figure 3-3.

The energy difference between **A** and **D** is about 5 kcal/mole, significantly larger than the rotational barrier in ethane (3 kcal/mole). This suggests that in **D**, where vicinal methyl groups are eclipsed, some repulsion is occurring between these methyl groups. The energy difference between conformers **A** and **C**, 0.9 kcal/mole, represents the increase in energy when two methyl groups move from a non-interacting arrangement to one in which they are separated by 60° (Newman projection). Since this is a *gauche conformer* (or *skewed* conformer), this repulsion is termed a *gauche* (or *skewed*) *butane interaction*.

Although butane could exist as either of two different conformers, the barrier between them is low enough for them to interconvert rapidly at room temperature. Thus, butane is an equilibrium mixture of conformers, the precise composition of the mixture being controlled by the free energy difference (ΔG) between the various conformers.

The equilibrium constant between two species (*e.g.*, $A \rightleftharpoons B$) can be calculated from the equation:

$$\Delta G = -RT \ln K_{eq}$$

where

ΔG = the free energy difference between species

R = the universal gas constant, 1.987 cal deg^{-1} $mole^{-1}$

T = the temperature (°K), 273 + °C

$\ln K_{eq}$ = natural logarithm of the equilibrium constant between two species (*e.g.*, $K_{eq} = [B]/[A]$)

ΔG is defined by the Gibbs-Helmholtz equation as equal to $\Delta H - T\Delta S$, where ΔH is the enthalpy or heat content difference and ΔS is the entropy difference between the product and the starting material. It is common, but not wholly accurate, to assume $T\Delta S$ to be small and to approximate ΔG by ΔH.

An enthalpy change (ΔH) may be defined as that heat change which accompanies any process carried out at a constant pressure, as are most reactions. Bond rotations and chemical transformations with $\Delta H < 0$ tend to be spontaneous. The entropy of a substance, like its enthalpy, is one of its characteristic properties. Qualitatively, entropy is a measure of disorder or randomness of a system. Substances which are highly disordered have high entropies; low entropy is associated with highly ordered substances. If the entropy change accompanying a process is positive ($\Delta S > 0$), the term $-T\Delta S$ makes a negative contribution to ΔG. Therefore, there will be a tendency for processes to be spontaneous if the products are less ordered than the reactants. For example, gasoline spilled on the ground tends to change from the liquid state, in which molecules are comparatively ordered, to the vapor state, in which molecules are much less ordered and dispersed through the atmosphere. Although enthalpy factors are involved, the pollution of the atmosphere by spilled volatile materials is partly a consequence of the entropy change which accompanies the evaporation.

A study of the distribution of conformations, the factors controlling the distribution, and the influence of conformational distribution upon chemical reactivity constitutes the area of research known as *conformational analysis*.

3. Build models of **C** and **E** in Figure 3–3 and see what relationship exists between them. What relationship exists between **C**, **E** and **D**?
4. What is the effect of increasing the temperature upon the population distribution of ethane conformers? of butane conformers?

A Structural Isomer of Butane. **Isomers** are compounds which have the same molecular formula but which differ in the arrangement of their atoms. Conformations are excluded from this definition by adding that isomers must be separated by significant energy barriers (*i.e.*, they cannot interconvert rapidly while they are being studied). "Isomers" may be subdivided into two classes: **constitutional isomers** or **structural isomers** (isomers with different atom-to-atom bonding sequences) and **stereoisomers** (isomers with the same atom-to-atom bonding sequence but with the atoms arranged differently in space). We will elaborate upon structural isomers in the following discussion, but we will wait until Chapter 4 to discuss stereoisomers.

There are two possible structural isomers corresponding to the formula C_4H_{10}. One of these, *n*-butane, already has been discussed. A second one, a compound which does not have a continuous chain of four carbons, bears the name *isobutane*.

$$\begin{array}{ccccc} & & H & & \\ & & | & & \\ H & H- & C & -H & \\ | & & | & & \\ H-C & \text{———} & C & -H & \\ | & & | & & \\ H & H- & C & -H & \\ & & | & & \\ & & H & & \end{array} \equiv CH_3-\boxed{\begin{array}{c} CH_3 \\ | \\ C-H \\ | \\ CH_3 \end{array}} \equiv CH_3-CH(CH_3)_2 \equiv CH(CH_3)_3 \qquad \text{isobutane}$$

The names "*n*-butane" and "isobutane" are part of one of two important schemes for naming organic compounds. In this "non-systematic" (or "common" or "trivial") method for naming structures, alkanes containing only a continuous ("straight") chain of carbons are called *n*-alkanes (*n* is an abbreviation for *normal*). That is why $CH_3—CH_2—CH_2—CH_3$ is called *n*-butane. Compounds whose carbon chains end in two methyl groups (*i.e.*, $—CH(CH_3)_2$) are, by way of contrast, called isoalkanes. Isobutane is the simplest alkane capable of having a non-continuous chain of carbons, a property termed *chain branching*. It is, of course, the simplest isoalkane.

Isomeric Pentanes. Compounds of the formula C_5H_{12} are *pentanes*. The three isomeric pentanes are given the non-systematic names *n*-pentane, isopentane, and neopentane. (The prefix *neo* means that the chain ends in a $—C(CH_3)_3$ group.)

$$\underset{n\text{-pentane}}{CH_3-CH_2-CH_2-CH_2-CH_3} \qquad \underset{\text{isopentane}}{CH_3-CH_2-\begin{array}{c} CH_3 \\ | \\ C \\ | \\ CH_3 \end{array}-H} \qquad \underset{\text{neopentane}}{CH_3-\begin{array}{c} CH_3 \\ | \\ C \\ | \\ CH_3 \end{array}-CH_3}$$

The number of isomers increases rapidly as the number of carbon atoms increases; when one considers isomers of eicosane ($C_{20}H_{42}$), one must deal with more than one-third of a million compounds!

3.4 NOMENCLATURE[a]

The complexity of organic chemistry is due to the variety of compounds which carbon can form. In order to avoid confusion arising from faulty communication, a systematic nomenclature scheme has evolved under the guidance of the International Union of Pure and Applied Chemistry, IUPAC. (Some older texts may refer to the IUC nomenclature, since the International Union of Chemistry evolved into the IUPAC.) Unfortunately, systematization began only after a number of commonly occurring compounds had received trivial names, and many of these are indelibly inscribed in the nomenclature of organic chemistry. Furthermore, the systematic names of some compounds are so complex that trivial names are justified. We must, therefore, learn several ways of naming compounds. Of these, the IUPAC scheme is the most important and will be discussed first. Many of the rules presented in this section will be used throughout the text and in every future contact you will have with organic chemistry.

The IUPAC Nomenclature. This system is occasionally called the *Geneva system* since the IUC adopted its initial set of rules in Geneva, Switzerland in 1892. A summary of the current rules for naming an alkane follows:

1. The ending of the name is *ane*.

2. The longest *continuous* chain of carbons in the molecule is the backbone upon which substituents are considered to be attached. The name of this backbone chain forms the root of the IUPAC name of this molecule. Although non-IUPAC (common) names of alkanes will be discussed shortly, these common names are presented in Table 3–2 because they are used to designate the backbone of an alkane in the IUPAC scheme.

---

[a] For a complete discussion of this subject, see "Nomenclature of Organic Compounds," edited by J.H. Fletcher, O.C. Dermer, and R.B. Fox, Advances In Chemistry #126, American Chemical Society, Washington, 1974.

TABLE 3-2 The Straight Chain Alkanes, $CH_3(CH_2)_{n-2}CH_3$

Number of Carbons in Chain (n)	Name	bp	mp
1	methane	−162	−183
2	ethane	−89	−172
3	propane	−42	−188
4	*n*-butane	−0.5	−135
5	*n*-pentane	36	−130
6	*n*-hexane	69	−95
7	*n*-heptane	98	−91
8	*n*-octane	126	−57
9	*n*-nonane	151	−54
10	*n*-decane	174	−30
11	undecane[a]	196	−26
12	dodecane[a]	216	−10
20	eicosane[a]	343	37
30	tricontane[a]	446	66

[a] The prefix *n* is rarely used beyond decane and is, of course, unnecessary for methane, ethane and propane. The IUPAC names for all of these straight chain hydrocarbons simply deletes the *n*.

For example, the compound shown below is named as a pentane, since the longest "straight" or continuous chain contains five carbons (shown in boldface).

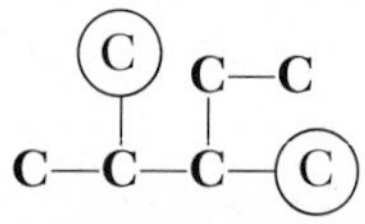

```
(C)  C—C
 |   |
C—C—C—(C)
```

Substituents are circled (only one possible set is shown)

If a compound contains two or more continuous chains of equal length, the more highly substituted one is used as the backbone.

3. The backbone carbons are numbered sequentially from the end of the chain which gives the pendent substituent the lowest number. If several alkyl substituents are present, the numbering order which gives the lowest sum is used. (More strictly, when two names are possible, one chooses the name that leads to the lower numbering on the *first* occasion of difference between the names.)

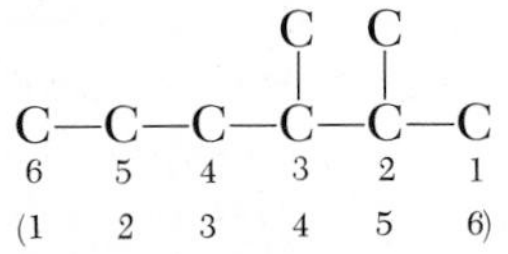

```
          C  C
          |  |
C—C—C—C—C—C
6  5  4  3  2  1     correct numbering
(1 2  3  4  5  6)    (incorrect numbering)
```

4. The names of alkyl substituents, also called *branches* or *side chains*, are attached as prefixes to the parent name (backbone). The position of a side chain is indicated by the number of the atom of the parent chain to which it is attached. If two branches are on the same carbon, the number is used twice. These numbers precede the names of the side chains to which they refer and are separated from them by a hyphen. Commas are used to separate consecutive numbers from one another. Several illustrations may help to clarify these rules. Hydrogens have been omitted in the structures which follow in order that you may concentrate on the carbon skeleton. However, remember that each carbon should have four bonds to it, and that any carbon with less than four bonds is lacking some hydrogens.

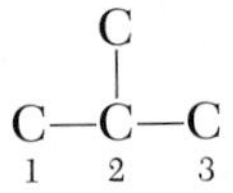

```
   C
   |
C—C—C          2-methylpropane
1  2  3

      C
      |
C—C—C—C        2-methylbutane
4  3  2  1

   C  C
   |  |
C—C—C—C        2,3-dimethylbutane
4  3  2  1

 5
 C     C
 |     |
4C—C—C—C       2-methylpentane
    3  2  1

9  8  7  6  5
C—C—C—C—C—C—C—C—C      5-butyl-2,4,4-trimethylnonane
            |           (refer back to this after you
          4 |            have read rule #6)
         C—C—C
            |   2  1
          3 C—C—C
                |
                C
```

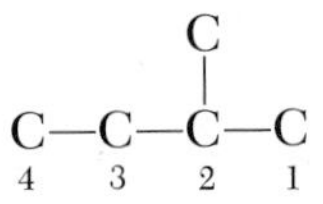

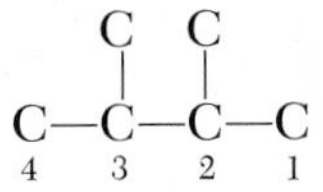

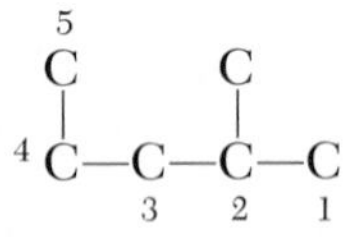

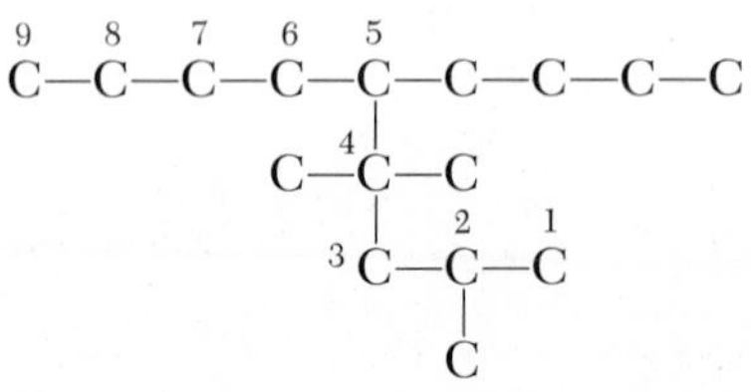

5. The IUPAC rules permit the use of common names for designating some side chains. These are summarized in Table 3–3. The compound shown below the table can benefit from this permissive rule.

TABLE 3-3 Alkyl Groups And Simple Alkanes

STRUCTURE	GROUP NAME	COMMON NAME OF CORRESPONDING HYDROCARBON[a]
$H_3C—$	methyl*	methane*
$CH_3CH_2—$	ethyl*	ethane*
$CH_3CH_2CH_2—$	*n*-propyl[b]	propane*
$(CH_3)_2CH—$	isopropyl*	propane*
$CH_3CH_2CH_2CH_2—$	*n*-butyl[b]	*n*-butane
H | $CH_3—C—CH_2—$ | CH_3	isobutyl*	isobutane*
$CH_3—CH—CH_2CH_3$ |	*sec*-butyl[c]* (*s*-butyl)	*n*-butane
CH_3 | $CH_3—C—CH_3$ |	*tert*-butyl[d]* (*t*-butyl)	isobutane*
$CH_3CH_2CH_2CH_2CH_2—$	*n*-pentyl[b] (*n*-amyl)	*n*-pentane
H | $CH_3—C—CH_2—CH_2—$ | CH_3	isopentyl* (isoamyl)	isopentane*
CH_3 | $CH_3—C—CH_2—$ | CH_3	neopentyl*	neopentane*
CH_3 | $CH_3—CH_2—C—$ | CH_3	*tert*-pentyl[d]* (*tert*-amyl) (*t*-pentyl) (*t*-amyl)	isopentane*

[a] Resulting from addition of hydrogen to the open bond. The names marked with an asterisk are accepted by the IUPAC.

[b] The prefix *n* must be dropped when using this to designate a side chain in the IUPAC system.

[c] *sec* = secondary; often abbreviated *s*.

[d] *tert* = tertiary; often abbreviated *t*.

```
1  2  3  4  5  6  7  8  9
C—C—C—C—C—C—C—C—C
   |        |
   C        C—C
            |
            C
            |
            C
```

5-*sec*-butyl-2-methylnonane

If an alkyl side chain is complex, and cannot be named using the terminology in Table 3-3, it is named as if it were an alkane except that the name ends in *yl*, rather than in *ane*. The atom in the side chain bonded to the backbone is numbered 1 and becomes the first carbon of the backbone *of the side chain*. The entire name of the side chain is enclosed in parentheses.

These serve to show which numbers are part of the name of the side chain and which refer to the parent chain.

```
1   2   3   4   5   6   7   8   9
C—C—C—C—C—C—C—C—C        side chains are circled
  |         |
 (C)      (1)C—C
             |
          (2)C—C
             |
          (3)C
```

2-methyl-5-(1,2-dimethylpropyl)nonane

6. What is the proper order in which to list substituents? There are two acceptable sequences, (a) alphabetical order and (b) in order of increasing complexity. The more common practice in the United States, and the one preferred by the American Chemical Society, is listing by alphabetical order. Substituents are not listed numerically unless this happens to correspond to either of the accepted methods.

```
          C
          |
      C   C
      |   |
C—C—C—C—C—C—C
1  2  3  4  5  6  7
```

4-ethyl-3-methylheptane *correct*
3-methyl-4-ethylheptane *less desirable*

The IUPAC name of a compound is based upon the length of the backbone. However, in discussing an alkane it usually is identified by the *total* number of carbons in the molecule. Thus, 4-ethyl-3-methylheptane is considered to be one of the isomeric *decanes*.

5. Provide IUPAC names for the following and identify the type of hydrocarbon represented by each (*i.e.*, butane, pentane, etc.).

```
                                                   H    CH3
                                                   |    |
    H3C—CH2             H    CH3          H   H—C———C—CH3
        |               |    |            |    |    |
(a) H3C—CH2      (b) H3C—C———C—H    (c) H3C—C————C—H H
                        |    |            |    |
                       CH3  CH3           H    |    H
                                               |    |
                                           H—C———C—CH3
                                               |    |
                                               H   CH3

                                                        H
                                                        |
                                                  H3C—C—CH3
    CH3 CH3 H              CH3 CH3 H               H    |
    |   |   |              |   |   |               |    |
(d) H3C—C———C———C—CH3 (e) H3C—C———C———C—CH3 (f) H3C—C———C—CH3
    |   |   |              |   |   |               |    |
    CH3 CH3 H              CH2 CH3 H               |    H
                           |                  H3C—C—CH3
                           CH3                     |
                                                   H
```

Many organic compounds can be viewed as derivatives of alkanes, and they are usually named that way. This most significant aspect of the IUPAC scheme is illustrated using halogenated alkanes.

$BrCH_2CH_3$	$Br—CH_2CH_2CH_3$	CH_3CHCH_3 (Br on C-2)
bromoethane ("1" is deleted if numbering is unambiguous)	1-bromopropane	2-bromopropane

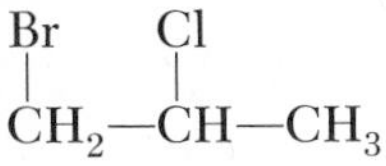

$$\begin{array}{ccccc} \text{Br} & & \text{Cl} & & \\ | & & | & & \\ CH_2 & — & CH & — & CH_3 \end{array}$$

1-bromo-2-chloropropane

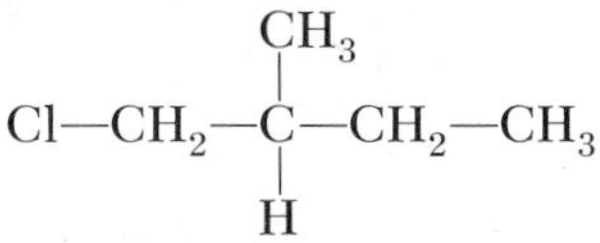

$$\begin{array}{ccccccccc} & & & & CH_3 & & & & \\ & & & & | & & & & \\ \text{Cl} & — & CH_2 & — & C & — & CH_2 & — & CH_3 \\ & & & & | & & & & \\ & & & & H & & & & \end{array}$$

1-chloro-2-methylbutane

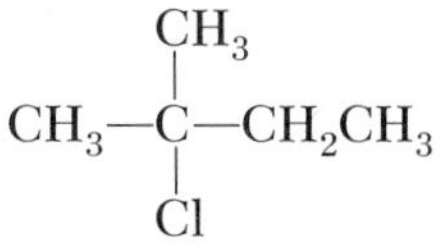

$$\begin{array}{ccccc} & & CH_3 & & \\ & & | & & \\ CH_3 & — & C & — & CH_2CH_3 \\ & & | & & \\ & & \text{Cl} & & \end{array}$$

2-chloro-2-methylbutane

Non-IUPAC Nomenclature. Non-IUPAC names are most often used for relatively simple alkanes. Common names of alkanes are based upon the total number of carbons in the alkane, employing the names in Table 3–3 with modification as necessary. In the common name system, unbranched chains are given the prefix *n* (*normal*), while branched chains have special prefixes. We have already used these common names and prefixes in discussing the isomers of butane and pentane.

One problem which perplexes some students is the use of a common name to indicate an *alkyl group* compared to its use in generating names of *alkanes*. Several representative pitfalls are illustrated below. In examining these, it is helpful to note that in naming alkanes bearing a functional group such as a halogen, the ordering of the first of the components differs in the IUPAC scheme and in the common scheme. Thus CH_3CH_2Cl is chloroethane in the IUPAC scheme but ethyl chloride in the common scheme.

1. 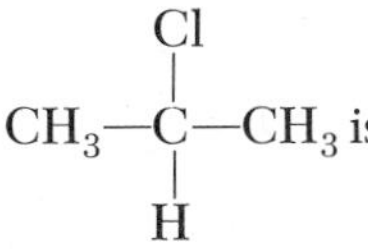 $\begin{array}{c} \text{Cl} \\ | \\ CH_3—C—CH_3 \\ | \\ H \end{array}$ is isopropyl chloride, but $\begin{array}{c} H \\ | \\ CH_3—C—CH_3 \\ | \\ H \end{array}$ is propane; there is no isopropane.

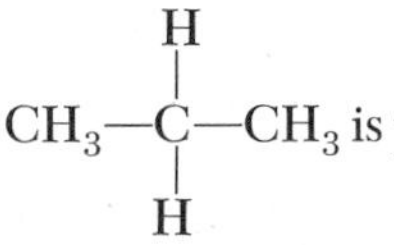

2. 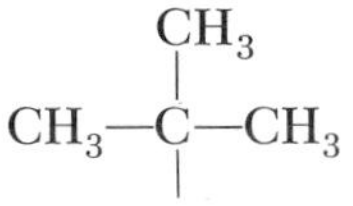 $\begin{array}{c} CH_3 \\ | \\ CH_3—C—CH_3 \\ | \end{array}$ is a *t*-butyl group, but $\begin{array}{c} CH_3 \\ | \\ CH_3—C—CH_3 \\ | \\ H \end{array}$ is isobutane; the chain ends with—$CH(CH_3)_2$.

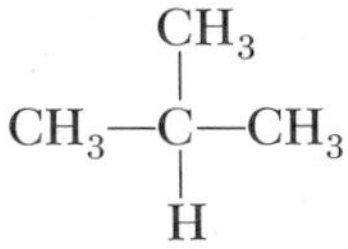

3. 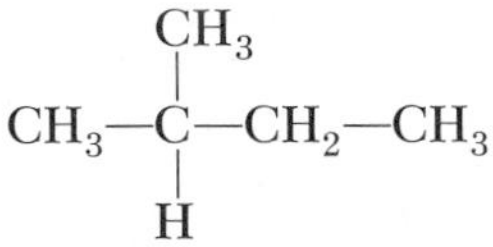 $\begin{array}{c} CH_3 \\ | \\ CH_3—C—CH_2—CH_3 \\ | \\ H \end{array}$ is *iso*pentane, but $\begin{array}{c} CH_3 \\ | \\ CH_3—C—CH_2—CH_3 \\ | \\ \text{Cl} \end{array}$ is *t*-pentyl chloride.

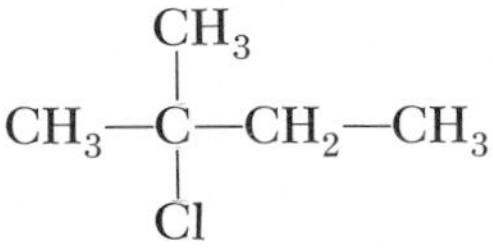

Perhaps a few more examples will further clarify the IUPAC and the non-systematic schemes. In those examples in which one is possible, the non-systematic name is presented in parentheses.

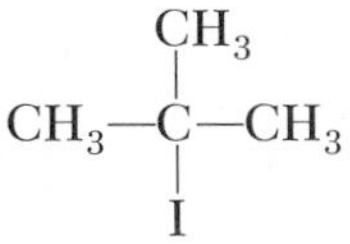

$$\begin{array}{c} CH_3 \\ | \\ CH_3—C—CH_3 \\ | \\ I \end{array}$$

2-iodo-2-methylpropane
(*t*-butyl iodide)

$$\text{Cl}—CH_2CH_2CH_2CH_2—\text{Cl}$$

1,4-dichlorobutane
(tetramethylene dichloride)

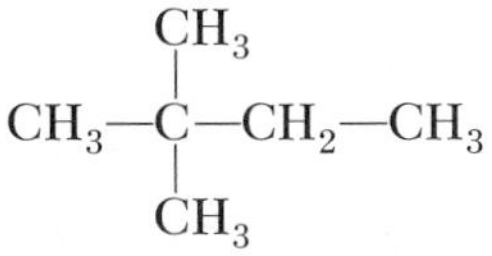

$$\begin{array}{c} CH_3 \\ | \\ CH_3—C—CH_2—CH_3 \\ | \\ CH_3 \end{array}$$

2,2-dimethylbutane
(neohexane)

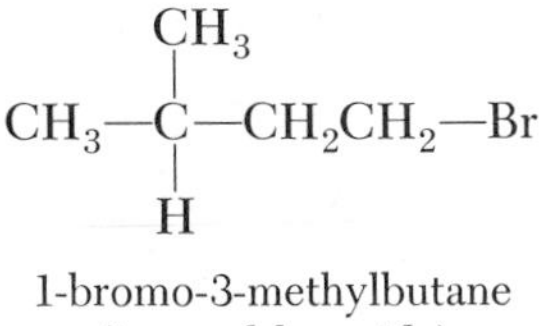

$$\begin{array}{c} CH_3 \\ | \\ CH_3—C—CH_2CH_2—\text{Br} \\ | \\ H \end{array}$$

1-bromo-3-methylbutane
(isoamyl bromide)

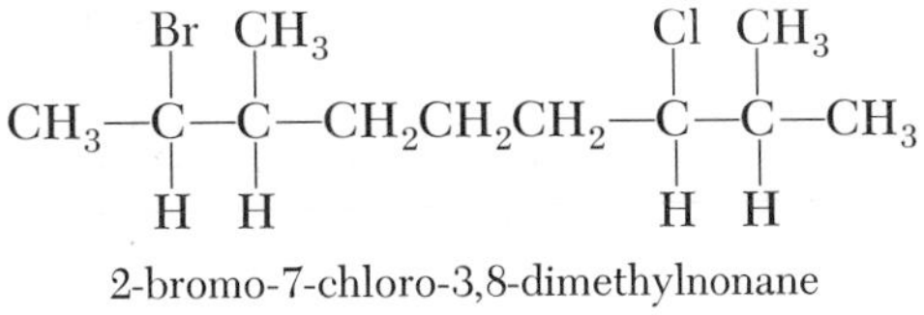

$$\begin{array}{ccccccccccc} & & \text{Br} & & CH_3 & & & & \text{Cl} & & CH_3 & & \\ & & | & & | & & & & | & & | & & \\ CH_3 & — & C & — & C & — & CH_2CH_2CH_2 & — & C & — & C & — & CH_3 \\ & & | & & | & & & & | & & | & & \\ & & H & & H & & & & H & & H & & \end{array}$$

2-bromo-7-chloro-3,8-dimethylnonane

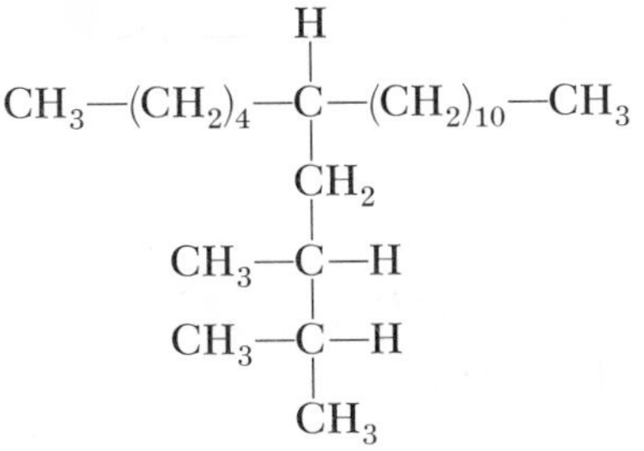

$$\begin{array}{c} H \\ | \\ CH_3—(CH_2)_4—C—(CH_2)_{10}—CH_3 \\ | \\ CH_2 \\ | \\ CH_3—C—H \\ | \\ CH_3—C—H \\ | \\ CH_3 \end{array}$$

6-(2,3-dimethylbutyl)heptadecane

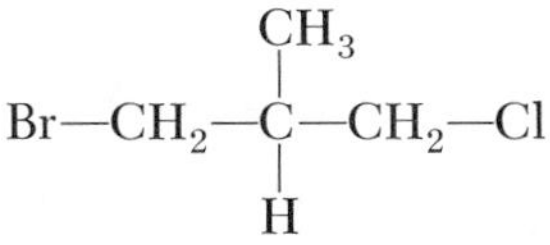

$$\begin{array}{c} CH_3 \\ | \\ \text{Br}—CH_2—C—CH_2—\text{Cl} \\ | \\ H \end{array}$$

1-bromo-3-chloro-2-methylpropane

Alkanes are sometimes named as alkylated methanes. In this scheme, the most highly branched carbon is considered the "methane carbon" unless a less-substituted carbon permits naming a smaller group. In the three examples that follow, the methane carbon is in boldface type. Once again, note that substituents are listed alphabetically. (Hydrogens are intentionally deleted.)

```
        C                    C   C                  C
        |                    |   |                  |
  C—C—C—C            C—C—C—C—C              C—C—C
    |   |                  |       |                |
    C   C                  C       C                C
```

trimethylisopropylmethane (*not* dimethyl-*t*-butyl-methane)

t-butyl-methylisopropylmethane

tetramethylmethane (preferred over *t*-butylmethane)

3.5 COMBUSTION OF ALKANES—A LOOK AT CHEMICAL ENERGETICS

Hydrocarbons are the largest single source of energy available to man (excluding sunlight, whose use as a practical energy source may be far off). Combustion of alkanes provides heat and light, and drives many of our mechanical devices. We shall, therefore, pause to consider some quantitative aspects of these reactions.

$$CH_4 + 2O_2 \rightarrow CO_2 + 2H_2O + \text{heat}$$

$$2C_8H_{18} + 25O_2 \rightarrow 16CO_2 + 18H_2O + \text{heat}$$

combustion of methane and octane

Heat of Formation. The **heat of formation** of a compound, ΔH_f, is the enthalpy change that occurs when one mole of compound is formed from its elements, *i.e.*, the difference between the enthalpy of the substance and the sum of the enthalpies of the elements of which it is made. Thus, the heat of formation of ethane, C_2H_6, is given by:

$$\Delta H_f \text{ ethane} = H \text{ ethane} - (2H \text{ carbon} + 3H \text{ hydrogen})$$

molar heat of formation of ethane = enthalpy of one mole of ethane − (enthalpy of two gram atoms of carbon + enthalpy of three moles of hydrogen gas)

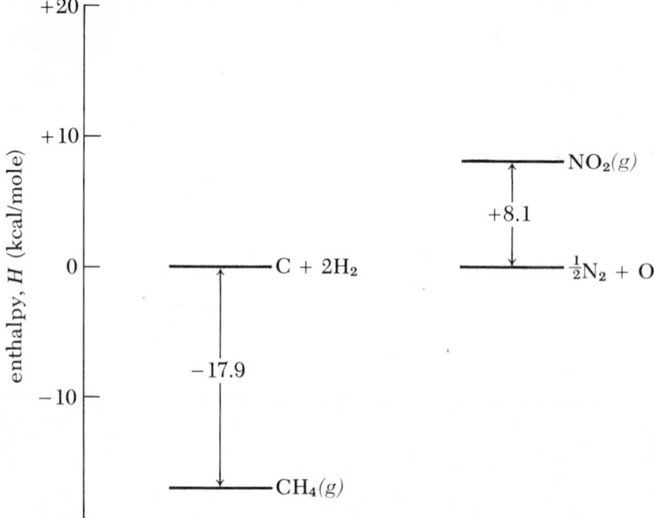

Figure 3-5 Heats of formation. Most compounds, like methane, have negative heats of formation; heat is evolved when they are formed from the elements. A few compounds, like NO_2, have positive heats of formation. (Adapted from Masterton, W. L., and Slowinski, E. J.: *Chemical Principles*, 3rd ed. Philadelphia, W. B. Saunders Company, 1973.)

For the sake of this definition, the elements are postulated to be in their so-called standard states, that is, the stable form at room temperature and atmospheric pressure. Since elements in their standard states are arbitrarily taken as having an enthalpy of zero, *the heat content of a compound equals its heat of formation.* A diagrammatic representation of the relationship of heat of formation to enthalpy change is shown in Figure 3–5, while a collection of heats of formation is found in Table 3–4.

TABLE 3-4 Heats Of Formation[a]

Compound	Heat of Formation[b]
$CH_{4(g)}$	−17.9
$CCl_{4(l)}$	−33.3
$CHCl_{3(l)}$	−31.5
$CH_3OH_{(l)}$	−57.0
$CO_{(g)}$	−26.4
$CO_{2(g)}$	−94.0
$C_2H_{2(g)}$	+54.2
$C_2H_{4(g)}$	+12.5
$C_2H_{6(g)}$	−20.2
$C_3H_{8(g)}$	−24.8
$n\text{-}C_4H_{10(g)}$	−29.8
$n\text{-}C_5H_{12(l)}$	−41.4
$H_2O_{(g)}$	−57.8
$H_2O_{(l)}$	−68.3
$H_2O_{2(l)}$	−44.8

[a] An extensive collection of heat of formation (ΔH_f), entropy of formation (S^0) and free energy of formation (ΔG_f) data for hydrocarbons can be found in the current edition of the *Handbook of Chemistry and Physics*, published by The Chemical Rubber Company, Cleveland, O.

[b] In kcal/mole at 25° and 1 atm.

Heat of Reaction. For reactions carried out under constant pressure (and this includes most laboratory reactions), the **heat of reaction** equals the enthalpy change (ΔH) associated with that reaction. In turn, the enthalpy change accompanying a reaction equals the heat of formation of the products minus the heat of formation of the starting materials. A reaction with a negative value for ΔH produces heat and is termed **exothermic.** A reaction with a positive value for ΔH absorbs heat and is termed **endothermic.**

$$\Delta H = \Sigma\Delta H_f \text{ products} - \Sigma\Delta H_f \text{ starting materials}$$

If, for example, we wished to calculate ΔH for the reaction of chlorine with hydrogen fluoride to produce hydrogen chloride and fluorine, we need only subtract the heat of formation of hydrogen fluoride (−64.2 kcal/mole) from that of hydrogen chloride (−22.1 kcal/mole). The chlorine and fluorine have, by definition, heats of formation equal to zero and they do not enter the calculation.

$$Cl_2 + 2HF \rightarrow 2HCl + F_2$$

$$\Delta H = 2\Delta H_f HCl + \Delta H_f F_2 - 2\Delta H_f HF - \Delta H_f Cl_2$$

$$\Delta H = 2\Delta H_f HCl + 0 - (2\Delta H_f HF + 0)$$

$$\Delta H = 2(-22.1\text{ kcal}) - 2(-64.2\text{ kcal}) = +84.2\text{ kcal}$$

We just have calculated that the reaction is endothermic, *i.e.*, that heat is absorbed as this reaction occurs. We now may re-write the equation as follows:

$$Cl_2 + 2HF \rightarrow 2HCl + F_2 - 84.2 \text{ kcal}$$

Heat of Combustion. The heat of combustion of a hydrocarbon compound is the amount of heat liberated during the conversion of one mole of material into carbon dioxide and water. Since all combustions are exothermic, the heat of combustion of a compound is defined as the negative of the enthalpy change (ΔH) accompanying the combustion. This permits heats of combustion to be positively signed numbers. Using methane as a representative hydrocarbon, the heats of combustion of carbon, hydrogen, and a hydrocarbon are related to the heat of formation of that hydrocarbon in Figure 3–6.

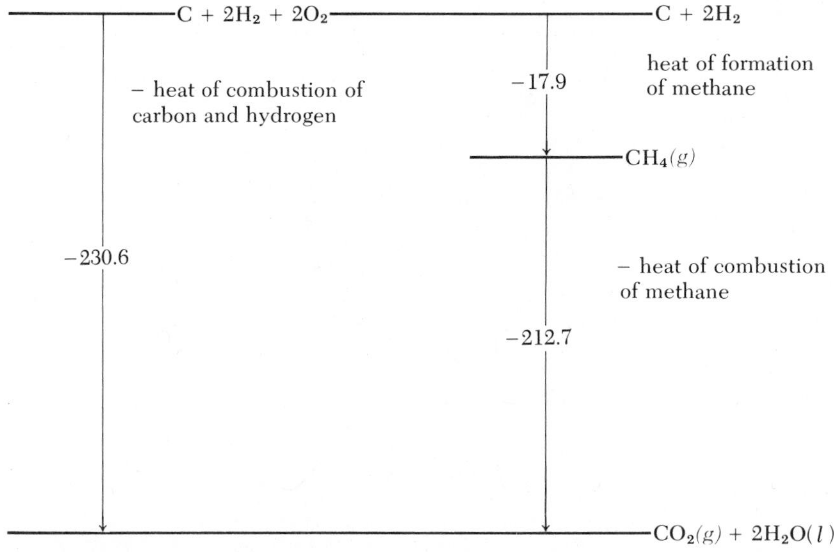

Figure 3-6 Relating the heats of formation and combustion of methane. The heat of formation of methane (−17.9 kcal/mole) plus the negative of the heat of combustion of methane (−212.7 kcal/mole) equals the negative of the heat of combustion of one gram-atom of carbon and two moles of hydrogen (−230.6 kcal/mole). Knowledge of any two of these values permits the determination of the third.

Given the equation for the combustion of methane, one can substitute the heats of formation (Table 3–4) of the various reagents and calculate the heat of combustion of methane.

$$CH_4(g) + 2O_2(g) \rightarrow CO_2(g) + 2H_2O(l) \qquad \text{initial equation}$$

$$-17.9 + 2(0) \rightarrow -94.0 + 2(-68.3) \qquad \text{insertion of heats of formation}$$

For this reaction $\Delta H = (-94.0 - 2(68.3)) - (-17.9 + 0) = -212.7$ kcal/mole. The heat of combustion is $-\Delta H$ or $+212.7$ kcal/mole.

$$CH_4(g) + 2O_2(g) \rightarrow CO_2(g) + 2H_2O(l) + 212.7 \text{ kcal} \qquad \text{complete equation}$$

This value was reached using liquid water as a product. If we wish to calculate the heat of combustion producing gaseous water instead of liquid water, we have to consider the amount of energy (the *heat of vaporization*) required to convert liquid water to gaseous water. This vaporization process is endothermic, with $\Delta H = +10.4$ kcal/mole. The corrected ΔH would then be:

$$-212.7 \text{ kcal/mole} + 2(10.4 \text{ kcal/mole}) = -191.9 \text{ kcal/mole};$$

the corrected heat of combustion is $+191.9$ kcal/mole.

This calculation does not require any knowledge of *how* the reaction occurs, only a knowledge of the heats of formation of starting materials and products. In fact, *any* calculations of *net* heat changes accompanying a reaction are independent of the number of steps involved in the conversion and depend only upon the enthalpy of starting material and the enthalpy of product. This is a statement of Hess' Law, also called the *law of constant heat summation.* The heats of combustion of several compounds are presented in Table 3–5.

TABLE 3–5 Heats Of Combustion[a]

FORMULA	NAME	HEAT OF COMBUSTION[b]
$CH_{4(g)}$	methane	212.7
$C_2H_{6(g)}$	ethane	372.8
$C_2H_{4(g)}$	ethylene	337.2
$C_2H_{2(g)}$	acetylene	310.6
$C_3H_{8(g)}$	propane	530.6
$CH_3OH_{(l)}$	methanol	173.6
$CH_3CH_2OH_{(l)}$	ethanol	326.7
$C_6H_{12}O_{6(s)}$	glucose or blood sugar	673
$C_{12}H_{22}O_{11(s)}$	sucrose, common sugar	1349

[a] Following the convention described in the text, these values are positive. Other sources may give them as negative values.
[b] In kcal/mole at 25°.

Problem: The reaction shown below is believed to be an important step in the combustion of natural gas. Calculate ΔH for the process.

$$CH_4(g) + \cdot OH(g) \rightarrow \cdot CH_3(g) + H_2O(g)$$

Solution: We cannot use Table 3–4 to calculate ΔH for this reaction, since values for $\cdot OH$ and $\cdot CH_3$ are not listed. However, looking at this reaction from the point of view of bond making and breaking, we see that all that is involved is breaking a C—H bond in the starting material and forming an O—H bond in the product. We can *estimate* ΔH from heats of formation of these bonds, *i.e.*, their bond strengths (Table 2–3):

$$\Delta H \approx \Delta H_f\text{O—H} - \Delta H_f\text{C—H}$$
$$\Delta H = -110\text{ kcal} - (-99\text{ kcal}) = -11\text{ kcal}$$

6. Calculate the equilibrium constant for the combustion of methane to carbon dioxide and water. Assume $\Delta G = \Delta H$ and $T = 25°C$. (Remember: $\Delta G = -RT \ln K_{eq}$)

3.6 CHEMICAL REACTIVITY AND ENERGY PROFILES

In order for two species to react they must first collide, converting kinetic energy present before collision to a minimum amount of potential energy after collision. This requires that the colliding species which are to react must possess a minimum kinetic energy. Since there is a distribution of molecular velocities at any temperature, only a fraction of colliding molecules possess the necessary kinetic energy. Those molecules which do not have enough kinetic energy simply bounce off one another. The minimum amount of energy necessary for a reaction to occur is the **energy of activation** of that reaction. This value varies from reaction to reaction.

Consider the three curves in Figure 3–7. Note that each pair of energy minima (ground states **1** and **2**) is separated by an energy maximum (the **transition state**). The species corresponding to ground states **1** and **2** and the transition state are "starting material," "product," and "activated complex," respectively. *The terms "transition state" and "activated complex" are often used interchangeably, although the transition state is a place along the energy profile while the activated complex is the species present at the transition state.* The energy of activation of the forward reaction is the energy required to go from ground state **1** to the transition state. (The energy of activation for the reverse process is the energy required to go from ground state **2** to the same transition state.)

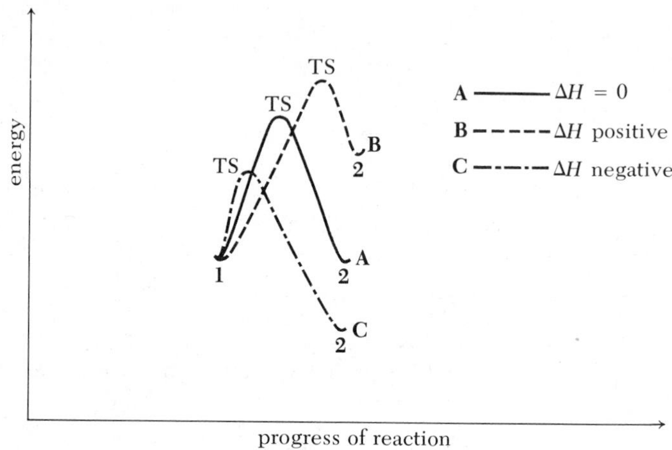

Figure 3-7 Representative energy profiles. The displacement of the transition state from the midpoint between starting material, **1**, and product, **2**, has been accentuated intentionally. It is assumed that ΔS equals zero for reactions **A**, **B**, and **C**. TS = transition state.

While the energy profiles in Figure 3–7 do not correspond to any specific reaction, they do convey the idea that an activated complex will resemble the ground state material that is closest to it in energy. For example, the more exothermic a reaction becomes, the more the transition state will be like the starting material. Curve **B** also shows that in an endothermic reaction the energy of activation must be equal to or greater than the energy difference between the two ground states.

The energy profile displays a system in equilibrium, for which K_{eq} is given (as noted on p. 59) by $\Delta G = -RT \ln K_{eq}$. Since ΔH often is used to approximate ΔG, K_{eq} is only *approximately* dictated by ΔH. A good rule of thumb is that K_{eq} will be >1 if ΔH is more negative than -15 kcal and <1 if ΔH is more positive than $+15$ kcal.

It is clear that it is the activation energy which controls the spontaneity of a reaction *and* that reactions can be extremely exothermic (*e.g.*, the combustion of methane) yet not be spontaneous. Those reactions which appear to be spontaneous (*i.e.*, those which require no energy input to begin) actually derive their energy of activation from collisions occurring at room temperature.

The Rate of Reaction. The study of the experimental property known as *reaction rate* falls within the province of **chemical kinetics.** Kinetics is treated in detail in courses in physical chemistry and in several monographs. Here we will examine only a classic kinetic relationship, the *Arrhenius equation.*

Reaction rates increase with an increase in temperature, approximately doubling with every increase of 10°C. Arrhenius is credited with the first quantitative relationship between rate and temperature,

$$k = Ae^{-E_a/RT}$$

where k = specific rate constant (a proportionality factor relating the actual rate of reaction to the concentration of the various reactants; when all concentrations are unity, the overall rate equals k)

E_a = the Arrhenius activation energy (a constant dependent upon the specific reaction)

A = the Arrhenius A factor (also called the pre-exponential factor)

T = temperature (°K)
R = universal gas constant, 1.987 cal deg^{-1} mole^{-1}

The Arrhenius equation is commonly used by organic chemists to interpret kinetic data, although the equation is empirical in origin and does not apply to multi-step processes—only to discrete, single-step reactions. A plot of ln *k versus* $1/T$ produces a straight line with a negative slope. The slope of this line can be used to calculate the Arrhenius energy of activation (E_a), as shown in Figure 3–8.

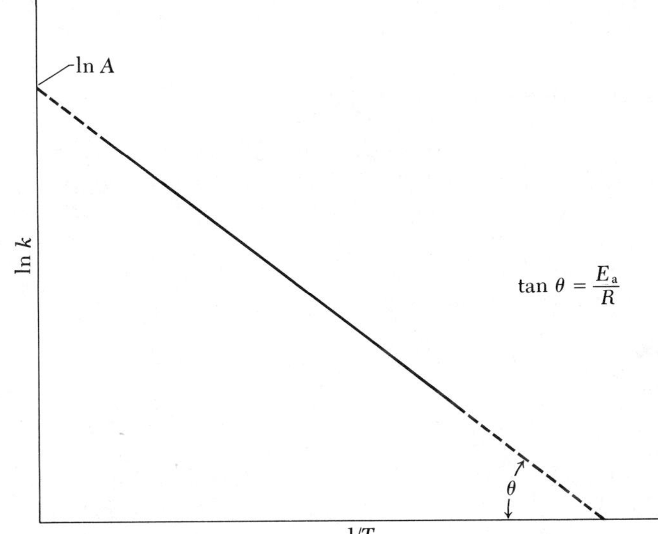

Figure 3-8 The Arrhenius plot, a graphical method for determining the Arrhenius activation energy, E_a, and the Arrhenius A factor.

The Arrhenius activation energy is not the only "energy of activation" associated with a given reaction. However, it is related to the theoretically derived activation energies which are implied in most energy profiles, and is similar to them in magnitude.

3.7 HALOGENATION OF ALKANES

Alkanes are important starting materials for the synthesis of a group of compounds called *alkyl halides,* alkanes in which one or more hydrogens have been replaced by a halogen. Next to alkane combustion, alkane halogenation is the most important reaction of alkanes and the source of compounds which are used as solvents, anesthetics, insecticides, and bacteriocides. We shall, therefore, examine the synthesis of alkyl halides from alkanes. Bear in mind that alkane halogenation, while it is extremely important to the chemical industry, rarely is performed in the synthetic chemistry laboratory.

$$CH_3—CH_2—CH_2—CH_3 + Cl_2 \rightarrow CH_3—\underset{}{\overset{Cl}{\overset{|}{C}}}H—CH_2—CH_3 + HCl$$ an alkane halogenation

butane — 2-chlorobutane (an alkyl halide)

Data—The Chlorination of Methane. While nothing appears to happen when methane and chlorine are mixed in a flask at room temperature in the dark, chlorine does react with methane under the proper conditions. The following statements apply to the chlorination of methane:

1. Reaction does not occur in the dark at low temperatures.
2. Reaction is induced by ultraviolet light or heat.

3. Reaction is quite rapid, even explosive, when it occurs.
4. Light-induced reactions produce many molecules of product for each quantum of light energy (photon) absorbed.
5. Reaction produces a variety of products, including HCl, CH_3Cl (chloromethane, methyl chloride), CH_2Cl_2 (dichloromethane, methylene chloride), $CHCl_3$ (trichloromethane, chloroform), CCl_4 (tetrachloromethane, carbon tetrachloride) and ethane. Other products are formed in lesser amounts.

$$CH_4 + Cl_2 \xrightarrow[\text{light}]{\text{heat or}} CH_3Cl + CH_2Cl_2 + CHCl_3 + CCl_4 + C_2H_6 + \text{other products}$$

6. Reaction is exothermic.

How Can We Explain These Observations? Postulation of a Mechanism. Armed with the above facts, we can construct a mechanism for this reaction. By **mechanism** we mean a detailed-as-possible account of how reactants become products. Since our knowledge increases with time, what was once an accepted mechanism may become but a part of the currently accepted mechanism or may be rejected completely. The following sequence of steps is the accepted mechanism for the chlorination of methane:

STEP	*REACTION*	*DESCRIPTION*
A	$Cl_2 \rightarrow 2Cl\cdot$ (chlorine radical)	initiation
B	$Cl\cdot + CH_4 \rightarrow HCl + \cdot CH_3$ (methyl radical)	propagation
C	$\cdot CH_3 + Cl_2 \rightarrow CH_3Cl + \cdot Cl$	propagation
D	$2Cl\cdot \rightarrow Cl_2$	termination

This type of reaction, in which an initiation step leads to two or more self-perpetuating reactions (propagation steps) which are then followed by some termination process, is called a *chain reaction*. This particular sequence involves species (radicals) with an odd number of electrons and is, therefore, a *radical chain reaction* or *free radical chain reaction*.

The methyl radical, $H_3C\cdot$, plays a critical role in the halogenation of methane. It is a nearly-planar, extremely energetic species. Its carbon is sp^2 hybridized, the odd electron being located in the unhybridized p-orbital. Being an electron-deficient species, *i.e.*, one which lacks an outer octet, its reactions usually are those which will provide it with eight outer electrons. It accomplishes this objective in free-radical chlorination by abstracting a chlorine atom from a chlorine molecule.

$\underset{1s}{\uparrow\downarrow}$ $\underset{2sp^2}{\overset{H}{\uparrow\downarrow}}$ $\underset{2sp^2}{\overset{H}{\uparrow\downarrow}}$ $\underset{2sp^2}{\overset{H}{\uparrow\downarrow}}$ $\underset{2p}{\uparrow}$ — methyl radical (sp^2)

methyl radical + :C̈l:C̈l: ⟶ methyl chloride ($H_3C{:}\ddot{C}l{:}$) + $\cdot\ddot{C}l{:}$ chlorine radical

methyl radical · chlorine molecule · methyl chloride · chlorine radical

methyl radical abstracts a chlorine radical

Given this mechanism, let us rationalize the facts summarized earlier. (To be acceptable, a mechanism must explain everything that is known about a reaction.) Step **A** involves the cleavage of the Cl—Cl bond into two radicals and consumes energy ($\Delta H = +58$ kcal/mole). Since each atom involved in the bond received one electron, this is an example of a homolysis. Both low temperature and darkness (which prevents the introduction of electromagnetic energy [light] to the Cl—Cl bond) act to prevent step **A** from occurring. When enough energy is provided, the Cl—Cl bond cleaves. Once initiated, the chain reaction proceeds to generate product.

7. Suggest a reason for the fact that ultraviolet light causes Cl_2 (gas) to cleave homolytically, to produce two chlorine radicals, rather than heterolytically, to produce $Cl^{\oplus}$ and $Cl^{\ominus}$.

Why does one photon produce so many molecules of product? The answer to this is at the heart of the definition of a chain reaction. Once initiated, the reaction cycles (**B** → **C** → **B** → **C** → . . .) and produces many product molecules, consuming CH_4 and Cl_2 along the way. The number of such cycles occurring before the termination step defines the *chain length* of the overall reaction. In this particular reaction the chain length may reach 10^4. The chain length depends upon the exact experimental conditions.

As the amount of methyl chloride that is present increases, it can replace CH_4 in step **B** of the original sequence, and now two new propagating steps, **B′** and **C′**, are introduced:

$$\mathbf{B'} \quad \cdot Cl + CH_3Cl \rightarrow HCl + \cdot CH_2Cl$$

$$\mathbf{C'} \quad \cdot CH_2Cl + Cl_2 \rightarrow CH_2Cl_2 + \cdot Cl$$

This sequence produces methylene chloride. Similar steps, using CH_2Cl_2 and $CHCl_3$ as reactants, eventually produce carbon tetrachloride. (Can you write out these propagation steps?)

Ethane formation is explained by noting that step **B** produces a methyl radical. This highly reactive species will, as the amount of Cl_2 diminishes, find itself more likely to encounter another methyl radical. Such collisions, which also are chain terminating steps because they consume radicals, produce ethane.

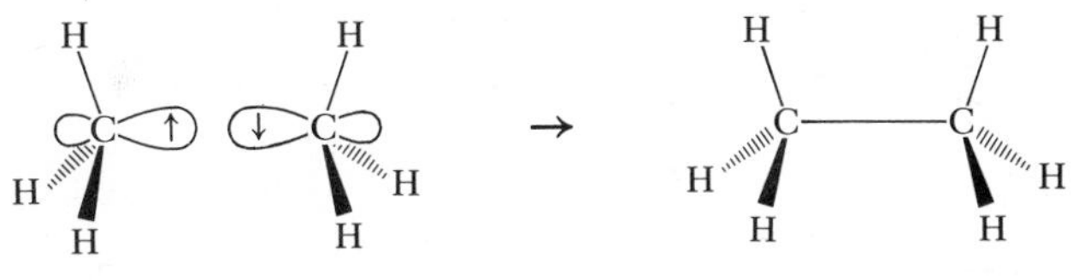

methyl radicals → ethane

Finally, let us explain the exothermicity of the reaction. Our approach will involve an examination of the thermochemistry of steps **A** through **D**, since these add up to a balanced equation for the reaction.

STEP **A**	$Cl_2 \rightarrow 2Cl\cdot$	$\Delta H = +58$ kcal/mole
STEP **B**	$\cdot Cl + H_3C{-}H \rightarrow H{-}Cl + \cdot CH_3$	$\Delta H = -1$ kcal/mole
STEP **C**	$\cdot CH_3 + Cl{-}Cl \rightarrow H_3C{-}Cl + \cdot Cl$	$\Delta H = -23$ kcal/mole
STEP **D**	$2Cl\cdot \rightarrow Cl_2$	$\Delta H = -58$ kcal/mole

Addition of these four equations, and the associated ΔH values, indicates a net ΔH of -24 kcal/mole of methyl chloride produced (or methane or chlorine consumed). This calculation was made for a chain length of 1. What would the net ΔH be for a chain length of 2?

Only a detailed analysis of the individual steps explains why energy must be supplied to get this exothermic reaction started.

Can We Test the Proposed Mechanism? To some degree the mechanism has already been tested, since it explains all of the observations presented earlier. However, questioning a proposed mechanism is a common procedure among chemists, either because they wish to substantiate it or because they favor an alternative mechanism.

1. Influence of Molecular Oxygen on the Reaction. Oxygen slows down many radical reactions by combining with a radical to form a less-reactive *peroxy radical.* Because of this, oxygen is termed a *radical inhibitor.*

$$\underset{\text{alkyl radical}}{R\cdot} \quad + \quad \underset{\text{molecular oxygen}}{O_2} \quad \rightarrow \quad \underset{\text{alkyl peroxy radical}}{R{-}\ddot{\underset{\cdot\cdot}{O}}{-}\ddot{\underset{\cdot\cdot}{O}}\cdot}$$

(Reviewing the m.o. picture of O_2 [Figure 2–13] may help you to explain oxygen's propensity for reacting with radicals.) If the chlorination of methane is attempted in a vessel containing a small amount of oxygen, the reaction will not proceed until all of the oxygen is consumed. The time elapsed before product formation begins, called the *induction period,* is explained by assuming that methyl radicals are produced but are *trapped* by O_2, thus interrupting the chain.

2. Influence of Lead Tetraethyl on the Reaction. Covalent bonds can be formed between carbon and some metallic elements. An illustration of the resultant *organometallic compound* is lead tetraethyl (or tetraethyl lead), $Pb(C_2H_5)_4$. When heated to approximately 150°, it decomposes into metallic lead and ethyl radicals. When even a small amount (about 0.1%) of lead tetraethyl is added to a mixture of chlorine and methane, the temperature required to initiate reaction drops from 400° to 150°. This observation supports the proposed mechanism, since ethyl radicals are expected to react with molecular chlorine and provide the chlorine radicals otherwise produced by initiation step **A.** Lead tetraethyl is a *radical initiator,* a compound which produces radicals under mild conditions.

$$H_5C_2{-}\overset{\displaystyle C_2H_5}{\underset{\displaystyle C_2H_5}{Pb}}{-}C_2H_5 \xrightarrow{\text{heat}} Pb + 4\ \cdot C_2H_5$$

$$CH_3CH_2\cdot + Cl_2 \rightarrow CH_3CH_2{-}Cl + Cl\cdot \qquad \text{alternate initiation step}$$

8. A commonly used free-radical initiator is AIBN (azobisisobutyronitrile). What are the products of its decomposition in an inert solvent?

$$CH_3{-}\overset{\displaystyle CN}{\underset{\displaystyle CH_3}{C}}{-}\ddot{N}{=}\ddot{N}{-}\overset{\displaystyle CN}{\underset{\displaystyle CH_3}{C}}{-}CH_3$$

AIBN

Reactions of Other Halogens with Methane. The observed ease of reaction of methane with the halogens is $F_2 > Cl_2 > Br_2 > I_2$. This pattern of relative rates follows the ΔH values of the overall reactions, as can be seen in Table 3–6. Although it is true in this case, we must not always expect such a parallel.

TABLE 3-6 Enthalpy Changes (ΔH) Associated with Methane Halogenation

REACTION	F	Cl	Br	I	
	HALOGEN				
$X_2 \rightarrow 2X\cdot$	+37	+58	+46	+36	$= \Delta H^a$
$X\cdot + CH_4 \rightarrow \cdot CH_3 + H{-}X$	−33	−1	+15	+31	$= \Delta H^a$
$\cdot CH_3 + X_2 \rightarrow CH_3X + X\cdot$	−71	−23	−21	−17	$= \Delta H^a$
$2X\cdot \rightarrow X_2$	−37	−58	−46	−36	$= \Delta H^a$
	−104	−24	−6	+14	$= \Delta H_{net}{}^a$

[a] In kcal/mole.

There are several things which are not apparent from the data in Table 3-6. For example, they do not tell you that fluorine reacts explosively with methane in the dark and at low temperatures. They do not inform you that iodine does not react with methane to produce methyl iodide and that only chlorine and bromine react at rates which can be studied conveniently. It should be added that when two species "do not react," this could mean that K_{eq} for the reaction is unfavorable, that the activation barrier is very high, or both. Moreover, if any step in a chain reaction does not occur readily, then the entire reaction will be slow.

Using the data in Table 3-6 (and other requisite information), we can construct energy profiles for the steps in the reaction of methane with chlorine and bromine. Composite energy profiles for the propagation steps in these two reactions are presented in Figure 3-9.

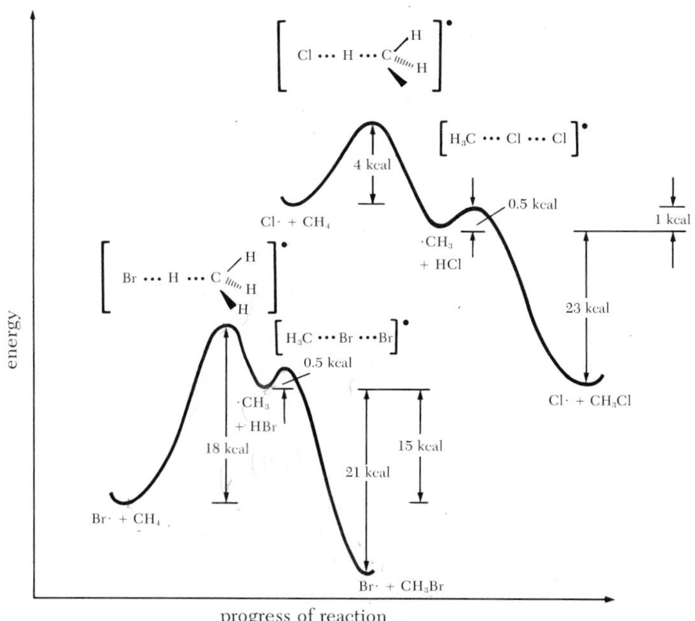

Figure 3-9 The reaction of chlorine and bromine radicals with methane. The initiation and termination steps have been omitted. One curve has been placed above the other for convenience; no energetic significance is associated with this placement. The slow step in both processes is the reaction of the halogen radical with methane, that is, the first step shown. The species in brackets represent the appropriate activated complexes. Energy is in kcal/mole.

Every step in fluorination, except the first, is exothermic. The difficulty that a system has in dissipating the heat generated in such reactions makes it all but impossible to control the reaction. The amount of energy liberated is large enough that all C—H bonds are quickly broken and replaced by C—F bonds. The direct fluorination of methane even produces some pure carbon.

Although direct fluorination of hydrocarbons generally is synthetically valueless, fluorine can be substituted for hydrogen *via* the use of a less reactive fluorinating agent, such as cobaltic fluoride (CoF_3). Cobaltic fluoride is prepared by the reaction of cobaltous fluoride (CoF_2) with F_2. After excess F_2 vapor has been removed, methane is passed over the CoF_3 bed. The methane is fluorinated and CoF_3 reduced to CoF_2. Regeneration of the CoF_3 by reaction of the cobaltous fluoride with F_2 allows this "batch process" to be repeated.

$$2CoF_2 + F_2 \rightarrow 2CoF_3$$

$$CH_4 + 8CoF_3 \rightarrow CF_4 + 8CoF_2 + 4HF$$

Chlorination of Higher Alkanes. Chlorination of ethane produces only one monochloro product, C_2H_5Cl (ethyl chloride or chloroethane), supporting the idea that, allowing for free rotation, all of the hydrogens in ethane are chemically equivalent to each other. The chlorination of ethane follows a mechanism similar to that presented for the chlorination of methane.

$$H_3C{-}CH_3 + Cl_2 \rightarrow H_3C{-}CH_2{-}Cl + HCl$$

It does, however, differ in that ethylene (C_2H_4) is a by-product, arising from the *disproportionation* of the ethyl radical ($CH_3CH_2\cdot$). A hydrogen atom is transferred from one ethyl radical to a second one, producing ethylene and ethane. ("Fish-hook" arrows are sometimes used to indicate the movement of *one* electron, the normal arrowhead then being reserved for two-electron movements.)

$$CH_3CH_2\cdot \;\; H{-}CH_2{-}\dot{C}H_2 \rightarrow CH_3{-}CH_3 + H_2C{=}CH_2$$

This reaction is not a valuable means of synthesizing ethylene; however, it is useful in that it suggests the transient existence of the ethyl radical and, thus, supports the accepted mechanism.

In propane, unlike ethane, not all of the hydrogens are equivalent. Six of them, those of the methyl groups, are primary (1°), while two of them, those of the methylene group, are secondary (2°). A *primary hydrogen* is one which is bonded to a carbon which, in turn, is bonded to only one other carbon. A *secondary hydrogen* is one which is bonded to a carbon which, in turn, is bonded to two other carbons. A *tertiary hydrogen* (3°) is bonded to a carbon which is bonded to three carbons. The carbons bearing these hydrogens also are called 1°, 2°, and 3°.

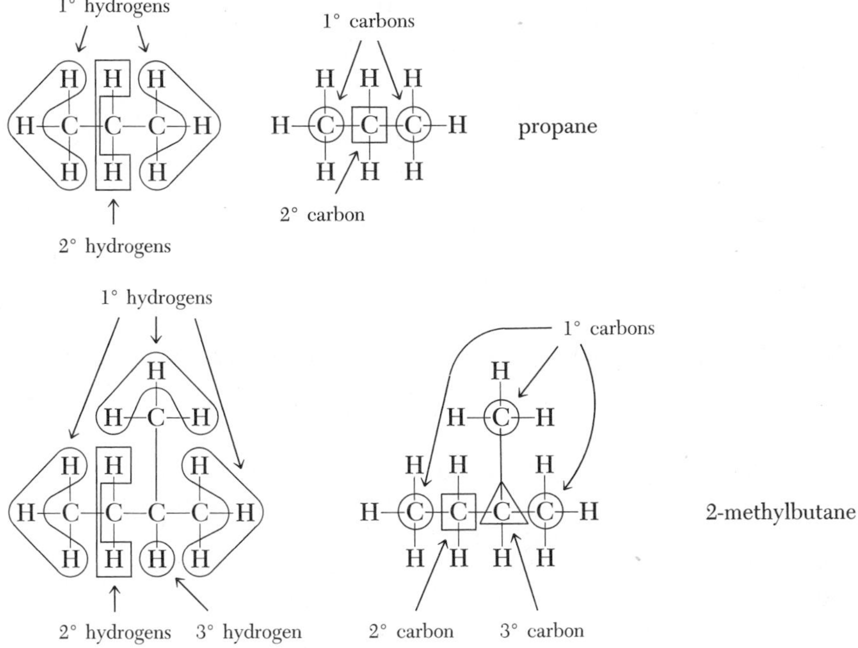

On purely statistical grounds, one would expect to replace the primary hydrogens of propane six times for every two replacements of the secondary hydrogens. If the reaction is carried out at high temperatures (>450°) one does observe the predicted product ratio, 3:1 (*n*-propylchloride: isopropyl chloride). If the reaction is carried out at low temperatures the reaction produces more isopropyl chloride (55%) than *n*-propyl chloride (45%)—the less populous 2° hydrogen accounting for the larger portion of product.

$$CH_3CH_2CH_3 + Cl_2 \xrightarrow[\text{(photon)}]{h\nu} H_3C{-}CH_2{-}CH_2Cl + CH_3{-}\underset{\displaystyle H}{\overset{\displaystyle Cl}{C}}{-}CH_3$$

propane — *n*-propyl chloride — isopropyl chloride

We must conclude that at high temperatures the chlorine radical is unselective and will abstract any hydrogen with which it collides. (How is this explained in view of our earlier discussion of "chemical reactivity"?) At low temperatures the chlorine radical becomes more selective (a general property of all reacting species) and will not abstract every hydrogen with which it collides. This suggests that it is inherently easier to abstract a 2° H than a 1° H. A number of studies have suggested that, for abstraction at room temperature by Cl·, 2° H's are about four times as reactive as 1° H's.

A statistical product distribution is not observed when 2-methylpropane (isobutane) reacts with Cl_2 under mild conditions. Rather, only 64% of the product results from 1° H abstraction, while attack at the single 3° H accounts for the remainder.

$$CH_3-\underset{\displaystyle H}{\overset{\displaystyle CH_3}{C}}-CH_3 + Cl_2 \xrightarrow{h\nu} CH_3-\underset{\displaystyle H}{\overset{\displaystyle CH_3}{C}}-CH_2-Cl + CH_3-\underset{\displaystyle Cl}{\overset{\displaystyle CH_3}{C}}-CH_3$$

isobutyl chloride (64%) — *t*-butyl chloride (36%)

On the basis of a variety of data, the relative reactivities of 3°, 2°, and 1° hydrogens toward Cl· at room temperature are 5: 4: 1 *per hydrogen*. For Br· the comparable values are 2000: 100: 1. What does this suggest concerning the "inherent" reactivity of Cl· *vs.* Br·?

9. An alkane, molecular weight 72, can form only one monochloro derivative. What is its structure? How many dichloro derivatives can this alkane form?

Why Are Different Protons Abstracted at Different Rates? The reaction of an alkane with a chlorine radical involves the breaking of a C—H bond and the formation of an H—Cl bond. Examine the bonds being made and broken within a group of alkanes: the energy contribution of the HCl bond will be constant but the strength of the C—H bond will vary, as will the energy content of the radical being produced.

$$R-\underset{|}{\overset{|}{C}}-H + Cl\cdot \rightarrow R-\underset{|}{\overset{|}{C}}\cdot + H-Cl$$

From the relative rates of abstraction, it is clear that it requires less energy to abstract a 3° H than a 2° H, and so on. This does not *demand* that a 3° radical[a] be more stable than a 2° radical (although this is most often true). The only thing such reactivity data require is that the 3° radical formed by abstraction of the 3° H be more stable compared to its parent hydrocarbon than is the 2° radical compared to its progenitor hydrocarbon.

[a] Radicals are identified by the number of groups residing on the carbon bearing the odd electrons: 3°, $R_3C\cdot$, 2°, $R_2HC\cdot$, 1°, $RH_2C\cdot$.

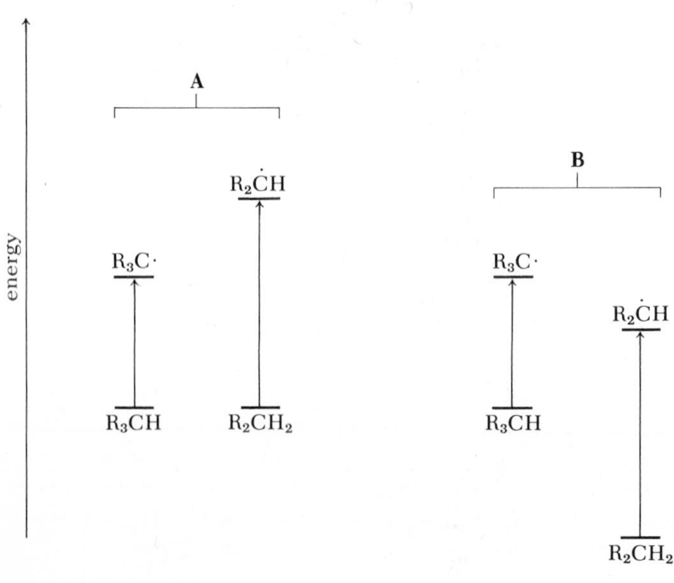

Figure 3-10 Alternate schemes for relating radical stability to ease of hydrogen abstraction. The stability order suggested in **A** is more commonly observed; *i.e.*, a 3° radical is more stable than a 2° radical. Each arrow corresponds to the loss of a hydrogen radical (*i.e.*, hydrogen atom) and, therefore, to the generation of an alkyl radical (upper species) from an alkane (lower species).

This point is illustrated in Figure 3–10. In both **A** and **B** less energy is required to abstract a 3° H than a 2° H. However, in **A** the 3° radical is more stable than the 2° radical while in **B** the 2° radical is more stable (lower in ground state energy). Quite obviously, E_a for H abstraction need not reflect the relative stabilities of the products.

The general order of radical stability is $3° > 2° > 1° > \cdot CH_3$. This order of stability appears to be due to the interaction of the odd electron with the C—H bonds on adjacent carbons. Since the number of such bonds increases in going from 1° to 2° to 3° radical, one expects this order of stability.

3.8 NITRATION OF ALKANES

Alkanes react with nitric acid or dinitrogen tetroxide, N_2O_4, in the vapor phase to form *nitro* derivatives ($R—NO_2$). Nitration of alkanes is a very significant industrial reaction but is of extremely limited laboratory utility.

Nitromethane (CH_3NO_2) and nitroethane ($CH_3CH_2NO_2$) are important solvents, and nitromethane has been used as an exotic fuel in automobile racing. The vapor phase nitration of propane produces 1-nitropropane ($CH_3CH_2CH_2NO_2$) in 25% yield. This is an unusual reaction since it also produces nitromethane and nitroethane, as well as 2-nitropropane ($CH_3CHNO_2CH_3$). Formation of these smaller nitroalkanes requires that C—C bond cleavage accompany vapor phase nitration.

$$CH_3CH_2CH_3 + N_2O_4 \rightarrow CH_3NO_2 + CH_3CH_2NO_2 + CH_3(CH_2)_2NO_2 + (CH_3)_2CHNO_2$$

The isomeric nitropropanes are important solvents for cellulose acetate, lacquers, synthetic rubbers and other organic materials. All of the simple nitroalkanes are flammable and quite toxic.

APPENDIX

DETERMINATION OF MOLECULAR FORMULA USING COMBUSTION ANALYSIS

Complete combustion of an organic compound, carried out in an atmosphere of pure oxygen and with the aid of a catalyst (*e.g.*, vanadium pentoxide, V_2O_5), is a valuable analytical procedure. By burning a known weight of compound and measuring the weight of oxides produced, one can calculate the compound's empirical formula. Since combustion analysis cannot determine the amount of oxygen present (Why?), oxygen is usually determined by difference. The *presence* of oxygen can usually be determined by spectral procedures.

Determination of an approximate molecular weight can be used to convert a compound's empirical formula into its exact molecular formula. In the problem which follows, the molecular weight has been determined by the Rast procedure. This method is based upon the lowering (depression) of the melting point (freezing point) of a pure substance (the solvent) by the addition of some compound (the solute) to that substance. The equation below is used to calculate approximate molecular weights from freezing point depression data. This equation contains a constant, the cryoscopic constant, which is unique for every solvent (Table 3–7). The larger the constant, the greater will be the depression of the freezing point of the solvent for a given number of moles of solute added to it. Consequently, the use of solvents with large cryoscopic constants leads to more accurate molecular weight determinations. Camphor, with a cryoscopic constant of 39.7, is the most useful solvent.

$$\text{molecular weight of solute} = K_f \frac{1000\, w_2}{\Delta T\, w_1}$$

where K_f = cryoscopic constant of the solvent (the pure compound, present in excess, whose mp is being lowered)
w_2 = grams of solute in mixture
w_1 = grams of solvent in mixture
ΔT = [mp of pure solvent] − [mp of mixture]

TABLE 3-7 Cryoscopic Constants

SOLVENT	MELTING POINT[a]	CRYOSCOPIC CONSTANT
water	0°	1.86
acetic acid	16°	3.86
benzene	5°	5.12
camphene	49°	31.0
camphor	178°	39.7

[a] For pure solvent.

Problem: Calculate the molecular formula of a compound, 0.1824 g of which produced 0.2681 g of carbon dioxide and 0.1090 g of water upon combustion. When 25 g of this compound was dissolved in 100 g of water, the freezing point of the solution was 2.2° lower than that of pure water. (Note that, as organic compounds go, this compound is extraordinarily soluble in water.)

Solution: The ratio of carbon in carbon dioxide is 12/44. Therefore,

$$\text{weight of carbon in sample} = 12/44 \times 0.2681\text{ g} = 0.07312\text{ g}$$

The ratio of hydrogen in water is 2/18. Therefore,

$$\text{weight of hydrogen in sample} = 2/18 \times 0.1090\text{ g} = 0.01211\text{ g}$$

The percentage of carbon in the sample = 0.07312/0.1824 × 100 = 40.09%.
The percentage of hydrogen in the sample = 0.01211/0.1824 × 100 = 6.64%.
The remainder, 100 − (40.09 + 6.64), must be oxygen = 53.27%.

Each weight percentage can be divided by the atomic weight of the particular element to give a ratio of atoms present in the sample:

carbon: atomic ratio = 40.09/12 = 3.34
hydrogen: atomic ratio = 6.64/1 = 6.64
oxygen: atomic ratio = 53.27/16 = 3.33

The atomic ratio is then divided through by the lowest common denominator, 3.33 in this instance, to give integral atomic ratios present in the compound.

carbon: 3.34/3.33 = 1
hydrogen: 6.64/3.33 = 2
oxygen: 3.33/3.33 = 1

Thus, the *empirical formula* of the substance is CH_2O. Calculation of the molecular formula requires a calculation of the molecular weight.

The molecular weight can be calculated from the data given about the freezing point depression of water. Using the equation given earlier:

$$\text{molecular weight} = \frac{1.86 \times 1000 \times 25}{2.2 \times 100}$$

$$= 2.1 \times 10^2$$

The empirical formula "weighed" 30; therefore, there are seven empirical formula units in the molecular formula, which must be

$$C_7H_{14}O_7.$$

IMPORTANT TERMS

Activated complex: The species present at the transition state of a reaction. It is the least abundant species present in a simple, one-step reaction.

Alkyl group: The fragment which remains when a hydrogen is removed from an alkane. Among the more common alkyl groups are

$—CH_3$	(methyl group)
$—CH_2CH_3$	(ethyl group)
$—CH(CH_3)_2$	(isopropyl group)
$—C(CH_3)_3$	(*t*-butyl group)

Chain reaction: A series of consecutive reactions which have a repetitive portion (the propagation sequence). Many molecules of product may result from one chain. A chain reaction need not involve free radicals, although those discussed in this chapter are free radical chain reactions.

Condensed structural formula: A structure (usually linear) which relates the sequence of atoms in the molecule but does not show any explicit bonds.

$$CH_3CHCH_3CH_2CH_2CH_3$$

condensed structural formula of 2-methylpentane

Conformation: A specific geometry of a molecule. Conformations of a single compound differ from one another by the extent of rotation around one or more σ bonds.

Conformer: A conformation corresponding to an energy minimum along an energy profile. A stable conformation.

Cryoscopic constant: A measure of how much the freezing point of a compound is depressed by the addition of impurity. A group of such constants is collected in Table 3–7.

Eclipsed conformation: That conformation in which groups on adjacent atoms reside one behind the other as seen in a Newman projection. The torsional angle is 0° in an eclipsed conformation. Usually a high energy species.

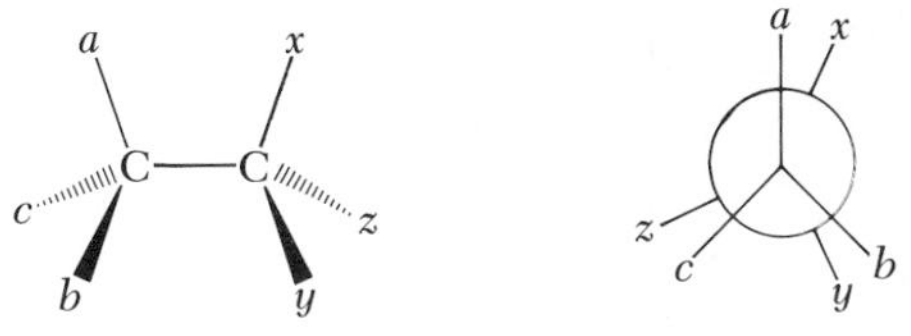

two views of an eclipsed conformation

Empirical formula: The lowest whole number ratio of atoms in a molecule, expressed as a formula. The empirical formula of mannose, $C_6H_{12}O_6$, is CH_2O.

Energy of activation: The minimum amount of energy required to convert a starting material into a specific activated complex (transition state).

Energy profile: A plot of the energy of a system (ordinate) against some parameter which measures the progress of the process being studied (*e.g.*, a chemical reaction or conformational change) (abscissa). The abscissa is generally labeled "progress of reaction" or "reaction coordinate."

Gauche conformation: A conformation with a torsional angle of 60° between two specific substituents.

Heat of combustion: The negative of the enthalpy or heat content change (ΔH) accompanying the complete conversion of one mole of a compound into carbon dioxide and water. A positively signed number.

Heat of formation: The enthalpy or heat content change accompanying the formation of a compound from its elements. Abbreviated ΔH_f, it is given by

$$\Delta H_f = [\text{heat content of compound}] - [\text{heat content of percursor elements}]$$

Most compounds have a negatively signed heat of formation.

Heat of reaction: The enthalpy or heat content change accompanying a reaction. Abbreviated ΔH, it is given by

$$\Delta H = [\text{heat content of products}] - [\text{heat content of starting materials}]$$

Isomers: Compounds with the same molecular formula. They may be divided into two varieties—structural isomers and stereoisomers.

Newman projection: A representation of the geometry around a bond, which is used to emphasize the torsional angle between various substituents. A Newman projection can be constructed by viewing along *any* bond and simply drawing what you see. For simplicity, the two atoms which reside at the bond termini are not shown explicity. However, the front terminus is indicated by the intersection of covalent bonds from it, while the back terminus is indicated by a circle.

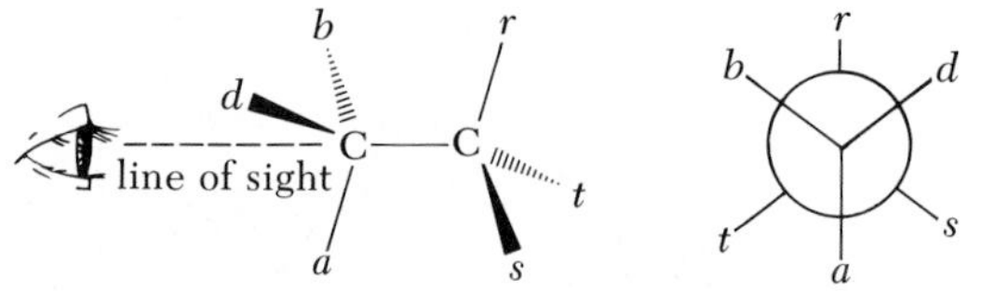

generation of a Newman projection of a staggered conformation

Radical: A species with an unpaired electron. Simple radicals usually are uncharged. Some atoms, *e.g.*, the chlorine atom, are radicals. The term "free radical" is sometimes used. (In older treatises the term "radical" was also used to mean "functional group.")

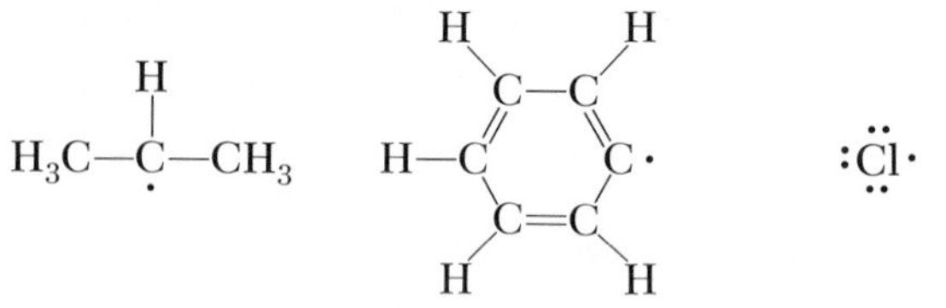

isopropyl radical   phenyl radical   chlorine radical

Radical initiator: A substance which decomposes under mild conditions to produce a radical. Lead tetraethyl is a typical radical initiator, making it useful as a gasoline additive.

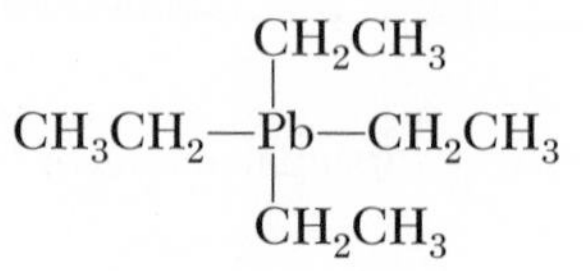

lead tetraethyl

Radical inhibitor: A substance which slows down radical reactions by combining with radicals involved in the reaction and preventing them from reacting or diverting them into other reaction paths. Oxygen is a common radical inhibitor.

Staggered conformation: A conformation with a torsional angle of 60° as seen in a Newman projection. Usually a conformer. The distinction between a staggered conformation and a gauche conformation is shown below.

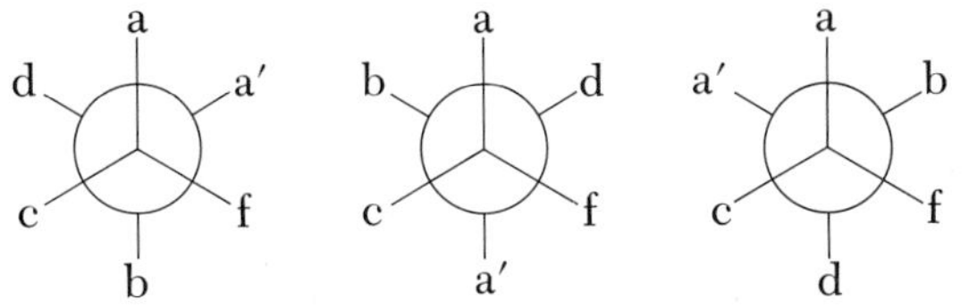

1	2	3
front: a, c, f; back: d, a′, b	front: a, c, f; back: b, d, a′	front: a, c, f; back: a′, b, d

1, **2** and **3** are all staggered, but only **1** and **2** are gauche with respect to groups a′ and f

Stereoisomers: Isomers which have the same bonding sequence (*i.e.*, not structural isomers) but which differ in how the atoms are arranged in space. There are several types of stereoisomers, two of which are discussed in great detail in Chapter 4.

Structural formula: An extended representation of a molecule, showing all covalent bonds. The structural formula below is that of α-alanine, an important amino acid found in proteins.

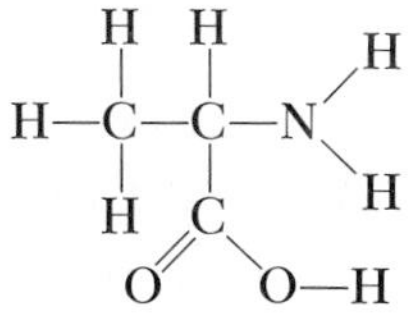

H–C(H)(H)–C(H)(C(=O)O–H)–N(H)H

α-alanine

Structural isomers: Isomers which have different atom-to-atom bonding sequences. α-Alanine and β-alanine are structural isomers of one another. Methyl ether and ethanol also are structural isomers of one another.

CH_3—$CH(NH_2)CO_2H$
α-alanine

$CH_2(NH_2)$—CH_2CO_2H
β-alanine

CH_3OCH_3
methyl ether

CH_3CH_2OH
ethanol

Transition state: An energy maximum on an energy profile. The species present at the transition state is the activated complex. The term is applicable whether the energy profile depicts a conformational change or a chemical reaction.

PROBLEMS

10. Provide an acceptable name for each of the following:
    (a) $(CH_3)_2CH(CH_2)_8CH_3$
    (b) $(CH_3)_2CHCH(CH_3)_2$
    (c) $CH_3(CH_2)_3C(CH_3)(C_2H_5)CH(CH_3)C_2H_5$
11. Identify the 1°, 2° and 3° hydrogens and carbons in the following:
    (a) *n*-heptane
    (b) 2,2,4-trimethylpentane (sometimes erroneously called "isooctane")

(c) tetramethylmethane
(d) isooctane

12. Indicate the mistake(s) in the following and provide a structure and an acceptable name for each.
(a) methylbutane
(b) 4-butylpentane
(c) 4-methylpentane
(d) 2-isopropylhexane

13. The following sequence is suggested as an alternative mechanism for the halogenation of methane. Discuss the relative merits of this mechanism and the one provided in the text; consider (a) product analysis and (b) energetics.

$$Cl_2 \rightarrow 2Cl\cdot \text{ (initiation)}$$
$$\left.\begin{array}{r}\cdot Cl + CH_4 \rightarrow CH_3Cl + H\cdot \\ \cdot H + Cl_2 \rightarrow HCl + Cl\cdot\end{array}\right\} \text{ (propagation)}$$
$$2Cl\cdot \rightarrow Cl_2 \text{ (termination)}$$

14. Although the iodine molecule is the easiest halogen to cleave homolytically, it is the only halogen which does not react with methane. How can you account for this?

15. An important industrial reaction of ethane is its conversion to ethylene and hydrogen. This reaction has $\Delta H = 32$ kcal/mole and $\Delta S = 30$ cal/deg mole (entropy units, e.u.). What is the free energy change for the reaction at 25°? What is the equilibrium constant at 25°?

$$\underset{\text{ethane}}{C_2H_6} \xrightarrow[\text{catalyst}]{\text{heat}} \underset{\text{ethylene}}{C_2H_4} + H_2$$

16. The free energy of formation for *n*-pentane is -2.00 kcal/mole, while that of isopentane is -3.50 kcal/mole. Assuming a method for equilibration, calculate the expected equilibrium constant at 25°. (Remember that $R = 1.987$ cal deg^{-1} mole^{-1} and that $\ln X = 2.3 \log X$.)

17. Why does food cook faster in a pressure cooker?

18. When gaseous methyl chloride and sodium vapor are reacted at very high temperatures in a stream of hydrogen, methane is a major product. When a mixture of iodine vapor and nitrogen is substituted for the hydrogen, methyl iodide is the major product. (Under these conditions methyl chloride and iodine alone do not react.) Suggest a mechanism for these reactions.

19. *t*-Butyl peroxide has a bond dissociation energy of 37 kcal/mole and is often used to initiate free radical reactions. Given an Arrhenius *A* factor of 10^{12} sec^{-1}, calculate the rate constant for dissociation at 100°C. (To answer this you must read the section in small type on p. 70.)

$$(CH_3)_3C\ddot{\underset{\cdot\cdot}{O}}-\ddot{\underset{\cdot\cdot}{O}}C(CH_3)_3 + 37\text{ kcal} \rightarrow 2(CH_3)_3C-\ddot{\underset{\cdot\cdot}{O}}\cdot$$

*20. Calculate the cryoscopic constant that a solvent must have in order to display a 0.1° drop in melting point when 0.1 mole of solute (MW = 300) is dissolved in 2 kg of solvent.

*21. Compound X, molecular weight = 72, depressed the freezing point of benzene by 1.58° when dissolved in 15.0 g of benzene. How much X was dissolved in the benzene?

*22. Calculate the empirical formula for compound M, 0.614 g of which produced 0.88 g of CO_2 and 0.55 g of H_2O upon complete combustion. Does this compound contain oxygen?

---

*Based upon material in the Appendix.

STEREOISOMERISM IN ALIPHATIC COMPOUNDS

4.1 INTRODUCTION

There is no such thing as a "left hand hammer"; left and right hand hammers are the same. Similarly, there is no such thing as a "left hand teaspoon"—a teaspoon is a teaspoon is a teaspoon. In contrast, there *is* a left hand glove and a right hand glove, just as there are left and right hands. What permits the existence of two types of glove but only one type of teaspoon?

Assuming your hands to be of equal dexterity, striking a nail with a hammer held in either hand will produce identical results. However, if you attempt to place first one and then another of your hands in one member of a pair of gloves, you will encounter a perfect fit in one instance and an impossibly difficult task in the other. Why do your hands interact with the hammer and nail with equal facility? Why do your hands interact unequally with a single glove? The answers to these and related questions depend upon the symmetry properties of the objects being discussed and upon the extent of interaction between objects possessing different symmetry characteristics.

These everyday examples have analogies at the molecular level. We encounter molecules which, like the hammer, cannot be said to be "left handed" or "right handed" and some which, like the glove, have "left handed" and "right handed" forms. This journey into the influence of symmetry properties upon organic molecules begins with an examination of "handedness," emphasizing those geometric properties which permit *or* prevent a compound's possession of this molecular characteristic. In future chapters we will learn that these "handed" molecules react with equal facility with molecules having no handedness but that they react unequally with a species which, itself, has handedness. This is especially important to those interested in the chemistry of living systems since, as we shall observe throughout this text, all biological reactions involved handed molecules.

4.2 ENANTIOMERS

The relationship between your left and right hands is the same as that exhibited between any object which has a non-superimposable mirror image and that mirror image. Your hands, and any other similarly related mirror image pairs of objects, are said to be **enantiomers** or **enantiomorphs** of one another. While it may be difficult to put into words just how your hands differ from one another, a chemist would say that they differ in **configuration.** Indeed, since an object can have only one mirror image, he would add that they have *opposite* configurations.

The configuration of a *molecule* refers to the arrangement of its atoms or functional groups in space. In order to keep down the number of possibilities, the various *conformations* of a single compound are excluded when considering the possible configurations of a given structure. By way of illustration, **A** and **B** (below) differ in configuration and are enantiomers, while **C** and **D** are considered simply to have different conformations. Put slightly differently, **A** and **B** have opposite configurations, while **C** and **D** have identical configurations but different conformations.

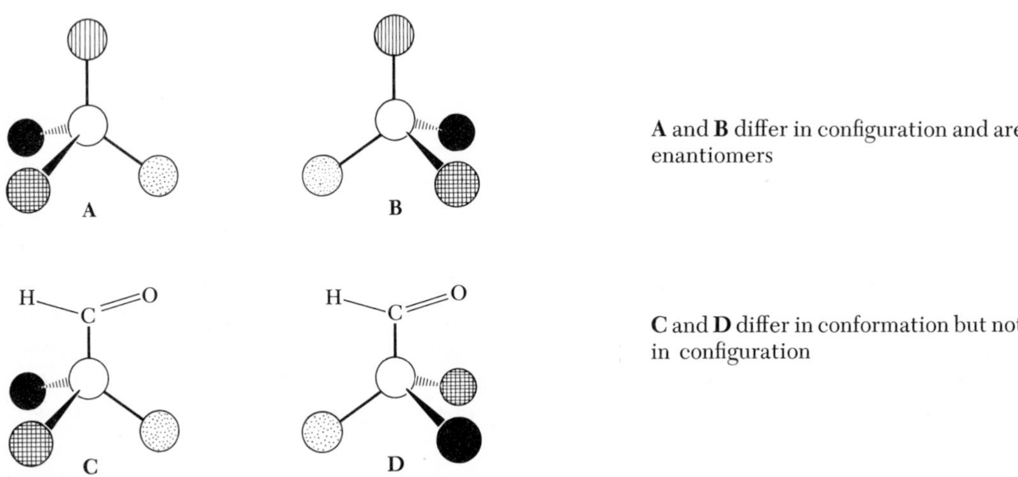

We shall soon realize that, whenever we wish to describe the opposite configurations of mirror image molecules, it will be necessary to speak in terms of our earliest learning experiences, using pairs of terms like *clockwise* and *counterclockwise*, *left handed* and *right handed*. The occurrence of such pairs of terms serves to emphasize that any object has one and only one mirror image.

Molecular Symmetry. The symmetry properties of any substance can be described in terms of the presence or absence of certain symmetry elements. The most important of these are (a) a plane of symmetry, (b) a point of symmetry, and (c) a simple axis of symmetry. A **plane of symmetry** is a plane which bisects an object in such a way as to make the pieces of the object on opposite sides of the plane enantiomeric (*i.e.*, mirror images). A teaspoon and a hammer both possess such a plane (Figure 4–1). Some of the planes of symmetry present

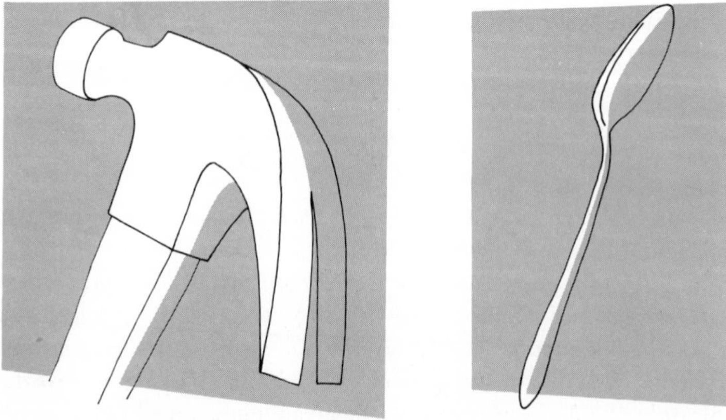

Figure 4–1 Planes of symmetry in a hammer and a teaspoon.

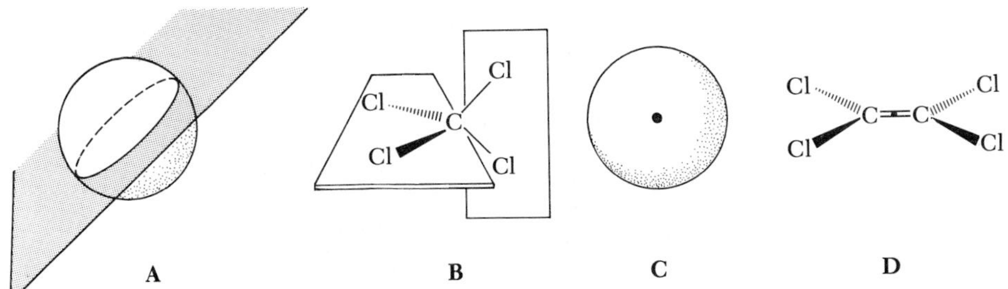

Figure 4-2 Planes and points of symmetry. Only one of an infinite number of planes of symmetry of the sphere is shown in **A.** Two of the six mutually perpendicular planes of symmetry in the carbon tetrachloride molecule are shown in **B.** All of the planes of symmetry of the sphere share the point (or center) of symmetry (**C**). How many planes of symmetry are present in **D**? Do they all contain the point of symmetry that is shown?

in a sphere and in carbon tetrachloride are indicated in Figure 4–2. A **point of symmetry** is a point from which two lines can be directed at an angle of 180° and always encounter the same environment at the same distance from that point. The center of a sphere corresponds to the only point of symmetry for that sphere (Figure 4–2). While some objects may possess more than one plane of symmetry, no object can possess more than one point of symmetry. Of course, some objects, such as your left hand, have neither a plane nor a point of symmetry.

A **simple axis of symmetry** is an axis which passes through a substance and produces a structure identical to the starting one after a rotation around that axis of $360°/n$. The "degree of foldedness" of the axis is given by the value of n. For example, the C—H bond in chloroform is coincident with the three-fold ($n = 3$) simple axis of symmetry of chloroform.

A three-fold simple axis of symmetry: - - - - - . The figure is replicated every 120° of rotation around the axis.

H
C
Cl Cl Cl

Generally, in order for a substance to exist as a pair of enantiomers, it must lack both a point and a plane of symmetry. Such substances are called **dissymmetric.** We can say that *if an object lacks both a point of symmetry and a plane of symmetry, then it will have a non-superimposable mirror image (enantiomer).* However, if an object has *either* a point *or* a plane of symmetry, then it will not be dissymmetric and will not have an enantiomer. An **asymmetric** compound is one which lacks all symmetry elements, including a point and a plane of symmetry. Therefore, all asymmetric compounds must be dissymmetric; since most simple dissymmetric compounds are asymmetric, it is common to find the terms "asymmetric" and "dissymmetric" used interchangeably.

1. Which of the following idealized objects is capable of existing as pairs of enantiomers?
   (a) a spool of thread
   (b) a spool of thread with all the thread and labels removed
   (c) a conch shell
   (d) a coiled cable connecting a telephone receiver to the instrument

(e) a donut (plain)
(f) a round-head screw
(g) a bicycle wheel
(h) a monkey-wrench
(i) a pair of scissors
(j) a baseball
(k) a propeller

2. Identify a plane of symmetry, point of symmetry, or simple axis of symmetry in each of the following objects.
(a) a bald truck tire (neglect labels)
(b) an apple
(c) a "ping-pong" ball
(d) a roll of adhesive tape (neglect labels)
(e) a piece of chalk
(f) a light bulb (glass portion only)
(g) a flashlight battery (neglect labels)
(h) the planet Saturn

This definition of dissymmetric is a bit simplified, since the point and plane of symmetry are but two special cases of a more general symmetry element called a **rotation-reflection axis** or the **alternating axis of symmetry.** This axis is identified when the following operations convert the initial object into an *identical* arrangement: (1) rotation of the molecule around an axis (going through the molecule) through an angle of $360°/n$; followed by (2) reflection of this new structure in a plane perpendicular to the rotational axis (Figure 4–3). Thus, a more correct definition of a dissymmetric compound is one which lacks a rotation-reflection axis.

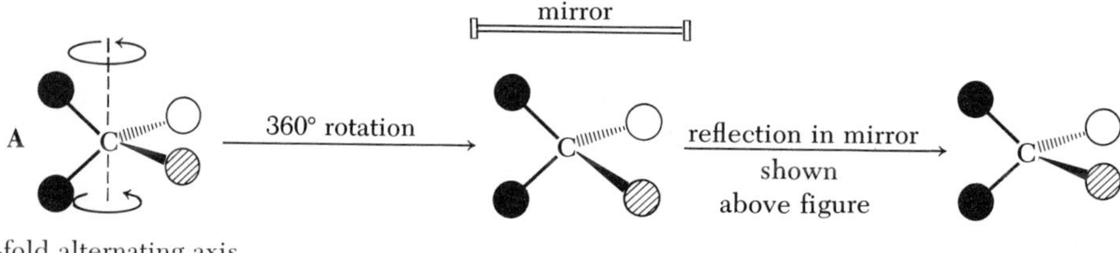

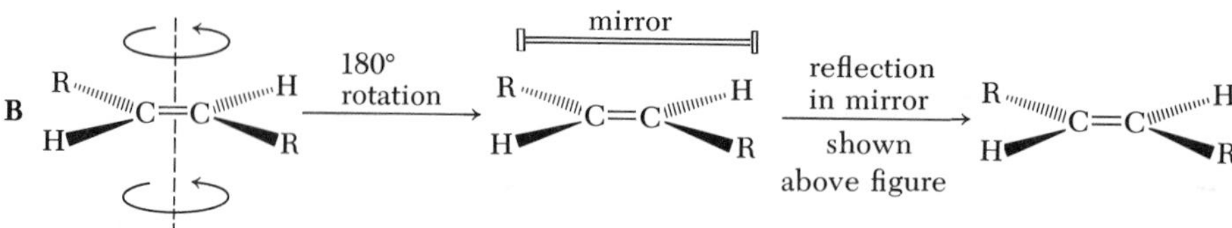

Figure 4–3 A one-fold (**A**) and a two-fold (**B**) alternating axis of symmetry. In **A** and **B** the reflected structure has been translated from a position on the other side of the mirror.

Because of confusion generated by the use (and misuse) of the terms "asymmetric" and "dissymmetric," another term has been introduced to describe whether or not a substance is capable of having an enantiomer. This term is **chiral.** Chemists now prefer to state that if a substance is chiral then it must have an enantiomer. Chirality, then, becomes the necessary and sufficient condition for the existence of enantiomers. The safest way to determine if a molecule is chiral is to build a model of it and of its mirror image and to see whether you can make exactly coincident all parts of one with the identical parts of the other. If they cannot be made to coincide completely, then the molecule is chiral.

With some practice you will be able to make sketches of three-dimensional structures and to use these sketches to determine whether two structures are enantiomers. But there is an even easier way to identify a chiral substance. In general, if a molecule lacks both a point of symmetry and a plane of symmetry, then it is chiral.

A **symmetric substance** is one which possesses any symmetry element other than the one-fold simple axis of symmetry. It most often is recognized by the presence of a plane of symmetry or a point of symmetry. A molecule that does not have an enantiomer is said to be **achiral.** The relationship between symmetry properties, chirality, and optical activity (a physical property to be discussed shortly) is summarized in Table 4–1.

TABLE 4-1 Relationship Between Symmetry, Chirality, and Optical Activity

Simple Axis	Alternating Axis[a]	Symmetry	Chirality	Optical Activity
present	present	symmetric	achiral	no
absent	present	symmetric	achiral	no
present	absent	dissymmetric	chiral	usually
absent	absent	asymmetric	chiral	usually

[a] The significance of the alternating axis of symmetry is discussed in the small print on p. 88.

3. Neglecting other "surface" features, does the presence of a "red spot" on the planet Jupiter render the planet chiral? (You may assume the red spot to be circular and fixed in position on the surface.) Is the Earth chiral? Explain.

Structures Capable of Existing as Enantiomers. The simplest means of constructing a chiral compound (a compound capable of existing as a pair of enantiomers) is by placing four different atoms or functional groups at the apices of the tetrahedron described by a central, sp^3 hybridized atom. It is common practice, although incorrect, to call this central atom such things as "asymmetric" or "the asymmetric center" or "the chiral center." Actually, it is the *environment* around the atom which is chiral; however, the convenience of these terms justifies their continued use. Examples of pairs of enantiomers having chiral centers are presented in Figure 4–4.

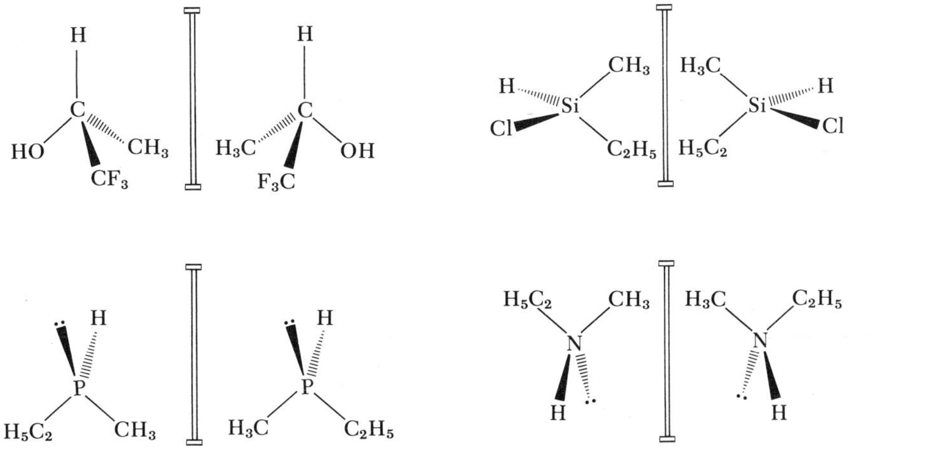

H C HO CH₃ CF₃ | H C H₃C OH F₃C

H Si Cl CH₃ C₂H₅ | H₃C H₅C₂ Si H Cl

H P H₅C₂ CH₃ | H P H₃C C₂H₅

H₅C₂ CH₃ N H | H₃C C₂H₅ N H

Figure 4-4 Pairs of enantiomers containing asymmetric atoms. The center bar in each pair represents a mirror.

If an atom possesses three identical substituents out of four, then that atom is achiral. For example, each plane containing the fragment Cl—C—H is a plane of symmetry for the chloroform molecule. Chloroform, since it possesses at least one plane of symmetry, is achiral and cannot have an enantiomer.

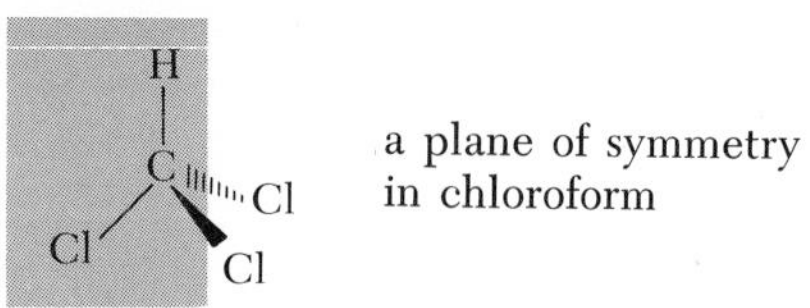

a plane of symmetry in chloroform

If only two groups around an sp^3 hybridized atom are identical, the molecule must have a plane of symmetry and, by definition, must be achiral. (In the illustration below, the methyl group of 1,1-dichloroethane is considered to be a spherical, symmetric "lump." It is common to treat all groups which lack a chiral center in this way; this does not alter the validity of any conclusions reached about molecular chirality.)

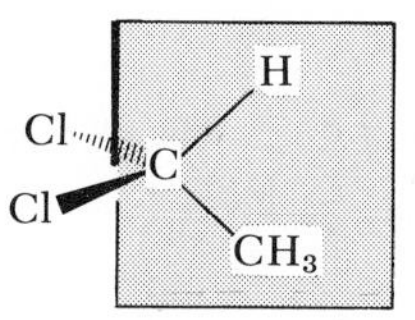

the plane of symmetry contains the central atom, hydrogen and the methyl group

1,1-dichloroethane (achiral)

Allene, $H_2C{=}C{=}CH_2$, is the simplest hydrocarbon containing a set of cumulative (contiguous) double bonds. The orthogonality of the p orbitals of the central, sp hybridized carbon, coupled with the need for each of these orbitals to be parallel to the p orbital of one of the outer, sp^2 hybridized carbons, causes the four hydrogens to reside at the apices of an *elongated* tetrahedron.

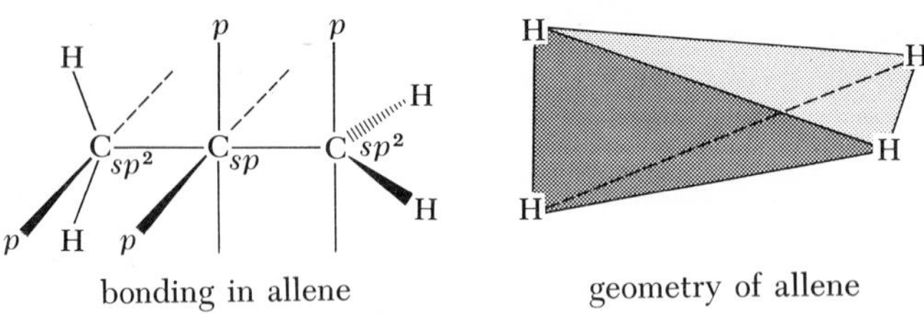

bonding in allene

geometry of allene

While allene itself is achiral, a pair of enantiomers is created when four different groups replace the hydrogens of allene. However, unlike the regular tetrahedron, the *elongated* tetrahedron can produce enantiomers using only two different groups, so long as neither carbon bears two identical groups. An allene (note how the name of the simplest member of a series becomes the generic name for a class of compounds) which has two identical groups on the same carbon possesses a plane of symmetry and must, therefore, be achiral.

enantiomers

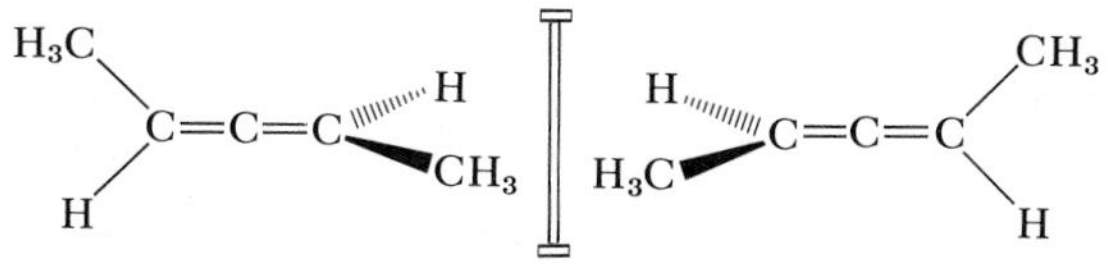

identical

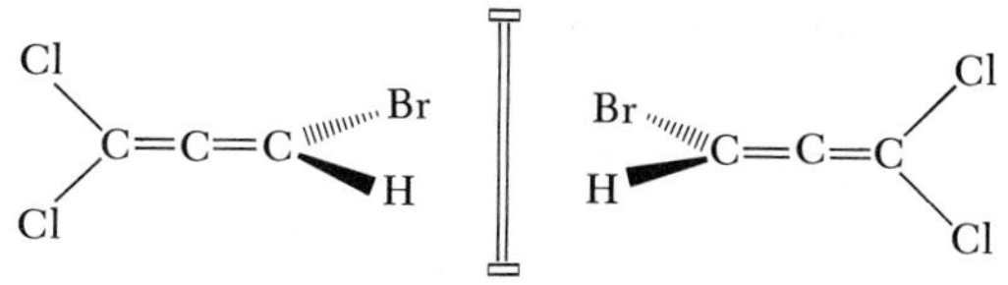

plane of symmetry

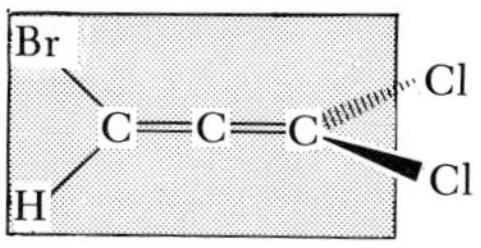

4. Identify those structures below which are capable of existing as enantiomers. Indicate a point or plane of symmetry present in the remaining structures.

(a) 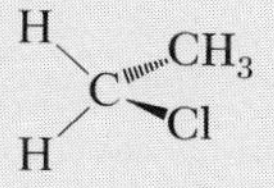 (f)

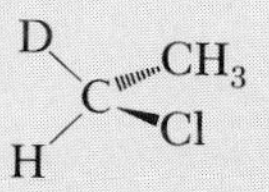

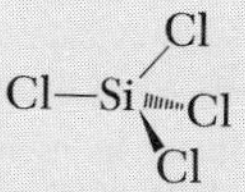

(b) (g)

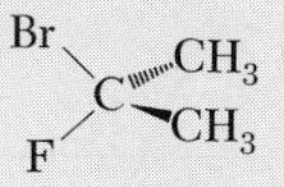

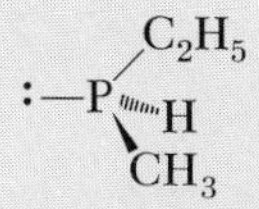

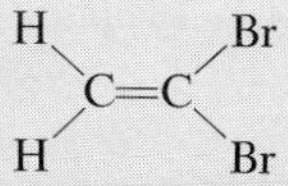

(c) (h)

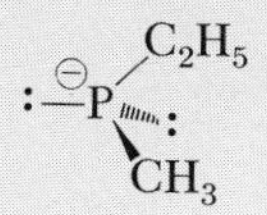

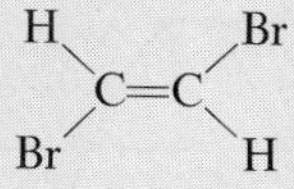

(d) (i)

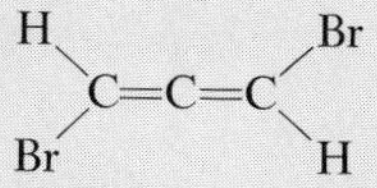

(e) (j)

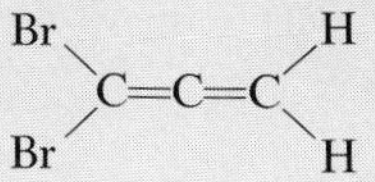

4.3 POLARIZED LIGHT AND OPTICAL ACTIVITY— THE INFLUENCE OF CHIRALITY UPON PHYSICAL PROPERTIES

Enantiomers will have identical physical and chemical properties if the system used to study them is achiral. For example, the enantiomers of 2-chlorobutane, $CH_3CHClCH_2CH_3$, have the same boiling point, density, melting point, solubility in water, and rate of reaction with chlorine radicals. *Only when the probe used to study a compound is chiral can enantiomers be distinguished one from the other.* Returning to our hand-and-glove analogy, we can view a right hand glove as a chiral "probe" to distinguish between a right and a left hand. One chiral probe useful in studying enantiomeric molecules is polarized light. Its use in studying naturally occurring enantiomers dates back to the 19th century.

Polarized Light. Ordinary light consists of electromagnetic waves which vibrate in all directions perpendicular to the direction of propagation. When ordinary light is passed through a Nicol prism or a piece of Polaroid, the emergent light is *linearly polarized* (also called *plane polarized*), as shown in Figure 4–5. When viewed from the side, the electric vector of plane polarized light describes a sine wave; the plane containing this sine wave is termed the *plane of polarization* (Figure 4–6).

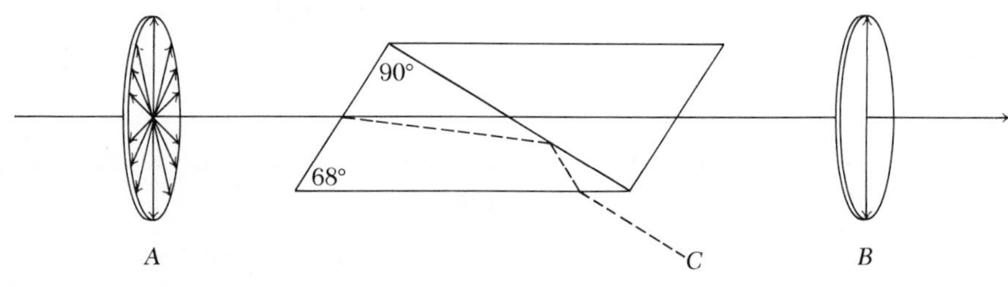

Figure 4-5 Linear polarization of ordinary light by a Nicol prism. The incoming beam (*A*) is converted into two beams, the extraordinary beam (*B*) and the ordinary beam (*C*). *B* and *C* are both polarized, but their planes of polarization are mutually perpendicular. The plane of the paper is the plane of polarization of *B*.

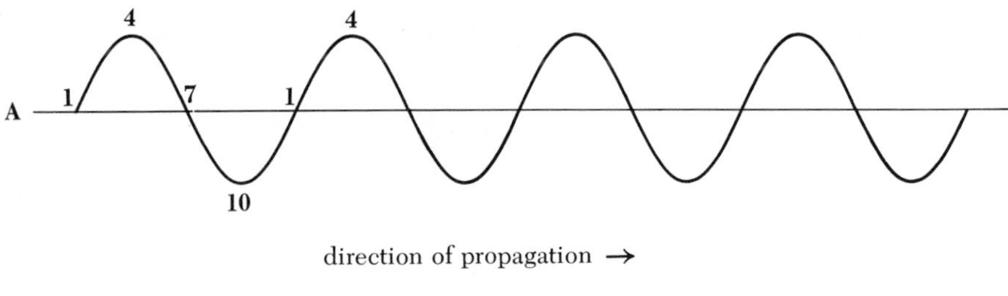

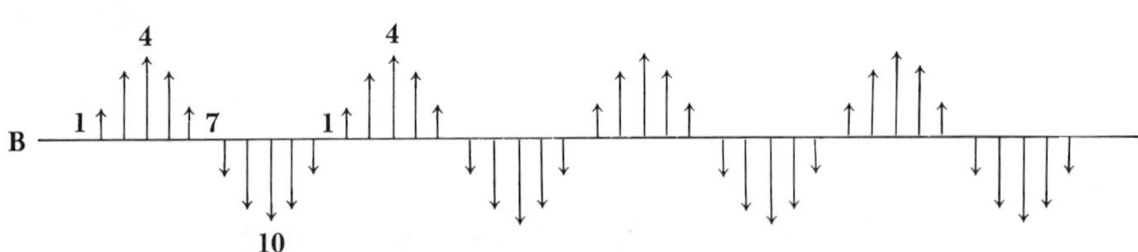

Figure 4-6 Linearly polarized light. (**A**) The sine wave generated by the oscillating electric vector. (**B**) The varying magnitude of the electric vector generating **A**. The plane of the paper is the plane of polarization. Numbers correspond to the resultant vectors of the diagrams in Figure 4–7(**A**).

Such an oscillating vector can be described as the resultant of two in-phase vectors, one moving in a clockwise direction and the other moving in a counterclockwise direction (Figure 4–7). Each rotating vector describes a helical path in space, quite similar to a partially elongated Slinky (a child's toy). However, while all Slinkies seem to have the twist in the same direction, such circularly polarized light vectors have opposite chirality or handedness or *helicity*. Linearly polarized light is, thus, made up of left and right handed circularly polarized light.

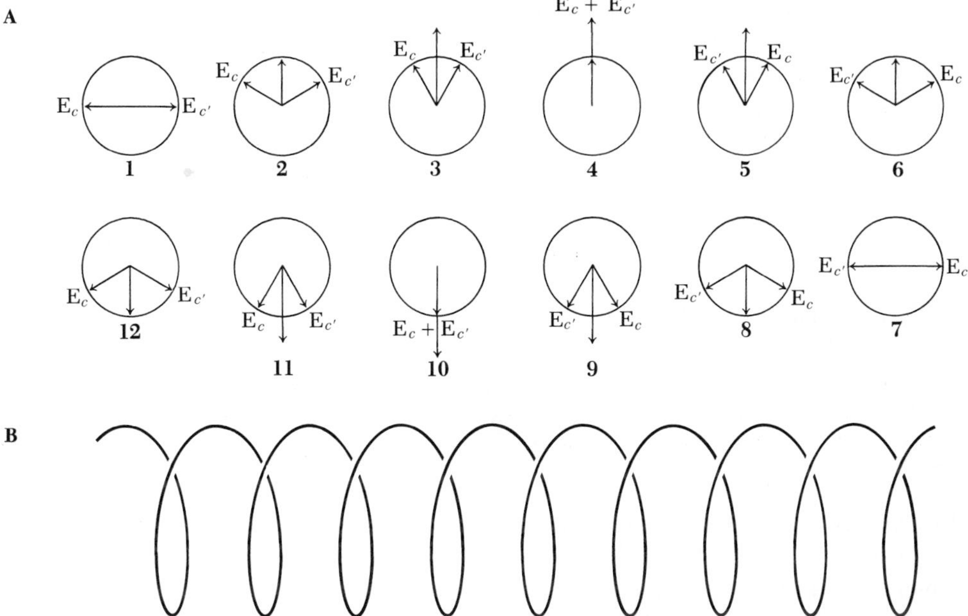

Figure 4-7 (**A**) The representation of linearly polarized light as circularly polarized components of opposite helicity. E_c travels in the clockwise direction, while $E_{c'}$ travels in the counterclockwise direction. The resultant is the electric vector shown in Figure 4–6. Horizontal components always cancel one another, so there is a net vector in the vertical direction. (**B**) An elongated "Slinky."

As linearly polarized light passes through a chiral medium, one of its circularly polarized components is slowed down relative to the other. This phenomenon, **circular birefringence,** places the left and right hand vectors out of phase and, consequently, changes the orientation of the plane of polarization. The plane is constantly changing its direction as it passes through the medium so that it appears as a twisted ribbon in the optically active medium (Figure 4–8).

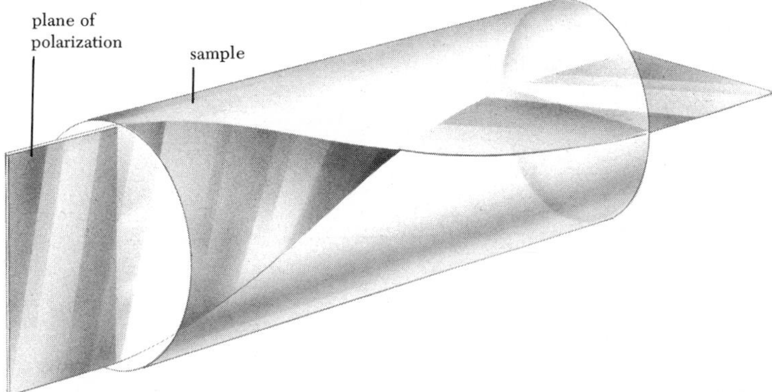

Figure 4-8 Rotation of polarized light by an optically active compound. As a beam of plane polarized light passes through a sample of an optically active compound (shown here as a cylinder), the plane of polarization is continuously rotated. The direction of rotation depends upon the molecular structure of the compound, and the amount of rotation per unit length of the sample is characteristic of the compound.

This twisting of the plane is known as **optical rotation.** The ability of a substance to cause optical rotation is **optical activity,** and the substance is said to be **optically active.**

THE POLARIMETER. The instrument used to measure the magnitude of optical rotation is the **polarimeter.** Its major components are presented schematically in Figure 4–9. If two Nicol prisms are oriented at right angles to one another, plane polarized light from the first ("polarizer") will not pass through the second ("analyzer"); an observer will see a dark viewing area within the instrument. When a solution of an optically active material—either one enantiomer or a non-equimolar mixture of enantiomers—is placed in the sample cell of the instrument, optical rotation occurs. The analyzer must be rotated in order to be perpendicular to the incoming plane of polarization and, thereby, maintain a dark viewing area.

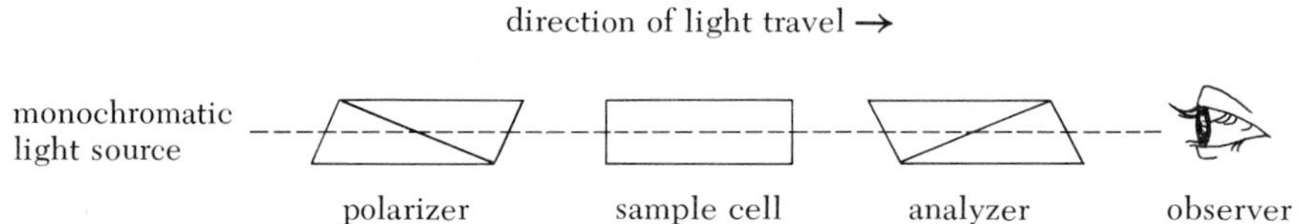

Figure 4-9 A schematic representation of a manual polarimeter. A monochromatic light beam consists of light of one wavelength. Both the analyzer and the polarizer are Nicol prisms in most polarimeters.

If a clockwise rotation (as viewed by the experimenter) of the analyzer is required, the substance is termed **dextrorotatory** (rotation to the right) and is labelled "(+)" or "*d.*" The enantiomer of a (+) substance rotates the plane of polarization an equal amount but in the opposite direction. It is **levorotatory,** abbreviated "(−)" or "*l.*" Some examples of this nomenclature which you may have already encountered include "*l*-menthol," "*d*-glucose," "*d*-penicillin," "*l*-amphetamine," "(+)-camphor," and "(+)-LSD." Modern usage prefers (+) and (−) over *d* and *l.*

An equimolar mixture of enantiomers has a net optical rotation of zero and is called a **racemic modification.** It is commonly indicated by "(±)" or "*d,l.*" The absence of any prefix before the name of a compound, *e.g.*, ". . . 2-chlorobutane . . .," implies a racemic modification unless the text indicates otherwise, *e.g.*, ". . . optically active 2-chlorobutane. . . ." Spatial restrictions in texts, scientific journals, and other published material prevent printers from showing both enantiomers even when a racemic modification is intended. For example, in the equation below, it is the intention of the writer to tell the student of the reaction of *racemic* 2-chlorobutane with chlorine radicals. However, using the soon-to-be-described Fischer convention, the novice might read this to mean the reaction of *one enantiomer* of this material with chlorine radicals. Unless there is evidence to the contrary, the student should always take the representation of one enantiomer in an equation or structure as implying a racemic modification.

$$CH_3\text{—}\underset{\underset{\textstyle Cl}{|}}{\overset{\overset{\textstyle H}{|}}{C}}\text{—}CH_2\text{—}CH_3 + Cl\cdot \rightarrow \text{products}$$

OPTICAL ROTATORY POWER. The magnitude of optical rotation under any set of conditions is expressed in degrees and is termed the **observed rotation,** α. The angle α is a function of the length of the sample cell, the structure and concentration of the optically active material, the solvent, the temperature of the solution, and the wavelength of the polarized light used in the study. The sodium D line (589 nm) is used as the source of light in most polarimeters.

The use of the **specific rotation,** $[\alpha]_{\lambda}^{t°}$, compensates for the effect of path length and concentration on α; it is given by

$$[\alpha]_{\lambda}^{t°} = \frac{100\,\alpha}{lc}$$

where α = measured rotation (degrees of arc)
$t°$ = temperature (°C)
λ = wavelength of light (generally 589 nm, abbreviated "D")
l = cell length in decimeters
c = concentration (g solute/100 ml solution)

For neat (*i.e.*, pure) liquids, $[\alpha]_{\lambda}^{t°} = \alpha/ld$, where d is the density in g/ml of the neat liquid.

Sometimes the molecular rotation, $[M]_{\lambda}^{t°}$, is used in place of $[\alpha]_{\lambda}^{t°}$:

$$[M]_{\lambda}^{t°} = \frac{[\alpha]_{\lambda}^{t°}M}{100}$$

where M = molecular weight of solute

Problem: Calculate the molecular rotation of **A**, given the following data. A solution of **A** (20.0 g of **A**/1000 ml of solution) has an observed rotation of 1.03° at 20°C in methanol in a 2 dm long cell. The molecular weight of **A** is 206. The wavelength of light used was 589 nm, the D line.

Solution:

$$[\alpha]_{\lambda}^{20°} = \frac{100\alpha}{lc}$$

$\alpha = 1.03°$
$l = 2$ dm
$c = 20$ g/1000 ml $= 2.0$ g/100 ml
$\lambda = 589$ nm (normally written simply as "D")

$$[\alpha]_{D}^{20°} = \frac{100 \times 1.03}{2 \times 2.00} = \frac{103}{4.00} = 25.8°$$

The specific rotation is 25.8°.

$$[M]_{D}^{20°} = \frac{[\alpha]M}{100}$$

$$= \frac{25.8 \times 206}{100} = 53.2°$$

The molecular rotation is 53.2°.

5. How many grams of compound (mol. wt. 150) with a specific rotation of +120° must be dissolved in sufficient water to make up 100 ml of a solution which will have an observed rotation of 0.10°? Assume the use of a 1 decimeter tube. What would your answer be with a 2 decimeter tube?

Optical Rotatory Dispersion. A plot of $[\alpha]$ or $[M]$ as a function of the wavelength of light is called an **ord** (optical rotatory dispersion) **curve.** An idealized ord curve is presented in Figure 4–10. The region in which the rotational magnitude changes rapidly with wavelength is the *Cotton effect* region; the gradually sloping portion is the *plain curve.* Many rotations of organic compounds, taken at 589 nm, occur in plain curve regions. As seen in Figure 4–10, enantiomers exhibit enantiomeric (mirror image) ord curves.

Ord curves are obtained using spectropolarimeters. Similar in principle to the polarimeter, the spectropolarimeter substitutes a light source and a *monochromator* for the sodium vapor lamp, and a photoelectric detector for the human eye. The application of ord to the determination of relative and absolute configurations will be discussed in Chapter 17.

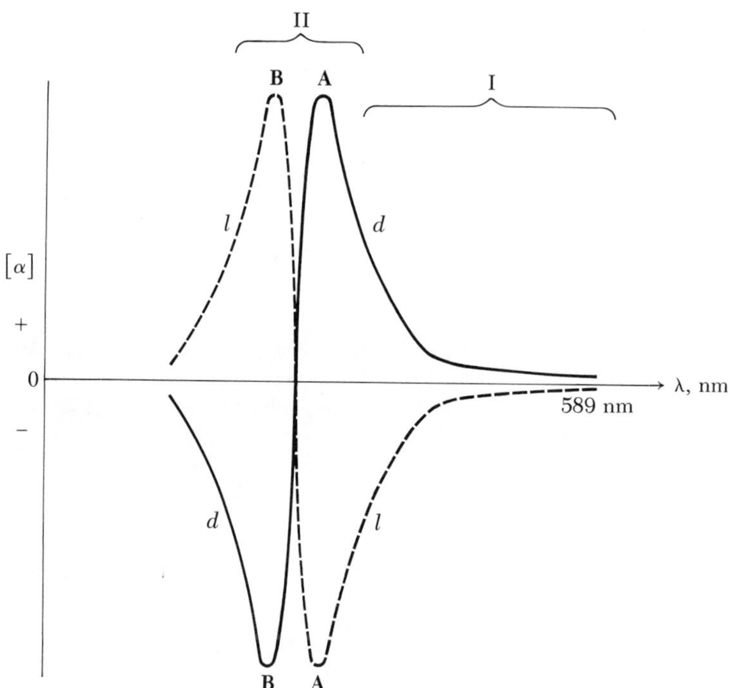

Figure 4–10 Idealized optical rotatory dispersion (ord) curve of a pair of enantiomers. The solid and dashed curves represent the two enantiomers. **A** is the "first extremum," and **B** is the "second extremum." For the *d* isomer (solid line), the first extremum is a *peak* and the second extremum is a *trough*. The *d* isomer shows a positive Cotton effect (*i.e.*, the peak occurs at a longer wavelength than does the trough). Region I is the *plain curve* region, and region II is the *Cotton effect* region. A plot that contains both a plain curve and a Cotton effect is dubbed an *anomalous curve*. The Cotton effect arises because E_c and $E_{c'}$ (described in Figure 4–7) are absorbed by a single enantiomer to different extents.

Is Chirality Sufficient for Optical Activity? In the next chapter you will learn one way to separate one enantiomer from another. However, let us assume that you already know how and that you go into the laboratory and do whatever you think must be done in order to *resolve* (*i.e.*, separate the enantiomers of) butylethylmethylamine, $CH_3(CH_3CH_2)\ddot{N}CH_2CH_2CH_2CH_3$. You would be disappointed to find that none of the materials which you prepared in the laboratory showed even the slightest trace of optical activity. But, you argue, the nitrogen has four different "things" around it ($—CH_3$; $—CH_2CH_3$; $—CH_2CH_2CH_2CH_3$; and a lone electron pair) which are arranged in a tetrahedron, and such a chiral array must be separable into enantiomers. Continued years of failure, coupled with the inability to find any report of anyone ever preparing an optically active amine of the type you want, might lead you to conclude that having an sp^3 hybridized atom with four different "things" around it is not enough. EUREKA!

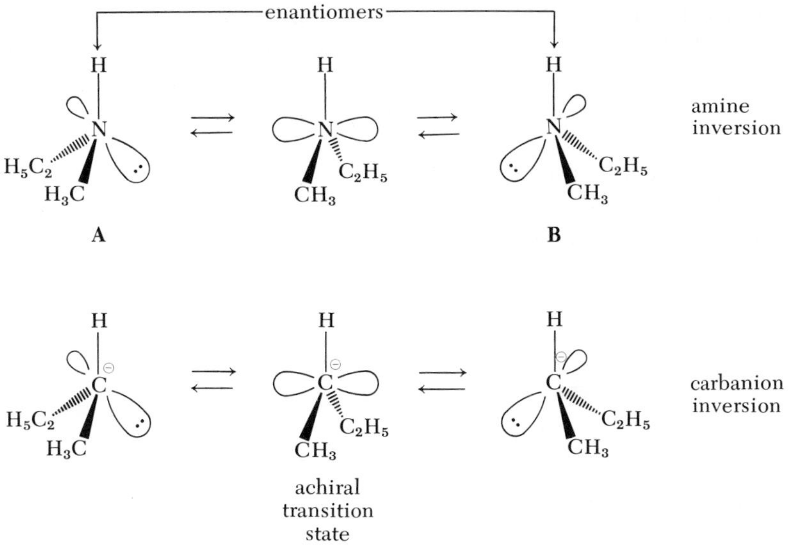

enantiomers

A   B

amine inversion

carbanion inversion

achiral transition state

Figure 4-11 Pyramidal inversion in simple substituted compounds of first period elements. The transition state for each reaction possesses a planar, archiral sp^2-hybridized central atom. Both equations illustrate the principle that *any* reaction beginning with an optically active starting material and involving an achiral transition state (or intermediate) *must* produce an optically inactive product.

The inability to prepare optically active butylethylmethylamine is due to the *configurational instability* of nitrogen in this compound. Put differently, one enantiomer rapidly interconverts into the other enantiomer. The net result of this is the conversion of one enantiomer into a racemic modification, a process dubbed *racemization*. So, as soon as you prepared one enantiomer in our little scenario, it spontaneously converted into a racemic modification. We must conclude that *optical activity is possible only if a substance is chiral* ***and*** *it is configurationally stable*. A material which does not rotate a plane of polarized light under any experimental conditions is said to be optically inactive. Butylethylmethylamine is optically inactive, as are all racemic modifications.

What types of compounds are configurationally stable or configurationally unstable? Simple, trisubstituted atoms of the first long period of the periodic table generally are configurationally unstable, and when these atoms are the chiral centers in molecules, all attempts to prepare one enantiomer free of the other lead to racemic product. This explains why simple optically active amines and carbanions, $RR'R''C{:}^{\ominus}$, have yet to be prepared. The racemization of these species is believed to occur through a planar, achiral transition state (Figure 4-11).

If the lone electron pair is coordinated, thus preventing its tunneling through the nucleus, the molecules should be configurationally stable and, if chiral, separable into enantiomers. This prediction is confirmed by the configurational stability of substituted carbon ($RR'R''R'''C$) and of substituted ammonium cations ($RR'R''R'''N^{\oplus}$).

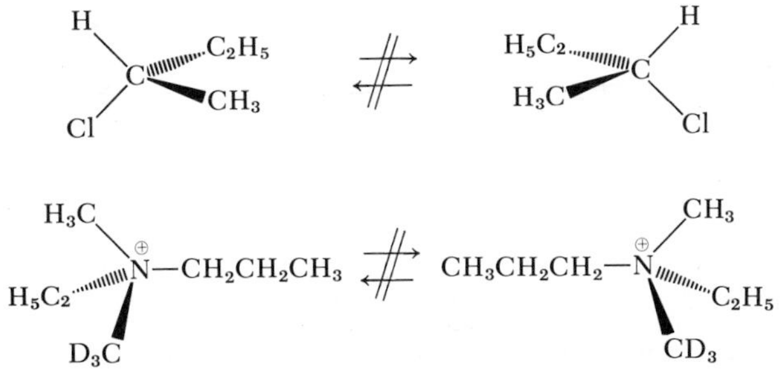

Configurational stability is sometimes called *optical stability* because of the change in optical rotation which accompanies inversion of configuration.

The Fischer Projection[a]. In this scheme for representing a three-dimensional object on a two-dimensional surface, a chiral center is drawn with its four bonds at right angles to one another, pointed at the major compass points. Vertical bonds are imagined to project behind the plane of the paper, while horizontal bonds are imagined to project toward the viewer; the central atom is in the plane of the paper.[b] This view may be said to resemble a bow-tie attached to the cervical vertebrae.

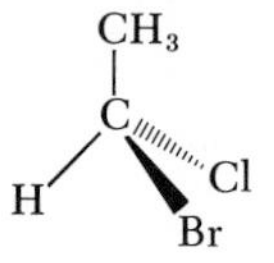

CH_3, C, H, Cl, Br — one view of a wedge projection

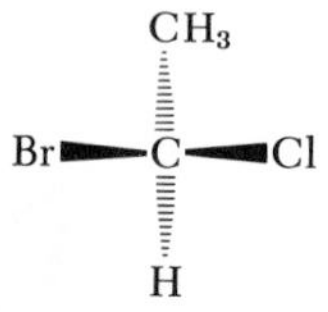

CH_3, Br, C, Cl, H — alternate view of wedge projection

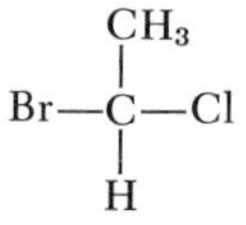

CH_3, Br—C—Cl, H — Fischer projection

1-bromo-1-chloroethane

One point which often escapes the novice is that a substance will have more than one completely acceptable Fischer projection. This is due to the fact that you can look at a given three-dimensional structure from different vantage points and, consequently, see various sets of bonds as "horizontal" and "vertical." To illustrate this, we have drawn three acceptable Fischer projections of one enantiomer of 1-bromo-1-chloroethane below. (Remember that the horizontal bonds in the projection are the ones sticking out toward you as you view the structure, *i.e.*, the "bow-tie.")

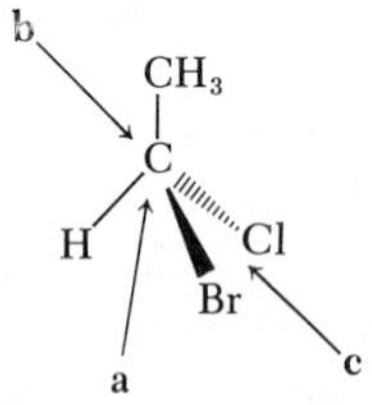

vantage point	Fischer projection
a (in H-C-Br plane)	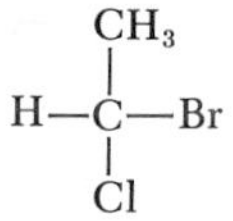 top CH_3; H—C—Br; bottom Cl
b (in H-C-CH_3 plane)	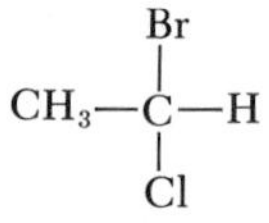 top Br; CH_3—C—H; bottom Cl
c (in Br-C-Cl plane)	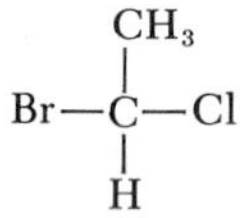 top CH_3; Br—C—Cl; bottom H

In the Fischer projection, the exchange of two adjacent groups produces an enantiomer of the starting material. Using a three-dimensional presentation such as the wedge projection, one produces an enantiomer of the starting material with *any* simple exchange around the chiral center. What happens when two substituents separated by 180° in a Fischer projection are exchanged?

[a] Initially presented on p. 54. It is suggested that you review that material.

[b] While some chemists exclude explicitly drawing the central atom, others include it. The American Chemical Society appears to favor inclusion: "In [the Fischer type projection] the asymmetric carbon atom is placed . . ." (*Nomenclature of Organic Compounds*, ed. J.H. Fletcher, O.C. Dermer, R.B. Fox, American Chemical Society, Washington, 1974, p. 110. In this monograph the chiral atom is explicitly included in Fischer projections.

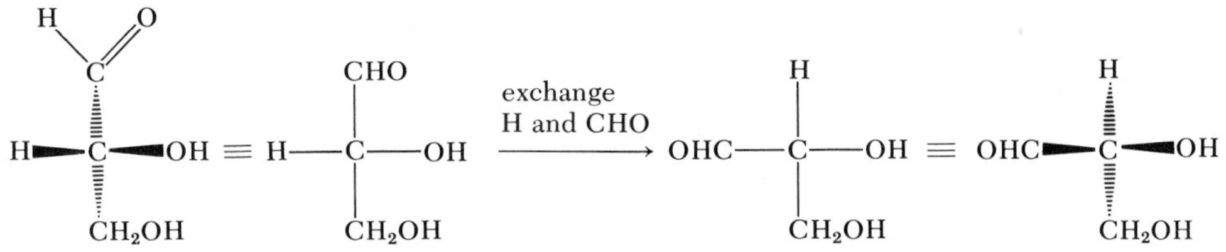

(+)-glyceraldehyde

exchange H and CHO

(−)-glyceraldehyde

The enantiomer of the starting material is created if a Fischer projection is rotated in the plane of the paper by either 90° or 270°, but not when rotation equals 180°.

6. Convert each of the following into a Fischer projection.

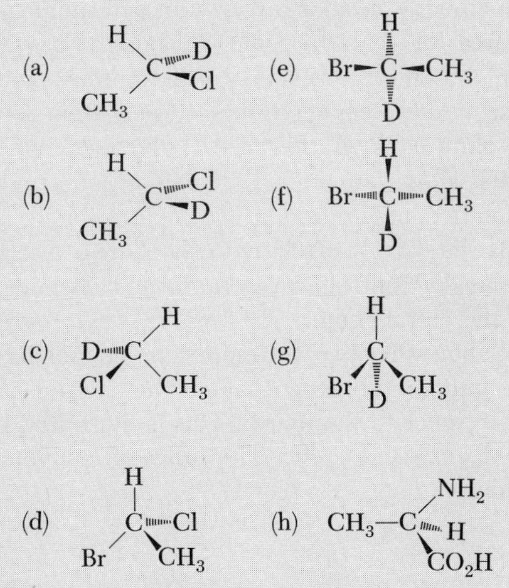

(a) (b) (c) (d) (e) (f) (g) (h)

7. Convert each of the following Fischer projections into a suitable wedge projection.

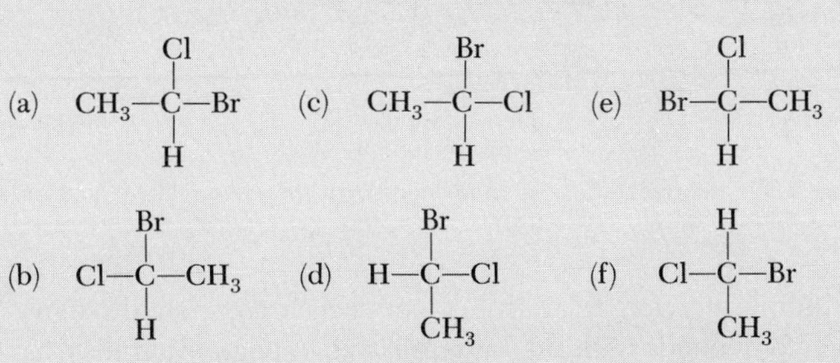

(a) CH_3—C(Cl)(H)—Br (c) CH_3—C(Br)(H)—Cl (e) Br—C(Cl)(H)—CH_3

(b) Cl—C(Br)(H)—CH_3 (d) H—C(Br)(CH_3)—Cl (f) Cl—C(H)(CH_3)—Br

4.4 CONFIGURATION

Relative Configuration vs. Absolute Configuration. A polarimeter determines the sign of rotation of a compound, but it cannot tell *which* enantiomer of a pair has *that* sign of rotation. For example, which of the two structures below, **A** or **B**, is *l*-2-butanol? The answer to this question would provide the *absolute configuration*, that is, the actual spatial arrangement around the chiral center, of *l*-2-butanol.

H_3C, H, OH, H_5C_2 — **A**

CH_3, HO, H, C_2H_5 — **B**

The determination of absolute configuration by X-ray crystallography is an active area of research. However, the amount of labor involved in determining an absolute configuration portends activity in the area for years to come. Amazingly, it was not until 1951 that the first absolute configuration of a chiral substance was determined. In that year J.M. Bijvoet and co-workers determined the absolute configuration of the sodium rubidium salt of (+)-tartaric acid. Before 1951 the absolute configuration of (+)-glyceraldehyde was arbitrarily taken to be that shown on page 99. However, after 1951 this assignment was confirmed and, with the knowledge of one absolute configuration, it became possible to correlate the configuration of other substances to this standard. Correlative procedures, described throughout this text, include ord studies, biologically-controlled reactions, and chemical reactions which do not influence the chirality of the chiral center.

Often it is enough to show that two (or more) compounds have the same **relative configuration,** *i.e.*, that three common substituents about the chiral center have the same relative orientations in space with respect to the fourth. This is illustrated in Figure 4–12, using two hypothetical compounds—C*abxy* and C*abxz*. The series of compounds shown below all have the same relative configuration.

Br, H, C_2H_5, H_3C | HO, H, C_2H_5, H_3C | CH_3O, H, C_2H_5, H_3C | D, H, C_2H_5, H_3C

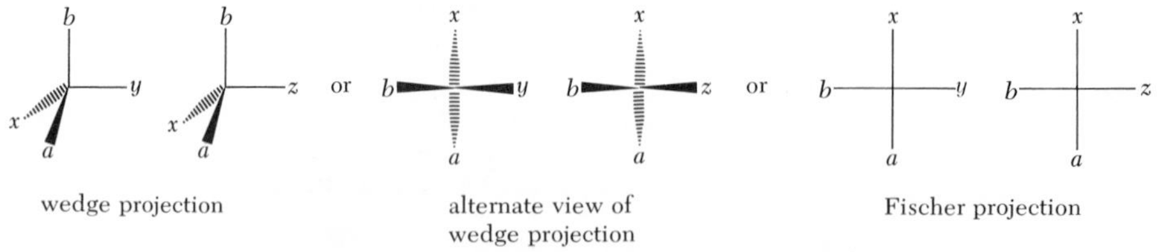

wedge projection — alternate view of wedge projection — Fischer projection

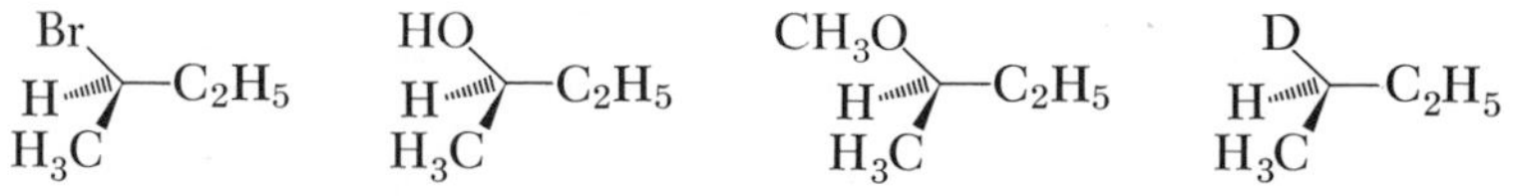

Figure 4–12 Representations of compounds with the same relative configurations.

There is no simple *a priori* relationship between a compound's sign of rotation and its configuration; two compounds with the same relative configuration may have opposite D line rotations, and seemingly similar compounds with the same D line rotation may have opposite relative configurations.

The Designation of Absolute Configuration Using the Cahn-Ingold-Prelog System. Since the sign of rotation does not convey absolute configuration, and drawing a structure in three dimensions is often inconvenient, it became necessary to devise a scheme to convey the information called "absolute configuration" whenever it was known. The method of greatest utility is that one devised by Sir Christopher Ingold, R.S. Cahn, and V. Prelog.

The **Cahn-Ingold-Prelog nomenclature system,** henceforth referred to as the "**R,S**" system, considers the chirality of a molecule to be describable in terms of the chirality of its chiral centers. The asymmetric atom is the most commonly occurring chiral center, and we will concentrate upon the designation of the absolute configuration around asymmetric atoms.[a]

The **R,S** system is based upon the *arbitrary* assignment of priorities to the substituents around the chiral center using the "sequence rule." This rule ranks a given set of four substituents as either "1", "2", "3", or "4" (1 having the highest priority). Configurational assignment is then made by mentally constructing a triangle using groups 1, 2 and 3 and viewing group 4 (which is behind the chiral center) through the triangle. If one travels in a clockwise fashion in going down the priorities ($1 \rightarrow 2 \rightarrow 3$) while viewing 4, the configuration of the chiral center is *defined* as **R** (Latin, *rectus* = right). If a jaunt down the priorities leads you in the counterclockwise direction, the configuration is **S** (Latin, *sinister* = left). Enantiomeric chiral centers have opposite designations, **R** and **S**, as shown in Figure 4–13. A racemic modification is designated "**R,S.**"

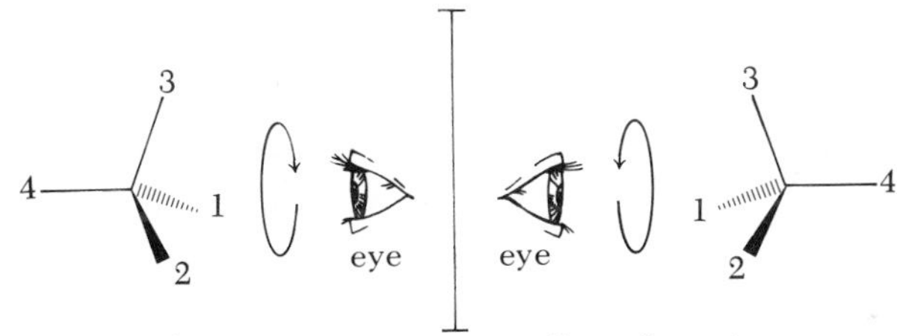

R configuration **S** configuration

Figure 4–13 Designation of absolute configuration about a chiral center after assignment of priorities.

The indicators **R** and **S** are added to the names of compounds as prefixes enclosed in parentheses. Thus, the enantiomers of 1-bromo-1-chloroethane are (**R**)-1-bromo-1-chloroethane and (**S**)-1-bromo-1-chloroethane. The corresponding racemic modification is (**R,S**)-1-bromo-1-chloroethane.

The **sequence rule** is composed of several sub-rules, of which the most pertinent are:

1. Higher atomic number precedes (*i.e.*, has a higher priority than) lower atomic number.
2. Higher mass number precedes lower mass number. This is applicable only when considering isotopic substituents.
3. Non-bonded electrons follow atoms. Thus, non-bonding electrons have a lower priority than does "H".

8. Arrange the following sets in order of decreasing priority.
   (a) —H, —Cl, —Br
   (b) —I, —S, —N
   (c) —^{3}H, —^{1}H, —^{2}H
   (d) —Br, —H, —Sn, —lone electron pair

[a] More detailed discussions are available: R.S. Cahn, *J. Chem. Ed.*, **41,** 116 (1964); E.L. Eliel, *J. Chem. Ed.*, **48,** 163 (1971).

Let us apply this to an enantiomer of bromochlorofluoromethane:

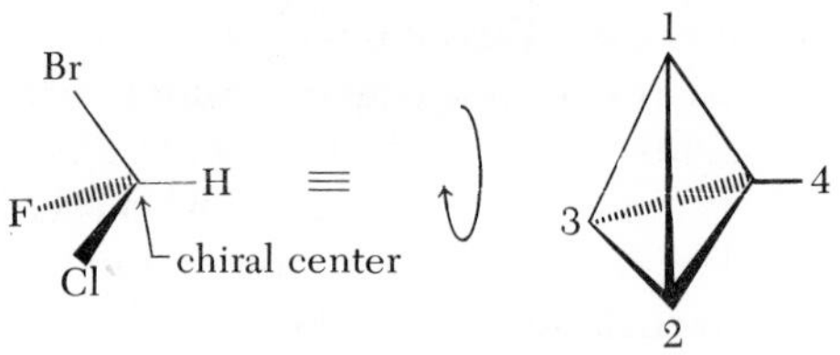

(**R**)-bromochlorofluoromethane

Beginning at the chiral center, we probe along the four bonds and assign priorities using sub-rule 1. This is translated into the drawing beside the original structure, showing that this enantiomer is (**R**)-bromochlorofluoromethane. Of course, this tells us nothing about the optical rotation of this enantiomer. The sign and magnitude of its optical rotation must be determined experimentally.

Most chiral centers will have at least two identical atoms, commonly carbon, bonded to them. When this occurs, the probe must be extended beyond the first "shell" of atoms to the second, and so on, until group priorities are established.

Consider the enantiomer of 3-bromo-2-methylpentane shown below. Going from the chiral carbon out to the first shell of atoms, one encounters Br, **C, C,** and H. The bromine has the highest priority, 1, while the hydrogen has the lowest, 4. But how do we distinguish between the ethyl group and the isopropyl group? Both are attached to the central atom by carbon, so we must move farther away from the chiral center in order to establish priorities. Every time we move out from a branch point, such as the carbons shown in boldface, we go out to the highest priority substituent and so on down in atom priorities until a distinction is made. Beginning at each carbon that is a branch point, we encounter, in order of descending priority, C, H, and H with the ethyl group and C, C, and H with the isopropyl group. The second atom attached to each branch point presents a hydrogen from the ethyl group but a higher priority carbon from the isopropyl group. Therefore, since this represents the first distinction between these groups, the ethyl group is given the lower priority as a group.

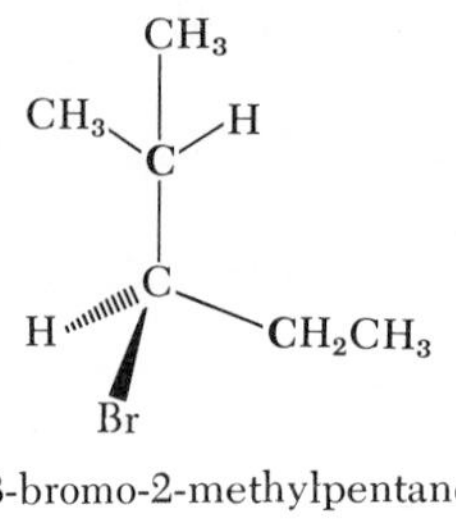

3-bromo-2-methylpentane

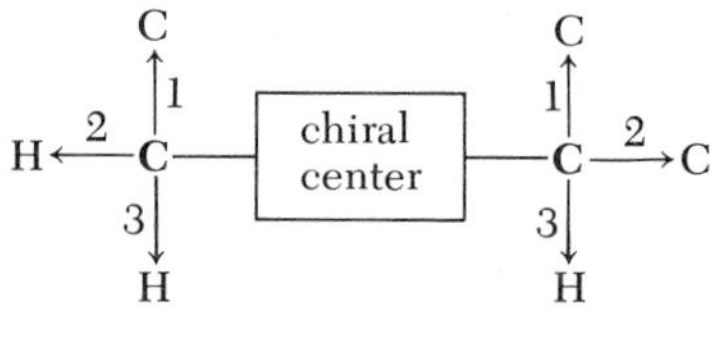

priority assignment to ethyl and isopropyl groups

Having now assigned priorities to all substituents around the chiral center, we assign the **R** absolute configuration to the chiral center and name the compound (**R**)-3-bromo-2-methylpentane.

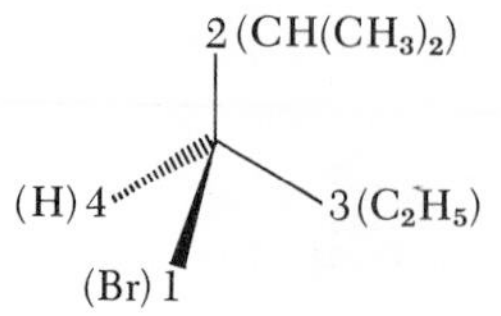

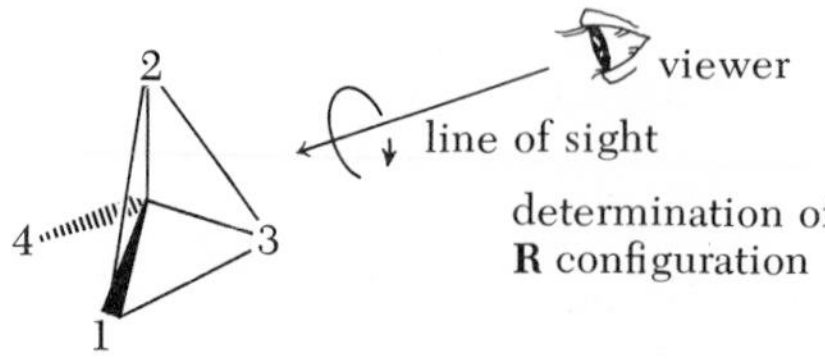

determination of **R** configuration

9. Arrange the following sets in order of decreasing priority.
   (a) $-H$, $-CH_3$, $-C_2H_5$, $-C(CH_3)_3$, $-CH(CH_3)_2$
   (b) $-CH_2CH_2CH_3$, $-CH_2CH(CH_3)_2$, $-CH(CH_2CH_3)_2$
   (c) $-CHCH_3CH_2CH_3$, $-C(CH_3)_2CH_2CH_2CH_3$, $-C(CH_3)_2CH_2CH_3$

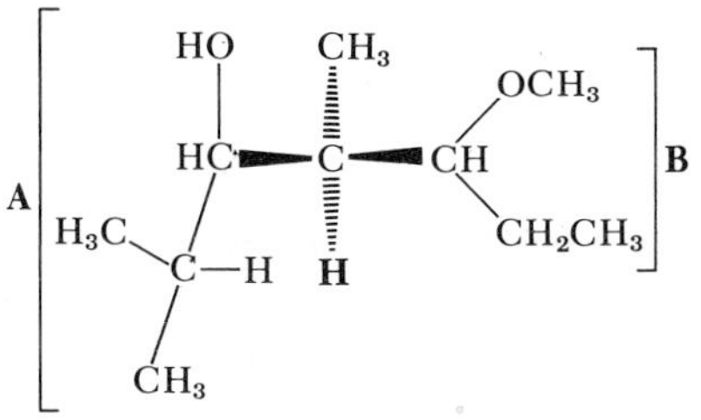

3-hydroxy-5-methoxy-2,4-dimethylheptane

Consider the stereoisomer of 3-hydroxy-5-methoxy-2,4-dimethylheptane shown here. While **H** has the lowest priority, we must travel out beyond the first shell of carbon atoms to establish the priorities of the remaining groups, $-CH_3$, **A**, and **B.**

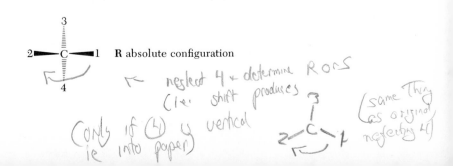

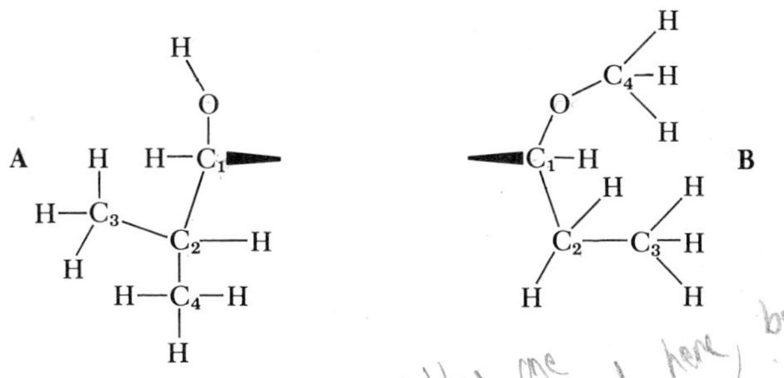

Referring to the enlargement of molecular fragments **A** and **B,** departing the central atoms, C_1(**A**) and C_1(**B**), can lead to either C_2(**A**) or C_2(**B**) *or* to O(**A**) or O(**B**). The paths to oxygen are followed, again illustrating the rule that at every branch point the road to follow first is the one containing the highest priority atom. Since both oxygens have identical priorities we continue, coming to H from O(**A**) but to C_4 from O(**B**). Therefore, since C has a higher priority than H, group **B** has a higher priority than group **A.** The methyl group has a lower priority than either **A** or **B.** The compound can now be assigned the **R** absolute configuration.

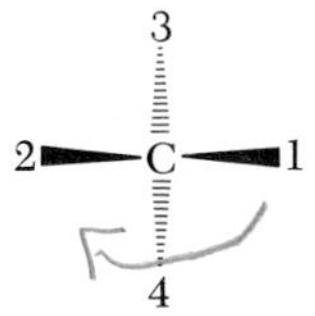

R absolute configuration

Multiple bonds must be "pretreated" in order to be able to assign priorities, using the rules already described, to functional groups containing multiple bonds. This pretreatment involves replicating the atoms involved in multiple bond formation until each of the atoms that originally formed a multiple bond bears four substituents, including lone electron pairs. In the three examples below we see how this replication is applied to a symmetric double bond, an unsymmetric double bond and an unsymmetric triple bond. In each illustration the atoms involved in multiple bond formation are shown in bold typeface, and the atoms that have been "created" as a result of the replication process are encircled.

$$-\mathbf{C}(R)=\mathbf{C}(R)-R \xrightarrow[\text{replication}]{\text{atom}} -\mathbf{C}(R)(\text{Ⓒ})-\mathbf{C}(R)(\text{Ⓒ})-R$$

original functional group — expanded version

$$R-\mathbf{C}=\ddot{\mathbf{O}}: \xrightarrow[\text{replication}]{\text{atom}} R-\mathbf{C}(\text{Ⓞ})-\ddot{\mathbf{O}}:(\text{Ⓒ})$$

$$-\mathbf{C}\equiv\mathbf{N}: \xrightarrow[\text{replication}]{\text{atom}} -\mathbf{C}(\text{Ⓝ})(\text{Ⓝ})-\mathbf{N}:(\text{Ⓒ})(\text{Ⓒ})$$

To establish the priority of the original substituent, then, one simply treats the expanded version of the functional group according to sub-rule 1. We can, for example, show that the order of priorities for the groups $—CO_2H$, $—CHO$, and $—CH_2OH$ is $—CO_2H > —CHO > —CH_2OH$ by expanding the first two. (The third need not be expanded since it lacks multiple bonds.)

```
  O—C            O—C            O—H
  |              |              |
 —C—O           —C—O           —C—H
  |              |              |
  O—H            H              H
expanded —CO2H  expanded —CHO  —CH2OH
```

10. Arrange the following sets in order of decreasing priority.

(a) $—CH{=}CH_2$, $—CCH_3{=}CH_2$, $—CH{=}CHCH_3$
(b) $—C{\equiv}CH$, $—C(CH_3)_3$, $—CH_2CH_2CH_2C(CH_3)_3$
(c) $—OCH_3$, $—CH_2OH$, $—CH_2OCH_3$, $—CH_2SH$
(d) $—C(=O)OH$, $—C(=O)OCH_3$, $—CH(OCH_3)_2$, $—CH(OCH_3)_2$

Since the absolute configuration of (+)-malic acid ($HO_2CCH_2CHOHCO_2H$) is known, we can describe it in the **R,S** system. The least prior substituent is H, while OH is the most prior. The carbon of the carbonyl group is bound to three oxygens (because of the duplication when C=O is expanded) while the highest priority atom attached to the methylene carbon is only a carbon. Applying sub-rule 1, the $—CO_2H$ group has a higher priority than does the $—CH_2CO_2H$ group (2 *vs.* 3). While either (**R**)-malic acid or (+)-malic acid is an acceptable name, (+)-(**R**)-malic acid is most desirable since it conveys both the experimental D-line rotation and the absolute configuration.

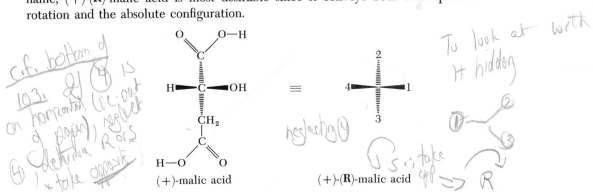

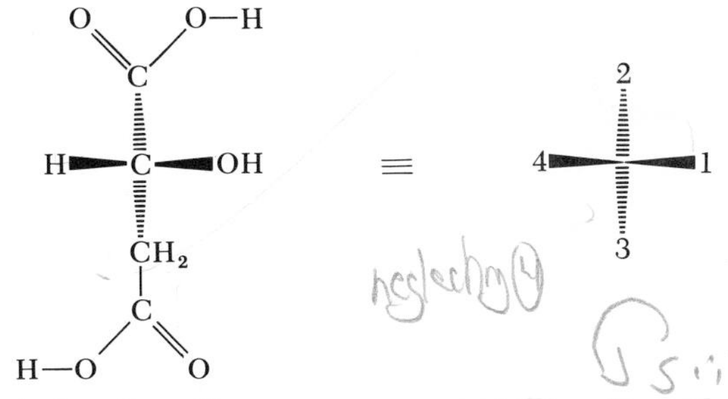

(+)-malic acid ≡ (+)-(**R**)-malic acid

> 11. The enantiomer of (+)-malic acid occurs in many fruits and is called "apple acid." Draw a three-dimensional structure of apple acid. What is its sign of rotation? What is its absolute configuration?

Serine, $HOCH_2CH(NH_2)CO_2H$, is a compound which contains an amino group ($—NH_2$) adjacent to a carboxy group and, therefore, is one member of that class of compounds called amino acids. Amino acids are very important and are discussed at length in Chapters 24 and 27. Let us simply note here that proteins of our body are nothing more than polymers of such amino acids and that special proteins, called enzymes, control all reactions in our body.

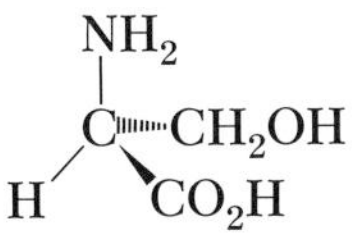

(−)-serine 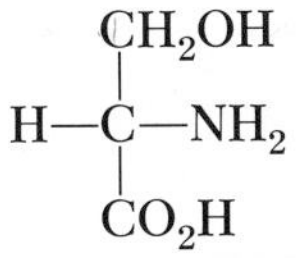 | (−)-serine (Fischer projection) 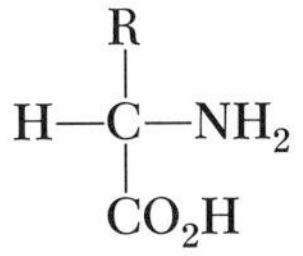 | general structure of a naturally occurring amino acid

We may assign an absolute configuration to (−)-serine by ordering the substituents about the chiral center according to their priority in the **R,S** system: $—NH_2 > —CO_2H > —CH_2OH > —H$. Thus (−)-serine has the **S** absolute configuration. Almost all R groups (see structure above) present in naturally occurring amino acids have a priority less than that of the carboxy group but greater than that of hydrogen. Since most naturally occurring amino acids are chiral and have the same relative configuration at this chiral center, they too are expected to have the **S** absolute configuration. That this appears to be correct can be seen from the structures of (+)-valine, (−)-leucine, and (−)-phenylalanine, all of which are amino acids that occur in nature in proteins.

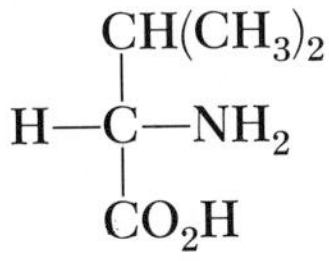

(+)-valine 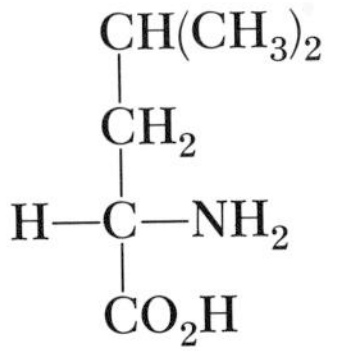 | (−)-leucine 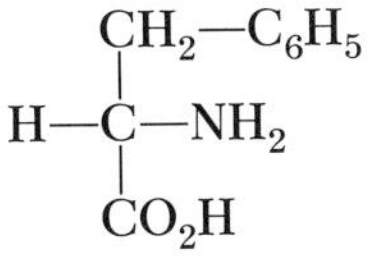 | (−)-phenylalanine

12. Assign absolute configurations to all of the chiral centers in problems 4.6 and 4.7 (p. 99).
13. Assign absolute configurations to the chiral centers below.

(a) 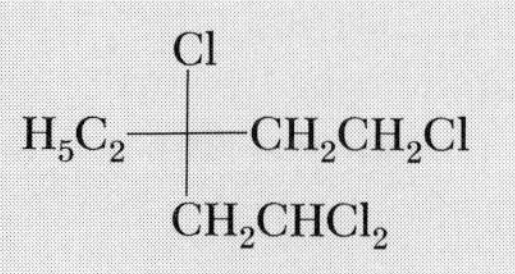 Cl above; H_5C_2—C—CH_2CH_2Cl; CH_2CHCl_2 below

(b) Cl above; H_5C_2—C—CCl_2CH_3; CH_2CHCl_2 below

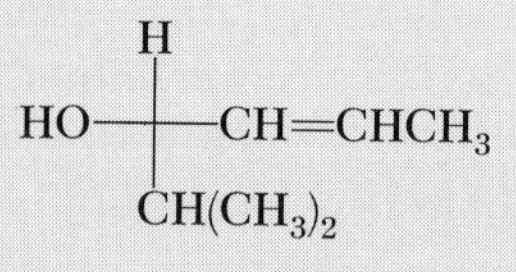

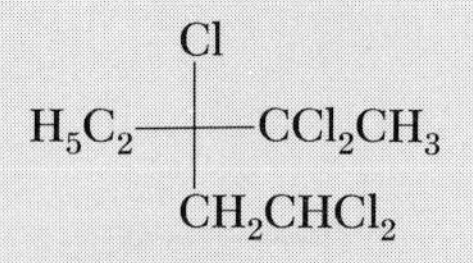

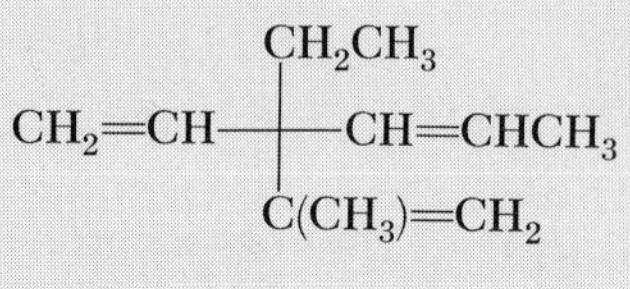

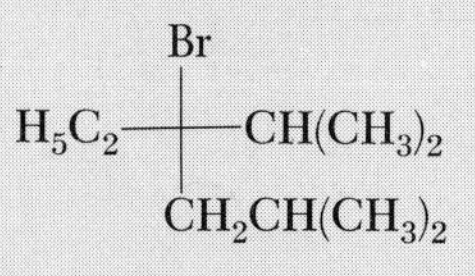

(c) Br above; H_5C_2—C—$CH(CH_3)_2$; $CH_2CH(CH_3)_2$ below

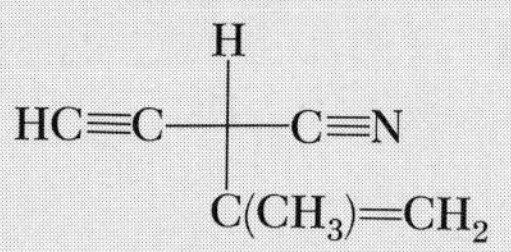

(d) 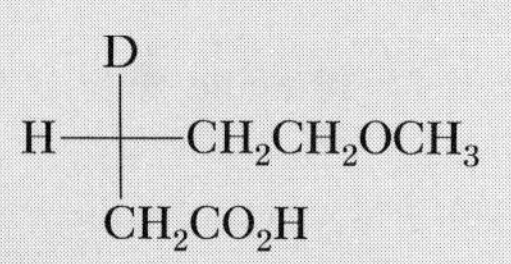 D above; H—C—$CH_2CH_2OCH_3$; CH_2CO_2H below

(e) 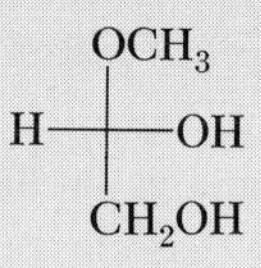 OCH_3 above; H—C—OH; CH_2OH below

(f) H above; HO—C—CH=$CHCH_3$; $CH(CH_3)_2$ below

(g) CH_2CH_3 above; CH_2=CH—C—CH=$CHCH_3$; $C(CH_3)$=CH_2 below

(h) H above; HC≡C—C—C≡N; $C(CH_3)$=CH_2 below

(i) 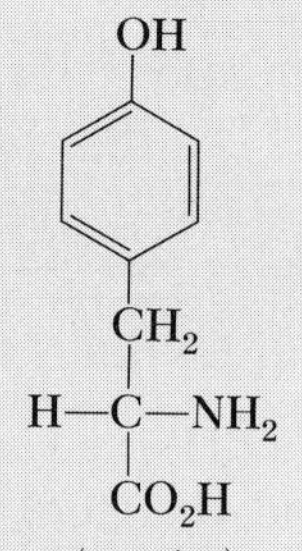 HO–C_6H_4–CH_2 above; H—C—NH_2; CO_2H below
(tyrosine)

4.5 DIASTEREOMERS—COMPOUNDS WITH SEVERAL CHIRAL CENTERS[a]

How many stereoisomers arise when two chiral centers, each bearing the same set of substituents, occur in the same molecule? Abbreviating the chiral groups (*i.e.*, the chiral centers) in such a substance as G_R and G_S (the subscript reflecting the absolute configuration of the group), we can generate the following four structures.

G_R	G_S	G_R	G_S
|	|	|	|
G_R	G_S	G_S	G_R
1	**2**	**3**	**4**

Compounds **1** and **2** are enantiomers, while **3** and **4** are identical; hence, we need consider only **1**, **2**, and **3**. Enantiomers **1** and **2** are stereoisomers of **3**, since they differ from **3** only

[a] Such compounds sometimes are called "classical diastereomers."

in the configuration at a chiral center. The relationship between **1** and **3** or between **2** and **3** is *diastereomeric;* **1** and **3** (or **2** and **3**) are diastereomers of one another. *Non-enantiomeric stereoisomers are* ***diastereomers.***

The set of enantiomers, **1** and **2**, is termed the *d,l* form, while **3** is termed the *meso* form. The *meso* form is optically inactive, while both **1** and **2** are optically active. These points are illustrated using tartaric acid.

HO(O=)C—CH(OH)—CH(OH)—C(=O)OH

tartaric acid

Below are drawn wedge projections of *d-*, *l-*, and *meso*-tartaric acid, each in an eclipsed conformation. These eclipsed conformations have been selected since one of the Fischer projections derived from them (shown immediately below them) often is used to represent these isomers. We have included the correct **R** and **S** designations for the chiral centers.

d-tartaric acid	*l*-tartaric acid	*meso*-tartaric
CO_2H (1)	CO_2H (1)	CO_2H (1)
H—C(2)—OH (R)	HO—C(2)—H (S)	H—C(2)—OH (R)
HO—C(3)—H (R)	H—C(3)—OH (S)	H—C(3)—OH (S)
CO_2H (4)	CO_2H (4)	CO_2H (4)

d-tartaric acid and *l*-tartaric acid: enantiomers

We may describe *d-*, *l-*, and *meso*-tartaric acid using the **R,S** system. All that is necessary is to number the carbon skeleton (as shown in the Fischer projections) and to use these numbers to indicate whether the chiral centers are **R** or **S.**

d-tartaric acid *is equivalent to* (2**R**,3**R**)-tartaric acid
l-tartaric acid *is equivalent to* (2**S**,3**S**)-tartaric acid
meso-tartaric acid *is equivalent to* (2**R**,3**S**)-tartaric acid

One way to recognize the *meso* form of tartaric acid (or any other *meso* form) is to note that it can be drawn in a wedge form (and in a Fischer projection) which has a plane of symmetry. This is not true for either *d-* or *l*-tartaric acid or the members of any other *d,l* form.

plane of symmetry

HO_2C, CO_2H, H, H, HO, OH

CO_2H
H—C—OH
H—C—OH
CO_2H

Be warned, however, that not every wedge or Fischer projection of a *meso* form must have a plane of symmetry. Again we can call upon *meso*-tartaric acid to illustrate this point.

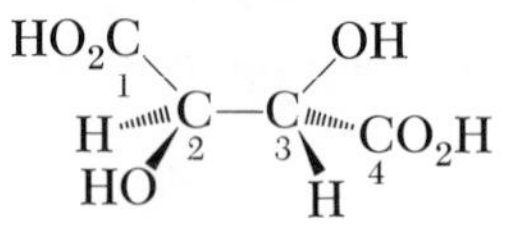

another conformation of *meso*-tartaric acid

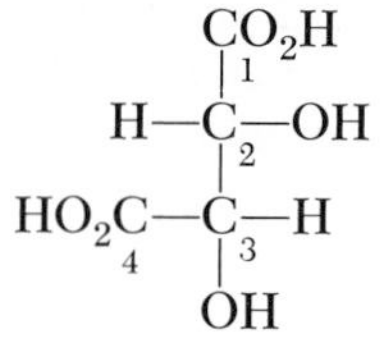

Fischer projection of *meso*-tartaric acid lacking a plane of symmetry

l-Tartaric acid ("unnatural tartaric acid") and *d*-tartaric acid ("natural tartaric acid," "Weinsaüre") are enantiomers and, therefore, are identical in achiral environments. *meso*-Tartaric acid ("unresolvable tartaric acid"), since it is a diastereomer of both of these, will have properties different from those of the other two forms (Table 4–2). Indeed, diastereomers always are expected to have different properties since they are different compounds!

TABLE 4–2 Comparison of Tartaric Acids[a]

Form	$[\alpha]_D^{20}(H_2O)$	m.p. (°C)
l	−11.98	168–170
d	+11.98	168–170
meso	0.00	140
"*d,l*"	0.00	206

[a] For more detail see "Merck Index," 8th Edition, 1968, p. 1014.

TABLE 4–3 Comparison of Lactic Acids

Form	$[\alpha]_D^{22}(H_2O)$	m.p. (°C)
l	−2.6	53
d	+2.6	53
"*d,l*"	0.00	17

An interesting feature of Table 4–2 is the appearance of a fourth entry, the so-called "*d,l*-tartaric acid" ("resolvable tartaric acid"). This "*d,l*" is the designation for the racemic modification, suggesting that here the racemic modification is somehow different from what one might have expected from a mixture of enantiomers. The sharp melting point and other data indicate that "*d,l*-tartaric acid" is a distinct substance *in the solid state*. Such a compound is called a *racemate*. It is impossible to predict whether a given racemic modification will be a racemate.

Racemate formation can also occur when a compound contains only one chiral center, as illustrated for lactic acid (Table 4–3). (Why is there no entry for "*meso*" lactic acid?)

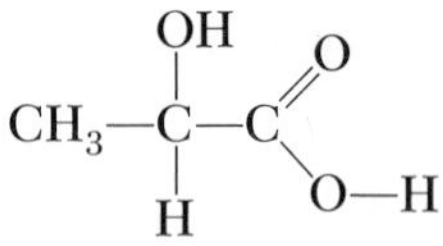

lactic acid

Why Is Meso-Tartaric Acid Optically Inactive? The best explanation for the optical inactivity of *meso*-tartaric acid is that the compound exists in solution as a mixture of sets of enantiomers, each set being equally populated by *d* and *l* forms. These *d* and *l* forms are enantiomeric conformations and illustrate the phenomenon of *conformational enantiomerism*. The diagram in Figure 4–14 illustrates how, beginning with an eclipsed conformation, one *must* generate equal amounts of conformational enantiomers from *meso*-tartaric acid.

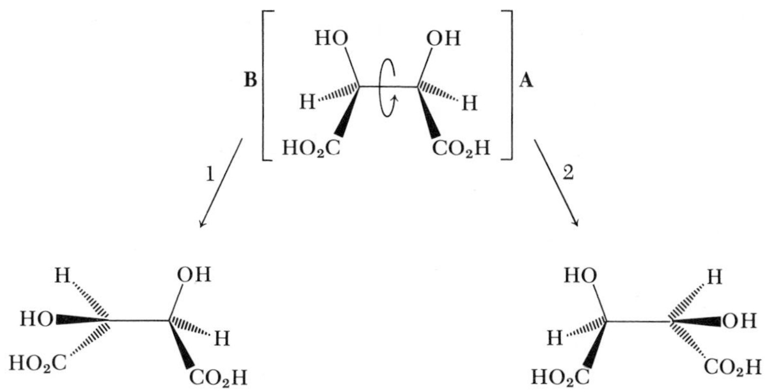

Figure 4-14 Conformational enantiomerism in *meso*-tartaric acid. In process 1, fragment **B** is rotated 90°; while in process 2, fragment **A** undergoes an equally probable 90° rotation. The products are **1** and **2**, a set of enantiomeric conformations. The same principle applies for any rotation around the central bond.

Threo and Erythro Forms. Each of the possible stereoisomers of 1-bromo-1,2-dichloropropane, $CHBrClCHClCH_3$, is presented in one wedge and one Fischer projection in Figure 4-15. Note that **3** and **4** are not identical but are enantiomers. This pair is called the *erythro* form. Compounds **1** and **2** also are enantiomers and constitute the *threo* form. Sets (**1** and **3**), (**1** and **4**), (**2** and **3**), *and* (**2** and **4**) represent sets of diastereomers. While both members of such a set (*e.g.*, **1** and **3**) are optically active, they are expected to have rotations of unequal magnitude. The signs of rotation may or may not be alike.

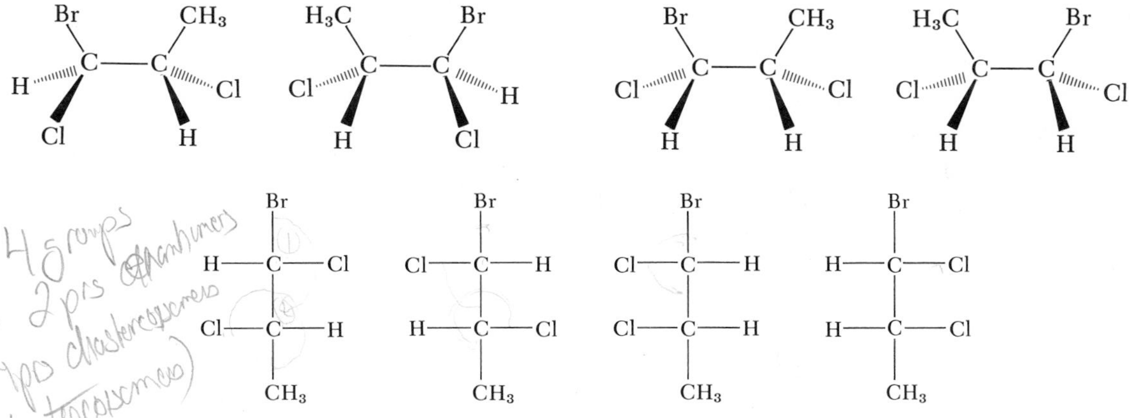

1 (1S,2S)-1-bromo-1,2-dichloropropane
2 (1R,2R)-1-bromo-1,2-dichloropropane
threo

3 (1R,2S)-1-bromo-1,2-dichloropropane
4 (1S,2R)-1-bromo-1,2-dichloropropane
erythro

Figure 4-15 The stereoisomers of 1-bromo-1,2-dichloropropane. Structures **1** and **2** are enantiomers, as are **3** and **4**. The *threo* form is represented by **1** and **2**, while the *erythro* form is represented by **3** and **4**.

In general, the *erythro* form of a compound is akin to the *meso* form except that, instead of having identical sets of substitutents on both chiral centers, the *erythro* form has only *two* substituents in common on the chiral centers. The *erythro* form can easily be recognized once you note that it can exist in a conformation in which similar groups are eclipsed (after allowance is made for the identity of only two of the three substituents on one chiral center with those on the other). Although it resembles the *meso* form, the replacement of one group with a second

(x for a in the example below) permits the *erythro* form to exist as a pair of enantiomers.

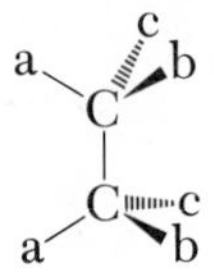

one conformation of a generalized *meso* form

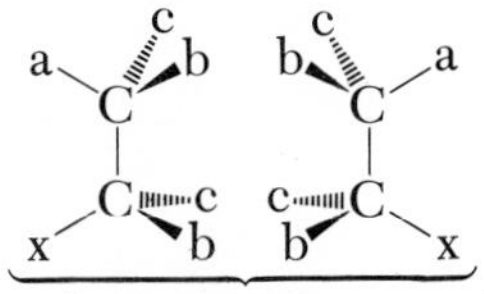

one conformation of a generalized pair of *erythro* isomers

The *threo* form resembles the *d,l* form except that, instead of having three substitutents in common on the chiral centers, the *threo* form has only two sets of identical substituents. Like the *d,l* form, the *threo* form exists as a pair of enantiomers.

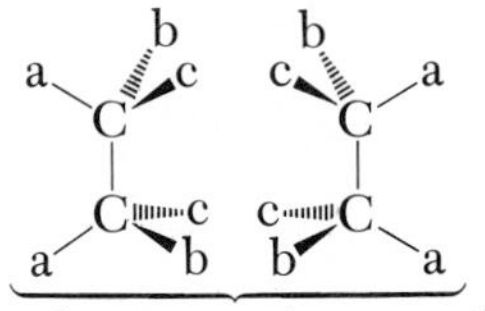

one conformation of a generalized pair of *d,l* isomers

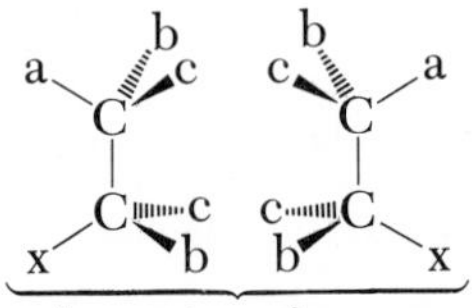

one conformation of a generalized pair of *threo* isomers

When a compound possesses two chiral centers, it can form as many as four stereoisomers. This illustrates the rule that a molecule containing n non-identical chiral centers may exist as a maximum of 2^n stereoisomers. For example, 2-bromo-3-chloro-4-fluoropentane should exist as 8 ($= 2^3$) stereoisomers.

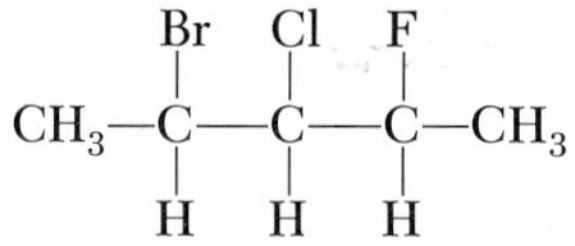

2-bromo-3-chloro-4-fluoropentane

Prochirality. An enzymic reaction (a reaction accelerated by a complex biological molecule known as an enzyme) is often capable of distinguishing between two apparently identical groups, as, for example, occurs in the hypothetical enzyme substrate C*aabe*. This arises because the enzyme, although a complex molecule containing many chiral centers, is enantiomerically pure. When the enzyme (abbreviated E(+)) attacks the two a groups, the central carbon atom becomes chiral and two diastereomeric enzyme-C*aabe* complexes are formed.

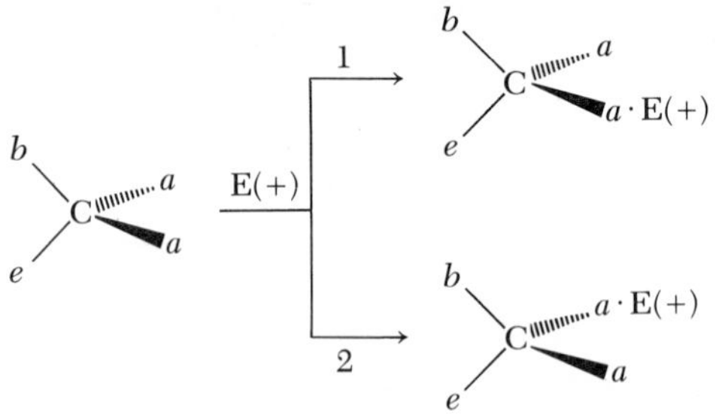

Diastereomers, having different chemical properties, will be formed at different rates and will react at different rates. One of these paths (*e.g.*, 1) will proceed much more rapidly than the other, and this permits the enzyme to distinguish between the two a groups.

The central atom in C*aabe* is said to be **prochiral** because it can become chiral by replacing one of the two "identical" groups with an achiral substituent.

The *a* groups in C*aabe* are said to be *enantiotopic,* meaning that the environment about one group is enantiomeric with the environment around the other group. A convenient test for enantiotopic groups is to replace first one, and then the other, with an achiral substituent (*e.g.*, D for H). Enantiotopic substituents will produce enantiomers.

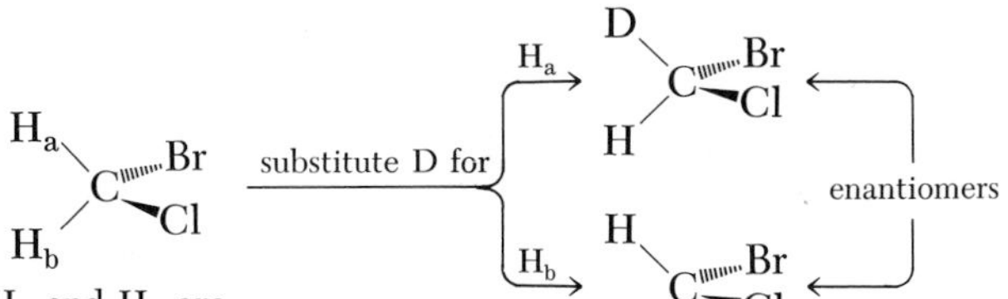

H_a and H_b are enantiotopic

substitute D for H_a → D, H, C, Br, Cl; substitute D for H_b → H, D, C, Br, Cl — enantiomers

14. Identify one pair of enantiotopic groups in each of the following:
    (a) CH_2ClBr
    (b) $CHCl(CH_3)_2$
    (c) $CHCl(CH_2CH_3)_2$
    (d) $ClCH_2CH_2Br$
    (e) $ClCH_2CH_2Cl$

4.6 THE SYNTHESIS OF DIASTEREOMERS BY FREE-RADICAL HALOGENATION

In discussing the free-radical halogenation of alkanes (Chapter 3), we carefully avoided an elaborate discussion of the preparation of dihalides; we limited ourselves to monohalogenation. Part of the reason for this, as you now may be able to guess, is that such dihalides often exist as diastereomers; we could not give intelligent consideration to the formation of diastereomeric dihalides if we did not know what diastereomers were! Now that we have introduced some basic stereochemical concepts, we can expand our picture of free-radical halogenation to include the halogenation of alkyl halides (*i.e.*, the preparation of dihalides). The *concepts* introduced in this discussion are valid for many reactions that produce diastereomers, and they will be employed on many occasions in this text. To keep the size of the current discussion reasonable, we will consider only the dichlorination of *n*-butane.

The Chlorination of *n*-Butane—A Further Look. If *n*-butane is chlorinated, the methylene hydrogens, being 2°, will react most rapidly. The two enantiotopic methylene hydrogens on a given carbon will be abstracted at the same rate to produce enantiomeric radicals which equilibrate very rapidly. The reactions of these radicals can be predicted on the basis of a single, planar structure.

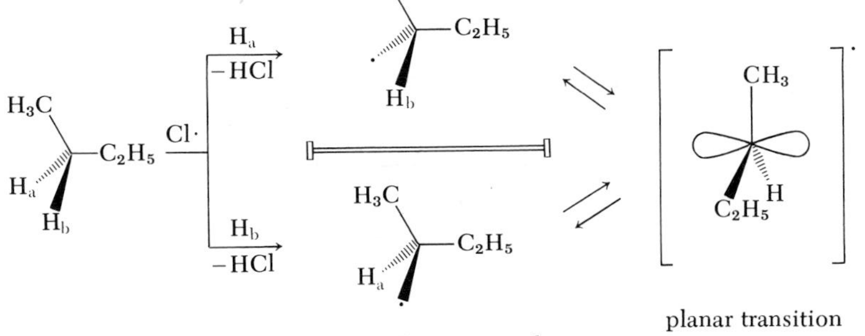

H_3C, C_2H_5, H_a, H_b; $Cl\cdot$; H_a $-HCl$; H_b $-HCl$; H_3C, C_2H_5, H_b; H_3C, C_2H_5, H_a; CH_3, H, C_2H_5

shallow pyramid

planar transition state

This planar radical will be attacked at equal rates from either face to produce equimolar quantities of two enantiomeric products. This reaction and, indeed, any reaction involving an achiral reagent and enantiotopic reaction sites must produce a racemic modification!

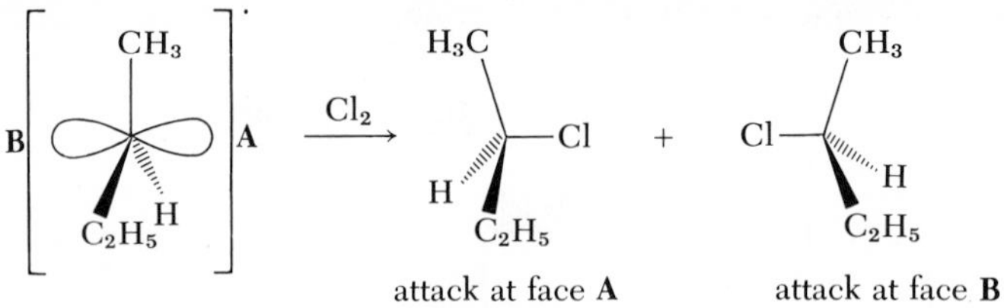

B, A; CH_3; H; C_2H_5; $\xrightarrow{Cl_2}$; H_3C, Cl, H, C_2H_5 + Cl, CH_3, H, C_2H_5

attack at face **A** attack at face **B**

Consider the further chlorination of 2-chlorobutane, focusing on the remaining methylene group. The two remaining methylene protons of (**R**)-2-chlorobutane are not identical. This is most readily seen if we label first one (designated H_a) and then the other (designated H_b) with deuterium. The two diastereomeric products have opposite configurations at the newly created asymmetric center, and these two hydrogens, H_a and H_b, are said to be **diastereotopic.**

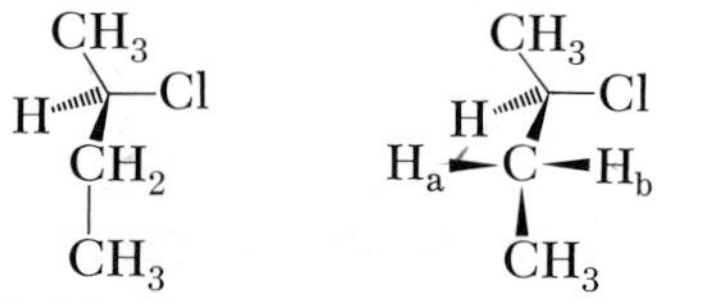

CH_3, H, Cl, CH_2, CH_3 — (**R**)-2-chlorobutane; CH_3, H, Cl, H_a—C—H_b, CH_3

H_a and H_b are diastereotopic protons

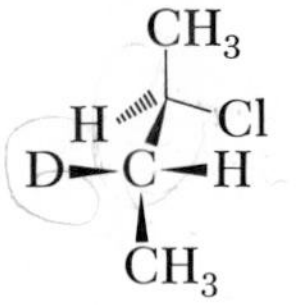

(2**R**,3**S**)-2-chloro-3-deuteriobutane 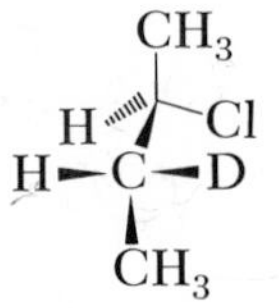 (2**R**,3**R**)-2-chloro-3-deuteriobutane

These diastereotopic hydrogens will be abstracted by Cl· at different rates through transition states which are diastereomeric. Rapid inversion of the configuration of the resultant radicals permits us to discuss the fate of these radicals in terms of a planar radical. However, this time the reactive faces are **diastereotopic.**

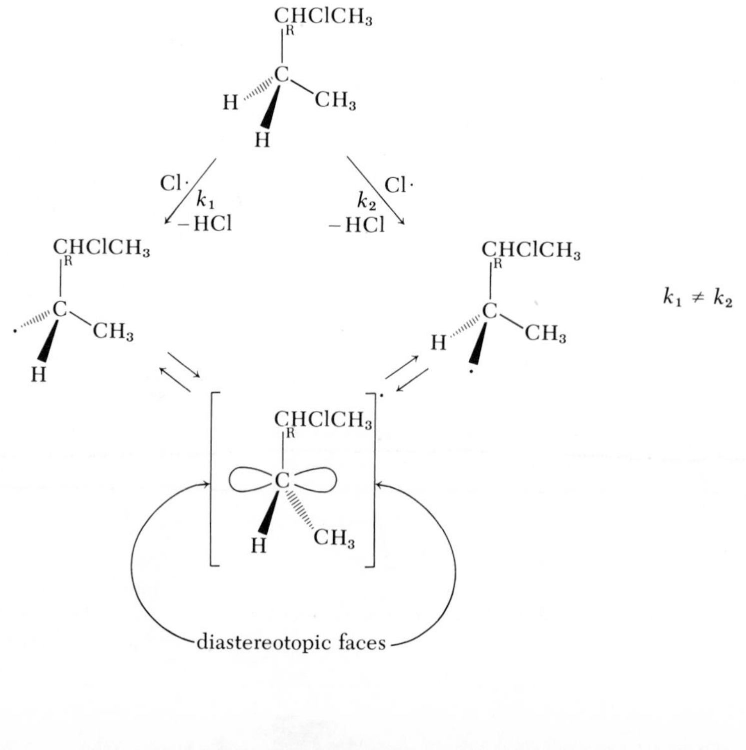

$CHClCH_3$ (R); C; H; CH_3; H

Cl·, k_1, −HCl; Cl·, k_2, −HCl

$CHClCH_3$ (R); C; CH_3; H $CHClCH_3$ (R); C; H; CH_3

$k_1 \neq k_2$

$CHClCH_3$ (R); C; H; CH_3

diastereotopic faces

Using a single conformation of this radical as an illustration, we see that attack at one face (**B′**) produces one of the *d,l* isomers, while attack at the other face (**A′**) produces the *meso* form. In addition to producing different products (*i.e.*, diastereomers), attack at **A′** and **B′** occurs at different rates.

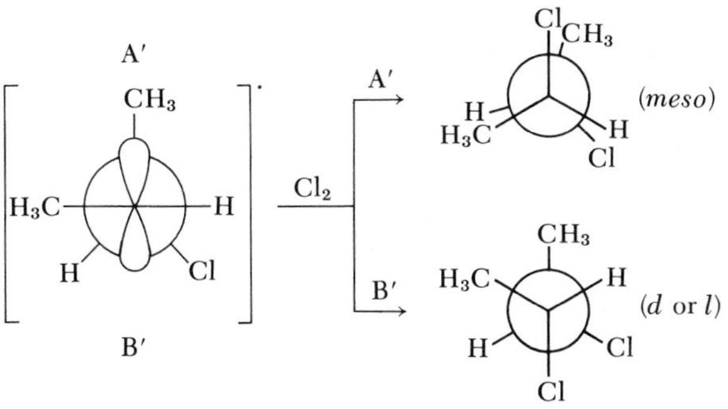

The chlorination of the monohalide enantiomer, *i.e.*, (**S**)-2-chlorobutane, produces the same amount of *meso*-2,3-dichlorobutane as does the chlorination of (**R**)-2-chlorobutane. Also, (**S**)-2-chlorobutane produces the *enantiomer* of the particular *d,l*-form of 2,3-dichlorobutane generated from (**R**)-2-chlorobutane. The ratio of *meso* to *d* (or *l*) produced from the **R** enantiomer equals the ratio of *meso* to *l* (or *d*) produced from the **S** enantiomer. Therefore, free radical chlorination of (**R,S**)-2-chlorobutane (the racemic modification) must produce equal amounts of *d*- and *l*-2,3-dichlorobutane. The amounts of *meso* and *d,l* forms produced are unequal, being controlled by the conformational preferences of the radical and by the relative reactivity of faces **A′** and **B′**.

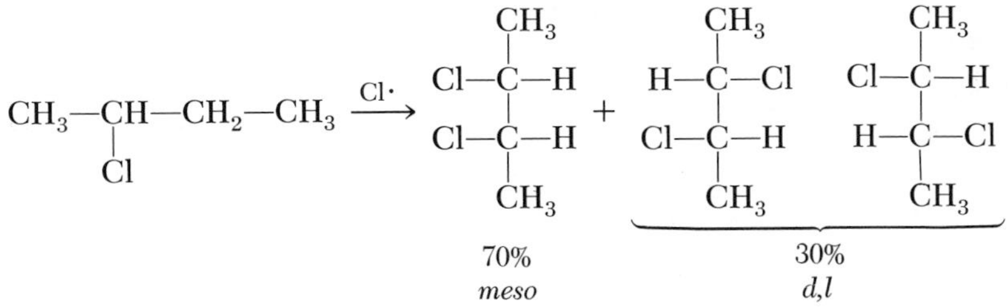

15. Assuming that the process of vapor phase chromatography, vpc (also called gas chromatography), can separate diastereomers but not enantiomers, how many dibromides would appear to be produced by the free radical bromination of each of the following if the reaction mixture were analyzed by vpc?
    (a) ethane
    (b) propane
    (c) butane
    (d) 2-methylpropane
    (e) 2,2-dimethylpropane
    (f) 2,2-dimethylbutane
    (g) pentane
16. If, in problem 4.15, you had used an analytical procedure which separated enantiomers from one another as well as diastereomers from one another, how many products would have been detected from each reaction?

IMPORTANT TERMS

Achiral Compound: Any substance which is superimposable upon its mirror image. Such a compound is symmetric. Most often a compound with a point or plane of symmetry.

Asymmetric substance: Any substance which lacks all symmetry elements. For practical purposes, the one-fold simple axis of symmetry, which must convert an object into itself, is not considered because *everything* possess such an axis.

Chiral compound: Any substance which is not superimposable upon its mirror image. Such a compound may be dissymmetric or asymmetric.

Conformational enantiomers: Substances which are enantiomers in one or more conformations. A compound must be chiral in the conformation in which it exhibits this property.

Br, H, H–C–C(H)(H)–Cl and Cl–C(H)(H)–C(Br)(H)–H are conformational enantiomers

Diastereomers: Two or more substances which are stereoisomers but which are not enantiomers. Included within this are *classical diastereomers,* stereoisomers which possess more than one chiral center and which differ in configuration at one or more than one of these; however, classical diastereomers cannot have completely opposite configurations at all chiral centers since this would make them enantiomers.

Diastereotopic groups: Groups which reside in diastereomeric environments. Replacement of first one and then another with an achiral group creates diastereomers.

Dissymmetric substance: Any substance lacking both a point and a plane of symmetry. More generally, the criterion for dissymmetry is the absence of an alternating axis of symmetry (see p. 88).

***d,l* form:** When dealing with a set of compounds of the type C*abx*—C*abc*, the *d,l* form is that set which constitutes a pair of enantiomers. Remember that, although the word "form" is singular, the "*d,l* form" consists of two substances! Each member of the *d,l* form has the *meso* form as a diastereomer. When dealing with a compound with only one chiral center, such as lactic acid, the prefix *d,l* implies a racemic modification.

```
        H
        |
   Cl—C—CH3
        |
 CH3—C—Cl
        |
        H
```

one enantiomer of a *d,l* form

Enantiomers: Stereoisomers which are non-superimposable mirror images.

Enantiotopic groups: Groups which reside in enantiomeric environments. Replacement of first one and then another with an achiral group creates enantiomers.

***erythro* form:** A diastereomer which exists in enantiomeric forms. Chiral centers have two of three substituents identical (C*abx*—C*aby*). Had all three substituents been the same, this would have been a *meso* form (C*abx*—C*abx*).

***meso* form:** The diastereomer which is achiral (does not have an enantiomer). Both chiral centers have identical substituents (C*abx*—C*abx*).

```
      H
      |
 Cl—C—CH3
      |
 Cl—C—CH3
      |
      H
```

meso form

Optical activity: The ability, shown by chiral molecules, to rotate the plane of plane polarized light.

Prochiral center: A center which contains two identical and two non-identical substituents. Replacement of one identical substituent with a completely different one converts this into a chiral center.

Racemic modification: A sample containing only equimolar amounts of both enantiomers.

Racemization: The conversion of one enantiomer into a racemic modification. Several asymmetic centers may be involved, but *all* must be inverted in configuration in order to have racemization of the molecule.

Specific rotation: A property of chiral substances which, for a given wavelength of light, temperature, and solvent, is considered a molecular constant. It is given by

$$[\alpha]_{\lambda}^{t^\circ} = \frac{100\alpha}{lc}$$

Symmetric substance: Any substance possessing a point or plane of symmetry. More correctly, it is any substance possessing an alternating axis of symmetry (see p. 88).

***threo* form:** A diastereomer which exists in enantiomeric forms. Chiral centers have two of three substituents identical (C*abx*—C*aby*). Had both sets of substituents been the same, this would have been a *d,l* form (C*abx*—C*abx*).

PROBLEMS

17. Provide a specific illustration of each of the following.
    - (a) a chiral molecule
    - (b) an achiral molecule
    - (c) a prochiral molecule
    - (d) a pair of enantiomers
    - (e) three diastereomers
    - (f) *threo* forms
    - (g) *erythro* forms
    - (h) a racemization reaction
    - (i) a *meso* form
    - (j) a *d,l* form
    - (k) a plane of symmetry
    - (l) a point of symmetry
    - (m) enantiotopic protons
    - (n) diastereotopic protons

18. Arrange the members of each part into sets of identical substances.

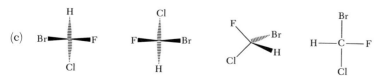

(a)

```
     Br            Cl            F             Br            H
     |             |             |             |             |
  H—C—Cl        H—C—Br        H—C—Cl        H—C—F        Br—C—F
     |             |             |             |             |
     F             F             Br            Cl            Cl
```

(b) H, Cl, F, Br   Cl, Br, F, H   H, F, Cl, Br   F, Br, H, Cl   Br, Cl, F, H

(c)

```
      H              Cl                              Br
      |              |           F                   |
 Br———+———F     F————+————Br       >—Br          H———C———F
      |              |         Cl    H               |
      Cl             H                               Cl
```

(d) 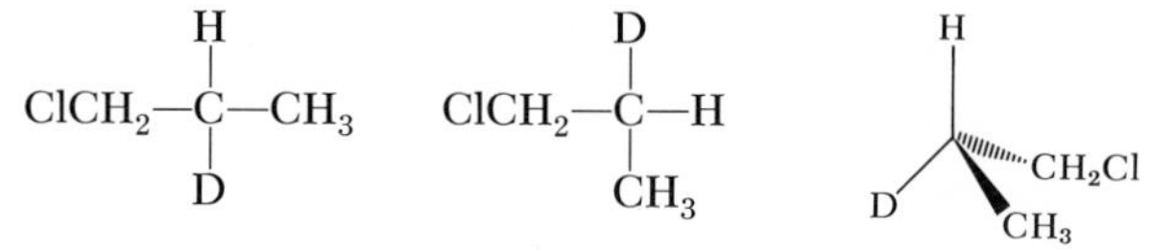 $ClCH_2$—C(H)(D)—CH_3 $ClCH_2$—C(D)(CH_3)—H H, D, CH_2Cl, CH_3 (wedge structure)

(e) 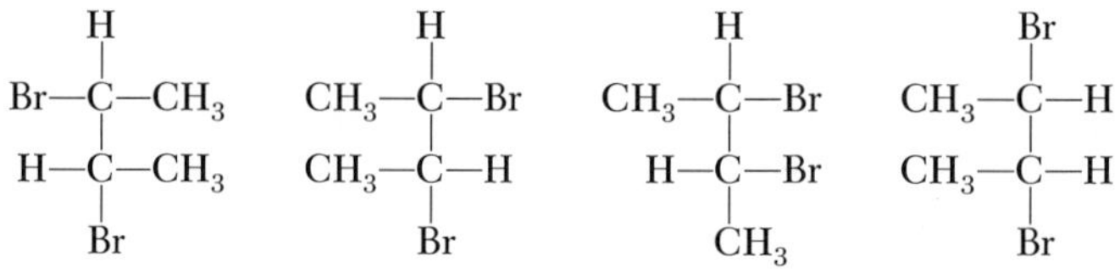 Br—C(H)—CH_3 / H—C(Br)—CH_3 CH_3—C(H)—Br / CH_3—C(Br)—H CH_3—C(H)—Br / H—C(CH_3)—Br CH_3—C(Br)—H / CH_3—C(Br)—H

19. Identify the following as *identical, enantiomers,* or *diastereomers.* (Assume free rotation around single bonds.)

(a) 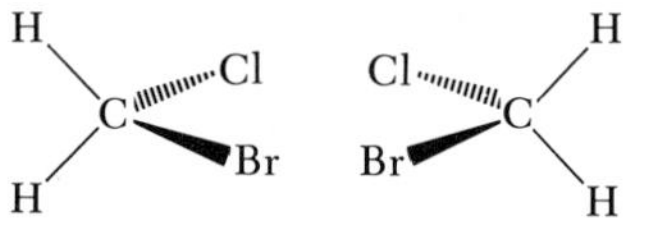 H, H, C, Cl, Br Cl, Br, C, H, H

(b) H—C(Br)(F)—Cl Br—C(Cl)(H)—F

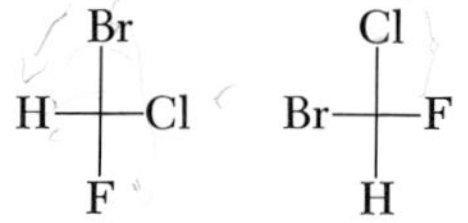

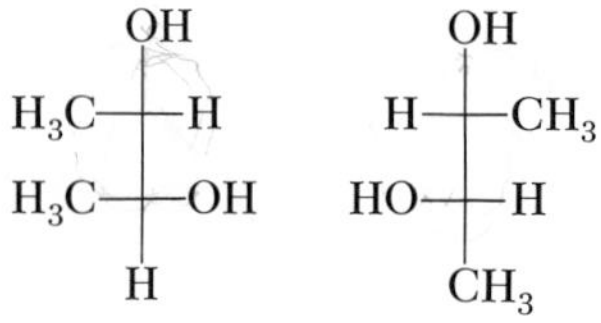

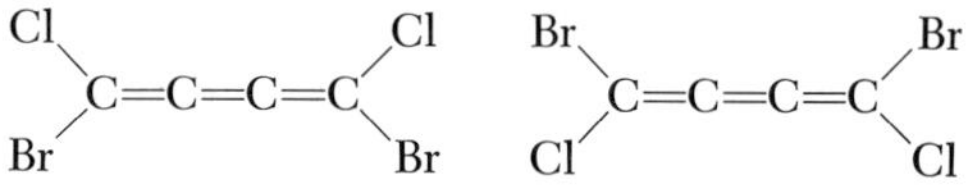

(c) OH / H_3C—H / H_3C—OH / H OH / H—CH_3 / HO—H / CH_3

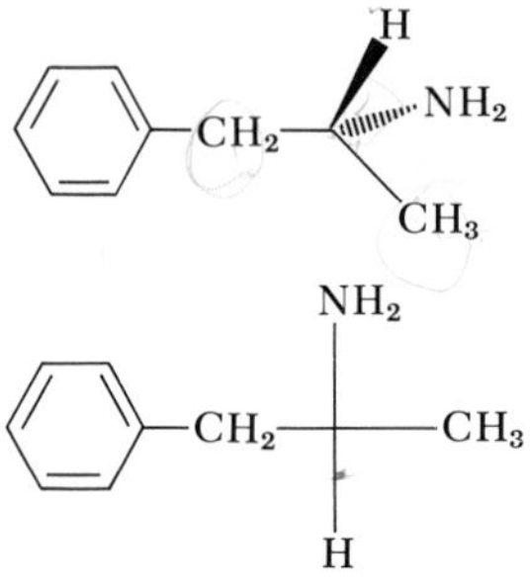

(d) Cl(Br)C=C=C=C(Cl)(Br) Br(Cl)C=C=C=C(Br)(Cl)

(e) C_6H_5—CH_2—C(H)(NH_2)(CH_3) C_6H_5—CH_2—C(NH_2)(H)—CH_3

(f) HO, H_3C, H — H, OH, CH_3 H_3C, HO, H — H, CH_3, OH

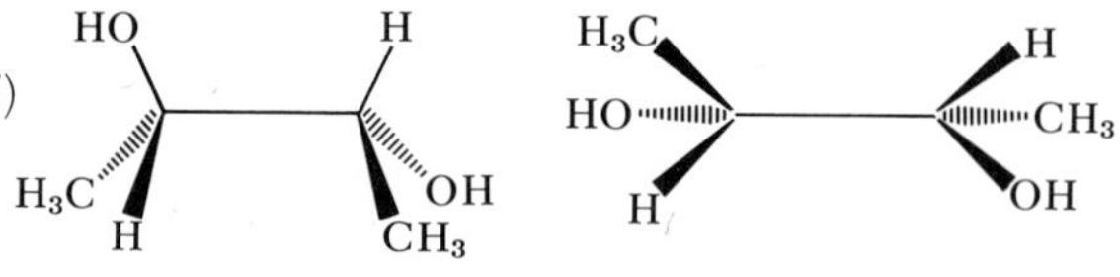

20. Some of the following compounds contain chiral centers. Assign absolute configurations to all of these.

(a) 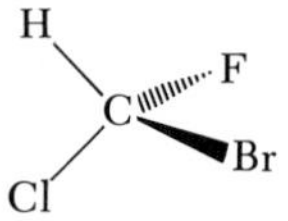 H, Cl, C, F, Br

(b) 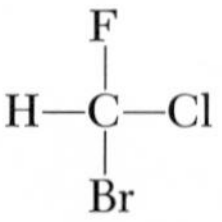 H—C(F)(Br)—Cl

(c) 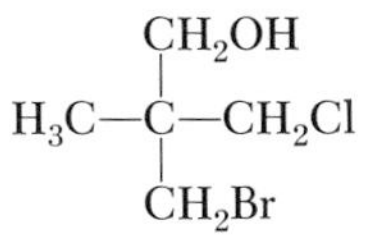 $H_3C—C(CH_2OH)(CH_2Br)—CH_2Cl$

(d) 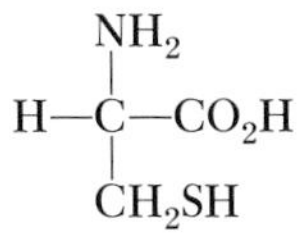 $H—C(NH_2)(CH_2SH)—CO_2H$

(e) 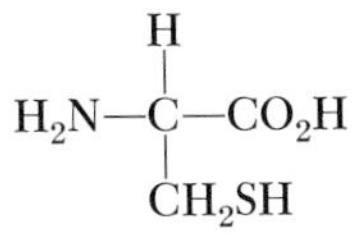 $H_2N—C(H)(CH_2SH)—CO_2H$

(f) 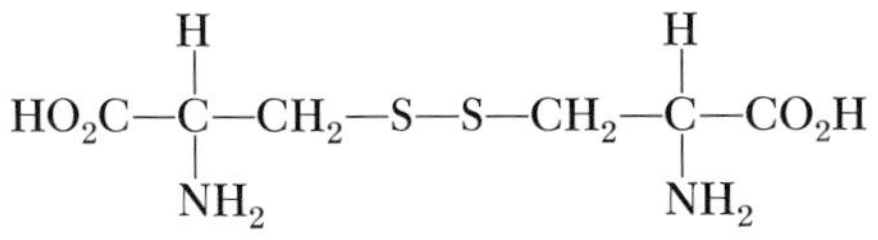 $HO_2C—C(H)(NH_2)—CH_2—S—S—CH_2—C(H)(NH_2)—CO_2H$

(g) 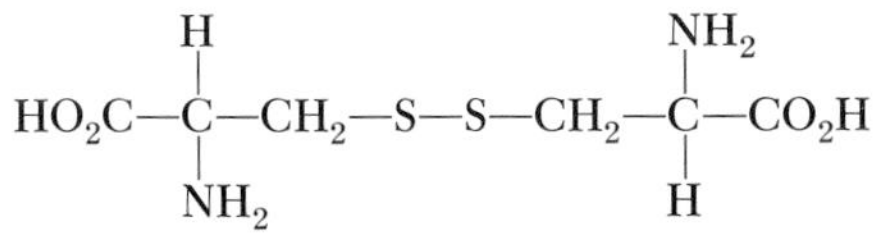 $HO_2C—C(H)(NH_2)—CH_2—S—S—CH_2—C(NH_2)(H)—CO_2H$

(h) $H—C(CH_3)(H)—C(Br)(CH_3)—CH_2—C(Br)(CH_3)—H$

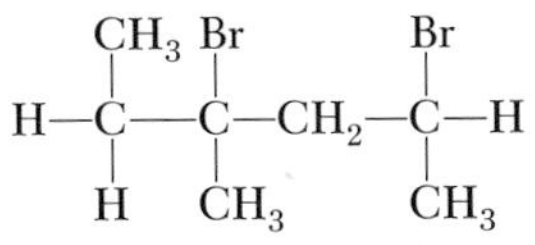

(i) 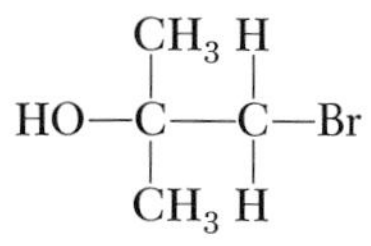 $HO—C(CH_3)(CH_3)—C(H)(H)—Br$

(j) 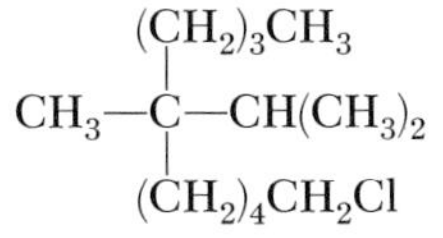 $CH_3—C((CH_2)_3CH_3)((CH_2)_4CH_2Cl)—CH(CH_3)_2$

(k) 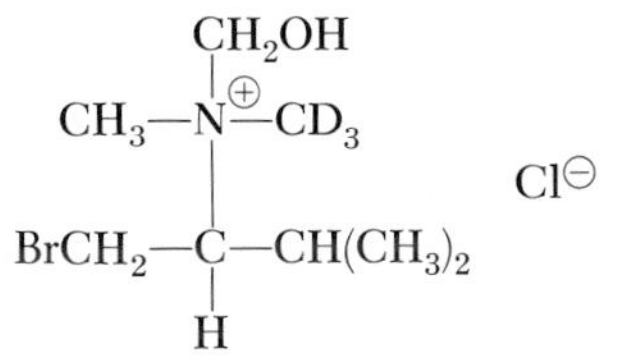 $CH_3—\overset{\oplus}{N}(CH_2OH)—CD_3$ attached to $BrCH_2—C(H)—CH(CH_3)_2$ $Cl^{\ominus}$

(l) 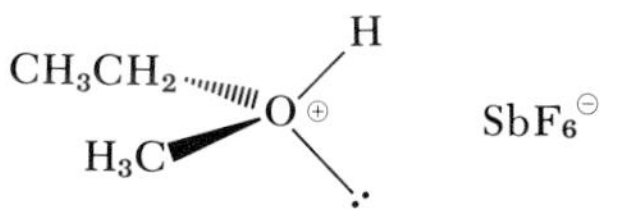 CH_3CH_2, H_3C, H on $\overset{\oplus}{O}$ (with lone pair) $SbF_6^{\ominus}$

21. Identify points and planes of symmetry in the following:
   (a) methane
   (b) allene
   (c) ethylene
   (d) acetylene
   (e) boron trifluoride
   (f) formaldehyde
   (g) 1,1-dichloro-3,3-diiodoallene
   (h) chloromethane
   (i) carbon tetrachloride

22. In which of the following ways should enantiomers differ, at least in principle?
   (a) melting point
   (b) boiling point
   (c) sign of specific rotation
   (d) magnitude of specific rotation
   (e) absolute configuration
   (f) solubility in water
   (g) solubility in chloroform
   (h) solubility in (+)-2-chloropentane
   (i) solubility in (−)-2-chloropentane
   (j) solubility in (±)-2-chloropentane
   (k) interaction with left handed circularly polarized light
   (l) toxicity to human beings (Note that all reactions in our bodies are controlled by enzymes.)

23. The **R** enantiomer of mevalonic acid is biologically active. Replacement of some of the four methylene hydrogens with deuterium converts **C**3 to the **S** configuration. (a) Identify all of the sets of prochiral hydrogens in (**R**)-mevalonic acid. (b) Which hydrogens can be replaced by deuterium in order to convert the configuration at **C**3 from **R** to **S**? (Structure on p. 118.)

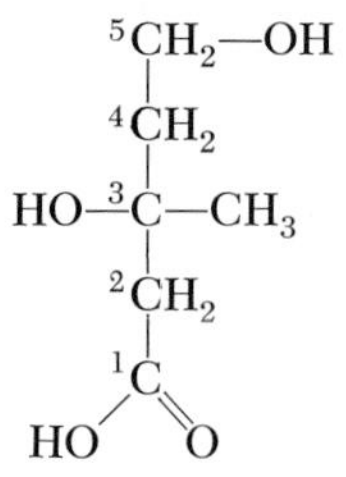

$$
\begin{array}{c}
{}^{5}CH_2\text{—}OH \\
| \\
{}^{4}CH_2 \\
| \\
HO\text{—}{}^{3}C\text{—}CH_3 \\
| \\
{}^{2}CH_2 \\
| \\
{}^{1}C(=O)OH
\end{array}
$$

mevalonic acid

24. Each of the following compounds is involved in biological chemistry or medicine. Identify all of the chiral centers and indicate how many stereoisomers of each compound are possible. (Note that the biological activity is usually associated with a particular stereoisomer.)

(a) general formula for a hexose sugar

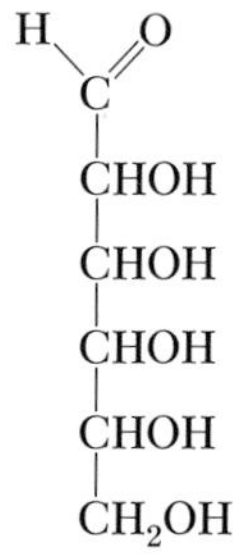

$$
\begin{array}{c}
H\text{—}C{=}O \\
| \\
CHOH \\
| \\
CHOH \\
| \\
CHOH \\
| \\
CHOH \\
| \\
CH_2OH
\end{array}
$$

(b) glucose (blood sugar)

$$
\begin{array}{c}
H\text{—}C{=}O \\
H\text{—}C\text{—}OH \\
HO\text{—}C\text{—}H \\
H\text{—}C\text{—}OH \\
H\text{—}C\text{—}OH \\
CH_2OH
\end{array}
$$

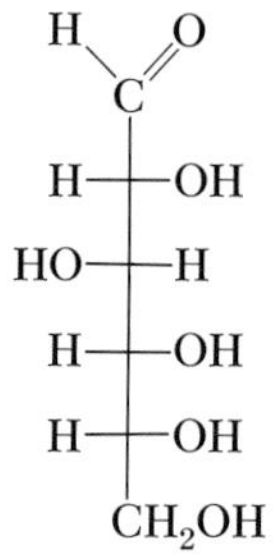

(c) ambucetamide (anti-spasmodic)

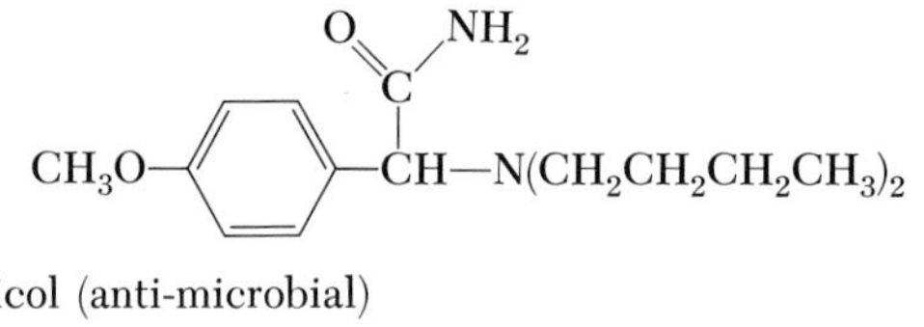

$$CH_3O\text{—}C_6H_4\text{—}CH(C(=O)NH_2)\text{—}N(CH_2CH_2CH_2CH_3)_2$$

(d) chloramphenicol (anti-microbial)

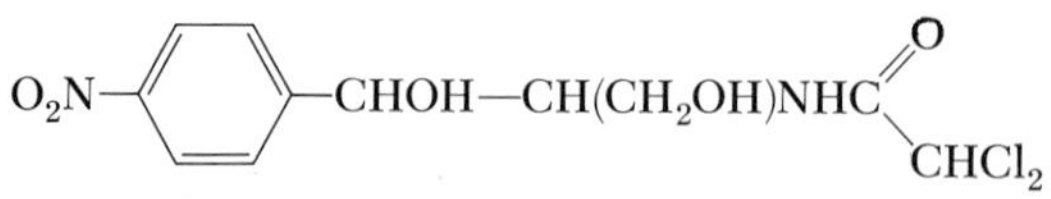

$$O_2N\text{—}C_6H_4\text{—}CHOH\text{—}CH(CH_2OH)NHC(=O)CHCl_2$$

(e) α-chloro-α-phenylacetylurea (anti-epileptic)

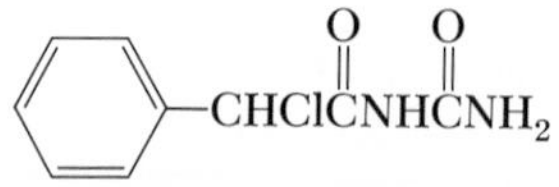

$$C_6H_5\text{—}CHCl\overset{O}{\overset{\|}{C}}NH\overset{O}{\overset{\|}{C}}NH_2$$

(f) pantothenic acid (human nutrition)

$$HO_2CCH_2CH_2NH\overset{O}{\overset{\|}{C}}CHOHC(CH_3)_2CH_2OH$$

(g) Ciodrin (ectoparasite control in livestock)

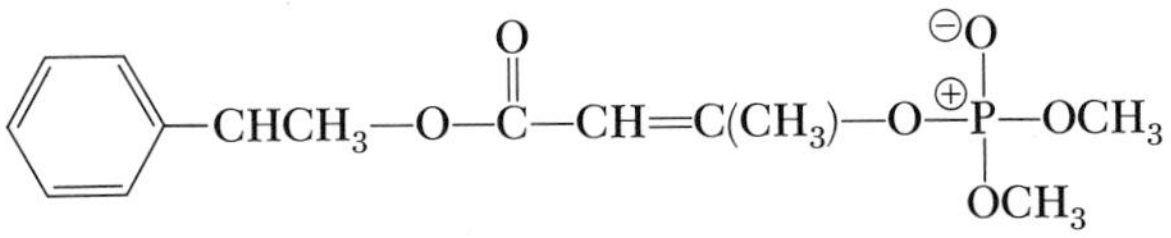

$$C_6H_5\text{—}CHCH_3\text{—}O\text{—}\overset{\overset{\displaystyle O}{\|}}{C}\text{—}CH{=}C(CH_3)\text{—}O\text{—}\overset{\ominus O}{\underset{OCH_3}{\overset{\oplus}{P}}}\text{—}OCH_3$$

25. Many biologically important compounds contain prochiral centers. Identify all of the prochiral centers in the following.

(a) nitroglycerin (coronary vasodilator)

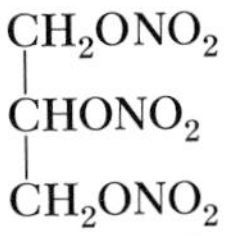

$$\begin{array}{l} CH_2ONO_2 \\ | \\ CHONO_2 \\ | \\ CH_2ONO_2 \end{array}$$

(b) Phorate (systemic insecticide)

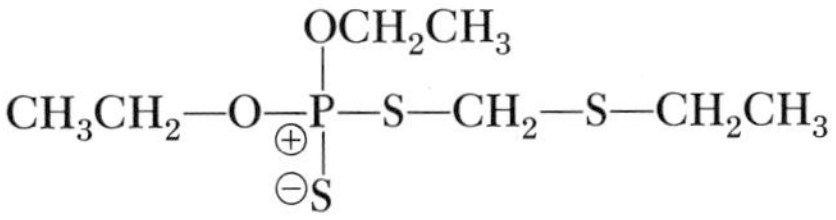

$$CH_3CH_2\text{—}O\text{—}\overset{\displaystyle OCH_2CH_3}{\underset{\ominus S}{\overset{\oplus}{P}}}\text{—}S\text{—}CH_2\text{—}S\text{—}CH_2CH_3$$

(c) sorbitol (humectant; candy manufacture)

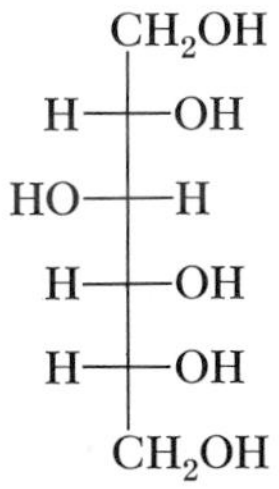

$$\begin{array}{rcl} & CH_2OH & \\ H & \text{—}\!\!+\!\!\text{—} & OH \\ HO & \text{—}\!\!+\!\!\text{—} & H \\ H & \text{—}\!\!+\!\!\text{—} & OH \\ H & \text{—}\!\!+\!\!\text{—} & OH \\ & CH_2OH & \end{array}$$

(d) tyrosine (amino acid)

$$HO\text{—}C_6H_4\text{—}CH_2CH(NH_2)CO_2H$$

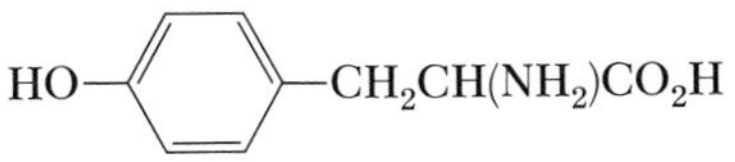

(e) tyramine (sympathomimetic)

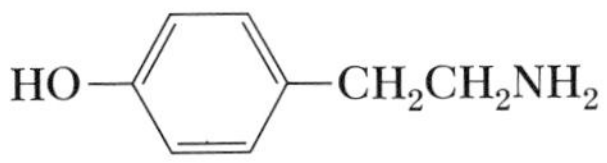

$$HO\text{—}C_6H_4\text{—}CH_2CH_2NH_2$$

(f) norcymethadol (narcotic analgesic)

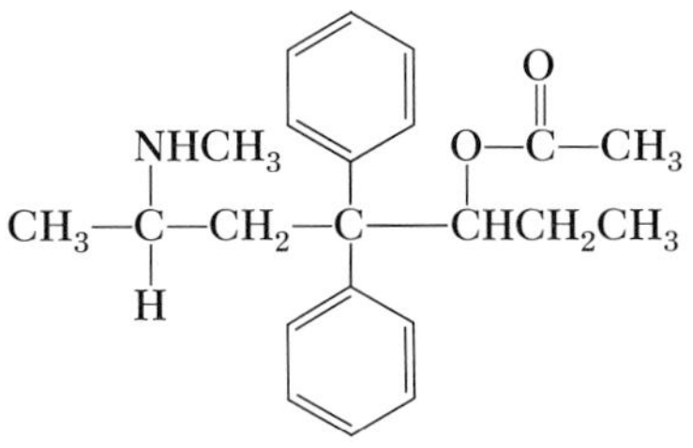

$$CH_3\text{—}\underset{H}{\overset{NHCH_3}{C}}\text{—}CH_2\text{—}\underset{C_6H_5}{\overset{C_6H_5}{C}}\text{—}\overset{O\text{—}\overset{O}{\overset{\|}{C}}\text{—}CH_3}{CHCH_2CH_3}$$

26. Amines (RNH_2) react with carboxylic acids ($R'CO_2H$) to form salts ($RNH_3^{\oplus}R'CO_2^{\ominus}$) by transfer of a proton to the lone electron pair of nitrogen. Ammonium acetate formation is a simple illustration.

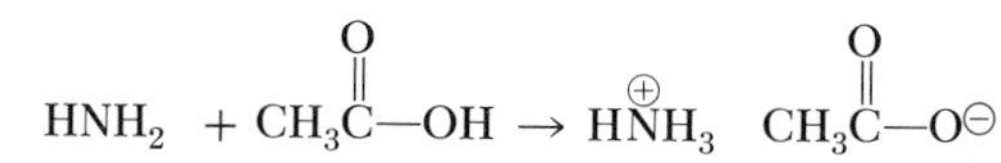

$$\underset{\text{ammonia}}{HNH_2} + \underset{\text{acetic acid}}{CH_3\overset{\overset{O}{\|}}{C}-OH} \rightarrow \underset{\text{ammonium acetate}}{H\overset{\oplus}{N}H_3 \quad CH_3\overset{\overset{O}{\|}}{C}-O^{\ominus}}$$

When 1-aminobutane reacts with (**R**)-4-chloropentanoic acid, only one type of salt is produced. However, when 2-aminobutane is used, two different salts are formed. Account for this difference. How are the salts produced from 2-aminobutane related to one another?

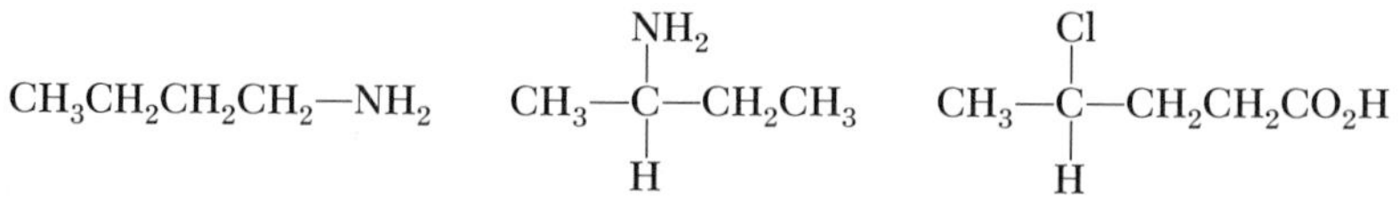

$$\underset{\text{1-aminobutane}}{CH_3CH_2CH_2CH_2-NH_2} \qquad \underset{\text{2-aminobutane}}{CH_3-\overset{\overset{NH_2}{|}}{\underset{\underset{H}{|}}{C}}-CH_2CH_3} \qquad \underset{\text{4-chloropentanoic acid}}{CH_3-\overset{\overset{Cl}{|}}{\underset{\underset{H}{|}}{C}}-CH_2CH_2CO_2H}$$

27. Neglecting conformers, draw all of the possible isomers of the following. Label every chiral center. Provide the IUPAC name for each. Identify those compounds incapable of optical activity.
    (a) C_4H_9Cl
    (b) $C_4H_8Cl_2$
    (c) C_4H_8ClBr

28. (a) When 5.678 g of cane sugar is dissolved in water and brought to a total volume of 20 ml at 20°C, the rotation of the solution in a 1 dm tube is 18.88°. What is the specific rotation of cane sugar? (b) The observed rotation of an aqueous solution of cane sugar in a 2 dm tube is 10.75°. What is the concentration of the sugar solution? (c) Can you calculate the molecular weight of cane sugar from these data?

29. (a) A sample of pure amyl alcohol having a density of 0.8 g/ml at 20°C in a 20 cm tube gives a rotation of 9.44°. Calculate the specific rotation of the alcohol. (b) A fraction of fusel oil having a density of 0.8 g/ml at 20°C has a rotation (observed) of 3.56° in a 4 dm tube. What percentage of optically active amyl alcohol is present in the fusel oil sample? Assume that amyl alcohol is the only optically active compound in fusel oil. (Fusel oil is a byproduct of fermentation processes which yield ethyl alcohol. It is rather toxic, 30 ml having caused death.)

30. Do you agree with the following statement? Explain your position. "Enantiotopic hydrogens participating in free radical substitution (such as chlorination) must produce *identical* products, while diastereotopic hydrogens must produce diastereomers."

31. "Even though a molecule contains a chiral center, that does not mean that the molecule is chiral." Provide an illustration which supports this point of view. Can a molecule which possesses a plane of symmetry have a chiral center (asymmetric atom) which resides in that plane? Provide an example to support your answer.

ALKYL HALIDES—NUCLEOPHILIC SUBSTITUTION REACTIONS

5.1 WHERE ARE WE GOING?

Most organic chemists realize the futility of trying to memorize the "... compound X is converted to compound Y ..." statement without an understanding of *how* it happens. One learns very quickly that there are a limited number of "hows," better known as *mechanisms*, and that learning the "what" without the "how" is similar to acquiring a wardrobe without a place to hang it—a frustrating, chaotic situation must follow. In this and following chapters we will see how stereochemistry, kinetics, and product studies can be used to determine reaction mechanisms.

This will be our first contact with the so-called *ionic aliphatic substitution reaction*, one in which an ion is used to achieve the replacement of one functional group by another on a substituted alkane. In contradistinction to free-radical substitution, ionic aliphatic substitution is an extremely important process in the practical, day-to-day synthesis of organic compounds. Because of the facile preparation of alkyl halides, often by routes other than free-radical halogenation, and because of the ease with which alkyl halides undergo ionic substitution, much of this chapter (and the one which follows) is built around the chemistry of alkyl halides. However, most of the ideas presented in these chapters are applicable to a host of compounds and will be called upon on many occasions throughout this text.

5.2 SYNTHESIS OF ALKYL HALIDES INVOLVING IONIC SPECIES—THE S_N2 REACTION

There are two major types of ionic aliphatic substitution reactions. We begin by discussing the simpler of these two, the S_N2 reaction.

Facts. If (+)-2-iodobutane is heated in acetone (CH_3COCH_3, a moderately polar solvent) and then recovered, it is found to be unchanged and, in particular, to be free of (−)-2-iodobutane. However, if (+)-2-iodobutane is heated for a protracted period in acetone *containing sodium iodide*, the recovered 2-iodobutane is racemic, that is, the product consists of an equimolar mixture of (+)- and (−)-2-iodobutane. If heating is stopped prior to complete racemization, the reaction mixture is richer in (+)- than (−)-2-iodobutane. Indeed, the reaction mixture never becomes richer in (−)-2-iodobutane than (+)-2-iodobutane. No other organic products (termed by-products) are formed during these reactions.

$(+)\text{-}C_2H_5CH(CH_3)\text{—I} \xrightarrow[\text{heat}]{\text{acetone}}$ no reaction

$(+)\text{-}C_2H_5CH(CH_3)\text{—I} \xrightarrow[\text{prolonged reflux}]{\text{acetone}/I^{\ominus}} (+)\text{-}C_2H_5CH(CH_3)\text{—I} + \text{I—}CH(CH_3)C_2H_5\text{-}(-)$ (1:1)

$(+)\text{-}C_2H_5CH(CH_3)\text{—I} \xrightarrow[\text{short reflux}]{\text{acetone}/I^{\ominus}} (+)\text{-}C_2H_5CH(CH_3)\text{—I} > \text{I—}CH(CH_3)C_2H_5\text{-}(-)$

If (+)-2-iodobutane is heated in acetone containing sodium iodide which is labeled (that is, if the sodium iodide contains ^{131}I, a radioactive isotope, abbreviated *I), some of the recovered 2-iodobutane is found to be radioactive. If the reaction mixture is heated for only a short time, then only (−)-2-iodobutane contains the radioactive isotope. Continued heating of this same mixture eventually yields a racemic product with the label equally distributed between both enantiomers of 2-iodobutane. Finally, if a large excess of Na*I is used, then (after protracted heating) the racemic product is completely labeled.

$(+)\text{-}C_2H_5CH(CH_3)\text{—I} \xrightarrow[\text{short heating}]{\text{acetone}/{}^{*}I^{\ominus}} (+)\text{-}C_2H_5CH(CH_3)\text{—I} > {}^{*}\text{I—}CH(CH_3)C_2H_5\text{-}(-)$

$(+)\text{-}C_2H_5CH(CH_3)\text{—I} \xrightarrow[\text{prolonged heating}]{\text{acetone/xs}\,{}^{*}I^{\ominus}} (+)\text{-}C_2H_5CH(CH_3)\text{—}{}^{*}\text{I} + {}^{*}\text{I—}CH(CH_3)C_2H_5\text{-}(-)$ (1:1)

A polarimeter usually cannot detect the difference between the optical rotation of a chiral substance and the rotation of the same chiral substance with one atom replaced by an isotope. For example, (+)-2-iodobutane and (+)-2-iodobutane bearing ^{131}I have identical D-line rotations (within experimental error). Therefore, an equimolar mixture of (+)-2-iodobutane and (−)-2-iodobutane-^{131}I will be optically inactive, even though these two compounds are not, strictly speaking, enantiomers. Armed with this information, let us now add one more fact: when (+)-2-iodobutane is reacted with $^{*}I^{\ominus}$, the rate of "racemization" (the rate of loss of optical activity) is twice the rate of incorporation of ^{131}I.

A Proposed Mechanism—Bimolecular Nucleophilic Substitution. All of the reactions described above have a common mechanism. Let us begin to understand it by explaining how (+)-2-iodobutane reacts with iodide ion to form (−)-2-iodobutane. It is proposed that iodide ion, acting as a **nucleophile** (an atom or group of atoms donating a pair of electrons to an element other than hydrogen), attacks the back lobe of the C—I bonding molecular orbital, thereby displacing the incumbent iodine (now called the **leaving group**). The incoming $I^{\ominus}$ becomes bonded to the carbon during this **concerted process.** The transition state of the reaction is reached when the two halogens are equally bonded to the central carbon. At the transition state the atom which has been attacked appears to violate the octet rule, in that it is at least in partial possession of ten outer electrons. The partial bonding between the nucleophile, the leaving group, and the atom attacked by the nucleophile is indicated by a dotted line. The entire process, as you can see from Figure 5–1, results in the inversion of relative configuration in going from starting material to product. This inversion sometimes is called a **Walden inversion.**

The reaction described above is a substitution process, the nucleophile displacing the

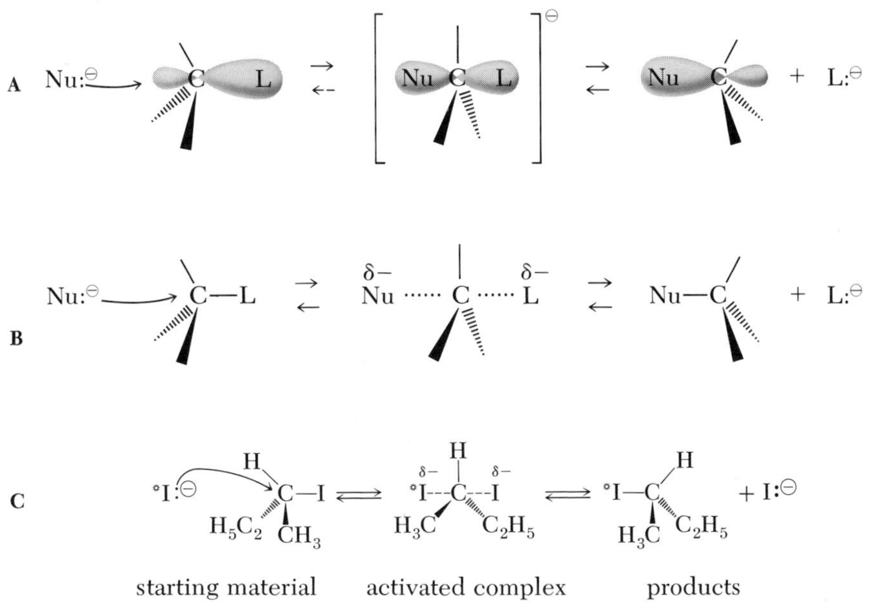

A Nu:⊖ → C L ⇌ [Nu C L]⊖ ⇌ Nu C + L:⊖

B Nu:⊖ → C—L ⇌ δ− Nu ······ C ······ L δ− ⇌ Nu—C + L:⊖

C *I:⊖ H C—I H_5C_2 CH_3 ⇌ H δ− *I---C---I δ− H_3C C_2H_5 ⇌ *I—C H H_3C C_2H_5 + I:⊖

starting material activated complex products

Figure 5-1 Mechanism for S_N2 substitution. The reaction begins with attack of an electron pair from the nucleophile, Nu⊖, upon the back lobe of the carbon-leaving group (C—L) bond (depicted by a curved arrow). Carbon is a trigonal bipyramid in the activated complex. **A** shows the orbital overlap in the activated complex, and **B** shows the bonds being made and broken (dotted lines). Scheme **C** depicts the specific reaction under discussion. The curved arrows describe the motion of the electron pairs involved in the reaction.

leaving group. When the leaving group departs, it takes with it the electron pair which bound it to the atom attacked by the nucleophile. Because it also is a bimolecular process (meaning that two species are involved in the activated complex), the mechanism is termed **S_N2,** meaning substitution-nucleophilic-bimolecular.

Why does the reaction of (+)-2-iodobutane with iodide ion eventually produce (±)-2-iodobutane instead of only (−)-2-iodobutane? After all, isn't the reaction supposed to proceed with inversion of configuration? Yes, it is true that every single S_N2 reaction proceeds to a product with the opposite relative configuration. At the inception of the reaction there only are (+)-2-iodobutane molecules present, and each S_N2 encounter can lead only to (−)-2-iodobutane. However, as the amount of (−)-2-iodobutane increases and the amount of (+)-2-iodobutane decreases, it becomes increasingly likely that an iodide ion will react with one of the (−)-2-iodobutane (product) molecules. Finally, when the reaction contains an equimolar amount of (+)- and (−)-2-iodobutane, it becomes just as likely that the (+) enantiomer will react as it is that the (−) enantiomer will react. From that point forward, the composition will remain unchanged, *i.e.*, it will remain racemic.

This general mechanism accounts for the incorporation of *I as well as for the relative rates of incorporation and racemization. Our definition of racemization is the conversion of one enantiomer into a racemic modification, *i.e.*, (+) → (±). Since every attack by *I⊖ produces an alkyl iodide with an inverted configuration, *two* molecules (one unreacted and one inverted) are "racemized" by each inversion. Since every attack by *I⊖ racemizes two molecules but introduces only one radioactive iodine, the rate of racemization must be twice the rate of label incorporation.

The reaction series just described is rather unique because it is completely reversible and, neglecting isotopic differences, the leaving group and the nucleophile are identical. The more common type of S_N2 reaction, in which nucleophile and leaving group are different, does lead to a product with a relative configuration opposite to that of starting material. In the example at the top of p. 124, the reverse reaction is restricted by the insolubility of sodium chloride in acetone.

$$Na^{\oplus}I^{\ominus} + \underset{C_2H_5}{\overset{H}{CH_3}}\!\!\!-\!C\!-\!Cl \xrightarrow{\text{acetone}} I\!-\!\underset{C_2H_5}{\overset{H}{C}}\!-\!CH_3 + Na^{\oplus}Cl^{\ominus}\ (\text{ppt})$$

1. A flask contains one molecule of (**R**)-2-iodobutane and one iodide ion. (a) Draw the starting material, showing its correct absolute configuration. Describe the product, including absolute configuration, of the first successful encounter between the alkyl halide and the anion. Does this lead to "racemization"? (b) If the original flask contained two molecules of (**R**)-2-iodobutane and one iodide ion, what would be the consequences of the first successful encounter between an alkyl halide and the anion? Does this lead to racemization? (c) Is there any assumption implicit in the statement that "(**R**)-2-iodobutane can be racemized by reaction with iodide ion"?
2. A flask contains two molecules of (**R**)-2-chloropentane and one iodide ion. Describe the minimum sequence of reactions, including absolute configurations, which could lead to a racemic product.
3. (a) A flask contains two molecules of (2**R**,3**R**)-2-chloro-3-methylpentane and one chloride ion. What is the product of the first successful encounter between an alkyl halide molecule and the halide anion? Is the starting material *threo* or *erythro*? (b) What is the configuration of the product? (c) Can an S_N2 reaction between a large amount of (2**R**,3**R**)-2-chloro-3-methylpentane and chloride ions lead to racemization? (d) Will the final product be optically active? Explain.
4. Draw the product of the reaction of one molecule of each of the following with a bromide ion. Give each a suitable name.
   (a) (**R**)-2-bromobutane (b) (**S**)-2-bromobutane
   (c) (**R**)-2-chloropentane (d) (**S**)-1-chloro-2-methylbutane

Testing the Mechanism. If the concerted process described above is the correct mechanism for the S_N2 reaction, then we should be able to analyze the influence of varying the reaction conditions upon the course of reaction.

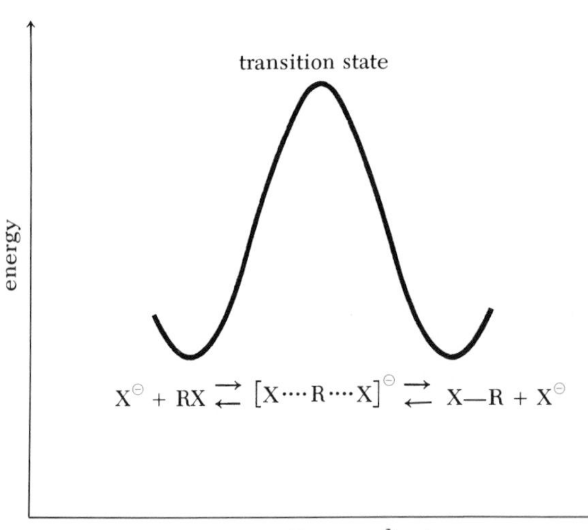

Figure 5-2 Energy profile for S_N2 substitution. The activated complex is formed at the transition state.

*1. **Kinetics.*** As seen in Figure 5–2, the rate-determining step is the same as the product-determining step in the S_N2 reaction. (Of course, this follows: The reaction is a one-step, *i.e.*, concerted, process.) The rate of the S_N2 reaction is known to be directly proportional to the concentrations of substrate and nucleophile, and it is said to be *first order* in substrate, *first order* in nucleophile, and *second order overall.* That the rate of the S_N2 reaction is directly related to the concentration of substrate and nucleophile can be seen from the following equation:

$$R = k[\mathrm{RI}][\mathrm{I}^{\ominus}]$$

where R = rate of reaction
k = specific rate constant (a proportionality factor)
[RI] and $[I^{\ominus}]$ = molar concentrations

As might be expected from this equation, a doubling of the concentration of either the substrate (alkyl iodide) or the nucleophile (iodide ion) doubles the rate of reaction. If both are doubled, then the rate of reaction increases by a factor of four.

A reaction's **kinetic order** in a given reactant is the exponent following that reactant's concentration term in the rate equation; the overall order is the sum of these exponents. One should be careful not to confuse "molecularity" with "order"; the former is derived from our mechanistic picture, while the latter is determined by experiment.

The kinetic order of a reaction with respect to a particular reagent is determined by varying the concentration of only that reactant and evaluating the effect of this change upon the overall rate of reaction. For example, if doubling the concentration of a particular reactant doubles the rate of reaction, that reaction is first order (2^1) in the particular reactant. Had doubling the concentration caused a four-fold increase in reaction rate, the rate would have been second order (2^2) in that reactant. The illustrative rate equation below is first order in A and second order in B. This means that keeping the concentration of A constant and trebling the concentration of B causes the rate, R, to increase nine-fold; but that keeping B constant and trebling the concentration of A causes R to treble.

$$R = k[A][B]^2$$

Any species which enters the reaction may appear in the kinetic equation so long as it does not become involved after the rate-determining step; thus, any number of terms may be in the equation. However, the *molecularity* of a multi-step process is conventionally defined as the number of molecules undergoing covalency change in the rate-determining step. Therefore, there is no reason to expect these two to be the same, even though they often are. One example of a difference between order and molecularity is the *pseudo first-order reaction.*

If the concentration of the nucleophile is very high, small changes in its concentration will not have a measurable effect on the rate of reaction. Instead of having a rate equation $R = k$[substrate][nucleophile], the reaction is governed by the rate equation $R = k$[substrate]. This type of reaction is dubbed a pseudo first-order reaction. For example, reactions in which the solvent is also the nucleophile are expected to exhibit pseudo first-order kinetics. The reaction between methyl iodide (substrate) and water (nucleophile) is an illustration, the concentration of the nucleophile being about 55M.

$$CH_3I + H_2O \rightarrow CH_3OH + HI$$

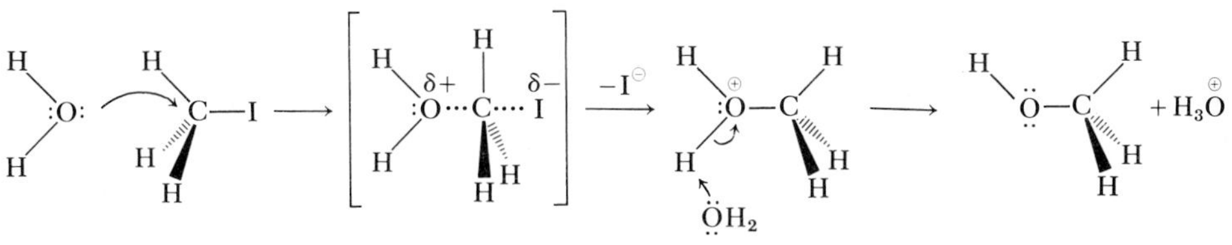

$$R = k'[CH_3I]$$

2. ***Nature of Substituents Bonded to the Atom Under Attack.*** S_N2 reactions occur readily only on sp^3 hybridized carbon compared to other states of carbon hybridization. Since reaction requires back-side attack on the C—L bond, the presence of a large number of groups bonded to the carbon bearing the leaving group, L, should hinder attack by the nucleophile. The existence of *steric hindrance* to S_N2 reactions is supported by the data in Table 5–1.

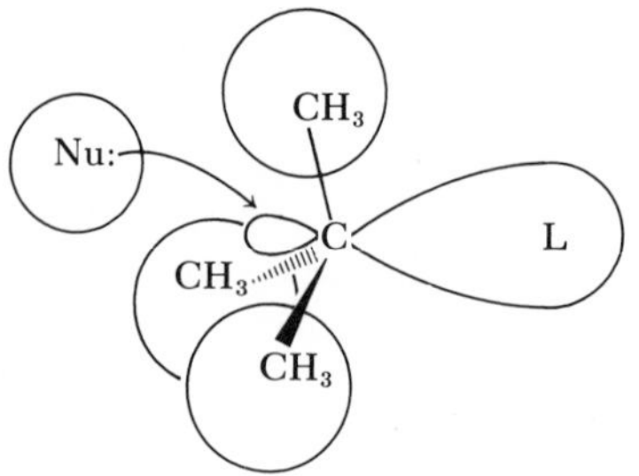

the presence of three methyl groups prevents attack by the nucleophile upon the C—L bond: an example of steric hindrance to the S_N2 reaction

TABLE 5-1 Average Comparative Effects of Alkyl Moieties Upon Rates of S_N2 Processes

$$Nu^{\ominus} + \text{Alk-L} \rightarrow \text{Alk-Nu} + L^{\ominus a}$$

Alkyl Moiety	Relative Rate of Substitution
$—CH_3$	30
$—CH_2CH_3$	1
$—CH(CH_3)_2$	0.03
$—C(CH_3)_3$	~0
$—CH_2C(CH_3)_3$	1×10^{-5}
$—CH_2—CH{=}CH_2$	50

[a] $Nu^{\ominus}$ represents a nucleophile, and L represents a leaving group.

While these data tell us, for example, that *t*-butyl bromide reacts more slowly with bromide ion than does methyl bromide, they give no information about the eventual yield of the product. Actually, S_N2 reactions proceed in high yield on 1° substrates (*e.g.*, ethyl bromide), moderate yield on 2° substrates (*e.g.*, isopropyl bromide), and extremely low yield (<1%) on 3° substrates (*e.g.*, *t*-butyl bromide). As the rate of an S_N2 reaction decreases, side reactions can become competitive with the "major" reaction. In Chapter 6 we will learn that elimination reactions (reactions which usually result in double bond formation) compete with substitution reactions and that, while tertiary halides do not undergo S_N2 reactions, they do undergo elimination.

$$\underset{\text{1-chloropentane}}{CH_3CH_2CH_2CH_2CH_2Cl} + NaI \xrightarrow[\text{heat}]{\text{acetone}} \underset{\text{1-iodopentane}}{CH_3CH_2CH_2CH_2CH_2I} + NaCl\ (\text{ppt.})$$

$$\underset{\text{3-chloropentane}}{CH_3CH_2CHClCH_2CH_3} + NaI \xrightarrow[\text{heat}]{\text{acetone}} \underset{\text{3-iodopentane}}{CH_3CH_2CHICH_2CH_3} + NaCl\ (\text{ppt.})$$

$$\underset{\text{3-chloro-3-methylpentane}}{CH_3CH_2C(CH_3)ClCH_2CH_3} + NaI \xrightarrow[\text{heat}]{\text{acetone}}\!\!\!\!\!\!/\!\!/\ \underset{\substack{\text{3-iodo-3-methylpentane}\\ \text{(DOES NOT OCCUR)}}}{CH_3CH_2C(CH_3)ICH_2CH_3} + NaCl\ (\text{ppt.})$$

$$\text{—}\overset{|}{\underset{|}{C}}\text{—}\overset{\overset{|}{\text{—}C\text{—H}}}{\underset{\underset{|}{\text{—}C\text{—}}}{C}}\text{—X} \xrightarrow{Nu^{\ominus}} \begin{cases} \xrightarrow{-X^{\ominus}}\!\!\!\!\!/\!\!/\ \text{—}\overset{|}{\underset{|}{C}}\text{—}\overset{\overset{|}{\text{—}C\text{—H}}}{\underset{\underset{|}{\text{—}C\text{—}}}{C}}\text{—Nu} & \text{bimolecular substitution (does not occur)} \\ \xrightarrow{-HX} \text{—}\overset{|}{\underset{|}{C}}\text{—}\overset{\overset{\diagdown\ \diagup}{C}}{\overset{\|}{C}}\text{—}\overset{|}{\underset{|}{C}}\text{—} & \text{elimination (does occur)} \end{cases}$$

Neopentyl substrates are exceptional in that, although 1°, they undergo substitution quite slowly under S_N2 conditions. This is due to steric hindrance to attack of the nucleophile by the β-methyl groups in the neopentyl group.

neopentyl chloride $H_3C—C(CH_3)_2—CH_2—Cl$ (β, α)

$^{\delta-}Cl \cdots C \cdots Cl^{\delta-}$

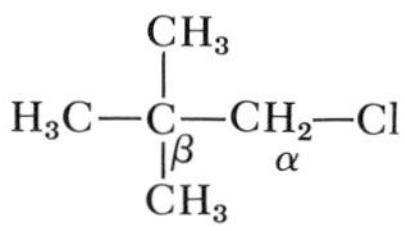

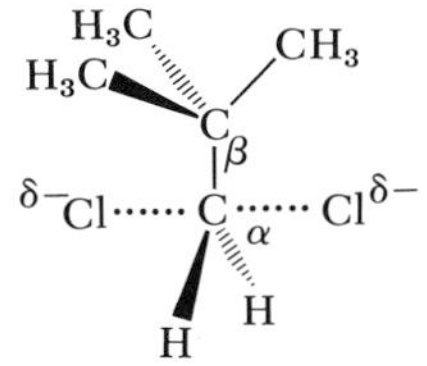

transition state for an S_N2 process at the "neopentyl carbon"

Allyl bromide, $CH_2{=}CH—CH_2—Br$, is a primary halide which undergoes S_N2-type substitution reactions slightly more rapidly than does ethyl bromide and much more rapidly than does neopentyl bromide (Table 5–1). Since ethyl bromide is our standard (it has a relative rate of reaction of 1 in Table 5–1), we must conclude that *allyl bromide* is accelerated in its S_N2 reactions. Why does this happen? The simplest answer, and one which nearly begs the question, is that the rate enhancement results from a stabilization of the transition state for the S_N2 process, *i.e.*, a decrease in the energy content of the activated complex, and a concomitant decrease in the energy of activation for the process. We now must go on and ask the more critical question: Why is the activated complex for S_N2 reactions involving allyl bromide of lower energy than is that for S_N2 reactions involving ethyl bromide? The stabilization of the activated complex involving allyl bromide is presumed due to the interaction ("overlap") of the π electron system with the bonds being made and broken in that complex. The precise reason that the overlap of the π orbital stabilizes this species cannot be given without recourse to detailed mathematical arguments.

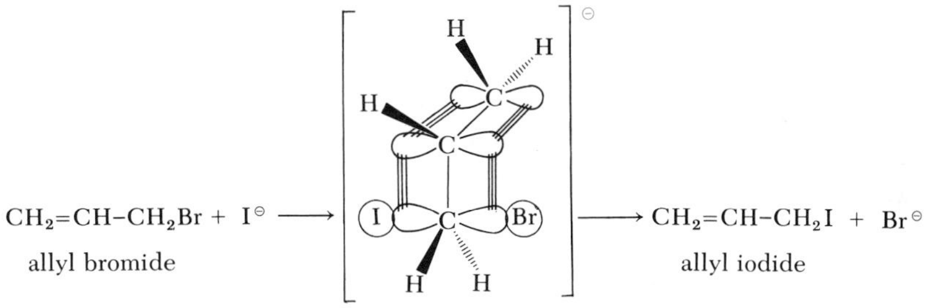

$$CH_2{=}CH{-}CH_2Br + I^{\ominus} \longrightarrow [\,I \cdots C \cdots Br\,]^{\ominus} \longrightarrow CH_2{=}CH{-}CH_2I + Br^{\ominus}$$

allyl bromide — allyl iodide

π overlap stabilizing the S_N2 activated complex

5. 6-Chloro-1-hexene, $CH_2{=}CH(CH_2)_3CH_2Cl$, reacts with methoxide ion, $CH_3O^{\ominus}$, more slowly than does allyl chloride (S_N2 conditions) but at about the same rate as does 1-chlorohexane. Explain.

6. The relative reactivity of a series of alkyl bromides in the reaction shown below is as follows: G = H, 100; G = CH_3, 3; G = C_2H_5, 1; G = $CH(CH_3)_2$, 0.1; G = $C(CH_3)_3$, 0.0001. Account for these results.

$$GCH_2Br + CH_3CH_2O^{\ominus} \rightarrow GCH_2OCH_2CH_3 + Br^{\ominus}$$

7. (+)-4-Iodo-3,3-dimethylhexane neither racemizes upon exposure to iodide ions (S_N2 conditions) nor incorporates any label if the iodide ions are labeled. Explain.

3. Solvent Effects. An increase in solvent polarity slightly decreases the rate of halogen-halogen exchange (the so-called Finkelstein reaction). To explain this, we must evaluate the effect of changing the solvent on both the starting material and the activated complex. The starting material has a full negative charge localized on the anion, and that charge is stabilized largely by interactions with the solvent. On the other hand, the negative charge in the activated complex is dispersed over both halogens. Since charge is dispersed in the activated complex, this complex is not as stabilized by interactions with the solvent as is the starting material. Increasing the solvent polarity stabilizes the starting material more than the activated complex. This increases the energy difference between the starting material and the activated complex and, therefore, increases the energy of activation. The net result is a decrease in the reaction rate (Figure 5–3).

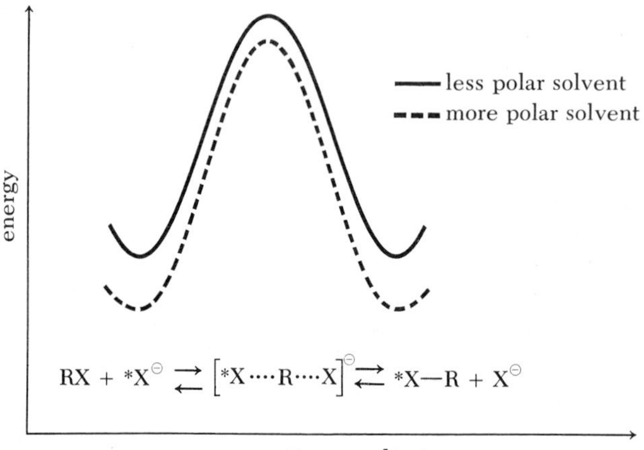

Figure 5-3 Effect of solvent polarity upon the S_N2 halogen-halogen exchange process. Increasing solvent polarity stabilizes the activated complex less than it does the starting materials and products. Thus, increasing the solvent polarity in this type of S_N2 reaction increases the energy of activation and decreases the rate of reaction slightly.

Early research concentrated upon S_N2 reactions carried out in solvents capable of hydrogen-bonding, so called *protic solvents*. Recently, the use of *dipolar aprotic solvents* (solvents of high dielectric constant but incapable of contributing hydrogen to a hydrogen bond) has opened up new areas of synthetic and mechanistic study. Where ethanol, dioxane, aqueous alcohol and acetone once were the "solvents of choice" for the S_N2 reaction, dimethyl sulfoxide (DMSO), sulfolane, hexamethylphosphoric triamide (HMPT) and dimethylformamide (DMF) are now dominating much of the solvent picture.

$:O(CH_2—CH_2)_2O:$
dioxane

$(H_3C)_2\overset{\oplus}{S}—\overset{\ominus}{O}:$
dimethyl sulfoxide (DMSO)

$H_2C—CH_2$ / $H_2C—CH_2$ ring with $\overset{+2}{S}(O^{\ominus})_2$
sulfolane

$((CH_3)_2N)_3\overset{\oplus}{P}—\overset{\ominus}{O}:$
hexamethylphosphoric triamide (HMPT)

$(CH_3)_2NC(=O)H$
dimethylformamide (DMF)

The great utility of dipolar aprotic solvents can be seen in the conversion of 1,3-dichloropropane to 1,3-dicyanopropane *via* the S_N2 process; the rate is about 1000 times greater in DMSO than in aqueous ethanol.

$$2N{\equiv}C{:}^{\ominus} + Cl{-}CH_2CH_2CH_2{-}Cl \rightarrow NC{-}CH_2CH_2CH_2{-}CN + 2Cl^{\ominus}$$

cyanide ion 1,3-dichloropropane 1,3-dicyanopropane

The use of dipolar aprotic solvents has shown that highly *polar* protic solvents can act in opposite directions—slowing a reaction down by deactivating a nucleophile through hydrogen-bonding *or* increasing the speed of reaction by stabilizing a charge-separated activated complex. The enhanced reactivity of anions in dipolar aprotic solvents, where the anion is not stabilized by hydrogen-bonding, suggests that the polarity of solvents may not be as important in influencing reactions as is their solvating ability.

5.3 A BROADER VIEW OF THE S_N2 PROCESS

Nothing in what has so far been presented limits the S_N2 process to halides, either as attacking groups or as leaving groups. Indeed, nothing has been stated which would limit the nucleophile to an anion; it only has to be capable of donating an electron pair. Put differently, the terms "nucleophile" and "Lewis base" are synonymous. Similarly, while most leaving groups depart as anions, there has been no requirement made that the departed group be an anion. Furthermore, the S_N2 scheme is not limited to carbon as the attacked atom; it has been observed on sulfur, bromine, silicon, tin, and other elements. One might even view the acid-base reaction between water and hydrogen chloride as representative of an S_N2 reaction on hydrogen. In this reaction it is, of course, impossible to speak of inversion of configuration, since hydrogen has no configuration.

$$H_2\ddot{O}{:} + H{-}\ddot{C}l{:} \rightleftarrows H_2\overset{\oplus}{O}{-}H + {:}\ddot{C}l{:}^{\ominus}$$

Nucleophiles and Nucleophilicity. Any electron-pair donor is a nucleophile when it reacts at a comparatively electron-deficient site. However, if the atom being attacked by the electron-pair of the nucleophile is hydrogen, the reaction is considered to be an acid-base reaction and the nucleophile then is called a *base*. All bases are nucleophiles, even though they are not of equivalent effectiveness in S_N2 reactions; all nucleophiles are bases, even though they may not be "good" bases, *i.e.*, even though they do not readily accept protons. The terms "nucleophile" and "base" may be used to describe exactly the same species participating in different reactions. For example, the $OH^{\ominus}$ which reacts (S_N2) with methyl bromide to form methyl alcohol is a *nucleophile*, while the same $OH^{\ominus}$ is a *base* when it reacts with methanol to form a methoxide ion.

$$H{-}\ddot{O}{:}^{\ominus} + H_3C{-}Br \longrightarrow H{-}O{-}CH_3 + Br^{\ominus}$$

methyl bromide methanol

$OH^{\ominus}$: nucleophile

$$H{-}\ddot{O}{:}^{\ominus} + H{-}O{-}CH_3 \rightleftarrows H{-}O{-}H + {}^{\ominus}O{-}CH_3$$

methanol methoxide ion

$OH^{\ominus}$: base

Proton transfers usually, but certainly not always, are rapid. Therefore, **basicity** is most commonly defined in terms of the position of an equilibrium reaction rather than upon how rapidly that equilibrium is established. Since the position of any equilibrium is determined by the difference in the energies of the product and of the starting material, basicity is often described as a "thermodynamic property."

$$\text{BASE} + \text{ACID} \rightarrow \text{“SALT”} \qquad \frac{[\text{“SALT”}]}{[\text{BASE}]\,[\text{ACID}]} \text{ defines “basicity”}$$

Nucleophilic substitutions normally are slower than proton transfers and are amenable to kinetic studies. The **nucleophilicity** of a species is a measure of how rapidly it reacts and is considered a "kinetic property."

Those species which are reactive nucleophiles in the common, protic solvents are characterized by an attacking electron pair located on a large atom. If the attacking electrons are *polarizable* (if the outer electron cloud is easily deformed), they will be able to respond better to the slight positive charge at the site of attack. Anions like $I^{\ominus}$, $RS^{\ominus}$ and $HS^{\ominus}$, potent nucleophiles but poor bases, possess polarizable outer electrons. Anions like $OH^{\ominus}$ and $H_2N^{\ominus}$ are potent bases but poor nucleophiles, reflecting the low polarizability of oxygen and nitrogen. In general, polarizability increases as one moves down a group of elements in the periodic table. (Why?)

Nucleophiles have to escape their solvent shells when attacking a substrate. Anions which are small, and highly dependent upon solvation for charge stabilization, should be less reactive than large anions because of this. The order of nucleophilicity for halide ions in protic solvents like ethanol ($I > Br > Cl \gg F$), which is the reverse order of their basicity in water, bears this out. The common ordering of nucleophilic reactivity derived from results in protic solvents is:

$$HS^{\ominus}, RS^{\ominus} > I^{\ominus} > Br^{\ominus} > RO^{\ominus} > Cl^{\ominus} > H_3C{-}C(=O)O^{\ominus} > {}^{\ominus}O{-}N^{\oplus}(=O)O^{\ominus}$$

mercaptide > iodide > bromide > alkoxide > chloride > acetate > nitrate

8. Predict the product(s) of the S_N2 reaction of each of the following nucleophiles with (±)-2-iodooctane.

(a) $Cl^{\ominus}$ (b) $I^{\ominus}$
(c) $HS^{\ominus}$ (d) NH_3
(e) $CH_3C(O)CH_2^{\ominus}$ (f) (**R**)-$CH_3CH(O^{\ominus})CH_2CH_3$
(g) (**S**)-$CH_3CH(O^{\ominus})CH_2CH_3$ (h) (**R,S**)-$CH_3CH(O^{\ominus})CH_2CH_3$

As implied in the discussion of dipolar aprotic solvents, it should be possible to alter relative nucleophilicities by passing from protic to aprotic solvents. This is borne out in the order of halide nucleophilicity in dimethylformamide: $Cl^{\ominus} > Br^{\ominus} > I^{\ominus}$. The removal of hydrogen bonds between anion and solvent has made the less polarizable chloride ion a better nucleophile than the more polarizable iodide ion.

Leaving Groups. While strong bases may act as nucleophiles, they are rarely leaving groups. Included in a list of *uncommon* leaving groups are: $OH^{\ominus}$, $NH_2^{\ominus}$, $RO^{\ominus}$, $R_2N^{\ominus}$, $CH_3^{\ominus}$, and $H^{\ominus}$. Thus, bromide ion will NOT react with methyl alcohol to produce methyl bromide, nor with ethane to produce methyl bromide or ethyl bromide. These three NON-reactions are shown below.

$$\left.\begin{array}{l} Br^{\ominus} + CH_3\text{—}OH \not\rightarrow CH_3\text{—}Br + OH^{\ominus} \\ Br^{\ominus} + CH_3\text{—}CH_3 \not\rightarrow CH_3\text{—}Br + CH_3^{\ominus} \\ Br^{\ominus} + CH_3CH_2\text{—}H \not\rightarrow CH_3CH_2\text{—}Br + H^{\ominus} \end{array}\right\} \text{ do NOT occur}$$

There is a fair *inverse* correlation between the basicity of a species and its leaving group ability. Thus iodide ion, a weaker base than chloride ion, is a better leaving group. The best common leaving groups, the sulfonates, are very weak bases.

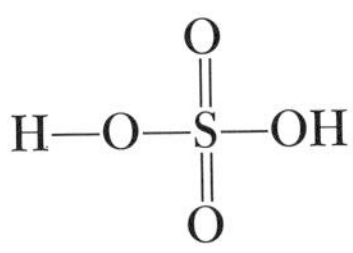

$H\text{—}O\text{—}S(=O)_2\text{—}OH$ sulfuric acid

$H_3C\text{—}S(=O)_2\text{—}OH$ methanesulfonic acid

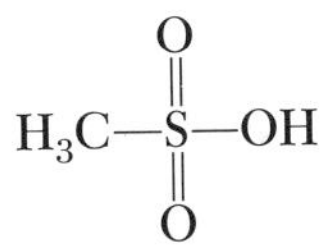

$H_3C\text{—}S(=O)_2\text{—}O^{\ominus}$ methanesulfonate ion (excellent leaving group)

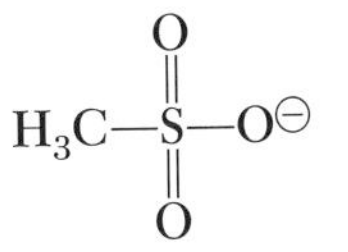

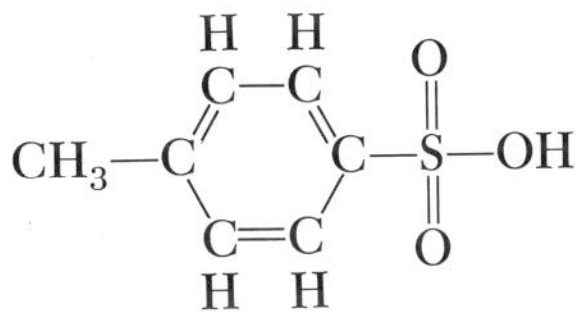

$CH_3\text{—}C_6H_4\text{—}S(=O)_2\text{—}OH$ *p*-toluenesulfonic acid (HOTs)

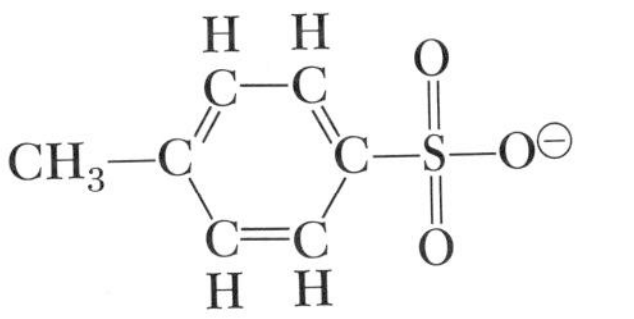

$CH_3\text{—}C_6H_4\text{—}S(=O)_2\text{—}O^{\ominus}$ *p*-toluenesulfonate ion *or* tosylate ion ($^{\ominus}$OTs) (excellent leaving group)

order of decreasing leaving group ability:

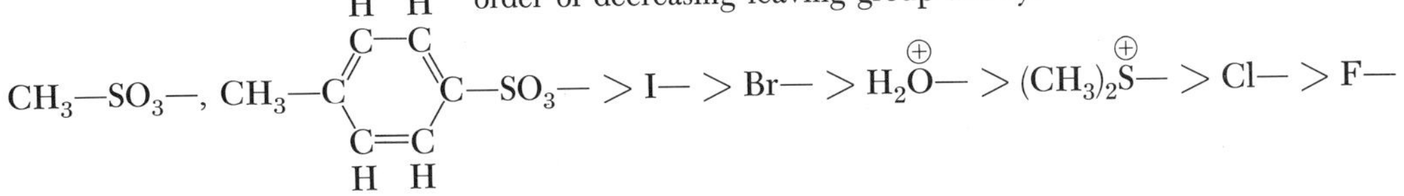

$$CH_3\text{—}SO_3\text{—},\ CH_3\text{—}C_6H_4\text{—}SO_3\text{—} > I\text{—} > Br\text{—} > H_2\overset{\oplus}{O}\text{—} > (CH_3)_2\overset{\oplus}{S}\text{—} > Cl\text{—} > F\text{—}$$

One question which students sometimes ask is, "How do I know which atom in a complex structure is going to be attacked by the nucleophile?" The easiest answer to this is, "Look for the carbon bearing the best leaving group, because, other things being equal, that carbon will be attacked most easily."

$$Br\text{—}CH_2CH_2CH_2\text{—}Cl + I^{\ominus} \rightarrow I\text{—}CH_2CH_2CH_2\text{—}Cl > Br\text{—}CH_2CH_2CH_2\text{—}I$$

9. Each of the following is a potential substrate for nucleophilic substitution. For each, indicate (1) the leaving group(s) and (2) the nature of this group once it has departed.

(a) $CH_3CH_2CH_2OS(O)_2CH_3$ (b) $CH_3OS(O)_2OCH_3$

(c) $CH_3CH_2\overset{\oplus}{O}H_2$ (d) $(CH_3)_2CHCH_2CH_2\overset{\oplus}{O}(CH_3)_2$

(e) $(CH_3)_3CCH_2CH_2\overset{\oplus}{O}(CH_3)C_2H_5$ (f) $(CH_3)_4N^{\oplus}$

10. Predict the major product(s) from the reaction of each compound in problem 9 with an excess of sodium iodide dissolved in acetone.

Functional Group Synthesis via S_N2 Reactions. The existence of a wide range of potential nucleophiles and leaving groups makes the S_N2 reaction an extremely valuable synthetic tool. Indeed, almost any functional group bearing a non-bonding electron pair can serve, under the proper conditions, as a nucleophile. Even if the functional group is of low nucleophilicity, as, for example, is the acetate ion ($CH_3CO_2^{\ominus}$), it may be used in an S_N2-type synthesis by selecting a substrate which bears a very good leaving group. This is illustrated below using the preparation of methyl acetate, $CH_3C(O)OCH_3$, by two alternate S_N2 routes. Here we also encounter the use of the very potent methylating agent dimethyl sulfate, $CH_3OS(O)_2OCH_3$, a compound which readily provides a methyl group to a nucleophile.

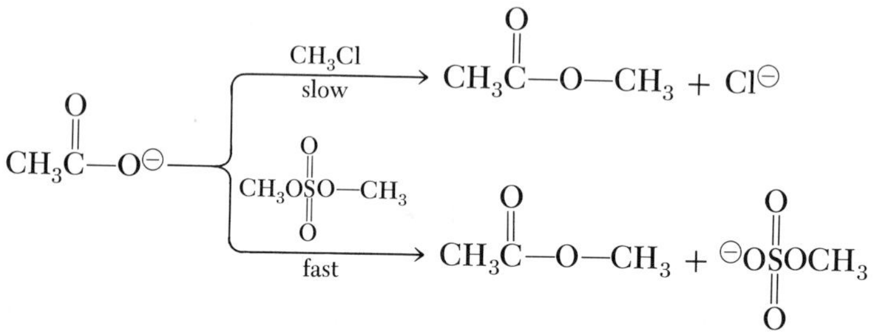

$$CH_3C(=O)O^{\ominus} \xrightarrow[\text{slow}]{CH_3Cl} CH_3C(=O){-}O{-}CH_3 + Cl^{\ominus}$$

$$CH_3C(=O)O^{\ominus} \xrightarrow[\text{fast}]{CH_3OS(=O)_2O{-}CH_3} CH_3C(=O){-}O{-}CH_3 + {}^{\ominus}OS(=O)_2OCH_3$$

The sweep of the S_N2 reaction may be gathered from the examples in Table 5–2. However, it is not just in such variety that the S_N2 reaction has major advantages. It is important, in synthesizing a compound, to be certain that a new functional group is going into a specific, known location in the molecule. Since the entering group becomes bonded to the same atom which bore the leaving group, the S_N2-type synthesis is excellent when unequivocal structural assignments are necessary (see, however, p. 146 and problem 33).

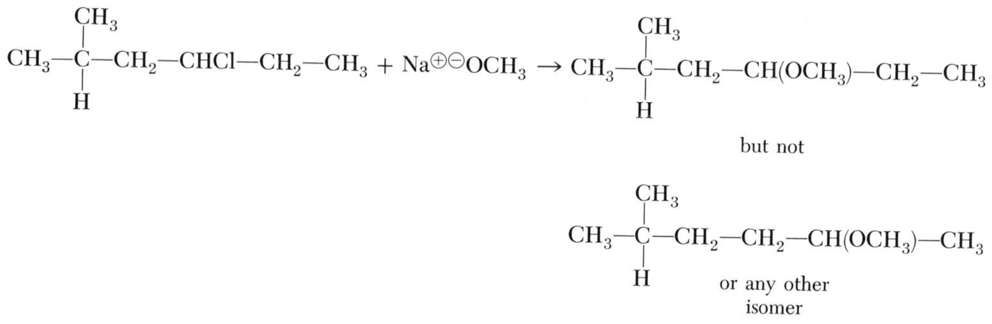

$$CH_3{-}CH(CH_3){-}CH_2{-}CHCl{-}CH_2{-}CH_3 + Na^{\oplus}{}^{\ominus}OCH_3 \rightarrow CH_3{-}CH(CH_3){-}CH_2{-}CH(OCH_3){-}CH_2{-}CH_3$$

but not

$$CH_3{-}CH(CH_3){-}CH_2{-}CH_2{-}CH(OCH_3){-}CH_3$$

or any other isomer

11. Each of the following might have been synthesized by an S_N2 reaction. Suggest a combination of substrate and nucleophile which could have led to their production.

(a) CH_3OCH_3

(b) 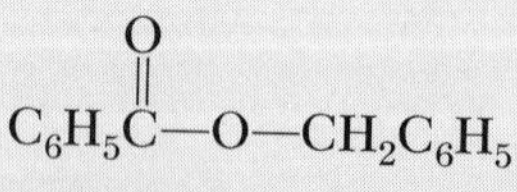 $C_6H_5C(=O){-}O{-}CH_2C_6H_5$

(c) (**R**)-$CH_3CH(OCH_3)CH_2CH_3$

(d) $(CH_3)_4N^{\oplus}Cl^{\ominus}$

(e) $CH_2{=}CHCH_2CH_2CH{=}CH_2$

(f) 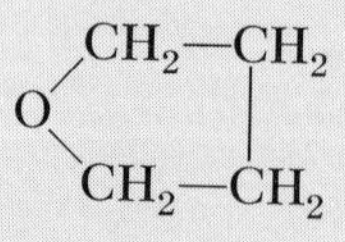 $O(CH_2{-}CH_2)(CH_2{-}CH_2)$ (ring: O bonded to $CH_2{-}CH_2$ and $CH_2{-}CH_2$, which are joined)

(g) 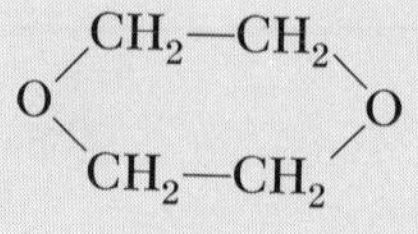 $O(CH_2{-}CH_2)_2O$ (ring: two O atoms bridged by two $CH_2{-}CH_2$ units)

TABLE 5-2 Functional Group Interconversion Via S_N2 Reactions[a]

NUCLEOPHILE + SUBSTRATE → PRODUCT + DEPARTED GROUP

A. $I^{\ominus} + \mathbf{C}H_3Cl \rightarrow CH_3{-}I + Cl^{\ominus}$

methyl chloride; methyl iodide

B. $N{\equiv}C^{\ominus} + CH_3OS(O)_2O\mathbf{C}H_3 \rightarrow CH_3CN + CH_3OS(O)_2O^{\ominus}$

dimethyl sulfate; acetonitrile

C. $HO^{\ominus} + H_2\mathbf{C}{-}\mathbf{C}H_2$ (ring with NH) $\rightarrow HO{-}CH_2CH_2{-}NH^{\ominus} \rightleftarrows {}^{\ominus}O{-}CH_2CH_2{-}NH_2$

aziridine; ethanolamine anion

D. ${}^{\ominus}N{=}\overset{\oplus}{N}{=}N^{\ominus} + \mathbf{C}H_3Cl \rightarrow CH_3{-}N{=}\overset{\oplus}{N}{=}N^{\ominus} + Cl^{\ominus}$

azide ion; methyl azide

E. $(C_6H_5)_3P + \mathbf{C}H_3I \rightarrow (C_6H_5)_3\overset{\oplus}{P}CH_3\ I^{\ominus}$

triphenyl phosphine; methyltriphenylphosphonium iodide

F. $CH_3{-}\overset{O}{\overset{\|}{C}}{-}O^{\ominus} + CH_3\mathbf{C}H_2I \rightarrow CH_3CH_2OC(O)CH_3 + I^{\ominus}$

acetate ion; ethyl iodide; ethyl acetate

G. $PH_2^{\ominus} + \mathbf{C}H_3Br \rightarrow PH_2CH_3 + Br^{\ominus}$

phosphide ion; methyl phosphine

H. $H{-}C{\equiv}C^{\ominus} + CH_3\mathbf{C}H_2Br \rightarrow H{-}C{\equiv}C{-}CH_2CH_3 + Br^{\ominus}$

acetylide ion; ethylacetylene

I. $CH_3{-}S{-}CH_3 + \mathbf{Cl}{-}\mathbf{Cl} \rightarrow CH_3{-}\overset{Cl}{\overset{|}{\underset{\oplus}{S}}}{-}CH_3 + Cl^{\ominus}$

methyl sulfide; chlorodimethylsulfonium chloride

J. $I^{\ominus} + \mathbf{C}H_3{-}\overset{\oplus}{O}(\mathbf{C}H_3){-}\mathbf{C}H_3 \rightarrow CH_3I + CH_3{-}O{-}CH_3$

trimethyloxonium ion; methyl ether

[a] The atom attacked by the nucleophile is printed in boldface. If equivalent sites of attack are present, all are indicated.

Let us now examine the "real thing," picking up the synthesis of *tyramine* at an early stage in order to see how substitution reactions can be used to interconvert functional groups. Tyramine has been selected since it is highly toxic, is found in many foods (*e.g.*, cheese), but is destroyed in our bodies by a complex protein (an enzyme) called MAO (*monoamine oxidase*). The synthesis is analyzed on p. 134; however, this is not the way in which a synthesis should be worked out initially. It is often advantageous to work backward in solving a synthetic problem. One needs to remember much less if, beginning with the final product, one asks, "How is that made in a one step reaction?" This, in turn, leads to a choice of several immediate precursors of the final product, and the same question is then asked about these. Usually, by the time you have gone back two or three steps in the synthesis, the problem has been solved *or* the synthetic route of choice has become self-evident.

$$\text{HO—C}_6\text{H}_4\text{—CH}_2\text{OH} + \text{HCl} \xrightarrow{-\text{H}_2\text{O}} \text{HO—C}_6\text{H}_4\text{—CH}_2\text{Cl} \xrightarrow[-\text{NaCl}]{\text{NaCN}}$$

$$\text{HO—C}_6\text{H}_4\text{—CH}_2\text{—CN} \xrightarrow[\text{Ni/pressure}]{\text{H}_2} \text{HO—C}_6\text{H}_4\text{—(CH}_2)_2\text{—NH}_2$$

tyramine

Tyramine Synthesis

1. $$\text{HO—C}_6\text{H}_4\text{—CH}_2\text{OH} + \text{HCl} \rightarrow \text{HO—C}_6\text{H}_4\text{—CH}_2\text{Cl} + \text{H}_2\text{O}$$

This conversion occurs by an S_N1 mechanism, a process which will be discussed later in this chapter. After you have finished reading about S_N1 reactions, return to this step and see if you can explain (a) the mechanism, (b) why the reaction is not S_N2, and (c) why only one of two —OH groups was replaced. The last question is the most difficult of these three and you may have to wait until Chapter 22 for a good answer.

2. $$\text{HO—C}_6\text{H}_4\text{—CH}_2\text{Cl} + \text{CN}^{\ominus} \rightarrow \text{HO—C}_6\text{H}_4\text{—CH}_2\text{—CN} + \text{Cl}^{\ominus}$$

This useful means of lengthening a carbon chain by one carbon is an S_N2 reaction. Cyanide ion is the nucleophile and chloride ion is the leaving group.

3. $$\text{HO—C}_6\text{H}_4\text{—CH}_2\text{—C}\equiv\text{N} + 2\text{H}_2 \xrightarrow{\text{Ni}} \text{HO—C}_6\text{H}_4\text{—CH}_2\text{—CH}_2\text{—NH}_2$$

This is a reduction reaction and is not a nucleophilic substitution. In this reaction, two moles of hydrogen convert the cyano group, —C≡N, to the aminomethyl group, $\text{—CH}_2\text{—NH}_2$. This useful synthesis of amines is discussed on p. 630. Like many other reactions which use a metal as a catalyst, its mechanism is still the subject *of* disagreement and *for* experimentation.

5.4 THE S_N1 REACTION

The second major type of ionic aliphatic substitution reaction is, like the S_N2 process, a nucleophilic substitution; hence, it also is an "S_N" reaction. However, it involves only *one* species in the more important of two activated complexes produced during the conversion of starting material to product. For this reason this second process is a *unimolecular* nucleophilic substitution reaction, abbreviated S_N1. The S_N1 reaction is not a concerted reaction (the S_N2 reaction is concerted) but involves two distinct steps. The first of these two steps, in which the substrate (*e.g.*, an alkyl halide) ionizes to produce an anion (*e.g.*, a halide anion) and a cation, $R^{\oplus}$, is the more important step. In the second, culminating step the cation reacts with the nucleophile to produce the substitution product.

$$-\overset{|}{\underset{|}{C}}-I \xrightarrow[\text{(slow)}]{\text{ionization}} -\overset{|}{\underset{|}{C}}^{\oplus} + I^{\ominus}$$

alkyl iodide (starting material) — cation (intermediate)

$$\xrightarrow[\text{ion combination (fast)}]{Br^{\ominus}} -\overset{|}{\underset{|}{C}}-Br$$

alkyl bromide (product)

an idealized S_N1 conversion of an iodide to a bromide

The energy profile (Figure 5–4) which describes the idealized alkyl iodide-to-alkyl bromide interconversion shown above is rather revealing. As you might imagine, of the two steps in the sequence (ionization and ion combination), *ionization* is the more difficult. After all, while it takes work to cleave a covalent bond and generate oppositely charged ions, energy should be released when oppositely charged ions are brought together to create a new covalent bond. The activated complex leading to ionization, which may be described as having a stretched bond between carbon and iodine, involves only one species (R—I) and, as noted earlier, it is for this reason that the overall reaction is termed *unimolecular*.

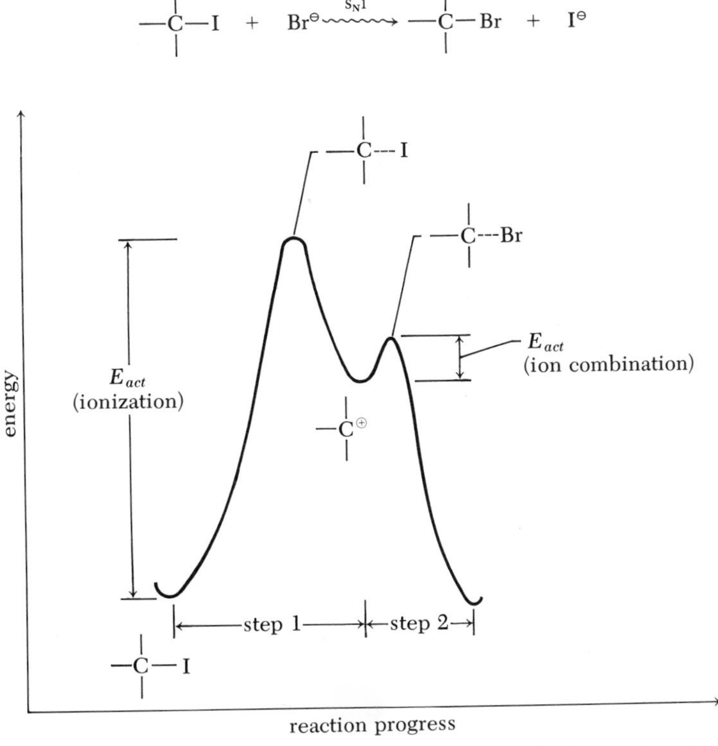

Figure 5-4 An energy profile for an S_N1 reaction. Step 1 is an ionization, producing a cation, while step 2 represents the combination of the cation with a nucleophile. Step 1 is the rate-determining step (it has the higher E_{act}), while step 2 is the product-determining step (it converts the cation into the product). Compare this to Figure 5–2, in which the rate- and product-determining steps are identical.

Rather than continuing in generalities, let us look at a representative S_N1 reaction, the synthesis of *t*-butyl bromide from *t*-butyl alcohol.

Reaction of *t*-Butyl Alcohol with Hydrobromic Acid. *t*-Butyl alcohol reacts rapidly with concentrated hydrobromic acid to produce *t*-butyl bromide, water, and by-products. The steric hindrance associated with the *t*-butyl group (Table 5–1) suggests that this reaction cannot occur by an S_N2 process. An S_N1 process is a reasonable alternative. However, because of the very poor leaving group ability of $OH^{\ominus}$, the rate determining step (the slow step) is postulated to be cleavage of *O*-protonated *t*-butyl alcohol to produce the *t*-butyl cation and water, rather than the cleavage of *t*-butyl alcohol to produce the *t*-butyl cation and hydroxide ion. The *t*-butyl cation then will undergo several reactions, including combination with $Br^{\ominus}$ to form *t*-butyl bromide.

net reaction: $(H_3C)_3C{-}OH + HBr \rightarrow (H_3C)_3C{-}Br + H_2O$
t-butyl alcohol → *t*-butyl bromide

mechanism: $CH_3{-}C(CH_3)({:}\ddot{O}{-}H){-}CH_3 \overset{H^{\oplus}}{\rightleftharpoons} CH_3{-}C(CH_3)({:}\overset{\oplus}{O}H_2){-}CH_3 \overset{\text{slow}}{\rightleftharpoons} CH_3{-}\overset{\oplus}{C}(CH_3){-}CH_3 + H_2O$ (*t*-butyl cation)

$CH_3{-}\overset{\oplus}{C}(CH_3){-}CH_3 \xrightarrow[\text{rapid}]{Br^{\ominus}} CH_3{-}C(CH_3)(Br){-}CH_3$

From the energy profile for this reaction (Figure 5–5), it is clear that the rate of *t*-butyl bromide formation is independent of the bromide ion concentration, the rate-determining step occurring before bromide ion becomes involved. The rate of this reaction is given by $R = k[(CH_3)_3COH_2]^{\oplus}$ and, like other S_N1 reactions, depends only upon the rate of cation formation.

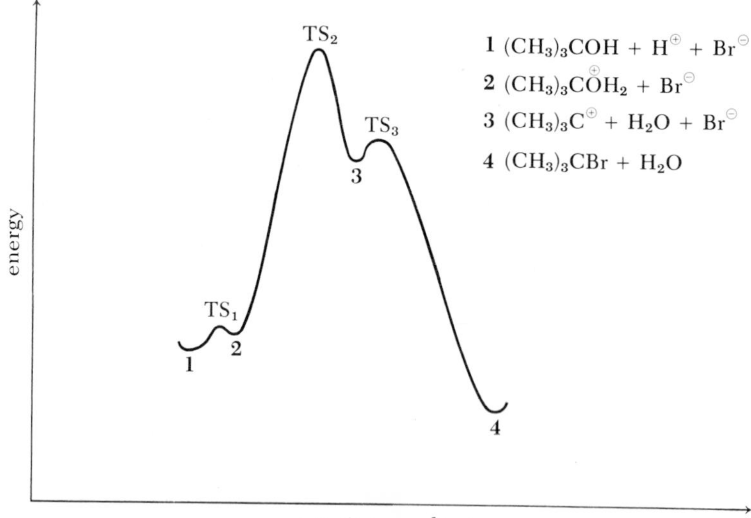

Figure 5-5 The reaction of *t*-butyl alcohol with hydrobromic acid. The rate-determining step is the conversion of **2** to **3**. TS_2 may be represented by $(H_3C)_3\overset{\delta+}{C}\cdots\overset{\delta+}{O}H_2$.

Stereochemistry of the S_N1 Reaction. The lowest energy geometry of a carbocation (see pp. 40 and 138) is one which is planar about its positively charged, sp^2 hybridized carbon. Assuming that there are no other chiral centers in the molecule, this cation will have a plane of symmetry and must be achiral. An S_N1 reaction occurring at a chiral center must, therefore, proceed through an achiral intermediate carbocation. Such an achiral cation will react with a nucleophile with equal facility at both faces to produce an equimolar mixture of enantiomers. Thus, an optically active starting material undergoing an S_N1 reaction must produce a racemic product.

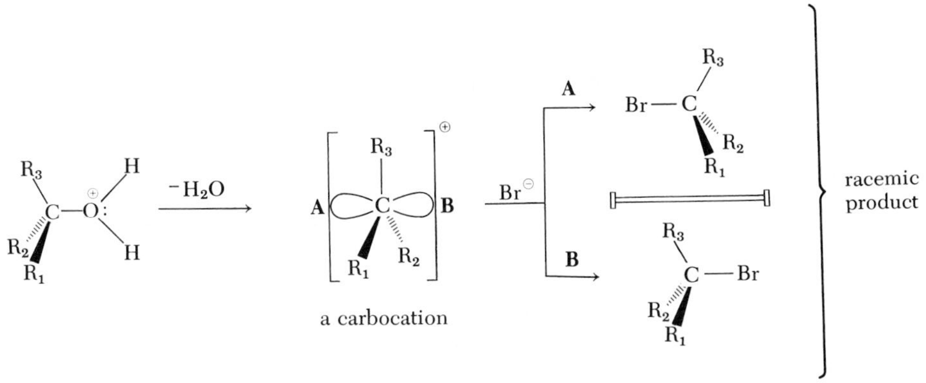

$-H_2O$

A B

a carbocation

$Br^{\ominus}$

A B

racemic product

$$(+)\text{—}CH_3\text{—}\underset{OH}{\overset{CH_2CH_3}{C}}\text{—}(CH_2)_4CH_3 + HBr \xrightarrow{S_N1} (\pm)\text{—}CH_3\text{—}\underset{Br}{\overset{CH_2CH_3}{C}}\text{—}(CH_2)_4CH_3 + H_2O$$

We see here a major difference between S_N2 and S_N1 reactions. Beginning with an optically active starting material, if the nucleophile and the leaving group are different, then the activated complex in an S_N2 reaction will be chiral and will go on to produce an optically active product (of inverted configuration). However, the comparable S_N1 process will lead to an optically inactive product (an equimolar mixture of enantiomers) via an achiral intermediate. It is quite common to generalize and say that *the S_N2 process occurs with inversion of configuration while the S_N1 process occurs with racemization.*

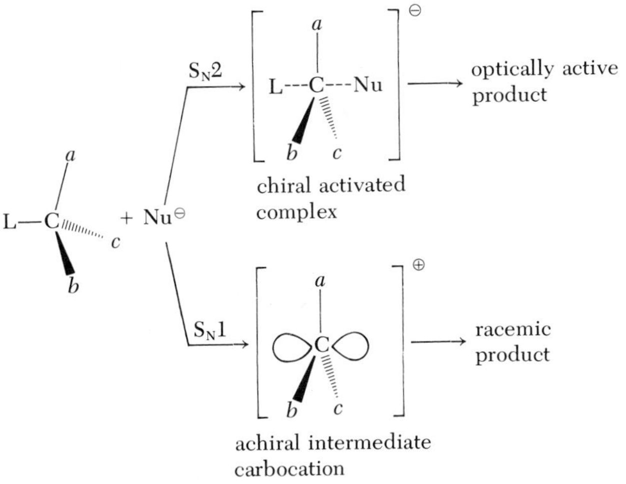

S_N2 → L---C---Nu → optically active product

chiral activated complex

L—C + $Nu^{\ominus}$

S_N1 → racemic product

achiral intermediate carbocation

While the S_N1 reaction usually results in racemic product, a very few S_N1 reactions are known in which the product is slightly richer in the enantiomer opposite in relative configuration to the starting material. The hydrolysis of 1-phenylethyl chloride (1-chloro-1-phenylethane) is a classic example (p. 138). In this reaction the departing chloride ion hinders attack by the nucleophile at side **B**, resulting in more product from attack at side **A**.

$$\mathrm{H_3C{-}\overset{H}{\underset{C_6H_5}{C}}{-}Cl} \xrightarrow{-\mathrm{Cl}^{\ominus}} \mathrm{H_3C{-}\overset{H}{\underset{C_6H_5}{C^{\oplus}}}} \longleftarrow :\ddot{\mathrm{O}}\mathrm{H_2} \longrightarrow \mathrm{H_3C{-}\overset{H}{\underset{C_6H_5}{C}}{-}\overset{\oplus}{O}H_2} \xrightarrow{-\mathrm{H}^{\oplus}} \mathrm{H_3C{-}\overset{H}{\underset{C_6H_5}{C}}{-}OH}$$

$$\mathrm{(H)(H_3C)(C_6H_5)C{-}Cl} \longrightarrow \mathrm{A}\ \text{(planar cation, H, CH}_3\text{, C}_6\text{H}_5\text{)}\ \mathrm{B}\ \ \mathrm{Cl}^{\ominus} \xrightarrow[\text{(b) } -\mathrm{H}^{\oplus}]{\text{(a) } \mathrm{H_2O}} \underset{51\%}{\mathrm{HO{-}C(H)(CH_3)(C_6H_5)}} + \underset{49\%}{\mathrm{(H)(H_3C)(C_6H_5)C{-}OH}}$$

12. Each of the following is heated in water (S_N1 conditions). What product(s) will each produce? (*Hint:* All will be alcohols from the reaction $R^{\oplus} + H_2O \rightarrow ROH$.)
   (a) *t*-butyl bromide
   (b) *t*-butyl iodide
   (c) (**R**)-3-chloro-3-methylhexane
   (d) (**S**)-3-chloro-3-methylhexane
   (e) (3**R**,5**S**)-3-chloro-3,5-dimethylheptane

5.5 CARBOCATIONS

Before continuing with our discussion of S_N1 reactions, we must learn more of these positively charged *reactive intermediates* (*i.e.*, species corresponding to an energy minimum between starting material and product) called "carbocations."

Nomenclature. The term "carbonium ion" has, in the past, been used to describe a positively charged species in which the positive charge is on carbon. For example, $CH_3^{\oplus}$ has often been called the "methyl carbonium ion." However, for the sake of consistency with inorganic nomenclature, in which $Na^{\oplus}$ is the sodium *cation*, $CH_3^{\oplus}$ is better called the "methyl cation." Using this approach, we can generate names such as:

$\mathrm{CH_3{-}\overset{\oplus}{C}H_2}$
ethyl cation

$\mathrm{CH_3{-}CH_2{-}\overset{\oplus}{C}H_2}$
propyl cation

$\mathrm{CH_3{-}\overset{\oplus}{C}H{-}CH_3}$
isopropyl cation
(2-propyl cation)

$\mathrm{CH_3{-}\overset{\oplus}{C}H{-}CH_2{-}CH_3}$
sec-butyl cation
(2-butyl cation)

$\mathrm{CH_3{-}\overset{CH_3}{C}H{-}CH_2^{\oplus}}$
isobutyl cation
(2-methyl-1-propyl cation)

$\mathrm{CH_3{-}\overset{CH_3}{C}H{-}CH_2{-}\overset{\oplus}{C}H_2}$
isoamyl cation
(3-methyl-1-butyl cation)

Along with the fall into disfavor of the term "carbonium ion," there has appeared a new generic term to describe a positively charged species with the positive charge on carbon, the *carbocation*. This represents a condensation of "carbon cation" and is, therefore, analogous to terms such as "sodium cation." We shall use the term "carbocation" to describe $RR'R''C^{\oplus}$, that species which has classically been called a "carbonium ion." [A more recent suggestion is that such species are rather similar to a carbene (R_2C:) which has gained a positively charged ligand (*e.g.*, $R^{\oplus}$) and should be called a *carbenium ion*. However, this suggestion has not yet withstood the test of time and will not be considered further.]

Stability and Reactivity. Anything which disperses positive charge tends to *stabilize*, that is, lower the ground state energy of, a carbocation. Dispersal of this charge may occur by internal or external factors, or both. In any single step of a reaction, the activated complex resembles the ground state closest to it in energy. In an S_N1 reaction, the activated complex of the ionization step is best described by the product cation. Therefore, we can estimate the influence of structural changes upon the activated complex leading to a carbocation by examining the effects of those structural changes upon the carbocation itself.

A compound's *reactivity* is measured by the energy of activation required by that species to form a particular activated complex. Therefore, one must speak of a specific reaction when discussing reactivity. Figure 5–6 contains two energy profiles to illustrate how one compound can be both more reactive and more stable than another.

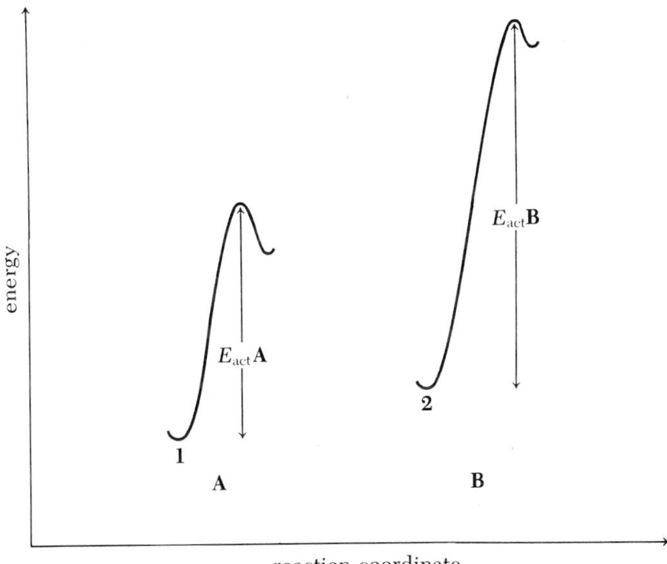

Figure 5–6 Reactivity *vs.* stability. Compound **1** is more stable than compound **2**; *i.e.*, it has a lower energy. However, **1** is more reactive than **2**; *i.e.*, E_{act}**A** < E_{act}**B**. Here, **A** and **B** represent two different reactions that pass through different transition states.

13. "Limestone is quite stable, degrading but slightly after centuries of exposure to the atmosphere." "Limestone is extremely reactive, dissolving very rapidly in hydrochloric acid to produce carbon dioxide and calcium chloride." While the factual material is correct, it seems paradoxical to have limestone described as both "quite stable" and "extremely reactive." Clarify this apparent paradox. Incidentally, limestone's destruction is accelerated by atmospheric sulfur dioxide because of the reaction of water (rain) with it to produce "acid rain" (H_2SO_3).

Inductive Stabilization of Carbocations. When compared to hydrogen, alkyl groups tend to release electron density along the σ-bond framework to centers which are electron deficient. In releasing electron density, the alkyl groups develop a slight positive charge and thereby disperse the charge originally concentrated on one atom. Thus, increasing alkylation at a cationic center stabilizes that cation. Halogens, being highly electronegative, destabilize a cation by withdrawing electron density through the σ-bond framework and increasing the positive charge on the cationic center. These distortions of electron density in bonds because of electronegativity differences are termed **inductive effects.**

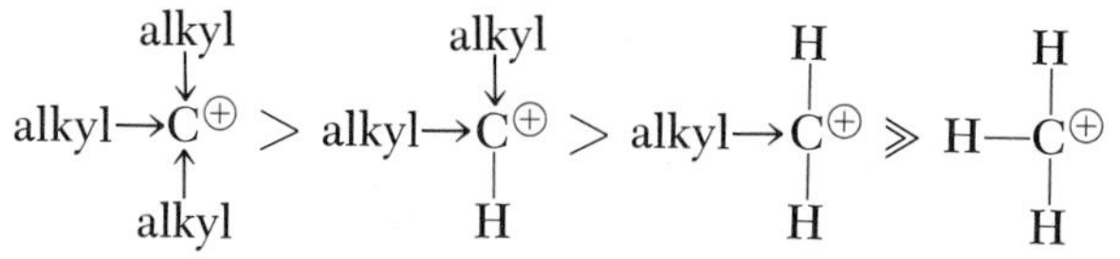

$$\text{alkyl}\rightarrow\overset{\overset{\text{alkyl}}{\downarrow}}{\underset{\underset{\text{alkyl}}{\uparrow}}{C}}{}^{\oplus} > \text{alkyl}\rightarrow\overset{\overset{\text{alkyl}}{\downarrow}}{\underset{\underset{H}{|}}{C}}{}^{\oplus} > \text{alkyl}\rightarrow\overset{\overset{H}{|}}{\underset{\underset{H}{|}}{C}}{}^{\oplus} \gg H-\overset{\overset{H}{|}}{\underset{\underset{H}{|}}{C}}{}^{\oplus}$$

order of stability based on inductive effect

The same element will have varying inductive effects, depending upon its state of hybridization. Since orbital electronegativity varies in the order $sp > sp^2 > sp^3$, the inductive stabilization of a carbocation should be least for C_{sp} bonded to the positive center and most for C_{sp^3} bonded to a positive center. The inductive effects of C_{sp^2} and C_{sp} *destabilize* carbocations when compared to H or C_{sp^3}.

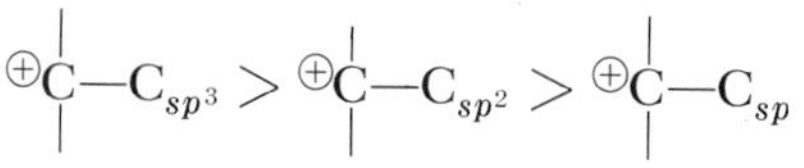

$$^{\oplus}\overset{|}{\underset{|}{C}}-C_{sp^3} > {}^{\oplus}\overset{|}{\underset{|}{C}}-C_{sp^2} > {}^{\oplus}\overset{|}{\underset{|}{C}}-C_{sp}$$

order of stability based on inductive effect

Inductive effects are comparatively weak and drop off rapidly with an increase in the number of intervening bonds. They are virtually non-existent beyond three σ bonds.

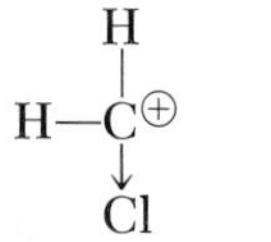

$$H-\overset{\overset{H}{|}}{\underset{\underset{Cl}{\downarrow}}{C}}{}^{\oplus} \qquad Cl\leftarrow CH_2\leftarrow CH_2\leftarrow CH_2-\overset{\oplus}{C}H_2$$

destabilized by inductive effect of chlorine

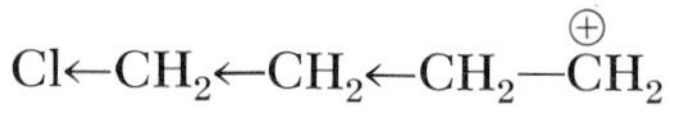

cation uninfluenced by inductive effect of chlorine

Resonance Stabilization of Carbocations. Allyl chloride, $H_2C{=}CH{-}CH_2{-}Cl$, is a primary halide that undergoes S_N1 reactions more rapidly than does *t*-butyl chloride. Since the inductive effect of C_{sp^2} should destabilize an adjacent positively charged carbon, another factor must outweigh this in order to make the allyl cation as stable as it is.

$$\underset{\text{allyl chloride}}{CH_2{=}CH{-}CH_2Cl} \xrightarrow{-Cl^{\ominus}} \underset{\text{allyl cation}}{CH_2{=}CH{-}\overset{\oplus}{C}H_2} \xrightarrow{X^{\ominus}} CH_2{=}CH{-}CH_2X$$

an S_N1 reaction of allyl chloride

This second factor is commonly called the **resonance effect** (review p. 42). The allyl cation can be drawn as two identical arrangements of atoms differing only in their electronic distribution. These two resonance contributing structures have no physical existence, and no time-dependent relationship between them is intended; *they are not in equilibrium.* (The double-headed "resonance arrow" emphasizes the non-equilibrium relationship between these structures.) Rather, the real allyl cation is a hybrid of the two contributing structures, having an electronic distribution between these degenerate canonical forms. Unfortunately, this hybrid cannot be adequately represented by a single Lewis structure.

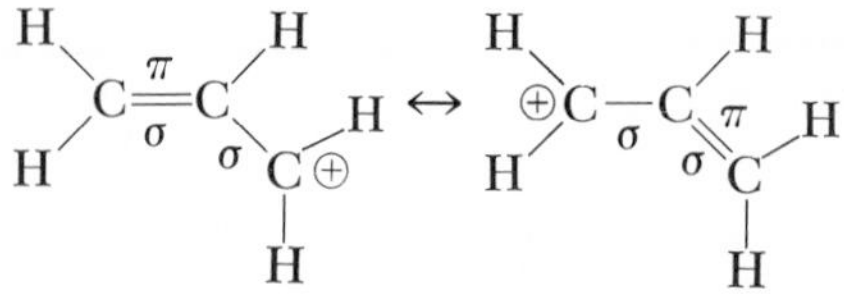

$$H_2C\overset{\pi}{\underset{\sigma}{=}}CH\overset{\sigma}{-}\overset{\oplus}{C}H_2 \leftrightarrow H_2\overset{\oplus}{C}\underset{\sigma}{-}CH\overset{\pi}{\underset{\sigma}{=}}CH_2$$

resonance contributing structures of the allyl cation—note that the charge is distributed over two carbons

Whenever one can draw resonance structures which neither create charge separation nor require atomic reorganization, the real molecule will be more stable than any of the contributing structures. (Charge-separated forms may be important if a basic atom, *e.g.*, nitrogen, donates the electron pair and becomes positively charged.) The hybrid resembles more closely the most stable contributing structure if the contributors are of unequal energy.

The allyl cation can be represented by a structure containing "partial bonds." If one views each partial bond as providing one half of a bond to each atom involved in that bond, each of the outer carbons carries half of the formal positive charge while the central carbon is neutral. This representation suggests, as do the resonance structures, that there is partial double bond character in each carbon-carbon linkage. (What are possible consequences of this?) The use of partial bonds permits the chemist to convey the concept of electron (and charge) delocalization without recourse to more than one structure. Unfortunately, one can easily confuse the significance of such dotted lines, and we will avoid them unless their meaning is unambiguous.

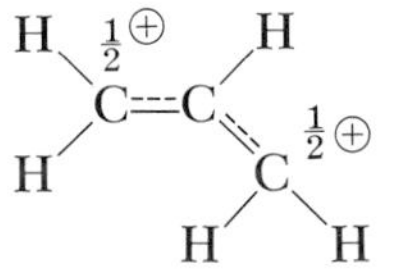

H, $\frac{1}{2}\oplus$, H, C---C, $\frac{1}{2}\oplus$, H, C, H, H

the allyl cation represented by a partial bond structure

One also may represent the allyl cation using atomic orbitals. The picture of the delocalized cation, in which electron density is distributed over all p orbitals, does not clearly demonstrate the lack of positive charge at the central carbon atom, and it makes the resonance picture more valuable because only those atoms bearing some positive charge will combine with the nucleophile in an S_N1 reaction.

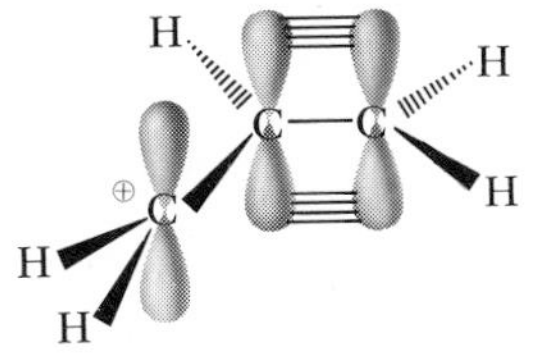

localized cation

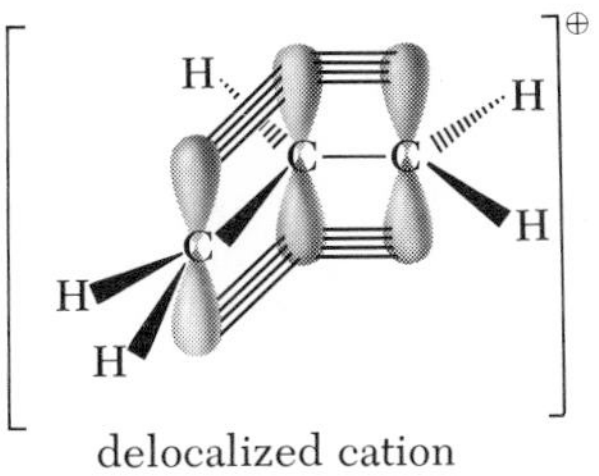

delocalized cation

$$CH_2{=}CH{-}CH_2^{\oplus} + Br \rightarrow \underbrace{CH_2{=}CH{-}CH_2Br + BrCH_2{-}CH{=}CH_2}_{\text{identical}}$$

but no product with Br bonded to the central carbon

Since the positive charge is distributed over two carbons, nucleophilic substitution upon allylic systems often is accompanied by an *allylic rearrangement.* In this rearrangement the incoming nucleophile becomes bonded to *either* of the carbons bearing the positive charge in the intermediate allylic cation. This is illustrated below using the hydrolysis of crotyl chloride.

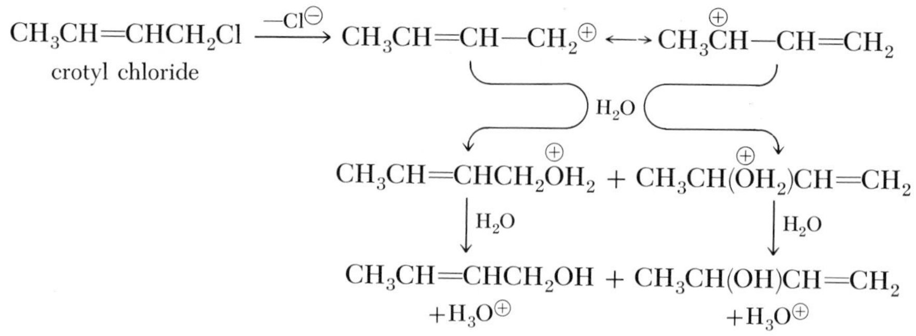

$$CH_3CH{=}CHCH_2Cl \xrightarrow{-Cl^{\ominus}} CH_3CH{=}CH{-}CH_2^{\oplus} \longleftrightarrow CH_3\overset{\oplus}{C}H{-}CH{=}CH_2$$

crotyl chloride

$$\xrightarrow{H_2O} CH_3CH{=}CHCH_2\overset{\oplus}{O}H_2 + CH_3CH(\overset{\oplus}{O}H_2)CH{=}CH_2$$

$$\xrightarrow{H_2O} CH_3CH{=}CHCH_2OH + H_3O^{\oplus} \quad + \quad CH_3CH(OH)CH{=}CH_2 + H_3O^{\oplus}$$

Benzylic halides, $C_6H_5CH_2{-}X$, produce highly charge-delocalized carbocations. One might almost view them as "super allyl cations." Consequently, the presence of a phenyl group ($C_6H_5{-}$) on a carbon bearing a leaving group encourages S_N1 substitution at that carbon. This explains why 1-chloro-1-phenylethane undergoes a significant amount of S_N1 substitution (p. 137).

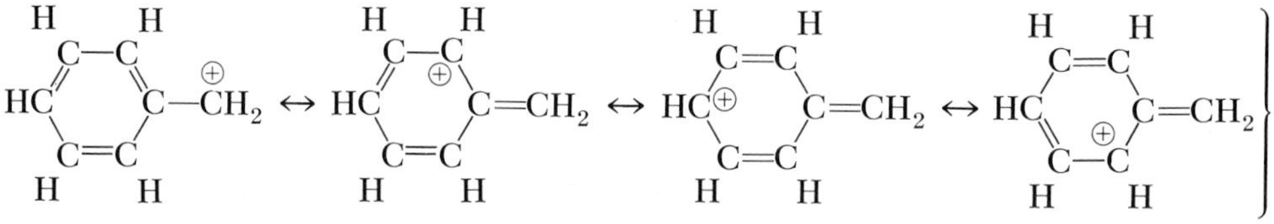

[Resonance structures: $C_6H_5\overset{\oplus}{C}H_2 \leftrightarrow$ ortho-cation $C{=}CH_2 \leftrightarrow$ para-cation $C{=}CH_2 \leftrightarrow$ ortho-cation $C{=}CH_2$] — the benzyl cation

> 14. Allyl chloride is not stabilized by resonance, the way the allyl cation is, yet the total energy of allyl chloride is less than is that of a separated allyl cation and a chloride anion. Clarify this apparent paradox.

Hyperconjugation. The inductive effect may not be of sufficient magnitude to explain the stability of highly alkylated carbocations such as the *t*-butyl cation. Another phenomenon, **hyperconjugation,** involving delocalization between a C—H σ-bond and an adjacent, vacant p orbital, has been suggested as a further stabilizing influence. The hyperconjugative process can be represented both by an orbital picture and with the aid of resonance contributing structures. The resonance picture suggests that in the resonance hybrid (the real species) the hydrogen adjacent to the positive charge has some protonic character.

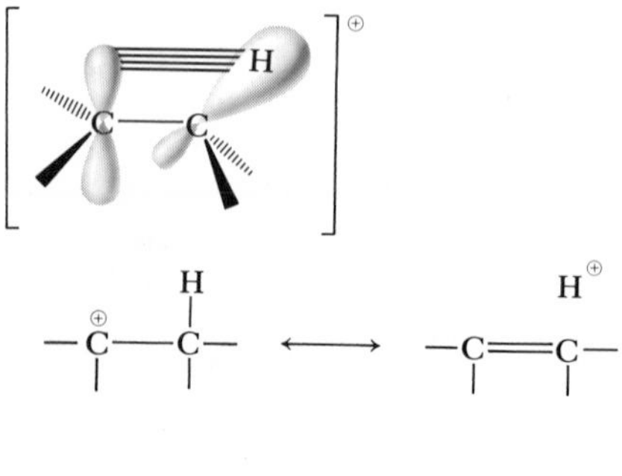

[Orbital picture: C—C with H overlapping vacant p orbital]$^{\oplus}$

$$-\overset{\oplus}{\underset{|}{C}}-\overset{H}{\underset{|}{\overset{|}{C}}}- \longleftrightarrow -\underset{|}{C}{=}\underset{|}{C}- \quad H^{\oplus}$$

5.6 CARBOCATION REARRANGEMENTS

The reaction of 2-methyl-2-butanol with hydrochloric acid produces 2-chloro-2-methylbutane by an S_N1 reaction. Under similar conditions, 3-methyl-2-butanol produces both the expected 2-chloro-3-methylbutane *and* 2-chloro-2-methylbutane.

$$H_3C-\underset{OH}{\overset{CH_3}{C}}-CH_2CH_3 + HCl \rightarrow H_3C-\underset{Cl}{\overset{CH_3}{C}}-CH_2CH_3$$

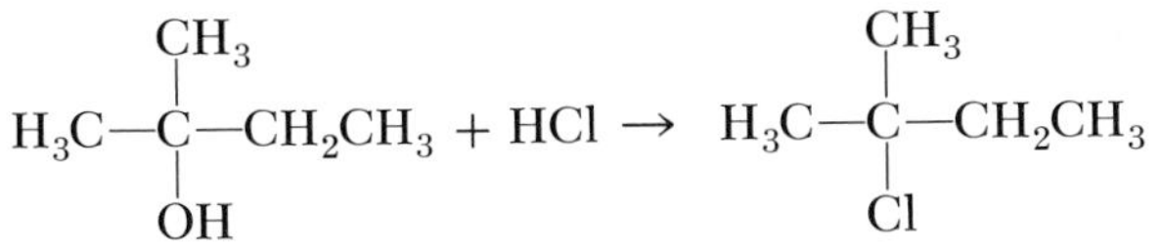

2-methyl-2-butanol   2-chloro-2-methylbutane

$$H_3C-\underset{H}{\overset{CH_3}{C}}-\underset{H}{\overset{OH}{C}}-CH_3 + HCl \rightarrow H_3C-\underset{H}{\overset{CH_3}{C}}-\underset{H}{\overset{Cl}{C}}-CH_3 + H_3C-\underset{Cl}{\overset{CH_3}{C}}-CH_2CH_3$$

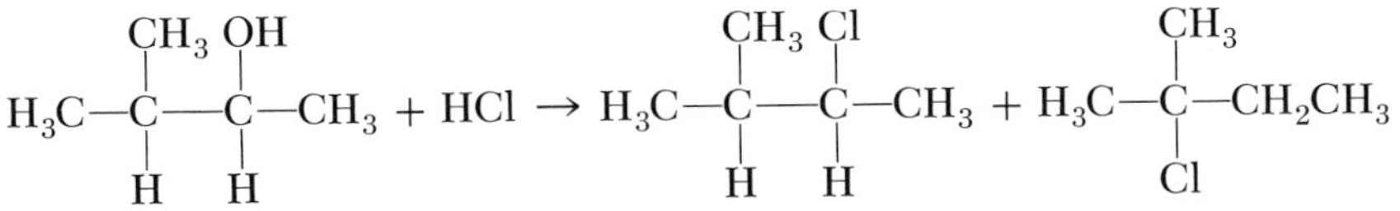

3-methyl-2-butanol   2-chloro-3-methylbutane

How do we account for the appearance of 2-chloro-2-methylbutane in this second reaction? This product results from a combination of at least two factors. First, carbocations are extremely energy-rich species and, second, there is a great difference in the energy content of 1°, 2°, and 3° carbocations—3° cations are more stable than 2° cations, and 1° cations are less stable than either. Like its precursor alcohol, the cation derived from 3-methyl-2-butanol undergoes rapid rotation around all of its single bonds. It can be seen from the diagrams below that one conformation of the cation has the vacant p orbital of the positively charged carbon parallel to an adjacent C—H bond, an arrangement which maximizes interaction between these two.

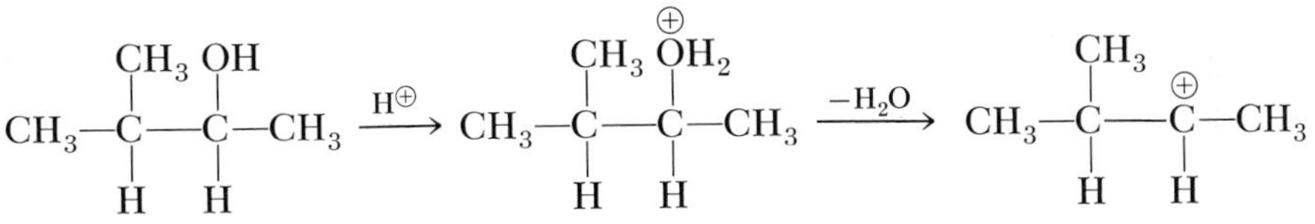

$$CH_3-\underset{H}{\overset{CH_3}{C}}-\underset{H}{\overset{OH}{C}}-CH_3 \xrightarrow{H^{\oplus}} CH_3-\underset{H}{\overset{CH_3}{C}}-\underset{H}{\overset{\overset{\oplus}{O}H_2}{C}}-CH_3 \xrightarrow{-H_2O} CH_3-\underset{H}{\overset{CH_3}{C}}-\underset{H}{\overset{\oplus}{C}}-CH_3$$

3-methyl-2-butyl cation

H, CH$_3$, CH$_3$, ⊕, H, CH$_3$ $\underset{\longleftarrow}{\overset{90°}{\longrightarrow}}$ H, CH$_3$, CH$_3$, H, ⊕, CH$_3$ $\rightleftarrows$ bond rotation in carbocation

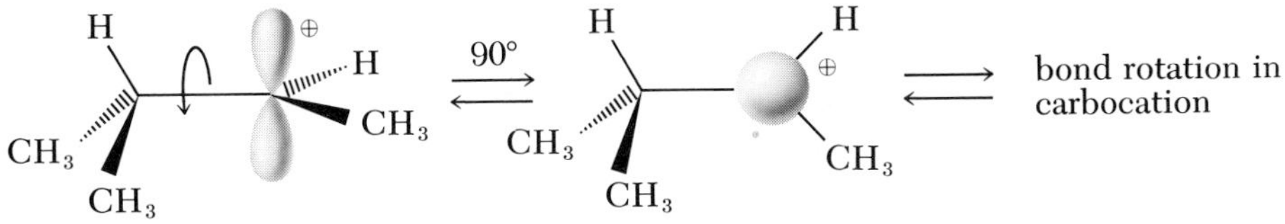

C—H bond parallel to vacant p orbital (maximum interaction) p orbital directed at viewer (minimum interaction)

It was Frank Whitmore who suggested that, in such conformations, the hydrogen of this C—H fragment can migrate, *with the electron pair which bonds it to carbon,* into the vacant p orbital on the adjacent carbon. Since both the hydrogen and the electron pair migrate, this is equivalent to a hydride ion migration or, as it is commonly called, a hydride shift. This, the so-called Whitmore mechanism,[a] is an example of a **1,2-shift,** a rearrangement in which the migrating group moves from one atom to the next atom.

---

[a] F.C. Whitmore, *J. Amer. Chem. Soc.,* **54,** 3274 (1932).

2° carbocation → 3° carbocation

The net result of this hydride migration is the conversion of a less stable 2° cation into a more stable 3° cation. In general, if a 1,2-shift can yield a more stable cation, such shifts, or rearrangements, can be expected to occur. The possibility that rearrangement may occur is but one reason that S_N1 reactions are sometimes shunned in favor of S_N2 reactions in synthetic organic chemistry. Of course, if rearrangement is extensive, and you desire the rearranged product, the S_N1 route may be the one to employ.

Can Only Hydrogens Participate in Rearrangements? No! The stabilization gained by isomerization of a carbocation permits the 1,2-shift of other groups, most frequently alkyl substituents. For example, the reaction of 3,3-dimethyl-2-butanol with hydriodic acid produces a fair amount of 2-iodo-2,3-dimethylbutane. As shown below, this is explained by a *methide ion*, $CH_3{:}^{\ominus}$, migration from C3 to C2. This type of rearrangement is important in organic chemistry, since it provides a method for converting one carbon skeleton into another.

$$H_3C{-}C(CH_3)_2{-}CH(OH){-}CH_3 \xrightarrow{HI} H_3C{-}C(CH_3)_2{-}CH(\overset{\oplus}{O}H_2){-}CH_3 \xrightarrow{-H_2O} H_3C{-}C(CH_3)_2{-}\overset{\oplus}{C}H{-}CH_3$$

3,3-dimethyl-2-butanol → 2° cation

2° cation → 3° cation $\xrightarrow{I^{\ominus}}$ 2-iodo-2,3-dimethylbutane

In both of these examples, rearrangement was pictured as producing a more stable cation. Although pathways existed for alternate 1,2-shifts, these were not used since they resulted in the production of less stable cations and should, therefore, have higher energies of activation than the ones producing more stable cations.

```
                                        H
                                      ⊕ |
                                a   ┌→ C—C—C—CH3      preferred
     H     H                        │    |  |
     | a ⊕ b|                       │    C  H
  C—C—→C←—C—H  ──────────────────────┤   3° cation
     |  H   |                       │
     C      H                       │    H  H
   2° cation                        │    |  |  ⊕
                                b   └→ C—C—C—CH2
                                         |  |
                                         C  H
                                       1° cation
```

```
                                        C
                                      ⊕ |
                                a   ┌→ C—C—C—CH3      preferred
     C     H                        │    |  |
     | a ⊕ b|                       │    C  H
  C—C—→C←—C—H  ──────────────────────┤   3° cation
     |  H   |                       │
     C      H                       │    C  H
   2° cation                        │    |  |  ⊕
                                b   └→ C—C—C—CH2
                                         |  |
                                         C  H
                                       1° cation
```

One of the great distinctions between organic chemistry's three important reactive intermediates (carbocations, carbanions, and free radicals) is that only carbocations are prone to undergo isomerization to a more stable species. Carbanions and free radicals apparently require a rather high-energy transition state for 1,2-shifts and, consequently, their involvement in a reaction sequence does not portend skeletal rearrangement even when a more stable species might result.

15. Each of the following cations is capable of rearranging to a more stable cation. Limiting yourself to a single 1,2-shift, suggest a structure for the rearranged cation.

(a) $CH_3CHCH_3\overset{\oplus}{C}HCH_3$ (b) $(CH_3)_3\overset{\oplus}{C}CHCH_3$

(c) $CH_3CH_2CH_2\overset{\oplus}{C}HCH(CH_3)C(CH_3)_3$ (d) $CH_2{=}CHCH_2\overset{\oplus}{C}HCH_2CH_3$

(e) $CH_3OCH_2\overset{\oplus}{C}HC(CH_3)_3$

5.7 WILL IT BE S_N1 OR S_N2?

Let us now examine some of the factors which determine whether a particular substitution will be S_N1 or S_N2.

1. Nature of the Carbon Skeleton. Only those substrates which produce extremely stable cations undergo S_N1 reactions. Typically, only compounds producing trialkyl or resonance-stabilized cations undergo S_N1 reactions in preference to S_N2 reactions. The effect of increasing alkylation upon these competitive reactions is illustrated in Table 5-3. It is clear from the data in this table that while increasing alkylation favors S_N1 reactions, it hinders S_N2 reactions. This reflects the contrast between the steric requirements of the S_N2 reaction and the electronic requirements of the S_N1 reaction.

TABLE 5-3 Relative Rates of S_N2 and S_N1 Competitive Reactions in the Series:

$$RBr + I^{\ominus} \rightarrow RI + Br^{\ominus} \quad (S_N2)$$
$$RBr + H_2O \rightarrow ROH + HBr \quad (S_N1)$$

R	Relative Rate S_N2	Relative Rate S_N1
$—CH_3$	145	1.05
$—CH_2CH_3$	1	1
$—CH(CH_3)_2$	8×10^{-3}	12
$—C(CH_3)_3$	5×10^{-4}	1.2×10^6

The generalization that "primary halides react by an S_N2 route while tertiary halides react by an S_N1 route" is subject to exceptions, particularly with regard to the reactivity of primary halides. We have, for example, already noted (p. 140 *et seq.*) that allyl and benzyl halides undergo S_N1 reactions quite easily. A related instance of exceptional behavior is provided by chloromethyl methyl ether, CH_3OCH_2Cl, which also undergoes rapid S_N1 substitution. The reason for this unusual behavior is the same as the reason for the unusual behavior of allyl chloride—chloromethyl methyl ether forms a resonance stabilized cation. However, rather than using a pair of electrons in a π bond to delocalize the positive charge, the cation derived from chloromethyl methyl ether employs a pair of non-bonding electrons on oxygen to delocalize the charge.

$$\underset{\text{chloromethyl methyl ether}}{CH_3—\ddot{O}—CH_2—Cl} \xrightarrow{-Cl^{\ominus}} \underset{\text{resonance stabilized cation}}{CH_3—\ddot{O}—\overset{\oplus}{C}H_2 \leftrightarrow CH_3—\overset{\oplus}{O}=CH_2} \xrightarrow{\overset{\ominus}{O}CH_3} CH_3OCH_2—OCH_3$$

If a good nucleophile (*e.g.*, O, S, or N) is in the same molecule but separated from the leaving group by two atoms, it may act to displace the leaving group. A three-membered ring is formed in this S_N2 process. The cyclic product is, in turn, highly strained and prone to ring-opening attack by a second nucleophile.

$$R—\ddot{Z}—CH_2—CH_2—Cl \rightarrow R—\overset{\oplus}{Z}(CH_2)_2 \quad Cl^{\ominus}$$

$$R—\overset{\oplus}{Z}(CH_2)_2 + X^{\ominus} \rightarrow R—\ddot{Z}—CH_2—CH_2—X$$

$$\underset{\text{mustard gas}}{Cl—CH_2—CH_2—\ddot{S}—CH_2—CH_2—Cl} \rightarrow Cl—CH_2—CH_2—\overset{\oplus}{S}(CH_2)_2 \quad Cl^{\ominus}$$

The war gas *bis*(2-chloroethyl)sulfide ("mustard gas" or "Yperite") is a deadly vesicant (blistering agent) which is toxic in concentrations of 1 ppm (part per million) in the air. Its reactivity is associated with this *neighboring group participation.* Either sodium hypochlorite or calcium hypochlorite is capable of deactivating the gas by oxidizing the nucleophilic sulfur to the *much* less nucleophilic sulfonyl group. (During World War II, air raid shelters in England sometimes were "painted" with calcium hypochlorite solutions.)

$$R{-}\ddot{\underset{\cdot\cdot}{S}}{-}R + Ca(OCl)_2 \rightarrow R{-}\overset{\overset{O}{\|}}{\underset{\underset{O}{\|}}{S}}{-}R + CaCl_2 \qquad R = Cl{-}CH_2{-}CH_2{-}$$

Rapid ring-opening reactions make these three-membered rings potent alkylating agents when attacked by a nucleophile. This is used to advantage in drugs such as chlorambucil, a nitrogen mustard used in treating chronic lymphocytic leukemia. The detailed action of such drugs is as yet unknown; however, the need for two chloroethyl ($Cl{-}CH_2CH_2{-}$) groups attached to each nitrogen suggests that these mustards serve to link cellular components together. Much of our knowledge of the *in vitro* behavior of these drugs is based upon earlier studies of mustard gas!

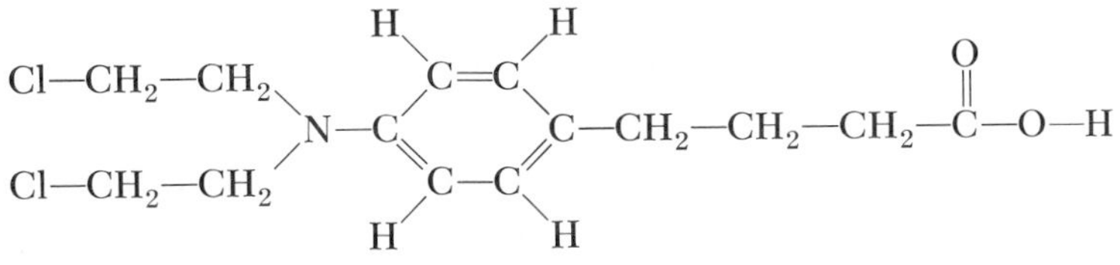

$$(Cl{-}CH_2{-}CH_2)_2N{-}C_6H_4{-}CH_2{-}CH_2{-}CH_2{-}\overset{O}{\overset{\|}{C}}{-}O{-}H$$

chlorambucil

Ethylene oxide, $\overline{CH_2CH_2{-}O}$, is used to sterilize hospital instruments and also has been used to sterilize spacecraft prior to their trips to the moon. It is quite effective against bacteria, fungi, and viruses. This sterilizing ability is believed due to the facile ring opening, by an S_N2 attack at the ring carbons, of this three-membered ring. The nucleophiles most likely are lone pairs of electrons on nitrogen, oxygen, or sulfur associated with proteins (Chapter 24) at or near the surface of the microorganism. (While we will say more about the ring opening of ethylene oxide in Chapter 11, we should note here that facile S_N2 substitution on the C—O—C linkage is not a common reaction.)

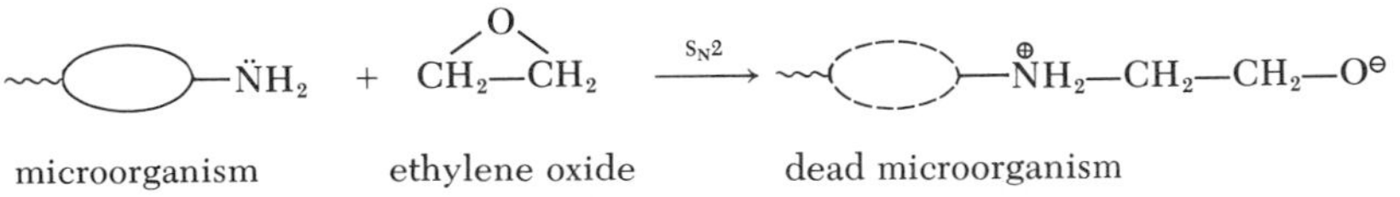

$$\text{microorganism}{-}\ddot{N}H_2 + \underset{\text{ethylene oxide}}{CH_2{-}CH_2 \text{ (O bridged)}} \xrightarrow{S_N2} \text{dead microorganism}{-}\overset{\oplus}{N}H_2{-}CH_2{-}CH_2{-}O^{\ominus}$$

microorganism — ethylene oxide — dead microorganism

Myleran (1,4-butanediol dimethanesulfonate) is an alkylating agent which is an antineoplastic agent (*neoplasm:* any abnormal tissue growth). Although it is a compound which has two points for attack by nucleophiles, this drug cannot invoke intramolecular participation by a nucleophile since it lacks a nucleophilic center. Its main use is in the treatment of acute myelogenous leukemia.

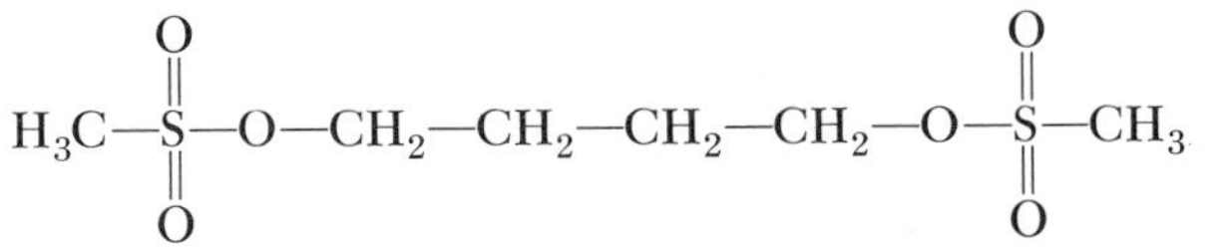

$$H_3C{-}\overset{\overset{O}{\|}}{\underset{\underset{O}{\|}}{S}}{-}O{-}CH_2{-}CH_2{-}CH_2{-}CH_2{-}O{-}\overset{\overset{O}{\|}}{\underset{\underset{O}{\|}}{S}}{-}CH_3$$

Myleran

2. *Solvent.* The more the solvent can stabilize the anion and cation, the more likely is it that an S_N1 reaction will occur. Those solvents favoring S_N1 reactions include water, mixed solvents containing large quantities of water, and acetic acid. Acetone and ethanol are common solvents favoring S_N2 reactions. Of course, regardless of the solvent, a highly hindered compound such as *t*-butyl bromide cannot react by S_N2 processes. Dipolar aprotic solvents are generally not as effective in promoting S_N1 reactions as are hydrogen-bonding solvents.

3. *Electrophilic Catalysis.* The C—F bond is the most stable carbon-halogen bond and is generally inert to most reactions, including S_N1 and S_N2 processes. This accounts, in part, for the practical value of TFE (Teflon, terafluoroethylene polymer) and other fluorinated hydrocarbons (*e.g.*, the Freons). However, S_N reactions of alkyl fluorides can be achieved in acidic media, in which the fluorine is strongly hydrogen-bonded to a proton. The leaving group has been modified so that it is HF rather than $F^{\ominus}$. This is an example of acceleration by an electron-seeking species (an *electrophile*) and is analogous to the reaction of *t*-butyl alcohol with HBr (p. 136).

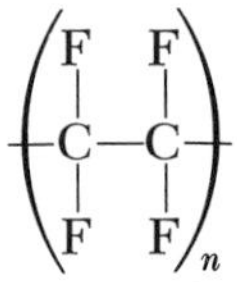

$$\left(-\overset{\displaystyle F}{\underset{\displaystyle F}{C}}-\overset{\displaystyle F}{\underset{\displaystyle F}{C}}-\right)_n \qquad\qquad F-\overset{\displaystyle Cl}{\underset{\displaystyle Cl}{C}}-F$$

Teflon polymer 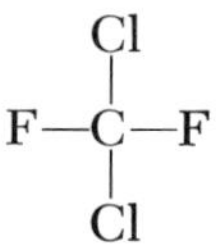   dichlorodifluoromethane (Freon 12)

The most common electrophiles are the Lewis acids, represented by $H^{\oplus}$, $AlCl_3$, $ZnCl_2$, BF_3, and $FeCl_3$. Aside from the proton, aluminum chloride is the most commonly used of these catalysts. The avidity with which aluminum chloride seeks out electron pairs is so great that it coordinates with most organic solvents and reacts violently with water. The lack of specificity which accompanies this great reactivity precludes the use of aluminum chloride as a catalyst in S_N reactions.

Silver(I) and mercury(II) ions are stable in most solvents and coordinate with halogens, thus increasing the latter's propensity to act as leaving groups. These reactions are S_N1-like in character, although there is evidence which suggests that such a view is an over-simplification.

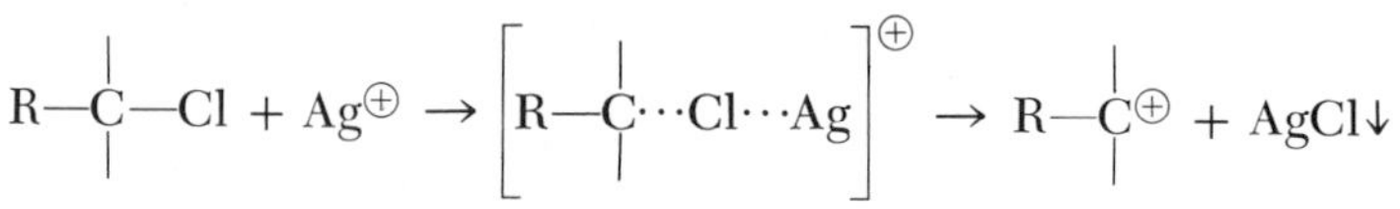

$$R-\overset{|}{\underset{|}{C}}-Cl + Ag^{\oplus} \rightarrow \left[R-\overset{|}{\underset{|}{C}}\cdots Cl\cdots Ag\right]^{\oplus} \rightarrow R-\overset{|}{\underset{|}{C}}^{\oplus} + AgCl\downarrow$$

The precipitation of silver halide (AgCl, AgBr, and AgI) from the reaction of alkyl halides with ethanolic silver nitrate follows the order: benzyl, allyl $> 3° > 2° > 1°$ halide and is a useful test for the presence of carbon-halogen bonds. Vinylic halides, $R_2C{=}CRX$, do not give positive tests because of the comparative instability of the vinyl cation ($R_2C{=}CR^{\oplus}$).

4. *Effect of Added Salts.* An S_N1 reaction is generally aided by the presence of a salt, since the ionic strength of the solution increases with increasing salt concentration. These electrolyte-rich solutions stabilize the transition state. If the salt contains the anion that is produced in the S_N1 reaction, one may expect rate retardation because of the common ion effect (mass-law effect). The more stable the carbocation, the more likely it is to show this rate retardation. The more reactive carbocations can't wait around for a "common ion"; they react with anything that can provide an electron pair for the vacant orbital. This often is the solvent.

5.8 NUCLEOPHILE, ELECTROPHILE—WHO ATTACKS WHOM?

The curved arrows used in describing mechanisms relate electron movements and *not* the movement of atoms. Therefore, *our mechanistic pictures invariably depict nucleophiles attacking electrophiles.* In spite of this, our verbal or written descriptions do not always reflect this universal approach to descriptions of mechanisms.

Verbal descriptions of mechanisms use varying criteria to identify the "attacking group" and the "substrate."

1. If both reactants are of comparable energy, the nucleophile is usually considered as the "attacking group."
2. If the two reagents are of grossly unequal energy, the more energetic species is considered as the "attacking group." Since ions are considered to be, on the average, more energetic than neutral molecules, the verbal description of a reaction between an ion and a molecule reflects the character of the ion. This explains why, for example, reactions **1** and **2** (below) both use arrows to represent electron-pair donation but **1** represents "nucleophilic attack" while **2** represents "electrophilic attack."

$$\mathbf{1}\quad H_3C{-}\overset{\displaystyle CH_3}{\underset{\displaystyle H}{N}}\!:\;\curvearrowright H_3C{-}Cl \rightarrow H_3C{-}\overset{\displaystyle CH_3}{\underset{\displaystyle H}{N^{\oplus}}}{-}CH_3\quad Cl^{\ominus}$$

$$\mathbf{2}\quad H_3C{-}CH_2{-}\ddot{\underset{\cdot\cdot}{Cl}}\!:\;\curvearrowright Ag^{\oplus} \rightarrow CH_3\overset{\oplus}{C}H_2 + AgCl\downarrow$$

5.9 HOW CAN YOU SEPARATE ENANTIOMERS?— OPTICAL RESOLUTION!

In this chapter and in Chapter 4 we made extensive use of isolated enantiomers; it is now time to describe one way to separate one enantiomer from another. We have waited this long to describe the process in order to be able to emphasize the care that must go into planning such a separation. Certainly, this chapter has shown us that if we use the wrong process we could inadvertently racemize the very enantiomer that we were attempting to isolate. It makes no sense to employ any chemical reaction which may convert the chiral center into a cation, a carbanion, or a free radical, since all of these will destroy its chirality.

The classic method for separating a mixture of enantiomers is to convert them into two diastereomers, separate the diastereomers by a common laboratory procedure (generally crystallization), and then regenerate the individual enantiomers (Figure 5–7). This process is called an *optical resolution.*

It is obvious from the resolution scheme that in order to conduct a successful resolution of this type, the resolving agent (the substance which forms two diastereomers upon reaction with a mixture of enantiomers) should be of high optical purity. Furthermore, the initial reaction should go largely to completion and the regeneration of the enantiomers should occur without side reactions. None of these reactions should alter the chirality of the enantiomer.

The one reaction type which fits these criteria rather well involves salt formation, the

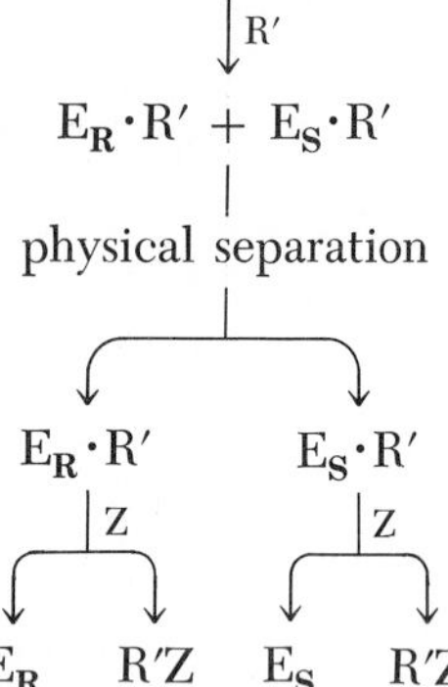

Figure 5-7 Classical optical resolution. The mixture of enantiomers (E_R and E_S) is allowed to react with one enantiomer (R′) to produce two diastereomers ($E_R \cdot R'$ and $E_S \cdot R'$). These diastereomers are separated by physical means and the individual diastereomers are then decomposed (using Z) to regenerate one enantiomer and R′Z.

reaction of an acid with a base. The most common organic bases are amines (the chiral centers are generally carbon atoms), while the most commonly employed acids are carboxylic acids (RCO_2H) and, less frequently, sulfonic acids (RSO_3H).

$$\underset{\text{base}}{R{-}NH_2} + \underset{\text{acid}}{R'CO_2H} \rightarrow \underset{\text{salt}}{R{-}\overset{\oplus}{N}H_3 \quad R'CO_2^{\ominus}}$$

The preparation of optically active *sec*-butylamine ($CH_3CHNH_2C_2H_5$) using (+)-tartaric acid and racemic *sec*-butylamine is outlined below.[a]

$$(\pm)\text{-}CH_3CH(NH_2)C_2H_5 + (+)\text{-}HO_2CCH(OH)CH(OH)CO_2H$$

$\downarrow$ H_2O; an acid-base reaction, with tartaric acid transferring a proton to the amine nitrogen

$$\underbrace{(+)\text{-}CH_3CH(\overset{\oplus}{N}H_3)C_2H_5\ (+)\text{-}HO_2CCH(OH)CH(OH)CO_2^{\ominus}}_{\text{less soluble salt}}$$

$$+\ (-)\text{-}CH_3CH(\overset{\oplus}{N}H_3)C_2H_5\ (+)\text{-}HO_2CCH(OH)CH(OH)CO_2^{\ominus}$$

$\downarrow$ recrystallization permits isolation of the less soluble salt

$$(+)\text{-}CH_3CH(\overset{\oplus}{N}H_3)C_2H_5\ (+)\text{-}HO_2CCH(OH)CH(OH)CO_2^{\ominus}$$

$\downarrow$ aqueous KOH (excess)

$$\underset{\text{(extracted into ether)}}{(+)\text{-}CH_3CH(NH_2)C_2H_5} + \underset{\text{(in aqueous phase)}}{(+)\text{-}K^{\oplus\ominus}O_2CCH(OH)CH(OH)CO_2^{\ominus}K^{\oplus}}$$

In this sequence a strong base is used to regenerate the amine from its tartrate salt. Had an amine been used to resolve a racemic carboxylic acid, hydrochloric acid (or another strong acid) would have been used to regenerate the resolved acid from its salt.

[a] The details of this separation can be found in *Selected Experiments In Organic Chemistry*, G.K. Helmkamp and H.W. Johnson, Jr., W.H. Freeman and Co., San Francisco, 1964.

Since schemes to separate a racemic modification into its individual enantiomers themselves require a separate enantiomer (the resolving agent), one is forced to ask about the source of such resolving agents. Ultimately, nature provides the chemist with the necessary resolving agents. How? Chemical syntheses in living systems employ catalysts called enzymes, and enzymes invariably produce one member of a pair of enantiomers in preference to the other. The enzyme-controlled biological syntheses in plants actually provide our most important resolving agents. Three of these are (+)-tartaric acid (already discussed), strychnine (a basic compound, famous because of its toxicity, which has an extremely complex structure), and (−)-ephedrine (used, medically, as a treatment for symptoms of allergies).

$$\begin{array}{c} CH_3 \\ | \\ H\text{—}C\text{—}NHCH_3 \\ | \\ H\text{—}C\text{—}OH \\ | \\ C_6H_5 \end{array}$$

(−)-ephedrine

Optical Purity. Sometimes one may attempt an optical resolution but, for a variety of reasons, fail to separate one enantiomer completely from the other. When this occurs, it becomes necessary to describe the extent of separation which was achieved. This is usually done by stating the **optical purity** of the product. Optical purity may be considered to be a measure of the amount of one enantiomer present in excess of the other enantiomer.

In calculating the optical purity of a substance composed of unequal amounts of enantiomers, the enantiomer present in lower concentration is considered to be part of a racemic modification. For example, a mixture of 70% of (+)-2-chlorobutane and 30% of (−)-2-chlorobutane is viewed as being 40% (+) and 60% (±)—an optical purity of 40%, not 70%.

$$\%\ \text{optical purity} = \frac{\text{specific rotation of mixture of enantiomers}}{\text{specific rotation of one pure enantiomer}} \times 100$$

16. Enantiomerically pure compound **A** has an observed rotation of 11° under a given set of conditions. Under identical conditions, a sample of **A** has a rotation of only 9°. If this is enough data, calculate the optical purity of this second sample of **A**. If this is not enough, what else must be provided?

IMPORTANT TERMS

Aprotic solvent: A solvent which lacks any hydrogens capable of hydrogen bond formation; a solvent which cannot function as a proton donor.

Bimolecular reaction: A reaction involving two species in the rate determining step. The S_N2 reaction is bimolecular. Molecularity is determined by our conceptualization of the mechanism, and not from kinetic data.

Canonical form: A given resonance contributing structure.

Carbenium ion: An alternate name for carbocation.

Carbocation: A positively charged ion of the type $-\overset{\oplus}{\underset{|}{C}}-$. The carbon is sp^2 hybridized and possesses a vacant p orbital (see p. 40).

Carbonium ion: An alternate, older name for carbocation.

Concerted process: A one-step reaction, *i.e.*, a reaction in which the starting material and product are separated by only one transition state.

Contributing structure: One of several acceptable molecular structures differing only in electron distribution. When taken together, such structures describe a resonance hybrid.

Dipolar aprotic solvent: An aprotic solvent with a high dipole moment. Examples include dimethyl sulfoxide and dimethylformamide.

Electrophile: Any species which coordinates (reacts) with electron pairs. Typical electrophiles include the proton and Lewis acids (*e.g.*, aluminum chloride).

Hydride shift: The translocation of a hydrogen with its bonding electron pair, *i.e.*, $H:^{\ominus}$, from one atom to another. Such shifts usually convert less stable carbocations to more stable carbocations.

Hyperconjugation: Stabilization of a partially filled or vacant orbital by overlap with a filled, bonding σ orbital.

Inductive effect: A complex, multi-faceted effect which may be viewed as being related to the distortion of electron density within bonds arising from electronegativity differences of atoms involved in those bonds. Halogens, oxygen, and nitrogen withdraw electron density from neighboring atoms *via* this effect.

Nucleophile: Any atom or functional group which can act as an electron-pair donor (Lewis base) to an element other than hydrogen. The most effective nucleophiles include $I^{\ominus}$, $HS^{\ominus}$, and $H_2P^{\ominus}$. In each of these the attacking electron pair is associated with a polarizable element. A species donating an electron pair to hydrogen commonly is termed a "base."

Optical purity: A measure of the enantiomeric composition of a non-equimolar mixture of enantiomers. The sample is considered to contain two entities, a certain amount of racemic modification and a certain amount of pure enantiomer. A racemic modification has an optical purity of 0%. A mixture of 10% of one enantiomer and 90% of the other has an optical purity of 80%.

Protic solvent: A solvent bearing a hydrogen which can participate in hydrogen bonding or be transferred to a base.

Racemization: The conversion of one enantiomer to an equimolar mixture of both enantiomers. If a molecule possesses more than one chiral center, each one must be converted into its enantiomeric configuration in order for racemization to occur.

Reactive intermediate: The species corresponding to an energy minimum between two energy maxima in an energy profile. The three common intermediates in organic chemistry are the carbocation, the carbanion, and the free radical.

Reactivity: A measure of the energy of activation for a particular process.

Resonance: A phenomenon constructed by scientists because of their inability to accurately represent the electron distribution within certain species using only one structure. It does not imply any dynamic process and, in this respect, differs from the way the term is used in physics (see Chapter 2).

Resonance hybrid: The atomic and electronic distribution characteristic of the "real molecule," in contrast to the distribution in a given contributing structure (see Chapter 2).

S_N1 Reaction: A unimolecular nucleophilic substitution. A reaction involving an alkyl carbocation. A non-concerted, two-step process.

S_N2 Reaction: A bimolecular nucleophilic substitution. A concerted process.

Stability: An indication of the ground state energy of a species. Used in this sense, stability is not directly related to reactivity. If a phenomenon stabilizes a species (resonance is such a phenomenon) then that species has a lower ground state energy when that phenomenon is operative than when it is not.

Unimolecular reaction: A reaction in which only one species is involved in bond making and/or breaking in the activated complex. One example is the ionization of a covalent bond, $R—L \rightarrow R^{\oplus}L^{\ominus}$.

Walden inversion: Inversion of configuration at a chiral center undergoing S_N2 substitution. If the substrate has only one chiral center, Walden inversion converts a substance into its enantiomer and must, eventually, lead to racemization. If a molecule contains several chiral centers and only one undergoes Walden inversion, the result is the generation of a diastereomer of the starting material.

PROBLEMS

17. Define the following terms *in your own words*, providing a specific illustration where possible.
    (a) nucleophile
    (b) electrophile
    (c) overall kinetic order
    (d) molecularity
    (e) bimolecular reaction
    (f) unimolecular reaction
    (g) dipolar aprotic solvent
    (h) optical purity
18. Predict the major product(s), if any, of the following:
    (a) chloromethane + excess ammonia
    (b) ammonia + excess chloromethane
    (c) neohexyl bromide + sodium methylmercaptide ($Na^{\oplus}CH_3S^{\ominus}$)
    (d) (**R**)-2-bromobutane + $Cl_2/h\nu$
    (e) (**S**)-2-chlorobutane + $Br_2/h\nu$
    (f) Myleran + sodium cyanide in dimethylformamide (solvent)
19. Any ionic reaction can be viewed as a reaction between an electron pair donor (the nucleophile) and an electron pair receiver (the electrophile). In the following reactions, identify (1) the nucleophile, (2) the electrophile, and (3) the leaving group.

(a) $I^{\ominus} + CH_3Cl \rightarrow CH_3I + Cl^{\ominus}$

(b) $Br^{\ominus} + CH_3I \rightarrow CH_3Br + I^{\ominus}$

(c) $HO^{\ominus} + NH_4^{\oplus} \rightarrow H_2O + NH_3$

(d) $NH_3 + Cl—S(O)_2—CH_3 \rightarrow H_3C—S(O)_2—\overset{\oplus}{N}H_3 \quad Cl^{\ominus}$

(e) $CH_3CHFCH_2CH_3 + Ag^{\oplus} \rightarrow CH_3\overset{\oplus}{C}HCH_2CH_3 + AgF$

(f) $CH_3S—SCH_3 + H^{\ominus} \rightarrow CH_3SH + CH_3S^{\ominus}$

(g) $CH_3—S—CH_2—CH_2—Cl \rightarrow CH_3—\overset{\oplus}{S}\langle(CH_2—CH_2)\rangle \quad Cl^{\ominus}$ (three-membered ring: S bonded to two CH_2 groups which are bonded to each other)

20. Suggest the nucleophile which would convert *n*-butyl bromide into the following:
    (a) *n*-butyl chloride
    (b) *n*-octane
    (c) tetra *n*-butylammonium bromide, $(n\text{-butyl})_4N^{\oplus}Br^{\ominus}$
    (d) *n*-butyl methyl ether ($CH_3OCH_2CH_2CH_2CH_3$)

(e) 2-methylhexane
(f) *n*-butyl acetate ($CH_3CH_2CH_2CH_2OC(O)CH_3$)
(g) *n*-butyl cyanide ($CH_3CH_2CH_2CH_2CN$)

21. For each of the following, suggest a means of converting the given starting material into the desired product. Several steps may be required. Use any necessary reagents.
(a) methane to methyl chloride
(b) methyl chloride to methyl iodide
(c) methyl iodide to methyl alcohol (CH_3OH)
(d) methyl iodide to methylamine (CH_3NH_2)
(e) 2-methylpropane to *t*-butyl chloride
(f) *t*-butyl chloride to *t*-butyl methyl ether ($(CH_3)_3COCH_3$)
(g) methane to acetonitrile (CH_3CN)
(h) 2,2-dimethylpropane to 1-chloro-2,2-dimethylpropane
(i) 2,2-dimethylpropane to 2-methyl-2-butanol ($CH_3C(OH)CH_3CH_2CH_3$)

22. "I view the reaction of ammonia with hydrogen chloride as a nucleophilic substitution that is bimolecular when carried out in an inert solvent such as *n*-hexane. It is, however, unfortunate that at least one of the major criteria used to identify an S_N2 reaction is absent from this process. May I be so bold as to suggest that when dry hydrogen chloride gas is dissolved in water, to produce what we call hydrochloric acid (but which actually contains $H_3O^{\oplus}$ and $Cl^{\ominus}$), the most significant reaction is again an S_N2 process. I make these statements because, among other things, hydrogen chloride gas (free of water) dissolves in hexane but no ions are produced." This statement is correct. Draw a representation of the transfer of a proton from hydrogen chloride to (a) ammonia and (b) water, viewing these as S_N2 processes. What criterion of an S_N2 reaction cannot be detected in these processes? The lack of ionization of dry hydrogen chloride in alkanes rules out an S_N1 reaction, in the view of the speaker. What is the basis of his point of view?

23. Treatment of neopentyl magnesium bromide with iodine produces neopentyl iodide in good yield. Suggest a mechanism for this reaction, which is carried out in ether. *Note:* Neopentyl magnesium bromide may be considered to be in equilibrium with its carbanion, *i.e.*,

$$\underset{\text{neopentyl magnesium bromide}}{CH_3-\overset{\overset{CH_3}{|}}{\underset{\underset{CH_3}{|}}{C}}-CH_2-MgBr} \rightleftarrows CH_3-\overset{\overset{CH_3}{|}}{\underset{\underset{CH_3}{|}}{C}}-\overset{\ominus}{CH_2}: \quad \overset{\oplus}{MgBr}$$

24. Predict the order of increasing reactivity of the following compounds to S_N2 displacement: (a) $CH_3CH_2CCl(CH_3)_2$, (b) $CH_3(CH_2)_4Cl$, (c) $CH_3(CH_2)_4I$, (d) $(CH_3)_3CCH_2Cl$, (e) $CH_3CH_2CH_2CH_2Cl$.

25. A graduate student hoped to improve the yield of a reaction in which methyl iodide was attacked by a nucleophile by employing a dipolar aprotic solvent instead of dioxane. He selected dimethylsulfoxide because of its high boiling point (his first mistake?). In working up the reaction he found that he had produced very little of what he had wanted. Instead, he produced a large quantity of a crystalline solid which had both sulfur and iodine in it. This compound reacted instantly with aqueous silver nitrate to produce a precipitate of silver iodide. Can you suggest an explanation for these observations?

26. While resonance has been considered important in explaining the facile S_N1 reactivity of allyl halides by lowering the energy of activation for the production of the allyl cation, resonance serves to *decrease* the nucleophilicity of various anions by delocalizing their charge, thereby stabilizing the species. Included in these resonance stabilized anions are nitrate, $NO_3^{\ominus}$, sulfate, $SO_4^{\,(-2)}$, and phosphate, $PO_4^{\,(-3)}$. Remembering that resonance contributing structures differ only in their electron distribution, draw two resonance structures for each of these ions, trying to keep the number of formal charges in each contributor to a minimum. (A review of some of pages 42 to 47 may be of help, since resonance in ions was discussed there.) What is the maximum number of equivalent resonance structures possible for each of these ions?

Just to help you get started, the important resonance contributing structures to the carbonate anion, $CO_3^{\,(-2)}$, are given below. Notice how each oxygen gets a chance to be part of a C=O unit. We could have created more contributing structures, such as the one labeled *yuck*, but that one requires more charge separation than the others and is considered to be unimportant. As you may imagine from those three contributing structures, carbonate is a poor nucleophile.

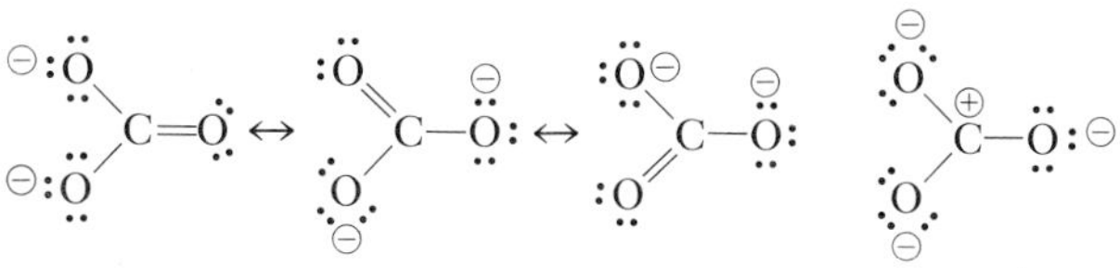

yuck

27. Benzyl chloride reacts with water to form benzyl alcohol at a rate slightly greater than the rate of reaction of allyl chloride with water to produce allyl alcohol. Both of these are S_N1 reactions. Explain.

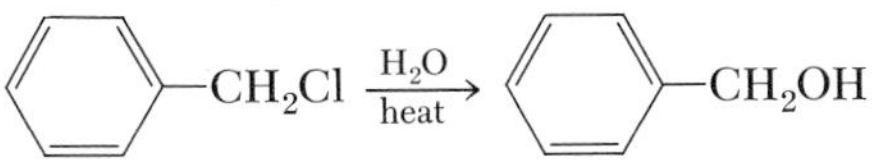

benzyl chloride   benzyl alcohol

$$CH_2{=}CHCH_2Cl \xrightarrow[\text{heat}]{H_2O} CH_2{=}CHCH_2OH$$

allyl chloride   allyl alcohol

28. After it is left standing in aqueous acid, (**R**)—$CH_3CH_2CHOHCH_3$ is observed to lose its optical activity. Account for this.
29. When either *meso-* or *d,l*-2,3-dichlorobutane is allowed to react with excess sodium iodide in acetone, the reaction produces a mixture of *meso-* and *d,l*-2,3-diiodobutane. Either starting material produces the same mixture. The amount of *meso* produced does not equal the amount of *d,l* produced. Explain these facts.
30. (2**R**,3**S**)-3-Methyl-2-pentanol will react with concentrated hydrobromic acid to produce rearranged and unrearranged alkyl bromides as products. Indicate the products, noting which will be optically active.
31. Suppose that we had defined the rate of racemization as the rate of *interconversion* of enantiomers, *i.e.*, $(+) \rightarrow (-)$, instead of defining it as the rate of conversion of an enantiomer into a racemic modification. How would this have altered the comparative rates of label incorporation and racemization in the reaction of radioactive iodide ion and 2-iodobutane? (See p. 123.) Explain your answer.
32. Chloroacetone, CH_3COCH_2Cl, undergoes bimolecular substitution faster than does *n*-propyl chloride, but it reacts with alcoholic silver nitrate (an S_N1 reaction) much more slowly. Account for these results.
33. The reaction shown below is an interesting displacement, particularly because the skeleton of the product differs from that of the starting material. Suggest a mechanism for this. (*Hint:* An intermediate, $C_8H_{18}NCl$, can, with care, be isolated from the reaction mixture. This intermediate is water soluble. If the starting material is optically active, the product also is optically active.)

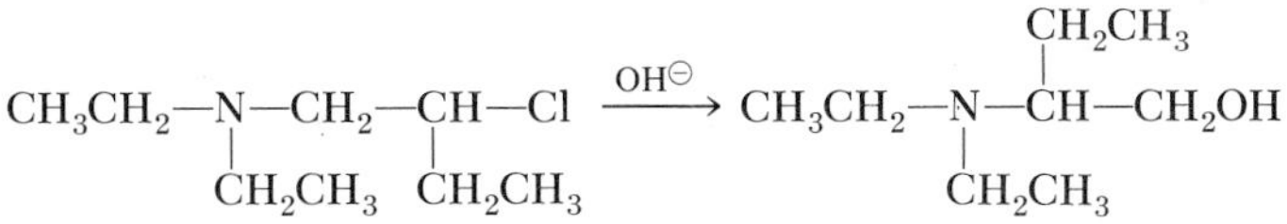

34. An increase in solvent polarity will increase the rate of S_N2 reaction of an amine, $R_3N\!:$, with an alkyl halide, R′X, to form an ammonium salt, $R_3\overset{\oplus}{N}R'\ X^{\ominus}$. Draw an energy profile for the reaction in a less polar solvent and in a more polar solvent. Using arguments built around the nature of the activated complex and the starting materials, account for this solvent effect.

$$R_3N + R'X \rightarrow R_3\overset{\oplus}{N}R'\ X^{\ominus} \qquad (S_N2)$$

35. Outline a scheme for isolating the enantiomers of **A** from a racemic modification of **A**. What complication(s) may accompany this optical resolution?

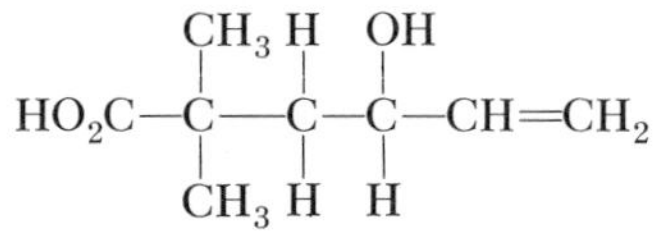

$$
\begin{array}{ccccccc}
 & & CH_3 & & H & & OH \\
 & & | & & | & & | \\
HO_2C & - & C & - & C & - & C - CH{=}CH_2 \\
 & & | & & | & & | \\
 & & CH_3 & & H & & H
\end{array}
$$

A

36. 3-Chloro-1-butene, $CH_3CHCl—CH{=}CH_2$, reacts with a solution of sodium ethoxide in ethanol ($Na^{\oplus}\ C_2H_5O^{\ominus}/C_2H_5OH$) to produce 3-ethoxy-1-butene, $CH_3CH(OC_2H_5)CH{=}CH_2$. The rate of reaction is second order overall, first order in 3-chloro-1-butene, and first order in ethoxide ion. In the absence of ethoxide ion, however, 3-chloro-1-butene reacts with ethanol to produce both 3-ethoxy-1-butene and 1-ethoxy-2-butene, $CH_3CH{=}CHCH_2OC_2H_5$. Explain these results.

37. Suggest a reason that pentacoordinated carbon is encountered only in energy-rich species, like the S_N2 transition state, while pentacoordinated phosphorus compounds, such as PCl_5, are fairly common, stable species. (*Hint:* Consider the orbitals available to phosphorus and carbon.)

The chemistry of organophosphorus compounds will be discussed in Chapter 23.

ALKYL HALIDES— ELIMINATION REACTIONS AND GRIGNARD REAGENTS

6

6.1 WHERE ARE WE GOING?

Few organic reactions proceed to a 100% yield, even after allowances are made for losses encountered during isolation, because it is uncommon to find a set of reaction conditions which will permit one reaction to occur to the total exclusion of another. The best that one can do is to select conditions which favor one reaction over competitive processes. Nucleophilic substitution reactions are no exception, having to compete for starting material with *elimination reactions*. Indeed, proper selection of reagents and reaction conditions may actually result in more starting material undergoing elimination than substitution. It is because of the intimate relationship between substitution and elimination that the bulk of this chapter is devoted to this second important ionic aliphatic reaction.

Once we have completed our study of elimination reactions, we shall take an introductory look at one of the most versatile types of reagents ever discovered—the Grignard reagent. These are introduced here in order to complete our study of the major uses of alkyl halides—as substrates for substitution, as substrates for elimination, and as substrates for the preparation of compounds containing a carbon-metal bond. Most of what might be called the "reaction chemistry" of Grignard reagents will be developed in future chapters.

6.2 ELIMINATION REACTIONS

The most common type of elimination reaction is one in which two fragments are removed from a substrate to produce a modified substrate and two smaller units. One of these fragments is usually recognizable as a "leaving group" of the type discussed in Chapter 5. Eliminations usually produce a new σ or π bond in the modified substrate.

$$\text{substrate} \rightarrow \text{modified substrate} + \text{fragment 1} + \text{fragment 2}$$

One means of classifying elimination reactions employs the Greek alphabet to designate the skeletal atoms bearing the groups which leave the substrate. The skeletal atom bearing the typical leaving group (for example, chlorine) is lettered $C\alpha$, and the one bearing the atypical leaving group (for example, hydrogen) is lettered sequentially from $C\alpha$. In those eliminations which involve two of the more usual leaving groups (for example, two chlorines), the designation of which atom is $C\alpha$ is arbitrary. Thus, elimination reactions may be considered to involve $C\alpha$ and $C\alpha$, $C\alpha$ and $C\beta$, $C\alpha$ and $C\gamma$, and so forth. For simplicity, such reactions are classified according to the higher letter designation and, using this approach, the reactions just mentioned would be called α-, β- and γ-eliminations, respectively.

If both departing fragments are bonded to the same skeletal atom, then the process, an α-elimination, produces a member of the electron deficient species known as *carbenes* (discussed in detail in Chapter 8). α-Elimination reactions are not very common (see problem 6.2).

$$R_2C_{\alpha}(Y)\text{—}L \rightarrow R\text{—}\ddot{C}\text{—}R + L + Y \qquad \alpha\text{-elimination}$$

a carbene

When the departing groups are on adjacent atoms, then the product, containing a new π bond, is formed *via* a β-elimination.

$$Y\text{—}C_{\beta}R_2\text{—}C_{\alpha}R_2\text{—}L \rightarrow R_2C{=}CR_2 + L + Y \qquad \beta\text{-elimination}$$

Elimination reactions involving more widely separated groups usually produce cyclic compounds or polymers. As the number of intervening atoms increases, the likelihood of polymer formation increases. (The polymer is a very large molecule made up of repeating units, each of which contains the same number of carbons as the starting material.)

$$\underset{\alpha}{L\text{—}CH_2}\text{—}\underset{\beta}{CH_2}\text{—}\underset{\gamma}{CH_2}\text{—}Y \rightarrow \underset{\text{cyclopropane}}{(CH_2)_3} + L + Y \qquad \gamma\text{-elimination}$$

$$nL\text{—}(CH_2)_{10}\text{—}Y \rightarrow \underset{\text{polymer}}{\text{—}(\text{—}(CH_2)_{10}\text{—})_n\text{—}} + nL + nY$$

1. Identify each of the following as an α, β, γ, or δ elimination.
   (a) $CH_3CH_2Br \rightarrow CH_2{=}CH_2$
   (b) $CH_3CH_2CHBrCH_3 \rightarrow CH_3CH{=}CHCH_3$
   (c) $BrCH_2CH_2CHBrCH_3 \rightarrow$ cyclo-$CH_2CH_2CHCH_3$ (three-membered ring: CH_2, CH_2, $CHCH_3$)
   (d) $CH_3CH{=}CHCH_2Br \rightarrow CH_2{=}CH\text{—}CH{=}CH_2$
   (e) $CH_2Br_2 \rightarrow CH_2\!:$
   (f) $CH_3CHBrCHClCH_3 \rightarrow CH_3CH{=}CHCH_3$

2. Considering only the number of bonds being made and broken, why should you expect an α-elimination leading to a carbene to be much slower (that is, have a higher energy of activation) than a β-elimination leading to an alkene?

$$\text{R—}\underset{\text{R}}{\overset{\text{H}}{\text{C}}}\text{—}\underset{\text{R}}{\overset{\text{H}}{\text{C}}}\text{—L} \xrightarrow[\text{(slower)}]{\alpha\text{-elimination}} \text{R—}\underset{\text{R}}{\overset{\text{H}}{\text{C}}}\text{—}\ddot{\text{C}}\text{—R} + \text{H—L}$$

$$\text{R—}\underset{\text{R}}{\overset{\text{H}}{\text{C}}}\text{—}\underset{\text{R}}{\overset{\text{H}}{\text{C}}}\text{—L} \xrightarrow[\text{(faster)}]{\beta\text{-elimination}} \text{R}_2\text{C=C(H)R} + \text{H—L}$$

Three types of mechanisms for achieving elimination will be discussed in this chapter: E2 (elimination, bimolecular); E1 (elimination, unimolecular); and E1cb (elimination, unimolecular, conjugate base). While they are treated separately, it will be shown that these three "different" pathways really represent three readily identifiable points along an entire spectrum of mechanistic types.

As you read the following pages, remember that elimination reactions are classified in two *independent* ways. One of them designates the relative positions of the two departing groups in the starting material. (This is the α, β designation.) The other does not consider the relative location of the two groups but, instead, focuses upon the mechanism of loss of these groups. (This is the E1, E2, E1cb designation.) It is possible, at least in principle, to encounter elimination reactions which are α-E1, α-E2, α-E1cb, β-E1, and so on. For a variety of reasons, some of which will be presented later in this chapter (see also problem 6.2), the β-elimination is *much* more common than is the α-elimination or higher eliminations. Furthermore, it will become apparent that the most prevalent mechanisms are those designated E1 and E2. Since this chapter will deal almost exclusively with β-eliminations, we will begin by discussing the β-E2 process of elimination.

6.3 E2 REACTIONS

When substitution (S_N2) is placed in an unfavorable competitive position, the *basicity* of the nucleophile can dictate the course of the reaction between an electron pair donor and an electron pair acceptor. Substitution at hydrogen becomes competitive with substitution at carbon, and a bimolecular β-elimination is the final result. The mechanism for this type of elimination is dubbed E2. While E2 reactions are, as a class, not limited to the loss of hydrogen (as $H^{\oplus}$ or a related species) and a halide ion, we shall concentrate upon these *dehydrohalogenations* in this chapter.

$$\text{R—}\underset{\text{R}}{\overset{\text{H}}{\text{C}_\beta}}\text{—}\underset{\text{X}}{\overset{\text{R}}{\text{C}_\alpha}}\text{—R} \xrightarrow[\text{base}]{-\text{HX}} \text{R}_2\text{C=CR}_2$$

dehydrohalogenation (a β-elimination reaction)

Instead of beginning with a set of vague generalities, let us develop a picture of this E2 process with a concrete example, the reaction of isopropyl bromide with ethoxide ion. When isopropyl bromide reacts with sodium ethoxide, the S_N2 product (ethyl isopropyl ether) accounts for only 20% of the *total* product. The remainder is propene, derived from an E2 reaction.

$$CH_3CH_2\ddot{\underset{..}{O}}{:}^{\ominus} + \text{H—}\underset{C_\beta H_3}{\overset{C_\beta H_3}{C_\alpha}}\text{—Br} \xrightarrow{C_2H_5OH}$$

isopropyl bromide

$$\xrightarrow[-\text{Br}^{\ominus}]{S_N2} CH_3CH_2OCH(CH_3)_2 \quad (20\%)$$

ethyl isopropyl ether

$$\xrightarrow[-\text{HBr}]{\text{E2}} CH_3CH{=}CH_2 \quad (80\%)$$

propene

Mechanism. In the E2 reaction between isopropyl bromide and ethoxide ion, the base ($C_2H_5O^{\ominus}$) attacks a hydrogen β to the leaving group ($Br^{\ominus}$) with an oxygen electron pair. As this is going on, the electron pair which constitutes the C_β—H bond "swings around" and attacks the carbon ($C\alpha$) adjacent to it. This carbon, unable to accommodate ten outer electrons, simultaneously begins to eject bromine along with the two electrons which constituted the C—Br bond. In this *concerted* process, a π bond is introduced between $C\alpha$ and $C\beta$, converting the bond between them from a single to a double bond.

As noted in Section 5.8, all electron pair movements are shown by curved arrows, as in the structures below. Recall that these curved arrows begin on a specific pair of electrons and point to where those electrons will go. When you see such arrows on a formula, remember that they depict the electron movement that will produce the next structure.

$$CH_3CH_2\ddot{\underset{\cdot\cdot}{O}}:^{\ominus} \;\; H-\underset{H}{\overset{H}{C_\beta}}-\underset{H}{\overset{CH_3}{C_\alpha}}-Br \rightarrow CH_3CH_2OH + Br^{\ominus} + H_2C{=}CHCH_3$$

The activated complex in this **concerted elimination** is pictured as having a partially formed O—H bond, a partially broken H—C_β bond, a partially formed C_α—C_β π bond, and a partially broken C_α—Br bond.

$$\overset{\delta-}{Et-\ddot{\underset{\cdot\cdot}{O}}}\cdots H\cdots\underset{H}{\overset{H}{C_\beta}}\overset{\cdots}{=}\underset{H}{\overset{CH_3}{C_\alpha}}\cdots\overset{\delta-}{Br}$$

activated complex for E2 loss of HBr from isopropyl bromide

The rate equation for the elimination is the same as for the S_N2 process, except that the specific rate constants are normally different. The rate of formation of propene from the elimination reaction is given by $R = k'[CH_3CHBrCH_3][CH_3CH_2O^{\ominus}]$. An energy profile for the reaction is given in Figure 6–1.

3. What is the rate equation for the formation of ethyl isopropyl ether from ethoxide anion and isopropyl bromide? What is the rate equation for the formation of propene from the same reagents? How would increasing the concentration of base two-fold influence these two competitive reactions?

Before moving on with our detailed discussion of the β-elimination which follows the E2 mechanism, let us examine a generalized presentation of this process. Although the electron-pair donor (base) may be neutral (*e.g.*, water) or negatively charged (*e.g.*, hydroxide ion), and although the group which departs may produce an anion (*e.g.*, chloride ion) or a neutral species (*e.g.*, water), the scheme below depicts only one possible combination (negatively charged base and anionic leaving group).

$$\underset{\text{base}}{B:^{\ominus}} \;\; \underset{\text{substrate}}{H-\overset{|}{\underset{|}{C}}-\overset{|}{\underset{|}{C}}-L} \rightarrow \underset{\text{transition state}}{\left[B\cdots H\cdots\overset{|}{\underset{|}{C}}\overset{\cdots}{=}\overset{|}{\underset{|}{C}}\cdots L\right]^{\ominus}} \rightarrow B-H + \underset{\text{alkene}}{\rangle C{=}C\langle} + L:^{\ominus}$$

The rate equation for this process is $R = k'[B^{\ominus}][H-\overset{|}{\underset{|}{C}}-\overset{|}{\underset{|}{C}}-L]$, where R is the rate of reaction and k' is the specific rate constant for a given combination of base and substrate.

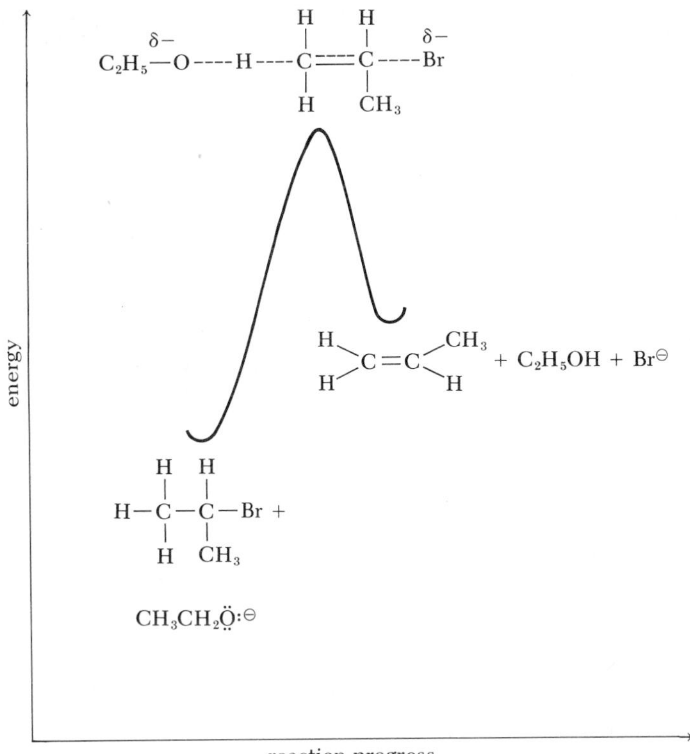

Figure 6–1 The E2 reaction. A considerable amount of double bond character has developed in the activated complex.

Isotope Effect. It is a tenet of all elementary approaches to chemical reactivity that reactions of a particular element are dictated by its electron configuration and that isotopes of the same element, since they have the same electron configuration, exhibit identical chemical properties. This extrapolation, from identical electronic structure to identical chemical reactivity, is incorrect.

When we speak of *chemical reactivity*, we must consider both the process going on and the rate at which it is occurring. While isotopes undergo similar reactions, they usually undergo them at different rates. A detailed explanation of this is beyond our needs; however, this difference is partly due to the lower energy (more stable) bond that is formed between one element and the heavier of two isotopes of another element. For example, the C—D bond is more stable than the C—H bond.

Bond strength differences persist in the transition states of reactions involving isotopes. However, in a linear, symmetric transfer of an isotope from one atom to another, a situation approximated by the proton transfer from $C\beta$ to the base in the E2 reaction, differences in bonding in the activated complex are unimportant. Thus, only the bond strength differences which occur in the ground state need be considered in discussing the effect of isotopic substitution on the rate of the E2 process. Since lowering the ground state energy, without significantly altering the transition state energy, should increase the energy of activation, it is anticipated that bonds to protium are broken faster than corresponding bonds to deuterium (Figure 6–2, p. 162).

base · · · H · · · C
base · · · D · · · C

activated complexes for H and D transfer in the E2 reaction

This, the **kinetic isotope effect,** can be used to demonstrate that a carbon-hydrogen bond is broken in the activated complex of an E2 reaction. When isopropyl bromide and its hexadeuterated analog are subjected to the E2 elimination, it is observed that $CD_3CHBrCD_3$ reacts at 0.15 times the rate of $CH_3CHBrCH_3$. The slower reaction rate of the deuterated compound is consistent only with breakage of the C—D bond in the activated complex.

D CD$_3$
δ^- | | δ^-
EtO···D···C⋯C···Br
| |
D H

1,1,1,3,3,3-hexadeuterioisopropyl bromide shows a kinetic isotope effect

While isotopes are quite useful in studying reaction mechanisms, isotopic substitution does not always lead to an *observable* isotope effect. This derives from the fact that the effect is greatest when a change in isotopes produces the largest percentage change in atomic mass. Clearly, the ^1H to ^3H substitution should always produce the largest isotope effect since this is the largest percentage change possible. Owing to the difficulties and expense of working with tritium, the organic chemist usually settles for the 100% mass change involving ^1H and ^2H. Of course, you also will not get an isotope effect if the bond to the isotopic atom is not broken in the rate-determining step.

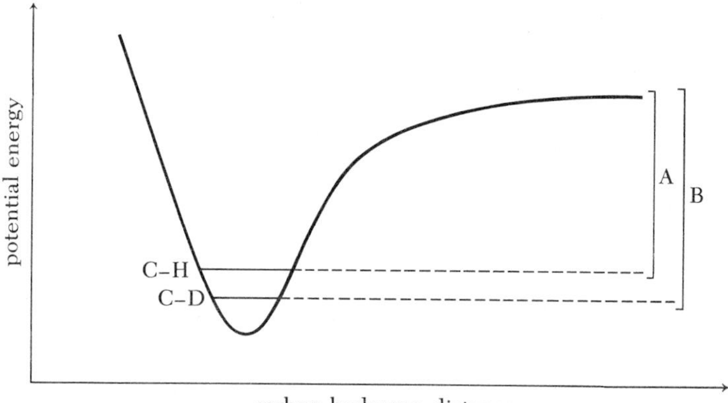

Figure 6-2 The isotope effect and bond dissociation energy. While both C—H and C—D cleavage follow the same profile, the C—D bond vibrates at a lower frequency; thus, the bond dissociation energy of the C—D bond is greater than that of the C—H bond. A similar situation makes the D—D bond more stable than the H—H bond by 2 kcal/mole. A = BDE of the C—H bond; B = BDE of the C—D bond. (You may wish to review the pertinent section in Chapter 2, and especially Figure 2-15.)

Stereoelectronic Constraints upon the E2 Reaction. The bonds being broken in the activated complex of an E2 reaction should be *anti* to one another. This geometry permits the electron pair being generated at one carbon (Cβ) to attack the backside of the bonding molecular orbital of the C_α-leaving bond. Coplanarity of the two bonds of the substrate which are broken in the activated complex also assures maximum π-type overlap of the resultant p orbitals during π bond formation. This added dimension of the E2 reaction is shown below, again using isopropyl bromide as the substrate, in both wedge and Newman-like projections.

+EtOH + Br$^\ominus$

(side view)

π-bond

(end view)

+EtOH+Br$^\ominus$

anti-elimination of HBr from isopropyl bromide

activated complex

propene

A **stereospecific reaction** is one in which a given stereoisomeric compound produces one stereoisomer in preference to another. For example, the S_N2 reaction of (2**R**,3**S**)-2-chloro-3-methylpentane with bromide ion produces only (2**S**,3**S**)-2-bromo-3-methylpentane and not the isomeric (2**R**,3**S**)-2-bromo-3-methylpentane. Therefore, this displacement is a stereospecific reaction.

(2**R**,3**S**)-2-chloro-3-methylpentane —[$Br^{\ominus}$ / $-Cl^{\ominus}$]→

(2**S**,3**S**)-2-bromo-3-methylpentane — observed product

(2**R**,3**S**)-2-bromo-3-methylpentane — not observed

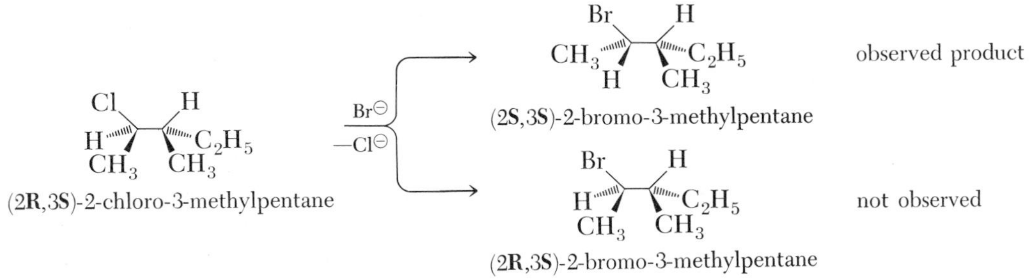

The proof of the existence of a stereoelectronic constraint upon the E2 reaction is its stereospecificity. For example, (1**R**,2**S**)-1-bromo-1,2-diphenylpropane undergoes the E2 loss of HBr to give only one alkene, *cis*-1,2-diphenylpropene; the *trans* isomer is not formed.

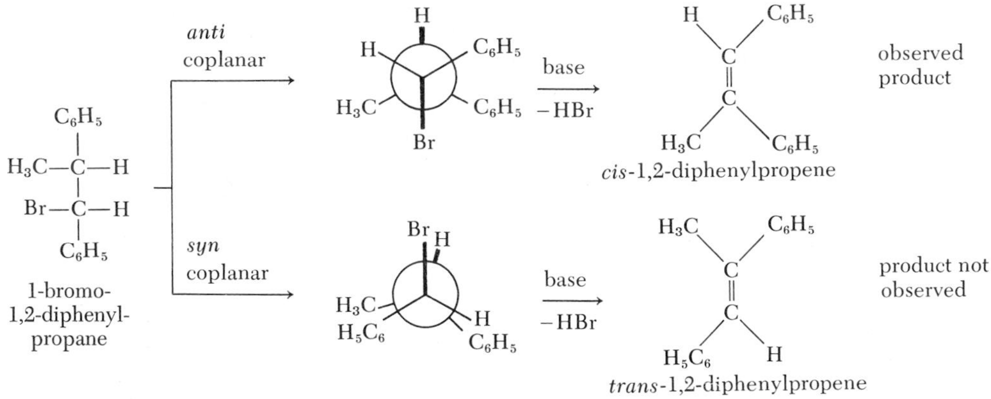

1-bromo-1,2-diphenylpropane —*anti* coplanar→ —base, −HBr→ *cis*-1,2-diphenylpropene — observed product

1-bromo-1,2-diphenylpropane —*syn* coplanar→ —base, −HBr→ *trans*-1,2-diphenylpropene — product not observed

Although the naming of alkenes is discussed in Chapter 8, it now is necessary for us to know that if two identical groups occur on the same side of the plane containing the π bond, they are said to be *cis* to one another. If the two identical groups are on opposite sides of the π bond, they are said to be *trans* to one another. (These prefixes are used in the example above.) The E2 reaction sometimes is described as being a *trans* elimination reaction. However, this is because of the relative position of the two groups being eliminated (*trans* and *anti* often are used interchangeably when discussing the conformation around a single bond) and NOT because of the geometry of the product alkene.

4. The reaction of (±)-2,3-dibromobutane with ethoxide ion produces *trans*-2-bromo-2-butene while, under the same conditions, *meso*-2,3-dibromobutane produces the corresponding *cis* isomer. With the aid of three-dimensional representations, determine whether this is an *anti* (*trans*) elimination. Explain how it is possible for a "*trans*" elimination to produce both *cis* and *trans* product.

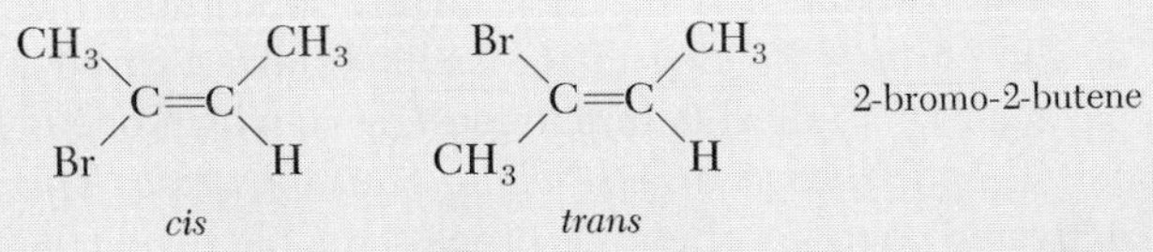

2-bromo-2-butene: *cis* and *trans*

Problem: An ϵ (*epsilon*)-E2 elimination should, in principle, compete with a β-E2 elimination. Such a competition is depicted below. Since the hydrogens β and ϵ to the leaving group (Br) are of about the same acidity, suggest a reason that ϵ-elimination, to form a five-membered ring, is much slower than is β-elimination.

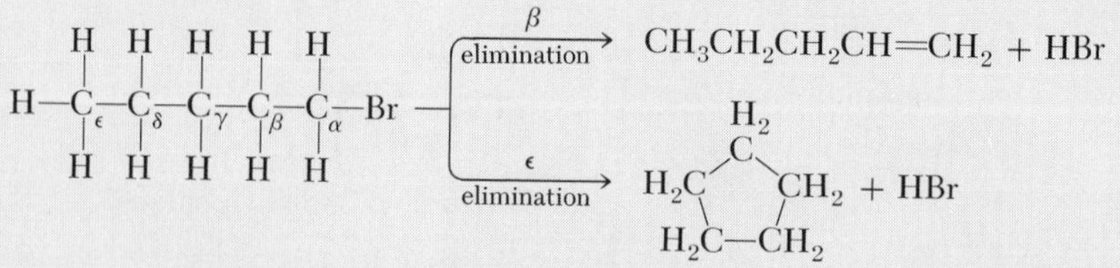

$$H\text{—}\underset{H}{\overset{H}{C}}_{\epsilon}\text{—}\underset{H}{\overset{H}{C}}_{\delta}\text{—}\underset{H}{\overset{H}{C}}_{\gamma}\text{—}\underset{H}{\overset{H}{C}}_{\beta}\text{—}\underset{H}{\overset{H}{C}}_{\alpha}\text{—}Br \xrightarrow[\text{elimination}]{\beta} CH_3CH_2CH_2CH{=}CH_2 + HBr$$

$$\xrightarrow[\text{elimination}]{\epsilon} \text{cyclo-}(CH_2)_5 \; (H_2C\text{—}CH_2\text{—}CH_2\text{—}CH_2\text{—}CH_2 \text{ ring}) + HBr$$

Solution: In order to achieve an ϵ-elimination, the molecule must restrict the conformation about a number of otherwise freely rotating carbon-carbon bonds. This necessarily increases the energy of activation for the process by increasing the entropy (orderliness) of the substrate in the activated complex ($\Delta G_{act} = \Delta H_{act} - T\Delta S_{act}$).

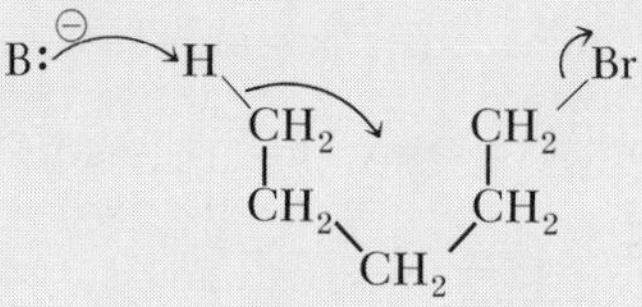

$$B:^{\ominus} \curvearrowright H\text{—}CH_2\text{—}CH_2\text{—}CH_2\text{—}CH_2\text{—}CH_2\text{—}Br$$

successful ϵ-elimination requires restriction of rotation about the four bonds marked with heavy lines

In order to achieve a β-elimination, the molecule has only to restrict rotation around one carbon-carbon bond. Therefore, the increase in the energy of activation due to entropy considerations is less for the β-elimination than for the ϵ-elimination.

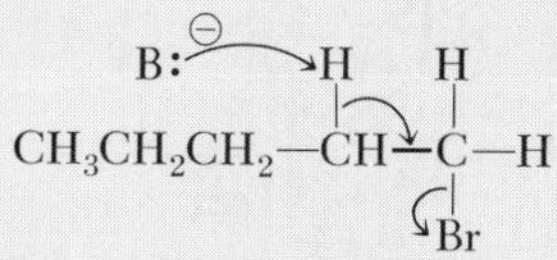

$$B:^{\ominus} \curvearrowright H \quad CH_3CH_2CH_2\text{—}\overset{H}{CH}\text{—}\underset{Br}{\overset{H}{C}}\text{—}H$$

Can you combine this information with your answer to problem 2 and explain why β-eliminations are the most common type of elimination?

Direction of Elimination—Saytzeff and Hofmann. When only one type of β hydrogen is present, as in isopropyl bromide, the direction of elimination is certain. Isopropyl bromide can produce only propene.

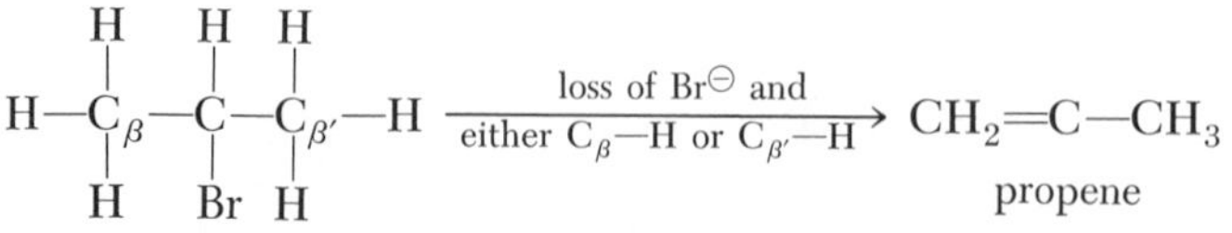

$$H\text{—}\underset{H}{\overset{H}{C}}_{\beta}\text{—}\underset{Br}{\overset{H}{C}}\text{—}\underset{H}{\overset{H}{C}}_{\beta'}\text{—}H \xrightarrow[\text{either } C_{\beta}\text{—}H \text{ or } C_{\beta'}\text{—}H]{\text{loss of } Br^{\ominus} \text{ and}} \underset{\text{propene}}{CH_2{=}C\text{—}CH_3}$$

When all β hydrogens cannot be made equally available by bond rotation, as in 2-bromobutane, the alkene normally produced is the most highly substituted one, *i.e.*, the one with the largest number of alkyl groups bonded to the sp^2 hybridized carbons. This, a statement of **Saytzeff's rule,** is particularly applicable to eliminations in which one of the leaving groups is an anion. Thus, *most dehydrohalogenations follow Saytzeff's rule.* This behavior supports the view that in the activated complex of the E2 process, double bond character ($C_{sp^3} \rightarrow C_{sp^2}$)

is already present, since increasing alkylation normally leads to more stable double bonds. Two illustrations of Saytzeff's rule are shown below.

$$CH_3\underset{\substack{|\\Br}}{C}HCH_2CH_3 \xrightarrow{C_2H_5O^{\ominus}} \underset{\substack{\text{2-butene}\\ \text{81\% of alkene product}\\ \text{Saytzeff product}}}{CH_3CH{=}CHCH_3} + \underset{\substack{\text{1-butene}\\ \text{19\% of alkene product}}}{CH_3CH_2CH{=}CH_2}$$

$$CH_3CH_2\overset{\substack{CH_3\\|}}{\underset{\substack{|\\Br}}{C}}CH_3 \xrightarrow{C_2H_5O^{\ominus}} \underset{\substack{\text{2-methyl-2-butene}\\ \text{62\% of alkene product}\\ \text{Saytzeff product}}}{CH_3CH{=}C(CH_3)_2} + \underset{\substack{\text{2-methyl-1-butene}\\ \text{38\% of alkene product}}}{CH_3CH_2C(CH_3){=}CH_2}$$

While many eliminations produce a preponderance of the most highly substituted alkene, there are some which follow **Hofmann's rule** and produce a preponderance of the least highly substituted alkene. For example, if a substrate were to produce more 1-butene than 2-butene, it would be described as following Hofmann's rule rather than Saytzeff's rule. *Hofmann's rule usually is operative when the leaving group is a positively charged species* (*e.g.*, $-\overset{\oplus}{S}(CH_3)_2$ or $-\overset{\oplus}{N}(CH_3)_3$), that is, when it departs as a neutral molecule.

net reaction:

$$(CH_3)_3\overset{\oplus}{N}{-}CHCH_3CH_2CH_3 \quad \overset{\ominus}{O}H \xrightarrow{\text{heat}} (CH_3)_3N + H_2O + \underset{\text{1-butene (95\%)}}{CH_2{=}CHCH_2CH_3} + \underset{\text{2-butene (5\%)}}{CH_3CH{=}CHCH_3}$$

mechanism:

$$(CH_3)_3\overset{\oplus}{N}{-}\underset{\substack{|\\CH_2\\|\\H\quad \overset{\ominus}{O}H}}{CH}{-}\overset{\substack{H\quad \overset{\ominus}{O}H\\|}}{C}HCH_3 \xrightarrow{E2} \begin{cases} \xrightarrow{\text{path a}} (CH_3)_3N + \underset{\text{1-butene}}{CH_2{=}CHCH_2CH_3} + H_2O \\ \xrightarrow{\text{path b}} (CH_3)_3N + \underset{\text{2-butene}}{CH_3CH{=}CHCH_3} + H_2O \end{cases}$$

$$(CH_3)_2CH{-}\underset{\substack{|\\ ^{\oplus}S(CH_3)_2}}{CH}{-}CH_3 \xrightarrow[\text{heat}]{^{\ominus}OH} (CH_3)_2CHCH{=}CH_2 > (CH_3)_2C{=}CHCH_3$$

The reason that positively charged leaving groups induce Hofmann-type elimination, rather than the Saytzeff-type elimination commonly associated with halide leaving groups, is complex and will not be discussed in detail. However, it is related to the fact that positively charged leaving groups are much larger than are simple halides. The effect of size upon the balance between Saytzeff and Hofmann elimination can be demonstrated in other ways. For example, increasing the bulk of the base or the size of the substituents at Cα and Cβ increases the amount of Hofmann product even if the leaving group is an anion. However, the examples below illustrate that *large* steric alterations are required in order to provide a significant amount of Hofmann product.

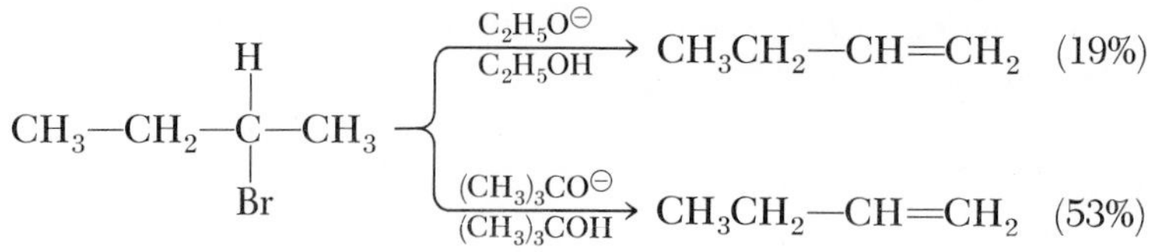

$$CH_3{-}CH_2{-}\underset{\large Br}{\overset{\large H}{C}}{-}CH_3 \begin{cases} \xrightarrow[C_2H_5OH]{C_2H_5O^{\ominus}} CH_3CH_2{-}CH{=}CH_2 \quad (19\%) \\ \xrightarrow[(CH_3)_3COH]{(CH_3)_3CO^{\ominus}} CH_3CH_2{-}CH{=}CH_2 \quad (53\%) \end{cases}$$

effect of varying base size

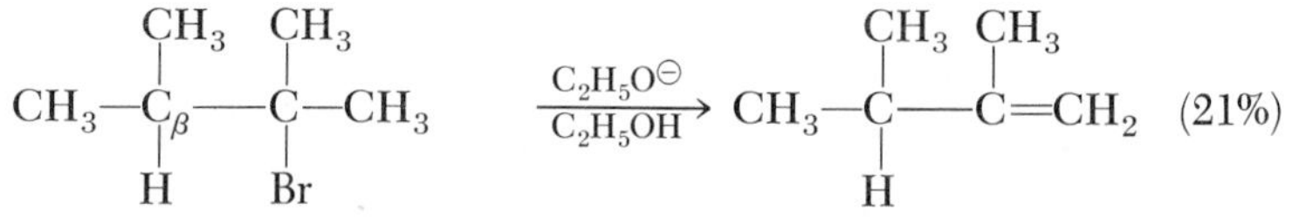

$$CH_3{-}\underset{\large H}{\overset{\large CH_3}{C_\beta}}{-}\underset{\large Br}{\overset{\large CH_3}{C}}{-}CH_3 \xrightarrow[C_2H_5OH]{C_2H_5O^{\ominus}} CH_3{-}\underset{\large H}{\overset{\large CH_3}{C}}{-}\overset{\large CH_3}{C}{=}CH_2 \quad (21\%)$$

(two methyl groups bonded to $C\beta$)

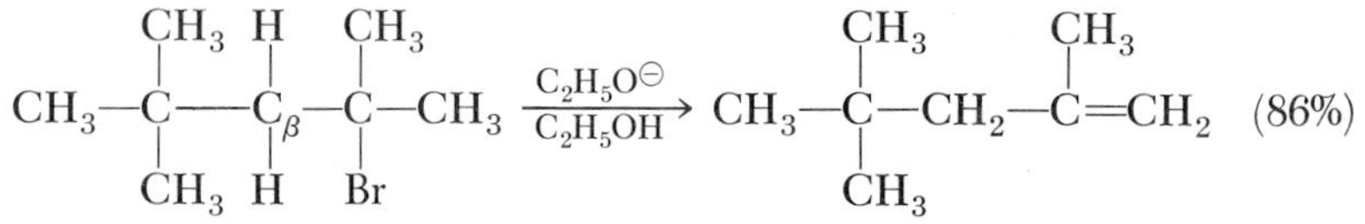

$$CH_3{-}\underset{\large CH_3}{\overset{\large CH_3}{C}}{-}\underset{\large H}{\overset{\large H}{C_\beta}}{-}\underset{\large Br}{\overset{\large CH_3}{C}}{-}CH_3 \xrightarrow[C_2H_5OH]{C_2H_5O^{\ominus}} CH_3{-}\underset{\large CH_3}{\overset{\large CH_3}{C}}{-}CH_2{-}\overset{\large CH_3}{C}{=}CH_2 \quad (86\%)$$

(t-butyl group bonded to $C\beta$)

effect of varying substituents at $C\beta$

5. For each of the following, indicate whether the substrate is following Saytzeff's or Hofmann's rule.
   (a) $CH_3CHBrCH_2CH_3 \rightarrow CH_2{=}CHCH_2CH_3$
   (b) $CH_3CHBrCH_2CH_3 \rightarrow CH_3CH{=}CHCH_3$
   (c) $CH_3CH_2CH(CH_3)\overset{\oplus}{S}(CH_3)_2 \rightarrow CH_3CH_2CH{=}CH_2\ (+CH_3SCH_3)$
   (d) $CH_3CH_2\overset{\oplus}{N}(CH_3)_2CH_2CH_2CH_3 \rightarrow CH_2{=}CH_2\ (+CH_3CH_2CH_2N(CH_3)_2)$
   (e) $CH_3CH_2C(CH_3)_2Br \rightarrow CH_3CH{=}C(CH_3)_2$

Which Geometric Isomer Is Formed? If both *cis* and *trans* alkenes can be formed in an E2 reaction, which isomer is expected to predominate? To answer this we must determine which process, formation of *cis* alkene or formation of *trans* alkene, has the lower energy of activation. The reaction with the lower E_{act} will go faster and, therefore, will produce the majority of the product. The activated complex for the ideal E2 reaction has both departing groups (*e.g.*, H and Cl in a dehydrochlorination) well on their way to being lost *and* possesses a substantial amount of double bond character between the two carbons forming the π bond. Consequently, it is believed that the activated complex will have a geometry resembling that of an alkene, with the non-leaving groups (Z^1 to Z^4, below) nearly coplanar (refer back to drawings on p. 162).

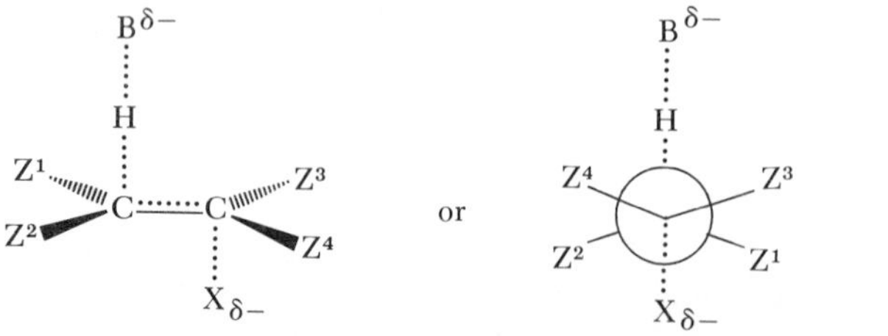

$B^{\delta-}$ ⋮ H ⋮ Z^1, Z^2 C······C Z^3, Z^4 ⋮ $X_{\delta-}$ or $B^{\delta-}$ ⋮ H ⋮ Z^4 Z^3 Z^2 Z^1 ⋮ $X_{\delta-}$

E2 activated complex has a geometry similar to that of the product alkene

Alkenes with larger substituents on the same side of the double bond usually are less stable than alkenes with larger substituents on opposite sides of the double bond (*e.g.*, *cis*-2-butene is less stable than *trans*-2-butene). The extra energy of the *cis* isomers arises from steric repulsion (van der Waals strain) between eclipsed substituents. Applying this idea to the transition state leading to isomeric alkenes, the "*cis* transition state" should be more energetic than the "*trans* transition state," and production of *cis* alkene should have a higher E_{act}. An illustration of this is the dehydrobromination of 2-bromopentane, an E2 reaction which produces more of the alkene that has an internal double bond. Only one-quarter of this Saytzeff product is the *cis* isomer; the majority of it is the *trans* alkene.

$$\underset{\text{2-bromopentane}}{CH_3CH_2CH_2\text{—}CHBr\text{—}CH_3} \xrightarrow[E2]{-HBr} \underset{\substack{\text{Saytzeff product}\\\text{(2-pentene)}}}{CH_3CH_2CH\text{=}CHCH_3} + \underset{\substack{\text{Hofmann product}\\\text{(1-pentene)}}}{CH_3CH_2CH_2CH\text{=}CH_2}$$

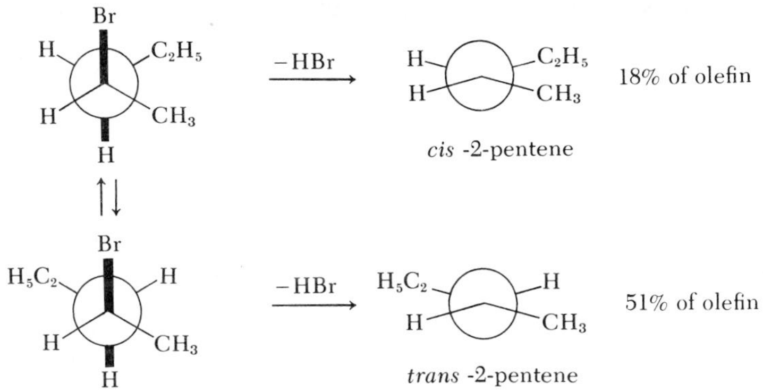

Br; H, C_2H_5; H, CH_3; H —HBr→ H, C_2H_5; H, CH_3 — 18% of olefin

cis -2-pentene

⇅

Br; H_5C_2, H; H, CH_3; H —HBr→ H_5C_2, H; H, CH_3 — 51% of olefin

trans -2-pentene

6. Treatment of 2,3-dibromobutane with iodide ion produces 2-butene, $CH_3CH\text{=}CHCH_3$. Compare the behavior of *meso*- and *d,l*-2,3-dibromobutane with iodide ion in this E2 process.

$$CH_3CHBrCHBrCH_3 + I^{\ominus} \rightarrow CH_3CH\text{=}CHCH_3 + IBr + Br^{\ominus}$$

7. Suggest a substrate-base combination which can lead to the given product by an E2 reaction:
   (a) $CH_2\text{=}CH_2$
   (b) $CH_3CH\text{=}CH_2$
   (c) $(CH_3)_2C\text{=}CH_2$
   (d) $CH_3CH\text{=}CHCH_2CH_3$
   (e) $CH_3CH\text{=}CHCH_3$
   (f) $(CH_3)_2CH\text{—}C(CH_3)\text{=}CH_2$
   (g) $CH_2\text{=}CH\text{—}CH_2CH_2CH_3$

8. Treatment of 2-bromo-2-methylbutane with potassium *t*-butoxide ($K^{\oplus}(CH_3)_3CO^{\ominus}$) dissolved in *t*-butyl alcohol produces a mixture of 69% 2-methyl-1-butene and 31% 2-methyl-2-butene. These product ratios are reversed when potassium ethoxide dissolved in ethanol replaces potassium *t*-butoxide in *t*-butanol. What conclusion can you reach from the effect of a change in base upon the course of these E2 elimination reactions?

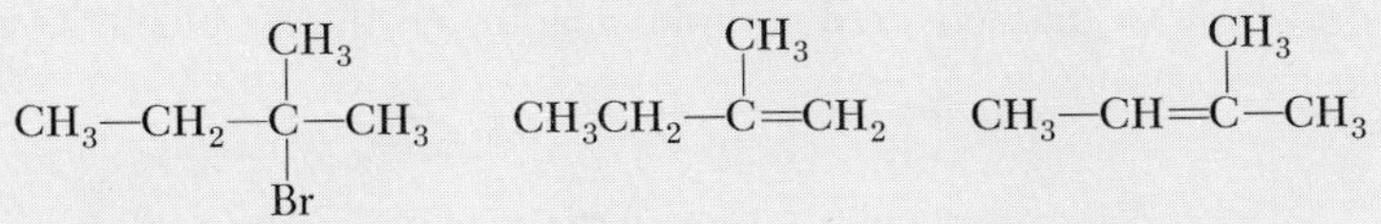

$$\underset{\text{2-bromo-2-methylbutane}}{CH_3\text{—}CH_2\text{—}\overset{\overset{CH_3}{|}}{\underset{\underset{Br}{|}}{C}}\text{—}CH_3} \qquad \underset{\text{2-methyl-1-butene}}{CH_3CH_2\text{—}\overset{\overset{CH_3}{|}}{C}\text{=}CH_2} \qquad \underset{\text{2-methyl-2-butene}}{CH_3\text{—}CH\text{=}\overset{\overset{CH_3}{|}}{C}\text{—}CH_3}$$

Before moving to a discussion of the β-E1 reaction, we will briefly introduce one more new term, *regiospecific* (pronounced rē·jē·ō·specific). A reaction is regiospecific when it gives exclusively one of several possible products from the viewpoint of orientation. For example, if an elimination gives exclusively Saytzeff product (and no Hofmann product), then that process is designated regiospecific. Should a process only give more of one possible product than of another, it is designated *regioselective*. Most of the eliminations discussed in this chapter are regioselective.

6.4 E1 REACTIONS

We learned in the previous chapter that alkyl halides, alkyl tosylates, protonated alcohols, and other alkane derivatives can ionize to carbocations, and that these cations can then combine with a nucleophile to complete the sequence known as the S_N1 reaction. We also saw that some carbocations could rearrange to a more stable cation prior to reacting with a nucleophile. The yields in such reactions often are not very high, with alkenes representing a major portion of the reaction by-products. These alkenes are produced by unimolecular elimination reactions, dubbed E1 reactions. Not only are E1 reactions competitive with S_N1 reactions, but they have the same rate determining step, the carbocation formation. The mechanism of formation of the alkene is straightforward, involving the loss of a proton from a carbon adjacent to the cationic center (Cβ) and simultaneous formation of a π bond between Cα and Cβ.

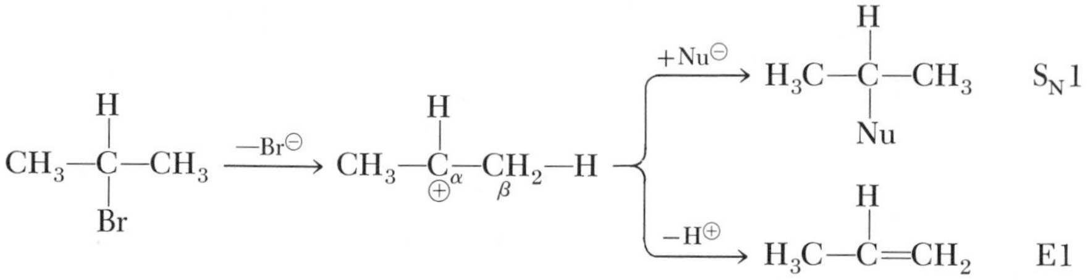

$$CH_3-CH(Br)-CH_3 \xrightarrow{-Br^{\ominus}} CH_3-\overset{\oplus}{C}H_{\alpha}-CH_{2\,\beta}-H \begin{cases} \xrightarrow{+Nu^{\ominus}} H_3C-CH(Nu)-CH_3 & S_N1 \\ \xrightarrow{-H^{\oplus}} H_3C-CH{=}CH_2 & E1 \end{cases}$$

mechanism of elimination:

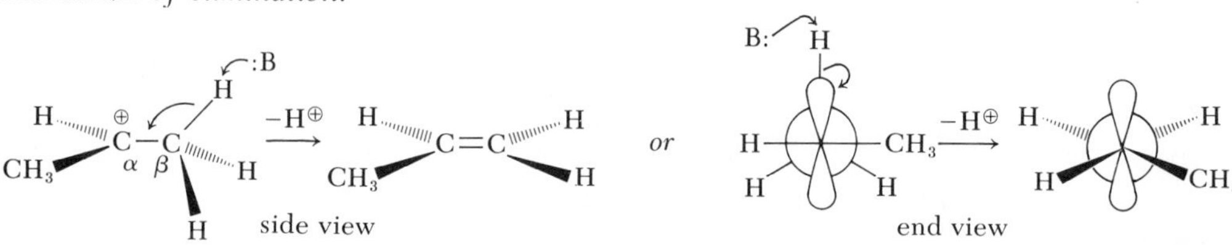

side view: H, CH_3 on $\overset{\oplus}{C}_\alpha$–$C_\beta$(H)(H)(H), B: removes H $\xrightarrow{-H^{\oplus}}$ $H(CH_3)C{=}CH_2$ *or* end view: B: removes H; H, H, CH_3 $\xrightarrow{-H^{\oplus}}$ H, CH_3, H, H alkene

If several β hydrogens are available, the one producing the most stable olefin usually is the one which is lost, *i.e., Saytzeff's rule is generally followed in E1 reactions.*

$$CH_3-CH_2-CH(Br)-CH_3 \xrightarrow[-HBr]{E1} \underset{\text{trans} > \text{cis}}{CH_3CH{=}CHCH_3} \gg CH_3CH_2CH{=}CH_2$$

Generally, *trans* substituted double bonds are formed more readily than are *cis* double bonds because of strain in the activated complex leading to the *cis* olefin. This parallels our commentary on E2 reactions.

$$CH_3CH_2CHBrCH_3 \longrightarrow CH_3CH_2\overset{\oplus}{C}HCH_3 \longrightarrow \textit{trans} > \textit{cis}$$

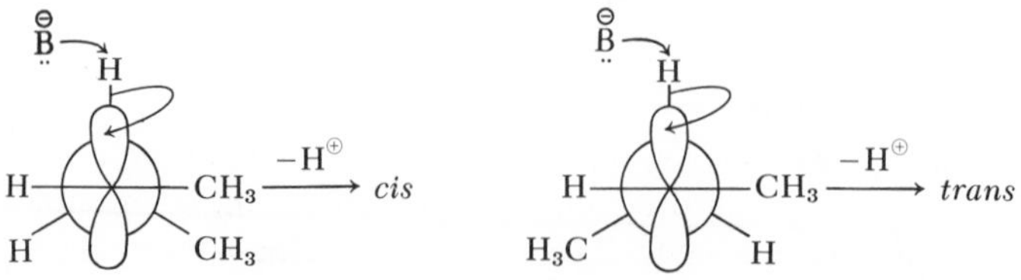

$\ddot{B}^{\ominus}$ removes H; H, H, CH_3, CH_3 $\xrightarrow{-H^{\oplus}}$ *cis*

$\ddot{B}^{\ominus}$ removes H; H, H_3C, CH_3, H $\xrightarrow{-H^{\oplus}}$ *trans*

The E1 reaction of 2-chloro-2,4,4-trimethylpentane is interesting because it produces an excess of the *less* substituted olefin. However, this still goes through the more stable activated complex. In this instance the steric repulsions are greatest in the transition state leading to the *most* highly substituted product. This is an example of a steric effect which leads to Hofmann elimination in an E1 process.

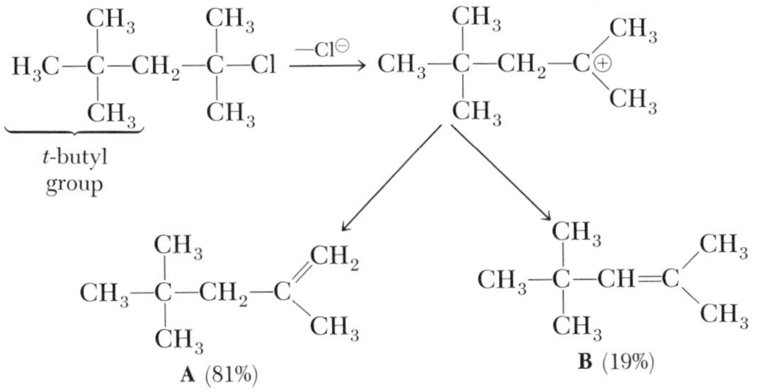

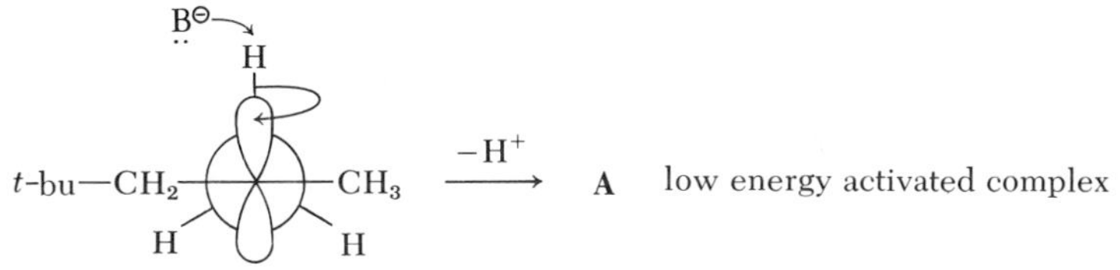

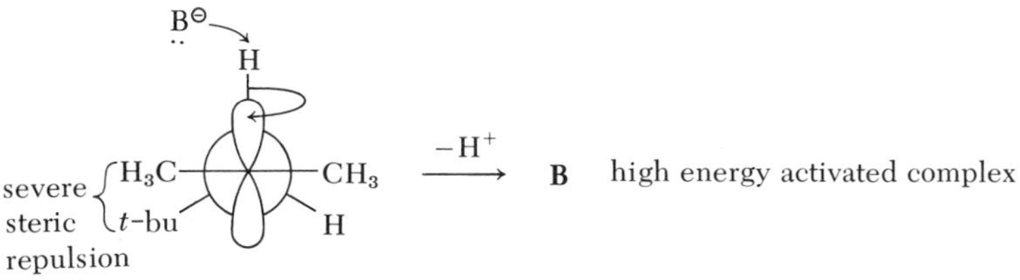

$$(CH_3)_3C\text{—}CH_2\text{—}C(CH_3)_2\text{—}Cl \xrightarrow{-Cl^{\ominus}} CH_3\text{—}C(CH_3)_2\text{—}CH_2\text{—}C^{\oplus}(CH_3)_2$$

t-butyl group

$$CH_3\text{—}C(CH_3)_2\text{—}CH_2\text{—}C(=CH_2)CH_3 \quad \textbf{A}\ (81\%)$$

$$CH_3\text{—}C(CH_3)_2\text{—}CH=C(CH_3)_2 \quad \textbf{B}\ (19\%)$$

B⊖ → H; *t*-bu—CH₂, CH₃, H, H $\xrightarrow{-H^+}$ **A** low energy activated complex

B⊖ → H; H₃C, *t*-bu (severe steric repulsion), CH₃, H $\xrightarrow{-H^+}$ **B** high energy activated complex

Other illustrations of the E1 reaction are shown below. This is a good time to note that E2 reactions usually give better yields of alkenes than do E1 reactions and, therefore, that E2 reactions are the so-called "method of choice" for synthesizing alkenes.

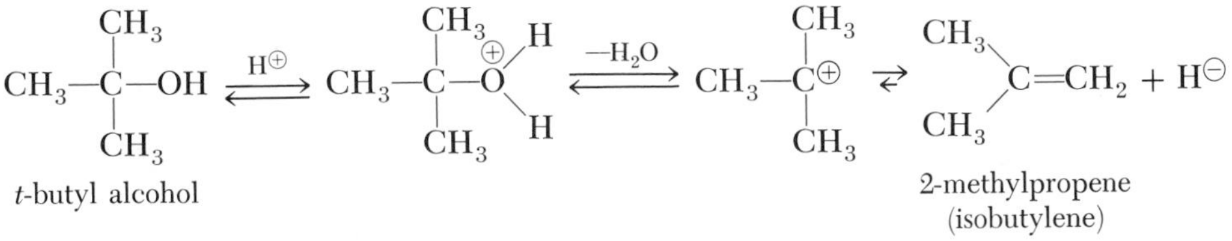

$$(CH_3)_3C\text{—}OH \underset{}{\overset{H^{\oplus}}{\rightleftharpoons}} (CH_3)_3C\text{—}\overset{\oplus}{O}H_2 \underset{}{\overset{-H_2O}{\rightleftharpoons}} (CH_3)_3C^{\oplus} \rightleftharpoons (CH_3)_2C=CH_2 + H^{\ominus}$$

t-butyl alcohol — 2-methylpropene (isobutylene)

$$(CH_3)_3CBr \xrightarrow[70°C]{80\%\ C_2H_5OH} (CH_3)_3\overset{\oplus}{C} \rightarrow (CH_3)_2C=CH_2$$

elimination represents only 35% of total product

$$CH_3CH_2CBr(CH_3)_2 \xrightarrow[35°C]{C_2H_5OH}$$

$$CH_3CH=C(CH_3)_2 > CH_3CH_2C(CH_3)=CH_2$$

2-methyl-2-butene : 2-methyl-1-butene = 4 : 1

elimination represents only 34% of total product

$$(CH_3)_3C-\overset{\oplus}{S}(CH_3)_2 \xrightarrow{80\%\ C_2H_5OH} (CH_3)_2C{=}CH_2$$ elimination represents only 35% of total products

$$CH_3CH_2-\underset{\oplus S(CH_3)_2}{\overset{CH_3}{\underset{|}{\overset{|}{C}}}}-CH_3 \xrightarrow{C_2H_5OH} \underset{\text{87\% of alkene product}}{CH_3CH{=}C(CH_3)_2} + \underset{\text{13\% of alkene product}}{CH_3CH_2C(CH_3){=}CH_2}$$

E1 reactions, like their S_N1 counterparts, are complicated by cationic rearrangements. For example, in the acid-catalyzed conversion of 3-methyl-2-butanol to alkenes, the formation of 2-methyl-1-butene is best accounted for by a 1,2-hydride shift prior to proton loss.

net reaction:

$$\underset{\text{3-methyl-2-butanol}}{CH_3-\underset{H}{\overset{CH_3}{C}}-\underset{H}{\overset{OH}{C}}-CH_3} \xrightarrow[-H_2O]{H^{\oplus}}$$

$$\underset{\text{2-methyl-2-butene}}{CH_3-\overset{CH_3}{C}{=}\overset{H}{C}-CH_3} + \underset{\text{3-methyl-1-butene}}{CH_3-\underset{H}{\overset{CH_3}{C}}-\overset{H}{C}{=}\overset{H}{C}-H} + \underset{\text{2-methyl-1-butene}}{H-\overset{H}{C}{=}\overset{CH_3}{C}-CH_2CH_3}$$

mechanism:

$$\underset{\text{3-methyl-2-butanol}}{CH_3-\underset{H}{\overset{CH_3}{C}}-\underset{OH}{\overset{H}{C}}-CH_3} \xrightarrow{H^{\oplus}} CH_3-\underset{H}{\overset{CH_3}{C}}-\underset{\oplus OH_2}{\overset{H}{C}}-CH_3 \xrightarrow{-H_2O} CH_3-\underset{H}{\overset{CH_3}{C}}-\underset{\oplus}{\overset{H}{C}}-CH_3$$

$$CH_3-\underset{H_A}{\overset{CH_3}{C}}-\underset{\oplus}{\overset{H}{C}}-CH_2-H_B \begin{cases} \xrightarrow{-H_A^{\oplus}} (CH_3)_2C{=}CHCH_3 & \text{(2-methyl-2-butene)} \\ \xrightarrow{-H_B^{\oplus}} CH_3-\underset{H}{\overset{CH_3}{C}}-CH{=}CH_2 & \text{(3-methyl-1-butene)} \end{cases}$$

$$\Big\downarrow \ H_A^{\ominus} \text{ shift and re-labeled } H_Y$$

$$H_X-CH_2-\underset{\oplus}{\overset{CH_3}{C}}-\underset{H_Y}{\overset{H}{C}}-CH_3 \begin{cases} \xrightarrow{-H_Y^{\oplus}} (CH_3)_2C{=}CHCH_3 & \text{(2-methyl-2-butene)} \\ \xrightarrow{-H_X^{\oplus}} CH_2{=}C(CH_3)CH_2CH_3 & \text{(2-methyl-1-butene)} \end{cases}$$

6.5 E1cb REACTIONS

A third mechanism for β-elimination of "HX" involves a rapid loss of a proton to a base, thus forming a stabilized carbanion, which is then converted, in the rate-determining step, to an alkene. This process, since it proceeds through the conjugate base of the starting material, is abbreviated E1cb. The E1cb reaction competes with the E2 process. However, due largely to the general instability of carbanions, E1cb reactions are much less common than are E2 reactions. Indeed, only a very small percentage of eliminations of any kind follow this pathway, and its very existence has been the subject of controversy.

$$\text{H}-\overset{|}{\underset{|}{C}}-\overset{|}{\underset{|}{C}}-X \underset{}{\overset{B^{\ominus}}{\rightleftharpoons}} \ominus{:}C-C-X \xrightarrow[\text{slow}]{-X^{\ominus}} C{=}C \qquad \text{an E1cb } \beta\text{-elimination}$$

The reaction is, like the E2, first order in base and first order in substrate. However, it is *unimolecular* because only one species is involved in the rate-determining step.

The use of deuterium labeling can help to distinguish the E1cb from the E2 pathway. If an E2 reaction is carried out in solvent which could act as a deuterium source, and the reaction is interrupted before it has gone to completion, recovered starting material is free of any deuterium. This is to be expected because there is no way, according to our description of the E2 process, by which deuterium could become incorporated into the starting material. Any reaction which involves a carbanion as an intermediate, on the other hand, should produce recovered starting material labeled with deuterium. It is precisely such deuterium incorporation which occurs in those reactions presumed to be E1cb.

E2 system

$$\text{H}-\overset{|}{\underset{|}{C}}-\overset{|}{\underset{|}{C}}-X \xrightarrow[C_2H_5OD]{KOD} \text{D}-\overset{|}{\underset{|}{C}}-\overset{|}{\underset{|}{C}}-X \qquad \text{this incorporation does NOT occur}$$

E1cb system

$$\text{H}-\overset{|}{\underset{|}{C}}-\overset{|}{\underset{|}{C}}-X \xrightarrow[D_2O]{OD^{\ominus}} \overset{\ominus}{\ddot{C}}-\overset{|}{\underset{|}{C}}-X \xrightarrow{D_2O} \text{D}-\overset{|}{\underset{|}{C}}-\overset{|}{\underset{|}{C}}-X + \overset{\ominus}{OD} \qquad \text{this incorporation DOES occur}$$

The reaction of 1,1-dichloro-1-deuterio-2,2,2-trifluoroethane illustrates a β-elimination which follows the E1cb pathway. It also shows how, by starting with a deuterated substrate, the carbanion can be trapped using $OH^{\ominus}/HOH$ rather than $OD^{\ominus}/DOD$.

$$\underset{\text{1,1-dichloro-1-deuterio-2,2,2-trifluoroethane}}{DCCl_2-CF_3} \xrightarrow[D_2O]{OD^{\ominus}} Cl-\overset{\ominus\,\cdot\cdot}{\underset{Cl}{C}}-CF_2-F \rightarrow (Cl)_2C{=}C(F)_2$$

$$\downarrow OH^{\ominus}/H_2O$$

$$Cl_2\overset{\cdot\cdot\ominus}{C}-CF_3 \xrightarrow[\text{fast}]{H_2O} Cl_2CH-CF_3$$

The reaction of chloroform with base, an α-elimination which produces dichlorocarbene, is considered to go *via* the E1cb pathway.

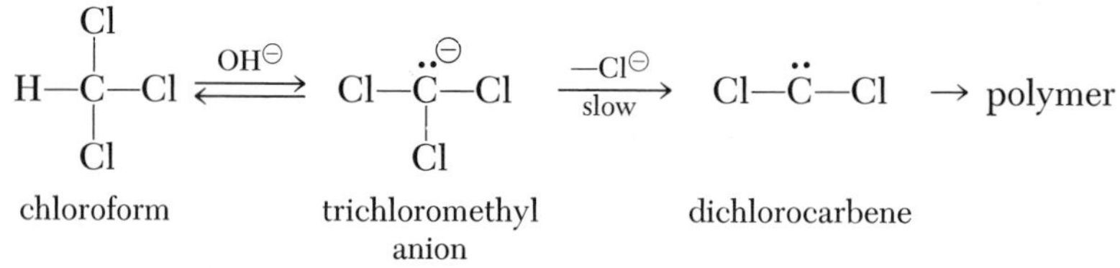

$$\underset{\text{chloroform}}{H\text{—}CCl_3} \underset{}{\overset{OH^{\ominus}}{\rightleftharpoons}} \underset{\text{trichloromethyl anion}}{Cl\text{—}\ddot{C}^{\ominus}Cl\text{—}Cl} \xrightarrow[\text{slow}]{-Cl^{\ominus}} \underset{\text{dichlorocarbene}}{Cl\text{—}\ddot{C}\text{—}Cl} \rightarrow \text{polymer}$$

The E1, E2, E1cb Spectrum. The E2 loss of HX is a concerted process, two bonds being made and two being broken at the same time. One can imagine variations in the "extent of concertedness," ranging from one extreme, at which the C—H bond is broken but the C—X bond is intact, to another, at which the C—X bond is broken but the C—H bond is undisturbed. These extremes, pictured in Figure 6–3, are clearly nothing more than E1cb and E1 mechanisms, respectively. Thus, while it is helpful to think of separate E1, E2, and E1cb reactions, it is clear that all three are related, differing mainly in the timing of various bond making and breaking processes.

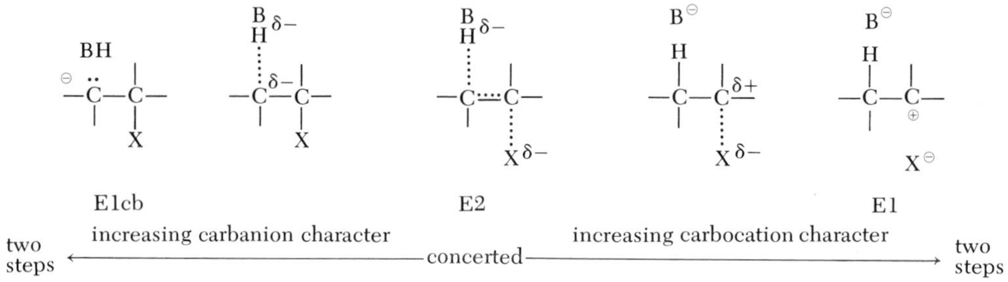

E1cb — E2 — E1

two steps ← increasing carbanion character — concerted — increasing carbocation character → two steps

Figure 6–3 The spectrum of elimination processes.

6.6 SUBSTITUTION AND ELIMINATION—SOME GUIDELINES

Because of the myriad possible combinations of substrates, nucleophiles, solvents, and reaction temperatures, and because of gaps in our understanding, it is impossible to always predict the outcome of the attack of a nucleophile-base upon an alkyl moiety bearing a potential leaving group. The "rules" which follow are generalizations and, therefore, are subject to exceptions; however, they can be used to predict the behavior of a variety of reacting systems with reasonable confidence.

Although much of this discussion has employed dehydrohalogenation, these guidelines are applicable to other types of eliminations and may be helpful in understanding reactions throughout this text.

1. Strong bases (*e.g.*, $NH_2^{\ominus}$) favor elimination over substitution, encouraging E2 (and E1cb) over E1.

2. While leaving group activity varies in the same way for both S_N and E reactions, certain groups favor E1 over E2. These include the better leaving groups $—OH_2^{\oplus}$ and $—N_2^{\oplus}$. The best common anionic leaving group, $RSO_3^{\ominus}$, acts well in both E2 and E1 reactions but favors S_N over E. The fluoride ion is a particularly good leaving group for E1cb processes. (Dehydro-

fluorinations by bases such as methoxide ion, $CH_3O^{\ominus}$, give unusually high amounts of Hofmann elimination product.)

3. More polar solvents usually favor S_N2 over E2 reactions. Thus, alcoholic KOH is used for E2 reactions but aqueous KOH is used for S_N2 reactions. Carbocations invariably produce more substitution than elimination ($S_N1 > E1$), and this cannot be counterbalanced by any change in solvent.

4. Substitution at Cα favors E2 over S_N2; the amount of E1 is increased but not enough to make S_N1 a minor reaction path compared to it. Substitution at Cβ increases the amount of elimination. Electron withdrawing groups (*e.g.*, Br and CN) in the β position increase the acidity of the hydrogen bonded to Cβ and favor eliminations of the E2 (and E1cb) variety.

5. Higher temperatures always increase the amount of elimination at the expense of substitution.

6. E1 reactions, because of product complexity brought about by isomerizations, are usually less useful in alkene synthesis than E2 reactions.

A slightly different way of presenting some of the same ideas, but one perhaps more useful to some, is found in Table 6–1.

TABLE 6–1 A Comparison of S_N And E Conditions

Substrate	Leaving Group	Nucleophile	Major Anticipated Reaction(s)
RCH_2—L	$X^{\ominus}$, $OSO_2R^{\ominus}$	$OH^{\ominus}$, $OR^{\ominus}$, $CN^{\ominus}$, NH_3	S_N2
(R = alkyl)	$X^{\ominus}$, $OSO_2R^{\ominus}$	HO—H, RO—H	[a]
	—$OH_2^{\oplus}$[b]	$X^{\ominus}$	S_N2
	—$NR_3^{\oplus}$	$OH^{\ominus}$, $OR^{\ominus}$, $CN^{\ominus}$, NH_3	E2
R_2CH—L	$X^{\ominus}$, $OSO_2R^{\ominus}$	$OH^{\ominus}$, $OR^{\ominus}$, $CN^{\ominus}$, NH_3	$S_N2 > E2$[c]
(R = alkyl)	$X^{\ominus}$, $OSO_2R^{\ominus}$	HO—H, RO—H	$S_N1 > S_N2$
	—$OH_2^{\oplus}$	$X^{\ominus}$	$S_N1 > S_N2$
	—$NR_3^{\oplus}$	$OH^{\ominus}$, $OR^{\ominus}$, $CN^{\ominus}$, NH_3	E2
R_3C—L	$X^{\ominus}$, $OSO_2R^{\ominus}$	$OH^{\ominus}$, $OR^{\ominus}$, $CN^{\ominus}$, NH_3	E2
(R = alkyl)	$X^{\ominus}$, $OSO_2R^{\ominus}$	HO—H, RO—H	S_N1
	—$OH_2^{\oplus}$	$X^{\ominus}$	S_N1
	—$NR_3^{\oplus}$	$OH^{\ominus}$, $OR^{\ominus}$, $CN^{\ominus}$, NH_3	E2

[a] This combination is too slow to be of much practical value.
[b] This represents a protonoted alcohol.
[c] Elevated temperature increases the amount of E2 component.

6.7 GRIGNARD REAGENTS

The two most valuable applications of alkyl halides are as substrates for nucleophilic substitutions and for eliminations. Having examined these two in detail we shall turn our attention, at an introductory level, to the use of alkyl halides to prepare organometallic compounds, *i.e.*, compounds containing a carbon-metal bond. Because of their importance, our discussion will be limited to Grignard reagents, compounds of the type R—Mg—X.

Bonding in organometallic compounds can vary from largely ionic, with extremely electropositive metals like potassium, to largely covalent, with less-active metals like tin and mercury. As with any other bonds, the balance between ionic character and covalent character in the carbon-metal bond can be represented in resonance terms. If the resonance contributing structure on the left (p. 174) is more important, then the bond is considered to be covalent.

If the resonance contributor on the right is more important, then the bond is considered to be ionic.

$$-\overset{|}{\underset{|}{C}}-\text{Metal} \leftrightarrow -\overset{|}{\underset{|}{C}}:^{\ominus}\ \text{Metal}^{\oplus}$$ resonance in a carbon-metal bond

The highly ionic compounds, such as alkylpotassiums, are very reactive, igniting in air and in contact with water. Organomercury compounds, by way of contrast, are stable in air, unreactive to water, soluble in most organic solvents, and quite toxic. When combined, these properties make organomercury compounds quite hazardous constituents of the environment. Organomagnesium compounds of the type RMgX are stable enough to be handled conveniently, yet reactive enough to make them potent reagents in organic synthesis. Victor Grignard, the discoverer of these substances, received the Nobel prize in chemistry (1912) for his efforts in this area.

$$-\overset{|}{\underset{|}{C}}-\text{MgX} \leftrightarrow -\overset{|}{\underset{|}{C}}:^{\ominus}\ \overset{\oplus}{\text{Mg}}-\text{X}$$ the Grignard reagent

Synthesis of Grignard Reagents. Alkyl Grignard reagents are prepared by the dropwise addition of an alkyl halide to a mixture of magnesium and ethyl ether. Great care must be taken to add the halide to the magnesium neither too rapidly nor in large excess. If the halide is added too rapidly, the reaction may "take off" (get out of hand). An excess of alkyl halide increases the probability of observing a coupling reaction, that is, the synthesis of a hydrocarbon by reaction between the Grignard reagent and the unchanged alkyl halide. This coupling may be viewed as an S_N2 reaction in which the Grignard reagent is the nucleophile and the unreacted alkyl halide is the substrate (however, see p. 176).

$$CH_3CH_2CH_2I + Mg \xrightarrow{\text{ether}} CH_3CH_2CH_2MgI$$

propyl magnesium iodide, a Grignard reagent

net reaction: $2CH_3CH_2CH_2I + Mg \rightarrow CH_3CH_2CH_2{-}CH_2CH_2CH_3 + MgI_2$

mechanism: $CH_3CH_2CH_2:^{\ominus}\ \ CH_2(I)CH_2CH_3 \rightarrow CH_3CH_2CH_2{-}CH_2CH_2CH_3$

coupling reaction

The apparatus used for synthesizing and carrying out reactions with Grignard reagents must be protected from the air by a drying tube. The tube prevents exposure of the reagents to water. The use of anhydrous solvents and oven-dried or "flamed-out"[a] glassware further reduces the chance of accidental reaction of the Grignard reagent with water. Here the carbanionic character of the reagent is exposed as it behaves like a strong base, reacting with the weak acid water to form a hydrocarbon. Of course, this destroys the organometallic compound. (This is used to advantage in the Zerewitnoff determination, described on p. 176.)

net reaction: $CH_3MgX + HOH \rightarrow CH_4 + MgXOH$ reaction of a Grignard reagent with water

mechanism: $H_3\overset{\ominus}{C}:\ \ H{-}OH \rightarrow H_3C{-}H + \overset{\ominus}{O}H$

Nitrogen often is used to displace air from the reaction system in order to prevent the

[a] Always done *before* the ether and other reagents are added!

oxidation of the Grignard reagent. The oxidized Grignard reagent is easily hydrolyzed to an organic hydroperoxide.

$$CH_3MgCl + O_2 \rightarrow CH_3\text{—}\ddot{\underset{..}{O}}\text{—}\ddot{\underset{..}{O}}\text{—}MgCl \xrightarrow{H_2O} \underset{\text{methyl hydroperoxide}}{CH_3\text{—}\ddot{\underset{..}{O}}\text{—}\ddot{\underset{..}{O}}\text{—}H} + MgClOH$$

Nature of the Grignard Reagent. Our knowledge of the nature of organometallic compounds, including Grignard reagents, is rather sparse. While the single crystal X-ray study of Grignard reagents has been accomplished, our knowledge of the substance in solution is deficient. Although it is commonplace to write the formula for a Grignard reagent as RMgX, it often reacts as if it were composed of an alkyl carbanion and an $MgX^{\oplus}$ counter ion. The inability to isolate stable Grignard reagents free of solvent suggests that they are *strongly* solvated in solution. One plausible structure for methyl magnesium iodide etherate, the species present in an ethereal solution of methyl magnesium iodide, is shown below. It would appear that a good Lewis base such as an ether ($R\ddot{\underset{..}{O}}R$) is required to stabilize the magnesium of the Grignard reagent.

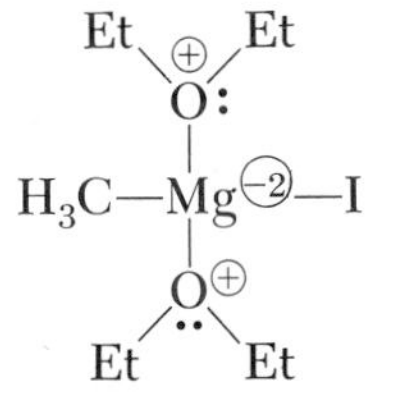

$$\begin{array}{c} Et_2\overset{\oplus}{O}: \\ | \\ H_3C\text{—}Mg^{\ominus 2}\text{—}I \\ | \\ \ddot{O}^{\oplus}Et_2 \end{array}$$

methyl magnesium iodide etherate

While ethyl ether is the usual solvent for preparing Grignard reagents, tetrahydrofuran (THF) is preferred in the preparation of Grignard reagents derived from compounds containing C_{sp^2}-halogen bonds. This may be due to the greater exposure of the electron pairs on the oxygen of THF, which results from the "tying back" of the alkyl residue, $\text{—}CH_2CH_2CH_2CH_2\text{—}$, in the cyclic ether, THF. (Enhanced exposure of the oxygen electron pairs should make it easier for THF to stabilize the magnesium.) In addition to THF, dimethoxyethane (DME) is another valuable solvent for the preparation of Grignard reagents.

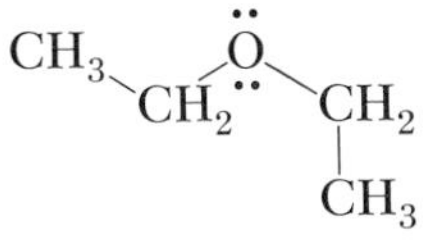

$CH_3CH_2\ddot{\underset{..}{O}}CH_2CH_3$ — ethyl ether ("ether")

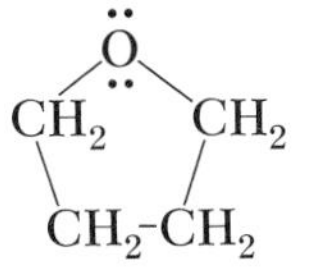

$\overset{..}{O}(CH_2CH_2CH_2CH_2)$ (cyclic) — tetrahydrofuran (THF)

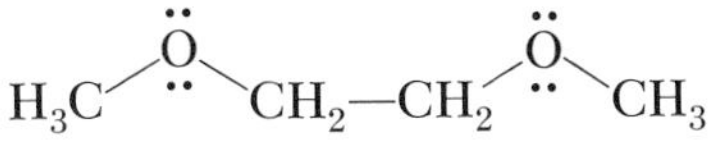

$H_3C\ddot{\underset{..}{O}}CH_2\text{—}CH_2\ddot{\underset{..}{O}}CH_3$ — dimethoxyethane (DME)

9. What advantage might there be to the use of dimethoxyethane, rather than ethyl ether, as a solvent for Grignard reagents?

Grignard Reagents as Nucleophiles. The carbanionic character of Grignard reagents renders them powerful bases but only fair nucleophiles. This strong basicity makes E2 reactions quite competitive with S_N2 reactions when the substrate is a simple alkyl halide. Fortunately, both S_N2 and E2 reactions between alkyl Grignard reagents and alkyl halides are sufficiently slow that alkyl Grignard reagents can be prepared from the majority of alkyl halides and magnesium in excellent yields.

$$CH_3MgX + CH_3X \xrightarrow{S_N2} CH_3{-}CH_3 + MgX_2 \quad \text{slow at } 25^\circ$$

$$CH_3CH_2MgX + CH_3CH_2X \xrightarrow{E2} CH_3{-}CH_3 + CH_2{=}CH_2 + MgX_2 \quad \text{slow at } 25^\circ$$

Only particularly reactive alkyl halides, such as allyl halides, undergo facile S_N2 reactions with Grignard reagents.

$$\underset{\text{allyl chloride}}{CH_2{=}CHCH_2Cl} + CH_3CH_2MgCl \xrightarrow{(C_2H_5)_2O} \underset{\text{1-pentene}}{CH_2{=}CHCH_2CH_2CH_3} + MgCl_2$$

10. Benzyl bromide, $C_6H_5CH_2Br$, reacts with magnesium to form a small amount of hydrocarbon, $C_{14}H_{14}$. (a) Suggest a structure for this compound. (b) How is it formed?

Grignard Reagents as Bases—The Zerewitnoff Determination. Methyl magnesium iodide reacts with water, and with other compounds containing acidic hydrogens, to form methane. The determination of the amount of methane liberated by known amounts of a substance forms the basis of the Zerewitnoff determination of "active" hydrogens. Protons bound to oxygen, nitrogen, and sulfur usually can be estimated by this procedure. Some of the functional groups that will react as acids toward methyl magnesium iodide (or other Grignard reagents) include —SH, —OH, $—NH_2$, $—CO_2H$, and $—SO_3H$. Hydrogens bonded to sp^2 and sp^3 hybridized carbon are not sufficiently acidic to react with Grignard reagents.

$$CH_3CH_2SH + CH_3MgI \rightarrow CH_4 + \overset{\oplus}{Mg}I(CH_3CH_2\overset{\ominus}{S})$$

$$HOCH_2CH_2OH + 2CH_3MgI \rightarrow 2CH_4 + (\overset{\oplus}{Mg}I)_2(\overset{\ominus}{O}CH_2CH_2\overset{\ominus}{O})$$

A variation of the acid-base reaction between a Grignard reagent and a compound containing acidic hydrogens can be used to prepare deuterium-labeled hydrocarbons. For example, racemic 2-deuteriobutane can be prepared from butane by the sequence of halogenation, Grignard formation, and hydrolysis with heavy water (D_2O).

$$CH_3{-}CH_2{-}CH_2{-}CH_3 \xrightarrow[h\nu]{Cl_2} CH_3{-}\underset{\displaystyle Cl}{\underset{|}{CH}}{-}CH_2{-}CH_3 \xrightarrow[\text{ether}]{Mg} CH_3{-}\underset{\displaystyle MgCl}{\underset{|}{CH}}{-}CH_2{-}CH_3$$

$$\xrightarrow{D_2O} \underset{\text{2-deuteriobutane}}{CH_3{-}CHD{-}CH_2{-}CH_3}$$

11. How much ethanol, C_2H_5OH, is required to produce one mole of methane by reaction with an excess of methyl magnesium chloride?

$$C_2H_5OH + CH_3MgCl \rightarrow CH_4 + C_2H_5OMgCl$$

Some Grignard Reagents Are Extremely Unstable. Grignard reagents with good leaving groups β to the C—Mg bond are unstable, rapidly decomposing to an alkene and a magnesium salt. A classic example of this is the dehalogenation of a *vic*-dibromide (*vic* is the abbreviation for *vicinal*, meaning "neighboring") to form an alkene and magnesium bromide. A mechanism for this is suggested below.

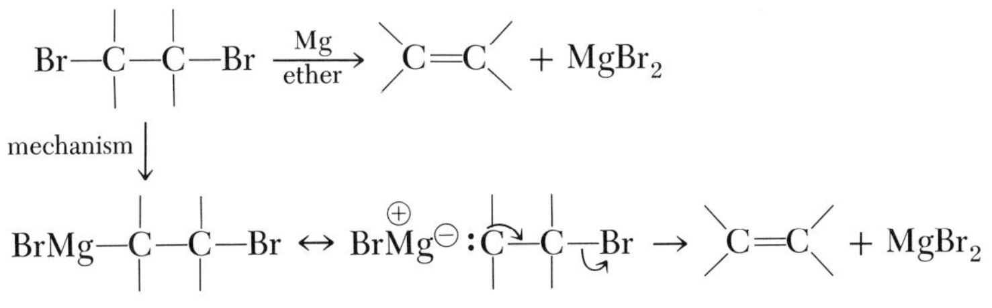

$$\text{Br—}\overset{|}{\underset{|}{\text{C}}}\text{—}\overset{|}{\underset{|}{\text{C}}}\text{—Br} \xrightarrow[\text{ether}]{\text{Mg}} \text{>C=C<} + \text{MgBr}_2$$

mechanism ↓

$$\text{BrMg—}\overset{|}{\underset{|}{\text{C}}}\text{—}\overset{|}{\underset{|}{\text{C}}}\text{—Br} \leftrightarrow \overset{\oplus}{\text{BrMg}}{}^{\ominus}\text{:}\overset{|}{\underset{|}{\text{C}}}\text{—}\overset{|}{\underset{|}{\text{C}}}\text{—Br} \rightarrow \text{>C=C<} + \text{MgBr}_2$$

12. Although depicted as an E1cb process, dehalogenation with magnesium tends to follow the geometry that one would expect from an E2 elimination. (The big difference is that such reactions are not stereospecific and produce minor amounts of the isomer *not* anticipated upon the basis of a concerted *anti* elimination.) Predict the major products of the reactions of magnesium with (a) (+)-2,3-dibromobutane, (b) (−)-2,3-dibromobutane, and (c) *meso*-2,3-dibromobutane.

Organolithium Compounds. Alkyl halides react with metallic lithium in ethereal solvents to form organolithium compounds. Their properties are similar, though not identical, to those of Grignard reagents. They often are used interchangeably in the laboratory. *n*-Butyl lithium is being used increasingly as a strong base.

$$CH_3CH_2CH_2CH_2Br + 2Li \xrightarrow{Et_2O} \underset{\text{n-butyl lithium}}{CH_3CH_2CH_2CH_2\text{—}Li} + LiBr$$

6.8 HALOGENATED COMPOUNDS OF SPECIAL INTEREST

With the advent of "truth-in-packaging" concepts, it became clear that telling the average consumer what was in the thing that he was buying would be of little help unless he had some knowledge of what the items on the label were. Although it is becoming trite to speak of "relevance," there is certainly no harm in being relevant. This section introduces organic halides that are important in their own right.

Aldrin: A potent insecticide, its toxicity to humans is felt only after ingestion of large quantities (>1 g). (However, see dieldrin, p. 178.)

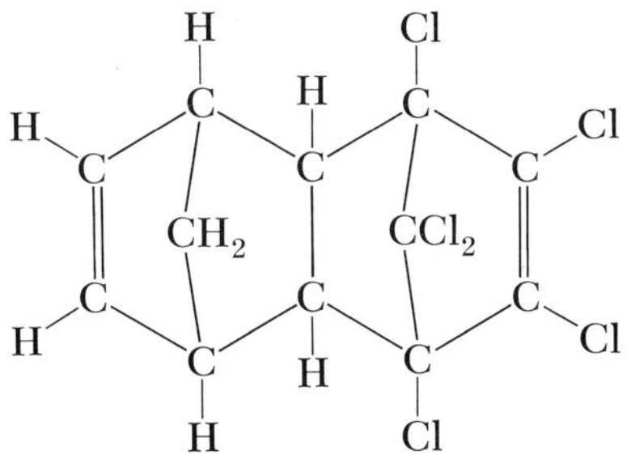

[Structure of aldrin: bicyclic hexachloro structure with CH$_2$ bridge and CCl$_2$ bridge; labels H, Cl, C, CH$_2$, CCl$_2$]

aldrin

Captan: A compound with several functional groups, including a trichloromethyl group; it is used as an agricultural fungicide.

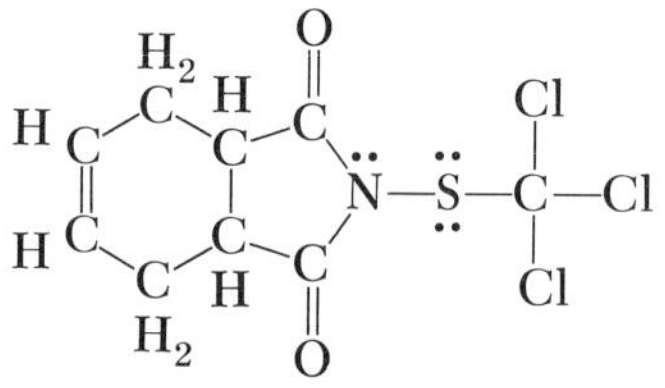

```
       H2       O
        C  H    ||             Cl
  H  C/   \C/-C\     ..        |
     ||    |     N—S—C—Cl
  H  C\   /C\-C/     ..        |
        C  H    ||             Cl
       H2       O
```

captan

Carbon tetrachloride: Carbon tetrachloride (CCl_4) is converted upon heating in air to the poisonous gas phosgene (carbonyl chloride, $COCl_2$). This has resulted in the almost complete abandonment of the use of CCl_4 in fire extinguishers, once an important use of this simplest of *chlorocarbons*.

Chloroform ($CHCl_3$)**:** At one time an inhalation anesthetic, its narrow safety margin has now led to the use of other anesthetics. Large doses cause death, and chloroform is sometimes used to "sacrifice" animals or insects for biological study.

Chlorophyll: There is no chlorine, or other halogen, in chlorophyll. This is one illustration of a name which is not only unrelated to the compound's structure but is, in fact, misleading.

DDT: Properly 1,1,1-trichloro-2,2-bis(*p*-chlorophenyl)ethane, this insecticide has single-handedly curtailed the most prevalent disease in the world, malaria, by acting against the mosquito (Anopheles) which carries the disease. Unfortunately, it is toxic to humans (estimated fatal dose 500 mg/kg) and plays havoc with the life cycles of avian species. DDT-resistant insects are becoming more prevalent, although "superbug" has still to make an appearance.

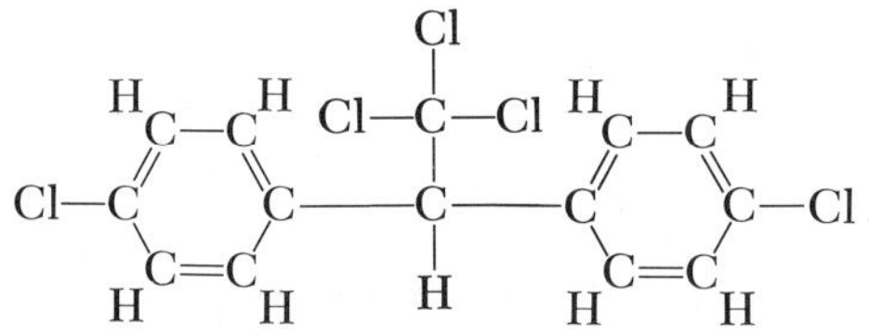

```
                          Cl
                          |
        H     H       Cl—C—Cl    H     H
         C—C              |        C—C
        //    \\          |       //    \\
   Cl—C        C————C————C        C—Cl
        \      /          |       \      /
         C=C              |        C=C
        H     H           H       H     H
```

DDT

Lindane, aldrin, dieldrin, and DDT represent the class of insecticides known as "chlorinated hydrocarbons." A sort of "bad word" in a society attempting to keep its world virile, these materials are threats to the environment because of their extreme stability (non-degradation in the biosphere) as much as because of their toxicity.

***sym*-Dichloroethyl ether:** Related to mustard gas, this material has been used as a soil fumigant *and* as a phytocide (plant killer).

$$Cl—CH_2CH_2—O—CH_2CH_2—Cl$$

sym-dichloroethyl ether

Dieldrin: Related to aldrin, dieldrin is another potent insecticide. It has been suggested that insects convert aldrin to the more toxic dieldrin *in vivo!* In mid-1972 the EPA (Environmental Protection Agency) began to take steps to ban the use of aldrin and dieldrin because of their potential for inducing sexual impotence and cancer in humans.

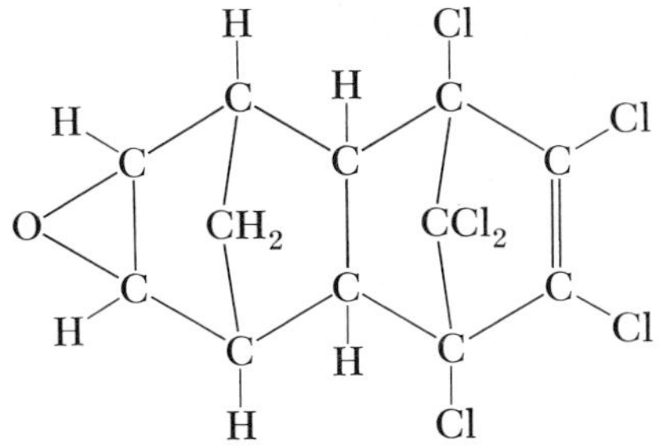

dieldrin

Ethyl chloride (C_2H_5Cl): Physicians planning to use a local anesthetic may spray the area with a material which lowers the temperature at the puncture site. That spray is ethyl chloride, cooling the skin as it boils away (b.p. 13°C).

Lindane ($C_6H_6Cl_6$): This particular isomer of 1,2,3,4,5,6-hexachlorocyclohexane is the only one that is a potent insecticide. Other names include "Gammophene" and "BHC." It is particularly effective against the boll weevil. (The structure of ring systems is discussed in the following chapter.)

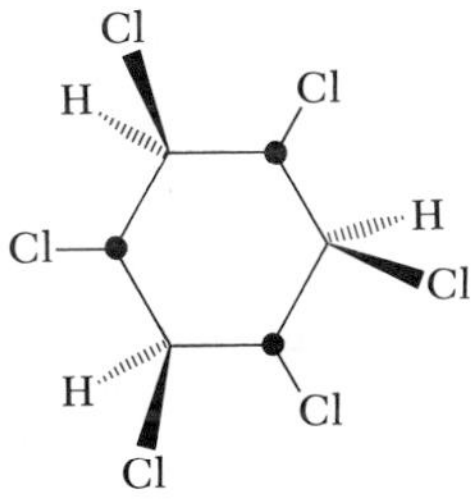

Methoxychlor: Fifteen times less toxic to humans than DDT, this insecticide is quite similar in structure to its big brother, DDT.

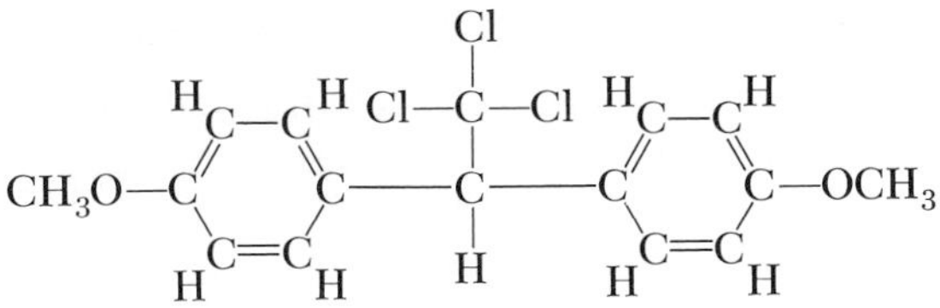

methoxychlor
(1,1,1-trichloro-2,2-*bis*(*p*-methoxyphenyl)ethane)

This list has not been intentionally weighted to favor chlorine over the other halogens. Rather, it reflects the lower cost of using materials made from the most abundant of the halogens.

6.9 SPECTRAL PROPERTIES OF ALKYL HALIDES

The material in this section is designed to be read only after the chapters on molecular spectroscopy have been covered. However, it is included *here* in order to increase the reference value of this text. Many of the subsequent chapters will end in a similar fashion.

Ultraviolet Spectroscopy. Ultraviolet spectroscopy is of little value in the identification of organic halides.

Infrared Spectroscopy. The absence or presence of halogen in organic compounds is difficult to determine by infrared spectroscopy. Both C—Br and C—I bonds may absorb outside the range of most instruments, while both C—F and C—Cl absorb over wide ranges (1350–950 cm^{-1} and 850–500 cm^{-1}, respectively).

In the cyclohexane ring, an *equatorial* substituent absorbs at a higher frequency than does the corresponding *axial* substituent. Therefore, ir can distinguish between *equatorial* C—Cl bonds (~800 cm^{-1}) and *axial* C—Cl bonds (~700 cm^{-1})!

NMR Spectroscopy. The presence of fluorine can be demonstrated by nmr spectroscopy, either by direct observation of the fluorine nuclei or by observation of the splitting of proton signals by fluorine nuclei. Fluorine resonance spectroscopy is beyond the scope of this text, although it is alluded to in Chapter 29. Some typical H,F coupling constants are presented in Table 6–2. The effect of halogens upon the chemical shift of protons is discussed in Chapter 29.

TABLE 6–2 Typical H,F Coupling Constants

Compound	$J_{H,F}$ (Hz)	Compound	$J_{H,F}$ (Hz)
$CH_3—CH_2—F$	47	(H)(F)C=C (H and F on the same carbon)	~20
$CH_3—CH_2—F$	25		
F—C—C—H (wedge/dash structure)	~4	H—C=C—F (H and F trans)	~50
		H—C=C, F on same side as H (cis)	~80
F—C—C—H (wedge/dash structure)	~20	$H_3C—CH=CH—F$	~3

IMPORTANT TERMS

α-Elimination: An elimination reaction in which both departing groups are bonded to the same atom in the substrate prior to their loss. If both departing groups are bonded to carbon, the product is an unstable, electron-deficient carbon species called a carbene. There is no relationship between the designation α for a carbon and its location on a carbon chain. Put differently, when used to designate a type of elimination, Cα need not refer to a terminal carbon of a backbone chain.

$$CH_3—CCl_2—CH_2CH_3 \xrightarrow{Mg} CH_3—\ddot{C}—CH_2CH_3 + MgCl_2$$

ethyl methyl carbene — an α elimination

***anti* Elimination:** An elimination reaction in which the departing groups must be antiparallel to one another. These groups, and the vicinal carbons connected to them, all reside in one plane. (The torsional angle between the two leaving groups is 180°.) This is the usual geometry for an E2 elimination. (This also is termed a *trans* elimination.)

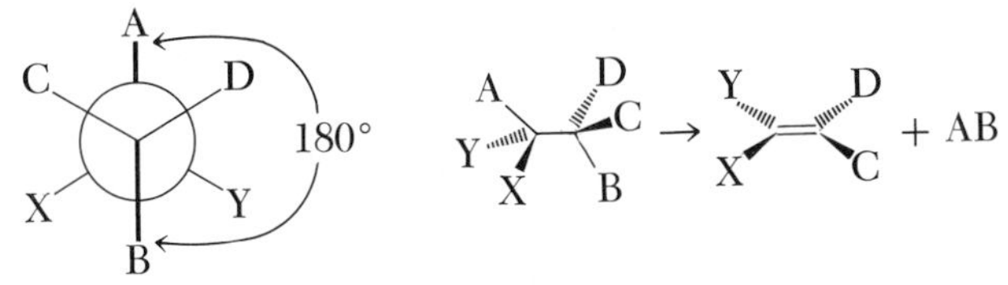

anti geometry *anti* elimination

β-Elimination: An elimination reaction in which the departing groups are bonded to vicinal atoms. This is sometimes called a "1,2-elimination." However, in this latter terminology the numbers "1" and "2" do not refer to the IUPAC numbers of backbone carbons; rather, they serve to indicate that departing groups are bonded to adjacent carbons.

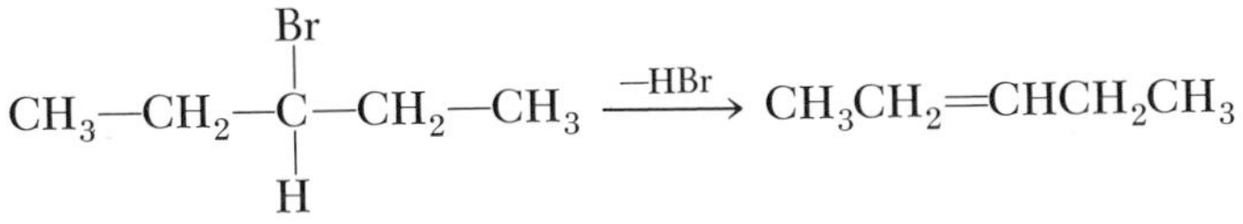

$$CH_3{-}CH_2{-}C(Br)(H){-}CH_2{-}CH_3 \xrightarrow{-HBr} CH_3CH_2{=}CHCH_2CH_3 \quad \text{a } \beta\text{-elimination}$$

Carbene: A highly reactive species containing a carbon with only six outer electrons (four involved in covalent bonding and two non-bonded). The product of an α elimination. (Carbenes will be discussed in much greater detail in Chapter 8.)

Dehydrohalogenation: Loss of hydrogen and a halogen in an elimination reaction.

E1 Reaction: An elimination involving loss of a proton from a carbocation. The proton is lost from a carbon adjacent to the cationic center. This reaction competes with the S_N1 and has the same rate determining step, *i.e.*, carbocation formation. The two steps of this non-concerted process are illustrated below.

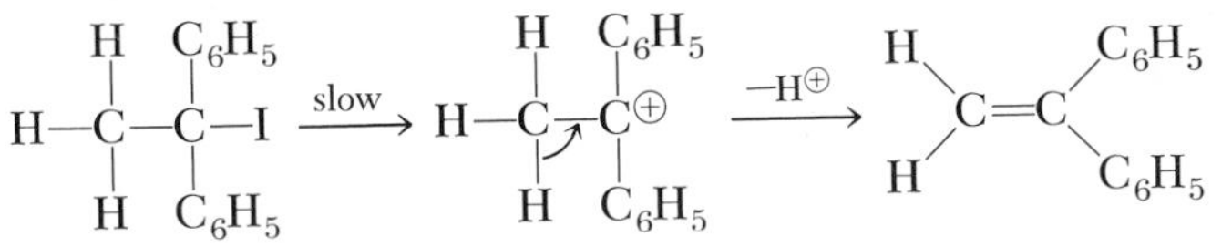

an E1 reaction

E1cb Reaction: An elimination involving loss of an anion from a comparatively stable carbanion. E1cb reactions are much less common than are E1 and E2 reactions.

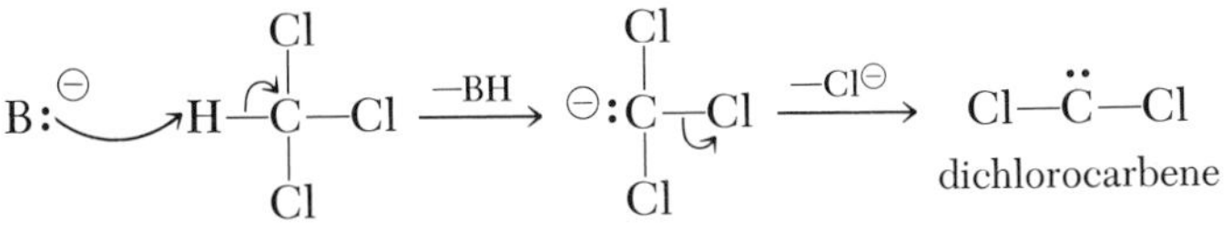

an E1cb reaction

dichlorocarbene

E2 Reaction: An elimination involving the concerted loss of two fragments, most commonly on adjacent atoms. This process competes with the S_N2 reaction. Remember that the terms E1, E2, and E1cb describe mechanisms along a spectrum of reaction types (refer to Figure 6–3). Most E2 reactions are *anti* eliminations.

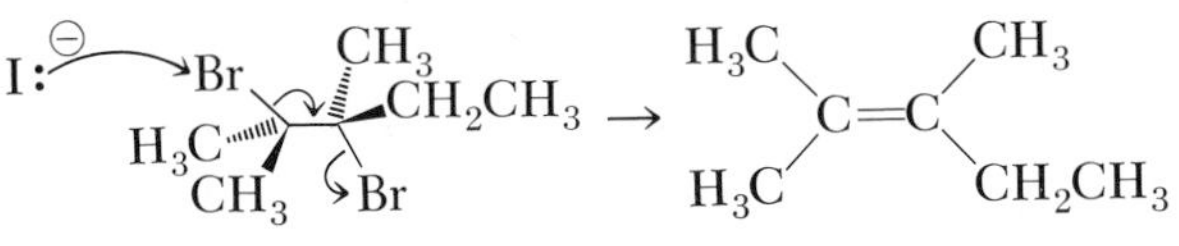

an E2 reaction

Grignard reagent: A compound of the type R—Mg—X, most often prepared by the reaction of a halide with magnesium metal in the presence of an ether solvent. This highly reactive type of compound behaves as if it were a carbanion, *i.e.*, $R:^{\ominus}MgX^{\oplus}$. Grignard reagents cannot be stable if the carbanionic center has a good leaving group adjacent to it, since this will lead to elimination (alkene formation).

$$\text{Br—C(CH}_3)_2\text{—CH}_2\text{—OCH}_3 \xrightarrow[\text{ether}]{\text{Mg}} \text{BrMg—C(CH}_3)_2\text{—CH}_2\text{—OCH}_3 \rightarrow$$

a Grignard reagent

$$(CH_3)_2C{=}CH_2 + MgBr(OCH_3)$$

formation and decomposition of an unstable Grignard reagent

Hofmann's rule: Elimination from a quaternary ammonium salt (see example below) or other substrate in which the leaving group is positively charged tends to produce the least highly substituted alkene. The decomposition of quaternary ammonium hydroxides to alkenes $\left(R_3\overset{\oplus}{N}\text{—}\overset{|}{\underset{|}{C}}\text{—}\overset{|}{\underset{|}{C}}\text{—H}\ \ OH^{\ominus} \xrightarrow{\text{heat}} R_3N + \overset{|}{\underset{|}{C}}{=}\overset{|}{\underset{|}{C}} + H_2O\right)$ is a specific E2 reaction called the *Hofmann elimination.* It is from a study of this type of elimination that the Hofmann rule evolved. In general, the least substituted alkene produced in an elimination often is termed the "Hofmann product."

$$R\text{—}CH_2\text{—}\underset{\oplus N(CH_3)_3}{CH}\text{—}CH_3 \xrightarrow[\text{heat}]{OH^{\ominus}}$$

$$R\text{—}CH_2\text{—}CH{=}CH_2 + N(CH_3)_3 + H_2O$$

an elimination following Hofmann's rule

Kinetic isotope effect: An alteration in the rate of a reaction caused by the substitution of one isotope of an element for another.

Saytzeff's rule: Elimination from a substrate in which the leaving group departs as an anion produces the most highly substituted alkene. The most common example of this is dehydrohalogenation. In general, the most highly substituted alkene product in an elimination often is termed the "Saytzeff product."

$$R\text{—}CH_2\text{—}\underset{X}{CH}\text{—}CH_3 \xrightarrow{OH^{\ominus}} R\text{—}CH{=}CH\text{—}CH_3 + X^{\ominus} + H_2O$$

an elimination following Saytzeff's rule

Stereospecific reaction: A reaction in which one particular stereoisomeric starting material produces only one stereoisomeric product. In a stereospecific process, two stereoisomeric starting materials will produce two diastereomeric products. The reaction shown below is stereospecific, since one starting material produces only *d,l* product while the other produces only *meso* product. If a reaction produces *more* than one diastereomeric product from one stereoisomeric starting material, then the reaction is not stereospecific. However, if one diastereomeric product predominates in this mixture, the process can be called *stereoselective.*

H H / C=C / CH_3 CH_3 $\xrightarrow{OsO_4}$ HO OH / H C—C H / H_3C CH_3

meso

H CH_3 / C=C / CH_3 H $\xrightarrow{OsO_4}$ HO OH / H C—C CH_3 / H_3C H + HO OH / H_3C C—C H / H CH_3

d,l

the mechanism of this type of stereospecific reaction is discussed in Chapter 8

PROBLEMS

13. Draw all of the optically active compounds with the formula $C_5H_{11}Cl$. Predict the major product(s), if any, of the reaction of each with (a) alcoholic KOH, (b) aqueous KOH, and (c) sodium amide ($NaNH_2$, strong base) in liquid ammonia.
14. Predict the major product(s), if any, of the following:
    (a) $BrCH_2CH_2SCH_3$ + Mg (*Note:* $RS^{\ominus}$ is a fair leaving group.)
    (b) $BrCH_2SCH_3$ + Mg
    (c) CH_3CH_2OH + NaOH
    (d) $(CH_3)_3CCH_2Br$ + KOH/ethanol
    (e) $(CH_3)_3CCH_2Br$ + KOH/water
    (f) $(CH_3)_3CCH_2CH_2Br$ + $NaSCH_3$/ethanol
    (g) $(CH_3)_3C(CH_2)_4Br$ + $NaSCH_3$/ethanol
    (h) $(CH_3)_3C(CH_2)_4Br$ + KOH/ethanol
    (i) $(CH_3)_3C(CH_2)_4Br$ + KOH/water
    (j) (**R**)-2-bromobutane + KOH/ethanol
    (k) (**R**)-2-bromobutane + $NaNH_2$
    (l) Myleran + CH_3NH_2
15. Indicate the major alkene produced by each of the following:
    (a) $CH_3CH_2CH_2CBr(CH_3)_2$ + $C_2H_5O^{\ominus}$
    (b) $CH_3CH_2CH_2C(CH_3)_2\overset{\oplus}{N}(CH_3)_3$ $OH^{\ominus}$ + heat
    (c) $CH_3CH_2CH(OTs)CH_3$ + heat (acetic acid solvent)
    (d) $(CH_3)_3\overset{\oplus}{N}C(CH_3)_2CH_2CH_3$ $OH^{\ominus}$ + heat
    (e) $(CH_3)_3CO^{\ominus}$ + $CH_3CH_2CH_2CHBrCH_3$
    (f) $(CH_3)_3CCH(OH)CH_3$ + $H^{\oplus}$
16. Suggest a synthesis, using the given starting material and anything else you require, for each of the given products. Some syntheses will require only one step, while others are multi-step processes. At this early stage of learning organic chemistry, you may be forced to use some reactions which are repetitive or wasteful; your synthetic "vocabulary" will expand in the following chapters.
    (a) methane *to* monodeuteriomethane (CH_3D)
    (b) ethane *to* monodeuterioethane
    (c) ethane *to* ethene (C_2H_4)
    (d) hexadeuterioethane (C_2D_6) *to* tetradeuterioethene (C_2D_4)
    (e) allyl bromide *to* $CH_2{=}CH{-}CH_2CH_2{-}CH{=}CH_2$
17. (a) How many grams of methane will be liberated when 100 g of an equimolar mixture of methanol and ethanol reacts with an excess of methyl magnesium iodide? (b) What volume would this be expected to occupy at STP? (*Remember:* One mole of gas occupies 22.4 l at STP.)
18. For each pair that follows, indicate which member will give the highest substitution-to-elimination ratio. Explain your answers.
    (a) $CH_3CH_2CClCH_3CH_2CH_3$ *or* $CH_3(CH_2)_2CBrCH_3CH_2CH_3$ heated in aqueous dioxane.
    (b) $(CH_3)_2CHC(CH_3)_2Cl$ *or* $CH_3CH_2CH_2CCl(CH_3)_2$ heated in aqueous dioxane.
    (c) $(CH_3)_2CHCH_2Cl$ *or* $(CH_3)_2CHCH_2\overset{\oplus}{N}(CH_3)_3$ $Cl^{\ominus}$ treated with sodium ethoxide in ethanol.
    (d) $CH_3CH_2CH_2Br$ reacted with 10% KOH in HOH *or* with 10% KOH in CH_3OH.

19. What possible conclusion(s) could you draw if you noted that changing a reaction solvent from ethanol to pyridine increased the amount of E1 product compared to E2 product?

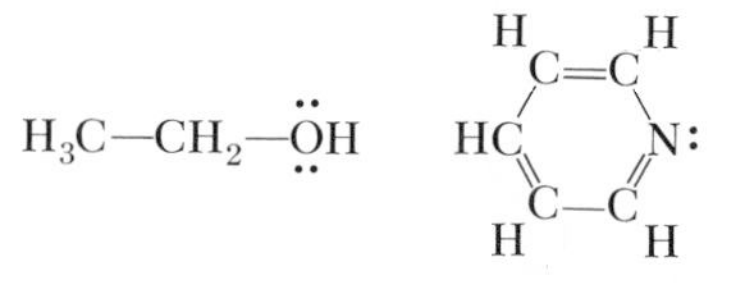

$H_3C{-}CH_2{-}\ddot{O}H$ (ethanol)

pyridine: HC=CH ring with N: (C₅H₅N)

20. As shown below, the product distribution in an E1 reaction is almost independent of the leaving group. Explain this.

$$(CH_3)_3CBr \xrightarrow[65°]{80\%\ \text{ethanol}/20\%\ HOH} \underset{36\%}{CH_2{=}C(CH_3)_2} + \text{substitution products}$$

$$(CH_3)_3\overset{\oplus}{S}(CH_3)_2 \xrightarrow[65°]{80\%\ \text{ethanol}/20\%\ HOH} \underset{36\%}{CH_2{=}C(CH_3)_2} + \text{substitution products}$$

What substitution products are expected to accompany the alkene in each of these reactions?

21. Iodine can be used to catalyze the dehydration of an alcohol to an alkene. A suggested hypoiodite intermediate is shown below. Does the product distribution suggest whether the reaction is E1 or E2? Design an experiment aimed at disproving the E2 process, including anticipated results.

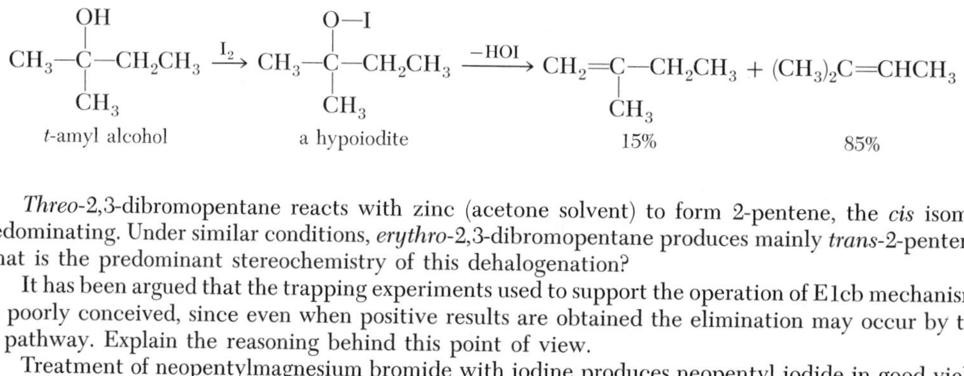

$$\underset{t\text{-amyl alcohol}}{CH_3{-}C(OH)(CH_3){-}CH_2CH_3} \xrightarrow{I_2} \underset{\text{a hypoiodite}}{CH_3{-}C(O{-}I)(CH_3){-}CH_2CH_3} \xrightarrow{-HOI} \underset{15\%}{CH_2{=}C(CH_3){-}CH_2CH_3} + \underset{85\%}{(CH_3)_2C{=}CHCH_3}$$

22. *Threo*-2,3-dibromopentane reacts with zinc (acetone solvent) to form 2-pentene, the *cis* isomer predominating. Under similar conditions, *erythro*-2,3-dibromopentane produces mainly *trans*-2-pentene. What is the predominant stereochemistry of this dehalogenation?

23. It has been argued that the trapping experiments used to support the operation of E1cb mechanisms are poorly conceived, since even when positive results are obtained the elimination may occur by the E2 pathway. Explain the reasoning behind this point of view.

24. Treatment of neopentylmagnesium bromide with iodine produces neopentyl iodide in good yield. Provide a mechanism for this conversion.

$$(CH_3)_3CCH_2MgBr + I_2 \rightarrow (CH_3)_3CCH_2I + MgBrI$$

25. The reactions shown below have similar mechanisms. Suggest a mechanism for both of these.

(a) $$BrCH_2C(CH_3)_2{-}C(=O)O^{\ominus} \xrightarrow{\text{heat}} CH_2{=}C(CH_3)_2 + CO_2 + Br^{\ominus}$$

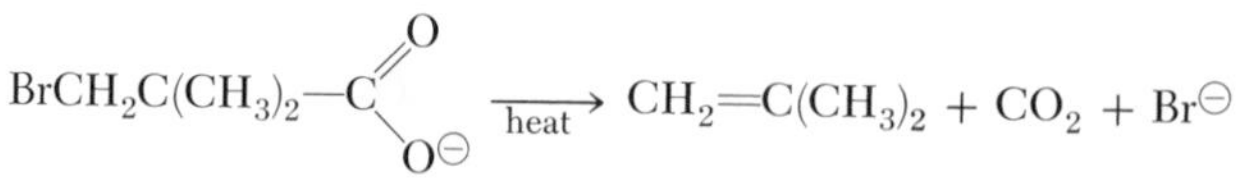

(b) $$C_6H_5{-}CH(Br){-}CH(Br){-}C(=O)O^{\ominus} \xrightarrow{\text{heat}} C_6H_5{-}CH{=}CHBr + CO_2 + Br^{\ominus}$$

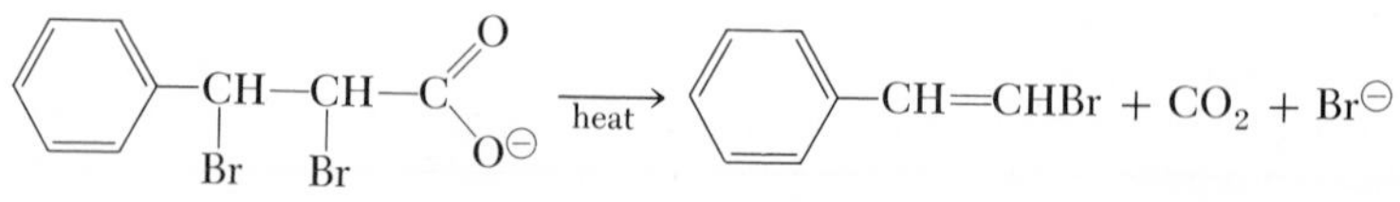

26. Do you agree with the following? "Enantiotopic hydrogens participating in E2 dehydrohalogenation reactions must produce identical products, while diastereotopic hydrogens must produce diastereomers." Explain your answer, using specific illustrations whenever possible.

THE STEREOCHEMISTRY OF RING SYSTEMS

7.1 INTRODUCTION

A **cycloalkane** is a hydrocarbon with a cyclic skeleton containing only sp^3 hybridized carbon. A **spiro** compound results when two rings share one atom. When two rings share two adjacent atoms, the system is said to be **fused.** When two rings share non-adjacent atoms, the system is **bridged** and the atoms common to both rings are called *bridgeheads.*

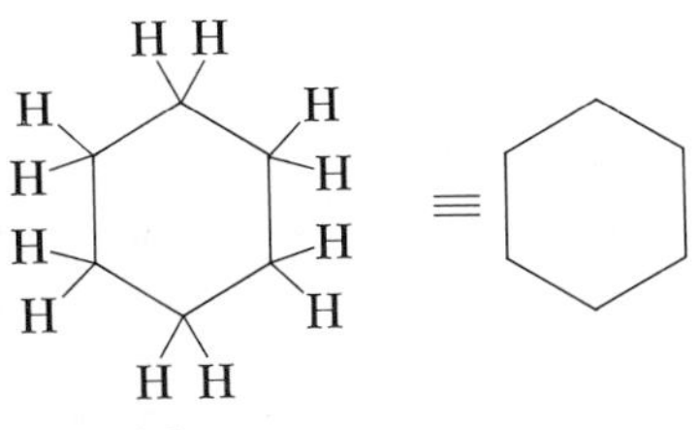

cyclohexane
(monocyclic carbocycle)

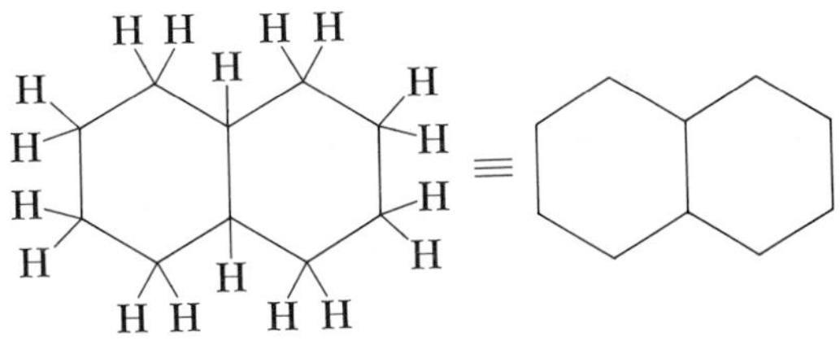

decalin
(fused bicyclic hydrocarbon)

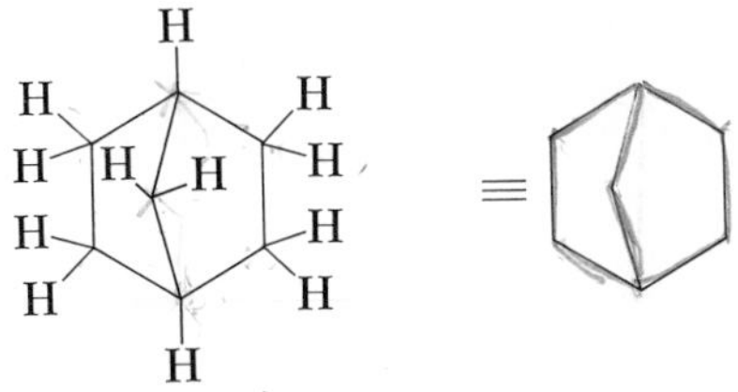

bicyclo[2.2.1]heptane
(bridged bicyclic hydrocarbon)

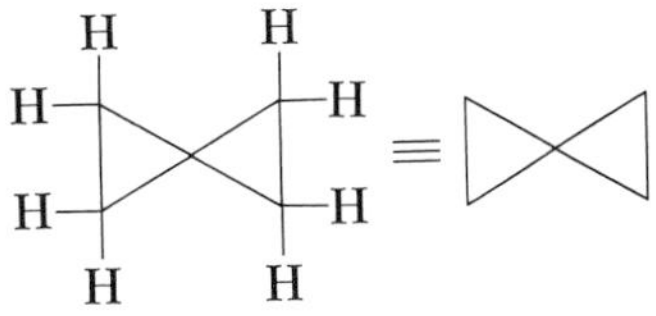

spiropentane
(spiroalkane)

Rings containing heteroatoms (atoms other than carbon) are **heterocycles,** a term distinguishing them from the **carbocycles** described above. It is not unusual to encounter polycyclic compounds in which some rings are heterocyclic and some are carbocyclic.

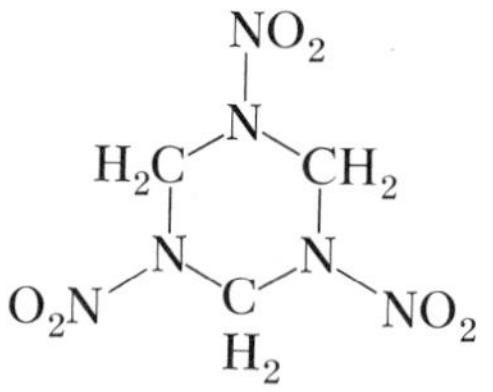

cyclonite
(high explosive)

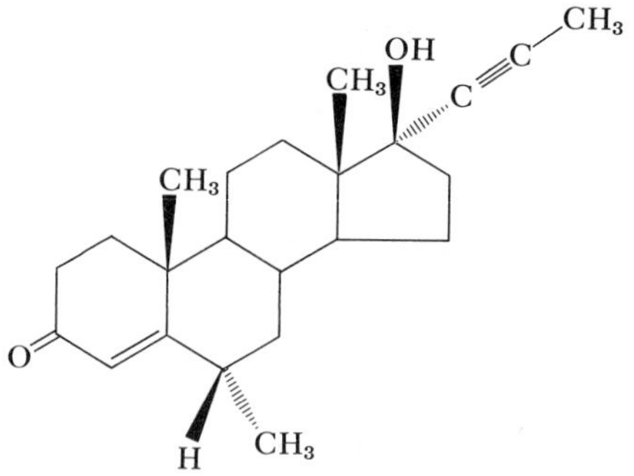

dimethisterone
(oral contraceptive)

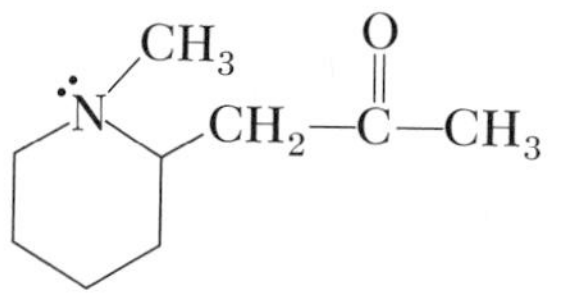

N-methylpelletierine
(treatment for dog tapeworm)

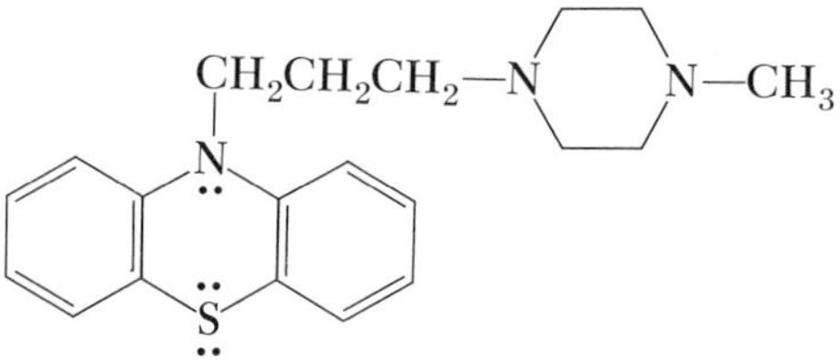

perazine
(tranquilizer)

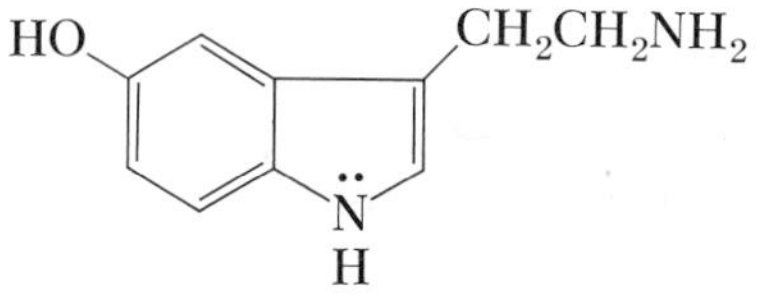

serotonin
(vasoconstrictor)

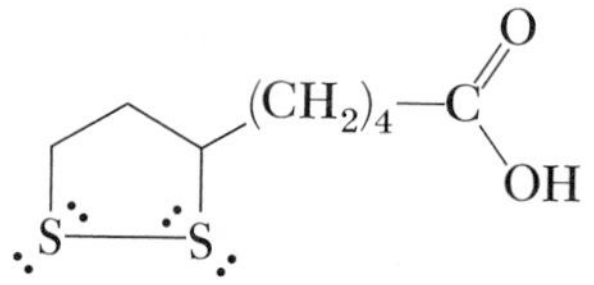

lipoic acid
(biological oxidative decarboxylation)

In the preceding discussion we introduced the very common procedure of abbreviating carbocyclic rings by idealized polygons. Thus, the simplest designation of a three-membered ring is an equilateral triangle, that of a six-membered ring is a hexagon, and so on. In this convention it is assumed that the intersection of two lines represents a $—CH_2—$ group. If such a methylene group is replaced by a heteroatom, then that heteroatom replaces the intersection. Other extensions of this shorthand representation include the use of a single line to represent a methyl group and the use of three intersecting lines to represent a methine group ($-\overset{|}{\underset{|}{C}}-H$). Several examples should help make this convention clearer and demonstrate how it can serve to abbreviate alkyl group structures as well as ring systems.

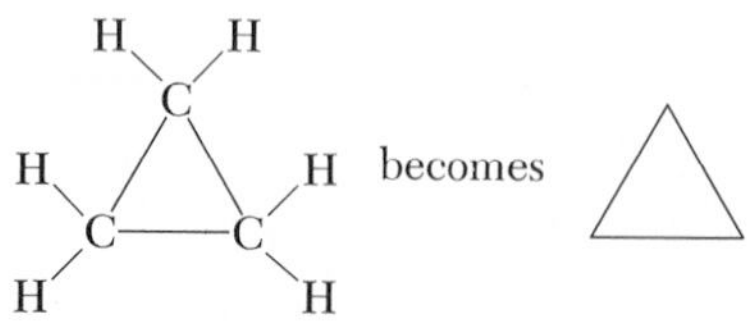

becomes

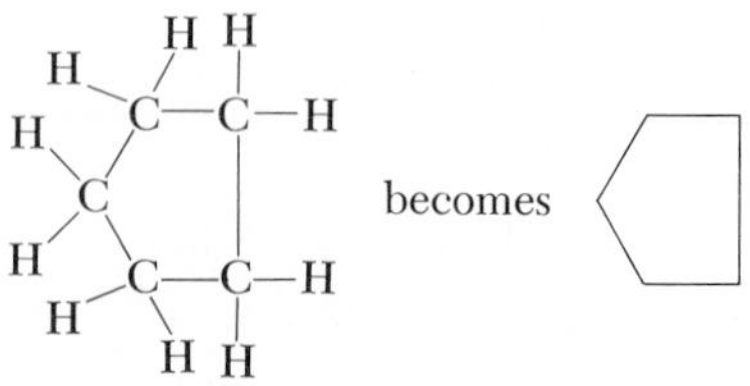

becomes

1. For each abbreviated structure below, draw the corresponding complete structure showing all atoms and bonds.

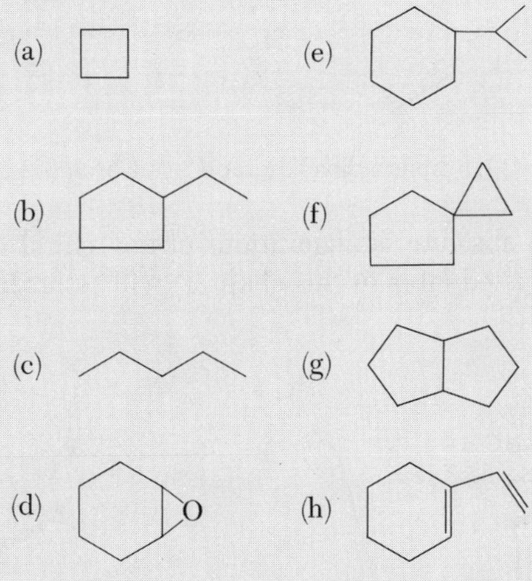

Identify each of these structures using one or more of the following terms: *acyclic, monocyclic, bicyclic, bridged ring system, spiro ring system, heterocycle, alkene, alkyne.*

7.2 NOMENCLATURE

There are too many ring systems known to organic chemists to discuss them all in this section, in this chapter, or even in this book. If you want some idea of how many ring systems are known, you can examine The Ring Index and its supplements.[a] We will limit this discussion to simple cycloalkanes and bridged bicyclic systems.

The **simple cycloalkanes,** $(CH_2)_n$, are named by adding the prefix *cyclo* to the name of

---

[a] "Ring Index," *2nd* ed. and supplements, A.M. Patterson, American Chemical Society, Washington, D.C., 1960.

TABLE 7-1 Cycloalkanes, $(CH_2)_n$

n	COMPOUND	BP (MP)	CLASSIFICATION
3	cyclopropane	−33	small ring
4	cyclobutane	13	small ring
5	cyclopentane	49	common ring
6	cyclohexane	81	common ring
7	cycloheptane	118	common ring
8	cyclooctane	149	medium membered ring
9	cyclononane	178	medium membered ring
10	cyclodecane	201	medium membered ring
11	cycloundecane	(4)	medium membered ring
12	cyclododecane	(63)	large ring[a]

[a] Rings larger than cyclododecane are all considered as *large rings*.

the corresponding straight chain hydrocarbon (Table 7–1). Substituted cycloalkanes are named and numbered in the same way as are their acyclic (non-cyclic) counterparts.

cyclopentane 1,1-dimethylcyclohexane 1-bromo-3-chlorocyclohexane

The designation of the absolute configurations of any chiral centers or the geometry of *cis,trans* isomers (to be discussed later in this chapter) will result in the appending of prefixes to these simple names.

(1R,2S)-1,2-dimethylcyclohexane (3S)-1,1-dichloro-3-ethylcycloheptane

Monosubstituted cycloalkanes are sometimes named as derivatives of the *cycloalkyl* unit. This is particularly common among, though not limited to, halogen-substituted cycloalkanes.

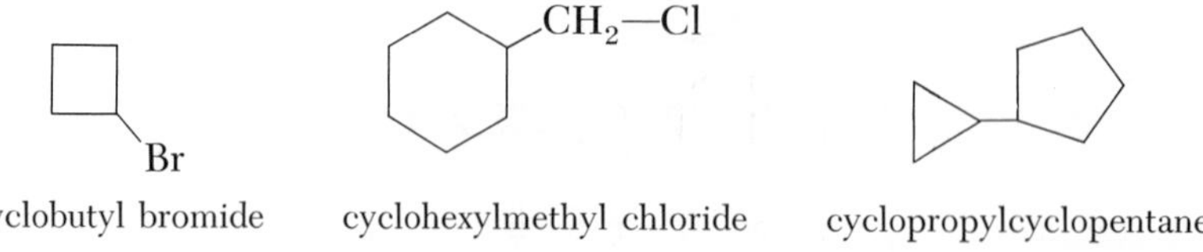

cyclobutyl bromide cyclohexylmethyl chloride cyclopropylcyclopentane

Bicyclic systems derive their basic names from the total number of carbons in the ring system, these being prefixed by the term *bicyclo*. The most highly substituted carbons of the ring system are identified as the bridgehead carbons, atoms which are treated as tying together polymethylene chains of various sizes. The lengths of these bridges are determined, beginning with the longest one, and the values given in brackets in order of decreasing length. While decalin can be viewed as a bridged system in which one bridge has a *zero* length, common fused systems are usually not named as bicyclic systems.

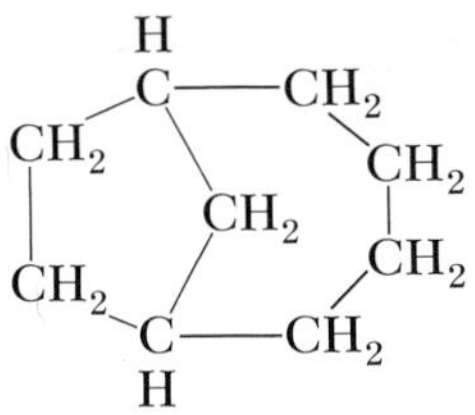

bicyclo[4.2.1]nonane 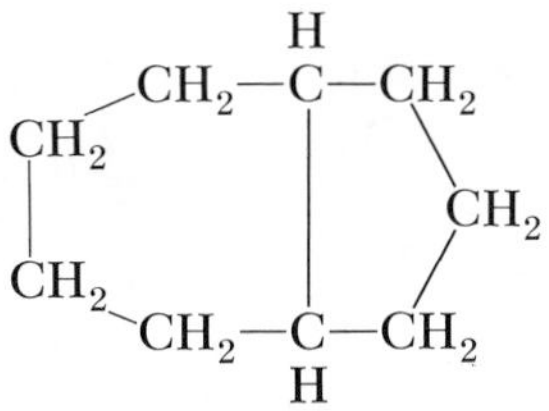 bicyclo[4.3.0]nonane

7.3 THE SHAPES OF RINGS

Cycloalkanes. *Cyclopropane,* an inhalation anesthetic (bp −33°), must have a planar skeleton because three points determine a plane and the ring contains only three carbons. Any of the three hydrogens on one face of the plane is *trans* to any hydrogen on the other face of the plane. Any two hydrogens on the same face of the ring are *cis* and eclipse one another. In general, the term *cis* may be equated with "on the same side," while *trans* may be equated with "on opposite sides."

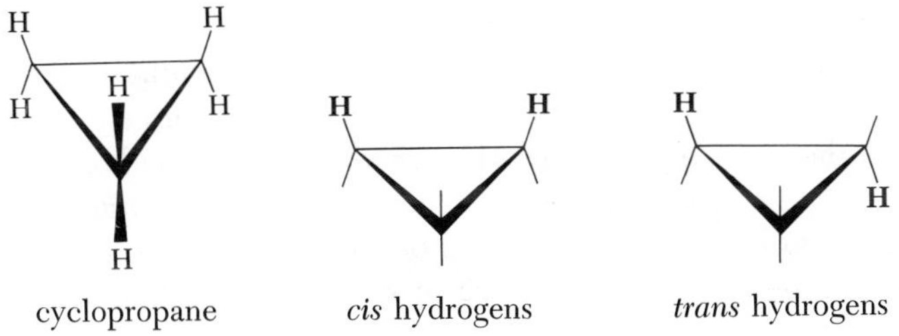

cyclopropane *cis* hydrogens *trans* hydrogens

Cyclobutane is a slightly puckered molecule, but some planar derivatives are known. Since the extent of folding is small, with a resultant low barrier to interconversion, cyclobutane derivatives can be considered to be planar.

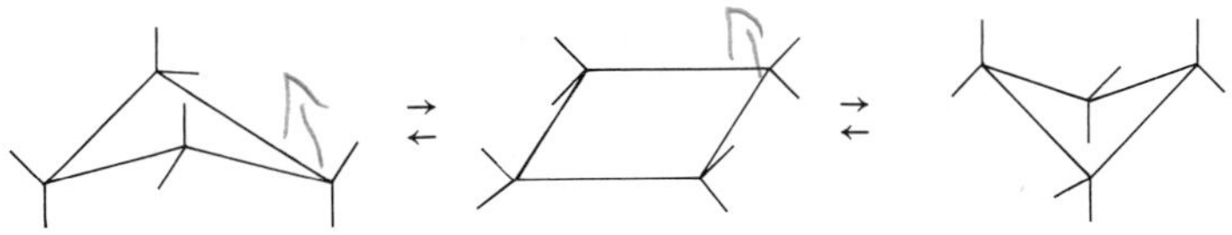

puckered ⇄ planar ⇄ puckered

cyclobutane

The *cis/trans* relationships of representative hydrogens on cyclobutane are presented below. These do not depend upon the shape (flat *vs.* puckered) of the ring.

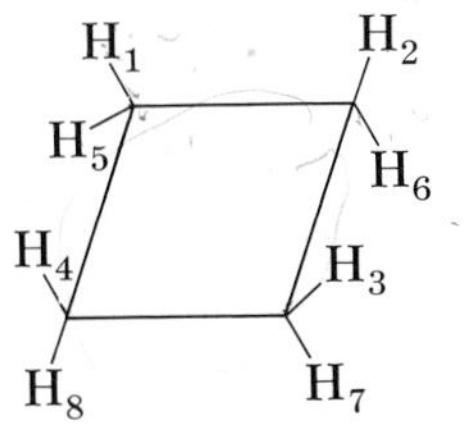

cis	*trans*
H_1, H_2	H_1, H_5
H_2, H_3	H_1, H_6
H_3, H_4	H_1, H_7
H_1, H_4	H_1, H_8
H_1, H_3	H_2, H_8

2. (a) How many pairs of eclipsed hydrogens are present in planar cyclobutane? (b) In puckered cyclobutane?

Cyclopentane is slightly puckered. This puckering travels around the ring by a series of C—C bond rotations (equivalent to one carbon going "up" and an adjoining one going "down"), the entire process being called *pseudo-rotation* (Figure 7–1).

Figure 7–1 Pseudo-rotation in a five-membered ring. The structure in the upper left is planar, while all others exist in the "envelope" form. The dashed line represents a fold; i.e., it marks the intersection of two planes.

Cyclohexane exists in several important conformations, the most stable of which is the **chair conformer** or **chair form.** The chair conformer has a three-fold simple axis of symmetry. Six C—H bonds are parallel to this axis, three pointing "up" and three pointing "down." These six hydrogens are in the *axial* (*a*) positions. The remaining six C—H bonds are nearly perpendicular to the symmetry axis, the hydrogens occupying the *equatorial* (*e*) positions. A given carbon is bonded to two hydrogens, one *a* and one *e*. These bonds are oriented in opposite directions, that is, one is pointing up relative to some idealized molecular plane while the other is pointing down.

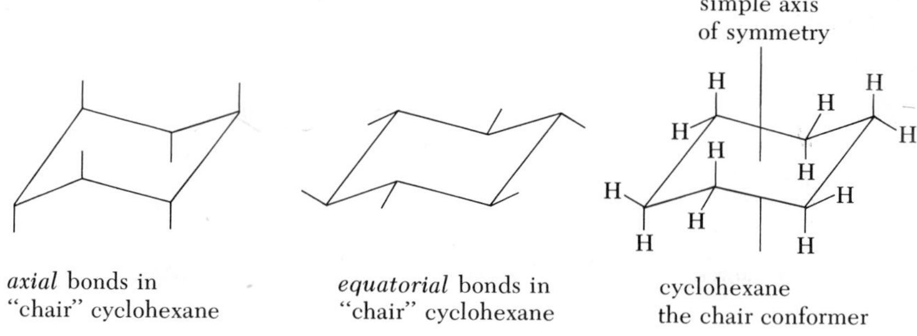

axial bonds in "chair" cyclohexane

equatorial bonds in "chair" cyclohexane

cyclohexane the chair conformer

The hydrogens which are *cis* and *trans* to the equatorial and axial hydrogens on Cl (H_1^e and H_1^a) are presented below. Note the repeating . . . *a, e, a, e* . . . patterns.

cis to H_1^a: H_2^e, H_3^a, H_4^e, H_5^a, H_6^e
trans to H_1^a: H_2^a, H_3^e, H_4^a, H_5^e, H_6^a
cis to H_1^e: H_2^a, H_3^e, H_4^a, H_5^e, H_6^a
trans to H_1^e: H_2^e, H_3^a, H_4^e, H_5^a, H_6^e

It is clear from the Newman projection of "chair" cyclohexane (Figure 7–2) that adjacent hydrogens are not eclipsed. Observe, also, that each four-carbon, butane-like fragment (*e.g.*, carbons bearing H_2, H_4, H_6, and H_8 in Figure 7–2) is staggered. These factors are important in making the chair form the more stable of cyclohexane's two conformers.

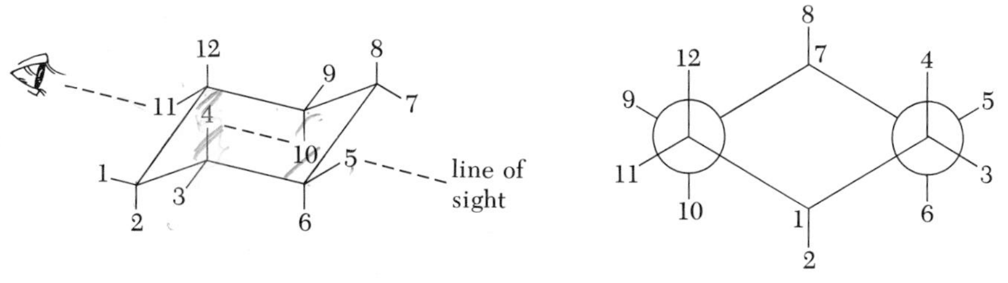

normal representation   Newman projection

Figure 7-2 The Newman projection of cyclohexane. Hydrogens 1 and 7 project directly toward and away from the observer, and thus cannot be seen in this representation. However, their positions are marked by their numbers.

In addition to the chair form, you must learn to recognize the *boat, half-chair* (or *half-twist*), and *twist* forms of cyclohexane. These forms are collectively known as the *flexible form*, although the term is more properly applied to the twist form. (Build a model.)

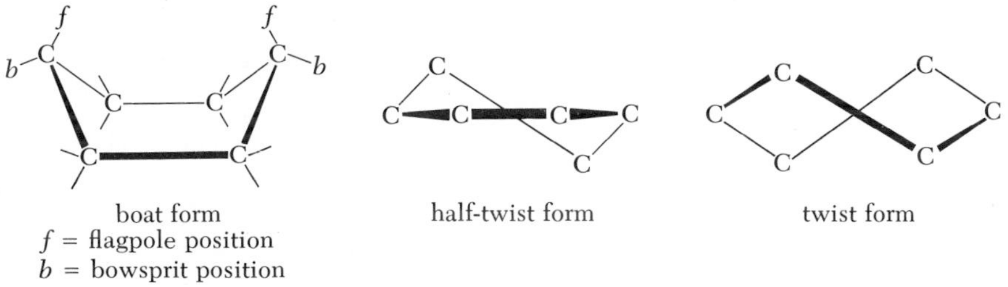

boat form   half-twist form   twist form

f = flagpole position
b = bowsprit position

The boat form represents the transition between various twist forms, while the half-twist form represents the transition between the chair and twist forms (Figure 7–3).

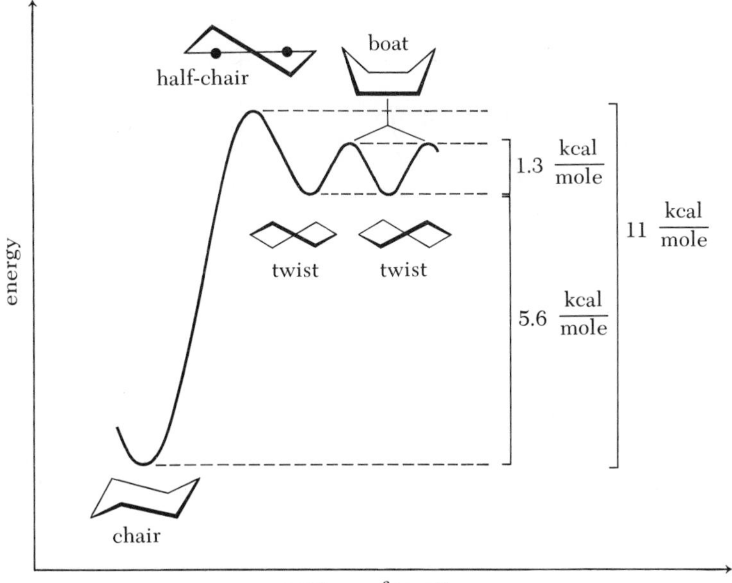

Figure 7-3 Conformational changes in cyclohexane. Only the energy minima (chair and twist forms) are conformers; however, *every* species shown is a conformation of cyclohexane.

The half-twist conformation is also involved in the important process of *chair-chair interconversion* ("ring flipping"). In this process all of those bonds which are *a* become *e*, while those which are *e* become *a* (*cis-trans* relationships do not change). Although the net process is simple, the actual sequence of steps (or "mechanism") is rather complex (Figures 7–4 and 7–5).

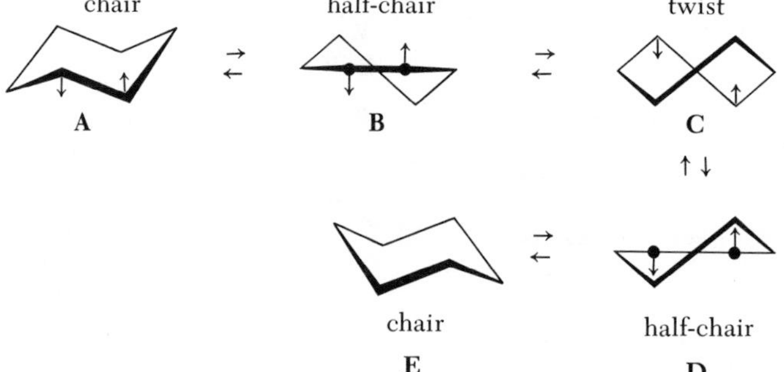

Figure 7–4 Cyclohexane ring inversion. The top line shows the net process, and the mechanism is given below. Vertical arrows indicate displacement of atoms.

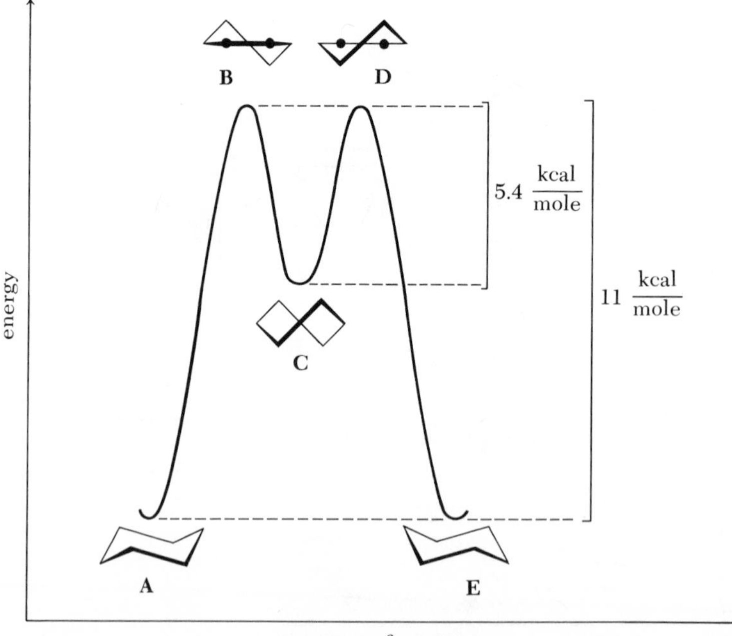

Figure 7–5 Ring inversion in cyclohexane—the energy profile. The E_{act} for inversion is 11 kcal/mole, the barrier between the chair and half-chair forms. The lettered structures correspond to those in Figure 7–4.

Cyclohexane is the smallest ring system encountered which cannot be dismissed as "nearly planar." Reactions of its derivatives are influenced by the orientation (*a vs. e*) of ring substituents and by the conformational mobility of the cyclohexane ring.

3. Identify any planes of symmetry in the (a) chair, (b) boat, (c) twist, and (d) half-twist forms of cyclohexane. Models may be of some help.

The number of conformers increases as the ring size increases. The conformations believed to be preferred in several larger ring systems are presented below.

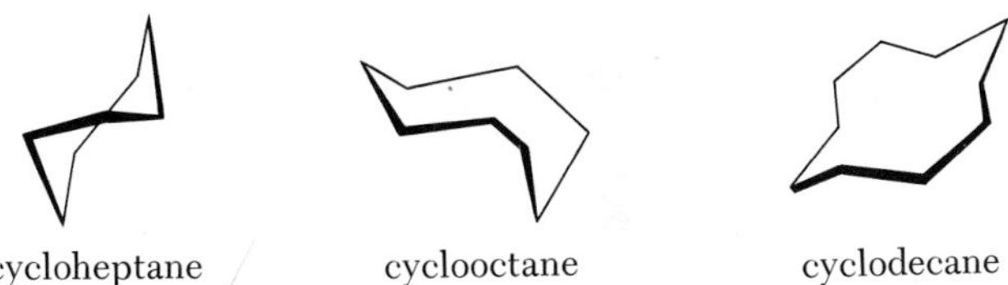

Heterocyclic Rings. Replacement of skeletal atoms of a carbocycle by heteroatoms (most commonly O, N, and S) does not significantly change the picture presented above, although structural details (*e.g.*, bond distances, bond angles, and conformational exchange barriers) will vary. When numbering the ring atoms of these heterocycles, the heteroatom is given the lowest number, "1". When two heteroatoms are present, the one with the higher atomic number is numbered "1" (see *morpholine* below).

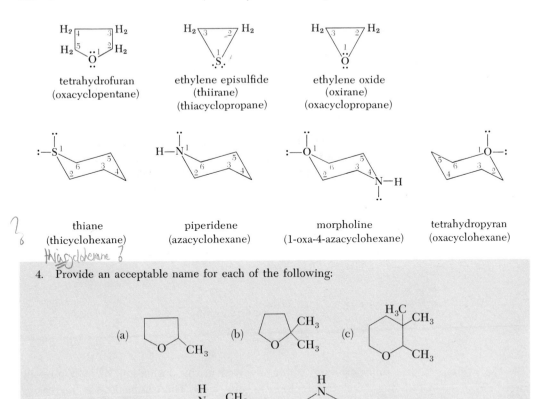

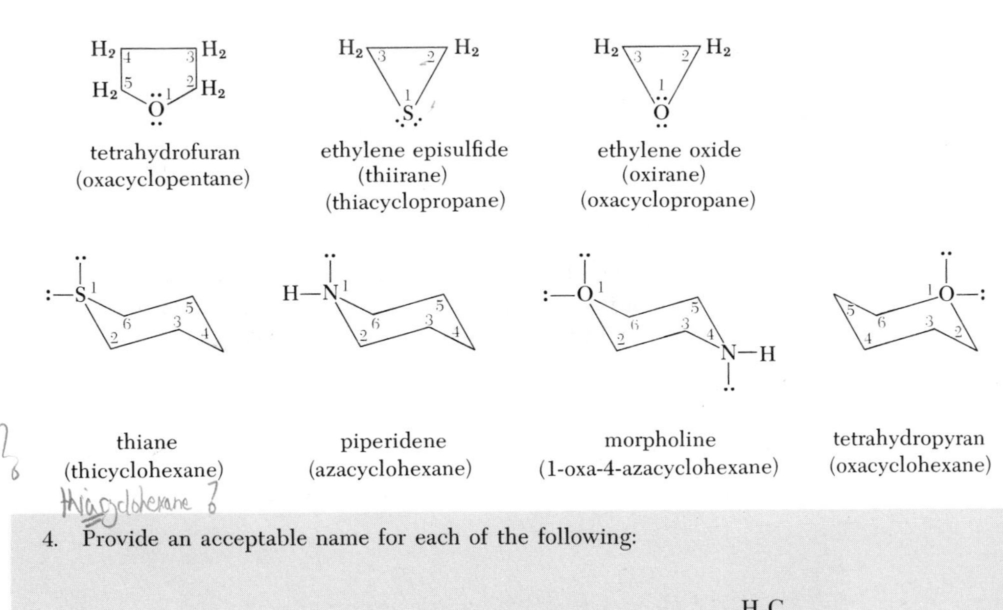

4. Provide an acceptable name for each of the following:

(a) (b) (c)

(d) (e)

7.4 CYCLOPROPANE—A BENT BOND

The small C—C—C interatomic angle in cyclopropane (60°), viewed in light of the sp^3 interorbital angle (109.5°), suggests that the C—C bonds of cyclopropane do not involve maximum a.o. overlap. These C—C bonds, which are σ-like but which lack the necessary cylindrical symmetry, have been called "banana bonds," "bent bonds," and "τ-bonds" (τ = *tau*, Greek). The bent bond picture of cyclopropane is presented in Figure 7–6.

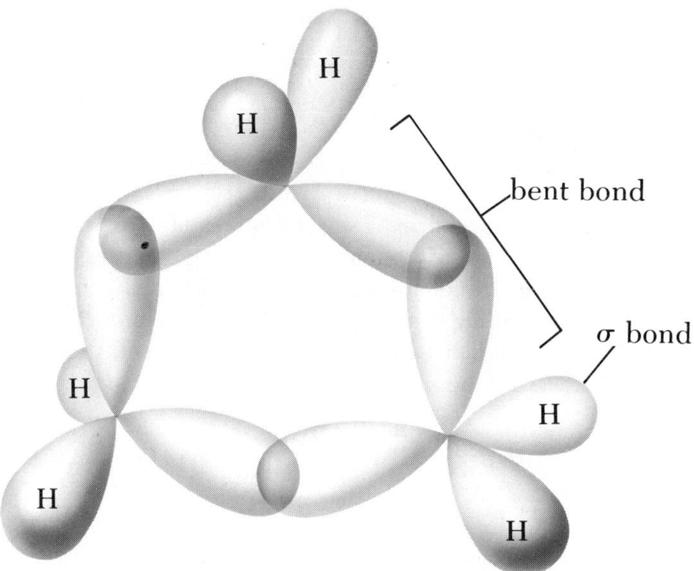

Figure 7–6 The bent bonds of cyclopropane.

The ring of cyclopropane is opened by hydrogen (in the presence of a catalyst such as finely divided platinum) under conditions which do not result in the cleavage of the C—C bonds in cyclopentane. The existence of this reaction attests to the weakness of the bent bond compared with the σ-bond.

$$\text{cyclo-}(CH_2)_3 \xrightarrow[\text{Pt}]{H_2} CH_3CH_2CH_3$$

$$\text{cyclo-}(CH_2)_5 \xrightarrow[\text{Pt}]{H_2} \text{no reaction}$$

The overlap of the "inner orbitals" (those involved in ring formation) of cyclopropane can be improved by bringing these orbitals closer to colinearity. This can be done by increasing the p character of these inner orbitals. In turn, the p character of the carbon orbitals used to bond C to H ("outer orbitals") must decrease, because the *total* number of atomic orbitals undergoing the mathematical manipulation called "hybridization" must remain one s and three p.

The inner orbitals are designated sp^{3+}, while the outer orbitals are sp^{3-} (or sp^{2+}). *Real* cyclopropane probably does a little of both—it violates the goal of maximum overlap but is hybridized to come closer to that goal.

sp^3 sp^3 C sp^3 sp^3 $\equiv$ sp^{2+} sp^{2+} C sp^{3+} sp^{3+}

$a = sp^{3+}$
$b = sp^{2+}$

Cyclopropane is the most acidic of the cycloalkanes, as predicted by the increased *s* character in the outer orbitals.

$$\text{cyclo-}(CH_2)_3 + B^{\ominus} \rightarrow \text{cyclo-}(CH_2CH_2CH^{\ominus}) + BH$$

This picture of modified sp^3 hybridization and bent bonds is not the only possible representation of bonding in cyclopropane. However, the others also predict the enhanced acidity of the cyclopropane hydrogens.

5. Explain the following order of acidity of C—H bonds.

$$Csp^3\text{—}H < Csp^2\text{—}H < Csp\text{—}H.$$

(*Hint:* Consider the electronegativity of the various hybrid orbitals.)

7.5 STRAIN

One might expect that the amount of heat liberated by burning one mole of cycloalkane would increase as the molecular weight increases *but* that the value per methylene group would be constant. The data in Table 7–2 indicate, however, that there is more energy *per methylene*

TABLE 7–2 Heats of Combustion of Cycloalkanes, $(CH_2)_n$

n	HEAT OF COMBUSTION PER METHYLENE GROUP (KCAL/MOLE)
3	166.6
4	164.0
5	158.7
6	157.4
7	158.3
8	158.6
9	158.8
10	158.6
11	158.4
12	157.6

group in cyclopropane than in any other cycloalkane. The heat of combustion per methylene group reaches a minimum at cyclohexane and then begins to rise again, reaching a maximum at cyclononane. This type of behavior, coupled with gradations in chemical reactivity, has led to the categorization of rings presented in Table 7–1, p. 188.

The factors which influence the heat of combustion per CH_2 in this series of cycloalkanes are collectively termed "strain," and the extra energy imparted to a given molecule by this strain is its **strain energy.** The standards of comparison are the cyclohexane ring or a long chain alkane, both of which are considered to be essentially strain-free.

What Types of Strain Are Present in Cycloalkanes? The classic explanation for the unusual reactivity of cyclopropane resides in the *Baeyer strain theory*. Adolf von Baeyer noted[a] that the normal tetrahedral angle is 109.5° but that cyclopropane would be forced to have internal (C—C—C) interatomic angles of 60°, planar cyclobutane would have angles of 90°, and planar cyclopentane would have internal angles of 108°. The strain caused by the deviation from 109.5° for the planar systems represents the *angle strain* (sometimes called *small angle strain*) or *Baeyer strain* in the ring.

Extrapolation of this approach to cyclohexane requires an internal angle of 120° (the internal angle of a hexagon) and would suggest more strain in cyclohexane than cyclopentane. The data (Table 7–2) do not bear this out. The anomaly is resolved when we realize that a critical component of Baeyer's approach is the assumption of a *planar* ring system. Only cyclopentane and smaller rings approximate planarity; thus, cyclohexane and larger rings cannot be expected to follow the predictions made *via* the Baeyer approach. However, these ring systems, as well as non-cyclic compounds, are subject to other forms of strain.

The existence of the barrier to rotation in ethane (p. 56) shows that there is an increase in energy whenever adjacent atoms are not perfectly staggered. Since the bonds are responsible for this interaction, and not the atoms, it is possible for substituents to be separated by a distance greater than the sum of their van der Waals radii and still interact. The effect is termed **bond opposition strain, torsional strain,** or **Pitzer strain.** Its maximal value for two adjacent bonds which are eclipsed is approximately 1 kcal/mole, one third of the ethane rotational barrier.

$1\ \frac{\text{kcal}}{\text{mole}}$

bond opposition strain

If the two substituent groups, *X* and *Y*, in the fragment *X*—C—C—*Y* are large enough to be separated by a distance less than the sum of their van der Waals radii, their mutual steric repulsion can be reduced by changing the torsional angle to something other than the ideal 60° found in the staggered array. The net result is the replacement of **van der Waals strain** with the less energetic torsional strain. Reduction of van der Waals strain may also be achieved by distortion of a molecule from its ideal valence angles. The concommitant increase in strain energy results from **bond angle strain** (also called **angle strain**).

---

[a] A. Baeyer, *Chemische Berichte*, **18**, 2269, 2277 (1885).

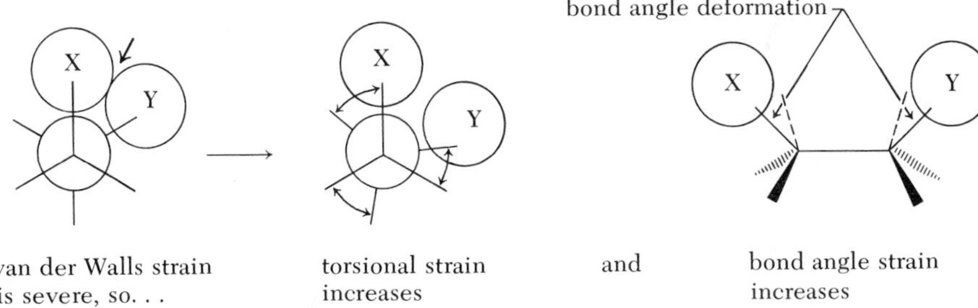

Finally, atoms on opposite ends of a medium-membered ring (see Table 7–1) can approach one another closely enough that they may be repelled by van der Waals interactions. The strain caused by this unusual juxtaposition of groups is called **transannular strain** or **Prelog strain.** A transannular interaction in cyclooctane is shown below.

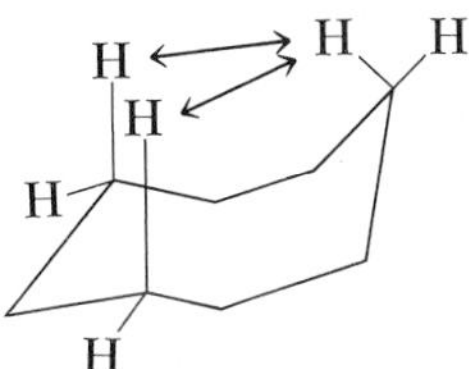

A real molecule, cyclic or acyclic, will acquire *that* geometry which will give it a minimum amount of energy. This explains, for example, why the cyclopentane ring is puckered and why the cyclohexane ring exists largely in the chair form. If these compounds did not adopt these geometries, their strain energy would increase.

7.6 ISOMERISM IN CYCLOALKANE DERIVATIVES

Cyclopropane. While there can be only one monosubstituted cyclopropane, a disubstituted cyclopropane bearing two identical substituents can exist as two diastereomers. Although analogous to *meso* and *d,l* forms, these are traditionally described as *cis* and *trans* isomers. If the substituents are not identical, both diastereomers exist as pairs of enantiomers, analogous to *erythro* and *threo* forms. In addition to these, a 1,1-disubstituted cyclopropane (a structural isomer) is also possible.

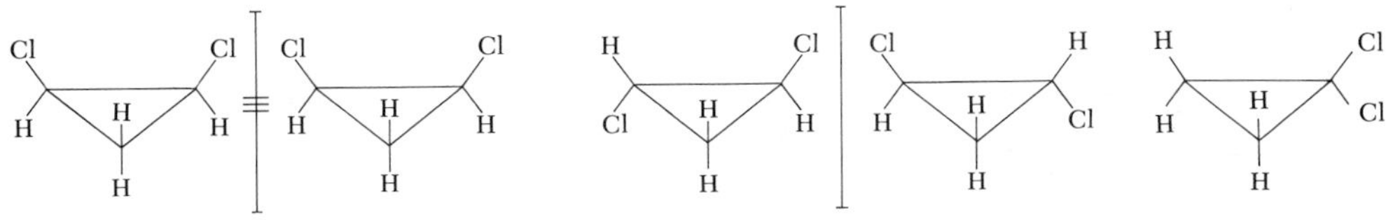

cis-1,2-dichlorocyclopropane

trans-1,2-dichlorocyclopropane

1,1-dichlorocyclopropane

Cyclobutane. Only one monosubstituted cyclobutane is possible. All of the possible dichlorocyclobutanes are shown on p. 198. Of these five, only **C** exhibits enantiomerism. If the two substituents are different, then three isomers are symmetric while two exhibit enantiomerism. Can you identify these five?

The dot convention is a useful method for abbreviating structures of isomers of cyclic compounds in which the ring is drawn as an idealized geometric shape. If a hydrogen on a carbon bearing a substituent faces you ("sticks up"), it is represented by a dot. The substituent in this case is behind the idealized molecular plane and is connected to the ring *via* a solid line. If the hydrogen is behind the idealized plane it is not shown; the substituent-to-ring bond is again shown as a solid line. As with most abbreviated representations, extraneous hydrogens (especially those of methylene groups) are deleted.

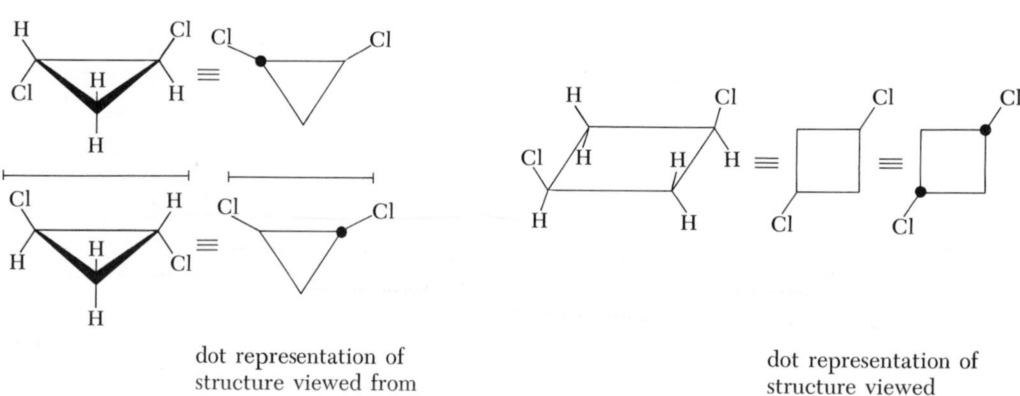

dot representation of structure viewed from the top

dot representation of structure viewed from the top and bottom

Cyclopentane. Five disubstituted cyclopentanes are possible. When the substituents are identical, **C** and **E** exhibit enantiomerism. With non-identical substituents, **B, C, D,** and **E** exhibit enantiomerism.

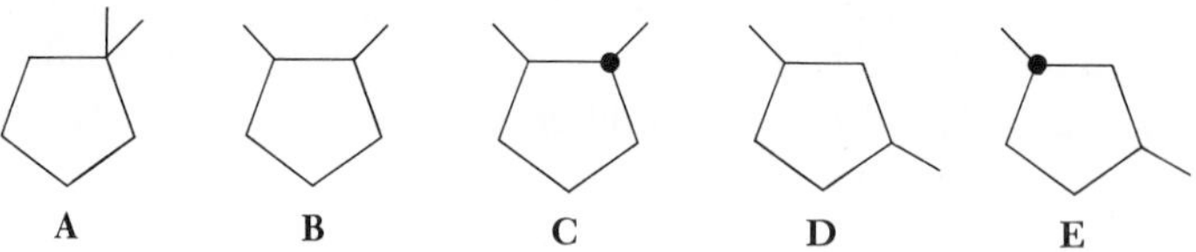

6. (a) Which of the dimethylcyclopentanes possess chiral centers? (b) Which of these compounds is chiral? (c) achiral?

Cyclohexane. Monosubstitued cyclohexanes are achiral if the substituent itself is achiral. However, the *a* and *e* conformers are diastereomers and a monosubstituted cyclohexane does, therefore, exhibit **conformational diastereoisomerism.** The barrier to interconversion of these diastereomers normally is much higher than that found in monosubstituted cyclobutanes and cyclopentanes possessing puckered rings. However, it is still much too low a barrier to permit the isolation of these diastereomeric conformations at room temperature.

conformational diastereomers

The preference of a given group for the *e* or *a* position is commonly expressed as the free energy difference between *equatorial* and *axial* isomers. The data in Table 7–3 indicate that most substituents prefer the equatorial array. The value for the *t*-butyl group shows that *t*-butylcyclohexane is virtually all *equatorial.*

TABLE 7-3 Conformational Distribution of Monosubstituted Cyclohexanes

SUBSTITUENT	$-\Delta G^a$	SUBSTITUENT	$-\Delta G^a$
$—CH_3$	1.7	—I	0.4
$—CH_2CH_3$	1.8	—OH	~0.8
$—CH(CH_3)_2$	2.1	$—OCH_3$	0.7
$—C(CH_3)_3$	>4.4	$—C_6H_5$	3.1
—F	0.2	—CN	0.2
—Cl	0.4	$—CO_2H$	1.2
—Br	0.4	$—NH_2$	~1.5

[a] The free energy difference between *equatorial* and *axial* substituents, in kcal/mole at 25°. The more negative the number, the more stable the *equatorial* conformer. A value of "0" represents an equal distribution of *axial* and *equatorial* conformers. Values are average values and may be influenced by solvent, especially where hydrogen bonding is important.

Why do substituents usually exist in the *equatorial* rather than the *axial* position? A question which provides the same answer, but which is more to the point, is "Why is an *equatorially*-substituted cyclohexane of lower energy than an *axially*-substituted cyclohexane?" The origin of this energy difference can be seen, from one perspective, by comparing the skew-butane-like interactions present in cyclohexane, *equatorial* methylcyclohexane, and *axial* methylcyclohexane. From the structures in Figure 7–7, we can see that cyclohexane and both monosubstituted derivatives possess six skew-butane-like interactions contained solely within the ring. While the *equatorial* methyl group does not generate any additional skew-butane-like interactions, the *axial* methyl group introduces two more. Therefore, one expects *axial* methylcyclohexane to be less stable than *equatorial* methylcyclohexane by 2 × 0.9 or 1.8 kcal/mole. In fact, this value is quite close to the experimentally determined value of 1.7 kcal/mole. A similar argument can be made for substituents other than methyl groups.

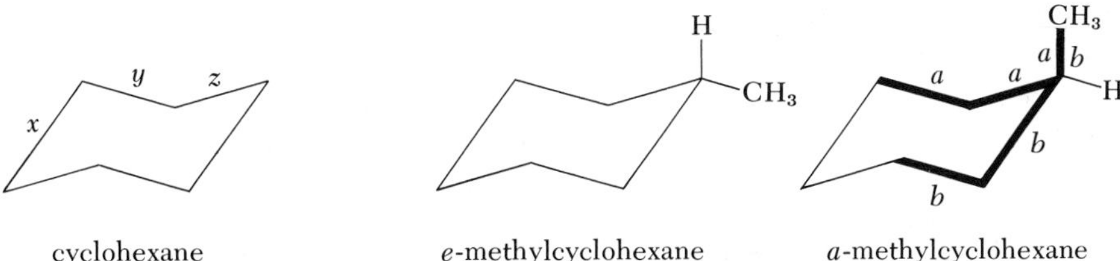

cyclohexane *e*-methylcyclohexane *a*-methylcyclohexane

Figure 7–7 Skew-butane interactions in substituted cyclohexanes. There are six skew-butane-like interactions in cyclohexane and in *e*-methylcyclohexane. These are all restricted to the ring and are typified by the fragment *x-y-z*. Introduction of an axial methyl group produces two additional interactions, marked *a-a-a* and *b-b-b*.

The destabilization of the axial conformer is presented, from a slightly different point of view, in the structure below. Here we emphasize the interaction between two *axial* hydrogens and one of the hydrogens on the *axial* methyl group.

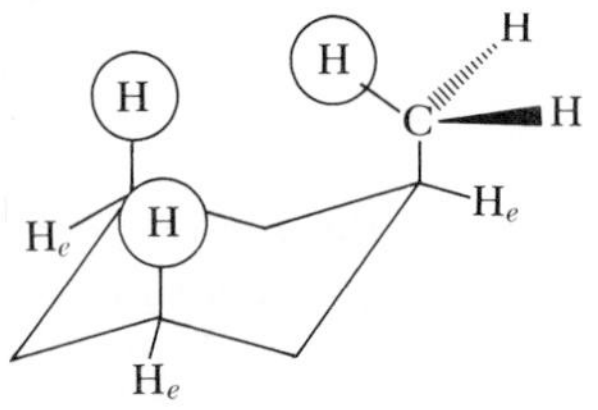

As the substituent increases in size, this type of repulsive interaction is expected to increase. (An apparent anomaly is discussed in problems 34 and 35.) Indeed, the repulsion between an axial *t*-butyl group and the axial hydrogens is so severe that the *t*-butyl group is never axial.

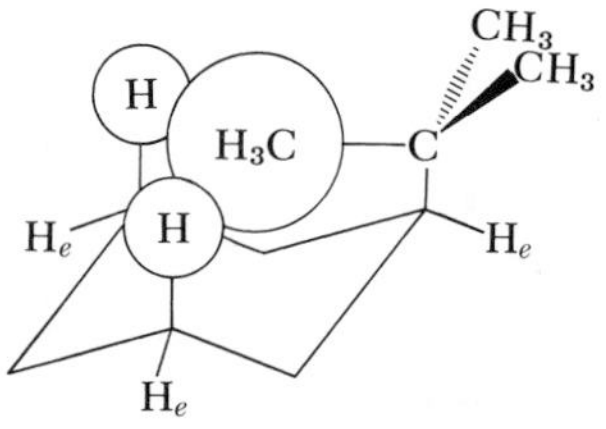

> It was noted in Chapter 2 that dispersion forces, which are attractive in nature, are responsible for our ability to liquefy helium. Dispersion forces between an axial substituent and the neighboring two axial hydrogens might be expected to stabilize an *axially*-substituted cyclohexane. However, in order that these dispersion forces become significant, the substituent must be highly polarizable *and* it must be a rather specific distance from the axial hydrogens. These conditions are rarely met in simple, monosubstituted cyclohexanes.

The possible disubstituted cyclohexanes are shown below. The conformational distributions which will result if the substituents are identical are presented in Chart 7–1.

1,1-	*cis*-1,2-	*trans*-1,2-	*cis*-1,3-	*trans*-1,3-	*cis*-1,4-	*trans*-1,4-
A	**B**	**C**	**D**	**E**	**F**	**G**

Let us refer to Chart 7–1: See opposite page.

While the two conformers of *cis*-1,2 (**B**) are enantiomers, those of *trans*-1,2 (**C**) are diastereomers. The diequatorial conformer of **C** is more stable than the diaxial one because of the larger number of skew-butane interactions in the latter. The *cis*-1,3 isomer (**D**) is analogous to the *trans*-1,2 isomer (**C**) in that it exists as a mixture of conformational diastereomers. The diaxial form of **D** is especially unstable because there are repulsions between substituents themselves (*1,3-diaxial interactions*), amounting to 3.7 kcal/mole when the substituents are methyl groups.

The *cis*- and *trans*-1,4 isomers (**F** and **G,** respectively) are pairs of identical and non-identical conformers, respectively. Compounds of this type have been used to study the behavior of a given group in either the *a* or the *e* position by choosing *t*-butyl as the other substituent.

CHART 7–1 The Conformational Analysis of Disubstituted Cyclohexanes

ISOMER DESIGNATION	TEXT FIGURE[a]	CONFORMATIONAL DISTRIBUTION X	⇄	Y	$K_{eq}\left(\frac{[Y]}{[X]}\right)$
1,1-	A	Cl, Cl	⇄	Cl, Cl	1
cis-1,2-	B		⇄		1
trans-1,2-	C		⇄		<1
cis-1,3-	D		⇄		<1
trans-1,3-	E		⇄		1
cis-1,4-	F		⇄		1
trans-1,4-	G		⇄		<1

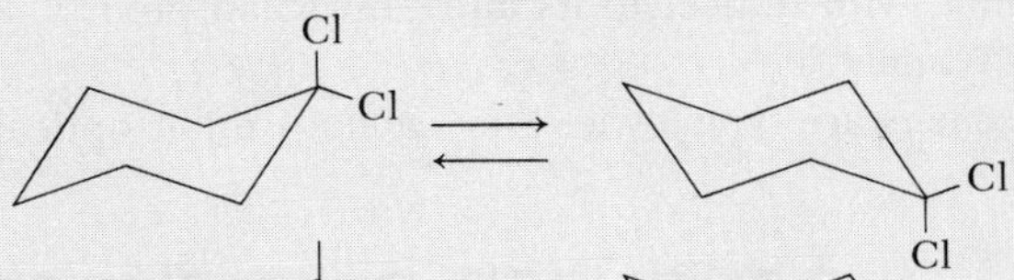

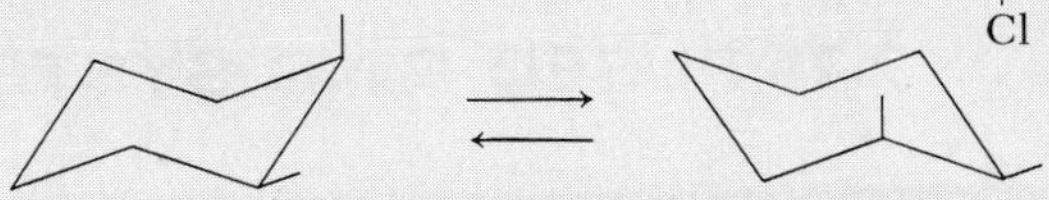

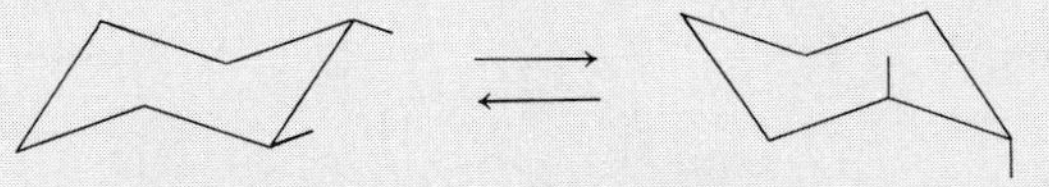

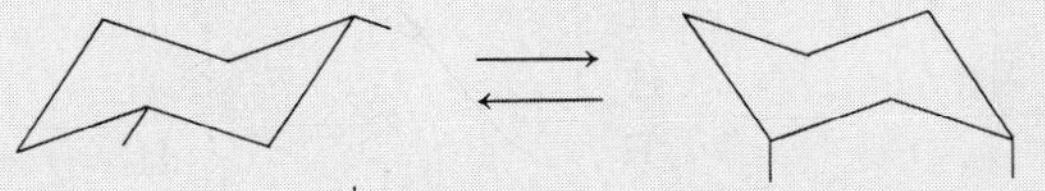

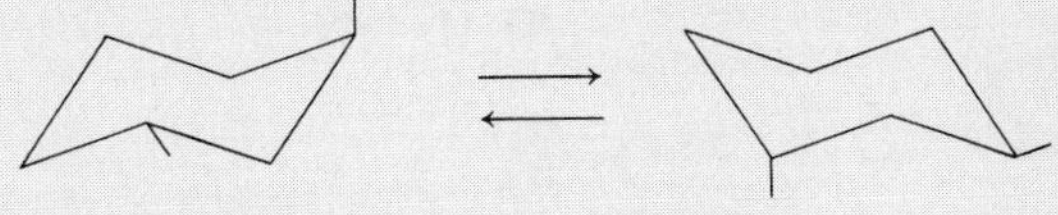

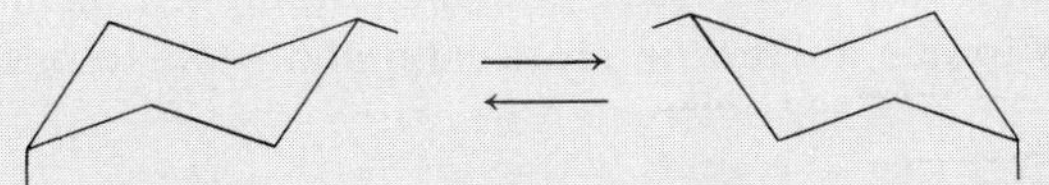

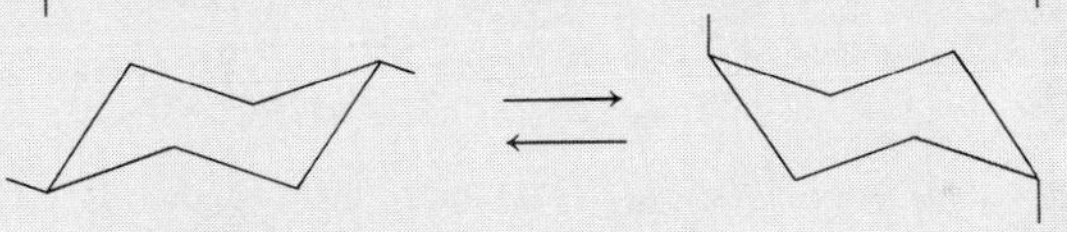

[a] Refers to structures on p. 200.

The high conformational preference of the *t*-butyl group makes it a **holding group,** fixing the conformation of a given stereoisomer into that one possessing an *equatorial t*-butyl group.

$(CH_3)_3C$, H, Br, H ⇄ $C(CH_3)_3$, H, Br, H — pure *axial* Br

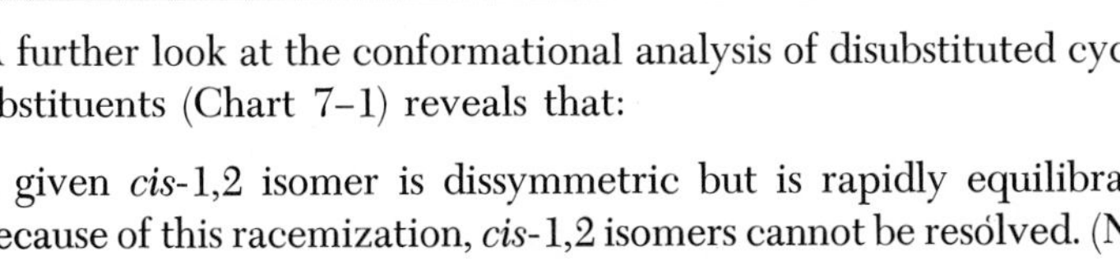

$(CH_3)_3C$, H, H, Br ⇄ $C(CH_3)_3$, H, H, Br — pure *equatorial* Br

7. One of the isomeric 1,3-di-*t*-butylcyclohexanes exists with the ring in a non-chair conformation. (a) Which? (b) Why?

A further look at the conformational analysis of disubstituted cyclohexanes bearing identical substituents (Chart 7–1) reveals that:

1. A given *cis*-1,2 isomer is dissymmetric but is rapidly equilibrated with its enantiomer. Because of this racemization, *cis*-1,2 isomers cannot be resolved. (Note how the dot structure

leads to this conclusion, although for the wrong reason.)

2. A given *trans*-1,2 isomer does not interconvert into its enantiomer by ring inversion and should be resolvable.
3. A given *cis*-1,3 isomer is like the *trans*-1,2 isomer, and the *trans*-1,3 isomer is like the *cis*-1,2 isomer, with respect to its interconversion process. However, both conformers of *cis*-1,3 are symmetric.
4. Both 1,4 isomers are symmetric; they cannot exhibit optical activity.

7.7 FUSED RING SYSTEMS

Decalin consists of two cyclohexane rings sharing a common edge.

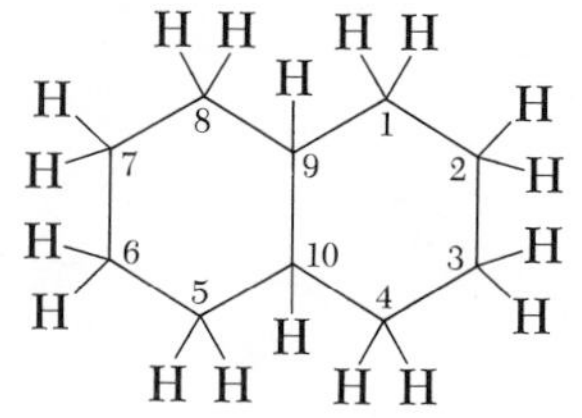

decalin

The two isomers of decalin differ in configuration at C9 and C10. The methine (tertiary) hydrogens are *cis* to one another in *cis*-decalin and *trans* to one another in *trans*-decalin.

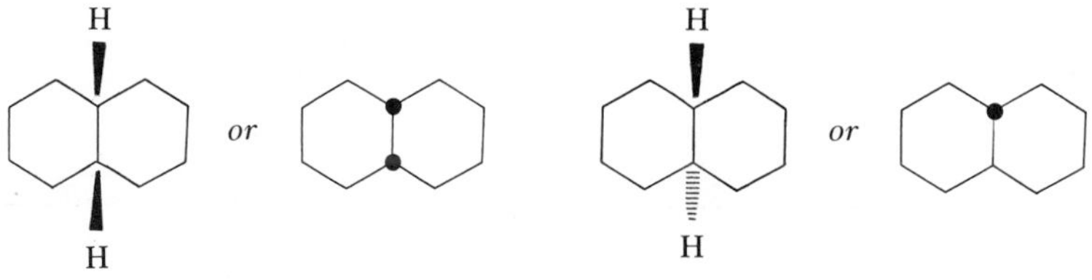

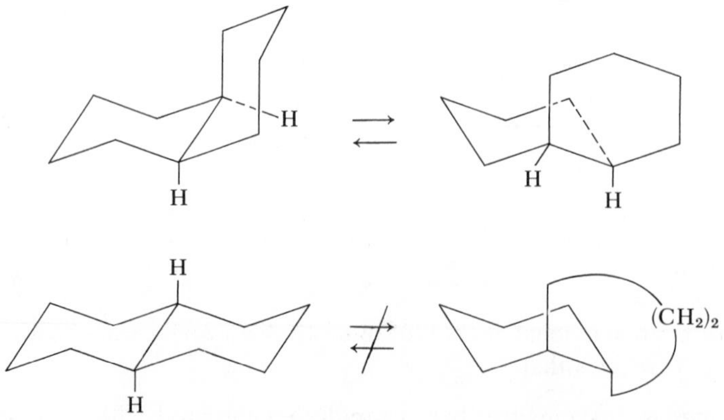

cis-decalin

trans-decalin

While the dot projections correctly suggest the proper number of stereoisomers, only a three-dimensional representation (or molecular models) can show that one conformer of *trans*-decalin is possible while two conformers of *cis*-decalin are possible.

cis-decalin

$(CH_2)_2$

trans-decalin

stable conformation

non-existent conformation

Cis-decalin has three more skew-butane interactions than does the *trans* isomer and is expected to be less stable by 2.7 kcal/mole (3 × 0.9 kcal/mole). This is close to the experimental value of approximately 3 kcal/mole.

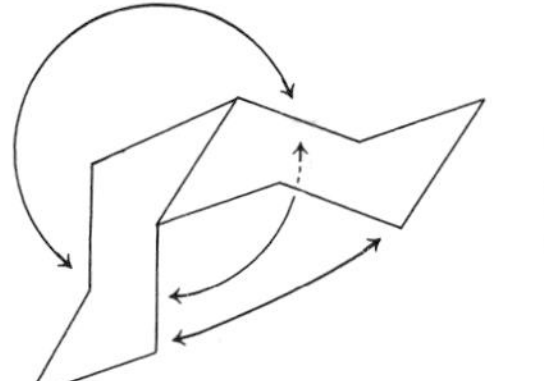

three skew-butane interactions present in *cis*-decalin but absent in *trans*-decalin

Hydrindane contains a five-membered ring fused to a six-membered ring. Like decalin, it can exist in *cis* and *trans* forms. However, these are of nearly equal stability.

cis-hydrindane

trans-hydrindane

Steroids are a group of compounds containing four rings and three ring fusions. The rings are usually indicated by letter designations:

Much of the variation in steroids is found in the stereochemistry of the A/B ring fusion. A "5α-steroid" contains a *trans*-fused A/B junction, while a "5β-steroid" contains a *cis*-fused junction. The designations "α" and "β" are used in ring systems (particularly steroids and sugars) to indicate substituents *below* and *above* the ring, respectively. For these purposes the ring is considered to be flat and is drawn with the A ring at the lower left and the D ring at the upper right. In these specific cases the designations α and β refer to the hydrogens at **C**5.

5α-steroid (A/B *trans*)

5β-steroid (A/B *cis*)

Most steroids have methyl groups at C10 and C13, and these are usually *axial* (*i.e.*, β). Because of their wide distribution, the carbons of these methyl groups are given standard numbers (19 and 18, respectively).

8. Androsterone, a male sex hormone of weaker activity than testosterone, is shown below. Draw this compound in three dimensions, assigning absolute configurations to all chiral centers. (*Note:* Androsterone is a 5α-steroid.)

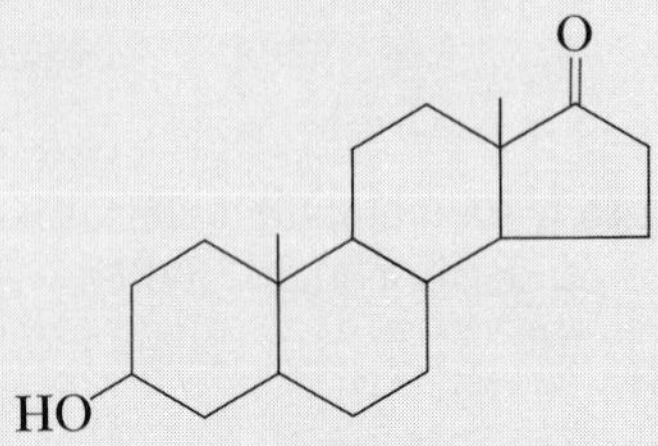

HO O

androsterone

9. The structure of methyl cholate is shown below. Assign an absolute configuration to each chiral center. Draw a planar representation of methyl cholate which conveys this stereochemical information. Is methyl cholate 5α or 5β?

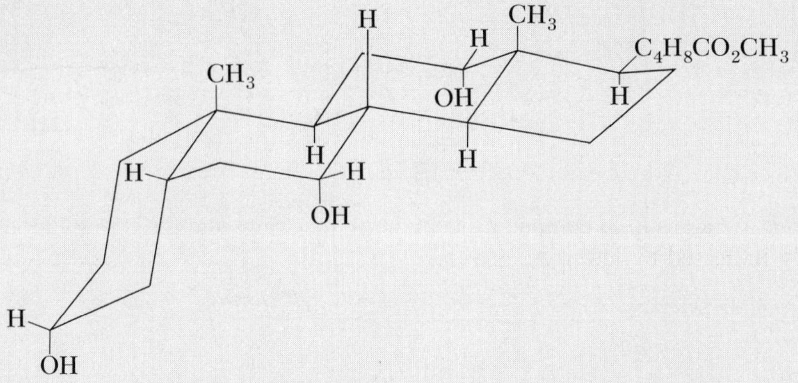

CH_3 H H CH_3 $C_4H_8CO_2CH_3$ OH H H H H H H OH H OH

methyl cholate

10. Identify each C—H bond in the basic 5α steroid skeleton as α or β.

Although the steroid structure resembles a decalin system fused to a hydrindane system, it is sometimes more valuable to view the steroid nucleus as being composed of a five-membered ring (D) fused to a *perhydrophenanthrene* ring system.

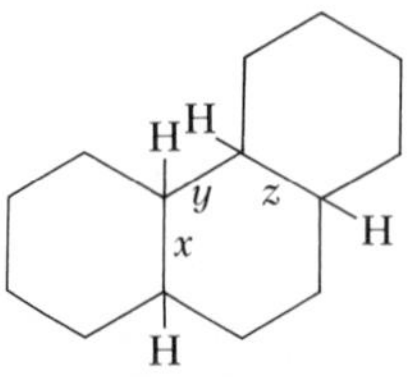

H H y z H x H

perhydrophenanthrene

A special nomenclature exists for representing the relative orientation of the methine hydrogens (hence the geometry of ring fusion) for such polycyclic systems. The relative configuration at bonds *x* and *z* is labeled *trans* or *cis,* while the configuration at y is *anti* or *syn*. The usage is indicated below. Note that the *cis, trans* nomenclature describes the stereochemistry in rings A and C but that *syn* and *anti* refer to the stereochemistry of ring B.

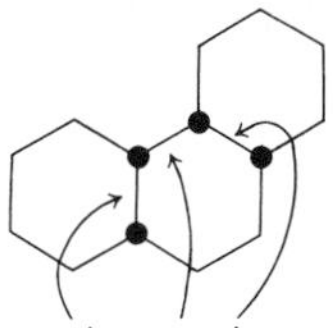

cis-syn-*cis* 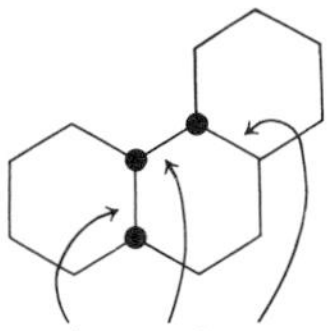 *cis*-syn-*trans* 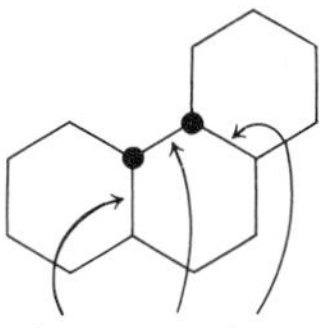 *trans*-syn-*trans*

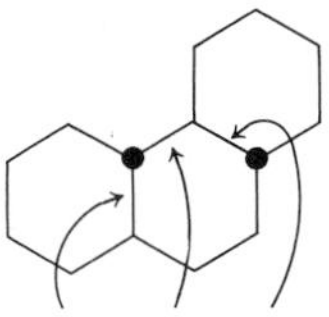

trans-anti-*trans* 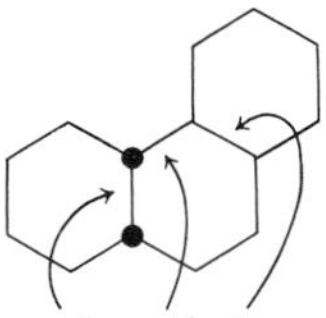 *cis*-anti-*cis*

7.8 POLYCYCLIC RING SYSTEMS

Interest has focused upon two types of polycyclic systems, (a) those which are highly strained and (b) those which may contain a boat-like six-membered ring. Excitement about the former results from the information about bonding which they may provide and because of the (emotional?) need to answer the question "How much strain can it take?" The latter are interesting because they occur extensively in Nature and because of the information they can provide about reaction mechanisms.

Presented below are several highly strained polycyclic alkanes. Their syntheses are unique and, in the vast scheme of chemistry, are probably no more than curiosa. Their existence is a measure of the ingenuity of organic chemists *and* an indication that atoms can take a lot of punishment. Their systematic names are complex and are presented below in order to illustrate how useful common names can be when dealing with very complex molecules.

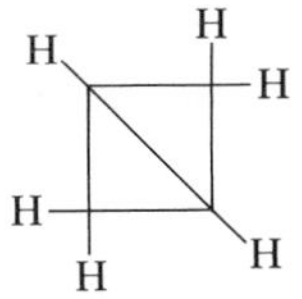

"bicyclobutane"
bicyclo[1.1.0]butane

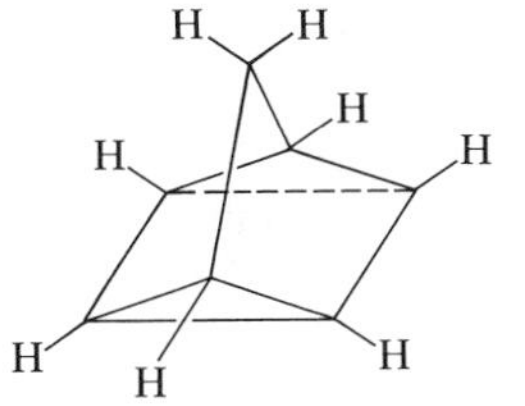

"quadricyclene"
tetracyclo[$3.2.0.0^{2.7}.0^{4.6}$] heptane

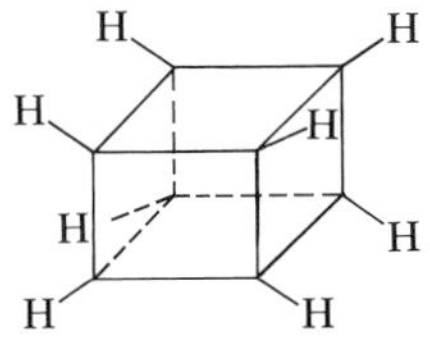

"cubane"
pentacyclo[$4.2.0.0^{2.5}.0^{3.8}.0^{4.7}$]octane

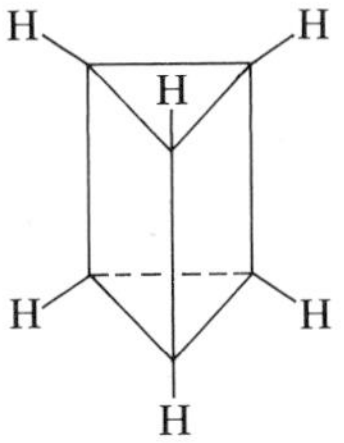

"prismane"
tetracyclo[$2.2.0.0^{2.6}.0.^{3.5}$]hexane

Terpenes are naturally-occurring compounds related to the volatile "essential oils" that can be isolated by steam distillation of plant parts. Found in both plants and animals, terpenes are compounds which are derivable (at least in principle) from *isoprene.*

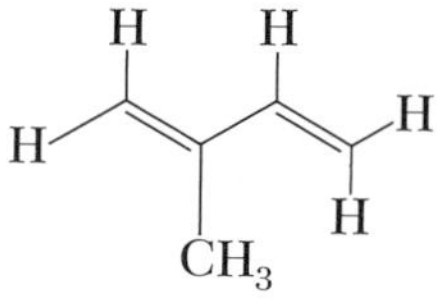

isoprene

While terpenes are discussed in detail elsewhere (Chapter 12), we present the structures of several bicyclic terpenes below. Five are carbocyclic, and one is heterocyclic, but all share the property of having characteristic but not unpleasant odors. This is consistent with the suggestion[a] that the shape of a compound has much to do with its odor.

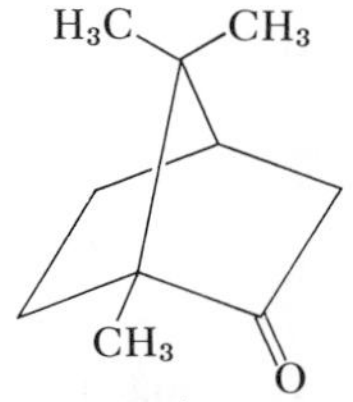

camphor
(oil of camphor)

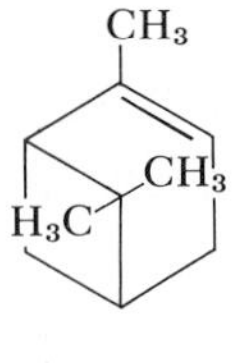

α-pinene
(turpentine)

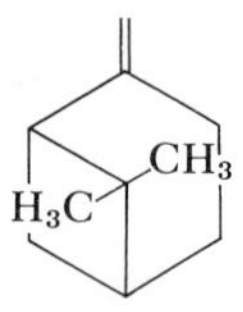

β-pinene
(turpentine)

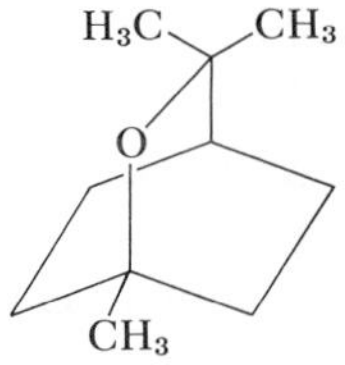

1,8-cineole
(oil of eucalyptus)

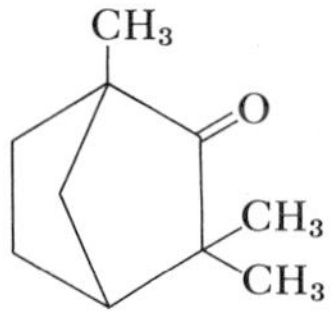

fenchone
(fennel oil)

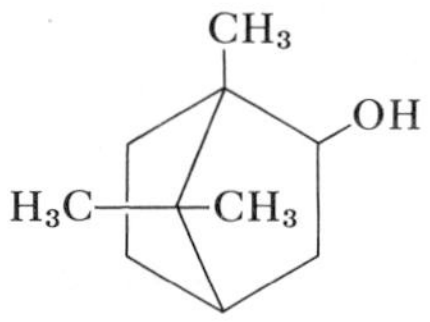

borneol
(oil of camphor)

Camphor, borneol, and fenchone represent the bicyclo[2.2.1]heptane system (also called "norbornane"), while 1,8-cineole represents the bicyclo[2.2.2]octane system. Both of these contain a *boat* six-membered ring; however, the amount of angle strain is greater in the former than in the latter.

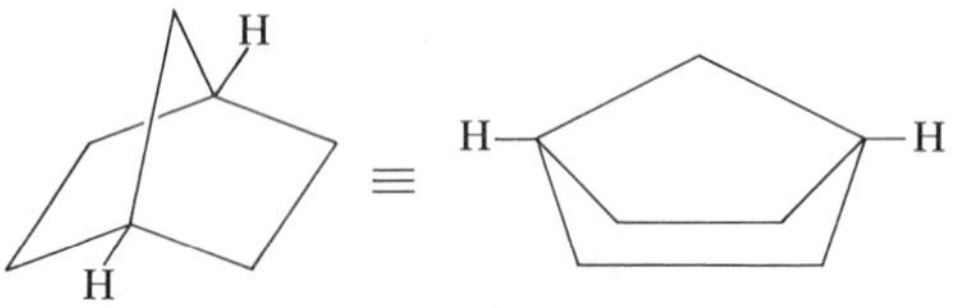

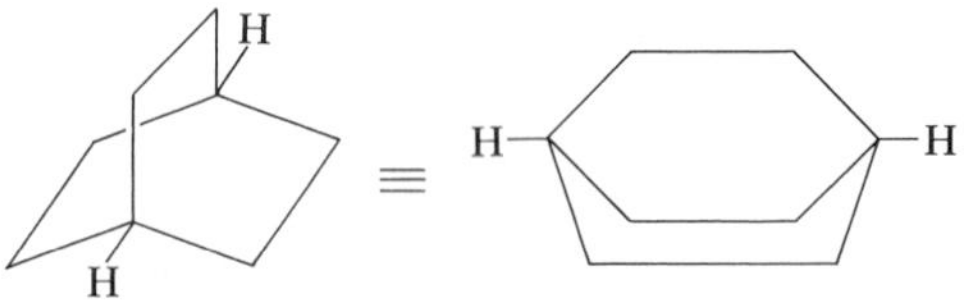

In the [2.2.1] system, a substituent on the boat ring which is directed toward the bridge (*i.e.*, "up") is called *exo,* while one pointed away (*i.e.*, "down") is called *endo.* (What can you say about the [2.2.2] system?) Models may help to show that every bond from the bridgehead

[a] "The Stereochemical Theory of Odor," J.E. Amoore, J.W. Johnston, Jr., and M. Rubin in *Organic Chemistry of Life,* ed. M. Calvin and W.A. Pryor, W.H. Freeman and Co., San Francisco, 1973.

that is in the six-membered ring (*e.g.*, bond *a* in the sketch below) is *anti* to an *exo* substituent (*e.g.*, R). Adjacent substituents (*exo-exo* or *endo-endo*) eclipse one another.

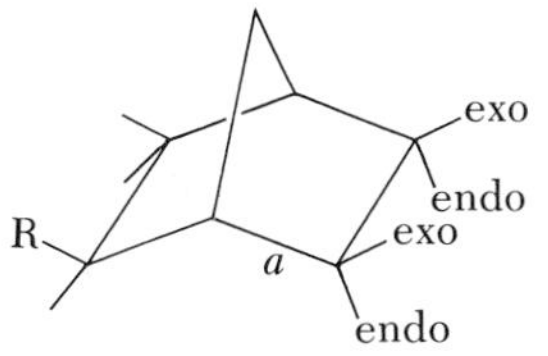

exo: on the same side as the bridge
endo: on the side away from the bridge

Because of the relative unimportance of the *spiro* system (see p. 185), this type of compound will not be considered.

11. (a) Why is it that the terms *exo* and *endo* have no significance in describing the bridged compounds **A** and **B** shown below? (b) What designation could be used to compare **A** and **B**?

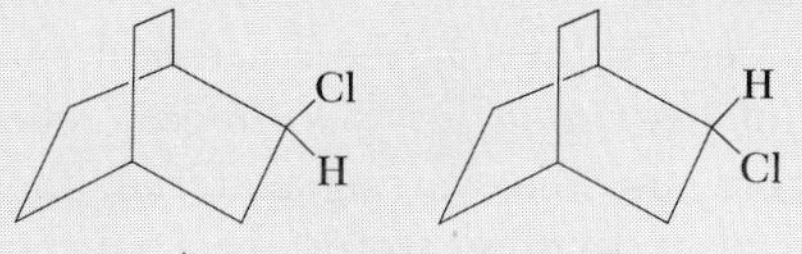

A **B**

7.9 RING STRUCTURE AND REACTIVITY

How does the behavior of a functional group placed on a ring system differ from its behavior in a non-cyclic environment? This question will be answered by looking briefly at the effect of ring structure on the course of S_N and E reactions. Because of their practical importance, much of this discussion will concentrate on substituted cyclohexanes.

S_N **Reactions.** S_N2 reactions proceed with inversion of configuration at the atom undergoing substitution. In an appropriately disubstituted ring system, the result of an S_N2 reaction is the conversion of a *cis* isomer into a *trans* isomer and *vice versa.* For example, *cis*-1-chloro-3-methylcyclopentane reacts with iodide ion to produce *trans*-1-iodo-3-methylcyclopentane; iodide ion reacting with *trans*-1-chloro-3-methylcyclopentane produces *cis*-1-iodo-3-methylcyclopentane.

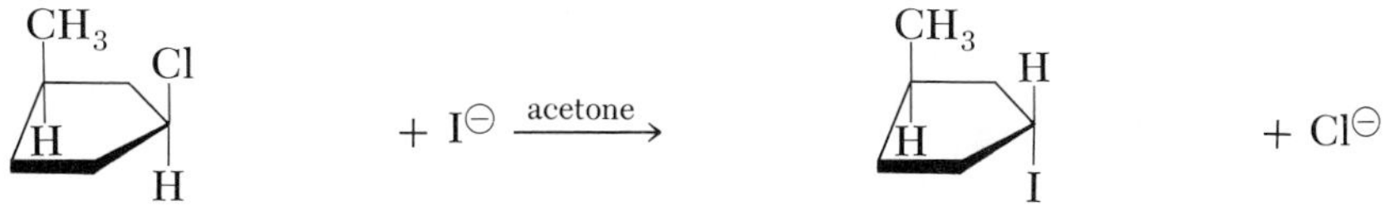

cis-1-chloro-3-methylcyclopentane $+ I^{\ominus} \xrightarrow{\text{acetone}}$ *trans*-1-iodo-3-methylcyclopentane $+ Cl^{\ominus}$

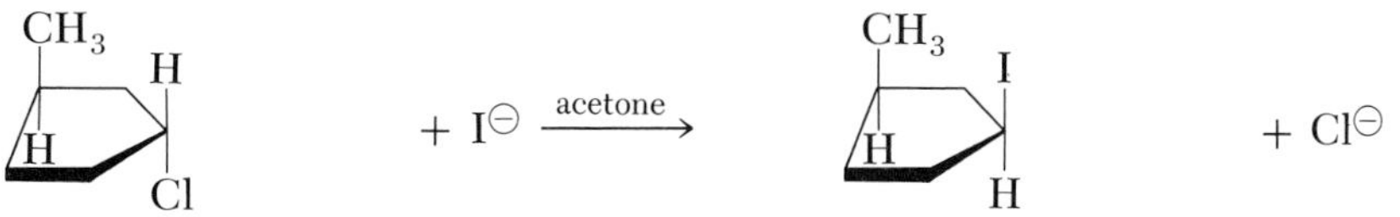

trans-1-chloro-3-methylcyclopentane $+ I^{\ominus} \xrightarrow{\text{acetone}}$ *cis*-1-iodo-3-methylcyclopentane $+ Cl^{\ominus}$

The same stereochemical result is obtained in the cyclohexyl systems, *e.g.*, *cis*-1-chloro-4-methylcyclohexane is converted to *trans*-1-iodo-4-methylcyclohexane by reaction with iodide ion. When the cyclohexane ring is fixed in conformation (for example, by a *t*-butyl group), S_N2 displacement on the ring converts the reactive substituent from one geometry (*e.g.*, *a*) to another (*e.g.*, *e*).

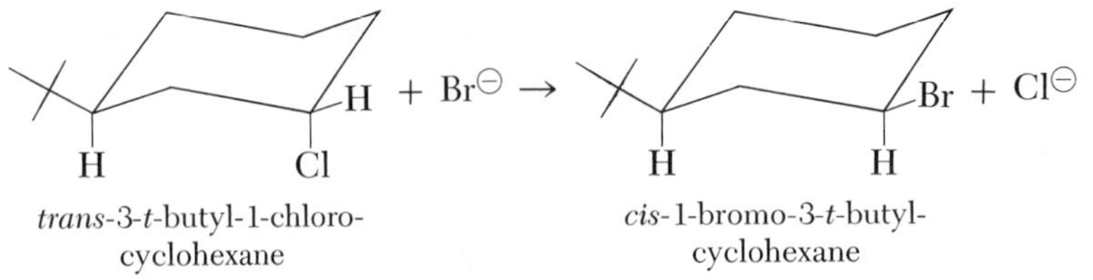

H Cl + Br⊖ → Br H H + Cl⊖

trans-3-*t*-butyl-1-chloro-cyclohexane

cis-1-bromo-3-*t*-butyl-cyclohexane

conversion from *axial* to *equatorial* substitution by an S_N2 reaction

Which substituents are displaced more readily, *axial* or *equatorial?* To answer this question we must examine the activated complexes for displacement of an axial and an equatorial substituent. This is done below for the reaction of *axial* cyclohexyl iodide and *equatorial* cyclohexyl iodide with *I⊖. What is immediately obvious is that the activated complex is the same for both reactions. Therefore, any differences in reaction rate must be due to differences in the energy of the starting materials. Since the *equatorial* conformation of cyclohexyl iodide is more stable than the *axial* (see Table 7–3), the equatorial conformer must have the lower energy. Because the energy of activation for this process reflects the difference in energy between the activated complex and the starting material, the process involving the axial conformer must have the lower E_{act}. In other words, the axial conformer must have the greater reactivity. This may be considered to be an example of *steric acceleration.* This type of argument becomes increasingly complex when the nucleophile and the leaving group are non-identical, and it will not be considered further.

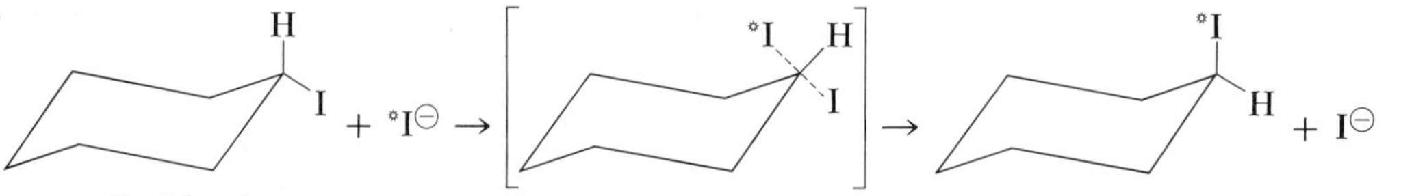

H I + *I⊖ → [*I H I] → *I H + I⊖

e C—I bond

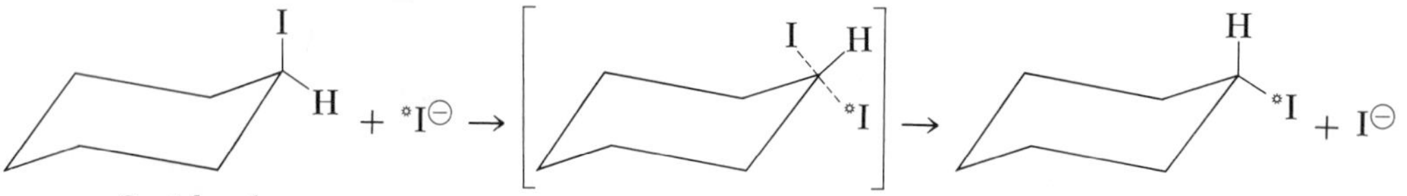

I H + *I⊖ → [I H *I] → H *I + I⊖

a C—I bond

If we consider that both equatorial and axial substituted cyclohexanes ionize to the same carbocation, we can construct an argument when discussing the S_N1 reactivity of substituted cyclohexanes which parallels that describing their S_N2 reactivity. To wit, since the *axially* substituted compound has the higher ground state energy, it should be the most reactive. This is borne out by the observation that *cis*-4-*t*-butylcyclohexyl tosylate is more reactive under S_N1 conditions than is the corresponding *trans* isomer by a factor of about four.

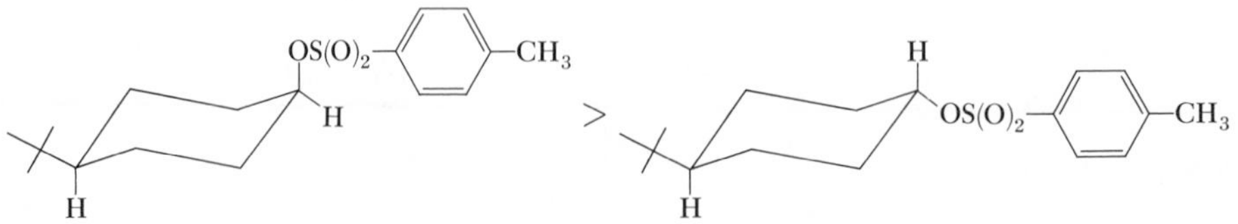

$OS(O)_2$—C₆H₄—CH_3 H H > H $OS(O)_2$—C₆H₄—CH_3 H

cis-4-*t*-butylcyclohexyl tosylate

trans-4-*t*-butylcyclohexyl tosylate

order of S_N1 reactivity

E Reactions. The normal stereoelectronic requirement of the E2 reaction is that the leaving groups be oriented *anti* and coplanar to one another. In the cyclohexane ring system only those substituents which are 1,2-diaxial (*e.g.*, **a** and **d** below) satisfy this constraint. Thus, a substituent which exists in an *equatorial* position cannot act as leaving group in the E2 reaction.

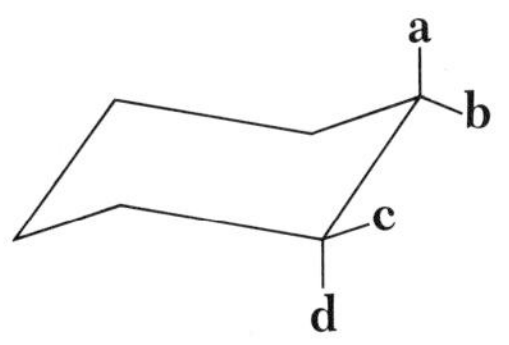

a
b
c
d

An interesting example of this is found in the E2 reactions of menthyl chloride and neomenthyl chloride with ethoxide ion. Menthyl chloride must adopt a conformation in which all substituents are axial in order to have the chlorine in the axial position (which is necessary for elimination). This is a high energy conformation and is therefore only a minor conformer; thus, the conversion of menthyl chloride to 2-methene is rather slow. Furthermore, the alkene which is formed is the less highly substituted product, a violation of Saytzeff's rule.

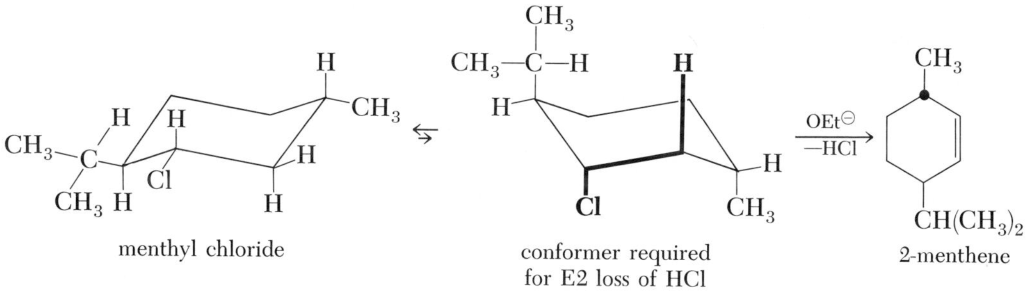

menthyl chloride

conformer required for E2 loss of HCl

$\xrightarrow[-HCl]{OEt^{\ominus}}$

2-menthene

Neomenthyl chloride, on the other hand, possesses an *axial* chlorine in its most stable conformation. Because of the availability of two axial hydrogens (**H**$_a$ and **H**$_b$), one of which is on the carbon bearing the isopropyl group, the E2 reaction of neomenthyl chloride produces both 3-methene (75%) and 2-menthene (25%). This product distribution indicates that the more highly substituted (hence, more stable) alkene is preferred and that the elimination follows Saytzeff's rule.

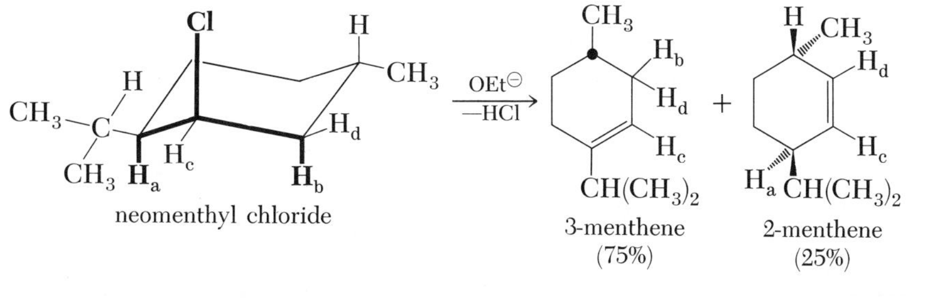

neomenthyl chloride

$\xrightarrow[-HCl]{OEt^{\ominus}}$

3-menthene (75%) + 2-menthene (25%)

12. The E2 reaction of neomenthyl chloride is about 200 times faster than is the E2 reaction of menthyl chloride. Explain.
13. When menthyl chloride is heated in ethanol in the absence of base, it produces 3-methene and 2-menthene. The product ratio is 2:1. Account for these observations.
14. Both *trans*- and *cis*-2-phenylcyclopentyl tosylate give 1-phenylcyclopentene as the only alkene when treated with $\overset{\oplus\ominus}{KO}t$-bu/$t$-buOH. The *cis* isomer reacts fifteen times as rapidly as does the *trans* isomer. Explain.

Bredt's Rule. Whenever a transition state leads to a product with extremely high energy, the anticipated reaction may be prevented from occurring by virtue of a prohibitively high energy of activation. One such product is the compound shown below, in which the would-be double bond has little, if any, *p-p* π overlap.

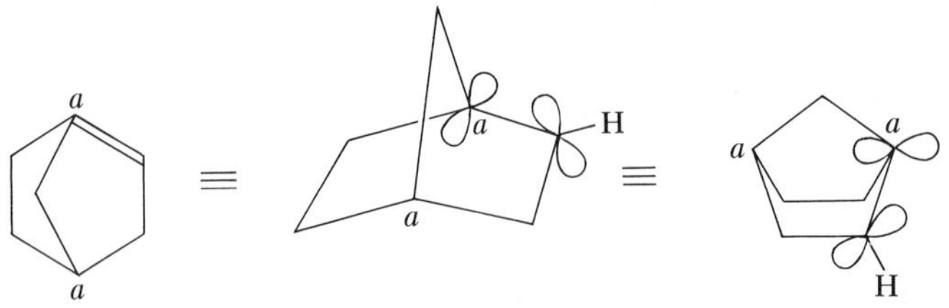

The inability of *p* orbitals on bridgehead positions (*a* in the diagram above) to overlap with *p* orbitals adjacent to them is a statement of Bredt's rule. It is often put, somewhat too simply, "You cannot have a double bond at a bridgehead." In the example below, elimination occurs only across the bond that does not include a bridgehead.

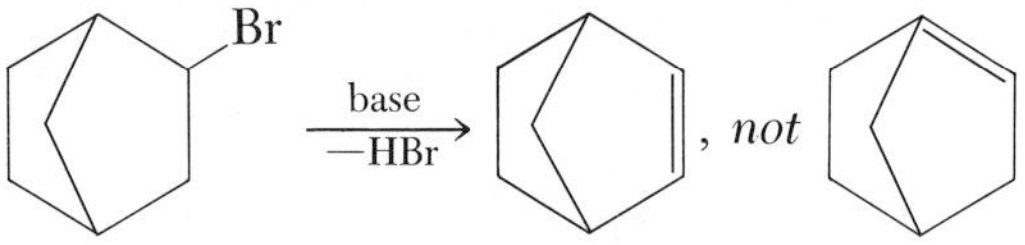

Bredt's rule is violated for very large ring systems, in which the *p* orbitals can adopt positions of maximum π-type overlap. (Build a model!)

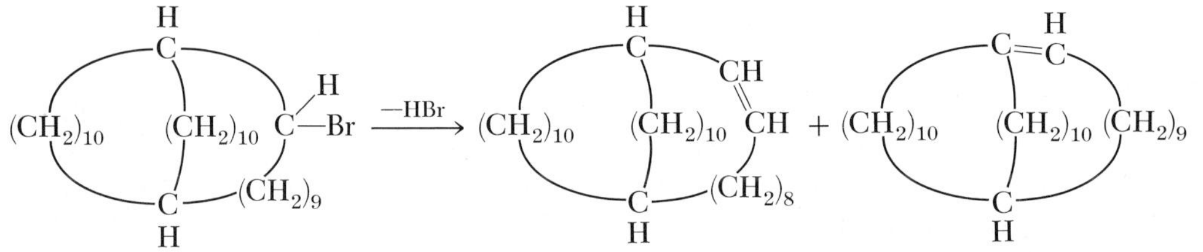

The recently reported (1973) synthesis of the compound shown below suggests that the limits of Bredt's rule may not be as broad as had previously been believed.

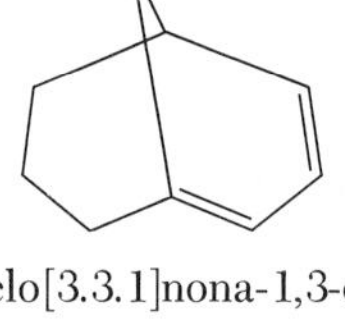

bicyclo[3.3.1]nona-1,3-diene

Cis-Eliminations. A growing number of concerted elimination reactions are being recognized in which departing groups are coplanar and eclipsed. Such reactions, called "*cis*-eliminations," occur only under forcing conditions *and* only if the alternative *anti*- coplanar pathway cannot operate. This latter condition is most often satisfied by the rigid geometry found in certain cyclic systems. The conversion of *exo*-norbornyl chloride to norbornene is illustrative:

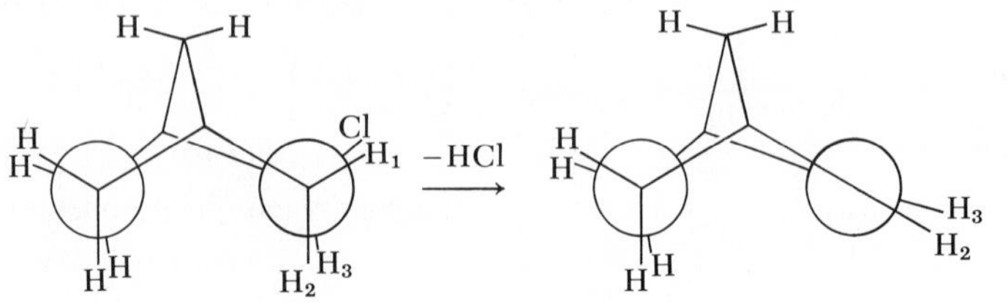

Cis elimination is also observed in one isomer of 11,12-dichloro-9,10-dihydro-9,10-ethanoanthracene, compound **A** below.

very slow → $-HCl$ ← slow $-HCl$

A

IMPORTANT TERMS

Angle strain: Increase in the energy of a molecule owing to deviations from ideal bond angles. Also called "bond angle strain."

Axial bond: A bond on the chair form of cyclohexane which is parallel to the three-fold simple axis of symmetry of the molecule.

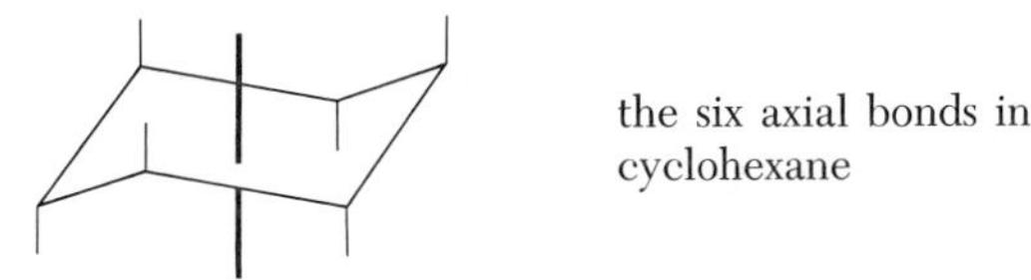

the six axial bonds in cyclohexane

Baeyer strain: Increase in the energy of a ring system owing to differences between the internal angles of regular polygons and the 109.5° angle of sp^3 hybridized carbon. Since this is most applicable to three, four, and five membered rings, it often is called "small angle" strain.

Bent bond: A σ-like bond in which maximum overlap is not attained. In this bond the component a.o.'s do not extend along the internuclear axis. Also called a "τ bond" or a "banana bond." This bond is most commonly associated with the carbon-carbon bond in cyclopropane.

Boat form: A high energy conformation of cyclohexane. The transition state between twist forms.

the boat form of cyclohexane

Bredt's rule: A statement indicating the impossibility of forming a double bond at a bridgehead of a moderately-sized bridged ring system. The impossibility arises because adjacent p orbitals, one of which is on a bridgehead atom, cannot overlap effectively.

Bridgehead positions: Those positions shared by rings in a bridged ring system.

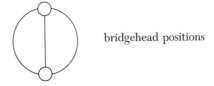

bridgehead positions

Carbocyclic compound: A compound possessing a ring which is made up solely of carbon.

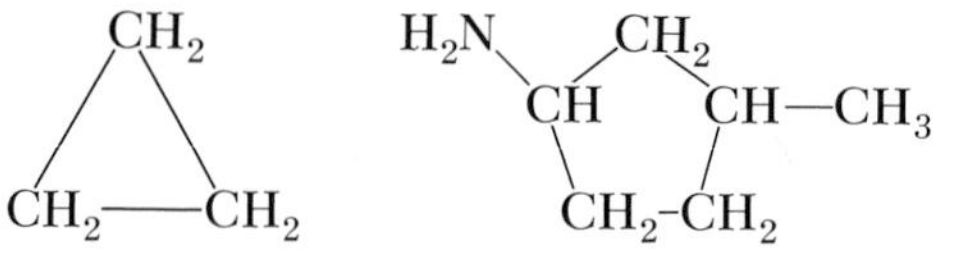

CH_2 / CH_2—CH_2 $\qquad$ H_2N—CH, CH_2, CH—CH_3, CH_2-CH_2

two carbocyclic molecules

Chair form: The most stable conformer of cyclohexane. It possesses a three-fold simple axis of symmetry, six axial bonds, and six equatorial bonds.

the chair form of cyclohexane

***Cis* elimination:** An elimination (most often β) in which the departing groups are eclipsed in the starting material. This pathway is followed only if the *anti* pathway is prohibited. This also is called a *syn* elimination. A relatively rare process.

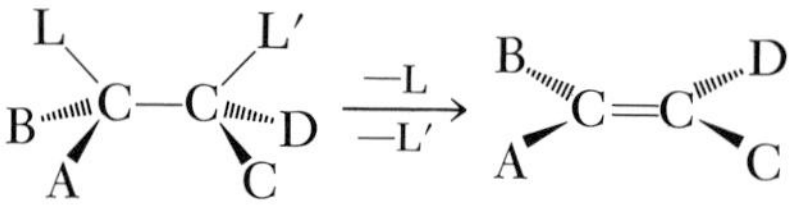

L, B, A—C—C—L′, D, C $\xrightarrow[-L']{-L}$ B, A—C=C—D, C

a *cis* elimination

Cycloalkane: A carbocyclic compound whose ring atoms are all sp^3 hybridized carbons.

***Endo* position:** In a bridged bicyclic molecule, a position pointing away from the shortest chain ("bridge").

←bridge

endo

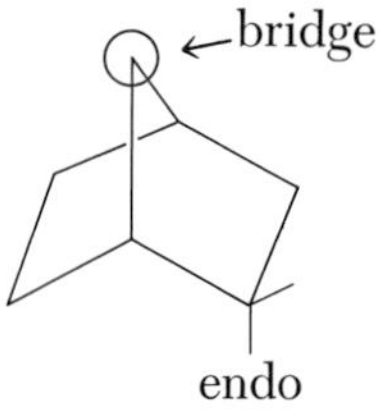

Equatorial bond: A bond nearly perpendicular to the three-fold symmetry axis of cyclohexane. It is tipped either up a bit or down a bit.

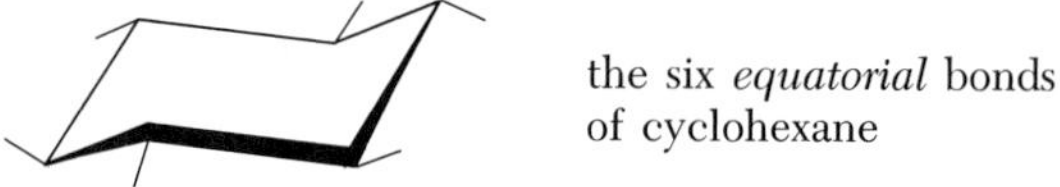

the six *equatorial* bonds of cyclohexane

***Exo* position:** In a bridged bicyclic molecule, a position pointing toward the shortest chain ("bridge").

←bridge

exo

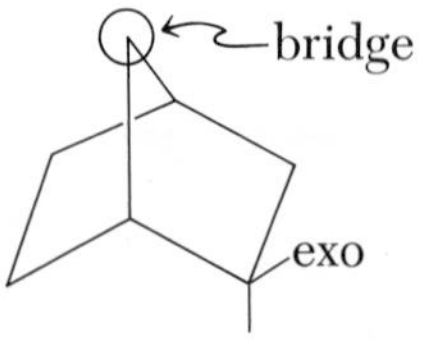

Fused ring system: A molecule containing two (or more) rings, in which rings share adjacent atoms.

fused ring systems

Heterocyclic compound: A molecule in which one or more rings contain at least one atom other than carbon.

heterocyclic ring systems

Torsional strain: The increase in energy of a molecule owing to the eclipsing of σ bonds. The rotational barrier in ethane is due to an increase in this strain in going from conformer to conformer *via* the eclipsed conformation. It also is called Pitzer strain (or bond opposition strain) and has an energy of about 1 kcal/mole.

Transannular strain: Increase in the energy of a molecule arising from repulsion between atoms at opposite ends of a ring. This strain, also called Prelog strain, is associated with the medium-membered rings (C_8–C_{11}).

Van der Waals strain: Increase in the energy of a molecule arising from the repulsion between two (or more) atoms. A result of these atoms being in the same space at the same time. Van der Waals' strain actually is due to the repulsion of the electrons on one atom by electrons on other atoms.

PROBLEMS

15. Using the numbering system below, indicate those hydrogens which are (a) *cis* and (b) *trans* to H_3.

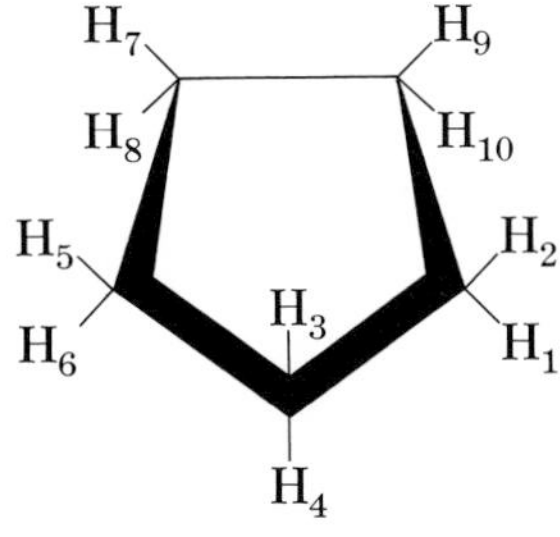

16. For each of the structures below, indicate whether the pairs of hydrogens (H_1 and H_2) are *identical, enantiotopic*, or *diastereotopic*.

(a) (b)

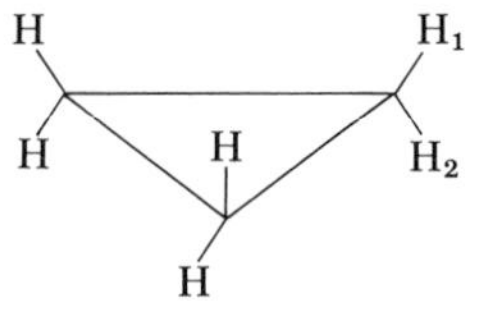

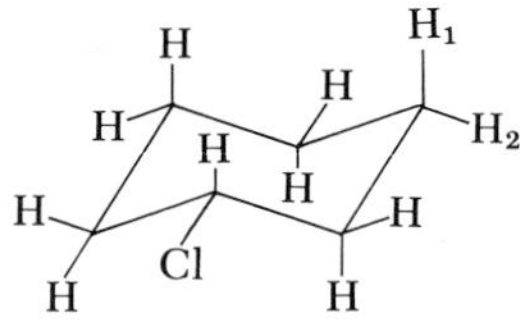

(c) 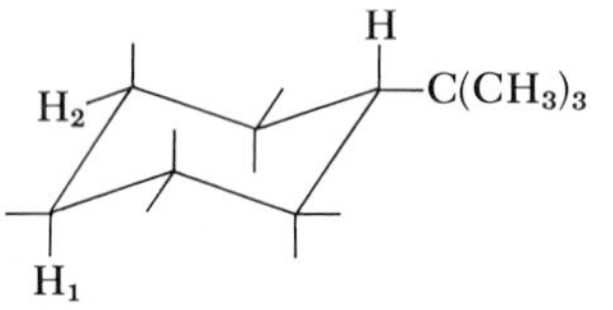 (d)

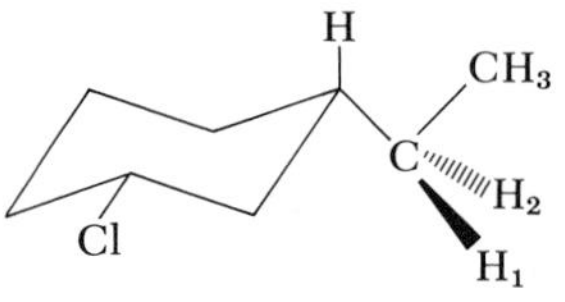

(e) 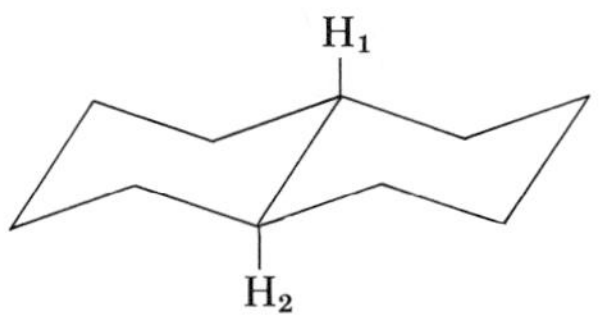 (f)

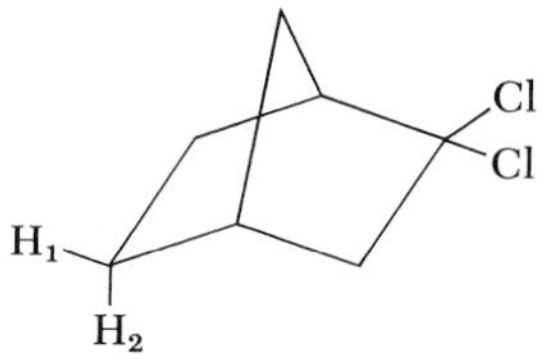

(g) 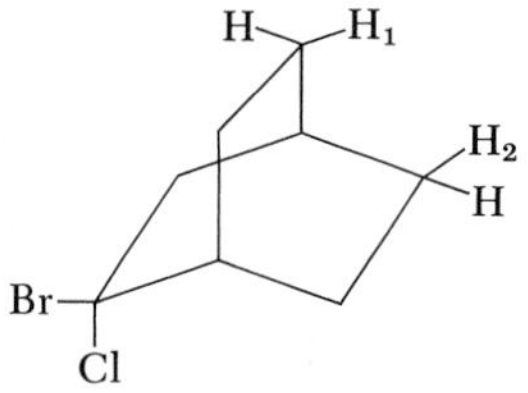 (h)

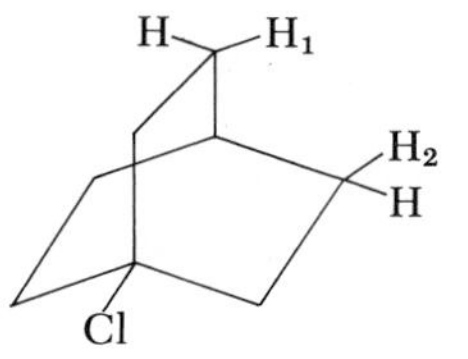

17. Chlorination of bromocyclohexane produces a number of monochlorinated products. (a) Draw *all* of the stereoisomers (neglecting conformations) produced by this reaction. (b) Which are symmetric? (c) Which should have different boiling points?

18. Inositols, natural products found in many living organisms, are hexahydroxycyclohexanes. Using the dot convention, draw all of the possible inositols. Which are incapable of optical activity?

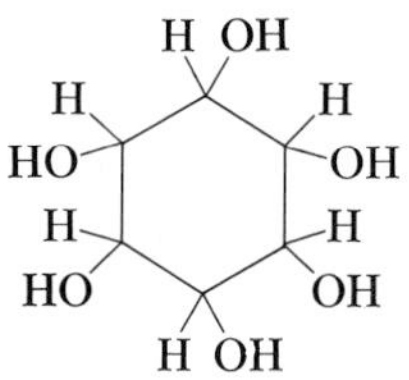

inositol

19. Mycaminose is related to the antibiotic magnamycin. On the basis of the representation below, draw its preferred conformation. Assign an absolute configuration to each chiral center.

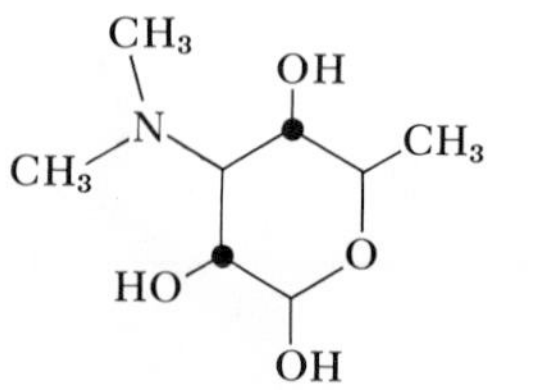

mycaminose

20. Bromination of optically active **A** produces several dibromides, including some which are optically

inactive. Draw all of the possible products of dibromination, indicating which ones are optically active. Assign absolute configurations to the chiral centers in the symmetric molecules.

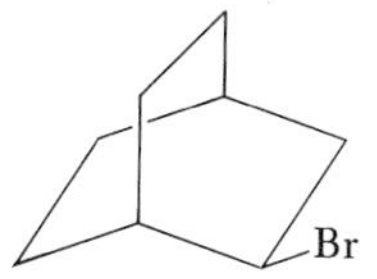

Br

A

21. Bromination of **B** produces several monobromo derivatives. Draw all of the possible products of monobromination of **B,** indicating which are resolvable.

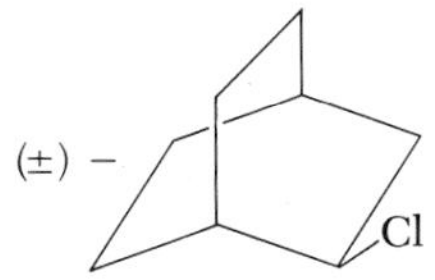

(±) –

Cl

B

22. The free radical chlorination of chlorocyclohexane produces a small amount of **C.** (a) Account for its formation. (b) How many stereoisomers are represented by structure **C?** (c) Draw each product in its most stable conformation.

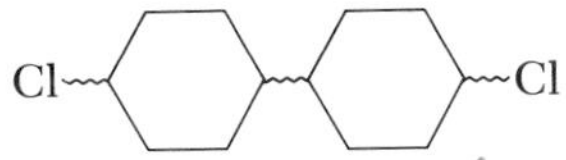

Cl Cl

C

23. (a) Neglecting enantiomers, are all of the stereoisomers of perhydrophenanthrene presented in this chapter? If not, complete the set. (b) Which of these are capable of enantiomerism? (c) Draw the various stereoisomers in their preferred conformations.

24. *Trans*-5-hydroxy-2-phenyl-1,3-dioxane (**D**) can exist in two conformations. Spectral evidence suggests a strong intramolecular hydrogen bond in the compound. Account for this.

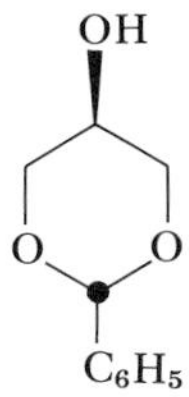

OH

O O

C_6H_5

D

25. With the aid of models, suggest sources of strain in the boat form of cyclohexane.

26. (a) Estimate the amount of torsional strain in planar cyclopentane. (b) Is the amount of torsional strain decreased in going from the planar to the puckered form? (c) Estimate the amount of torsional strain in planar cyclobutane and in planar cyclohexane.

27. It has been argued that since 1,4-dioxane has no dipole moment, it *must* exist in the chair form. Do you agree? Explain.

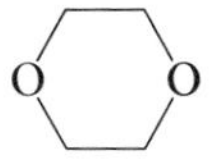

O O

1,4-dioxane

28. *cis*-1,3-Dimethylcyclopentane is more stable than the *trans* isomer. (a) Suggest a reason for this. (b) Could gas chromatographic analysis of the monobrominated derivatives of these two compounds help to identify which stereoisomer is *cis* and which is *trans*? Explain.

29. Phenothiazine (I), acridane (II), thioxanthene (III), and thianthrene (IV) all form the basic part of many biologically important materials. All have geometries similar to that of 9,10-dihydroanthracene (V). What is the shape of the central ring in I through V? Suggest the preferred conformation for 9-ethyl-9,10-dihydroanthracene (VI). Models will help.

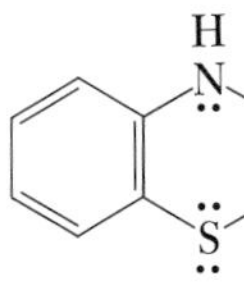

I 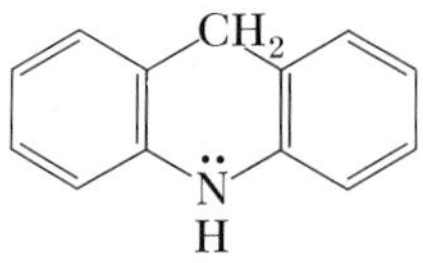 II 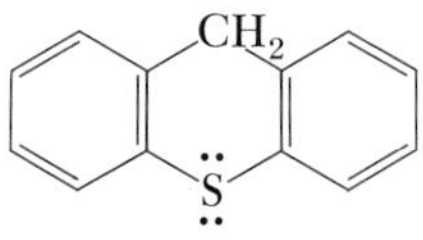 III

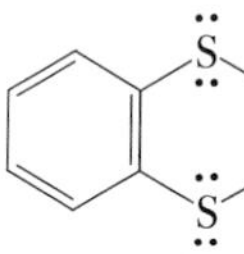

IV 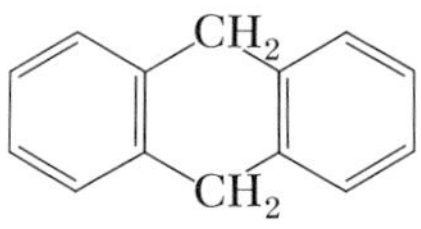 V 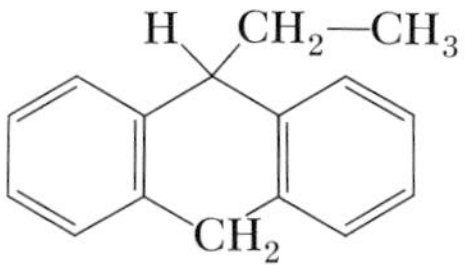 VI

30. Suggest a means of converting the given starting material to the desired product for each of the following.

(a) cyclohexyl bromide to cyclohexene
(b) *cis*-1-bromo-4-*t*-butylcyclohexane to 4-*t*-butylcyclohexene
(c) *trans*-1-bromo-4-*t*-butylcyclohexane to 4-*t*-butylcyclohexene
(d) cyclohexanol, $C_6H_{11}OH$, to cyclohexene

31. One of the isomers of 1,2,3,4,5,6-hexachlorocyclohexane is inert to E2 dehydrohalogenation. Draw its structure. Explain this lack of reactivity.

32. The acetolysis (solvolysis in which acetic acid is solvent) of *trans*-4-methoxycyclohexyl tosylate produces a large amount of *trans*-4-methoxycyclohexyl acetate. (a) Suggest a mechanism for this. (b) How might you test your answer? (Assume access to any reagents you require.)

33. Although the iodine atom is larger than the chlorine atom, the data in Table 7–3 indicate that conformational differences (*e vs. a*) for these species in cyclohexane are about the same. Explain this.

34. According to the data in Table 7–3, —CH_3, —C_2H_5, and —$CH(CH_3)_2$ have nearly identical conformational preferences in cyclohexane. Suggest a reason for this.

35. From the data in Table 7–3, and using the equation $\Delta G = -RT \ln K$, calculate the percentage of (a) methylcyclohexane, (b) isopropylcyclohexane, (c) fluorocyclohexane, and (d) iodocyclohexane that is *equatorial*. What percentage is *axial*?

36. Account for the following order of S_N1 reactivity.

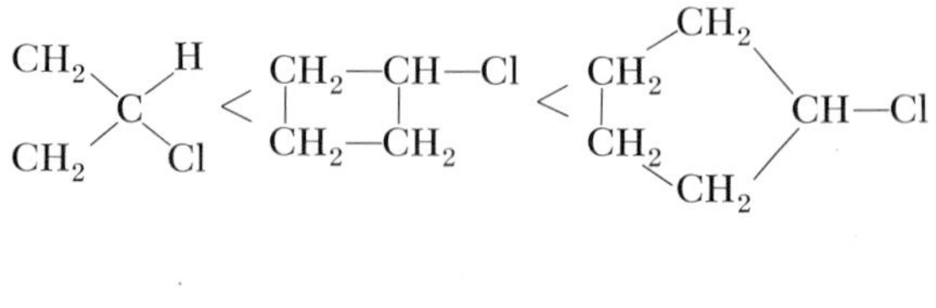

37. *cis*-4-*t*-Butylcyclohexyl iodide reacts more rapidly with $I^{\ominus}$ than does the *trans* isomer, although both are S_N2 halogen displacements. Account for this.

38. Cyclopropyl bromide undergoes the S_N2 reaction much more slowly than does cyclopentyl bromide. Explain this difference.

39. Compound **A** reacts with magnesium to form a stable Grignard reagent, while **B** does not. Explain.

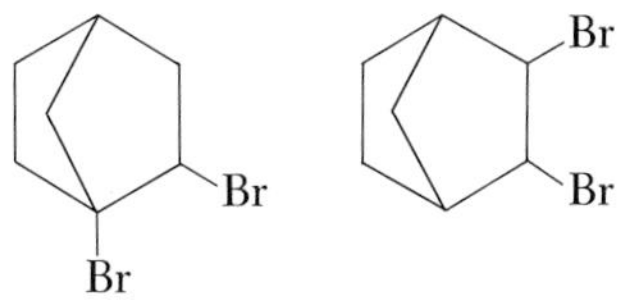

Br Br Br Br

A **B**

40. Compare the effect of increasing methylation on the relative stabilities of *cis*- and *trans*-decalin systems by calculating the number of skew-butane interactions in (a) *cis*-9-methyldecalin, (b) *trans*-9-methyldecalin, (c) *cis*-9,10-dimethyldecalin, and (d) *trans*-9,10-dimethyldecalin.

8 ALKENES

8.1 INTRODUCTION

All compounds containing double or triple bonds are said to be *unsaturated* because the atoms forming such bonds could (at least in principle) each acquire one or two more substituents respectively. Those unsaturated compounds containing the C=C group are specifically called *alkenes* or *olefins*. The identical electronegativity of the two atoms involved in the bond, and the absence of non-bonding electron pairs, result in a relatively simple chemical behavior for the carbon-carbon double bond compared to other double bonds, *e.g.*, C=O and C=N.

Since the carbon-carbon π bond is weaker than the carbon-carbon σ bond (~60 kcal/mole *vs.* ~80 kcal/mole), and since it is more exposed, reactions of the double bond usually involve initial attack by a reagent upon the π system. Moreover, because the π bond serves as a readily accessible electron source, most assaults upon alkenes involve an initial attack by an electron-deficient species such as a cation, a radical, or a carbene.

Once the electron-deficient species has attacked, the reaction may be consummated by cleavage of the π bond and concomitant generation of two new σ bonds (*i.e.*, each C_{sp^2} becomes C_{sp^3}) *or* by cleavage of the π and σ bonds, usually producing two new double bonds. However, one also may encounter reactions in which the double bond is regenerated. Specific illustrations of the first two follow. The bulk of this chapter will deal with reactions similar to these.

$$H_2C=CH_2 + Br_2 \rightarrow Br-CH_2-CH_2-Br$$

conversion of a π bond into two σ bonds

$$(CH_3)_2C=C(CH_3)_2 + KMnO_4 \xrightarrow[\text{heat}]{H_3O^{\oplus}} (CH_3)_2C=O \quad O=C(CH_3)_2$$

cleavage of π and σ bond with creation of a new set of double bonds

Because alkenes can serve as their precursors, this chapter presents the first of several glimpses into that vast class of compounds known as *polymers*. Our initial view is limited to members of the group of polymers commonly called "plastics."

A decade or more ago people tended to view this period in man's history as the "atomic age." While the atom and the atomic bomb have played very important roles in modern history, it may be even more correct to call this the "synthetic age." And of the panorama of ersatz items which have come to represent our society, no substance looms more important than the "plastic." Is it really surprising that the age of man which developed the atomic weapon also developed the plastic flower?

8.2 NOMENCLATURE

The skeleton selected in naming an alkene by the IUPAC scheme is the longest continuous chain containing the double bond. The skeletal atoms are numbered so that the sum of the numbers corresponding to the position of the double bond has the lowest possible value. The chain receives the name of the corresponding alkane, but the *ane* ending becomes *ene* (or *diene*, *triene*, and so forth, as needed).

$$H_3C{-}CH{=}CH{-}CH_2{-}CH_3$$
2-pentene

$$H_2C{=}CH_2$$
ethene

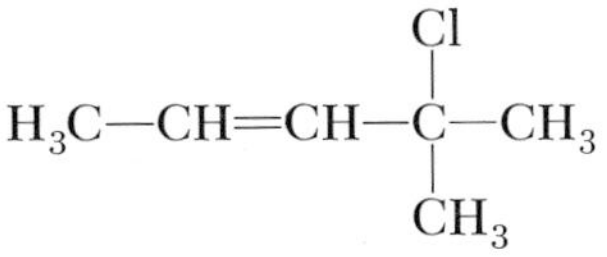

$$H_3C{-}CH{=}CH{-}C(Cl)(CH_3){-}CH_3$$
4-chloro-4-methyl-2-pentene (*not* 2-chloro-2-methyl-3-pentene)

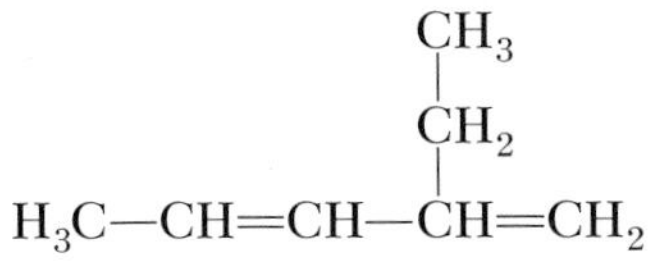

$$H_3C{-}CH{=}CH{-}C(CH_2CH_3){=}CH_2$$
2-ethyl-1,3-pentadiene

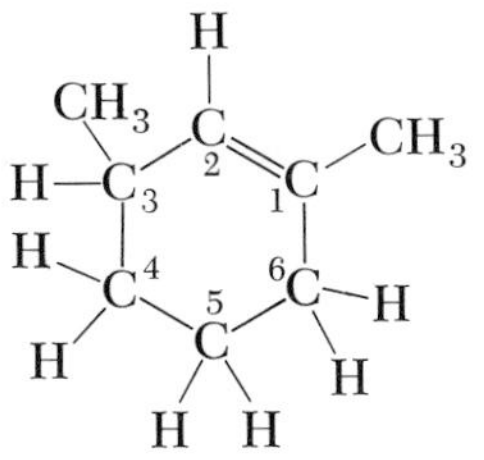

1,3-dimethylcyclohexene

$$H_2C{=}C{=}CH_2$$
propadiene

Simpler alkenes usually are named by unique, trivial names or as substituted *ethylenes* ("ethylene" is the common name for ethene).

$H_2C{=}CHCl$ chloroethylene

$(NC)_2C{=}C(CN)_2$ tetracyanoethylene ("TCNE")

$H_2C{=}C(CH_3)CH{=}CH_2$ isoprene

$H_2C{=}C{=}CH_2$ allene

Functional groups containing double bonds are given common names, which often find their way into names of compounds.

$H_2C{=}CH{-}$ the vinyl group	$H_2C{=}CHCl$ vinyl chloride
$H_2C{=}CH{-}CH_2{-}$ the allyl group	$H_2C{=}CH{-}CH_2{-}OH$ allyl alcohol
$H_2C{=}$ the "methylene" group	$H_2C{=}C(CH_2)_2$ (ring) methylenecyclopropane
$(H)(H_3C)C{=}$ the ethylidene group	ethylidenecyclohexane
$(H_3C)_2C{=}$ the isopropylidene group	$(H_3C)_2C{=}C(C_6H_5)_2$ isopropylidenediphenylmethane

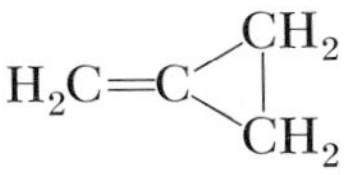

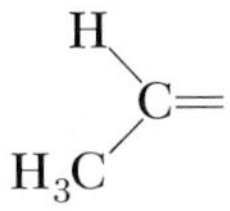

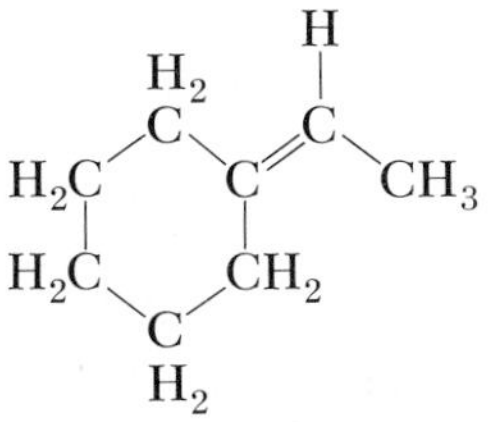

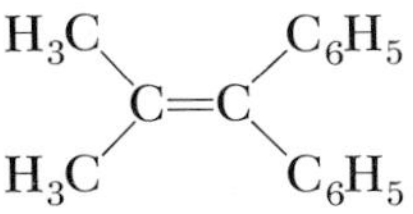

If both C_{sp^2} atoms of a double bond are within a ring system, the double bond is said to be *endocyclic*. If only one is within the ring, the double bond is *exocyclic*.

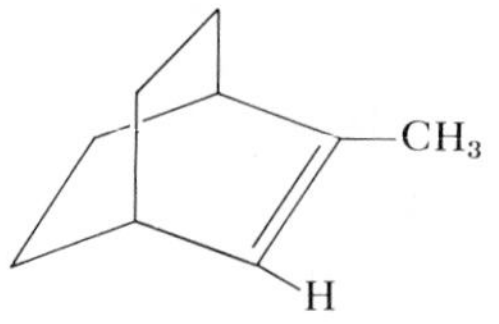

an endocyclic double bond 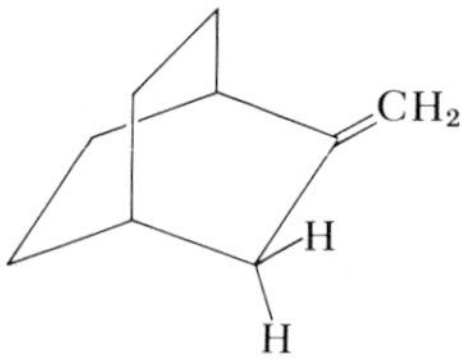 an exocyclic double bond

The fragments shown below illustrate the three ways in which double bonds occur in molecules.

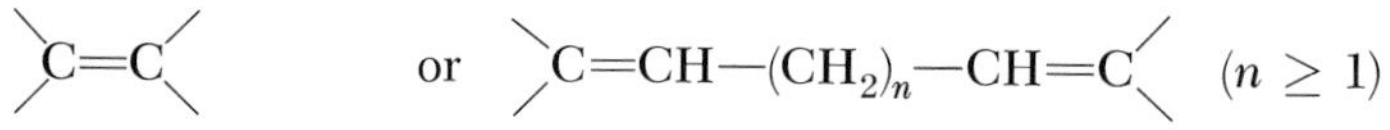

$>C=C<$ or $>C=CH-(CH_2)_n-CH=C<$ $(n \geq 1)$ *isolated* double bond

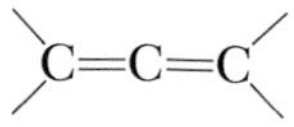

$>C=C=C<$ *cumulative* double bonds

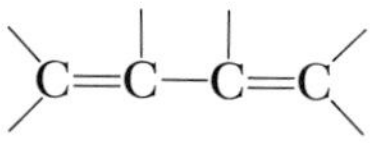

$>C=\overset{|}{C}-\overset{|}{C}=C<$ *conjugated* double bonds

8.3 ISOMERISM IN SIMPLE ALKENES

Three types of isomerism can be observed in simple alkenes. In one, the ordering of the skeletal atoms is changed. In a second, the skeletal arrangement is the same but the position of the double bond within the skeleton is changed. In the third, both the skeleton and the position of the double bond are unchanged but the distribution of groups around the double bond is varied. Even a C_4 alkene system can serve to illustrate these three possibilities.

Butene. Calling a substance "butene" says no more than that it is an alkene with the formula C_4H_8; all of the structures below meet this simple requirement.

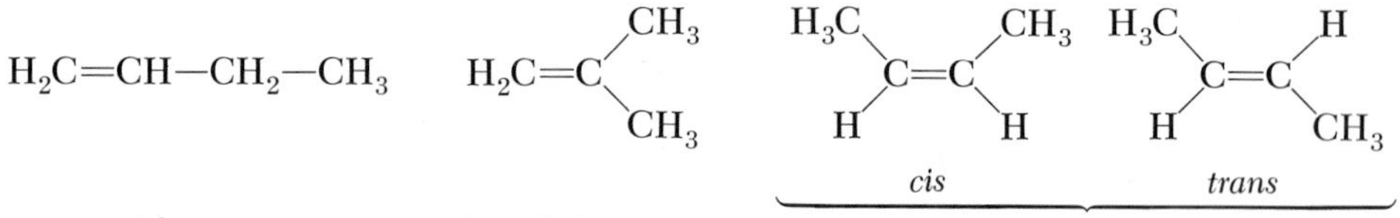

$H_2C=CH-CH_2-CH_3$ — 1-butene

$H_2C=C(CH_3)_2$ — 2-methylpropene (isobutylene)

$H_3C(H)C=C(H)CH_3$ — *cis* 2-butene; $H_3C(H)C=C(CH_3)H$ — *trans* 2-butene

1-Butene and 2-methylpropene are **constitutional isomers** of one another, that is, they have the same molecular formula but different atom-to-atom bonding sequences. Since 1-butene and the 2-butenes have the same skeleton but differ in the position of the double bond, these are dubbed **positional isomers** of one another. Thus, 1-butene and 2-methylpropene illustrate the first type of isomerism alluded to at the beginning of this section, while 1-butene and 2-butene illustrate the second type. The two forms of 2-butene are **geometric isomers** of one another and represent the third type of isomerism.[a] The independent existence of two geometric

[a] As non-enantiomeric stereoisomers, they are also diastereomers.

isomers, which differ in the way in which groups are distributed on either side of the π system, is assured because of the restricted rotation around the double bond (if necessary, review material on p. 35, *et seq.*).

Positions on the same side of the double bond are said to be *cis* to one another. Positions on opposite sides and at opposite ends of a π system are *trans* to one another.

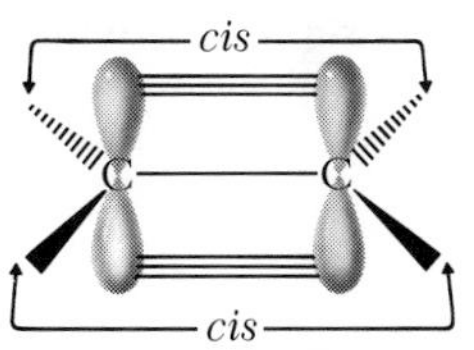

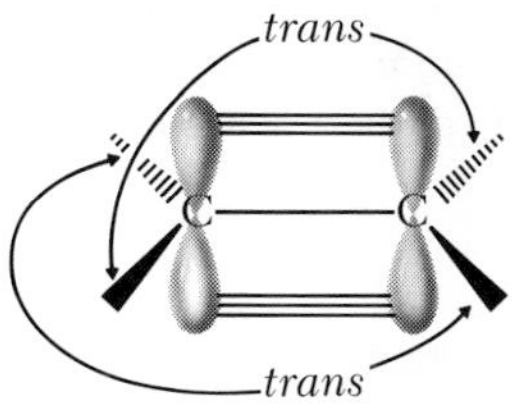

The terms *cis* and *trans* also are used as a part of the nomenclature of alkenes to indicate any stereochemistry about the double bond. For this application, one identifies the longest chain containing the double bond; if the elements of the chain are *cis* to one another, then the term *cis* is attached to the name of the alkene. If the elements of the chain are *trans* to one another, then the term *trans* is prefixed to the name. Some examples follow:

H_3C, H / C=C / CH_3, H	H_3C, H / C=C / H, CH_3	H_3C, H / C=C / H, CH_2CH_3
cis-2-butene	*trans*-2-butene	*trans*-2-pentene

H_3C, H / C=C / CH_3, CH_2CH_3	H_3C, H / C=C / CH_2CH_3, CH_3
trans-3-methyl-2-pentene	*cis*-3-methyl-2-pentene

If an alkene is disubstituted, then the prefixes *cis* and *trans* are applied even when the chain is only two carbons long, *i.e.*, for an ethene derivative.

H, Cl / C=C / H, Cl	Cl, H / C=C / H, Cl
cis-1,2-dichloroethene	*trans*-1,2-dichloroethene

H, O_2N / C=C / H, Cl	H, O_2N / C=C / Cl, H
cis-1-chloro-2-nitroethene	*trans*-1-chloro-2-nitroethene

Unfortunately, this simple scheme breaks down for tri- and tetrasubstituted alkenes in which the main chain is in doubt or nonexistent. For example, which of the following is *cis*?

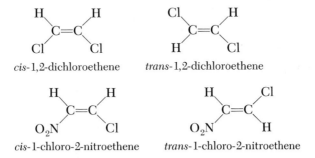

H_3C, Br / C=C / Cl, NO_2	H_3C, Br / C=C / NO_2, Cl

The "**E,Z**" nomenclature system for identifying geometric isomers has been introduced to alleviate this type of problem. The rules for this scheme are:

1. Using the Cahn-Ingold-Prelog system, assign relative priorities to the groups attached

to each end of the double bond; that is, assign each substituent "1" or "2," depending on whether it has the higher or lower priority.

2. If the two groups of higher priority are on the same side of the plane defined by the π bond, the configuration is designated **Z.** If the higher priority groups are on opposite sides of the plane the configuration is designated **E.**

In the examples below, the numbers **1** and **2** refer to priorities established under the **R,S** system rules. Plane *a* separates the two carbons for priority assignment, while plane *b* is the reference plane for **E,Z** assignment.

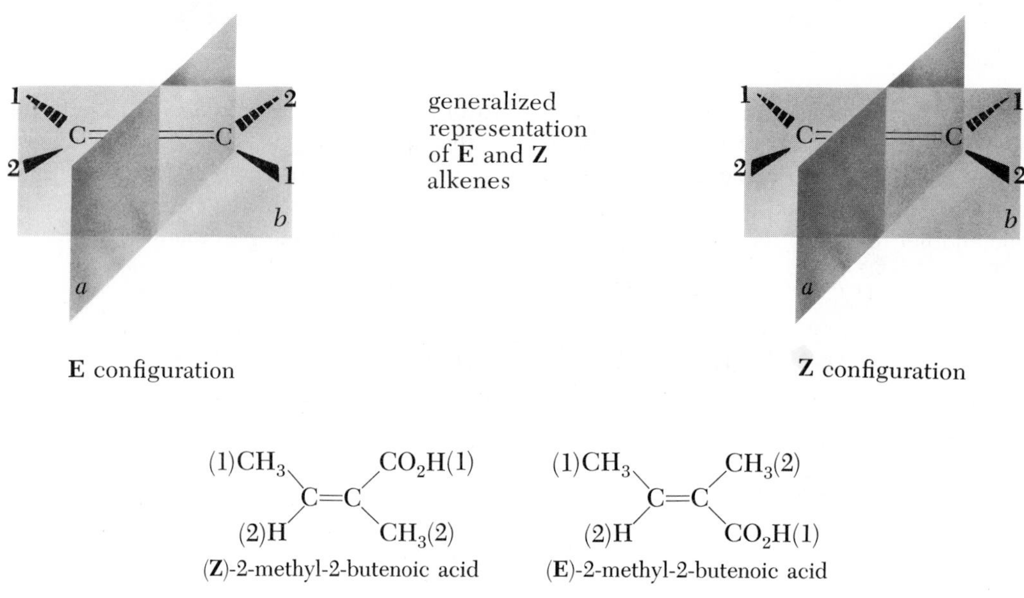

This system can be applied to alkenes which also are describable in the *cis,trans* terminology, and because of its universal applicability the "**E,Z**" system gradually is replacing the "*cis,trans*" system.

Br, H / C=C / H_3C, NO_2

(**E**)-2-bromo-1-nitropropene

Br, NO_2 / C=C / H_3C, H

(**Z**)-2-bromo-1-nitropropene

H_3C, H / C=C / H; C=C / H, $(CH_2)_3CH_3$, H

(**E,Z**)-2,4-nonadiene
(or *trans,cis*-2,4-nonadiene)

H, H / C=C / H_3C; C=C / H, $(CH_2)_3CH_3$, H

(**Z,Z**)-2,4-nonadiene
(or *cis,cis*-2,4-nonadiene)

1. Provide an acceptable name for each of the following. Use the **E,Z** convention to specify configuration.

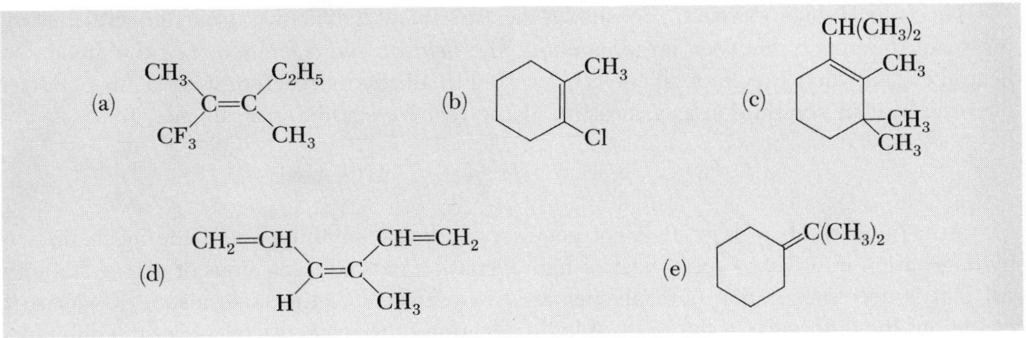

(a) $CH_3(CF_3)C{=}C(C_2H_5)CH_3$ (b) 1-methyl-2-chlorocyclohexene (c) cyclohexene with $CH(CH_3)_2$, CH_3, and $C(CH_3)_2$ substituents

(d) $CH_2{=}CH(H)C{=}C(CH{=}CH_2)CH_3$ (e) cyclohexylidene $=C(CH_3)_2$

8.4 RELATIVE STABILITY OF ISOMERIC ALKENES

Even though a simple explanation is not currently available, we have already noted (p. 164) that the stability of an alkene usually increases with increasing alkylation at the double bond:

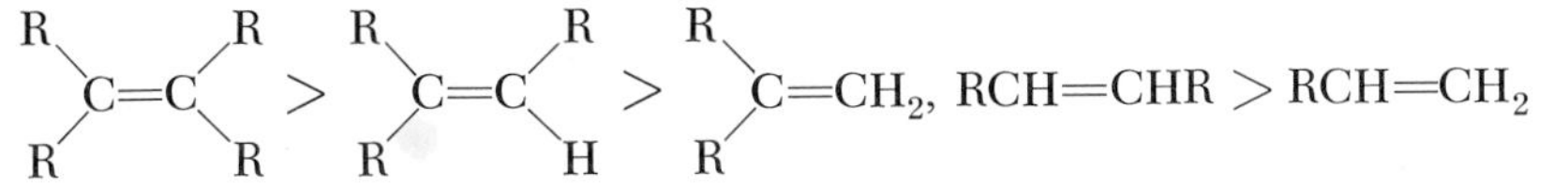

$$R_2C{=}CR_2 > R_2C{=}CRH > R_2C{=}CH_2,\ RCH{=}CHR > RCH{=}CH_2 \qquad \text{alkene stability}$$

What about the relative stability of geometric isomers, in which the extent of substitution around the double bond is constant? The answer to this can be determined quantitatively in the following ways.

Heat of Combustion. Since *cis-* and *trans-*2-butene are diastereomers, they must have different chemical and physical properties, including different ground state energies. One way to determine their relative stabilities is to oxidize them to CO_2 and H_2O; the more stable alkene should have the smaller heat of combustion (Figure 8–1). The heat of combustion of *cis-*2-butene is 647.81 kcal/mole, while that of the *trans* isomer is 646.81 kcal/mole. Although they are nearly identical, these values suggest that the *trans* isomer is the more stable by approximately 1 kcal/mole.

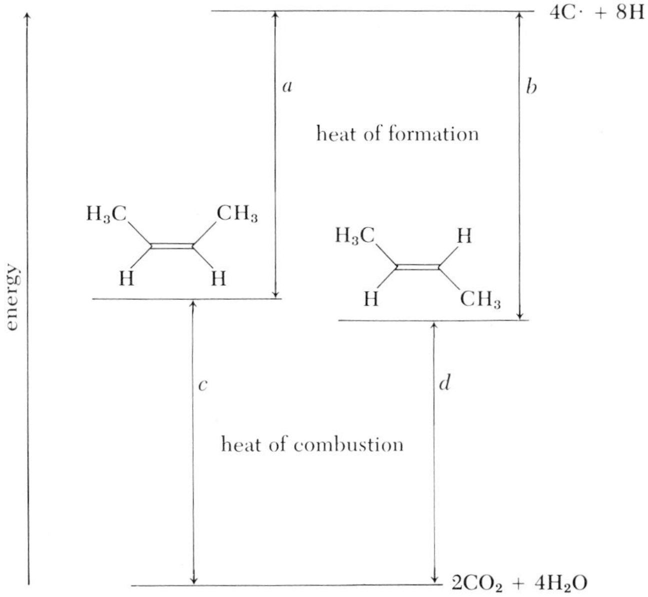

Figure 8-1 The heats of combustion of *cis-* and *trans-*2-butene. Values *a*, *b*, *c*, and *d* are positive since they equal $-\Delta H$ for the appropriate reactions and both reactions (formation from atoms *and* combustion) are exothermic (*i.e.*, they have negative values for ΔH). $a = +1.39$ kcal/mole; $b = +2.34$ kcal/mole.

Heats of Hydrogenation. Because of the experimental difficulty associated with measuring small differences between large numbers, the *heat of hydrogenation* (*i.e.*, the amount of heat liberated when 1 mole of alkene is converted to alkane by reaction with H_2) is preferred over the heat of combustion as a measure of the relative stabilities of alkenes.

$$\rangle C{=}C\langle + H_2 \rightarrow H{-}\overset{|}{\underset{|}{C}}{-}\overset{|}{\underset{|}{C}}{-}H + \text{heat}$$

As a result of Hess' Law, it is not necessary to know anything about the mechanism of hydrogenation in order to apply heat of hydrogenation data to discussions of alkene stability. All that is necessary is that both alkenes are hydrogenated to the same alkane. Under such conditions the difference in the heats of hydrogenation of two alkenes measures the difference in the energy content of the alkenes (Figure 8–2).

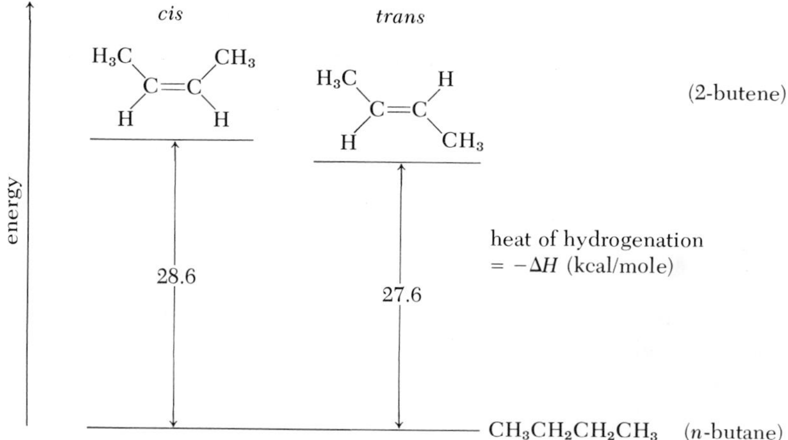

Figure 8–2 The heats of hydrogenation for *cis*- and *trans*-2-butene. Upon reduction, *trans*-2-butene releases one less kilocalorie per mole than does the *cis* isomer; thus, the *trans* isomer is more stable by 1 kcal/mole.

The data in Table 8–1 support the hypothesis of greater stability of *trans*-2-butene than of *cis*-2-butene and, moreover, demonstrate that we are now dealing with small changes in rather small numbers, ensuring greater reliability of the data. With few exceptions, *trans* alkenes are more stable than the corresponding *cis* isomers. This is ascribed to unfavorable steric interactions in the latter.

cis *trans*

One group of exceptions to this generalization is the smaller cycloalkenes, of which only the *cis* alkene is known. Cycloöctene is the smallest cycloalkene that can exist in both *cis* and *trans* forms, and the *trans* isomer is much less stable than the *cis* isomer (about 10 kcal/mole difference in their heats of hydrogenation).

cis-cycloöctene *trans*-cycloöctene

TABLE 8-1 Heats of Hydrogenation of Alkenes

ALKENE	$-\Delta H$ (KCAL/MOLE)
$CH_3{-}CH{=}CH_2$	30.1
$CH_3CH_2{-}CH{=}CH_2$	30.3
$CH_3{-}CH{=}CH{-}CH_3$	28.6 (*cis*) 27.6 (*trans*)
$(CH_3)_2C{=}CH_2$	28.4
$(CH_3)_2C{=}CHCH_3$	26.9
cyclopentene	26.9
cyclohexene	28.6

A second type of exception is exemplified by *cis*- and *trans*-1,2-dichloroethene, the *cis* isomer being more stable by 0.25 kcal/mole. This turn of events is ascribed to *stabilization* of the *cis* isomer by dispersion forces between the two chlorine atoms. Such stabilization is relatively rare, since it requires a rather specific distance between the atoms involved in the attractive interaction.

$$\text{ClHC}{=}\text{CHCl}\ (cis) > \text{ClHC}{=}\text{CHCl}\ (trans) \qquad \text{order of stability}$$

2. Why can't the heats of hydrogenation of cyclopentene and cyclohexene be used as a direct measure of the relative stability of these two alkenes?

DIRECT EQUILIBRATION. The most direct method for determining relative stabilities of alkenes (or of any other isomers, for that matter) involves establishing the equilibrium constant between the two species.

$$\underset{cis}{\text{abC}{=}\text{Cab}} \rightleftarrows \underset{trans}{\text{abC}{=}\text{Cba}} \qquad K_{eq} = \frac{[trans]}{[cis]}$$

Although K_{eq} reflects the free energy difference (ΔG) between the isomers, neglect of ΔS (a sometimes reasonable approximation) permits K_{eq} to reflect the enthalpy difference (ΔH) between the isomers.

$$\Delta G = -RT \ln K_{eq}$$

$$\Delta G = \Delta H - T\Delta S$$

Heating is the most direct means of interconverting geometric isomers; however, decomposition may occur during equilibration. If decomposition does occur, a valid equilibrium constant is obtained only if both alkenes decompose at known rates.

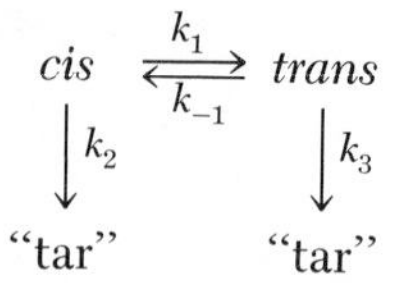

$$\begin{array}{ccc} cis & \underset{k_{-1}}{\overset{k_1}{\rightleftarrows}} & trans \\ \downarrow k_2 & & \downarrow k_3 \\ \text{"tar"} & & \text{"tar"} \end{array}$$

k_2/k_3 must be known to evaluate K_{eq}

$$K_{eq} = \frac{k_1}{k_{-1}} = \frac{[trans]}{[cis]}$$

The direct equilibration procedure has been applied to the isomeric 1,2-dichloroethenes, and the results support the view that *cis* is the more stable isomer; at 185° (the equilibration temperature) the equilibrium mixture contains 63.5% *cis* and 36.5% *trans*.

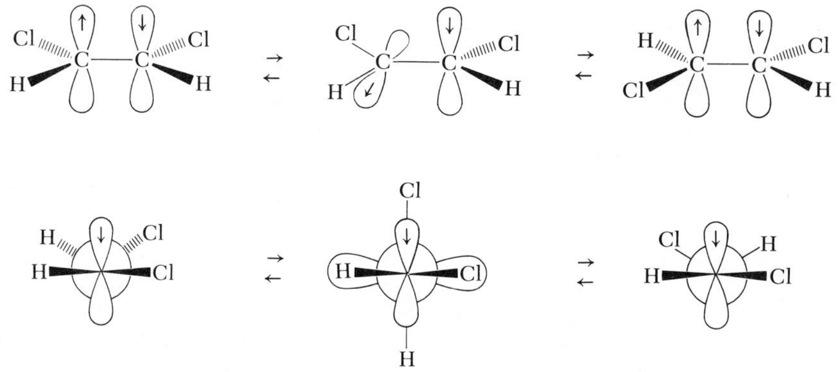

uncoupled electrons maintain opposite spins

8.5 REDUCTION OF DOUBLE BONDS

The Concept of Reduction and Oxidation in Organic Compounds. In your previous chemistry courses you probably have seen the term "oxidation" used to mean "the loss of electrons" and the term "reduction" used to mean "the gain of electrons." The change in the degree of oxidation or reduction was easily recognized by a change in the oxidation number, *e.g.*,

$$Mg \rightarrow Mg^{\oplus 2} + 2e^{\ominus} \qquad \text{(oxidation)}$$

$$F_2 + 2e^{\ominus} \rightarrow 2F^{\ominus} \qquad \text{(reduction)}$$

This approach is impossible when dealing with most organic compounds because the changes accompanying reaction involve covalent bonds and there is no clear "ownership" of bonding electrons. To surmount this difficulty, we assign "ownership" of electrons based upon the electronegativity of the elements.

Since most heteroatoms (*e.g.*, O, N, S, Br) are more electronegative than carbon, each heteroatom bonded to a carbon is considered to make a +1 contribution to the oxidation state of that carbon. Each hydrogen, being less electronegative than carbon, makes a −1 contribution to the carbon to which it is attached. A carbon bonded to another carbon makes a *zero* contribution to the oxidation state of the carbon to which it is bonded. Multiple bonds are counted according to the multiplicity of the bond (*e.g.*, in C=O the oxygen contributes +2). Remember that in the final molecule the algebraic sum of the oxidation numbers equals *zero.*

$\overset{-4}{\mathrm{C}}H_4$ (H—C—H with H above and below) $CH_3\overset{-3}{\mathrm{C}}H_3$ $H\overset{-1}{\mathrm{C}}(CH_3)_3$ selected oxidation states

$H_3\overset{-2}{\mathrm{C}}—O—H$ $H—\overset{0}{\mathrm{C}}H{=}O$ $H_2\overset{-2}{\mathrm{C}}{=}\overset{-1}{\mathrm{C}}H—\overset{-3}{\mathrm{C}}H_3$

If we apply the method for determining changes in oxidation states to the reaction of an alkene with hydrogen, we see that a net *reduction* of carbon takes place. The molecular hydrogen is oxidized.

$$H_2\overset{-2}{C}{=}\overset{-2}{C}H_2 + H_2 \rightarrow H_3\overset{-3}{C}—\overset{-3}{C}H_3$$

3. Determine the oxidation number of each atom in the following.
(a) propene
(b) 1-butene
(c) *cis*-2-butene
(d) 1-chloro-2-butene
(e) (**E**)-2-chloro-2-butene
(f) 1,1-dichloropropane
(g) acetic acid, CH_3CO_2H
(h) methanethiol, CH_3SH

Heterogeneously Catalyzed Hydrogenation. Ethylene and other alkenes do not react readily with hydrogen gas because of the high E_{act} associated with the addition. However, through the use of a *catalyst*, which provides a comparatively low energy pathway for the addition, hydrogen can be used to convert an alkene to an alkane. The heat of hydrogenation of the alkene in the presence of the catalyst is the same as it would be had hydrogenation occurred in the absence of the catalyst; only the E_{act} has changed (see Figure 8–3).

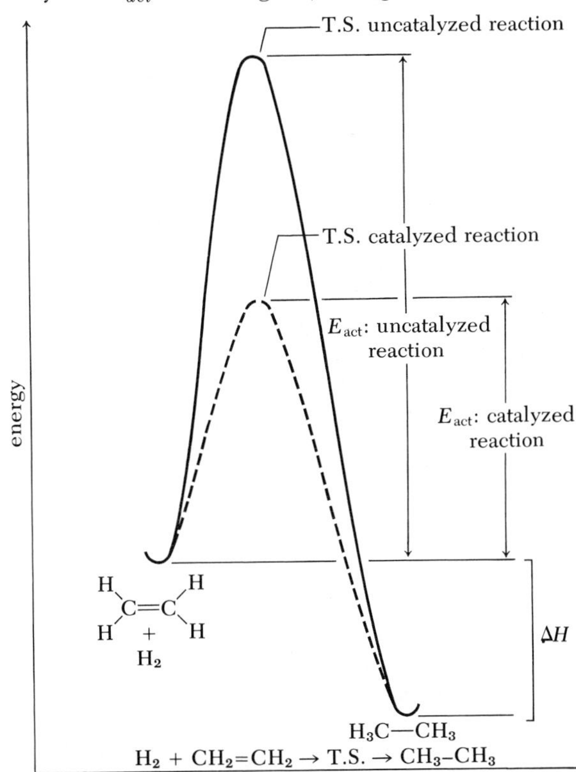

Figure 8-3 Energy profile for the hydrogenation of a double bond—the effect of a catalyst.

As with all true catalysts, lowering the E_{act} increases the rates of both the forward and the reverse reaction to the same extent. *Catalysis simply provides a faster way to establish an equilibrium between two species.*

Catalysts suitable for hydrogenation include finely divided platinum, palladium, ruthenium, rhodium, and nickel. These metals, by virtue of their insolubility in organic solvent, are *heterogeneous* catalysts. The most famous of these include *Adam's catalyst* and *Raney nickel.* The former consists of finely divided platinum metal, called *platinum black,* prepared by the reduction of platinum oxide by H_2 *in situ* (*i.e.*, directly in the reaction vessel). It is used in low pressure (1 to 2 atm) hydrogenations. Raney nickel is prepared by treating a Ni—Al alloy with aqueous sodium hydroxide (which reacts with the Al to form H_2) and consists of finely divided nickel saturated with H_2. It is used for low pressure hydrogenations, although it is slightly less active than either Pt or Pd.

$$(H_3C)_2C{=}CH_2 \xrightarrow[\text{1 atm } H_2]{\text{Pt, 25°}} CH_3{-}CH(CH_3){-}CH_3$$

Adkins catalyst (also called "copper chromite") is considerably cheaper than any of the catalysts mentioned above, but requires much more vigorous conditions (*e.g.*, 300 atm H_2 pressure). This catalyst, as well as other non-platinum metals, is usually used in industrial hydrogenation reactions because of its low cost.

"$CuO \cdot CuCr_2O_4$" formula of copper chromite

In heterogeneously catalyzed reactions, the H_2 associates with the catalyst's surface and proceeds to add across the double bond. This is shown schematically in Figure 8–4.

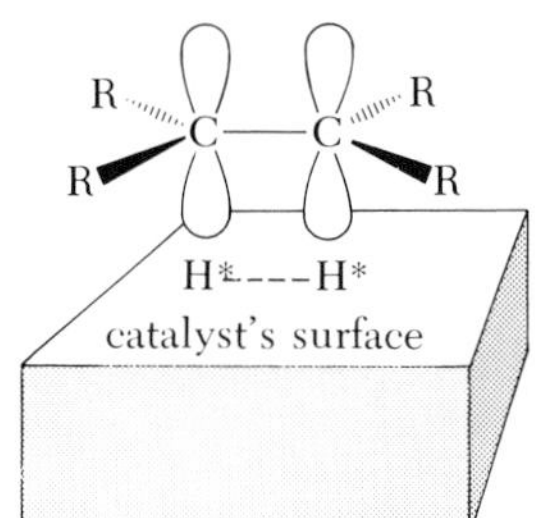

Figure 8–4 Schematic representation of catalytic hydrogenation. H°---H° represents H_2 adsorbed on the catalyst's surface; as a consequence of adsorption, the hydrogen is "activated" as a reductant.

The mechanism is such that the incoming hydrogens add on the same face of the π system, a so-called *cis addition* (or *syn* addition). Thus, the catalytic reduction of 1,2-dimethylcyclohexene produces mainly *cis*-1,2-dimethylcyclohexane (see problem 48).

$$\text{1,2-dimethylcyclohexene} \xrightarrow[\text{1 atm, } Pt^0]{H_2\text{, 25°}} \textit{cis}\text{-1,2-dimethylcyclohexane}$$

Catalytic hydrogenations usually are performed in an excess of hydrogen in order to repress the reverse reaction, the *de*hydrogenation of an alkane. However, dehydrogenations are important in the processing of petroleum. Certain petroleums are rich in cycloalkanes, especially methylcyclopentane, 1,2-dimethylcyclopentane, cyclohexane, and methylcyclohexane. These cycloalkanes, called *naphthenes* in the petroleum industry, are intentionally isomerized and dehydrogenated to aromatic hydrocarbons during petroleum refining. Aromatic hydrocarbons, discussed in detail in Chapters 14 and 15, are important commercial substances and, therefore, this process of *catalytic reforming* is a major industrial reaction.

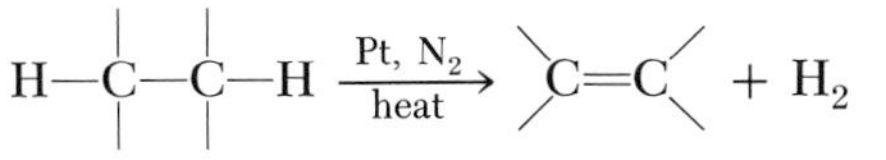

$$\text{H—}\overset{|}{\underset{|}{\text{C}}}\text{—}\overset{|}{\underset{|}{\text{C}}}\text{—H} \xrightarrow[\text{heat}]{\text{Pt, N}_2} \text{>C=C<} + \text{H}_2 \qquad \text{dehydrogenation}$$

$$\text{methylcyclohexane (aliphatic)} \xrightarrow[560°;\ 300\ \text{psi}]{\text{Mo}_2\text{O}_3\cdot\text{Al}_2\text{O}_3} \text{toluene (aromatic)} + 3\text{H}_2$$

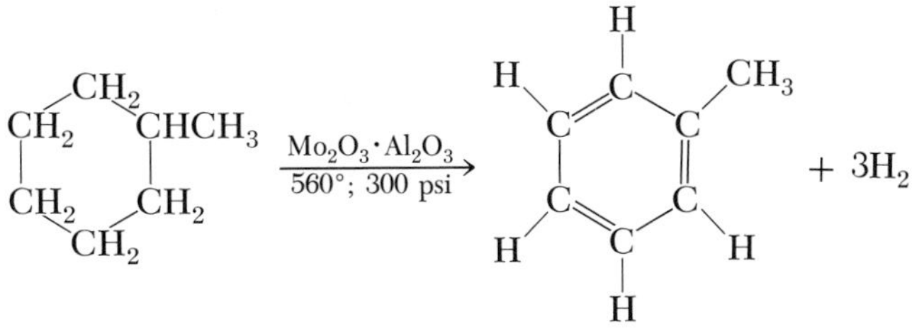

methylcyclohexane
(aliphatic)

toluene
(aromatic)

4. If cyclohexene is treated with deuterium and a platinum catalyst for a long period of time, there is obtained a small quantity of perdeuteriocyclohexane, C_6D_{12}. How could this be formed? (*Note:* The expression "perdeuterio" signals replacement of all protons with deuterons.)
5. (a) What is the product of the catalytic hydrogenation of each of the following? (b) What is the product if hydrogen is replaced by D_2?

(a) 1-butene (d) (**E**)-2-pentene
(b) *trans*-2-butene (e) (**Z**)-2-pentene
(c) *cis*-2-butene (f) cyclohexene

Homogeneously Catalyzed Reductions. Catalytic hydrogenation using metallic catalysts suffers from several drawbacks. Among these is the fact that these heterogeneous catalysts sometimes cause isomerization of alkenes and/or rupture of carbon-carbon single bonds. During the past decade there have been developed, largely through the efforts of Wilkinson, a group of hydrogenation catalysts which are (a) soluble in organic solvents and (b) less likely to cause rearrangement or decomposition of the starting alkene. This development, which represents a major advance in synthetic *organic* chemistry, is the outgrowth of a research program in *inorganic* chemistry.

The most useful of these **homogeneous** hydrogenation catalysts are complexes between rhodium or ruthenium chloride and triphenylphosphine, $(C_6H_5)_3P$. These complexes are called tris(triphenylphosphine)chlororhodium and hydridochlorotris(triphenylphosphine)ruthenium, $((C_6H_5)_3P)_3RhCl$ and $((C_6H_5)_3P)_3RuClH$, respectively. The more familiar rhodium complex is easily synthesized by reacting rhodium chloride with triphenylphosphine.

$$RhCl_3 + 3(C_6H_5)_3P \xrightarrow[\text{heat}]{\text{ethanol}} ((C_6H_5)_3P)_3RhCl$$

The rhodium complex can be used to hydrogenate isolated carbon-carbon double bonds at ordinary pressures and temperatures. Cyanides (RC≡N) and ketones (R_2C=O) are not hydrogenated under these conditions. Variations in the substitution pattern around the double bond alter its ease of reduction, with tri- and tetrasubstituted double bonds being much more slowly reduced than mono- or disubstituted double bonds. This permits the selective reduction of compounds containing both types of double bond, as shown below for carvone and linalool, two compounds isolated from plants.

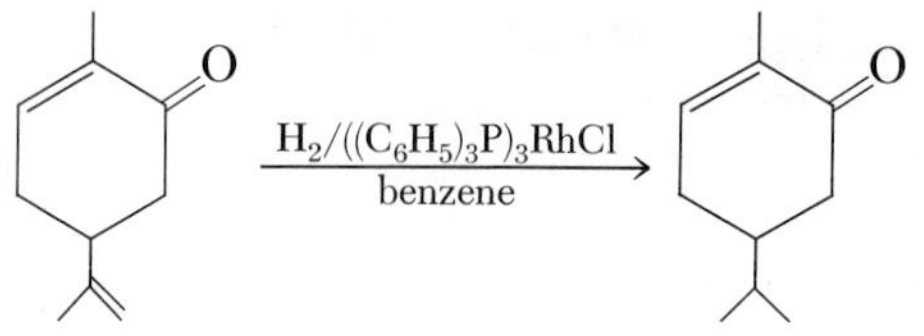

$$\text{carvone} \xrightarrow[\text{benzene}]{H_2/((C_6H_5)_3P)_3RhCl}$$

(+)-carvone and (−)-carvone are responsible for the odors of caraway and spearmint, respectively

carvone

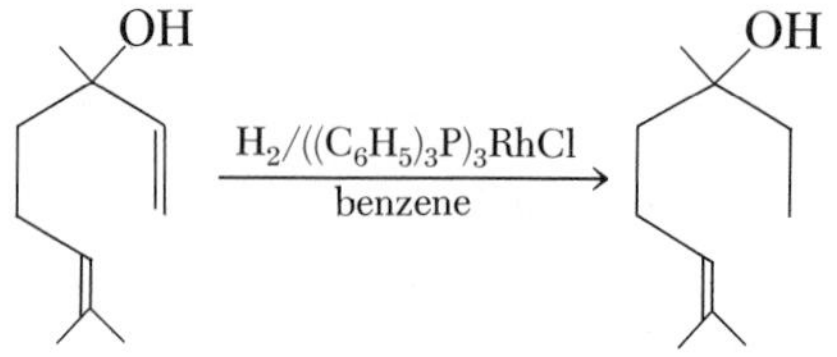

$$\text{linalool} \xrightarrow[\text{benzene}]{H_2/((C_6H_5)_3P)_3RhCl}$$

linalool

The addition of hydrogen follows a *cis* path; reduction of *cis*-2-butene with D_2 produces *meso*-2,3-dideuteriobutane.

$$\text{cis-2-butene} \xrightarrow[\text{benzene}]{D_2/((C_6H_5)_3P)_3RhCl} \text{meso-2,3-dideuteriobutane}$$

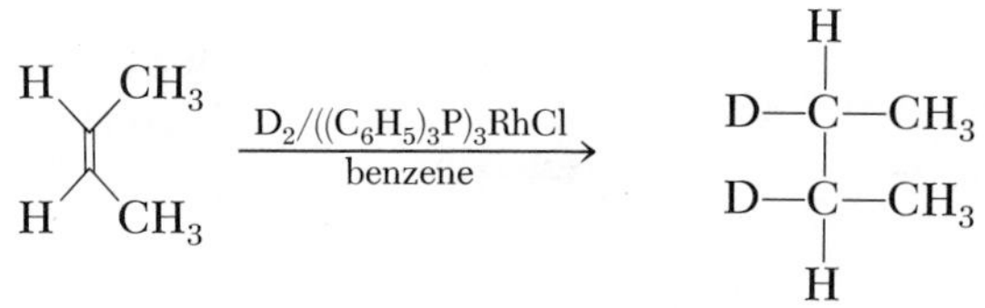

cis-2-butene *meso*-2,3-dideuteriobutane

The rhodium catalyst suffers from a drawback in that it *decarbonylates* aldehydes, that is, it converts RCHO to RH + CO.

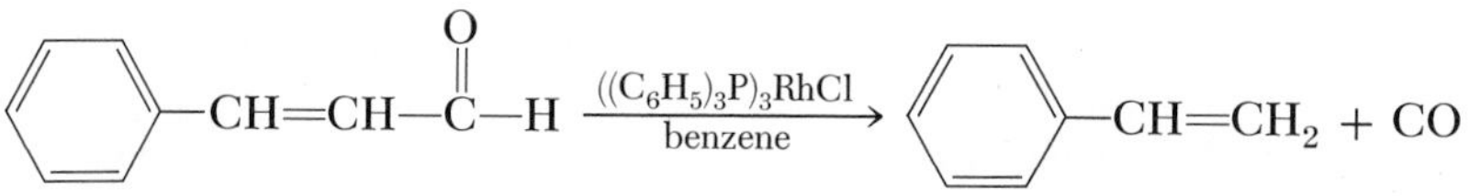

$$C_6H_5{-}CH{=}CH{-}\overset{O}{\overset{\|}{C}}{-}H \xrightarrow[\text{benzene}]{((C_6H_5)_3P)_3RhCl} C_6H_5{-}CH{=}CH_2 + CO$$

cinnamaldehyde styrene

The mechanism of these reductions, and of related reactions, is still under investigation. It currently is believed that the rhodium complex dissociates in solvents to produce a solvated rhodium complex, the solvent replacing one $(C_6H_5)_3P$ unit. In turn, this solvated complex forms an adduct with the hydrogen that will eventually add to the π bond. The alkene to be reduced displaces the solvent to form a new adduct. Reduction occurs at this stage, with the hydrogen that is bonded to rhodium being transferred to the alkene. This sequence is outlined below:

1. $((C_6H_5)_3P)_3RhCl + Sol \rightleftarrows ((C_6H_5)_3P)_2Rh(Sol)Cl + (C_6H_5)_3P$
2. $((C_6H_5)_3P)_2Rh(Sol)Cl + H_2 \rightleftarrows ((C_6H_5)_3P)_2Rh(Sol)ClH_2$
3. $((C_6H_5)_3P)_2Rh(Sol)ClH_2 + {>}C{=}C{<} \rightleftarrows ((C_6H_5)_3P)_2RhClH_2({>}C{=}C{<}) + Sol$
4. $((C_6H_5)_3P)_2RhClH_2({>}C{=}C{<}) \rightarrow {>}\underset{H}{C}{-}\underset{H}{C}{<} + ((C_6H_5)_3P)_2Rh(Sol)Cl$

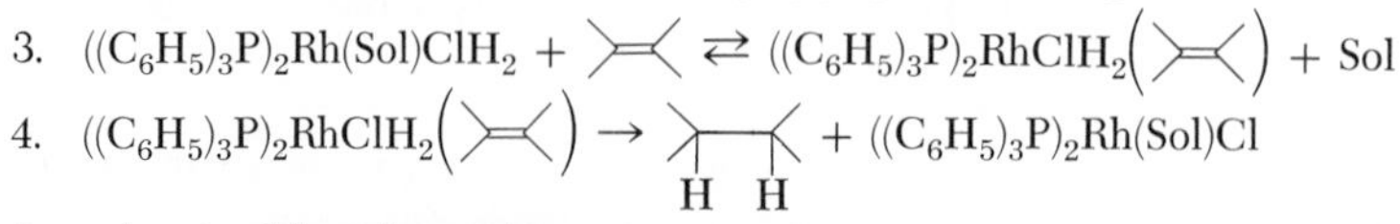

1. solvent adduct formation
2. complexation with hydrogen
3. complexation with alkene
4. hydrogenation of alkene

Note that this sequence regenerates the rhodium complex so that the complex is, in fact, a true catalyst.

6. Suggest a compound which, when subjected to homogeneous catalytic hydrogenation, will give the desired product.

(a) $CH_3CH_2CH_3$
(b) $CH_3CH_2CH_2CH_3$
(c) $(CH_3)_2CHCH_2CH_3$
(d) $(CH_3)_3CCH_2CH_2CH_3$
(e) *meso*-$CH_3CHDCHDCH_3$
(f) *d,l*-$CH_3CHDCHDCH_3$
(g) methylcyclohexane
(h) ethylcyclohexane

Diimide Reduction. Hydrazine (N_2H_4, known principally as a component of early liquid rocket fuels) is oxidized by hydrogen peroxide in the presence of Cu^{+2} to diimide, N_2H_2.

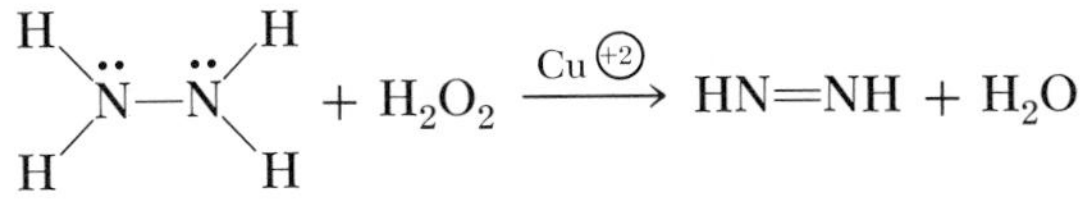

$$H_2\ddot{N}-\ddot{N}H_2 + H_2O_2 \xrightarrow{Cu^{(+2)}} HN{=}NH + H_2O$$

hydrazine — diimide

Diimide is unstable and, in the absence of an alkene, decomposes to H_2 and N_2. However, if diimide is prepared in the presence of an alkene, the double bond of the alkene is reduced and nitrogen is produced. The reduction proceeds through a cyclic activated complex and, like catalytic hydrogenation, adds incoming hydrogens in a *cis* fashion.

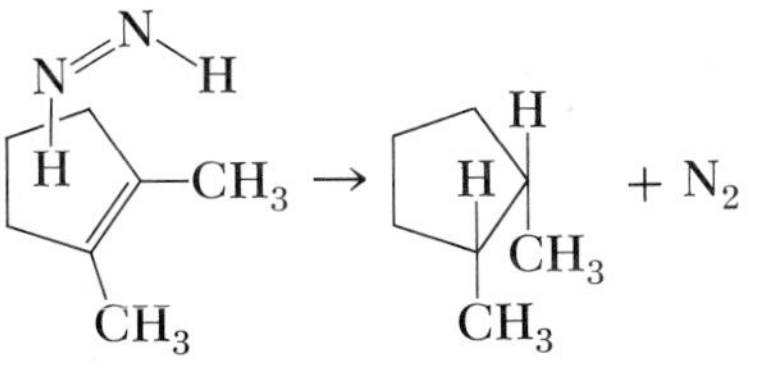

[Scheme: 1,2-dimethylcyclopentene with HN=NH → cis-1,2-dimethylcyclopentane (H, H, CH_3, CH_3) $+ N_2$]

diimide reduction of 1,2-dimethylcyclopentene

7. Suggest an experiment to establish that diimide reduction requires only one molecule of diimide to reduce an isolated double bond.

8. Draw an orbital representation of diimide, showing bond angles, non-bonding electrons, and so forth. What is the hybridization of nitrogen? It has been suggested that there are two isomers of diimide but that only one is capable of reducing alkenes. Explain the idea(s) behind this statement.

Reduction via Organoboranes. When boron trifluoride etherate (a solution of BF_3 in diethyl ether, containing the BF_3 complexed as $F_3\overset{\ominus}{B}-\overset{\oplus}{O}Et_2$) reacts with a complex metal hydride, the product is the highly toxic gas diborane, B_2H_6. Two of the more commonly employed complex hydrides are lithium aluminum hydride, $LiAlH_4$, and sodium borohydride, $NaBH_4$.

$$3LiAlH_4 + 4BF_3 \rightarrow 2B_2H_6 + 3LiAlF_4$$

$$B_2H_6 + 2CH_3CH_2OCH_2CH_3 \rightarrow 2H_3\overset{\ominus}{B}-\overset{\oplus}{O}Et_2$$

Diborane readily adds to alkenes to form alkylboranes in a process called *hydroboration*. For example, ethylene reacts with diborane to produce triethylborane.

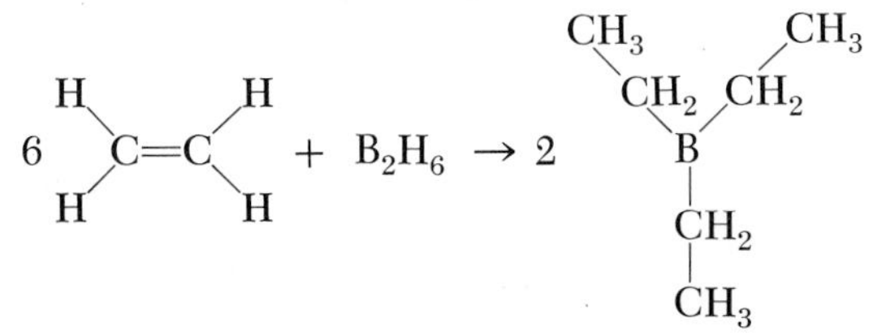

$$6\ H_2C{=}CH_2 + B_2H_6 \rightarrow 2\ (CH_3CH_2)_3B$$

alkene — diborane — triethylborane

This immensely valuable intermediate, the alkylborane, undergoes a number of reactions (see the Index), including conversion to alkanes by reaction with a carboxylic acid (RCO_2H). Since carboxylic acids are more effective than are simple mineral acids (*e.g.*, HCl), the reaction is presumed to involve a cyclic transition state.

net reaction:

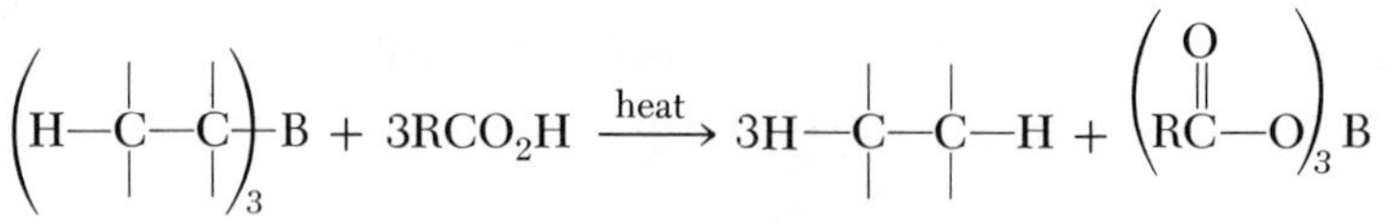

$$\left(\text{H—}\overset{|}{\underset{|}{\text{C}}}\text{—}\overset{|}{\underset{|}{\text{C}}}\right)_3\text{B} + 3\text{RCO}_2\text{H} \xrightarrow{\text{heat}} 3\text{H—}\overset{|}{\underset{|}{\text{C}}}\text{—}\overset{|}{\underset{|}{\text{C}}}\text{—H} + \left(\text{R}\overset{\text{O}}{\overset{\|}{\text{C}}}\text{—O}\right)_3\text{B}$$

mechanism:

$$\rightarrow \text{H—}\overset{|}{\underset{|}{\text{C}}}\text{—}\overset{|}{\underset{|}{\text{C}}}\text{—H} + \text{O=C(R)—O—B}$$

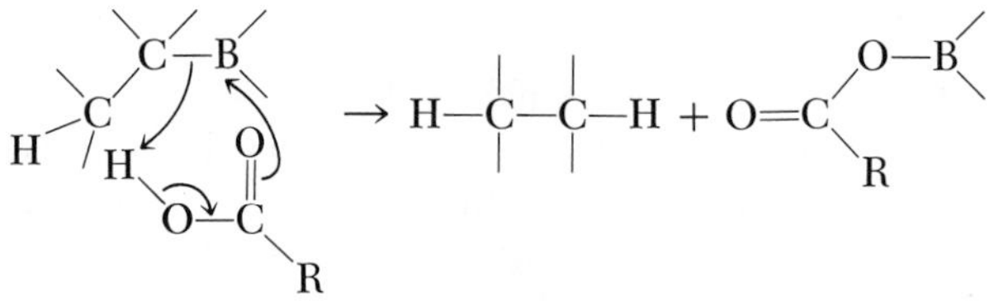

Diborane adds *cis* to a double bond, and the conversion of the C—B bond to the C—H bond proceeds with retention of configuration. Hence, hydroboration followed by acid decomposition produces a net *cis* addition of H_2.

$$\xrightarrow{B_2H_6} \xrightarrow{CH_3CH_2CO_2H}$$

1,2-dideuteriocyclopentene *cis*-1,2-dideuteriocyclopentane

9. How could you establish that, in the diborane/carboxylic acid reduction of alkenes, one hydrogen comes from the diborane and one from the acid?

8.6 CATIONIC ADDITIONS TO DOUBLE BONDS

Many additions to alkenes are initiated by the attack of a cation or incipient cation upon the π electrons of the double bond and are culminated by an attack of a nucleophile upon the resulting carbocation (this would be a good time to review pp. 138–145).

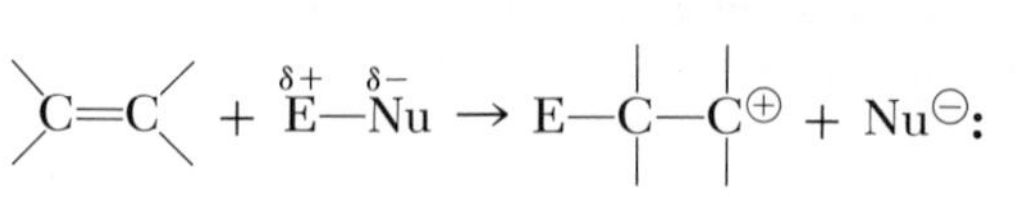

$$\text{C=C} + \overset{\delta+}{\text{E}}\text{—}\overset{\delta-}{\text{Nu}} \rightarrow \text{E—}\overset{|}{\underset{|}{\text{C}}}\text{—}\overset{|}{\underset{|}{\text{C}}}^{\oplus} + \text{Nu}^{\ominus}\text{:}$$ attack of electrophile, $E^{\oplus}$

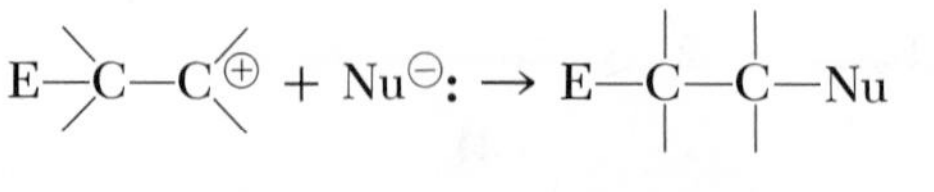

$$\text{E—C—C}^{\oplus} + \text{Nu}^{\ominus}\text{:} \rightarrow \text{E—}\overset{|}{\underset{|}{\text{C}}}\text{—}\overset{|}{\underset{|}{\text{C}}}\text{—Nu}$$ attack of nucleophile, $Nu^{\ominus}$

On the next several pages we will examine a number of reactions which follow this basic pattern of *electrophilic addition*. The best understood of these is the halogenation of alkenes.

ADDITION OF HALOGEN ACROSS THE DOUBLE BOND—THE POSTULATE. Alkenes react with solutions of Br_2 or Cl_2 (I_2 is much less reactive) in an inert solvent (*e.g.*, carbon tetrachloride) to produce *vic*-dihalides. The rate of halogenation increases with increasing alkylation at the double bond and is normally first order in halogen and in alkene.

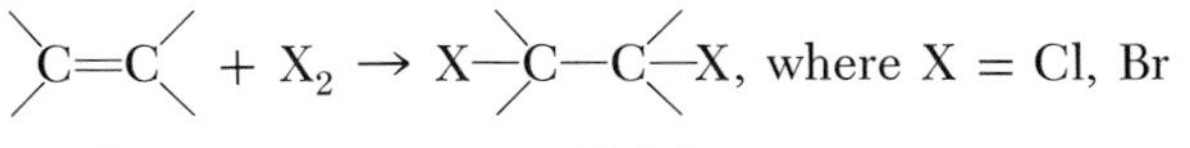

$$\rangle C{=}C\langle + X_2 \rightarrow X{-}\overset{|}{\underset{|}{C}}{-}\overset{|}{\underset{|}{C}}{-}X, \text{ where } X = Cl, Br$$

an alkene　　a *vic*-dihalide

$$R = k[\text{alkene}][X_2]$$

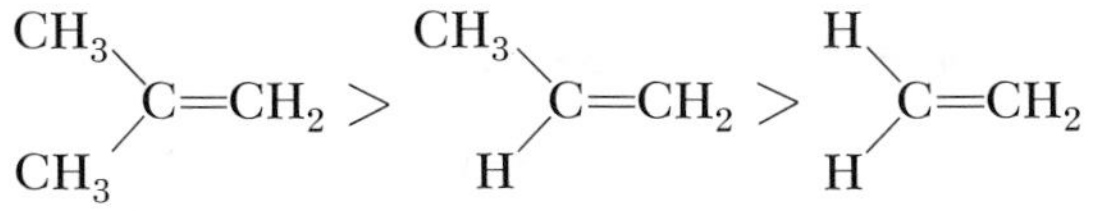

$(CH_3)_2C{=}CH_2 > (CH_3)(H)C{=}CH_2 > (H)_2C{=}CH_2$　order of increasing reactivity

While halogens do add to alkenes in the presence of light, the reaction now being discussed occurs in the dark. The implication of this is that while alkenes can undergo free radical addition of halogen, in the absence of a free radical initiator the mechanism for addition is ionic. The scheme used to explain ionic halogenation is detailed below, using bromine as the halogen.

1. The halogen molecule is temporarily polarized (by the electron-rich carbon-carbon π bond) along its σ bond, and the positive halogen reacts with the π system to form a weak complex called a "π-*complex*."

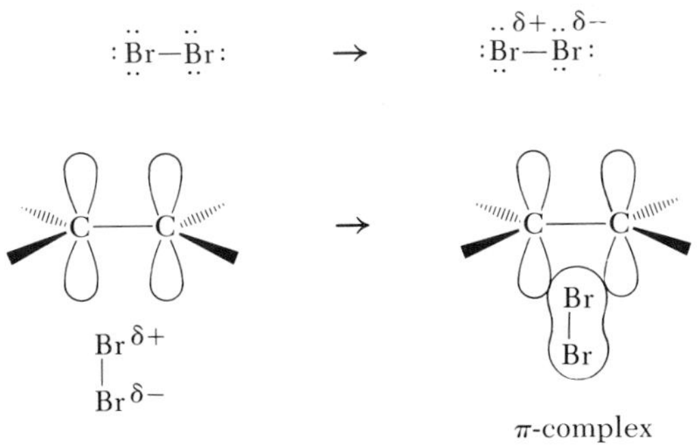

$$:\ddot{Br}{-}\ddot{Br}: \rightarrow :\ddot{Br}^{\delta+}{-}\ddot{Br}^{\delta-}:$$

$Br^{\delta+}$—$Br^{\delta-}$ → π-complex

2. The π-complex breaks down to form a σ-*complex*.

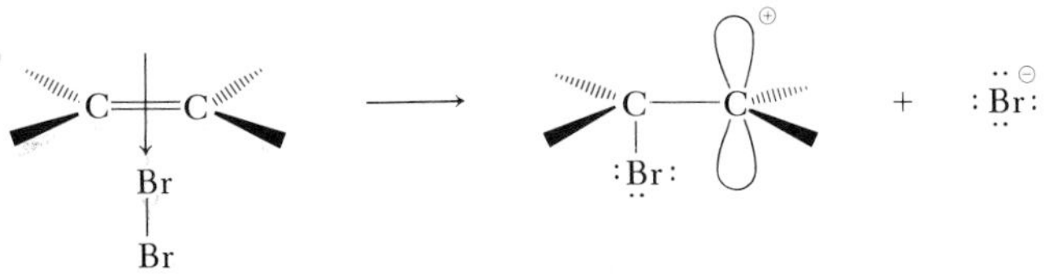

π-complex → σ-complex + $:\ddot{Br}:^{\ominus}$

3. The vacant p orbital on C_{sp^2} overlaps with a non-bonding electron pair on Br to form a cyclic *bromonium ion* (a cation in which the bulk of the positive charge is on bromine).

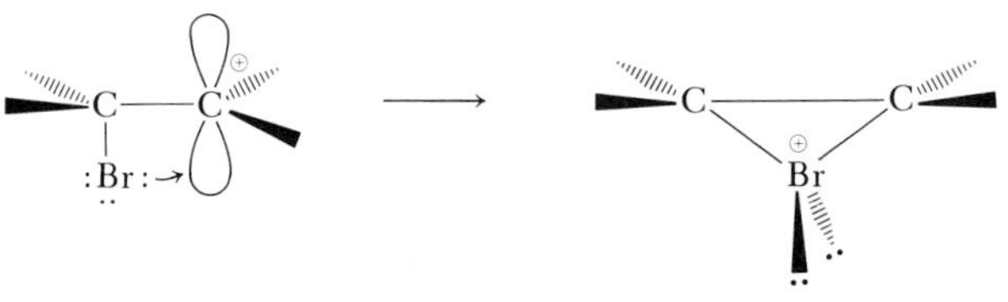

σ-complex → cyclic bromonium ion

4. The cyclic bromonium ion is attacked by a nucleophile at the carbons which are part of the three-membered ring (the Br⊕ makes them slightly positive). This can be considered an S_N2 reaction where the leaving group is —Br⊕—.

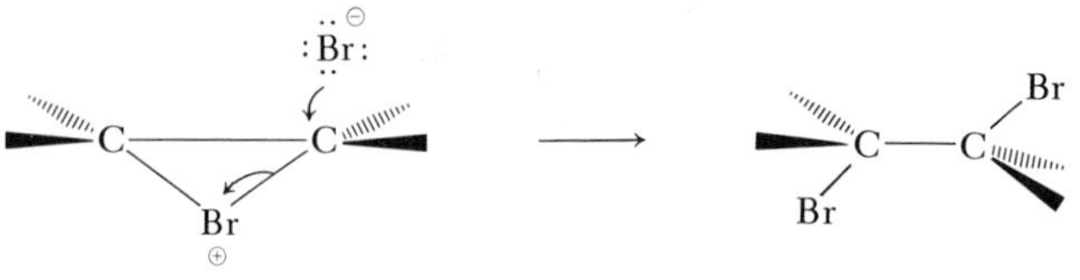

:Br:⊖ → C—C (Br⊕ bridge) ⟶ Br—C—C—Br

10. "The closure of the σ-complex to the cyclic bromonium ion (step 3 in the sequence) is akin to the S_N1 reaction." Explain.
11. (a) Which of the bromonium ions shown below should be more susceptible to conversion to a dibromide by reaction with Br⊖? (b) What product will each form? Explain.

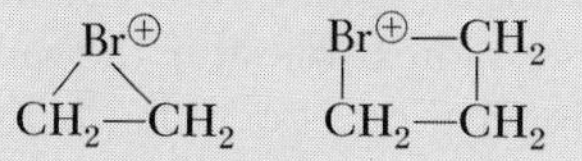

A: Br⊕ bridging CH_2—CH_2

B: Br⊕—CH_2, CH_2—CH_2 (four-membered ring)

What Is the Evidence for the Mechanism?

A. If bromination is carried out in a solution containing a nucleophile other than Br⊖ (for example, Cl⊖), the product includes a mixed halogen product but does not include a dichloride.

$$H_2C{=}CH_2 \xrightarrow[Cl^{\ominus}]{Br_2} H_2CBr{-}CH_2Br + H_2CBr{-}CH_2Cl \quad (\text{no } H_2CCl{-}CH_2Cl)$$

These data are consistent with the existence of a discrete, positively charged intermediate (since it was trapped by Cl⊖) and, more importantly, rules out a concerted addition of Br_2.

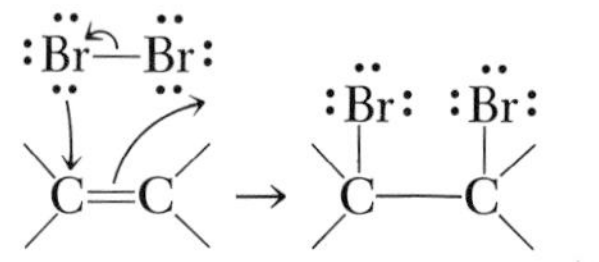

:Br—Br: + C=C → :Br: :Br: on C—C

path EXCLUDED by incorporation of Cl

B. Step 4 requires that the electrophile and nucleophile add to opposite sides of the molecular plane, *i.e.*, add in a *trans* fashion. Using ethylene as a substrate cannot help substantiate the existence of the cyclic halonium ion, because *any* addition (*i.e.*, *cis* or *trans*) across the π bond gives the same product mixture because of free rotation in the resultant C—C σ bond. However, the stereochemistry of addition can be decided by employing a cyclic olefin, because the stereochemistry of the initial reaction then will be reflected in the product. As seen below, bromine adds *trans* to cyclohexene and to cyclopentene, firmly establishing the stereochemistry of addition.

$$H_2C{=}CH_2 \xrightarrow[CCl_4]{Br_2} Br{-}CH_2{-}CH_2{-}Br$$

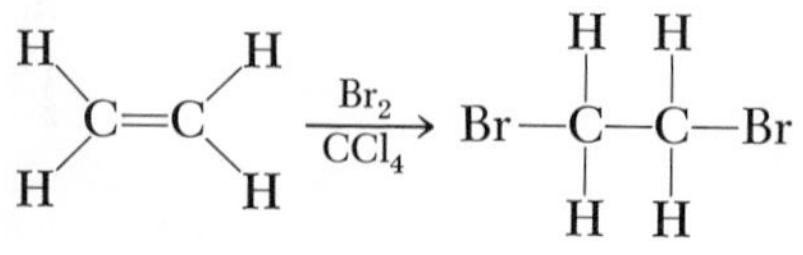

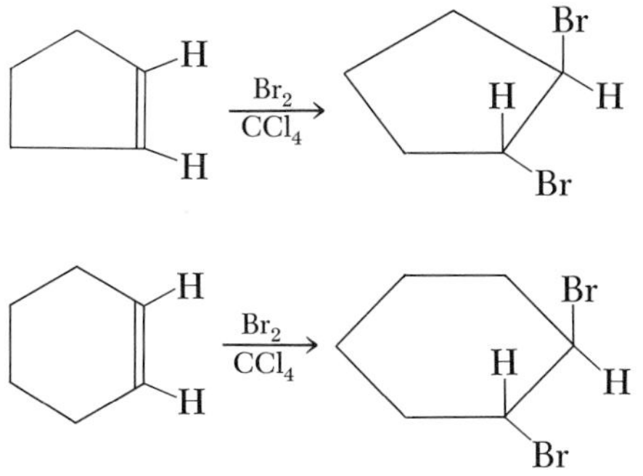

$$\text{cyclopentene} \xrightarrow[CCl_4]{Br_2} \text{trans-1,2-dibromocyclopentane}$$

$$\text{cyclohexene} \xrightarrow[CCl_4]{Br_2} \text{trans-1,2-dibromocyclohexane}$$

An "open" or "free" carbocation similar to that postulated in step 2 is inconsistent with this stereochemistry, since the vacant p orbital should be subject to attack from either side—producing both the *cis* and the *trans* adduct. (Since attack at b is more hindered than is attack at a, a free carbocation would be expected to yield more *trans* than *cis* product.)

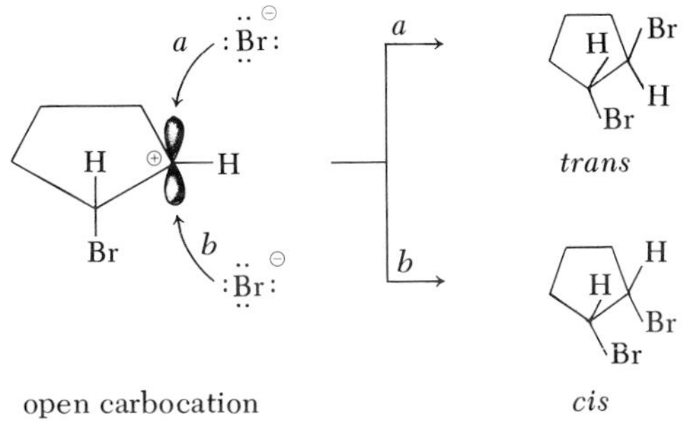

open carbocation — a → *trans*; b → *cis*

A concerted addition, already ruled out by product studies, would have produced *cis* addition.

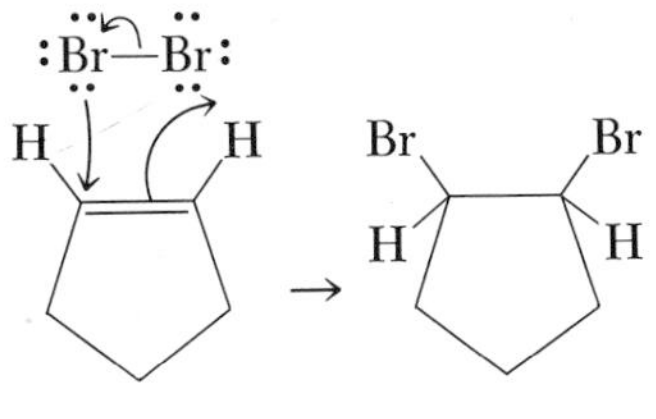

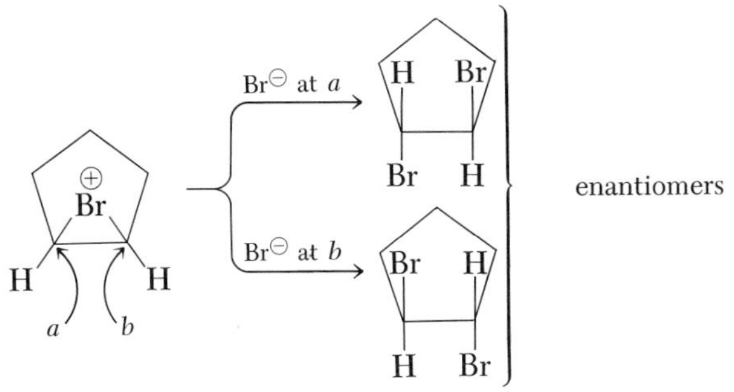

enantiomers — $Br^{\ominus}$ at a; $Br^{\ominus}$ at b

The exclusively *trans* addition is consistent only with the formation of some type of cyclic intermediate. *Therefore, either step 2 does not occur (in which case the π-complex is converted directly to the cyclic bromonium ion), or the σ-complex is extremely short-lived and is converted*

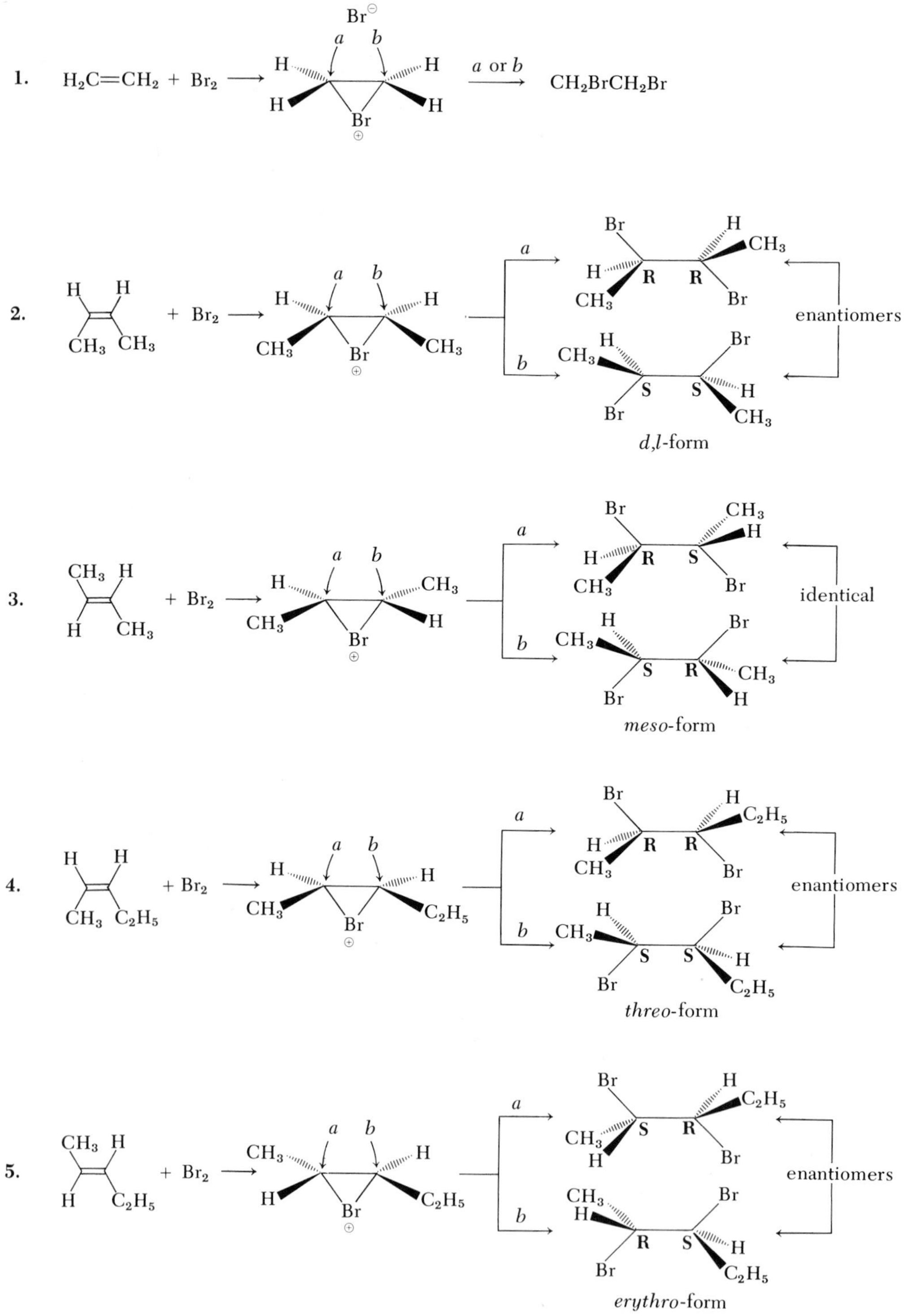

1. $H_2C{=}CH_2 + Br_2 \longrightarrow$ [bromonium ion, attack at a or b by $Br^\ominus$] $\xrightarrow{a \text{ or } b}$ CH_2BrCH_2Br

2. cis-2-butene (CH_3, CH_3 cis) $+ Br_2 \longrightarrow$ bromonium ion; a → (R,R); b → (S,S) — enantiomers — *d,l*-form

3. trans-2-butene $+ Br_2 \longrightarrow$ bromonium ion; a → (R,S); b → (S,R) — identical — *meso*-form

4. cis-2-pentene (CH_3, C_2H_5 cis) $+ Br_2 \longrightarrow$ bromonium ion; a → (R,R); b → (S,S) — enantiomers — *threo*-form

5. trans-2-pentene $+ Br_2 \longrightarrow$ bromonium ion; a → (S,R); b → (R,S) — enantiomers — *erythro*-form

Figure 8-5 Reaction of various alkenes with bromine. In each example, attacks at a and b occur with equal probability. Had the bromonium ions been drawn with $Br^\oplus$ on top of each planar system, rather than on the bottom, the *net* result would have been the same. It is suggested that you construct a chart of your own, using a bromonium ion with $Br^\oplus$ "on top," *i.e.*, [bromonium ion with $Br^\oplus$ on top], in order to confirm your understanding and to develop one further point.

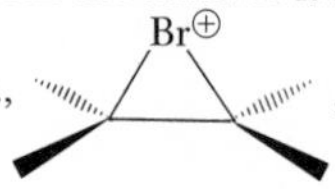

to the cyclic bromonium ion before rotation can occur around the Br—C—C$^{\oplus}$ *σ bond.* A symmetric, bridged ion explains the formation of *equimolar* amounts of the enantiomeric *trans*-1,2-dibromocyclopentanes, since attack at carbons *a* and *b* is equally probable. Figure 8–5 presents a detailed stereochemical analysis of bromine addition to a variety of alkenes.

12. By itself, does the isolation of $BrCH_2CH_2Cl$ from the reaction of ethylene with Br_2 and $Cl^{\ominus}$ prove that reaction is initiated by an electrophilic attack on the alkene? Explain.
13. How could you use *cis-* and *trans*-2-butene to prove that bromine adds *trans* (sometimes described as *anti*) to a double bond?

A Deeper Look. One of the products of the addition of bromine to propene, carried out in the presence of chloride ion, is 1-bromo-2-chloropropane; however, 2-bromo-1-chloropropane is not produced in this reaction.

$$CH_2{=}CHCH_3 + Br_2 + Cl^{\ominus} \rightarrow BrCH_2CHClCH_3 \text{ but no } ClCH_2CHBrCH_3$$

To account for this we must modify our picture of the cyclic bromonium ion so as to strengthen the C1—Br bond, weaken the C2—Br bond and, in so doing, place a larger amount of the positive charge on C2 than on C1. This is not unreasonable, since the polarizability of the methyl group should stabilize some positive charge at C2.

The incoming nucleophile ($Cl^{\ominus}$) preferentially attacks the unsymmetric ion at C2, producing 1-bromo-2-chloropropane rather than 2-bromo-1-chloropropane. These results suggest the existence of a continuum of intermediates that can be involved in electrophilic halogen addition to alkenes. One extreme is a symmetric, cyclic halonium ion, while the other extreme is the unsymmetric, open carbocation. The intermediate type of ion, the unsymmetric halonium ion, is the one which is responsible for the product formed when propene reacts with bromine in the presence of added nucleophile.

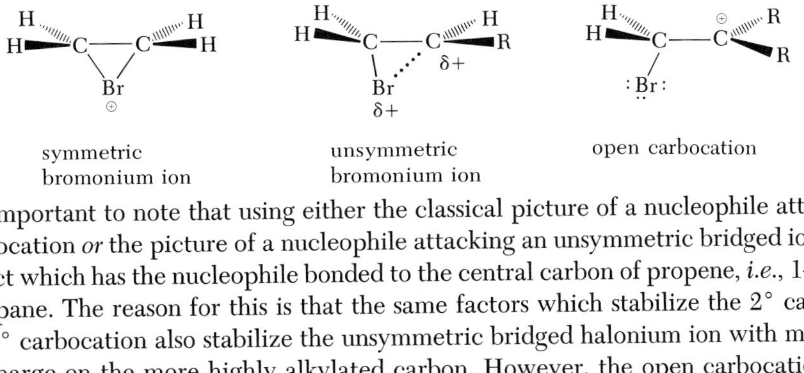

symmetric bromonium ion | unsymmetric bromonium ion | open carbocation

It is important to note that using either the classical picture of a nucleophile attacking an open carbocation *or* the picture of a nucleophile attacking an unsymmetric bridged ion leads to the product which has the nucleophile bonded to the central carbon of propene, *i.e.*, 1-bromo-2-chloropropane. The reason for this is that the same factors which stabilize the 2° carbocation over the 1° carbocation also stabilize the unsymmetric bridged halonium ion with more of the positive charge on the more highly alkylated carbon. However, the open carbocation cannot explain the *trans* stereochemistry that is observed in many of these additions. Thus, although the direction of addition is that predicted from an open carbocation, the stereochemistry of the product requires a bridged ion as an intermediate.

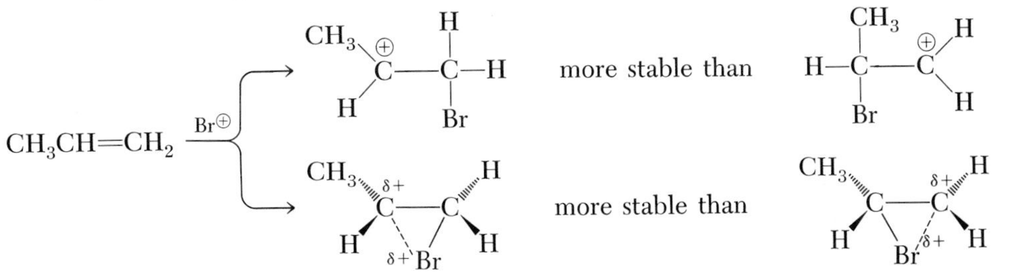

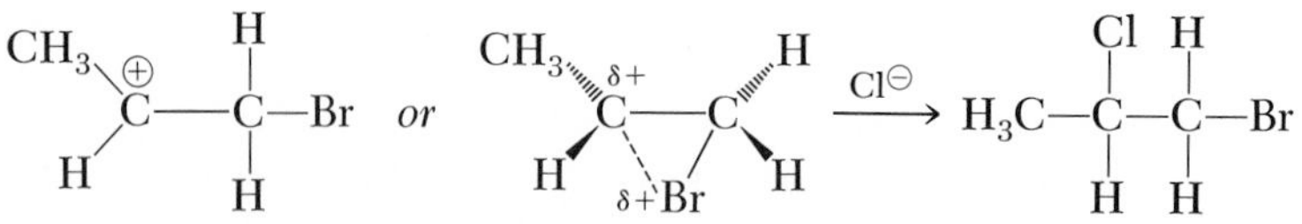

$$\underset{\text{H}}{\overset{\text{CH}_3}{}}\overset{\oplus}{\text{C}}-\underset{\text{H}}{\overset{\text{H}}{\text{C}}}-\text{Br} \quad or \quad \text{CH}_3\overset{\delta+}{\text{C}}(\text{H})-\text{C}(\text{H})(\text{H})\cdots\overset{\delta+}{\text{Br}} \xrightarrow{\text{Cl}^{\ominus}} \text{H}_3\text{C}-\underset{\text{H}}{\overset{\text{Cl}}{\text{C}}}-\underset{\text{H}}{\overset{\text{H}}{\text{C}}}-\text{Br}$$

14. Compare the symmetric bromonium ion, the unsymmetric bromonium ion, and the open carbocation with regard to:
   (a) free rotation around the C—C bond.
   (b) extent of positive charge on carbons.
   (c) extent of positive charge on bromine.
   (d) possibility that they lead to *cis* (sometimes described as *syn*) addition.
   (e) distance between bromine and both carbons.

Addition of Acids to Olefins. When ethylene is bubbled into concentrated sulfuric acid, the solution warms up and the bubbles disappear. Dissolution signals the reaction of the alkene with the acid by the mechanism outlined below:

net reaction:

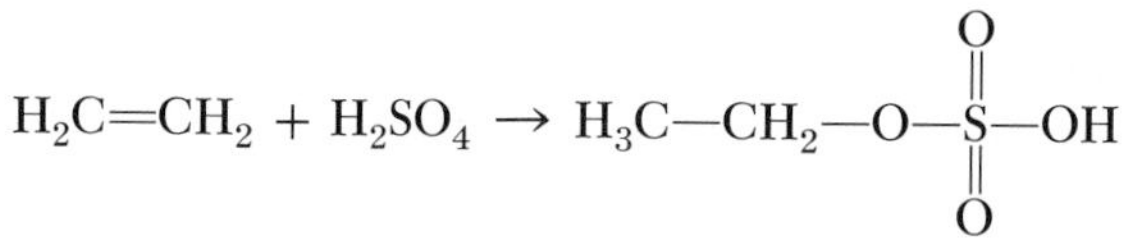

$$H_2C{=}CH_2 + H_2SO_4 \rightarrow H_3C{-}CH_2{-}O{-}\overset{\overset{O}{\|}}{\underset{\underset{O}{\|}}{S}}{-}OH$$

alkyl hydrogen sulfate

mechanism:

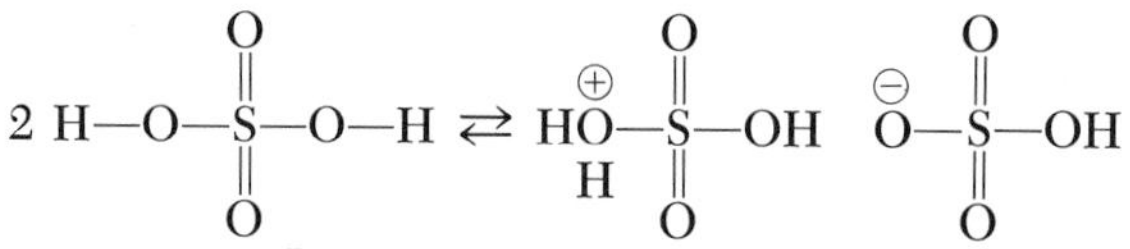

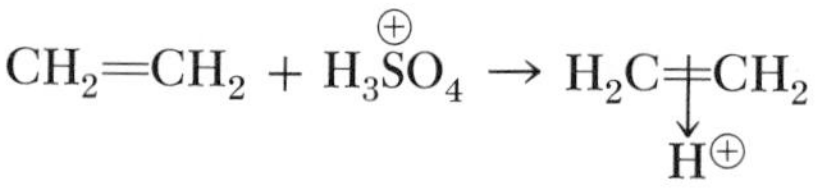

$$2\ H{-}O{-}\overset{\overset{O}{\|}}{\underset{\underset{O}{\|}}{S}}{-}O{-}H \rightleftarrows H\overset{\oplus}{O}(H){-}\overset{\overset{O}{\|}}{\underset{\underset{O}{\|}}{S}}{-}OH \quad \overset{\ominus}{O}{-}\overset{\overset{O}{\|}}{\underset{\underset{O}{\|}}{S}}{-}OH$$

$$CH_2{=}CH_2 + H_3\overset{\oplus}{S}O_4 \rightarrow \underset{\downarrow\,H^{\oplus}}{H_2C{=}CH_2} \qquad \pi \text{ complex formation}$$

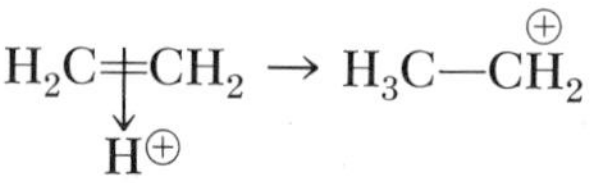

$$\underset{\downarrow\,H^{\oplus}}{H_2C{=}CH_2} \rightarrow H_3C{-}\overset{\oplus}{C}H_2 \qquad \sigma \text{ complex formation}$$

$$H_3C{-}\overset{\oplus}{C}(H)(H) \leftarrow :\!\ddot{O}(H){-}S(=O)(=O){-}O{-}H \rightarrow H_3C{-}CH_2{-}\overset{\oplus}{O}(H)(SO_3H) \xrightarrow{-H^{\oplus}} H_3C{-}CH_2{-}OSO_3H$$

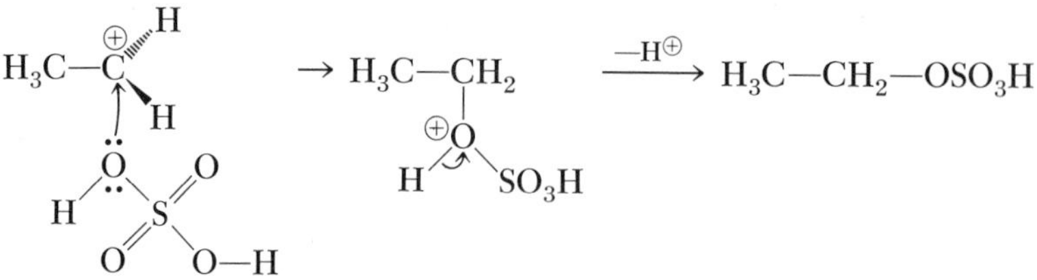

Other acids, including the hydrohalic acids, behave similarly. The ease of addition of various hydrohalic acids follows the order of their acidity, *i.e.*, HI > HBr > HCl > HF, and supports an initial, rate-determining attack by $H^{\oplus}$ on the alkene.

$$H_2C{=}CH_2 + HBr \rightarrow CH_3CH_2Br$$

Markownikoff's Rule. When HCl is added to propene, the major product is 2-chloropropane. This indicates that the π-complex collapses more rapidly to form a 2° cation than it does to form a 1° cation (Figure 8–6).

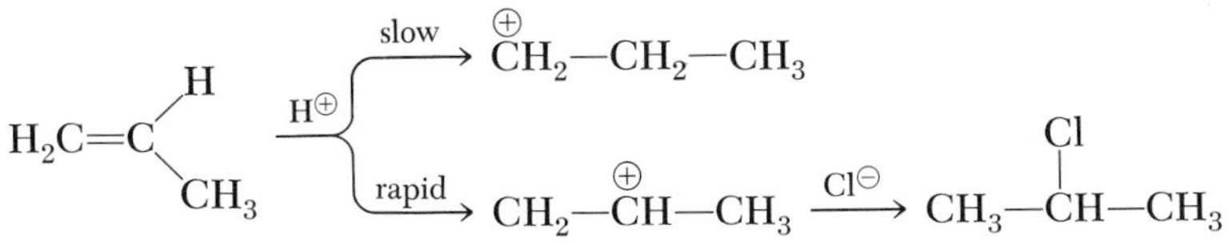

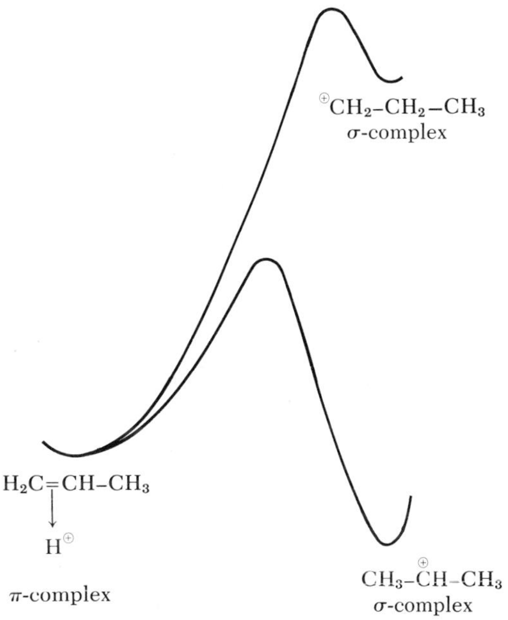

$$H_2C{=}C(H)(CH_3) \xrightarrow{H^{\oplus}} \begin{cases} \xrightarrow{\text{slow}} \overset{\oplus}{C}H_2{-}CH_2{-}CH_3 \\ \xrightarrow{\text{rapid}} CH_2{-}\overset{\oplus}{C}H{-}CH_3 \xrightarrow{Cl^{\ominus}} CH_3{-}\underset{}{\overset{Cl}{C}H}{-}CH_3 \end{cases}$$

$^{\oplus}CH_2-CH_2-CH_3$
σ-complex

$H_2C{=}CH{-}CH_3$ ↓ $H^{\oplus}$

π-complex

$CH_3{-}\overset{\oplus}{C}H{-}CH_3$
σ-complex

Figure 8-6 Decomposition of the π-complex protonated propene. Formation of $CH_3\overset{\oplus}{C}HCH_3$ occurs more rapidly than does formation of $\overset{\oplus}{C}H_2CH_2CH_3$.

The observation that an acid will add to an unsymmetric alkene in such a way as to place the incoming proton on the carbon bearing the largest number of hydrogens is called **Markownikoff's rule.** With the benefit of mechanistic insight, Markownikoff's rule can be rationalized by stating that *addition of a proton proceeds to form the most stable cation.* Several examples of "Markownikoff addition" follow:

(a) 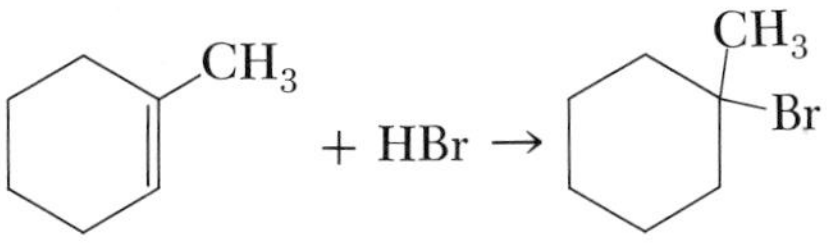 1-methylcyclohexene (CH_3) + HBr → 1-bromo-1-methylcyclohexane (CH_3, Br)

(b) $CH_3{-}CH{=}CH_2 + HOSO_3H \rightarrow CH_3{-}\underset{OSO_3H}{CH}{-}CH_3$

(c) 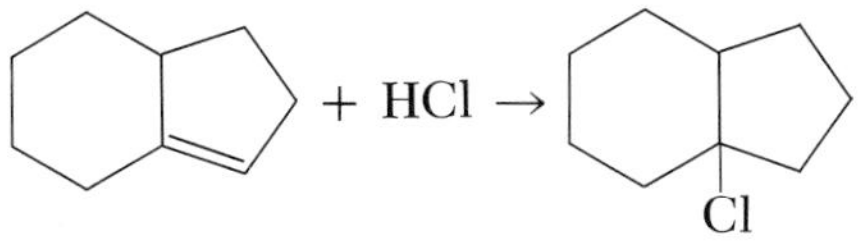 bicyclic alkene + HCl → product with Cl at ring-fusion carbon

If both possible cations are equally stable, then one expects comparable amounts of both isomeric products.

$$CH_3CH{=}CHCH_2CH_3 + HI \rightarrow CH_3CH_2CHICH_2CH_3 + CH_3CHICH_2CH_2CH_3$$

Anti-Markownikoff addition is observed if the alkene bears a powerful electron-withdrawing group attached directly to the double bond.

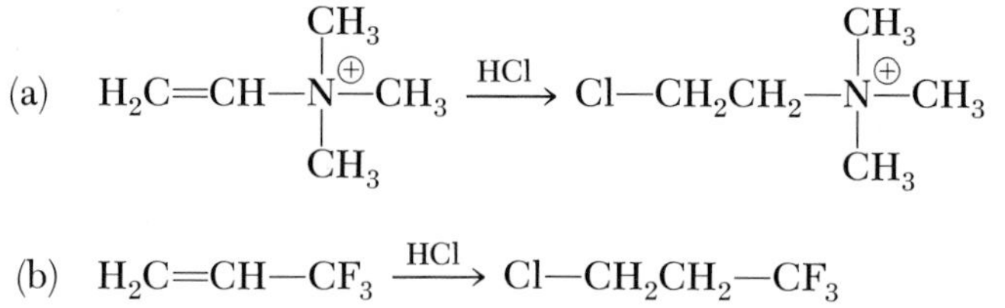

$$\text{(a)}\quad H_2C{=}CH{-}\overset{\overset{\displaystyle CH_3}{|}\;\oplus}{\underset{\underset{\displaystyle CH_3}{|}}{N}}{-}CH_3 \xrightarrow{HCl} Cl{-}CH_2CH_2{-}\overset{\overset{\displaystyle CH_3}{|}\;\oplus}{\underset{\underset{\displaystyle CH_3}{|}}{N}}{-}CH_3$$

anti-Markownikoff additions

$$\text{(b)}\quad H_2C{=}CH{-}CF_3 \xrightarrow{HCl} Cl{-}CH_2CH_2{-}CF_3$$

These anti-Markownikoff additions *do* proceed through the most stable cations. However, the substituents at the double bond destabilize the higher order cation relative to the lower order cation. Thus, while they violate the empiricism called "Markownikoff's rule," they follow the theoretical basis for the rule (see problem 31).

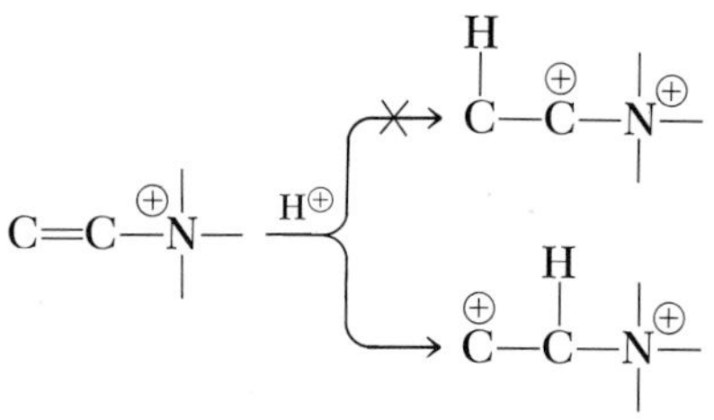

$$C{=}C{-}\overset{\oplus}{N}{-} \xrightarrow{H^{\oplus}} \begin{cases} \not\to\ \overset{H}{C}{-}\overset{\oplus}{C}{-}\overset{\oplus}{N}{-} & \text{positive charges juxtaposed} \\ \to\ \overset{\oplus}{C}{-}\overset{H}{C}{-}\overset{\oplus}{N}{-} & \text{more stable cation} \end{cases}$$

The Stereochemistry of HX Addition. The "normal" *trans* addition of HX is postulated as proceeding through a bridged ion in which H is the "bridging atom." However, since H does not have available non-bonding electron pairs (as does Br), the cyclic *protonium* ion cannot have the same type of structure as does the cyclic halonium ion. Instead, the cyclic protonium ion is viewed as a resonance hybrid of two contributing structures. This ion, like the halonium ion, will be opened by an S_N2 attack. (An unsymmetric ion, necessary to explain Markownikoff's rule, is derived from *unequally* weighted contributing structures.)

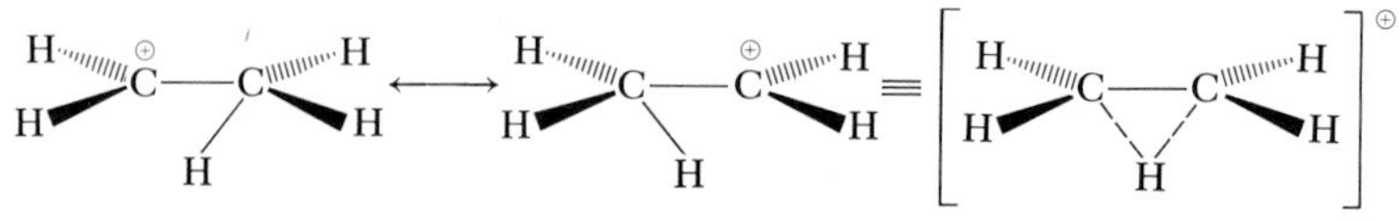

symmetric protonium ion

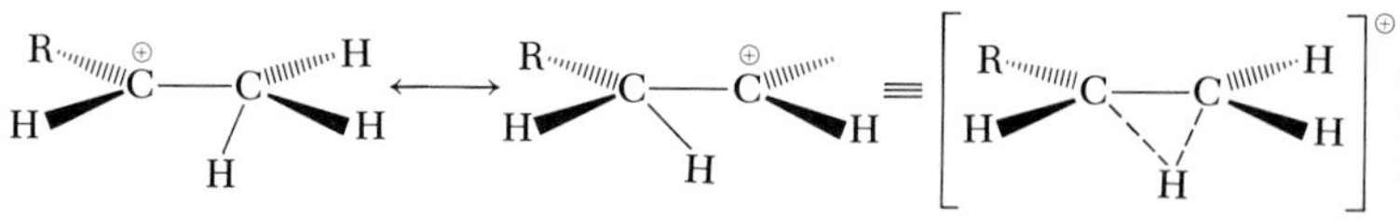

unsymmetric protonium ion

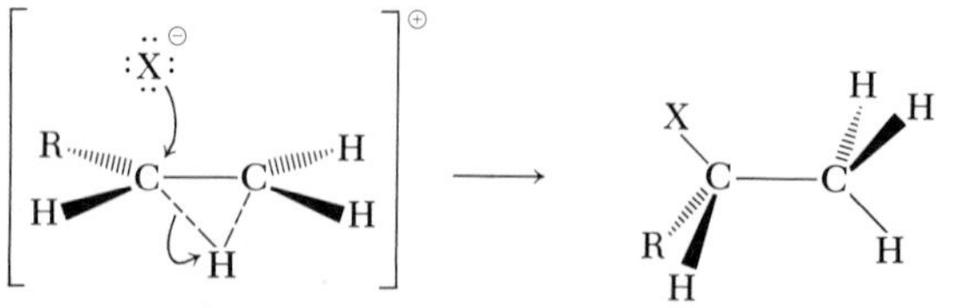

A number of chemists do not agree with this picture and prefer to view HX addition to alkenes as proceeding through an essentially open carbocation. Regardless of the correct answer, assuming that there even is a single, correct answer to fit all cases, electrophilic addition to a double bond usually produces much more *trans* than *cis* adduct.

15. Why *must* the bonding in the cyclic bromonium ion and that in the cyclic protonium ion be different? Which would you expect to be more stable?

Until recently it had been assumed that *all* additions produce more *trans* than *cis* product. However, within the past decade some additions that proceed by an exclusively *cis* pathway have been discovered. Although the number of such reactions represents no more than 1% of all known additions, there are now examples of the *cis* electrophilic addition of both HX and X_2 to alkenes.

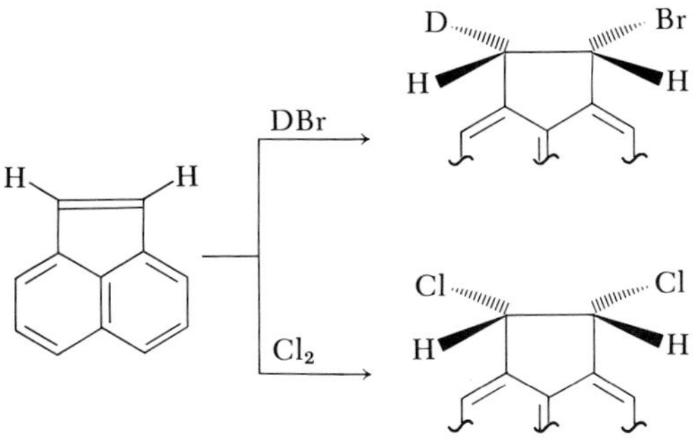

DBr

Cl_2

examples of *cis* addition to alkenes

It is believed that *cis* addition is favored when the cation is free (*i.e.*, highly stabilized by resonance) and rapidly collapses to product. Such a rapid association with the anion prevents the anion from migrating to the other side of the π system *and* prevents a second molecule from providing the consummating nucleophile from the other side of the π system.

HBr

π-complex

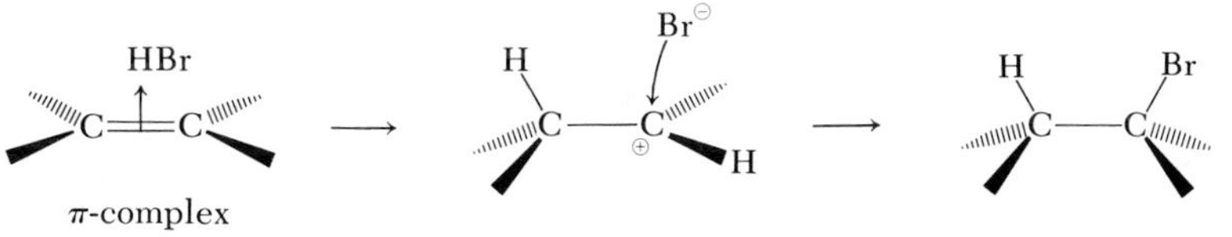

cis addition

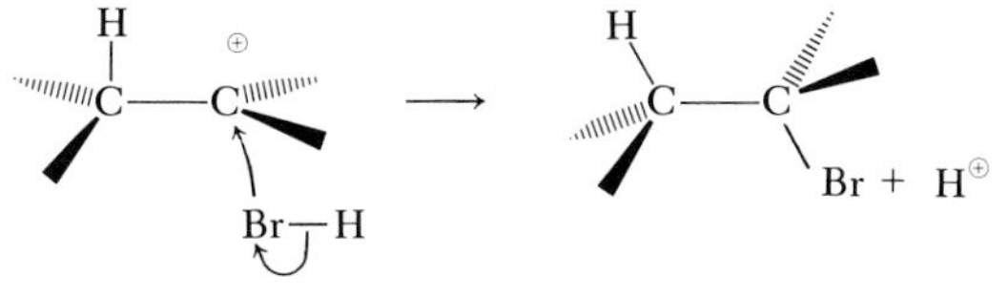

trans addition is prevented by rapid collapse of ion pair

Hydration of Alkenes. If an alkene is treated with dilute aqueous acid, the major product is an alcohol, corresponding to the addition of water to the double bond. The reaction follows Markownikoff's rule and gives predominately *trans* addition. The reaction is analogous to the addition of HX, except that H_2O successfully competes with $X^{\ominus}$ for the intermediate cation.

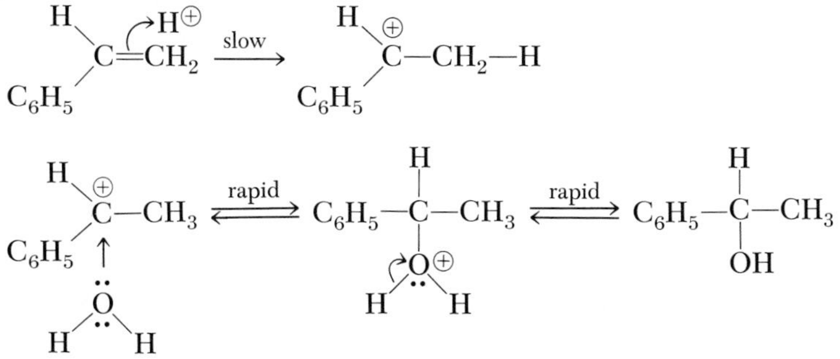

$$C_6H_5(H)C{=}CH_2 + H^{\oplus} \xrightarrow{\text{slow}} C_6H_5(H)\overset{\oplus}{C}{-}CH_2{-}H$$

$$C_6H_5(H)\overset{\oplus}{C}{-}CH_3 + H_2\ddot{O} \underset{}{\overset{\text{rapid}}{\rightleftharpoons}} C_6H_5{-}CH(CH_3){-}\overset{\oplus}{O}H_2 \underset{}{\overset{\text{rapid}}{\rightleftharpoons}} C_6H_5{-}CH(CH_3){-}OH$$

Hydration of an alkene also can be accomplished by addition of cold, concentrated sulfuric acid followed by hydrolysis of the product alkyl hydrogen sulfate. (Be sure to try problem 33.)

$$\text{cyclohexene} \xrightarrow{H_2SO_4} \text{cyclohexyl hydrogen sulfate} \xrightarrow[(S_N2)]{OH_2} \text{(oxonium ion)} \xrightarrow{-H^{\oplus}} \text{cyclohexyl alcohol (cyclohexanol)}$$

cyclohexene   cyclohexyl hydrogen sulfate   cyclohexyl alcohol (cyclohexanol)

$$\text{1-methylcyclohexene} \xrightarrow{H_2SO_4} \text{(1-methylcyclohexyl hydrogen sulfate)} \xrightarrow[(S_N1)]{OH_2} \text{(oxonium ion)} \xrightarrow{-H^{\oplus}} \text{1-methylcyclohexanol}$$

1-methylcyclohexene   1-methylcyclohexanol

16. What product would you expect to get from the acid-catalyzed hydration of the following?
(a) $CH_2{=}CHC_2H_5$
(b) $CH_3CH{=}CHCH_3$ (*cis* and *trans*)
(c) $C_6H_5CH{=}CH_2$
(d) 1-methylcyclohexene
(e) ethylidenecyclohexane

17. The acid-catalyzed hydration of **A** produces some **B** as well as **C**, but not **D**. Account for this.

$$\underset{\textbf{A}}{(CH_3)_3CCH{=}CH_2} \xrightarrow[H_2O]{H_3O^{\oplus}} \underset{\textbf{B}}{(CH_3)_2C(OH)CH(CH_3)_2} + \underset{\textbf{C}}{(CH_3)_3CCH(OH)CH_3}$$

$$\underset{\textbf{D}}{(CH_3)_3CCH_2CH_2OH}$$

8.7 FREE RADICAL ADDITION REACTIONS

Radicals, like cations, are electron-deficient species and will add across the π system of a double bond.

Hydrobromination. When propylene reacts with hydrogen bromide in the presence of a radical initiator, such as dibenzoyl peroxide, the product is *n*-propyl bromide. This anti-Markownikoff addition follows the free-radical chain mechanism presented below.

initiation

$$\underset{\text{dibenzoyl peroxide}}{C_6H_5\overset{O}{\overset{\|}{C}}{-}O{-}O{-}\overset{O}{\overset{\|}{C}}C_6H_5} \rightarrow \underset{\text{benzoylperoxy radical}}{2C_6H_5\overset{O}{\overset{\|}{C}}{-}\ddot{\underset{\cdot\cdot}{O}}\cdot}$$

the weak —O—O— bond makes peroxides good radical initiators

$$C_6H_5CO_2\cdot + HBr \rightarrow C_6H_5CO_2H + Br\cdot$$

propagation

$$\cdot Br + CH_3CH{=}CH_2 \rightarrow CH_3{-}\dot{C}H{-}CH_2Br \quad \text{(addition step)}$$

$$CH_3\dot{C}HCH_2Br + HBr \rightarrow CH_3CH_2CH_2Br + Br\cdot \quad (\textit{chain transfer}\text{ step})$$

Termination occurs by any (or all) of the reactions which consume species involved in the propagation steps.

termination steps

$$2\ Br\cdot \rightarrow Br_2$$

$$2\ CH_3{-}\dot{C}H{-}CH_2Br \rightarrow CH_3{-}\underset{H}{\overset{CH_2Br}{C}}{-}\underset{H}{\overset{CH_2Br}{C}}{-}CH_3$$

$$Br\cdot + CH_3{-}\dot{C}H{-}CH_2Br \rightarrow CH_3{-}CHBr{-}CH_2Br$$

Thus, anti-Markownikoff addition of HBr arises because (a) Br· (and not H·) is the first to react with the π system, and (b) the most stable radical is formed most rapidly.

$$CH_3{-}\underset{CH_3}{\overset{CH_3}{C}}\cdot > CH_3{-}\underset{CH_3}{\overset{H}{C}}\cdot > H{-}\underset{CH_3}{\overset{H}{C}}\cdot > H{-}\underset{H}{\overset{H}{C}}\cdot \qquad \text{order of radical stability}$$

18. What do cationic Markownikoff addition, cationic anti-Markownikoff addition, and free radical anti-Markownikoff addition have in common?

Neither HCl nor HI undergoes this "abnormal" addition. The former is energetically unfavorable because of the strength of the H—Cl bond (preventing the chain transfer step), while the latter is unfavorable because of the high E_{act} associated with addition of I· to the double bond. (Remember that in an endothermic process E_{act} must be greater than ΔH.)

$$R\cdot + H{-}Cl \xrightarrow{slow} R{-}H + Cl\cdot \qquad \Delta H = +5\ \text{kcal/mole}$$

$$I\cdot + \gt C{=}C\lt \xrightarrow{slow} I{-}\overset{|}{\underset{|}{C}}{-}\overset{|}{\underset{|}{C}}\cdot \qquad \Delta H = +7\ \text{kcal/mole}$$

Halogenation. Both Cl_2 and Br_2 can add across the π bond by a free radical chain process. These reactions are much more rapid than are the corresponding ionic additions but require a free-radical initiator in order to occur.

initiation $\quad X_2 \rightarrow 2X\cdot \quad (X = Cl\ \text{or}\ Br)$

propagation

$$X\cdot + \gt C{=}C\lt \rightarrow X{-}\overset{|}{\underset{|}{C}}{-}\overset{|}{\underset{|}{C}}\cdot$$

$$X_2 + \cdot C{-}C{-}X \rightarrow X{-}C{-}C{-}X + X\cdot \qquad \text{(chain transfer step)}$$

termination steps

$$2X\cdot \rightarrow X_2$$

$$X\cdot + X{-}\overset{|}{\underset{|}{C}}{-}\overset{|}{\underset{|}{C}}\cdot \rightarrow X{-}C{-}C{-}X$$

$$2X{-}\overset{|}{\underset{|}{C}}{-}\overset{|}{\underset{|}{C}}\cdot \rightarrow X{-}\left(\overset{|}{\underset{|}{C}}\right)_4{-}X$$

At very high temperatures (about 300°C) propene reacts with chlorine to give allyl chloride. This is *not* an addition reaction but, in fact, is a substitution reaction.

$$CH_2{=}CHCH_3 + Cl_2 \xrightarrow{300°} CH_2{=}CHCH_2Cl + HCl$$

How is this allyl chloride formed? After initiation, abstraction of a methyl hydrogen forms a stable, allylic radical. This radical survives long enough to be able to collide with a chlorine molecule, abstract Cl· from Cl_2, and form the allyl halide. When a Cl· adds to the π system, it is ejected because of the reversibility of the addition under the reaction conditions. Thus, the double bond survives the reaction conditions while the allylic position is halogenated. (This reaction works well with Br_2 replacing Cl_2.)

$$Cl_2 \rightarrow 2Cl\cdot$$

$$Cl\cdot + H_2C{=}CH{-}CH_3 \rightarrow CH_2{=}CH{-}\dot{C}H_2 + HCl$$

$$Cl_2 + CH_2{=}CH{-}\dot{C}H_2 \rightarrow CH_2{=}CH{-}CH_2Cl + Cl\cdot$$

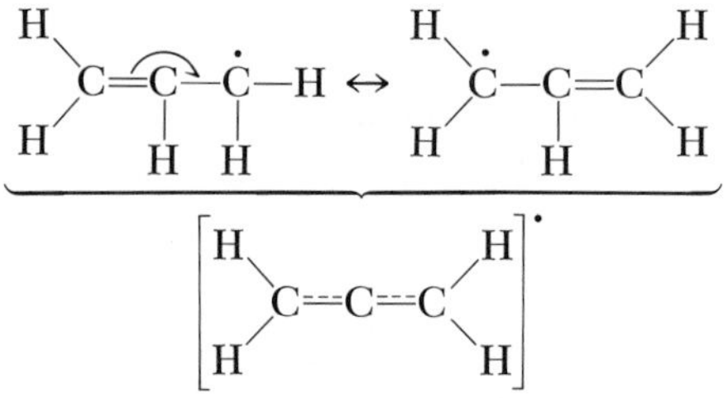

$$H_2C{=}CH{-}\dot{C}H_2 \leftrightarrow H_2\dot{C}{-}CH{=}CH_2 \qquad \left[H_2C{\cdots}CH{\cdots}CH_2\right]^{\cdot}$$

stabilization of the allyl radical by resonance delocalization of the odd electron

Allylic bromination is accomplished in the laboratory through the use of *N*-bromosuccinimide (NBS) and dibenzoyl peroxide (see problem 43).

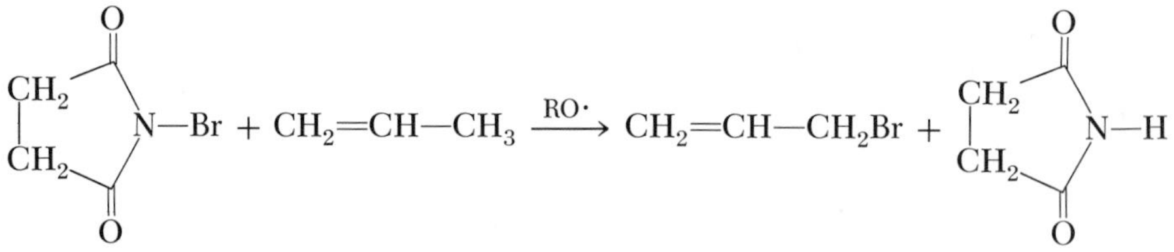

$$\text{NBS} + CH_2{=}CH{-}CH_3 \xrightarrow{RO\cdot} CH_2{=}CH{-}CH_2Br + \text{succinimide}$$

NBS

succinimide

$$\text{cyclohexene} \xrightarrow[RO\cdot]{NBS} \text{3-bromocyclohexene}$$

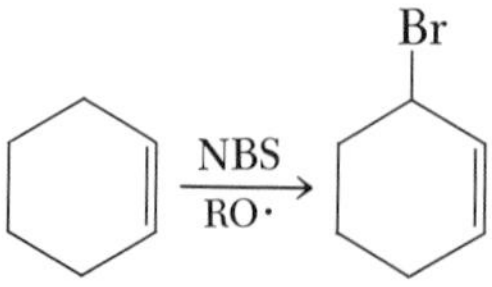

allylic bromination using NBS

Other Free Radical Additions. A number of other reagents also can add across the double bond under free radical conditions. It is suggested that the student construct reasonable chain mechanisms to account for the reactions summarized in the following diagram.

$$CH_3{-}CH{=}CH_2 \xrightarrow{CH_3CO_2\cdot} \begin{cases} \xrightarrow{RSH} CH_3CH_2CH_2SR \\ \xrightarrow{CCl_4} CH_3CHClCH_2CCl_3 \\ \xrightarrow{CBr_4} CH_3CHBrCH_2CBr_3 \\ \xrightarrow{CCl_3Br} CH_3CHBrCH_2CCl_3 \\ \xrightarrow{(C_6H_5)_3SiH} CH_3CH_2CH_2Si(C_6H_5)_3 \\ \xrightarrow{ClCOCl} CH_3CHClCH_2COCl \end{cases}$$

It is necessary to carry out free radical addition reactions in the presence of an excess of addend in order to prevent the reaction of the initially formed radical with a second molecule of alkene. If unchecked, such a side-reaction would eventually produce a *polymer*. Polymerization also is favored if A—B, the addend, does not easily provide "A" when attacked by the radical (step 3, below).

homolysis $A{-}B \xrightarrow[\text{heat}]{} A\cdot + B\cdot$ step 1

addition $>C{=}C< + B\cdot \rightarrow B{-}\overset{|}{\underset{|}{C}}{-}\overset{|}{\underset{|}{C}}\cdot$ step 2

chain transfer $B{-}\overset{|}{\underset{|}{C}}{-}\overset{|}{\underset{|}{C}}\cdot + A{-}B \rightarrow B{-}C{-}C{-}A + B\cdot$ step 3

polymerization $>C{=}C< + B{-}\overset{|}{\underset{|}{C}}{-}\overset{|}{\underset{|}{C}}\cdot \rightarrow B{-}\overset{|}{\underset{|}{C}}{-}\overset{|}{\underset{|}{C}}{-}\overset{|}{\underset{|}{C}}{-}\overset{|}{\underset{|}{C}}\cdot \xrightarrow{\text{etc.}} B{\left(\overset{|}{\underset{|}{C}}{-}\overset{|}{\underset{|}{C}}\right)_n}\cdot$

$$\xrightarrow{AB} B\cdot + B{\left(\overset{|}{\underset{|}{C}}{-}\overset{|}{\underset{|}{C}}\right)_n}A$$

The importance of the chain transfer process (hence, the nature of AB) in controlling the yield of a radical addition can be seen in Table 8–2. Halomethanes decrease in ability to act as chain transfer agents in the order $CBr_4 > CBrCl_3 > CCl_4 > CHCl_3$—and this is also the order in which the yield of 1:1 adduct between 1-octene and halomethane decreases.

TABLE 8–2 Reaction of 1-Octene With Halomethanes[a]

Halomethane	Adduct	% Yield
$CHCl_3$	$Cl_3C{-}CH_2{-}CHH{-}C_6H_{13}$	22
CCl_4	$Cl_3C{-}CH_2{-}CHCl{-}C_6H_{13}$	85
$CBrCl_3$	$Cl_3C{-}CH_2{-}CHBr{-}C_6H_{13}$	88
CBr_4	$Br_3C{-}CH_2{-}CHBr{-}C_6H_{13}$	96

[a] $A{-}B + \underset{\text{1-octene}}{CH_2{=}CH{-}C_6H_{13}} \rightarrow \underset{\text{adduct}}{A{-}CH_2{-}CHB{-}C_6H_{13}}$

8.8 POLYMERIZATION OF ALKENES

A polymer is a molecule composed of many (perhaps several hundred thousand) repeating units called "mers." It is made by the successive addition (polymerization) of smaller molecules called *monomers*. A polymer synthesized by the polymerization of two different monomers is known as a *copolymer* or *heteropolymer*. It must be emphasized that while any given molecule of polymer has very discrete properties (*e.g.*, molecular weight), the "polymer" as produced in the laboratory or in industry is a mixture with a range of molecular weights. The range will be narrow or broad, depending upon reaction conditions and the specific reagents involved in the polymerization.

$$\underset{\text{monomer}}{nA} \rightarrow \underset{\text{polymer}}{A_n}$$

$$\underset{\text{monomer B}}{mB} + \underset{\text{monomer A}}{nA} \rightarrow$$

$$\underset{\substack{\text{copolymer}\\ \text{(random arrangement)}}}{\text{—A—A—B—A—B—A—B—B—B—A—A—B—A—A—B—A—B—A—}}$$

Polymers have "end groups" which are quite different from their repeating units. However, since end groups represent a very small portion of the polymer, it is common to discuss the polymer without considering the specific end groups.

$$\underset{\substack{\text{polymerization}\\ \text{initiator}}}{X\text{—}Y} + \underset{\text{monomer}}{nA} \rightarrow \underset{\text{end groups}}{X\text{—}(A)_n\text{—}Y}$$

Most of the polymerizations encountered in the research laboratory produce the "tars," "gunk," and "goop" which are the hallmark of organic chemistry, particularly to non-organic chemists. Polymerization is, however, a very important industrial process. The majority of chemists are employed in some phase of polymer synthesis, characterization, or fabrication. The household use of terms such as "polyethylene," "polypropylene," "PVC," "Teflon," and "Saran," all of which are polymers, support this point of view. In other portions of this text we will encounter other familiar polymers including "Nylon," "Dacron," "Bakelite," "polyurethane," "starch," and "DNA." From safety devices in automobiles to physical contraceptive devices, polymer chemistry is one aspect of organic chemistry that touches us all.

Polymerization Reactions. There are three ways to produce a polymer. The simplest of these, and the one implied in the earlier illustrations, is **addition polymerization,** a process in which a monomer *adds* to a second monomer, and so on, until a long chain is produced. Addition polymerization can be initiated by cations, anions, and radicals. A second type of polymerization, one initiated by coordination of an alkene with complex inorganic materials, is termed **coordination polymerization.** This will also be discussed in this chapter. The final type, **condensation polymerization,** will be discussed in future chapters. This does not relegate condensation polymerization to a position of lesser importance, for our bodies are provided their proteins, carbohydrates, enzymes, and genes by such reactions.

$$A + C{=}C \rightarrow A\text{—}C\text{—}C \xrightarrow{C=C} A\text{—}C\text{—}C\text{—}C\text{—}C \xrightarrow{C=C} \textit{etc.} \qquad \text{addition polymerization}$$

Cationic Polymerization. A Lewis acid (*e.g.*, $H^{\oplus}$, BF_3 or $AlCl_3$) can initiate the polymerization of many alkenes in a reaction analogous to electrophilic addition to a double bond.

$$>C{=}C< + H^{\oplus} \rightarrow H\text{—}\overset{|}{\underset{|}{C}}\text{—}\overset{|}{\underset{|}{C}}^{\oplus}$$

So long as the resultant cation is reasonably stable, and alkene remains available, continued reaction of the cation with more alkene will form a growing chain. This chain can be terminated by any process which destroys the cationic center, *e.g.*, alkene formation by proton loss.

$$H^{\oplus} + \rangle C{=}C\langle \rightarrow H{-}\overset{|}{\underset{|}{C}}{-}\overset{|}{\underset{|}{C}}^{\oplus} \xrightarrow{C=C} H{-}\overset{|}{\underset{|}{C}}{-}\overset{|}{\underset{|}{C}}{-}\overset{|}{\underset{|}{C}}{-}\overset{|}{\underset{|}{C}}^{\oplus} \rightsquigarrow H\left(\overset{|}{\underset{|}{C}}{-}\overset{|}{\underset{|}{C}}\right)_n \overset{|}{\underset{|}{C}}{-}\overset{H}{\underset{|}{C}}^{\oplus}$$

$$\downarrow -H^{\oplus}$$

$$H\left(\overset{|}{\underset{|}{C}}{-}\overset{|}{\underset{|}{C}}\right)_n \overset{|}{\underset{|}{C}}{=}C\langle$$

This reaction is not used to polymerize ethylene (Why?), but it is important in the preparation of polyisobutylene and poly-α-methylstyrene.

$$\left(\overset{H}{\underset{H}{C}}{-}\overset{CH_3}{\underset{CH_3}{C}}\right)_n \qquad \left(\overset{H}{\underset{H}{C}}{-}\overset{C_6H_5}{\underset{CH_3}{C}}\right)_n$$

polyisobutylene poly-α-methylstyrene

Note how the structure of polyisobutylene reflects a chain growing with the most stable cation at its growth terminus. This kind of consistent addition is known as a "head-to-tail" addition.

$$E^{\oplus} + CH_2{=}C(CH_3)_2 \rightarrow E{-}CH_2{-}\overset{CH_3}{\underset{CH_3}{C}}{}^{\oplus} + \overset{H}{\underset{H}{C}}{=}C(CH_3)_2 \rightarrow$$

$$E{-}CH_2{-}\overset{CH_3}{\underset{CH_3}{C}}{-}CH_2{-}\overset{CH_3}{\underset{CH_3}{C}}{}^{\oplus} \rightarrow \text{etc.}$$

head-to-tail polymerization

If the concentration of alkene is low or if the concentration of a chain-ending reagent is high, the polymerization may stop with the formation of an **oligomer** (a "polymer" of 2 to 8 monomers). Isobutylene in 60% sulfuric acid is only dimerized because the water concentration is sufficiently high to remove the proton adjacent to the cationic site after two molecules of monomer have united. The resultant mixture of alkenes is known as "diisobutylene."

Anionic Polymerization. Simple alkenes do not react with bases to form stable carbanions and, consequently, most alkenes do not polymerize under basic conditions. An important exception is acrylonitrile; its polymer is soluble in dimethylformamide and can be *spun* from this solvent to form the fiber Orlon.

$$\underset{\text{acrylonitrile}}{CH_2{=}CHCN} + NH_2^{\ominus} \rightarrow H_2N{-}CH_2{-}\overset{CN}{\underset{H}{C}}{:}^{\ominus} \xrightarrow{CH_2=CHCN} \xrightarrow{\text{etc.}} \underset{\substack{\text{polyacrylonitrile}\\ \text{(Orlon)}}}{{-}(CH_2{-}CHCN)_n{-}}$$

Free Radical Polymerization. Many of the polymeric materials used in the United States are prepared by free radical polymerizations. Teflon (polytetrafluoroethylene) is made by reacting tetrafluoroethylene (TFE) with a radical (*e.g.*, acetoxy radical).

$$\underset{\text{acetyl peroxide}}{CH_3-\overset{O}{\overset{\|}{C}}-O-O-\overset{O}{\overset{\|}{C}}-CH_3} \xrightarrow{\text{heat}} \underset{\text{acetoxy radical}}{2\underbrace{CH_3-\overset{O}{\overset{\|}{C}}-O\cdot}_{R}}$$

$$R\cdot \; \underset{\text{TFE}}{F_2C{=}CF_2} \rightarrow R-\overset{F}{\underset{F}{C}}-\overset{F}{\underset{F}{C}}\cdot \xrightarrow{nCF_2CF_2} \underset{\text{Teflon}}{\left(\overset{F_2}{C}-\overset{F_2}{C}\right)_{n+1}}$$

With a molecular weight as high as 2×10^6, Teflon is inert to concentrated acids, aqueous alkalies, organic solvents, and most oxidants. Its useful temperature range is -70 to 250°; at 600 to 800° it depolymerizes to tetrafluoroethylene. The inertness and nonsticking properties of Teflon are largely responsible for the annual production of more than 15 million pounds of this material. The ecological impact of our increasing usage of Teflon has yet to be determined.

Ethylene is polymerized to polyethylene under extreme conditions (1000 psi, 100°C) with dibenzoyl peroxide as the initiator.

$$C_6H_5-\overset{O}{\overset{\|}{C}}-O-O-\overset{O}{\overset{\|}{C}}-C_6H_5 \xrightarrow{\text{heat}} 2C_6H_5-\overset{O}{\overset{\|}{C}}-\ddot{\underset{..}{O}}\cdot$$

$$C_6H_5CO_2\cdot + nCH_2{=}CH_2 \rightarrow \underset{\text{polyethylene}}{\left(CH_2-CH_2\right)_n}$$

The termination step can involve the dimerization of a large radical:

$$2R(CH_2CH_2)_nCH_2CH_2\cdot \rightarrow R(CH_2CH_2)_{2n+2}R$$

In an alternate termination step, the growing chain disproportionates to a mixture of alkene and alkane. This amounts to the transfer of a hydrogen atom from one radical to another.

$$2R-(CH_2CH_2)_n-CH_2CH_2\cdot \rightarrow R-(CH_2CH_2)_nCH{=}CH_2 + R-(CH_2CH_2)_n-CH_2CH_3$$

Such disproportionation reactions are common in radical reactions and are not limited to polymerization processes.

$$2CH_3-CH_2-CH_2\cdot \rightarrow CH_3-CH{=}CH_2 + CH_3CH_2CH_3$$ disproportionation of the propyl radical

When an unsymmetric alkene is polymerized in a free radical process, the growing end of the polymer represents the more stable possible radical. This produces head-to-tail polymerization, such as for styrene:

$$R\cdot + \underset{\text{styrene}}{C_6H_5CH{=}CH_2} \begin{cases} \not\rightarrow C_6H_5CHR-CH_2\cdot \\ \rightarrow R-CH_2-\dot{C}HC_6H_5 \xrightarrow{CH_2=CHC_6H_5} \end{cases}$$

$$R-CH_2\underset{C_6H_5}{\underset{|}{C}H}-CH_2-\underset{C_6H_5}{\underset{|}{\dot{C}H}} \xrightarrow{C_6H_5CH=CH_2} \text{polymer}$$

Polyvinyl chloride (PVC), made by radical polymerization of vinyl chloride ($CH_2{=}CHCl$), is used to construct low pressure water pipe and phonograph records. In these uses the polymer has a molecular weight of about 1.5×10^6.

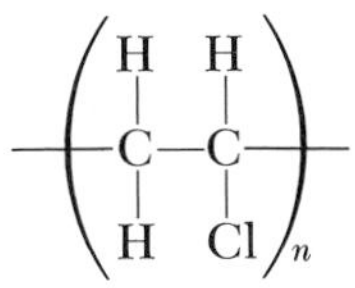

$$\left(\begin{array}{cc} H & H \\ | & | \\ -C & -C- \\ | & | \\ H & Cl \end{array} \right)_n$$

polyvinyl chloride

This hard polymer can be made softer by mixing it with a *plasticizer*, generally a low molecular weight ester such as a dialkyl phthalate.

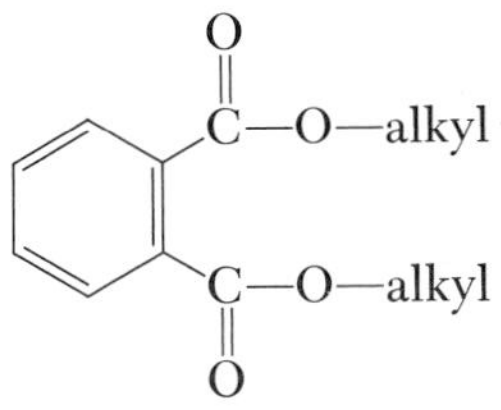

$$C_6H_4(C(=O){-}O{-}\text{alkyl})_2$$

dialkyl phthalate

In this softened condition the polymer is used to make plastic raincoats, simulated leather, electrical insulation, and the laboratory tubing known as "Tygon." "Saran" is made by copolymerizing vinyl chloride and 1,1-dichloroethene.

Coordination Polymerization. In 1963, K. Ziegler and G. Natta shared the Nobel prize in chemistry for their discovery of catalysts that would control the polymerization of alkenes such as propylene. These "Ziegler-Natta catalysts" are complex substances composed of a reductant and a transition metal salt. The most common one is probably $Al(C_2H_5)_3 \cdot TiCl_4$. With these catalysts the monomer adds to the growing polymer chain by *inserting* itself between the members of a carbon-to-metal bond.

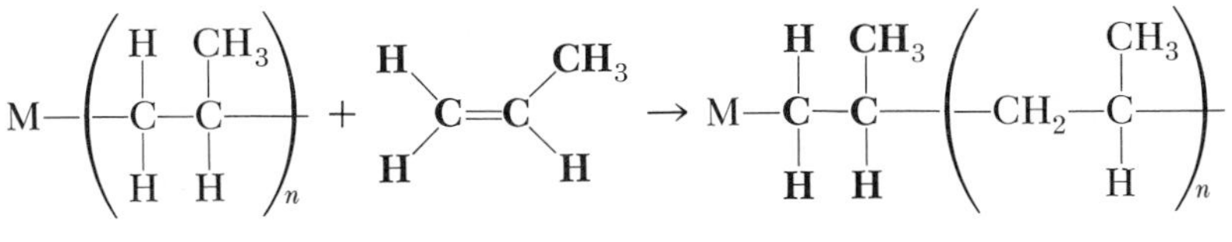

$$M{-}\left(CH_2{-}CH(CH_3)\right)_n + \begin{array}{c} H \\ H \end{array}\!\!C{=}C\!\!\begin{array}{c} CH_3 \\ H \end{array} \rightarrow M{-}CH_2{-}CH(CH_3){-}\left(CH_2{-}CH(CH_3)\right)_n$$

The use of these catalysts permits control of the stereochemistry of the polymer. Every time a propylene monomer is incorporated into a growing polymer chain, a new chiral center is created. Polypropylene prepared by older methods had a random arrangement of methyl groups around the backbone of the polymer. Such an **atactic polymer** contains chains which, because of the random geometry, cannot approach one another closely. The result is an amorphous, rubbery polymer of limited practical value.

Through the use of Ziegler-Natta catalysts, one now may prepare polypropylene in which the methyl groups alternate *regularly* from one side of the polymer backbone to another, *or* polypropylene in which all of the methyl groups are on the same side of the chain. The former is an example of a **syndiotactic polymer,** while the latter is an **isotactic polymer.** These three are represented in Figure 8–7.

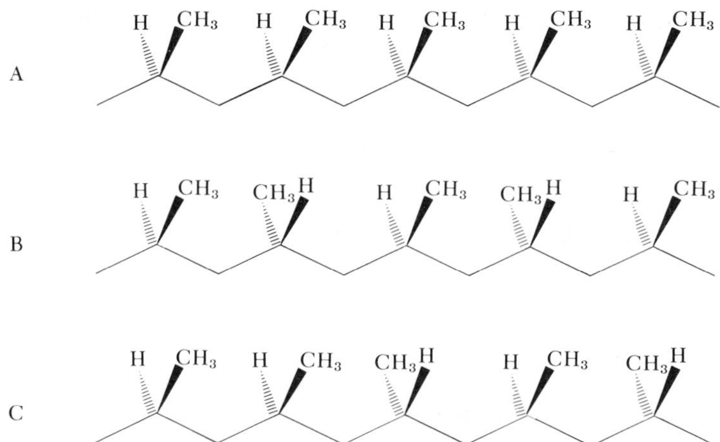

Figure 8-7 Three different polypropylene structures: A = isotactic; B = syndiotactic; C = atactic.

How physical properties are based upon structure and how this can influence the practical applications of polymers can be seen from a brief look at isotactic polypropylene. This isomer of polypropylene adopts a helical chain structure because repulsions between the methyl groups are present in a non-helical array. This helical geometry gives isotactic polypropylene a fairly high melting point (170°) and allows it to be formed into fibers (Figure 8–8).

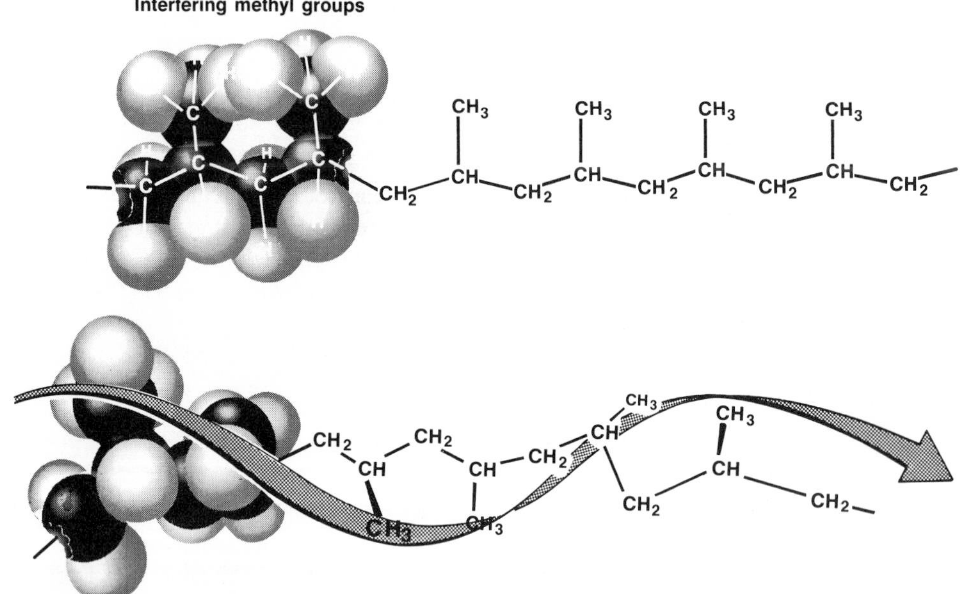

Figure 8-8 Isotactic polypropylene. The bulky methyl groups are close to each other in the top arrangement. To eliminate crowding of the methyl groups on one side of the chain, the chain flexes to produce the helical arrangement shown in the bottom view. (From Jones, M.M., Netterville, J.T., Johnston, D.O., and Wood, J.L.: Chemistry, Man and Society. W.B. Saunders Company, Philadelphia, 1972.)

8.9 OXIDATION OF DOUBLE BONDS

The term "oxidation," when applied to double bonds, is usually limited to reactions which cleave the π bond or both the σ and π bonds, and substitute bonds to oxygen for these bonds.

Extensive oxidation may also cleave C=**C—H** bonds. Although it changes the oxidation state of the double-bonded carbons, a reaction such as halogen addition to a double bond is normally not called an oxidation.

The possible oxidation products from alkenes include *epoxides, vic-diols, aldehydes, ketones, carboxylic acids,* and carbon dioxide.

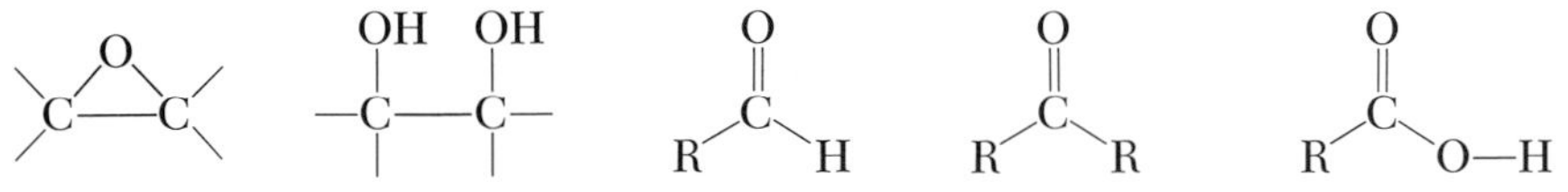

EPOXIDE *vic*-DIOL ALDEHYDE KETONE CARBOXYLIC ACID

The specific products of an oxidation are a function of the alkene, the oxidants, and the reaction conditions.

Epoxidation. When an alkene is treated with a peroxy compound (*e.g.*, peroxybenzoic acid) in the absence of potent nucleophiles, the product is an epoxide. (The preparation and reactions of epoxides are discussed in detail in Chapter 11.)

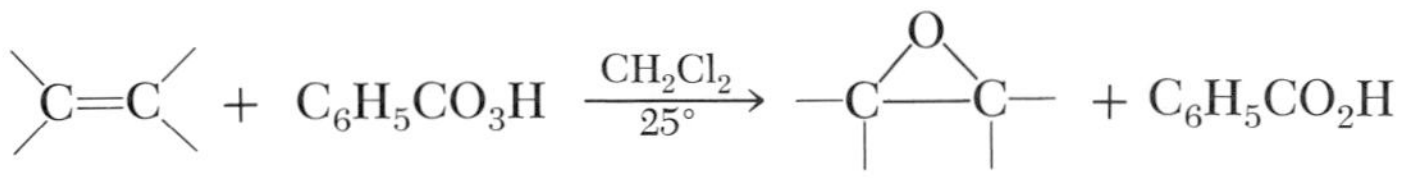

$$>C{=}C< + C_6H_5CO_3H \xrightarrow[25^\circ]{CH_2Cl_2} \text{epoxide} + C_6H_5CO_2H$$

alkene, peroxybenzoic acid, epoxide, benzoic acid

***vic*-Diol Formation.** The best known *vic*-diol is ethylene glycol, commonly used as "permanent" anti-freeze in automobile engines. It is prepared industrially by the reaction of ethylene oxide with water.

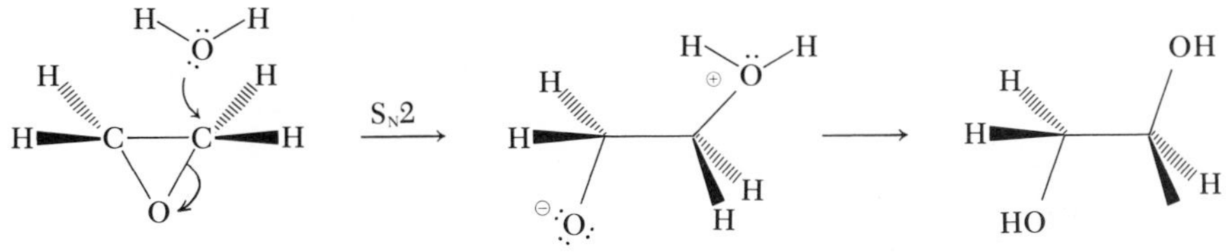

S_N2

Although the ring opening of epoxides is discussed in great detail in Chapter 11, you should recognize that this process (a) is an S_N2 reaction in which water is the nucleophile, (b) is analogous to the opening of the cyclic bromonium ion, *and* (c) results in *trans* dihydroxylation.

19. Compare and contrast the mechanisms of the addition of bromine to ethene and the reaction of ethylene oxide with hydroxide ion.

When an alkene reacts with either $MnO_4^{\ominus}$ (permanganate ion) or OsO_4 (osmium tetroxide), the product is an inorganic ester (permanganate and osmate, respectively). The ester is then converted to a *cis* diol, the stereochemistry of the product being dictated by the cyclic nature of the intermediate ester. Osmium tetroxide is the reagent of choice because, although it is highly toxic and expensive, it generally produces diols in yields approaching 100%.

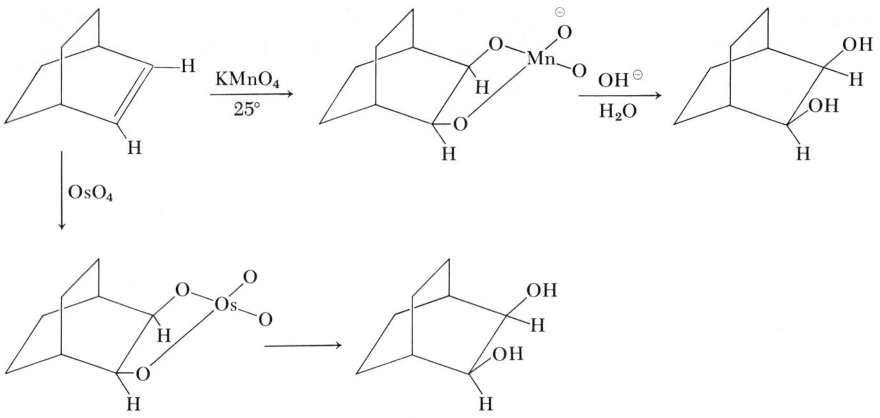

$\xrightarrow[25^\circ]{KMnO_4}$ $\xrightarrow[H_2O]{OH^\ominus}$

$\downarrow OsO_4$

It is important that the alkene be oxidized with cold, alkaline permanganate solutions, because heat and acidity produce the more vigorous oxidation conditions which lead to over-oxidation. This over-oxidation will result in the rupture of the carbon-carbon σ bond. The ultimate products of over-oxidation include ketones, carboxylic acids, and carbon dioxide, depending upon the starting alkene.

$$RCH{=}CH_2 \xrightarrow[\text{heat}]{MnO_4^\ominus/H^\oplus} \left[\underset{\substack{\text{an aldehyde}\\\text{(not isolated)}}}{RCHO} + \underset{\substack{\text{formaldehyde}\\\text{(not isolated)}}}{CH_2O}\right]^a \xrightarrow[\text{heat}]{MnO_4^\ominus/H^\oplus} \underset{\substack{\text{a carboxylic}\\\text{acid}}}{RCO_2H} + CO_2$$

$$R_2C{=}CH_2 \xrightarrow[\text{heat}]{MnO_4^\ominus/H^\oplus} \underset{\substack{\text{a ketone}\\\text{(isolated)}}}{R_2C{=}O} + \left[\underset{\substack{\text{formaldehyde}\\\text{(not isolated)}}}{CH_2O}\right]^a \xrightarrow[\text{heat}]{MnO_4^\ominus/H^\oplus} CO_2$$

$$RCH{=}CHR \xrightarrow[\text{heat}]{MnO_4^\ominus/H^\oplus} \left[\underset{\substack{\text{an aldehyde}\\\text{(not isolated)}}}{2RCHO}\right]^a \xrightarrow[\text{heat}]{MnO_4^\ominus/H^\oplus} \underset{\substack{\text{a carboxylic}\\\text{acid}}}{2RCO_2H}$$

$$R_2C{=}CHR' \xrightarrow[\text{heat}]{MnO_4^\ominus/H^\oplus} \underset{\substack{\text{a ketone}\\\text{(isolated)}}}{R_2C{=}O} + \left[\underset{\substack{\text{an aldehyde}\\\text{(not isolated)}}}{R'CHO}\right]^a \xrightarrow[\text{heat}]{MnO_4^\ominus/H^\oplus} R'CO_2H$$

$$R_2C{=}CR'_2 \xrightarrow[\text{heat}]{MnO_4^\ominus/H^\oplus} \underset{\substack{\text{a ketone}\\\text{(isolated)}}}{R_2C{=}O} + \underset{\substack{\text{a ketone}\\\text{(isolated)}}}{R'_2C{=}O}$$

20. Predict the product(s) formed by reacting each of the following with (a) osmium tetroxide and (b) permanganate ion (mild conditions).
   (a) propene
   (b) 1-butene
   (c) (**R**)-3-chloro-1-butene
   (d) (**S**)-3-chloro-1-butene
   (e) cyclohexene
   (f) (**R**)-3-methylcyclohexene
   (g) (**S**)-3-methylcyclohexene

21. Predict the product(s) from the reaction of each compound in problem 20 with warm, acidic potassium permanganate.

Alkenes are oxidized fairly rapidly by cold, neutral solutions of potassium permanganate. During the oxidation the purple color of the permanganate ion disappears and is replaced

[a] A variety of complex intermediates, deleted for simplicity, actually are formed.

by a brown precipitate of manganese dioxide. (This is due to the reduction of the manganese from Mn^{VII} to Mn^{II}.) This loss of purple color, coupled with the appearance of the brown precipitate, constitutes the "Baeyer test for unsaturation."

$$\text{>C=C<} + MnO_4^{\ominus} \rightarrow \text{—}\underset{OH}{\overset{|}{C}}\text{—}\underset{OH}{\overset{|}{C}}\text{—} + MnO_2\downarrow$$

colorless — purple — colorless — brown

$MnO_4^{\ominus}$— a chemical test for alkenes

Since permanganate ion actually can be reduced by several functional groups (*e.g.*, —C≡C— and R_2CHOH), a positive Baeyer test is not absolute proof that a double bond is present. However, it is one of the preliminary chemical (so-called "wet") tests used to determine whether a sample may contain a double bond. A second such test is the decolorization of solutions of bromine in carbon tetrachloride by alkenes.

$$\text{>C=C<} + Br_2 \xrightarrow{CCl_4} Br\text{—}\underset{|}{\overset{|}{C}}\text{—}\underset{|}{\overset{|}{C}}\text{—}Br$$

colorless — brown-orange — colorless

Br_2/CCl_4— a chemical test for alkenes

Ozonolysis. Alkenes react with ozone, O_3, by a mechanism not fully understood, to cleave the σ and π bonds of the double bond.

$$\text{>C=C<} + \overset{\oplus}{\ddot{O}}\text{—}\ddot{O}\text{—}\overset{\ominus}{\ddot{O}}\text{:} \rightarrow \text{molozonide (five-membered ring: C—C, O—O—O)}$$

a molozonide

$$\text{molozonide} \xrightarrow{\text{spontaneous}} \text{ozonide (ring: C—O—O—C, C—O—C)}$$

an ozonide

$$\text{ozonide} \xrightarrow[H^{\oplus}]{Zn} \text{>C=O O=C<}$$

If an oxidizing work-up is used instead of the reductive one employed above ($Zn/H^{\oplus}$), then carboxylic acids are isolated whenever aldehydes were present initially. These are formed by oxidation of the initially formed aldehyde. Ketones are not affected by such a change in work-up.

$$R'HC{=}CR_2 \xrightarrow{O_3} \xrightarrow{H_2O_2, H^{\oplus}} R'CO_2H + R\text{—}\overset{\overset{O}{\|}}{C}\text{—}R \qquad \text{oxidative work-up}$$

$$H_2C{=}CR_2 \xrightarrow{O_3} \xrightarrow{H_2O_2, H^{\oplus}} HCO_2H + R\text{—}\overset{\overset{O}{\|}}{C}\text{—}R \qquad \text{oxidative work-up}$$

Ethylidenecyclohexane (H, CH_3 on C=cyclohexane) $\xrightarrow{O_3}$

$$\xrightarrow[H^{\oplus}]{Zn} \text{cyclohexanone} + CH_3\text{—}\overset{\overset{O}{\|}}{C}\text{—}H \qquad \text{reductive work-up}$$

$$\xrightarrow[H^{\oplus}]{H_2O_2} \text{cyclohexanone} + CH_3\text{—}\overset{\overset{O}{\|}}{C}OH \qquad \text{oxidative work-up}$$

The major application of ozonolysis is not to synthesis but to analysis. Cleavage of the carbon skeleton and identification of the resultant fragments aids in determining the positions of double bonds within a molecule. Often just knowing the number of different products can be useful, as in the example below.

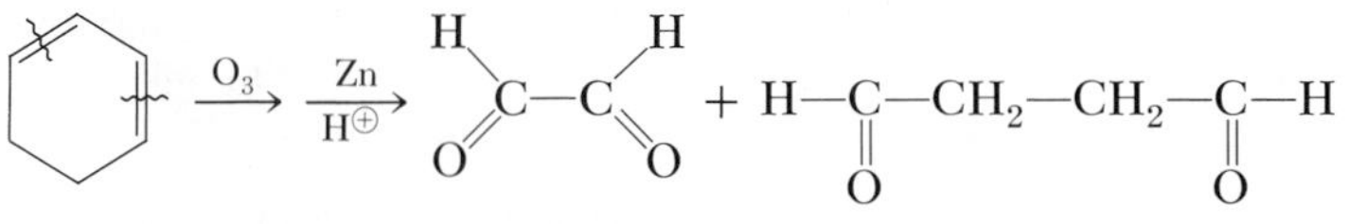

O_3 ; Zn, $H^{\oplus}$ → H–C(=O)–C(=O)–H + H–C(=O)–CH_2–CH_2–C(=O)–H

glyoxal butanedial

O_3 ; Zn, $H^{\oplus}$ → 2H–C(=O)–CH_2–C(=O)–H

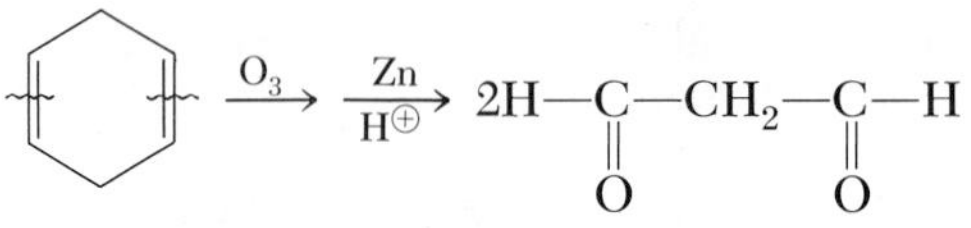

propanedial

22. What would result from the ozonolysis (reductive work-up) of (a) cyclohexene, (b) 1-methylcyclohexene, (c) 1,3-cyclohexadiene?

A solution of chromic acid (CrO_3) is much more vigorous (and, consequently, much less selective) an oxidant than is ozone and will oxidize a number of functional groups. However, CrO_3 does oxidize alkenes to produce those products normally expected from an oxidative work-up of an ozonolysis reaction. Because of the occurrence of side reactions, ozonolysis or permanganate oxidation usually is preferred to chromic acid oxidation.

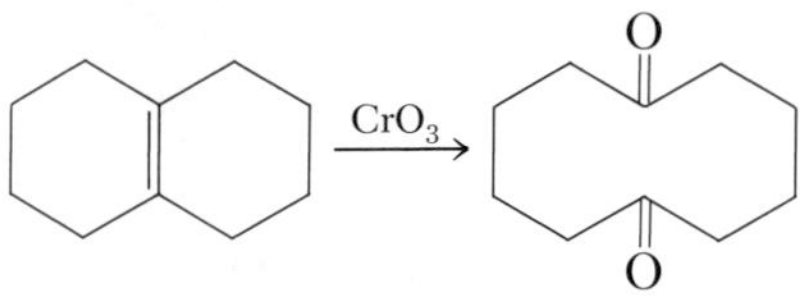

$\xrightarrow{CrO_3}$

O O

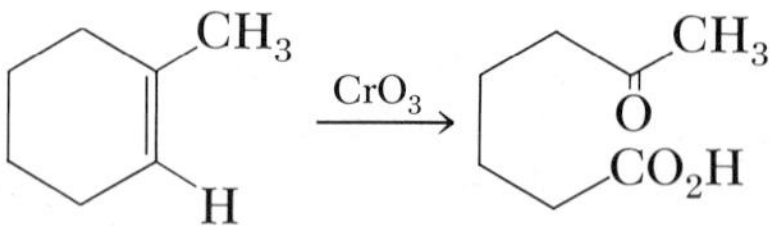

CH_3 H $\xrightarrow{CrO_3}$ CH_3 O CO_2H

8.10 CARBENES—THEIR STRUCTURE AND REACTIONS WITH ALKENES

In our introduction to the discussion of carbon hybridization (Chapter 2) we noted that the species $:CH_2$ has only six outer electrons, that it was well characterized about a decade ago, and that there are actually two molecules with the formula CH_2. One of the chemical characteristics of species containing only two bonds and six outer electrons on carbon, called **carbenes,** involves reaction with double bonds.

Methylene. The carbene that has the formula CH_2 is called *methylene.* This neutral molecule exists in two different forms. In one, the two outer electrons that are not involved in bonding are spin-paired and exist in a common orbital. This species is termed *singlet* methylene. In the other, these outer electrons are not spin-paired and each electron is found in a separate orbital. This second form, which is a *diradical,* is called *triplet* methylene.

The singlet is the less stable form of methylene. It is produced when diazomethane, CH_2N_2, is photolyzed (decomposed by the action of light). The singlet form can transfer energy (and spin) to its surroundings by collisions with neighboring molecules and is converted to the more stable triplet form by this process.

$$:\overset{\ominus}{C}H_2{-}\overset{\oplus}{N}{\equiv}N: \xrightarrow{h\nu} :CH_2 + N_2$$

diazomethane → methylene

The carbon of singlet methylene may be *approximated* by an sp^2 hybridized carbon, while the carbon of triplet methylene may be *approximated* by an *sp* hybridized carbon. Deviations from the ideal bond angles (120° and 180°, respectively) are due to valence shell electron pair repulsion and other, complex, phenomena.

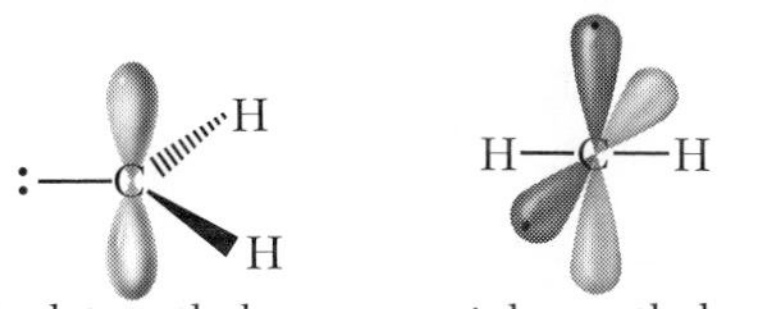

singlet methylene triplet methylene

Addition to Alkenes. The π bond of alkenes can provide a pair of electrons in order to complete the octet of the electron-deficient carbenes. Both singlet and triplet methylene react with alkenes to form cyclopropane derivatives.

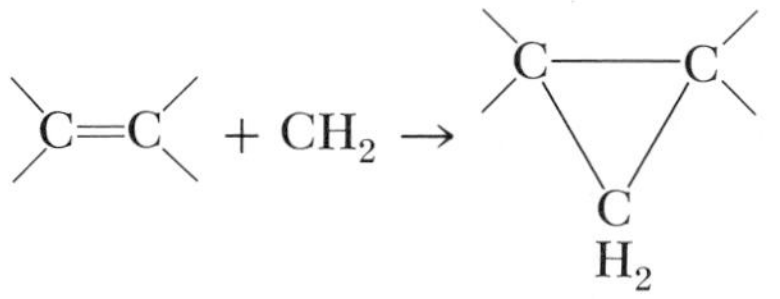

$$>C{=}C< + CH_2 \rightarrow \text{cyclopropane } (>C{-}C<, \text{ bridged by } CH_2)$$

One important difference between the behavior of singlet carbenes and that of triplet carbenes toward alkenes is the stereochemistry of these additions. The singlet addition is stereospecific, while the triplet addition is non-stereospecific. For example, when a gaseous mixture of either *cis*- or *trans*-2-butene and diazomethane is photolyzed, the product contains both *cis*- and *trans*-1,2-dimethylcyclopropane. In this reaction the first formed singlet carbene has a chance to convert to the triplet before it adds to the alkene. When photolysis is carried out in the more condensed liquid state, the first formed singlet rapidly encounters, and adds to, the alkene. For example, in the liquid state *cis*-2-butene produces only *cis*-1,2-dimethylcyclopropane, while the *trans* alkene produces only *trans*-1,2-dimethylcyclopropane.

triplet cis-$CH_3CH{=}CHCH_3$ or trans-$CH_3CH{=}CHCH_3$ $+ CH_2N_2 \xrightarrow[\text{gas phase}]{h\nu}$ cis-1,2-dimethylcyclopropane + trans-1,2-dimethylcyclopropane

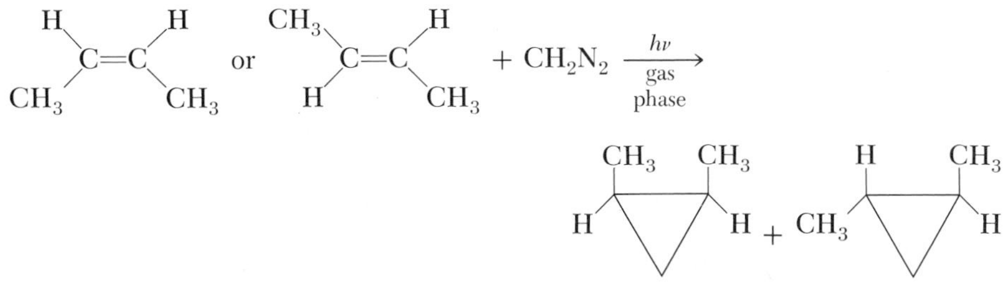

$$\text{singlet}\quad \underset{CH_3}{\overset{H}{}}\!\!C{=}C\!\!\underset{CH_3}{\overset{H}{}} + CH_2N_2 \xrightarrow[\text{liquid phase}]{h\nu} \text{cis-1,2-dimethylcyclopropane (}CH_3, CH_3\text{ cis; H, H)}$$

$$\underset{H}{\overset{CH_3}{}}\!\!C{=}C\!\!\underset{CH_3}{\overset{H}{}} + CH_2N_2 \xrightarrow[\text{liquid phase}]{h\nu} \text{trans-1,2-dimethylcyclopropane (}CH_3, H; H, CH_3\text{)}$$

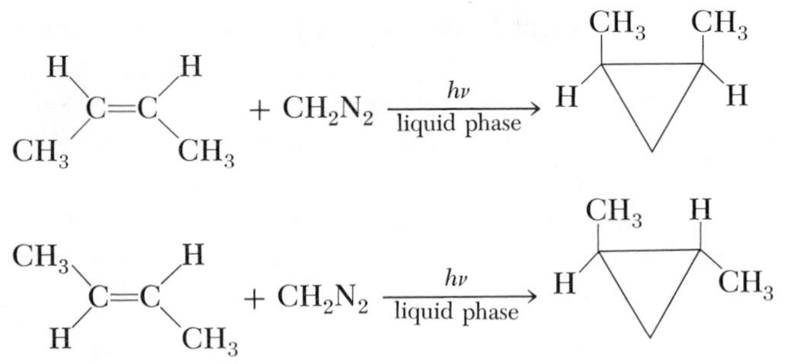

The non-stereospecificity of the reaction of triplet methylene results from its *stepwise* addition to the alkene. The diradical intermediate (shown below) lacks a carbon-carbon π bond, can undergo free rotation, and, therefore, leads to isomer interconversion.

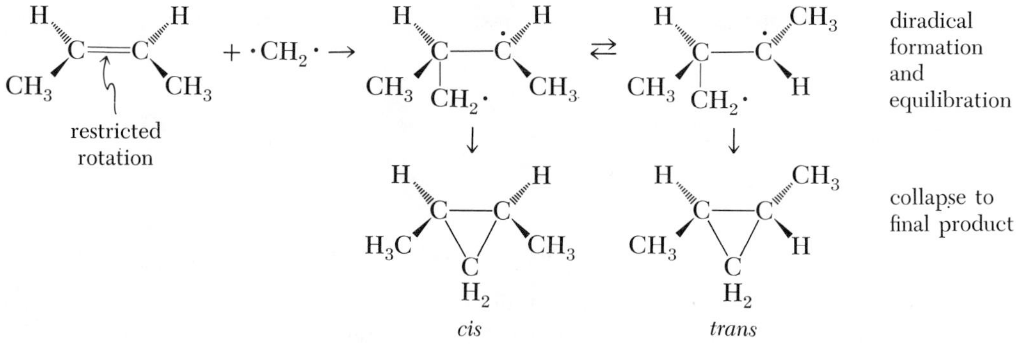

$$\text{cis-}CH_3CH{=}CHCH_3 \text{ (restricted rotation)} + \cdot CH_2 \cdot \rightarrow CH_3(H)\overset{}{C}(CH_2\cdot)\text{—}\dot{C}(H)CH_3 \rightleftarrows CH_3(H)C(CH_2\cdot)\text{—}\dot{C}(CH_3)H \quad \text{diradical formation and equilibration}$$

$$\downarrow \qquad\qquad \downarrow$$

$$\text{cis (}H_3C, CH_3\text{ on ring C–C–}C H_2\text{)} \qquad \text{trans (}CH_3, H\text{; }H, CH_3\text{)} \quad \text{collapse to final product}$$

restricted rotation

cis *trans*

The stereospecificity of the singlet addition arises because its addition is a *concerted* process rather than a stepwise one.

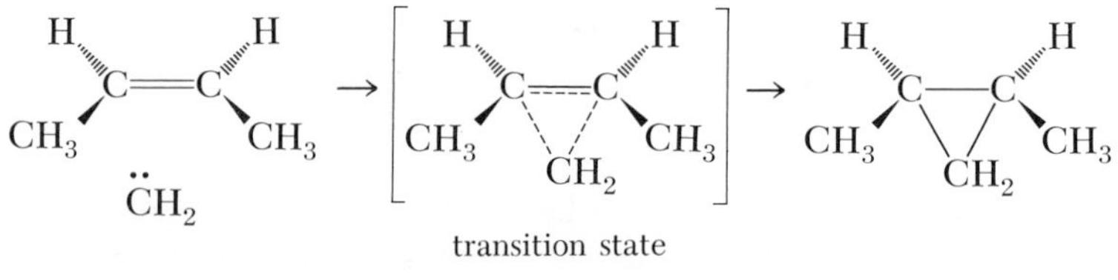

$$CH_3(H)C{=}C(H)CH_3 + \ddot{C}H_2 \rightarrow \left[CH_3(H)C\overset{\cdots}{=}C(H)CH_3 \;\; \overset{\text{\textbackslash\ /}}{CH_2} \right] \rightarrow CH_3(H)C\text{—}C(H)CH_3 \text{ (ring with }CH_2\text{)}$$

transition state

In our discussion of elimination reactions (Chapter 6) we noted the existence of α-elimination reactions and how these lead to carbenes. A classic example is the conversion of chloroform to dichlorocarbene. As with other carbenes, dichlorocarbene adds to alkenes.

$$CHCl_3 + t\text{-buO}^{\ominus} \rightleftarrows t\text{-buOH} + :\overset{\ominus}{C}Cl_3$$

$$:\overset{\ominus}{C}Cl_3 \rightarrow \underset{\text{dichlorocarbene}}{:CCl_2} + Cl^{\ominus}$$

$$CH_3CH{=}CH_2 + CCl_2 \rightarrow \underset{\text{1,1-dichloro-2-methylcyclopropane}}{CH_3CH\text{—}CH_2 \text{ (ring with } CCl_2\text{)}}$$

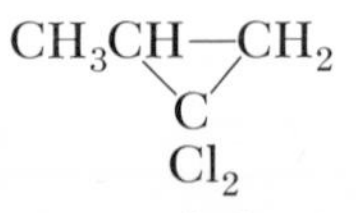

Unlike singlet methylene, singlet dichlorocarbene is more stable than its triplet isomer. (The reason for this is still not perfectly clear, although donation of an electron pair from chlorine to the vacant orbital on carbon may be part of the answer.) When dichlorocarbene is formed in the dehydrochlorination of chloroform, it is produced in its singlet state and will add *stereospecifically* to an alkene.

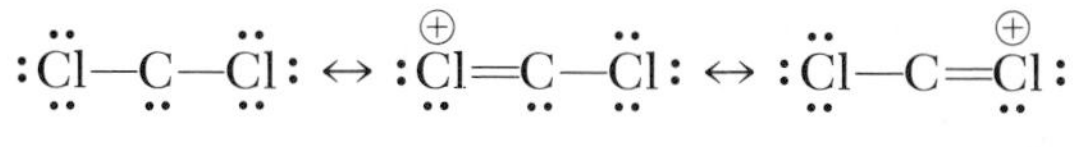

$$:\ddot{Cl}-\ddot{C}-\ddot{Cl}: \leftrightarrow :\overset{\oplus}{Cl}=\ddot{C}-\ddot{Cl}: \leftrightarrow :\ddot{Cl}-\ddot{C}=\overset{\oplus}{Cl}:$$ stabilization of dichlorocarbene

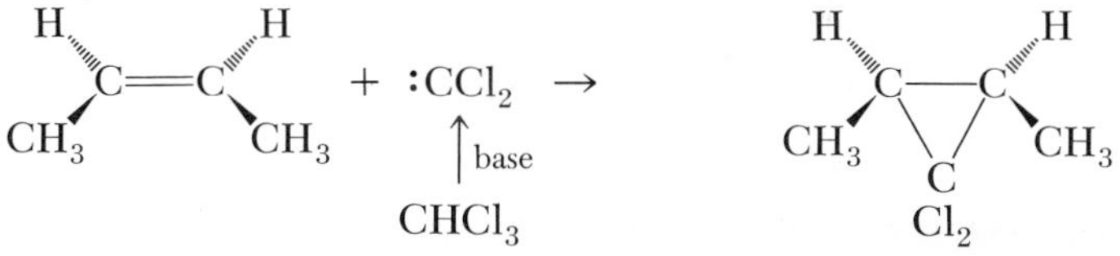

$$cis\text{-}CH_3CH=CHCH_3 + :CCl_2 \rightarrow \text{(}CCl_2 \text{ from } CHCl_3 \xrightarrow{base}\text{)}$$

cis-1,1-dichloro-2,3-dimethylcyclopropane

Insertion Reactions. Either state of methylene is, in general, much more reactive than are the dihalocarbenes, CX_2. This is due to some (as yet unclear) stabilization arising from the presence of the halogens. This diminished reactivity explains why CH_2 will *insert* itself between the members of a C—H bond, to produce the next higher homolog of the starting material, while dichlorocarbene generally does not undergo insertion reactions.

$$CH_3-CH_2-H + \mathbf{CH_2} \rightarrow CH_3-CH_2-\mathbf{CH_2}-H$$ insertion into a C—H bond

$$CH_3-CH_2-H + CCl_2 \rightarrow \text{no reaction}$$

Carbenoids. We might note, in closing, that there are some materials that behave like carbenes but, in fact, are not free carbenes. These compounds are referred to as *carbenoids*. The best known example of a carbenoid is the Simmons-Smith reagent, made from methylene iodide and a zinc-copper couple (mixture). While it does not produce a free methylene, the organozinc compound (which is the actual reagent in this mixture) does lead to addition to alkenes without competitive insertion. It is most useful in the synthesis of cyclopropanes.

$$CH_2I_2 + Zn(Cu) \rightarrow ICH_2ZnI$$

Simmons-Smith reagent

$$cis\text{-}CH_3CH=CHCH_3 + IZnCH_2I \rightarrow cis\text{-}1,2\text{-dimethylcyclopropane}$$

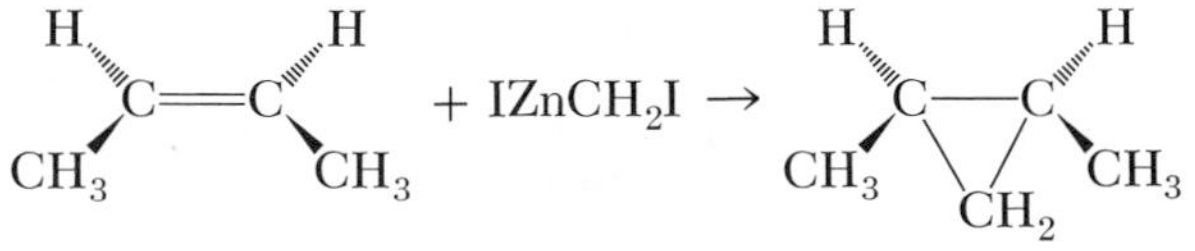

In Chapters 10 and 11 we will consider still more additions to alkenes.

IMPORTANT TERMS

Addition polymer: A polymer made by the continuing addition of an unsaturated compound to the growing end of the polymer. The monomer is almost identical to the repeating unit of the polymer. Polyethylene and other polymeric alkenes are addition polymers.

$$n\,CH_2=CH_2 \rightarrow -(CH_2-CH_2)_n-$$

polyethylene

Alkylborane: A compound possessing an alkyl group bonded to boron. The term has been used to describe compounds which have one, two, or three alkyl groups bonded to boron, although these also are termed mono-, di-, and trialkylboranes, respectively.

alkyl—B(H)(H)	alkyl—B(alkyl)(H)	alkyl—B(alkyl)(alkyl)
monoalkylborane	dialkylborane	trialkylborane

Anti-Markownikoff addition: Addition of HZ to an unsymmetric alkene such that the carbon bearing the fewer hydrogens is bonded to the hydrogen of HZ.

$$\mathrm{HRC{=}CR_2 + \mathbf{H}Z \rightarrow HRZC{-}CR_2\mathbf{H}}$$

While the classic example is the free radical addition of hydrogen bromide to an alkene, ionic additions also may lead to anti-Markownikoff addition. See the text and problem 31 for examples.

Atactic polymer: A polymer in which the groups bonded to the backbone have a random stereochemistry with respect to a given face of the backbone.

Baeyer test: A test for the presence of groups oxidized by cold, dilute, neutral aqueous potassium permanganate. A positive test requires both loss of the pink color of the permanganate ion and production of a brown precipitate of manganese dioxide, MnO_2. A positive result is given by carbon-carbon double and triple bonds and by aldehydes. Sometimes this is called the Baeyer test for unsaturation.

Bromonium ion: An ion that carries a positive charge on bromine. A cyclic bromonium ion is formed during the addition of bromine to alkenes.

Carbene: The general name for a class of compounds possessing neutral, divalent carbon. Carbenes may exist as singlets (in which non-bonding electrons on carbon have opposite spins) or triplets (in which non-bonding electrons have similar spins).

Catalytic reforming: Conversion of one kind of hydrocarbon into another using various catalysts and reaction conditions. This is especially important in the conversion of petroleum into benzene and its derivatives. *Hydroforming* reactions are those in which petroleum and hydrogen react to form hydrogenated cyclic compounds. When the catalyst is platinum on clay, the process is called platforming.

Chain transfer agent: A reagent that accomplishes a chain transfer reaction.

Chain transfer reaction: The simultaneous termination of one chain and initiation of another. In the example below, the termination of a growing polystyrene chain is accomplished by reacting the chain with carbon tetrachloride. This termination produces the trichloromethyl radical, which, by reacting with a styrene molecule, begins another polymer molecule. One can view the transfer agent and the monomer as competing with one another for the growing radical polymer chain.

$$\boxed{\text{polystyrene}}\sim CH_2{-}\underset{C_6H_5}{\overset{H}{C}}\cdot + CH_2{=}CHC_6H_5 \xrightarrow{\text{polymerization}}$$

$$\boxed{\text{polystyrene}}\sim CH_2{-}\underset{C_6H_5}{\overset{H}{C}}{-}CH_2{-}\underset{C_6H_5}{\overset{H}{C}}\cdot$$

$$\boxed{\text{polystyrene}}\sim CH_2-\underset{C_6H_5}{\overset{H}{\underset{|}{\overset{|}{C}}}}\cdot + CCl_4 \xrightarrow{\text{chain transfer}} \boxed{\text{polystyrene}}\sim CH_2-\underset{C_6H_5}{\overset{H}{\underset{|}{\overset{|}{C}}}}-Cl + \cdot CCl_3$$

$$\cdot CCl_3 + CH_2{=}CHC_6H_5 \xrightarrow[\text{polymer chain}]{\text{beginning a new}} Cl_3C-CH_2-\underset{C_6H_5}{\overset{H}{\underset{|}{\overset{|}{C}}}}\cdot$$

***cis* addition:** Addition in which both fragments end up bonded on the same face of a (used-to-be) π bond. A classic example of a *cis* addition is the catalytic hydrogenation of a double bond. Also called *syn* addition.

Condensation polymer: A polymer prepared by the removal of a small molecule (*e.g.*, hydrogen chloride) from two monomers. The beginnings of such a polymer are shown below. The mechanism of this type of reaction will be discussed later.

$$Cl-\overset{O}{\overset{\|}{C}}-Cl \quad HO-(CH_2)_3-OH \quad Cl-\overset{O}{\overset{\|}{C}}-Cl \quad HO-(CH_2)_3-OH \quad Cl-\overset{O}{\overset{\|}{C}}-Cl$$

$$\downarrow$$

$$Cl-\overset{O}{\overset{\|}{C}}-O-(CH_2)_3-O-\overset{O}{\overset{\|}{C}}-O-(CH_2)_3-O-\overset{O}{\overset{\|}{C}}-Cl + 4HCl$$

Conjugated double bonds: An arrangement of alternating single and double bonds. The simplest molecule containing two conjugated double bonds is 1,3-butadiene. (Further elaboration will be presented in Chapter 12. A definition of "conjugated system" is found on p. 366.)

Coordination polymerization: A polymerization in which the growth terminus (*i.e.*, the growing end of the polymer) is a carbon-metal bond.

Cumulative double bonds: An arrangement in which two double bonds share a common, *sp* hybridized carbon. The simplest molecule containing a pair of cumulative double bonds is allene.

Decarbonylation: A reaction in which carbon monoxide is lost from a compound.

Electrophilic addition: Addition which begins by the reaction of a positively charged species with a π bond.

Endocyclic double bond: A double bond in which both sp^2 hybridized carbons are found within a ring.

Exocyclic double bond: A double bond in which only one sp^2 hybridized carbon is found within a ring.

Halonium ion: A cation in which the positive charge is on a halogen.

Heterogeneous catalyst: A catalyst which is insoluble in a given reaction medium. The classic example is the metal used in catalytic hydrogenation.

Homogeneous catalyst: A catalyst which is soluble in a given reaction mixture.

Hydroboration: Addition of diborane to a π bond to form an organoborane (*i.e.*, a compound containing a carbon-boron bond).

Isotactic polymer: A polymer in which the substituents on the backbone all are on one face of the backbone.

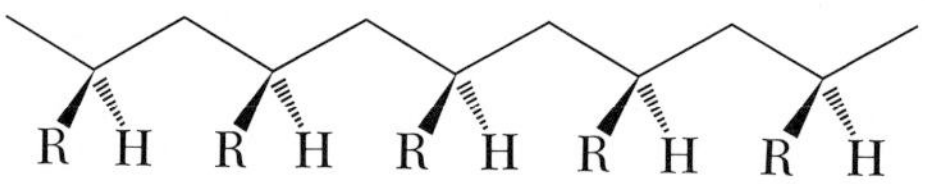

R H R H R H R H R H

Insertion reaction: A reaction in which a carbene inserts itself into a C—H bond. This type of reaction is unique to carbenes.

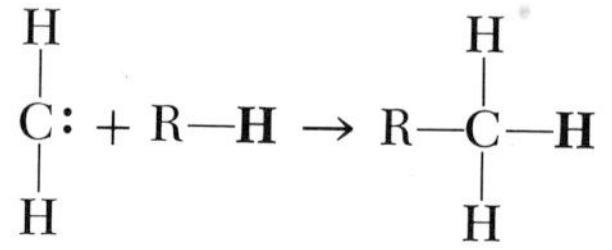

$$H_2C\!: + R\text{—}\mathbf{H} \rightarrow R\text{—}CH_2\text{—}\mathbf{H}$$

Markownikoff addition: An addition reaction that follows Markownikoff's rule.

Markownikoff's rule: A rule dictating the course of addition of an unsymmetric reagent (H—Z) to an unsymmetric alkene ($HRC{=}CR_2$). Simply stated, the hydrogen of HZ becomes bonded to the carbon of the double bond that already bears more hydrogens.

$$HRC{=}CR_2 + HZ \rightarrow H_2RC\text{—}CR_2Z$$

The mechanistic basis of the rule is that if two reactions are competing, then the one that goes *via* the lower energy activated complex will proceed more rapidly. The addition of HZ involves a carbocation, and it is the more stable carbocation that is formed more rapidly.

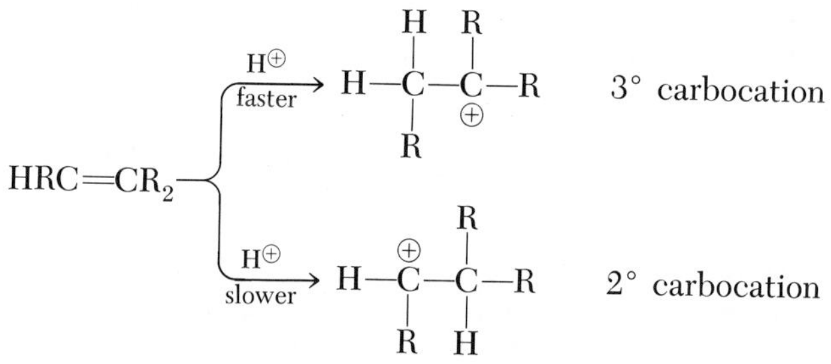

$$HRC{=}CR_2 \xrightarrow[\text{faster}]{H^{\oplus}} H\text{—}CHR\text{—}\overset{\oplus}{C}R\text{—}R \quad 3° \text{ carbocation}$$

$$HRC{=}CR_2 \xrightarrow[\text{slower}]{H^{\oplus}} H\text{—}\overset{\oplus}{C}R\text{—}CHR\text{—}R \quad 2° \text{ carbocation}$$

Even those reactions which violate Markownikoff's rule (*i.e.*, give anti-Markownikoff addition) follow the theoretical basis for the rule and go through the most stable intermediate.

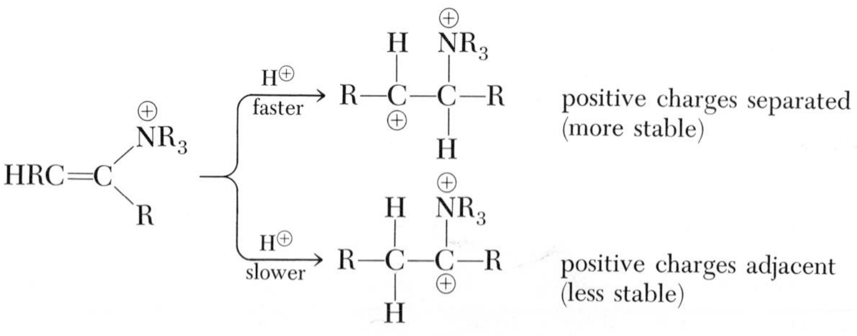

$$HRC{=}C(\overset{\oplus}{N}R_3)R \xrightarrow[\text{faster}]{H^{\oplus}} R\text{—}\overset{\oplus}{C}H\text{—}CH(\overset{\oplus}{N}R_3)\text{—}R \quad \text{positive charges separated (more stable)}$$

$$HRC{=}C(\overset{\oplus}{N}R_3)R \xrightarrow[\text{slower}]{H^{\oplus}} R\text{—}CH_2\text{—}\overset{\oplus}{C}(\overset{\oplus}{N}R_3)\text{—}R \quad \text{positive charges adjacent (less stable)}$$

Olefin: An older name for alkene. This name originated because ethylene reacts with chlorine to give a liquid (1,2-dichloroethane, $ClCH_2CH_2Cl$); ethylene, consequently, was termed an *olefiant gas* (meaning *oil-making gas*). The term olefiant gas then was contracted to olefin and became a general term for unsaturated hydrocarbons containing a double bond.

Pi complex: A weak adduct formed between a π bond, acting as an electron donor, and some electron-deficient species. The electron-deficient species may carry either a formal positive charge or a partial positive charge. Since electron density is transferred from the π bond to the positively charged species, this sometimes is called a charge-transfer complex.

Polymer: A large molecule made up of one type or of several different types of repeating units. Molecular weights in the hundreds of thousands are not uncommon. While there are alternate means of classification, polymers can be grouped into addition, coordination, and condensation polymers.

Protonium ion: A bridged ion in which a proton bridges two centers. It may be viewed as a hybrid of two resonance contributing structures, neither of which places a full positive charge on hydrogen. The existence of this species is a moot point.

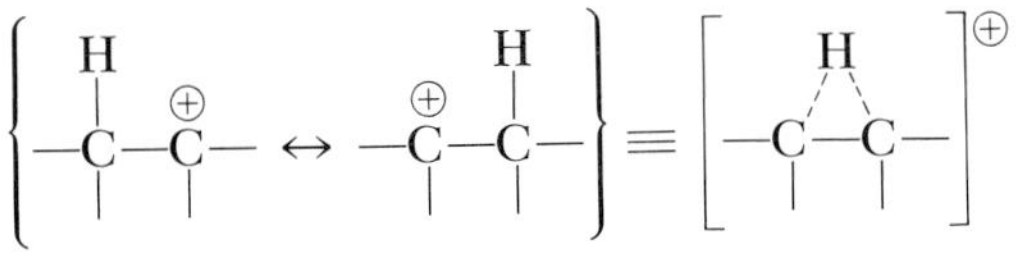

Sigma complex: A species possessing a σ bond between an electrophile and what used to be a π bond. A σ complex can be viewed as nothing more than a carbocation which originated by addition of a cation to a π bond. (This term will be expanded upon in Chapter 15.)

Syndiotactic polymer: A polymer in which the substituents on the backbone regularly alternate their position on the front and back faces of the backbone.

Unsaturated compound: A compound in which not all of the carbons are sp^3 hybridized. Such compounds have carbons which, at least in principle, are capable of acquiring at least one more substituent.

PROBLEMS

23. Draw all of the isomeric alkenes with the formula C_6H_{12}. Provide a suitable name, including the specification of stereochemistry, for each.
24. Assign stereochemistry, using **E,Z** nomenclature, to the following:

(a) 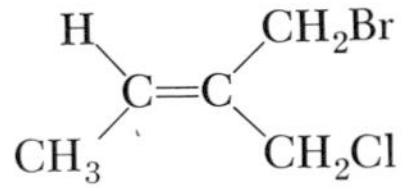 (b) 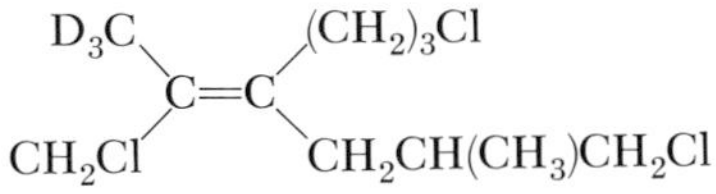 (c)

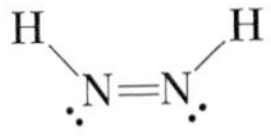

(d) 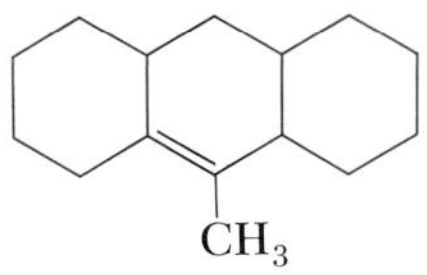 (e) 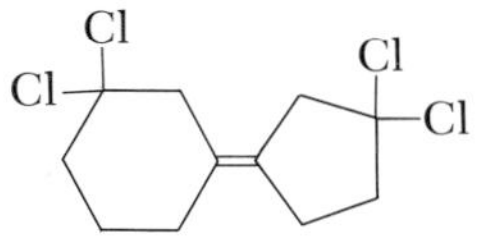 (f)

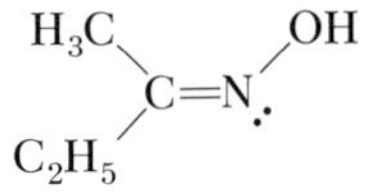

(g)

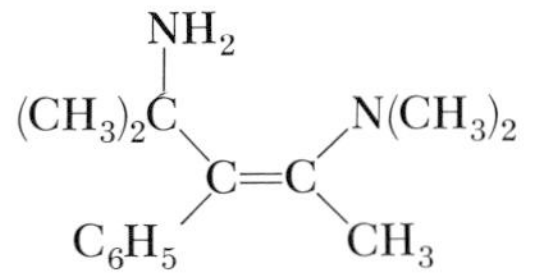

25. Predict the major product(s) from the reaction of (a) *cis*-2-butene and (b) *trans*-2-butene with each of the following:

(a) H_2O, $H^{\oplus}$
(b) CH_3OH, HCl (dry)
(c) ICl
(d) BrCl
(e) O_3/oxidative work-up
(f) $KMnO_4$ (cold, dilute, alkaline)
(g) OsO_4
(h) $Br_2 + Cl^{\ominus}$
(i) B_2H_6 followed by CH_3CO_2H
(j) B_2H_6 followed by CH_3CO_2D
(k) B_2D_6 followed by CH_3CO_2H
(l) H_2SO_4 (concentrated, cold)
(m) Cl_2/H_2O
(n) $Pt/D_2/CH_3OH$ solvent
(o) DCl
(p) CH_2 (singlet)
(q) CH_2 (triplet)
(r) $CHCH_3$ (singlet)
(s) $CHCH_3$ (triplet)

26. Suggest a means of converting cyclohexene to the following:

(a) cyclohexane
(b) *cis*-1,2-dideuteriocyclohexane
(c) adipic acid ($HO_2C(CH_2)_4CO_2H$)
(d) adipaldehyde ($OHC(CH_2)_4CHO$)
(e) *trans*-1,2-dibromocyclohexane
(f) 3-chlorocyclohexene
(g) 3-bromocyclohexene
(h) chlorocyclohexane
(i) deuteriocyclohexane ($C_6H_{11}D$)—two ways
(j) cyclohexyl hydrogen sulfate
(k) cyclohexyl alcohol ($C_6H_{11}OH$)
(l) 1,2-dibromo-3-chlorocyclohexane (mixed isomers)
(m) 1,2,3-trideuteriocyclohexane

27. Starting with propane, and using any other needed reagents, suggest a synthesis for each of the following. Provide acceptable names for these products.

(a) $CH_3CHBrCH_3$
(b) CH_3CHICH_3
(c) $CH_3CH{=}CH_2$
(d) $CH_3CH_2CH_2Br$
(e) $CH_3CHBrCH_2Br$
(f) $CH_3CH_2CH_2Cl$
(g) $ClCH_2CH{=}CH_2$
(h) $CH_2DCH{=}CH_2$
(i) CH_3CHDCH_2D
(j) $CH_2DCHDCH_2D$
(k) $CH_3CH_2CH_2D$
(l) $CH_3CH_2CH_2CN$
(m) CH_3CHDCH_3
(n) $CH_3CH_2CH_2OH$
(o) $CH_2{=}CHCH_2OH$
(p) $CH_3CHCNCH_3$
(q) CH_3CHO
(r) CH_3CO_2H
(s) $CH_3CHOHCH_2OH$

28. Identify the error(s) in each of the syntheses proposed below. Given the particular starting material, suggest corrections to these syntheses which will yield the desired final product.

(a) $C_2H_6 \xrightarrow[h\nu]{F_2} CH_3CH_2F \xrightarrow[\text{alcohol}]{KOH} CH_2{=}CH_2$

(b) cyclohexene $\xrightarrow[400^\circ]{Cl_2}$ 3-chlorocyclohexene (Cl) $\xrightarrow[H_2O]{Mg}$ (MgCl) $\xrightarrow{CD_4}$ (D)

(c) bicyclic diene $+ H_2 \xrightarrow[\text{benzene}]{((C_6H_5)_3P)_3RhCl}$ decalin $\xrightarrow[h\nu]{Br_2}$ (Br, Br) $\xrightarrow{Mg/Et_2O}$ (MgBr, MgBr) $\xrightarrow{D_2O}$ (D, D)

(stereochemistry unspecified)

(d) $CH_2{=}CH{-}\overset{H}{\underset{(CH_2)_4CH_3}{\underset{(S)}{C}}}{-}CH_3 \xrightarrow{HCl} CH_3{-}\overset{Cl}{\underset{H}{C}}{-}\overset{H}{\underset{(CH_2)_4CH_3}{C}}{-}CH_3 \xrightarrow[H_2O]{KOH} CH_3{-}CH{=}\overset{CH_3}{C}{-}(CH_2)_4CH_3$

$\downarrow N_2H_2$

$CH_3CH_2{-}\overset{H}{\underset{(CH_2)_4CH_3}{\underset{(R)}{C}}}{-}CH_3$

29. (a) Indicate the product(s) expected from ozonolysis, followed by reductive work-up, of the following. (b) What changes would result if you employed an oxidative work-up? Assume that all double bonds react. (c) Provide acceptable names for compounds a through g.

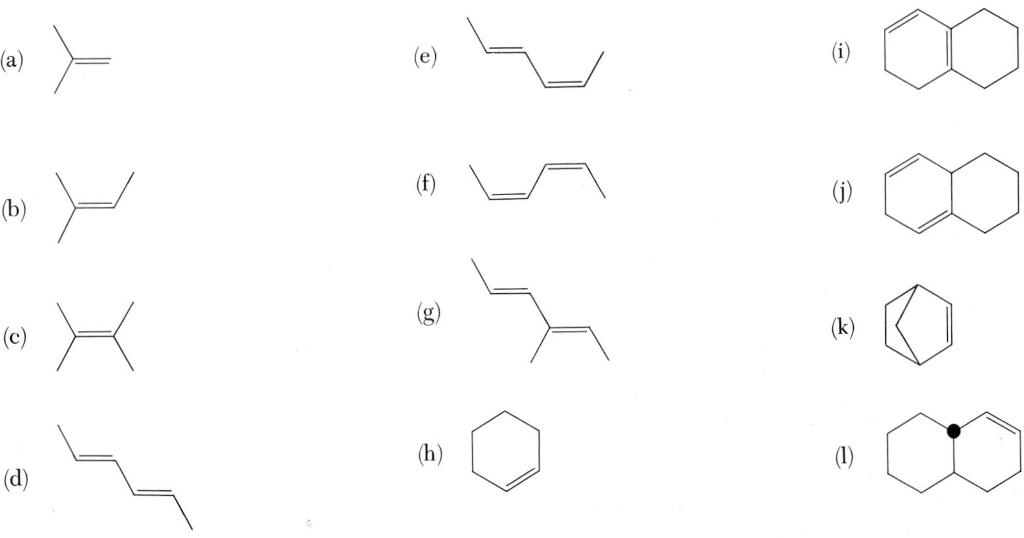

30. Account for the following: $CH_3CH{=}CHBr \xrightarrow{HCl} CH_3CH_2CHBrCl$

31. Acrylic acid ($H_2C{=}CH{-}CO_2H$) reacts with HCl to produce **A**; it does not produce **B**. Account for this.

$ClCH_2CH_2CO_2H$ (**A**) $\quad CH_3CHClCO_2H$ (**B**)

32. Under ionic conditions, methylenecyclobutane reacts with HCl to produce **A**, while methylenecyclohexane produces **B**. Account for this.

A: cyclobutane carbon bearing H and CH_2Cl **B**: cyclohexane carbon bearing CH_3 and Cl

33. Alcohol **A** reacts with cold sulfuric acid to produce several products, including **B** and **C.** Account for their formation.

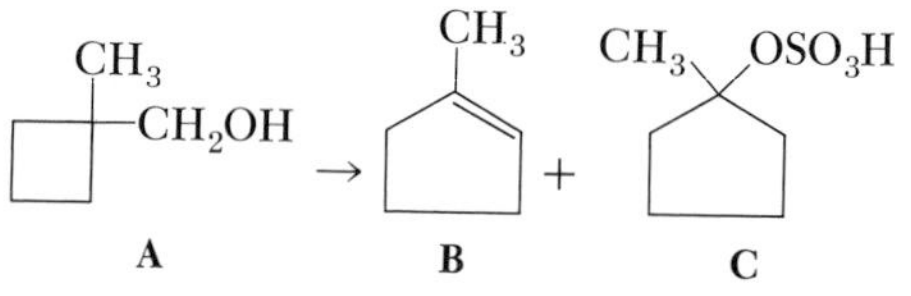

CH$_3$ CH$_2$OH → CH$_3$ + CH$_3$ OSO$_3$H

A **B** **C**

34. The ionic additions of I_2, IBr, and ICl to ethylene occur with the relative rates of $1:3 \times 10^3:1 \times 10^5$. Suggest a reason for this trend.

35. An alternate explanation for the *trans* addition of HCl to 1,2-dimethylcyclopentene involves the simultaneous reaction of alkene with two molecules of HCl:

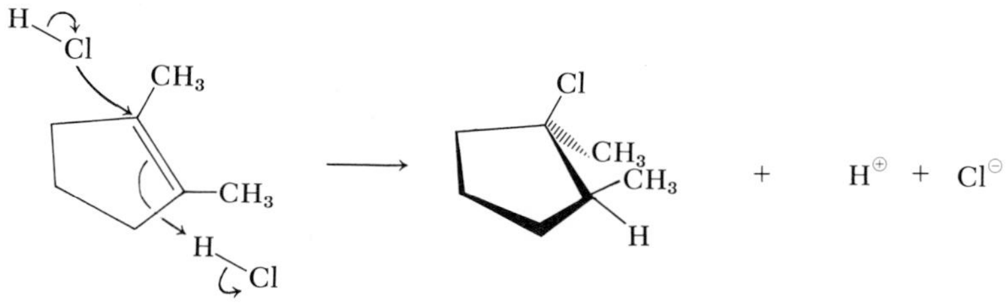

H–Cl CH$_3$ CH$_3$ H–Cl ⟶ Cl CH$_3$ CH$_3$ H + $H^{\oplus}$ + $Cl^{\ominus}$

Suggest an experimental means of distinguishing between this path and the mechanism presented in the text.

36. When diborane reacts with propene and the adduct is decomposed by refluxing with CH_3CO_2D, the product is 1-deuteriopropane. This suggests that the diborane adds to unsymmetric alkenes in an anti-Markownikoff fashion. Account for this, assuming that the reaction is ionic; Table 1–6 may be helpful.

37. Propene recovered from the incomplete reaction of propene with DCl/D_2O does not contain any deuterium. What does this suggest about the mechanism of hydration?

*38. Methylenecyclohexane reacts with trifluoroacetic acid (CF_3CO_2H) to produce two compounds (**D** and **C**), both with the formula $C_9H_{13}O_2F_3$. Compound **D** shows a sharp singlet (3**H**) in the nmr, while **C** does not. (a) Assign structures to **D** and **C**. (b) Which would be produced to the larger extent? Explain.

39. Ozonolysis (oxidative work-up) led a chemist to conclude that she was working with 1,5-cycloöctadiene when, in fact, she had a sample of 1,5-hexadiene. (a) Can you suggest a likely reason for the error? (b) How might it have been avoided?

40. Ethylene can be polymerized simply by heating the alkene in the presence of oxygen under high pressure. Suggest a function of O_2 in this process.

41. What intermolecular force, or forces, acts to hold the molecules of a polymer such as polyvinyl chloride together? With this in mind, suggest how a plasticizer may function in altering the physical properties of polymers.

42. The addition of BH_3 to a double bond is kinetically controlled, *i.e.*, the initial product of reaction is that one which is produced most easily. If the adduct formed between a straight-chain alkene and BH_3 is heated until decomposition occurs (decomposition here entails loss of BH_3 and regeneration of the alkene), and the resultant product is subjected to ozonolysis (reductive work-up), one product obtained from all alkenes is formaldehyde. (a) What are the possible structures of the precursor alkylboranes? (b) If the boron occupies the position that produces the least crowding, which is the actual precursor? (c) How might you prove the structure of this adduct?

43. When cyclohexene reacts with Br_2 under free-radical conditions, the major product is 1,2-dibromocyclohexane if the concentration of Br_2 is high *but* 3-bromocyclohexene if the concentration of Br_2 is very low. (a) How could the concentration of Br_2 influence the competition presented on the following page. (b) With the aid of this information, suggest a mechanism for the conversion of cyclohexene to 3-bromocyclohexene using *N*-bromosuccinimide.

---

*This problem requires a knowledge of nmr spectroscopy; see Chapter 29.

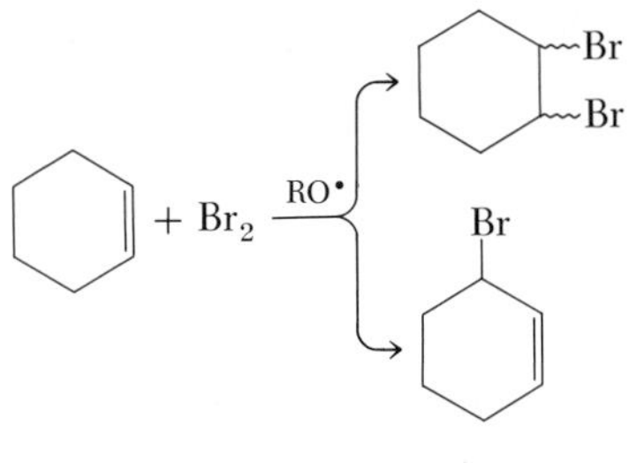

44. While ethylene reacts with diborane to produce trialkylborane, 2-methyl-2-butene reacts to produce a dialkylborane and 2,4,4-trimethyl-2-pentene produces a monoalkylborane. (a) What are the structures of these adducts (neglect stereoisomers)? (b) Account for the variation in the extent of alkylation.

45. It is known that, all other things being equal, an endocyclic double bond in a six-membered ring is usually more stable than is an exocyclic double bond. However, the heats of hydrogenation in set **A** cannot be accepted as proof of this, while the heats of hydrogenation in set **B** do prove this. Why is this so?

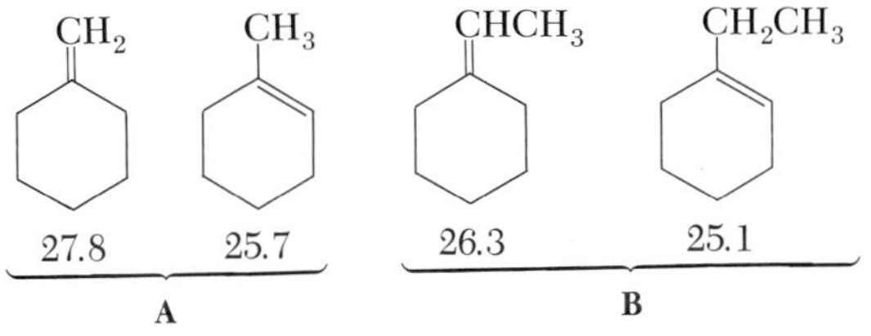

heats of hydrogenation in kcal/mole

	CH_2 (exocyclic)	CH_3 (endocyclic)	$CHCH_3$ (exocyclic)	CH_2CH_3 (endocyclic)
	27.8	25.7	26.3	25.1
Set	**A**	**A**	**B**	**B**

46. In general, stereospecific radical additions are much rarer than are stereospecific ionic additions; indeed, they are the *exception* rather than the rule. However, at low temperatures ($\sim -70°$), *cis*-2-butene adds DBr under radical conditions to give *threo*-3-deuterio-2-bromobutane, while *trans*-2-butene gives the corresponding *erythro* isomer. What is the stereochemistry of DBr addition under these conditions? What explanations would account for the formation of *racemic* product in each instance?

47. In the presence of dilute acid, 1-methoxycyclohexene (a member of that class of compounds called *vinyl ethers*) is converted to cyclohexanone. Suggest a mechanism for this conversion. (*Hint*: Two distinct processes are required—one is an addition, the other an elimination.)

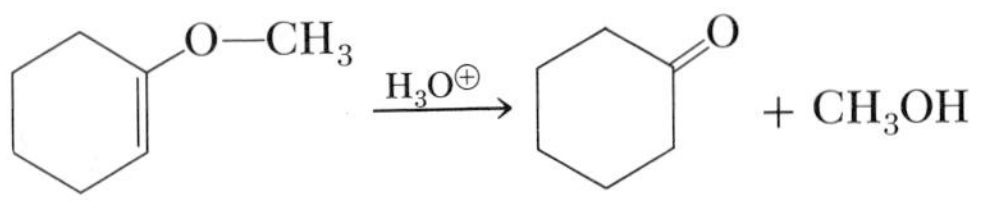

$$\text{1-methoxycyclohexene (O—CH}_3) \xrightarrow{H_3O^{\oplus}} \text{cyclohexanone} + CH_3OH$$

48. The catalytic reduction of 1,2-dimethylcyclohexene (H_2/Pt; 25°; CH_3CO_2H solvent) produces 86% *cis*- and 14% *trans*-1,2-dimethylcyclohexane. The *trans* isomer may be produced by (1) the non-stereospecific reduction of the double bond (*i.e.*, both isomers are formed by the initial reduction) or (2) the isomerization of the less stable *cis* isomer to the more stable *trans* isomer. (a) How might these two alternatives be distinguished? (b) Suggest a mechanism for the isomerization of the kinetically favored *cis* isomer to the thermodynamically favored *trans* isomer. (c) Sketch an energy profile to express this observation.

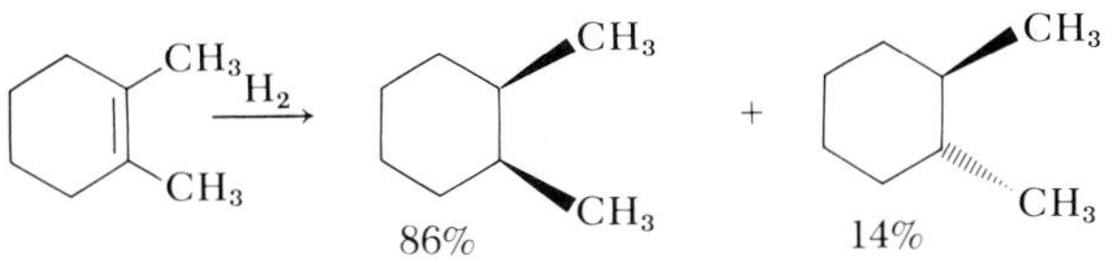

$$\text{1,2-dimethylcyclohexene} \xrightarrow{H_2} \textit{cis}\text{-1,2-dimethylcyclohexane (86\%)} + \textit{trans}\text{-1,2-dimethylcyclohexane (14\%)}$$

49. If singlet carbene is added to the π bond of *cis*-2-butene, the product is a non-resolvable cyclopropane derivative. However, the adduct between triplet carbene and *cis*-2-butene can, at least in principle, be resolved. (a) Account for this difference. (b) Would you expect similar behavior for cyclobutene? cyclodecene?

50. One of the two isomers of cycloöctene has been prepared in optically active form. Suggest, with the aid of models, which isomer could be capable of optical activity and indicate the "center" of dissymmetry.

51. Addition polymerizations are invariably exothermic. How can you account for this?

52. Is it possible to devise an alternate description for *atactic, isotactic,* and *syndiotactic* polymers based upon the **R,S**-nomenclature system? Explain.

53. When tetramethylammonium chloride ($(CH_3)_4\overset{\oplus}{N}Cl^{\ominus}$) reacts with phenyl sodium ($C_6H_5^{\ominus}Na^{\oplus}$), a strong base, in the presence of cyclohexene, one of the products is *norcarane*. Account for its formation.

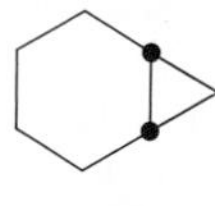

norcarane

54. Tetracyanoethylene (TCNE) is prepared by the reaction of dibromomalononitrile with a metal, *e.g.*, Zn or Cu. Suggest how TCNE can be formed under these conditions. (*Hint*: An *alpha* elimination is involved.)

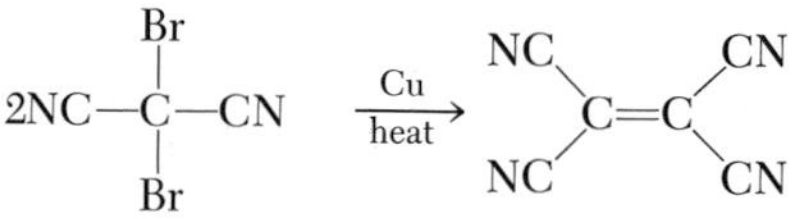

$$2NC{-}CBr_2{-}CN \xrightarrow[\text{heat}]{\text{Cu}} (NC)_2C{=}C(CN)_2$$

dibromomalononitrile TCNE

*55. Suggest means, chemical or spectral, for distinguishing between individual members of the following sets. Be sure to include your observations.

(a) 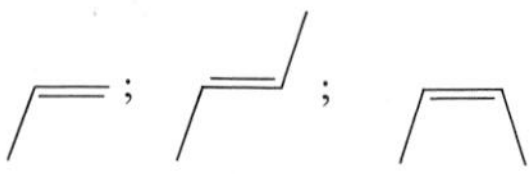 1; 2; 3

(b) 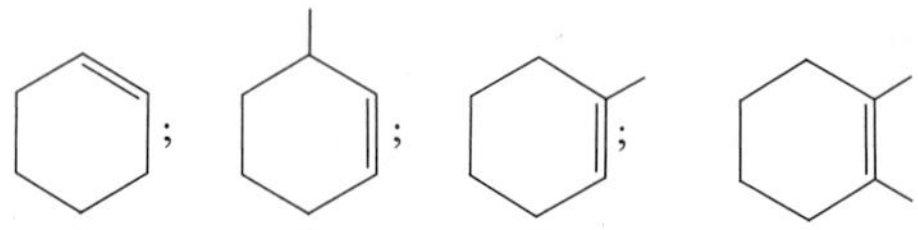 1; 2; 3; 4

(c) 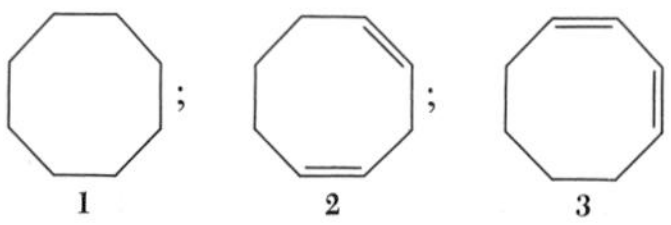 1; 2; 3

(d) 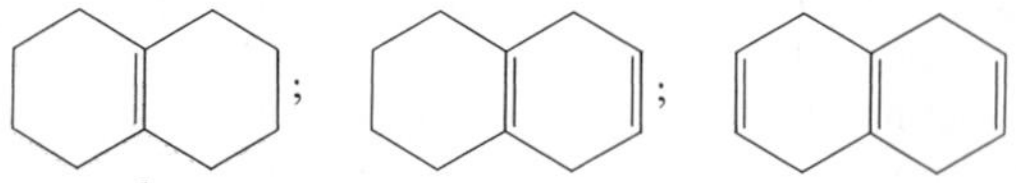 1; 2; 3

56. (a) In what ways does $CH_3CH_2CH_2^{\ominus}$ differ from $CH_3\overset{\ominus}{C}HCN$? (b) How does this relate to the synthesis of Orlon given on p. 247?

* This problem requires a knowledge of spectroscopy.

ALKYNES

9.1 INTRODUCTION

Alkynes are hydrocarbons containing the —C≡C— bond[a]; simple alkynes have the general formula C_nH_{2n-2}. Acetylene, C_2H_2, is the best known member of this class of compounds, and as a result alkynes are often called "acetylenes," or "substituted acetylenes."

Since both alkenes and alkynes are unsaturated (*i.e.*, they both contain carbon-carbon π bonds), our emphasis in this chapter will be upon a comparison of these two groups of compounds. We should be aware that the reactions of alkenes have been much more thoroughly studied than have those of alkynes; however, alkene chemistry and alkyne chemistry are active research areas in both industrial and academic laboratories.

9.2 NOMENCLATURE

In the IUPAC scheme, the procedure for naming *alkenes* is followed in naming *alkynes*. The most significant difference is the replacement of the suffix *ene* by *yne*.

H—C≡C—H	H_3C—C≡C—H	H_3C—C≡C—CH_3
ethyne	propyne	2-butyne

H_3C—C≡C—CH_2—CH_3	H_2C—C≡C—CH_2 (ring closed by) H_2C—$(CH_2)_4$—CH_2	H_2C=CH—CH=CH—C≡C—H
2-pentyne	cyclodecyne	1,3-hexadien-5-yne

Applications of the common, or "acetylene," nomenclature are presented below.

H—C≡C—H	H_3C—C≡C—H	H_3C—C≡C—CH_3
acetylene	methylacetylene	dimethylacetylene

H_3C—C≡C—C_6H_5	CH_2—C≡C—CH_2 (ring closed by) CH_2—$(CH_2)_4$—CH_2	H_2C=CH—C≡CH
methylphenylacetylene	octamethyleneacetylene	vinylacetylene

The names "ethynyl," H—C≡C—, and "propargyl," H—C≡C—CH_2—, are sometimes encountered.

H—C≡C—Cl	H—C≡C—CH_2—CN
ethynyl chloride	propargyl cyanide

---

[a] Bonding in the functional group was discussed on p. 37.

1. Provide an acceptable name for each of the following:
(a) $HC{\equiv}CCH_3$ (e) $C_6H_5C{\equiv}CH$
(b) $DC{\equiv}CCH_3$ (f) $H_2C{=}CHC{\equiv}CH$
(c) $DC{\equiv}CCD_3$ (g) $H_2C{=}C(CH_3)C{\equiv}CH$
(d) $HC{\equiv}CCH_2Cl$ (h) $CH_3CH_2C{\equiv}CC(CH_3)_2$

9.3 ACIDITY OF ALKYNES

It is not difficult to explain the high proton-donating propensity of hydrochloric acid or sulfuric acid; indeed, calling them "acid" prepares us to accept this idea. The acidity of aqueous hydrochloric acid is a result of the stabilization of the chloride ion by hydrogen-bonding to water, and of the electronegativity of chlorine; resonance-stabilization of the bisulfate anion, $HOSO_3^{\ominus}$, is largely responsible for sulfuric acid's acidity. As a rule of thumb, the acidity of a given hydrogen increases with the increasing electronegativity of the element to which it is attached. Since the order of electronegativity is $F > O > N > C$, it is not surprising that **carbon acids** (compounds possessing an acidic hydrogen bonded to carbon) rarely are encountered in freshman chemistry courses.

Hydrocarbons do not behave like the usual acids; they do not turn blue litmus red, nor do they have the tart taste commonly associated with acids. However, this "non-acidity" can be viewed as a result of the *comparatively* poor basic properties of water! In fact, if a much stronger base than water is used, hydrocarbons *do* begin to show acidic properties. The acidity of a spectrum of compounds is included in Table 9–1. From these data it can be seen, for example, that hydrochloric acid is about 10^{12} times as strong an acid as is acetic acid (found in vinegar), but that acetic acid is 10^{21} times as acidic as is acetylene. That makes hydrochloric acid about 10^{33} times as strong an acid as acetylene!

TABLE 9–1 Acidity of Common Acidic Compounds

COMPOUND	FORMULA[a]	K_a[b]
perchloric acid	$\mathbf{H}ClO_4$	10^{10}
hydriodic acid	**H**I	10^{10}
hydrochloric acid	**H**Cl	10^{7}
nitric acid	$\mathbf{H}NO_3$	$10^{2.3}$
acetic acid	$CH_3CO_2\mathbf{H}$	$10^{-4.8}$
ethanol	$CH_3CH_2O\mathbf{H}$	10^{-16}
acetylene	$HC{\equiv}C\mathbf{H}$	10^{-26}
ammonia	$N\mathbf{H}H_2$	10^{-35}

[a]The acidic hydrogen is shown in boldface.
[b]K_a measures the position of the equilibrium $HA \rightleftharpoons H^{\oplus}\,A^{\ominus}$ and equals $[H^{\oplus}]\,[A^{\ominus}]/[HA]$. Larger values of K_a imply stronger acids. These values are all related to water as a standard base. Acids that are very weak (e.g., acetylene) do not protonate water and, therefore, their values are calculated for other base/solvent systems and then are extrapolated to water solutions.

Terminal acetylenes, compounds of the type $R{-}C{\equiv}C{-}H$, are weak acids, as is acetylene itself. The reason for this acidity is the high degree of *s* character in the *sp* orbital used to bond carbon to hydrogen. [You might want to review the discussions of orbitals and electronegativity (p. 39) and of bond polarity (pp. 29 and 39).] These acetylenes react with strong bases to form salts of the type $M^{\oplus n}(RC{\equiv}C^{\ominus})_n$, called *acetylides*. One restriction placed upon acetylide formation is that the base, in accepting a proton from the acetylene, cannot become an acid stronger than the acetylene whose salt is being prepared. If *that* happened, the newly formed acid (which used to be the base before it acquired a proton) would immediately re-protonate the acetylide anion. Ammonia is a weaker acid than is acetylene (see Table 9–1), and this permits the use of the base derived from ammonia, the amide anion ($H_2N^{\ominus}$), to form acetylides. The requisite sodium amide can be made by the reaction of sodium with liquid ammonia in the presence of iron. Acetylide formation plays an important role in both the analysis and the synthesis of alkynes, as we shall see in the following sections.

$$Na + NH_3 \xrightarrow[NH_3(l)]{Fe} NaNH_2 + \tfrac{1}{2}H_2$$

$$\underset{\text{a terminal acetylene}}{R{-}C{\equiv}C{-}H} + \overset{\oplus}{Na}\overset{\ominus}{NH_2} \rightleftarrows \underset{\text{an acetylide}}{R{-}C{\equiv}\overset{\ominus}{C}{:}\ \overset{\oplus}{Na}} + NH_3$$

2. Identify the acids and bases in the reactions shown. If you have more than one of each classification, list them in order of increasing potency.
   (a) $HCl + H_2O \rightleftarrows H_3O^{\oplus} + Cl^{\ominus}$
   (b) $HCl + OH^{\ominus} \rightleftarrows H_2O + Cl^{\ominus}$
   (c) $HC{\equiv}CH + NH_2^{\ominus} \rightleftarrows HC{\equiv}C^{\ominus} + NH_3$
   (d) $HC{\equiv}C^{\ominus} + NH_4^{\oplus} \rightleftarrows HC{\equiv}CH + NH_3$
3. What is wrong with the following? "Ammonia is not a base. After all, it does not react with ethane (to form $CH_3CH_2^{\ominus}$ and $NH_4^{\oplus}$)."

Alkenes and alkanes are not sufficiently acidic (or ammonia is too acidic!) to form salts by reaction with sodium amide.

$$R_2C{=}CH_2 + NaNH_2 \rightarrow \text{no reaction}$$

$$R_3C{-}CH_3 + NaNH_2 \rightarrow \text{no reaction}$$

4. You are given a mixture of 1-butyne (bp 8°) and 2-butyne (bp 27°). How could you separate, and then recover, each of these compounds using chemical means?
5. You are given separate samples of 1-butyne, 1-butene, and octane. How could you identify each of these compounds by chemical means?
6. Ethane does not react with amide ion. What would you expect to happen when you mix $CH_3CH_2^{\ominus}$ and NH_3? Why?

Analysis. The heavy metal "salts" of terminal alkynes actually are covalent compounds, insoluble in water; they precipitate when produced in aqueous media. Thus, precipitation of, for example, a silver acetylide (R—C≡C—Ag) can serve to distinguish between a *terminal* and an *internal* triple bond. Ammoniacal silver nitrate is used, because NH_3 prevents precipitation of $Ag(OH)_2$ from the alkaline medium that is necessary to produce the acetylide.

$$R{-}C{\equiv}C{-}H + Ag(NH_3)_2^{\oplus}NO_3^{\ominus} \rightarrow R{-}C{\equiv}C{-}Ag + NH_4^{\oplus}NO_3^{\ominus} + NH_3$$

$$R{-}C{\equiv}C{-}R + Ag(NH_3)_2^{\oplus}NO_3^{\ominus} \rightarrow \text{no reaction}$$

Since silver(I) forms a very stable complex with cyanide ions, silver acetylides regenerate acetylenes upon treatment with aqueous sodium cyanide. [*Note:* Heavy metal acetylides are sensitive to shock and are highly explosive when dry. Therefore, these acetylides usually are precipitated and discarded or decomposed with nitric acid without ever being dried.]

$$R{-}C{\equiv}C{-}Ag + 2CN^{\ominus} + H_2O \rightarrow R{-}C{\equiv}C{-}H + Ag(CN)_2^{\ominus} + OH^{\ominus}$$

Terminal alkynes are sufficiently acidic to react with alkyl Grignard reagents, forming an acetylenic Grignard reagent and an alkane. This reaction, an unusual Zerewitnoff determination (see p. 176), serves to prepare the synthetically valuable R—C≡C—MgX.

$$CH_3C{\equiv}C{-}H + CH_3CH_2{-}MgBr \rightarrow C_2H_6 + CH_3C{\equiv}C{-}MgBr$$

Since acetylides are salts of weak acids, they react rapidly and completely with water or dilute acids to regenerate the corresponding alkynes. This can be used to introduce deuterium into a molecule.

$$CH_3{-}C{\equiv}C{-}MgBr + D_2O \rightarrow \underset{\text{1-deuteriopropyne}}{CH_3{-}C{\equiv}C{-}D} + MgOHBr$$

$$C_6H_5{-}CH_2{-}C{\equiv}\overset{\ominus}{C}{:}\,Na^{\oplus} + D_2O \rightarrow \underset{\text{1-deuterio-3-phenylpropyne}}{C_6H_5{-}CH_2{-}C{\equiv}C{-}D} + NaOD$$

$$H{-}C{\equiv}C{-}Ag + 2CN^{\ominus} + D_2O \rightarrow \underset{\text{deuterioethyne}}{H{-}C{\equiv}C{-}D} + Ag(CN)_2^{\ominus} + OD^{\ominus}$$

7. You are given the following sets of samples; each member of a given set is in its own container. How could you distinguish one member from the other by chemical means?
   (a) 1-butyne and 2-butyne
   (b) 1-butyne and 1-butene
   (c) 1-butene and butane
   (d) 1-butyne and butane

9.4 SYNTHESIS OF ALKYNES

Acetylene. Acetylene is prepared by the reaction of calcium carbide with water, the culmination of a reaction sequence that begins with limestone.

$$\underset{\text{limestone}}{CaCO_3} \xrightarrow[\text{heat}]{\text{kiln}} \underset{\text{lime}}{CaO} + CO_2$$

$$CaO + 3C \xrightarrow{2500^\circ} \underset{\text{calcium carbide}}{CaC_2} + CO$$

$$CaC_2 + 2H_2O \xrightarrow{25^\circ} Ca(OH)_2 + HC{\equiv}CH$$

Acetylene also is produced by the oxidation of methane at very high temperatures ($\sim 1500^\circ$) for a very short period of time (*e.g.*, 0.1 sec). Perhaps one third of the acetylene manufactured in the United States has its ultimate origin in methane from natural gas.

$$2CH_4 \xrightarrow[1500^\circ]{0.1\text{ sec}} H{-}C{\equiv}C{-}H + 3H_2$$

Terminal Alkynes. The acetylide anion, conveniently prepared by the reaction of acetylene with sodamide (another name for sodium amide), can be used to prepare terminal alkynes by a number of reactions, including the S_N2 displacement.

$$H{-}C{\equiv}C^{\ominus}Na^{\oplus} + CH_3(CH_2)_3Br \rightarrow \underset{(77\%)}{H{-}C{\equiv}C{-}(CH_2)_3CH_3}$$

Lithium acetylide forms a stable complex with ethylenediamine, EDA, which serves as a convenient source of acetylide anions in solvents other than liquid ammonia.

$$H{-}C{\equiv}C{-}Li \cdot NH_2CH_2CH_2NH_2 \qquad \text{lithium acetylide-EDA complex}$$

$$F{-}(CH_2)_5{-}Cl + LiC{\equiv}CH{\cdot}EDA \xrightarrow[25^\circ]{DMSO} \underset{(92\%)}{F{-}(CH_2)_5C{\equiv}C{-}H} + LiCl + EDA$$

$$LiC{\equiv}C{-}H \cdot EDA + \overset{O}{CH_2{-}CH_2} \xrightarrow{THF} \xrightarrow{HOH} H{-}C{\equiv}C{-}CH_2CH_2OH$$

Both terminal and internal acetylenes can be prepared by elimination reactions, *e.g.*, two consecutive dehydrohalogenations.

$$-\underset{H}{\overset{Br}{C}}-\underset{H}{\overset{Br}{C}}- \xrightarrow[-HBr]{E2} \underset{\text{a vinyl halide}}{\overset{}{C}(H){=}C(Br)} \xrightarrow[-HBr]{E2} -C{\equiv}C-$$

The necessary starting material can be prepared by the halogenation of an alkene. Because of the comparatively high E_{act} for dehydrohalogenation of a vinyl halide, a *strong* base (*e.g.*, $NH_2^{\ominus}$) is usually used to convert a dihalide directly to an alkyne. The use of potassium hydroxide produces only the vinyl halide except under very vigorous conditions, when several products, including the alkyne, are formed.

$$\underset{H\quad H}{\overset{CH_3\quad CH_3}{C{=}C}} + Br_2 \rightarrow CH_3{-}\underset{Br}{\overset{H}{C}}{-}\underset{H}{\overset{Br}{C}}{-}CH_3 \xrightarrow[NH_3(l)]{NaNH_2} CH_3{-}C{\equiv}C{-}CH_3$$

$$CH_3{-}\underset{Br}{\overset{H}{C}}{-}\underset{H}{\overset{Br}{C}}{-}CH_3 \xrightarrow[C_2H_5OH]{KOH} \underset{CH_3\quad Br}{\overset{H\quad CH_3}{C{=}C}} \xrightarrow[NH_3(l)]{NaNH_2} CH_3{-}C{\equiv}C{-}CH_3$$

As with most other E2 processes, dehydrohalogenation of a vinyl halide usually requires a *trans* arrangement of departing groups.

$$\overset{\delta-}{B}\cdots H,\ R \;\; \underset{R\qquad X^{\delta-}}{C{=}C} \qquad \text{transition state for E2 reaction of a vinyl halide}$$

8. Suggest a reaction sequence that will convert the given starting material into the desired product for each of the following. Use any other necessary reagents.
   (a) ethane *to* ethene
   (b) ethane *to* ethyne
   (c) ethane *to* acetylene
   (d) phenyl magnesium bromide (C_6H_5MgBr) *to* 3-phenyl-1-propyne
9. Why does the reaction shown below yield **A** but not **B**?

$$F(CH_2)_5Cl + LiC{\equiv}CH \xrightarrow{DMSO} \underset{\mathbf{A}}{F(CH_2)_5C{\equiv}CH} \qquad \underset{\mathbf{B}}{Cl(CH_2)_5C{\equiv}CH}$$

Internal Alkynes. When possible, internal alkynes are prepared from S_N2 reactions in which the nucleophile is the acetylide of a terminal acetylene. When this is impossible, or if starting materials are unavailable, the elimination route already described is most useful.

(**A**) $CH_3C{\equiv}C^{\ominus} + CH_3Br \rightarrow CH_3C{\equiv}CCH_3$ acetylene synthesis by S_N2

(**B**) $CH_3{-}\overset{CH_3}{\underset{CH_3}{C}}{-}C{\equiv}C{:}^{\ominus} + CH_3Br \rightarrow CH_3{-}\overset{CH_3}{\underset{CH_3}{C}}{-}C{\equiv}C{-}CH_3$ acetylene synthesis by S_N2

(Why not $CH_3C{\equiv}C{:}^{\ominus} + (CH_3)_3CBr$?)

(**C**) $C_6H_5{-}CH_2CH_3 \xrightarrow[h\nu]{Br_2} C_6H_5CHBrCH_3 \xrightarrow{OH^{\ominus}} C_6H_5{-}CH{=}CH_2$ acetylene synthesis by E2

$C_6H_5{-}CH{=}CH_2 \xrightarrow[CCl_4]{Br_2} C_6H_5{-}CHBr{-}CH_2Br \xrightarrow[NH_3(l)]{NaNH_2} C_6H_5{-}C{\equiv}C{-}H$

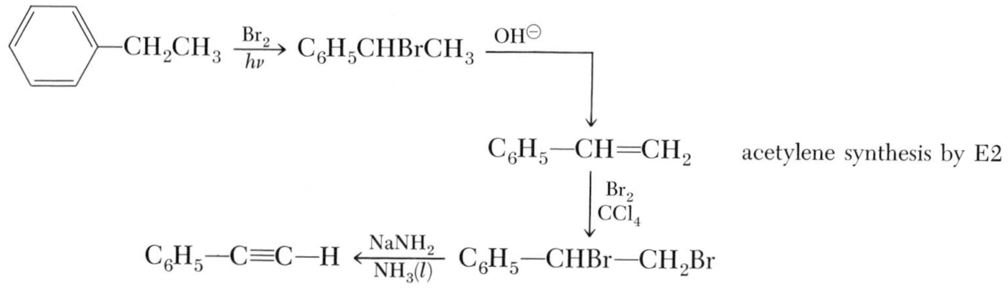

Sequence **C** must be used to prepare phenylacetylene because S_N2 reactions on simple phenyl halides have a prohibitively high E_{act} (see p. 669).

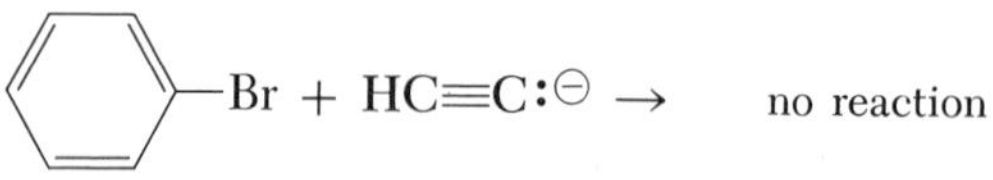

$C_6H_5{-}Br + HC{\equiv}C{:}^{\ominus} \rightarrow$ no reaction

9.5 HYDRATION OF ALKYNES

Alkynes undergo the same electrophilic addition reactions that are undergone by alkenes. The major difference is that alkynes usually react more slowly than do alkenes. For example, while alkenes are brominated "instantly" with Br_2/CCl_4, several minutes are required for alkynes to decolorize this reagent. When unsymmetric reagents add to alkynes, under electrophilic conditions, the reaction follows Markownikoff's rule and usually leads to *trans* addition. This is similar to the behavior of alkenes (p. 239).

$$CH_3-C\equiv C-H \xrightarrow[CCl_4]{Br_2} (CH_3)(Br)C{=}C(Br)(H) \xrightarrow[CCl_4]{Br_2} CH_3-CBr_2-CBr_2-H$$

$$CH_3-C\equiv C-H \xrightarrow{DBr} (CH_3)(Br)C{=}C(D)(H) \xrightarrow{DBr} CH_3-CBr_2-CD_2-H$$

10. Why doesn't the reaction of propyne with two moles of DBr produce 1,2-dideuterio-1,2-dibromopropane instead of 2,2-dibromo-1,1-dideuteriopropane? (Hint: Consider the possibility of resonance stabilization of the intermediate cation.)

In the next section we shall survey several addition reactions of alkynes, including free-radical additions. This section is devoted to reactions that add water across a triple bond. This type of reaction has been singled out because alkyne hydration also introduces us to the phenomenon of *tautomerism.*

Alkenes hydrate more readily in dilute aqueous acid than do alkynes. Mercuric ion, Hg^{+2}, which is presumed to form a π complex with the triple bond, is used to catalyze the addition of water to alkynes. The initial adduct, an *enol,* is unstable and rapidly isomerizes to the more stable carbonyl compound. The carbonyl form is called the *keto* form whether it is a ketone (RC(O)R) or an aldehyde (RC(O)H).

$$H-C\equiv C-H \xrightarrow[Hg^{+2}]{H_2O,H^{\oplus}} \left[(H)(H)C{=}C(OH)(H)\right] \leftrightharpoons H-CH_2-C({=}O)H$$

"enol form" "keto form" (an aldehyde)

$$CH_3-C\equiv C-H \xrightarrow[Hg^{+2}]{H_2O,H^{\oplus}} \left[(CH_3)(HO)C{=}C(H)(H)\right] \leftrightharpoons CH_3-C({=}O)-CH_3$$

"enol form" "keto form" (a ketone)

What is the mechanism of the conversion of enol to keto form? In the acidic medium used to prepare the enol, a proton attacks the π bond (electrophilic attack) to produce the keto form.

$$H^{\oplus} + (H)(H)C{=}C(\ddot{O}-H)(R) \rightleftarrows \left[H-CH_2-\overset{\oplus}{C}(\ddot{O}-H)(R)\right] \rightleftarrows H-CH_2-C({=}O)(R) + H^{\oplus}$$

"enol" "keto"

The equations presented above do not specifically employ the mercuric ion but were, for years, the way in which the mercury-catalyzed Markownikoff hydration of alkynes was presented. Mounting evidence, in a still active area of research, suggests that mercury actually is involved in several discrete steps in the hydration of alkynes. As shown below, Hg(II) forms a cyclic complex with the alkyne, which is ring-opened (an S_N reaction) to produce an organomercury compound that is also an enol. This enol then reacts with aqueous acid to produce an α-mercury ketone. In the last step of the conversion, this keto form suffers C—Hg bond rupture and yields the final product.

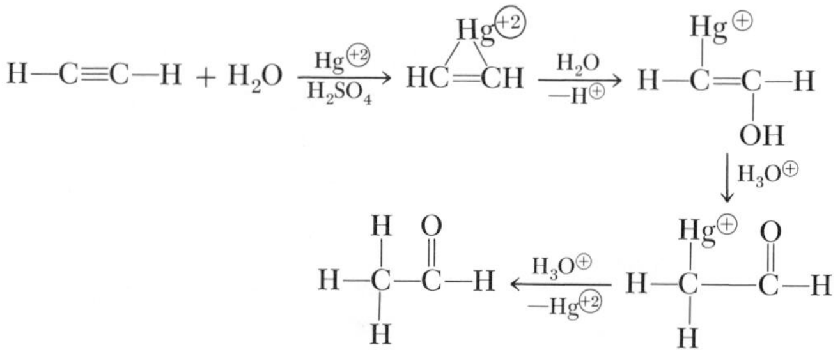

$$\text{H—C}\equiv\text{C—H} + H_2O \xrightarrow[H_2SO_4]{Hg^{+2}} \text{HC}\overset{Hg^{+2}}{=}\text{CH} \xrightarrow[-H^{\oplus}]{H_2O} \text{H—C(Hg}^{\oplus})\text{=C(OH)—H} \xrightarrow{H_3O^{\oplus}} \text{H—CH(Hg}^{\oplus})\text{—C(=O)—H} \xrightarrow[-Hg^{+2}]{H_3O^{\oplus}} \text{H—CH}_2\text{—C(=O)—H}$$

This "keto-enol equilibrium," as it is called, is a specific kind of *prototropic* equilibrium (*i.e.*, an equilibrium between species which differ only in the position of H and electrons). Another term used to describe this type of reaction is **tautomerism** or **keto-enol tautomerism.**

Enols are unstable in neutral or alkaline media as well as in acidic solution. In alkaline media an electrophile (*e.g.*, $H^{\oplus}$) is not present in sufficiently high concentrations to initiate the equilibration. Instead, the reaction proceeds through a resonance-stabilized enolate anion. (Enols and enolate anions are discussed in greater detail in Chapter 17.)

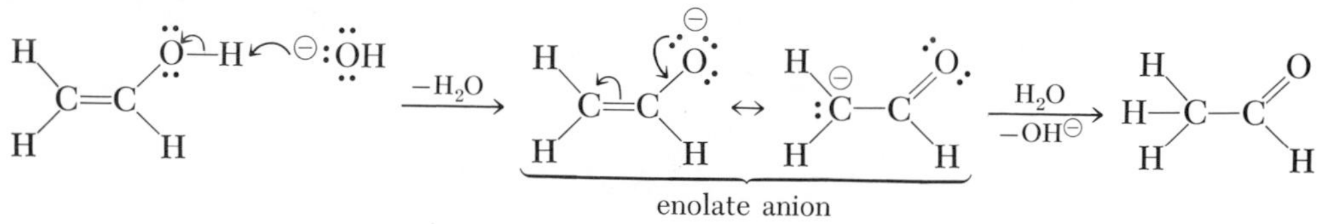

$$H_2C\text{=CH—}\ddot{O}\text{—H} \;\; {}^{\ominus}\!:\!\ddot{O}H \xrightarrow{-H_2O} \underbrace{H_2C\text{=CH—}\ddot{O}\!:^{\ominus} \leftrightarrow \;{}^{\ominus}\!:\!CH_2\text{—CH=O}}_{\text{enolate anion}} \xrightarrow[-OH^{\ominus}]{H_2O} \text{H—CH}_2\text{—CH=O}$$

Alkynes undergo hydroboration to produce *cis* vinylboranes, the addition being anti-Markownikoff (similar to the direction observed for alkenes). When the vinylborane is oxidized with alkaline hydrogen peroxide, the resultant enol tautomerizes to the ketone or aldehyde.[a]

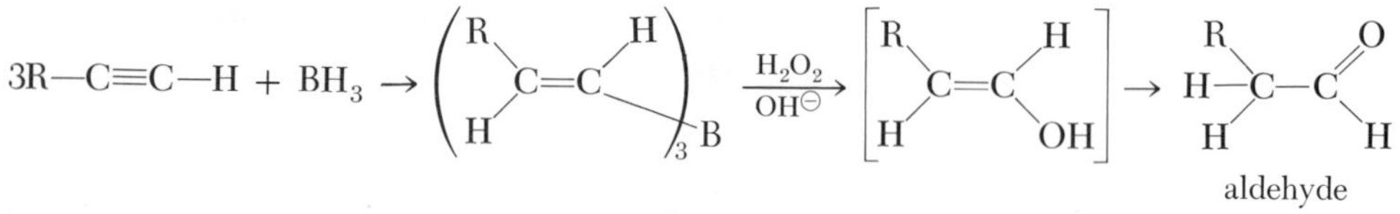

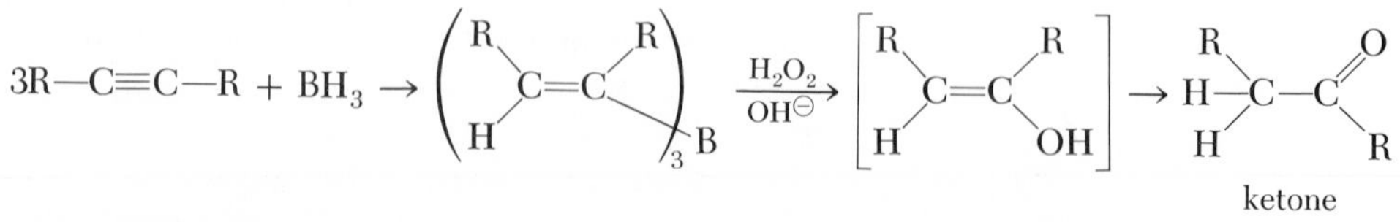

$$3\text{R—C}\equiv\text{C—H} + BH_3 \rightarrow (\text{R(H)C=CH—})_3B \xrightarrow[OH^{\ominus}]{H_2O_2} [\text{R(H)C=C(H)OH}] \rightarrow \underset{\text{aldehyde}}{\text{H—CH(R)—C(=O)H}}$$

$$3\text{R—C}\equiv\text{C—R} + BH_3 \rightarrow (\text{R(H)C=CR—})_3B \xrightarrow[OH^{\ominus}]{H_2O_2} [\text{R(H)C=C(R)OH}] \rightarrow \underset{\text{ketone}}{\text{H—CH(R)—C(=O)R}}$$

---

[a] This sequence of reactions, which results in addition of water across a π bond, is discussed on p. 292.

9.6 ASSORTED ALKYNE ADDITIONS

Nucleophilic Addition. For reasons which are still subject to debate, alkynes are *more easily* attacked by nucleophiles than are alkenes and, consequently, certain additions to carbon-carbon triple bonds can be base-catalyzed. For example, alcohols will add to an alkyne to form, in two distinct steps, *enol ethers* and *gem-diethers* (also called *acetals* or *ketals*—see Chapter 16).

$$\text{R—C}\equiv\text{C—R} + \text{EtOH} \xrightarrow{\text{OH}^\ominus} \underset{\text{enol ether}}{\text{RCH=CROEt}} \xrightarrow[\text{OH}^\ominus]{\text{EtOH}} \underset{\textit{gem}\text{-diether}}{\text{RCH}_2\text{CR(OEt)}_2}$$

The sulfur analogs of alcohols, *thiols*, react similarly.

$$\text{R—C}\equiv\text{C—R} + \text{EtSH} \xrightarrow{^\ominus\text{OH}} \text{RCH=CRSEt} \xrightarrow[\text{OH}^\ominus]{\text{EtSH}} \text{RCH}_2\text{CR(SEt)}_2$$

Enamines, nitrogen analogs of enols, are produced when amines add to alkynes. These compounds, discussed in more detail later in this text, are mentioned here because they can provide an illustration of a type of keto-enol tautomerism, enamine-imine tautomerism.

$$\text{R—C}\equiv\text{C—R} + \text{R'NH}_2 \rightarrow \underset{\text{an enamine}}{\text{RCH=C(NHR')—R}} \rightleftarrows \underset{\text{an imine}}{\text{RCH}_2\text{—C(=NR')—R}}$$

11. Reaction of 4-octyne with *n*-propylamine, $CH_3CH_2CH_2NH_2$, produces *two* enamines in unequal amounts. Explain.

An important example of this type of addition is the addition of ammonia and primary amines (compounds of the type RNH_2) to compounds with adjacent triple bonds to produce the *pyrrole* ring system. Pyrroles are important since the ring system is found in many natural products, including hemoglobin. The reaction shown below is suggested to occur by two consecutive 1,2-additions.

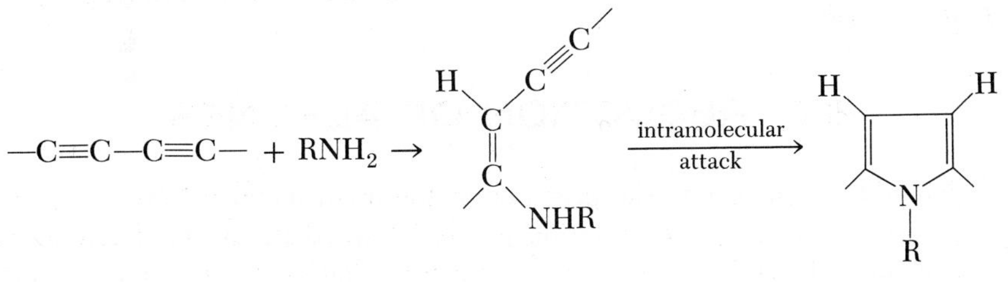

$$\text{—C}\equiv\text{C—C}\equiv\text{C—} + \text{RNH}_2 \rightarrow \text{(H)(—C}\equiv\text{C—)C=C(NHR)—} \xrightarrow[\text{attack}]{\text{intramolecular}} \text{a pyrrole (N—R)}$$

a pyrrole

Electrophilic Addition. Electrophilic catalysis can be used to add species other than water across the carbon-carbon triple bond. We have already alluded (p. 273) to the addition of Br_2 and DBr. An interesting example, and one not often discussed at the elementary level, is the Hg(II) catalyzed, stepwise addition of carboxylic acids (R—C(O)—OH). This reaction produces, depending upon reaction conditions, either an *enol ester* or a *gem-diester*.

$$\text{R—C}\equiv\text{C—R} + \text{R}'\text{—}\overset{\overset{\displaystyle\text{O}}{\|}}{\text{C}}\text{—OH} \xrightarrow{\text{Hg}^{+2}} \underset{\text{enol ester}}{\text{RCH}=\overset{\overset{\displaystyle \text{O—}\overset{\overset{\text{O}}{\|}}{\text{C}}\text{—R}'}{|}}{\text{C}}\text{—R}} \xrightarrow[\text{Hg}^{+2}]{\text{R}'\text{CO}_2\text{H}} \underset{gem\text{-diester}}{\text{RCH}_2\text{CR}\left(\text{O—}\overset{\overset{\text{O}}{\|}}{\text{C}}\text{R}'\right)_2}$$

Hydrogen sulfide and thiols add to alkynes under electrophilic catalysis. Boron trifluoride often is used as the catalyst for this type of reaction.

$$\text{R—C}\equiv\text{C—R} + \text{R}'\text{SH} \xrightarrow{\text{BF}_3} \text{RCH}=\text{CR(SR}') \xrightarrow[\text{BF}_3]{\text{R}'\text{SH}} \text{RCH}_2\text{CR(SR}')_2$$

Free-Radical Reactions. The S—H bond is easily cleaved under free-radical conditions; as a result, hydrogen sulfide and thiols add to alkynes in the presence of a radical initiator. Free-radical additions to alkynes differ from electrophilic additions in that they yield the anti-Markownikoff product (review p. 242).

$$\text{CH}_3\text{C}\equiv\text{CH} + \text{R}'\text{SH} \xrightarrow{\text{R}\cdot} \text{CH}_3\text{CH}=\text{CH(SR}')$$

Beyond this reaction, there are those free-radical processes which, by now, should be expected to occur—halogen *and* hydrogen bromide addition brought about by a free-radical initiator.

$$\text{CH}_3\equiv\text{C—H} + \text{Br}_2 \xrightarrow{\text{R}\cdot} \overset{\text{CH}_3}{\underset{\text{Br}}{}}\!\!\text{C}=\text{C}\!\!\overset{\text{Br}}{\underset{\text{H}}{}} \xrightarrow[\text{R}\cdot]{\text{Br}_2} \text{CH}_3\text{—}\underset{\underset{\text{Br}}{|}}{\overset{\overset{\text{Br}}{|}}{\text{C}}}\text{—}\underset{\underset{\text{Br}}{|}}{\overset{\overset{\text{Br}}{|}}{\text{C}}}\text{—H}$$

$$\text{CH}_3\text{C}\equiv\text{C—H} + \text{DBr} \xrightarrow{\text{R}\cdot} \overset{\text{CH}_3}{\underset{\text{D}}{}}\!\!\text{C}=\text{C}\!\!\overset{\text{Br}}{\underset{\text{H}}{}}$$

12. Compare the final products for the four sequences below.
 (a) 1-propyne and HBr (radical conditions) followed by HBr (ionic conditions).
 (b) 1-propyne and HBr (ionic conditions) followed by HBr (radical conditions).
 (c) 1-propyne and 2 equivalents of HBr (ionic conditions).
 (d) 1-propyne and 2 equivalents of HBr (radical conditions).

 For the sake of this exercise, assume that the stabilizing influence of the methyl group controls the radical additions.

9.7 REDUCTION OF ALKYNES

Hydrogen adds to alkynes in the presence of the usual metal catalysts (*e.g.*, Pt, Pd) to produce the corresponding alkane. Isolation of the intermediate alkene is made difficult by its facile reduction to the alkane. However, the use of Lindlar's catalyst (Pd, PbO, and $CaCO_3$) allows the reaction to be stopped, after the addition of only one mole of hydrogen to the triple bond, at the alkene stage. This reduction is a *cis* addition.

$$\text{RC}\equiv\text{CR} + 2\text{H}_2 \xrightarrow{\text{Pt}} \text{RCH}_2\text{CH}_2\text{R} \quad \text{(alkene usually not isolated)}$$

$$\text{RC}\equiv\text{CR} + \text{H}_2 \xrightarrow[\text{catalyst}]{\text{Lindlar's}} \underset{(cis)}{\text{RCH}=\text{CHR}}$$

The synthesis of *cis* alkenes also may be achieved by reducing alkynes in a pyridine solution, using as a catalyst Pd supported on barium sulfate.

$$H_2 + CH_3O-\overset{O}{\overset{\|}{C}}-C\equiv C-\overset{O}{\overset{\|}{C}}-OCH_3 \xrightarrow[\text{pyridine}]{Pd/BaSO_4} CH_3O-\overset{O}{\overset{\|}{C}}(H)C=C(H)\overset{O}{\overset{\|}{C}}-OCH_3$$

pyridine—a nitrogen analog of benzene

Alkynes also are reduced by sodium and liquid ammonia. This step-wise process produces the *trans* rather than the *cis* alkene.

$$R-C\equiv C-R + 2Na + 2NH_3 \xrightarrow{NH_3(l)} (R)(H)C=C(H)(R) + 2NaNH_2$$

Finally, vinylboranes react with carboxylic acids to produce *cis* alkenes, another reaction that has an analog in alkene chemistry. One interesting difference between the two, however, is that vinylboranes are reduced under much milder conditions than are alkylboranes.

$$3CH_3C\equiv CH + BH_3 \rightarrow (CH_3CH=CH)_3B \xrightarrow[0^\circ]{CH_3CO_2H} CH_3CH=CH_2$$

While the synthetic implications of this chapter may be self-evident, the interconversion of the geometric isomers of an alkene *via* alkynes provides an interesting summary of some of the reactions discussed here.

$$(CH_3)(H)C=C(CH_3)(H) \xrightarrow[CCl_4]{Br_2} Br-C(CH_3)(H)-C(CH_3)(H)-Br \xrightarrow{NH_2^\ominus} CH_3-C\equiv C-CH_3$$

$$CH_3-C\equiv C-CH_3 \xrightarrow{Na/NH_3(l)} (CH_3)(H)C=C(H)(CH_3)$$

$$(CH_3)(H)C=C(H)(CH_3) \xrightarrow[CCl_4]{Br_2} Br-C(CH_3)(H)-C(CH_3)(H)-Br \xrightarrow[NH_3]{NH_2^\ominus} CH_3-C\equiv C-CH_3$$

$$CH_3-C\equiv C-CH_3 \xrightarrow{H_2,\ \text{Lindlar's catalyst}} (CH_3)(H)C=C(CH_3)(H)$$

9.8 OXIDATION OF ALKYNES

The same reagents that oxidize alkenes also oxidize alkynes. Both ozone and potassium permanganate are capable of cleaving an alkyne at the carbon-carbon triple bond and producing two molecules of carboxylic acid.

$$CH_3(CH_2)_4-C\equiv C-(CH_2)_3CH_3 \xrightarrow[CCl_4]{O_3} \xrightarrow{H_2O} CH_3(CH_2)_4CO_2H + HO_2C(CH_2)_3CH_3$$

$$CH_3(CH_2)_4-C\equiv C-(CH_2)_3CH_3 \xrightarrow[OH^\ominus,\ 25^\circ]{KMnO_4} \xrightarrow{H^\oplus} CH_3(CH_2)_4CO_2H + HO_2C(CH_2)_3CH_3$$

hexanoic acid — pentanoic acid

α-Diketones, prepared in moderate yield by carefully controlling the pH during the oxidation, are believed to be intermediates in the oxidation.

$$CH_3(CH_2)_4{-}C{\equiv}C{-}(CH_2)_3CH_3 \xrightarrow[\text{pH7}]{\text{dil. KMnO}_4} CH_3(CH_2)_4{-}\overset{\overset{\displaystyle O}{\|}}{C}{-}\overset{\overset{\displaystyle O}{\|}}{C}{-}(CH_2)_3CH_3$$

an α-diketone

9.9 POLYACETYLENES—NATURAL OCCURRENCE AND SYNTHESIS

While alkanes and alkenes abound in nature, naturally occurring alkynes are scarce; in fact, before 1950 it was believed that they were virtually nonexistent. However, over the past three decades chemists have begun to reveal the existence in nature of a respectable number of alkynes. Some of these compounds bear other functional groups (*e.g.*, —CO_2H, —OH and —C(O)—) and some are hydrocarbons. However, they all share the properties of being rather unstable and being present in very small amounts even in their prime sources. Interestingly, among the major sources of these alkynes are the fungi (Basidiomycetes).

Most of the natural acetylenic compounds are polyacetylenes and contain from two to five carbon-carbon triple bonds. These triple bonds normally are conjugated to one another, to double bonds, or to both. This pattern of extensive conjugation is evident in the polyacetylenes below, all of which are found in the plant kingdom.

$$CH_3CH{=}CH{-}C{\equiv}C{-}C{\equiv}C{-}CH{=}CH{-}CH{=}CH{-}CH{=}CH_2 \quad \textbf{(E,E,E)}$$
$$CH_3CH{=}CH{-}C{\equiv}C{-}C{\equiv}C{-}C{\equiv}C{-}CH{=}CH{-}CH{=}CH_2 \quad \textbf{(E,E)}$$
$$C_6H_5C{\equiv}C{-}C{\equiv}C{-}C{\equiv}CCH_3$$

One of the methods that chemists employ to prove the structure of a natural product is the so-called "unequivocal synthesis." In the case of conjugated polyacetylenes, the most common *in vitro* (outside a living system) synthesis is the oxidative coupling of terminal alkynes.

Oxidative Coupling of Alkynes. Although we have described the use of heavy metal acetylides in organic analysis, the most important use of copper acetylides is in the synthesis of conjugated acetylenes *via* the oxidative coupling procedure. The coupling can be accomplished by shaking the terminal alkyne(s) in a solution of cuprous chloride in methanol/pyridine (a base) in the presence of oxygen.

$$2CH_3{-}\underset{\underset{\displaystyle OH}{|}}{\overset{\overset{\displaystyle CH_3}{|}}{C}}{-}C{\equiv}C{-}H \xrightarrow[O_2]{\text{Cu(I)}} CH_3{-}\underset{\underset{\displaystyle OH}{|}}{\overset{\overset{\displaystyle CH_3}{|}}{C}}{-}C{\equiv}C{-}C{\equiv}C{-}\underset{\underset{\displaystyle OH}{|}}{\overset{\overset{\displaystyle CH_3}{|}}{C}}{-}CH_3$$

75%

$$CH_2{=}CH(CH_2)_4C{\equiv}CH + HC{\equiv}C(CH_2)_7CO_2H \xrightarrow[O_2]{\text{Cu(I)}}$$

$$CH_2{=}CH(CH_2)_4C{\equiv}C{-}C{\equiv}C(CH_2)_7CO_2H$$

erythrogenic acid
~30%

This process also can be used to prepare cyclic alkynes, albeit in poor yields.

$$2HC{\equiv}C(CH_2)_4C{\equiv}CH \xrightarrow[C_2H_5OH/H_2O/O_2]{Cu(I)/NH_4Cl} \text{cyclic dimer: } (CH_2)_4 \text{ and } (CH_2)_4 \text{ joined by two } C{\equiv}C{-}C{\equiv}C \text{ units}$$

10%

One possible mechanism for this as yet incompletely understood reaction is shown below. Note that the actual coupling step is presented as a radical dimerization.

$$R{-}C{\equiv}C{-}H \xrightarrow{base} R{-}C{\equiv}C^{\ominus}$$

$$R{-}C{\equiv}C^{\ominus} + Cu(I) \rightarrow R{-}C{\equiv}C{-}Cu$$

$$R{-}C{\equiv}C{-}Cu + O_2 \rightsquigarrow R{-}C{\equiv}C{-}Cu^{\oplus}$$

$$R{-}C{\equiv}C{-}Cu^{\oplus} \rightarrow R{-}C{\equiv}C\cdot + Cu^{\oplus}$$

$$2R{-}C{\equiv}C\cdot \rightarrow R{-}C{\equiv}C{-}C{\equiv}C{-}R$$

A valuable modification of this coupling reaction employs a two-reagent system, consisting of an ethynyl bromide and a terminal alkyne. Very little oxidative coupling of the terminal alkyne accompanies this crossed coupling.

$$CH_3{-}C(CH_3)(OH){-}C{\equiv}C{-}Br + HC{\equiv}C(CH_2)_8CO_2H \xrightarrow[base]{Cu(I)/H_2O} CH_3{-}C(CH_3)(OH){-}C{\equiv}C{-}C{\equiv}C(CH_2)_8CO_2H$$

98%

We should note that diacetylene, H—C≡C—C≡C—H, the simplest conjugated diacetylene, is conveniently prepared by something other than an oxidative coupling. 1,4-Butynediol (commercially available) can be converted to the corresponding dichloride using phosphorus trichloride,[a] PCl_3, and this can be dehydrohalogenated to diacetylene with dilute sodium hydroxide solution. Use of a strong base like sodium amide leads directly to the disodium salt of diacetylene, which, in turn, can be dialkylated to an internal disubstituted diacetylene.

$$HOCH_2{-}C{\equiv}C{-}CH_2OH \xrightarrow{PCl_3} ClCH_2{-}C{\equiv}C{-}CH_2Cl \xrightarrow{4NaNH_2} Na^{\oplus}\ {}^{\ominus}C{\equiv}C{-}C{\equiv}C^{\ominus}\ Na^{\oplus}$$

$$ClCH_2{-}C{\equiv}C{-}CH_2Cl \xrightarrow{OH^{\ominus}/H_2O} HC{\equiv}C{-}C{\equiv}CH \quad \text{diacetylene } 60\%$$

$$Na^{\oplus}\ {}^{\ominus}C{\equiv}C{-}C{\equiv}C^{\ominus}\ Na^{\oplus} \xrightarrow{2RX} R{-}C{\equiv}C{-}C{\equiv}C{-}R$$

9.10 SPECTRAL PROPERTIES OF ALKYNES[b]

Since simple alkynes do not absorb in the ultraviolet, we shall comment only upon their ir and nmr spectra.

---

[a] This type of reaction is described in detail on p. 305.

[b] This material is meant to be discussed along with (or after) the detailed chapters on spectroscopy.

Infrared Spectra. Simple alkynes show a characteristically sharp, moderate-to-weak absorption in the 2255 to 2100 cm^{-1} region. While terminal acetylenes show a prominent C≡C stretch, internal alkynes usually show a weak absorption. This absorption may be completely absent from the spectra of symmetric alkynes (R—C≡C—R).

Terminal alkynes exhibit a rather strong, sharp **≡C—H** stretching vibration near 3300 cm^{-1}. The infrared spectra of 5-phenyl-1-pentyne (Figure 9–1) and 6-phenyl-2-hexyne (Figure 9–2) illustrate some of these spectral features.

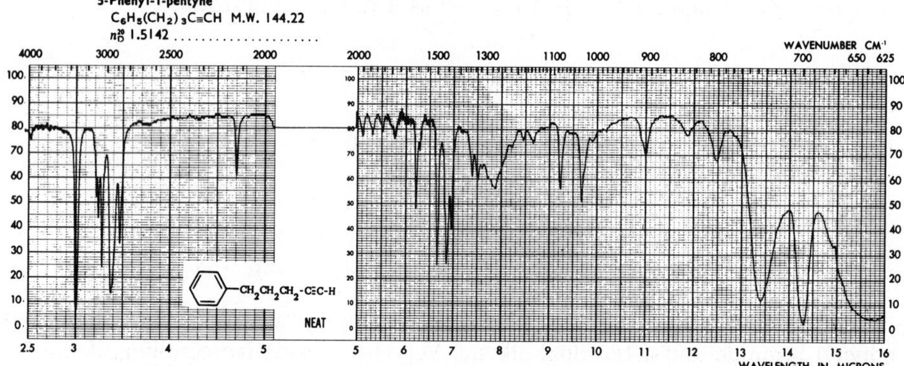

Figure 9–1 Infrared spectrum of 5-phenyl-1-pentyne.

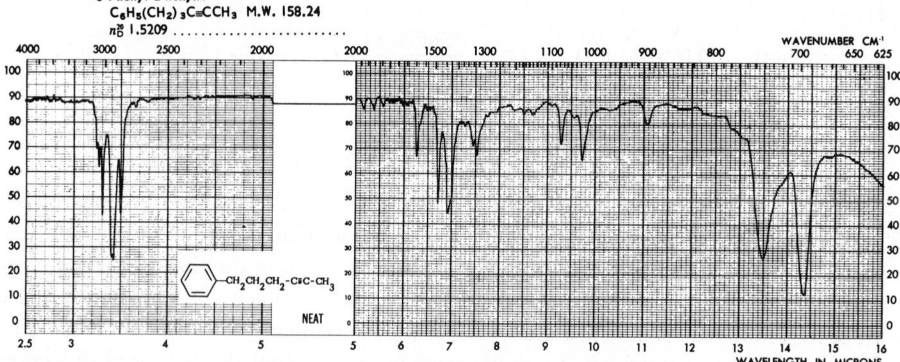

Figure 9–2 Infrared spectrum of 6-phenyl-2-hexyne.

NMR Spectra. Terminal alkynes usually show a **≡C—H** resonance in the range from 1.8 to 3.1δ. This signal occurs at unusually high field owing to the anisotropy of the carbon-carbon triple bond. (Chapter 29 contains a detailed discussion of anisotropic effects.)

The π system of the triple bond transmits sufficient "spin information" that the terminal acetylenic proton may be split by a nearest-neighbor proton:

$$\begin{array}{c} \mathbf{H} \\ | \\ -\mathrm{C}-\mathrm{C}{\equiv}\mathrm{C}-\mathbf{H} \\ | \end{array} \qquad J \approx 2\text{–}3\,\mathrm{Hz}$$

The nmr spectrum of 1-methoxy-1-buten-3-yne (Figure 9–3) is most informative. The methoxy group appears as a singlet, shifted downfield by the inductive effect of the oxygen and the combined anisotropic effects of the multiple bonds (see Chapter 29). The methine protons show a complex series of multiplets. $H_{(3)}$ (3.08δ) is split into a doublet by $H_{(2)}$ (4.52δ). Each signal of the resultant $H_{(3)}$ doublet is further split by weak coupling between $H_{(3)}$ and $H_{(1)}$ (6.35δ). $H_{(2)}$ is split into a doublet by $H_{(1)}$ (large J), and each signal of the $H_{(2)}$ doublet is further split by the coupling to $H_{(3)}$ already noted. Finally, $H_{(1)}$ is split into a doublet by $H_{(2)}$ and then further split by the weak coupling to $H_{(3)}$. This is presented in Figure 9–4.

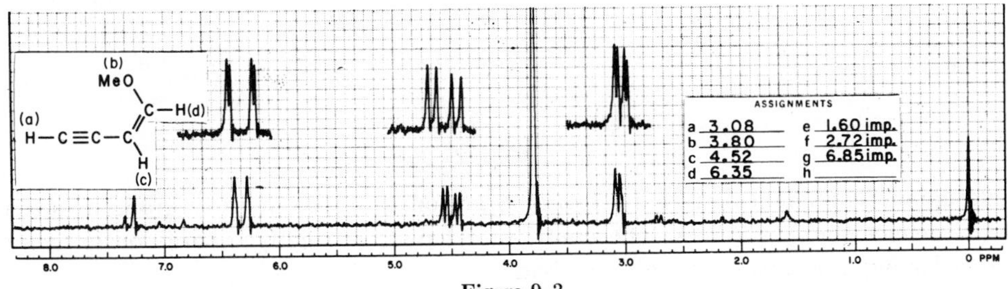

Figure 9-3

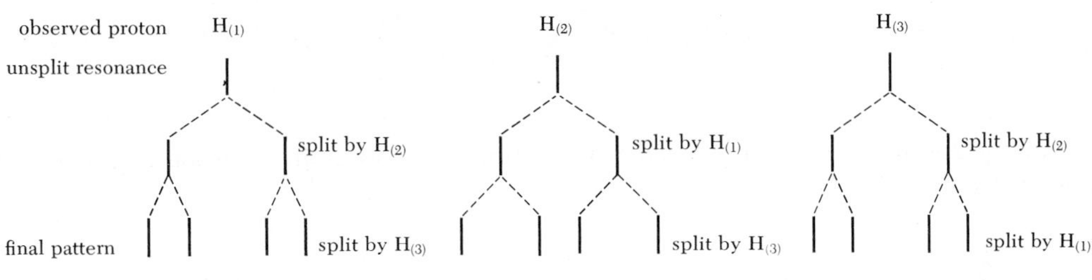

$$H_{(3)}—C≡C—\underset{H_{(2)}}{C}=\underset{H_{(1)}}{C}—O—CH_3$$

Figure 9-4 Analysis of the splitting pattern of the methine protons of 1-methoxy-1-buten-3-yne. The indicated splittings are not drawn to scale. The ordering of the splitting sequence does not alter the final pattern. Thus, $H_{(3)}$ could be shown splitting $H_{(2)}$ before $H_{(1)}$ splits $H_{(2)}$, and the final appearance would be the same.

These long-range couplings (couplings extending beyond three bonds) clearly demonstrate how the presence of multiple bonds may increase the complexity of an nmr spectrum.

IMPORTANT TERMS

Acetylide: A salt of a terminal acetylene (alkyne). The term also is used to describe compounds that have a covalent bond between a metal and a terminal carbon-carbon triple bond.

$$Na^{\oplus}\ominus{:}C≡C—H \qquad Ag—C≡C—C_6H_5$$

sodium acetylide   silver phenylacetylide

Carbon acid: A compound that can donate, to an appropriate base, a proton which is bonded to carbon. The resultant anion is, of course, a carbanion. The most common carbon acids are the terminal alkynes. Alkanes are very weak carbon acids.

Enamine: A compound possessing a nitrogen bonded to a carbon-carbon double bond. The nitrogen analog of an enol. Enamines are tautomers of imines.

$$\rangle C=C\langle^{NH_2} \rightleftarrows H—\overset{|}{\underset{|}{C}}—C\langle^{\,NH}$$

an enamine   an imine

Enol: A compound possessing a hydroxy group bonded to a carbon-carbon double bond. It may be considered to be a vinyl alcohol. Enols are normally unstable and isomerize to their tautomers, compounds containing a carbonyl group. This tautomer is called the keto form of the compound.

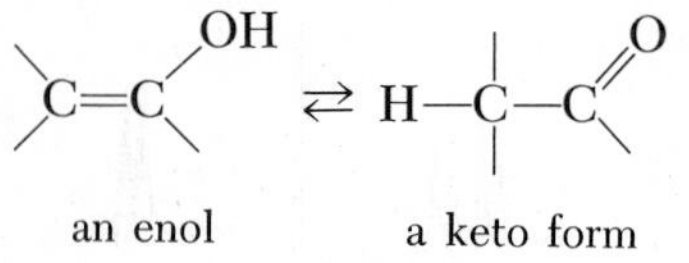

$$\mathrm{>C{=}C(OH)<} \rightleftarrows \mathrm{H{-}\overset{|}{\underset{|}{C}}{-}C({=}O)<}$$

an enol a keto form

Enolate anion: The anion derived by loss of a proton from the hydroxy group of an enol. It is a resonance stabilized anion, one contributor having a negative charge on oxygen and the other having a negative charge on carbon.

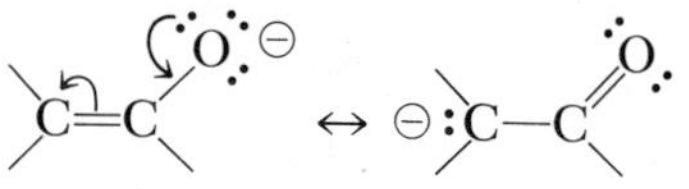

$$\mathrm{>C{=}C({:}\ddot{O}{:}^{\ominus})<} \leftrightarrow \mathrm{{}^{\ominus}{:}C{<}{-}C({=}\ddot{O}{:})<}$$

An enolate anion also is produced when a proton is lost from a carbon adjacent to a carbonyl group (—C=O).

$$\mathrm{B{:}\ H{-}\overset{|}{\underset{|}{C}}{-}C({=}\ddot{O}{:})<} \rightarrow \mathrm{BH} + \mathrm{{}^{\ominus}{:}C{<}{-}C({=}\ddot{O}{:})<} \leftrightarrow \mathrm{>C{=}C({:}\ddot{O}{:}^{\ominus})<}$$

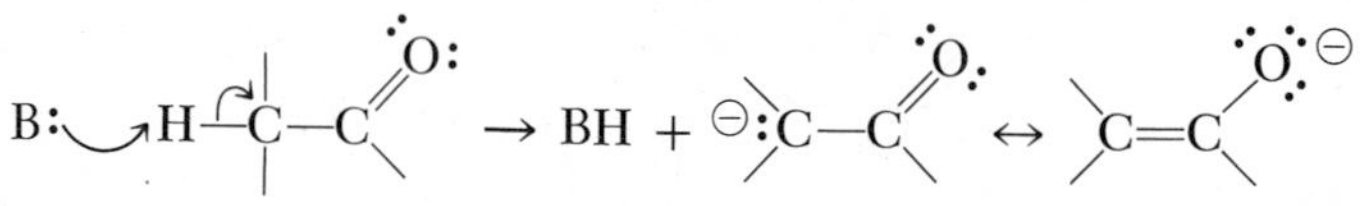

Enol ether: An ether (compound of the type R—O—R) where one of the groups bonded to the oxygen is a double bond.

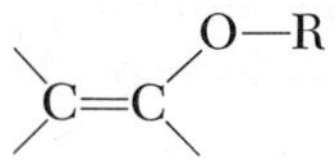

$$\mathrm{>C{=}C(O{-}R)<}$$

an enol ether

Keto (form): A compound containing a carbon-hydrogen bond adjacent to a carbonyl group (—C=O). The tautomer of an enol. The term is usually used only when one is discussing tautomeric structures.

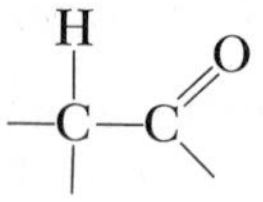

$$\mathrm{{-}\overset{H}{\underset{|}{C}}{-}C({=}O)<}$$

Lindlar's catalyst: A heterogeneous hydrogenation catalyst that converts an alkyne to a *cis* alkene. It consists of palladium, lead oxide (PbO), and calcium carbonate.

Prototropic equilibrium: An equilibrium between species which differ in the location of a proton. More usually, they differ in the position of a proton *and* of a single and a multiple bond. Examples of this include the keto-enol tautomerism and the equilibrium between enamine and imine. The base-catalyzed equilibration of an alkyne and an allene, one example of which is shown below, is another such equilibrium.

$$CH_3{-}C{\equiv}C{-}CH_3 \rightleftarrows CH_2{=}C{=}CH{-}CH_3$$

Tautomers: Species that can interconvert *via* a prototropic equilibrium. The phenomenon is termed *tautomerism*.

PROBLEMS

13. Draw out the structures and give acceptable names for all isomeric alkynes of C_6H_{10}. Neglect enantiomers.

14. Compare the behavior of 1-butyne and 2-butyne toward the following reagents.
   (a) Na, $NH_3(l)$
   (b) $HgSO_4$, H_2SO_4, H_2O
   (c) 2 moles of DBr
   (d) 2 moles of Cl_2, acetyl peroxide
   (e) $NaNH_2$, $NH_3(l)$
   (f) $Ag(I)/NH_4OH$
   (g) H_2, Pd, $BaSO_4$, pyridine
   (h) CH_3MgCl
   (i) B_2H_6 followed by H_2O_2, $OH^{\ominus}$
   (j) CH_3SH, acetyl peroxide

15. Starting with acetylene, and using anything else required, suggest a rational synthesis for each of the following.
   (a) $CH_2{=}CH_2$
   (b) C_2H_6
   (c) $CH_3{-}C{\equiv}C{-}H$
   (d) $CH_3{-}C{\equiv}C{-}D$
   (e) $CH_3CHCHCH_3$ (*cis*)
   (f) $CH_3CHCHCH_3$ (*trans*)
   (g) CH_3CClCH_2
   (h) $CH_3CHCHCl$ (*cis*)
   (i) $CH_3CHCHCl$ (*trans*)
   (j) C_2D_2
   (k) CH_3CH_2OH
   (l) CH_3CHO
   (m) CH_3CDO
   (n) $CH_3C(O)CH_3$
   (o) CCl_2CCl_2
   (p) $HC{\equiv}CCH_2CH{=}CH_2$
   (q) $CH_3C(O)CH_2CH_3$
   (r) polyacrylonitrile (see p. 247)
   (s) polyvinylchloride (see p. 249)

16. Predict the major product(s) from the addition of HBr to each of the following under (a) ionic and (b) radical conditions.
   (a) methylacetylene
   (b) chloroacetylene
   (c) trifluoromethylacetylene
   (d) cyclodecyne

17. Although dehalogenation of the type shown below is, in principle, a good synthetic route to alkynes, it is not very often employed. Can you suggest why?

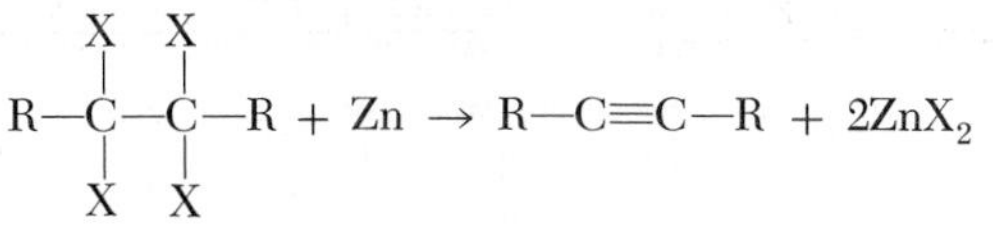

$$R{-}\underset{X}{\overset{X}{\underset{|}{\overset{|}{C}}}}{-}\underset{X}{\overset{X}{\underset{|}{\overset{|}{C}}}}{-}R + Zn \rightarrow R{-}C{\equiv}C{-}R + 2ZnX_2$$

18. When allene reacts with lithium amide, $LiNH_2$, in liquid ammonia, and the reaction mixture is treated with D_2O, $CH_3{-}C{\equiv}C{-}D$ is produced. Provide a mechanism for this conversion.

19. $^{13}CH_3{-}C{\equiv}C{-}H$ can be converted to a mixture of $^{13}CH_3{-}C{\equiv}C{-}H$ and $CH_3{-}C{\equiv}{}^{13}C{-}H$ by treatment with amide anion dissolved in liquid ammonia. (a) Provide a mechanism for this reaction. (b) How is this similar to the base-catalyzed conversion of an enol to a keto form? (c) How do they differ?

20. Suggest simple chemical tests to distinguish between the following.
   (a) methylacetylene and dimethylacetylene
   (b) 1-chloropropyne and 1-chloropropane
   (c) propyne and propene
   (d) cyclodecane, cyclodecene, and cyclodecyne

21. The heats of formation of ethane, ethene, and ethyne are -20, $+12$, and $+54$ kcal/mole, respectively. What is the significance of these data? The heats of formation of carbon dioxide and water are -94 and -58 kcal/mole, respectively. Using these data, calculate which of the hydrocarbons provides the largest amount of heat upon complete combustion to carbon dioxide and water.

22. The reaction of *t*-butylacetylene with hydrochloric acid produces a small amount of 3-chloro-2,3-dimethyl-1-butene. Account for its formation.

23. Suggest a mechanism for each of the following reactions.

(a) 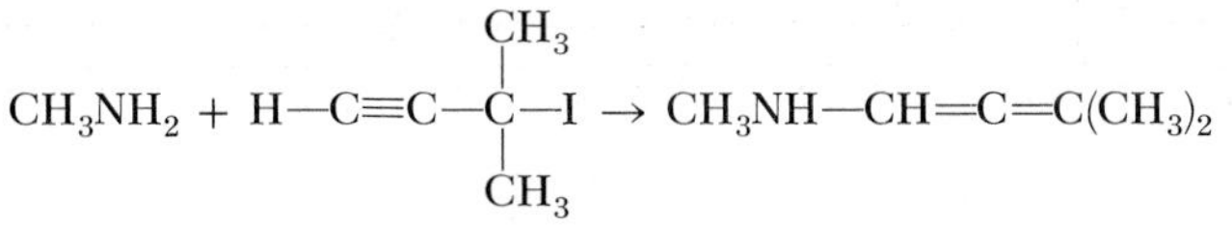 $$CH_3NH_2 + H{-}C{\equiv}C{-}\underset{CH_3}{\overset{CH_3}{\underset{|}{\overset{|}{C}}}}{-}I \rightarrow CH_3NH{-}CH{=}C{=}C(CH_3)_2$$

(b) $$CH_3NH_2 + (CH_3)_2CH{-}C{\equiv}C{-}C(CH_3)_2Br \rightarrow C(CH_3)_2{=}C{=}C{=}C(CH_3)_2$$

24. Using bond strength data (Tables 2–3 and 2–5), determine whether one would expect acetaldehyde (CH_3CHO) to be more stable than the corresponding enol.

*25. Suggest a spectral method for distinguishing between the following.

(a) ⬡—C≡C—H *and* ⬡—C≡C—⬡

(b) CH_3—C≡C—CH_2CH_3 *and* CH_3—C≡C—CH_2CD_3

(c) CH_3—C≡C—CH_3 *and* CH_3—C≡C—CH_2CH_3

(d) CHCl=CH—CH_2—C≡C—H *and* H_2C=CH—CH_2—C≡C—Cl

(e) *cis-* and *trans-*CH_3C≡CCH=CHCH_3

*26. The nmr spectra below are those of allyl bromide and 2-bromopropene. Which spectrum belongs to which compound? Explain your answer.

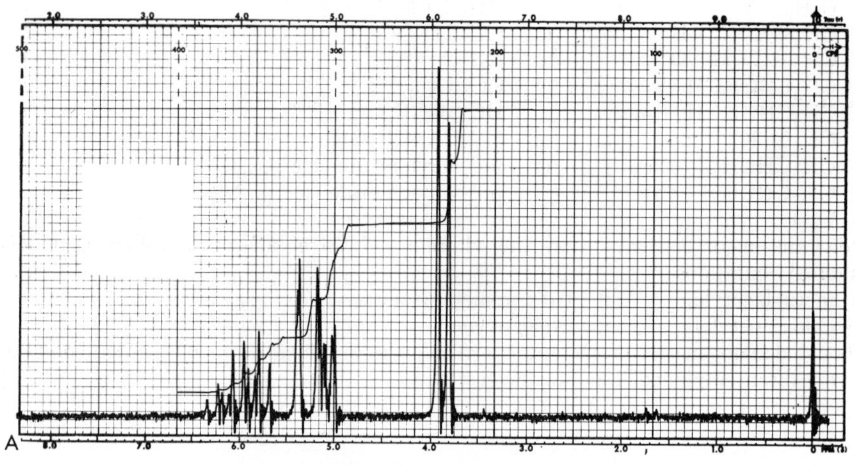

© Sadtler Research Laboratories, Inc., 1976.

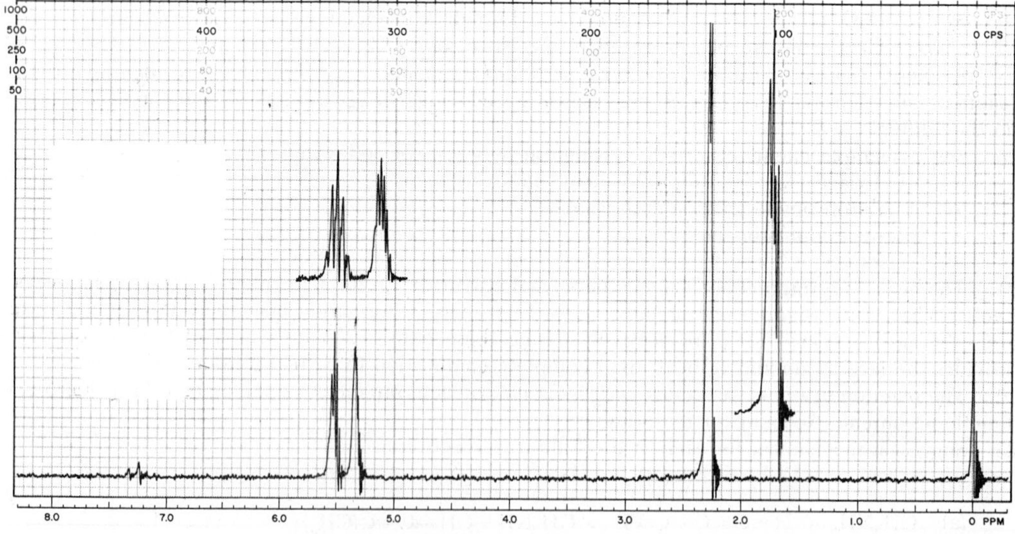

*This problem requires a knowledge of spectroscopy.

*27. Sketch the nmr spectra of the following.

(a) $CH_3{-}C{\equiv}C{-}CH_3$

(b) $CH_3CH_2C{\equiv}C{-}H$

(c) $CH_2{=}C{=}CHCH_3$

*28. How would the ir spectra of 1-butyne and 2-butyne differ?

*29. Account for the differences indicated by the arrows.

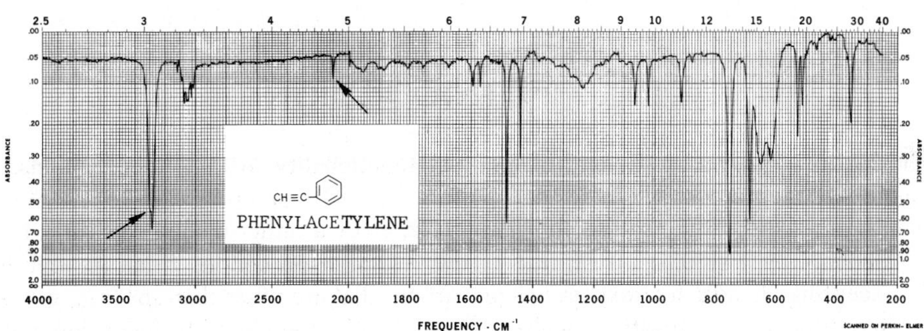

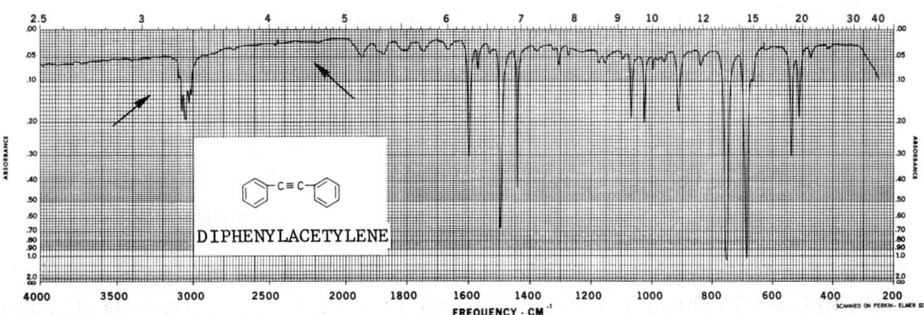

---

*This problem requires a knowledge of spectroscopy.

10 ALCOHOLS

10.1 INTRODUCTION

The hydroxy group, $-\ddot{\underset{\cdot\cdot}{O}}H$, is the principal functionality in two important classes of compounds—*phenols* and *alcohols*. Phenols possess the —OH group bonded to an sp^2 carbon of a benzene ring and, as we shall see in Chapter 22, enjoy a rather unique chemistry because of this. Alcohols, by way of contrast, possess the —OH group bonded to an sp^3 hybridized carbon. Excepting phenols, hydroxy groups on carbons that are other than sp^3 (*e.g.*, sp^2) either are unstable relative to enolization or else are more aptly considered to be an integral part of another functional group. This latter type of "OH" is incorporated into the carboxy group, —COOH.

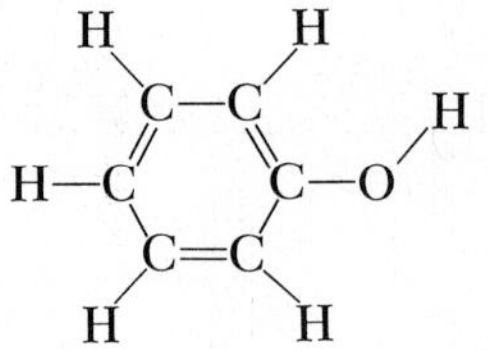

phenol

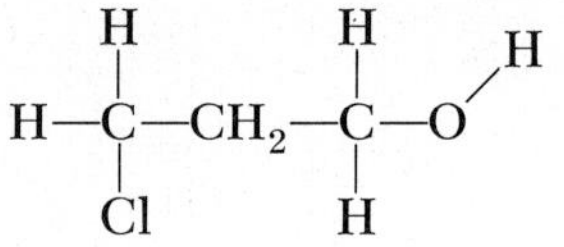

an alcohol

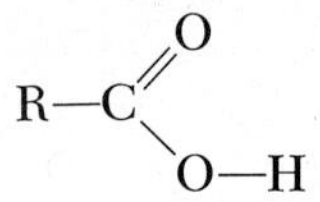

a carboxylic acid

"Alcohols" may be divided, somewhat arbitrarily, into three major categories—*simple alcohols*, *sterols*, and *carbohydrates*. This list, though not all-inclusive, covers the bulk of compounds possessing the C_{sp^3}—OH linkage. Sterols and carbohydrates are not considered here since, although both groups have biological importance, the former possess a characteristic polycyclic ring system (see Chapter 7), while the latter have a large number of hydroxy groups (usually at least three) per molecule.

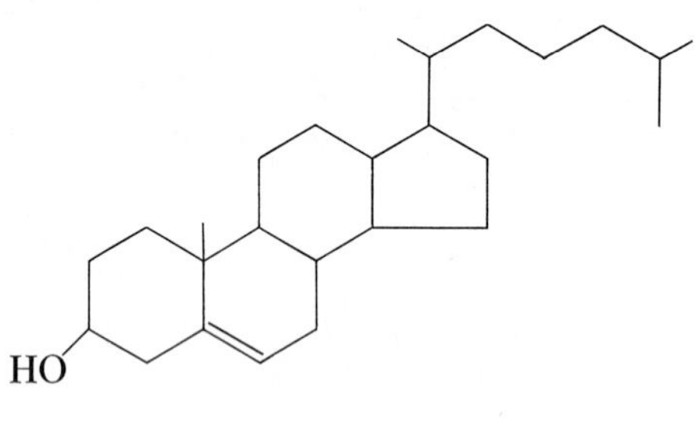

cholesterol
(a sterol)

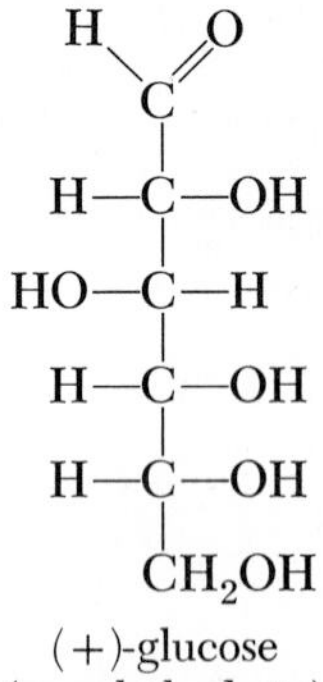

(+)-glucose
(a carbohydrate)

10.2 PHYSICAL PROPERTIES OF ALCOHOLS

The presence of the hydroxy group is responsible for many of the atypical physical properties exhibited by low molecular weight alcohols. For example, the simplest members of several families of compounds, including the alkanes, alkenes, alkynes, alkyl halides, and ethers, all are low boiling in comparison to the simplest alcohol, methanol (see Table 10–1). Ethanol, the next higher homolog of methanol, has a much higher boiling point (+78°) than does dimethyl ether (bp −24°), even though both of them have the same molecular formula, C_2H_6O. Moreover, methanol and ethanol are completely miscible with water, unlike any of the non-alcohols listed in Table 10–1.

TABLE 10–1 Physical Properties of Alcohols And Simple Molecules

COMPOUND			
Formula	*Name*	BP	WATER SOLUBILITY[a]
CH_4	methane	−161	3.5 ml/100 ml
C_2H_6	ethane	−88	4.7 ml/100 ml
$CH_3CH_2CH_3$	propane	−42	6.5 ml/100 ml
$CH_3CH_2CH_2CH_3$	butane	−0.50	15 ml/100 ml
$CH_3(CH_2)_3CH_3$	pentane	+36	0.034 g/100 ml
$C(CH_3)_4$	2,2-dimethylpropane	+9.5	"insoluble"
$CH_3(CH_2)_6CH_3$	octane	+126	"insoluble"
$CH_2{=}CH_2$	ethene	−102	11 ml/100 ml
$CH_2{=}CHCH_3$	propene	−48	
$HC{\equiv}CH$	ethyne	−75	100 ml/100 ml
$HC{\equiv}CCH_3$	propyne	−23	
CH_3Cl	chloromethane	+24	303 ml/100 ml
CH_3CH_2Cl	chloroethane	+13	
$CH_3CH_2CH_2Cl$	1-chloropropane	+47	
$CH_3CH_2CH_2CH_2Cl$	1-chlorobutane	+79	
CH_3OCH_3	methyl ether	−24	37 ml/100 ml
$CH_3CH_2OCH_2CH_3$	ethyl ether	+35	
CH_3OH	methanol	+65	infinite
C_2H_5OH	ethanol	+78	infinite
$CH_3CH_2CH_2OH$	1-propanol	+98	infinite
$CH_3(CH_2)_4OH$	1-pentanol	+138	2.7 g/100 ml
$CH_3(CH_2)_7OH$	1-octanol	+195	"insoluble"

[a] Where quantities of solute are given in ml, the reference is to ml of gaseous solute at 17 to 20°.

All of these differences can be explained by invoking hydrogen bonding to or from the hydroxy group of the alcohol molecule. Intermolecular hydrogen bonding between alcohol molecules is responsible for the higher boiling points of alcohols compared to those of alkyl halides and similar compounds, while intermolecular hydrogen bonding between alcohol molecules and water molecules is responsible for the miscibility of simple alcohols with water. As the hydrocarbon portion of the molecule increases, the relative influence of hydrogen bonding decreases. This latter idea explains why the physical properties of large monohydric alcohols (*i.e.*, alcohols containing only one hydroxy group) approach those of the corresponding hydrocarbons (Table 10–1).

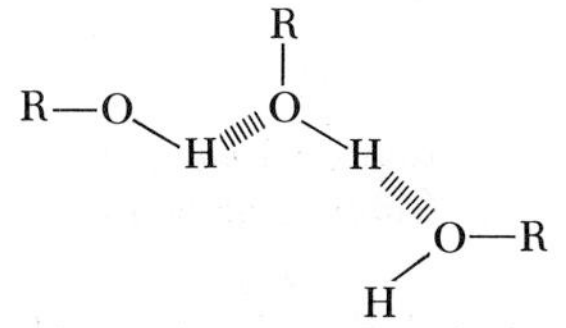

intermolecular hydrogen-bonding raises boiling point

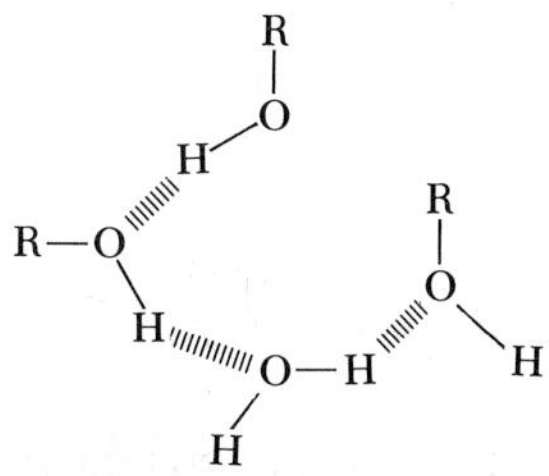

intermolecular hydrogen-bonding increases water miscibility

The spectral properties of alcohols, including the influence of hydrogen bonding upon molecular spectra, are discussed at the end of this chapter.

10.3 NOMENCLATURE

There are three methods available for naming alcohols—the *IUPAC* system, the *carbinol* system, and the *alcohol* system. The carbinol system is the one that is the least used of these three. However, it is included because it will be encountered when reading the literature of organic chemistry. In the IUPAC system, simple alcohols are considered to be *alkanols* and are named accordingly.

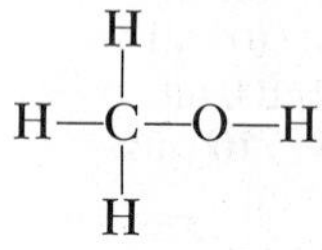

$$H_3C{-}OH$$

methanol

$$H_3C{-}CH_2{-}CH_2{-}CH_2{-}CH_2{-}CH(OH){-}CH_3$$

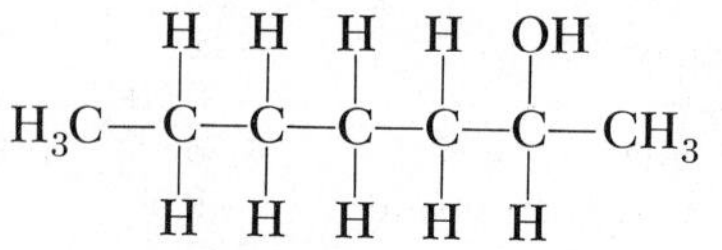

2-heptanol

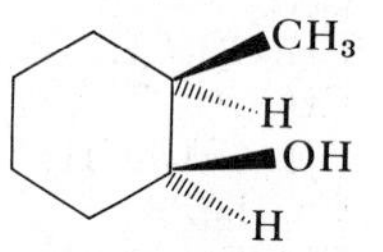

cis-2-methylcyclohexanol

In the *carbinol system* for naming alcohols, methanol is called "carbinol" and alcohols are named as derivatives of "carbinol." The so-called "carbinol carbon" is indicated in **boldface** in the examples below.

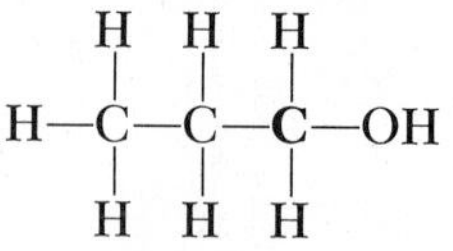

$$CH_3{-}CH_2{-}\mathbf{C}H_2{-}OH$$

ethylcarbinol

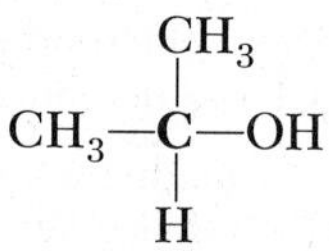

$$CH_3{-}\mathbf{C}H(CH_3){-}OH$$

dimethylcarbinol

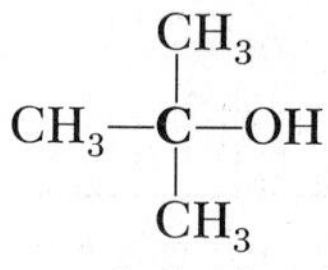

$$(CH_3)_3\mathbf{C}{-}OH$$

trimethylcarbinol

$$C_6H_5{-}\mathbf{C}H(CH_3){-}OH$$

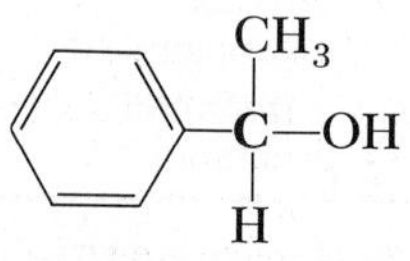

methylphenylcarbinol

In the last scheme for naming alcohols, and the only one of the three to actually use the word "alcohol," compound names are obtained by identifying the alkyl moiety bearing the hydroxy group and then following its name with the word "alcohol." The common alcohols are usually called by something other than their IUPAC names (Table 10–2).

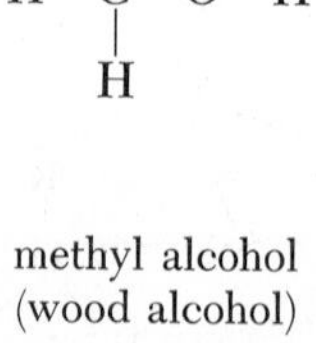

$$H_3C{-}O{-}H$$

methyl alcohol
(wood alcohol)

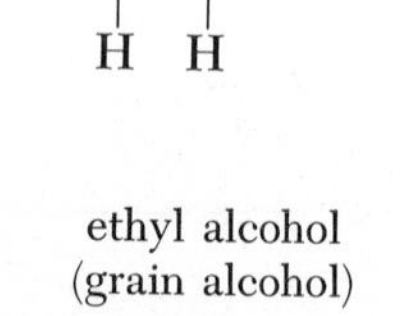

$$H_3C{-}CH_2{-}O{-}H$$

ethyl alcohol
(grain alcohol)

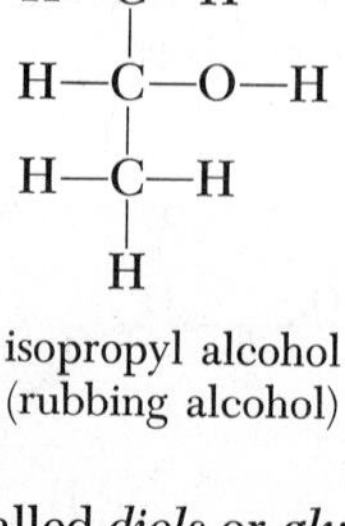

$$H_3C{-}CH(O{-}H){-}CH_3$$

isopropyl alcohol
(rubbing alcohol)

Simple compounds containing two hydroxy groups are called *diols* or *glycols*. Compounds containing two hydroxy groups on the same carbon, *i.e.*, 1,1-diols or *gem*-diols, exist in equilib-

TABLE 10-2 Alcohols

NAME	FORMULA	MP	BP
methyl	CH_3OH	−97	65
ethyl	CH_3CH_2OH	−115	78
n-propyl	$CH_3(CH_2)_2OH$	−126	97
n-butyl	$CH_3(CH_2)_3OH$	−90	118
n-pentyl (or *n*-amyl)	$CH_3(CH_2)_4OH$	−79	138
n-hexyl	$CH_3(CH_2)_5OH$	−52	157
n-heptyl	$CH_3(CH_2)_6OH$	−34	176
n-octyl	$CH_3(CH_2)_7OH$	−15	195
n-nonyl	$CH_3(CH_2)_8OH$	−8	215
n-decyl	$CH_3(CH_2)_9OH$	6	228
n-dodecyl (or lauryl)	$CH_3(CH_2)_{11}OH$	24	262
n-tetradecyl (or myristyl)	$CH_3(CH_2)_{13}OH$	38	289
n-hexadecyl (or cetyl)	$CH_3(CH_2)_{15}OH$	49	
isopropyl	$(CH_3)_2CHOH$	−86	83
isobutyl	$(CH_3)_2CHCH_2OH$	−108	108
sec-butyl	$CH_3CH_2CH(CH_3)OH$	−114	100
t-butyl	$(CH_3)_3COH$	26	83
isopentyl (or isoamyl)	$(CH_3)_2CHCH_2CH_2OH$	−117	132
t-pentyl (or *t*-amyl)	$CH_3CH_2C(OH)(CH_3)_2$	−12	102
allyl	$CH_2{=}CHCH_2OH$	−129	97
α-phenylethyl	$C_6H_5CH(OH)CH_3$	20	205
β-phenylethyl	$C_6H_5CH_2CH_2OH$	−27	221

rium with a carbonyl compound and water. This equilibrium usually favors the carbonyl compound and, as a result, *gem*-diols do not constitute an important class of diols.

$HOCH_2CH_2OH$ 1,2-ethanediol (ethylene glycol)

$HOCH_2COHHCH_3$ 1,2-propanediol (propylene glycol)

$HOCH_2CH_2CH_2OH$ 1,3-propanediol (trimethylene glycol)

$$CH_3C(=O)CH_3 + H_2O \rightleftarrows CH_3{-}C(OH)_2{-}CH_3$$

acetone (carbonyl compound) — 2,2-propanediol (*gem*-diol)

As usual, you can expect to find alcohols with names that do not seem to relate to anything. While the origin of the term "wood alcohol" will become obvious shortly, the name "pinacol" for 2,3-dimethyl-2,3-butanediol has a more obscure origin.

Regardless of its specific name, an alcohol may be classified according to the number of hydrogens bonded to the carbinol carbon. A *primary alcohol* (1°) has two hydrogens attached to the carbinol carbon, while a *secondary alcohol* (2°) has one and a *tertiary alcohol* (3°) has none. For convenience, methyl alcohol is considered along with primary alcohols.

$C_6H_5CH_2OH$	$C_6H_5CHOHC_6H_5$	$(C_6H_5)_2COHCH_2C_6H_5$
benzyl alcohol (1° alcohol)	benzhydrol (2° alcohol)	benzyldiphenylcarbinol (3° alcohol)

1. Provide an acceptable name for each of the following and identify the alcohols as 1°, 2°, or 3°.
   (a) $CH_3CH{=}CHCH_2OH$
   (b) $CH_3CH_2CH_2CHCH_3OH$
   (c) 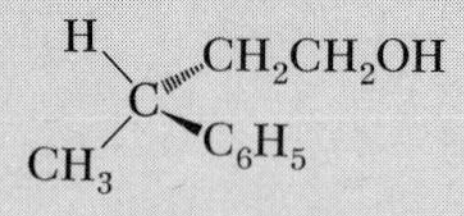 C bearing H, CH_2CH_2OH (hashed bond), C_6H_5 (wedge), and CH_3
   (d) 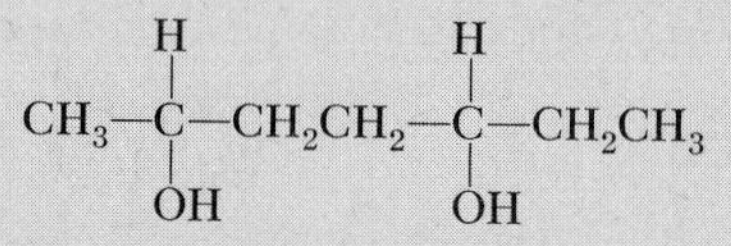 $CH_3-\underset{OH}{\overset{H}{C}}-CH_2CH_2-\underset{OH}{\overset{H}{C}}-CH_2CH_3$
   (e) 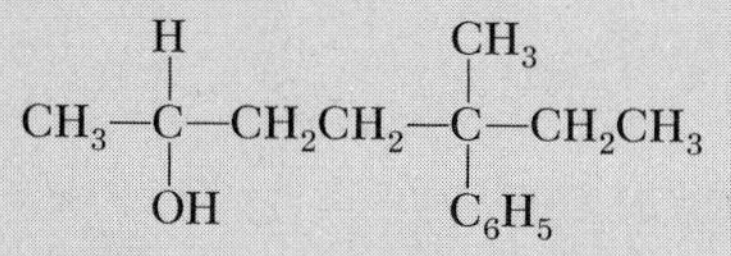 $CH_3-\underset{OH}{\overset{H}{C}}-CH_2CH_2-\underset{C_6H_5}{\overset{CH_3}{C}}-CH_2CH_3$
   (f) 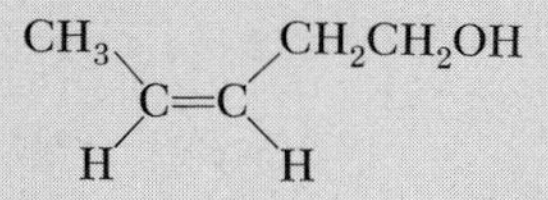 $(CH_3)(H)C{=}C(H)(CH_2CH_2OH)$, with CH_3 and CH_2CH_2OH on the same side

2. Draw the structure and provide the IUPAC name for a 1°, a 2°, and a 3° alcohol, each of which contains seven carbons.

10.4 SYNTHESIS OF ALCOHOLS

After a review of some methods that have appeared in earlier chapters, we will examine six general routes to alcohols. During this discussion we will look in detail at the use of Grignard reagents for the synthesis of alcohols, and at some limitations on these reactions. (You might want to review the discussion of Grignard reagents beginning on p. 173.) A presentation of the industrial preparations of several alcohols will end this section.

A Brief Review. Without specifically labeling them as such, we have already examined several routes to alcohols. On pages 129 and 241 we encountered reactions that could be used to prepare alcohols by S_N reactions and by the direct hydration of alkenes. When using these reactions, you must remember the limitations placed upon their synthetic utility by the nature of their mechanisms. The examples below review these types of reactions and, in example *b*, re-emphasize how a carbocation may rearrange to give an unexpected product.

$$\text{(a)}\quad (+)\text{-}C_6H_5CHClCH_3 + H_2O \xrightarrow[20\%\ H_2O]{80\%\ CH_3\overset{O}{\overset{\|}{C}}-CH_3} (\pm)\text{-}C_6H_5CHOHCH_3$$

$$\text{(b)}\quad C_6H_5CHClCH{=}CH_2 \xrightarrow[\text{heat}]{H_2O} C_6H_5CH{=}CHCH_2OH$$

(c) $CH_3SCH_2CH_2Cl \xrightarrow[\text{heat}]{H_2O} CH_3SCH_2CH_2OH$

(d) $NO_2CH{=}CHCO_2H \xrightarrow[\text{heat}]{HCO_2H^a/H_2O} NO_2CH_2CH(OH)CO_2H$

3-nitropropenoic acid → 2-hydroxy-3-nitropropanoic acid

Another important example of rearrangement accompanying solvolysis (S_N reaction in which the solvent is the nucleophile) is the reaction of neopentyl tosylate with water. When it is hydrolyzed, the alcohol that is produced is 2-methyl-2-butanol and *not* neopentyl alcohol. This observation, along with many others, suggests that the neopentyl cation is not produced under most reaction conditions. It is believed, instead, that methide migration occurs simultaneously with the departure of the tosylate ion in the example above. This produces the 2-methyl-2-butyl cation without the intermediacy of the 1°, neopentyl cation.

neopentyl tosylate → ionization begins → activated complex → a 3° cation

Ts: $-SO_2-C_6H_4-CH_3$

> 3. Suggest a mechanism for the conversion of $C_6H_5CHClCH{=}CH_2$ to $C_6H_5CH{=}CHCH_2OH$ in water.

Why does rearrangement accompany ionization? Why doesn't it occur after ionization? The answer to this must, of course, be that the energy of activation is lower if rearrangement accompanies ionization than if a primary cation is formed and rearrangement follows ionization. The lower energy of activation is a result of the distribution of the developing positive charge over two carbons, including one that is trisubstituted. One can actually view the methide migration as assisting ionization. Such reactions, in which migration is synchronous with, and increases the rate of, ionization, are said to be **anchimerically assisted.**

These and other data suggest that the simple 1° cation, $RCH_2^{\oplus}$, is formed in solution only under very unusual conditions. Even when a reaction appears to involve a 1° cation, one can probably construct an alternate mechanism that is S_N2 in nature. An example of such a reaction is the conversion of primary alkyl halides to ethers with the aid of silver ions. This reaction was presented on page 148 as being S_N1-like; however, it actually goes through an S_N2 process in which the solvent is the nucleophile.

---

[a] HCO_2H is formic acid.

$$CH_3(CH_2)_2{-}Br + Ag^{\oplus} \rightarrow CH_3CH_2\overset{H}{\underset{H}{C}}{}^{\delta+}\cdots Br\cdots \overset{\delta+}{Ag}$$

$$CH_3CH_2\overset{\delta+}{C}H_2\cdots Br\cdots \overset{\delta+}{Ag} \xrightarrow[-AgBr]{S_N2} CH_3CH_2CH_2{-}\overset{\oplus}{\underset{H}{O}}{-}C_2H_5 \leftrightarrows$$

$C_2H_5{-}O{-}H$ (attacks C) H—O(C_2H_5)H

$$CH_3CH_2CH_2OC_2H_5 + C_2H_5\overset{\oplus}{O}H_2$$

an ether

Hydroboration-Oxidation. Recall that diborane adds to alkenes to form trialkylboranes (p. 231). If such an adduct is oxidized with an alkaline solution of hydrogen peroxide, the C—B bond is converted to the C—OH bond with retention of configuration. Since diborane adds *cis* and anti-Markownikoff to an alkene, the net result of hydroboration followed by oxidation is *cis* anti-Markownikoff hydration of an alkene. This forms, then, a valuable counter-point to acid-catalyzed hydration of an alkene, which produces *trans* > *cis* Markownikoff addition of water.

$$H_2C{=}CHCH_2CH_3 \xrightarrow{B_2H_6} (H_2\underset{B}{C}CH_2CH_2CH_3)_3 \xrightarrow[OH^{\ominus}]{H_2O_2} HOCH_2CH_2CH_2CH_3$$

norbornene $\xrightarrow{B_2H_6}$ (norbornyl)$_3$B $\xrightarrow[OH^{\ominus}]{H_2O_2}$ *exo*-norbornyl alcohol

One mechanism proposed for the alkaline hydrogen peroxide oxidation of the organoboron compound involves an initial addition of the hydroperoxide anion to boron. This intermediate then rearranges, with loss of hydroxide ion, to eventually form a borate ester. Hydrolysis of the borate ester yields the final alcohol.

$$H_2O_2 + OH^{\ominus} \rightleftarrows \underset{\text{hydroperoxide anion}}{HO{-}O^{\ominus}} + H_2O$$

$$R_2B{-}R + HO_2^{\ominus} \xrightarrow[\text{formation}]{\text{adduct}} R_2\overset{\ominus}{B}(R){-}O{-}O{-}H \xrightarrow[-OH^{\ominus}]{\text{R migrates}} R_2B{-}OR$$

$$R_2B{-}OR \xrightarrow{HO_2^{\ominus}} \xrightarrow{-OH^{\ominus}} RB(OR)_2 \xrightarrow{HO_2^{\ominus}} \xrightarrow{-OH^{\ominus}} \underset{\text{borate ester}}{B(OR)_3}$$

$$B(OR)_3 + 3H_2O \rightarrow 3ROH + H_3BO_3$$

4. Identify the alcohol that will be produced by the hydroboration-oxidation of the following:
(a) $D_2C{=}CHCH_3$
(b) $CH_3CH{=}CHCH_3$ (**Z** isomer)
(c) $CH_3CH{=}CHCH_3$ (**E** isomer)
(d) 1-methylcyclohexene (CH_3 on ring double bond)
(e) methylenecyclohexane ($=CH_2$)

Hydroboration-Carbonylation-Oxidation. Trialkylboranes form adducts with carbon monoxide which are, in turn, oxidized by alkaline hydrogen peroxide to trialkylcarbinols. In this reaction the carbon of the carbon monoxide becomes the carbinol carbon. Since the reaction with carbon monoxide is rapid at atmospheric pressure, alkylboranes are excellent precursors to symmetric trialkylcarbinols. The carbonylation of the organoboron compound involves a complex series of three separate migrations of alkyl groups from boron to the carbon atom of carbon monoxide, and it will not be discussed further.

$$\text{cyclohexene} \xrightarrow{B_2H_6} (C_6H_{11})_3B \xrightarrow[100^\circ]{CO} \xrightarrow[OH^\ominus]{H_2O_2} (C_6H_{11})_3COH$$

tricyclohexylcarbinol

$$H_2C{=}CHCH_3 \xrightarrow{B_2H_6} (CH_3CH_2CH_2)_3B \xrightarrow[100^\circ]{CO} \xrightarrow[OH^\ominus]{H_2O_2} (CH_3CH_2CH_2)_3COH$$

tri-*n*-propylcarbinol

$$\text{norbornene} \xrightarrow{B_2H_6} (\text{norbornyl})_3B \xrightarrow[100^\circ]{CO} \xrightarrow[OH^\ominus]{H_2O_2} (\text{norbornyl})_3COH$$

norbornene — tri-*exo*-norbornylcarbinol

Synthesis via Organometallics—General Considerations. Grignard reagents and organolithium compounds usually do not add to the carbon-carbon double bond because the olefinic carbons are not sufficiently positive to attract the incipient carbanions and, perhaps more importantly, because addition of a carbanion to an alkene would produce a high-energy intermediate, the carbanion. The carbonyl group, on the other hand, has an sp^2 carbon that is rather positive (oxygen is more electronegative than carbon), and the anion that is produced by addition of a carbanion to a carbonyl carbon is the relatively stable alkoxide ion. Therefore, the reactions of Grignard reagents and organolithium compounds with carbonyl compounds produce the salts of alcohols. These *alkoxide* salts can be converted to the corresponding alcohols simply by addition of water or dilute acid.

$$-\overset{|}{\underset{|}{C}}{:}^{\ominus}\ {}^{\oplus}MgX \longrightarrow \ >C{=}C< \ \not\longrightarrow\ -\overset{|}{\underset{|}{C}}-\overset{|}{\underset{|}{C}}-\overset{|}{\underset{|}{C}}{:}^{\ominus}\ \overset{\oplus}{MgX} \qquad \text{reaction does NOT occur}$$

$$-\overset{|}{\underset{|}{C}}{:}^{\ominus}\ {}^{\oplus}MgX \longrightarrow \ >C{=}\ddot{O}{:} \ \rightarrow\ -\overset{|}{\underset{|}{C}}-\overset{|}{\underset{|}{C}}-\ddot{\underset{..}{O}}{:}^{\ominus}\ \overset{\oplus}{MgX} \xrightarrow{H_2O} -\overset{|}{\underset{|}{C}}-\overset{|}{\underset{|}{C}}-OH + MgXOH$$

carbonyl compound — alkoxide salt — alcohol

As seen from the scheme below, the reaction of Grignard reagents with formaldehyde, higher aldehydes, or ketones serves as an excellent route to 1°, 2°, and 3° alcohols, respectively.

$$CH_3MgI \xrightarrow[\text{formaldehyde}]{H_2C{=}O} CH_3CH_2\overset{\ominus\oplus}{O}MgI \xrightarrow{H_2O} CH_3{-}\underset{H}{\overset{OH}{C}}{-}H \quad (1°\ \text{alcohol})$$

$$CH_3MgI \xrightarrow[\substack{\text{acetaldehyde}\\ \text{(an aldehyde)}}]{CH_3\overset{O}{\overset{\|}{C}}H} CH_3{-}\underset{CH_3}{\overset{\overset{\ominus\oplus}{O}MgI}{C}}{-}H \xrightarrow{H_2O} CH_3{-}\underset{CH_3}{\overset{OH}{C}}{-}H \quad (2°\ \text{alcohol})$$

$$CH_3MgI \xrightarrow[\substack{\text{acetone}\\ \text{(a ketone)}}]{CH_3\overset{O}{\overset{\|}{C}}CH_3} CH_3{-}\underset{CH_3}{\overset{\overset{\ominus\oplus}{O}MgI}{C}}{-}CH_3 \xrightarrow{H_2O} CH_3{-}\underset{CH_3}{\overset{OH}{C}}{-}CH_3 \quad (3°\ \text{alcohol})$$

These reactions, which are *nucleophilic additions* to the carbonyl group, share features with S_N2 reactions. The carbanion may be viewed as a nucleophile that attacks a carbon p orbital, thus displacing a pair of electrons from the just-attacked carbon. However, since it is a π bonding electron pair which is being displaced, and since a σ bond still holds carbon and oxygen together, addition occurs rather than substitution.

In a related reaction, a Grignard reagent or an organolithium compound attacks a carbonyl group bound to a potential leaving group (*e.g.*, Cl, Br, OCH_3, O—$COCH_3$) rather than to hydrogen or to hydrocarbon fragments. This type of reaction eventually produces a 3° alcohol in good yield.

$$2RMgX + R'{-}\overset{O}{\overset{\|}{C}}{-}L \rightsquigarrow R'{-}\underset{R}{\overset{OH}{C}}{-}R \qquad L = \text{leaving group}$$

For the sake of simplicity, at this stage we will consider only the reaction of a Grignard reagent with an ester (L=OR). The initial step in the reaction of a Grignard reagent with an ester is an addition similar to that which was just discussed.

$$\overset{\ominus}{R}{:}\ {}^{\oplus}MgX \quad R'{-}\overset{:\ddot{O}}{\overset{\|}{C}}{-}\ddot{O}R'' \rightarrow R'{-}\underset{R}{\overset{:\ddot{O}:^{\ominus}\ \ \overset{\oplus}{MgX}}{C}}{-}\ddot{O}R'' \qquad \text{addition of a Grignard reagent to an ester}$$

Once the adduct is produced, a pair of electrons on oxygen attacks the central carbon and ejects the leaving group in a step analogous to a step in the E1cb reaction (p. 171). The ketone that is produced is more reactive toward the Grignard reagent than was the original ester, and it reacts immediately to form the salt of a 3° alcohol. The key step in this sequence, the conversion of the ester to the ketone, is an example of an *addition-elimination* process. The *net* result of an addition-elimination sequence is the substitution of a nucleophile for a leaving group.

$$R'—C(O^{\ominus})(R)—\ddot{O}R'' \rightarrow R'—C(=O)—R + \overset{\ominus}{O}R''$$

leaving group (L) — $\ddot{O}R''$; a ketone

$$R'—C(=O)—R + RMgX \rightarrow R'—C(O^{\ominus}\overset{\oplus}{M}gX)(R)—R \xrightarrow{H_2O} R'—C(OH)(R)—R$$

Since compounds of the type R—C(O)L, and esters in particular, are easily prepared (see Chapter 18), this sequence offers a valuable route to simple 3° alcohols. For example, triphenylcarbinol is prepared in yields of better than 90% by the scheme below.

$$C_6H_5—C(=O)—\underbrace{OCH_2CH_3}_{L} + 2\ C_6H_5—MgBr \xrightarrow{THF} \xrightarrow{H_2O} (C_6H_5)_3COH$$

ethyl benzoate (ester); phenyl magnesium bromide (Grignard reagent); triphenylcarbinol (3° alcohol)

In our discussion of alkenes (Chapter 8) we noted that ethylene oxide, because of strain present in the ring, is easily ring-opened *via* nucleophilic attack by water. In a related process, ethylene oxide is ring-opened by Grignard reagents to produce 1° alcohols; the alcohol chain is two carbon atoms longer than the precursor to the Grignard reagent. Trimethylene oxide, also called *oxetane*, can be ring-opened by an S_N2 process to produce an alcohol three carbons longer than the Grignard reagent. This reaction usually requires more vigorous conditions than does oxirane ring opening. Can you suggest a reason for this?

$$RMgX + \underset{O}{CH_2—CH_2} \xrightarrow[\text{ether}]{S_N2} RCH_2CH_2\overset{\ominus}{O}\overset{\oplus}{M}gX \xrightarrow{H_2O} RCH_2CH_2OH$$

$$RC{\equiv}C—MgX + \underset{CH_2—CH_2}{CH_2—O} \xrightarrow[\text{THF}]{S_N2} RC{\equiv}C—(CH_2)_3—\overset{\ominus}{O}\overset{\oplus}{M}gX \xrightarrow{H_2O} RC{\equiv}C—(CH_2)_3OH$$

oxetane

Synthesis via Organometallics—Approaches and Limitations. In planning a Grignard synthesis of an alcohol, it is important that (a) the Grignard reagent be stable and (b) the substrate for the Grignard reagent contain only one functional group capable of facile reaction with the reagent. Let us begin by considering the limitations upon the Grignard reagent.

Since the reagent is a strong base, it will react rapidly with even modest proton donors, converting the C—Mg bond into a C—H bond. Therefore, the synthesis of 1,4-butanediol proposed below is impractical because the desired Grignard reagent will react with itself faster than it will react with ethylene oxide.

POORLY PLANNED SYNTHESIS:

$$BrCH_2CH_2OH + Mg \xrightarrow{ether} BrMgCH_2CH_2OH$$

$$BrMgCH_2CH_2OH + \underset{\diagdown O \diagup}{CH_2{-}CH_2} \xrightarrow{ether} \xrightarrow{H_2O} HOCH_2CH_2CH_2CH_2OH + MgBrOH$$

WHAT ACTUALLY OCCURS:

$$BrMgCH_2CH_2OH \rightarrow HCH_2CH_2\overset{\ominus\ \oplus}{OMg}Br$$

It should be clear that a stable Grignard reagent must lack all of those functional groups which normally react with Grignard reagents! This "obviousism" precludes preparing Grignard reagents possessing any of the following groups: —OH, —SH, —NH_2, —CHO, —CO_2H, —C(O)OR, —CN, —C≡C—H, —NO_2, and —C(O)X (X=Cl, Br, F).

Even though iodides appear to form Grignard reagents more rapidly than do bromides, and bromides more rapidly than chlorides, the difference in reactivity of alkyl halides is normally not great enough to permit the selective reaction of one halogen over another. This reactivity difference *is* important for halobenzenes; bromobenzene rapidly forms a Grignard reagent in ether, while chlorobenzene is much less reactive.

$$Br\text{-}C_6H_{10}\text{-}CH_2CH(Cl)CH_3 \xrightarrow[ether]{Mg} BrMg\text{-}C_6H_{10}\text{-}CH_2CH(Cl)CH_3 + Br\text{-}C_6H_{10}\text{-}CH_2CH(MgCl)CH_3$$

$$\underset{\text{bromobenzene}}{C_6H_5Br} + Mg \xrightarrow[rapid]{THF} C_6H_5MgBr$$

$$\underset{\text{chlorobenzene}}{C_6H_5Cl} + Mg \xrightarrow[slow]{THF} C_6H_5MgCl$$

$$C_6H_5MgBr \text{ or } C_6H_5MgCl \xrightarrow{CH_2O} \xrightarrow{H_3O^{\oplus}} \underset{\text{benzyl alcohol}}{C_6H_5CH_2OH}$$

In discussing limitations placed upon the potential Grignard reagents, we have inadvertently hinted at the limitations placed upon any potential substrate for Grignard attack. If, for example, we wish to synthesize a 2° alcohol by the reaction of a Grignard reagent with an aldehyde, that aldehyde must be free of any other aldehyde groups as well as any other groups that might react with the Grignard reagent. This latter limitation can be divided into two sub-categories, based upon the nature of the reaction between the Grignard reagent and the "other" group.

One kind of interfering functional group does little more than destroy the Grignard reagent by a process that does not materially alter the group itself. This difficulty can be overcome by using an "excess" of Grignard reagent. The other type of interfering functional group also destroys the Grignard reagent but, in so doing, is itself destroyed. Using an excess of the Grignard reagent does not help this second type of problem and may, in fact, make it worse.

By way of illustration of the first type of complication, let us consider the reaction of 4-hydroxybutanal ($HOCH_2CH_2CH_2CHO$) with methylmagnesium chloride. As seen from the equations below, two moles of Grignard reagent are required to form the final dialkoxide.

However, once the dialkoxide is formed, reaction with water produces the desired product, 1,4-pentanediol, by regenerating the hydroxy group which initially consumed one mole of Grignard reagent.

$$\underset{\text{4-hydroxybutanal}}{HOCH_2CH_2CH_2CHO} + CH_3MgCl \xrightarrow{\text{ether}} CH_4 + Cl\overset{\oplus}{Mg}\overset{\ominus}{O}CH_2CH_2CH_2CHO$$

$$Cl\overset{\oplus}{Mg}\overset{\ominus}{O}CH_2CH_2CH_2CHO + CH_3MgCl \xrightarrow{\text{ether}} Cl\overset{\oplus}{Mg}\overset{\ominus}{O}CH_2CH_2CH_2{-}\underset{\underset{CH_3}{|}}{\overset{\overset{\ominus\oplus}{OMgCl}}{\overset{|}{C}}}{-}H$$

$$\xrightarrow{H_2O} \underset{\text{1,4-pentanediol}}{HOCH_2CH_2CH_2CH(OH)CH_3}$$

In order to illustrate the second type of problem, let us consider the reaction of 4-carbomethoxybutanal ($CH_3OC(O)CH_2CH_2CH_2CHO$) with the same Grignard reagent. From the equations below we note a profusion of possible reactions between these two. Initially, the organometallic may react either with the ester functional group (—$C(O)OCH_3$) or with the aldehyde functional group (—CHO). If we assume that both of these groups show *about* the same reactivity toward Grignard reagents, once the *desired salt* is produced it can still react at the ester site with the remainder of the Grignard reagent. Thus, a fraction of the desired first-formed product will be destroyed by the continued addition of just one mole of methylmagnesium chloride.

$$\underset{\text{4-carbomethoxybutanal}}{CH_3O{-}\overset{\overset{O}{\|}}{C}(CH_2)_3CHO} + CH_3MgCl \xrightarrow{\text{ether}} CH_3{-}\overset{\overset{O}{\|}}{C}(CH_2)_3CHO + \underset{\text{desired salt}}{CH_3O{-}\overset{\overset{O}{\|}}{C}(CH_2)_3\underset{\underset{CH_3}{|}}{\overset{\overset{\ominus\oplus}{OMgCl}}{\overset{|}{C}}}{-}H}$$

produced by addition of a portion of one mole of CH_3MgCl

$$CH_3{-}\overset{\overset{O}{\|}}{C}(CH_2)_3CHO + \underset{\text{desired salt}}{CH_3O{-}\overset{\overset{O}{\|}}{C}(CH_2)_3\underset{\underset{CH_3}{|}}{\overset{\overset{\ominus\oplus}{OMgCl}}{\overset{|}{C}}}{-}H} \xrightarrow{CH_3MgCl}$$

$$CH_3{-}\underset{\underset{CH_3}{|}}{\overset{\overset{\ominus\oplus}{OMgCl}}{\overset{|}{C}}}(CH_2)_3CHO + CH_3{-}\overset{\overset{O}{\|}}{C}(CH_2)_3\underset{\underset{CH_3}{|}}{\overset{\overset{\ominus\oplus}{OMgCl}}{\overset{|}{C}}}{-}H + CH_3{-}\overset{\overset{O}{\|}}{C}(CH_2)_3\underset{\underset{CH_3}{|}}{\overset{\overset{\ominus\oplus}{OMgCl}}{\overset{|}{C}}}{-}H$$

Addition of an *excess* of methylmagnesium chloride to *this* mixture would serve to further destroy any of the desired salt ($CH_3OC(O){-}CH_2CH_2CH_2CH(CH_3){-}O^{\ominus}\overset{\oplus}{Mg}Cl$) formed initially! Finally, addition of water would not regenerate the —$C(O)OCH_3$ group in any molecule in which it had been destroyed by reaction with the Grignard reagent. Quite clearly, one might do well to consider an alternative synthesis of $CH_3OC(O)CH_2CH_2CH_2CHCH_3OH$.

But how do we plan the synthesis of an alcohol by the Grignard route? If we apply the "working-in-reverse" approach to the planning of an alcohol synthesis by the Grignard route, we recognize that several different paths may lead to the same product. For example, ethylmethylphenylcarbinol can be prepared by any of the three routes below; the method of choice is dictated by the availability of starting materials.

$$CH_3\overset{O}{\overset{\|}{C}}C_6H_5 + CH_3CH_2MgCl \xrightarrow{\text{ether}} \xrightarrow{H_3O^{\oplus}} CH_3-\underset{C_2H_5}{\underset{|}{\overset{OH}{\overset{|}{C}}}}-C_6H_5$$

ethylmethylphenylcarbinol

$$CH_3CH_2\overset{O}{\overset{\|}{C}}C_6H_5 + CH_3MgCl \xrightarrow{\text{ether}} \xrightarrow{H_3O^{\oplus}} CH_3-\underset{C_2H_5}{\underset{|}{\overset{OH}{\overset{|}{C}}}}-C_6H_5$$

$$CH_3\overset{O}{\overset{\|}{C}}CH_2CH_3 + C_6H_5MgCl \xrightarrow{\text{THF}} \xrightarrow{H_3O^{\oplus}} CH_3-\underset{C_2H_5}{\underset{|}{\overset{OH}{\overset{|}{C}}}}-C_6H_5$$

In general, we can approach the Grignard synthesis of an alcohol by hypothetically removing one substituent from the carbinol carbon, imagining it to be derived from a Grignard reagent, and considering the residue as derived from a carbonyl compound. Several examples of this procedure follow. In the reconstruction of the synthesis of 1-methylcyclohexanol included below, we also see the most common route used to synthesize alcohols which contain the carbinol carbon as part of a ring system.

$$C_6H_5-\underset{CH_3}{\underset{|}{\overset{OH}{\overset{|}{C}}}}-CH_3$$ was made from

$$C_6H_5-\overset{O}{\overset{\|}{C}}-CH_3$$ and CH_3MgCl *or* $CH_3\overset{O}{\overset{\|}{C}}CH_3$ and C_6H_5MgCl

$$H-\underset{H}{\underset{|}{\overset{OH}{\overset{|}{C}}}}-CH_3$$ was made from $H-\overset{O}{\overset{\|}{C}}-H$ and CH_3MgCl

$HOCH_2(CH_2)_3CH_2OH$ was made from 2 moles of $H-\overset{O}{\overset{\|}{C}}-H$ and $ClMg(CH_2)_3MgCl$

1-methylcyclohexanol (cyclohexane ring bearing $\overset{OH}{\overset{|}{C}}-CH_3$) was made from cyclohexanone (ring $C{=}O$) and CH_3MgCl

5. Suggest a Grignard reaction which will produce the following:

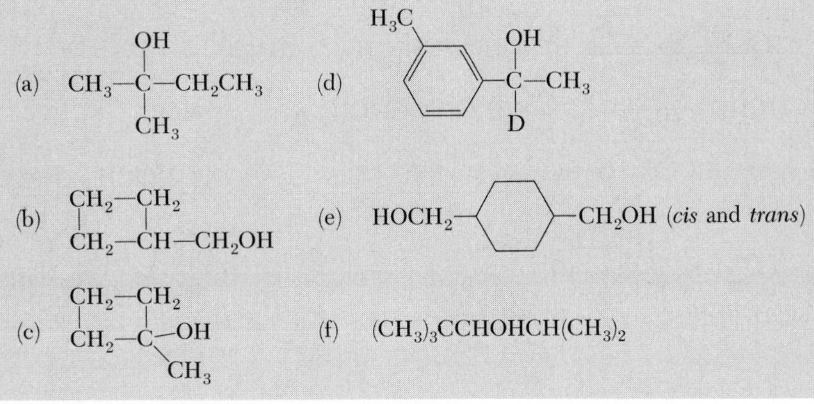

(a) $CH_3-C(OH)(CH_3)-CH_2CH_3$

(b) cyclobutane ring: CH_2-CH_2 / $CH_2-CH-CH_2OH$

(c) cyclobutane ring: CH_2-CH_2 / $CH_2-C(CH_3)-OH$

(d) 3-methylphenyl (H_3C on ring)$-C(OH)(D)-CH_3$

(e) $HOCH_2$–cyclohexane–CH_2OH (*cis* and *trans*)

(f) $(CH_3)_3CCHOHCH(CH_3)_2$

Reduction of Aldehydes and Ketones. The polar carbonyl group can be attacked at the carbon by anions other than carbanions. One such anion is the hydride ion, H:$^{\ominus}$. Actually, a simple hydride such as sodium hydride, NaH, will not reduce ketones or aldehydes. Successful attack by the hydride requires the coordination of the carbonyl oxygen with a Lewis acid. Such coordination enhances the ability of the carbonyl carbon to act as a hydride acceptor by increasing the positive charge on carbon, and may be viewed as a type of electrophilic catalysis.

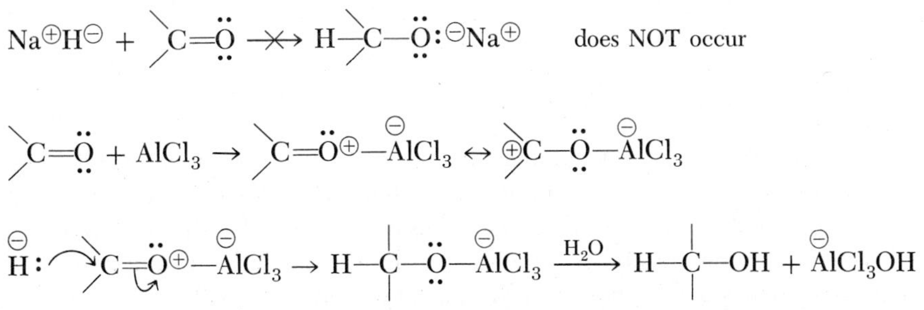

$$Na^{\oplus}H^{\ominus} + \rangle C{=}\ddot{O} \not\rightarrow H{-}\rangle C{-}\ddot{O}{:}^{\ominus}Na^{\oplus} \qquad \text{does NOT occur}$$

$$\rangle C{=}\ddot{O} + AlCl_3 \rightarrow \rangle C{=}\ddot{O}^{\oplus}{-}\overset{\ominus}{Al}Cl_3 \leftrightarrow {}^{\oplus}\rangle C{-}\ddot{O}{-}\overset{\ominus}{Al}Cl_3$$

$$\overset{\ominus}{H}{:}\ \rangle C{=}\ddot{O}^{\oplus}{-}\overset{\ominus}{Al}Cl_3 \rightarrow H{-}\overset{|}{\underset{|}{C}}{-}\ddot{O}{-}\overset{\ominus}{Al}Cl_3 \xrightarrow{H_2O} H{-}\overset{|}{\underset{|}{C}}{-}OH + \overset{\ominus}{Al}Cl_3OH$$

While it is possible to reduce aldehydes and ketones with a mixture of a simple hydride and a Lewis acid, *e.g.*, aluminum chloride, it is much more convenient to use a *complex metal hydride* such as sodium borohydride, $NaBH_4$, or lithium aluminum hydride, $LiAlH_4$. The complex anion coordinates with the carbonyl group and enhances its reactivity as a hydride acceptor. Sodium borohydride has an advantage in that, being less basic than lithium aluminum hydride, it can be used in aqueous or alcoholic media with only slow solvent-induced decomposition. Lithium aluminum hydride is the more potent reductant and reacts violently and instantly with alcohols and water. However, it is soluble in ethereal solvents and is normally used in tetrahydrofuran or ethyl ether.

$$Na^{\oplus}BH_4^{\ominus} + 4CH_3OH \xrightarrow[25°]{\text{slow at}} Na^{\oplus}B(OCH_3)_4^{\ominus} + 4H_2\uparrow$$

$$Li^{\oplus}AlH_4^{\ominus} + 4CH_3OH \xrightarrow[25°]{\text{very rapid at}} Li^{\oplus}Al(OCH_3)_4^{\ominus} + 4H_2\uparrow$$

The examples below point up both the high yields obtained from hydride reduction *and* the fact that one can reduce carbonyl groups in the presence of double bonds without necessarily reducing the double bond.

$$CH_3(CH_2)CHO \xrightarrow[\text{ether}]{LiAlH_4} \xrightarrow{H_2O} CH_3(CH_2)_5CH_2OH \qquad (86\%)$$

$$CH_3\overset{\overset{\displaystyle O}{\|}}{C}CH_2CH_3 \xrightarrow[\text{ether}]{LiAlH_4} \xrightarrow{H_2O} CH_3CHOHCH_2CH_3 \qquad (80\%)$$

$$CH_3CH{=}CHCHO \xrightarrow[\text{ether}]{LiAlH_4} \xrightarrow{H_2O} CH_3CH{=}CHCH_2OH \qquad (70\%)$$

HOCH$_2$ O— … CH$_3$ … OH … O … CH$_3$ … O … O $\xrightarrow[CH_3OH/H_2O]{NaBH_4}$ HOCH$_2$ O— C … O … OH … CH$_3$ … OH … H … CH$_3$ … O … O (85%)

The reduction of 2-butenal, $CH_3CH{=}CHCHO$, must be carried out at low temperatures in order to avoid the reduction of the double bond. Unless care is taken, 2-butenal will react with lithium aluminum hydride to produce 1-butanol. This behavior is typical of double bonds conjugated to carbonyl groups, *e.g.*, α,β-unsaturated aldehydes.

$$\underset{\text{an }\alpha,\beta\text{-unsaturated aldehyde}}{>C{=}C<^{CHO}} \xrightarrow[70^\circ]{LiAlH_4} \xrightarrow{H_2O} \underset{\text{a saturated alcohol}}{>\overset{H}{\overset{|}{C}}-\overset{H}{\overset{|}{C}}<^{CH_2OH}}$$

At this stage in your study of organic chemistry, you are probably better off using sodium borohydride as a reductant in your "graphite and cellulose" syntheses of alcohols, since it will reduce little else beyond aldehydes and ketones. This is in contrast to lithium aluminum hydride, which will reduce a wide variety of functional groups; using lithium aluminum hydride to reduce simple aldehydes and ketones is a kind of chemical "over-kill."

Diborane reduces aldehydes and ketones to the corresponding alcohols. For small quantities of starting material this procedure has the advantage of a simple work-up; the only by-product is boric acid, and that can easily be removed with an alkaline wash.

$$R-\overset{\overset{\displaystyle O}{\|}}{C}-R + BH_3 \rightarrow \left[R-\underset{\oplus}{\overset{\overset{\displaystyle R}{|}}{C}}-O-\underset{\underset{\displaystyle H}{|}}{\overset{\overset{\displaystyle H}{|}}{B^{\ominus}}}-H\right] \rightarrow \left[R-\underset{\underset{\displaystyle H}{|}}{\overset{\overset{\displaystyle R}{|}}{C}}-O-B\begin{smallmatrix}H\\ H\end{smallmatrix}\right] \xrightarrow{R_2CO}$$

$$\left[\left(R-\underset{\underset{\displaystyle R}{|}}{\overset{\overset{\displaystyle H}{|}}{C}}-O\right)_2 BH\right] \xrightarrow{H_2O} R-\underset{\underset{\displaystyle R}{|}}{\overset{\overset{\displaystyle OH}{|}}{C}}-H$$

A summary of those common functional groups which are reduced by lithium aluminum hydride, sodium borohydride, and diborane is found in Table 10–3.

TABLE 10-3 Common Functional Groups Reduced By Lithium Aluminum Hydride, Sodium Borohydride and Diborane

Reductant	Functional Group	Product
$LiAlH_4$	$—CO_2H$	$—CH_2OH$
$LiAlH_4$	—C(O)H (—C(O)R)	$—CH_2OH$ (—CHROH)
$LiAlH_4$	$—CO_2R$	$—CH_2OH + ROH$
$LiAlH_4$	—CONHR	$—CH_2NHR$
$LiAlH_4$	$—CONR_2$	$—CH_2NR_2$
$LiAlH_4$	—CN	$—CH_2NH_2$
$LiAlH_4$	alkyl-NO_2	alkyl-NH_2
$LiAlH_4$	$—CH_2OS(O)_2C_6H_5$	$—CH_3$
$LiAlH_4$	$—CH—CR_2$ (epoxide, O bridging)	$—CH_2—CR_2OH$
$NaBH_4$	—C(O)H (—C(O)R)	$—CH_2OH$ (—CHROH)
B_2H_6	$—CO_2H$	$—CH_2OH$
B_2H_6	$—CO_2R$	$—CH_2OH + ROH$
B_2H_6 (followed by acid)	—CH=CH—	$—CH_2CH_2—$
B_2H_6	$—CH—CR_2$ (epoxide, O bridging)	$—CH(OH)—CHR_2$
B_2H_6	—C(O)H (—C(O)R)	$—CH_2OH$ (—CHROH)

Catalytic Hydrogenation of the Carbonyl Group. Aldehydes and ketones are easily reduced with hydrogen and a catalyst (*e.g.*, Pt) to the corresponding 1° and 2° alcohols. Reduction conditions are similar to those used to reduce carbon-carbon double bonds, and care must be taken if it is necessary to reduce one without reducing another when both are present in the same molecule. Here again, the reaction is a *cis* addition.

$$RC{\equiv}CR > RCHO > RCH{=}CHR > R_2C{=}O$$ order of ease of reduction by H_2 + catalyst

$$\text{cyclohexanone} \xrightarrow[25°;\ C_2H_5OH]{H_2(3\text{ atm});\ Pt} \text{cyclohexanol (H, OH)} \quad (98\%)$$

Alcohols are not *hydrogenolyzed* to hydrocarbons under conditions which reduce the carbonyl group.

$$\text{cyclohexanol (H, OH)} \underset{25°;\ C_2H_5OH}{\overset{H_2\ (3\text{ atm});\ Pt}{\not\longrightarrow}} \text{cyclohexane (H, H)} + H_2O$$ this hydrogenolysis does NOT occur

Oxymercuriation. Alkenes react with mercuric acetate to produce an organomercury compound. The C—Hg bond of organomercurials can be cleaved with sodium borohydride to yield mercury and a C—H bond. The net result of **oxymercuriation** followed by borohydride reduction is the conversion of an alkene to an alcohol in yields that are often 80% or better. Oxymercuriation-demercuriation leads to the Markownikoff hydration of an alkene and minimizes the amount of rearranged products.

$$>C{=}C< + \underset{\text{mercuric acetate}}{Hg(OAc)_2} + H_2O \rightarrow HO—\overset{|}{\underset{|}{C}}—\overset{|}{\underset{|}{C}}—Hg—OAc + HOAc$$ oxymercuriation

$$\xrightarrow[H_2O]{NaBH_4} HO—\overset{|}{\underset{|}{C}}—\overset{|}{\underset{|}{C}}—H + Hg$$ demercuriation

$$CH_3CH_2CH{=}CH_2 + Hg(OAc)_2 \xrightarrow{H_2O} \underset{\substack{| \\ OH}}{CH_3CH_2CH}{-}\underset{\substack{| \\ HgOAc}}{CH_2} \xrightarrow[H_2O]{NaBH_4} \underset{\substack{| \\ OH}}{CH_3CH_2CH}{-}CH_3$$

$$\text{(norbornene)} + Hg(OAc)_2 + H_2O \rightarrow \xrightarrow[H_2O]{NaBH_4} \text{(exo-norborneol; OH, H, H, H shown)}$$

Commercial Synthesis of Alcohols. It is interesting to compare these laboratory syntheses with the processes presented below, each of which is used to turn out millions of pounds of product per year at the lowest possible cost.

Methanol carries the pseudonym *wood alcohol* because until *ca.* 1925 it was prepared by the destructive distillation of wood. (*Destructive distillation* implies heating to high temperatures in the absence of an oxidizing atmosphere, *i.e.*, air.) It is now made largely by the catalytic hydrogenation of carbon monoxide.

$$2H_2 + CO \xrightarrow[400°;\ cat.]{3000\ psi} CH_3OH$$

Ethanol can be prepared by the fermentation of carbohydrates. The grape, the fig, and the date have been harvested since antiquity, and ethanol production by the fermentation of these fruits is undoubtedly the oldest large-scale organic synthesis known to man.

Commercial alcohol is usually a constant boiling mixture (*i.e.*, one that cannot be separated by distillation into fractions with different boiling points) of ethanol and water (95.57/4.43 weight percent). Since this mixture has a lower boiling point than does pure ethanol (also called *absolute* or *200 proof alcohol*) (78.2° *vs.* 78.3°), 100% ethanol cannot be prepared by simple distillation of a mixture composed largely of water and ethanol; it never gets beyond 95.57% ethanol. The removal of that last 4.43% of water is not necessary for many uses but, if required, it can be accomplished by treating the wet alcohol with quicklime (calcium oxide, CaO) or by other means. Denatured alcohol is ethyl alcohol that has been made unfit for use as a beverage by the addition of some contaminant (*e.g.*, methanol).

$$CH_3CH_2OH + HOH + CaO \rightarrow CH_3CH_2OH + Ca(OH)_2$$

The large demand for ethanol has now led to the preparation of significant quantities by the hydration of ethylene.

$$CH_2{=}CH_2 \xrightarrow[1000\ psi]{H_3O^{\oplus},\ 300°} CH_3CH_2OH$$

Isopropanol is made largely by the hydration of propene under conditions similar to those used to synthesize ethanol.

$$CH_2{=}CHCH_3 \xrightarrow[1000\ psi]{H_3O^{\oplus},\ 300°} CH_3CHOHCH_3$$

n-Butanol is made, along with acetone, by the action of *Clostridium acetobutylicum* Weizmann (a bacterium) upon carbohydrates—another fermentation process. This synthesis was developed by Chaim Weizmann, a man well known for his role in the development of the state of Israel as well as for his significant contributions to chemistry.

10.5 ALCOHOLS AS ACIDS

Since alcohols have a proton bonded to an electronegative element (*i.e.*, oxygen) they can serve as proton donors to even modestly strong bases. Moreover, the anion produced after the proton is lost is sufficiently stable to permit the isolation and storage of metal alkoxides. Today a number of salts of alcohols, *e.g.*, sodium ethoxide, sodium methoxide, potassium *t*-butoxide, and thallium ethoxide, are commercially available. The first three of these, at least, are conveniently made in the laboratory by "dissolving" the appropriate metal in the alcohol of choice.

$$2CH_3CH_2OH + 2Na \rightarrow H_2 + 2Na^{\oplus}CH_3CH_2O^{\ominus}$$

(sodium ethoxide *or* sodium ethylate)

For many synthetic purposes, an *excess* of alcohol is treated with metal, and the alcoholic solution of the salt is used "as is." This is why one often hears of "sodium ethoxide in ethanol" but rarely of "sodium ethoxide in methanol." (What would be another reason for not using a solution of "sodium ethoxide in methanol"?)

One never encounters reactions that call for the use of "sodium methoxide in water," or a similar alkoxide in water, because the water is a sufficiently strong acid (compared to the alcohol) to convert the alkoxide almost completely to the alcohol. Such solutions would, in fact, be largely solutions of sodium hydroxide in alcohol/water mixtures.

$$M^{\oplus}OR^{\ominus} + H_2O \rightleftarrows M^{\oplus}OH^{\ominus} + ROH$$

Is there a variation in the acidity of solutions of alcohols? Indeed there is, with tertiary alcohols usually being the least acidic. The decreased acidity of *t*-butyl alcohol compared to that of ethyl alcohol is due to the hindrance caused by the alkyl group to the solvation of the alkoxide anion. This lack of solvation increases the energy content of the anion, raises the E_{act} for proton removal, and increases the energy difference between alcohol and salt (Figure 10–1). We should note that the order of alcohol acidity in the vapor phase, where solvation is unimportant, is the reverse of what it is in solution.

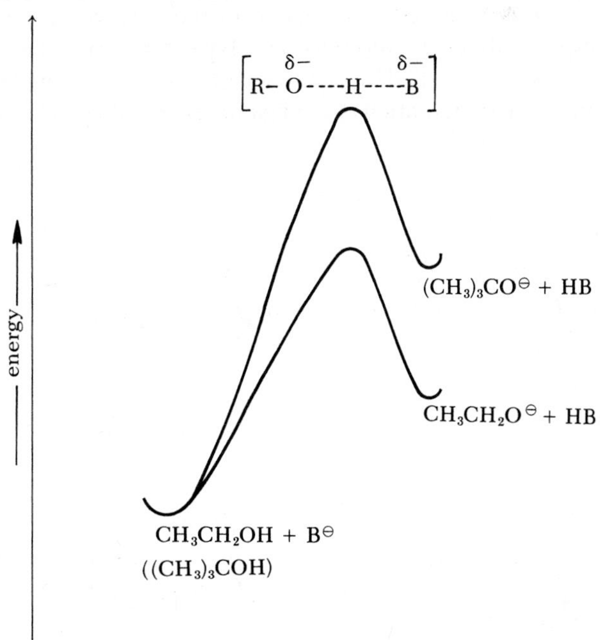

Figure 10-1 Effect of increasing alkylation upon alcohol acidity. These results apply to reactions in solution.

10.6 CONVERSION OF C—OH TO C-LEAVING GROUP

One of the most important chemical transformations of alcohols involves conversion of the hydroxy group to a good leaving group. We have already seen (p. 136) that protonation of the hydroxy group enhances its leaving ability; in this chapter we will focus attention upon the conversion of alcohols to (i) halides and (ii) esters.

Halide Formation. In Chapter 5 we examined the conversion of alcohols to alkyl halides by the action of hydrohalic acids. This reaction, which may follow either an S_N1 or an S_N2 path, is reviewed below.

$$CH_3CH_2OH + HBr \rightarrow CH_3CH_2\overset{\oplus}{O}H_2\ Br^{\ominus} \xrightarrow{-H_2O} CH_3CH_2Br \qquad S_N2$$

$$(CH_3)_3COH + HX \rightarrow (CH_3)_3C\overset{\oplus}{O}H_2\ X^{\ominus} \xrightarrow{-H_2O} \left[(CH_3)_3C^{\oplus}\right] \xrightarrow{X^{\ominus}} (CH_3)_3CX \qquad S_N1$$

$HI > HBr > HCl > HF$ order of reactivity of HX toward alcohols

While these approaches to alkyl halides have sound teaching value, they are often not the best route to alkyl halides. Many chemists prefer the reaction of an alcohol with thionyl chloride, $SOCl_2$, as a method for preparing alkyl chlorides. This reaction is rapid and relatively free of side reactions, and the by-products are gases under the conditions of the reaction. Because sulfur dioxide and hydrogen chloride are gases and escape during the reaction, the reaction is driven forward (*i.e.*, toward product) by application of the ideas summarized in LeChatelier's principle.

$$R{-}OH + SOCl_2 \xrightarrow[\text{heat}]{\text{ether}} R{-}Cl + SO_2\uparrow + HCl\uparrow$$

The reaction often proceeds with retention of configuration at the carbinol carbon. Decomposition of the first-formed chlorosulfite to the observed product is the accepted reason for the stereospecificity of the reaction. This type of reaction, *i.e.*, collapse of an ion pair (oppositely charged ions surrounded by a common group of solvent molecules or "solvent shell") to produce substitution with retention of configuration, is called an S_Ni reaction (Substitution *N*ucleophilic *i*nternal).

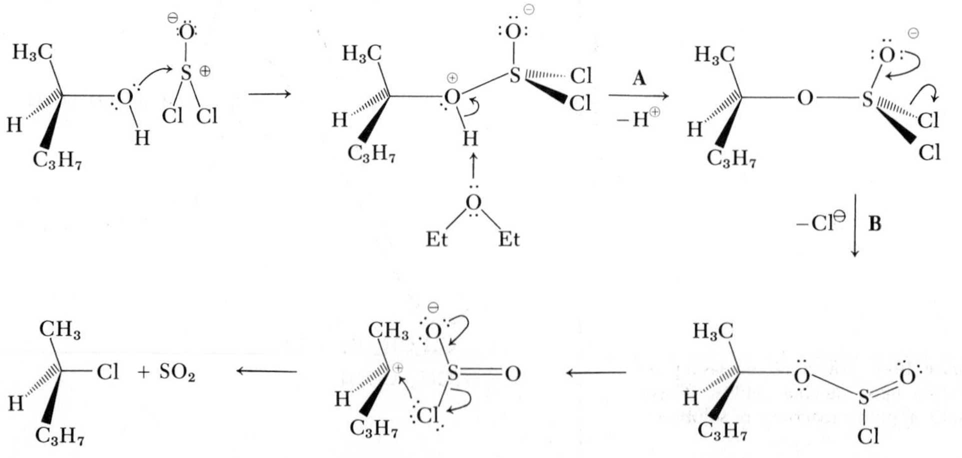

an ion pair

a chlorosulfite

Whether **A** occurs before **B**, or vice versa, and the exact amount of time separating these two steps, are matters of conjecture; but the answer does not alter the basic scheme. The intermediacy of the chlorosulfite has been established by isolating it from a mixture of alcohol and thionyl chloride held at low temperature, and then showing that the chlorosulfite decomposes, upon warming, to the chloride.

6. Reaction of (**R**)-2-butanol with thionyl chloride actually produces *two* chlorosulfites. These have the same molecular formula. When placed in a polarimeter, both of them are found to be optically active, but they have different rotations. Both of them decompose to the same product, (**R**)-2-chlorobutane. What is the significance of these data? Why are there two chlorosulfites?

An interesting modification of this alkyl chloride synthesis is achieved when a base that is a poor nucleophile (*e.g.*, pyridine) is present during the reaction. Under these circumstances the hydrogen chloride is trapped by the base, and the chloride ion so produced participates in an S_N2 displacement of the O—SO—Cl moiety to produce an alkyl chloride with an *inverted* configuration.

pyridine N: + H—Cl ⇌ pyridinium N⊕—H Cl⊖

:Cl:⊖ + (CH_3)(H)(C_3H_7)C—O—S⊕(O⊖)—Cl ⟶ Cl—C(CH_3)(H)(C_3H_7) + SO_2 + Cl⊖

Because of the short shelf life of thionyl bromide, $SOBr_2$, alkyl bromides are usually made by reacting an alcohol with phosphorus tribromide, PBr_3. Unlike the thionyl chloride reaction, this process produces a non-volatile by-product (phosphorous acid, H_3PO_3), which must be removed by extraction with aqueous base. Nonetheless, the reaction of an alcohol with PBr_3 is among the best methods for synthesizing alkyl bromides (however, see problems 23–38 for some more detail).

$$3ROH + PBr_3 \rightarrow 3RBr + H_3PO_3$$

The mechanism for the reaction involves an initial attack by the alcohol upon the phosphorus tribromide, followed by either an S_N1 or an S_N2 displacement of a phosphorus-containing moiety by the bromide ion. Primary alcohols react by the S_N2 process, while tertiary alcohols react by the S_N1 process.

C—Ö(H): + :PBr₃ ⟿ —C—O⊕(H)—P(Br)₂: Br⊖

:Br:⊖ + C—O⊕(H)—PBr₂ ⟶ Br—C + $HOPBr_2$ S_N2 decomposition

C—O⊕(H)—PBr₂ $\xrightarrow{-HOPBr_2}$ —C⊕ $\xrightarrow{Br^{\ominus}}$ —C—Br S_N1 decomposition

The use of phosphorus triiodide, PI_3, converts an alcohol to an alkyl iodide. The mechanism is similar to that followed by phosphorus tribromide.

$$3ROH + PI_3 \rightarrow 3RI + H_3PO_3$$

Ester Formation. An ester is the adduct formed by the reaction of an oxyacid (*e.g.*, $HONO_2$, nitric acid) with an alcohol; loss of water accompanies this reaction. Methyl nitrate is an *inorganic* ester, since it is formed from an alcohol and an *inorganic* acid.

$$CH_3\ddot{O}H + HO\text{—}N^{\oplus}(=\ddot{O})\text{—}O^{\ominus} \rightarrow CH_3\overset{\oplus}{O}H\text{—}\overset{\oplus}{N}(OH)(O^{\ominus})\text{—}\ddot{O}^{\ominus} \rightarrow CH_3\ddot{O}\text{—}\overset{\oplus}{N}(OH)(O^{\ominus})\text{—}\ddot{O}\text{—}H \xrightarrow{HNO_3}$$

$$CH_3\ddot{O}\text{—}\overset{\oplus}{N}(OH)(O^{\ominus})\text{—}\overset{\oplus}{O}H_2 \rightarrow CH_3\text{—}O\text{—}N^{\oplus}(=O)\text{—}O^{\ominus}$$

methyl nitrate

While esters can be made by the reaction of an alcohol with an acid, they are more often prepared by reactions of an alcohol with a derivative of an acid, for example, an acid chloride. The synthesis of trimethyl phosphate, using the acid chloride (actually a *tri*chloride) of phosphoric acid, illustrates this approach.

$$CH_3\ddot{O}H + \ddot{O}^{\ominus}\text{—}P^{\oplus}Cl_3 \xrightarrow{-Cl^{\ominus}} CH_3\overset{\oplus}{O}H\text{—}P^{\oplus}(O^{\ominus})Cl_2 \xrightarrow{-H^{\oplus}} CH_3O\text{—}P^{\oplus}(\ddot{O}^{\ominus})Cl_2$$

phosphorus oxychloride

$$\downarrow \text{etc.}$$

$$(CH_3O)_3\overset{\oplus}{P}\text{—}\ddot{O}\text{:}^{\ominus}$$

trimethyl phosphate

$OPCl_3$ is the acid chloride of phosphoric acid, H_3PO_4

Just as phosphoric acid and its halogen derivatives form phosphate esters, so sulfuric acid and its halide derivatives form *sulfate esters.*

$$CH_3OH + HO\text{—}SO_2\text{—}Cl \rightarrow CH_3O\text{—}SO_2\text{—}OH + HCl\uparrow$$

chlorosulfonic acid; methyl bisulfate (methyl hydrogen sulfate)

$$2CH_3\text{—}O\text{—}H + H_2SO_4 \text{ (or } SO_2Cl_2) \rightarrow CH_3\text{—}O\text{—}SO_2\text{—}O\text{—}CH_3 + 2H_2O \text{ (or } 2HCl)$$

dimethyl sulfate

In addition to these sulfates, we should note the existence of the *sulfonates*, esters of organic analogs of sulfuric acid which possess a C—S bond. We already have alluded to the fact that the tosylate (*p*-toluenesulfonate) group is a good leaving group (p. 131).

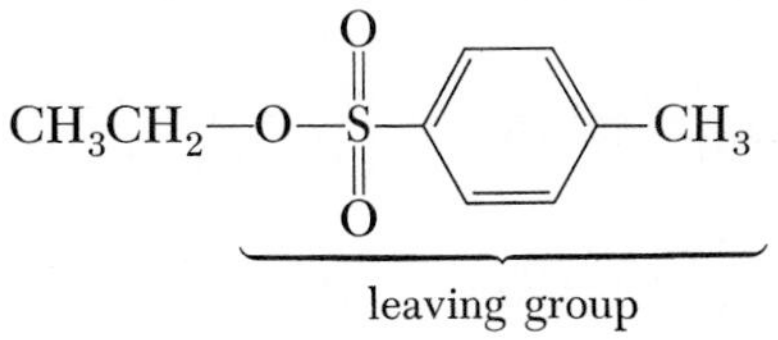

$$CH_3CH_2\!-\!O\!-\!\overset{O}{\underset{O}{\overset{\|}{\underset{\|}{S}}}}\!-\!C_6H_4\!-\!CH_3$$

ethyl *p*-toluenesulfonate (ethyl tosylate), a sulfonate ester

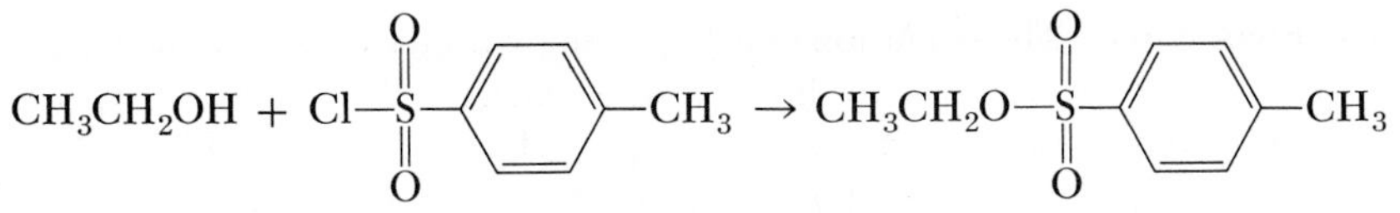

leaving group (the $-O-SO_2-C_6H_4-CH_3$ portion)

$$CH_3CH_2OH + Cl\!-\!\overset{O}{\underset{O}{\overset{\|}{\underset{\|}{S}}}}\!-\!C_6H_4\!-\!CH_3 \rightarrow CH_3CH_2O\!-\!\overset{O}{\underset{O}{\overset{\|}{\underset{\|}{S}}}}\!-\!C_6H_4\!-\!CH_3$$

p-toluenesulfonyl chloride (tosyl chloride)

The inorganic portions of most inorganic esters are good leaving groups, especially when compared to OH$^{\ominus}$. This is due mainly to the resonance stabilization of the resultant anion.

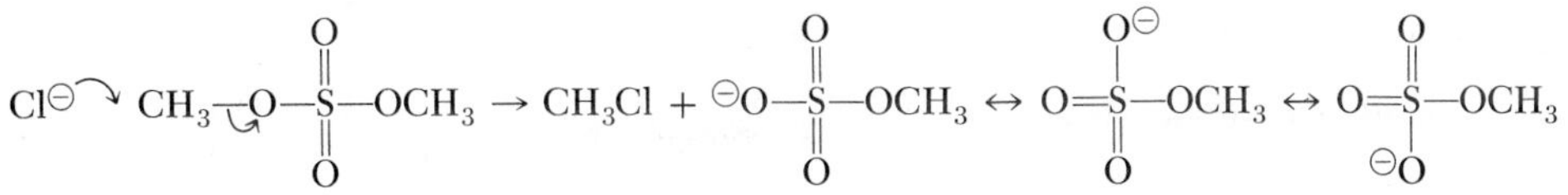

$$Cl^{\ominus} \curvearrowright CH_3\!-\!O\!-\!\overset{O}{\underset{O}{\overset{\|}{\underset{\|}{S}}}}\!-\!OCH_3 \rightarrow CH_3Cl + {}^{\ominus}O\!-\!\overset{O}{\underset{O}{\overset{\|}{\underset{\|}{S}}}}\!-\!OCH_3 \leftrightarrow O\!=\!\overset{O^{\ominus}}{\underset{O}{\overset{|}{\underset{\|}{S}}}}\!-\!OCH_3 \leftrightarrow O\!=\!\overset{O}{\underset{{}^{\ominus}O}{\overset{\|}{\underset{|}{S}}}}\!-\!OCH_3$$

Alcohols can be converted to *carboxylate esters* by refluxing a mixture of the alcohol, a carboxylic acid, and a trace of acid (commonly sulfuric acid); provision usually is made for the removal of water as it is formed during the reaction. (How is this an application of LeChatelier's principle?)

$$CH_3CH_2OH + CH_3CO_2H \underset{H^{\oplus}}{\overset{H^{\oplus}}{\rightleftharpoons}} CH_3CH_2O\!-\!\overset{O}{\overset{\|}{C}}CH_3 + H_2O$$

a carboxylate ester

esterification of ethyl alcohol

The C=O linkage of carboxylate esters usually is more susceptible to addition by a Lewis base than the alkyl group is susceptible to the S_N reaction. (This has already been used to advantage in the synthesis of 3° alcohols *via* the reaction of Grignard reagents with carboxylate esters.) As a result of this, and because of the facile departure of phosphates and tosylates, carboxylates are not used when R—OH must be converted to R-Leaving.

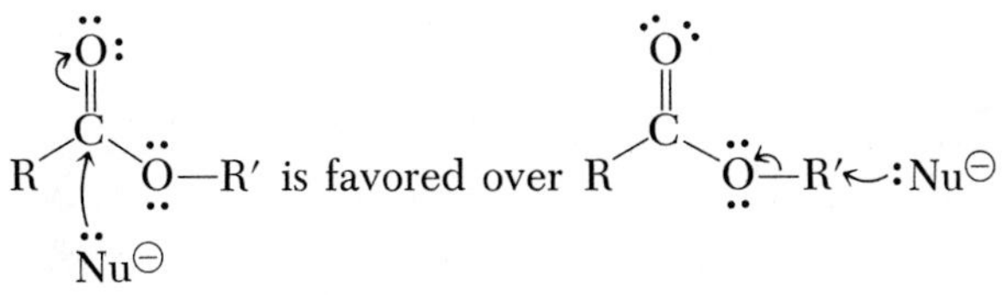

$$R\!-\!\overset{:\ddot{O}:}{\overset{\|}{C}}\!-\!\ddot{O}\!-\!R' \text{ (attacked at C by } :Nu^{\ominus}) \text{ is favored over } R\!-\!\overset{:\ddot{O}:}{\overset{\|}{C}}\!-\!\ddot{O}\!-\!R' \leftarrow :Nu^{\ominus}$$

10.7 OXIDATION OF ALCOHOLS

Just as aldehydes and ketones may be reduced to 1° and 2° alcohols, respectively, so may 1° and 2° alcohols be oxidized to aldehydes and ketones, respectively. Tertiary alcohols, lacking a hydrogen on the carbinol carbon, are inert to the common oxidants under mild conditions.

$$RCH_2OH \underset{[H]}{\overset{[O]}{\rightleftharpoons}} RCHO \qquad \begin{array}{l}\text{[O]: unspecified oxidant}\\ \text{[H]: unspecified reductant}\end{array}$$

$$R_2CHOH \underset{[H]}{\overset{[O]}{\rightleftharpoons}} R_2CO$$

Aldehyde Formation. The oxidation of a 1° alcohol with the most commonly employed oxidants, Cr^{VI}-containing species, usually involves the reaction of the alcohol with chromic acid, H_2CrO_4. The reaction begins with formation of a chromate ester, which subsequently loses $HCrO_4^{\ominus}$ in a rate-determining step.

The aldehyde produced by this oxidation is hydrated to form, in low concentration, the 1,1-diol. This forms another chromate ester which, after another E2 reaction, is converted to a carboxylic acid. Thus, unless steps are taken to prevent over-oxidation, 1° alcohols are quickly converted to carboxylic acids. The amount of over-oxidation can be reduced by carrying out the oxidation at a temperature above the boiling point of the intermediate aldehyde but below the boiling point of the alcohol. (Intermolecular hydrogen bonding raises the boiling point of the alcohol relative that of the aldehyde.) Under these conditions, the aldehyde distills out of the reaction vessel before over-oxidation can occur.

net equation: $RCH_2OH + H_2CrO_4 \rightarrow RCHO + H_2CrO_3 + H_2O$

mechanism:

$$R\!-\!CH_2\!-\!OH + HCrO_4^{\ominus} + H^{\oplus} \rightleftharpoons R\!-\!CH_2\!-\!O\!-\!Cr(=O)_2\!-\!OH + H_2O$$

$$H_2O: + R\!-\!CH_2\!-\!O\!-\!Cr(=O)_2\!-\!OH \xrightarrow{E2} R\!-\!CH\!=\!O + HCrO_3^{\ominus} + H_3O^{\oplus}$$

an aldehyde

over-oxidation:

$$H_2O + RCHO \rightleftharpoons RCH(OH)_2 \xrightarrow{H_2CrO_4} R\!-\!CH(OH)\!-\!O\!-\!Cr(=O)_2\!-\!OH$$

$$H_2O: + R\!-\!CH(OH)\!-\!O\!-\!Cr(O)_2OH \rightarrow R\!-\!C(=O)\!-\!OH + H_3O^{\oplus} + HCrO_3^{\ominus}$$

carboxylic acid

A relatively mild oxidation employs a dilute solution of CrO_3 (chromium trioxide) in dilute sulfuric acid as the oxidant. The alcohol is dissolved in acetone and treated with the stoichiometric amount of the oxidant, the temperature being maintained at 15 to 20°. This procedure is called a *Jones oxidation*.

Aldehydes also can be produced in excellent yield by the oxidation of a 1° alcohol with the complex formed by reacting CrO_3 with pyridine (a combination called *Sarett's reagent*). The oxidation is usually performed in methylene chloride solvent at 25°. Sarett's reagent is quite valuable in oxidizing acid-sensitive compounds, since pyridine is a base.

$$\underset{}{CrO_3} + \underset{\text{pyridine}}{2C_5H_5N} \rightarrow \underset{\text{Sarett's reagent}}{CrO_3 \cdot (C_5H_5N)_2}$$

$$CH_3(CH_2)_5{-}CH_2OH \xrightarrow[CH_2Cl_2/25^\circ]{\text{Sarett's reagent}} CH_3(CH_2)_5{-}CHO \quad (93\%)$$

$$C_6H_5CH{=}CH{-}CH_2OH \xrightarrow[\text{pyridine}/25^\circ]{CrO_3} C_6H_5CH{=}CH{-}CHO \quad (81\%)$$

A special procedure, using freshly prepared manganese dioxide, is capable of oxidizing a 1° alcohol *adjacent to a double bond* to an aldehyde without oxidizing ordinary 1° alcohols.

$$2KMnO_4 + 3MnSO_4 + 4NaOH \rightarrow \underset{\text{manganese dioxide}}{5MnO_2} \downarrow + K_2SO_4 + 2Na_2SO_4$$

$$CH_2{=}CH{-}CH_2OH \xrightarrow[25^\circ]{MnO_2} \underset{\text{acrolein}}{CH_2{=}CH{-}CHO}$$

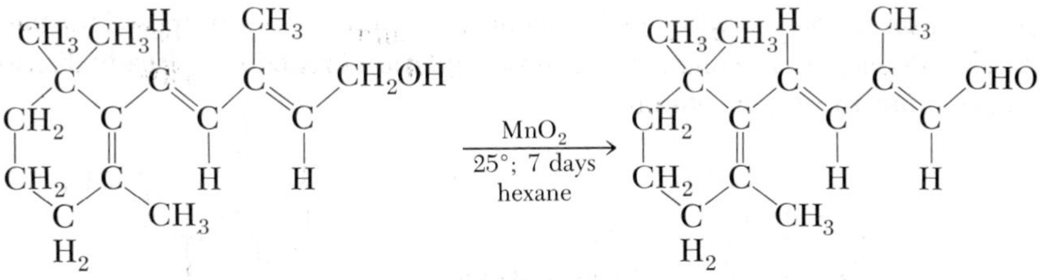

[Structure: vitamin A (ring with CH_3, CH_3, CH_2 groups and side chain $-C(H)=C(CH_3)-C(H)=C(H)-CH_2OH$)] $\xrightarrow[\substack{25^\circ;\ 7\ \text{days} \\ \text{hexane}}]{MnO_2}$ [same structure with CHO in place of CH_2OH]

vitamin A

An alcohol may also be oxidized by first converting it to a chloride or tosylate and then oxidizing this derivative with dimethyl sulfoxide. The sequence is completed by an E2 elimination of dimethyl sulfide, CH_3SCH_3, and concomitant formation of the carbonyl group.

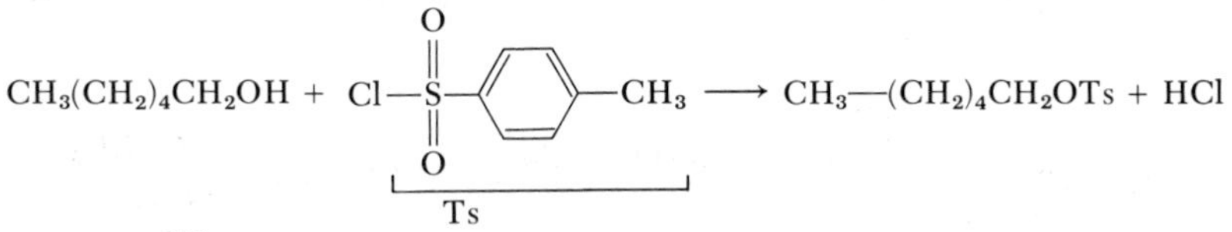

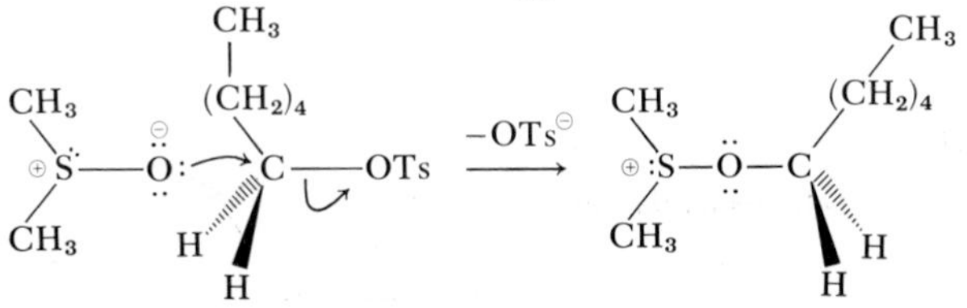

$$CH_3(CH_2)_4CH_2OH + \underbrace{Cl{-}SO_2{-}C_6H_4{-}CH_3}_{Ts} \longrightarrow CH_3{-}(CH_2)_4CH_2OTs + HCl$$

$$(CH_3)_2\overset{\oplus}{S}{-}\ddot{\underset{..}{O}}{:}^{\ominus} + CH_3(CH_2)_4{-}CH_2{-}OTs \xrightarrow{-OTs^{\ominus}} (CH_3)_2\overset{\oplus}{:S}{-}\ddot{\underset{..}{O}}{-}CH_2(CH_2)_4CH_3$$

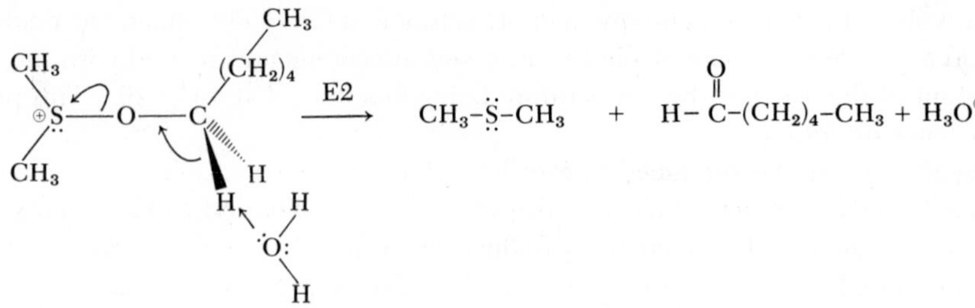

$$CH_3-\ddot{S}-CH_3 + H-C(=O)-(CH_2)_4-CH_3 + H_3O^{\oplus}$$

Ketones. Since ketones can be oxidized only with difficulty, over-oxidation of 2° alcohols (*i.e.*, oxidation to something other than a ketone) is usually not a problem. Because of this, simple 2° alcohols are readily converted to ketones by oxidation with (a) $K_2Cr_2O_7/H_2SO_4/H_2O$ or (b) $CrO_3/CH_3CO_2H/H_2O$.

$$3R_2CHOH + K_2Cr_2O_7 + 4H_2SO_4 \rightarrow 3R_2CO + Na_2SO_4 + Cr_2(SO_4)_3 + 7H_2O$$

$$3R_2CHOH + 2CrO_3 + 6CH_3CO_2H \rightarrow 3R_2CO + 2Cr(CH_3CO_2)_3 + 6H_2O$$

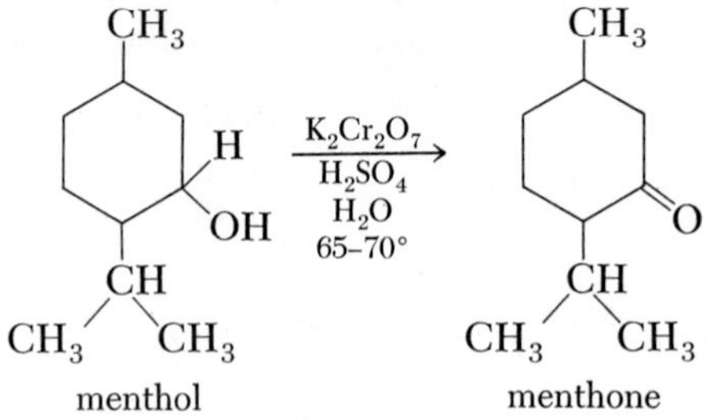

menthol $\xrightarrow[H_2O,\ 65\text{–}70°]{K_2Cr_2O_7,\ H_2SO_4}$ menthone

The mechanism for the oxidation of 2° alcohols is similar to that proposed for the oxidation of 1° alcohols to aldehydes, and one can use those oxidation procedures already discussed under **Aldehyde Formation** to prepare ketones.

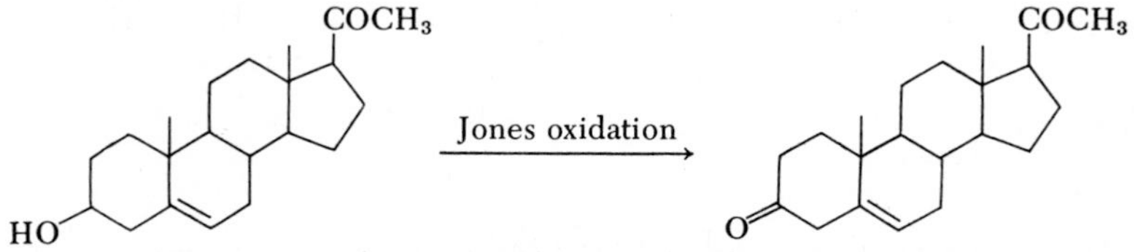

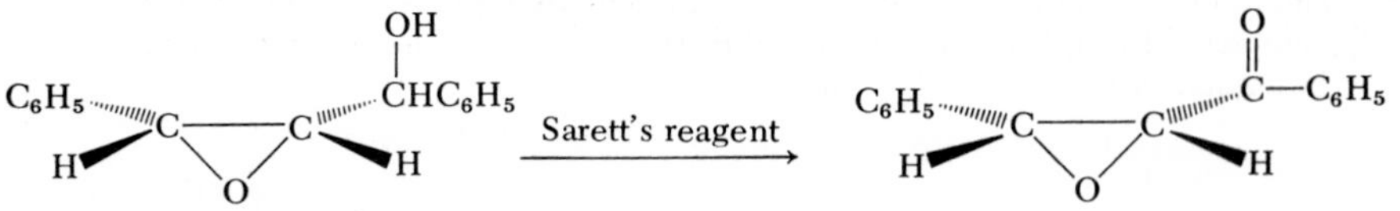

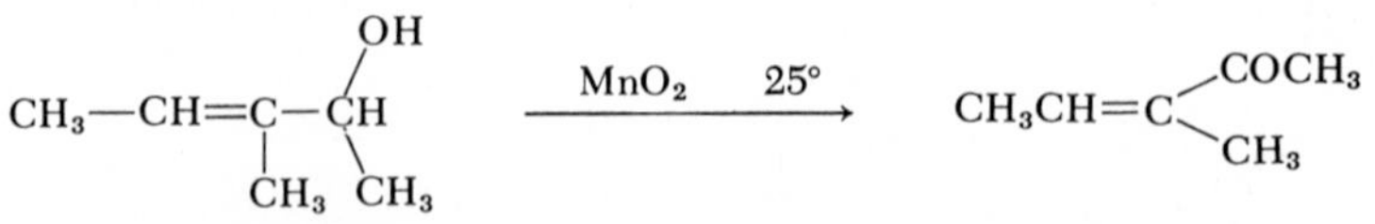

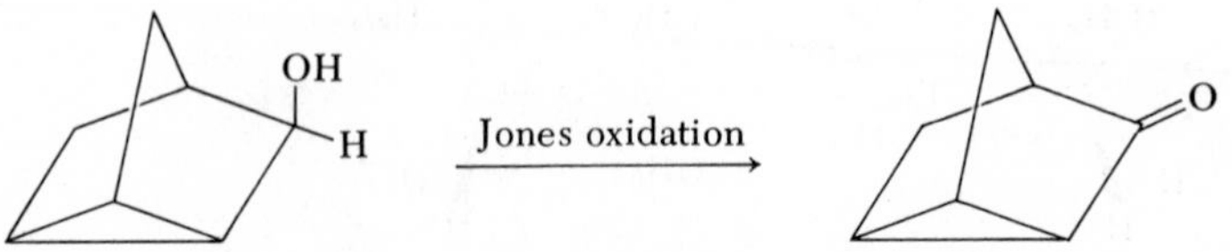

Jones oxidation

$$C_6H_5-CH(O)CH-CH(OH)C_6H_5 \xrightarrow{\text{Sarett's reagent}} C_6H_5-CH(O)CH-C(=O)-C_6H_5$$

$$CH_3-CH=C(CH_3)-CH(OH)CH_3 \xrightarrow{MnO_2 \quad 25°} CH_3CH=C(CH_3)COCH_3$$

Jones oxidation

$$CH_3-\underset{H}{\overset{OH}{C}}-C{\equiv}C(CH_2)_4CH_3 \xrightarrow[5°;\ acetone]{CrO_3;\ H_3O^{\oplus}} CH_3-\overset{O}{\overset{\|}{C}}-C{\equiv}C(CH_2)_4CH_3$$

Potassium permanganate oxidizes 1° alcohols to carboxylic acids, *via* the aldehyde, and 2° alcohols to the ketones, with ease. These reactions also proceed through inorganic esters (this time they are permanganates), but they may be more complex than Cr^{VI} oxidations and will not be discussed further.

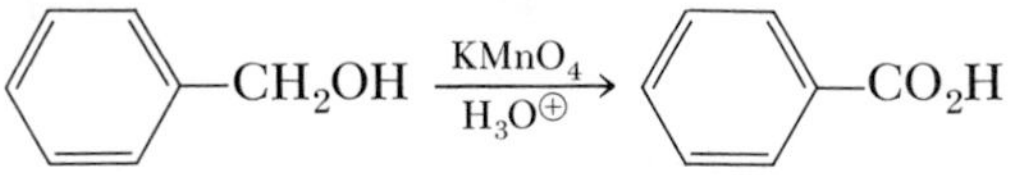

$$C_6H_5-CH_2OH \xrightarrow[H_3O^{\oplus}]{KMnO_4} C_6H_5-CO_2H$$

benzyl alcohol — benzoic acid

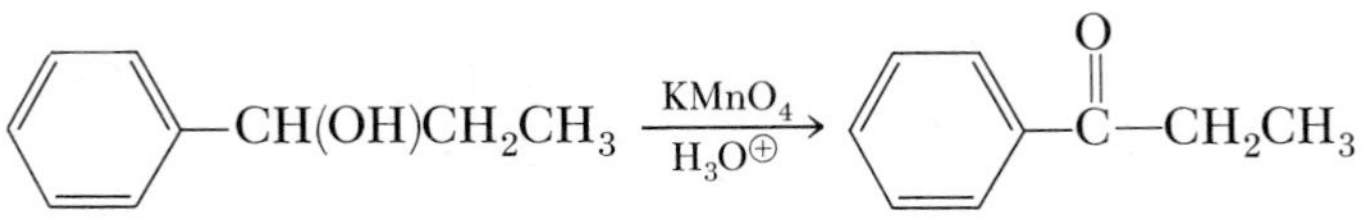

$$C_6H_5-CH(OH)CH_2CH_3 \xrightarrow[H_3O^{\oplus}]{KMnO_4} C_6H_5-\overset{O}{\overset{\|}{C}}-CH_2CH_3$$

1-phenyl-1-propanol — ethyl phenyl ketone

It is important for the experimentalist to realize that *many organic compounds will explode violently when placed in contact with potent oxidants.* For this reason, oxidations that employ potassium permanganate, perchloric acid, and similar reagents are never carried out in the absence of solvent. The solvent moderates the reaction by dissipating heat as it is evolved.

7. Suggest a substrate-oxidant combination that will produce:
   (a) CH_3COCH_3
   (b) $CH_3CH{=}CHCHO$
   (c) 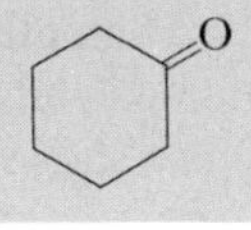 [cyclohexanone]
   (d) 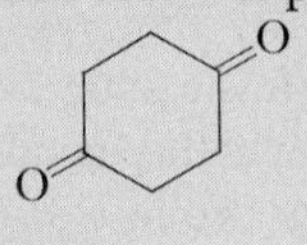 [1,4-cyclohexanedione]
   (e) 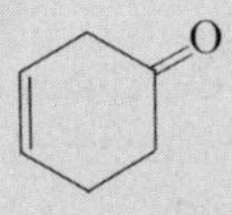 [2-cyclohexen-1-one]

10.8 HOW CAN ONE DISTINGUISH AMONG 1°, 2°, AND 3° ALCOHOLS BY CHEMICAL MEANS? THE LUCAS TEST

Both HI and HBr react rather rapidly with alcohols. This high reactivity reflects (a) the strength of these acids and (b) the nucleophilicity of their anions. HCl is less acidic than either HI or HBr, *and* $Cl^{\ominus}$ is less nucleophilic than either $I^{\ominus}$ or $Br^{\ominus}$. This combination permits the use of hydrochloric acid in a facile determination of the character of an alcohol, *i.e.*, whether it is 1°, 2°, or 3°. If a mixture of concentrated hydrochloric acid and anhydrous zinc chloride, a combination known as the **Lucas reagent,** is added to an alcohol, the following observations are made:

3° alcohol: immediate reaction; heat evolved; oil produced immediately (*i.e.*, a two-layer system is formed)

2° alcohol: reaction within five minutes but not immediate; evolution of heat not obvious; oil produced on standing for no more than five minutes

1° alcohol: no reaction even after standing at room temperature for one hour

This order of reactivity mirrors the order of carbocation stability (*i.e.*, 3° > 2° > 1°) and reflects the ease of conversion of an alcohol to its chloride *via the corresponding carbocation.*

$$CH_3OH + HCl \xrightarrow{ZnCl_2} N.R.$$

$$CH_3\text{—}\underset{H}{\overset{CH_3}{C}}\text{—}OH + HCl \xrightarrow{ZnCl_2} CH_3\text{—}\underset{H}{\overset{CH_3}{C}}{}^{\oplus} \xrightarrow{Cl^{\ominus}} CH_3\text{—}\underset{H}{\overset{CH_3}{C}}\text{—}Cl$$

2° cation

$$CH_3\text{—}\underset{CH_3}{\overset{CH_3}{C}}\text{—}OH + HCl \xrightarrow{ZnCl_2} CH_3\text{—}\underset{CH_3}{\overset{CH_3}{C}}{}^{\oplus} \xrightarrow{Cl^{\ominus}} CH_3\text{—}\underset{CH_3}{\overset{CH_3}{C}}\text{—}Cl$$

3° cation

The oil produced in these reactions is, of course, the alkyl chloride. Since "oil formation" is the most recognizable sign of reaction, it follows that the alcohol must be soluble in the Lucas reagent in order to detect a positive reaction. Actually, most simple alcohols are soluble in the Lucas reagent, probably because of the formation of an oxonium salt.

$$\underset{\text{alcohol}}{ROH} + HCl \rightarrow \underset{\text{oxonium salt}}{R\overset{\oplus}{O}H_2\, Cl^{\ominus}}$$

The results of the Lucas test may be misleading, because it is really based upon how rapidly a carbocation is produced and not upon whether an alcohol is 1°, 2°, or 3°. Thus, allyl alcohol, which is a 1° alcohol that forms a very stable cation (p. 140), reacts immediately with the Lucas reagent. The same is true of benzyl alcohol, which also forms a very stable cation.

$$H_2C{=}CHCH_2OH + \text{Lucas reagent} \xrightarrow[\text{fast}]{\text{very}} H_2C{=}CHCH_2Cl$$

$$C_6H_5CH_2OH,\ H_2C{=}CHCH_2OH > CH_3\text{—}\underset{CH_3}{\overset{CH_3}{C}}\text{—}OH > CH_3\text{—}\underset{H}{\overset{CH_3}{C}}\text{—}OH > CH_3\text{—}\underset{H}{\overset{H}{C}}\text{—}OH$$

order of reactivity toward Lucas reagent

10.9 SPECTRAL PROPERTIES OF SIMPLE ALCOHOLS

Since simple alcohols are transparent in the ultraviolet, only their ir and nmr spectra will be considered.

Infrared Spectra. The most characteristic spectral features of alcohols are the *O—H stretching vibration* (3650 to 3000 cm^{-1}) and the *C—OH stretching vibration* (1200 to 1000 cm^{-1}).

Several types of O—H absorptions may be found in the "OH region." Because of intermolecular hydrogen bonding, the spectra of alcohols obtained from either pure compounds or concentrated solutions exhibit a rather broad absorption at 3550 to 3200 cm^{-1}. Alcohols in the vapor state or in dilute solution (usually less than 0.05 *M*) exhibit a sharper and weaker absorption ascribed to non-hydrogen-bonded —OH groups. These "free" OH vibrations are found within the 3650 to 3550 cm^{-1} region. The infrared spectra of most alcohols, under commonly used conditions (approximately 10% solutions in carbon tetrachloride), exhibit both free and bonded OH absorptions.

Some alcohols show OH absorptions that are broad but that *do not* decrease in intensity with increasing dilution. These are due to OH groups participating in *intra*molecular hydrogen bonding; since such bonds are sometimes stronger than *inter*molecular hydrogen bonds, these absorptions may be found below 3200 cm^{-1}.

RO—H	RO—H···O(R)H	intramolecular C–O–H···O–C ring
$\nu_{O-H} \approx 3650\ cm^{-1}$	$\nu_{O-H} \approx 3550\text{–}3200\ cm^{-1}$	$\nu_{O-H} \approx 3500\text{–}3000\ cm^{-1}$
"free" —OH	intermolecular H-bond	intramolecular H-bond

The ir spectrum of pure 2,3-dimethyl-1-pentanol exhibits only hydrogen bonded OH absorption, while that of pure 3-ethyl-3-pentanol exhibits both free and bonded OH (Figure 10–2). Can you suggest a reason for this difference?

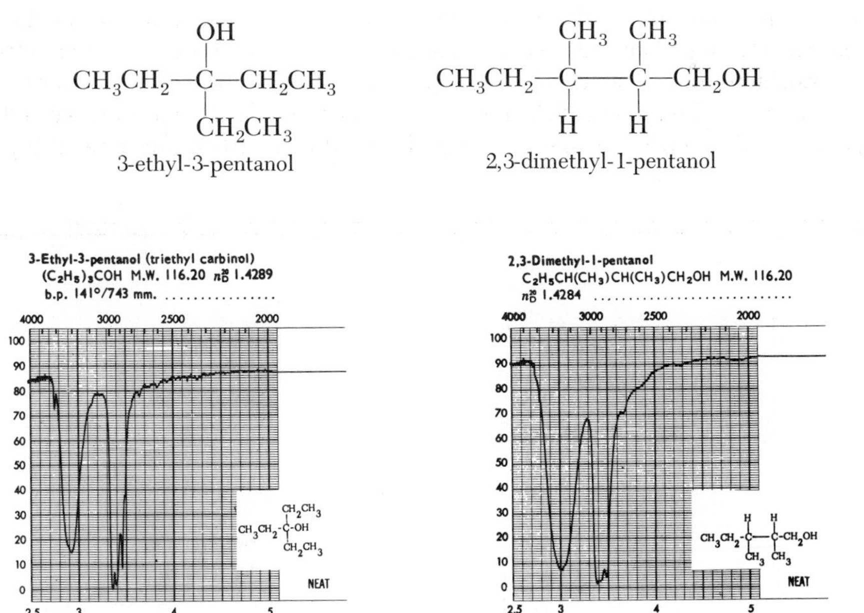

Figure 10–2 The 4000 to 2000 cm^{-1} region of the infrared spectra of 3-ethyl-3-pentanol and 2,3-dimethyl-1-pentanol. Both compounds are liquids. These spectra are from "neat" samples (i.e., samples free of solvent).

The C—O stretching vibration near 1050 cm^{-1} can sometimes help to determine whether an alcohol is 1°, 2°, or 3°; see Table 10–4.

TABLE 10-4 Characteristic C—O Vibration Of Alcohols

ALCOHOL	ABSORPTION RANGE (cm^{-1})
$R_3C—OH$	1200–1125
$R_2CH—OH$ $C{=}C—CR_2—OH$ $(CH_2)_nC(R)(OH)$ (ring)	1125–1085
$C{=}C—CHR—OH$ $RCH_2—OH$ $(CH_2)_nC(H)(OH)$ (ring) $n = 5,6$	1085–1050
others	below 1050

NMR SPECTRA. The hydroxy group of alcohols normally appears as a singlet at 2 to 4δ. The lack of splitting of the O—H signal of 1° and 2° alcohols by protons bonded to the carbinol carbon is due to the rapid exchange of various ionizable protons in the media (an effect called *chemical exchange*), *and* to the fact that, once a proton departs from a hydroxy group, the replacement may have either the same spin or the opposite spin from that of the one that departed. The exchange-induced decoupling of alcohols in solvents such as carbon tetrachloride and deuteriochloroform is catalyzed by the presence of traces of hydrogen chloride in these solvents. Therefore, observation of alcohols in most common nmr solvents produces a decoupled **OH** resonance. The nmr spectrum of benzyl alcohol (Figure 10–3) illustrates this point.

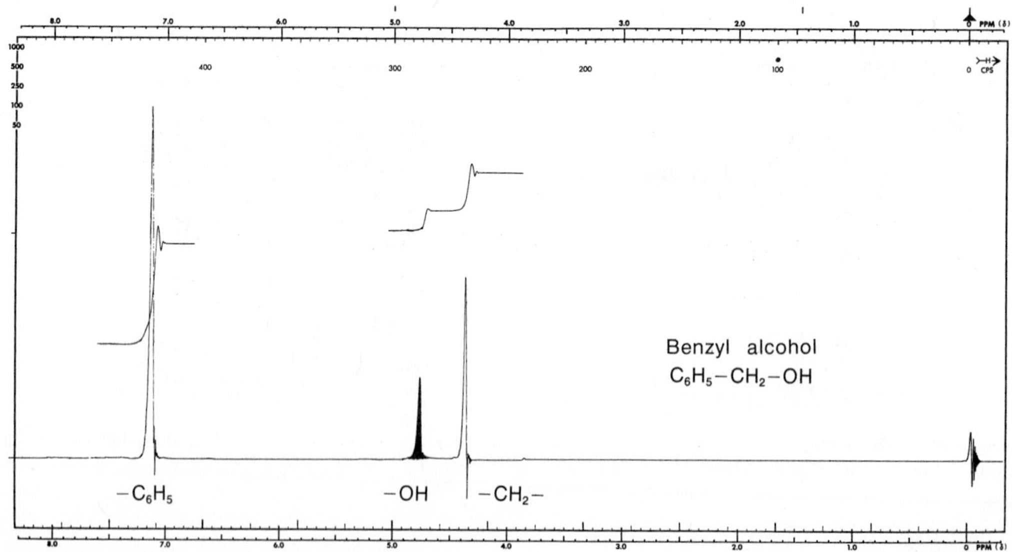

Figure 10-3 The nmr spectrum of benzyl alcohol. (From J. A. Moore and D. L. Dalrymple, *NMR Spectra of Unknowns,* W. B. Saunders Company, 1971.)

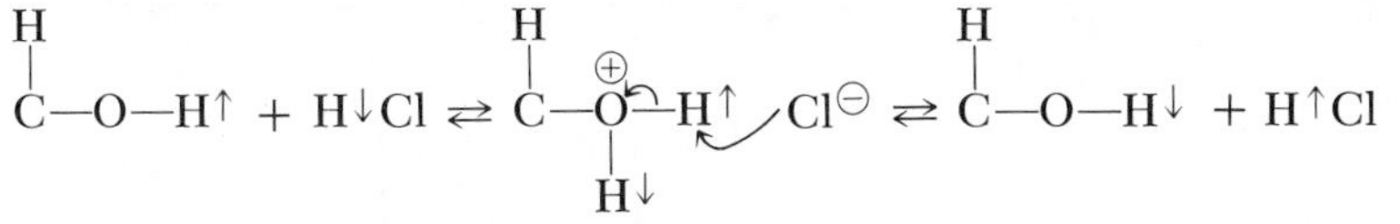

$$\overset{\displaystyle H}{\overset{|}{C}}\text{—O—H}\uparrow + \text{H}\downarrow\text{Cl} \rightleftarrows \overset{\displaystyle H}{\overset{|}{C}}\text{—}\underset{\displaystyle \underset{\text{H}\downarrow}{|}}{\overset{\oplus}{\text{O}}}\text{—H}\uparrow \; \text{Cl}^{\ominus} \rightleftarrows \overset{\displaystyle H}{\overset{|}{C}}\text{—O—H}\downarrow + \text{H}\uparrow\text{Cl} \quad \text{exchange decoupling of an OH resonance}$$

If fully deuterated dimethyl sulfoxide (d_6-DMSO) is used as the solvent, the strong hydrogen bond between the alcohol and the solvent slows the exchange rate sufficiently that coupling between OH and vicinal hydrogens is observed. Moreover, the formation of the hydrogen bond shifts the OH resonance downfield to approximately 5 δ. The combination of (a) observation of coupling and (b) relocation of the OH resonance permits the unequivocal assignment of the nature of the alcohol (*i.e.*, 1°, 2°, or 3°) by nmr spectroscopy. The nmr spectra of *n*-propyl, isopropyl, and *t*-butyl alcohol in DMSO (Figure 10–4) are illustrative.

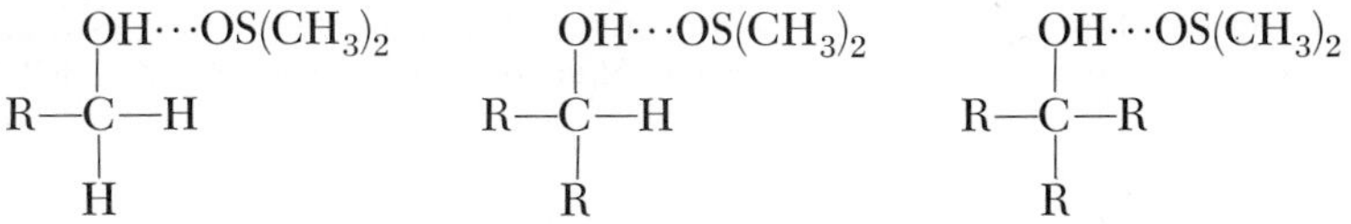

$$\text{R—}\underset{\displaystyle H}{\underset{|}{\overset{\displaystyle \text{OH}\cdots\text{OS(CH}_3)_2}{\overset{|}{C}}}}\text{—H} \qquad \text{R—}\underset{\displaystyle R}{\underset{|}{\overset{\displaystyle \text{OH}\cdots\text{OS(CH}_3)_2}{\overset{|}{C}}}}\text{—H} \qquad \text{R—}\underset{\displaystyle R}{\underset{|}{\overset{\displaystyle \text{OH}\cdots\text{OS(CH}_3)_2}{\overset{|}{C}}}}\text{—R}$$

—**OH** triplet
—CH_2—O— doublet

—**OH** doublet
—CH—O— doublet

—**OH** singlet

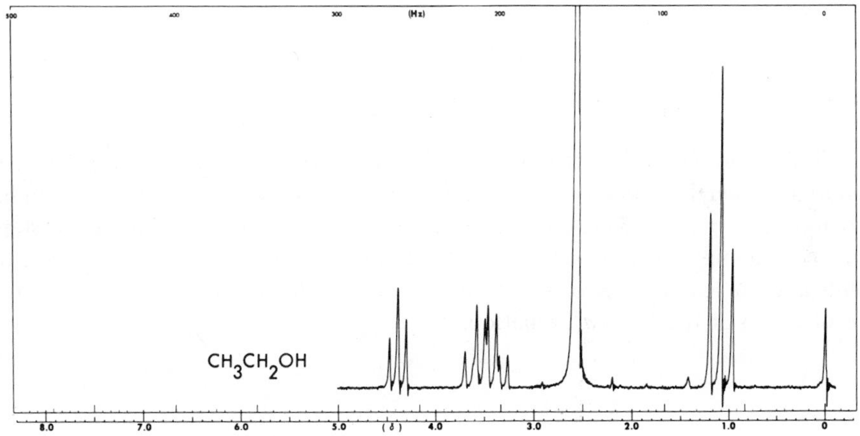

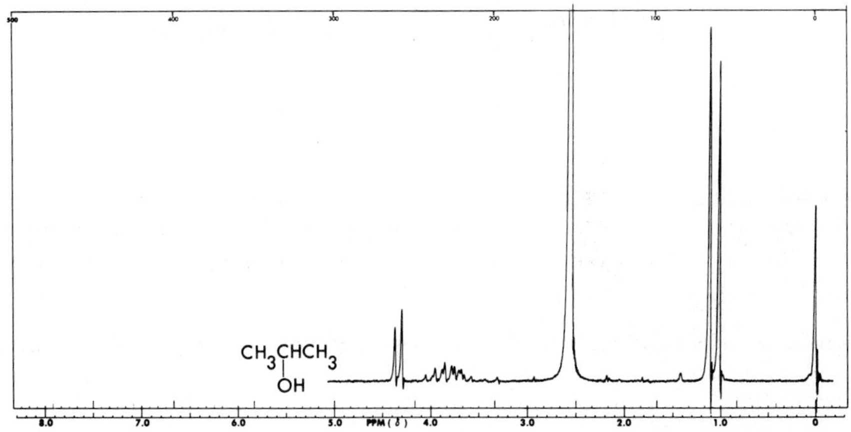

Illustration continued on following page.

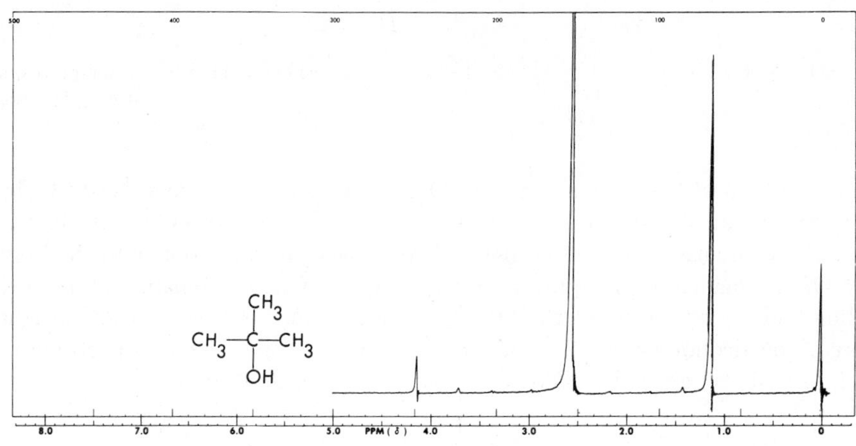

Figure 10-4 Nmr spectra of *n*-propyl, isopropyl, and *t*-butyl alcohol in dimethyl sulfoxide. Absorption at 2.6δ is that of DMSO. (From D. J. Pasto and C. R. Johnson, *Organic Structure Determination*. Prentice-Hall, Inc., Englewood Cliffs, 1969.)

*8. Predict the structure(s) of the alcohols produced by reducing the compounds listed in problem 7 with sodium borohydride. What will their OH resonances look like in d_6-DMSO?

IMPORTANT TERMS

Addition-elimination reaction: A reaction in which a nucleophile replaces a leaving group that is bonded to a carbonyl carbon, *via* a two-step sequence. In the first step the nucleophile adds to a carbon-oxygen π bond. This creates an alkoxide anion ($RO^{\ominus}$). In the second step a pair of electrons on the negatively charged oxygen attacks the carbon, which now bears both the nucleophile and the leaving group, and displaces the leaving group. This last step is analogous to the second step of the E1cb elimination.

$$\overset{:O:}{\underset{\overset{\ominus}{Nu:}}{-\overset{\|}{C}-L}} \xrightarrow[\text{step}]{\text{addition}} -\overset{:\overset{..}{O}:^{\ominus}}{\underset{Nu}{\overset{|}{\underset{|}{C}}}}-L \xrightarrow[\text{step}]{\text{elimination}} -\overset{:O:}{\overset{\|}{C}}-Nu + L:^{\ominus}$$

While the leaving group is bonded to a carbonyl group in this example (and in the definition above), this type of process can occur when a leaving group is bonded to other multiply bonded carbons, *e.g.*, C=N. Addition-elimination reactions are common in organic chemistry and will be discussed in several places in this text.

Anchimeric assistance: Participation by a neighboring group in the departure of a leaving group

*This problem requires knowledge of spectroscopy.

during an S_N1 type process. Ionizations in which anchimeric assistance takes place occur more rapidly than they would if this participation were not present. Migration of the neighboring group during the ionization lowers the energy of the transition state by dispersing the charge in the activated complex. This leads to a decrease in the energy of activation for the process, and to a resultant rate increase.

Carbinol carbon: That carbon of an alcohol which bears the hydroxy group.

Carbohydrate: A group of compounds that consists of polyhydroxy aldehydes, polyhydroxy ketones, and their derivatives. The most familiar carbohydrates include sugar, starch, and glycogen. These compounds are discussed in detail in Chapters 25A and 25B.

Ester: A compound that consists of two readily identifiable fragments, one of which is derived from an alcohol and the other from either an inorganic oxyacid (an acid whose acidic hydrogen is bonded to oxygen) or an organic acid. The chemistry of the esters of carboxylic acids is presented in Chapters 18 and 19.

Hydrogenolysis: Cleavage of a σ bond by molecular hydrogen.

$$A{-}B + H_2 \rightarrow A{-}H + H{-}B$$

Intermolecular hydrogen bond: A hydrogen bond between two molecules. The amount of intermolecular hydrogen bonding decreases with decreasing concentration.

Intramolecular hydrogen bond: A hydrogen bond in which both the donor and the acceptor are found within one molecule. The amount of intramolecular hydrogen bonding is constant with decreasing concentration.

Lucas test: A test to determine whether an alcohol is primary, secondary, or tertiary. Based upon the conversion of an alcohol to an alkyl chloride, the test actually measures the ease with which an alcohol is converted into a carbocation. The reagent (Lucas reagent) consists of hydrochloric acid and anhydrous zinc chloride.

PROBLEMS

9. Draw all of the alcohols with molecular formula $C_5H_{12}O$, neglecting enantiomers. Provide *two* acceptable names for each compound. Indicate whether each alcohol is 1°, 2°, or 3°.

10. Which of the following do not belong to the group of isomeric octyl alcohols?
    - (a) isobutyl-*n*-propylcarbinol
    - (b) 2,6-dimethyl-4-heptanol
    - (c) 4-ethyl-3-hexanol
    - (d) 2,3,4-trimethyl-2-pentanol
    - (e) diethylisopropylcarbinol
    - (f) 2-methyl-1-octanol

11. Suggest how each of the following products could be synthesized by using a Grignard reagent and the given starting material. Give the IUPAC name for each product.
    - (a) $(CH_3)_2C{=}O$ *to* $(CH_3)_3COH$
    - (b) $(CH_3)_2C{=}O$ *to* $(CH_3)_2C(OH)CH_2CH_3$
    - (c) $CH_3CO_2CH_3$ *to* $(CH_3)_3COH$
    - (d) $CH_3CO_2C_2H_5$ *to* $(CH_3)_3COH$
    - (e) cyclohexanone (ring with =O) *to* cyclohexane ring bearing OH and CH_2CH_3

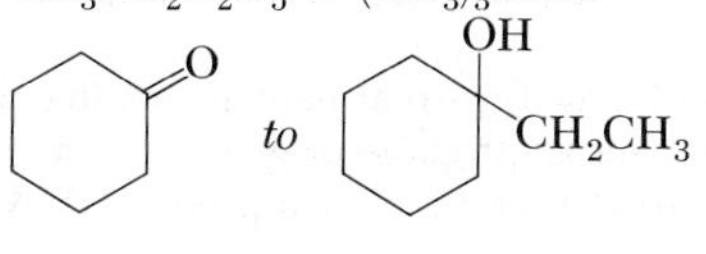

(f) 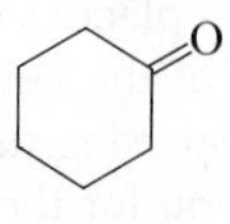 [cyclohexanone] *to* [1-cyclohexylcyclohexanol]

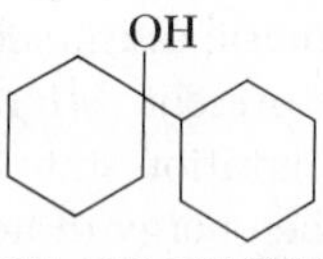

(g) $(CH_3)_2C{=}O$ *to* $CH_3CH_2CH(OH)CH_3$ (several steps)

12. Give reactions illustrating the synthesis of the following alcohols from any suitable alkenes.
    (a) 2-pentanol
    (b) 1-pentanol
    (c) 3-pentanol
    (d) dimethyl-*n*-propylcarbinol
    (e) *t*-amyl alcohol
    (f) triethylcarbinol

13. Give a reaction sequence useful for the preparation of:
    (a) an alkoxide from *n*-octyl alcohol
    (b) an alkane from isobutylmethylcarbinol
    (c) an inorganic ester from 2,3-dimethyl-1-pentanol
    (d) a trialkylcarbinol from propylene

14. Suggest a route to convert propene to each of the following. Use any other needed reagents.
    (a) propane
    (b) 1-propanol
    (c) 2-propanol
    (d) 1-chloropropane
    (e) 2-chloropropane
    (f) 1,2-propanediol
    (g) 1-deuterio-2-propanol
    (h) 2-deuterio-1-propanol
    (i) propanal (CH_3CH_2CHO)
    (j) 2-propanone (acetone, CH_3COCH_3)
    (k) propylene oxide $CH_3CH{-}CH_2$ (with O bridging the two carbons)
    (l) 1-butene
    (m) 1-butyne
    (n) methyl ethyl ketone ($CH_3COCH_2CH_3$)
    (o) isobutyl alcohol
    (p) isobutylene
    (q) isoamyl alcohol
    (r) 3-methyl-1-butene
    (s) 3-methyl-2-butanol
    (t) 2-chloro-3-methylbutane
    (u) 2-bromo-3-methylbutane
    (v) 2-deuterio-3-methylbutane

15. Suggest a chemical test that could be used to distinguish between members of the following pairs. Include expected observations.
    (a) 1-propanol and propane
    (b) 1-propanol and propene
    (c) 1-propanol and 2-propanol
    (d) 2,3,4-trimethyl-2-pentanol and 4-ethyl-3-hexanol
    (e) *t*-butyl alcohol and *t*-butylcarbinol
    (f) *t*-butyl alcohol and allyl alcohol
    (g) propyne and propanol
    (h) cyclohexanol and 1-methylcyclohexanol
    (i) allyl alcohol and propargyl alcohol

16. Piperylene can be prepared by the following sequence in an overall yield of about 65%. When subjected to ozonolysis and reductive work-up, piperylene produces CH_3CHO, CH_2O, and $(CHO)_2$ in equimolar amounts. What structures for piperylene are consistent with these data?

$$CH_3CH{=}CHCHO \xrightarrow{CH_3MgCl} \xrightarrow{H_2O} \xrightarrow[\text{heat}]{Al_2O_3} \text{piperylene}$$

17. Benzyl alcohol reacts instantly with the Lucas reagent even though it is only a primary alcohol. Explain.

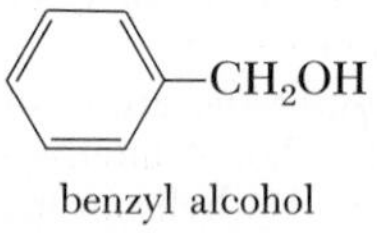

$C_6H_5{-}CH_2OH$

benzyl alcohol

18. The following represents the order of increasing ease of acid-catalyzed dehydration. Account for this trend.

1-butanol < 2-methyl-1-propanol < 2-butanol < 3-methyl-2-butanol < 2-methyl-2-propanol

19. "Although no 2-methylpropene has been found in the products from 2-butanol, the acid-catalyzed dehydration of 3,3-dimethyl-2-butanol yields very little *t*-butylethylene." What point is the author attempting to make?

20. Nitroglycerin is used extensively in explosives and also in the treatment of angina pectoris. It is made by the complete esterification of glycerin (1,2,3-trihydroxypropane *or* glycerol) using a mixture of nitric and sulfuric acids (**A**). The need for glycerin is so great that, in 1939, a process (**B**) was devised

for converting propylene to glycerin. Provide structures for all of the intermediates in **B** and mechanisms for their formation.

$$\text{A: } HOCH_2CH(OH)CH_2OH \xrightarrow[H_2SO_4]{HNO_3} O_2NOCH_2CH(ONO_2)CH_2ONO_2$$

glycerin — nitroglycerin *or* "nitro"

$$\text{B: propene} + Cl_2 \xrightarrow{h\nu} \mathbf{A}\ (C_3H_5Cl) \xrightarrow{OH^{\ominus}} \mathbf{B}\ (C_3H_6O) \xrightarrow{Cl_2/H_2O} \mathbf{C}\ (C_3H_7ClO_2) \xrightarrow[\text{heat}]{Ca(OH)_2} \mathbf{D}\ (C_3H_6O_2) \xrightarrow{H_2O} \text{glycerin}$$

21. Suggest a mechanism for the following reaction.

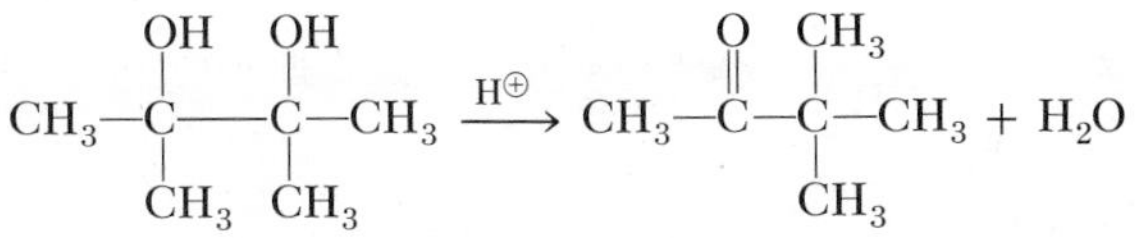

$$CH_3-C(OH)(CH_3)-C(OH)(CH_3)-CH_3 \xrightarrow{H^{\oplus}} CH_3-C(=O)-C(CH_3)_2-CH_3 + H_2O$$

22. Dihydropyran, **A**, can be made by reacting tetrahydrofurfuryl alcohol, **B**, with acid. (a) Suggest a mechanism for this. (b) What product would you expect from the hydroboration-oxidation of **A**? Explain.

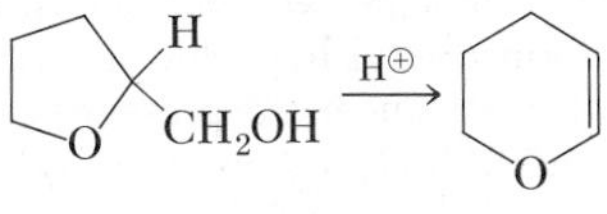

$$\mathbf{B} \xrightarrow{H^{\oplus}} \mathbf{A}$$

23. Suggest a reason for the preponderance of 2-butanol over 2-methyl-1-propanol in the product of the reaction of propylene oxide with methyl magnesium iodide.

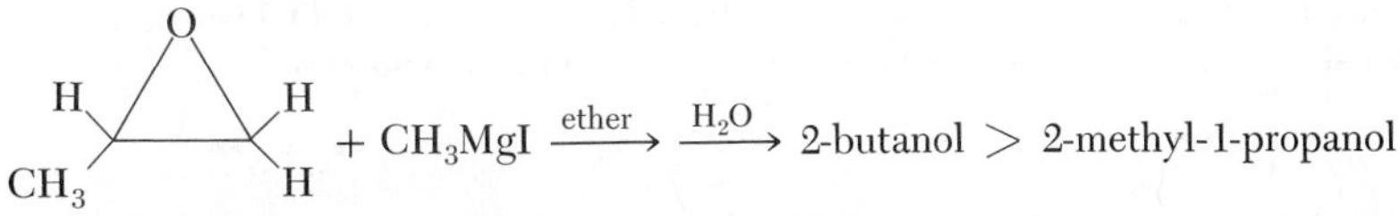

$$\text{propylene oxide} + CH_3MgI \xrightarrow{\text{ether}} \xrightarrow{H_2O} \text{2-butanol} > \text{2-methyl-1-propanol}$$

24. The following represents the order of increasing ease of oxidation with chromium trioxide. What conclusions, if any, can be reached from this information?

2-methyl-2-propanol < 3-methyl-2-butanol < 2-butanol < 2-methyl-1-propanol < 1-butanol

25. Account for the fact that **A** (below) reacts with diborane to produce a mixture containing 37% of the boron bonded to C2 (the remainder is bonded to Cl) while **B** (below) produces a mixture with only 6% of the boron bonded to C2. What products do you expect when these alkenes are subjected to (a) hydroboration-oxidation and (b) hydroboration-carbonylation-oxidation?

$$\underset{\mathbf{A}}{H_2C{=}CH{-}Si(CH_3)_3} \qquad \underset{\mathbf{B}}{H_2C{=}CH{-}C(CH_3)_3}$$

26. Oxetane can be synthesized, in moderate yield, by the following sequence. Provide a mechanism for each step.

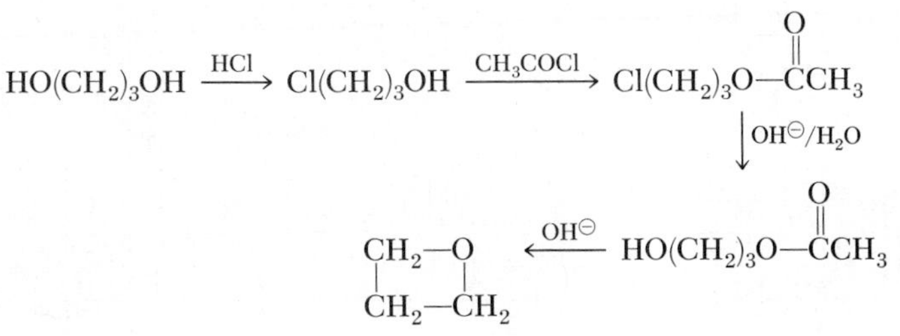

$$HO(CH_2)_3OH \xrightarrow{HCl} Cl(CH_2)_3OH \xrightarrow{CH_3COCl} Cl(CH_2)_3O-\overset{O}{\overset{\|}{C}}CH_3 \xrightarrow{OH^{\ominus}/H_2O} HO(CH_2)_3O-\overset{O}{\overset{\|}{C}}CH_3 \xrightarrow{OH^{\ominus}} \underset{\text{oxetane}}{\text{CH}_2{-}\text{O} / \text{CH}_2{-}\text{CH}_2}$$

27. Silicate esters, $(RO)_4Si$, are hydrolyzed by water to form silica and alcohol. This reaction could proceed by an S_N2 reaction of water upon silicon or upon the R group. How might you distinguish between these two paths?

$$\underset{\text{silicate ester}}{(RO)_4Si} + 4H_2O \rightarrow 4ROH + Si(OH)_4$$

$$Si(OH)_4 \longrightarrow SiO_2 + 2H_2O$$

28. Trimethylamine oxide reacts with *sec*-butyl tosylate to produce 2-butanone and trimethylamine. Propose a mechanism for this reaction, which is shown below.

$$\underset{sec\text{-butyl tosylate}}{CH_3\underset{\underset{OSO_2-C_6H_4-CH_3}{|}}{C}HCH_2CH_3} + \underset{\text{trimethylamine oxide}}{(CH_3)_3\overset{\oplus}{N}-\overset{\ominus}{O}} \rightarrow \underset{\text{2-butanone}}{CH_3\overset{O}{\overset{\|}{C}}CH_2CH_3} + \underset{\text{trimethylamine}}{(CH_3)_3N}$$

°29. When *t*-butyl alcohol reacts with chlorine in the presence of sodium hydroxide, there is produced *t*-butyl hypochlorite, $(CH_3)_3COCl$, a valuable chlorinating agent. How could it be shown that the product is not 1-chloro-2-methyl-2-propanol? The strength of the RO—Cl bond is approximately 52 kcal/mole. From bond strength data, indicate whether chlorination at carbon (to form 1-chloro-2-methyl-2-propanol) or at oxygen (to form *t*-butyl hypochlorite) is energetically more favorable.

°30. Acetone, $(CH_3)_2C{=}O$, reacts with acetylene in the presence of sodamide to produce, after work-up, a compound **A** which reacts with Br_2 and $KMnO_4$ but not with $NaBH_4$. Further, **A** reacts with a mixture of mercuric sulfate, water, and sulfuric acid to produce **B,** whose formula is $C_5H_{10}O_2$. Compound **B** reacts with $NaBH_4$ but not with $KMnO_4$. The nmr spectrum of **B** in d_6-DMSO consists of three singlets in the ratio of 6H:3H:1H. Identify **A** and **B.**

°31. Sketch the nmr spectra of ethanol (a) in pure d_6-DMSO and (b) in d_6-DMSO containing D_2O. Why are they different?

°32. The infrared spectra below are those of CH_3OH, CH_3OD, and CD_3OD. Determine which spectrum is associated with each of these, making as many assignments as you can.

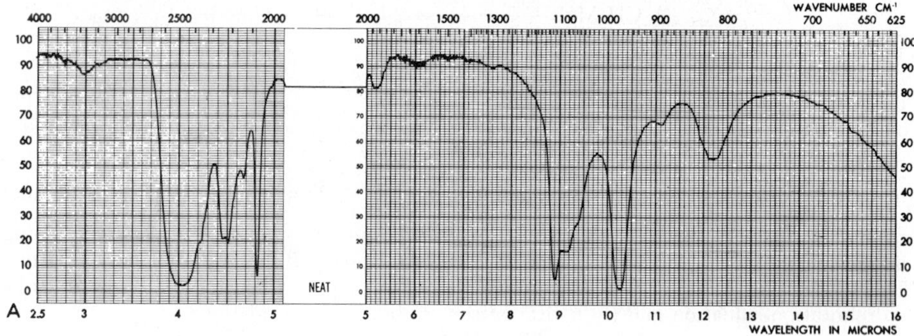

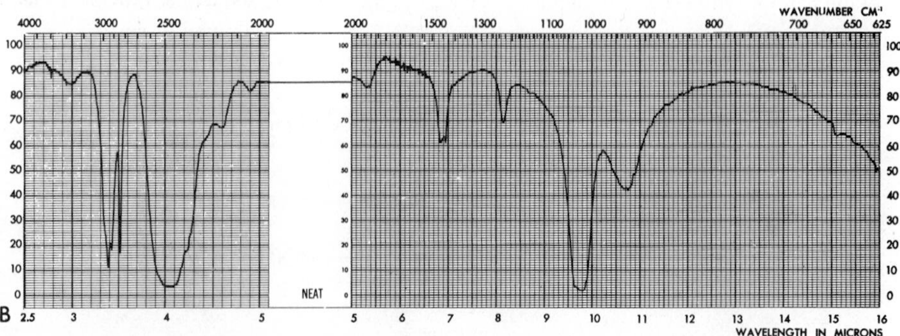

°These problems require knowledge of spectroscopy.

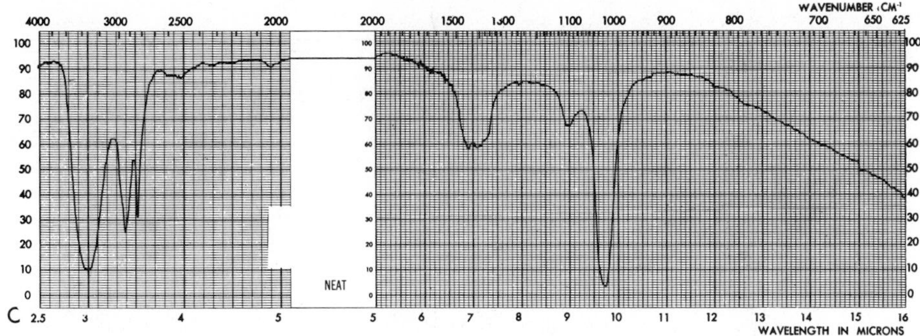

*33. The fermentation of grain to produce ethanol also yields *fusel oil,* a by-product of variable composition but usually rich in C_5 alcohols. Since fusel oil is moderately toxic, it must be absent from all alcohol destined for internal use. The nmr and ir spectra of the two major constituents of fusel oil are shown below. Identify these two C_5 alcohols.

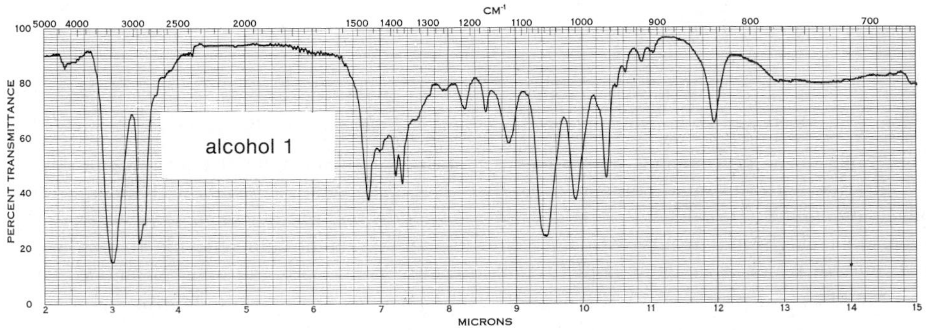

alcohol 1

alcohol 2

alcohol 1

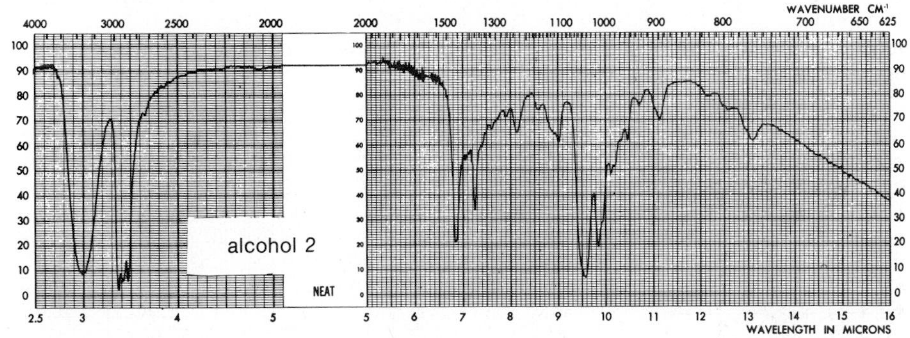

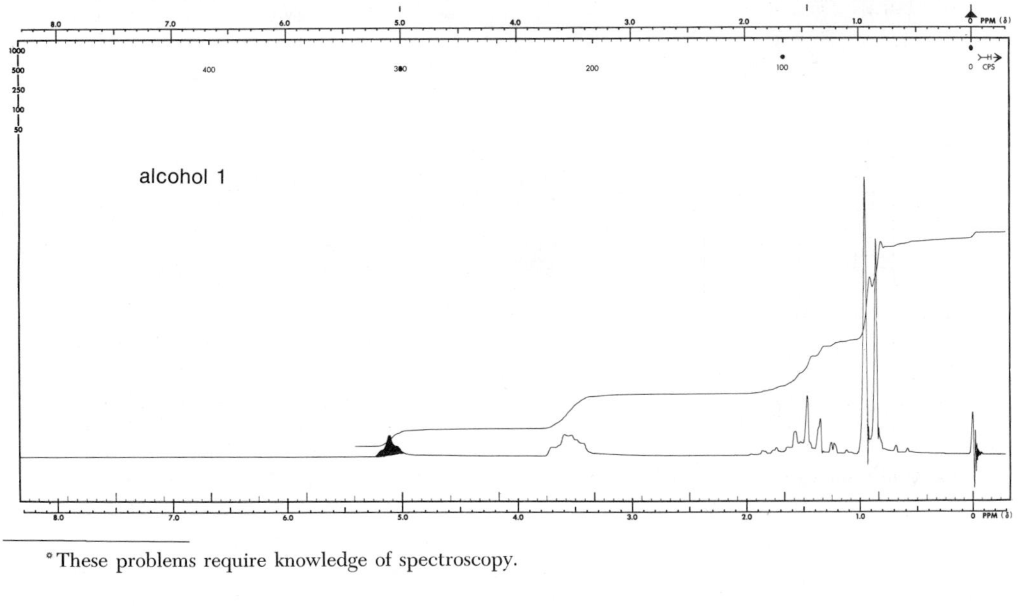

*These problems require knowledge of spectroscopy.

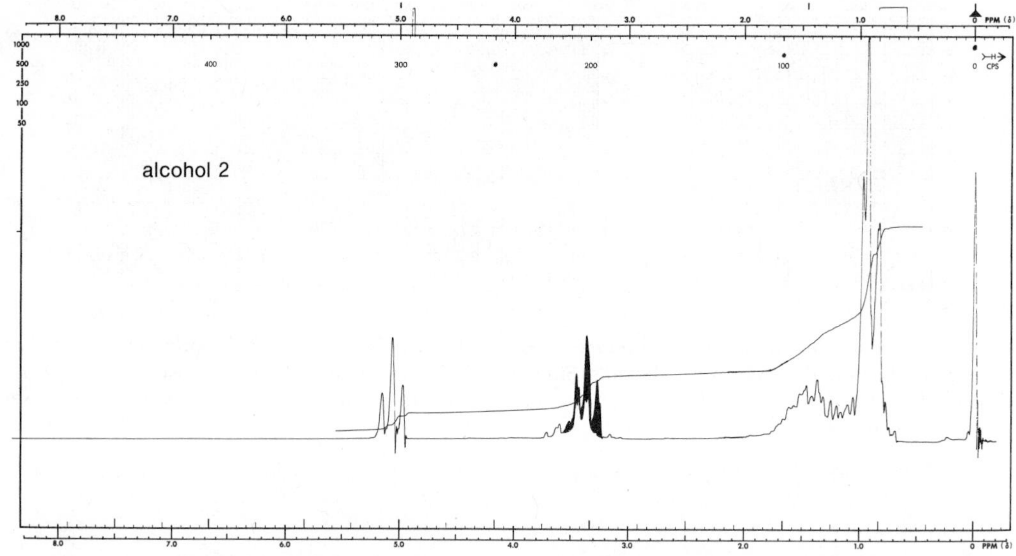

(From J. A. Moore and D. L. Dalrymple, *NMR Spectra of Unknowns*, W. B. Saunders Company, 1971.)

°34. The reaction of allyl alcohol with aqueous chlorine produces **A** ($C_3H_7ClO_2$), while the reaction of allyl chloride with the same reagent forms **B** ($C_3H_6Cl_2O$). When **A** reacts with sodium hydroxide, it produces *glycidol* ($C_3H_6O_2$); while **B**, under the same conditions, produces *epichlorohydrin* (C_3H_5ClO).

One mole of glycidol reacts with one mole of methyl magnesium chloride to produce several products, one of which is methane. Epichlorohydrin also reacts with one mole of methyl magnesium chloride but does not produce any methane. Neither glycidol nor epichlorohydrin contains a $\rangle C{=}O$ linkage. (a) Suggest structures for **A, B,** glycidol, and epichlorohydrin consistent with these data. (b) Explain the difference between the behavior of allyl chloride and that of allyl alcohol toward aqueous chlorine.

The infrared spectra of glycidol and epichlorohydrin are presented below. (c) Are they consistent with your conclusions?

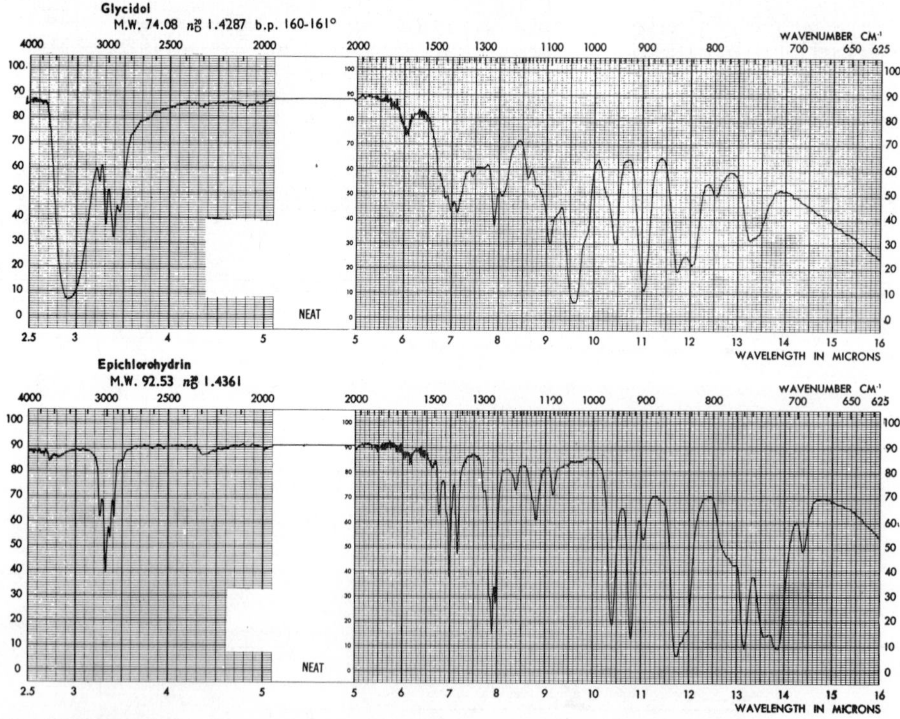

(d) What would be the product of the oxidation of glycidol with Sarett's reagent?

ETHERS, EPOXIDES AND DIOLS

11.1 INTRODUCTION

Nowhere is there as broad an example of the effects of ground state strain upon reactivity as is found in the chemistry of ethers, compounds of the type R—O—R. Ethers are, by and large, a rather dull, unreactive lot, and their chemistry offers very little that is new and exciting. If, however, the ethereal oxygen is part of a three-membered ring system, a so-called **epoxide,** the picture changes to include a wealth of interesting and important reactions. The contrast in behavior is so great that one normally does not think of epoxides when the word "ether" is mentioned.

In addition to ethers and epoxides, this chapter also examines some of the chemistry of *vic*-diols, a group of compounds whose chemistry is intertwined with that of epoxides.

$H_3C—O—CH_3$
methyl ether
(a symmetric ether)

$CH_3CH_2—O—CH_2CH_3$
ethyl ether
("ether")

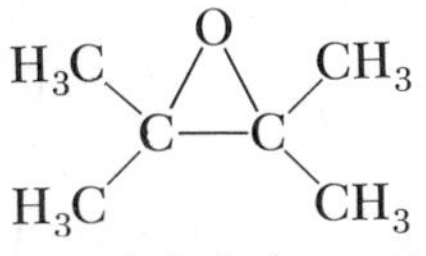

$CH_3CH_2—O—CH_3$
ethyl methyl ether
(an unsymmetric ether)

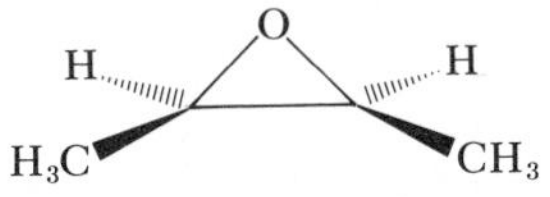

tetramethylethylene oxide
(an epoxide)

cis-2,3-epoxybutane
(*cis*-2-butene oxide)
(an epoxide)

1. Provide an acceptable name for each of the following:
   (a) $CH_3OCH_2CH_2CH_3$
   (b) $H_2C=CH—O—CH=CH_2$
   (c) $H_2C=CH—CH_2OC\equiv CH$
   (d) 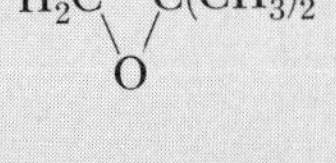 $H_2C—C(CH_3)_2$ (epoxide ring with O)
   (e) 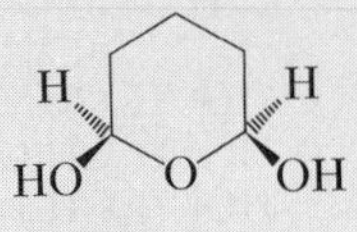 (cyclic ether structure with H, HO, O, H, OH)
   (f) 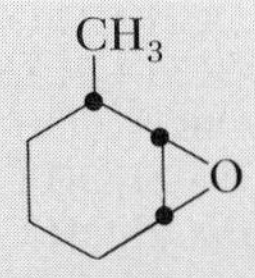 (methylcyclohexene oxide structure with CH_3 and O)

11.2 PHYSICAL PROPERTIES OF DIOLS AND ETHERS

Diols. The differences between the physical properties of alkanes and those of alcohols were ascribed, in Chapter 10, to the hydrogen bonding present in monohydric alcohols but absent in alkanes. Increasing the number of hydroxy groups should magnify the differences between alkanes and alcohols by increasing the extent of intermolecular hydrogen bonding, and this is, in fact, observed. For example, while ethane boils at $-89°$, ethanol boils at $+78°$ and 1,2-ethanediol (ethylene glycol) boils at 198° (Table 11–1). Paralleling this increase in boiling point is an increase in viscosity; ethane "flows" extremely easily and ethanol pours as easily as water, but ethylene glycol is rather viscous. The elevated boiling point and increased viscosity of ethylene glycol result from an extensive network of intermolecular hydrogen bonding made possible by the presence of two hydroxy groups per molecule. One might imagine pure ethylene glycol to be a polymer-like substance, with several ethylene glycol molecules held together by hydrogen bonds.

TABLE 11–1 Physical Properties of Ethers, Epoxides and Diols

Compound	Structure	bp(°C)	mp(°C)
methyl ether	CH_3OCH_3	−24	−140
ethyl ether	$C_2H_5OC_2H_5$	35	−116
n-propyl ether	$CH_3CH_2CH_2OCH_2CH_2CH_3$	90	−122
isopropyl ether	$(CH_3)_2CHOCH(CH_3)_2$	69	−60
n-butyl ether	$CH_3(CH_2)_3O(CH_2)_3CH_3$	143	−98
tetrahydrofuran	$(CH_2)_4O$	66	−108
epoxyethane	$\underset{O}{CH_2—CH_2}$	11	−112
epoxypropane	$\underset{O}{CH_2—CHCH_3}$	34	
1,2-epoxybutane	$\underset{O}{CH_2—CHCH_2CH_3}$	63	
cis-2,3-epoxybutane	$\overset{O}{(H)(CH_3)C—C(H)(CH_3)}$	59	−80
trans-2,3-epoxybutane	$\overset{O}{(CH_3)(H)C—C(H)(CH_3)}$	54	−85
1,2-ethanediol	$HOCH_2CH_2OH$	198	−13
1,2-propanediol	$HOCH_2CH(OH)CH_3$	187	
1,3-propanediol	$HOCH_2CH_2CH_2OH$	214	−30
1,2-butanediol	$HOCH_2CH(OH)CH_2CH_3$	192	
1,3-butanediol	$HOCH_2CH_2CH(OH)CH_3$	215	
1,4-butanediol	$HO(CH_2)_4OH$	230	19
meso-2,3-butanediol	$CH_3CH(OH)CH(CH_3)OH$	182	34
d,l-2,3-butanediol	$CH_3CH(OH)CH(CH_3)OH$	177	8

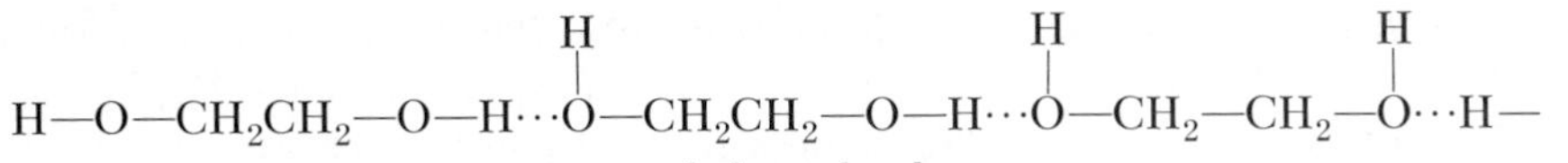

$$\text{H—O—CH}_2\text{CH}_2\text{—O—H}\cdots\underset{|}{\overset{\text{H}}{}}\text{O—CH}_2\text{CH}_2\text{—O—H}\cdots\text{O(H)—CH}_2\text{—CH}_2\text{—O(H)}\cdots\text{H—}$$

ethylene glycol

Diols are much more soluble in water than are similar monohydric alcohols because of the greater opportunity for hydrogen bonding between dihydric alcohols (diols) and water.

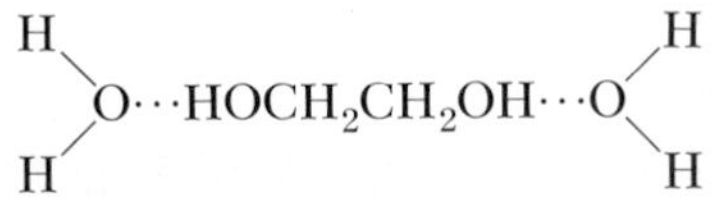

$$\text{H}_2\text{O}\cdots\text{HOCH}_2\text{CH}_2\text{OH}\cdots\text{OH}_2$$

Ethers. The students in an organic chemistry laboratory quickly get several impressions about "ether," the common name for ethyl ether, including the following: (a) water is insoluble in ether, (b) ether is a good solvent for organic compounds, and (c) ether is extremely flammable. While the last two statements are correct, the first is, in fact, somewhat in error. If you shake together ether and water, as in an extraction experiment, and recover the ether, it will be found to contain about 7% water. It is for this reason that the ether extract of an aqueous phase is dried prior to removal of the ether by distillation (a warning is presented on p. 329). For perspective, we might note that n-butyl alcohol and ether are equally soluble in water.

Why Are Ether and Water Partially Miscible? The non-bonding electrons on the oxygen of an ether are good hydrogen bond acceptors, and it is hydrogen bonding between ethyl ether and water that causes the partial miscibility of these two. Dioxane, more correctly 1,4-dioxane, contains two ethereal oxygens for four carbons, and it hydrogen bonds with water sufficiently to make these two completely miscible. Tetrahydrofuran, THF, has only one oxygen per four carbons, just like ethyl ether. However, the oxygen of THF is more exposed than is the oxygen of ethyl ether because of the "tying back" of the alkyl groups of THF into a ring structure. This makes it easier for THF to undergo hydrogen bonding with water; thus, THF, like dioxane, is completely miscible with water. The increase in water miscibility of these ethers parallels their increased ability to stabilize Grignard reagents (see p. 175).

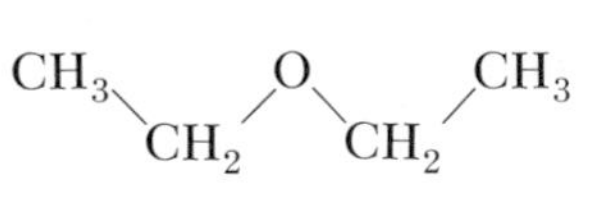

$CH_3CH_2OCH_2CH_3$ — ethyl ether (3-oxapentane)

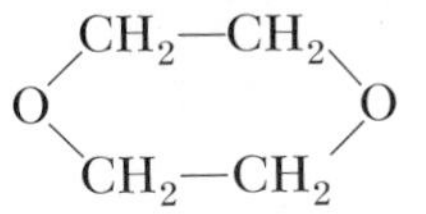

$\text{O(CH}_2\text{—CH}_2)_2\text{O}$ — 1,4-dioxane (1,4-dioxacyclohexane)

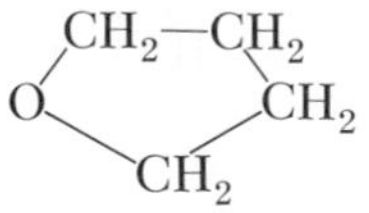

$\text{O(CH}_2\text{—CH}_2\text{—CH}_2\text{—CH}_2)$ — tetrahydrofuran (oxacyclopentane)

Why Is Ether a Good Solvent for Covalent Compounds? First of all, ethyl ether does not form hydrogen bonds with itself. Therefore, a great deal of energy is not needed to free ether molecules from one another before they can solvate another compound. Secondly, the C—O bonds in ethers are polar, permitting ethers to solvate polar functional groups effectively. Finally, the ethereal oxygen is a good electron pair donor (Lewis base) and can, by electron pair donation, coordinate and solubilize electron-deficient species such as boron halides and aluminum halides (Lewis acids).

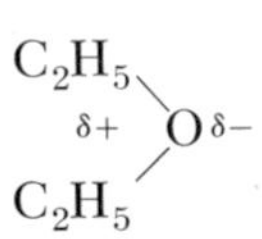

$$(C_2H_5)_2\overset{\delta+\;\;\delta-}{O}$$

dipolar nature of ether

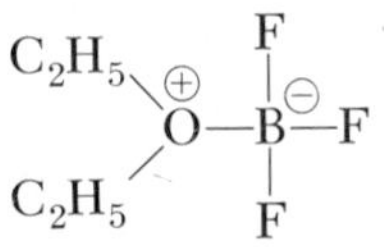

$$(C_2H_5)_2\overset{\oplus}{O}\text{—}\overset{\ominus}{B}F_3$$

coordination complex between ether (a Lewis base) and boron trifluoride (a Lewis acid)

Why Is Ether So Hazardous? This is an extremely difficult question to answer because so little is known about flame chemistry, the chemistry of explosions, and related areas. Suffice it to note the existence of a property of substances, the *flash point,* which measures the minimum temperature that a material must attain in order to commence a self-sustained fire when subjected to an outside flame, and the fact that the flash point of ether is lower than that of gasoline and among the lowest of the commonly used laboratory solvents. For example, the flash points of ether, 100 octane gasoline, benzene, and ethanol are $-49°$, $-36°$, $+12°$, and $+55°$, respectively.

11.3 SYNTHESIS OF ETHERS

There are only three useful, general syntheses of ethers. One is the **Williamson ether synthesis,** an S_N2 reaction involving an oxyanion and an alkylating agent. The utility of this synthesis, like that of all S_N2 reactions, is limited by competitive E2 eliminations. Many ethers can be made by either of two alternative S_N2 routes. However, if the nucleophile is also a potent base (*e.g.*, $(CH_3)_3CO^{\ominus}$) *or* if the substrate is quite hindered (*e.g.*, $(CH_3)_3CBr$), the synthetic approaches become limited.

$$\left.\begin{array}{l} CH_3I + CH_3CH_2O^{\ominus}Na^{\oplus} \\ CH_3O^{\ominus}Na^{\oplus} + CH_3CH_2I \end{array}\right\} \rightarrow CH_3OCH_2CH_3 + NaI \qquad S_N2$$

$$CH_3I + (CH_3)_3CO^{\ominus}K^{\oplus} \rightarrow \underset{t\text{-butyl methyl ether}}{(CH_3)_3COCH_3} + KI \qquad S_N2$$

$$CH_3O^{\ominus}K^{\oplus} + (CH_3)_3CBr \rightarrow CH_3OH + (CH_3)_2C{=}CH_2 + KBr \qquad E2$$

Since S_N2 reactions do not occur readily on sp^2 hybridized carbons, ethers of the type $Csp^3{-}O{-}Csp^2$ can be made in only one way.

$$CH_3O^{\ominus}Na^{\oplus} + C_6H_5I \xrightarrow{75°} \text{No reaction}$$

$$CH_3I + C_6H_5O^{\ominus}Na^{\oplus} \xrightarrow{25°} C_6H_5O{-}CH_3 + NaI$$

methyl phenyl ether
(anisole)

The second, and perhaps most valuable, route to ethers is **alkoxymercuriation-demercuriation.** This reaction sequence is analogous to the Markownikoff hydration of alkenes achieved by oxymercuriation-demercuriation (see p. 301). The major difference between these two schemes, and one that is quite logical once you think of an alcohol as an organic derivative of water, is that in alkoxymercuriation the mercury salt reacts with the alkene in an alcoholic solution rather than in an aqueous solution. Another difference is that mercuric trifluoroacetate is used instead of mercuric acetate.

$$>C{=}C< + Hg^{\oplus 2}(CF_3CO_2^{\ominus})_2 + ROH \rightarrow -\underset{OR}{\overset{|}{\underset{|}{C}}}-\underset{HgOC(O)CF_3}{\overset{|}{\underset{|}{C}}}- \xrightarrow{NaBH_4} -\underset{OR}{\overset{|}{\underset{|}{C}}}-\underset{H}{\overset{|}{\underset{|}{C}}}-$$

This type of process is more useful than is the Williamson synthesis because there is no elimination reaction competing with ether formation. Unfortunately, like the Williamson synthesis, alkoxymercuriation-demercuriation cannot be used to prepare di-*t*-alkyl ethers (*e.g.*, $R_3C{-}O{-}CR_3$). The reason for this is, as yet, unclear but is probably steric in origin.

$$CH_3{-}\underset{CH_3}{\overset{CH_3}{C}}{-}CH{=}CH_2 + CH_3CH_2OH \xrightarrow{Hg(CF_3CO_2)_2} \xrightarrow{NaBH_4} CH_3{-}\underset{CH_3}{\overset{CH_3}{C}}{-}\underset{OC_2H_5}{CH}{-}CH_3$$

3-ethoxy-2,2-dimethylbutane

The third common route to ethers, one that is useful only for preparing symmetric ethers, is the acid-catalyzed intermolecular dehydration of alcohols.

$$2ROH \xrightarrow{acid} R{-}O{-}R + H_2O$$

The formation of ether by intermolecular dehydration is an example of a nucleophilic displacement upon a protonated alcohol by an alcohol molecule.

$$CH_3CH_2OH + H_2SO_4 \rightleftarrows CH_3CH_2\overset{\oplus}{O}H_2 \quad HSO_4^{\ominus}$$

$$CH_3CH_2{-}\overset{\oplus}{O}H_2 \rightarrow CH_3CH_2{-}\underset{H}{\overset{\oplus}{O}}{-}CH_2CH_3 + H_2O \rightleftarrows (CH_3CH_2)_2O + H_3O^{\oplus}$$

(attacking nucleophile: $H{-}O{-}CH_2CH_3$)

t-Butyl ether cannot be isolated from the reaction of *t*-butyl alcohol with sulfuric acid. This does not mean that the ether is not produced, *but rather that it isn't isolated!* The ether probably is produced, as its protonated salt, but this is unstable relative to the *t*-butyl cation. The channeling off of the *t*-butyl cation to *t*-butyl hydrogen sulfate and poly(isobutylene) becomes responsible for the majority of the observed reaction product.

$$(CH_3)_3COH + H_2SO_4 \rightleftarrows (CH_3)_3C\overset{\oplus}{O}H_2 \quad HSO_4^{\ominus}$$

$$\rightleftarrows (CH_3)_3C^{\oplus} + H_2O$$

$$(CH_3)_3C^{\oplus} + O\begin{matrix}H\\C(CH_3)_3\end{matrix} \rightleftarrows (CH_3)_3C{-}\overset{H}{O^{\oplus}}{-}C(CH_3)_3 \xrightleftharpoons{-H^{\oplus}} ((CH_3)_3C)_2O$$

O-protonated *t*-butyl ether — *t*-butyl ether

$$(CH_3)_3C^{\oplus} \underset{}{\overset{-H^{\oplus}}{\rightleftarrows}} CH_2{=}C(CH_3)_2 \xrightarrow{H_2SO_4} \text{poly(isobutylene)}$$

$$(CH_3)_3C^{\oplus} \overset{H_2SO_4}{\rightleftarrows} (CH_3)_3C{-}OSO_2OH + H^{\oplus}$$

t-butyl hydrogen sulfate

2. Using alkenes as the only organic starting materials, suggest a synthesis for each of the following:
   (a) ethyl ether
   (b) 2-hexyl isopropyl ether
   (c) methyl ether
   (d) cyclohexyl cyclopentyl ether
   (e) *t*-butyl ethyl ether

The intermolecular dehydration of alcohols can also be accomplished by heating them in the presence of alumina, elimination again being a competitive process at higher temperatures.

$$CH_3CH_2OH \xrightarrow{Al_2O_3} \begin{cases} \xrightarrow{375^\circ} H_2C{=}CH_2 + H_2O \\ \xrightarrow{300^\circ} CH_3CH_2OCH_2CH_3 + H_2O \end{cases}$$

11.4 REACTIONS OF ETHERS

Ethers usually are quite unreactive to ionic processes, such as S_N1, S_N2, E1, and E2 reactions. Perhaps the most significant reaction undergone by ethers is their acid-catalyzed cleavage; in these cleavages the *reactive* starting material is the *O*-protonated ether. The reagent most often used to achieve ether cleavage is hydriodic acid (aqueous hydrogen iodide).

$$CH_3OCH_3 + HI \rightleftarrows CH_3{-}\overset{\overset{\displaystyle H}{|}}{O}{}^{\oplus}{-}CH_3 \;\; I^{\ominus} \xrightarrow{S_N2} CH_3OH + CH_3I \qquad \text{ether cleavage}$$

While the cleavage of methyl ether, depicted above, is an S_N2 reaction, the nature of the ether dictates whether the cleavage of other ethers will be S_N1 or S_N2. In turn, this dictates, for an unsymmetric ether, which portion becomes the halide and which the alcohol. Several examples should help to make this clear.

$$CH_3{-}O{-}C_6H_5 \xrightarrow[S_N2]{HI} CH_3{-}I + HO{-}C_6H_5, \text{ not } I{-}C_6H_5 \text{ and } CH_3{-}OH$$

$$(CH_3)_3C{-}O{-}C_6H_5 \xrightarrow[S_N1]{HI} (CH_3)_3C{-}I + HO{-}C_6H_5, \text{ not } I{-}C_6H_5 \text{ and } (CH_3)_3C{-}OH$$

$$(CH_3)_3C{-}O{-}CH_2CH_3 \xrightarrow[S_N1]{HBr} (CH_3)_3C{-}Br + HO{-}CH_2CH_3,$$

$$\text{not } Br{-}CH_2CH_3 \text{ and } (CH_3)_3C{-}OH$$

3. Indicate the product(s) from the reaction of the compounds in problem 2 with hydriodic acid.

Peroxides. Ethers are easily air oxidized to peroxides and higher oxides. Of the common ethers, isopropyl ether is particularly prone to peroxide formation. These oxides are *explosive, unstable, insidious* materials. Since they are less volatile than the parent ethers, they are concentrated in the residue during the distillation of an ether. Ethers, therefore, must never be "distilled to dryness," because an explosion may result. The purification of ethers is often accomplished by distillation from lithium aluminum hydride, a reductant capable of destroying any water and peroxide that may be present.[a]

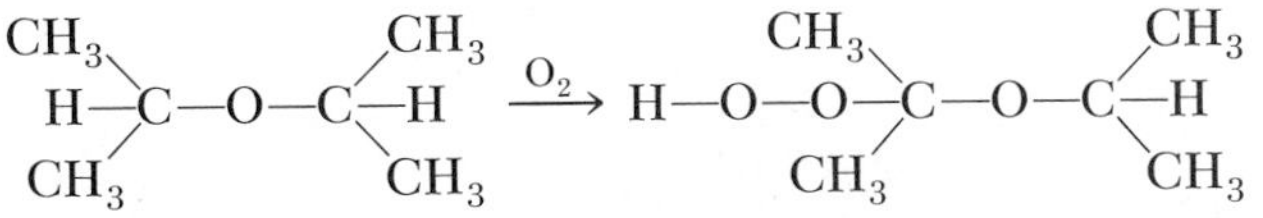

$$(CH_3)_2CH{-}O{-}CH(CH_3)_2 \xrightarrow{O_2} H{-}O{-}O{-}C(CH_3)_2{-}O{-}CH(CH_3)_2$$

isopropyl ether — peroxide of isopropyl ether

11.5 SYNTHESIS OF EPOXIDES

Epoxidation. The most direct synthesis of an epoxide is by the oxidation of an alkene. Peroxyacids are the most common oxidants used in this *epoxidation* reaction.

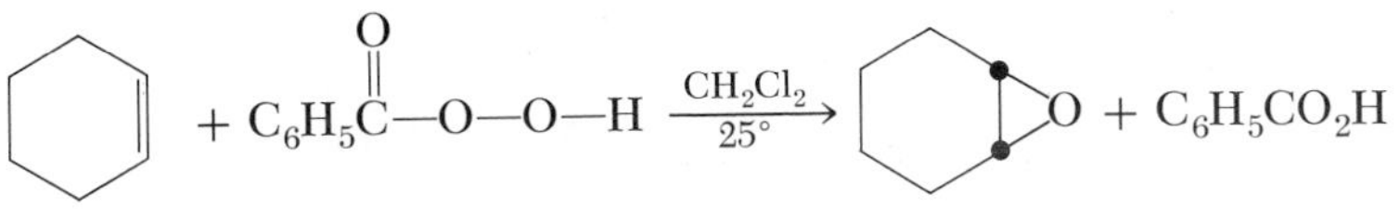

$$\text{cyclohexene} + C_6H_5\overset{O}{\overset{\|}{C}}{-}O{-}O{-}H \xrightarrow[25°]{CH_2Cl_2} \text{cyclohexene oxide} + C_6H_5CO_2H \qquad \text{epoxidation}$$

cyclohexene — peroxybenzoic acid — cyclohexene oxide — benzoic acid

There have been several mechanisms suggested for epoxidation, and it is possible that any one of several different mechanisms may be operative, depending upon the particular peracid and the alkene. The one shown below is believed to be operative in most epoxidation reactions.

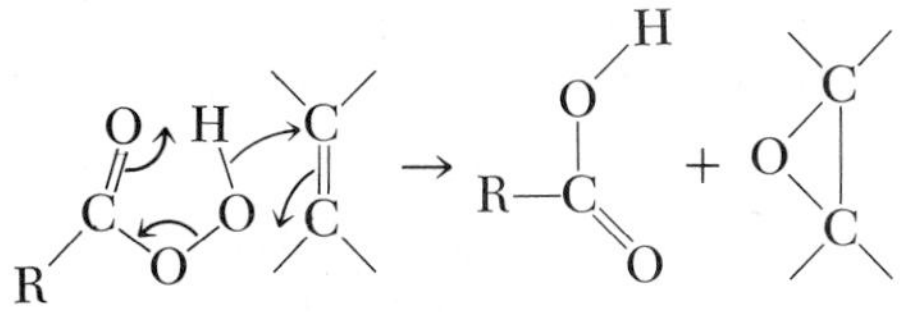

$$R{-}C(=O){-}O{-}O{-}H + C{=}C \rightarrow R{-}C(=O){-}O{-}H + \text{epoxide}$$

peroxyacid — epoxide

Because both carbon-oxygen bonds of the epoxide are formed simultaneously, as indicated in the above mechanism, the stereochemistry of the alkene is retained in the epoxide. Thus, *cis* alkenes produce *cis* epoxides and *trans* alkenes produce *trans* epoxides.

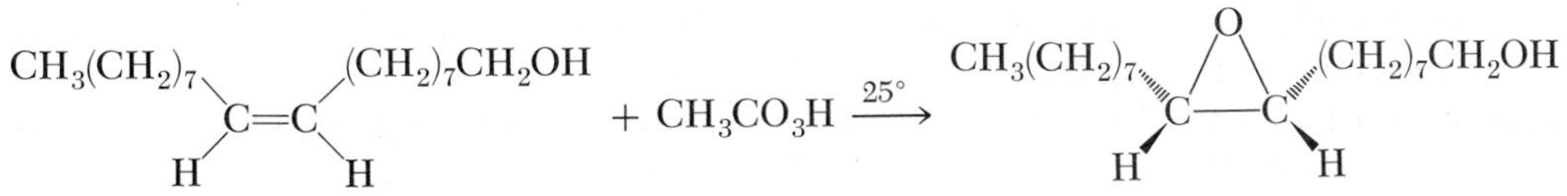

$$cis\text{-}CH_3(CH_2)_7CH{=}CH(CH_2)_7CH_2OH + CH_3CO_3H \xrightarrow{25°} cis\text{-}CH_3(CH_2)_7\overset{O}{CH{-}CH}(CH_2)_7CH_2OH$$

cis-9,10-epoxy-1-octadecanol

$$trans\text{-}CH_3(CH_2)_7CH{=}CH(CH_2)_7CH_2OH + CH_3CO_3H \xrightarrow{25°} trans\text{-}CH_3(CH_2)_7\overset{O}{CH{-}CH}(CH_2)_7CH_2OH$$

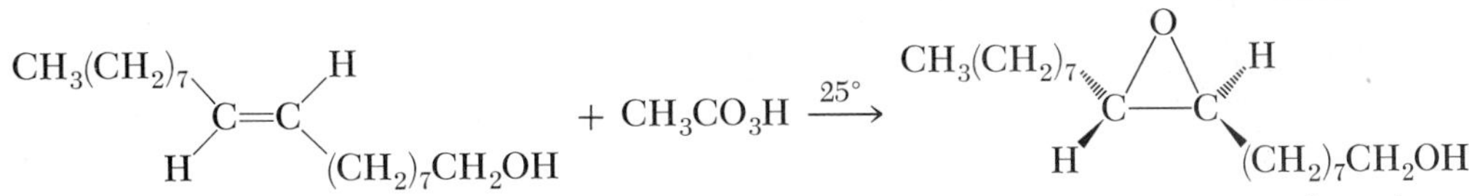

trans-9,10-epoxy-1-octadecanol

---

[a] Peroxide-free ethers usually are stored over metallic sodium or calcium hydride.

Epoxidation is favored by electron-releasing groups attached to the double bond, while electron-attracting groups decrease the rate of reaction. For example, tri- and tetraalkylated double bonds epoxidize more rapidly than do mono- or dialkylated double bonds. Unsymmetric cycloalkenes usually are oxidized preferentially at the less hindered face of the double bond.

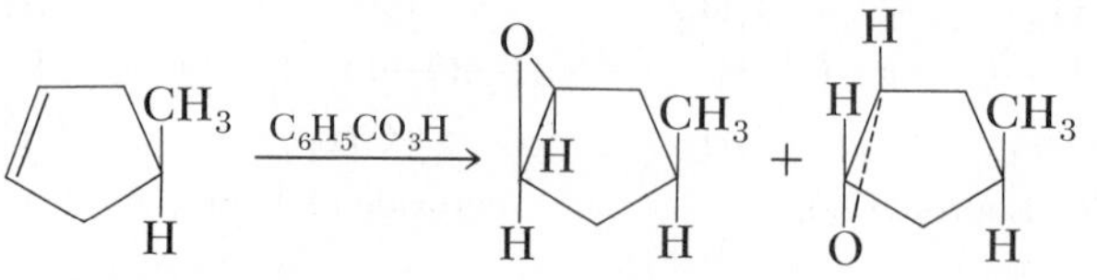

4-methylcyclopentene $\xrightarrow{C_6H_5CO_3H}$ (24%) + (76%)

Ethylene and propylene oxides, the most important industrial epoxides, are now prepared by the catalytic air-oxidation of the corresponding alkenes.

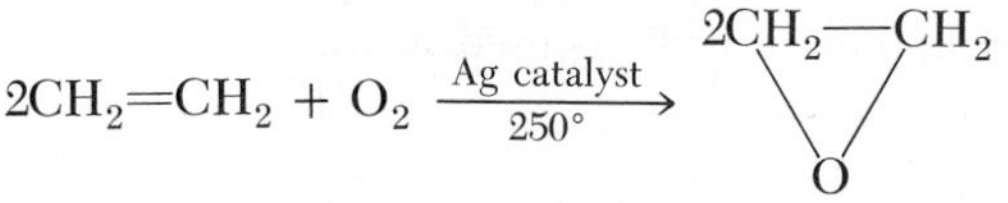

$$2CH_2{=}CH_2 + O_2 \xrightarrow[250°]{\text{Ag catalyst}} 2CH_2{-}CH_2 \text{ (with O bridge)}$$

ethylene oxide

Dehydrohalogenation of a Halohydrin. An older commercial synthesis of propylene oxide employs a chlorohydrin as an intermediate. This is a straightforward application of the Williamson synthesis in which the addition of a hypohalous acid (HOX) across a double bond, followed by reaction of the resultant halohydrin with base, converts an alkene into an epoxide. This type of scheme, while declining in industrial importance, is still an important laboratory procedure.

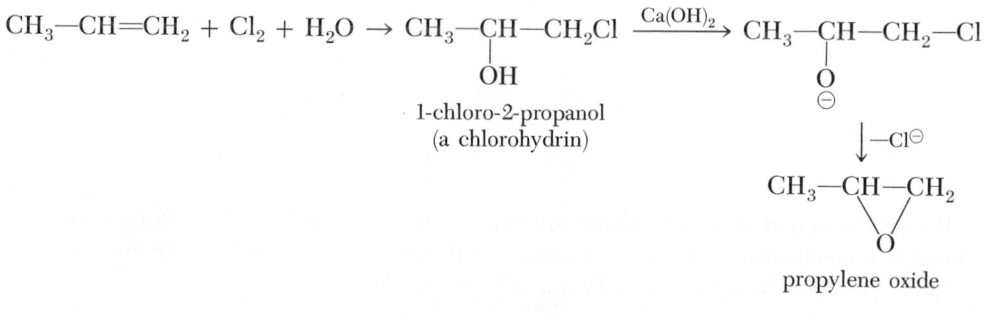

$$CH_3{-}CH{=}CH_2 + Cl_2 + H_2O \rightarrow CH_3{-}CH(OH){-}CH_2Cl \xrightarrow{Ca(OH)_2} CH_3{-}CH(O^{\ominus}){-}CH_2{-}Cl$$

1-chloro-2-propanol (a chlorohydrin)

$$\downarrow -Cl^{\ominus}$$

$$CH_3{-}CH{-}CH_2 \text{ (with O bridge)}$$

propylene oxide

The stereochemical constraints upon the *chlorohydrin-to-epoxide* conversion are illustrated by the behavior of *cis*- and *trans*-2-chlorocyclohexanol toward base—only the *trans* isomer produces epoxide. A diaxial geometry is required in order to achieve the necessary backside attack, and only the *trans* isomer can attain the necessary geometry.

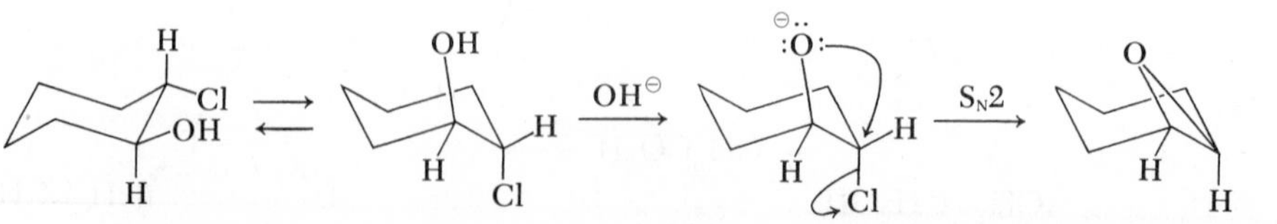

trans-2-chlorocyclohexanol $\rightleftharpoons$ (diaxial conformer) $\xrightarrow{OH^{\ominus}}$ (alkoxide) $\xrightarrow{S_N2}$ cyclohexene oxide

The diastereomeric *cis* isomer reacts, *via* an intramolecular hydride transfer, to form cyclohexanone.

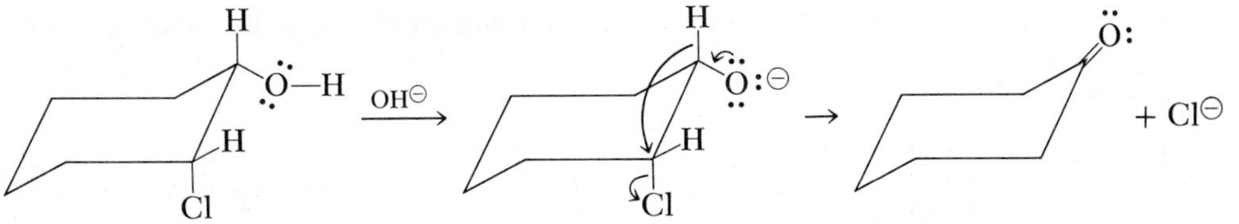

cis-2-chlorocyclohexanol $\xrightarrow{OH^{\ominus}}$ → cyclohexanone + $Cl^{\ominus}$

Synthesis of an epoxide with oxygen on the *more* hindered face of the double bond can be achieved by chlorohydrin formation followed by the Williamson synthesis. The stereochemistry of the final epoxide differs from that produced by direct epoxidation because it is the initially formed halonium ion that has the less hindered geometry.

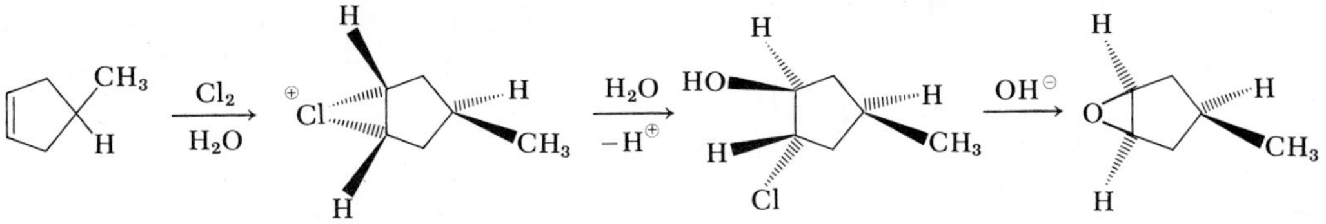

$\xrightarrow[H_2O]{Cl_2}$ $\xrightarrow[-H^{\oplus}]{H_2O}$ $\xrightarrow{OH^{\ominus}}$

A more recently discovered synthesis of epoxides, the reaction of carbonyl compounds with sulfur ylids, will be discussed in Chapter 23.

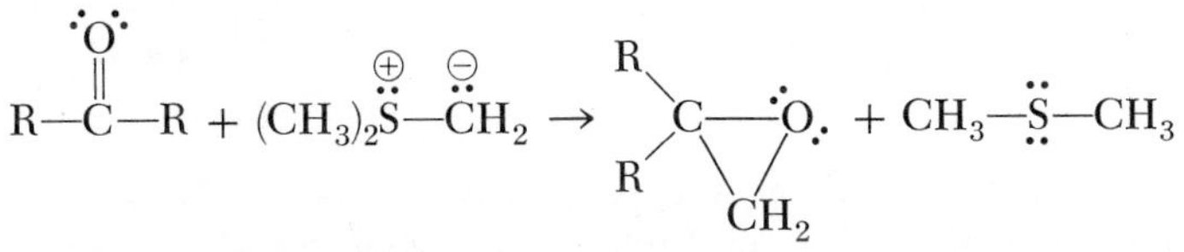

$$R{-}\overset{\overset{\textstyle O}{\|}}{C}{-}R + (CH_3)_2\overset{\oplus}{S}{-}\overset{\ominus}{C}H_2 \rightarrow R_2C{-}O{-}CH_2 + CH_3{-}S{-}CH_3$$

ketone | a sulfur ylid | an epoxide | a sulfide

4. Draw the structure of the epoxide(s) produced by epoxidation of:
(a) 1-hexene (b) 2-hexene (c) 3-hexene
(d) (**R**)-3-methylcyclopentene (e) (**Z**)-(2**R**,5**R**)-2,5-dichloro-3-hexene

11.6 REACTIONS OF EPOXIDES

This section concentrates upon ring opening reactions and reductions of epoxides.

Ring Opening Reactions. Unlike ordinary ethers, epoxides readily undergo carbon-oxygen bond cleavage by S_N reactions. Under neutral or alkaline conditions the reaction is invariably S_N2, as evidenced both by the kinetics of the ring opening and by the stereospecificity of the process. For example, the ring opening of cyclopentene oxide by methoxide ion produces only *trans*-2-methoxycyclopentanol.

cyclopentene oxide $\xrightarrow[S_N2]{^{\ominus}OCH_3}$ (a or b) $\xrightarrow[-OH^{\ominus}]{H_2O}$ *trans*-2-methoxy cyclopentanol

enantiomeric products

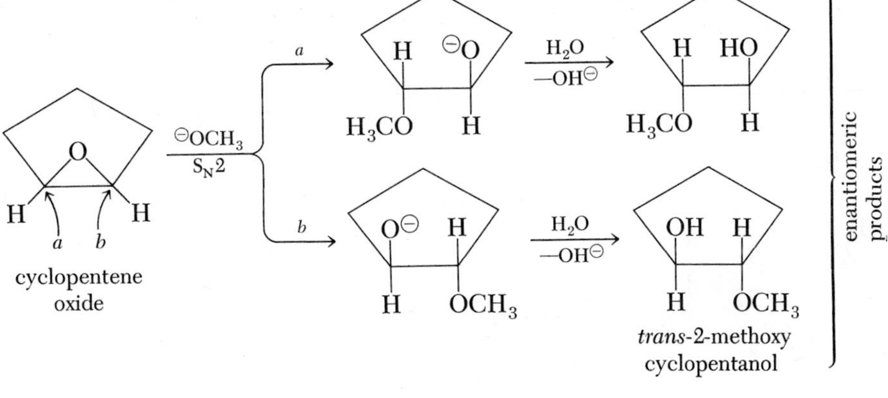

The rate of S_N2 opening can be enhanced by using an electrophilic catalyst, the most common being a proton. This procedure is used, for example, in the commercial preparation of ethylene glycol from ethylene oxide.

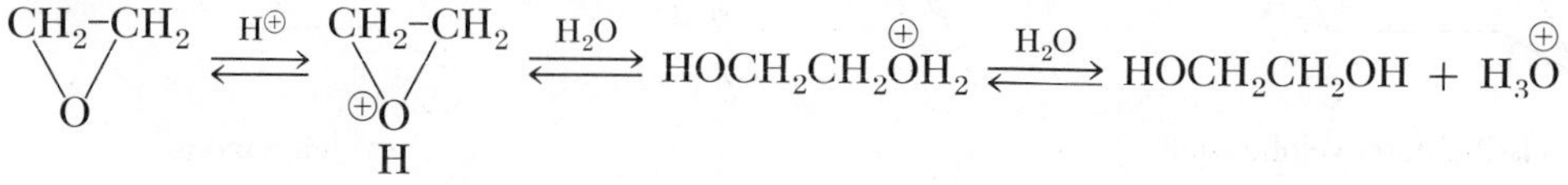

$$\underset{\text{O}}{\text{CH}_2\text{-CH}_2} \underset{}{\overset{\text{H}^\oplus}{\rightleftharpoons}} \underset{\oplus\text{O-H}}{\text{CH}_2\text{-CH}_2} \overset{\text{H}_2\text{O}}{\rightleftharpoons} \text{HOCH}_2\text{CH}_2\overset{\oplus}{\text{O}}\text{H}_2 \overset{\text{H}_2\text{O}}{\rightleftharpoons} \text{HOCH}_2\text{CH}_2\text{OH} + \text{H}_3\overset{\oplus}{\text{O}}$$

The reaction of propylene oxide with methanol in the presence of sulfuric acid produces 2-methoxypropanol, a product most easily accounted for by capture of a 2° cation with methanol. *However, there is evidence to suggest that this pathway, outlined below, is not taken.*

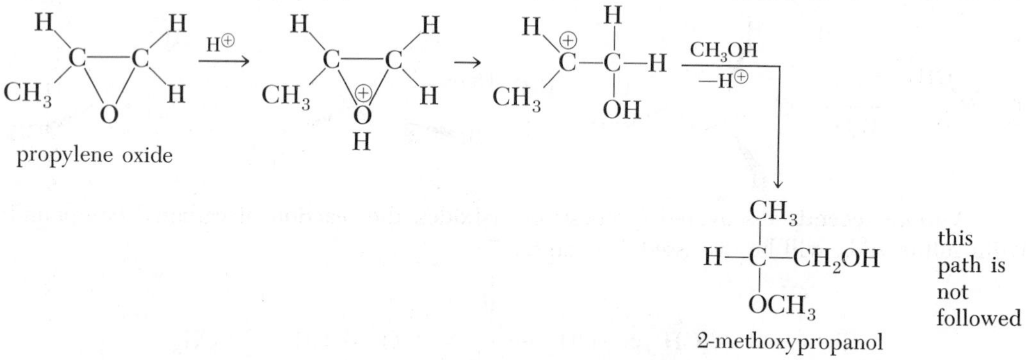

propylene oxide $\xrightarrow{\text{H}^\oplus}$ protonated epoxide $\rightarrow$ $\text{CH}_3\overset{\oplus}{\text{C}}\text{H}-\text{C}(\text{H})(\text{OH})-\text{H}$ $\xrightarrow[-\text{H}^\oplus]{\text{CH}_3\text{OH}}$ $\text{H}-\text{C}(\text{CH}_3)(\text{OCH}_3)-\text{CH}_2\text{OH}$

2-methoxypropanol

this path is not followed

Instead of such an S_N1 reaction, the process is believed to be an S_N2 reaction, with methanol attacking the more highly substituted ring carbon. Since we have previously noted (Chapter 5) the existence of steric hindrance towards S_N2 processes, we must explain this anomaly before we can accept this mechanism. To do this we envision the protonated epoxide as a resonance hybrid similar to the cyclic bromonium ion discussed in Chapter 8. This picture places more positive charge on the ring carbon bearing the methyl group than on the ring carbon bearing two hydrogens. It is this "excess" positive charge on the more highly substituted ring carbon that attracts the attacking electron pair on the oxygen of methanol, resulting in attack at C2 of propylene oxide. Thus, the reaction is an S_N2 process with considerable "S_N1 character."

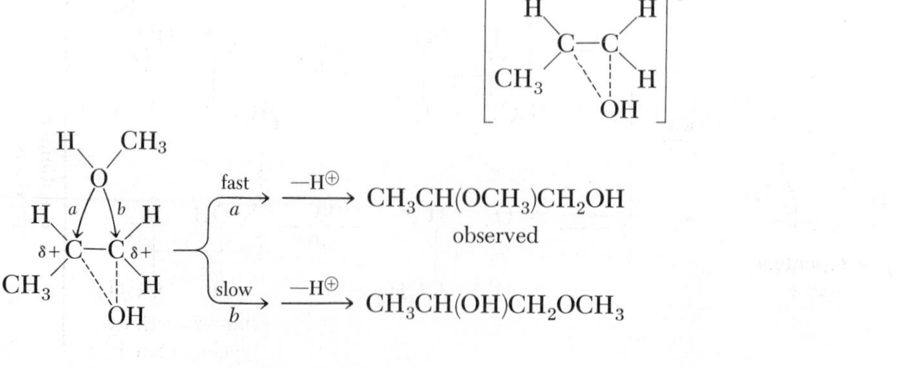

propylene oxide $\overset{\text{H}^\oplus}{\rightleftharpoons}$ [protonated epoxide $\leftrightarrow$ $\text{CH}_3\overset{\oplus}{\text{C}}\text{H}-\text{CH}_2\text{OH}$ $\leftrightarrow$ $\text{CH}_3\text{CH}(\text{OH})-\overset{\oplus}{\text{C}}\text{H}_2$] (more important contributors)

$$\left[\text{CH}_3\text{CH}\overset{\text{OH}}{\cdots}\text{CH}_2\right]^\oplus$$

CH_3OH attacking $\delta+$C (path a) or $\delta+$C (path b) of the hybrid:

fast, a $\xrightarrow{-\text{H}^\oplus}$ $\text{CH}_3\text{CH}(\text{OCH}_3)\text{CH}_2\text{OH}$

observed

slow, b $\xrightarrow{-\text{H}^\oplus}$ $\text{CH}_3\text{CH}(\text{OH})\text{CH}_2\text{OCH}_3$

Epoxides are cleaved by organometallic reagents (*e.g.*, Grignard reagents) to form alcohols. The reaction proceeds by inversion of configuration and is S_N2-like; coordination between the "ethereal" oxygen and the organometallic undoubtedly facilitates the reaction. Ring opening is not limited to organomagnesium or organolithium compounds, as evidenced by the action of lithium dimethylcopper (an organocopper compound) upon epoxycyclohexane.

$$CH_3Br + Li \rightarrow \underset{\text{methyl lithium}}{CH_3Li} \xrightarrow{Cu_2I_2} \underset{\text{lithium dimethylcopper}}{(CH_3)_2CuLi}$$

$$\text{epoxycyclohexane} + Li(CH_3)_2Cu \longrightarrow \xrightarrow{H_2O} \text{trans-2-methylcyclohexanol}$$

5. Predict the product(s) of the reaction of ethylene oxide with:
(a) $H_3O^{\oplus}$
(b) $OH^{\ominus}/H_2O$, then $H_3O^{\oplus}$
(c) HCl
(d) $CH_3S^{\ominus}Na^{\oplus}$
(e) $CH_3CH_2CH_2MgBr$, then $H_3O^{\oplus}$
(f) $Li(CH_3)_2Cu$, then $H_3O^{\oplus}$
(g) $K^{\oplus\ominus}SCN$
(h) conc. H_2SO_4
(i) $CH_3C{\equiv}C^{\ominus}Na^{\oplus}$

Reduction and Deoxygenation. Expoxides are conveniently reduced to alcohols by lithium aluminum hydride. Preferential formation of the more highly substituted carbinol suggests an S_N2 attack by $H^{\ominus}$, or its equivalent, upon the less hindered carbon. The less highly substituted carbinol is produced by reduction with diborane (review Table 10–3).

$$(H_3C)(C_2H_5)C\overset{O}{—}C(H)(CH_3) \xrightarrow{LiAlD_4} \xrightarrow{H_2O} HO(H_3C)(C_2H_5)C—C(H)(CH_3)(D)$$

$$(H_3C)(CH_3)C\overset{O}{—}C(H)(CH_3) \xrightarrow{B_2H_6} \xrightarrow{H_2O} H_3C—CH(CH_3)—CH(OH)—CH_3$$

Epoxides can be *deoxygenated* to alkenes by organophosphines, including triphenylphosphine, $(C_6H_5)_3P$. This reaction is initiated by an S_N2 attack by the phosphine upon the oxirane ring. Bond rotation followed by a *cis* elimination produces the final alkene with a configuration opposite to that of the starting epoxide.

$$(R_1)(R_2)C\overset{O}{—}C(R_4)(R_3) + \ddot{P}(C_6H_5)_3 \longrightarrow (C_6H_5)_3\overset{\oplus}{P}—C(R_1)(R_2)—C(\overset{\ominus}{O})(R_4)(R_3) \longrightarrow (C_6H_5)_3\overset{\oplus}{P}—C(R_1)(R_2)—C(R_3)(R_4)—O^{\ominus} \longrightarrow (R_1)(R_2)C{=}C(R_3)(R_4) + (C_6H_5)_3\overset{\oplus}{P}—O^{\ominus}$$

This reaction, which illustrates the propensity for triligant phosphorus compounds to be oxidized (see Chapter 23), can be employed to interconvert (**E,Z**) isomers.

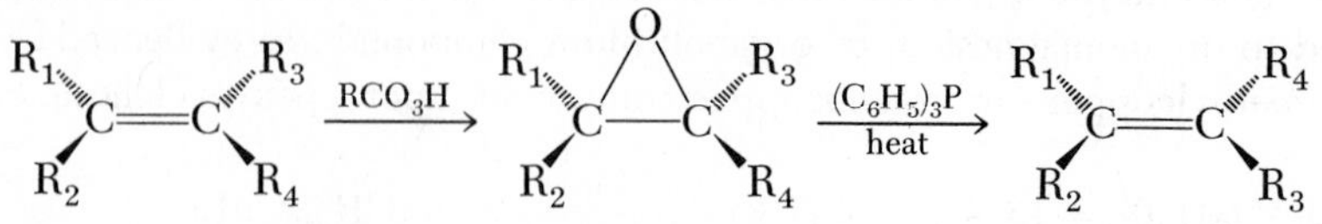

$$R_1R_2C{=}CR_3R_4 \xrightarrow{RCO_3H} R_1R_2C{-}CR_3R_4\ (\text{epoxide, O}) \xrightarrow[\text{heat}]{(C_6H_5)_3P} R_1R_2C{=}CR_4R_3$$

6. "Cyclobutene oxide cannot be deoxygenated with triphenylphosphine *via* the same mechanism by which ethylene oxide is deoxygenated." Explain.

11.7 THE OCCURRENCE OF EPOXIDES IN NATURE

Simple ethers do not appear to be of great biochemical importance. However, it is becoming apparent that epoxides do play some critical role in the "grand design" of living systems. One of the more interesting of these epoxides is a complex substance called a *juvenile hormone.*

Two hormones necessary for the normal development of insects are *ecdysone* and *juvenile hormone.* The former causes differentiation during the adult stage of life, while the latter causes growth but not maturation. Either the absence of ecdysone or the presence of excess juvenile hormone prevents the insect from becoming an adult. Since insects are thus rendered incapable of reproduction, control of their ecdysone–juvenile hormone balance may be a useful way of controlling insect pests. It is not in wide use today because juvenile hormone is extremely difficult to obtain in even modest quantities and because it is non-descriminating and will interfere with the maturation of all insects, including those considered helpful.

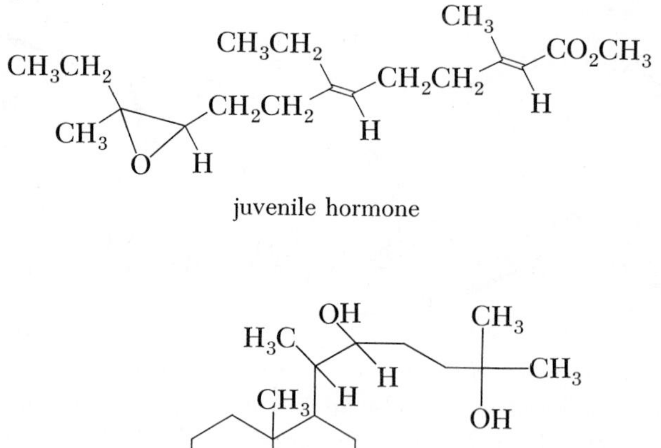

CH_3CH_2, CH_3, O, H, CH_2CH_2, CH_3CH_2, H, CH_2CH_2, CH_3, CO_2CH_3, H

juvenile hormone

OH, H_3C, H, H, CH_3, CH_3, CH_3, OH, CH_3, HO, HO, OH, H, O

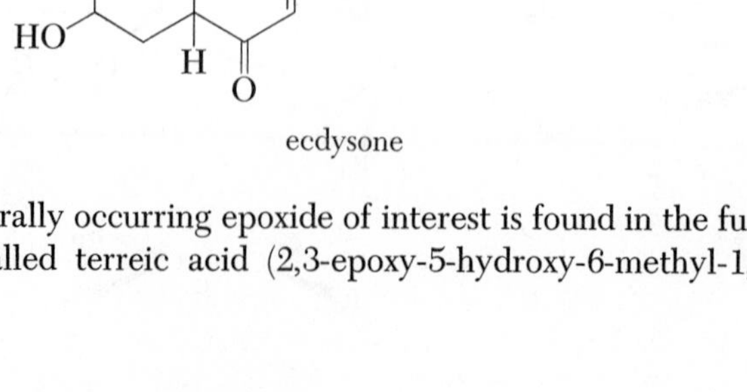

ecdysone

A second naturally occurring epoxide of interest is found in the fungus *Aspergillus terreus.* This substance, called terreic acid (2,3-epoxy-5-hydroxy-6-methyl-1,4-benzoquinone), is an antibiotic.

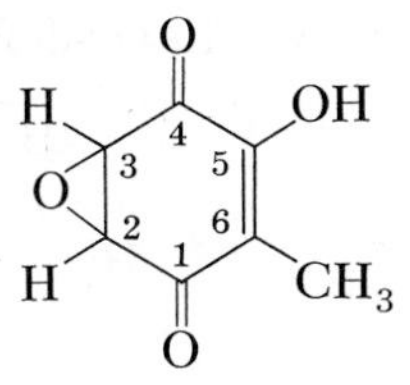

terreic acid

Just so that you will not get the impression that all natural epoxides are "good things," note that tutin, a diepoxide whose structure is shown below, is a poison. It is found in several places, including the seed of the bush *Coriaria ruscifolia,* also called the tutu.

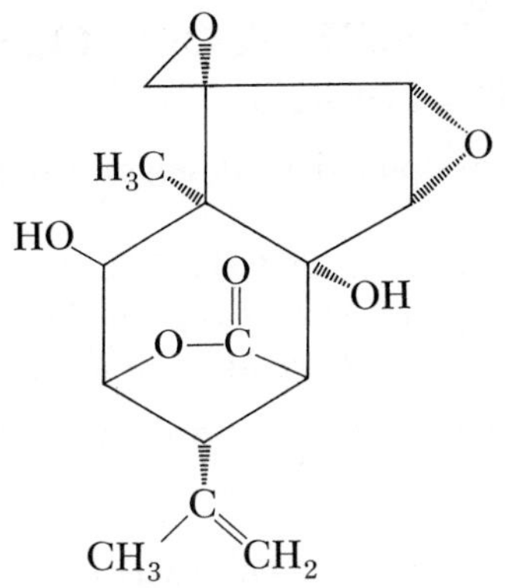

tutin

In the following chapter we will learn how an epoxide plays a critical role in the synthesis of steroids (sterols) in living systems.

11.8 SYNTHESIS OF *VIC*-DIOLS

The most important routes to *vic*-diols have already been encountered. Oxidation of an alkene with permanganate ion or osmium tetroxide produces *cis* dihydroxylation, while base-catalyzed hydrolysis of an epoxide produces *trans* dihydroxylation of the double bond.

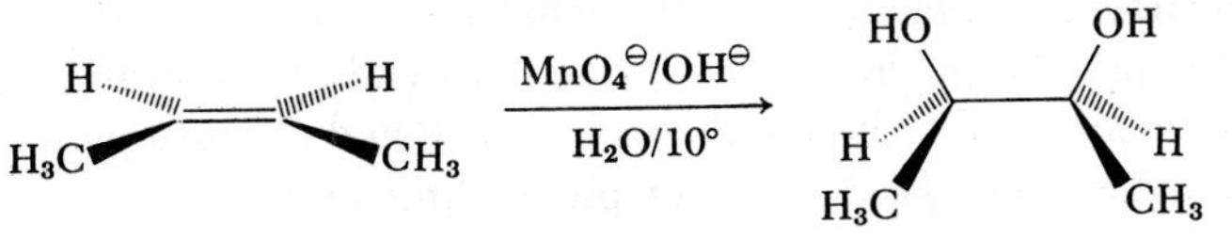

(**Z**)-2-butene $\xrightarrow[H_2O/10°]{MnO_4^{\ominus}/OH^{\ominus}}$ *meso*-2,3-butanediol (*cis* dihydroxylation)

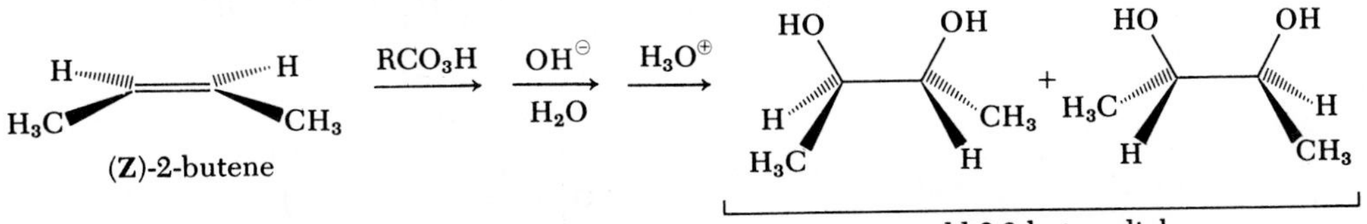

(**Z**)-2-butene $\xrightarrow{RCO_3H}$ $\xrightarrow[H_2O]{OH^{\ominus}}$ $\xrightarrow{H_3O^{\oplus}}$ *d,l*-2,3-butanediol (*trans* dihydroxylation)

Halohydrins undergo alkaline hydrolysis to *vic*-diols *via* an intermediate epoxide.

$$\text{cyclopentene} \xrightarrow[H_2O]{Cl_2} \text{trans-2-chlorocyclopentanol} \xrightarrow[H_2O]{OH^{\ominus}} [\text{cyclopentene oxide}] \xrightarrow[H_2O]{OH^{\ominus}} \text{trans-1,2-cyclopentanediol}$$

trans-1,2-cyclopentanediol
(*trans* dihydroxylation)

The synthesis of *vic*-diols starting with aldehydes or ketones, a process called the *pinacol reduction*, is discussed in Chapter 16.

11.9 ETHER-ALCOHOLS AND *VIC*-DIOLS

Three important classes of compounds, *cellosolves, carbitols,* and *carbowaxes,* are made from ethylene oxide.

Cellosolves, compounds of the type $ROCH_2CH_2OH$, are made by the reaction of an alkoxide with ethylene oxide. *Butyl cellosolve* has good solvent characteristics, and is used in hydraulic brake fluids and as an anti-icing additive in aviation fuels.

$$\underset{\text{butanol}}{CH_3(CH_2)_3OH} + NaOH \rightleftarrows \underset{\text{sodium butoxide}}{CH_3(CH_2)_3O^{\ominus}\ Na^{\oplus}} + H_2O$$

$$CH_3(CH_2)_3O^{\ominus} + (CH_2)_2O \rightarrow CH_3(CH_2)_3OCH_2CH_2O^{\ominus} \xrightarrow{H_2O} \underset{\text{butyl cellosolve}}{CH_3(CH_2)_3O(CH_2)_2OH}$$

Carbitols, used as solvents and in lacquer formulation, are monoalkyl ethers of di(ethylene glycol), $HOCH_2CH_2OCH_2CH_2OH$, and are made by reacting a cellosolve with an equivalent of ethylene oxide.

$$\underset{\text{methyl cellosolve}}{CH_3OCH_2CH_2OH} + (CH_2)_2O \rightarrow \underset{\text{methyl carbitol}}{CH_3OCH_2CH_2OCH_2CH_2OH}$$

Methylation of methyl carbitol produces an important high boiling (162°) solvent, *diglyme.*

$$\underset{\text{diglyme}}{CH_3O(CH_2)_2O(CH_2)_2OCH_3}$$

When ethylene oxide reacts with a *vic*-diol (*e.g.*, ethylene glycol), the product is a polymer of varying molecular weight known as a *carbowax.* Since the ethylene glycol is bifunctional (*i.e.*, it contains two functional groups), the polymer grows from both ends. Carbowaxes are used as lubricants, as bases for ointments, and as liquid phases for gas liquid chromatography (also called gas liquid phase chromatography, *glpc,* or vapor phase chromatography, *vpc*).

$$HO^{\ominus} + (CH_2)_2O \rightarrow HOCH_2CH_2O^{\ominus} \rightleftarrows {}^{\ominus}OCH_2CH_2OH$$

$$\downarrow (CH_2)_2O \qquad \downarrow (CH_2)_2O$$

$$HO(CH_2)_2O(CH_2)_2O^{\ominus} \rightleftarrows {}^{\ominus}O(CH_2)_2O(CH_2)_2OH$$

$$\downarrow n(CH_2)_2O$$

$$\downarrow H_2O,\ -OH^{\ominus}$$

$$\underset{\text{a carbowax}}{HO\text{—}(CH_2CHO)_{n+2}\text{—}H}$$

7. Suggest a synthesis for each of the following, using any starting material containing less than four carbons.
   (a) $CH_3SCH_2CH_2OCH_2CH_2OH$
   (b) $HOCH_2CH_2OCH_2CH_2OCH_2CH_2OCH_2CH_2OCH_3$
   (c) $CH_3(CH_2)_3OCH_2CH_2CH_2OCH_2CH_2OH$

vic-Diols react with oxidants to form aldehydes (RCHO), ketones (R_2CO), and/or carboxylic acids (RCO_2H), depending upon the diol and the oxidant. For example, warm solutions of potassium permanganate or chromic acid ($CrO_3 + H_2SO_4$) oxidize diols to ketones and carboxylic acids, the latter arising from intermediate aldehydes.

$$CH_3CHOHCOH(CH_3)C_6H_5 \xrightarrow[H_2O/heat]{MnO_4^{\ominus}} \underset{\text{acetaldehyde}}{[CH_3CHO]} + \underset{\text{acetophenone}}{H_3C\overset{O}{\overset{\|}{C}}C_6H_5}$$

$$[CH_3CHO] \xrightarrow[H_2O/heat]{MnO_4^{\ominus}} CH_3CO_2H$$

$$CH_3CHOHCH_2OH \xrightarrow[H_2O/heat]{MnO_4^{\ominus}} [CH_3CHO + \underset{\text{formaldehyde}}{CH_2O}] \xrightarrow[H_2O/heat]{MnO_4^{\ominus}} CH_3CO_2H + [HOCO_2H] \rightarrow H_2O + CO_2$$

Periodic acid (HIO_4) and lead tetracetate ($Pb(OAc)_4$) oxidize *vic*-diols to aldehydes and/or ketones, but normally do not lead to further oxidation. The mechanism involves a cyclic intermediate.

$$H_3C-\underset{OH}{\overset{CH_3}{C}}-\underset{OH}{\overset{H}{C}}-CH_3 + Pb(OAc)_4 \xrightarrow{-2HOAc} \text{[cyclic intermediate: } H_3C-\overset{CH_3}{C}-\overset{H}{C}-CH_3 \text{ with both O bonded to } Pb(OAc)_2\text{]} \rightarrow$$

$$CH_3\overset{O}{\overset{\|}{C}}CH_3 + CH_3\overset{O}{\overset{\|}{C}}H + Pb(OAc)_2$$

Because ketones and aldehydes are easily characterized and identified, mild oxidation has found extensive use in the study of polyhydroxy compounds (see Chapter 25).

11.10 CHARACTERIZATION OF ETHERS, EPOXIDES, AND *VIC*-DIOLS

The presence of an ether linkage is often established by negative evidence, in that many of the ordinary analytical reagents do not react with the sample. Most ethers are sufficiently basic to dissolve in cold, concentrated sulfuric acid; the resulting solutions often regenerate the ether upon dilution with water. This permits their distinction from alkanes.

$$R{-}O{-}R + H_2SO_4 \rightleftarrows R{-}\overset{H}{\overset{|\oplus}{O}}{-}R \quad HSO_4^{\ominus} \xrightarrow{H_2O\ (excess)} R{-}O{-}R + H_3O^{\oplus} + HSO_4^{\ominus}$$

Epoxides are often characterized by a multi-step procedure that involves hydrolysis to a *vic*-diol, oxidation of the diol to carbonyl compounds, and characterization of the carbonyl compounds, as described in Chapter 17.

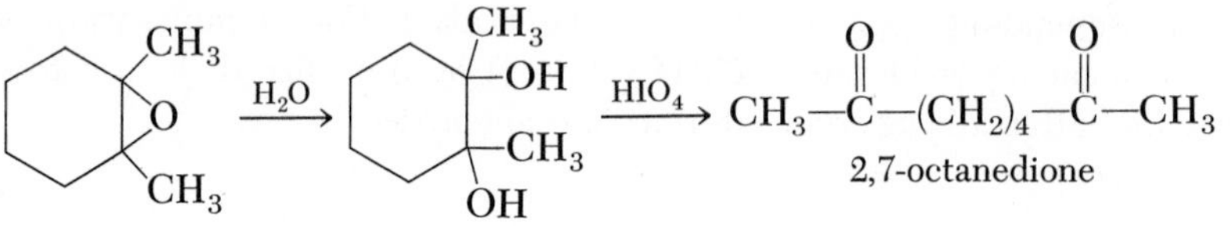

$$\text{1,2-dimethylcyclohexene oxide} \xrightarrow{H_2O} \text{1,2-dimethyl-1,2-cyclohexanediol} \xrightarrow{HIO_4} CH_3{-}\overset{O}{\overset{\|}{C}}{-}(CH_2)_4{-}\overset{O}{\overset{\|}{C}}{-}CH_3$$

2,7-octanedione

The presence of a *vic*-diol can be established by a variation of the periodic acid oxidation called the *Malaprade reaction.* In this test the diol is treated with a mixture of aqueous nitric acid and periodic acid; after approximately 15 seconds, aqueous silver nitrate is added to the entire mixture. A white precipitate of silver iodate, $AgIO_3$, forms immediately. (Silver periodate is soluble in dilute nitric acid, while silver iodate is not.)

$$RCHOHCHOHR' + HIO_4 \rightarrow RCHO + R'CHO + H_3O^{\oplus} + IO_3^{\ominus}$$

$$IO_3^{\ominus} + Ag^{\oplus} \xrightarrow[H_2O]{HNO_3} AgIO_3\downarrow$$

silver iodate

As with many other "wet tests," the Malaprade reaction is based upon the occurrence of a reaction (*i.e.*, oxidation) rather than on the presence of an explicit functional group. Therefore, *any* compound that either is converted to a *vic*-diol or is directly oxidized by periodic acid will give a positive test. This explains why positive tests can sometimes be obtained with epoxides and α-hydroxy carbonyl compounds.

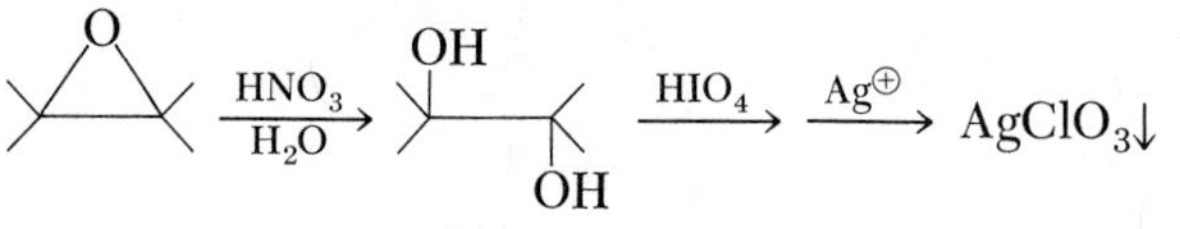

$$\text{tetramethyloxirane} \xrightarrow[H_2O]{HNO_3} \text{2,3-dimethyl-2,3-butanediol} \xrightarrow{HIO_4} \xrightarrow{Ag^{\oplus}} AgClO_3\downarrow$$

$$RCHOHC(O)R' + HIO_4 \rightarrow RCHO + R'CO_2H + HIO_3 \xrightarrow{Ag^{\oplus}} AgClO_3\downarrow$$

Spectral Characterization. Simple ethers show a C—O stretching vibration near 1100 cm^{-1}, as do alcohols; but, unlike alcohols, the ethers lack the broad, characteristic O—H absorption near 3300 cm^{-1}. Epoxides often exhibit an absorption near 1250 cm^{-1}, ascribed to a C—O stretching mode. However, the absence of this absorption does not preclude the presence of an epoxide group. Absorptions near 850 cm^{-1} in the ir spectra of a number of epoxides may be analogous to the C—H bending modes in alkenes.

vic-Diols are recognized in the ir by characteristic OH absorptions and, if geometry permits, by intramolecular hydrogen bond formation. Simple distillation can often serve to distinguish between a simple alcohol and a diol; extensive intermolecular hydrogen bonding gives a diol a much higher boiling point than that of a comparable monohydric alcohol (Table 11-1).

The nmr spectra of a diol can show characteristic **OH** resonances, especially in d_6-DMSO. The utility of nmr spectroscopy in studies of ethers and epoxides is limited to the effects that the oxygen may have upon proximal C—H signals. The infrared and nmr spectra of ethanol, ethylene glycol, and propylene oxide are presented in Figures 11-1 and 11-2.

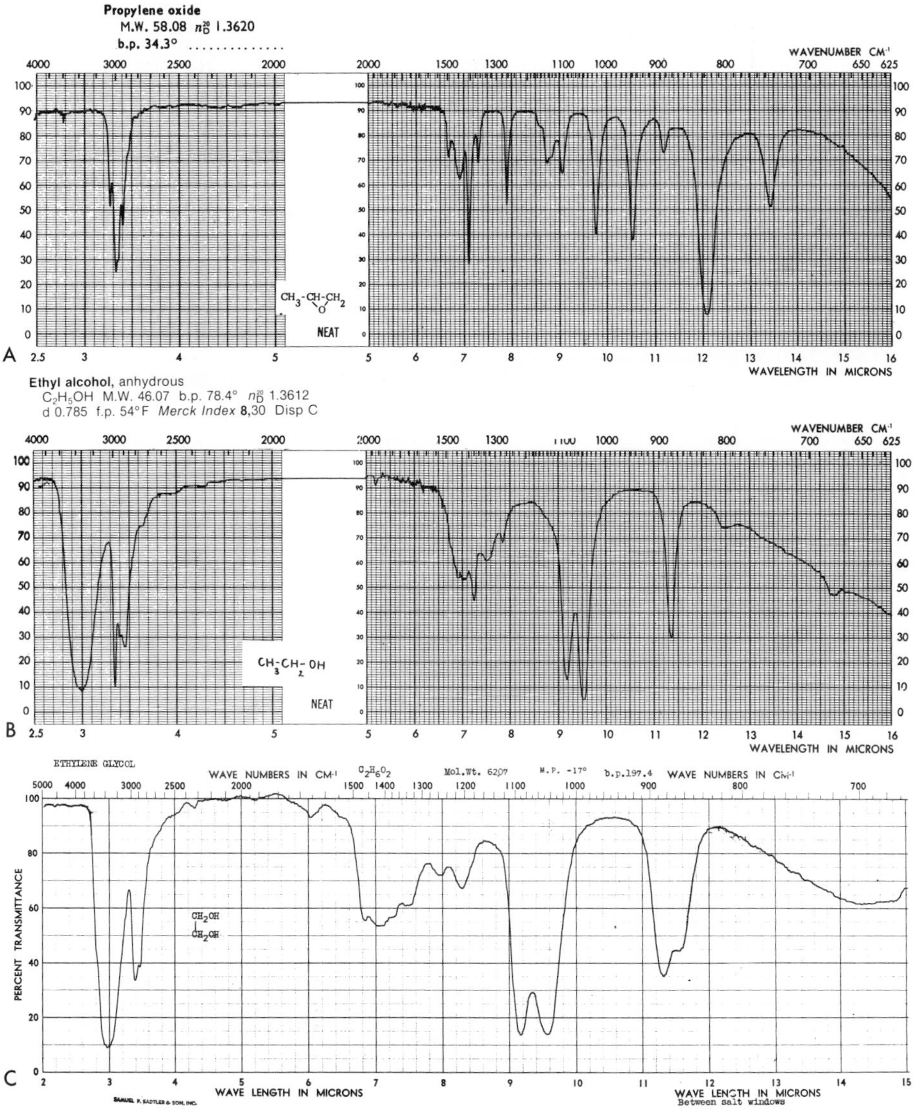

Propylene oxide
M.W. 58.08 n_D^{20} 1.3620
b.p. 34.3°
WAVENUMBER CM⁻¹
CH3-CH-CH2 O
NEAT
A
WAVELENGTH IN MICRONS
Ethyl alcohol, anhydrous
C_2H_5OH M.W. 46.07 b.p. 78.4° n_D^{20} 1.3612
d 0.785 f.p. 54°F Merck Index 8,30 Disp C
WAVENUMBER CM⁻¹
CH3-CH2-OH
NEAT
B
WAVELENGTH IN MICRONS
ETHYLENE GLYCOL
WAVE NUMBERS IN CM⁻¹
$C_2H_6O_2$
Mol.Wt. 62.07
M.P. -17°
b.p.197.4
PERCENT TRANSMITTANCE
CH2OH
CH2OH
C
WAVE LENGTH IN MICRONS
WAVE LENGTH IN MICRONS
Between salt windows

© Sadtler Research Laboratories, Inc., 1976.

Figure 11-1 Infrared spectra of ethyl alcohol, propylene oxide, and ethylene glycol.

a b c
CH3CH2-OH
ASSIGNMENTS
a 1.17
b 3.58
c 4.40
d 3.32 imp.
A

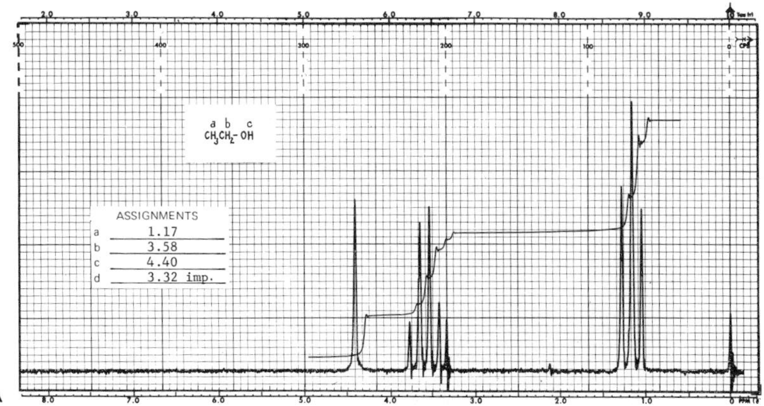

Figure 11-2 *See legend on page 340.*

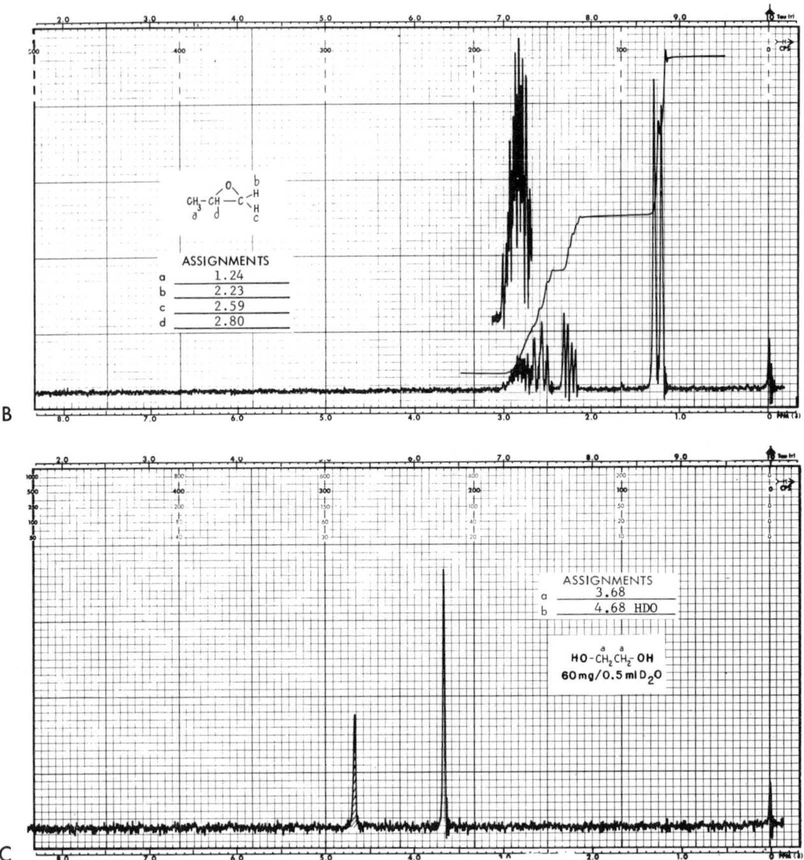

Figure 11-2 NMR spectra of ethyl alcohol, propylene oxide, and ethylene glycol. © Sadtler Research Laboratories, Inc., 1976.

IMPORTANT TERMS

Epoxide: A compound containing the $>\overset{O}{C{-}C}<$ functional group.

Ether: A compound containing the R—O—R linkage.

***gem*:** An abbreviation for *geminal*, meaning bonded to a common carbon. Methylene chloride, CH_2Cl_2, is a geminal dihalide.

Halohydrin: A compound containing a halogen and a hydroxy group, most often bonded to adjacent carbons. If the halogen is chlorine, the halohydrin is a *chlorohydrin*.

Monohydric: Containing one hydroxy group.

Peroxide: A compound containing the —O—O— linkage. Organic peroxides are unstable and may explode when heated. The facile cleavage of the oxygen-oxygen bond (bond strength $\approx$33 kcal/mole) makes peroxides good radical initiators.

Polyhydric: Containing several hydroxy groups.

***vic*:** An abbreviation for *vicinal*, meaning bonded to adjacent carbons. Cyclohexane-1,2-diol is a vicinal diol.

Viscosity: A measure of resistance to flow. The more viscous a material is, the more slowly it flows. The opposite of viscosity is fluidity.

PROBLEMS

8. Draw the structural formulas for:
   (a) methyl ether
   (b) ethyl ether
   (c) ethyl methyl ether
   (d) isobutyl *sec*-butyl ether
   (e) allyl ether
   (f) vinyl ether
   (g) (**R**)-2-chloropropyl ether
   (h) bicyclo[2.2.2]octene oxide
   (i) *cis*-2,3-epoxypentane
   (j) *trans*-2,3-epoxypentane
9. Suggest a test to distinguish between:
   (a) ethanol and ethyl ether
   (b) ethanol and ethyl iodide
   (c) cyclohexene and cyclohexene oxide
   (d) vinyl ether and ethyl ether
   (e) ethanol, isopropyl alcohol, and isopropyl ether
   (f) ethynyl ether and vinyl ether
   (g) ethylene oxide and cyclohexane
   (h) 1,2-octanediol and 1,6-octanediol
   (i) *n*-butyl chloride, *t*-butyl chloride, and *n*-propyl ether
10. Predict the product(s), if any, obtained by reacting the following with methyl magnesium chloride (one mole).
   (a) ethylene oxide
   (b) 2-chloroethanol
   (c) 3-chloropropanol
   (d) epoxycyclohexane
   (e) *cis*-2-butene oxide
   (f) ethyl ether
   (g) ethylene glycol
   (h) (**R**)-propylene oxide
11. What product(s), if any, should be produced by the reactions of the following pairs?
   (a) sodium ethoxide and ethyl iodide
   (b) sodium ethoxide and *t*-butyl iodide
   (c) ethylene oxide and potassium *t*-butoxide
   (d) propyl ether and cold, conc. sulfuric acid
   (e) propyl ether and hot, conc. sulfuric acid
   (f) methyl isopropyl ether and hot, conc. hydriodic acid
12. What product(s) are expected from the periodic acid oxidation of the following?
   (a) ethylene glycol
   (b) 1,2-propanediol
   (c) $CHO(CHOH)_3CH_2OH$
   (d) *cis*-1,2-cyclopentanediol
13. Draw all of the stereoisomers of insect juvenile hormone. Assign a configuration (**R, S, E,** and **Z**) to each appropriate center.

C_2H_5, CH_3 / epoxide (O) / H, CH_2CH_2—C(CH_2CH_3)=C(H)—CH_2CH_2—C(CH_3)=C(H)—C(=O)—O—CH_3

insect juvenile hormone

14. Grignard reagents do not react readily with alkyl halides, but they do react (couple) with α-chloroethers ($ROCHR'Cl$). Rationalize this difference.
15. Account for the following pH dependence:

(a)
$$Cl^{\ominus} + CH_3CH\overset{O}{—}CH_2 \xrightarrow[H_2O]{pH\ 7} \underset{(86\%)}{CH_3CHOH—CH_2Cl} + \underset{(14\%)}{CH_3CH(Cl)—CH_2OH}$$

(b)
$$Cl^{\ominus} + CH_3CH\overset{O}{—}CH_2 \xrightarrow[H_2O]{pH\ 4} \underset{(64\%)}{CH_3CH(OH)—CH_2Cl} + \underset{(36\%)}{CH_3CH(Cl)—CH_2OH}$$

16. One method for protecting hydroxy groups involves the reaction of an alcohol with an α,β-unsaturated ether. Provide mechanisms to account for the following:

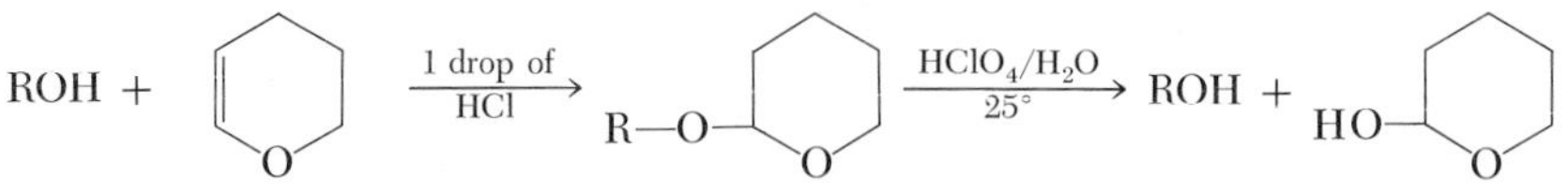

ROH + dihydropyran $\xrightarrow[\text{HCl}]{\text{1 drop of}}$ R—O—(tetrahydropyranyl) (protected alcohol) $\xrightarrow[25°]{HClO_4/H_2O}$ ROH + HO—(tetrahydropyranyl)

When one mole of (**R**)-2-butanol reacts with less than one mole of dihydropyran, and the resultant mixture is separated by distillation, three optically active fractions are obtained. Account for this. What are the structures of these three compounds?

17. In an attempt to prove that epoxidation involves the concerted *cis* addition of oxygen to the π bond, both *cis*-2-butene and cyclopentene were shown to undergo stereospecific formation of epoxide. Is either of these observations better proof than the other of the *cis* nature of the epoxidation?

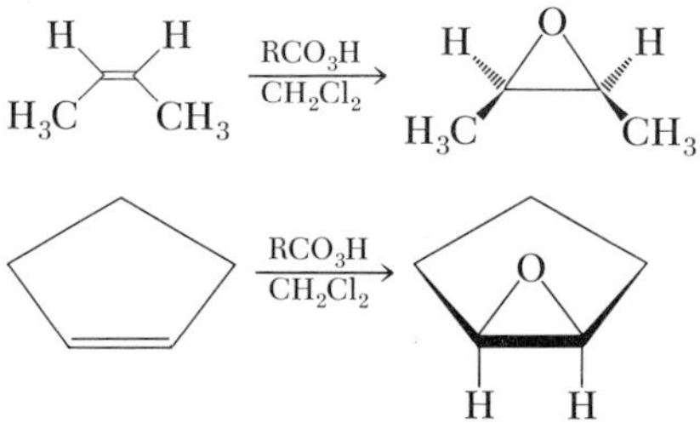

cis-2-butene $\xrightarrow[CH_2Cl_2]{RCO_3H}$ *cis*-2,3-epoxybutane

cyclopentene $\xrightarrow[CH_2Cl_2]{RCO_3H}$ cyclopentene oxide

18. While compound **A**, below, has never been prepared (and may never be), **B** is well known. Suggest a reason for this difference. (*Hint:* Build a model of each.)

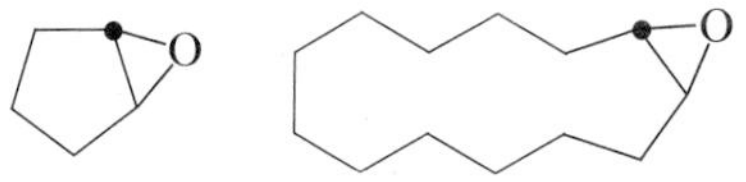

A **B**

19. Two compounds have the molecular formula $C_5H_{12}O$. One reacts with sodium, with the evolution of hydrogen; the other does not react. Both react rapidly with concentrated hydrochloric acid to give a water-insoluble product. Identify these compounds.

20. Two compounds have the molecular formula $C_4H_{10}O$. The boiling point of one is above 80°, and that of the other is under 50°. Each gives a single product when heated with hydrogen iodide. When passed over alumina (at 350°), one compound gives a gas while the other gives a liquid at these temperatures. Ozonolysis of the liquid gives a one-carbon aldehyde and a three-carbon aldehyde. Identify the two starting materials.

*21. The nmr spectra on p. 343 are those of anisole, butyl vinyl ether, *p*-methoxyanisole, and *p*-methylanisole. Assign the correct structure to each of these spectra. Identify as many spectral features as possible.

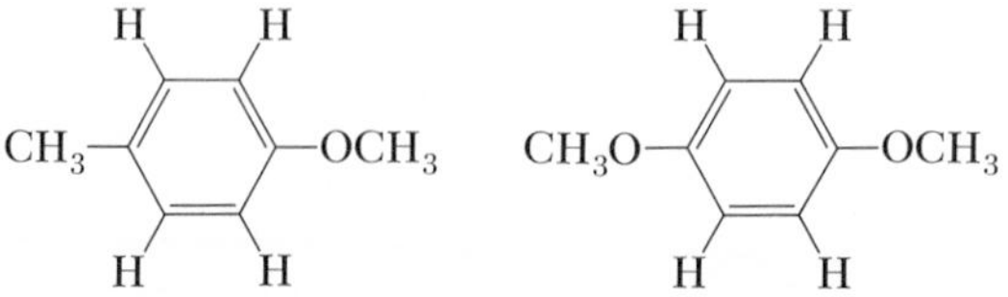

CH_3—C_6H_4—OCH_3 *p*-methylanisole

CH_3O—C_6H_4—OCH_3 *p*-methoxyanisole

*This problem requires a knowledge of spectroscopy.

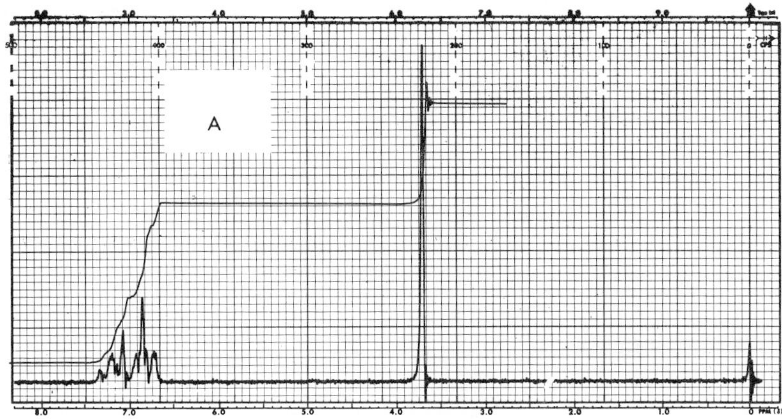
A

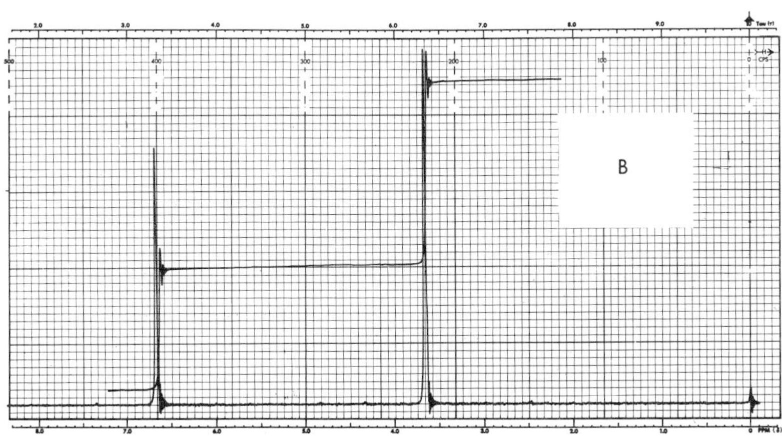
B

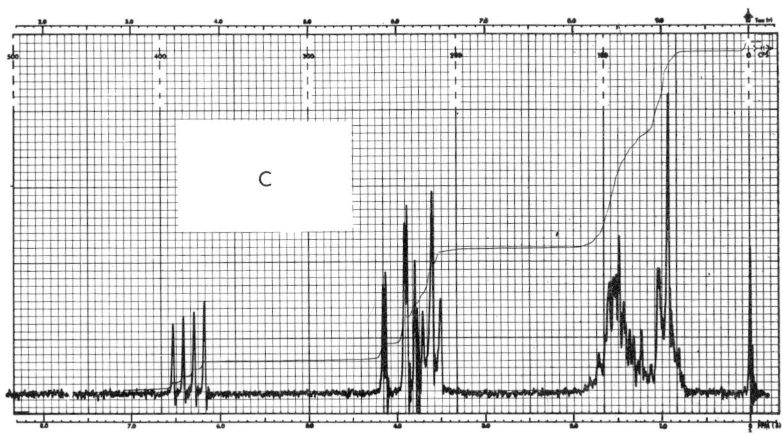
C

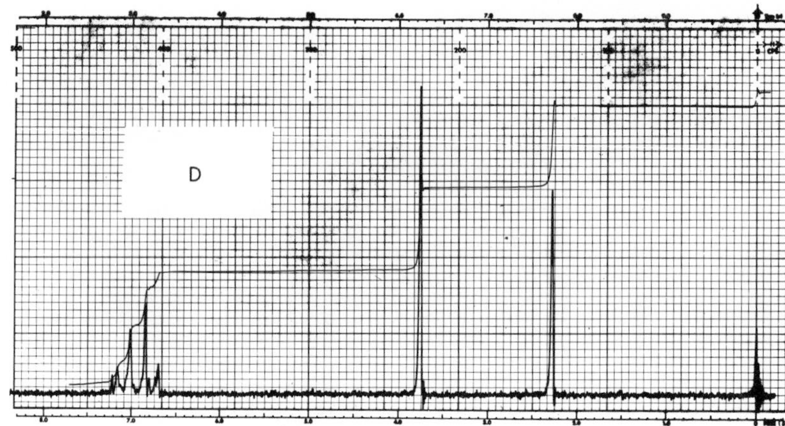

*22. Using either nmr or ir for each member of a pair below, explain how one could distinguish one member of a pair from another. If you use, for example, nmr for the first member of a pair, you must use nmr for the second member of the pair as well.

(a) ethyl ether and ethyl alcohol
(b) ethyl alcohol and ethylene glycol
(c) diethylene glycol and diethylene glycol dimethyl ether
(d) *n*-hexane and *n*-hexyl ether
(e) 1-hexene and 1-hexene oxide
(f) propene and propene oxide
(g) propene oxide and propylene chlorohydrin

*23. The infrared spectra of concentrated solutions (CCl_4 solvent) of *cis*- and *trans*-1,2-cyclopentanediol are rather similar in the O—H stretching region. However, as these solutions are diluted with CCl_4 the spectra begin to change. At approximately 10^{-5} *M*, the *trans* isomer shows only free —OH absorption while the *cis* isomer shows both free and hydrogen-bonded —OH absorption. Further dilution does not alter either spectrum. Account for the effect of dilution upon the ir spectra of these isomers.

---

*This problem requires a knowledge of spectroscopy.

DIENES AND TERPENES

12.1 WHERE ARE WE GOING?

The previous chapters dealt largely with isolated functional groups, *i.e.*, groups whose reactions are essentially independent of adjoining functionalities. This and the following chapter will, in contrast, concentrate upon the effect that mobile electron pairs separated by a single bond have upon one another. Such an arrangement, in which electron pairs (and the groups bearing them) are **conjugated,** raises a number of important questions: (a) To what extent may conjugated groups influence one another? (b) How can such interactions be quantified? (c) What is the theoretical basis for such interaction? (d) Is it ever valid to consider a conjugated system as a new functional group?

The existence of an almost limitless combination of conjugated systems, some of which are presented below, precludes a detailed discussion of all conjugated systems. Instead, we will concentrate upon conjugated carbon-carbon double bonds, since conjugated dienes are both extremely important and rather well studied. The topics covered in this and the following chapter constitute one unit of work divided, somewhat artificially, for convenience.

$H_2C{=}CH{-}CH{=}CH_2$ 1,3-butadiene

$H_2C{=}CH{-}CH{=}\ddot{O}{:}$ acrolein

representative conjugated systems

$H_2C{=}CH{-}C{\equiv}N{:}$ acrylonitrile

$H_2C{=}CH{-}\ddot{Cl}{:}$ vinyl chloride

$H_2C{=}CH{-}\ddot{O}{-}CH_3$ methyl vinyl ether

$C_6H_5{-}CH{=}CH_2$ styrene

12.2 THE STRUCTURE OF 1,3-BUTADIENE

Because 1,3-butadiene represents the simplest π-π conjugated system, it will receive the bulk of our attention. If interaction between the π bonds in 1,3-butadiene does occur, it should be detectable by *careful* analysis of the properties associated with either the central single bond or the outer double bonds.

Bond Distance. The accurate determination of the C2-C3 (*i.e.*, single bond) distance in 1,3-butadiene created a flurry among organic chemists and theoreticians because the bond appeared to be "too short." It is, in fact, shorter than the single bond distances in ethane, propene, *n*-butane, and 2-butene.

1,3-butadiene	ethane	propene	*n*-butane	2-butene
$C_1{=}C_2{-}C_3{=}C_4$	C—C	$C_1{=}C_2{-}C_3$	$C_1{-}C_2{-}C_3{-}C_4$	$C_1{-}C_2{=}C_3{-}C_4$
$C_1{-}C_2$: 1.33 Å	C—C: 1.54 Å	$C_1{-}C_2$: 1.34 Å	C—C: 1.51 Å	$C_1{-}C_2$: 1.54 Å
$C_2{-}C_3$: 1.48 Å		$C_2{-}C_3$: 1.50 Å		$C_2{-}C_3$: 1.39 Å

For a number of years the "bond shortening" was taken as proof that something was occurring between the two π bonds of 1,3-butadiene. Since the central bond was "too short," and since double bonds are shorter than single bonds, it was concluded that π-π interaction made the C2-C3 bond more double-bond–like than is suggested by the normal Lewis structure. The most direct way to account for the presumed increase in double bond character between C2 and C3 is by invoking a π electron distribution that is different from that already given—one in which the terminal carbons carry opposite charges and the C2-C3 bond is a double bond.

$$\underset{\mathbf{A}}{H_2C{=}\overset{H}{C}{-}\overset{H}{C}{=}CH_2} \leftrightarrow \underset{\mathbf{B}}{H_2\overset{\oplus}{C}{-}\overset{H}{C}{=}\overset{H}{C}{-}\overset{\ominus}{C}H_2} \leftrightarrow \underset{\mathbf{B'}}{H_2\overset{\ominus}{C}{-}\overset{H}{C}{=}\overset{H}{C}{-}\overset{\oplus}{C}H_2}$$

The statement that 1,3-butadiene is inadequately represented by **A** but, instead, should be pictured as having qualities of **A, B,** and **B′** is equivalent to saying that 1,3-butadiene is a resonance hybrid of **A, B,** and **B′**. (Review the discussions of resonance beginning on p. 42 and p. 140, if necessary.) This does *not* mean that 1,3-butadiene is a mixture of **A, B,** and **B′**; it does *not* mean that **A, B,** and **B′** are discrete species in equilibrium; indeed, it does *not* mean that **A, B,** and **B′** are real! Instead, this is the chemist's way of stating that the molecule is a *hybrid* of different amounts of **A** and **B**/**B′**. As long as we draw structures that show electrons as dots, it will be impossible to describe a smeared-out electron cloud (which is what we really are trying to express) with one classical structure.

Structures **B** and **B′** are equally important in describing *real* 1,3-butadiene, but much less important than **A,** since **B** and **B′** involve charge separation and, as a corollary, fewer covalent bonds.

A Day of Reckoning. While resonance theory accounts for the short "single bond" in 1,3-butadiene, in the late 1950's M.J.S. Dewar pointed out that the bond under such intense study was not *unusually* short. Indeed, he claimed that the bond distance is what is expected—if one knows what to expect! Dewar suggested that the single bond distances in ethane, propene, butane, and 2-butene are simply the wrong models to use because they represent single bonds between C_{sp^2}-C_{sp^3} or C_{sp^3}-C_{sp^3} *and not single bonds between* C_{sp^2}-C_{sp^2}. Since sp^2 orbitals do not extend quite as far as do sp^3 orbitals, one should expect single bond distances to follow the order: C_{sp^2}-$C_{sp^2} < C_{sp^2}$-$C_{sp^3} < C_{sp^3}$-C_{sp^3}—precisely what is observed.

Today it is accepted by most that some small amount of π electron delocalization does occur in 1,3-butadiene, although the debate about magnitude continues. However, since the work of Dewar, the bond distance argument has lost much of its appeal.

Heat of Hydrogenation. Another property germane to this discussion is the heat of hydrogenation of dienes. The data in Table 12–1 indicate a heat of hydrogenation of ~30 kcal/mole for monoalkenes bearing one alkyl group on the double bond. Extrapolation to 1,3-butadiene suggests that it should have a heat of hydrogenation of about 60 kcal/mole. However, the observed heat of hydrogenation for this and other conjugated dienes is somewhat less, indicating that conjugated dienes are more stable than expected. The energy that 1,3-butadiene has lost by conjugation is called the *conjugation energy*[a] of 1,3-butadiene (Figure 12–1).

TABLE 12-1 Heats Of Hydrogenation

Compound	Heat of Hydrogenation (kcal/mole)
propene	30.1
1-pentene	30.3
1,4-pentadiene	60.6
1,3-pentadiene	54.1
1,3-butadiene	57.0

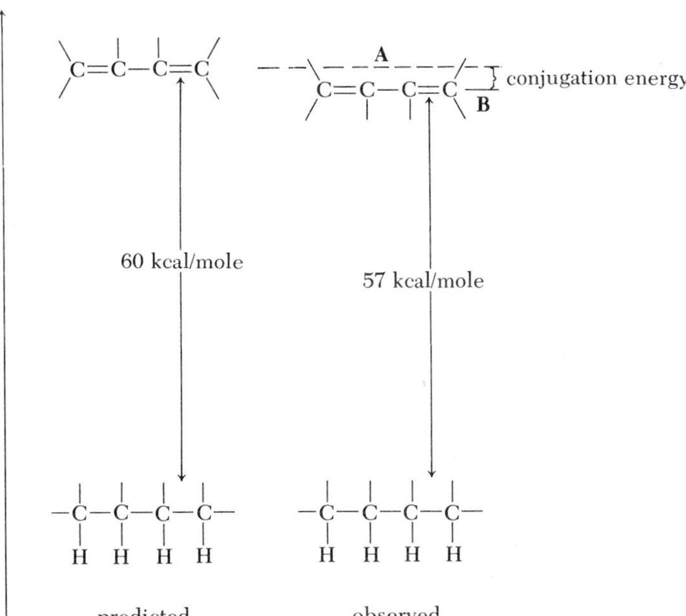

Figure 12–1 The effect of conjugation upon the heat of hydrogenation of 1,3-butadiene. Energy level **A** is that calculated by doubling the propene heat of hydrogenation, while level **B** is that determined from the observed heat of hydrogenation of 1,3-butadiene. The *conjugation energy* equals **A** − **B**.

Since resonance must make the real molecule more stable than any individual contributing structure, some argue that the existence of this conjugation energy is proof that π electron delocalization occurs in 1,3-butadiene. Undoubtedly mindful of the "too short bond" controversy, others contend that this 3 kcal/mole energy difference may reflect yet another, unrecognized error in our expectations. The answer is probably somewhere between these two points of view—some of the 3 kcal/mole being due to delocalization and some to "other effects."

Molecular Orbital Approach to Butadiene. An alternate approach to the description of the π electron distribution in 1,3-butadiene is found in the molecular orbital method. The initial molecular orbital picture of 1,3-butadiene is that of a σ framework, containing electron pairs and four p orbitals. These four p orbitals then are allowed to interact (*i.e.*, the wave equations describing the orbitals are combined), producing four new molecular orbitals, the π system of interest. Each of these molecular orbitals involves all four carbons.

The energies of the four π m.o.'s of 1,3-butadiene relative to those of the isolated a.o.'s,

[a] Other terms sometimes used for this quantity are *delocalization energy* and *resonance energy;* see p. 366.

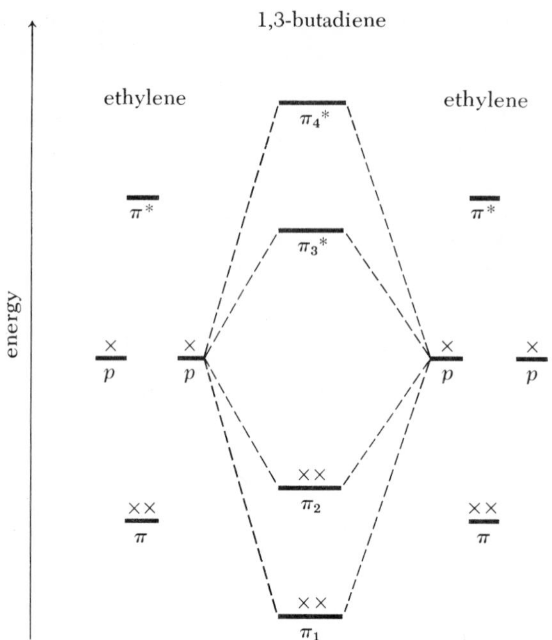

Figure 12–2 The molecular orbitals of 1,3-butadiene. The levels labeled "p" are those of the isolated orbitals. Those labeled π and π^* are those of ethylene and are included for comparison with 1,3-butadiene's orbitals, which are labeled π_1, π_2, π_3^*, and π_4^*.

and the energies of the m.o.'s of ethylene, are shown in Figure 12–2. The two bonding orbitals, π_1 and π_2, are occupied; the anti-bonding orbitals, π_3^* and π_4^*, are unoccupied. The complete filling of the bonding m.o.'s, and the non-occupancy of the anti-bonding m.o.'s, suggests that 1,3-butadiene should be a stable molecule with all electrons paired.

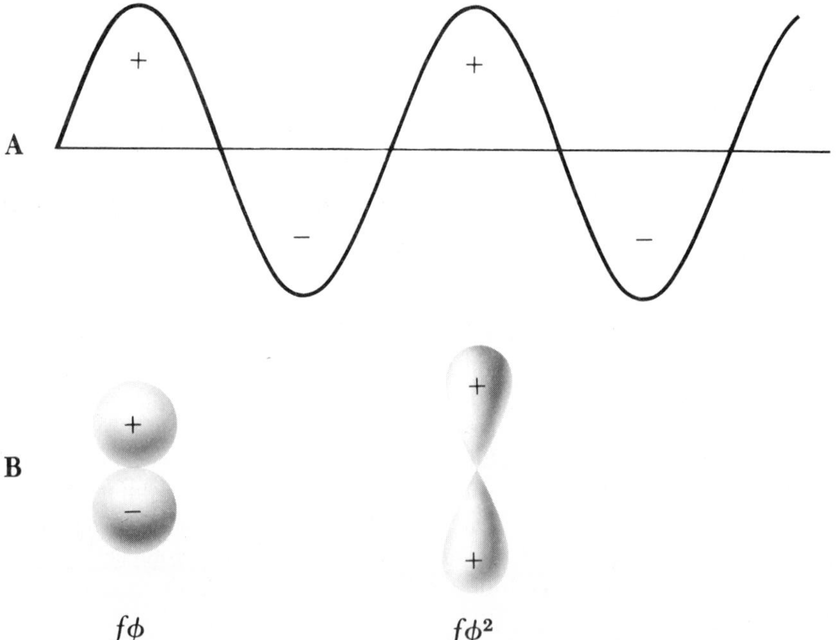

Figure 12–3 Signs associated with (**A**) a sine wave and (**B**) the 2p orbital. While $f\phi$ results in oppositely signed lobes, both lobes are positively signed for $f\phi^2$. However, the *shape* of the orbital is described by $f\phi^2$, not by $f\phi$.

Much more information about π_1, π_2, π_3^*, and π_4^*, and the importance of Figure 12–2, can be obtained if we backtrack to our original picture of a *p* orbital. In Chapter 1 we described the shape of the *p* orbital in terms of the probability of finding the electron at a given distance from the nucleus. While the wave function, ψ, has no simple physical significance, ψ^2 does—as the determining factor in defining the orbital's probability surface. In the same way that a wave may be thought to have positive and negative amplitude, so the *p* orbital wave function, ψ, will be signed. This makes one lobe positive and one negative.[a] (The lobes drawn as ψ, rather than ψ^2, are spherical and oppositely signed, while the lobes drawn as probability contours, ψ^2, are both positive and elongated; see Figure 12–3.) While it is useful to draw ψ^2, we shall soon see that it is helpful to think of the sign associated with ψ. Consequently, we shall commit the error of drawing ψ^2 while assigning signs to the lobes. At our level this will cause no harm and will facilitate discussion.

With this new perspective, let us now take a second look (review Figure 2–12) at π type overlap between adjacent *p* orbitals. The lobes will be signed and can take up two alternate arrangements (**A** and **B**, below). Wave theory indicates that while overlap can occur when similarly signed orbitals are proximal, a node (*i.e.*, no overlap) occurs when two oppositely signed lobes are juxtaposed. Thus, **A** and **B** actually represent the bonding and anti-bonding π orbitals, respectively.

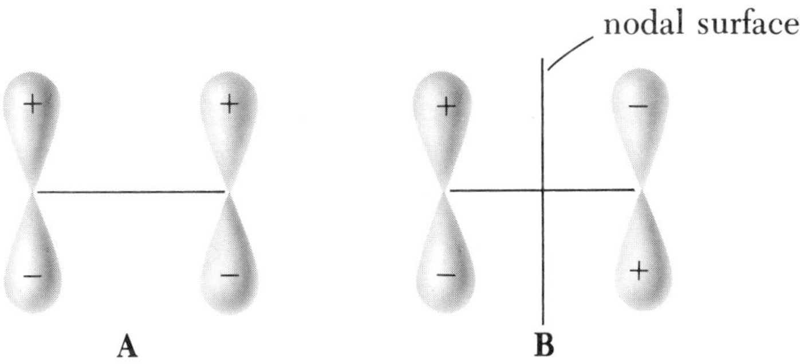

With these ideas as a guide, we construct the four m.o.'s of 1,3-butadiene (Figure 12–4). The lowest energy orbital, π_1, extends over all four carbons, *i.e.*, it does not contain any nodes between carbons. The other occupied orbital, π_2, contains a node between C2 and C3. Only one stabilizing overlap (between C2 and C3) is present in π_3^*, while π_4^* lacks any semblence of a π bond. Note that the energy of a molecular orbital increases as the number of nodes it possesses increases.

The molecular orbital picture of 1,3-butadiene clearly requires some double bond character between C2 and C3. Indeed, a simple examination of π_1 and π_2 suggests a substantial amount of π bond in the central bond. A more detailed analysis would reveal that the simple molecular orbital picture of 1,3-butadiene has over-emphasized the extent of π bond formation, but it would support the existence of some delocalization of π electrons through conjugated double bonds.

---

[a] This is *not* electrical positive and negative, but the mathematical sign of the electron wave function in that area of space.

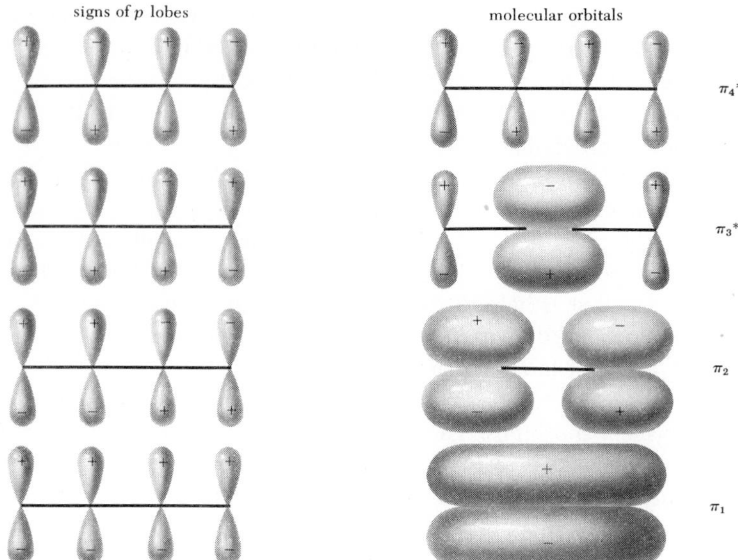

Figure 12-4 The molecular orbitals of 1,3-butadiene. The signs of the various atomic orbitals are shown on the left, and the resultant molecular orbitals are shown on the right. Note that nodes result when orbital lobes of opposite sign are juxtaposed.

12.3 PROPERTIES OF 1,3-BUTADIENE

In this section we shall see how three properties, bond order, electronic spectra, and bond rotation, are consistent with some π-π interaction between C2 and C3 of 1,3-butadiene.

Bond Order. Accepting the conclusion that the central bond in 1,3-butadiene is neither single nor double, we must now come up with a convenient way of suggesting what it is. That is, we must generate a means of indicating how much "double bond character" is present in this "single" bond. While one may approach this problem theoretically or experimentally, the experimental approach provides a simple, straightforward answer and will be taken here.

The bond order (defined as *the number of covalent bonds, or equivalents, between two nuclei*) is 1 for ethane, 2 for ethene, and 3 for ethyne. Pauling has suggested that a plot of *bond order vs. bond length,* similar to that in Figure 12–5, could be used to determine the bond order of bonds of known length.

The curve in Figure 12–5 is approximated by the following equation:

$$\%\,\text{DBC} = \frac{100(r_s - r)}{2r - 3r_d + r_s}$$

where % DBC = per cent double bond character

r = observed bond distance
r_s = normal single bond distance (1.54 Å)
r_d = normal double bond distance (1.33 Å)

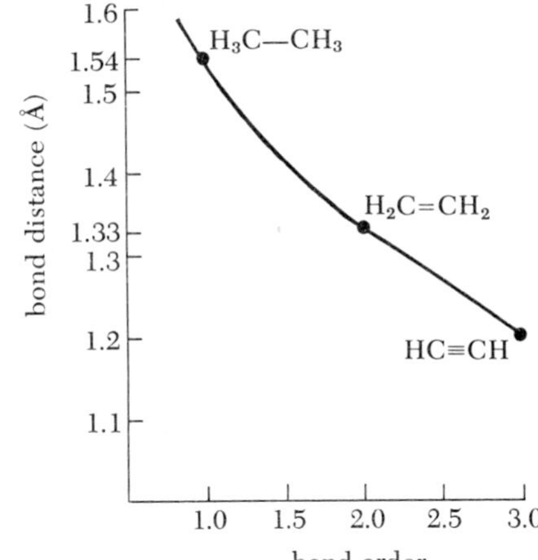

Figure 12-5 An experimental definition of bond order. Bond order can be defined in terms of an experimentally determined bond distance.

Let us apply this equation to 1,3-butadiene, recalling that the central bond distance is 1.48 Å.

$$\% \text{ DBC} = \frac{100(1.54 - 1.48)}{(2 \times 1.48) - (3 \times 1.33) + 1.54} \approx 12$$

Having calculated that the central bond has about 12% double bond character, we indicate its bond order as 1.12.

1. The π *bond order* is another measure of the amount of double bond character between two atoms. A single bond has a π bond order of zero, while a double bond has a π bond order of 1. What is the π bond order of the C2-C3 bond in 1,3-butadiene?
2. Phenanthrene reacts with bromine to produce 9,10-dibromophenanthrene. *On this basis,* which bond in phenanthrene has the highest π bond order?

H H
Br Br

+ Br_2 →

phenanthrene 9,10-dibromophenanthrene

Ultraviolet Spectra.[a] The longest wavelength (lowest energy) transition in the electronic spectrum of 1,3-butadiene occurs at 217 nm. It is ascribed to the excitation of one electron from π_2 into π_3^*. Ethylene, on the other hand, absorbs at 187 nm. The basis for this difference in absorption maxima is found in Figure 12–2. Here we see that conjugation in 1,3-butadiene decreases the separation between the highest occupied m.o. (HOMO), π_2, and

[a] It is assumed, for this discussion, that you have read Chapter 28.

the lowest unoccupied m.o. (LUMO), π_3, and reduces the energy necessary for excitation. This diminution of the energy required for excitation actually continues to increase with increasing conjugation; highly conjugated molecules absorb in the visible region, *i.e.*, they are colored (Table 12–2).

TABLE 12–2 Electronic Spectra Of Conjugated Polyenes, H—(CH=CH)$_n$—H[a]

n	$\lambda_{max}^{iso\text{-}octane}$ (nm)
2	217
3	268
4	304
5	335
6	365
7	390
8	410
9	446

[a] Where geometric isomers are possible, the all-*trans* isomer is presumed to be the major contributor to the spectral data. As the wavelength, λ, increases, the energy separating HOMO and LUMO decreases.

The spectrum of 1,3-butadiene and the evidence in Figure 12–2 clearly indicate that the conjugated polyene cannot be viewed simply as "a lot of ethylenes." The conjugated diene represents a new chromophore with its own characteristic electronic transitions.

3. Why is the electronic spectrum of the compound shown below more like that of tetramethylethylene than that of hexamethyl-1,3-butadiene? (*Hint:* Build a model.)

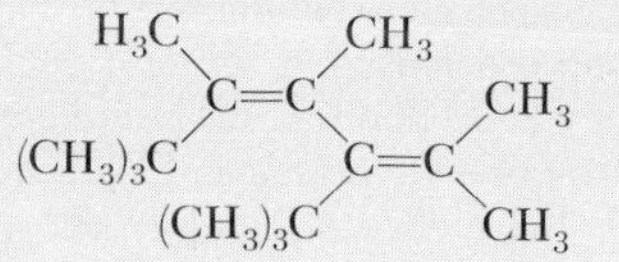

$$\begin{array}{c} H_3C \quad\quad CH_3 \\ C{=}C \quad\quad CH_3 \\ (CH_3)_3C \quad\quad C{=}C \\ (CH_3)_3C \quad\quad CH_3 \end{array}$$

Restricted Rotation about C2-C3. Spectroscopic evidence indicates a barrier to rotation about the C2-C3 bond in 1,3-butadiene of approximately 5 kcal/mole, a value significantly higher than the ethane rotational barrier of 3 kcal/mole (Figure 3–1).

This rotational barrier is consistent with a small degree of double bond character at C2-C3. The two conformers of 1,3-butadiene, which interconvert rapidly at room temperature, are called "s-*cis*" and "s-*trans*" forms. These conformational diastereomers differ in geometry about the "single" bond (hence, "s") possessing double bond character.

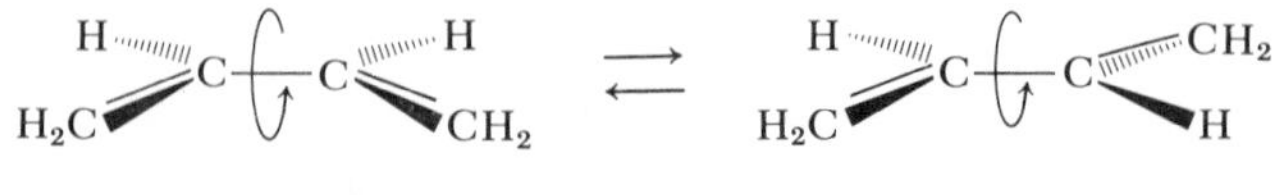

$$H_2C{=}C(H){-}C(H){=}CH_2 \ (\text{s-}cis) \rightleftharpoons H_2C{=}C(H){-}C(H){=}CH_2 \ (\text{s-}trans)$$

s-*cis* s-*trans*

4. 1,3-Cyclohexadiene exists only in the s-*cis* form, while 1,3-cyclohexadecadiene exists in both s-*cis* and s-*trans* forms. Account for this.

12.4 ADDITION REACTIONS OF 1,3-BUTADIENE

While there may be some debate about the extent of interaction in the π system of conjugated dienes, there is no doubt about the interdependence of the conjugated bonds in a host of reactions.

Ionic Bromination of 1,3-Butadiene. When 1,3-butadiene reacts with an excess of bromine (in carbon tetrachloride solvent), the product is a mixture of *meso*- and *d,l*-1,2,3,4-tetrabromobutane.

$$H_2C{=}CH{-}CH{=}CH_2 + 2Br_2 \rightarrow \underset{(meso \text{ and } d,l)}{BrCH_2{-}CHBr{-}CHBr{-}CH_2Br}$$

When the addition is carried out under conditions favoring addition of one mole of Br_2 (what would these conditions be?), the products include 1,4-dibromo-2-butene (*trans* > *cis*) as well as the expected 3,4-dibromo-1-butene.

$$H_2C{=}CH{-}CH{=}CH_2 + Br_2 \rightarrow \underset{\text{3,4-dibromo-1-butene}}{BrCH_2{-}CHBr{-}CH{=}CH_2} + \underset{\substack{\text{1,4-dibromo-2-butene} \\ (trans>cis)}}{BrCH_2{-}CH{=}CH{-}CH_2Br}$$

If the reaction is carried out at low temperatures (*e.g.*, $-80°$), the major product is the anticipated "1,2" adduct, 3,4-dibromo-1-butene. However, if the reaction is carried out at $+40°$, the "1,4" adduct, 1,4-dibromo-2-butene, becomes the major product.

$$H_2C{=}CH{-}CH{=}CH_2 + Br_2 \begin{cases} \xrightarrow{+40°} \underset{\sim 20\%}{\overset{\text{"1,2" adduct}}{BrCH_2{-}CHBr{-}CH{=}CH_2}} \quad \underset{\sim 80\%}{\overset{\text{"1,4" adduct}}{BrCH_2{-}CH{=}CH{-}CH_2Br}} \\ \xrightarrow{-80°} \underset{\sim 80\%}{BrCH_2{-}CHBr{-}CH{=}CH_2} \quad \underset{\sim 20\%}{BrCH_2{-}CH{=}CH{-}CH_2Br} \end{cases}$$

Either the "1,2" adduct or the "1,4" adduct is converted upon standing in an ionizing solvent at $+40°$ to the same mixture produced by addition of one mole of Br_2 at that temperature.

$$\substack{BrCH_2CHBrCH{=}CH_2 \\ (or\ BrCH_2CH{=}CHCH_2Br)} \xrightarrow{+40°} \underset{\sim 20\%}{BrCH_2CHBrCH{=}CH_2} + \underset{\sim 80\%}{BrCH_2CH{=}CHCH_2Br}$$

In contrast, both "1,2" and "1,4" adducts are stable at $-80°$; they do not convert to a mixture of products.

$$\substack{BrCH_2CHBrCH{=}CH_2 \\ (or\ BrCH_2CH{=}CHCH_2Br)} \xrightarrow{-80°} \text{no change}$$

These data are rationalized by assuming that the reaction begins, at any temperature, by formation of a π complex between Br_2 and one double bond—a process strictly analogous to the first step in addition of Br_2 to an isolated double bond (p. 233). However, this collapses to the resonance-stabilized allylic carbocation (p. 140), rather than to the cyclic bromonium ion, because of the lower energy of the former compared to that of the latter.

$$CH_2{=}CH{-}CH{=}CH_2 + Br_2 \rightarrow CH_2\overset{Br_2}{\overset{\uparrow}{\mp}}CH{-}CH{=}CH_2 \rightarrow$$

$$BrCH_2{-}\overset{\oplus}{C}H{-}CH{=}CH_2 \leftrightarrow BrCH_2{-}CH{=}CH{-}\overset{\oplus}{C}H_2$$

The π complex does not collapse to the 1° carbocation, ${}^{\oplus}CH_2{-}CHBr{-}CH{=}CH_2$, because this has a higher energy than does the 2° (resonance-stabilized) carbocation.

$$HC{=}CH{-}CH{=}CH_2 \xrightarrow{Br_2} \begin{cases} \longrightarrow [BrCH_2{-}CH{\cdots}CH{\cdots}CH_2]^{\oplus} \\ \not\longrightarrow {}^{\oplus}CH_2{-}CHBr{-}CH{=}CH_2 \end{cases}$$

At $-80°$ the halide ion, $Br^{\ominus}$, probably does not move away rapidly from the just-generated carbocation, and the bromide ion becomes bonded to C2, producing 3,4-dibromo-1-butene. It is clear that this is the easiest product to form, *i.e.*, it forms most rapidly, and is called the *kinetically controlled product.*

When the reaction is carried out at $+40°$, the speed of "1,2" addition is enhanced. However, the product, being an allylic halide, can ionize to produce halide ion and resonance-stabilized allylic cation. Furthermore, at $+40°$ collisions become energetic enough that some bromide ions are moved along to the terminal carbon, where they are captured by the positively charged carbon.

$$BrCH_2{-}CHBr{-}CH{=}CH_2 \xrightarrow[-Br^{\ominus}]{40°} BrCH_2{-}\overset{\oplus}{C}H{-}CH{=}CH_2 \leftrightarrow BrCH_2{-}CH{=}CH{-}\overset{\oplus}{C}H_2$$

Thus, while elevated temperatures increase the rate of formation of "1,2" adduct, they also increase its rate of decomposition to "1,4" adduct. This "1,4" adduct also is an allylic halide and should ionize to an allylic cation. While this does happen at $+40°$, it is harder to remove (by ionization) bromide ion from the "1,4" adduct than from the "1,2" adduct because the "1,4" adduct has a lower energy than does the "1,2" adduct (remember that an internal double bond is more stable than a terminal double bond). Since the "1,4" adduct is more stable, it is called the *thermodynamically controlled product.*

$$\underset{\text{"1,2" adduct}}{BrCH_2{-}CHBr{-}CH{=}CH_2} \overset{40°}{\rightleftharpoons} \underset{Br^{\ominus}}{\left[BrCH_2{-}\overset{H}{C}{\cdots}\overset{H}{C}{\cdots}CH_2\right]^{\oplus}} \rightleftarrows \underset{\text{"1,4" adduct}}{BrCH_2{-}CH{=}CH{-}CH_2Br}$$

Out of this explanation can be created an energy profile depicting "1,2 *vs.* 1,4 addition" (a specific example of the more general phenomenon of *kinetic vs. thermodynamic control*). Such a profile, Figure 12–6, is entered in the middle, at the allylic cation formed by the initial reaction.

One further question may be asked. Why does *any* "1,4" adduct form at $-80°$? To answer this, we must remember that at any given temperature molecules possess a distribution of velocities. If one of the more energetic solvent molecules strikes the $Br^{\ominus}$, it can move the ion over to C4 and permit "1,4" adduct formation even at $-80°$.

When passage over all energy barriers occurs at a reasonable rate, the product mixture is an equilibrium distribution whose composition is dictated by the energy difference between the species in equilibrium. While the free energy difference, ΔG, between the "1,2" and "1,4" adduct defines the equilibrium constant, ΔG is often approximated by ΔH, as noted in Figure 12–6.

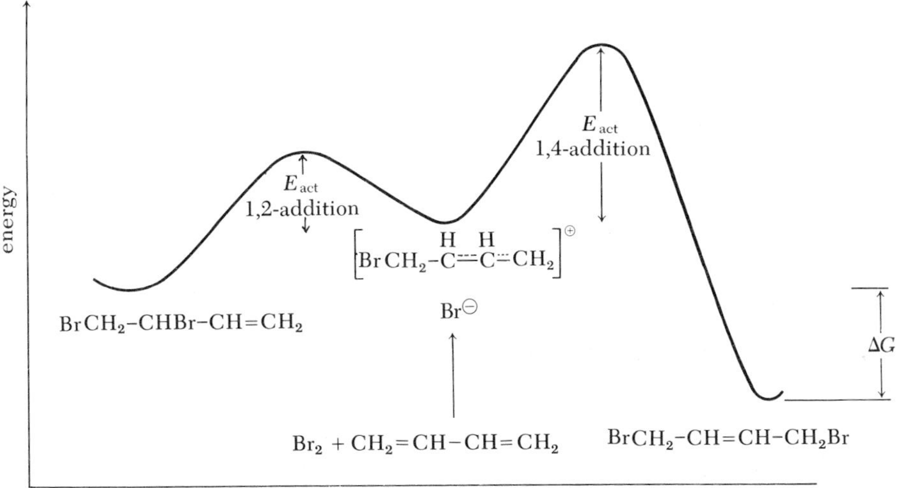

Figure 12-6 Kinetic versus thermodynamic control in the ionic addition of Br_2 to 1,3-butadiene.

Substituted Dienes. Under thermodynamically controlled conditions, most conjugated dienes produce an excess of "1,4" adduct. If the diene is unsymmetric, Markownikoff's rule can be used to predict the site of initial electrophilic attack.

$$H_2C{=}CH{-}CH{=}CH_2 \xrightarrow[+60^\circ]{HBr} CH_3{-}CH{=}CH{-}CH_2Br > CH_3{-}CHBr{-}CH{=}CH_2$$

$$H_2C{=}\overset{\overset{\large CH_3}{|}}{C}{-}CH{=}CH_2 \xrightarrow[+60^\circ]{HBr} CH_3{-}\overset{\overset{\large CH_3}{|}}{C}{=}CH{-}CH_2Br$$

5. Suggest two mechanisms to account for the following ionic addition. How might you distinguish between them?

cyclopentadiene + HCl → 3-chlorocyclopentene

Free Radical Bromination of Dienes. A similar, though not identical, situation to that just described occurs in the free radical addition of bromine to dienes. A stable allylic radical is the initial product, and it can abstract a bromine radical from Br_2 to produce either 3,4-dibromo-2-butene or 1,4-dibromo-2-butene. As with ionic addition, 1,4-addition predominates. This chain reaction is presented below.

(1) $Br_2 \xrightarrow{heat} 2Br\cdot$

(2) $H_2C{=}CH{-}CH{=}CH_2 + Br\cdot \rightarrow$

$$\underbrace{Br{-}CH_2{-}\dot{C}H{-}CH{=}CH_2 \leftrightarrow Br{-}CH_2{-}CH{=}CH{-}\dot{C}H_2}$$
$$\equiv [Br{-}CH_2{-}CH{\because}CH{\because}CH_2]\cdot$$

(3) $[Br{-}CH_2{-}CH{\because}CH{\because}CH_2]\cdot + Br_2 \rightarrow$

$$Br{-}CH_2{-}CH{=}CH{-}CH_2{-}Br > Br{-}CH_2{-}CHBr{-}CH{=}CH_2 + Br\cdot$$

(4) $2Br\cdot \rightarrow Br_2$

The major difference between radical and ionic addition is that radical addition is not subject to rapid equilibration by a radical process (only 46 kcal/mole are required to cleave Br_2 into radicals, while approximately 67 kcal/mole are required for homolytic cleavage of the C—Br bond). Predominance of 1,4-addition occurs because the allylic radical must exist for a brief period before it collides with a bromine molecule, and the kinetic advantage of proximity that existed in ionic addition is lost in radical addition.

Where Is Resonance More Important—1,3-Butadiene or its Allylic Intermediate? If a mixture of a conjugated diene and an isolated alkene are subjected to addition, the diene reacts almost completely before any isolated alkene has reacted. Since the diene has a lower ground state energy than does the isolated alkene, but is also more reactive than the alkene, it follows that the transition state of the reaction involving the diene must be more resonance-stabilized than is the diene itself. Were this not so, resonance stabilization of the diene would make it less reactive than the alkene. Because the transition state resembles the ground state nearest to it in energy, one must conclude that *delocalization is more important in stabilizing radicals and cations than in stabilizing adjacent double bonds* (Figure 12–7).

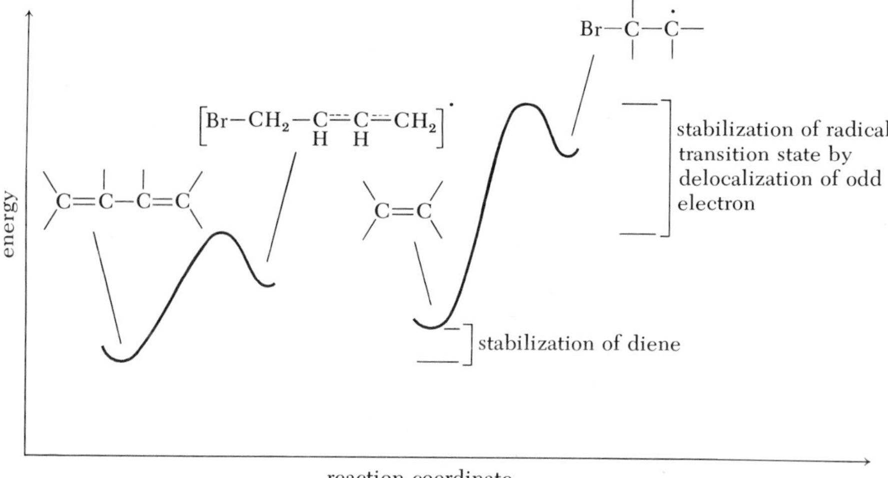

Figure 12-7 Stabilization of 1,3-diene versus allyl radical.

12.5 POLYMERIZATION OF CONJUGATED DIENES

The major difference between diene polymerization and alkene polymerization is that the former produces a product with many isolated double bonds, while the latter produces an alkane. Like simple addition, both ionic and free radical polymerization of dienes favor the 1,4-addition pictured below:

$$n\ \mathrm{>C{=}C<} \xrightarrow{\mathrm{R\cdot}} \left(\mathrm{-\overset{|}{\underset{|}{C}}-\overset{|}{\underset{|}{C}}-}\right)_n \qquad \text{alkene linear polymer}$$

$$n\ \mathrm{>C{=}\overset{|}{C}-\overset{|}{C}{=}C<} \xrightarrow{\mathrm{R\cdot}} \left(\mathrm{-\overset{|}{\underset{|}{C}}-\overset{|}{C}{=}\overset{|}{C}-\overset{|}{\underset{|}{C}}-}\right)_n \qquad \text{1,4-addition; diene linear polymer}$$

Polymerization of a substituted diene proceeds by attachment of a new diene in the same fashion as the previous one, a process called **head-to-tail** addition. This is simply a consequence of the relative stabilities of the alternative intermediates (pp. 247, 248).

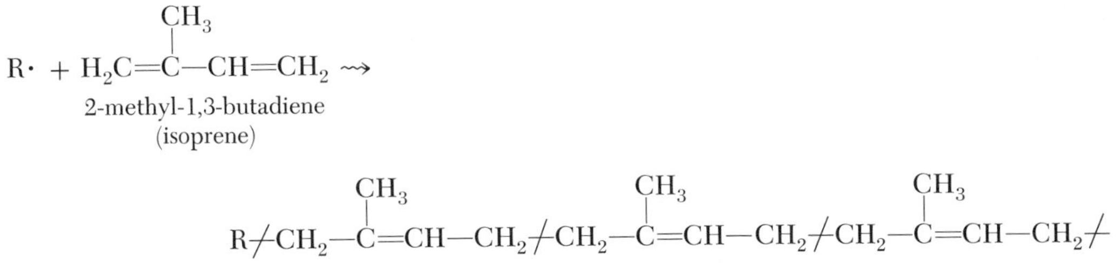

$$R\cdot + H_2C{=}\overset{\overset{\displaystyle CH_3}{|}}{C}{-}CH{=}CH_2 \rightsquigarrow$$

2-methyl-1,3-butadiene
(isoprene)

$$R\!\not{}\,CH_2{-}\overset{\overset{\displaystyle CH_3}{|}}{C}{=}CH{-}CH_2\!\not{}\,CH_2{-}\overset{\overset{\displaystyle CH_3}{|}}{C}{=}CH{-}CH_2\!\not{}\,CH_2{-}\overset{\overset{\displaystyle CH_3}{|}}{C}{=}CH{-}CH_2\!\not{}$$

head-to-tail polymer

In the polymerization of isoprene (pictured above) and other unsymmetric conjugated dienes, the possibility of *cis-trans* isomerism around the product double bonds must be considered. Consistent with the data for simple 1,4-addition, the *trans* isomer (*i.e.*, the isomer in which the linkages to the polymer backbone are *trans* to one another) predominates.

Gutta percha is a naturally occurring all-*trans* polymer of isoprene, familiar to most as the hard "rubber" covering on golf balls.

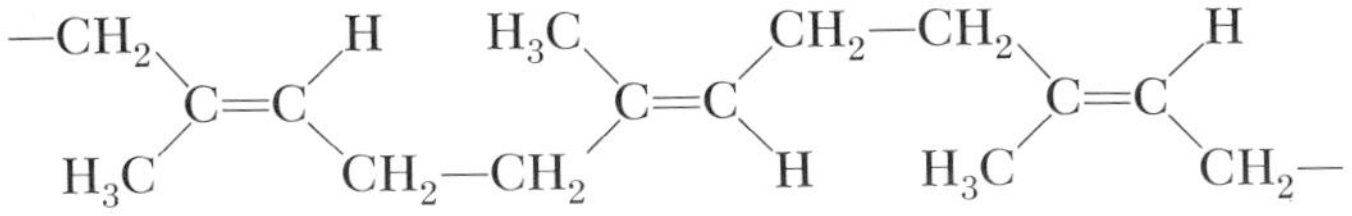

$$\begin{array}{ccccccccccc} -CH_2 & & & & H & H_3C & & & & CH_2{-}CH_2 & & & & H \\ & \diagdown & C{=}C & \diagup & & & \diagdown & C{=}C & \diagup & & & \diagdown & C{=}C & \diagup \\ H_3C & \diagup & & \diagdown & CH_2{-}CH_2 & & \diagup & & \diagdown & H & H_3C & \diagup & & \diagdown & CH_2- \end{array}$$

poly-*trans*-isoprene
(gutta percha)

Certain trees native to the Eastern hemisphere produce, when their bark is damaged, an exudate known as **latex.** Latex is a suspension of the all-*cis* polymer of isoprene, also called "natural rubber," in water. When latex is processed, the natural rubber that is isolated is a sticky, gummy substance of moderate elasticity.

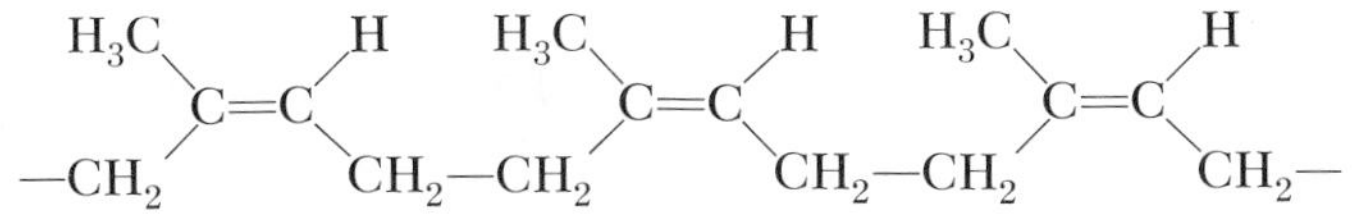

$$\begin{array}{ccccccccccc} H_3C & & & & H & H_3C & & & & H & H_3C & & & & H \\ & \diagdown & C{=}C & \diagup & & & \diagdown & C{=}C & \diagup & & & \diagdown & C{=}C & \diagup \\ -CH_2 & \diagup & & \diagdown & CH_2{-}CH_2 & & \diagup & & \diagdown & CH_2{-}CH_2 & & \diagup & & \diagdown & CH_2- \end{array}$$

poly-*cis*-isoprene
(natural rubber)

6. Natural rubber is readily decomposed by ozone (oxidative work up) to form levulinic acid, $C_5H_8O_3$. (a) Suggest a structure for levulinic acid. (b) What product(s) would be formed by reduction of levulinic acid with sodium borohydride? (c) How many moles of methane would be produced by reacting levulinic acid with an excess of methyl magnesium chloride?

In 1839 Charles Goodyear discovered that if this tacky natural rubber were heated with sulfur (as much as 8%), it would be converted to a non-sticky substance that was still elastic and resilient. This *elastomer* has become known as *vulcanized rubber*, and the process of producing it is known as vulcanization. Still more sulfur in the vulcanization process leads to a hard polymer known as *hard rubber*.

Until 1955 trees were better synthetic chemists than were men, at least to judge by their ability to produce poly-*cis*-isoprene. Human attempts, largely *via* free radical polymerization of isoprene, produced polymers of variable stereochemistry and little utility as substitute rubbers. In 1955 both the Goodyear and Firestone companies announced that isoprene could be stereospecifically polymerized to poly-*cis*-isoprene using Ziegler-Natta type catalysts. The rubber industry will never again be the same.

Neoprene. When acetylene is dimerized, using cuprous chloride as a catalyst, the product is vinylacetylene, $H_2C{=}CH{-}C{\equiv}CH$. The addition of hydrogen chloride to vinylacetylene, again in the presence of cuprous chloride, produces 2-chloro-1,3-butadiene, commonly called *chloroprene.*

$$2H{-}C{\equiv}C{-}H \xrightarrow{Cu_2Cl_2} \underset{\text{vinylacetylene}}{H_2C{=}CH{-}C{\equiv}CH}$$

$$H_2C{=}CH{-}C{\equiv}CH + HCl \xrightarrow{Cu_2Cl_2} \underset{\text{chloroprene}}{H_2C{=}CH{-}\underset{\displaystyle Cl}{\underset{|}{C}}{=}CH_2}$$

The similarity between chloroprene and isoprene is obvious and, in fact, chloroprene can be polymerized by a free-radical process to produce an essentially all-*trans* 1,4-addition polymer called **neoprene.** Neoprene was the first synthetic American rubber (1932) and was developed under the supervision of W.H. Carothers of DuPont. Neoprene can be vulcanized by heating with metallic oxides (ZnO, MgO) and, though too expensive for use in tires, is extremely valuable because of its high resistance to organic solvents and air oxidation. Consequently, neoprene is used in washers, laboratory tubing, and other applications in which oil-resistant rubber is necessary.

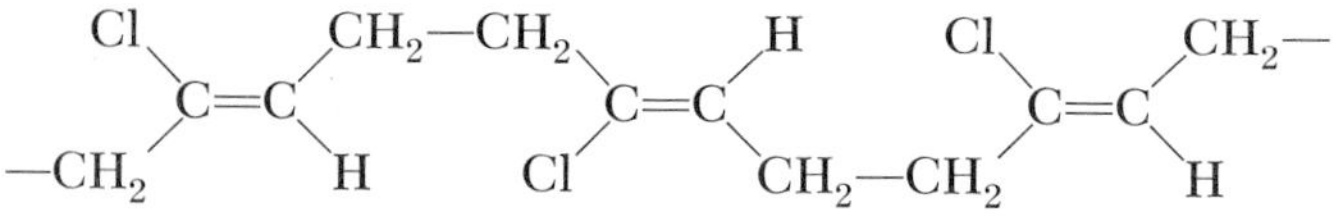

$$\text{—}CH_2(Cl)C{=}C(H)CH_2{-}CH_2(Cl)C{=}C(H)CH_2{-}CH_2(Cl)C{=}C(H)CH_2\text{—}$$

neoprene

Another group of rubbers of some importance are the *Buna rubbers,* obtained by polymerizing butadiene with metallic sodium (*bu*tadiene-*na*trium). Buna was developed in Germany in the late 1920's but, unfortunately, is not a good rubber substitute. During World War II the Allies could not rely upon Asian rubber plantations for latex; they were not welcome there. The tires produced during WWII were made largely of SBR (also known as Buna S rubber). SBR is made by co-polymerizing butadiene (3 parts) and styrene (1 part) under free-radical conditions. Like Buna itself, SBR was first introduced in Germany; but, unlike Buna, SBR can be vulcanized to produce an elastomer that is at least as good as natural rubber in some applications.

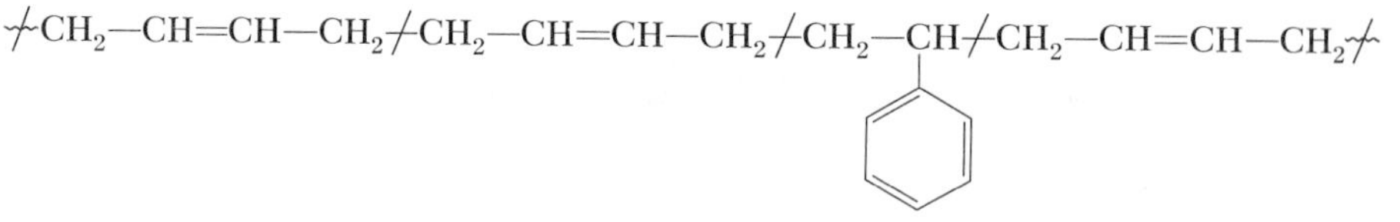

$$\nmid CH_2{-}CH{=}CH{-}CH_2 \nmid CH_2{-}CH{=}CH{-}CH_2 \nmid CH_2{-}\underset{C_6H_5}{CH} \nmid CH_2{-}CH{=}CH{-}CH_2 \nmid$$

SBR synthetic rubber

12.6 TERPENES

> "And the rose herself has got
> Perfume which on earth is not."
>
> Keats, *Bards of Passion and of Mirth*

For over four hundred years the odorific components, usually part of the "essential oils," of flowers have been the source of speculation, consternation, and occasional income for a small group of individuals, the perfumers. In the same way that mountaineers climb mountains "because they are there," so chemists, even alchemists, were interested in the substances responsible for the odors of various plant parts. (The monetary aspects were, to be sure, not neglected.)

By 1920 it was known that most of a plant's pleasant odors arose from a group of compounds called **terpenes.**[a] These terpenes share the characteristic of possessing carbons in multiples of 5, *e.g.*, 5, 10, or 15. While most terpenes are unsaturated hydrocarbons, alcohols, aldehydes, ketones, carboxylic acids, and even peroxides are included in this broad classification of natural products.

In 1920, L. Ruzicka began a classic series of experiments that culminated in the *isoprene rule.* This axiom, which unified most of the preceding structural analyses of terpenes, stated that *terpenes are formed as if by head-to-tail polymerization of isoprene.* We shall see shortly that isoprene itself is not the material that is converted to terpenes by the plant. Nonetheless, it is helpful to think of terpenes as addition polymers of isoprene. The types of known terpenes are summarized in Table 12–3. The non-existence of C_{25} and C_{35} terpenoids will be explained later in this chapter.

TABLE 12–3 Classification Of Terpenes

TYPE	NUMBER OF ISOPRENE UNITS	NUMBER OF CARBONS
monoterpene	2	10
sesquiterpene	3	15
diterpene	4	20
triterpene	6	30
tetraterpene	8	40

MONOTERPENES. The variety of terpene structures is quickly made apparent by considering the ways in which Mother Nature can disguise even the two isoprene units found in monoterpenes. Such methods include translocation of double bonds and σ bonding that results in ring formation. Some of the monoterpenes presented below contain heavy bonds to make it easier to recognize the isoprene units in these molecules. Of course, not all isoprene units are shown in this way in order to give you the chance to find a few yourself.

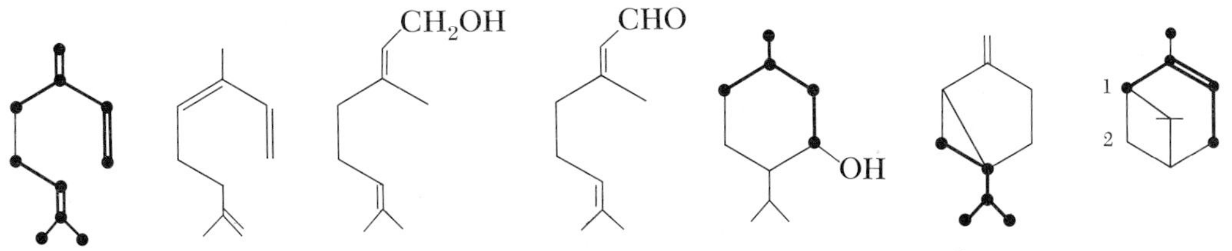

CH_2OH CHO OH 1 2

myrcene (bayberry) ocimene (basil) geraniol (rose) geranial (lemon grass) menthol (peppermint) sabinene (savin) α-pinene (turpentine)

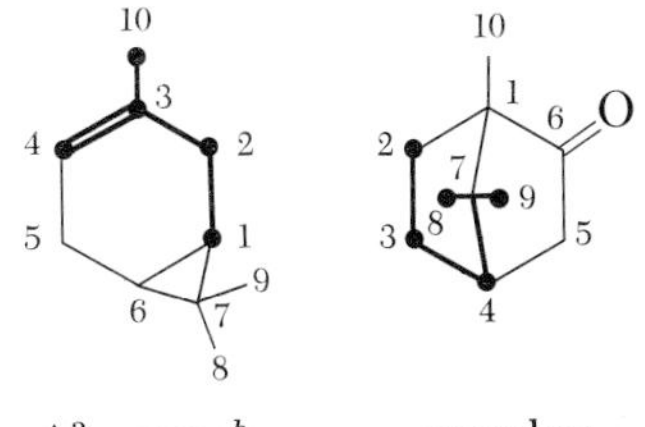

10 3 4 2 5 1 9 6 7 8 10 1 6 O 2 7 9 3 8 5 4

Δ^3-carene[b] (turpentine) camphor (camphor tree)

---

[a] Also called *terpenoids* or *isoprenoids.*

[b] The Greek delta, Δ, symbolizes the double bond; the superscript identifies the lower of the two carbons forming the double bond. This nomenclature is antiquated but may still be found in current literature.

The sources listed above are not exclusive; these compounds are usually found in several plants and geraniol, for example, is found in almost every plant. Moreover, while we worry about hydrocarbon contamination from the automobile engine, it is important to know that *trees exude isoprene* (and other hydrocarbons) *through their leaves*.

Sesquiterpenes. Sesquiterpenes commonly exist as acyclic, monocyclic, bicyclic, and tricyclic structures.

Farnesol, the odorific component of lily-of-the-valley, is an acyclic terpenoid alcohol. Farnesol is an attractant to certain male insects and may, therefore, assist in the plant pollination process. Perfumes based upon lily-of-the-valley have also been used to attract the male of *H. sapiens.*

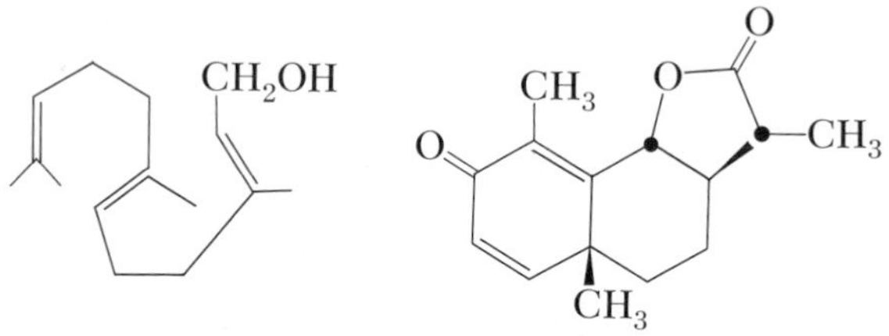

farnesol santonin

Santonin, a tricyclic sesquiterpene, has a rather complex structure. Indeed, superficial inspection might not even suggest that it is a terpene. It is found in plants of the Artemesia variety and is an active ingredient in an herb remedy used on the Asian subcontinent for centuries. It is currently used as a specific for *Ascaris lumbricoides*[a] in humans; isolation of the compound is performed by extraction of the unexpanded flower heads of the plant.

Diterpenes. When chlorophyll is hydrolyzed in alkaline media it produces, among other things, an alcohol called *phytol,* $C_{20}H_{39}OH$. First discovered by Willstatter (1909), phytol is currently used as a precursor in the commercial synthesis of vitamins E and K_1.

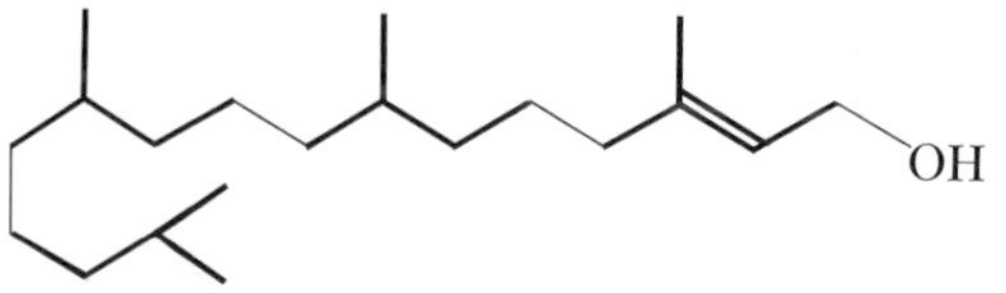

isoprene units are emphasized

phytol

The material that exudes from the damaged bark of the pine tree is called "naval stores" because of its use in caulking wooden ships' hulls and in weatherproofing rope. Steam distillation separates naval stores into a volatile *gum turpentine* and a non-volatile *rosin.* Rosin is used extensively in the preparation of varnishes and paper sizing and, from a different perspective, is the largest natural source of organic acids. Abietic acid is the most abundant of these acids. It is present as a primary plant constituent and is also readily formed by isomerization of levopimaric acid.

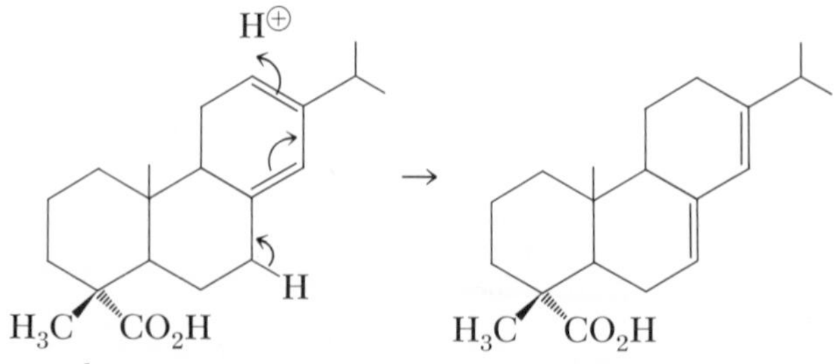

levopimaric acid → abietic acid

[a] A nematode. The nematodes include roundworms and pinworms.

Triterpenes. *Squalene,* $C_{30}H_{50}$, is a very important triterpene. It occurs in yeast, wheat germ, and olive oil and makes up as much as 40%, by weight, of shark's liver oil. While of limited commercial value, squalene is a fundamental triterpene and is the biological precursor of steroids. The sequence in Figure 12-8 shows how squalene, by a series of carbocation reactions, is converted to the steroid *lanosterol* (found in *lanolin* or *wool fat*). Lanosterol is, in turn, biologically demethylated to *cholesterol.*

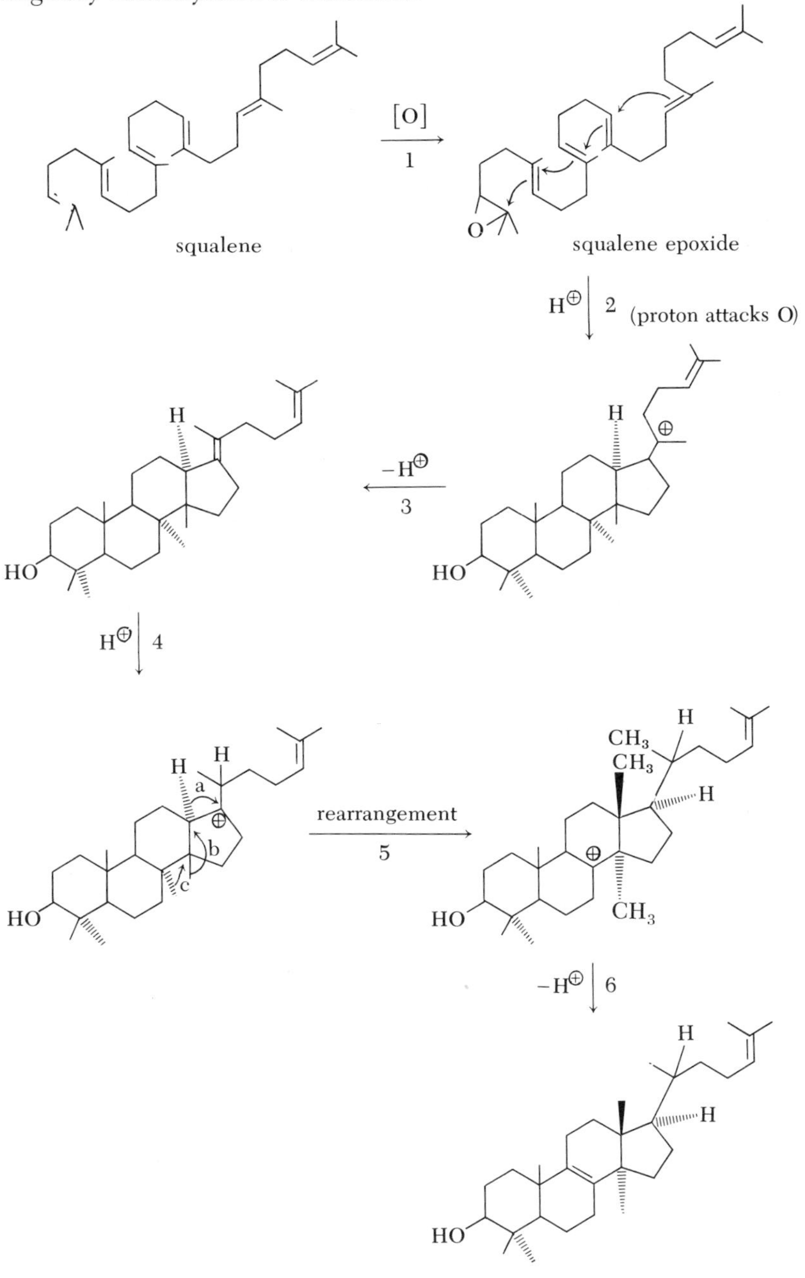

squalene

[O]
1

squalene epoxide

H⊕ 2 (proton attacks O)

−H⊕
3

H⊕ 4

rearrangement
5

−H⊕ 6

lanosterol

Figure 12-8 The biosynthesis of lanosterol from squalene. *Step 1:* Enzymatic epoxidation of the C2-C3 double bond. *Step 2:* Protonation of epoxide oxygen non-bonding pair, followed by attack by a proximal bond. About this time, a series of electrons migrate; positive charge ends up on a 3° carbon. *Step 3:* Loss of a proton adjacent to the positive charge, an E1 reaction, produces a double bond. *Step 4:* Protonation of the newly created double bond. *Step 5:* Three sequential (perhaps synchronous) migrations: (a) hydrogen with its bonding pair; (b) methyl with its electron pair; and (c) another methyl with its electron pair. The net result of interconversion of 3° cations is production of the steroid nucleus. *Step 6:* Loss of a proton produces lanosterol.

Tetraterpenes. The tetraterpenes occur widely in nature and, because of extensive conjugation of double bonds, are often found in mixtures of plant pigments. A standard experiment in some laboratory manuals employs column chromatography to separate lycopene from tomato paste. β-Carotene, an isomer of lycopene, is found in carrots.

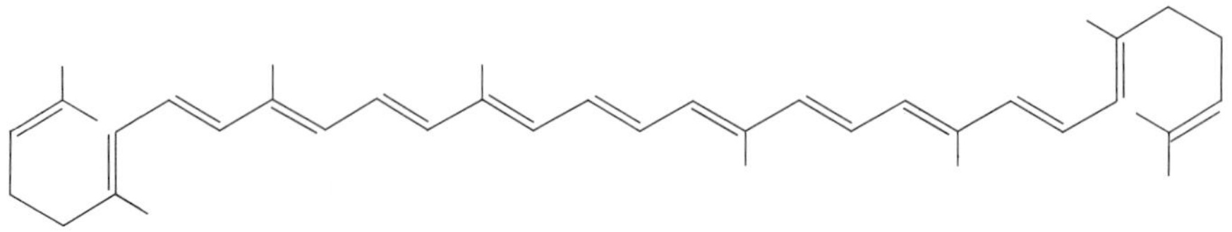

lycopene

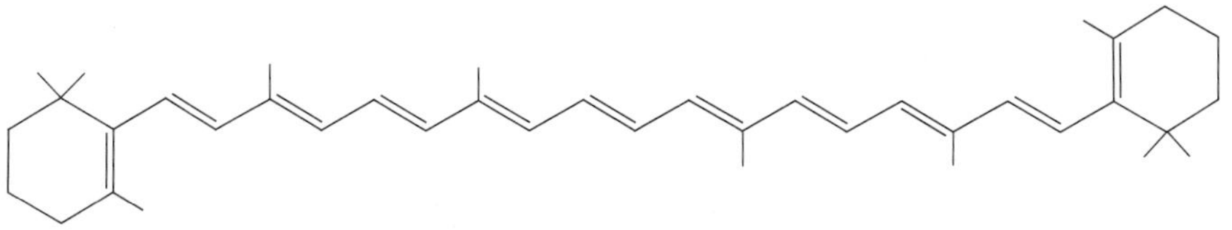

β-carotene

β-Carotene can be converted *in vivo* to two molecules of vitamin A, also called *retinol.* This alcohol is important to normal health because it is converted *in vivo* to 11-*cis*-retinal, which plays a critical role in the visual process. The role of vitamin A in maintaining good health clearly extends beyond the visual process, since deprivation of vitamin A in experimental animals leads to death.

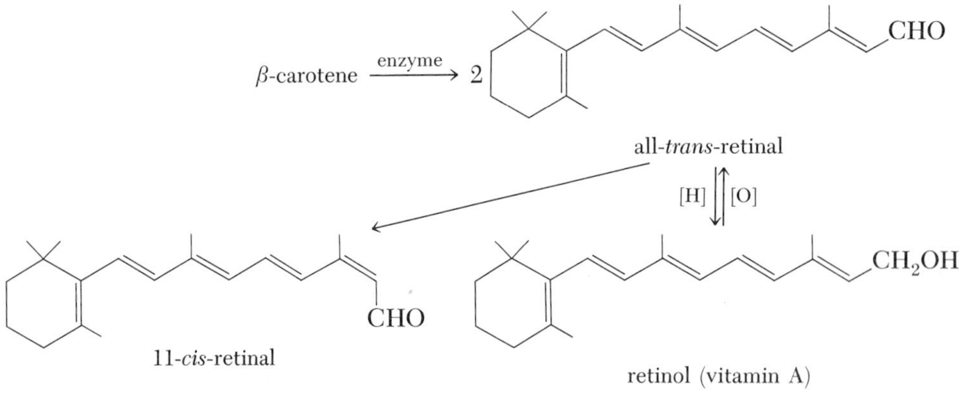

β-carotene $\xrightarrow{\text{enzyme}}$ 2 [CHO]

all-*trans*-retinal

[H] [O]

CHO

11-*cis*-retinal

CH_2OH

retinol (vitamin A)

Lycopenemia is an apparently harmless condition brought about by eating too many tomatoes; the lycopene turns the tomato fancier a bright red-orange. Not to be outdone, the carrot lover can sometimes consume large quantities of carotene and turn yellow. This *carotenemia,* as it is called, is also harmless. Both of these conditions are reversible.

Biogenesis of Terpenes and Steroids. Acetic acid, CH_3CO_2H, is the starting material for biological synthesis of terpenes and, necessarily, steroids. It is converted, with the assistance of coenzyme A (CoA), to mevalonic acid. (Coenzyme A is discussed in detail later in this text.) Further enzymic action converts mevalonic acid to isopentyl pyrophosphate (IPP). IPP is enzymically equilibrated to the isomeric 3,3-dimethylallyl pyrophosphate (DMAP).

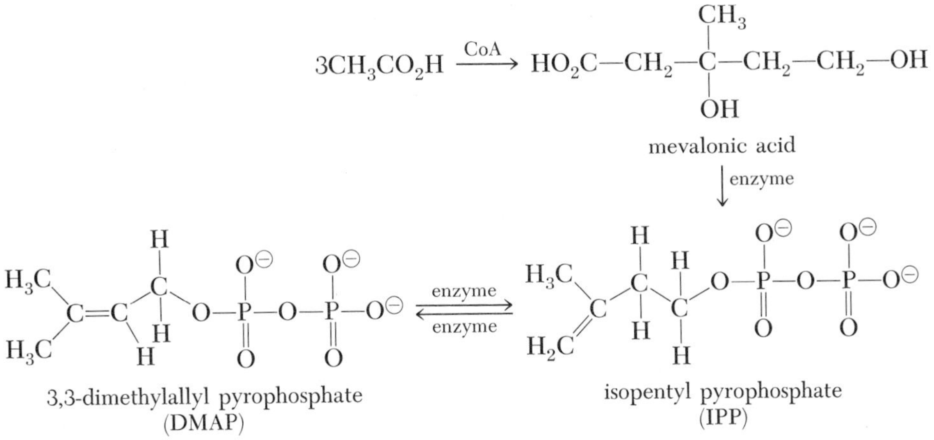

$$3CH_3CO_2H \xrightarrow{CoA} HO_2C\text{—}CH_2\text{—}\overset{\displaystyle CH_3}{\underset{\displaystyle OH}{C}}\text{—}CH_2\text{—}CH_2\text{—}OH$$

mevalonic acid

↓ enzyme

3,3-dimethylallyl pyrophosphate (DMAP) ⇌ (enzyme / enzyme) isopentyl pyrophosphate (IPP)

DMAP ionizes to form a resonance-stabilized allylic cation and the relatively stable pyrophosphate anion, $P_2O_7^{\ominus 4}$. (Pyrophosphate is a good leaving group.) This cation can attack the π system of IPP to produce the pyrophosphate ester of the monoterpene geraniol, geranyl pyrophosphate. Geranyl pyrophosphate can be hydrolyzed *via* an S_N1 process to geraniol; or it can be converted, *via* another carbocation alkylation, to the sesquiterpene derivative farnesyl pyrophosphate. A further alkylation of the sesquiterpene, by a similar mechanism, forms a C_{20} pyrophosphate.

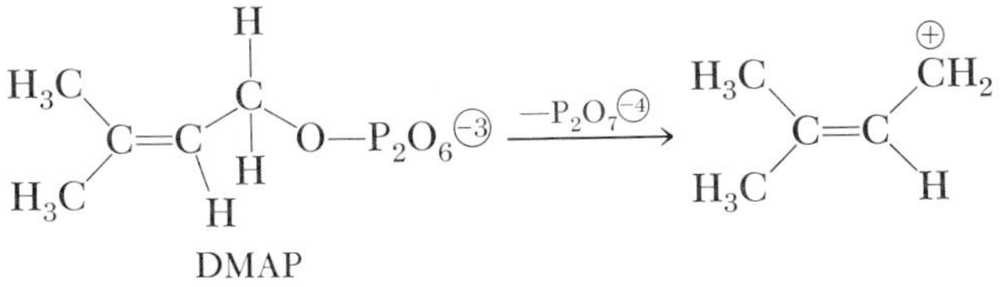

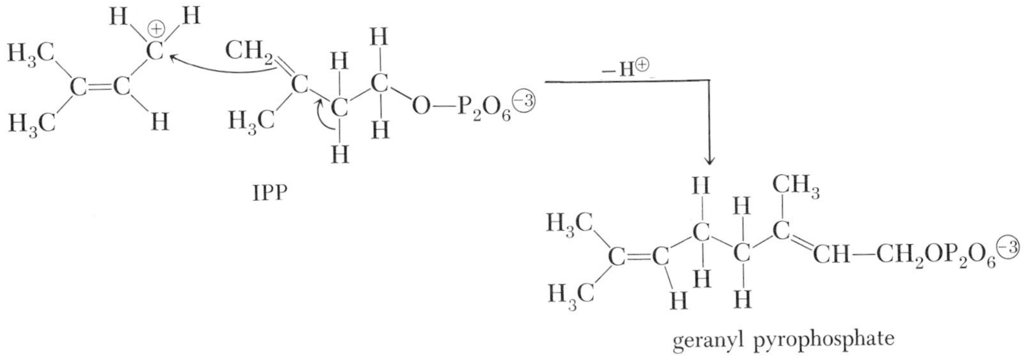

$$(H_3C)_2C{=}CH\text{—}CH_2\text{—}O\text{—}P_2O_6^{\ominus 3} \xrightarrow{-P_2O_7^{\ominus 4}} (H_3C)_2C{=}CH\text{—}\overset{\oplus}{C}H_2$$

DMAP

IPP $\xrightarrow{-H^{\oplus}}$

$$(H_3C)_2C{=}CH\text{—}CH_2\text{—}CH_2\text{—}C(CH_3){=}CH\text{—}CH_2OP_2O_6^{\ominus 3}$$

geranyl pyrophosphate

The tri- and tetraterpenes are not produced by continued alkylation but, instead, are formed by "dimerization" of sesqui- and diterpenes, respectively (Figure 12–9). This explains, at least to some degree, why there are no C_{25} or C_{35} terpenoids (Table 12–3).

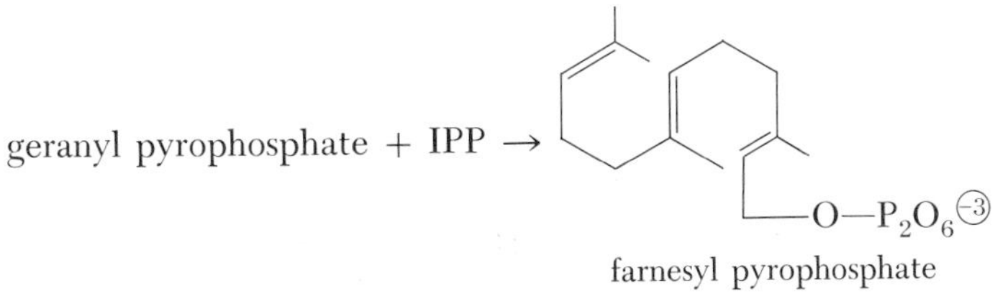

geranyl pyrophosphate + IPP → —O—$P_2O_6^{\ominus 3}$

farnesyl pyrophosphate

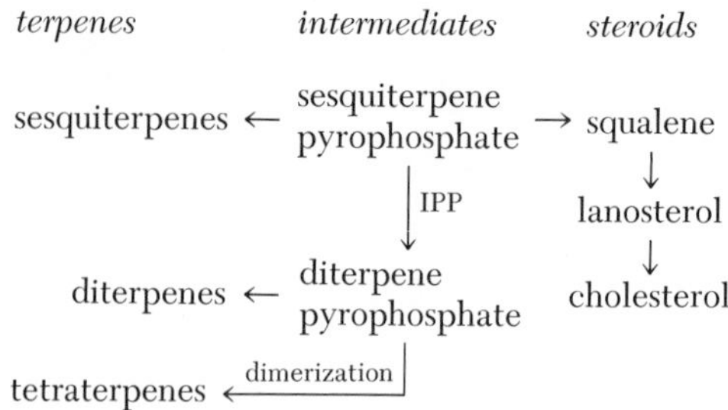

Figure 12-9 The biological route to steroids and tetraterpenes.

12.7 PHEROMONES

The major means of communication among humans involve visual or acoustical contact. However, among some other mammals and especially among simpler animals, the major method of communication appears to be chemical in nature. Most animals are capable of excreting compounds into the environment that influence the behavior of other animals. Once called *ectohormones,* these compounds now are termed **pheromones.** Two different types of pheromones are recognized—*releaser* pheromones and *primer* pheromones. Releaser pheromones cause rapid, reversible changes in animal behavior while primer pheromones produce protracted physiological changes in the recipient.

Pheromones are not very complex molecules, many having molecular weights under 300. Those that have been studied in detail do not employ a host of functional groups to achieve their remarkable results. Among the functional groups found in pheromones are carboxaldehyde (—CHO), ester (—CO_2R), hydroxy, and the double bond. A simple example of a pheromone is geraniol. Having already acknowledged that it is a monoterpene found in roses, we may now add that geraniol is a *recruiting* pheromone of the honey bee. When a honey bee encounters a source of food, it excretes geraniol which, in turn, attracts other bees to that site. Citral,* another monoterpene, is used by the honey bee for the same purpose.

Another type of releaser pheromone is called an *alarm* pheromone. To see an alarm pheromone at work, all one has to do is to disturb an ant that is working near an ant hill. Once disturbed, the ant will excrete a pheromone that causes other ants in its environs to become excited. While different species of ants appear to have different alarm pheromones, the two shown below are typical. Again, we note the amazing fact that these are both terpenes. More interesting is the observation that citral, a recruiting pheromone for the honey bee, is an alarm pheromone for the ant *Atta sexdens.*

$(CH_3)_2C{=}CHCH_2CH_2C(CH_3){=}CHCHO$ — citral

3-substituted furan, side chain $(CH_2)_2CH{=}C(CH_3)(CH_2)_2CH{=}C(CH_3)_2$ — dendrolasin

alarm pheromones

Not all pheromones are terpenes. (Since we know comparatively little about pheromones, it is still difficult to generalize effectively about their structure.) An excellent illustration of a non-terpene pheromone is the alarm pheromone of the honey bee, the ester called isoamyl acetate, $(CH_3)_2CHCH_2CH_2OC(O)CH_3$. If a bee keeper is stung by a bee, it does not take long for other bees to come and attack the spot previously stung. The reason is that the first bee released the alarm pheromone when it stung the keeper, and that served to attract other bees.

---

* What relationship exists between citral and geranial (p. 359)?

Humans smell isoamyl acetate and detect the odor of bananas. (Appropriate puns about the reactions of bees to isoamyl acetate are left to the interested reader.)

Of the various types of pheromones, none has received as much attention as have the so-called "sex attractants." While such *sex* pheromones serve to inform the male of a species of the availability of a female, they serve other functions as well. For example, a sex pheromone causes the male cabbage looper to exhibit copulatory behavior in the absence of a female. This compound is (**Z**)-7-dodecenyl acetate.

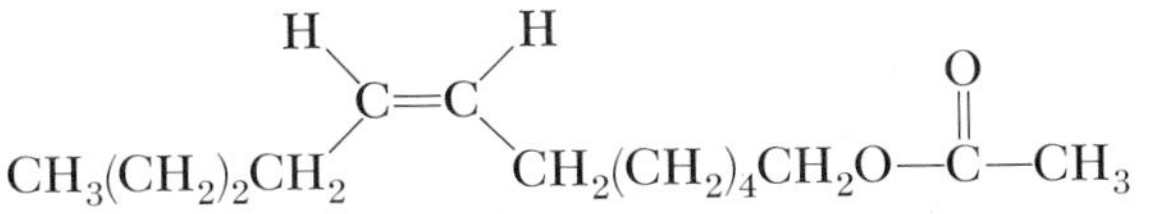

$$\begin{matrix} H & & H \\ & C{=}C & \\ CH_3(CH_2)_2CH_2 & & CH_2(CH_2)_4CH_2O{-}\overset{\displaystyle O}{\overset{\|}{C}}{-}CH_3 \end{matrix}$$

(**Z**)-7-dodecenyl acetate

The potency of some sex pheromones is astounding. For example, the sex pheromone of the silkworm moth, bombykol, can attract the male even in concentrations of approximately 200 molecules per milliliter of air! Put differently, one female silkworm moth contains enough sex attractant to stimulate over one *billion* males. Bombykol is an alkadienol and, as can be seen from the structure below, is not *that* different from the sex attractant of the cabbage looper.

$$CH_3{-}CH_2{-}CH_2{-}CH{=}CH{-}CH{=}CH{-}(CH_2)_8{-}CH_2OH$$

bombykol

For many, the big question is, "Do humans excrete pheromones?" The answer, at this early stage, is that we do not yet know. However, the fact that sexually mature males and females frequently respond differently to some odorific compounds suggests that sex-related pheromones for humans may exist. But if they do exist, the problem of human behavior is so complex that it may be a long while before they can be detected and evaluated.

7. Ecdysone is the molting *hormone* (not a pheromone) that encourages the maturation of a larval stage of the silkworm into the pupal stage. It can kill insects by causing premature molting. (a) What general class of compounds is represented by ecdysone? (b) Identify the chiral centers and assign their absolute configurations where possible. (c) How many products would be produced if ecdysone were reduced by catalytic hydrogenation under conditions that resulted in the uptake of two equivalents of hydrogen? (d) What, if anything, would be the relationship among these compounds? (Review the discussion of ecdysone and juvenile hormone in Chapter 11 for more perspective.)

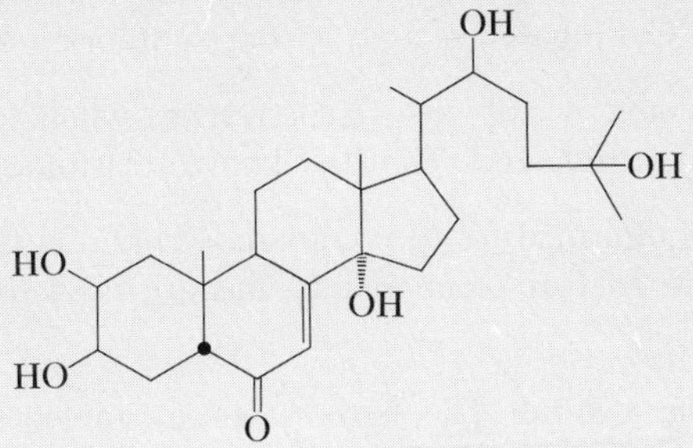

OH

OH

HO

OH

HO

O

ecdysone

IMPORTANT TERMS

Bond order: The number of covalent bonds (or equivalents) holding two atoms together. The carbon-carbon bond orders in ethane, ethene, and ethyne are 1, 2, and 3, respectively. If a bond participates in resonance, then its bond order will not be an integer (whole number).

Chromophore: A functional group that absorbs energy in the electromagnetic spectrum, most frequently the ultraviolet and the visible region.

Conjugated system: An arrangement of electrons in which a set of potentially mobile electrons (usually non-bonding electrons or π bonds) is separated from an orbital that can accept them by a single bond. A system which, if in the proper geometry, may produce electron delocalization. (Even if a system is conjugated, delocalization will not occur if the appropriate orbitals do not overlap. Thus, conjugation and delocalization really are not the same thing.)

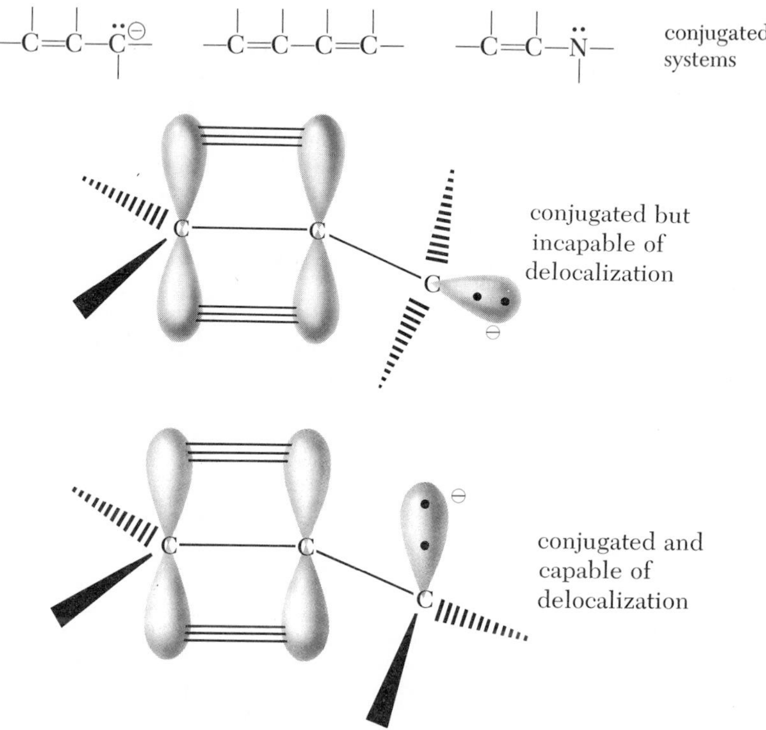

Conjugation energy: Energy lost by a molecule by virtue of electron delocalization. Also called resonance energy or delocalization energy. Because of the difference between "conjugation" and "delocalization" (see "**Conjugated system**"), some authors prefer these latter terms.

Elastomer: An *elastic* poly*mer*. A polymer that stretches when tension is applied, but that contracts when the tension is removed. The best known elastomer is rubber.

Essential oil: A volatile (liquid) mixture of compounds obtained from plants. These mixtures contain the odorific constituents of many plants and are used in medicine, flavorings, and perfumes.

Highest occupied molecular orbital: The highest energy molecular orbital to contain any electrons. Abbreviated HOMO.

Kinetically controlled product: Product from that reaction, of a set of competing reactions, which has the lowest energy of activation. The kinetically controlled product is formed most rapidly in any reaction. The kinetically controlled product may or may not be the same as the thermodynamically controlled product.

Lowest unoccupied molecular orbital: The lowest energy molecular orbital to be free of any electrons. Abbreviated LUMO.

Pheromone: A compound released by one species of animal in order to influence the behavior of other animals, most often other members of the same species. The materials used in "chemical communication."

Terpene: A low molecular weight polymer that may be thought of as derived from isoprene. A derivative of the polymer just mentioned.

Thermodynamically controlled product: The most stable product produced by a series of competing reactions. It may or may not be the same as the kinetically controlled product.

Vulcanization: Treatment of polymers to reduce their tackiness and to produce other desirable properties. This is accomplished by cross-linking the polymers (binding a divalent atom to one carbon on each of two adjacent backbones), thereby converting a series of linear molecules into a semi-rigid three-dimensional network. Sulfur usually is used to vulcanize rubber. (The rubber in tires is not black because of the vulcanization process, *per se*. Carbon is added to the rubber, and this gives it the black color.)

PROBLEMS

8. Draw structures for all dienes with the formula C_5H_8, indicating the geometry of the molecules and the hybridization of each carbon. Provide an acceptable name for each compound. (*Hint:* Remember that allene is the simplest diene.)
9. Identify the conjugated system in each of the following:
   (a) $CH_2{=}CHCH_3CH{=}CH_2$
   (b) $CH_2{=}CHCH_2CH{=}CHCH{=}CHCH_3$
   (c) $CH_3OCH{=}C(CH_3)_2$
   (d) $ClCH_2CH{=}CCH_3C{\equiv}CH$
   (e) $H_2NCH{=}CH_2$
   (f) $H_2NCH{=}CHCH{=}CH_2$
   (g) $H_3\overset{\oplus}{N}CH{=}CHCH{=}CH_2$
10. Give the names and structures of products expected from the reaction, if any, of (i) 1,4-pentadiene and (ii) 1,3-butadiene with:
    (a) 1 mole of $Br_2/CCl_4(+40°)$
    (b) excess Br_2/CCl_4
    (c) 1 mole H_2/Pt
    (d) 1 mole D_2/Pt
    (e) excess H_2/Pt
    (f) 1 mole DCl (+40°)
    (g) excess DCl
    (h) excess CH_3OH/mercuric trifluoroacetate, followed by $BH_4^{\ominus}$
    (i) excess ozone (reductive work-up)
    (j) excess ozone (oxidative work-up)
    (k) excess B_2H_6, followed by $CH_3CH_2CO_2D$ (reflux)
11. Write the structures of the intermediate cation produced by the addition of $Br^{\oplus}$ to:
    (a) 1-hexene
    (b) 1,3-hexadiene
    (c) 1,5-hexadiene
    (d) cyclohexene
    (e) 1,3-cyclohexadiene
12. *vic*-Dibromides usually react with bases to produce more alkyne than conjugated diene. What factor(s) may be controlling this?

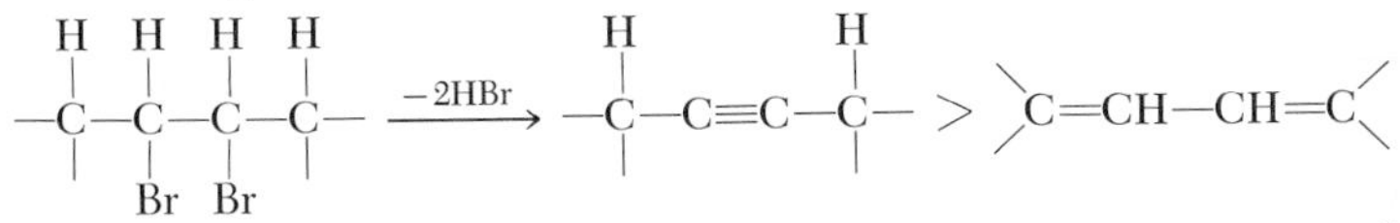

$$-\overset{H}{\underset{|}{\overset{|}{C}}}-\overset{H}{\underset{Br}{\overset{|}{C}}}-\overset{H}{\underset{Br}{\overset{|}{C}}}-\overset{H}{\underset{|}{\overset{|}{C}}}- \xrightarrow{-2HBr} -\overset{H}{\underset{|}{\overset{|}{C}}}-C{\equiv}C-\overset{H}{\underset{|}{\overset{|}{C}}}- \; > \; {>}C{=}CH-CH{=}C{<}$$

Two exceptions to this behavior are 1,2-dibromocyclohexane and 2,3-dichloro-2,3-dimethylbutane. How do you account for these exceptions?
13. Geraniol is converted by acid to α-terpineol. Suggest a mechanism for this reaction.

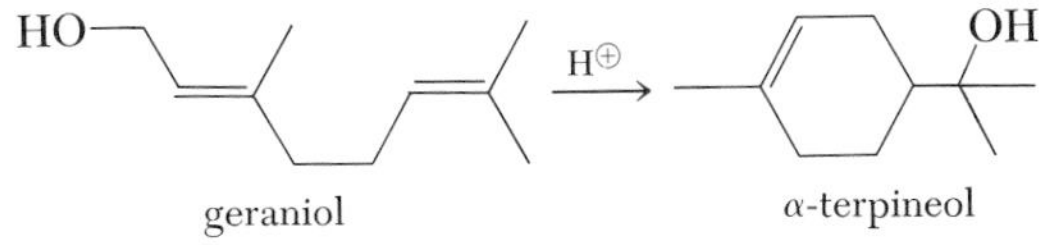

HO— (geraniol structure) $\xrightarrow{H^{\oplus}}$ (α-terpineol structure, OH)

geraniol α-terpineol

14. What products would you expect from the ozonolysis (reductive work-up) of the following compounds?

(a) 1-pentene
(b) 2-pentene
(c) 1-pentyne
(d) 1,3-butadiene
(e) 2-methyl-1,3-butadiene
(f) myrcene
(g) ocimene
(h) α-pinene
(i) farnesol
(j) geraniol
(k) geranial

15. Dihydromyrcene, $C_{10}H_{18}$, is formed by the addition of one mole of hydrogen to myrcene. Catalytic hydrogenation of dihydromyrcene produces $C_{10}H_{22}$. Upon oxidation with potassium permanganate, dihydromyrcene produces:

$$CH_3C(O)CH_2CH_2CO_2H,\ CH_3CO_2H,\ \text{and}\ (CH_3)_2CO$$

Suggest a structure for dihydromyrcene.

16. Predict the major product(s) of the reaction of 2-methyl-1,3-butadiene with (a) HCl and (b) Br_2. Explain your answer.

17. Predict the major product(s) from the free radical addition of one mole of $BrCCl_3$ to the following:

(a) 1-butene
(b) 1,3-butadiene
(c) 1,3-pentadiene
(d) 1,4-pentadiene
(e) 2-methyl-1,3-butadiene

18. One mole of bromine adds to 1,3,5-hexatriene to produce only 1,6-dibromo-2,4-hexadiene and 5,6-dibromo-1,3-hexadiene. Suggest a mechanism that explains these as the only products of addition.

19. The bark beetle (*Ips confusus*) is an insect pest that destroys billions of board feet of timber every year. The trees are attacked by a few beetles, which excrete a material ("frass") that contains feces, wood, and an aggregating pheromone. This pheromone, 2-methyl-6-methylidene-7-octen-4-ol, serves to attract massive amounts of other bark beetles. (a) Draw the structure of this pheromone. (b) Is it a terpene? (c) Is there any relationship between it and geraniol? (d) Catalytic hydrogenation of this material with excess hydrogen produces two substances, both of which are optically active. Explain.

20. Pure samples of either 1-chloro-3-methyl-2-butene or 3-chloro-3-methyl-1-butene react with a dilute solution of sodium carbonate to produce the same mixture of **A** and **B** (below). In turn, pure samples of either **A** or **B** react with acid to produce the same mixture of **A** and **B.** Account for these results.

$$\underset{\mathbf{A}}{(CH_3)_2C(OH)CH{=}CH_2} \qquad \underset{\mathbf{B}}{(CH_3)_2C{=}CHCH_2OH}$$

21. *Chicle* is a resin obtained by concentrating the latex of the sapodilla tree (*Acras zapota*), which grows only on the Yucatan peninsula. It is a mixture, containing about 5% *cis-* and 12% *trans*-polyisoprene, whose major use is in the manufacture of chewing gum. Can you suggest a means of determining the relative amounts of these two polymers present in chicle?

*22. Two moles of acetone react with an alloy of mercury and magnesium to produce **A** ($C_6H_{14}O_2$). Compound **A** reacts with acid to produce **B** (C_6H_{10}). Compound **A** does not react with potassium permanganate or bromine, but it does react with sodium amide (to generate ammonia). Compound **B** reacts with both potassium permanganate and bromine. Ozonolysis (reductive work-up) of **B** produces two moles of formaldehyde and one mole of 2,3-butanedione, $CH_3C(O)C(O)CH_3$. Provide structures for **A** and **B** consistent with these data. (*Hint:* The infrared spectra of **A** and **B** are presented below and on p. 369.)

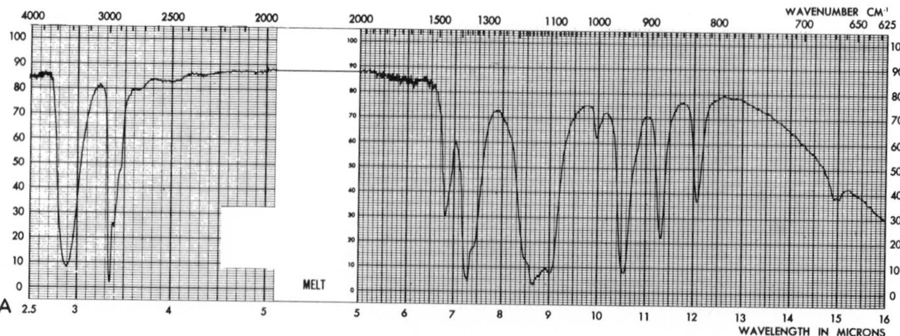

---

*This problem may require a knowledge of infrared spectroscopy.

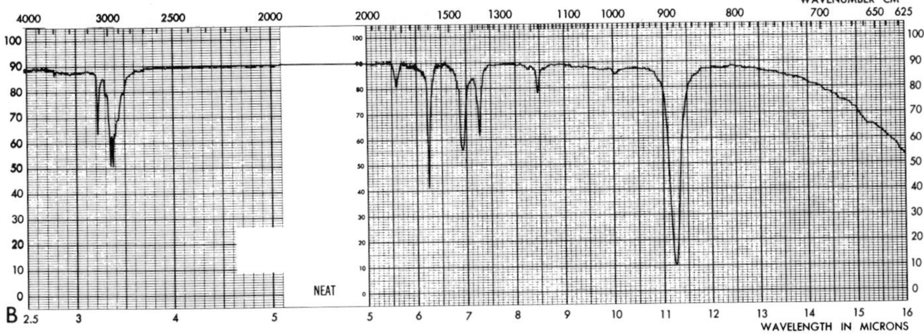

23. Camphene reacts with acetic acid, in the presence of sulfuric acid, to produce isobornyl acetate. Suggest a mechanism for this reaction. (*Hint:* Acetic acid may react as a nucleophile in the ways shown below.)

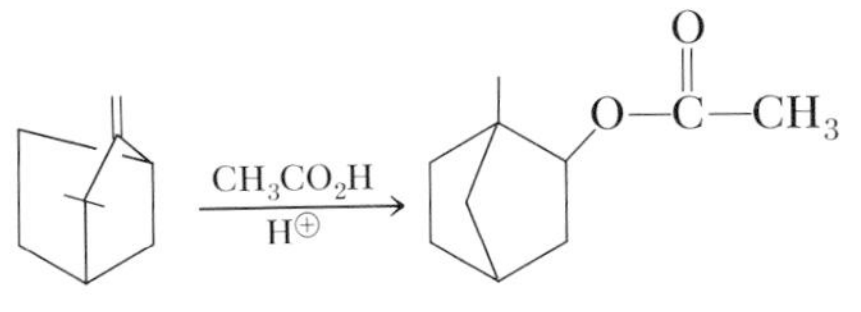

camphene $\xrightarrow[H^{\oplus}]{CH_3CO_2H}$ isobornyl acetate ($O—C(=O)—CH_3$)

$CH_3—C(=O)—\ddot{O}(H)\!:\curvearrowright Z—L$ *or* $CH_3—C(=O)—\ddot{O}(H)\!:\curvearrowright Z^{\oplus}\ L^{\ominus}$

acetic acid

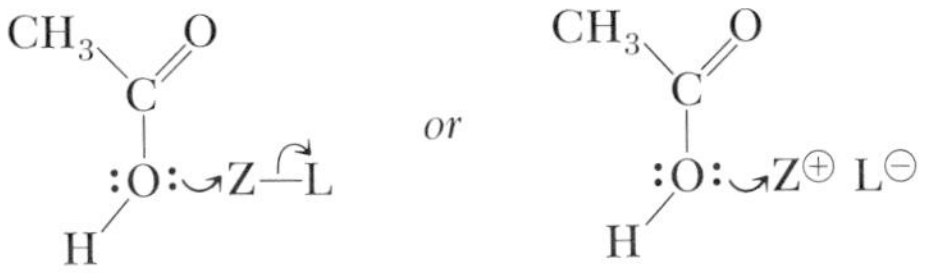

acetic acid as a nucleophile

°24. The nmr spectrum of the 2-methylallyl cation at $-60°$, in a solution of antimony pentafluoride and sulfur dioxide, consists of one signal (3**H**) at 3.85 δ and one signal (4**H**) at 8.95 δ. Is this consistent with the structure shown below? Explain. If your answer is negative, what changes and assumptions must be made in order to explain this spectrum?

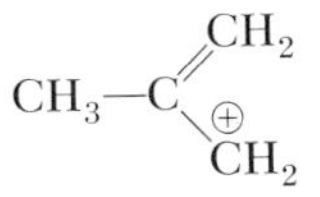

$CH_3—C(=CH_2)—\overset{\oplus}{C}H_2$

2-methylallyl cation

°25. Acetylene reacts with acetone, $(CH_3)_2C(O)$, in the presence of a base to form **A** (C_5H_8O). **A** reacts with one mole of hydrogen to form **B** ($C_5H_{10}O$) which, in turn, reacts with acid to form **C** (C_5H_8). Compound **C** reacts with ozone to produce CH_2O and $CH_3C(O)CHO$. (a) What are the structures of **A, B,** and **C**? (b) What is the common name for **C**?

(c) What differences would you expect in the (a) ir and (b) nmr spectra of **A, B,** and **C**? (d) Describe a simple chemical test to distinguish (i) **A** from **B** and (ii) **A** from **C**; include anticipated observations.

26. The equilibrium constant for the reaction shown below is approximately 50. Which of the following approximates the free energy difference between *s-cis-* and *s-trans-*1,3-butadiene? Explain. (a) 0.5; (b) 1.6; (c) 2.4; (d) 5.0 kcal/mole.

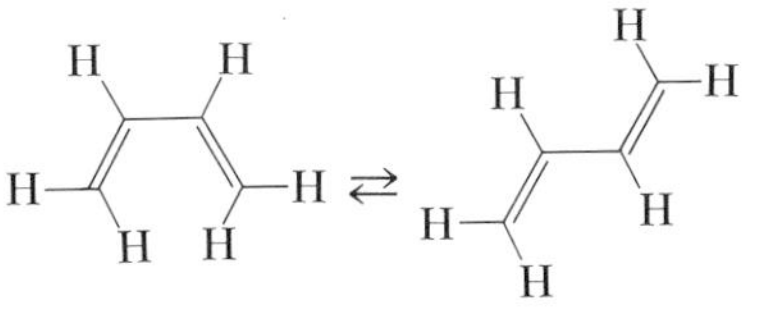

s-cis $\rightleftarrows$ *s-trans*

° These problems require a knowledge of spectroscopy.

27. Construct a picture similar to Figure 12–2 for $CH_2{=}CH{-}CH{=}CH_2$ plus one electron, *i.e.*, $[CH_2{=}CH{-}CH{=}CH_2]^-$. This species is a "radical anion" (or anion radical), that is, a species which bears a negative charge and carries an odd electron. (This beast has five π electrons.) Would you expect this radical anion to be more or less stable than 1,3-butadiene itself? Explain.

28. β-Ionone is an important chemical in the synthesis of vitamin A. It is prepared by the reaction of pseudo-ionone with sulfuric acid, *d,l*-α-iononc being produced in the same reaction. Suggest a mechanism for the acid-catalyzed conversion of pseudo-ionone to α- and β-ionone.

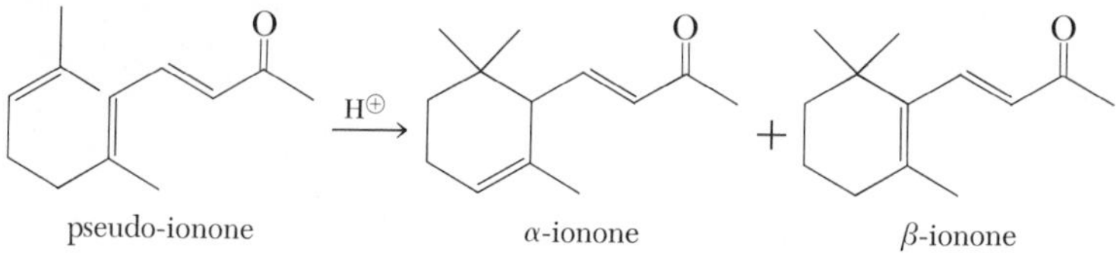

$H^{\oplus}$

pseudo-ionone $\longrightarrow$ α-ionone + β-ionone

Why might you anticipate that the cyclization produces more β- than α-ionone?

29. The insecticide aldrin is made by a sequence called the *Diels-Alder reaction.* This involves the condensation of a 1,3-diene with a *dienophile, e.g.*, an isolated double bond, but the reaction does not display many of the characteristics of either ionic or free-radical reactions. Attempt to come up with a non-ionic *and* non-free-radical reaction mechanism for this conversion. (If you become frustrated, peek into the next chapter!)

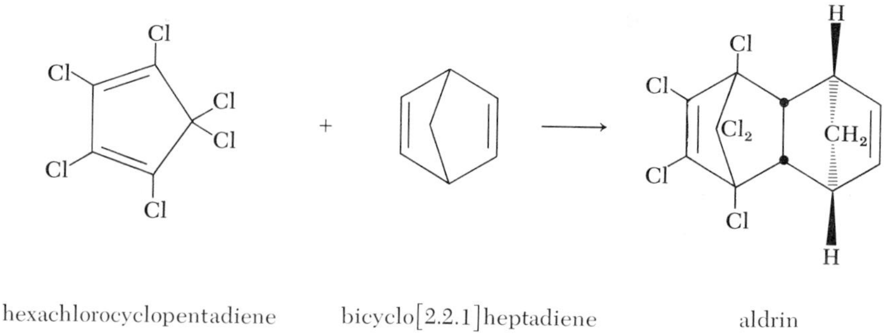

Cl Cl Cl Cl Cl Cl + $\longrightarrow$ Cl Cl Cl_2 Cl Cl H CH_2 H

hexachlorocyclopentadiene bicyclo[2.2.1]heptadiene aldrin

ELECTROCYCLIC AND CYCLOADDITION REACTIONS

13.1 INTRODUCTION

It is helpful to group reactions according to the types of mechanisms they follow or the types of intermediates they involve. For example, we speak of reactions that are nucleophilic displacements, or that employ carbocations or carbanions. However, there exists a group of reactions that cannot be described in either of these ways; they occur by a concerted process, with a set of bonds being made and broken simultaneously. In the past such reactions have been said to have the "no mechanism" mechanism, this statement conveying the idea that they do not involve the common reactive intermediates of organic chemistry and *not* that they lack a mechanism. Perhaps the most famous and most controversial example of this group is the Diels-Alder reaction, in which an appropriately substituted diene reacts with an alkene to form a new ring system. An early way of presenting its mechanism (shown below) suggests the absence of any of the classical intermediates.

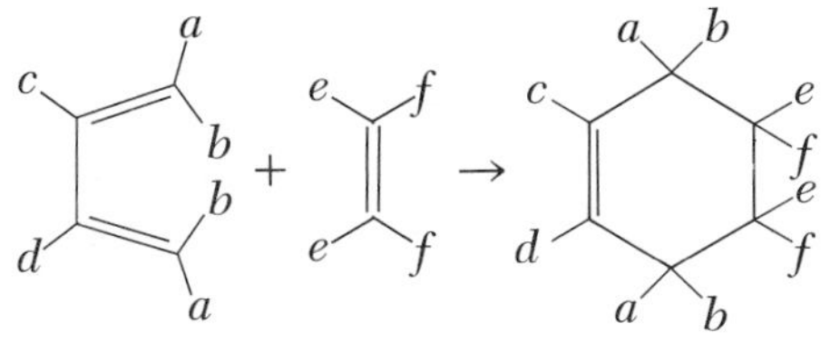

the Diels-Alder reaction

activated complex

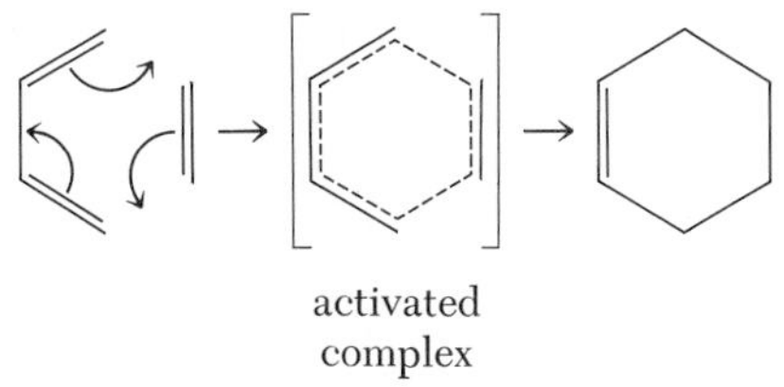

one "mechanism" of the Diels-Alder reaction

In 1965, R.B. Woodward and R. Hoffmann presented a series of papers that contained what have since become known as the *Woodward-Hoffmann rules*. These rules are used to predict whether reactions that are concerted will have low or high energies of activation under the influence of (a) increased temperature and (b) ultraviolet light. Concerted reactions that are predicted to have a low energy of activation are described as *symmetry allowed,* while those that are predicted to have a high energy of activation are described as *symmetry forbidden.*

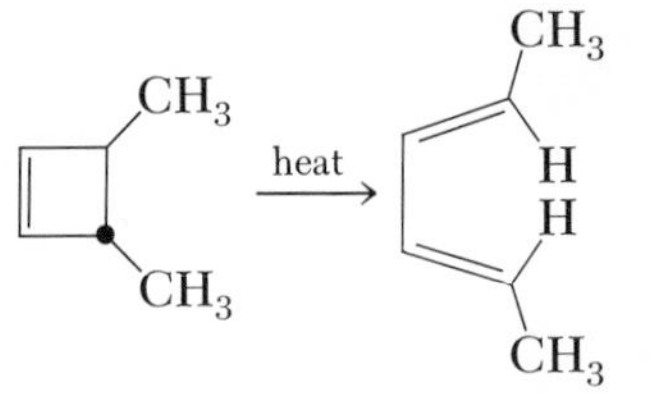

heat → a symmetry allowed reaction

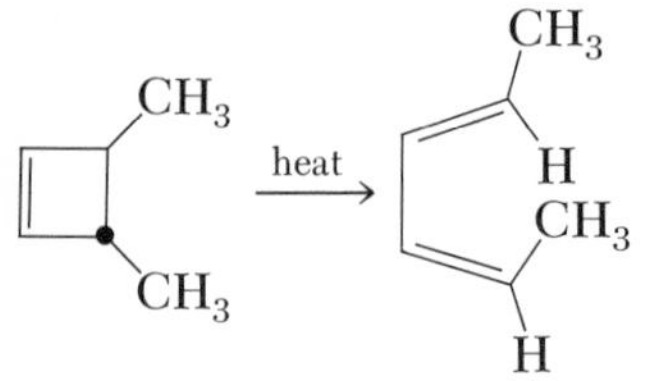

heat → a symmetry forbidden reaction—does *not* occur by a concerted process

The basic idea behind the Woodward-Hoffmann rules is intuitively quite obvious: when considering several alternate mechanisms for a reaction, the process that occurs most readily is that one which maintains the maximum amount of bonding during the reaction sequence. Of course, this suggests that a concerted process with a low energy of activation maintains a considerable amount of bonding in the activated complex. Many reactions that are concerted processes require the participation of orbitals that have two lobes (*e.g.*, a *p* orbital). We noted on p. 349 that two atomic orbitals may be aligned in a bonding or an anti-bonding array, depending upon whether or not lobes that are juxtaposed are of the same sign or of opposite sign. Those concerted processes that can arrange for adjacent lobes of orbitals involved in the reaction to maintain similar signs can be expected to have low energies of activation.

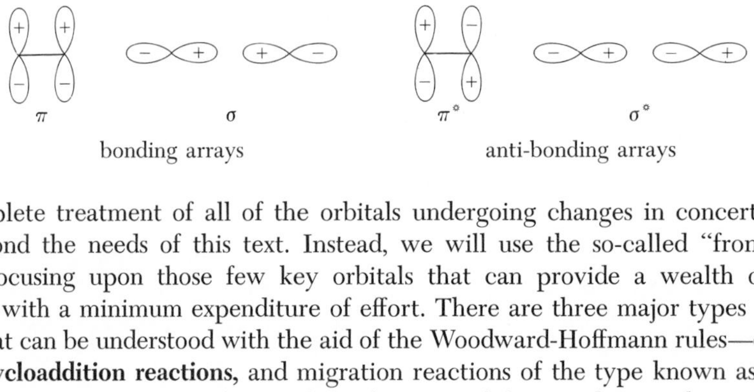

π σ π^* σ^*

bonding arrays anti-bonding arrays

A complete treatment of all of the orbitals undergoing changes in concerted reactions is well beyond the needs of this text. Instead, we will use the so-called "frontier orbital" approach, focusing upon those few key orbitals that can provide a wealth of significant information with a minimum expenditure of effort. There are three major types of concerted reactions that can be understood with the aid of the Woodward-Hoffmann rules—**electrocyclic reactions, cycloaddition reactions,** and migration reactions of the type known as **sigmatropic shifts.** Our effort will be directed toward the first two of these and, even then, will be rather superficial. The intent is to demonstrate the potential of the method and not to cover it in detail.

13.2 ELECTROCYCLIC REACTIONS

Depending on the starting material, conjugated polyenes undergo concerted cyclization under the influence of *either* light *or* heat. When the reaction is over, a double bond has disappeared, other double bonds have altered location, and a new single bond has appeared between the reactive ends of the conjugated system. This type of reaction and the reverse reaction (in which a cyclic compound is converted to an acyclic polyene) are called **electrocyclic reactions.**

an electrocyclic reaction

(**Z**)-1,3,5-hexatriene → 1,3-cyclohexadiene

The most amazing attribute of electrocyclic reactions is their high stereospecificity. Thus, *trans*-3,4-dimethylcyclobutene is converted exclusively to (2**E**, 4**E**)-2,4-hexadiene, while *cis*-3,4-dimethylcyclobutene is converted to (2**E**, 4**Z**)-2,4-hexadiene upon heating.

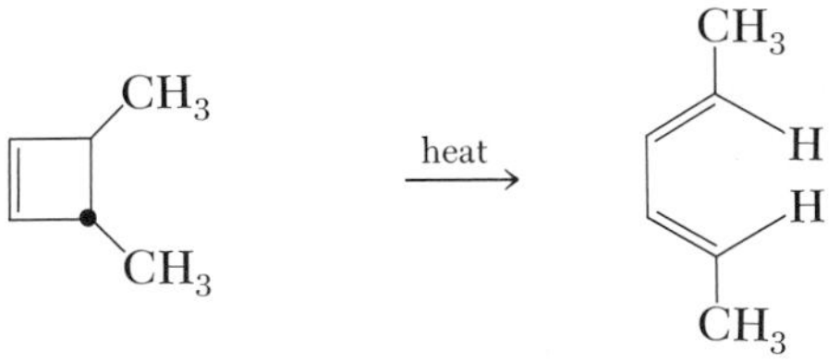

CH_3 CH_3 —heat→ CH_3 H H CH_3

trans-3,4-dimethylcyclobutene (2**E**,4**E**)-2,4-hexadiene

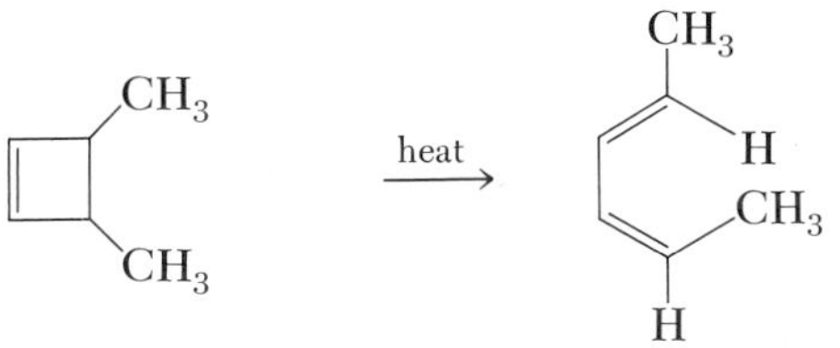

CH_3 CH_3 —heat→ CH_3 H CH_3 H

cis-3,4-dimethylcyclobutene (2**E**,4**Z**)-2,4-hexadiene

Although electrocyclic reactions are, as noted earlier, reversible, we shall concentrate on the "cyclization" process rather than upon the "ring opening" process, since this will permit a simpler presentation of the fundamental mechanism controlling such reactions. Furthermore, the principle of microscopic reversibility assures us that the mechanisms of the forward and reverse reactions are simply the reverse of one another. These reactions, while they are reversible, may favor one side over another because of thermodynamic considerations. Therefore, these reactions will be written so as to indicate the product that is favored under specific experimental conditions.

In our discussion of the ultraviolet spectrum of 1,3-butadiene (p. 351) we focused attention on the excitation of an electron from the highest occupied molecular orbital (HOMO) to the lowest unoccupied molecular orbital (LUMO) (thus producing the first excited state of 1,3-butadiene). Here again we will concentrate upon the behavior of the HOMO of the polyene, since its electrons may be viewed as the outer or "valence" electrons of the molecule's π system. The HOMO of a 1,3-butadiene is π_2 (see Figures 12–2 and 12–4) and may be represented as shown below. Since these are the valence electrons of the molecule, we can imagine these as the electrons that form the new σ bond when a 1,3-butadiene closes to a cyclobutene.

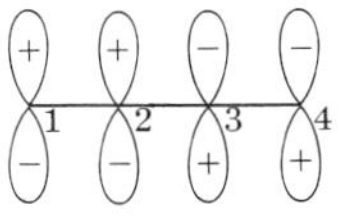

+ + − −
1 2 3 4
− − + +

π_2 of 1,3-butadiene

1. How many electrons are in the π_1, π_2, π_3^* and π_4^* orbitals of the ground state of 1,3-butadiene? In the first excited state of 1,3-butadiene?

In order to get overlap when ring closure occurs, the lobes that will form the new σ bond (lobes on Cl and C4 of π_2) must have the same sign. This can be attained only if the bonds between Cl and C2 and between C3 and C4 are rotated in the same direction. This type of motion is described as **conrotatory.** When such a conrotatory transformation occurs, the cyclization process has a low energy of activation and is described as symmetry allowed.

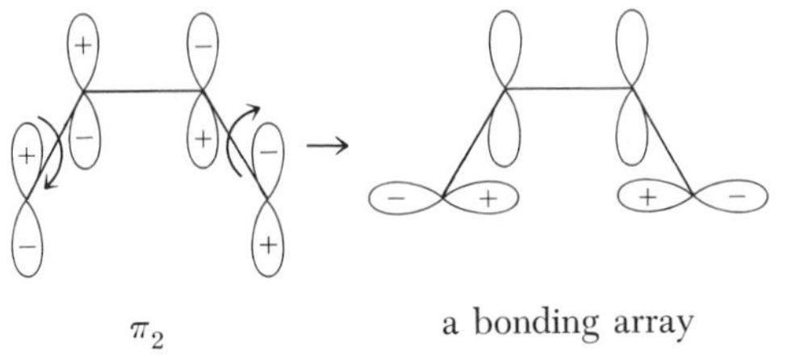

conrotatory motion (symmetry allowed)

π_2 → a bonding array

If rotation around these two bonds had occurred in opposite directions (so-called **disrotatory** motion), the lobes that moved together to form the new σ bond would have had *opposite* signs. As noted earlier, this type of arrangement is anti-bonding, and any process leading to it would have a high energy of activation. Such a process, therefore, is described as symmetry forbidden. Consequently, we expect the ring closure of 1,3-butadiene to follow a conrotatory pathway when heat is used to initiate reaction. (The heat provides the energy for the necessary bond rotations, among other things.)

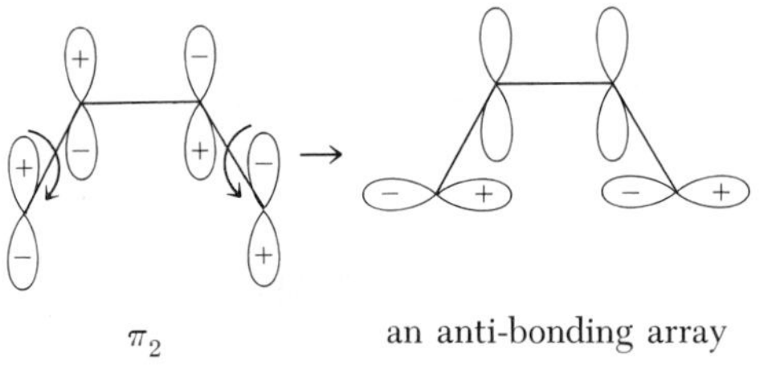

disrotatory motion (symmetry forbidden)

π_2 → an anti-bonding array

When we add to these simple 1,3-butadiene skeletons the methyl groups that convert butadiene to the isomeric 2,4-hexadienes, we can explain the high degree of stereospecificity noted earlier.

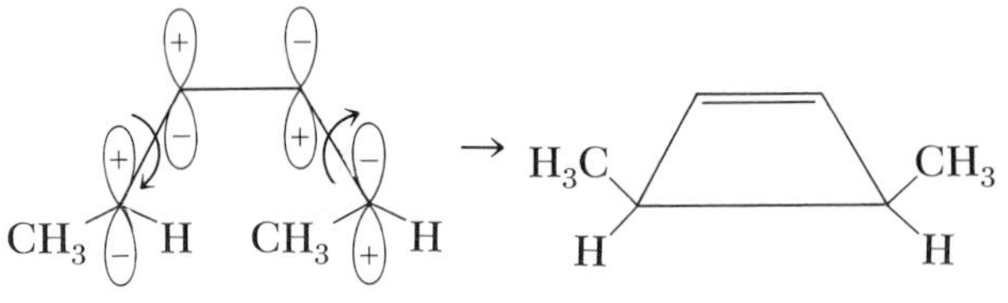

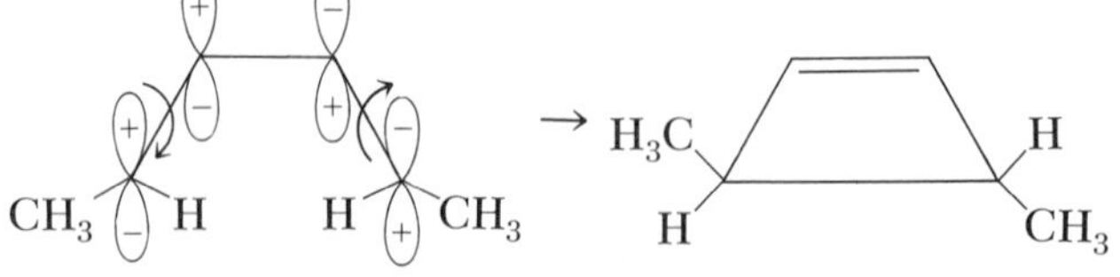

CH_3 H CH_3 H → H_3C H CH_3 H

conrotatory motion (symmetry allowed)

CH_3 H H CH_3 → H_3C H H CH_3

conrotatory motion (symmetry allowed)

We can expand this approach to encompass the thermal cyclization of trienes to cyclohexadienes. For example, heating converts (**2E, 4Z,** 6**E**)-2,4,6-octatriene into 5,6-*cis*-dimethyl-1,3-cyclohexadiene, but converts (**2E, 4Z,** 6**Z**)-2,4,6-octatriene into 5,6-*trans*-dimethyl-1,3-cyclohexadiene. This stereospecificity is accounted for by examining the highest occupied molecular orbital, π_3, of the 1,3,5-hexatriene system (Figure 13–1) and noting orbital alignment

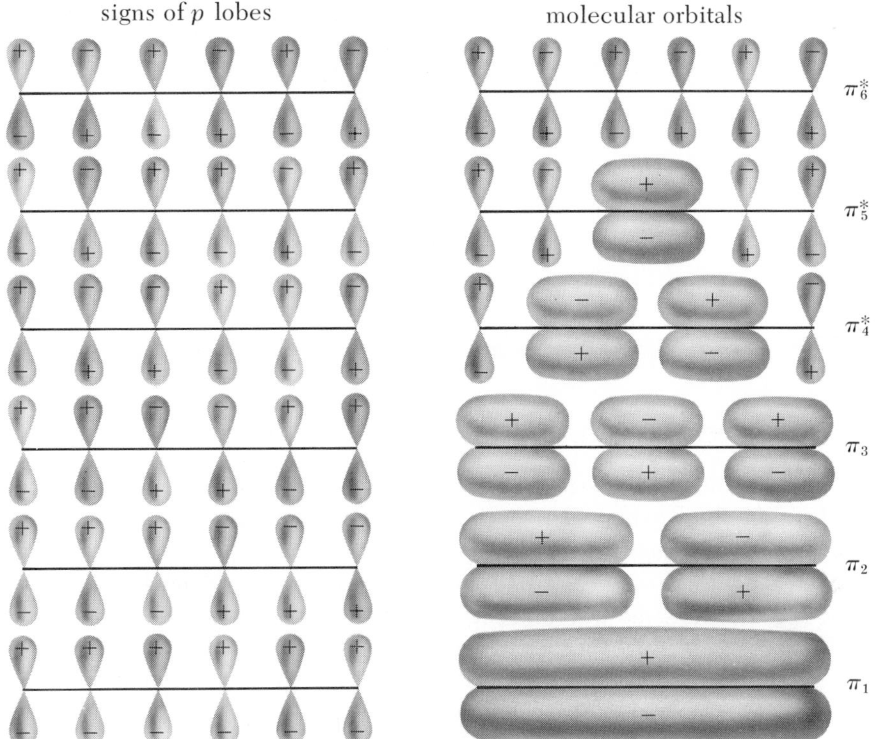

Figure 13-1 Molecular orbitals of 1,3,5-hexatriene.

after conrotatory and disrotatory motion. As shown below, in order to bring the outer orbitals of π_3 into a σ bonding array, they must be brought together by a disrotatory motion. A conrotatory motion leads to an anti-bonding array. Here, therefore, the disrotatory process is the one that is symmetry allowed and has a low energy of activation.

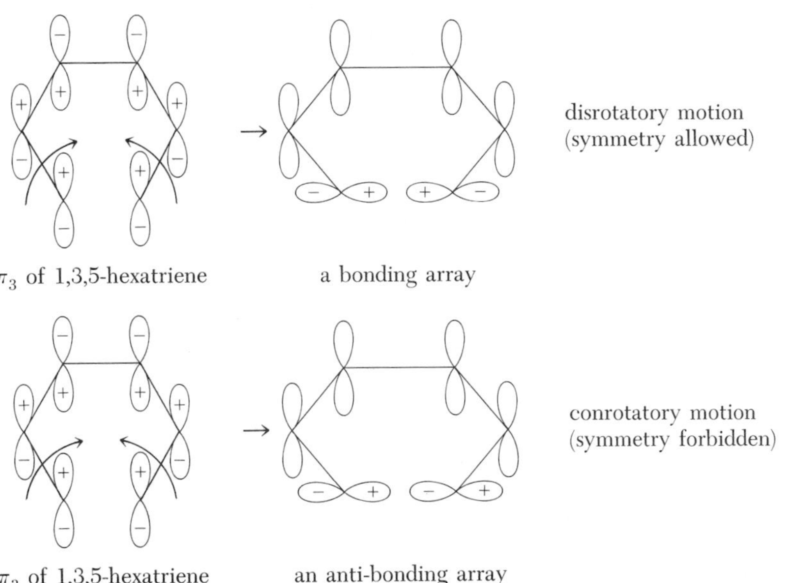

(2**E**,4**Z**,6**E**)-2,4,6-octatriene → 5,6-*cis*-dimethyl-1,3-cyclohexadiene

(2**E**,4**Z**,6**Z**)-2,4,6-octatriene → 5,6-*trans*-dimethyl-1,3-cyclohexadiene

This behavior can be summarized by noting that *thermally induced electrocyclic reactions involving 4n π electrons* (where *n* is an integer) *require conrotatory motion. Under similar conditions, electrocyclic reactions involving 4n + 2 π electrons follow disrotatory motion.*

Conjugated polyenes can react, under the influence of ultraviolet light, to form cyclic compounds in much the same way that they cyclize upon heating. The significant difference between thermal and photochemical cyclization is in the stereochemistries of the products. For example, (2**E**, 4**E**)-2,4-hexadiene is converted to *cis*-3,4-dimethylcyclobutene by ultraviolet light. How can we account for this difference between the thermal and the photochemical reactions?

(2**E**,4**E**)-2,4-hexadiene $\underset{}{\overset{h\nu}{\rightleftarrows}}$ *cis*-3,4-dimethylcyclobutene — photochemical cyclization

We must remember that the absorption of ultraviolet light promotes an electron from π_2 of 1,3-butadiene to π_3^*. Thus, in the photoexcited state of 1,3-butadiene, the highest occupied molecular orbital is π_3^* and not π_2. As shown below, a disrotatory motion of the terminal orbitals of π_3^* will produce a bonding array (a new σ bond). Conrotatory motion, on the other hand, produces an anti-bonding array. Thus, it is the disrotatory motion in π_3^* which is responsible for the observed stereospecificity of the photochemical ring closure of (2**E**, 4**E**)-2,4-hexadiene.

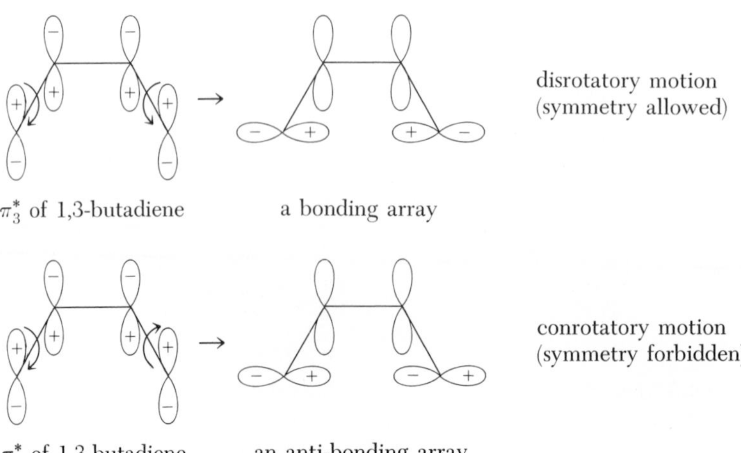

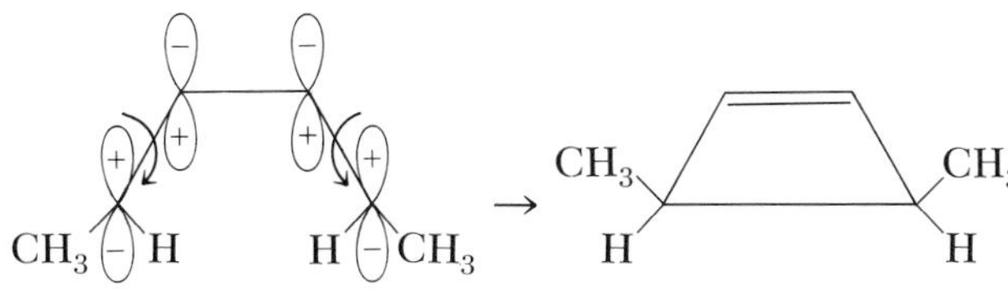

CH$_3$ H H CH$_3$ → CH$_3$ CH$_3$ H H

π_3^* of (2E,4E)-2,4-hexadiene

cis-3,4-dimethylcyclobutene

In general, *photochemical cyclizations are symmetry allowed for disrotatory motion with 4n π electron systems and for conrotatory motion with 4n + 2 π electron systems.* This, of course, is the *opposite* of what is observed for thermal cyclization.

2. You have just prepared the product shown below. Should you be concerned about protecting your product from heat or from light? Was the compound prepared by heating or by illuminating the starting material? Explain your answer.

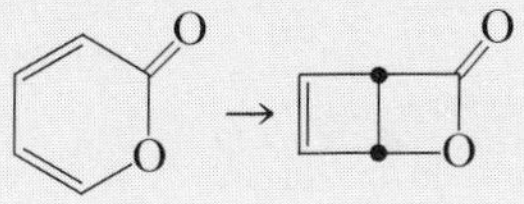

3. Show why the cyclization of 1,3,5-hexatriene to 1,3-cyclohexadiene follows a conrotatory pathway under the influence of light but a disrotatory pathway under the influence of heat. Suggest an appropriately substituted 1,3,5-hexatriene that can prove that the reaction follows the pathway suggested. Justify your selection.

13.3 CYCLOADDITION REACTIONS

A **cycloaddition reaction** is one in which two molecules containing unsaturated functional groups combine to form a cyclic molecule, two new σ bonds being formed from π electrons. Cycloaddition reactions are often spontaneous or, at most, require modest amounts of energy for inception. One of the simplest cycloaddition reactions is that involving two molecules of ethylene to produce cyclobutane. This process does *not* occur upon heating ethylene to moderate temperatures, but does occur when ethylene is exposed to ultraviolet light. How can we account for this difference?

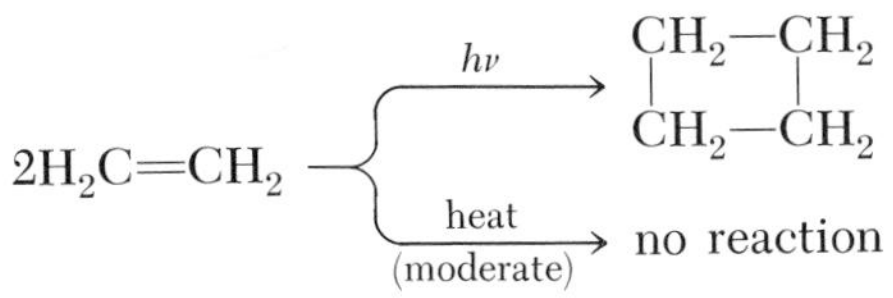

$$2H_2C{=}CH_2 \xrightarrow{h\nu} \begin{array}{c} CH_2{-}CH_2 \\ | \quad\;\; | \\ CH_2{-}CH_2 \end{array}$$

$$2H_2C{=}CH_2 \xrightarrow[\text{(moderate)}]{\text{heat}} \text{no reaction}$$

Ethylene Cyclodimerization. Ethylene's inability to undergo facile thermal cyclization may be explained after recourse to a picture of the orbitals used in this cycloaddition. When the new σ bonds are formed, they can be pictured as resulting from the interaction of the HOMO of one ethylene molecule with the lowest unoccupied molecular orbital (LUMO) of the other ethylene molecule. The reason that the interaction occurs with the LUMO of the second ethylene molecule is that the HOMO of the first molecule must provide electrons to the second molecule, and only an *unoccupied* molecular orbital of the second molecule may accept them. (The HOMO of the second ethylene molecule already has its complement of two electrons.)

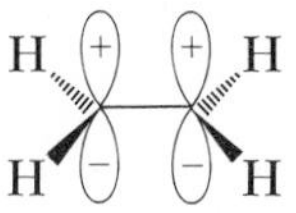

π orbital of ethylene ("HOMO")

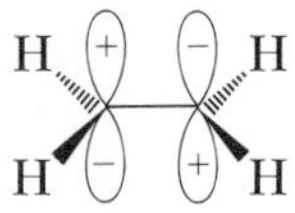

π^* orbital of ethylene ("LUMO")

If we attempt to overlap the HOMO of one ethylene molecule and the LUMO of another, then, as seen below, we generate one bonding and one anti-bonding orbital. There is, therefore, no *net* bonding in this process. Expanding the terminology used in our discussion of electrocyclic reactions, such a process will have a high energy of activation and will be symmetry forbidden. That is why thermal cyclodimerization of ethylene does not readily occur by a concerted process.

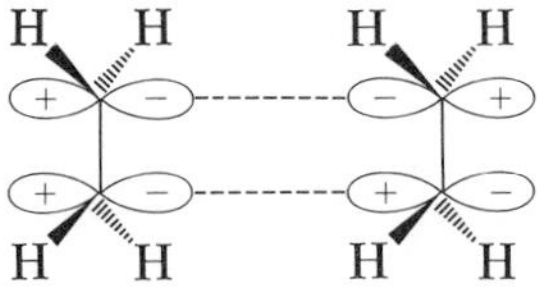

HOMO of ethylene (π) LUMO of ethylene (π^*)

symmetry forbidden

Photochemical cycloaddition of ethylene to ethylene involves the overlap of the HOMO of a photoexcited ethylene molecule with the LUMO of a second molecule in the ground state. The HOMO of the photoexcited state is the π^* orbital; as seen below, this interaction *does* produce a net bonding array and is symmetry allowed. This is why photochemical dimerization of ethylene has a low energy of activation.

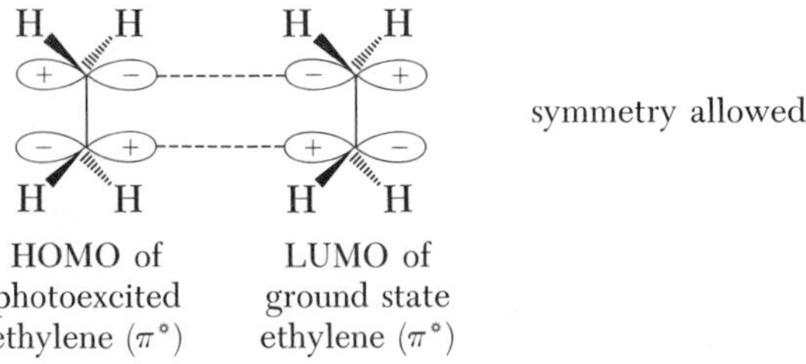
HOMO of photoexcited ethylene (π^*) LUMO of ground state ethylene (π^*)

symmetry allowed

The photochemical cyclodimerization of ethylene is described as a suprafacial process, both bonds to one ethylene being formed from the same face of the molecule. Processes in which bonds are formed (or broken) on opposite faces of a molecule are described as antarafacial. This type of terminology can be used to describe the course of some familiar reactions. Thus, the epoxidation of ethylene (p. 329) is a suprafacial process. It is important to bear in mind the idea that these new terms reflect the stereochemistry of the *process* and not the stereochemistry of the product.

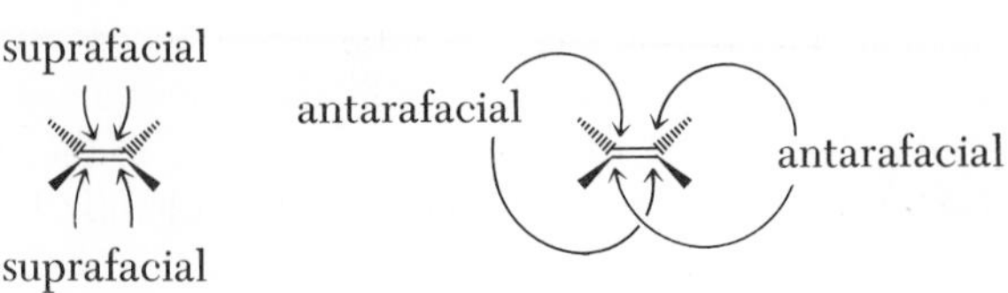

a suprafacial process

The Diels-Alder Reaction—The Mechanism. The cyclodimerization of ethylene is classified as a [2 + 2] cycloaddition reaction because it involves a molecule with two π electrons reacting with another molecule containing two π electrons. A much more common type of cycloaddition reaction is the thermally induced [4 + 2] process, in which a system of four π electrons reacts with a system of two π electrons. The most famous example of this is the **Diels-Alder reaction,** the reaction of a conjugated diene with an alkene (dubbed the *dienophile*) to form a cyclohexene derivative.

the Diels-Alder reaction—
a [4+2] cycloaddition

The Diels-Alder reaction is a very useful process for building complex ring systems, as seen from the examples below.

As with other reactions controlled by orbital symmetry, the Diels-Alder reaction possesses some rigid stereochemical requirements. First, the diene must react in the s-*cis* conformation, since only this one permits both termini of the diene to react with both ends of the dienophile simultaneously. Second, the reaction is stereospecific in that the addition to the dienophile is a suprafacial process. Finally, the addition of the diene to a dienophile bearing an unsaturated group (used to enhance the reactivity of the dienophile) occurs in an *endo,* rather than an *exo,* fashion. This means that the unsaturated group of the dienophile resides near the developing double bond in the diene fragment.

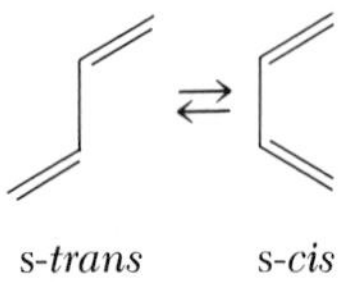

s-*trans* ⇄ s-*cis*

s-*cis* geometry required for the Diels-Alder reaction

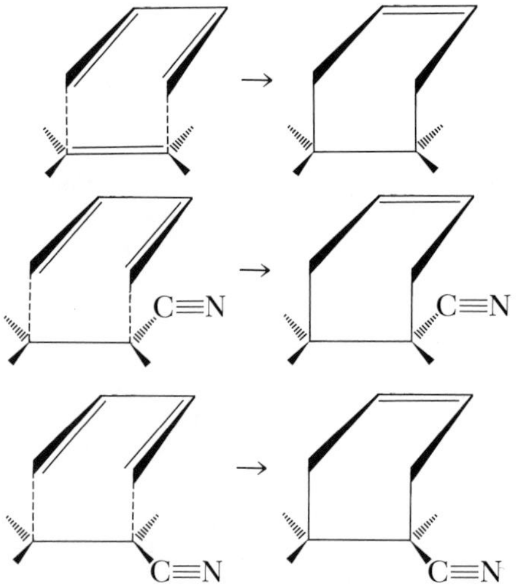

suprafacial addition to the dienophile

C≡N → C≡N

endo addition, preferred in the Diels-Alder reaction

C≡N → C≡N

exo addition, dis-favored in the Diels-Alder reaction

4. Would you anticipate that vinylacetylene, $H_2C{=}CH{-}C{\equiv}CH$, could function as a diene in the Diels-Alder reaction? Explain your answer.

All of these stereochemical constraints, as well as the reversibility of the Diels-Alder reaction, are illustrated in the dimerization of cyclopentadiene to "dicyclopentadiene" and the thermal "cracking" of this dimer back to the starting material. In this example one compound acts as both diene and dienophile.

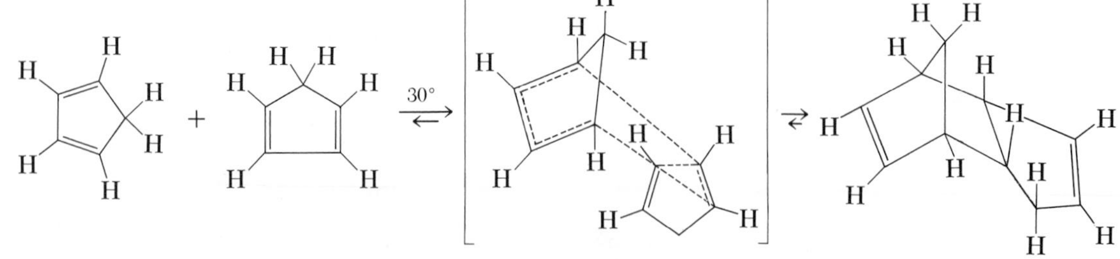

diene (cyclopentadiene) + dienophile (cyclopentadiene) $\xrightarrow{30°}$ [Activated complex] ⇄ dicyclopentadiene

Is the ease of this thermal cycloaddition reaction consistent with the molecular orbital approach to the study of concerted reactions? Indeed, it is! As with the [2 + 2] cycloaddition, we must worry about combining the HOMO of one molecule with the LUMO of the other. Should we use the LUMO of the diene or that of the dienophile? It makes no difference since, as shown below, either combination (LUMO diene, HOMO dienophile *or* HOMO diene, LUMO dienophile) is symmetry allowed and produces a bonding array. In other words, the thermal cycloaddition should have a low energy of activation regardless of which way it is presented. This is consistent with the observation that 1,3-cyclopentadiene is usually stored at $-78°$ in order to prevent rapid dimerization. The normal preparation of 1,3-cyclopentadiene involves the thermal cracking of the dimer.

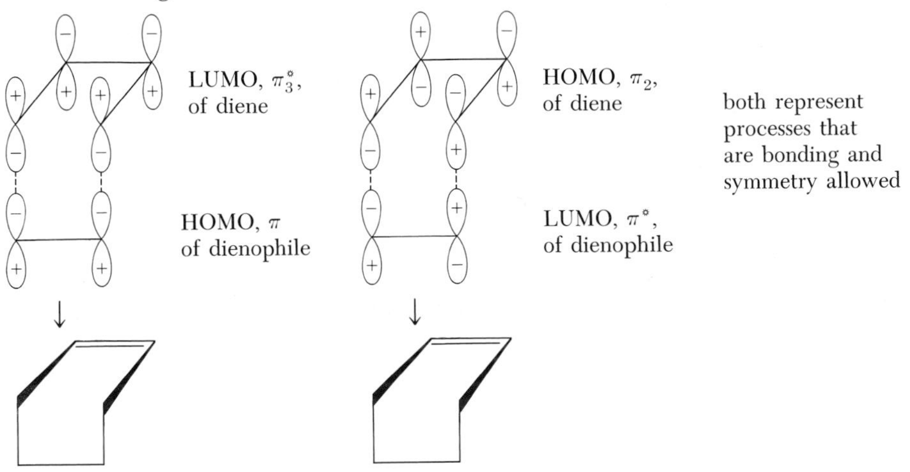

We should note, in summary, that cycloaddition reactions involving a total of $4n$ π electrons are accelerated by the action of light but not by the action of heat. In contrast, cycloadditions in which the total number of π electrons is $4n + 2$ proceed readily by the action of heat but are not accelerated by light.

The Diels-Alder Reaction—Applications. In general, the most reactive dienophiles are those that have the alkene double bond conjugated to some electron withdrawing group. The most common such dienophiles include the following:

tetracyanoethylene	$(NC)_2C{=}C(CN)_2$
crotonaldehyde	$CH_3CH{=}CHCHO$
maleic anhydride	
1-nitropropene	$CH_3CH{=}CHNO_2$
acrylonitrile	$CH_2{=}CHCN$
ethyl acrylate	$CH_2{=}CHCO_2C_2H_5$
benzoquinone	

By way of contrast, alkylation of the diene usually accelerates the Diels-Alder reaction. While most Diels-Alder reactions appear to involve dienes that are electron-rich relative to the dienophile, it is now believed that what is essential for adduction to occur is that the two reacting molecules have complementary electronic natures. Thus, some Diels-Alder reactions occur between electron-rich dienophiles and electron-poor dienes. A good example of this is the reaction used to produce the insecticide aldrin, in which hexachlorocyclopentadiene serves as an electron-poor diene (review problem 29, p. 370).

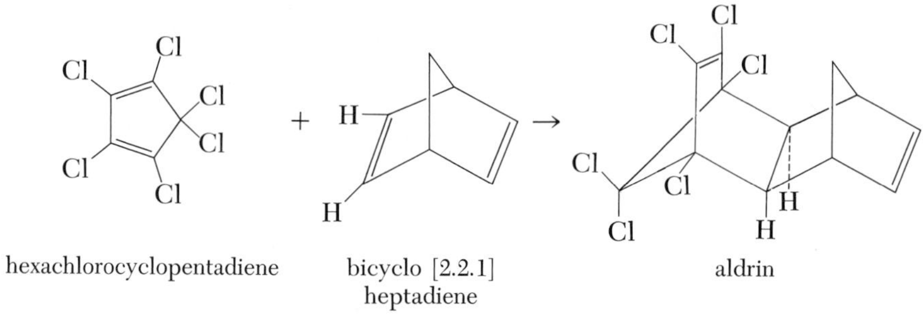

hexachlorocyclopentadiene + bicyclo [2.2.1] heptadiene → aldrin

In addition to simple double bonds, the reactive dienophile may also be an allene or a triple bond. However, allenes normally are not useful as dienophiles.

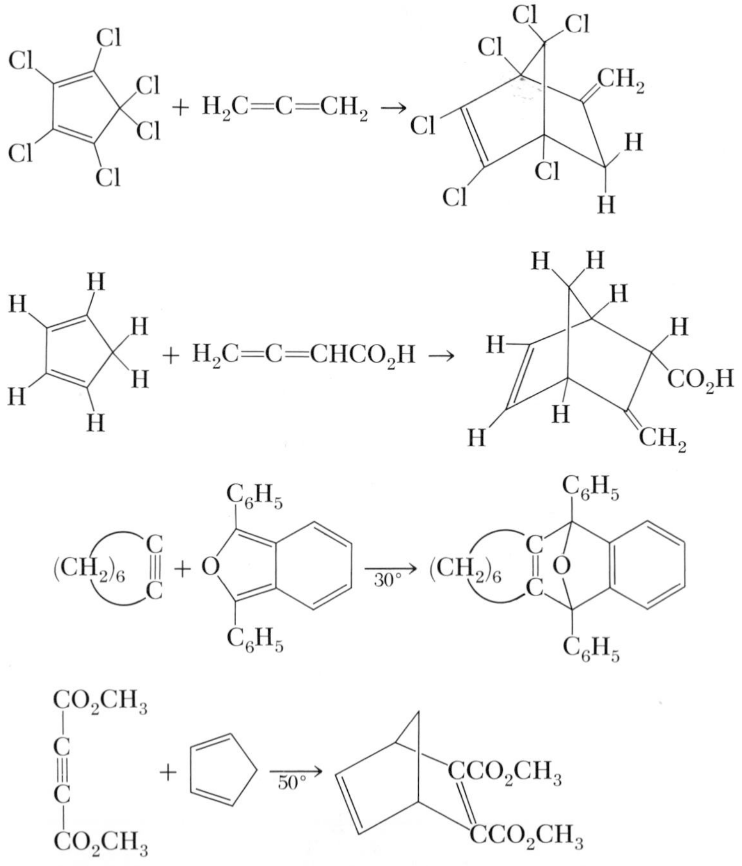

$$+ H_2C{=}C{=}CH_2 \rightarrow$$

$$+ H_2C{=}C{=}CHCO_2H \rightarrow$$

$$(CH_2)_6\,C{\equiv}C + \text{1,3-diphenylisobenzofuran} \xrightarrow{30^\circ}$$

$$CH_3O_2C{-}C{\equiv}C{-}CO_2CH_3 + \text{cyclopentadiene} \xrightarrow{50^\circ}$$

While the most common reactants with dienes in the Diels-Alder reaction are alkenes and alkynes, other combinations of [4 + 2] π electrons can lead to rapid and synthetically valuable cycloadditions. The most important application of such variations is in the synthesis of heterocycles, several examples of which are shown below. We see from these illustrations that the diene can be replaced by C=C—C=O and that the alkene can be replaced by C=O, N=O and N=N. Other substitutions also are possible.

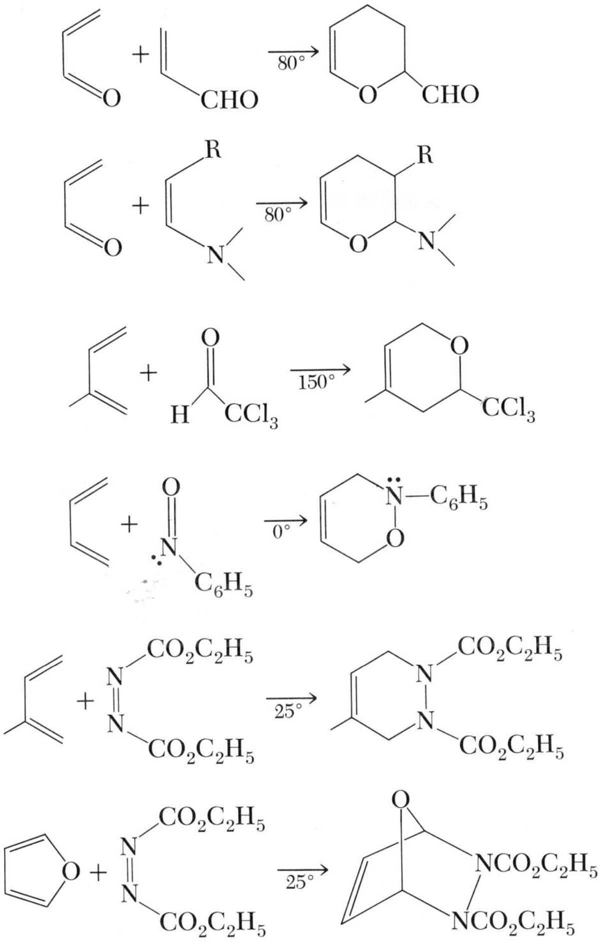

In this chapter we have examined two major contributions to organic chemistry. One, the Woodward-Hoffmann rules, is quite new and represents a major theoretical advance. The other, the Diels-Alder reaction, is much older but is a most important synthetic tool. It is interesting to note that both R.B. Woodward *and* O. Diels and K. Alder have won Nobel prizes in chemistry. The former received his (1965) for "organic synthesis," while the latter received theirs (1950) for "development of the diene synthesis." While some people have questioned the wisdom associated with the awarding of some Nobel prizes, there is no doubt of the soundness of these particular awards.

IMPORTANT TERMS

Antarafacial process: A process in which bonds are being made or broken on opposite sides (faces) of a π bond. We have not discussed any reactions like this in this chapter. Since a process must be concerted in order to be termed antarafacial, the *trans* addition of a halogen to a double bond is *not* antarafacial.

Conrotatory motion: Rotation of bonds in the same direction, i.e., both clockwise or both counterclockwise.

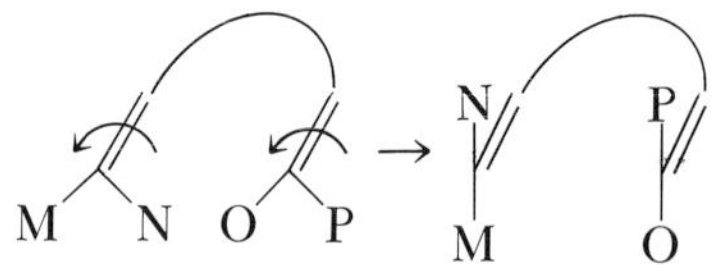

M N O P → N M P O conrotatory motion

Cycloaddition: A reaction in which two unsaturated molecules react to produce a cyclic compound. During the process π bonds are converted to σ bonds, and new π bonds may appear. The classic example of a cycloaddition reaction is the Diels-Alder reaction, a concerted process useful in synthesizing cyclohexenes and related systems. The Diels-Alder reaction is thermally allowed (see below).

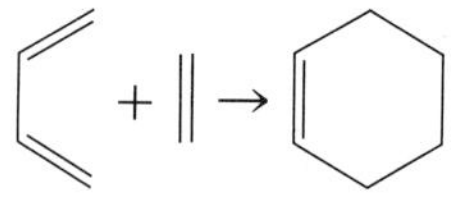

a Diels-Alder cycloaddition reaction

Dienophile: A compound that, by employing a π bond, will undergo a Diels-Alder reaction with a diene.

Disrotatory motion: Rotation of bonds in opposite directions, *i.e.*, one clockwise and one counterclockwise.

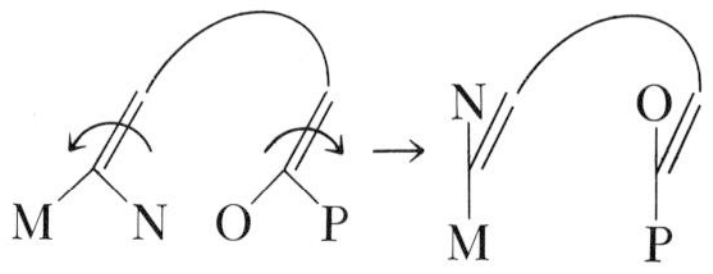

M N O P → N M O P disrotatory motion

Electrocyclic reaction: Conversion of a conjugated polyunsaturated compound into an isomeric cyclic compound. The bond responsible for ring formation occurs between the two ends of the conjugated system of the starting material.

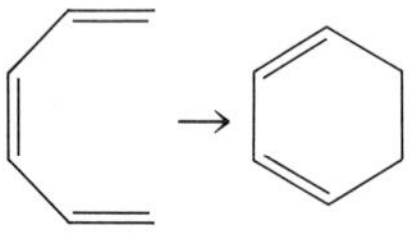

an electrocyclic reaction

Photochemically allowed reaction: A reaction that will be concerted *and* will possess a comparatively low energy of activation when initiated by light.

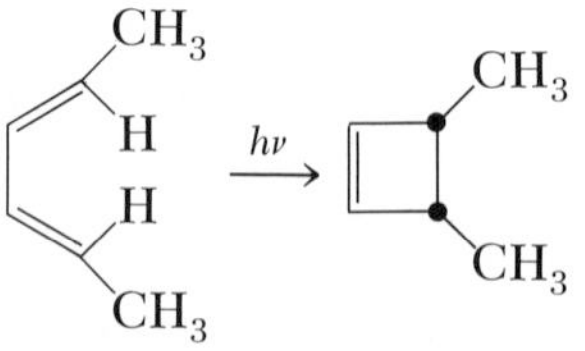

CH_3 H H CH_3 $\xrightarrow{h\nu}$ CH_3 CH_3 a photochemically allowed reaction

Suprafacial process: A process in which bonds are simultaneously being made (or broken) on the same side (face) of a π bond. The Diels-Alder reaction is an example of this.

Symmetry allowed reaction: A reaction that will occur by a concerted process and will not involve radical or ionic intermediates. A symmetry allowed process may be accelerated either by heat or by light.

Symmetry forbidden reaction: A reaction that cannot occur by a *concerted* process under a given set of conditions (*e.g.*, light). Reactions that are "symmetry forbidden" do occur. However, when they do occur they are not concerted but, instead, involve classical intermediates (often radicals). They usually have high energies of activation.

Thermally allowed reaction: A reaction that will be concerted and will possess a comparatively low energy of activation when initiated by heat. The Diels-Alder reaction is one example of a thermally allowed reaction.

PROBLEMS

5. Predict the product of the Diels-Alder reaction of 1,3-butadiene with each of the following:
   (a) ethylene (e) maleic anhydride
   (b) acrylonitrile (f) tetracyanoethylene
   (c) benzoquinone (g) crotonaldehyde
   (d) ethyl acrylate
6. Would you expect reaction **A** to have a lower energy of activation than reaction **B**? Explain your answer.

A. + ‖ —heat→

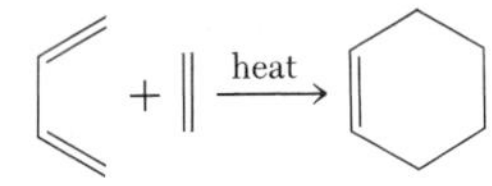

B. + ‖ —light→

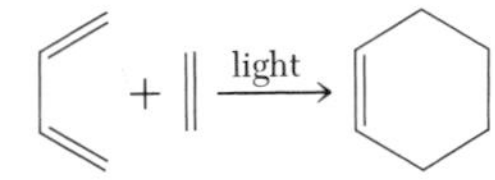

7. Suggest a diene-dienophile combination that will produce each of the following:

(a) O O

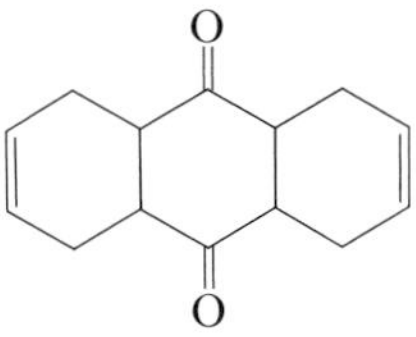

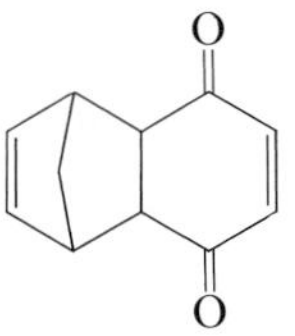

(f) CHO C_2H_5

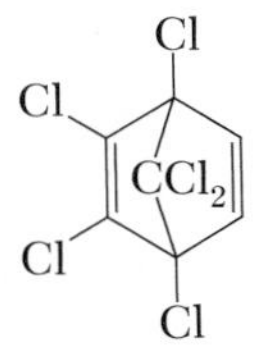

(k) $OCOCH_3$ H CO_2CH_3 $OCOCH_3$

(b) O

(g) O

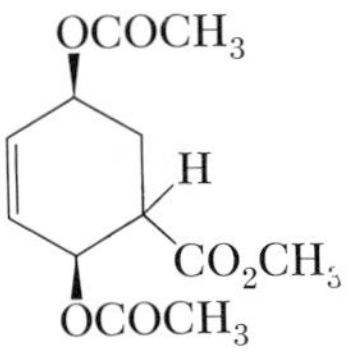

(l) CH_3 O O O O

(h) O OCH_3

(c)

(i) Cl Cl CCl_2 Cl Cl

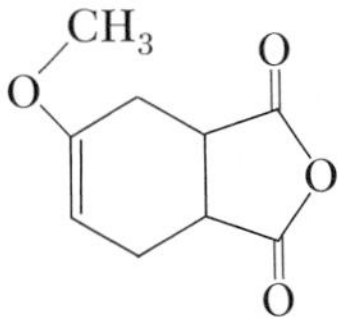

(m) Cl CH_3—C—N CH_3 CH_3 O CH_3

(d) O O

(n) $OCOCH_3$ $CH_3C(O)O$ O O O

(j) Cl Cl CCl_2 Cl Cl

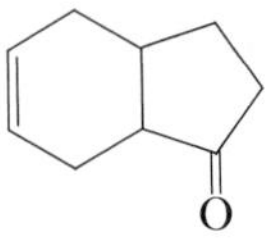

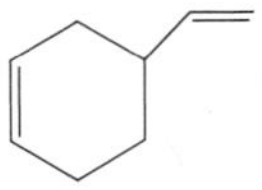

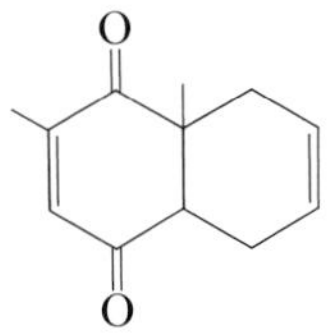

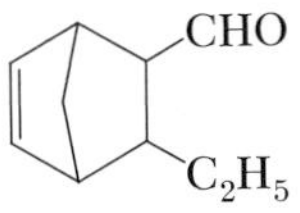

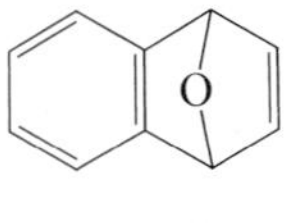

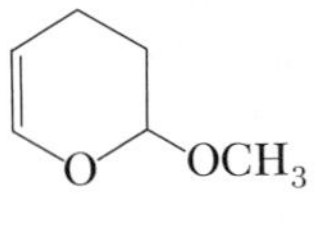

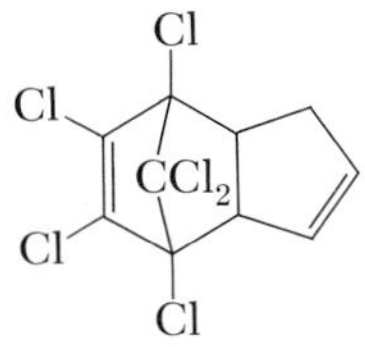

(e) O O

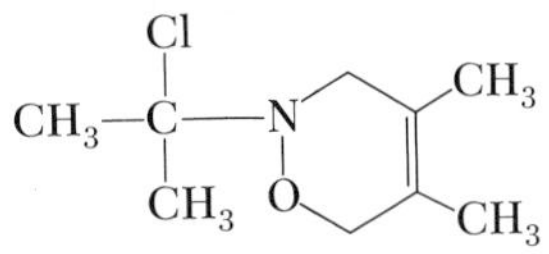

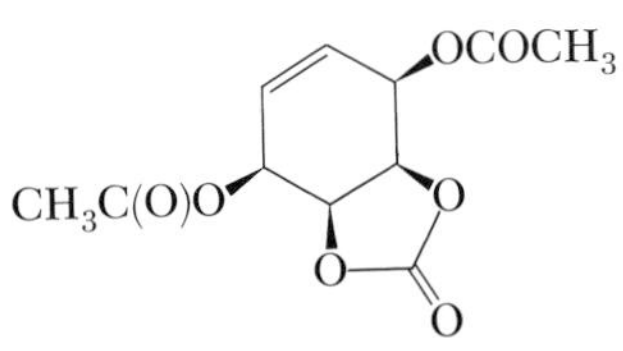

(o) CH_3, CH_3, CH_3, CH_3, CH, CH, O, O, O

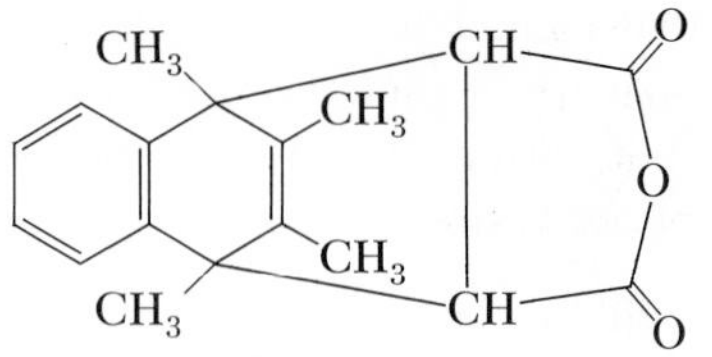

(q) 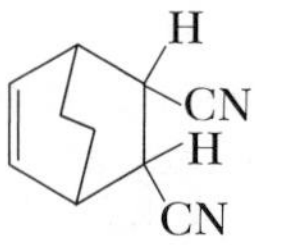 H, CN, H, CN

(r) 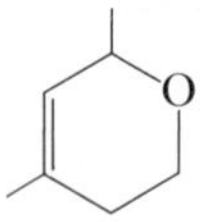 O

(p) 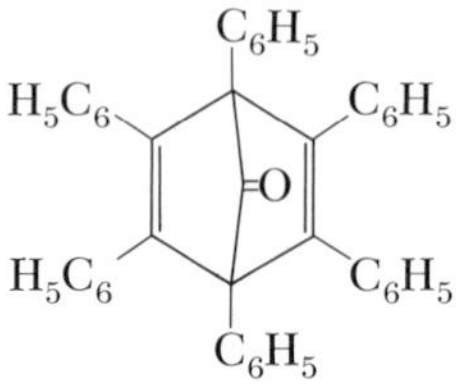 C_6H_5, H_5C_6, C_6H_5, O, H_5C_6, C_6H_5, C_6H_5

(s) 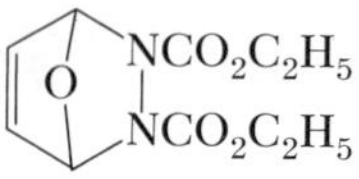 O, $NCO_2C_2H_5$, $NCO_2C_2H_5$

8. Suggest a reason for the greater reactivity of cyclopentadiene compared to (**2E**, **4E**)-2,4-hexadiene as a diene in the Diels-Alder reaction.

9. *Trans*-piperylene ($CH_2{=}CH{-}CH{=}CHCH_3$) reacts with maleic anhydride at 0° (benzene solvent) to give a quantitative yield of the Diels-Alder adduct. However, *cis*-piperylene produces less than 5% of a Diels-Alder adduct with maleic anhydride even after 8 hours at 100°. Account for this difference.

10. Identify the optically active compounds produced by the reaction of (**R**)-3-methyl-1-pentene with cyclopentadiene.

11. What is the importance, if any, of the following in understanding the mechanism of the Diels-Alder reaction?

A. 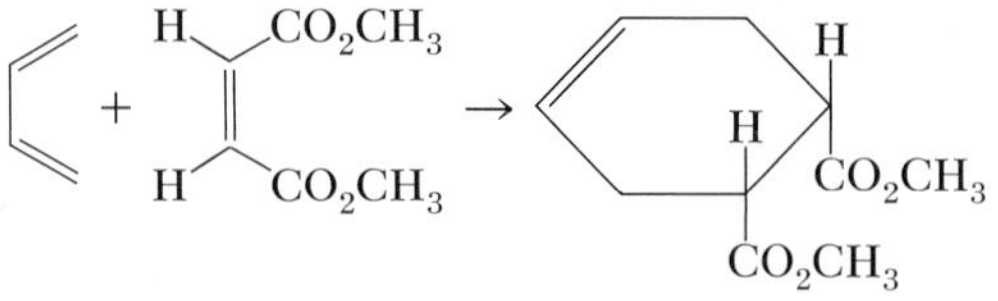 + H, CO_2CH_3, H, CO_2CH_3 → H, H, CO_2CH_3, CO_2CH_3

B. 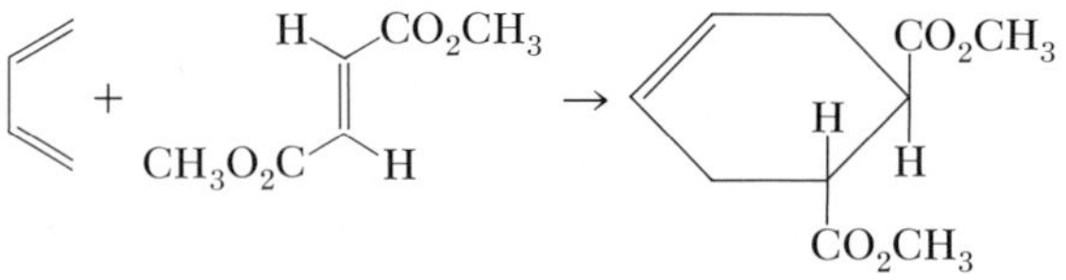 + H, CO_2CH_3, CH_3O_2C, H → CO_2CH_3, H, H, CO_2CH_3

12. Predict the structure of the 2,4-heptadiene produced by the thermal ring opening of *cis*-3-ethyl-4-methylcyclobutene.

13. Although the *endo* product is formed most rapidly in the Diels-Alder reaction (see p. 380), reactions that are carried out at higher temperatures may actually produce more *exo* than *endo* product. Explain this in terms of "kinetic *vs.* thermodynamic control" and suggest an experiment that might test this explanation.

14. The reaction of **A** with base, in the presence of 1,3-cyclopentadiene, produced **B** and **C.** Account for their formation.

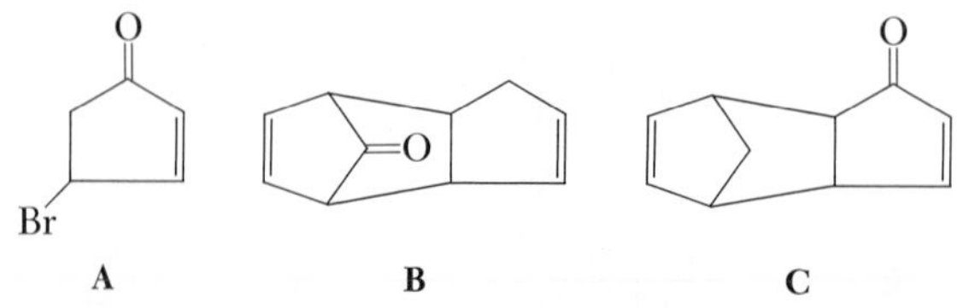

15. The thermal ring closure of **A, B,** and **C** (below) follows the Woodward-Hoffmann rules. Predict the structure of the dimethylcyclooctatriene produced by each of these. (*Hint:* Two of them give the same product.)

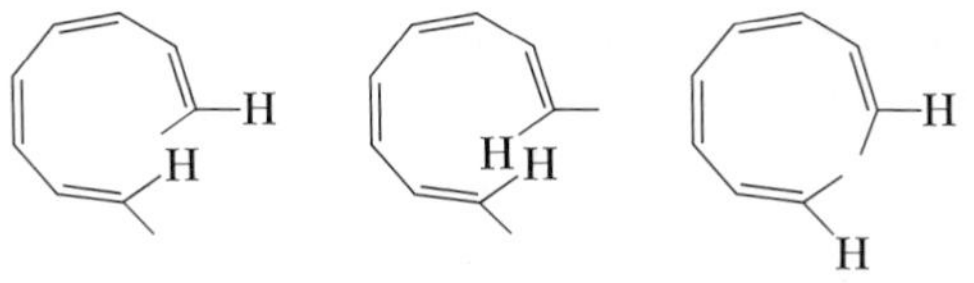

A B C

16. Predict the products of the photochemical ring closure of **A**, **B**, and **C** in problem 15.
17. What, if anything, do the following have in common?

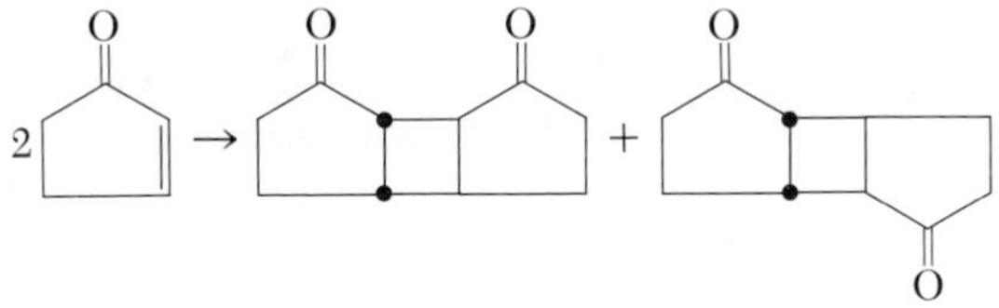

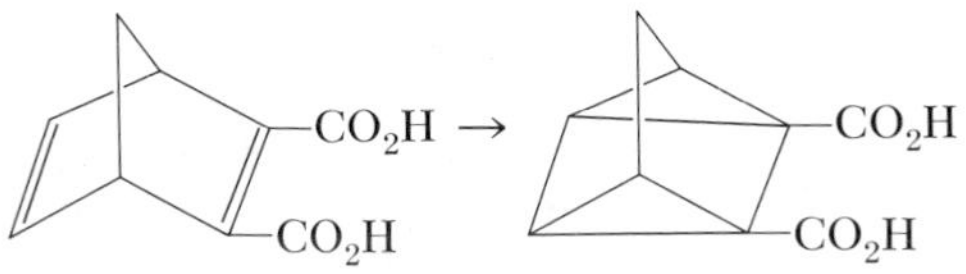

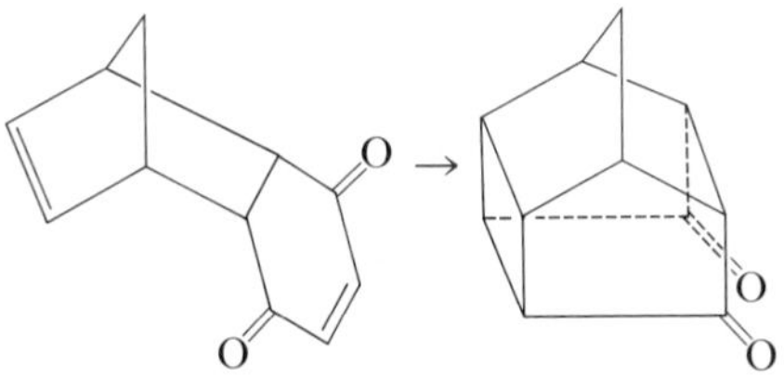

18. Would you expect the reactions shown below to be accelerated by heat or by light? Explain.

(a)

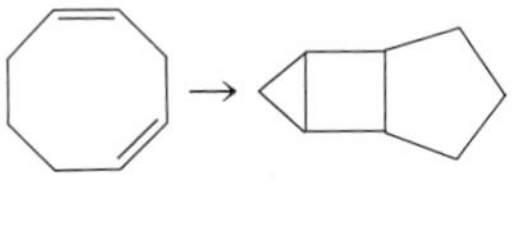

(b) 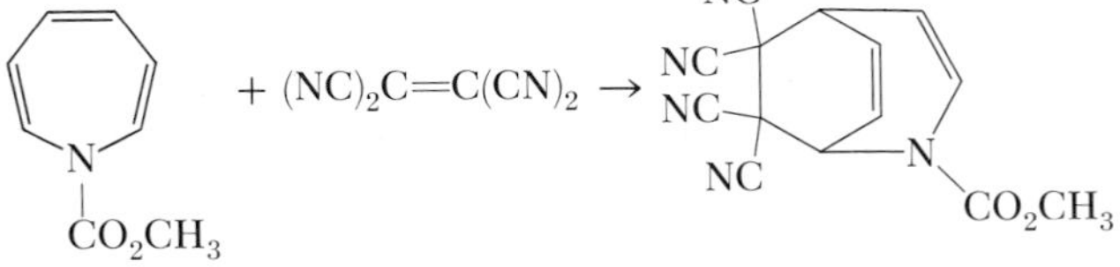 $+ (NC)_2C{=}C(CN)_2 \rightarrow$

(c)

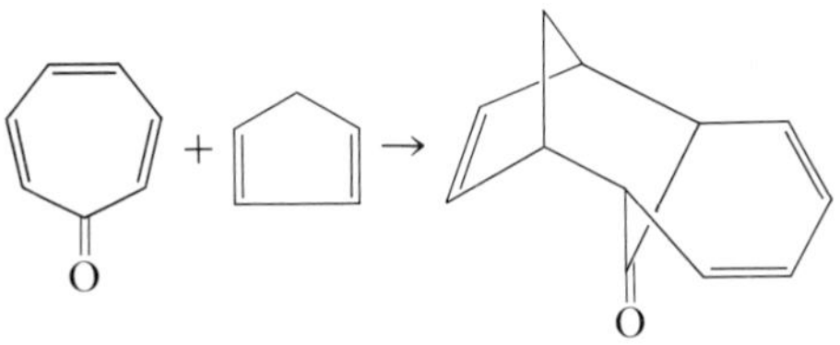

(d)

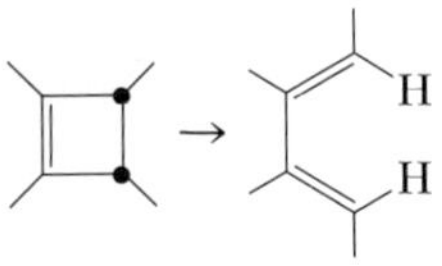

19. The interconversion shown below can be explained by a combination of an *intra*molecular Diels-Alder reaction and a *retro* or reverse Diels-Alder reaction. Suggest a sequence that accounts for this isomerization.

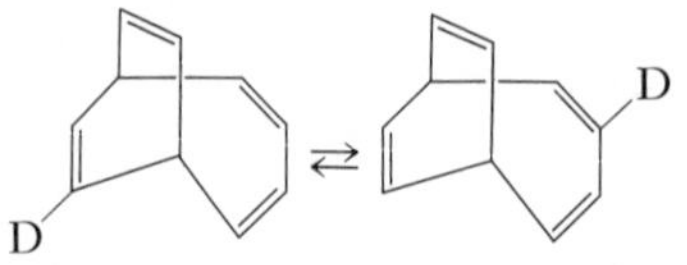

D ⇄ D

20. The interconversion shown below is one example of a **Cope rearrangement,** in which a given diallylic system is converted to a new diallylic system. Suggest a "no mechanism" mechanism for this reaction. How might you test whether mechanism **A** (below) is operative? (By the way, it is an *incorrect* suggestion.)

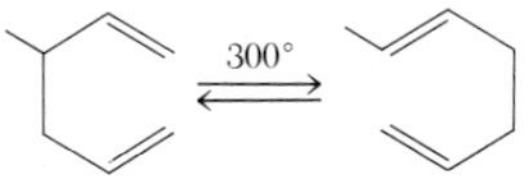

300° Cope rearrangement

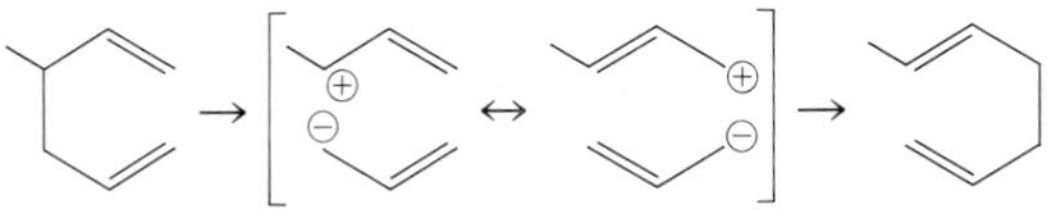

mechanism **A**

21. Thermal [2 + 2] "cycloaddition" reactions occur at elevated temperatures. However, unlike [4 + 2] cycloadditions, they are believed to occur in two steps, *i.e.*, to be non-concerted. A diradical is formed in the first step and cyclization is completed in the second. When two diradicals are possible, the more stable one is responsible for product formation. For the non-concerted reactions presented below (**A** and **B**), indicate the intermediate diradical that would lead to product. (*Note:* Fluorine is less effective in stabilizing radicals than is chlorine.)

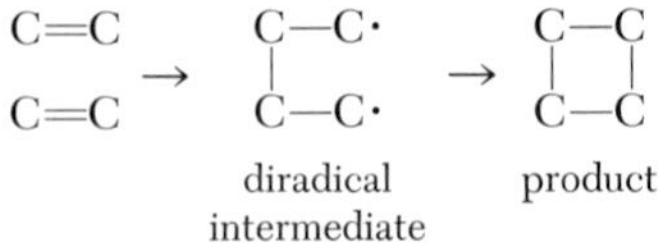

C=C / C=C → C—C· / C—C· (diradical intermediate) → C—C / C—C (product) non-concerted reaction

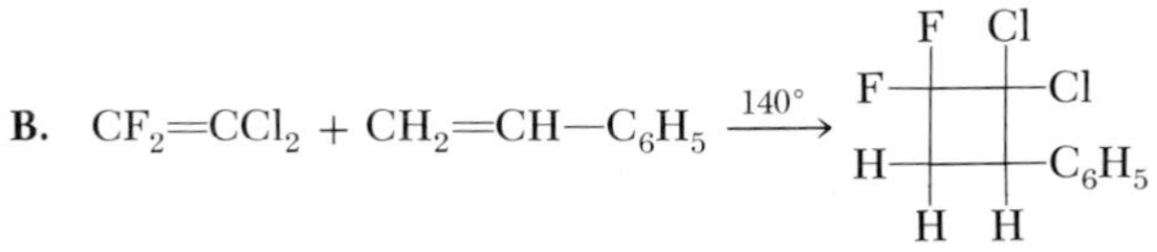

A. $2CF_2{=}CCl_2 \xrightarrow{200^\circ}$ cyclobutane (F, F, F, F / Cl, Cl, Cl, Cl)

B. $CF_2{=}CCl_2 + CH_2{=}CH{-}C_6H_5 \xrightarrow{140^\circ}$ cyclobutane (F, F, H, H / Cl, Cl, C_6H_5, H)

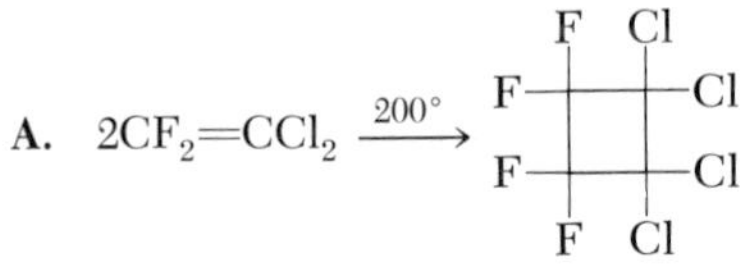

22. 7-Dehydrocholesterol is converted (by the action of sunlight) to vitamin D_3, which, by regulating calcium metabolism, prevents the bone disease termed *rickets*. A laboratory analysis of the sequence indicates that an intermediate, pre-vitamin D_3, is the compound actually formed by the action of light. (a) Provide a mechanism for its formation.

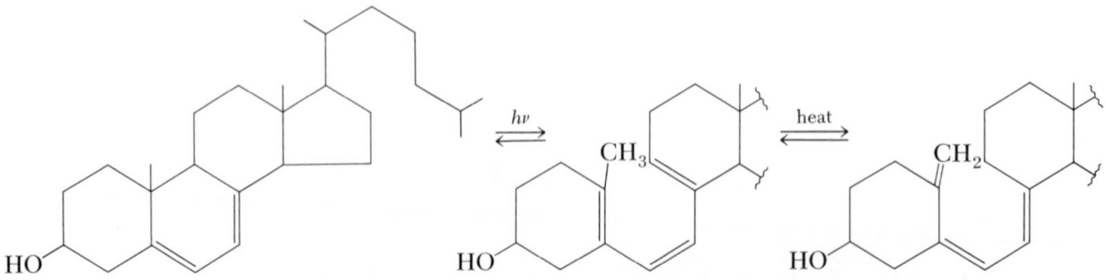

HO — 7-dehydrocholesterol $\underset{}{\overset{h\nu}{\rightleftarrows}}$ HO — CH_3 — pre-vitamin D_3 $\overset{\text{heat}}{\rightleftarrows}$ HO — CH_2 — vitamin D_3

The conversion of pre-vitamin D_3 to vitamin D_3 can be "explained" with the aid of a "no mechanism" mechanism. (b) Suggest a "mechanism" for this interconversion that lacks any of the classical intermediates of organic chemistry.

AROMATICITY

14.1 WHERE ARE WE GOING?

There exists a group of compounds, called *annulenes*, which have the empirical formula CH. They share the common characteristics of being *cyclic* and *highly unsaturated*. Within this class of hydrocarbons are found a compound so reactive as to have only a fleeting existence, a compound with little chemical similarity to an alkene, compounds with unusually low heats of combustion and hydrogenation, *and* compounds with unusual molecular spectra.

[4]-annulene (cyclobutadiene) | [6]-annulene (benzene) | [8]-annulene (cyclooctatetraene)

The purpose of this chapter is to explore some of the factors that can make a superficially similar group of compounds so different in physical and chemical properties. At the heart of our search is the word "aromaticity." Since there are several different definitions for this term, we will not define it here. Instead, we shall develop these sometimes-overlapping definitions during this chapter. This chapter closes with a presentation of the most usually accepted method for experimentally determining whether a compound is "aromatic." A fuller appreciation of this last section requires some knowledge of nmr spectroscopy (Chapter 29).

14.2 BENZENE—AN ARCHETYPE

Until relatively recent times it was considered injurious to the health to bathe too frequently. To mask some of the obvious consequences, the wealthy covered themselves with oils and essences, which were called "aromatic" because of their pleasant, persistent odors. One such aromatic essence was obtained from crushed vanilla beans, the active ingredient being vanillin. Vanillin could be related to another compound, benzoic acid, and it is in this relationship, and others like it, that we see the first link between "aromatic" and "benzene."

OH, OCH_3, CHO — vanillin | CO_2H — benzoic acid

In 1825, Michael Faraday reported the isolation of a mildly odorific compound during the thermal decomposition of whale oil. This substance, named *bicarburet of hydrogen* by Faraday, had the formula C_6H_6. Shortly thereafter, E. Mitscherlich prepared the same C_6H_6 by the decomposition of benzoic acid, and called it *benzin*. This compound, now called *benzene*, is the parent hydrocarbon of many of the odorous compounds alluded to above; it was said to be "aromatic," as were all compounds structurally related to it.

The inadequacy of the "odor criterion" for establishing the presence of a benzene ring is two-fold: (a) fragrant compounds like the terpenes bear no relationship to benzene, and (b) many benzenoid compounds (*i.e.*, compounds containing benzene rings) have either no odor, because of low vapor pressure, or horrid odors, because of substituents on the benzene ring.

Benzene—A Non-olefin. Since saturated cyclic compounds have the general formula C_nH_{2n} (for example, cyclohexane is C_6H_{12}), a compound with the formula C_6H_6 (C_nH_n) is quite *un*saturated and is expected to be prone to undergo typical addition reactions. Benzene, however, is inert toward bromine except in the presence of certain specific catalysts, and when it does react the product is not $C_6H_6Br_6$. This reluctance of benzene and its derivatives to undergo halogen addition and other reactions typical of alkenes, in spite of their high degree of unsaturation, became a second way to identify (and, later, define) aromatic compounds.

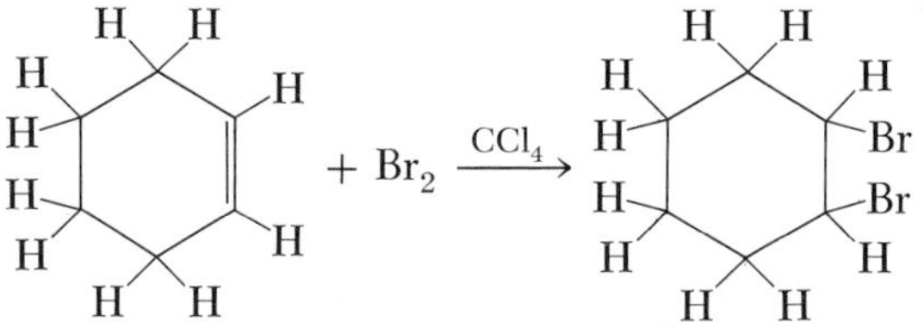

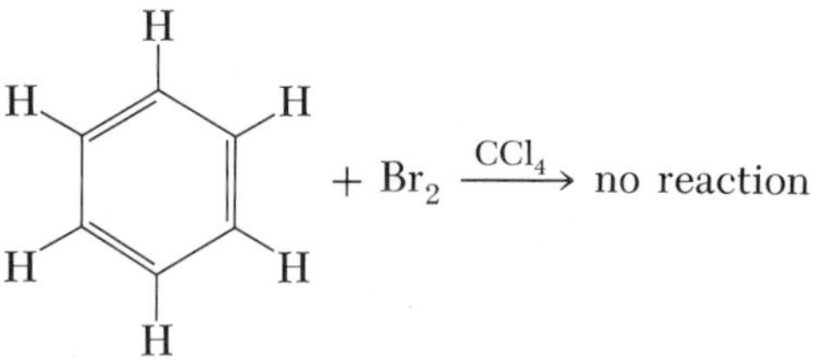

$$C_6H_{10} + Br_2 \xrightarrow{CCl_4} C_6H_{10}Br_2$$

$$\text{benzene} + Br_2 \xrightarrow{CCl_4} \text{no reaction}$$

benzene

The debate over the structure and bonding in this C_6H_6 raged for about thirty years, and the development of this subject makes fascinating reading. Suffice it to say that it was Kekulé who, in 1865, made the most salient contribution by suggesting an ill-defined equilibrium between two alternative 1,3,5-cyclohexatrienes, **A** and **B** (below).

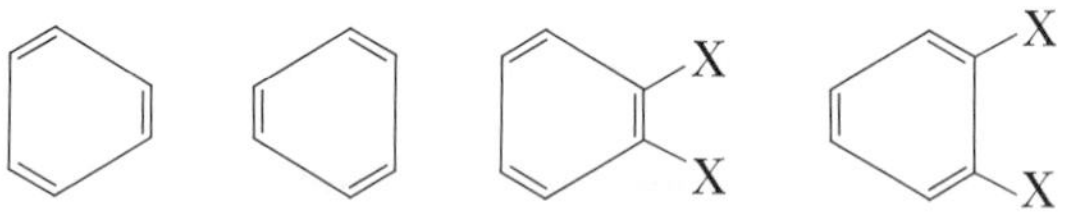

A **B** **C** **D**

One obvious inadequacy of this view is that it requires the existence of two distinct disubstituted benzene derivatives bearing adjacent substituents, **C** and **D**; such a pair of isomers (technically, *valence bond isomers*) has never been detected. Moreover, any true equilibrium between two species does not, *a priori*, destroy the uniqueness of the species in equilibrium. Therefore, any equilibration between **A** and **B** would still permit these two to react as if they were 1,3,5-cyclohexatriene.

Through many theoretical efforts, capped by the works of Coulson, Ingold, Pauling and Wheland, Kekulé's picture matured until it was suggested that benzene was not cyclohexatriene *but* that one could devise a structure for benzene beginning with cyclohexatriene. To do this, one starts with 1,3,5-cyclohexatriene and provides sufficient energy to shorten the single bonds and lengthen the double bonds until all six carbon-carbon bonds are of equal length. (X-ray analysis of crystalline benzene has shown that all carbons and hydrogens of benzene reside in one plane and that the carbon-carbon bond distances are *all* 1.39 Å.) Maximum overlap could then occur between the parallel p orbitals of adjacent sp^2 carbons in either of two equivalent ways. Thus, benzene has been pictured as a resonance hybrid with the two *Kekulé structures* as the major contributors.

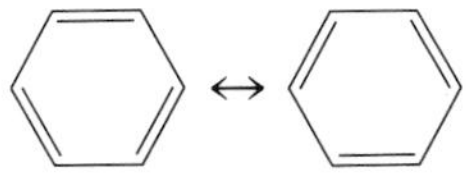

↔ resonance between Kekulé structures

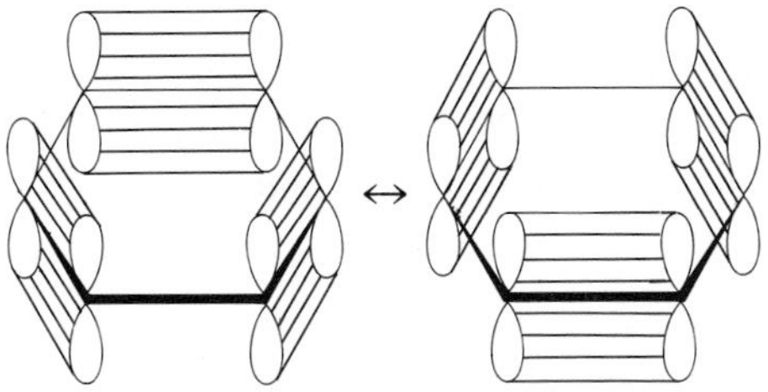

↔ a.o. overlap in Kekulé structures

Some of the other, less important, contributing structures for benzene include those shown below. (How many more can you devise?) The three forms with the long π bonds are called *Dewar structures.*

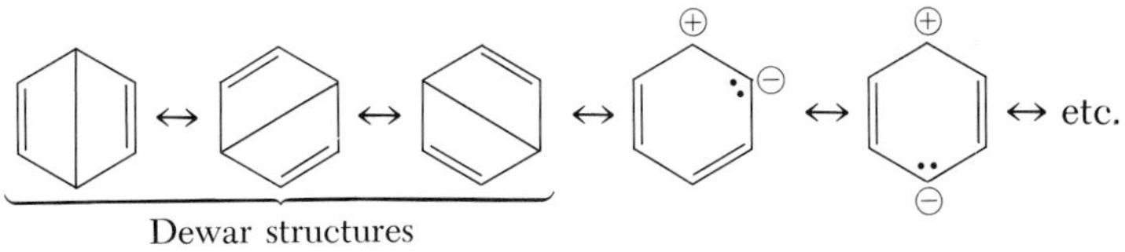

↔ ↔ ↔ ↔ ↔ etc.

Dewar structures

To facilitate picturing the resultant equality of all carbon-carbon bonds, the hybrid (*i.e.*, *real* benzene) is sometimes drawn with its σ frame surrounded by two "donuts" of π electron density.

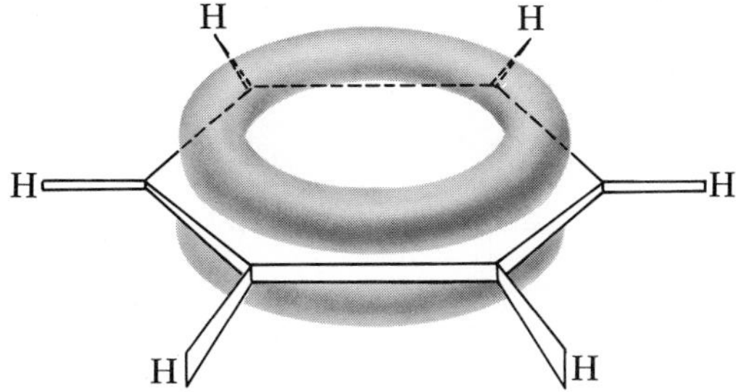

How does one conveniently represent "real" benzene, a compound with six *equivalent* carbons, each bonded to hydrogen and each participating in the aromatic bonding? The best way probably is to draw one of the Kekulé structures, leaving it to the reader to recognize the role of resonance. Another way is to draw a circle, or broken circle, to represent the completely conjugated circuit of six π electrons. In this text we will use Kekulé structures whenever possible for reasons to be made clear later in this chapter.

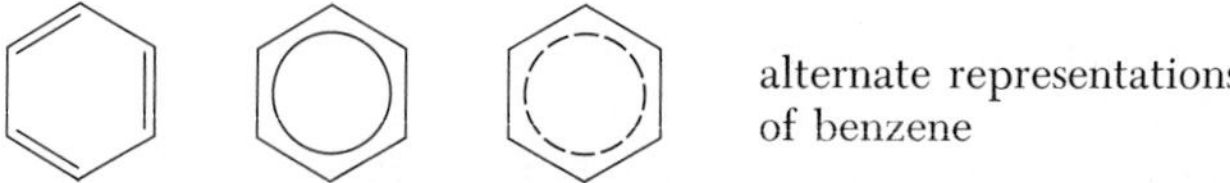

1. How many disubstituted benzenes of the type $C_6H_4X_2$ are possible? Draw their structures.

Resonance Energy of Benzene. In Chapter 12 we saw how the ~3 kcal/mole difference in the heat of hydrogenation of 1,3-butadiene and non-conjugated dienes was used to support the presence of some π electron delocalization in the former. Is there similar evidence for π electron delocalization in benzene?

The hydrogenation of one mole of cyclohexene to cyclohexane releases 28.8 kcal. The heat of hydrogenation of benzene should be thrice this value or 86.4 kcal/mole; however, it is only 49.8 kcal/mole. This means that benzene is more stable than predicted by 36.6 kcal/mole. This energy, energy which benzene does not possess, is its **resonance energy** (Figure 14–1).

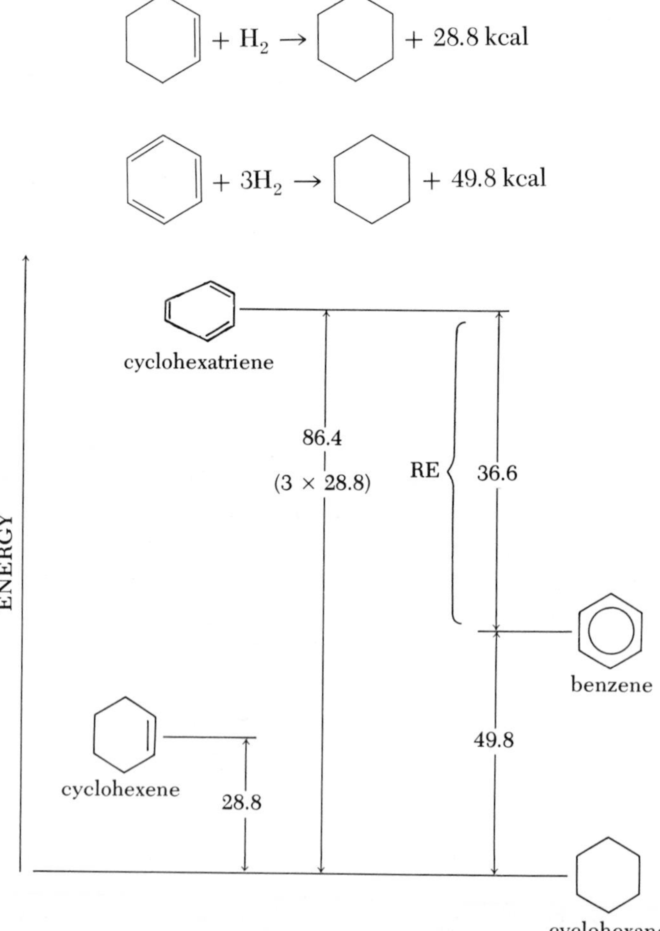

Figure 14–1 The resonance energy (RE) of benzene, based upon heats of hydrogenation. "Cyclohexatriene" is a hypothetical substance whose heat of hydrogenation is assumed to be three times that of cyclohexene. All values are in kcal/mole.

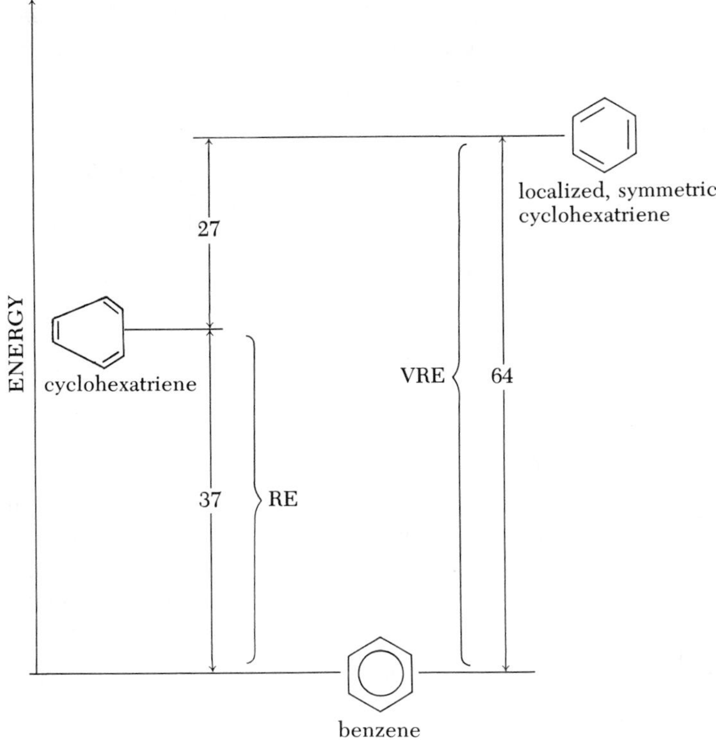

Figure 14–2 The vertical resonance energy (VRE) of benzene. Energy differences are in kcal/mole. Of the three structures pictured, only *benzene* is a real molecule.

We should not equate the resonance energy of benzene with the stabilization that it gains from π electron delocalization. Recall that we had to provide energy to convert "1,3,5-cyclohexatriene" to a substance without delocalization but with all carbon-carbon bond distances equal (Figure 14–2). Thus, the stabilization gained by delocalization equals the resonance energy *plus* the energy required to deform 1,3,5-cyclohexatriene. Calculations suggest that the latter is about 27 kcal/mole. Therefore, the *vertical resonance energy* (*i.e.*, the energy between "localized benzene" and "delocalized benzene") is 27 + 37, or 64 kcal/mole.

Whether one elects to speak of the resonance energy or the vertical resonance energy, it is clear that benzene possesses considerably less energy than expected. This type of "stability" is called *thermodynamic stability*. Another use of the term "stability" is to convey the facility with which a compound undergoes a specific reaction. This "stability" is termed *kinetic stability* and is related to the magnitude of the activation barrier for the reaction; the greater the kinetic stability, the higher the energy of activation and the slower the reaction. Statements defining aromaticity in terms of kinetic stability, *i.e.*, the ease or difficulty with which certain reactions occur, are less desirable than those based upon thermodynamic stability, because the former must depend upon the specific reaction under study. The most satisfactory, simple conceptual definition of aromaticity is *unusual stability ascribed to π electron delocalization.*

Let us leave this section with an understanding of the difference between *conjugated* double bonds and *delocalized* double bonds. (Refer back to the definition of *conjugated system* on p. 366.) The former term indicates that a molecule possesses a pattern of *alternating* single and double bonds, —C=C—C=C—; it implies nothing about the π electron distribution within that unit. The latter term indicates that the π electron density is not localized within one pair of adjacent p orbitals but is distributed over the π orbital network. For example, while both fragments below are conjugated, the conformation around the central single bond permits delocalization only in the fragment on the right.

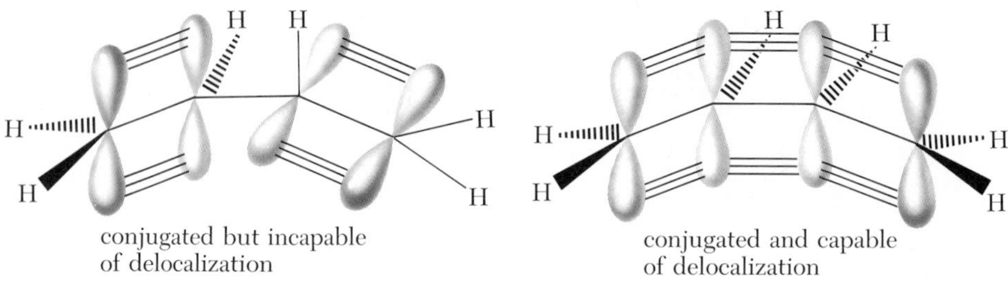

The Molecular Orbital Picture of Benzene. In the m.o. picture of benzene we imagine a localized σ framework composed of six carbon-carbon bonds and six carbon-hydrogen bonds. The six $2p$ atomic orbitals of carbon are then combined to produce six molecular orbitals, three bonding and three anti-bonding (Figure 14–3). Both the two higher energy bonding m.o.'s and the two lower energy anti-bonding m.o.'s are degenerate. That the energy of a molecular orbital increases with an increasing number of nodes is consistent with the observation that π_1 is free of nodes while π_6^* has six; the degenerate orbitals, being of equal energy, have the same number of nodes (Figure 14–4).

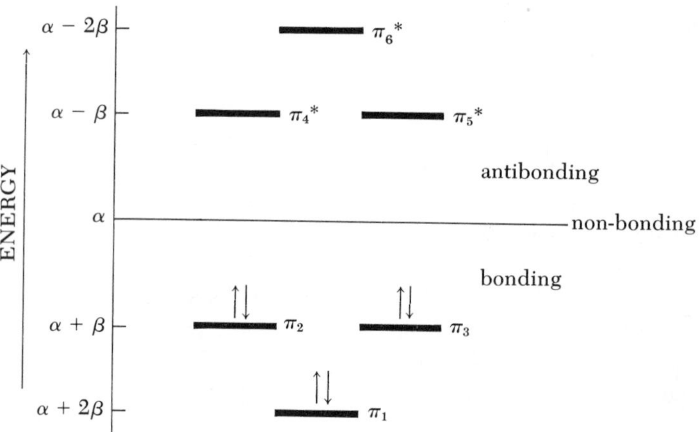

Figure 14-3 The molecular orbitals of benzene. Molecular orbitals π_1, π_2, and π_3 are bonding, while π_4^*, π_5^*, and π_6^* are antibonding. The energy difference between π_1 and π_2 (or π_3) and between π_4^* (or π_5^*) and π_6^* is sometimes expressed in units of β. If the energy of the system without delocalization is assigned the value α, then the m.o.'s are of the following energies: π_1: $\alpha + 2\beta$; π_2 and π_3: $\alpha + \beta$; π_4^* and π_5^*: $\alpha - \beta$; π_6^*: $\alpha - 2\beta$. Why must you conclude that β is a negative number?

It has been common to state that benzene is unusually stable because it possesses a torus of six π electrons above and below the molecular plane. This picture is readily reconciled with the resonance account of benzene, which shows six π electrons moving in concert. The m.o. picture, on the other hand, shows that only π_1 completely enshrouds all six carbons. These seemingly divergent pictures are reconciled if the two higher bonding m.o.'s of benzene are considered along with π_1. In the end the total symmetry of benzene's occupied molecular orbitals presents a uniform distribution of π electron density between the various carbons.

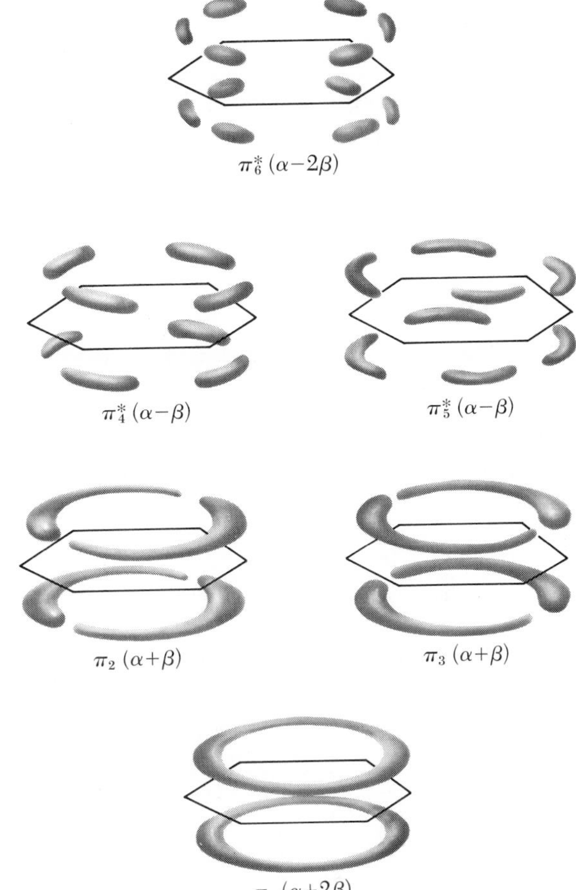

Figure 14-4 The π molecular orbitals of benzene. π_1, π_2, and π_3 are occupied. (From D.S. Urch, *Orbitals and Symmetry*, Penguin Books, Inc., Baltimore, 1970.)

2. Biphenyl, $C_6H_5—C_6H_5$, has a conjugation energy of 71 kcal/mole. Draw an atomic orbital picture of biphenyl. What are its most important resonance contributing structures? *Estimate* the heat of hydrogenation of biphenyl.

14.3 ANNULENES AND HÜCKEL'S RULE

At this point one might imagine that any compound for which one could write several reasonable contributing structures should be aromatic. Quite the opposite! Only those structures containing 2, 6, 10, . . . [*i.e.*, $4n + 2$ (where n is an integer)] π electrons are capable of enhanced thermodynamic stability. This is formalized in **Hückel's rule:** *Only monocyclic carbocycles containing $4n + 2$ π electrons in a closed, planar, conjugated circuit are capable of enhanced thermodynamic stability.* Hückel's rule suggests that [2]-, [6]-, [10]-, . . . annulenes should be aromatic. A theoretical limit on the size of an annulene that may be aromatic has been set at [22]- or [26]-annulene.

Leaving [2]-annulene aside (it is ethylene, the basis for many comparisons), [6]-annulene is benzene, aromatic by any set of criteria. For this reason, the six π electrons of benzene have been called the "aromatic sextet." [10]-Annulene is an extremely reactive molecule and may be non-aromatic because crowding within the molecule causes it to be non-planar. In turn, this non-planarity prevents significant delocalization of the ten π electrons over the p orbitals, because the orbitals' major axes are no longer parallel. However, [14]-, [18]-, and [22]-annulene have been shown to be aromatic by the application of the nmr criterion described in the last section of this chapter. Refined m.o. calculations suggest that annulenes containing $4n$ electrons actually are destabilized by delocalization, and these are now called *anti-aromatic* systems.

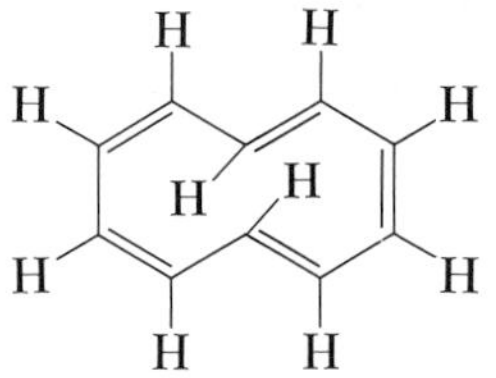

[10]-annulene 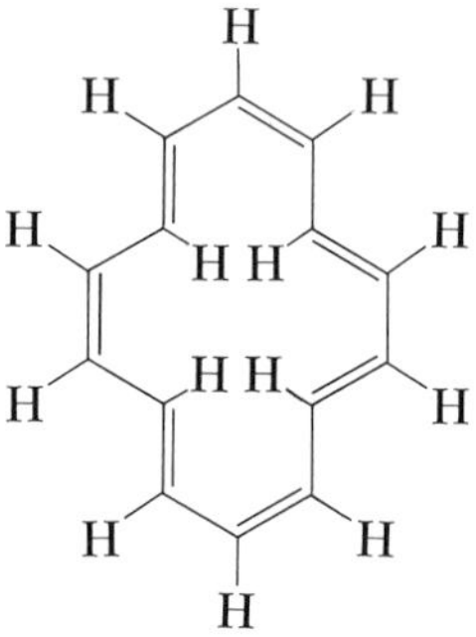 [14]-annulene

3. The anion of benzene, $C_6H_5^{\ominus}$, is aromatic although it contains a total of eight "free" electrons. How do you account for this?

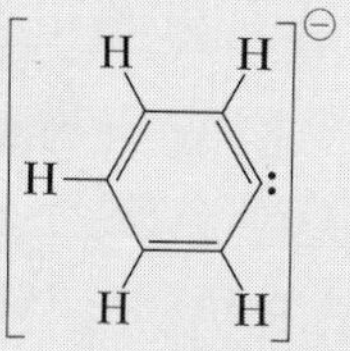

phenyl anion

4. "Benzyne is aromatic yet is extremely reactive." (a) Why is it aromatic? (b) Why would you expect it to be extremely reactive? (c) Are these two ideas inconsistent with each other? Explain.

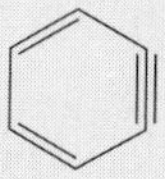

benzyne

We can, using a simple graphic presentation, suggest the nature of various *planar*, fully conjugated electronic systems. A symmetric polygon representing the annulene is rested on one apex, and energy levels are constructed according to the locations of the apices. The line representing the non-bonding energy level passes through the center of the geometric figure; bonding orbitals are below this line, of course. When this device is applied to benzene, a hexagon generates the orbitals already shown to be correct. Since six electrons enter the π system, the resultant picture depicts benzene with all electrons spin-paired, all bonding orbitals filled, and all anti-bonding orbitals vacant (compare to Figure 14–3).

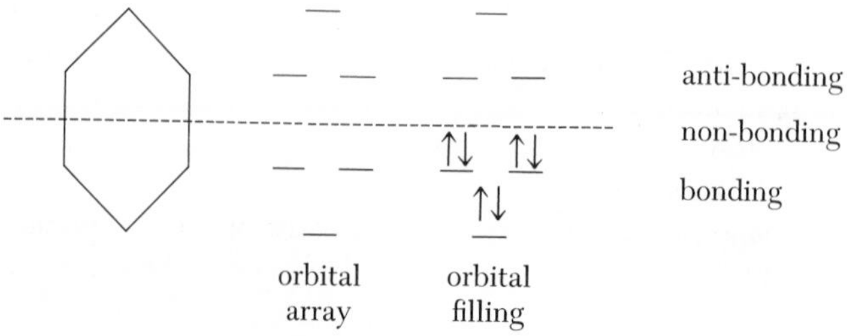

anti-bonding
non-bonding
bonding

orbital array orbital filling

Application of this method to cyclobutadiene leads to the prediction that it should be a diradical, in accord with the known high reactivity of this compound. The suggestion has been made that cyclobutadiene is probably not perfectly square planar, precisely in order to avoid the symmetry that leads to this high-energy electronic arrangement. (This distortion of geometry to reduce electronic energy is called Jahn-Teller distortion.) A recent (1973) report, based upon spectral observations, has cast doubt upon this long cherished picture by announcing that triplet (diradical with spins unpaired) cyclobutadiene is square planar. At this writing the nature of cyclobutadiene and the applicability of this device to cyclobutadiene remain moot points (see problem 30).

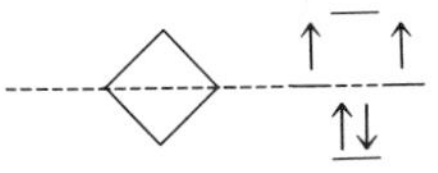

π electron distribution in square planar cyclobutadiene

Applying this device to cyclooctatetraene leads to a prediction of a (highly reactive) diradical ground state. However, cyclooctatetraene is stable, behaving like a typical alkene. The reason for this discrepancy is that cyclooctatetraene is non-planar (a result of Jahn-Teller distortion) and this scheme is not applicable to it. Planar cyclooctatetraene has never been prepared.

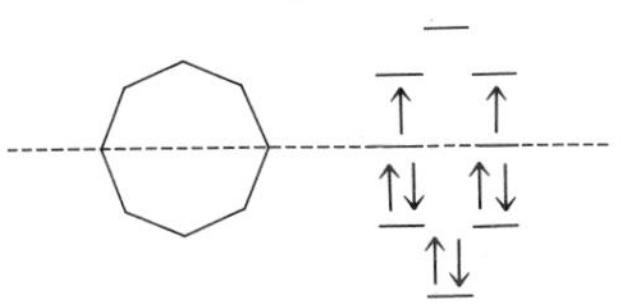

π electron distribution in planar cyclooctatetraene

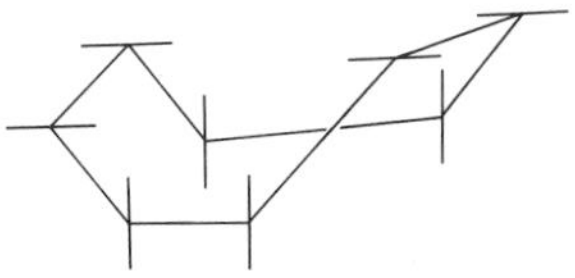

actual shape of cyclooctatetraene

Can an Ion Be Aromatic? There is no constraint present in Hückel's rule that limits the existence of aromaticity to neutral species. The purpose of this section is to introduce you to several carbocations and carbanions that are aromatic.

The synthesis presented below has produced the aromatic triphenylcyclopropenium perchlorate, a derivative of the cyclopropenium ion, $C_3H_3^{\oplus}$, which is an "$n = 0$" aromatic. This salt is sufficiently stable to have been bottled, and its structure has been determined by X-ray crystallography.

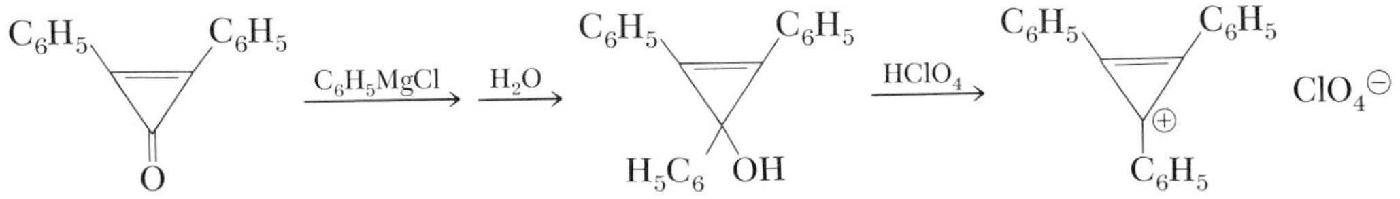

C_6H_5 C_6H_5 O $\xrightarrow{C_6H_5MgCl}$ $\xrightarrow{H_2O}$ C_6H_5 C_6H_5 H_5C_6 OH $\xrightarrow{HClO_4}$ C_6H_5 C_6H_5 C_6H_5 $ClO_4^{\ominus}$

diphenylcyclopropenone

triphenylcyclopropenium perchlorate

$\equiv$

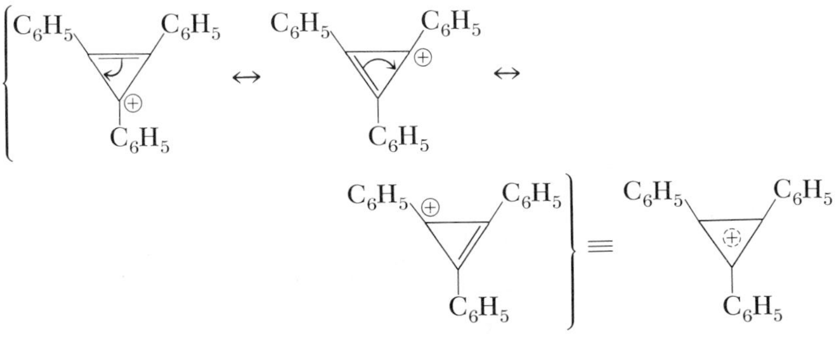

here the circle represents two π electrons

One could imagine that the extreme stability of this cation is due to the existence of nine contributing structures that place the positive charge on the three phenyl groups. However, tri-*n*-propylcyclopropenium ion is even *more* stable than the triphenyl compound, suggesting that the inductive effect of the phenyl groups actually destabilizes the triphenyl cation. Tri-*n*-propylcyclopropenium ion is so stable that its equilibrium reaction with water favors the cation!

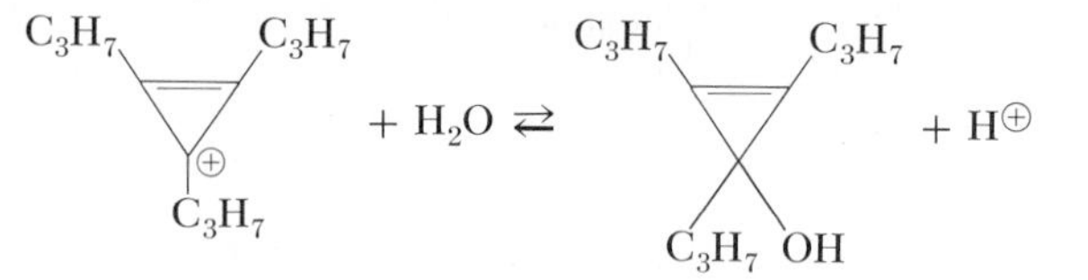

C_3H_7 C_3H_7 C_3H_7 (cation) $+ H_2O \rightleftarrows$ C_3H_7 C_3H_7 C_3H_7 OH $+ H^{\oplus}$

This behavior can be contrasted with that of the cyclopropenyl carbanion, a $4n$ system, no derivative of which has ever been prepared. It is analogous to the cyclobutadiene system and should be anti-aromatic.

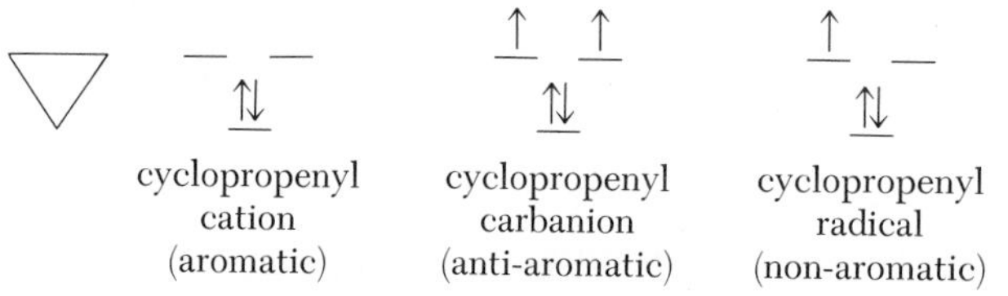

cyclopropenyl cation (aromatic) — cyclopropenyl carbanion (anti-aromatic) — cyclopropenyl radical (non-aromatic)

1,3-Cyclopentadiene is more acidic ($K_a = 10^{-15}$) than acetylene, even though the proton lost by 1,3-cyclopentadiene is bonded to an sp^3 carbon and the proton lost by acetylene is bonded to an sp carbon. In the same way that we rationalized the greater acidity of acetylene than of ethane by arguing that the acetylide ion is more stable than the ethide ion ($CH_3CH_2^{\ominus}$), so we explain *this* difference by arguing that the 1,3-cyclopentadienide ion ($C_5H_5^{\ominus}$) is considerably more stable than the acetylide ion. However, instead of attributing this stabilization to the state of hybridization of the carbon losing the proton, as we did with acetylene, we argue that the 1,3-cyclopentadienide ion is of unusually low energy content because it is aromatic. A look at the resonance contributing structures for this aromatic carbanion shows uniform distribution of electron density and negative charge over all five p orbitals of this planar, six π electron system.

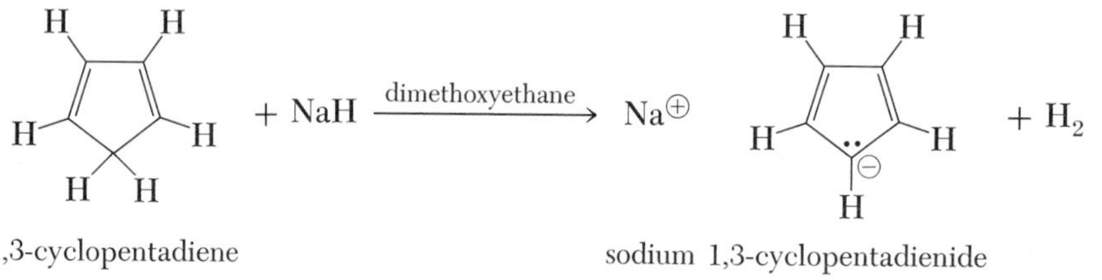

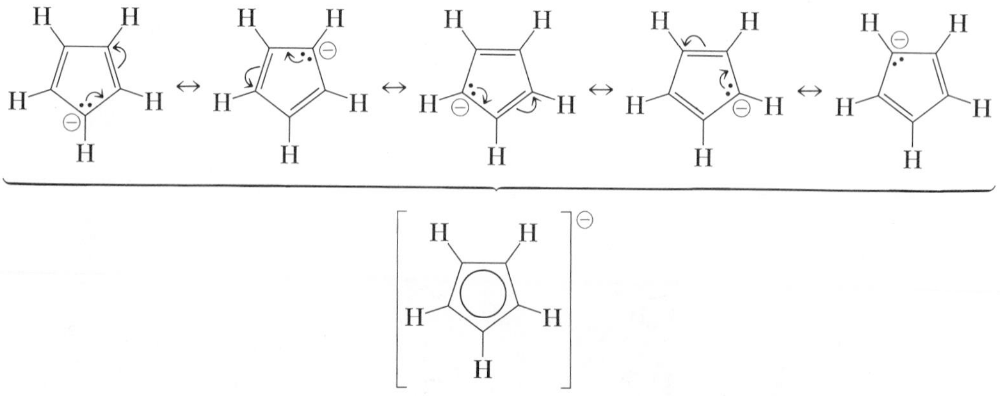

1,3-cyclopentadiene $+$ NaH $\xrightarrow{\text{dimethoxyethane}}$ $Na^{\oplus}$ sodium 1,3-cyclopentadienide $+ H_2$

resonance contributing structures for 1,3-cyclopentadienide ion

5. Would you expect the cyclopentadienyl cation, $C_5H_5^{\oplus}$, to be as stable as the cyclopentadienide anion, $C_5H_5^{\ominus}$? Explain.

By way of contrast, 1,3,5-cycloheptatriene is much *less* acidic ($K_a = 10^{-45}$) than acetylene, even though *seven* contributing structures can be drawn for the 1,3,5-cycloheptatrienide anion. The explanation for this is that *it is not the number of resonance structures that can be drawn but, rather, the number of mobile π electrons that are delocalized, that determines whether a species is aromatic.* Since the cycloheptatrienide anion contains eight π electrons, it is not aromatic. Indeed, the *planar* anion contains 4*n* electrons, is anti-aromatic, and is destabilized by delocalization.

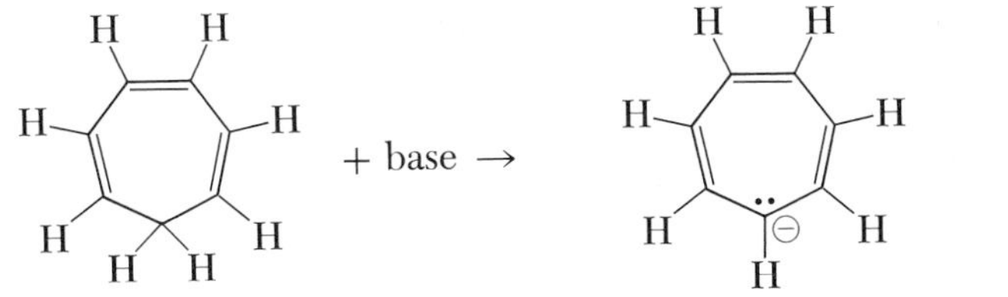

+ base →

1,3,5-cycloheptatriene

1,3,5-cycloheptatrienide anion, a 4*n* π electron system

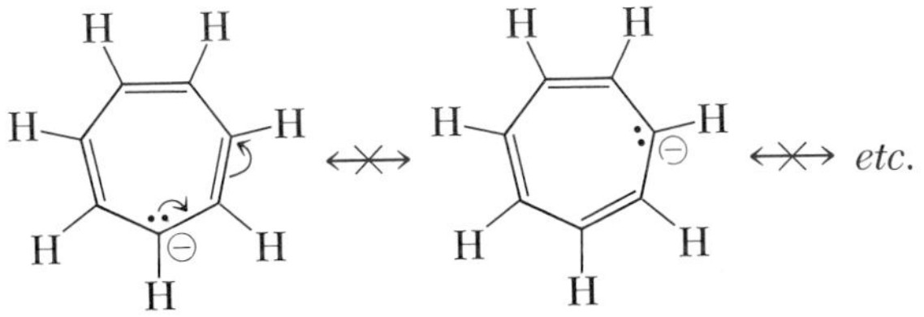

etc.

this delocalization is *not* stabilizing and does not occur

6. Would you expect the cycloheptatrienide anion to be planar? Explain.

Tropone, 1,3,5-cycloheptatrienone, is a rather strong base, forming salts with acids such as hydrochloric acid. This unusual basicity of an electron pair on a carbonyl group is ascribed to the stability of the aromatic cation produced by protonation. Here the aromatic species contains seven *p* orbitals and six delocalized π electrons.

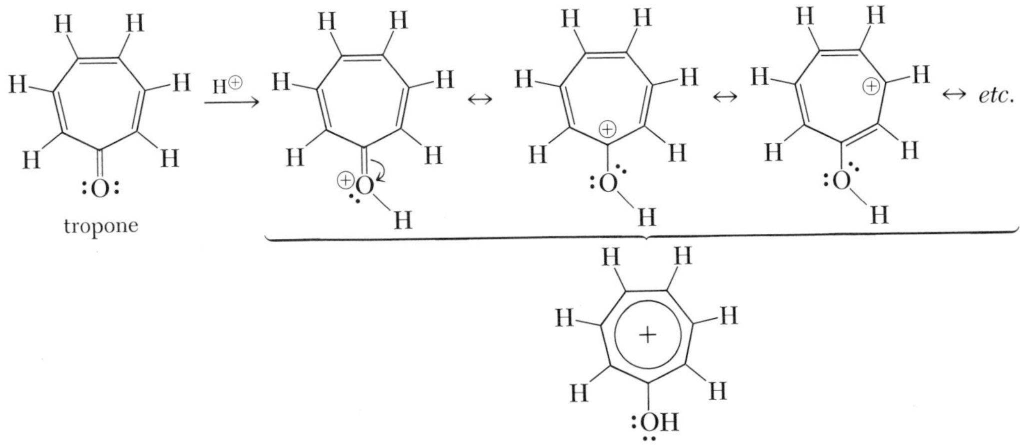

tropone

etc.

7. All of the carbon-carbon bond lengths in tropolone are 1.40 Å. Heat of combustion experiments suggest that it has a conjugation energy of about 20 kcal/mole. Interpret these data in terms of resonance theory and Hückel's rule.

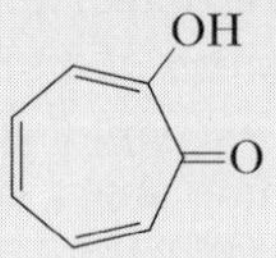

OH

=O tropolone

While [10]-annulene is probably non-aromatic for reasons already presented, one can prepare a monocyclic ten π electron system from cyclooctatetraene. Two gram atoms of potassium will react with one mole of cyclooctatetraene, each gram atom donating one equivalent of electrons to the hydrocarbon, converting it to a planar, aromatic (ten π electrons) dianion.

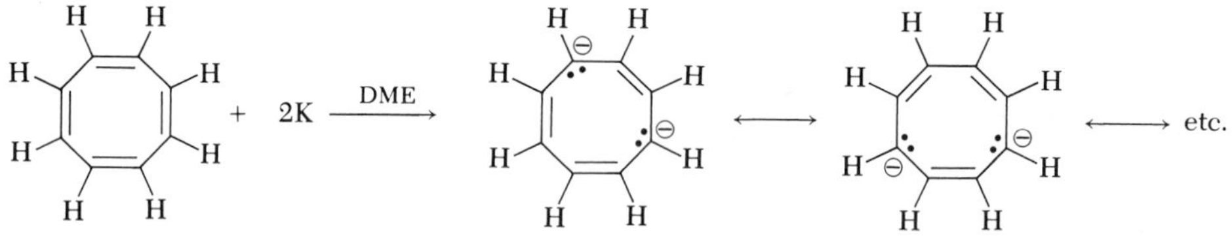

+ 2K $\xrightarrow{\text{DME}}$ ⟷ ⟷ etc.

cyclooctatetraene dianion

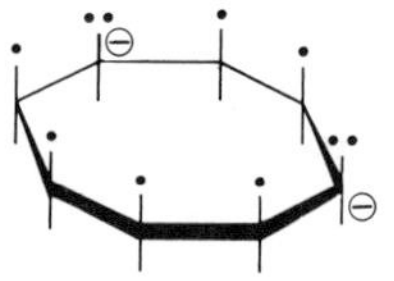

eight p orbitals (vertical lines)
ten π electrons

one contributor of the cyclooctatetraene dianion

Aromatic Polycycles. The Hückel rule has been found to be applicable to *polynuclear* hydrocarbons (*i.e.*, hydrocarbons with fused ring systems) that possess a continuous circuit of π electron density. Thus, naphthalene can be viewed as a ten π electron system that is bicyclic with respect to the carbon skeleton but monocyclic with respect to the π electron distribution.

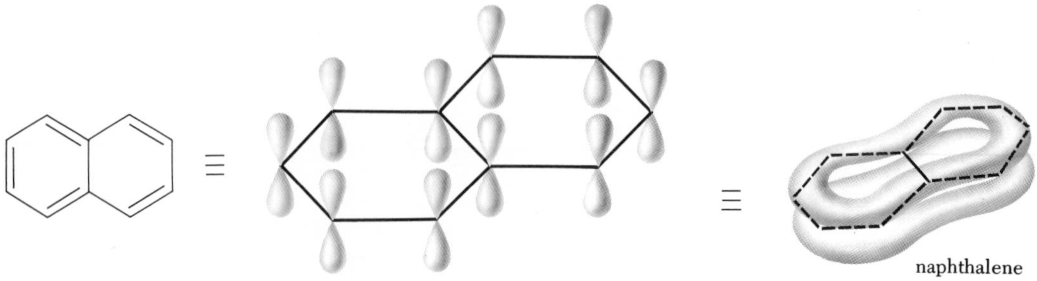

≡ ≡

naphthalene

Three Kekulé resonance contributing structures can be constructed for naphthalene.

A ↔ B ↔ B′

The existence of polynuclear aromatics like naphthalene forces us to examine two important points. First, it is misleading to represent naphthalene's π electrons with two circles. In benzene the circle represents six π electrons, while in naphthalene it represents only five. It is to avoid such confusion that we shall avoid the use of the circle to represent aromatic systems unless its significance is clear.

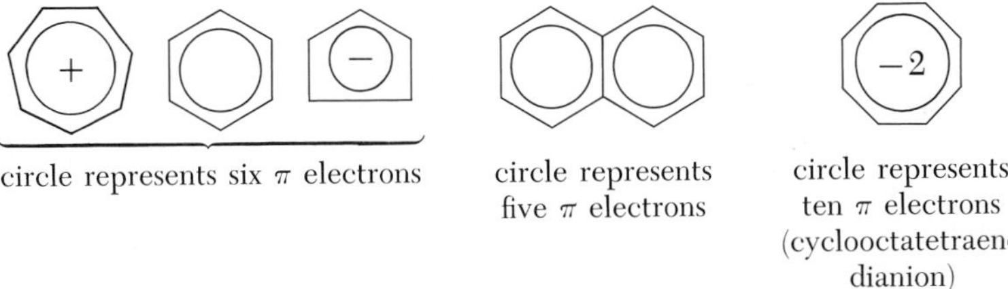

circle represents six π electrons — circle represents five π electrons — circle represents ten π electrons (cyclooctatetraene dianion)

Second, when comparing aromatic compounds, we can estimate their relative resonance energies by determining the number of Kekulé structures that can be drawn for each. In general, the more Kekulé structures possible, the greater will be the resonance energy. This approach correctly suggests that benzene, with only two Kekulé structures, is less stabilized by aromaticity than is naphthalene, for which one can draw three Kekulé structures (Table 14–1). However, naphthalene is *not* more aromatic than is benzene because, on a "per electron" basis, the resonance energy of benzene is greater than that of naphthalene.

TABLE 14–1 Resonance Energy of Aromatic Hydrocarbons

COMPOUND	STRUCTURE[a]	RESONANCE ENERGY (KCAL/MOLE)
benzene		37
naphthalene		61
anthracene		84
phenanthrene		92

[a] Only one Kekulé form is shown.

Azulene, an isomer of naphthalene, is a blue hydrocarbon (a remarkable property for a hydrocarbon) that is aromatic. In support of its aromaticity are its planar geometry and a heat of combustion approximately 40 kcal/mole lower than expected. The dipole moments of azulene (1.08 D) and 1-chloroazulene (2.69 D) suggest that the "aromatic contributing structure," **B,** is an important contributor to the hybrid that is azulene. (Why?)

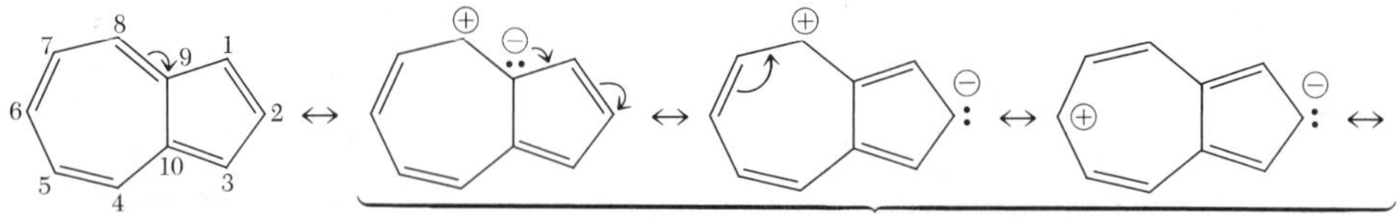

A
azulene

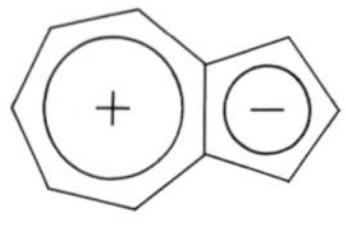

in this representation each circle represents 5 π electrons

B

Aromatic Heterocycles. Although originally limited to carbocycles, Hückel's rule seems equally applicable to many heterocyclic systems. Compounds like furan, pyrrole, thiophene, imidazole, and thiazole, whose structures are shown below, are planar and use an electron pair in a p orbital of an sp^2 hybridized heteroatom to complete the aromatic sextet. The electrons on the heteroatom used to complete the sextet are shown as dots; these are perpendicular to the molecular plane. Other (non-bonding) electron pairs are perpendicular to the π system (*i.e.*, in the molecular plane) and are shown, for clarity, as ×'s.

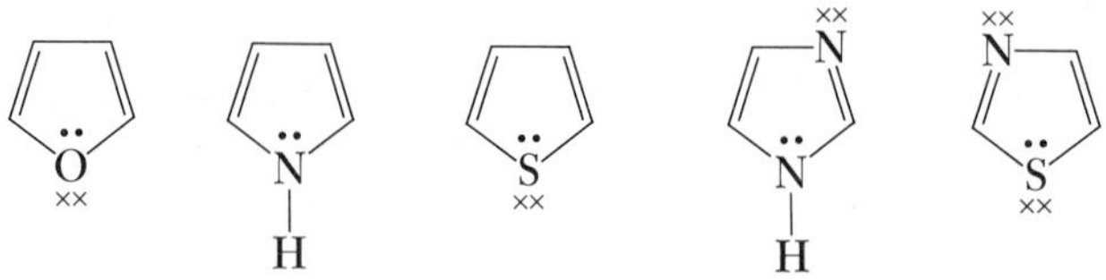

furan pyrrole thiophene imidazole thiazole

Pyridine and pyrimidine are aromatic heterologs of benzene, each containing six π electrons.

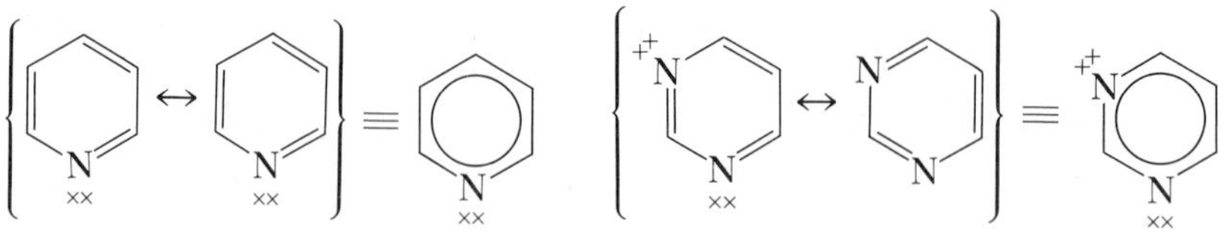

pyridine pyrimidine

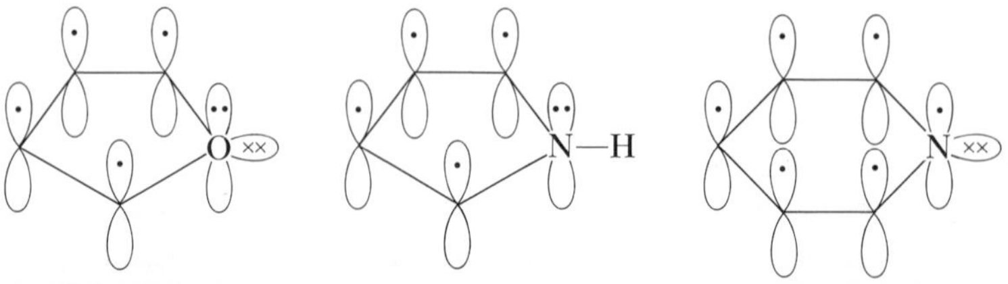

furan pyrrole pyridine atomic orbital representations

Quinoline and isoquinoline are aromatic ten π electron systems that are nitrogen analogs of naphthalene.

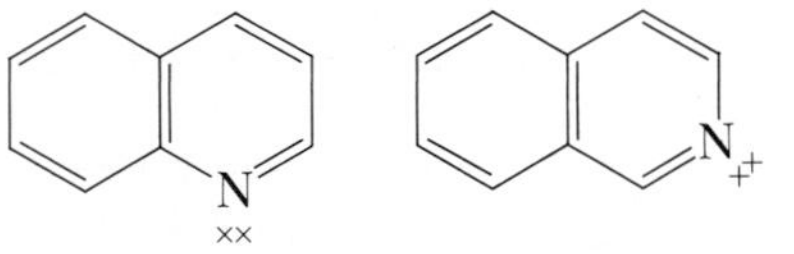

quinoline isoquinoline

Purine is an aromatic ten π electron system that contains an imidazole ring fused to a pyrimidine ring.

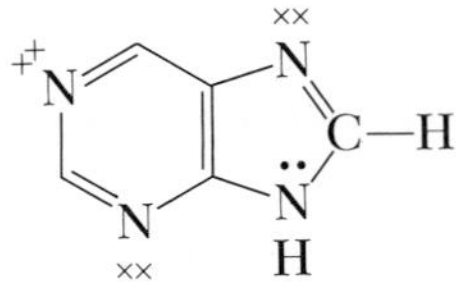

purine

8. Both of the following are aromatic. Suggest the geometry of these molecules and identify the aromatic electronic array.

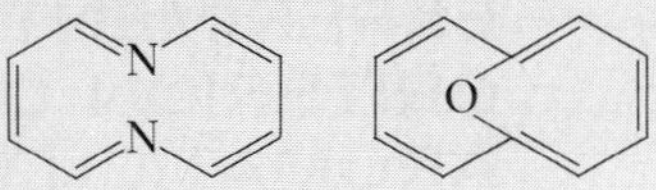

Many of these nitrogen-containing heterocycles can act as electron pair donors (*i.e.*, bases), without destroying their aromaticity, by donating those electron pairs designated by ×× to an acid. Pyridine, for example, reacts with hydrogen chloride to form the salt *pyridinium chloride* (also called pyridine hydrochloride), which still has an aromatic sextet.

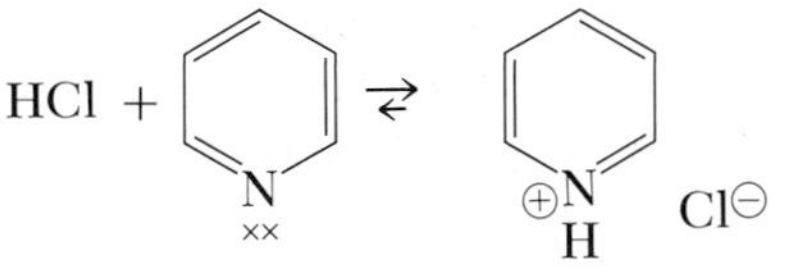

$$HCl + \text{pyridine} \rightleftarrows \text{pyridinium chloride}$$

pyridine pyridinium chloride

Purine and pyrimidine can behave in the same manner; this is fortunate, indeed, since these heterocycles are found in DNA and RNA (Chapter 26), and hydrogen-bonding to these basic sites is important in maintaining the integrity of biopolymers (*e.g.*, genes). (You should review Figure 2–6.)

9. (a) What is the hybridization of the heteroatoms in furan, pyrrole, pyridine, and pyrimidine? (b) What is the hybridization of nitrogen in pyridinium chloride?
10. γ-Pyrone is more basic than either acetone or vinyl ether. (a) Account for this. (b) Which oxygen of γ-pyrone is the more basic? (c) Why?

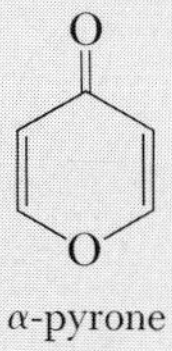

α-pyrone

11. Many of the beautiful colors of butterflies' wings, autumn leaves, and flowers are due to the presence of flavone derivatives called *anthocyanins*. When anthocyanins are hydrolyzed in hydrochloric acid (to remove sugar residues), they produce aromatic oxygen-containing salts. One of these is delphinidin chloride. Explain why delphinidin chloride is aromatic. What is the hybridization of the heterocyclic oxygen in this compound?

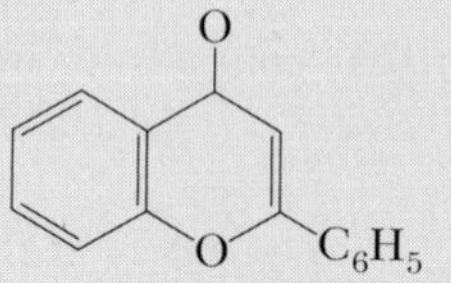

flavone 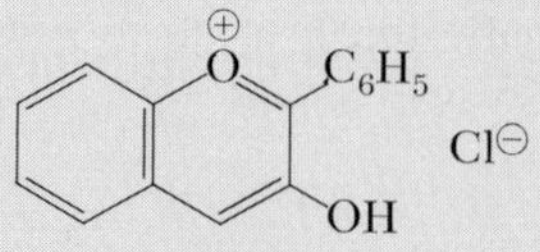　　　delphinidin chloride

In summary, a broad view of Hückel's rule has made it clear that aromaticity is not limited to benzenoid molecules but includes many so-called "non-benzenoid" substances. Furthermore, aromatic compounds can be heterocyclic substances and even ions.

14.4 WHAT IS THE EXPERIMENTAL CRITERION OF AROMATICITY?[a]

When a new compound is prepared and it is necessary to establish whether it is aromatic, one rarely employs a determination of the heat of combustion or hydrogenation because such experiments are time-consuming and fraught with experimental difficulty. Instead, one usually measures the compound's nmr spectrum.

Benzene shows a single, sharp singlet at 7.37 δ. The multiplicity is not surprising, since all of the hydrogens are chemical shift equivalent and do not appear to split one another. However, the chemical shift is much farther downfield than is normally observed for alkenes. When the π system is subjected to the applied field, H_0, a flow of π electrons is induced around the ring. This *ring current* generates an auxiliary magnetic field that augments H_0 in the vicinity of the benzene protons and opposes H_0 in the center of the ring. This deshields the benzene protons and brings them into resonance at lower values of H_0, *i.e.*, downfield (Figure 14–5).

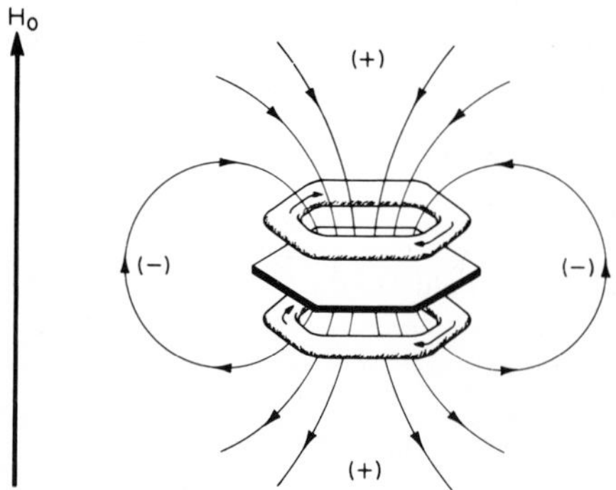

Figure 14-5 Electron density, ring currents, and magnetic lines of force about a benzene ring. (From F.A. Bovey, *Nuclear Magnetic Resonance Spectroscopy*, Academic Press, New York, 1968.)

[a] This section requires some familiarity with nmr spectroscopy (Chapter 29).

It appears that only an aromatic molecule can produce a ring current of this magnitude and direction; its presence, indicated by unusual chemical shifts, is the most common experimental criterion of aromaticity. The nmr spectra of benzene (aromatic carbocycle), furan (aromatic heterocycle), and cyclooctatetraene (non-aromatic annulene) illustrate this phenomenon and its constraint (Figure 14–6).

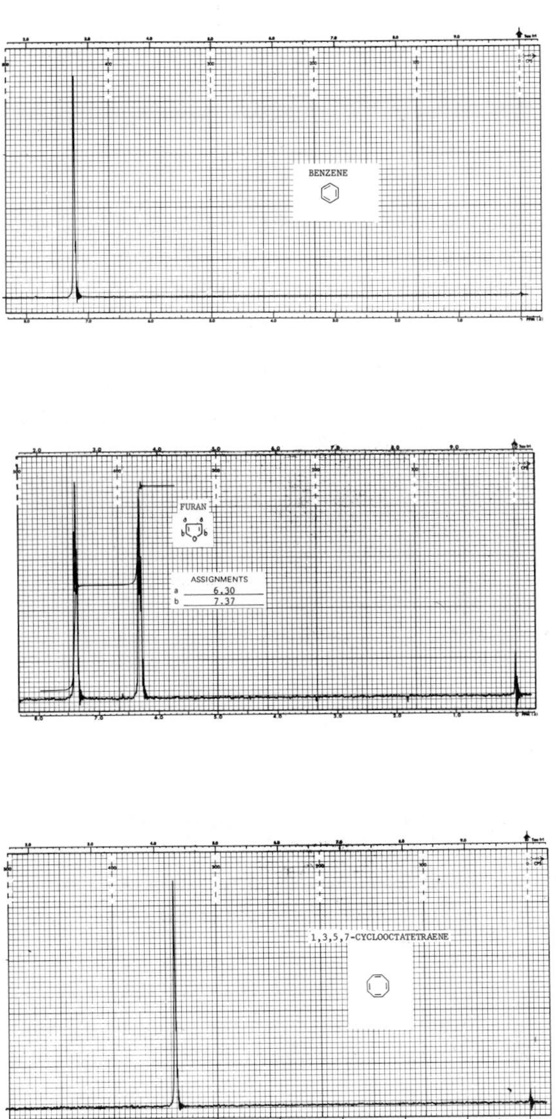

Figure 14–6 Nmr spectra of benzene, furan, and cyclooctatetraene. © Sadtler Research Laboratories, Inc., 1976.

The ring current occasionally produces somewhat bizarre spectra, especially in non-benzenoid aromatic compounds. The nmr spectrum of 1,6-methanocyclodecapentaene, consisting of a group of signals (8**H**) centered at 7.2 δ and a sharp singlet (2**H**) at $-0.5\ \delta$ (that is,

0.5 ppm *upfield* from TMS), is a case in point. The deshielded signals are ascribed to the protons around the perimeter of the π system; the methylene group, however, being located in the middle of the induced field, is shielded and occurs upfield of TMS. We should note that this compound is similar to [10]-annulene but that the methano (—CH_2—) bridge forces the sp^2 hybridized carbons into one plane, thereby permitting π electron delocalization through all ten parallel p orbitals.

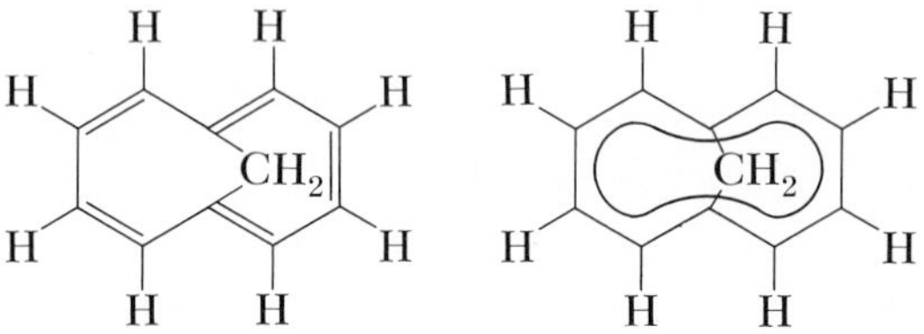

two representations of a ten π electron non-benzenoid aromatic compound: 1,6-methanocyclodecapentaene

The nmr spectrum of [18]-annulene consists of two groups of signals. The downfield set, $\sim 9\ \delta$, 12**H**, represents the outer protons, while the upfield set, $-3\ \delta$, 6**H**, corresponds to the inner protons, *i.e.*, those subjected to the shielding influence of the induced field. Once again, the presence of aromaticity is proved by the ring current effect.

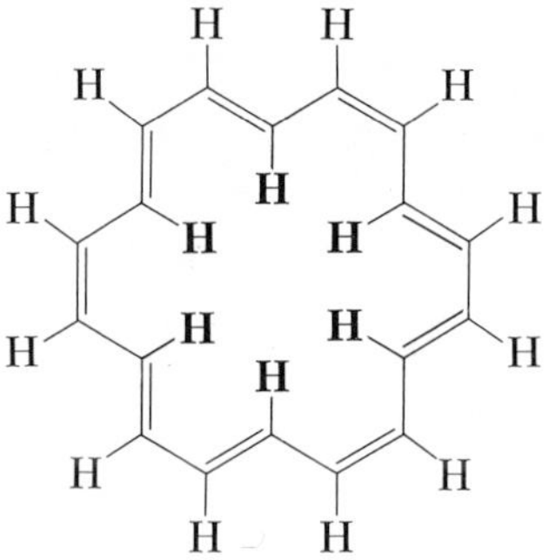

H occurs at $-3\ \delta$
H occurs at $\sim 9\ \delta$

[18]-annulene

IMPORTANT TERMS

Anti-aromaticity: A marked increase in the amount of energy possessed by a species when that species delocalizes its π electrons. Thus, anti-aromatic compounds are more stable (lower energy) when their π electrons are localized. In order to be anti-aromatic a species must have $4n$ π electrons. *Planar* cyclooctatetraene, with eight π electrons, is an anti-aromatic molecule.

Aromaticity: A marked decrease in the energy of a species caused by π electron delocalization. The standard of comparison for calculating the amount of the "decrease" is the same species *without* π electron delocalization. The energy that the species lacks as a result of delocalization is its *resonance energy*. Defined in this way, aromaticity is a reflection of unusual thermodynamic stability (see below).

The most common experimental criterion for aromaticity is the presence of a ring current (see below).

Hückel's rule: A statement describing the π electron requirements for the presence of aromaticity. It relates that a species containing 6, 10, 14 . . . (*i.e.*, $4n + 2$) π electrons in a fully conjugated π network will exhibit enhanced thermodynamic stability. (Note that the definition does not consider the number of p orbitals or the number of π bonds involved, just the number of π electrons!) Although the rule originally focused upon monocyclic carbocycles, it is now recognized as applicable to polycyclic and heterocyclic compounds.

Kekulé structure: A Lewis structure for an aromatic compound. A Kekulé structure is an important contributing structure to the resonance hybrid that describes the real molecule. Aromatic compounds are often represented by one of their Kekulé structures. Each of the structures shown below is a Kekulé structure. Note that in each of them the bond lengths are shown as being equal.

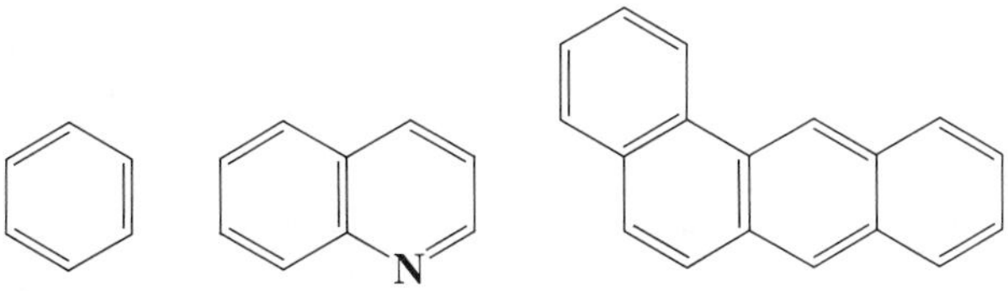

N

Kinetic stability: The existence of a comparatively high energy of activation for a specific reaction. When a compound undergoes a reaction only very slowly (or not at all), it is exhibiting kinetic stability (with regard to that particular reaction). You may recall that in Chapter 6 we noted that the E2 reaction almost always requires an *anti* geometry. Other conformations, therefore, may be said to be kinetically stable with regard to the E2 elimination.

The importance of kinetic stability to this chapter stems from the fact that aromatic compounds used to be defined as a group of substances that did not undergo additions to their "double bonds." They were, therefore, described as being quite "stable." The point that was overlooked, of course, is that aromatic compounds may be quite stable to some reactions yet quite reactive under other conditions. Thus, kinetic stability is a less desirable criterion of aromaticity than is thermodynamic stability (see below).

Non-aromatic compound: A compound that does not change its energy *markedly* as a result of π electron delocalization. The radical derived by abstraction of a hydrogen from cyclopentadiene is non-aromatic.

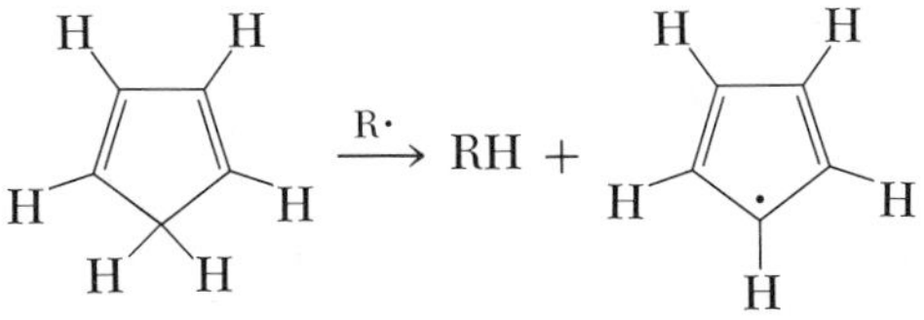

H H H H H H $\xrightarrow{R\cdot}$ RH + H H H H H

5 π electrons

Ring current: Considered as the motion of π electrons in a closed loop, it generates magnetic fields that may move protons away from their "normal" resonance frequency. Most commonly, a ring current moves protons on aromatic compounds (*e.g.*, benzene derivatives) downfield, that is, to the left as you view the spectrum, from the position that would be expected if the ring current were absent (see Figures 14–5 and 14–6). Its presence is an experimental criterion of aromaticity.

Thermodynamic stability: Possession of a low ground state energy. Compounds that are aromatic show enhanced thermodynamic stability as a result of π electron delocalization. (The basis of comparison is a structure with similar geometry but without delocalization of π electrons.)

PROBLEMS

12. Define the following terms in your own words:
    (a) Hückel's rule
    (b) annulene
    (c) diradical
    (d) conjugated
    (e) ring current
    (f) resonance energy (RE)
    (g) vertical resonance energy
    (h) aromatic
    (i) anti-aromatic

13. Each of the following is a particular contributing structure of a compound. Which of these compounds are aromatic?

(a) 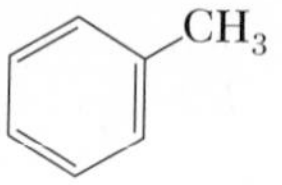 CH_3

(b)

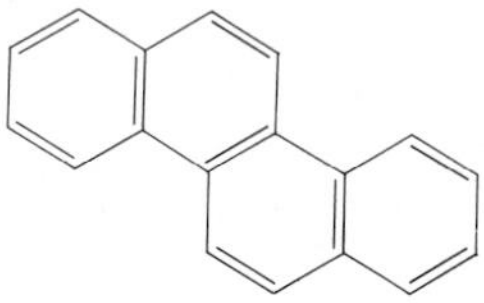

(c)

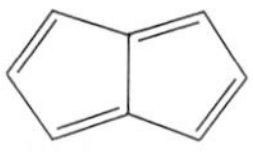

(d)

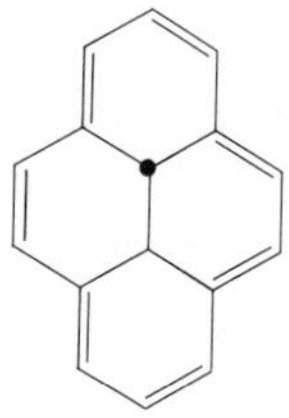

(e)

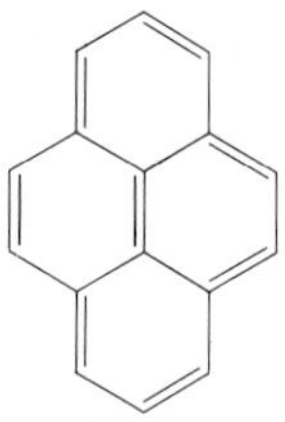

(f) 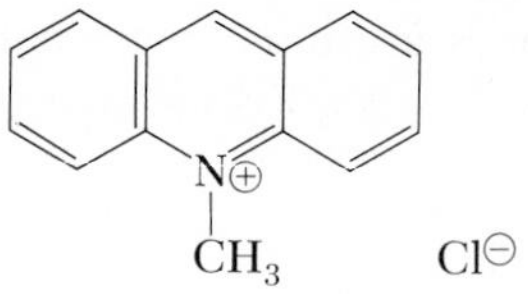 $N^{\oplus}$ CH_3 $Cl^{\ominus}$

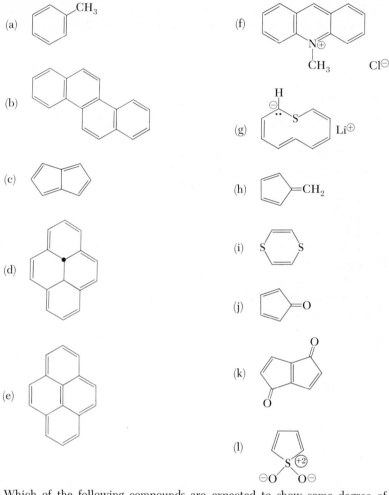

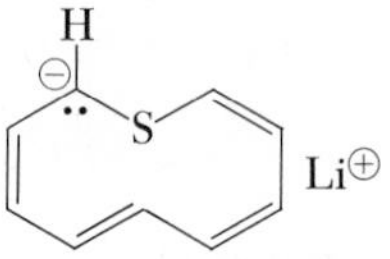

(g) H ⊖ S $Li^{\oplus}$

(h) $=CH_2$

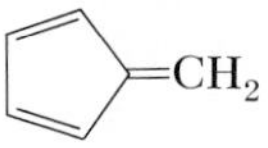

(i) 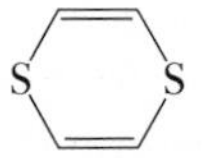 S S

(j) 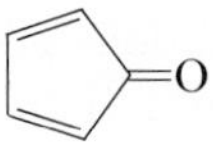 $=O$

(k) 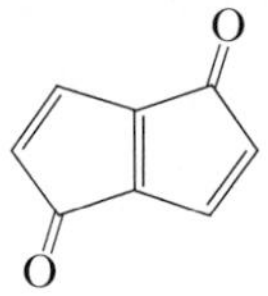 O O

(l) 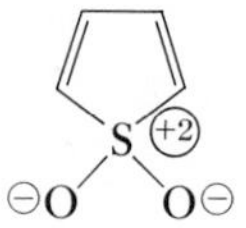 S (+2) ⊖O O⊖

14. Which of the following compounds are expected to show some degree of aromaticity? What physical method other than nmr spectroscopy might be used to support this view?

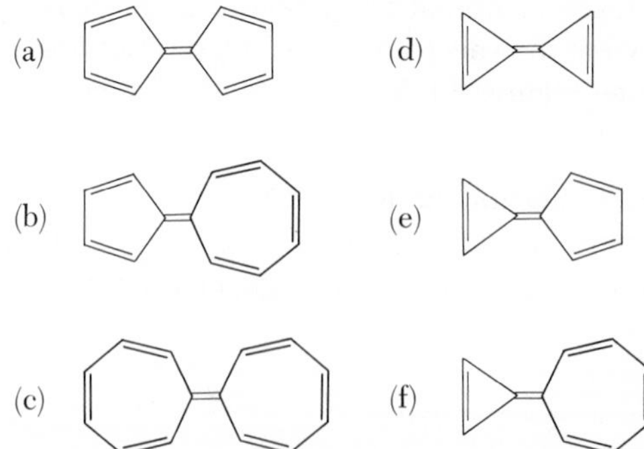

15. Suggest a reason for the greater resonance stabilization of thiophene than of furan and pyrrole. (*Hint:* Consider the electronic structure of sulfuric acid.)

16. When the methylcyclopentadiene shown below (**A**) reacts with sodium hydride and the resultant anion is quenched with methyl iodide, there are produced a number of dimethylcyclopentadienes. (a) Account for the formation of these isomers. (b) How many dimethylcyclopentadienes can result? (c) Which are capable of optical activity?

(d) Would your answers be different if the starting compound had been **B** instead of **A**? Explain.

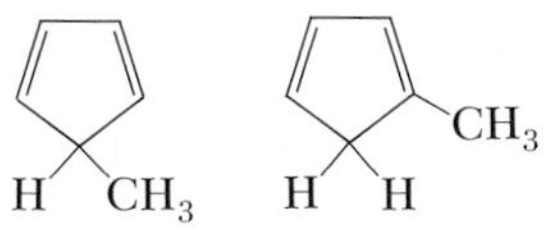

H CH_3 H H CH_3

A **B**

17. Do you agree with the following? "Cycloheptatrienyl cation, $C_7H_7^{\oplus}$, cannot be aromatic. After all, it is so unstable that it reacts instantly with water." Explain your answer. What is the product of this reaction?

18. When 9,10-dihydroanthracene is refluxed in sodium methoxide/methanol for several hours and then recovered, the product is identical to the starting material. However, anthrone rapidly exchanges its methylene protons for deuterons under the same conditions. Account for this difference. (*Hint:* Consider the structure of the intermediate anion.)

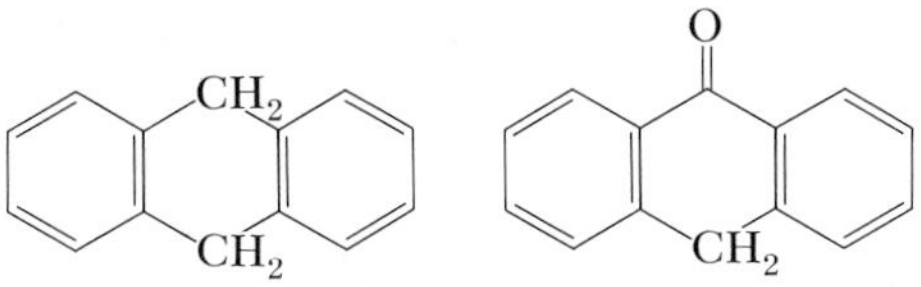

CH_2 CH_2 O CH_2

9,10-dihydroanthracene anthrone

19. "Sand is extremely stable and is, therefore, aromatic." Do you agree? Explain.

20. In 1869, Ladenburg suggested that benzene had the structure shown below. If you were a 19th century chemist, how would you test this hypothesis?

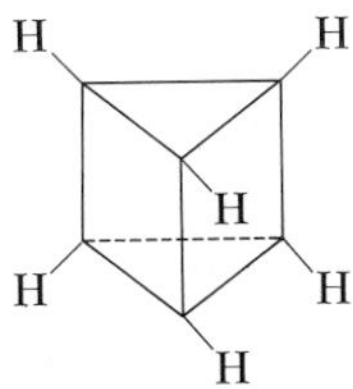

Ladenburg benzene

21. The suggestion has been made that ethylene is aromatic, that is, it is a $4n + 2$ system in which $n = 0$. Why couldn't the value of the heat of hydrogenation of ethylene be used to support this view?

22. The graphic representation of Hückel m.o.'s described in the text can be made more quantitative by inscribing the regular polygon in a circle of radius 2β. (The value of β is approximately -18 kcal/mole.) The intersection of the polygon with the circle defines the energies of the molecular orbitals. This is done below for benzene.

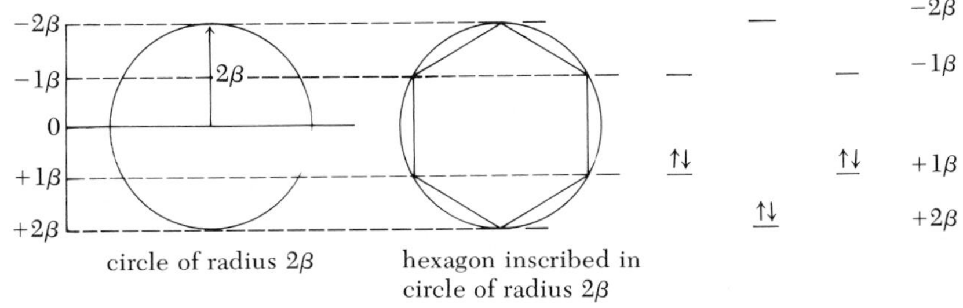

The calculated π electron energy using this scheme is $(2 \times 2\beta) + (4 \times 1\beta)$ or 8β. Cyclohexatriene, possessing three sets of double bonds, has a π energy of 6β (that is, 2β for each isolated double bond). The delocalization energy of benzene (calculated additional bond energy resulting from the delocalization of electrons originally treated as isolated π bonds) is, therefore, $8\beta - 6\beta$ or 2β.

Using this same scheme, can you verify a calculated delocalization energy of planar cyclooctatetraene of $1.657\,\beta$?

23. With the aid of models or a good scale drawing, account for the non-planarity of [10]-annulene.

24. Borazole, $B_3N_3H_6$, is believed to be an aromatic heterocycle. What are the major contributing structures to the resonance hybrid? Discuss the possibility of borazole having a molecular dipole moment.
25. Di-*n*-propylcyclopropenone has a very high dipole moment (5 D) compared to simple ketones such as acetone (3 D). Moreover, this compound reacts with perchloric acid to form a stable salt ($C_9H_{15}ClO_5$). Account for these facts.

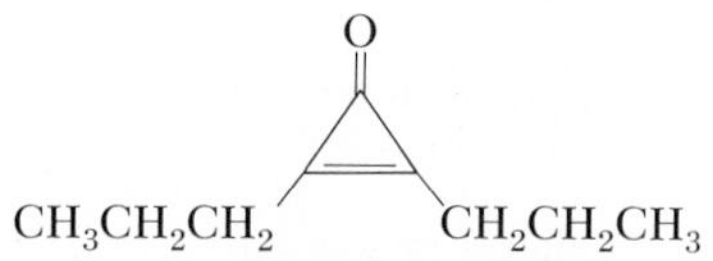

O

$CH_3CH_2CH_2$ $CH_2CH_2CH_3$

di-*n*-propylcyclopropenone

°26. Pyridazine and pyrazine are aromatic isomers of pyrimidine. The nmr spectrum of pyrazine consists of a single line. Suggest structures for pyridazine and pyrazine.
27. Hexahelicene is capable of existing in enantiomeric forms. How do you account for this? (*Hint:* Build a model.)

hexahelicene

28. Which of the carbon-carbon bonds in phenanthrene should have the shortest bond length? Explain your answer. (*Hint:* Examine the five Kekulé forms.)
29. Is the reaction shown below one that might be explicable in terms of orbital symmetry arguments? If your answer is "yes," would you expect this reaction to be accelerated by heat or by light? Explain. Why does the equilibrium favor the dihydronaphthalene?

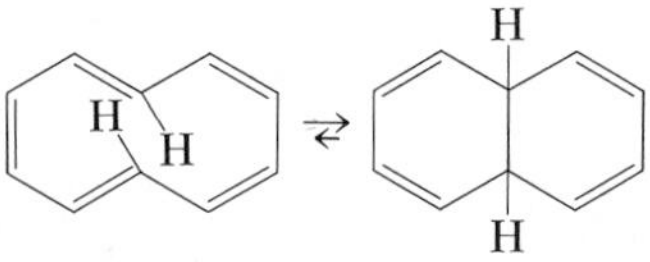

H H ⇄ H H

30. The orbital energies of *rectangular,* planar cyclobutadiene are shown below in comparison to the orbital energies of *square,* planar cyclobutadiene. Should rectangular cyclobutadiene be a diradical? Explain. What could be wrong with picturing rectangular, planar cyclobutadiene as a hybrid of the two structures shown below?

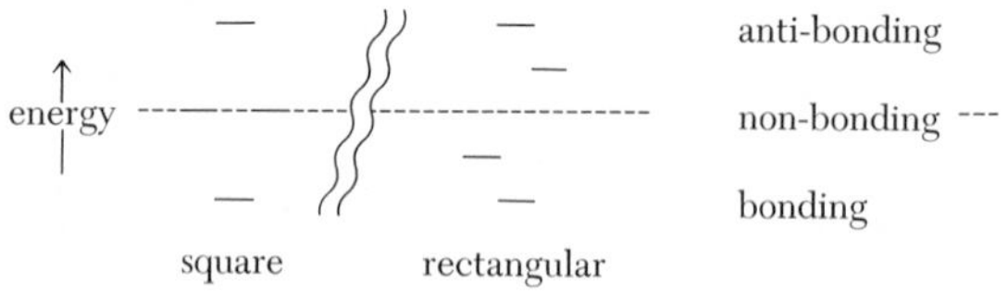

energy

anti-bonding

non-bonding

bonding

square rectangular

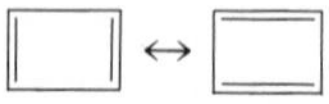

↔

°31. 3-Chlorocyclopropene reacts with antimony pentachloride, $SbCl_5$, to produce a crystalline solid, $C_3H_3SbCl_6$, which is insoluble in nonpolar solvents but soluble in polar solvents such as sulfur dioxide. The nmr spectrum of this solid consists of a single line. Account for these facts.

°These problems require knowledge of nmr spectroscopy.

[*] 32. Compound **A** reacts with two moles of antimony pentafluoride to produce a salt, **B**. The nmr spectrum of **B** consists of a single line. Suggest a structure for **B** and account for its stability.

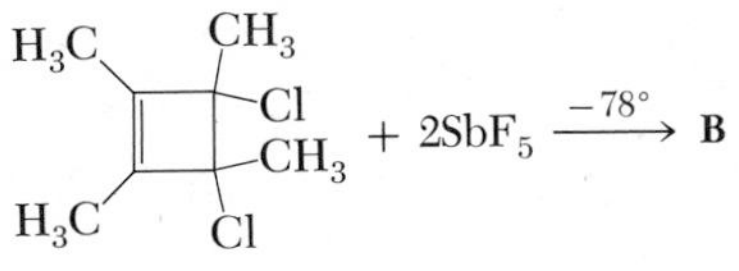

H_3C CH_3 Cl CH_3 H_3C Cl $+ 2SbF_5 \xrightarrow{-78^\circ}$ **B**

A

33. One excellent means of defining aromaticity is by calculating the energy of a linear π electron system and comparing it to that of the corresponding cyclic π electron system. If "cyclization" decreases the energy of the π system, the cyclic compound is described as aromatic; while if cyclization increases the energy of the π system, the cyclic compound is anti-aromatic. (In each conceptual "cyclization" the linear molecule loses a mole of hydrogen.) What is the linear species that is used as the basis for determining whether (a) cyclooctatetraene, (b) the cyclopropenium cation ($C_3H_3^{\oplus}$), and (c) the cyclopropenyl anion ($C_3H_3^{\ominus}$) are aromatic? (*Hint:* Two examples of this method are shown below.)

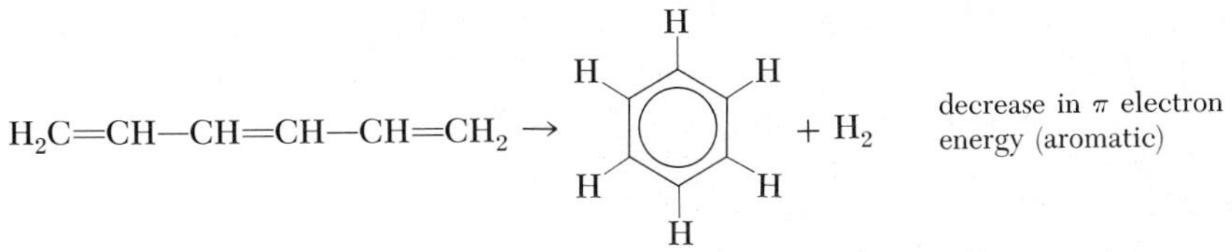

$$H_2C{=}CH{-}CH{=}CH{-}CH{=}CH_2 \rightarrow C_6H_6 + H_2$$

decrease in π electron energy (aromatic)

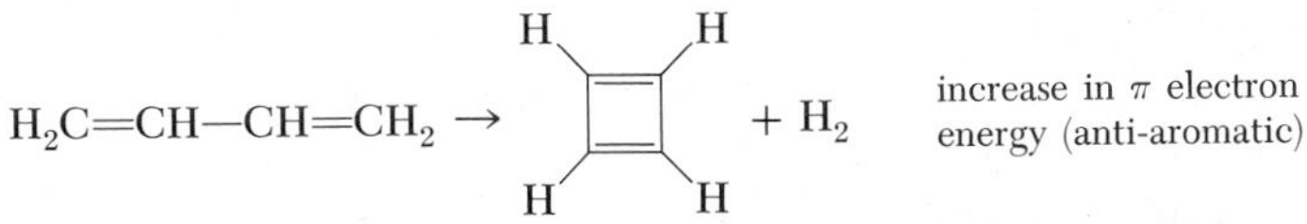

$$H_2C{=}CH{-}CH{=}CH_2 \rightarrow C_4H_4 + H_2$$

increase in π electron energy (anti-aromatic)

[*] This problem requires knowledge of nmr spectroscopy.

ELECTROPHILIC AROMATIC SUBSTITUTION

15.1 INTRODUCTION

Before the dawn of the twentieth century, Charles Friedel and James Mason Crafts had begun a classic series of studies on the replacement of the hydrogens bonded to the benzene ring by a wide variety of reagents that we now recognize as electrophiles. Over fifty years later the details of the mechanisms of these **electrophilic aromatic substitution** reactions were still wrapped in mystery. Much of the credit for beginning the unraveling of these secrets belongs to Lars Melander and his elegant studies of the influence of isotopic substitution upon the rate of aromatic substitution. However, it has taken the efforts of many leading chemists, including G. Wheland, H. Zollinger, E.A. Halevi, P.B.D. de la Mare, G. Hammond, H.C. Brown, J.H. Ridd, G. Olah, M.J.S. Dewar, and V. Gold,[a] to provide us with a detailed, although still incomplete, picture of these processes.

We can only survey the more qualitative aspects of electrophilic aromatic substitution in this chapter. Citations in the bibliography, especially the monograph by L. Stock, will help interested students to pursue this subject further.

15.2 NOMENCLATURE OF BENZENE DERIVATIVES

The term *arene* describes any aromatic hydrocarbon or any hydrocarbon containing a fragment that is aromatic. This includes benzene and its polynuclear analogs (*e.g.*, naphthalene) as well as their alkylated derivatives. From this nomenclature is derived the term *aryl group*, a general name for any aromatic substituent directly bonded to a carbon skeleton. The discussion in this section is limited to arenes and aryl groups in which the benzene ring is the focus of aromaticity.

Monosubstituted Benzenes. Simple benzene derivatives are named conveniently as substituted benzenes. Thus, the compound in which an ethyl group has replaced a hydrogen on benzene is *ethylbenzene*, while *chlorobenzene* has a chlorine replacing one of benzene's hydrogens. The monosubstituted benzene ring, C_6H_5-, is called the *phenyl group*, and monosubstituted derivatives of benzene also can be named by treating the phenyl group as a substituent. For example, ethylbenzene and chlorobenzene may be called phenylethane and

---

[a] This list is not comprehensive, nor is it arranged in any particular order.

[b] Alkylaryl groups are of the type R—Ar—, while arylalkyl groups are of the type Ar—R—. Ar: aryl group. R: alkyl group.

phenyl chloride, respectively. Because of its common occurrence, it is convenient to abbreviate the phenyl group as Ph.

C_2H_5 Cl

phenyl group (Ph or C_6H_5—) ethylbenzene (phenylethane *or* PhEt) chlorobenzene (phenyl chloride *or* PhCl)

Many simple benzene derivatives have unique names, common names being much more prevalent in aromatic chemistry than in aliphatic chemistry. Among the more important of these common names, and ones which should be memorized, are *toluene* (methylbenzene), *aniline* (aminobenzene), *phenol* (hydroxybenzene), *anisole* (methoxybenzene), and *cumene* (isopropylbenzene).

CH_3 NH_2 OH OCH_3 $CH(CH_3)_2$

toluene aniline phenol anisole cumene

Polysubstituted Benzenes. When only two substituents are present on benzene, they can be separated by a minimum of *zero, one,* or *two* ring carbons. Such substituents are said to be *ortho* (*o*), *meta* (*m*), or *para* (*p*) to one another, respectively.

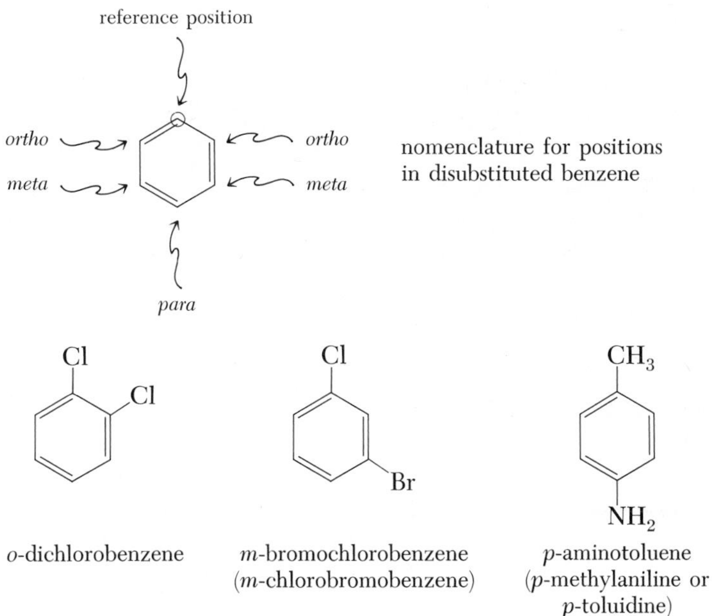

nomenclature for positions in disubstituted benzene

o-dichlorobenzene *m*-bromochlorobenzene (*m*-chlorobromobenzene) *p*-aminotoluene (*p*-methylaniline or *p*-toluidine)

When two or more substituents are present on one ring, their positions can be indicated by numbers. Usually, but certainly not always, the formula is written with one group at the top ring carbon and this becomes C1 of the ring. The remaining ring carbons are numbered C2 through C6 in the direction around the ring that gives the substituents the smaller numbers.

It is desirable to list substituents in alphabetical order. In numbering, it is sometimes helpful to place the substituent bonded to C1 at the end of the list of substituents and to delete its number. If a polyfunctional benzene derivative bears a methyl, hydroxy, or amino group, these normally are considered to be bonded to C1. The order of preference in designation is $OH > NH_2 > CH_3$ if two or more of these appear on the same ring.

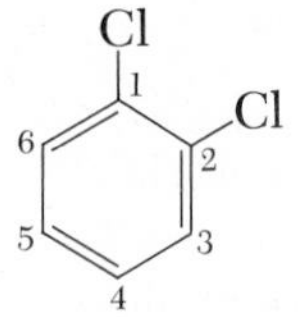

1,2-dichlorobenzene
(2-chlorochlorobenzene)

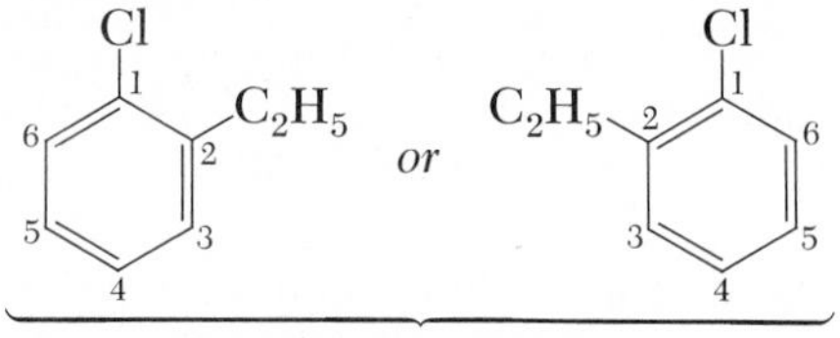

or

1-chloro-2-ethylbenzene
(2-ethylchlorobenzene)

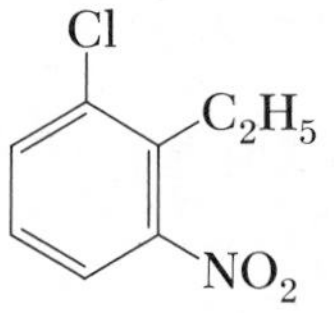

1-chloro-2-ethyl-3-nitrobenzene
(2-ethyl-3-nitrochlorobenzene)

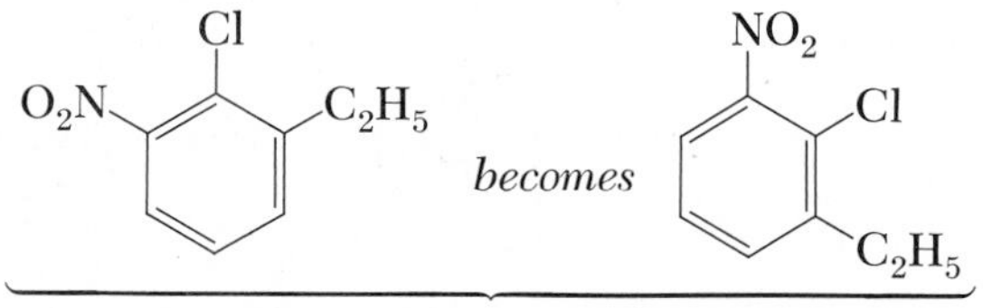

becomes

2-chloro-1-nitro-3-ethylbenzene
(2-chloro-3-ethylnitrobenzene)

Proper numbering also permits the naming of substituted benzene rings as substituents on carbon skeletons.

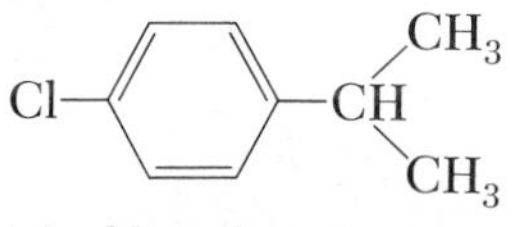

2-(4-chlorophenyl)propane

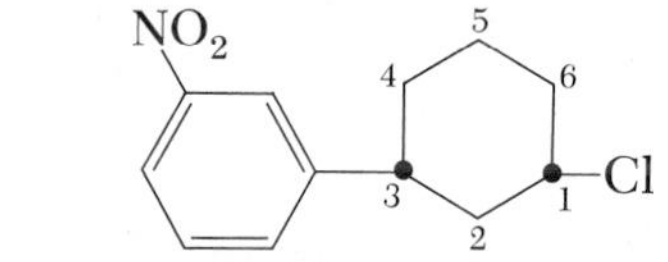

cis-3-(3-nitrophenyl)chlorocyclohexane

Isomeric dimethylbenzenes usually are called *xylenes*. Thus, 1,2-, 1,3-, and 1,4-dimethylbenzene are *o*-, *m*-, and *p*-xylene, respectively. These and other common polysubstituted arenes are shown below.

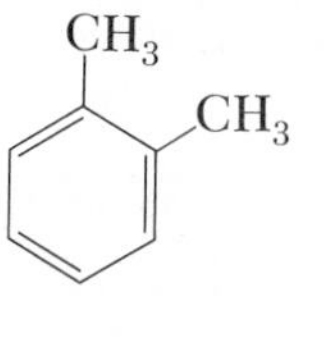

o-xylene

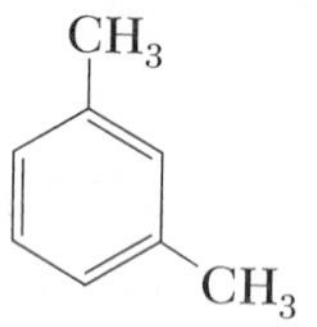

m-xylene

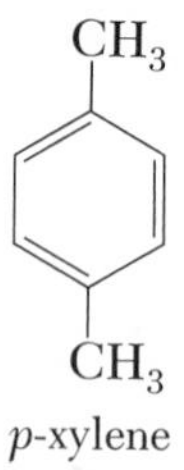

p-xylene

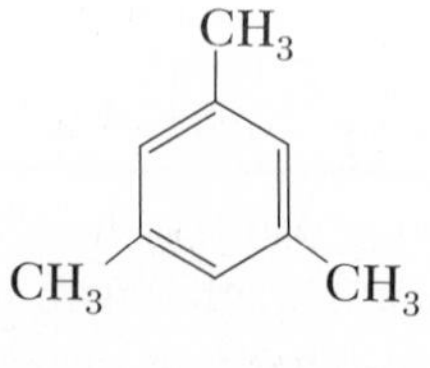

mesitylene

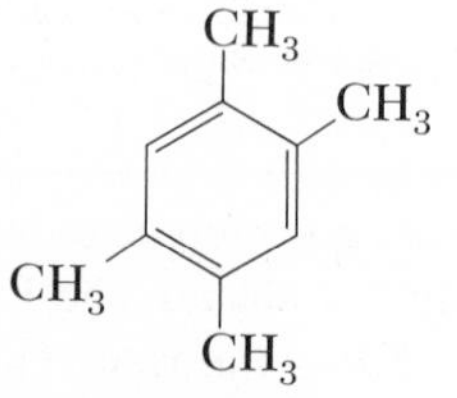

durene

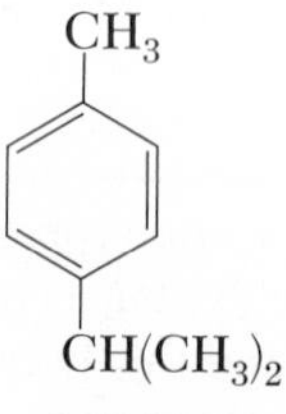

p-cymene

Two important types of functional groups are derivable from toluene. The *benzyl group* is produced by removing a methyl hydrogen, while the *tolyl group* is produced by removing a phenyl hydrogen. Since phenyl hydrogens are disposed *o*-, *m*-, and *p*- to the methyl group, there are three distinct tolyl groups: *o*-tolyl, *m*-tolyl, and *p*-tolyl.

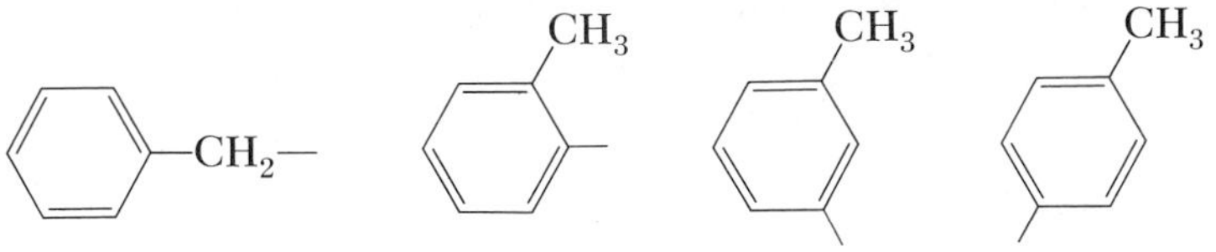

CH_2—

benzyl group

CH_3 CH_3 CH_3

ortho *meta* *para*

tolyl groups

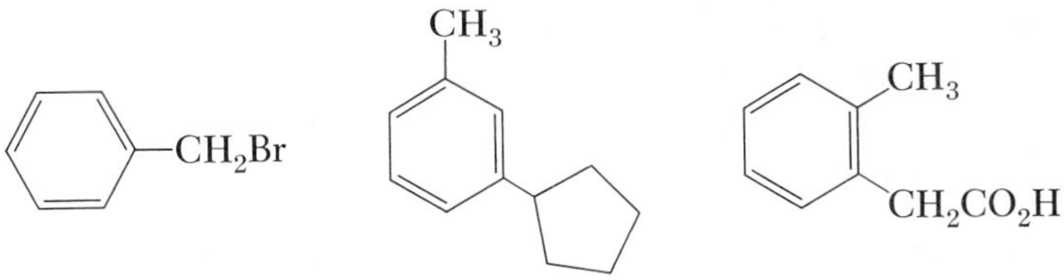

CH_2Br

benzyl bromide

CH_3

m-tolylcyclopentane

CH_3 CH_2CO_2H

2-(*o*-tolyl) acetic acid

1. Draw the structure of each of the following.
   (a) 1,2-dibromobenzene
   (b) 3-ethylnitrobenzene
   (c) 1,2-dibromo-3-phenylbenzene
   (d) 3-bromotoluene
   (e) *m*-nitroanisole
   (f) (**R**)-2-(4-bromophenyl)butane
   (g) 1-ethyl-2-(2-hexyl)benzene

15.3 THE GENERAL MECHANISM OF ELECTROPHILIC AROMATIC SUBSTITUTION

In this section we will not examine any specific electrophilic aromatic substitution reactions. Instead, we will develop a very general picture of the mechanism of such substitutions using $E^{\oplus}$ to designate any electrophile that is reacting with a benzene ring. In the next section we shall see how this scheme can, with minor modification, be used to explain the behavior of benzene toward a number of specific electrophiles.

All electrophilic aromatic substitutions follow the same basic pattern, beginning with attack of an electrophile (either a cation or the positive end of a highly polar bond) upon the hydrocarbon's π electron system to form a resonance-stabilized, non-aromatic σ *complex* or *benzenonium ion*. This is followed by proton loss and concomitant re-aromatization, producing a compound in which the electrophile has replaced the hydrogen on the carbon that was attacked initially.

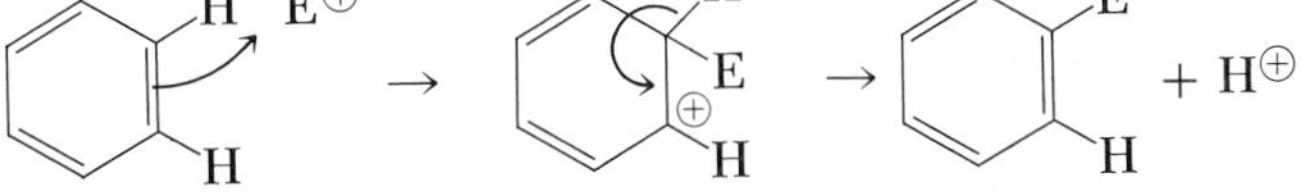

H $E^{\oplus}$ H → H E ⊕ H → E H $+ H^{\oplus}$

benzenonium ion
(σ complex)

The intermediate σ complex has several contributing structures and is rather like a "super allyl cation," distributing positive charge over three of five available *p* orbitals. These involve two equivalent carbons *ortho* to the sp^3 carbon and one *para* to that reference carbon. Two equivalent positions *meta* to the sp^3 carbon carry no formal charge but undoubtedly are somewhat positive because of the adjacent, positively charged carbons.

alternate representations of a benzenonium ion

While the first step in electrophilic aromatic substitution is analogous to the first step in electrophilic addition to alkenes, the benzenonium ion does not complete the reaction by adding a nucleophile. The pathway leading to the re-formation of the benzene ring, *i.e.*, proton loss, has a lower activation energy than does the one leading to addition, and is the one which occurs most rapidly (Figure 15–1).

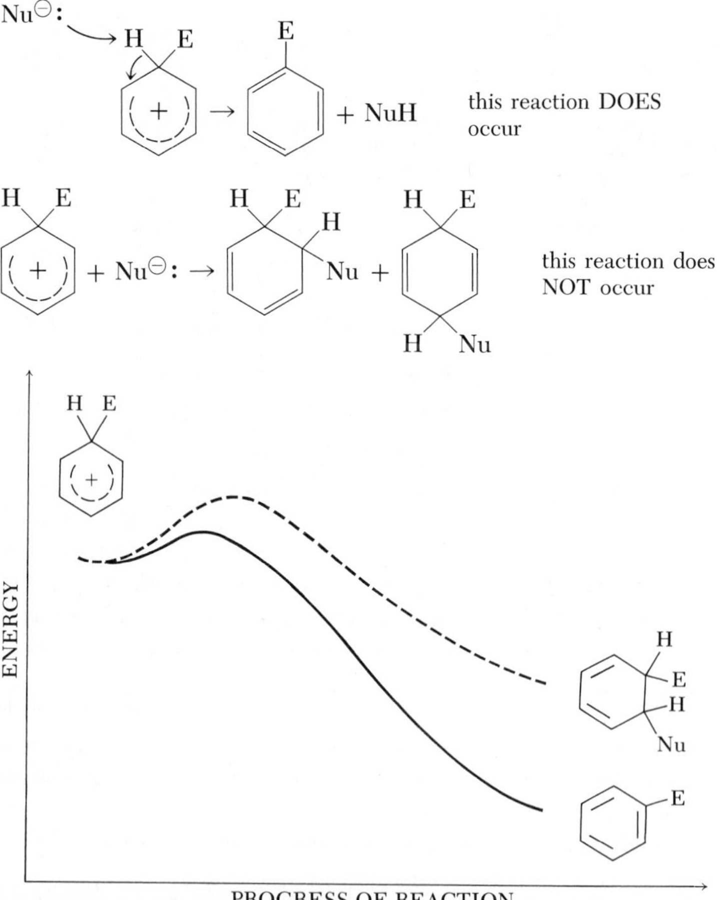

Figure 15-1 Competition between addition of a nucleophile and proton loss in the reaction of a benzenonium ion. The energy if activation for proton loss (solid line) is less than that for nucleophile addition (dashed line).

This scheme for substitution is an oversimplification; the aromatic compound and the electrophile may form a weak complex prior to benzenonium ion production. In this *π complex* the benzene π system acts as electron donor and the electrophile acts as electron acceptor.

However, π complex formation and dissociation is extremely rapid and π complexation is not significant in determining either the rate of reaction or the nature of the product for the vast majority of substitutions (Figure 15–2). Unless there is a special need, we will not consider the formation of the π complex in future substitution schemes.

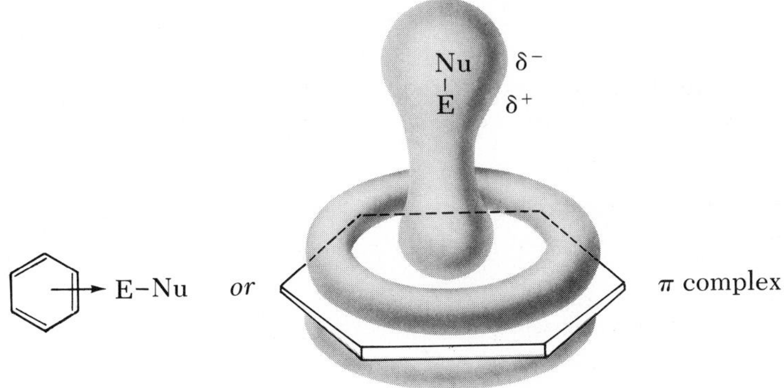

$$\text{Ar–H} + \text{E}^{\oplus} \rightleftharpoons \pi\ \text{complex}_1 \rightleftharpoons \sigma\ \text{complex} \rightleftharpoons \pi\ \text{complex}_2 \rightleftharpoons \text{Ar–E} + \text{H}^{\oplus}$$

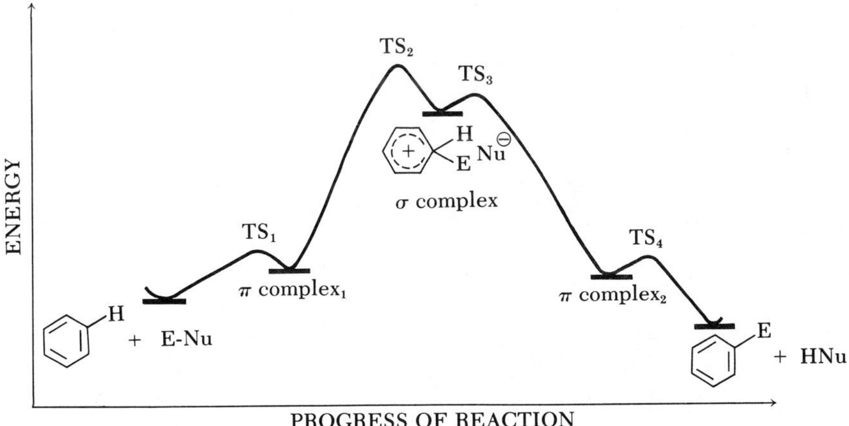

Figure 15-2 Idealized energy profile for aromatic substitution showing π complexation. This profile suggests that σ complex formation is rate determining. TS is the transition state.

The relative energy content of the transition states leading *to* and *from* the σ complex (TS_2 and TS_3, respectively, in Figure 15–2) can be determined by testing for the presence of a kinetic isotope effect. If formation of the σ complex is rate determining, then C_6D_6 should react at about the same rate as C_6H_6. However, if proton loss is slow (*i.e.*, $E_{TS_3} > E_{TS_2}$), substitution of D for H should produce a diminution in the rate of reaction. In fact, most electrophilic aromatic substitutions do not exhibit an isotope effect, proving that σ complex decomposition is not rate determining.

15.4 VARIATIONS ON A THEME

In this section we will see how *nitration* (substitution of —NO_2 for —H), *halogenation* (substitution of —Br or —Cl for —H), *alkylation* (substitution of -alkyl for —H), *acylation* (substitution of —C(O)R for —H), *sulfonation* (substitution of $—SO_3H$ for —H), and *isotope exchange* (substitution of —D for —H) are mechanistically similar and follow the pattern presented in the previous section.

Nitration. Benzene and concentrated nitric acid react to produce nitrobenzene at a *very* slow rate. The rate is greatly accelerated by the addition of concentrated sulfuric acid; aromatic nitrations are, as a result, often carried out in a mixture of $HNO_3:H_2SO_4$ (1:2) called *mixed acid.* The function of the sulfuric acid is to protonate the nitric acid (which acts as a base toward sulfuric acid) and to increase the concentration of the actual electrophile, the *nitronium ion,* $NO_2^{\oplus}$.

$$C_6H_6 + HNO_3 \xrightarrow[\text{slow}]{\text{very}} C_6H_5NO_2 + H_2O \qquad \text{absence of } H_2SO_4$$

nitrobenzene

$$C_6H_6 + HNO_3 \xrightarrow{H_2SO_4} C_6H_5NO_2 + H_2O \qquad \text{nitration in mixed acid}$$

$$2HONO_2 \rightleftharpoons HO^{\oplus}(H)—N^{\oplus}(=O)—O^{\ominus}\ NO_3^{\ominus} \underset{}{\overset{\text{slow}}{\rightleftharpoons}} NO_2^{\oplus} + H_2O + NO_3^{\ominus} \qquad \text{absence of } H_2SO_4$$

nitronium ion

$$HONO_2 + 2H_2SO_4 \rightleftharpoons HO^{\oplus}(H)—N^{\oplus}(=O)—O^{\ominus}\ HSO_4^{\ominus} + H_2SO_4 \overset{\text{slow}}{\rightleftharpoons} NO_2^{\oplus} + H_3O^{\oplus} + 2HSO_4^{\ominus} \qquad H_2SO_4 \text{ added}$$

$$C_6H_6 + NO_2^{\oplus} \rightarrow [C_6H_6NO_2]^{\oplus} \rightarrow C_6H_5NO_2 + H^{\oplus}$$

There is much evidence that favors the existence and activity of the nitronium ion in mixed acid. Moreover, nitronium tetrafluoroborate, a direct source of nitronium ion, readily nitrates benzene.

$$\underset{\text{nitryl chloride}}{NO_2Cl} + AgBF_4 \quad \underset{\text{nitronium tetrafluoroborate}}{NO_2^{\oplus}\ BF_4^{\ominus}} + AgCl$$

$$C_6H_6 + NO_2^{\oplus}\ BF_4^{\ominus} \rightarrow C_6H_5NO_2 + HBF_4$$

Nitrations also can be performed using other sources of nitronium ion, *e.g.*, acetyl nitrate and dinitrogen pentoxide.

$$H_3C{-}\overset{\overset{\displaystyle O}{\|}}{C}{-}O{-}NO_2 + H^{\oplus} \rightleftarrows H_3C{-}\overset{\overset{\displaystyle O}{\|}}{C}{-}\overset{\oplus}{O}(H){-}NO_2 \rightarrow CH_3{-}\overset{\overset{\displaystyle O}{\|}}{C}{-}OH + NO_2^{\oplus}$$

acetyl nitrate

$$O_2N{-}O{-}NO_2 + H^{\oplus} \rightleftarrows (HO)(O)\overset{\oplus}{N}{-}O{-}\overset{\oplus}{N}(O)O^{\ominus} \rightarrow (HO)(O)\overset{\oplus}{N}{-}O^{\ominus} + NO_2^{\oplus}$$

2. Draw the principal resonance forms of the nitronium ion.

Halogenation. Spectral and X-ray studies have proved that halogens form π complexes with aromatic compounds. Although this complexation undoubtedly promotes substitution by juxtaposing the reactants, it does not by itself usually lead to halogenation. Halogenation of benzene requires a catalytic amount of Lewis acid (*e.g.*, anhydrous aluminum chloride or anhydrous ferric halide), which coordinates with the halogen. This acid-base reaction increases the positive character of the halogen closest to the benzene ring and, hence, the reactivity of the electrophile.

$$C_6H_6 + X_2 \rightleftarrows C_6H_6 \rightarrow X{-}X \xrightarrow{AlCl_3} C_6H_6 \rightarrow \overset{\delta+}{X}\cdots\overset{\delta-}{X}AlCl_3$$

π complex

$$\rightarrow [C_6H_6X]^{+}\ (\sigma\ \text{complex}) + \overset{\ominus}{AlCl_3X} \rightarrow C_6H_5X + HAlCl_3X$$

σ complex

Under the reaction conditions the complex acid $HAlCl_3X$ dissociates and regenerates the catalyst.

$$H^{\oplus}AlCl_3X^{\ominus} \rightarrow AlCl_3 + HX\uparrow$$

Bromine and chlorine are the most commonly employed halogens in this type of scheme. Fluorine is not introduced directly because the ensuing violent reaction produces a variety of products, among which is relatively little fluorobenzene. Iodine, on the other hand, is too unreactive; iodination must often be accomplished by using I_2 in the presence of an oxidant such as nitric acid. Under these conditions the electrophile is believed to be $IOH_2^{\oplus}$. The more reactive mixed halogen iodine monochloride, ICl, also can be used to produce iodobenzene.

$$C_6H_6 + I_2 \xrightarrow{HNO_3} C_6H_5I + HI$$

$$C_6H_6 + ICl \xrightarrow{AlCl_3} C_6H_5I + HCl$$

Alkylation. The simplest method of preparing alkylated benzenes is by the reaction of benzene with an alkyl halide in the presence of aluminum chloride. This *Friedel-Crafts alkylation* is sufficiently common that the expressions "Lewis acid" and "Friedel-Crafts catalyst" often are used interchangeably.

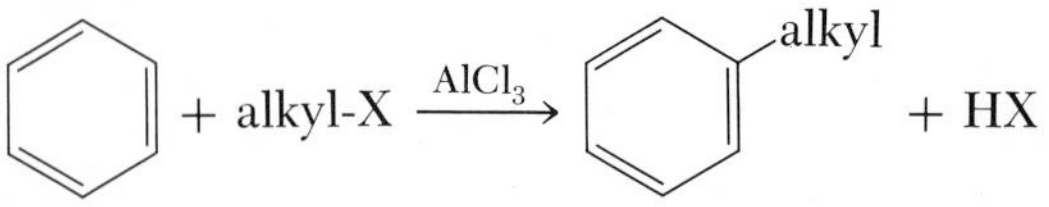

$$C_6H_6 + \text{alkyl-X} \xrightarrow{AlCl_3} C_6H_5\text{-alkyl} + HX$$

The initial step in the reaction is formation of a complex between aluminum chloride and the halide. This complex possesses a highly polarized C—X bond.

$$-\overset{|}{\underset{|}{C}}-\ddot{\underset{..}{X}}: + AlCl_3 \rightleftarrows -\overset{|}{\underset{|}{C}}^{\delta+}\cdots\ddot{\underset{..}{X}}\cdots AlCl_3^{\delta-}$$

Primary halide complexes undergo displacement of $XAlCl_3^{\ominus}$ by the benzene π system, while tertiary halide complexes ionize and react by an S_N1-like attack of the cation upon the π electrons.

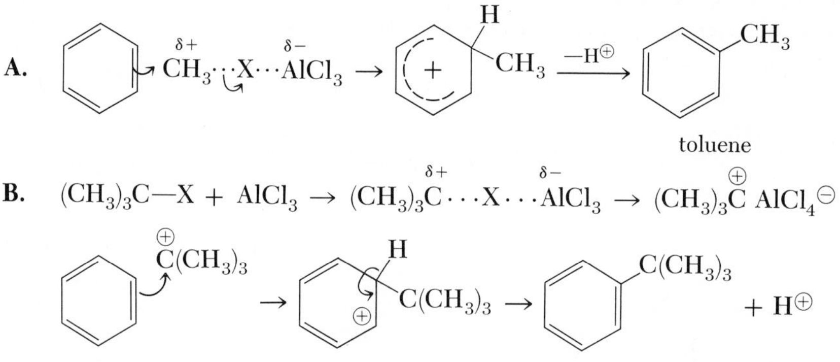

A. $$C_6H_6 + \overset{\delta+}{CH_3}\cdots X\cdots \overset{\delta-}{AlCl_3} \rightarrow [C_6H_6(H)(CH_3)]^{+} \xrightarrow{-H^{\oplus}} C_6H_5CH_3$$

toluene

B. $$(CH_3)_3C-X + AlCl_3 \rightarrow \overset{\delta+}{(CH_3)_3C}\cdots X\cdots \overset{\delta-}{AlCl_3} \rightarrow (CH_3)_3\overset{\oplus}{C}\ AlCl_4^{\ominus}$$

$$C_6H_6 + \overset{\oplus}{C}(CH_3)_3 \rightarrow [C_6H_6(H)(C(CH_3)_3)]^{\oplus} \rightarrow C_6H_5C(CH_3)_3 + H^{\oplus}$$

Rearrangements often plague carbocation reactions, and aromatic substitution is no exception. Thus, benzene reacts with either *n*-propyl chloride or isopropyl chloride, in the presence of a catalyst, to produce isopropylbenzene (cumene) as the major product. The conversion of *n*-propyl chloride to isopropyl cation probably proceeds by an assisted ionization rather than by ionization followed by rearrangement, as shown below.

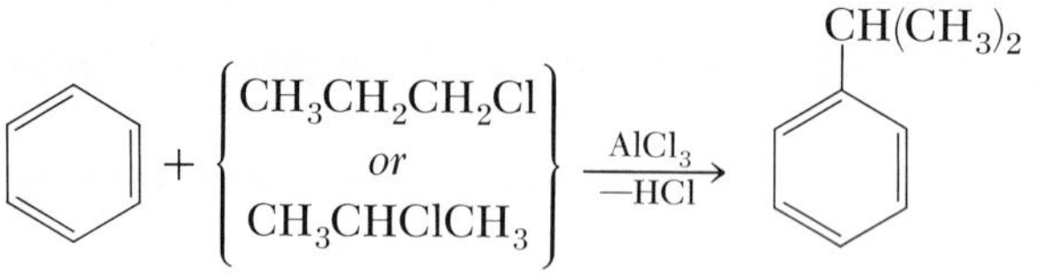

$$C_6H_6 + \left\{\begin{matrix} CH_3CH_2CH_2Cl \\ or \\ CH_3CHClCH_3 \end{matrix}\right\} \xrightarrow[-HCl]{AlCl_3} C_6H_5CH(CH_3)_2$$

isopropylbenzene
(cumene)

NO $$CH_3-\overset{H}{\underset{H}{C}}-\overset{\delta+}{CH_2}\cdots Cl\cdots \overset{\delta-}{AlCl_3} \xrightarrow{-AlCl_4^{\ominus}} CH_3-\overset{H}{CH}-\overset{\oplus}{C}H_2 \rightarrow CH_3-\overset{\oplus}{C}H-CH_2H$$

YES $$CH_3-\overset{H}{\underset{H}{C}}-\overset{\delta+}{CH_2}\cdots Cl\cdots \overset{\delta-}{AlCl_3} \xrightarrow{-AlCl_4^{\ominus}} CH_3-\overset{\oplus}{C}HCH_3$$

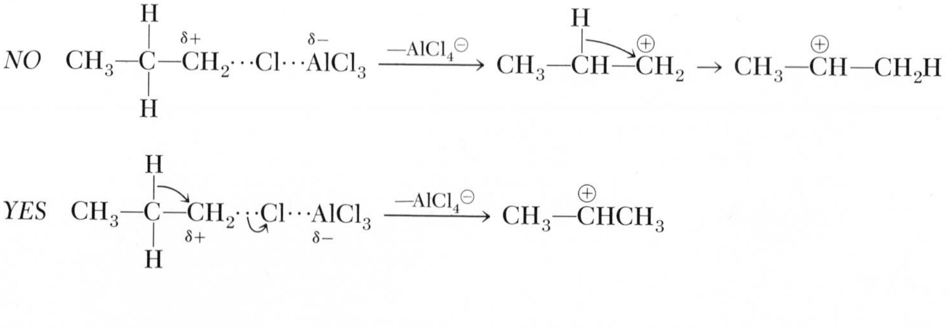

Any compound or mixture capable of forming a carbocation can alkylate benzene. The most commonly employed alternate sources of carbocation are alkenes and alcohols, the cations being generated by reaction with strong acids.

$$C_6H_6 \xrightarrow[HF(l)]{H_2C{=}CH_2} C_6H_5CH_2CH_3 \quad \text{ethylbenzene}$$

$$C_6H_6 \xrightarrow[HF(l)]{H_2C{=}CHCH_3} C_6H_5CH(CH_3)_2 \quad \text{isopropylbenzene}$$

$$C_6H_6 \xrightarrow[H_3PO_4]{(CH_3)_3COH} C_6H_5C(CH_3)_3 \quad t\text{-butylbenzene}$$

Friedel-Crafts alkylation and related reactions have two serious drawbacks. First, carbocation rearrangements make it impossible to prepare isomer-free straight-chain alkyl benzenes. Second, it is difficult to stop alkylation at the monoalkylation stage (for reasons to be discussed shortly); polyalkylated benzenes can represent a major source of wasted starting materials. Fortunately, use of an excess of the aromatic substrate represses the formation of polyalkylated products. (Why?)

3. Suggest two alternate routes to each of the following, starting with benzene and using anything else required.
   (a) toluene (c) *sec*-butylbenzene
   (b) *t*-butylbenzene (d) ethylbenzene
4. Suggest a reason why liquid hydrogen fluoride is preferable to concentrated, aqueous hydrobromic acid in inducing propene to react with benzene to produce isopropylbenzene.

Acylation. In a related reaction, acyl halides (usually the chloride, RC(O)Cl) react with benzene in the presence of excess anhydrous aluminum chloride to produce a ketone (R—C(O)—C_6H_5). The usual solvent for these **acylations** is either nitrobenzene or carbon disulfide, CS_2.

$$C_6H_6 + R{-}C(=O)Cl \xrightarrow[CS_2]{AlCl_3} C_6H_5{-}C(=O){-}R + HCl$$

an acyl chloride (R is alkyl or aryl) — a ketone

The initial step in this acylation reaction involves the formation of a complex between the acyl halide and aluminum chloride. This then ionizes to produce the *acylium ion,* $R\overset{\oplus}{C}O$. Attack of this acylium ion upon the benzene ring leads to the final product.

$$R{-}C(=O)Cl + AlCl_3 \rightarrow R{-}\overset{O}{\overset{\|}{C}}{}^{\delta+}\cdots Cl\cdots \overset{\delta-}{AlCl_3} \rightarrow R{-}\overset{\oplus}{C}{=}\ddot{O} + AlCl_4^{\ominus}$$

acylium ion

$$C_6H_6 + R{-}\overset{\oplus}{C}{=}O \rightarrow [C_6H_6(H)(C(=O)R)]^{\oplus} \rightarrow C_6H_5{-}C(=O){-}R + H^{\oplus}$$

ketone

The product ketone is sufficiently basic to form a stable adduct with aluminum chloride. This removes aluminum chloride from the reaction mixture and necessitates the use of more than one equivalent of "catalyst." Coordination of the carbonyl group in the ketone with aluminum chloride also has the effect of reducing the reactivity of the aromatic π system sufficiently to prevent further acylation. (Thus, while polyalkylation of benzene is quite common, polyacylation is quite rare.) When the reaction is completed, water must be added to the reaction mixture in order to liberate the desired ketone from the complex.

$$R{-}C(=O){-}C_6H_5 + AlCl_3 \rightleftarrows R{-}C(=\overset{\oplus}{O}{-}\overset{\ominus}{Al}Cl_3){-}C_6H_5$$

ketone-aluminum chloride complex

$$R{-}C(=\overset{\oplus}{O}{-}\overset{\ominus}{Al}Cl_3){-}C_6H_5 + 3H_2O \rightarrow RC(O)C_6H_5 + 3HCl + Al(OH)_3$$

Unlike many alkyl cations, the acylium ion does not rearrange—in one of its resonance forms, each atom has a complete octet. Therefore, we can add a single, straight chain alkyl group to a benzene ring *via* a two-step sequence; the first step is acylation, and the second is reduction of the ketonic carbonyl group. One method for reducing ketones to hydrocarbons ($\rangle C{=}O \rightarrow \rangle CH_2$) involves the use of a mixture of zinc amalgam and hydrochloric acid, a reaction called the **Clemmensen reduction.**

$$C_6H_6 + CH_3(CH_2)_5{-}C(=O){-}Cl \xrightarrow{AlCl_3} \xrightarrow{H_2O} C_6H_5{-}C(=O){-}(CH_2)_5CH_3$$

heptanoyl chloride — *n*-heptyl phenyl ketone

$$C_6H_5{-}C(=O){-}(CH_2)_5CH_3 \xrightarrow[\text{reflux}]{Zn \cdot Hg/HCl} C_6H_5{-}(CH_2)_6CH_3$$

n-heptylbenzene

Clemmensen reduction

Acylation of arenes also can be achieved by using other sources of acylium ions, including carboxylic acid anhydrides, R—C(O)—O—C(O)—R, and carboxylic acids, R—C(O)—OH. The equations below indicate that *more than two* moles of aluminum chloride must be used when acylating with an anhydride; one mole is complexed with the ketonic product and a second mole reacts with the carboxylate anion.

$$R{-}C(=O){-}O{-}C(=O){-}R + AlCl_3 \rightarrow R{-}C(=O){-}O{-}C(=\overset{\oplus}{O}{-}\overset{\ominus}{Al}Cl_3){-}R \rightarrow R{-}\overset{\oplus}{C}{=}O + R{-}C(=O){-}O{-}\overset{\ominus}{Al}Cl_3$$

carboxylic acid anhydride

$$R{-}C(=O){-}O{-}\overset{\ominus}{Al}Cl_3\; H^{\oplus} \rightarrow R{-}C(=O){-}O{-}AlCl_2 + HCl$$

$$C_6H_6 + R{-}\overset{\oplus}{C}{=}O \rightarrow \overset{\oplus}{[C_6H_6{-}C(=O)R]} \rightarrow C_6H_5{-}C(=O){-}R + H^{\oplus}$$

5. Draw the principal resonance forms of the $CH_3CO^{\oplus}$ ion.

A carboxylic acid can acylate arenes if the reaction is carried out in the presence of a sufficiently strong acid. Experimentally, this usually is accomplished by dissolving the arene and the carboxylic acid in liquid hydrogen fluoride.

$$C_6H_6 + R{-}\overset{O}{\overset{\|}{C}}{-}OH \xrightarrow{HF(l)} C_6H_5{-}\overset{O}{\overset{\|}{C}}{-}R + H_2O$$

carboxylic acid

One can use the fact that the Clemmensen reaction reduces ketones but not carboxylic acids to devise a means of fusing a new ring system onto a benzene ring. We have already encountered all of the reactions required to do this; all that we need do is begin our acylation with a *cyclic* acid anhydride rather than an acyclic anhydride. The sequence below describes the conversion of benzene to naphthalene. The last step in this sequence is a *catalytic dehydrogenation* that shows the aromatization of a six-membered ring. Catalytic dehydrogenation is encouraged by sweeping away hydrogen, as it is formed, with a stream of inert gas such as nitrogen.

$$C_6H_6 + \text{succinic anhydride} \xrightarrow[CS_2]{AlCl_3} \xrightarrow{H_2O} C_6H_5{-}\underset{\underset{O}{\|}}{C}{-}(CH_2)_2\overset{O}{\overset{\|}{C}}{-}OH \xrightarrow[HCl]{Zn\cdot Hg} C_6H_5{-}CH_2CH_2CH_2{-}\overset{O}{\overset{\|}{C}}{-}OH \xrightarrow{HF(l)} \alpha\text{-tetralone} \xrightarrow[HCl]{Zn\cdot Hg} \text{tetralin}$$

succinic anhydride

tetralin α-tetralone

$$\text{tetralin} \xrightarrow[heat]{Pt} \text{naphthalene} + 2H_2$$

tetralin naphthalene

Sulfonation. The sulfonation of benzene can be accomplished by the reaction of benzene with fuming sulfuric acid (sulfuric acid containing dissolved sulfur trioxide, SO_3). While the mechanism for this is not as well understood as are the mechanisms of nitration and chlorination, it seems likely that the most important sulfonating agent in fuming sulfuric acid is sulfur trioxide.

$$C_6H_6 + SO_3 \rightleftarrows [C_6H_6SO_3^{\ominus}]^{\oplus} \rightleftarrows C_6H_5SO_3^{\ominus}H^{\oplus} \rightleftarrows C_6H_5SO_3H$$

benzenesulfonic acid

Sulfonation is different from most other substitutions in that (a) it is readily reversed and (b) it shows a small isotope effect. This is best explained with the aid of the energy profile in Figure 15–3. The barrier to loss of sulfur trioxide from the sulfur trioxide–benzene adduct must be relatively low (see Figure 15–3), since the reaction is reversible, and the energy barriers on both sides of the σ complex must be of approximately the same height.

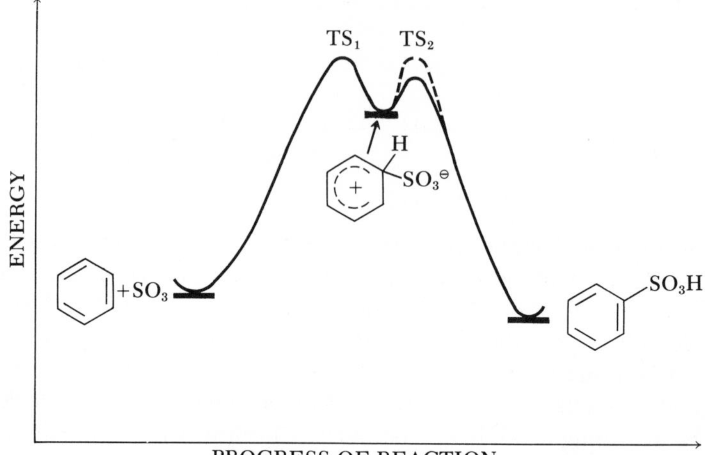

Figure 15–3 Energy profile for sulfonation of benzene (solid line) and perdeuteriobenzene (dashed line). Because of the similar energy content of TS_1 and TS_2, the reaction is reversible and the rate of the forward reaction is decreased when deuterium is substituted for hydrogen.

Since it requires more energy to cleave a C—D bond than a C—H bond, every time a molecule passes over the final transition state more energy will be required if the bond being broken is C—D rather than C—H. This explains the k_H/k_D of ~2 that is observed for sulfonation. (This argument, of course, is after the fact; one could not have predicted that sulfonation would be unusual.)

The desulfonation of benzenesulfonic acid is achieved in the laboratory by heating the acid in superheated steam, *i.e.*, steam above 100°.

$$C_6H_5SO_3H + H_2O \underset{\text{heat}}{\rightleftarrows} C_6H_6 + H_2SO_4$$

Aromatic sulfonic acids are similar to sulfuric acid in being highly water soluble (because of hydrogen bonding), very corrosive, and generally difficult to work with. Because work-up procedures are simpler, an organic chemist will often perform a *chloro*sulfonation rather than a sulfonation. Chlorosulfonation produces a sulfonyl chloride, and most aryl sulfonyl chlorides are solids that are easy to manipulate and that can be converted to sulfonic acids by refluxing in water. Chlorosulfonic acid is used for chlorosulfonation reactions; as with sulfonation, a catalyst such as aluminum chloride is unnecessary.

$$C_6H_6 + 2Cl{-}SO_2{-}OH \rightarrow C_6H_5SO_2Cl + HCl + H_2SO_4 \qquad \text{chlorosulfonation}$$

chlorosulfonic acid — benzenesulfonyl chloride

$$C_6H_5SO_2Cl \xrightarrow[\text{heat}]{H_2O} C_6H_5SO_3H + HCl$$

6. Draw the principal resonance forms of (a) sulfuric acid, (b) benzenesulfonic acid, (c) the sulfate anion, and (d) the benzenesulfonate anion.

Hydrogen Exchange and Protonation. When benzene is mixed with deuterated sulfuric acid, D_2SO_4, it is slowly converted to perdeuteriobenzene, C_6D_6. This reaction is much faster than sulfonation, and the formation of benzenesulfonic acid is not a major side reaction.

H, H + D_2SO_4 ⇄ H, D, ⊕ H, $DSO_4^{\ominus}$ ⇄ D, H + $HDSO_4$

The ease of hydrogen exchange between various acids and benzene follows the order of acidity of those acids: $D_2SO_4 > D_3\overset{\oplus}{O} > C_6H_5OD \gg D_2O$. The rate determining step in these isotope exchanges is σ complex formation.

Evidence that an ionic intermediate is involved comes, among other things, from the observation that a solution of toluene in liquid hydrogen fluoride conducts electricity better than either does separately. More convincing, perhaps, is the isolation of a salt of hydrofluoric acid and toluene, which contains the electrophile (**H**) *para* to the methyl group.

CH_3, H + HF ⇌ (−80°) CH_3, (+), H **H** $F^{\ominus}$

p-protonated toluene

If such salts are in equilibrium with hydrocarbons, and if increasing the concentration of salt increases a solution's electrical conductivity, then varying the number and location of methyl groups (which stabilize positive charges) on a benzene ring should influence the conductivity of methylated benzenes in liquid hydrogen fluoride. The data in Table 15–1 bear this out, hexamethylbenzene being approximately 10^5 times as conductive as *p*-xylene.

TABLE 15–1 Relative Conductance of Methylated Benzenes in Hydrogen Fluoride

Compound	Relative Conductance
methylbenzene	<1
1,4-dimethylbenzene	1
1,2-dimethylbenzene	1
1,3-dimethylbenzene	26
1,2,4,5-tetramethylbenzene	140
1,2,3,4-tetramethylbenzene	400
1,3,5-trimethylbenzene	13,000
1,2,3,4,5,6-hexamethylbenzene	97,000

Can we explain, for example, why a solution of 1,3,5-trimethylbenzene (mesitylene) in hydrogen fluoride is a better electrical conductor than is a solution of 1,4-dimethylbenzene (*p*-xylene) in the same solvent? This is equivalent to asking why the cation produced by protonating mesitylene is more stable than the one produced by protonating *p*-xylene. In answering this question we will also learn *where* to expect the protonation of alkyl benzenes to occur.

Mesitylene can, in principle, be protonated either at C1 (a carbon bearing a methyl group) or at C2 (a carbon bearing a hydrogen). Protonation at C2 produces a cation that is stabilized by charge delocalization within the ring and by polarization and hyperconjugative effects of all three methyl groups (see the resonance structures below). In contradistinction, protonation at C1 produces a cation that can be stabilized by resonance within the ring but cannot be directly stabilized by the three methyl groups. Therefore, we predict that mesitylene will be protonated at C2 and, thereby, form the more stable cation.

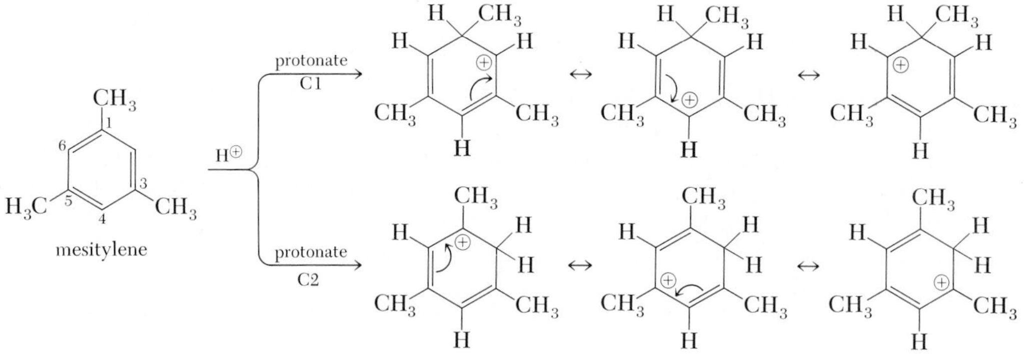

mesitylene
H⊕
protonate C1
protonate C2

Protonation of *p*-xylene can occur either at C1 or at C2. The cation produced by protonation at C1 is stabilized by delocalization of charge within the ring and by the C4 methyl group. The cation produced by protonation at C2 will be stabilized by charge delocalization within the ring and by the C1 methyl group. Therefore, considering that the significant stabilization gained by ring delocalization and methyl group interaction is the same for protonation at C1 as it is at C2, it is difficult to predict whether protonation will occur at C1 or at C2 in *p*-xylene.

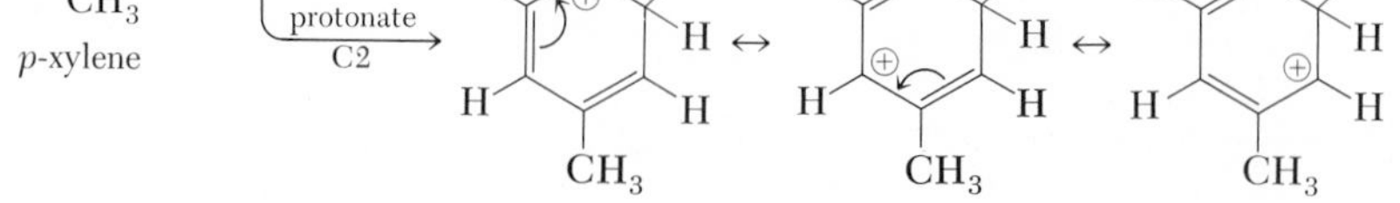

p-xylene
H⊕
protonate C1
protonate C2

Comparing the results of the preceding two paragraphs, protonation of *p*-xylene at either C1 or C2 produces a cation that is stabilized by delocalization within the ring and by *one* methyl group, while protonation of mesitylene at C2 (which is the preferred site of protonation) produces a cation that is stabilized by delocalization within the ring and by *three* methyl groups.

Therefore, C2 protonated mesitylene is more stable than either C1 or C2 protonated *p*-xylene; and mesitylene will produce a larger concentration of salt in liquid hydrogen fluoride than will *p*-xylene. Similar arguments can be used to explain other data in Table 15–1. (These ideas are extrapolated to a slightly more complex situation in problem 42.)

7. Recalling that hyperconjugative stabilization of a positive charge may be represented by $\overset{\ominus}{C}—CH_3 \leftrightarrow C{=}CH_2\ H^{\oplus}$, explain why *six* resonance contributing structures can be drawn for *p*-protonated toluene. Draw these six.
8. *m*-Xylene may be protonated in four different positions. Two protonated *m*-xylenes are comparatively stable, while two are comparatively unstable. Draw these four protonated species and suggest which should be the more stable ones. Explain your reasoning.
9. Draw the possible protonated *o*-xylenes and the protonated *m*-xylenes (see question 8). Using these structures, explain why *m*-xylene is a stronger base than is *o*-xylene.
10. How many different monoprotonated 1,2,3,4,5,6-hexamethylbenzenes are possible? Why is diprotonation of this compound less likely than monoprotonation?

15.5 THE EFFECTS OF EXISTING SUBSTITUENTS ON ELECTROPHILIC AROMATIC SUBSTITUTION

In the same way that methyl groups determine the extent and position of protonation of methylated benzenes, so every existing substituent influences the course of further electrophilic aromatic substitution. This influence is expressed in two ways: (a) a general enhancement or diminution of the rate of substitution, and (b) direction of the incoming electrophile toward particular positions relative to the existing substituent. For the more ambitious, a quantitative treatment that considers *both* of these ideas at the same time is presented in an appendix to this chapter (p. 453).

Reactivity. The activated complex leading to formation of a benzenonium ion (TS_2 in Figure 15–2) possesses some positive charge on the π system of the benzene ring. If we approximate this activated complex by the benzenonium ion (σ complex) that it is about to produce, it is reasonable to assume that anything that stabilizes this benzenonium ion should stabilize the activated complex leading to it and, therefore, increase the rate of that particular substitution reaction. Conversely, anything that destabilizes this benzenonium ion should destabilize the activated complex leading to it and decrease the rate of that particular substitution. We must conclude, therefore, that substituents already present on a benzene ring may either activate or deactivate the ring toward further substitution. The influence of such substituents varies from mild to severe, depending largely upon the substituent but, to a smaller degree, upon the specific electrophile attacking the ring. Table 15–2 (p. 428) summarizes common deactivating and activating groups.

What is the rationale for the division presented in Table 15–2? In a nutshell, any substituent that is on the ring and that places a positive charge (full or partial) on that ring will deactivate the ring. This deactivation reflects an increase in the energy of the activated complex leading to the σ complex, brought about by the juxtaposition of two similar charges on the ring (one from the electrophile and one from the substituent). Groups that can deactivate a benzene ring include $—NO_2$, $—CF_3$, $—SO_3H$, $—\overset{\oplus}{N}R_3$, and $—CN$. Even the inductive effect of the

TABLE 15-2 Influence of Existing Groups on Electrophilic Substitution

ortho,para DIRECTORS	*meta* DIRECTORS
Activating[a]	*Deactivating*
—OH	—CN
$—NH_2$, —NHR, $—NR_2$	$—CO_2H$
—OR	$—CO_2R$
—NHC(O)R	—CHO
—alkyl	—COR
—aryl	$—NO_2$
Deactivating	$—\overset{\oplus}{N}R_3$
—halogen	$—SO_3H$
	$—SO_2OR$

[a] Within this group, substituents near the top are strongly activating, while substituents near the bottom are weakly activating.

halogens is sufficient to reduce the rate of substitution on halobenzenes relative to that on benzene.

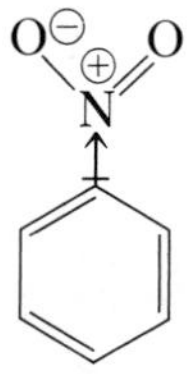

Nitrobenzene is attacked more slowly by $\overset{\oplus}{N}O_2$ than is benzene

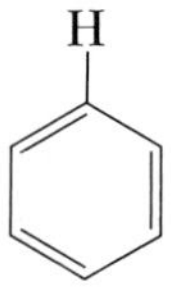

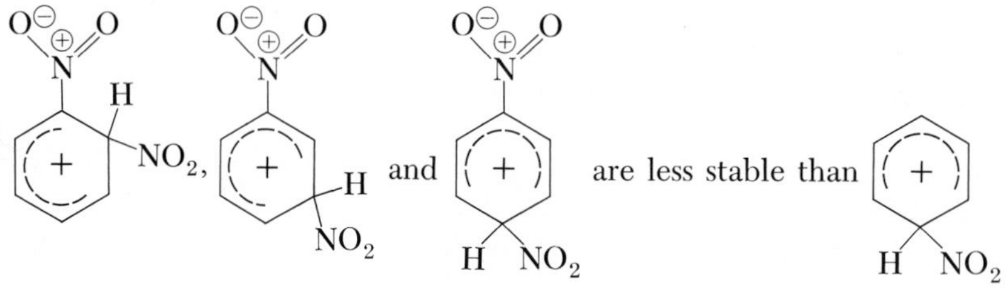

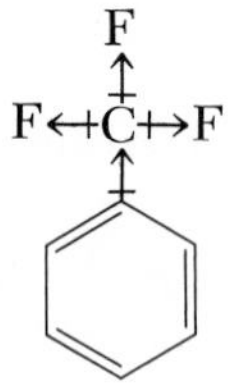

(ortho, meta and para arenium ions of nitrobenzene bearing H and NO_2 at the attacked carbon) NO_2, H and ... are less stable than (benzene arenium ion with H and NO_2)

(Trifluoromethyl)benzene is attacked more slowly by $(CH_3)_3CCl/AlCl_3$ than is benzene

(ortho, meta and para arenium ions of CF_3-benzene bearing H and $C(CH_3)_3$ at the attacked carbon) $C(CH_3)_3$, H and ... are less stable than (benzene arenium ion with H and $C(CH_3)_3$)

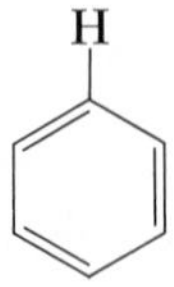

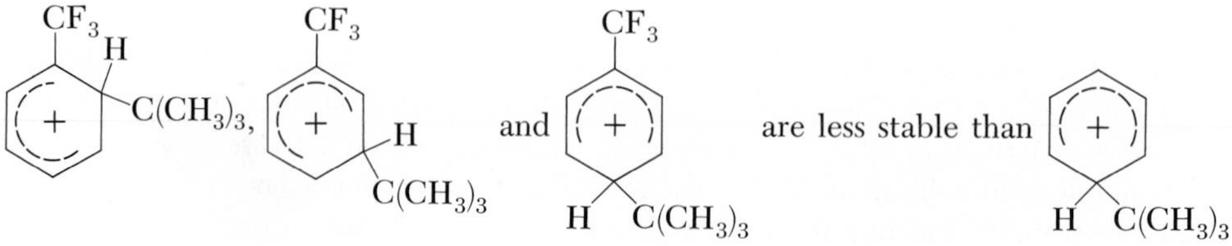

The effect may be quite severe. For example, Friedel-Crafts reactions usually will not occur on aromatics more deactivated than aryl halides.

NO_2 $\xrightarrow[AlCl_3]{RC(O)Cl}$ no reaction

OCH_3 NO_2 $\xrightarrow[AlCl_3]{RC(O)Cl}$ OCH_3 NO_2 C O R + HCl

On the other hand, groups that stabilize the developing charge in the activated complex, or in the resulting σ complex, activate the ring. Aside from the alkyl groups, the common activating groups possess at least one non-bonding electron pair on that atom which bonds the substituent to the ring. Such groups include —NH_2, —OH, —OR, —NHR, —NR_2, and —NH—C(O)R. While a substituent such as —OCH_3 is inductively electron-withdrawing, it is electron-donating by resonance; since the methoxy group is a potent activator, it is clear that the resonance effect outweighs the inductive effect in this instance.

OCH_3 + H E

inductive effect of —OCH_3 is deactivating but unimportant

$\ddot{O}$—CH_3 ⊕ H R ↔ ⊕ :O—CH_3 H R ↔ etc.

resonance effect of —OCH_3 is activating and important

OCH_3 is attacked more rapidly by $Br_2/FeBr_3$ than is H

[O—CH_3 H E]⊕ is more stable than H + H E

One can draw resonance structures for the halobenzenes that are similar to those presented for a substituted anisole; yet substitution in halobenzenes is at least partly controlled by (deactivating) inductive effects. Why are halogens deactivating substituents?

inductive effect of —Cl is deactivating

resonance effect of —Cl is activating

One explanation is that resonance of the type shown above requires overlap of a $2p$ orbital on carbon with a $3p$ orbital on chlorine. Since different main quantum levels are involved, this overlap will not be as effective as carbon-nitrogen or carbon-oxygen overlap. Thus, the inductive effect outweighs the resonance effect.

Orientation. Existing substituents usually are classified as either *o,p*-directors (causing an incoming substituent to attack *ortho* or *para* to the existing substituent) or *m*-directors (causing further substitution at the *meta* position). Excepting halogens, the common *o,p*-directors are also ring activators; all *m*-directors are ring deactivators. The halogens are *o,p*-directors but ring deactivators. Most electrophilic aromatic substitutions are carried out under conditions that restrict the amount of isomerization undergone by the initial reaction product. Therefore, the observed product distribution (*i.e.*, the $o:m:p$ ratio) is determined by the comparative ease of attack of the electrophile at positions *o*, *m*, and *p* to an existing substituent. Stated differently, *electrophilic aromatic substitution usually is kinetically controlled* and not thermodynamically controlled. Let us now explain how a typical *meta*-director, the nitro group, and a typical *o,p*-director, the methoxy group, can control the attack on a substituted benzene ring.

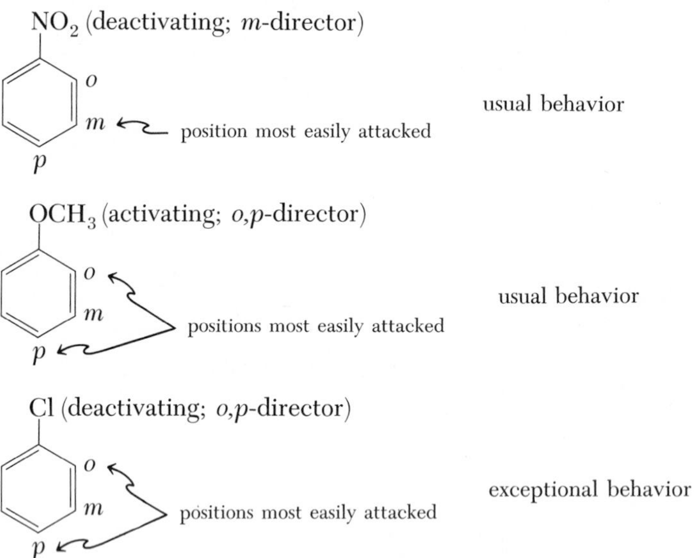

Again approximating activated complexes by the corresponding benzenonium ions, we see that the presence on the ring of a nitro group places a positive charge directly on a carbon bearing a positive charge when the electrophile attacks *o* or *p* to $—NO_2$. However, the two positive charges are never juxtaposed when the electrophile attacks *m* to $—NO_2$. Thus, the *m*-substituted intermediate is most stable (the activated complex leading to it is most stable and, therefore, it is formed most rapidly), not because the *meta* position is activated but because it is deactivated less than are the *o* and *p* positions. Deactivating *meta* directors deactivate all ring positions relative to unsubstituted benzene!

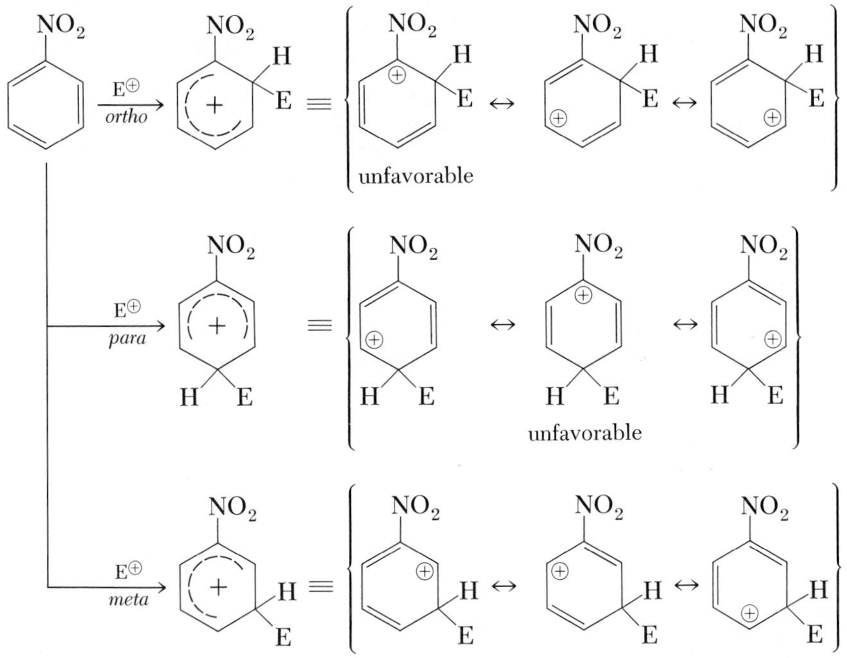

NO$_2$ E$^{\oplus}$ ortho; para; meta; unfavorable

An activating group, such as methoxy (—OCH_3), stabilizes the positive charge developing in the activated complex whenever that charge can be placed on the carbon bearing the methoxy group. Since these extra resonance structures are possible only for *o* or *p* attack, —OCH_3 and other electron pair donors are *o,p*-directors. (This was implicit in our discussion of the reactivity of anisole.)

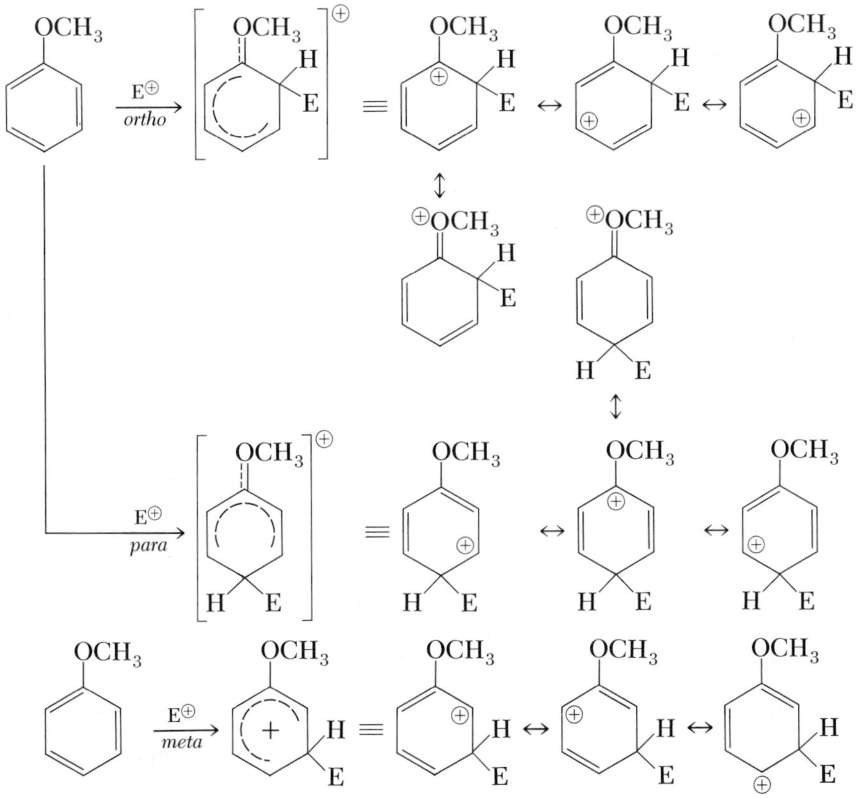

OCH$_3$ E$^{\oplus}$ ortho; para; meta

Experimental *facts* lead to the conclusion that while it is the inductive effect of the halogens that dictates the deactivation of the ring, it is their ability to act as electron-pair donors that renders them *o,p*-directors.

We can explain why an alkyl group is an *o,p*-director by referring back to our discussion of the protonation of mesitylene. Only when the incoming electrophile attacks *o* or *p* to an extant —CH_3 will the methyl group be capable of dispersing the positive charge.

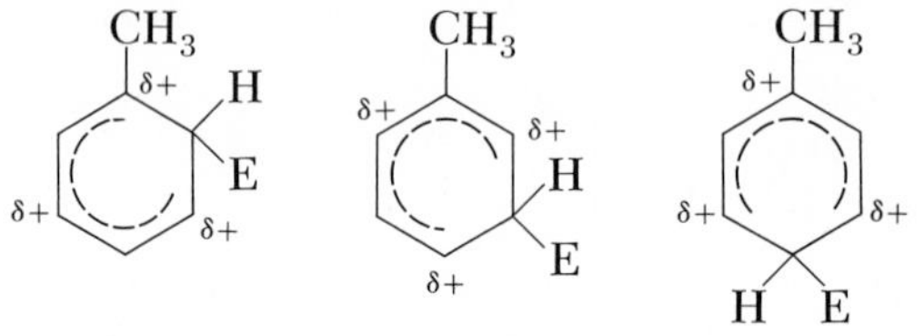

CH_3 δ+ H E δ+ δ+ — *ortho*

CH_3 δ+ δ+ H E δ+ — *meta*

CH_3 δ+ δ+ δ+ H E — *para*

Several examples of this "directive effect" follow:

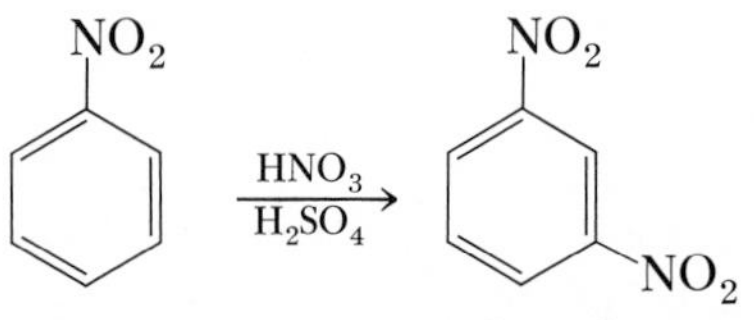

nitrobenzene $\xrightarrow[H_2SO_4]{HNO_3}$ *m*-dinitrobenzene (93%)

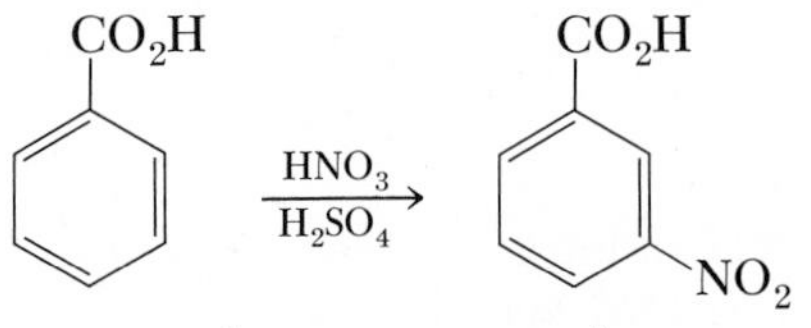

benzoic acid $\xrightarrow[H_2SO_4]{HNO_3}$ *m*-nitrobenzoic acid (80%)

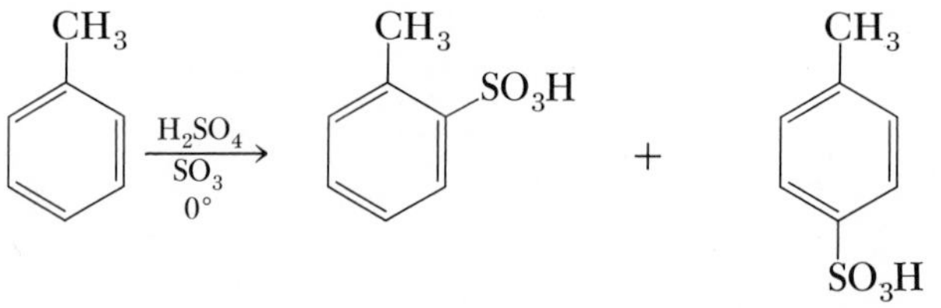

toluene $\xrightarrow[SO_3,\ 0°]{H_2SO_4}$ *o*-toluenesulfonic acid (43%) + *p*-toluenesulfonic acid (53%)

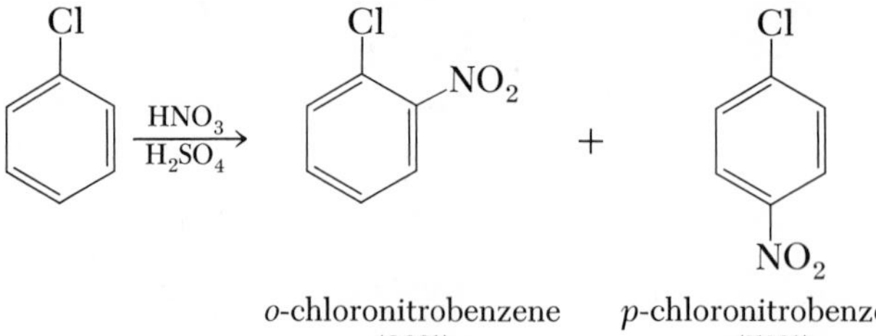

chlorobenzene $\xrightarrow[H_2SO_4]{HNO_3}$ *o*-chloronitrobenzene (30%) + *p*-chloronitrobenzene (70%)

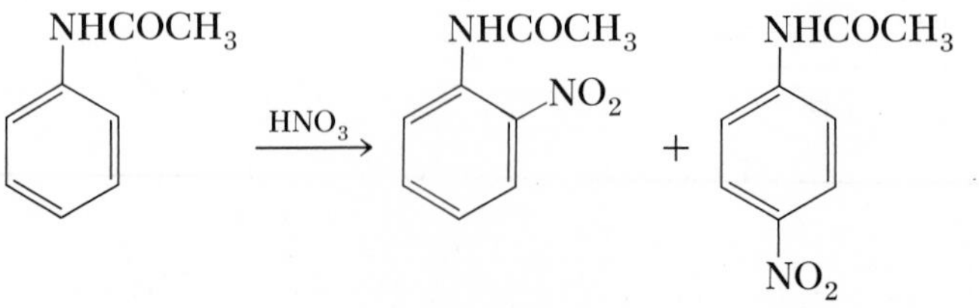

acetanilide ($NHCOCH_3$) $\xrightarrow{HNO_3}$ *o*-nitroacetanilide (5%) + *p*-nitroacetanilide (95%)

11. Explain, with the aid of structures, why —OH is an *ortho,para* director. With the aid of similar structures, determine whether the —O$^{\ominus}$ group should be an *o,p*- or an *m*-director.
12. Even in dilute acidic solution, aniline, which exists mainly as the anilinium ion ($C_6H_5NH_3^{\oplus}$), undergoes *o,p*-substitution. Account for this.

ALL OR NOTHING-AT-ALL? While groups are classified, for convenience, as *o,p*- or *m*-directors, there are very few reactions that produce only *o, p*- or only *m*-substituted product. Most reactions produce all three, the "directing influence" of the substituent already on the ring simply indicating which is produced in excess. Moreover, the composition of the *ortho,para* portion of the product rarely is exactly 67% *ortho* and 33% *para*, the statistical distribution. For example, nitration of bromobenzene produces 62% *p*-nitrobromobenzene, 37% *o*-nitrobromobenzene, and 1% *m*-nitrobromobenzene.

Br — $[NO_2^{\oplus}]$ → Br, NO_2 (37%) + Br, NO_2 (62%) + Br, NO_2 (1%)

The complete distinction between *o,p*- and *m*-direction rarely is observed because there always are a few energetic collisions that provide sufficient energy to surmount the higher barrier and proceed through the more energetic activated complex. Data in Table 15–3 bear this out and also show us that most substituents produce more *para* than *ortho* substitution. The reason for this is largely steric in origin—there is less hindrance in the activated complex leading to *para* attack than in the one leading to *ortho* attack. This may be seen clearly if one approx-

TABLE 15–3 Isomer Distribution in Nitration of Benzene Derivatives

G — $[NO_2^{\oplus}]$ → G, NO_2

G	PRODUCT DISTRIBUTION		
	% *ortho*	% *para*	% *meta*
$—CH_3$	58	38	4
—F	12	88	—
—Cl	30	70	—
—Br	37	62	1
—I	38	60	2
$—CO_2H$	19	1	80
$—CONH_2$	27	<3	70
$—\overset{\oplus}{N}(CH_3)_3$	—	11	89

imates the geometry of the activated complex leading to benzenonium ion formation by the geometry of the benzenonium ion itself.

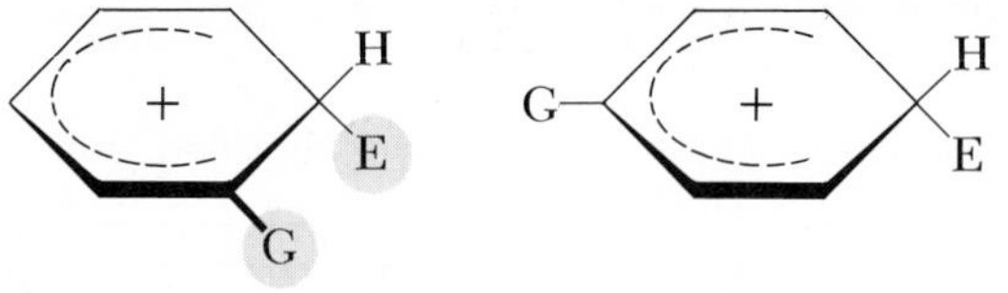

ortho attack *para* attack

15.6 SYNTHESIS OF SUBSTITUTED BENZENES

In designing the synthesis of substituted benzenes, it is essential to consider the directive influence of a substituent already on the ring. For example, chlorination of nitrobenzene produces *m*-chloronitrobenzene, while nitration of chlorobenzene produces *o*- and *p*-chloronitrobenzene. Thus, starting with benzene, nitration must precede chlorination to yield the *meta* isomer but must follow it to yield *ortho* or *para* isomer.

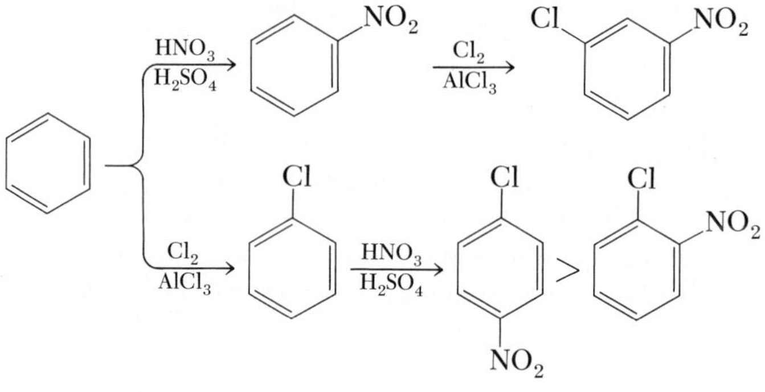

The syntheses that follow illustrate several points, including:

(1) When a benzene ring contains two activating groups, the more powerful activator controls the entry of a third substituent.
(2) If a ring contains an activator and a deactivator, the activating group controls the entry of a third group.
(3) It is quite unlikely that two groups *meta* to one another will direct a group between them, regardless of their directing influence.
(4) The reversibility of the sulfonation reaction provides a useful means of temporarily blocking a position and of introducing a deuterium onto a benzene ring.

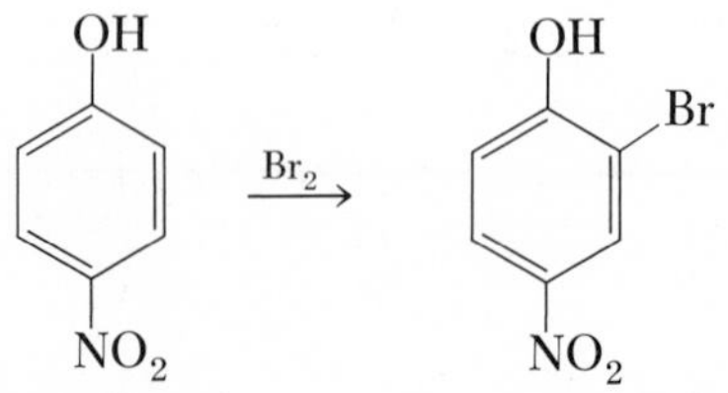

4-nitrophenol 2-bromo-4-nitrophenol

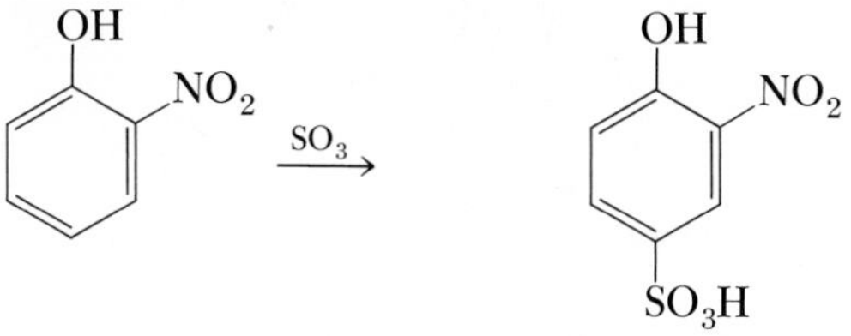

$\xrightarrow{SO_3}$

2-nitrophenol → 4-hydroxy-3-nitrobenzenesulfonic acid

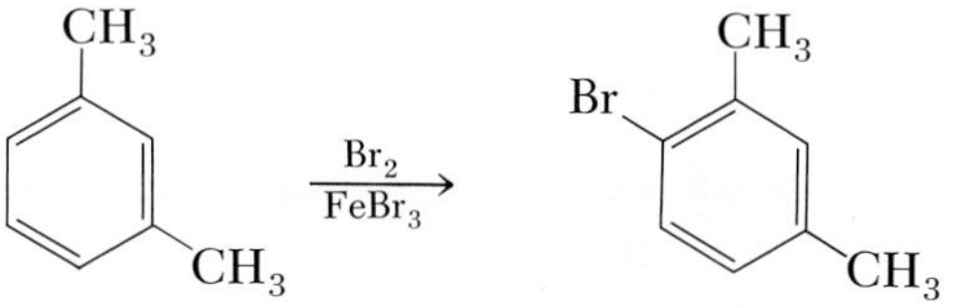

$\xrightarrow[FeBr_3]{Br_2}$

1-bromo-2,4-dimethylbenzene

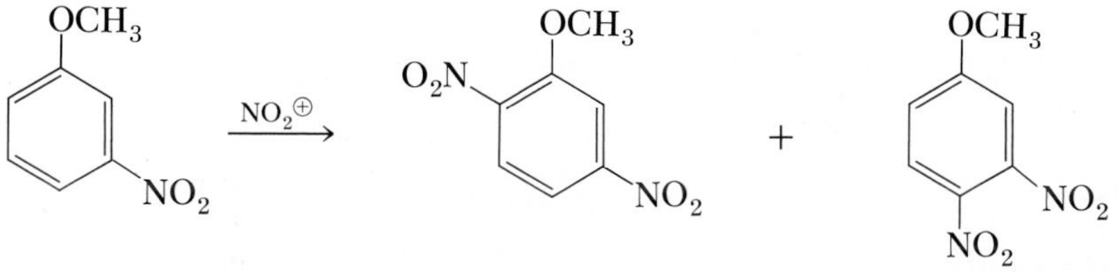

$\xrightarrow{NO_2^{\oplus}}$

1-methoxy-2,5-dinitrobenzene + 1-methoxy-3,4-dinitrobenzene

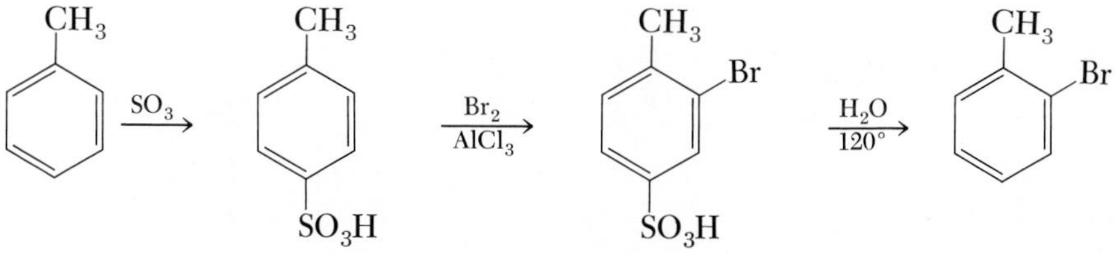

$\xrightarrow{SO_3}$ $\xrightarrow[AlCl_3]{Br_2}$ $\xrightarrow[120°]{H_2O}$

toluene → *p*-toluenesulfonic acid → 3-bromo-4-methylbenzenesulfonic acid → *o*-bromotoluene (free of *p*-isomer)

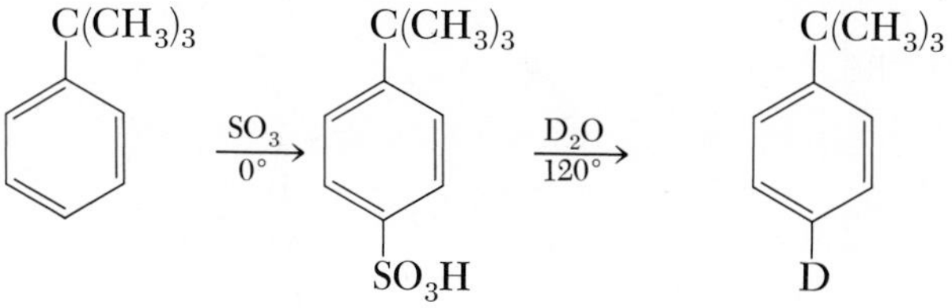

$C(CH_3)_3$ $\xrightarrow[0°]{SO_3}$ $\xrightarrow[120°]{D_2O}$

p-deuterio-*t*-butylbenzene

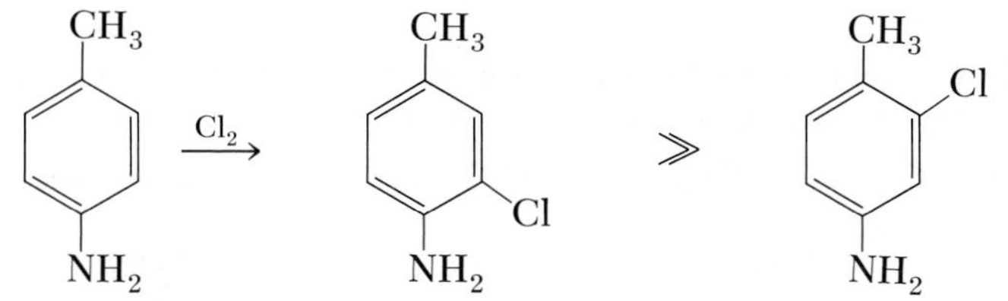

$\xrightarrow{Cl_2}$

p-toluidine → 2-chloro-4-methylaniline ≫ 3-chloro-4-methylaniline

$\xrightarrow[FeBr_3]{Br_2}$

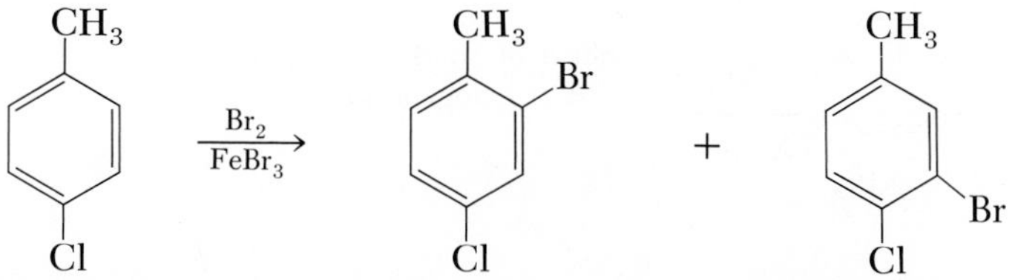

p-chlorotoluene → 2-bromo-4-chlorotoluene + 3-bromo-4-chlorotoluene

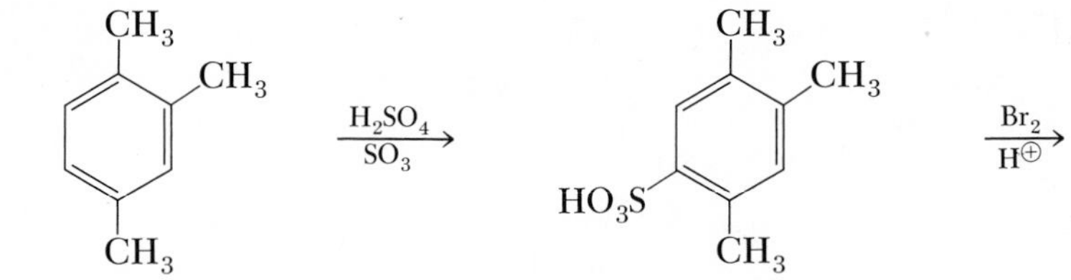

1,2,4-trimethylbenzene $\xrightarrow[SO_3]{H_2SO_4}$ 2,4,5-trimethylbenzenesulfonic acid $\xrightarrow[H^{\oplus}]{Br_2}$

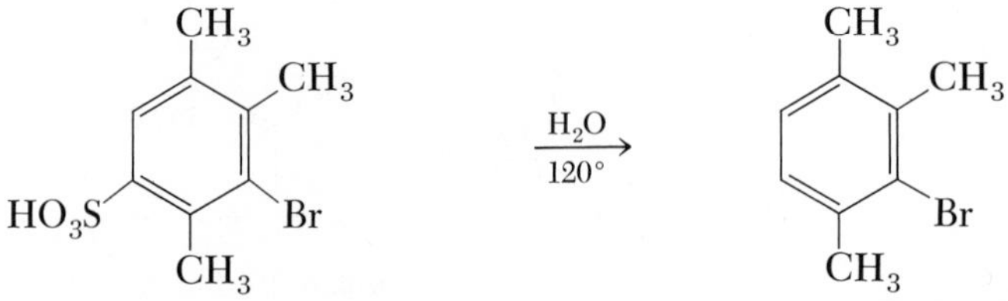

3-bromo-2,4,5-trimethylbenzenesulfonic acid $\xrightarrow[120°]{H_2O}$ 3-bromo-1,2,4-trimethylbenzene

13. Suggest the major product(s) from the following reactions.

(a) 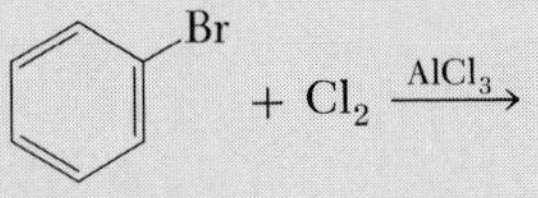 $C_6H_5Br + Cl_2 \xrightarrow{AlCl_3}$

(b) $C_6H_5CH_3 + CH_3CH_2C(O)Cl \xrightarrow{AlCl_3}$

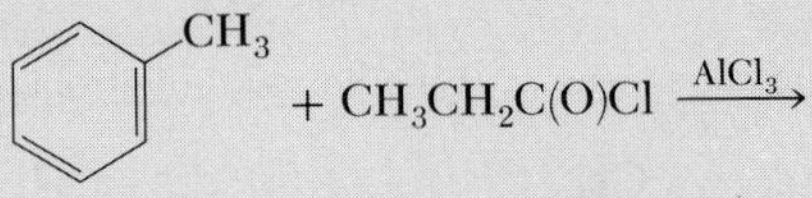

(c) $C_6H_5C(CH_3)_3 + I_2 \xrightarrow{HNO_3}$

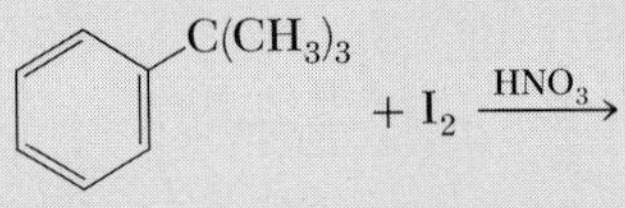

(d) 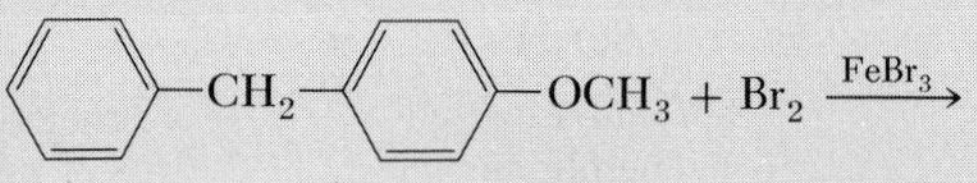 $C_6H_5\text{—}CH_2\text{—}C_6H_4\text{—}OCH_3 + Br_2 \xrightarrow{FeBr_3}$

(e) 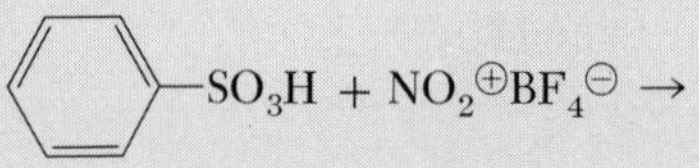 $C_6H_5\text{—}SO_3H + NO_2^{\oplus}BF_4^{\ominus} \rightarrow$

15.7 A COUPLE OF GOODIES

In this section we will examine two reactions that are not electrophilic aromatic substitutions. They both deal with the reactivity of carbons adjacent to a benzene ring, so-called *benzylic* carbons. These carbons, by virtue of being adjacent to a benzene ring, are more reactive than are other alkyl side chain carbons.

Halogenation of Alkyl Side Chains. While halogenation of an alkylbenzene in the presence of aluminum chloride (ionic conditions) results in ring substitution, free radical conditions result in side chain substitution.

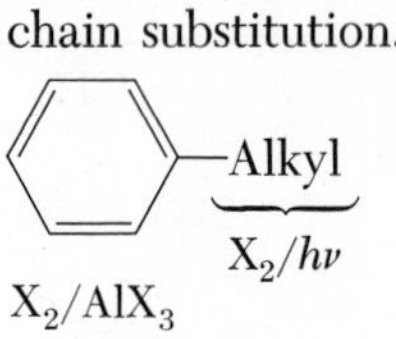

$C_6H_5\text{—Alkyl}$ (ring: X_2/AlX_3; side chain: $X_2/h\nu$) effect of conditions upon site of halogenation

In this free radical chain process the halogen radical abstracts the *benzylic* hydrogen, *i.e.*, the hydrogen on the carbon adjacent to the benzene ring, to produce a resonance stabilized benzylic radical.

$$X_2 \xrightarrow{h\nu} 2X\cdot$$

$$X\cdot + C_6H_5CH_2CH_2CH_3 \rightarrow C_6H_5\dot{C}HCH_2CH_3 + HX$$

$$X_2 + C_6H_5\dot{C}HCH_2CH_3 \rightarrow C_6H_5CHXCH_2CH_3 + X\cdot$$

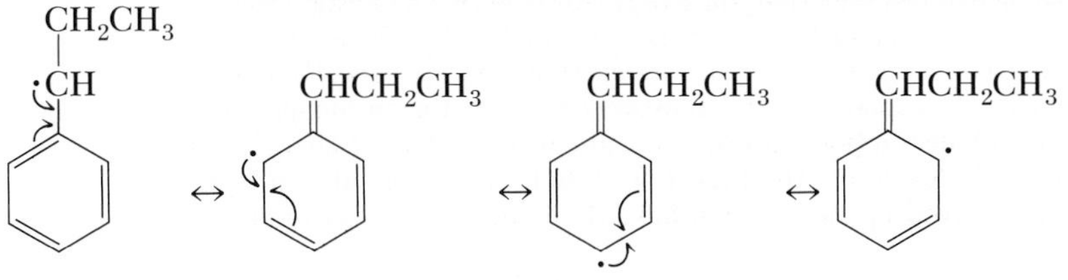

resonance stabilized benzylic radical—the fish-hook arrow indicates delocalization of one electron

The scheme in Figure 15–4 reveals how this may be used to synthesize a number of benzene derivatives.

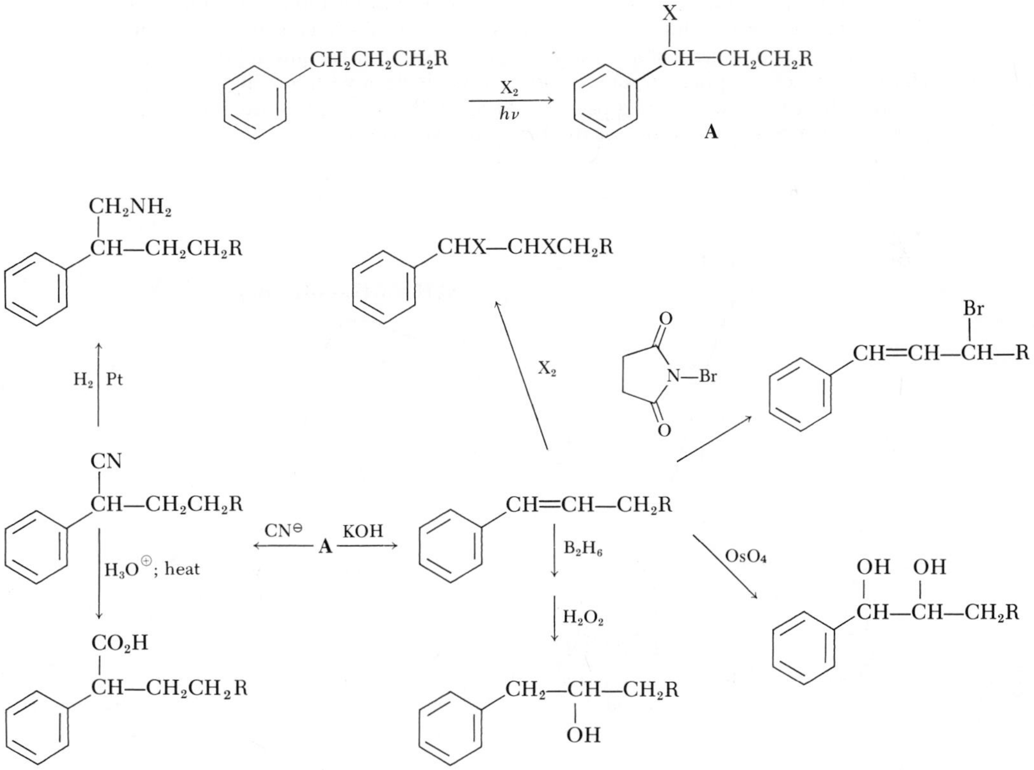

Figure 15-4 A series of syntheses beginning with a benzylic halide. Suggest mechanisms for those conversions on the right side of the scheme.

When ethylbenzene reacts with $Br_2/h\nu$, it produces only 1-bromo-1-phenylethane; however, reaction with $Cl_2/h\nu$ produces both 1-chloro-1-phenylethane and 2-chloro-1-phenylethane. Why?

CH_2CH_3 (benzene ring) $\xrightarrow{Br_2/h\nu}$ (benzene ring)$CHBr—CH_3$ 100%

$\xrightarrow{Cl_2/h\nu}$ (benzene ring)$CHClCH_3$ (91%) + (benzene ring)CH_2CH_2Cl (9%)

This difference is consistent with our earlier observations (p. 78) that Cl· is more reactive, hence less selective, than Br·. However, we had established the ratio of reactivity of 3°, 2°, and 1° hydrogens to abstraction by Cl· as 5.0:3.8:1.0. On this scale the benzylic hydrogen has a reactivity of 1.3 *even though it is resonance stabilized.* To explain this, we must recognize that even though a bond is cleaved by radical abstraction, and the radical carries no formal charge, some charge character will develop in the transition state because the halogen, hydrogen, and carbon involved in the reaction have different electronegativities.

$$\text{R—H} + \text{X}\cdot \rightarrow [\overset{\delta+}{\text{R}}\cdots\text{H}\cdots\overset{\delta-}{\text{X}}]^{\cdot} \rightarrow \text{R}\cdot + \text{H—X}$$

transition state

We must conclude that with the highly reactive chlorine radical the transition state is reached early along the energy profile (Figure 15–5). This means that the C—H bond is still largely intact and the organic residue has acquired little radical character. Since the phenyl group is inductively electron-withdrawing, and since this destabilizes any developing positive charge in the transition state, the phenyl group decreases the reactivity of the benzylic hydrogens.

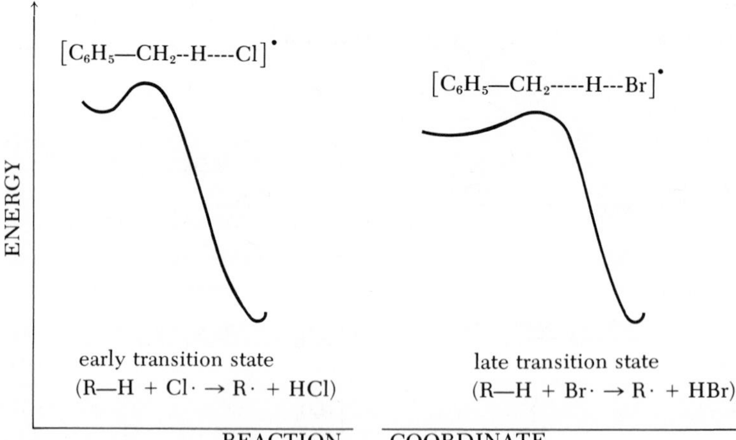

Figure 15-5 The effect of the radical on the transition state for hydrogen abstraction. The more reactive chlorine radical promotes a transition state with less radical character at the benzylic carbon than does the less reactive bromine radical.

The transition state for bromination, involving the less reactive bromine radical, is reached later along the energy profile (Figure 15–5). At the transition state considerable C—H bond cleavage has occurred and the ability of the phenyl group to delocalize the odd electron becomes the dominating factor.

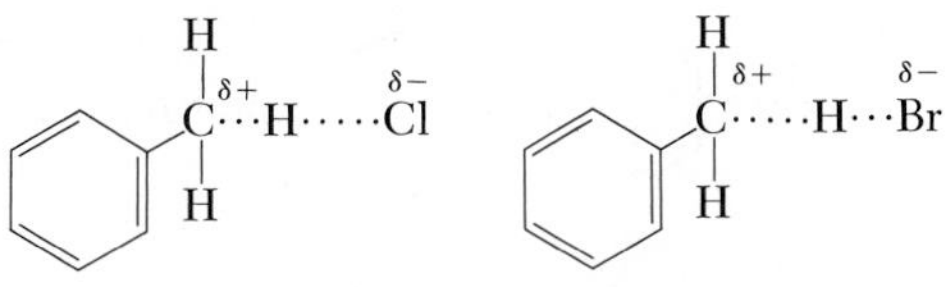

C_6H_5–$\overset{H}{\underset{H}{C}}$ $\delta+$ ···H·····Cl $\delta-$

transition state for chlorination

C_6H_5–$\overset{H}{\underset{H}{C}}$ $\delta+$ ·····H···Br $\delta-$

transition state for bromination

Evidence that the free-radical chlorination of toluene is subject to polar effects is found in the following order of reactivity toward free radical chlorination. Note how increasing the electronegativity of the substituent decreases the ease of chlorination.

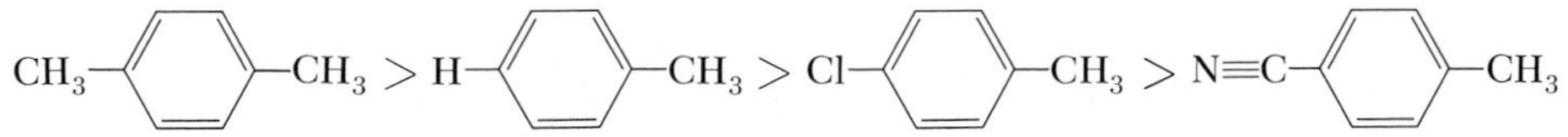

CH_3–C_6H_4–CH_3 > H–C_6H_4–CH_3 > Cl–C_6H_4–CH_3 > N≡C–C_6H_4–CH_3

rate of radical chlorination of —CH_3

Side Chain Oxidation. While alkanes and benzene are relatively inert to oxidation with potassium permanganate in aqueous alkali, alkyl groups attached to benzene rings are oxidized to carboxy groups under forcing conditions. (The oxidation requires several hours at the reflux temperature.) However, alkyl groups that lack a benzylic hydrogen, *e.g.*, *t*-butylbenzene, are not readily oxidized under these conditions.

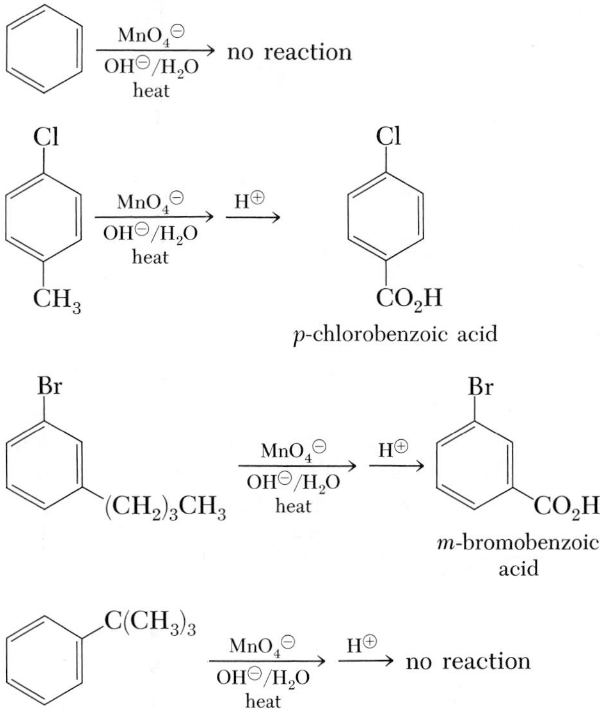

$$C_6H_6 \xrightarrow[\text{heat}]{MnO_4^{\ominus},\ OH^{\ominus}/H_2O} \text{no reaction}$$

$$p\text{-}Cl\text{–}C_6H_4\text{–}CH_3 \xrightarrow[\text{heat}]{MnO_4^{\ominus},\ OH^{\ominus}/H_2O} \xrightarrow{H^{\oplus}} p\text{-}Cl\text{–}C_6H_4\text{–}CO_2H$$

p-chlorobenzoic acid

$$m\text{-}Br\text{–}C_6H_4\text{–}(CH_2)_3CH_3 \xrightarrow[\text{heat}]{MnO_4^{\ominus},\ OH^{\ominus}/H_2O} \xrightarrow{H^{\oplus}} m\text{-}Br\text{–}C_6H_4\text{–}CO_2H$$

m-bromobenzoic acid

$$C_6H_5\text{–}C(CH_3)_3 \xrightarrow[\text{heat}]{MnO_4^{\ominus},\ OH^{\ominus}/H_2O} \xrightarrow{H^{\oplus}} \text{no reaction}$$

Since carboxylic acid derivatives of benzene are all solids, this procedure has the advantage of being able to convert alkylated benzenes, which are often liquids, to solid derivatives.

o-xylene (bp 144°) $\xrightarrow{[O]}$ phthalic acid, mp 231°

m-xylene (bp 139°) $\xrightarrow{[O]}$ isophthalic acid, mp 348°

p-xylene (bp 138°) $\xrightarrow{[O]}$ terephthalic acid, mp 300°

Functional groups that are subject to oxidation under mild conditions suffer simultaneous oxidation under these forcing conditions.

$$p\text{-}CH_3C_6H_4CHO \xrightarrow[OH^{\ominus},\ heat]{MnO_4^{\ominus}} \xrightarrow{H^{\oplus}} p\text{-}HO_2CC_6H_4CO_2H \xleftarrow{H^{\oplus}} \xleftarrow[OH^{\ominus},\ heat]{MnO_4^{\ominus}} p\text{-}CH_3C_6H_4CH_2OH$$

$$p\text{-}H_3CC_6H_4CH{=}CH_2 \xrightarrow[OH^{\ominus},\ heat]{MnO_4^{\ominus}} \xrightarrow{H^{\oplus}} p\text{-}HO_2CC_6H_4CO_2H$$

Aromatic carboxylic acids will lose carbon dioxide upon heating to high temperatures in the presence of base ($RCO_2H \rightarrow R{-}H + CO_2$). This process is termed a decarboxylation and, while it will be discussed further in Chapter 18, it will be used here to demonstrate its usefulness in the synthesis of arenes. The sequence of alkyl side chain oxidation followed by decarboxylation can completely remove an alkyl side chain from a benzene ring that does not bear any other easily oxidized functional groups. One application of this idea is the synthesis of 1,3,5-trinitrobenzene (TNB) from 2,4,6-trinitrotoluene (TNT).

$$C_6H_5CH_3 \xrightarrow[H_2SO_4\ (excess)]{HNO_3} \underset{TNT}{2,4,6\text{-}(O_2N)_3C_6H_2CH_3} \xrightarrow{[O]} 2,4,6\text{-}(O_2N)_3C_6H_2CO_2H \xrightarrow[heat]{base} \underset{TNB}{1,3,5\text{-}(O_2N)_3C_6H_3}$$

1,3,5-Trinitrobenzene cannot be prepared by the direct nitration of benzene because of the deactivating effect of two nitro groups present on an intermediate in the proposed sequence.

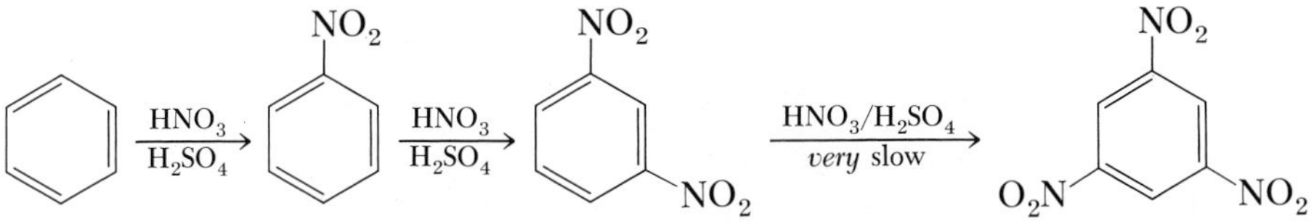

NO_2 NO_2 NO_2

$\xrightarrow[H_2SO_4]{HNO_3}$ $\xrightarrow[H_2SO_4]{HNO_3}$ NO_2 $\xrightarrow[\text{very slow}]{HNO_3/H_2SO_4}$ O_2N NO_2

TNB

14. What arene derivatives are expected from the oxidation of the following with hot, alkaline potassium permanganate followed by acidification?

(a) 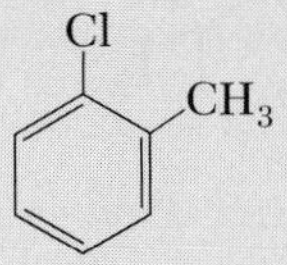 Cl, CH_3

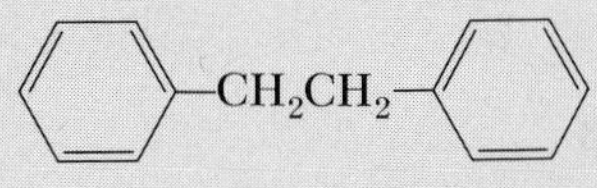

(b) 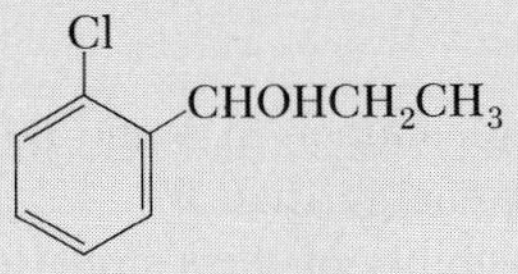 Cl, $CHOHCH_2CH_3$

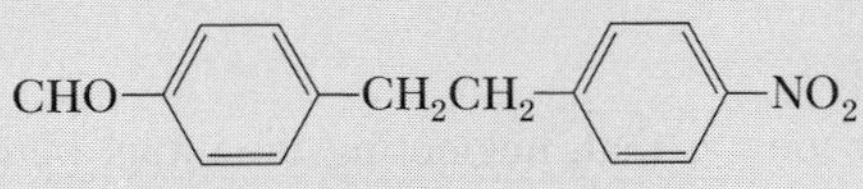

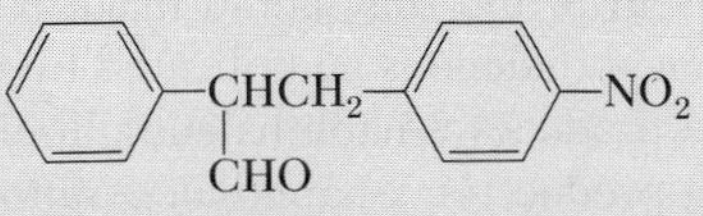

(c) 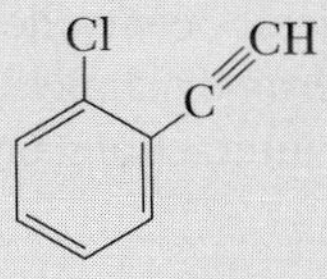 Cl, C≡CH

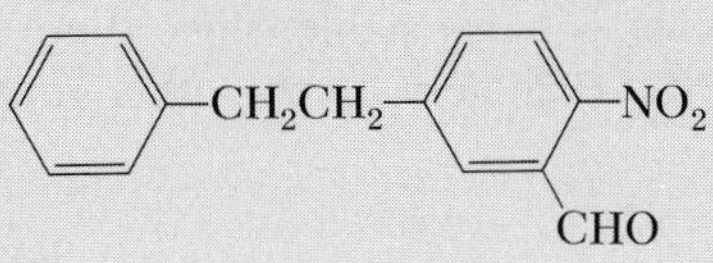

(d)

(e) $-CH_2CH_2-$

(f) $CHO-$ $-CH_2CH_2-$ $-NO_2$

(g) $-CHCH_2-$ $-NO_2$, CHO

(h) $-CH_2CH_2-$ $-NO_2$, CHO

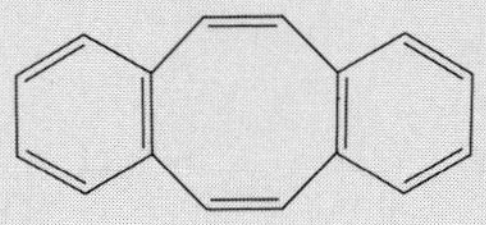

15.8 NAPHTHALENE

Compounds possessing two or more benzene rings fused together are termed polynuclear aromatic hydrocarbons. This section will consider the most abundant of these, naphthalene, $C_{10}H_{10}$.

Reactions. When subjected to attack by an electrophile, naphthalene produces a 1- or α-substituted product.

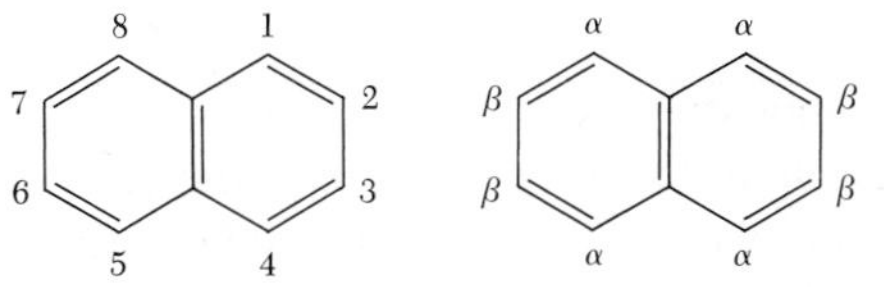

8 1 7 2 6 3 5 4 — α α β β β β α α — positional nomenclature of naphthalene

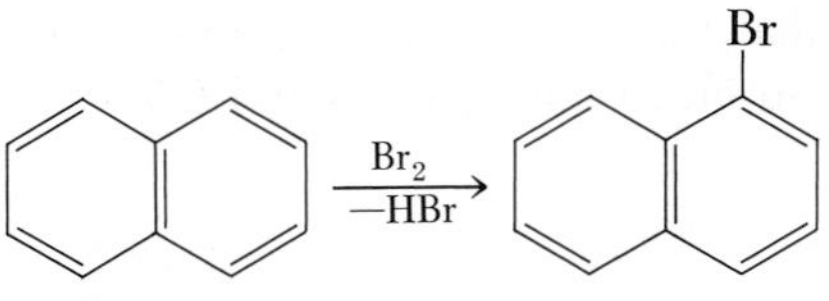

$\xrightarrow[-HBr]{Br_2}$ Br — catalyst not required for halogenation

This behavior is rationalized by comparing the two cations produced by α- and β-substitution and noting that more Kekulé forms *containing a benzene ring* can be drawn for the cation leading to α-substitution.

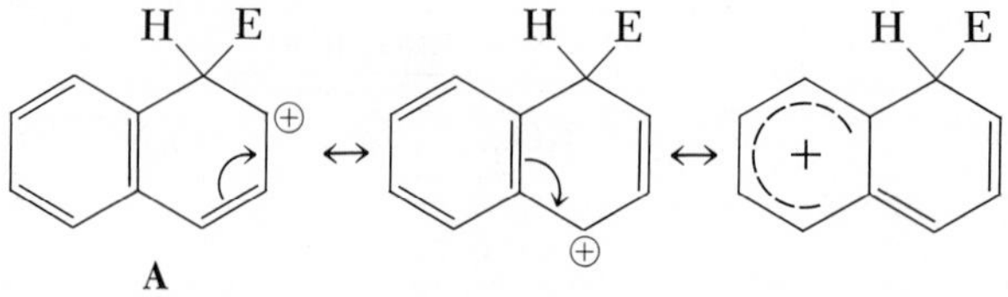

H E ⊕ ↔ H E ⊕ ↔ H E +

A

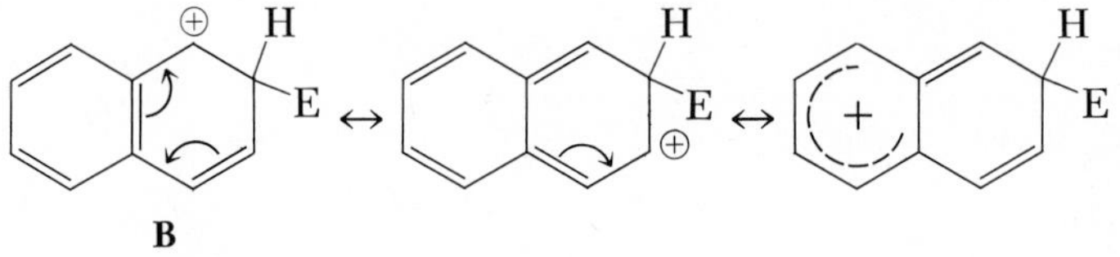

⊕ H E ↔ H E ⊕ ↔ + H E

B

Another view is that, neglecting the intact benzene ring common to **A** and **B,** the carbocation is allylic only in **A** and this should make **A** of lower energy than **B.**

Sulfonation of naphthalene is complicated by the reversibility of the sulfonation process. At 50° naphthalene produces 1-naphthalenesulfonic acid, the kinetically controlled product. At 150° the major product is 2-naphthalenesulfonic acid, the thermodynamically favored product. The β-isomer is more stable, since there are no repulsive interactions between the substituent and the C_α-H moiety on the other ring.

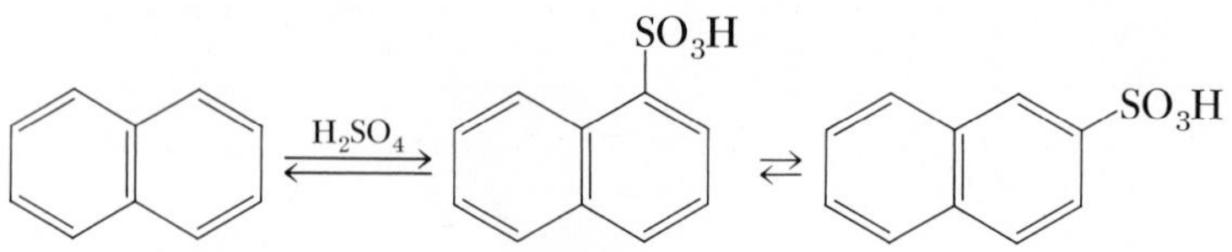

SO_3H $\xrightleftharpoons{H_2SO_4}$ ⇄ SO_3H

kinetic product thermodynamic product

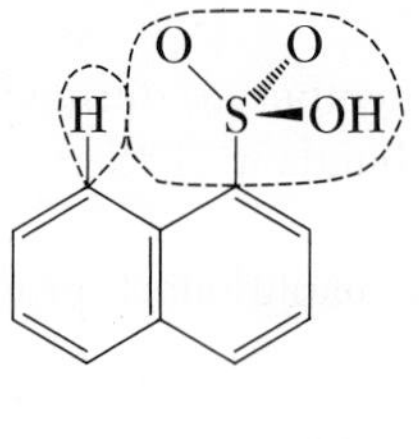

O O H S OH

destabilizing interaction present in α isomer

If the naphthalene bears an activating substituent on C1, an attacking electrophile enters at C4 and C2 (C4 > C2). A deactivating group at C1 directs attack to the other ring, favoring C8 and C5 (C8 > C5). Surprisingly, an activating group at C2 favors C1, C6, and C3 (C1 > C6 ≫ C3). While there is no simple reason for C6 being preferred over C3, the preference surely is not steric, considering that the major product arises from attack at C1. Finally, a deactivating group at C2 directs an electrophile to C8 and C5 (C8 > C5). These patterns are summarized in Figure 15–6.

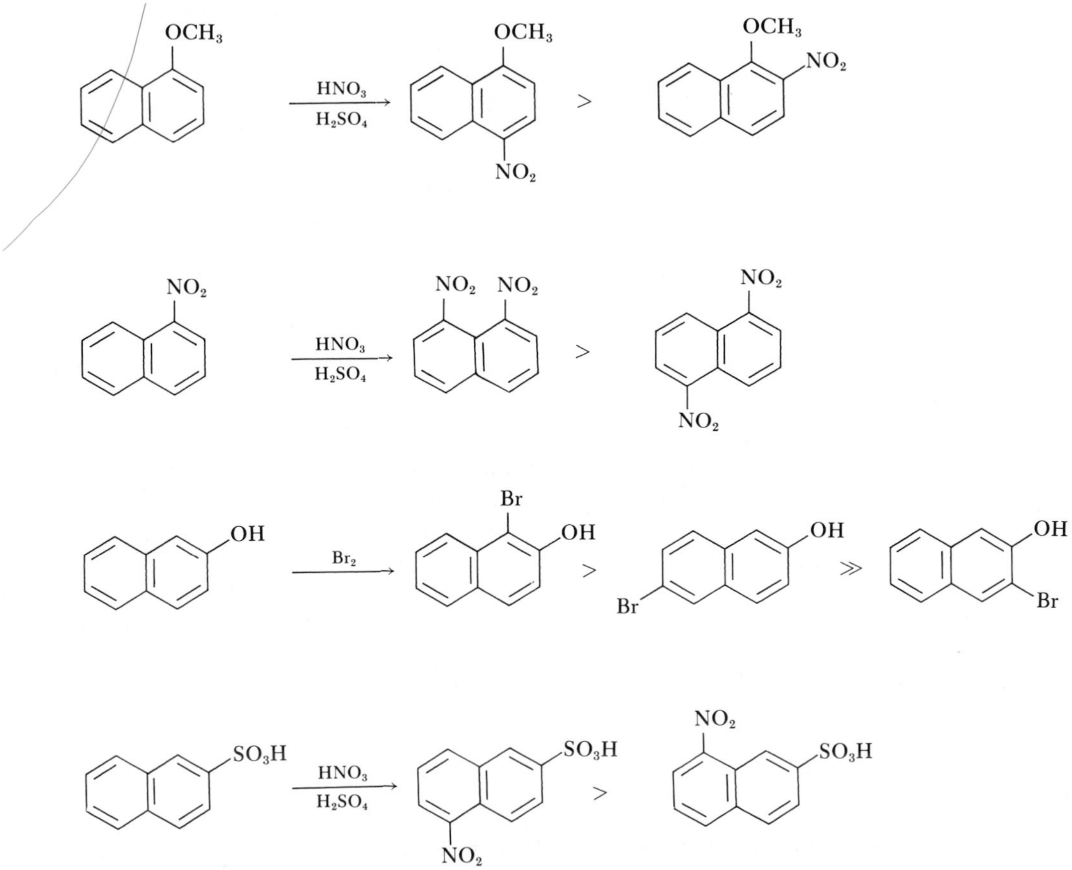

Figure 15-6 The effect of substituents upon further substitution in naphthalene.

If the naphthalene ring contains both a powerful activator and a deactivator, an incoming electrophile is led by the activator.

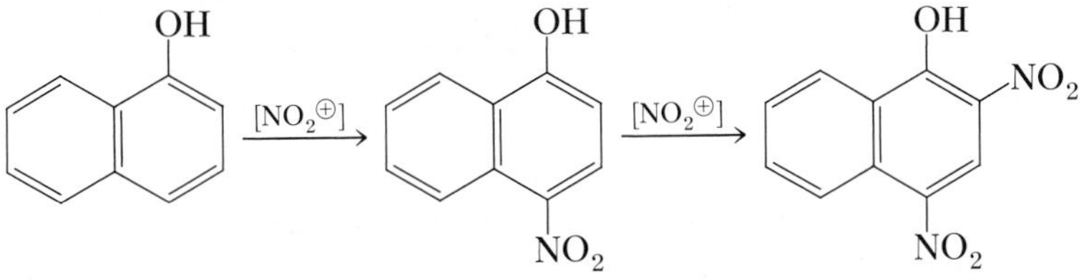

Martius Yellow
(a dye)

The complexities associated with naphthalene substitution are evidenced in the acylation of naphthalene with acetyl chloride and $AlCl_3$. In nitrobenzene solvent the major product is 2-acetylnaphthalene, while in carbon disulfide the major product is 1-acetylnaphthalene. This is presumably due to the larger bulk of the $CH_3COCl{:}AlCl_3{:}C_6H_5NO_2$ complex compared to the $CH_3COCl{:}AlCl_3{:}CS_2$ complex. (What effect should size have?)

$\xrightarrow[AlCl_3]{CH_3COCl}$ ($C_6H_5NO_2$) → 2-acetylnaphthalene; (CS_2) → 1-acetylnaphthalene

Polynuclear aromatic compounds are, in general, more reactive than benzene. This is ascribed to the lower resonance energy *per ring* in the polycyclic system. Thus, naphthalene is oxidized by $K_2Cr_2O_7/H_2SO_4$ to phthalic acid, *via* a class of compounds called *quinones* (discussed in Chapter 22).

$\xrightarrow[H_2SO_4]{Cr_2O_7^{-2}}$ 1,4-naphthoquinone $\xrightarrow[H_2SO_4]{Cr_2O_7^{-2}}$ phthalic acid (CO_2H, CO_2H)

If the ring is substituted, oxidation occurs at the more activated ring.

1-nitronaphthalene (NO_2) $\xrightarrow[H_2SO_4]{Cr_2O_7^{-2}}$ 3-nitrophthalic acid (NO_2, CO_2H, CO_2H)

1-methoxynaphthalene (OCH_3) $\xrightarrow[H_2SO_4]{Cr_2O_7^{-2}}$ phthalic acid (CO_2H, CO_2H)

As with benzene, catalytic reduction of naphthalene occurs only under strenuous conditions. The partial reduction of naphthalene, to tetralin, is achieved by reduction with a sodium amalgam using ethanol as a hydrogen source.

$\xrightarrow[\text{heat; pressure}]{H_2/Ni}$ decalin

$\xrightarrow[EtOH]{Na \cdot Hg}$ tetralin

When we proceed to still more complex polynuclear hydrocarbons, the number of possible and observed products can increase sharply. Phenanthrene produces five different mononitration products; the relative rates of their formation are shown below. Because of the multiplicity of products, polynuclear hydrocarbons will not be considered further.

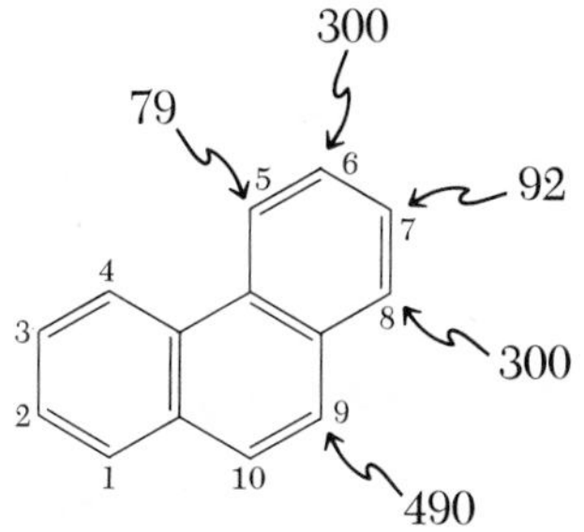

300
79
92
300
490

relative reactivity in phenanthrene nitration; a value of 1 represents the reactivity of a single position on benzene

Naphthalene in the Body. We have noted that our environment is "contaminated" with hydrocarbons. Although we are properly concerned with man's propensity for spoiling the environment, you should now be aware that the presence of hydrocarbons in the environment is a normal situation. (The problem is in the *amount* that is present.) Natural oil spills, gas evolution from caves, mines, and natural fissures, and decaying animal and vegetable matter all supply hydrocarbons to our biosphere.

One piece of evidence has suggested to *some* scientists that members of the animal kingdom may have been exposed to these hydrocarbons for hundreds of thousands of years. This is the ability of animals to metabolize some hydrocarbons when they are intentionally introduced into those animals. A specific example of this biological processing is the ability of rats and rabbits to metabolize naphthalene to *trans*-1,2-dihydroxy-1,2-dihydronaphthalene. This glycol can exist as (+)- and (−)-forms, and it is an interesting comment on the stereochemical specificity of enzymes that rats produce the levorotatory glycol, while rabbits produce the dextrorotatory glycol.[a]

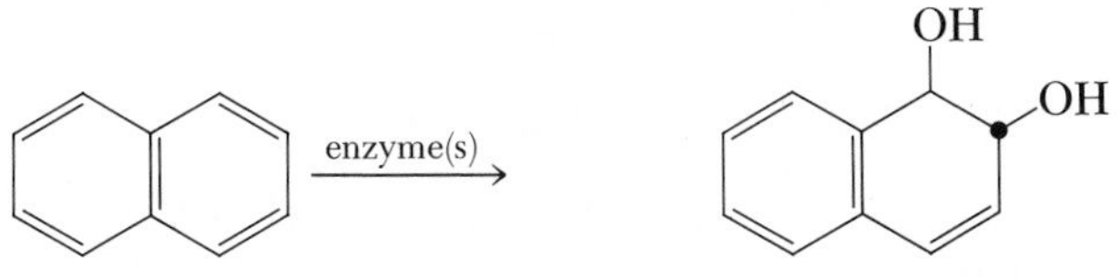

enzyme(s) →

OH
OH

trans-1,2-dihydroxy-1,2-dihydronaphthalene

biological oxidation of naphthalene

15.9 HETEROCYCLES

Five-membered aromatic heterocycles such as furan, thiophene, and pyrrole undergo halogenation, nitration, and sulfonation just like other aromatics. They are, as a rule, much more reactive than benzene, having the reactivity normally found in phenol (hydroxybenzene) and aniline (aminobenzene) (see Chapters 21 and 22), and often do not need the potent catalysts required for benzene substitution. Since both pyrrole and furan decompose in the presence of protic acids, some modifications of the standard reactions are necessary. Instead of fuming sulfuric acid, the complex formed between pyridine and SO_3 serves as the SO_3 source in the sulfonation reaction; acetyl nitrate may be used as a nitrating agent.

---

[a] For a different, but cogent, argument, see "How the Liver Metabolizes Foreign Substances," A. Kappas and A.P. Alvares, *Sci. Amer.*, **232**(6):22, 1975.

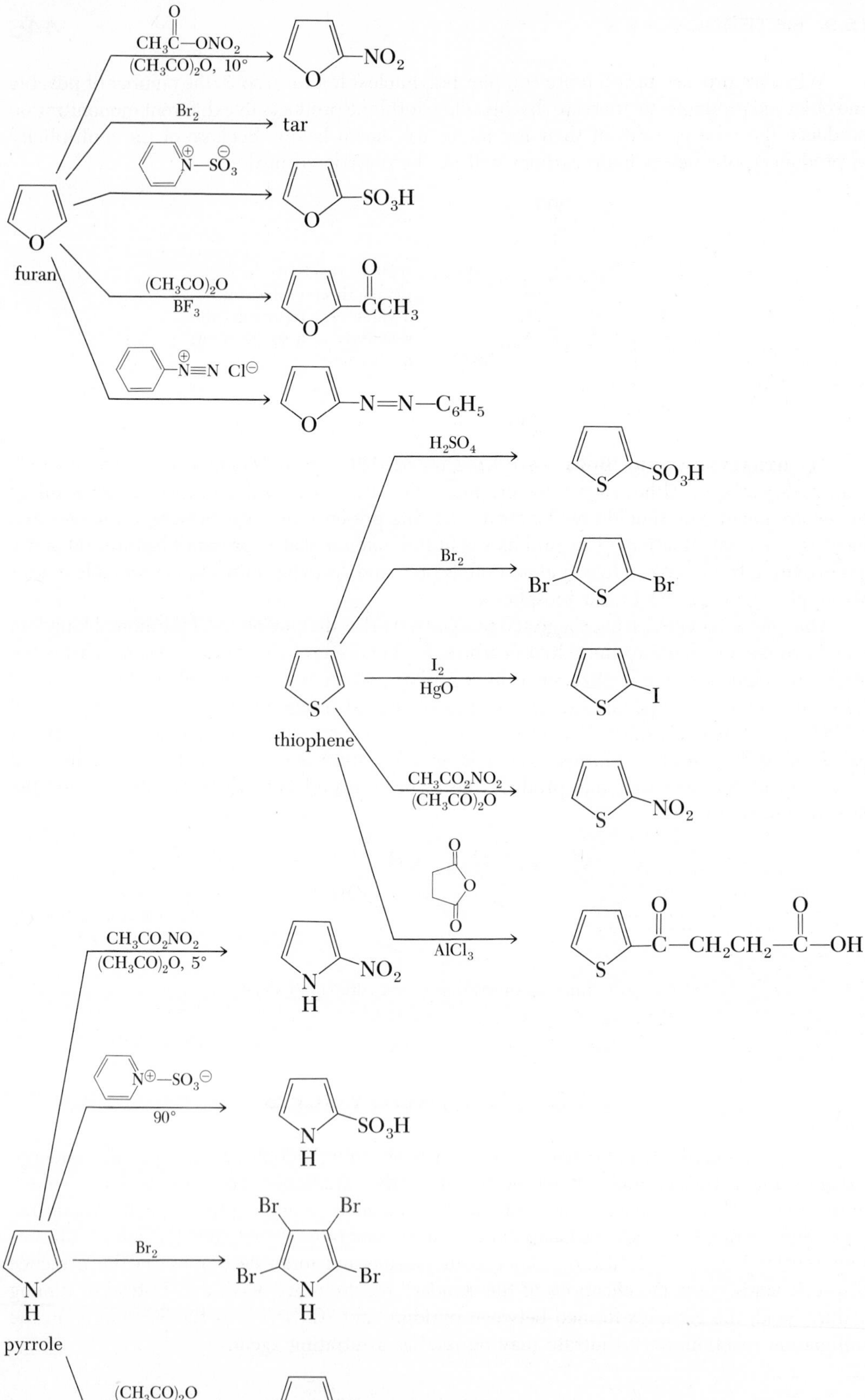

furan

$\xrightarrow[(CH_3CO)_2O,\ 10°]{CH_3C(=O)-ONO_2}$ 2-nitrofuran (NO_2)

$\xrightarrow{Br_2}$ tar

$\xrightarrow{C_5H_5\overset{\oplus}{N}-\overset{\ominus}{S}O_3}$ furan-2-SO_3H

$\xrightarrow[BF_3]{(CH_3CO)_2O}$ furan-2-$\overset{O}{\overset{\|}{C}}CH_3$

$\xrightarrow{C_6H_5-\overset{\oplus}{N}{\equiv}N\ Cl^{\ominus}}$ furan-2-$N{=}N-C_6H_5$

thiophene

$\xrightarrow{H_2SO_4}$ thiophene-2-SO_3H

$\xrightarrow{Br_2}$ 2,5-dibromothiophene (Br, Br)

$\xrightarrow[HgO]{I_2}$ 2-iodothiophene (I)

$\xrightarrow[(CH_3CO)_2O]{CH_3CO_2NO_2}$ 2-nitrothiophene (NO_2)

$\xrightarrow[AlCl_3]{\text{succinic anhydride}}$ thiophene-2-$\overset{O}{\overset{\|}{C}}-CH_2CH_2-\overset{O}{\overset{\|}{C}}-OH$

pyrrole

$\xrightarrow[(CH_3CO)_2O,\ 5°]{CH_3CO_2NO_2}$ 2-nitropyrrole (NO_2, N–H)

$\xrightarrow[90°]{C_5H_5N^{\oplus}-SO_3^{\ominus}}$ pyrrole-2-SO_3H (N–H)

$\xrightarrow{Br_2}$ 2,3,4,5-tetrabromopyrrole (Br, Br, Br, Br; N–H)

$\xrightarrow[200°]{(CH_3CO)_2O}$ pyrrole-2-$\underset{\|}{\overset{}{C}}-CH_3$ (C=O) (N–H)

15. Technical grade benzene is normally contaminated with thiophene because of their similar boiling points (80.10° *vs.* 84.12°, respectively). Thiophene-free benzene can be prepared from this technical grade by shaking the impure benzene with sulfuric acid, discarding the sulfuric acid, and washing the organic phase with water. After the aqueous phase is discarded, the organic phase is dried and, finally, distilled. The distillate is free of thiophene.

Explain the function(s) of sulfuric acid and water in this purification. Suggest a reason why substitution of fuming sulfuric acid for concentrated sulfuric acid may actually reduce the amount of thiophene-free benzene produced.

Substitution usually occurs at C2 because of the greater stability of the cation and the concomitant stabilization of the transition state leading to it.

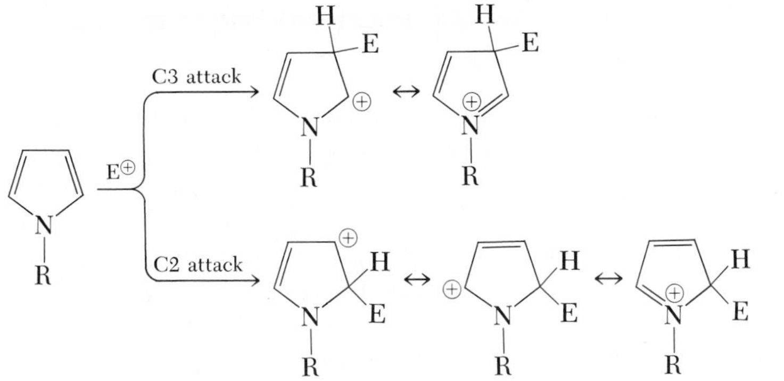

C3 attack

$E^{\oplus}$

C2 attack

Indole is an aromatic, ten π electron heterocycle that undergoes substitution at C3. Suggest a reason why pyrrole reacts at its C2 position, while indole reacts at its C3 position. Is there any relationship between substitution in naphthalene and in indole?

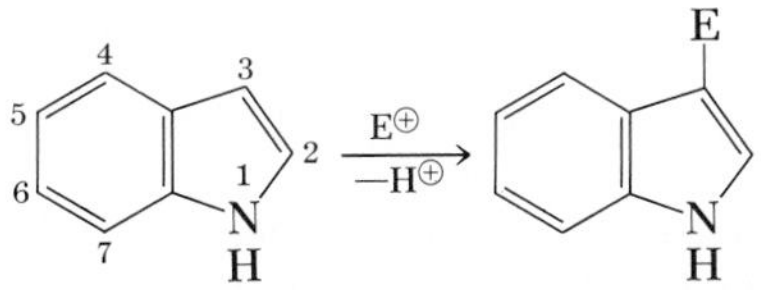

$\xrightarrow[-H^{\oplus}]{E^{\oplus}}$

indole

Pyridine. Electrophiles react toward pyridine as they do toward nitrobenzene, *i.e.*, reluctantly! Pyridine is substituted under vigorous conditions in the C3 position, the heterocyclic nitrogen behaving like the *meta*-directing, deactivating nitro group.

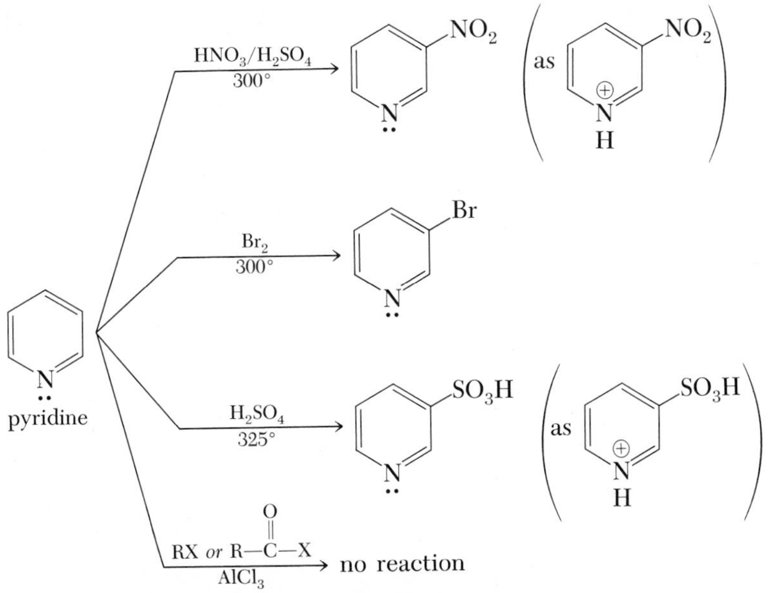

HNO_3/H_2SO_4, 300° → 3-nitropyridine (as the pyridinium ion)

Br_2, 300° → 3-bromopyridine

H_2SO_4, 325° → pyridine-3-sulfonic acid (as the pyridinium ion)

RX *or* R—C(=O)—X, $AlCl_3$ → no reaction

pyridine

Attack of the electrophile at C2 or C4 in pyridine leads to only two important resonance contributing structures, while three contributing structures are possible for attack at C3. The decreased rate of reaction of pyridines is ascribed to the electronegativity of nitrogen, making the π electrons less available for attack.

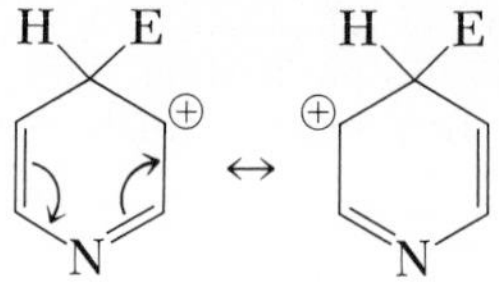

$E^{\oplus}$ attacks C4

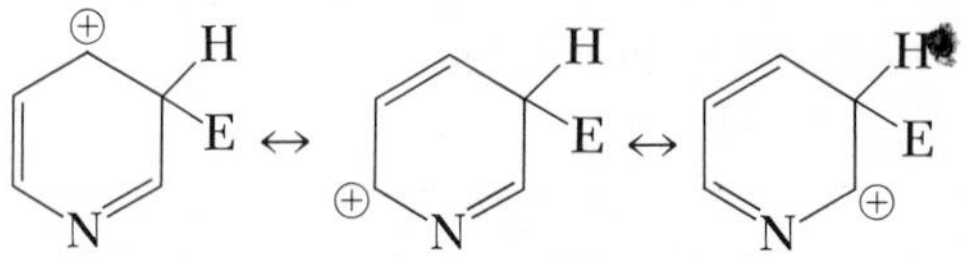

$E^{\oplus}$ attacks C3

Pyridine derivatives occur in nature, and some of them will be discussed in Chapter 27. However, we might note that a very important biological oxidation-reduction reaction involves a quaternary salt of the amide of nicotinic acid (nicotinamide). Biochemists refer to this complex salt as NAD (short for *n*icotinamide *a*denine *d*inucleotide) and it, along with structurally similar compounds, plays a role in cell respiration, photosynthesis, synthesis of straight chain carboxylic acids ("fatty acids"), and vision. A generalized reaction, showing the conversion of NAD to its reduced form, is presented below. Note that the redox reaction involves the interconversion of a pyridinium ring and a 1,4-dihydropyridine ring.

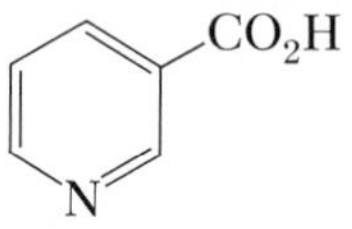

nicotinic acid

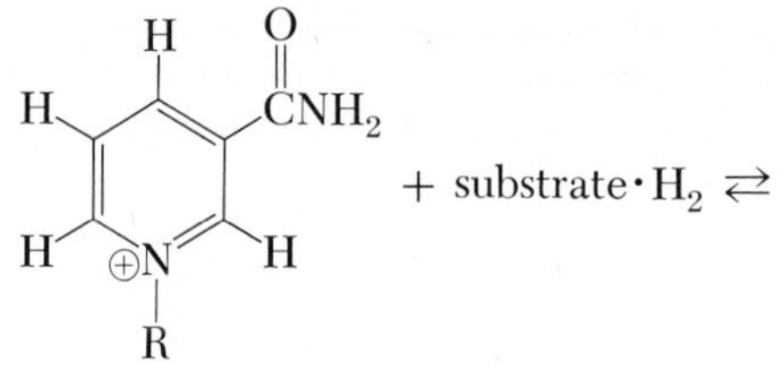

a quaternary nicotinamide salt (NAD) + substrate·H_2 ⇄

reduced form of substrate

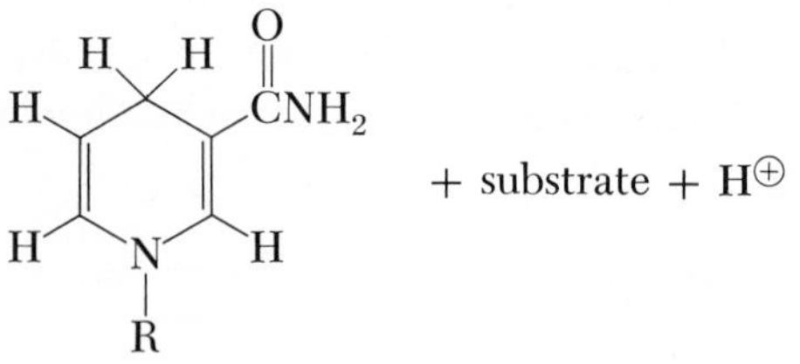

a 1,4-dihydronicotinamide (reduced NAD) + substrate + $H^{\oplus}$

oxidized form of substrate

Quinoline is a nitrogen heterolog of naphthalene. It undergoes electrophilic substitution at C5 and C8. (Compare this to the behavior of 1- and 2-nitronaphthalene.)

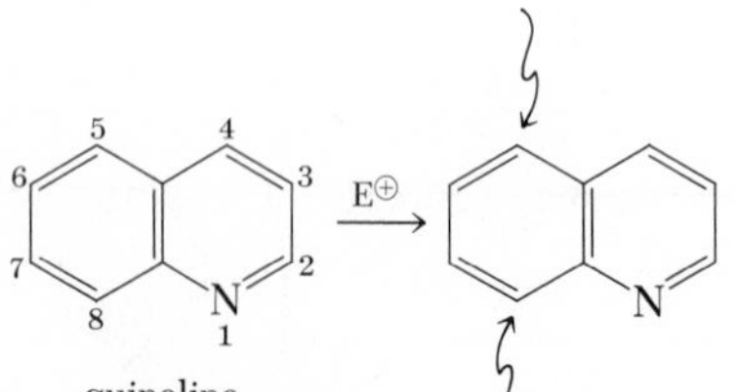

quinoline $\xrightarrow{E^{\oplus}}$

positions attacked by an electrophile

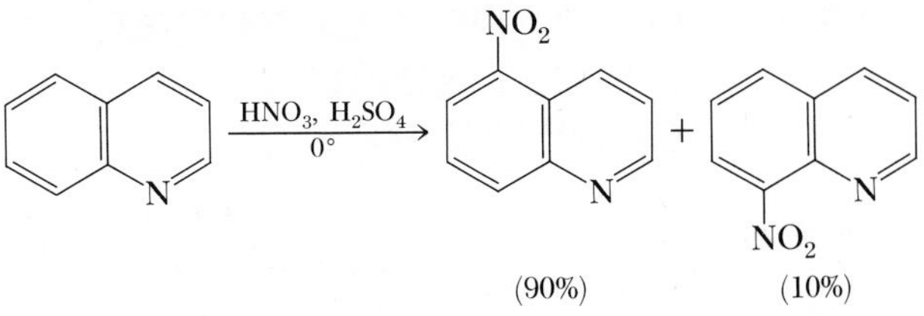

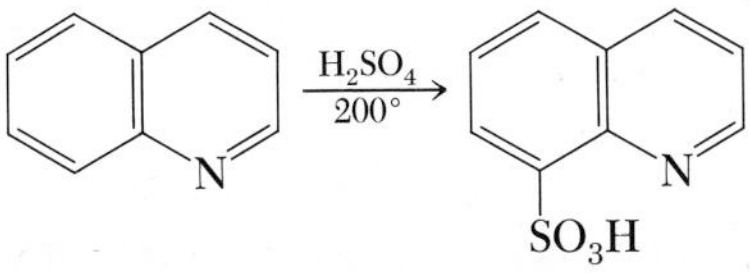

$\xrightarrow[0°]{HNO_3,\ H_2SO_4}$ NO_2 (5-nitroquinoline) + NO_2 (8-nitroquinoline)

(90%) (10%)

$\xrightarrow[200°]{H_2SO_4}$ SO_3H

15.10 ADDITION-ELIMINATION EQUALS SUBSTITUTION

Benzene does not show the typical addition reactions of alkenes. However, as the size of the polynuclear hydrocarbon increases, it often becomes easier for it to undergo typical alkene additions. Thus phenanthrene and anthracene form monobromo derivatives by addition of Br_2 followed by loss of HBr. While anthracene is more reactive than phenanthrene, and actually adds across the 9,10-positions more rapidly than it undergoes substitution, the 9,10-adducts readily decompose to give the 9-substitution products. It should be noted that the 9,10 double bond of phenanthrene shows almost normal isolated alkene reactivity.

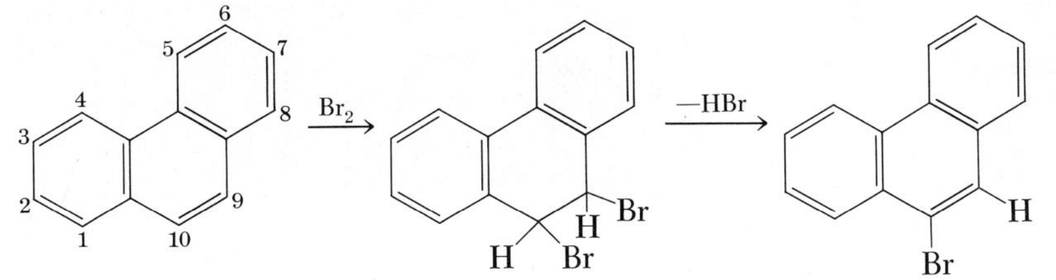

phenanthrene $\xrightarrow{Br_2}$ (H, Br, H, Br) $\xrightarrow{-HBr}$ (H, Br)

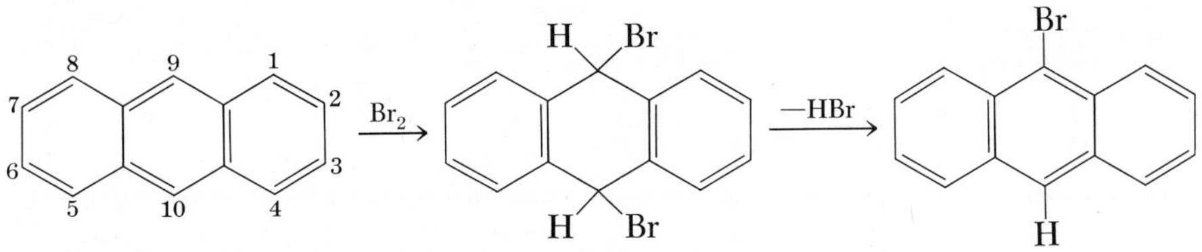

anthracene $\xrightarrow{Br_2}$ (H, Br, H, Br) $\xrightarrow{-HBr}$ (Br, H)

16. Draw the structures and provide names for all of the possible dichloro derivatives of (a) naphthalene, (b) phenanthrene, and (c) anthracene.

15.11 INFRARED SPECTRA OF BENZENE DERIVATIVES[a]

The effect of the ring current upon the nmr spectra of benzene (and, by extrapolation, its derivatives) is presented on page 405. Furthermore, the nmr spectra of several benzene derivatives are shown in Chapter 29. Therefore, this discussion will be limited to the infrared spectra of derivatives of benzene and will focus upon three important regions: $>3000\ cm^{-1}$, 2000 to 1400 cm^{-1}, and 900 to 675 cm^{-1}.

The $>3000\ cm^{-1}$ Region. While C_{sp^3}—H bonds usually exhibit their stretching vibration at frequencies less than ~3000 cm^{-1}, C_{sp^2}—H bonds absorb above 3000 cm^{-1}. Benzene derivatives commonly exhibit their stretching vibrations from 3080 to 3030 cm^{-1} (moderate-weak intensity). Since alkenes also absorb in this region, an absorption above 3000 cm^{-1} can only be taken as suggestive of the presence of a benzene ring. Only a benzene derivative lacking ring hydrogens will fail to show some absorption above 3000 cm^{-1}.

The 2000 to 1400 cm^{-1} Region. Benzene derivatives normally show four absorptions between 1650 and 1400 cm^{-1}. These absorptions, due to C=C vibrations, occur near 1600, 1585, 1500, and 1450 cm^{-1}. The region from 2000 to 1650 cm^{-1} contains a group of combination and overtone bands that are rather weak. However, when they can be observed (*e.g.*, in concentrated solutions), they are quite useful in determining the substitution pattern of the benzene derivative. The patterns that may be expected for variously substituted benzenes are shown in Figure 15–7.

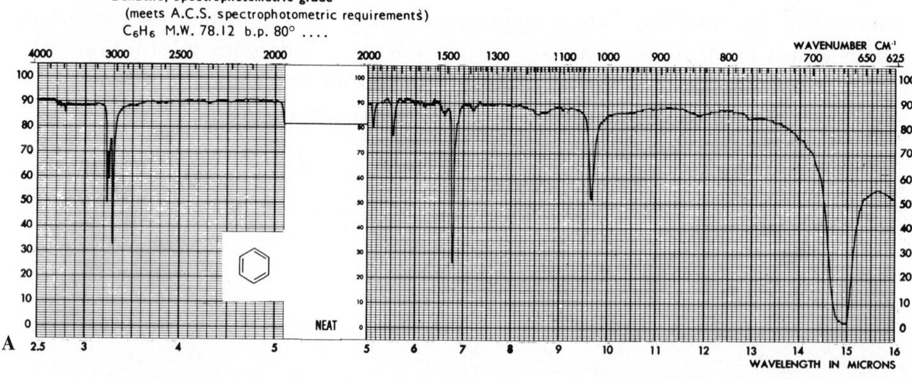

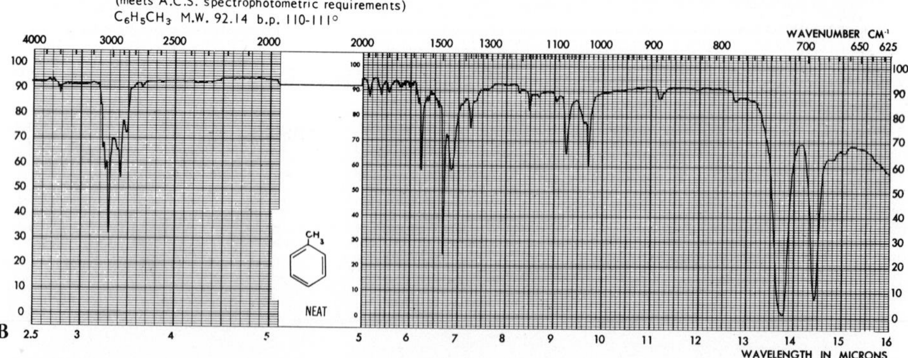

Figure 15–7 *Illustration continued on opposite page.*

---

[a] A knowledge of the infrared discussion in Chapter 28 is assumed.

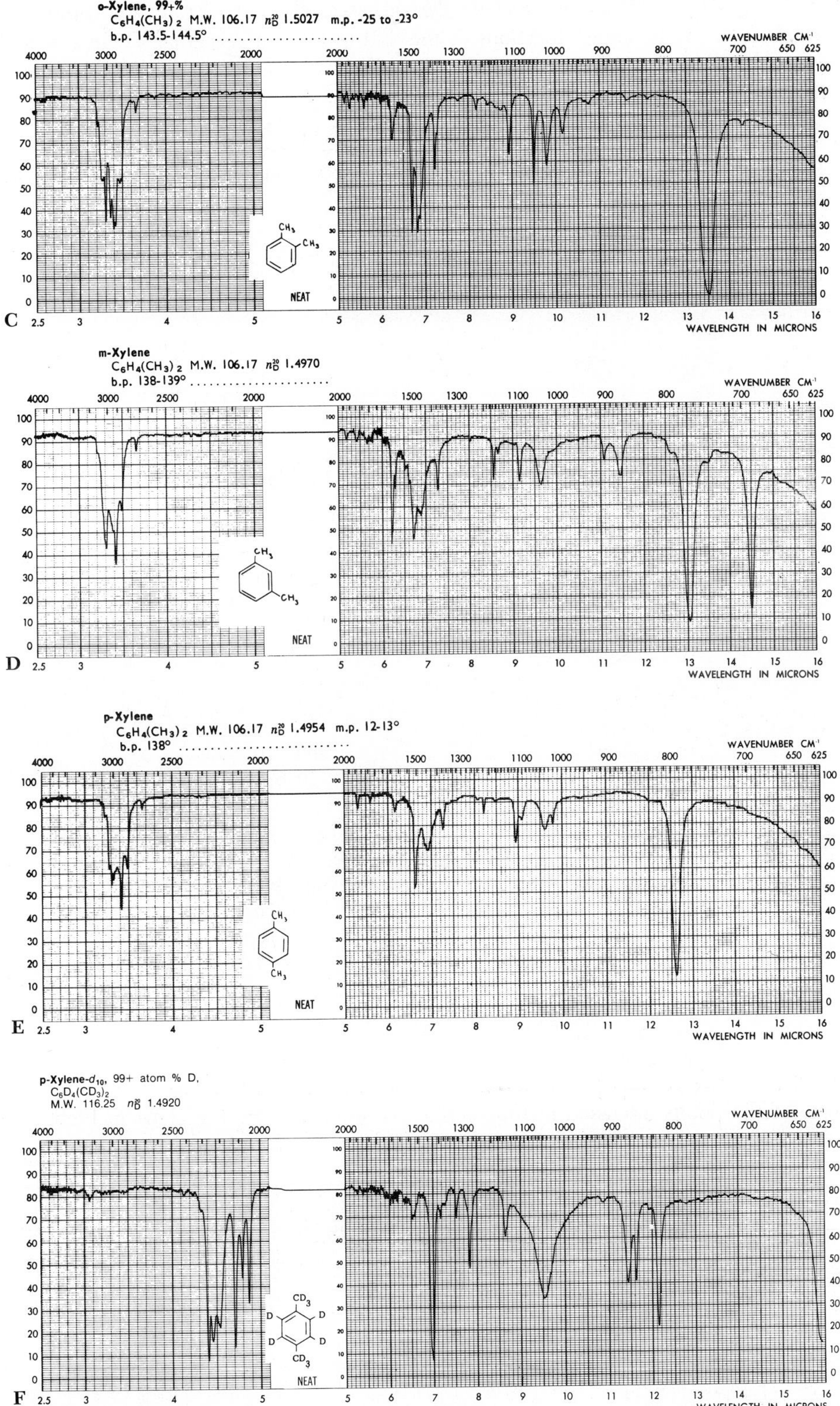

Figure 15-7 Infrared spectra of some benzenoid compounds. In *p*-xylene-d_{10}, all of the hydrogen atoms have been replaced with deuterium.

The 900 to 675 cm^{-1} Region. Benzene derivatives exhibit fairly strong absorptions caused by out-of-plane C—H bending vibrations in the 900 to 675 cm^{-1} region. These absorptions, like those in the 2000 to 1650 cm^{-1} region, are useful in determining the substitution pattern of the derivative and are summarized in Table 15–4.

TABLE 15-4 Absorption in the 900 to 675 cm^{-1} Region

Substitution Pattern	Absorption Bands (cm^{-1})
	770–735 710–685
	760–745
	900–860 790–770 725–685
	830–800
	800–770 720–685
	900–860 860–800
	900–860 865–810 730–675

Some of the points discussed in this section are illustrated by the chart in Figure 15–8.

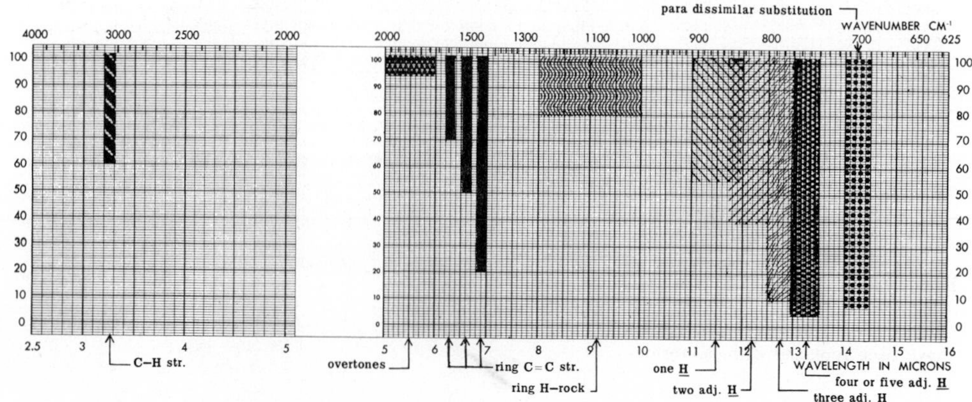

Figure 15-8 Characteristic aromatic absorptions.

APPENDIX

How Can We Express Reactivity and Selectivity Simultaneously? The overall rate of attack upon a benzene ring bearing substituent G, *i.e.*, C_6H_5G, can be expressed as the sum of the rates of attack at the positions *ortho, meta,* and *para* to G. Designating the *overall* rate of reaction as $k_{C_6H_5G}$ and the rates of attack *o, m,* and *p* to G as k_oG, k_mG, and k_pG, respectively, it follows that $k_{C_6H_5G} = 2k_oG + 2k_mG + k_pG$. The coefficient "2" appears because the rate constants refer to attack at single positions but C_6H_5G has two *o* and two *m* positions.

The fraction of product formed by attack *ortho* to G is related to k_oG and $k_{C_6H_5G}$ by the following:

$$\text{fraction of } \textit{ortho} \text{ product} = \frac{\%\ o \text{ isomer formed}}{100} = \frac{2k_0G}{k_{C_6H_5G}}$$

Similarly, the fractions of *m*- and *p*-substituted products resulting from attack upon C_6H_5G equal $2k_mG/k_{C_6H_5G}$ and $k_pG/k_{C_6H_5G}$, respectively.

Consider substitution upon benzene itself. The *overall* rate of reaction of benzene, $k_{C_6H_6}$, must equal six times the rate of substitution at any of the six equivalent positions of benzene. Designating the rate of substitution at one position of benzene as k_H, we find $k_{C_6H_6} = 6k_H$.

Let us now introduce a new term, the **partial rate factor,** which compares the rate of substitution at a single position of C_6H_5G to the rate of substitution at any single position of benzene. The partial rate factors for the positions *ortho, meta,* and *para* to G are abbreviated o_f^G, m_f^G, and p_f^G, respectively, and are defined by:

$$o_f^G = \frac{k_oG}{k_H} \qquad m_f^G = \frac{k_mG}{k_H} \qquad p_f^G = \frac{k_pG}{k_H}$$

Remembering that the *overall* rate of reaction of benzene equals $6k_H$, these partial rate factors can also be expressed as

$$o_f^G = \frac{6k_oG}{k_{C_6H_6}} \qquad m_f^G = \frac{6k_mG}{k_{C_6H_6}} \qquad p_f^G = \frac{6k_pG}{k_{C_6H_6}}$$

We can now substitute these latter relationships into our initial expression for the fractions of various isomers produced in the substitution of C_6H_5G. Thus, for *ortho* substitution:

$$\frac{\%\ o \text{ isomer formed}}{100} = \frac{2k_oG}{k_{C_6H_5G}}$$

becomes

$$\frac{\%\ o \text{ isomer formed}}{100} = \frac{2\left(\dfrac{k_{C_6H_6} \times o_f^G}{6}\right)}{k_{C_6H_5G}}$$

Solving for the partial rate factor then produces

$$o_f^G = \frac{\%\ o \text{ isomer formed}}{100} \times \frac{3k_{C_6H_5G}}{k_{C_6H_6}}$$

Using a similar procedure, we can arrive at the *operational* definitions of m_f^G and p_f^G shown below. The importance of these equations is that they permit the calculation of partial rate factors from the amounts of *ortho, meta,* and *para* isomers produced and from a knowledge of the *overall* rates of reaction of C_6H_5G and C_6H_6.

$$m_f^G = \frac{\%\ m \text{ isomer formed}}{100} \times \frac{3k_{C_6H_5G}}{k_{C_6H_6}}$$

$$p_f^G = \frac{\%\ p \text{ isomer formed}}{100} \times \frac{6k_{C_6H_5G}}{k_{C_6H_6}}$$

If a given position in C_6H_5G is deactivated by G, substitution at that position will have a partial rate factor <1, while activation will produce a partial rate factor >1. If the substituent G has the same effect as does hydrogen, the partial rate factor will equal 1.

We can illustrate the use of the partial rate factor by noting that when toluene is brominated in acetic acid, the position *para* to the methyl group is brominated 2420 times as rapidly as is a single position in benzene; hence, $p_f^{CH_3} = 2420$ *for this reaction.* This and the other partial rate factors for the bromination of toluene are shown below. Note how the values of the partial rate factors indicate both the directive influence of the methyl group ($o_f^{CH_3}$ and $p_f^{CH_3} \gg m_f^{CH_3}$) *and* the activation of all positions on the ring (all partial rate factors greater than unity).

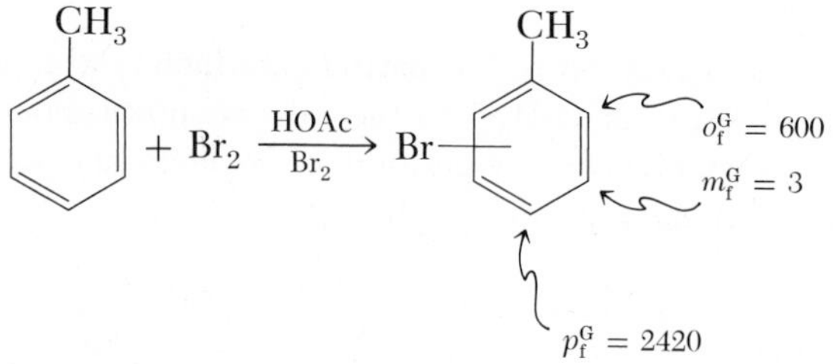

CH$_3$–C$_6$H$_5$ + Br_2 $\xrightarrow[Br_2]{HOAc}$ Br–C$_6$H$_4$–CH$_3$

$o_f^G = 600$
$m_f^G = 3$
$p_f^G = 2420$

The partial rate factors for chlorination of toluene are presented below:

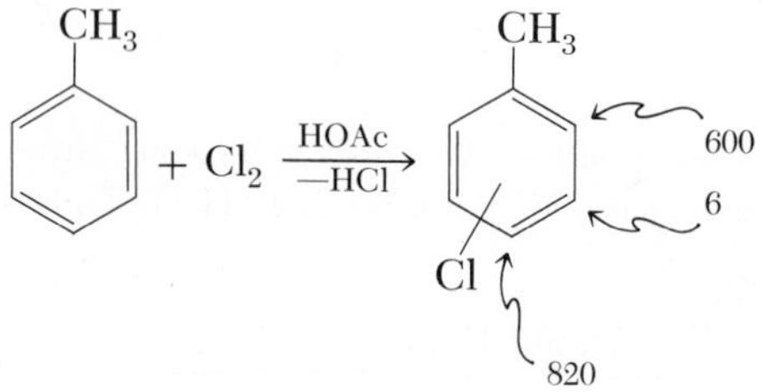

CH$_3$–C$_6$H$_5$ + Cl_2 $\xrightarrow[-HCl]{HOAc}$ Cl–C$_6$H$_4$–CH$_3$

600
6
820

The lower value of $p_f^{CH_3}$ for chlorination (820) compared to that for bromination (2420) implies that the methyl group has less of an effect upon the stabilization of the transition state for chlorination than for bromination. In turn, this suggests that there is less positive charge developed in the transition state for p chlorination than for p bromination. This indicates that the transition state leading to the chlorinated σ complex does not involve as much destruction of the aromatic sextet as does the transition state leading to the brominated σ complex. In other words, the transition state for chlorination is reached earlier along the energy profile.

The values of $o_f^{CH_3}$ for chlorination and bromination are both about 600. This apparent inconsistency (one might have expected a lower value for chlorination) arises because the decreased size of chlorine compared to bromine decreases the steric hindrance to o attack and compensates for the decrease in positive charge present on the ring in the transition state.

Representative partial rate factors are collected in Table 15–5.

TABLE 15-5 Partial Rate Factors for Selected Benzene Derivatives, C_6H_5—G

G	Chlorination			Bromination			Nitration		
	o_f	m_f	p_f	o_f	m_f	p_f	o_f	m_f	p_f
$-OCH_3$	6×10^6	—	5×10^7	—	—	—	—	—	—
$-C_6H_5$	310	0.7	670	11	0.3	16	36	—	36
—F	0.2	0.006	4	—	—	—	0.4	—	0.8
—Cl	0.01	0.002	0.4	—	—	—	0.03	8×10^{-4}	0.13
$-NO_2$	—	—	—	—	5×10^{-5}	—	—	—	—

IMPORTANT TERMS

Activating group: A substituent that makes a benzene ring more reactive toward electrophilic substitution than is benzene itself.

Arene: The class name for "aromatic hydrocarbon."

Benzenonium ion: A cation formed by attack of an electrophile upon the π system of benzene. Since there exists a formal σ bond between the electrophile and the ring, the benzenonium ion is a σ complex. The benzenonium ion does not have an aromatic sextet and is non-aromatic. It is the important intermediate in electrophilic aromatic substitution.

Benzylic position: The carbon atom of an alkyl group which bonds that group to a benzene ring. The term may be used to describe similar positions in other aromatic systems.

Clemmensen reduction: The reduction of a carbonyl group to a methylene group by zinc amalgam in the presence of hydrochloric acid.

Deactivating group: A substituent that makes a benzene ring less reactive toward electrophilic aromatic substitution than is benzene.

Electrophilic aromatic substitution: Replacement of a hydrogen that is bonded to a benzene ring, by initial reaction of an electrophile with the benzene ring. The reaction has two major steps, formation and cleavage of a benzenonium ion. It is the most common way of preparing many derivatives of benzene and other aromatic hydrocarbons. (Those of you who go on in organic chemistry will learn that groups other than hydrogen can be replaced by an electrophile.)

Friedel-Crafts acylation: Formation of an aryl ketone by reaction of an acylium ion, $R-\overset{\oplus}{C}=O$, with an aromatic compound. The acylium ion may be formed from several combinations of acylating agents and Friedel-Crafts catalysts. Unlike Friedel-Crafts alkylation, acylation is not accompanied by skeletal rearrangement or over-reaction.

Friedel-Crafts alkylation: Formation of an alkyl-substituted aromatic compound by reaction of an arene with an alkyl halide in the presence of a Friedel-Crafts catalyst. Modifications employ the acid-catalyzed reaction of arenes with alkenes or alcohol. The reaction is accompanied by the usual carbocation rearrangements and by over-alkylation. The latter occurs because the initially produced alkylbenzene undergoes substitution more rapidly than does the starting benzene.

Friedel-Crafts catalyst: A synonymn for Lewis acid. The most common one is anhydrous aluminum chloride.

***meta* Director:** A substituent that directs incoming electrophiles preferentially to positions *meta* to it.

***meta* Position:** A position on a benzene ring one atom removed from a reference position. A monosubstituted benzene has two *meta* positions.

NAD: Nicotinamide adenine dinucleotide. A naturally occurring complex organic substance that is involved in biological oxidation-reduction reactions. Its most salient chemical feature is a pyridine ring, which, by virtue of coordination at the heterocyclic nitrogen, has been converted to a pyridinium salt. (You may recall from Chapter 14 that such coordination will not destroy the aromaticity of the heterocyclic ring.)

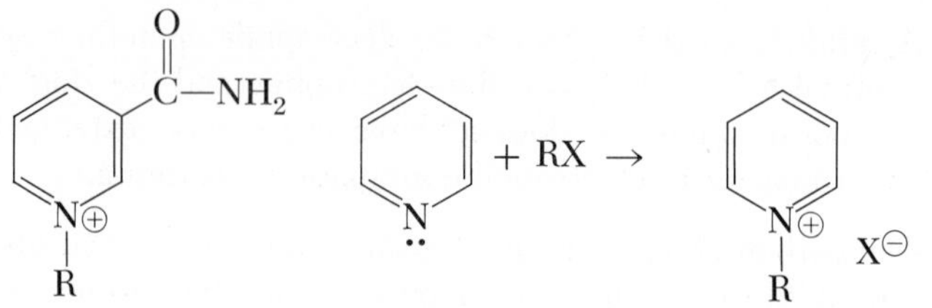

$C{-}NH_2$ (O) ... $N^{\oplus}$—R   pyridine $+ RX \rightarrow$ $N^{\oplus}$—R $X^{\ominus}$

NAD   pyridine   a pyridinium salt

This type of salt is sometimes called a quarternary salt in order to indicate that the nitrogen has become bonded to four organic groups. The complete structure of NAD is presented on p. 891. However, it should be noted here that the "R" of NAD is extremely complex and chiral.

Nitration: Replacement of a hydrogen on an aromatic hydrocarbon with the nitro group ($—NO_2$). The usual reagent is the nitronium ion, $NO_2^{\oplus}$.

***ortho,para* Director:** A substituent that directs incoming electrophiles preferentially to positions *ortho* and *para* to it.

***ortho* Position:** A position on a benzene ring adjacent to a reference position. A monosubstituted benzene has two *ortho* positions.

***para* Position:** A position on a benzene ring two atoms removed from a reference position. A monosubstituted benzene has one *para* position.

Partial rate factor: A quantitative comparison of the reactivity of a specific position on a substituted benzene to that of a single position on benzene. Partial rate factors are given with reference to a specific reaction and particular reaction conditions. A partial rate factor greater than 1 indicates that the position in question is more reactive than is a single position in benzene. A partial rate factor of less than 1 indicates that the position in question is less reactive than is a single position in benzene.

Sigma complex: A species possessing a sigma (σ) bond between an electrophile and a (used-to-be) π system. When the electrophile attacks a benzene ring, the intermediate σ complex is non-aromatic; it reverts to a substituted benzene by loss of hydrogen from the carbon bearing the electrophile. In reality, the σ complexes discussed in this chapter are just highly delocalized carbocations. The term σ complex serves to tell us something about how they were formed.

Sulfonation: Replacement of a hydrogen on an aromatic hydrocarbon with a sulfo group (—SO_2OH).

PROBLEMS

17. Draw structures for the following:
   (a) *m*-dinitrobenzene
   (b) *p*-chloronitrobenzene
   (c) *p*-nitrochlorobenzene
   (d) 1,2-dimethyl-4-nitrobenzene
   (e) 1,3,5-trideuteriobenzene
   (f) 3-hydroxynaphthalene
   (g) α-bromonaphthalene
   (h) 2-aminopyridine
   (i) 3-carboxypyridine (niacin)
   (j) 2-methyl-3-hydroxy-4,5-dihydroxymethylpyridine (pyridoxine)
   (k) 9,10-dimethylanthracene
   (l) 2,4,6-trinitrotoluene (TNT)
   (m) 2-acetylthiophene

18. Write structures (one or more may be possible) for aromatic compounds that satisfy the following restrictions:
   (a) $C_6H_3Br_3$ that can give three possible monochlorination products.
   (b) $C_6H_3Cl_2Br$ that can give two possible mononitration products.
   (c) C_8H_{10} that can give only one mononitration product.

19. A student was given an unknown and asked to determine whether it was an aromatic compound or an alkene. He wisely treated it with a mixture of bromine and aluminum chloride and observed that dense white "smoke" appeared at the top of the test tube. (It was a typical, humid Cleveland afternoon.) The sight of the "smoke" convinced him that the material was aromatic. Explain his reasoning.

20. Identify, by name and structure, the principal product(s) obtained from the reaction of cumene with each of the following. Where no reaction occurs, write *N.R.*
   (a) Cl_2
   (b) $Cl_2/h\nu$
   (c) Cl_2/Fe
   (d) $Cl_2/200°$
   (e) Mg/ether
   (f) conc. HCl
   (g) conc. H_2SO_4
   (h) $NO_2^{\oplus}BF_4^{\ominus}$
   (i) *t*-butyl alcohol/H_3PO_4
   (j) $KMnO_4/OH^{\ominus}$/heat
   (k) O_3
   (l) $H_2/Pt/25°/1$ atm
   (m) $H_2/Ni/200°/75$ atm
   (n) $C_6H_5Br/AlBr_3$
   (o) $(CH_3)_2C{=}CH_2/H_3PO_4$
   (p) B_2H_6 followed by H_2O_2
   (q) $AgNO_3$/ethanol
   (r) acetyl nitrate/acid
   (s) *n*-propyl chloride/$AlCl_3$

21. Starting with benzene as the only aromatic compound, and using any other reagents, provide rational syntheses for the following. Assume that you can separate *ortho* from *para* isomers.
   (a) $C_6H_5C_2H_5$
   (b) 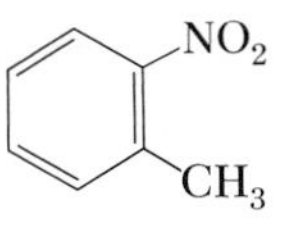 [benzene ring with NO_2 and CH_3 ortho]
   (c) 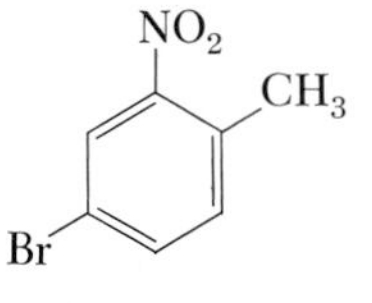 [benzene ring with NO_2, CH_3, and Br]
   (d) C_6H_5D
   (e) 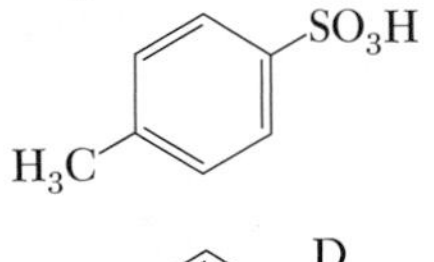 [benzene ring with H_3C and SO_3H para]
   (f) [benzene ring with H_3C and D para]
   (g) [benzene ring with CO_2H]
   (h) Cl—[benzene ring]—CO_2H

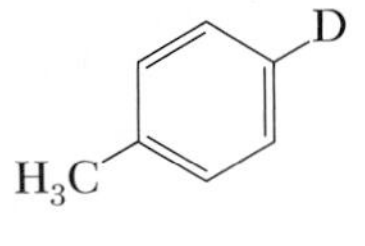

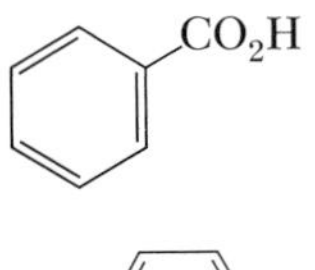

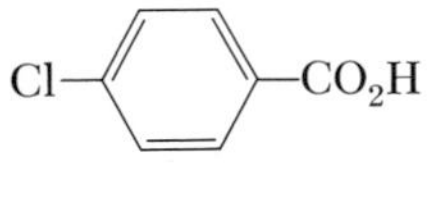

(i) $C_6H_5CH_2Cl$
(j) $C_6H_5CH_2I$
(k) $C_6H_5CH_2D$
(l) (1,2-benzenedicarboxylic acid: CO_2H, CO_2H)
(m) $C_6H_5—CH_2—C_6H_4—Cl$
(n) $C_6H_5CH(CH_3)C_6H_5$
(o) (anthracene)
(p) (H_3C-substituted tetralone, C=O)
(q) (H_3C-substituted dihydronaphthalene)
(r) $H_3C—C_6H_3—(CH_2)_2CO_2H$, with CO_2H

22. X-ray analysis of N_2O_5 reveals the presence of $NO_2^{\oplus}$ and $NO_3^{\ominus}$. Knowing that these ions are both planar, what is the state of hybridization of nitrogen in these ions?
23. Although tropylium bromide, $C_7H_7^{\oplus}Br^{\ominus}$, is aromatic, it does not undergo Friedel-Crafts alkylation. Suggest a reason for this.
24. Aniline reacts instantly with bromine to produce 2,4,6-tribromoaniline (also called *sym*-tribromoaniline). However, if aniline is first acetylated (—NH_2 converted to —$NHC(O)CH_3$), the monobromination product is prepared without any danger of "over-bromination." Account for the effect of acetylation upon the rate of bromination.

$$3Br_2 + C_6H_5NH_2 \xrightarrow[\text{fast}]{\text{very}} \text{2,4,6-}Br_3C_6H_2NH_2$$

$$Br_2 + C_6H_5—\underset{|}{\overset{H}{N}}—\underset{\|}{C}(O)—CH_3 \xrightarrow{\text{moderate}} Br—C_6H_4—NH—C(O)—CH_3$$

(*p*, *o*)

25. Why does the halogenation of benzene with ICl produce iodobenzene instead of chlorobenzene?
26. (a) Suggest why the *nitroso* group, —N=O, is an *o,p*-director although the nitro group is a *m*-director. (b) Why is the phenyl group an *o,p*-director?
27. Explain the difference between the conductances of 1,2,3,4- and 1,2,4,5-tetramethylbenzene in liquid hydrogen fluoride (Table 15-1).
28. The insecticide DDT is prepared by the following route. Suggest a mechanism for the reaction.

$$Cl_3C—CHO + 2C_6H_5Cl \xrightarrow{H_2SO_4} Cl_3C—CH(C_6H_4—Cl)_2$$

DDT

29. The potential utility of the *t*-butyl group as a removable blocking group is illustrated in the sequence below. Provide a mechanism for each step in the sequence. What is the fate of the *t*-butyl group?

$$C_6H_5CH_3 + (CH_3)_3COH \xrightarrow{H^{\oplus}} (CH_3)_3C—C_6H_4—CH_3 \xrightarrow[AlCl_3]{C_6H_5COCl}$$

$$(CH_3)_3C—C_6H_3(CH_3)—C(O)—C_6H_5 \xrightarrow[C_6H_6]{AlCl_3} o\text{-}CH_3C_6H_4—C(O)—C_6H_5$$

30. In the *chloromethylation reaction*, an aromatic compound reacts with formaldehyde and hydrogen chloride in the presence of anhydrous zinc chloride (a mild Lewis acid) to form an aryl chloromethane. Suggest a mechanism for this reaction.

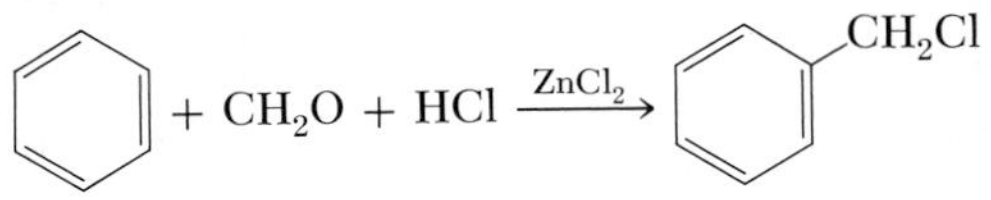

$$+ \ CH_2O + HCl \xrightarrow{ZnCl_2} \quad CH_2Cl$$

31. Account for the relative ease of bromination *para* to the methoxy group that is shown in the series below.

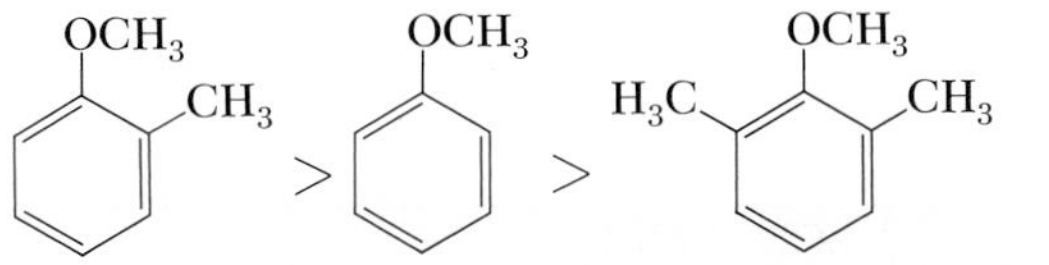

OCH_3 CH_3 > OCH_3 > OCH_3 H_3C CH_3

32. The chlorination of *p*-*t*-butyltoluene produces a large quantity (>90%) of 1-methyl-2-chloro-4-*t*-butylbenzene and equimolar amounts (<10%) of *p*-chlorotoluene and *t*-butyl chloride. Neither *p*-chloro-*t*-butylbenzene nor 1-methyl-3-chloro-4-*t*-butylbenzene is formed in the reaction. Account for these observations.

33. Suggest a reason for the faster substitution at C3 than at C2 in indole.

34. Consider the nitration of aromatic hydrocarbons using $NO_2^{\oplus}BF_4^{\ominus}$ in sulfolane. Relative to benzene, toluene has a reactivity of 1.67 and ethylbenzene has a reactivity of 2.71. Is this consistent with rate-determining σ complex formation? (Similar behavior has been observed in mixed acid.)

35. Why are the following product distributions inconsistent with a mechanism for methylation that depicts a free methyl cation, $CH_3^{\oplus}$, attacking toluene in the product determining step?

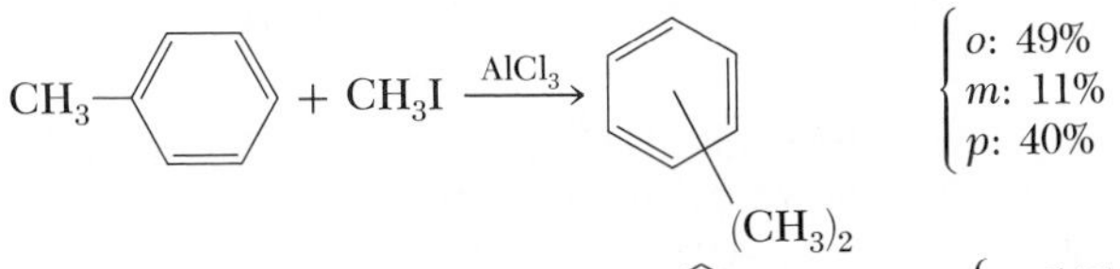

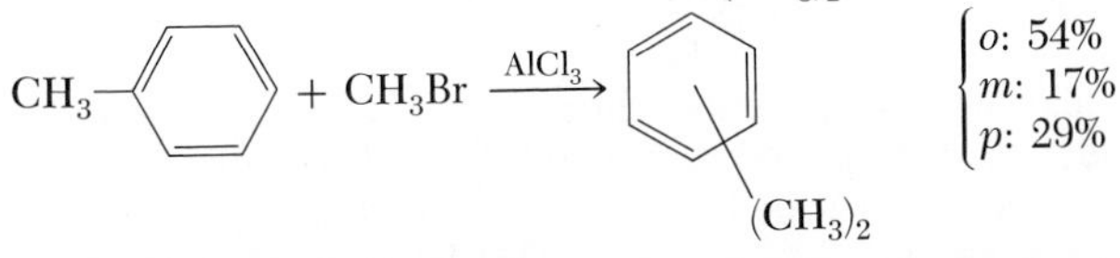

$$CH_3-C_6H_5 + CH_3I \xrightarrow{AlCl_3} C_6H_4(CH_3)_2 \quad \begin{cases} o\text{: } 49\% \\ m\text{: } 11\% \\ p\text{: } 40\% \end{cases}$$

$$CH_3-C_6H_5 + CH_3Br \xrightarrow{AlCl_3} C_6H_4(CH_3)_2 \quad \begin{cases} o\text{: } 54\% \\ m\text{: } 17\% \\ p\text{: } 29\% \end{cases}$$

36. In an attempt to establish that anthracene forms 9-bromoanthracene by first forming a 9,10-dibromo adduct, a student elected to study the reaction of 9,10-dimethylanthracene with bromine. He believed that this would give a more stable adduct than would anthracene. Why did he feel this way? To his surprise, he succeeded in isolating *two* compounds, which both appeared to be 9,10-dibromo-9,10-dimethylanthracene. Explain.

37. Nitration of the ether **A** with N_2O_5 gives almost all *o*-nitro product. Suggest an explanation for this in light of the fact that *n*-butylbenzene gives much less *o* than *p* product under similar conditions.

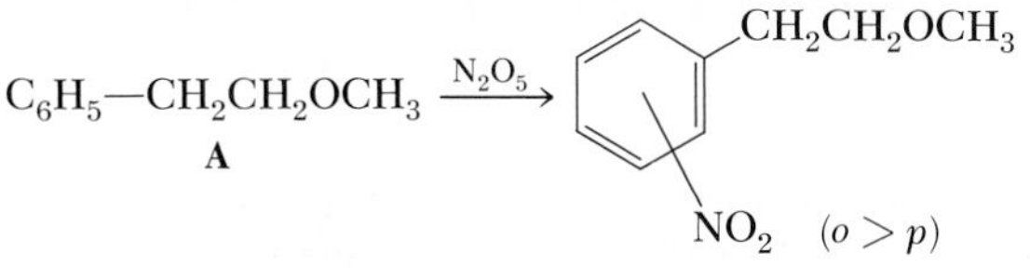

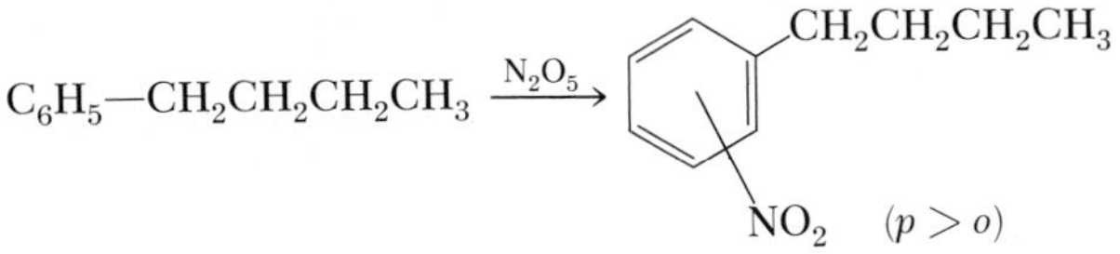

$$\underset{\mathbf{A}}{C_6H_5-CH_2CH_2OCH_3} \xrightarrow{N_2O_5} C_6H_4(CH_2CH_2OCH_3)NO_2 \quad (o > p)$$

$$C_6H_5-CH_2CH_2CH_2CH_3 \xrightarrow{N_2O_5} C_6H_4(CH_2CH_2CH_2CH_3)NO_2 \quad (p > o)$$

38. Many foot powders contain 8-hydroxyquinoline, a mild bacteriocide and fungicide. What is its structure? When oxidized with alkaline potassium permanganate, it is converted to quinolinic acid, $C_7H_5NO_4$. Suggest a structure for quinolinic acid.

39. *Indene* is a hydrocarbon, C_9H_8, that occurs in coal tar and that reacts rapidly with Br_2/CCl_4. It is catalytically hydrogenated to C_9H_{10} under mild conditions and to C_9H_{16} under vigorous conditions. Prolonged refluxing of indane, C_9H_{10}, with alkaline potassium permanganate produces phthalic acid. What is the structure of (a) indane and (b) indene?

40. Suggest a detailed mechanism for each of the following.

(a) 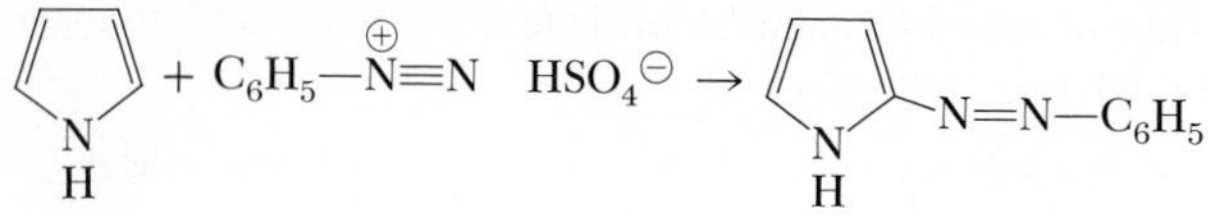 pyrrole + $C_6H_5—\overset{\oplus}{N}≡N\ \ HSO_4^{\ominus}$ → 2-phenylazopyrrole ($—N=N—C_6H_5$)

(b) 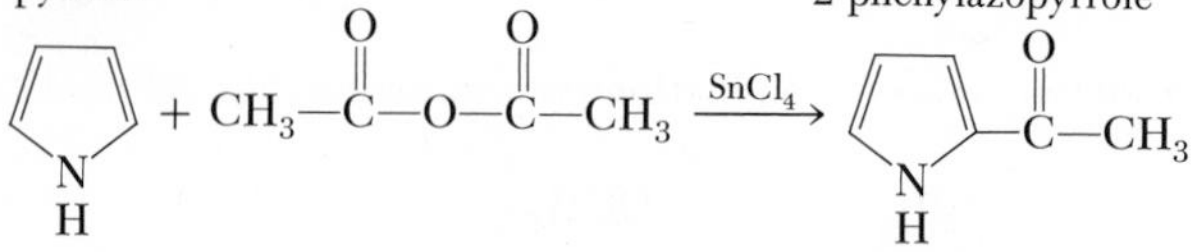 pyrrole + $CH_3—\overset{O}{\overset{\|}{C}}—O—\overset{O}{\overset{\|}{C}}—CH_3 \xrightarrow{SnCl_4}$ 2-acetylpyrrole ($—\overset{O}{\overset{\|}{C}}—CH_3$)

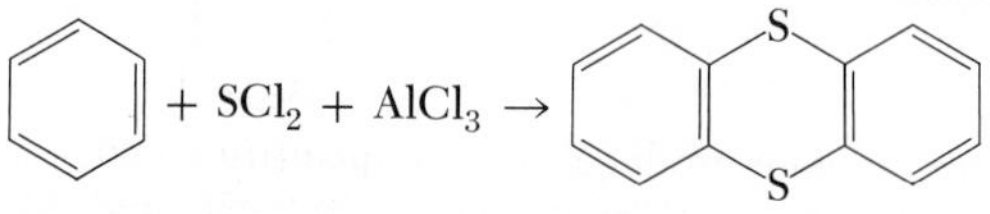

(c) benzene + SCl_2 + $AlCl_3$ → thianthrene

(d) 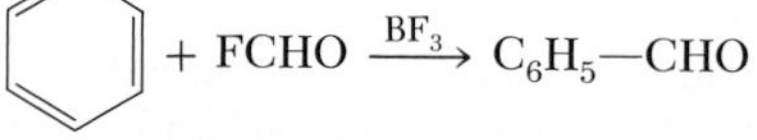 benzene + FCHO $\xrightarrow{BF_3}$ $C_6H_5—CHO$

(e) 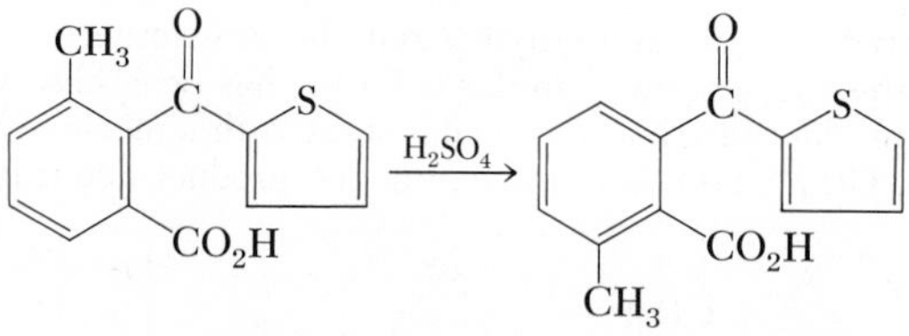 $\xrightarrow{H_2SO_4}$ (substituents: CH_3, $\overset{O}{\overset{\|}{C}}$, S, CO_2H; product: CO_2H, CH_3)

41. The relative reactivities toward free-radical chlorination, per hydrogen, of the hydrogens in *n*-butyl chloride are shown below. Explain the variation for the three methylene groups.

$$\underset{5.6}{CH_3}—\underset{17}{CH_2}—\underset{5.6}{CH_2}—\underset{1.0}{CH_2}—Cl$$

42. When toluene is alkylated with methyl chloride and aluminum chloride, and the reaction is quenched with water after a short period (1 minute), the xylene distribution is 55% *o*, 10% *m*, and 35% *p*. If the reaction is permitted to go on for several hours and then quenched with water, the product distribution is 20% *o*-, 60% *m*-, and 20% *p*-xylene. This latter product mixture also is formed when *any* pure isomeric xylene is allowed to stand in contact with anhydrous hydrogen chloride and a trace of aluminum chloride. Account for these results.

43. A somewhat antiquated test for alkylbenzenes involves the formation of a red or orange color when an alkylbenzene is treated with chloroform and aluminum chloride. This color is presumed to be due to the presence of a triarylmethyl cation. Account for its formation.

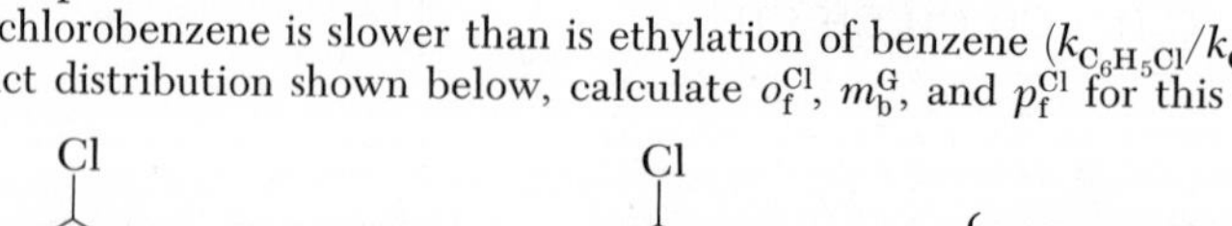

$$\text{alkyl}—C_6H_5 + CHCl_3 + AlCl_3 \rightarrow (\text{alkyl}—C_6H_4)_3\overset{\oplus}{C}\ \ \overset{\ominus}{AlCl_4}$$

red-orange

[a] 44. Mesitylene reacts (−80°) with ethyl fluoride and boron trifluoride to form a salt. Upon warming to −15°, this salt is converted to a hydrocarbon whose nmr spectrum exhibits three different types of methyl resonances of relative intensities 2:1:1. Account for this, suggesting a structure for the salt. What would its nmr spectrum be like?

[b] 45. Ethylation of chlorobenzene is slower than is ethylation of benzene ($k_{C_6H_5Cl}/k_{C_6H_6} = 0.20$). Using this, and the product distribution shown below, calculate o_f^{Cl}, m_b^{G}, and p_f^{Cl} for this alkylation.

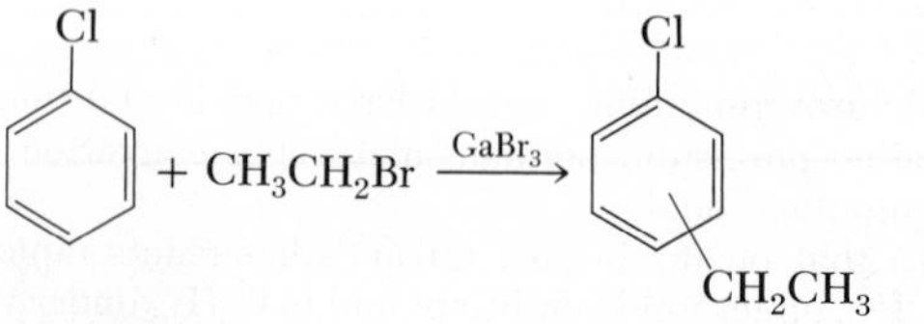

$$C_6H_5Cl + CH_3CH_2Br \xrightarrow{GaBr_3} ClC_6H_4CH_2CH_3 \begin{cases} 42.4\%\ ortho \\ 15.9\%\ meta \\ 41.9\%\ para \end{cases}$$

[b] 46. Starting with the definition of partial rate factor, derive an expression relating p_f^{G} to the rate of substitution *para* to G in C_6H_5G.

---

[a] Requires a knowledge of nmr spectroscopy.
[b] Requires a knowledge of material found in the chapter's Appendix.

SYNTHESIS AND SIMPLE ADDITION REACTIONS OF ALDEHYDES AND KETONES

16.1 WHERE ARE WE GOING?

The carbonyl group, $\rangle C{=}O$, is usually found in two different molecular environments. If only alkyl groups, aryl groups, or hydrogen are bonded to the carbonyl carbon, the resultant compound is likely to be either an *aldehyde* or a *ketone*. If at least one of the atoms attached to the carbonyl carbon contains a non-bonding electron pair, the resultant compounds are carboxylic acids or their derivatives. The presence of such non-bonding electron pairs can dramatically alter the characteristics of the C=O linkage.

This and the following chapter will be limited to a discussion of aldehydes and ketones. The chemistry of these carbonyl compounds will be divided, for convenience, into two areas—reactions involving the carbonyl group alone, and reactions involving groups that are activated by the presence of a carbonyl group. This chapter focuses upon the synthesis and "simple" reactions of aldehydes and ketones. Chapter 17 will deal with more complex reactions of aldehydes and ketones and with their spectral properties.

ALDEHYDES: $R{-}\overset{O}{\overset{\|}{C}}{-}H$ *or* RCHO *KETONES:* $R{-}\overset{O}{\overset{\|}{C}}{-}R$ *or* RC(O)R

SOME CARBOXYLIC ACID DERIVATIVES: $R{-}\overset{O}{\overset{\|}{C}}{-}\ddot{\underset{\cdot\cdot}{O}}{-}H$ (carboxylic acid) $R{-}\overset{O}{\overset{\|}{C}}{-}\ddot{\underset{\cdot\cdot}{O}}{-}R$ (ester)

$R{-}\overset{O}{\overset{\|}{C}}{-}\ddot{\underset{\cdot\cdot}{X}}{:}$ (acyl halide) $R{-}\overset{O}{\overset{\|}{C}}{-}\ddot{N}R_2$ (amide) $R{-}\overset{O}{\overset{\|}{C}}{-}\ddot{\underset{\cdot\cdot}{O}}{-}\overset{O}{\overset{\|}{C}}{-}R$ (anhydride)

R is alkyl or aryl

As you read this chapter, compare the reactions of the carbon-oxygen double bond to those of the carbon-carbon double bond. Bear in mind that additions to alkenes usually involve attack of an electrophile upon the π system, while additions to carbonyl groups involve attack of a nucleophile at the carbonyl carbon. The tendency for the carbonyl group to be attacked by nucleophiles stems, in part, from the existence of two resonance forms for the carbonyl group, one of which places a positive charge on the carbonyl carbon. In addition to this resonance effect, the electronegativity of oxygen creates a permanent polarization of the carbonyl group, with the carbon bearing a partial positive charge.

$$\underset{\text{resonance effect}}{\geq C{=}\ddot{O}: \leftrightarrow \geq \overset{\oplus}{C}{-}\ddot{O}:^{\ominus}} \qquad \underset{\text{permanent polarization}}{\geq \overset{\delta+}{C}{=}\overset{\delta-}{O}}$$

What things should you look for in comparing C=C bonds to C=O bonds? One *similarity* is in the effect of a catalyst upon reactivity. In the same way that electrophilic attack upon a carbon-carbon double bond sometimes requires the presence of a catalyst (often to increase the concentration of electrophile), so nucleophilic addition to a carbonyl group is accelerated by coordination of a Lewis acid to the carbonyl oxygen. A striking *difference* between addition to C=C bonds and addition to C=O bonds is the greater tendency of carbonyl adducts to undergo reversal of the initial addition, thus regenerating the carbonyl group. Many additions across the carbonyl group can be treated as true equilibria.

$$>C{=}C< + E^{\oplus} \rightarrow >\overset{\oplus}{C}{-}C(E)< \xrightarrow{Nu:^{\ominus}} >C(Nu){-}C(E)<$$

$$(CH_3)_2C{=}C(CH_3)_2 + HCl \rightarrow (CH_3)_2C(H){-}\overset{\oplus}{C}(CH_3)_2 \rightarrow (CH_3)_2C(H){-}C(Cl)(CH_3)_2 \qquad \text{electrophilic addition}$$

$$Nu:^{\ominus}\ >C{=}O \rightleftarrows Nu{-}C{-}O^{\ominus} \xrightleftharpoons{E^{\oplus}} Nu{-}C{-}O{-}E$$

$$N{\equiv}C:^{\ominus} + (CH_3)_2C{=}O \rightleftarrows NC{-}C(CH_3)_2{-}O^{\ominus} \xrightleftharpoons{H^{\oplus}} NC{-}C(CH_3)_2{-}OH \qquad \text{nucleophilic addition}$$

In Chapter 10 the term "addition-elimination" was used to describe the mechanism for the conversion of an ester to a ketone by the action of a Grignard reagent. In this chapter we re-introduce this term, this time to describe a group of reactions of aldehydes and ketones. Re-examining pertinent sections of Chapter 10 will help you to get a clear picture of the differences between these two "addition-elimination" processes.

$$R{-}\overset{O}{\overset{\|}{C}}{-}O{-}R + R'MgX \rightarrow R{-}\overset{O}{\overset{\|}{C}}{-}R' + RMgX \qquad \textit{mechanism: } \text{addition-elimination}$$

$$R{-}\overset{O}{\overset{\|}{C}}{-}R + NH_3 \rightleftarrows R{-}\overset{NH}{\overset{\|}{C}}{-}R + H_2O \qquad \textit{mechanism: } \text{addition-elimination}$$

1. Explain why acetone, $CH_3C(O)CH_3$, is completely miscible with water while propene is not.

16.2 NOMENCLATURE

The IUPAC name for an aldehyde is formed by taking the longest straight chain containing the aldehyde group, —CHO, dropping the final "*e*" from its IUPAC name, and adding the ending "*al*." It is understood that the aldehydic carbon $\left(i.e., \text{—}\overset{\text{O}}{\overset{\|}{\text{C}}}\text{—H}\right)$ is C1. Ketones are named similarly except that the appropriate ending is "*one*." While a ketonic carbon $\left(\text{R—}\overset{\text{O}}{\overset{\|}{\text{C}}}\text{—R}\right)$ is never C1 (Why?), it is given the lowest possible number. Several illustrations may help you to understand this scheme.

ALDEHYDES

$\text{H—}\overset{\text{O}}{\overset{\|}{\text{C}}}\text{—H}$ methanal

CH_3CHO ethanal

$CH_3CHBrCHO$ 2-bromopropanal

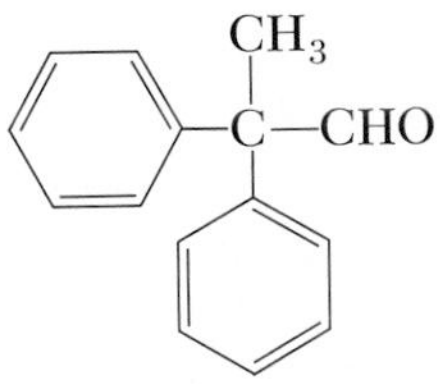

$BrCH_2CH_2CHO$ 3-bromopropanal

$(C_6H_5)_2C(CH_3)\text{—}CHO$ 2,2-diphenylpropanal

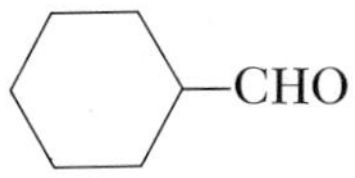

$C_6H_{11}\text{—}CHO$ cyclohexylmethanal

$H_2C\text{=}CH\text{—}CHO$ propenal

KETONES

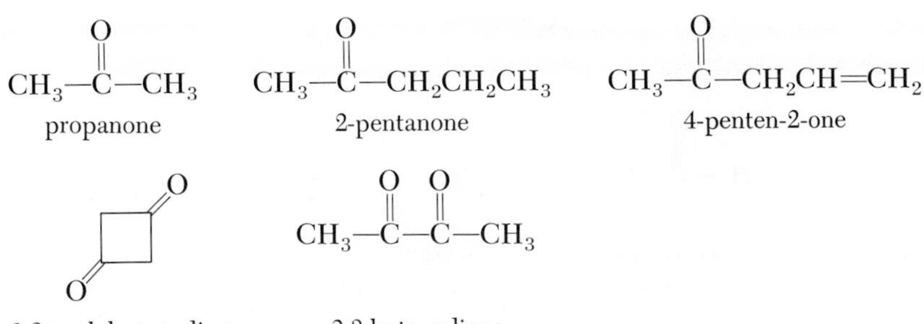

Some schemes for naming aldehydes emphasize the relationship between aldehydes and carboxylic acids. These may use the name of the carboxylic acid as the root and add the suffix "aldehyde," or they may simply call the —CHO group *carboxaldehyde*. The term "formyl," used as a prefix, also indicates the presence of the —CHO group.

$\text{H—}\overset{\text{O}}{\overset{\|}{\text{C}}}\text{—OH}$ formic acid

$\text{H—}\overset{\text{O}}{\overset{\|}{\text{C}}}\text{—H}$ formaldehyde

$C_6H_5\text{—}CH_2CH_2\text{—}CO_2H$ β-phenylpropionic acid

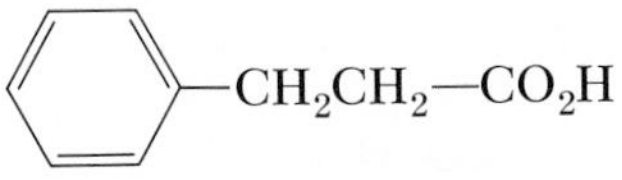

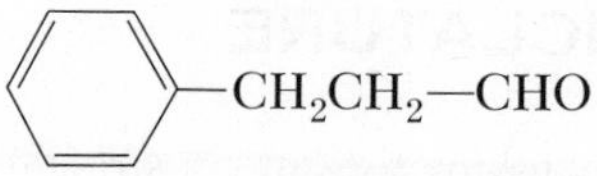

C_6H_5—CH_2CH_2—CHO β-phenylpropionaldehyde (*not* 2-phenylpropionaldehyde)

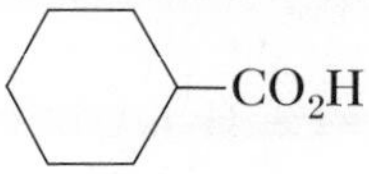

C_6H_{11}—CO_2H cyclohexanecarboxylic acid

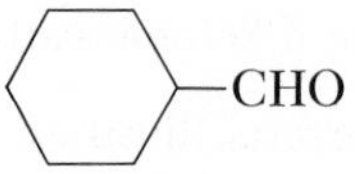

C_6H_{11}—CHO cyclohexanecarboxaldehyde

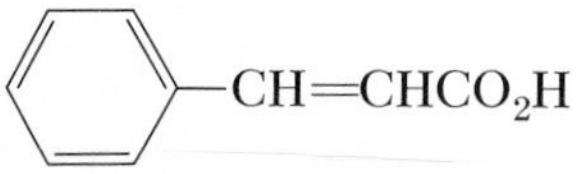

C_6H_5—CH=CHCO$_2$H cinnamic acid

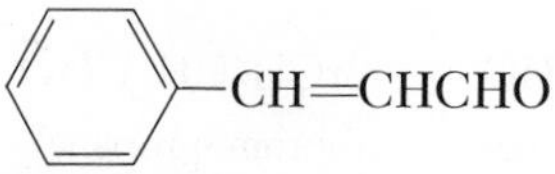

C_6H_5—CH=CHCHO cinnamaldehyde

HO_3S—C_6H_4—CHO 4-formylbenzenesulfonic acid

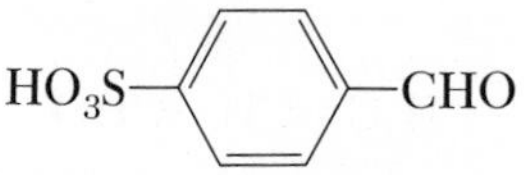

There are always some compounds whose names are not easily related to any scheme.

CH=CH—CHO acrolein

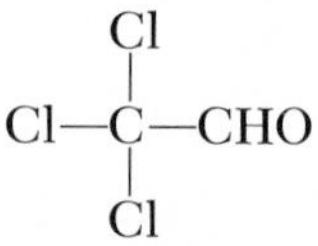

Cl_3C—CHO chloral

In addition to their IUPAC names, ketones can be named as derivatives of a hypothetical compound called "ketone." The following examples illustrate this method.

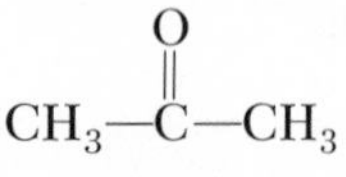

CH_3—CO—CH_3 dimethyl ketone

CH_3—CO—C_6H_5 methyl phenyl ketone

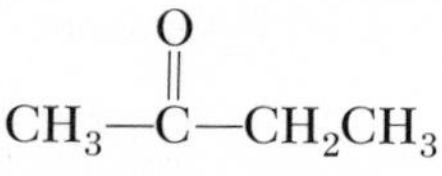

CH_3—CO—CH_2CH_3 methyl ethyl ketone (MEK)

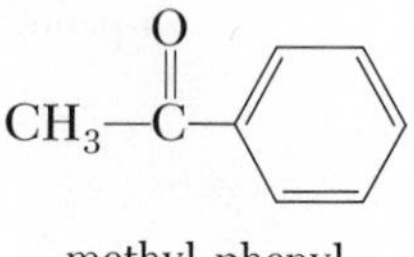

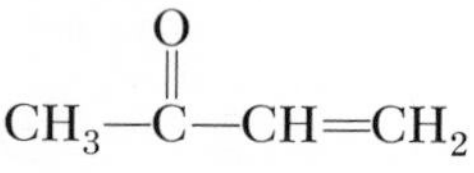

CH_3—CO—CH=CH_2 methyl vinyl ketone (MVK)

Some very common ketones are given names that give no hint of their structure, while some names, such as "mesityl oxide," are downright misleading.

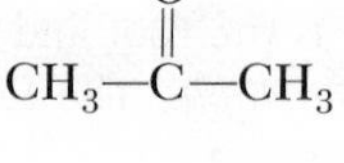

CH_3—CO—CH_3 acetone

CH_3—CO—C_6H_5 acetophenone

C_6H_5—CO—C_6H_5 benzophenone

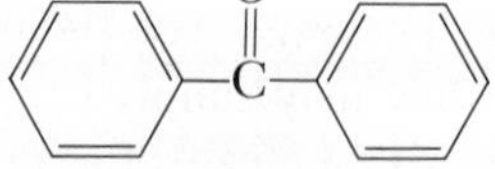

isophorone (3,5,5-trimethylcyclohex-2-enone: ring with CH_3, CH_3, CH_3 and O)

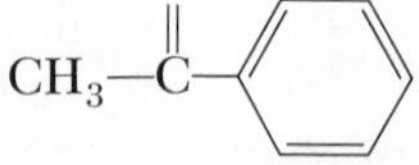

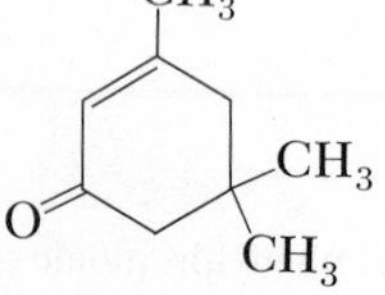

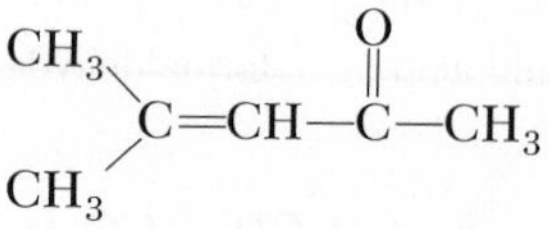

$(CH_3)_2C$=CH—CO—CH_3 mesityl oxide

2. Provide an alternate acceptable name for each of the following.
   (a) ethanal
   (b) acrolein
   (c) cinnamaldehyde
   (d) trichloroacetaldehyde
   (e) acetone
   (f) methyl ethyl ketone
   (g) mesityl oxide

16.3 SYNTHESIS OF ALDEHYDES AND KETONES

A number of methods of synthesizing both aldehydes and ketones were presented in earlier chapters. These will not be discussed again but, for review, the more important ones are summarized in Figure 16–1. It would be wise to re-examine any that you may have forgotten.

ALDEHYDE:

1. $RCH{=}CHR \xrightarrow{O_3} 2RCHO$ (p. 253)
2. $RCH_2OH \xrightarrow[\text{acetone, 15°}]{CrO_3/H_3O^{\oplus}} RCHO$ (p. 309)
3. $RCH_2OH \xrightarrow[25°]{CrO_3/\text{pyridine}} RCHO$ (p. 309)
4. $RCH{=}CH{-}CH_2OH \xrightarrow[25°]{MnO_2} RCH{=}CHCHO$ (p. 309)
5. $RCH_2OH \rightarrow RCH_2OTs \xrightarrow[\text{heat}]{DMSO} RCHO$ (p. 309)
6. $RCH(OH)CH(OH)R \xrightarrow{Pb(OAc)_4} 2RCHO$ (p. 337)
7. $RC{\equiv}CH \xrightarrow{BH_3} \xrightarrow[OH^{\ominus}]{H_2O_2} RCH_2CHO$ (p. 274)

KETONE:

8. $R_2C{=}CR_2 \rightarrow 2R_2CO$ via $MnO_4^{\ominus}$, $OH^{\ominus}$/heat or via O_3 (pp. 252, 253)
9. $RC{\equiv}CR + H_2O \xrightarrow{H^{\oplus}/Hg^{+2}} RC(O)CH_2R$ (p. 273)
10. $RC{\equiv}CR \xrightarrow{BH_3} \xrightarrow[OH^{\ominus}]{H_2O_2} RC(O)CH_2R$ (p. 274)
11. $R_2CHOH \rightarrow R_2CO$ via $Cr_2O_7^{-2}/H^{\oplus}$, or CrO_3/pyridine, or CrO_3/CH_3CO_2H (p. 310)
12. $R_2CHOH \rightarrow R_2CHOTs \xrightarrow[\text{heat}]{DMSO} R_2CO$ (pp. 309, 310)
13. $R_2CHOH \xrightarrow[\text{acetone/15°}]{CrO_3/H^{\oplus}} R_2CO$ (p. 310)
14. $R_2C{=}CHOCH(OH)R \xrightarrow[25°]{MnO_2} R_2C{=}CHC(O)R$ (pp. 309, 310)
15. $R_2C(OH)C(OH)R_2 \xrightarrow{Pb(OAc)_4} 2R_2CO$ (p. 337)
16. $ArH + RC(O)Cl \xrightarrow{AlCl_3} ArC(O)R$ (p. 421)
17. $ArH + RC(O)OC(O)R \xrightarrow{AlCl_3} ArC(O)R$ (p. 422)

(Friedel-Crafts acylation using acyl halides (16) and acid anhydrides (17))

Figure 16–1 Synthesis of aldehydes and ketones. These reactions have appeared in earlier chapters. R usually can be an alkyl or aryl group. Reactions 16 and 17 are electrophilic aromatic substitutions, and the requisite aryl hydrocarbon is designated ArH. Unusual restrictions or limited reactions have been deleted.

Oxidation of Aryl Methyl Groups. Aryl methyl groups can be oxidized to aldehydes by a mixture of chromium trioxide and acetic anhydride. Over-oxidation is prevented by the formation of a *gem*-diacetate, which is later hydrolyzed to the aldehyde. Yields are rarely more than 50%.

$$p\text{-}CH_3C_6H_4CH_3 \xrightarrow[H_2SO_4]{CrO_3/(CH_3CO)_2O} p\text{-}(AcO)_2CH\text{-}C_6H_4\text{-}CH(OAc)_2 \xrightarrow[H^{\oplus}]{H_2O} p\text{-}OHC\text{-}C_6H_4\text{-}CHO \quad (50\%)$$

gem-diacetate terephthalaldehyde

$$OAc \equiv -O-\overset{O}{\overset{\|}{C}}-CH_3$$

A related reaction, the *Étard oxidation*, uses chromyl chloride (CrO_2Cl_2) as the oxidant. This reagent is a powerful oxidant and can cause some organic compounds to ignite. It must, therefore, be used with caution. Fortunately, both carbon tetrachloride and carbon disulfide are inert to it and can serve as solvents. As with the chromium trioxide, over-oxidation is prevented by complexation.

$$p\text{-}I\text{-}C_6H_4\text{-}CH_3 \xrightarrow[CCl_4]{CrO_2Cl_2} \left[p\text{-}I\text{-}C_6H_4\text{-}CH(OCrCl_2OH_2)_2\right] \xrightarrow{H_3O^{\oplus}} p\text{-}I\text{-}C_6H_4\text{-}CHO \quad (60\%)$$

p-iodobenzaldehyde

Reduction of Acid Chlorides. Carboxylic acids, RCO_2H, are difficult to reduce, while aldehydes are easily reduced to 1° alcohols. Thus, direct reduction of a carboxylic acid with lithium aluminum hydride or with diborane produces a 1° alcohol *via* an intermediate aldehyde or its equivalent. On the other hand, acid chlorides, RC(O)Cl, are reduced to aldehydes in excellent yield by the action of lithium tri-*t*-butoxyaluminum hydride at −78°.

$$LiAlH_4 + 3\underbrace{(CH_3)_3C}_{t\text{-bu}}{-}OH \longrightarrow LiAlH(O\text{-}t\text{-bu})_3 + H_2$$

lithium tri-*t*-butoxyaluminum hydride

$$LiAlH(O\text{-}t\text{-bu})_3 + RC(O)Cl \xrightarrow[\text{diglyme}]{-78^\circ} RCHO \quad (50\text{–}90\%)$$

$$p\text{-}ClOC\text{-}C_6H_4\text{-}COCl + LiAlH(O\text{-}t\text{-bu})_3 \xrightarrow[\text{diglyme}]{-78^\circ} p\text{-}OHC\text{-}C_6H_4\text{-}CHO \quad (77\%)$$

Gattermann Hydrogen Cyanide Synthesis of Aryl Aldehydes. When an aromatic compound bearing an activating substituent is treated with a mixture of hydrogen cyanide and hydrogen chloride in the presence of anhydrous zinc chloride, the product isolated after hydrolysis is an aldehyde. Yields in the reaction can be high (~90%), but even a recent modification, using zinc cyanide instead of hydrogen cyanide, has not stopped the slow fall into oblivion of this preparation, the **Gattermann reaction.** The Gattermann reaction involves an intermediate aldiminium salt (*i.e.*, $RCH{=}\overset{\oplus}{N}H_2\ Cl^{\ominus}$), which is formed by an electrophilic substitution reaction. Can you suggest a mechanism for its formation?

$$\text{1,3-}(HO)_2C_6H_4 + HCN + HCl \xrightarrow[30^\circ]{ZnCl_2} \left[(HO)_2C_6H_3{-}CH{=}\overset{\oplus}{N}H_2\ Cl^{\ominus} \right] \xrightarrow{H_2O} (HO)_2C_6H_3{-}CHO$$

$$C_6H_5CH_3 + HCN + HCl \xrightarrow[100^\circ]{ZnCl_2} \xrightarrow{H_2O} p\text{-}CH_3C_6H_4CHO$$

Reduction of Nitriles. While the Gattermann reaction is being used less in spite of modernization, the reduction of aromatic nitriles, known since 1925, has been improved by modern synthetic chemists and remains an often-used route to aldehydes. In the **Stephen's reduction** an aromatic nitrile is reduced with stannous chloride to an aldiminium salt, which is then hydrolyzed to the aldehyde.

$$R{-}C{\equiv}N + HCl \rightarrow \left[R{-}\underset{Cl}{\underset{|}{C}}{=}NH \xrightarrow[HCl]{SnCl_2} (RCH{=}NH)_2SnCl_4 \right] \xrightarrow{H_2O} 2RCHO \qquad \text{Stephen's reduction}$$

$$\beta\text{-cyanonaphthalene} \xrightarrow[SnCl_2]{HCl} \xrightarrow{H_2O} \beta\text{-naphthaldehyde} \quad (95\%)$$

β-cyanonaphthalene *β*-naphthaldehyde

If a solution of one equivalent of lithium aluminum hydride in tetrahydrofuran is added to a nitrile, and the reaction mixture is decomposed with dilute aqueous acid, the corresponding aldehyde is isolated in yields that may sometimes approach 90%.

$$C_3H_5{-}CN \xrightarrow[\text{ether, } -50^\circ]{LiAlH_4} \xrightarrow{H_3O} C_3H_5{-}CHO \qquad LiAlH_4 \text{ added to nitrile}$$

cyclopropyl cyanide cyclopropylcarboxaldehyde (70%)

$$R{-}C{\equiv}N + LiAlH_4 \rightarrow RCH{=}N{-}Li + AlH_3$$

$$RCH{=}N{-}Li + H_2O \rightarrow RCH{=}NH + LiOH$$

$$RCH{=}NH + H_3O^{\oplus} \rightarrow RCHO + NH_4^{\oplus}$$

The order of addition of reagents is important; if the nitrile is *added to* the lithium aluminum hydride, over-reduction occurs and the main product is an amine (R—NH_2).

$$\triangleright\!-C{\equiv}N \xrightarrow[\text{ether}]{LiAlH_4} \triangleright\!-CH_2NH_2 \qquad \text{nitrile added to } LiAlH_4$$

cyclopropylmethylamine

A valuable modification of the hydride reduction of nitriles uses lithium tri-ethoxyaluminum hydride in place of lithium aluminum hydride.

$$CH_3(CH_2)_4CN \xrightarrow[(b)\ H_3O^{\oplus}]{(a)\ LiAlH(OEt)_3} CH_3(CH_2)_4CHO \qquad (55\%)$$

capronitrile → caproaldehyde

3. Suggest a synthesis, starting with toluene, for each of the following.
   (a) benzaldehyde
   (b) *p*-tolualdehyde (*p*-$CH_3C_6H_4CHO$)
   (c) *p*-tolylcarbinol
   (d) *p*-tolylphenylcarbinol
   (e) *p*-xylene
   (f) terephthalaldehyde
   (g) benzyl bromide

Rearrangement of *vic*-Diols. In acidic solutions, 1,2-diols containing a —CH_2OH unit undergo a 1,2-hydride ion migration (*i.e.*, a carbocation rearrangement) to form aldehydes. Isobutyraldehyde is prepared commercially by this route.

$$(CH_3)_2C{=}CH_2 \xrightarrow{[O]} (CH_3)_2COHCH_2OH \xrightarrow[-H_2O]{H^{\oplus}} CH_3-\overset{CH_3}{\underset{\oplus}{C}}-\overset{H}{\underset{OH}{C}}-H$$

$$\downarrow$$

$$(CH_3)_2CHCHO \xleftarrow{-H^{\oplus}} CH_3-\overset{CH_3}{\underset{H}{C}}-\overset{\oplus}{C}(H)-\ddot{O}-H$$

isobutyraldehyde

If the *vic*-diol does not include a 1° alcohol, rearrangement produces a ketone. The classic example of this is the conversion of 2,3-dimethyl-2,3-butanediol to 3,3-dimethyl-2-butanone. Since the starting material and product are called *pinacol* and *pinacolone,* respectively, this type of reaction is dubbed a "pinacol" or "pinacol-pinacolone" rearrangement.

$$CH_3-\overset{CH_3}{\underset{OH}{C}}-\overset{CH_3}{\underset{OH}{C}}-CH_3 \xrightarrow{H^{\oplus}} CH_3-\overset{CH_3}{\underset{\overset{\oplus}{O}H_2}{C}}-\overset{CH_3}{\underset{OH}{C}}-CH_3 \xrightarrow{-H_2O} CH_3-\overset{CH_3}{\underset{\oplus}{C}}-\overset{CH_3}{\underset{OH}{C}}-CH_3$$

pinacol

$$\downarrow$$

$$CH_3-\overset{CH_3}{\underset{CH_3}{C}}-\overset{O}{\overset{\|}{C}}-CH_3 \xleftarrow{-H^{\oplus}} CH_3-\overset{CH_3}{\underset{CH_3}{C}}-\overset{\oplus}{C}(CH_3)-\ddot{O}-H$$

pinacolone

The necessary pinacol, $R_2COHCOHR_2$, is usually made by a bimolecular reduction of a ketone with magnesium amalgam (Mg·Hg).

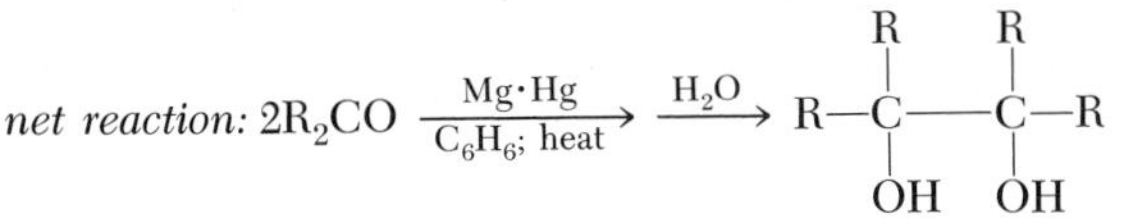

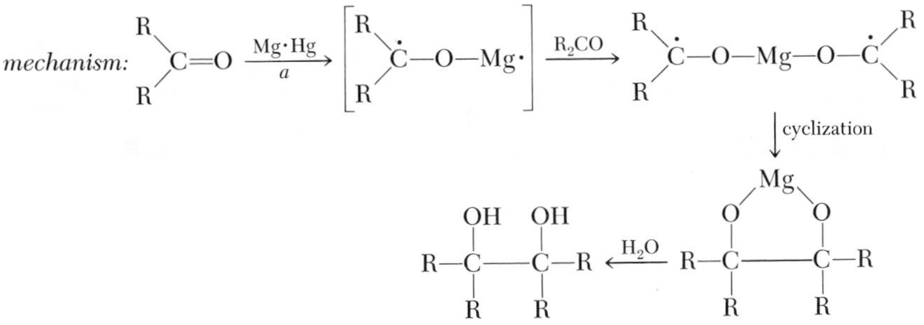

$$\textit{net reaction: } 2R_2CO \xrightarrow[C_6H_6;\ heat]{Mg\cdot Hg} \xrightarrow{H_2O} R-\underset{OH}{\overset{R}{C}}-\underset{OH}{\overset{R}{C}}-R$$

$$\textit{mechanism: } R_2C{=}O \xrightarrow[a]{Mg\cdot Hg} \left[R_2\dot{C}-O-Mg\cdot\right] \xrightarrow{R_2CO} R_2\dot{C}-O-Mg-O-\dot{C}R_2$$

$$\xrightarrow{\text{cyclization}} R-\underset{R}{C}(-O-Mg-O-)\underset{R}{C}-R \xrightarrow{H_2O} R-\underset{R}{\overset{OH}{C}}-\underset{R}{\overset{OH}{C}}-R$$

In the mechanism suggested above, the magnesium is seen (step *a*) as giving an electron to the carbonyl group to form a radical. Two such radicals, bound to the same magnesium (remember that Mg is "divalent" and can part with two electrons), cyclize by radical dimerization to form the pinacol.

4. The reduction of 2-butanone with magnesium produces, after hydrolysis, two isomeric glycols. What are their structures?

Grignard Addition to Nitriles. Grignard reagents add across the triple bond of the C≡N group to form a ketimine salt, which precipitates out of ether at low temperatures, *e.g.*, −60°. Once the salt is thus removed from further reaction, water is added to it to produce a ketone in reasonable yield. The use of halonitriles possessing halogens that cannot react with Grignard reagents permits the synthesis of haloketones. One example is the synthesis of *o*-bromoacetophenone shown below.

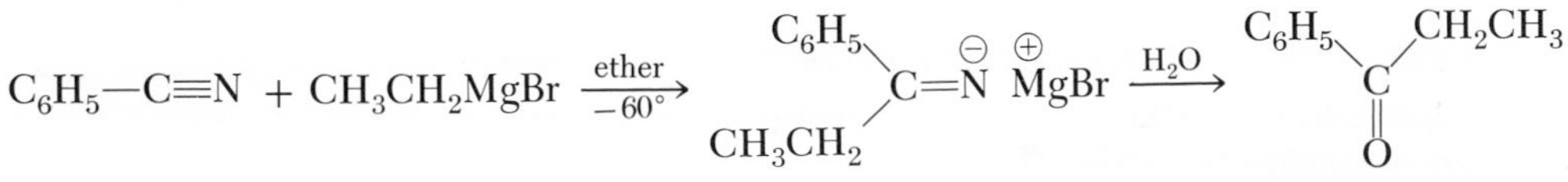

$$C_6H_5-C{\equiv}N + CH_3CH_2MgBr \xrightarrow[-60^\circ]{ether} (C_6H_5)(CH_3CH_2)C{=}\overset{\ominus}{N}\ \overset{\oplus}{Mg}Br \xrightarrow{H_2O} (C_6H_5)(CH_2CH_3)C{=}O$$

benzonitrile (cyanobenzene) — a ketimine salt — ethyl phenyl ketone

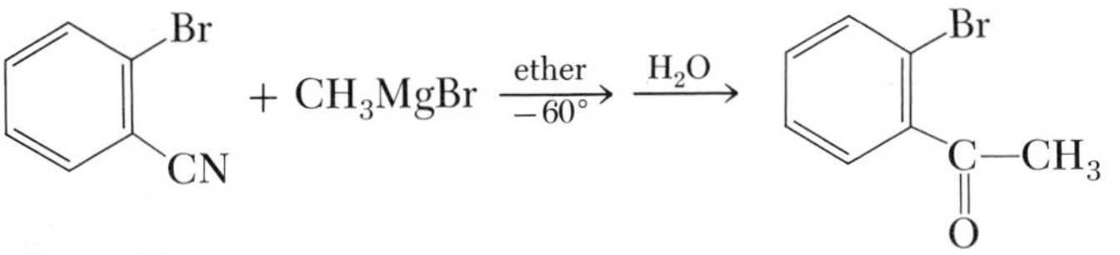

$$o\text{-}BrC_6H_4CN + CH_3MgBr \xrightarrow[-60^\circ]{ether} \xrightarrow{H_2O} o\text{-}BrC_6H_4-\underset{\|}{C}(=O)-CH_3 \quad (80\%)$$

o-bromoacetophenone

5. Predict the product from each of the following, assuming a normal work-up.
   (a) $CH_3MgCl + C_6H_5CN$
   (b) *p*-$NCC_6H_4CN + CH_3MgCl$ (2 moles)
   (c) $CH_3CN + LiAlH_4$ (excess)
   (d) CH_3CN (excess) + $LiAlH_4$

(e) $p\text{-}CH_2{=}CHC_6H_4CN + C_6H_5MgBr$
(f) CH_3MgCl (excess) + $C_6H_5C(O)OCH_3$

6. When cyanoethane, CH_3CH_2CN, reacts with ethyl magnesium iodide, ethane can be detected before the reaction mixture is quenched with water. If the reaction mixture is quenched with D_2O rather than H_2O, one of the products is CH_3CHDCN. Account for these data. Since propane does not react with ethyl magnesium iodide, what must you conclude about the CN group? How would you rationalize this effect?

Dialkylketones from Trialkylboranes. On page 293 we noted that trialkylcarbinols result from the oxidation of the product formed in the reaction of trialkylboranes with carbon monoxide. If water is present during the initial addition of carbon monoxide, the reaction produces a dialkyl ketone. By this procedure, dicyclopentyl ketone can be prepared from cyclopentene in 90% yield.

$$\text{cyclopentene} \xrightarrow{B_2H_6} (C_5H_9)_3B \xrightarrow[100^\circ]{CO/H_2O} \xrightarrow[OH^\ominus]{H_2O_2} (C_5H_9)_2C{=}O + C_5H_9OH$$

(90%)

$$CH_3{-}CH{=}CH{-}CH_3 \xrightarrow{B_2H_6} \xrightarrow[100^\circ]{CO/H_2O} \xrightarrow[OH^\ominus]{H_2O_2} (CH_3{-}CH_2{-}CH(CH_3){-})_2C{=}O$$

16.4 NUCLEOPHILIC ADDITION REACTIONS

Hydration. Most carbonyl compounds of fewer than six carbons exhibit some water solubility. This is due, in large measure, to hydrogen bond formation between the carbonyl oxygen and the water hydrogens.

$$(CH_3)_2C{=}O\cdots H{-}O{-}H \quad (\text{two water molecules hydrogen bonded to the carbonyl oxygen})$$

acetone is completely miscible with water

A limited number of carbonyl compounds react with water to form *gem*-diols. The position of equilibrium between a carbonyl compound and water on the one hand, and the hydrate (*i.e.*, *gem*-diol) on the other, is controlled by steric and electronic factors. For example, the inductive effect of a trichloromethyl group destabilizes the charge separation that is inherent in carbonyl groups; this effect is responsible for the formation of a stable hydrate of trichloroacetaldehyde (also called *chloral*). The product, *chloral hydrate*, has received notoriety as the active ingredient in a Mickey Finn. More conventionally, it is used as a sedative and hypnotic (sleep-inducer) and in veterinary medicine as an anesthetic for large animals (*e.g.*, horses and swine).

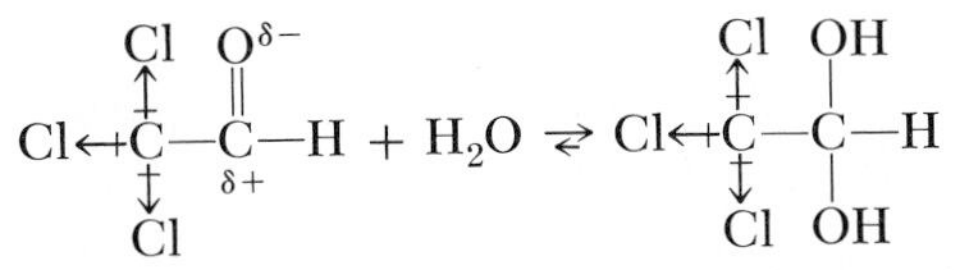

$$Cl_3C{-}C(=O^{\delta-})^{\delta+}{-}H + H_2O \rightleftarrows Cl_3C{-}C(OH)_2{-}H$$

chloral　　　chloral hydrate

Formaldehyde, CH_2O, does not have any alkyl groups bonded to its carbonyl group. Since alkyl groups stabilize positively charged centers to which they are bonded, formaldehyde is destabilized, relative to most aldehydes, by the absence of these alkyl groups. Consequently, formaldehyde exists completely as its hydrate in aqueous solution. A 40% solution of formaldehyde in water, usually called formalin, is used as a preservative of biological specimens.

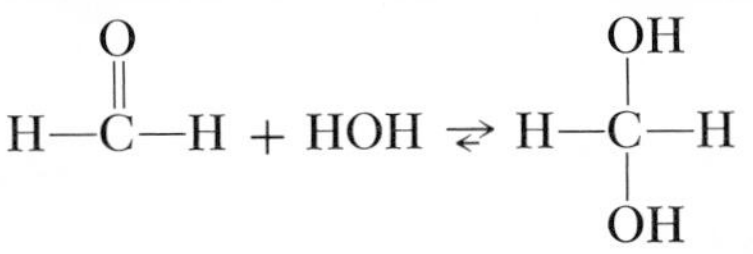

$$H{-}C(=O){-}H + HOH \rightleftarrows H{-}C(OH)_2{-}H \qquad \text{formalin solution}$$

If the rate of dehydration of a *gem*-diol is greater than the rate of hydration of the corresponding carbonyl compound, the equilibrium between them will favor the carbonyl compound over the *gem*-diol. Indeed, K_{eq} can be expressed as $k_{\text{forward reaction}}/k_{\text{reverse reaction}}$. Therefore, the inability to detect a *gem*-diol does not mean that it isn't produced; it simply means that it decomposes faster than it is formed. One can establish that a reaction, presumably *gem*-diol formation, is occurring by observing the incorporation of ^{18}O into the carbonyl group of an aldehyde or ketone that has been dissolved in $H_2{}^{18}O$.

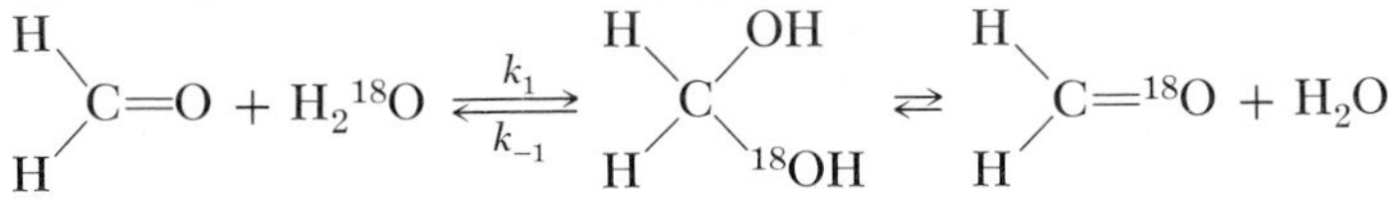

$$H_2C{=}O + H_2{}^{18}O \underset{k_{-1}}{\overset{k_1}{\rightleftarrows}} H_2C(OH)(^{18}OH) \rightleftarrows H_2C{=}^{18}O + H_2O$$

$$K_{eq} = \frac{[CH_2(OH)_2]}{[CH_2O][H_2O]} = \frac{k_1}{k_{-1}}$$

Incorporation of ^{18}O is subject to acid or base catalysis. Mechanisms for these processes are presented in Figure 16–2.

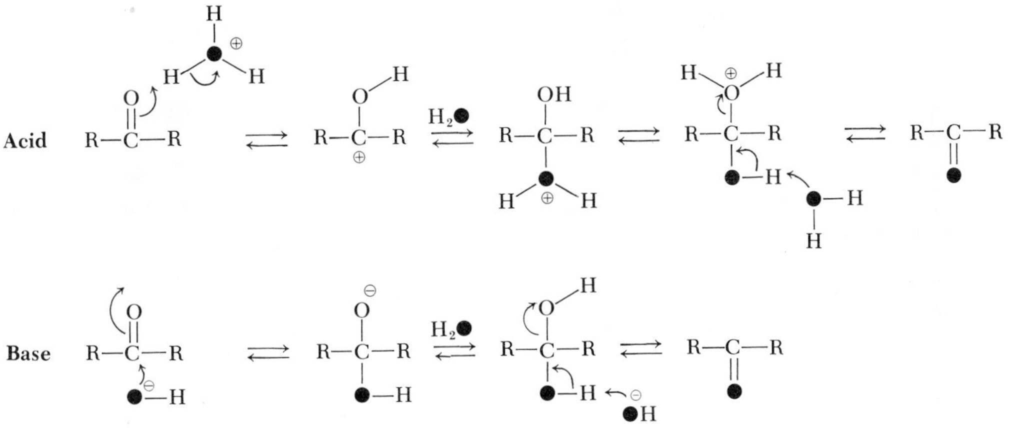

Acid R—C(=O)—R ⇌ R—C⊕(OH)—R ⇌(H₂●) R—C(OH)(●H₂⊕)—R ⇌ R—C(O⊕H₂)(●—H)—R ⇌ R—C(=●)—R

Base R—C(=O)—R + ⊖●—H ⇌ R—C(O⊖)(●—H)—R ⇌(H₂●) R—C(OH)(●—H)—R + ⊖●H ⇌ R—C(=●)—R

Figure 16–2 The mechanism of acid- and base-catalyzed ^{18}O incorporation in carbonyl compounds. ^{16}O is indicated by "O", while ^{18}O is indicated by "●".

7. Using the mechanism in Figure 16–2 as a model, suggest a mechanism for the acid-catalyzed hydrolysis of the following aldimines and ketimines to aldehydes and ketones, respectively (p. 472).

$$RCH{=}NH + H_2O \xrightarrow{H^{\oplus}} RCHO + NH_4^{\oplus}$$
aldimine

$$RCH{=}NR' + H_2O \xrightarrow{H^{\oplus}} RCHO + R'NH_3^{\oplus}$$
N-substituted aldimine

$$R_2C{=}NH + H_2O \xrightarrow{H^{\oplus}} R_2CO + NH_4^{\oplus}$$
ketimine

$$R_2C{=}NR' + H_2O \xrightarrow{H^{\oplus}} R_2CO + R'NH_3^{\oplus}$$
N-substituted ketimine

While many carbonyl compounds rapidly form hydrates, the hydrate is normally unstable relative to the carbonyl compound (Table 16–1). Compare this with the hydration of an alkene, which is very slow in the absence of a catalyst, but which forms a product (an alcohol) that *does not* spontaneously dehydrate to the starting material. This can be related, in part, to the *exo*thermicity of the alkene hydration and the *endo*thermicity of the carbonyl hydration.

$$H_2O + \;\rangle C{=}O \underset{\text{faster}}{\overset{\text{fast}}{\rightleftharpoons}} HO{-}\overset{|}{\underset{|}{C}}{-}OH \; - \sim 5\,\text{kcal/mole}$$

$$H_2O + \;\rangle C{=}C\langle \; \underset{\text{slower}}{\overset{\text{slow}}{\rightleftharpoons}} H{-}\overset{|}{\underset{|}{C}}{-}\overset{|}{\underset{|}{C}}{-}OH \; + \sim 15\,\text{kcal/mole}$$

TABLE 16–1 Stability of Carbonyl Hydrates

CARBONYL COMPOUND	% HYDRATE AT pH 7
H_2CO	100
CH_3CHO	58
$(CH_3)_2CO$	very small
$(CF_3)_2CO$	100
CCl_3CHO	100
C_6H_5CHO	very small

ADDITION OF ALCOHOLS. Alcohols react with carbonyl compounds to form adducts containing either one or two alkoxy residues (—OR). When the carbonyl compound is an aldehyde, these adducts are dubbed *hemiacetals* and *acetals,* respectively; ketones form *hemiketals* and *ketals.* (It is common to find both acetals and ketals categorized as "acetals" because of their similarity.)

$$R{-}\overset{O}{\overset{\|}{C}}{-}H + R'OH \rightleftharpoons \underset{\text{hemiacetal}}{R{-}\overset{OH}{\underset{OR'}{C}}{-}H} \xrightleftharpoons{R'OH} \underset{\text{acetal}}{R{-}\overset{OR'}{\underset{OR'}{C}}{-}H} + H_2O$$

$$R{-}\overset{O}{\overset{\|}{C}}{-}R + R'OH \rightleftharpoons \underset{\text{hemiketal}}{R{-}\overset{OH}{\underset{R}{C}}{-}OR'} \xrightleftharpoons{R'OH} \underset{\text{ketal}}{R{-}\overset{OR'}{\underset{R}{C}}{-}OR'} + H_2O$$

Hemiacetal formation, like hydration, may be acid- or base-catalyzed:

acid catalyzed:

$$R-\overset{\displaystyle O}{\overset{\|}{C}}-H \underset{}{\overset{H^{\oplus}}{\rightleftharpoons}} R-\overset{\displaystyle {}^{\oplus}OH}{\overset{\|}{C}}-H \leftrightarrow R-\overset{\displaystyle OH}{\overset{|}{\overset{\oplus}{C}}}-H$$

$$R-\underset{\oplus}{\overset{\displaystyle OH}{\overset{|}{C}}}-H + R'OH \rightleftharpoons R-\underset{\underset{\displaystyle \underset{\oplus}{R'-O-H}}{|}}{\overset{\displaystyle OH}{\overset{|}{C}}}-H \rightleftharpoons R-\underset{\underset{\displaystyle OR'}{|}}{\overset{\displaystyle OH}{\overset{|}{C}}}-H + H^{\oplus}$$

base catalyzed:

$$R'OH + B\colon \rightleftharpoons R'O\colon^{\ominus} + \overset{\oplus}{B}H$$

$$R-\overset{\displaystyle O}{\overset{\|}{C}}-H + R'O^{\ominus} \rightleftharpoons R-\underset{\underset{\displaystyle OR}{|}}{\overset{\displaystyle O^{\ominus}}{\overset{|}{C}}}-H$$

$$R-\underset{\underset{\displaystyle OR'}{|}}{\overset{\displaystyle O^{\ominus}}{\overset{|}{C}}}-H + \overset{\oplus}{B}H \rightleftharpoons R-\underset{\underset{\displaystyle OR'}{|}}{\overset{\displaystyle OH}{\overset{|}{C}}}-H + B\colon$$

While hemiacetal formation can be either acid- or base-catalyzed, conversion of an aldehyde or hemiacetal to an acetal is only acid-catalyzed. Base-catalyzed acetal formation would require the loss of hydroxide ion, an unlikely process because of the poor leaving-group ability of that ion. Acetals are hydrolyzed by aqueous acid to form the aldehyde by the reverse of the mechanism of its formation. Acetals are, however, stable to hydrolysis in basic solution. (Why?)

$$\left.\left\{\begin{array}{l} R'OH + B\colon \rightleftharpoons RO^{\ominus} + \overset{\oplus}{B}H \\[2ex] R-\underset{\underset{\displaystyle OR'}{|}}{\overset{\displaystyle OH}{\overset{|}{C}}}-H + R'O^{\ominus} \rightleftharpoons R-\underset{\underset{\displaystyle OR'}{|}}{\overset{\displaystyle OR'}{\overset{|}{C}}}-H + OH^{\ominus} \end{array}\right\}\right. \text{does } \textit{not} \text{ occur}$$

$$\underset{\text{hemiacetal}}{R-\underset{\underset{\displaystyle OR'}{|}}{\overset{\displaystyle OH}{\overset{|}{C}}}-H} + H^{\oplus} \rightleftharpoons R-\underset{\underset{\displaystyle OR'}{|}}{\overset{\displaystyle \overset{\oplus}{O}H_2}{\overset{|}{C}}}-H \overset{-H_2O}{\rightleftharpoons} \underbrace{R-\underset{\underset{\displaystyle OR'}{|}}{\overset{\oplus}{C}}-H \leftrightarrow R-\underset{\underset{\displaystyle {}^{\oplus}OR}{\|}}{C}-H}$$

$$\Big\Updownarrow R'OH$$

$$\underset{\text{acetal}}{R-\underset{\underset{\displaystyle OR'}{|}}{\overset{\displaystyle OR'}{\overset{|}{C}}}-H} \rightleftharpoons R-\underset{\underset{\displaystyle OR'}{|}}{\overset{\displaystyle H\overset{\oplus}{O}R'}{\overset{|}{C}}}-H$$

In the laboratory, acetal formation is used to protect the carboxaldehyde group in alkaline media; most unprotected aldehydes slowly polymerize in alkaline solution (Chapter 17). Protection also prevents the otherwise facile oxidation of aldehydes to carboxylic acids. The conversion of acrolein to glyceraldehyde, shown below, is possible only because of the protection of the carbonyl group.

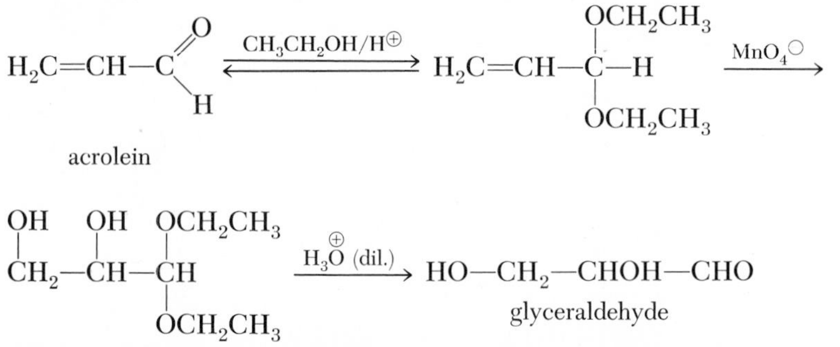

$$\underset{\text{acrolein}}{H_2C{=}CH{-}CHO} \xrightleftharpoons{CH_3CH_2OH/H^{\oplus}} H_2C{=}CH{-}CH(OCH_2CH_3)_2 \xrightarrow{MnO_4^{\ominus}}$$

$$HOCH_2{-}CH(OH){-}CH(OCH_2CH_3)_2 \xrightarrow{H_3O^{\oplus}\text{ (dil.)}} \underset{\text{glyceraldehyde}}{HO{-}CH_2{-}CHOH{-}CHO}$$

Five-membered cyclic acetals called 1,3-dioxolanes are prepared from an aldehyde and a *vic*-diol (such as ethylene glycol). Formation of the five-membered ring rather than a polymer is favored by the proximity of the second —OH group to the hemiacetal carbon.

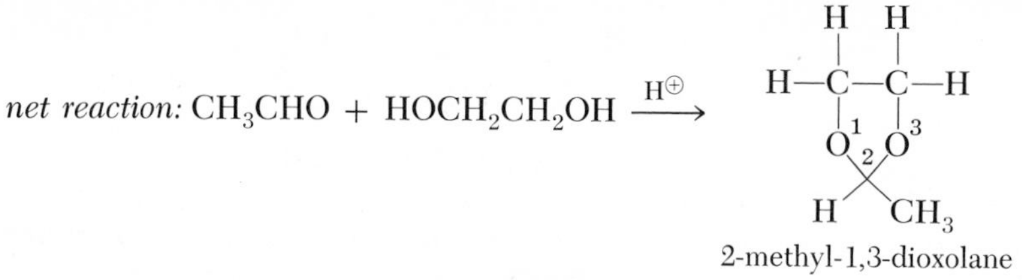

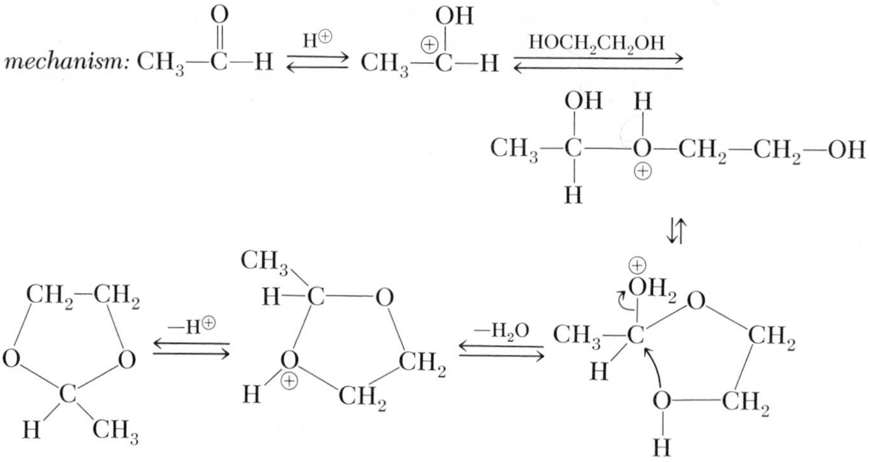

net reaction: $CH_3CHO + HOCH_2CH_2OH \xrightarrow{H^{\oplus}}$ 2-methyl-1,3-dioxolane

2-methyl-1,3-dioxolane

mechanism: $CH_3{-}\overset{O}{\overset{\|}{C}}{-}H \xrightleftharpoons{H^{\oplus}} CH_3{-}\overset{\oplus}{C}(OH){-}H \xrightleftharpoons{HOCH_2CH_2OH} CH_3{-}CH(OH){-}\overset{\oplus}{O}H{-}CH_2{-}CH_2{-}OH$

$\rightleftharpoons$ (cyclic intermediate with $\overset{\oplus}{O}H_2$) $\xrightleftharpoons{-H_2O}$ (protonated 2-methyl-1,3-dioxolane) $\xrightleftharpoons{-H^{\oplus}}$ 2-methyl-1,3-dioxolane

If a molecule contains an aldehyde group and a hydroxy group separated by either three or four carbons, these will react to form a cyclic hemiacetal containing a five- or six-membered ring.

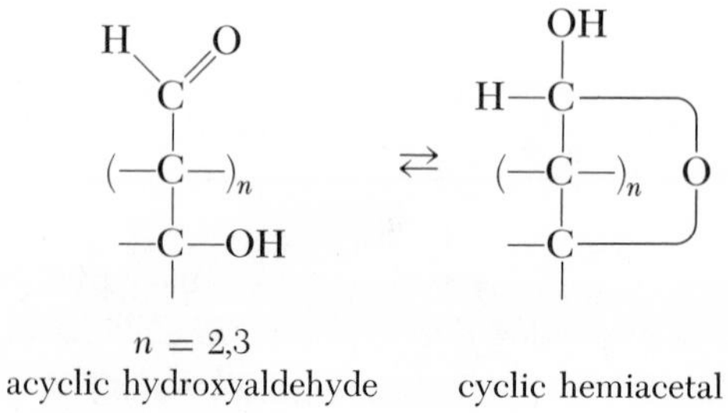

$$\underset{\substack{n\,=\,2,3\\ \text{acyclic hydroxyaldehyde}}}{HC(=O){-}(C)_n{-}C{-}OH} \rightleftharpoons \underset{\text{cyclic hemiacetal}}{H{-}C(OH){-}(C)_n{-}C{-}O\ (\text{ring})}$$

This type of reaction is extremely important in understanding the chemistry of carbohydrates (discussed in detail in Chapter 25). For example, glucose, the most important simple carbohydrate, exists almost completely in a cyclic hemiacetal form.

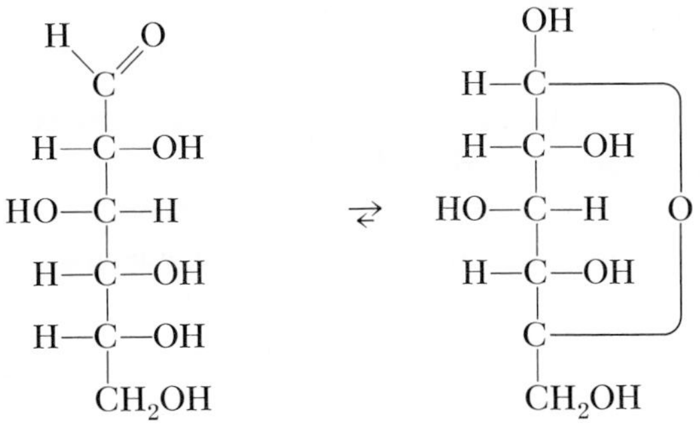

glucose (acyclic hydroxyaldehyde) ⇌ glucose (cyclic hemiacetal)

Carbohydrates such as glucose form acetals as well as hemiacetals. Indeed, most of the complex carbohydrates found in nature, including cellulose, starch, and glycogen, are actually polymeric acetals. Even the carbohydrate components of things as complex as the gene (Chapter 26) and the bacterial cell wall (Chapter 27) are polymeric acetals. Since these will be detailed later, we shall indicate only in a schematic fashion the nature of such biologically important acetals. Note that a molecule that forms such a polymer must contain at least two hydroxy groups and one aldehyde group. One hydroxy group reacts *intra*molecularly to form the cyclic hemiacetal, while the second reacts *inter*molecularly to form the polymeric acetal.

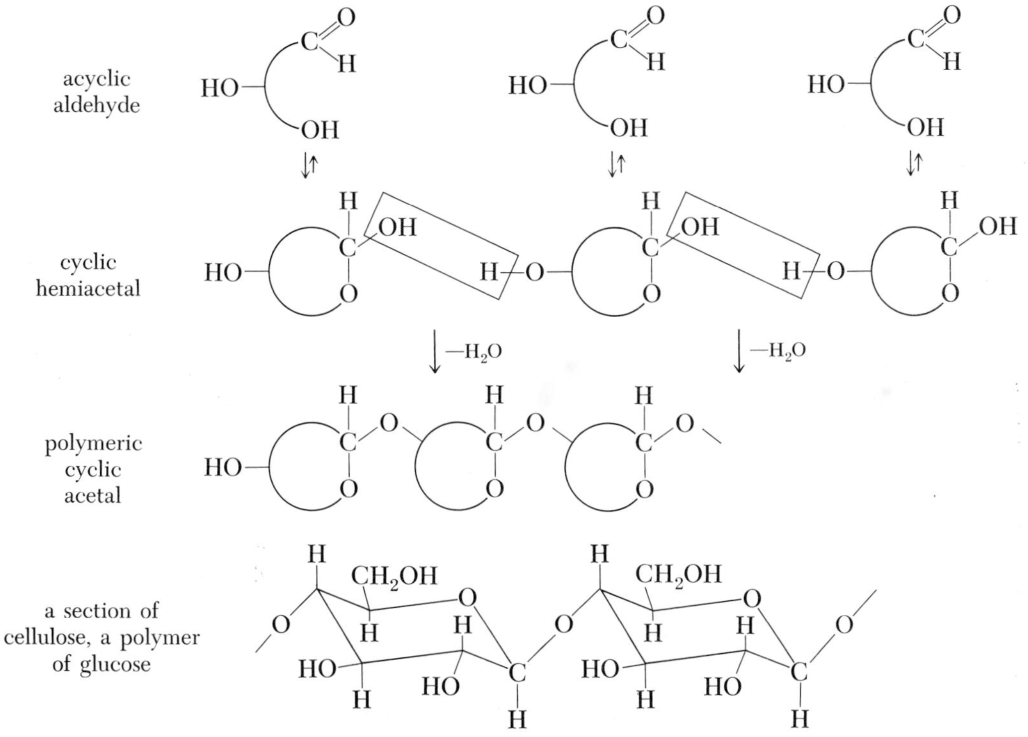

acyclic aldehyde

cyclic hemiacetal

$-H_2O$

polymeric cyclic acetal

a section of cellulose, a polymer of glucose

8. Glucose actually forms two cyclic hemiacetals, each containing a six-membered ring. These are isomers, have different melting points and different specific rotations. When either of them is dissolved in water, there results a mixture of three compounds, that is, both of these isomers and a trace of the acyclic form. Account for these facts. (The solution to this problem is discussed in detail in Chapter 25A.)

Most of this discussion has centered upon hemiacetals and acetals because, at room temperature, the position of equilibrium in the reaction between alcohols and ketones does not favor hemiketals or ketals. When a ketal is required (hemiketals are not important except as non-isolated intermediates in ketal hydrolysis), it is often prepared by the reaction of a ketone with a *vic*-diol (however, the method used in problem 9 is also applicable to ketones). The single most important ketal is probably 2,2-dimethoxypropane, synthesized commercially by the reaction of acetone with methanol in the presence of acid.

$$CH_3-\overset{\overset{\displaystyle O}{\|}}{C}-CH_3 + 2CH_3OH \xrightleftharpoons{-27°,\ H^{\oplus}} CH_3-\underset{\underset{\displaystyle OCH_3}{|}}{\overset{\overset{\displaystyle OCH_3}{|}}{C}}-CH_3 + H_2O$$

2,2-dimethoxypropane

The extremely rapid hydrolysis of 2,2-dimethoxypropane, and the concomitant production of two volatile end products (*i.e.*, acetone and methanol), makes it a valuable reagent for removing water formed in other reactions and for the chemical drying of compounds.

9. Suggest a mechanism for the acid-catalyzed conversion of an aldehyde into an acetal by reaction with an *orthoester*, *i.e.*, a compound of the type $RC(OR)_3$.

$$CH_3CHO + \underset{\text{methyl orthoformate}}{HC(OCH_3)_3} \xrightarrow{H^{\oplus}} \underset{\text{acetaldehyde dimethylacetal}}{CH_3CH(OCH_3)_2} + \underset{\text{methyl formate}}{H-\overset{\overset{\displaystyle O}{\|}}{C}-OCH_3}$$

10. While 2,2-dimethoxypropane can be viewed as an ether, it is rapidly hydrolyzed to methanol and acetone in dilute aqueous acid, although diethyl ether and 1,2-dimethoxypropane are stable in this medium. Account for this difference, providing a mechanism for the acid-catalyzed hydrolysis of 2,2-dimethoxypropane.

11. The acetals produced from glucose are called glucosides. Suggest a mechanism for the acid-catalyzed hydrolysis of methyl glucoside.

$$\begin{array}{c} OCH_3 \\ | \\ H-C- \\ | \\ H-C-OH \\ | \\ HO-C-H \quad O \\ | \\ H-C-OH \\ | \\ H-C- \\ | \\ CH_2OH \end{array} \xrightarrow{H_3O^{\oplus}} \begin{array}{c} H \quad O \\ \backslash \ /\!/ \\ C \\ | \\ H-C-OH \\ | \\ HO-C-H \\ | \\ H-C-OH \\ | \\ H-C-OH \\ | \\ CH_2OH \end{array} + CH_3OH$$

methyl glucoside — glucose

One of the many applications of ketalization in steroid chemistry is found in the synthesis of norethisterone (used in oral contraceptive preparations), 17α-ethynyl-19-nortestosterone. Key steps of this synthesis, showing how one of two carbonyl groups is protected, are presented below.

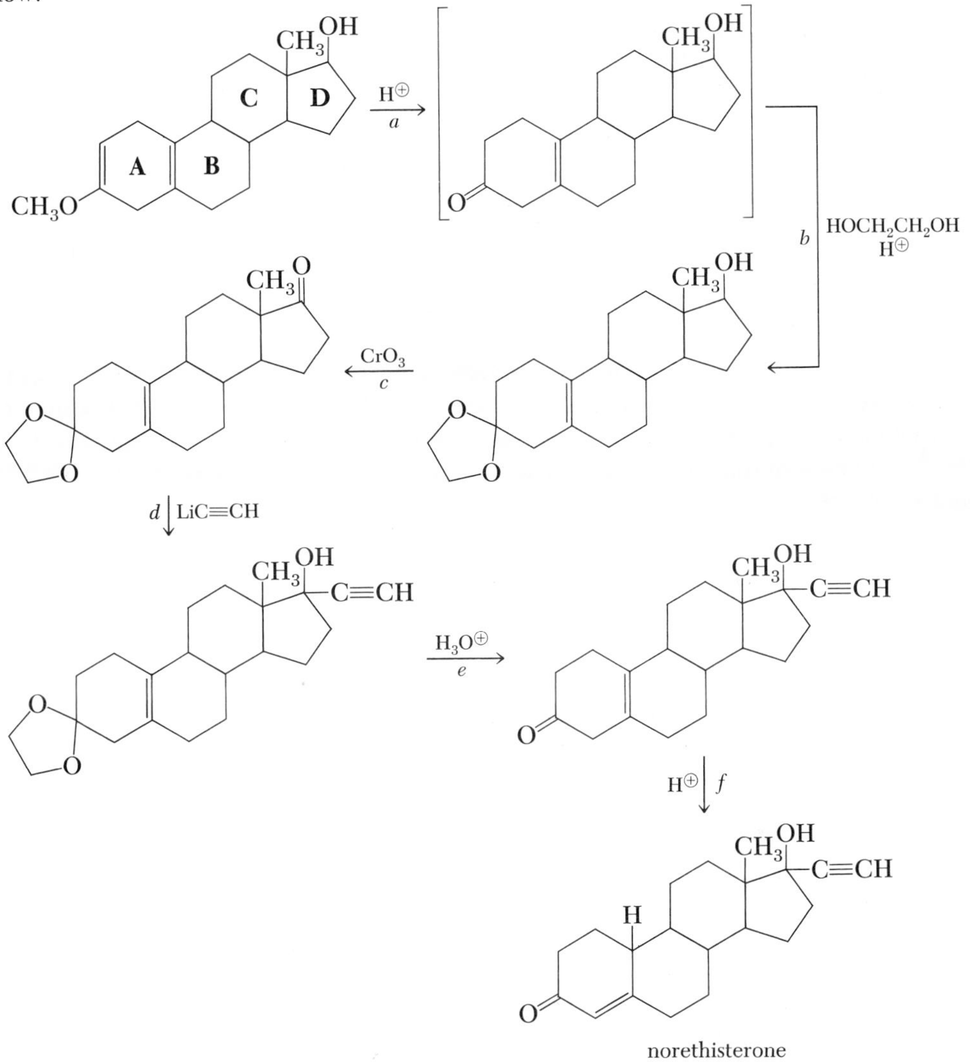

CH_3 OH, C, D, A, B, CH_3O $\xrightarrow[a]{H^{\oplus}}$ [CH_3 OH, O] $\xrightarrow[b]{HOCH_2CH_2OH,\ H^{\oplus}}$ CH_3 OH $\xrightarrow[c]{CrO_3}$ CH_3 O $\xrightarrow[d]{LiC{\equiv}CH}$ CH_3 OH, $-C{\equiv}CH$ $\xrightarrow[e]{H_3O^{\oplus}}$ CH_3 OH, $-C{\equiv}CH$, O $\xrightarrow[f]{H^{\oplus}}$ CH_3 OH, $-C{\equiv}CH$, H, O

norethisterone

12. Below are presented alternate mechanisms for step *a* in the synthesis of norethisterone. Which seems more reasonable? Explain.

(A) H, CH_3O $\xrightarrow{H^{\oplus}}$ H, $H{-}\overset{\oplus}{O}$, CH_3, O, H, H $\rightarrow$ H, HO $+ \overset{\oplus}{C}H_3OH_2$

$H^{\oplus}$, H, O, H, OH_2 $\rightarrow$ O $+ H_3O^{\oplus}$

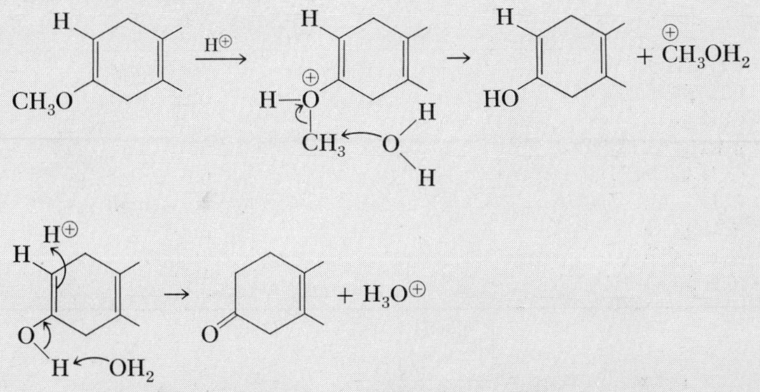

(B) [reaction scheme: protonation of 1-methoxycyclohexene derivative; $H^{\oplus}$; CH_3O; $\xrightarrow{H_2O}$; CH_3O, $\overset{\oplus}{O}H_2$; $\xrightarrow{-CH_3OH}$ ketone $+ H_3O^{\oplus}$; H—OH_2]

Cyanohydrin Formation. Addition of hydrogen cyanide to a carbonyl compound produces a *gem*-hydroxycyanide, more usually called a *cyanohydrin*. The reaction is base-catalyzed, the function of the base being to generate the active nucleophile, $CN^{\ominus}$, from the weak acid HCN. It can be seen from the data in Table 16–2 that cyanohydrin formation is more favored than is hydration.

net process: $R—\overset{O}{\overset{\|}{C}}—R + HCN \rightleftharpoons R—\underset{CN}{\overset{OH}{C}}—R$ (cyanohydrin)

mechanism: $H—CN + \underset{\text{base}}{B:} \rightleftharpoons BH^{\oplus} + CN^{\ominus}$ (fast)

$RC(O)R + CN^{\ominus} \rightleftharpoons R—\underset{CN}{\overset{O^{\ominus}}{C}}—R$ (slow)

$R—\underset{CN}{\overset{O^{\ominus}}{C}}—R + BH^{\oplus} \rightleftharpoons R—\underset{CN}{\overset{OH}{C}}—R + B:$ (fast)

TABLE 16-2 Stability of Cyanohydrins

Carbonyl Compound	Formation Constant of Cyanohydrin K_{form}[a]
CH_3CHO	very large
$CH_3C(O)CH_2CH_3$	38
C_6H_5CHO	210
p-$CH_3OC_6H_4CHO$	32
$C_6H_5C(O)CH_3$	0.8
$C_6H_5C(O)C_6H_5$	very small
$(CH_2)_4CO$	500
$(CH_2)_5CO$	10,000
$(CH_2)_6CO$	75

[a] Defined by: $K_{form} = \dfrac{\left[R\begin{smallmatrix}OH\\CN\end{smallmatrix}\right]}{[RCHO][HCN]}$

Let us examine some of the data in Table 16–2. The difference between benzaldehyde and *p*-methoxybenzaldehyde reflects the decreased double bond character in the carbonyl group of the latter because of resonance interaction between the methoxy and carbonyl groups. Put differently, *p*-methoxybenzaldehyde is stabilized by resonance to a greater extent than is benzaldehyde; this decreases the energy difference between *p*-methoxybenzaldehyde and its cyanohydrin compared to benzaldehyde and its cyanohydrin.

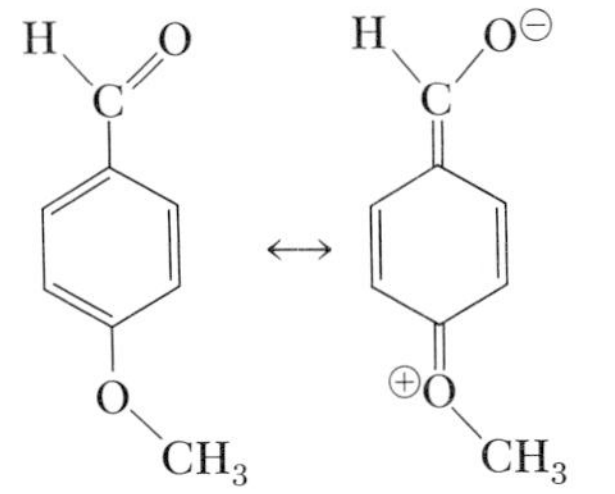

resonance in *p*-methoxybenzaldehyde

Cyclohexanone cyanohydrin is much more stable, relative to its progenitor ketone, than is cyclopentanone cyanohydrin. This is ascribed to introduction of eclipsing interactions when the sp^2 carbon in cyclopentanone is rehybridized to sp^3. Unfavorable steric interactions are probably *reduced* in going from cyclohexanone to its cyanohydrin, since the lowest energy conformation of the six-membered ring is the chair form (*i.e.*, all carbons sp^3).

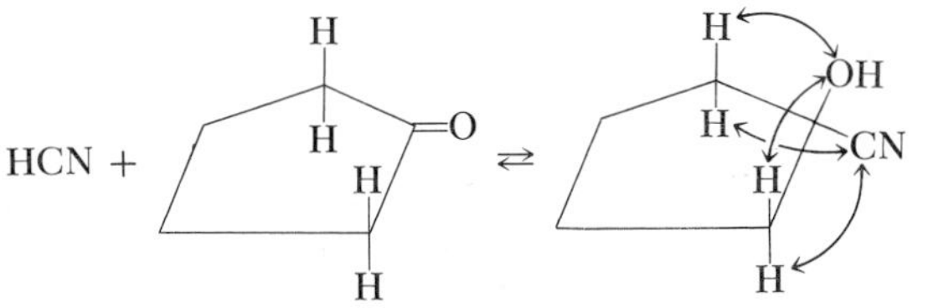

HCN + cyclopentanone ⇄ cyclopentanone cyanohydrin (eclipsing interactions) $K_{eq}^{HCN} \approx 500$

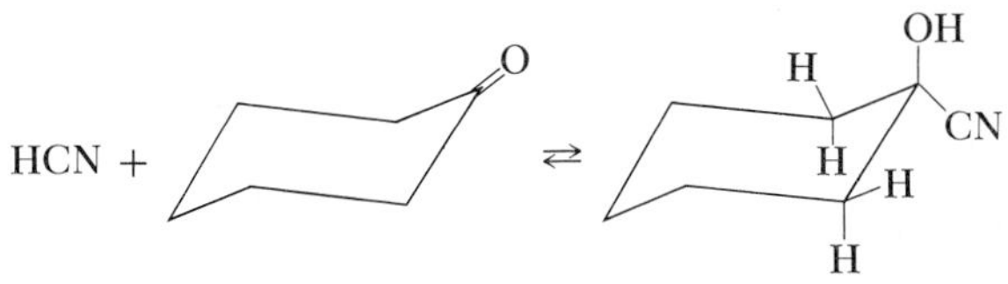

HCN + cyclohexanone ⇄ cyclohexanone cyanohydrin $K_{eq}^{HCN} \approx 10{,}000$

Plants are capable of synthesizing cyanohydrins, some of which are common in Nature. This gives rise to the serious possibility of fatalities arising from inadvertent ingestion of a naturally occurring cyanohydrin followed by liberation of hydrogen cyanide within the body. A classic example of this danger is found among plants of the genus *Prunus*, the plants that give us such fruits as plums, cherries, peaches, and almonds. There are, for example, two types of almond which we will consider: *Prunus amygdalus dulcis* and *Prunus amygdalus amara*. While the last part of a three-part name might not cause the average almond lover to pause in wonder, he had better! The former is sometimes called "sweet almond," while the latter is usually called "bitter almond." Both plants produce kernels that can be crushed to provide oils, yielding *oleum amygdalae dulcis* and *oleum amygdalae amarae*, respectively. The latter is better known as "oil of bitter almonds" and produces, upon hydrolysis, between 2 and 4% hydrogen cyanide due to a cyanohydrin component. *Oleum amygdalae dulcis* is used in confectionery.

Bisulfite Addition. When aldehydes, methyl ketones, and some cyclic ketones are shaken with a concentrated solution of sodium bisulfite, $NaHSO_3$, they react to form *bisulfite addition compounds*. These adducts are water-soluble salts that are insoluble in common organic solvents.

$$RCHO + NaHSO_3 \rightarrow RCH(OH)SO_3^{\ominus}Na^{\oplus}$$

sodium bisulfite

The main application of this salt formation is in the purification of mixtures containing carbonyl compounds that form these adducts. For example, one can prepare diphenylcarbinol by reacting benzaldehyde with phenyl magnesium bromide, followed by the normal work-up. A convenient way to remove small amounts of unreacted benzaldehyde is by shaking the crude reaction product with aqueous sodium bisulfite, thereby converting it to its bisulfite adduct. If the resultant mixture is extracted with ether, the desired carbinol will dissolve in the ether while the benzaldehyde, now as its bisulfite adduct, remains in the aqueous phase. Should recovery of the unreacted benzaldehyde be necessary, it can be achieved simply by treating the bisulfite addition compound with dilute aqueous acid or base or with formaldehyde.

$$C_6H_5\text{—}CHO + C_6H_5\text{—}MgBr \xrightarrow{THF} \xrightarrow{H_2O}$$

benzaldehyde

$$C_6H_5\text{—}CH(OH)\text{—}C_6H_5 + C_6H_5\text{—}CHO$$

diphenylcarbinol — unreacted

(a) $NaHSO_3/H_2O$
(b) ether

$$C_6H_5\text{—}CH(OH)\text{—}C_6H_5 + C_6H_5\text{—}CH(OH)SO_3^{\ominus}Na^{\oplus}$$

in ether layer — bisulfite adduct of C_6H_5CHO in water layer

$$C_6H_5\text{—}CH(OH)SO_3^{\ominus}Na^{\oplus} \xrightarrow{HCl/H_2O} C_6H_5\text{—}CHO + NaCl + H_2O + SO_2\uparrow$$

$$C_6H_5\text{—}CH(OH)SO_3^{\ominus}Na^{\oplus} \xrightarrow[H_2O]{Na_2CO_3} C_6H_5\text{—}CHO + Na_2SO_3 + NaHCO_3$$

$$C_6H_5\text{—}CH(OH)SO_3^{\ominus}Na^{\oplus} \xrightarrow[H_2O]{CH_2O} C_6H_5\text{—}CHO + H_2C(OH)SO_3^{\ominus}Na^{\oplus}$$

For many years the structure of these adducts was a subject of intense debate. Now, however, there is no doubt that sodium bisulfite reacts as a nucleophile to form a carbon-sulfur bond.

$$Na^{\oplus}\ HO\text{—}S(=O)(O^{\ominus})\text{:} + RHC\text{=}O \rightarrow Na^{\oplus}\ HO\text{—}S(=O)_2\text{—}CHR\text{—}O^{\ominus} \rightleftarrows Na^{\oplus}\ {}^{\ominus}O\text{—}S(=O)_2\text{—}CHR\text{—}OH$$

aldehyde-bisulfite adduct

Steric hindrance prevents most ketones from forming bisulfite adducts. This reaction is analogous to carbonyl hydration or cyanohydrin formation; changing the carbonyl carbon from sp^2 to sp^3 in the adduct increases the number of skew butane interactions and destabilizes the adduct if other factors (*e.g.*, relief of angle strain) do not compensate for this.

Aldehyde Polymerization. Simple aldehydes, but not ketones, undergo head-to-tail polymerization to form both cyclic trimers and linear polymers. The tendency for simple aldehydes to polymerize is so great that pure formaldehyde is not commercially available. Instead, one can purchase only paraformaldehyde (a linear polymer made by dissolving formaldehyde in water) or 1,3,5-trioxane (a cyclic trimer made by heating paraformaldehyde in dilute acid).

Formaldehyde is prepared in the laboratory by heating 1,3,5-trioxane and sweeping out the formaldehyde produced by depolymerization with a stream of nitrogen gas. The nitrogen also serves to dilute the formaldehyde, thus preventing rapid repolymerization.

$$\underset{\text{formaldehyde}}{nCH_2O} + H_2O \rightarrow \underset{\text{paraformaldehyde}}{H(OCH_2)_n\text{—}OH}$$

$$3CH_2O + H^{\oplus} \rightarrow \underset{\substack{\text{1,3,5-trioxane} \\ \text{(mp 61°)}}}{(CH_2O)_3} \xrightarrow[\text{heat}]{N_2} 3CH_2O$$

Acetaldehyde, available as the monomer (bp 20°), forms a cyclic trimer called *paraldehyde.*

$$3CH_3CHO \xrightarrow{H^{\oplus}} \underset{\text{paraldehyde}}{(CH_3CHO)_3}$$

Both paraformaldehyde and paraldehyde can be found in many hospitals. The former is used to sterilize walls, instruments, and other items, while the latter is a sedative-hypnotic. Paraformaldehyde also has been used as a spermicide in vaginal contraceptive cream.

16.5 ADDITION-ELIMINATION REACTIONS

Most addition-elimination reactions of aldehydes and ketones involve the condensation of the carbonyl group with a reagent of the type H_2Z, where Z represents the remainder of the molecule. The products include water and an organic compound in which C═O has been replaced by C═Z. As suggested by the title of this section, these reactions proceed by an initial addition across the carbonyl group followed by a 1,2-elimination. The largest class of compounds that react with carbonyl compounds in this fashion are derivatives of ammonia, and we will, therefore, restrict the subsequent discussion to these adducts.

$$\gt C{=}O + H_2NR \xrightarrow{\text{addition}} \left[HO{-}\overset{|}{\underset{|}{C}}{-}\overset{H}{\overset{|}{N}}{-}R \right] \xrightarrow{\text{elimination}} \gt C{=}NR + H_2O$$

Ammonia and its Derivatives. Ammonia behaves like many other nucleophiles in forming adducts with carbonyl compounds. Except for the adduct with acetaldehyde, these are usually unstable and dehydrate to form *imines*, also called *Schiff's bases*. Most unsubstituted imines, *i.e.*, compounds of the type $\gt C{=}NH$, are unstable and polymerize upon standing. The use of substituted amines, RNH_2, leads to the comparatively stable *N*-substituted imines. Imines revert to their progenitor carbonyl compounds upon hydrolysis (see problem 7).

$$CH_3CHO + NH_3 \rightleftarrows CH_3{-}\overset{OH}{\overset{|}{\underset{H}{\underset{|}{C}}}}{-}NH_2 \qquad \text{the exception}$$

$$R{-}\overset{O}{\overset{\|}{C}}{-}H(R) + NH_3 \rightleftarrows R{-}\overset{OH}{\overset{|}{\underset{NH_2}{\underset{|}{C}}}}{-}H(R) \rightleftarrows \underset{\text{imine}}{R{-}\overset{NH}{\overset{\|}{C}}{-}H(R)} + H_2O \qquad \text{the rule}$$

$$R{-}\overset{O}{\overset{\|}{C}}H(R) + R'NH_2 \rightleftarrows R{-}\overset{OH}{\overset{|}{\underset{NHR}{\underset{|}{C}}}}{-}H(R) \rightleftarrows \underset{N\text{-substituted imine}}{R{-}\overset{NR}{\overset{\|}{C}}{-}H(R)} + H_2O$$

$$R{-}\overset{NH(R')}{\overset{\|}{C}}{-}H(R) + H_2O \rightarrow R{-}\overset{O}{\overset{\|}{C}}{-}H(R) + H_2NH(R') \qquad \text{imine hydrolysis}$$

As with many other reactions, ketones are less reactive than are aldehydes, although both usually undergo these condensation reactions.

aldehyde + amine ↓ *aldimine*

$$CH_3CHO + CH_3CH_2NH_2 \xrightarrow{-H_2O} \underset{\text{ethylidene ethylamine}}{CH_3CH{=}NCH_2CH_3} \qquad (77\%)$$

$$C_6H_5CHO + CH_3NH_2 \xrightarrow{-H_2O} \underset{N\text{-benzylidene methylamine}}{C_6H_5C{=}NCH_3} \qquad (70\%)$$

ketone + amine ↓ *ketimine*

$$\underset{\text{fluorenone}}{\text{(fluorenone structure)}} + NH_3 \xrightarrow{-H_2O} \underset{\text{fluorenimine}}{\text{(fluorenimine structure)}} \qquad (66\%)$$

A host of ammonia derivatives other than amines also form similar condensation products with aldehydes and ketones. Included in this group are *hydroxylamine* ($H_2N{-}OH$), *hydrazine* ($H_2N{-}NH_2$), *phenylhydrazine* ($H_2N{-}NHC_6H_5$), and *semicarbazide* ($H_2N{-}NHC(O)NH_2$). The products that these form are called *oximes, hydrazones, phenylhydrazones,* and *semicarbazones,* respectively. Also, the *2,4-dinitrophenylhydrazones* of carbonyl compounds are encountered frequently; these require *2,4-dinitrophenylhydrazine* (2,4-DNPH or DNPH) as the condensing reagent.

$$R_2C{=}O \xrightarrow[\text{(hydroxylamine)}]{H_2NOH} R_2C{=}N{-}OH \quad \text{(oxime)}$$

$$R_2C{=}O \xrightarrow[\text{(hydrazine)}]{H_2N{-}NH_2} R_2C{=}N{-}NH_2 \quad \text{(hydrazone)} \xrightarrow{R_2C{=}O} R_2C{=}N{-}N{=}CR_2 \quad \text{(azine)}$$

$$R_2C{=}O \xrightarrow[\text{(phenylhydrazine)}]{H_2N{-}NHC_6H_5} R_2C{=}N{-}NHC_6H_5 \quad \text{(phenylhydrazone)}$$

$$R_2C{=}O \xrightarrow[\text{(semicarbazide)}]{H_2N{-}NH{-}C(O){-}NH_2} R_2C{=}N{-}NH{-}C(O){-}NH_2 \quad \text{(semicarbazone)}$$

$$R_2C{=}O \xrightarrow[\text{(2,4-dinitrophenylhydrazine)}]{H_2N{-}NH{-}C_6H_3(NO_2)_2} R_2C{=}N{-}NH{-}C_6H_3(NO_2)_2 \quad \text{(2,4-dinitrophenylhydrazone)}$$

While many aldehydes and ketones are liquids, most of the derivatives just cited are solids with sharp melting points. Therefore, these derivatives are often used as aids in identifying carbonyl compounds. Since the condensation products can be hydrolyzed to yield the original carbonyl compound, such reactions also can serve to isolate an aldehyde or ketone from a reaction mixture.

The condensation of carbonyl compounds with these reagents is normally carried out with the assistance of an electrophilic catalyst, most commonly a proton. The function of the catalyst is to coordinate with the carbonyl group, thus making it more subject to attack by a nucleophile. Attack by the most nucleophilic nitrogen, proton transfer and, finally, 1,2-elimination produce the desired product. This entire sequence is outlined below for the reaction of acetone with semicarbazide.

$$(CH_3)_2C{=}O + H^{\oplus} \rightleftarrows (CH_3)_2C{=}\overset{\oplus}{O}H \leftrightarrow (CH_3)_2\overset{\oplus}{C}{-}OH$$

$$CH_3{-}\overset{OH}{\underset{\overset{\oplus}{N}H_2NHC(O)NH_2}{C}}{-}CH_3 \rightleftarrows CH_3{-}\overset{OH}{\underset{H-\overset{\oplus}{N}(NHC(O)NH_2)-H}{C}}{-}CH_3 \rightleftarrows CH_3{-}\overset{\overset{\oplus}{O}H_2}{\underset{N(H)NHC(O)NH_2}{C}}{-}CH_3 \xrightarrow{\text{fast}} (CH_3)_2C{=}N{-}NHC(O)NH_2$$

It is noteworthy that the use of a large amount of acid may actually diminish the rate of reaction. This is due to the fact that the nucleophilic nitrogens also are bases and, in acidic solution, will be converted to non-nucleophilic cations.

$$H_2\ddot{N}{-}OH + H_3\overset{\oplus}{O} \rightleftarrows H_3\overset{\oplus}{N}{-}OH + H_2O$$

nucleophilic nitrogen — non-nucleophilic nitrogen

The pH that provides the maximum rate of reaction will depend on the basicity of the nucleophilic nitrogen and, therefore, upon the specific condensing reagent. For example, 2,4-dinitrophenylhydrazine is a comparatively poor base and can be used in strongly acidic solutions. By way of contrast, hydroxylamine is a stronger base and its reactions are favored by a more weakly acidic solution.

A Biological Role for Imine Formation and Hydrolysis. Protein ingestion is essential to good nutrition, and this has triggered a worldwide hunt for better and cheaper sources of protein. Some of the results of this are "hamburger" containing plant protein, and "bacon bits" made with soy flour and containing no bacon. Proteins, as we shall learn in Chapter 24, are complex polymers made from amino acids, $RCH(NH_2)CO_2H$. Two of the major needs of the body are to convert various compounds into amino acids and to convert certain amino acids, present in abundance, into compounds that may be in short supply. One way in which the cell does this employs imine formation and hydrolysis to convert an amino acid and an α-keto acid into a keto acid derived from the original amino acid and an amino acid derived from the original keto acid.

$$R{-}CH(NH_2){-}CO_2H + R'{-}C(=O){-}CO_2H \xrightleftharpoons{\text{enzyme}} R{-}C(=O){-}CO_2H + R'{-}CH(NH_2){-}CO_2H$$

$$\text{amino acid I} + \alpha\text{-keto acid II} \xrightleftharpoons{\text{enzyme}} \alpha\text{-keto acid I} + \text{amino acid II}$$

This interconversion requires pyridoxal phosphate, a member of the vitamin B_6 family, and uses the carboxaldehyde group of this pyridoxal phosphate to (a) form an imine with amino acid I, (b) "store" the NH_2 group when amino acid I is converted to the corresponding α-keto acid I, and (c) form an imine with α-keto acid II. What is remarkable is that, after this complex series of reactions is completed, the pyridoxal phosphate is present in its original form, ready to begin another series of interconversions. While the reaction sequence outlined below shows pyridoxal phosphate as *the* active compound in this *transamination*, it is actually an enzyme-pyridoxal phosphate complex which catalyzes this reaction.

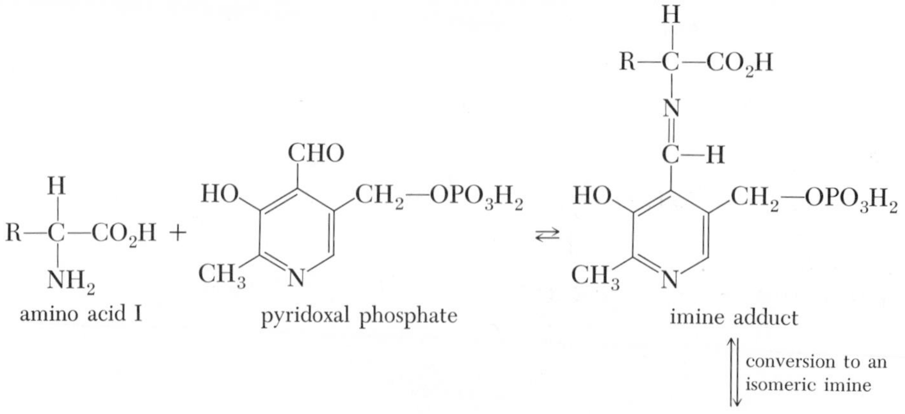

$$R{-}CH(NH_2){-}CO_2H + \text{(pyridine ring: CHO, HO, } CH_2{-}OPO_3H_2\text{, } CH_3\text{)} \rightleftarrows \text{(pyridine ring: } C(H){=}N{-}CH(R){-}CO_2H\text{, HO, } CH_2{-}OPO_3H_2\text{, } CH_3\text{)}$$

amino acid I — pyridoxal phosphate — imine adduct

⇅ conversion to an isomeric imine

Formula continued on opposite page.

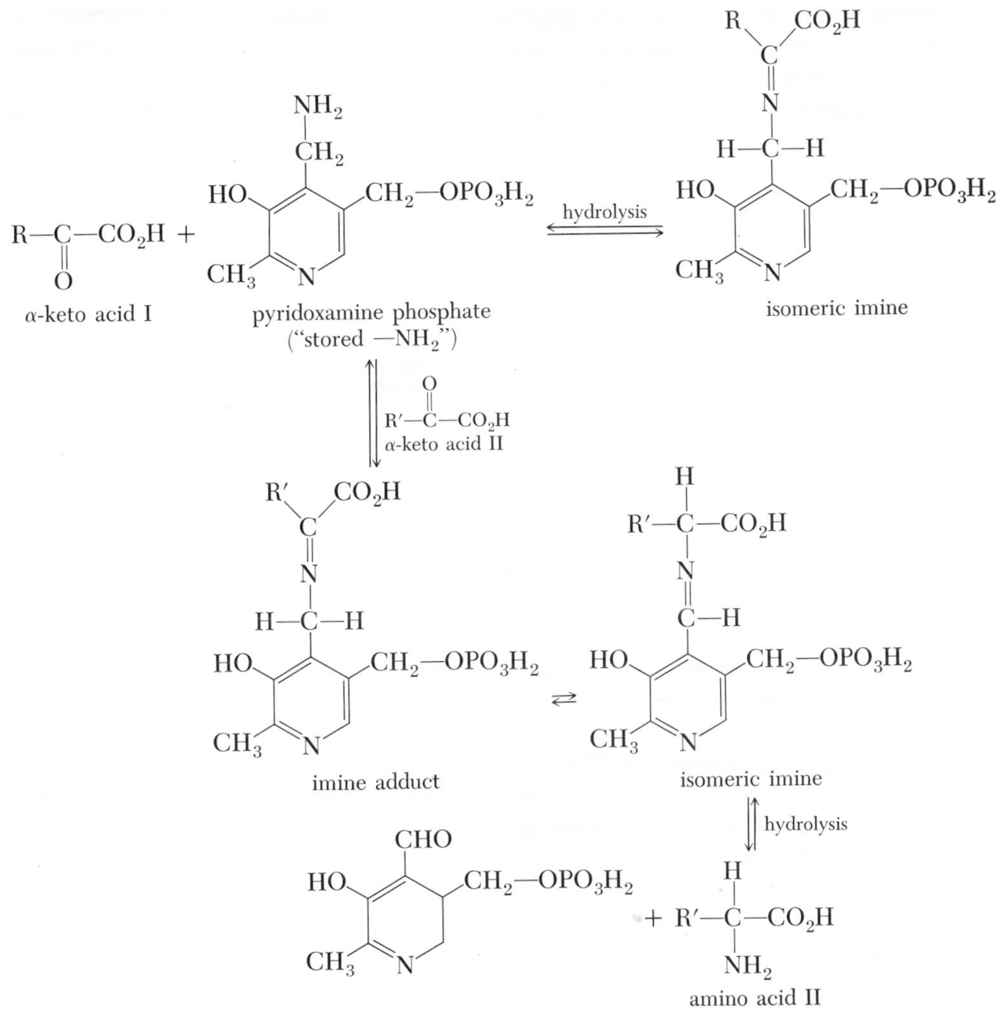

R—C(=O)—CO_2H + pyridoxamine phosphate ("stored —NH_2") $\underset{\text{hydrolysis}}{\rightleftharpoons}$ isomeric imine

α-keto acid I

pyridoxamine phosphate
("stored —NH_2")

⇅ R′—C(=O)—CO_2H
α-keto acid II

imine adduct ⇄ isomeric imine

⇅ hydrolysis

(pyridoxal phosphate: CHO, HO, CH_3, CH_2—OPO_3H_2) + R′—CH(NH_2)—CO_2H

amino acid II

16.6 REDUCTION OF CARBONYL COMPOUNDS TO ALCOHOLS—STEREOCHEMISTRY

The reduction of aldehydes and ketones to alcohols has been discussed in Chapter 10. Here we shall elaborate briefly upon the stereochemistry of these reductions.

Reduction of Conformationally Restricted Ketones. When an unsymmetric ketone (R—CO—R′) is reduced to a 2° alcohol, the carbinol carbon becomes chiral. If the ketone already possesses a chiral center, *two* diastereomeric products are formed and questions about product distribution must be raised.

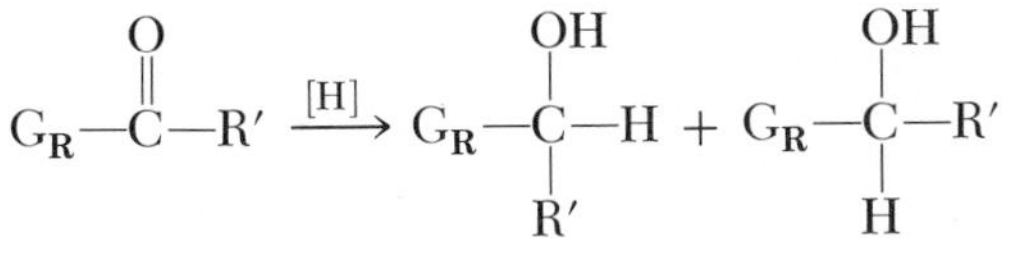

$$G_R\text{—C(=O)—R}' \xrightarrow{[H]} G_R\text{—C(OH)(R}'\text{)—H} + G_R\text{—C(OH)(H)—R}'$$

G_R is a group with a chiral center

When reduction of a ketone by a metal hydride involves two faces of the carbonyl group that are different, as in an unsymmetrically substituted ketone, the hydride usually attacks from the less hindered side. This, in turn, reflects coordination of the reductant to the carbonyl group at the less hindered side. Such a reaction, being irreversible, must be kinetically controlled.

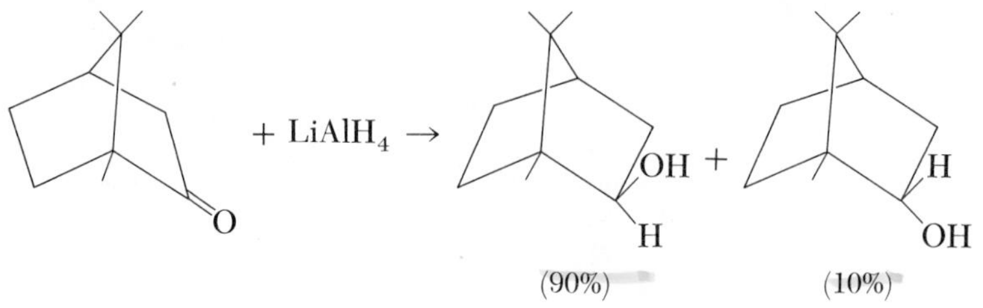

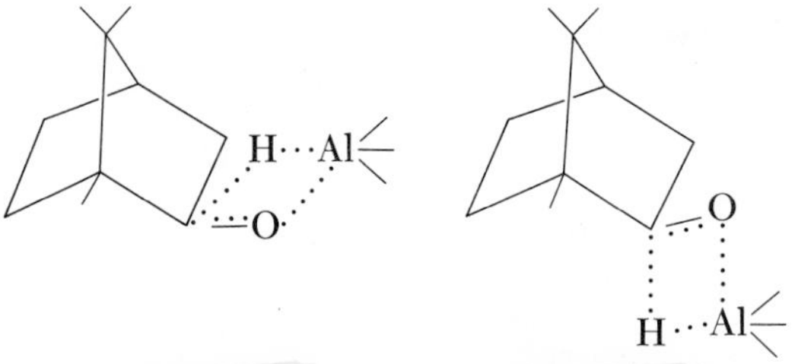

+ $LiAlH_4$ →

OH + H

H OH

(90%) (10%)

H···Al

O

O

H···Al

more hindered transition state

less hindered transition state

The reduction of 4-*t*-butylcyclohexanone forms much more *trans*- than *cis*-4-*t*-butylcyclohexanol. Here the kinetic and thermodynamic products happen to be the same, but the reaction is still controlled kinetically (*i.e.*, it passes through the lower energy transition state).

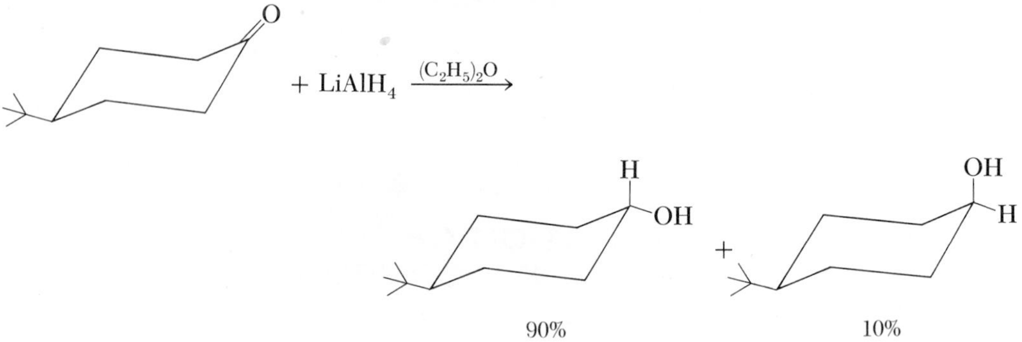

O

+ $LiAlH_4$ $\xrightarrow{(C_2H_5)_2O}$

H OH + OH H

90% 10%

Something can be learned about the transition state for reduction of 4-*t*-butylcyclohexanone by examining the effect of introduction of a substituent, R, at C3. There are, of course, two such derivatives, one with R *axial* and one with R *equatorial*. The activated complex for reduction when R is *equatorial* may be approximated by structure **E** (below). In this geometry the $AlH_4^{\ominus}$ anion attacks from the "top" (*axial* position) and leads to an *equatorial* hydroxy group. If this pictorialization is correct, then placing R in the *axial* position (structure **A**) should introduce diaxial repulsion between R and $AlH_4^{\ominus}$ and cause the transition state leading to reduction to increase in energy. In turn, this should favor attack of $AlH_4^{\ominus}$ from the "bottom" and production of an *axial* hydroxy group. This last geometry, which removes interaction between R and $AlH_4^{\ominus}$, is shown as structure **A′**. The fact that the introduction of an *axial* substituent at C3 does, indeed, increase the amount of *axial* hydroxy group produced in this reduction strongly supports this mechanistic picture.

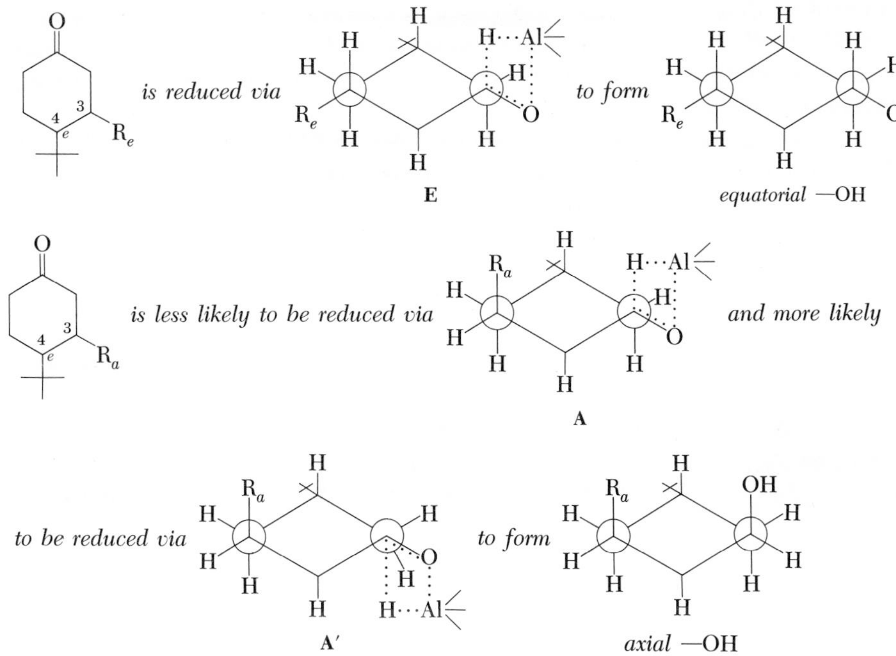

is reduced via

E

to form

equatorial —OH

is less likely to be reduced via

A

and more likely

to be reduced via

A′

to form

axial —OH

Reduction of Acyclic Carbonyl Compounds. The course of a kinetically controlled nucleophilic addition to a ketone adjacent to a chiral center can be predicted with the aid of what is usually called *Cram's rule*. To apply Cram's rule one draws the ketone in Newman projection with the carbonyl oxygen between the small and medium-sized substituents on the adjacent chiral center. The nucleophile is then pictured as attacking from the less hindered face of the carbonyl group. Application of this rule to the lithium aluminum hydride reduction of 3-phenyl-2-pentanone is shown below.

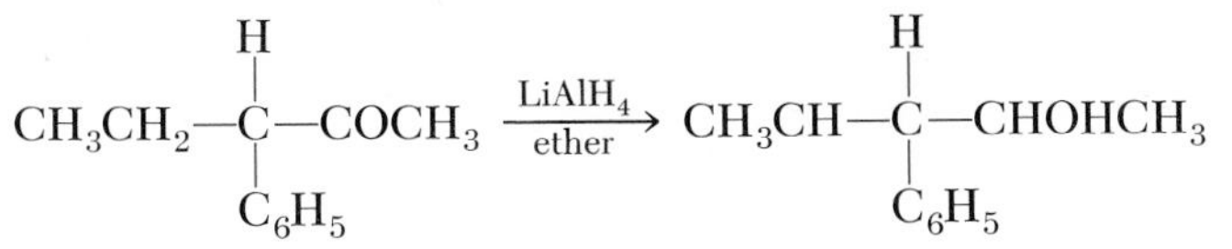

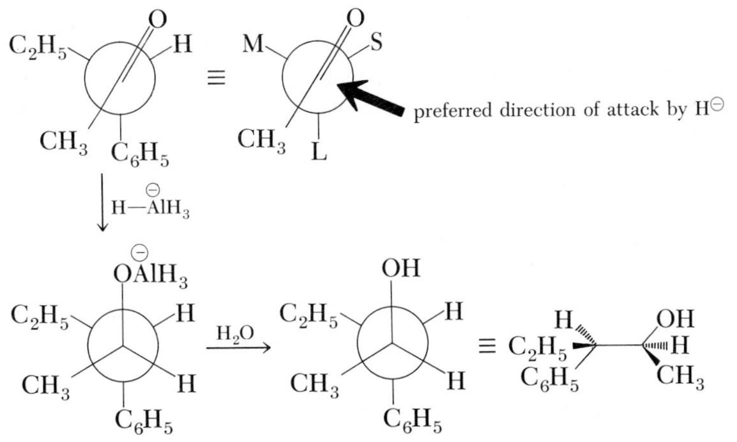

$$CH_3CH_2-\underset{C_6H_5}{\overset{H}{C}}-COCH_3 \xrightarrow[\text{ether}]{LiAlH_4} CH_3CH-\underset{C_6H_5}{\overset{H}{C}}-CHOHCH_3$$

preferred direction of attack by $H^{\ominus}$

$H-\overset{\ominus}{A}lH_3$

$\overset{\ominus}{O}AlH_3$

H_2O

This rule is best applied to compounds that lack a basic site (*e.g.*, —OH, $—NH_2$) at the chiral carbon, since these may coordinate with the reductant and produce violations of Cram's rule. Moreover, the rule does not suggest a stereospecific process. Instead, it is used to predict the isomer produced in excess.

A hindered ketone will react with a hindered Grignard reagent to produce a 2° rather than a 3° alcohol, with simultaneous conversion of the Grignard reagent into an alkene. For example, the reaction of cyclohexanone with *t*-butyl magnesium chloride produces a large amount of cyclohexanol and only a small amount of the expected 1-*t*-butylcyclohexanol.

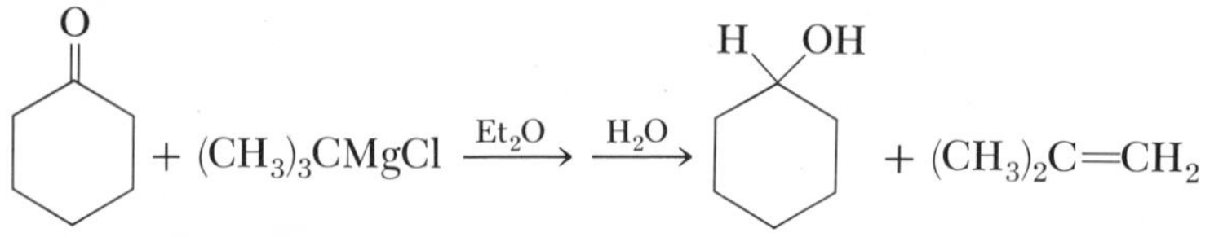

$$\text{cyclohexanone} + (CH_3)_3CMgCl \xrightarrow{Et_2O} \xrightarrow{H_2O} \text{cyclohexanol} + (CH_3)_2C{=}CH_2$$

reduction of a ketone, using a Grignard reagent

The mechanism for this reduction, which is known to require the presence of a β-hydrogen on the Grignard reagent, involves the cyclic transition state shown. In this transition state a hydride ion is transferred from Cβ of the Grignard reagent to the carbonyl carbon.

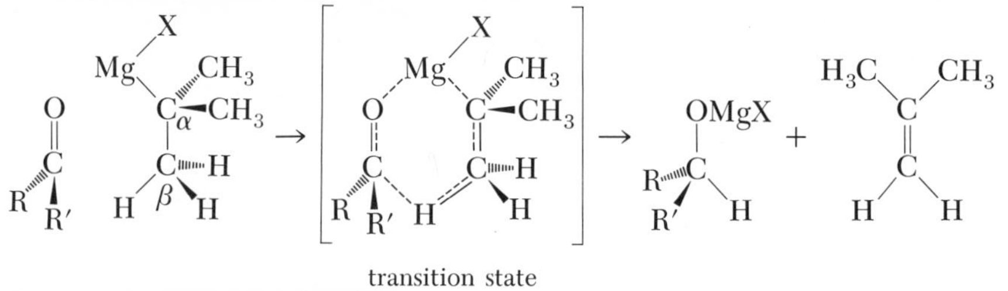

$$RR'C{=}O + XMg{-}C_\alpha(CH_3)_2{-}C_\beta H_3 \rightarrow [\text{transition state}] \rightarrow RR'CH{-}OMgX + (H_3C)_2C{=}CH_2$$

transition state

13. (a) When a ketone is dissolved in a solution of isopropanol containing aluminum isopropoxide, the ketone is reduced to the corresponding alcohol and an equimolar amount of isopropanol is converted to acetone. This procedure, which is extremely selective in that it reduces only carbonyl compounds, is called the Meerwein-Pondorf-Verley (MPV) reduction. If, as in the example below, the carbinol carbon bears a deuteron, the alcohol formed by reduction is found to possess deuterium. Suggest a mechanism for this reduction. (*Hint:* The process is akin to the reduction of carbonyl compounds by Grignard reagents.)

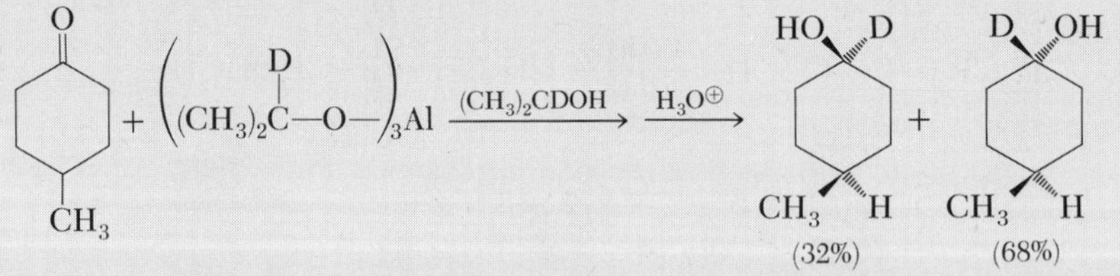

$$\text{4-methylcyclohexanone} + ((CH_3)_2CD{-}O{-})_3Al \xrightarrow{(CH_3)_2CDOH} \xrightarrow{H_3O^\oplus} \text{HO, D / CH}_3\text{, H (32\%)} + \text{D, OH / CH}_3\text{, H (68\%)}$$

(b) The reverse of this reaction, called the Oppenauer oxidation, converts an alcohol into a ketone. This reaction requires the use of a small amount of aluminum isopropoxide and a large excess of acetone; only alcohols are oxidized. Suggest a mechanism for the Oppenauer oxidation. A specific example and the first step in the reaction are shown below and on the following page.

net reaction:

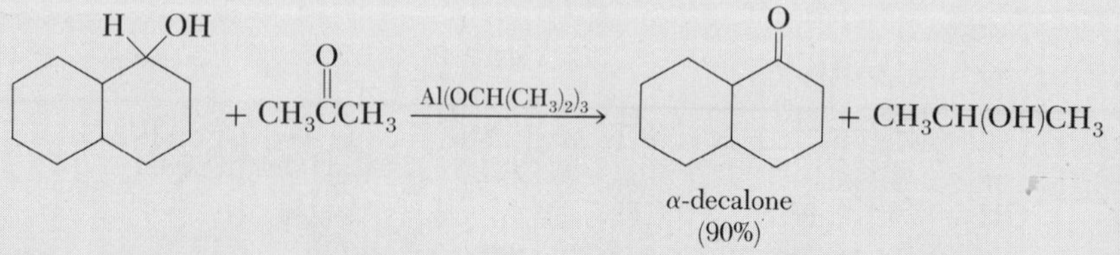

$$\text{1-decalol} + CH_3\overset{O}{\overset{\|}{C}}CH_3 \xrightarrow{Al(OCH(CH_3)_2)_3} \text{α-decalone (90\%)} + CH_3CH(OH)CH_3$$

α-decalone (90%)

first step:

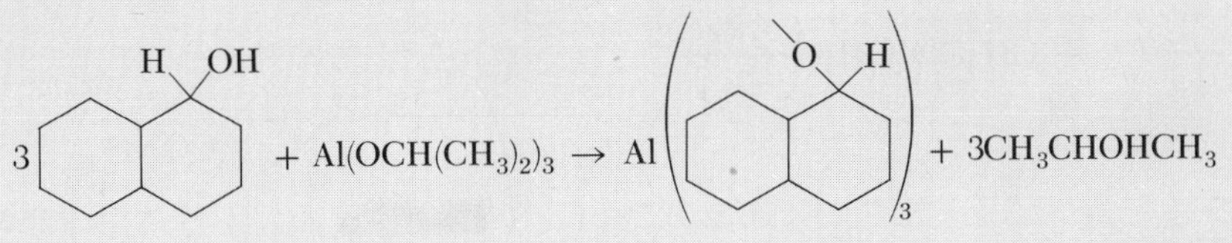

$$3\ \text{(decalol: H, OH)} + Al(OCH(CH_3)_2)_3 \rightarrow Al\left(\text{(decalyl-O, H)}\right)_3 + 3CH_3CHOHCH_3$$

Unlike organomagnesium compounds, organolithium compounds do *not* show a marked tendency to act as hydride donors. Therefore, syntheses of hindered alcohols are best accomplished with organolithium (or organosodium) compounds.

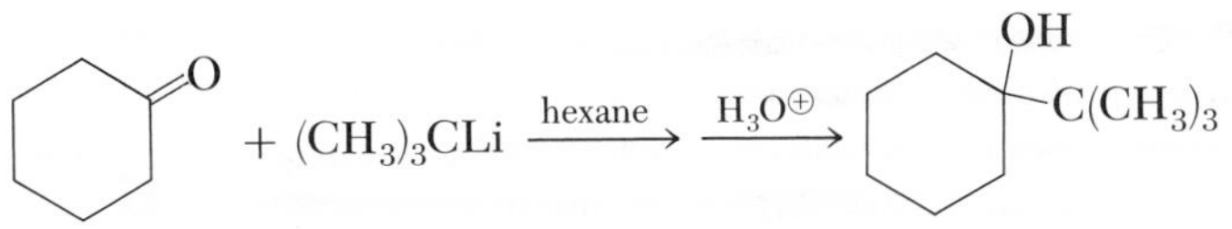

$$\text{cyclohexanone} + (CH_3)_3CLi \xrightarrow{\text{hexane}} \xrightarrow{H_3O^{\oplus}} \text{1-}t\text{-butylcyclohexanol (OH, } C(CH_3)_3)$$

$$(CH_3)_3C{-}\overset{\displaystyle O}{\overset{\|}{C}}{-}C(CH_3)_3 + (CH_3)_3C^{\ominus}Na^{\oplus} \rightarrow \xrightarrow{H_2O} ((CH_3)_3C)_3COH$$

tri-*t*-butylcarbinol

Biological Reduction. On page 448 we noted that a pyridinium salt derived from nicotinamide, called NAD, could serve as a hydrogen acceptor and oxidize an appropriate substrate. At that time we indicated that the reaction was reversible and that a 1,4-dihydropyridine derivative (reduced NAD) could reduce an oxidized substrate. We will now return to this and show that reduced NAD, in consort with an enzyme, is an excellent biochemical reductant of aldehydes. Furthermore, we shall see that the reactions of this combination of reduced NAD and enzyme are extremely stereospecific. In the reaction shown below, reduced NAD (associated with the enzyme *yeast alcohol dehydrogenase*) reduces acetaldehyde to ethanol quantitatively.

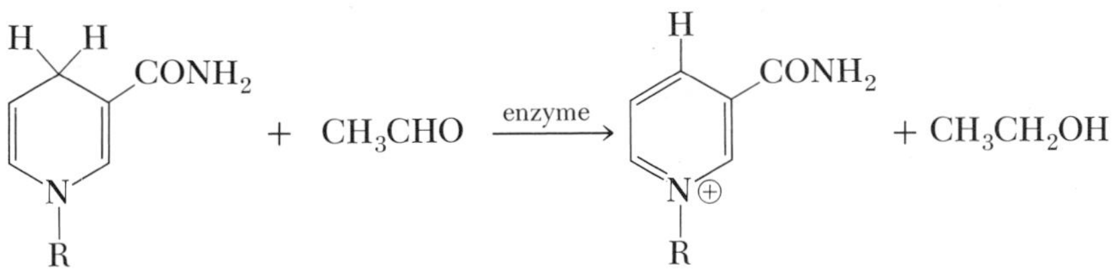

$$\text{reduced NAD (H, H; } CONH_2\text{; N–R)} + CH_3CHO \xrightarrow{\text{enzyme}} \text{NAD (H; } CONH_2\text{; } N^{\oplus}\text{–R)} + CH_3CH_2OH$$

reduced NAD — acetaldehyde — NAD — ethanol

in NAD, R is chiral

If the reverse of this reaction is carried out using deuterated ethanol as the compound being oxidized, then the reduced NAD that is formed is found to have deuterium incorporated on only one side of the 1,4-dihydropyridine ring. Put slightly differently, a new chiral center in the reduced NAD has been created (C4), but only one of its enantiomers has been formed.

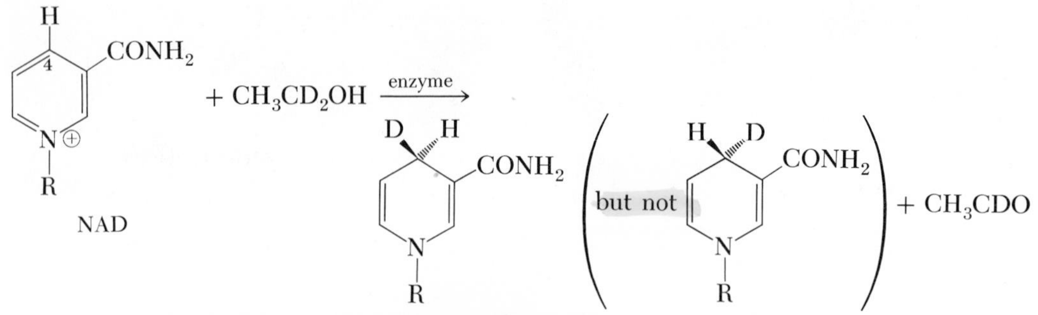

H CONH$_2$ + CH_3CD_2OH $\xrightarrow{\text{enzyme}}$ D H CONH$_2$ (but not H D CONH$_2$) + CH_3CDO

NAD

reduced NAD (labeled)

What happens if this reduced NAD, which is now stereospecifically labeled with deuterium, is used to reduce acetaldehyde in the presence of an enzyme? First, all of the ethanol that is formed contains deuterium. This means that the enzyme can distinguish between the two diastereotopic "hydrogens" at C4. Second, only one of the two enantiomeric alcohols is produced. This means that the enzyme–reduced NAD complex is capable of distinguishing between the two faces of the carbonyl group. This great specificity is not surprising in enzyme-controlled reactions, as we shall see throughout the remainder of this text.

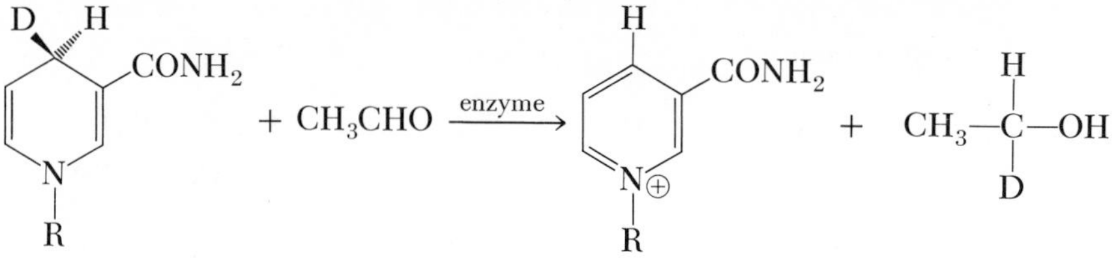

D H CONH$_2$ + CH_3CHO $\xrightarrow{\text{enzyme}}$ H CONH$_2$ + CH_3—C(H)(D)—OH

reduced NAD (labeled)

(only one enantiomer)

The general basis for this stereospecificity may be understood if we recognize that there are four possible activated complexes for the reduction of acetaldehyde by reduced NAD *but that all four are diastereomers of one another.* The four reactions described by these four activated complexes will proceed at different rates, since the chances that even two diastereomeric activated complexes will have the same energy are extremely small. The reaction with the most stable activated complex will proceed most rapidly and will yield the ultimate product. These four activated complexes are shown schematically below. Note that the two depicting the transfer of protium (**A** and **A′**) are discounted by product studies, *i.e.*, no CH_3CH_2OH is formed in the reaction.

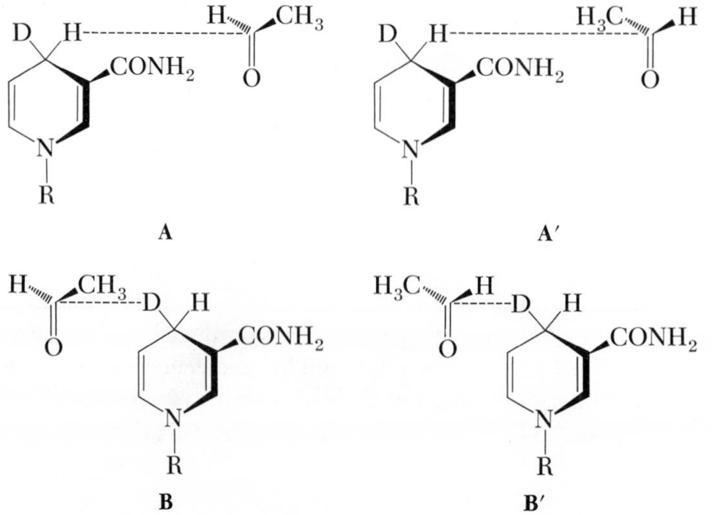

A **A′**

B **B′**

four possible activated complexes in the reduction of acetaldehyde with labeled, reduced NAD

This type of enzymic reduction has been used to prepare optically active neopentyl-*d* alcohol.

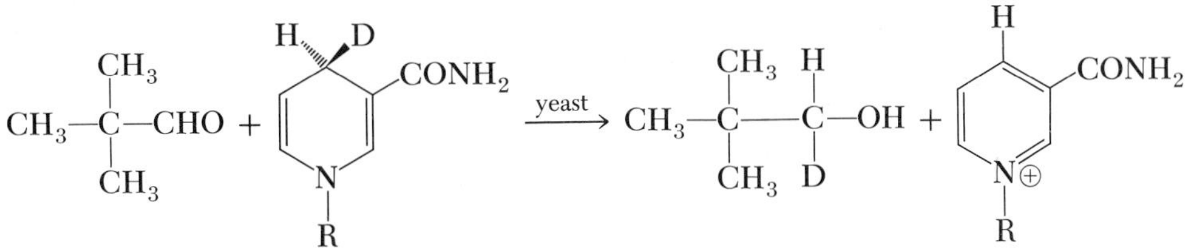

$$(CH_3)_3C\text{—}CHO + \text{(1-R-4-H,4-D-1,4-dihydropyridine-3-}CONH_2) \xrightarrow{\text{yeast}} CH_3\text{—}C(CH_3)_2\text{—}CHD\text{—}OH + \text{(1-R-pyridinium-3-}CONH_2)^{\oplus}$$

one enantiomer of neopentyl-*d* alcohol

16.7 REDUCTION OF A CARBONYL GROUP TO A METHYLENE GROUP

The Clemmensen reduction was used earlier (p. 422) to convert the products of Friedel-Crafts acylation to hydrocarbons. It is suitable for the reduction of aldehydes as well.

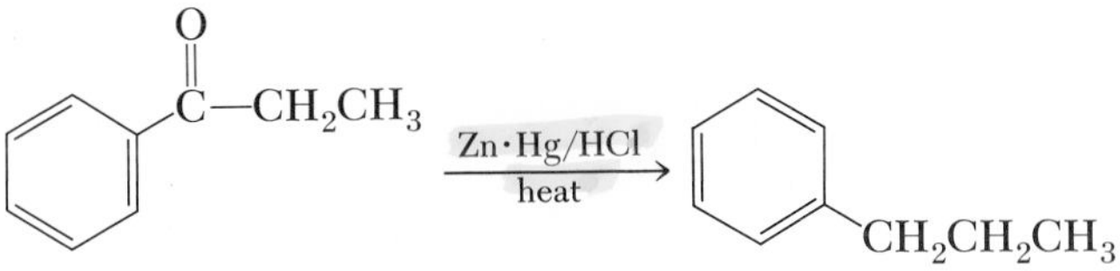

$$C_6H_5\text{—}\overset{O}{\overset{\|}{C}}\text{—}CH_2CH_3 \xrightarrow[\text{heat}]{Zn\cdot Hg/HCl} C_6H_5\text{—}CH_2CH_2CH_3$$

the Clemmensen reduction

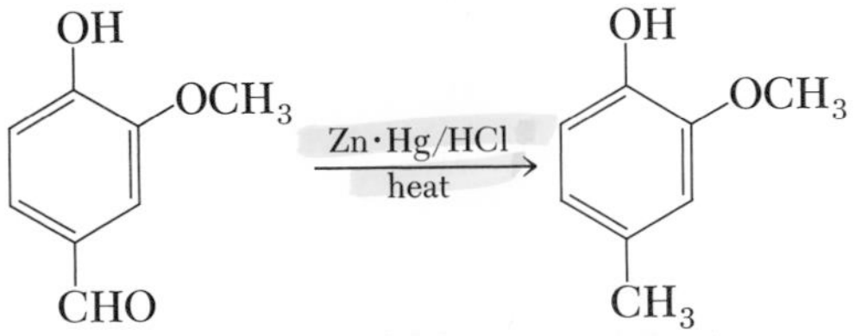

$$\text{4-hydroxy-3-methoxybenzaldehyde (OH, }OCH_3\text{, CHO)} \xrightarrow[\text{heat}]{Zn\cdot Hg/HCl} \text{2-methoxy-4-methylphenol (OH, }OCH_3\text{, }CH_3) \quad (65\%)$$

While the mechanism is still under investigation, it is known that the best yields are obtained when the reaction is carried out in a three-phase system of zinc, aqueous acid, and toluene. The toluene, a fair solvent for ketones, dissolves most of the ketone and permits only a low ketone concentration in the aqueous phase. If this is not done, ketone concentration near the zinc becomes high enough to permit bimolecular reduction.

$$C_6H_5\text{—}\overset{O}{\overset{\|}{C}}\text{—}CH_2CH_3 \xrightarrow{H^{\oplus}} C_6H_5\text{—}\overset{\overset{\oplus}{O}H}{\overset{\|}{C}}\text{—}CH_2CH_3 \xrightarrow{Zn} C_6H_5\text{—}\overset{OH}{\overset{|}{\dot{C}}}\text{—}C_2H_5$$

(radical)

$$2C_6H_5\text{—}\overset{OH}{\overset{|}{\underset{C_2H_5}{\underset{|}{C}}}}\cdot \rightarrow C_6H_5\text{—}\overset{OH}{\overset{|}{\underset{C_2H_5}{\underset{|}{C}}}}\text{—}\overset{OH}{\overset{|}{\underset{C_2H_5}{\underset{|}{C}}}}\text{—}C_6H_5$$

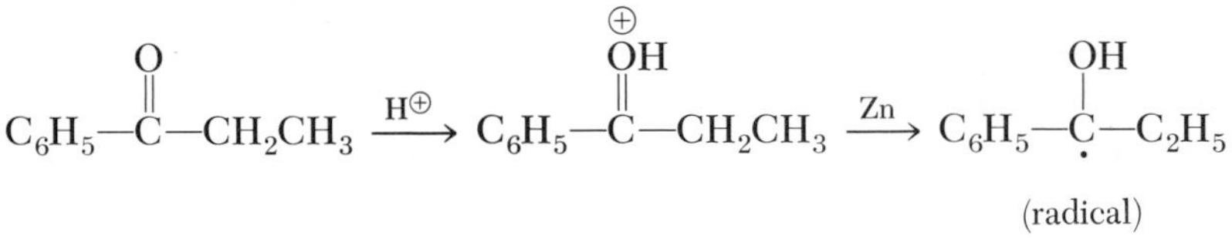

probable cause of bimolecular reduction

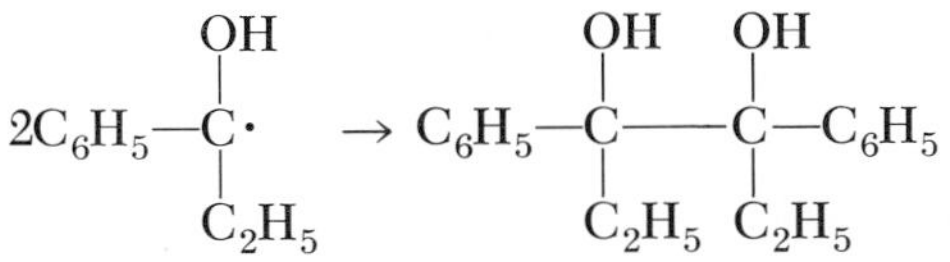

When a molecule contains two keto groups held in close proximity, Clemmensen reduction produces a high yield of *vic*-diol.

$$\text{(cage diketone)} \xrightarrow[80^\circ]{Zn\cdot Hg/HCl} \text{(cage vic-diol, OH, OH)} \quad 98\%$$

bimolecular reduction

The **Wolff-Kishner reduction** is useful for reduction of carbonyl groups in compounds that are stable to base but unstable to acid. The reduction involves the reaction of a carbonyl group with hydrazine and the subsequent decomposition of the hydrazone with base. While the original procedure called for the isolation of the hydrazone prior to decomposition, the more popular **Huang-Minlon modification** accomplishes the entire reaction in one "pot."

$$R\text{—}\overset{O}{\overset{\|}{C}}\text{—}R' + H_2N\text{—}NH_2 \xrightarrow[180^\circ]{KOH;\ (HOCH_2CH_2)_2O} RCH_2R'$$

Huang-Minlon modification of the Wolff-Kishner reduction

One possible mechanism for the reaction follows:

$$\begin{matrix}R\\ \;\;C{=}N{-}NH_2\\ R'\end{matrix} \overset{OH^\ominus}{\rightleftharpoons} H{-}\underset{R'}{\overset{R}{C}}{-}N{=}N{-}H \xrightarrow[-H_2O]{OH^\ominus} H{-}\underset{R'}{\overset{R}{C}}{-}N{=}\ddot{N}{:}^\ominus$$

$$\downarrow -N_2$$

$$H{-}\underset{R'}{\overset{R}{C}}{-}H \xleftarrow[-Sv^\ominus]{SvH} H{-}\underset{R'}{\overset{R}{C}}{:}^\ominus$$

Sv: solvent

16.8 DISPROPORTIONATION REACTIONS

Since the carbonyl group is midway in oxidation state between an alcohol and an acid, one might inquire whether there are reactions in which disproportionation occurs between two carbonyl groups (*i.e.*, reactions in which one carbonyl group is oxidized while the other is reduced). Two examples of this kind of reaction follow. The first reaction is *inter*molecular, disproportionation being the result of hydride transfer. An *intra*molecular carbanion migration results in disproportionation in the second reaction.

Cannizzaro Reaction. Aldehydes that lack an α hydrogen, *i.e.*, a hydrogen adjacent to the carbonyl group, react with concentrated aqueous sodium hydroxide to produce one mole of alcohol and one mole of acid.

$$2\ C_6H_5\text{—}\overset{O}{\overset{\|}{C}}\text{—}H + NaOH \xrightarrow[heat]{H_2O} C_6H_5\text{—}CH_2OH + C_6H_5\text{—}\overset{O}{\overset{\|}{C}}\text{—}O^\ominus\ Na^\oplus$$

benzaldehyde   benzyl alcohol   sodium benzoate

The mechanism for the reaction is shown below, using formaldehyde as the substrate.

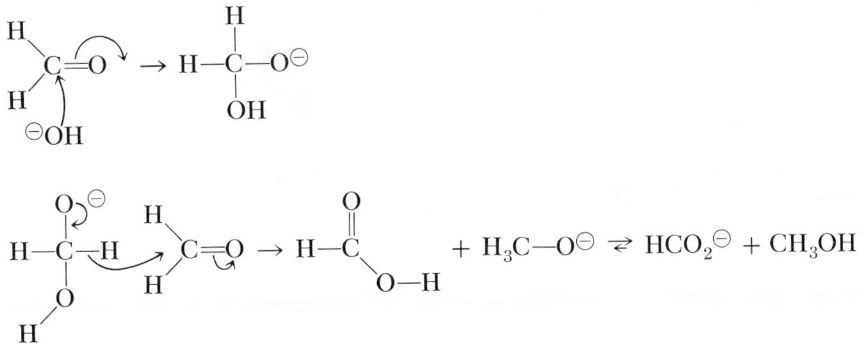

$$H_2C{=}O + {}^{\ominus}OH \rightarrow H{-}\underset{OH}{\overset{H}{C}}{-}O^{\ominus}$$

$$H{-}\underset{\underset{H}{O}}{\overset{O^{\ominus}}{C}}{-}H + H_2C{=}O \rightarrow H{-}\overset{O}{\overset{\|}{C}}{-}O{-}H + H_3C{-}O^{\ominus} \rightleftarrows HCO_2^{\ominus} + CH_3OH$$

Because formaldehyde is more easily oxidized than are aromatic aldehydes, a *crossed Cannizzaro* is possible, with the formaldehyde being oxidized to formic acid and the aromatic aldehyde being reduced to an alcohol.

$$C_6H_5CHO + H_2CO \xrightarrow[\text{heat}]{\text{conc. NaOH}} C_6H_5CH_2OH + HCO_2^{\ominus}\ Na^{\oplus}$$

Benzilic Acid Rearrangement. When benzil, $C_6H_5COCOC_6H_5$, is heated with aqueous sodium hydroxide, a phenyl anion migrates to a neighboring carbon to produce benzilic acid (as its sodium salt) and, coincidentally, reduce one carbonyl while oxidizing another.

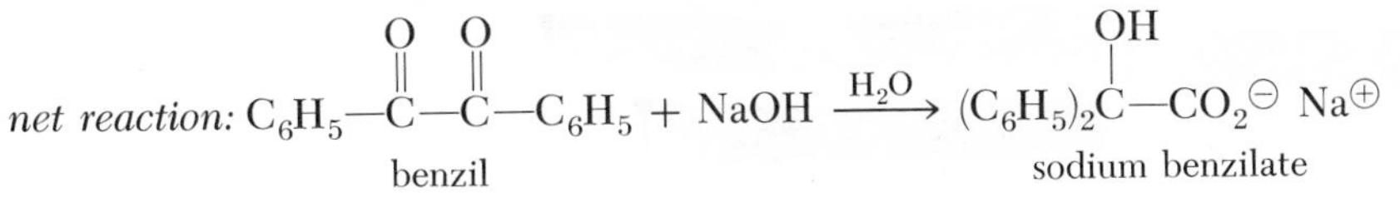

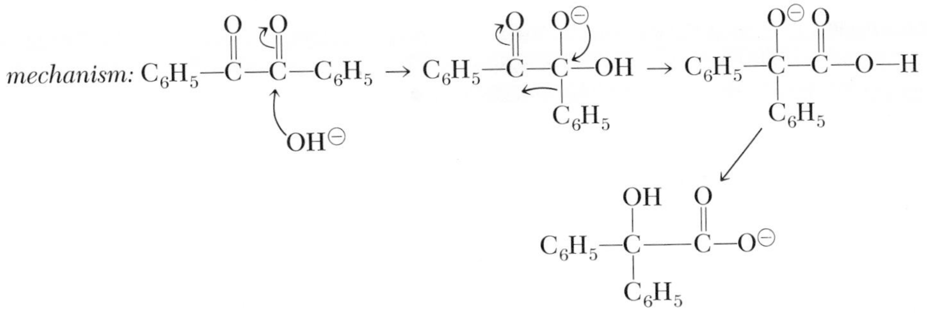

net reaction: $$\underset{\text{benzil}}{C_6H_5{-}\overset{O}{\overset{\|}{C}}{-}\overset{O}{\overset{\|}{C}}{-}C_6H_5} + NaOH \xrightarrow{H_2O} \underset{\text{sodium benzilate}}{(C_6H_5)_2\overset{OH}{\overset{|}{C}}{-}CO_2^{\ominus}\ Na^{\oplus}}$$

mechanism: $$C_6H_5{-}\overset{O}{\overset{\|}{C}}{-}\overset{O}{\overset{\|}{C}}{-}C_6H_5 + OH^{\ominus} \rightarrow C_6H_5{-}\overset{O}{\overset{\|}{C}}{-}\underset{C_6H_5}{\overset{O^{\ominus}}{C}}{-}OH \rightarrow C_6H_5{-}\underset{C_6H_5}{\overset{O^{\ominus}}{C}}{-}\overset{O}{\overset{\|}{C}}{-}O{-}H \rightarrow C_6H_5{-}\underset{C_6H_5}{\overset{OH}{C}}{-}\overset{O}{\overset{\|}{C}}{-}O^{\ominus}$$

This reaction is something of an oddity, since it is one of the few *base*-catalyzed skeletal rearrangements that are known. Most, of course, are *acid*-catalyzed and involve carbocations.

16.9 CONJUGATE ADDITION

When a double bond and a carbonyl group are conjugated, *i.e.*, C=C—C=O, the tendency for electrophilic addition to occur at the C=C is reduced. This can be explained either by invoking the inductive effect of the C=O dipole, which places a partial positive charge adjacent to the double bond, or by constructing a resonance picture that shows the π electron density of the alkene group displaced toward the carbonyl oxygen.

The resonance picture, on the other hand, suggests that the C=C bond should be more susceptible to attack by a nucleophile than should an isolated carbon-carbon double bond. (Why?)

Electrophilic Addition. The product of addition of HCl to acrolein indicates that electrophilic addition to α,β-unsaturated carbonyl systems occurs in an "anti-Markownikoff" fashion.

$$H_2C{=}CH{-}CHO + HCl(g) \xrightarrow{-10^\circ} Cl{-}CH_2{-}CH_2{-}CHO$$

The mechanism for this addition is related to the problem of 1,2-*vs.* 1,4-addition in dienes (p. 353) and begins with protonation of the carbonyl oxygen to form an allylic cation. This cation then is attacked by a nucleophile at two possible positions; one eventually leads to product (*i.e.*, addition across the C=C linkage), while the other leads back to starting material.

$$H_2C{=}CH{-}CH{=}O \xrightarrow{H^\oplus} CH_2{=}CH{-}\overset{\oplus}{C}H{-}O{-}H \leftrightarrow \overset{\oplus}{C}H_2{-}CH{=}CH{-}OH$$

$$\overset{\delta+}{C}H_2{\overset{---}{}}CH{\overset{---}{}}\overset{\delta+}{C}H{-}OH$$

$$\overset{\delta+}{C}{\cdots}C{\cdots}\overset{\delta+}{C}{-}O{-}H \;\; (\text{attack by } {:}Nu^\ominus \text{ at } ① \text{ or } ②) \begin{cases} \xrightarrow{1} Nu{-}CH_2{-}CH{=}CH{-}OH \\ \xrightarrow{2} CH_2{=}CH{-}CH(Nu){-}OH \end{cases}$$

The compound formed *via* path **2** is an adduct which, like the HCl adduct of simple aldehydes, reverts to the original carbonyl compound. Attack along path **1**, on the other hand, produces the enol form (p. 273) of the final product.

$$CH_2{=}CH{-}CH(Nu){-}O{-}H \rightleftarrows CH_2{=}CH{-}\overset{O}{\overset{\|}{C}}{-}H + HNu$$

$$Nu{-}CH_2CH{=}CH{-}OH \rightleftarrows Nu{-}CH_2CH_2{-}C(=O)H$$

While initial protonation of the carbonyl oxygen makes the "1,2- *vs.* 1,4-addition" mechanism attractive for HNu addition, Br_2 addition may proceed *via* an unsymmetric bromonium ion.

Examples of Electrophilic Addition

$$H_2C{=}CHCO_2H + H_2O \xrightarrow{H^\oplus} HOCH_2CHCO_2H$$

$$CH_3CH{=}CHCO_2H + HCl \rightarrow CH_3CHClCH_2CO_2H$$

$$\underset{\text{mesityl oxide}}{(CH_3)_2C{=}CHC(O)CH_3} + CH_3OH \xrightarrow{H^\oplus} \underset{\text{4-methoxy-4-methyl-2-pentanone}}{(CH_3)_2C(OCH_3)CH_2C(O)CH_3}$$

Nucleophilic Addition. When a nucleophile adds to an unsaturated carbonyl compound, usually called an "enone," it may attack the carbonyl carbon or the remote end of the double bond.

possible sites of nucleophilic attack

charge is *not* resonance stabilized

charge *is* resonance stabilized

One might imagine that resonance delocalization of charge should make the transition state leading to attack at C=C more stable than the transition state leading to attack at the carbonyl carbon. While this is very often the case (*i.e.*, addition occurs preferentially at the C=C linkage), very reactive anions such as Grignard reagents give a fair amount of "direct addition" product. It is possible, however, to alter this preference of the Grignard reagent so that "conjugate addition" is observed—all that is required is the presence of Cu(I) during the addition.

$(CH_3)_2C{=}CH{-}C(=O){-}CH_3 \xrightarrow[\text{conjugate addition}]{CH_3NH_2} (CH_3)_2C(\overset{\oplus}{N}H_2CH_3){-}CH{=}C(O^{\ominus})CH_3 \rightarrow (CH_3)_2C(NHCH_3){-}CH{=}C(OH)CH_3 \rightarrow (CH_3)_2C(NHCH_3){-}CH_2{-}C(=O){-}CH_3$

direct addition, CH_3CH_2MgCl:

$(CH_3)_2C{=}CH{-}C(OMgBr)(CH_2CH_3){-}CH_3 \xrightarrow[H^{\oplus}]{H_2O} (CH_3)_2C{=}CH{-}C(OH)(CH_2CH_3){-}CH_3$

CH_3MgI, $(CH_3CH_2)_2O$ → product of direct addition

Cu(I), CH_3MgI → product of conjugate addition

The following rules may be useful in determining whether a reaction will be a conjugate addition or a direct addition.

1. Conjugate addition is favored by increased steric hindrance at the carbonyl group.
2. Direct addition is favored by increased crowding at the position β to the carbonyl group.
3. Conjugate addition is favored by a decrease in the double bond character of the carbonyl group.
4. Direct addition is favored by increased reactivity of the attacking nucleophile.

When an α,β-unsaturated carbonyl compound is reduced with $LiAlH_4$ or $NaBH_4$, the product usually is that from direct addition. Which, if any, of these rules is consistent with that result?

$$\text{2-cyclohexenone} \xrightarrow[\text{(b) } H_2O]{\text{(a) } LiAlH_4 \text{ or } NaBH_4} \text{2-cyclohexen-1-ol (H, OH)}$$

Catalytic reduction with one mole of hydrogen, on the other hand, reduces the C=C unit selectively.

$$\text{2-cyclohexenone} \xrightarrow[\text{1 atm}]{H_2/Pt} \text{cyclohexanone}$$

Conjugate Addition of Trialkylboranes. α,β-Unsaturated carbonyl compounds produce aldehydes or ketones that are alkylated at the double bond upon reaction with trialkylboranes.

$$R_3B + CH_2{=}CH{-}\overset{O}{\overset{\|}{C}}{-}CH_3 \xrightarrow[\text{1 hr}]{25°} \left[RCH_2CH{=}C\begin{matrix}CH_3 \\ OBR_2\end{matrix} \right] \xrightarrow{H_2O}$$

$$RCH_2CH{=}\overset{CH_3}{\overset{|}{C}}{-}OH \rightleftarrows RCH_2CH_2{-}\overset{O}{\overset{\|}{C}}{-}CH_3$$

$$(\text{C}_6\text{H}_{11})_3B + CH_2{=}CH{-}\overset{O}{\overset{\|}{C}}{-}H \rightarrow \xrightarrow{H_2O} \text{C}_6\text{H}_{11}{-}CH_2CH_2{-}\overset{O}{\overset{\|}{C}}{-}H$$

While the reaction is wasteful, in that only one of the groups bonded to boron is used in the alkylation, the simplicity of operation makes this a highly desirable procedure.

IMPORTANT TERMS

Acetal: A *gem*-diether, most often prepared by the reaction of an aldehyde with two moles of an alcohol in the presence of an acid. The product of the reaction of a ketone with two moles of an alcohol also may be called an acetal, although the term "ketal" is more descriptive.

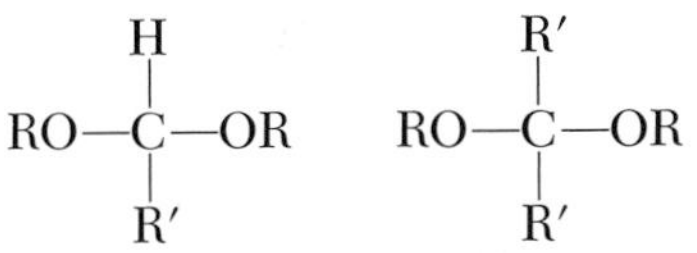

$$\underset{\text{acetal}}{RO{-}\overset{\overset{\large H}{|}}{\underset{\underset{\large R'}{|}}{C}}{-}OR} \qquad \underset{\text{ketal}}{RO{-}\overset{\overset{\large R'}{|}}{\underset{\underset{\large R'}{|}}{C}}{-}OR}$$

both of these may be called "acetal"

Anesthetic: A substance that produces loss of sensitivity to pain or other feeling.

Cannizzaro reaction: A base-induced intermolecular disproportionation between two aldehydes, both of which lack hydrogens on the carbon adjacent to the carbonyl group. The reaction produces an alcohol and the salt of a carboxylic acid.

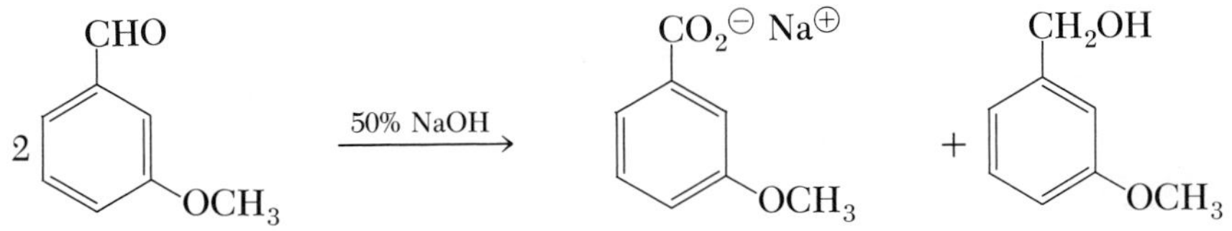

$$2\ \underset{m\text{-methoxybenzaldehyde}}{m\text{-}CH_3OC_6H_4CHO} \xrightarrow{50\%\ NaOH} \underset{\text{sodium } m\text{-methoxybenzoate}}{m\text{-}CH_3OC_6H_4CO_2^{\ominus}\ Na^{\oplus}} + \underset{m\text{-methoxybenzyl alcohol}}{m\text{-}CH_3OC_6H_4CH_2OH}$$

Cram's rule: A rule for predicting the effect of a chiral center adjacent to a carbonyl group on the course of nucleophilic addition to that group.

Enone: An α,β-unsaturated carbonyl compound, *i.e.*, $-C{=}\overset{|}{C}{-}\overset{|}{C}{=}O$.

Hemiacetal: A *gem*-hydroxy ether, most often prepared by the reaction of an aldehyde with one mole of alcohol. The product of the reaction of a ketone with one mole of alcohol also may be called a hemiacetal, although the term "hemiketal" is more descriptive.

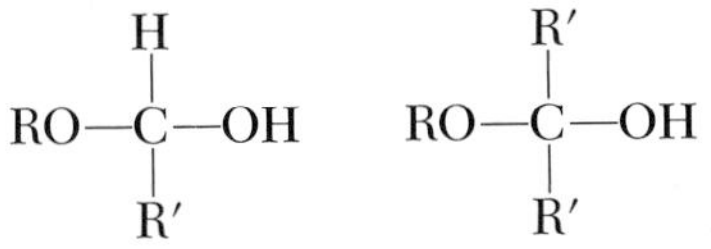

$$\underset{\text{hemiacetal}}{RO{-}\overset{\overset{\large H}{|}}{\underset{\underset{\large R'}{|}}{C}}{-}OH} \qquad \underset{\text{hemiketal}}{RO{-}\overset{\overset{\large R'}{|}}{\underset{\underset{\large R'}{|}}{C}}{-}OH}$$

both of these may be called hemiacetal

Hypnotic: A substance capable of inducing sleep.

Mickey Finn: A beverage, usually alcoholic, which renders the drinker helpless.

Schiff's base: A compound of the type $R_1R_2C{=}NR_3$. (Most compounds of the type in which R_3 is hydrogen are unstable and polymerize spontaneously.) These compounds are also called **imines.**

Sedative: A compound capable of reducing an individual's functional activity.

Stephen's reduction: The reduction of an aromatic nitrile (Ar—CN) to an aromatic aldehyde (Ar—CHO) through the action of stannous chloride ($SnCl_2$) and hydrochloric acid.

Transamination: The transfer of an amino group ($-NH_2$) from one molecule to another. The usual transamination reactions occur in living systems and involve the synthesis of one amino acid ($RCH(NH_2)CO_2H$) from another amino acid. An example of such an enzyme-controlled exchange is shown below.

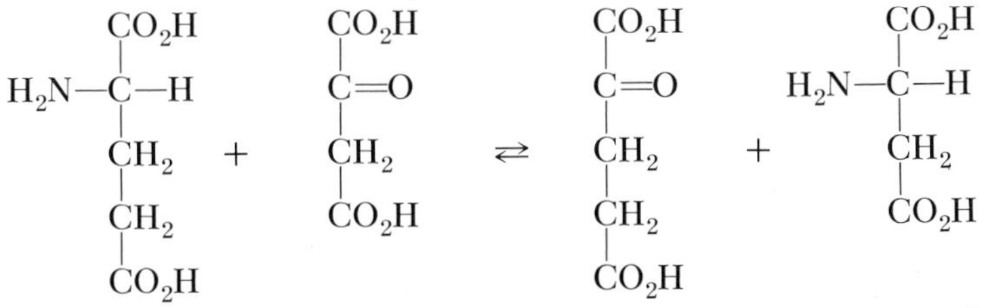

$$\underset{\text{L-glutamic acid}}{HO_2C{-}CH(NH_2){-}CH_2{-}CH_2{-}CO_2H} + \underset{\text{oxaloacetic acid}}{HO_2C{-}C({=}O){-}CH_2{-}CO_2H} \rightleftarrows \underset{\alpha\text{-ketoglutaric acid}}{HO_2C{-}C({=}O){-}CH_2{-}CH_2{-}CO_2H} + \underset{\text{L-aspartic acid}}{HO_2C{-}CH(NH_2){-}CH_2{-}CO_2H}$$

Biological transaminations are not simple, one-step processes but, as noted in the text (p. 484), are a series of complex reactions. Imines are critical intermediates in these reactions.

Wolff-Kishner reaction: The reduction of a carbonyl group by alkaline decomposition of a hydrazone at high temperatures. In the *Huang-Minlon* modification of this reaction, the hydrazone is not isolated but is decomposed in the same flask in which it is prepared. This modification is used much more often than is the original procedure.

$$R_2C{=}O + H_2NNH_2 \rightarrow R_2C{=}NNH_2 \xrightarrow[200^\circ]{OH^\ominus} R_2CH_2 + N_2$$

PROBLEMS

14. Provide acceptable names for each of the following.
   (a) CH_3CHO
   (b) CH_2O
   (c) C_6H_5CHO
   (d) CO
   (e) $COCH_3$
   (f) Cl— —CHO
   (g) $H_2C{=}NH$
   (h) O
   (i) $CH_3CH_2COCH_2C_6H_5$
   (j) $CH_3CH{=}CHCOCH_2Cl$
   (k) Cl_3CCHO
   (l) O O
   (m) $CH_3{-}C({=}NH{-}NHC_6H_5){-}CH_3$
   (n) $(CH_3)_2C{=}CHCOCH_3$
   (o) $C_6H_5CH{=}CHCHO$

15. Suggest a substrate-oxidant combination that, after an appropriate work-up, will produce each of the following.
   (a) CH_3COCH_3
   (b) $CH_3COCH_2CH_3$
   (c) $CH_2{=}CHCHO$
   (d) $CH_3C(O)(CH_2)_4C(O)CH_3$
   (e) $OHC(CH_2)_4CHO$
   (f) O
   (g) O
   (h) 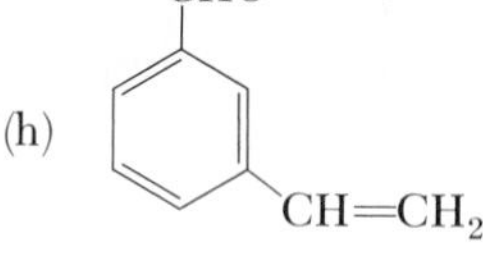

16. Provide equations for the following. If no reaction occurs, write "NR."
   (a) acetone + Mg·Hg
   (b) acetophenone + Mg·Hg
   (c) di-*t*-butyl ketone + $NaHSO_3$(*aq.*)
   (d) cyclohexanone + hydroxylamine (NH_2OH) + dil. HCl
   (e) 1-propyne + $LiAlH_4$
   (f) 2,2-dimethoxypropane + $OH^\ominus/H_2O$
   (g) formaldehyde + H_2O
   (h) acrolein + H_2/Pt (1 mole)
   (i) acrolein + $LiAlH_4$
   (j) 1,3-cyclopentanedione + hydrazine
   (k) 1-propyne + $H_2SO_4/HgSO_4/H_2O$

(l) 2-hexyne + $H_2SO_4/HgSO_4/H_2O$
(m) cyclohexanecarboxaldehyde + $NaBH_4/H_2O$
(n) acetone + CO/H_2O + heat followed by $H_2O_2/OH^{\ominus}$
(o) 3-hexanone + semicarbazide + dil. HCl
(p) chloral hydrate + sodium bisulfite
(q) methyl vinyl ketone + $LiAlH_4$
(r) $NC{-}(CH_2)_7{-}CN$ + $SnCl_2/HCl$ + heat
(s) ethanal + ethanol (excess) + $H^{\oplus}$
(t) ethanal + ethanol (excess) + $OH^{\ominus}$
(u) pivaldehyde (2,2-dimethylpropanal) + $OH^{\ominus}$
(v) methanal diethylacetal + methanol (excess) + $H^{\oplus}$
(w) acrolein + cyclopentadiene
(x) $CH_3CH_2CO_2H$ + $LiAlH_4$
(y) CH_3CN + $LiAl(OEt)_3H$
(z) CH_3CN + cyclohexylmagnesium bromide ($-70°$) followed by H_2O
(aa) CH_3CN + $LiAlH_4$ (excess) followed by H_2O
(bb) $LiAlH_4$ + CH_3CN (excess) followed by H_2O
(cc) norbornene + B_2H_6 followed by CO/H_2O + heat followed by $H_2O_2/OH^{\ominus}$
(dd) tri-*iso*-butylborane + 3-penten-2-one
(ee) triphenylacetaldehyde + formaldehyde + $OH^{\ominus}$
(ff) 3,3-dimethyl-2-butanol + dil. H_2SO_4
(gg) 2,3-dimethoxypropane + $H_2O/H^{\oplus}$
(hh) *cis*-4,5-dimethyl-1,3-dioxolane + dil. H_2SO_4
(ii) *trans*-4,5-dimethyl-1,3-dioxolane + dil. H_2SO_4
(jj) ethylene oxide (excess) + dil. H_2SO_4
(kk) trioxane + heat

17. Suggest a substrate-Grignard reagent combination that, after an appropriate work-up, will produce each of the following.

(a) ethanol
(b) isopropanol
(c) diphenylcarbinol
(d) *t*-butyl alcohol
(e) 1-methylcyclohexanol
(f) cyclohexylcarbinol
(g) 3-buten-2-ol

Which of these will be oxidized by (a) MnO_2 at 25° and (b) Sarrett's reagent? Indicate the expected products.

18. Starting with cyclohexanone, and anything else you require, provide rational syntheses for the following.

(a) 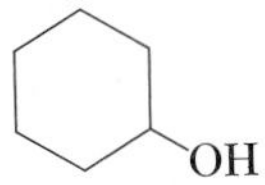 OH

(b)

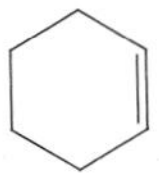

(c) 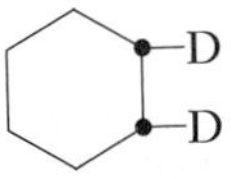 D 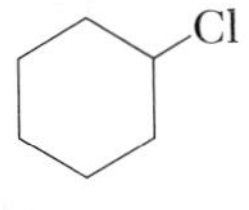 D

(d) Cl

(e) 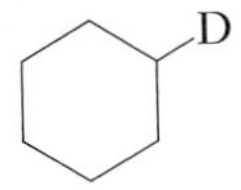 D

(f) 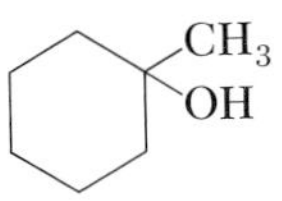 CH_3 OH

(g) 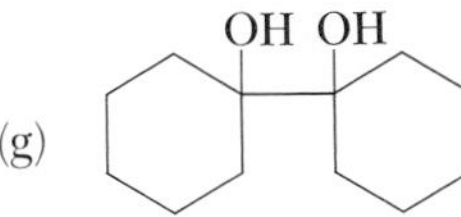

(h) 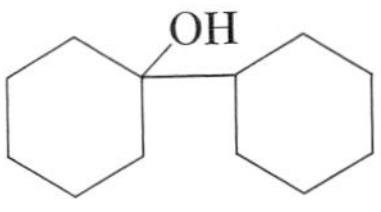 OH

(i) 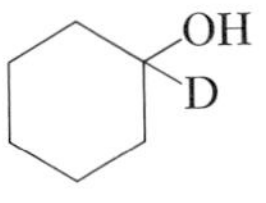 OH D

(j)

(k) 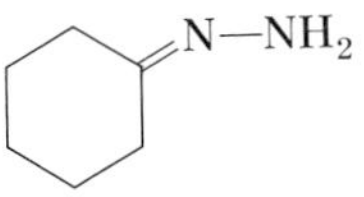 $N{-}NH_2$

(l) 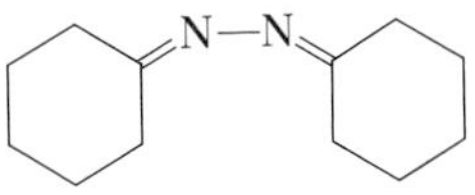 N—N

(m) 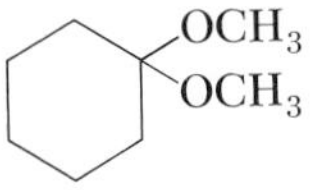 OCH_3 OCH_3

(n)

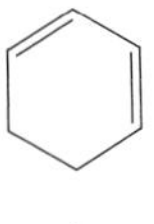

(o)

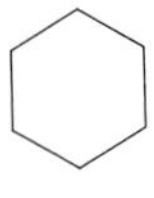

(p) $HO_2C(CH_2)_4CO_2H$

(q) 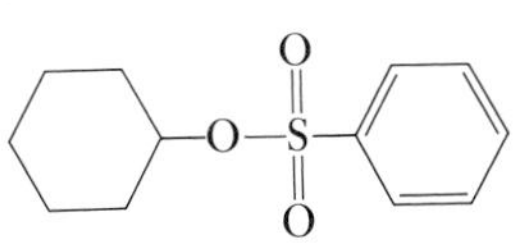 O—S(=O)(=O)—

(r) 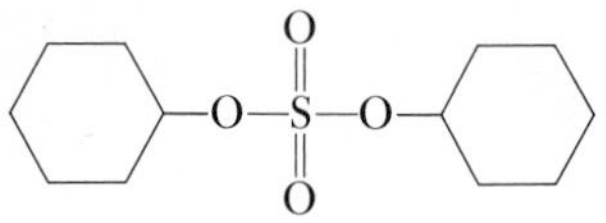 (s) 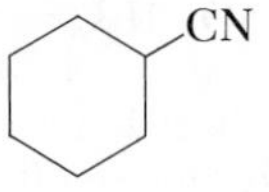 CN (t) 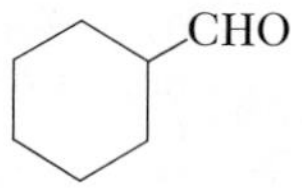 CHO

19. Prepare a chart showing the ways in which organoboranes can be used to synthesize (a) alcohols, (b) aldehydes, and (c) ketones. Do not restrict yourself to reactions introduced in this chapter.
20. *Dybenal* (2,4-dichlorobenzyl alcohol) is an anti-bacterial agent that has been used as an antiseptic. Suggest a synthesis for it beginning (a) with benzene, (b) with benzaldehyde, and (c) with benzoic acid.

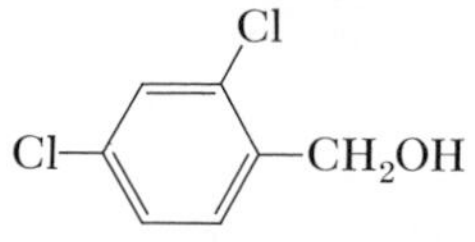

Cl Cl CH_2OH

Dybenal

21. A student was preparing the oxime of phenyl ethyl ketone and, upon checking the purity of his product, found that he had prepared *two* distinct materials that both corresponded to an oxime of phenyl ethyl ketone. Explain what the compounds were.
22. The labels have fallen off of four bottles that contain isopropyl bromide, methyl ethyl ketone, butanal, and hexane. What would you do to help re-label the bottles correctly? Assume that you do not have authentic samples available.
23. Puberulic acid, an antibiotic isolated from *Penicillium puberulum*, is soluble in aqueous acid. Suggest a reason for this.

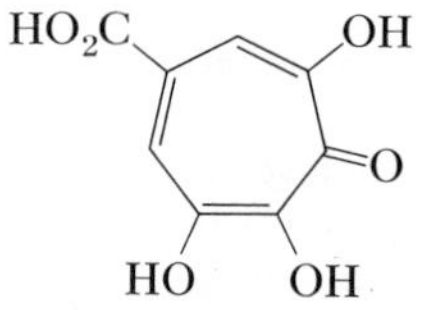

HO_2C OH O HO OH

24. Cyclohexene is oxidized with dilute HNO_3 to adipic acid. Concentrated nitric acid will, with heating, oxidize cyclohexanone to adipic acid. Does this preclude cyclohexanone as an intermediate in the oxidation of cyclohexene? Explain.
25. Predict the absolute stereochemistry of the newly created chiral center for each of the following. Assume that Grignard reagent additions follow Cram's rule.
   (a) (**R**)-3-methyl-3-phenylpropanal + CH_3MgX
   (b) (**S**)-3-phenyl-2-butanone + $LiAlH_4$
   (c) (**R**)-3-methyl-3-phenylpropanal + $LiAlD_4$
   (d) (**S**)-2-phenyl-3-pentanone + CH_3MgX
   (e) (**R**)-2-methylcyclopentanone + $LiAlH_4$
   (f) (**R**)-2-methylcyclopentanone + $LiAlD_4$
26. In an (illegal) attempt to improve the yield in his laboratory synthesis of diphenylcarbinol from benzaldehyde and phenyl magnesium bromide, a student started out with more benzaldehyde than was called for. After the reaction was completed, the reaction mixture was worked up by a procedure that included washing with an aqueous sodium bisulfite solution. When this student carried out this procedure, he alone got a white, crystalline precipitate. Astounded, he took this to his laboratory instructor. The laboratory instructor, after being told of everything except of the use of an excess of benzaldehyde, informed the student that he had used more benzaldehyde than called for. How did Professor Holmes figure this out?
27. Aqueous solutions of glutaraldehyde are potent biocides, used in hospitals to destroy bacteria, viruses, and spores on environmental surfaces. An nmr investigation has shown that the major component of a 25% solution of glutaraldehyde is a cyclic hydrate. Suggest a mechanism for its formation and reversion to glutaraldehyde.

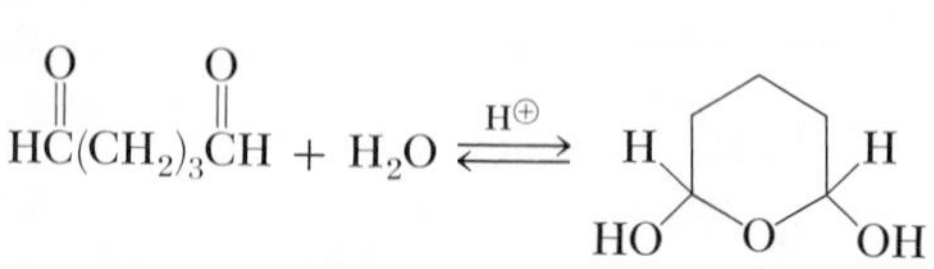

$$HC(=O)(CH_2)_3CH(=O) + H_2O \underset{}{\overset{H^{\oplus}}{\rightleftharpoons}} \text{cyclic hydrate}$$

glutaraldehyde — cyclic hydrate (H, H, HO, O, OH)

28. Suggest mechanisms for the steps of the following synthesis of α,β-unsaturated methyl ketones.

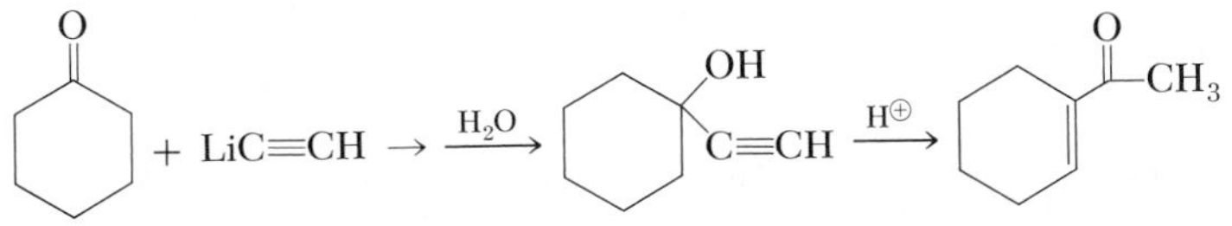

cyclohexanone + $LiC{\equiv}CH$ $\xrightarrow{H_2O}$ 1-ethynylcyclohexanol (OH, $C{\equiv}CH$) $\xrightarrow{H^{\oplus}}$ 1-acetylcyclohexene (C(=O)CH_3)

29. Account for the following order of stability toward dehydration of 1,1-diols.

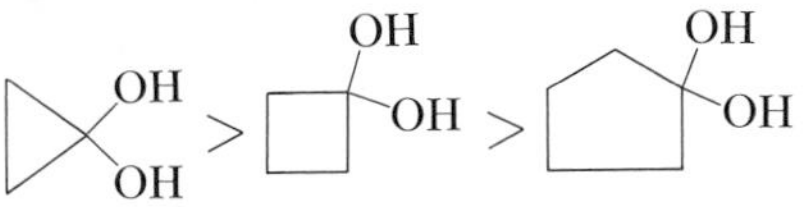

cyclopropane-1,1-diol (OH, OH) > cyclobutane-1,1-diol (OH, OH) > cyclopentane-1,1-diol (OH, OH)

30. Ninhydrin, a reagent used to detect amino acids by a color reaction, is the hydrate of hydrindantrione. Why is the hydrate stable? Why is this the most likely structure for the hydrate?

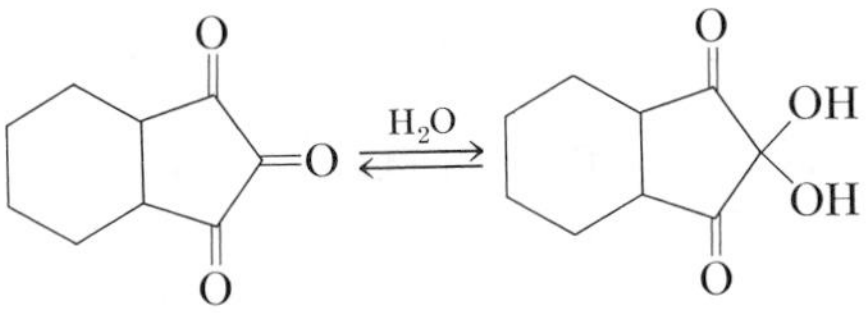

hydrindantrione (O, =O, O) $\underset{}{\overset{H_2O}{\rightleftharpoons}}$ ninhydrin (O, OH, OH, O)

hydrindantrione ninhydrin

31. (a) Suggest a mechanism for the conversion of glucose to the dithioacetal shown below.

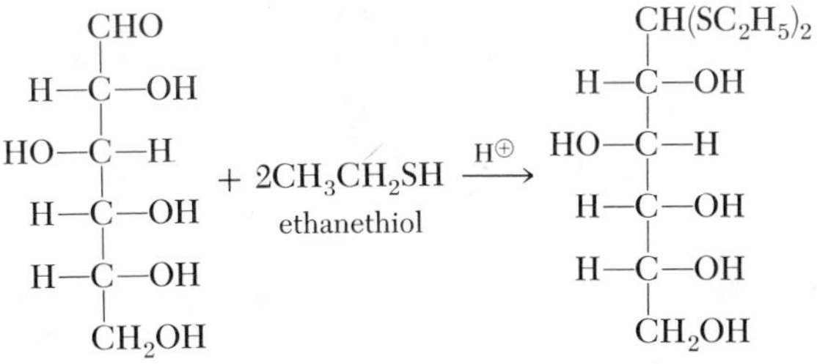

CHO H—C—OH HO—C—H H—C—OH H—C—OH CH_2OH	+ $2CH_3CH_2SH$ ethanethiol	$\xrightarrow{H^{\oplus}}$	$CH(SC_2H_5)_2$ H—C—OH HO—C—H H—C—OH H—C—OH CH_2OH

A related biochemical process involves the conversion of glyceraldehyde 3-phosphate to an enzyme-bound thiohemiacetal. (b) Suggest a mechanism for this reaction.

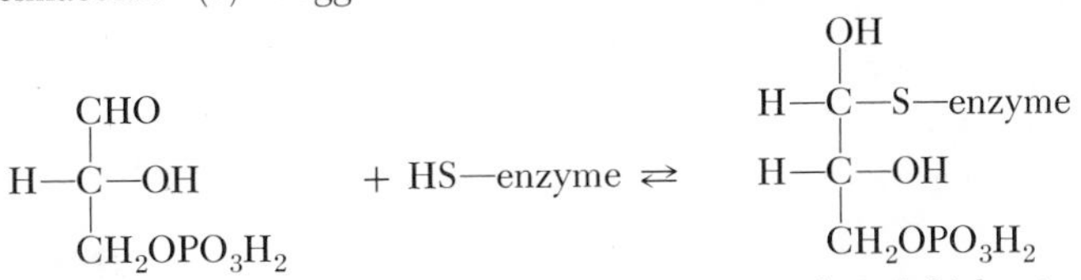

CHO H—C—OH $CH_2OPO_3H_2$ glyceraldehyde 3-phosphate	+ HS—enzyme	⇄	OH H—C—S—enzyme H—C—OH $CH_2OPO_3H_2$ enzyme-bound thiohemiacetal

32. Suggest a mechanism for the useful synthesis of *yneones* shown below. (*Hint:* Look back into Chapter 10.)

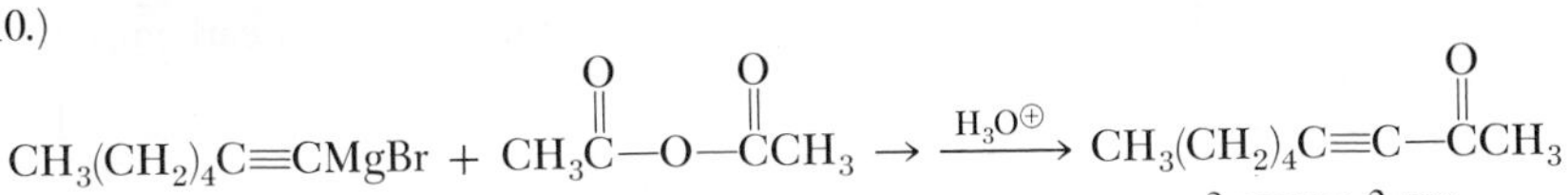

$$CH_3(CH_2)_4C{\equiv}CMgBr + CH_3\overset{O}{\overset{\|}{C}}{-}O{-}\overset{O}{\overset{\|}{C}}CH_3 \rightarrow \xrightarrow{H_3O^{\oplus}} CH_3(CH_2)_4C{\equiv}C{-}\overset{O}{\overset{\|}{C}}CH_3$$

3-nonyne-2-one

33. The Strecker synthesis of amino acids involves the conversion of an aldehyde to an amino acid *via* an amino nitrile. Suggest a mechanism for the formation of the amino nitrile. (Syntheses of amino acids are discussed in more detail in Chapter 24.)

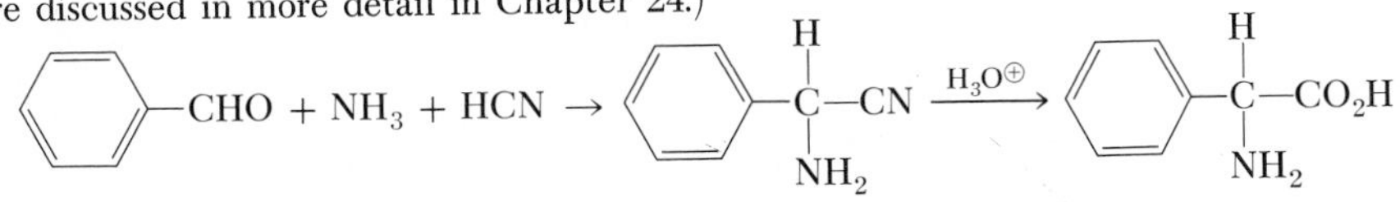

C_6H_5—CHO + NH_3 + HCN → C_6H_5—CH(NH_2)—CN $\xrightarrow{H_3O^{\oplus}}$ C_6H_5—CH(NH_2)—CO_2H

amino nitrile amino acid

34. The Amadori rearrangement, which converts a six-membered nitrogen analog of a hemiacetal into a five-membered hemiacetal, is shown below. Provide mechanisms for as many steps as you can. The reaction occurs in aqueous media.

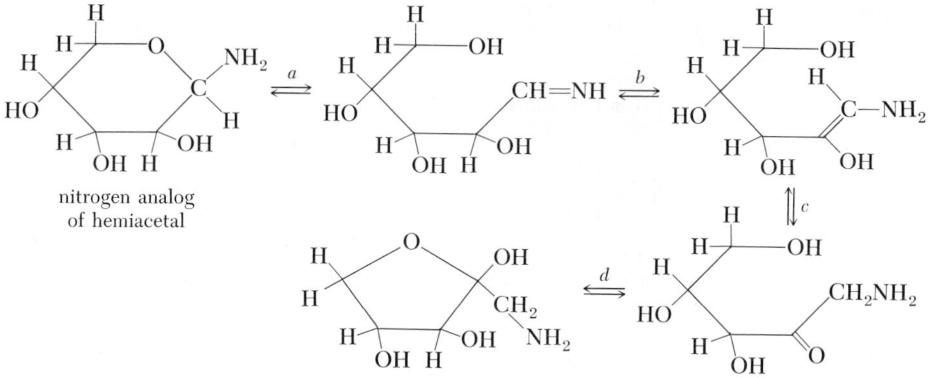

nitrogen analog of hemiacetal

35. Provide structures for compounds **A** through **D**, and mechanisms leading to their formation.

$$\text{acrolein} + HCl \xrightarrow{EtOH} \mathbf{A}(C_3H_5ClO) \xrightarrow[EtOH]{HCl} \mathbf{B}(C_7H_{15}ClO_2) \xrightarrow{OH^{\ominus}} \mathbf{C}(C_7H_{14}O_2) \xrightarrow[OH^{\ominus}]{KMnO_4} \mathbf{D}(C_7H_{16}O_4) \xrightarrow[\text{dil.}]{H_3O^{\oplus}} \text{glyceraldehyde}$$

36. Carbonyl compounds react with PCl_5 to produce *gem*-dihalides. Suggest a mechanism for this reaction. (PCl_5 may be considered to be $PCl_4^{\oplus}PCl_6^{\ominus}$.)

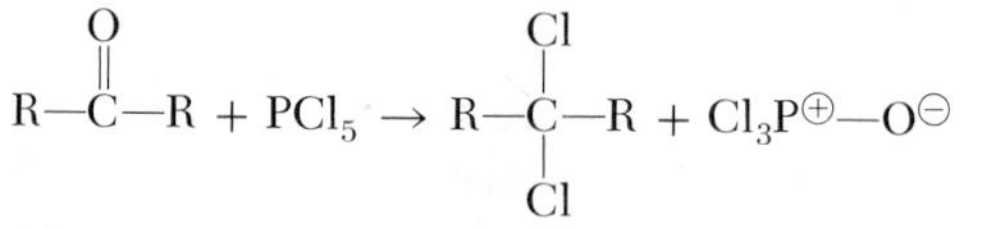

$$R-\overset{O}{\overset{\|}{C}}-R + PCl_5 \rightarrow R-\underset{Cl}{\underset{|}{\overset{Cl}{\overset{|}{C}}}}-R + Cl_3P^{\oplus}-O^{\ominus}$$

37. Methyl β-ribofuranoside reacts with acetone to give an acid-labile, base-stable adduct. Suggest a mechanism for this reaction. Why were *these* two hydroxy groups the ones that reacted?

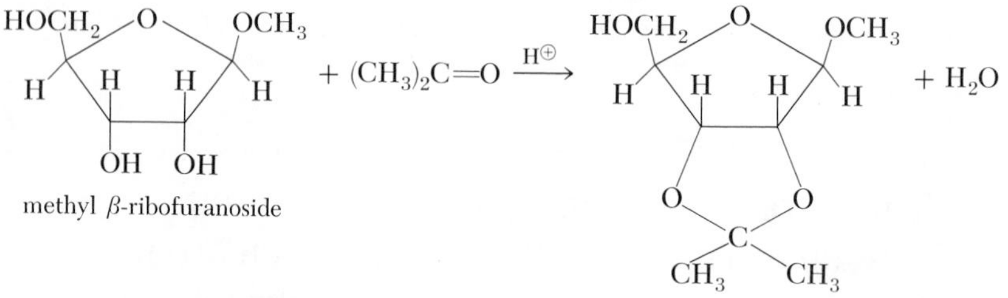

$$\text{methyl } \beta\text{-ribofuranoside} + (CH_3)_2C{=}O \xrightarrow{H^{\oplus}} \text{adduct} + H_2O$$

methyl β-ribofuranoside

38. The structure shown below is that of uridine, an important constituent of nucleic acids. (Nucleic acids are the materials that transfer genetic information from generation to generation.) This compound contains a carbonyl group "hidden" as a derivative. Identify this carbonyl's carbon. How might this carbonyl group be liberated?

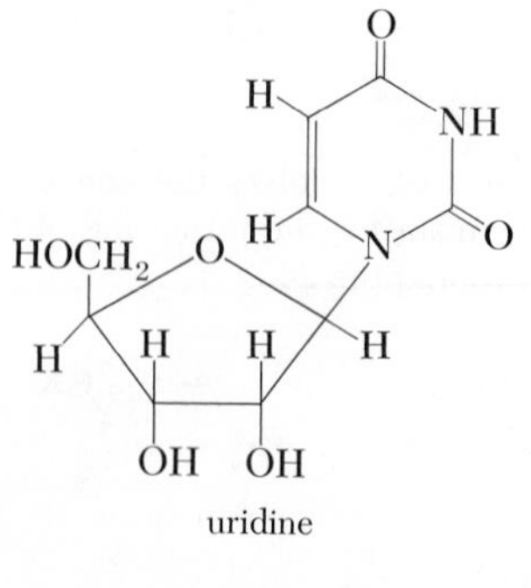

uridine

39. A spectroscopic analysis of an aqueous solution of 4-hydroxybutanal reveals that only 6% of the anticipated aldehyde groups are present. How might this be explained? In what form is the remaining 94%?

40. Hemiacetals can be oxidized to esters by halogen in aqueous base. Suggest a mechanism for this reaction. (This is an important process in carbohydrate chemistry.)

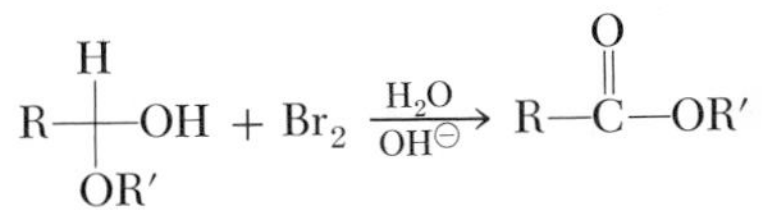

$$\underset{\text{hemiacetal}}{R-\underset{OR'}{\overset{H}{C}}-OH} + Br_2 \xrightarrow[OH^{\ominus}]{H_2O} \underset{\text{ester}}{R-\overset{O}{\overset{\|}{C}}-OR'}$$

41. The rate of reduction of cycloalkanones to the corresponding alcohols by sodium borohydride is cyclopentanone < cyclohexanone < cyclobutanone. Account for this order.

42. Do you agree with this statement? "Any reaction that is not reversible *must* be kinetically controlled." Explain your stand.

43. When formaldehyde is polymerized to a linear polymer in heptane solvent, with $(C_6H_5)_3P$ or a proton as a catalyst, it yields a thermally unstable material (**A**) that unzips readily to re-form formaldehyde. However, if the end of the polymer is "capped," the product is a highly crystalline, tough polymer. The example below, **B,** is called *Delrin* and is already replacing nylon as a molding resin. Suggest a reason for the stability gained by capping the polymer.

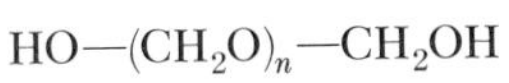

$$\underset{\mathbf{A}}{HO-(CH_2O)_n-CH_2OH} \qquad \underset{\mathbf{B}}{CH_3-\overset{O}{\overset{\|}{C}}-O-(CH_2O)_n-CH_2-O-\overset{O}{\overset{\|}{C}}-CH_3}$$

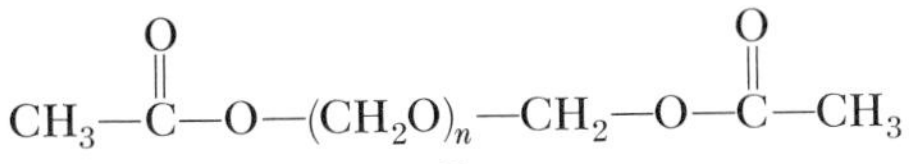

ALDEHYDES AND KETONES— THEIR CARBANIONS AND SPECTRA

17.1 INTRODUCTION

The carbonyl group is a polar functional group, the carbonyl carbon bearing some positive charge. This enhances the acidity of a hydrogen bonded to a carbon adjacent (or α) to the carbonyl group. Because of this acidity, because of the acidity of a hydrogen bonded to oxygen, and because of the comparable strengths of the carbon-carbon and carbon-oxygen double bonds, carbonyl compounds possessing a C—H group α to the carbonyl group can exist in an isomeric form. This isomer is referred to as the *enol* form and, while it is often less stable than the isomeric *keto* form, it accounts for many of the interesting reactions of carbonyl compounds. The following section is dedicated to these enol forms, to their relationship to the keto form, and to *enolization*, the process that converts the keto form to the enol form. Enolization is important in its own right but also is a specific type of *tautomerism, i.e.*, an equilibrium between species that differ in the location of a hydrogen and the translocation of a double bond and a single bond. (Tautomerism was introduced on p. 273, and that discussion should be reviewed.)

$$\underset{\text{keto form}}{-\overset{\overset{\displaystyle H}{|}}{\underset{|\,\alpha}{C}}-\overset{\overset{\displaystyle O}{\|}}{C}-} \rightleftarrows \underset{\text{enol form}}{-\overset{|}{\underset{\alpha}{C}}=\overset{\overset{\displaystyle O-H}{|}}{C}-}$$

keto-enol tautomerism

17.2 ENOLIZATION

When 3-phenyl-2-butanone is dissolved in a deuterated acidic or basic solution (*e.g.*, D_2O, CH_3OD, $OD^{\ominus}$) and then isolated, the recovered ketone contains deuterium. If enough time has elapsed, all of the hydrogens α to the carbonyl group will have been exchanged for deuterium. However, if the reaction is stopped after isotope exchange has just begun, the only hydrogen found to be exchanged is the one on the carbon bearing the phenyl group.

$$C_6H_5-\underset{CH_3}{\overset{H}{C}}-\overset{O}{\overset{\|}{C}}-CH_3 + D_2O \xrightarrow[OD^{\ominus}]{D^{\oplus}\text{ or}} C_6H_5-\underset{CH_3}{\overset{D}{C}}-\overset{O}{\overset{\|}{C}}-CD_3 \quad \text{longer exposure}$$

3-phenyl-2-butanone

$$C_6H_5-\underset{CH_3}{\overset{H}{C}}-\overset{O}{\overset{\|}{C}}-CH_3 + D_2O \xrightarrow[OD^{\ominus}]{D^{\oplus}\text{ or}} C_6H_5-\underset{CH_3}{\overset{D}{C}}-\overset{O}{\overset{\|}{C}}-CH_3 \quad \text{shorter exposure}$$

These isotopic exchanges are explained by the **enolization of the carbonyl group,** a concept first introduced while discussing the hydration of triple bonds (p. 272).

$$H-\underset{CH_3}{\overset{H_5C_6}{C}}-\overset{O}{\overset{\|}{C}}-CH_3 \;(+\; D-\overset{\oplus}{O}D_2) \rightleftarrows H-\underset{CH_3}{\overset{H_5C_6}{C}}-\overset{\overset{\oplus}{O}-D}{\overset{\|}{C}}-CH_3 \;(+\; D_2O) \rightleftarrows (H_5C_6)(H_3C)C{=}C(OD)(CH_3)$$

enol form

$$D_2\overset{\oplus}{O}-D \leftarrow (H_5C_6)(H_3C)C{=}C(O-D)(CH_3) \leftarrow D_2O \rightleftarrows D-\underset{CH_3}{\overset{H_5C_6}{C}}-\overset{O}{\overset{\|}{C}}-CH_3$$

keto form

acid-catalyzed enolization causes deuterium incorporation

$$DO^{\ominus}: \rightarrow H-\underset{CH_3}{\overset{H_5C_6}{C}}-\overset{:O:}{\overset{\|}{C}}-CH_3 \underset{}{\overset{-DOH}{\rightleftarrows}} (H_5C_6)(H_3C)\overset{\ominus}{:}C-\overset{O}{\overset{\|}{C}}-CH_3 \leftrightarrow (H_5C_6)(H_3C)C{=}C(O^{\ominus})(CH_3)$$

enolate anion

$$D-O-D \leftarrow \overset{\ominus}{:}\underset{H_3C}{\overset{H_5C_6}{C}}-\overset{O}{\overset{\|}{C}}-CH_3 \rightleftarrows D-\underset{H_3C}{\overset{H_5C_6}{C}}-\overset{O}{\overset{\|}{C}}-CH_3 + OD^{\ominus}$$

base-catalyzed enolization causes deuterium incorporation

If the starting ketone had been optically active, it would have racemized under the conditions that led to exchange, and the rate constants for racemization and for deuterium incorporation would be equal. Since the first symmetric species produced in the schemes above are the *enol* and the *enolate anion,* it is presumed that their formation is rate-determining in acidic and basic media, respectively.

3-Phenyl-2-butanone reacts rapidly with halogens (chlorine, bromine, or iodine), the reaction being catalyzed by acid or base. The rate of incorporation of the first halogen (see below) is the same as is the rate of deuterium incorporation. This requires that halogenation have the same rate-determining step as in isotope exchange and racemization. The accepted mechanisms for these halogenations are as shown:

base-catalyzed halogenation

$$C_6H_5-\underset{CH_3}{\overset{H}{C}}-\overset{O}{\overset{\|}{C}}-CH_3 \xrightarrow[-H_2O]{OH^{\ominus}} C_6H_5-\underset{CH_3}{\overset{\ominus}{\ddot{C}}}-\overset{O}{\overset{\|}{C}}-CH_3 \quad \text{(rate-determining)}$$

$$C_6H_5-\underset{C(=O)CH_3}{\overset{CH_3}{C}}:^{\ominus} \longrightarrow Br-Br \xrightarrow{-Br^{\ominus}} C_6H_5-\underset{C(=O)CH_3}{\overset{CH_3}{C}}-Br \quad \text{(rapid)}$$

acid-catalyzed halogenation

$$C_6H_5-\underset{CH_3}{\overset{H}{C}}-\overset{O}{\overset{\|}{C}}-CH_3 \xrightarrow{H^{\oplus}} (H_5C_6)(H_3C)C=C(OH)(CH_3) \quad \text{(rate-determining)}$$

$$(H_5C_6)(H_3C)C=C(O-H)(CH_3) + Br-Br \xrightarrow{-HBr} CH_3-\underset{Br}{\overset{H_5C_6}{C}}-\overset{O}{\overset{\|}{C}}-CH_3 \quad \text{(rapid)}$$

1. If acetone is dissolved in $D_2{}^{18}O$ containing a trace of DCl and then recovered, some of the acetone is found to be converted to $CD_3C^{18}OCD_3$. Provide a mechanism for this interconversion.
2. Ketene, $CH_2{=}C{=}O$, reacts with water to form acetic acid, CH_3CO_2H. Suggest a mechanism for this reaction. When ketene reacts with acetone containing a trace of sulfuric acid, there is produced 1-methylvinyl acetate (isopropenyl acetate). Using the reaction of water with ketene as a model, suggest a mechanism for this latter reaction.

$$CH_3-\overset{O}{\overset{\|}{C}}-CH_3 + CH_2{=}C{=}O \xrightarrow{H^{\oplus}} (CH_3)(H_2C{=})C-O-\overset{O}{\overset{\|}{C}}-CH_3$$

isopropenyl acetate

3. What is the significance of the following representation of the enolate anion?

$$-\overset{\ominus}{\underset{|}{C}}\cdots\underset{|}{\overset{O}{\overset{|}{C}}}-$$

4. Explain why the rates of acid-catalyzed bromination and iodination of acetophenone, $C_6H_5COCH_3$, are identical.

How Rapidly Does Enolization Occur? The conversion of a keto form into the corresponding enol form is quite rapid in water. The data in Table 17–1 indicate that increasing the electron-withdrawing capability of a substituent α to a carbonyl group increases the rate

TABLE 17-1 Relative Rate of Carbanion Formation

$$-\overset{\overset{\text{O}}{\|}}{\text{C}}-\overset{\overset{\text{H}}{|}}{\underset{|}{\text{C}}}- + H_2O \rightarrow -\overset{\overset{\text{O}}{\|}}{\text{C}}-\overset{\ominus}{\underset{|}{\text{C}}}- + \overset{\oplus}{H_3O}$$

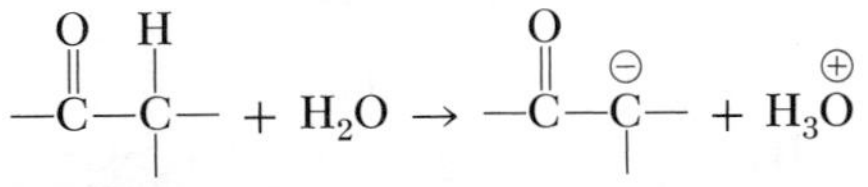

COMPOUND	RELATIVE RATE[a]
CH_3COCH_3 (italic H)	1
CH_3COCH_2Cl	1.2×10^2
$CH_3COCHCl_2$	1.6×10^3
$CH_3COCH_2COCH_3$	3.6×10^7
$CH_3COCH_2CO_2Et$	2.6×10^6

[a] Reaction at 50°. The hydrogen lost in enolate formation is italicized.

of carbanion formation at Cα. This is ascribed to the dispersion, by the Cα substituent, of developing negative charge in the transition state leading to the carbanion.

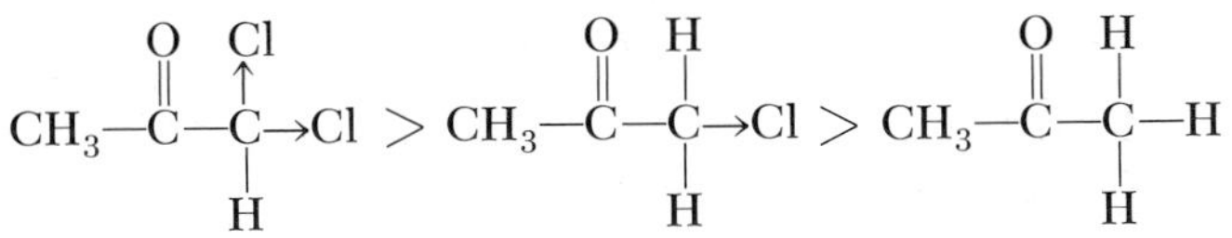

$$CH_3-\overset{\overset{\text{O}}{\|}}{\text{C}}-\overset{\overset{\text{Cl}}{\uparrow}}{\underset{\text{H}}{\underset{|}{\text{C}}}}\rightarrow Cl > CH_3-\overset{\overset{\text{O}}{\|}}{\text{C}}-\overset{\overset{\text{H}}{|}}{\underset{\text{H}}{\underset{|}{\text{C}}}}\rightarrow Cl > CH_3-\overset{\overset{\text{O}}{\|}}{\text{C}}-\overset{\overset{\text{H}}{|}}{\underset{\text{H}}{\underset{|}{\text{C}}}}-H$$

electron withdrawal enhances acidity

Carbanions that are resonance stabilized by two carbonyl groups (*e.g.*, the anion from 2,4-pentanedione) usually are more easily formed than are carbanions from monoketones. However, if one of the carbonyl groups is already participating in delocalization, as in the ketoester ethyl acetoacetate, the enhancement of proton loss is diminished (see Table 17–1).

TABLE 17-2 Keto-Enol Composition at Equilibrium[a]

COMPOUND	STRUCTURE	% ENOL FORM
acetone	CH_3COCH_3	0.00025
biacetyl	$CH_3COCOCH_3$	0.0056
cyclohexanone	$(CH_2)_5CO$	0.020
acetylacetone	$CH_3COCH_2COCH_3$	80
1,2-cyclohexanedione	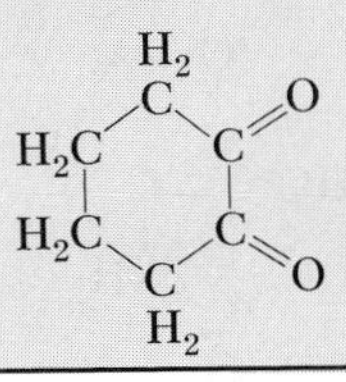 six-membered ring: H_2C, H_2C, CH_2, CH_2, C=O, C=O	100

[a] Data are for pure liquid samples.

anion of 2,4-pentanedione (acetylacetone)

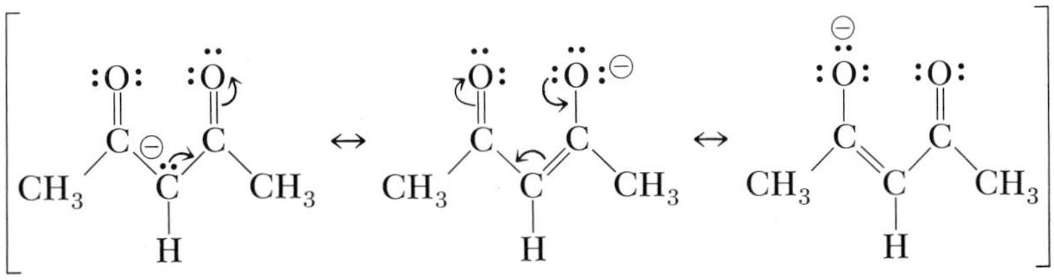

$$\left[CH_3-\overset{\overset{:\text{O}:}{\|}}{\text{C}}-\overset{\ominus}{\underset{\underset{\text{H}}{|}}{\text{C}}}-\overset{\overset{:\ddot{\text{O}}}{\|}}{\text{C}}-CH_3 \leftrightarrow CH_3-\overset{\overset{:\ddot{\text{O}}:}{\|}}{\text{C}}-\underset{\underset{\text{H}}{|}}{\text{C}}=\overset{\overset{:\ddot{\text{O}}:^{\ominus}}{|}}{\text{C}}-CH_3 \leftrightarrow CH_3-\overset{\overset{:\ddot{\text{O}}:^{\ominus}}{|}}{\text{C}}=\underset{\underset{\text{H}}{|}}{\text{C}}-\overset{\overset{:\text{O}:}{\|}}{\text{C}}-CH_3 \right]$$

anion of ethyl acetoacetate (acetoacetic ester)

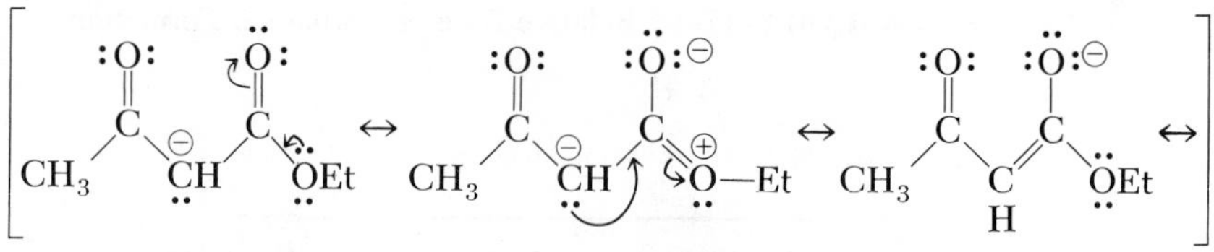

How Much Keto, How Much Enol? Most simple aldehydes and ketones exist mainly in their keto (*i.e.*, carbonyl) forms. Thus, both acetone and cyclohexanone (Table 17–2, p. 507) contain less than 1% enol as the pure liquid. However, the considerable difference between even these two arises because single-bond rotation becomes restricted in going from keto to enol forms; this unfavorable entropy change is already partly accomplished in the cyclic structure of cyclohexanone, but not in the acyclic structure of acetone.

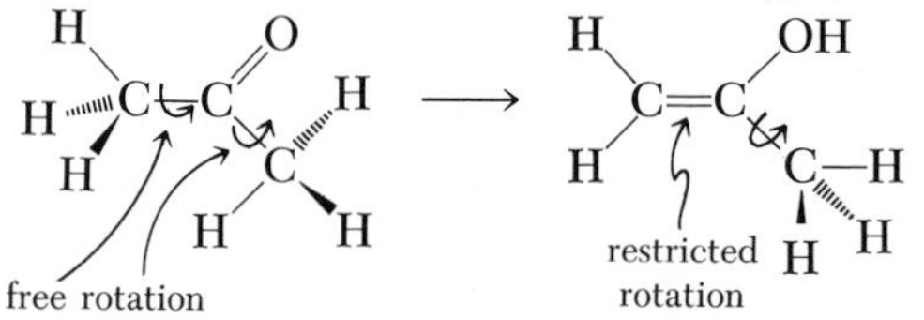

free rotation

acetone (keto form)

restricted rotation

acetone (enol form)

partly restricted

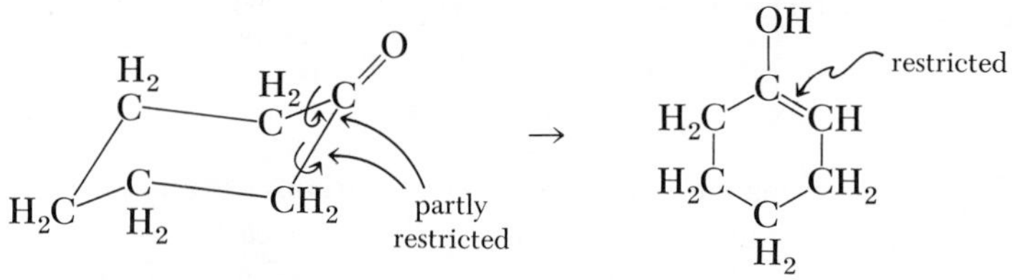

cyclohexanone (keto form)

restricted

cyclohexanone (enol form)

Biacetyl, an α- or 1,2-diketone, has a slightly greater enolic content than does acetone; but it has a much smaller enol content than does 1,2-cyclohexanedione, another α-diketone. The difference between acetone and biacetyl is probably not indicative of any single effect. However, 1,2-cyclohexanedione possesses two carbonyl groups with dipoles directed in similar directions. Repulsion between these dipoles can be effectively reduced only by enolization. Biacetyl, on the other hand, can reduce similar repulsion by a simple rotation around the σ bond connecting the carbonyl groups.

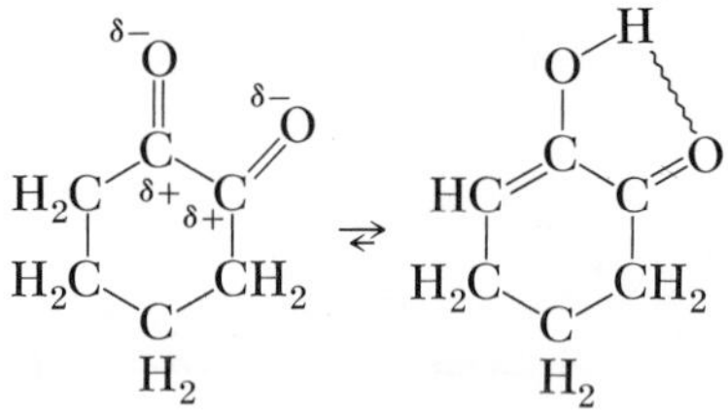

1,2-cyclohexanedione

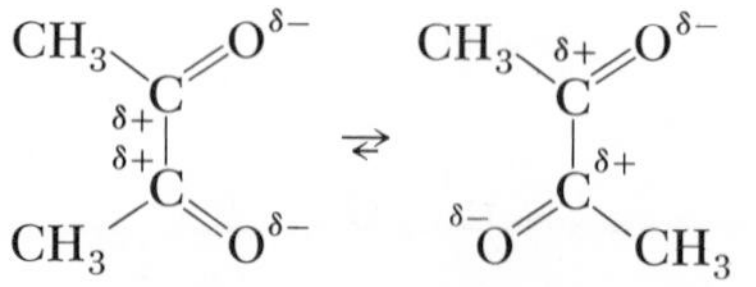

biacetyl

While it seems likely that intramolecular hydrogen bonding assists in stabilizing the enols of diketones and related compounds, intramolecular hydrogen bonding is not a necessary condition for high enol content. For example, the β- or 1,3-diketone, 5,5-dimethyl-1,3-cyclohexanedione (also called *methone*), exists nearly exclusively as the enol. There may be several factors that cause this—*e.g.*, severe conformational restriction in the diketo form—but *internal* hydrogen-bonding cannot be one of them because the hydroxy group and the carbonyl group are too far apart to permit such an interaction. Incidentally, the enol form of methone that is formed is the one which results in conjugation of the double bond rather than in an isolated double bond, *i.e.*, **A** rather than **B.**

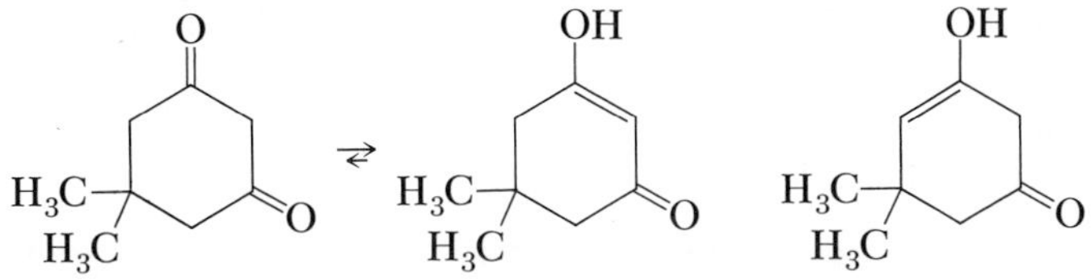

methone ⇄ **A** **B**

There are some ketones that cannot enolize because of the impossibility of achieving 2*p*-2*p* π overlap in the enolic double bond. For example, the bridgehead hydrogen (shown in boldface) in the structure below is non-enolizable, while the others shown are enolizable. This is yet another application of Bredt's rule (p. 210).

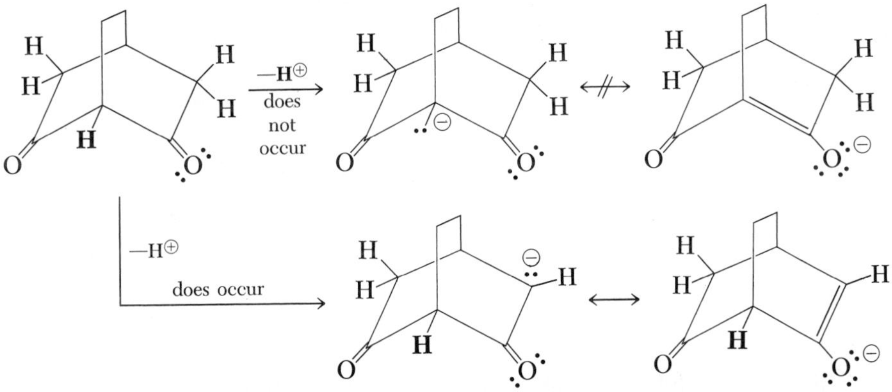

—H⊕ does not occur

—H⊕ does occur

In addition to enolizations prohibited by stereoelectronic constraints such as those implied in Bredt's rule, enolizations that might produce unstable conjugated systems are not observed.

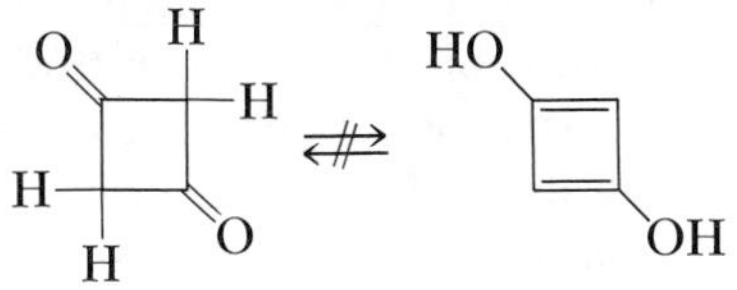

an anti-aromatic structure

5. Indicate the preferred enol form of each of the following, explaining your choice.
   (a) 2-methylcyclohexanone
   (b) 1,3-cyclohexanedione
   (c) methyl ethyl ketone
   (d) methyl *t*-butyl ketone
   (e) methyl allyl ketone
6. (a) Explain why **A,** below, is less stable than its enol. (b) How can **A** and **B** form the same enol in acidic solution? (c) What is its structure?

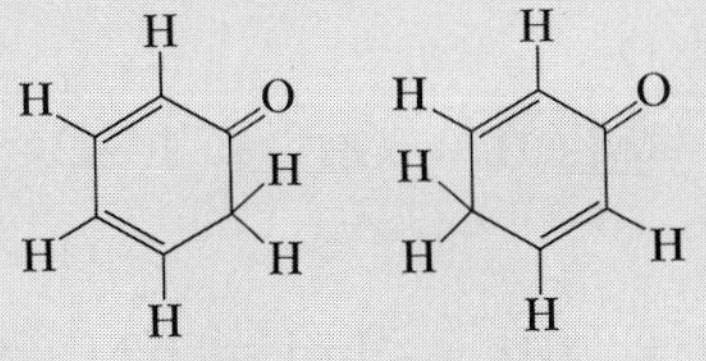

A **B**

How Is The Enol Content Determined? Spectral procedures, including ir and nmr, can be used to assess the composition of a keto-enol mixture. Since these techniques will be discussed in subsequent portions of this chapter, we will limit this discussion to a classical "wet" procedure.

The classical method for determining the amount of enol present in a keto-enol mixture is based upon the work of Kurt Meyer (1911), who showed that only the *enol* form (and not the *keto* form) reacts with bromine to give bromoketones. The enol content is determined by (a) rapid over-titration of the enol with bromine, (b) rapid destruction of excess bromine with β-naphthol, (c) oxidation of iodide ion by the bromoketone to iodine, and (d) titrimetric determination of I_2 with thiosulfate anion, $S_2O_3^{2-}$.

$$CH_3-\overset{\overset{O}{\|}}{C}-CH_2-\overset{\overset{O}{\|}}{C}-CH_3 \rightleftharpoons CH_3-\overset{\overset{O}{\|}}{C}-CH=\overset{\overset{OH}{|}}{C}-CH_3 \xrightarrow[\text{rapid}]{\text{excess } Br_2} CH_3-\overset{\overset{O}{\|}}{C}-\underset{\underset{Br}{|}}{CH}-\overset{\overset{O}{\|}}{C}-CH_3$$

$$Br_2(\text{unreacted}) + \underset{\beta\text{-naphthol}}{C_{10}H_7OH} \xrightarrow{\text{rapid}} \text{1-bromo-2-naphthol} + HBr$$

$$CH_3-\overset{\overset{O}{\|}}{C}-\underset{\underset{Br}{|}}{CH}-\overset{\overset{O}{\|}}{C}-CH_3 + H^{\oplus} + 2I^{\ominus} \rightarrow I_2 + Br^{\ominus} + CH_3-\overset{\overset{O}{\|}}{C}-CH_2-\overset{\overset{O}{\|}}{C}-CH_3$$

$$I_2 + \underset{\text{thiosulfate}}{2S_2O_3^{2-}} \rightarrow 2I^{\ominus} + S_4O_6^{2-}$$

All of the reagents for this analysis are used in excess (an excess of bromine over enol, and an excess of β-naphthol over bromine). Success of the analysis hinges upon the rapid addition of bromine followed by rapid addition of β-naphthol. (This minimizes the generation of more enol during the analysis.) Analysis of the equations reveals that each mole of thiosulfate consumed is equivalent to 0.5 mole of enol in the sample titrated.

While halogenation can be used, as a tool of quantitative analysis, to determine the enol content of a carbonyl compound, it also is important, in qualitative analysis, in identifying certain types of ketones, as seen from the following discussion.

The Haloform Reaction. A standard wet-test for the presence of a CH_3—C(O)— group is the appearance of a light yellow precipitate, m.p. 119 to 121°, when an unknown is mixed with iodine in alkaline solution ($OH^{\ominus}/H_2O$). The test is based upon the rapid *tri*halogenation of the methyl group and the subsequent loss of a rather stable trihalomethide ion in an addition-elimination sequence.

net reaction: $$R-\overset{\overset{O}{\|}}{C}-CH_3 + I_2 \xrightarrow[OH^{\ominus}]{H_2O} R-\overset{\overset{O}{\|}}{C}-O^{\ominus} + CHI_3$$

mechanism: $$R-\overset{\overset{O}{\|}}{C}-CH_3 + OH^{\ominus} \rightleftharpoons R-\overset{\overset{O}{\|}}{C}-CH_2^{\ominus} + H_2O$$

$$R-\overset{\overset{O}{\|}}{C}-\overset{\ominus}{C}H_2 \curvearrowright I-I \rightleftharpoons R-\overset{\overset{O}{\|}}{C}-CH_2I + I^{\ominus}$$

$$R\text{—}\overset{O}{\overset{\|}{C}}\text{—}CH_2I + OH^{\ominus} \underset{}{\overset{-HOH}{\rightleftharpoons}} R\text{—}\overset{O}{\overset{\|}{C}}\text{—}\overset{\ominus}{C}HI \xrightarrow{I_2} R\text{—}\overset{O}{\overset{\|}{C}}\text{—}CHI_2$$

$$R\text{—}\overset{O}{\overset{\|}{C}}\text{—}CI_3 \overset{I_2}{\rightleftharpoons} R\text{—}\overset{O}{\overset{\|}{C}}\text{—}\overset{\ominus}{C}I_2 \overset{OH^{\ominus}}{\rightleftharpoons} R\text{—}\overset{O}{\overset{\|}{C}}\text{—}CHI_2$$

$$R\text{—}\overset{\ddot{O}:}{\overset{\|}{C}}\text{—}CI_3 + :\ddot{O}H^{\ominus} \rightleftharpoons R\text{—}\overset{:\ddot{O}:^{\ominus}}{\underset{OH}{C}}\text{—}CI_3 \rightarrow R\text{—}\overset{O}{\overset{\|}{C}}\text{—}OH + :CI_3^{\ominus} \rightarrow RCO_2^{\ominus} + \underset{\text{iodoform}}{HCI_3}$$

The yellow precipitate formed in the test is *iodoform*, CHI_3. For this reason the reaction is also called the *iodoform test*. That old "hospital smell" was largely due to the liberal use of iodoform as an antiseptic. While it has been replaced by other bactericides in hospitals, iodoform is still used in veterinary medicine (*e.g.*, for treatment of fissured heels and thrush in horses).

The procedure works equally well for bromine and chlorine, but since bromoform, $CHBr_3$, and chloroform, $CHCl_3$, are liquids, the analytical advantage rests with iodine. The haloform reaction is, in addition to being a test for methyl ketones, a valuable means of synthesizing carboxylic acid from methyl ketones.

$$\underset{\text{cyclopropyl methyl ketone}}{\triangleright\text{—}\overset{O}{\overset{\|}{C}}\text{—}CH_3} \xrightarrow[OH^{\ominus}]{Br_2} \xrightarrow{H^{\oplus}} \underset{\text{cyclopropanecarboxylic acid}}{\triangleright\text{—}CO_2H} \quad (85\%)$$

$$\text{(steroid)}\text{—}\overset{O}{\overset{\|}{C}}\text{—}CH_3 \xrightarrow[H_2O/\text{dioxane}]{Br_2/OH^{\ominus}} \xrightarrow{H^{\oplus}} \text{(steroid)}\text{—}CO_2H \quad (\sim 60\%)$$

$$(CH_3)_2C{=}CH\text{—}\overset{O}{\overset{\|}{C}}\text{—}CH_3 \xrightarrow[H_2O/\text{dioxane}]{Cl_2/OH^{\ominus}} \xrightarrow{H^{\oplus}} (CH_3)_2C{=}CH\text{—}CO_2H \quad (\sim 50\%)$$

A lack of total specificity of the test results, in part, from the ability of halogens to oxidize 2° methyl carbinols, $RCH(OH)CH_3$, to methyl ketones, $RC(O)CH_3$ (see also problem 25). We will close with one further observation on the scope of the haloform reaction. Methyl and methylene groups attached to benzene rings bearing an acetyl group also are oxidized by alkaline iodine. Can you suggest why?

$$CH_3\text{—}C_6H_4\text{—}\overset{O}{\overset{\|}{C}}\text{—}CH_3 \xrightarrow{I_2/OH^{\ominus}} \xrightarrow{H^{\oplus}} HO_2C\text{—}C_6H_4\text{—}CO_2H \quad (47\%)$$

$$\text{(dihydrophenanthryl)}\text{—}\overset{O}{\overset{\|}{C}}\text{—}CH_3 \xrightarrow{I_2/OH^{\ominus}} \xrightarrow{H^{\oplus}} \text{(biphenyl-2,2',4'-tricarboxylic acid: } CO_2H,\ CO_2H,\ CO_2H) \quad (49\%)$$

7. Identify the carboxylic acid, if any, produced by treating the following with alkaline, aqueous chlorine, followed by acidification.
   (a) $C_6H_5COCH_3$
   (b) $CH_3C(O)(CH_2)_5C(O)CH_3$
   (c) $(CH_3)_2C{=}CHC(O)CH_3$
   (d) methyl *o*-tolyl ketone
   (e) methyl *m*-tolyl ketone

Enolization in Biological Systems. Enolization and other tautomerization processes are encountered in many biological reactions. For example, transfer of an amino group between an amino acid and an α-keto acid, described in the previous chapter (p. 484), requires tautomerism in a derivative of pyridoxamine. We will now examine how enolization enters into the biochemical process termed *alcoholic fermentation*, a process that ultimately converts glucose (a sugar) into ethanol and that forms the basis of the alcoholic beverage industry. For the sake of completeness we will begin near the beginning.

$$\underset{\text{glucose}}{C_6H_{12}O_6} \rightarrow 2C_2H_5OH + 2CO_2 \qquad \text{alcoholic fermentation}$$

In an early step (better follow along using Figure 17–1), glucose is converted (with the aid of an enzyme and a complex molecule called adenosine triphosphate, "ATP") into glucose 6-phosphate. Glucose 6-phosphate then is converted, by an enolization reaction, into fructose 6-phosphate (fructose is another sugar). Following this, fructose 6-phosphate is converted, by a new enzyme and ATP, into fructose 1,6-diphosphate. This diphosphate is cleaved into two smaller molecules, glyceraldehyde 3-phosphate and dihydroxyacetone phosphate. These two small molecules are equilibrated, with the aid of yet another enzyme, by what is believed to be another enolization reaction. Glyceraldehyde 3-phosphate is then converted, by a complex process, into 1,3-diphosphoglyceric acid. Loss of a phosphate group converts this 1,3-diphosphoglyceric acid into 3-phosphoglyceric acid which, in turn, is equilibrated with the isomeric 2-phosphoglyceric acid. This complex sequence continues with the dehydration of 2-phosphoglyceric acid to produce phosphoenolpyruvic acid. After loss of a phosphate group, which produces enolpyruvic acid, the sequence moves along to yield pyruvic acid. Pyruvic acid is decarboxylated (loses carbon dioxide) to form acetaldehyde and, in the last step, acetaldehyde is reduced to ethanol. (We discussed this last step in the previous chapter, p. 489.)

Although this complex process is presented in Figure 17–1, the enolization steps are singled out below. We should note that the phosphate group appears so frequently in this and other biochemical schemes because living systems employ the phosphate group as, among other things, an excellent leaving group.

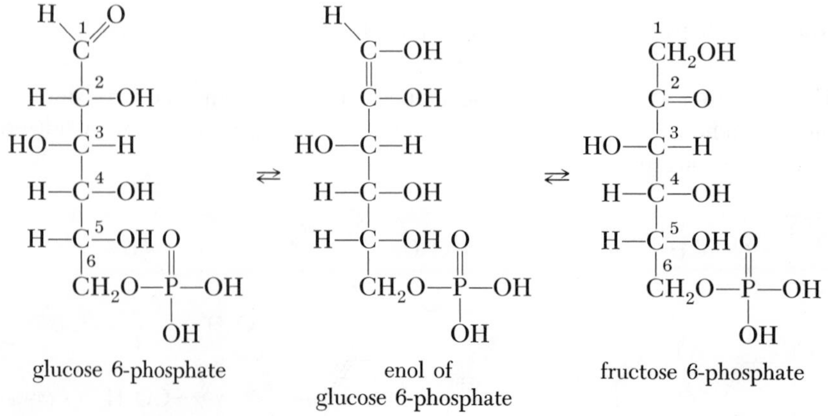

```
  H 1 O                 H                            1
   \ //                  \                          CH2OH
    C                     C—OH                       | 2
    | 2                   ||                         C=O
 H—C—OH                   C—OH                       | 3
    | 3                   |                     HO—C—H
HO—C—H          ⇄    HO—C—H              ⇄         | 4
    | 4                   |                      H—C—OH              step 2
 H—C—OH                H—C—OH                        | 5
    | 5                   |                      H—C—OH O
 H—C—OH O              H—C—OH O                     |6    ||
    |6    ||              |     ||                 CH2O—P—OH
   CH2O—P—OH             CH2O—P—OH                      |
        |                     |                         OH
        OH                    OH

glucose 6-phosphate    enol of               fructose 6-phosphate
                       glucose 6-phosphate
```

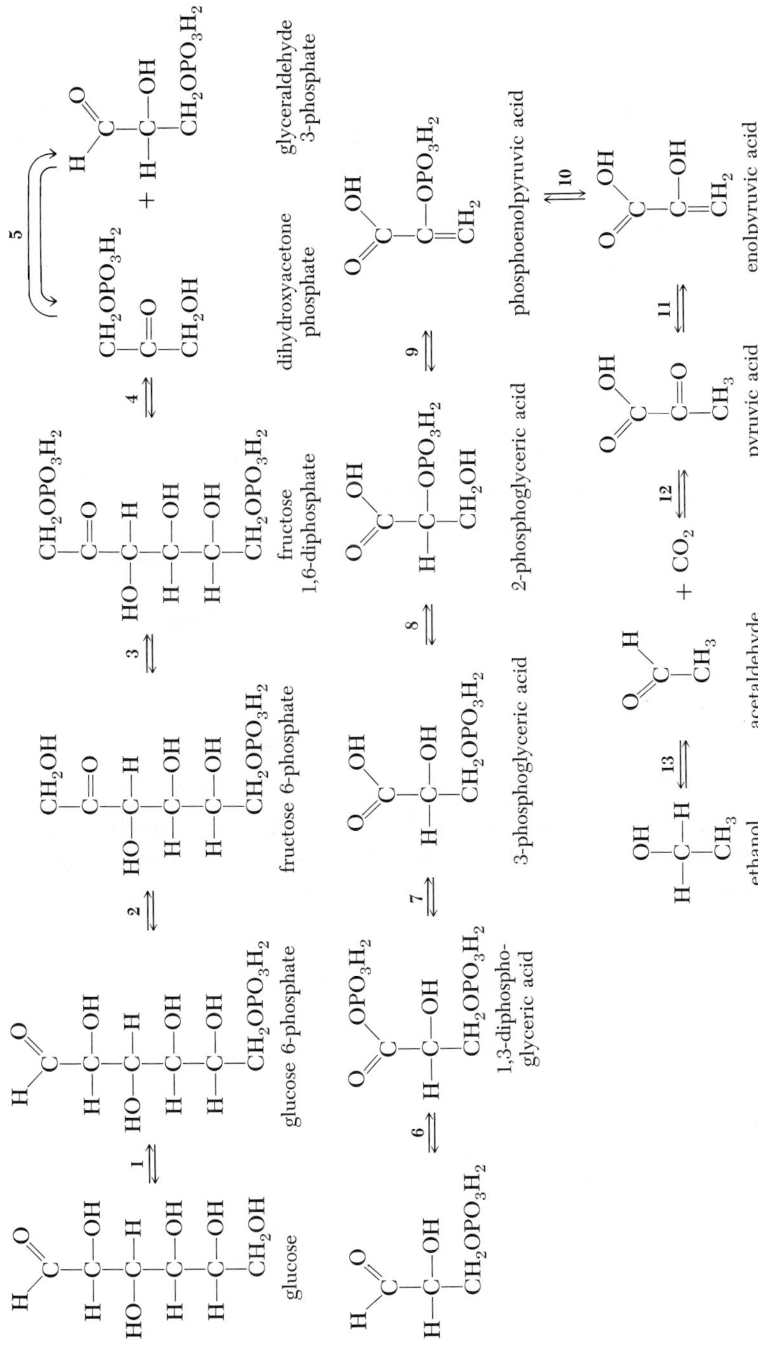

Figure 17-1 The alcoholic fermentation of glucose. The glucose originates from an enzyme-catalyzed decomposition of glucose polymers. Reactions such as step **2** will be examined in more detail in Chapter 25. Steps **2**, **5**, and **11** involve tautomerism. Other enolizations also may be involved.

$$\underset{\text{dihydroxyacetone phosphate}}{\mathrm{H_2C(OPO_3H_2)-C(=O)-CH_2OH}} \rightleftarrows \underset{\text{enol of dihydroxyacetone phosphate}}{\mathrm{H_2C(OPO_3H_2)-C(OH)=CH(OH)}} \rightleftarrows \underset{\text{glyceraldehyde 3-phosphate}}{\mathrm{H_2C(OPO_3H_2)-CH(OH)-CHO}} \qquad \text{step } \mathbf{5}$$

$$\underset{\text{enolpyruvic acid}}{\mathrm{CH_2=C(OH)-CO_2H}} \rightleftarrows \underset{\text{pyruvic acid}}{\mathrm{CH_3-C(=O)-CO_2H}} \qquad \text{step } \mathbf{11}$$

17.3 α-HALOCARBONYL COMPOUNDS

Synthesis. The usual method for synthesizing α-halocarbonyl compounds involves an acid-catalyzed halogenation of a carbonyl compound, because the yields from base-catalyzed halogenation are considerably less than from acid-catalyzed halogenation.

$$\underset{p\text{-bromoacetophenone}}{\mathrm{Br-C_6H_4-C(=O)-CH_3}} \xrightarrow[25^\circ]{\mathrm{Br_2/CH_3CO_2H}} \underset{p\text{-bromophenacyl bromide}}{\mathrm{Br-C_6H_4-C(=O)-CH_2Br}}$$

Another useful procedure for the selective halogenation of unsymmetric ketones at the more highly substituted α-carbon employs copper(II) halide to (a) catalyze enolization and (b) transfer halogen to the enol. The technique also is valuable for synthesizing α-haloaldehydes.

$$\mathrm{HO-C_6H_4-C(=O)-CH_3} + \mathrm{CuBr_2} \xrightarrow[\text{chloroform, heat}]{\text{ethyl acetate}} \underset{p\text{-hydroxyphenacyl bromide (100\%)}}{\mathrm{HO-C_6H_4-C(=O)-CH_2Br}} + \mathrm{CuBr}$$

$$\mathrm{CH_3-C(=O)-CH(CH_3)-CH_3} + \mathrm{CuCl_2} \xrightarrow[90^\circ]{\mathrm{LiCl/DMF}} \underset{\text{3-chloro-3-methyl-2-butanone (85\%)}}{\mathrm{CH_3-C(=O)-CCl(CH_3)-CH_3}} + \mathrm{CuCl}$$

$$\mathrm{H-C(=O)-CH(CH_3)-CH_3} + \mathrm{CuCl_2} \xrightarrow[\text{water, heat}]{\text{acetone}} \underset{\text{2-chloro-2-methylpropanal (96\%)}}{\mathrm{H-C(=O)-CCl(CH_3)-CH_3}} + \mathrm{CuCl}$$

Reduction. α-Haloalcohols can be prepared in moderate yield by the aluminum hydride (not lithium aluminum hydride) reduction of α-haloketones. If lithium aluminum hydride is employed as a reductant, then both the carbon-halogen bond and the carbonyl group are reduced.

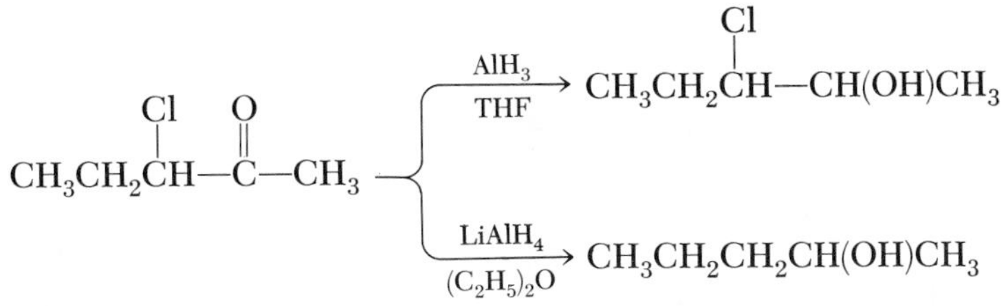

$$CH_3CH_2CH(Cl)-C(=O)-CH_3 \xrightarrow[THF]{AlH_3} CH_3CH_2CH(Cl)-CH(OH)CH_3$$

$$CH_3CH_2CH(Cl)-C(=O)-CH_3 \xrightarrow[(C_2H_5)_2O]{LiAlH_4} CH_3CH_2CH_2CH(OH)CH_3$$

Selective reduction of α-haloketones to ketones may be achieved by reaction with zinc and acetic acid, as illustrated by the synthesis of *cis*- and *trans*-α-decalone.

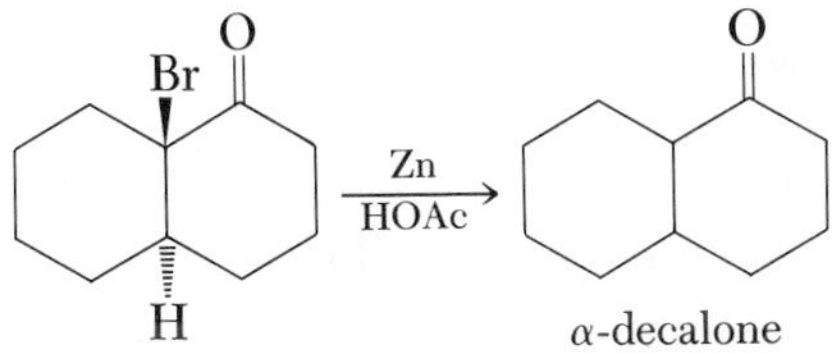

$$\text{(bromo-α-decalone: Br, O, H)} \xrightarrow[HOAc]{Zn} \text{α-decalone}$$

α-decalone
(*cis* and *trans*)

Dehydrohalogenation. Attempts to reduce α-haloketones with hydrazine (a variation of the Wolff-Kishner reduction) produce alkenes, in moderate yield, rather than the expected alkyl halide.

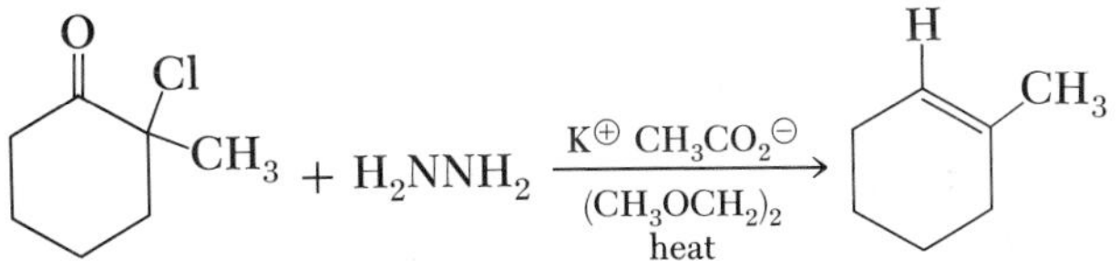

$$\text{2-chloro-2-methylcyclohexanone} + H_2NNH_2 \xrightarrow[(CH_3OCH_2)_2,\ \text{heat}]{K^{\oplus}\ CH_3CO_2^{\ominus}} \text{1-methylcyclohexene}$$

(~40%)

The dehydrohalogenation of α-haloketones, forming α,β-unsaturated ketones, can be accomplished by heating with a base such as γ-collidine (2,4,6-trimethylpyridine).

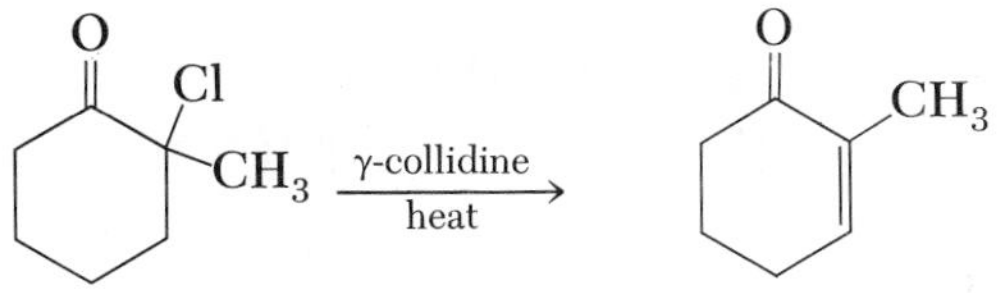

$$\text{2-chloro-2-methylcyclohexanone} \xrightarrow[\text{heat}]{\gamma\text{-collidine}} \text{2-methyl-2-cyclohexenone}$$

2-methyl-2-cyclohexenone
(50%)

This simple base-catalyzed dehydrohalogenation can be complicated by an interesting reaction, the *Favorskii rearrangement,* which converts an α-haloketone into a carboxylate salt. One base commonly used to *encourage* the Favorskii rearrangement is hydroxide ion.

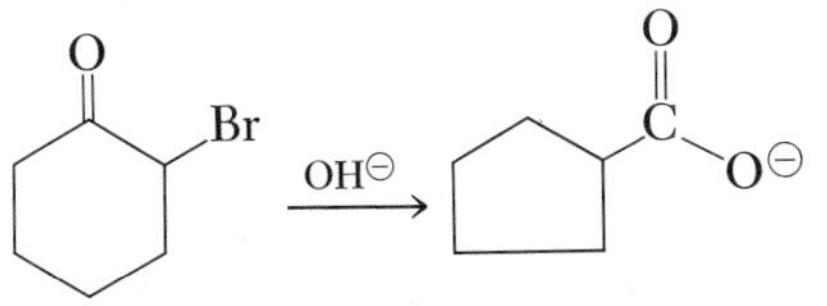

$$\text{2-bromocyclohexanone} \xrightarrow{OH^{\ominus}} \text{cyclopentyl}-C(=O)-O^{\ominus}$$

a Favorskii rearrangement

The mechanism of the rearrangement, a subject of considerable controversy a decade or so ago, is now believed to involve a transient cyclopropanone intermediate formed by a γ-elimination reaction. This cyclopropanone is ring-opened by hydroxide ion to form, after proton transfer, the final product.

$$\text{2-bromocyclohexanone} \xrightarrow[H_2O]{OH^{\ominus}} \text{(carbanion at C-2)} + \text{(carbanion at C-6)}$$

$$\xrightarrow{-Br^{\ominus}} \text{bicyclic cyclopropanone} \xrightarrow[\text{rapid}]{OH^{\ominus}} \rightarrow \rightarrow \text{cyclopentanecarboxylate } (C{-}O^{\ominus})$$

Nucleophilic Substitution. As noted in Chapter 5, allylic halides are particularly reactive toward S_N2 (and S_N1) displacements. α-Halocarbonyl compounds are, by way of contrast, reactive toward S_N2 but not S_N1 reactions. An excellent example of this enhanced S_N2 reactivity is the use of *p*-bromophenacyl bromide (a lachrymator!) to convert liquid carboxylic acids to solid esters. These solid esters are useful in characterizing unknown acids.

$$\underset{\substack{\text{carboxylic acid}\\\text{(solid or liquid)}}}{R{-}\overset{O}{\overset{\|}{C}}{-}OH} \xrightarrow{\text{base}} \underset{\text{carboxylate anion}}{R{-}\overset{O}{\overset{\|}{C}}{-}O^{\ominus}}$$

$$R{-}\overset{O}{\overset{\|}{C}}{-}O^{\ominus} + \underset{p\text{-bromophenacyl bromide}}{Br{-}CH_2{-}\overset{O}{\overset{\|}{C}}{-}C_6H_4{-}Br} \xrightarrow{S_N2} \underset{\substack{p\text{-bromophenacyl ester}\\\text{(a solid)}}}{R{-}\overset{O}{\overset{\|}{C}}{-}O{-}CH_2{-}\overset{O}{\overset{\|}{C}}{-}C_6H_4{-}Br}$$

17.4 ALKYLATION OF CARBONYL COMPOUNDS

The reaction of an aldehyde or ketone with an equivalent of strong base, preferably under homogeneous conditions, can result in the nearly complete conversion of the compound into its enolate salt. [It is important that salt formation be accomplished rapidly in order to reduce the amount of aldol-condensation product (see section 17.5).] These salts, $RCO\overset{\ominus}{C}R_2\ M^{\oplus}$, can act as nucleophiles to produce alkylated ketones or aldehydes. (Because of complications with side reactions, ketones are more common substrates than are aldehydes.) The alkyl group donor is usually an alkyl halide or tosylate.

$$R{-}\overset{O}{\overset{\|}{C}}{-}\overset{H}{\overset{|}{C}}{-} + \overset{\ominus}{N}H_2 \rightarrow \underset{\text{enolate anion}}{\left\{R{-}\overset{O}{\overset{\|}{C}}{-}\overset{\ominus}{C}{-} \leftrightarrow R{-}\overset{O^{\ominus}}{\overset{|}{C}}{=}C{-}\right\}} \xrightarrow{R'X} R{-}\overset{O}{\overset{\|}{C}}{-}\overset{R'}{\overset{|}{C}}{-} + X^{\ominus}$$

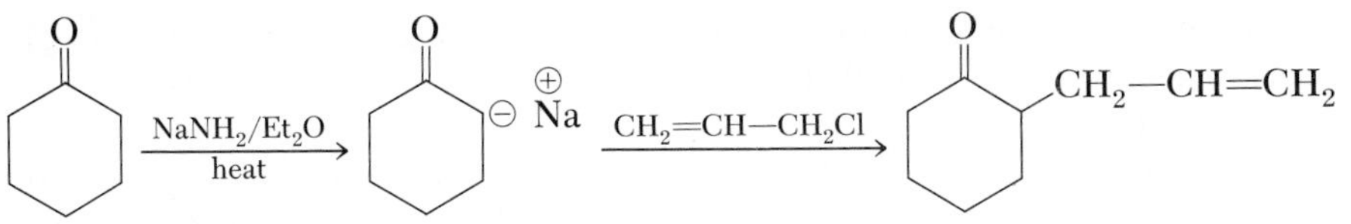

$\xrightarrow[\text{heat}]{NaNH_2/Et_2O}$ $\ominus$ Na$^{\oplus}$ $\xrightarrow{CH_2{=}CH{-}CH_2Cl}$ $CH_2{-}CH{=}CH_2$

(60%)

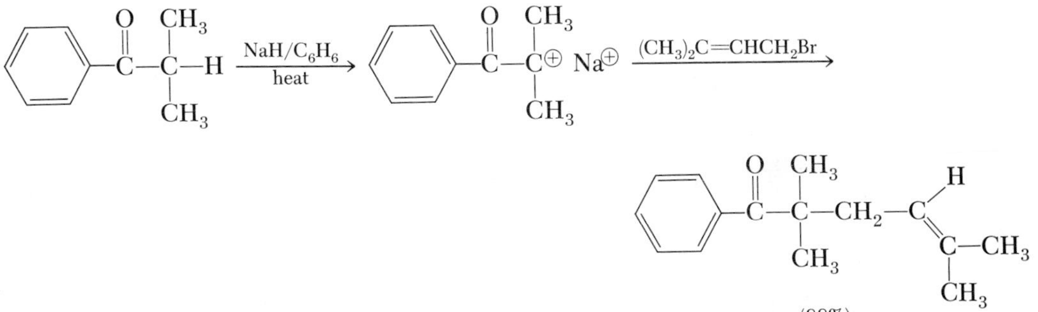

$\xrightarrow[\text{heat}]{NaH/C_6H_6}$ $\xrightarrow{(CH_3)_2C{=}CHCH_2Br}$

(88%)

$\xrightarrow[C_6H_6;\ \text{heat}]{NaOCH_2C(CH_3)_3}$

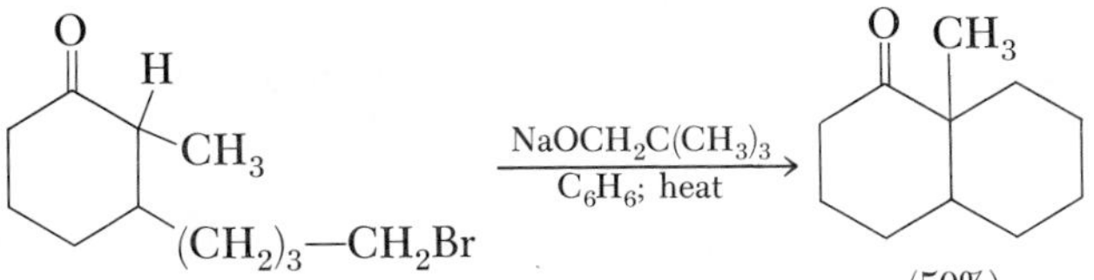

(50%)

Over-alkylation, *i.e.*, attachment of more than one alkyl group, is a common side reaction; it can be minimized by adding a solution of the enolate salt to a large excess of the alkylating agent. This permits consumption of the original enolate anion before proton exchange between that enolate anion and the alkylated product, the step prerequisite to dialkylation, can occur.

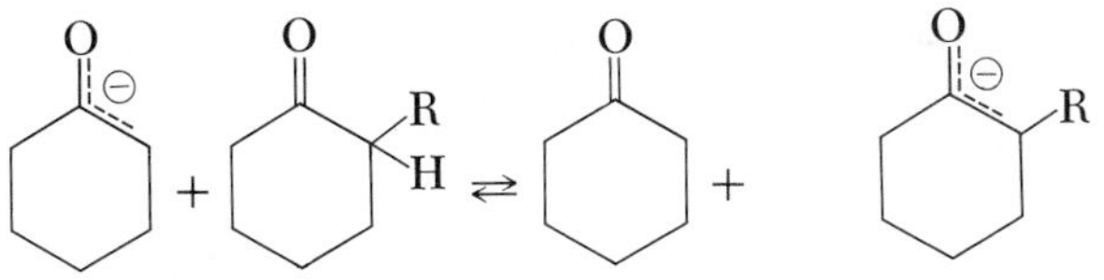

alkylation of this species forms dialkylated by-product

forward reaction minimized by adding enolate anion to an excess of alkylating agent

If the ketone is unsymmetric, and if one of the possible anions can be stabilized by further conjugation, the expected product is the one derived from the more stable anion.

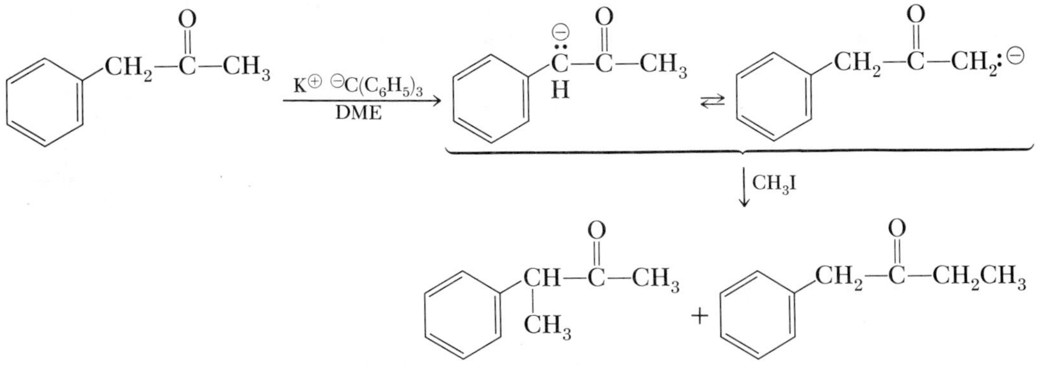

$\xrightarrow[DME]{K^{\oplus}\ {}^{\ominus}C(C_6H_5)_3}$ $\rightleftarrows$

$\downarrow CH_3I$

(93%) (1%)

If neither substituent on Cα is particularly capable of stabilizing negative charge (*e.g.*, if both are alkyl groups), alkylation at both α carbons is expected, the specific distribution being a function of experimental conditions.

An interesting procedure used to control the position of alkylation is to block the preferred position, alkylate the less preferred position, and then remove the blocking group. For example, the following synthesis uses ethyl formate followed by 1-propanethiol to block methylation at the preferred C2 position.

(*a*) KO*t*-bu/*t*-buOH
(*b*) CH_3I

(major product)

A: $H-\overset{O}{\overset{\|}{C}}-OEt$, $OEt^{\ominus}/HOEt$

B: $CH_3CH_2CH_2SH/H^{\oplus}$

$CH(SCH_2CH_2CH_3)$

C: (*a*) $K^{\oplus}$ $O^{\ominus}$*t*-bu/*t*-buOH
(*b*) CH_3I

H_2O/KOH, $HOCH_2CH_2OH$, heat

(80%)
(*cis* and *trans*)

Suggest a mechanism for steps **A** and **B.**

Alkylation of β-Dicarbonyl Compounds. Alkylation of β-dicarbonyl compounds usually occurs at the carbon flanked by the carbonyl groups, *i.e.*, $-CO-CH_2-CO-$. Because of the enhanced acidity of such methylene groups compared to that of a methylene adjacent to an isolated carbonyl, enolate anions of β-dicarbonyl compounds are prepared with moderate bases such as solutions of metal hydroxides in acetone or ethanol.

$$-\overset{O}{\overset{\|}{C}}-\overset{H}{\overset{|}{\underset{|}{C}}}-\overset{O}{\overset{\|}{C}}- \xrightarrow{\text{base}} \left\{ -\overset{O}{\overset{\|}{C}}-\overset{\ominus}{\underset{|}{C}}-\overset{O}{\overset{\|}{C}}- \leftrightarrow -\overset{O^{\ominus}}{\overset{|}{C}}=\underset{|}{C}-\overset{O}{\overset{\|}{C}}- \leftrightarrow -\overset{O}{\overset{\|}{C}}-\underset{|}{C}=\overset{O^{\ominus}}{\overset{|}{C}}- \right\} \xrightarrow[-X^{\ominus}]{RX} -\overset{O}{\overset{\|}{C}}-\overset{R}{\overset{|}{\underset{H}{\underset{|}{C}}}}-\overset{O}{\overset{\|}{C}}-$$

$$CH_3-\overset{O}{\overset{\|}{C}}-CH_2-\overset{O}{\overset{\|}{C}}-CH_3 \xrightarrow[CH_3COCH_3/\text{heat}]{K_2CO_3/CH_3I} CH_3-\overset{O}{\overset{\|}{C}}-\overset{CH_3}{\overset{|}{\underset{H}{\underset{|}{C}}}}-\overset{O}{\overset{\|}{C}}-CH_3$$

3-methyl-2,4-pentanedione
(75%)

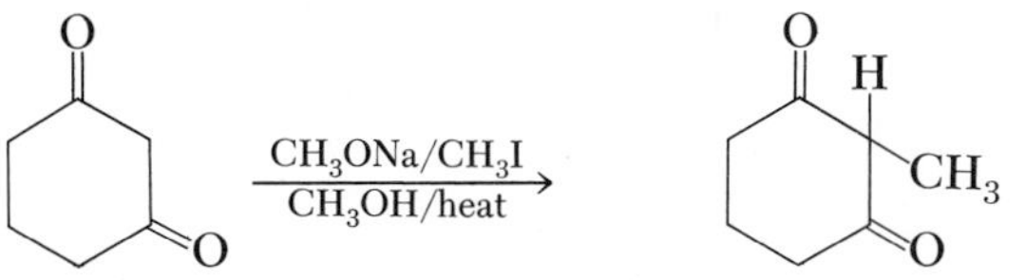

CH_3ONa/CH_3I / CH_3OH/heat

2-methyl-1,3-cyclohexanedione (60%)

As with monoketones, over-alkylation can be a major problem. The use of thallium ethoxide, $TlOCH_2CH_3$, as the base affords extremely high yields of the monoalkylated product. Dialkylated product is prepared by two consecutive alkylations.

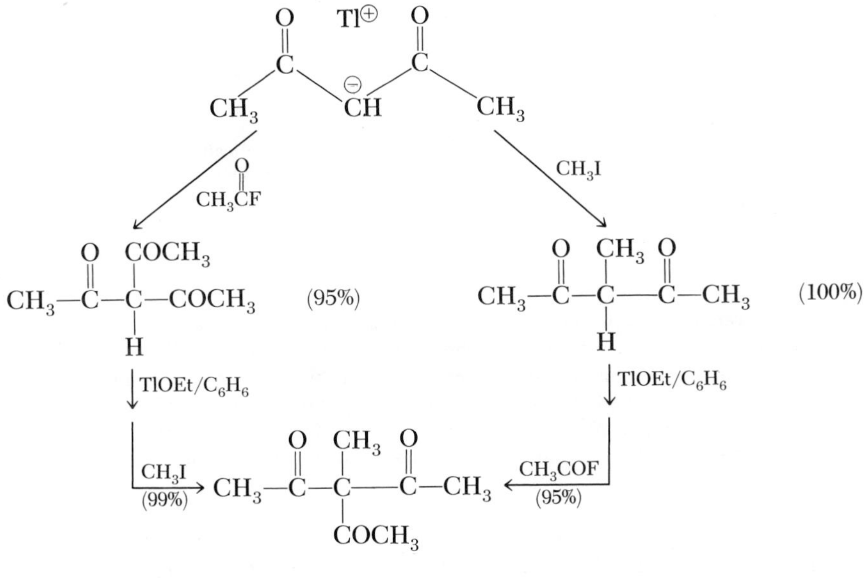

$CH_3COCH^{\ominus}COCH_3$ $Tl^{\oplus}$

CH_3CF (O) → $CH_3C(O)CH(COCH_3)COCH_3$ (95%)

CH_3I → $CH_3C(O)CH(CH_3)C(O)CH_3$ (100%)

↓ $TlOEt/C_6H_6$; CH_3I (99%) → $CH_3C(O)C(CH_3)(COCH_3)C(O)CH_3$ ← CH_3COF (95%); $TlOEt/C_6H_6$

Alkylation of α,β-Unsaturated Ketones. Alkali metal salts of 3° alcohols, acting as bases, produce two different alkylated products from α,β-unsaturated ketones. While product distribution can be controlled to some extent, alkylation usually produces some of both products.

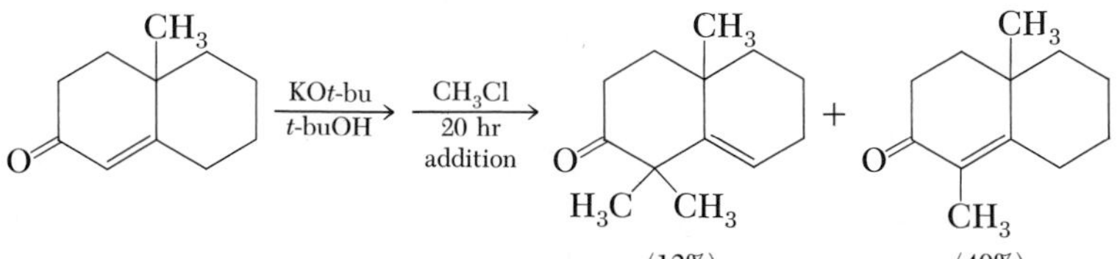

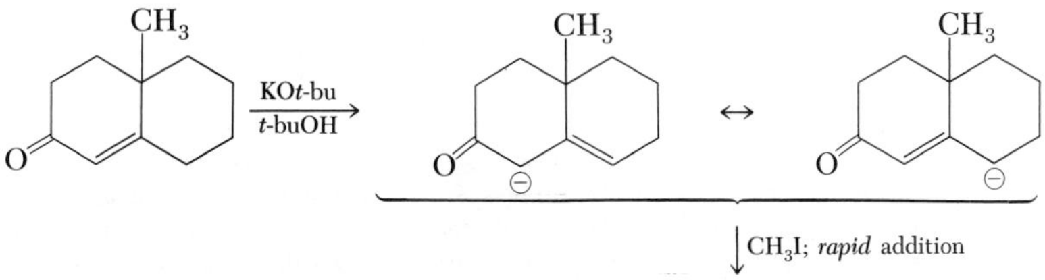

KO*t*-bu / *t*-buOH → CH_3Cl / 20 hr addition → (12%) + (40%)

KO*t*-bu / *t*-buOH → ↔

↓ CH_3I; *rapid* addition

Forward to p. 520

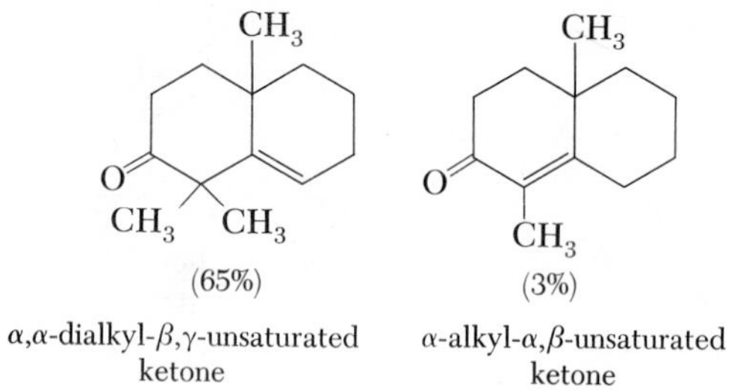

CH_3 CH_3

O CH_3 CH_3 O CH_3

(65%) (3%)

α,α-dialkyl-β,γ-unsaturated ketone α-alkyl-α,β-unsaturated ketone

This dependence of product distribution upon the nature and rate of addition of methyl halide that is implied in the two reactions just presented is rationalized by considering an initial, kinetically controlled alkylation of Cα of the enolate anion to give α-alkyl-β,γ-unsaturated ketone.

CH_3 $\xrightarrow{CH_3X}$ CH_3 kinetically controlled product

O ⊖ O (H) X

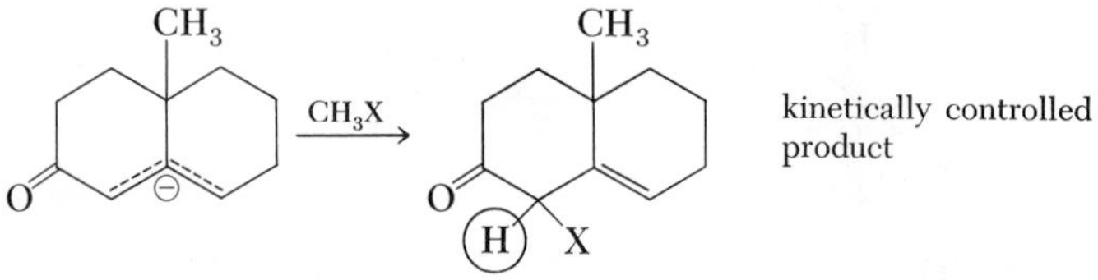

The remaining α proton, encircled in the structure, is especially acidic (Why?) and is removed by base to give, in the presence of alkyl halide, the α,α-dialkyl-β,γ-unsaturated ketone. If alkylation is slow, as with methyl chloride in the illustration above, then the thermodynamically favored α,β-unsaturated ketone is formed from the β,γ-unsaturated ketone. A generalized scheme depicting various carbanion interconversions in this type of system is presented in Figure 17-2.

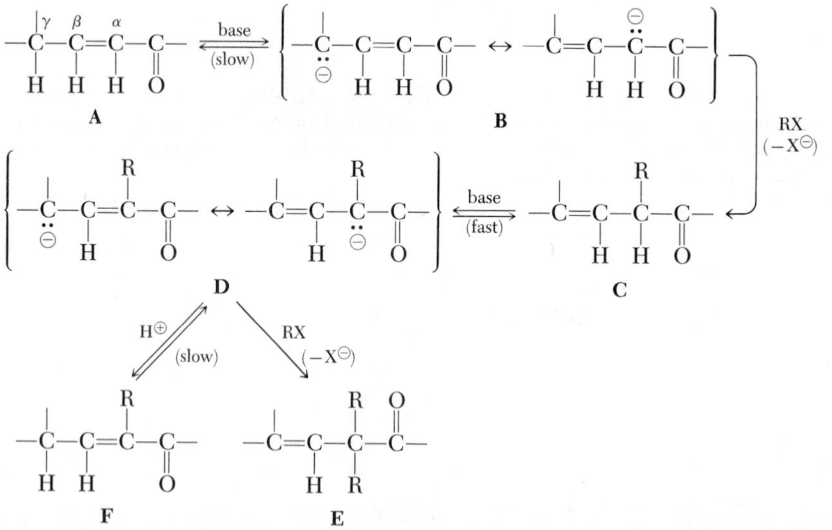

γ β α

A base (slow) B RX ($-X^{\ominus}$)

D base (fast) C

$H^{\oplus}$ (slow) RX ($-X^{\ominus}$)

F E

Figure 17-2 A general scheme for alkylation of α,β-unsaturated ketones. The anion **B,** produced by loss of a proton from Cγ in **A,** is alkylated preferentially at Cα to produce the monoalkylated derivative **C.** In the presence of base, **C** is converted to **D.** If a low concentration of alkylating agent is present, or if the alkylating agent is not very reactive, **D** will be able to abstract a proton from the solvent (or other proton source) and form product **F.** If, however, the alkylating agent is in excess, or quite reactive, carbanion **D** will be alkylated at Cα to form **E.** Thus, a cause of dialkylation is that **C,** a β,γ-unsaturated ketone, loses a proton faster than either **A** or **F,** both of which are α,β-unsaturated ketones.

17.5 ALDOL AND BENZOIN CONDENSATIONS

Aldehydes undergo a spontaneous conversion to β-hydroxyaldehydes when treated with small amounts of aqueous acid or base. This type of reaction is termed an **aldol condensation** although, in the strictest sense of the word, it is not a condensation but an *addition*. (In a condensation reaction a small molecule, *e.g.*, water, is usually produced along with the organic product.) The simplest product of this type, $CH_3CH(OH)CH_2CHO$, bears the name "aldol" and gives its name to the reaction.

$$2\ -\overset{H}{\underset{|}{\overset{|}{C}}}-\overset{O}{\overset{\|}{C}}-H \xrightarrow{H^{\oplus} \text{ or } OH^{\ominus}} -\overset{H}{\underset{|}{\overset{|}{C}}}-\overset{OH}{\underset{H}{\overset{|}{C}}}-\underset{|}{\overset{|}{C}}-\overset{O}{\overset{\|}{C}}-H$$

an aldol condensation

$$\underset{\text{acetaldehyde}}{2CH_3CHO} \xrightarrow{H^{\oplus} \text{ or } OH^{\ominus}} \underset{\text{aldol}}{CH_3-\overset{OH}{\underset{H}{\overset{|}{C}}}-CH_2-\overset{O}{\overset{\|}{C}}-H}$$

In the reactions below we will see how base-catalyzed aldol condensations proceed by attack of the anion of an enol (*enolate anion*) upon a carbonyl group. Acid-catalyzed condensations involve nucleophilic addition of the π electrons of the enol to a protonated carbonyl group. Aldols lose water, to form α,β-unsaturated aldehydes, either upon heating with base or upon treatment with dilute aqueous acid at room temperature; thus, it is nearly impossible to stop at the β-hydroxy stage in the acid-catalyzed reaction. Because of this we will, after presentation of the acid-catalyzed synthesis of aldol, focus future discussions upon base-catalyzed aldol condensations.

base-catalyzed condensation

$$CH_3CHO \overset{OH^{\ominus}}{\rightleftharpoons} \underset{\text{enolate anion}}{{}^{\ominus}CH_2-\overset{O}{\overset{\|}{C}}-H \leftrightarrow CH_2=\overset{O^{\ominus}}{\overset{|}{C}}-H} + H_2O$$

$$CH_3-\overset{O}{\overset{\|}{C}}-H \ \ (\text{attacked by } {}^{\ominus}CH_2-CHO) \rightleftharpoons CH_3-\overset{O^{\ominus}}{\underset{CH_2-CHO}{\overset{|}{C}}}-H \overset{H_2O}{\rightleftharpoons} \underset{\text{aldol}}{CH_3-\overset{OH}{\underset{CH_2-CHO}{\overset{|}{C}}}-H} + OH^{\ominus}$$

acid-catalyzed condensation

$$H-CH_2-\overset{O}{\overset{\|}{C}}-H \overset{H^{\oplus}}{\rightleftharpoons} H-CH_2-\overset{{}^{\oplus}OH}{\overset{\|}{C}}-H \ \ (H_2O:) \rightleftharpoons \underset{\text{enol}}{CH_2=\overset{OH}{\overset{|}{C}}-H} + H_3O^{\oplus}$$

$$CH_3-\overset{O}{\overset{\|}{C}}-H \overset{H^{\oplus}}{\rightleftharpoons} CH_3-\overset{{}^{\oplus}OH}{\overset{\|}{C}}-H \ \ \ CH_2=C(H)-O-H \leftarrow OH_2 \rightleftharpoons$$

$$\underset{\text{aldol}}{CH_3-\overset{OH}{\underset{H}{\overset{|}{C}}}-CH_2-\overset{O}{\overset{\|}{C}}-H} + H_3O^{\oplus}$$

$$CH_3-\underset{H}{\overset{OH}{C}}-\underset{H}{\overset{H}{C}}-\overset{O}{\overset{\|}{C}}-H \xrightarrow{-H_2O} CH_3-CH{=}CH-CHO \quad \text{(crotonaldehyde)}$$

base-catalyzed dehydration

$$CH_3\underset{H}{\overset{OH}{C}}-\underset{H}{\overset{H}{C}}CHO \overset{H^{\oplus}}{\rightleftharpoons} CH_3\underset{H}{\overset{\overset{\oplus}{O}H_2}{C}}-\underset{H}{\overset{H}{C}}CHO \xrightarrow{-H_2O} CH_3CH{=}CHCHO$$

acid-catalyzed dehydration

Attempts to carry out aldol condensations on mixtures of two aldehydes, both of which possess α hydrogens, are not synthetically useful because such reactions produce all possible products.

$$CH_3CHO + CH_3CH_2CHO \xrightarrow{OH^{\ominus}} CH_3-\underset{H}{\overset{OH}{C}}-CH_2-\overset{O}{\overset{\|}{C}}-H +$$

$$CH_3CH_2-\underset{H}{\overset{OH}{C}}-\underset{H}{\overset{CH_3}{C}}-\overset{O}{\overset{\|}{C}}-H + CH_3-\underset{H}{\overset{OH}{C}}-\underset{CH_3}{\overset{H}{C}}-\overset{O}{\overset{\|}{C}}-H + CH_3CH_2-\underset{H}{\overset{OH}{C}}-CH_2-\overset{O}{\overset{\|}{C}}-H$$

An effective *crossed aldol condensation* can be carried out if an excess of an aldehyde *lacking* an α hydrogen is treated with an aldehyde *possessing* an α hydrogen. Under such conditions the former serves as the carbanion acceptor, while the latter serves as a carbanion source.

$$(CH_3)_3CCHO + CH_3CHO \xrightarrow[\text{heat}]{OH^{\ominus}} \left[(CH_3)_3C\underset{H}{\overset{OH}{C}}-\underset{H}{\overset{H}{C}}CHO\right] \xrightarrow{-H_2O} (CH_3)_3CCH{=}CHCHO$$

Because ketones are poorer carbanion acceptors than are aldehydes, they can be used in crossed condensations (also called "aldol condensations") with aldehydes lacking α hydrogens. Conditions must, of course, be kept at the minimum required for condensation in order to prevent ketone-ketone condensation. This means, among other things, the maintenance of a low ketone concentration in the reaction mixture. The preparation of benzalacetone (below) brings out yet another point—β-hydroxycarbonyl compounds bearing a benzene ring or other unsaturated group on Cβ undergo dehydration so rapidly as to be non-isolable, the enone being isolated instead.

$$C_6H_5CHO \text{ (benzaldehyde)} + CH_3\overset{O}{\overset{\|}{C}}CH_3 \xrightarrow[\text{dilute}]{OH^\ominus} \left[C_6H_5\overset{OH}{\overset{|}{\underset{H}{\underset{|}{C}}}}-CH_2-\overset{O}{\overset{\|}{C}}-CH_3\right] \rightarrow C_6H_5CH{=}CH-\overset{O}{\overset{\|}{C}}-CH_3 \text{ (benzalacetone)}$$

Although both the aldol condensation and the Cannizzaro reaction (p. 492) are base-catalyzed aldehyde reactions, the Cannizzaro reaction requires a high concentration of base and, therefore, normally does not compete with the aldol condensation.

Ketonic Aldol Condensations. Ketones are much less prone to undergo self-aldol condensations than are aldehydes. For example, while acetone can be completely converted to mesityl oxide by following the aldol condensation of acetone with an acid-catalyzed dehydration, very little 4-hydroxy-4-methyl-2-pentanone (the aldol product) is formed from acetone and base under equilibrium conditions. However, ketone condensations that form five- and six-membered rings are favored by entropy considerations and are synthetically significant.

$$2CH_3\overset{O}{\overset{\|}{C}}CH_3 \underset{}{\overset{OH^\ominus}{\rightleftharpoons}} CH_3-\overset{OH}{\overset{|}{\underset{CH_3}{\underset{|}{C}}}}-CH_2-\overset{O}{\overset{\|}{C}}-CH_3 \xrightarrow{H^\oplus} (CH_3)_2C{=}CH-\overset{O}{\overset{\|}{C}}-CH_3 \text{ (mesityl oxide)}$$

$$CH_3-\overset{O}{\overset{\|}{C}}-CH_2-CH_2-\overset{O}{\overset{\|}{C}}-CH_3 \xrightarrow[\text{dioxane, heat}]{OH^\ominus} \text{3-methyl-2-cyclopentenone } (H_3C \text{ on ring C=C, ring C=O})$$

8. Acetone is produced, in low yield, when 4-methyl-4-hydroxy-2-pentanone is allowed to remain in alkaline solution. Show all the steps in a mechanism to account for its formation.

9. Provide mechanisms for the acid- and base-catalyzed aldol condensations of each of the following. Which should be more valuable in synthesizing the desired product? Which may not give an isolable β-hydroxycarbonyl compound?

(a) propanal (c) cyclopentanone
(b) acetophenone (d) (**R**)-2-methylbutanal

10. When a concentrated solution of acetaldehyde underwent an aldol condensation in $D_2O/OD^\ominus$, the product contained very few C—D bonds. (a) How is this consistent with carbanion formation being the rate-determining step in this condensation? A more dilute solution of acetaldehyde in $D_2O/OD^\ominus$ did produce a product with many more C—D bonds. (b) Explain this difference. Finally, at similar concentrations, the aldol condensation product of acetone is richer in deuterium than is the aldol condensation product of acetaldehyde. In fact, acetone undergoes base-induced deuterium incorporation faster than it undergoes condensation. (c) Explain.

11. One of the major synthetic uses of the aldol condensation is in the preparation of α,β-unsaturated carbonyl compounds for further catalytic reduction to alcohols. With this in mind, suggest a synthesis that employs a condensation reaction for each of the following.

(a) 4-phenyl-2-butanol (e) 2-ethyl-1-hexanol
(b) 3-phenyl-1-propanol (f) 4-methyl-2-pentanol
(c) 1,3-diphenyl-1-propanol (g) 1,3-diphenyl-2-butanol
(d) 1-butanol

Aldol Condensations in Biological Systems. Earlier in this chapter we described the conversion of glucose to ethanol and carbon dioxide (Figure 17–1). One of the critical steps in this process is the cleavage of fructose 1,6-diphosphate into dihydroxyacetone phosphate and glyceraldehyde 3-phosphate. This reaction is reversible and can, with the aid of enzymes, provide a route *to* fructose 1,6-diphosphate.

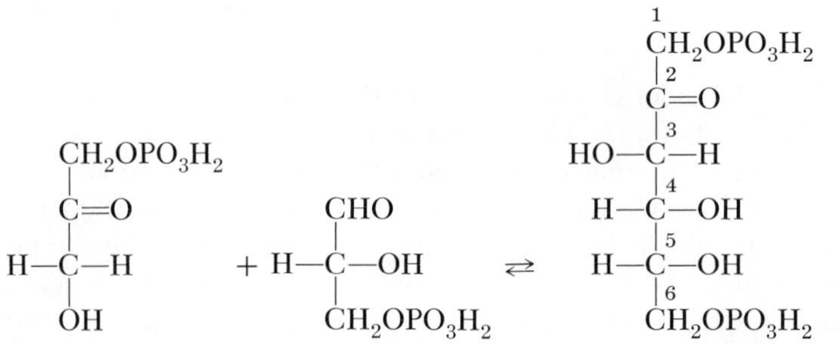

$$\underset{\text{dihydroxyacetone phosphate}}{\begin{array}{c} CH_2OPO_3H_2 \\ | \\ C{=}O \\ | \\ H{-}C{-}H \\ | \\ OH \end{array}} + \underset{\text{glyceraldehyde 3-phosphate}}{\begin{array}{c} CHO \\ | \\ H{-}C{-}OH \\ | \\ CH_2OPO_3H_2 \end{array}} \rightleftarrows \underset{\text{fructose 1,6-diphosphate}}{\begin{array}{c} {}^{1}CH_2OPO_3H_2 \\ |\,{}^{2} \\ C{=}O \\ |\,{}^{3} \\ HO{-}C{-}H \\ |\,{}^{4} \\ H{-}C{-}OH \\ |\,{}^{5} \\ H{-}C{-}OH \\ |\,{}^{6} \\ CH_2OPO_3H_2 \end{array}}$$

While the mechanism is a bit more complex than that shown below, this synthesis of fructose 1,6-diphosphate is presented here as a simple, base-catalyzed aldol condensation. Although an approximation, it does reveal how an aldol condensation can produce this six-carbon sugar derivative from two three-carbon precursors.

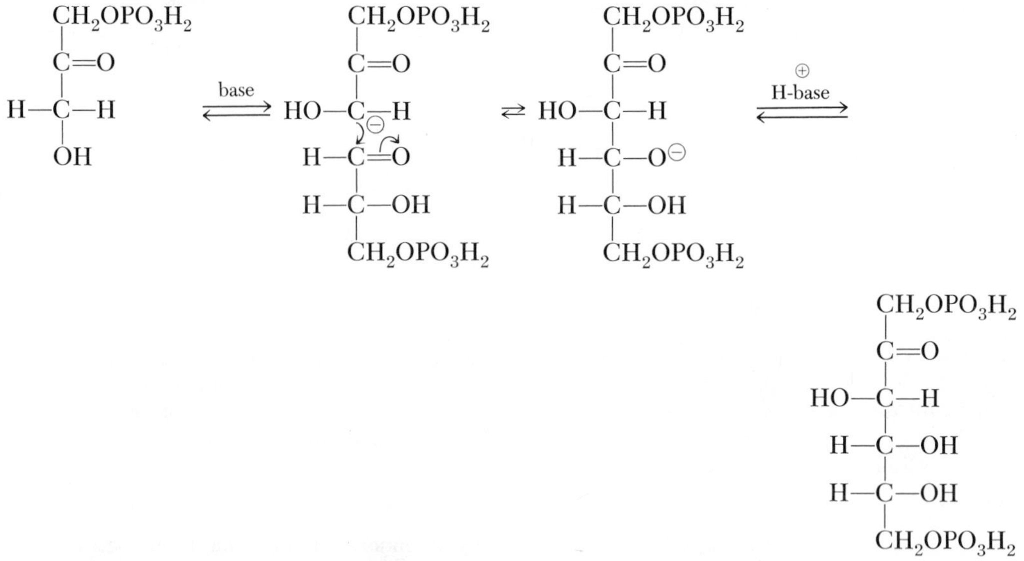

$$\begin{array}{c} CH_2OPO_3H_2 \\ | \\ C{=}O \\ | \\ H{-}C{-}H \\ | \\ OH \end{array} \xrightleftharpoons{\text{base}} \begin{array}{c} CH_2OPO_3H_2 \\ | \\ C{=}O \\ | \\ HO{-}\overset{\ominus}{C}{-}H \\ \\ H{-}C{=}O \\ | \\ H{-}C{-}OH \\ | \\ CH_2OPO_3H_2 \end{array} \rightleftarrows \begin{array}{c} CH_2OPO_3H_2 \\ | \\ C{=}O \\ | \\ HO{-}C{-}H \\ | \\ H{-}C{-}O^{\ominus} \\ | \\ H{-}C{-}OH \\ | \\ CH_2OPO_3H_2 \end{array} \xrightleftharpoons{\overset{\oplus}{\text{H-base}}} \begin{array}{c} CH_2OPO_3H_2 \\ | \\ C{=}O \\ | \\ HO{-}C{-}H \\ | \\ H{-}C{-}OH \\ | \\ H{-}C{-}OH \\ | \\ CH_2OPO_3H_2 \end{array}$$

The cleavage of fructose 1,6-diphosphate is a *retro-aldol condensation*, that is, a reversion of an aldol condensation product into the starting materials. In general, heating an aldol with base will result in partial reversion to the starting materials, which, being more volatile than the original condensation product, will distill from the reaction mixture.

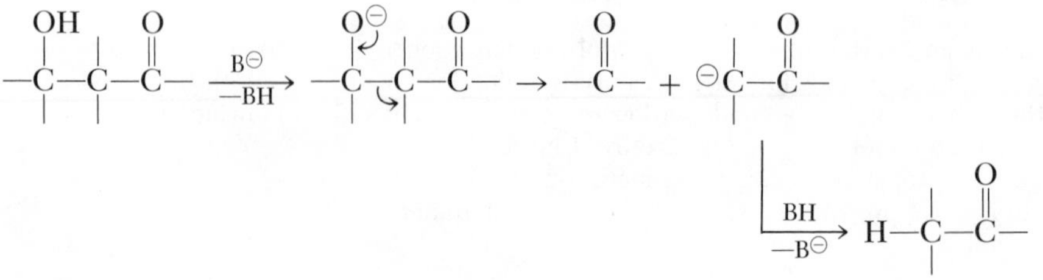

$$\begin{array}{c} OH \;\;\;\;\; O \\ | \;\;\;\;\;\; \| \\ {-}C{-}C{-}C{-} \\ | \;\;\; | \end{array} \xrightarrow[{-BH}]{B^{\ominus}} \begin{array}{c} O^{\ominus} \;\;\; O \\ | \;\;\;\;\;\; \| \\ {-}C{-}C{-}C{-} \\ | \;\;\; | \end{array} \rightarrow \begin{array}{c} O \\ \| \\ {-}C{-} \end{array} + \begin{array}{c} \;\;\;\;\; O \\ | \;\;\; \| \\ {}^{\ominus}C{-}C{-} \\ | \end{array} \xrightarrow[{-B^{\ominus}}]{BH} \begin{array}{c} \;\;\;\;\;\; O \\ | \;\;\; \| \\ H{-}C{-}C{-} \\ | \end{array}$$

retro-aldol condensation

Benzoin Condensation. Two molecules of aromatic aldehyde may be condensed by the action of cyanide ion to produce an α-hydroxyketone. Since the simplest such compound, produced by the condensation of benzaldehyde, is called *benzoin*, the reaction sequence is known as the **benzoin condensation.**

$$2\ C_6H_5\text{—CHO} \xrightarrow[\text{heat}]{CN^{\ominus}/H_2O/EtOH} C_6H_5\text{—CH(OH)—C(=O)—}C_6H_5 \qquad (90\%)$$

benzoin

Cyanide ion is a unique catalyst for this condensation because it (a) is sufficiently nucleophilic to produce **A** (see below), (b) stabilizes **B** by resonance, and (c) is a good enough leaving group to permit formation of the final product.

$$C_6H_5\text{—CH=O} + CN^{\ominus} \rightleftarrows \underset{\mathbf{A}}{C_6H_5\text{—C(O}^{\ominus}\text{)(CN)—H}} \rightleftarrows \underset{\mathbf{B}}{C_6H_5\text{—C}^{\ominus}\text{(OH)—CN}} \leftrightarrow C_6H_5\text{—C(OH)=C=N}^{\ominus}$$

$$C_6H_5\text{—C}^{\ominus}\text{(OH)(CN)} + \text{O=CH—}C_6H_5 \rightleftarrows C_6H_5\text{—C(OH)(CN)—CH(O}^{\ominus}\text{)—}C_6H_5$$

$$\Updownarrow$$

$$C_6H_5\text{—C(=O)—CH(OH)—}C_6H_5 \xleftarrow{-CN^{\ominus}} C_6H_5\text{—C(O}^{\ominus}\text{)(CN)—CH(OH)—}C_6H_5$$

The benzoin condensation does not occur with aliphatic aldehydes, although α-hydroxyketones flanked by alkyl groups, known as *acyloins*, can be prepared by other routes (see Chapter 19). Heterocyclic aldehydes can undergo the reaction, as seen from the conversion of furfural[a] to furoin.

$$\text{2-furyl—CHO} \xrightarrow{CN^{\ominus}} \text{2-furyl—CH(OH)—C(=O)—2-furyl}$$

furfural furoin

The benzoin condensation of two identical aldehydes is not readily accomplished with hydroxy- and amino-substituted aromatic aldehydes. However, these do participate in *crossed benzoin condensations*, in which they end up as the "ketone" fragment.

$$(CH_3)_2N\text{—}C_6H_4\text{—CHO} + C_6H_5\text{—CHO} \xrightarrow{CN^{\ominus}} (CH_3)_2N\text{—}C_6H_4\text{—C(=O)—CH(OH)—}C_6H_5$$

---

[a] Prepared in large quantities from corn cobs.

17.6 ULTRAVIOLET SPECTRA AND OPTICAL ROTATORY DISPERSION OF KETONES[a]

Simple Carbonyl Compounds. Simple alkyl ketones show two absorptions in the near ultraviolet region, a weak transition (ϵ 10^{-3}) near 400 nm and a stronger transition (ϵ 10) near 290 nm. These are due to triplet and singlet transitions of an oxygen non-bonding electron to the carbonyl anti-bonding π orbital, respectively. Because of the extremely low intensity of the triplet transition, the 290 nm $n \rightarrow \pi^*$ absorption is the most studied of the carbonyl absorptions. Table 17–3 summarizes the absorption maxima for selected simple and conjugated carbonyl compounds.

TABLE 17-3 The Long Wavelength Ultraviolet Absorption for Selected Carbonyl Compounds[a]

Compound	Structure	Wavelength	ϵ
acetone	CH_3COCH_3	279 nm	13
cyclohexanone	$(CH_2)_5CO$	285	14
acetaldehyde	CH_3CHO	292	12
mesityl oxide	$(CH_3)_2C{=}CHCOCH_3$	327	98
acrolein	$CH_2{=}CHCHO$	328	20

[a] Data are from solutions in hexane or heptane. All of these transitions are $n \rightarrow \pi^*$ excitations.

The $n \rightarrow \pi^*$ transition of simple carbonyl compounds moves to shorter wavelengths (*i.e.*, the transition requires more energy) as the solvent used in obtaining the spectrum is changed from heptane to water (Table 17–4) because hydrogen bonds are broken in going from the electronic ground state to the electronic excited state in protic solvents.

Unsaturated Carbonyl Compounds.[b] α,β-Unsaturated carbonyl compounds exhibit an intense ($\epsilon \approx 20{,}000$) absorption at about 220 nm, which is often called the *K* band (German: *konjugierte*). It can be considered as the $\pi \rightarrow \pi^*$ transition of the enone chromophore. A weak transition, analogous to the carbonyl $n \rightarrow \pi^*$ transition, occurs at about 320 nm (ϵ 50).

TABLE 17-4 Solvent Dependence of the $n \rightarrow \pi^*$ Transition of Acetone

Solvent	λ_{max}(nm)
heptane	279
chloroform	277
ethanol	272
water	265

[a] The appropriate material in Chapter 28 is essential for an appreciation of this section. An introduction to optical rotatory dispersion was presented on p. 95 and should be reviewed.

[b] An m.o. diagram of an α,β-unsaturated carbonyl group is included in the problems in Chapter 28.

TABLE 17-5 Effect of Solvent upon the Ultraviolet Spectrum of Mesityl Oxide, $(CH_3)_2C{=}CHC(O)CH_3$

	$\pi \rightarrow \pi^*$		$n \rightarrow \pi^*$	
SOLVENT	λ_{max}	ϵ	λ_{max}	ϵ
hexane	229 nm	12,600	327 nm	98
ether	230	12,600	326	96
ethanol	238	10,700	312	74
water	245	10,000	305	60

Solvent changes influence these two transitions in opposite ways. As the hydrogen-bonding ability of the solvent increases, the $n \rightarrow \pi^*$ absorption goes to shorter wavelengths while the *K* band moves to longer wavelengths. (What does this suggest about the excited state of the *K* band transition?) The effect of solvent upon the spectrum of mesityl oxide is summarized in Table 17–5. Substitution at Cα and Cβ, and ring formation, also can influence the ultraviolet spectra of α,β-unsaturated ketones and aldehydes. A presentation of these effects is included in Chapter 28 and should be reviewed at this time.

TRANSANNULAR EFFECTS. If their *p* orbitals are properly oriented, double bonds and carbonyl groups that are not conjugated (*i.e.*, are not connected by one σ bond) can interact with one another. These transitions, which may be viewed as **charge transfer transitions** (*i.e.*, electron density moves from one multiple bond to another), are conveniently pictured in classical resonance terms.

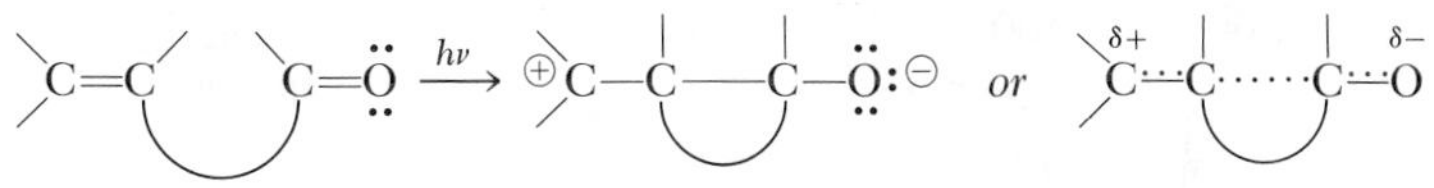

$$\text{C=C} \quad \text{C=\ddot{O}} \xrightarrow{h\nu} {}^{\oplus}\text{C—C—C—\ddot{O}:}^{\ominus} \quad or \quad \overset{\delta+}{\text{C}}\text{---C}\cdots\cdots\text{C---}\overset{\delta-}{\text{O}}$$

charge transfer complex

One group of compounds that show extensive interactions of this type are the medium-membered rings (Chapter 7). The proximity of groups at "opposite ends" of such rings has yielded many interesting examples, including the one that follows.

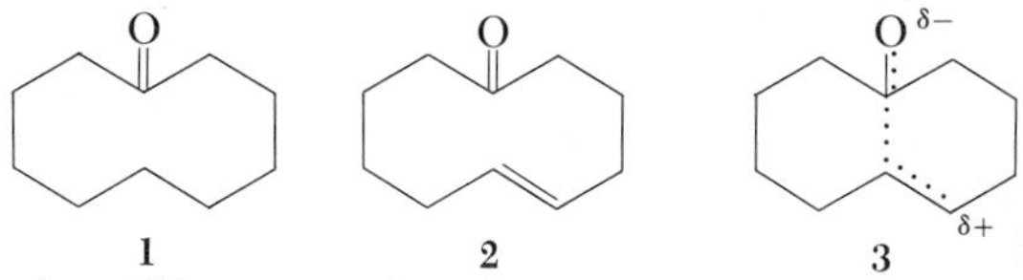

1	2	3
λ_{max} 288 nm	λ_{max} 302 nm	charge transfer band
ϵ 16	ϵ 73	λ_{max} 260 nm
		$\epsilon \approx 425$

The enhancement of ϵ and the shift to longer wavelength of the $n \rightarrow \pi^*$ transition in going from **1** to **2** indicate some unusual electronic interaction in **2.** Proof of this comes from the appearance of an electronic transition in **2** that is absent from the spectrum of **1.** This absorption at 260 nm (ϵ 425) is ascribed to a transannular interaction between the π electrons of the carbon-carbon double bond and the π electrons of the carbonyl group; it is pictured in **3.**

OPTICAL ROTATORY DISPERSION.[a] The dependence of $[\alpha]$, the specific rotation, upon λ, wavelength, is reflected in the optical rotatory dispersion (ord) curve of an optically active

[a] Now be honest—have you reviewed Chapter 4 yet?

compound. While an ord curve is *formally* a plot of rotation *vs.* wavelength, it also is an indicator that tells which electronic transitions are, so to speak, optically active. Thus, the place at which, on a simple ord curve, the Cotton effect region crosses the x-axis (*i.e.*, where α equals zero) or has an inflection point represents the wavelength maximum of an optically active transition in the ultraviolet (Figure 17–3).

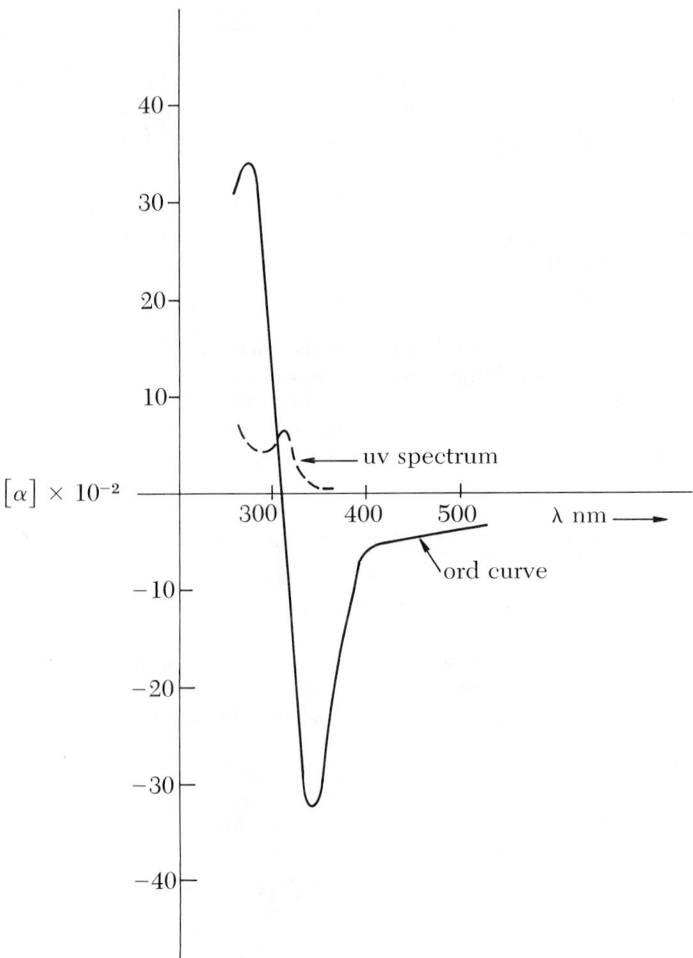

Figure 17–3 The relationship between uv and ord. The solid line represents the ord curve of some hypothetical compound, and the axes are labeled for the presentation of ord data, *i.e.*, $[\alpha]$ is plotted against λ. The dashed line represents a segment of the uv spectrum of the same compound. While the ordinate given in the figure cannot be related to the uv curve, which requires an ordinate based upon A or ϵ, the wavelength scales are the same for both. It is clear, therefore, that the ultraviolet absorption maximum and the place where the ord curve crosses the abscissa occur at the same wavelength. For this compound that value is 310 nm. More complex structures may lead to an inflection in the ord curve rather than to a crossing of the abscissa. An example of this is given in problem 42.

There are two types of optically active transitions that can be discerned by ord. One has an extraordinarily large amplitude, arising from an allowed transition, and corresponds to what has been called an *inherently dissymmetric chromophore* (idc). Such chromophores are normally twisted, elongated π systems. The second type of optically active transition arises from a simple symmetric chromophore that exists in a dissymmetric environment. Such *dissymmetrically perturbed symmetric chromophores* (dpsc's) usually have weak to moderately intense ord active transitions. The distinction between *idc* and *dpsc* is largely one of convenience, since it is a phenomenological one; both systems can be treated similarly at the theoretical level. A superpositioning of the ord curves (Figure 17–4) of bicycloheptenone and bicycloheptanone clearly shows the experimental basis for distinguishing idc's from dpsc's.

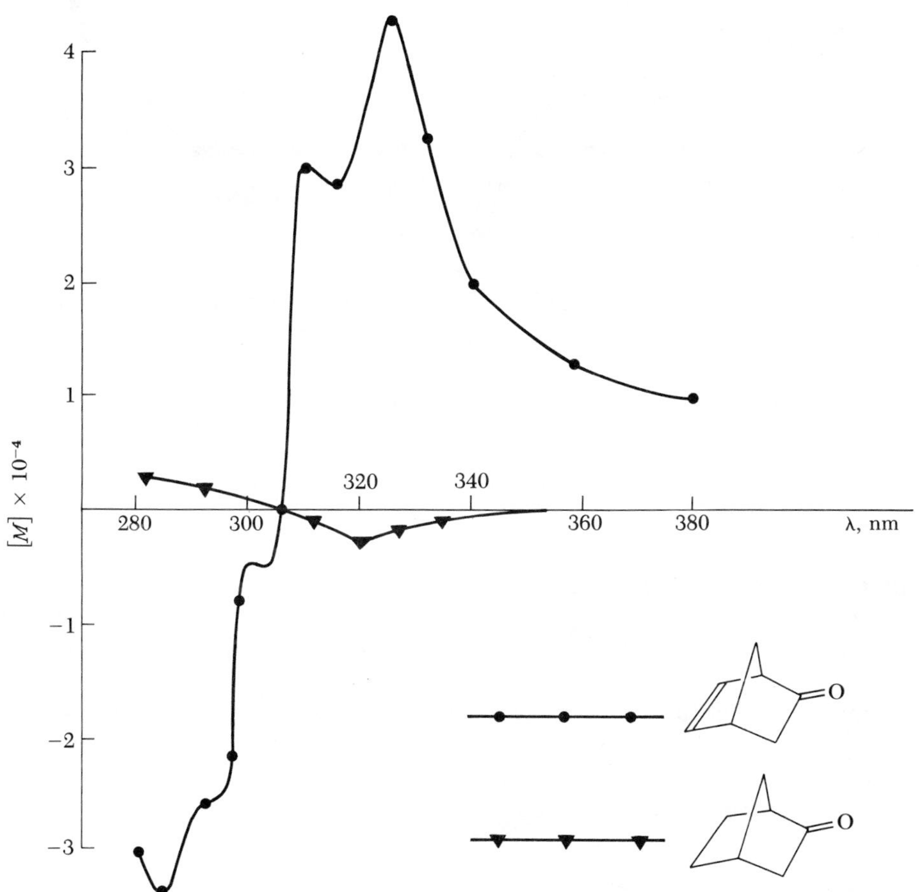

Figure 17-4 The ord curves of bicycloheptenone and bicycloheptanone, as specific rotation $[M]$ *vs.* λ. Bicycloheptenone contains an inherently dissymmetric chromophore which consists of a helical array of four π electrons. This *idc* shows a very intense ord transition. Although bicycloheptanone also is a dissymmetric ketone, its carbonyl group is perturbed only by dissymmetric sp^3 hybridized carbons. This perturbation is not great, since π electrons are more polarizable and delocalizable than σ electrons, and the result is a comparatively weak ord transition. It is suggested that you build a model of bicycloheptenone if you have difficulty envisioning the twisted π electron system.

Just as uv spectroscopy was used to demonstrate the presence of a transannular electronic interaction in cyclodecen-6-one, here we see the use of ord, in the form of an enhanced ord transition, to demonstrate a transannular π electron interaction in a bicyclic molecule.

After much experimental and theoretical study it has become possible to determine the *absolute* sense of twist, *i.e.*, the chirality, of an unsaturated ketone from its ord curve. The two orientations shown in Figure 17–5 give rise to enantiomeric Cotton effects. These two arrangements are actually different conformers of the same bonding sequence; however, conformational restraints imposed by ring formation can prevent their interconversion.

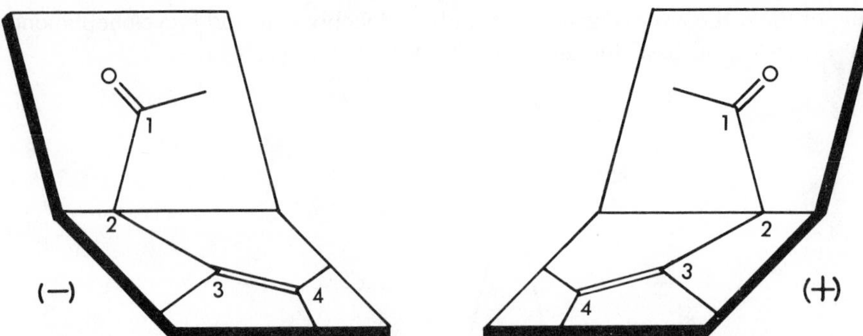

Figure 17-5 Enantiomeric arrangements of the inherently dissymmetric chromophores associated with some β, γ-unsaturated ketones. The signs refer to the relevant Cotton effect. (From Kurt Mislow, *Introduction to Stereochemistry*, copyright © 1966, W. A. Benjamin, Inc., Menlo Park, California.)

A similar arrangement is also valid for dienes, which permits the determination of the absolute configuration of many steroidal and terpenoid structures.

O $\quad$ CH_2 $\quad$ these conjugated systems display positive Cotton effects

The Octant Rule. A large amount of experimental and theoretical work has been summarized in an empiricism called **the octant rule.** The octant rule is used to predict the absolute configuration of a chiral center that perturbs the "symmetric" carbonyl chromophore in a cyclohexanone system.

To apply the octant rule to a cyclohexanone, one divides the space around the carbonyl group into eight unequal volumes, or *octants*. Each octant is assigned a sign (+ *or* −) based upon how a substituent in that octant would affect the sign of the Cotton effect of the carbonyl group's $n \rightarrow \pi^*$ transition.

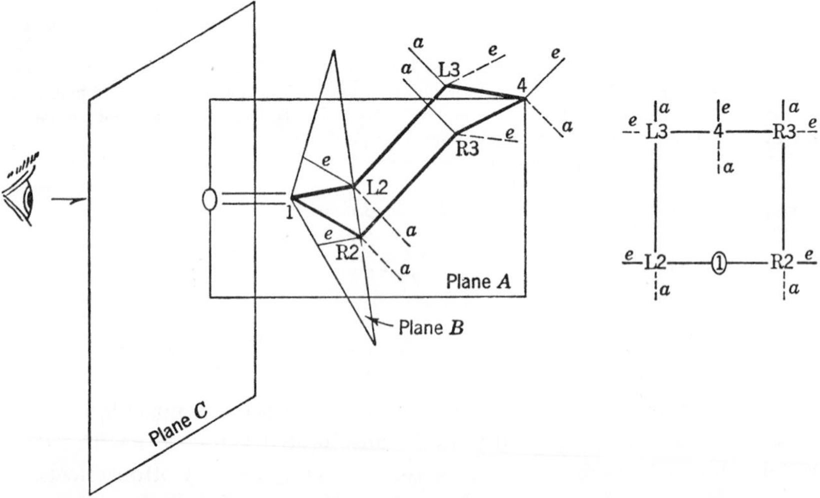

Figure 17-6 Geometry of cyclohexanone illustrating three planes creating eight octants. Abbreviations: *a*, axial; *e*, equatorial; L, left; R, right. (From Carl Djerassi, *Optical Rotatory Dispersion*, copyright © 1960 by McGraw-Hill, Inc. Used by permission of McGraw-Hill Book Co., New York.)

The theoretical basis of the octant rule, and related rules, is not the mere *existence* of groups in certain locations. Instead, it is the presence of electrons, and their polarizability, around the carbonyl group that influences the $n \rightarrow \pi^*$ Cotton effect.

The octants are displayed in Figure 17–6; the planes used to generate the octants are:

Plane 1: contains C1, C2, and C6, and the carbonyl O
Plane 2: contains C1 and C4 and bisects the ring
Plane 3: passes through the C=O bond and is perpendicular to the C=O bond.

Looking down the carbonyl group, we may designate the octants as near (*n*) or far (*f*)–upper (*u*) or lower (*l*)–left (*l*) or right (*r*). With this terminology, in this sequence, the *dextro* and *levo* contributing octants are summarized as follows:

dextro (+) octants	*levo* (−) octants
n-l-l	n-l-r
n-u-r	n-u-l
f-l-r	f-l-l
f-u-l	f-u-r

Substituents that lie on the planes used to define the octants make no contribution to the sign of the Cotton effect. For reference, the signs of various substituent contributions are summarized in Table 17–6.

TABLE 17–6 Substituent Contributions to Cyclohexanone ORD

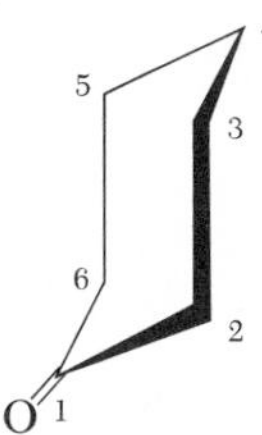

Carbon Number (Following Numbering In Figure Above)	Axial or Equatorial	Sign of Contribution[a]
2	axial	positive
	equatorial	zero
3	axial	negative
	equatorial	negative
4	axial	zero
	equatorial	zero
5	axial	positive
	equatorial	positive
6	axial	negative
	equatorial	zero

[a] All contributions that are "zero" represent substituents that reside on one of the planes used to define the various octants.

There are myriad applications of the octant rule, but we will consider only one now, the determination of the absolute configuration of (+)-*trans*-10-methyl-2-decalone. This compound, with both a positive D-line rotation and a positive Cotton effect, must have either structure **A** or structure **B.**

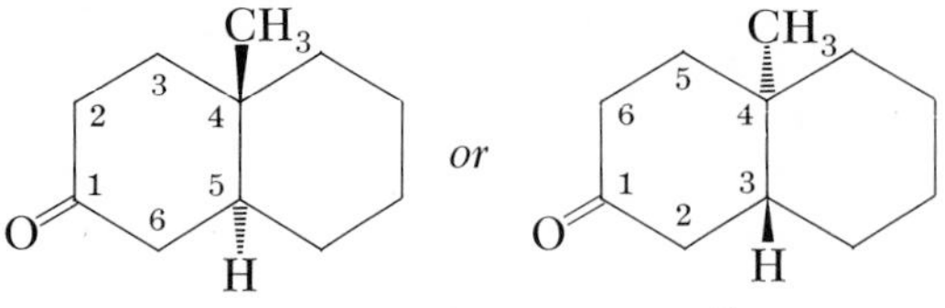

structures numbered for octant rule analysis

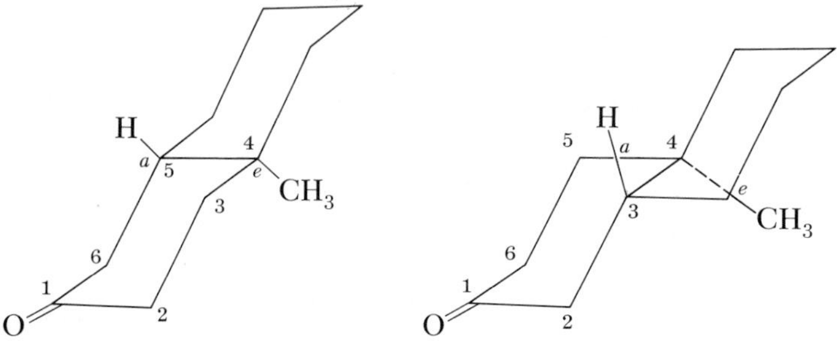

structures drawn and numbered for octant rule analysis

Structure **A** has contributions of zero from 4*a* and 4*e*, and a positive contribution from the 5*e* group. Structure **B** has zero contributions from 4*a* and 4*e*, and a negative contribution from 3*e*. Therefore, **A,** and not **B,** must represent the correct configuration of (+)-*trans*-10-methyl-2-decalone.

A Footnote. Our theoretical knowledge has not yet progressed to the point where we can predict the ord curve of an optically active compound as complex as those routinely found in living systems. What has happened during the past fifteen years, however, is a mushrooming of empirical observations, using compounds of known relative and/or absolute configuration, which has produced valuable analytical procedures for practicing organic and bio-organic chemists. Ord and its sister phenomenon, *circular dichroism*, are very powerful tools for correlating relative configurations.

17.7 INFRARED SPECTRA

The >C=O linkage is the most studied chromophore in the ir, partly because of its great intensity and partly because of its sensitivity to relatively minor changes in its environment. In turn, this makes the >C=O absorption a very useful one when trying to assign structures to compounds.

The Ballpark. All >C=O absorptions occur in the range from ~1900 to ~1550 cm^{-1}. Setting 1715 cm^{-1} as a reference point (we'll see why in a moment), aldehydes absorb at a slightly higher frequency (~1725 cm^{-1}), esters at still higher frequency (~1735 cm^{-1}), and carboxylic acids at yet higher frequency (~1760 cm^{-1}). On the other side of 1715 cm^{-1} are the amides, absorbing at about 1685 cm^{-1}.

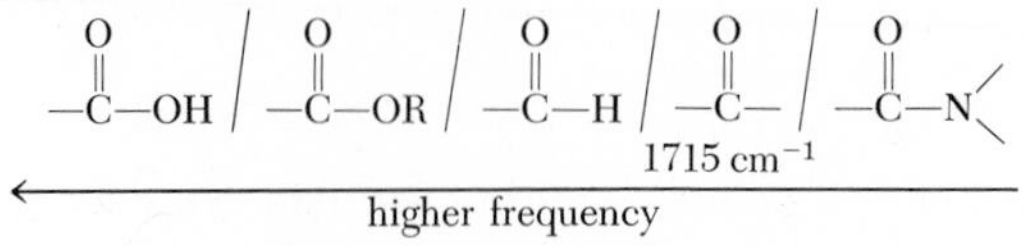

—C(=O)—OH / —C(=O)—OR / —C(=O)—H / —C(=O)— / —C(=O)—N<

1715 cm^{-1}

← higher frequency

trend in >C=O infrared vibrations

The discrepancy between the original 1900 to 1550 cm^{-1} range and these values, extending *from* 1760 *to* 1685 cm^{-1}, arises because the specific values are for simple, unsubstituted compounds while the range portends the consequences of substitution around a C=O linkage. A frequency of 1715 cm^{-1} was chosen as the reference because it is the stretching frequency of a simple dialkyl ketone such as acetone.

Aldehydes. Aldehydes exhibit their carbonyl absorption at ~1725 cm^{-1}. If the aldehydic carbonyl group is conjugated, the stretching frequency of the carbonyl group is reduced because of decreased double bond character in the carbonyl linkage.

$$\text{alkyl}-\overset{\overset{\displaystyle O}{\|}}{C}-H \qquad C_6H_5-\overset{\overset{\displaystyle O}{\|}}{C}-H \qquad -\overset{|}{C}=\overset{|}{C}-\overset{\overset{\displaystyle O}{\|}}{C}-H \leftrightarrow -\overset{|}{\overset{\oplus}{C}}-\overset{|}{C}=\overset{\overset{\displaystyle O^{\ominus}}{|}}{C}-H$$

$$\nu_{C=O}\ 1725\ cm^{-1} \qquad \nu_{C=O}\ 1700\ cm^{-1} \qquad \nu_{C=O}\ 1685\ cm^{-1}$$

While the carbonyl stretching frequency is helpful in suggesting the presence of —CHO, it is a doublet, of complex origin, at 2720 cm^{-1} and 2820 cm^{-1} that is most useful in pinning down the group as "aldehyde."

The spectra of several representative aldehydes constitute Figures 17–7 to 17–10.

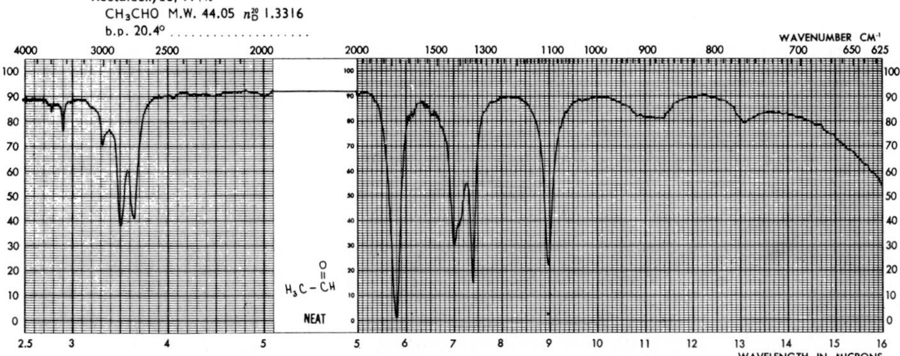

Figure 17–7 The infrared spectrum of acetaldehyde.

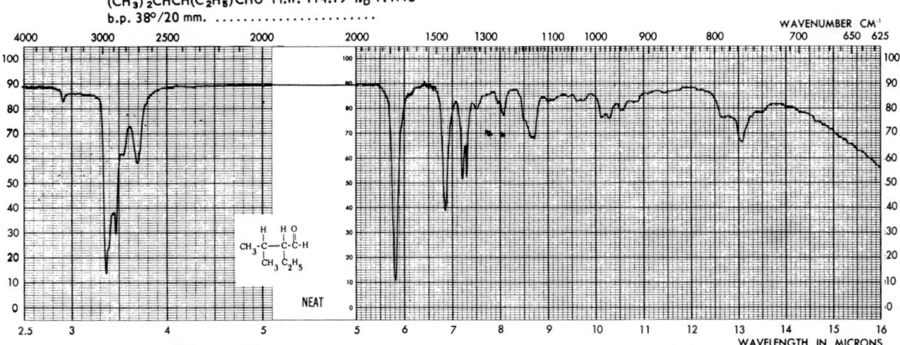

Figure 17–8 The infrared spectrum of 2-ethylisovaleraldehyde.

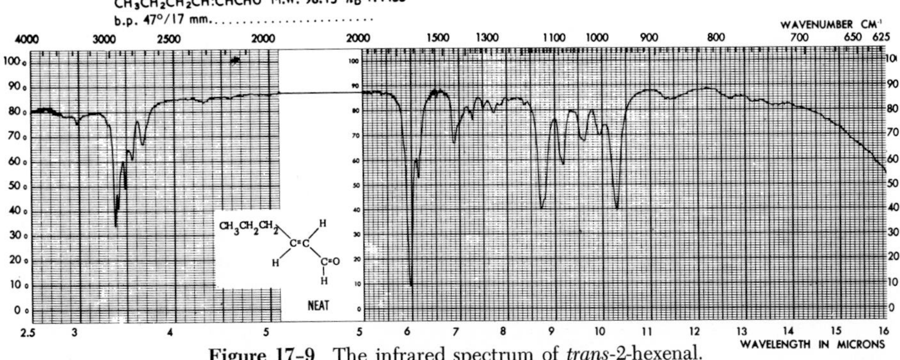

Figure 17–9 The infrared spectrum of *trans*-2-hexenal.

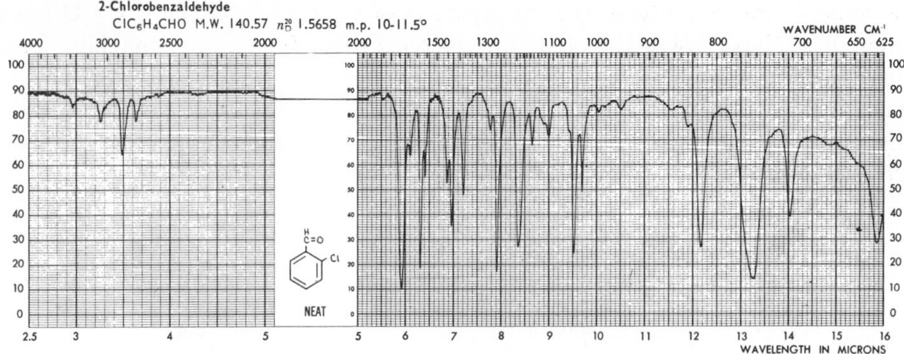

Figure 17-10 The infrared spectrum of 2-chlorobenzaldehyde.

Ketones. The "normal" $\nu_{\mathrm{C\;O}}$ for ketones, found in acyclic dialkyl ketones and simple cyclohexanones, is 1715 cm^{-1}. As with aldehydes, and for the same reason, conjugation reduces the C═O stretching frequency.

alkyl—C(═O)—alkyl $\nu_{C=O}$ 1715 cm^{-1}

C_6H_5—C(═O)—alkyl $\nu_{C=O}$ 1690 cm^{-1}

—C═C—C(═O)—alkyl $\nu_{C=O}$ 1675 cm^{-1}

C_6H_5—C(═O)—C_6H_5 $\nu_{C=O}$ 1665 cm^{-1}

The idea that the cyclopropane ring can behave as does a double bond toward conjugation has reappeared in organic chemistry for a number of years. One piece of evidence used to argue this point is the $\nu_{C=O}$ for cyclopropyl ketones; compounds of the type $(CH_2)_2CHC(O)alk$ usually absorb at about 1695 cm^{-1}.

Ring size can greatly influence the carbonyl stretching frequency of a cyclic ketone. Cyclohexanone, containing a keto group in a comparatively strain-free environment, absorbs at 1715 cm^{-1}. As the size of the ring decreases, and the ring strain increases, the $\nu_{\mathrm{C\;O}}$ of a cycloalkanone increases.

1715 cm^{-1} 1745 cm^{-1} 1780 cm^{-1} 1838 cm^{-1}

effect of ring size on $\nu_{C=O}$

Highly strained polycyclic ketones, such as the two compounds below, also exhibit $\nu_{\mathrm{C\;O}}$ near 1800 cm^{-1}.

7-ketobicyclo[2.2.1]heptane 6-ketobicyclo[2.1.1]hexane

Electron-withdrawing groups (*e.g.*, halogens) attached to the α carbon increase $\nu_{C=O}$. The variation in $\nu_{C=O}$ is a function of the torsional angle around the Z—C—C═O group, the greatest increase in $\nu_{C=O}$ coming when the torsional angle is near 0°. It has been suggested that in this planar array, the negative end of the C—Z dipole reduces the extent of charge separation in the carbonyl group, thus giving the carbonyl group more double bond character and thereby increasing $\nu_{C=O}$.

$Z \leftrightarrow C—C=O$ $C—C=O$ with Z below

torsional angle $\approx 0°$ torsional angle $\approx 90°$

This is best seen in the α-halocyclohexanones, in which an *axial* haloketone absorbs at lower frequency than does the corresponding *equatorial* haloketone.

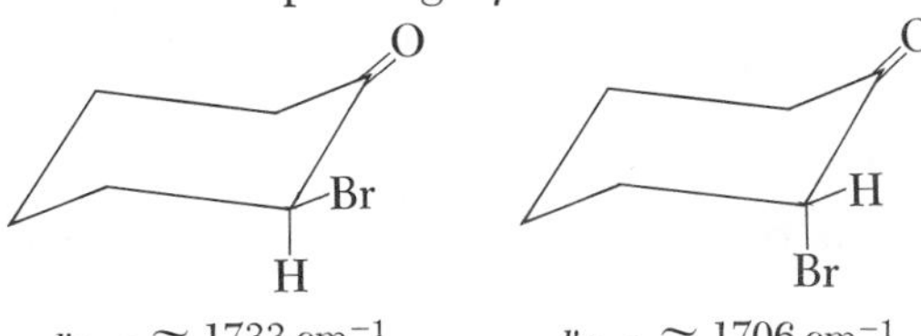

$\nu_{C=O} \approx 1733\ cm^{-1}$ $\nu_{C=O} \approx 1706\ cm^{-1}$

The spectra of several representative ketones constitute Figures 17–11 to 17–14.

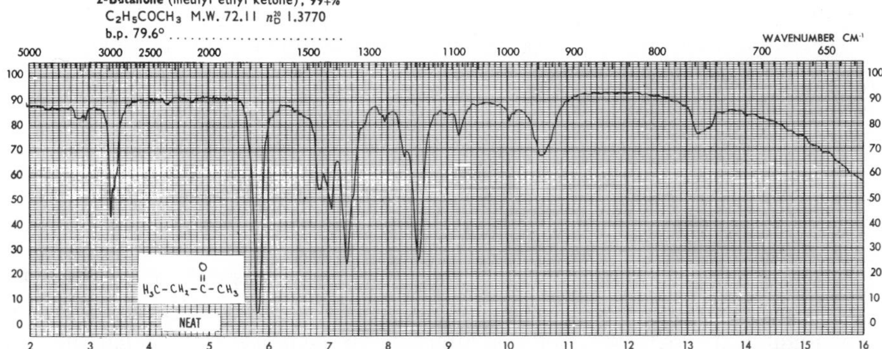

Figure 17–11 The infrared spectrum of 2-butanone.

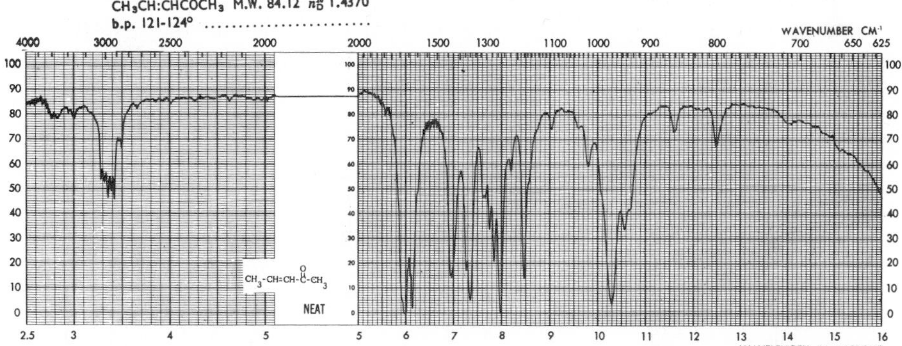

Figure 17–12 The infrared spectrum of 3-penten-2-one.

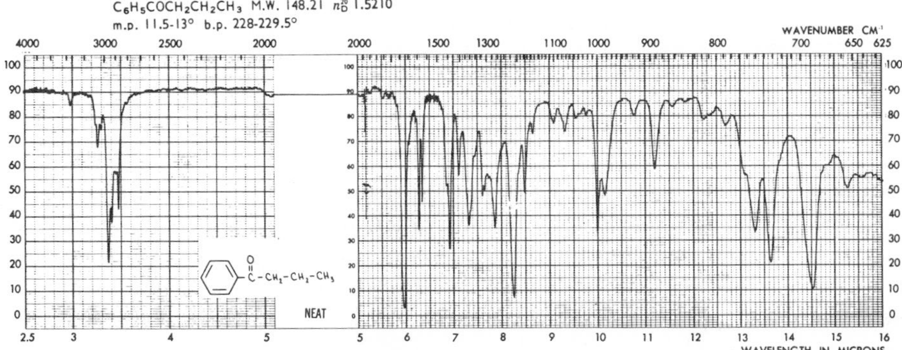

Figure 17–13 The infrared spectrum of *n*-butyrophenone.

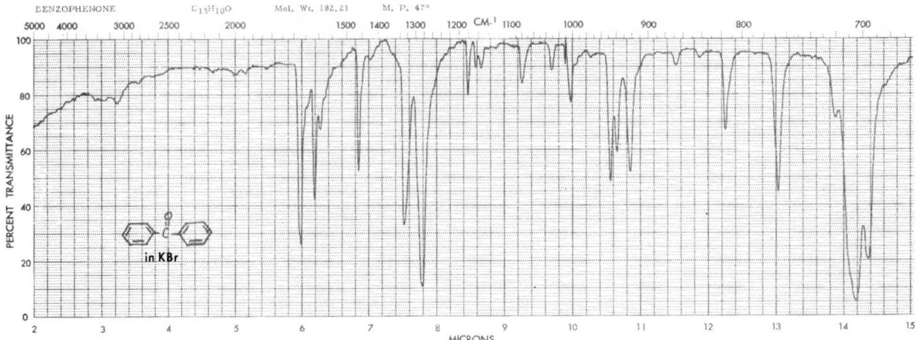

Figure 17-14 The infrared spectrum of benzophenone.

Keto-Enol Equilibria. A carbonyl compound containing a significant amount of enol will exhibit the broad OH stretching frequency characteristic of alcohols in addition to the carbonyl stretching vibration. If, as in the case of β-diketones, the extent of enolization is great, and if intramolecular hydrogen bonding is extensive and strong, $\upsilon_{C=O}$ will appear at lower frequency (~1650 to 1550 cm^{-1}). If enolization is nearly complete, as in methone for example, then only a very weak carbonyl absorption will be observed—it may even be absent from the spectrum!

17.8 NUCLEAR MAGNETIC RESONANCE SPECTRA[a]

Aldehydes are not easily distinguished from ketones on the basis of the >C=O vibration; it is the aldehydic hydrogen, vibrationally coupled with the carbonyl group, that causes the analytically useful 2720–2820 cm^{-1} doublet. Similarly, in the nmr it is the aldehydic hydrogen, by virtue of its appearance downfield of even aromatic hydrogens, that indicates the presence of the carboxaldehyde group.

The nmr spectrum of butyraldehyde (Figure 17–15) reveals that the —CHO resonance not only is far downfield (~10 ppm from TMS) but also is coupled only weakly (J = 1 to 3) to adjacent protons.

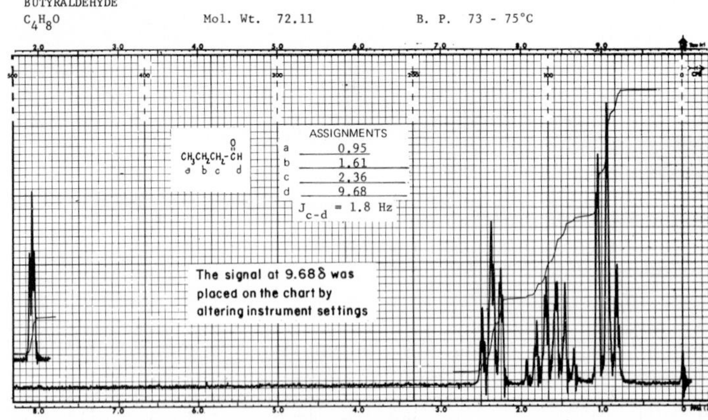

Figure 17-15 Nmr spectrum of butyraldehyde. © Sadtler Research Laboratories, Inc., 1976.

Because it lacks any protons of its own, the presence of a keto group must be inferred from its influence upon proximal protons. The deshielding effect of the keto group is evident in Figure 17–16, which compares the nmr spectra of acetophenone and ethylbenzene.

If a compound exists as an *enol,* one can expect to observe the enolic hydrogen as far downfield as 15 to 16 ppm. This deshielding is due to (a) the electronegativity of the oxygen

[a] A figure showing the magnetic environment around the >C=O group is included in Chapter 29 (Figure 29–5).

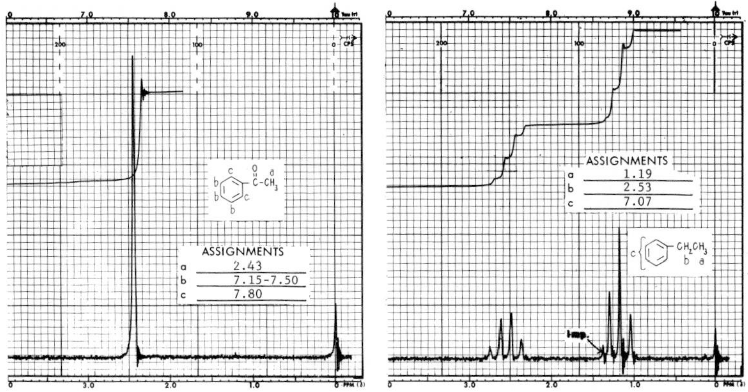

Figure 17-16 Nmr spectra of acetophenone and ethylbenzene. © Sadtler Research Laboratories, Inc., 1976.

bearing the enolic hydrogen and (b) the powerful intramolecular hydrogen bond associated with most enols.

IMPORTANT TERMS

Alcoholic fermentation: A complex sequence of enzyme-controlled reactions that converts a sugar (glucose) into ethanol and carbon dioxide.

$$C_6H_{12}O_6 \xrightarrow{\text{enzymes}} 2C_2H_5OH + CO_2$$

Aldol condensation: Most commonly, the acid- or base-catalyzed conversion of two molecules of an aldehyde bearing a proton on the carbon adjacent to the carbonyl group into a β-hydroxy aldehyde. The products are unstable if they bear a hydrogen on the carbon between the hydroxy and the carbonyl groups and will dehydrate either upon heating with base or by treatment with dilute acid at room temperature.

$$2R-\underset{H}{\overset{H}{C}}-CHO \xrightarrow[OH^{\ominus}]{H^{\oplus}\text{ or}} R-\underset{H}{\overset{H}{C}}-\underset{H}{\overset{OH}{C}}-\underset{R}{\overset{H}{C}}-CHO \xrightarrow[OH^{\ominus}+\text{ heat}]{H^{\oplus}\text{ or}} R-\underset{H}{\overset{H}{C}}-\overset{H}{C}=\overset{R}{C}-CHO$$

The reverse reaction, the cleavage of a β-hydroxy carbonyl compound, is called the retro-aldol condensation.

The reaction forms all possible aldol condensation products if two different aldehydes, each of which contains an (acidic) α hydrogen, are used. Since ketones react more slowly with carbanions than do aldehydes, a successful reaction can be achieved by treating an aldehyde lacking an α hydrogen with a ketone possessing an (acidic) α hydrogen.

$$R_3C\text{—}CHO + R'\text{—}CH_2\text{—}\overset{\overset{\large O}{\|}}{C}\text{—}R'' \xrightarrow{OH^{\ominus}} R_3C\text{—}\underset{H}{\overset{OH}{C}}\text{—}\underset{R'}{\overset{H}{C}}\text{—}\overset{\overset{\large O}{\|}}{C}\text{—}R''$$

Reactions using two different carbonyl compounds are termed crossed or mixed aldol condensations, even when ketones are employed as starting materials.

Benzoin condensation: The reaction between two molecules of an aryl aldehyde to produce an α-hydroxy ketone (ArCHOHCOAr). Such compounds are termed "benzoins." The reaction is specifically catalyzed by cyanide ion.

$$2Ar\text{—}CHO \xrightarrow[\text{heat}]{^{\ominus}CN/H_2O/EtOH} Ar\text{—}CHOH\overset{\overset{\large O}{\|}}{C}\text{—}Ar$$

a benzoin

Favorskii reaction: The conversion of an α-halo ketone to the salt of a carboxylic acid by the action of a base, usually hydroxide ion. It is believed to involve a cyclopropanone intermediate.

$$\text{2-halocyclohexanone (C=O, H, X)} \xrightarrow{OH^{\ominus}} \text{cyclopentane-}CH\text{—}CO_2^{\ominus} + X^{\ominus}$$

Haloform reaction: The conversion of a methyl ketone into a carboxylic acid and a haloform (CHX_3) by treatment with a mixture of halogen and base. (Since increasing halogenation stabilizes the intermediate carbanions produced in the reaction, the process rapidly leads to a haloform *via* a trihalide ion.)

$$R\text{—}\overset{\overset{\large O}{\|}}{C}\text{—}CH_3 \xrightarrow[OH^{\ominus}]{X_2} \left[R\text{—}\overset{\overset{\large O}{\|}}{C}\text{—}\overset{\ominus}{C}H_2\right] \rightarrow R\text{—}\overset{\overset{\large O}{\|}}{C}\text{—}CH_2X \xrightarrow{OH^{\ominus}} R\text{—}\overset{\overset{\large O}{\|}}{C}\text{—}\overset{\ominus}{C}HX \xrightarrow[OH^{\ominus}]{X_2} RCO_2^{\ominus} + HCX_3$$

Octant rule: An empiricism useful for predicting the sign of rotation of chiral cyclohexanone derivatives. The practical consequences of the rule are summarized in Table 17–6.

Optical rotatory dispersion (ord): A plot of the optical rotation of a sample as a function of the wavelength of polarized light passing through the sample. A simple ord curve can be broken down into two regions. In one region, the rotation changes only slightly with wavelength changes; in the other, the rotation changes rapidly with a small wavelength change. The overall shape of the ord curve is related to the absolute configuration of the sample. Enantiomers have enantiomeric (mirror image) ord curves. (Refer to Figure 4–10 and the related discussion for a bit more detail.)

PROBLEMS

12. When norbornanone is exposed to NaOD/D_2O, both protons at C3 are exchanged with deuterium. While this is not so surprising, it was found that the *exo* proton was exchanged more rapidly than the *endo* proton. Account for this difference.

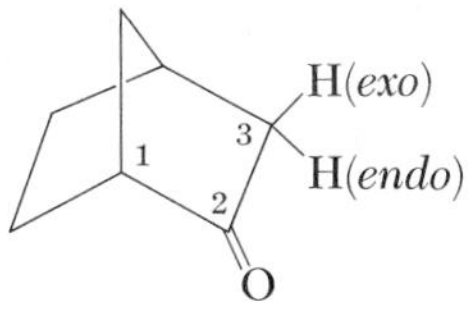

H(*exo*)
H(*endo*)
norbornanone

13. (a) Explain why (**R**)-2-methylbutanal reacts with base to form several aldol condensation products, none of which is optically active. (b) What would you expect from the base-catalyzed aldol condensation of (**R**)-3-methylpentanal?

14. When fructose is exposed to aqueous alkali, it is converted to two isomeric sugars. One of these is glucose, while the other is mannose. Suggest a mechanism for this conversion of fructose to glucose. Considering this, suggest (a) a structure for mannose and (b) a means of preparing some mannose from glucose.

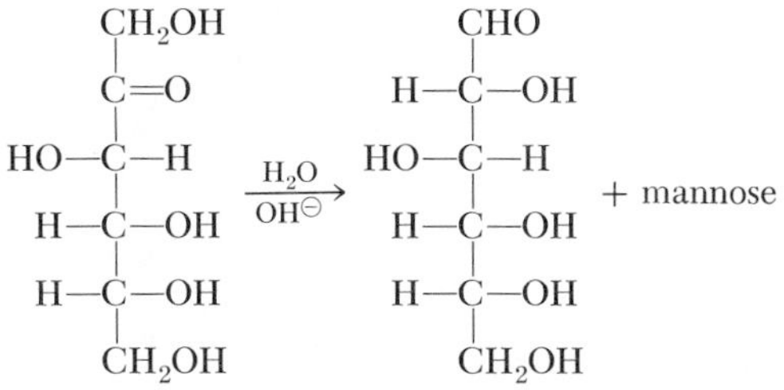

fructose		glucose	
CH_2OH		CHO	
C=O		H—C—OH	
HO—C—H	$\xrightarrow[OH^{\ominus}]{H_2O}$	HO—C—H	+ mannose
H—C—OH		H—C—OH	
H—C—OH		H—C—OH	
CH_2OH		CH_2OH	

15. The compounds shown below do *not* enolize, as shown. Account for this behavior.

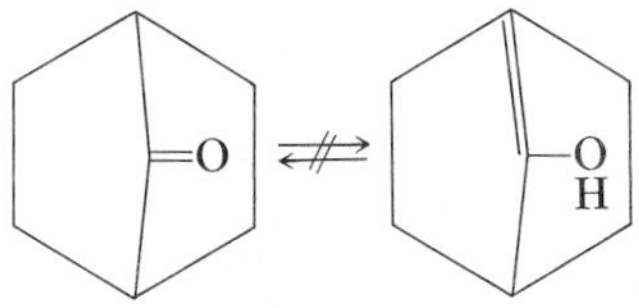

(a) (b) (c) (d)

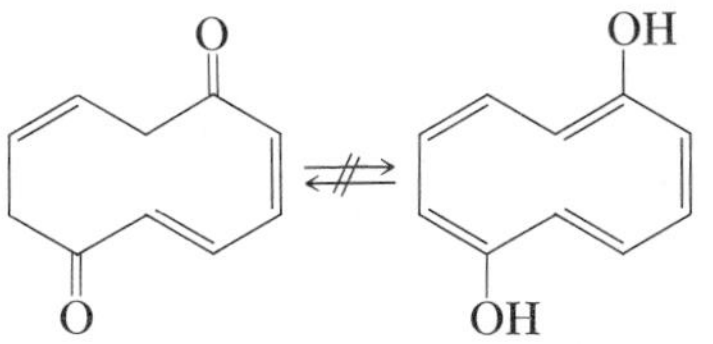

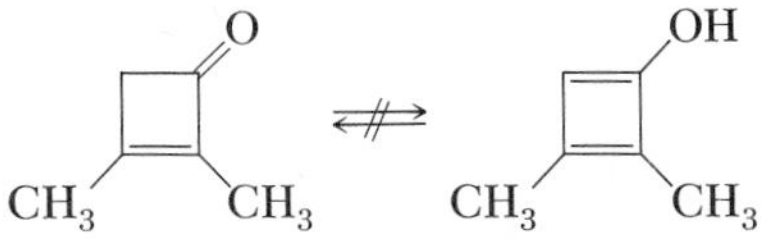

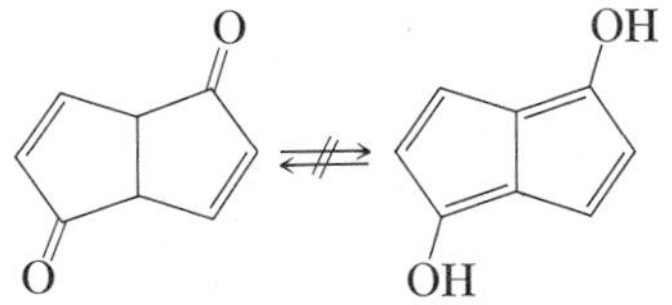

16. Prototropic equilibria (tautomerisms) are not limited to aldehydes and ketones. Draw another "tautomer" for each of the following.

(a)

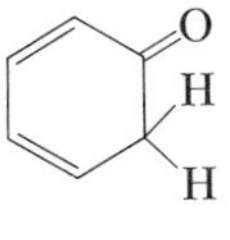

(b)

(c) $R_2CHN{=}O$
(d) $R_2C{=}N{-}NHR$
(e) CH_3NO_2
(f) $R_2CHCR{=}NR$

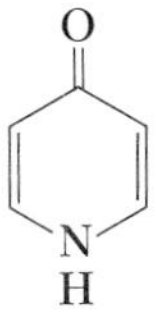

17. Acetoacetic ester has an enol content of 0.4% in water and 20% in toluene. Account for this solvent dependence. (*Hint:* Consider hydrogen bonding.)

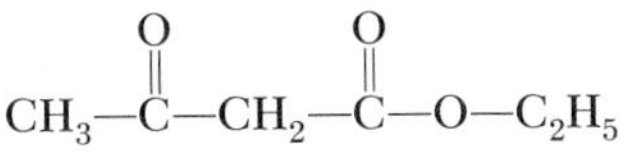

$$CH_3{-}\overset{O}{\overset{\|}{C}}{-}CH_2{-}\overset{O}{\overset{\|}{C}}{-}O{-}C_2H_5$$

acetoacetic ester

18. Starting with cyclohexanone, and using any other necessary reagents, suggest reasonable syntheses for the following.

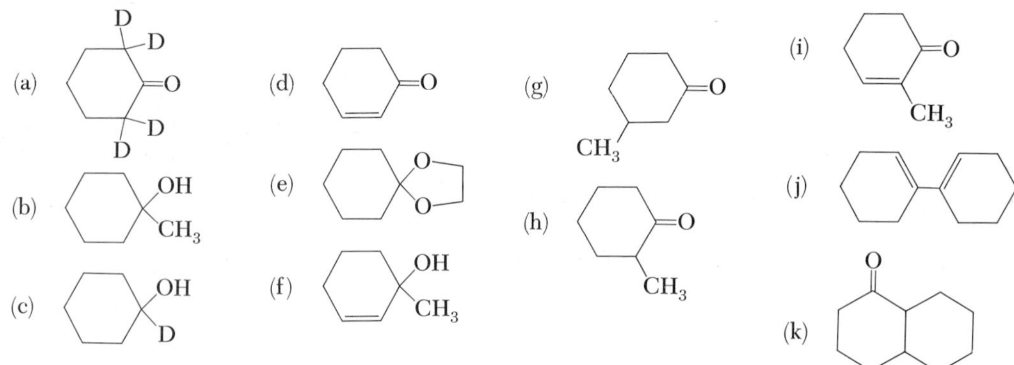

19. Write an equation, indicating all organic products, for the reaction of *p*-chlorobenzaldehyde with each of the following. Indicate the absence of reaction by "NR."

(a) dilute NaOH
(b) conc. NaOH
(c) acetaldehyde, dilute NaOH
(d) acetone, dilute NaOH
(e) acetophenone, dilute NaOH
(f) formaldehyde, conc. NaOH
(g) sodium salt of 2,4-pentandione
(h) benzaldehyde, dilute NaOH

20. Starting with methyl phenyl ketone (acetophenone), and using any other needed reagents, suggest syntheses for the following.

(a) methylphenylcarbinol
(b) 1-phenylethanol
(c) ethylbenzene
(d) dimethylphenylcarbinol
(e) benzoic acid
(f) phenyl ethyl ketone
(g) 2,3-diphenyl-2,3-butanediol
(h) 3,3-diphenyl-2-butanone
(i) 2,2-diphenylpropanoic acid ($(C_6H_5)_2C(CH_3)CO_2H$)

21. Starting with anything you need, so long as starting organic compounds are benzene or contain no more than four carbons, offer rational syntheses for the following.

(a) $(CH_3CH_2)_3COH$
(b) pivalic acid, $(CH_3)_3CCO_2H$
(c) pivaldehyde, $(CH_3)_3CCHO$
(d) benzalacetone
(e) 4-methyl-2-pentanone
(f) 2-phenyl-1,3-dioxolane
(g) hexanedial
(h) *p*-chlorobenzaldehyde
(i) 4-(*p*-chlorophenyl)-3-buten-2-one
(j) dimethylvinylcarbinol
(k) dicyclobutyl ketone

22. In the iodoform oxidation of acetone to acetic acid, no iodoacetic acid is produced. How do you explain this lack of halogenation on both sides of the carbonyl group?

$$CH_3-\overset{\overset{\displaystyle O}{\|}}{C}-CH_3 + I_2 + OH^{\ominus} \rightarrow \xrightarrow{H_3O^{\oplus}} \underset{\text{not produced}}{I-CH_2-\overset{\overset{\displaystyle O}{\|}}{C}-OH}$$

*23. You are given tubes **A** and **B.** Each contains one of the compounds in one of the following pairs. Describe, including anticipated observations, a spectral method for properly identifying members of the following pairs.

(a) methylcyclopentane; 2-methylcyclohexanone
(b) 2-cyclohexenone; 3-cyclohexenone
(c) [bicyclic ketone structures] =O; O=
(d) acetone; 1,1,1-trichloroacetone
(e) 2,4-pentanedione; 2,3-butanedione
(f) [4-*tert*-butyl-2-chlorocyclohexanone structure] =O, Cl (*cis* and *trans*)
(g) [hydroxycyclohexanone structures] O, H, OH; O, HO H
(h) [cyclohexanedione structures] O, O; O, O

*This problem requires knowledge of spectroscopy.

(i) $CH_3CH{=}CH{-}CHO$; $(CH_3)_2C{=}CH{-}CHO$

(j) 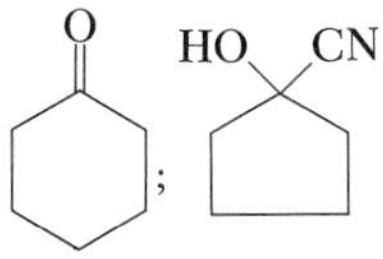 cyclohexanone ; 1-hydroxycyclopentanecarbonitrile (HO, CN)

(k) 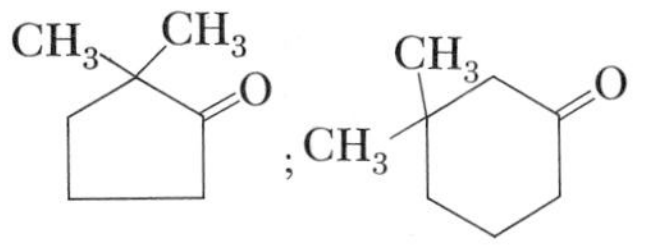 2,2-dimethylcyclopentanone (CH_3, CH_3, O) ; 2,2-dimethylcyclohexanone (CH_3, CH_3, O)

*24. (a) Repeat problem 23, restricting yourself to ir. (b) Repeat problem 23, restricting yourself to nmr. Are there any pairs that cannot be identified after the imposition of these restrictions?

25. Solutions of the following compounds in ethanol all give positive iodoform tests. What do you conclude?

o-chlorobenzaldehyde (Cl, CHO); benzyl alcohol (CH_2OH); cyclohexanone (O); toluene (CH_3)

26. Ethyl carbonate reacts with cyclooctanone, in the presence of sodium hydride (a strong base), to produce an α-acylated ketone, that is, a ketone bonded to a —C(O)—OR group at Cα. Suggest a mechanism for this. (*Hint:* Consider the course of reaction of a Grignard reagent with ethyl carbonate.)

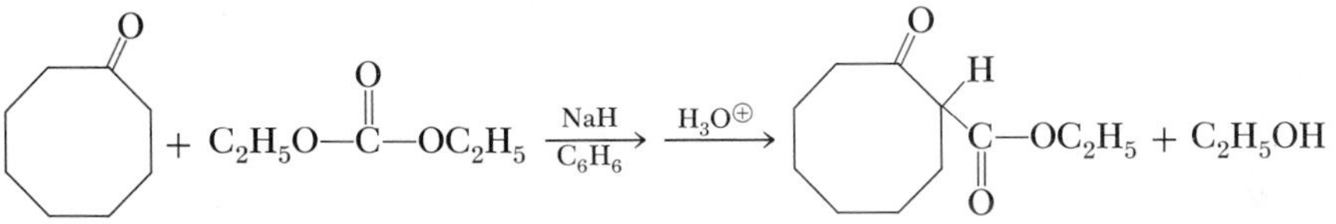

$$\text{cyclooctanone} + C_2H_5O{-}\overset{O}{\overset{\|}{C}}{-}OC_2H_5 \xrightarrow[C_6H_6]{NaH} \xrightarrow{H_3O^{\oplus}} \text{2-(}\overset{O}{\overset{\|}{C}}{-}OC_2H_5\text{)cyclooctanone (H at C}\alpha\text{)} + C_2H_5OH$$

ethyl carbonate

*27. Deuterated trifluoroacetic acid (TFA-d) catalyzes the self-condensation of acetone. Unfortunately, the kinetics of the reaction cannot be followed by nmr because all of the signals, except one near 10 δ, disappear. Explain this.

*28. You are given a sample of ketone that is approximately 25% enol. How would you determine the exact composition of the mixture by spectral means? Is there any advantage to an nmr analysis instead of an ir study?

29. Dihydrocitral, $C_{10}H_{14}O$, is produced in the reaction of β-methylcrotonaldehyde, $(CH_3)_2C{=}CHCHO$, with dilute sodium hydroxide. Suggest a structure for dihydrocitral and a mechanism for its formation.

30. Benzaldehyde reacts with malononitrile, $NC{-}CH_2{-}CN$, in base, to produce $C_6H_5CH{=}C(CN)_2$ after acidification. Suggest a mechanism for the reaction.

31. Cyclohexanone reacts with diazomethane, CH_2N_2, to produce **A** and **B**. Suggest mechanisms for their formation. (*Hint:* Draw the resonance structures for diazomethane.)

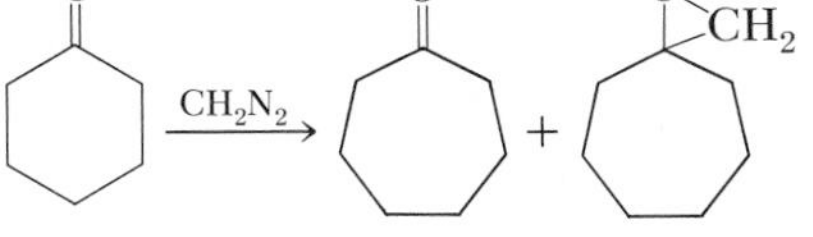

$$\text{cyclohexanone} \xrightarrow{CH_2N_2} \underset{\mathbf{A}}{\text{cycloheptanone}} + \underset{\mathbf{B}}{\text{spiro-epoxide (O, }CH_2\text{) of cycloheptane}}$$

*32. Suggest a reason why hydrogen-bonding solvents shift $\nu_{C=O}$ to a lower frequency (by as much as 50 cm^{-1}) compared to spectra taken in carbon tetrachloride.

33. Suggest a mechanism for the following reaction.

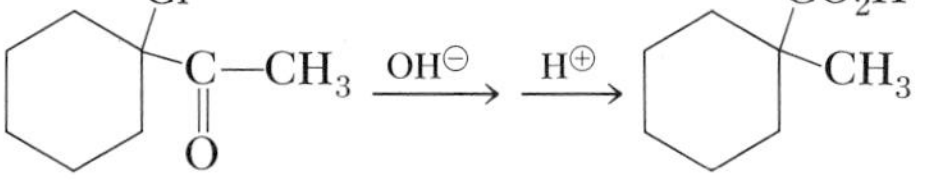

$$\text{1-chlorocyclohexyl}{-}\overset{O}{\overset{\|}{C}}{-}CH_3 \xrightarrow{OH^{\ominus}} \xrightarrow{H^{\oplus}} \text{1-methylcyclohexane-}CO_2H$$

34. In a future chapter we shall discuss the synthesis and alkylation of β-ketoesters, $RC(O)CH_2C(O)OR$. (a) However, with the information in this chapter, attempt to suggest a reason for the facile alkylation shown below.

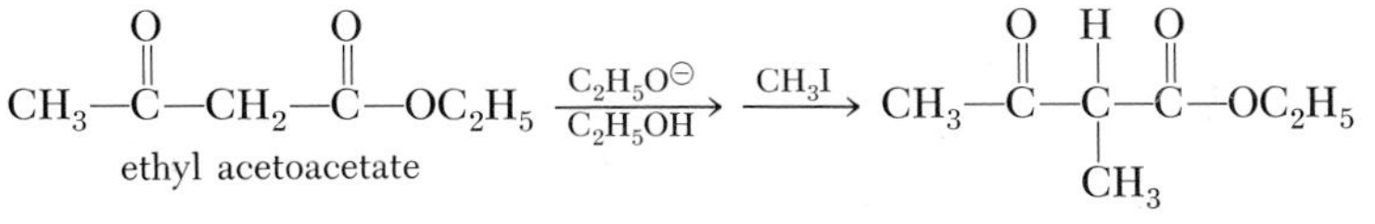

$$CH_3{-}\overset{O}{\overset{\|}{C}}{-}CH_2{-}\overset{O}{\overset{\|}{C}}{-}OC_2H_5 \xrightarrow[C_2H_5OH]{C_2H_5O^{\ominus}} \xrightarrow{CH_3I} CH_3{-}\overset{O}{\overset{\|}{C}}{-}\underset{CH_3}{\underset{|}{\overset{H}{\overset{|}{C}}}}{-}\overset{O}{\overset{\|}{C}}{-}OC_2H_5$$

ethyl acetoacetate

---

*This problem requires knowledge of spectroscopy.

(b) Suggest a mechanism for the conversion of ethyl acetate to ethyl acetoacetate. (c) What does this suggest about the —C(O)OC$_2$H$_5$ group?

$$2CH_3-\overset{O}{\overset{\|}{C}}-OC_2H_5 \xrightarrow{C_2H_5O^{\ominus}} CH_3-\overset{O}{\overset{\|}{C}}-CH_2-\overset{O}{\overset{\|}{C}}-OC_2H_5 + C_2H_5O^{\ominus}$$

35. Provide a mechanism for each of the following.

(a) 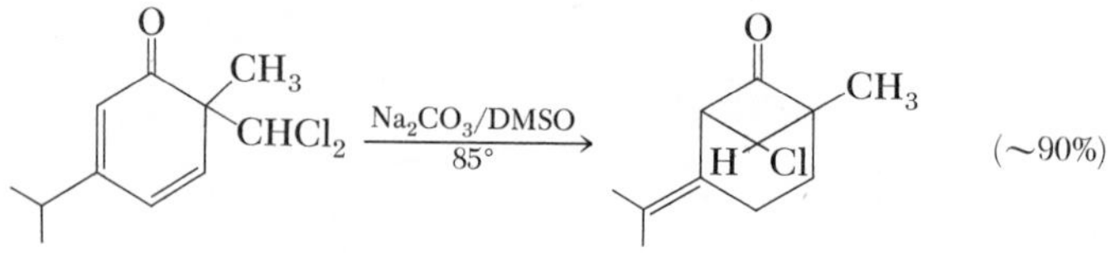 [structure: CH$_3$, CHCl$_2$] $\xrightarrow[85°]{Na_2CO_3/DMSO}$ [structure: CH$_3$, H, Cl] (~90%)

(b) 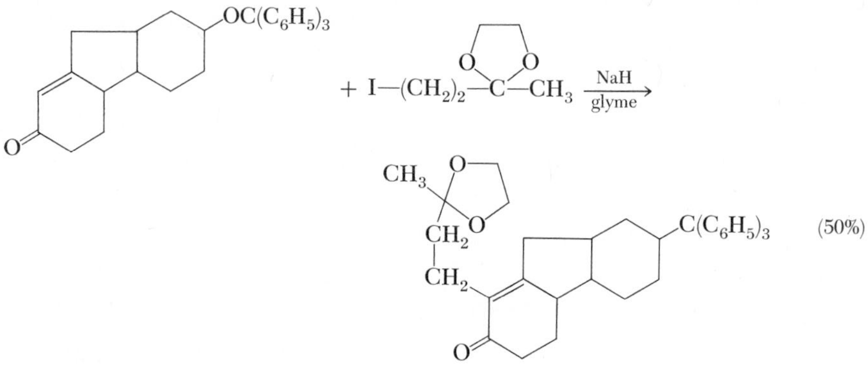 [structure: OC(C$_6$H$_5$)$_3$] + I—(CH$_2$)$_2$—C(O,O)—CH$_3$ $\xrightarrow[glyme]{NaH}$ [structure: CH$_3$, CH$_2$, CH$_2$, C(C$_6$H$_5$)$_3$] (50%)

(c) [structure: CO$_2$CH$_3$, —CH$_2$—CH$_2$—C(=O)—CH$_3$] $\xrightarrow[(b)\ dil.\ H_3O^{\oplus}]{(a)\ base}$ [structure: CO$_2$CH$_3$] (81%)

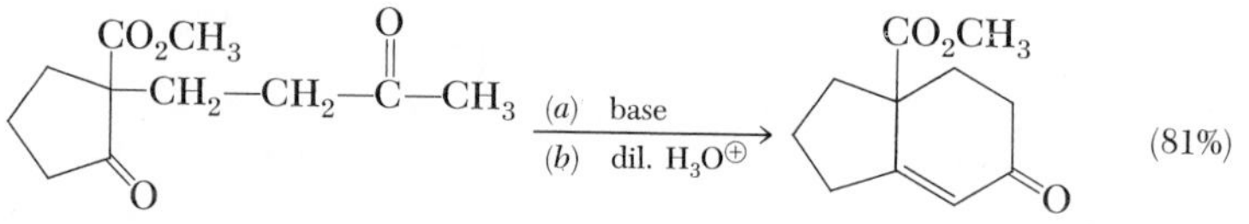

(d) 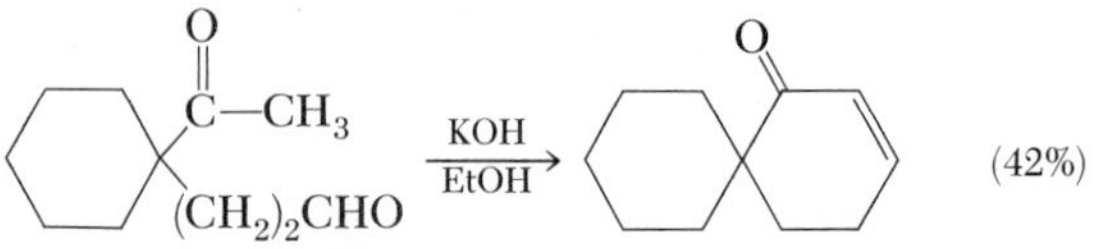 [structure: C(=O)—CH$_3$, (CH$_2$)$_2$CHO] $\xrightarrow[EtOH]{KOH}$ [structure] (42%)

(e) 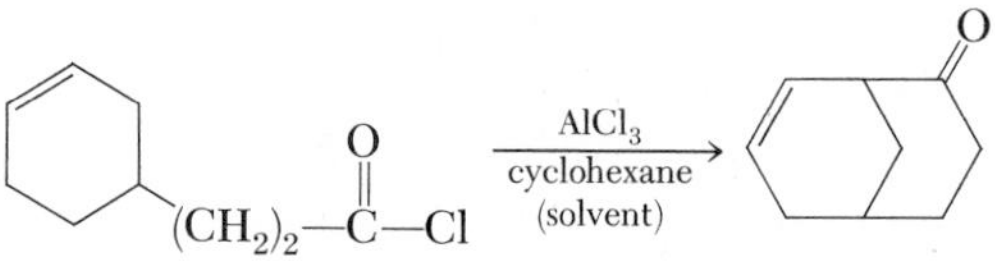 [structure: (CH$_2$)$_2$—C(=O)—Cl] $\xrightarrow[\text{cyclohexane (solvent)}]{AlCl_3}$ [structure]

(f) 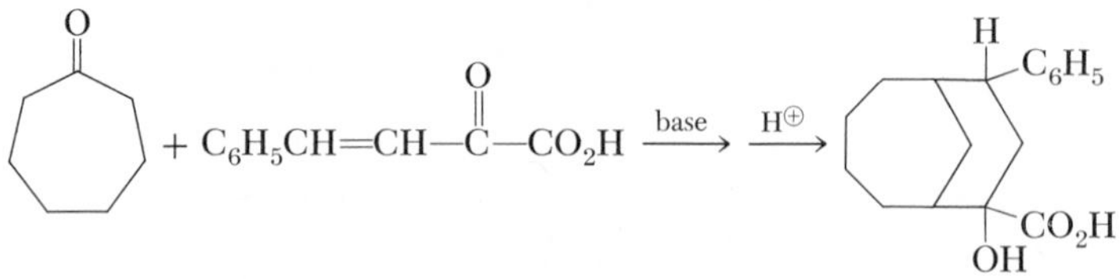 [structure] + C$_6$H$_5$CH=CH—C(=O)—CO$_2$H $\xrightarrow{base}$ $\xrightarrow{H^{\oplus}}$ [structure: H, C$_6$H$_5$, CO$_2$H, OH]

(g) $C_6H_5-\overset{O}{\overset{\|}{C}}-CH=CH-OC_2H_5 + C_6H_5MgBr \xrightarrow{ether} C_6H_5-\overset{O}{\overset{\|}{C}}-CH_2CH(C_6H_5)_2$ (90%)

(h) 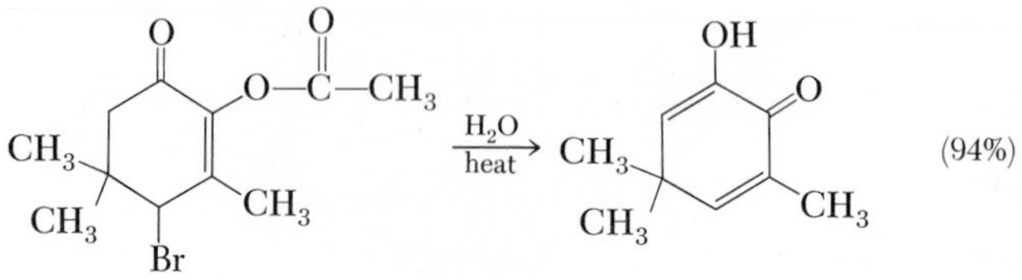 [structure: O—C(=O)—CH$_3$, CH$_3$, CH$_3$, CH$_3$, Br] $\xrightarrow[heat]{H_2O}$ [structure: OH, O, CH$_3$, CH$_3$, CH$_3$] (94%)

(i) 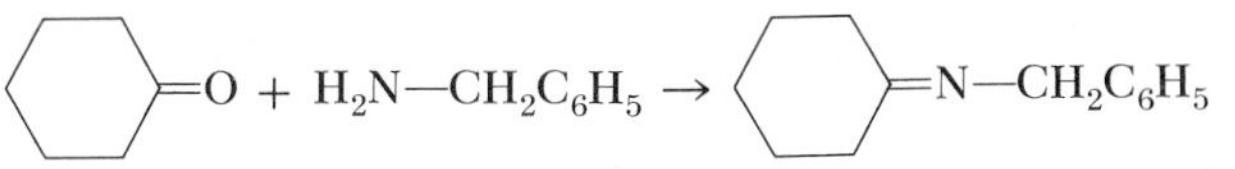 cyclohexanone$=O + H_2N{-}CH_2C_6H_5 \rightarrow$ cyclohexylidene$=N{-}CH_2C_6H_5$

(j) 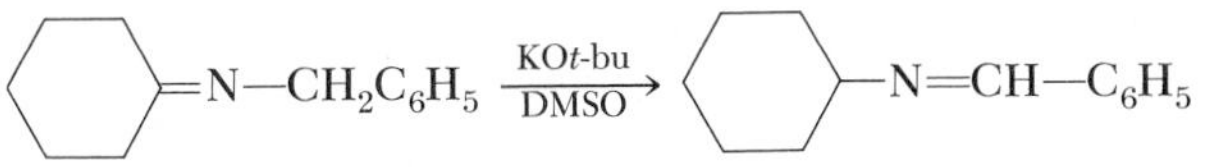 cyclohexylidene$=N{-}CH_2C_6H_5 \xrightarrow[\text{DMSO}]{\text{KO}t\text{-bu}}$ cyclohexyl$-N{=}CH{-}C_6H_5$

(k) 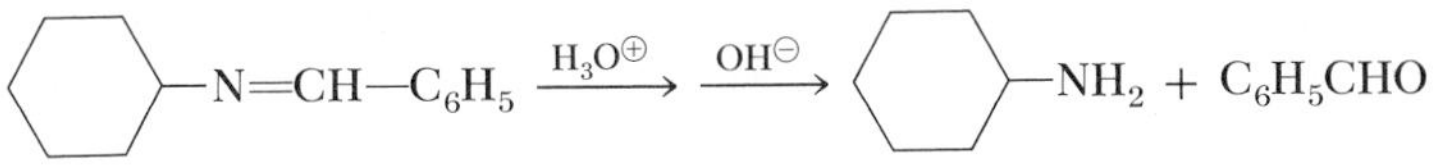 cyclohexyl$-N{=}CH{-}C_6H_5 \xrightarrow{H_3O^{\oplus}} \xrightarrow{OH^{\ominus}}$ cyclohexyl$-NH_2 + C_6H_5CHO$

(l) 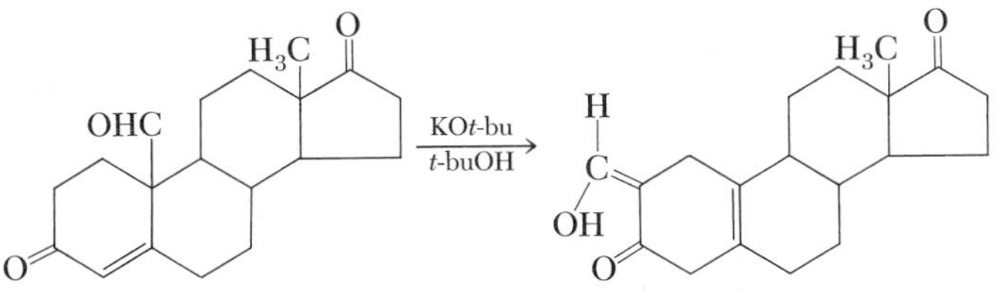 steroid with H_3C, O, OHC, O $\xrightarrow[t\text{-buOH}]{\text{KO}t\text{-bu}}$ steroid with H_3C, O, H–C(OH)=, O (63%)

36. How might one combine a condensation, an electrocyclic reaction, and a couple of other things to convert benzaldehyde to the cyclobutane derivatives shown below? You will need to use other reagents in addition to benzaldehyde.

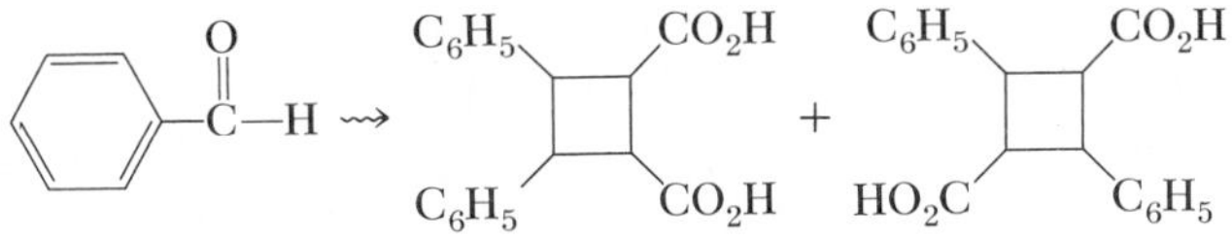

$C_6H_5CHO \rightsquigarrow$ cyclobutane (C_6H_5, CO_2H, C_6H_5, CO_2H) + cyclobutane (C_6H_5, CO_2H, HO_2C, C_6H_5)

37. What conclusion(s) might you draw from the following?

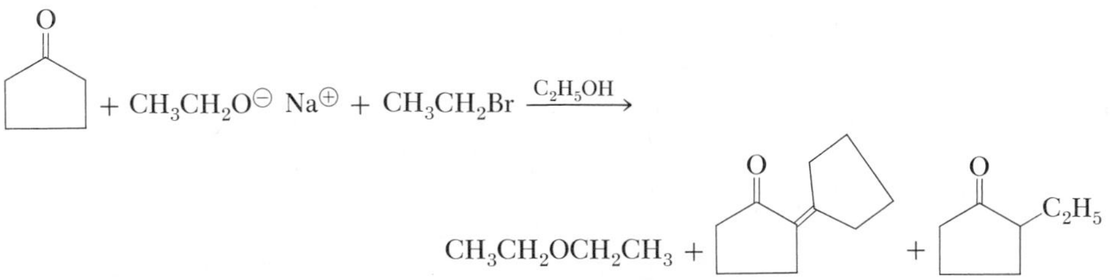

cyclopentanone $+ CH_3CH_2O^{\ominus}\ Na^{\oplus} + CH_3CH_2Br \xrightarrow{C_2H_5OH}$

$CH_3CH_2OCH_2CH_3$ + 2-cyclopentylidenecyclopentanone + 2-ethylcyclopentanone (C_2H_5)

major product | major product | minor product

38. Using an aldol condensation somewhere in each scheme, suggest a synthesis for each of the following. Provide an acceptable name for each product.

(a) $(CH_3)_2CDCHDCH(OH)CH_3$

(b) $(CH_3)_2CHCH_2CD(OH)CH_3$

(c) 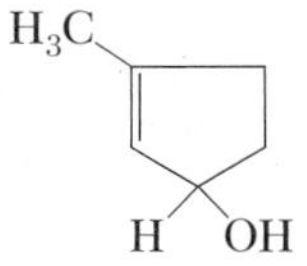 cyclopentene with H_3C and H, OH

(d) 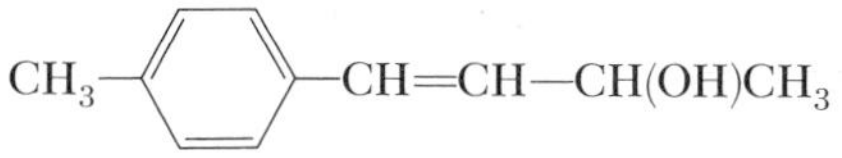 $CH_3{-}C_6H_4{-}CH{=}CH{-}CH(OH)CH_3$

(e) $CH_3CH(OH)CH_2CH_2OH$

(f) $CH_3CH_2CH(OH)CH(CH_3)CH_2OH$

(g) $(CH_3)_3CCH_2CH_2CH_2OH$

(h) 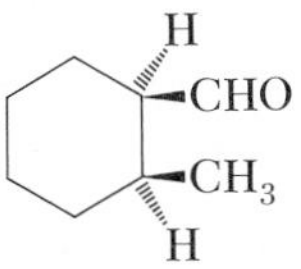 cyclohexane with H, CHO and H, CH_3 (*Hint:* cycloaddition)

39. The reaction of cyclopentanone with isopropylmagnesium chloride, followed by the usual acid work-up, produces only **A**. Suggest a mechanism for its formation.

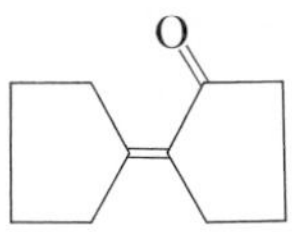

2-cyclopentylidenecyclopentanone (O)

A

40. (+)-3-Methylcyclohexanone has both a positive D-line rotation and a positive Cotton effect. It is also known to have the **R** absolute configuration. Assign a preferred conformation to it, with the aid of the octant rule.

41. *Trans*-2-chloro-5-methylcyclohexanone can be either diaxial or diequatorial. In methanol the compound is dextrorotatory and shows a positive Cotton effect, while in octane the compound is levorotatory and shows a negative Cotton effect. Rationalize this behavior, first by assigning conformations and then by examining the role of solvent in dictating (or not dictating) these conformations.

42. The ord curve below is that of a compound with a positive D-line rotation but a negative Cotton effect. (Such situations are not as unusual as that may seem.) With the aid of this curve, explain how a compound may have a D-line rotation with one sign but a nearby Cotton effect with another sign.

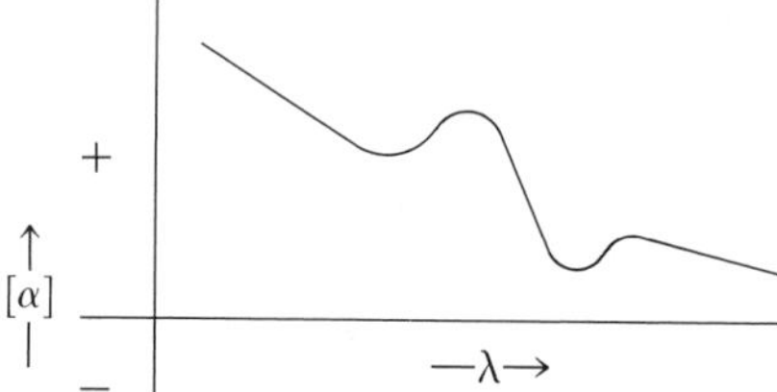

43. Ketones **A** and **B** both show positive Cotton effects, although the amplitude of the curve for **A** is much smaller than that of **B.** However, these can be unambiguously distinguished by adding a solution of hydrogen chloride in methanol to each sample and re-running the ord curves. The amplitude of the curve for **A** becomes almost zero, while that of **B** is virtually unchanged. Explain this last observation.

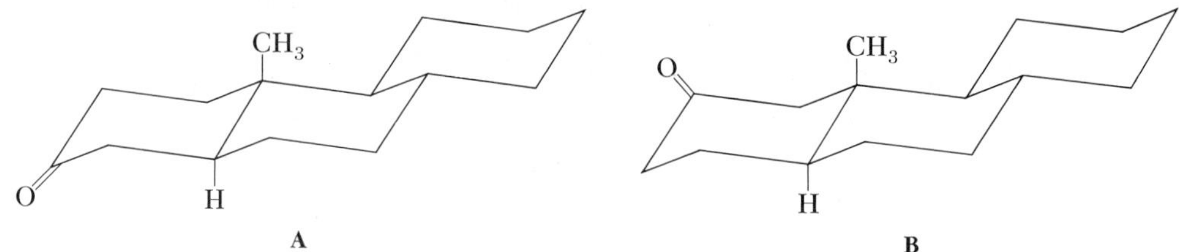

*44. The infrared spectra shown below are those of 1,1,1-trichloroacetone, pentachloroacetone, and 1,3-dichloro-1,1,3,3-tetrafluoroacetone. Match each spectrum with its compound.

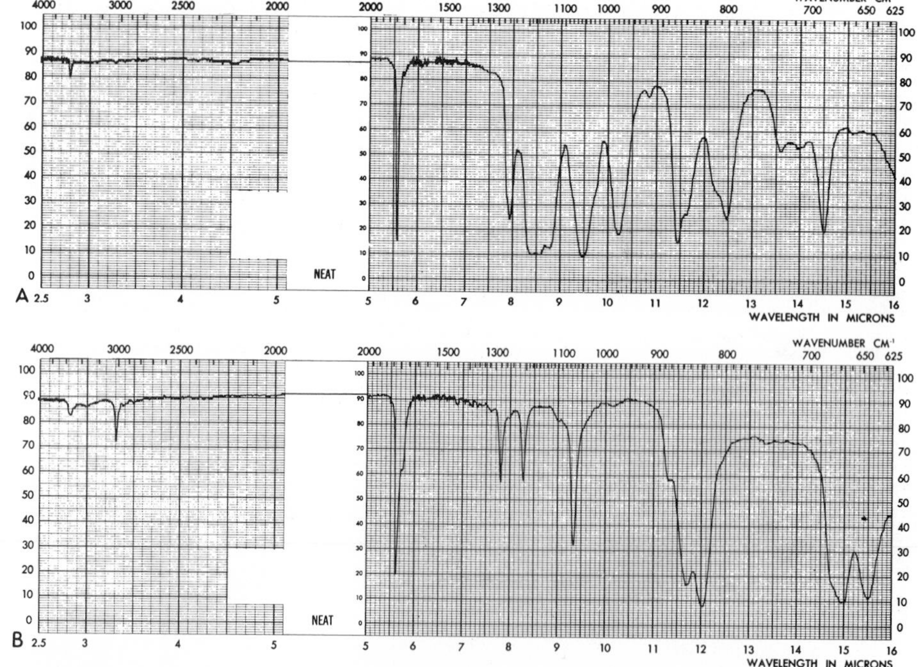

*This problem requires knowledge of ir spectroscopy.

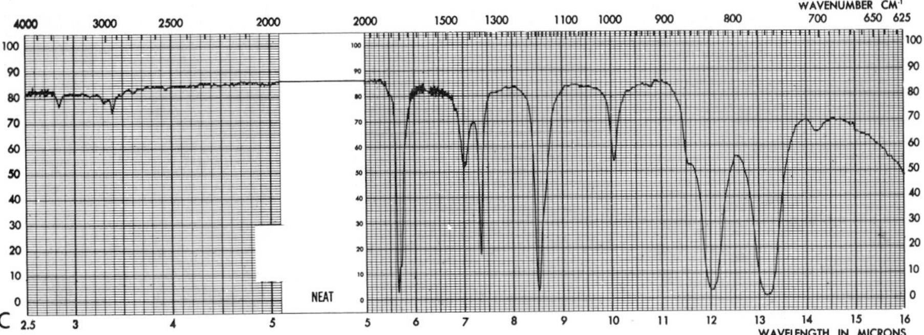

*45. The infrared spectra shown below are those of 5-hexen-2-one, 1-penten-3-one, and 3-pentanone. Match each spectrum with its compound.

A

B

C

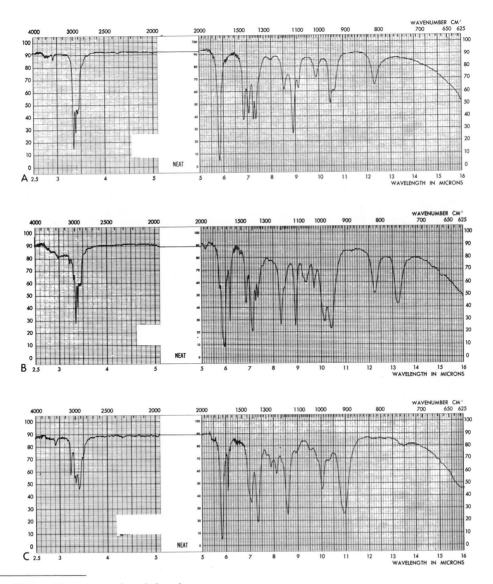

* This problem requires knowledge of ir spectroscopy.

*46. The infrared spectra shown below are those of acrolein, butyraldehyde, chloral, and 10-undecenal. Match each spectrum with its compound.

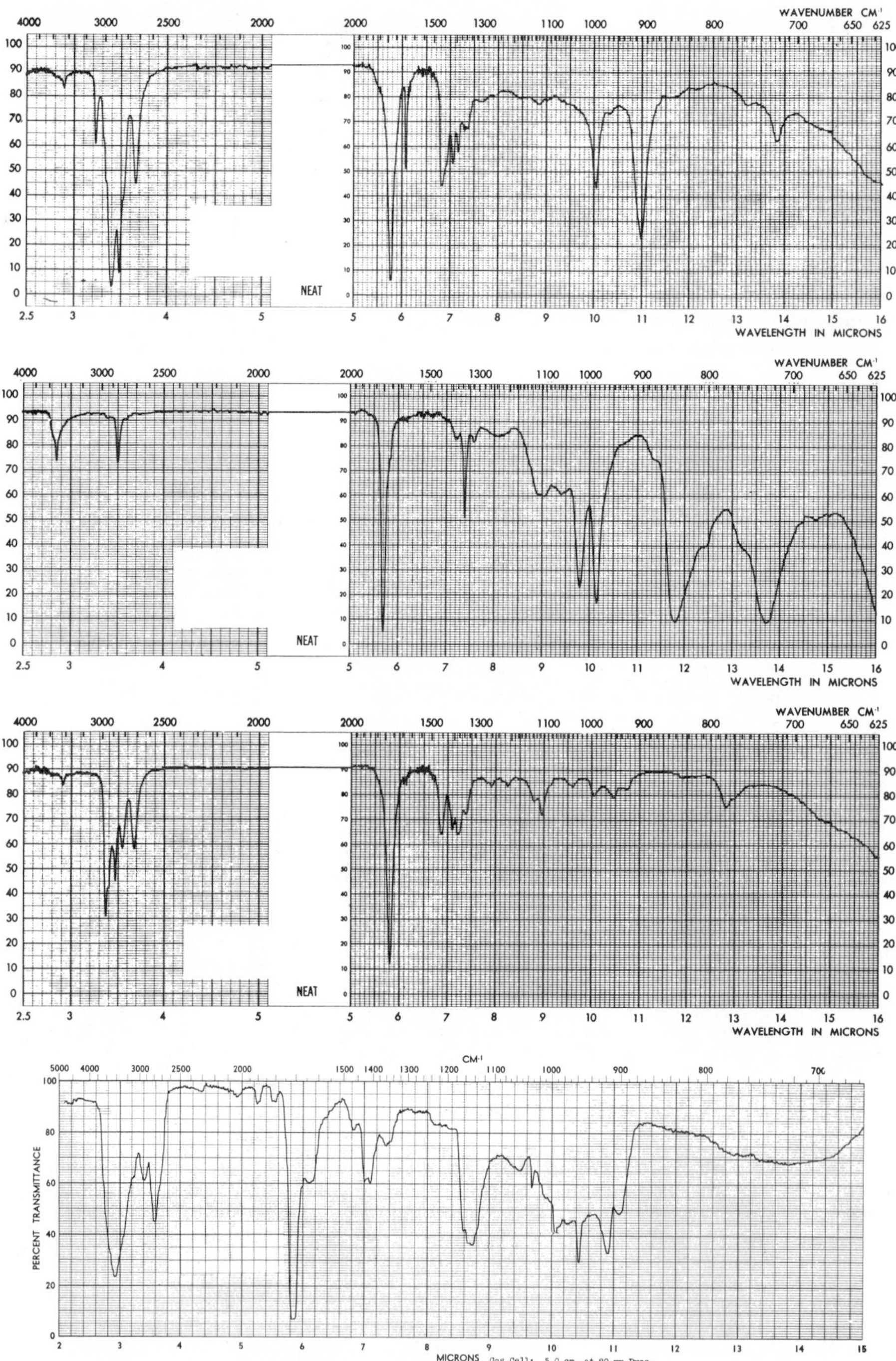

47. Benzaldehyde reacts with bromine water (Br_2/H_2O) to produce benzoic acid. This oxidation is typical of aldehydes. Propose a mechanism for this oxidation.

---

*This problem requires knowledge of ir spectroscopy.

CARBOXYLIC ACIDS AND THEIR DERIVATIVES

18.1 WHERE ARE WE GOING?

The carboxy group, —C(O)OH, is formally composed of a carbonyl group bonded to a hydroxy group. However, since the π bond of the carbonyl group and a lone pair of electrons on the hydroxy group are conjugated, the carboxy group is best treated as a new functionality. Some reactions of carboxylic acids, R—CO_2H, have analogs in ketone and alcohol chemistry; but conjugation within the carboxy group has an effect upon these as well.

Proton donation, *i.e.*, the ability to act as an acid, is the hallmark of the carboxy group. However, the importance of carboxylic acids extends well beyond this simple property because of the diversity of acid derivatives that are known and the multitude of valuable reactions that these derivatives undergo. The common derivatives of carboxylic acids are presented in Figure 18–1.

R—C(=O)—OH carboxylic acid	R—C(=O)—OR ester	R—C(=O)—O—C(=O)—R acid anhydride
R—C(=O)—Cl acid chloride (acyl chloride)	R—C(=O)—NH_2 amide	R—C≡N nitrile
R—C(=O)—O—OH peroxyacid	R—C(=O)—O—O—C(=O)—R peroxyanhydride (acyl peroxide)	RR′C=C=O ketene
cyclic —O—C=O lactone	cyclic —NR—C=O lactam	R—C(=O)—N(R′)—C(=O)—R″ or cyclic C(=O)—N—R′—C(=O) imide (acyclic or cyclic)
RO—C(=O)—OR carbonate	R_2N—C(=O)—NR_2 a urea	R_2N—C(=O)—OR urethane

Figure 18-1 Carboxylic acid derivatives. The circle is used to symbolize a cyclic structure. The nitrile also is called a *cyanide*, proper usage being described in section 18.2. Although ketenes may be viewed as intramolecular anhydrides, the term "anhydride" is restricted to compounds possessing the —C(=O)—O—C(=O)— group.

This chapter begins with a discussion of the nomenclature of carboxylic acids and their more common derivatives. After an examination of factors influencing the acidity of carboxylic acids, we shall turn our attention toward several means of synthesizing these compounds. In Section 18.5 we will survey the more important reactions of simple carboxylic acids, while in Section 18.6 we will examine some of the consequences of having two carboxy groups in one molecule. Because of their importance in biological chemistry, we will look in detail at some of the chemistry of esters (RCO_2R) and amides ($RCONH_2$) in Sections 18.8 and 18.9. After a brief look at esters from a more biological slant, the chapter ends with a presentation of some of the most important synthetic polymers—polyesters and polyamides.

18.2 NOMENCLATURE

Carboxylic Acids. Carboxylic acids are named in the IUPAC system by selecting the longest chain containing the carboxy group, dropping the terminal *e* from the name, and adding *oic acid.* The carboxy carbon is numbered **1** in this system.

$CH_3CH_2CO_2H$ — propanoic acid

$CH_3C(CH_3)_2CO_2H$ — dimethylpropanoic acid

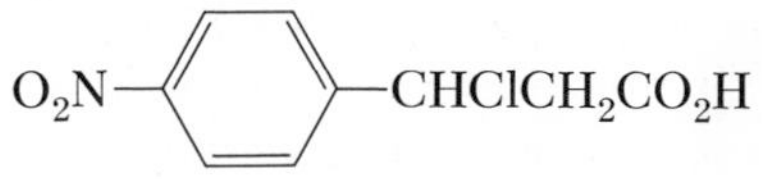

$O_2N{-}C_6H_4{-}CHClCH_2CO_2H$ — 3-chloro-3-(*p*-nitrophenyl)-propanoic acid

Salts of acids are named by stating the cation and then the name of the acid with the *ic acid* altered to *ate.*

$NH_4^{\oplus}\ CH_3CO_2^{\ominus}$ — ammonium ethanoate

$Ca^{(+2)}\ (CH_3CH_2CO_2^{\ominus})_2$ — calcium propanoate (a common preservative in bread)

Carboxylic acids were among the earliest known organic compounds and, consequently, have frequently used common names. These common names usually reflect a prime natural source of the acid or of some derivative. Table 18–1 includes the common names of some important acids. At least the first twelve entries should be committed to memory.

When substituents are located on a carboxylic acid identified by a common name, their positions are designated by Greek letters rather than by numbers. One never mixes numbers with common names *or* Greek letters with IUPAC names.

common: $\delta\ \ \gamma\ \ \beta\ \ \alpha$

$C{-}C{-}C{-}C{-}CO_2H$ — substituent designation

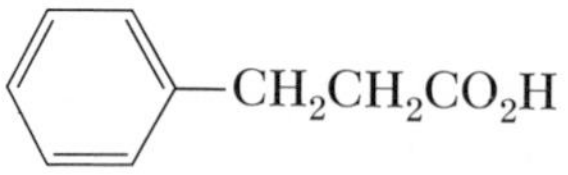

IUPAC: 5 4 3 2 1

$C_6H_5{-}CH_2CH_2CO_2H$ — *β*-phenylpropionic acid

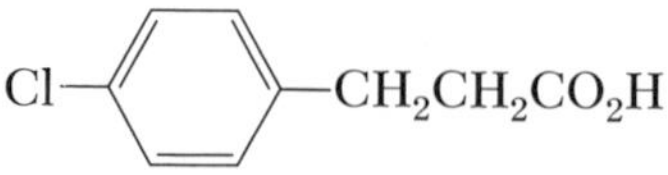

$Cl{-}C_6H_4{-}CH_2CH_2CO_2H$ — *β*-(*p*-chlorophenyl)propionic acid

Aromatic carboxylic acids usually are given names based upon the aromatic system bearing the carboxy group.

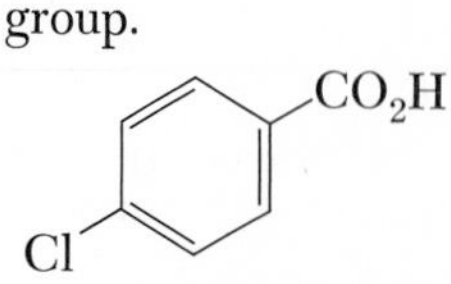

p-chlorobenzoic acid

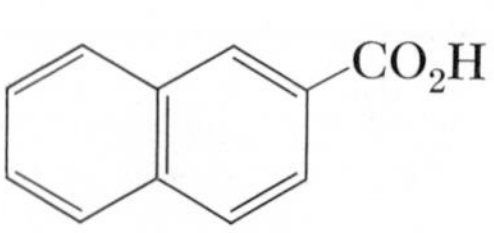

β-naphthoic acid

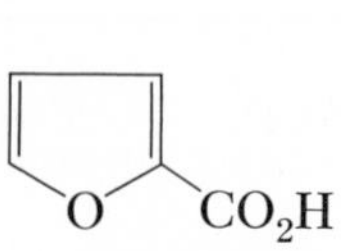

α-furoic acid

TABLE 18-1 Common Names of Normal Carboxylic Acids[a]

No. of Carbon Atoms	Name of Acid	Derivation of Name	Boiling Point	Melting Point
1	Formic	L. *formica* ant	100.7	8.4
2	Acetic	L. *acetum* vinegar	118.2	16.6
3	Propionic	Gr. *proto* first, *pion* fat	141.4	−20.8
4	Butyric	L. *butyrum* butter	164.1	−5.5
5	Valeric	valerian root (L. *valere* to be strong)	186.4	−34.5
6	Caproic	L. *caper* goat	205.4	−3.9
7	Enanthic	Gr. *oenanthe* vine blossom	223.0	−7.5
8	Caprylic	L. *caper* goat	239.3	16.3
9	Pelargonic	pelargonium	253.0	12.0
10	Capric	L. *caper* goat	268.7	31.3
11	Undecanoic		280	28.5
12	Lauric	laurel		43.2
13	Tridecanoic			41.6
14	Myristic	*Myristica fragrans* (nutmeg)		54.4
15	Pentadecanoic			52.3
16	Palmitic	palm oil		62.8
17	Margaric	Gr. *margaron* pearl		61.2
18	Stearic	Gr. *stear* tallow		69.6
19	Nonadecanoic			68.7
20	Arachidic	*Arachis hypogaea* (peanut)		75.4
21	Heneicosanoic			74.3
22	Behenic	behen oil		79.9
23	Tricosanoic			79.1
24	Tetracosanoic			84.2
25	Pentacosanoic			83.5
26	Cerotic	L. *cera* wax		87.7

[a] From Noller, C.R., *Chemistry of Organic Compounds*, 3rd edition. Philadelphia, W.B. Saunders Company, 1965.

The peroxy acids, compounds of the type RC(O)—OOH, are named similarly.

$CH_3C(=O)\text{—O—O—H}$ peroxyacetic acid

m-chloroperoxybenzoic acid (Cl and CO_3H on benzene ring)

On occasion one needs to refer to various segments of a carboxy group; appropriate terminology includes the following:

R—C(=O)—O— : acyl group (boxed R—C=O); acyloxy group (boxed R—C(=O)—O); acyl oxygen (boxed carbonyl O); alkyl oxygen (ether oxygen) (boxed —O—)

1. Provide as many acceptable names as possible for the following.
   (a) $CH_3CH_2CHCH_3CO_2H$
   (b) $CH_3CHCH_3CH_2CH_2CO_2H$
   (c) $CH_3CH(CO_2H)CH_2CH_3$
   (d) $BrCH_2CH_2CHCH_3CO_2H$
   (e) $BrCH_2CH_2CH(CH_2CH_2CH_3)CH_2CO_2H$
   (f) $p\text{-}O_2NC_6H_4CO_2H$
   (g) $CH_3CHCH_3CH(p\text{-}ClC_6H_4)CH_2CO_2H$

Derivatives. The names of acid derivatives are usually simple extensions of the name of the acid. Thus **acyl halides** are named by dropping the ending *ic acid* and replacing it with *yl halide.* **Acid anhydrides** are named by dropping the ending *acid* and adding *anhydride.* **Amides** are named in the IUPAC system by dropping *oic acid* and adding *amide.* In the common scheme amides are named by dropping the ending *ic acid* and adding *amide.* Finally, **esters** are named as derivatives of the acid portion of the molecule, not as derivatives of the alcohol portion. The first part of the name reflects the alkyl or aryl group of the alcohol, while the second portion reflects the acid. In naming the acid part of an ester, the ending *ic acid* is replaced by *ate.* We might note that the suffix *ate,* therefore, is used to indicate both esters and salts.

In the examples that follow, the common name of a compound is placed above its IUPAC name.

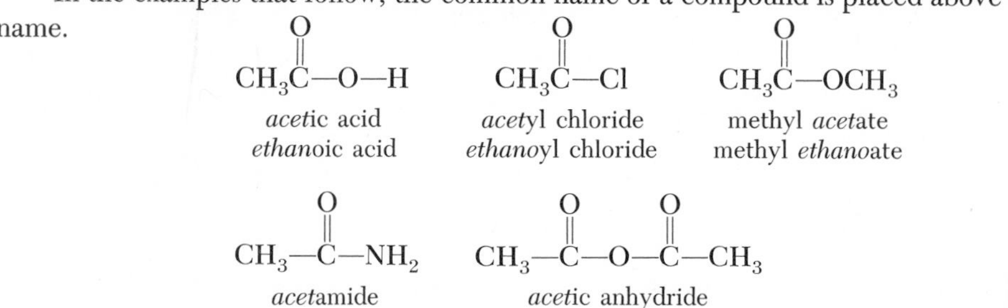

*acet*ic acid / *ethan*oic acid — *acet*yl chloride / *ethano*yl chloride — methyl *acet*ate / methyl *ethano*ate

*acet*amide / *ethan*amide — *acet*ic anhydride / *ethan*oic anhydride

Another group of compounds that can be considered as derivatives of acids are the *nitriles* or *cyanides,* R—CN. Their nomenclature presents an interesting twist in the numbering scheme. If a compound is named as a *nitrile* then, being named as a carboxylic acid derivative, C1 is the carbon of the cyano group, —CN. The name of the nitrile then reflects the total number of carbons in the backbone, including C1. Naming the same compound as a *cyanide* means that the carbon of the cyano group is not being considered as a part of the backbone but, instead, as part of the substituent.

CH_3CH_2CN

propanenitrile (IUPAC)
propionitrile (common)
cyanoethane (IUPAC)
ethyl cyanide (common)

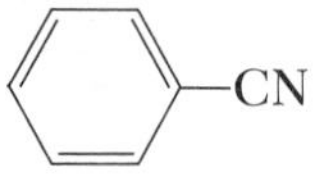

C_6H_5—CN

cyanobenzene (IUPAC)
phenyl cyanide (common)

The naming of ureas, carbonates, and urethanes is illustrated below.

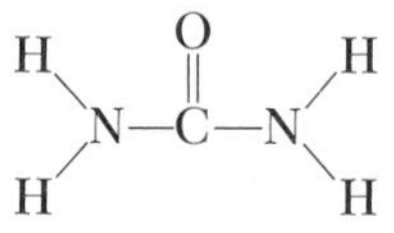

H_2N—C(=O)—NH_2
urea

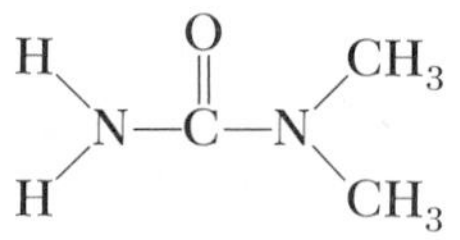

H_2N—C(=O)—$N(CH_3)_2$
N,N-dimethylurea

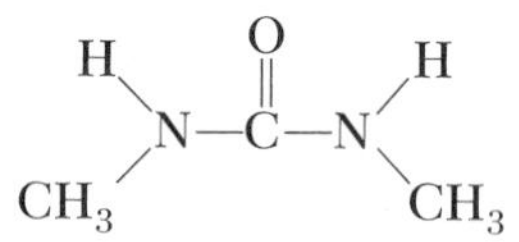

CH_3NH—C(=O)—$NHCH_3$
N,N′-dimethylurea[a]

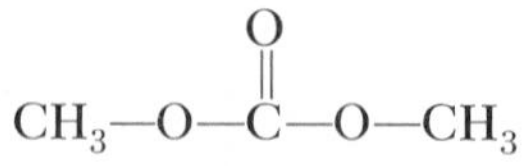

CH_3—O—C(=O)—O—CH_3
dimethyl carbonate

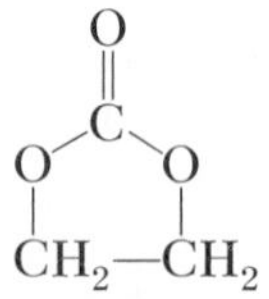

O=C(—O—CH_2—CH_2—O—) (cyclic)
ethylene carbonate

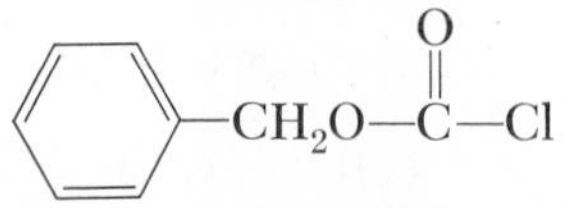

C_6H_5—CH_2O—C(=O)—Cl
benzylchlorocarbonate

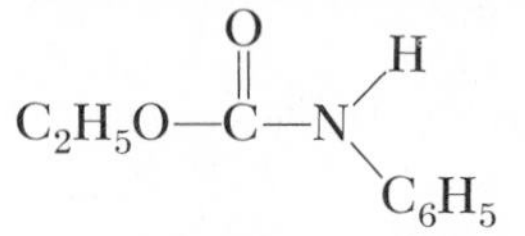

C_2H_5O—C(=O)—NH(C_6H_5)
ethyl *N*-phenylurethane

[a] The presence of a substituent on each nitrogen requires the "*N,N′*" designation.

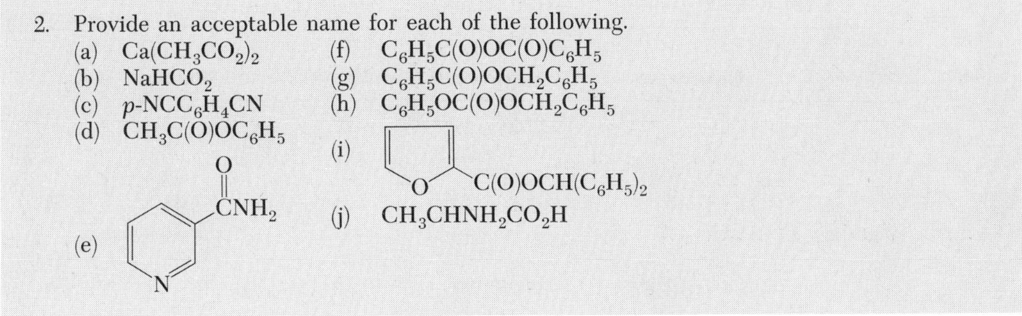

2. Provide an acceptable name for each of the following.

(a) $Ca(CH_3CO_2)_2$
(b) $NaHCO_2$
(c) p-NCC_6H_4CN
(d) $CH_3C(O)OC_6H_5$
(e) [pyridine ring with $\overset{O}{\|}CNH_2$ substituent]
(f) $C_6H_5C(O)OC(O)C_6H_5$
(g) $C_6H_5C(O)OCH_2C_6H_5$
(h) $C_6H_5OC(O)OCH_2C_6H_5$
(i) [furan ring with $C(O)OCH(C_6H_5)_2$ substituent]
(j) $CH_3CHNH_2CO_2H$

18.3 HYDROGEN BONDING AND ACIDITY

Carboxylic acids share with enols and alcohols the ability to act as hydrogen bond donors. If the acceptor is a sufficiently strong base, hydrogen bonding simply precedes complete proton transfer to the base. When this happens, the hydrogen bond donor is considered to be an "acid." What is interesting, of course, is that the classification of the compound either as a "hydrogen bond donor" or as an "acid" is largely in the hands of the "hydrogen bond acceptor" or "base," as the case may be. The stronger the base, the more likely it is that a given compound will behave as an acid toward it.

$$R{-}C(=\ddot{O})\ddot{O}{-}H \;+\; :B \;\rightarrow\; \overset{\oplus}{B}{-}H \;\; R{-}C(=\ddot{O})\ddot{O}:^{\ominus}$$

acid base salt

Even when proton *transfer* is not achieved, hydrogen bonding strongly influences the properties of carboxylic acids. For example, hydrogen bonding causes both 1-pentanol and pentanoic acid to have about the same solubility in water (~3 g/100 g of water) and to be much more soluble in water than are either pentane or pentanal.

$$R{-}C\begin{matrix}\diagup O\sim H \diagdown \\ \diagdown O{-}H \diagup\end{matrix} O{-}H$$

hydrogen bonding between water and an acid

Intermolecular hydrogen bonding between carboxylic acid molecules is sufficiently strong that these acids exist to a considerable degree as dimers even in the vapor state. This intermolecular association also makes carboxylic acids much less volatile than the corresponding alcohols and aldehydes. A comparison of the boiling points of related acids, alcohols, and aldehydes is presented in Table 18–2.

$$R{-}C\begin{matrix}\diagup O\sim H{-}O \diagdown \\ \diagdown O{-}H\sim O \diagup\end{matrix}C{-}R$$

hydrogen bonding in an acid dimer

TABLE 18-2 Boiling Points of Aldehydes, Alcohols, and Acids

Type	Compound	Functionality	BP (MP)	Molecular Weight
1 carbon	CH_2O	aldehyde	−21	30
	CH_3OH	1° alcohol	65	32
	HCO_2H	acid	101	46
2 carbon	CH_3CHO	aldehyde	21	44
	CH_3CH_2OH	1° alcohol	79	46
	CH_3CO_2H	acid	118	60
3 carbon	CH_3CH_2CHO	aldehyde	50	58
	$CH_3CH_2CH_2OH$	1° alcohol	97	60
	$CH_3CHOHCH_3$	2° alcohol	83	60
	$CH_3CH_2CO_2H$	acid	141	74
aromatic	C_6H_5CHO	aldehyde	185	106
	$C_6H_5CH_2OH$	1° alcohol	205	108
	$C_6H_5CO_2H$	acid	249 (123)	122

Carboxylic acids are more acidic than are alcohols because the carboxylate anion is stabilized by delocalization of the negative charge, while the alcoholate anion is not (Figure 18–2).

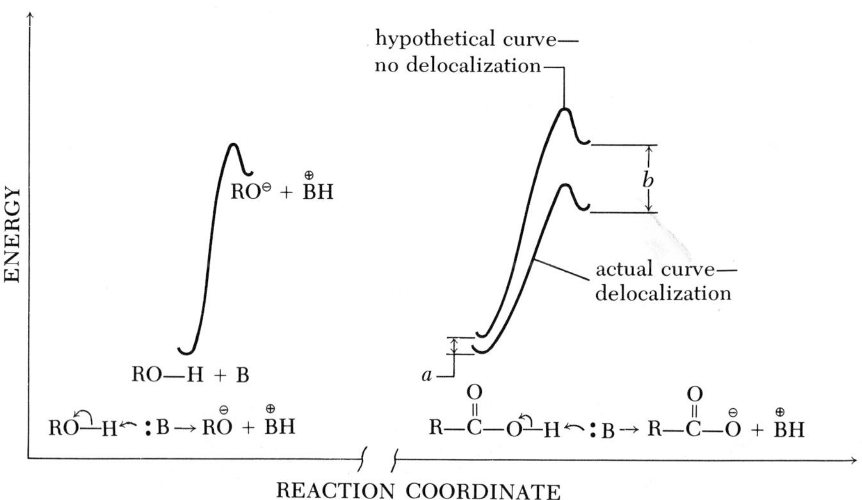

Figure 18-2 The effect of delocalization upon acidity. The extent of delocalization is $RCO_2^{\ominus} \gg RCO_2H \gg ROH$, $RO^{\ominus}$. The energy *a* is the stabilization due to delocalization in RCO_2H; *b* is the stabilization due to delocalization in $RCO_2^{\ominus}$.

What is the evidence that π electron delocalization occurs in the carboxylate anion? If delocalization did *not* occur, one would expect the carboxylate anion to have two different types of carbon-oxygen bonds. However, the X-ray analysis of carboxylate salts shows that both carbon-oxygen bonds have the same length, an observation consistent only with a resonance-stabilized anion. While resonance contributing structures also can be constructed for the carboxy group itself, resonance here is unimportant because delocalization produces charge separation and, moreover, is not between two identical contributors.

R—C(=O)—O—H ← different bond lengths

carboxylic acid

R—C(O)(O)⊖ ← identical bond lengths

carboxylate anion

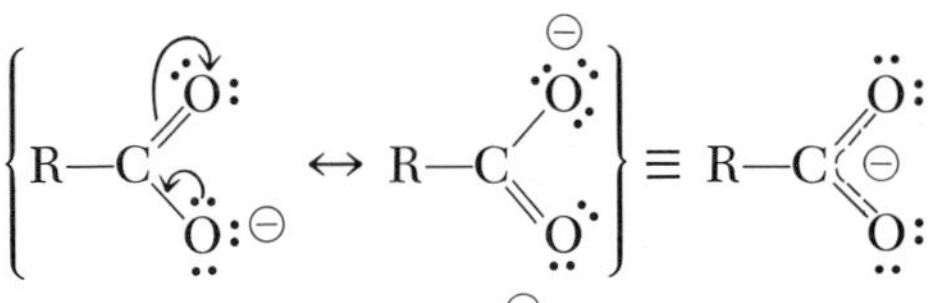

delocalization in the carboxylate anion

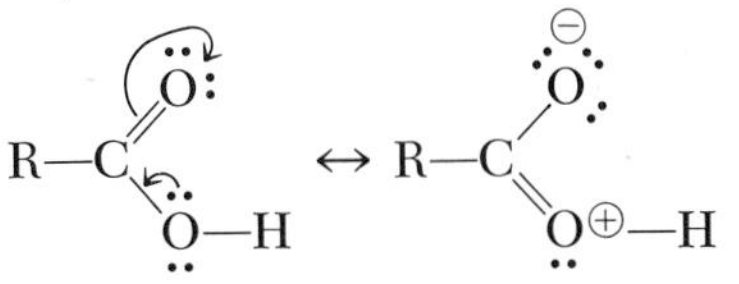

this delocalization is unimportant

A quantitative statement about the acidity of a compound can be made in terms of its **acidity constant** (also called **dissociation constant**), K_a, or the negative logarithmn of K_a, symbolized by pK_a. The acidity constants of most carboxylic acids are about 10^{-5} (*e.g.*, K_a of acetic acid is 1.8×10^{-5} at 20°), reflecting a relatively high concentration of un-ionized acid, $[RCO_2H]$. This is the reason that most carboxylic acids are classified as *weak acids.*

$$RCO_2H \rightleftharpoons RCO_2^{\ominus} + H^{\oplus}$$

$$K_a = \frac{[H^{\oplus}][RCO_2^{\ominus}]}{[RCO_2H]}$$

$$pK_a = -\log K_a$$

The extent of protonation of the carboxylate anion by water, a reaction that produces a carboxylic acid, is an indirect measure of a compound's acidity. This is quantitated in terms of the pK_b of the carboxylate anion.

$$RCO_2^{\ominus} + H_2O \rightleftharpoons RCO_2H + OH^{\ominus}$$

$$K_b = \frac{[RCO_2H][OH^{\ominus}]}{[RCO_2^{\ominus}]}$$

$$pK_b = -\log K_b$$

In dilute, aqueous solutions K_a and K_b are related by

$$K_w = K_a K_b$$

where $$K_w = [H^{\oplus}][OH^{\ominus}] \approx 10^{-14}$$

Therefore, $$pK_a + pK_b \approx 14$$

Although most simple carboxylic acids have pK_a's of about 5, any substituent that stabilizes a carboxylate anion relative to its precursor acid will enhance that acid's acidity (the value of the acid's pK_a will decrease). One of the simplest means of stabilizing the carboxylate anion is the presence of halogens, or other electronegative substituents, that help to disperse the negative charge. The data in Table 18–3 indicate the effect of halogenation upon the acidity of acetic acid. These data also indicate that when the electronegative substituent is removed from the vicinity of the carboxy group, the acidity returns to its normal value. This is consistent with the rapid decrease of the inductive effect of one group upon another as the number of intervening bonds increases.

Resonance effects do not have a great influence upon the acidity of aromatic carboxylic acids (Table 18–3). This reflects the inability to draw contributing structures that place the negative charge of the carboxylate anion onto the aromatic ring. Any attempt to do this cleaves the bond between the ring and the carboxylate group; moreover, the resulting phenyl anion is incapable of charge delocalization around the ring. (Why?) The most that a ring substituent

TABLE 18-3 Effect Of Substituents Upon Acidity

Compound	pK_a[a]
CH_3CO_2H	4.8
$ClCH_2CO_2H$	2.9
Cl_2CHCO_2H	1.3
Cl_3CCO_2H	0.7
FCH_2CO_2H	2.6
F_3CCO_2H	0.2
$Cl—CH_2CH_2CH_2CO_2H$	4.6
$CH_3CHClCH_2CO_2H$	4.0
$CH_3CH_2CHClCO_2H$	2.9
benzoic acid	4.2
p-methoxybenzoic acid[b]	4.5
p-chlorobenzoic acid	4.0
p-bromobenzoic acid	4.0
p-nitrobenzoic acid	3.4

[a] Remember, the smaller the value of pK_a, the stronger the acid.

[b] Benzoic acids bearing substituents *ortho* to the carboxy group usually show enhanced acidities for reasons that are complex and probably not completely understood. For example, the pK_a of *o*-toluic acid, *o*-$CH_3C_6H_4CO_2H$, is 3.9 while that of *p*-toluic acid, *p*-$CH_3C_6H_4CO_2H$, is 4.4.

can do to change the acidity of benzoic acid, $C_6H_5CO_2H$, is to alter the inductive effect of the ring or the extent of solvation of the acid and the anion. It is not surprising, therefore, that aliphatic acids, which can have substituents bonded directly to Cα, are more susceptible to inductive effects than are aromatic acids.

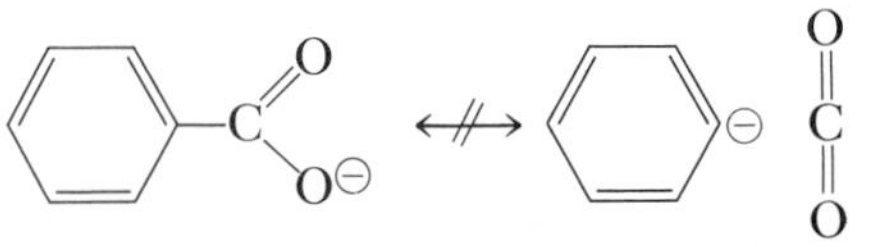

$C_6H_5-C(=O)O^{\ominus}$ ↮ $C_6H_5^{\ominus}$ O=C=O does *not* occur

If the carboxylate anion can be stabilized by intramolecular hydrogen bonding, one expects the acidity of the parent acid to be enhanced. This effect can be seen in a comparison of the acidities of salicylic acid (*o*-hydroxybenzoic acid), in which hydrogen bonding is possible, and *o*-methoxybenzoic acid, in which it is impossible.

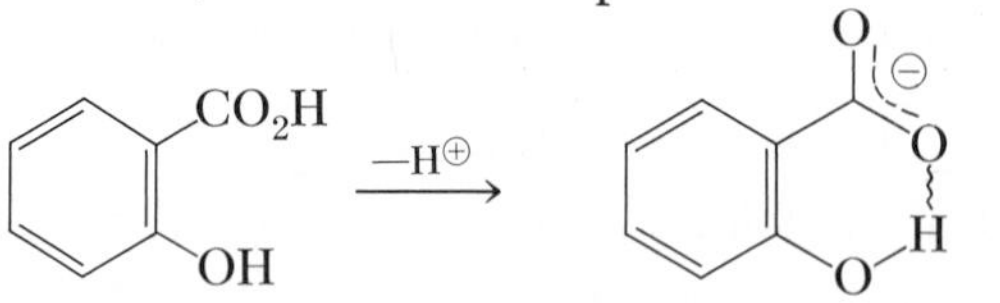

salicylic acid (CO_2H, OH) $\xrightarrow{-H^{\oplus}}$ salicylate anion — hydrogen bond anion stabilization

salicylic acid
$K_a = 105 \times 10^{-5}$

salicylate anion
(exact geometry unknown)

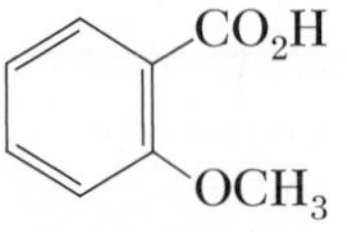

CO_2H, OCH_3

o-methoxybenzoic acid
$K_a = 8.2 \times 10^{-5}$

3. On the basis of data in Table 18–3, determine the direction of the inductive effect of the phenyl group.
4. *o*-Nitrobenzoic acid and *m*-nitrobenzoic acid have K_a's of 670×10^{-5} and 32×10^{-5}, respectively. Calculate their pK_a's. Using the data in Table 18–3, calculate the K_a of *p*-nitrobenzoic acid. Compare the acidities of *p*- and *m*-nitrobenzoic acids. To what conclusion does this lead you?

18.4 SYNTHESIS OF CARBOXYLIC ACIDS

A number of important routes to carboxylic acids have already been presented and are summarized in Figure 18–3.

$$R\text{—}CH_2OH \xrightarrow[OH^{\ominus},\ heat]{MnO_4^{\ominus}} \xrightarrow{H_3O^{\oplus}} RCO_2H \qquad \text{(p. 311)}$$

$$R\text{—}CHO \xrightarrow[H^{\oplus},\ heat]{Cr_2O_7^{\ominus 2}} RCO_2H \qquad \text{(p. 308)}$$

$$R\text{—}CH\text{=}CH\text{—}R' \xrightarrow{O_3} \xrightarrow{H_2O_2} RCO_2H + R'CO_2H \qquad \text{(p. 253)}$$

$$R\text{—}CH\text{=}CH\text{—}R' \xrightarrow[OH^{\ominus},\ heat]{MnO_4^{\ominus}} \xrightarrow{H_3O^{\oplus}} RCO_2H + R'CO_2H \qquad \text{(pp. 252, 337)}$$

$$R\text{—}C\equiv C\text{—}R' \xrightarrow{O_3} \xrightarrow{H_2O} RCO_2H + R'CO_2H \qquad \text{(p. 277)}$$

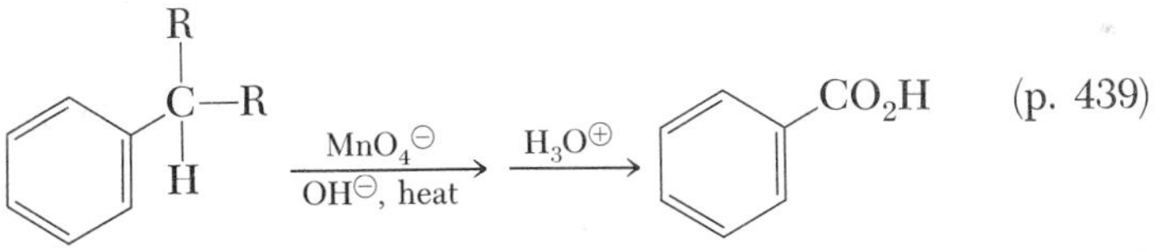

$$C_6H_5\text{—}C(R)(R)H \xrightarrow[OH^{\ominus},\ heat]{MnO_4^{\ominus}} \xrightarrow{H_3O^{\oplus}} C_6H_5\text{—}CO_2H \qquad \text{(p. 439)}$$

Figure 18-3 Selected carboxylic acid syntheses—a review. R and R′ may be alkyl or aryl.

Grignard Synthesis. Organomagnesium compounds react with carbon dioxide to form carboxylate salts. Upon acidification with an acid stronger than the desired carboxylic acids, these salts are converted to carboxylic acids.

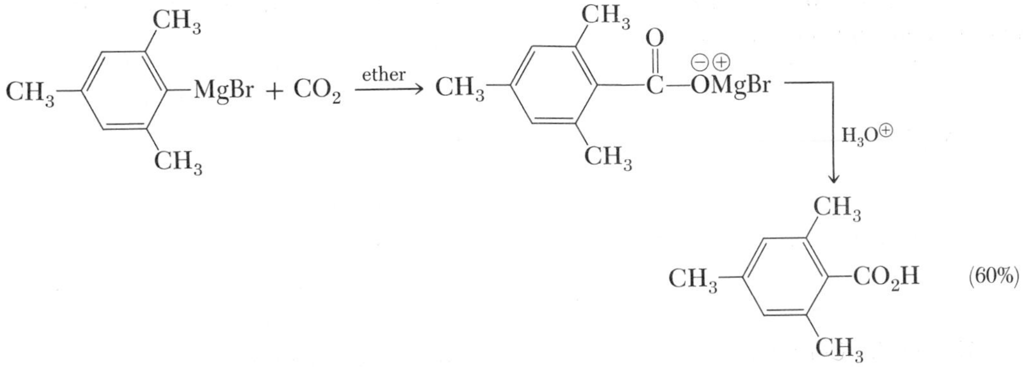

$$2,4,6\text{-}(CH_3)_3C_6H_2\text{—}MgBr + CO_2 \xrightarrow{ether} 2,4,6\text{-}(CH_3)_3C_6H_2\text{—}\overset{O}{\overset{\|}{C}}\text{—}\overset{\ominus\oplus}{O}MgBr \xrightarrow{H_3O^{\oplus}} 2,4,6\text{-}(CH_3)_3C_6H_2\text{—}CO_2H \quad (60\%)$$

mesitoic acid

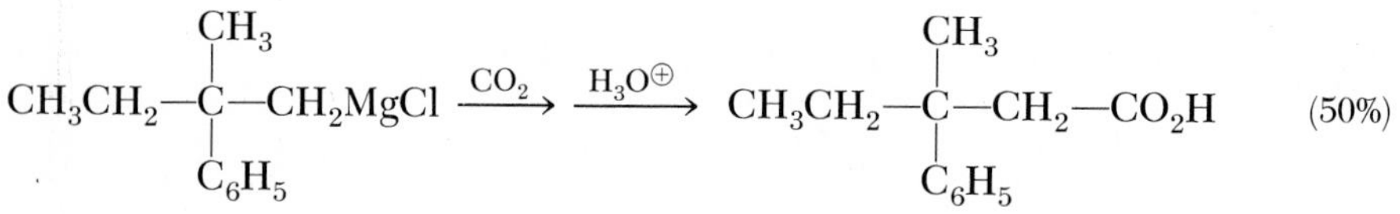

$$CH_3CH_2-\underset{C_6H_5}{\overset{CH_3}{C}}-CH_2MgCl \xrightarrow{CO_2} \xrightarrow{H_3O^{\oplus}} CH_3CH_2-\underset{C_6H_5}{\overset{CH_3}{C}}-CH_2-CO_2H \quad (50\%)$$

3-methyl-3-phenylpentanoic acid

5. Starting with any alkene, and using any other necessary reagents, outline the synthesis of each of the following.
   (a) formic acid
   (b) ethanoic acid
   (c) 2,2-dimethylbutanoic acid
   (d) benzoic acid
   (e) $HO_2C(CH_2)_4CO_2H$ (adipic acid)

Carboxylation of the Aromatic Nucleus. Aromatic carboxylic acids can be prepared in good yields by the direct introduction of the carboxy group *via* a Friedel-Crafts reaction. One useful reagent for this process is *oxalyl chloride*, ClC(O)C(O)Cl.

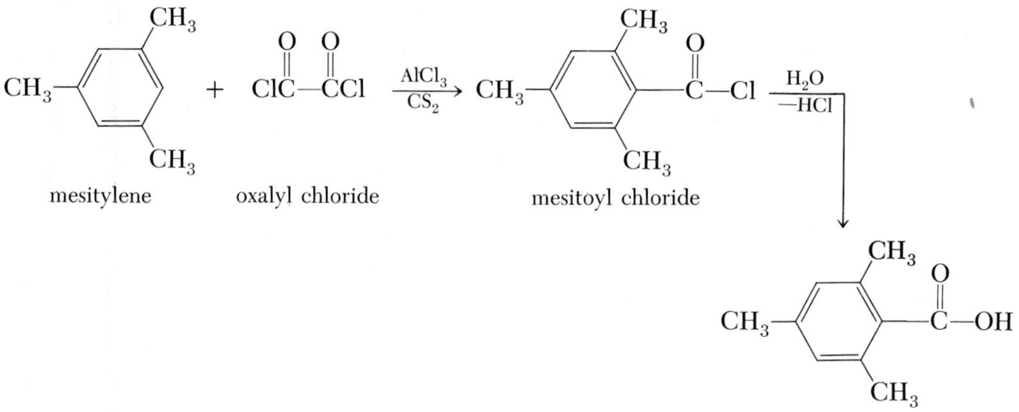

$$\text{mesitylene} + ClC(=O)-C(=O)Cl \xrightarrow[CS_2]{AlCl_3} \text{mesitoyl chloride } (2,4,6\text{-}(CH_3)_3C_6H_2-C(=O)-Cl) \xrightarrow[-HCl]{H_2O} 2,4,6\text{-}(CH_3)_3C_6H_2-C(=O)-OH$$

mesitylene — oxalyl chloride — mesitoyl chloride

6. Suggest a mechanism for the synthesis of mesitoyl chloride from mesitylene, oxalyl chloride, and aluminum chloride.

Nitrile Hydrolysis. Both aliphatic and aromatic nitriles hydrolyze, upon refluxing in aqueous mineral acid, to the corresponding carboxylic acid.

$$C_6H_5CH_2CN + H_2O \xrightarrow[\text{reflux}]{H_2SO_4} C_6H_5CH_2CO_2H \quad (77\%)$$

phenylacetic acid

$$CH_2{=}CH-CH_2-CN + H_2O \xrightarrow[\text{15 min. heat}]{HCl} CH_2{=}CH-CH_2-CO_2H \quad (55\%)$$

vinylacetic acid

The mechanism for this hydrolysis is shown on the following page.

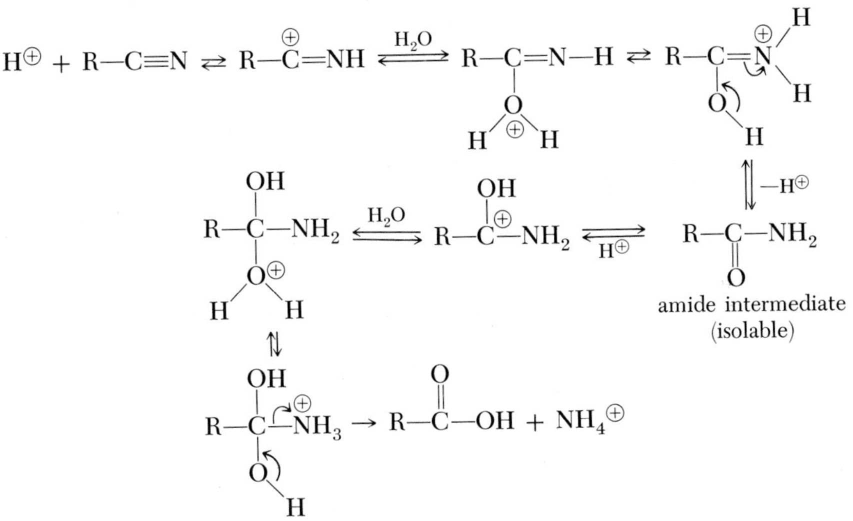

$$H^{\oplus} + R{-}C{\equiv}N \rightleftarrows R{-}\overset{\oplus}{C}{=}NH \xrightleftharpoons{H_2O} R{-}\underset{\underset{H\ \ \oplus\ \ H}{O}}{C}{=}N{-}H \rightleftarrows R{-}\underset{\underset{H}{O}}{C}{=}\overset{\oplus}{N}H_2$$

$$\xrightleftharpoons{-H^{\oplus}} R{-}\underset{\underset{}{O}}{\overset{}{C}}{-}NH_2 \text{ (C=O)}$$

amide intermediate
(isolable)

$$R{-}\underset{\|}{C}(=O){-}NH_2 \xrightleftharpoons{H^{\oplus}} R{-}\overset{OH}{\overset{|}{C}}{}^{\oplus}{-}NH_2 \xrightleftharpoons{H_2O} R{-}\overset{OH}{\underset{\underset{H\ \ \ H}{O^{\oplus}}}{C}}{-}NH_2$$

$$\rightleftarrows R{-}\overset{OH}{\underset{\underset{H}{O}}{C}}{-}\overset{\oplus}{N}H_3 \rightarrow R{-}\overset{O}{\overset{\|}{C}}{-}OH + NH_4^{\oplus}$$

7. Suggest a means of converting 1,4-dibromobutane to 1,6-dibromohexane. (*Hint:* One possible precursor is 1,6-hexanediol.)

While the necessary aliphatic nitriles usually are prepared by S_N2 processes, the aromatic nitriles are prepared by a specific sequence discussed in Chapter 21. However, if an amide were available, it could be dehydrated, using phosphorus pentoxide (P_2O_5) or thionyl chloride ($SOCl_2$), to the corresponding aliphatic or aromatic nitrile. Unfortunately, the conversion *amide → nitrile → carboxylic acid* is not of value in the preparation of carboxylic acids, since the amide is actually an intermediate in the conversion of a nitrile to the corresponding carboxylic acid (see mechanism above).

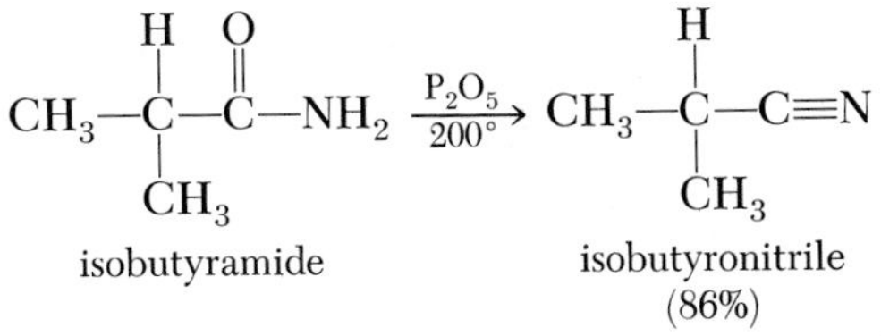

$$CH_3{-}\overset{H}{\underset{CH_3}{C}}{-}\overset{O}{\overset{\|}{C}}{-}NH_2 \xrightarrow[200^\circ]{P_2O_5} CH_3{-}\overset{H}{\underset{CH_3}{C}}{-}C{\equiv}N$$

isobutyramide → isobutyronitrile (86%)

In a related reaction, *aldoximes* (compounds of the type RCH=N—OH) are dehydrated to cyanides by acetic anhydride. The mechanism of this dehydration is shown below. Steps **1** and **2** of this mechanism show how anhydrides and other compounds of the type RC(O)—L (L = Cl, Br, OC(O)R, OR or other leaving group) can act as acylating agents, *i.e.*, reagents that introduce the RC(O)— group into a molecule. The most common acylation process is *acetylation*, introduction of the $CH_3C(O)$— or acetyl group.

net reaction:

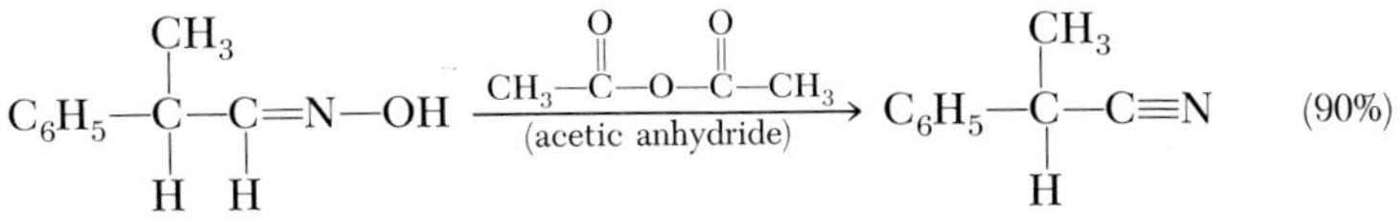

$$C_6H_5{-}\overset{CH_3}{\underset{H}{C}}{-}\underset{H}{C}{=}N{-}OH \xrightarrow[\text{(acetic anhydride)}]{CH_3-\overset{O}{\overset{\|}{C}}-O-\overset{O}{\overset{\|}{C}}-CH_3} C_6H_5{-}\overset{CH_3}{\underset{H}{C}}{-}C{\equiv}N \quad (90\%)$$

an aldoxime dehydration

mechanism:

$$\underbrace{C_6H_5-\underset{H}{\overset{CH_3}{C}}-\underset{H}{C}}_{R}=N-OH + (CH_3CO)_2O \xrightarrow{1} \underset{H}{\overset{R}{\;}}C=N-\underset{H}{\overset{\oplus}{O}}-\overset{CH_3}{C}(O^{\ominus})-O-C(=O)CH_3$$

$$\downarrow 2$$

$$\underset{H}{\overset{R}{\;}}C=N-\underset{H}{\overset{\oplus}{O}}-\overset{CH_3}{C}=O \;\; {}^{\ominus}O-\underset{O}{\overset{\|}{C}}-CH_3$$

$$\underset{H}{\overset{R}{\;}}C=N-O-\overset{O}{\overset{\|}{C}}-CH_3 \xleftarrow{-HOAc}$$

$$\downarrow -HOAc$$

$$R-C\equiv N$$

Chain-Lengthening of a Carboxylic Acid. The **Arndt-Eistert synthesis** provides a convenient means of converting a carboxylic acid into its next higher homolog. The overall sequence and a step-by-step analysis follow.

$$\underset{1}{R-CO_2H} \xrightarrow{A} \underset{2}{R-COCl} \xrightarrow{B} \underset{3}{RC(O)-CHN_2} \xrightarrow{C} \underset{4}{RCH_2CO_2H}$$

Step **A:** Carboxylic acids are readily converted to their acid chlorides by reaction with thionyl chloride in ether solution.

Step **B:** Conversion of the acid chloride (**2**) to a *diazoketone* (**3**) involves the acylation of diazomethane, CH_2N_2. (The preparation of this extremely reactive small molecule is described in Chapter 23.)

$$^{\ominus}{:}CH_2-\overset{\oplus}{N}\equiv N{:} \leftrightarrow CH_2=\overset{\oplus}{N}=\ddot{N}{:}^{\ominus}$$ resonance contributors of diazomethane

$$RC(O)Cl + CH_2N_2 \rightarrow \underset{\text{diazoketone}}{RC(O)-CHN_2} + HCl$$

The diazoketone is much more stable than is diazomethane, or other diazoalkanes, because of its ability to delocalize the negative charge into the carbonyl group

$$R-\overset{O}{\overset{\|}{C}}-\underset{H}{\overset{\ominus}{\ddot{C}}}-\overset{\oplus}{N}\equiv N \leftrightarrow R-\overset{O}{\overset{\|}{C}}-\underset{H}{C}=\overset{\oplus}{N}=\ddot{N}{:}^{\ominus} \leftrightarrow R-\overset{^{\ominus}O}{\overset{|}{C}}=\underset{H}{C}-\overset{\oplus}{N}\equiv N$$ resonance contributors of diazoketones

Step **C:** Silver ion is used to catalyze the decomposition of the diazoketone to a carbene or some equivalent; molecular nitrogen also is produced. The R group migrates to the electron-deficient carbon to form a ketene. This ketene reacts immediately with water to form the desired acid. (There is some question about whether a free carbene actually is produced or whether migration of R accompanies the loss of nitrogen. If the latter is correct, then steps *a* and *b* (opposite page) occur simultaneously, a process represented in *c*.)

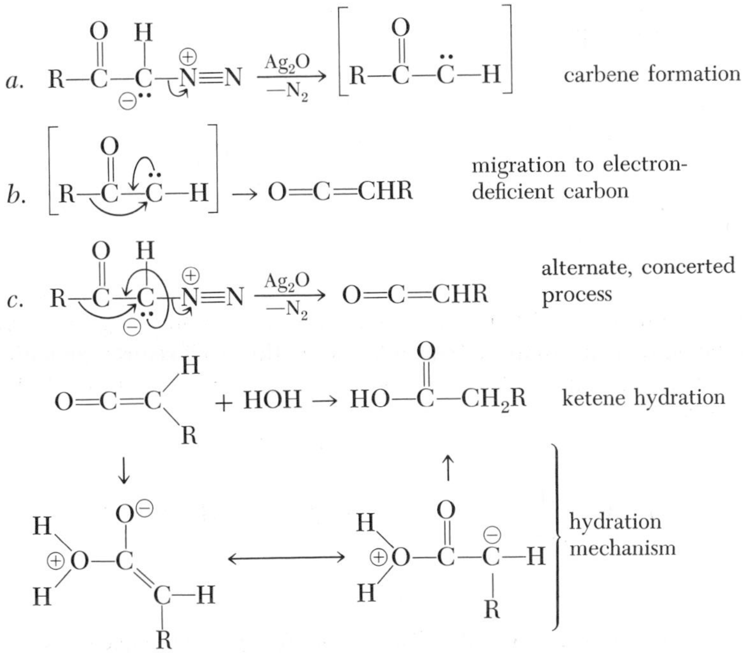

$$a.\quad R-\overset{O}{\overset{\|}{C}}-\overset{H}{\overset{|}{\underset{\ominus}{\ddot{C}}}}-\overset{\oplus}{N}\equiv N \xrightarrow[-N_2]{Ag_2O} \left[R-\overset{O}{\overset{\|}{C}}-\ddot{C}-H\right]$$ carbene formation

$$b.\quad \left[R-\overset{O}{\overset{\|}{C}}-\ddot{C}-H\right] \rightarrow O{=}C{=}CHR$$ migration to electron-deficient carbon

$$c.\quad R-\overset{O}{\overset{\|}{C}}-\overset{H}{\overset{|}{\underset{\ominus}{C}}}-\overset{\oplus}{N}\equiv N \xrightarrow[-N_2]{Ag_2O} O{=}C{=}CHR$$ alternate, concerted process

$$O{=}C{=}C(H)(R) + HOH \rightarrow HO-\overset{O}{\overset{\|}{C}}-CH_2R$$ ketene hydration

$$H_2\overset{\oplus}{O}-C(O^{\ominus}){=}C(H)R \longleftrightarrow H_2\overset{\oplus}{O}-\overset{O}{\overset{\|}{C}}-\overset{\ominus}{C}(R)-H$$ hydration mechanism

If the R group of the acid chloride is bound to the sp^2 carbon, C1, with a chiral carbon, the final product is produced with retention of configuration. This means that R cannot migrate as a free carbanion, since free carbanions are not configurationally stable. *Rearrangement with retention of configuration is always observed in migration of alkyl groups to electron-deficient atoms.*

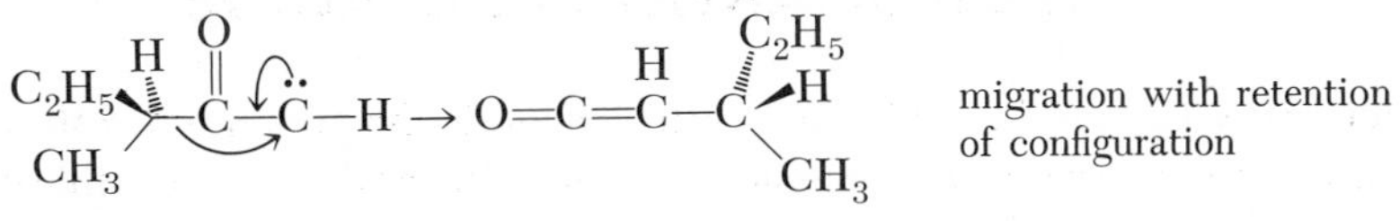

$$(C_2H_5)(CH_3)HC-\overset{O}{\overset{\|}{C}}-\ddot{C}-H \rightarrow O{=}C{=}C(H)-C(C_2H_5)(H)(CH_3)$$ migration with retention of configuration

The entire sequence often can be carried out in one day with overall yields varying from 50% to 80%. This is, therefore, a useful reaction scheme.

Cyclic diazoketones, prepared as shown below, undergo ring contraction as a result of R group migration. The net conversion is a valuable synthesis of cycloalkyl carboxylic acids from cycloalkanones.

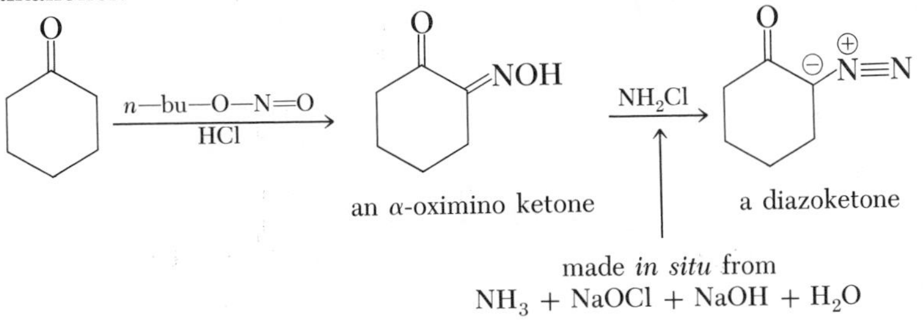

cyclohexanone $\xrightarrow[HCl]{n\text{—bu—O—N=O}}$ 2-(=NOH)cyclohexanone (an α-oximino ketone) $\xrightarrow{NH_2Cl}$ 2-($\ominus$ C—$\overset{\oplus}{N}\equiv N$)cyclohexanone (a diazoketone)

NH_2Cl made *in situ* from $NH_3 + NaOCl + NaOH + H_2O$

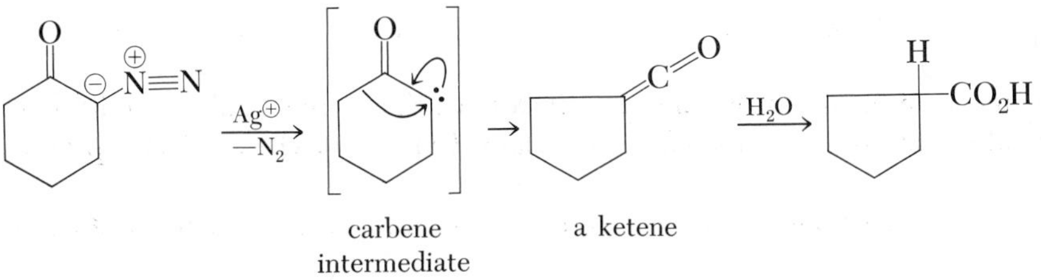

diazoketone $\xrightarrow[-N_2]{Ag^{\oplus}}$ [carbene intermediate] $\rightarrow$ cyclopentyl—C=O (a ketene) $\xrightarrow{H_2O}$ cyclopentyl(H)—CO_2H

carbene intermediate

a ketene

8. Suggest a mechanism for

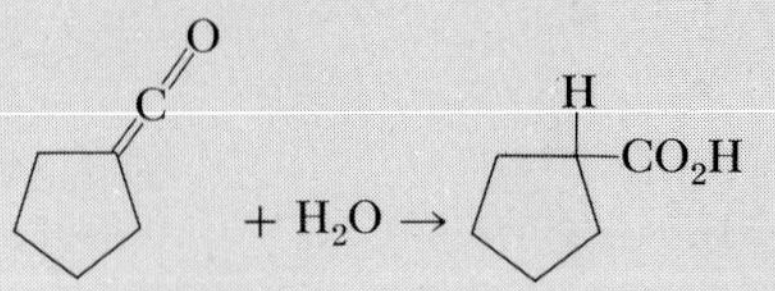

$$\text{(cyclopentylidene ketene, } C_5H_8{=}C{=}O) + H_2O \rightarrow \text{cyclopentane-}CH\text{-}CO_2H$$

The critical step in the Arndt-Eistert synthesis and in the ring contraction, *i.e.*, the conversion of the diazoalkane to the ketene, is known as the **Wolff rearrangement.** If the catalyst (silver ion) is not present, the Wolff rearrangement does not occur. Instead, the diazoketone is hydrolyzed to an α-ketocarbinol.

$$R{-}C(=O){-}\overset{\ominus}{C}H{-}\overset{\oplus}{N}{\equiv}N + H_2O \rightarrow R{-}C(=O){-}CH_2OH + N_2$$

9. What product would you expect if a diazoketone ($RC(O)CHNH_2$) decomposes in methanol in the absence of a catalyst?

18.5 REACTIONS OF ACIDS AT THE CARBOXY GROUP—MONOCARBOXYLIC ACIDS

Esterification. Carboxylic acids react with alcohols in the presence of a strong acid (*e.g.*, sulfuric acid) to produce esters. This **Fischer esterification** produces an equilibrium mixture, and high yields of product are ensured only by using an excess of the less valuable starting material and by removing water as it is formed.[a]

$$RCO_2H + R'OH \overset{H^\oplus}{\rightleftharpoons} R{-}C(O){-}OR' + H_2O \qquad \text{Fischer esterification}$$

$$CH_3CO_2H + C_2H_5OH \overset{H_2SO_4}{\rightleftharpoons} CH_3C(O){-}OC_2H_5 + H_2O$$

ethyl acetate (65%)

$$m\text{-}CH_3C_6H_4CO_2H + CH_3OH \overset{H_2SO_4}{\rightleftharpoons} m\text{-}CH_3C_6H_4C(=O){-}OCH_3$$

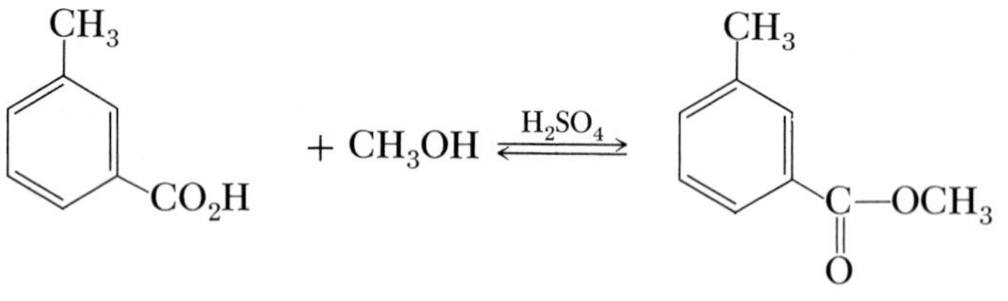

m-methylbenzoic acid — methyl *m*-methylbenzoate

The mechanism of the Fischer esterification follows the familiar addition-elimination pathway (often referred to as a *nucleophilic acyl substitution*). The catalyst, a proton, functions

[a] The water can be removed by adding benzene to the reaction mixture and distilling off the benzene-water azeotrope.

by increasing the susceptibility of the carbonyl group to attack by a nucleophile. (The synthesis of esters and their conversion back to alcohols and acids will be discussed in detail in Section 18.8.)

$$R{-}\overset{O}{\overset{\|}{C}}{-}OH \xrightleftharpoons{H^{\oplus}} R{-}\underset{\oplus}{\overset{OH}{\overset{|}{C}}}{-}OH \xrightleftharpoons{R'OH} R{-}\overset{O{-}H}{\overset{|}{C}}{-}O{-}H\ (\text{with } {-}\overset{\oplus}{O}(H)R') \rightleftarrows R{-}\overset{\overset{\oplus}{O}H_2}{\overset{|}{C}}{-}O{-}H\ (\text{with } {-}OR') \rightleftarrows$$

$$R{-}\overset{O}{\overset{\|}{C}}{-}OR' + H_3O^{\oplus}$$

Because of the inductive effect of the chlorine, which enhances the electron-deficient nature of the carbonyl carbon, and because of the good departing qualities of chlorine, acyl chlorides react vigorously with most nucleophiles even in the absence of a catalyst. Therefore, esters can be made by simply reacting an acyl halide with an alcohol. Esters could, of course, also be prepared by reacting an alcohol with an acid anhydride—another acylating agent.

$$C_6H_5{-}\overset{O}{\overset{\|}{C}}{-}Cl + CH_3OH \rightarrow C_6H_5{-}\overset{O}{\overset{\|}{C}}{-}O{-}CH_3 + HCl$$

benzoyl chloride → methyl benzoate

$$C_6H_5{-}\overset{O}{\overset{\|}{C}}{-}O{-}\overset{O}{\overset{\|}{C}}{-}C_6H_5 + CH_3OH \xrightarrow[\text{heat}]{} C_6H_5{-}\overset{O}{\overset{\|}{C}}{-}O{-}CH_3 + C_6H_5{-}\overset{O}{\overset{\|}{C}}{-}OH$$

benzoic anhydride → methyl benzoate

10. Provide a mechanism for the formation of methyl acetate from the reaction of (a) acetyl chloride and (b) acetic anhydride with methanol. While it is good practice to run *all* syntheses in a hood, which of these reactions *must* be run in a hood? Explain. In these mechanisms, identify the nucleophile and the electrophile.

Methyl esters are often made, in excellent yields, by reacting carboxylic acids with diazomethane. This procedure is of special value when working with small quantities, since the process is nearly quantitative and usually does not produce by-products.

$$\text{Thianthrene}{-}CO_2H \xrightarrow[\text{ether}]{CH_2N_2} \text{Thianthrene}{-}CO_2CH_3 \quad (98\%)$$

2-thianthrenecarboxylic acid → methyl 2-thianthrenecarboxylate

$$\text{Ar}{-}\overset{O}{\overset{\|}{C}}{-}\ddot{O}{-}H\ \ :\overset{\ominus}{C}H_2{-}\overset{\oplus}{N_2} \rightarrow \text{Ar}{-}\overset{O}{\overset{\|}{C}}{-}\ddot{O}{:}^{\ominus}\ \ CH_3{-}\overset{\oplus}{N_2} \rightarrow \text{Ar}{-}\overset{O}{\overset{\|}{C}}{-}OCH_3 + N_2$$

11. Suggest a synthesis of the following, using an inorganic compound as the deuterium source. Use any necessary starting materials.
   (a) $CH_3C(O)OCH_2D$
   (b) $CH_3C(O)OCD_2H$
   (c) CH_2DCO_2H

Acyl Halide Formation. Carboxylic acids react with phosphorus trichloride, PCl_3, thionyl chloride, $SOCl_2$, phosphorus pentachloride, PCl_5, and phosphorus tribromide, PBr_3, to produce the corresponding acyl halide in good yield. However, because of their lower cost, acid chlorides are used far more commonly in addition-elimination reactions than are the corresponding bromides.

$$RCO_2H \xrightarrow{SOCl_2} RCOCl$$
$$RCO_2H \xrightarrow{PCl_5} RCOCl$$
$$RCO_2H \xrightarrow{PCl_3} RCOCl$$
$$RCO_2H \xrightarrow{PBr_3} RCOBr$$

$$m\text{-}O_2NC_6H_4CO_2H \xrightarrow[\text{heat}]{SOCl_2} m\text{-}O_2NC_6H_4COCl \quad (90\%)$$

m-nitrobenzoic acid → *m*-nitrobenzoyl chloride

Amide Formation. We have already noted that amides can be made by the partial hydrolysis of nitriles. However, they are more often prepared by a nucleophilic acyl substitution.

When a carboxylic acid and ammonia are mixed together, the product is an ammonium carboxylate, *not* an amide.

$$NH_3 + RCO_2H \longrightarrow \overset{\oplus}{N}H_4\ RCO_2^{\ominus}$$
$$NH_3 + RCO_2H \not\longrightarrow RC(O){-}NH_2 + H_2O$$

The reason that amides are not formed this way is that transfer of a proton from the acid (RCO_2H) to the base (NH_3) is much faster than is the nucleophilic attack by the ammonia on the carbonyl group of the acid. Once this proton transfer is accomplished, the negative charge on the carboxylate ion hinders attack by any nucleophile. Moreover, if excess nucleophile were present; and if it did attack the carboxylate anion, amide formation would require the loss of hydroxide ion, a poor leaving group.

Preparation of an amide requires a substituent on the acyl carbon that cannot donate a proton to a base and that is a good leaving group. Acid chlorides are the usual substrates, but their use requires two equivalents of ammonia because one equivalent will be consumed by reaction with the product hydrogen chloride to form a salt.

$$RC(O)Cl + 2NH_3 \rightarrow RC(O)NH_2 + NH_4Cl$$

Ammonia is sufficiently nucleophilic that concentrated aqueous ammonia reacts with acid chlorides to form amides that are virtually uncontaminated by carboxylic acid salts.

$$(CH_3)_2CHC(O)Cl + \text{conc. } NH_3(aq.) \rightarrow (CH_3)_2CHCONH_2$$

isobutyramide (80%)

Substituted amides are made similarly, by reacting amines with acid chlorides. Here, again, two equivalents of amine are required in order to compensate for the hydrogen chloride that is produced. The synthesis of such amines is discussed in Chapter 20, while the mechanism of amide formation and hydrolysis is covered in more detail in Section 18.9.

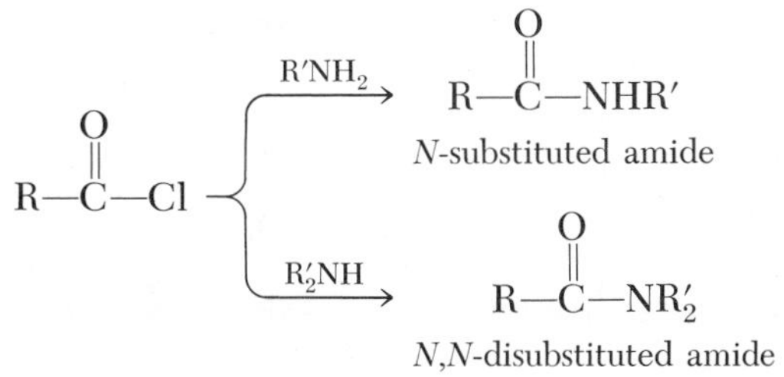

$$R-\overset{O}{\overset{\|}{C}}-Cl \xrightarrow{R'NH_2} R-\overset{O}{\overset{\|}{C}}-NHR'$$

N-substituted amide

$$R-\overset{O}{\overset{\|}{C}}-Cl \xrightarrow{R'_2NH} R-\overset{O}{\overset{\|}{C}}-NR'_2$$

N,N-disubstituted amide

Acid Anhydride Formation. Acid anhydrides can be made by a reaction analogous to ester formation, *i.e.*, nucleophilic "displacement" on an acyl halide. The attacking reagent (nucleophile) is most often the salt of a carboxylic acid. Simple heating of a monocarboxylic acid normally will *not* lead to formation of the corresponding anhydride. One of the reactions illustrated below produces a *mixed* anhydride, that is, an anhydride of two different acids. While compounds like this can be made, they tend to revert to *symmetric* anhydrides or, at best, to mixtures of the mixed anhydride and both appropriate symmetric anhydrides.

$$CH_3-\overset{O}{\overset{\|}{C}}-Cl + CH_3-\overset{O}{\overset{\|}{C}}-O^{\ominus}\ Na^{\oplus} \rightarrow CH_3-\overset{O}{\overset{\|}{C}}-O-\overset{O}{\overset{\|}{C}}-CH_3 + NaCl$$

$$C_6H_5-\overset{O}{\overset{\|}{C}}-Cl + CH_3-\overset{O}{\overset{\|}{C}}-O^{\ominus}\ Na^{\oplus} \rightarrow CH_3-\overset{O}{\overset{\|}{C}}-O-\overset{O}{\overset{\|}{C}}-C_6H_5 + NaCl$$

acetic benzoic anhydride (mixed anhydride)

$$2CH_3-\overset{O}{\overset{\|}{C}}-O-\overset{O}{\overset{\|}{C}}-C_6H_5 \rightleftarrows CH_3-\overset{O}{\overset{\|}{C}}-O-\overset{O}{\overset{\|}{C}}-CH_3 + C_6H_5-\overset{O}{\overset{\|}{C}}-O-\overset{O}{\overset{\|}{C}}-C_6H_5$$

mixed anhydride — symmetric anhydride — symmetric anhydride

$$2CH_3CH_2CO_2H \underset{heat}{\not\longrightarrow} CH_3CH_2-\overset{O}{\overset{\|}{C}}-O-\overset{O}{\overset{\|}{C}}-CH_2CH_3$$

this reaction does not occur

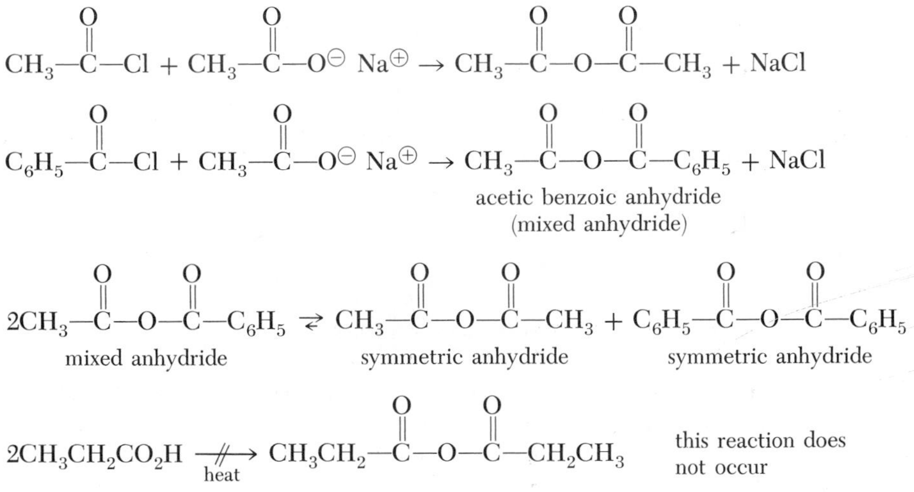

12. Write a mechanism for the reaction of acetic acid with acetyl chloride to produce acetic anhydride. What drawback(s) can be envisioned in this route to anhydrides?

18.6 REACTIONS OF CARBOXYLIC ACIDS—DICARBOXYLIC ACIDS

Structures of some of the more common dicarboxylic acids are presented in Table 18–4. (Note that the first letters of the names of the first six correspond to the first letters in the words, "*O*h *m*y, *s*uch *g*ood *a*pple *p*ie!") As one might expect from an increase in the number of polar groups attached to the skeleton, dicarboxylic acids are more soluble in water than are analogous monocarboxylic acids. For reasons that include increased intermolecular hydro-

TABLE 18-4 Dicarboxylic Acids

NAME	STRUCTURE	MP	K_a[a]	K'_a[a]
oxalic	$HO_2C{-}CO_2H$	189	5.4×10^{-2}	5.2×10^{-5}
malonic	$HO_2C{-}CH_2{-}CO_2H$	136	1.4×10^{-3}	2.0×10^{-6}
succinic	$HO_2C{-}(CH_2)_2{-}CO_2H$	185	6.4×10^{-5}	2.3×10^{-6}
glutaric	$HO_2C{-}(CH_2)_3{-}CO_2H$	98	4.5×10^{-5}	3.8×10^{-6}
adipic	$HO_2C{-}(CH_2)_4{-}CO_2H$	151	3.7×10^{-5}	3.9×10^{-6}
pimelic	$HO_2C{-}(CH_2)_5{-}CO_2H$	105	3.1×10^{-5}	3.7×10^{-6}
suberic	$HO_2C{-}(CH_2)_6{-}CO_2H$	144	3.0×10^{-5}	3.9×10^{-6}
azelaic	$HO_2C{-}(CH_2)_7{-}CO_2H$	106	2.9×10^{-5}	3.9×10^{-6}
sebacic	$HO_2C{-}(CH_2)_8{-}CO_2H$	134	2.6×10^{-5}	4.0×10^{-6}
fumaric	$HO_2CCH{=}CHCO_2H$ (*trans*)	302	9.6×10^{-4}	4.1×10^{-5}
maleic	$HO_2CCH{=}CHCO_2H$ (*cis*)	131	1.0×10^{-2}	5.5×10^{-7}

[a] K_a represents the ionization of the first carboxy group, while K'_a represents loss of a proton from the second (less acidic) carboxy group after the first (more acidic) has lost a proton.

$$HO_2C{------}CO_2H \rightarrow HO_2C{------}CO_2^{\ominus} + H^{\oplus};\ K_a$$
$$HO_2C{------}CO_2^{\ominus} \rightarrow {}^{\ominus}O_2C{------}CO_2^{\ominus} + H^{\oplus};\ K'_a$$

K_a and K'_a are often referred to as K_1 and K_2, respectively.

gen bonding, dicarboxylic acids usually have higher melting points than do analogous monocarboxylic acids. This increased hydrogen bonding also decreases the vapor pressure of dicarboxylic acids and, as a result, they lack the stench that is associated with low molecular weight monocarboxylic acids.

In the main, there is almost no difference between the reactions of dicarboxylic acids and those of their monocarboxylic analogs. Thus, diacids form diamides, diesters, and so forth by the same mechanisms as do monoacids. The only exceptions are diacids with three or fewer atoms between the carboxy groups. These diacids, capable of having two carboxy groups react with a single functional group, or with one another, show unusual characteristics in reactions involving five- or six-membered ring activated complexes and/or products.

THE EFFECT OF HEAT UPON DICARBOXYLIC ACIDS. The simplest example of the anomalous behavior of dicarboxylic acids is their behavior toward heat. For example, oxalic acid decomposes at 150° to formic acid and carbon dioxide. Oxalic acid is toxic and occurs in plants such as rhubarb and green potatoes (portion of the potato exposed to the sun). It is important to destroy this acid by adequate cooking before these vegetables are eaten.

$$\underset{\substack{\text{oxalic acid}\\\text{(ethanedioic acid)}}}{HO_2C{-}CO_2H} \xrightarrow{150^\circ} \underset{\text{formic acid}}{HCO_2H} + CO_2$$

Malonic acid, a β-dicarboxylic acid, loses carbon dioxide to form acetic acid, a monocarboxylic acid, when it is heated to 150°. Since the loss of carbon dioxide is quite difficult (requires much higher temperatures and catalysts) if more than one atom separates the two carboxy groups of a dicarboxylic acid, a cyclic activated complex is postulated for this reaction. In the next chapter we will see how the loss of carbon dioxide from substituted malonic acids is a key feature of many chain-lengthening processes.

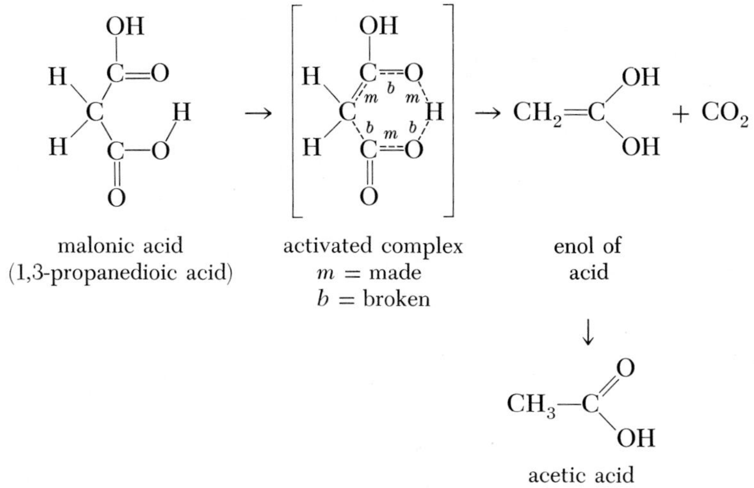

malonic acid (1,3-propanedioic acid) → activated complex (m = made, b = broken) → $CH_2{=}C(OH)_2 + CO_2$ (enol of acid) → $CH_3{-}COOH$ (acetic acid)

$$HO_2C{-}CR_2CO_2H \xrightarrow{\text{heat}} HCR_2CO_2H + CO_2$$

a disubstituted malonic acid → a disubstituted acetic acid

γ-Dicarboxylic acids, *i.e.*, compounds with carboxy groups separated by two atoms, undergo an entropy-favored *cyclodehydration* on heating.

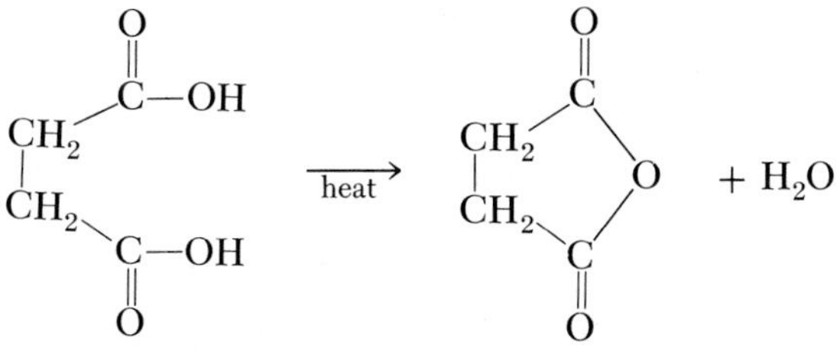

succinic acid (1,4-butanedioic acid) $\xrightarrow{\text{heat}}$ succinic anhydride $+ H_2O$

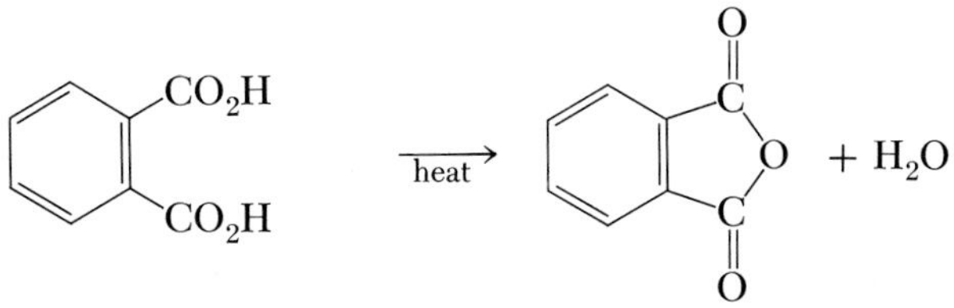

phthalic acid (1,2-benzenedicarboxylic acid) $\xrightarrow{\text{heat}}$ phthalic anhydride $+ H_2O$

Glutaric acid, a δ-dicarboxylic acid, is the largest of the simple dicarboxylic acids that forms a cyclic anhydride upon heating. Larger dicarboxylic acids usually require the addition of a dehydrating agent, such as phosphorus pentoxide, to form an anhydride; and even then the anhydride is a linear polymer.

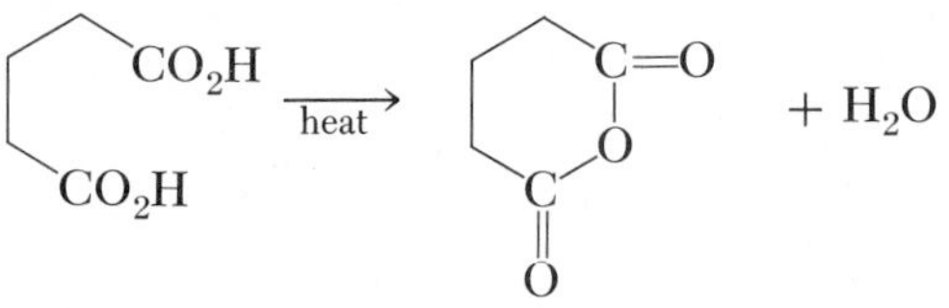

$$\text{glutaric acid} \xrightarrow{\text{heat}} \text{glutaric anhydride} + H_2O$$

glutaric acid glutaric anhydride

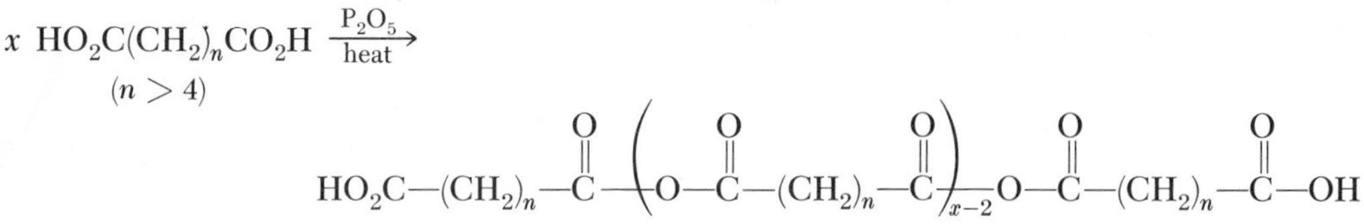

$$x\ HO_2C(CH_2)_nCO_2H \xrightarrow[\text{heat}]{P_2O_5} HO_2C{-}(CH_2)_n{-}\overset{O}{\overset{\|}{C}}{-}\left(O{-}\overset{O}{\overset{\|}{C}}{-}(CH_2)_n{-}\overset{O}{\overset{\|}{C}}\right)_{x-2}{-}O{-}\overset{O}{\overset{\|}{C}}{-}(CH_2)_n{-}\overset{O}{\overset{\|}{C}}{-}OH$$

$(n > 4)$

13. What product is expected from the reaction of glutaric anhydride with each of the following?
(a) water (d) ammonia
(b) methanol (e) acetic acid
(c) (**R**)-2-butanol (f) phthalic acid

18.7 DECARBOXYLATION REACTIONS

One reason for the utility of carboxylic acids in organic synthesis is the ease with which some acids lose carbon dioxide. Under most experimental conditions this reaction produces a hydrocarbon derived by "simple" loss of carbon dioxide and conversion of a C—CO_2H group into a C—H unit. It is this type of reaction which normally is implied by the term **decarboxylation.** However, under the proper conditions the loss of carbon dioxide can produce a halide (R—CO_2H → R—X) or a higher molecular weight hydrocarbon ($2RCO_2H$ → R—R). Since these reactions also require an initial loss of carbon dioxide, they will be discussed here.

β-Keto Acids. By a mechanism similar to that suggested for the decarboxylation of malonic acid, β-keto acids undergo facile thermal decarboxylation to form ketones.

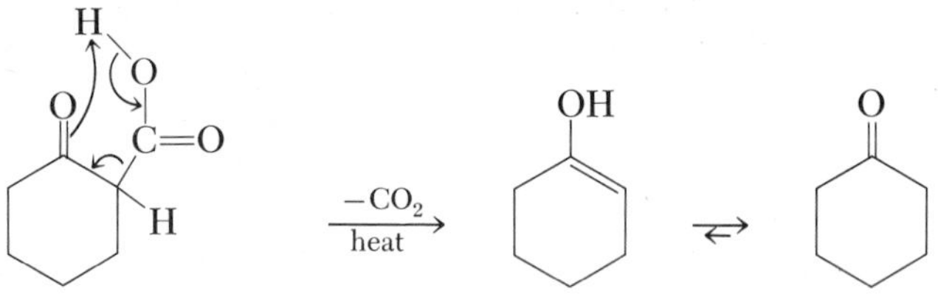

$$\text{2-oxocyclohexanecarboxylic acid} \xrightarrow[\text{heat}]{-CO_2} \text{1-cyclohexenol} \rightleftarrows \text{cyclohexanone}$$

2-oxocyclohexanecarboxylic acid 1-cyclohexenol cyclohexanone

β,γ-Unsaturated Acids. The result of replacing the carbonyl group of a β-keto acid with a carbon-carbon double bond is the formation of a β,γ-unsaturated acid. Such an acid undergoes a thermal decarboxylation similar to that observed for β-keto acids.

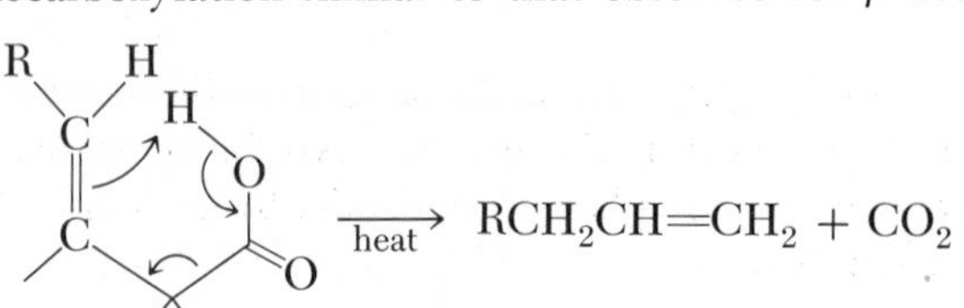

$$\xrightarrow{\text{heat}} RCH_2CH{=}CH_2 + CO_2$$

Simple Carboxylic Acids. Several copper-containing mixtures can be used to decarboxylate aliphatic and aromatic carboxylic acids. Such reactions proceed in moderate yield when carried out in refluxing quinoline.

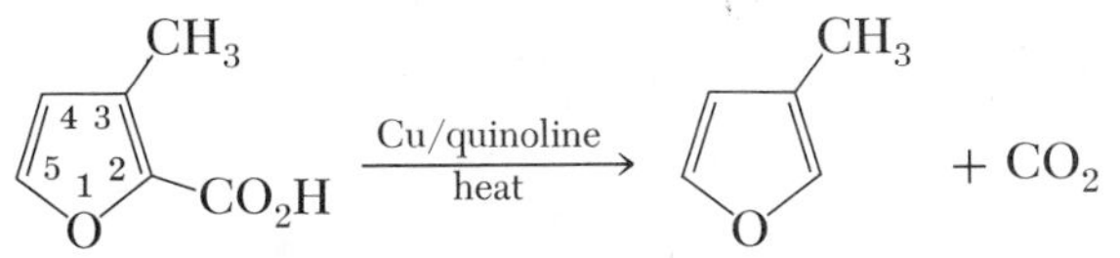

3-methyl-2-furoic acid $\xrightarrow[\text{heat}]{\text{Cu/quinoline}}$ 3-methylfuran $+ CO_2$

$$(C_6H_5)_2C{=}CHCO_2H \xrightarrow[\text{quinoline, heat}]{\text{copper chromite}} (C_6H_5)_2C{=}CH_2 + CO_2$$

3,3-diphenyl-2-propenoic acid → 1,1-diphenylethene

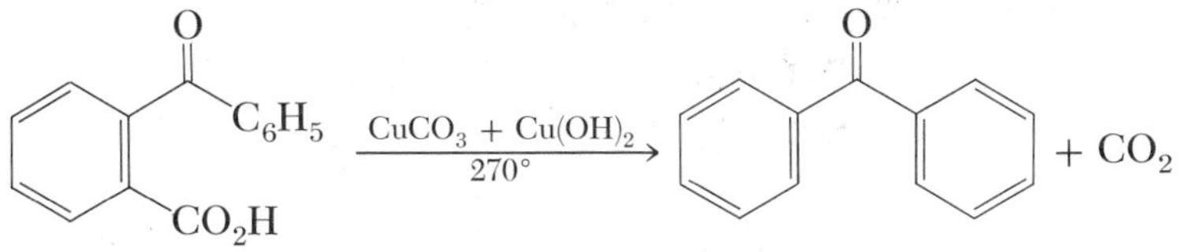

o-benzoylbenzoic acid $\xrightarrow[270°]{CuCO_3 + Cu(OH)_2}$ diphenyl ketone (80%) $+ CO_2$

Modified Decarboxylations. The above reactions have been "decarboxylations" in the strictest sense of the word. In the reactions that follow, we will see how intermediates derived from carbon dioxide loss can be diverted to give something other than a simple hydrocarbon. The first of these sequences produces an alkyl bromide from the silver salt of a carboxylic acid and is termed the **Hunsdiecker reaction.**

A critical step in the Hunsdiecker reaction, pictured below, is the homolytic cleavage of an oxygen-bromine bond to produce a carboxy radical. This, then, loses carbon dioxide to produce a radical that is trapped by bromine in the product-determining step.

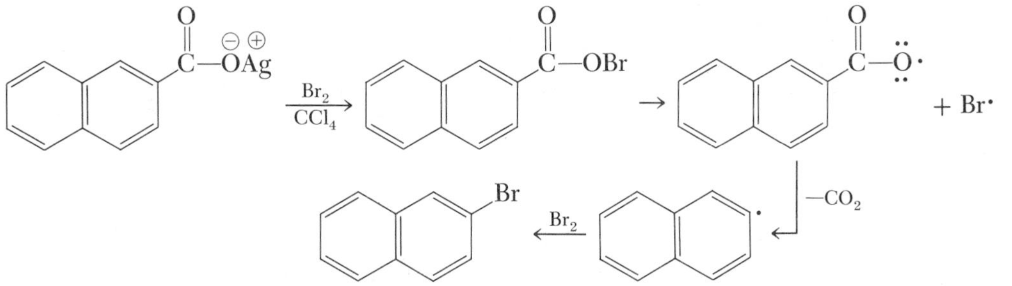

$C_{10}H_7$–C(=O)–$O^{\ominus}Ag^{\oplus}$ $\xrightarrow[CCl_4]{Br_2}$ $C_{10}H_7$–C(=O)–OBr → $C_{10}H_7$–C(=O)–Ö· + Br·

$\xrightarrow{-CO_2}$ $C_{10}H_7$· $\xrightarrow{Br_2}$ $C_{10}H_7$Br

One useful modification of the Hunsdiecker reaction utilizes mercury (II) oxide, HgO, in place of the silver salt.

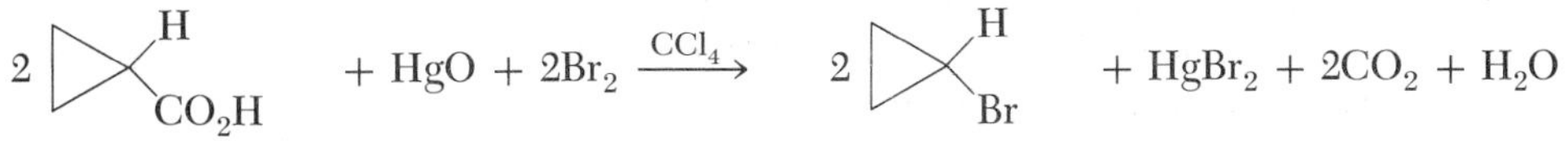

$$2\ \text{cyclopropanecarboxylic acid} + HgO + 2Br_2 \xrightarrow{CCl_4} 2\ \text{bromocyclopropane} + HgBr_2 + 2CO_2 + H_2O$$

cyclopropanecarboxylic acid; bromocyclopropane (45%)

In a different type of reaction, the **Kolbe electrolysis,** a carboxylate anion is oxidized electrolytically to form a radical intermediate similar to the one involved in the Hunsdiecker reaction. In the absence of an efficient radical trap, the resultant radicals dimerize to form a new carbon-carbon single bond. Unlike the Hunsdiecker reaction, the Kolbe electrolysis is carried out in water; therefore, sodium hydroxide and hydrogen are formed at the cathode.

$$\text{R—C(=O)—O}^{\ominus} \xrightarrow[\text{oxidation}]{\text{anodic}} \text{R—C(=O)—\ddot{O}\cdot} + e^{\ominus} \qquad \text{Kolbe electrolysis}$$

$$\text{R—C(=O)—\ddot{O}\cdot} \rightarrow \text{R}\cdot + CO_2$$

$$2\text{R}\cdot \rightarrow \text{R—R}$$

18.8 ESTERIFICATION, TRANSESTERIFICATION, AND ESTER HYDROLYSIS

The ester is among the most commonly encountered derivatives of acids. A great deal of time and effort have been devoted to the study of the mechanisms of ester formation and hydrolysis. Indeed, it would not be wrong to say that one of the cornerstones of modern organic chemistry arose from the study of esters. These experiments demonstrate, among other things, the value of isotopic substitution in unraveling a reaction mechanism and, because of this, will be covered in some detail. Furthermore, many of the compounds used by living systems are esters, and an understanding of the behavior of esters is important to biochemists. We will begin with an examination of how acids catalyze ester hydrolysis.

Acid-Catalyzed Hydrolysis. The *principle of microscopic reversibility* states that if a certain sequence of steps is favored for a given forward reaction, the reverse sequence of steps is favored for the reverse reaction. Thus, we can gain insight into the mechanism of acid-catalyzed *esterification* by studying acid-catalyzed *hydrolysis*.

$$\text{R—C(=O)—O—R'} + H_2O \xrightarrow{H^{\oplus}} \text{R—C(=O)—OH} + \text{R'OH} \qquad \text{acid-catalyzed ester hydrolysis}$$

$$\text{R—C(=O)—O—R'} + H_2O \xleftarrow{H^{\oplus}} \text{R—C(=O)—OH} + \text{R'OH} \qquad \text{acid-catalyzed ester formation}$$

If one can establish that acid-catalyzed ester hydrolysis occurs by acyl-oxygen fission (see below), then, by the principle of microscopic reversibility, one also establishes that acyl-oxygen bond formation occurs during the Fischer esterification procedure (p. 560). This has been done by hydrolyzing an ester with ^{18}O-labeled water in the presence of an acid such as hydrochloric acid or sulfuric acid. A useful example of this approach is the hydrolysis of γ-butyrolactone. (A **lactone** is an *intra*molecular ester, that is, one formed by reaction of a carboxy group and a hydroxy group within the same molecule.)

$$\text{R—C(=O)—O—R'}$$

acyl-oxygen fission breaks this bond (C—O)

alkyl-oxygen fission breaks this bond (O—R′)

The products from the acid-catalyzed hydrolysis of the cyclic ester γ-butyrolactone in $H_2{}^{18}O$ prove that this ester hydrolyzes by acyl-oxygen fission—the ^{18}O is found in the carboxy group of the product acid and *not* in the hydroxy group of the product alcohol. The accepted mechanism is presented in scheme **A** (below). An S_N2 displacement (scheme **B**) or an ionization (scheme **C**) would have produced product labeled in the hydroxy group. (Put differently, the formation of labeled alcohol would have established that the mechanism involved alkyl-oxygen fission—schemes **B** or **C**.)

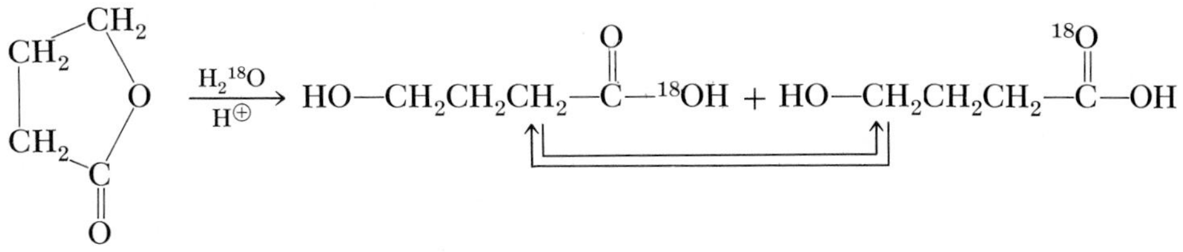

$$\text{γ-butyrolactone} \xrightarrow[H^{\oplus}]{H_2{}^{18}O} HO{-}CH_2CH_2CH_2{-}\overset{O}{\overset{\|}{C}}{-}{}^{18}OH + HO{-}CH_2CH_2CH_2{-}\overset{^{18}O}{\overset{\|}{C}}{-}OH$$

γ-butyrolactone

A: (scheme) usual mechanism of ester hydrolysis
O: oxygen-16
●: oxygen-18

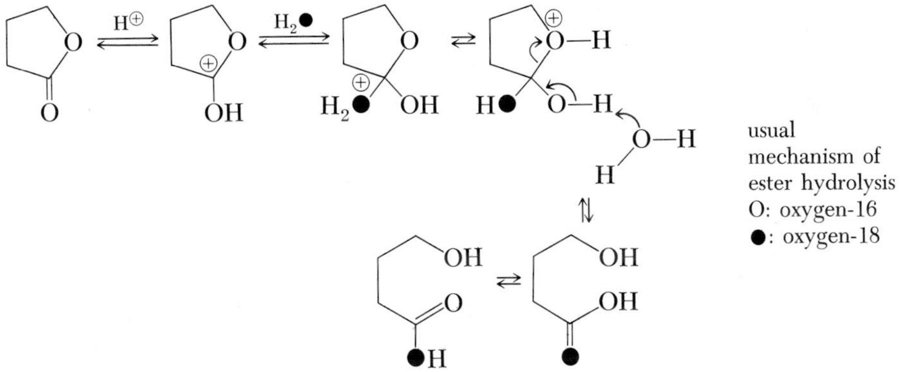

B: (scheme)

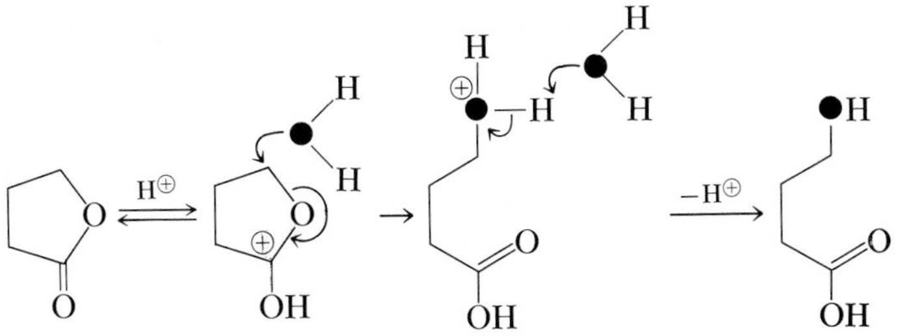

C: (scheme)

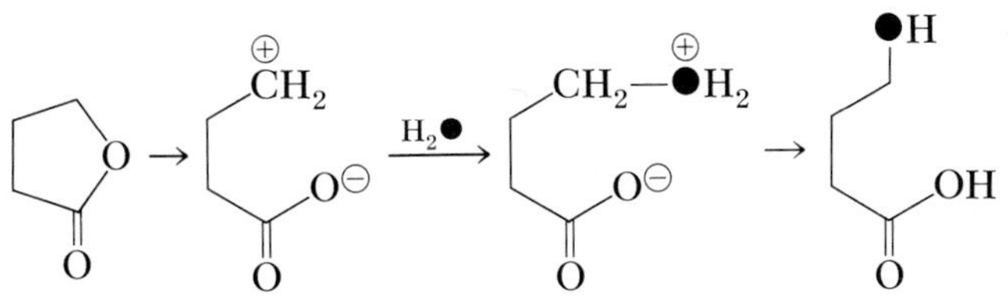

Another "proof" that mechanism **A** is operative (this is sometimes called the $A_{AC}2$ mechanism[a]) is the observation that hydrolysis of an optically active ester with a chiral center bonded to the alkyl oxygen produces an alcohol of the same

[a] Acid-catalyzed; *AC*yl fission; bimolecular.

configuration as was present in the ester. An S_N2 process (scheme **B**) would have produced inversion of configuration, while ionization (scheme **C**) would have led to racemization.

$$\text{R—C(=O)—O—C}abc \xrightarrow[A_{AC}2]{H_3O^{\oplus}} \text{R—C(=O)—OH} + \text{HO—C}abc$$

The reversibility of the initial hydration step, and further support for this mechanism, was established by examining "unreacted" starting material and noting ^{18}O incorporation in the carbonyl group.

$$\gamma\text{-butyrolactone (C=O)} \xrightarrow[H^{\oplus}]{H_2\bullet} \gamma\text{-butyrolactone (C=O)} + \gamma\text{-butyrolactone (C=}\bullet) \text{ but } NO \text{ ring-}\bullet \text{ lactone (C=O)}$$

If an ester is dissolved in *alcoholic* acid, rather than aqueous acid, an *ester interchange* results. The mechanism for these **transesterifications** is similar to the mechanism of hydrolysis.

$$\gamma\text{-butyrolactone} + CH_3OH \underset{}{\overset{H^{\oplus}}{\rightleftharpoons}} HOCH_2CH_2CH_2C(=O)OCH_3$$

Steric hindrance in the transition state that converts the sp^2 hybridized carbonyl carbon into an sp^3 hybridized carbon in the intermediate requires that highly hindered carboxylic acids be esterified under unusual conditions. The most common method involves dissolving the acid in concentrated sulfuric acid, thus forming the acylium ion corresponding to the acid, and then quenching the ion with an alcohol.

$$\text{2,4,6-(CH}_3)_3C_6H_2CO_2H \xrightarrow[H^{\oplus}]{CH_3OH} \text{no reaction}$$

$$\text{2,4,6-(CH}_3)_3C_6H_2CO_2H \xrightarrow[-H_2O]{H_2SO_4} \text{2,4,6-(CH}_3)_3C_6H_2C^{\oplus}{=}O \xrightarrow[-H^{\oplus}]{CH_3OH} \text{2,4,6-(CH}_3)_3C_6H_2C(=O)O{-}CH_3$$

an acylium ion

Base-Catalyzed Hydrolysis. **Saponification** (this is the term most often used to designate a base-catalyzed ester hydrolysis) results in acyl-oxygen fission and involves an addition-elimination sequence. Evidence for this mechanism comes from ^{18}O labeling and stereochemical studies.

$$\textit{net reaction: } CH_3-\overset{\overset{O}{\|}}{C}-O-CH_2CH_3 + OH^{\ominus} \rightarrow CH_3-\overset{\overset{O}{\|}}{C}-O^{\ominus} + CH_3CH_2OH$$

$$\textit{mechanism: } H_3C-\overset{\overset{\ddot{O}:}{\|}}{C}-OCH_2CH_3 \rightleftarrows CH_3-\overset{\overset{:\ddot{O}:^{\ominus}}{|}}{\underset{\underset{:\bullet H}{|}}{C}}-O-CH_2CH_3$$
$$\quad :\bullet H^{\ominus}$$

$$CH_3C\overset{\nearrow O}{\searrow \bullet^{\ominus}} + CH_3CH_2OH \leftarrow CH_3-\overset{\overset{O}{\|}}{C}-\bullet H + CH_3CH_2O^{\ominus}$$

The mechanism for this saponification is sometimes designated as $B_{AC}2$. (Why?)

While hydrolyses and transesterifications can be base-catalyzed, esterifications are *never* based-catalyzed. *Why?* Because the acid quickly reacts with the basic catalyst and is converted into an anion. This anion is not subject to attack by a nucleophile, such as an alcohol molecule or an alcoholate anion.

$$RCO_2H + R'OH \xrightarrow{OH^{\ominus}} RC(O)OR' \qquad \text{does } \textit{not} \text{ occur}$$

Alkyl-Oxygen Fission. Although the $A_{AC}2$ and $B_{AC}2$ mechanisms are followed in the majority of ester hydrolyses, there are at least two types of esters that hydrolyze by mechanisms involving alkyl-oxygen fission. These two differ from one another; one of them involves carbocation formation, and is a variety of S_N1 reaction, while the other is an S_N2 reaction.

Esters of 3° carbinols usually hydrolyze by an alkyl-oxygen fission because of the stability of the resultant 3° carbocation. This reaction is accelerated by protonation of the acyl oxygen, since this weakens the bond destined for S_N1 cleavage.

$$R-\overset{\overset{O}{\|}}{C}-O-\underset{\underset{R}{|}}{\overset{\overset{R}{|}}{C}}-R \xrightarrow{H_2O} R-\overset{\overset{O}{\|}}{C}-O^{\ominus} + {}^{\oplus}CR_3 \xrightarrow{H_2O} HOCR_3 + H^{\oplus}$$

14. Suggest a test of the S_N1-type mechanism of ester hydrolysis that (a) does not depend upon the identity of the products and (b) does not depend upon the stereochemistry of the products.

This mechanism also is used to account for the pH dependence of the transesterification of *t*-butyl benzoate. In methanol containing sodium methoxide, *t*-butyl benzoate forms *t*-butanol and methyl benzoate, the anticipated transesterification products. In acidic methanol, however, the products include benzoic acid, methyl benzoate, and methyl *t*-butyl ether. Ether formation is best explained by an ionization of the ester to the *t*-butyl cation followed by attack of methanol upon the carbocation.

$$C_6H_5-\overset{\overset{O}{\|}}{C}-OC(CH_3)_3 \xrightarrow{CH_3O^{\ominus}} C_6H_5-\underset{\underset{OCH_3}{|}}{\overset{\overset{O^{\ominus}}{|}}{C}}-OC(CH_3)_3 \rightarrow C_6H_5-\overset{\overset{O}{\|}}{C}-OCH_3 + (CH_3)_3CO^{\ominus}$$

$$C_6H_5-\overset{\overset{O}{\|}}{C}-O-C(CH_3)_3 \xrightarrow{H^{\oplus}} C_6H_5-\overset{\overset{^{\oplus}OH}{\|}}{C}-O-C(CH_3)_3 \rightarrow C_6H_5-\overset{\overset{O}{\|}}{C}-OH + (CH_3)_3C^{\oplus} \xrightarrow[CH_3OH]{-H^{\oplus}} CH_3OC(CH_3)_3$$

15. What oxygen-free compound might be expected to be formed in the acid-catalyzed reaction of *t*-butyl benzoate with methanol to produce *t*-butyl methyl ether? Suggest its origin. If there is more than one possible mechanism, suggest how they might be distinguished experimentally.

While S_N2 displacements are very common reactions, they are encountered only under unusual circumstances when dealing with ester hydrolysis. This is due to the great susceptibility of the carbonyl group to attack by a nucleophile (p. 307). Perhaps the only esters that routinely undergo S_N2 hydrolysis are β-lactones. The reasons for this are complex and will not be presented here.

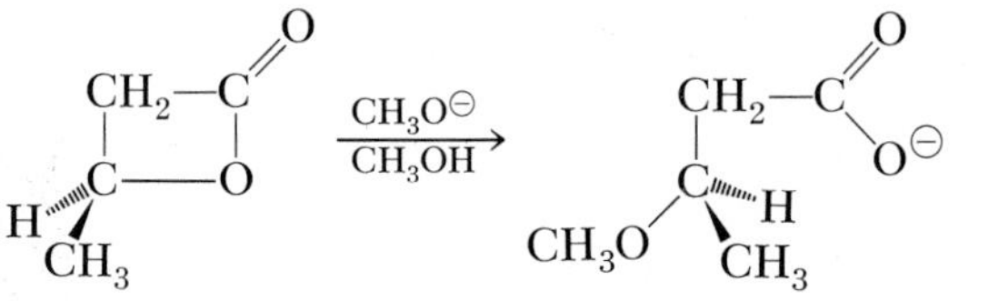

$$\beta\text{-lactone } (CH_2\text{—}C(=O)\text{—}O\text{—}CH(CH_3)\text{, ring}) \xrightarrow[CH_3OH]{CH_3O^{\ominus}} CH_3O\text{—}CH(CH_3)\text{—}CH_2\text{—}C(=O)O^{\ominus}$$

18.9 AMIDES—FORMATION AND HYDROLYSIS

No other single group of compounds is as interesting to the biochemist at this moment as are the proteins, complex polymers whose individual units (called amino acids) are held together by amide bonds. When one wishes to synthesize or degrade such molecules (yes, some proteins have been synthesized outside of living systems) one often employs rather special techniques. Chapter 24 is devoted to the chemistry of amino acids, and these polymers and the special techniques of the protein chemist will be discussed there. However, in order to better appreciate that material, we will look a bit more closely at a classical synthesis of amides and the hydrolysis of the amide bond in this section.

$$\text{—}\overset{O}{\overset{\|}{C}}\text{—}\overset{H}{\overset{|}{N}}\text{—CHR—}\overset{O}{\overset{\|}{C}}\text{—}\overset{H}{\overset{|}{N}}\text{—CHR}'\text{—}\overset{O}{\overset{\|}{C}}\text{—}\overset{H}{\overset{|}{N}}\text{—} \quad \text{a section of protein}$$

$$\downarrow H_2O$$

$$\text{—}\overset{O}{\overset{\|}{C}}\text{—OH} + \underset{\text{an amino acid}}{H_2N\text{—CHR—}CO_2H} + \underset{\text{an amino acid}}{H_2N\text{—CHR}'\text{—}CO_2H} + H_2N\text{—} \quad \text{hydrolysis products}$$

16. One reason that polyamides, rather than polyesters, are so prevalent in nature may be that the amide bond is harder to hydrolyze ("destroy") than is the ester bond. Suggest a reason for this difference. (*Hint:* The C=O bond of esters absorbs at a higher frequency than does the C=O bond of amides.)

Because of the nucleophilicity of ammonia and because of the poor leaving group characteristics of $NH_2^{\ominus}$, esters react with ammonia (or amines) to produce high yields of amides. The violence that can accompany acylations using acyl halides (especially acetyl chloride) makes the ammonolysis of esters a valuable synthetic route to amides.

$$\underset{\text{ethyl cyanoacetate}}{NC\text{—}CH_2\text{—}\overset{O}{\overset{\|}{C}}\text{—}OC_2H_5} + NH_3 \rightarrow \underset{\substack{\text{cyanoacetamide}\\ 88\%}}{NC\text{—}CH_2\text{—}\overset{O}{\overset{\|}{C}}\text{—}NH_2} + C_2H_5OH$$

$$\text{R—}\overset{\text{O}}{\overset{\|}{\text{C}}}\text{—OR}' + \text{NH}_3 \rightleftarrows \text{R—}\underset{\oplus\text{NH}_3}{\overset{\text{O}^\ominus}{\text{C}}}\text{—OR}' \rightleftarrows \text{R—}\underset{\text{NH}_2}{\overset{\text{OH}}{\text{C}}}\text{—OR}'$$

mechanism of ester ammonolysis

$$\text{R—}\underset{\text{NH}_2}{\overset{\text{O—H}\cdots\text{:NH}_3}{\text{C}}}\text{—OR}' \rightarrow \text{R—}\overset{\text{O}}{\overset{\|}{\text{C}}}\text{—NH}_2 + \overset{\oplus}{\text{N}}\text{H}_4\ \text{R}'\text{O}^\ominus$$

amide

Ketones are converted to amides in good yield by reaction with hydrazoic acid, HN_3, and a strong acid to serve as a catalyst. The proposed mechanism is as follows:

$$\text{R—}\overset{\text{O}}{\overset{\|}{\text{C}}}\text{—R} \xrightleftharpoons{\text{H}^\oplus} \text{R—}\underset{\oplus}{\overset{\text{OH}}{\text{C}}}\text{—R} \xrightarrow{\text{HN}_3} \text{R—}\underset{\text{R}}{\overset{\text{OH}}{\text{C}}}\text{—NH—}\overset{\oplus}{\text{N}}\text{≡N} \xrightarrow{-\text{H}_2\text{O}} \text{R}_2\text{C=N—}\overset{\oplus}{\text{N}}_2$$

$$\text{R}_2\text{C=}\ddot{\text{N}}\text{—}\overset{\oplus}{\text{N}}\text{≡N} \rightarrow \left[\begin{array}{c}\text{R—}\overset{\oplus}{\text{C}}\text{=}\ddot{\text{N}}\text{—R}\\ \updownarrow \\ \text{R—C≡}\overset{\oplus}{\text{N}}\text{—R}\end{array}\right] \xrightarrow[-\text{H}^\oplus]{\text{H}_2\text{O}} \text{R—}\overset{\text{O}}{\overset{\|}{\text{C}}}\text{—NHR}$$

Evidence supporting this mechanism includes the isolation of a tetrazole if excess hydrazoic acid is present.

$$\text{R—N=}\overset{\oplus}{\text{C}}\text{R} \xrightarrow{\text{HN}_3} \text{R—}\ddot{\text{N}}\text{=C(R)—NH—}\overset{\oplus}{\text{N}}\text{≡N} \rightarrow \text{(cyclic) R—}\overset{\oplus}{\text{N}}\text{=C(R)—N(H)—N=N—} \xrightarrow{-\text{H}^\oplus} \text{R—N—C—R (tetrazole ring: N=N—N=)}$$

a tetrazole

This ketone-amide conversion is commonly called the **Schmidt reaction.**

A simplified view of base-catalyzed hydrolysis of amides is presented below. Although the mechanism of acid-catalyzed hydrolysis was presented earlier in this chapter, it is repeated here in order to provide a ready comparison.

$$\text{R—}\overset{:\text{O}:}{\overset{\|}{\text{C}}}\text{—}\ddot{\text{N}}\text{H}_2 + :\ddot{\text{O}}\text{H}^\ominus \rightleftarrows \text{R—}\underset{:\ddot{\text{O}}\text{H}}{\overset{:\ddot{\text{O}}:^\ominus}{\text{C}}}\text{—}\ddot{\text{N}}\text{H}_2 \rightarrow \text{R—}\overset{:\text{O}:}{\overset{\|}{\text{C}}}\text{—}\ddot{\text{O}}\text{H} + :\ddot{\text{N}}\text{H}_2^\ominus \rightarrow \text{R—}\overset{:\text{O}:}{\overset{\|}{\text{C}}}\text{—}\ddot{\text{O}}:^\ominus + :\text{NH}_3$$

base-catalyzed amide hydrolysis

$$R{-}\overset{\overset{\displaystyle :O:}{\|}}{C}{-}\ddot{N}H_2 \overset{H^{\oplus}}{\rightleftharpoons} R{-}\overset{\overset{\displaystyle :\ddot{O}{-}H}{|}}{\underset{\oplus}{C}}{-}\ddot{N}H_2 \overset{H_2O}{\rightleftharpoons} R{-}\overset{\overset{\displaystyle :\ddot{O}{-}H}{|}}{\underset{\underset{\displaystyle H\overset{\oplus}{O}H}{|}}{C}}{-}\ddot{N}H_2 \rightleftharpoons R{-}\overset{\overset{\displaystyle :\ddot{O}{-}H}{|}}{\underset{\underset{\displaystyle :\ddot{O}{-}H}{|}}{C}}{-}\overset{\oplus}{N}H_3 \longrightarrow R{-}\overset{\overset{\displaystyle :O:}{\|}}{C}{-}\ddot{\underset{\cdot\cdot}{O}}{-}H + NH_4^{\oplus}$$

acid-catalyzed amide hydrolysis

18.10 BIOLOGICAL IMPORTANCE OF ESTERS—A PREVIEW

This will be the first of several discussions of the biological importance of esters and will provide more of an overview than a detailed analysis. We will limit ourselves to a peek at the triesters known as *fats* and *oils* and to the very important *coenzyme A*.

Fats. Fats are solid carboxylate esters of glycerin, 1,2,3-trihydroxypropane. These triesters are more properly called *triacylglycerols,* although they are also called *triglycerides*. If a triacylglycerol is a liquid at room temperature, it is called an *oil*. With few exceptions the carboxylic acids found in fats and oils may vary in length from three to eighteen carbons. Beginning with C_6 acids, most of the acids found in fats and oils contain even numbers of carbons. Some of the more common sources of fats and oils include butter, corn, olives, peanuts, soybeans, and lard.

When fats or oils are boiled with aqueous sodium or potassium hydroxide, they are saponified just as any other esters would be. The products of this hydrolysis are glycerin (sometimes called glycerol) and the sodium or potassium salts of various long-chain (so-called "fatty") acids. These salts are commonly called *soaps*.

$$\begin{matrix} CH_2{-}O{-}\overset{\overset{\displaystyle O}{\|}}{C}{-}R \\ | \\ CH{-}O{-}\overset{\overset{\displaystyle O}{\|}}{C}{-}R' \\ | \\ CH_2{-}O{-}\overset{\overset{\displaystyle O}{\|}}{C}{-}R'' \end{matrix} \xrightarrow[\text{heat}]{NaOH/H_2O} \begin{matrix} CH_2OH \\ | \\ CHOH \\ | \\ CH_2OH \end{matrix} + \left\{\begin{matrix} Na^{\oplus}\ {}^{\ominus}O_2CR \\ + \\ Na^{\oplus}\ {}^{\ominus}O_2CR' \\ + \\ Na^{\oplus}\ {}^{\ominus}O_2CR'' \end{matrix}\right\} \text{soaps}$$

(a fat or oil R and/or R′ may or may not be the same as R″)

glycerin (glycerol)

Today soap comes perfumed, filled with air (so that the bar floats), laced with alcohol (so that it is translucent), dyed (so that it matches the "decor"), and so on. In the proverbial "good ole days" soap was made by boiling together lard or other available fat, wood ashes (which contain a crude potassium carbonate called *potash*), and water in an iron kettle. Whether it's one of today's "beauty soaps" or yesterday's "lye soap," the cleaning action of soap is still its important property. In Chapter 23 we will look at some detergents and discuss how soaps and detergents differ in structure but are similar in function.

Fats have several biochemical functions, the most important of which is the storage of energy. When the body has taken in more food than it needs immediately, the excess is converted into fat and stored until it is needed. This serves as a useful mechanism for animals that hibernate to survive long periods without ingestion. Beyond this, compounds very similar to fats also are important in maintaining the structure of the cell membrane. Both the biosynthesis of fatty acids and the structure of the cell membrane will be discussed later in this text.

Thiol Esters. Thiols, compounds of the type R—SH, can produce *thiol esters*, R′C(O)—SR, by reaction with carboxylic acids or appropriate derivatives. Thiol esters, while superficially similar to ordinary esters, actually are much more prone to undergo substitution by a nucleophile. This enhanced reactivity stems from the fact that thiol esters are not stabilized by resonance as are ordinary esters, and from the good departing characteristics of the RS$^{\ominus}$ group.

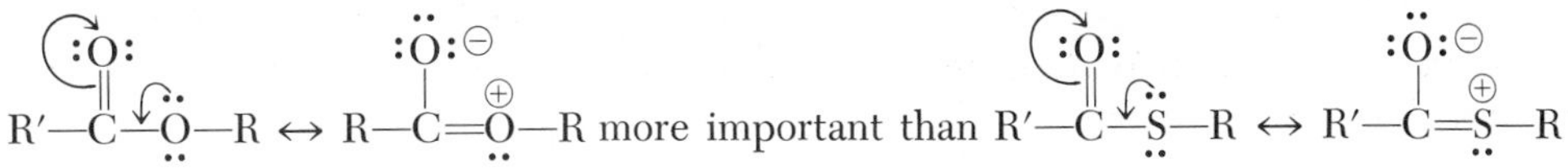

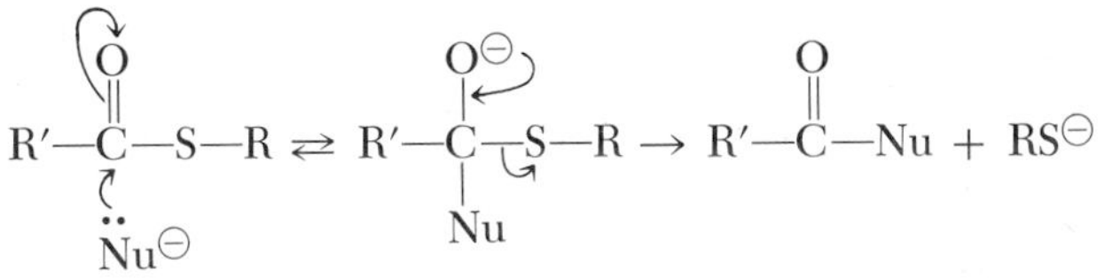

$$R'-C(=O)-\ddot{O}-R \leftrightarrow R-C(-O^{\ominus})=\overset{\oplus}{O}-R \text{ more important than } R'-C(=O)-\ddot{S}-R \leftrightarrow R'-C(-O^{\ominus})=\overset{\oplus}{S}-R$$

$$R'-C(=O)-S-R + Nu^{\ominus} \rightleftarrows R'-C(O^{\ominus})(Nu)-S-R \rightarrow R'-C(=O)-Nu + RS^{\ominus}$$

acylation of Nu$^{\ominus}$ using a thiol ester

The body uses thiol esters to string together (or uncouple) acyl groups and so construct (or degrade) long carbon chains, including those of carboxylic acids. This process will be examined in some detail in Chapter 27, but we should note here that the thiol used by the body is a complex molecule called *coenzyme* A. To emphasize that coenzyme A functions as a thiol, it is often abbreviated as CoASH rather than CoA. The most famous thiol ester of coenzyme A is the one that is derived from acetic acid, *acetyl coenzyme* A (abbreviated AcCoA or CoASC(O)CH_3).

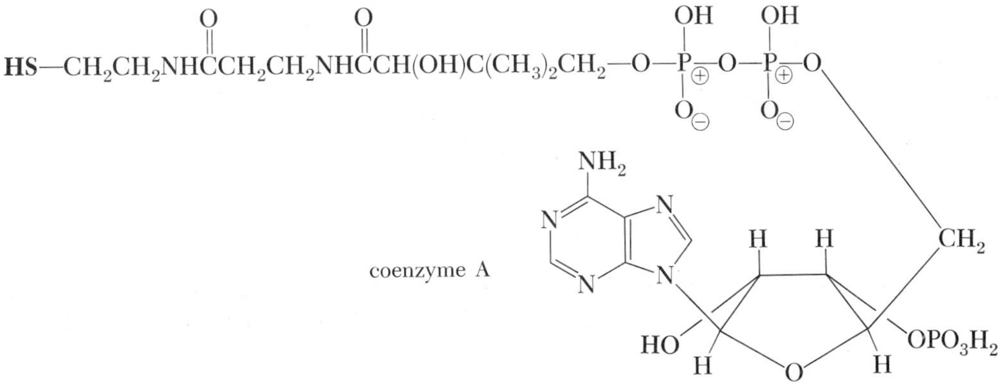

$$HS-CH_2CH_2NHC(O)CH_2CH_2NHC(O)CH(OH)C(CH_3)_2CH_2-O-\overset{\oplus}{P}(OH)(O^{\ominus})-O-\overset{\oplus}{P}(OH)(O^{\ominus})-O-CH_2-\text{(ribose: HO, H, H, H, H, } OPO_3H_2\text{; adenine with } NH_2\text{)}$$

coenzyme A

18.11 POLYMERIC ESTERS AND AMIDES

If a molecule has functional groups at two sites, it can react with another bifunctional compound to form, after a series of reactions, a polymer.

$$RCO_2H + HOR' \xrightarrow{-H_2O} RCO_2R' \quad \text{a "dead end" monoester}$$

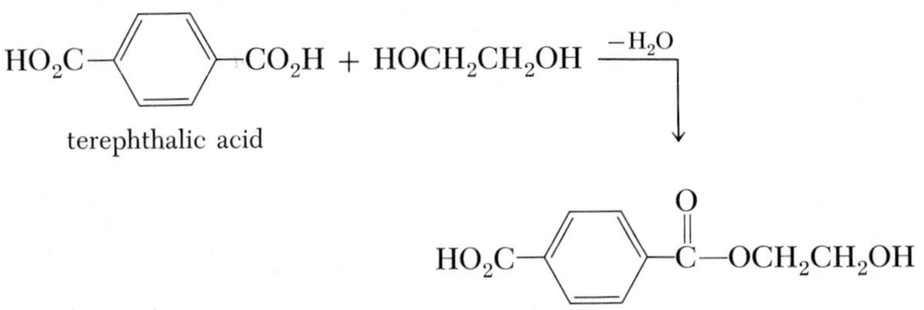

$$HO_2C{-}C_6H_4{-}CO_2H + HOCH_2CH_2OH \xrightarrow{-H_2O} HO_2C{-}C_6H_4{-}C(=O){-}OCH_2CH_2OH$$

terephthalic acid

monoester capable of further esterification

In the example above, ethylene glycol reacts with terephthalic acid to produce an ester that is not a "dead end"; it retains functionalities that can be further esterified. The resulting *polyester* can be spun into fibers of *Dacron* or *Fortrel* or processed into the highly tear-resistant film called *Mylar.* One important use for Mylar is as a backing for magnetic tapes used in tape recorders and computers.

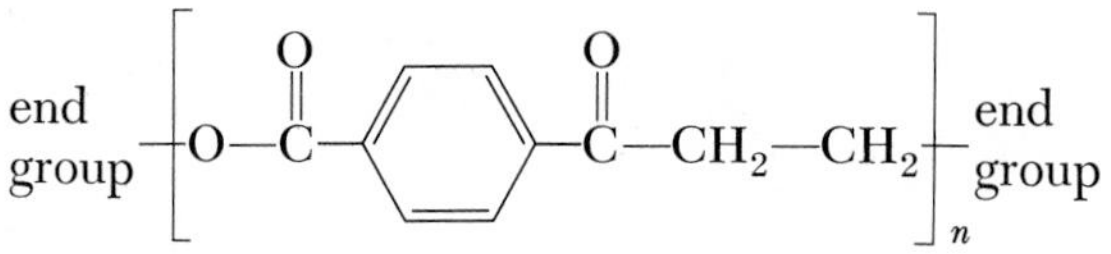

$$\text{end group}{-}\left[O{-}C(=O){-}C_6H_4{-}C(=O){-}CH_2{-}CH_2\right]_n{-}\text{end group}$$

Dacron or Fortrel

By changing the diol in the above polymerization process, fibers with different properties can be obtained. For example, catalytic reduction of dimethyl terephthalate produces 1,4-dihydroxymethylcyclohexane, a diol. If this is now substituted for ethylene glycol in the polymerization just presented, a polyester is produced that is spun into fiber known as *Kodel.* This polyester is used in clothing fabrics, carpeting, and furniture fabric.

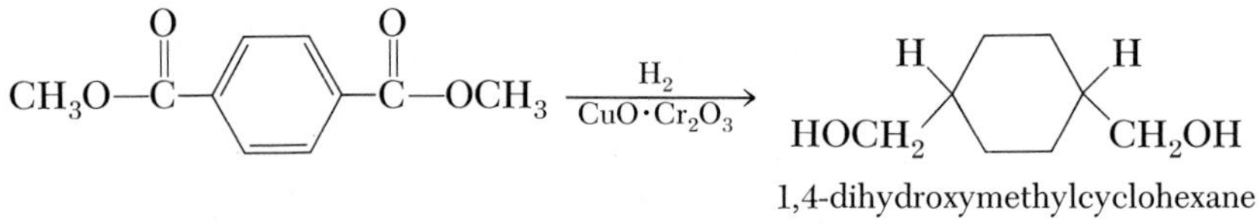

$$CH_3O{-}C(=O){-}C_6H_4{-}C(=O){-}OCH_3 \xrightarrow[CuO\cdot Cr_2O_3]{H_2} HOCH_2{-}C_6H_{10}{-}CH_2OH$$

1,4-dihydroxymethylcyclohexane

$$\text{end group}{-}\left[O{-}C(=O){-}C_6H_4{-}C(=O){-}O{-}CH_2{-}C_6H_{10}{-}CH_2\right]_n{-}\text{end group}$$

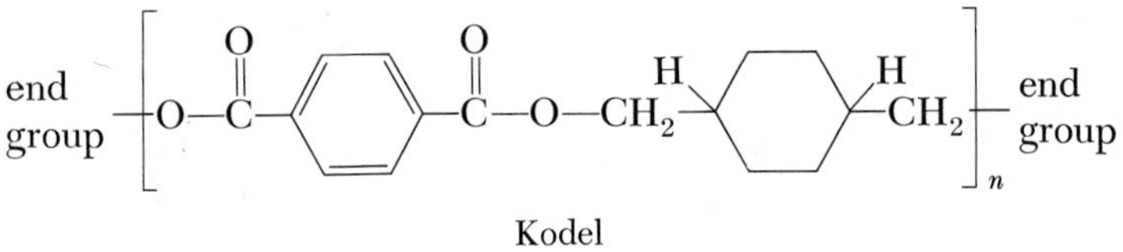

Kodel

Phosgene, also called *carbonyl chloride,* is the acid dichloride of carbonic acid. This deadly gas, used with impunity during WWI, reacts with alcohols to produce organic *carbonates,* diesters of carbonic acid.

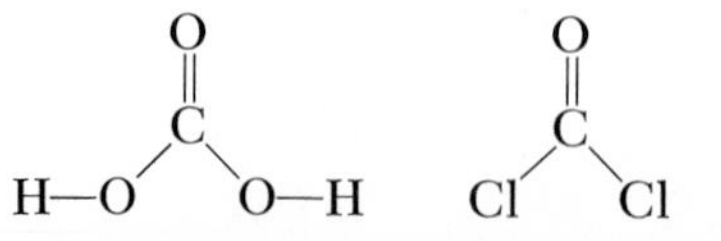

$$H{-}O{-}C(=O){-}O{-}H \qquad Cl{-}C(=O){-}Cl$$

carbonic acid phosgene

$$2CH_3CH_2OH + ClCOCl \rightarrow C_2H_5O{-}C(O)OC_2H_5$$

diethyl carbonate

If the alcohol were a diol, phosgene could react to produce a polymer.

$$Cl-\overset{O}{\overset{\|}{C}}-Cl + HO(CH_2)_4OH \rightarrow Cl-\overset{O}{\overset{\|}{C}}-O(CH_2)_4OH \xrightarrow{HO(CH_2)_4OH} HO(CH_2)_4O-\overset{O}{\overset{\|}{C}}-O(CH_2)_4OH$$

$$HO(CH_2)_4O-\overset{O}{\overset{\|}{C}}-O(CH_2)_4OH \xrightarrow{ClCOCl} Cl\overset{O}{\overset{\|}{C}}-(CH_2)_4O-\overset{O}{\overset{\|}{C}}-O(CH_2)_4OH \xrightarrow{etc.} \text{polycarbonate polymer}$$

The tough, clear plastics made by reacting phosgene with diols are known as *polycarbonates*. Of several important polycarbonates, none is more unusual than poly[2,2-*bis*(4-phenylene)propane carbonate], commonly called *Lexan* or *Merlon*. This polymer, first synthesized in Germany in 1953, is as clear as glass and nearly as tough as steel. A sheet can stop a .38-caliber bullet fired from 12 feet away. A quarter-inch thick sheet of Lexan bends harmlessly under the blows of a five pound hammer. Such unusual properties have resulted in Lexan's use in "burglar proof" windows and as visors in astronauts' space helmets.

$$Cl-\overset{O}{\overset{\|}{C}}-Cl + HO-C_6H_4-C(CH_3)_2-C_6H_4-OH \xrightarrow[\text{pyridine}]{25^\circ} \left[-O-C_6H_4-C(CH_3)_2-C_6H_4-O-\overset{O}{\overset{\|}{C}}-\right]_{80\text{ to }100}$$

Lexan

When glycerol, $HOCH_2CHOHCH_2OH$, is polymerized with phthalic acid, the product contains unreacted —OH groups.

$$HOCH_2-\underset{OH}{\underset{|}{CH}}-CH_2OH + C_6H_4(CO_2H)_2 \rightarrow -OCH_2-\underset{\boxed{OH}}{\underset{|}{CH}}-CH_2O-\overset{O}{\overset{\|}{C}}-C_6H_4-\overset{O}{\overset{\|}{C}}-OCH_2-\underset{\boxed{OH}}{\underset{|}{CH}}-CH_2O-$$

free OH groups

If the reaction mixture contains three moles of phthalic acid for every two moles of glycerin, these free hydroxy groups can react with "unused" carboxy groups to form cross-links between growing chains. Cross-linked polymers of this type are known as *alkyd resins*[a] and

[a] A resin is a polymer capable of being molded.

are used to make enamel paints and dentures. The specific type of alkyd resin described above is a *glyptal*. With some modification these resins are used in "water-base" paints, *e.g.*, "Lucite."

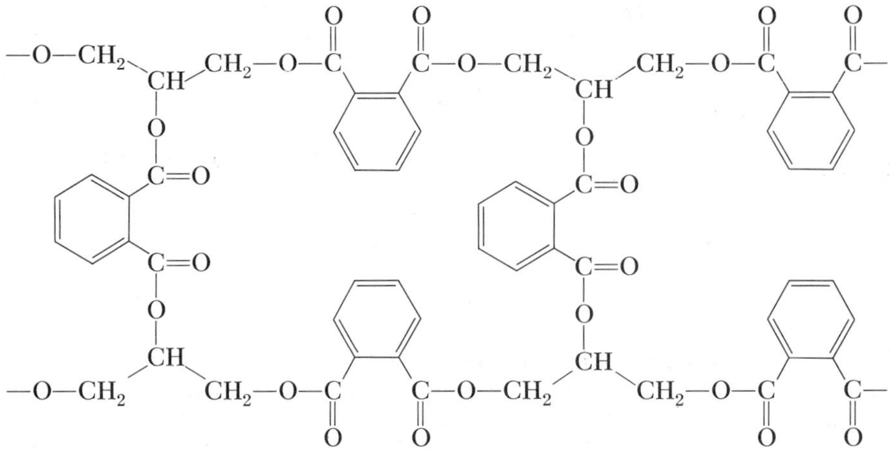

glyptal

The same condensation can be achieved using phthalic anhydride rather than phthalic acid. More than half of the approximately 600 million pounds of phthalic anhydride produced annually goes into the production of alkyd resins. The major sources of this anhydride are naphthalene and *o*-xylene, conversion being achieved by catalytic oxidation.

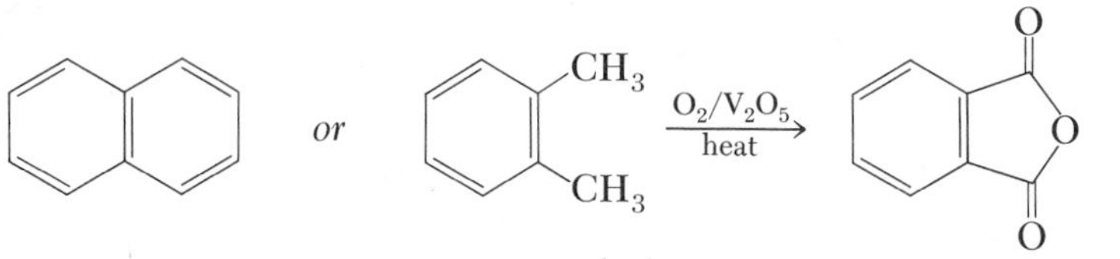

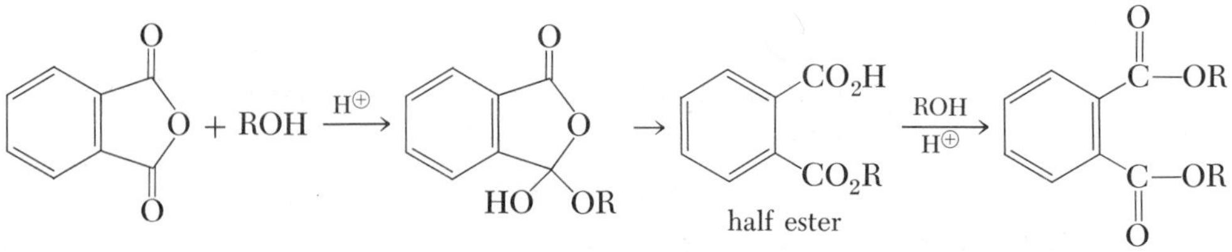

$$\text{naphthalene} \quad \textit{or} \quad o\text{-}C_6H_4(CH_3)_2 \xrightarrow[\text{heat}]{O_2/V_2O_5} \text{phthalic anhydride}$$

phthalic anhydride

$$\text{phthalic anhydride} + ROH \xrightarrow{H^{\oplus}} \text{(HO)(OR) adduct} \rightarrow o\text{-}C_6H_4(CO_2H)(CO_2R) \xrightarrow[H^{\oplus}]{ROH} o\text{-}C_6H_4(COOR)_2$$

half ester

Polyamides, like polyesters, have had a major impact upon our lives. One polyamide in particular, discovered at the DuPont laboratories, has had a most unique effect—that polymer is *nylon*.

Nylon is made by the polymerization of adipic acid and hexamethylene diamine.

$$\underset{\text{adipic acid}}{HO_2C(CH_2)_4CO_2H} + \underset{\text{hexamethylene diamine}}{H_2N(CH_2)_6NH_2} \rightarrow \underset{\text{Nylon 66}}{\left[\overset{O}{\overset{\|}{C}}-(CH_2)_4-\overset{O}{\overset{\|}{C}}-NH(CH_2)_6-NH \right]_n}$$

This particular "nylon" is specifically Nylon 66; the two 6's reflect the number of carbons in the starting diacid and in the diamine. This is the "nylon" used in fabrics, stockings, *etc.*

Another nylon, Nylon 6, is made by polymerizing the cyclic amide, or *lactam*, caprolactam. Nylon 6 is not as suitable for fabric use as is Nylon 66, but it is cheaper and is used, for example, in tire cord and carpeting.

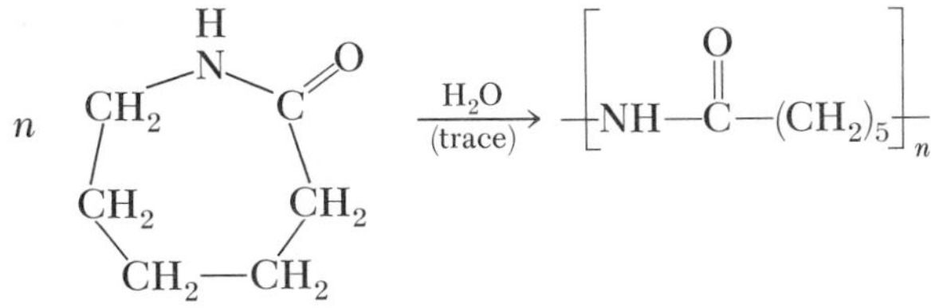

$$n \; \text{caprolactam (cyclic } \text{—NH—C(=O)—(CH}_2)_5\text{—)} \xrightarrow[\text{(trace)}]{H_2O} \left[NH-\overset{O}{\overset{\|}{C}}-(CH_2)_5 \right]_n$$

caprolactam — Nylon 66

Polyesters and polyamides are unlike polymers produced by the reactions of radicals with, for example, alkenes. Polyesters and polyamides are **condensation polymers,** produced by splitting out a small molecule as two units come together. Alkenes, on the other hand, produced addition polymers (see pp. 245–246). One practical difference between these two types is that the condensation reaction is often reversible by a mild hydrolysis, while the addition reaction is not. This accounts, at least in part, for the greater durability associated with polyalkenes than with polyesters or polyamides.

IMPORTANT TERMS

Acylium ion: A carbocation in which the bulk of the positive charge is on the carbon of a carbonyl group, $R-\overset{\oplus}{C}=O$. The reactive intermediate in the Friedel-Crafts acylation reaction.

Acyl-oxygen fission: A bond cleavage reaction in which an ester group is cleaved between the carbonyl carbon and the single-bonded oxygen.

$$R-\overset{O}{\overset{\|}{C}}-OR' + Nu^{\ominus} \rightarrow R-\overset{O}{\overset{\|}{C}}-Nu + OR^{\ominus}$$

Alkyl-oxygen fission: A bond cleavage reaction in which an ester group is cleaved between the single-bonded oxygen and the non-carbonyl carbon.

$$R-\overset{O}{\overset{\|}{C}}-OR' + Nu^{\ominus} \rightarrow R-\overset{O}{\overset{\|}{C}}-O^{\ominus} + R'Nu$$

Arndt-Eistert synthesis: A procedure for converting a carboxylic acid into its next higher homolog. A key step in the process is the conversion of a diazoketone to a ketene. This single step is the *Wolff rearrangement.*

$$R-CO_2H \rightsquigarrow R-CH_2-CO_2H \qquad \text{Arndt-Eistert synthesis}$$

Coenzyme A: A complex, naturally occurring compound possessing a thiol group (—SH). Its biological function centers on the transfer of acetyl groups ($CH_3C(O)$—) from one substance to another. The compound is unstable in air and is converted into a biologically inactive material.

Decarboxylation: A reaction in which a compound, usually a carboxylic acid, loses carbon dioxide.

Dissociation constant: A number that characterizes the strength of an acid (or base). For an acid, HA, ionizing in water to produce $H_3O^{\oplus}$ and $A^{\ominus}$, the dissociation constant, K_a, is given by

$$K_a = \frac{[H_3O^{\oplus}][A^{\ominus}]}{[HA]}$$

where [HA] = concentration of undissociated acid

For a base, B, which reacts with water to form hydroxide ion and protonated base, $BH^{\oplus}$, the dissociation constant, K_b, is given by

$$K_b = \frac{[BH^{\oplus}][OH^{\ominus}]}{[B]}$$

where [B] = concentration of unprotonated base

Some typical values of K_a and K_b are:

HCO_2H (formic acid) $K_a = 1.8 \times 10^{-4}$	CH_3CO_2H (acetic acid) 1.8×10^{-5}	H_3BO_3 (boric acid) 5.8×10^{-10}
NH_3 (ammonia) $K_b = 1.8 \times 10^{-5}$	C_5H_5N (pyridine) 1.6×10^{-9}	$(C_6H_5)_2NH$ (diphenylamine) 6.9×10^{-14}

Fat: A solid triester whose alcohol is glycerol (1,2,3-trihydroxypropane) and whose acid is one (or more) straight chain saturated or unsaturated carboxylic acid(s). The most commonly encountered acids are palmitic acid, stearic acid, oleic acid [(**Z**)-9-octadecenoic acid] and linoleic acid [(**Z,Z**)-9,12-octadecenedioic acid].

$CH_3(CH_2)_{14}CO_2H$ palmitic acid
$CH_3(CH_2)_{16}CO_2H$ stearic acid
$CH_3(CH_2)_7CH{=}CH(CH_2)_7CO_2H$ oleic acid
$CH_3(CH_2)_4CH{=}CHCH_2CH{=}CH(CH_2)_7CO_2H$ linoleic acid

Fischer esterification: The reaction of a carboxylic acid and an alcohol, in the presence of a trace of strong acid, to form an ester and water. The reaction may be driven to completion by the addition of an excess of acid (or alcohol) or by removal of water as it is formed.

Lactam: An amide in which the amide linkage is part of a cyclic structure.

H
N
H_2C C=O
a lactam
H_2C—CH_2
azacyclopentan-2-one

Lactone: An ester in which the ester linkage is part of a cyclic structure.

O
H_2C C=O
a lactone
H_2C—CH_2
oxacyclopentan-2-one

Oil: A triacylglycerol that melts below room temperature. Unsaturation lowers the melting point of triacylglycerols, so that saturated acids predominate in fats, while unsaturated acids predominate in oils. Oils become rancid chiefly through the oxidation of their double bonds.

Peroxy acid: A compound that contains the —C(O)O_2H group. Sometimes called a "peracid." Peroxy acids are useful as oxidants.

Principle of microscopic reversibility: If a forward reaction follows one pathway, the reverse reaction will follow the reverse pathway. In other words, the path of lowest energy over the barrier separating products and starting materials is the same, and involves the same transition state, for both the forward and the reverse reactions. This principle permits a single energy profile to describe both a given reaction and the reverse of that reaction.

Saponification: The base-catalyzed hydrolysis of an ester.

Soap: The sodium (or potassium) salt of a long-chain carboxylic acid.

Transesterification: The conversion of one ester into another that possesses a different alcohol fragment. Transesterification usually is accomplished by refluxing a solution of an ester with an excess of alcohol in the presence of a trace of strong acid.

$$RCO_2R' + R''OH \xrightarrow{H^{\oplus}} RCO_2R'' + R'OH$$

Triacylglycerol: A triester of glycerin (glycerol). The term encompasses both "fat" and "oil." An older, less desirable term is "triglyceride."

Triglyceride: An old term for "triacylglycerol."

Wolff rearrangement: The conversion of a diazoketone to a ketene and nitrogen. The process usually is consummated by reaction of the ketene with an alcohol (or amine) to form an ester (or amide).

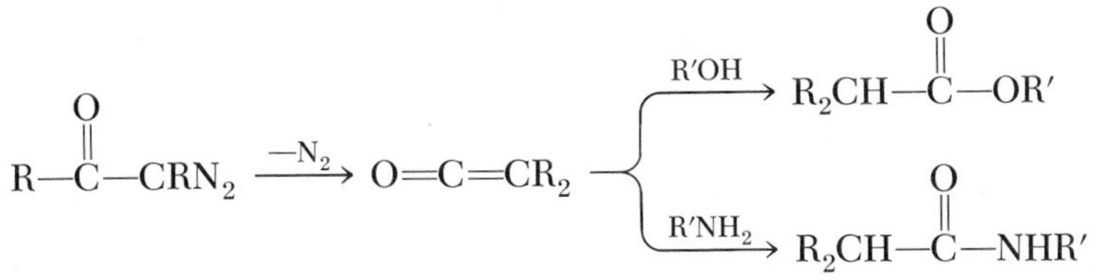

$$R{-}\overset{O}{\overset{\|}{C}}{-}CRN_2 \xrightarrow{-N_2} O{=}C{=}CR_2 \begin{cases} \xrightarrow{R'OH} R_2CH{-}\overset{O}{\overset{\|}{C}}{-}OR' \\ \xrightarrow{R'NH_2} R_2CH{-}\overset{O}{\overset{\|}{C}}{-}NHR' \end{cases}$$

PROBLEMS

17. Write the skeletal structures for the acyclic saturated six-carbon monocarboxylic acids. Provide an IUPAC name for each.

18. Provide acceptable names for the following compounds.

(a) CF_3CO_2H

(b) $CF_3C(O)OC(O)CF_3$

(c) $CF_3C(O)NH_2$

(d) CF_3CN

(e) 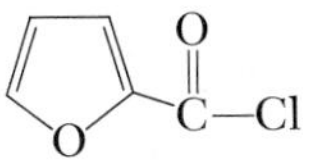 2-furyl–C(=O)–Cl (furan ring bearing $\overset{O}{\overset{\|}{C}}{-}Cl$)

(f) 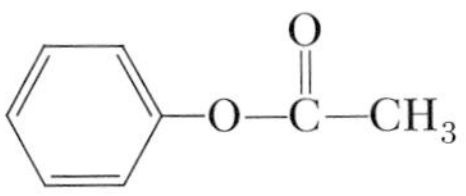 $C_6H_5{-}O{-}\overset{O}{\overset{\|}{C}}{-}CH_3$

(g) $H_3C{-}O{-}\overset{O}{\overset{\|}{C}}{-}C_6H_5$

(h) $C_6H_5{-}CH_2{-}O{-}\overset{O}{\overset{\|}{C}}{-}C_6H_5$

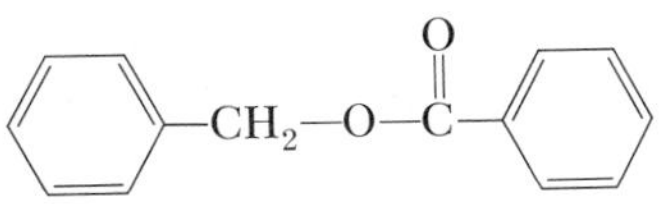

(i) 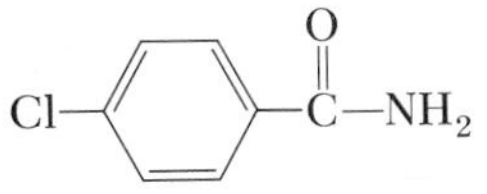 $Cl{-}C_6H_4{-}\overset{O}{\overset{\|}{C}}{-}NH_2$

(j) 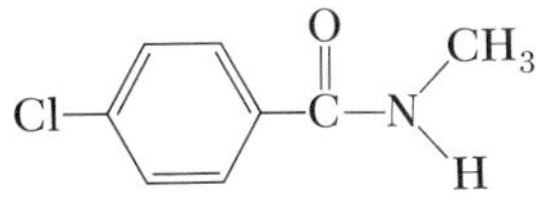 $Cl{-}C_6H_4{-}\overset{O}{\overset{\|}{C}}{-}N(CH_3)(H)$

(k) 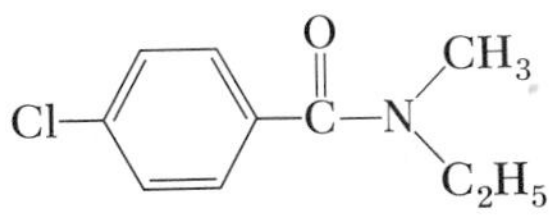 $Cl{-}C_6H_4{-}\overset{O}{\overset{\|}{C}}{-}N(CH_3)(C_2H_5)$

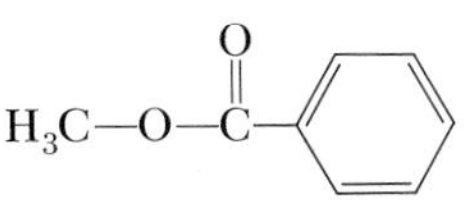

(l) $(CH_3)_2C(CO_2H)_2$

(m) $H_2C(CO_2C_2H_5)_2$

(n) $H-\underset{C_6H_5}{\overset{Cl}{C}}-CH_2CO_2H$

(o) CO_2H, CO_2H

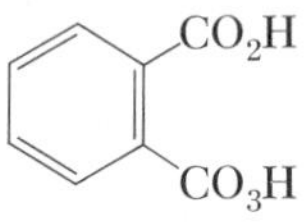

(p) O, =O, O

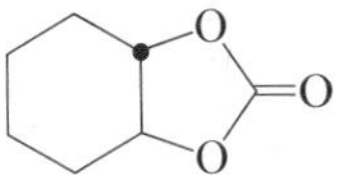

(q) O, =O, O

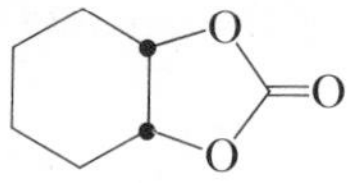

(r) CO_2H, OH

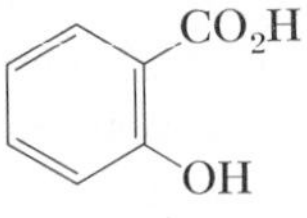

(s) O, $\overset{\|}{C}-OCH_3$, OH

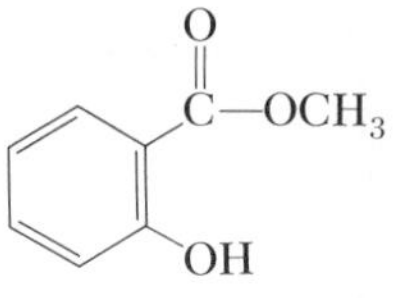

(t) CO_2H, Cl

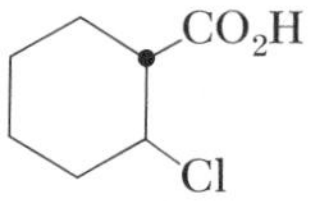

(u) CN

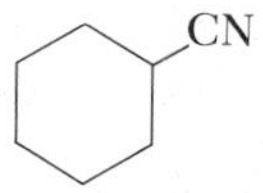

(v) HCO_3H

(w) CH_3, Cl, C=C, HO_2C, C_2H_5

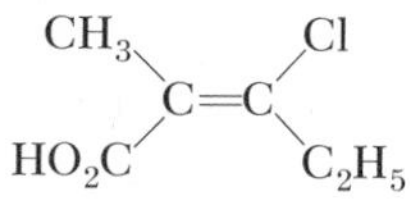

(x) $CH_3-\overset{O}{\overset{\|}{C}}-CH-N_2$

(y) $CH_2=C=O$

(z) CO_2H, CO_2H

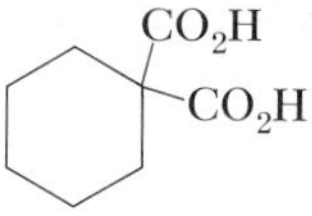

19. List the following in order of increasing acidity: (a) propanoic acid, (b) hydrochloric acid, (c) acetylene, (d) 1-propanol, (e) pentane, (f) 2-chloropropanoic acid, (g) 3-chloropropanoic acid.

20. List the following in order of increasing ability to confer water solubility when they replace a hydrogen of an alkane: (a) chloro, (b) hydroxy, (c) methoxy, (d) carboxy, (e) deuterium.

21. (a) The procedure presented below was used to separate a carboxylic acid from an ester. Explain the significance of each step. Does this process contain any unnecessary steps?

The mixture of acid and ester was a paste and was, therefore, dissolved in chloroform. The chloroform solution was washed with a solution of 10% aqueous sodium bicarbonate. The aqueous phase was separated and then treated with dilute hydrochloric acid. The resulting precipitate was washed with water and air dried to afford the desired acid. The chloroform layer was dried with anhydrous sodium sulfate, the drying agent was removed by filtration, and the dried solution was concentrated to afford the crude ester. This crude ester was distilled to afford the pure product.

(b) Why would it have been pointless to have substituted sodium chloride for sodium bicarbonate in the extraction step? Why was it unnecessary to use sodium hydroxide instead of sodium bicarbonate?

(c) Anhydrous sodium carbonate is a good drying agent. Why wasn't it used instead of anhydrous sodium sulfate?

22. Complete the following table by providing the missing entries. The first one has been done for you.

ACID +	*BASE*	→ *CONJUGATE BASE OF* + *ACID*	*CONJUGATE ACID OF BASE*
H_2O	H_2O	$OH^{\ominus}$	$H_3O^{\oplus}$
H_2O	$H^{\ominus}$	$OH^{\ominus}$	*a*
H_2SO_4	H_2O	*b*	$H_3O^{\oplus}$
c	$OH^{\ominus}$	$CF_3CO_2^{\ominus}$	HOD
H_2SO_4	CH_3OH	*d*	*e*
Cl_2CO_2H	CH_3CO_2H	*f*	*g*
h	*i*	$Cl^{\ominus}$	$NH_4^{\oplus}$
j	*k*	H_2O	H_2O

23. Benzamidine is about 10^{14} times as potent a base as is benzamide. How do you account for this? If the pK_b of benzamidine is 2.4, what is the pK_b of benzamide?

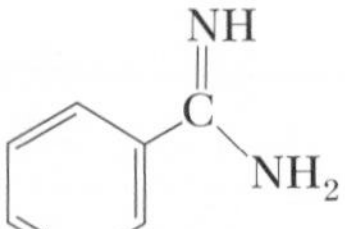

NH, C, NH_2

benzamidine

24. Sodium benzoate is frequently added to foods (*e.g.*, carbonated soft drinks) to inhibit the growth of microorganisms and, thereby, act as a preservative. However, to be most effective, the food must have a pH of less than 4.5. What does this suggest about the nature of the actual preservative?

25. Starting with acetic acid or benzoic acid, and anything else required, suggest a rational synthesis for each of the following.

(a) acetyl chloride
(b) acetic anhydride
(c) acetophenone
(d) benzaldehyde
(e) benzoyl chloride
(f) benzamide
(g) propanoic acid
(h) propionitrile
(i) phenylacetic acid
(j) ethanol
(k) dibenzyl carbonate
(l) benzyl acetate
(m) ethyl acetate
(n) sodium propionate
(o) cyanobenzene (benzonitrile)
(p) cyclohexylcarbinol
(q) bromomethane
(r) benzyl bromide
(s) benzyl phenyl ketone

26. When ammonia reacts with a simple acid anhydride, such as acetic anhydride, the product is acetamide, CH_3CONH_2. Very little of the corresponding imide, $CH_3C(O)—NH—C(O)CH_3$, is formed. However, if a cyclic anhydride is used the major product is a cyclic imide, *e.g.*,

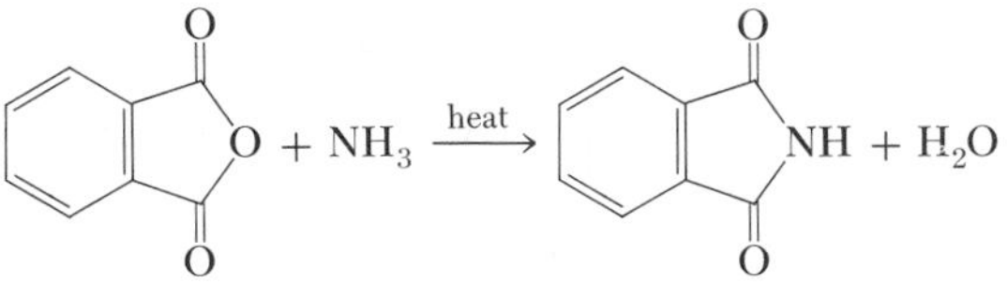

$$\text{(phthalic anhydride)} + NH_3 \xrightarrow{\text{heat}} \text{(phthalimide)} + H_2O$$

phthalimide

What factor(s) might cause this difference in behavior?

27. The behavior of various hydroxy acids toward heating is summarized below. Suggest a structure for each of the products, rationalizing the variation in behavior.

(a) 4-hydroxybutanoic acid → $C_4H_6O_2$ (insoluble in dilute base)
(b) 3-hydroxybutanoic acid → $C_4H_6O_2$ (soluble in dilute base)
(c) 2-hydroxypropanoic acid (lactic acid) → $C_6H_8O_4$ (insoluble in dilute base)

28. While ammonia is not very acidic, amides are weakly acidic and imides are even more acidic. How do you account for this order?

29. While most elements do not form covalent bonds to themselves, a few compounds containing three contiguous nitrogens are known, *e.g.*, hydrazoic acid. How, then, can you explain the stability of the tetrazole ring system, which contains a "chain" of four contiguous nitrogens?

30. Ketene, $H_2C{=}C{=}O$, can be made by the pyrolysis of acetic acid at 700°. This internal acid anhydride undergoes the reactions outlined below. Present a mechanism for each of these.

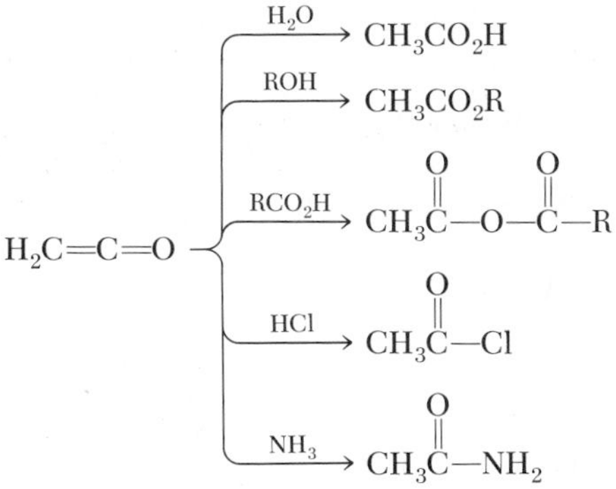

$$H_2C{=}C{=}O \xrightarrow{H_2O} CH_3CO_2H$$
$$H_2C{=}C{=}O \xrightarrow{ROH} CH_3CO_2R$$
$$H_2C{=}C{=}O \xrightarrow{RCO_2H} CH_3C({=}O){-}O{-}C({=}O){-}R$$
$$H_2C{=}C{=}O \xrightarrow{HCl} CH_3C({=}O){-}Cl$$
$$H_2C{=}C{=}O \xrightarrow{NH_3} CH_3C({=}O){-}NH_2$$

31. Thalidomide, the sedative and sleep-producer that led to many deformed babies in the 1960's, is synthesized as shown below. Suggest mechanisms for these reactions.

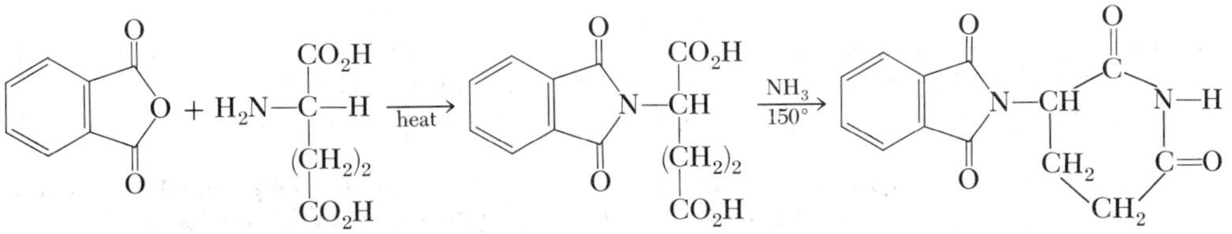

$$\text{(phthalic anhydride)} + H_2N{-}CH(CO_2H){-}(CH_2)_2{-}CO_2H \xrightarrow{\text{heat}} \text{(phthalimido)}N{-}CH(CO_2H){-}(CH_2)_2{-}CO_2H \xrightarrow[150^\circ]{NH_3} \text{thalidomide}$$

glutamic acid

thalidomide

32. Suggest a mechanism for the acylation of diazomethane with acetyl chloride to produce diazoacetone.

$$CH_3{-}\overset{\overset{\displaystyle O}{\|}}{C}{-}Cl + CH_2N_2 \rightarrow CH_3{-}\overset{\overset{\displaystyle O}{\|}}{C}{-}CHN_2$$

diazoacetone

33. Caprolactam, the starting material for Nylon 6, can be prepared in the laboratory by starting with cyclohexanone. Suggest how this might be done.

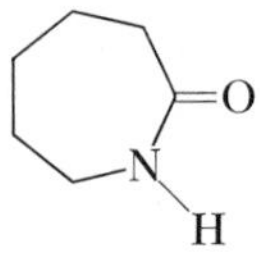

caprolactam

34. (a) While many α,β-unsaturated carboxylic acids undergo decarboxylation upon heating, the α,β-unsaturated acid shown below is immune to such a reaction. What does this suggest about the mechanism of decarboxylation of α,β-unsaturated acids?

$$CH_3{-}\underset{\underset{\displaystyle CH_3}{|}}{\overset{\overset{\displaystyle CH_3}{|}}{C}}{-}CH{=}CH{-}CO_2H$$ does not decarboxylate readily

(b) Explain why there should be a difference in the ease of decarboxylation of the following acids, both of which are β,γ-unsaturated.

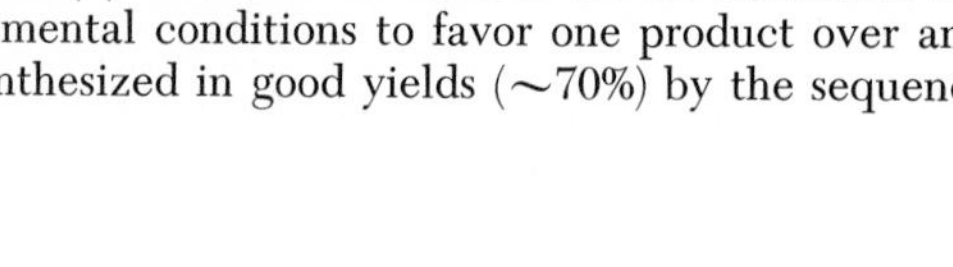

CO_2H and $(CH_2)_{10}$ CO_2H

35. Cyclic carbonates (*e.g.*, ethylene carbonate) can be prepared by reacting phosgene with a diol (*e.g.*, ethylene glycol). These same reagents, however, can also react to form a polymer. (a) What is the structure of the polymer? (b) What is the mechanism for formation of the cyclic carbonate? (c) How could one vary experimental conditions to favor one product over another?

36. Thiochromanone is synthesized in good yields (~70%) by the sequence outlined below. Provide mechanisms for both steps.

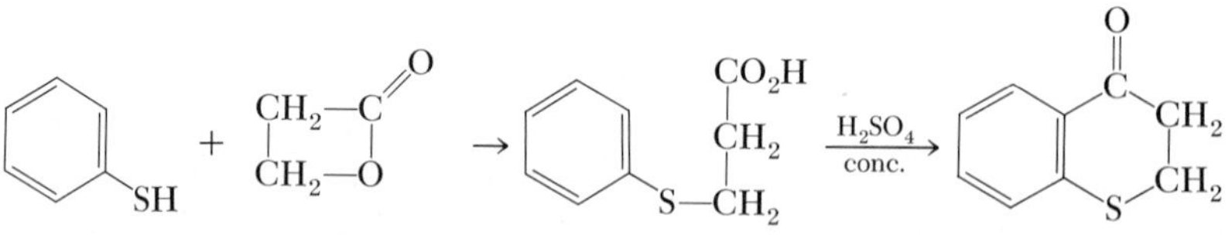

thiophenol + β-propiolactone (carcinogen!) → $C_6H_5{-}S{-}CH_2{-}CH_2{-}CO_2H$ $\xrightarrow[\text{conc.}]{H_2SO_4}$ thiochromanone

37. While aqueous ammonia reacts with acid chlorides to form the corresponding amides, aqueous trimethylamine, $:N(CH_3)_3$ (which is quite water-soluble), reacts with acid chlorides to form carboxylic acids. Working through the mechanism, explain why this difference is observed. (Several suggestions are, in principle, possible.)

38. The hydrolysis of triphenylmethyl benzoate in water does not require the presence of added acid or base and, in fact, is not influenced significantly by small amounts of added acid or base. Account for these results by suggesting a mechanism for the reaction.

39. The equilibrium constant for ionization of an acid is related to the free energy difference between product and starting material by

$$\Delta G = -2.3\,RT \log K_a$$

Using this equation, show that making ionization more favorable by 1.4 kcal/mole is equivalent to lowering the pK_a by one unit at 25°C.

40. Assuming an equilibrium constant of about 4 for an esterification, calculate the free energy change accompanying the reaction. (See problem 39.)

41. Phthalic anhydride is converted to a diacid dichloride, *o*-phthaloyl dichloride, upon reaction with phosphorus pentachloride. *o*-Phthaloyl dichloride reacts with anhydrous aluminum chloride to form the isomer **A.** (a) Suggest a mechanism for this isomerization.

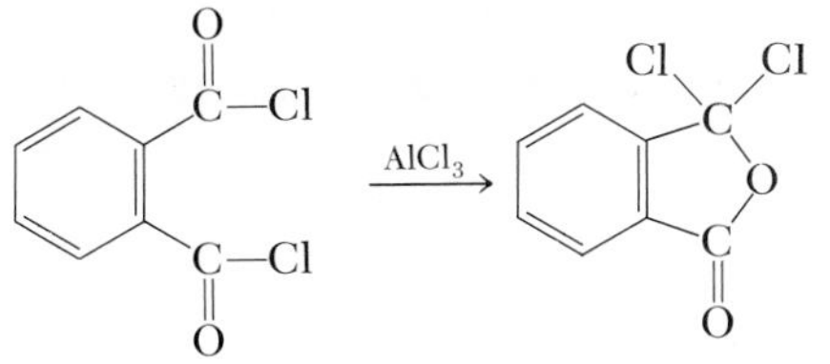

o-phthaloyl dichloride $\xrightarrow{AlCl_3}$ **A**

Compound **A** reacts with ammonia to produce, after acidification, *o*-cyanobenzoic acid. (b) Suggest a mechanism for this reaction. (*Hint:* An intermediate, $C_8H_5NO_2$, is formed in the reaction.)

$$\mathbf{A} \xrightarrow{NH_3} C_8H_5NO_2 \xrightarrow{H_3O^{\oplus}} o\text{-cyanobenzoic acid}$$

Compound **A** is reduced with a mixture of zinc and hydrochloric acid to phthalide. Upon heating with potassium cyanide (KCN), phthalide is converted to compound **B** ($C_9H_6NO_2^{\ominus}K^{\oplus}$). The salt **B** produces, after refluxing with sulfuric acid, compound **C** (molecular weight 180). (c) Suggest structures for **B** and **C** and mechanisms for the reactions leading to their formation. The nmr spectrum of **C** shows two types of protons.

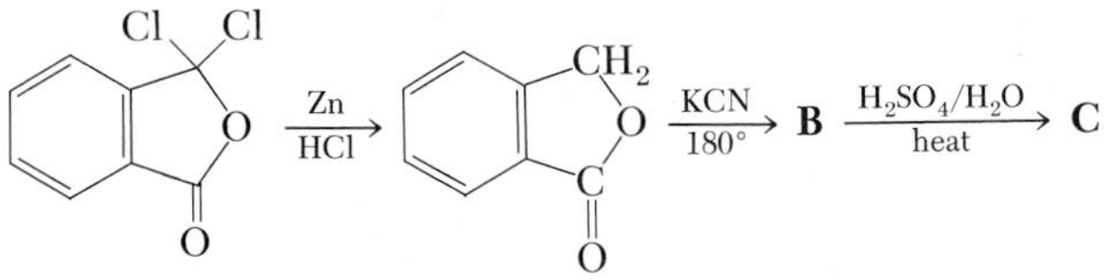

A $\xrightarrow[HCl]{Zn}$ phthalide $\xrightarrow[180°]{KCN}$ **B** $\xrightarrow[heat]{H_2SO_4/H_2O}$ **C**

42. Carboxylic acids can be prepared by the hydrolysis of the trichloromethyl group. Suggest mechanisms for the sequence shown below, which at one time was an important industrial synthesis of benzoic acid.

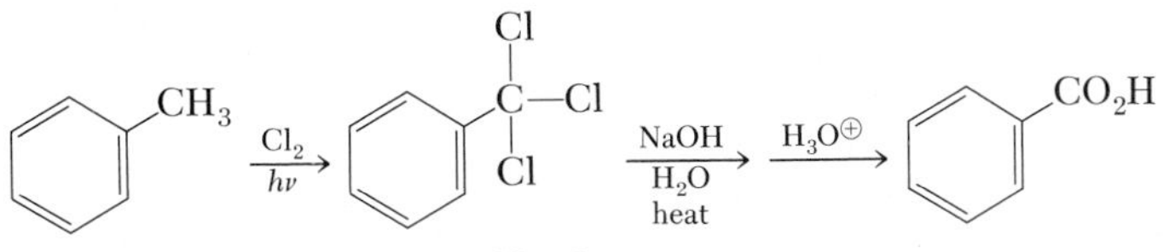

$C_6H_5CH_3 \xrightarrow[h\nu]{Cl_2} C_6H_5CCl_3$ (benzotrichloride) $\xrightarrow[\text{heat}]{NaOH,\ H_2O} \xrightarrow{H_3O^{\oplus}} C_6H_5CO_2H$

43. Diphenylketene, $(C_6H_5)_2C{=}C{=}O$, is prepared by the sequence outlined below. Suggest a mechanism for steps **1** and **3.**

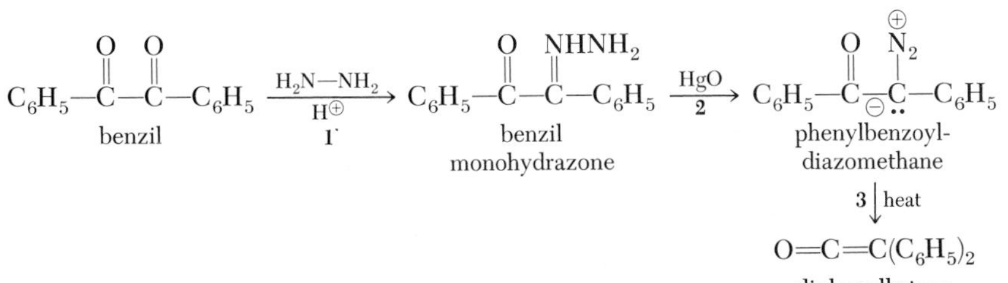

$$C_6H_5\text{—}\overset{O}{\overset{\|}{C}}\text{—}\overset{O}{\overset{\|}{C}}\text{—}C_6H_5 \xrightarrow[1]{H_2N\text{—}NH_2,\ H^{\oplus}} C_6H_5\text{—}\overset{O}{\overset{\|}{C}}\text{—}\overset{NHNH_2}{\overset{\|}{C}}\text{—}C_6H_5 \xrightarrow[2]{HgO} C_6H_5\text{—}\overset{O}{\overset{\|}{C}}\text{—}\overset{\overset{\oplus}{N_2}}{\overset{|}{\underset{\ominus}{C}}}\text{—}C_6H_5$$

benzil; benzil monohydrazone; phenylbenzoyl-diazomethane

$\xrightarrow[3]{heat}$ $O{=}C{=}C(C_6H_5)_2$

diphenylketene

°44. When graphite is oxidized with fuming nitric acid, a potent oxidant, it is converted to mellitic anhydride, $C_{12}O_9$. This oxide of carbon reacts with boiling water to produce mellitic acid, $C_{12}H_{16}O_{12}$. The infrared spectrum of mellitic anhydride exhibits strong "anhydride" absorptions, while the spectrum of mellitic acid is shown below. Assign structures to these two compounds.

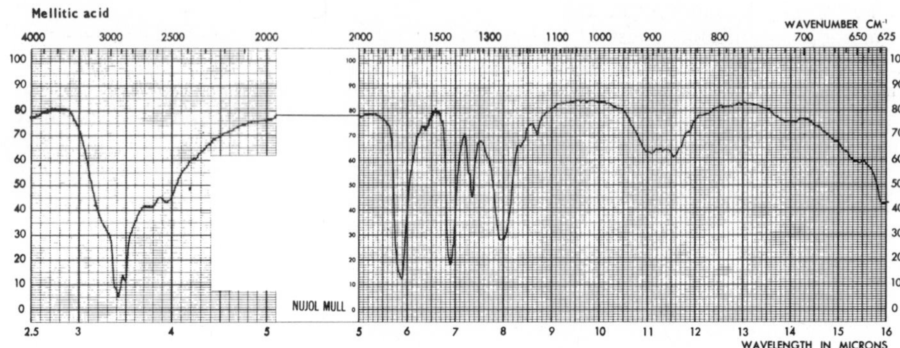

Problem 44: Infrared spectrum of mellitic acid.

†45. When 4-pentenoic acid is treated with an acid, there results a compound, $C_5H_8O_2$, which is insoluble in acidic or alkaline aqueous media. The compound's nmr spectrum is virtually unchanged in going from $CDCl_3$ to $(CD_3)_2SO$, while its ir spectrum shows no significant absorption above 3300 cm^{-1}. Suggest a structure for this compound. How is it formed?

46. The reaction of phthalic anhydride with hydrogen peroxide in the presence of sodium carbonate, followed by acidification, produces monoperoxyphthalic acid. Suggest a mechanism for this reaction.

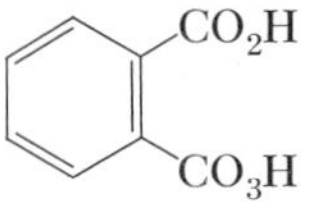

CO_2H

CO_3H

monoperoxyphthalic acid

47. Nitriles are converted to amides by hydrogen peroxide and dilute base (sodium hydroxide).

$$\text{R—C}\equiv\text{N} + 2\text{H}_2\text{O}_2 \xrightarrow{\text{OH}^\ominus} \text{R—CONH}_2 + \text{O}_2 + \text{H}_2\text{O}$$

If the reaction is carried out with labeled hydrogen peroxide, $H_2{}^{18}O_2$, dissolved in unlabeled water, the amide's acyl oxygen is labeled with ^{18}O. Considering that dilute sodium hydroxide, alone, will not cause this nitrile hydrolysis and that the rate equation for the reaction is as shown below, suggest a mechanism for the reaction.

$$R = k[\text{RCN}][\text{H}_2\text{O}_2][\text{OH}^\ominus]$$

°This problem requires knowledge of infrared spectroscopy.

†This problem requires knowledge of ir and nmr spectroscopy.

SYNTHESIS OF CARBON-CARBON BONDS EMPLOYING ESTERS AND OTHER ACID DERIVATIVES

19

19.1 INTRODUCTION

A parallel in the chemical behavior of aldehydes, ketones, and carboxylic acid derivatives already has been demonstrated. For instance, it has been shown that carbon-oxygen double bonds in all three classes are prone to attack by nucleophiles. *Differences* between them arise because the acid derivatives contain leaving groups bonded to the acyl carbon, while the ketones and aldehydes do not possess leaving groups bonded to the carbonyl carbon. Reactions with acid derivatives normally result in substitution *via* addition-elimination, while reactions involving carbonyl compounds usually result in addition and, on occasion, the conversion of the C=O bond to a C=Z bond (where Z is commonly nitrogen).

$$\underset{\text{aldehyde (ketone)}}{R{-}\overset{\overset{\displaystyle O}{\|}}{C}{-}H(R)} + NH_3 \rightarrow \left[R{-}\underset{\underset{\displaystyle NH_2}{|}}{\overset{\overset{\displaystyle OH}{|}}{C}}{-}H(R)\right] \rightarrow \underset{\text{imine}}{R{-}\overset{\overset{\displaystyle NH}{\|}}{C}{-}H(R)}$$

$$\underset{\text{ester}}{R{-}\overset{\overset{\displaystyle O}{\|}}{C}{-}OR} + NH_3 \rightarrow \left[R{-}\underset{\underset{\displaystyle NH_2}{|}}{\overset{\overset{\displaystyle OH}{|}}{C}}{-}OR\right] \rightarrow \underset{\text{amide}}{R{-}\overset{\overset{\displaystyle O}{\|}}{C}{-}NH_2}$$

This chapter will extend the parallel between aldehydes, ketones, and acids and their derivatives to include reactions occurring at the carbon α to the carbon-oxygen double bond. Much of this similarity between carbonyl and acyl compounds is due to the acidifying effect that an acyl group exerts upon hydrogens bonded to Cα. Most of the compounds discussed in this chapter will be esters, for while the acidity of esters is orders of magnitude less than that of similar ketones, it is sufficient for carbanion production at Cα.

As noted earlier in this text, the decreased acidity of esters, compared to carbonyl compounds, is a result of less resonance stabilization of the Cα carbanion by the acyl group compared to the carbonyl group. This, in turn, stems from the fact that any stabilization of a carbanion by an ester group will result in a decreased resonance interaction between the alkyl oxygen and the acyl group of the ester moiety.

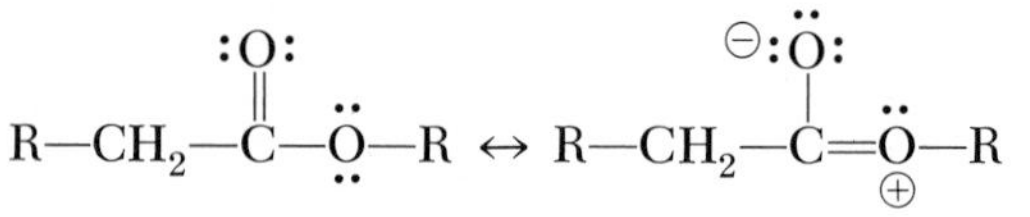

$$R{-}CH_2{-}\overset{:O:}{\overset{\|}{C}}{-}\ddot{\underset{..}{O}}{-}R \leftrightarrow R{-}CH_2{-}\overset{\ominus:\ddot{O}:}{\overset{|}{C}}{=}\underset{\oplus}{\ddot{O}}{-}R$$

resonance within the ester group

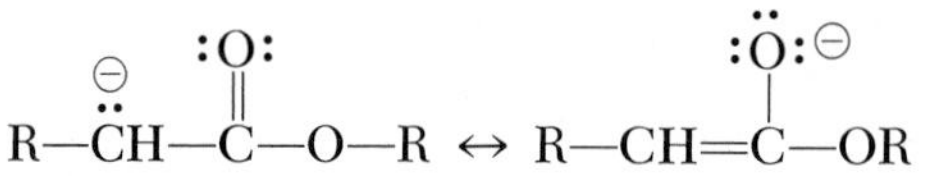

$$R{-}\overset{\ominus}{\ddot{C}}H{-}\overset{:O:}{\overset{\|}{C}}{-}O{-}R \leftrightarrow R{-}CH{=}\overset{:\ddot{O}:\ominus}{\overset{|}{C}}{-}OR$$

carbanion stabilized by an ester group

A great deal of carbanion chemistry will be presented (with mechanistic interpretation) in this chapter because carbanions derived from esters and other acid derivatives are extremely important in synthetic organic chemistry. Indeed, some of the most valuable ways to create new carbon-carbon bonds will be discussed here. The significance of the material in this chapter will be more apparent if you recall the analogous reactions of aldehydes and ketones at every opportunity.

In Chapter 17 (Figure 17–1) we described the alcoholic fermentation of glucose. At the end of this chapter we will examine an alternate fate for the pyruvic acid produced by this scheme, a pathway usually called the Krebs cycle or the citric acid cycle.

19.2 SYNTHESIS OF α-HALO ACIDS AND ESTERS

The most often used entry into a series of α-halo acids and their derivatives involves the reaction of an aliphatic acid with a halogen (bromine or chlorine) in the presence of a catalytic amount of phosphorus trihalide. One possible mechanism for this **Hell-Volhard-Zelinsky reaction** is shown below.

$$RCH_2CO_2H + Cl_2 \xrightarrow{PX_3} RCHClCO_2H + HCl$$

Hell-Volhard-Zelinsky reaction

a possible mechanism:

$$3RCH_2CO_2H + PX_3 \rightarrow \underset{\text{acyl halide}}{3RCH_2COX} + H_3PO_3$$

$$RCH_2COX \rightleftarrows \underset{\text{enol}}{RCH{=}CX(OH)} \xrightarrow{Cl_2} \underset{\alpha\text{-chloro acyl halide}}{RCHCl{-}COX} + HCl$$

$$RCH_2CO_2H + RCHClCOX \rightleftarrows RCHClCO_2H + RCH_2COX$$

The function of the phosphorus trihalide is to convert the acid, which enolizes with difficulty, into the more readily enolized acyl halide. It is the enol derived from the acyl halide which then reacts with the chlorine (or other halogen) to form an α-chloro acyl halide. Equilibration of the α-chloro acyl halide and the unreacted acid generates the desired α-halo acid and more acyl halide. It is this last equilibration, between acid and α-halo acyl halide, that makes it unnecessary to use a large amount of phosphorus trihalide. Overall yields in the sequence may approach 100%.

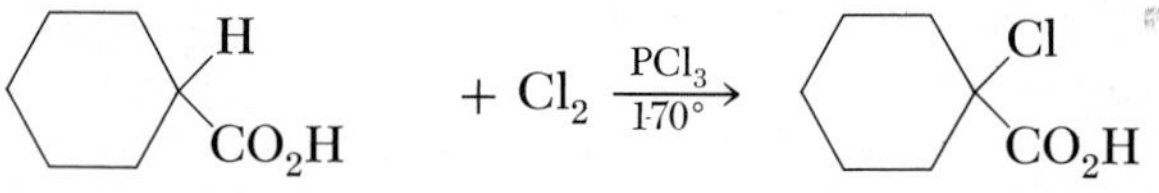

$$\underset{\text{cyclohexanecarboxylic acid}}{C_6H_{11}CO_2H} + Cl_2 \xrightarrow[170^\circ]{PCl_3} \underset{\alpha\text{-chlorocyclohexanecarboxylic acid}}{C_6H_{10}(Cl)CO_2H} \quad (94\%)$$

$$C_6H_5-CH_2CO_2H + Br_2 \xrightarrow[C_6H_6;\ heat]{PCl_3} C_6H_5-\underset{Br}{\overset{H}{C}}-CO_2H \quad (60\%)$$

phenylacetic acid → α-bromophenylacetic acid

$$CH_3(CH_2)_4CO_2H \xrightarrow[Br_2;\ 100^\circ]{PCl_3} CH_3(CH_2)_3CHBrCO_2H \quad (83\%)$$

caproic acid → α-bromocaproic acid

1. Cyclohexanecarboxylic acid reacts with chlorine at 270° (in the absence of phosphorus trichloride) to form all of the possible monochloro cyclohexanecarboxylic acids. Suggest a mechanism for these reactions. Draw all of the possible products, indicating which can exist as enantiomers.
2. Suggest a reason why acyl halides enolize more readily than do acids. (*Hint:* Consider the effect of resonance upon the stability of acids and acid chlorides.)

A second valuable, and apparently general, synthesis of α-bromo acids involves the reaction of *N*-bromosuccinimide with an acid chloride in refluxing carbon tetrachloride. Some questions still remain about the mechanism of this reaction.

The importance of α-halo acids stems from their facile conversion to other carboxylic acid derivatives. Some typical reactions of α-halo acids are presented in Figure 19–1, pages 590 and 591. Since simple monoesters do not undergo direct halogenation cleanly (see below), simple α-halo esters usually are prepared by esterification of α-halo acids.

$$RCHCl-CO_2H \xrightarrow[ether]{SOCl_2} RCHCl-COCl \xrightarrow{MeOH} RCHCl-CO_2CH_3$$

(directly: $RCHCl-CO_2H \xrightarrow{CH_2N_2/ether} RCHCl-CO_2CH_3$)

Aliphatic carboxylic acids are not readily halogenated by iodine, bromine, or chlorine. Indeed, they are sufficiently inert to the halogens that acetic acid commonly is used as a solvent for ionic brominations. Attempts to halogenate *esters* usually are of little preparative value because simple esters often react to form a variety of products. This lack of reactivity or selectivity is to be contrasted to the behavior of β-diesters, β-keto esters, and β-diketones. These compounds are rather acidic, their carbanions being stabilized by two functional groups, and they readily undergo base-catalyzed halogenation at the acidic or *activated* methylene group. Thus, beginning with a β-diester, it is possible to synthesize α-halo acids, compounds of some synthetic importance, in reasonable yields.

$$R-\overset{O}{\overset{\|}{C}}-OH(R') + X_2 \rightarrow \text{N.R. } (\textit{or} \text{ variety of products})$$

$$\underset{\text{a }\beta\text{-diester}}{C_2H_5O-\overset{O}{\overset{\|}{C}}-CH_2-\overset{O}{\overset{\|}{C}}-OC_2H_5} \xrightarrow{B^\ominus} C_2H_5O-\overset{O}{\overset{\|}{C}}-\overset{\ominus}{\underset{\cdot\cdot}{C}H}-\overset{O}{\overset{\|}{C}}-OC_2H_5$$

$$\downarrow X_2 \quad (X = Cl,\ Br,\ I)$$

$$\left[HO\overset{O}{\overset{\|}{C}}-CHX-\overset{O}{\overset{\|}{C}}OH\right] \xleftarrow[heat]{H_3O^\oplus} C_2H_5O-\overset{O}{\overset{\|}{C}}-CHX-\overset{O}{\overset{\|}{C}}-OC_2H_5$$

$$\xrightarrow{-CO_2} H-CHX-\overset{O}{\overset{\|}{C}}-OH$$

an α-halo acid

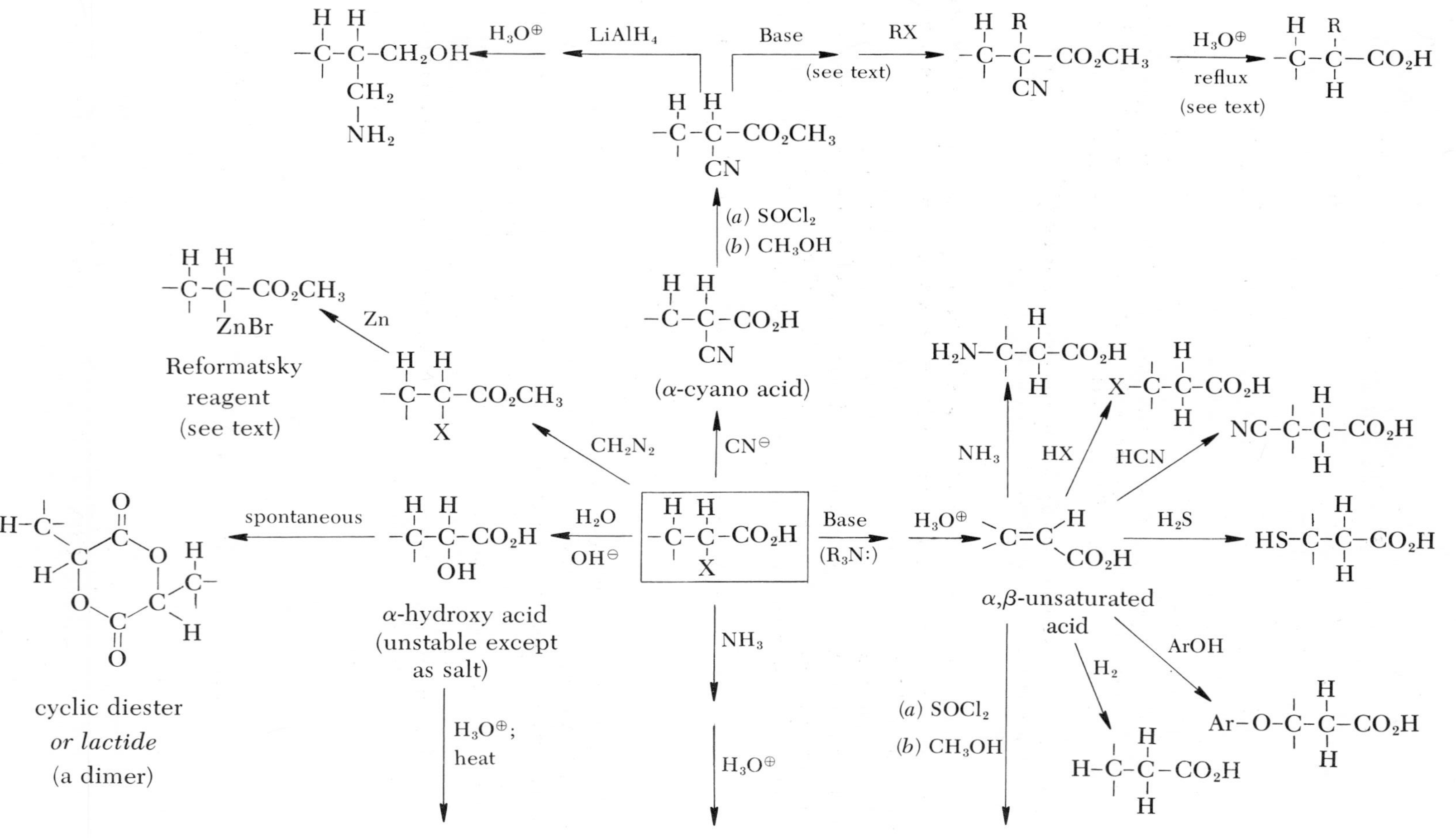

Illustration continued on opposite page.

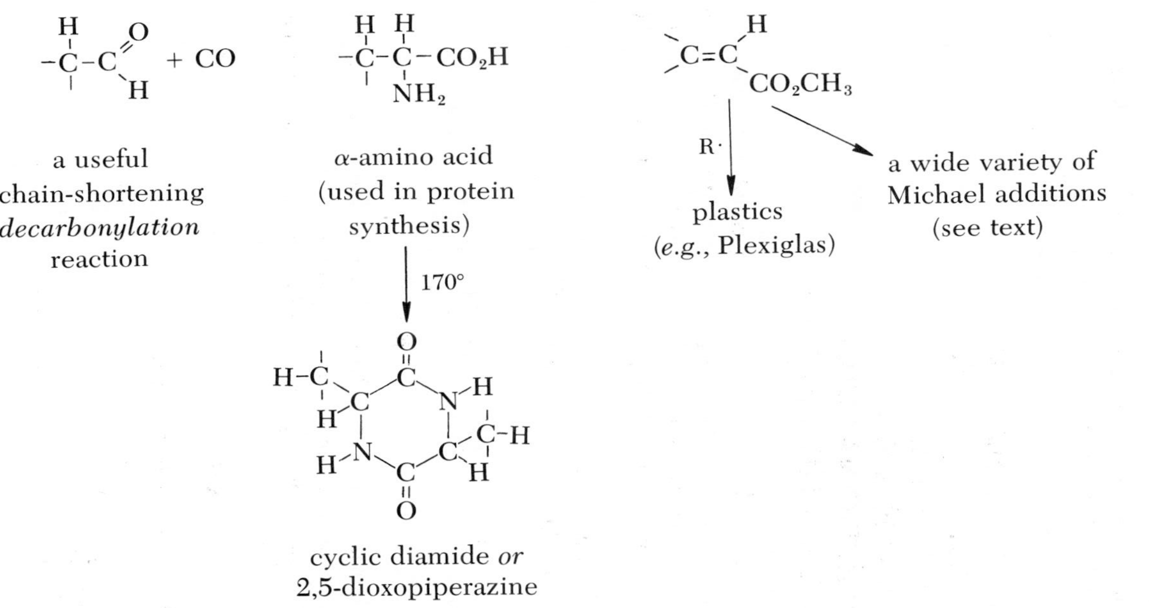

Figure 19-1 Reactions of α-haloacids and derivatives.

3. Write six specific transformations of α-halo acids into other compounds. Suggest a mechanism for at least two of these.
4. Starting with ethylene oxide and methyl iodide as the only organic materials, suggest a synthesis of alanine, $CH_3CH(NH_2)CO_2H$, an important amino acid.

19.3 BASE-PROMOTED CONDENSATION REACTIONS—SYNTHESIS OF FUNCTIONALIZED ACIDS AND ESTERS

The alkylation of anions derived from simple esters is not often encountered because of the low acidity of simple esters. The enhanced acidity typical of β-keto esters and β-diesters is necessary to produce a significant anion concentration and, thereby, a synthetically valuable substrate for alkylation. However, before we examine such alkylations we must learn several routes to these difunctional compounds.

Claisen Ester Condensation. Ethyl acetoacetate (also called *acetoacetic ester*) is prepared by the base-induced self-condensation of ethyl acetate.

$$2CH_3CO_2C_2H_5 \xrightarrow[C_2H_5OH]{C_2H_5O^{\ominus}} \xrightarrow{\text{dil. HCl}} CH_3\overset{\overset{\displaystyle O}{\|}}{C}-CH_2-\overset{\overset{\displaystyle O}{\|}}{C}OC_2H_5$$

acetoacetic ester
(80%)

The mechanism of this condensation is quite similar to that of the aldol condensation, any differences arising because the attacked acyl carbon of an ester possesses a potential leaving group. Stated slightly differently, this reaction is nothing more than the acylation of a carbanion that is (resonance) stabilized by an ester group.

$$C_2H_5O-\overset{\overset{\displaystyle :O:}{\|}}{C}-CH_3 \underset{}{\overset{Na^{\oplus}\ C_2H_5O^{\ominus}}{\rightleftharpoons}} C_2H_5O-\overset{\overset{\displaystyle :O:\ Na^{\oplus}}{\|}}{C}-\overset{\ominus}{\ddot{C}}H_2 \leftrightarrow C_2H_5O-\overset{\overset{\displaystyle :\ddot{O}:^{\ominus}\ Na^{\oplus}}{|}}{C}=CH_2$$

$$CH_3-\overset{\overset{\displaystyle :O:}{\|}}{C}-OC_2H_5 + C_2H_5O-\overset{\overset{\displaystyle :O:}{\|}}{C}-\ddot{C}H_2^{\ominus} \rightleftharpoons CH_3-\overset{\overset{\displaystyle :\ddot{O}:^{\ominus}}{|}}{\underset{\underset{\displaystyle CH_2-\underset{\underset{\displaystyle :O:}{\|}}{C}-OC_2H_5}{|}}{C}}-\ddot{O}C_2H_5 \xrightarrow{-\overset{\ominus}{O}C_2H_5} CH_3-\overset{\overset{\displaystyle :O:}{\|}}{C}-CH_2-\overset{\overset{\displaystyle :O:}{\|}}{C}-OC_2H_5$$

$$\Big\Updownarrow \overset{\ominus}{O}C_2H_5\ Na^{\oplus}$$

$$\left[CH_3\overset{\overset{\displaystyle :\ddot{O}:^{\ominus}\ Na^{\oplus}}{|}}{C}=CH-\overset{\overset{\displaystyle :O:}{\|}}{C}OC_2H_5 \leftrightarrow CH_3\overset{\overset{\displaystyle :O:\ Na^{\oplus}}{\|}}{C}-CH=\overset{\overset{\displaystyle :\ddot{O}:^{\ominus}}{|}}{C}-OC_2H_5 \leftrightarrow CH_3-\overset{\overset{\displaystyle :O:\ Na^{\oplus}}{\|}}{C}-\overset{\ominus}{\ddot{C}}H-\overset{\overset{\displaystyle :O:}{\|}}{C}-C_2H_5\right]$$

sodioacetoacetic ester

$$\Big\downarrow \text{dil. HCl}$$

$$CH_3-\overset{\overset{\displaystyle O}{\|}}{C}-CH_2-\overset{\overset{\displaystyle O}{\|}}{C}-OC_2H_5$$

acetoacetic ester

The need to add dilute acid in order to produce the final product is a consequence of the acidity of the hydrogens of the active methylene group. The acid ensures the conversion of any sodioacetoacetic ester to the final product.

The reaction we have just studied is one example of the **Claisen ester condensation,** a reaction in which an ester with sufficiently acidic protons on the carbon α to the ester acyl group undergoes a base-induced condensation to form a β-keto ester. While acetoacetic ester is the most important β-keto ester, any appropriately substituted β-keto ester can be produced in this way as long as the Cα hydrogens of the prerequisite ester are sufficiently acidic.

$$2\text{R—CH}_2\text{—}\overset{\overset{\text{O}}{\|}}{\text{C}}\text{—OC}_2\text{H}_5 \xrightarrow[\text{C}_2\text{H}_5\text{OH}]{\text{C}_2\text{H}_5\text{O}^\ominus} \text{R—CH}_2\text{—}\overset{\overset{\text{O}}{\|}}{\text{C}}\text{—}\underset{\underset{\text{R}}{|}}{\overset{\overset{\text{H}}{|}}{\text{C}}}\text{—}\overset{\overset{\text{O}}{\|}}{\text{C}}\text{—OC}_2\text{H}_5$$

general Claisen condensation

β-keto ester

$$\text{C}_6\text{H}_5\text{CH}_2\overset{\overset{\text{O}}{\|}}{\text{C}}\text{OC}_2\text{H}_5 \xrightarrow[\text{C}_2\text{H}_5\text{OH}]{\text{C}_2\text{H}_5\text{O}^\ominus} \text{C}_6\text{H}_5\text{CH}_2\text{—}\overset{\overset{\text{O}}{\|}}{\text{C}}\text{—CH(C}_6\text{H}_5\text{)—}\overset{\overset{\text{O}}{\|}}{\text{C}}\text{—OC}_2\text{H}_5$$

ethyl phenylacetate — ethyl 2,4-diphenyl-3-oxobutanoate[a]

Crossed-Claisen condensations, *i.e.*, condensation reactions involving two different esters, are subject to the same types of limitations that apply to crossed-aldol condensations—such reactions are not synthetically significant if both esters contain α hydrogens of comparable acidity. Attempts to carry out crossed-Claisen condensations involving two different esters, both of which contain acidic hydrogens, will produce all of the four possible products.

$$\text{R—CH}_2\text{—}\overset{\overset{\text{O}}{\|}}{\text{C}}\text{—OC}_2\text{H}_5 + \text{R}'\text{—CH}_2\text{—}\overset{\overset{\text{O}}{\|}}{\text{C}}\text{—OC}_2\text{H}_5 \xrightarrow{\text{OEt}^\ominus}$$

$$\text{RCH}_2\text{—}\overset{\overset{\text{O}}{\|}}{\text{C}}\text{—CHR}'\text{—}\overset{\overset{\text{O}}{\|}}{\text{C}}\text{—OC}_2\text{H}_5$$

$$\text{R}'\text{CH}_2\text{—}\overset{\overset{\text{O}}{\|}}{\text{C}}\text{—CHR—}\overset{\overset{\text{O}}{\|}}{\text{C}}\text{—OC}_2\text{H}_5$$

$$\text{RCH}_2\text{—}\overset{\overset{\text{O}}{\|}}{\text{C}}\text{—CHR—}\overset{\overset{\text{O}}{\|}}{\text{C}}\text{—OC}_2\text{H}_5$$

$$\text{R}'\text{CH}_2\text{—}\overset{\overset{\text{O}}{\|}}{\text{C}}\text{—CHR}'\text{—}\overset{\overset{\text{O}}{\|}}{\text{C}}\text{—OC}_2\text{H}_5$$

5. Suggest a reason why ethyl isobutyrate does not lead to ethyl α,α,γ-trimethylacetoacetate when treated with ethoxide anion, but does form this compound in the presence of sodium triphenylmethide ($Na^{\oplus}$ $(C_6H_5)_3C^{\ominus}$), a very potent base. (*Hint:* Consider the stability of the product as it is formed in the reaction mixture.)
6. Ordinary esters (RCO_2R) are not as reactive as are thiol esters (R(O)SR) in the Claisen condensation. Explain.

Successful crossed-Claisen condensations make use of one ester which, lacking any α hydrogens, can serve only as a carbanion acceptor and a second ester, possessing an α hydrogen,

[a] Complex compounds containing keto groups may have these named as *oxo* groups. The oxo groups are numbered as are other substituents.

to serve as a carbanion source. In order to minimize self-condensation of the latter, it is added *to* a mixture of the incipient acylating agent (*i.e.*, carbanion acceptor) and the base.

$$C_6H_5{-}\overset{O}{\overset{\|}{C}}{-}OC_2H_5 + CH_3CH_2\overset{O}{\overset{\|}{C}}{-}OC_2H_5 \xrightarrow{OEt^{\ominus}} C_6H_5{-}\overset{O}{\overset{\|}{C}}{-}\underset{CH_3}{\underset{|}{CH}}{-}\overset{O}{\overset{\|}{C}}{-}OC_2H_5$$

ethyl 2-methyl-3-phenyl-3-oxopropanoate

$$\text{(ethylene carbonate)} + CH_3CH_2{-}\overset{O}{\overset{\|}{C}}{-}OCH_3 \xrightarrow{OCH_3^{\ominus}} HO{-}CH_2CH_2O{-}\overset{O}{\overset{\|}{C}}{-}\underset{CH_3}{\underset{|}{CH}}{-}\overset{O}{\overset{\|}{C}}{-}OCH_3$$

ethylene carbonate

methyl 2-methyl-3-(2-hydroxyethyloxy)-3-oxopropanoate

$$CH_3O{-}\overset{O}{\overset{\|}{C}}{-}\overset{O}{\overset{\|}{C}}{-}OCH_3 + CH_3{-}\overset{O}{\overset{\|}{C}}{-}OC_2H_5 \xrightarrow{OCH_3^{\ominus}} CH_3O{-}\overset{O}{\overset{\|}{C}}{-}\overset{O}{\overset{\|}{C}}{-}CH_2{-}\overset{O}{\overset{\|}{C}}{-}OCH_3$$

dimethyl oxalate

dimethyl oxaloacetate

The propensity for esters to act as acylating agents, when combined with the moderate acidity of ketones, permits a successful crossed-Claisen condensation between esters and ketones.

$$CH_3{-}\overset{O}{\overset{\|}{C}}{-}CH_3 + CH_3{-}\overset{O}{\overset{\|}{C}}{-}O{-}C_2H_5 \xrightarrow{B:^{\ominus}} CH_3{-}\overset{O}{\overset{\|}{C}}{-}CH_2{-}\overset{O}{\overset{\|}{C}}{-}CH_3$$

acetylacetone

7. Suggest a crossed-condensation that will produce each of the following.
   (a) $HC(O)CH(CH_2CO_2C_2H_5)CO_2C_2H_5$
   (b) $C_6H_5C(O)CH_2CO_2C_2H_5$

The **Dieckmann condensation** is a very valuable variant of the simple Claisen condensation, whereby a diester is converted to a *cyclic* keto ester. Entropic considerations (*i.e.*, the chances of bringing two ends of a molecule together) and strain effects make this *cyclization* most favorable for five- and six-membered ring formation.

$$C_2H_5O{-}\overset{O}{\overset{\|}{C}}{-}(CH_2)_4{-}\overset{O}{\overset{\|}{C}}{-}OC_2H_5 \xrightarrow{C_2H_5O^{\ominus}} \text{(cyclopentanone ring with H and } CO_2C_2H_5)$$

2-carbethoxycyclopentanone

8. In a Dieckmann condensation involving a diester with non-equivalent α-positions, acylation occurs at the less substituted α-carbon because this produces the more stable enolate anion of the keto ester. With this in mind, predict the product of the reaction of $H_5C_2O_2C(CH_2)_5CH(CH_3)CO_2C_2H_5$ and base followed by acidification.

Perkin and Knoevenagel Condensations. While the Claisen-type condensation reactions involve the attack of an acyl-stabilized carbanion upon an acylating agent, two other important syntheses of carbon-carbon bonds employ such enolate anions in addition reactions with aldehydes. These reactions are the **Perkin condensation** and the **Knoevenagel condensation.** Both of these usually are used to prepare α,β-unsaturated acids and esters rather than β-diesters or β-keto esters. However, they are considered at this point because of the mechanistic similarity between these two and the Claisen condensation, and because α,β-unsaturated esters will be used in Section 19.5.

In the **Perkin condensation,** the enolate anion of an acid anhydride adds to an aromatic aldehyde to produce, in the end, an α,β-unsaturated acid. In order to prevent side reactions, the base that is employed is usually the sodium salt of the acid corresponding to the anhydride.

$$\text{ArCHO} + \text{RCH}_2\overset{\overset{\text{O}}{\|}}{\text{C}}\text{—O—}\overset{\overset{\text{O}}{\|}}{\text{C}}\text{CH}_2\text{R} \xrightarrow[\text{heat}]{\text{RCH}_2\text{CO}_2^{\ominus}} \text{ArCH=CR—}\overset{\overset{\text{O}}{\|}}{\text{C}}\text{OH} \qquad \text{Perkin condensation}$$

$$\text{C}_6\text{H}_5\text{CHO} + \text{CH}_3\overset{\overset{\text{O}}{\|}}{\text{C}}\text{—O—}\overset{\overset{\text{O}}{\|}}{\text{C}}\text{CH}_3 \xrightarrow[\text{heat}]{\text{CH}_3\text{CO}_2^{\ominus}} \underset{\text{cinnamic acid}}{\text{C}_6\text{H}_5\text{CH=CHCO}_2\text{H}}$$

Mechanistic detail for the Perkin synthesis of cinnamic acid follows:

$$\underset{(\text{Ac}_2\text{O})}{\text{CH}_3\text{—}\overset{\overset{:\text{O}:}{\|}}{\text{C}}\text{—O—}\overset{\overset{:\text{O}:}{\|}}{\text{C}}\text{—CH}_3} \xrightarrow[1]{\text{NaOAc; heat}} {\ominus}\!:\text{CH}_2\text{—}\overset{\overset{:\text{O}:}{\|}}{\text{C}}\text{—O—}\overset{\overset{:\text{O}:}{\|}}{\text{C}}\text{—CH}_3 \leftrightarrow \text{CH}_2\text{=}\overset{\overset{:\ddot{\text{O}}:^{\ominus}}{|}}{\text{C}}\text{—O—}\overset{\overset{:\text{O}:}{\|}}{\text{C}}\text{—CH}_3$$

$$\text{C}_6\text{H}_5\overset{\overset{:\ddot{\text{O}}:}{\|}}{\text{C}}\text{H} + {\ominus}\!:\text{CH}_2\text{—}\underset{\underset{\text{O}}{\|}}{\text{C}}\text{—O—}\underset{\underset{\text{O}}{\|}}{\text{C}}\text{—CH}_3 \xrightarrow{2} \text{C}_6\text{H}_5\text{—}\overset{\overset{:\ddot{\text{O}}:^{\ominus}}{|}}{\underset{\underset{\text{H}}{|}}{\text{C}}}\text{—CH}_2\text{—}\overset{\overset{\text{O}}{\|}}{\text{C}}\text{—O—}\overset{\overset{\text{O}}{\|}}{\text{C}}\text{—CH}_3$$

Step 3: the alkoxide oxygen of $C_6H_5CH(O^{\ominus})CH_2C(=O)OC(=O)CH_3$ attacks the terminal carbonyl carbon (cyclic six-membered intermediate) to give a tetrahedral intermediate with $:\ddot{O}:^{\ominus}$ and CH₃ on that carbon; step 4: this intermediate breaks down to

$$\text{C}_6\text{H}_5\text{CH}(\text{O—}\overset{\overset{:\text{O}:}{\|}}{\text{C}}\text{—CH}_3)\text{—CH}_2\text{—C(=O)—}:\ddot{\text{O}}:^{\ominus}$$

$$\xrightarrow[5]{\text{Ac}_2\text{O};\ -\text{OAc}^{\ominus}} \text{C}_6\text{H}_5\text{CH}(\text{O—}\overset{\overset{\text{O}}{\|}}{\text{C}}\text{—CH}_3)\text{—CH}_2\text{—C(=O)—O—}\overset{\overset{\text{O}}{\|}}{\text{C}}\text{—CH}_3 \quad (\text{H removed by } {}^{\ominus}\text{OAc})$$

$$\xrightarrow{6} \text{C}_6\text{H}_5(\text{H})\text{C=C}(\text{H})\text{—}\underset{\underset{\text{O}}{\|}}{\text{C}}\text{—O—}\underset{\underset{\text{O}}{\|}}{\text{C}}\text{—CH}_3 \xrightarrow[7]{\text{H}_2\text{O}} \underset{(\textbf{E})\text{-cinnamic acid}}{\text{C}_6\text{H}_5(\text{H})\text{C=C}(\text{H})\text{CO}_2\text{H}}$$

Step 1: Formation of the enolate anion of acetic anhydride, using sodium acetate as the base.
Step 2: Addition of the enolate anion to the aldehyde.
Steps 3 & 4: Intramolecular attack by the newly created alkoxide anion upon the anhydride end of the molecule—an intramolecular acetylation.
Step 5: Carboxylate anion produced in step **4** is acetylated by acetic anhydride.
Step 6: Acetate ion, produced in step **5,** acts as a base and induces an E2 reaction.
Step 7: The anhydride is hydrolyzed to the final product.

The **Knoevenagel reaction** employs the attack of the enolate anion of a reactive methylene group, such as that produced from a β-diester, upon an aldehyde to produce an unsaturated acid. Piperidine often is used as the basic catalyst. The reaction is driven to completion by distilling off water as it is produced.

$$ROC(O)CH_2C(O)OR + R'CHO \xrightarrow{\text{piperidine}} R'(H)C{=}C(CO_2R)_2 \qquad \text{Knoevenagel reaction}$$

$$\underset{\text{ethyl malonate}}{EtO{-}\overset{O}{\overset{\|}{C}}{-}CH_2{-}\overset{O}{\overset{\|}{C}}{-}OEt} + \underset{\text{piperidine}}{C_5H_{10}\ddot{N}H} \rightleftarrows \underset{\text{piperidinium}}{C_5H_{10}\overset{\oplus}{N}H_2} + \underset{\text{diethyl malonate}}{EtO{-}\overset{O}{\overset{\|}{C}}{-}\overset{\ominus}{\underset{\cdot\cdot}{C}H}{-}\overset{O}{\overset{\|}{C}}{-}OEt}$$

$$CH_3CH_2CHO + EtO{-}\overset{O}{\overset{\|}{C}}{-}\overset{\ominus}{\underset{\cdot\cdot}{C}H}{-}\overset{O}{\overset{\|}{C}}{-}OEt \rightarrow CH_3CH_2{-}\underset{H}{\overset{O^{\ominus}}{C}}{-}CH(CO_2Et)_2$$

$$\xrightarrow{C_5H_{10}\overset{\oplus}{N}H_2} CH_3CH_2{-}\underset{H}{\overset{OH}{C}}{-}CH(CO_2Et)_2 \xrightarrow{-H_2O} \overset{CH_3CH_2}{\underset{H}{}}C{=}C\overset{CO_2Et}{\underset{CO_2Et}{}}$$

If the aldehyde is aliphatic, and if an excess of ethyl malonate is used, the initial Knoevenagel reaction usually is followed by a Michael addition (see the following section). The end product of such a process is a tetracarboxylic ester.

$$CH_3CH_2CH{=}C(CO_2C_2H_5)_2 + CH_2(CO_2C_2H_5) \xrightarrow{B:} \xrightarrow{H^{\oplus}} CH_3CH_2CH[CH(CO_2C_2H_5)_2]_2$$

9. Suggest a combination of reagents that will produce each of the following.
(a) $CH_3CH_2CH{=}C(CO_2C_2H_5)_2$
(b) $C_6H_5CH{=}C(C(O)CH_3)CO_2C_2H_5$

19.4 BASE-PROMOTED CONDENSATION REACTIONS—ALKYLATIONS

Now that we have described four routes to functionalized acids or their derivatives, we can return to the problem (p. 592) of constructing complex carbon skeletons through the application of various alkylation reactions. Two of the carbanion sources most often used are acetoacetic ester and malonic ester. Their popularity is due to the ease with which their respective carbanions can be formed and the ease with which the products of their alkylation can be converted (by hydrolysis followed by decarboxylation) to ketones and acids, respectively.

Acetoacetic Ester Syntheses. Acetoacetic ester anion, usually prepared by the reaction of acetoacetic ester with sodium ethoxide, can be alkylated at Cα by reaction with an alkyl halide or tosylate. Resonance stabilization of the anion is sufficiently great that the anion of a monoalkyl acetoacetic ester can be formed and further alkylated.

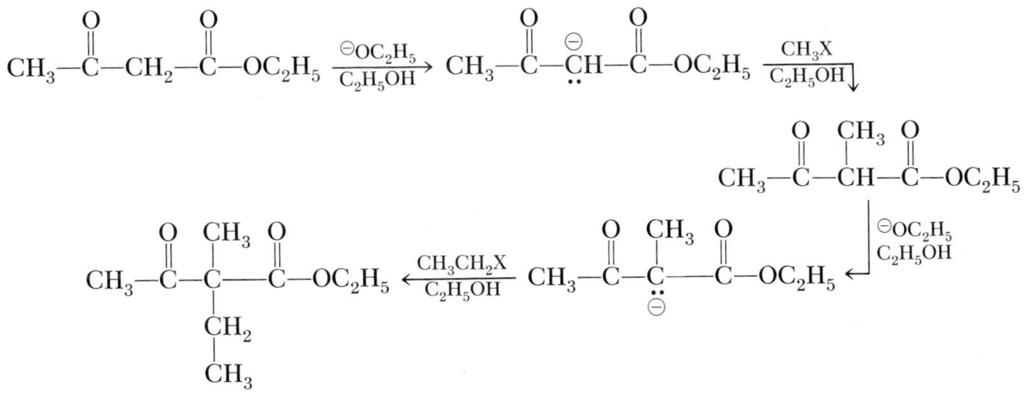

$$CH_3-\overset{O}{\overset{\|}{C}}-CH_2-\overset{O}{\overset{\|}{C}}-OC_2H_5 \xrightarrow[C_2H_5OH]{^{\ominus}OC_2H_5} CH_3-\overset{O}{\overset{\|}{C}}-\overset{\ominus}{\underset{\cdot\cdot}{C}}H-\overset{O}{\overset{\|}{C}}-OC_2H_5 \xrightarrow[C_2H_5OH]{CH_3X} CH_3-\overset{O}{\overset{\|}{C}}-\overset{CH_3}{\overset{|}{C}}H-\overset{O}{\overset{\|}{C}}-OC_2H_5$$

$$CH_3-\overset{O}{\overset{\|}{C}}-\overset{CH_3}{\overset{|}{C}}H-\overset{O}{\overset{\|}{C}}-OC_2H_5 \xrightarrow[C_2H_5OH]{^{\ominus}OC_2H_5} CH_3-\overset{O}{\overset{\|}{C}}-\underset{\ominus}{\underset{\cdot\cdot}{\overset{CH_3}{\overset{|}{C}}}}-\overset{O}{\overset{\|}{C}}-OC_2H_5 \xrightarrow[C_2H_5OH]{CH_3CH_2X} CH_3-\overset{O}{\overset{\|}{C}}-\underset{\underset{CH_3}{|}}{\underset{CH_2}{|}}{\overset{CH_3}{\overset{|}{C}}}-\overset{O}{\overset{\|}{C}}-OC_2H_5$$

ethyl 2-ethyl-2-methyl-3-oxobutanoate

Either the mono- or dialkylacetoacetic ester can be hydrolyzed in dilute acid (*e.g.*, sulfuric acid) to yield the corresponding β-keto acid which, in turn, may be decarboxylated to form a ketone.

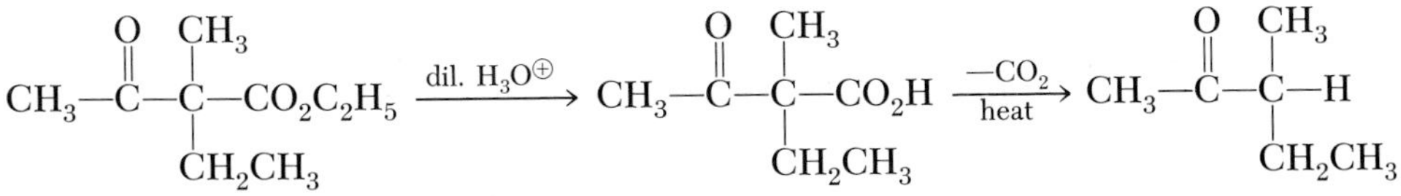

$$CH_3-\overset{O}{\overset{\|}{C}}-\underset{CH_2CH_3}{\underset{|}{\overset{CH_3}{\overset{|}{C}}}}-CO_2C_2H_5 \xrightarrow{\text{dil. } H_3O^{\oplus}} CH_3-\overset{O}{\overset{\|}{C}}-\underset{CH_2CH_3}{\underset{|}{\overset{CH_3}{\overset{|}{C}}}}-CO_2H \xrightarrow[\text{heat}]{-CO_2} CH_3-\overset{O}{\overset{\|}{C}}-\underset{CH_2CH_3}{\underset{|}{\overset{CH_3}{\overset{|}{C}}}}-H$$

β-keto ester → β-keto acid → ketone

This sequence, which can use almost any β-keto ester as a starting material, is an extremely important synthetic method because it can create a variety of ketones which, in turn, are a cornerstone in synthetic organic chemistry. The following scheme combines a crossed-Claisen condensation, an acetoacetic ester alkylation, and a hydrolysis with decarboxylation to synthesize a complex ketone.

$$C_6H_5-\overset{O}{\overset{\|}{C}}-OC_2H_5 + CH_3-\overset{O}{\overset{\|}{C}}-OC_2H_5 \xrightarrow{^{\ominus}OC_2H_5} C_6H_5-\overset{O}{\overset{\|}{C}}-CH_2-\overset{O}{\overset{\|}{C}}-OC_2H_5$$

$$\downarrow (a)\ ^{\ominus}OC_2H_5 \quad (b)\ C_6H_5CH_2Br$$

$$C_6H_5-\overset{O}{\overset{\|}{C}}-(CH_2)_2C_6H_5 \xleftarrow[\text{heat}]{H_3O^{\oplus}} C_6H_5-\overset{O}{\overset{\|}{C}}-\underset{CH_2C_6H_5}{\overset{H}{C}}-\overset{O}{\overset{\|}{C}}-OC_2H_5$$

The Claisen condensation is reversed in the presence of strong alkali, the β-keto ester being cleaved to two molecules of acid (as the salt). Since substituted acetoacetic ester derivatives behave similarly, they can be used to prepare α-substituted or α,α-disubstituted acetic acids. This type of reaction is often called a **retrograde Claisen condensation.**

$$\textbf{A:}\ CH_3-\overset{O}{\overset{\|}{C}}-CH_2-\overset{O}{\overset{\|}{C}}-OC_2H_5 \xrightarrow{\text{conc. } OH^{\ominus}} 2CH_3CO_2^{\ominus} + CH_3CH_2OH$$

$$\textbf{B:}\ CH_3-\overset{O}{\overset{\|}{C}}-\underset{R'}{\overset{R}{C}}-\overset{O}{\overset{\|}{C}}-OC_2H_5 \xrightarrow{\text{conc. } OH^{\ominus}} \xrightarrow{\text{dil. } H_3O^{\oplus}}$$

$$CH_3CO_2H + CHRR'CO_2H + CH_3CH_2OH$$

The mixture of two acids produced in sequence **B** can easily be separated, since acetic acid is completely miscible with water while the monocarboxylic acids containing five or more carbons show reduced water solubility.

Malonic Ester Syntheses. The hydrogens of the methylene group between the acyl groups of diethyl malonate (usually called *malonic ester*) are acidic, forming a sodium salt with sodium ethoxide. This salt, sodiomalonic ester, is alkylated by a sequence of S_N2 reactions analogous to those involved in the alkylation of acetoacetic ester and related to the illustration of the Knoevenagel reaction.

$$C_2H_5O-\overset{O}{\overset{\|}{C}}-\underset{H}{\overset{H}{C}}-\overset{O}{\overset{\|}{C}}-OC_2H_5 \xrightarrow[C_2H_5OH]{NaOC_2H_5} C_2H_5O-\overset{O}{\overset{\|}{C}}-\underset{H}{\overset{\ominus}{\ddot{C}}}-\overset{O}{\overset{\|}{C}}-OC_2H_5 \ Na^{\oplus} \xrightarrow{RX}$$

$$\underset{\alpha,\alpha\text{-dialkyl malonic ester}}{C_2H_5O-\overset{O}{\overset{\|}{C}}-\underset{R}{\overset{R'}{C}}-\overset{O}{\overset{\|}{C}}-OC_2H_5} \xleftarrow[(b)\ R'X]{(a)\ NaOC_2H_5} C_2H_5O-\overset{O}{\overset{\|}{C}}-\underset{H}{\overset{R}{C}}-\overset{O}{\overset{\|}{C}}-OC_2H_5$$

The malonic ester synthesis is now the preferred route to substituted acetic acids; the only additional steps that are required are (a) hydrolysis and (b) decarboxylation. Neither the acetoacetic ester route nor the malonic ester route to carboxylic acids ever can lead directly to a substituted acetic acid with all three α hydrogens replaced by alkyl groups; decarboxylation always results in one residual α hydrogen.

$$CH_3CH_2O\overset{O}{\overset{\|}{C}}-\underset{R'}{\overset{R}{C}}-\overset{O}{\overset{\|}{C}}OC_2H_5 \xrightarrow[\text{heat}]{\text{dil. } H_2SO_4} HO_2C-\underset{R'}{\overset{R}{C}}-CO_2H \xrightarrow{160^\circ} H-\underset{R'}{\overset{R}{C}}-CO_2H + CO_2$$

The facile decomposition of β-dicarboxylic acids can be used to prepare other dicarboxylic acids by the malonic ester route. The selective decarboxylation of a tricarboxylic acid, for example, is used below in the preparation of methylsuccinic acid.

$$\mathrm{EtO{-}\overset{\overset{O}{\|}}{C}{-}CH_2{-}\overset{\overset{O}{\|}}{C}{-}OEt \xrightarrow[(b)\ CH_3CHBrCO_2Et]{(a)\ OEt^{\ominus}} EtO{-}\overset{\overset{O}{\|}}{C}{-}\overset{\overset{H}{|}}{C}{-}\overset{\overset{O}{\|}}{C}{-}OEt}$$
$$\mathrm{\quad\quad CH_3{-}\overset{|}{C}{-}H,\ \overset{|}{C}O_2Et}$$
$$\xrightarrow[(b)\ H_3O^{\oplus}]{(a)\ \text{dil. } OH^{\ominus}} \mathrm{HO_2C{-}CH{-}CO_2H\ (CH_3{-}\overset{|}{C}{-}H,\ \overset{|}{C}O_2H)} \xrightarrow[\text{heat}]{-CO_2} \mathrm{HO_2CCH_2CHCH_3CO_2H}$$

methylsuccinic acid

19.5 MICHAEL ADDITIONS

Strictly speaking, a **Michael reaction** is the 1,4-addition of enolate anion to a conjugated double bond, such as the ones that occur in α,β-unsaturated esters or nitriles. The more general definition, and one based upon mechanistic similarity, encompasses the conjugate addition of any nucleophile to an α,β-unsaturated system. The expression will be used in its broader context in this text unless otherwise specified.

$$\mathrm{EtO_2C{-}CH_2{:}^{\ominus} + CH_2{=}CHCN \rightsquigarrow EtO_2C{-}CH_2{-}CH_2{-}CH_2{-}CN} \quad \text{Michael reaction}$$

$$\mathrm{H_3C\ddot{\underset{..}{S}}{:}^{\ominus} + CH_2{=}CH{-}CN \rightsquigarrow CH_3S{-}CH_2{-}CH_2{-}CN} \quad \text{Michael-like addition}$$

The competition between 1,4-addition, also called **conjugate addition,** and 1,2-addition, called **direct addition,** was touched upon in a discussion of the reactions of Grignard reagents with carbonyl compounds (page 495). That material should be reviewed if necessary. The general pattern of Michael-like additions, the pathway normally anticipated when a nucleophile adds to an α,β-unsaturated system, is clear from the reactions in Figure 19–2.

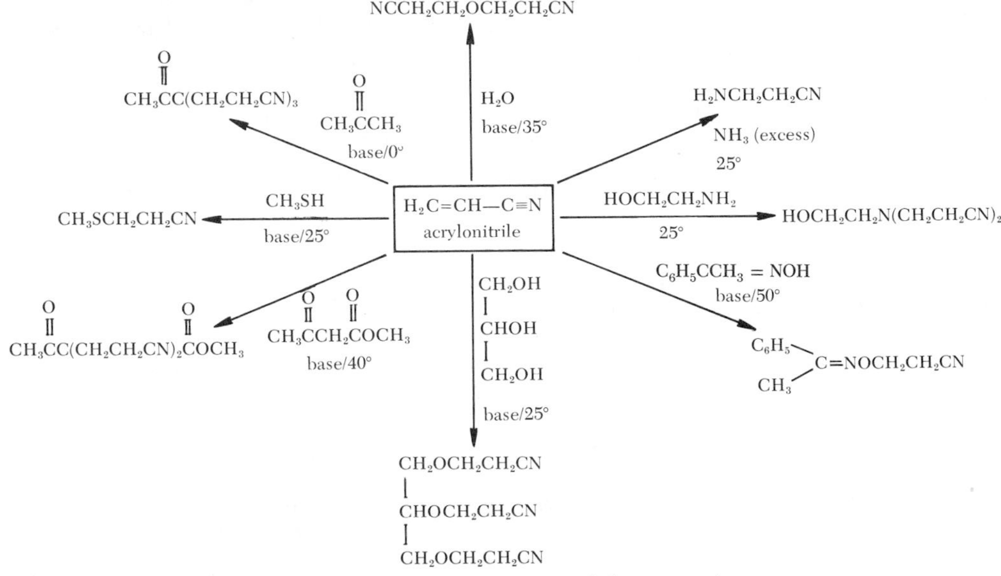

Figure 19-2 Representative cyanoethylation reactions.

All of the processes depicted in Figure 19–2 are called cyanoethylation reactions because they introduce the $—CH_2CH_2CN$ group into a molecule. Cyanoethylation of carbanions is an important method of lengthening a chain by three carbons. All of the reactions in Figure 19–2 share a common mechanism, and the cyanoethylation of methanethiol, CH_3SH, may be considered representative.

$$CH_3\ddot{S}H + B:^{\ominus} \rightleftarrows CH_3\ddot{S}:^{\ominus} + BH$$

$$CH_3\ddot{S}:^{\ominus} + H_2C{=}CH{-}CN \rightarrow CH_3SCH_2{-}\overset{\ominus}{\ddot{C}}H{-}C{\equiv}N: \leftrightarrow CH_3S{-}CH_2{-}CH{=}C{=}\ddot{N}^{\ominus}$$

$$CH_3SCH_2\overset{\ominus}{C}HCN + BH \rightarrow CH_3SCH_2CH_2CN + B:^{\ominus}$$

The Michael reaction, unlike some other base-induced reactions, involves the regeneration of a base in the last step; therefore, base is required only in catalytic amounts. The need to use only a small amount of base is fortunate, since the use of high concentrations of base and protracted reaction times may cause reversal of the Michael addition. This **retrograde Michael reaction** does not always regenerate the compounds used in the original Michael condensation. An example of this follows:

$$C_6H_5{-}CH{=}CH{-}\overset{O}{\overset{\|}{C}}{-}C_6H_5 + CH_2(CO_2Et)_2 \xrightarrow{\text{trace OEt}^{\ominus}} C_6H_5{-}\overset{O}{\overset{\|}{C}}{-}CH_2{-}\underset{}{\overset{C_6H_5}{\overset{|}{C}H}}{-}CH(CO_2Et)_2$$

$$\xrightarrow[\text{heat}]{\text{conc. OEt}^{\ominus}} C_6H_5{-}\overset{O}{\overset{\|}{C}}{-}CH_3 + C_6H_5{-}CH{=}C(CO_2Et)_2$$

A fundamental difference between the Claisen condensation and the Michael reaction is that in the former the product is a stronger acid than the reactant, while in the latter the product is a weaker acid than the reactant. This means that while an excess of base can drive the Claisen condensation to completion, a low concentration of base is necessary to produce a high yield from a Michael reaction.

A compound having two or more double bonds in conjugation with a carbonyl group may, in principle, undergo several different Michael additions. The product that may be expected to predominate is the one that retains the maximum amount of conjugation in the product.

$$\underset{\delta}{H_2C}{=}\underset{\gamma}{CH}{-}\underset{\beta}{CH}{=}\underset{\alpha}{CH}{-}\overset{O}{\overset{\|}{C}}{-}OCH_3 + CH_2(CO_2CH_3)_2 \xrightarrow{^{\ominus}OCH_3} (CH_3O_2C)_2CH{-}CH_2{-}CH_2{-}CH{=}CH{-}CO_2CH_3$$

δ-attack
major product

Predominance of the most conjugated product is expected because this should be the thermodynamically favored (*i.e.*, most stable) product, and the Michael reaction is usually carried out under conditions that permit equilibration to the most stable adduct.

Robinson Annelation Reaction. The Michael addition, used in conjunction with an aldol condensation, has been used to fuse a six-membered ring to a carbon skeleton. This sequence, the Robinson ring-forming reaction or **Robinson annelation reaction,** can be very important in the synthesis of polycyclic molecules. A simple illustration is found in part **A** of Figure 19–3, while a more complex illustration is found in sequence **B**. The latter sequence represents the beginning of the *in vitro* (in the laboratory) synthesis of the steroid epiandrosterone.

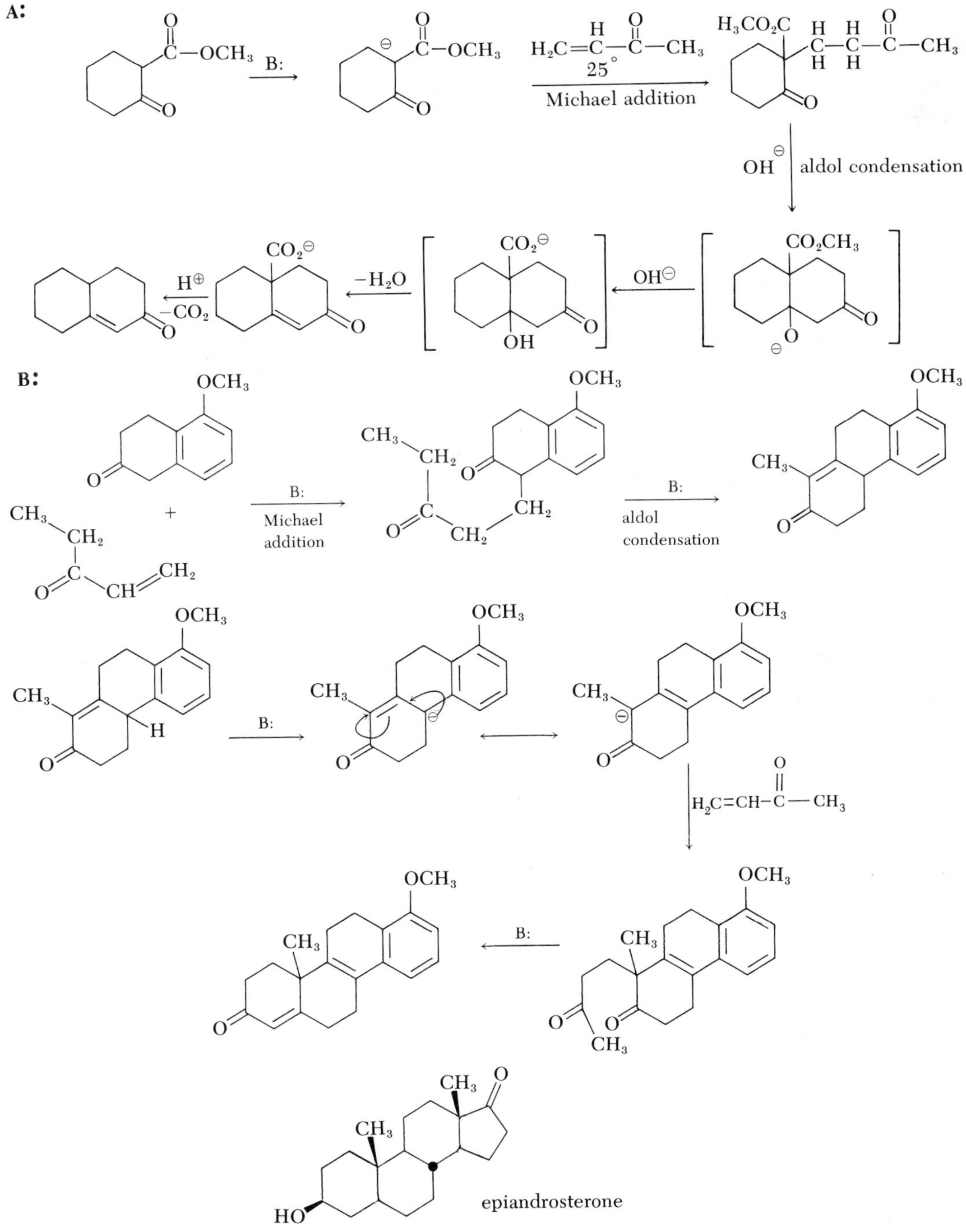

A:

B:

$H_2C{=}CH{-}C({=}O){-}CH_3$, 25°

Michael addition

$OH^{\ominus}$ aldol condensation

$OH^{\ominus}$

$-H_2O$

$H^{\oplus}$, $-CO_2$

B:

B: Michael addition

B: aldol condensation

B:

$H_2C{=}CH{-}C({=}O){-}CH_3$

B:

epiandrosterone

Figure 19-3 Robinson annelation reaction.

10. Identify the lettered compounds in the sequence below.

$$\text{acetone} \xrightarrow{H^{\oplus}} \mathbf{A}\ (C_6H_{10}O)$$

$$\mathbf{A} + CH_2(CO_2C_2H_5)_2 \xrightarrow{EtO^{\ominus}} \mathbf{B}\ (C_{13}H_{22}O_5) \xrightarrow{EtO^{\ominus}} \mathbf{C}\ (C_{11}H_{16}O_4)$$

$$\mathbf{C} \xrightarrow[\text{heat}]{H_2O/KOH} \mathbf{D}\ (C_9H_{11}O_4K) \xrightarrow[\text{heat}]{H_3O^{\oplus}} \text{5,5-dimethyl-1,3-cyclohexanedione}$$

5,5-dimethyl-1,3-cyclohexanedione

19.6 REFORMATSKY REACTION

Organomagnesium compounds rapidly add to the carbon-oxygen double bonds of esters, ketones, anhydrides, and similar derivatives. It is impossible, therefore, to prepare a Grignard reagent that also possesses one of these reactive functional groups. Organozinc compounds are less reactive nucleophiles than are the corresponding magnesium compounds; while they will react with aldehydes and ketones, they are unreactive toward esters. This is used to advantage in the **Reformatsky reaction,** the addition of an organozinc compound derived from an α-halo ester to aldehydes and ketones to produce β-hydroxy esters. β-Hydroxy acids and esters are easily dehydrated upon warming with acid (*e.g.*, sulfuric acid) to give α,β-unsaturated acids and esters. Since these are of great synthetic importance (see Figure 19-1), this makes the Reformatsky reaction an important preparative procedure.

The Reformatsky reagent, that is, the organozinc compound, is prepared in a manner similar to the preparation of Grignard reagents. The major difference is that the solvent is usually benzene-ether in the Reformatsky reaction and ether in the Grignard reaction.

$$Zn + BrCH_2\overset{O}{\overset{\|}{C}}{-}OC_2H_5 \xrightarrow[\text{benzene}]{\text{ether-}} BrZnCH_2CO_2C_2H_5$$

a Reformatsky reagent

$$BrZnCH_2CO_2C_2H_5 + CH_3\overset{O}{\overset{\|}{C}}CH_2CH_3 \rightarrow CH_3{-}\underset{C_2H_5}{\underset{|}{\overset{OZnBr}{\overset{|}{C}}}}{-}CH_2CO_2C_2H_5$$

$$\downarrow H_3O^{\oplus}$$

$$CH_3{-}\underset{C_2H_5}{\underset{|}{\overset{OH}{\overset{|}{C}}}}{-}CH_2{-}CO_2C_2H_5$$

ethyl 3-hydroxy-3-methylpentanoate

$$CH_3{-}C_6H_4{-}CHO + Br\overset{CH_3}{\overset{|}{C}H}{-}CO_2C_2H_5 \xrightarrow[\text{ether-benzene}]{Zn} \xrightarrow{H_3O^{\oplus}}$$

$$CH_3{-}C_6H_4{-}\underset{OH}{\underset{|}{\overset{H}{\overset{|}{C}}}}{-}\overset{CH_3}{\overset{|}{C}H}{-}CO_2C_2H_5$$

ethyl 3-hydroxy-2-methyl-3-(4-methylphenyl)propanoate

19.7 WITTIG REACTION

One of the most important routes to alkenes was reported in 1942 by Georg Wittig. Now known as the **Wittig reaction,** this process succeeds in converting a carbonyl group into a carbon-carbon double bond, *i.e.*, $>C{=}O \rightarrow >C{=}CRR'$. The reagent that accomplishes this task is an *ylid* (see Chapter 23) containing a positively charged phosphorus atom and an adjacent carbanionic center. The reaction of this *Wittig reagent* with the carbonyl group can be pictured as the attack of a carbanion upon the carbonyl carbon. This attack leads to a *betaine* (a neutral molecule with non-adjacent opposite charges), which collapses to the final alkene and a phosphine oxide, R_3PO.

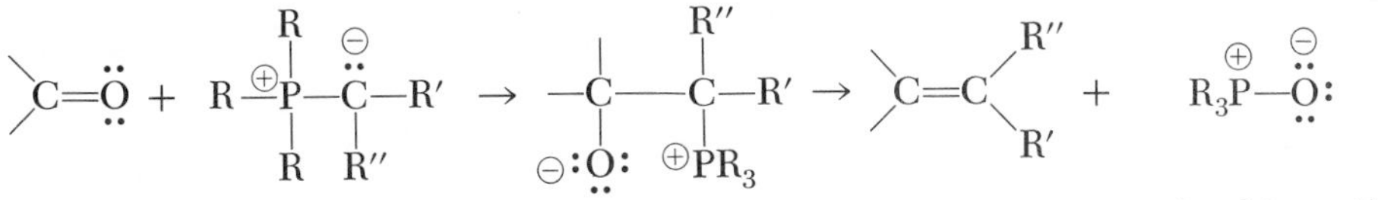

$$>C{=}\ddot{O} + R{-}\overset{\oplus}{P}(R)_2{-}\overset{\ominus}{\ddot{C}}(R'')\text{—}R' \rightarrow {-}C(\ominus{:}\ddot{O}{:}){-}C(R'')(\oplus PR_3){-}R' \rightarrow >C{=}C(R'')R' + R_3\overset{\oplus}{P}{-}\overset{\ominus}{\ddot{O}}{:}$$

a phosphorus ylid (Wittig reagent) — betaine — a phosphine oxide

The phosphorus ylid, whose nature is discussed in more detail in Chapter 23, is prepared in a two step process. The first involves an S_N2 displacement by a phosphine (R_3P) on an alkyl halide, and the second is a proton abstraction by a base from the resultant salt. Since the proton to be abstracted is on the carbon that bore the halogen which was displaced, the starting halide may be 1° or 2° but not 3°. In order that only one type of acid hydrogen be available for abstraction, the most commonly employed phosphine is triphenylphosphine, $(C_6H_5)_3P$.

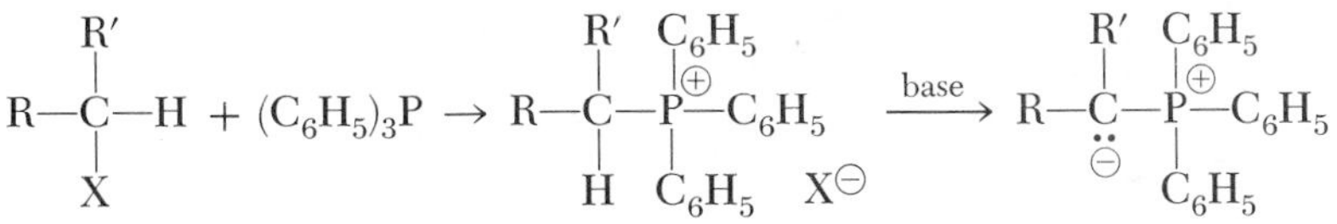

$$R{-}C(R')(X){-}H + (C_6H_5)_3P \rightarrow R{-}C(R')(H){-}\overset{\oplus}{P}(C_6H_5)_2{-}C_6H_5\ X^{\ominus} \xrightarrow{\text{base}} R{-}\overset{\ominus}{\ddot{C}}(R'){-}\overset{\oplus}{P}(C_6H_5)_2{-}C_6H_5$$

a phosphonium salt — a Wittig reagent

Representative syntheses of simple alkenes are shown below. When the possibility for *cis-trans* isomerism exists, the *trans* isomer usually predominates, although changes in experimental conditions can readily produce an excess of the *cis* alkene.

A: $CH_3Br + (C_6H_5)_3P \rightarrow (C_6H_5)_3\overset{\oplus}{P}{-}CH_3\ Br^{\ominus} \xrightarrow[\text{THF}]{C_6H_5Li} (C_6H_5)_3\overset{\oplus}{P}{-}\overset{\ominus}{\ddot{C}}H_2 + C_6H_5{-}H + LiBr$

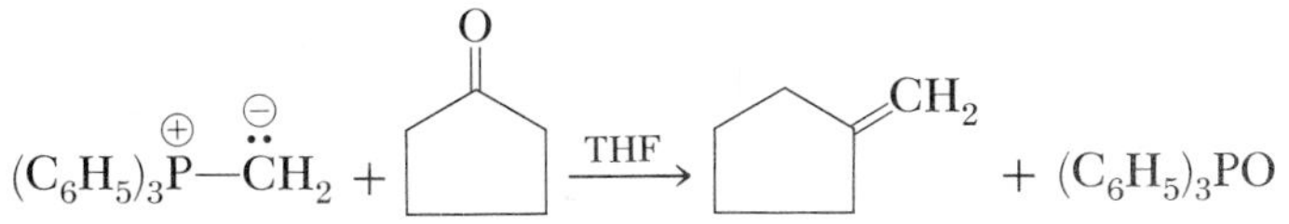

$$(C_6H_5)_3\overset{\oplus}{P}{-}\overset{\ominus}{\ddot{C}}H_2 + \text{cyclohexanone} \xrightarrow{\text{THF}} \text{methylenecyclohexane }(=CH_2) + (C_6H_5)_3PO$$

B: $CH_2{=}CH{-}CH_2Cl + (C_6H_5)_3P \rightarrow (C_6H_5)_3\overset{\oplus}{P}{-}CH_2CH{=}CH_2\ Cl^{\ominus} \xrightarrow[\text{DMF}]{NaOCH_3} (C_6H_5)_3\overset{\oplus}{P}{-}\overset{\ominus}{\ddot{C}}H{-}CH{=}CH_2$

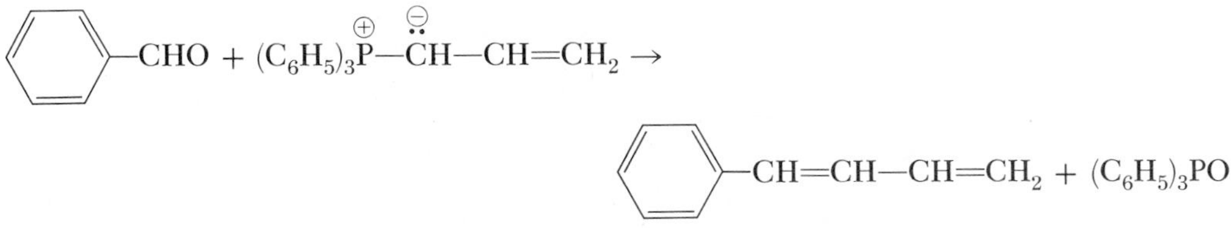

$$C_6H_5{-}CHO + (C_6H_5)_3\overset{\oplus}{P}{-}\overset{\ominus}{\ddot{C}}H{-}CH{=}CH_2 \rightarrow C_6H_5{-}CH{=}CH{-}CH{=}CH_2 + (C_6H_5)_3PO$$

If the negative charge of the ylid is delocalized by resonance to a very large degree or if the carbanionic center is highly hindered, or both, the ylid may not attack a given carbonyl group. For example, ylid **A** (below) normally does not react with ketones, and ylid **B** is unreactive even toward aldehydes.

$$(C_6H_5)_3\overset{\oplus}{P}-\overset{\ominus}{\underset{\cdot\cdot}{C}}H-C(=O)H \qquad (C_6H_5)_3\overset{\oplus}{P}-\overset{\ominus}{\underset{\cdot\cdot}{C}}(C_6H_5)_2$$

A **B**

A useful variation of the Wittig reaction employs phosphite esters (compounds of the types $(RO)_3P$) instead of phosphines to prepare the reactive nucleophile. These phosphite esters react with α-halo esters to produce phosphonates, $(RO)_2P(O)R'$. (Note the difference between this behavior and the original Wittig reaction, in which a phosphine and a halide react to form a salt.) If the phosphonate contains a hydrogen on the carbon of R′ that is adjacent to phosphorus, this hydrogen will be abstracted by a base to form a nucleophilic carbanion. (At this stage the classical Wittig reaction produced an ylid.)

$$\underset{\text{triethyl phosphite}}{(EtO)_3P} + BrCHRCO_2CH_3 \rightarrow \underset{\text{a phosphonate}}{EtO-\overset{\oplus}{P}(\overset{\ominus}{:}\ddot{O}:)(OEt)-CHR-CO_2CH_3} + C_2H_5Br$$

$$EtO-\overset{\oplus}{P}(\overset{\ominus}{:}\ddot{O}:)(OEt)-CHR-CO_2CH_3 \xrightarrow[\text{DME}]{\text{NaH}} EtO-\overset{\oplus}{P}(\overset{\ominus}{:}\ddot{O}:)(OEt)-\overset{\ominus}{\underset{\cdot\cdot}{C}}R-CO_2CH_3$$

The carbanion derived from a phosphonate ester reacts with a carbonyl compound, in much the same way that an ylid does, to produce an α,β-unsaturated ester. In the sequence below, cyclohexanone is converted to ethyl cyclohexylideneacetate by this modification of the Wittig reaction.

$$(EtO)_3P + BrCH_2CO_2C_2H_5 \rightarrow (EtO)_2\overset{\oplus}{P}(\overset{\ominus}{:}\ddot{O}:)-CH_2-CO_2C_2H_5 + C_2H_5Br$$

$$\downarrow \text{NaH/DME}$$

$$\underset{\substack{\text{ethyl cyclohexylideneacetate}\\(70\%)}}{C_6H_{10}{=}CHCO_2C_2H_5} \xleftarrow[-(EtO)_2PO_2^{\ominus}]{\text{cyclohexanone}} (EtO)_2\overset{\oplus}{P}(\overset{\ominus}{:}\ddot{O}:)-\overset{\ominus}{\underset{\cdot\cdot}{C}}H-CO_2C_2H_5$$

Thus, the Wittig reaction is an extremely valuable synthesis of alkenes; in addition, one modification of it provides an excellent route to compounds that can be subjected to Michael addition.

There are many other variations of the original Wittig reaction, but spatial restrictions prevent us from giving a detailed treatment of them. We will note only that ketones can be converted to aldehydes containing one more carbon by reaction with an ylid derived from chloromethyl methyl ether. The synthesis of formylcyclohexane (cyclohexanecarboxaldehyde) shown below illustrates this procedure.

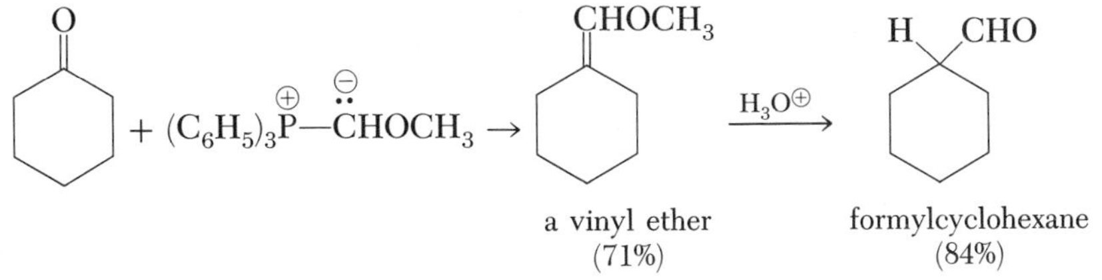

$$\text{cyclohexanone} + (C_6H_5)_3\overset{\oplus}{P}-\overset{\ominus}{\ddot{C}}HOCH_3 \rightarrow \underset{\substack{\text{a vinyl ether} \\ (71\%)}}{\text{cyclohexylidene}=CHOCH_3} \xrightarrow{H_3O^{\oplus}} \underset{\substack{\text{formylcyclohexane} \\ (84\%)}}{\text{cyclohexane(H)(CHO)}}$$

11. Suggest a carbonyl compound/ylid combination that will yield:
   (a) 1-butene
   (b) 2-butene
   (c) ethylidenecyclohexane
   (d) 1,1-diphenylethene
   (e) 1,2-diphenylethene
   (f) methyl 3-methylbutenoate

19.8 ACYLOIN CONDENSATION

The **acyloin condensation** does not involve the carbon that is α to an acyl carbon and, one might argue, does not belong with a discussion of carbanion acylation and alkylation. However, because this reaction is a very important route to new carbon-carbon bonds, and since the reactive groups are the ester groups themselves, we will discuss this reaction briefly.

In the absence of a proton donor, an ester will react with metallic sodium to form a radical anion. If a diester is used, the two radicals can dimerize to form a dialkyl α-hydroxy ketone, or *acyloin*. (Similar diaryl compounds are termed *benzoins*.) Because radical anion formation occurs on the metal's surface, diesters tend to have their radical sites in proximity to one another. This favors cyclization (over polymerization) and makes the acyloin condensation useful for synthesizing rings larger than C_6.

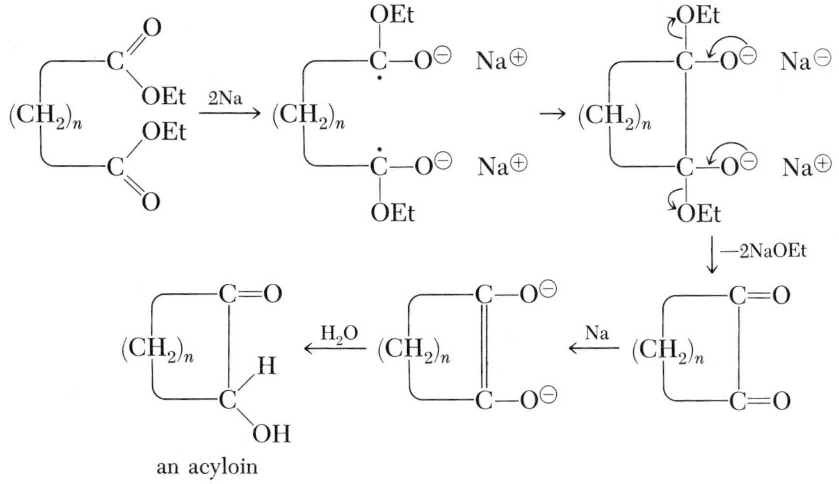

$$(CH_2)_n(COOEt)_2 \xrightarrow{2Na} (CH_2)_n[\dot{C}(OEt)-O^{\ominus}\ Na^{\oplus}]_2 \rightarrow (CH_2)_n[C(OEt)-O^{\ominus}\ Na^{\oplus}]_2 \text{ (C–C bonded)} \xrightarrow{-2NaOEt} (CH_2)_n[C{=}O]_2$$

$$(CH_2)_n[C{=}O]_2 \xrightarrow{Na} (CH_2)_n[C(-O^{\ominus}){=}C(-O^{\ominus})] \xrightarrow{H_2O} \underset{\text{an acyloin}}{(CH_2)_n[C{=}O][CH(OH)]}$$

The absence of a proton source is imperative for the success of the acyloin condensation. In the presence of water (achieved by using ether containing water), for example, the ester group is reduced by metallic sodium to an alcohol. This is related to the Bouvault-Blanc reduction, a procedure for reducing esters to alcohols by reaction with sodium and alcohol. At present the Bouvault-Blanc reduction is largely of historic interest, having been superseded by lithium aluminum hydride reduction.

side reactions accompanying the acyloin condensation:

$$\text{R—}\overset{\overset{\text{O}}{\|}}{\text{C}}\text{—OR}' + \text{Na} \xrightarrow{\text{wet ether}} \text{RCH}_2\text{OH} + \text{R}'\text{OH}$$

$$(\text{CH}_2)_n(\text{CO}_2\text{R})_2 + \text{Na} \xrightarrow{\text{wet ether}} (\text{CH}_2)_n(\text{CH}_2\text{OH})_2 + 2\text{ROH}$$

Bouvault-Blanc reduction:

$$\text{CH}_3\text{CH}_2\text{CH}_2\text{—}\overset{\overset{\text{O}}{\|}}{\text{C}}\text{—O—CH}_2\text{C}_6\text{H}_5 + \text{Na} \xrightarrow[\text{heat}]{\text{EtOH}} \text{CH}_3\text{CH}_2\text{CH}_2\text{CH}_2\text{OH} + \text{C}_6\text{H}_5\text{CH}_2\text{OH}$$

19.9 THE KREBS CYCLE

It has been suggested for about a century that the biological processing of food resembles the combustion of organic molecules. In fact, it is true that the *net* results of "biological oxidation" and "combustion" are the same—the production of carbon dioxide, water, and energy along with the consumption of oxygen. However, while the uncontrolled combustion of compounds such as glucose produces very little useful energy, the biological oxidation of glucose occurs in a number of steps and the body stores the energy released in these steps in so-called "high energy" compounds. The most famous of these high energy compounds is adenosine triphosphate, commonly called ATP. In this section we will examine a most important group of reactions that, in the end, lead to ATP synthesis. They do this by playing a role in the oxidation of fats, sugars, and, if necessary, proteins. This sequence is called the **Krebs cycle,** the **tricarboxylic acid cycle,** or the **citric acid cycle,** and it represents a common terminal phase for the aerobic oxidation of biological molecules.

The ultimate result of the Krebs cycle is to supply electrons into the respiratory chain (presented in more detail in Chapter 22). The respiratory chain and the Krebs cycle occur in the subcellular structures called mitochondria and together have, as their final result, the reduction of oxygen to water. The relationship of the Krebs cycle to other biological transformations is schematized in Figure 19–4.

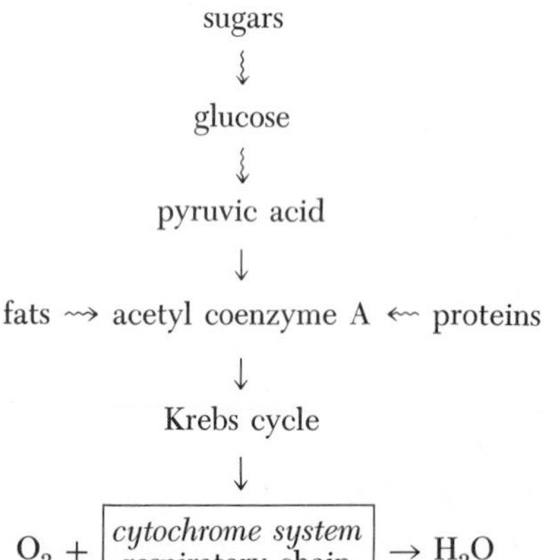

Figure 19-4 The role of the Krebs cycle in the metabolism of fats, proteins, and sugars.

The gross transformations involved in the Krebs cycle are shown, as a biochemist might present them, in Figure 19–5. If we begin at the top, we see that the cycle "begins" with the reaction of acetyl coenzyme A with oxaloacetic acid to form the first tricarboxylic acid in the cycle, citric acid. A dehydration-rehydration converts citric acid to isocitric acid, which

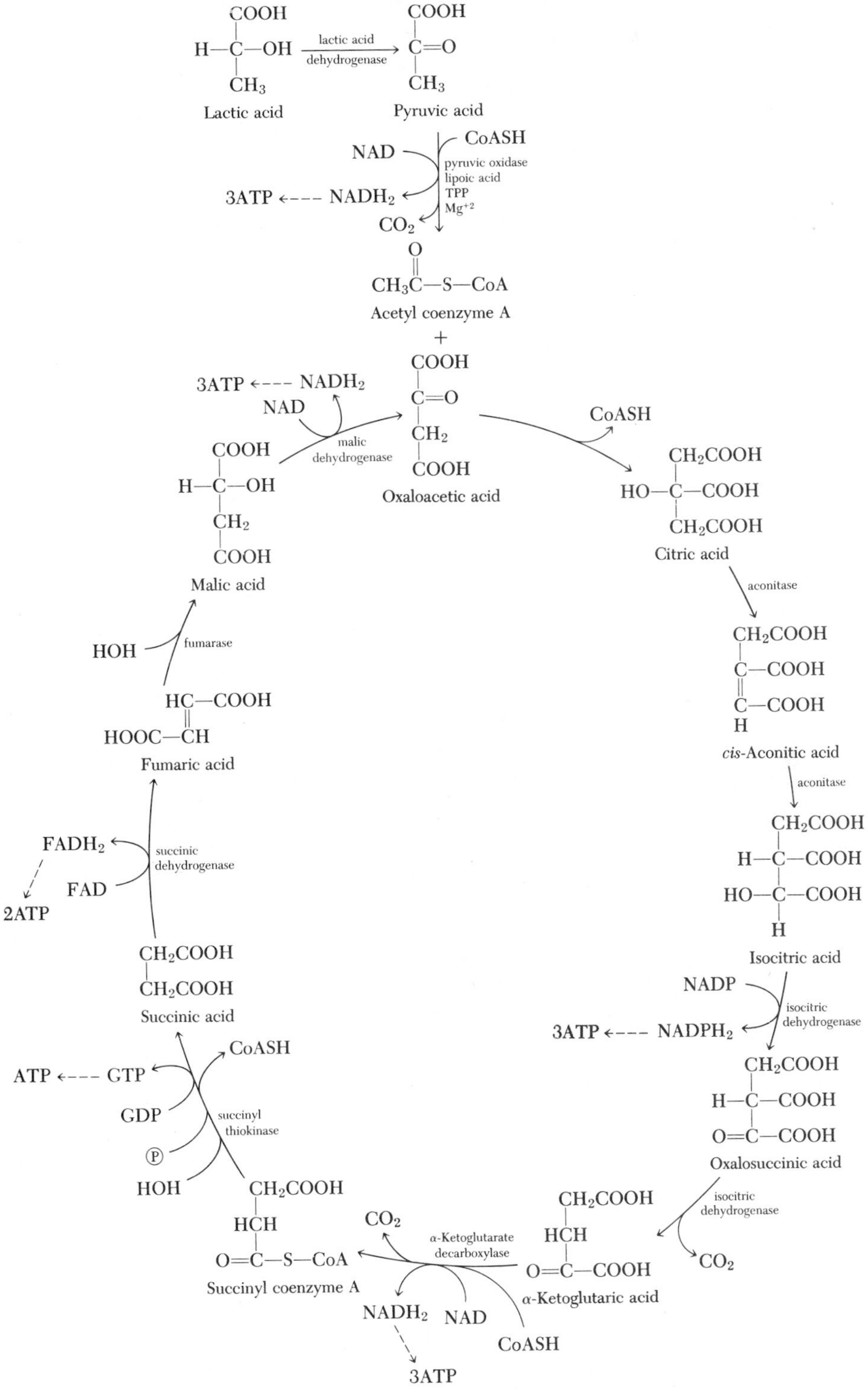

$H-C(OH)(COOH)-CH_3$ —lactic acid dehydrogenase→ $COOH-C(=O)-CH_3$

Lactic acid → Pyruvic acid

NAD, CoASH → pyruvic oxidase, lipoic acid, TPP, Mg^{+2} → 3ATP ←--- $NADH_2$, CO_2

$CH_3C(=O)-S-CoA$

Acetyl coenzyme A

\+

$COOH-C(=O)-CH_2-COOH$

Oxaloacetic acid

CoASH

$CH_2COOH-C(OH)(COOH)-CH_2COOH$ (HO—C—COOH)

Citric acid

aconitase

$CH_2COOH-C(COOH)=C(COOH)H$

cis-Aconitic acid

aconitase

$CH_2COOH-CH(COOH)-CH(OH)(COOH)$ (H—C—COOH, HO—C—COOH, H)

Isocitric acid

NADP → isocitric dehydrogenase → 3ATP ←--- $NADPH_2$

$CH_2COOH-CH(COOH)-C(=O)COOH$ (H—C—COOH, O=C—COOH)

Oxalosuccinic acid

isocitric dehydrogenase, CO_2

$CH_2COOH-HCH-O=C-COOH$

α-Ketoglutaric acid

α-Ketoglutarate decarboxylase: NAD, CoASH → $NADH_2$ ---→ 3ATP, CO_2

$CH_2COOH-HCH-O=C-S-CoA$

Succinyl coenzyme A

succinyl thiokinase: GDP, Ⓟ, HOH → ATP ←--- GTP, CoASH

$CH_2COOH-CH_2COOH$

Succinic acid

succinic dehydrogenase: FAD → $FADH_2$ ---→ 2ATP

$HC-COOH \; = \; HOOC-CH$

Fumaric acid

fumarase: HOH

$COOH-C(H)(OH)-CH_2-COOH$ (H—C—OH)

Malic acid

malic dehydrogenase: NAD → 3ATP ←--- $NADH_2$

Figure 19–5 The Krebs cycle. This also is called the tricarboxylic acid cycle.

is then oxidized to oxalosuccinic acid. Decarboxylation, followed by an oxidative decarboxylation, converts oxalosuccinic acid to succinic acid. Succinic acid is oxidized to fumaric acid, which is then converted to malic acid. Oxidation of the malic acid to oxaloacetic acid completes the cycle and permits the processing of another molecule of acetyl coenzyme A. In the long run, the Krebs cycle converts acetyl coenzyme A into energy and may be characterized by the equation

$$CH_3CO_2H + 2H_2O \rightarrow 2CO_2 + 8H^{\oplus} + 8e^{\ominus}$$

While much of the detail of the Krebs cycle is understood, its full analysis is beyond the scope of this text. We will, however, look at some steps that can be explained in terms of reactions presented in this and the previous chapter. In a few instances we will dip into material from future chapters.

Origin of Acetyl CoA. The conversion of pyruvic acid to acetyl CoA is a complex process, involving several enzymes and discrete intermediates. Indeed, its formation is not a part of the Krebs cycle but is a step necessary in order for sugars (carbohydrates) to enter the cycle.

The process begins by formation of a reactive α-hydroxy derivative of pyruvic acid with a complex molecule, thiamine pyrophosphate (TPP). This adduct loses carbon dioxide to form the equivalent of an adduct of acetaldehyde with thiamine pyrophosphate. (Some details of the process are shown in Figure 27–9.) This adduct, called "active aldehyde," reacts with lipoic acid to form acetyl dihydrolipoic acid, a thiol ester of acetic acid.

$$\underset{\text{pyruvic acid}}{CH_3{-}\overset{\overset{\displaystyle O}{\|}}{C}{-}CO_2H} + TPP \rightsquigarrow \underset{\text{active aldehyde}}{\left[CH_3{-}\overset{\overset{\displaystyle OH}{|}}{\underset{\underset{\displaystyle H}{|}}{C}}{-}TPP\right]} + CO_2$$

$$\Big\downarrow \; \text{lipoic acid: } H_2C\langle S{-}S\rangle CH(CH_2)_4CO_2H \text{ (ring closed by } CH_2\text{)}$$

$$\underset{\text{acetyl dihydrolipoic acid}}{CH_3\overset{\overset{\displaystyle O}{\|}}{C}{-}SCH_2CH_2CH(SH)(CH_2)_4CO_2H}$$

Acetyl dihydrolipoic acid then reacts with coenzyme A *via* a transacylation (ester interchange *or* transesterification) to produce acetyl coenzyme A and dihydrolipoic acid.

$$\underset{\text{acetyl dihydrolipoic acid}}{CH_3C(O)S{-}CH_2CH_2CH(SH)(CH_2)_4CO_2H} + CoASH \rightarrow$$

$$\underset{\text{acetyl CoA}}{CoASC(O)CH_3} + \underset{\text{dihydrolipoic acid}}{HSCH_2CH_2CH(SH)(CH_2)_4CO_2H}$$

Citric Acid Synthesis. The reaction of acetyl coenzyme A with oxaloacetic acid to form citric acid requires an enzyme (*citric synthase*), but it is nothing more than a mixed condensation in which acetyl CoA provides the carbanion (or equivalent). Coenzyme A is regenerated in the same sequence.

$$CoASC(O)CH_3 + HO_2CC(O)CH_2CO_2H \xrightarrow[\text{synthase}]{\text{citric}}$$

$$\underset{\text{citric acid}}{HO_2CC(OH)(CH_2CO_2H)CH_2CO_2H} + CoASH$$

This reaction illustrates the fact that thiol esters, such as acetyl coenzyme A, are more reactive than ordinary esters toward condensation reactions. The enhanced reactivity stems from the fact that thiol esters have a lower resonance energy than do ordinary esters and, therefore, can more readily form enols and enolates (see problem 6).

The Isomerization of Citric Acid. The conversion of citric acid to isocitric acid, *via* *cis*-aconitic acid, is a dehydration-hydration sequence. While all three species are present at equilibrium, the isocitric acid is continuously removed in the cell and this drives the cycle forward.

Isocitric Acid Oxidation. Isocitric acid may be viewed as a complex alcohol, and its oxidation to oxalosuccinic acid as nothing more than the oxidation of a 2° alcohol to a ketone. As implied in Chapter 16, such reactions involve $NAD^{\oplus}$[a] as an oxidant.

$$\underset{\text{isocitric acid}}{HO_2C\underset{\displaystyle CH_2CO_2H}{\underset{|}{C}}H\overset{\displaystyle OH}{\overset{|}{C}}HCO_2H} + NAD^{\oplus} \xrightarrow{\text{enzyme}} \underset{\text{oxalosuccinic acid}}{HO_2C\underset{\displaystyle CH_2CO_2H}{\underset{|}{C}}H\overset{\displaystyle O}{\overset{\|}{C}}CO_2H} + NADH + H^{\oplus}$$

Toward Succinic Acid. Oxalosuccinic acid is a β-keto acid and, as such, is readily decarboxylated (to α-ketoglutaric acid). Like pyruvic acid, another α-keto acid, α-ketoglutaric acid undergoes decarboxylation in the cell to form an acylated coenzyme A derivative. In this instance, however, the derivative is succinyl coenzyme A. Hydrolysis of succinyl CoA yields succinic acid.

$$\underset{\alpha\text{-ketoglutaric acid}}{HO_2CCH_2CH_2C(O)CO_2H} + CoASH \rightsquigarrow \underset{\text{succinyl coenzyme A}}{HO_2CCH_2CH_2C(O){-}SCoA} + CO_2$$

Back Again to Citric Acid. Succinic acid is oxidized enzymically to fumaric acid which then, by conjugated addition of water, is converted to malic acid. Once again $NAD^{\oplus}$ is called upon to oxidize an alcohol, this time converting malic acid into oxaloacetic acid. This may then condense with a second molecule of acetyl coenzyme A and begin the entire cycle again.

$$\underset{\text{succinic acid}}{HO_2CCH_2CH_2CO_2H} \xrightarrow{[O]} \underset{\text{fumaric acid}}{HO_2CCH{=}CHCO_2H} \xrightarrow{HOH} \underset{\text{malic acid}}{HO_2CCH(OH)CH_2CO_2H}$$

In Retrospect. It is important to recognize that if we begin with a molecule of the sugar called glucose ($C_6H_{12}O_6$) and oxidize it biologically, four molecules of carbon dioxide are produced in the Krebs cycle and two are produced in converting pyruvic acid into acetyl CoA. (Remember that each molecule of glucose produces two molecules of pyruvic acid; review Figure 17–1.) However, only 10% of the energy produced in the respiratory (*i.e.*, "oxygen requiring") breakdown of glucose occurs in the Krebs cycle; the remainder is produced in the *respiratory chain*, where NADH and O_2 react to form $NAD^{\oplus}$ and H_2O.

$$\underset{\text{glucose}}{C_6H_{12}O_6} \rightsquigarrow \underset{\text{pyruvic acid}}{2CH_3C(O)CO_2H}$$

$$2CH_3C(O)CO_2H \rightsquigarrow 2\text{ acetyl coenzyme A} + 2CO_2$$

$$\xrightarrow{\text{Krebs cycle}} 4CO_2 + 2CoASH$$

---

[a] Since it carries a unit positive charge on nitrogen (p. 448), NAD sometimes is abbreviated $NAD^{\oplus}$. The reduced form of this compound, since it carries no positive charge, is always written NADH.

The Krebs cycle demonstrates catalysis in biological systems at two different levels. On the more usual level, we find a group of enzymes that catalyze a series of discrete reactions. On the larger scale, we see how the entire process is catalytic, with one molecule of oxaloacetic acid being able to "process" many molecules of acetyl CoA.

19.10 PROSTAGLANDINS

Prostaglandins are compounds that were discovered about forty-five years ago but that were recognized as being of vast biological importance only during the past two decades. This corresponds roughly, and accidentally, to the first determination of the structure of a prostaglandin in 1962. It was not until 1968 that prostaglandins were synthesized in the laboratory. The soaring interest in prostaglandins is reflected in the fact that, while only 68 publications in 1965 dealt with prostaglandins, 521 publications dealing with these compounds were published in 1971.

What are these prostaglandins (abbreviated PG)? Chemically, they are derivatives of eicosanoic acid (C_{20} carboxylic acid) that contain a cyclopentane ring. While there appear to be many prostaglandins in nature, they are all believed to be derived from the so-called *primary* prostaglandins, PGE_1, PGF_{1a}, PGE_2, PGF_{2a}, PGE_3, and PGF_{3a}. (The PG's are classified as A, B, E, and F, depending upon the nature of the functional groups present.) The primary prostaglandins are derived, in pairs, from three unsaturated fatty acids: (8**Z**, 11**Z**, 14**Z**)-eicosatrienoic acid, (5**Z**, 8**Z**, 11**Z**, 14**Z**)-eicosatetraenoic acid, and (5**Z**, 8**Z**, 11**Z**, 14**Z**, 17**Z**)-eicosapentaenoic acid. The need for *unsaturated* fatty acids in the diet stems, in part, from their role in the biosynthesis of prostaglandins.

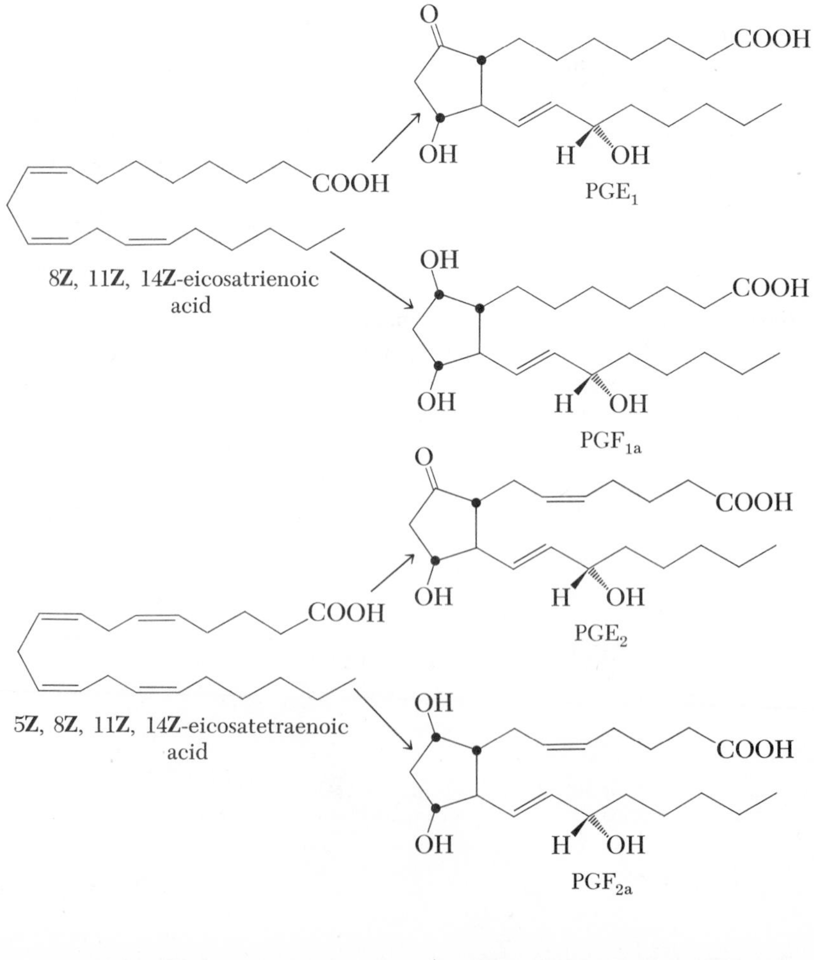

8**Z**, 11**Z**, 14**Z**-eicosatrienoic acid

PGE_1

PGF_{1a}

5**Z**, 8**Z**, 11**Z**, 14**Z**-eicosatetraenoic acid

PGE_2

PGF_{2a}

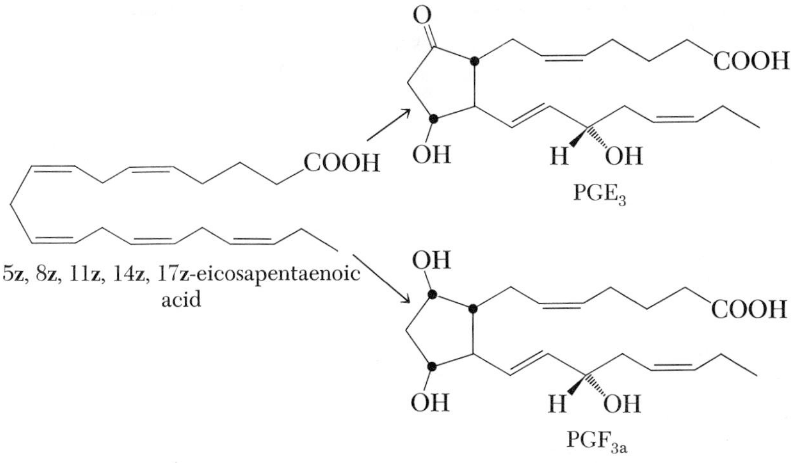

The scheme below is an *abbreviated* presentation of the synthesis of one pair of primary prostaglandins, PGE$_1$ and PGF$_{1a}$. In the scheme the doubly allylic radical, formed by abstraction of a hydrogen atom, reacts with RO· and molecular oxygen to produce a PG precursor that contains the cyclopentane ring. Oxidation of the C9—H bond in this precursor leads to PGE$_1$, while reduction of the O—O bond leads to PGF$_{1a}$.

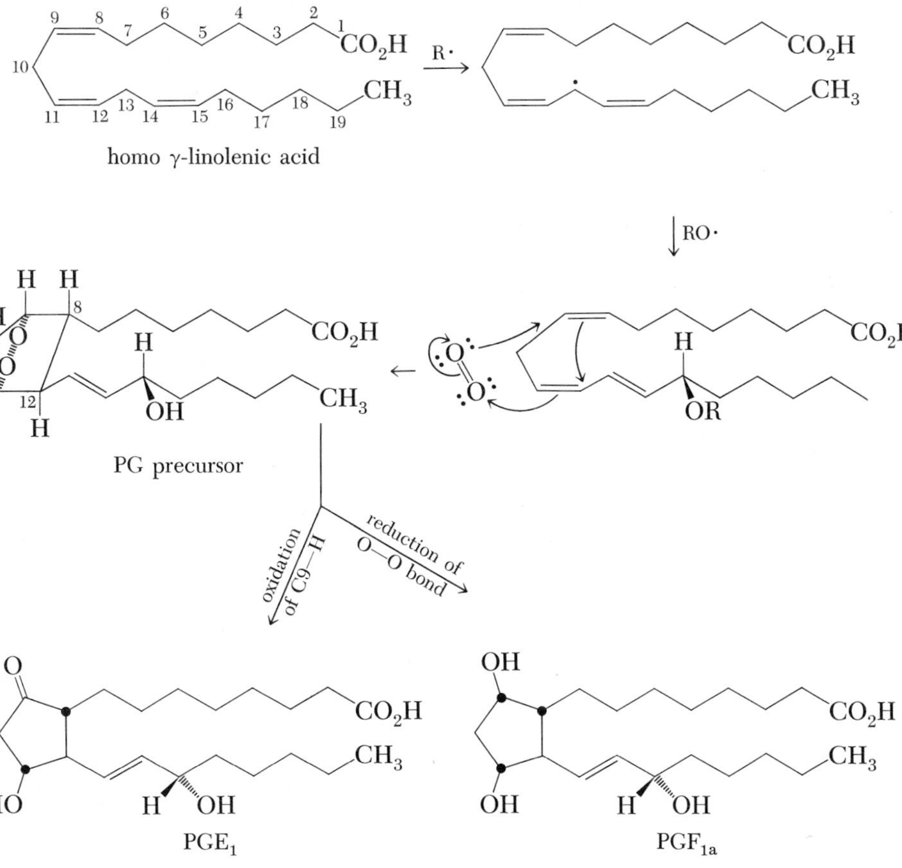

The need for prostaglandins for research is severe,[a] and syntheses employing relatively simple starting materials are of major interest in many laboratories. Some of the important steps in a laboratory synthesis of PGE_1 are shown below:

The physiological effects of prostaglandins are varied and often complex. However, one of their more general characteristics is the ability to induce smooth muscle contraction. This property has led to the use of prostaglandins for abortion. Unfortunately, side effects (nausea, diarrhea) often accompany its use. It also is believed that the IUD (intrauterine device) may

---

[a] One of the more recently discovered natural sources of prostaglandins is the sea whip coral (*Plexaura homomalla*). This coral, which is abundant in the Carribean, is approximately 0.75% PG_{2a}. It is presumed to play some defensive role in the coral.

derive its contraceptive capability, at least in part, from its ability to increase the amount of PG present locally.

In addition to their use in birth control, prostaglandins also are being examined for their role in inflammation, blood clotting, hypertension, menstrual regulation, asthma, and stomach ulcer generation. The highest concentration of PG in humans is found in semen, and it has been suggested that prostaglandins are necessary for erection and ejaculation in the male. (One type of infertility may be due to a lack of type E prostaglandins.) It has been revealed that sickle cell anemia crises (periods accompanied by severe pain) have been induced by prostaglandins. Even aspirins, our most consumed drug, may act by influencing the prostaglandins in the body. The mode of action of prostaglandins is a long way from being crystal clear. However, it is suggested that prostaglandins play a critical role in the moderation of hormone activity. What is most fascinating is their potency—solutions of as little as 1 nanogram (one billionth of a gram) per milliliter are able to induce smooth muscle contraction. Clearly, prostaglandin research is on the frontier of organic and biochemistry as well as that of medicine.

12. Examining the structures in the text, determine the meaning of the subscripts 1, 2, and 3 in the names of the various prostaglandins.

IMPORTANT TERMS

Activated methylene group: A methylene group whose hydrogens are unusually acidic. When these hydrogens are lost to a base, the resulting anions are resonance stabilized. Several moieties containing activated methylene groups are shown below.

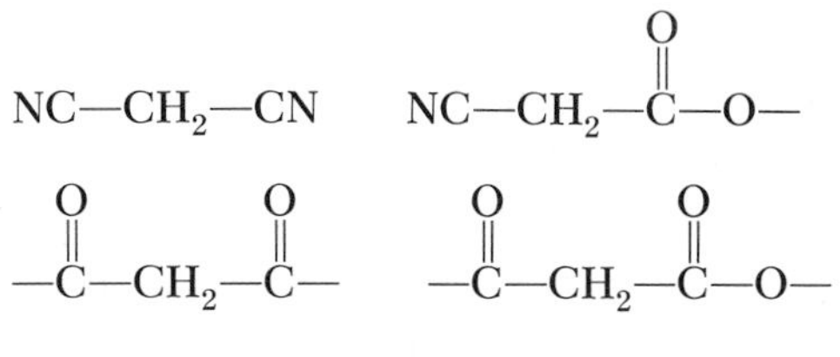

$$NC—CH_2—CN \qquad NC—CH_2—\overset{\overset{\displaystyle O}{\|}}{C}—O—$$

$$—\overset{\overset{\displaystyle O}{\|}}{C}—CH_2—\overset{\overset{\displaystyle O}{\|}}{C}— \qquad —\overset{\overset{\displaystyle O}{\|}}{C}—CH_2—\overset{\overset{\displaystyle O}{\|}}{C}—O—$$

Acyloin condensation: A reductive dimerization of esters to produce α-hydroxy ketones. The reaction is carried out in the presence of metallic sodium and is presumed to occur on the surface of the metal. A key intermediate is the enediol.

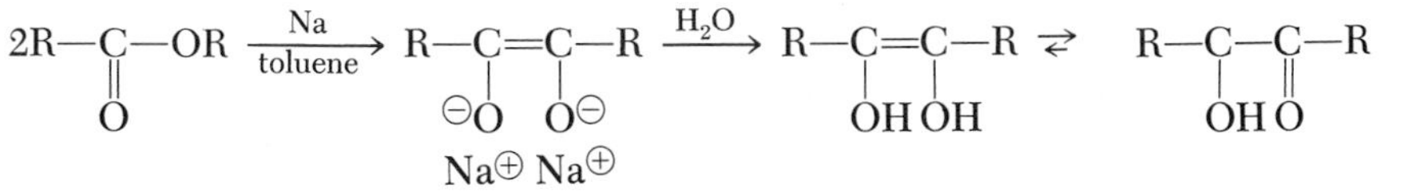

$$2R—\underset{\underset{\displaystyle O}{\|}}{C}—OR \xrightarrow[\text{toluene}]{\text{Na}} R—\underset{\underset{\displaystyle \ominus O\ Na^{\oplus}}{|}}{C}=\underset{\underset{\displaystyle O^{\ominus}\ Na^{\oplus}}{|}}{C}—R \xrightarrow{H_2O} \underset{\text{an enediol}}{R—\underset{\underset{\displaystyle OH}{|}}{C}=\underset{\underset{\displaystyle OH}{|}}{C}—R} \rightleftarrows \underset{\text{an }\alpha\text{-hydroxy ketone}}{R—\underset{\underset{\displaystyle OH}{|}}{C}—\underset{\underset{\displaystyle O}{\|}}{C}—R}$$

Adenosine triphosphate: A complex compound used, among other things, to transfer phosphate groups ("phosphorylation"). It usually is called ATP. The exothermic hydrolysis of ATP often is coupled, in living systems, to a second, endothermic reaction. The energy released by ATP hydrolysis may be said to "drive" the endothermic process. For this reason ATP is considered a "high energy compound" and may be viewed as a carrier of chemical energy in biologically useful form.

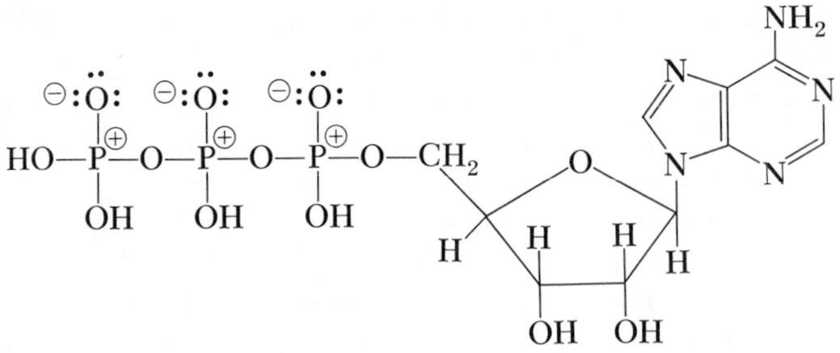

$$\text{HO}-\overset{\ominus:\ddot{O}:}{\underset{OH}{\overset{|\oplus}{\underset{|}{P}}}}-\text{O}-\overset{\ominus:\ddot{O}:}{\underset{OH}{\overset{|\oplus}{\underset{|}{P}}}}-\text{O}-\overset{\ominus:\ddot{O}:}{\underset{OH}{\overset{|\oplus}{\underset{|}{P}}}}-\text{O}-\text{CH}_2-(\text{ribose; H, H, H, H; OH, OH})-\text{adenine }(\text{NH}_2)$$

ATP

Aerobic oxidation: The conversion of the acetyl group of acetyl coenzyme A into carbon dioxide and water by a complex process that has as a major component the tricarboxylic acid (Krebs) cycle. It is termed "aerobic" because oxygen is ultimately reduced to water (see Figure 19–4).

Citric acid cycle: A reaction cycle that begins with the acetyl CoA–assisted conversion of oxaloacetic acid into citric acid and that ultimately produces two moles of carbon dioxide (corresponding to the two carbons transferred to oxaloacetic acid). It is viewed as a cycle because the last step regenerates oxaloacetic acid. The process is presented in Figure 19–5. It also is called the Krebs cycle and the tricarboxylic acid cycle.

Claisen ester condensation: The reaction of two moles of ester with base (usually alkoxide ion) to form a β-keto ester. The starting material must have a proton α to the ester group.

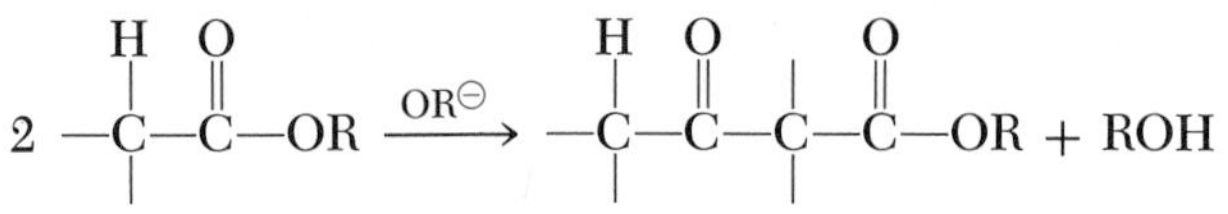

$$2\ -\overset{H}{\underset{|}{\overset{|}{C}}}-\overset{O}{\overset{\|}{C}}-OR \xrightarrow{OR^{\ominus}} -\overset{H}{\underset{|}{\overset{|}{C}}}-\overset{O}{\overset{\|}{C}}-\underset{|}{\overset{|}{C}}-\overset{O}{\overset{\|}{C}}-OR + ROH$$

Cytochromes: A group of complex proteins, found in mitochondria, that are of major importance in biological oxidations. The best characterized of these is called *cytochrome c* (molecular weight 12,400). The cytochromes carry out the terminal steps in the oxidative metabolism of food.

Hell-Volhard-Zelinsky reaction: The conversion of a carboxylic acid to an α-halo carboxylic acid by halogen and a catalytic amount of phosphorus trihalide.

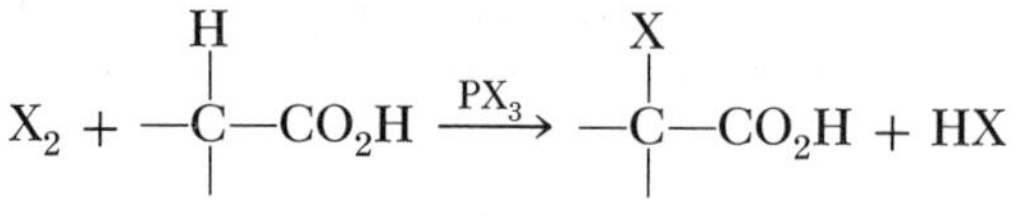

$$X_2 + -\overset{H}{\underset{|}{\overset{|}{C}}}-CO_2H \xrightarrow{PX_3} -\overset{X}{\underset{|}{\overset{|}{C}}}-CO_2H + HX$$

Michael addition: In the broadest sense, a Michael addition is any 1,4- or conjugate addition reaction. In a stricter sense, it refers to the 1,4-addition of a resonance-stabilized carbanion to a conjugated double bond (*e.g.*, an enone or an α,β-unsaturated ester).

Perkin condensation: The reaction of an anhydride, an aryl aldehyde, and the sodium (or potassium) salt of the acid corresponding to the anhydride. The ultimate product usually is an α,β-unsaturated acid.

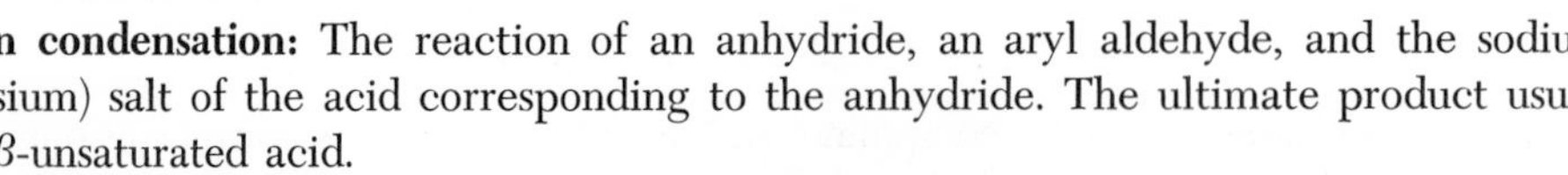

$$Ar-CHO + RCH_2-\overset{O}{\overset{\|}{C}}-O-\overset{O}{\overset{\|}{C}}-CH_2R \xrightarrow[180°]{RCH_2CO_2^{\ominus}} \xrightarrow{H_3O^{\oplus}} ArCH{=}CRCO_2H$$

Reformatsky reaction: The reaction of an aldehyde or ketone with a Reformatsky reagent to produce a β-hydroxy ester. The Reformatsky reagent has the general structure $R_2C(ZnBr)CO_2R'$ and is made by the reaction of an α-halo ester with metallic zinc in a mixture of ether and benzene. The product β-hydroxy ester may be dehydrated in acid to an α,β-unsaturated ester.

Respiratory chain: A series of reversible oxidation-reduction reactions that ultimately reduce molecular oxygen. The electrons for this reduction come from the citric acid cycle. The energy released during the operation of the respiratory chain is used to synthesize ATP. The respiratory chain will be re-introduced in Chapter 22.

Thiamine pyrophosphate: Thiamin pyrophosphate. A derivative of vitamin B1 (thiamine). Its structure and utility are presented on p. 891.

Wittig reaction: The reaction of an aldehyde or ketone with a phosphorus ylid to produce an alkene. One of the most important synthetic routes to alkenes.

Ylid: A molecule containing adjacent, opposite charges in which delocalization of a pair of electrons from the negatively charged center onto the positively charged center requires expansion of the octet of the latter. This electron pair is delocalized by interaction with a *d* orbital on the positively charged center. Ylids, *d* orbitals, and other such things will be discussed in more detail in Chapter 23.

$$-\overset{\oplus}{P}-\overset{\ominus}{\ddot{C}}- \leftrightarrow -P{=}C- \quad \text{a phosphorus ylid}$$

PROBLEMS

13. Provide a *specific* illustration of each of the following.
    (a) Claisen condensation
    (b) Perkin condensation
    (c) aldol condensation
    (d) Knoevenagel reaction
    (e) Michael reaction
    (f) crossed-Claisen condensation
    (g) acetoacetic ester synthesis
    (h) malonic ester synthesis
    (i) Dieckmann condensation
    (j) acyloin condensation
    (k) Robinson annelation reaction
    (l) Wittig reaction
    (m) Reformatsky reaction
14. Starting with ethyl acetate as the only organic compound, and any needed inorganic reagents, provide a rational synthesis for each of the following.
    (a) acetic acid
    (b) ethanol
    (c) ethylene
    (d) methanol
    (e) propanoic acid
    (f) acetoacetic ester
    (g) ethyl 3-hydroxybutanoate
    (h) α-bromopropionic acid
    (i) 1,3-butanediol
    (j) diethyl malonate
    (k) ethyl 4-cyanobutanoate
    (l) ethyl 4-cyano-2-methylbutanoate
    (m) 2-methylpentanedioic acid
    (n) ethyl α-deuterioacetate
    (o) $CH_3CH_2CH_2C(CH_3)C(CO_2Et)_2$
    (p) butanoic acid
    (q) acetone
15. Give the reaction sequence that will accomplish each of the following conversions. You must use the given starting material, but you can employ anything else you need.
    (a) $ClCH_2CO_2H$ to $CH_3CHOHCH_2CO_2CH_3$
    (b) $(EtO_2C)_2CH_2$ to 2,5-dimethylhexanoic acid
    (c) malonic acid to 4-methyl-2-pentenoic acid
    (d) acetylene to adipic acid
    (e) benzaldehyde to phenylsuccinic acid

(f) malonic ester to 1,5-pentanedioic acid
(g) diethyl adipate to cyclopentanol
(h) diethyl octanedioate to cyclo-octanol
(i) 1,2-dibromoethane to 1,1-cyclopropanedicarboxylic acid
(j) ethyl cyanoacetate to $NCCH(CO_2Et)CH_2CH_2CHO$
(k) propanoic acid to propenoic acid
(l) ethyl acetate to ethyl 3-hydroxy-3-phenylpropanoate
(m) ethylene carbonate to $CH_3C(O)OCH_2CH_2OC(O)CH_2C(O)CH_3$

* 16. Suggest a specific procedure, including observations, that would allow distinction between members of each of the following pairs.
(a) acetic acid and ethyl acetate
(b) acetoacetic ester and ethyl acetate
(c) acetic acid and acetic anhydride
(d) sodium acetate and sodioacetoacetic ester
(e) cyclohexanone and 2-carbethoxycyclohexanone
(f) caproic acid and caprylic acid
(g) acetic anhydride and succinic anhydride
(h) ethylene carbonate and propylene carbonate
(i) 3,3-dimethylbutanal and pivalaldehyde

(j) 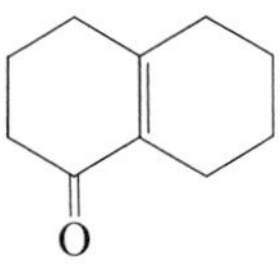 [bicyclic enone, O] and [bicyclic enone, O]

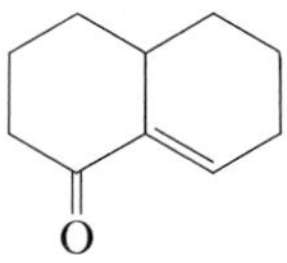

(k) 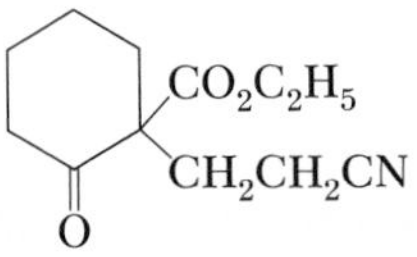 [cyclohexanone with $CO_2C_2H_5$ and CH_2CH_2CN; O] and [cyclohexanone with H and CH_2CH_2CN; O]

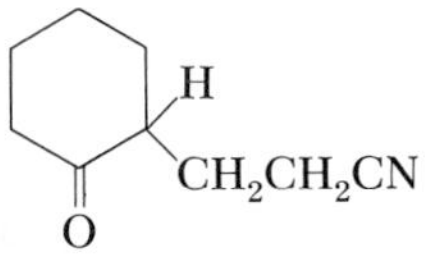

17. Tetracyanoethylene is attacked by water to form HCN and a compound, **A**, which shows a strong OH absorption in the ir. **A** has the formula C_5HN_3O. Suggest a structure for **A** and account for its spectrum.

18. *Malathion*, an insecticide of low toxicity to humans, is made by a Michael addition. Given the structure of malathion, suggest the immediate precursors for its synthesis. (For comments on the biological action of such compounds, see p. 757.)

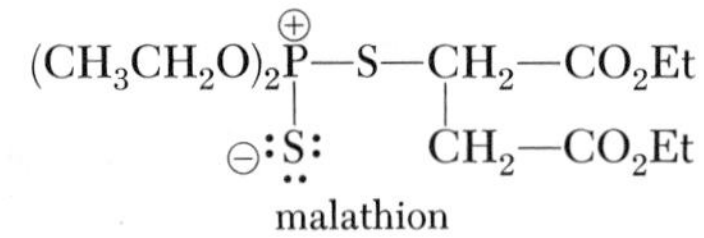

$$(CH_3CH_2O)_2\overset{\oplus}{P}(\text{—}S^{\ominus})\text{—}S\text{—}CH(CH_2\text{—}CO_2Et)\text{—}CO_2Et$$

malathion

19. Provide a mechanism for each of the following:

(a) $RCOCH_3 + NaH \rightarrow A \xrightarrow{EtO-C(=O)-OEt} RCOCH_2CO_2Et$

(b) $HCO_2Et + CH_3CO_2Et \xrightarrow{OEt^{\ominus}} HCOCH_2CO_2Et$
ethyl formylacetate

(c) $BrCH_2CH_2Br + EtO_2CCH_2CO_2Et \xrightarrow{OEt^{\ominus}}$ [cyclopropane with CO_2Et, CO_2Et]

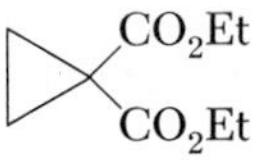

(d) 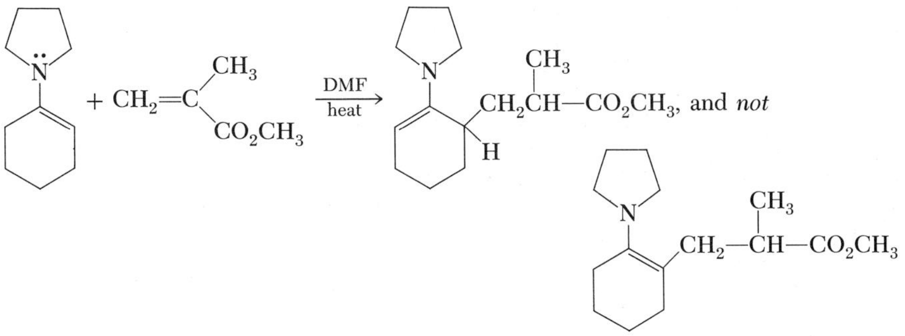 [pyrrolidine enamine of cyclohexanone] + $CH_2{=}C(CH_3)CO_2CH_3$ $\xrightarrow[\text{heat}]{\text{DMF}}$ [enamine product] $CH_2CH(CH_3)-CO_2CH_3$, H, and *not* [enamine product] $CH_2-CH(CH_3)-CO_2CH_3$

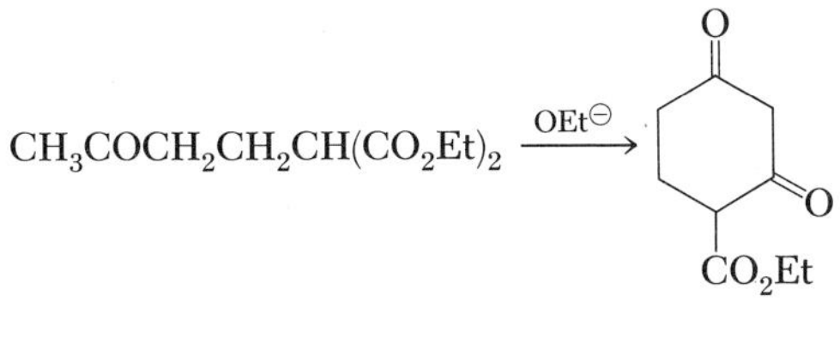

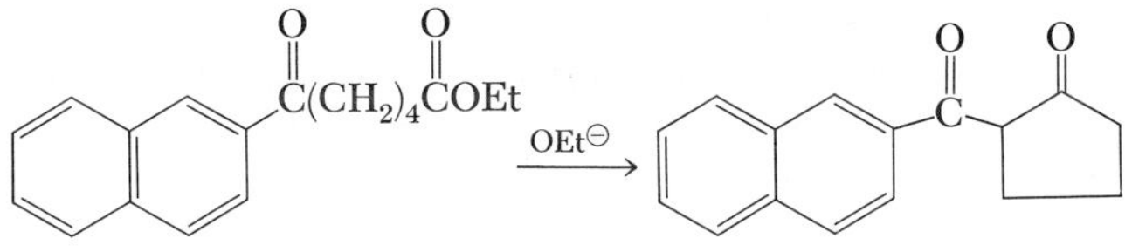

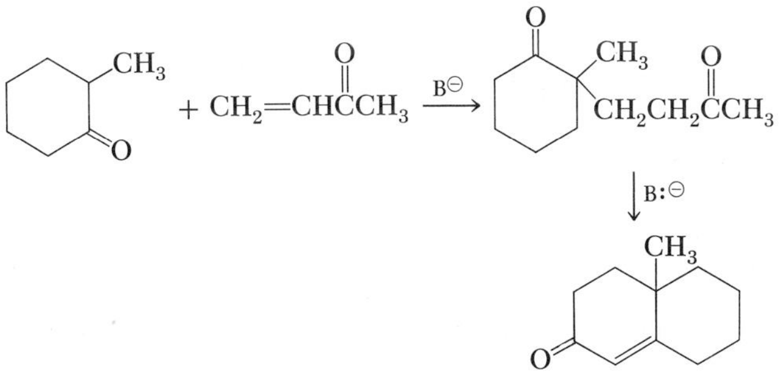

(e) $CH_3COCH_2CH_2CH(CO_2Et)_2 \xrightarrow{OEt^{\ominus}}$ [cyclohexane-1,3-dione] CO_2Et

(f) [2-naphthyl] $\overset{O}{\overset{\|}{C}}(CH_2)_4\overset{O}{\overset{\|}{C}}OEt \xrightarrow{OEt^{\ominus}}$ [2-naphthyl] $\overset{O}{\overset{\|}{C}}$-[cyclopentanone]

(g) [2-methylcyclohexanone] CH_3, O + $CH_2{=}CH\overset{O}{\overset{\|}{C}}CH_3 \xrightarrow{B^{\ominus}}$ [cyclohexanone] CH_3, $CH_2CH_2\overset{O}{\overset{\|}{C}}CH_3$ $\xrightarrow{B:^{\ominus}}$ [bicyclic enone] CH_3, O

(h) [cyclobutane] CH_2Br, CH_2Br + $EtO\overset{O}{\overset{\|}{C}}CH_2\overset{O}{\overset{\|}{C}}OEt \xrightarrow{OEt^{\ominus}}$ **A** $\xrightarrow{H_3O^{\oplus}} \xrightarrow{\text{heat}}$ [spiro[3.3]heptane]$-CO_2H$

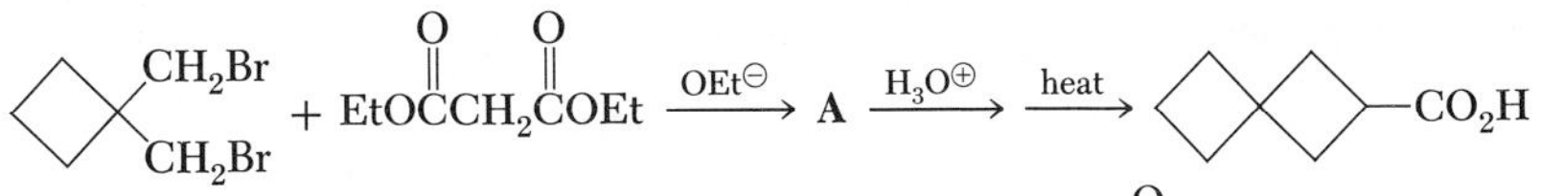

(i) 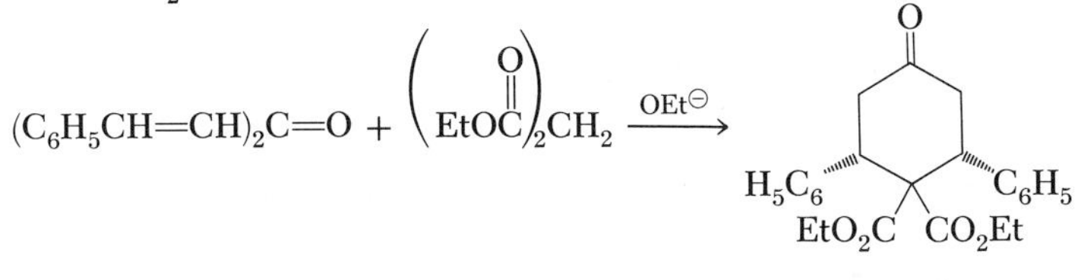 $(C_6H_5CH{=}CH)_2C{=}O + \left(EtO\overset{O}{\overset{\|}{C}}\right)_2CH_2 \xrightarrow{OEt^{\ominus}}$ [cyclohexanone] O, H_5C_6, C_6H_5, EtO_2C, CO_2Et

20. Nitromethane, H_3CNO_2, is sufficiently acidic to be soluble in aqueous sodium hydroxide. Furthermore, if the solvent is deuterated (*i.e.*, $D_2O/OD^{\ominus}$), when the nitromethane is recovered it is now CD_3NO_2. (By way of contrast, CH_3CF_3 is insoluble in aqueous alkali.) Infrared spectral evidence indicates that nitromethane can be converted to an isomeric *aci* form. (a) Construct a consistent explanation for all of these data. (b) With the aid of your answer to (a), provide a mechanism for each of the following reactions.

(i) 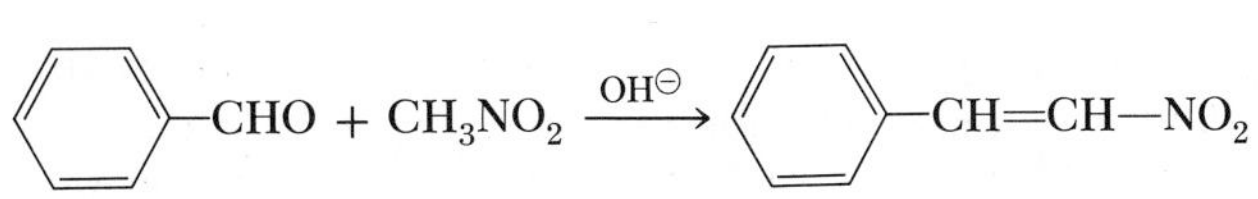 [benzene]$-CHO + CH_3NO_2 \xrightarrow{OH^{\ominus}}$ [benzene]$-CH{=}CH-NO_2$

β-nitrostyrene

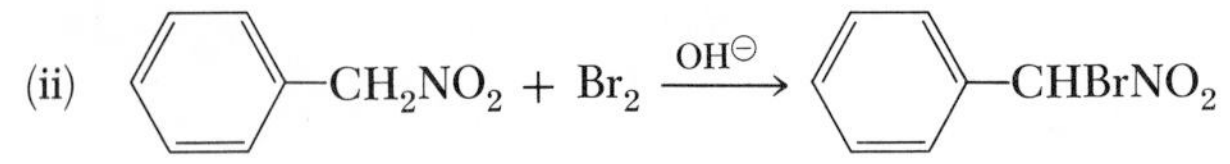

(ii) $C_6H_5CH_2NO_2 + Br_2 \xrightarrow{OH^\ominus} C_6H_5CHBrNO_2$

(iii) $(CH_3)_2CHNO_2 + CH_2{=}CHCN \xrightarrow[25^\circ]{OH^\ominus} (CH_3)_2C(NO_2){-}CH_2CH_2CN$

21. The intramolecular attack of a nitrile-stabilized carbanion upon a second nitrile group to form a five- or six-membered ring is called a **Thorpe cyclization**; an example follows. Suggest a mechanism for this reaction. What is the structure of **A**?

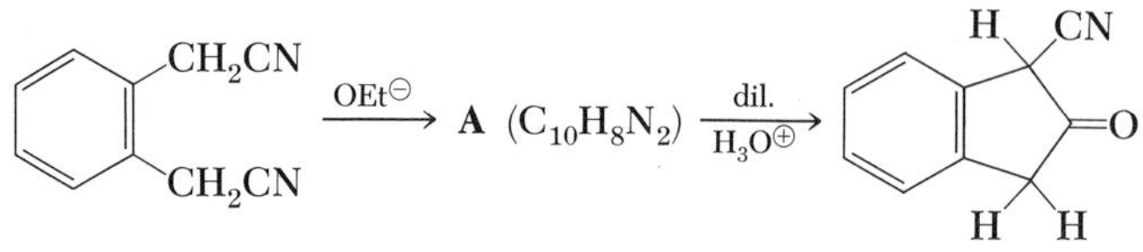

$$o\text{-}C_6H_4(CH_2CN)_2 \xrightarrow{OEt^\ominus} \mathbf{A}\ (C_{10}H_8N_2) \xrightarrow[H_3O^\oplus]{\text{dil.}} \text{2-indanone with H, CN at C-1 and H, H at C-3}$$

What experimental precautions would be required in order to form large rings by this reaction or by the Dieckmann reaction?

22. The following isomerization illustrates the reversibility of the Dieckmann condensation. What starting material will lead, upon reaction under Dieckmann conditions, to compound **A**? Suggest a mechanism for the conversion of **A** to **B**.

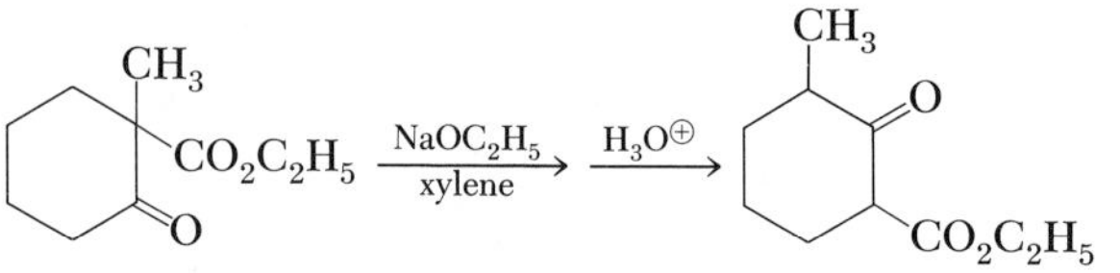

$$\underset{\mathbf{A}}{\text{2-methyl-2-}CO_2C_2H_5\text{-cyclohexanone}} \xrightarrow[\text{xylene}]{NaOC_2H_5} \xrightarrow{H_3O^\oplus} \underset{\mathbf{B}}{\text{2-methyl-6-}CO_2C_2H_5\text{-cyclohexanone}}$$

23. Reaction of α,α,β-triphenylpropionitrile with potassium amide, a strong base, leads to triphenylethylene in 94% yield. Suggest a mechanism for this reaction.

24. Benzaldehyde and benzyl cyanide react in the presence of sodium cyanide to produce 1,2-dicyano-1,2-diphenylethane. Suggest a mechanism for this.

25. Refluxing a mixture of sodium ethoxide and diethyl succinate in ethanol for one day yields 2,5-dicarbethoxy-1,4-cyclohexanedione in 65% yield. Suggest a mechanism for this reaction.

$$2C_2H_5O{-}\overset{O}{\overset{\|}{C}}{-}CH_2CH_2{-}\overset{O}{\overset{\|}{C}}{-}OC_2H_5 \xrightarrow[\text{EtOH}]{\text{NaOEt}} \text{1,4-cyclohexanedione with H, }CO_2C_2H_5\text{ at C-2 and H, }H_5C_2O_2C\text{ at C-5}$$

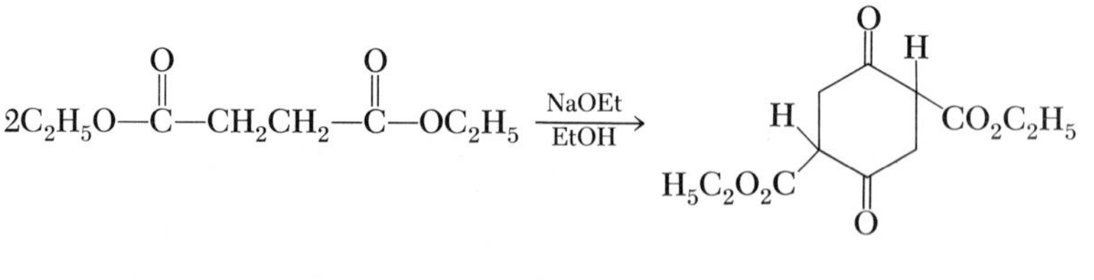

26. Malononitrile condenses with aldehydes or ketones to give unsaturated nitriles in yields from 60 to 90%. Suggest a mechanism for the example below.

$$C_6H_5CHO + NC{-}CH_2{-}CN \xrightarrow{\text{pyridine}} C_6H_5CH{=}C(CN)_2$$

27. Diphenylacetonitrile, $(C_6H_5)_2CHCN$, reacts with benzyl chloride in the presence of potassium amide to form α,α,β-triphenylpropionitrile in 95% yield. Suggest a mechanism for this reaction.

28. Suggest a method of synthesizing triphenylphosphine, using PCl_3 as your source of phosphorus. (*Hint:* Friedel-Crafts.)

29. With the aid of a periodic table, can you suggest a reason why triphenylphosphine is useful for preparing Wittig reagents (ylids) while triphenylamine is not? (This is discussed in detail in Chapter 23.)

30. Suggest a combination of Wittig reagents and carbonyl compounds that will produce all of the isomeric monoalkenes that have the isobutane carbon skeleton.

*31. The nmr spectra below are those of diethyl ethylmalonate ($C_2H_5O_2CH(C_2H_5)CO_2C_2H_5$), diethyl ethylidenemalonate ($CH_3CH{=}C(CO_2C_2H_5)_2$), acetoacetic ester ($CH_3COCH_2CO_2C_2H_5$), and ethyl acetate ($CH_3CO_2C_2H_5$). Assign each spectrum to one of these compounds. How might you distinguish among these compounds by chemical means?

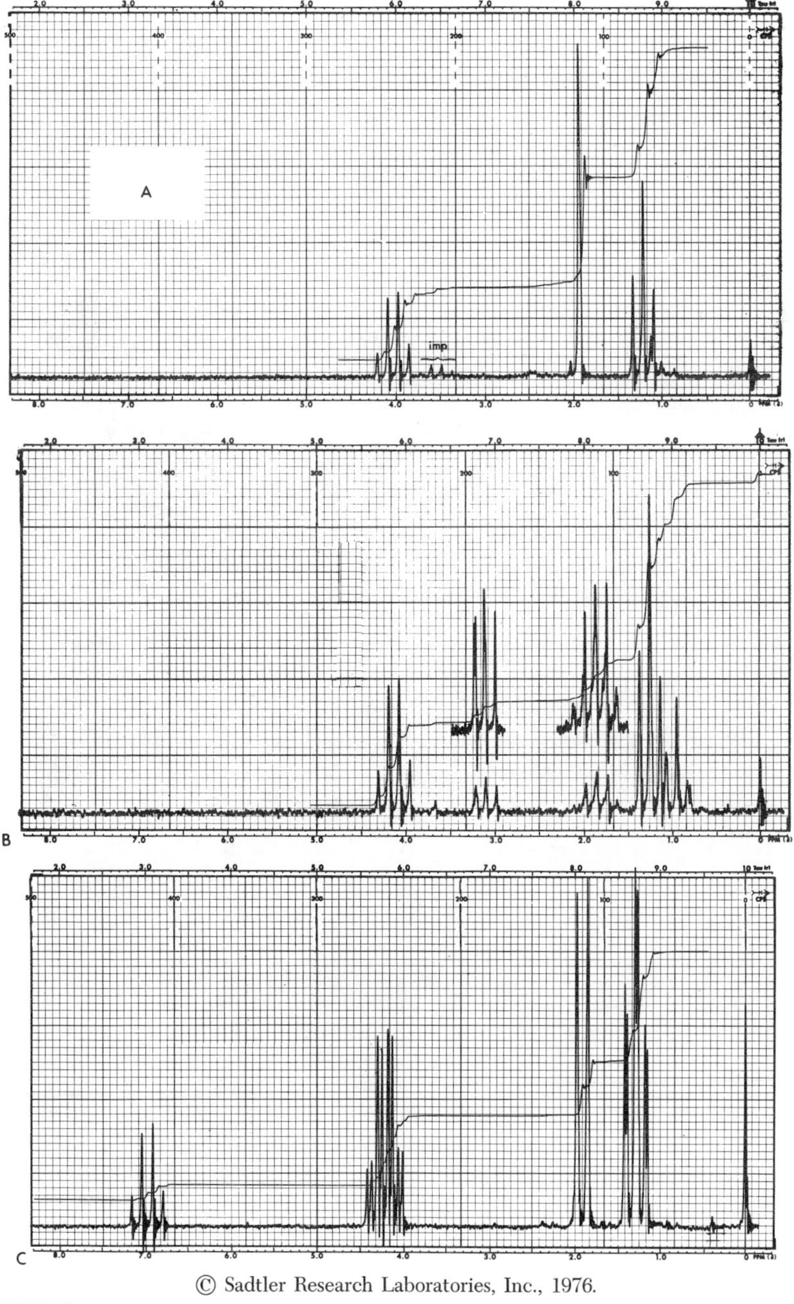

* This problem requires a knowledge of spectroscopy.

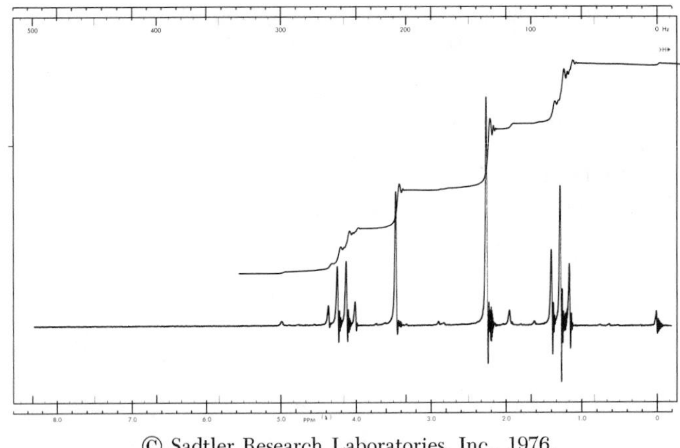

°32. The infrared and nmr spectra below are those of acetylacetone. Account for as many spectral features as possible.

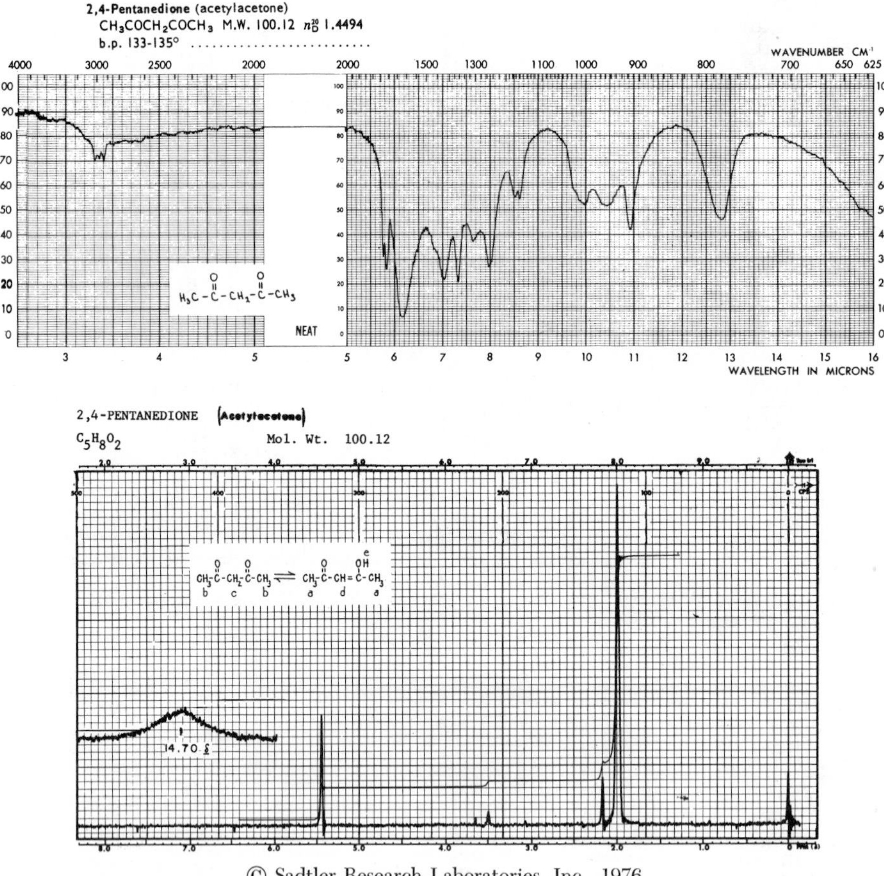

°33. Compound **A,** $C_5H_8O_3$, is soluble in concentrated aqueous sodium hydroxide solution. The nmr spectrum of **A** consists of three singlets (3H, 2.12δ; 2H, 3.35δ; 3H, 3.62δ). Assign a structure to **A** that is consistent with these data.

°This problem requires a knowledge of spectroscopy.

AMINES

20.1 INTRODUCTION

Amines are the most important of the monofunctional organic nitrogen compounds and, as we shall see in future chapters, the amino group is essential to life as we know it. The chemistry of rather uncomplicated amines forms the core of the material in this and the following chapter. This chapter deals largely with the synthesis and reactions of alkylamines, *i.e.*, amines bearing only alkyl groups on nitrogen. The following chapter concentrates upon arylamines, *i.e.*, amines bearing at least one aromatic group on nitrogen. The artificiality of dividing the chemistry of amines into "alkylamine chemistry" and "arylamine chemistry" will become apparent when, early in this chapter, synthesis of aliphatic and aromatic amines are discussed in the same breath. However, it will help us if we group selected reactions into those that involve arylamines and those that involve alkylamines.

20.2 NOMENCLATURE

Amines are classified as *primary* (1°), *secondary* (2°), or *tertiary* (3°) *according to the number of groups bonded to nitrogen*. Thus, an amine with only one of the ammonia hydrogens replaced by an organic moiety is a primary amine. This is to be contrasted to the naming of alcohols and alkyl halides, in which the number of substituents *on the carbon bearing the functional group* determines the designation of the compound as primary, secondary, or tertiary.

RNH_2 1° amine	R_2NH 2° amine	R_3N 3° amine
RCH_2OH 1° alcohol	R_2CHOH 2° alcohol	R_3COH 3° alcohol
RCH_2X 1° halide	R_2CHX 2° halide	R_3CX 3° halide

A fourth type of compound in this series is the quaternary ammonium salt, $R_4N^{\oplus}X^{\ominus}$. The non-specific term "ammonium salt" identifies any of a series of five different types of compounds—$NH_4^{\oplus}X^{\ominus}$, $RNH_3^{\oplus}X^{\ominus}$, $R_2NH_2^{\oplus}X^{\ominus}$, $R_3NH^{\oplus}X^{\ominus}$, and $R_4N^{\oplus}X^{\ominus}$.

Alkylamines commonly are named by identifying the alkyl group(s) bound to nitrogen and adding the suffix "amine," the entire compilation being treated as one word. The ordering of substituents is alphabetical.

$CH_3{-}\ddot{N}H{-}H$	$(CH_3)_3C{-}\ddot{N}H{-}CH_3$	$CH_3CH_2{-}\ddot{N}(CH_2CH_3){-}CH_2CH_3$	$CH_3{-}N^{\oplus}(CH_3)_2{-}CH_3\ Cl^{\ominus}$
methylamine (1° amine)	*t*-butylmethylamine (2° amine)	triethylamine (3° amine)	tetramethylammonium chloride (quaternary ammonium salt)

Arylamines may be named similarly, although they generally are treated as derivatives of aniline.

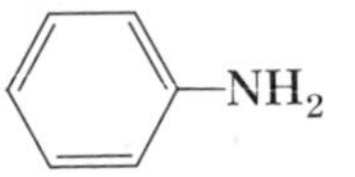

aniline — C_6H_5—NH_2

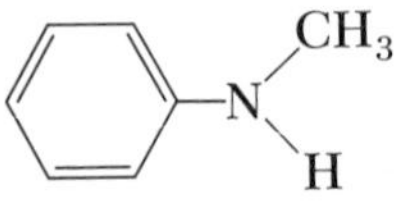

N-methylaniline — C_6H_5—N(CH_3)H

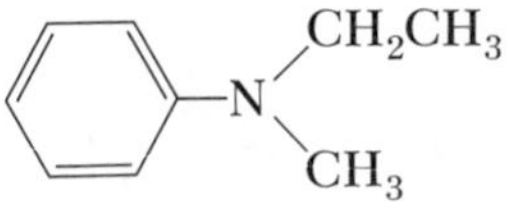

N-ethyl-*N*-methylaniline — C_6H_5—N(CH_2CH_3)(CH_3)

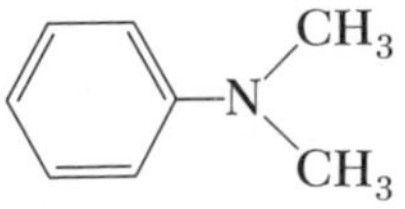

N,*N*-dimethylaniline — C_6H_5—N(CH_3)(CH_3)

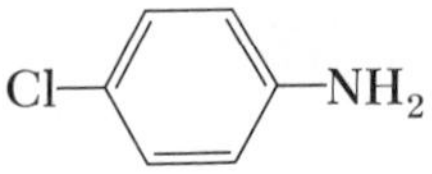

p-chloroaniline — Cl—C_6H_4—NH_2

p-toluidine (special name) — H_3C—C_6H_4—NH_2

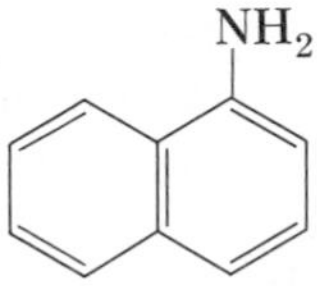

α-naphthylamine — NH_2

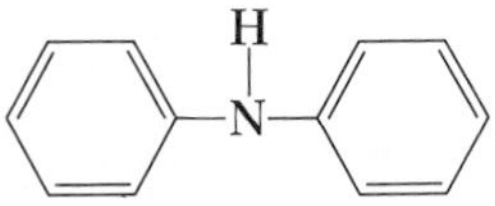

diphenylamine (*not N*-phenylaniline)

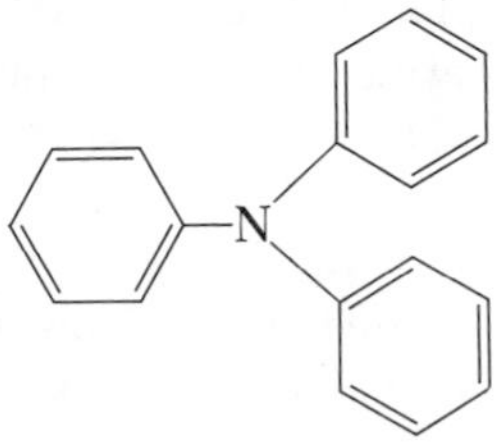

triphenylamine (*not N,N*-diphenylaniline)

Heterocyclic amines, some of which already have been encountered, usually carry special names.

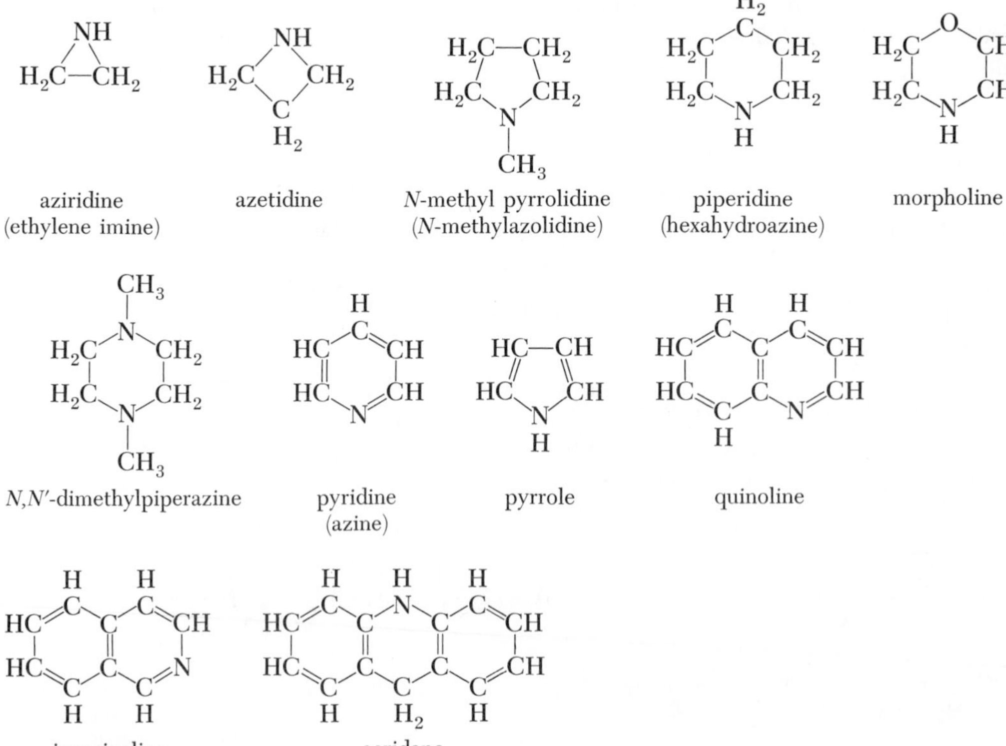

aziridine (ethylene imine)

azetidine

N-methyl pyrrolidine (*N*-methylazolidine)

piperidine (hexahydroazine)

morpholine

N,*N*′-dimethylpiperazine

pyridine (azine)

pyrrole

quinoline

isoquinoline

acridane

More complex amines sometimes can be named by including "aza" in a hydrocarbon name, thereby indicating replacement of a $—CH_2—$ (or $—\overset{|}{C}H—$) by $—NH—$ (or $—\overset{|}{N}—$).

$CH_3CH_2NHCH_2CH_3$ — 3-azapentane

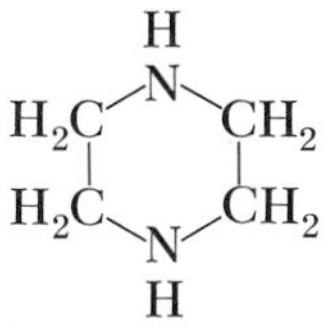

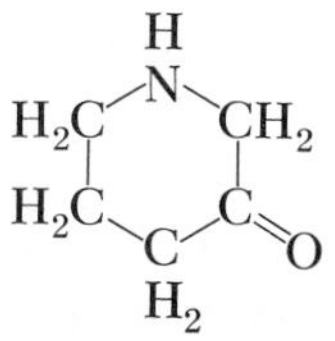

1,4-diazacyclohexane

3-azacyclohexanone

1. Provide an acceptable name for each of the following.

(a) CH_3NH_2

(b) $(CH_3)_2\overset{\oplus}{N}H_2\ Cl^{\ominus}$

(c) $C_6H_5NHC_2H_5$

(d) $(C_6H_5)_2\overset{\oplus}{N}(CH_3)_2\ OH^{\ominus}$

(e)

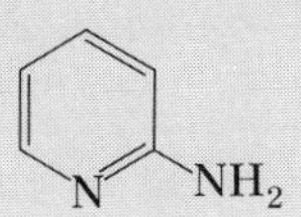

(f)

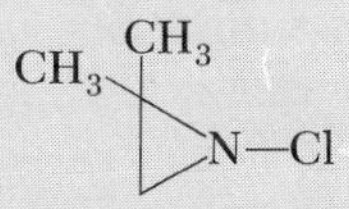

(g)

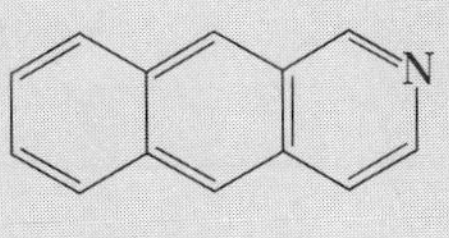

20.3 PHYSICAL PROPERTIES OF AMINES

Hydrogen Bonding. Most amines are potent hydrogen bond acceptors, the hydrogen bonding ability generally increasing with increasing alkylation. In addition, the N—H moieties of 1° and 2° amines act as good hydrogen bond donors. Intermolecular hydrogen bonding is responsible for the comparatively high boiling points of 1° and 2° amines, while the absence of a donor hydrogen permits 3° amines to have more normal boiling points (Table 20–1). The unlimited solubility of low molecular weight amines in water is ascribed to extensive hydrogen bonding between these amines and the solvent.

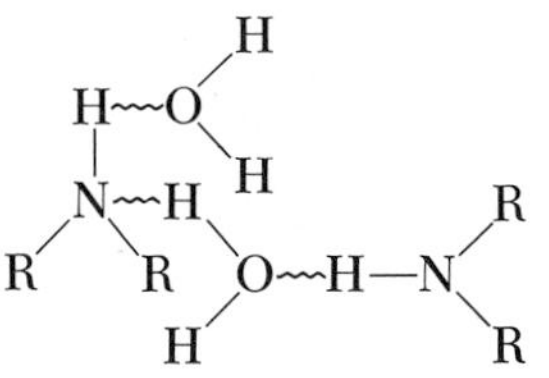

hydrogen bonding involving a 2° amine and water

TABLE 20-1 Amines

NAME	MP (°C)	BP (°C)	K_b
methylamine	−92	−7.5	4.5×10^{-4}
dimethylamine	−96	+7.5	5.4×10^{-4}
trimethylamine	−117	3	0.6×10^{-4}
ethylamine	−80	17	5.1×10^{-4}
diethylamine	−39	55	10.0×10^{-4}
triethylamine	−115	89	5.6×10^{-4}
n-propylamine	−83	49	4.1×10^{-4}
n-butylamine	−50	78	4.8×10^{-4}
isobutylamine	−85	68	3.0×10^{-4}
sec-butylamine	−104	63	4.0×10^{-4}
t-butylamine	−67	46	5.0×10^{-4}
cyclohexylamine	−17	134	5.1×10^{-4}
aniline	−6	184	4.2×10^{-10}
N-methylaniline	−57	196	7.1×10^{-10}
N,N-dimethylaniline	+3	194	11.7×10^{-10}
diphenylamine	53	302	6.0×10^{-14}
triphenylamine	127	365	
o-toluidine	−28	200	2.6×10^{-10}
m-toluidine	−30	203	5.0×10^{-10}
p-toluidine	44	200	1.2×10^{-9}
p-chloroaniline	70	232	1.0×10^{-10}
p-nitroaniline	148	332	1.0×10^{-13}
pyridine	−42	115	2.0×10^{-9}
quinoline	−15	237	1.3×10^{-10}
piperidine	−13	106	1.6×10^{-3}
morpholine	−6	129	2.5×10^{-6}

Basicity. Ammonia has a pK_b[a] of 4.75 at 25°. As might be expected from the electron donating character of alkyl groups, replacing the hydrogens of ammonia with alkyl groups increases the basicity of nitrogen. However, replacement of all three hydrogens results in a decrease in basicity, a result often considered to be a "solvation" effect.

$$CH_3CH_2NH_2 \quad K_b = 5.1 \times 10^{-4} \qquad (CH_3CH_2)_2NH \quad K_b = 10.0 \times 10^{-4} \qquad (CH_3CH_2)_3N \quad K_b = 5.6 \times 10^{-4}$$

Arylamines, on the other hand, are less basic than ammonia or alkylamines because both the inductive and resonance effects of the aryl group withdraw electron density from the nitrogen. The influence of resonance upon aniline's basicity is illustrated in Figure 20–1.

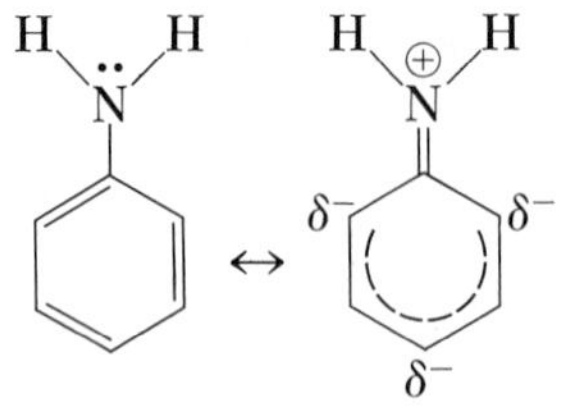

H H N ↔ H ⊕ H N δ− δ− δ−

resonance withdrawal of electron density from nitrogen

Heterocyclic amines, such as pyridine, that can donate an electron pair without losing aromaticity exhibit moderate basic character and, therefore, are useful as organic bases. On

[a] You may wish to review the discussion of K_a and K_b in Chapter 18.

$$Ar—\ddot{N}H_2 + HB \rightarrow Ar—\underset{H}{\overset{H}{N}} \overset{\delta+}{\cdots} H \cdots \overset{\delta-}{B} \rightarrow Ar\overset{\oplus}{N}H_3 \quad B^{\ominus}$$

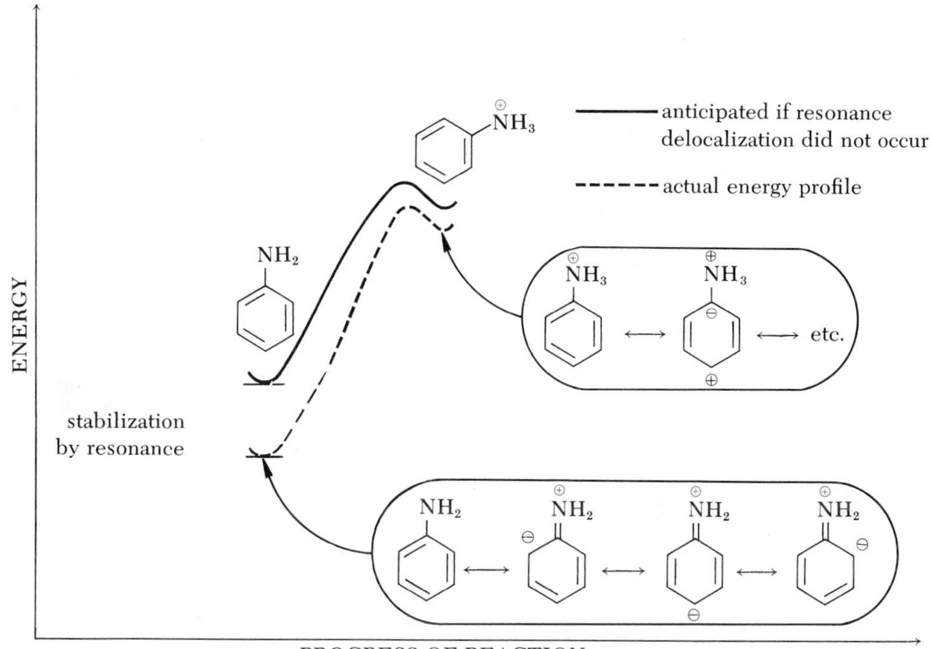

Figure 20-1 The effect of resonance on the basicity of aniline. Resonance stabilizes the free amine, relative to the salt, and makes arylamines weaker bases than alkylamines. The behavior of alkylamines, in which resonance does not occur, is approximated by the solid line.

the other hand, pyrrole derivatives, which lose their aromaticity upon protonation, tend to decompose in acidic media.

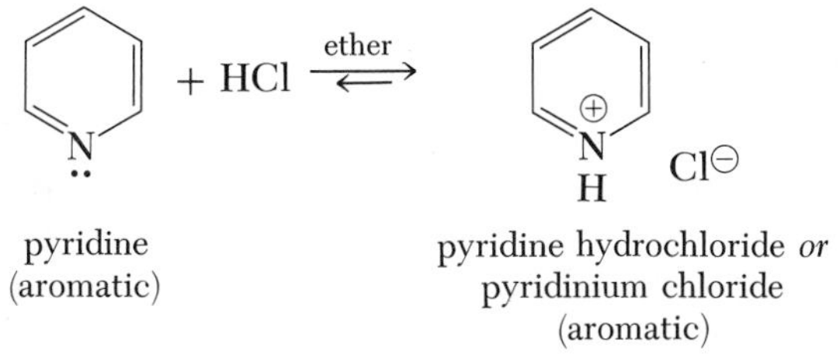

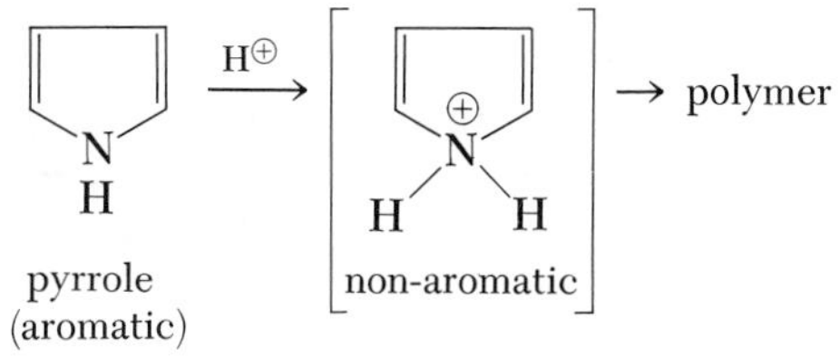

Most amines are converted to the corresponding ammonium salts by reaction with mineral or other acids. In turn, these salts liberate the amine from which they were produced when the salts are treated with a strong base such as sodium hydroxide. Salt formation followed by decomposition of the salt with strong base provides an excellent means of freeing amines from other organic compounds. This procedure works well because only amines, of the common types of organic compounds, are insoluble in water, soluble in dilute acid, and regenerated by basification with hydroxide ion.

$$\left.\begin{matrix} RNH_2 \\ R_2NH \\ R_3N \end{matrix}\right\} \underset{OH^{\ominus}}{\overset{H^{\oplus}A^{\ominus}}{\rightleftarrows}} \left\{\begin{matrix} R\overset{\oplus}{N}H_3\ A^{\ominus} \\ R_2\overset{\oplus}{N}H_2\ A^{\ominus} \\ R_3\overset{\oplus}{N}H\ A^{\ominus} \end{matrix}\right.$$

free amine (often water insoluble) — ammonium salt (water soluble)

2. While azanonatetraene is comparatively stable at room temperature, *N*-carbethoxyazanonatetraene rapidly isomerizes under the same conditions. Account for this difference, suggesting a mechanism for this isomerization.

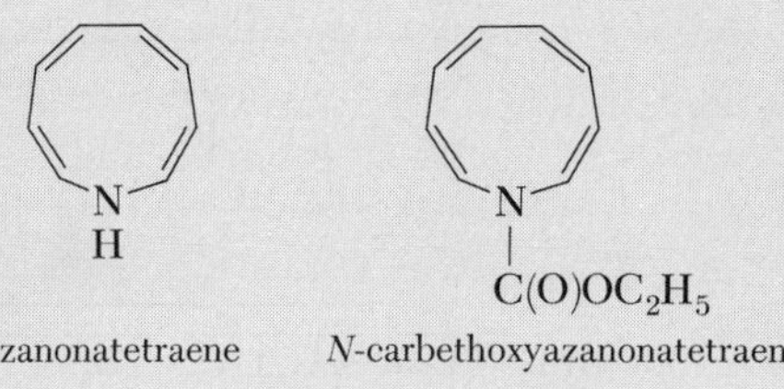

azanonatetraene — *N*-carbethoxyazanonatetraene

Odor. The methylamines are flammable gases that smell much like ammonia. The "fishy" odor of higher molecular weight alkylamines is not surprising, since alkylamines are responsible for the odor of fish brines. The Gothic "odor of death and decay" of flesh is due in large measure to the presence of two diamines with the descriptive names *putrescine* and *cadaverine.*

$$H_2N{-}CH_2CH_2CH_2CH_2{-}NH_2 \qquad H_2N{-}CH_2CH_2CH_2CH_2CH_2{-}NH_2$$

putrescine (1,4-butanediamine) — cadaverine (1,5-pentanediamine)

Arylamines are not as malodorous as are alkylamines, but they are, as a rule, toxic. Danger of poisoning is enhanced by the facile absorption of amines through the skin. Motivated in part by the carcinogenicity of several arylamines, such as β-naphthylamine, federal law now requires a cautionary label on all substances containing more than 5% of an arylamine.

Configurational Stability. As noted in Chapter 4, simple alkylamines and arylamines are configurationally unstable and undergo rapid pyramidal inversion at room temperature.[a] By way of contrast, there is spectral evidence that indicates that the nitrogen of the aziridine ring system does not invert configuration very rapidly (nmr time scale). Thus, the room temperature nmr spectrum of tetramethylaziridine shows two signals of equal intensity, assignable to those methyl groups which are *cis* and *trans* to the hydrogen on nitrogen. Had nitrogen inversion been rapid on the nmr time scale, all four methyl groups would have appeared as one time-averaged signal.

Why, we must ask, is nitrogen inversion slower in aziridine derivatives (at least in this one) than in simple acyclic amines? The answer to this question is found in the angle strain

[a] The rate of inversion of ammonia is approximately 10^{11}/sec.

that is generated when the nitrogen in the three-membered ring passes from sp^3 to sp^3 hydbridization *via* an sp^2 hybridized, planar transition state. The increased strain in going to such a transition state leads to a *comparatively* high activation energy for pyramidal inversion. Acyclic amines, on the other hand, do not suffer any major increase in strain energy in going from a pyramidal nitrogen to a planar nitrogen and, therefore, invert quite easily.

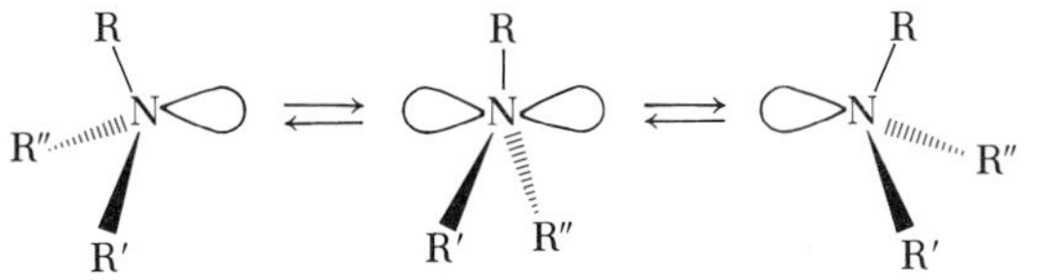

racemization by pyramidal inversion is rapid at 25°

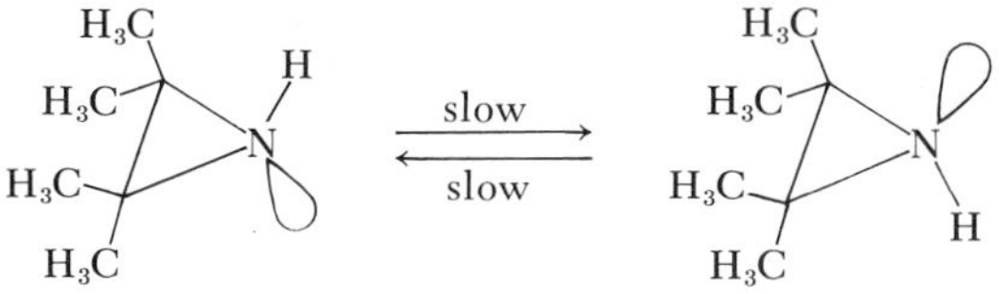

planar, achiral transition state

tetramethylaziridine: slow inversion at 25°

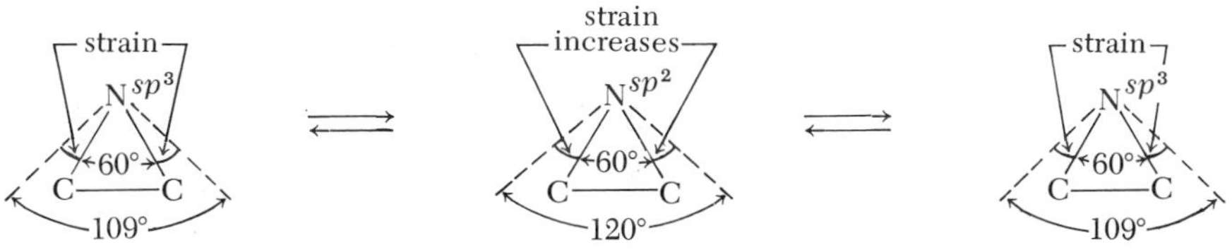

strain in ground state of aziridine

strain present when nitrogen is planar in transition state

strain in ground state of aziridine

Inversion of nitrogen is prevented if the lone electron pair is coordinated, since this prevents that electron pair from "tunneling" from one "side" of nitrogen to the other. This explains why allylbenzylmethylphenylammonium cation is separable into enantiomers.

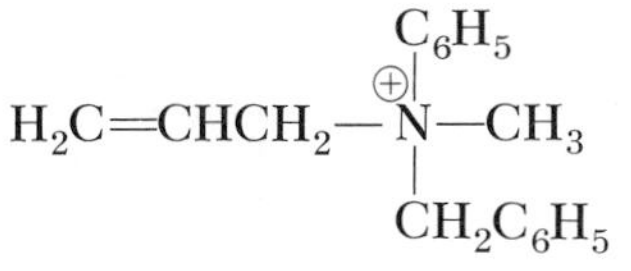

$$H_2C{=}CHCH_2{-}\overset{\oplus}{N}(C_6H_5)(CH_2C_6H_5){-}CH_3$$

configurationally stable at room temperature

Unlike the *quaternary* ammonium salt, the other ammonium salts racemize rapidly by ionizing to a configurationally unstable amine. Proton recapture by the racemized amine leads to racemic ammonium salt.

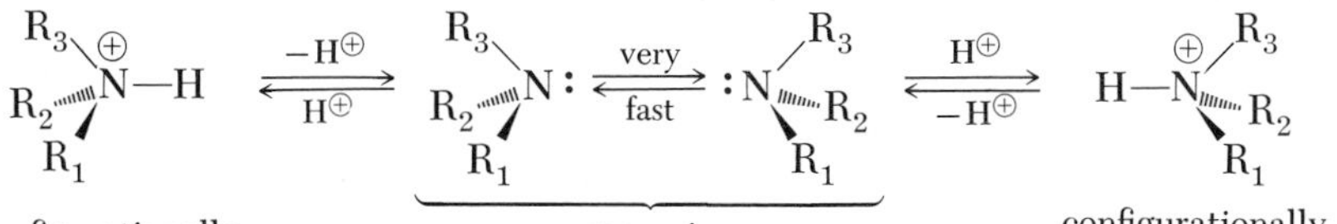

configurationally stable

racemic modification

configurationally stable

Aziridines are not the only triligant nitrogen compounds that exhibit restricted configurational inversion. The methylene hydrogens of benzylmethoxymethylamine appear in the nmr spectrum as an AB quartet at $-40°$, reflecting a decreased rate of nitrogen inversion at this low temperature. (Recall that the hydrogens of a methylene group adjacent to a chiral center are diastereotopic. Therefore, they appear as an AB quartet.) This compound is not a simple amine. Rather, it is a substituted *hydroxylamine*, and it is the presence of the electronegative $—OCH_3$ group on nitrogen which is responsible for the increased configurational stability at nitrogen.

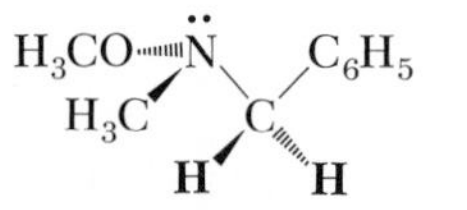

H_3CO—$\ddot{N}$(CH_3)—$C(H)(H)$—C_6H_5

benzylmethoxymethylamine

H's constitute an AB spin system if rate of *N* inversion is slow

3. Suggest a means of demonstrating that benzylmethylphenylamine is not configurationally stable at room temperature.
4. Predict all of the possible products in the following reactions.
   (a) (**R,S**)-$CH_3CH(C_2H_5)CH_2NH_2$ + (**R**)-$CH_3CH(OH)CO_2H$
   (b) (**R,S**)-$CH_3CH(C_2H_5)CH_2N(CH_3)C_2H_5$ + (**R**)-$CH_3CH(OH)CO_2H$

20.4 SYNTHESIS OF AMINES

Ammonolysis of Alkyl Halides. Ammonia reacts with alkyl halides to produce 1° alkylamines. Monoalkylamines are better nucleophiles than is ammonia and, unless the alkyl halide is maintained in low concentration, they will react with additional alkyl halide to produce significant amounts of higher alkylated amines and even a quaternary ammonium salt. Considering the availability of alternate syntheses (see below), direct ammonolysis is *not* a "method of choice" for preparing 1° amines.

monoalkylation:

$$CH_3CH_2I + NH_3 \xrightarrow{S_N2} CH_3CH_2\overset{\oplus}{N}H_3\ I^{\ominus}$$

$$CH_3CH_2\overset{\oplus}{N}H_3\ I^{\ominus} + NH_3(\text{excess}) \rightleftarrows CH_3CH_2NH_2 + \overset{\oplus}{N}H_4\ I^{\ominus}$$

over-alkylation:

$$CH_3CH_2NH_2 + CH_3CH_2I \rightarrow (CH_3CH_2)_2\overset{\oplus}{N}H_2\ I^{\ominus}$$

$$(CH_3CH_2)_2\overset{\oplus}{N}H_2\ I^{\ominus} \underset{}{\overset{NH_3(-\overset{\oplus}{N}H_4\ I^{\ominus})}{\rightleftarrows}} (CH_3CH_2)_2NH \ (2°\ \text{amine})$$

$$(CH_3CH_2)_2NH \xrightarrow{C_2H_5I} (CH_3CH_2)_3\overset{\oplus}{N}H\ I^{\ominus}$$

$$(CH_3CH_2)_3\overset{\oplus}{N}H\ I^{\ominus} \overset{NH_3(-\overset{\oplus}{N}H_4\ I^{\ominus})}{\rightleftarrows} (CH_3CH_2)_3N\ (3°\ \text{amine})$$

$$(CH_3CH_2)_3N \xrightarrow{C_2H_5I} (CH_3CH_2)_4N^{\oplus}\ I^{\ominus}\ (\text{quaternary salt})$$

Gabriel Synthesis. Phthalimide, produced by heating phthalic anhydride with ammonia, is acidic because of extensive, symmetric delocalization of negative charge in the imide anion.

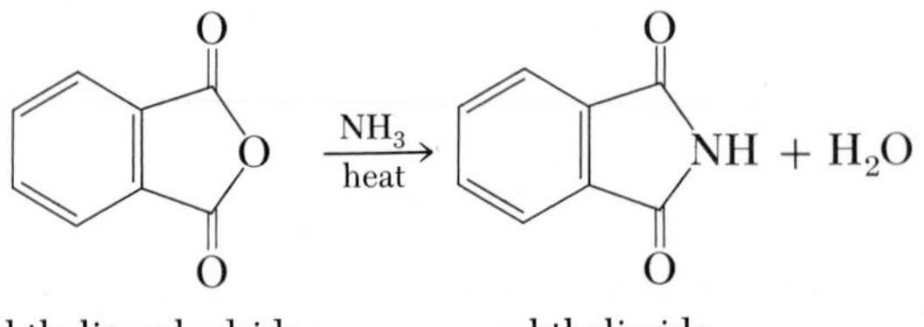

$$\text{phthalic anhydride} \xrightarrow[\text{heat}]{NH_3} \text{phthalimide} + H_2O$$

phthalic anhydride phthalimide

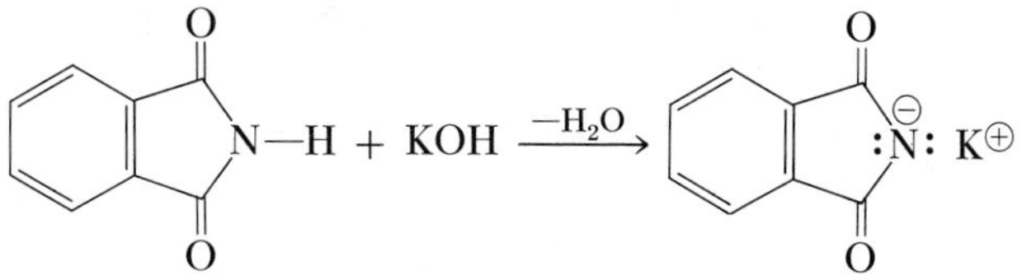

potassium phthalimide

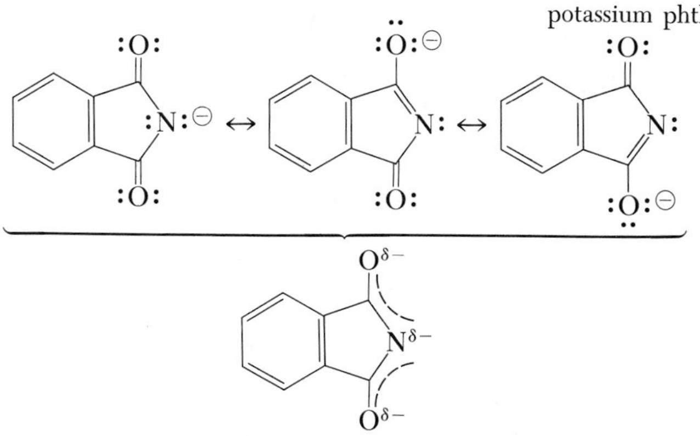

delocalization in phthalimide anion

5. Is the potassium salt of the imide of maleic acid aromatic? Explain.

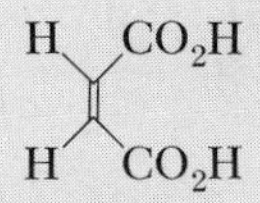

maleic acid

Alkylation of potassium phthalimide, by an S_N2 reaction, produces an *N*-alkyl phthalimide which is easily hydrolyzed to the corresponding amine. This, the **Gabriel synthesis,** produces 1° amines free of higher alkylated products.

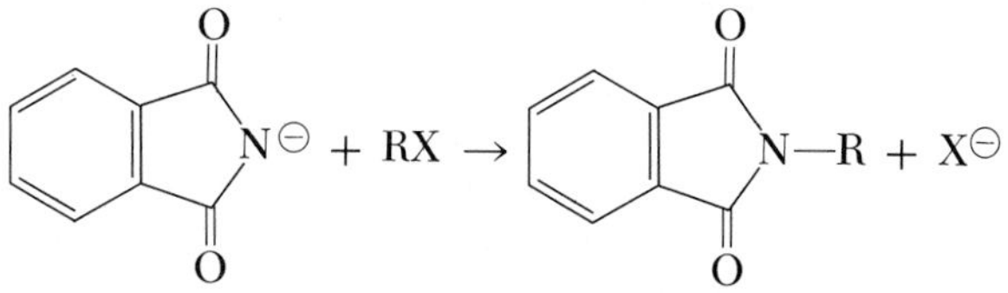

N-alkyl phthalimide

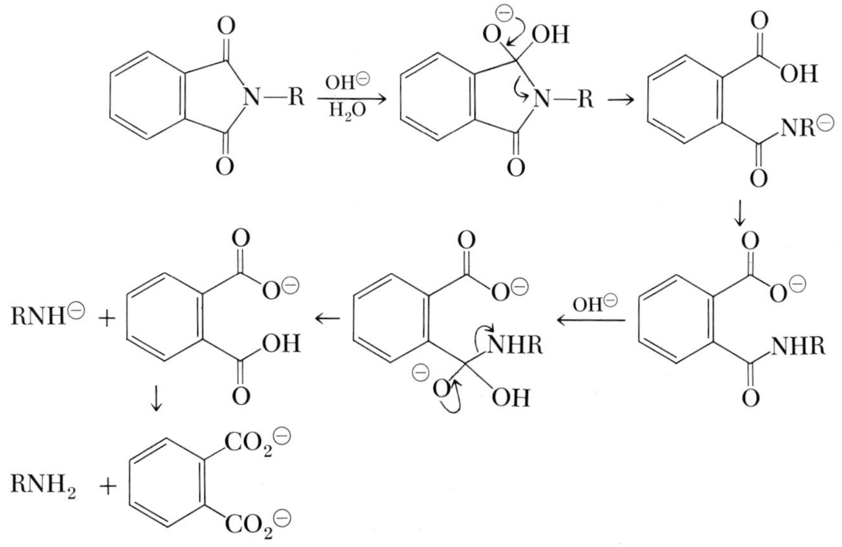

1° amine

Nitrile Reduction. The reduction of nitriles can produce 1° amines of the types RCH_2NH_2 in excellent yields. One of the older reduction procedures employs hydrogen in the presence of a catalyst (*e.g.*, Raney nickel) as the reductant. This procedure is presumed to involve an intermediate *imine*, RCH=NH. (It is known that *N*-alkylimines, RCH=NR, are easily hydrogenated to 1° amines under these conditions.) Unfortunately, this reduction is complicated by the tendency of the desired product to react with the first-formed imine to yield, after hydrogenation, a symmetric 2° amine.

$$\underset{\text{nitrile}}{R{-}C{\equiv}N} + H_2 \xrightarrow{Ni} \underset{\text{imine}}{[RCH{=}NH]} \xrightarrow[Ni]{H_2} \underset{1^\circ\text{ amine}}{RCH_2{-}NH_2} \qquad \text{nitrile hydrogenation}$$

$$RCH{=}NH + RCH_2NH_2 \rightleftharpoons RCH_2NH{-}\underset{NH_2}{CHR} \rightleftharpoons RCH_2{-}N{=}CHR + NH_3$$

$$RCH_2{-}N{=}CHR \xrightarrow{H_2/Ni} \underset{2^\circ\text{ amine}}{RCH_2{-}NH{-}CH_2R} \qquad \text{by-product formation}$$

This side reaction can be avoided in several ways, including carrying out the reduction in a solution of sodium acetate in acetic anhydride. Such a medium protects the 1° amine by converting it to a mixture of an amide and a salt. After the reduction is completed, the desired amine is liberated by alkaline hydrolysis.

$$2RCH_2NH_2 + \underset{Ac_2O}{CH_3C(O)OC(O)CH_3} \rightarrow \underbrace{RCH_2NHC(O)CH_3 + RCH_2\overset{\oplus}{N}H_3\ \overset{\ominus}{O}C(O)CH_3}$$

$$\underbrace{RCH_2NHC(O)CH_3 + RCH_2\overset{\oplus}{N}H_3\ \overset{\ominus}{O}C(O)CH_3} \xrightarrow[\text{heat}]{H_2O/OH^{\ominus}} 2RCH_2NH_2$$

$$\underset{\text{benzyl cyanide}}{C_6H_5CH_2CN} + 2H_2 \xrightarrow[NaOAc/50^\circ/50\text{ psi}]{Ni/Ac_2O} \xrightarrow[\text{heat}]{HO^{\ominus}/H_2O} \underset{\beta\text{-phenylethylamine (97\%)}}{C_6H_5CH_2CH_2NH_2}$$

Both lithium aluminum hydride and diborane reduce nitriles to 1° amines. Of these two, diborane has the advantage that it will not reduce ester or nitro groups that also may be present in the molecule. The yields resulting from lithium aluminum hydride reduction are improved by adding aluminum chloride to the reaction mixture.

$$C_6H_5{-}\underset{C_6H_5}{\overset{H}{C}}{-}CN \xrightarrow[\substack{AlCl_3\\ \text{ether}}]{LiAlH_4} C_6H_5{-}\underset{C_6H_5}{\overset{H}{C}}{-}CH_2NH_2 \qquad (91\%)$$

$$m\text{-}O_2NC_6H_4CN + B_2H_6 \xrightarrow{THF} m\text{-}O_2NC_6H_4CH_2NH_2 \qquad (88\%)$$

Hofmann Hypohalite Reaction. Alkyl and aryl carboxamides react with alkaline solutions of iodine, bromine, or chlorine to form 1° amines. This, the **Hofmann hypohalite reaction,**[a] is not only a good synthesis of 1° amines but also a means of degrading a carbon chain by one carbon.

[a] This is also called the Hofmann rearrangement.

net reaction: $Br_2 + H_2O + RCONH_2 \xrightarrow{OH^\ominus} RNH_2 + CO_2 + 2HBr$

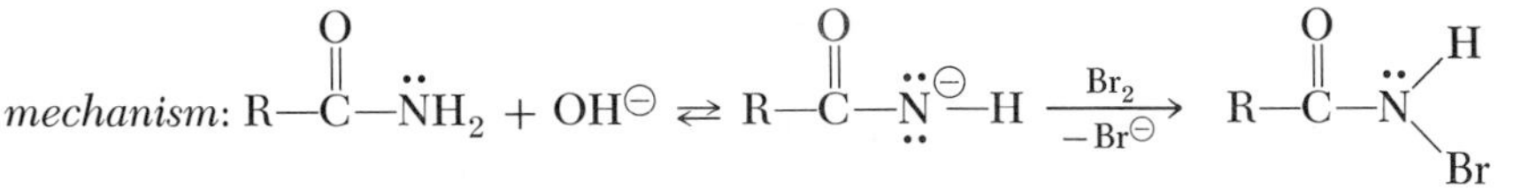

mechanism: $R-\overset{O}{\overset{\|}{C}}-\ddot{N}H_2 + OH^\ominus \rightleftarrows R-\overset{O}{\overset{\|}{C}}-\ddot{\underset{\cdot\cdot}{N}}^\ominus-H \xrightarrow[-Br^\ominus]{Br_2} R-\overset{O}{\overset{\|}{C}}-\ddot{N}(H)(Br)$

an *N*-bromoamide

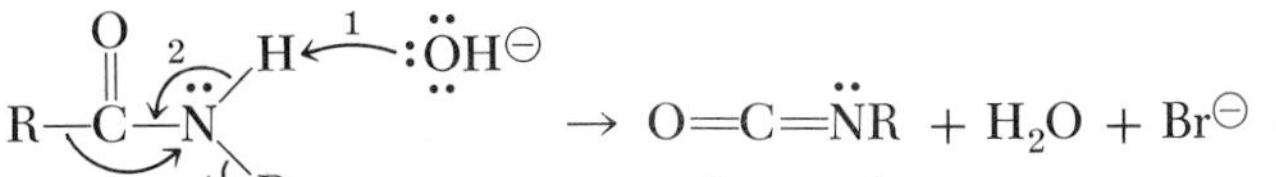

$R-\overset{O}{\overset{\|}{C}}-\ddot{N}(H)(Br) + :\ddot{O}H^\ominus \rightarrow O{=}C{=}\ddot{N}R + H_2O + Br^\ominus$

an isocyanate

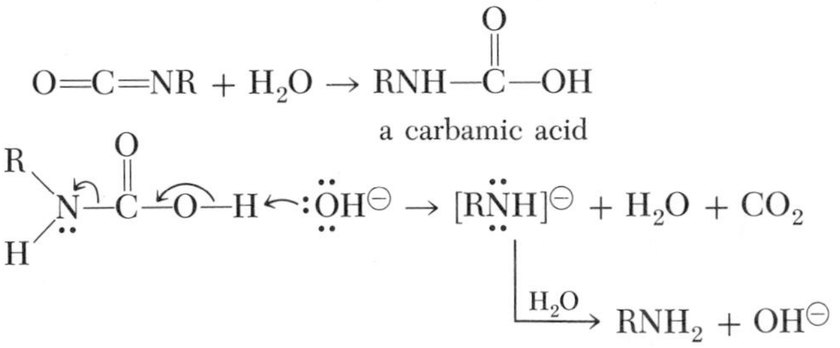

$O{=}C{=}NR + H_2O \rightarrow RNH-\overset{O}{\overset{\|}{C}}-OH$

a carbamic acid

$R(H)\ddot{N}-\overset{O}{\overset{\|}{C}}-O-H + :\ddot{O}H^\ominus \rightarrow [R\ddot{N}H]^\ominus + H_2O + CO_2$

$[R\ddot{N}H]^\ominus \xrightarrow{H_2O} RNH_2 + OH^\ominus$

The initial sequence forms an *N*-haloamide from the resonance-stabilized amide anion. The haloamide is unstable in alkaline solution, being converted to an isocyanate. Isocyanates, like their carbon analogs (ketenes), react rapidly with water. This hydration product, a carbamic acid, readily decarboxylates to an amine.

There is some question about the mechanism of isocyanate formation. While one school of thought favors the concerted process pictured above, another suggests formation of an intermediate *nitrene*, the nitrogen analog of a carbene.

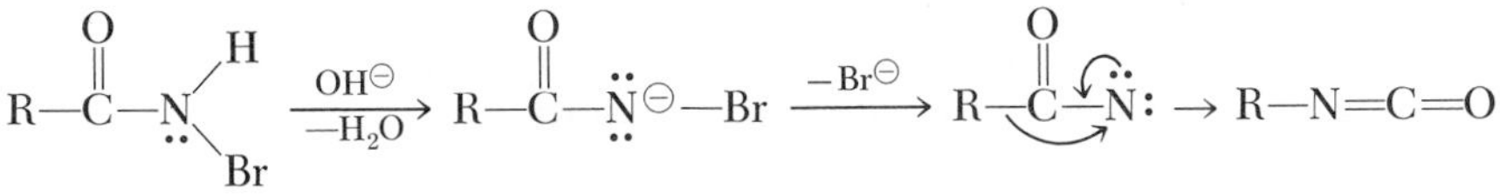

$R-\overset{O}{\overset{\|}{C}}-\ddot{N}(H)(Br) \xrightarrow[-H_2O]{OH^\ominus} R-\overset{O}{\overset{\|}{C}}-\ddot{\underset{\cdot\cdot}{N}}^\ominus-Br \xrightarrow{-Br^\ominus} R-\overset{O}{\overset{\|}{C}}-\ddot{N}: \rightarrow R-N{=}C{=}O$

a nitrene

Whether or not the rearrangement is concerted, it *is* known that the migrating group retains configuration at the migrating carbon. Thus, like the Wolff rearrangement (p. 559), the Hofmann hypohalite reaction does *not* involve migration of a free carbanion.

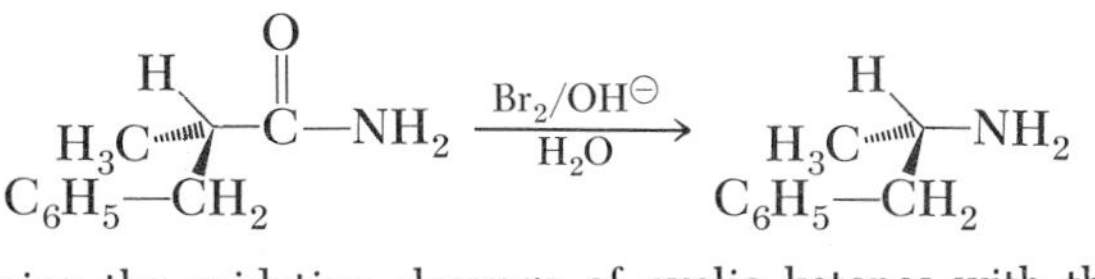

$(H)(H_3C)(C_6H_5CH_2)C-\overset{O}{\overset{\|}{C}}-NH_2 \xrightarrow[H_2O]{Br_2/OH^\ominus} (H)(H_3C)(C_6H_5CH_2)C-NH_2$

By following the oxidative cleavage of cyclic ketones with this Hofmann amide degradation, one can synthesize terminal or "α,ω" diamines.

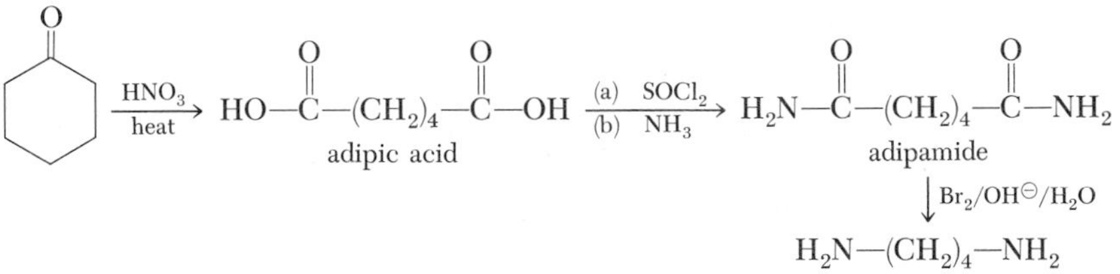

$\text{cyclohexanone} \xrightarrow[heat]{HNO_3} HO-\overset{O}{\overset{\|}{C}}-(CH_2)_4-\overset{O}{\overset{\|}{C}}-OH \xrightarrow[(b)\ NH_3]{(a)\ SOCl_2} H_2N-\overset{O}{\overset{\|}{C}}-(CH_2)_4-\overset{O}{\overset{\|}{C}}-NH_2$

adipic acid; adipamide

$\xrightarrow{Br_2/OH^\ominus/H_2O} H_2N-(CH_2)_4-NH_2$

1,4-butanediamine (putrescine)

The **Curtius rearrangement** is the thermally induced conversion of an acyl azide to an isocyanate. Reaction of the isocyanate with water then produces an amine. The azide is prepared by the reaction of an acyl halide with sodium azide, NaN_3.

$$\underset{}{R-\overset{\overset{\large O}{\|}}{C}-Cl} + \underset{\text{sodium azide}}{Na^{\oplus}N_3^{\ominus}} \rightarrow R-\overset{\overset{\large O}{\|}}{C}-\ddot{N}^{\ominus}-\overset{\oplus}{N}\equiv N: \xrightarrow[\text{heat}]{-N_2} R-N=C=O \xrightarrow{H_2O} RNH_2 + H_2O$$

As with the Hofmann hypohalite reaction, there is some question about the exact mechanism of isocyanate formation, but there is no doubt that if R is chiral it migrates with retention of configuration.

Reduction of Amides. Amides can be reduced to amines by lithium aluminum hydride. This scheme provides a valuable route to 1°, 2°, and 3° amines.

$$R'COCl \begin{cases} \xrightarrow{NH_3} R'CONH_2 \xrightarrow{LiAlH_4} \xrightarrow{H_2O} R'CH_2NH_2 \\ \xrightarrow{RNH_2} R'CONHR \xrightarrow{LiAlH_4} \xrightarrow{H_2O} R'CH_2NHR \\ \xrightarrow{R_2NH} R'CONR_2 \xrightarrow{LiAlH_4} \xrightarrow{H_2O} R'CH_2NR_2 \end{cases}$$

6. Which, if any, of the following could not be made by reduction of an amide?
   (a) $NH(CH_3)_2$
   (b) $N(CH_3)_2C_2H_5$
   (c) $NH(CH_3)C_6H_5$
   (d) $NH(C_6H_5)_2$
   (e) $N(C_6H_5)_2C_2H_5$

Reduction of Nitro Compounds. Nitro compounds are converted to primary amines by potent reductants. The greater availability of aryl nitro compounds, compared to that of alkyl nitro compounds, limits the utility of this procedure to the synthesis of primary arylamines. The most commonly employed reductant is a mixture of tin and hydrochloric acid; the conversion of nitrobenzene to aniline by this method has appeared in undergraduate lab manuals for decades.

$$\underset{\text{nitrobenzene}}{C_6H_5NO_2} \xrightarrow[HCl/H_2O]{Sn} C_6H_5\overset{\oplus}{N}H_3\ Cl^{\ominus} \xrightarrow[H_2O]{OH^{\ominus}} \underset{\text{aniline}}{C_6H_5NH_2}$$

20.5 SYNTHESIS OF HETEROCYCLIC AMINES

Individual steps of the reactions used to synthesize heterocyclic amines are no different from individual steps found in reactions discussed throughout this text. However, since condensations are usually involved, and since many syntheses of heterocycles employ complex reaction mixtures, their syntheses are treated separately.

Pyridine. There is no commercially valuable synthesis of pyridine; coal tar is the major source of this important heteroaromatic compound.

The most important preparation of substituted pyridines employs electrophilic aromatic substitution on pyridine or pyridine *N*-oxide (review Chapter 15 if necessary).

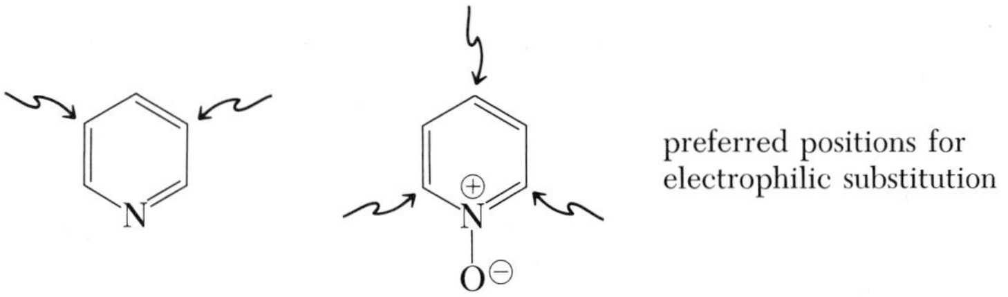

preferred positions for electrophilic substitution

If direct substitution is impossible, a condensation reaction involving ammonia, an aldehyde, and an unsaturated carbonyl compound may be employed to synthesize the substituted ring system directly. This reaction, the **Hantzsch pyridine synthesis,** actually produces a dihydropyridine which is aromatized to the corresponding pyridine by oxidation. (Compare this to the NADH/NAD$^{\oplus}$ interconversion described in Chapter 16.) In the illustration below, the "unsaturated carbonyl compound" is the enol of acetoacetic ester. The aldehydic carbon, here the carbon of formaldehyde, becomes C4 in the final product. Details of this rather complex-looking conversion are presented in Figure 20–2.

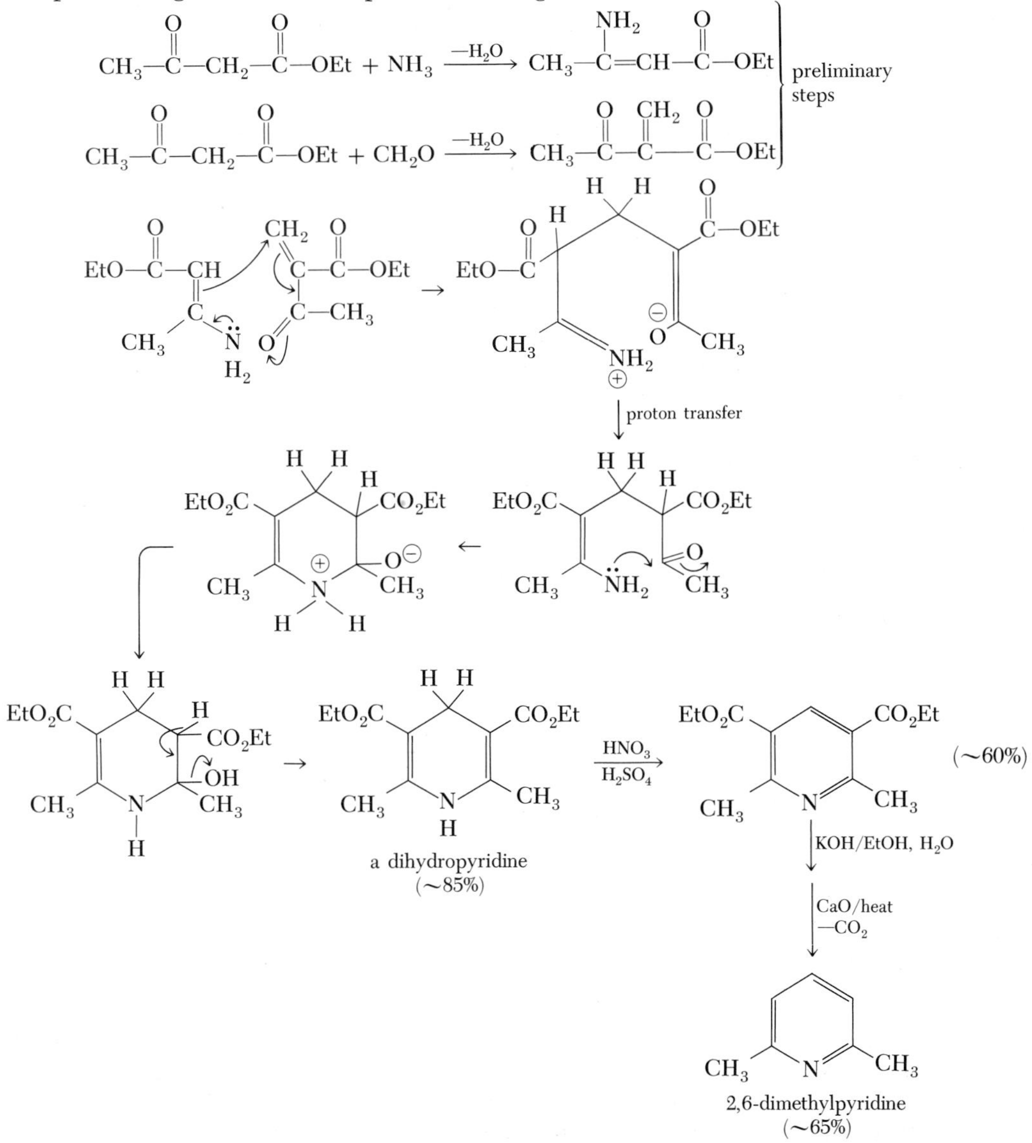

$$CH_3-\overset{O}{\overset{\|}{C}}-CH_2-\overset{O}{\overset{\|}{C}}-OEt + NH_3 \xrightarrow{-H_2O} CH_3-\overset{NH_2}{\overset{|}{C}}=CH-\overset{O}{\overset{\|}{C}}-OEt$$

$$CH_3-\overset{O}{\overset{\|}{C}}-CH_2-\overset{O}{\overset{\|}{C}}-OEt + CH_2O \xrightarrow{-H_2O} CH_3-\overset{O}{\overset{\|}{C}}-\overset{CH_2}{\overset{\|}{C}}-\overset{O}{\overset{\|}{C}}-OEt$$

preliminary steps

proton transfer

a dihydropyridine (~85%)

HNO_3, H_2SO_4 (~60%)

KOH/EtOH, H_2O

CaO/heat, $-CO_2$

2,6-dimethylpyridine (~65%)

Figure 20-2 Mechanism of the Hantzsch pyridine synthesis. After oxidation, the diester is hydrolyzed to the acid salt. Heating the salt in the presence of calcium oxide causes decarboxylation and production of 2,6-dimethylpyridine.

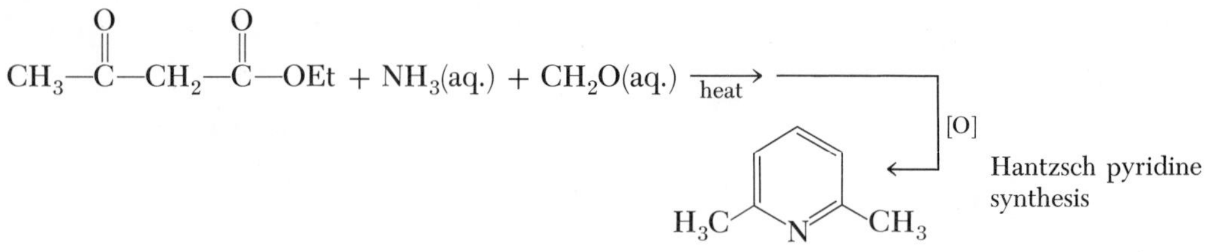

$$CH_3-\overset{O}{\overset{\|}{C}}-CH_2-\overset{O}{\overset{\|}{C}}-OEt + NH_3(aq.) + CH_2O(aq.) \xrightarrow[heat]{} \xrightarrow{[O]}$$ 2,6-dimethylpyridine (H_3C, N, CH_3)

Hantzsch pyridine synthesis

This basic scheme can be modified by including the heterocyclic nitrogen in the starting material as something other than ammonia, and by employing aldehydes other than formaldehyde. For example,

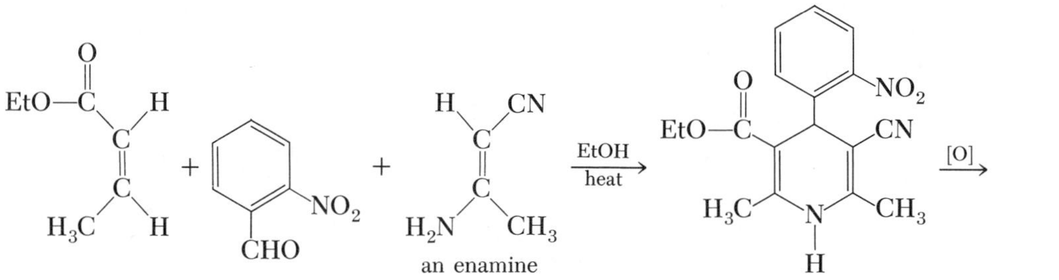

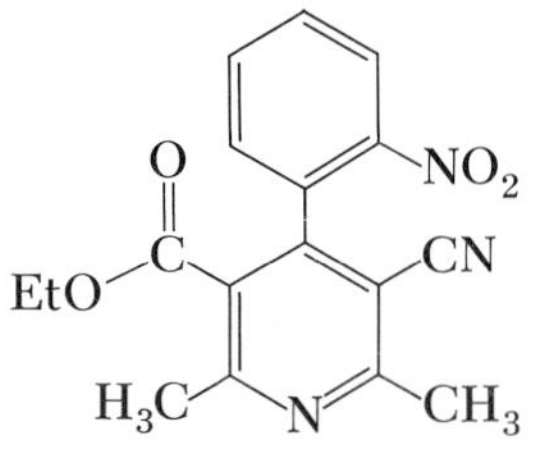

EtO—C(=O)(H)C=C(H)(H₃C) + o-nitrobenzaldehyde (CHO, NO_2) + H(CN)C=C(H₂N)(CH₃) (an enamine) $\xrightarrow[heat]{EtOH}$ 1,4-dihydropyridine (EtO—C(O), NO_2, CN, H_3C, N—H, CH_3) $\xrightarrow{[O]}$ pyridine (EtO—C(O), NO_2, CN, H_3C, N, CH_3)

7. Suggest a mechanism for step **A** in the modification of the Hantzsch pyridine synthesis shown below.

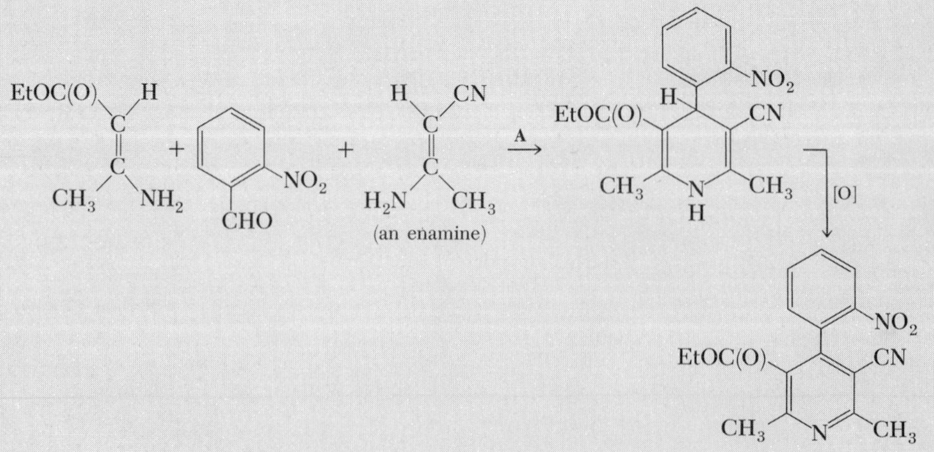

EtOC(O)(H)C=C(CH_3)(NH_2) + o-nitrobenzaldehyde (CHO, NO_2) + H(CN)C=C(H_2N)(CH_3) (an enamine) $\xrightarrow{A}$ 1,4-dihydropyridine (H, EtOC(O), NO_2, CN, CH_3, N—H, CH_3) $\xrightarrow{[O]}$ pyridine (EtOC(O), NO_2, CN, CH_3, N, CH_3)

While pyridine and its derivatives are quite important in "pure" organic chemistry, even some very simple derivatives of pyridine play vital roles in maintaining our health and well-being. *Niacin* (3-pyridinecarboxylic acid) is a vitamin, while Isoniazid is a drug used in the chemotherapy of tuberculosis.

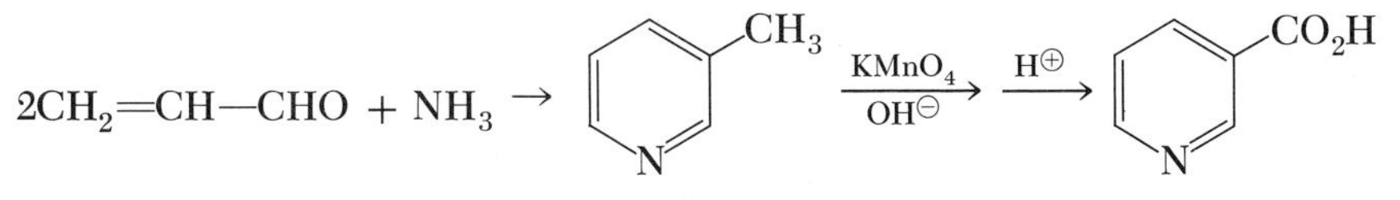

$$2CH_2{=}CH{-}CHO + NH_3 \rightarrow \beta\text{-picoline} \xrightarrow[OH^{\ominus}]{KMnO_4} \xrightarrow{H^{\oplus}} \text{niacin}$$

β-picoline niacin

$$\gamma\text{-picoline} \xrightarrow[OH^{\ominus}]{KMnO_4} \xrightarrow{H^{\oplus}} (CO_2H) \xrightarrow{SOCl_2} (COCl) \xrightarrow[\text{(excess)}]{H_2N{-}NH_2} \text{Isoniazid}$$

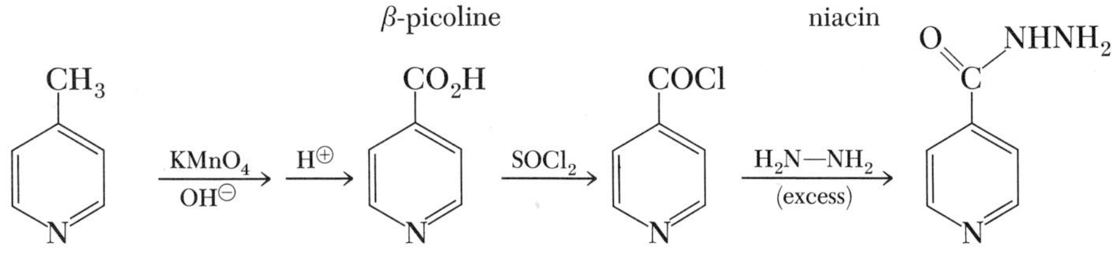

γ-picoline (from coal tar) Isoniazid

Isoniazid is an interesting chemotherapeutic agent, usually used with a second drug to reduce the possibility of genetic mutation of the bacilli into a strain that is resistant to the chemotherapeutic regime to which the patient is subjected. We note, for the more biochemically oriented reader, that Isoniazid is one of the few drugs known to be subject to a genetic variation in metabolism. Very elegant studies[a] have demonstrated that the unusually slow metabolism of Isoniazid is an autosomal recessive trait. Studies of the genetic control of drug metabolism offer an exciting area of research for chemists, biochemists, and molecular biologists.

8. Isoniazid synthesis requires the use of γ-picolinoyl chloride. Why doesn't this compound, which contains both a moderately nucleophilic nitrogen and a reactive acyl halide group, decompose by the polymerization shown below?

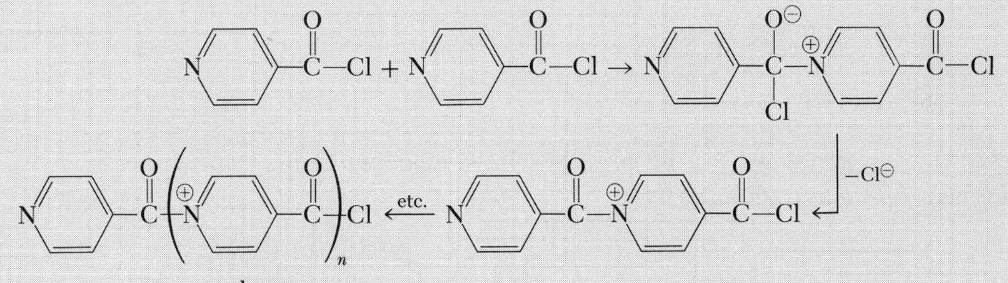

$$\text{N}C_5H_4{-}C(=O){-}Cl + \text{N}C_5H_4{-}C(=O){-}Cl \rightarrow \text{N}C_5H_4{-}C(O^{\ominus})(Cl){-}N^{\oplus}C_5H_4{-}C(=O){-}Cl \xrightarrow{-Cl^{\ominus}} \text{N}C_5H_4{-}C(=O){-}N^{\oplus}C_5H_4{-}C(=O){-}Cl \xrightarrow{\text{etc.}} \text{N}C_5H_4{-}C(=O){-}\left(N^{\oplus}C_5H_4{-}C(=O)\right)_n{-}Cl$$

polymer

Quinoline. Quinoline occurs in coal tar but is easily prepared by means of a classic "pot boiler," the **Skraup synthesis.** The simplest example, including mechanism, is shown below. Note that aniline appears on both sides of the equation, since the aniline that you start with is not the aniline with which you finish.

Skraup synthesis:

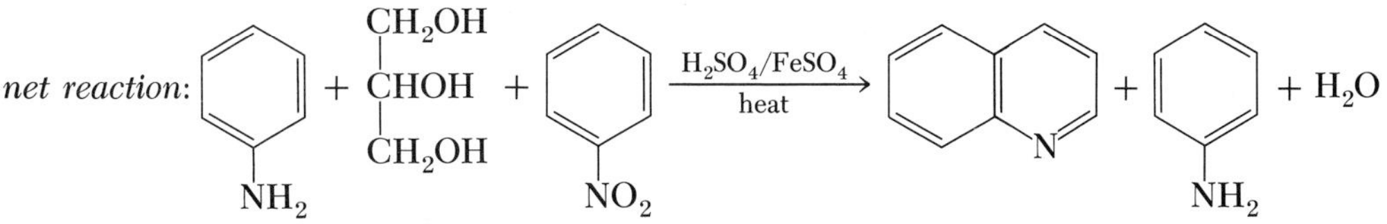

net reaction:

$$C_6H_5NH_2 + \underset{\text{glycerol}}{CH_2OH{-}CHOH{-}CH_2OH} + C_6H_5NO_2 \xrightarrow[\text{heat}]{H_2SO_4/FeSO_4} \underset{\text{quinoline}}{C_9H_7N} + C_6H_5NH_2 + H_2O$$

---

[a] D.A.P. Evans, K.A. Manley, and V.A. McKusick: Genetic Control of Isoniazid Metabolism In Man, *Brit. Med. J.*, 2:485 (1960).

mechanism: $CH_2OH—CHOH—CH_2OH \xrightarrow[\text{heat}]{H_2SO_4} CH_2{=}CH—CHO + 2H_2O$
acrolein

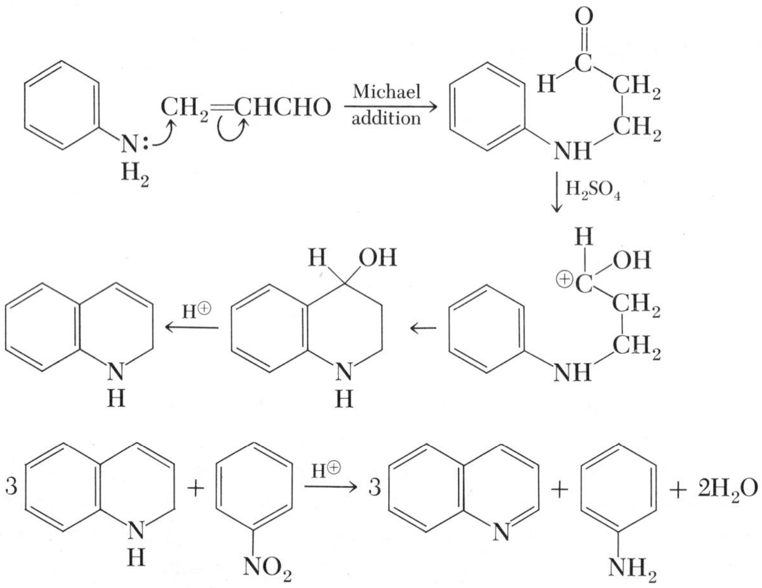

In the last step of this process, nitrobenzene displays its ability to act as an oxidant, being reduced to aniline along the way. This role for nitrobenzene is confirmed by the fact that this quinoline synthesis can be carried out using an oxidant such as arsenic acid, H_3AsO_4, instead of nitrobenzene.

9. Suggest a mechanism for the conversion of glycerol to acrolein in the presence of sulfuric acid.

Substituted quinolines can be prepared by similar reactions, using either glycerol or a pre-formed α,β-unsaturated aldehyde.

$C_6H_5NH_2 + CH_3—CH{=}CH—CHO \rightsquigarrow$

2-methylquinoline

$+ CH_2OHCHOHCH_2OH \rightsquigarrow$

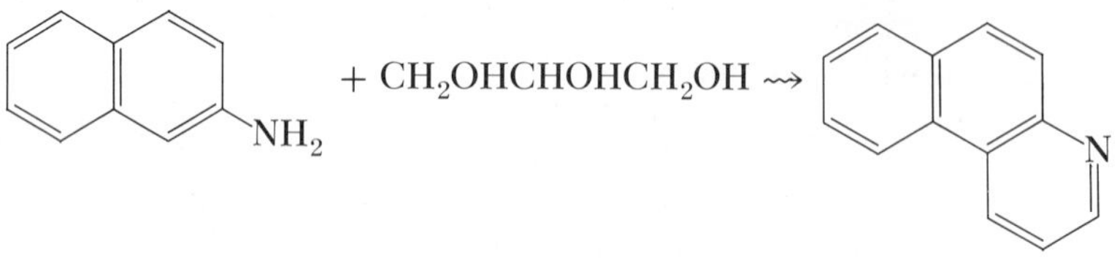

β-naphthylamine
(carcinogen!)

Since the function of glycerol in the Skraup synthesis is to provide the α,β-unsaturated aldehyde, it can be replaced by aldehydes or ketones that contain α hydrogens; these carbonyl compounds undergo aldol condensations under the reaction conditions to form α,β-unsaturated carbonyl compounds. This modification of the Skraup synthesis is called the **Doebner-Miller reaction.**

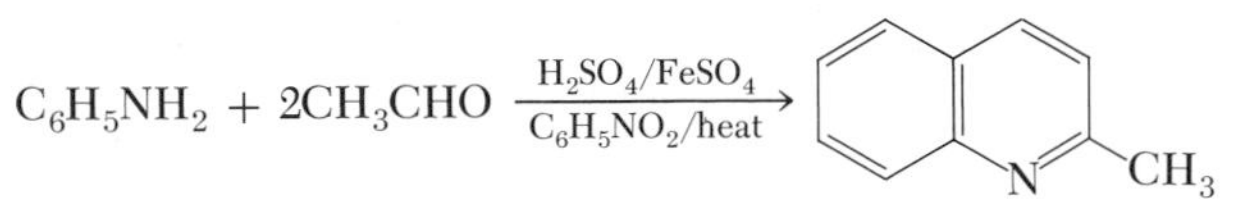

$$C_6H_5NH_2 + 2CH_3CHO \xrightarrow[C_6H_5NO_2/\text{heat}]{H_2SO_4/FeSO_4} \text{2-methylquinoline}$$

Compounds that produce reasonable concentrations of α,β-unsaturated carbonyl compounds by enolization also function in the Doebner-Miller reaction. In the following example, the enol form of acetylacetone is used to prepare 2,4-dimethylquinoline.

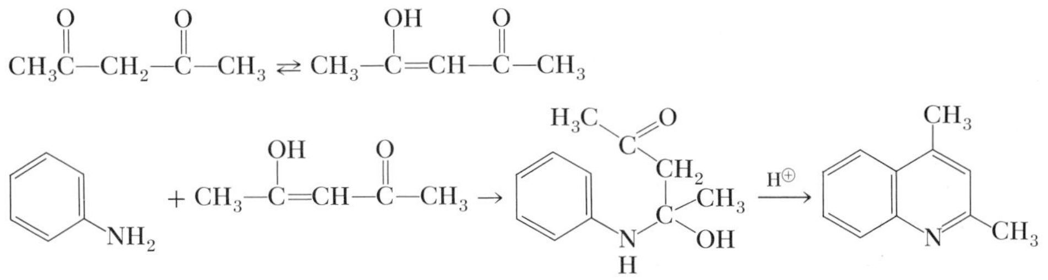

$$CH_3\overset{O}{\overset{\|}{C}}-CH_2-\overset{O}{\overset{\|}{C}}-CH_3 \rightleftarrows CH_3-\overset{OH}{\overset{|}{C}}=CH-\overset{O}{\overset{\|}{C}}-CH_3$$

$$C_6H_5NH_2 + CH_3-\overset{OH}{\overset{|}{C}}=CH-\overset{O}{\overset{\|}{C}}-CH_3 \rightarrow \text{intermediate} \xrightarrow{H^{\oplus}} \text{2,4-dimethylquinoline}$$

Isoquinoline. The **Bischler-Napieralski synthesis** provides a route to isoquinoline *via* the acid-catalyzed cyclization of amides of β-phenylethylamine. Completion of the synthesis requires aromatization of the intermediate dihydroisoquinoline. However, the aromatization usually is accomplished by catalytic dehydrogenation, rather than by chemical oxidants as in the Hantzsch pyridine synthesis.

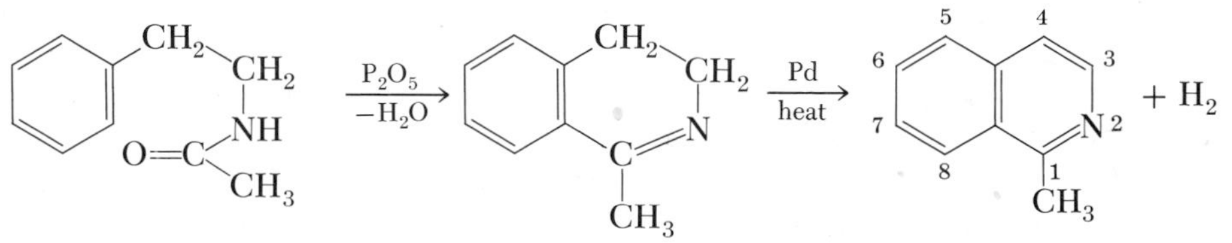

$$\text{N-(2-phenylethyl)acetamide} \xrightarrow[-H_2O]{P_2O_5} \text{1-methyl-3,4-dihydroisoquinoline} \xrightarrow[\text{heat}]{Pd} \text{1-methylisoquinoline} + H_2$$

N-(2-phenylethyl)acetamide 1-methylisoquinoline

10. Starting with benzene and any organic compounds containing less than three carbons, suggest a synthesis of *N*-(2-phenylethyl)acetamide.

20.6 REACTIONS OF AMINES

The ability to act both as bases and as nucleophiles are two characteristic reactions of amines. Given a potent enough base, however, some amines can act as proton donors, *i.e.*, acids.

Amines as Acids. Ammonia, primary amines, and secondary amines can donate protons to bases to form amide anions. (Take care to distinguish this from an amide, $RCONH_2$.) Since these amide anions are strong bases, the proton acceptor must be an even stronger base. The "bases" most commonly used to convert amines to amide anions are metallic sodium and metallic lithium.

Dissolving sodium in liquid ammonia produces a blue solution that contains sodium cations and solvated electrons. If iron(III) chloride is added to this solution, hydrogen is evolved and the blue color fades. Cessation of hydrogen evolution marks the completion of the conversion to sodium amide (also called sodamide).

$$2Na + 2NH_3(l) \xrightarrow[NH_3(l)]{Fe(III)} 2Na^{\oplus}NH_2^{\ominus} + H_2$$

sodamide

Acylation of Amines. In discussing carboxylic acids and their derivatives, we noted the conversion of acid chlorides to amides by reaction with ammonia. This same process can be used to prepare amides of 1° and 2° amines and constitutes an excellent way to convert a liquid amine into a solid derivative.

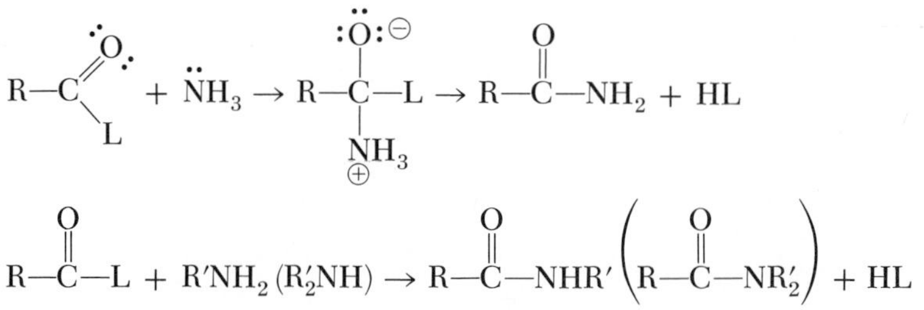

$$R{-}C(=\ddot{O}{:}){-}L + \ddot{N}H_3 \rightarrow R{-}C(:\ddot{O}:^{\ominus})(\overset{\oplus}{N}H_3){-}L \rightarrow R{-}\overset{O}{\overset{\|}{C}}{-}NH_2 + HL$$

$$R{-}\overset{O}{\overset{\|}{C}}{-}L + R'NH_2\ (R'_2NH) \rightarrow R{-}\overset{O}{\overset{\|}{C}}{-}NHR' \left(R{-}\overset{O}{\overset{\|}{C}}{-}NR'_2\right) + HL$$

The production of the acid, HL, will consume an equivalent amount of unreacted amine (How?), and if the amine is scarce or expensive this becomes especially wasteful. Because of this, amine acylation often is carried out by the so-called **Schotten-Baumann procedure.** This calls for the reaction of the amine and the acylating reagent in the presence of aqueous sodium hydroxide. In this two-phase reaction the amine successfully competes with hydroxide ion for the acylating agent, while the hydroxide ion consumes the liberated acid.

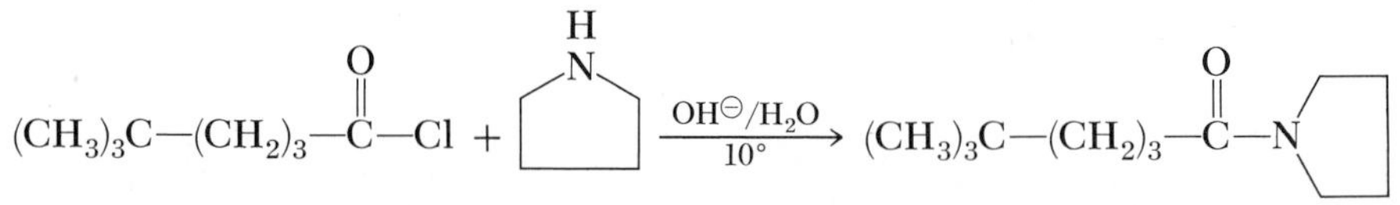

$$(CH_3)_3C{-}(CH_2)_3{-}\overset{O}{\overset{\|}{C}}{-}Cl + \text{pyrrolidine (N–H)} \xrightarrow[10^{\circ}]{OH^{\ominus}/H_2O} (CH_3)_3C{-}(CH_2)_3{-}\overset{O}{\overset{\|}{C}}{-}N\text{(pyrrolidinyl)}$$

Tertiary amines react with acyl halides to form salts which are, themselves, acylating agents. In the illustration below, such a salt is used to acylate water. While the overall process is the hydrolysis of an acyl halide, it is mediated by pyridine.

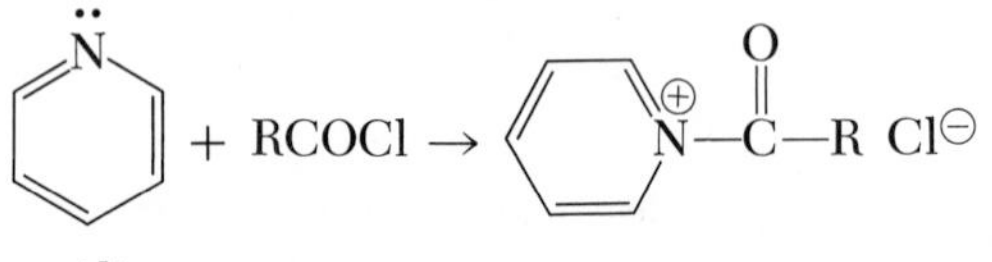

$$\text{pyridine} + RCOCl \rightarrow C_5H_5\overset{\oplus}{N}{-}\overset{O}{\overset{\|}{C}}{-}R\ Cl^{\ominus}$$

pyridine

$$C_5H_5\overset{\oplus}{N}{-}\overset{:\ddot{O}}{\overset{\|}{C}}{-}R + H\ddot{O}H \rightarrow C_5H_5\overset{\oplus}{N}{-}\overset{:\ddot{O}:^{\ominus}}{\overset{|}{C}}(\overset{\oplus}{O}H_2){-}R \rightarrow C_5H_5N + R{-}\overset{:O:}{\overset{\|}{C}}{-}\overset{\oplus}{\ddot{O}}H_2 \rightarrow C_5H_5\overset{\oplus}{N}{-}H + R{-}\overset{O}{\overset{\|}{C}}{-}OH$$

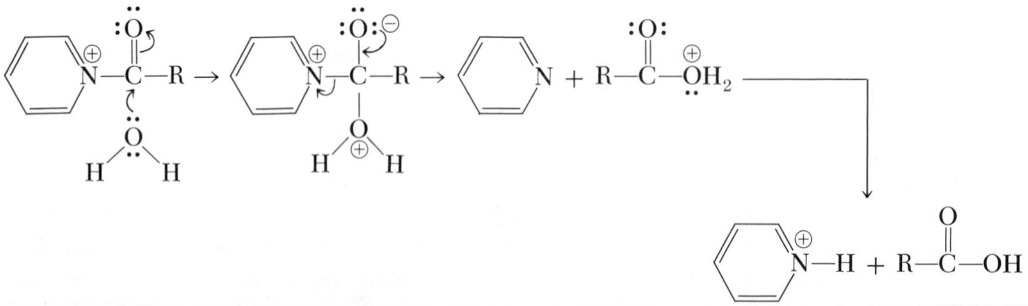

In Chapter 27 this type of process will be shown to play an important role in some enzyme-catalyzed reactions.

Nitrosation. While nitrous acid, HONO, is not stable, it can be prepared in aqueous solution by dissolving sodium nitrite in dilute aqueous acid (*e.g.*, hydrochloric acid). Primary *aliphatic* amines react with cold aqueous nitrous acid to produce alkyl *diazonium* salts, which decompose to a morass of products. For example, *n*-propylamine undergoes the following reaction sequence.

$$CH_3CH_2CH_2NH_2 \xrightarrow[H_2O]{[HONO]} [CH_3CH_2CH_2N_2^{\oplus}] \xrightarrow[H_2O]{-N_2} \begin{cases} CH_3CH_2CH_2OH \\ CH_3CHOHCH_3 \\ CH_3CH{=}CH_2 \\ (CH_2)_3 \end{cases}$$

Primary arylamines react with nitrous acid to produce the much more important and more stable aryl diazonium salts. These will be discussed in the next chapter.

$$C_6H_5NH_2 + [HONO] \xrightarrow[H_2O]{0^\circ} \underset{\text{phenyldiazonium cation}}{C_6H_5N_2^{\oplus}}$$

Secondary amines, both alkyl and aryl, react with nitrous acid to produce yellow *N*-nitrosamines. These compounds, amides of nitrous acid, are very weak bases. Inorganic nitrites, long used as food preservatives and in the meat processing industry, are believed to be mutagens. A possible mode of action of these agents involves formation of unstable *N*-nitrosamines after nitrous acid is produced from nitrite ions at the pH of the stomach and of cells.

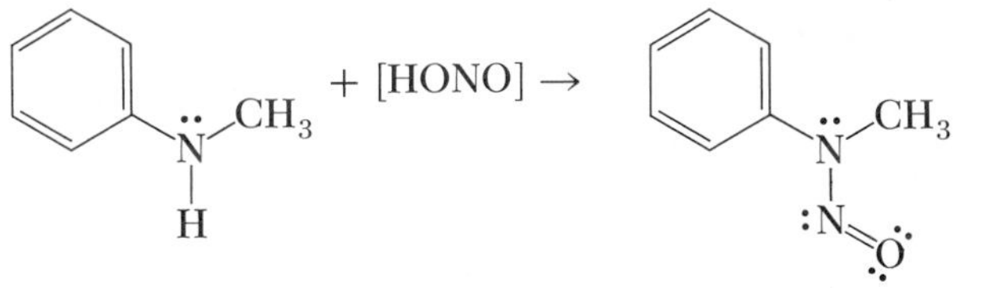

$$C_6H_5-\ddot{N}(CH_3)-H + [HONO] \rightarrow C_6H_5-\ddot{N}(CH_3)-\ddot{N}{=}\ddot{O}$$

N-methylaniline | *N*-nitroso-*N*-methylaniline

Tertiary alkylamines react with nitrous acid to produce complex mixtures; therefore, this reaction will not be discussed further. Tertiary arylamines react rapidly with nitrous acid to produce a *p*-nitroso arylamine. This electrophilic aromatic substitution involves attack of $NO^{\oplus}$ upon the aromatic ring.

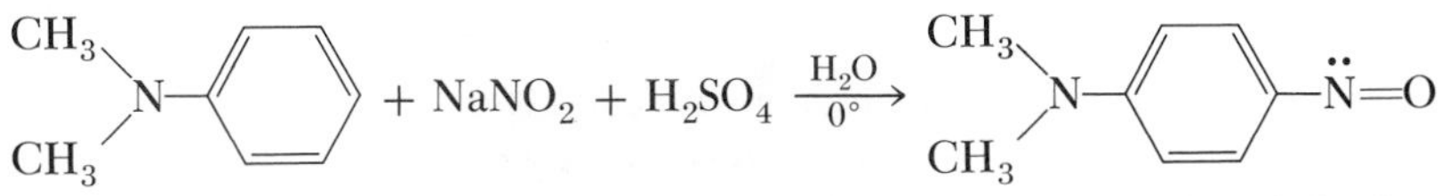

$$(CH_3)_2N-C_6H_5 + NaNO_2 + H_2SO_4 \xrightarrow[0^\circ]{H_2O} (CH_3)_2N-C_6H_4-\ddot{N}{=}O$$

N,N-dimethylaniline | *p*-nitroso-*N,N*-dimethylaniline

Excess nitrous acid can be removed from reaction mixtures by the addition of urea. This diamide, like simpler monoamides, reacts with nitrous acid to produce a carboxylic acid and nitrogen. However, with urea the eventual products are H_2O, CO_2 and N_2. The reaction's key step is the conversion of the amide to a diazonium ion.

$$R-\overset{O}{\overset{\|}{C}}-NH_2 \xrightarrow{[HONO]} \left[R-\overset{O}{\overset{\|}{C}}-N_2\right]^{\oplus} \xrightarrow{H_2O} RCO_2H + N_2 + H^{\oplus}$$

$$\underset{\text{urea}}{H_2N-\overset{O}{\overset{\|}{C}}-NH_2} + 2[HONO] \rightarrow 2N_2 + 2H_2O + \left[HO-\overset{O}{\overset{\|}{C}}-OH\right] \longrightarrow CO_2 + H_2O$$

N-alkylamides react with nitrous acid to form *N*-nitroso derivatives, which are precursors to diazoalkanes. The best known example of this is an *N*-nitrososulfonamide that is commonly called *Diazald*. This is a widely used precursor to diazomethane, CH_2N_2. *N*-nitroso-*N*-methylurea, whose synthesis is shown below, is another useful diazomethane precursor.

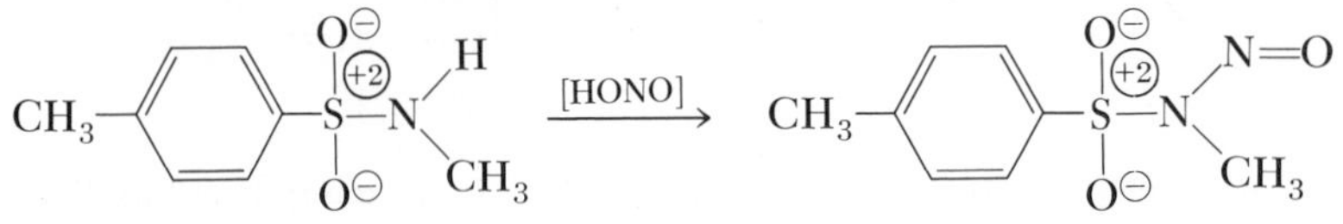

$$CH_3{-}C_6H_4{-}S^{+2}(O^-)_2{-}N(H)(CH_3) \xrightarrow{[HONO]} CH_3{-}C_6H_4{-}S^{+2}(O^-)_2{-}N(N{=}O)(CH_3)$$

N-nitroso-*N*-methyl-*p*-toluenesulfonamide
(Diazald)

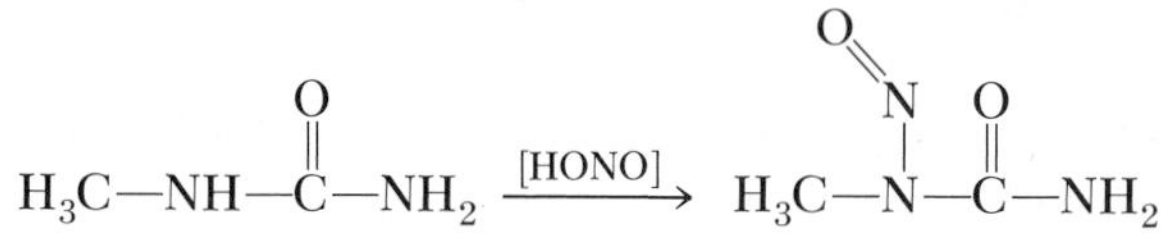

$$H_3C{-}NH{-}C(=O){-}NH_2 \xrightarrow{[HONO]} H_3C{-}N(N{=}O){-}C(=O){-}NH_2$$

N-nitroso-*N*-methylurea

$$CH_3{-}N(N{=}O){-}C(=O){-}NH_2 \xrightarrow[H_2O]{OH^-} CH_3{-}N(N{=}O){-}C(O^-)(OH){-}NH_2 \rightarrow CH_3{-}N{=}N{-}O^- + HO{-}C(=O){-}NH_2$$

methyldiazotate anion

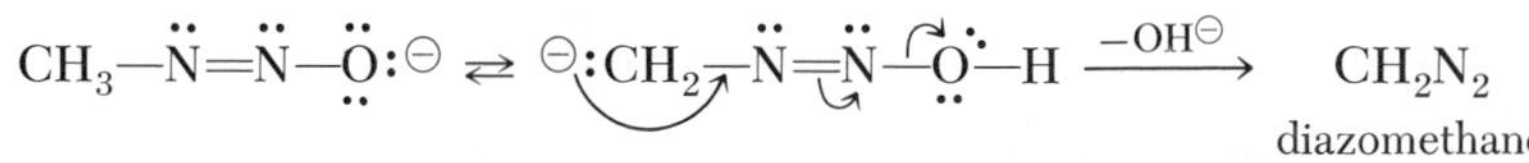

$$CH_3{-}N{=}N{-}O^- \rightleftarrows {}^-{:}CH_2{-}N{=}N{-}O{-}H \xrightarrow{-OH^-} CH_2N_2$$

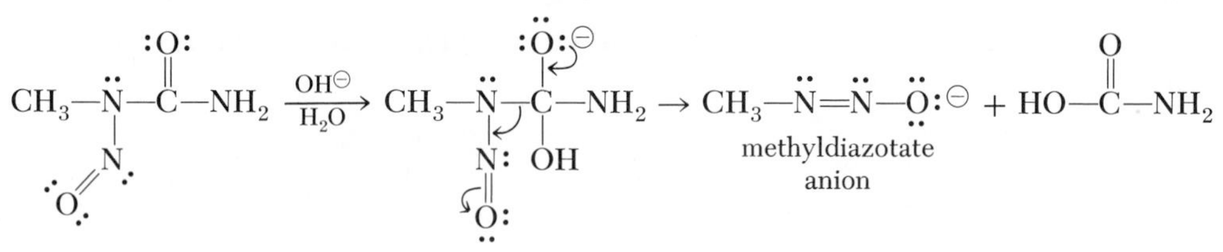

diazomethane

11. Provide two specific, different uses of diazomethane in organic synthesis. Draw the two principal resonance structures of diazomethane.

Imine Formation. The reaction of aldehydes and ketones with ammonia, to form imines, has already been presented in Chapter 16. Primary and secondary amines react with carbonyl compounds to produce *carbinolamines*, which undergo spontaneous dehydration to imines, also called *Schiff's bases*. Schiff's bases are important in a number of biological processes, including the Kreb's cycle, described in the previous chapter, and in the transaminations presented in Chapter 16.

$$R'NH_2 + R{-}C(=O){-}R \rightleftarrows R'{-}NH{-}C(OH)(R){-}R \xrightarrow{-H_2O} R'{-}N{=}CR_2$$

a carbinolamine — a ketimine

$$R'NH_2 \xrightarrow{R{-}C(=O){-}H} \text{a carbinolamine} \rightarrow R'{-}N{=}C(H)(R)$$

an aldimine

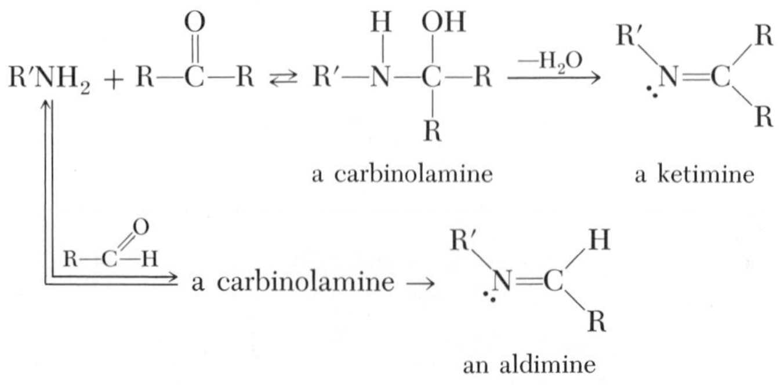

Carbonyl compounds containing an α hydrogen react with secondary amines, if water is removed as it is produced, to form enamines. Enamines are nitrogen analogs of enols.

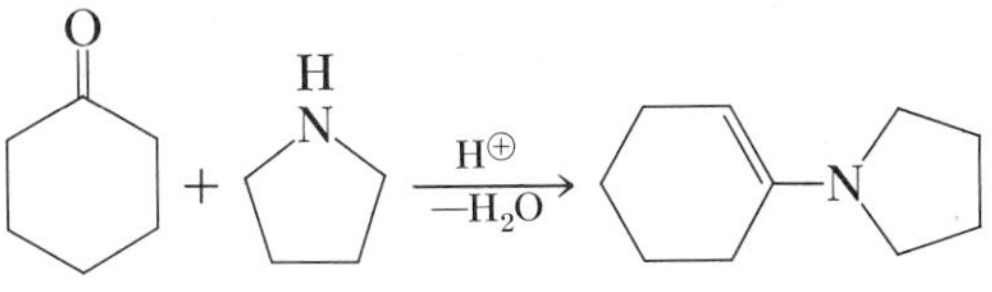

$$\text{cyclohexanone} + \text{pyrrolidine} \xrightarrow[-H_2O]{H^{\oplus}} \text{1-pyrrolidinylcyclohexene}$$

an enamine

Enamines also are produced by the mercuric acetate oxidation of 3° amines possessing a hydrogen α to nitrogen. The overall conversion involves coordination at nitrogen, a β-elimination, and proton loss:

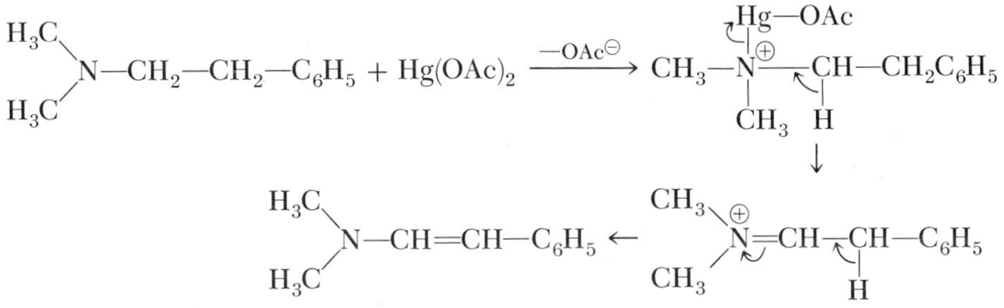

$$(H_3C)_2N{-}CH_2{-}CH_2{-}C_6H_5 + Hg(OAc)_2 \xrightarrow{-OAc^{\ominus}} CH_3{-}\overset{\oplus}{N}(CH_3)(HgOAc){-}CH_2{-}CH_2C_6H_5$$

$$\downarrow$$

$$(H_3C)_2N{-}CH{=}CH{-}C_6H_5 \leftarrow (CH_3)_2\overset{\oplus}{N}{=}CH{-}CH_2{-}C_6H_5$$

an enamine — an iminium ion

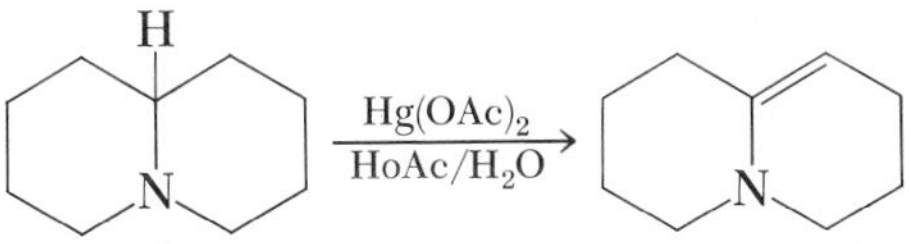

$$\text{quinolizidine} \xrightarrow[HoAc/H_2O]{Hg(OAc)_2} \text{enamine (60\%)}$$

quinolizidine (60%)

An interesting side reaction between the desired enamine and the prerequisite iminium ion can become a serious deterrent to the use of mercuric acetate oxidation of 3° cyclic amines. For example, mercuric acetate reacts with *N*-methylpiperidine to produce a dimeric product in greater than 65% yield.

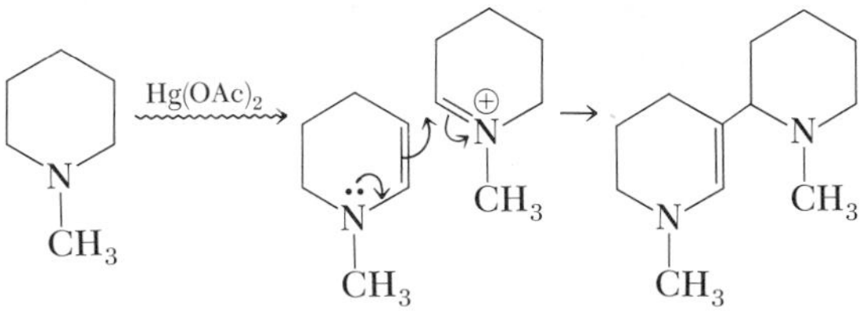

$$N\text{-methylpiperidine} \xrightarrow{Hg(OAc)_2} \text{enamine} + \text{iminium ion} \rightarrow \text{dimer}$$

One of many uses of enamines is found in their alkylation, the so-called **Stork reaction.** As seen from the examples that follow, α-alkylation of an unactivated ketone is made quite feasible by first converting it to an enamine.

A. $$\text{cyclohexanone} + \text{pyrrolidine (NH)} \xrightarrow{-H_2O} \left[\text{enamine} \leftrightarrow \text{iminium carbanion}\right] \xrightarrow{C_6H_5CH_2Cl} \text{2-}CH_2C_6H_5\text{ iminium } Cl^{\ominus}$$

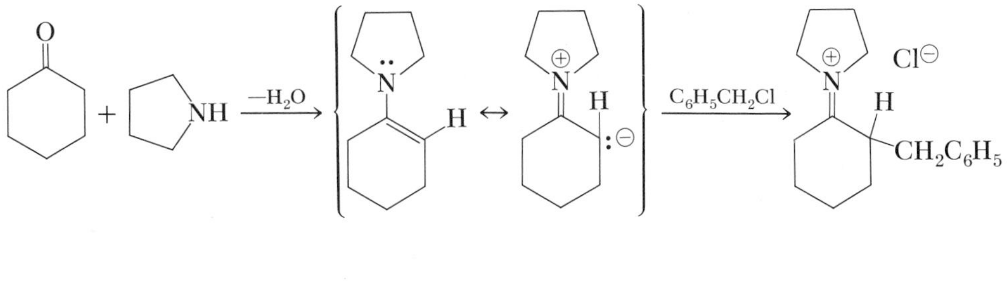

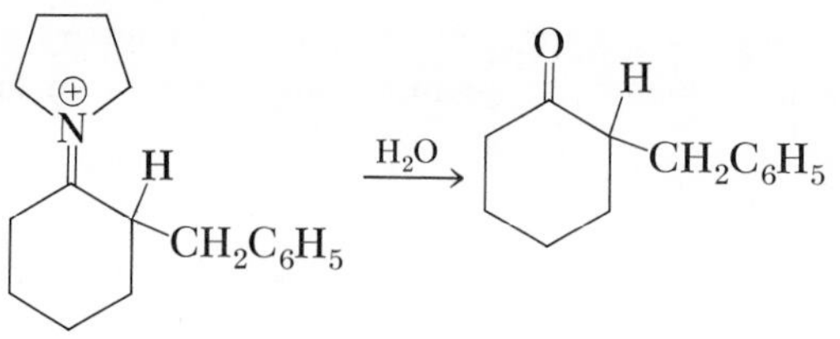

$\xrightarrow{H_2O}$ (55%)

B. $\xrightarrow{-H_2O}$ $\xrightarrow{CH_3I}$ $\xrightarrow{H_2O}$ (50%)

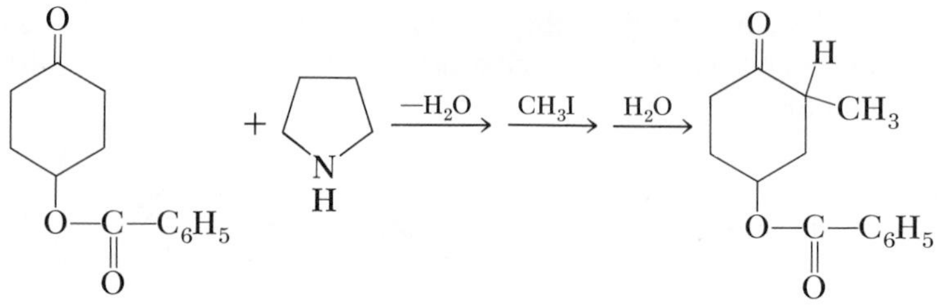

Alkylation may occur at C or at N; these *irreversible* processes ultimately yield ketones or quaternary ammonium salts, respectively.

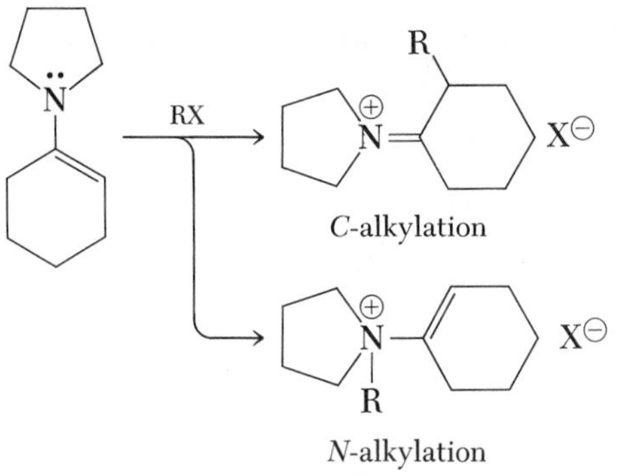

RX

C-alkylation

N-alkylation

As shown below, enamines may also be acylated, generally to produce β-diketones (which, as you should recall, are themselves readily alkylated).

Suggest a reason for postulating the desirability of step **A** *and* the mechanism for step **B** in the sequence below.

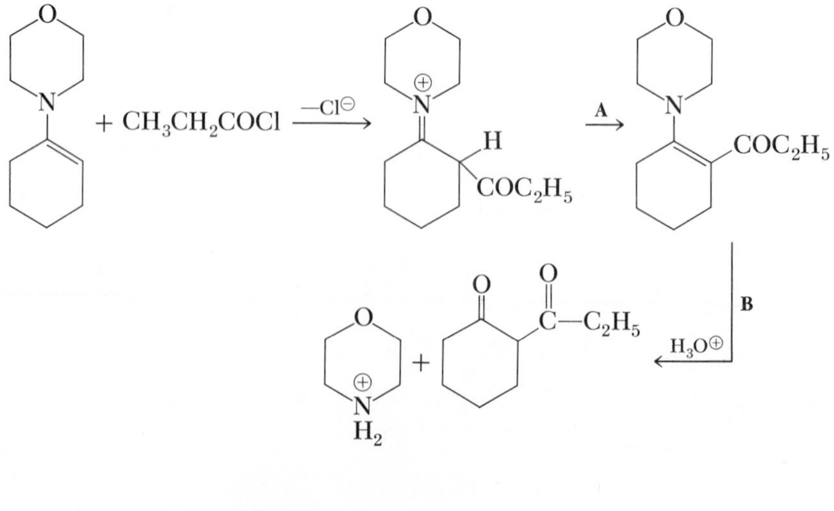

$+ CH_3CH_2COCl \xrightarrow{-Cl^{\ominus}}$ $\xrightarrow{A}$ $\xrightarrow{B,\ H_3O^{\oplus}}$

12. It has been suggested that the real advantage to the use of enamines is that they contain *nucleophilic carbon.* Identify this carbon in any simple enamine. What is the ultimate source of electrons that it uses in acting as a nucleophile?

Oxidation. Although all types of amines are easily oxidized, only the oxidation of 3° amines is of importance. One of the major uses of amine oxides is in the synthesis of alkenes of known stereochemistry.

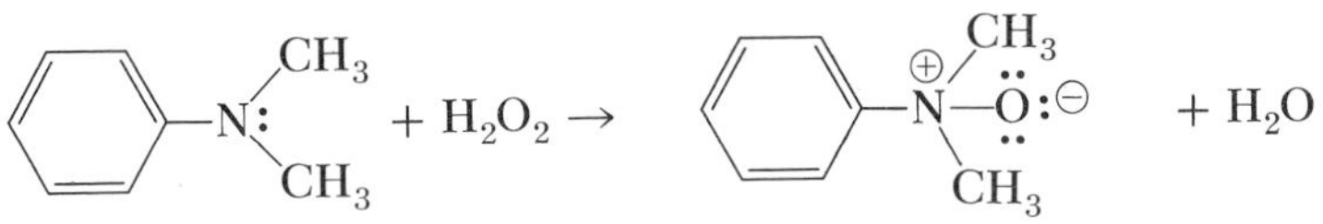

$$C_6H_5N(CH_3)_2 + H_2O_2 \rightarrow C_6H_5\overset{\oplus}{N}(CH_3)_2{-}\ddot{O}{:}^{\ominus} + H_2O$$

N,N-dimethylaniline *N*-oxide

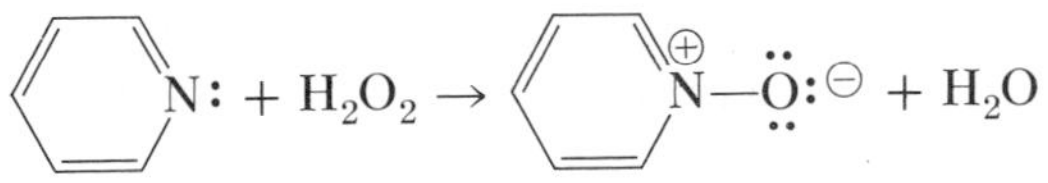

$$C_5H_5N{:} + H_2O_2 \rightarrow C_5H_5\overset{\oplus}{N}{-}\ddot{O}{:}^{\ominus} + H_2O$$

pyridine *N*-oxide

When an amine oxide containing a suitable β hydrogen is heated to about 150°, it undergoes decomposition to form an alkene and a derivative of hydroxylamine. Pyrolysis of the *threo* isomer of dimethyl(3-phenyl-2-butyl)amine oxide produces mainly *cis*-2-phenyl-2-butene, while the *erythro* isomer yields mainly the *trans* alkene. This specificity is interpreted in terms of a concerted, cyclic transition state in which the β hydrogen and the amine residue depart from the same face of the double bond. The five atoms of greatest importance to the transition state (H—C_β—C_α—N—O) reside in one plane in the activated complex. This is one example of a *cis* or *syn* elimination and is specifically called a *Cope elimination.*

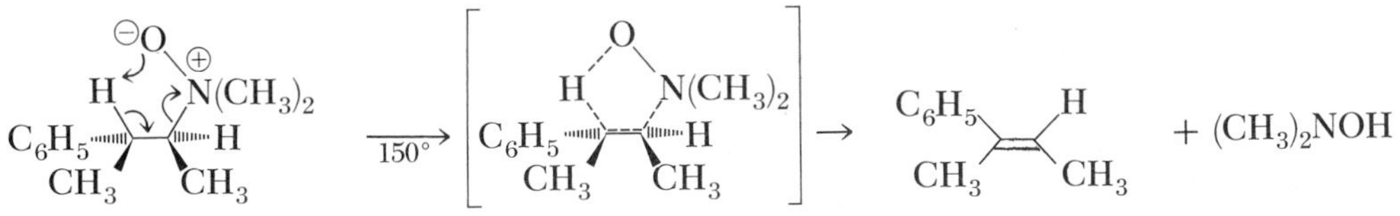

threo-dimethyl(3-phenyl-2-butyl)amine oxide $\xrightarrow{150°}$ [transition state] $\rightarrow$ *cis*-2-phenyl-2-butene $+ (CH_3)_2NOH$

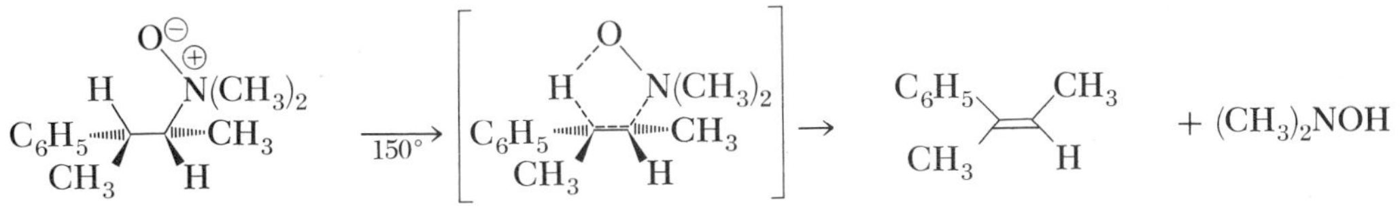

erythro-dimethyl(3-phenyl-2-butyl)amine oxide $\xrightarrow{150°}$ [transition state] $\rightarrow$ *trans*-2-phenyl-2-butene $+ (CH_3)_2NOH$

One advantage of this route to alkenes is that it forms a double bond under conditions that will not lead to isomerization or conjugation to an existing double bond.

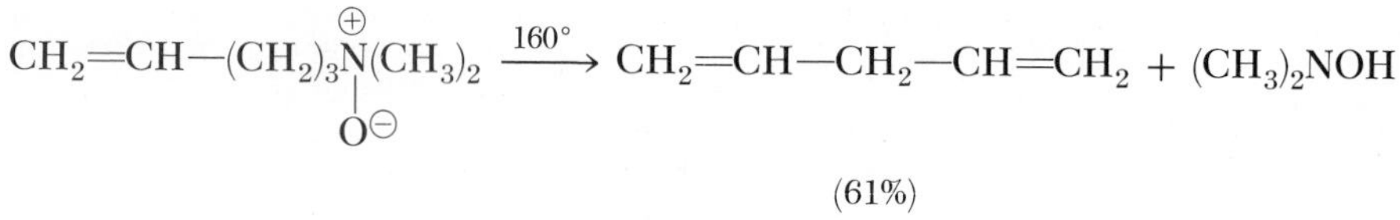

$$CH_2{=}CH{-}(CH_2)_3\overset{\oplus}{N}(CH_3)_2(O^{\ominus}) \xrightarrow{160^\circ} CH_2{=}CH{-}CH_2{-}CH{=}CH_2 + (CH_3)_2NOH$$

(61%)

If two different types of β hydrogens are available, the product distribution often will be that predicted on a statistical basis. Furthermore, if both *cis* and *trans* alkenes are possible, the *trans* isomer normally predominates. These points are illustrated in the example below.

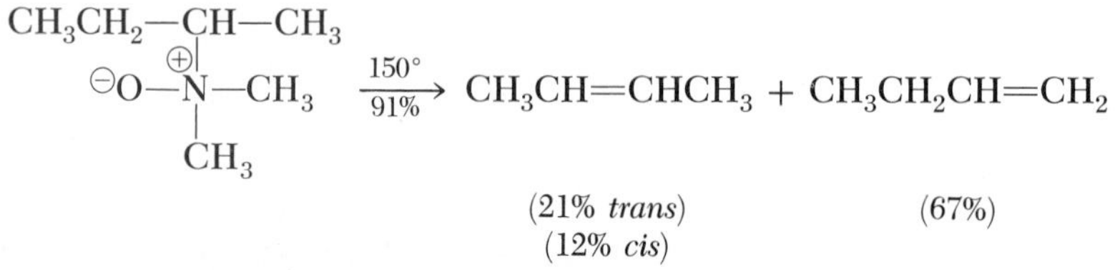

$$CH_3CH_2{-}CH(CH_3){-}\overset{\oplus}{N}(CH_3)_2{-}O^{\ominus} \xrightarrow[91\%]{150^\circ} CH_3CH{=}CHCH_3 + CH_3CH_2CH{=}CH_2$$

(21% *trans*) (12% *cis*) (67%)

13. Why is the Cope elimination called a *syn* elimination?
14. High temperatures lead to the conversion of esters to alkenes. Suggest a mechanism consistent with the following data.

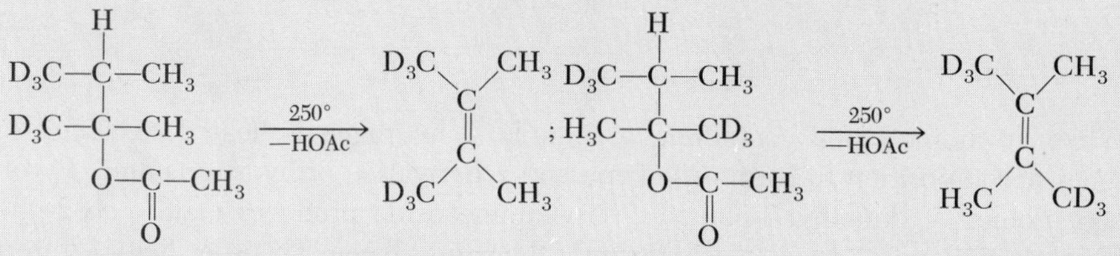

$$(D_3C)(CH_3)CH{-}C(CD_3)(CH_3){-}O{-}C(=O){-}CH_3 \xrightarrow[-HOAc]{250^\circ} (D_3C)(CH_3)C{=}C(CD_3)(CH_3)$$

(product: D_3C and CH_3 on the upper carbon; D_3C and CH_3 on the lower carbon)

$$; (D_3C)(CH_3)CH{-}C(H_3C)(CD_3){-}O{-}C(=O){-}CH_3 \xrightarrow[-HOAc]{250^\circ} (D_3C)(CH_3)C{=}C(H_3C)(CD_3)$$

What advantage does this have over taking the alcohol that is the ester precursor and dehydrating it with, for example, sulfuric acid?

20.7 ALKALOIDS

The word "alkaloid" is a rather ill-defined term used to designate basic nitrogen-containing compounds that originate in plants. These natural products usually possess at least one heterocyclic ring and often exhibit marked biological activity in humans. Although the function of these complex compounds in their plant sources is unclear, they have provided our society with some potent pharmaceutical agents and some severe social and moral problems.

The impossibility of covering even the more significant alkaloids forces our attention to no more than a handful of the most interesting members of several alkaloid families.

Coniine. One of history's most famous alkaloids, **coniine,** is known for its physiological properties but is virtually unknown by name. Coniine occurs in all parts of the poisonous hemlock (*Conium maculatum*), an extract of which is reputed to have killed Socrates.

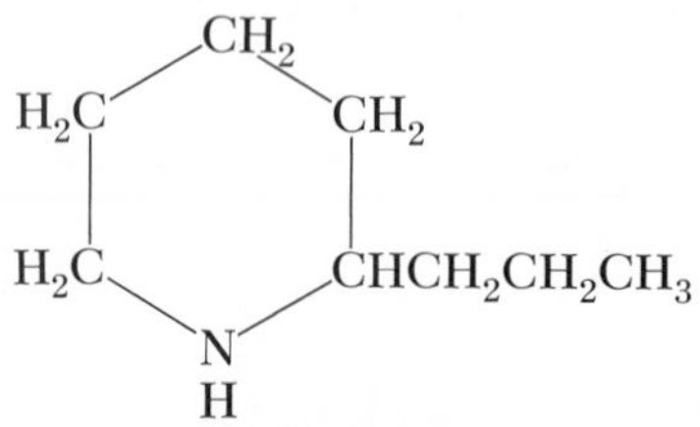

(ring structure: CH_2, H_2C, CH_2, H_2C, $CHCH_2CH_2CH_3$, N–H)

coniine

Nicotine. Nicotine represents 75% of the total alkaloid content in tobacco. In low concentrations nicotine is extremely toxic to humans. As a water-soluble salt, nicotine sulfate, this alkaloid is used as a potent contact insecticide.

Nicotine is interesting because it initially excites the autonomic ganglia and then blocks them so that they no longer respond to *any* agonists[a], including nicotine itself. Nicotine acts on the nerve by depolarizing a membrane and triggering an "action potential." In addition to its stimulant activity and toxicity, nicotine is known to be mutagenic and teratogenic[b] in some lower animals.

One possible laboratory synthesis of nicotine is presented below:

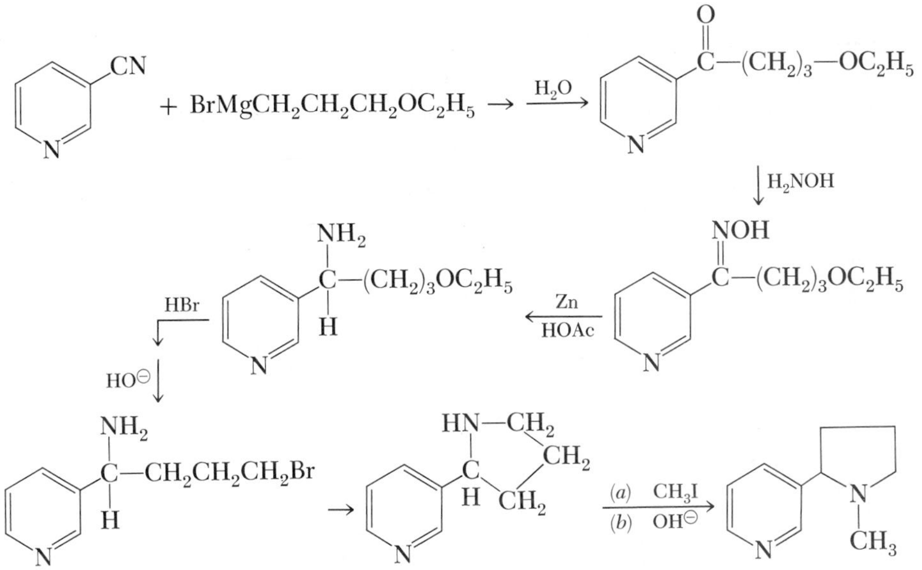

3-pyridyl–CN + $BrMgCH_2CH_2CH_2OC_2H_5$ → $\xrightarrow{H_2O}$ 3-pyridyl–C(=O)–$(CH_2)_3$–OC_2H_5

↓ H_2NOH

3-pyridyl–C(=NOH)–$(CH_2)_3OC_2H_5$ $\xrightarrow[HOAc]{Zn}$ 3-pyridyl–CH(NH_2)–$(CH_2)_3OC_2H_5$

↓ HBr, ↓ $HO^{\ominus}$

3-pyridyl–CH(NH_2)–$CH_2CH_2CH_2Br$ → 3-pyridyl–CH(HN–CH_2–CH_2–CH_2) $\xrightarrow[(b)\ OH^{\ominus}]{(a)\ CH_3I}$ 3-pyridyl–(N–CH_3 pyrrolidine)

nicotine

Capsaicin. Lovers of "hot" foods owe a great deal to **capsaicin,** the pungent ingredient in tabasco, cayenne and paprika. This compound, although considered an alkaloid because of its origin and physiological activity, lacks a heterocyclic ring and a basic nitrogen.

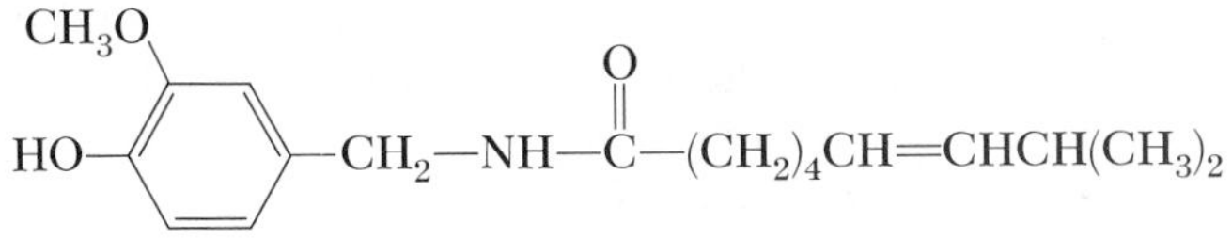

CH_3O, HO–C_6H_3–CH_2–NH–C(=O)–$(CH_2)_4CH{=}CHCH(CH_3)_2$

capsaicin

Hyoscyamine, Scopolamine, and Cocaine. These three alkaloids are part of the family of *tropane* alkaloids; they all possess the tropane ring system.

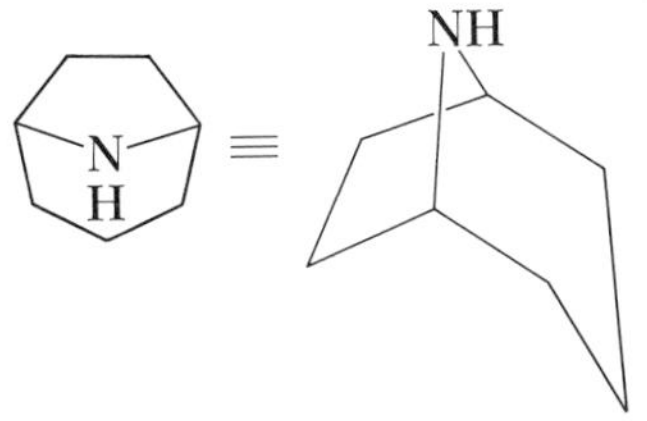

NH ≡ NH tropane

---

[a] An *agonist* is an agent that induces a certain excitation.

[b] Defined in Chapter 23.

Hyoscyamine is the major alkaloid of several infamous plants, including henbane and the deadly nightshade. It racemizes readily to the much more familiar *atropine.* Atropine is used in medicine as an anticholinergic, acting upon receptors at the parasympathetic effector organs and in the central nervous system (CNS). It exerts little effect at autonomic ganglia and the neuromuscular junction at low concentrations. However, its ability to act as an antidote for DFP poisoning (see Important Terms) is related to its effect upon ganglia at high concentrations.

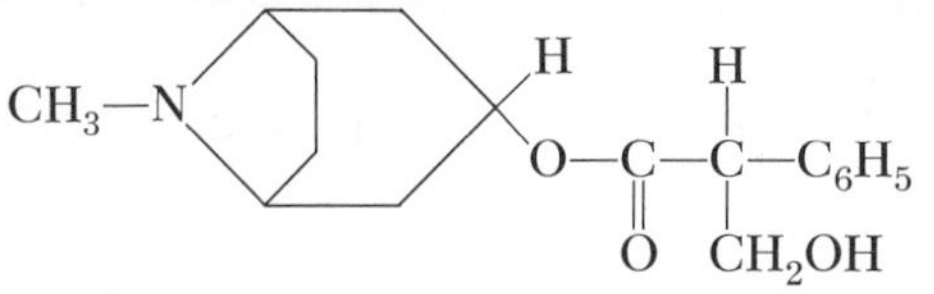

(−)-hyoscyamine and atropine

Scopolamine has received notoriety as a "truth drug," although it is also used as a CNS depressant, an anti-Parkinsonism agent, and a motion-sickness preventative.

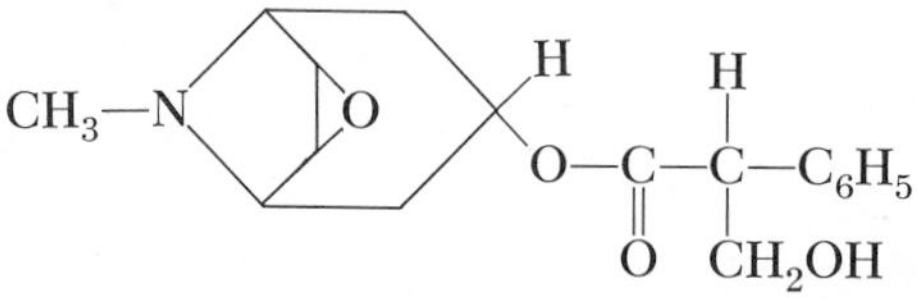

scopolamine

Cocaine, because of its ability to induce dependence, is limited to use as a topical anesthetic. An important source of cocaine is the leaves of the coca bush found in Peru and Bolivia.

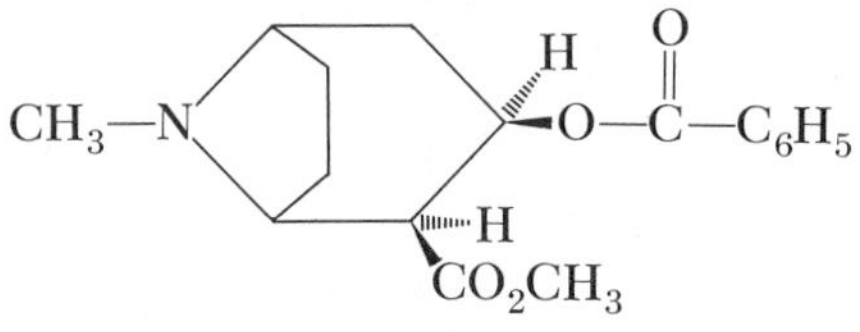

cocaine

Papaverine and Morphine. Opium, the dried latex of the "opium poppy" (*Papaver somniferum*), is a source of two families of alkaloids—*benzylisoquinoline* and *phenanthrene* alkaloids. The former includes **papaverine** and **narcotine,** while the latter includes **morphine** and **codeine.** Over two dozen individual alkaloids have been isolated from *Papaver somniferum.*

Papaverine is used as an antispasmodic; morphine, which constitutes about 10% of opium, is still used as a pain reliever although its use is severely limited because of its potential for causing addiction. Codeine differs from morphine in having a hydroxy group on one benzene ring replaced by a methoxy group. Codeine is most often encountered in cough "remedies," *e.g.,* elixir of terpin hydrate with codeine, but its use is monitored by the government because it is a narcotic.

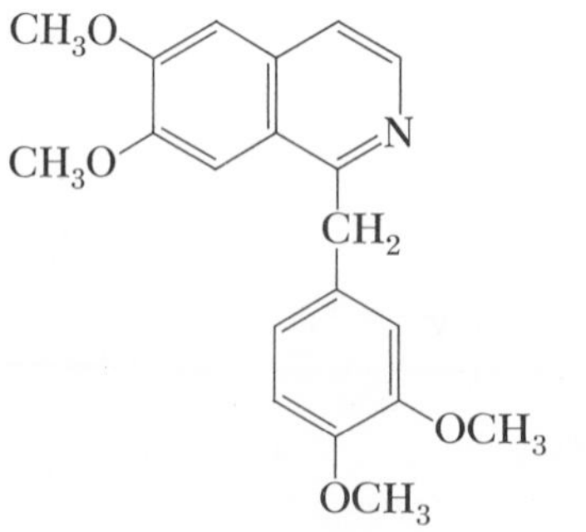

papaverine

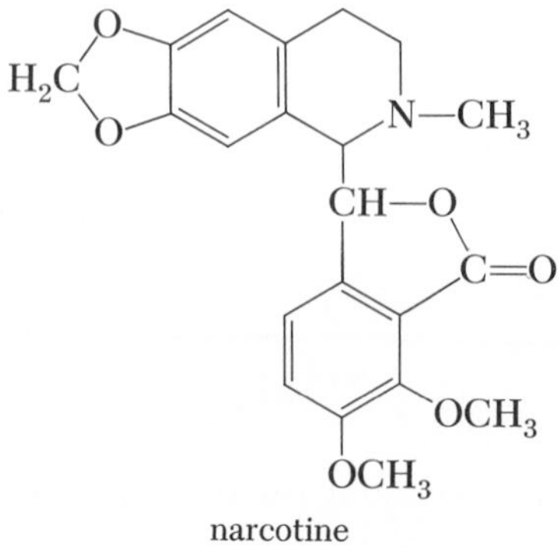

narcotine

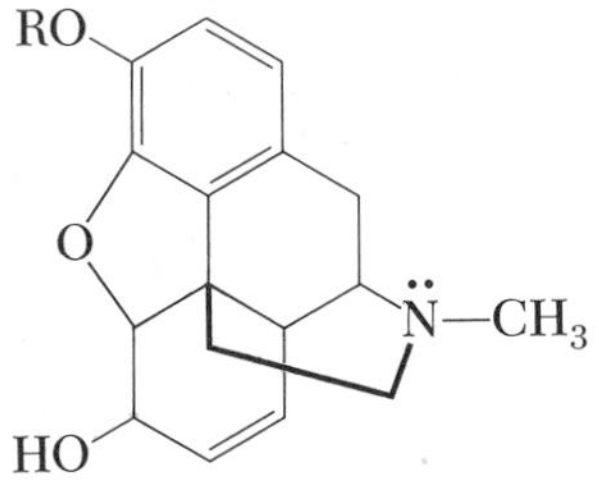

morphine R = H
codeine R = CH_3

Heroin, which is not a naturally occurring substance, is the diacetate of morphine.

Tubocurarine Chloride. Curare appears to have been used for centuries by the natives of South America as a poison transmitted to the victim by arrow. Three types of curare have been recognized, distinction being made by the type of container (bamboo tubes, clay pots, or dried gourds) used to pack the curare. The curare available for medical use is a standardized extract from the plant *Chondodendron tomentosum* R. & P., of the family *Menispermaceae,* which contains **(+)-tubocurarine chloride.**

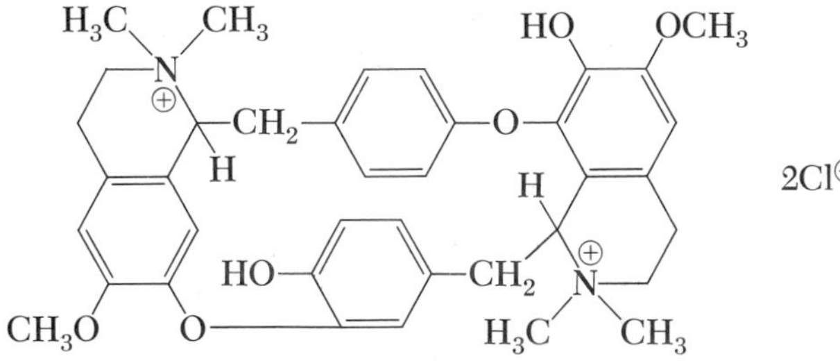

(+)-tubocurarine chloride

Curare has little effect upon the central nervous system. Deaths that result from contact with this potent muscle relaxant[a] usually stem from respiratory paralysis. Curare has been used as an adjunct to surgical anesthesia, since its use permits muscle relaxation without the dangers associated with deep anesthesia. While curare is being replaced by a synthetic diquaternary ammonium salt, **succinylcholine iodide,** in anesthesia, it is still used as a diagnostic agent for myasthenia gravis.

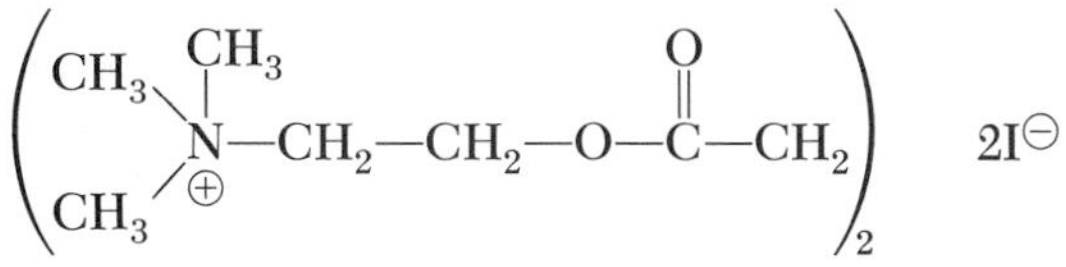

succinylcholine iodide

Quinine. While the average American thinks of heart disease or cancer as the major medical problem of the age, it is *malaria* which is the world's number one medical problem. Hundreds of millions of people have this disease, and over one million succumb to it every year: It is no wonder, then, that the *cinchona tree* has been among the world's most important resources. The bark of this tree produces *quinine,* the first drug available for the treatment of malaria. The need continually outstrips the supply; fortunately, several synthetic drugs have become available for the treatment of malaria. *Atabrine,* the first synthetic anti-malaria drug, was introduced in 1930. It, in turn, has been replaced by other drugs because it turns the skin yellow.

---

[a] A brief discussion of the chemical reactions that go on at the neuromuscular junction is found beginning on p. 757.

Quinine and atabrine contain two interesting heterocyclic systems. The former has a *quinuclidine* side chain, while the parent ring of the latter is *acridine.*

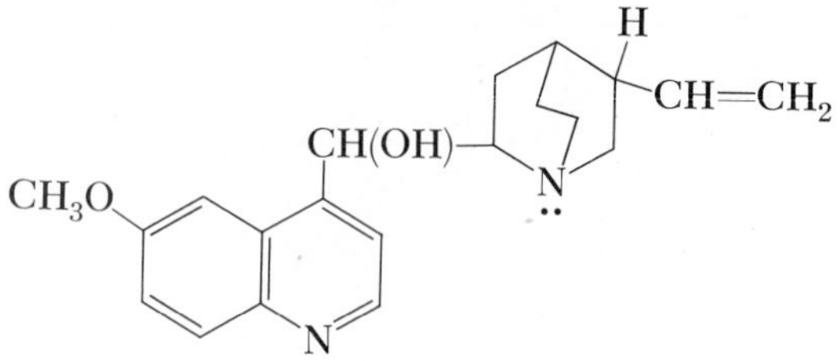

quinine

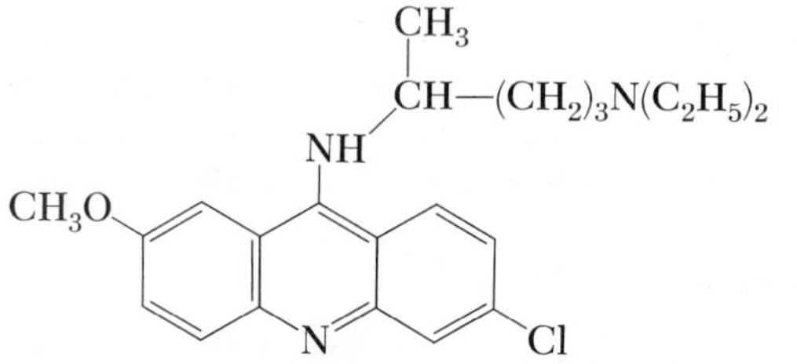

Atabrine
(quinacrine)

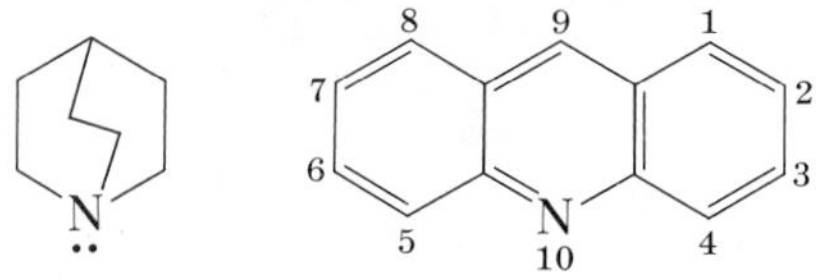

quinuclidine

acridine

LSD. Ergot is a fungus that grows on cereals, especially rye. St. Anthony's fire, a disease known for centuries, results from ergot poisoning. All of the alkaloids found in ergot are derivatives of lysergic acid, and it is to one of lysergic acid's derivatives that we now turn our attention.

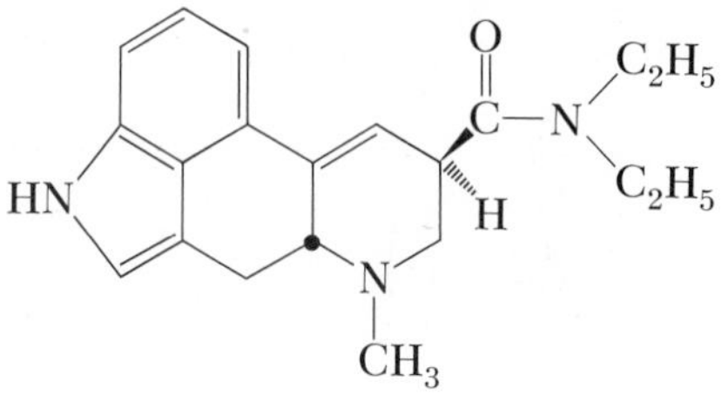

LSD

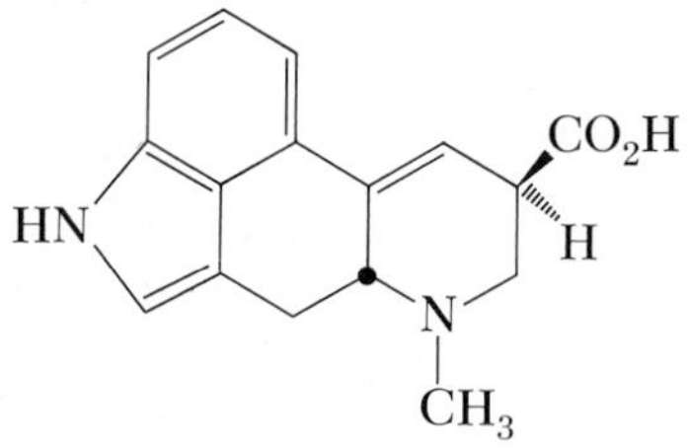

lysergic acid

The diethyl amide of lysergic acid usually is called LSD, from its German name, *Lyserg-säure diethylamid.* This compound is an hallucinogen, producing psychoses resembling schizophrenia. LSD is among the most powerful hallucinogens known; as little as 50 micrograms (50 millionths of a gram) can produce a psychotic state. It is currently believed that LSD disturbs the serotonin balance in the brain and that it is this imbalance which induces the psychotic state. In addition, LSD has been implicated in fetal abnormalities in experimental animals and in humans.

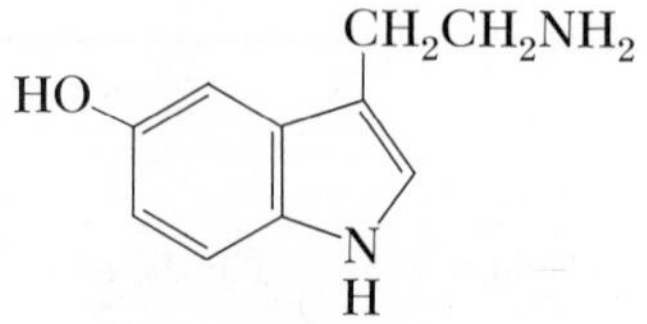

serotonin
(5-hydroxytryptamine, 5HT)

20.8 CHARACTERIZATION OF AMINES[a]

Hinsberg Test. The most common chemical test for distinguishing among 1°, 2°, and 3° amines involves the reaction of the amine with benzenesulfonyl chloride. The behavior expected for these amines is outlined below.

1° Amines. Primary amines react with benzenesulfonyl chloride to produce a sulfonamide. This sulfonamide contains an acidic hydrogen and will, therefore, dissolve in aqueous alkali and reprecipitate upon acidification.

$$C_6H_5NH_2 + C_6H_5SO_2Cl \rightarrow C_6H_5-\overset{O^{\ominus}}{\underset{O^{\ominus}}{\overset{|}{\underset{|}{S^{(+2)}}}}}-NH-C_6H_5$$

$$C_6H_5-\overset{O^{\ominus}}{\underset{O^{\ominus}}{\overset{|}{\underset{|}{S^{(+2)}}}}}-N\begin{matrix}H\\C_6H_5\end{matrix} + NaOH \xrightarrow{H_2O} \underset{\text{water-soluble salt}}{C_6H_5-\overset{O^{\ominus}}{\underset{O^{\ominus}}{\overset{|}{\underset{|}{S^{(+2)}}}}}-N^{\ominus}-C_6H_5\;Na^{\oplus}} \xrightarrow{H_3O^{\oplus}} C_6H_5-\overset{O^{\ominus}}{\underset{O^{\ominus}}{\overset{|}{\underset{|}{S^{(+2)}}}}}-N\begin{matrix}H\\C_6H_5\end{matrix}$$

2° Amines. Secondary amines also react with benzenesulfonyl chloride to produce a sulfonamide. Unlike the sulfonamide from a 1° amine, these sulfonamides lack an acidic hydrogen on nitrogen and are insoluble in base. (Why?)

$$(CH_3CH_2CH_2)_2NH + C_6H_5SO_2Cl \rightarrow C_6H_5-\overset{O^{\ominus}}{\underset{O^{\ominus}}{\overset{|}{\underset{|}{S^{(+2)}}}}}-N(CH_2CH_2CH_3)_2 \xrightarrow{OH^{\ominus}} \text{N.R.}$$

3° Alkylamines. Tertiary alkylamines react with benzenesulfonyl chloride to form a salt, **A.** This salt reacts with aqueous alkali to regenerate the amine. Hence, it appears as if tertiary alkylamines do *not* react when, in fact, they do. Addition of hydrochloric acid at the end of the experiment causes the free amine to dissolve as the hydrochloride salt.

$$C_6H_5SO_2Cl + N(CH_3)_3 \rightarrow C_6H_5-\overset{O^{\ominus}}{\underset{O^{\ominus}}{\overset{|}{\underset{|}{S^{(+2)}}}}}-\overset{CH_3}{\underset{CH_3}{\overset{|}{\underset{|}{N^{\oplus}}}}}-CH_3\;Cl^{\ominus}$$

$$C_6H_5-\overset{O^{\ominus}}{\underset{O^{\ominus}}{\overset{|}{\underset{|}{S^{(+2)}}}}}-\overset{CH_3}{\underset{CH_3}{\overset{|}{\underset{|}{N^{\oplus}}}}}-CH_3 + 2OH^{\ominus} \rightarrow C_6H_5-\overset{O^{\ominus}}{\underset{O^{\ominus}}{\overset{|}{\underset{|}{S^{(+2)}}}}}-O^{\ominus} + N(CH_3)_3 + H_2O$$

$$N(CH_3)_3 \xrightarrow{HCl} (CH_3)_3\overset{\oplus}{N}H\;Cl^{\ominus}$$

[a] A number of nmr spectra are presented in problem 35.

3° Arylamines. Tertiary anilines react with benzenesulfonyl chloride to produce a myriad of products. The scheme shown below is typical.

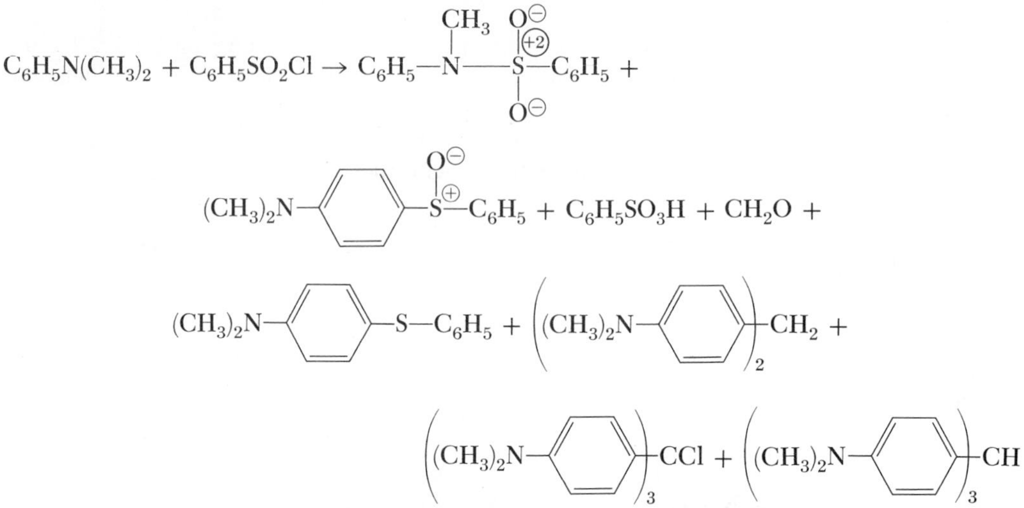

$$C_6H_5N(CH_3)_2 + C_6H_5SO_2Cl \rightarrow C_6H_5-\overset{CH_3}{\overset{|}{N}}-\overset{O^{\ominus}}{\underset{O^{\ominus}}{\overset{|}{\underset{|}{S}}}}{}^{(+2)}-C_6H_5 +$$

$$(CH_3)_2N-C_6H_4-\overset{O^{\ominus}}{\overset{|}{S^{\oplus}}}-C_6H_5 + C_6H_5SO_3H + CH_2O +$$

$$(CH_3)_2N-C_6H_4-S-C_6H_5 + \left((CH_3)_2N-C_6H_4\right)_2CH_2 +$$

$$\left((CH_3)_2N-C_6H_4\right)_3CCl + \left((CH_3)_2N-C_6H_4\right)_3CH$$

Reaction of benzenesulfonyl chloride with a 3° arylamine is slower than reaction with $OH^{\ominus}$, and in basic media it appears as though 3° arylamines do not react with $C_6H_5SO_2Cl$.

$$C_6H_5SO_2Cl + C_6H_5N(CH_3)_2 \rightarrow \text{slow reaction}$$
$$C_6H_5SO_2Cl + OH^{\ominus} \rightarrow \text{not as slow}$$

In summary, both alkyl and aryl 3° amines appear *not* to react with benzenesulfonyl chloride when, in fact, they can.

The observations associated with the Hinsberg test are summarized in Table 20–2.

TABLE 20–2 The Hinsberg Test

$$RSO_2Cl + \text{amine} \xrightarrow[H_2O]{OH^{\ominus}} \mathbf{A} \xrightarrow[H_2O]{HCl} \mathbf{B}$$

AMINE	OBSERVATIONS	
	A	**B**
1°	Reaction occurs and product dissolves in base.	Product precipitates.
2°	Reaction occurs but product does not dissolve in base.	No reaction.
3°	Reaction may or may not occur, depending upon type. However, product decomposes to regenerate amine.	Amine dissolves.

INFRARED SPECTRA. The most characteristic vibrations in the spectra of 1° and 2° amines are those associated with the N—H bond. Both alkyl and aryl 1° amines exhibit two N—H vibrations—an asymmetric stretching band at 3490 cm^{-1} and a symmetric stretching band near 3400 cm^{-1}. (Why are there two stretching vibrations?) Absorption in this region, while subject to hydrogen bonding effects, is not as influenced by hydrogen bonding as are O—H vibrations. This reflects, in part, the relative inability of the N—H moiety to serve as a hydrogen bond acceptor. When intermolecular hydrogen bonding does occur, it produces a complex, moderately intense set of absorptions in the region from 3300 to 3000 cm^{-1}.

Secondary amines show a single N—H vibration in the 3450 to 3300 cm^{-1} region. A high-frequency absorption usually indicates an aryl 2° amine, while an absorption in the 3350 to 3300 cm^{-1} range suggests an alkyl 2° amine. Tertiary amines, of course, do not show any N—H vibration.

The C—N vibrations of amines occur in the same region as do C—C and C—O vibrations (approximately 1350 to 1200 cm^{-1}) and are not very valuable in structural assignments.

A convenient technique for helping to distinguish among various types of amines involves preparation and determination of the ir spectrum of a hydrochloride salt of the amine. Salts of the type $RNH_3^{\oplus}$ will exhibit a broad, strong absorption near 3000 cm^{-1}. Both $R_2NH_2^{\oplus}$ and $R_3NH^{\oplus}$ exhibit a strong, broad absorption near 2700 to 2200 cm^{-1} but, of course, only the secondary amine would have shown an N—H vibration originally.

N-Methylamines exhibit a vibration near 2750 $\pm$ 50 cm^{-1}, and the absence of an absorption here precludes the presence of such a group. Unfortunately, since other absorptions may occur in this region, the presence of an absorption here may, at best, "strongly suggest" the presence of the N—CH_3 group. A number of representative spectra are collected in Figure 20–3.

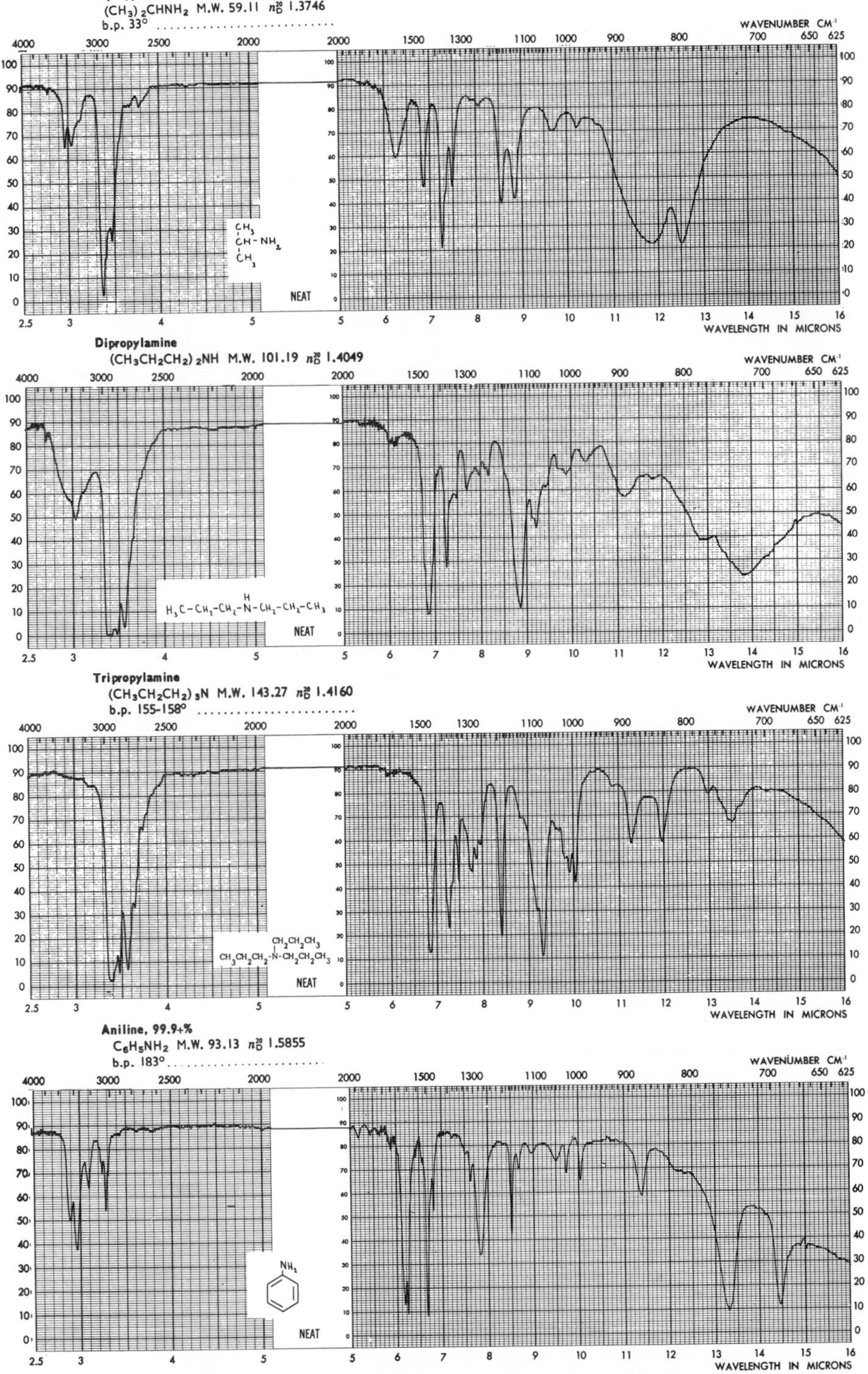

Figure 20–3 The infrared spectra of amines. *Illustration continued on following page.*

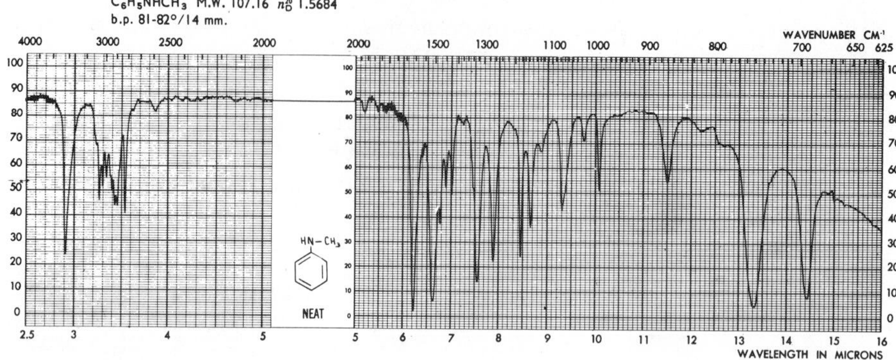

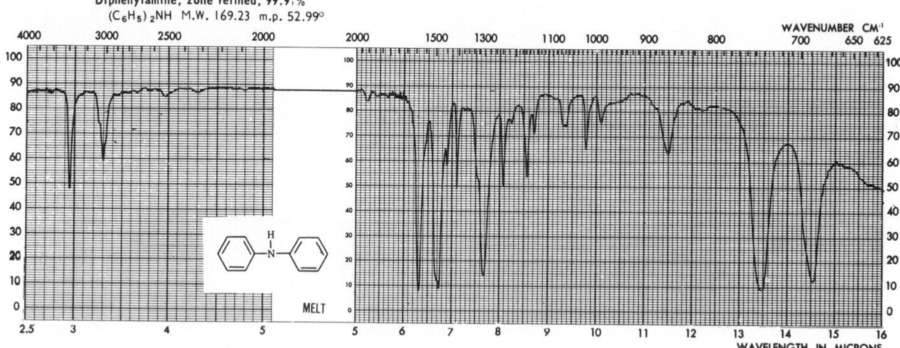

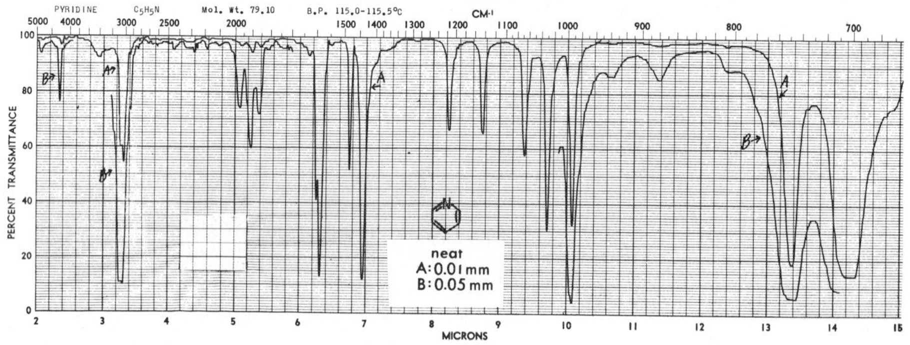

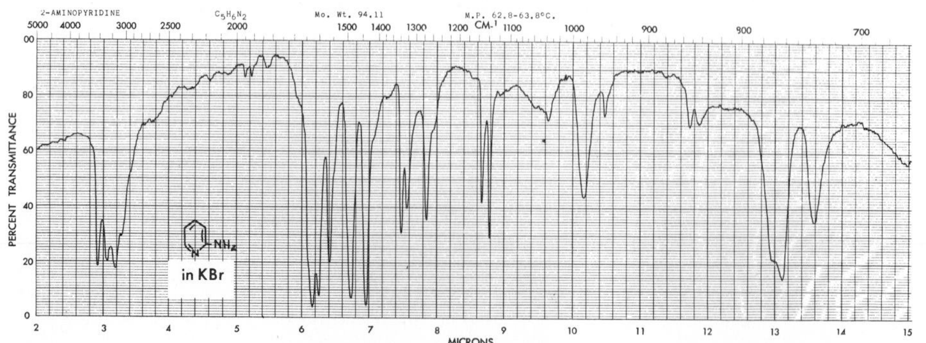

Figure 20-3 Continued.

IMPORTANT TERMS

Action potential: The potential resulting from rapid influx of sodium into a cell following membrane depolarization. An action potential initiates a propagated nerve impulse.

Alkaloid: A compound that is derived from a plant and that contains nitrogen. Alkaloids usually are basic and physiologically active.

Alkylamine: A compound containing an alkyl group bonded to an amine nitrogen. If the compound contains two or three alkyl groups bonded to this nitrogen, then it may be called a di- or trialkylamine, respectively.

Ammonium salt: A salt in which the cationic center is a nitrogen bonded to four groups, including protons. If none of the moieties bonded to the nitrogen are protons, the salt is a *quaternary* ammonium salt.

Ammonolysis: Cleavage of a bond by the action of ammonia. [Note the similarity to the terms *hydrolysis* (cleavage by water) and *solvolysis* (cleavage by solvent).]

Anticholinergic: A compound that acts as an antagonist to acetylcholine. Non-prescription sleep-inducing compounds (*e.g.*, Sominex) contain anticholinergic compounds.

Arylamine: A compound containing an alkyl group bonded to an amine nitrogen. If the compound contains two or three aryl groups bonded to this nitrogen, then it may be called a di- or triarylamine, respectively.

Autonomic ganglion: A cluster of interconnected nerve cells is termed a ganglion (plural: ganglia). Autonomic ganglia are ganglia that are part of the autonomic nervous system. The autonomic nervous system is loosely defined as the portion of the peripheral nervous system that is independent of willful control. [For the more biologically inclined, it is better defined as the efferent path to the viscera (except striated muscle) from the central nervous system.]

Autosomal: Pertaining to any chromosome that is not a sex chromosome.

Carbamic acid: The half-amide of carbonic acid, *i.e.*, H_2NCO_2H. In a general sense, a carbamic acid is a compound of the type $RNHCO_2H$ or $RNR'CO_2H$. An ester of a carbamic acid, *e.g.*, $RNHCO_2R'$, is a *urethane*. While carbamic acids are unstable, decomposing to carbon dioxide and amines, urethanes are rather stable.

Carcinogen: A material that causes cancer.

Central nervous system (CNS): The central portion of the nervous system consisting, in higher animals, of the spinal cord and the brain. The remainder of the nervous system is the *peripheral nervous system*.

Chemotherapy: The treatment of diseases with specific chemicals.

Curtius rearrangement: The conversion of an acyl azide, $RC(O)N_3$, to an isocyanate *via* a nitrene.

DFP: Diisopropyl fluorophosphonate. A highly toxic compound.

$$F-\overset{\overset{\displaystyle O^{\ominus}}{|}}{\underset{\underset{\displaystyle OCH(CH_3)_2}{|}}{P^{\oplus}}}-OCH(CH_3)_2$$

Diazonium salt: A salt in which the cation is of the type $R-\overset{\oplus}{N}\equiv N:$.

Gabriel synthesis: A synthesis of primary alkylamines based upon the hydrolysis of *N*-alkyl phthalimides.

Hallucinogen: A material that produces states resembling psychoses.

Myasthenia gravis: A degenerative muscle disorder.

Mutagen: A substance capable of inducing biological mutation.

Narcotic: A substance that reduces sensibility or induces sleep.

Parasympathetic: Referring to that portion of the nervous system which originates from the cranial or sacral region. It is one portion of the autonomic nervous system, the other being the sympathetic nervous system. Many organs of the body are influenced by both the sympathetic and parasympathetic nervous systems, usually in opposite directions. However, the difference between these two is *not* related to the changes they may produce but, instead, to origin (*i.e.*, to anatomical factors). A *parasympathetic effector organ* is an organ that reacts to a stimulus from the parasympathetic nervous system.

PROBLEMS

15. Provide an acceptable name for each of the following. Indicate whether the amino groups in these compounds are primary, secondary, or tertiary.

(a) $CH_2{=}CHCH_2NH_2$

(b) $CH_2{=}CHCH_2NHCH_3$

(c) $CH_3CH_2CHNHCH_3$

(d) $C_6H_5NH_2$

(e) $H_2NCH_2CO_2H$

(f) $(CH_3)_2CHN(CH_3)_2$

(g) $((CH_3)_3C)_3N$

(h) 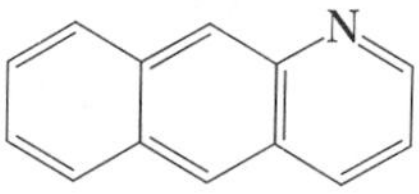 N

(i) $NCCH_2CH_2N(CH_3)C_6H_5$

(j) 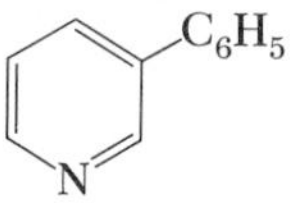 C_6H_5 N

(k) 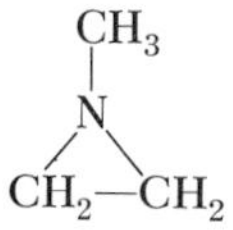 CH_3 N $CH_2{-}CH_2$

(l) 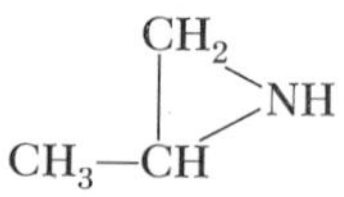 CH_2 NH $CH_3{-}CH$

(m) 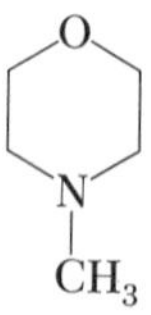 O N CH_3

(n) NO_2 $NHCH_3$

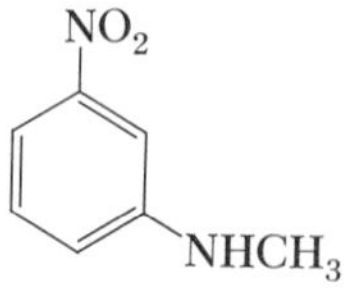

(o) $(CH_2)_5NCH_3$

(p) 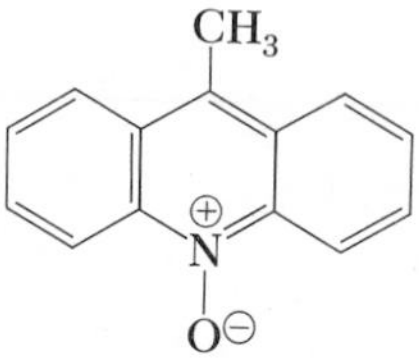 CH_3 $\oplus$ N $O^{\ominus}$

16. Provide a structure for each of the following.
(a) isoquinoline
(b) aniline
(c) quinoline *N*-oxide
(d) 1,4-dihydropyridine
(e) β-dimethylbutyronitrile
(f) *N*-benzylphthalimide
(g) ethyl 5-diphenylamino-2-pentenoate
(h) phenyl isocyanate
(i) *t*-butylamine
(j) 3-aminopentane
(k) 2-aza-3-ethylpentane
(l) triethylamine
(m) ethylenediamine
(n) ethyldimethylphenylammonium hydroxide

17. Describe the Hinsberg test and show how it can distinguish between 1°, 2°, and 3° amines. Be sure to include the anticipated observations.

18. Clearly identify each of the following.
(a) Skraup synthesis
(b) Hofmann hypohalite reaction
(c) Hantzsch synthesis
(d) Bischler-Napieralski reaction
(e) Gabriel synthesis
(f) Cope elimination

19. Would you expect the reaction of hydrazine, $H_2N—NH_2$, with two moles of methyl iodide to produce more *N,N*-dimethylhydrazine, $(CH_3)_2NNH_2$, or more *N,N'*-dimethylhydrazine, $CH_3NHNHCH_3$? Explain your answer.

20. Suggest procedures for separating the following mixtures and recovering the individual components.
(a) ether; ethylamine
(b) ether; ethylamine; ethanol
(c) ether; ethylamine; ethanol; ethylene glycol
(d) heptanol; heptylamine; heptanoic acid
(e) butylamine; dibutylamine; triethylamine
(f) 2-hexanol; 2-hexanone; 2-aminohexane

21. Suggest a chemical test or a spectral test to distinguish between members of the following pairs.
(a) ethylamine and ethanol
(b) ethylamine and diethylamine
(c) ethylamine and triethylamine
(d) triethylamine and tetraethylammonium chloride
(e) isoamylamine and dihexylamine
(f) dipropylamine and aniline
(g) hexylamine and 2-hexene
(h) pyridine and morpholine
(i) pyridine and piperidine

22. Starting with acetic acid as the only organic material, and using necessary inorganic reagents, outline the synthesis of each of the following.
(a) acetamide
(b) ethanol
(c) ethylamine
(d) methylamine
(e) ethyl chloride
(f) propanoic acid
(g) 2-aminopropane
(h) ethylene oxide
(i) ethanolamine, $HOCH_2CH_2NH_2$
(j) butanol
(k) butanoic acid
(l) methylethylamine
(m) ethyl acetate
(n) acetoacetic ester
(o) ethylene
(p) formaldehyde
(q) 2,6-dimethylpyridine

23. What problem(s) might you encounter in attempting to prepare a Grignard reagent from *p*-bromoaniline? Explain.

24. Suggest a synthesis for aziridine starting with ethanol. What product(s) would aziridine form with the following?
(a) NaOEt/EtOH
(b) H_2O
(c) H_2S
(d) CH_3NH_2
(e) $CH_2{=}CHCN$
(f) $CH_3CH{=}CHCO_2Et$

25. Suggest a means of converting the given starting material into the desired product. Use any other needed reagents.
(a) acteone to isopropylamine
(b) acetone to neopentylamine
(c) acetone to 2-isopropylaminoethanol
(d) ethyl acetate to 3-amino-3-methyl-1,4-pentadiyne
(e) ethylene to 3-amino-1-propanol
(f) ethylamine to ethylene
(g) ethanolamine ($HOCH_2CH_2NH_2$) to morpholine
(h) ethane to ethylamine (free of other amines)
(i) propylene to isobutylamine
(j) propylene to *n*-butylamine
(k) malonic ester to 3-aminopentane

26. Suggest a means of converting 2-hexanone to amylamine, free of other amines.
27. When a quaternary ammonium hydroxide is heated, it decomposes to form a tertiary amine, water, and an alkene. This reaction is called a *Hofmann elimination reaction.* Elimination will occur only if at least one of the substituents on nitrogen possesses a hydrogen on the carbon β to nitrogen; *e.g.*, tetraneopentylammonium hydroxide is comparatively stable to heat. (a) Suggest a mechanism for the observed elimination, using ethyltrimethylammonium hydroxide as the substrate.

(b) The elimination forms the basis for the *Hofmann degradation* of amines, sometimes called exhaustive methylation. In the early history of chemistry, exhaustive methylation played a critical role in determining the structures of various alkaloids. The scheme below is representative. Suggest a structure for the starting amine **A**, and identify the lettered intermediates.

$$\mathbf{A} + CH_3I \rightarrow \mathbf{B}\,(C_9H_{20}\overset{\oplus}{N}\,\overset{\ominus}{I})$$

$$\mathbf{B} + Ag_2O \xrightarrow{H_2O} \mathbf{C}\,(C_9H_{20}\overset{\oplus}{N}\,\overset{\ominus}{O}H)$$

$$\mathbf{C} \xrightarrow{\text{heat}} \text{ethylene} + \mathbf{D}\,(C_7H_{15}N)$$

$$\mathbf{D} + CH_3I \rightarrow \mathbf{E}\,(C_8H_{18}\overset{\oplus}{N}\,\overset{\ominus}{I})$$

$$\mathbf{E} + Ag_2O \xrightarrow{H_2O} \mathbf{F}\,(C_8H_{18}\overset{\oplus}{N}\,\overset{\ominus}{O}H)$$

$$\mathbf{F} \xrightarrow{\text{heat}} \mathbf{G}\,(C_8H_{17}N)$$

$$\mathbf{G} + CH_3I \rightarrow \mathbf{H}\,(C_9H_{20}\overset{\oplus}{N}\,\overset{\ominus}{I})$$

$$\mathbf{H} + Ag_2O \xrightarrow{H_2O} \mathbf{I}\,(C_9H_{20}\overset{\oplus}{N}\,\overset{\ominus}{O}H)$$

$$\mathbf{I} \xrightarrow{\text{heat}} \mathbf{J}\,(C_6H_{10}) + \text{trimethylamine}$$

$$\mathbf{J} \xrightarrow{\text{ozone}} \xrightarrow[\text{acid}]{\text{zinc}} 2CH_2O + CH_3C(O)C(O)CH_3$$

(*Hint:* There are two possible structures for **A** that fit these data.)

28. *N*-Bromosuccinimide, NBS, can be prepared by treating succinimide with a mixture of NaOH, Br_2, and H_2O at 0°. At higher temperatures the product that is isolated is 3-aminopropanoic acid (as the salt). Account for these results, providing mechanisms for both reactions.

$$\text{succinimide} + Br_2 \xrightarrow[0^\circ]{OH^\ominus/H_2O} \text{NBS} + Br_2 \xrightarrow[40^\circ]{OH^\ominus/H_2O} HO_2C{-}CH_2CH_2NH_2$$

succinimide NBS 3-aminopropanoic acid

29. The text describes the conversion of acetylacetone to 2,4-dimethylquinoline, a variation of the Doebner-Miller reaction. Suggest a mechanism for the acid-catalyzed cyclodehydration that leads from the intermediate carbinolamine into the final product. Suggest a reasonable consequence of replacing acetylacetone with (a) malonic ester and (b) acetoacetic ester in this sequence.
30. The Diels-Alder reaction of benzyne (dehydrobenzene) with pyrrole produces an intermediate that isomerizes in acid to α-naphthylamine (as the salt). Suggest a mechanism for this isomerization. Would the use of deuterated acid (*e.g.*, DCl) in the last step lead to deuterium incorporation in the molecule? If your answer is affirmative, where would the deuterium be?

$CO_2^\ominus$, $\overset{\oplus}{N_2}$ → [benzyne] —pyrrole→ (NH bridged adduct, H, H) $\xrightarrow{H^\oplus}$ NH_2 (α-naphthylamine)

31. (a) Suggest structures for the lettered compounds in the sequence below. (b) What is the function of morpholine? An alternate route to **B** involves the reaction of α-picoline with benzaldehyde in the presence of morpholine to produce **C**, and the reaction of **C** with methyl iodide. (c) Why might this be less desirable?

$$\alpha\text{-picoline} + CH_3I \rightarrow \mathbf{A}\,(C_7H_{10}N^\oplus\ I^\ominus) \xrightarrow[\text{trace of morpholine}]{C_6H_5CHO} \mathbf{B}\,(C_{14}H_{14}N^\oplus\ I^\ominus) + H_2O$$

α-picoline

32. Treatment of quinoline with mixed acid (a mixture of sulfuric acid and nitric acid) affords 5- and 8-nitroquinoline, while nitration of isoquinoline affords mainly 5-nitroisoquinoline. Treatment of quinoline with amide ion, $NH_2^{\ominus}$, produces 2-aminoquinoline, while treatment of isoquinoline with amide ion produces 1-aminoisoquinoline. Compare and contrast the behavior of quinoline and isoquinoline in these reactions, accounting for the observed products.

33. In the **Beckmann rearrangement,** a ketoxime reacts with acid (*e.g.*, H_2SO_4) to produce an amide: $R_2C{=}NOH \rightarrow RC(O)NHR$. The group R may be either aryl or alkyl. (a) What general type of mechanism could this follow? (b) Suggest a specific mechanism for this reaction. (c) Show how one could use the Beckmann rearrangement to synthesize aniline from benzophenone, $(C_6H_5)_2CO$. (d) What ketoxime would yield ε-caprolactam by the Beckmann rearrangement?

H O
N

ε-caprolactam

34. When pyridine reacts with sodamide at elevated temperatures, the major product is 2-aminopyridine. This is called the **Chichibabin reaction.** Use of lithium amide leads to the formation of an intermediate, **A,** $C_5H_7N_2Li$, which converts to 2-aminopyridine upon hydrolysis. (a) Suggest a structure for **A** and a mechanism for the Chichibabin reaction. (b) Why would you not expect 3-aminopyridine to be formed in this process? (c) Predict the ultimate product from the reaction of pyridine with phenyl lithium.

35. This chapter does not have an explicit discussion of the nmr spectra of amines. However, you should be able to determine some of the characteristics of these compounds on your own. In order to give you this opportunity, the following spectra are given with their proper identification. The absorptions that are shaded in black will disappear if D_2O is added to the sample. Analyze these spectra, making as many assignments as possible. What conclusions may be drawn about the nmr spectra of amines? (Chapter 29 presents some aspects of the nmr of amines.)

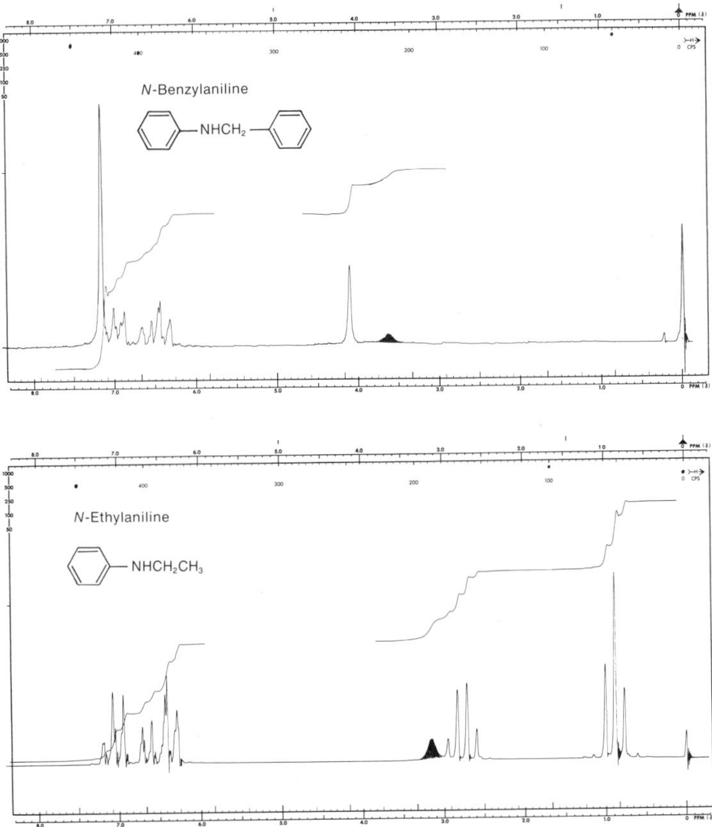

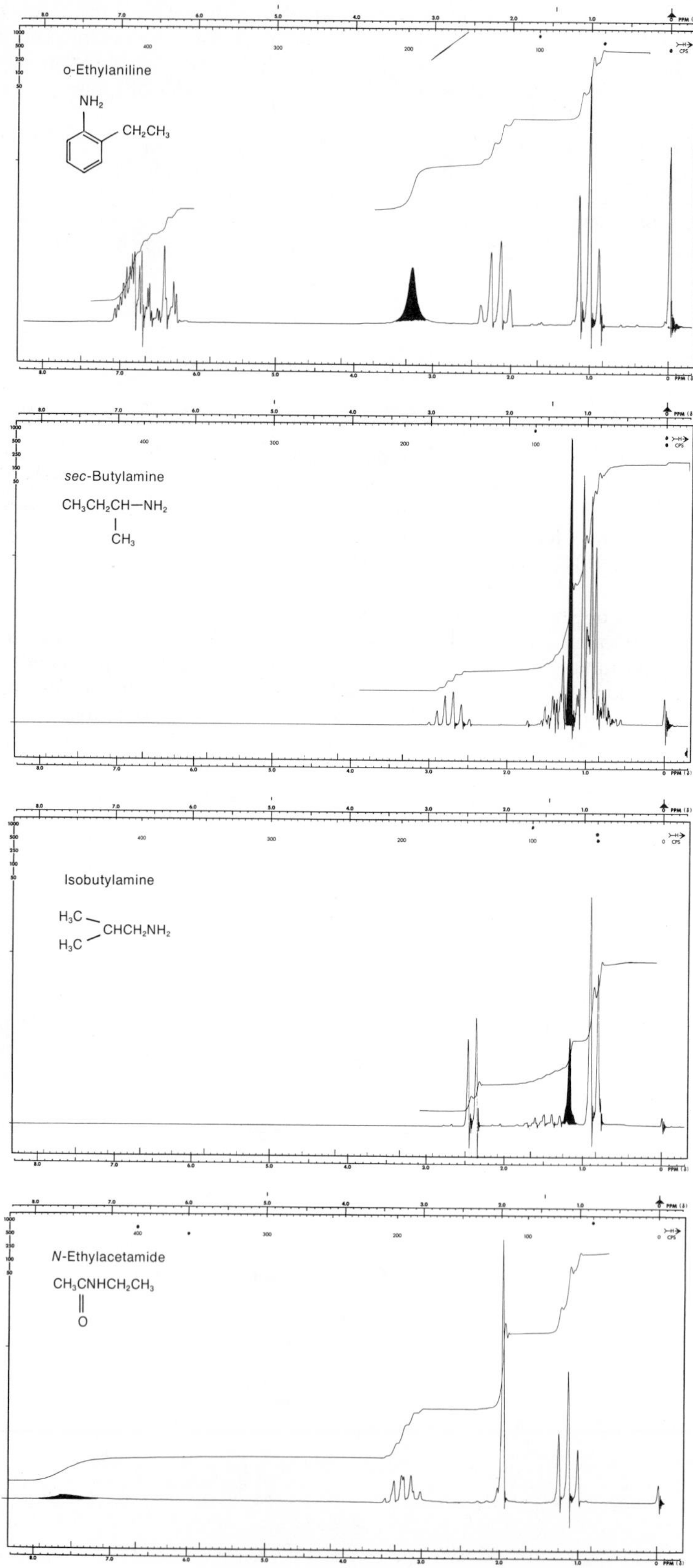
o-Ethylaniline
NH_2
CH_2CH_3
sec-Butylamine
$CH_3CH_2CH—NH_2$
CH_3
Isobutylamine
H_3C
$CHCH_2NH_2$
H_3C
N-Ethylacetamide
$CH_3CNHCH_2CH_3$
O

ARYL NITROGEN COMPOUNDS

21.1 BETWEEN NITROBENZENE AND ANILINE

Millions of pounds of nitrobenzene are produced annually, but very little nitrobenzene, as such, is used in the chemical industry. Most of it is immediately converted to its most important reduction product, aniline, $C_6H_5NH_2$. Depending upon the exact reaction conditions, reduction of nitrobenzene and other aryl nitro compounds produces nitroso, hydroxylamino, amino, azoxy, azo, and/or hydrazo compounds. All of these contain nitrogenous functional groups which, except for the amino compounds themselves, are intermediate in oxidation state between the nitro and amino groups. The influence of the pH of the reaction medium upon the product distribution during reductions with metals is summarized below, using nitrobenzene as a typical aryl nitro compound.

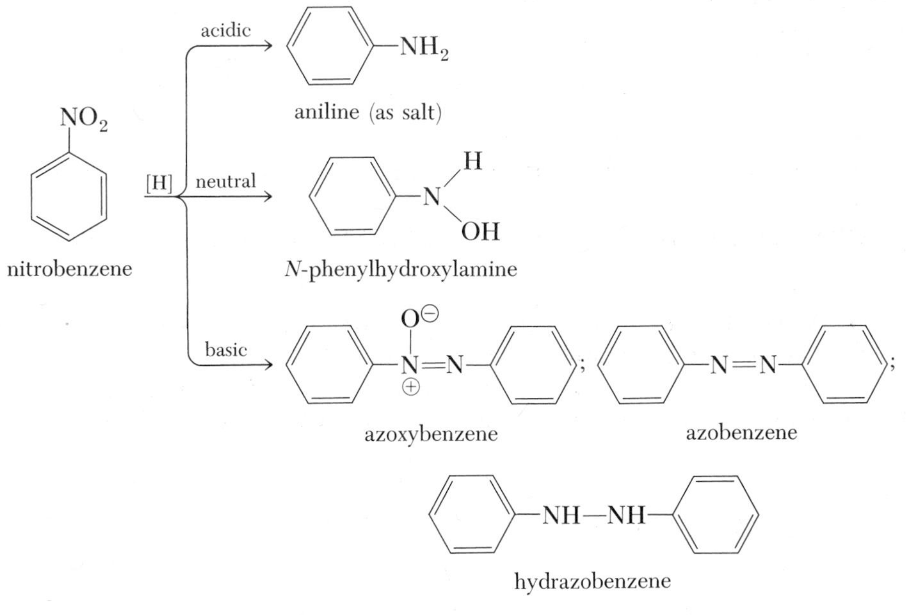

NO_2

nitrobenzene

[H] acidic → C_6H_5—NH_2 aniline (as salt)

[H] neutral → C_6H_5—N(H)OH *N*-phenylhydroxylamine

[H] basic → C_6H_5—$\overset{\oplus}{N}(O^{\ominus})$=N—$C_6H_5$ azoxybenzene; C_6H_5—N=N—C_6H_5 azobenzene; C_6H_5—NH—NH—C_6H_5 hydrazobenzene

Metal reductions in acid media are more potent than are reductions by the same metals in basic media. Therefore, *N*-phenylhydroxylamine, azoxybenzene, and hydrazobenzene all can be reduced to aniline by the action of tin and hydrochloric acid. Aniline is also the major product when nitrobenzene is reduced with tin and hydrochloric acid directly.

$$\left.\begin{array}{l} C_6H_5NHOH \\ C_6H_5N(O){=}NC_6H_5 \\ C_6H_5N{=}NC_6H_5 \\ C_6H_5NH{-}NHC_6H_5 \end{array}\right\} \xrightarrow[HCl]{Sn} \xrightarrow{OH^{\ominus}} C_6H_5{-}NH_2 \xleftarrow{OH^{\ominus}} \xleftarrow[HCl]{Sn} C_6H_5{-}NO_2$$

Nitrosobenzene, C_6H_5NO, plays a critical role in the formation of the "dimeric" reduction products (azoxy-, azo-, and hydrazobenzene), but it is so easily reduced that it cannot be isolated from a mixture of nitrobenzene, tin, and hydrochloric acid. Let us now examine these various reduction products individually.

Aniline. The most common route to 1° arylamines is the reduction of nitro aromatics, the usual reductant being a mixture of an active metal and acid. Catalytic hydrogenation achieves the same conversion but is much less frequently used because of the exothermicity of the reaction.

$$C_6H_5NO_2 + 3Zn + 7HCl + H_2O \rightarrow C_6H_5\overset{\oplus}{N}H_3\ Cl^{\ominus} + 3ZnCl_2 + 3H_2O$$

$$\xrightarrow{OH^{\ominus}} C_6H_5NH_2 + H_2O + Cl^{\ominus}$$

$$C_6H_5NO_2 + 3H_2 \xrightarrow[30\ atm]{Ni/25^\circ} C_6H_5NH_2 + 2H_2O + 85\ kcal/mole$$

The commercial preparation of aniline involves the reduction of nitrobenzene with a mixture of scrap iron, water, and only about 0.02% of the calculated amount of hydrochloric acid. This reduction is therefore brought about essentially by iron and water. The iron is oxidized to Fe_3O_4.

$$4C_6H_5NO_2 + 9Fe + 4H_2O \xrightarrow{H^{\oplus}} 4C_6H_5NH_2 + 3Fe_3O_4$$

Nitrosobenzene. Phenylhydroxylamine is oxidized by chromic acid to nitrosobenzene.

$$3C_6H_5NHOH + Na_2Cr_2O_7 + 4H_2SO_4 \xrightarrow{0^\circ} \underset{\text{nitrosobenzene}}{3C_6H_5NO} + Na_2SO_4 + Cr_2(SO_4)_3 + 7H_2O$$

Nitrosobenzene is green in the liquid and vapor states but, due to dimerization, is colorless in the solid state. In contrast to nitrosobenzene, *p*-nitroso-*N*,*N*-dimethylaniline is a green solid, since the solid consists only of monomer. Here resonance involving the dimethylamino group and the nitroso group, an interaction impossible in nitrosobenzene, stabilizes the monomer.

$$\underset{\text{green}}{2C_6H_5{-}N{=}O} \rightleftarrows \underset{\text{colorless}}{C_6H_5{-}\overset{\oplus}{N}(=O){-}N(O^{\ominus}){-}C_6H_5}$$

$$(CH_3)_2\ddot{N}{-}C_6H_4{-}\ddot{N}{=}\ddot{O}: \leftrightarrow (CH_3)_2\overset{\oplus}{N}{=}C_6H_4{=}\ddot{N}{-}\ddot{O}:^{\ominus}$$

p-nitroso-*N*,*N*-dimethylaniline

N-PHENYLHYDROXYLAMINE. Reduction of nitrobenzene with zinc dust and aqueous ammonium chloride (a weakly acidic solution) produces *N*-phenylhydroxylamine.

$$C_6H_5NO_2 + Zn \xrightarrow[H_2O]{NH_4^{\oplus}\ Cl^{\ominus}} C_6H_5N(H)OH$$

N-phenylhydroxylamine

This interesting compound is unstable in dilute sulfuric acid, isomerizing to *p*-hydroxyaniline.

$$H\text{—}C_6H_4\text{—}N(H)OH \xrightarrow{H^{\oplus}} H\text{—}C_6H_4\text{—}N(H)\overset{\oplus}{O}H_2 \xrightarrow{-H_2O} H\text{—}C_6H_4\text{—}\overset{\oplus}{\ddot{N}}H \leftrightarrow H\text{—}\overset{\oplus}{C_6H_4}\text{=}\ddot{N}H \quad etc.$$

$$H_2\ddot{O}: + H\text{—}\overset{\oplus}{C_6H_4}\text{=}NH \xrightarrow{-H^{\oplus}} HO(H)C_6H_4\text{=}NH \rightleftarrows HO\text{—}C_6H_4\text{—}NH_2$$

p-hydroxyaniline

The intermediate cation can be trapped by replacing the sulfuric acid with the hydrochloric acid. This leads to formation of *p*-chloroaniline.

$$H\text{—}\overset{\oplus}{C_6H_4}\text{=}NH \xrightarrow{Cl^{\ominus}} H(Cl)C_6H_4\text{=}NH \xrightarrow{H^{\oplus}} H(Cl)C_6H_4\text{=}\overset{\oplus}{N}H_2 \rightarrow Cl\text{—}C_6H_4\text{—}NH_2$$

p-chloroaniline

AZOXYBENZENE. Nitrobenzene is a sufficiently potent oxidant in alkaline solution to oxidize methanol to formaldehyde; the product of the accompanying reduction is azoxybenzene. The azoxybenzene arises by a dehydrative addition reaction between nitrosobenzene and phenylhydroxylamine, both of which are produced under these mild conditions. This mechanism of formation of azoxybenzene is supported by the fact that azoxybenzene derivatives can be prepared by the direct combination of the appropriate nitrosobenzene and phenylhydroxylamine derivatives.

$$C_6H_5NO_2 + 3CH_3OH \xrightarrow{NaOH} C_6H_5\text{—}\overset{\oplus}{N}(O^{\ominus})\text{=}N\text{—}C_6H_5 + 3CH_2O + 3H_2O$$

azoxybenzene

$$C_6H_5N\text{=}O + C_6H_5NHOH \rightarrow C_6H_5\overset{\oplus}{N}(O^{\ominus})\text{=}NC_6H_5 + H_2O$$

green colorless yellow

$$CH_3\text{—}C_6H_4\text{—}N\text{=}O + m\text{-}Cl\text{—}C_6H_4\text{—}NHOH \xrightarrow{-H_2O} CH_3\text{—}C_6H_4\text{—}N\text{=}\overset{\oplus}{N}(O^{\ominus})\text{—}C_6H_4\text{—}Cl(m)$$

Hydrazobenzene. Azobenzene is reduced by heating with zinc dust and aqueous sodium hydroxide to hydrazobenzene, $C_6H_5—NH—NH—C_6H_5$. This organic derivative of hydrazine, $H_2N—NH_2$, undergoes several interesting reactions, including (a) oxidation to azobenzene and (b) acid-catalyzed rearrangement to *benzidine, p,p'*-diaminobiphenyl. The mechanism for the latter reaction, the **benzidine rearrangement,** has been the subject of much investigation. While it is known that the rearrangement is intramolecular, details of the process are still subjects of controversy.

$C_6H_5—NH—NH—C_6H_5$

air / $Zn/OH^\ominus$; $H^\oplus$

$C_6H_5N{=}NC_6H_5$ $\qquad$ $H_2N—C_6H_4—C_6H_4—NH_2$

benzidine
(a carcinogen!)

Azobenzene. Distillation of azoxybenzene from iron powder, or reaction of azoxybenzene with an equivalent amount of zinc and methanolic sodium hydroxide, produces the orange solid, azobenzene, $C_6H_5—N{=}N—C_6H_5$. (Azobenzene may be viewed as an organic derivative of the unstable compound diimide, H—N=N—H.) An excess of zinc must be avoided in order to prevent over-reduction to hydrazobenzene.

$$C_6H_5—\overset{\oplus}{N}(—O^\ominus){=}N—C_6H_5 \xrightarrow[Zn/OH^\ominus/CH_3OH]{Fe,\ distill} C_6H_5—N{=}N—C_6H_5$$

"Azobenzene" is actually two compounds, *cis*- and *trans*-azobenzene. The *trans* isomer, being thermodynamically favored, is produced by the reductions described above. Irradiation of the *trans* isomer with ultraviolet light converts it to the *cis* isomer. (It is not unusual for light-induced isomerizations to produce the product disfavored by thermal isomerization.)

$$\text{trans-azobenzene} \underset{heat}{\overset{h\nu}{\rightleftharpoons}} \text{cis-azobenzene}$$

trans-azobenzene $\qquad$ *cis*-azobenzene

Unsymmetrical azobenzenes can be prepared by the reaction of a nitroso compound with a 1° amine.

$$CH_3—C_6H_4—N{=}O + C_{10}H_7—NH_2 \rightarrow C_{10}H_7—N{=}N—C_6H_4—CH_3$$

Aryl azo compounds are not very basic and are relatively stable, highly colored materials. The intensity of their colors makes them quite valuable as dyes. Alkyl azo compounds are colorless and much less stable, decomposing homolytically to form hydrocarbons and nitrogen. This propensity toward radical formation has led to the use of aliphatic azo compounds as initiators in radical processes. A classic example of a free radical generator is "Vazo," also called azobisisobutyronitrile or AIBN (p. 74).

$$2CH_3{-}\overset{\overset{\displaystyle O}{\|}}{C}{-}CH_3 + \underset{\text{hydrazine}}{N_2H_4} + 2HCN \rightarrow (CH_3)_2\overset{\overset{\displaystyle CN}{|}}{C}{-}NH{-}NH{-}\overset{\overset{\displaystyle CN}{|}}{C}(CH_3)_2 \qquad \text{synthesis of AIBN}$$

$$\downarrow HgO$$

$$\underset{\text{AIBN}}{(CH_3)_2\overset{\overset{\displaystyle CN}{|}}{C}{-}N{=}N{-}\overset{\overset{\displaystyle CN}{|}}{C}(CH_3)_2}$$

$$(CH_3)_2\underset{\underset{\displaystyle CN}{|}}{C}{-}N{=}N{-}\underset{\underset{\displaystyle CN}{|}}{C}(CH_3)_2 \xrightarrow[100°]{} 2(CH_3)_2\underset{\underset{\displaystyle CN}{|}}{C}\cdot + N_2 \qquad \text{homolysis of AIBN}$$

A scheme that relates the various reduction products of nitrobenzene is presented in Figure 21-1. Use this figure to help put the previous discussion in perspective.

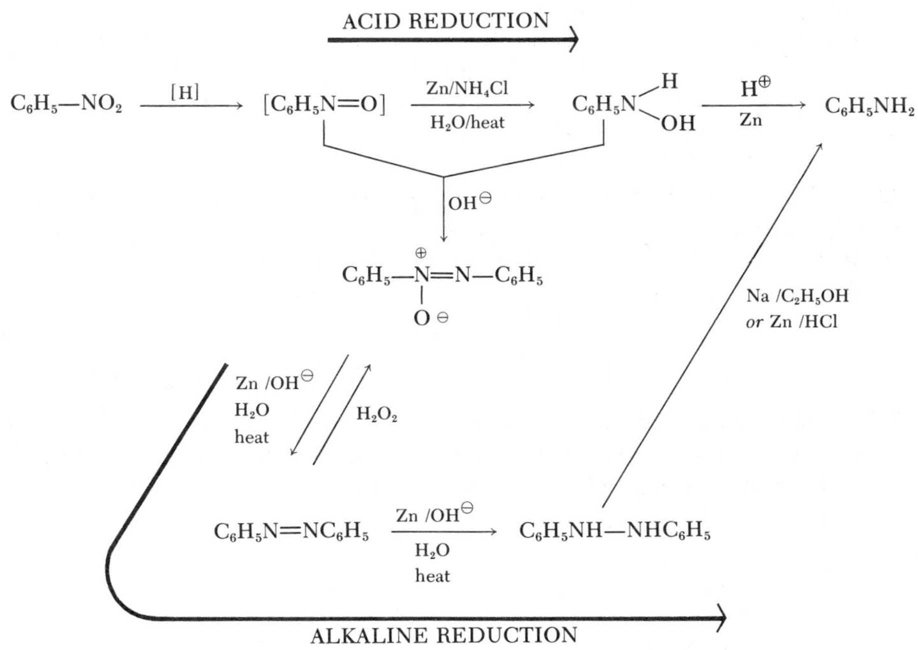

Figure 21-1 The reduction of nitrobenzene.

1. Suggest a synthesis for each of the following, using nitrobenzene and any other necessary reagents.
   (a) $C_6H_5NH_2$ (d) C_6H_5NHOH
   (b) $(C_6H_5NH)_2$ (e) $p\text{-}CH_3C_6H_4NH_2$
   (c) $C_6H_5N{=}NC_6H_5$ (f) $C_6H_5ND_2$
2. What organic compound(s) is (are) produced by the decomposition of Vazo?

21.2 DIAZOTIZATION AND DIAZONIUM SALTS

Primary arylamines such as aniline react with nitrous acid at 0° to 5° to produce solutions of diazonium salts. A reasonable mechanism for this reaction involves the *N*-nitrosation of the 1° amine.

$$H{-}\ddot{O}{-}\ddot{N}{=}\ddot{O} \xrightleftharpoons{H^{\oplus}} H{-}\overset{H}{\overset{|}{O^{\oplus}}}{-}\ddot{N}{=}\ddot{O} \rightarrow H_2O + [\ddot{N}{=}\ddot{O}]^{\oplus}$$

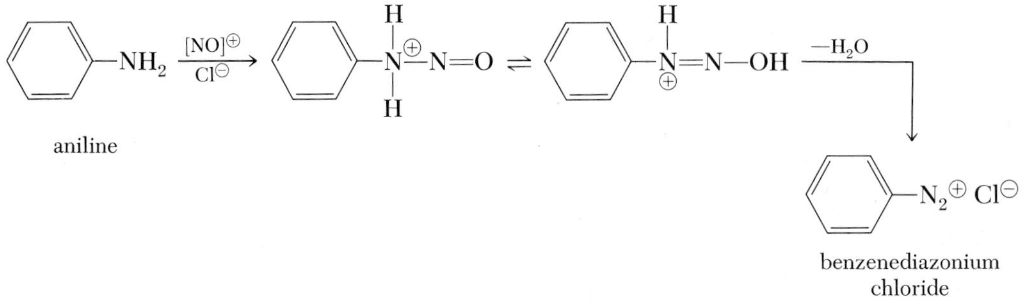

$$C_6H_5{-}NH_2 \xrightarrow[Cl^{\ominus}]{[NO]^{\oplus}} C_6H_5{-}\overset{H}{\underset{H}{N^{\oplus}}}{-}N{=}O \rightleftharpoons C_6H_5{-}\overset{H}{\underset{\oplus}{N}}{=}N{-}OH \xrightarrow{-H_2O} C_6H_5{-}N_2^{\oplus}\ Cl^{\ominus}$$

aniline

benzenediazonium chloride

Diazonium salts are unstable and, with few exceptions, tend to explode when they are dry. If they are required, they can be formed by the reaction of a 1° ammonium salt with an alkyl nitrite, RONO. The diazonium salt then precipitates upon addition of ether to the reaction mixture.

$$C_6H_5\overset{\oplus}{N}H_3Cl^{\ominus} + EtONO \rightarrow C_6H_5\overset{\oplus}{N}_2Cl^{\ominus} + EtOH + H_2O$$

aniline hydrochloride; ethyl nitrite

If the acid used to generate the nitrous acid produces a potent nucleophilic anion, it may react with the diazonium cation to form a covalent diazo compound. This is observed, for example, when HCN replaces HCl or H_2SO_4 in the diazotization mixture.

$$o\text{-}CH_3C_6H_4{-}NH_2 \xrightarrow[H_2O/0°]{NaNO_2/HCN} o\text{-}CH_3C_6H_4{-}N{=}N{-}CN$$

When a cold solution of a diazonium salt is treated with aqueous sodium hydroxide, a series of equilibria containing several new species is established. These are summarized in Figure 21-2. It is noteworthy that alkalinity destroys the electrophilic diazonium ion.

$$Ar{-}N{=}N{-}O^{\ominus}\ K^{\oplus}$$

potassium diazotate (basic)

$$\Big\Updownarrow KOH$$

$$Ar{-}\overset{\oplus}{N}{\equiv}N{:}\ Cl^{\ominus} \xrightarrow{KOH} Ar{-}\overset{\oplus}{N}{\equiv}N\ OH^{\ominus} \rightleftharpoons Ar{-}N{=}N{-}OH$$

diazonium chloride (acidic); diazonium hydroxide pH ≈ 7; diazotic acid (pH ≈ 7)

$$\Updownarrow$$

$$Ar{-}\underset{H}{N}{-}N{=}O$$

N-nitrosamine (pH ≈ 7)

Figure 21-2 The effect of pH upon solutions of diazonium salts.

Schiemann Reaction. The most important synthetic application of diazonium salts involves replacement of the diazo group with one of a number of anions. A straightforward illustration of this is the thermal decomposition of solid benzenediazonium tetrafluoroborate (one of the few insoluble diazonium salts) to produce fluorobenzene. This, the **Schiemann reaction,** is one of the few reactions believed to involve the phenyl cation.

$$C_6H_5NH_2 \xrightarrow[H_2SO_4/0°]{NaNO_2/H_2O} \xrightarrow[BF_3]{HF} C_6H_5N_2^{\oplus}\ BF_4^{\ominus} \xrightarrow[heat]{-N_2} C_6H_5^{\oplus} \xrightarrow{BF_4^{\ominus}} C_6H_5F + BF_3$$

phenyl cation

$$\text{4-bromo-1-naphthylamine} \xrightarrow[H_2SO_4/0°]{NaNO_2/H_2O} \xrightarrow{HBF_4} \text{4-bromo-1-naphthalenediazonium } BF_4^{\ominus} \xrightarrow[heat]{-N_2;\ -BF_3} \text{1-bromo-4-fluoronaphthalene}$$

(95%)

Hydrolysis. Once prepared, solutions of diazonium salts must be used without delay and must be maintained at 0° until their use. The reason for this is their facile hydrolysis to *phenols,* Ar—OH. Indeed, deliberate warming of a diazonium salt solution produces phenols in good yield. Like the Schiemann reaction, this also is believed to involve an aryl cation.

$$C_6H_5NH_2 \xrightarrow[H_2SO_4/0°]{NaNO_2/H_2O} C_6H_5N_2^{\oplus}\ HSO_4^{\ominus} \xrightarrow[50°,\ \text{rapid}]{H_2O} C_6H_5OH + N_2 + H_2SO_4$$

(diazonium salt → phenol: $H_2O/0°$, slow)

3. (a) Suggest a mechanism for the reaction of the phenyl cation with (i) $BF_4^{\ominus}$ and (ii) H_2O. (b) Why might you not expect a reaction between $C_6H_5^{\oplus}$ and CH_4?

Sandmeyer and Gattermann Reactions. The replacement of the $N_2^{\oplus}$ group of aryl diazonium salts with chlorine or bromine is usually accomplished by the **Sandmeyer reaction** or by the related **Gattermann reaction.** In the Sandmeyer procedure the diazonium salt reacts with a cuprous halide in the presence of the corresponding halogen acid; while in the Gattermann procedure the cuprous halide is replaced by metallic copper and halogen acid. The mechanism of these reactions still is under investigation; but the need for Cu(I), when considered along with the fact that copper exists in its compounds with two oxidation states that differ by only one electron [Cu(I) and Cu(II)], suggests that these reactions are free radical in nature.

$$C_6H_5NH_2 \xrightarrow[HX,\ 0°]{NaNO_2} \underset{\text{not isolated}}{C_6H_5N_2^{\oplus}\ X^{\ominus}} \xrightarrow{Cu_2X_2} C_6H_5—X \qquad \text{Sandmeyer reaction}$$

$$C_6H_5NH_2 \xrightarrow[HX,\ 0°]{NaNO_2} \underset{\text{not isolated}}{C_6H_5N_2^{\oplus}\ X^{\ominus}} \xrightarrow[HX]{Cu} C_6H_5—X \qquad \text{Gattermann reaction}$$

Iodination ($C—N_2^{\oplus} \rightarrow C—I$) does not require a cuprous ion, but can be accomplished just through the action of hydriodic acid upon a diazonium salt. There appears to be some question whether the mechanism of iodination is free radical or involves a nucleophilic aromatic substitution (see Section 21.3). In any event, this Sandmeyer-like reaction offers the only convenient route to aryl iodides. Aryl fluorides are prepared by the Schiemann reaction, as described earlier.

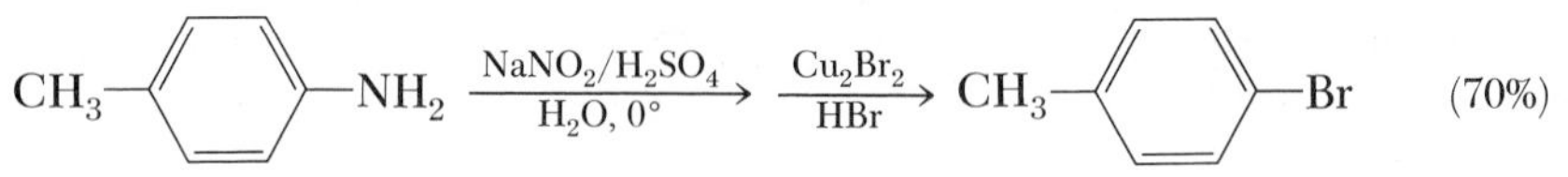

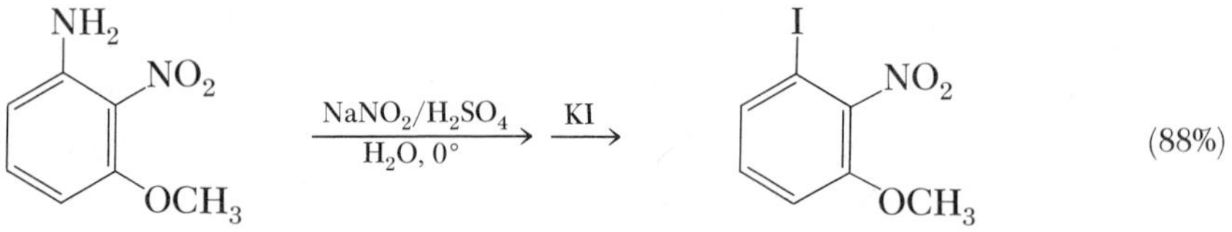

$$CH_3-C_6H_4-NH_2 \xrightarrow[H_2O,\ 0^\circ]{NaNO_2/H_2SO_4} \xrightarrow[HBr]{Cu_2Br_2} CH_3-C_6H_4-Br \quad (70\%)$$

$$\text{3-methoxy-2-nitroaniline} \xrightarrow[H_2O,\ 0^\circ]{NaNO_2/H_2SO_4} \xrightarrow{KI} \text{3-iodo-2-nitromethoxybenzene} \quad (88\%)$$

3-methoxy-2-nitroaniline — 3-iodo-2-nitromethoxybenzene

In addition to halide incorporation, the Sandmeyer procedure is also useful for the synthesis of aryl nitro and cyano compounds. These require the reaction of the diazonium salt with Cu(I) and sodium nitrite or cuprous cyanide, respectively. In the formation of the cyano compound it is important to neutralize the diazonium salt mixture prior to its addition to the cyanide solution in order to prevent the generation of large amounts of hydrogen cyanide.

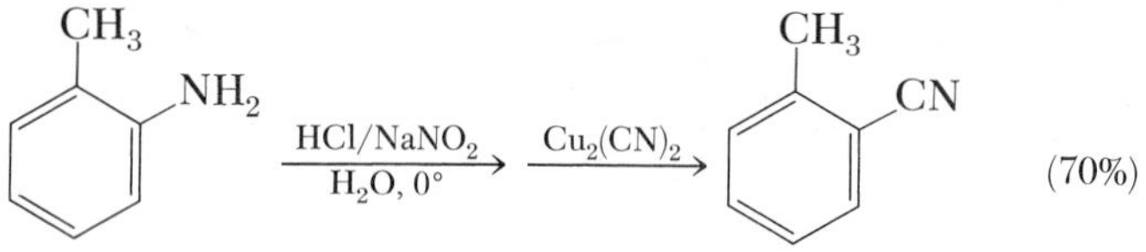

$$o\text{-}CH_3C_6H_4NH_2 \xrightarrow[H_2O,\ 0^\circ]{HCl/NaNO_2} \xrightarrow{Cu_2(CN)_2} o\text{-}CH_3C_6H_4CN \quad (70\%)$$

o-tolunitrile

$$H_2N-C_6H_4-C_6H_4-NH_2 \xrightarrow[0^\circ]{NaNO_2/H_2SO_4} \xrightarrow{KCN} NC-C_6H_4-C_6H_4-CN$$

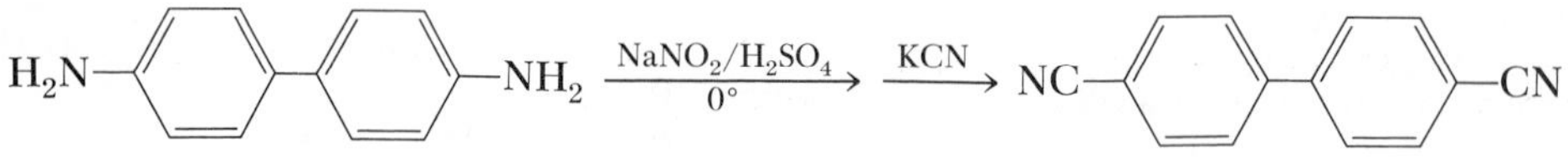

benzidine — 4,4′-dicyanobiphenyl (65%)

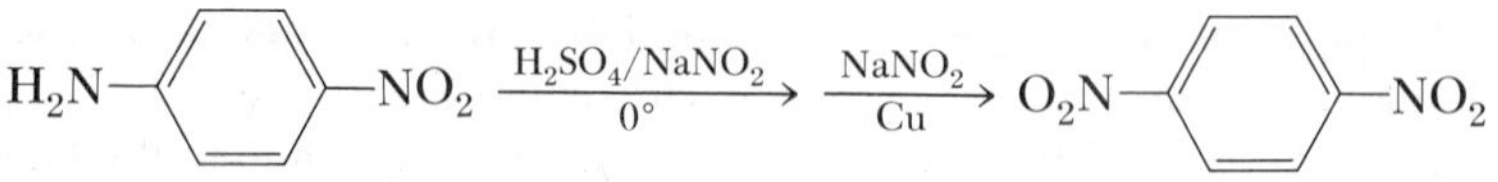

$$H_2N-C_6H_4-NO_2 \xrightarrow[0^\circ]{H_2SO_4/NaNO_2} \xrightarrow[Cu]{NaNO_2} O_2N-C_6H_4-NO_2$$

(70%)

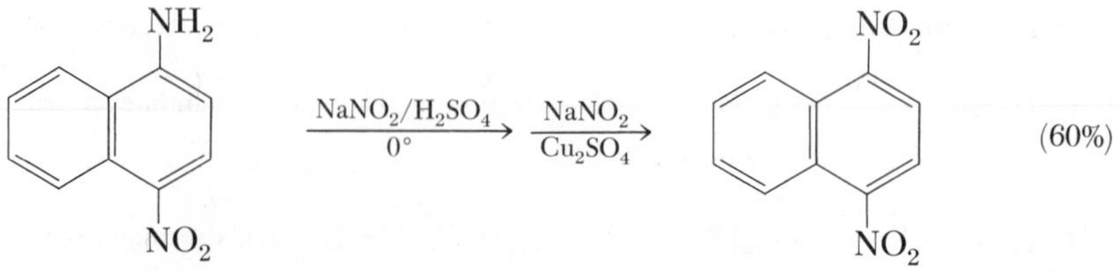

$$\text{1-amino-4-nitronaphthalene} \xrightarrow[0^\circ]{NaNO_2/H_2SO_4} \xrightarrow[Cu_2SO_4]{NaNO_2} \text{1,4-dinitronaphthalene} \quad (60\%)$$

1-amino-4-nitronaphthalene — 1,4-dinitronaphthalene

An interesting extension of the Sandmeyer type of transformation can be used to convert alkylamines to alkyl bromides. In this process a diazonium salt reacts with an amine to form an alkylaminoazobenzene. This, in turn, reacts with hydrogen bromide to produce an aniline and an alkyl bromide.

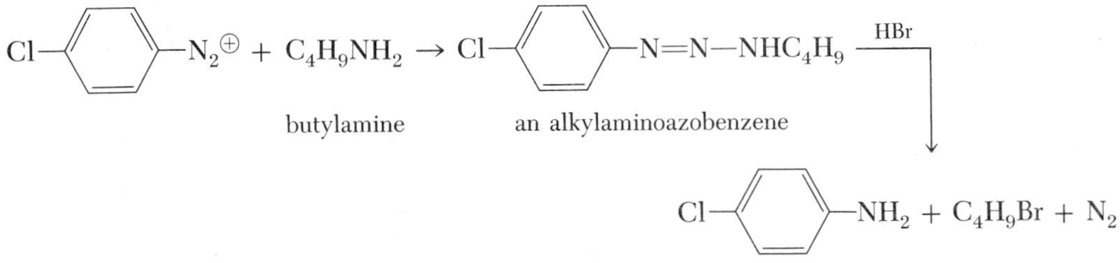

$$\text{Cl—}C_6H_4\text{—}N_2^{\oplus} + \underset{\text{butylamine}}{C_4H_9NH_2} \rightarrow \underset{\text{an alkylaminoazobenzene}}{\text{Cl—}C_6H_4\text{—N=N—NHC}_4H_9} \xrightarrow{\text{HBr}} \text{Cl—}C_6H_4\text{—}NH_2 + C_4H_9Br + N_2$$

4. Suggest a mechanism for the cleavage of alkylaminoazobenzenes with hydrogen bromide. According to your mechanism could aniline be substituted for the butylamine?

Replacement by Hydrogen. The diazo group can be replaced by hydrogen through the action of several agents, including sodium stannite, hypophosphorous acid (H_3PO_2), and ethanol. The mechanism of the reaction of a diazonium acetate with ethanol, which is believed to involve a free radical chain process, is shown below.

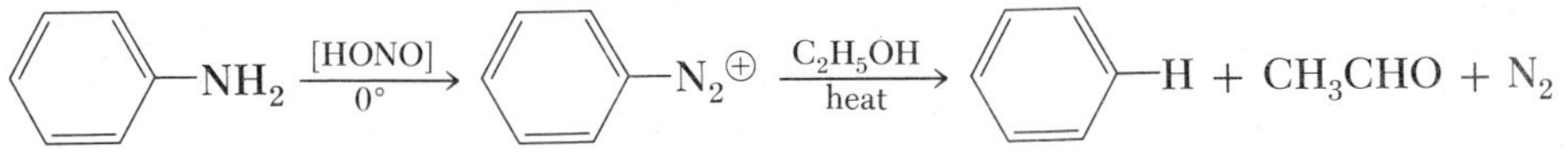

$$C_6H_5\text{—}NH_2 \xrightarrow[0^\circ]{[\text{HONO}]} C_6H_5\text{—}N_2^{\oplus} \xrightarrow[\text{heat}]{C_2H_5OH} C_6H_5\text{—H} + CH_3CHO + N_2$$

$$C_6H_5N_2^{\oplus}\ Cl^{\ominus} + Na^{\oplus}\ OAc^{\ominus} \rightarrow Na^{\oplus}\ Cl^{\ominus} + C_6H_5\text{—N=N—OAc}$$

$$C_6H_5\text{—N=N—OAc} \rightarrow C_6H_5\cdot + N_2 + \cdot OAc \qquad \text{initiation}$$

$$\left.\begin{array}{l} C_6H_5\cdot + HCH_2CH_2OH \rightarrow C_6H_6 + HCH_2\dot{C}HOH \\ C_6H_5N_2^{\oplus} + HCH_2\dot{C}HOH \rightarrow C_6H_5\cdot + HCH_2CHO + H^{\oplus} \end{array}\right\} \text{propagation}$$

5. (a) Suggest an experiment (or observation) that will establish the reduction of diazonium salts with ethanol as a free radical process. (b) Suggest a Lewis structure for H_3PO_2.

The net result of diazotization followed by hydrogen replacement is *deamination* (R—NH_2 R—H). This deamination procedure can be useful in syntheses that require the *o,p*-directing influence of amino or amido groups but that do not require these groups in the final product. A classic problem requiring this sort of solution is the synthesis of a disubstituted benzene bearing two *o,p*-directors *meta* to one another. The synthesis of *m*-chloropropylbenzene is exemplary and, *for the sake of review*, is presented on the following page using benzene as the starting material.

$$C_6H_6 + CH_3CH_2COCl \xrightarrow[CS_2]{AlCl_3} C_6H_5COCH_2CH_3 \xrightarrow[HCl]{Zn\cdot Hg} C_6H_5CH_2CH_2CH_3 \xrightarrow{HNO_3/H_2SO_4} CH_3CH_2CH_2C_6H_4NO_2\ (o,p)$$

$$\xrightarrow[para]{isolate} \xrightarrow[HCl]{Zn/H_2O} \xrightarrow{OH^{\ominus}} p\text{-}CH_3CH_2CH_2C_6H_4NH_2 \xrightarrow{Ac_2O} p\text{-}CH_3CH_2CH_2C_6H_4NHAc$$

$$\xrightarrow[HOAc]{Cl_2} \text{(3-Cl, 4-NHAc)}\ CH_3CH_2CH_2C_6H_3(Cl)NHAc \xrightarrow{H_3O^{\oplus}} CH_3CH_2CH_2C_6H_3(Cl)NH_2 \xrightarrow[0^\circ]{[HONO]} \xrightarrow{H_3PO_2} m\text{-}CH_3CH_2CH_2C_6H_4Cl$$

Reduction to Hydrazines. Aryl hydrazines, Ar—$NHNH_2$, are produced when diazonium salts are reduced with stannous chloride or sodium sulfite. However, since stannous chloride reduces nitro groups to azo groups, nitrophenyl diazonium salts must be reduced with sodium sulfite. It was the discovery (1875) of phenylhydrazine that led Emil Fischer to his work on the structure of sugars (see Chapter 25). Nitro substituted phenylhydrazines are reagents used to prepare solid derivatives of carbonyl compounds, the best known being 2,4-dinitrophenylhydrazine (p. 483).

$$Ar—\overset{\oplus}{N}\equiv N\ X^{\ominus} \xrightarrow[Na_2SO_3]{SnCl_2\ or} Ar—NHNH_2$$

$$O_2N—C_6H_4—\overset{\oplus}{N}_2\ HSO_4^{\ominus} \xrightarrow[H_2O]{Na_2SO_3} \xrightarrow{OH^{\ominus}} O_2N—C_6H_4—NHNH_2$$

Diazo Coupling. Highly activating substituents (*e.g.*, —$O^{\ominus}$, —$N(CH_3)_2$, and —OH) on a benzene ring permit it to be attacked by a diazonium salt, thus producing an azo compound (an electrophilic aromatic substitution). This reaction is termed a **diazo coupling.** The compound called Butter Yellow is produced in this manner. Once used as a coloring agent for oleomargarine, Butter Yellow's importance in food has diminished since it was discovered to be a carcinogen.

$$C_6H_5N_2^{\oplus}Cl^{\ominus} + C_6H_5—N(CH_3)_2 \rightarrow [\text{H}(N_2C_6H_5)C_6H_4^{\oplus}—N(CH_3)_2] \rightarrow C_6H_5—N{=}N—C_6H_4—N(CH_3)_2$$

Butter Yellow

If the diazotization of an amine is carried out in an insufficient amount of acid, the diazonium ion will react with the free amine to form a diazoamino compound. It is to prevent this that diazotizations are carried out in strongly acidic media since, in strong acids, the concentration of free amine will be negligible. The same result can be obtained by adding an excess of amine to a solution of diazonium salt.

$$C_6H_5N_2^{\oplus}X^{\ominus} + C_6H_5NH_2 \rightarrow \underset{\text{diazoaminobenzene}}{C_6H_5N{=}N{-}NHC_6H_5}$$

In the presence of acid, usually provided by addition of an amine hydrochloride, this condensation is reversed and a substituted azobenzene results from a coupling reaction.

$$C_6H_5N{=}N{-}NHC_6H_5 \xrightleftharpoons{C_6H_5\overset{\oplus}{N}H_3\ Cl^{\ominus}} C_6H_5{-}N{=}N{-}\overset{\oplus}{N}H_2C_6H_5$$

$$\rightleftharpoons C_6H_5N_2^{\oplus} + C_6H_5NH_2$$

$$\rightleftharpoons H_2N{-}C_6H_4{-}N{=}N{-}C_6H_5$$

p-aminoazobenzene

If the hydrochloride of a different amine is used, the final product results from attack of $ArN_2^{\oplus}$ upon the more reactive amine.

$$C_6H_5N{=}N{-}NHC_6H_5 \xrightarrow{C_6H_5\overset{\oplus}{N}H(CH_3)_2\ Cl^{\ominus}} (CH_3)_2N{-}C_6H_4{-}N{=}N{-}C_6H_5 + C_6H_5\overset{\oplus}{N}H_3\ Cl^{\ominus}$$

p-(dimethylamino)azobenzene

6. The reaction of $C_6H_5N_2^{\oplus}Cl^{\ominus}$ with $C_6D_5NH_2$ produces four isomers. What are they?

21.3 NUCLEOPHILIC AROMATIC SUBSTITUTION

While 2° alkyl halides can undergo substitution by an S_N2 process, there is no evidence that simple aryl halides react with nucleophiles under moderate reaction conditions. Direct displacement reactions on aryl substrates are extremely rare because of the difficulty in attaining the trigonal bipyramidal transition state *via* the backside attack characteristic of the S_N2 process.

$$I^{\ominus} + (H_3C)_2CHCl \rightleftharpoons \left[I{-}{-}{-}\underset{CH_3\ CH_3}{\overset{H}{C}}{-}{-}{-}Cl\right]^{\ominus} \rightleftharpoons I{-}CH(CH_3)_2 + Cl^{\ominus}$$

low energy

$I^{\ominus}$ + Cl ⇄ [I Cl]$^{\ominus}$ direct displacement very rare in aryl halides

very high energy

Substitution of nitro or nitroso groups *para* and *ortho* to the intended leaving group enhances the displacement rate sufficiently that reaction occurs at a reasonable rate even at room temperature. The reaction is kinetically second order overall, first order in aryl halide and first order in nucleophile.

Cl, NO_2, NO_2 + $CH_3O^{\ominus}$ $\xrightarrow{25°}$ OCH_3, NO_2, NO_2 + $Cl^{\ominus}$

$$R = k[\text{Ar—X}][\text{Nu}^{\ominus}]$$

These data are taken to indicate that the reaction proceeds by attack at the carbon bearing the leaving group to form a resonance-stabilized anionic *intermediate.* (Remember that there is no discrete intermediate in the S_N2 reaction.)

L + $\overset{\ominus}{\text{Nu}}$: → {L Nu ↔ L Nu ↔ L Nu ↕ L Nu} $\xrightarrow{-L^{\ominus}}$ Nu

The intermediate *benzenanion* is the anionic analog of the benzenonium ion and may be similarly represented.

L, NO_2 + $\overset{\ominus}{\text{Nu}}$: → L Nu, NO_2 ↔ L Nu

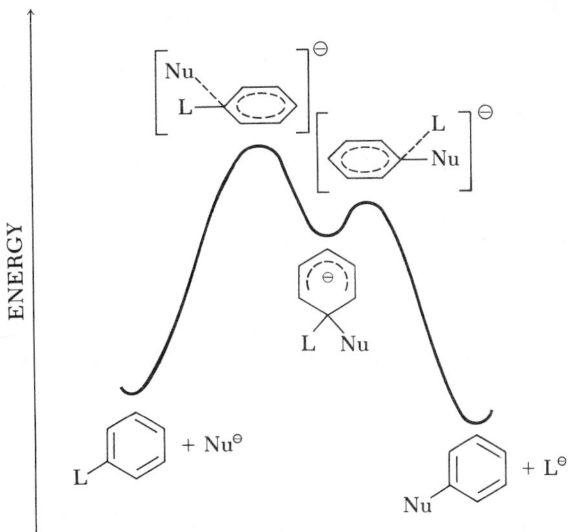

Figure 21-3 Energy profile of an idealized nucleophilic aromatic substitution. $\overset{\ominus}{\text{Nu}}$ is the attacking nucleophile and L is the leaving group.

The energy profile for an idealized nucleophilic aromatic substitution is presented in Figure 21–3. As with most electrophilic aromatic substitutions, the destruction of the aromatic π system is considered to be rate-determining. This mechanism has received support from the isolation of highly stabilized benzenanions, often called *Meisenheimer complexes*. These complexes can, after isolation, proceed on to the substitution products when re-subjected to the reaction conditions.

OC_2H_5, O_2N, NO_2, NO_2 + $K^{\oplus}$ $CH_3O^{\ominus}$ $\rightleftarrows$ $K^{\oplus}$ [H_3CO OC_2H_5, O_2N, NO_2, NO_2] $\rightleftarrows$

Meisenheimer complex
(red solid)

OCH_3, O_2N, NO_2, NO_2 + $K^{\oplus}$ $C_2H_5O^{\ominus}$

Nitro groups *meta* to the proposed leaving group have only a minor effect upon the reaction rate because of the impossibility of resonance delocalization of the negative charge onto these nitro groups.

7. What is the hybridization of each carbon atom in the following species?

CH_3O OCH_3, O_2N, NO_2, NO_2

The Nature of the Leaving Group. Since the bond between the benzene ring and the leaving group is not ruptured in the rate-determining step, all of the aryl halides react at about the same rate. This is quite different from S_N2 displacements, in which an alkyl iodide may be 10^6 times as reactive as the corresponding fluoride. Indeed, aryl fluorides are slightly ($\sim$10 times) *more* reactive than are the corresponding aryl iodides. This is presumed to be due to the greater inductive stabilization of the negative charge in the transition state by the more electronegative fluorine. While iodine is a better leaving group than is fluorine, even in aromatic substitutions, departure of the anion occurs after the rate-determining step and does not influence the rate of displacement (Figure 21–4).

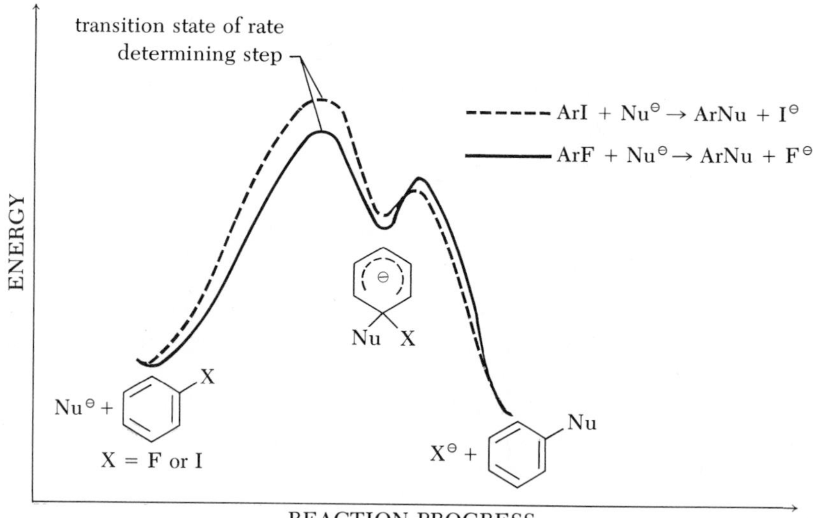

Figure 21-4 Effect of changing the leaving group on the rate of nucleophilic aromatic substitution.

21.4 DEHYDROBENZENE

While it is known that aryl halides do not normally undergo facile substitution, there are a group of reactions in which "displacement" is *remarkably* facile. For example, while chlorobenzene does not react with a boiling solution of aqueous sodium hydroxide, it is converted to aniline by potassium amide in liquid ammonia.

$$C_6H_5\text{—}Cl \xrightarrow[H_2O,\ 100^\circ]{NaOH} \text{N.R.}$$

$$C_6H_5\text{—}Cl \xrightarrow[NH_3(l)]{K^{\oplus}\ NH_2^{\ominus}} C_6H_5NH_2 + KCl$$

In 1953, J.D. Roberts carried out a beautiful experiment wherein he examined the reaction of 1-^{14}C-chlorobenzene with potassium amide in liquid ammonia. Degradation of the resultant aniline showed the product to be a 1:1 mixture of two anilines, 1-^{14}C-aniline and 2-^{14}C-aniline.

Cl — $\xrightarrow[NH_3(l)]{NH_2^{\ominus}}$ — NH_2 ($\sim$50%) + NH_2 ($\sim$50%)

$\bullet \equiv {}^{14}C$

This experiment, the capstone to a suggestion made several decades earlier, established the existence of a symmetric (neglecting the isotopic label) intermediate in this ammonolysis. This intermediate, pictured below, was termed *benzyne* by Roberts.

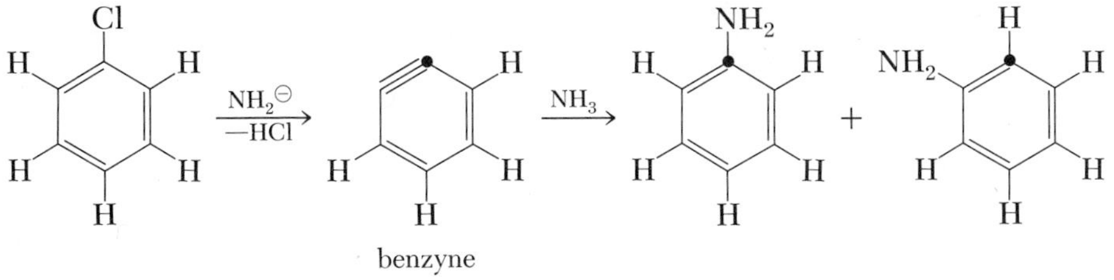

$$C_6H_5Cl \xrightarrow[-HCl]{NH_2^{\ominus}} \text{benzyne} \xrightarrow{NH_3} C_6H_5NH_2 + C_6H_5NH_2$$

benzyne

Although this term and the more general expression *aryne* are in common use, they are somewhat misleading. In one Kekulé form the generation of the "extra" bond leads to three bonds between two carbons, hence the *yne* suffix. However, in the other Kekulé form the "extra" bond generates a highly cumulated structure. Since this "extra" bond does not interact with the aromatic sextet (see below), it is hard to justify any preference in descriptive names. For this reason, and since the formula for benzyne is C_6H_4, this species is often called *dehydrobenzene.*

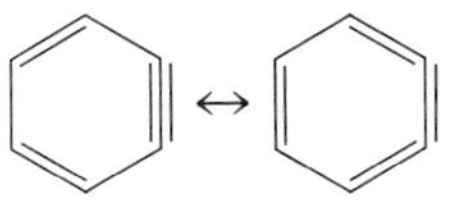

benzyne contributing structures

The C—H and C—Cl bonds involved in the generation of dehydrobenzene are perpendicular to the aryl π electron system (Figure 21–5). After HX elimination has occurred, the dehydrobenzene has its extra bond similarly oriented, *i.e.*, perpendicular to the aromatic sextet. Because of this orthogonality, there must be little, if any, interaction between this bond and the π system. To the first approximation, then, dehydrobenzene is aromatic. However, the presence of a long bond formed by the less-than-ideal overlap of two $2sp^2$ orbitals makes dehydrobenzene extremely reactive.

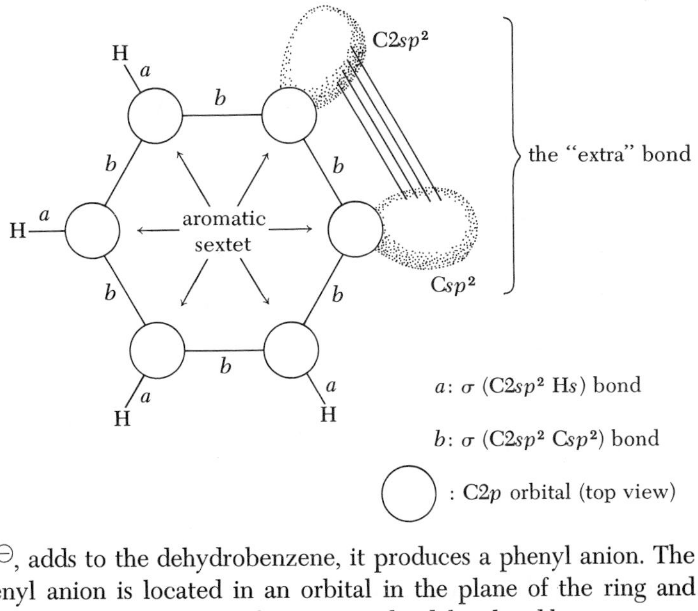

H a b b H a b b b a H a H C2sp^2 the "extra" bond Csp^2 aromatic sextet

a: σ (C2sp^2 Hs) bond

b: σ (C2sp^2 Csp^2) bond

: C2p orbital (top view)

Figure 21–5 Dehydrobenzene.

When a nucleophile, $Nu^{\ominus}$, adds to the dehydrobenzene, it produces a phenyl anion. The lone electron pair of the phenyl anion is located in an orbital in the plane of the ring and perpendicular to the aryl π system. The charge, therefore, cannot be delocalized by resonance.

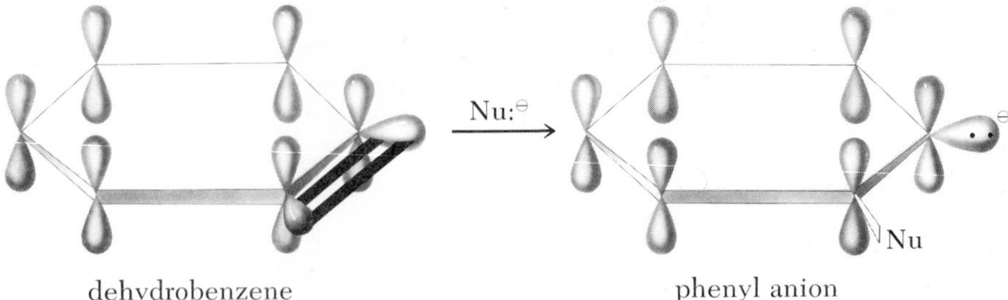

8. Why isn't the phenyl anion, $C_6H_5^{\ominus}$, anti-aromatic?
9. How is it that dehydrobenzene can be both aromatic and extremely reactive?

Because the electron pair of a phenyl anion is localized, the only influence that a substituent on the ring can exert in determining which carbon is attacked by the nucleophile will be an inductive effect. This explains why *p*-bromoanisole reacts with amide ion and ammonia to produce more *p*- than *m*-aminoanisole.

OCH_3 (Br) $\xrightarrow[NH_3]{NH_2^{\ominus}}$ OCH_3 (NH_2) > OCH_3 (NH_2)

OCH_3 (Br) $\xrightarrow{NH_2^{\ominus}}$ OCH_3

$\xrightarrow{:NH_3}$ OCH_3, $\overset{\oplus}{N}H_3$, $:\ominus$ — negative charge closer to electron-withdrawing —OCH_3

$\xrightarrow{:NH_3}$ OCH_3, $\overset{\oplus}{N}H_3$, $\ominus:$ — negative charge farther from electron-withdrawing —OCH_3

With nucleophiles that are weaker bases than amide ion, benzyne formation must compete with nucleophilic aromatic substitution. Under these circumstances, lower temperatures favor the nucleophilic displacement.

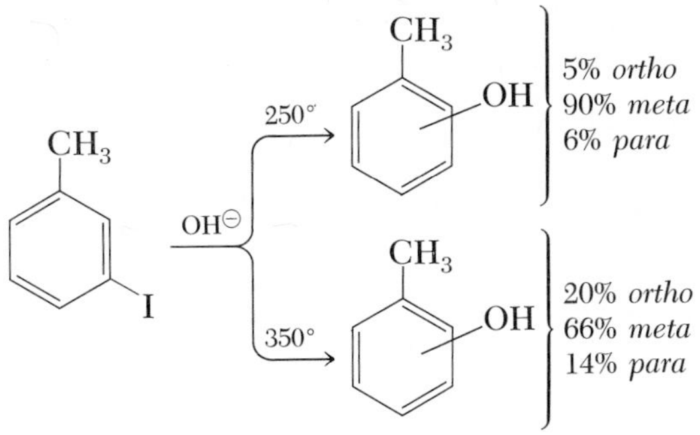

250° → 5% *ortho*, 90% *meta*, 6% *para*

$OH^{\ominus}$

350° → 20% *ortho*, 66% *meta*, 14% *para*

Dehydrobenzene and similar compounds are potent dienophiles. For such Diels-Alder reactions, benzenediazonium carboxylate serves as a useful dehydrobenzene precursor.

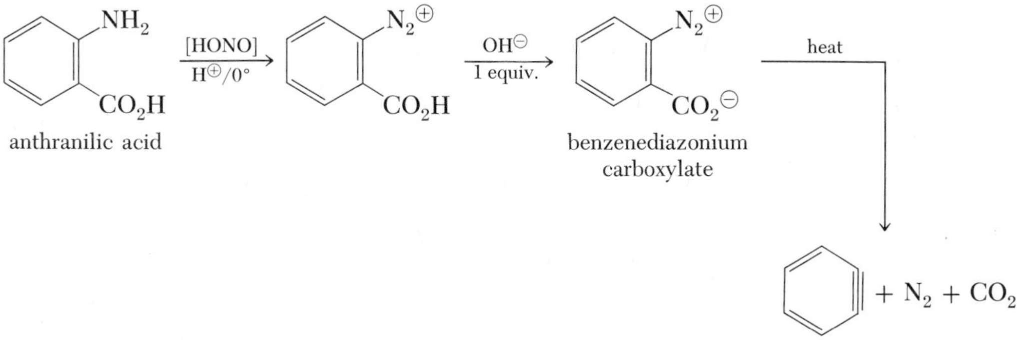

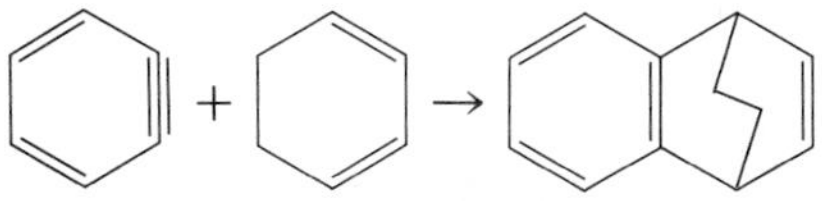

[HONO] $H^{\oplus}/0°$

$OH^{\ominus}$ 1 equiv.

heat

anthranilic acid

benzenediazonium carboxylate

$+ N_2 + CO_2$

a Diels-Alder reaction of dehydrobenzene

If dehydrobenzene is generated in the presence of *N*-methylpyrrole, a Diels-Alder reaction ensues. The first-formed adduct isomerizes in acid to a naphthylamine derivate, a reaction that is commonly observed in compounds possessing this type of bridged structure.

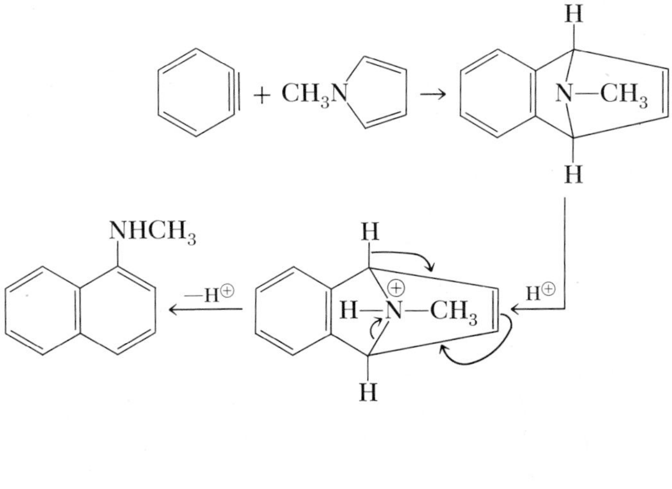

$+ CH_3N$ →

$-H^{\oplus}$ ← $H^{\oplus}$

$NHCH_3$

Arynes can also be formed by the elimination of two adjacent halogens with the aid of an active metal such as magnesium. In the example below, dehydrobenzene is used to prepare the hydrocarbon called triptycene.

Br, F + Mg $\xrightarrow{THF}$ [dehydrobenzene] $\xrightarrow[\text{Diels-Alder reaction}]{\text{anthracene}}$ triptycene

triptycene

10. 2-Methylanthracene reacts with 2-bromo-3-fluorotoluene, in the presence of magnesium, to form two compounds, both of which are capable of enantiomerism. How many products capable of enantiomerism are formed from the reaction of 2-methylanthracene with 3-bromo-4-fluorotoluene in the presence of magnesium? Are any products formed that are incapable of enantiomerism? Explain.

IMPORTANT TERMS

Aryne: A general term used to describe a highly reactive aromatic compound that, in one canonical form, contains a triple bond. The simplest aryne is dehydrobenzene (benzyne). A *heteroaryne* is an aryne derived from a heterocyclic system. *Pyridyne* is an aryne.

3-chloropyridine $\xrightarrow[NH_3(l)]{NH_2^\ominus}$ [a pyridyne] $\xrightarrow{NH_3}$ 3-aminopyridine (25%) + 4-aminopyridine (45%)

a pyridyne (25%) (45%)

Benzenanion: The anionic analog of a benzenonium ion. The product of addition of an anion to the π system of benzene or one of its derivatives. The benzenanion is not aromatic, although it has a highly delocalized negative charge. Stable benzenanions contain potent electron-withdrawing groups on the benzene ring, and sometimes are termed Meisenheimer complexes.

Benzyne: An alternate name for dehydrobenzene.

Deamination: The loss of an amino group ($-NH_2$) and concomitant conversion of a $C-NH_2$ moiety into a C—H moiety.

Dehydrobenzene: The specific aryne derived from benzene. It is less desirable to call it benzyne, since it contains a formal triple bond (hence the ending *yne*) in only one contributing structure.

fluorobenzene $\xrightarrow[-HF]{C_6H_5Li}$ [dehydrobenzene ↔ dehydrobenzene] $\xrightarrow{C_6H_5Li}$ 2-lithiobiphenyl $\xrightarrow{H_2O}$ biphenyl + LiOH

dehydrobenzene

biphenyl

Other contributing structures, whose importance remains a moot point, possess charge separation.

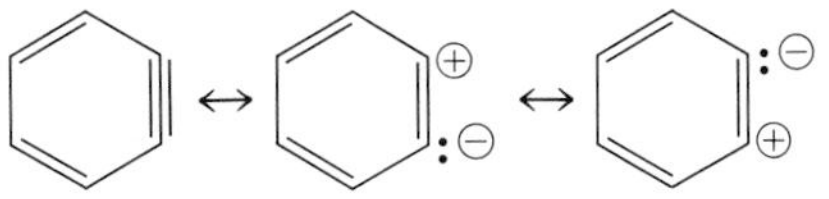

Diazo coupling: The action of an aryl diazonium salt upon an aromatic ring that is activated by either a hydroxy or an amino group, to form an azo compound (also called a *diazene*).

Meisenheimer complex: A stable benzenanion.

Nucleophilic aromatic substitution: Replacement of a leaving group (*e.g.*, halogen) on an aromatic ring by the action of a nucleophile. The two important nucleophilic aromatic substitution reactions discussed in this chapter are (a) the *addition-elimination process* (involving a benzenanion) and (b) the *elimination-addition process* (involving an aryne). The addition-elimination process sometimes is called the S_NAr mechanism (Substitution-*N*ucleophilic-*Ar*omatic).

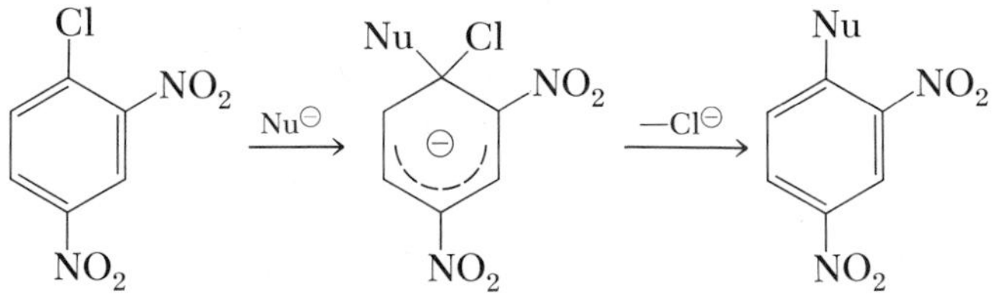

substitution by addition-elimination

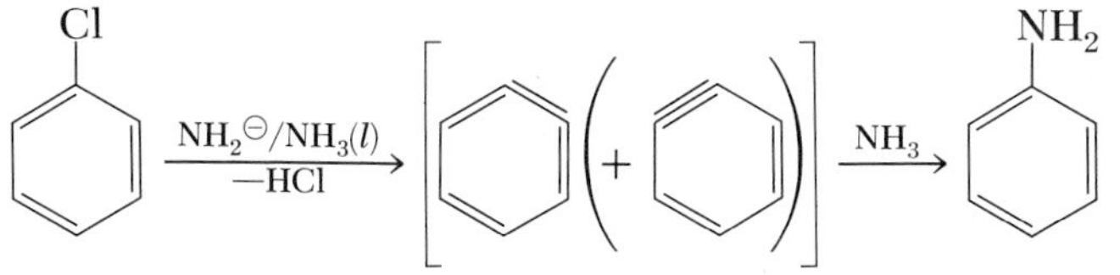

substitution by elimination-addition—the Lewis base acts first as a base and then as a nucleophile

PROBLEMS

11. Suggest a reagent or sequence of reagents to achieve the following interconversions.

(a) $C_6H_5NHOH \rightarrow C_6H_5-\overset{\oplus}{N}(O^{\ominus})=N-C_6H_4CH_3$ (o-tolyl)

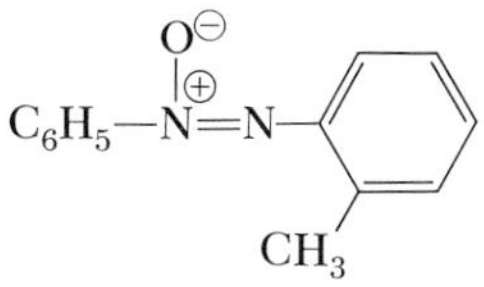

(b) $C_6H_5NHOH \rightarrow C_6H_5N{=}O$

(c) $C_6H_5NO_2 \rightarrow$ (3-ClC_6H_4)$-\overset{\oplus}{N}(O^{\ominus})=N-$(3-$ClC_6H_4$)

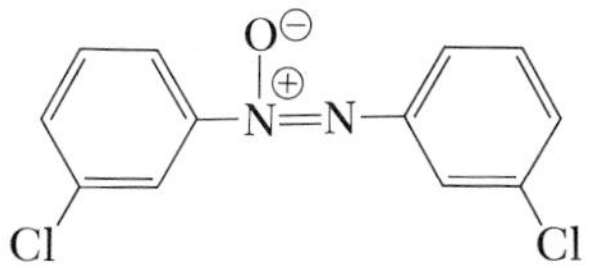

(d) $C_6H_5NO_2 \rightarrow C_6H_5NH_2$

(e) $C_6H_5N{=}O \rightarrow$ CH_3 —N=N—C_6H_5 (*cis*)

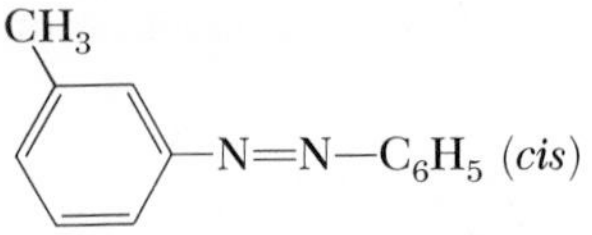

(f) $C_6H_5N{=}O \rightarrow$ CH_3 —N=N—C_6H_5 (*trans*)

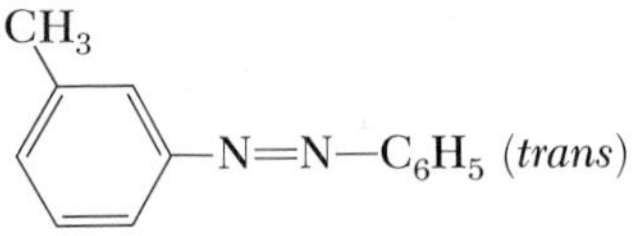

(g) $C_6H_5N{=}NC_6H_5 \rightarrow C_6H_5NHNHC_6H_5$
(h) $C_6H_5NHNHC_6H_5 \rightarrow C_6H_5NH_2$
(i) $C_6H_5NHNHC_6H_5 \rightarrow C_6H_5N{=}NC_6H_5$
(j) $C_6H_5NH_2 \rightarrow C_6H_5N{=}NCN$
(k) $C_6H_5NH_2 \rightarrow C_6H_5NHNH_2$

12. Starting with either toluene or nitrobenzene, and using any other needed reagents, suggest rational syntheses for the following.

(a) C_6H_5CN

(b) $C_6H_5CH_2NH_2$

(c) $C_6H_5COCH_3$

(d) 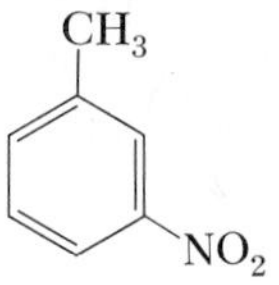 CH_3, NO_2

(e) 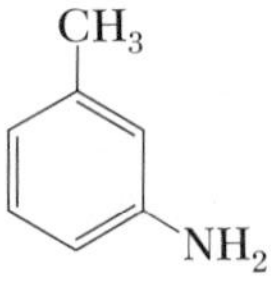 CH_3, NH_2

(f) 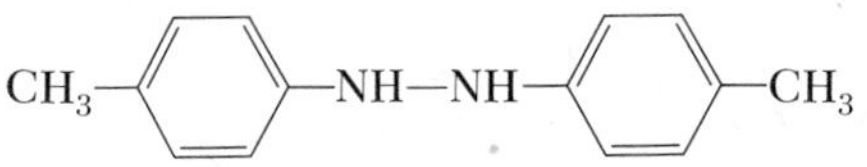 CH_3— —NH—NH— —CH_3

(g) H_2N— —NH_2

(h)

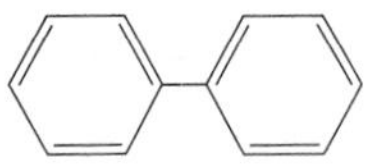

(i) 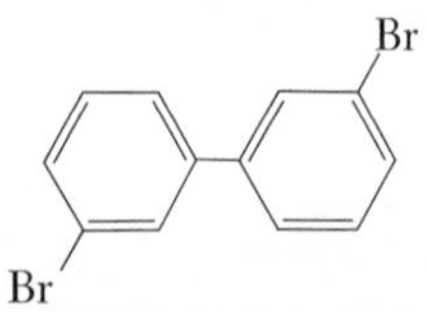 Br, Br

(j) 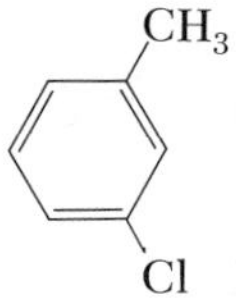 CH_3, Cl

(k) 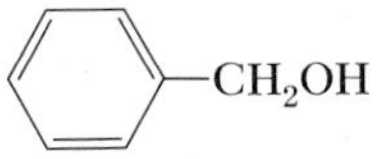 —CH_2OH

(l) 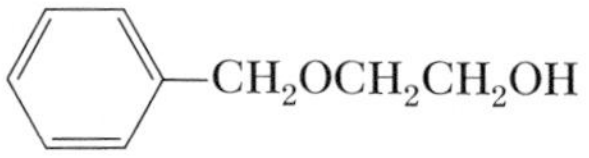 —$CH_2OCH_2CH_2OH$

(m) 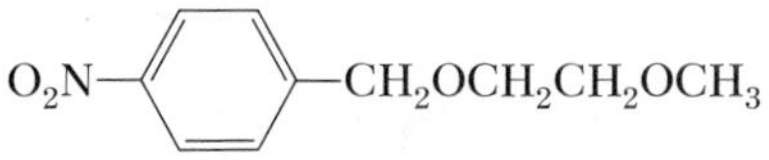 O_2N— —$CH_2OCH_2CH_2OCH_3$

(n) Br

(o) Br, Br

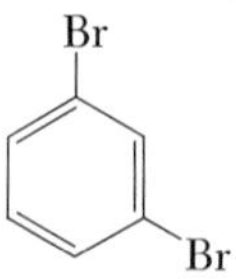

(p) 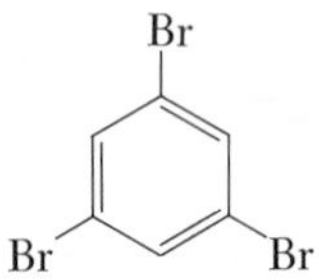 Br, Br, Br

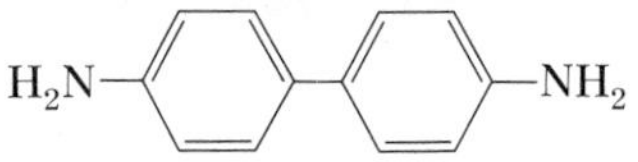

(q) C_6H_5D

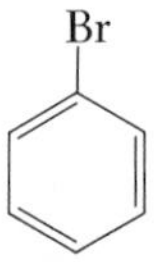

(r) 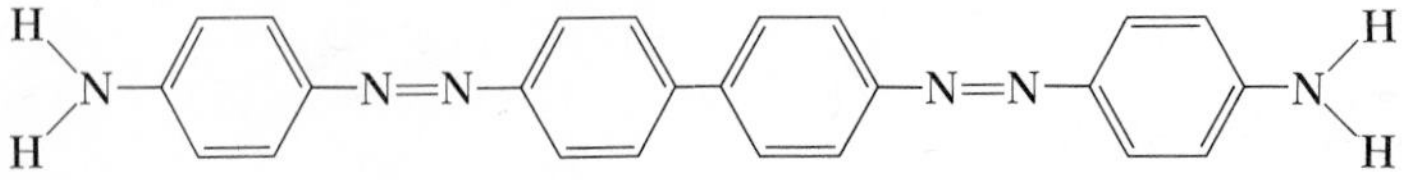 H, H N— —N=N— —N=N— —N H, H

13. Starting with benzene or toluene, and using any other needed reagents, suggest a synthesis for each of the following. Use a diazonium reaction where possible.

(a) *o*-chlorobenzoic acid
(b) 2-chloro-5-fluorobenzoic acid
(c) 1,3-benzenedicarboxylic acid
(d) *p*-fluoroaniline
(e) *o*-chlorophenol
(f) *m*-chlorophenol
(g) 4,4′-dinitrobiphenyl, $(4\text{-}O_2NC_6H_4)_2$

14. Aliphatic nitroso compounds of the type R_2CHNO tend to isomerize to a rather common type of compound, one often used in quantitative organic analysis. Suggest a structure for this isomer and a reason that this isomerization does not occur with nitrosobenzene.

15. Benzenediazonium bisulfate is hydrolyzed more rapidly than is *p*-nitrobenzenediazonium bisulfate. Account for this result.

16. The *Gomberg-Bachman reaction* involves the conversion of an aryl diazonium salt to a biaryl in alkaline solution; an example is presented below. Yields in this reaction may be as high as 60%. Bearing in mind that the reaction between benzenediazonium bisulfate and toluene produces nearly equal amounts of **A** and **B**, suggest a mechanism for the Gomberg-Bachman reaction.

$$CH_3\text{—}C_6H_5 + C_6H_5N_2^{\oplus} \xrightarrow[\text{heat}]{OH^{\ominus}/H_2O} CH_3\text{—}C_6H_4\text{—}C_6H_5 \ (\mathbf{A}) + CH_3\text{—}C_6H_4\text{—}C_6H_5 \ (\mathbf{B}) + CH_3\text{—}C_6H_4\text{—}C_6H_5$$

A B

17. Using benzene as the only organic starting material, and with the information produced by answering problem 16, suggest a synthesis of *p*-terphenyl.

p-terphenyl

18. Assign structures to compounds **A**, **B**, **C**, and **D**. Provide mechanisms for the steps producing **A**, **B**, and **C**.

$$C_6H_5C(=O)Cl + C_6H_6 \xrightarrow[CS_2]{AlCl_3} \mathbf{A}(C_{13}H_{10}O) \xrightarrow{C_6H_5MgCl} \xrightarrow{H_2O} \mathbf{B}(C_{19}H_{16}O)$$

$$\mathbf{B} + \text{aniline} \xrightarrow[(b)\ H_2O]{(a)\ HCl} \mathbf{C}(C_{25}H_{21}N) \xrightarrow[H_3PO_2]{NaNO_2} \mathbf{D}(C_{25}H_{20})$$

19. Aryl diazonium salts react with cuprous thiocyanate ($Cu_2(SCN)_2$) to give aryl thiocyanates, Ar—SCN, rather than aryl isothiocyanates, Ar—NCS. Draw the contributing structures for the thiocyanate anion and for both possible products. Assuming that the reaction is ionic rather than free radical, account for this result. What would you expect to be the product of the complete hydrolysis of C_6H_5—SCN in aqueous acid?

20. When *o*-fluorobromobenzene reacts with magnesium in the presence of 1,3-butadiene, the product, **A**, is found to react with Br_2/CCl_4 to produce **B**, $C_{10}H_{10}Br_2$. **A** reacts with refluxing alkaline $KMnO_4$ to produce a dicarboxylic acid, **C**, $C_8H_6O_4$. Account for the formation of **A** and provide structures for compounds **A**, **B**, and **C**.

21. While benzenediazonium chloride couples readily with *N,N*-dimethylaniline, it does not couple with *N,N*-2,6-tetramethylaniline. Account for this difference.

22. When the diazonium salt of anthranilic acid and one equivalent of iodine are refluxed in chloroform, *o*-diiodobenzene is produced in 65% yield. Suggest a mechanism for this reaction.

23. Benzenediazonium chloride reacts with 2,4-pentanedione to form a phenylhydrazone, **A**, $C_{11}H_{12}O_2N_2$. This phenylhydrazone is a rearrangement product of an azo compound. Suggest a structure for **A** and account for its formation.
24. The two compounds shown below are both azo dyes that are useful as *p*H indicators. Each can be made by two alternative coupling procedures. What are they? Discuss the practical utility of each approach.

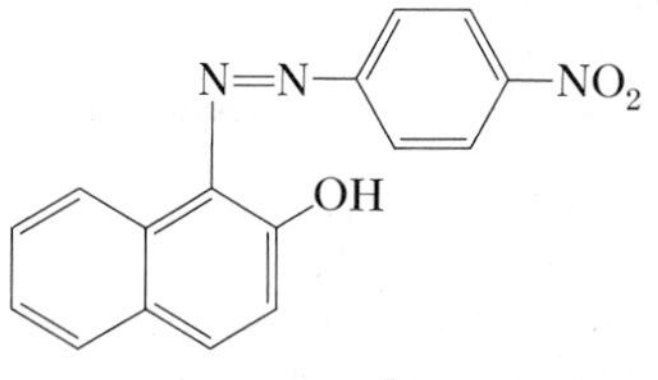

N=N—C_6H_4—NO_2 (2-hydroxynaphthyl)

para red

$Na^{\oplus}\ \overset{\ominus}{O}_3S$—$C_6H_4$—N=N—$C_6H_4$—$N(CH_3)_2$

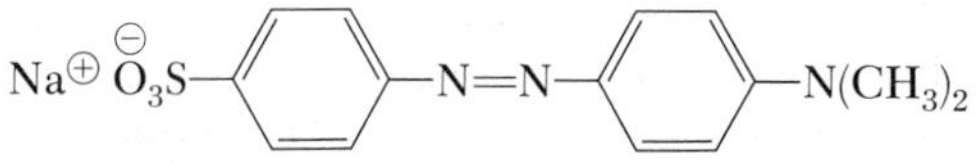

methyl orange

25. Detailed studies of the effect of *p*H upon the rate of coupling between diazonium salts and (a) phenols and (b) amines indicated that amines couple most rapidly in mildly acidic solutions, while phenols couple most rapidly in mildly alkaline solutions. Account for this difference.
26. Benzenediazonium chloride couples with phenol but not with anisole. However, 2,4-dinitrobenzenediazonium chloride does couple with anisole. Account for this difference.
27. Provide mechanisms for the following reactions.

(a) m-Cl—C_6H_4—$(CH_2)_3NHCH_3$ $\xrightarrow{NH_2^{\ominus}}$ N-methyl-1,2,3,4-tetrahydroquinoline (H_2C, CH_2, CH_2, N—CH_3)

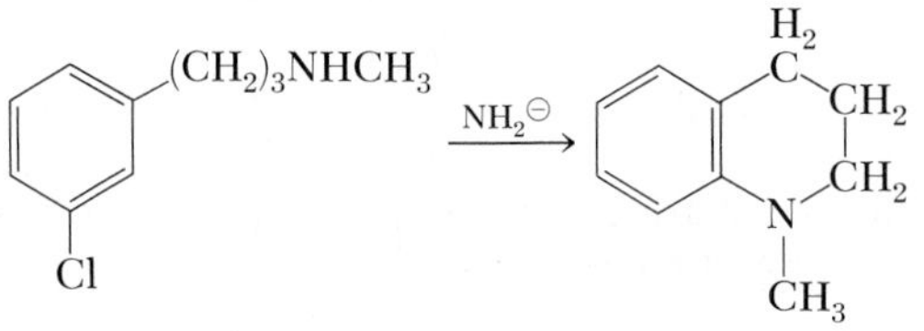

(b) 2 (benzothiadiazole dioxide, $\oplus 2$ S with two $O^{\ominus}$) $\xrightarrow{\text{heat}}$ biphenylene $+ 2N_2 + 2SO_2$

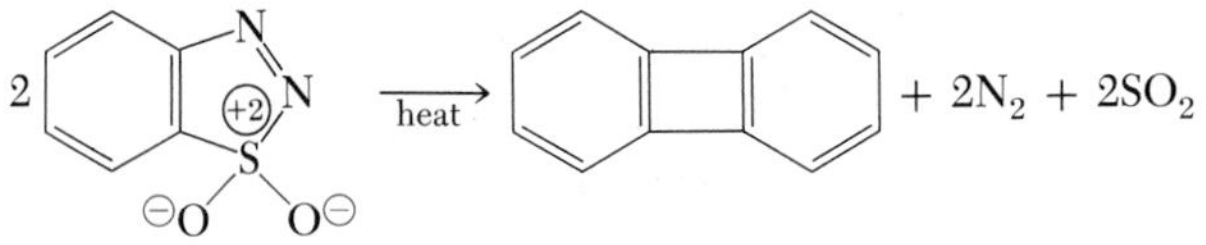

biphenylene

(c) $[C_6H_5—N{\equiv}^{15}N]^{\oplus} \rightarrow [C_6H_5—^{15}N{\equiv}N]^{\oplus}$

(d) 2 o-CO_2H—C_6H_4—$N_2^{\oplus}$ + 2Cu(I) $\xrightarrow{NH_3}$ (CO_2H-C_6H_4—C_6H_4-CO_2H) + 2Cu(II) + $2N_2$

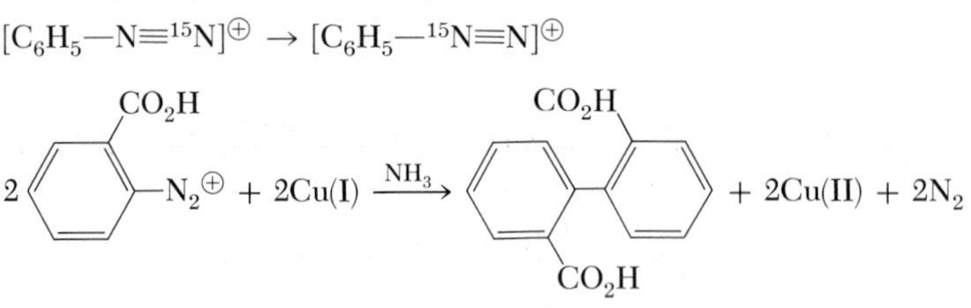

diphenic acid

(e) $C_6H_5N_2^{\oplus}\ BF_4^{\ominus} + C_6H_5Br \xrightarrow[\text{heat}]{C_6H_5Br(\text{solv.})}$ C_6H_5—$\overset{\oplus}{Br}$—C_6H_5 $BF_4^{\ominus}$

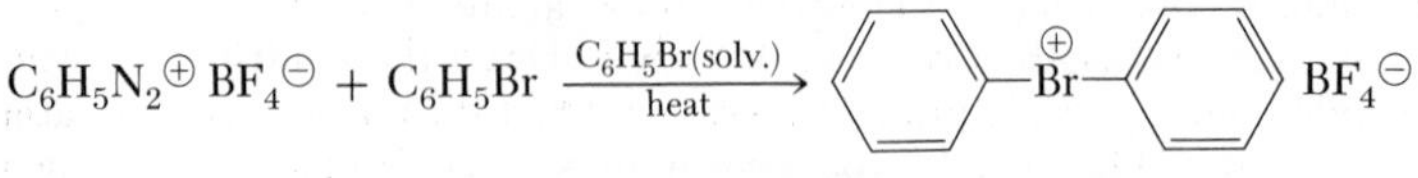

(trace)

(f) C_6H_5—$N_2^{\oplus}$ + $N_3^{\ominus}$ $\rightarrow$ C_6H_5—$N_3 + N_2$

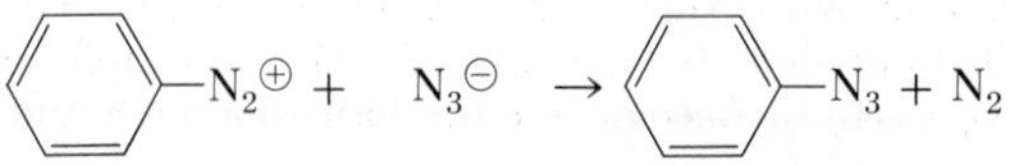

azide ion phenyl azide

ARYL OXYGEN COMPOUNDS

22.1 INTRODUCTION

There are two important types of aryl oxygen compounds—**phenols** (Ar—OH) and **aryl ethers** (Ar—OR). Of these two, the chemistry of phenols is much more interesting and important than is that of aryl ethers; therefore, phenols will be the focal point of this chapter. Phenols are readily oxidized to cyclic, unsaturated diketones known as **quinones,** and the relationship between the chemistries of phenols and quinones requires that quinones be introduced in this chapter. However, before we examine the chemistry of phenols and quinones, we will look briefly at a few uses of these compounds in order that we may better answer the question, "Why learn this material?" The brief discussion of biological oxidation-reduction at the end of this chapter should put the chemistry of quinones into an even more interesting perspective.

Phenols are important compounds to the chemical industry because they are, among other things, a prime raw material in the synthesis of plastics such as *Bakelite.* Anyone who has used "Lysol" liquid disinfectant has taken advantage of the bactericidal activity of *o-phenylphenol* and *2-benzyl-4-chlorophenol.* Even the simplest phenol, *phenol,* is a potent disinfectant; this may account for its use in the dentist's office under the name *carbolic acid.* And surely every microbiology student has encountered the "phenol coefficient," a means of comparing the antimicrobial activity of a compound to the activity of phenol.

—OH | —OH, C_6H_5 | Cl— —OH, $CH_2C_6H_5$

phenol | *o*-phenylphenol | 2-benzyl-4-chlorophenol

Creosote, the fraction of coal tar that is rich in phenol and *o*-, *m*-, and *p*-cresols, is used as a wood preservative, as is pentachlorophenol ("Pentachlor").

CH_3, —OH | CH_3, OH | H_3C— —OH | Cl, Cl, Cl— —OH, Cl, Cl

o-cresol mp 30°; bp 192° | *m*-cresol mp 11°; bp 203° | *p*-cresol mp 35°; bp 202° | Pentachlor mp 190°; bp 310°

Phenolic compounds are found in certain plants, including the creosote bush (*Larrea mexicana*). The toxic irritants of poison ivy and poison oak are phenols. This is consistent with the observation that phenols are corrosive and can cause serious damage to the skin, mucosa, and other membranes.

Quinones and related compounds find their major industrial application as dyes. It is the extended π electron system that is responsible for the color of these dyes—the greater the conjugation, the deeper the color. Even the simplest stable quinone, *p*-benzoquinone, is colored. In addition to being colored in the visible spectrum, most quinones fluoresce under ultraviolet light, making them suitable as pigments that "glow" in so-called "black light." Quinones and quinone-like molecules also play an essential role in the functioning of the normal cell.

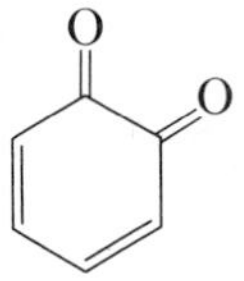

1,2-benzoquinone (*o*-benzoquinone) (red)

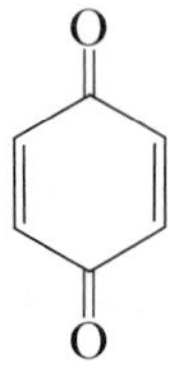

1,4-benzoquinone (*p*-benzoquinone) (yellow)

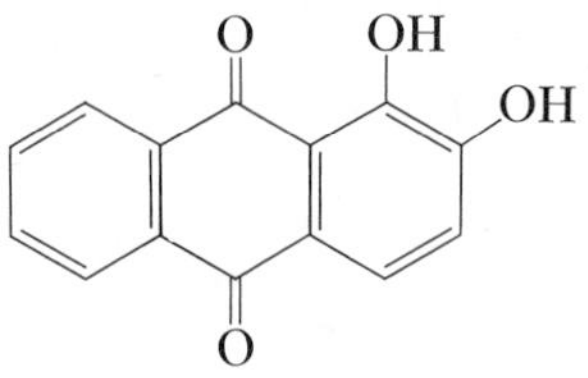

alizarin, a quinone dye found in madder root (red-purple)

PHENOLS

22.2 PROPERTIES OF PHENOLS

As the name *carbolic acid* implies, phenols are acidic substances. The facile loss of a proton from phenol stems from the delocalization of the negative charge in the resultant phenoxide anion. Alcohols, which cannot form resonance-stabilized anions, are considerably less acidic than phenols (phenols $pK_a \approx 10$; methanol $pK_a = 18$).

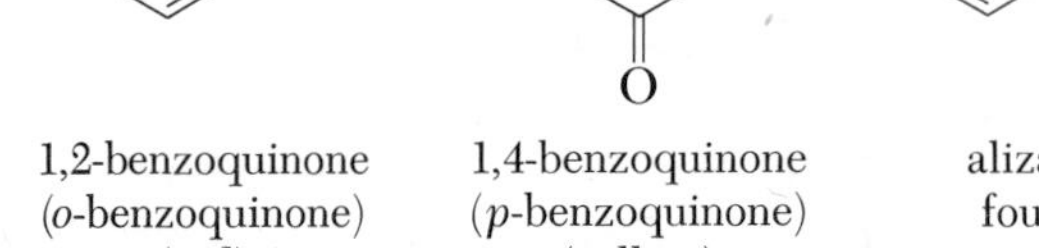

$$B{:} + H{-}\ddot{O}{-}C_6H_5 \rightleftarrows \overset{\oplus}{B}H + {:}\ddot{O}^{\ominus}{-}C_6H_5$$

phenoxide anion

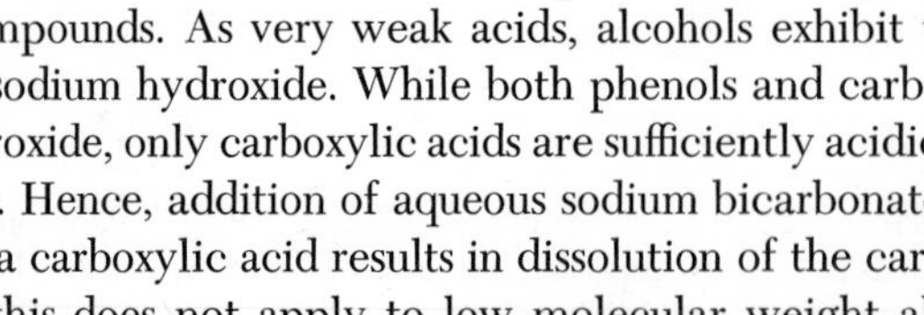

delocalization in the phenoxide anion

The difference in the acidity of carboxylic acids (benzoic acid $pK_a = 4.2$) on the one hand and phenols and alcohols on the other permits the convenient recognition and purification of these compounds. As very weak acids, alcohols exhibit the same (in)solubility in water as in aqueous sodium hydroxide. While both phenols and carboxylic acids are soluble in aqueous sodium hydroxide, only carboxylic acids are sufficiently acidic to form salts with aqueous sodium bicarbonate. Hence, addition of aqueous sodium bicarbonate to a mixture of a water-insoluble phenol and a carboxylic acid results in dissolution of the carboxylic acid but not of the phenol. Of course, this does not apply to low molecular weight alcohols and carboxylic acids, both of which are water-soluble.

$$ArOH + NaHCO_3 \leftrightharpoons ArO^{\ominus}\,Na^{\oplus} + H_2CO_3$$

$$ArCO_2H + NaHCO_3 \rightleftarrows ArCO_2^{\ominus}\,Na^{\oplus} + H_2CO_3$$

Simple phenols are either liquids or low-melting solids. Phenol (mp 43°) liquefies rapidly upon contact with water because of the ensuing depression in its melting point. The enhanced water solubility that is observed for phenols is the result of a combination of effects, including facile hydrogen bond formation and ionization to the phenoxide anion. An increase in intermolecular hydrogen bonding may be invoked to explain the slightly higher boiling point of phenol compared to cyclohexanol (182° *vs.* 161°).

1. Outline a procedure for separating components of the following mixtures.
   (a) hexane, acetic acid, cyclohexanol
   (b) hexane, *p*-chlorobenzoic acid, phenol
   (c) hexane, *p*-chlorobenzoic acid, acetic acid

Most simple phenols develop some color after standing exposed to the air. This is due, in part, to the oxidation of phenols to quinones.

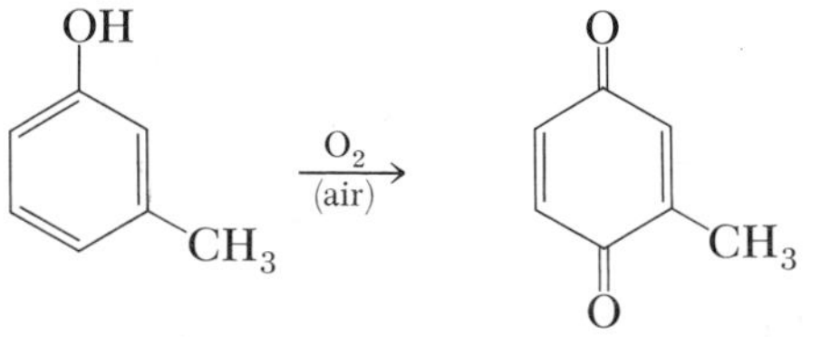

$$m\text{-cresol} \xrightarrow[\text{(air)}]{O_2} \text{2-methyl-1,4-benzoquinone}$$

m-cresol   2-methyl-1,4-benzoquinone

22.3 SYNTHESIS OF PHENOLS

As with other classes of compounds, the reactions that are suitable for large-scale, industrial syntheses often are not employed as laboratory syntheses, and *vice versa.*

Industrial Syntheses. While one might imagine that the direct (*i.e.,* air) oxidation of aromatic hydrocarbons should be an inexpensive route to phenols, it is of no practical value because phenols are more easily oxidized than are the prerequisite hydrocarbons.

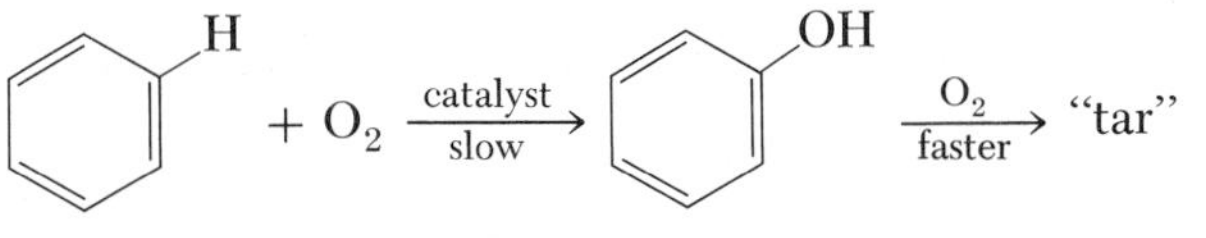

$$C_6H_5\text{–H} + O_2 \xrightarrow[\text{slow}]{\text{catalyst}} C_6H_5\text{–OH} \xrightarrow[\text{faster}]{O_2} \text{“tar”}$$

One of the two standard industrial routes to phenols involves the replacement of a chloro or sulfonic acid group on an aromatic ring by a hydroxide ion. The reaction is carried out under forcing conditions, and rearrangement may occur because of the intervention of a dehydrobenzene-type intermediate. Replacement of the sulfonic acid group was the basis of a process introduced in the early twentieth century, while replacement of the chloro group was introduced in 1928 by the Dow Chemical Company. The latter route is called the *Dow phenol process.* The interest in large-scale phenol synthesis revolved around the explosive power of picric acid, 2,4,6-trinitrophenol, and of its ammonium salt, ammonium picrate. This salt is at least equal to TNT in explosive power and has been used as the charge in armor-piercing shells.

$$C_6H_6 \xrightarrow[H_2SO_4]{\text{conc.}} C_6H_5SO_3H \xrightarrow[\text{fuse}]{NaOH} \xrightarrow{H_3O^{\oplus}} C_6H_5OH$$

$$C_6H_6 \xrightarrow[Fe]{Cl_2} C_6H_5Cl \xrightarrow[300°/3000\ \text{psi}]{8\%\ \text{aq. NaOH}} C_6H_5OH$$

Dow phenol process

$$\underset{\text{picric acid}}{2,4,6\text{-}(O_2N)_3C_6H_2OH} + NH_3 \rightarrow \underset{\text{ammonium picrate}}{2,4,6\text{-}(O_2N)_3C_6H_2O^{\ominus}\ \overset{\oplus}{N}H_4}$$

A more recent route to phenol involves the air oxidation of cumene and the acid-catalyzed decomposition of the resultant cumene hydroperoxide. A major advantage of this procedure is that it produces not only phenol but also acetone, another major industrial chemical.

$$C_6H_6 + CH_2{=}CH{-}CH_3 \xrightarrow{H_3PO_4} \underset{\text{cumene}}{C_6H_5{-}\overset{H}{\overset{|}{C}}(CH_3)_2} \xrightarrow[\text{cat.}]{O_2} \underset{\text{cumene hydroperoxide}}{C_6H_5{-}\overset{O-OH}{\overset{|}{C}}(CH_3)_2} \xrightarrow{H_3O^{\oplus}} C_6H_5OH + (CH_3)_2C{=}O$$

2. Suggest a mechanism for the acid-catalyzed conversion of cumene hydroperoxide to phenol and acetone.

Laboratory Syntheses. The laboratory synthesis of phenols can take any of several routes. For example, fusion of an arylsulfonate with a mixture of sodium and potassium hydroxide at 300°, followed by acidification, can produce a phenol in reasonable yield. While it is rarely the most desirable route to a phenol, fusion of β-naphthylsulfonic acid with sodium hydroxide provides the most convenient route to β-naphthol.

$$CH_3{-}C_6H_4{-}SO_3H \xrightarrow[300°]{NaOH/KOH} \xrightarrow{H_3O^{\oplus}} CH_3{-}C_6H_4{-}OH \quad (65\%)$$

$$C_{10}H_8 + H_2SO_4 \xrightarrow{150°} \beta\text{-}C_{10}H_7SO_3H \xrightarrow[\text{fuse}]{NaOH} \xrightarrow{H_3O^{\oplus}} \underset{\beta\text{-naphthol}}{\beta\text{-}C_{10}H_7OH}$$

Although simple aryl halides are rather unreactive toward aqueous alkali, aryl halides bearing electron-withdrawing groups *ortho* and/or *para* to the halide can be hydrolyzed to the corresponding phenols under moderate conditions. Such reactions proceed by nucleophilic displacement rather than by aryne formation.

$$O_2N\text{-}C_6H_4\text{-}Cl \xrightarrow[160°]{15\%\ aq.\ NaOH} \left[O_2N\text{-}C_6H_4^{\ominus}(Cl)(OH) \right] \xrightarrow{-Cl^{\ominus}} O_2N\text{-}C_6H_4\text{-}OH$$

not isolated

An aryl halide can be converted to a phenol *via* a Grignard reagent rather than through hydrolysis. Conversion of an aryl Grignard reagent to a phenol is accomplished by reaction with trimethylborate ($B(OCH_3)_3$) followed by oxidation of the resultant arylboronic acid with dilute hydrogen peroxide.

$$ClMg\text{—}C_6H_5 \xrightarrow[-80°]{(CH_3O)_3B} (CH_3O)_2B\text{—}C_6H_5 \xrightarrow{H_3O^{\oplus}} (HO)_2B\text{—}C_6H_5 \xrightarrow[H_3O^{\oplus}]{15\%\ H_2O_2} HO\text{—}C_6H_5$$

phenylboronic acid (70%)

$$\text{6-}CH_3O\text{-2-bromonaphthalene} \xrightarrow[THF]{Mg} \xrightarrow[-80°]{(CH_3O)_3B} \xrightarrow{H_3O^{\oplus}} \xrightarrow[H_3O^{\oplus}]{15\%\ H_2O_2} \text{6-}CH_3O\text{-2-naphthol}$$

(80%)

3. Which of the following would not be expected to give a good yield of phenol using the Grignard reagent–trimethyl borate sequence? Why?
   (a) $p\text{-}CH_3C_6H_4Br$
   (b) $p\text{-}BrC_6H_4Cl$
   (c) $p\text{-}NCC_6H_4Br$
   (d) $m\text{-}H_2NC_6H_4Br$
   (e) $p\text{-}CH_3OC(O)C_6H_4Br$

Another general synthesis of phenols, the hydrolysis of diazonium salts, was discussed in the previous chapter. We might note that yields can sometimes be improved by converting the diazonium salt to an ester and then hydrolyzing that ester.

$$ArN_2^{\oplus}\ HSO_4^{\ominus} \xrightarrow{HBF_4} ArN_2^{\oplus}\ BF_4^{\ominus} \xrightarrow{CH_3CO_2H} ArOC(=O)CH_3 \xrightarrow[heat]{H_3O^{\oplus}} ArOH + CH_3CO_2H$$

1-Naphthol is readily prepared by the acid-catalyzed rearrangement of the adduct formed by trapping dehydrobenzene with furan. The mechanism is similar to that of the reaction used to produce 1-naphthylamine from pyrrole (Chapter 20, problem 30). This represents a rather unique approach, and one that is to be contrasted with the synthesis of β-naphthol (2-naphthol) presented earlier (p. 684).

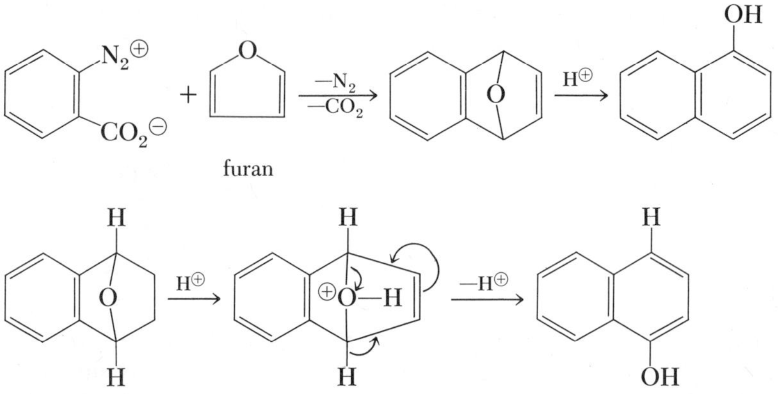

$N_2^{\oplus}$ $CO_2^{\ominus}$ + furan $\xrightarrow[-CO_2]{-N_2}$ (O-bridged adduct) $\xrightarrow{H^{\oplus}}$ (1-naphthol, OH)

furan

(H, O, H) $\xrightarrow{H^{\oplus}}$ (H, $^{\oplus}O$—H, H) $\xrightarrow{-H^{\oplus}}$ (H, OH)

> 4. Based upon the mechanism presented in the text, should 1-naphthol prepared from benzenediazonium carboxylate and furan contain any deuterium if the reaction is worked up using $D_3O^{\oplus}$ rather than $H_3O^{\oplus}$? Explain.

22.4 REACTIONS OF PHENOLS

"Reactions of phenols" may be divided into reactions at O—H, at C—O, and at the aryl ring. Since cleavage of the phenolic C—O bond is uncommon, we will restrict our attention to the OH group and the aryl ring.

Reactions at the Hydroxy Group. Like carboxylic acids, phenols are sufficiently acidic to react with diazoalkanes.[a] Of course, while carboxylic acids yield esters, phenols react with diazoalkanes to produce aryl alkyl ethers. Because of the importance of methyl ethers, the most commonly employed diazoalkane is diazomethane.

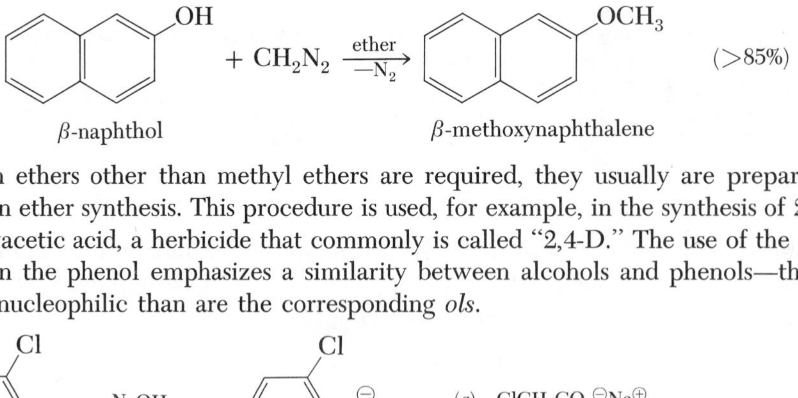

β-naphthol (OH) + CH_2N_2 $\xrightarrow[-N_2]{\text{ether}}$ β-methoxynaphthalene (OCH_3) (>85%)

When ethers other than methyl ethers are required, they usually are prepared by the Williamson ether synthesis. This procedure is used, for example, in the synthesis of 2,4-dichlorophenoxyacetic acid, a herbicide that commonly is called "2,4-D." The use of the phenoxide rather than the phenol emphasizes a similarity between alcohols and phenols—their anions are more nucleophilic than are the corresponding *ols*.

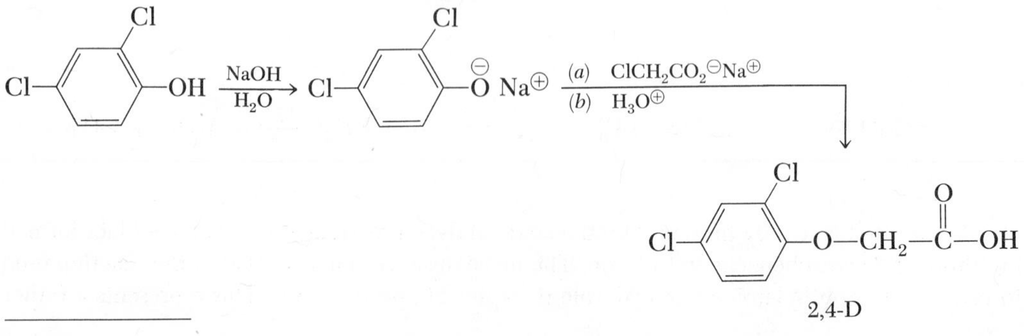

Cl—C_6H_3(Cl)—OH $\xrightarrow[H_2O]{\text{NaOH}}$ Cl—C_6H_3(Cl)—$O^{\ominus}$ $Na^{\oplus}$ $\xrightarrow[(b)\ H_3O^{\oplus}]{(a)\ ClCH_2CO_2^{\ominus}Na^{\oplus}}$ Cl—C_6H_3(Cl)—O—CH_2—C(=O)—OH

2,4-D

[a] Alcohols can be etherified with diazoalkanes only in the presence of boron trifluoride or other Lewis acids.

5. Suggest a synthesis for this isomer of 2,4-D, using 2,4-dichlorophenol as one starting material.

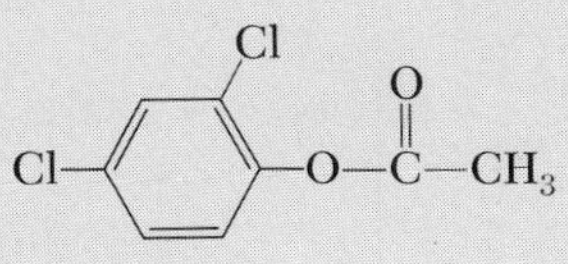
Cl
O
Cl
O—C—CH3

Etherification, which can be called *O-alkylation,* is the usual result of the reaction of an alkylating agent with a phenoxide. However, phenoxides, like enolate anions (see Chapter 19), are capable of *C*-alkylation, that is, alkylation on the aromatic ring. The explanation for the occurrence of *O*-alkylation in preference to *C*-alkylation is that *O*-alkylation is kinetically favored and generally irreversible.

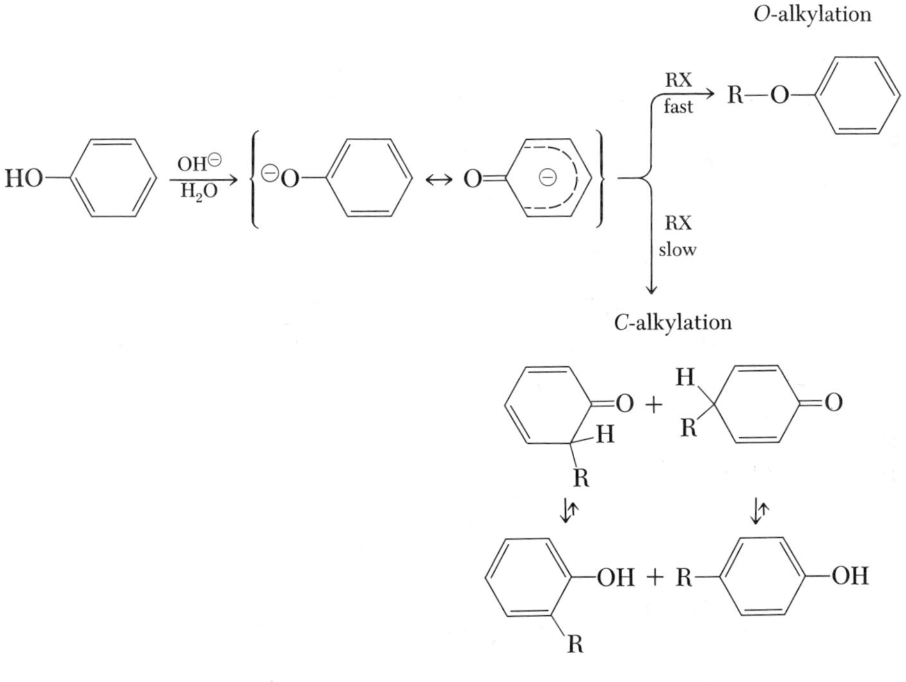
O-alkylation
RX
fast
R—O
HO
OH⊖
H2O
⊖O
O
RX
slow
C-alkylation
O +
H
R
H
R
O
OH + R
OH
R

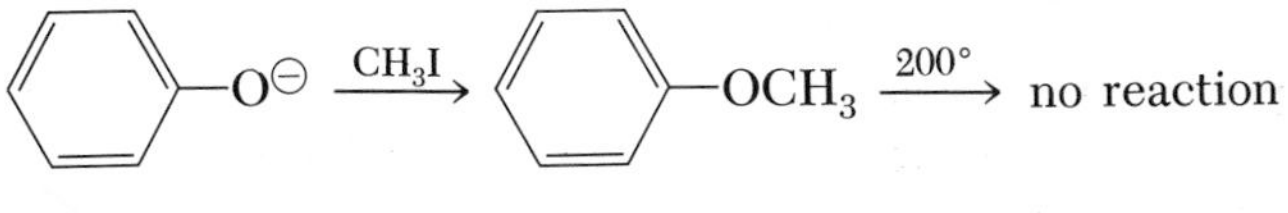
O⊖
CH3I
OCH3
200°
no reaction

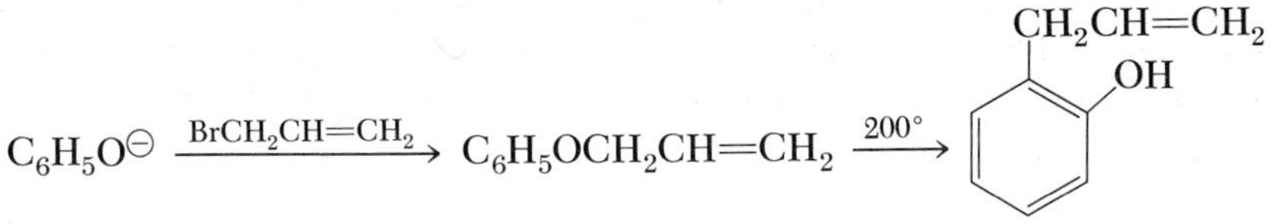
C6H5O⊖
BrCH2CH=CH2
C6H5OCH2CH=CH2
200°
CH2CH=CH2
OH

Claisen rearrangement

phenyl allyl ether

Unlike simple *O*-alkylated phenols, phenyl allyl ether is unstable at high temperatures, rearranging to *o*-allylphenol. This isomerization, the **Claisen rearrangement,** occurs *via* a cyclic transition state which, as seen below, provides a route for equilibration of *O*- and *C*-alkylated phenols. Since the rearrangement produces *C*-alkylation, it is construed that *C*-alkylation of phenols is thermodynamically favored over *O*-alkylation.

transition state
m: bond being made
b: bond being broken

$CH_2—CH=CH_2$

6. Predict the product of the Claisen rearrangement of each of the following.

(a) $O—CH_2—CH=CH_2$ (2,6-dideuterio)

(b) $O—CH_2—CH=CD_2$

(c) $O—CH_2—CD=CH_2$

(d) $O—CD_2—CH=CH_2$

7. Describe an experiment that can prove that the Claisen rearrangement does *not* follow the path shown below.

$OCH_2CH=CH_2$ ⇄ $O^{\ominus}$ + $\overset{\oplus}{C}H_2—CH=CH_2$ → $CH_2—CH=CH_2$ → *etc.*

In addition to etherification, phenols and phenoxides can participate in the nucleophilic addition and addition-elimination reactions characteristic of other nucleophiles. For example, phenols react with acid anhydrides to produce phenolic esters (a typical addition-elimination sequence), and they react with isocyanates, RNCO, to produce urethanes of the type RNHC(O)OAr (a typical addition reaction). Urethanes have high melting points, making them useful as derivatives of phenols.

$$C_6H_5—OH \xrightarrow{CH_3—C(=O)—O—C(=O)—CH_3} C_6H_5—O—C(=O)—CH_3 + CH_3CO_2H$$

phenyl acetate

$$C_6H_5—OH \xrightarrow{O_2N—C_6H_4—N=C=O} O_2N—C_6H_4—NH—C(=O)—O—C_6H_5$$

a urethane

Phenolic esters undergo the **Fries rearrangement** in the presence of aluminum chloride. The product of this isomerization is a mixture of *o*- and *p*-substituted phenol. The mechanism of the reaction is still subject to debate, in part because the product distribution is dependent upon temperature, solvent, and the ratio of catalyst to substrate. Regardless of the details, it seems likely that this internal acylation involves an acylium ion or its equivalent.

$$C_6H_5O{-}\overset{O}{\overset{\|}{C}}{-}CH_3 + AlCl_3 \rightarrow C_6H_5\overset{\oplus}{O}(\overset{\ominus}{Al}Cl_3){-}\overset{O}{\overset{\|}{C}}{-}CH_3 \rightarrow C_6H_5O{-}\overset{\ominus}{Al}Cl_3 + CH_3\overset{\oplus}{C}{=}O$$

acylium ion

$$C_6H_5O{-}\overset{\ominus}{Al}Cl_3 \xrightarrow{CH_3\overset{\oplus}{C}=O} \text{Cl}_3\overset{\ominus}{\text{Al}}{-}O{-}C_6H_5^{\oplus}(H)(COCH_3) \xrightarrow{-H^{\oplus}} \text{Cl}_3\overset{\ominus}{\text{Al}}{-}O{-}C_6H_4{-}\overset{O}{\overset{\|}{C}}{-}CH_3 \xrightarrow{H_3O^{\oplus}} o\text{-}HO{-}C_6H_4{-}\overset{O}{\overset{\|}{C}}{-}CH_3$$

(*p*-attack not shown)

$$m\text{-}CH_3C_6H_4O{-}\overset{O}{\overset{\|}{C}}{-}CH_3 \xrightarrow[\text{(b) } H_2O]{\text{(a) } AlCl_3/25^\circ} \text{4-hydroxy-2-methylacetophenone (OH, CH}_3\text{, C(=O)CH}_3\text{)}$$

$$m\text{-}CH_3C_6H_4O{-}\overset{O}{\overset{\|}{C}}{-}CH_3 \xrightarrow[\text{(b) } H_2O]{\text{(a) } AlCl_3/160^\circ} \text{2-hydroxy-4-methylacetophenone (OH, C(=O)CH}_3\text{, CH}_3\text{)}$$

Fries rearrangement temperature dependence

Reduction. Normal reducing agents (*e.g.*, zinc and acid) are incapable of converting phenol to benzene. However, catalytic hydrogenation does reduce the aromatic ring, thereby converting phenols to alicyclic alcohols.

$$\underset{\text{phenol}}{C_6H_5OH} \xrightarrow[Ni]{H_2} \underset{\text{cyclohexanol}}{C_6H_{11}OH}$$

Oxidation. Phenols are readily oxidized to a variety of products, including comparatively stable radicals, dimers of radicals, compounds derived from radical disproportionation, and quinones. Chromic acid oxidation of phenols leads, *via* an intermediate chromate ester, to *p*-benzoquinone.

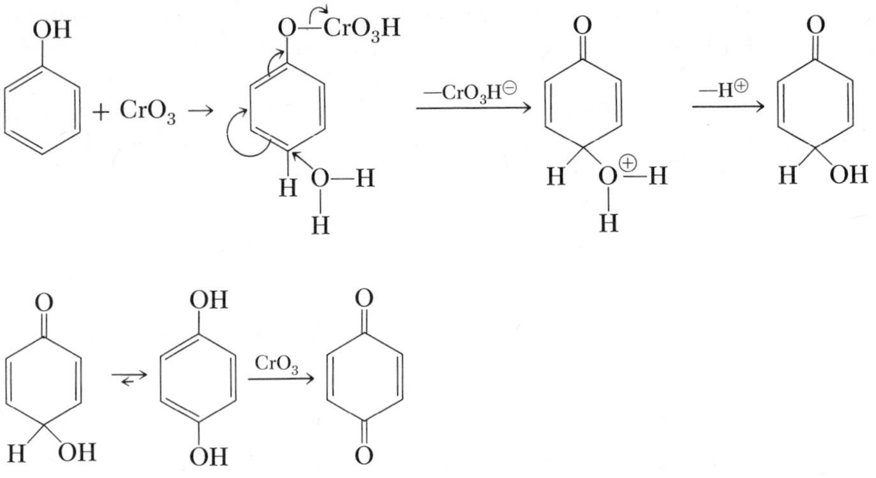

OH, + CrO_3 →, O—CrO_3H, H, O—H, H, $-CrO_3H^{\ominus}$ →, O, H, $O^{\oplus}$—H, H, $-H^{\oplus}$ →, O, H, OH

O, H, OH ⇄ OH, OH $\xrightarrow{CrO_3}$ O, O

Quinones also are produced by the oxidation of arylamines by dichromate ion.

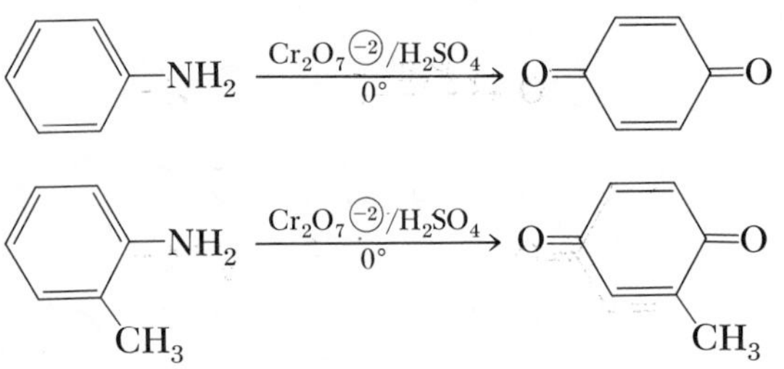

—NH_2 $\xrightarrow[0°]{Cr_2O_7^{\ominus 2}/H_2SO_4}$ O= =O

—NH_2, CH_3 $\xrightarrow[0°]{Cr_2O_7^{\ominus 2}/H_2SO_4}$ O= =O, CH_3

Oxidants that can produce one-electron transfers (*e.g.*, ferric chloride and potassium ferricyanide) convert phenols to radicals. This propensity for radical formation is due to the resonance stabilization of the resultant phenoxy radical. Particularly unreactive radicals are formed if the *ortho* and *para* positions are blocked by bulky substituents such as *t*-butyl groups.

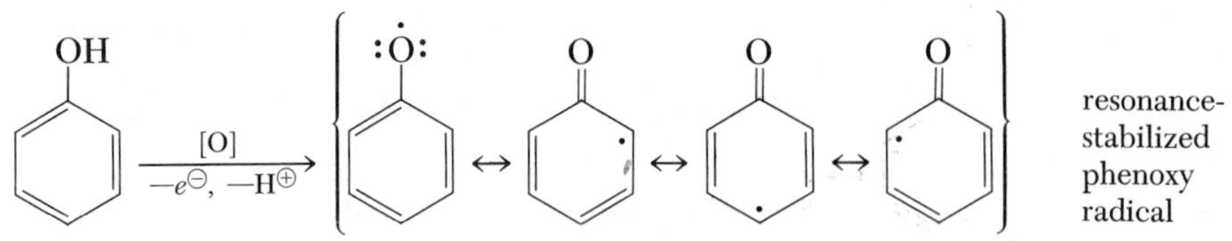

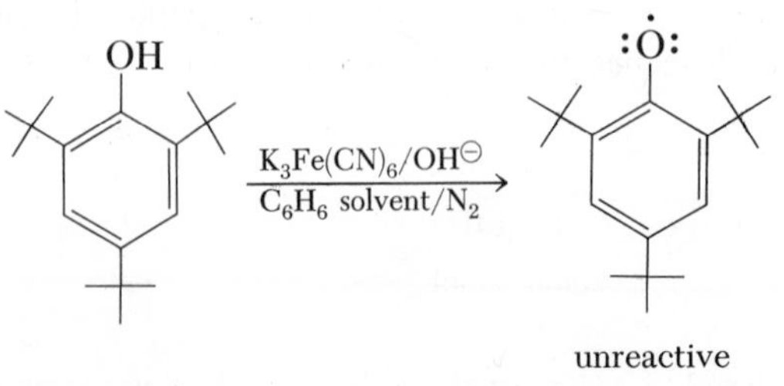

OH $\xrightarrow[-e^{\ominus},\ -H^{\oplus}]{[O]}$ {:Ȯ: ↔ O ↔ O ↔ O} resonance-stabilized phenoxy radical

OH $\xrightarrow[C_6H_6\ \text{solvent}/N_2]{K_3Fe(CN)_6/OH^{\ominus}}$:Ȯ:

unreactive

If an *ortho* or *para* position is not blocked, radical dimerization may occur.

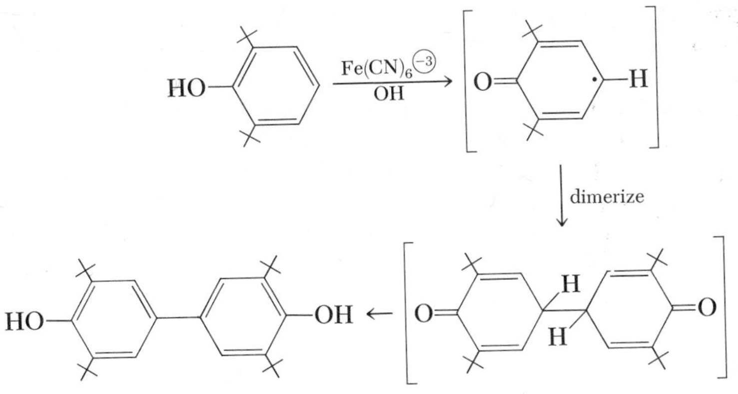

Fe(CN)$_6^{-3}$ / OH

dimerize

dimerization

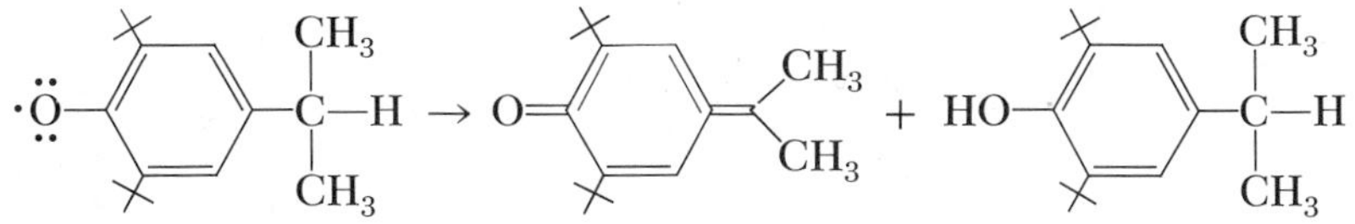

disproportionation

Many products occurring in nature have carbon skeletons that appear to have been synthesized by oxidative free radical coupling of phenols, including morphine (an alkaloid), griseofulvin (an anti-fungal antibiotic), and usnic acid (an anti-bacterial found in lichens).

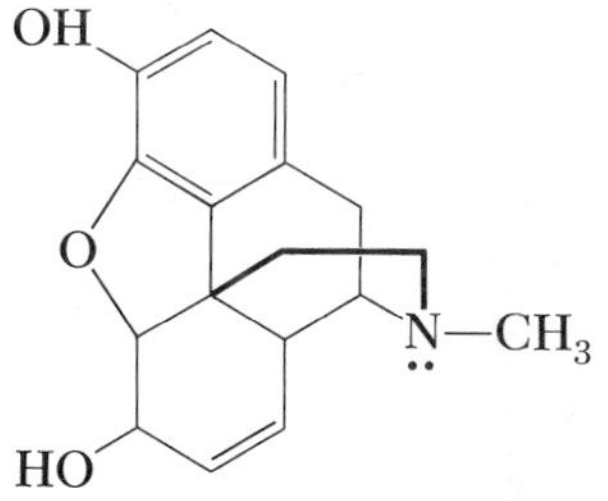

morphine

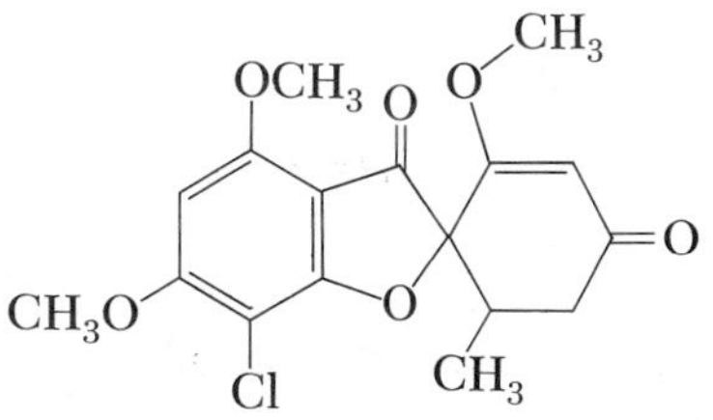

griseofulvin

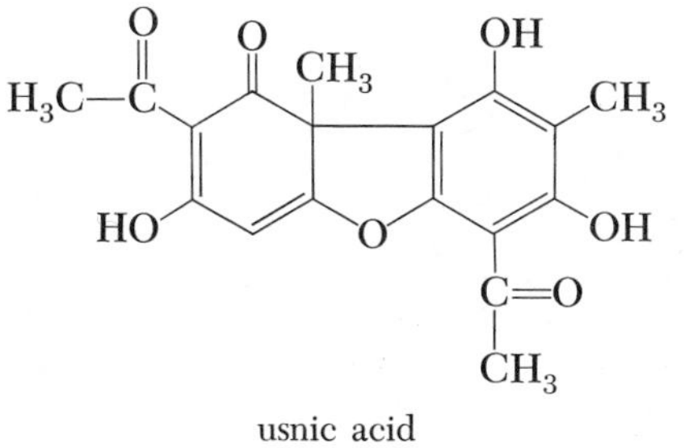

usnic acid

Because of this, and for many other reasons, there is a great deal of interest in the mechanism of oxidative coupling of phenols and in the synthetic utility of oxidative coupling. One application of ferric chloride and potassium ferricyanide to *in vitro* couplings is seen in the heterocyclic synthesis shown below.

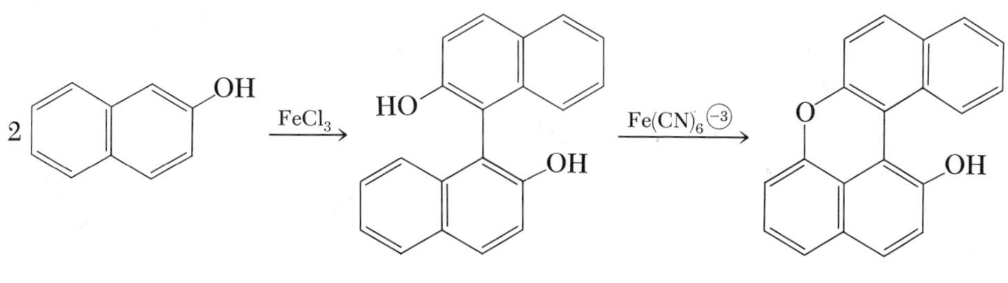

2 [2-naphthol] $\xrightarrow{FeCl_3}$ [binaphthol] $\xrightarrow{Fe(CN)_6^{-3}}$

Reactions at the Ring. The hydroxy group provides a source of electron density for the aromatic ring of phenols and, as a result, phenols easily undergo electrophilic aromatic substitution. All of the processes discussed in this section are facilitated by this enhanced electron density; some, like the Kolbe reaction, require such high electron density that they occur on the phenoxide anion but not on the corresponding phenol.

A simple example of the enhanced reactivity of the phenolic ring is the rapid reaction of phenol with aqueous bromine to produce 2,4,6-tribromophenol. While it is virtually impossible to isolate only monobrominated phenols under these conditions, the use of a non-polar solvent can limit the reaction to the monobromination stage. This diminished reactivity stems from a reduced concentration of the more nucleophilic phenoxide anion in the aprotic medium.

$$C_6H_5OH + 3Br_2 \xrightarrow{H_2O} \text{2,4,6-}Br_3C_6H_2OH + 3HBr$$

$$C_6H_5OH + Br_2 \xrightarrow{CS_2} BrC_6H_4OH + HBr$$

o and *p*

In addition to halogenation, another typical electrophilic substitution is the Friedel-Crafts reaction. Actually, Friedel-Crafts reactions usually are carried out on phenolic ethers rather than on the progenitor phenols (Can you suggest a reason for this?), even though they may be accompanied by de-alkylation of the ether.

$$1,3\text{-}(CH_3O)_2C_6H_4 + CH_3COCl \xrightarrow{AlCl_3} \xrightarrow{H_2O} \text{(4-}OCH_3\text{, 2-}OH\text{)}C_6H_3C(O)CH_3$$

de-alkylation accompanying Friedel-Crafts acylation

The **Gattermann** and **Hoesch** reactions are used to prepare phenolic aldehydes and ketones, respectively. Yields in the Hoesch reaction are greatest with extremely reactive phenols. A serious side reaction, particularly with monohydric phenols, involves ester formation at the phenolic oxygen. What are the mechanisms for these reactions?

$$C_6H_5OH \xrightarrow[AlCl_3]{HCl/HCN} \xrightarrow[heat]{H_2O} HOC_6H_4CHO$$

o and *p*

Gattermann formylation reaction

$$1,3,5\text{-}C_6H_3(OH)_3 \xrightarrow[ZnCl_2]{CH_3CN/HCl} \xrightarrow[heat]{H_2O} 2,4,6\text{-}(HO)_3C_6H_2C(O)CH_3$$

(87%)

Hoesch reaction

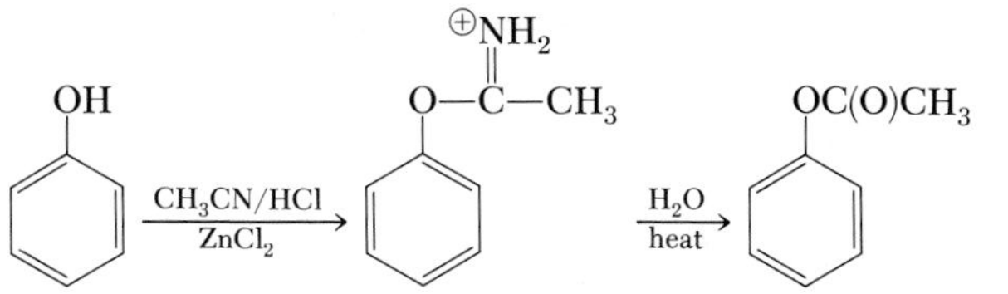

$$C_6H_5OH \xrightarrow[ZnCl_2]{CH_3CN/HCl} C_6H_5\text{—O—C}(=\overset{\oplus}{N}H_2)\text{—}CH_3 \xrightarrow[heat]{H_2O} C_6H_5OC(O)CH_3$$

side-reaction accompanying Hoesch reaction

Phenols, and particularly phenoxides, are sufficiently reactive that they undergo facile **diazo coupling** (review p. 668). The pH of the reaction mixture must be carefully controlled in order to maximize the amount of diazonium and phenoxide ions. Excess acid reduces the concentration of the phenoxide, while excess base converts the diazonium salt to the non-electrophilic diazotate salt. One common way to carry out diazo coupling reactions on phenols is to add, dropwise, a solution of the phenoxide salt (alkaline) to an acidic solution of the diazonium salt.

$$C_6H_5\text{—}O^{\ominus} + H_3O^{\oplus} \rightarrow C_6H_5\text{—}OH + H_2O$$

increased acidity reduces concentration of reactive substrate

$$C_6H_5\text{—}N_2^{\oplus} + 2OH^{\ominus} \rightarrow C_6H_5\text{—}N{=}N\text{—}O^{\ominus} + H_2O$$

increased basicity reduces concentration of electrophile

Phenolphthalein, used as a pH indicator and as an over-the-counter (non-prescription) laxative (*e.g.*, Ex-Lax), is prepared by the reaction of phenol with phthalic anhydride. In alkaline solution, phenolphthalein is converted to a resonance-stabilized, highly conjugated anion that is pink.

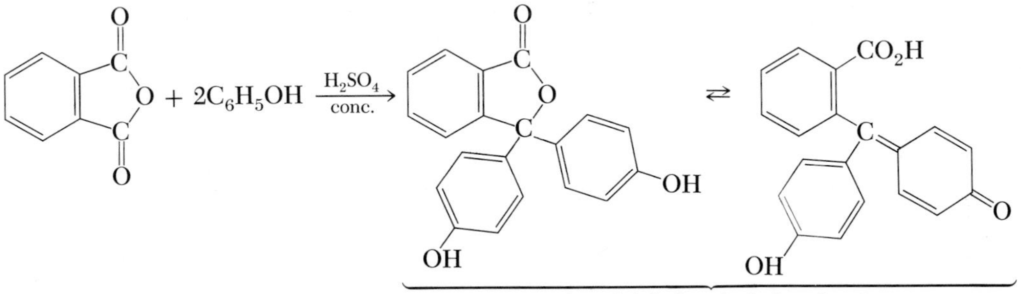

$$\text{phthalic anhydride} + 2C_6H_5OH \xrightarrow[conc.]{H_2SO_4} \text{(lactone form, OH, OH)} \rightleftarrows \text{(quinoid form, } CO_2H\text{, OH, O)}$$

phenolphthalein (colorless form)

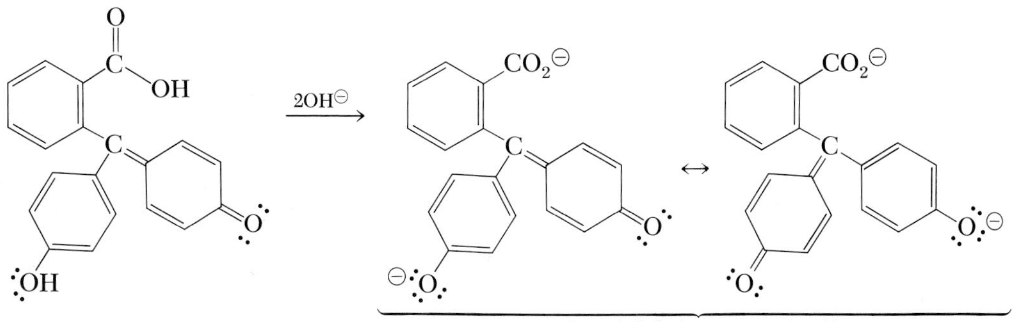

$$\text{(quinoid acid form, C(=O)OH, OH, O)} \xrightarrow{2OH^{\ominus}} \text{(}CO_2^{\ominus}\text{, }O^{\ominus}\text{, O)} \leftrightarrow \text{(}CO_2^{\ominus}\text{, O, }O^{\ominus}\text{)}$$

phenolphthalein anion (pink)

8. Suggest a mechanism for the steps in the synthesis of phenolphthalein.

The reactions discussed thus far are easily recognized as electrophilic aromatic substitutions, since they involve the attack of a cation or an incipient cation upon an aromatic ring. In the remaining two reactions we will encounter two rather atypical aromatic substitutions. The surprising thing about both is the nature of the electrophile—in one it is carbon dioxide, while in the other it is dichlorocarbene.

In the **Kolbe reaction,** a phenoxide anion reacts with carbon dioxide (a weak electrophile) to form a carboxylate salt. This process is an example of a carbanion adding to CO_2 and is similar to the carbonation of a Grignard reagent.

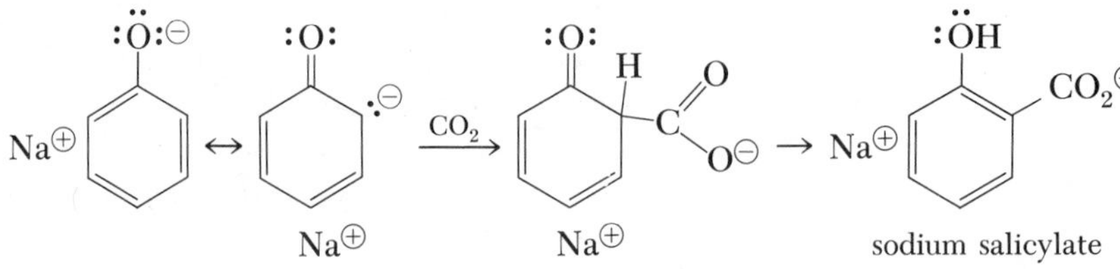

CO_2

$Na^{\oplus}$ $Na^{\oplus}$ $Na^{\oplus}$ $Na^{\oplus}$ $CO_2^{\ominus}$

sodium salicylate

9. Both aspirin (acetylsalicylic acid) and oil of wintergreen (methyl salicylate) are made from salicylic acid. Explain why the two routes shown below do *not* give significant amounts of the by-products shown.

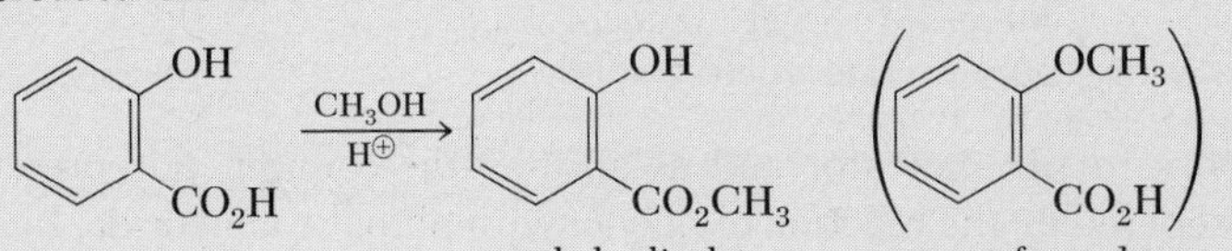

CH_3OH, $H^{\oplus}$

methyl salicylate — not formed

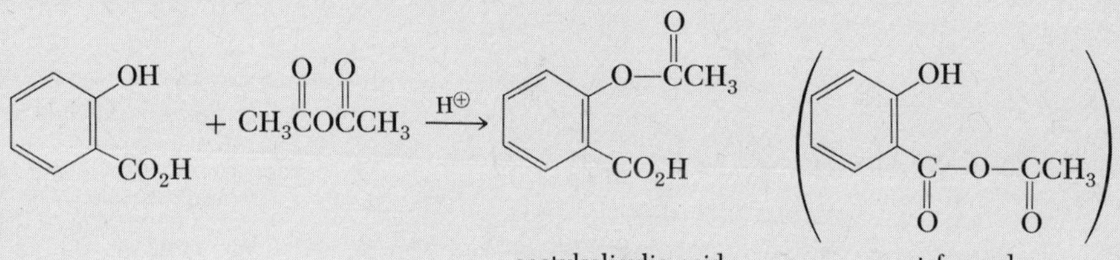

$+ CH_3COCCH_3$ ($H^{\oplus}$)

acetylsalicylic acid — not formed

Chloroform reacts with sodium hydroxide to produce dichlorocarbene by an α-elimination. In the **Reimer-Tiemann reaction,** the electron-deficient carbene reacts with sodium phenoxide to form a dihalide, which is then hydrolyzed to form an aldehyde.

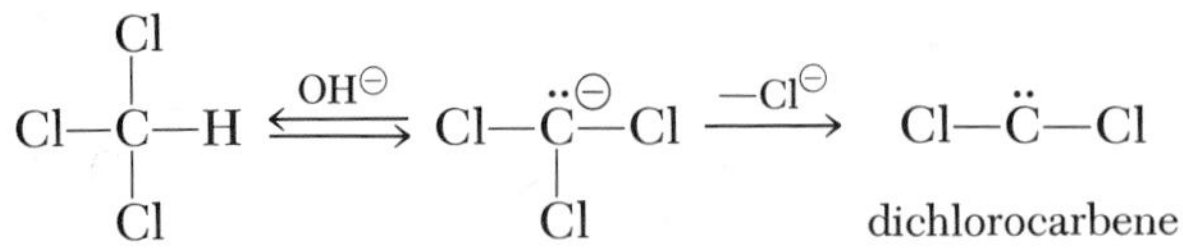

$$CHCl_3 \underset{}{\overset{OH^{\ominus}}{\rightleftharpoons}} {}^{\ominus}CCl_3 \xrightarrow{-Cl^{\ominus}} :CCl_2$$

dichlorocarbene

$:CCl_2$, H_2O, CCl_2H, $OH^{\ominus}$/H_2O, CHO, $H_3O^{\oplus}$

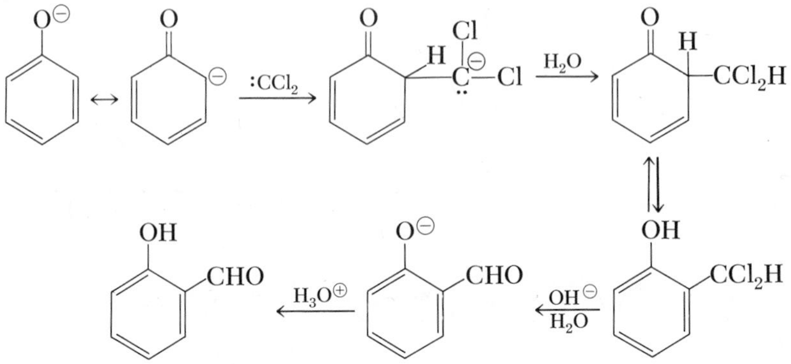

salicylaldehyde

22.5 POLYHYDRIC PHENOLS

A number of important, naturally occurring phenols possess two or more hydroxy groups, although some of them may be disguised as ethers or esters. The polyhydric phenols usually are identified by their common names rather than by systematic names. Thus, *o*-, *m*- and *p*-hydroxyphenol are much more commonly referred to as *catechol, resorcinol,* and *hydroquinone,* respectively. While sometimes found in the animal kingdom, these compounds are especially prevalent in plants.

$$\text{o-}C_6H_4Cl_2 \xrightarrow[OH^{\ominus}/200^\circ]{Cu(II),\ H_2O} \xrightarrow{H_3O^{\oplus}} \text{o-}C_6H_4(OH)_2$$

catechol

$$\text{m-}C_6H_4(SO_3H)_2 \xrightarrow[\text{fuse}]{NaOH} \xrightarrow{H_3O^{\oplus}} \text{m-}C_6H_4(OH)_2$$

resorcinol

$$H_2N{-}C_6H_5 \xrightarrow[H_2SO_4]{Cr_2O_7^{\,(-2)}} O{=}C_6H_4{=}O \rightsquigarrow HO{-}C_6H_4{-}OH$$

hydroquinone

The galls of trees (a *gall* is a swelling due to attack by a parasite) are rich in a material called *tannin.* Oak gall, for example, is nearly 50% tannin. Tannins are colored and astringent. The former property is responsible for the color of such things as tea and nut shells, while the latter makes tannin useful for converting animal hides to leather. Indeed, the conversion of hide to leather is called "tanning." (We might note that tannins are not the only materials now used to "tan" hides.) When tannin is hydrolyzed, it produces a large amount of 3,4,5-trihydroxybenzoic acid (also called, aptly enough, *gallic acid*).

Benzoic acids bearing hydroxy groups *ortho* and/or *para* to the carboxy group are decarboxylated upon heating. The product of thermal decarboxylation of gallic acid is 3,4,5-trihydroxybenzene, called *pyrogallol.* Pyrogallol is easily oxidized (that is, it is a potent reductant), and this makes it useful as a photographic developer. Since strongly alkaline solutions of pyrogallol are oxidized by oxygen gas, such solutions have been used in gas analysis and to remove traces of oxygen from gas streams.

$$(HO)_3C_6H_2CO_2H \xrightarrow[\text{heat}]{-CO_2} (HO)_3C_6H_3$$

gallic acid pyrogallol

The last of the polyhydric phenols to be discussed is 1,3,5-trihydroxybenzene, usually called *phloroglucinol.* This interesting compound is used in the diazo-printing industry and, in microscopy, as a decalcifier for bone specimens. Its commercial synthesis, presented below, is interesting because it begins with TNT.

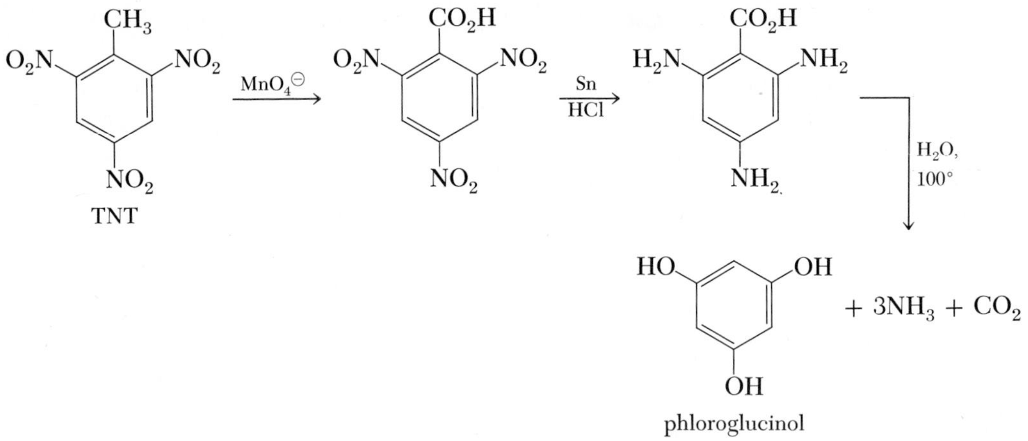

CH_3 O_2N NO_2 NO_2 TNT $\xrightarrow{MnO_4^{\ominus}}$ CO_2H O_2N NO_2 NO_2 $\xrightarrow[HCl]{Sn}$ CO_2H H_2N NH_2 NH_2 $\xrightarrow{H_2O,\ 100°}$ HO OH OH $+ 3NH_3 + CO_2$

phloroglucinol

22.6 BAKELITE RESINS

Phenol reacts with formaldehyde, in the presence of base, to produce a thermosetting resin named **Bakelite,** after its discoverer, Leo Baekeland. This plastic, because of its high thermal and electrical resistance, has found many uses including pot handles, drinking glasses, telephone bodies, and the bodies and handles of electrical switches.

The first step in the synthesis of Bakelite is mechanistically similar to an aldol condensation in which the phenoxide anion, acting as an ambident anion (an anion that can act as a nucleophile at two different atoms), provides the carbanion.

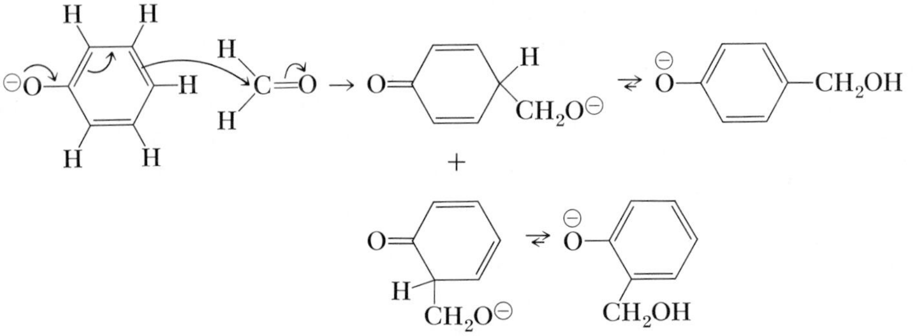

$\ominus O$–C_6H_5 + $H_2C{=}O$ → O= (H, $CH_2O^{\ominus}$) ⇄ $\overset{\ominus}{O}$–C_6H_4–CH_2OH

+

O= (H, $CH_2O^{\ominus}$) ⇄ $\overset{\ominus}{O}$–C_6H_4 (CH_2OH)

The next step involves addition of a phenoxide to a quinonmethide, formed by the dehydration of a phenolic alcohol.

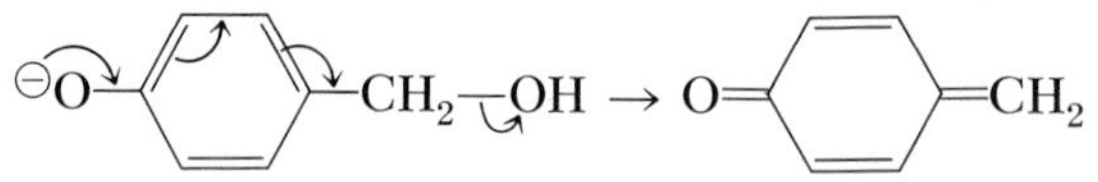

$\ominus O$–C_6H_4–CH_2–OH → O=C_6H_4=CH_2

a quinonmethide

$\ominus O$–C_6H_5 + $H_2C{=}C_6H_4{=}O$ → O= (H, CH_2–C_6H_4–$O^{\ominus}$)

↓

HO–C_6H_4–CH_2–C_6H_4–$O^{\ominus}$

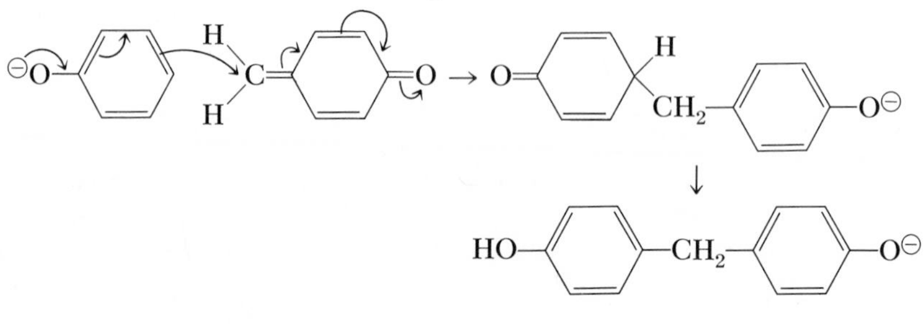

When this process is continued at all available *ortho* and *para* positions, the result is Bakelite, a cross-linked, three-dimensional polymer. A thermosetting resin like Bakelite undergoes a permanent chemical change when it is first formed (and melted), and then sets to a solid that cannot be re-melted because re-melting would require the breaking of very strong bonds. In the case of Bakelite, these bonds are the bonds between methylene groups and the benzene rings.

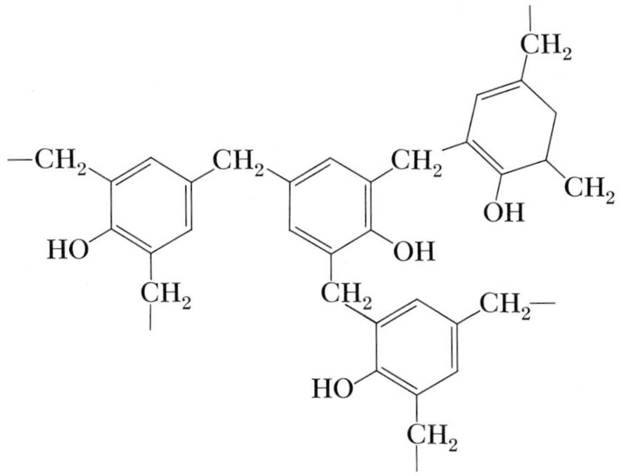

a section of phenol-formaldehyde (Bakelite) resin

QUINONES

Several decades ago the chemistry of quinones was of great interest because quinoidal compounds were used in the dye and photographic industries. After a long hiatus, interest in reactions of quinones has begun to grow again. The impetus is the recognition that quinones play a vital role in cellular oxidative phosphorylation and in the clotting of blood.

22.7 SYNTHESIS

While 1,4-quinones (*p*-quinones) are easily prepared and often rather stable, 1,2-quinones (*o*-quinones) are difficult to prepare and very reactive. The syntheses and numbering schemes for the more important quinones are summarized below.

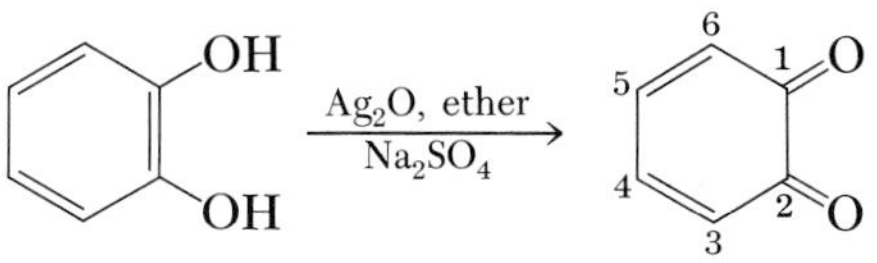

Ag_2O, ether / Na_2SO_4

o-benzoquinone

$Cr_2O_7^{-2}$ / $H_2SO_4/0°$

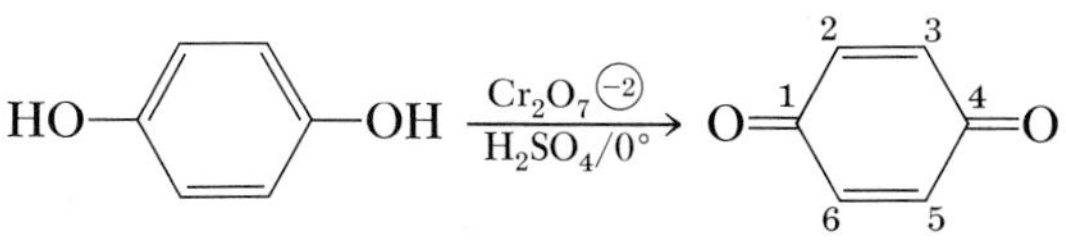

p-benzoquinone

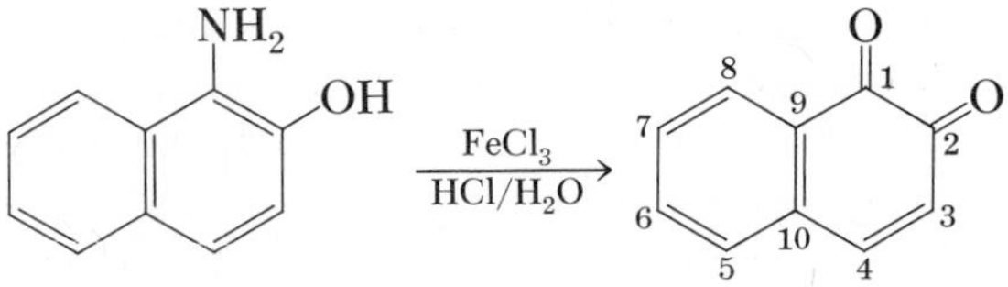

NH_2 OH $\xrightarrow[HCl/H_2O]{FeCl_3}$

1,2-naphthoquinone
(β-naphthoquinone)

OH $\xrightarrow[H_2SO_4]{Cr_2O_7^{-2}}$

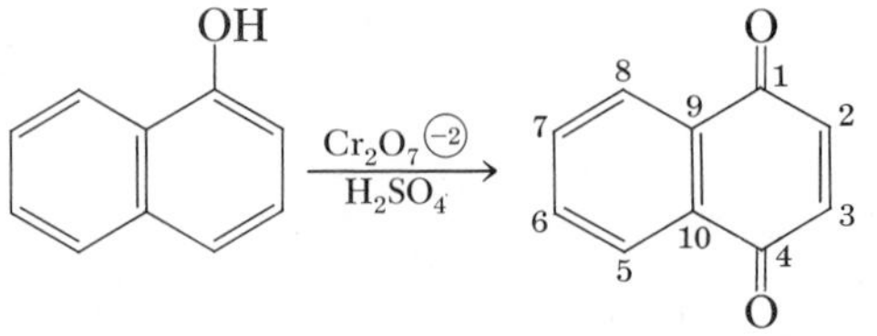

1,4-naphthoquinone
(α-naphthoquinone)

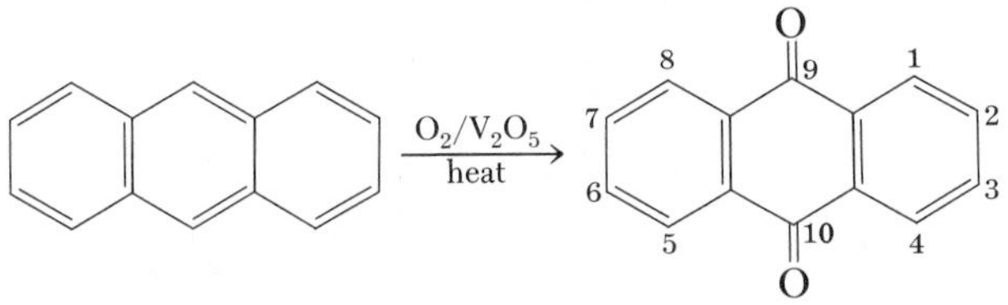

$\xrightarrow[heat]{O_2/V_2O_5}$

9,10-anthraquinone

$\xrightarrow[heat]{O_2/V_2O_5}$

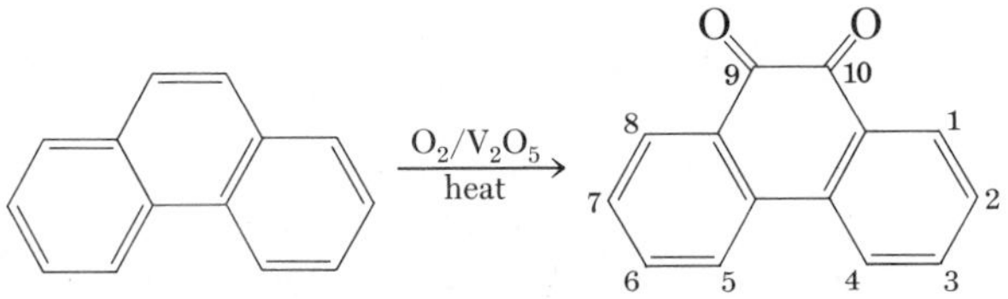

9,10-phenanthroquinone

22.8 REDUCTION OF QUINONES—QUINONES AS OXIDANTS

The reduction of quinones proceeds in two stages, the intermediate being a stabilized radical-anion called a **semiquinone.**

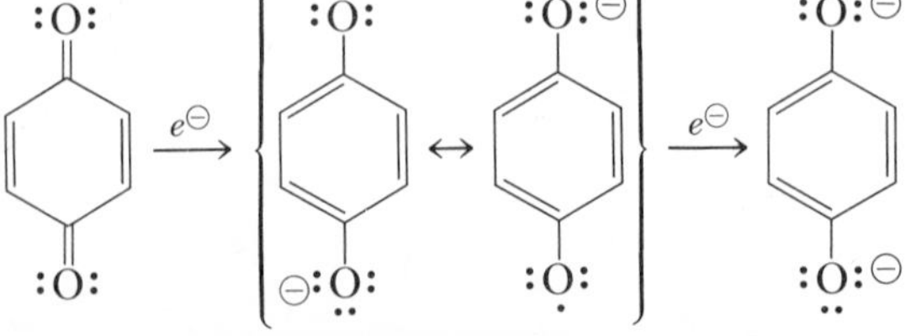

$\xrightarrow{e^{\ominus}}$ $\leftrightarrow$ $\xrightarrow{e^{\ominus}}$

semiquinone

Formation of a stable 1:1 adduct between quinone and hydroquinone, called **quinhydrone,** complicates the course of reduction. This adduct is a charge-transfer complex, with the hydroquinone acting as an electron donor and the quinone as electron acceptor.

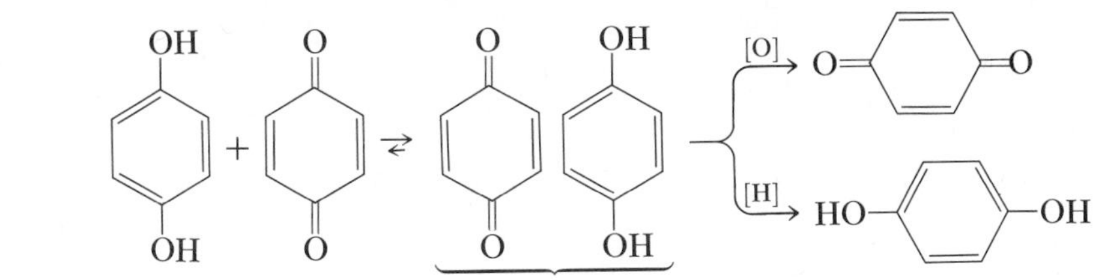

[O]

[H]

quinhydrone
(green-black)

Benzoquinones have positive reduction potentials in aqueous solutions. Benzoquinones bearing powerful electron-withdrawing groups have a higher potential and are potent oxidants. This explains the frequent use of DDQ (2,3-dichloro-5,6-dicyano-1,4-benzoquinone) as an oxidant in organic syntheses.

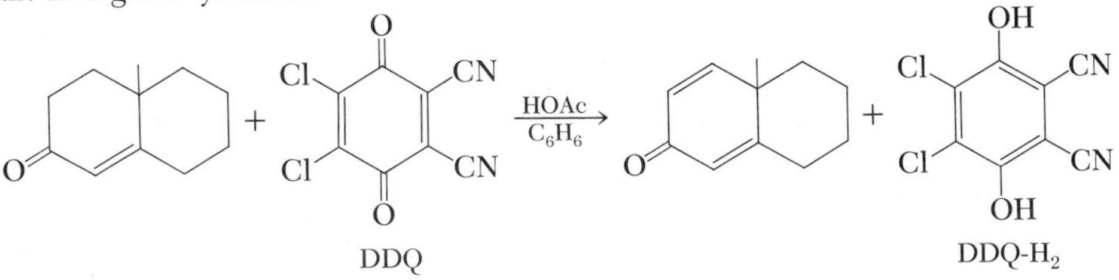

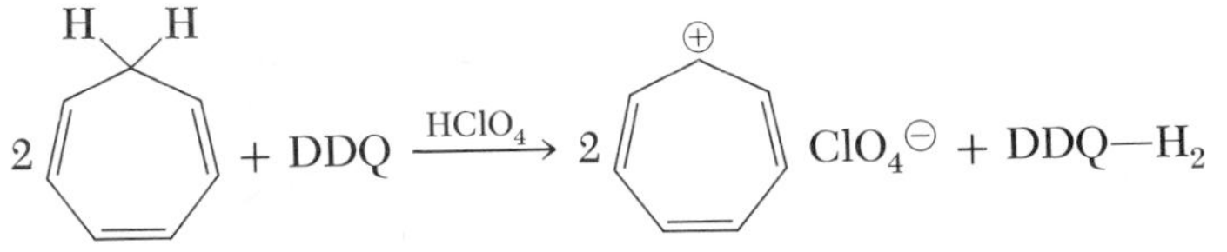

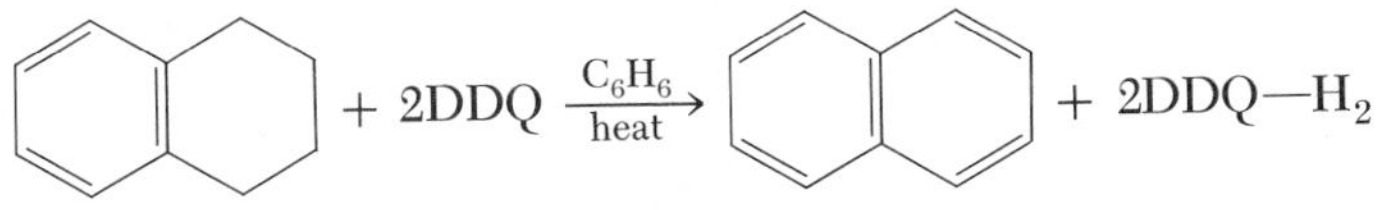

$\xrightarrow[C_6H_6]{HOAc}$

DDQ

DDQ-H_2

2 + DDQ $\xrightarrow{HClO_4}$ 2 $ClO_4^{\ominus}$ + DDQ—H_2

+ 2DDQ $\xrightarrow[\text{heat}]{C_6H_6}$ + 2DDQ—H_2

22.9 QUINONES AS UNSATURATED KETONES

The 1,4-quinones are α,β-unsaturated ketones and will undergo 1,2- and 1,4-addition reactions much like more ordinary unsaturated ketones. One example of this is the addition of hydrogen chloride to benzoquinone to form chlorohydroquinone. In turn, this can be oxidized by the starting quinone to form chloro-*p*-benzoquinone, which, upon addition of hydrogen chloride, yields 2,3-dichlorohydroquinone.

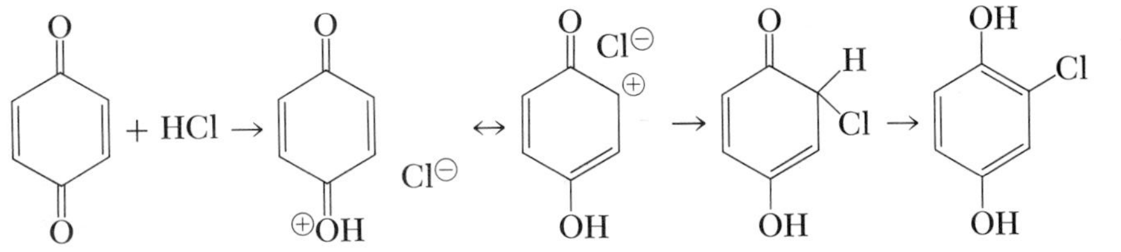

+ HCl →

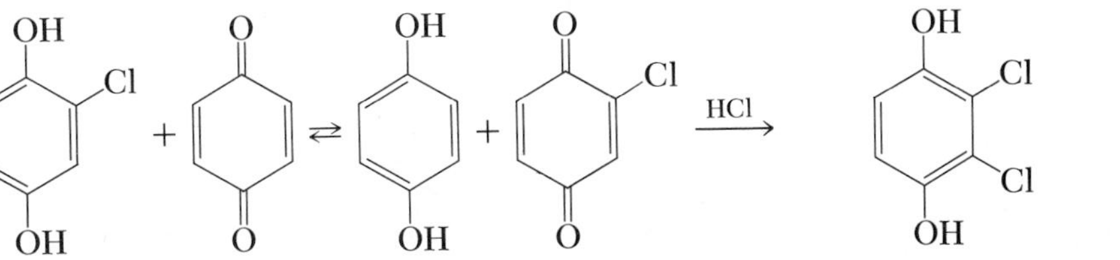

$\xrightarrow{HCl}$

2,3-dichlorohydroquinone

This same approach has been used to prepare DDQ.

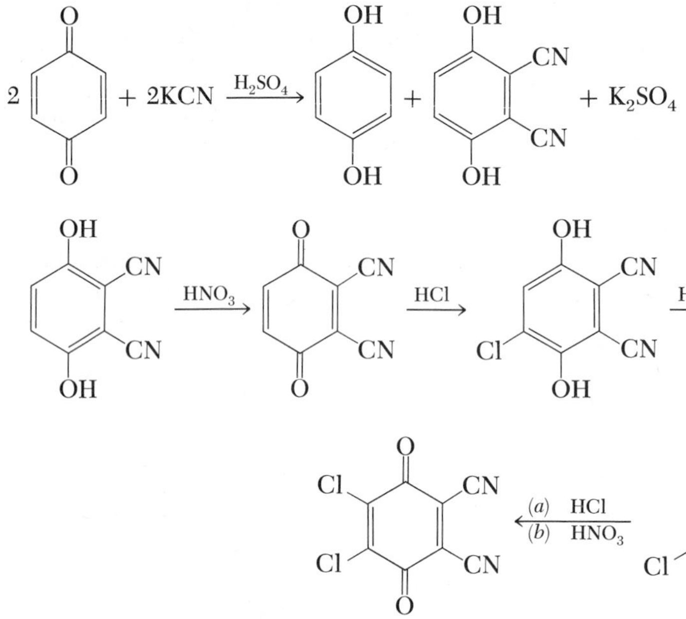

These ionic additions are not the only reactions of the quinone ring. If a quinone contains a double bond that is not part of an aromatic ring, *that* double bond will undergo a facile Diels-Alder reaction. This type of sequence permits the convenient build-up of polycyclic systems.

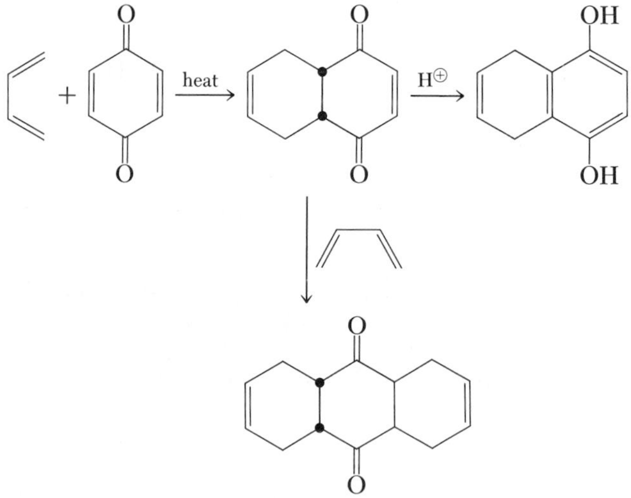

10. Explain the stereochemistry of the following reaction.

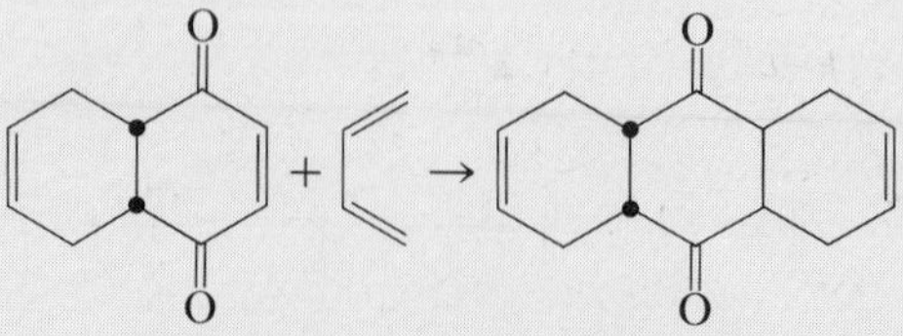

11. Suggest a mechanism for the following reaction.

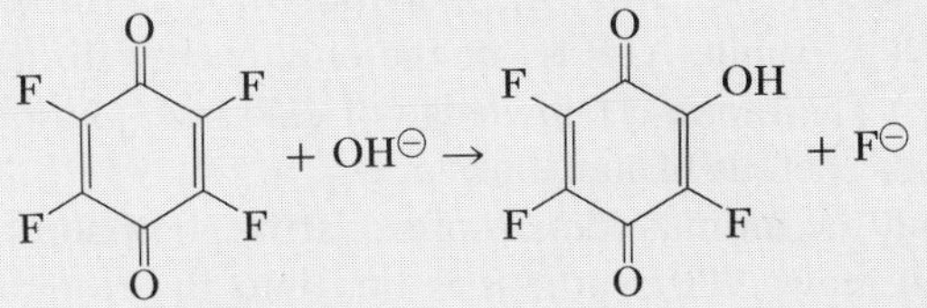

$$\text{tetrafluoro-p-benzoquinone} + OH^{\ominus} \rightarrow \text{2-hydroxy-3,5,6-trifluoro-p-benzoquinone} + F^{\ominus}$$

12. Lapachol occurs in the grain of some woods, including the South American lapacho. This yellow material can be prepared, *in vitro*, by the sequence shown below. Suggest a mechanism for this synthesis and for the interconversions of lapachol and α- and β-lapachone. Which of the latter two is more stable? How would you rationalize this?

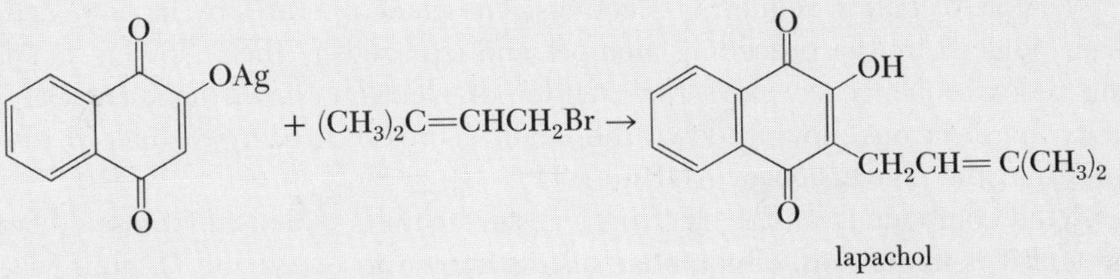

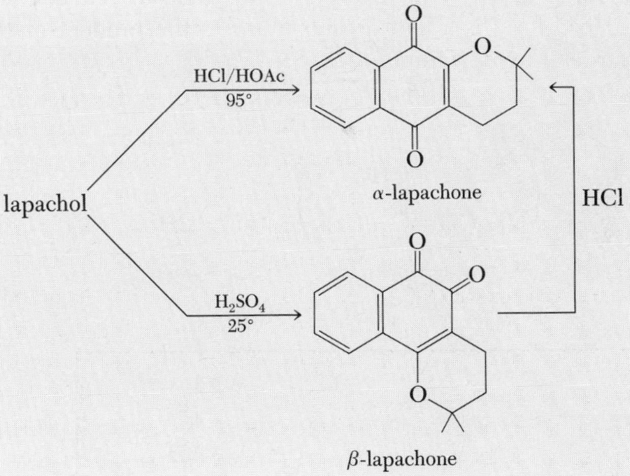

$$\text{(2-OAg-1,4-naphthoquinone)} + (CH_3)_2C{=}CHCH_2Br \rightarrow \text{(2-OH-3-}CH_2CH{=}C(CH_3)_2\text{-1,4-naphthoquinone)}$$

lapachol

lapachol —HCl/HOAc, 95°→ α-lapachone

lapachol —H_2SO_4, 25°→ β-lapachone

β-lapachone —HCl→ α-lapachone

22.10 COENZYME Q (UBIQUINONE)

Virtually all living creatures are aerobic, that is, they require oxygen in order to survive. The function of this oxygen is to oxidize various organic compounds that have been provided to the cell by ingestion or metabolism. However, unlike the "ordinary" oxidations carried out in the laboratory or in industry, biological oxidations involve a series of compounds whose function is to transport the electrons from the substrate (which, losing electrons, is oxidized) to the oxygen. This process is called *electron transport* and is promulgated by a group of compounds that constitute the *respiratory chain*. In the cell this electron transport method of oxidation is always accompanied by the conversion of adenosine diphosphate (ADP) to adenosine triphosphate (ATP)[a]; the combination is called *oxidative phosphorylation.*

The energy that a cell derives from the oxidation of small molecules is "stored" in the phosphate linkages of ATP. (While ATP is not the only energy storage facility in the cell, it is a very important one.) Therefore, a knowledge of oxidative phosphorylation must be part of an understanding of the details of metabolism.

We know little about the intimate details of oxidative phosphorylation. What may be the most important question, how ATP synthesis is tied into the respiratory chain, is far from answered. The reason for this ignorance is partly the fact that the process is limited largely to those organelles called mitochondria, and attempts to duplicate it outside a mitochondrion have, thus far, been unsuccessful. The systems controlling oxidative phosphorylation appear to require the surface of the mitochondrion in order to act properly.

It is known that the electron transfer chain is made up of a series of molecules that can undergo reversible oxidation-reduction reactions. The chain operates by having each member of the chain reduced by the preceding member and oxidized by the following member. This alternating redox sequence employs the transfer of electrons down an electrical potential gradient. As the electrons moving down the chain decrease in energy, some of the released energy goes into the conversion of ADP to ATP.

One of the compounds in the electron transfer chain is called *ubiquinone* because it is a quinone and it is ubiquitous. Ubiquinone also is known as coenzyme Q, abbreviated CoQ. Actually, several "coenzyme Q's" occur in nature. They all possess a benzoquinone ring and differ only in the number of isoprene units bound to it. In humans, the important CoQ contains ten isoprene units and is termed CoQ_{10}. CoQ_{10} and its reduced form are shown below; note that the reduced form is a hydroquinone and is the product of the reaction of CoQ_{10} with two electrons and two protons.

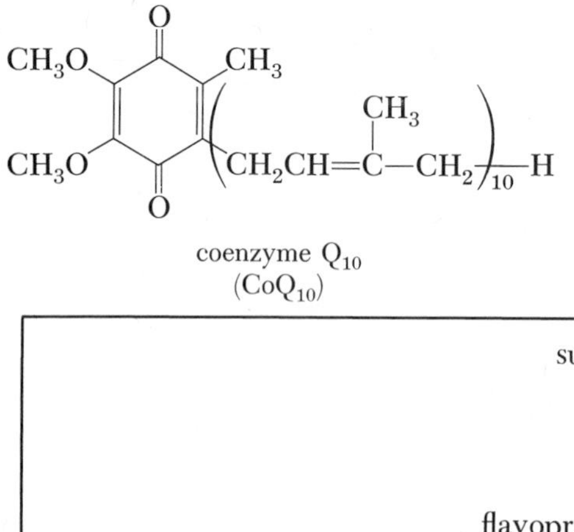

coenzyme Q_{10} (CoQ_{10}): ring substituents CH_3O, CH_3O, CH_3, two C=O, and side chain $\left(CH_2CH{=}C(CH_3){-}CH_2\right)_{10}{-}H$

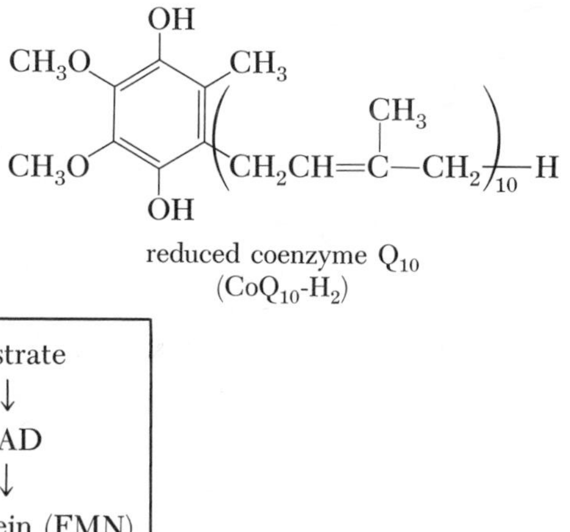

reduced coenzyme Q_{10} (CoQ_{10}-H_2): ring substituents CH_3O, CH_3O, CH_3, two OH, and side chain $\left(CH_2CH{=}C(CH_3){-}CH_2\right)_{10}{-}H$

substrate
↓
NAD
↓
flavoprotein (FMN)
↓
substrate → flavoprotein (FAD) → coenzyme Q
↓
cytochrome system
↓
oxygen

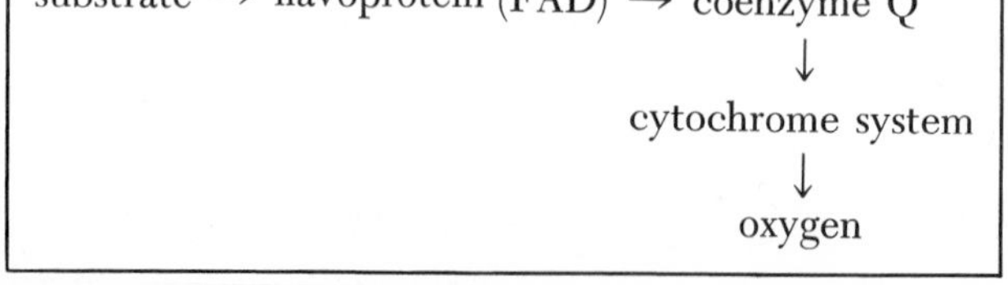

Figure 22-1 The role of coenzyme Q in the flow of electrons in mitochondria. The oxidation of one compound is accompanied by the reduction of the next along the line of arrows. The ultimate products are oxidized substrate and reduced oxygen, found as water. Note that the two types of systems (one directly oxidized by a flavoprotein, the other requiring NAD) join at coenzyme Q, from which electrons are fed to oxygen *via* the cytochrome system.

[a] The structures of ADP and ATP are discussed on page 854.

What makes CoQ_{10} all the more important in this electron transport process is that it appears to represent a branching point in the chain. As seen from Figure 22–1, CoQ_{10} funnels electrons to oxygen from two different redox systems, one based upon FMN (flavin mononucleotide) and the other upon FAD (flavin adenine dinucleotide). Even these complex molecules are quinone-like. To emphasize this similarity, the quinoidal array in these compounds is shown in boldface in the structures on page 704. Thus, the redox properties of quinones, which may pervade the behavior of complex heterocyclic molecules, is essential to living systems.

22.11 CHARACTERIZATION OF PHENOLS AND QUINONES

At this point we will abandon our separate treatment of phenols and quinones, ending this chapter with a brief look at the means of characterizing these classes of compounds.

Chemical Characterization. Many phenols give a color upon addition of a drop of aqueous ferric chloride to a solution of the phenol in water or alcohol. This test is negative with *m*- and *p*-hydroxybenzoic acids and with nitrophenols. Furthermore, compounds with moderate enol concentrations also give positive ferric chloride tests. Even though this test is very old, the origin of the color is not entirely clear.

Sodium hypochlorite oxidizes 2,6-dichloro-4-aminophenol to the corresponding *quinone chlorimine* (a quinonimine). This reagent reacts with alkaline solutions of phenols to produce intensely colored *indophenols* and their salts. This becomes, then, a useful color test for phenols with available *ortho* and/or *para* positions.

2,6-dichloro-4-aminophenol (OH, Cl, Cl, NH_2) + 2NaOCl → quinone chlorimine (O, Cl, Cl, N–Cl) + 2NaCl + $2H_2O$

$O{=}C_6H_2Cl_2{=}NCl$ + C_6H_5–OH $\xrightarrow{OH^{\ominus}}$ $O{=}C_6H_2Cl_2{=}N{-}C_6H_4{-}O^{\ominus}$ (blue)

↓ $H_3O^{\oplus}$

$O{=}C_6H_2Cl_2{=}N{-}C_6H_4{-}OH$ (red)

an indophenol

The color of a compound or the observation of fluorescence (or both) often is a clue to the presence of a quinone. Since quinones are capable of oxidizing iodide ion to iodine,

flavin adenine dinucleotide
(FAD)

flavin adenine dinucleotide (reduced)
(FAD—H_2)

flavin mononucleotide
(FMN)

flavin mononucleotide (reduced)
(FMN—H_2)

immediate generation of a purple-brown color from a sample dissolved in hydriodic acid can be used to test for quinones. However, since other oxidants may accomplish the same result, the test is non-specific.

Quinones, like simple ketones, can form derivatives with 2,4-DNPH. However, the product is not always the anticipated hydrazone.

$$O_2N\text{—}C_6H_3(NO_2)\text{—}NHNH_2 + O{=}C_6H_4{=}O \xrightarrow{-H_2O} O_2N\text{—}C_6H_3(NO_2)\text{—}N{=}N\text{—}C_6H_4\text{—}OH$$

2,4-DNPH
(2,4-dinitrophenylhydrazine)

The unusual *quinoxaline* system is formed when *o*-quinones (and other 1,2-diketones) react with *o*-phenylenediamine.

$$C_6H_4(NH_2)_2 + \text{phenanthrenequinone} \rightarrow \text{quinoxaline}$$

o-phenylenediamine   a quinoxaline

Spectral Characterization. The **infrared spectra** of phenols and alcohols are sufficiently similar that they cannot be *easily* distinguished by ir spectroscopy. The C—O stretching frequency of phenols occurs at about 1230 cm^{-1}, while the corresponding absorption in alcohols occurs from 1200 to 1050 cm^{-1}. Phenols exhibit characteristic aromatic absorption (see Chapter 15). The ir spectra of several phenols are presented in Figure 22–2.

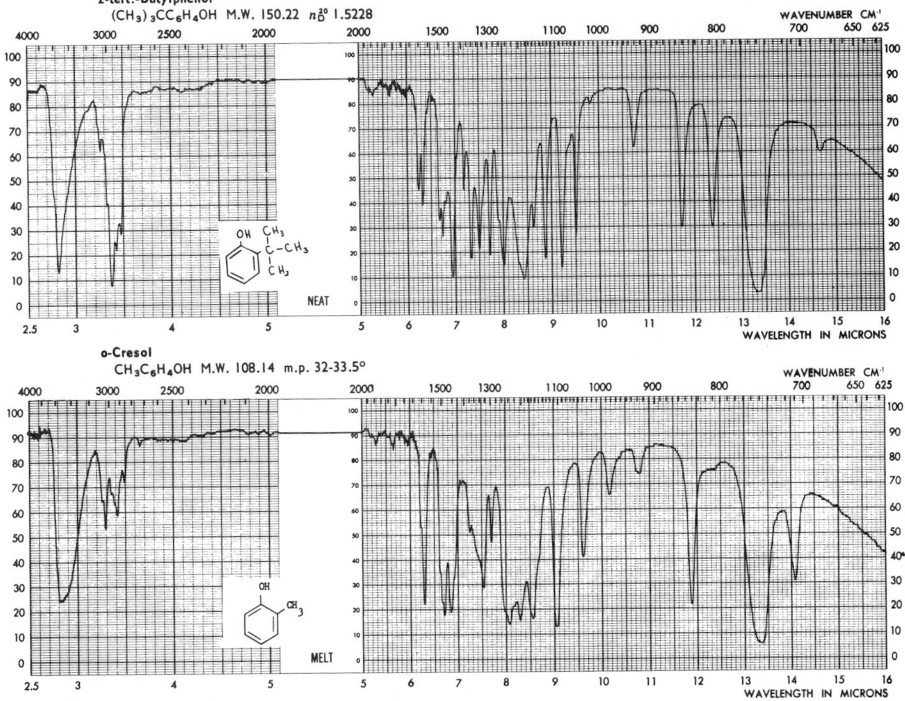

Figure 22–2 The infrared spectra of phenols. *Illustration continued on following page.*

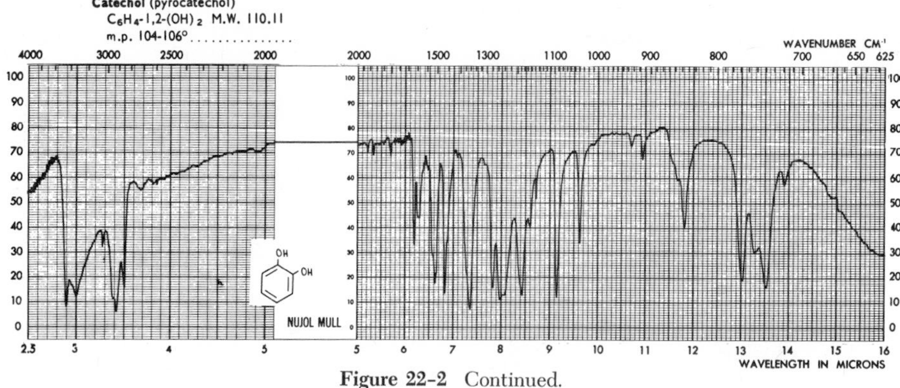

Figure 22-2 Continued.

Quinones usually show the carbonyl absorption near 1675 cm^{-1}, although the intramolecular hydrogen bonding can reduce the frequency to ~1630 cm^{-1} with a concomitant diminution in the intensity of the 1675 cm^{-1} band. Representative quinone spectra are shown in Figure 22-3.

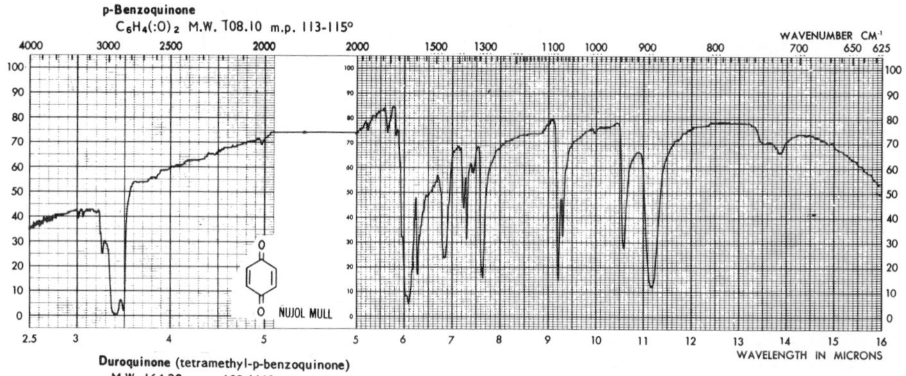

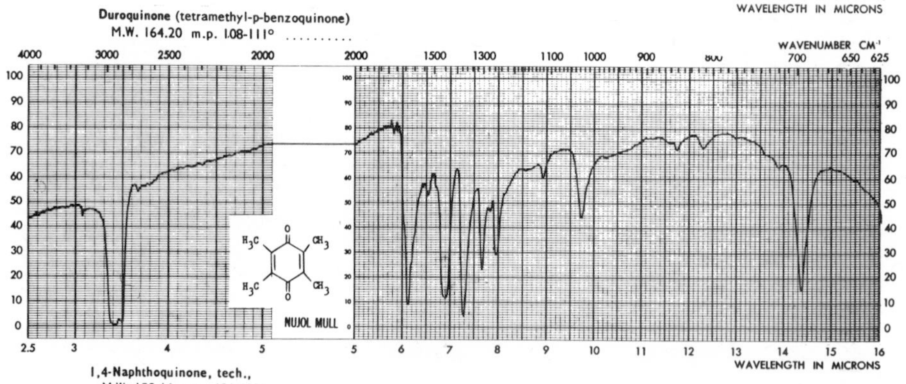

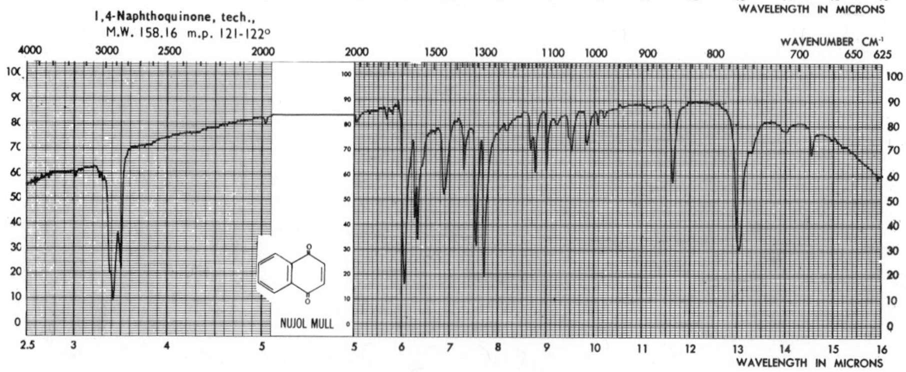

p-Benzoquinone
$C_6H_4(:O)_2$ M.W. 108.10 m.p. 113-115°
NUJOL MULL

Duroquinone (tetramethyl-p-benzoquinone)
M.W. 164.20 m.p. 108-111°
NUJOL MULL

1,4-Naphthoquinone, tech.,
M.W. 158.16 m.p. 121-122°
NUJOL MULL

Figure 22-3 The infrared spectra of quinones. *Illustration continued on opposite page.*

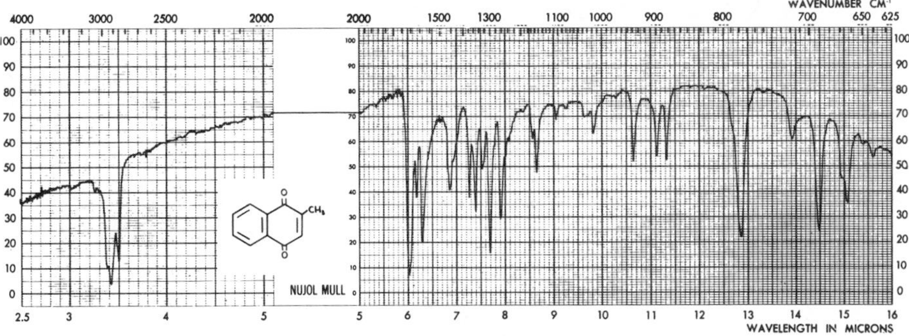

Figure 22-3 Continued.

The **nmr absorption** of the O—H proton of phenols may occur over a range from 4 to 12δ, depending upon solvent, temperature, concentration, and the presence (or absence) of hydrogen-bonding. Those absorptions which occur very far downfield usually represent protons experiencing intramolecular hydrogen bonding. The nmr spectra of several phenols and quinones are presented in Figures 22–4 and 22–5, respectively.

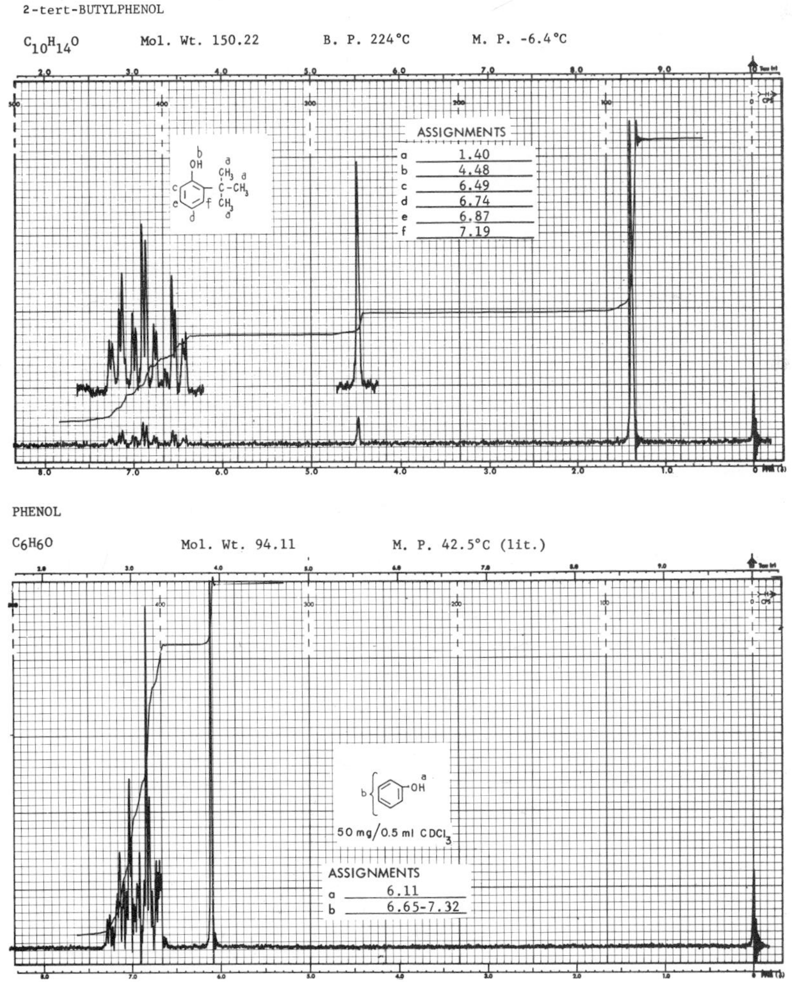

Figure 22-4 The nmr spectra of phenols. © Sadtler Research Laboratories, Inc., 1976. *Illustration continued on following page.*

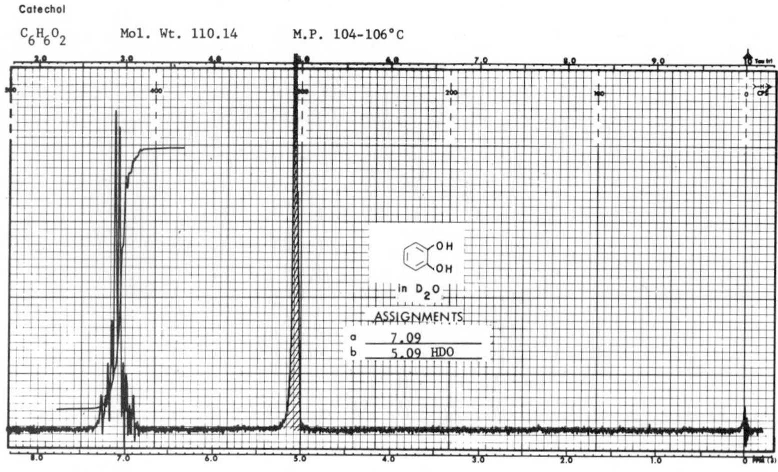

Figure 22-4 Continued.

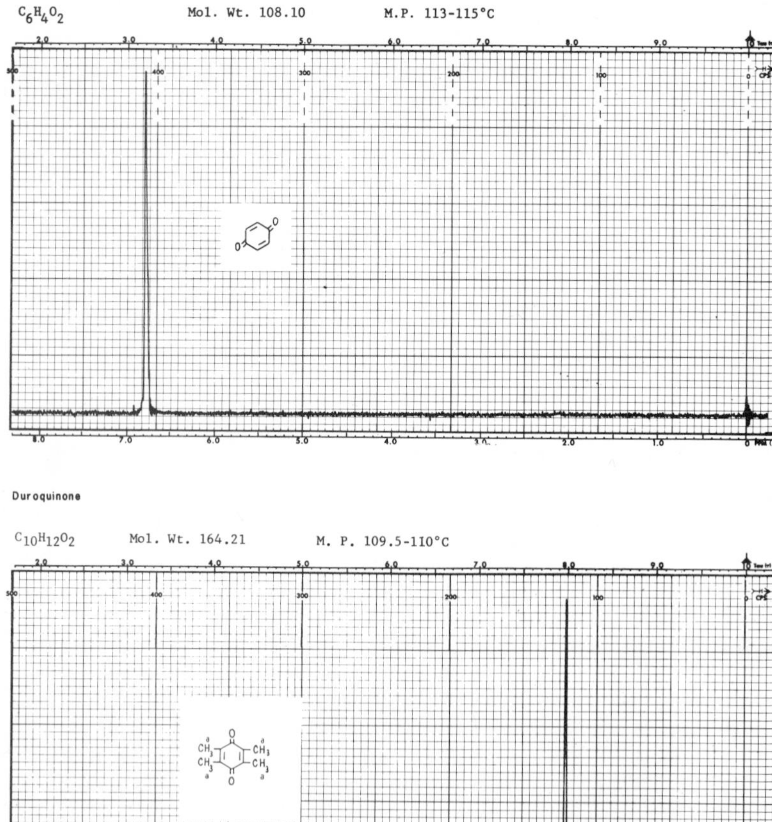

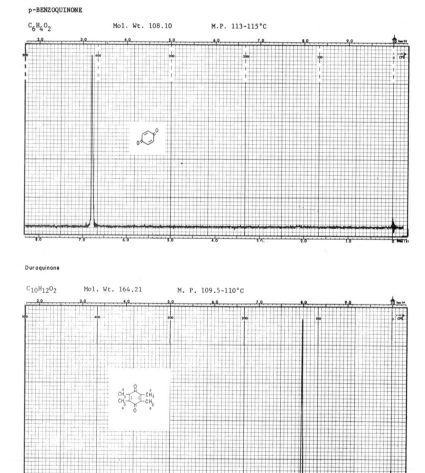

Figure 22-5 The nmr spectra of quinones. © Sadtler Research Laboratories, Inc., 1976. *Illustration continued on facing page.*

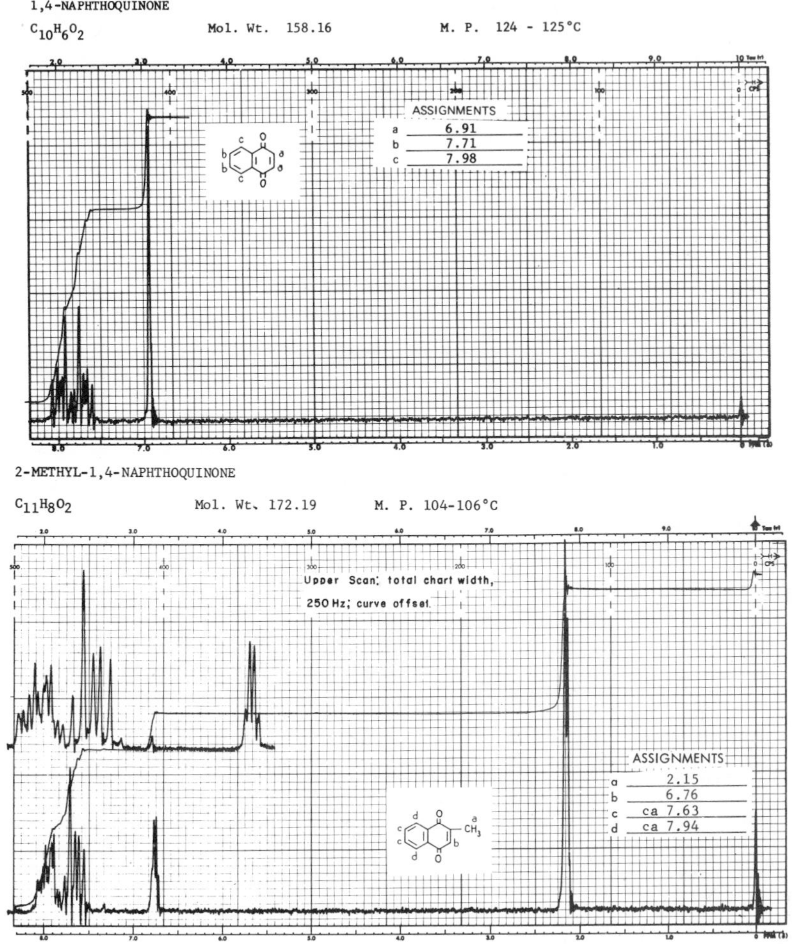

Figure 22-5 Continued.

IMPORTANT TERMS

Arylboronic acid: A compound of the type $ArB(OH)_2$. Oxidation of arylboronic acids produces phenols.

$$\text{Ar—MgX} + (\text{RO})_3\text{B} \rightarrow \text{Ar—B(OR)}_2 \xrightarrow{\text{H}_3\text{O}^\oplus} \underset{\text{arylboronic acid}}{\text{Ar—B(OH)}_2}$$

$$\text{Ar—B(OH)}_2 \xrightarrow[\text{H}_3\text{O}^\oplus]{15\%\ \text{H}_2\text{O}_2} \text{Ar—OH}$$

Claisen rearrangement: The thermal isomerization of an allyl phenyl ether to an *o*-allylphenol.

$$\text{allyl phenyl ether} \xrightarrow{200^\circ} \text{dienone intermediate} \xrightarrow{\text{tautomerize}} o\text{-allylphenol}$$

allyl phenyl ether (O—CH_2—CH=CH_2 on benzene ring); dienone intermediate (ring C=O, CH_2—CH=CH_2, H); *o*-allylphenol (OH, CH_2—CH=CH_2)

If the *ortho* position is blocked, that is, if tautomerization to a phenol is prevented by the absence of an ortho hydrogen, isomerization continues and the allyl group ends up on the *para* position. In the example below, step *b*, like step *a*, involves a six-membered cyclic transition state.

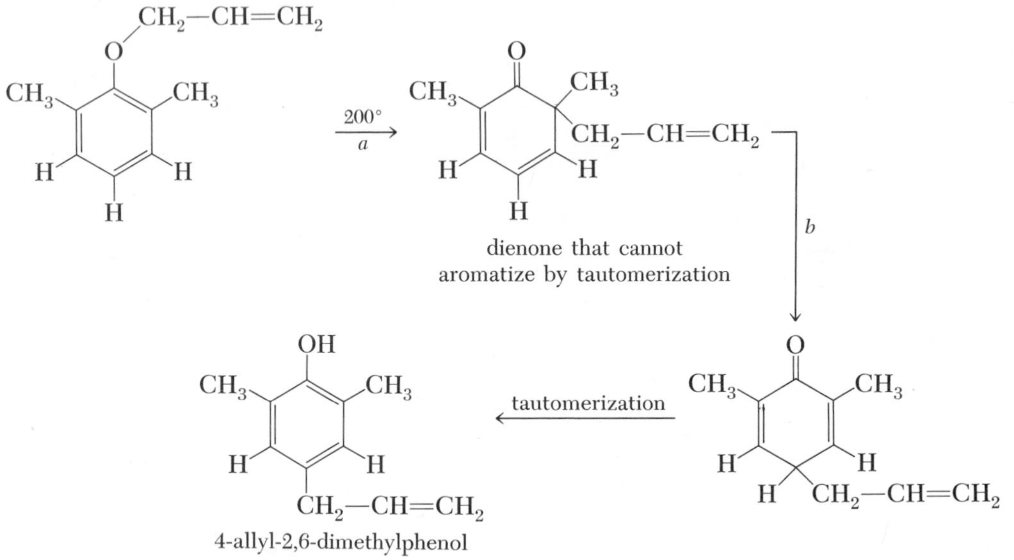

$CH_2—CH{=}CH_2$ O CH_3 CH_3 H H H $\xrightarrow[a]{200°}$ O CH_3 CH_3 $CH_2—CH{=}CH_2$ H H H

dienone that cannot aromatize by tautomerization

$\downarrow b$

O CH_3 CH_3 H H H $CH_2—CH{=}CH_2$ $\xrightarrow{\text{tautomerization}}$ OH CH_3 CH_3 H H $CH_2—CH{=}CH_2$

4-allyl-2,6-dimethylphenol

The Claisen rearrangement, like the Diels-Alder reaction and other processes discussed in Chapter 13, does not involve any of the standard reactive intermediates (anion, cation, or radical) of organic chemistry.

Herbicide: A material that will kill plants.

Oxidative phosphorylation: The coupling of biological oxidation (which releases energy) with a phosphorylation (which converts adenosine *di*phosphate to adenosine *tri*phosphate, ADP → ATP). These processes occur simultaneously.

coupled reactions { oxidation-reduction reaction → energy
adenosine diphosphate + phosphate + energy → adenosine triphosphate

Phenol coefficient: A number that reflects the disinfectant ability of a compound compared to that of phenol, which is arbitrarily given a value of 1.

Quinhydrone: A 1:1 complex between quinone and hydroquinone. Quinhydrone, a dark green solid that precipitates when alcoholic solutions of quinone and hydroquinone are mixed, dissociates slightly in solution.

$$\text{quinhydrone} \rightleftarrows \text{quinone} + \text{hydroquinone}$$

Quinonimine: An imine analog of a quinone. Quinonimines are much less common (and stable) than are quinones.

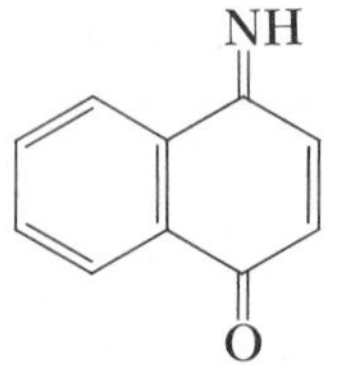

NH

O

Quinonmethide: An alkylidene analog of a quinone. Quinonmethides are much less common (and stable) than are quinones.

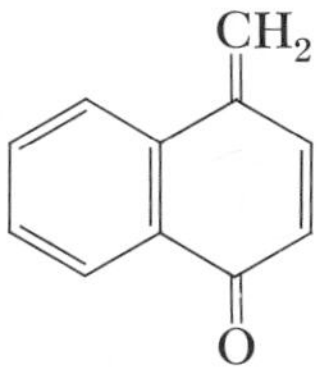

CH_2

O

PROBLEMS

13. Provide a reasonable name for each of the following:

(a) O, CN, CN, O

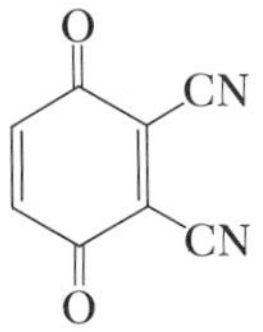

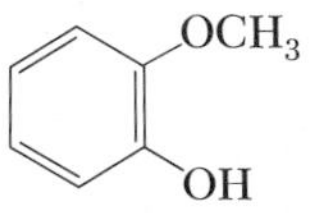

(f) O, C—H, O—C—CH_3, O

(b) O, Cl, Cl, O

(g) OH, CHO, Cl

(c) Cl, O, Cl, O

(h) OH, O

(d) OH, OH, C_6H_5

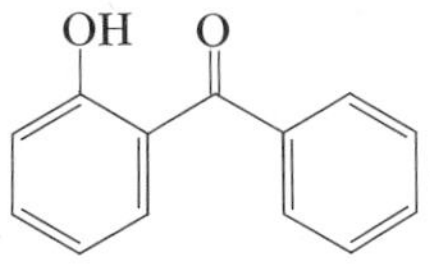

(i) O, O

(e) OCH_3, OH

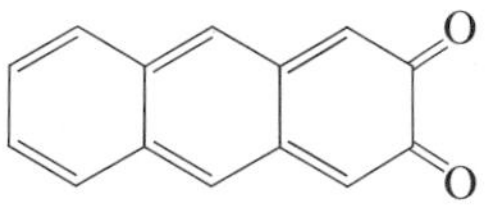

(j) OH, $SO_3^{\ominus}$ $Na^{\oplus}$

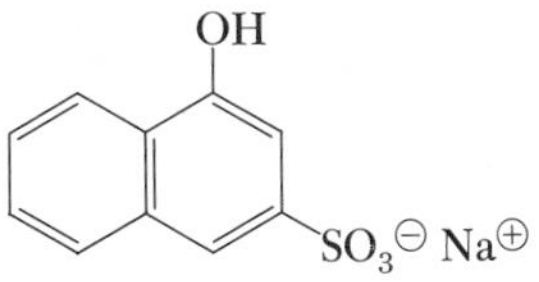

14. Predict the product(s) of the following reactions. If no reaction occurs, indicate with "NR."

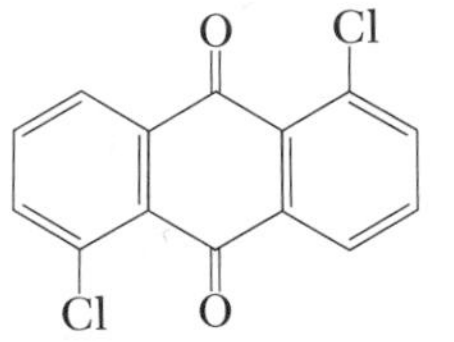

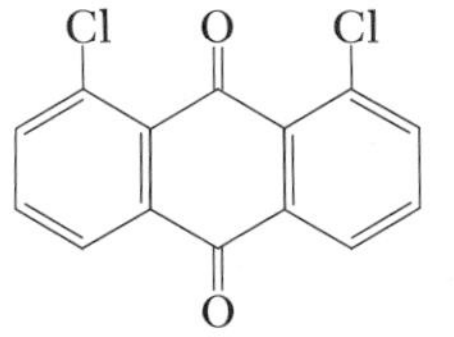

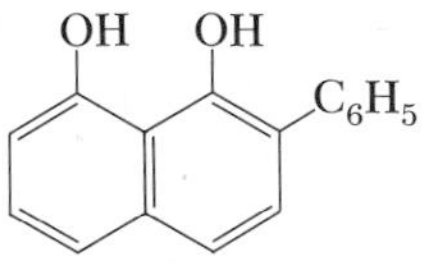

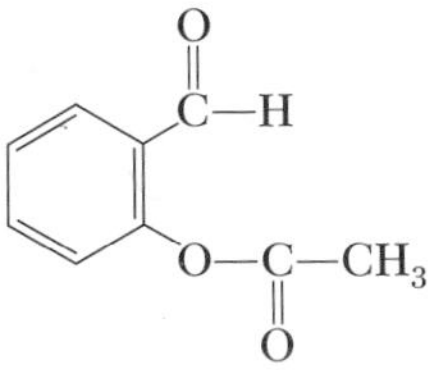

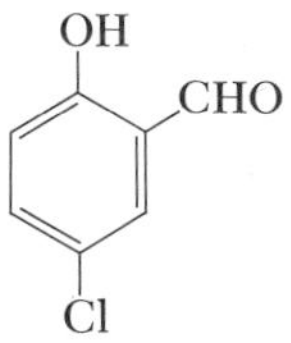

(a) OH $\xrightarrow[H_2O]{OH^{\ominus}}$ $\xrightarrow{(CH_3)_3CBr}$

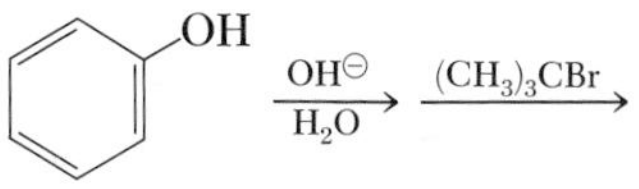

(b) —OH $\xrightarrow[H_2O]{OH^{\ominus}}$ $\xrightarrow{C_6H_5Br}$

(c) OH $\xrightarrow[H_2O]{OH^{\ominus}}$ $\xrightarrow{C_6H_5N_2^{\oplus}\ HSO_4^{\ominus}}$

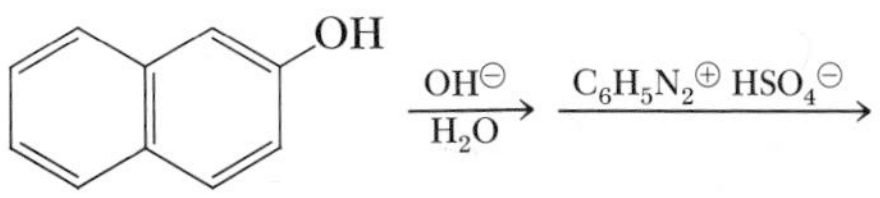

(d) OH + Cl_2(excess) $\xrightarrow{H_2O}$

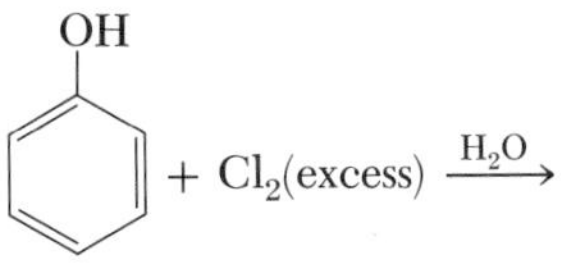

(e) CH_3, —O—C(=O)— $\xrightarrow{AlCl_3}$

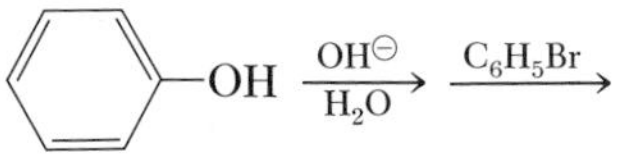

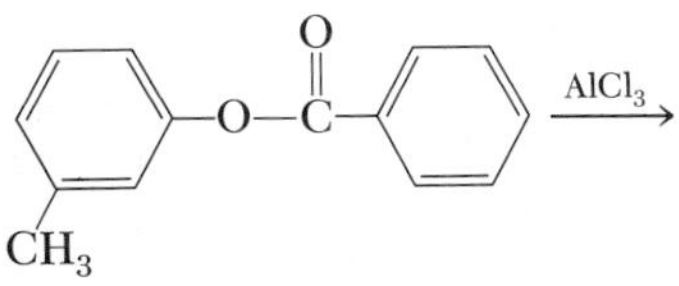

(f) 8-hydroxy-1-naphthoic acid (OH, CO_2H) $\xrightarrow{P_2O_5}$

(g) $C_6H_5{-}O{-}C(=O){-}C_6H_4{-}CH_3$ (m-) $\xrightarrow{AlCl_3}$

(h) 1,2,4,5-tetrahydroxybenzene (HO, OH, HO, OH) $\xrightarrow[H_2O]{OH^{\ominus}\text{ (excess)}}$ $\xrightarrow{CH_2I_2}$

(i) $C_6H_5OH \xrightarrow[\text{heat}]{D_2O/OD^{\ominus}}$

(j) $C_6H_5{-}O{-}C(CH_3)_3 \xrightarrow{BF_3}$

(k) $CH_3{-}C_6H_4{-}O{-}CH(CH_3){-}CH{=}CH_2 \xrightarrow{200^\circ}$

(l) m-cresol (CH_3, OH) $+ NH_3 \xrightarrow{C_6H_6}$

(m) H_3C-substituted 1-naphthol (OH) $\xrightarrow[AlCl_3]{HCN/HCl}$ $\xrightarrow{H_2O}$

(n) 1-naphthol (OH) $\xrightarrow[H_2SO_4,\ 0^\circ]{K_2Cr_2O_7}$

(o) $HO{-}C_6H_4{-}OH \xrightarrow[H_2SO_4,\ 0^\circ]{Cr_2O_7^{2-}}$

(p) o-$C_6H_4(N_2^{\oplus})(CO_2^{\ominus})$ + p-benzoquinone (O, O) $\rightarrow$

*15. When 2-methylnaphthalene is oxidized with $K_2Cr_2O_7/H_2SO_4$, the product possesses the ir and nmr spectra shown below. On the basis of these data, suggest a structure for this oxidation product. Suggest an alternate synthesis of this compound, beginning with naphthalene.

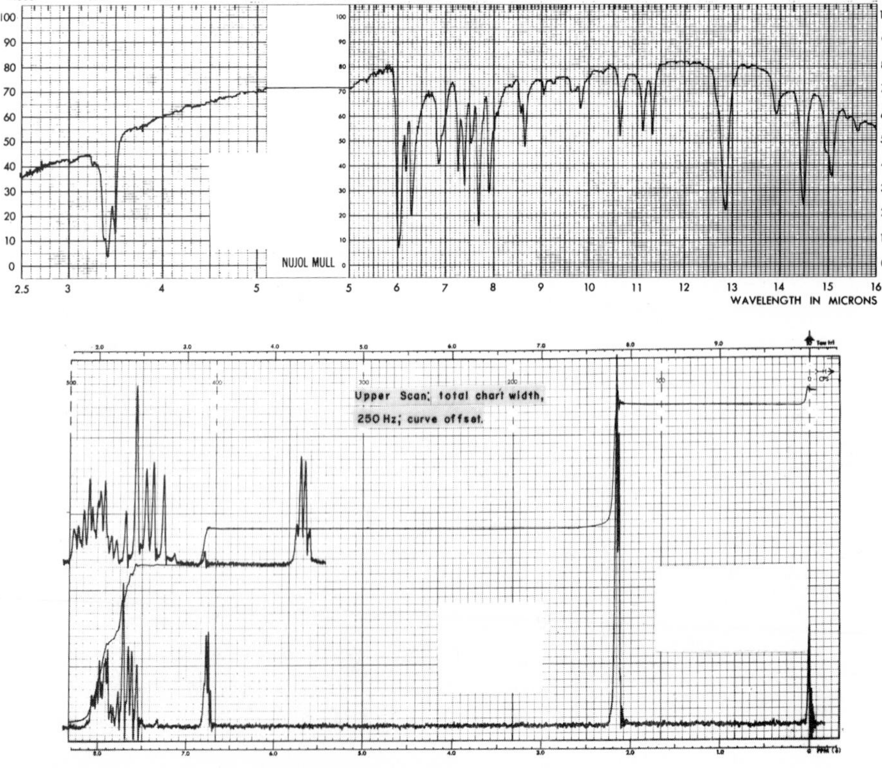

*This problem requires knowledge of spectroscopy.

16. When compound **A** is warmed in a mixture of $CH_3OD/NaOCH_3$, the product **B** is formed. Anthracene, on the other hand, is not converted to **C** under comparable conditions. Account for the conversion of **A** to **B**, bearing in mind the difference between **A** and **C**.

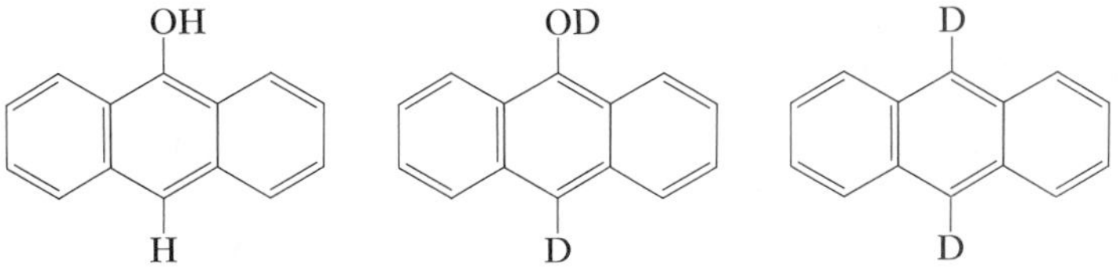

17. Arrange the following in the order of increasing acidity.

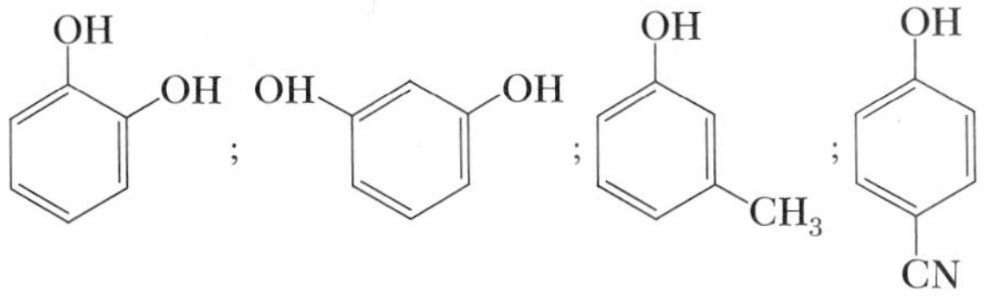

18. Gallic acid is easily thermolyzed (decomposed by heat) to pyrogallol, while benzoic acid is not easily thermolyzed to benzene. Account for this difference.

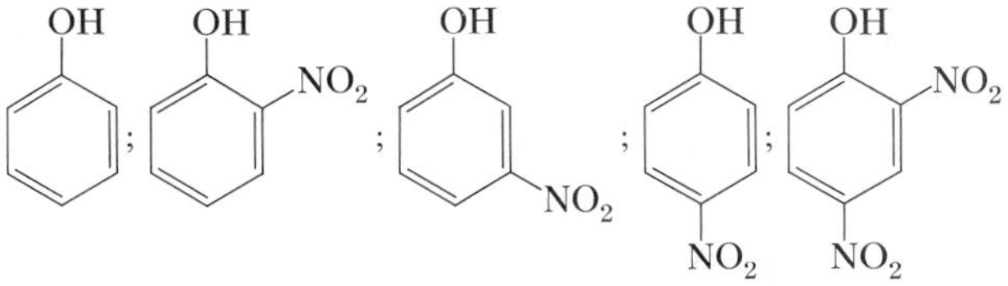

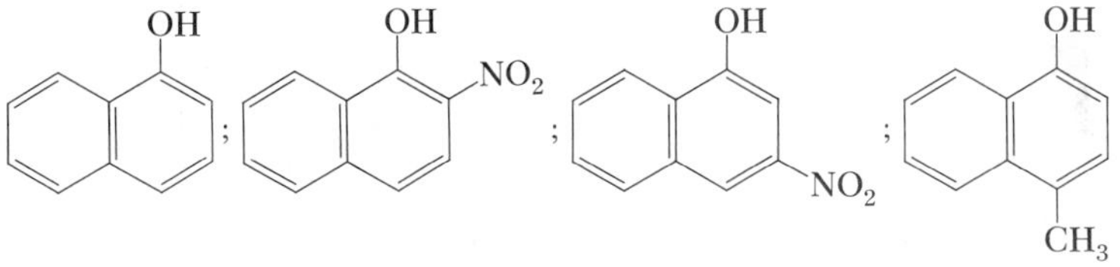

19. When an equimolar mixture of 2-cyano-1,4-benzoquinone (**A**) and hydroquinone (**B**) are mixed together, there is produced an equilibrium mixture containing these two compounds *and* 1,4-benzoquinone (**C**) and 2-cyano-1,4-dihydroxybenzene (**D**). Equal amounts of **C** and **D** are present in the equilibrium mixture. Equal amounts of **A** and **B** are present at equilibrium, but this is less than the amount of **C** and **D**. Account for these observations.

20. Suggest reactions to achieve the following conversions.
   (a) phenol to cyclohexanol
   (b) phenol to cyclohexanone
   (c) phenol to adipic acid
   (d) cyclohexanone to phenol
   (e) benzene to *m*-nitroanisole
   (f) phenol to *o*-propylphenol
   (g) chlorobenzene to 2,4-dinitrophenol
   (h) benzoic acid to 3,5-dinitroanisole
   (i) phenol to *p*-aminophenol
   (j) nitrobenzene to *p*-aminophenol
   (k) resorcinol to 4-ethyl-1,3-dihydroxybenzene
   (l) nitrobenzene to 4,4′-dimethoxybiphenyl
   (m) azobenzene to *o*-hydroxybenzaldehyde

21. The following sequence can be used to synthesize adrenaline (epinephrine). Identify the intermediates **A, B,** and **C.**

$$o\text{-nitrophenol} \xrightarrow[\text{Pt}]{H_2} \xrightarrow[\text{HCl}]{\text{HONO}} \xrightarrow[\text{heat}]{H_2O} \mathbf{A} \xrightarrow[PCl_3]{ClCH_2COCl} \mathbf{B} \xrightarrow[\text{1 eq.}]{H_2NCH_3} \mathbf{C} \xrightarrow[\text{Pt}]{H_2} \text{adrenaline}$$

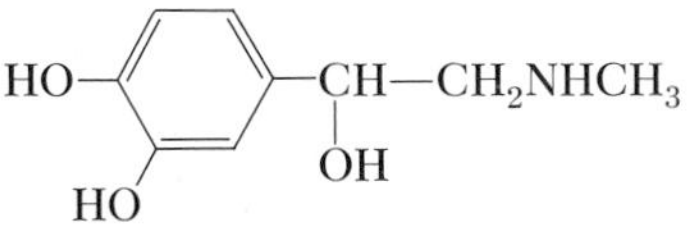

adrenaline (epinephrine)

Adrenaline, noradrenaline (also called norepinephrine), and dopamine are collectively called *catecholamines*. Suggest a reason for this name. Suggest syntheses of norepinephrine and dopamine, starting with any disubstituted benzene derivative.

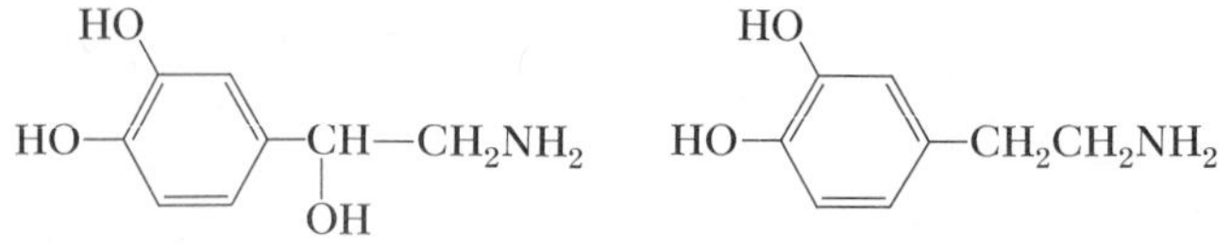

HO, HO— (ring) —CH—CH$_2$NH$_2$ with OH on CH

norepinephrine

HO, HO— (ring) —CH$_2$CH$_2$NH$_2$

dopamine

*22. Identify **A, B, C,** and **D** in this sequence.

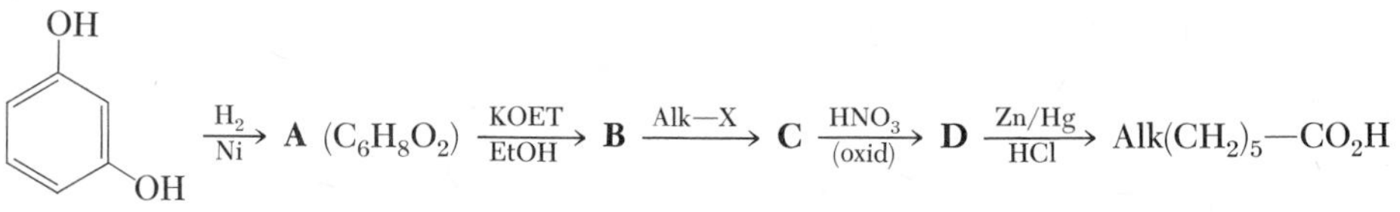

$$\text{(1,3-dihydroxybenzene: OH, OH)} \xrightarrow[\text{Ni}]{H_2} \textbf{A}\ (C_6H_8O_2) \xrightarrow[\text{EtOH}]{\text{KOET}} \textbf{B} \xrightarrow{\text{Alk—X}} \textbf{C} \xrightarrow[\text{(oxid)}]{HNO_3} \textbf{D} \xrightarrow[\text{HCl}]{\text{Zn/Hg}} \text{Alk}(CH_2)_5\text{—}CO_2H$$

The ir spectrum of **A** is shown below. Compound **D** reacts with 2,4-dinitrophenylhydrazine.

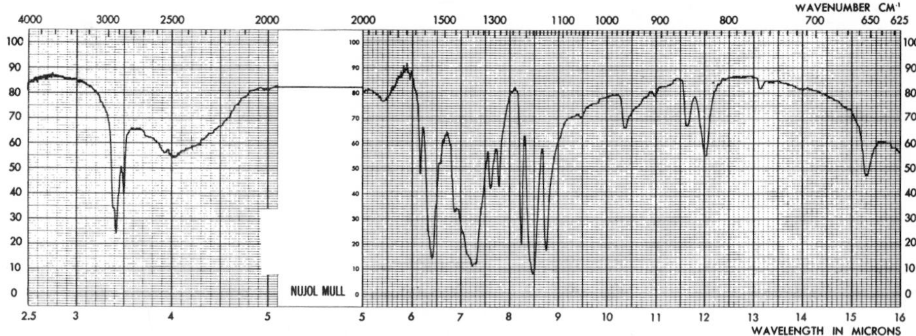

23. Suggest a simple means, preferably a test tube reaction, that would distinguish between members of the following pairs.

(a) aniline and phenol
(b) aniline and benzoic acid
(c) phenol and *p*-hydroxybiphenyl
(d) *cis*-4-*t*-butylcyclohexanol and 2-naphthol
(e) 2,4,6-trimethylphenol and *p*-cresol
(f) benzoic acid and salicylic acid
(g) 2,4,6-trinitrophenol (picric acid) and *m*-nitrophenol

24. Under certain circumstances phenol can be nitrosated to produce the same compound that arises from the reaction of *p*-benzoquinone with 1 mole of hydroxylamine (H_2NOH). Account for this.

25. Why can't the reaction shown below be used to prepare salicylaldehyde?

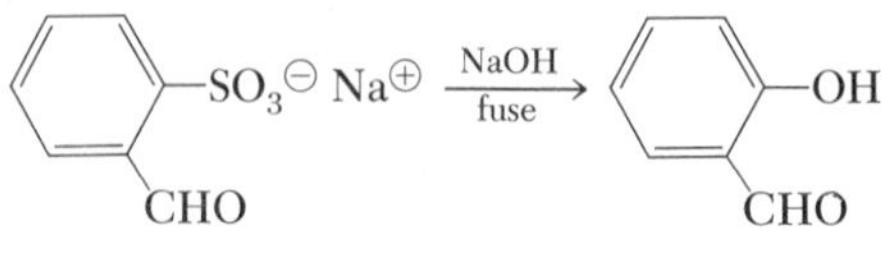

$$\text{(ring, CHO)}\text{—}SO_3^{\ominus}\ Na^{\oplus} \xrightarrow[\text{fuse}]{\text{NaOH}} \text{(ring, CHO)}\text{—OH}$$

26. BHA (*b*utylated *h*ydroxy*a*nisole) is a mixture of 2- and 3-*t*-butyl-4-methoxyphenol. This mixture is an anti-oxidant and is used to prevent rancidity in foods. Suggest a synthesis for either isomer of BHA.

27. *Guaiacol*, produced commercially from wood tar, is used to produce expectorants found in cough remedies. This compound is the isomeric methoxyphenol, which exhibits hydrogen-bonding (evidence from ir) even in very dilute solutions. What is its structure?

---

*This problem requires knowledge of spectroscopy.

28. A number of phenols and phenolic derivatives cause the fragrances of many natural products. Some of them, and their sources, are presented below. Suggest syntheses for these, starting with any monosubstituted benzene derivative.

(a) 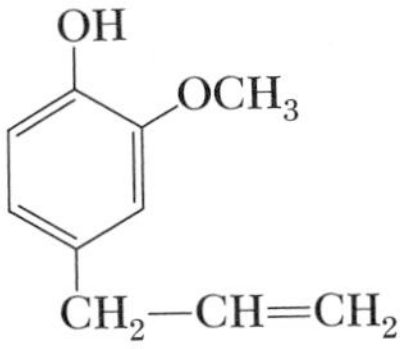 eugenol (oil of cloves): OH, OCH_3, $CH_2—CH=CH_2$

(b) 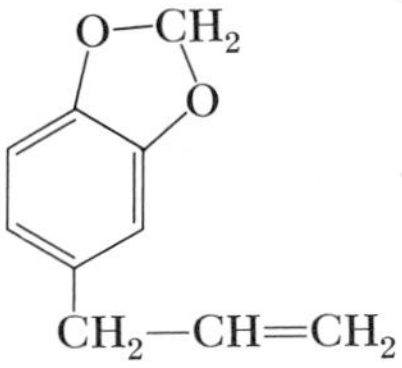 safrole (oil of sassafras): $O—CH_2—O$, $CH_2—CH=CH_2$

(c) 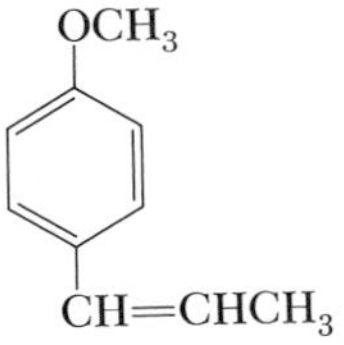 anethole (oil of anise): OCH_3, $CH=CHCH_3$

(d) 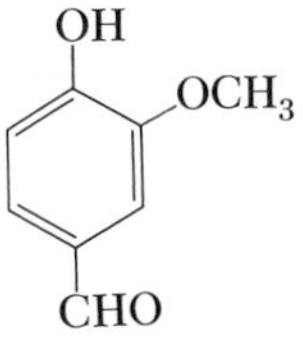 vanillin (oil of vanilla bean): OH, OCH_3, CHO

29. When 1-bromo-2-(*p*-hydroxyphenyl)ethane is solvolyzed to 1-hydroxy-2-(*p*-hydroxyphenyl)ethane, an isolable intermediate ketone, C_8H_8O, is formed. Give a structure for this, accounting for its formation. If it were not isolable, suggest an experiment using ^{14}C that might help establish its intermediacy.

30. Bromine is immediately decolorized upon addition to **A**, sodium 3,5-dibromo-4-hydroxybenzenesulfonate. The spectrum of a new substance (non-aromatic) forms and, after several hours, the spectrum of the solution becomes that of 2,4,6-tribromophenol. Account for these observations.

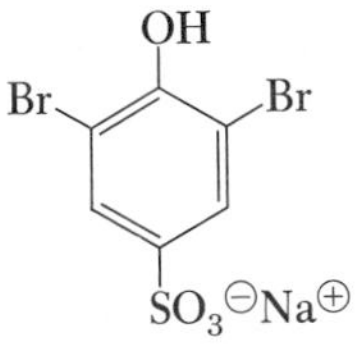

OH, Br, Br, $SO_3^{\ominus}Na^{\oplus}$

A

31. While quinones are attacked by nucleophiles at carbon, compound **A** (a quinonimmonium salt) reacts with a nucleophile (Z) to form **C** rather than **B**. Suggest why a nucleophilic addition to **A**, by attack at nitrogen, is more likely than is the related addition to a quinone, by attack at oxygen.

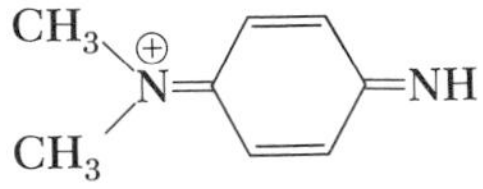

A: $(CH_3)_2\overset{\oplus}{N}=C_6H_4=NH$ [drawn as a ring structure]

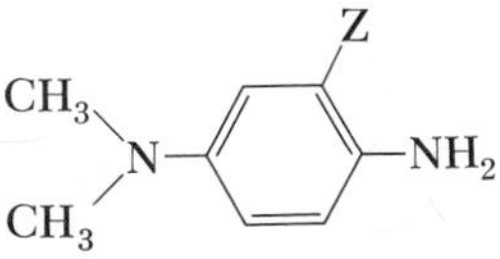

B: $(CH_3)_2N—C_6H_3(Z)—NH_2$ [drawn as a ring structure]

C: $(CH_3)_2N—C_6H_4—NHZ$ [drawn as a ring structure]

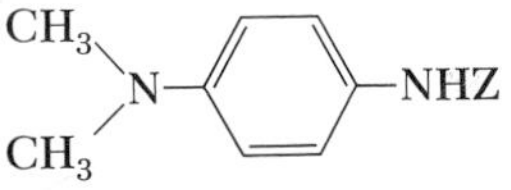

ORGANIC COMPOUNDS OF SULFUR AND PHOSPHORUS

INTRODUCTION

In its infancy, "organic chemistry" meant the study of compounds containing only carbon, hydrogen, nitrogen, and oxygen. The number of "organic" chemists working on compounds of the "inorganic elements" was small, and they often represented a fringe group. However, as theories advanced, as the role of heteroatoms in influencing life processes became clearer, and as the number of "organic" chemists increased, more and more studies bridging organic and inorganic chemistry appeared in the literature.

In response to the very real need to understand the chemistry of heteroatomic compounds, and in order to demonstrate that compounds containing heteroatoms are not *that* unusual, this chapter is devoted to sulfur and phosphorus chemistry. The student will note that, while the chapter is divided into two independent portions, problems for both are found at the end of the chapter.

SULFUR COMPOUNDS

23.1 SULFUR-CONTAINING FUNCTIONAL GROUPS

Many organosulfur compounds (*i.e.*, compounds containing a carbon-sulfur bond) are analogs of oxygen-containing compounds. Sulfur's position in the periodic table, immediately below oxygen, accounts for this similarity. However, while stable compounds of oxygen never involve more than three bonds to oxygen, stable sulfur compounds may possess as many as six bonds to sulfur. The ability to form six bonds, as in SF_6, results from sulfur's ability to employ 3*d* orbitals in its bonding, an attribute lacking in oxygen. One effect of this ability is to create two kinds of sulfur compounds, those that have oxygen analogs and those that do not.

The Oxygen Analogs. Just as alcohols and ethers are organic analogs of water, so *thiols* (also called *mercaptans*) and *sulfides* are organic analogs of hydrogen sulfide. The analogy extends to hydrogen peroxide, with *disulfides* being much more common than *hydrodisulfides*. The hypohalites are compounds of limited stability as are their sulfur analogs, *sulfenyl halides*.

$CH_3—S—H$ methanethiol	$CH_3—S—C_6H_5$ methyl phenyl sulfide	$CH_3—S—S—H$ methylhydrodisulfide
$CH_3—S—S—CH_3$ dimethyl disulfide	$CH_3—S—S—C(CH_3)_3$ *t*-butyl methyl disulfide	$C_6H_5—S—Cl$ phenylsulfenyl chloride

While the number of known aldehydes and ketones extends into the thousands, the number of *thioaldehydes* and *thioketones* that are known is quite limited. The only stable thioketones are diaryl thioketones containing potent electron donors on the aryl groups. This stability is due to the resonance interaction pictured below. It also is resonance stabilization that is responsible for the more-than-fleeting existence of the sulfur analogs of carboxylic acids and derivatives.

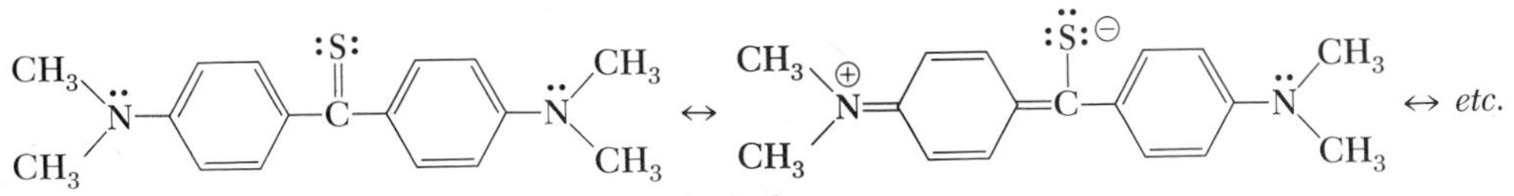

$$(CH_3)_2\ddot{N}-C_6H_4-C(=S)-C_6H_4-\ddot{N}(CH_3)_2 \leftrightarrow (CH_3)_2\overset{\oplus}{N}=C_6H_4=C(-S^{\ominus})-C_6H_4-\ddot{N}(CH_3)_2 \leftrightarrow etc.$$

a stable thioketone

$$R-C(=S)-\ddot{N}R_2 \leftrightarrow R-C(-S^{\ominus})=\overset{\oplus}{N}R_2$$

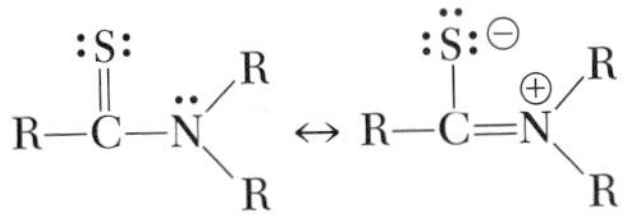

a stable thionamide

Oxygen Derivatives of Sulfides. Table 23–1 lists the more common types of organosulfur compounds, including those that have just been discussed, and illustrates the proper use of the nomenclature. The existence of the assortment of functional groups presented in Table 23–1 is due to the ability of sulfur to achieve ligancies higher than two, and to the existence of two kinds of sulfur-oxygen bonds, the nature of which will be discussed in Section 23.2. We will, in the interim, distinguish among them by writing one kind as an "ordinary" σ bond (S—O) and the other as a dative bond ($\overset{\oplus}{S}-\overset{\ominus}{O}$). It is interesting to see how, by starting with a simple compound (*e.g.*, diphenyl sulfide), one can generate a total of six stable organosulfur compounds just by adding and inserting oxygen. If the sulfide had been unsymmetric, *e.g.*, phenyl *p*-tolyl sulfide, several more products would have been included. (What would they be?)

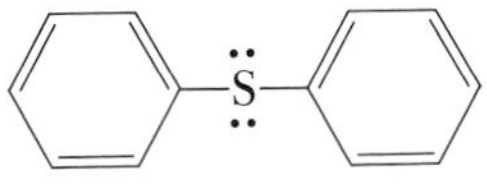

$C_6H_5-\ddot{S}-C_6H_5$ — diphenyl sulfide

$C_6H_5-\overset{\oplus}{S}(-O^{\ominus})-C_6H_5$ — diphenyl sulfoxide

$C_6H_5-\ddot{S}-\ddot{O}-C_6H_5$ — phenyl phenylsulfenate

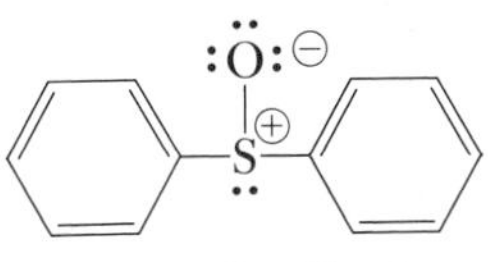

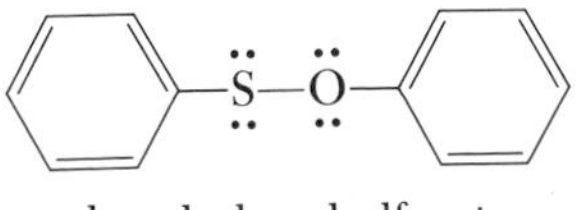

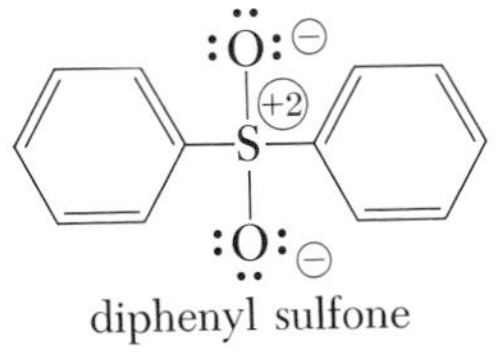

$C_6H_5-\overset{+2}{S}(-O^{\ominus})_2-C_6H_5$ — diphenyl sulfone

$C_6H_5-\overset{\oplus}{S}(-O^{\ominus})-\ddot{O}-C_6H_5$ — phenyl phenylsulfinate

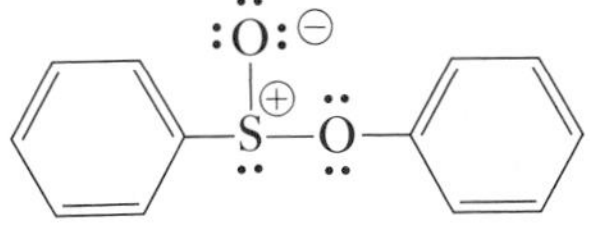

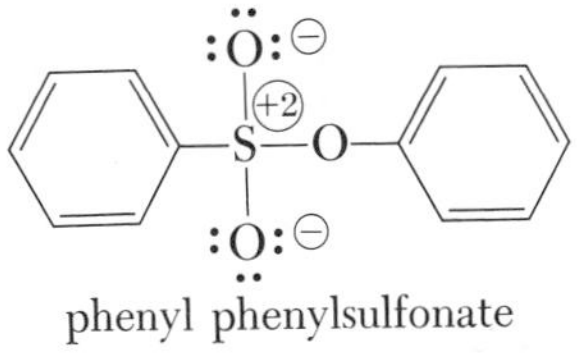

$C_6H_5-\overset{+2}{S}(-O^{\ominus})_2-O-C_6H_5$ — phenyl phenylsulfonate

The Oganosulfur Oxyacids. Another way to bring some order to the data in Table 23–1 is to recognize that a thiol can be treated in exactly the same way that diphenyl sulfide was treated. From a thiol one can generate three oxyacids and from these, in turn, can spring a host of derivatives. All of the types of compounds pictured below and on page 719 have been reported.

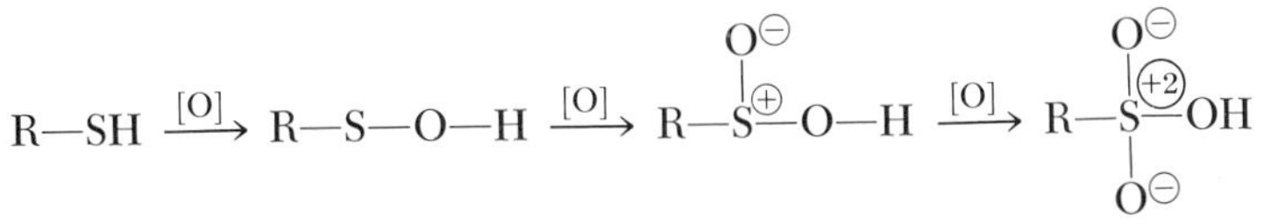

$$R-SH \xrightarrow{[O]} R-S-O-H \xrightarrow{[O]} R-\overset{\oplus}{S}(-O^{\ominus})-O-H \xrightarrow{[O]} R-\overset{+2}{S}(-O^{\ominus})_2-OH$$

thiol — sulfenic acid — sulfinic acid — sulfonic acid

TABLE 23-1 **Organosulfur Compounds**[a]

Compound	IUPAC Name	Alternate Names
CH_3-S-H	methanethiol	methyl mercaptan
CH_3-S-CH_3	methylthiomethane	**dimethyl sulfide**; dimethyl thioether; 2-thiapropane[b]
$CH_3-S-S-CH_3$	methyldithiomethane	**dimethyl disulfide**
$CH_3-\overset{CH_3}{\overset{\vert}{S}}{}^{\oplus}-CH_3\ Cl^{\ominus}$	trimethylsulfonium chloride	
$CH_3-S-O-CH_3$	methyl methanesulfenate	
$CH_3-S-O-H$	methanesulfenic acid	
$CH_3-\overset{O^{\ominus}}{\overset{\vert}{S}}{}^{\oplus}-CH_3$	methylsulfinylmethane	**dimethyl sulfoxide**
$CH_3-\overset{O^{\ominus}}{\overset{\vert}{S}}{}^{\oplus}-OH$	methanesulfinic acid	
$CH_3-\overset{O^{\ominus}}{\overset{\vert}{S}}{}^{\oplus}-Cl$	methanesulfinyl chloride	
$CH_3-\overset{O^{\ominus}}{\overset{\vert}{S}}{}^{\oplus}-OCH_3$	methyl methanesulfinate	
$CH_3-\underset{O^{\ominus}}{\underset{\vert}{\overset{O^{\ominus}}{\overset{\vert}{S}}}}{}^{(+2)}-CH_3$	methylsulfonylmethane	**dimethyl sulfone**
$CH_3-\underset{O^{\ominus}}{\underset{\vert}{\overset{O^{\ominus}}{\overset{\vert}{S}}}}{}^{(+2)}-Cl$	methanesulfonyl chloride	
$CH_3O-\overset{O^{\ominus}}{\overset{\vert}{S}}{}^{\oplus}-OCH_3$	dimethyl sulfite[c]	
$CH_3-\underset{O^{\ominus}}{\underset{\vert}{\overset{O^{\ominus}}{\overset{\vert}{S}}}}{}^{(+2)}-OH$	methanesulfonic acid	
$CH_3-\underset{O^{\ominus}}{\underset{\vert}{\overset{O^{\ominus}}{\overset{\vert}{S}}}}{}^{(+2)}-O-CH_3$	methyl methanesulfonate	
$CH_3O-\underset{O^{\ominus}}{\underset{\vert}{\overset{O^{\ominus}}{\overset{\vert}{S}}}}{}^{(+2)}-OCH_3$	dimethyl sulfate[c]	

[a] If an option is available, the more commonly used name is in boldface.
[b] The fragment "thia" reflects replacement of a hydrocarbon fragment by sulfur.
[c] This compound, although lacking C—S bonds, is included because of its importance.

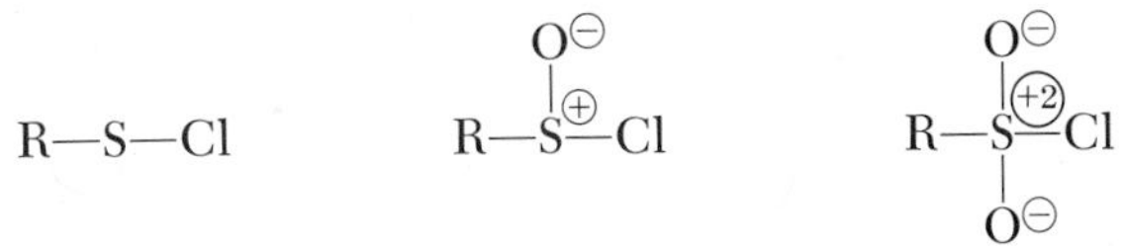

R—S—Cl	$R-\overset{O^{\ominus}}{\overset{\mid}{S^{\oplus}}}-Cl$	$R-\underset{O^{\ominus}}{\underset{\mid}{\overset{O^{\ominus}}{\overset{\mid}{S^{(+2)}}}}}-Cl$
sulfenyl chloride	sulfinyl chloride	sulfonyl chloride
$R-S-NR'_2$	$R-\overset{O^{\ominus}}{\overset{\mid}{S^{\oplus}}}-NR'_2$	$R-\underset{O^{\ominus}}{\underset{\mid}{\overset{O^{\ominus}}{\overset{\mid}{S^{(+2)}}}}}-NR'_2$
sulfenamide	sulfinamide	sulfonamide
R—S—O—R′	$R-\overset{O^{\ominus}}{\overset{\mid}{S^{\oplus}}}-OR'$	$R-\underset{O^{\ominus}}{\underset{\mid}{\overset{O^{\ominus}}{\overset{\mid}{S^{(+2)}}}}}-OR'$
sulfenate ester	sulfinate ester	sulfonate ester

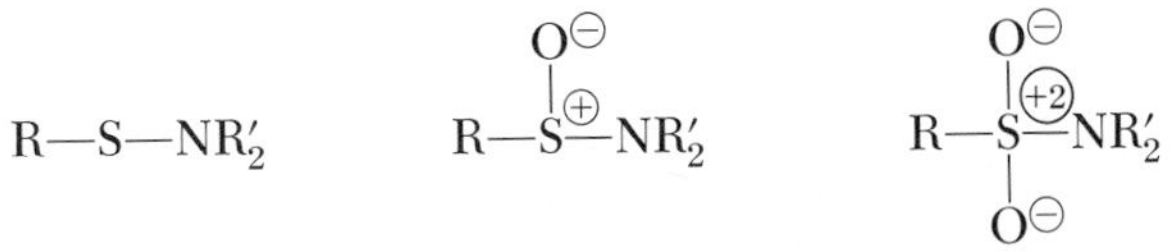

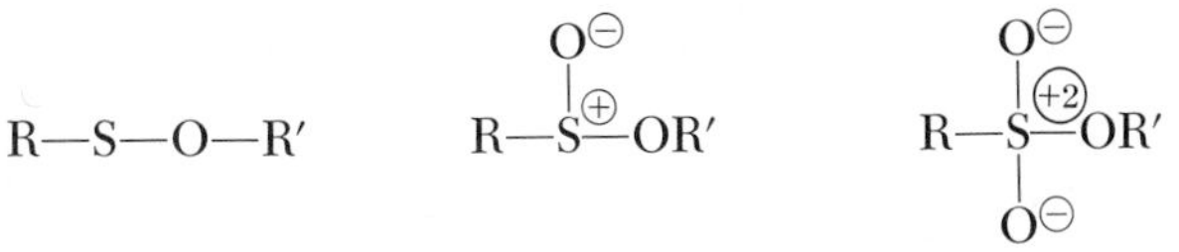

Occasionally you may encounter names of organosulfur compounds that are based upon the names of the corresponding oxygen compounds. In these instances the replacement of oxygen with sulfur is most commonly indicated by the inclusion of the term *thio* (or *thia*) in the name. For example, the sulfur analog of anisole ($C_6H_5OCH_3$) is called thioanisole ($C_6H_5SCH_3$). The sulfur analogs of 1,3-dioxolane and 1,4-dioxane are 1,3-dithiolane and 1,4-dithiane, respectively.

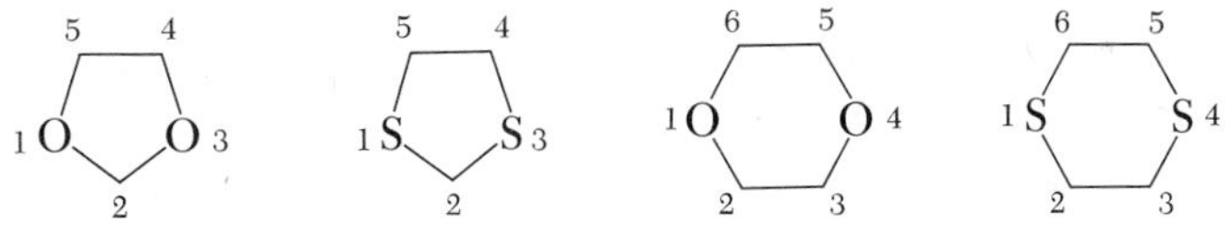

1,3-dioxolane	1,3-dithiolane	1,4-dioxane	1,4-dithiane
five-membered ring: O at 1 and 3; C at 2, 4, 5	five-membered ring: S at 1 and 3; C at 2, 4, 5	six-membered ring: O at 1 and 4; C at 2, 3, 5, 6	six-membered ring: S at 1 and 4; C at 2, 3, 5, 6

The replacement of a methylene or methine group with sulfur also may be signalled by the term "thia" as a means of modifying the name of the corresponding hydrocarbon. Several examples of this approach to the naming of sulfides are shown below.

$CH_3CH_2SCH_3$	$CH_3SCH_2CH_2CH(CH_3)_2$	$CH_3SCH_2SCH_3$
2-thiabutane	5-methyl-2-thiahexane	2,4-dithiapentane

With minor modification, to indicate the number and location of the oxygens, this scheme can be expanded to cover sulfoxides and sulfones. Three oxides of 2,5-dithiahexane are used to illustrate this approach.

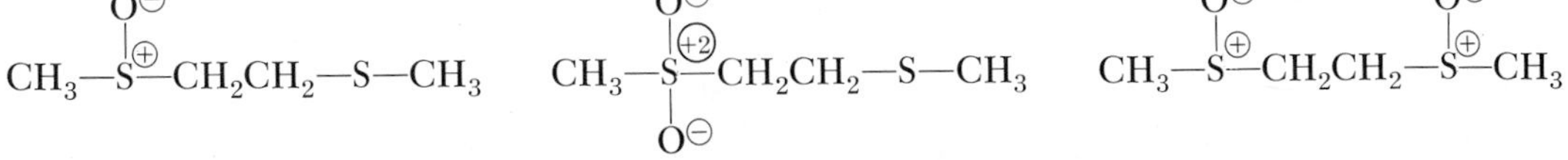

$CH_3-\overset{O^{\ominus}}{\overset{\mid}{S^{\oplus}}}-CH_2CH_2-S-CH_3$	$CH_3-\underset{O^{\ominus}}{\underset{\mid}{\overset{O^{\ominus}}{\overset{\mid}{S^{(+2)}}}}}-CH_2CH_2-S-CH_3$	$CH_3-\overset{O^{\ominus}}{\overset{\mid}{S^{\oplus}}}-CH_2CH_2-\overset{O^{\ominus}}{\overset{\mid}{S^{\oplus}}}-CH_3$
2,5-dithiahexane 2-oxide	2,5-dithiahexane 2,2-dioxide	2,5-dithiahexane 2,5-dioxide

1. Identify the types of functional groups in the following compounds.

(a) 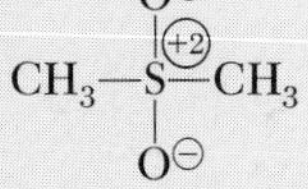 $CH_3-\overset{O^{\ominus}}{\underset{O^{\ominus}}{S^{+2}}}-CH_3$

(b) 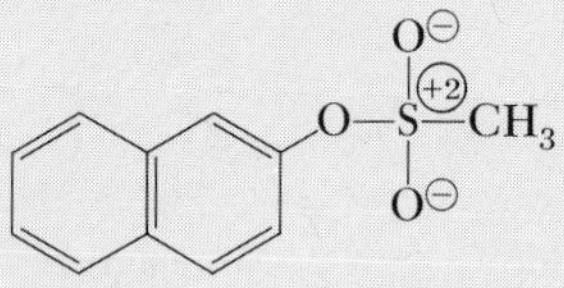 2-naphthyl–$O-\overset{O^{\ominus}}{\underset{O^{\ominus}}{S^{+2}}}-CH_3$

(c) 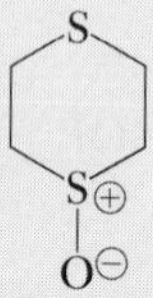 six-membered ring containing S and $S^{\oplus}-O^{\ominus}$

(d) 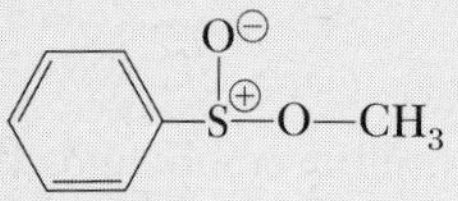 $C_6H_5-\overset{O^{\ominus}}{S^{\oplus}}-O-CH_3$

(e) 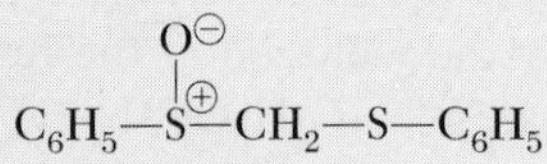 $C_6H_5-\overset{O^{\ominus}}{S^{\oplus}}-CH_2-S-C_6H_5$

(f) $C_6H_5-S-O-C_6H_5$

(g) $CH_3-S-S-C_6H_5$

(h) 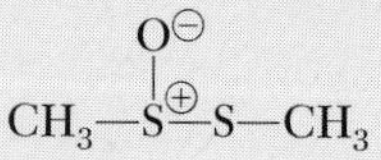 $CH_3-\overset{O^{\ominus}}{S^{\oplus}}-S-CH_3$

(i)  thiophene ring (S)

(j) 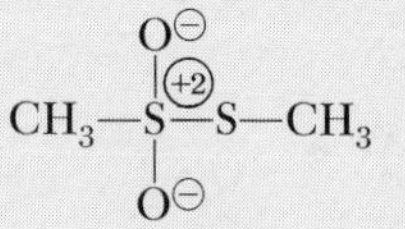 $CH_3-\overset{O^{\ominus}}{\underset{O^{\ominus}}{S^{+2}}}-S-CH_3$ (guess)

(k) 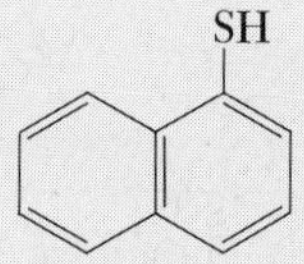 1-naphthyl–SH

(l) 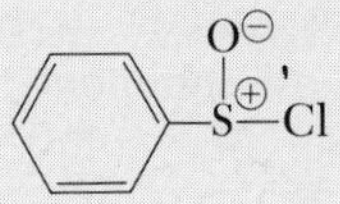 $C_6H_5-\overset{O^{\ominus}}{S^{\oplus}}-Cl$

2. Provide an acceptable name for each of the following.

(a) $CH_3-S-S-C_6H_5$

(b) 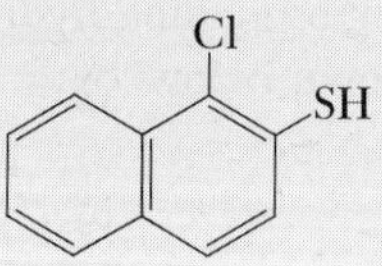 naphthalene ring with Cl and SH

(c) 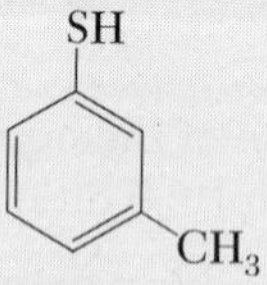 benzene ring with SH and CH_3

(d) 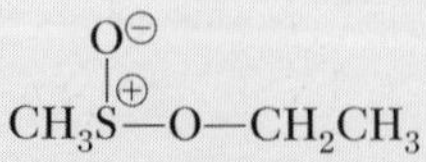 $CH_3\overset{O^{\ominus}}{S^{\oplus}}-O-CH_2CH_3$

(e) 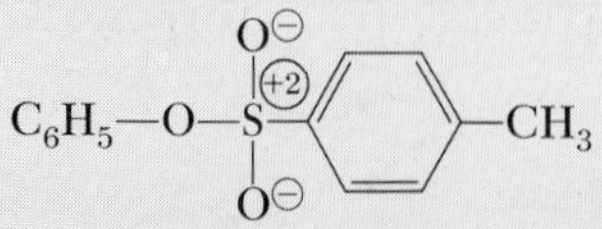 $C_6H_5-O-\overset{O^{\ominus}}{\underset{O^{\ominus}}{S^{+2}}}-C_6H_4-CH_3$

(f) 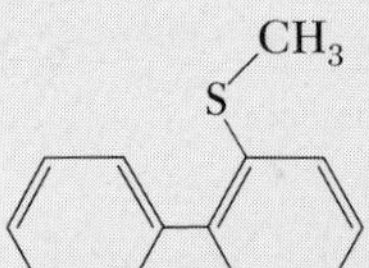 biphenyl with $S-CH_3$

(g) H_3C, H, C, $S-CH_2-CH_2-CH_3$, $S-CH_2-CH_2-CH_3$

(h) H, C_6H_5 on C of five-membered ring with S, S

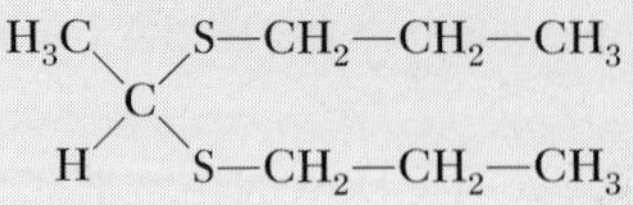

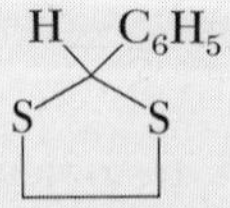

(i) 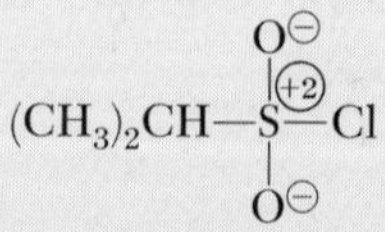 $(CH_3)_2CH-\overset{O^{\ominus}}{\underset{O^{\ominus}}{S^{+2}}}-Cl$

(j) $HO-CH_2CH_2-S-CH_3$

23.2 BONDING IN ORGANOSULFUR COMPOUNDS

The bond angle in hydrogen sulfide, 92°, suggests that sulfur is unhybridized in this compound. As the degree of alkylation around sulfur increases, the change in bond angle suggests a hybridization approaching $3sp^3$. The X-ray study of sulfonium salts reveals that they are not planar but pyramidal at sulfur, as would be required by sp^3 hybridization of sulfur.

H_2S	H_3C–S–H	$(H_3C)_2S$	$(CH_3)_3S^{\oplus}$ $Cl^{\ominus}$
92°	100°	105°	104°
hydrogen sulfide	methanethiol	dimethyl sulfide	trimethylsulfonium chloride

Multiple Bonding—Weak π Bonds. While single bonds formed by sulfur are usually not of great theoretical interest, the nature of multiple bonds to sulfur continues to challenge the chemist. The high reactivity of thioketones is but one proof that the *p,p* π bonds between first and second row elements are very weak. *Why is the C=S bond so weak?*

The comparative shapes of the 2*s*, 2*p*, 3*s*, and 3*p* orbitals, along with the signs of the lobes, are presented in Figure 23–1. A thiocarbonyl group requires π overlap between C2*p* and S3*p* orbitals. Such a bond will be weakened because one lobe of a 2*p* orbital interacts with a lobe of opposite sign on each "side" of a 3*p* orbital (Figure 23–2). (Recall that favorable overlap requires the interaction of lobes of similar sign.)

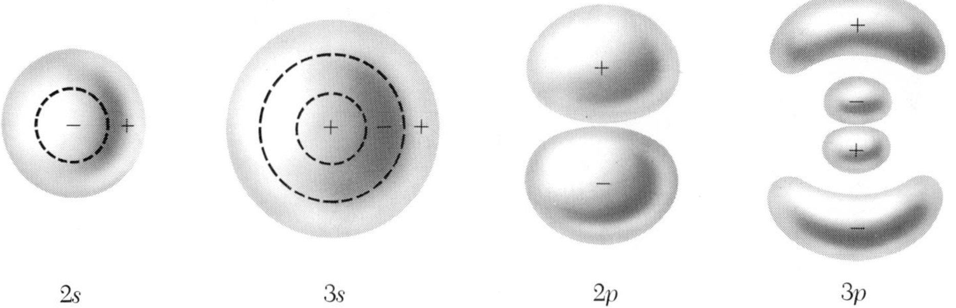

Figure 23–1 Shapes of *s* and *p* orbitals of the second and third main quantum levels. Contours shown are 95% probability lines.

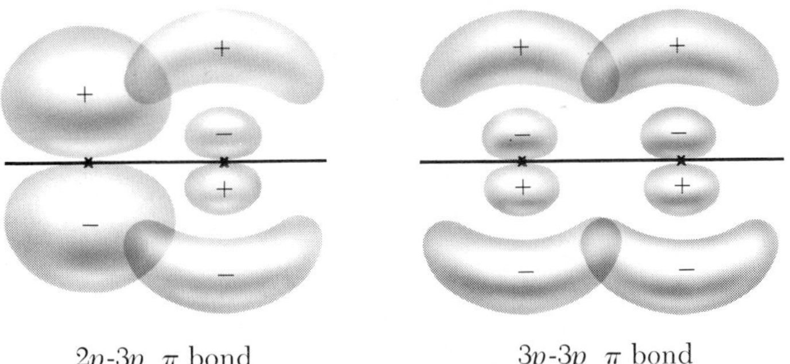

Figure 23–2 Pi bonding between 2*p* and 3*p* orbitals (left) and between a pair of 3*p* orbitals (right). The internuclear axis is shown, the nuclei being indicated by **x**.

A $3p,3p$ π bond is weak because the shape of the "outer" lobes diminishes the amount of overlap between orbitals. Furthermore, the increased internuclear distance diminishes the extent of interaction between both nuclei and the π electron pair (Figure 23–2).

The weak $3p,3p$ π bond helps to explain why the most stable form of oxygen contains π bonds (see Chapter 2), while the most stable form of sulfur is an eight-membered ring free of π bonds.

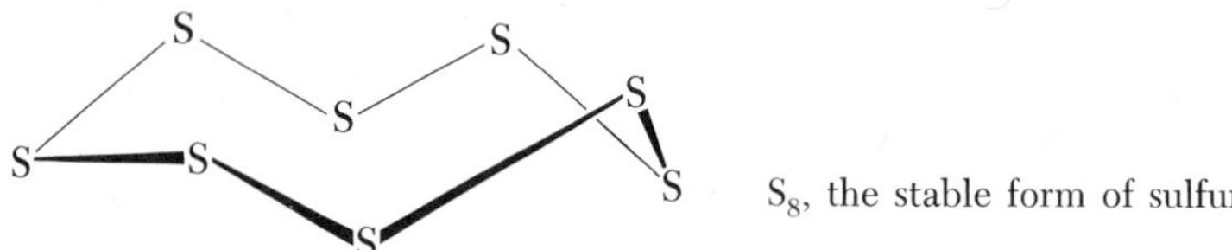

S_8, the stable form of sulfur

MULTIPLE BONDS USING *d* ORBITALS. Sulfur contains five accessible $3d$ orbitals, and these are useful in formation of multiple bonds to sulfur (Figure 23–3). Of the five $3d$ orbitals, two (d_{z^2} and $d_{x^2-y^2}$) are most often used in σ bond formation, while three (d_{xy}, d_{yz}, and d_{xy}) are usually employed in π bond formation.

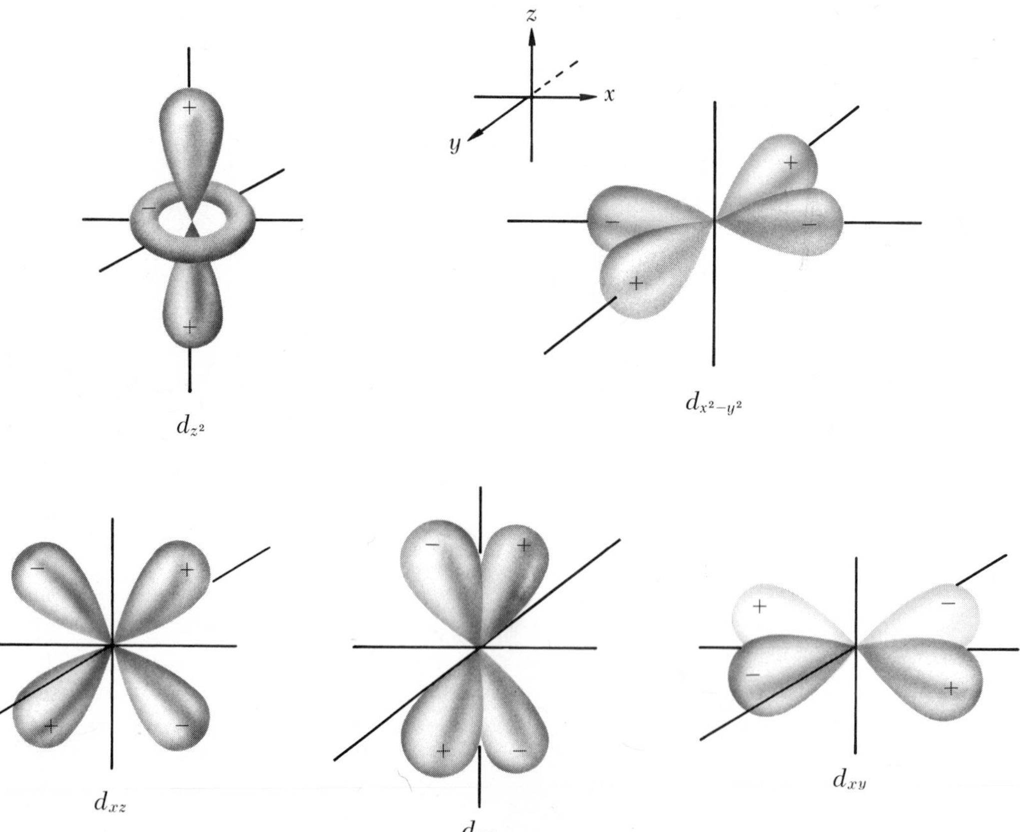

Figure 23-3 The five $3d$ orbitals. Note that the lobes of the $d_{x^2-y^2}$ orbital extend along the x and y axes, while the lobes of the d_{xz}, d_{yz}, and d_{xy} orbitals extend between the axes. It is the latter three that are most often used in π bonding.

It becomes clear why **A** (below) is not a bonding array while **B** (below) is a *p,d* π bond when one recalls that the signs of interacting lobes must be the same in order to achieve a bonding array.

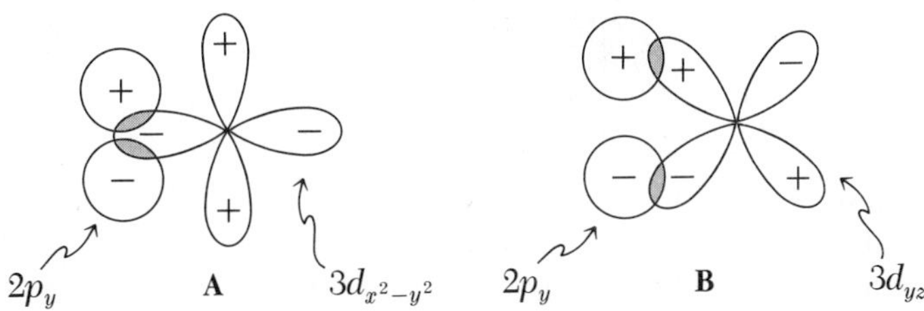

The sulfinate ester group contains two types of sulfur-to-oxygen bonds. The simpler, single bond results from overlap of sp^3 orbitals on sulfur and oxygen.

$$\begin{array}{c} \ominus\!:\ddot{\text{O}}: \\ | \\ \text{R}-\overset{\oplus}{\ddot{\text{S}}}-\ddot{\text{O}}-\text{R}' \end{array} \qquad \sigma_{3sp^3-2sp^3}$$

a sulfinate ester

The other bond is written as a charge-separated entity. Closer scrutiny suggests that this bond is actually made up of a σ bond and two π bonds. The latter involve donation of two electron pairs on oxygen into vacant, properly oriented *d* orbitals on sulfur. This phenomenon is sometimes termed *back-bonding.* The σ and π bonds involved in the sulfur-oxygen multiple bond are presented in Figure 23–4. The sulfur-oxygen multiple bond can be pictured as a

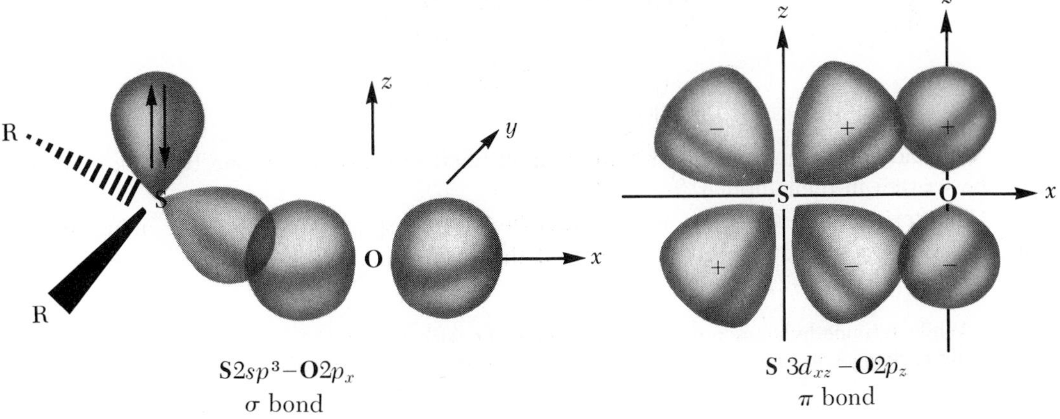

Figure 23-4 Bonding in the sulfinyl (sulfoxide) group. For clarity, the $2p_y$ and $2p_z$ orbitals of oxygen have been deleted in the representation of the σ bond, since they are not involved in σ bond formation. One π bond is shown. The $2p_y$ orbital of oxygen overlaps with the d_{xy} orbital of sulfur to form a second π bond. This has not been drawn because all that is required to represent it is to relabel the vertical axes in the drawing of the π bond as "y" rather than "z". Therefore, while the $2p_y$ and $2p_z$ orbitals of oxygen are not involved in σ bond formation, they are both used in π bond formation. It is these two orbitals that feed electron density into the vacant $3d$ orbitals, the "back-bonding" discussed in the text.

resonance hybrid of three contributing structures, dipole moment data suggesting that form **1** is the most important of the three.

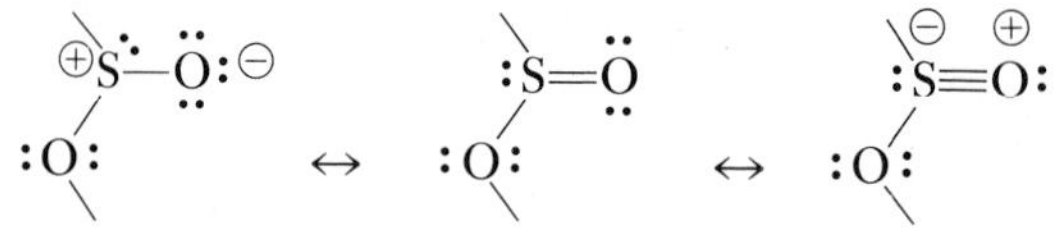

$$\overset{\oplus}{S}-\ddot{O}:^{\ominus} \;(\text{1, no } p,d\ \pi \text{ bond}) \leftrightarrow :S=\ddot{O} \;(\text{2, one } p,d\ \pi \text{ bond}) \leftrightarrow :\overset{\ominus}{S}\equiv\overset{\oplus}{O}: \;(\text{3, two } p,d\ \pi \text{ bonds})$$

1 no p,d π bond 2 one p,d π bond 3 two p,d π bonds

Back-bonding also explains the enhanced acidity of hydrogens on carbon adjacent to sulfur. Carbanions α to sulfur are sufficiently stable (and easily formed) that trimethylsulfonium ions rapidly incorporate deuterium under conditions in which tetramethylammonium ions are inert.

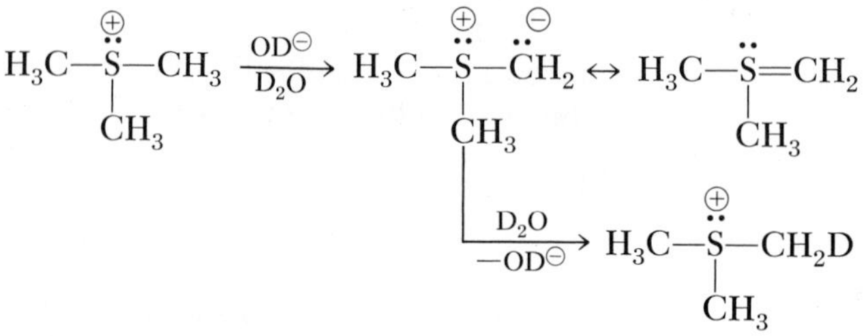

$$H_3C-\overset{\oplus}{S}(CH_3)-CH_3 \xrightarrow[D_2O]{OD^{\ominus}} H_3C-\overset{\oplus}{S}(CH_3)-\overset{\ominus}{C}H_2 \leftrightarrow H_3C-S(CH_3)=CH_2$$

$$\xrightarrow[-OD^{\ominus}]{D_2O} H_3C-\overset{\oplus}{S}(CH_3)-CH_2D$$

$$\overset{\oplus}{N}(CH_3)_4 \xrightarrow[D_2O]{OD^{\ominus}} \text{no reaction}$$

The double bonds between first row elements require a planar geometry, as is found in ethylene. Multiple bonding involving d orbitals is not so restricted, and compounds possessing multiple bonds to sulfur need not be planar. A comparison of acetone and dimethyl sulfoxide illustrates this point.

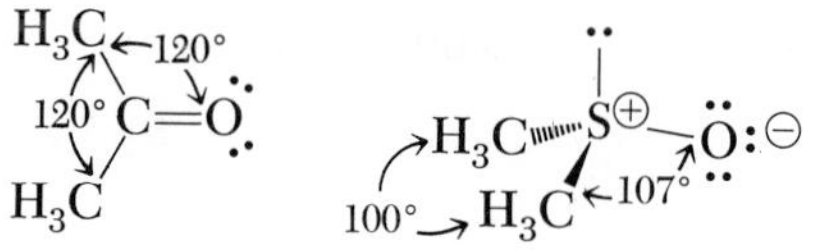

$(H_3C)_2C=\ddot{O}$: 120°, 120°
acetone (planar)

$(H_3C)_2\overset{\oplus}{S}-\ddot{O}:^{\ominus}$: 100°, 107°
dimethyl sulfoxide (pyramidal)

To avoid confusion, bonds that achieve multiplicities greater than one by back-bonding will be written in their dipolar form.

Complexities increase when we consider bonding in sulfones and sulfonic acids, and a discussion of this subject exceeds the needs of this text.

3. While tetramethylammonium ion does not react with $D_2O/OD^{\ominus}$, trimethylsulfonium ion rapidly exchanges its protons for deuterons in the mixture. Account for this difference.

23.3 SYNTHESIS AND REACTIONS—AN OVERVIEW

In this section we will discuss general approaches to the synthesis of the more common types of organosulfur compounds.

Thiols. *Aryl thiols,* often called **thiophenols,** are readily synthesized in a two-step process that begins with an aromatic hydrocarbon. The success of this method, since it involves an initial electrophilic aromatic substitution, depends upon the presence of the properly orienting substituent(s) in the starting material. In this scheme a first-formed sulfonyl chloride is reduced to the desired thiol with zinc and acid.

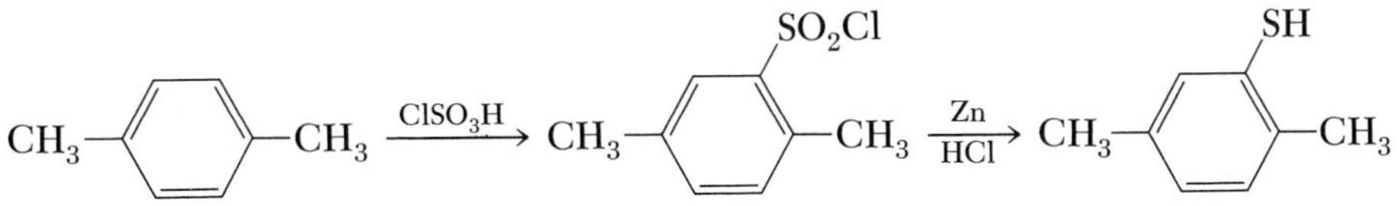

$$CH_3-C_6H_4-CH_3 \xrightarrow{ClSO_3H} CH_3-C_6H_3(SO_2Cl)-CH_3 \xrightarrow[HCl]{Zn} CH_3-C_6H_3(SH)-CH_3$$

2,4-dimethylthiophenol

While this synthesis is useful, it is restricted to those substrates otherwise inert to chlorosulfonic acid and the reducing medium, *e.g.,* alkyl or aryl substituted benzenes. A somewhat more general reaction involves the trapping of an aryl cation, produced by the decomposition of a diazonium salt, with potassium ethyl xanthate, $K^{\oplus}C_2H_5OCS_2^{\ominus}$, to form an aryl ethyl xanthate. Upon heating in aqueous base, this is hydrolyzed to a thiolate salt; acidification generates the desired thiol.

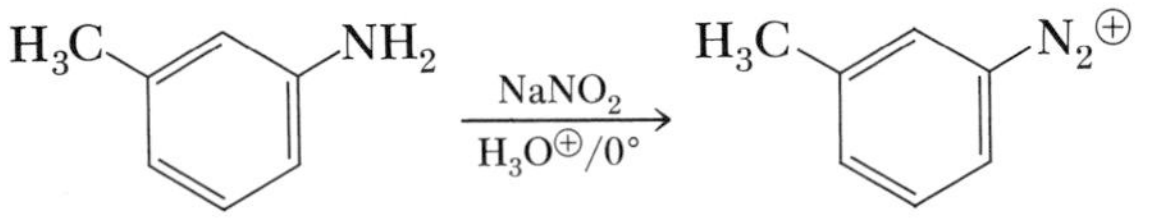

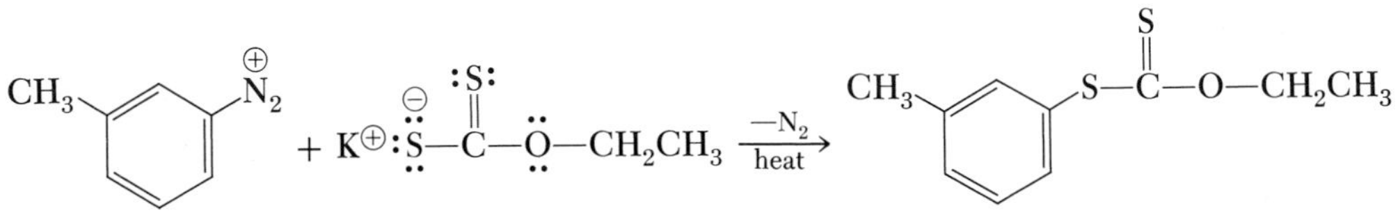

$$m\text{-}H_3C-C_6H_4-NH_2 \xrightarrow[H_3O^{\oplus}/0^\circ]{NaNO_2} m\text{-}H_3C-C_6H_4-N_2^{\oplus}$$

$$m\text{-}CH_3-C_6H_4-N_2^{\oplus} + K^{\oplus}\,{}^{\ominus}S-C(=S)-O-CH_2CH_3 \xrightarrow[heat]{-N_2} m\text{-}CH_3-C_6H_4-S-C(=S)-O-CH_2CH_3$$

potassium ethyl xanthate

m-tolyl-*O*-ethyldithiocarbonate (*m*-tolyl ethyl xanthate)

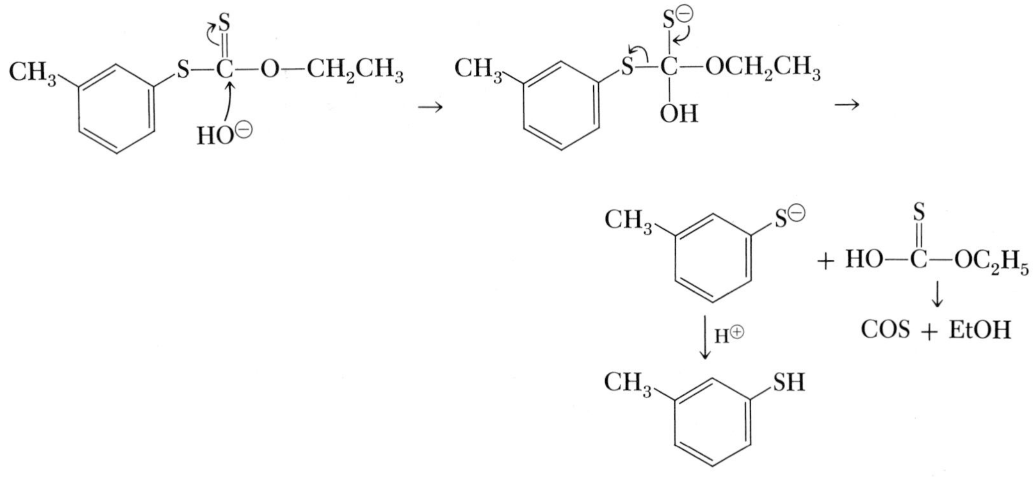

$$m\text{-}CH_3-C_6H_4-S-C(=S)-O-CH_2CH_3 + HO^{\ominus} \rightarrow m\text{-}CH_3-C_6H_4-S-C(S^{\ominus})(OH)-OCH_2CH_3 \rightarrow$$

$$m\text{-}CH_3-C_6H_4-S^{\ominus} + HO-C(=S)-OC_2H_5$$

$$HO-C(=S)-OC_2H_5 \rightarrow COS + EtOH$$

$$m\text{-}CH_3-C_6H_4-S^{\ominus} \xrightarrow{H^{\oplus}} m\text{-}CH_3-C_6H_4-SH$$

m-thiocresol (75%)

Aryl Grignard reagents react with sulfur to form, after acidification, aryl thiols. However, this approach is not a common one because the yields are generally unpredictable and by-product formation may be significant.

$$C_6H_5\text{—MgBr} \xrightarrow[\text{THF}]{S_8} \xrightarrow{H_3O^{\oplus}} C_6H_5\text{—SH} + (C_6H_5)_2S + (C_6H_5\text{—S})_2$$

Alkyl thiols can be prepared by the reaction of an alkyl Grignard reagent with sulfur. However, this reaction often produces low yields. One might also envision the conversion of an alkyl halide to an alkyl thiol by an S_N displacement of the halide with hydrosulfide (mercaptide) ion, $HS^{\ominus}$. This approach suffers from the need to carry out the reaction under stringently controlled conditions in order to avoid alkylation of the desired thiol.

$$C_6H_5\text{—}CH_2\text{—O—}\overset{O^{\ominus}}{\underset{O^{\ominus}}{\overset{|}{\underset{|}{S^{(+2)}}}}}\text{—}C_6H_4\text{—}CH_3 + NaSH \xrightarrow{\text{EtOH}} C_6H_5\text{—}CH_2SH + CH_3\text{—}C_6H_4\text{—}SO_3^{\ominus}\ Na^{\oplus}$$

α-toluenethiol
(benzylmercaptan)

side reactions:

$$C_6H_5\text{—}CH_2SH + HS^{\ominus} \rightleftarrows C_6H_5\text{—}CH_2S^{\ominus} + HSH \quad \textbf{A}$$

$$C_6H_5\text{—}CH_2S^{\ominus} + C_6H_5\text{—}CH_2OTs \rightarrow C_6H_5\text{—}CH_2\text{—S—}CH_2\text{—}C_6H_5 \quad \textbf{B}$$

dibenzyl sulfide

In spite of these drawbacks, alkyl halides can be used to prepare alkyl thiols in good yield. What is required is the conversion of the halide into an S-alkyl isothiouronium salt[a] and the subsequent hydrolysis of this salt into the desired thiol. In addition to simple alkyl thiols, this process can also be used to produce thiols from such compounds as halo-alcohols and halo-acids. This type of salt, prepared by the reaction of the alkyl halide with thiourea, is highly crystalline and has been used to form derivatives of alkyl halides.

$$(H_2\ddot{N})_2C{=}\ddot{S}\text{:} + R\text{—}X \rightarrow H_2\overset{\oplus}{N}{=}\underset{\underset{:NH_2}{|}}{C}\text{—}\ddot{\underset{\cdot\cdot}{S}}\text{—}R\ X^{\ominus}$$

thiourea — S-alkyl isothiouronium salt

[a] These also have been called thiouronium salts, thiuronium salts, and isothiuronium salts.

$$H_2N{-}\overset{SR}{S}{=}\overset{\oplus}{N}H_2 \;(:\ddot{O}H^{\ominus}) \xrightarrow{-H_2O} H_2N{-}\overset{SR}{C}{=}N{-}H \;\; {}^{\ominus}:\ddot{O}H \xrightarrow{-H_2O} H_2N{-}C{\equiv}N + RS^{\ominus}$$

cyanamide → polymer; $RS^{\ominus} \xrightarrow{H_3O^{\oplus}} RSH$

$$BrCH_2CH_2Br + (H_2N)_2C{=}S \xrightarrow{KOH} \xrightarrow{H^{\oplus}} HSCH_2CH_2SH \quad (60\%)$$

1,2-ethanedithiol

$$HO(CH_2)_3Cl + (H_2N)_2C{=}S \xrightarrow{KOH} \xrightarrow{H^{\oplus}} HO(CH_2)_3SH \quad (45\%)$$

3-hydroxypropanethiol

$$ClCH_2CH_2CO_2H + (H_2N)_2C{=}S \xrightarrow{KOH} \xrightarrow{H^{\oplus}} HSCH_2CH_2CO_2H \quad (70\%)$$

3-mercaptopropanoic acid

These S_N2 processes (alkyl halide + mercaptide ion; alkyl halide + thiourea) are useless for preparing 3° thiols (*e.g.*, *t*-butyl mercaptan), since elimination successfully competes with substitution.

$$(CH_3)_3CBr + KSH \xrightarrow{C_2H_5OH} (CH_3)_2C{=}CH_2 + H_2S + KBr$$

Fortunately, the production of thiols from the reaction of Grignard reagents with sulfur can be used to prepare such 3° thiols.

$$(CH_3)_3CBr + Mg \xrightarrow{THF} \xrightarrow{S_8} \xrightarrow{HOH} (CH_3)_3CSH$$

Disulfides. The thiol group, often called the *sulfhydryl group* by biochemists, is oxidized by mild oxidants to disulfides. In the laboratory the most common oxidant is iodine. This reaction is rapid, nearly quantitative, and suited for forming derivatives of both alkyl and aryl thiols. Potent oxidants, such as potassium permanganate, and forcing conditions must be avoided because they result in the oxidation of thiols to sulfonic acids.

$$2RSH + I_2 \xrightarrow[25^\circ]{C_2H_5OH/H_2O} RS{-}SR + 2HI$$

$$RSH \xrightarrow[H_2O/heat]{KMnO_4} RSO_3H$$

Unsymmetric disulfides can be prepared by the reaction of thiolate salts with a sulfenyl halide or with a symmetric disulfide.

$$C_6H_5{-}\ddot{S}:^{\ominus} Na^{\oplus} + Cl{-}S{-}C_6H_4{-}NO_2 \xrightarrow[S_N2 \text{ on sulfur}]{-Cl^{\ominus}} C_6H_5{-}S{-}S{-}C_6H_4{-}NO_2$$

sodium thiophenoxide; *p*-nitrophenyl sulfenyl chloride; *p*-nitrophenyl phenyl disulfide

$$CH_3S^{\ominus} Na^{\oplus} + C_6H_5{-}S{-}S{-}C_6H_5 \rightleftarrows CH_3{-}S{-}S{-}C_6H_5 + C_6H_5{-}S^{\ominus} Na^{\oplus}$$

diphenyl disulfide; methyl phenyl disulfide

The disulfide linkage can be reduced with lithium aluminum hydride or with a mixture of zinc and acid to produce two molecules of thiol. This is not a valuable thiol synthesis, since disulfides are usually made from thiols.

$$RS{-}SR \xrightarrow[\text{ether}]{LiAlH_4} \xrightarrow{H_3O^{\oplus}} 2RSH$$

Sulfides. The reaction between thiolate salts (*i.e.*, $RS^{\ominus}$) and alkyl halides, an analog of the Williamson ether synthesis, is a valuable synthesis of sulfides. It is subject to the usual restrictions placed upon S_N2 reactions and, for example, is generally useless in preparing diaryl sulfides. These can, however, be prepared by the reaction of a Grignard reagent with a diaryl disulfide, a process which involves a nucleophilic displacement by a carbanion on the sulfur of the disulfide bond.

$$C_6H_5S^{\ominus}\,Na^{\oplus} \begin{cases} \xrightarrow{CH_3Cl} C_6H_5SCH_3 + NaCl \\ \xrightarrow{C_6H_5Cl} \text{no reaction} \end{cases}$$

$$C_6H_5MgCl + (o\text{-}CH_3C_6H_4)S{-}S(C_6H_4CH_3\text{-}o) \xrightarrow{THF} C_6H_5{-}S{-}C_6H_4CH_3\text{-}o \quad (80\%)$$

$$(CH_3)_3CMgBr + C_6H_5{-}S{-}S{-}C_6H_5 \xrightarrow{THF} (CH_3)_3C{-}S{-}C_6H_5 \quad (65\%)$$

Episulfides. Episulfides, sulfur analogs of epoxides, cannot be prepared by the direct sulfuration of an alkene. Instead, they can be made by treating epoxides (or carbonates) with thiocyanate ion, $NCS^{\ominus}$. The reaction is stereospecific, *cis*-epoxide producing *cis*-episulfide and *trans*-epoxide producing *trans*-episulfide.

net equation: $RHC\overset{O}{\frown}CRH + NCS^{\ominus} \rightarrow RHC\overset{S}{\frown}CRH + NCO^{\ominus}$

mechanism:

$$\text{H}\cdots\text{C(R)}\overset{\ddot{O}}{\frown}\text{C(R)}\cdots\text{H} + :N{\equiv}C{-}\ddot{S}:^{\ominus} \rightarrow \text{H}\cdots\underset{R}{\overset{:\ddot{O}:^{\ominus}}{C}}{-}\underset{:S:-C\equiv N:}{\overset{H}{C}}{-}R \rightarrow \text{H}\cdots\underset{R}{\overset{^{\ominus}:\ddot{O}:}{C}}{-}\underset{H}{\overset{:S:-C\equiv N}{C}}\cdots R$$

$$\rightarrow \text{(cyclic: O–C(=N:}^{\ominus}\text{)–S)}\;\text{H}\cdots\underset{R}{C}{-}\underset{H}{C}\cdots R \rightarrow \text{H}\cdots\underset{R}{\overset{:\ddot{O}{-}C\equiv\ddot{N}}{C}}{-}\underset{H}{\overset{:\ddot{S}:^{\ominus}}{C}}\cdots R \rightarrow \text{H}\cdots\underset{R}{\overset{:\ddot{O}{-}C\equiv\ddot{N}}{C}}{-}\underset{:\ddot{S}:^{\ominus}}{\overset{H}{C}}{-}R$$

$$\xrightarrow{-NCO^{\ominus}} R\overset{H}{-}C\overset{\ddot{S}}{\frown}C\overset{H}{-}R$$

Episulfides react with nucleophiles to form ring-opened products and are, in this respect, quite similar to epoxides.

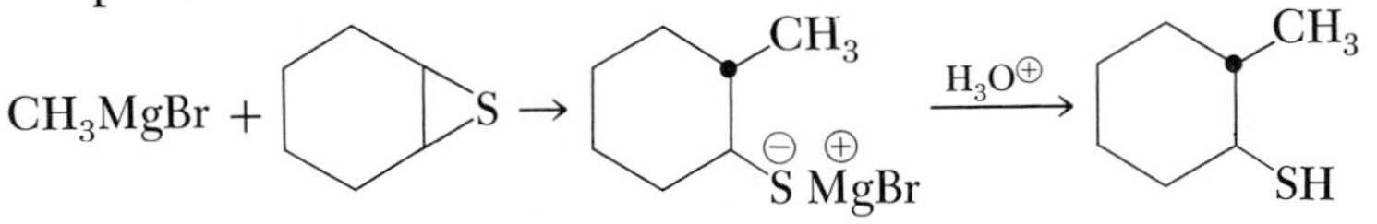

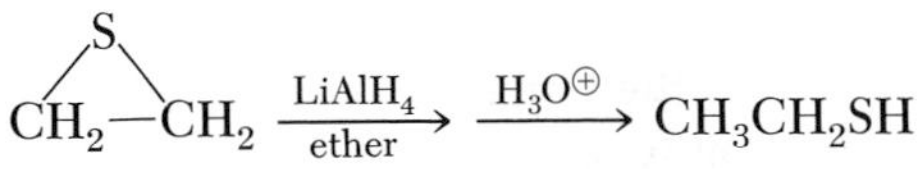

$$CH_3MgBr + \text{(cyclohexene episulfide)} \rightarrow \text{2-methylcyclohexyl-}\overset{\ominus}{S}\,\overset{\oplus}{Mg}Br \xrightarrow{H_3O^{\oplus}} \text{2-methylcyclohexanethiol (SH)}$$

$$\overset{S}{CH_2\text{—}CH_2} \xrightarrow[\text{ether}]{LiAlH_4} \xrightarrow{H_3O^{\oplus}} CH_3CH_2SH$$

Sulfonium Salts. Sulfides react as nucleophiles with alkyl halides and tosylates to form sulfonium salts (**A,** below). Sulfonium halides are stable, crystalline compounds that have served in analytical schemes as derivatives of sulfides. (The stability of sulfonium salts is to be contrasted to the instability and high reactivity of the oxygen analogs.) Sulfonium halides react with moist silver oxide to produce sulfonium hydroxides. These decompose upon heating to produce an alkene, a sulfide, and water. This *E2 reaction* is strictly analogous to the thermal decomposition of alkylammonium hydroxides (the Hofmann elimination, pp. 165 and 656) and, as shown below (**B**), can be used to convert one sulfide into another.

A: $CH_3CH_2Cl + CH_3S(CH_2)_4CH_3 \rightarrow CH_3CH_2\text{—}\overset{CH_3}{\underset{\oplus}{S}}\text{—}(CH_2)_4CH_3\ Cl^{\ominus}$

ethylmethylpentylsulfonium chloride

B: $C_6H_5SC_2H_5 + CH_3I \rightarrow C_6H_5\overset{\oplus}{S}(CH_3)C_2H_5\ I^{\ominus}$

$$C_6H_5\overset{\oplus}{S}(CH_3)C_2H_5\ OH^{\ominus} \xleftarrow{-AgI} (Ag_2O,\ H_2O)$$

$$C_6H_5\text{—}\overset{CH_3}{\underset{\oplus}{S}}\text{—}CH_2\text{—}CH_2\text{—}H\ \ {}^{\ominus}OH \xrightarrow{150°} C_6H_5\text{—}S\text{—}CH_3 + H_2C\text{=}CH_2 + H_2O$$

Sulfoxides. The most direct method for preparing sulfoxides is *via* the oxidation of the corresponding sulfide.

$$R\text{—}S\text{—}R\ (\text{or } R') \xrightarrow{[O]} R\text{—}\overset{O^{\ominus}}{\overset{|}{S^{\oplus}}}\text{—}R\ (\text{or } R')$$

While 30% hydrogen peroxide dissolved in glacial acetic acid is still a popular oxidant for sulfoxide preparation, emphasis is switching to dinitrogen tetroxide, N_2O_4, sodium metaperiodate, $NaIO_4$, and *m*-chloroperoxybenzoic acid. One reason for this is that over-oxidation, to sulfones, is more easily controlled with these oxidants.

$$CH_3CH_2\text{—}S\text{—}CH_2CH_3 \xrightarrow[0°]{N_2O_4/CHCl_3} CH_3CH_2\text{—}\overset{O^{\ominus}}{\overset{|}{S^{\oplus}}}\text{—}CH_2CH_3 \quad (98\%)$$

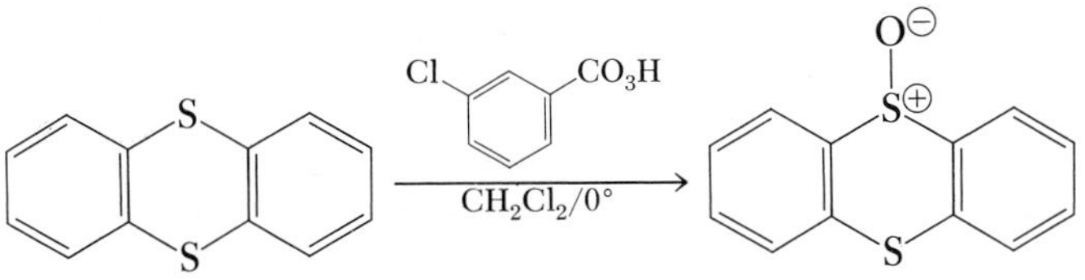

diethyl sulfide → diethyl sulfoxide

$$\text{thianthrene} \xrightarrow[CH_2Cl_2/0°]{m\text{-}ClC_6H_4CO_3H} \text{thianthrene monoxide} \quad (80\%)$$

thianthrene → thianthrene monoxide

A second important synthesis of sulfoxides employs the attack of a Grignard reagent upon a sulfinate ester.

$$\underset{\text{sulfinate ester}}{R-\overset{O^{\ominus}}{\overset{|}{S^{\oplus}}}-O-R'} + R''MgX \xrightarrow[10°]{\text{ether}} \underset{\text{sulfoxide}}{R-\overset{O^{\ominus}}{\overset{|}{S^{\oplus}}}-R''} + R'OMgX$$

This sequence actually makes double use of the Grignard reagent, the one shown above and an earlier one used to prepare the prerequisite sulfinic acid.

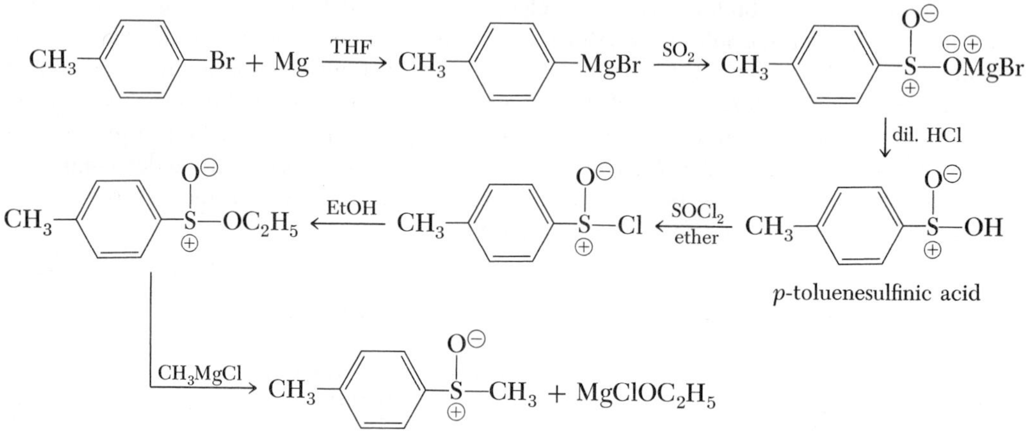

$$CH_3-C_6H_4-Br + Mg \xrightarrow{\text{THF}} CH_3-C_6H_4-MgBr \xrightarrow{SO_2} CH_3-C_6H_4-\overset{O^{\ominus}}{\overset{|}{\underset{\oplus}{S}}}-\overset{\ominus\oplus}{O}MgBr$$

$$\downarrow \text{dil. HCl}$$

$$CH_3-C_6H_4-\overset{O^{\ominus}}{\overset{|}{\underset{\oplus}{S}}}-OC_2H_5 \xleftarrow{\text{EtOH}} CH_3-C_6H_4-\overset{O^{\ominus}}{\overset{|}{\underset{\oplus}{S}}}-Cl \xleftarrow[\text{ether}]{SOCl_2} \underset{p\text{-toluenesulfinic acid}}{CH_3-C_6H_4-\overset{O^{\ominus}}{\overset{|}{\underset{\oplus}{S}}}-OH}$$

$$\xrightarrow{CH_3MgCl} \underset{\text{methyl } p\text{-tolylsulfoxide}}{CH_3-C_6H_4-\overset{O^{\ominus}}{\overset{|}{\underset{\oplus}{S}}}-CH_3} + MgClOC_2H_5$$

Sulfones. Sulfones usually are prepared by oxidizing the corresponding sulfide or sulfoxide with hot 30% hydrogen peroxide in glacial acetic acid.

$$CH_3SCH_3 \text{ or } CH_3\overset{O^{\ominus}}{\overset{|}{\underset{\oplus}{S}}}CH_3 \xrightarrow[\text{reflux}]{H_2O_2/HOAc} \underset{\text{dimethyl sulfone}}{CH_3\overset{O^{\ominus}}{\overset{|}{\underset{\underset{O^{\ominus}}{|}}{S^{(+2)}}}}CH_3} \qquad (100\%)$$

Friedel-Crafts condensation of an arylsulfonyl chloride with an aromatic nucleus provides a useful route to diaryl sulfones.

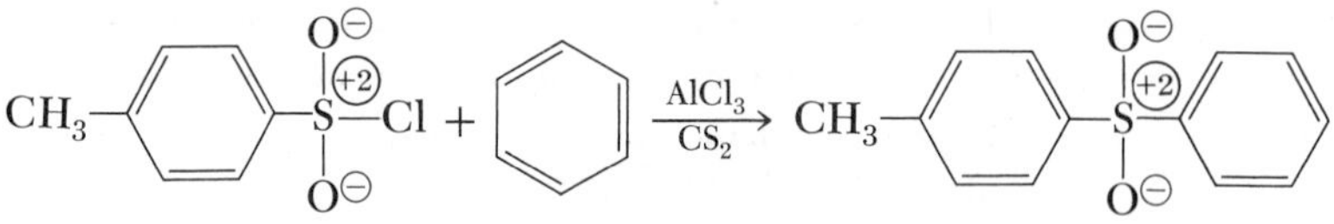

$$CH_3-C_6H_4-\overset{O^{\ominus}}{\overset{|}{\underset{\underset{O^{\ominus}}{|}}{S^{(+2)}}}}-Cl + C_6H_6 \xrightarrow[CS_2]{AlCl_3} \underset{\text{phenyl } p\text{-tolyl sulfone}}{CH_3-C_6H_4-\overset{O^{\ominus}}{\overset{|}{\underset{\underset{O^{\ominus}}{|}}{S^{(+2)}}}}-C_6H_5}$$

Unlike sulfoxides, which are subject to oxidation to sulfones and to reduction (using zinc and acid, or lithium aluminum hydride) to sulfides, sulfones are rather stable. They are very difficult to oxidize and are reduced only after protracted exposure to lithium aluminum hydride at elevated temperatures (~100°).

Sulfonic Acids. The importance of sulfonic acids and their derivatives is attested by the fact that they have already been discussed at various places in this text, even though they are organosulfur compounds. In Chapter 15 we emphasized the synthesis of aryl sulfonic acids by electrophilic aromatic substitution. In Chapter 22 we noted how such sulfonic acids could, upon fusion with sodium hydroxide, be converted to phenols. We have utilized sulfonate esters as substrates for nucleophilic substitution in several places. The synthesis of sulfonamides and their utility in identifying various amines *via* the Hinsberg test was described in Chapter 20. In spite of all of this coverage, there is more to tell. In the next two paragraphs we will examine the use of sulfonic acid salts as cleansing agents, while later in the chapter we will look closely at the behavior of antimicrobial sulfonamides. It is, perhaps, amusing that this first discussion hinges upon man's attempts to keep bacteria alive and feeding, while the latter will dwell upon his attempts to destroy bacteria.

Aliphatic sulfonic acids, Alk—SO_3H, usually are prepared by oxidizing a thiol or, industrially, by the direct reaction of sulfur trioxide or sulfuric acid with an alkene. The most common way to prepare aryl sulfonic acids is by sulfonation of the appropriate aromatic hydrocarbon (see Chapter 15).

The most common commercial use of sulfonic acids, particularly aryl sulfonic acids, is as detergents. Actually, sulfonic acids are quite acidic, and so their sodium salts are used; this gives the molecule a water-soluble ionic end in addition to its grease-soluble neutral organic portion. Such a combination produces the necessary solubilization of the oils and greases in water (this is illustrated, using a conceptually similar soap molecule, in Figure 23–5). Increasing

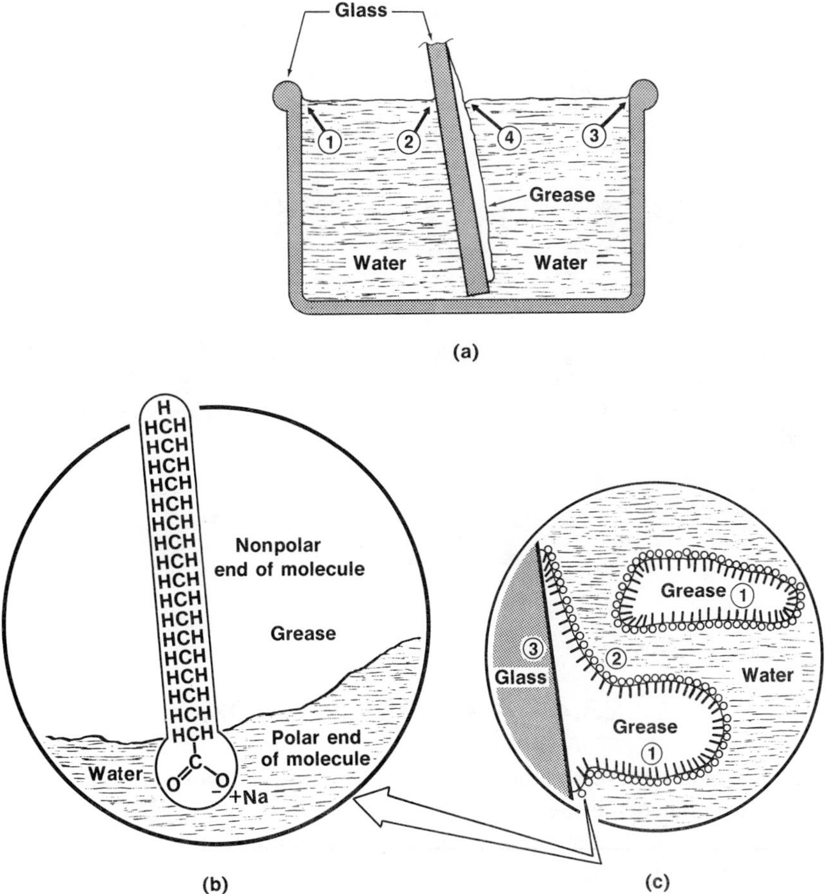

Figure 23-5 The cleansing action of soap. (*a*) A piece of glass coated with grease inserted in water gives evidence for the strong adhesion between water and glass at 1, 2 and 3. The water curves up against the pull of gravity to wet the glass. The relatively weak adhesion between oil and water is indicated at 4 by the curvature of the water away from the grease against the force tending to level the water. (*b*) A soap molecule, having oil soluble and water soluble ends, will orient at an oil-water interface such that the hydrocarbon chain is in the oil (with molecules that are electrically similar-nonpolar) and the COO^-Na^+ group is in the water (highly charged polar groups interacting electrically). (*c*) In an idealized molecular view, a grease particle, 1, is surrounded by soap molecules which in turn are strongly attracted to the water. At 2 another droplet is about to break away. At 3 the grease and clean glass interact before the water moves between. (From M. M. Jones, J. T. Netterville, D. O. Johnston, and J. L. Wood, *Chemistry, Man and Society*, copyright © 1972, W. B. Saunders Company, Philadelphia.)

concern about biodegradability has led detergent manufacturers away from the use of highly branched alkyl groups attached to the benzenesulfonate moiety and toward the use of alkyl groups that lack quaternary carbons. Bacteria have a difficult time completely degrading ("eating") alkyl groups if they possess carbons that lack hydrogens bonded to them.

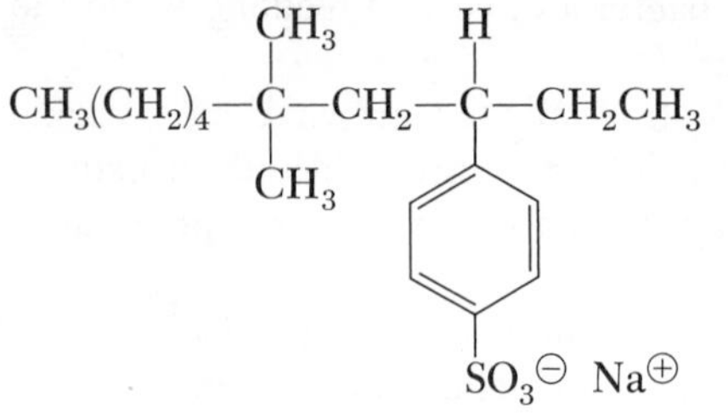

$$CH_3(CH_2)_4-\overset{\overset{CH_3}{|}}{\underset{\underset{CH_3}{|}}{C}}-CH_2-\overset{\overset{H}{|}}{\underset{\underset{C_6H_4SO_3^{\ominus}\ Na^{\oplus}}{|}}{C}}-CH_2CH_3$$

non-biodegradable detergent

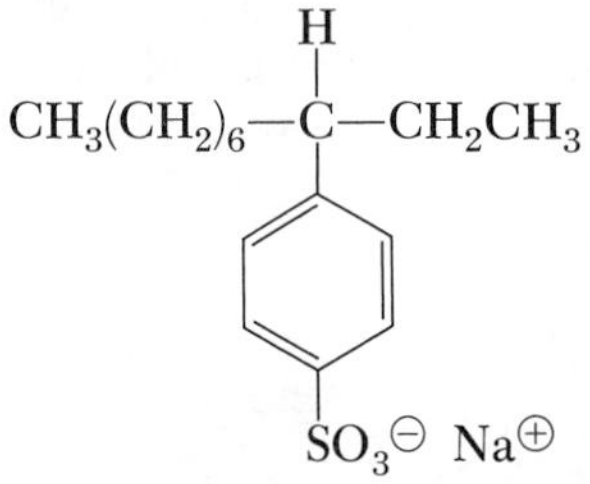

$$CH_3(CH_2)_6-\overset{\overset{H}{|}}{\underset{\underset{C_6H_4SO_3^{\ominus}\ Na^{\oplus}}{|}}{C}}-CH_2CH_3$$

biodegradable detergent

4. Suggest a synthesis for each of the following, beginning with benzenethiol and any other needed reagents.

(a) $C_6H_5SCH_3$ (e) $C_6H_5SSCH_3$
(b) $C_6H_5SSC_6H_5$ (f) $C_6H_5SO_3H$
(c) $C_6H_5\overset{\overset{O^{\ominus}}{|}}{S^{\oplus}}CH_3$ (g) $C_6H_5SSC(CH_3)_3$
(d) $C_6H_5S^{\oplus}(CH_3)_2\ Br^{\ominus}$ (h) $C_6H_5SC(CH_3)_3$

5. Would you expect **A** to form an optically active episulfide after reaction with thiocyanate ion? Explain.

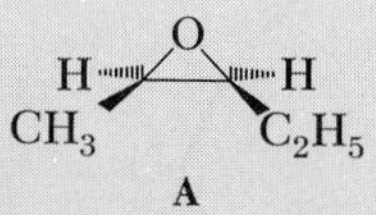

(Epoxide **A**: ring carbons bearing H and CH_3, and H and C_2H_5)

A

23.4 A CLOSER LOOK

Having presented a survey of the chemistry of certain organosulfur compounds, we will now examine a few functionalities in some detail.

***Gem*-Dithioethers.** Just as ketones and aldehydes produce 1,3-dioxolanes upon reaction with *vic*-diols, so *vic*-dithiols react to form 1,3-dithiolanes.

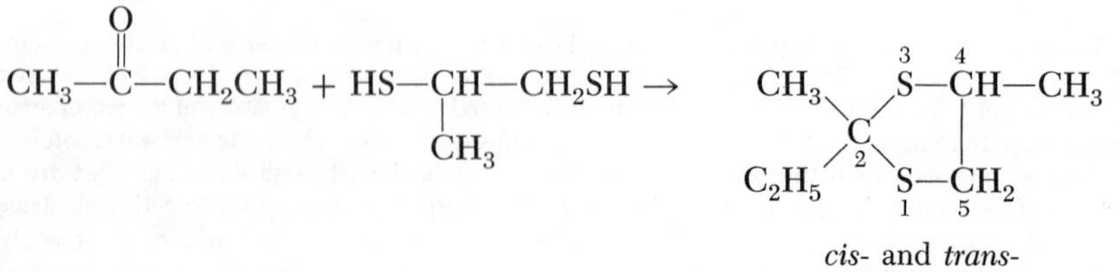

$$CH_3-\overset{\overset{O}{\|}}{C}-CH_2CH_3 + HS-\underset{\underset{CH_3}{|}}{CH}-CH_2SH \rightarrow \text{(ring: } C_2(CH_3)(C_2H_5)-S_3-C_4H(CH_3)-C_5H_2-S_1\text{)}$$

cis- and *trans*-
2-ethyl-2,4-dimethyl-1,3-dithiolane

1,3-Dithiolanes are often used as intermediates in the conversion of carbonyl groups to methylene groups. The reduction is actually a *reductive desulfurization* and is effective upon sulfides, disulfides, sulfoxides, and sulfones. This reaction, the *Mozingo reduction*, utilizes hydrogen that is adsorbed on the surface of the catalyst, Raney nickel. This process is often called "Raney nickel desulfurization."

$$C_6H_5-CH_2-S-CH_2-C_6H_4-CH_3 \xrightarrow[{[H_2]}]{\text{Raney Ni}} C_6H_5-CH_3 + CH_3-C_6H_4-CH_3$$

$$\text{(steroid 1,3-dithiolane, } H_3C, CH_3, O, H) \xrightarrow[{[D_2]}]{\text{Raney Ni}} \text{(gem-dideuterio steroid, D, D, } H_3C, CH_3, O, H)$$

Carbanions at C2 of dithiolanes are easily formed. In turn, these are efficiently alkylated and acylated, the products being hydrolyzed to aldehydes or ketones. Yields often approach 80%.

$$\text{1,3-dithiane (H, H)} \xrightarrow[-30°]{n\text{-buLi/THF}} \text{2-H-1,3-dithian-2-ide} \xrightarrow[0°]{CH_3CH_2Br} \text{2-}C_2H_5\text{-1,3-dithiane (H)} \xrightarrow{n\text{-buLi/THF}} \text{2-}C_2H_5\text{-1,3-dithian-2-ide}$$

$$\xrightarrow[-I^{\ominus}]{(CH_3)_2CHI} \text{2-}CH(CH_3)_2\text{-2-}C_2H_5\text{-1,3-dithiane} \xrightarrow[\sim 50°]{H_2O/CH_3OH,\ HgO/HgCl_2} CH_3CH_2-\overset{O}{\overset{\|}{C}}-\underset{CH_3}{\overset{CH_3}{\overset{|}{\underset{|}{C}}}}-H + (HS-CH_2)_2$$

$(HS-CH_2)_2$: ppt. as Hg(+2) salt

Applications of this type of reaction are summarized below.

Starting from $(-S)_2C(R)Li$:

$$\xrightarrow{\text{ethylene oxide}} (-S)_2C(R)CH_2CH_2OH \xrightarrow[{[H_2]}]{\text{Raney Ni}} RCH_2CH_2CH_2OH$$

$$\xrightarrow{R'C\equiv N} (-S)_2C(R)C(=NH)R \xrightarrow{H_3O^{\oplus}} (-S)_2C(R)C(=O)R'$$

$$\xrightarrow{R'COCl} (-S)_2C(R)C(=O)R'$$

$$\xrightarrow{CH_2=CHCHO} (-S)_2C(R)CH_2-CH_2-C(=O)H$$

The convenient preparation of CH_3SCH_2Li suggests that carbanions are adequately stabilized by even one thioether, —S—, linkage.

$$CH_3SCH_3 + CH_3CH_2CH_2CH_2Li \xrightarrow[20°]{THF} CH_3SCH_2Li + C_4H_{10}$$

$$CH_3SCH_2Li + C_6H_5CHO \xrightarrow[0°]{THF} \xrightarrow[H_3O^{\oplus}]{dil.} C_6H_5CH(OH)CH_2SCH_3 \ (84\%)$$

Sulfur Ylid[a] Reactions. Hydrogens on carbons α to a positively charged sulfur are abstracted by strong base to form ylids, a reaction analogous to the formation of the Wittig reagent (p. 603). As already noted, the stability of such sulfur ylids is due to back bonding of the electron pair on carbon into sulfur d orbitals.

Trimethylsulfonium iodide, $(CH_3)_3S^{\oplus}I^{\ominus}$, reacts with sodium hydride to form dimethylsulfonium methylide. This ylid is useful for converting carbonyl compounds into epoxides; it reacts, for example, with cyclohexanone to produce an epoxide with an *equatorial* C—O bond.

$$(CH_3)_3S^{\oplus}\ I^{\ominus} \xrightarrow[DMSO]{NaH} CH_3-\overset{\overset{CH_3}{|}}{\underset{\oplus}{S}}-\overset{..\ominus}{C}H_2 + H_2 + NaI$$

dimethylsulfonium methylide

$$(CH_3)_2\overset{\oplus}{S}-\overset{\ominus}{C}H_2 + C_6H_5-CH{=}CH-COCH_3 \rightarrow C_6H_5-CH{=}CH-\underset{\underset{CH_3}{|}}{C}(-O-)CH_2$$

4-*tert*-butylcyclohexanone $+ (CH_3)_2\overset{\oplus}{S}-\overset{\ominus}{C}H_2 \rightarrow$ epoxide (spiro CH_2–O)

(O is e)

When dimethyl sulfoxide is warmed with methyl iodide, and the resulting salt is treated with sodium hydride, another ylid, dimethylsulfoxonium methylide, is produced. This ylid reacts with cyclohexanone to produce an epoxide having an *axial* C—O bond.

$$CH_3-\overset{\overset{O^{\ominus}}{|}}{\underset{\oplus}{S}}-CH_3 + CH_3I \xrightarrow{heat} CH_3-\overset{\overset{O^{\ominus}}{|}}{\underset{\underset{CH_3}{|}}{S^{(+2)}}}-CH_3\ I^{\ominus} \xrightarrow{NaH} CH_3-\overset{\overset{:\ddot{O}:^{\ominus}}{|}}{\underset{\underset{CH_3}{|}}{S^{(+2)}}}-\overset{..\ominus}{C}H_2$$

dimethylsulfoxonium methylide

4-*tert*-butylcyclohexanone $+ (CH_3)_2\overset{\overset{O^{\ominus}}{|}}{\underset{(+2)}{S}}-\overset{\ominus}{C}H_2 \rightarrow$ epoxide (spiro O–CH_2)

(O is a)

[a] Ylids were defined on p. 615.

The difference in the stereochemistry of epoxide formation by these two ylids has been rationalized by assuming that the more stable dimethylsulfoxonium methylide adds with partial (or complete) equilibration of the intermediate zwitterion (see below) to produce the zwitterion with the larger substituent on the *equatorial* position. The more reactive dimethylsulfonium methylide produces the kinetically controlled adduct.

Another difference between these two ylids is that conjugated unsaturated carbonyl compounds react with dimethylsulfoxonium methylide to form cyclopropanes rather than epoxides.

$$C_6H_5{-}CH{=}COCH_3 + (CH_3)_2\overset{O^{\ominus}}{\underset{(+2)}{S}}{-}\overset{\ominus}{C}H_2 \rightarrow C_6H_5{-}\underset{\diagdown CH_2 \diagup}{CH{-}CH}{-}\overset{O}{\overset{\|}{C}}{-}CH_3 \quad (95\%)$$

Mechanisms for such reactions are similar to types that have already been discussed.

$$(CH_3)_2\overset{O^{\ominus}}{\underset{(+2)}{S}}{-}\overset{\ominus}{C}H_2 + \text{cyclohexanone} \rightarrow (CH_3)_2\overset{O^{\ominus}}{\underset{(+2)}{S}}{-}CH_2{-}C(O^{\ominus})(\text{cyclohexyl}) \rightarrow \text{cyclohexane spiro-epoxide } (O{-}CH_2) + \text{DMSO}$$

$$C_6H_5{-}CH{=}\overset{H}{C}{-}\overset{O}{\overset{\|}{C}}{-}C_6H_5 + {}^{\ominus}CH_2{-}\overset{(+2)}{S}(CH_3)_2{-}O^{\ominus} \rightarrow C_6H_5{-}\overset{H}{C}(CH_2{-}\overset{(+2)}{S}(CH_3)_2{-}O^{\ominus}){-}\overset{H}{C}{=}C(O^{\ominus}){-}C_6H_5 \xrightarrow{-\text{DMSO}} C_6H_5{-}\underset{\diagdown CH_2 \diagup}{\overset{H}{C}{-}\overset{H}{C}}{-}\overset{O}{\overset{\|}{C}}{-}C_6H_5$$

(*cis* and *trans*)

Dimsyl sodium, the sodium salt of DMSO, is prepared by the reaction of sodium hydride with DMSO. Though it is not an ylid, its reactions are profitably considered at this point.

$$CH_3{-}\overset{O^{\ominus}}{\overset{|}{S^{\oplus}}}{-}CH_3 + NaH \rightarrow CH_3{-}\overset{:\ddot{O}:^{\ominus}}{\overset{|}{\underset{\cdot\cdot}{S^{\oplus}}}}{-}\ddot{C}H_2^{\ominus}\ Na^{\oplus}$$

dimsyl sodium

This carbanion reacts with alkyl halides to produce higher sulfoxides.

$$CH_3{-}\overset{O^{\ominus}}{\overset{|}{S^{\oplus}}}{-}\ddot{C}H_2^{\ominus} + RX \xrightarrow{S_N2} CH_3{-}\overset{O^{\ominus}}{\overset{|}{S^{\oplus}}}{-}CH_2{-}R + X^{\ominus}$$

Its reaction with esters produces β-ketosulfoxides, which, in turn, are easily reduced to methyl ketones by aluminum amalgam.

$$\text{2-furyl}-\overset{O}{\overset{\|}{C}}-OCH_3 + CH_3-\overset{O^{\ominus}}{\overset{|}{S^{\oplus}}}-\ddot{C}H_2^{\ominus} \rightarrow \text{2-furyl}-\overset{O}{\overset{\|}{C}}-CH_2-\overset{O^{\ominus}}{\overset{|}{S^{\oplus}}}-CH_3 \xrightarrow{Al\cdot Hg/THF/0^{\circ}} \text{2-furyl}-\overset{O}{\overset{\|}{C}}-CH_3 \quad (70\%)$$

Since the methylene group between the carbonyl and sulfinyl groups is acidic, β-ketosulfoxides can be alkylated prior to reduction.

$$C_6H_5-\overset{O}{\overset{\|}{C}}-CH_2-\overset{O^{\ominus}}{\overset{|}{S^{\oplus}}}-CH_3 \xrightarrow{NaH} C_6H_5-\overset{O}{\overset{\|}{C}}-\underset{H}{\overset{\ominus}{C}}-\overset{O^{\ominus}}{\overset{|}{S^{\oplus}}}-CH_3 \xrightarrow{C_2H_5Br}$$

$$C_6H_5-\overset{O}{\overset{\|}{C}}-\underset{C_2H_5}{\overset{H}{C}}-\overset{O^{\ominus}}{\overset{|}{S^{\oplus}}}-CH_3 \xrightarrow[0^{\circ}]{Al\cdot Hg} C_6H_5\overset{O}{\overset{\|}{C}}CH_2CH_2CH_3$$

6. How might the product(s) of the reaction of 1,4-cyclohexadione with (a) dimethylsulfonium methylide and (b) dimethylsulfoxonium methylide differ? Explain.
7. What would you expect as the major sulfur-containing product from the following sequence? Explain.

$$CH_3\underset{O^{\ominus}}{\overset{\oplus}{S}}CH_2^{\ominus} + (CH_3)_3CBr \rightarrow \xrightarrow[\text{reflux}]{H_2O_2/HOAc} ?$$

Stereochemistry of Organosulfur Compounds. Unlike amines, triligant compounds of sulfur (such as sulfoxides, sulfinate esters, and sulfonium salts) are configurationally stable at room temperature.

Sulfinate Esters. The reaction of *p*-toluenesulfinyl chloride with (±)-2-butanol yields two isomeric sulfinate esters. The independent existence of these two diastereomers at room temperature *requires* that both the chiral carbon and the chiral sulfur be configurationally stable. That both products are sulf*in*ates is confirmed by the oxidation of both to the same racemic sulf*on*ate.

$$(\pm)-CH_3-C_6H_4-\overset{O^{\ominus}}{\overset{|}{S^{\oplus}}}-Cl + (\pm)-CH_3CH(OH)CH_2CH_3 \longrightarrow CH_3-C_6H_4-\overset{O^{\ominus}}{\overset{|}{\overset{*}{S}^{\oplus}}}-O-\overset{*}{C}H(CH_3)CH_2CH_3$$

$$\xrightarrow{H_2O_2} \underbrace{CH_3-C_6H_4}_{C_7H_7-}-\underset{O^{\ominus}}{\overset{O^{\ominus}}{S^{+2}}}-O-\underset{CH_3}{\overset{*}{C}}HCH_2CH_3 \quad \text{racemic}$$

asterisk denotes chiral centers

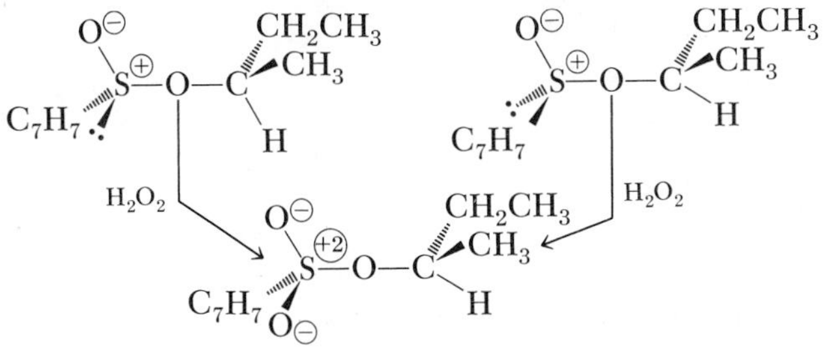

conversion of diastereomeric sulfinates into identical sulfonates

Sulfoxides. The most useful synthesis of optically active sulfoxides involves the elegant sequence presented below. What is particularly astounding is that this type of scheme was known for about fifty years before being applied by Andersen, in the early 1960's, to the preparation of optically active sulfoxides. A period of intense investigation of the stereochemistry of organosulfur compounds followed the report by K. Andersen. The Andersen synthesis (an example follows), coupled with detailed studies by K. Mislow, established that the sulfinate-to-sulfoxide conversion proceeds by *inversion of configuration* at sulfur.

l-menthol

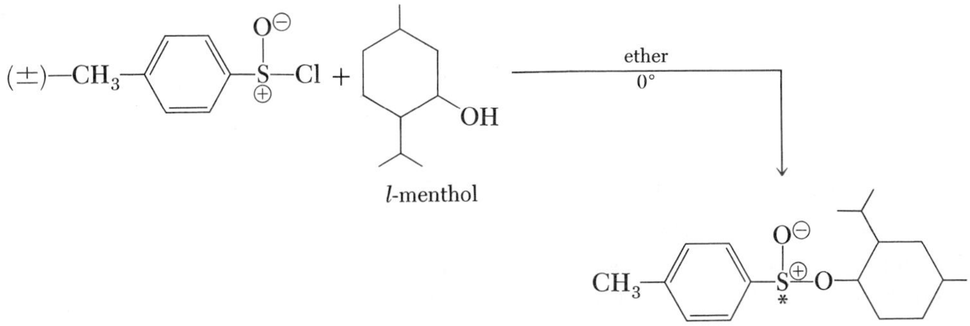

two diastereomers, differing in configuration at sulfur and separable by crystallization

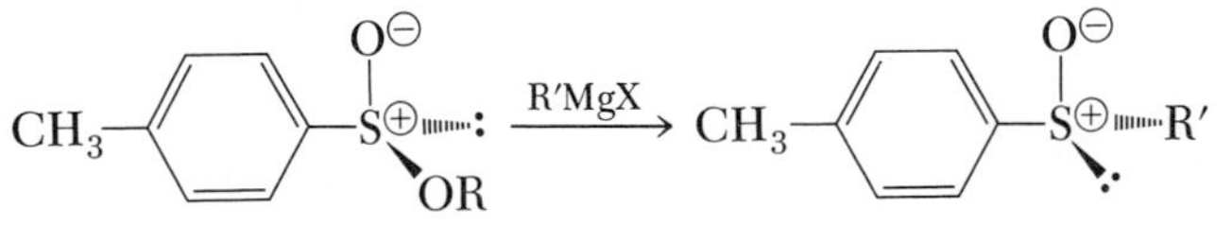

R = *l*-menthyl
sulfinate, one diastereomer

sulfoxide, one enantiomer

configurational inversion at sulfur

8. "All attempts to prepare and study optically active 2-naphthalenesulfinic acid are doomed to failure." What possible justification could be made for this prediction?

Sulfa Drugs. The earliest class of synthetics to receive fame as antimicrobial agents are the so-called *sulfa drugs.* A major step in modern chemotherapy was taken in 1933 when a youngster in Europe was cured of staphylococcal septicemia by a drug, patented in 1932, called *Prontosil.* A year later the wide range of activity of Prontosil against "strep" and "staph" infections was recognized and, in 1935, it was discovered that Prontosil was metabolized in the body of animals to sulfanilamide and that *this,* and not Prontosil, was active against bacteria *in vivo.*

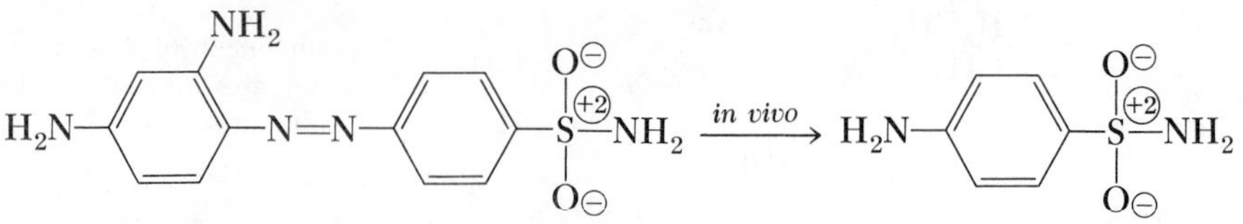

Prontosil

$\xrightarrow{\text{in vivo}}$

p-aminobenzenesulfonamide (sulfanilamide)

The laboratory synthesis of sulfanilamide provides a classic example of deactivation of an amino group by acetylation in order to be able to control the extent of electrophilic aromatic substitution on aniline.

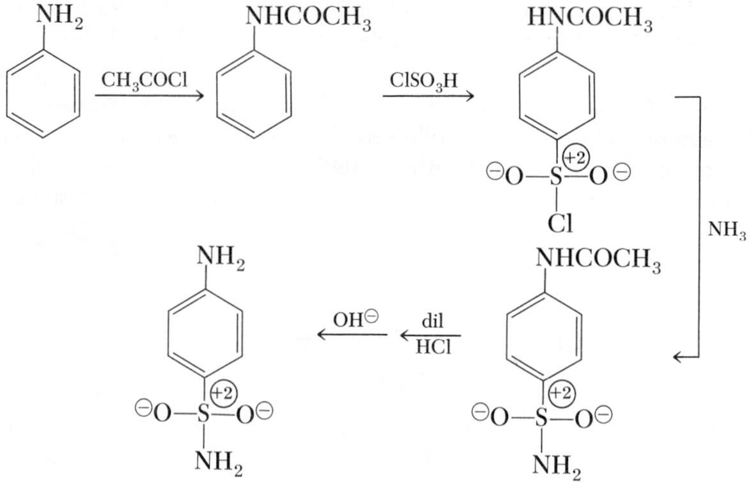

$\xrightarrow{CH_3COCl}$ $\xrightarrow{ClSO_3H}$ NH_3 $\xleftarrow{\text{dil HCl}}$ $\xleftarrow{OH^{\ominus}}$

sulfanilamide

Sulfanilamide acts against microorganisms by preventing the action of a metabolic chain that requires folic acid (also called pteroylglutamic acid).

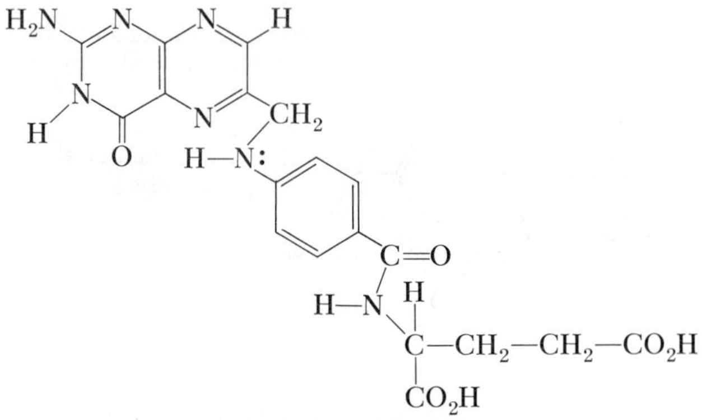

folic acid

Bacteria synthesize folic acid by reacting (2-amino-4-oxo-7,8-dihydro-6-pteridyl)methyl pyrophosphate, **A**, with *p*-aminobenzoic acid, abbreviated as PABA.

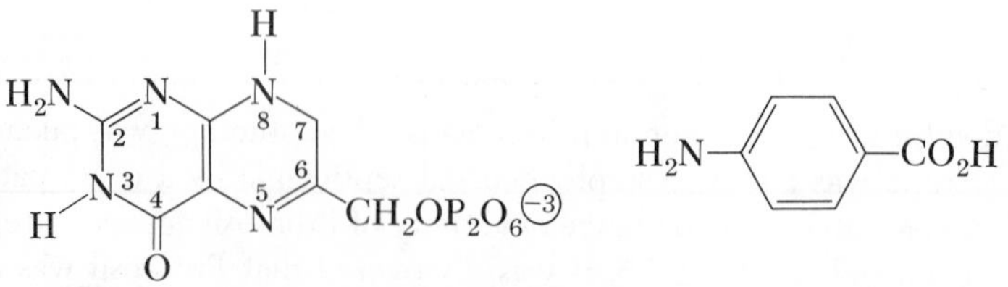

A PABA

The sulfanilamide competes effectively with PABA for **A**; but unfortunately for the bacteria, the end result is *not* folic acid and the bacteria die. Humans do not synthesize their own folic acid, but get it from external sources and from micro-organisms in the digestive tract; thus, people usually are invulnerable, at low concentrations, to the effects of sulfanilamide.

While over 5000 analogs of sulfanilamide have been synthesized and screened for biological activity, only perhaps a dozen are actively employed. Indeed, sulfanilamide's utility is now limited largely to veterinary medicine. The sulfa drugs still used in treating humans include sulfadiazine, sulfapyrazine, sulfathiazole, and sulfaethidole.

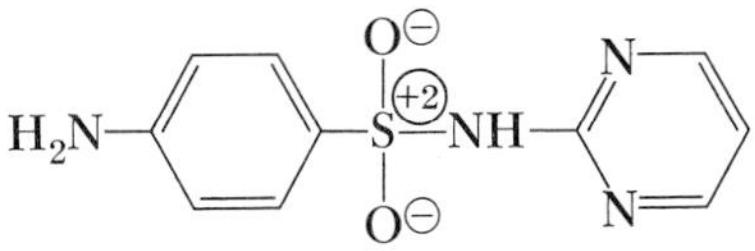

sulfadiazine

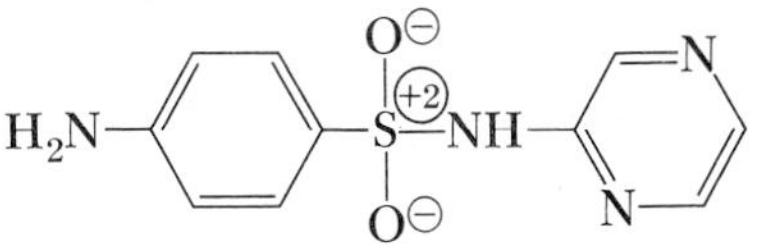

sulfapyrazine

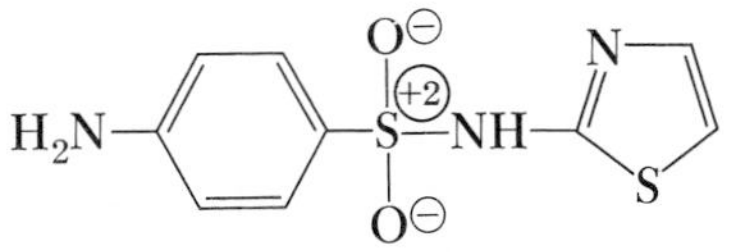

sulfathiazole

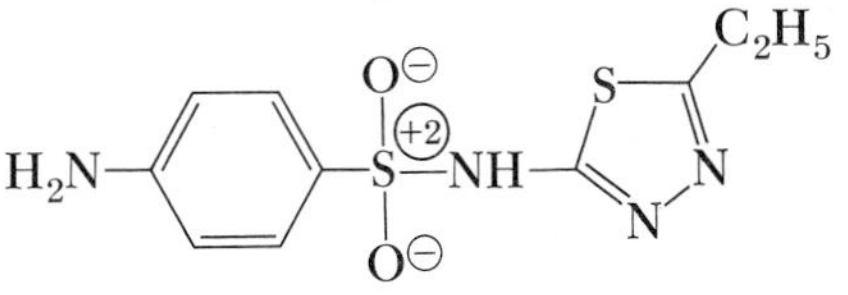

sulfaethidole

A slight departure from the fundamental sulfanilamide structure resulted in 4-homosulfanilamide, sometimes used in the treatment of gangrene.

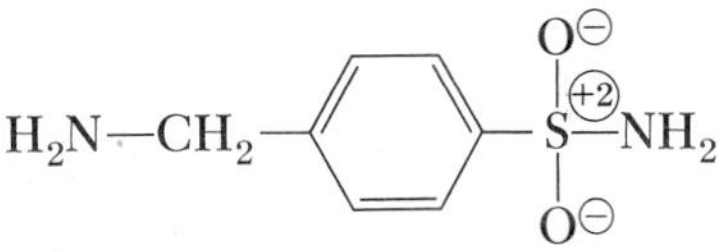

4-homosulfanilamide

"If they're so good, why aren't they used more often?" This question, if asked about sulfa drugs, would have to be answered at several different levels. The simplest answer is that some sulfa drugs have been replaced by more effective drugs, while the hazards associated with sulfa drug use have discouraged their continued use. Also, some bacterial strains are resistant to these drugs.

When sulfanilamide is administered to children under three years of age, and especially to premature babies, it displaces *bilirubin* from plasma proteins and causes *kernicterus,* a potentially fatal brain condition. Kernicterus may lead to mental retardation in surviving infants.

A teratogen is *any agent that can induce ANY abnormality in the somatic cells of a developing embryo.* (If the embryonic germ cells are undamaged, only the individual himself will be influenced, not his offspring.) The tragedies that ensued because of the use of thalidomide (1960–1962) point up the massive effects that an unchecked teratogen might have on the population. Because of this, most drugs, food additives, cosmetics and similar items are now checked for teratogenic potential. While the problem is quite complex, and the hazards have been adequately demonstrated only in small animals, sulfanilamide is a teratogen. The significance of this is unclear, however, since other known teratogens include penicillin, vitamins A, D, and E, carbon dioxide, nicotine, and salts of mercury.

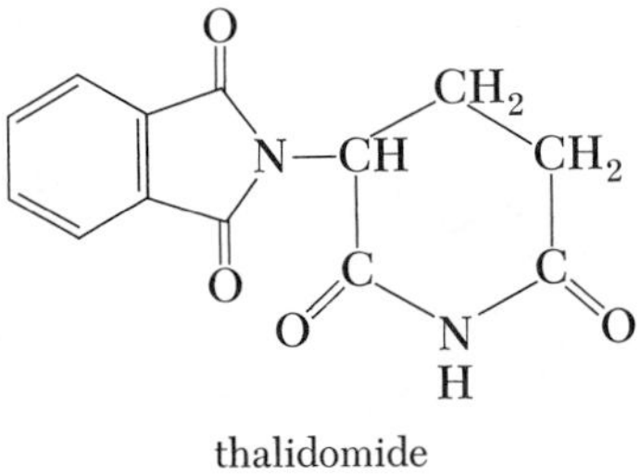

thalidomide

Finally, at least six sulfa drugs, including sulfanilamide, are known to induce *hemolytic anemia* in people with a genetically determined idiosyncratic susceptibility.

23.5 SPECTRAL PROPERTIES OF SULFUR COMPOUNDS

Infrared spectra are, in general, more valuable for identifying organosulfur compounds than are any other type of spectra.

Infrared Spectra. The **S—H** stretching mode occurs as a weak band in the 2600 to 2550 cm^{-1} region. Compounds containing the SH group do not form strong intermolecular hydrogen bonds, so spectra of pure liquids and of dilute solutions are similar in the SH region.

The **C—S** stretching frequency is weak and occurs over a range (720 to 550 cm^{-1}). Similarly, the **S—S** stretching mode occurs as a weak band in the 500 to 400 cm^{-1} region. Thus, *the presence of sulfides and disulfides cannot be established by ir.* The infrared spectra of several organosulfur compounds are presented in Figure 23–6.

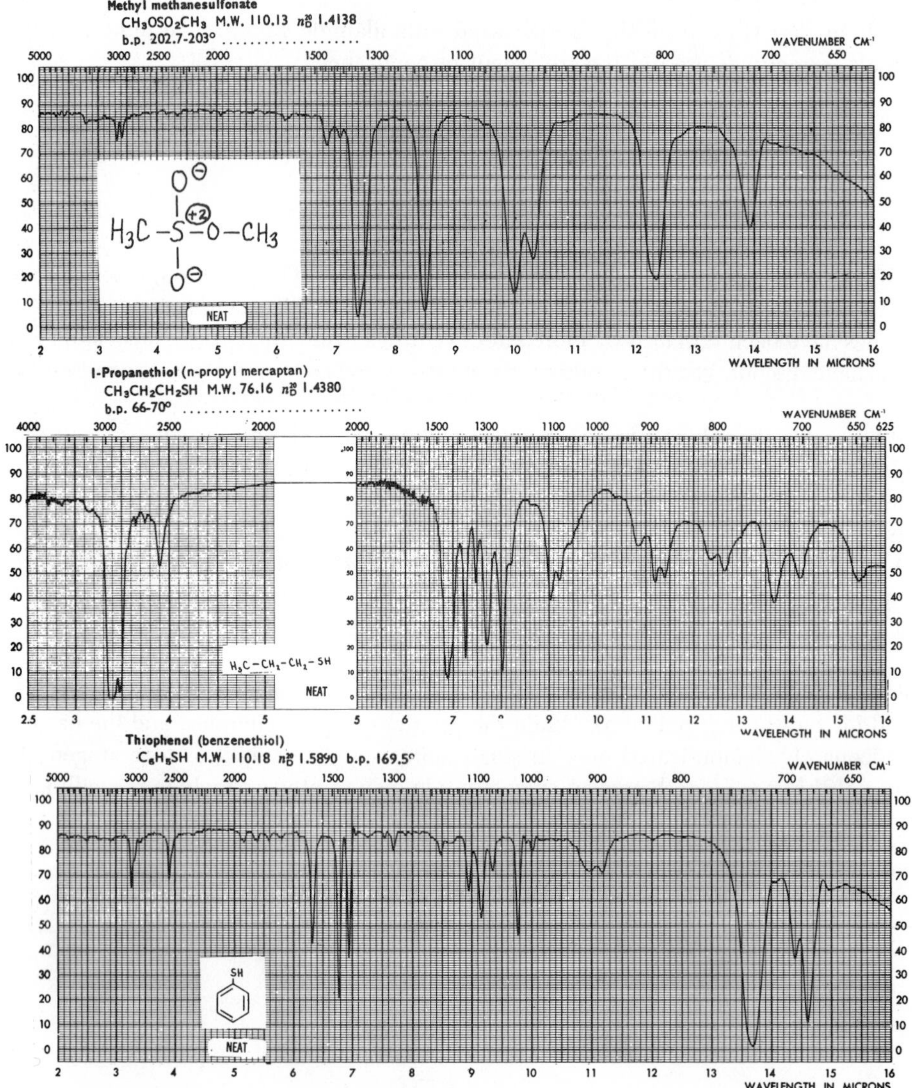

Figure 23-6 The infrared spectra of organosulfur compounds. *Illustration continued on opposite page.*

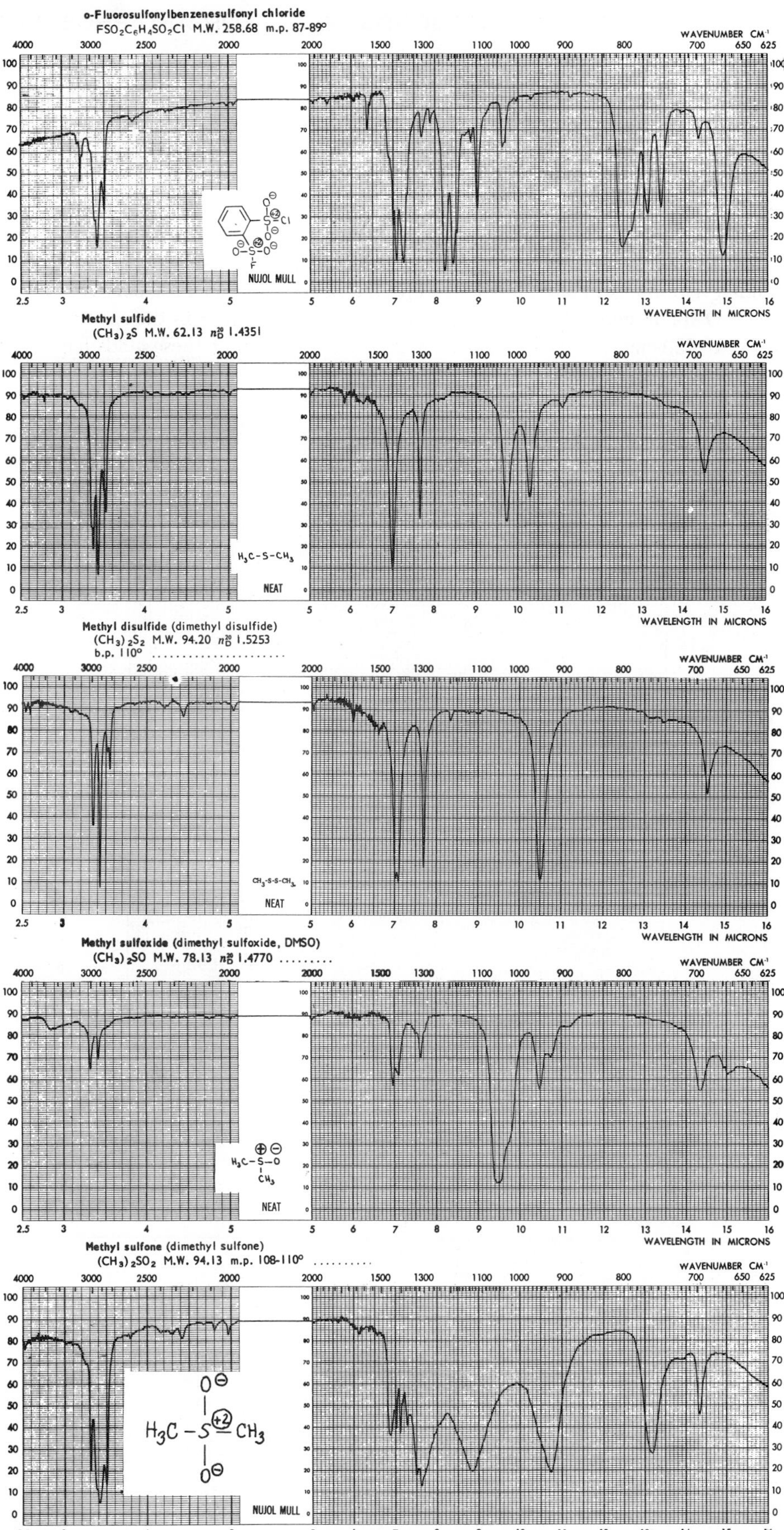
o-Fluorosulfonylbenzenesulfonyl chloride
FSO2C6H4SO2Cl M.W. 258.68 m.p. 87-89°
NUJOL MULL
Methyl sulfide
(CH3)2S M.W. 62.13 n20D 1.4351
H3C-S-CH3
NEAT
Methyl disulfide (dimethyl disulfide)
(CH3)2S2 M.W. 94.20 n20D 1.5253
b.p. 110°
CH3-S-S-CH3
NEAT
Methyl sulfoxide (dimethyl sulfoxide, DMSO)
(CH3)2SO M.W. 78.13 n20D 1.4770
NEAT
Methyl sulfone (dimethyl sulfone)
(CH3)2SO2 M.W. 94.13 m.p. 108-110°
NUJOL MULL
WAVENUMBER CM-1
WAVELENGTH IN MICRONS

In contrast to this behavior, $\overset{\oplus}{S}$—$\overset{\ominus}{O}$ bonds show fairly intense, characteristic bands near 1050 cm^{-1}. The infrared spectra of several compounds containing the polar S—O bond are included in Figure 23–6. Note that while sulfoxides appear to have only a single ν_{S-O}, sulfones possess two absorptions (1350 to 1300 cm^{-1} and 1185 to 1140 cm^{-1}). These two vibrations are asymmetric and symmetric modes, respectively.

Aromatic sulfonyl chlorides, the most common type of sulfonyl halide, exhibit asymmetric and symmetric modes at 1380 and 1180 cm^{-1}, respectively. These values are close to the corresponding vibrations in sulfonic acids, reported near 1350 and 1160 cm^{-1}.

The **S—O single bond** appears as a strong band near 900 cm^{-1} in the spectra of sulfonic acids and near 850 cm^{-1} in the spectra of sulfinic acids. Esters of these acids have an S—O single bond stretching frequency at about 750 cm^{-1}.

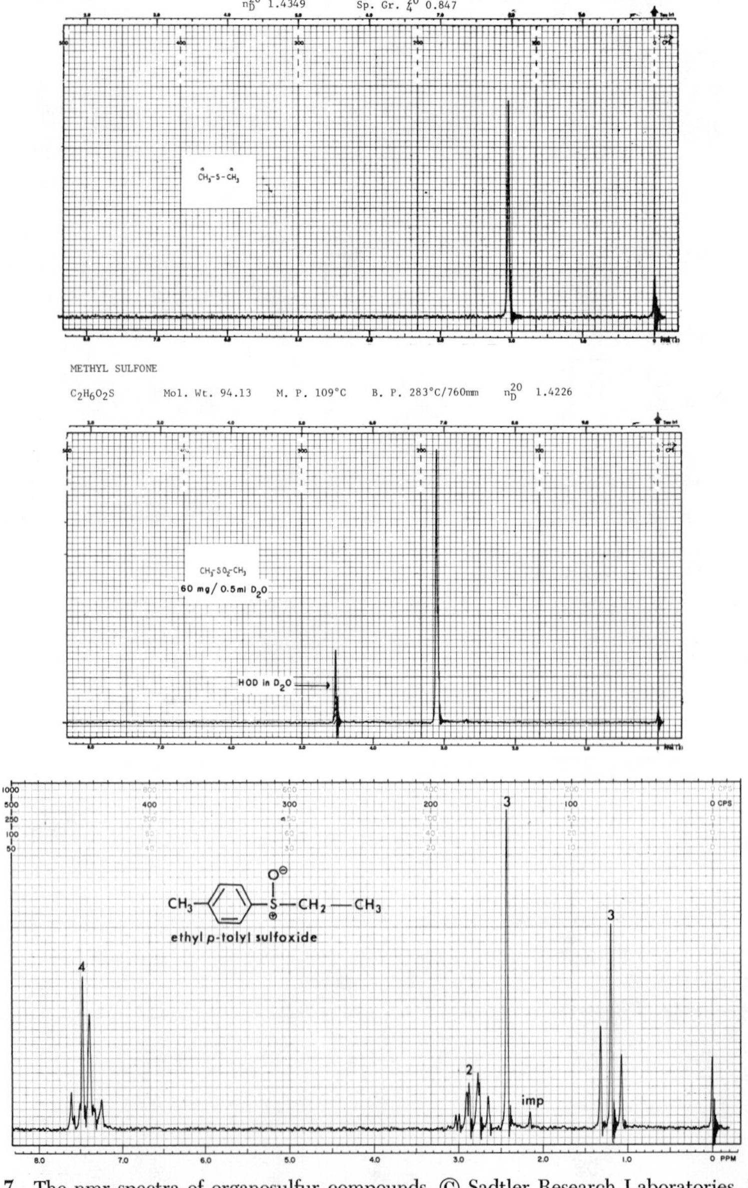

Figure 23-7 The nmr spectra of organosulfur compounds. © Sadtler Research Laboratories, Inc., 1976.

NMR Spectra. Protons on carbon adjacent to a sulfur are usually shifted downfield because of the inductive effect of the heteroatom. Increasing the electron-withdrawing capacity of sulfur by oxidation shifts those protons even farther downfield (Figure 23–7). Since the sulfinyl group is chiral, protons on a methylene group adjacent to a sulfinyl group are non-equivalent with respect to chemical shift and, if unsplit by other nuclei, will appear as an "AB quartet." This phenomenon accounts for the multiplicity of the —CH_2— signal in the spectrum of ethyl *p*-tolyl sulfoxide in Figure 23–7.

PHOSPHORUS COMPOUNDS

23.6 ORGANIC COMPOUNDS CONTAINING PHOSPHORUS

Any discussion of the organic chemistry of phosphorus will be complicated by the variety of organic compounds that contain phosphorus and by the existence of several different ways of naming many of them. The abundance of substances containing phosphorus stems, in part, from the ability of phosphorus to employ *d* orbitals in bonding, a capability that leads to compounds in which phosphorus bears as many as five or six substituents. In addition, phosphorus forms a strong dative bond to oxygen, $\overset{\oplus}{P}—\overset{\ominus}{O}$; consequently, there exist compounds containing the "normal" P—O bond, compounds containing the dative bond, and many containing both.

Complexities that have arisen in naming phosphorus compounds have been met by considering the compounds to be derivatives of various phosphorus oxyacids. These acids include *phosphinous acid,* H_2POH, *phosphonous acid,* $HP(OH)_2$, *phosphorous acid,* $P(OH)_3$, and *phosphoric acid,* $OP(OH)_3$. The esters of phosphinous acid, H_2POR, are *phosphinites;* the diesters of phosphonous acid are *phosphonites,* $HP(OR)_2$; the triesters of phosphorous acid are *phosphites,* $P(OR)_3$; and the triesters of phosphoric acid are *phosphates,* $OP(OR)_3$. Several examples may serve to clarify the naming of the organic analogs of these acids and their esters.[a]

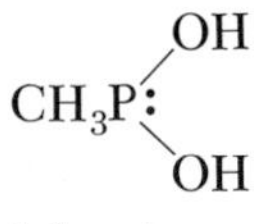

$CH_3\ddot{P}(OH)_2$
methylphosphonous acid (unstable)

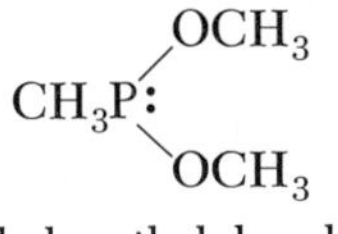

$CH_3\ddot{P}(OCH_3)_2$
dimethyl methylphosphonite

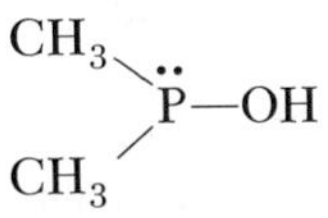

$(CH_3)_2\ddot{P}—OH$
dimethylphosphinous acid (unstable)

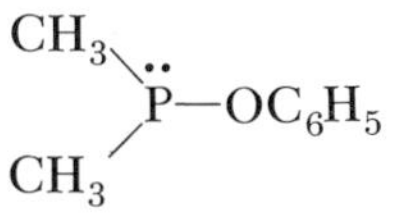

$(CH_3)_2\ddot{P}—OC_6H_5$
phenyl dimethylphosphinite

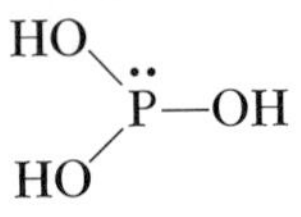

$(HO)_2\ddot{P}—OH$
phosphorous acid (unstable)

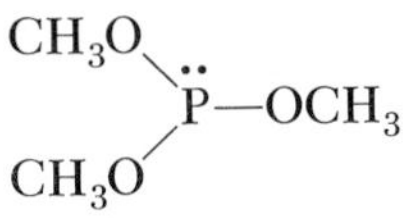

$(CH_3O)_2\ddot{P}—OCH_3$
trimethyl phosphite

[a] The unshared electron pair on phosphorus is included in the structures in this section but normally will not be shown in structures in subsequent sections.

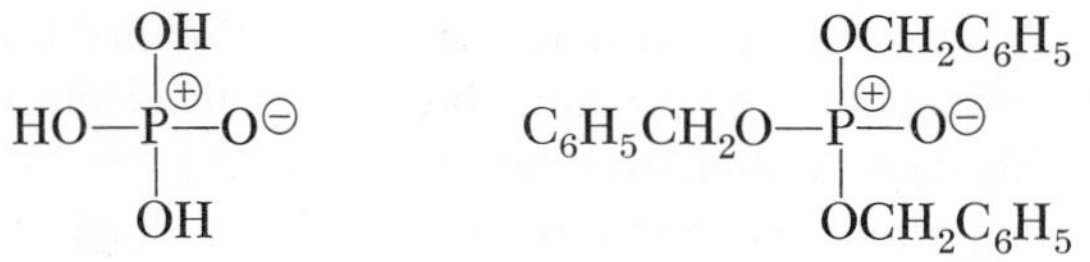

$$HO-\overset{\overset{OH}{|}}{\underset{\underset{OH}{|}}{\overset{\oplus}{P}}}-O^{\ominus}$$
phosphoric acid

$$C_6H_5CH_2O-\overset{\overset{OCH_2C_6H_5}{|}}{\underset{\underset{OCH_2C_6H_5}{|}}{\overset{\oplus}{P}}}-O^{\ominus}$$
tribenzyl phosphate

Three of the four oxyacids listed above are labeled "unstable" because they isomerize to a tautomer that contains a phosphoryl group. This facile equilibration is a direct consequence of the greater stability of the $\overset{\oplus}{P}-\overset{\ominus}{O}$ bond (~140 kcal/mole) than of the P—O bond (~95 kcal/mole). These stable tautomers all contain a phosphorus-hydrogen bond; one can generate a new series of phosphorus compounds simply by replacing those hydrogens with an alkyl or aryl group.

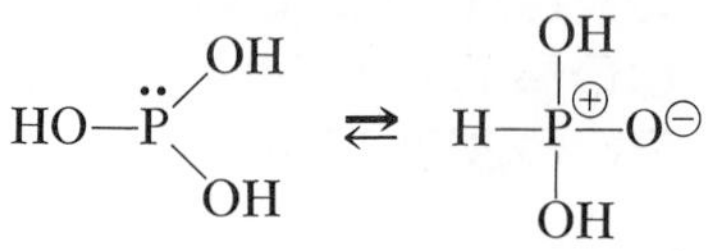

$$HO-\ddot{P}(OH)_2 \rightleftarrows H-\overset{\overset{OH}{|}}{\underset{\underset{OH}{|}}{\overset{\oplus}{P}}}-O^{\ominus}$$
phosphorous acid — phosphonic acid

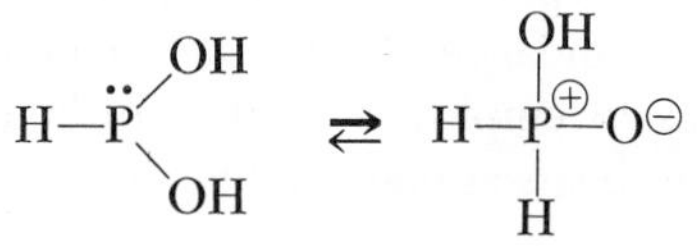

$$H-\ddot{P}(OH)_2 \rightleftarrows H-\overset{\overset{OH}{|}}{\underset{\underset{H}{|}}{\overset{\oplus}{P}}}-O^{\ominus}$$
phosphonous acid — phosphinic acid

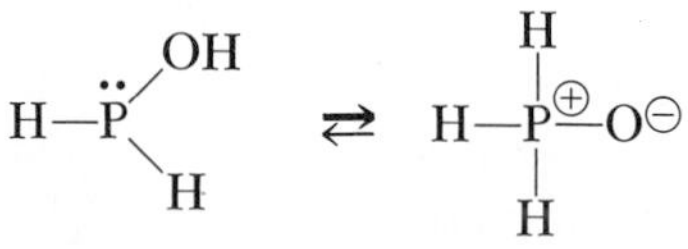

$$H-\ddot{P}(OH)(H) \rightleftarrows H-\overset{\overset{H}{|}}{\underset{\underset{H}{|}}{\overset{\oplus}{P}}}-O^{\ominus}$$
phosphinous acid — phosphine oxide

$$CH_3-\overset{\overset{OH}{|}}{\underset{\underset{OH}{|}}{\overset{\oplus}{P}}}-O^{\ominus}$$
methylphosphonic acid

$$CH_3-\overset{\overset{OH}{|}}{\underset{\underset{CH_3}{|}}{\overset{\oplus}{P}}}-O^{\ominus}$$
dimethylphosphinic acid

$$CH_3-\overset{\overset{OCH_3}{|}}{\underset{\underset{OCH_3}{|}}{\overset{\oplus}{P}}}-O^{\ominus}$$
dimethyl methylphosphonate

$$CH_3-\overset{\overset{CH_3}{|}}{\underset{\underset{CH_3}{|}}{\overset{\oplus}{P}}}-O^{\ominus}$$
trimethylphosphine oxide

The most important *acid chlorides* of these acids are the phosphinous chlorides, R_2PCl, the phosphonous dichlorides, $RPCl_2$, phosphorus trichloride, PCl_3, and phosphorus oxychloride (also called phosphoryl trichloride), $OPCl_3$. The naming of acid chlorides and related amides is illustrated below.

$$C_6H_5-\ddot{P}(C_6H_5)(Cl)$$
diphenylphosphinous chloride

$$C_6H_5-\ddot{P}(Cl)_2$$
phenylphosphonous dichloride

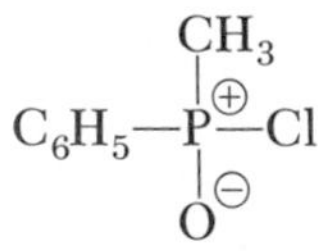

$C_6H_5{-}\overset{\oplus}{P}(CH_3)(O^{\ominus}){-}Cl$
methylphenylphosphinic chloride

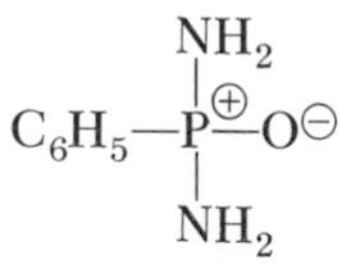

$C_6H_5{-}\overset{\oplus}{P}(NH_2)_2{-}O^{\ominus}$
phenylphosphonic diamide

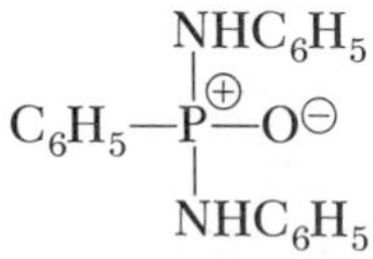

$C_6H_5{-}\overset{\oplus}{P}(NHC_6H_5)_2{-}O^{\ominus}$
N,N′-diphenyl phenylphosphonic diamide

Two of the most important types of organic phosphorus compounds are related to phosphine oxide. The first of these is the *phosphines*, and the second is the *phosphonium salts*. Such compounds are the structural analogs of amines and ammonium salts and are named similarly. Phosphine, PH_3, is the stable hydride of phosphorus and provides the root in the naming of compounds of the type RPH_2, R_2PH and R_3P. While phosphorane, PH_5, is an unknown substance, it provides the root for naming of compounds containing five bonds from carbon to phosphorus. However, further complications result from the existence of two varieties of phosphoranes. The simple variety (*e.g.*, R_5P) is called a "phosphorane," while those that achieve pentacovalency *via* back-bonding are termed *alkylidenephosphoranes*. One group of the latter contains the Wittig reagents derived from phosphonium salts (Chapter 19).

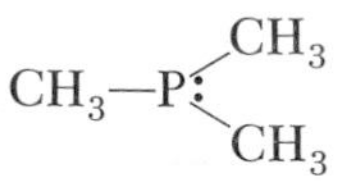

$CH_3{-}\ddot{P}(CH_3)_2$
trimethylphosphine

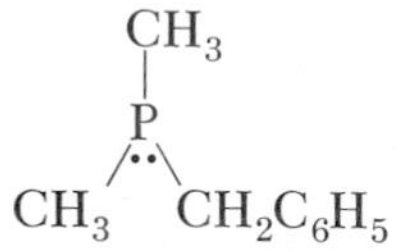

$\ddot{P}(CH_3)_2(CH_2C_6H_5)$
benzyldimethylphosphine

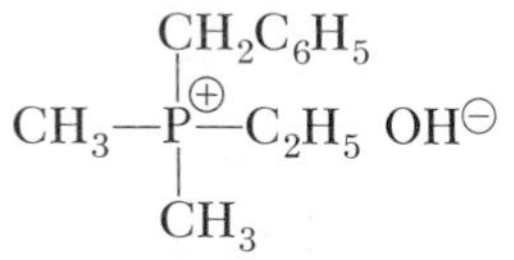

$CH_3{-}\overset{\oplus}{P}(CH_2C_6H_5)(CH_3){-}C_2H_5\ OH^{\ominus}$
benzylethyldimethylphosphonium hydroxide

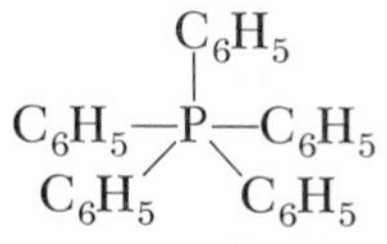

$P(C_6H_5)_5$
pentaphenylphosphorane
(a phosphorane)

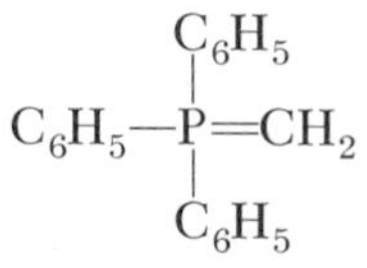

$(C_6H_5)_3P{=}CH_2$
methylenetriphenylphosphorane
(an alkylidenephosphorane)

In summary, the naming of organic compounds of phosphorus is based upon the group of compounds, some of which are unstable, that is presented below.

$:PH_3$ phosphine — PH_5 phosphorane — PH_3O phosphine oxide

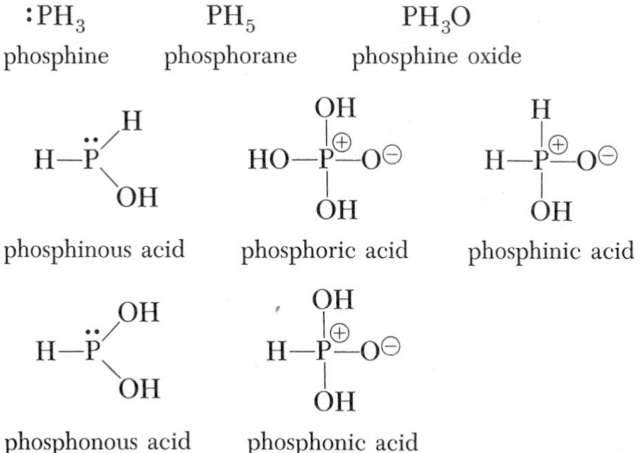

9. Provide an acceptable name for each of the following.

(a) PH_3 (e) $(HO)_2PH$
(b) PCl_3 (f) $OP(OH)_2H$
(c) $(HO)_3P$ (g) $(C_6H_5O)_3P$
(d) $^{\ominus}O\overset{\oplus}{P}H_3$ (h) $(C_6H_5O)_3\overset{\oplus\ominus}{PO}$

23.7 REACTIVITY, BONDING, AND STRUCTURE

Bond Strength and Reactivity. Since "bond energy" measures the energy required to cleave a bond homolytically but most organic reactions do not involve radicals, it is incorrect to draw strict correlations between bond energy and general chemical reactivity. However, some useful information can be gleaned by comparing the bond strengths of analogous bonds to carbon, nitrogen, and phosphorus, For example, phosphorus-hydrogen bonds are weaker than either nitrogen-hydrogen bonds or carbon-hydrogen bonds (Table 23–2) and are readily cleaved under both free-radical and ionic conditions. However, carbon-carbon, carbon-nitrogen, and carbon-phosphorus bonds are of comparable strength and, as might be expected, cleavage of the phosphorus-carbon bond is rather rare. One group of exceptions consists of reactions that produce the phosphoryl group as a result of phosphorus-carbon cleavage, *e.g.*, the Wittig reaction (p. 603).

$$R_2CO + (C_6H_5)_3\overset{\oplus}{P}—\overset{\ominus}{C}R_2' \rightarrow R_2C{=}CR_2' + (C_6H_5)_3\overset{\oplus}{P}—O^{\ominus} \qquad \text{Wittig reaction}$$

TABLE 23-2 Bond Energies in Phosphorus, Nitrogen, and Carbon Compounds[a]

Bond	Bond Energy (kcal/mole)	Bond	Bond Energy (kcal/mole)	Bond	Bond Energy (kcal/mole)
P—C	65	N—C	80	C—C	83
P—H	77	N—H	93	C—H	98
P—F	117	N—F	65	C—F	116
P—Cl	76	N—Cl	37	C—Cl	81
P—Br	62			C—Br	68
P—O	95	N—O	53	C—O	86
$P^{\oplus}—O^{\ominus}$	140				
P—P	50	N—N	40		

[a] Values are average bond energies and will vary for specific compounds.

Phosphorus-chlorine bonds are comparable in strength to carbon-chlorine bonds, and both are stronger than nitrogen-chlorine bonds. Thus, while nitrogen halides are unstable, sometimes explosive compounds, phosphorus halides and carbon halides are fairly stable and of comparable reactivity.

***d*-Orbital Participation.** The outer electronic configurations of nitrogen, $1s^22s^22p^3$, and of phosphorus, $1s^22s^22p^63s^23p^3$, are similar; thus, it is not surprising that there are parallels in their chemistries. However, the parallel is not strict, since there are many stable functional groups containing nitrogen for which there are no stable phosphorus analogs. Some of these include *amide*, $RCONH_2$, *azo*, RN=NR, *nitroso*, R—N=O, and *nitrile*, RCN.

On the other side of the coin, phosphorus, but not nitrogen, can possess ten outer electrons because phosphorus has available, low-lying *d* orbitals (3*d*). This explains the unusual stability of alkylidenephosphoranes and the phosphoryl group, both of which are described by resonance hybrids that include contributions containing a *p,d* π bond. We shall assume that *d*-orbital participation in phosphorus is similar to that in sulfur (p. 722).

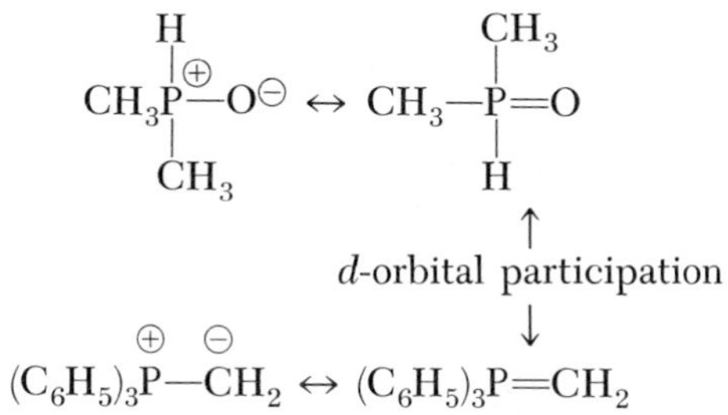

$$CH_3\overset{\overset{H}{|}}{\underset{\underset{CH_3}{|}}{P^{\oplus}}}—O^{\ominus} \leftrightarrow CH_3—\overset{\overset{CH_3}{|}}{\underset{\underset{H}{|}}{P}}=O$$

↑
d-orbital participation
↓

$$(C_6H_5)_3\overset{\oplus}{P}—\overset{\ominus}{C}H_2 \leftrightarrow (C_6H_5)_3P=CH_2$$

10. The free energy change of the reaction shown is −1.12 kcal at 25°. Assuming a P—H bond strength of 76 kcal/mole, a P—P bond strength of 55 kcal/mole, and an N—H bond strength of 93 kcal/mole, and $\Delta S = 0$, calculate the strength of the N—P bond.

$$(CH_3)_2NP(CH_3)_2 + (CH_3)_2PH \rightleftharpoons (CH_3)_2PP(CH_3)_2 + (CH_3)_2NH$$

11. Suggest a reason why compounds such as $CH_3N{=}NCH_3$, C_6H_5NO, and C_6H_5CN do not have phosphorus analogs (phosphorus replacing nitrogen).

Stereochemistry. Trisubstituted derivatives of phosphorus are pyramidal. Bond angles are not much larger than 90°, suggesting that the phosphorus is fairly close to being unhybridized. Tetrasubstituted derivatives of phosphorus have a tetrahedral geometry, the phosphorus being sp^3 hybridized in these compounds.

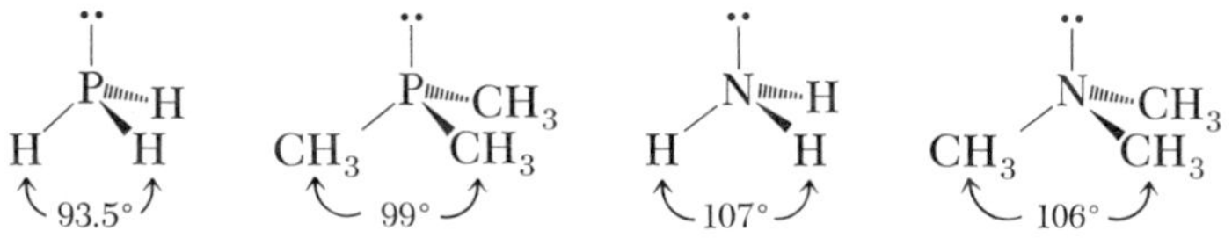

PH_3 (93.5°) $P(CH_3)_3$ (99°) NH_3 (107°) $N(CH_3)_3$ (106°)

For reasons that remain unclear, trisubstituted phosphorus derivatives are configurationally stable at room temperature, even though amines racemize rapidly at much lower temperatures. Thus, the pyramidal stability of phosphines resembles that of sulfonium salts, enantiomers of both being racemized only at elevated temperatures.

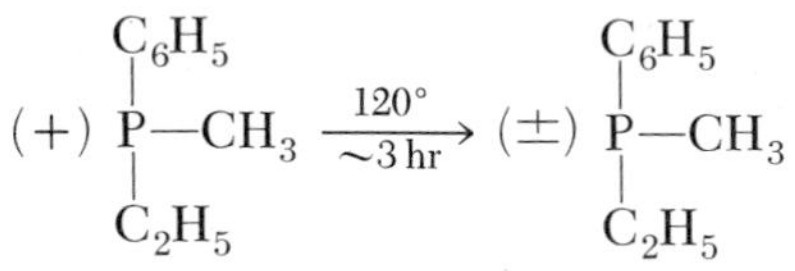

$$(+)\ C_6H_5(C_2H_5)P—CH_3 \xrightarrow[\sim 3\,hr]{120°} (\pm)\ C_6H_5(C_2H_5)P—CH_3$$

In light of the optical stability of phosphines, which possess a non-bonding electron pair on phosphorus, it should come as no surprise that phosphonium salts and phosphine oxides, which lack a non-bonding pair of electrons, also are configurationally stable.

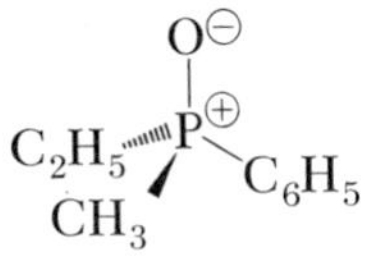

$C_2H_5(CH_3)(C_6H_5)P^{\oplus}—O^{\ominus}$ ethylmethylphenylphosphine oxide
$[\alpha]_D = \pm 22.8°$

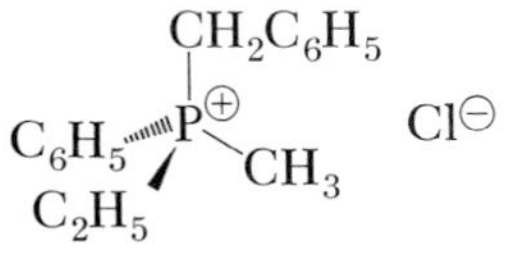

$C_6H_5(C_2H_5)(CH_3)(CH_2C_6H_5)P^{\oplus}\ Cl^{\ominus}$ benzylethylmethylphenylphosphonium chloride
$[\alpha]_D = \pm 21°$

Two geometries are possible for pentaligant phosphorus, trigonal bipyramid and square pyramid. The former is found in phosphorus pentachloride, PCl_5, while the latter is found in pentaphenylphosphorane. Phosphorus is dsp^3 hybridized in both of these. The difference in geometries is due to the use of a d_{z^2} orbital in the former and a $d_{x^2-y^2}$ orbital in the latter.

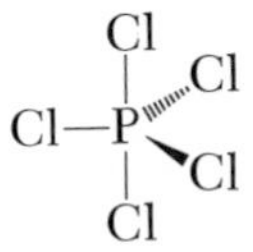

phosphorus pentachloride (trigonal bipyramid)

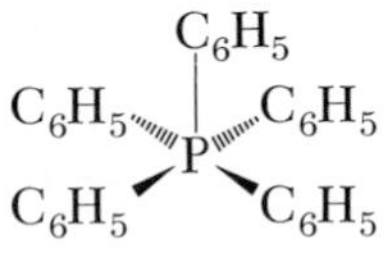

pentaphenylphosphorane (square pyramid)

23.8 SYNTHESIS OF SIMPLE PHOSPHINES

Phosphorus reacts with chlorine to produce phosphorus trichloride and phosphorus pentachloride, two compounds that are important starting points in the synthesis of organic phosphorus compounds. For example, phosphorus trichloride reacts with benzene, in the presence of aluminum chloride, to form phenylphosphonous dichloride, also called dichlorophenylphosphine.

$$P_4 + Cl_2(\text{excess}) \rightarrow PCl_3 + PCl_5$$

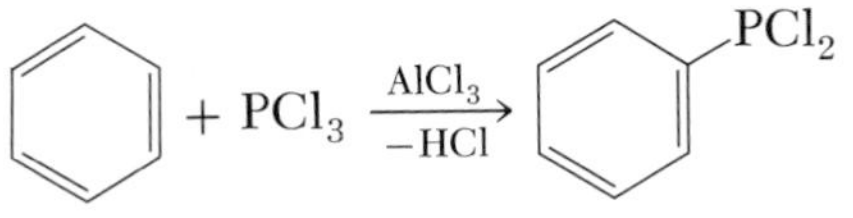

$$C_6H_6 + PCl_3 \xrightarrow[-HCl]{AlCl_3} C_6H_5PCl_2$$

Phenylphosphonous dichloride can be used to synthesize a wide variety of compounds containing a C_6H_5—P fragment. The usual point of departure is the reduction of the dichloride to phenylphosphine by lithium aluminum hydride. Hydrogens bonded to phosphorus are sufficiently acidic to react with metallic sodium, producing a salt (a sodium phosphide) and hydrogen. Thus, reaction of phenylphosphine with sodium produces the very nucleophilic phenylphosphide anion, $C_6H_5PH^{\ominus}$. Lithium aluminum hydride also will reduce phosphorus trichloride to phosphine, and, upon reaction with sodium, this will yield sodium phosphide, $NaPH_2$.

$$C_6H_5PCl_2 \xrightarrow[THF]{LiAlH_4} C_6H_5PH_2$$

$$2C_6H_5PH_2 + 2Na \rightarrow \underset{\text{sodium phenylphosphide}}{C_6H_5\overset{\ominus}{P}H\ Na^{\oplus}} + H_2$$

$$PCl_3 \xrightarrow{LiAlH_4} PH_3 \xrightarrow{Na} \underset{\text{sodium phosphide}}{NaPH_2}$$

The availability of these two nucleophiles sets the stage for the preparation of a variety of substituted phosphines, as outlined in Figure 23–8. The astonishing nucleophilicity of phosphide anions can be seen in the sequence that converts *m*-tolyl iodide into phenyl *m*-tolylphosphine. This reaction proceeds by a nucleophilic substitution and not by aryne formation (as do reactions of aryl halides with amide ions).

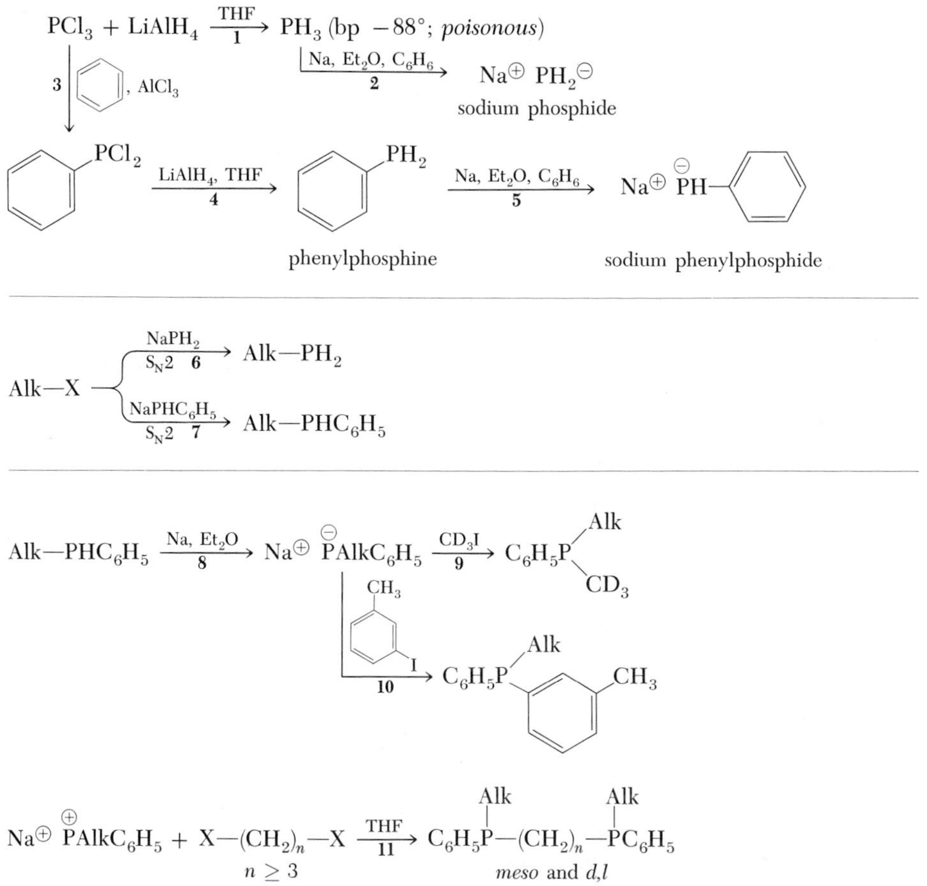

$$PCl_3 + LiAlH_4 \xrightarrow[1]{THF} PH_3 \text{ (bp } -88°\text{; } \textit{poisonous}\text{)} \xrightarrow[2]{Na,\ Et_2O,\ C_6H_6} Na^{\oplus}\ PH_2^{\ominus} \text{ (sodium phosphide)}$$

$$PCl_3 \xrightarrow[3]{C_6H_6,\ AlCl_3} C_6H_5PCl_2 \xrightarrow[4]{LiAlH_4,\ THF} C_6H_5PH_2 \text{ (phenylphosphine)} \xrightarrow[5]{Na,\ Et_2O,\ C_6H_6} Na^{\oplus}\ \overset{\ominus}{P}H{-}C_6H_5 \text{ (sodium phenylphosphide)}$$

$$Alk{-}X \xrightarrow[S_N2\ 6]{NaPH_2} Alk{-}PH_2$$

$$Alk{-}X \xrightarrow[S_N2\ 7]{NaPHC_6H_5} Alk{-}PHC_6H_5$$

$$Alk{-}PHC_6H_5 \xrightarrow[8]{Na,\ Et_2O} Na^{\oplus}\ \overset{\ominus}{P}AlkC_6H_5 \xrightarrow[9]{CD_3I} C_6H_5P(Alk)(CD_3)$$

$$Na^{\oplus}\ \overset{\ominus}{P}AlkC_6H_5 \xrightarrow[10]{m\text{-}CH_3C_6H_4I} C_6H_5P(Alk)(C_6H_4\text{-}m\text{-}CH_3)$$

$$Na^{\oplus}\ \overset{\oplus}{P}AlkC_6H_5 + X{-}(CH_2)_n{-}X \xrightarrow[11]{THF} C_6H_5\underset{}{\overset{Alk}{P}}{-}(CH_2)_n{-}\overset{Alk}{P}C_6H_5$$

$n \geq 3$ *meso* and *d,l*

Figure 23-8 Phosphine synthesis by nucleophilic substitution.

Although the reaction of organolithium or Grignard reagents with alkyl halides is not a "method of choice" for synthesizing carbon-carbon bonds, phosphorus-carbon bonds are prepared in good yield by treating the appropriate phosphorus chloride with either of these organometallic compounds. This suggests that trisubstituted phosphorus is more prone to attack *by* nucleophiles than is carbon.

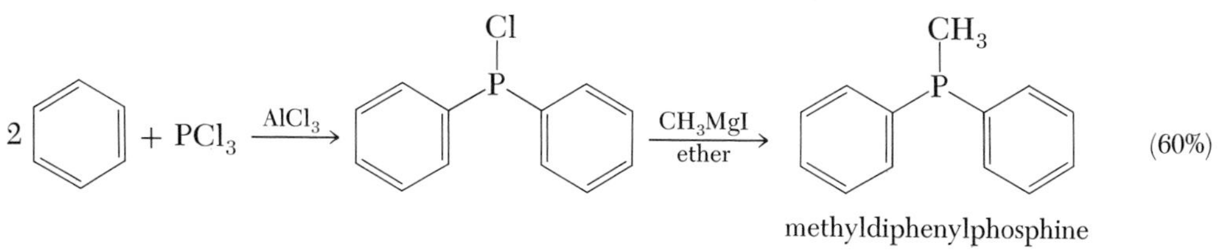

$$2\,C_6H_6 + PCl_3 \xrightarrow{AlCl_3} (C_6H_5)_2PCl \xrightarrow[ether]{CH_3MgI} (C_6H_5)_2PCH_3 \quad (60\%)$$

methyldiphenylphosphine

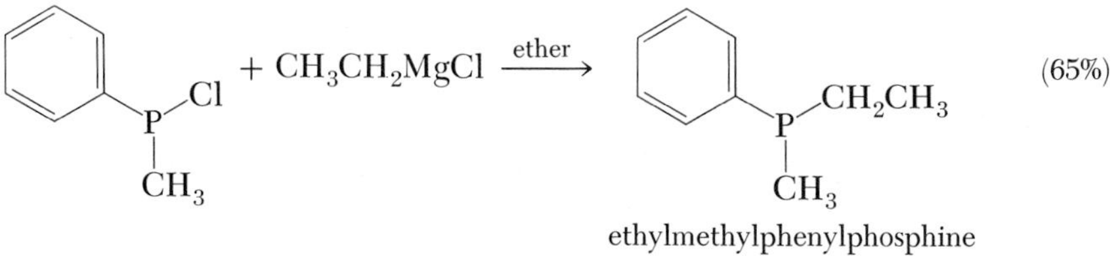

$$C_6H_5P(Cl)CH_3 + CH_3CH_2MgCl \xrightarrow{ether} C_6H_5P(CH_2CH_3)CH_3 \quad (65\%)$$

ethylmethylphenylphosphine

The synthesis of phosphines is complicated by their facile air oxidation to phosphine oxides. (Triphenylphosphine is an important exception.) This propensity toward air oxidation, coupled with their extremely unpleasant odor (once described by a student as "dead garlic") and moderate toxicity (phosphine is ten times as toxic as hydrogen cyanide) has made research with low molecular weight phosphines difficult.

23.9 REACTIONS OF PHOSPHORUS COMPOUNDS

Phosphorus as a Nucleophile. Phosphorus atoms are larger than nitrogen atoms, and one would anticipate that the outer orbitals of phosphorus should be deformed more readily than those of nitrogen. This is consistent with the observation that phosphines are more nucleophilic than are amines and explains, for example, why triphenylamine does not react with methyl iodide while triphenylphosphine forms a quaternary salt under the same conditions.

$$(C_6H_5)_3N + CH_3I \nrightarrow (C_6H_5)_3\overset{\oplus}{N}CH_3\ I^{\ominus} \qquad \text{not observed}$$

$$(C_6H_5)_3P + CH_3I \rightarrow (C_6H_5)_3\overset{\oplus}{P}CH_3\ I^{\ominus}$$

Alkyl groups are electron donors, and their introduction at a heteroatom should be expected to increase the nucleophilicity of that atom. However, increasing alkylation generally decreases the nucleophilicity of nitrogen although it increases the nucleophilicity of phosphorus. This reversal arises because nitrogen is smaller than phosphorus; increasing substitution introduces steric hindrance around nitrogen, which offsets the inductive effects of the alkyl groups. Such steric effects are absent in alkylated phosphines.

$$R_3P > R_2PH > RPH_2 > PH_3$$

increasing nucleophilicity

$$RNH_2 > R_2NH > R_3N$$

The propensity for phosphorus to act as a nucleophile will be better appreciated after we consider a few more examples, most of which will involve additions of phosphines to multiple bonds. Phosphine adds to simple alkenes, in the presence of acid, to produce a monosubstituted phosphine. It is believed that the increased basicity of phosphorus in the product results in its protonation and, by removing the electron pair at phosphorus, limitation of the alkylation to a single stage.

$$(CH_3)_2C{=}CH_2 + H^{\oplus} \rightleftarrows (CH_3)_2\overset{\oplus}{C}{-}CH_3 \xrightarrow{PH_3} H_2P{-}C(CH_3)_2{-}CH_3 + H^{\oplus} \rightleftarrows H_3\overset{\oplus}{P}{-}C(CH_3)_2{-}CH_3$$

Tertiary phosphines normally do not add to alkenes to form stable 1:1 adducts in the absence of acid. However, in the presence of concentrated acids, alkenes will react to form quaternary phosphonium salts.

$$(C_6H_5)_3P + CH_2{=}CHCO_2H \xrightarrow{48\%\ HBr} (C_6H_5)_3\overset{\oplus}{P}CH_2CH_2CO_2H\ Br^{\ominus}$$

$$(C_6H_5)_3P + C_6H_5{-}C{\equiv}C{-}CO_2H \xrightarrow{\text{conc. HCl}} (C_6H_5)_3\overset{\oplus}{P}C(C_6H_5){=}CHCO_2H\ Cl^{\ominus}$$

The adducts that are produced when tertiary phosphines react with alkenes can induce polymerization of the alkene by an anionic mechanism. As might be expected, such polymerizations are important only if the double bond is "activated," that is, if the resultant carbanion is stabilized by a substituent on the double bond.

$$R_3P + CH_2{=}CHZ \rightleftarrows \left[R_3\overset{\oplus}{P}{-}CH_2{-}\overset{\ominus}{C}H{-}Z \right]_{\text{unstable}} \xrightarrow{n\,CH_2=CHZ} R_3\overset{\oplus}{P}{-}(CH_2CHZ)_n{-}CH_2\overset{\ominus}{C}HZ$$

$Z = CN, CO_2R$

Evidence that *un*stable adducts are formed between phosphines and otherwise non-reactive alkenes includes the observation that alkenes may isomerize in the presence of phosphines.

$$\text{cis-}CH_3O{-}\overset{O}{\overset{\|}{C}}{-}CH{=}CH{-}\overset{O}{\overset{\|}{C}}{-}OCH_3 \xrightarrow[\text{1 hr}]{(C_6H_5)_3P;\ 150°} \text{trans-}CH_3O{-}\overset{O}{\overset{\|}{C}}{-}CH{=}CH{-}\overset{O}{\overset{\|}{C}}{-}OCH_3$$

The addition of triphenylphosphine to tetracyanoethylene (TCNE) produces an anion which, after addition of a second TCNE molecule, cyclizes to a phosphorane.

$$(C_6H_5)_3P + (NC)_2C{=}C(CN)_2 \rightarrow (C_6H_5)_3\overset{\oplus}{P}C(CN)_2\overset{\ominus}{C}(CN)_2 \xrightarrow{(NC)_2C=C(CN)_2} (C_6H_5)_3\overset{\oplus}{P}(C(CN)_2)_3\overset{\ominus}{C}(CN)_2 \rightarrow$$

(cyclic phosphorane: $(C_6H_5)_3P$ in a five-membered ring with four $C(CN)_2$ groups)

As is expected, phosphide anions will add to alkenes, particularly those alkenes that can stabilize a carbanion. In the example below, we see how this type of Michael addition (p. 599), when coupled with a Thorpe cyclization (see problem 19.21), can lead to the synthesis of a phosphorus heterocycle.

$$Na^{\oplus}\ C_6H_5\overset{\ominus}{P}H + CH_2{=}CHCN \rightarrow C_6H_5PH{-}CH_2\overset{\ominus}{C}HCN \xrightarrow{H_2O} C_6H_5PHCH_2CH_2CN$$

$$C_6H_5PHCH_2CH_2CN \xrightarrow{CH_2=CHCN,\ OH^{\ominus}} \xrightarrow{H_2O} C_6H_5P(CH_2CH_2CN)_2 \xrightarrow[\text{heat}]{OH^{\ominus}} \xrightarrow[\text{heat}]{H_3O^{\oplus}} C_6H_5{-}P\langle\text{ring}\rangle{=}O$$

Nucleophilic Reactions at Phosphorus. Compounds that possess phosphorus bonded to a potential leaving group are subject to attack by nucleophiles. This is presumed to be the mechanism for the hydrolysis and reduction of the phosphorus-halogen bond and for the reaction of Grignard reagents with phosphorus halides. These, and several other nucleophilic displacements at phosphorus, are presented below.

$$R_2PCl \xrightarrow[-Cl^{\ominus}]{OH^{\ominus}} R_2POH \rightleftarrows R_2\overset{\overset{O^{\ominus}}{|}}{P^{\oplus}}{-}H$$

$$R_2PCl \xrightarrow{H^{\ominus}} R_2PH + Cl^{\ominus}$$

$$R_2PCl \xrightarrow{R'MgX} R_2PR' + MgXCl$$

$$\underset{\text{phosphorous trichloride}}{PCl_3} + 3CH_3MgBr \rightarrow \underset{\text{trimethylphosphine}}{(CH_3)_3P} + 3MgBrCl$$

$$\underset{\text{diphenylphosphinic chloride}}{{}^{\ominus}O{-}\overset{\oplus}{P}Cl(C_6H_5)_2} + CH_3MgBr \rightarrow \underset{\text{methyldiphenylphosphine oxide}}{{}^{\ominus}O{-}\overset{\oplus}{P}(C_6H_5)_2CH_3} + MgBrCl$$

$$\underset{\text{phosphoryl trichloride}}{{}^{\ominus}O{-}\overset{\oplus}{P}Cl_3} + 3CH_3CH_2MgBr \rightarrow \underset{\text{triethylphosphine oxide}}{{}^{\ominus}O{-}\overset{\oplus}{P}(CH_2CH_3)_3} + 3MgBrCl$$

$$\underset{\text{dimethyl phosphonate}}{{}^{\ominus}O{-}\overset{\oplus}{P}H(OCH_3)_2} + 2CH_3MgBr \rightarrow \underset{\text{dimethylphosphine oxide}}{{}^{\ominus}O{-}\overset{\oplus}{P}H(CH_3)_2} + 2MgBrOCH_3$$

$$\underset{\text{methylphenylphosphinous chloride}}{CH_3P(C_6H_5)Cl} + NaOC_2H_5 \rightarrow \underset{\text{ethyl methylphenylphosphonite}}{CH_3P(C_6H_5)OC_2H_5} + NaCl$$

$$\underset{\text{methyl(2-naphthyl)phosphinous chloride}}{C_{10}H_7{-}P(CH_3){-}Cl} + \overset{\oplus}{Na}\ CH_3CO_2^{\ominus} \rightarrow \underset{\text{methyl(2-naphthyl)phosphinous acetate}}{C_{10}H_7{-}P(CH_3){-}O{-}\overset{O}{\overset{\|}{C}}{-}CH_3}$$

12. Phosphine reacts with formaldehyde in the presence of hydrochloric acid to produce tetrakis(hydroxymethyl)phosphonium chloride, THPC. This material is used in flame-retardant finishes for cotton fabric. Suggest a mechanism for its formation.

$$\underset{\text{THPC}}{P^{\oplus}(CH_2OH)_4\ Cl^{\ominus}}$$

13. When sodium diphenylphosphide reacts with an acetylene bromide, one main product is a trisubstituted phosphine. Suggest a mechanism for this reaction.

$$RC{\equiv}CBr + Na^{\oplus}(C_6H_5)_2P^{\ominus} \rightarrow RC{\equiv}C{-}P(C_6H_5)_2 + NaBr$$

The synthesis of phenylphosphonous dichloride, one of the earliest reactions discussed in Section 23.8, also is an example of a nucleophilic attack on phosphorus. What makes this less spectacular is that in the presence of the requisite catalyst, carbon also becomes a good electrophile (as in the Friedel-Crafts reaction).

$$C_6H_6 + PCl_3 + AlCl_3 \rightarrow \left[(C_6H_6^{+})(H)\text{—}P(Cl)(Cl)\cdots Cl\cdots \overset{\delta-}{AlCl_3} \right] \rightarrow C_6H_5\text{—}PCl_2 + HCl + AlCl_3$$

Alkaline Decomposition of Quaternary Phosphonium Salts. When an ammonium salt bearing a β hydrogen is heated with sodium hydroxide, the result is *alkene* formation by an E2 process (see problem 20.27).

$$(CH_3)_3\overset{\oplus}{N}\text{—}CH_2CH_3\ OH^{\ominus} \xrightarrow[\text{heat}]{} (CH_3)_3N + CH_2{=}CH_2 + H_2O$$

Under similar conditions, phosphonium salts produce *alkanes* and phosphine oxides. The difference arises because the phosphonium salt forms a phosphorane, which can decompose, in a E2 process, to form a $\overset{\oplus}{P}\text{—}\overset{\ominus}{O}$ bond and a carbanion. (Recall how rare it is for a carbanion to be a leaving group in an E2 process!)

$$CH_3\overset{\oplus}{P}(CH_3)_2\text{—}CH_2C_6H_5 + OH^{\ominus} \rightleftarrows (CH_3)_3P(OH)(CH_2C_6H_5)$$

a phosphorane

$$(CH_3)_3P(O\text{—}H)(CH_2C_6H_5) \xrightarrow{OH^{\ominus}} (CH_3)_3\overset{\oplus}{P}\text{—}\overset{\ominus}{O} + :\overset{\ominus}{C}H_2C_6H_5 \xrightarrow{H_2O} CH_3C_6H_5 + OH^{\ominus}$$

This could be viewed as a very unusual, specific way of reducing alkyl halides to alkanes.

$$P(CH_3)_3 + ClCH_2C_6H_5 \rightarrow (CH_3)_3\overset{\oplus}{P}CH_2C_6H_5\ \overset{\ominus}{Cl} \xrightarrow[\text{heat}]{H_2O/OH^{\ominus}} (CH_3)_3\overset{\oplus}{P}\text{—}\overset{\ominus}{O} + CH_3C_6H_5$$

The Arbuzov Reaction. The Arbuzov reaction combines two important aspects of organophosphorus chemistry—the high nucleophilicity of triligant phosphorus and the tendency to form the phosphoryl group—into an excellent route from alcohols to alkyl halides. In this reaction, a trialkyl phosphite reacts with an alkyl halide to produce a new alkyl halide and a dialkyl phosphonate.

$$\underset{\text{trialkyl phosphite}}{(AlkO)_3P} + Alk'\text{—}X \rightsquigarrow \underset{\text{dialkyl alkylphosphonate}}{(AlkO)_2\overset{\oplus}{P}(O^{\ominus})Alk'} + Alk\text{—}X \qquad \text{Arbuzov reaction}$$

This reaction proceeds by an initial nucleophilic attack by the phosphite upon the halide. The intermediate cation is then attacked by the displaced halide to form the observed products. It is the strength of the $\overset{\oplus}{P}\text{—}\overset{\ominus}{O}$ linkage which is responsible for this facile displacement.

$$3CH_3CH_2OH + PCl_3 \rightarrow \underset{\text{triethyl phosphite}}{P(OCH_2CH_3)_3}$$

$$(C_2H_5O)_3P: + CH_3CH_2CH_2Cl \xrightarrow{S_N2} CH_3CH_2CH_2\overset{\oplus}{P}(OC_2H_5)_3\ Cl^{\ominus}$$

$$CH_3\text{—}CH_2\text{—}O\text{—}\overset{\oplus}{P}(OC_2H_5)(O\text{—}C_2H_5)\text{—}CH_2CH_2CH_3\ \ Cl^{\ominus} \xrightarrow{S_N2} CH_3CH_2Cl + \overset{\ominus}{O}\text{—}\overset{\oplus}{P}(OC_2H_5)_2CH_2CH_2CH_3 \qquad \text{Arbuzov reaction}$$

14. Suggest a mechanism for the reaction shown below, which is a convenient route to acyl chlorides.

$$3RCO_2H + PCl_3 \rightarrow 3RCOCl + OP(OH)_2H$$

What argument(s) may be used to explain why this is a good route to acyl halides while the reaction of a carboxylic acid with hydrochloric acid is not a good route to acyl halides?

23.10 BIOLOGICAL CHEMISTRY OF PHOSPHORUS

Thus far, our attention has focused on phosphines and, to a lesser degree, on simple derivatives of phosphines. While phosphorus is found in all living systems, it usually does not occur as organophosphorus compounds. Instead, living systems employ esters of phosphoric acid and two other oxyacids of phosphorus, *diphosphoric acid* and *triphosphoric acid*. The former is sometimes called pyrophosphoric acid, since it can be envisioned as having arisen by heating ("*pyro*lysis") of phosphoric acid, with accompanying loss of water.

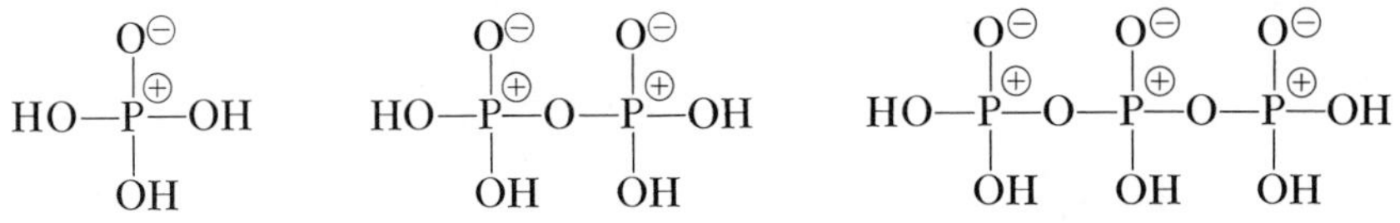

$$HO-\underset{OH}{\overset{O^{\ominus}}{P^{\oplus}}}-OH \qquad HO-\underset{OH}{\overset{O^{\ominus}}{P^{\oplus}}}-O-\underset{OH}{\overset{O^{\ominus}}{P^{\oplus}}}-OH \qquad HO-\underset{OH}{\overset{O^{\ominus}}{P^{\oplus}}}-O-\underset{OH}{\overset{O^{\ominus}}{P^{\oplus}}}-O-\underset{OH}{\overset{O^{\ominus}}{P^{\oplus}}}-OH$$

phosphoric acid — diphosphoric acid ("pyrophosphoric acid") — triphosphoric acid

Most naturally occurring esters of these acids are monoesters (that is, they bear only one alkoxy group); their names reflect this point. Thus, the terms methyl phosphate, methyl diphosphate, and methyl triphosphate refer to the monomethyl esters of three different acids, not to three different esters of the same acid. Since only one of the acidic hydrogens has been replaced in these monoesters, biological phosphates are acidic and, under physiological conditions, exist primarily as their anions. Because of this, biological phosphates sometimes are written as acids and sometimes as anions, depending upon the field of training of the author.

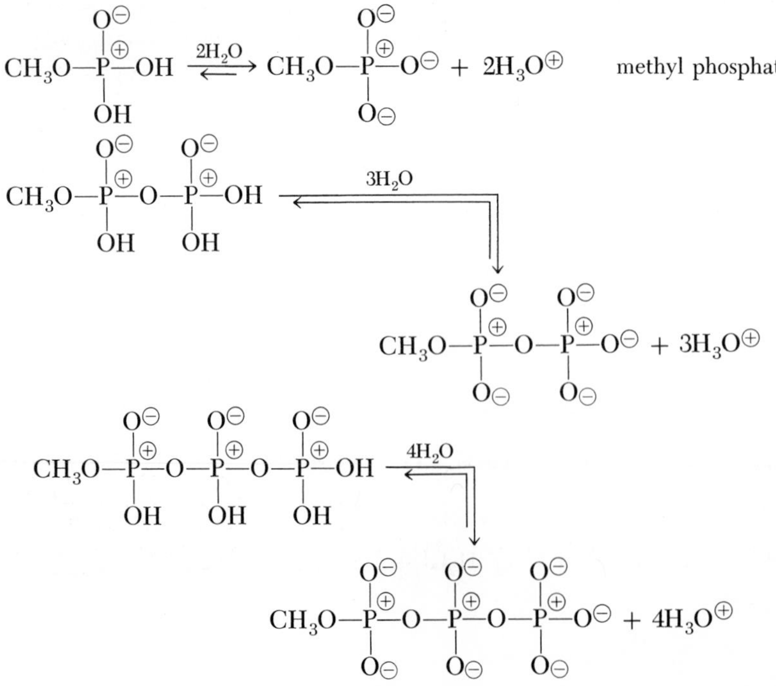

$$CH_3O-\underset{OH}{\overset{O^{\ominus}}{P^{\oplus}}}-OH \xrightleftharpoons{2H_2O} CH_3O-\underset{O^{\ominus}}{\overset{O^{\ominus}}{P^{\oplus}}}-O^{\ominus} + 2H_3O^{\oplus} \qquad \text{methyl phosphate}$$

$$CH_3O-\underset{OH}{\overset{O^{\ominus}}{P^{\oplus}}}-O-\underset{OH}{\overset{O^{\ominus}}{P^{\oplus}}}-OH \xrightleftharpoons{3H_2O} CH_3O-\underset{O^{\ominus}}{\overset{O^{\ominus}}{P^{\oplus}}}-O-\underset{O^{\ominus}}{\overset{O^{\ominus}}{P^{\oplus}}}-O^{\ominus} + 3H_3O^{\oplus} \qquad \text{methyl diphosphate}$$

$$CH_3O-\underset{OH}{\overset{O^{\ominus}}{P^{\oplus}}}-O-\underset{OH}{\overset{O^{\ominus}}{P^{\oplus}}}-O-\underset{OH}{\overset{O^{\ominus}}{P^{\oplus}}}-OH \xrightleftharpoons{4H_2O} CH_3O-\underset{O^{\ominus}}{\overset{O^{\ominus}}{P^{\oplus}}}-O-\underset{O^{\ominus}}{\overset{O^{\ominus}}{P^{\oplus}}}-O-\underset{O^{\ominus}}{\overset{O^{\ominus}}{P^{\oplus}}}-O^{\ominus} + 4H_3O^{\oplus} \qquad \text{methyl triphosphate}$$

The most important roles of biological phosphates include the interconversion of mono- *to* di- and triphosphates, and the reverse reactions. A typical reaction is the interconversion of a triphosphate into a monophosphate and a pyrophosphate anion, $P_2O_7^{\ominus 4}$. The latter is often abbreviated "PP" in the biochemical literature.

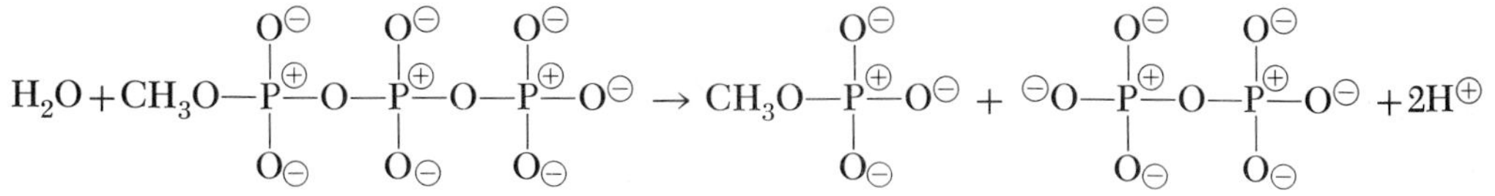

$$H_2O + CH_3O{-}\overset{O^{\ominus}}{\underset{O^{\ominus}}{\overset{|}{\underset{|}{P^{\oplus}}}}}{-}O{-}\overset{O^{\ominus}}{\underset{O^{\ominus}}{\overset{|}{\underset{|}{P^{\oplus}}}}}{-}O{-}\overset{O^{\ominus}}{\underset{O^{\ominus}}{\overset{|}{\underset{|}{P^{\oplus}}}}}{-}O^{\ominus} \rightarrow CH_3O{-}\overset{O^{\ominus}}{\underset{O^{\ominus}}{\overset{|}{\underset{|}{P^{\oplus}}}}}{-}O^{\ominus} + {}^{\ominus}O{-}\overset{O^{\ominus}}{\underset{O^{\ominus}}{\overset{|}{\underset{|}{P^{\oplus}}}}}{-}O{-}\overset{O^{\ominus}}{\underset{O^{\ominus}}{\overset{|}{\underset{|}{P^{\oplus}}}}}{-}O^{\ominus} + 2H^{\oplus}$$

pyrophosphate anion
"PP"

The "High-Energy" Phosphate Bond. When a physical chemist speaks of a high-energy bond, he usually means a rather stable bond, that is, one with a high bond energy or bond strength. Biochemists, on the other hand, often use the term to designate a bond that is rather weak ("energetic") and that is prone to release energy when involved in some reaction. Using the latter definition, there are a number of biological molecules that have high-energy phosphate bonds, *i.e.*, a negative ΔG of hydrolysis. These are included in Table 23–3.

$$ATP + H_2O \rightleftarrows ADP + H_3PO_4 + \text{energy}$$

TABLE 23-3 Standard Free Energy of Hydrolysis of Phosphylated Compounds[a]

Compound: Name	Compound: Structure	$\Delta G°$ (kcal/mole)
phosphoenolpyruvic acid	$CH_2{=}\underset{O\sim PO(OH)_2}{\underset{|}{C}}{-}CO_2H$	−14.80
1,3-diphosphoglyceric acid	$(HO)_2OPOCH_2CHOH\overset{O}{\overset{\|}{C}}{-}O\sim PO(OH)_2$	−11.80
acetylphosphoric acid	$CH_3\overset{O}{\overset{\|}{C}}{-}O\sim PO(OH)_2$	−10.10
ATP	see p. 854	−7.30
glucose 1-phosphoric acid	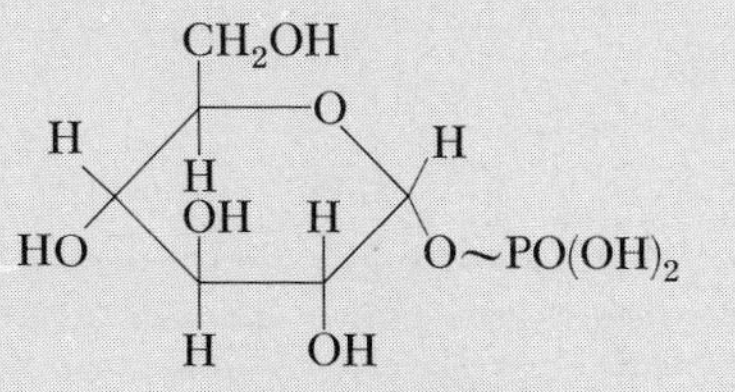 pyranose ring: CH_2OH; O; H; H; H; OH; H; HO; $O\sim PO(OH)_2$; H; OH	−5.00
glucose 6-phosphoric acid	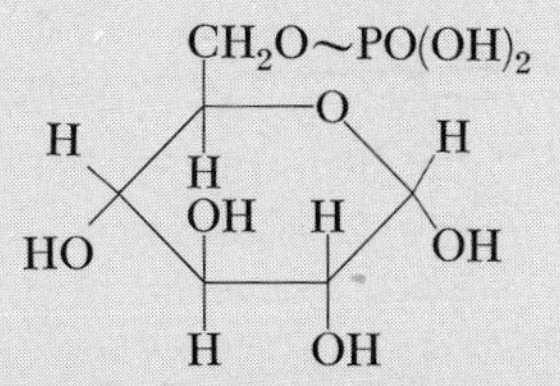 pyranose ring: $CH_2O\sim PO(OH)_2$; O; H; H; H; OH; H; HO; OH; H; OH	−3.30
glycerol 1-phosphoric acid	$HOCH_2CH(OH)CH_2O\sim PO(OH)_2$	−2.20

[a] Compounds with "high-energy" phosphate bonds are found at the top of the table.

The information in Table 23–3 explains how adenosine triphosphate (ATP) can be so important in providing "energy" for biological processes. The compounds with large, negative values of ΔG tend to hydrolyze completely at equilibrium, while those with small, negative values tend to hydrolyze only incompletely. Put differently, a compound with a large, negative ΔG easily loses a phosphate group. It is clear that ATP has an *intermediate* value for ΔG, indicating that it can readily release a phosphate ion (to form adenosine diphosphate, ADP) and that ADP can readily acquire a phosphate group and form ATP.

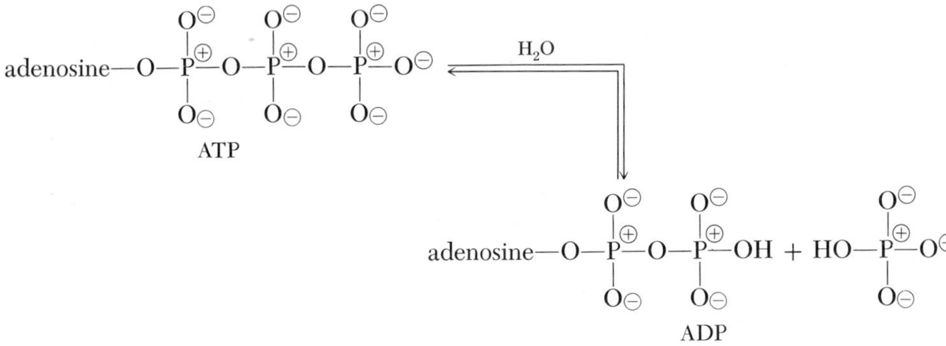

$$\text{adenosine—O—}\overset{\overset{\text{O}^{\ominus}}{|}}{\underset{\underset{\text{O}^{\ominus}}{|}}{\text{P}^{\oplus}}}\text{—O—}\overset{\overset{\text{O}^{\ominus}}{|}}{\underset{\underset{\text{O}^{\ominus}}{|}}{\text{P}^{\oplus}}}\text{—O—}\overset{\overset{\text{O}^{\ominus}}{|}}{\underset{\underset{\text{O}^{\ominus}}{|}}{\text{P}^{\oplus}}}\text{—O}^{\ominus} \underset{\text{ATP}}{} \xrightleftharpoons{H_2O}$$

$$\text{adenosine—O—}\overset{\overset{\text{O}^{\ominus}}{|}}{\underset{\underset{\text{O}^{\ominus}}{|}}{\text{P}^{\oplus}}}\text{—O—}\overset{\overset{\text{O}^{\ominus}}{|}}{\underset{\underset{\text{O}^{\ominus}}{|}}{\text{P}^{\oplus}}}\text{—OH} + \text{HO—}\overset{\overset{\text{O}^{\ominus}}{|}}{\underset{\underset{\text{O}^{\ominus}}{|}}{\text{P}^{\oplus}}}\text{—O}^{\ominus}$$

ATP → ADP

This equilibrium permits the ATP-ADP system to acquire a phosphate unit from a higher energy compound (such as acetyl phosphate), converting ADP to ATP, and then to transfer it to an appropriate acceptor and form a lower energy compound (such as glycerol 1-phosphate), converting the ATP back to ADP. Therefore, while ATP is considered the "grand-daddy" of all high-energy compounds, its biological importance stems from its ability to function as a phosphate shuttle, a property that results from an *intermediate* energy compared to other phosphates!

Adenosine triphosphate contains three phosphate units, but only two of them are removed with unusual ease; therefore, only two of them have high-energy bonds. ADP has only one high-energy bond. Using the convention that high-energy bonds are indicated by a "squiggle," ATP and ADP can be represented as shown below.

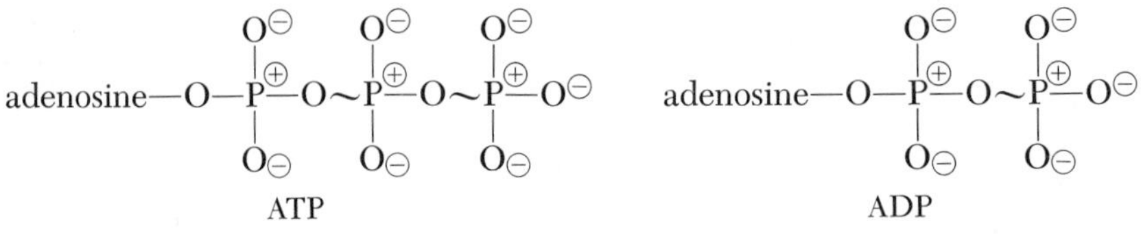

$$\underset{\text{ATP}}{\text{adenosine—O—}\overset{\overset{\text{O}^{\ominus}}{|}}{\underset{\underset{\text{O}^{\ominus}}{|}}{\text{P}^{\oplus}}}\text{—O}\sim\overset{\overset{\text{O}^{\ominus}}{|}}{\underset{\underset{\text{O}^{\ominus}}{|}}{\text{P}^{\oplus}}}\text{—O}\sim\overset{\overset{\text{O}^{\ominus}}{|}}{\underset{\underset{\text{O}^{\ominus}}{|}}{\text{P}^{\oplus}}}\text{—O}^{\ominus}} \qquad \underset{\text{ADP}}{\text{adenosine—O—}\overset{\overset{\text{O}^{\ominus}}{|}}{\underset{\underset{\text{O}^{\ominus}}{|}}{\text{P}^{\oplus}}}\text{—O}\sim\overset{\overset{\text{O}^{\ominus}}{|}}{\underset{\underset{\text{O}^{\ominus}}{|}}{\text{P}^{\oplus}}}\text{—O}^{\ominus}}$$

What is so unusual about ATP, or any other high-energy phosphate, that makes hydrolysis so favorable? Put differently, why is the energy of the hydrolysis products significantly lower than the energy of the starting material? There are several answers to this question. At pH 7, ATP molecules possess nearly four unit negative charges close to one another; this causes a certain increase in what has been termed "electrical stress." Upon hydrolysis, some of this charge repulsion is removed by the formation of two distinct, negatively charged ions. These ions, since they bear similar charges, show little tendency to recombine spontaneously.

Secondly, one can draw more resonance contributing structures for the two resultant anions than for the initial triphosphate, and this will also result in products that are more stable than the starting material.

Other Biological Phosphates. The occurrence of organic phosphates in biological systems is not limited to "high-energy" systems. The biological polymers commonly known as DNA and RNA possess a backbone made up of alternating units of polyhydroxylic materials (known as "sugars") and phosphate units. The structure of DNA and RNA will be discussed in detail in Chapter 26, so we will note here only that these compounds, which play an essential role in cell duplication and in inheritance, are "nothing more" than high molecular weight polymeric phosphate esters bearing unusual amines on the backbone of the polymer.

$$\cdots H\text{]}\!-\!O\!-\!\boxed{\text{SUGAR}}\!-\!O\!-\!\boxed{H\ \ HO}\!-\!\overset{\overset{OH}{|}\ \oplus}{\underset{\underset{O^{\ominus}}{|}}{P}}\!-\!\boxed{OH\ \ H}\!-\!O\!-\!\boxed{\text{SUGAR}}\!-\!O\!-\!\boxed{H\ \ HO}\!-\!\overset{\overset{OH}{|}\ \oplus}{\underset{\underset{O}{|}}{P}}\!-\!\text{[}OH\cdots$$

$$\downarrow -H_2O \qquad \downarrow -H_2O \qquad \downarrow -H_2O$$

$$-O\!-\!\boxed{\text{SUGAR}}\!-\!O\!-\!\overset{\overset{OH}{|}\ \oplus}{\underset{\underset{O^{\ominus}}{|}}{P}}\!-\!O\!-\!\boxed{\text{SUGAR}}\!-\!O\!-\!\overset{\overset{OH}{|}\ \oplus}{\underset{\underset{O^{\ominus}}{|}}{P}}\!-$$

portion of DNA or RNA backbone

23.11 TOXIC PHOSPHORUS COMPOUNDS

While phosphorus compounds are important in organic syntheses and essential to living systems, some compounds containing phosphorus are extremely toxic and are used as war gases and as insecticides. These poisons act by interfering with the transmission of nerve impulses; thus, before we can discuss the mode of action of a toxic phosphorus compound, we must present a brief review of how the interface (*synapse*) between nerves and muscles (the *neuromuscular junction*) functions.

The Neuromuscular Junction.[a] The neuromuscular junction (Figure 23–9) consists of a branched tip of a nerve axon resting in a groove on the muscle cell's surface. Normal transmission of an impulse from a nerve to a muscle involves three critical steps: (1) release of *acetylcholine* (a neurotransmitter) by the nerve endings into the synapse; (2) binding of this

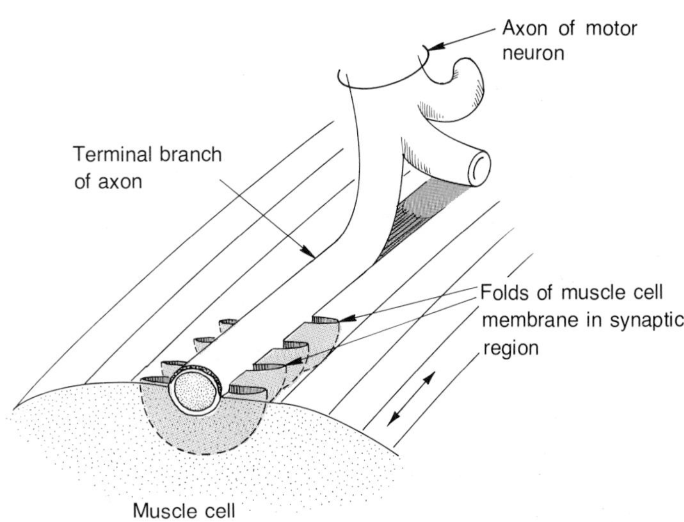

Figure 23-9 Schematic diagram of nerve ending on a muscle cell. Note the branched end of the axon buried in a trough of the muscle cell membrane. The latter has many folds in the synaptic region, greatly increasing the area available for synaptic binding sites. (After Birks, Huxley and Katz from M.E. Clark, *Contemporary Biology*, W.B. Saunders Company, Philadelphia, 1973.)

[a] A number of relevant terms were introduced in Chapter 20.

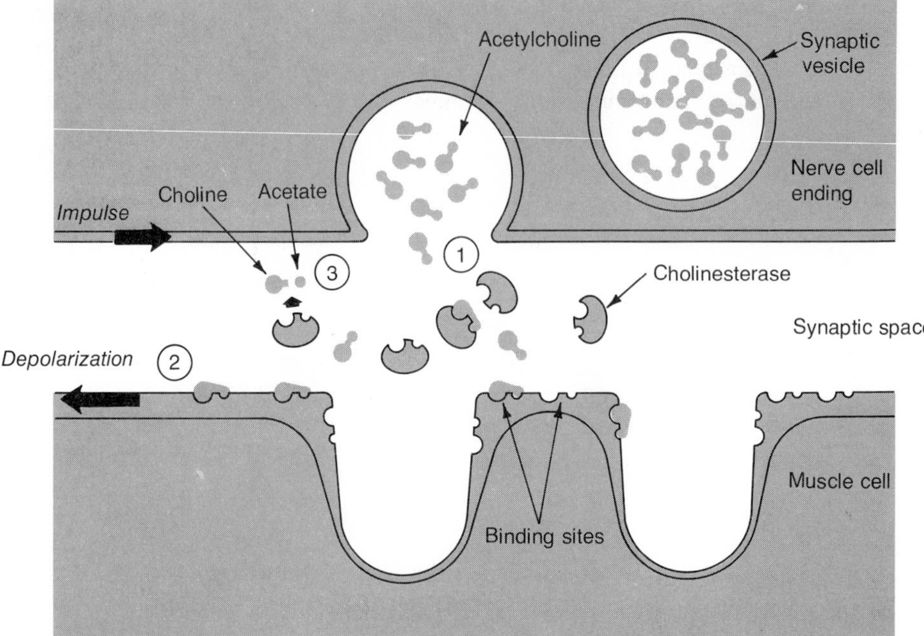

Figure 23-10 The steps in transmission of an impulse at a neuromuscular junction, shown in schematic diagram.

1. The nerve impulse releases acetylcholine from the synaptic vesicle into the synaptic space.
2. Acetylcholine binds in a lock-and-key fashion with binding sites on the muscle cell membrane, causing depolarization.
3. Acetylcholine is finally destroyed by the enzyme cholinesterase. This enzyme either acts in the synaptic space, as shown here, or else may be attached to the muscle cell membrane. In either case, the end result is destruction of the neurotransmitter. (From M.E. Clark, *Contemporary Biology*, copyright © 1973, W.B. Saunders Company, Philadelphia.)

acetylcholine to specific receptor sites on the membrane of the muscle cell; and (3) destruction of acetylcholine by hydrolysis, a process catalyzed by the enzyme *cholinesterase* (Figure 23–10). The second step is directly responsible for muscle contraction, but interference with any of these three steps will have an effect upon the reaction of a muscle cell to a nerve stimulus.

Neurotoxins. We already have mentioned (p. 647) that curare is a neurotoxin that, in large doses, leads to respiratory collapse and death. It acts by competing with acetylcholine

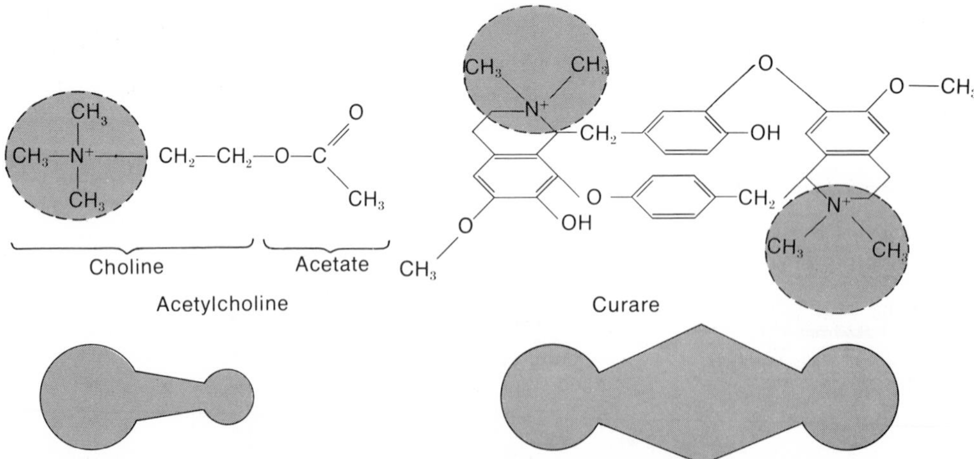

Figure 23-11 Chemical structures of the neurotransmitter acetylcholine and the postsynaptic blocking agent curare. Regions of similar structure which may compete for specific binding sites on the muscle cell membrane are indicated by shaded ovals. Underneath are shown simplified models. (From M.E. Clark, *Contemporary Biology*, copyright © 1973, W.B. Saunders Company, Philadelphia.)

for receptor sites on the muscle membrane, thereby preventing the acetylcholine from binding to the muscle and activating it. Through the use of radioactive curare (*i.e.*, curare labeled with ^{14}C), it has been shown that a single neuromuscular synapse may bind up to 2.6×10^6 molecules of curare and, therefore, that the synapse may have as many as 2.6×10^6 receptor sites.

The drawings in Figure 23–11 suggest that it is, at least in part, the shape of the curare molecule that permits it to compete with acetylcholine for the muscle receptor sites. A schematic view of how curare competes with acetylcholine is shown in Figure 23–12.

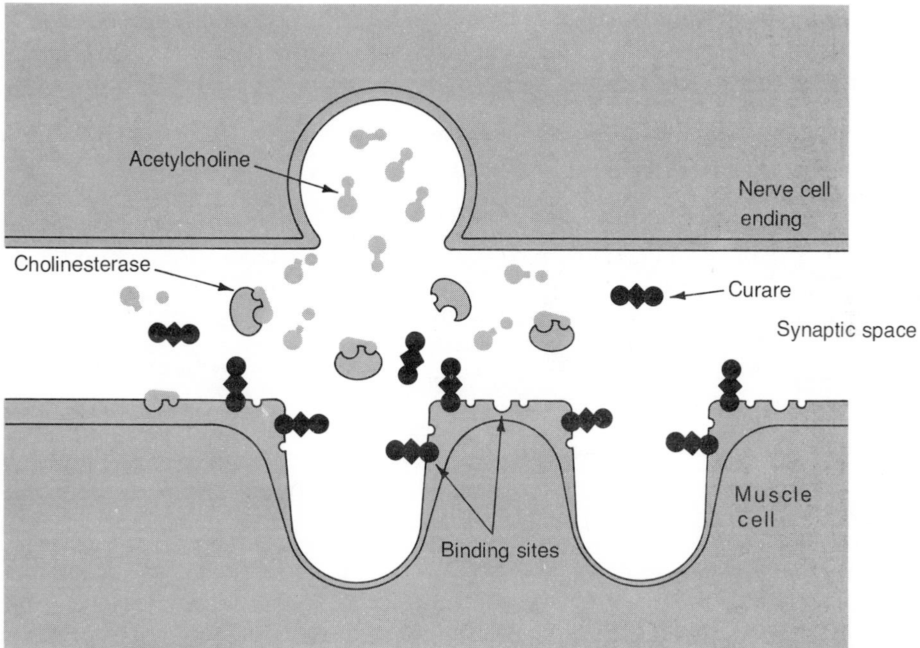

Figure 23-12 A schematic diagram illustrating how curare competes with acetylcholine for binding sites on the postsynaptic membrane. Since curare does not depolarize the muscle membrane, contraction cannot occur and the muscle is paralyzed. Note that curare does *not* bind with the enzyme cholinesterase, which continues to destroy acetylcholine. (From M.E. Clark, *Contemporary Biology*, copyright © 1973, W.B. Saunders Company, Philadelphia.)

Physostigmine and neostigmine bromide are used, among other things, to treat an overdose of curare. They act by binding to cholinesterase, thus preventing the rapid hydrolysis of acetylcholine. In turn, this increases the number of molecules of acetylcholine available at the synapse. When their concentration becomes sufficiently high, the acetylcholine molecules displace enough of the curare from the muscle cell membrane to develop what physiologists call the "action potential" of the muscle. Thus, while curare acts to produce paralysis by blocking the action of acetylcholine, neostigmine and physostigmine act in an opposite manner, potentiating the transmission of an impulse across the neuromuscular junction.

physostigmine

neostigmine bromide

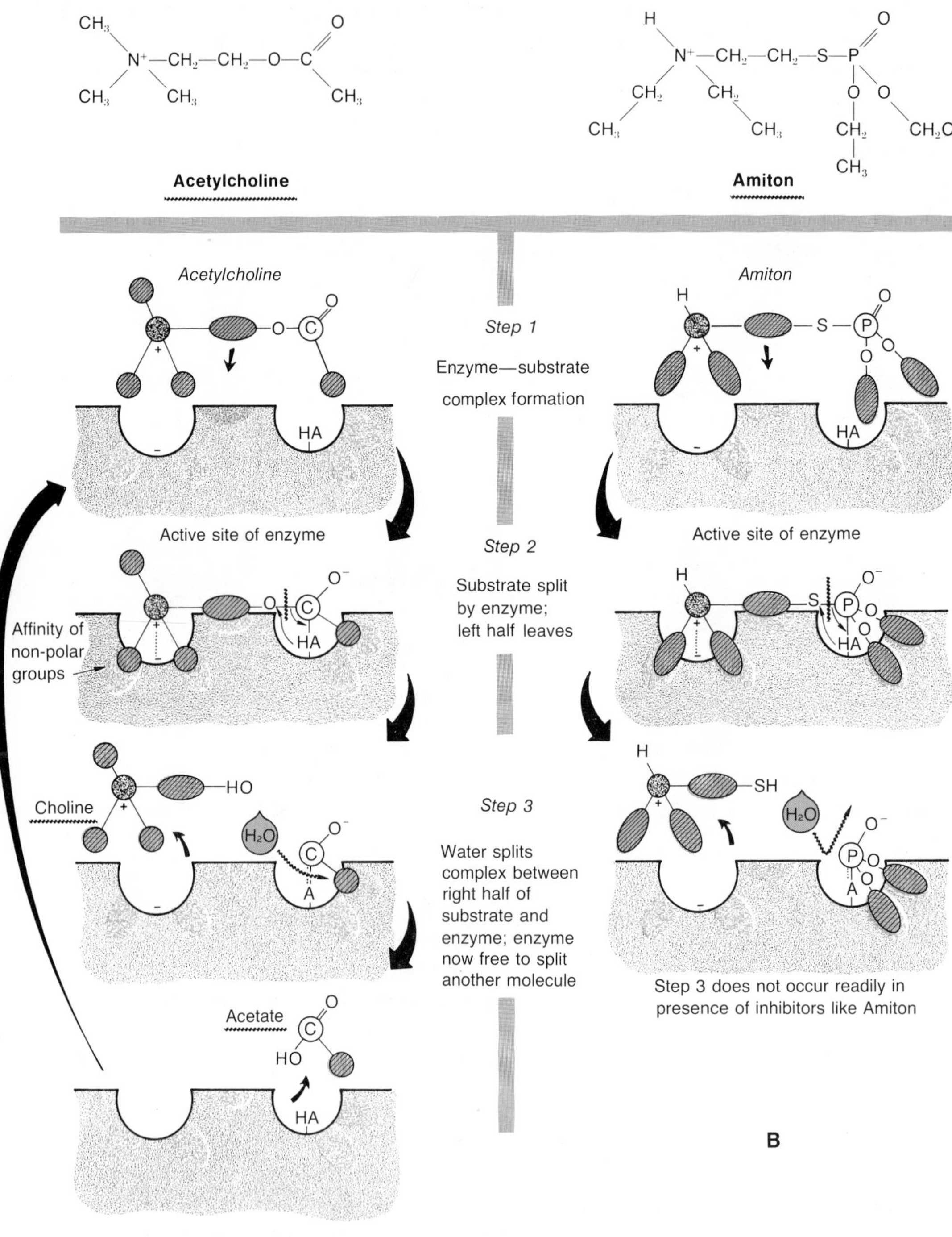

Figure 23-13 The binding and splitting of the natural substrate acetylcholine by the active site on the enzyme cholinesterase (A), compared with the binding and splitting of the organophosphorus inhibitor Amiton (B). Note the structural similarity of acetylcholine and Amiton, as well as the lock-and-key fit both between nonpolar regions of enzyme and substrate and between charged regions of both molecules. (From M.E. Clark, *Contemporary Biology*, copyright © 1973, W.B. Saunders Company, Philadelphia.)

War gases like Sarin, Soman, and Tabun are *cholinesterase inhibitors;* in very low amounts (*e.g.*, 9×10^{-3} g) they cause continuous muscle contraction which produces convulsions and, very quickly, death. Many insecticides are cholinesterase inhibitors that are esters of phosphorus. The most common of these are Amiton, Diazinon, Malathion, and Parathion.

$$C_2H_5O-\overset{\overset{\displaystyle O^{\ominus}}{|}}{\underset{\underset{\displaystyle OC_2H_5}{|}}{P^{\oplus}}}-SCH_2CH_2N(C_2H_5)_2$$

Amiton (Tetram, Inferno)

$$CH_3O-\overset{\overset{\displaystyle S^{\ominus}}{|}}{\underset{\underset{\displaystyle CH_3O}{|}}{P^{\oplus}}}-S-\underset{\underset{\displaystyle CH_2CO_2C_2H_5}{|}}{C}HCO_2C_2H_5$$

Malathion

$$C_2H_5O-\overset{\overset{\displaystyle S^{\ominus}}{|}}{\underset{\underset{\displaystyle C_2H_5O}{|}}{P^{\oplus}}}-O-C_4N_2H(CH_3)-CH(CH_3)_2$$

Diazinon

$$C_2H_5O-\overset{\overset{\displaystyle S^{\ominus}}{|}}{\underset{\underset{\displaystyle OC_2H_5}{|}}{P^{\oplus}}}-O-C_6H_4-NO_2$$

Parathion

Let us briefly examine how Amiton, with a salt whose structure is similar to acetylcholine, blocks the action of cholinesterase. To begin with, we imagine that there is a specific location on the enzyme (the *active site*) that must bond to the acetylcholine in order to catalyze its hydrolysis. When Amiton binds to this active site, hydrolysis of the sulfur-phosphorus bond produces a cation $((CH_3CH_2)_2\overset{\oplus}{N}HCH_2CH_2SH)$ and a phosphorus-containing anion $((CH_3CH_2O)_2P(O)O^{\ominus})$. While both portions produced by acetylcholine hydrolysis can readily leave the active site of the enzyme, the phosphorus-containing anion leaves the active site only with great difficulty. The strong bond between a portion of the active site and the anionic fragment prevents the enzyme from being used again (to hydrolyze acetylcholine). The enzyme is said to be *inhibited!* These two processes, hydrolysis of acetylcholine and of Amiton, are compared in Figure 23–13 (opposite page), while a closer look at uninhibited and inhibited cholinesterase is shown in Figure 23–14.

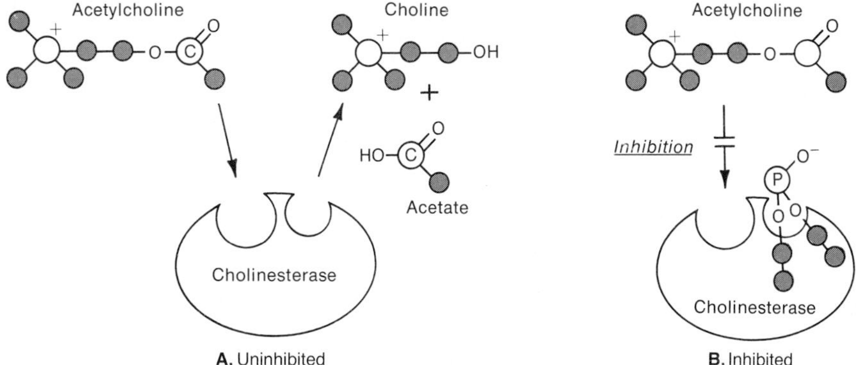

Figure 23-14 Comparison of uninhibited and inhibited cholinesterase. (A) Uninhibited enzyme quickly splits its normal substrate, acetylcholine, acting on 300,000 molecules per second. (B) Enzyme inhibited by P-containing half of organophosphorus pesticide cannot bind with acetylcholine, which accumulates in the synapse, causing convulsions. (From M.E. Clark, *Contemporary Biology*, copyright © 1973, W.B. Saunders Company, Philadelphia.)

Before leaving this discussion, we should ask why curare and Amiton, both of which are quaternary ammonium salts (as is acetylcholine), act in opposite ways at the synapse. The answer lies in the *specificity* of the active site on the cholinesterase, on the one hand, and the specificity of the receptor site on the muscle cell membrane, on the other. Thus Amiton can bind to the enzyme's active site but not the muscle's receptor site, while curare can bind to the muscle's receptor site but not to the enzyme's active site. Acetylcholine, by way of contrast, can bind to both. One may view acetylcholine as a sort of chemical "master key," capable of interacting with sites ("locks") on both the muscle cell membrane and the enzyme. Amiton and curare, to extend this analogy, may be viewed as separate "sub-master keys"—the former reacting only with the "lock" on the enzyme, and the latter reacting only with the "lock" on the muscle cell membrane. A detailed discussion of the nature of enzymes is given in Chapter 27; however, we may suggest at this stage that acetylcholine, Amiton, and curare differ in their activity because of variations in their size, geometry, and polarity.

IMPORTANT TERMS

Back bonding: Donation of a "non-bonding" electron pair on one atom into a vacant *d*-orbital on an adjacent atom to form a π bond. Back bonding is responsible for multiple bond formation in ylids.

Bilirubin: A compound formed by the biological degradation of red blood cells (specifically from hemoglobin). It, and related compounds, are (among other things) responsible for the color of feces.

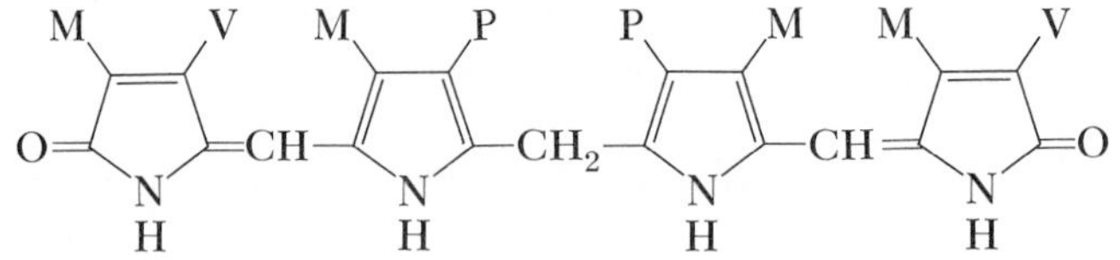

bilirubin

M: CH_3; V: $CH{=}CH_2$; P: $CH_2CH_2CO_2H$

Folate: An alternate name for folic acid, one that suggests ionization.

Hydrodisulfide: A compound containing the —S—S—H group. Such compounds are relatively rare.

Phosphorane: A compound containing a five-bonded phosphorus. If two of these bonds constitute a double bond, as in an ylid, the compound is an *alkylidenephosphorane*. The term "phosphorane" also refers to the hypothetical compound PH_5.

Phosphoryl group: The group with phosphorus bound to oxygen by a dative bond, $-\overset{|}{\underset{|}{P}}{}^{\oplus}-O^{\ominus}$.

Septicemia: A bacterial infection of the blood.

Sulfa drug: An antibiotic that is a derivative of sulfanilamide.

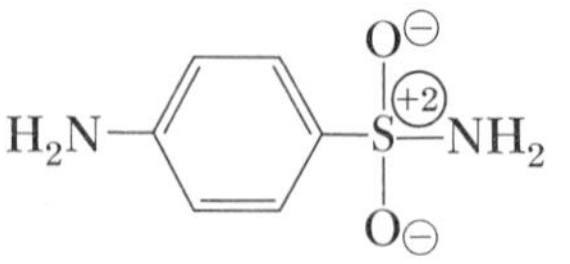

sulfanilamide

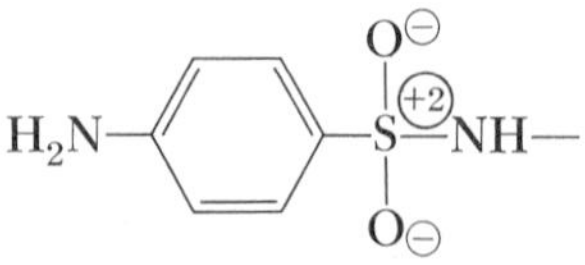

sulfanilamido group

Not all sulfanilamide derivatives are antimicrobial. Tolbutamide, for example, is a compound that is used to treat diabetes. (Of course, some sulfanilamides show no biological activity.)

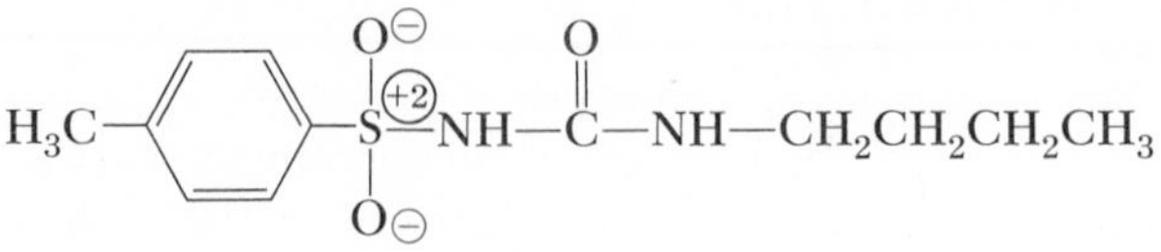

tolbutamide

Teratogen: From the Greek *teratos,* "monster," and *genés,* "born." A substance that alters cell differentiation (or development) in such a way that various organs or structures will not develop properly as the embryo develops. The result is sometimes called a "birth defect."

Thiocarbonyl group: The carbon-sulfur double bond, $>C{=}S$. In this group both carbon and sulfur are sp^2 hybridized. Thiocarbonyl compounds (*e.g.*, thioaldehydes) are not ylids, since *d*-orbitals are not involved in their bonding scheme. Most thiocarbonyl compounds are very reactive. Thioacetic acid, a thioacid, exists almost exclusively as its thiol tautomer.

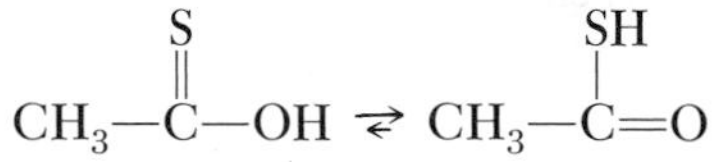

$$CH_3{-}\overset{\displaystyle S}{\overset{\|}{C}}{-}OH \rightleftharpoons CH_3{-}\overset{\displaystyle SH}{\overset{|}{C}}{=}O$$

thioacetic acid

The thiocarbonyl derivatives of alcohols known as *xanthates* are useful in organic synthesis. Thus, alkoxide ions react with carbon disulfide to form, after reaction with methyl iodide, *O*-alkyl methyl xanthates.

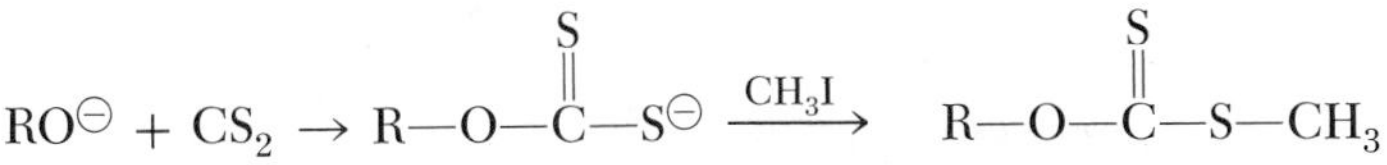

$$RO^{\ominus} + CS_2 \rightarrow R{-}O{-}\overset{\displaystyle S}{\overset{\|}{C}}{-}S^{\ominus} \xrightarrow{CH_3I} R{-}O{-}\overset{\displaystyle S}{\overset{\|}{C}}{-}S{-}CH_3$$

an *O*-alkyl methyl xanthate

Derivatives of this type undergo a rapid pyrolysis (thermal decomposition) to form alkenes. This, the **Chugaev reaction,** is depicted below. It is a means of converting alcohols to alkenes without the use of acid [and hence, a means of (ultimately) dehydrating alcohols while avoiding isomerizations and other side reactions].

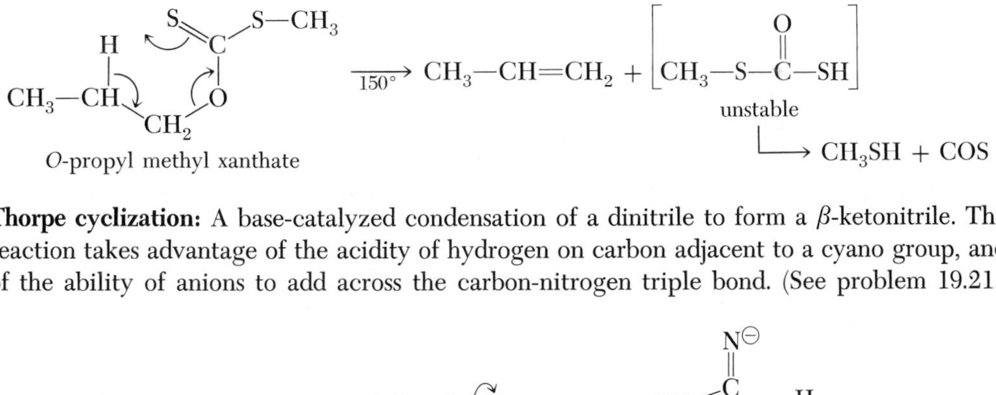

$$CH_3{-}CH(H){-}CH_2{-}O{-}C(=S){-}S{-}CH_3 \xrightarrow[150^{\circ}]{} CH_3{-}CH{=}CH_2 + \left[CH_3{-}S{-}\overset{\displaystyle O}{\overset{\|}{C}}{-}SH\right]$$

O-propyl methyl xanthate

unstable

$$\longrightarrow CH_3SH + COS$$

Thorpe cyclization: A base-catalyzed condensation of a dinitrile to form a β-ketonitrile. The reaction takes advantage of the acidity of hydrogen on carbon adjacent to a cyano group, and of the ability of anions to add across the carbon-nitrogen triple bond. (See problem 19.21.)

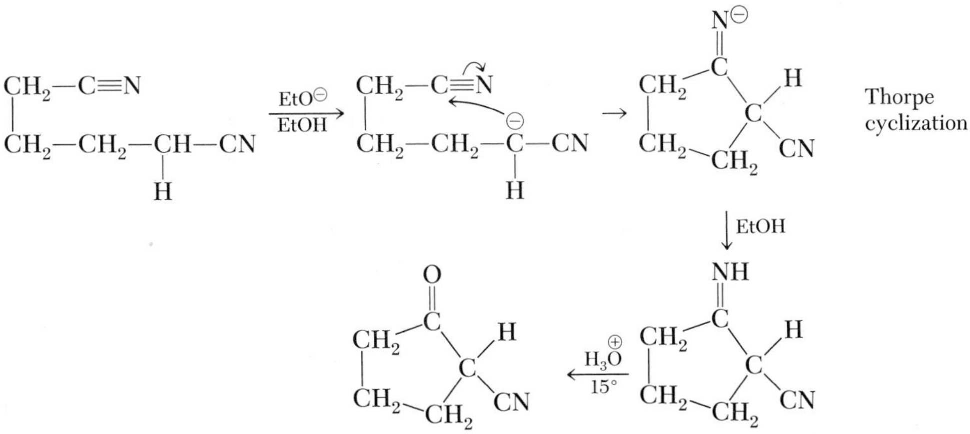

$$\begin{array}{l}CH_2{-}C{\equiv}N\\|\\CH_2{-}CH_2{-}CH(H){-}CN\end{array} \xrightarrow[EtOH]{EtO^{\ominus}} \begin{array}{l}CH_2{-}C{\equiv}N\\|\\CH_2{-}CH_2{-}\overset{\ominus}{C}(H){-}CN\end{array} \rightarrow \text{cyclic } C{=}N^{\ominus} \text{ (Thorpe cyclization)} \xrightarrow{EtOH} \text{cyclic } C{=}NH$$

Thorpe cyclization

$$\text{cyclic } C{=}NH \xrightarrow[15^{\circ}]{H_3O^{\oplus}} \text{2-cyanocyclopentanone}$$

2-cyanocyclopentanone
(a β-ketonitrile)

PROBLEMS

15. If dimethyl sulfide is treated with Br_2, at low temperatures, it forms a crystalline adduct. Hydrolysis of this adduct affords, among other things, dimethyl sulfoxide. Provide a mechanism for this conversion.

°16. When dimethyl sulfoxide reacts with methyl iodide, it forms a 1:1 adduct that isomerizes upon warming. The nmr spectrum of the first-formed adduct consists of two singlets with relative intensities of 2:1. When this adduct is dissolved in D_2O containing a trace of base, it eventually incorporates six deuterons. If this hexadeuterio adduct is heated in the absence of solvent, it decomposes to give $CD_3S(O)CD_3$ and CH_3I. Account for these observations.

17. The sulfinyl chlorides and sulfonyl chlorides that contain an α hydrogen react with trialkylamines to form the sulfoxides and sulfones, respectively, of thioketones.

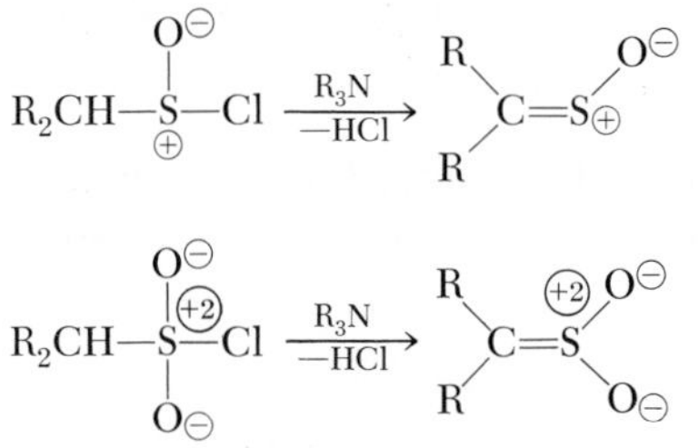

$$R_2CH{-}\overset{\overset{O^{\ominus}}{|}}{\underset{\oplus}{S}}{-}Cl \xrightarrow[-HCl]{R_3N} R_2C{=}\overset{\oplus}{S}{-}O^{\ominus}$$

$$R_2CH{-}\overset{O^{\ominus}}{\underset{O^{\ominus}}{\overset{|}{\underset{|}{S^{(+2)}}}}}{-}Cl \xrightarrow[-HCl]{R_3N} R_2C{=}\overset{(+2)}{S}(O^{\ominus})_2$$

Explain why reaction **A** produces two products while **B** produces only one.

A: (9-H-9-fluorenyl sulfinyl chloride, with CH_3 on ring; $\overset{\ominus}{O}{-}\overset{\oplus}{S}{-}Cl$ and H at C-9) $\xrightarrow[-HCl]{R_3N:}$

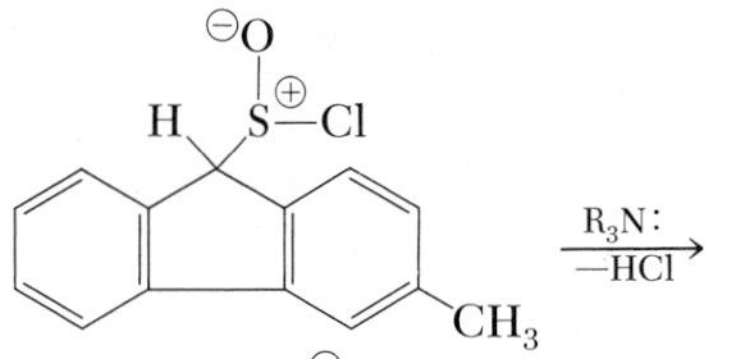

B: (9-H-9-fluorenyl sulfonyl chloride, with CH_3 on ring; $O^{\ominus}{-}\overset{(+2)}{S}(O^{\ominus}){-}Cl$ and H at C-9) $\xrightarrow[-HCl]{R_3N:}$

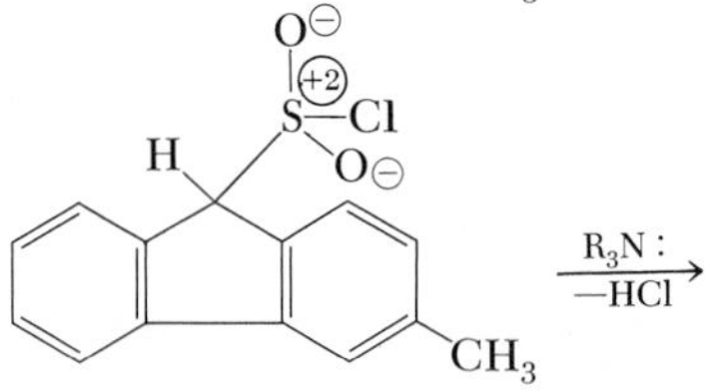

18. *p*-Nitrobenzenesulfenyl chloride has been used to form derivatives of alkenes; a representative reaction is shown below. Suggest a mechanism for this derivatization.

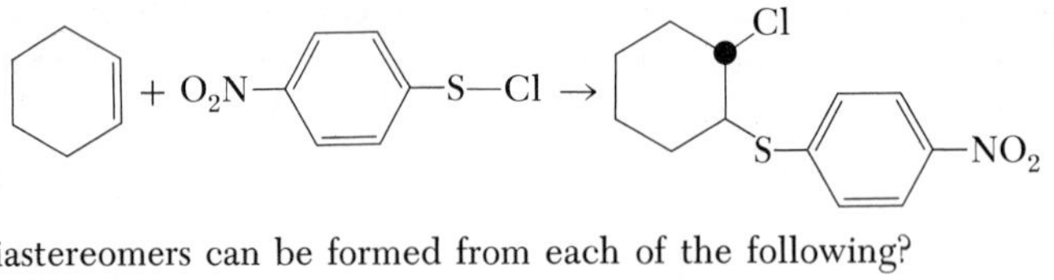

cyclohexene + $O_2N{-}C_6H_4{-}S{-}Cl$ → 2-chlorocyclohexyl $S{-}C_6H_4{-}NO_2$

19. How many diastereomers can be formed from each of the following?

(a) 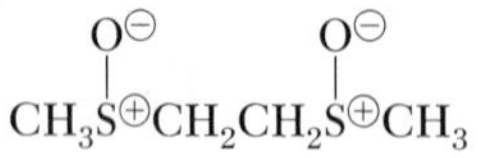 $CH_3\overset{O^{\ominus}}{\overset{|}{S^{\oplus}}}CH_2CH_2\overset{O^{\ominus}}{\overset{|}{S^{\oplus}}}CH_3$

(b)  $CH_3\overset{O^{\ominus}}{\overset{|}{S^{\oplus}}}OCH_2CH_2\overset{O^{\ominus}}{\overset{|}{S^{\oplus}}}CH_3$

(c) 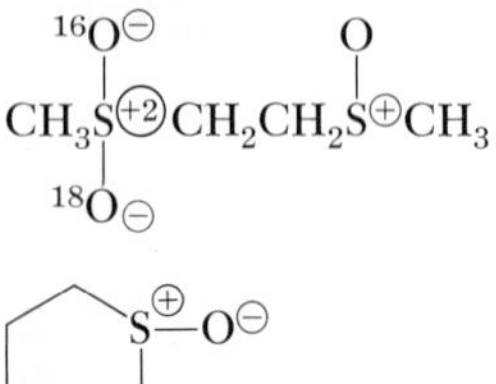 $CH_3\overset{^{16}O^{\ominus}}{\underset{^{18}O^{\ominus}}{\overset{|}{\underset{|}{S^{(+2)}}}}}CH_2CH_2\overset{O}{\overset{|}{S^{\oplus}}}CH_3$

(d) 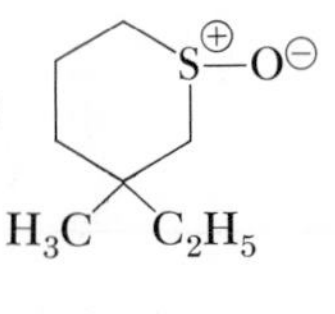 (thiane S-oxide, $\overset{\oplus}{S}{-}O^{\ominus}$, with H_3C and C_2H_5 on C-4)

°This problem requires knowledge of nmr spectroscopy.

20. Sulfoxides are reduced to the corresponding sulfides by prolonged heating with HCl or, much more rapidly, with HI. Suggest a mechanism for this reaction.

$$R{-}\overset{O^{\ominus}}{\overset{|}{S^{\oplus}}}{-}R + 2HI \rightarrow R{-}S{-}R + H_2O + I_2$$

21. The nitration of methyl phenyl sulfoxide produces substantial amounts of methyl *p*-nitrophenyl sulfoxide, while nitration of methyl phenyl sulfone produces methyl *m*-nitrophenyl sulfone. Account for this difference.

22. Sulfoxides containing a β hydrogen can undergo a pyrolytic elimination reaction to form alkenes. Suggest a mechanism for the reaction in light of the stereochemistry shown below.

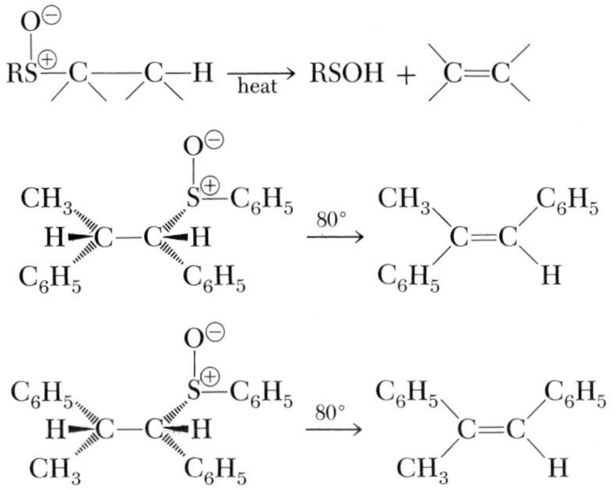

$$R\overset{O^{\ominus}}{\overset{|}{S^{\oplus}}}{-}C{-}C{-}H \xrightarrow{\text{heat}} RSOH + C{=}C$$

CH_3(H)(C_6H_5)C—C(H)(C_6H_5)$S^{\oplus}(O^{\ominus})C_6H_5$ $\xrightarrow{80°}$ $(CH_3)(C_6H_5)C{=}C(C_6H_5)(H)$ [CH_3 trans to C_6H_5]

C_6H_5(H)(CH_3)C—C(H)(C_6H_5)$S^{\oplus}(O^{\ominus})C_6H_5$ $\xrightarrow{80°}$ $(C_6H_5)(CH_3)C{=}C(C_6H_5)(H)$ [C_6H_5 trans to C_6H_5]

23. Using any needed starting materials, synthesize the following. You must use an organosulfur compound in some non-trivial step(s).

(a) $CH_3CH_2CHD_2$
(b) $CH_3CD_2CH_3$
(c) $CH_2DCH_2CH_2D$
(d) $CH_3COCH_2CH_3$
(e)  [spiro-oxirane on cyclobutane]
(f) $CH_2{=}CH{-}CH{=}CH_2$
(g) 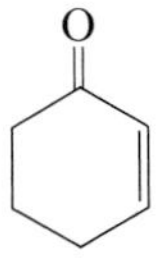 [2-cyclohexenone]
(h) 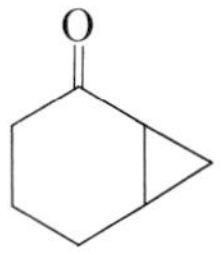 [bicyclo[4.1.0]heptan-2-one]
(i)  [bicyclic sulfide with S bridge]

*24. The nmr spectra below are those of ethyl *p*-tolyl sulfoxide (**A**), ethyl *p*-tolyl sulfide (**B**), and ethyl *p*-tolyl sulfone (**C**). Account for the differences in the spectra.

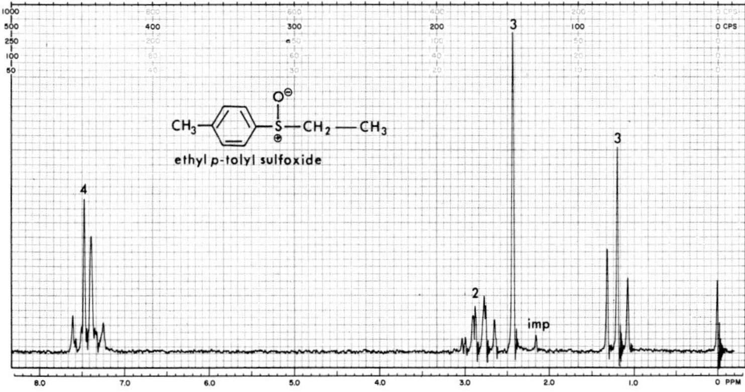

* This problem requires knowledge of nmr spectroscopy.

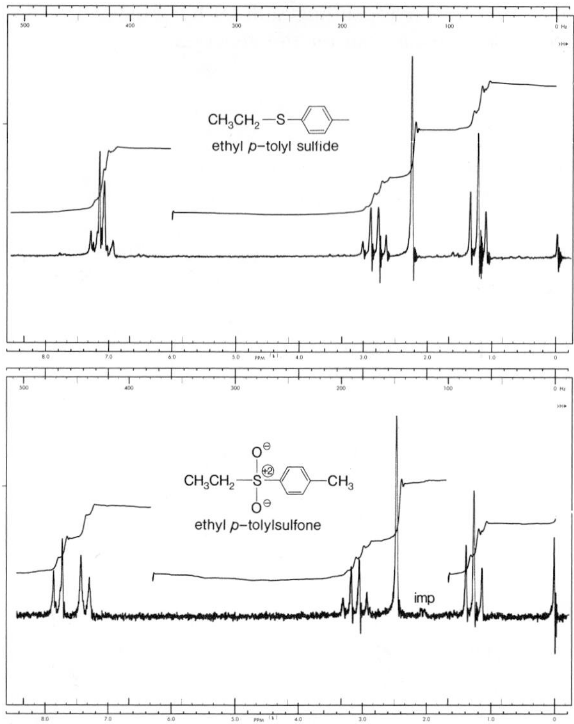

25. Thianthrene is prepared by the reaction of benzene with sulfur dichloride in the presence of $AlCl_3$. Provide a mechanism for its formation.

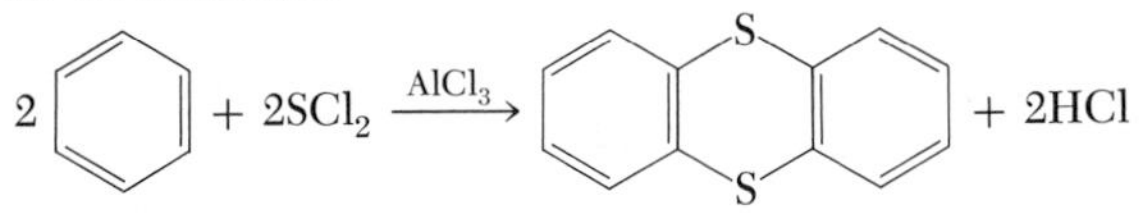

$$2\ C_6H_6 + 2SCl_2 \xrightarrow{AlCl_3} \text{thianthrene} + 2HCl$$

thianthrene

Can you suggest a synthesis for phenoxathiin?

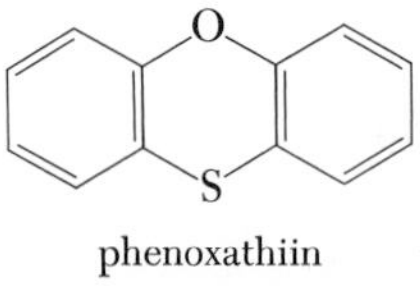

phenoxathiin

26. Thioxanthone is prepared in excellent yield by dissolving thiosalicylic acid in concentrated sulfuric acid and excess benzene, and then diluting the solution with water. Suggest a mechanism for this reaction. (*Two hints:* The odor of SO_2 fills the lab when the reaction is performed outside the hood. The disulfide shown below can replace thiosalicylic acid in this reaction.)

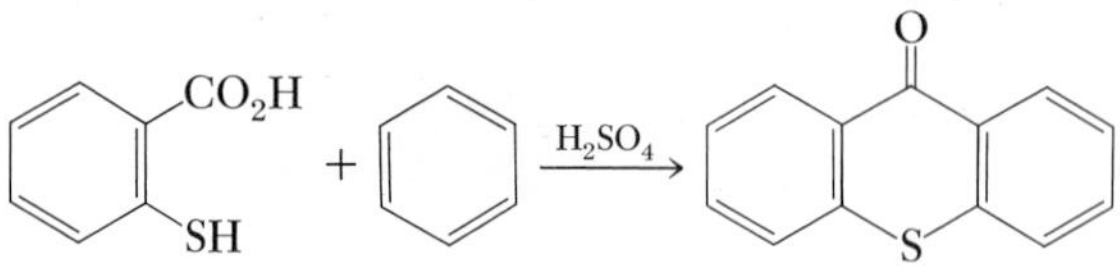

$$\text{thiosalicylic acid (}CO_2H, SH\text{)} + C_6H_6 \xrightarrow{H_2SO_4} \text{thioxanthone}$$

thiosalicylic acid thioxanthone

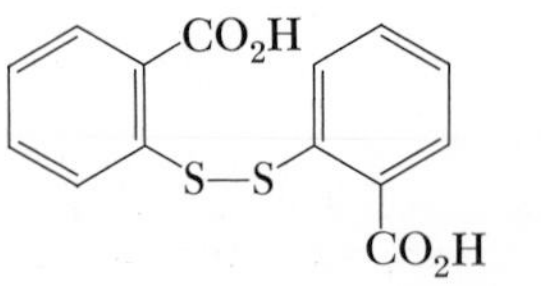

(alternate starting material)

27. Recently, compounds of the type $R_2S(OR)_2$ have been synthesized and studied by J. C. Martin and his students. A typical example is compound **B**, whose synthesis is shown at the top of p. 767.

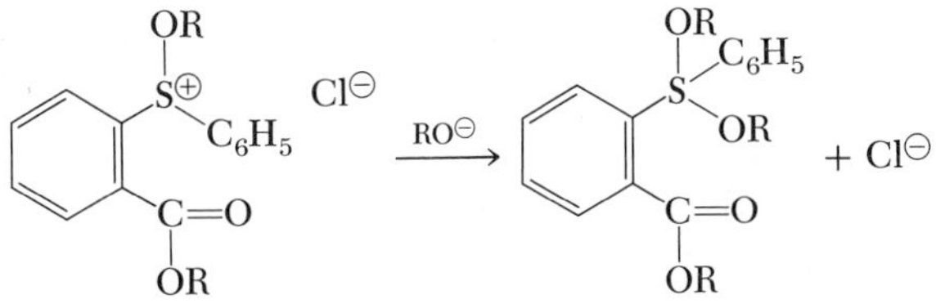

OR | S⊕ — C_6H_5 | C=O | OR $Cl^\ominus$ $\xrightarrow{RO^\ominus}$ OR | S — C_6H_5, OR | C=O | OR + $Cl^\ominus$

A **B**

where R ≡ C_6H_5—C(CF_3)(CF_3)—

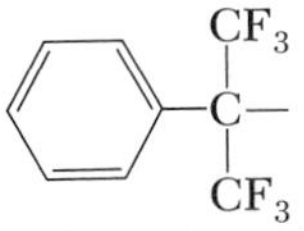

These workers reported that ^{19}F nmr spectral data support the structure assigned to **B,** since the trifluoromethyl groups of the S-alkoxy ligands appear as two quartets. (a) Why do they appear as quartets? (b) What is the significance of this observation and, in particular, why does it favor static structure **B** over the equilibrium, **C,** shown below?

C OR | S⊕ — C_6H_5 | C=O | OR $^\ominus$OR $\rightleftharpoons$ $RO^\ominus$ C_6H_5 | S⊕ — OR | C=O | OR

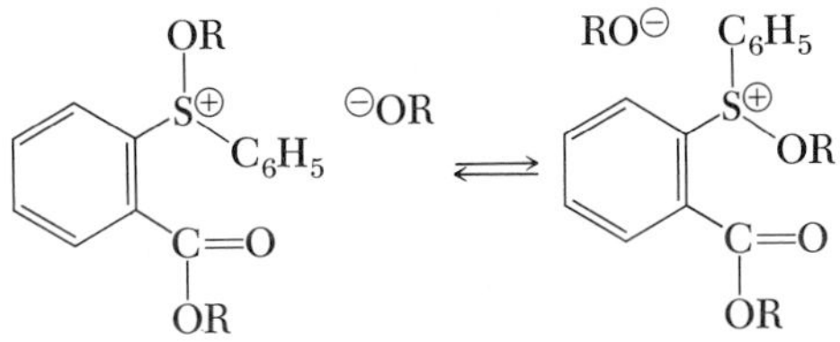

28. Provide an acceptable name for each of the following.

(a) $(C_6H_5CH_2O)_2P(OH)$ (e) $(CH_3)_2POCH_3$ (i) $(p\text{-}CH_3C_6H_4)_2PNH_2$

(b) $(C_6H_5CH_2O)_2H\overset{\oplus}{P}O^\ominus$ (f) $(CH_3)_3\overset{\oplus}{P}O^\ominus$ (j) $((CH_3)_2N)_3P$

(c) $(C_6H_5)_3PCH_2$ (g) $^\ominus O\overset{\oplus}{P}(CH_3)_2C_2H_5$ (k) $((CH_3)_2N)_3\overset{\oplus}{P}O^\ominus$

(d) $C_6H_5P(CH_2C_6H_5)Cl$ (h) $C_6H_5P(NH_2)_2$

29. Hypophosphorous acid, H_3PO_2, is another name for one of the stable oxyacids of phosphorus. Identify its other name. What is hypophosphorus acid used for in the synthetic organic chemistry laboratory? (Yes, it was discussed in a chapter dealing with amines.)

30. While trimethylphosphite boils at 111°, trimethylphosphate boils at 198°. Suggest a reason for this difference.

31. Explain why ammonia is about two thousand times more soluble in water than is phosphine. Which would you expect to be higher boiling, phosphine or ammonia? Explain how these two answers are related.

32. The synthesis of alkyl phosphites from phosphorus trichloride and an alcohol is carried out in the presence of a tertiary amine. The amine serves to react with the hydrochloric acid produced in the reaction. If the amine is not included in the original reaction mixture, a considerable amount of dialkyl phosphonate and alkyl chloride may result, and the yield of phosphite will be reduced by a similar amount. Account for the effect of excluding amine upon the reaction.

$$ROH + PCl_3 \rightarrow (RO)_3P \qquad \text{(with amine)}$$
$$ROH + PCl_3 \rightarrow (RO)_3P + RCl + (RO)_2P(O)H \qquad \text{(without amine)}$$

33. Suggest a reaction sequence that will convert the given starting material into the desired product.
(a) phosphorus trichloride *to* ethyl di-*n*-propylphosphinite
(b) *n*-butyl phosphonate *to* *n*-butylethylphosphonate
(c) *n*-propyl alcohol *to* *n*-propyl *n*-propylphosphonate
(d) triethyl phosphite *to* diethylphosphine oxide

34. Dialkylphosphines react with molecular oxygen to produce the corresponding oxide. Provide a mechanism for this reaction. (*Hint:* Re-examine the m.o. picture of oxygen.)

35. While sodium amide can be used to convert a monoalkylphosphine into its sodium salt, it usually will not convert a dialkylphosphine into its sodium salt. Suggest a reason for this difference.

36. Monoalkyl- and dialkylphosphines are conveniently prepared by the free-radical chain addition of phosphine, or a monoalkylphosphine, to an alkene. Suggest a reasonable mechanism for these reactions, including initiation, propagation, and termination steps.

$$PH_3 + CH_2{=}CHR \xrightarrow{R\cdot} H_2PCH_2CH_2R$$

$$R'PH_2 + CH_2{=}CHR \xrightarrow{R\cdot} R'PHCH_2CH_2R$$

*37. The reaction of trimethyl phosphite with ethyl bromoacetate produces **A.** Compound **A** reacts with sodium hydride to form a salt which, in turn, reacts with acetone to produce **B.** Account for the formation of **A** and **B.** What is the structure of the intermediate sodium salt? (*Hint:* The nmr spectrum of **B** ($C_7H_{12}O_2$) contains three methyl resonances.)

38. Triphenylphosphine reacts with iodobenzene to produce **A** ($C_{24}H_{20}PI$), which is water-soluble and which reacts with aqueous silver nitrate to form a precipitate. **A** reacts with phenyl lithium to afford two products, of which one was found to be insoluble in water (compound **B**) and the other (compound **C**, mw = 134) was soluble in water, had a rather high melting point, and reacted with aqueous silver nitrate to form a precipitate. Identify **A, B,** and **C.**

39. In our discussion of alcohols (p. 305) we noted that phosphorus trihalides can be used to convert alcohols to alkyl halides, while in this chapter we have noted that alcohols react with phosphorus trihalides (especially PCl_3) to produce a trialkylphosphite. How can these two views be reconciled? (*Hint:* The ease of formation of alkyl halides is $PI_3 > PBr_3 > PCl_3$. While primary alcohols react with PCl_3 under a given set of conditions to form mainly phosphite ester, PI_3 forms mainly alkyl iodides under the same conditions.)

40. The addition of triethylphosphine to **A** leads to a stable 1:1 adduct rather than to a polymer. Suggest a reason for this observation and a structure for the adduct.

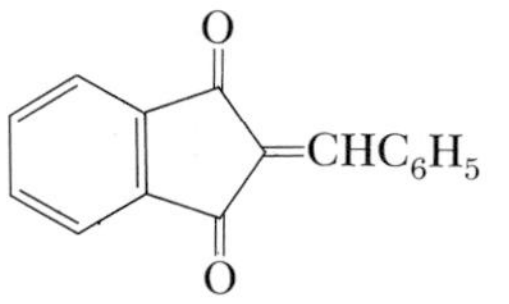

O

=CHC_6H_5

O

A

41. Dimethylphenylphosphine reacts with *cis*-2-butene episulfide to form dimethylphenylphosphine sulfide and *cis*-2-butene. Suggest (a) a mechanism that is consistent and (b) one that is inconsistent with this observation.

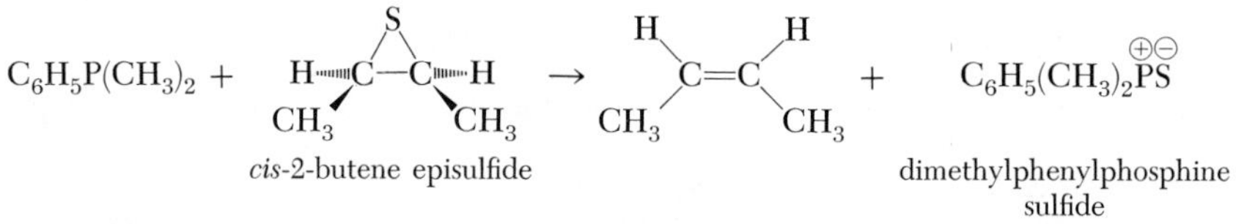

$C_6H_5P(CH_3)_2$ + H—C—C—H (S; CH_3, CH_3) → H, H C=C CH_3, CH_3 + $C_6H_5(CH_3)_2\overset{\oplus\ominus}{PS}$

cis-2-butene episulfide — dimethylphenylphosphine sulfide

42. Allenic derivatives of phosphine oxides can be prepared by the reaction shown. Suggest a mechanism for this reaction.

$$(C_6H_5)_2P{-}O{-}C(CH_3)_2C{\equiv}CH \xrightarrow{25^\circ} (C_6H_5)_2P(O)CH{=}C{=}C(CH_3)_2$$

43. Solutions of phosphines in halogenated solvents often are unstable and lead to ylid formation. The reaction of triphenylphosphine with carbon tetrachloride is typical. Account for the products. (*Hint:* When the reaction is carried out in the presence of an alkene, no cyclopropane derivatives are formed.)

$$(C_6H_5)_3P + CCl_4 \rightarrow (C_6H_5)_3P{=}CCl_2 + (C_6H_5)_3PCl_2$$

44. Triphenylphosphine reacts with chloromethyl methyl ether (CH_3OCH_2Cl) to produce a salt, which can be converted to an ylid by reaction with a strong base. This ylid can be used to convert ketones to aldehydes containing one more carbon, as shown in the sequence below. Identify the ylid and compound **A,** showing how **A** is converted to the final product.

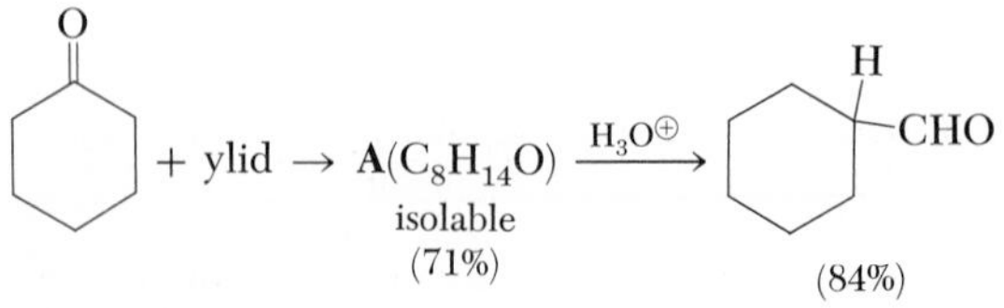

(cyclohexanone) + ylid → **A**($C_8H_{14}O$) isolable (71%) $\xrightarrow{H_3O^{\oplus}}$ (cyclohexane-H, CHO) (84%)

*This problem requires knowledge of nmr spectroscopy.

AMINO ACIDS, PEPTIDES, AND PROTEINS

24.1 INTRODUCTION

An *amino acid* is any compound containing both an acidic functionality and an amino group; however, the term usually is employed to mean a carboxylic acid bearing an amino group α to the carboxy group. These compounds are essential to all life as we know it. In spite of their importance in living systems, *free* amino acids are found rather rarely in nature. Instead, amino acids usually exist as polymers, known as *proteins*, which are involved in all critical functions of living systems. Proteins serve as nutrients, regulate metabolism, assist in the absorption of oxygen, play important roles in the functioning of the nervous system, provide the mechanical basis for muscle contraction, represent a major support material for the body, assist in the transfer of genetic information, and on, and on, and on.

$$RCH(NH_2)CO_2H \quad \textit{or} \quad R{-}\underset{\underset{NH_2}{|}}{\overset{\overset{H}{|}}{C}}{-}CO_2H \qquad \text{an } \alpha\text{-amino acid}$$

24.2 NOMENCLATURE

Although over seventy naturally occurring amino acids are known, there are about two dozen that play vital roles in most living systems. The common names and structures of these amino acids are listed in Table 24–1. The abbreviations for the names of these amino acids, which also are widely used, are included in the table.

All amino acids share a common structural unit, $R{-}CH(NH_2)CO_2H$,[a] and it is the nature of R that determines the differences between various amino acids. Some contain R groups that differ only slightly; for example, glutamine and asparagine, themselves α-amino acids, are the monoamides of glutamic acid and aspartic acid, respectively. Cysteine and cystine are related by a simple redox reaction, as are any thiol and the corresponding disulfide (p. 727).

$$\underset{\text{glutamine}}{H_2N{-}\overset{\overset{O}{\|}}{C}{-}CH_2CH_2{-}\underset{\underset{NH_2}{|}}{\overset{\overset{H}{|}}{C}}{-}CO_2H} \qquad \underset{\text{glutamic acid}}{HO{-}\overset{\overset{O}{\|}}{C}{-}CH_2CH_2{-}\underset{\underset{NH_2}{|}}{\overset{\overset{H}{|}}{C}}{-}CO_2H}$$

[a] Proline and hydroxyproline are exceptions (Table 24–1).

TABLE 24-1 Common α-Amino Acids, R—CHNH$_2$CO$_2$H

NAME (ABBREVIATION)	R	ISOELECTRIC POINT (pI)
glycine (gly)	H—	5.97
alanine (ala)	CH_3—	6.02
valine (val)	$(CH_3)_2CH$—	5.97
leucine (leu)	$(CH_3)_3C$—	5.98
isoleucine (ile)	$CH_3CH_2CH(CH_3)$—	6.02
proline[a] (pro)	H N —CO_2H	6.10
hydroxyproline[a] (hypro)	H N —CO_2H HO	
phenylalanine (phe)	$C_6H_5CH_2$—	5.98
tryptophan (try)	CH_2— N H	5.88
methionine (met)	$CH_3SCH_2CH_2$—	5.75
aspartic acid (asp)	HO_2CCH_2—	2.87
asparagine (asp(NH$_2$))	$H_2NC(O)CH_2$—	5.41
glutamic acid (glu)	$HO_2CCH_2CH_2$—	3.22
glutamine (glu(NH$_2$))	$H_2NC(O)CH_2CH_2$—	5.65
lysine (lys)	$H_2NCH_2CH_2CH_2CH_2$—	9.74
arginine (arg)	HN C—NH—$CH_2CH_2CH_2$— H_2N	10.76
histidine (his)	N N CH_2— H	7.58
serine (ser)	$HOCH_2$—	5.68
threonine (thr)	$HOCHCH_3$—	6.53
tyrosine (tyr)	HO—⟨ring⟩—CH_2—	5.65
cysteine (cySH)	$HSCH_2$—	5.02
cystine (cyS-Scy)	—CH_2S—SCH_2—	5.06

[a] Since this does not follow the general structure, the complete formula is given.

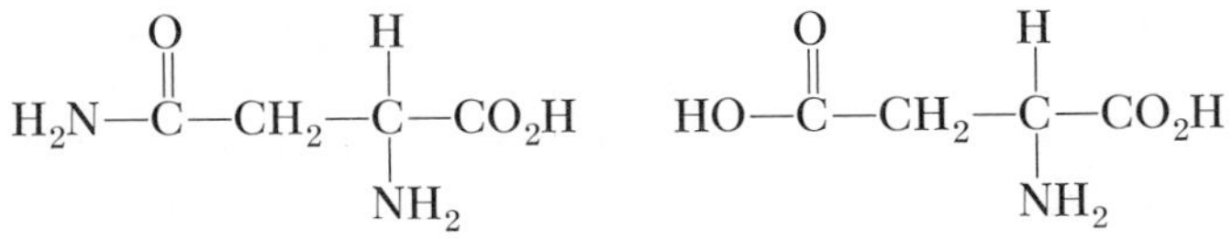

$$H_2N-\overset{O}{\overset{\|}{C}}-CH_2-\underset{NH_2}{\underset{|}{\overset{H}{\overset{|}{C}}}}-CO_2H \qquad HO-\overset{O}{\overset{\|}{C}}-CH_2-\underset{NH_2}{\underset{|}{\overset{H}{\overset{|}{C}}}}-CO_2H$$

asparagine — aspartic acid

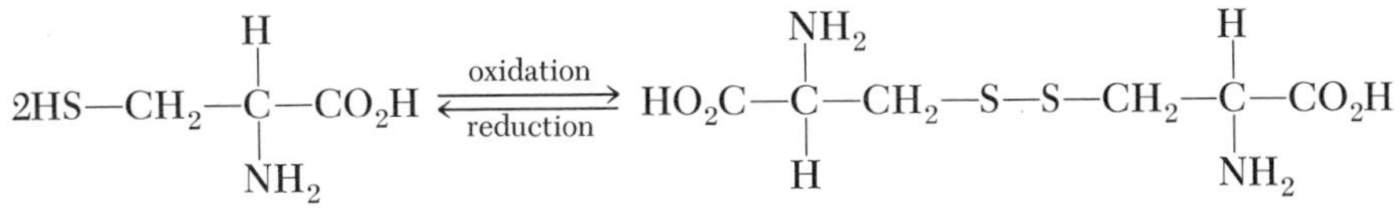

$$2HS-CH_2-\underset{NH_2}{\underset{|}{\overset{H}{\overset{|}{C}}}}-CO_2H \underset{\text{reduction}}{\overset{\text{oxidation}}{\rightleftharpoons}} HO_2C-\underset{H}{\underset{|}{\overset{NH_2}{\overset{|}{C}}}}-CH_2-S-S-CH_2-\underset{NH_2}{\underset{|}{\overset{H}{\overset{|}{C}}}}-CO_2H$$

cysteine — cystine

From such structural similarities, one would anticipate that the names usually used to identify amino acids would clearly indicate the nature of R. Unfortunately, common names tell us nothing about the nature of R (unless we already know which structure corresponds to which name), and we are forced to memorize each common name and the corresponding structure. While this situation might be construed as an excellent reason for identifying amino acids by their IUPAC names, common names are indelibly inscribed in the literature of organic chemistry and, more significantly, in that of biochemistry.

Essential amino acids are those that cannot be synthesized by an organism at a rate equal to physiological requirements from substances ordinarily present in the diet. The amino acids believed to be essential in man are: *isoleucine, leucine, lysine, methionine, phenylalanine, threonine, tryptophan* and *valine*. This list applies only to normal, healthy individuals. Certain metabolic disorders may render other amino acids "essential." Phenylketonuria (also called *phenylpyruvic oligophrenia* or *imbecillitas pyruvica*) is an inherited metabolic disorder that is related, in some unknown way, to certain mental disorders. Phenylketonuric individuals require *tyrosine* as an essential amino acid, since their bodies cannot convert phenylalanine to tyrosine as normal individuals do.

It is noteworthy that an organism must receive all of its essential amino acids simultaneously. If even a single member of a set of essential amino acids is withheld, only to be administered after several hours, the nutritional value of the entire group is diminished.

24.3 STEREOCHEMISTRY

All α-amino acids except glycine possess a chiral α carbon and can exist, in principle, as enantiomers. If, as in threonine, more than one chiral center is present, the possible stereoisomers include diastereomers as well as enantiomers.

Chemical correlations, one example of which is shown in Figure 24-1, have shown that

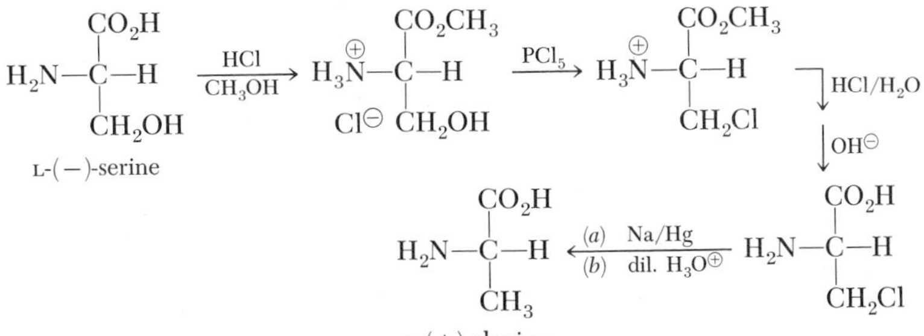

$$H_2N-\underset{CH_2OH}{\underset{|}{\overset{CO_2H}{\overset{|}{C}}}}-H \xrightarrow[CH_3OH]{HCl} \underset{Cl^{\ominus}}{H_3\overset{\oplus}{N}}-\underset{CH_2OH}{\underset{|}{\overset{CO_2CH_3}{\overset{|}{C}}}}-H \xrightarrow{PCl_5} H_3\overset{\oplus}{N}-\underset{CH_2Cl}{\underset{|}{\overset{CO_2CH_3}{\overset{|}{C}}}}-H$$

L-(−)-serine

$$\xrightarrow{HCl/H_2O} \xrightarrow{OH^{\ominus}} H_2N-\underset{CH_2Cl}{\underset{|}{\overset{CO_2H}{\overset{|}{C}}}}-H \xrightarrow[(b)\ \text{dil. } H_3O^{\oplus}]{(a)\ Na/Hg} H_2N-\underset{CH_3}{\underset{|}{\overset{CO_2H}{\overset{|}{C}}}}-H$$

L-(+)-alanine

Figure 24-1 Establishment of the relative configurations of L-serine and L-alanine.

almost all naturally occurring α-amino acids have the same relative configuration at Cα. Before elegant physical procedures were available, Emil Fischer arbitrarily assigned Cα of (−)-serine the "L" configuration and Cα of (+)-serine the "D" configuration. (By this convention, when the Fischer projection of an α-amino acid is written with the carboxy group on top and the R group on the bottom, an L-amino acid has the amino group on the left while a D-amino acid has it on the right.) Fischer's scheme for designating the configuration of an amino acid is, of course, applicable to all α-amino acids having a chiral α-carbon.

$$\begin{array}{cccc} \begin{array}{c} CO_2H \\ | \\ H_2N-C-H \\ | \\ CH_2OH \end{array} & \begin{array}{c} CO_2H \\ | \\ H-C-NH_2 \\ | \\ CH_2OH \end{array} & \begin{array}{c} CO_2H \\ | \\ H_2N-C-H \\ | \\ R \end{array} & \begin{array}{c} CO_2H \\ | \\ H-C-NH_2 \\ | \\ R \end{array} \\ \text{L-(−)-serine} & \text{D-(+)-serine} & \text{any L-amino acid} & \text{any D-amino acid} \end{array}$$

The data in Figure 24–1 reveal that an L-amino acid may be dextrorotatory, +, or levorotatory, −, depending upon the nature of R. This should not be surprising, since "L" refers to the configuration at Cα while the optical rotation is an experimentally determined property of the entire molecule. The vast majority of α-amino acids found in nature are L-amino acids. Their enantiomorphs usually are found only when dealing with microorganisms and are termed "unnatural" amino acids.

Using the (**R,S**) nomenclature, one can ascribe the **S** configuration to most of the "natural" or L-amino acids.[a] L-Isoleucine and L-threonine, with two chiral centers, could be either member of a pair of diastereomers, depending upon the configuration at Cβ. The correct absolute configurations for these are:

$$\begin{array}{cc} \begin{array}{c} CO_2H \\ | \\ H_2N-C-H \\ | \\ H_3C-C-H \\ | \\ C_2H_5 \end{array} & \begin{array}{c} CO_2H \\ | \\ H_2N-C-H \\ | \\ H-C-OH \\ | \\ CH_3 \end{array} \\ \text{L-isoleucine} & \text{L-threonine} \\ \text{((2S, 3S)-2-amino-3-} & \text{((2S, 3R)-2-amino-3-} \\ \text{methylpentanoic acid)} & \text{hydroxybutanoic acid)} \end{array}$$

1. Draw wedge representations for leucine, serine, valine, and alanine. Show the correct absolute configuration.

24.4 ACID-BASE PROPERTIES

Amino acids are amphoteric substances, capable of existing as cations or anions. In the solid state amino acids exist as **zwitterions** (also called **dipolar ions**), the carboxy group having transferred a proton to the α-amino group. The species distribution in an aqueous solution is pH-dependent. In strongly acidic solutions, amino acids exist as cations; while in basic solutions, they exist as anions.

[a] Since $HSCH_2$— has a higher priority than HO_2C—, L-cysteine is an exception; it has the **R** configuration.

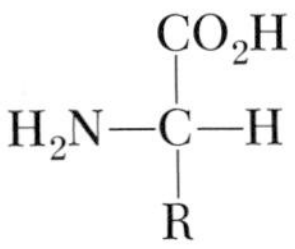

$$\underset{\substack{\text{non-ionic form;}\\ \text{idealized amino acid}}}{H_2N{-}\underset{\underset{R}{|}}{\overset{\overset{CO_2H}{|}}{C}}{-}H} \qquad \underset{\substack{\text{zwitterion;}\\ \text{amino acid as found}\\ \text{in the solid state}}}{H_3\overset{\oplus}{N}{-}\underset{\underset{R}{|}}{\overset{\overset{CO_2^{\ominus}}{|}}{C}}{-}H}$$

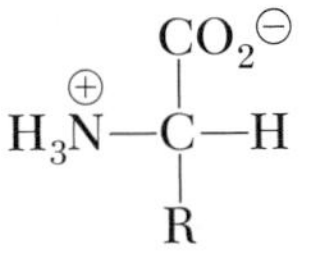

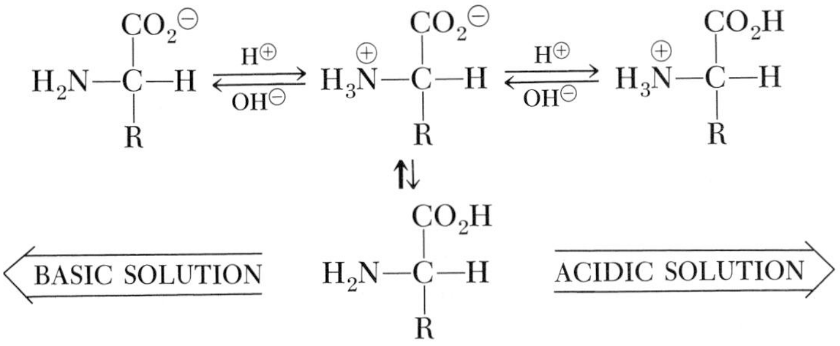

$$H_2N{-}\underset{\underset{R}{|}}{\overset{\overset{CO_2^{\ominus}}{|}}{C}}{-}H \underset{OH^{\ominus}}{\overset{H^{\oplus}}{\rightleftharpoons}} H_3\overset{\oplus}{N}{-}\underset{\underset{R}{|}}{\overset{\overset{CO_2^{\ominus}}{|}}{C}}{-}H \underset{OH^{\ominus}}{\overset{H^{\oplus}}{\rightleftharpoons}} H_3\overset{\oplus}{N}{-}\underset{\underset{R}{|}}{\overset{\overset{CO_2H}{|}}{C}}{-}H \quad \text{species distribution in solution}$$

$$\upharpoonleft\downharpoonright$$

$$\text{BASIC SOLUTION} \longleftarrow \quad H_2N{-}\underset{\underset{R}{|}}{\overset{\overset{CO_2H}{|}}{C}}{-}H \quad \longrightarrow \text{ACIDIC SOLUTION}$$

2. In extremely alkaline solutions an amino acid possesses two basic sites, $—NH_2$ and $CO_2^{\ominus}$. Which of these is more basic? Why? What would be the product of simple protonation of $H_2N—CHR—CO_2^{\ominus}$?

If an amino acid is placed in a conducting medium containing a pair of electrodes, it will migrate to the cathode in acidic solutions and to the anode in basic solutions. At some characteristic pH the amino acid will not migrate toward either electrode; this particular pH is termed the **isoelectric point** (pI) of the amino acid (see Table 24–1). The pI of an amino acid is rarely 7, since it depends upon the acidity of the alkylammonium ion, the basicity of the carboxylate anion, the effect of R upon both of these, and the presence of any additional basic or acidic groups.

We can better understand why the pI of an amino acid usually is not exactly 7 by considering what happens when a simple monoamino monocarboxylic acid such as glycine is dissolved in water. An aqueous solution of glycine, $H_3\overset{\oplus}{N}CH_2CO_2^{\ominus}$, is slightly acidic because the acidic group of the zwitterion ($—NH_3^{\oplus}$) is more acidic than the basic group ($—CO_2^{\ominus}$) is basic. Put slightly differently, this means that a solution of glycine in water contains more $H_2NCH_2CO_2^{\ominus}$ than $H_3\overset{\oplus}{N}CH_2CO_2H$ ($K_b = 2.5 \times 10^{-12}$ while $K_a = 1.6 \times 10^{-10}$ for glycine). In order to attain glycine's isoelectric point, a solution of pure glycine in water must have acid added to it until the resulting solution has a pH of 6.1. These added protons serve to repress the ionization of $H_3\overset{\oplus}{N}CH_2CO_2^{\ominus}$ to $H_2NCH_2CO_2^{\ominus}$ and, thereby, maximize the concentration of the zwitterion.

$$H_3\overset{\oplus}{N}—CH_2CO_2^{\ominus} + H_2O \rightarrow H_2N—CH_2CO_2^{\ominus} + H_3O^{\oplus} \qquad \text{reaction of glycine with water}$$

When a direct current is applied to a solution of a mixture of several amino acids, each individual amino acid will migrate to one of the two electrodes at a rate dependent upon the specific amino acid and upon the pH of the medium. Separation and analysis of amino acid mixtures by means of this phenomenon is called *electrophoresis*.

Ion-Exchange Chromatography—The Amino Acid Analyzer. Treatment of an insoluble polymer, or resin, bearing free sulfonic acid groups with sodium hydroxide converts the acidic sites to anions.

$$\text{(Resin)}—SO_3H + NaOH\ (aq) \rightarrow \text{(Resin)}—SO_3^{\ominus}\ Na^{\oplus} + H_2O$$

In ion-exchange chromatography a resin of this type is packed into a column and treated with an aqueous solution of a mixture of amino acids. The various zwitterions form salts with the basic sites on the resin and adhere to it.

$$\text{(Resin)}\text{—}SO_3^{\ominus} + H_3\overset{\oplus}{N}CHRCO_2^{\ominus} \rightarrow \text{(Resin)}\text{—}SO_3^{\ominus}\ H_3\overset{\oplus}{N}CHRCO_2^{\ominus}$$

If the amino acid mixture is put on the column as a solution of cationic salts, the attraction between the resin and the amino acids is even greater because no carboxylate anions are present.

$$H_3\overset{\oplus}{N}\text{—}CHR\text{—}CO_2^{\ominus} + HCl \rightarrow H_3\overset{\oplus}{N}\text{—}CHR\text{—}CO_2H\ Cl^{\ominus}$$

$$\text{(Resin)}\text{—}SO_3^{\ominus}\ Na^{\oplus} + H_3\overset{\oplus}{N}\text{—}CHRCO_2H\ Cl^{\ominus} \xrightarrow{-NaCl} \text{(Resin)}\text{—}SO_3^{\ominus}\ H_3\overset{\oplus}{N}\text{—}CHRCO_2H$$

To achieve separation of the amino acids in the mixture, the column of resin bearing the amino acid cations, $H_3N^{\oplus}CHRCO_2H$, is slowly eluted with buffers of gradually increasing pH. As the alkalinity of the buffer increases, different amino acids are converted to their anionic forms and washed from the column.

$$\text{(Resin)}\text{—}SO_3^{\ominus}\ H_3\overset{\oplus}{N}\text{—}CHRCO_2H + 2OH^{\ominus} \longrightarrow \text{(Resin)}\text{—}SO_3^{\ominus} + H_2N\text{—}CHRCO_2^{\ominus} + 2H_2O$$

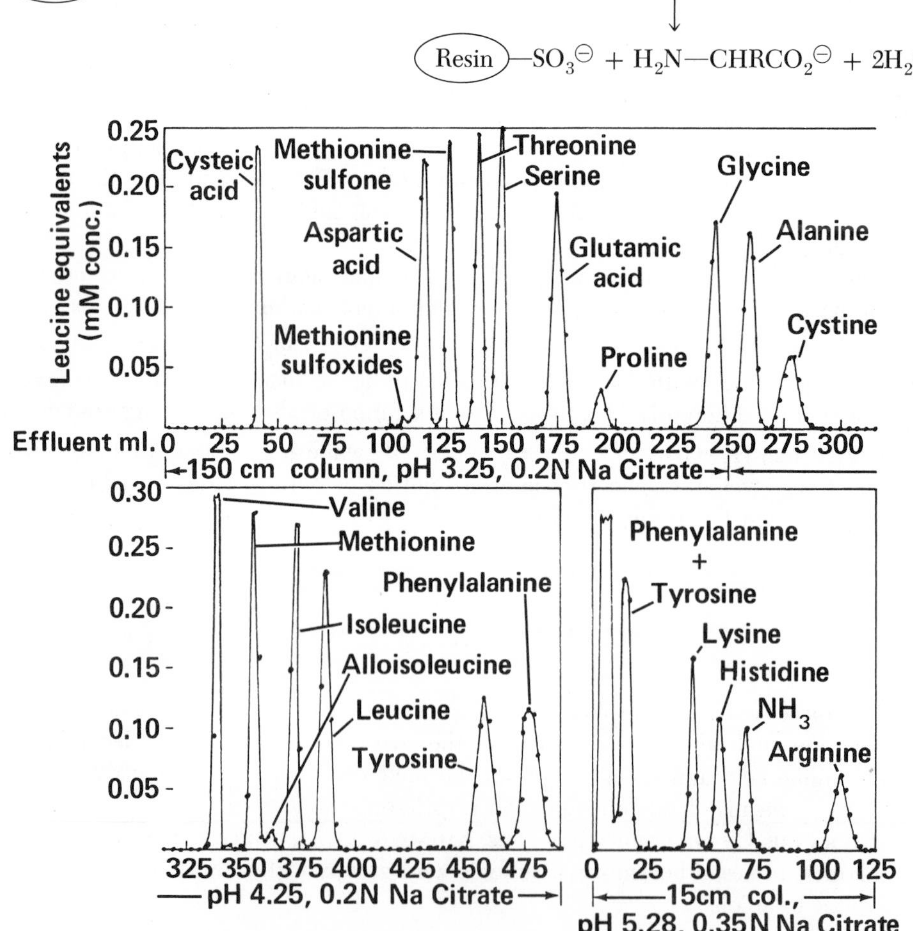

Figure 24-2 Chromatographic fractionation of an amino acid mixture by an ion exchange resin. (Reprinted with permission from Moore *et al.*, Anal. Chem., *30*:1186, 1958. Copyright by the American Chemical Society.)

The eluent bearing the now-separated amino acids can be passed through a detection system, which quantitates the amount of amino acid present and establishes an elution pattern for that particular amino acid mixture. The specific amino acids present in the unknown mixture can be determined by comparing this elution pattern to those of various mixtures of known amino acid composition. A typical elution pattern is presented in Figure 24–2.

Successful operation of the detection system requires that all amino acids have some readily detected property, or undergo some easily identifiable change. The property of amino acids most often employed is their ability to react with *ninhydrin* to form a characteristic blue color. The reactions involved in this **ninhydrin test** for amino acids are presented in Figure 24–3.

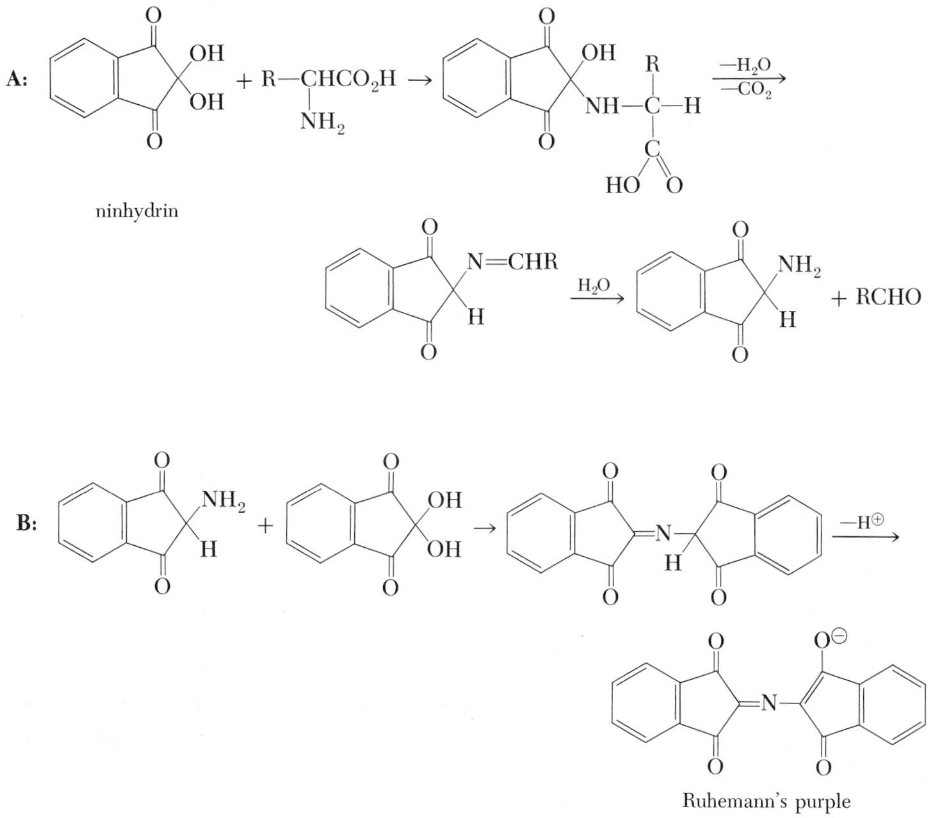

ninhydrin

Ruhemann's purple

Figure 24-3 Reactions responsible for a positive ninhydrin test. The product is responsible for the observed color (λ_{max} = 570 nm). Proline and hydroxyproline form a different product, which is yellow-brown in color (λ_{max} = 440 nm).

Instruments are available that completely automate this separation and identification procedure, and such *amino acid analyzers* can provide a quantitative analysis of a mixture of amino acids in two or three hours.

24.5 *IN VITRO* SYNTHESIS OF RACEMIC AMINO ACIDS

One of several useful syntheses of α-amino acids involves the ammonolysis of an α-halo acid, the halo acid often being made by a Hell-Volhard-Zelinsky reaction (p. 588).

$$RCH_2CO_2H + Br_2 \xrightarrow{PBr_3} RCHBrCO_2H \xrightarrow[\text{(b) dil. acid}]{\text{(a) } NH_3 \text{ (excess)}} RCH(NH_2)CO_2H$$

This route can be modified by preparing the α-bromo acid *via* a malonic ester-type synthesis or by amination *via* the Gabriel synthesis.

$$\left(EtO\overset{O}{\overset{\|}{C}}\right)_2\overset{\ominus}{C}H\ Na^{\oplus} \xrightarrow[(S_N2)]{RBr} \left(EtO\overset{O}{\overset{\|}{C}}\right)_2CHR \xrightarrow[\text{heat}]{KOH/H_2O} \xrightarrow{H^{\oplus}} \left(HO\overset{O}{\overset{\|}{C}}\right)_2CHR$$

sodiomalonic ester

$$\left(HO\overset{O}{\overset{\|}{C}}\right)_2CHR \xrightarrow{\text{heat} \mid Br_2/THF} \left(HO\overset{O}{\overset{\|}{C}}\right)_2CBrR \xrightarrow[\text{heat}]{-CO_2} HO-\overset{O}{\overset{\|}{C}}-\overset{H}{\overset{|}{\underset{Br}{\underset{|}{C}}}}-R$$

bromination *via* the malonic ester route

$$\text{(potassium phthalimide)}\ :\overset{\ominus}{N}:K^{\oplus} + ClCH_2-\overset{O}{\overset{\|}{C}}-OEt \rightarrow \text{(phthalimido)}N-CH_2-\overset{O}{\overset{\|}{C}}-OEt \xrightarrow{HCl/H_2O} H_3\overset{\oplus}{N}-CH_2CO_2H\ Cl^{\ominus}$$

glycine hydrochloride (90% overall)

Gabriel synthesis of α-amino acids

The *Strecker synthesis* of α-amino acids involves the reaction of a carbonyl compound with a mixture of ammonium chloride and sodium cyanide. These inorganic reagents react to form ammonia and hydrogen cyanide, the "active ingredients" in this process. An addition-elimination sequence involving ammonia and the carbonyl compound results in the formation of an imine, which then adds hydrogen cyanide to produce an α-amino nitrile. Hydrolysis converts this to the desired α-amino acid.

$$NH_4^{\oplus}Cl^{\ominus} + Na^{\oplus}CN^{\ominus} \rightleftarrows NH_4^{\oplus}CN^{\ominus} + Na^{\oplus}Cl^{\ominus}$$

$$NH_4^{\oplus}CN^{\ominus} \rightleftarrows NH_3 + HCN$$

$$RCHO \xrightleftharpoons{NH_3} \underset{NH_2}{\underset{|}{RCH}}-OH \xrightleftharpoons{-H_2O} RCH{=}NH \xrightarrow{HCN} \underset{NH_2}{\underset{|}{RCH}}-CN \xrightarrow{H_3O^{\oplus}} \underset{NH_2}{\underset{|}{RCH}}-CO_2H$$

α-amino nitrile

The *Bucherer synthesis* of α-amino acids, a modification of the Strecker synthesis, employs ammonium carbonate and sodium cyanide as the inorganic reagents. These react with the carbonyl compound to form an intermediate *hydantoin*, which is hydrolyzed to the amino acid. In the example below, the first step is not part of the Bucherer synthesis but is included for completeness.

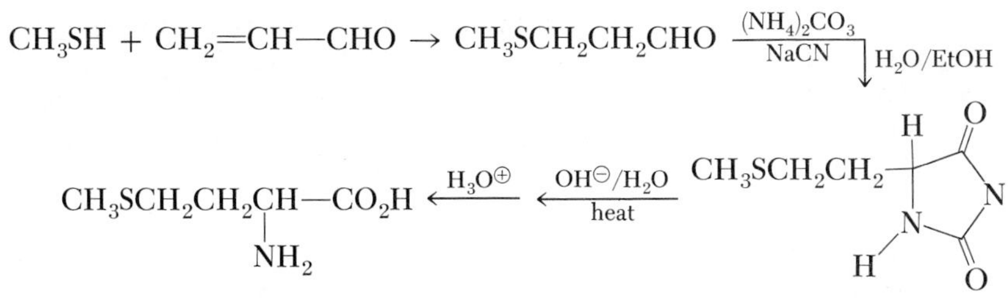

$$CH_3SH + CH_2{=}CH{-}CHO \rightarrow CH_3SCH_2CH_2CHO \xrightarrow[NaCN]{(NH_4)_2CO_3} \xrightarrow{H_2O/EtOH} \text{5-(2-methylthioethyl)-hydantoin}$$

$$\text{5-(2-methylthioethyl)-hydantoin} \xrightarrow[\text{heat}]{OH^{\ominus}/H_2O} \xrightarrow{H_3O^{\oplus}} CH_3SCH_2CH_2CH(NH_2){-}CO_2H$$

methionine

5-(2-methylthioethyl)-hydantoin

3. Suggest a mechanism for the hydrolysis of a hydantoin to an α-amino acid.

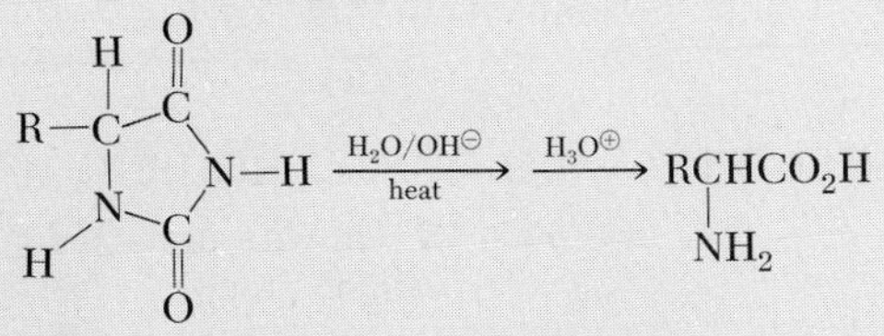

$$\text{a hydantoin} \xrightarrow[\text{heat}]{H_2O/OH^{\ominus}} \xrightarrow{H_3O^{\oplus}} RCH(NH_2)CO_2H$$

a hydantoin

4. Which of the two N—H hydrogens of a hydantoin should be the more acidic? Why?

Both the amination of α-halo acids and the Strecker synthesis produce racemic α-amino acids. Since the classical resolution of such racemic modifications is difficult and expensive, natural starting materials usually are used when large quantities of L-amino acids are needed. For example, monosodium glutamate (MSG, "Accent") is prepared from glutamic acid obtained by hydrolysis of wheat gluten. The annual world production of MSG is several hundred million pounds. Because of economic factors, there actually are several commercial routes to glutamic acid; however, they all have one common characteristic—Mother Nature is responsible for the enantiomeric homogeneity of the final product.

24.6 *IN VITRO* REACTIONS OF AMINO ACIDS

Most reactions that amino acids undergo in the laboratory are those expected from any amine or carboxylic acid. Of these, only amide formation at the carboxy group and arylation of the amino group will be singled out at this time.

Amide Formation. It would be difficult for the carboxy group of an amino acid to react with the amino group of any amine to form an amide, without simultaneous polymerization of the amino acid.

$$n H_2NCHR{-}\overset{O}{\overset{\|}{C}}OH \rightarrow H_2NCHR{-}\overset{O}{\overset{\|}{C}}{-}\left(NHCHR{-}\overset{O}{\overset{\|}{C}}\right)_{n-2}{-}NHCHR\overset{O}{\overset{\|}{C}}OH$$

poly(α-amino acid)

$$H_2N{-}CHR{-}\overset{O}{\overset{\|}{C}}{-}OH + R'NH_2 \rightarrow H_2N{-}CHR{-}\overset{O}{\overset{\|}{C}}{-}NHR' + H_2O$$

competing reactions

One way to prevent this amino acid polymerization is by blocking the amino group of the amino acid, thereby leaving only the amino group of the amine free to react. This can be done using *carbobenzoxy chloride* (also called *carbobenzyloxy chloride* and *benzyl chloroformate*[a]), a reagent that forms a carbamate with amino groups, including those of amino acids. This *protecting group* can be removed, without disturbing the new amide linkage or racemizing Cα, by catalytic hydrogenolysis or by treatment with a cold solution of hydrogen bromide in acetic acid. Since sulfur compounds poison the metal catalyst used in hydrogenation/hydrogenolysis, triethylsilane (Et_3SiH) and palladium chloride may be used to remove this protecting group from sulfur-containing amino acids.

$$\underset{\text{benzyl alcohol}}{C_6H_5CH_2OH} + \underset{\text{phosgene}}{ClC(=O)Cl} \xrightarrow{-20^\circ} \underset{\text{carbobenzoxy chloride}}{C_6H_5CH_2O{-}C(=O){-}Cl}$$

$$C_6H_5CH_2OC(=O)Cl + H_2NCHRCO_2H \rightarrow \underset{\alpha\text{-amino acid with protected amino group}}{C_6H_5CH_2OC(=O)NHCHRCO_2H}$$

$$C_6H_5CH_2O{-}C(=O){-}NHR \xrightarrow{H_2/Pt} C_6H_5CH_3 + \underset{\text{carbamic acid (unstable)}}{[HOC(O)NHR]} \rightarrow CO_2 + H_2NR$$

$$C_6H_5CH_2O{-}C(=O){-}NHR \xrightarrow[\text{cold}]{HBr/HOAc} C_6H_5CH_2Br + [HOC(O)NHR] \rightarrow CO_2 + H_2NR$$

In the early syntheses of amides of amino acids, the free carboxy group was "activated" for reaction with the amine. This was often done by converting the —C(O)—OH group to the —C(O)—O—C(O)OEt group with ethyl chloroformate.

$$C_6H_5CH_2O{-}C(=O){-}NHCHRCO_2H + C_2H_5O{-}C(=O){-}Cl \rightarrow C_6H_5CH_2O{-}C(=O){-}NHCHR{-}C(=O){-}O{-}C(=O){-}OEt$$

$$\xrightarrow[-C_2H_5OH,\ -CO_2]{R'NH_2} C_6H_5CH_2{-}O{-}C(=O){-}NHCHR{-}C(=O){-}NHR' \xrightarrow[H_2]{Pt} C_6H_5CH_3 + CO_2 + H_2N{-}CHR{-}C(=O){-}NHR'$$

The amidation of the carboxy group of amino acids has been improved by two synthetic advances. The *first improvement* involves a change in the group used to protect the amino group of the amino acid. *t*-Butoxycarboxazide, $(CH_3)_3COC(O)N_3$, converts an amino group into its *t*-butoxycarbonyl (BOC) derivative, but the BOC group can be easily removed by reaction with a mixture of hydrochloric and acetic acids. Because of this comparatively simple deblocking step, *t*-butoxycarboxazide is preferred to carbobenzoxy chloride for protecting amino groups. The *second improvement* has removed the need to activate the carboxy group prior to amidation. This has been accomplished by treating the amine-protected amino acid with an amine in the presence of dicyclohexylcarbodiimide (DCC), a potent, non-acidic dehydrating agent. The reactions that employ these reagents in the synthesis of an amide of an amino acid are presented on p. 779.

[a] The —OC(O)Cl group also is called *chlorocarbonate.*

$$\underset{\substack{t\text{-butoxycarboxazide} \\ (BOCN_3)}}{(CH_3)_3CO{-}\overset{\overset{\displaystyle O}{\|}}{C}{-}N_3} + H_2NCHRCO_2H \xrightarrow[(b)\ \text{acid}]{(a)\ \text{base}} BOC{-}NHCHRCO_2H$$

$$BOC{-}NHCHRCO_2H + H_2NR' + C_6H_{11}N{=}C{=}NC_6H_{11} \xrightarrow{CH_2Cl_2}$$

$$\underset{N,N'\text{-dicyclohexylurea}}{C_6H_{11}NH{-}\overset{\overset{\displaystyle O}{\|}}{C}{-}NHC_6H_{11}} + BOC{-}NHCHR\overset{\overset{\displaystyle O}{\|}}{C}{-}NHR'$$

$$BOC{-}NHCHR\overset{\overset{\displaystyle O}{\|}}{C}{-}NHR' \xrightarrow[HOAc]{HCl} (CH_3)_2C{=}CH_2 + CO_2 + H_3\overset{\oplus}{N}{-}CHR{-}\overset{\overset{\displaystyle O}{\|}}{C}{-}NHR'\ Cl^{\ominus}$$

5. The sequence below presents the synthesis of *t*-butoxycarboxazide. Identify the lettered intermediates.

$$\text{thiophenol} \xrightarrow{COCl_2} \mathbf{A}(C_7H_5ClOS) \xrightarrow[\text{base}]{(CH_3)_3COH} \mathbf{B}(C_{11}H_{14}O_2S) \xrightarrow{H_2NNH_2} \mathbf{C}(C_5H_{12}N_2O_2) \xrightarrow[H_3O^{\oplus}]{NaNO_2} (CH_3)_3COC(O)N_3$$

Arylation. Amino acids will react with 2,4-dinitrofluorobenzene, **Sanger's reagent,** in slightly alkaline solution to form a substituted dinitroaniline. These reactions proceed *via* nucleophilic aromatic substitution. The role of the reagent in biochemistry is explained on p. 784.

$$\underset{\text{Sanger's reagent}}{O_2N{-}C_6H_3(NO_2){-}F} + H_2N{-}CHRCO_2H \rightarrow O_2N{-}C_6H_3(NO_2){-}NH{-}CHRCO_2H$$

6. (a) Suggest a synthesis for Sanger's reagent, beginning with any monosubstituted benzene. (b) How would you prepare 3,5-dinitrofluorobenzene?

24.7 *IN VIVO* REACTIONS OF AMINO ACIDS

Simple amino acids, like many simple "biological" molecules, are not stored in the cell; excess amino acids normally are destroyed in processes that provide energy for the living system. Three reactions that are important in this cellular processing of amino acids are **deamination, transamination,** and **decarboxylation.** These processes, described below, are catalyzed by enzymes and do not occur without these catalysts.

Deamination-Transamination. Deamination may occur by either a *nonoxidative* or an *oxidative* route. The former, a reaction limited mainly to bacteria and fungi, is illustrated by the conversion of aspartic acid to fumaric acid and ammonia by the action of the enzyme *aspartase.*

$$\underset{\text{aspartic acid}}{HO_2C{-}CH_2{-}\underset{\displaystyle NH_2}{\underset{|}{C}H}{-}CO_2H} \xrightleftharpoons{\text{aspartase}} \underset{\text{fumaric acid}}{\begin{matrix} HO_2C & & H \\ & C & \\ & \| & \\ & C & \\ H & & CO_2H \end{matrix}} + NH_3 \qquad \text{deamination of aspartic acid}$$

In an oxidative deamination, such as the conversion of alanine to pyruvic acid, the enzyme catalyst requires an oxidant (dehydrogenator) to complete the conversion. A common hydrogen acceptor in such systems is FAD (flavine adenine dinucleotide), which is converted to a reduced form abbreviated $FADH_2$ (see p. 704). This oxidative deamination proceeds *via* an intermediate imine and is shown, below the more usual type of equation, as a biochemist might present it.

$$\underset{\text{alanine}}{\begin{matrix} CH_3 \\ | \\ H{-}C{-}NH_2 \\ | \\ CO_2H \end{matrix}} \xrightarrow{[O]} \begin{matrix} CH_3 \\ | \\ C{=}NH \\ | \\ CO_2H \end{matrix} \xrightarrow{H_2O} \underset{\text{pyruvic acid}}{\begin{matrix} CH_3 \\ | \\ C{=}O \\ | \\ CO_2H \end{matrix}} + NH_3$$

$$\begin{matrix} CH_3 \\ | \\ H{-}C{-}NH_2 \\ | \\ CO_2H \end{matrix} \xrightarrow[\text{FAD} \curvearrowright \text{FADH}_2]{\text{oxidase enzyme}} \begin{matrix} CH_3 \\ | \\ C{=}NH \\ | \\ CO_2H \end{matrix} \xrightarrow[NH_3]{H_2O} \begin{matrix} CH_3 \\ | \\ C{=}O \\ | \\ CO_2H \end{matrix}$$

Those enzymes classified as *transaminases* (also called *aminotransferases*) catalyze the interconversion of amino and carbonyl groups by a process of *transamination.* In addition to being a method for destroying a given amino acid, transamination also provides a route for amino acid biosynthesis. In the example below, *aspartate-α-ketoglutarate transaminase*[a] is shown to catalyze the interconversion of aspartic acid and α-ketoglutaric acid to oxaloacetic acid and glutamic acid. A mechanism for this type of reaction was given in Chapter 16 (p. 484).

$$\underset{\text{aspartic acid}}{\begin{matrix} CO_2H \\ | \\ CH_2 \\ | \\ H{-}C{-}NH_2 \\ | \\ CO_2H \end{matrix}} + \underset{\alpha\text{-ketoglutaric acid}}{\begin{matrix} CO_2H \\ | \\ CH_2 \\ | \\ CH_2 \\ | \\ C{=}O \\ | \\ CO_2H \end{matrix}} \xrightleftharpoons[\text{pyridoxal phosphate}]{\text{transaminase +}} \underset{\text{oxaloacetic acid}}{\begin{matrix} CO_2H \\ | \\ CH_2 \\ | \\ C{=}O \\ | \\ CO_2H \end{matrix}} + \underset{\text{glutamic acid}}{\begin{matrix} CO_2H \\ | \\ CH_2 \\ | \\ CH_2 \\ | \\ H{-}C{-}NH_2 \\ | \\ CO_2H \end{matrix}}$$

Decarboxylation. Another general biological reaction of α-amino acids is decarboxylation. The enzymes that catalyze these reactions are commonly called *decarboxylases* and are distributed throughout nature. Some amines show profound biological activity, and the decarboxylation of amino acids is an important source of them. The generation of *dopamine* by

[a] The systematic name of this enzyme, according to the *Recommendations (1964) of the International Union of Biochemistry on the Nomenclature and Classification of Enzymes,* is L-aspartate: 2-oxoglutarate aminotransferase.

the decarboxylation of *dopa* is especially significant, since dopamine is a biological precursor of *adrenaline.*

$$\text{HO(HO)C}_6\text{H}_3\text{—CH}_2\text{—C(NH}_2\text{)(H)—CO}_2\text{H} \xrightarrow[\textit{decarboxylase}]{\textit{dopa}} \text{HO(HO)C}_6\text{H}_3\text{—CH}_2\text{—CH}_2\text{NH}_2 + \text{CO}_2$$

dopa   dopamine

$$\text{HO}_2\text{C—CH(NH}_2\text{)—CH}_2\text{—CH}_2\text{—CO}_2\text{H} \xrightarrow[\alpha\textit{-decarboxylase}]{\textit{glutamic}} \text{H}_2\text{NCH}_2\text{CH}_2\text{CH}_2\text{CO}_2\text{H} + \text{CO}_2$$

γ-aminobutyric acid

The decarboxylation of glutamic acid is important because the resulting γ-aminobutyric acid is believed to act as an inhibitor of synaptic transmission in the central nervous system; it may be considered to be a natural "tranquilizer." *Glutamic decarboxylase* requires the presence of pyridoxal phosphate in order to convert glutamic acid to γ-aminobutyric acid. When pyridoxine (a precursor to pyridoxal phosphate) is removed from the diet of animals, they develop convulsions similar to epilepsy. The biological requirement for "vitamin B_6," a group of compounds that includes pyridoxal, pyridoxine, and pyridoxamine, is related to the fact that about twenty biological reactions of amino acids require pyridoxal phosphate as a "coenzyme" (coenzymes are discussed in detail in Chapter 27).

Pyridine ring with HO, CH_3 (as H_3C), and CH_2OH substituents; 4-substituent: CHO — pyridoxal; CH_2OH — pyridoxine; CH_2NH_2 — pyridoxamine.

pyridoxal   pyridoxine   pyridoxamine

Pyridinium ring (N⊕–H) with CHO, HO, H_3C, and $CH_2OPO_3H^{\ominus}$ substituents.

pyridoxal phosphate

Getting Rid of Nitrogen. While deaminations provide an organism with a means of removing excess amino acids, they increase the amount of unwanted nitrogenous material in the system. The toxicity associated with high concentrations of ammonia demands that organisms be able to excrete ammonia or a derivative of ammonia. Creatures that live in aqueous media (*e.g.*, the oceans) excrete ammonia directly into their environment, where diffusion removes it from its source. Land-locked organisms cannot do this, so they excrete their unwanted nitrogen as either of two solids, *urea* or *uric acid.*

The mammalian embryo excretes its excess nitrogen as urea, since urea is extremely water-soluble and can be disposed of in the maternal circulatory system. The embryos of birds and reptiles live in hard shells and cannot get rid of nitrogen this way. Instead, they convert their excess nitrogen into uric acid, which is insoluble in water and precipitates on the internal surface of the egg. After these various embryos develop, they continue to excrete the same compound.

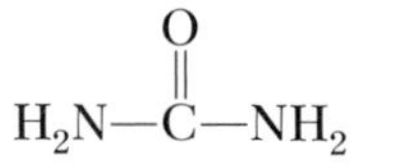

$$H_2N-\overset{O}{\overset{\|}{C}}-NH_2$$

urea

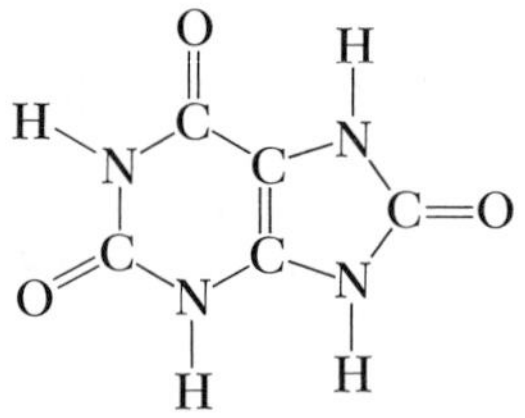

uric acid

Uric acid does show up in adult humans as a waste product, being present in urine in quantities of about 0.5 g/24 hr. The presence of larger amounts usually is associated with some medical problem. For example, kidney stones (urinary calculi) and bladder stones are rich in uric acid, and the painful symptoms of gout are due to the deposition of crystals of the monosodium salt of uric acid in joints. Gout affects nearly one million people in the United States, most of whom are sexually mature males. The presence of uric acid in the urine of humans may be important to evolutionists, since most animals completely degrade uric acid prior to excretion. It has been suggested, in part on the qualitative observation that many gouty people are more irritable than non-gouty people even before the onset of pain, that the presence of uric acid in the system may offer some evolutionary advantage to humans by providing them with an unusually aggressive personality. This point of view is a long way from being proved, but it may provide an interesting link between biochemistry and behavior.

7. What is a reasonable basis for the acidity of uric acid?

24.8 PEPTIDES

An amino acid can be made to polymerize to a polyamide. Polyamides derived from α-amino acids are commonly called **peptides** or **polypeptides,** and the amido linkage in this type of polymer is called a **peptide bond** or **peptide linkage.**

$$nH_2NCH_2CO_2H \xrightarrow{-(n-1)\,H_2O} H_2N-CH_2-\overset{O}{\overset{\|}{C}}\left(NH-CH_2-\overset{O}{\overset{\|}{C}}\right)_{n-2}NH-CH_2-\overset{O}{\overset{\|}{C}}-OH$$

glycine — polyglycine, a polypeptide

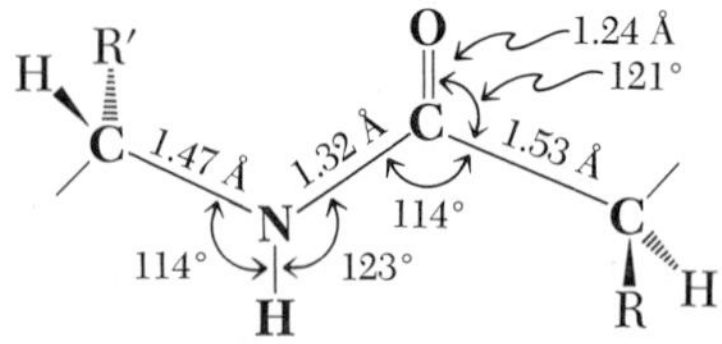

1.24 Å, 121°, 1.47 Å, 1.32 Å, 1.53 Å, 114°, 114°, 123°

the peptide bond; atoms in boldface reside in the plane of the paper

While homopolymers such as polyglycine and poly(L-valine) have been used as models for study, naturally occurring polypeptides are heteropolymers with very explicit amino acid sequences. Since very minor alterations in its amino acid sequence will alter or destroy the biological activity of a polypeptide, the chemist interested in life processes must know a great deal about the detailed structure of these heteropolymers. To aid in the transmission of this information, certain standards of nomenclature have been created. For example, in a heteropolyamide the amino acid at the end of the polymer bearing a free NH_2 group is termed the "*N*-terminal amino acid," while the end amino acid bearing the free carboxyl group is the "*C*-terminal amino acid." To facilitate communication, amino acid residues in a peptide chain can be given sequential numbers, beginning with the *N*-terminal amino acid, which is numbered "1".

Polypeptides are named as derivatives of the *C*-terminal amino acid, the *N*-terminal amino acid being listed first and the *C*-terminal amino acid last. This ordering is followed whether the name is made up of the full names of the constituent amino acids or from the abbreviations of those names. For example, glycylalanine is a dipeptide with glycine as the *N*-terminal amino acid and alanine as the *C*-terminal amino acid. This name can be abbreviated *gly-ala*. The name alanylglycine, abbreviated ala-gly, refers to the isomeric dipeptide that contains alanine as the *N*-terminal amino acid and glycine as the *C*-terminal amino acid.

$$H_2N-CH_2-\overset{\overset{\displaystyle O}{\|}}{C}-NH-\underset{\underset{\displaystyle CH_3}{|}}{CH}-CO_2H$$

glycylalanine
(gly-ala)
(gly(1)-ala(2))

$$H_2N-\underset{\underset{\displaystyle CH_3}{|}}{CH}-\overset{\overset{\displaystyle O}{\|}}{C}-NH-CH_2-CO_2H$$

alanylglycine
(ala-gly)
(ala(1)-gly(2))

The tetrapeptide *alanylserylglycylvaline* (ala-ser-gly-val) has alanine as the *N*-terminal amino acid and valine as the *C*-terminal amino acid. It is one of 24 tetramers that could be constructed using these four L-amino acids.

$$\underbrace{H_2N-\underset{\underset{\displaystyle CH_3}{|}}{CH}}_{N\text{-terminus}}-\overset{\overset{\displaystyle O}{\|}}{C}-NH-\underset{\underset{\underset{\underset{\displaystyle OH}{|}}{\displaystyle CH_2}}{|}}{CH}-\overset{\overset{\displaystyle O}{\|}}{C}-NH-\underset{\underset{\displaystyle H}{|}}{CH}-\overset{\overset{\displaystyle O}{\|}}{C}-\underbrace{NH-\underset{\underset{\displaystyle CH(CH_3)_2}{|}}{CH}-CO_2H}_{C\text{-terminus}}$$

ala-ser-gly-val
(ala(1)-ser(2)-gly(3)-val(4))

8. Twenty-four tetrapeptides are possible if four different amino acids are linked to one another, no single amino acid appearing more than once in a given tetrapeptide. Using alanine, serine, glycine, and valine, construct the abbreviated formulas for all twenty-four tetrapeptides. Indicate the *N*- and *C*-terminal amino acids for each molecule. Indicate those structures in which valine bears the number "2." Write out the structures for those compounds in which glycine carries the number "2."
9. A group of *n* different objects can be arranged *n*! ways, assuming that each is used once and only once in each array. How many different ways can seven different amino acids be arranged? Eight? Nine? [*Note: n*! (read "*n* factorial") equals $n(n-1)(n-2)\ldots(1)$.]
10. A protein contains 0.35% glycine. What is the minimum molecular weight of the protein?

The difference between a *polypeptide* and a *protein* is not distinct. Proteins have been defined as polypeptides possessing some arbitrary minimum molecular weight, *e.g.*, 5000. A more useful distinction involves the structure of the polymer at a level which is more subtle than its *primary structure, i.e.*, at a level which is concerned with something more than just the polymer's amino acid sequence. While a *polypeptide* is usually a linear, rather flexible molecule, proteins have long chains that may be coiled or folded into specific arrangements, often with clearly defined "holes" on the surface or within the body of the structure. Moreover, many of the proteins that act as enzymes have other moieties, called prosthetic groups, attached to the polyamide backbone. These prosthetic groups will be considered in more detail when we discuss how enzymes function (Chapter 27).

Before turning our attention to methods used by chemists to degrade ("analyze") and create ("synthesize") biologically active polypeptides, let us take a look at three medically important polyamides—oxytocin, vasopressin, and insulin. In doing this, be sure to note how two of them have remarkably similar amino acid sequences yet quite different physiological functions. All three possess a disulfide bond (—S—S—) and lose biological activity when the bond is reduced (—S—S— → —S—H + H—S—).

Oxytocin and vasopressin are polypeptide neurohormones that are released by discrete cells in the posterior lobe of the pituitary gland. The former is involved in uterine contraction and milk ejection, while the latter helps maintain the body's fluid balance by acting as an antidiuretic. Since oxytocin causes contraction of smooth muscle, particularly of the uterus, it is used in medicine to induce labor. Vasopressin and oxytocin, so unlike in physiological activity, have in common six of eight amino acids *and* a large ring formed by the presence of a disulfide linkage. If this disulfide linkage is destroyed (by reduction) the resultant non-cyclic structure has no biological activity.

The 1955 Nobel prize in chemistry was awarded to V. du Vigneaud for his ". . . work on biologically important sulfur compounds and particularly for the first synthesis of a polypeptidic hormone [vasopressin]."

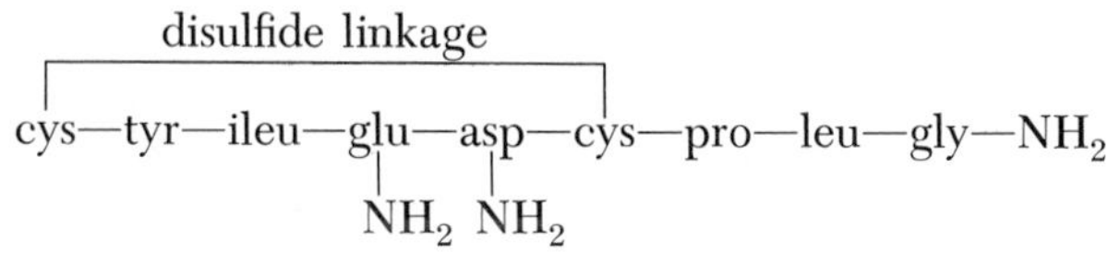

disulfide linkage (cys ... cys)

cys—tyr—ileu—glu—asp—cys—pro—leu—gly—NH_2
(glu: NH_2; asp: NH_2)

oxytocin

disulfide linkage (cys ... cys)

cys—tyr—phe—glu—asp—cys—pro—arg—gly—NH_2
(glu: NH_2; asp: NH_2)

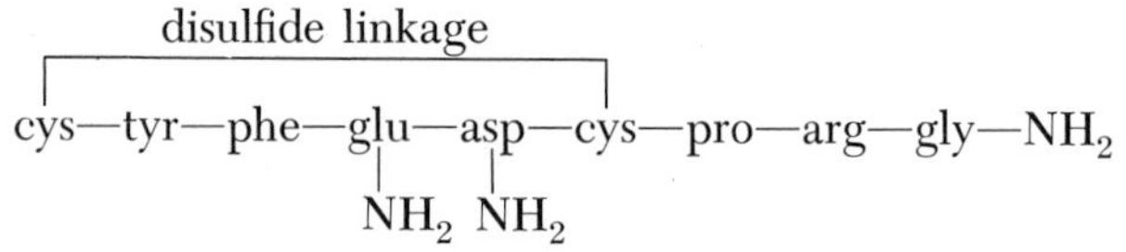

vasopressin

Insulin, like vasopressin and oxytocin, is a hormone. It is synthesized by the beta cells of the islets of Langerhans (in the pancreas), and its physiological role is in the control of glucose metabolism. Of course, it is most often associated with the treatment of diabetes mellitus. Like vasopressin and oxytocin, insulin possesses a disulfide linkage. However, in addition to simply forming rings along a peptide chain, in insulin (Figure 24–4, opposite page) disulfide linkages hold two chains together.

Analysis of Polypeptides. Polypeptides, like other amides, can be hydrolyzed with aqueous acid or base. If a polypeptide is completely hydrolyzed, the nature and proportions of its constituent amine acids can be determined through the use of the automated amino acid analyzer; but no information can be obtained through this procedure about its exact amino acid sequence. If the polypeptide is treated with Sanger's reagent before hydrolysis, the *N*-terminal amino acid can be identified because it will be converted into a stable, colored aniline derivative that is not decomposed under the hydrolysis conditions (p. 786).

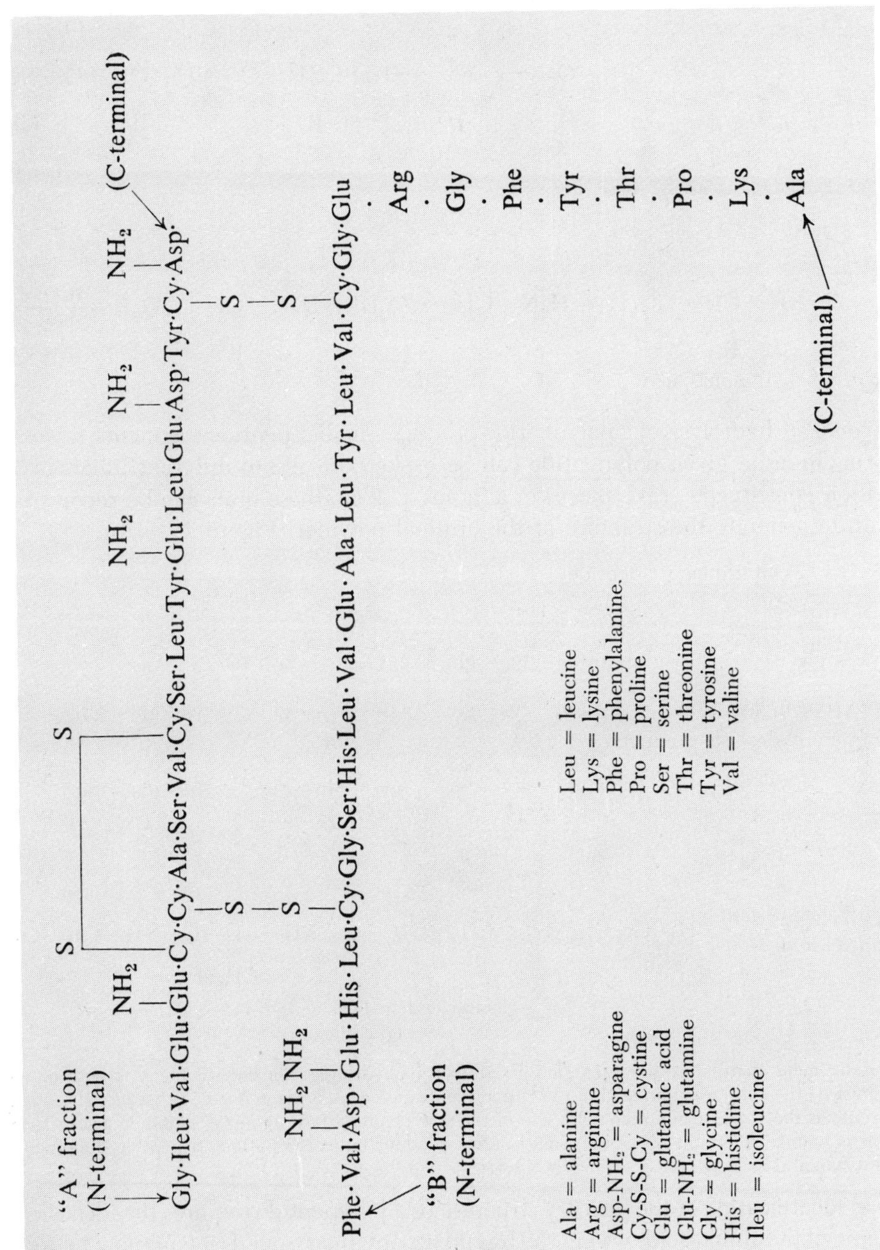

Figure 24-4 The structure of beef insulin. (From A. Mazur and B. Harrow, *Textbook of Biochemistry*, 10th ed., W.B. Saunders Company, 1971.)

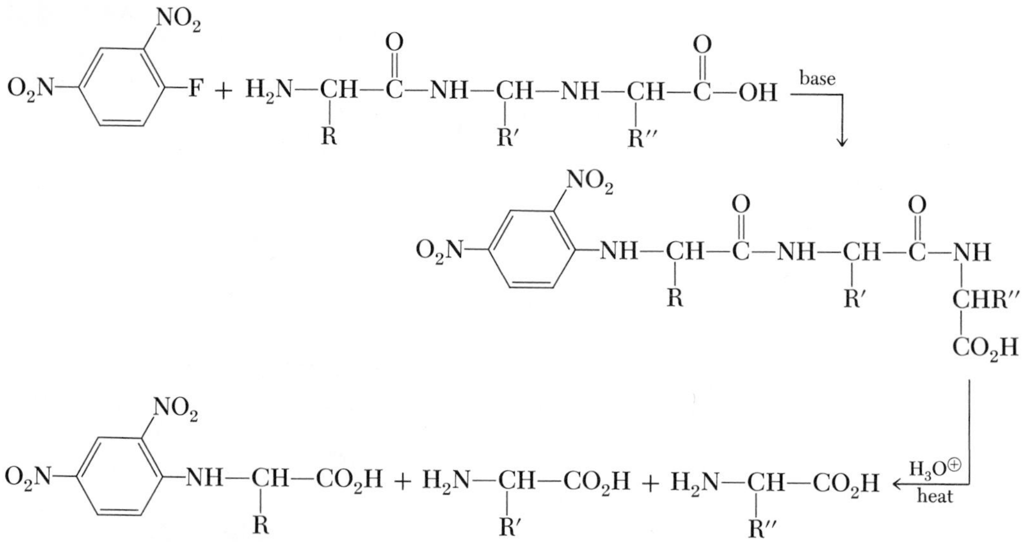

$$O_2N\text{-}C_6H_3(NO_2)\text{-}F + H_2N\text{—}\underset{R}{CH}\text{—}\overset{O}{\overset{\|}{C}}\text{—NH—}\underset{R'}{CH}\text{—NH—}\underset{R''}{CH}\text{—}\overset{O}{\overset{\|}{C}}\text{—OH} \xrightarrow{\text{base}}$$

$$O_2N\text{-}C_6H_3(NO_2)\text{—NH—}\underset{R}{CH}\text{—}\overset{O}{\overset{\|}{C}}\text{—NH—}\underset{R'}{CH}\text{—}\overset{O}{\overset{\|}{C}}\text{—NH—CHR''—}CO_2H$$

$$\xrightarrow[\text{heat}]{H_3O^{\oplus}} O_2N\text{-}C_6H_3(NO_2)\text{—NH—}\underset{R}{CH}\text{—}CO_2H + H_2N\text{—}\underset{R'}{CH}\text{—}CO_2H + H_2N\text{—}\underset{R''}{CH}\text{—}CO_2H$$

N-(2,4-dinitrophenyl) amino acid

The *incomplete* hydrolysis of a polypeptide produces smaller peptide fragments; by varying hydrolysis conditions, a given polypeptide can be broken down into different fragments that overlap in their constituents. Like pieces of a jigsaw puzzle, these units can be reconstructed (in the mind) to generate the structure of the original polymer (Figure 24–5).

glutathione (tripeptide)

complete hydrolysis → glu + cys + gly

partial hydrolysis → glu—cys + gly + glu + cys—gly + cys

All possible tripeptides:
(from complete hydrolysis products)

glu—cys—gly	gly—cys—glu	cys—gly—glu
A	**B**	**C**
glu—gly—cys	cys—glu—gly	gly—glu—cys
D	**E**	**F**

Only tripeptide consistent with partial hydrolysis:

$$HO_2C\text{—CH(NH}_2\text{)—}CH_2\text{—}CH_2\text{—}\overset{O}{\overset{\|}{C}}\text{—NH—}\underset{CH_2SH}{CH}\text{—}\overset{O}{\overset{\|}{C}}\text{—NH—}CH_2CO_2H$$

γ-glutamylcysteinylglycine (glu—cys—gly)
(glutathione)

Figure 24–5 Assignment of structure of glutathione based upon its hydrolysis. Since cys is found in both dipeptide fragments, it must be the central amino acid of the tripeptide. Structure **A** is elected over **B** because the fragment cys-gly has glycine as the *C*-terminal amino acid. Use of the side chain carboxy group of glutamic acid in peptide bond formation is unusual. Glutathione is not a degradation product but an important tripeptide, functioning as a cofactor (see Chapter 27).

The final identification of the primary structure of a polypeptide requires the identification and sequencing of the amino acids within the fragments obtained from its incomplete hydrolysis. This can be done by complete hydrolysis of the fragments, identification of the constituent amino acids, and chemical synthesis of the fragments; or it can be done by a selective hydrolysis, which removes one amino acid at a time from the peptide fragment. This latter approach, a stepwise degradation of a small polypeptide, can be accomplished using a group of pancreatic enzymes called **carboxypeptidases.** These hydrolyze off only *C*-terminal amino acids and, therefore, degrade a peptide fragment sequentially. Often a simple analysis of the varying concentrations of amino acids produced by a carboxypeptidase acting over increasing time periods can provide significant information about the sequencing of amino acids.

A major advance in the chemical analysis of polypeptides was described by P. Edman in 1950. He demonstrated that the *N*-terminal amino acid could be peeled away, through the use of phenyl isothiocyanate, to generate a new *N*-terminal amino acid that could be further degraded by the same type of sequence. This **Edman degradation** is described in Figure 24–6.

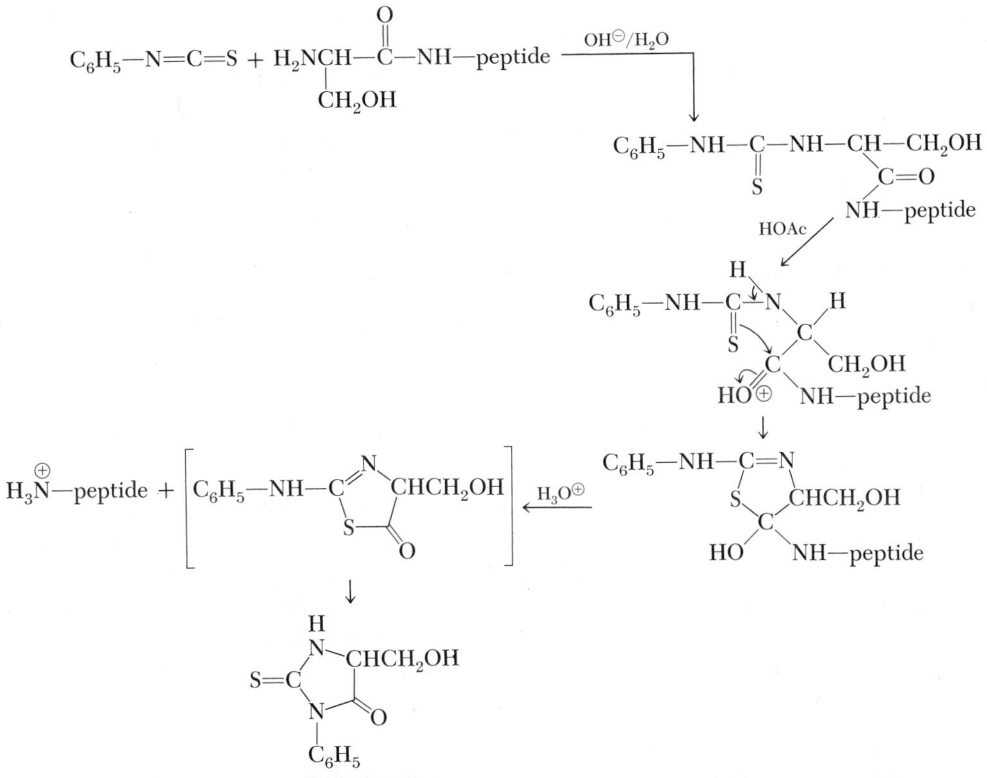

$$C_6H_5{-}N{=}C{=}S + H_2NCH(CH_2OH){-}C({=}O){-}NH{-}\text{peptide} \xrightarrow{OH^{\ominus}/H_2O} C_6H_5{-}NH{-}C({=}S){-}NH{-}CH(CH_2OH){-}C({=}O){-}NH{-}\text{peptide}$$

HOAc

H₃N⊕—peptide + [C₆H₅—NH—C(=N—CHCH₂OH—C(=O)—S) ring] ← H₃O⊕ ← C₆H₅—NH—C=N, S, CHCH₂OH, C(HO)(NH—peptide)

a thiohydantoin (S=C, N—H, CHCH₂OH, C=O, N—C₆H₅ ring)

Figure 24-6 Edman degradation. Phenyl isothiocyanate reacts with the *N*-terminal amino acid to produce an adduct (thioketone), which cyclizes in acetic acid. Acid hydrolysis cleaves the adduct to produce a new *N*-terminal amino acid ($H_3\overset{\oplus}{N}$-peptide) and an unstable intermediate that isomerizes to a thiohydantoin. The specific thiohydantoin serves to identify the first *N*-terminal amino acid. Repetition of the sequence can identify the next amino acid, since the degradation has converted it to the new *N*-terminal amino acid.

No matter how one attempts it, the chemical identification of the primary structure of even a simple polypeptide is a huge undertaking. In 1958, F. Sanger received the Nobel prize in chemistry for his unravelling of the primary structure of *insulin*—a compound containing "only" fifty-one amino acids (Figure 24–4).

11. A peptide produces the following dipeptides upon hydrolysis: glu-his; asp-glu; phe-val; val-asp. What is the structure of the peptide?

Classical Peptide Synthesis. We have already presented all of the reactions that would be needed to synthesize a peptide. After all, the reactions that we have indicated as being able to convert the carboxy group of an amino acid into the corresponding amide group could just as easily be employed to synthesize a peptide. To illustrate this, the classical approach to peptide synthesis, we have outlined the synthesis of gly-ala in Figure 24–7.

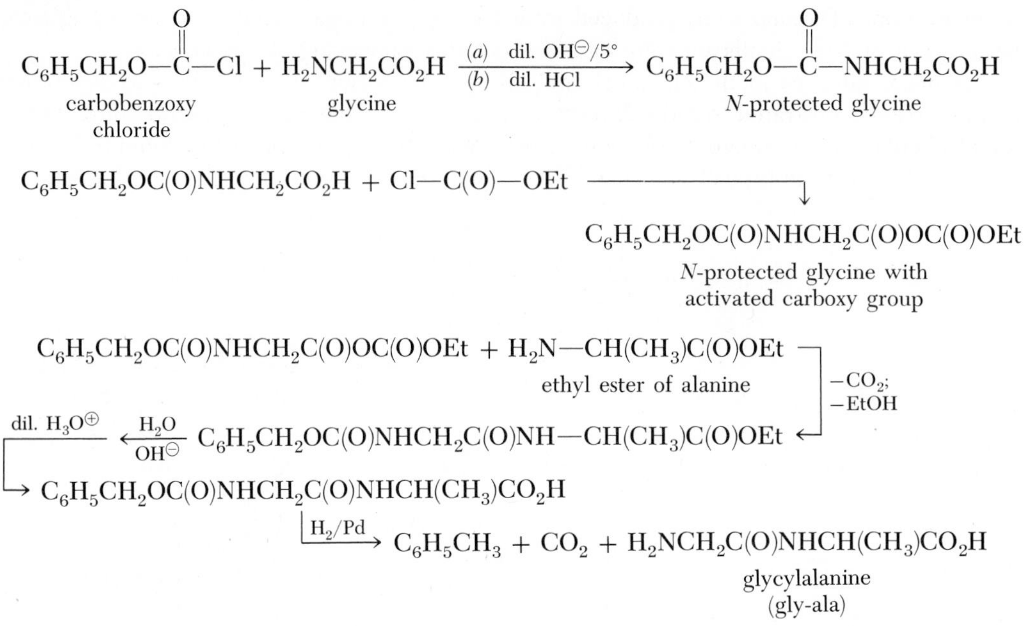

$$C_6H_5CH_2O{-}\overset{\overset{\displaystyle O}{\|}}{C}{-}Cl + H_2NCH_2CO_2H \xrightarrow[(b)\ \text{dil. HCl}]{(a)\ \text{dil. OH}^{\ominus}/5^{\circ}} C_6H_5CH_2O{-}\overset{\overset{\displaystyle O}{\|}}{C}{-}NHCH_2CO_2H$$

carbobenzoxy chloride — glycine — *N*-protected glycine

$$C_6H_5CH_2OC(O)NHCH_2CO_2H + Cl{-}C(O){-}OEt \longrightarrow C_6H_5CH_2OC(O)NHCH_2C(O)OC(O)OEt$$

N-protected glycine with activated carboxy group

$$C_6H_5CH_2OC(O)NHCH_2C(O)OC(O)OEt + H_2N{-}CH(CH_3)C(O)OEt \xrightarrow{-CO_2;\ -EtOH}$$

ethyl ester of alanine

$$C_6H_5CH_2OC(O)NHCH_2C(O)NH{-}CH(CH_3)C(O)OEt \xrightarrow[\text{OH}^{\ominus}]{H_2O} \xrightarrow{\text{dil. } H_3O^{\oplus}} C_6H_5CH_2OC(O)NHCH_2C(O)NHCH(CH_3)CO_2H$$

$$\xrightarrow{H_2/Pd} C_6H_5CH_3 + CO_2 + H_2NCH_2C(O)NHCH(CH_3)CO_2H$$

glycylalanine (gly-ala)

Figure 24-7 A classical synthesis of glycylalanine.

The scheme in Figure 24-7 begins with the protection of glycine, the *N*-terminal amino acid, with carbobenzoxy chloride. After blockage of the amino group, the carboxy group of glycine is activated by reaction with ethyl chloroformate. This activated carboxy group reacts with the methyl ester of alanine to produce the first intermediate containing a gly-ala bond. (The ester of alanine was used in preference to alanine itself in order to minimize side reactions involving the free carboxy group.) With the formation of the gly-ala bond, the ester linkage is hydrolyzed to liberate the carboxy group of the C-terminal amino acid. In the final step of the sequence, the original protecting group is removed by catalytic hydrogenolysis.

12. Outline the classical syntheses of ala-gly and gly-ala-gly.

The synthesis just outlined could be varied in several ways including, for example, substituting the use of dicyclohexylcarbodiimide for the activation with ethyl chloroformate. However, no matter how one attempts to vary the basic scheme, it clearly requires several steps to create a dipeptide. It should be obvious that the synthesis of a peptide containing fifty amino acid residues will require very many steps—and this proliferation of steps is at the root of the major difficulty with this approach to peptide preparation.

Each step in a reaction scheme such as that in Figure 24-7 generates a product that, before being used in a subsequent step, must be isolated and, perhaps, purified. Your exposure to the organic chemistry laboratory has, by now, convinced you of the difficulties often associated with the isolation and purification of organic compounds. Certainly, it is quite *un*common to obtain yields exceeding 90% or 95%. Stringing together a large number of reactions, each one providing a 90% yield, will result in a very small overall yield. For example, a sequence containing one hundred steps, each of which has a 90% yield, will afford an overall yield of $0.90^{100} \times 100\%$ or 0.003%! Thus, the synthesis of even a moderately large peptide may require a huge amount of starting material in order to produce an amount of product that is something more than "visible." A method has been developed for synthesizing peptides without suffering the extreme losses associated with continued isolation and purification. This method, described in the following section, is already of the utmost significance to biochemists.

Solid-Phase Peptide Synthesis. Recent research in peptide synthesis has focused on the solid-phase procedure pioneered by R.B. Merrifield. In this **solid-phase peptide synthesis** (SPPS), the amino acid that will become the *C*-terminal amino acid of the peptide is bound to a resin *via* its carboxy group. Quite often this involves the esterification of the carboxy group, using a chloromethyl group on the resin.

$$H_2N\text{—}CHRCO_2^{\ominus} + ClCH_2\text{—}(\text{Resin}) \xrightarrow{-Cl^{\ominus}} H_2N\text{—}CHR\text{—}\overset{\overset{\displaystyle O}{\|}}{C}\text{—}O\text{—}CH_2\text{—}(\text{Resin})$$

The free amino group then is acylated, to form a peptide bond, by reaction with an amino acid possessing a protected amino group and with dicyclohexylcarbodiimide.

$$(CH_3)_3COC(O)NHCHRCO_2H + H_2NCHRC(O)OCH_2\text{—}(\text{Resin}) + C_6H_{11}\text{—}N{=}C{=}N\text{—}C_6H_{11} \longrightarrow (CH_3)_3CO\text{—}C(O)NHCHRC(O)NHCHRC(O)OCH_2\text{—}(\text{Resin})$$

The *t*-butoxycarbonyl group is removed and, after neutralization, the peptide-containing resin is ready for another acylation.

$$(CH_3)_3COC(O)NHCHRC(O)NHCHRC(O)OCH_2\text{—}(\text{Resin}) \xrightarrow[(b)\ OH^{\ominus}]{(a)\ HCl} H_2NCHRC(O)NHCHRC(O)OCH_2\text{—}(\text{Resin})$$

After the last amino acid residue is added, the polypeptide-resin adduct is treated with a mixture of hydrogen bromide and trifluoroacetic acid, TFA. This cleaves the polypeptide from the resin and also removes the protecting group from the *N*-terminal amino acid.

$$(CH_3)_3COC(O)NHCHRC(O)NHCHRC(O)\sim\!\!/\!\!/\!\!\sim NHCHRC(O)\text{—}OCH_2\text{—}(\text{Resin}) \xrightarrow{HBr,\ TFA}$$

$$H_3\overset{\oplus}{N}CHRC(O)NHCHRC(O)\sim\!\!/\!\!/\!\!\sim NHCHRCO_2H + BrCH_2\text{—}(\text{Resin}) + (CH_3)_2C{=}CH_2 + CO_2$$

The unique advantage of this procedure over more classical methods is that the growing peptide chain never has to be isolated. Instead, the peptide-resin adduct, being extremely insoluble, is washed free of by-products, solvents, excess reagents, and so forth after one reaction and is ready for another without any loss of peptide. This technique has been automated, and programmed amino acid synthesizers now can easily add as many as six amino acids to a growing peptide chain over a 24 hour period. These machines add the reagents in their proper sequence, vary reaction conditions, wait an appropriate time period, wash away by-products, and then begin again. Both insulin and the enzyme *ribonuclease* (126 amino acids) have been prepared using this SPPS method.

One of the very interesting interfaces between polymer chemistry and biochemistry involves the development of new, improved resin supports for solid-phase peptide synthesis.

Our discussion of the synthesis of polypeptides and proteins has been limited to *in vitro* syntheses, that is, to syntheses carried on outside of living systems. In Chapter 26, which introduces us to some chemical aspects of the gene, we will learn about the biosynthesis of proteins and some consequences of improper protein biosynthesis in man and in bacteria.

24.9 THE STRUCTURE OF PROTEINS

The analysis of the "structure" of a protein does not end with the elucidation of its amino acid sequence (its **primary structure**). A protein has a complex "gross anatomy" that is controlled by interactions which are much more varied and numerous than the peptide bonds that make up its primary structure. The occurrence of regions possessing characteristic shapes within a protein implies the existence of an intermediate level of structural organization. The "gross anatomy" of a protein is spoken of as its **tertiary structure,** while this intermediate level of organization is its **secondary structure.**

Secondary Structure of Proteins. Hydrogen bonding plays a major role in dictating the conformation adopted by a polypeptide chain. In its most highly organized conformation, a single polypeptide chain of L-amino acids adopts a helical arrangement. The helix is defined by the spatial arrangement of (a) the carbonyl carbon, (b) the α carbon, and (c) the α-amino nitrogen of the constituent α-amino acids. The most stable of several helical arrangements is a right-handed α-helix, first suggested on theoretical grounds by Pauling and Corey in 1950. Hydrogen bonding in the α-helix occurs between a carbonyl oxygen and an amide hydrogen that are separated by three amino acid residues. The axis of the helix is nearly parallel to these hydrogen bonds (Figure 24–8).

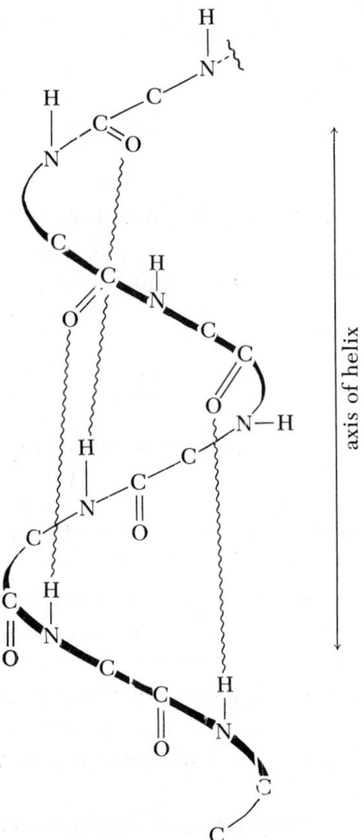

Figure 24-8 Hydrogen bonding in the α-helix. Hydrogen bonds are indicated by wavy lines.

The α-helix, being made up of only one type of repeating unit, *i.e.*, $-\overset{O}{\overset{\|}{C}}-{}_{(S)}\overset{|}{\underset{|}{C}}_{\alpha}-\overset{H}{\overset{|}{N}}-$, is rather constant in its dimensions. One pitch of the helix, that is, the linear distance along the helical axis that separates two similar atoms, is 1.5 Å. The angle between the normal to the helical axis and the plane described by a given amino acid residue is 26°. These dimensions are summarized in Figure 24–9.

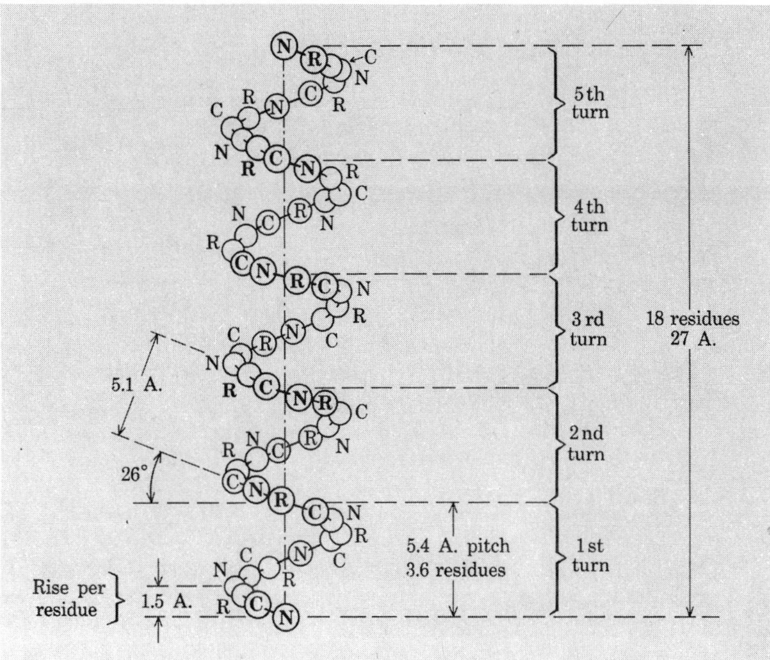

Figure 24–9 Molecular dimensions in the α-helix. (From R.F. Steiner, *Life Chemistry*, Van Nostrand, 1968.)

If the α-helix were the only type of secondary structure found in a protein, all proteins would be rather rigid, rod-like entities. Since they are not, it is to be concluded that only portions of polypeptide chains are α-helices. The reasons for deviations from the α-helical array are complex but include the existence of proline, hydroxyproline, and/or valine along the peptide chain. Proline and hydroxyproline lack an amide hydrogen once a peptide bond is formed and cannot participate in hydrogen bonding in the α-helix. Valine's isopropyl group appears to weaken the α-helix through steric repulsions.

A second organized arrangement of the polypeptide chain is the so-called *pleated sheet.* The pleated sheet is stabilized by hydrogen bonding between extended *adjacent* polypeptide chains and can involve chains that run in the same direction ("parallel pleated sheet") or in opposite directions ("antiparallel pleated sheet"); see Figure 24–10.

Tertiary Structure of Proteins. Proteins are categorized as either *fibrous* or *globular.* For classification purposes, proteins with a length/width ratio greater than 10 are considered fibrous, while those with a ratio less than 10 are globular. Silk fibroin and the β (extended) forms of keratin and synthetic polypeptides are fibrous proteins that are characterized by an almost completely extended polypeptide chain arranged in the pleated sheet array.

Collagen, another fibrous protein, is rich in glycine and proline. Because of this, collagen is incapable of forming either a pleated sheet or an α-helix. Instead, this most important protein is made up of three left-handed helices interwoven into a right-handed "super-helix" (Figure 24–11). Two of the three polypeptide chains which constitute collagen are identical in their primary structure.

A

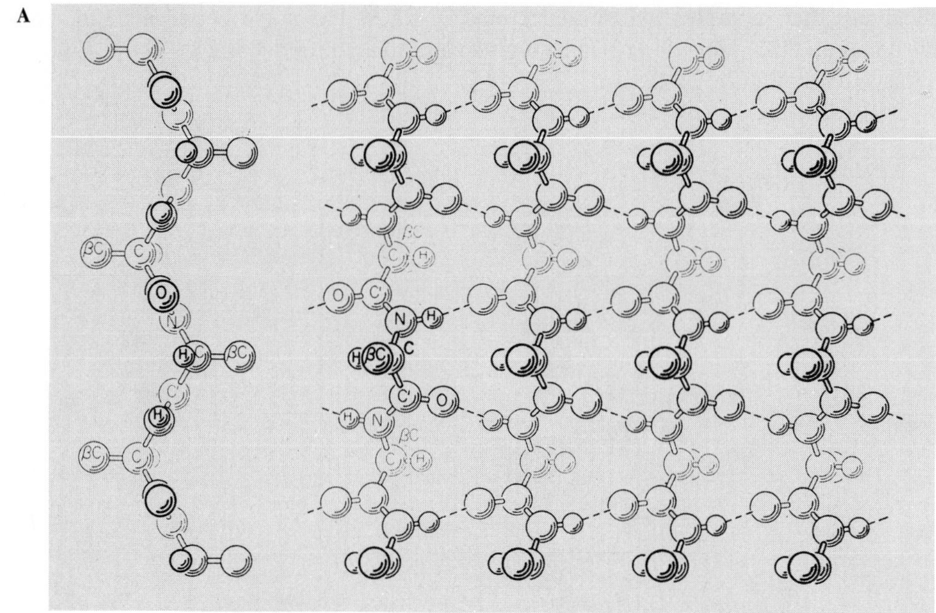

B

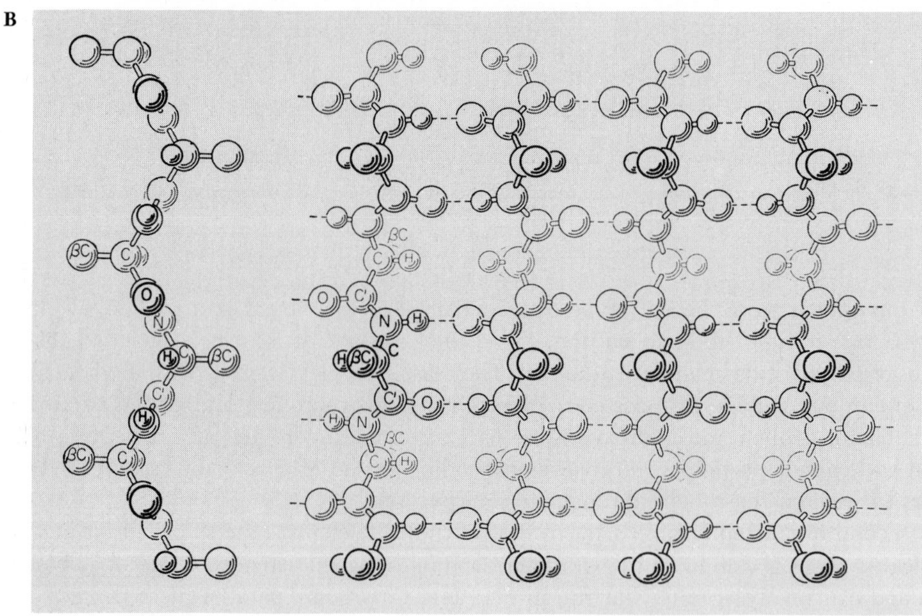

Figure 24-10 The pleated sheet structure of proteins. (A) Parallel pleated sheet. (B) Antiparallel pleated sheet. (From A. Mazur and B. Harrow, *Textbook of Biochemistry*, 10th ed., W.B. Saunders Company, 1971.)

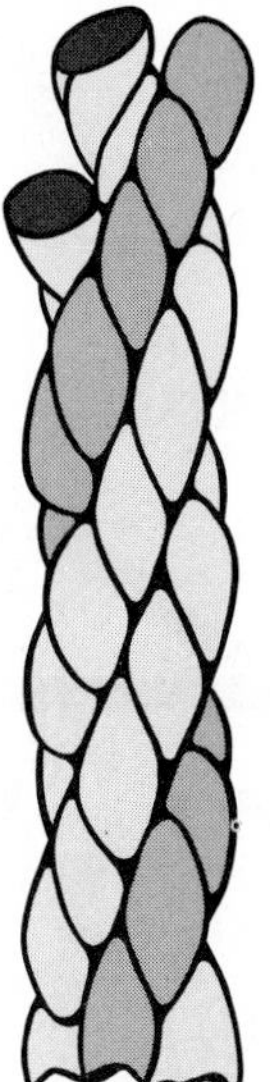

Figure 24-11 The strands of collagen are in a triple helix. Each of three peptide chains is individually coiled in a left-handed helix, and the three together are twisted in a right-handed helix. There are 10 turns in the individual chains per turn of the triple helix. (From R.W. McGilvery, *Biochemistry*, W.B. Saunders Company, 1970.)

Collagen is the predominant protein of vertebrates, representing nearly 50% of the dry weight of cartilage and about 30% of the solid matter in bone. It occurs in biological systems as bundles of linear fibrils, which have a tensile strength close to that of steel wire. Since collagen is so important, it is not surprising that there are a number of important diseases involving it. Perhaps the most famous of these is *scurvy,* caused by a deficiency of vitamin C. This condition stems from the disruption of normal collagen synthesis that occurs when the absence of vitamin C prevents the oxidation of proline to 3- and 4-hydroxyproline. Because the hydroxyprolines occur naturally only in collagen, analysis of tissues for their hydroxyproline content reflects the collagen content of those tissues.

$$\text{L-proline} \xrightarrow{\text{vitamin C}} \text{3-hydroxy-L-proline} + \text{4-hydroxy-L-proline}$$

Most of the regulatory systems in the plant and animal world rely upon those *globular proteins* called "enzymes" for catalysis. The high chemical specificity of enzymes is associated, in part, with the unique gross anatomy of these polymers. An appreciation of the complexity of the overall structure of proteins can be gained by examining the structure of the enzyme *ribonuclease,* shown in Figure 24–12. While hydrogen bonding is the single force responsible for the secondary structure of proteins, all bonds other than the peptide bond and the hydrogen bond between amide groups can be considered responsible for the great convolutions that give globular proteins their tertiary structure. These other bonds include (a) the disulfide bonds in cystine, (b) ionic bonds involving "extra" amino or carboxy groups, (c) hydrogen bonding, and (d) hydrophobic interactions (Figure 24–13).

Hydrophobic interactions arise, in part, from van der Waals attractive interactions between non-polar side chains on amino acid residues. Undoubtedly, they also are due to the greater attraction that water has for itself than for the non-polar organic residues of polypeptides.

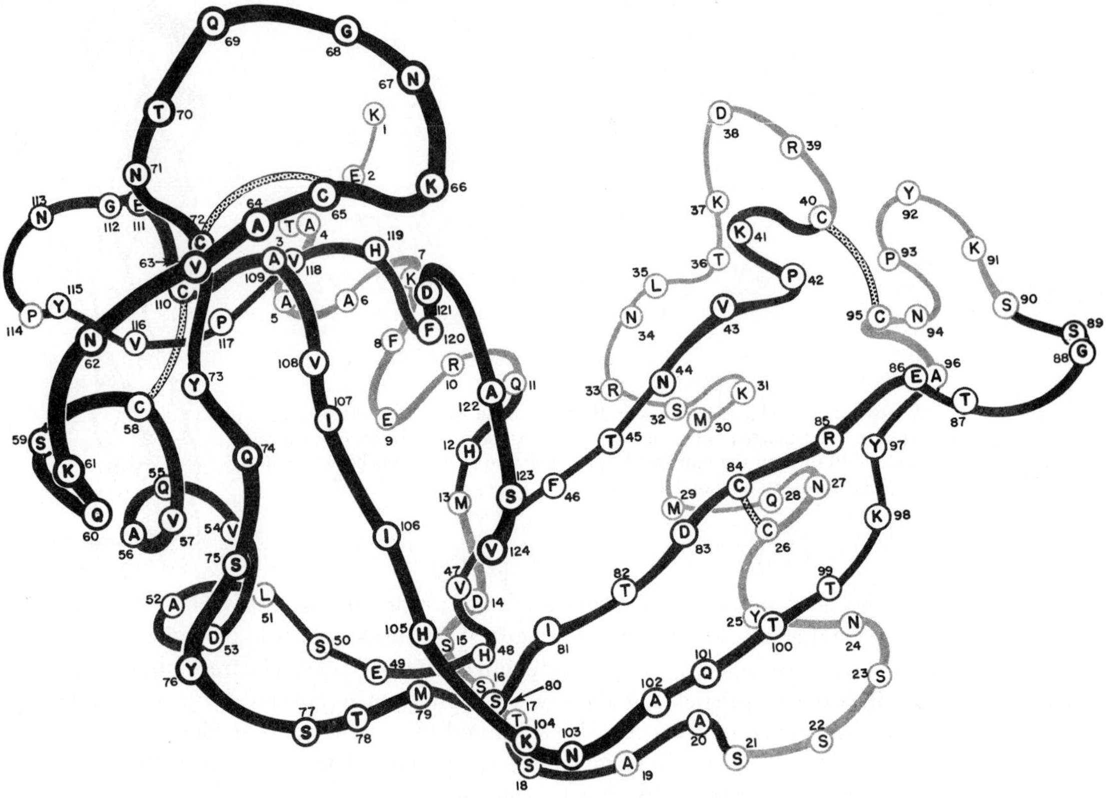

Figure 24-12 A three-dimensional illustration of the carbon atoms of the enzyme ribonuclease, as determined by x-ray crystallography. Three strands form an antiparallel pleated sheet structure (residues 42–48, 75–90, and 95–110), and there are three α-helical regions (residues 2–12, 26–33, and 50–58). (From A. Mazur and B. Harrow, *Textbook of Biochemistry*, 10th ed., W.B. Saunders Company, 1971.)

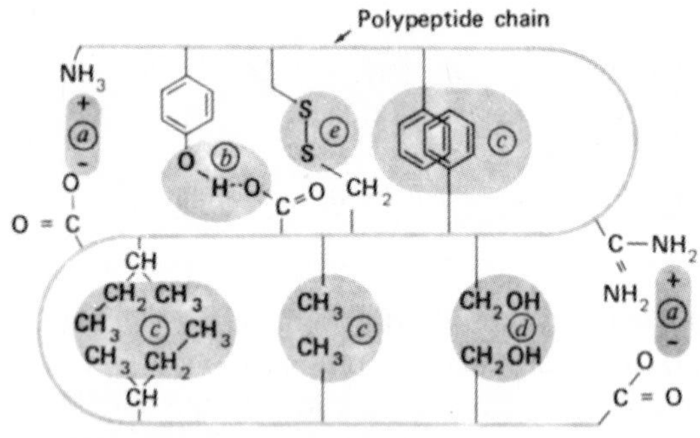

Figure 24-13 Non-covalent bonds stabilizing protein structure: (a) electrostatic interaction; (b) hydrogen bonding; (c) hydrophobic interaction; (d) dipole-dipole interaction; (e) disulfide bridge. (From J.I. Routh, *Introduction to Biochemistry*, W.B. Saunders Company, 1971.)

One might view hydrophobic interactions as a very important illustration of the adage "Like dissolves like."

Quaternary Structure. Some proteins must come together to form a macro-complex ("oligoprotein") consisting of several complete, proteinaceous sub-units before they can exhibit biological activity. In effect, each entire protein is a monomer, the quaternary structure defining the degree of association of such monomers in the biologically active material.

It is sometimes helpful to discuss the quaternary structure from a slightly different point of view. If identical sub-units come together, the macro-complex is said to have a *homogeneous* quaternary structure. If non-identical sub-units come together, the quaternary structure is *heterogeneous*. An interesting example of the latter is the tobacco mosaic virus, which requires RNA units (discussed in Chapter 26) complexed with protein units in order to exhibit viral activity. The virus may be viewed as a "super-molecule" made up of about 2,200 polypeptide chains!

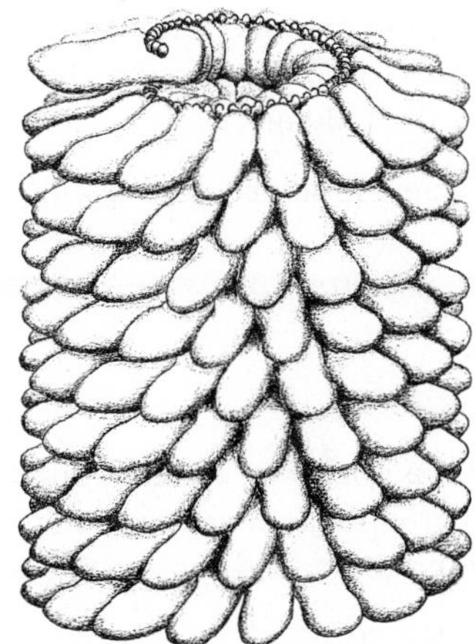

Tobacco mosaic virus. (From C.R. Goodheart, *Introduction to Virology*, W.B. Saunders Company, 1969.)

DENATURATION. The gross anatomy of a protein represents a very delicate balance among various attractive and repulsive forces between the biopolymer and its aqueous environment. Anything that upsets this balance may destroy all but the primary structure of the polypeptide. Such a loss of structural detail is called **denaturation;** depending on the extent of decomposition and the particular protein involved, denaturation may be either reversible or irreversible. A classic example of the latter is the coagulation of egg white during the cooking of an egg, the egg albumin (a protein) being denatured thermally.

Denaturation can be achieved by (a) an increase in temperature, (b) variation in pH of the medium, (c) addition of oxidizing or reducing agents (which destroy —S—S— bonds), (d) contact with detergents (which disturb hydrophobic interactions between water and the protein), (e) addition of strong hydrogen bond acceptors (*e.g.*, urea), or (f) physical abuse (*e.g.*, ultrasonic vibration).

This list should be considered a collection of "what-not-to-do's," since there is much more reason at present to study proteins in their *native state* than their denatured counterparts. It is fortunate that globular proteins are crystallizable in the native state (albeit with difficulty) while denatured proteins are non-crystalline, for single crystal X-ray analysis has been the tool used to deduce most of what we know about the secondary and tertiary structures of proteins.

IMPORTANT TERMS

α-Helix: A conformation of a peptide in which the repeating $-\overset{|}{\underset{|}{C_\alpha}}-\overset{O}{\overset{\|}{C}}-\underset{H}{\underset{|}{N}}-$ units form a helix.

In the α-helix every C=O and N—H residue in the helix is involved in intramolecular hydrogen bonding. This extensive hydrogen bonding makes this a very stable conformation for peptides.

Amino acid: In the most often used sense, a compound possessing an amino group on the carbon α to a carboxy group; $R{-}CHNH_2CO_2H$. In a more general but less often used sense, an amino acid is any compound possessing both an amine group and an acidic residue (*e.g.*, $-CO_2H$ or $-SO_3H$).

Collagen: A fibrous protein that is the major component of connective tissue. It consists of three peptide chains, of which two are identical and the third is different in amino acid composition. Each chain is a left-handed helix, the three helices being wound about each other in a right-handed "super-helix" of shallow pitch. Upon boiling in water, collagen is converted to gelatin.

C-Terminal amino acid: The amino acid in a polypeptide chain that (a) occurs at one end of the chain and (b) has a free carboxy group. The amino group of the *C*-terminal amino acid binds it to the polymer.

D-Amino acid: An amino acid having the following configuration at the α carbon:

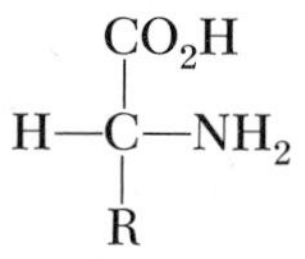

$$\begin{array}{c} CO_2H \\ | \\ H-C-NH_2 \\ | \\ R \end{array}$$

Deamination: Loss of an amino group from an amino acid. In living systems the carbon bearing the amino group may undergo either of the deamination routes shown below. The first is the *non-oxidative* route, and the second is the *oxidative* route.

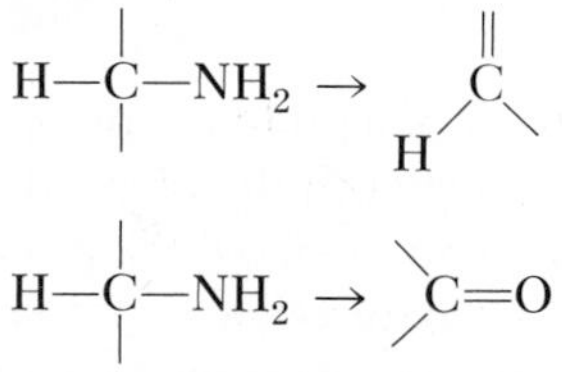

$$\begin{array}{c} | \\ H-C-NH_2 \\ | \end{array} \rightarrow \begin{array}{c} \| \\ C \\ H \diagup \quad \diagdown \end{array}$$

$$\begin{array}{c} | \\ H-C-NH_2 \\ | \end{array} \rightarrow \;\rangle C{=}O$$

Denaturation: The alteration of the conformation of a protein from that found in the native (natural) state. Such changes are often accompanied by loss of the biological function of that protein (*e.g.*, enzyme activity). Denaturation may, or may not, be a reversible process. Coagulation of egg white (albumin) when the egg is heated is an example of an irreversible denaturation.

Edman degradation: A method for removing (and identifying) *N*-terminal amino acids from proteins. It involves the reaction of the protein with phenyl isothiocyanate, ultimately producing a phenylthiohydantoin. The reaction involves a first-formed *N*-phenylthiocarbamyl derivative of the protein (called "thioketone" in Figure 24–6) which is cleaved, in acid, to produce an unstable thiazolone derivative (the substance in brackets in Figure 24–6). This derivative isomerizes to the phenylthiohydantoin.

Electrophoresis: An analytical procedure for analyzing amino acid or protein mixtures that depends upon the differing mobilities of the components of the mixture in an electric field (D.C.).

Fibrous protein: A protein that is thread-like in its dimensions. The protein in hair (keratin) is fibrous.

Globular protein: A relatively spherical protein (compared to fibrous proteins). Usually having lower molecular weights than fibrous proteins, most globular proteins can be crystallized from solution. Because of this potential crystallinity, they are suitable for precise X-ray diffraction analysis. Enzymes are globular proteins.

Gluten: The residue formed when flour is washed free of starch.

Gout: A disease whose most apparent symptoms are painful inflammations of joints, particularly of the hands and feet.

Isoelectric point: The pH at which the concentration of the zwitterionic form of an amino acid is maximized. For an amino acid containing only one carboxy and one amino group, it is the numeric average of the pK_a of $—CO_2H$ and $—NH_3^{\oplus}$ for that amino acid.

L-Amino acid: An amino acid having the following configuration. Most naturally occurring amino acids have the L configuration.

$$\begin{array}{c} CO_2H \\ | \\ H_2N—C—H \\ | \\ R \end{array}$$

***N*-Terminal amino acid:** The amino acid in a polypeptide chain that (a) occurs at one end of the chain and (b) has a free amino group. The carboxy group of the *N*-terminal amino acid binds it to the polymer.

Peptide: A general term used to describe polyamides made from α-amino acids. Depending upon the number of amino acids, the terms "dipeptide," "tripeptide,". . . , "polypeptide" may be used.

Peptide bond: The amide linkage between two amino acids. The four atoms usually considered in discussions of the peptide bond (—C(O)—NH—) reside in one plane in a fully extended peptide chain.

Proteins: High molecular weight polypeptides possessing, as a class, a host of biological functions. Some authors distinguish between "polypeptide" and "protein" on the basis of the natural occurrence and biological activity of the latter.

Sanger's reagent: 2,4-Dinitrofluorobenzene. Used in the identification of the *N*-terminal amino acid in a peptide.

Strecker synthesis: Synthesis of α-amino acids by the reaction of an aldehyde with ammonium chloride and sodium cyanide. An α-amino nitrile is an intermediate in the reaction.

Zwitterion: A charge-separated form of an otherwise covalent compound.

$$\begin{array}{ccc} H & & H \\ | & & | \\ H_3\overset{\oplus}{N}—C—CO_2^{\ominus} & \rightleftarrows & H_2N—C—CO_2H \\ | & & | \\ CH_3 & & CH_3 \end{array}$$

alanine (zwitterion) — alanine (covalent form)

PROBLEMS

13. Using starting materials available from either the Aldrich Chemical Company or the Eastman Kodak Company, outline reasonable syntheses for the following:
    - (a) valine
    - (b) alanine
    - (c) phenylalanine
    - (d) lysine
    - (e) serine
    - (f) proline
    - (g) aspartic acid
14. For each mixture of the following amino acids, suggest the order of elution from an ion exchange resin as the pH of the eluent increases.
    - (a) alanine, glycine, phenylalanine, leucine
    - (b) leucine, isoleucine, serine, proline
    - (c) asparagine, aspartic acid, histidine, lysine
    - (d) threonine, cysteine, arginine, methionine
    - (e) What types of acids are eluted early? Late? Explain.

15. Bradykinin is a physiologically active nonapeptide. It produces a lowering of blood pressure, increases capillary permeability, and causes pain. It has the sequence arg-pro-pro-gly-phe-ser-pro-phe-arg. What tripeptides containing phe would form upon incomplete hydrolysis of this peptide? What tripeptides containing phe would be formed if pro(3) and phe(5) were exchanged in bradykinin? How could these two nonapeptides be distinguished from arg-phe-pro-ser-phe-gly-pro-pro-arg?

16. Suggest a means of achieving the following interconversions:
   (a) isocaproic acid to leucine
   (b) isobutyraldehyde to valine
   (c) nitroethane to alanine
   (d) *sec*-butyl bromide to isoleucine

17. Amino acids that possess nucleophilic centers other than the α-amino group must have them protected before these amino acids can be used in a peptide synthesis. If not protected, these groups might compete with the α-amino group as shown below.

A: $C_6H_5CH_2O{-}\overset{O}{\overset{\|}{C}}{-}NHCHR{-}\overset{O}{\overset{\|}{C}}{-}O{-}\overset{O}{\overset{\|}{C}}{-}OC_2H_5 + \underset{\text{cysteine}}{HSCH_2CH(NH_2)CO_2H}$

$$\downarrow$$

$$\underset{\text{undesired product}}{C_6H_5CH_2O{-}\overset{O}{\overset{\|}{C}}{-}NHCHR{-}\overset{O}{\overset{\|}{C}}{-}SCH_2CH(NH_2)CO_2H} + CO_2 + C_2H_5OH$$

B: $\text{(Resin)}{-}CH_2O{-}\overset{O}{\overset{\|}{C}}{-}CH(NH_2)CH_2SH + (CH_3)_3CO{-}\overset{O}{\overset{\|}{C}}{-}NHCH_2CO_2H$

$$\downarrow \text{DCC}$$

$$\underset{\text{undesired product}}{\text{(Resin)}{-}CH_2O{-}\overset{O}{\overset{\|}{C}}{-}CH(NH_2)CH_2S{-}\overset{O}{\overset{\|}{C}}{-}CH_2NH{-}\overset{O}{\overset{\|}{C}}{-}OC(CH_3)_3}$$

Protecting groups should also be removable without disturbing the structure of the peptide.

When synthesizing peptides containing cysteine, the thiol group can be protected by treating cysteine hydrochloride with triphenylmethyl chloride. (a) Suggest a structure for the protected cysteine and a mechanism for its formation.

The amino group is not protected under these conditions. (b) Suggest a reason for this.

The protected cysteine residue can be converted back to a thiol by reaction with silver nitrate followed by reaction of the resultant silver mercaptide with hydrochloric acid. (c) Provide equations for the regeneration of cysteine from its S-protected derivative.

18. Refer to problem 17 for a discussion of the need for "protecting" certain amino acids. Dihydropyran can be used to protect the hydroxyl group of serine. The protecting group is readily hydrolyzed off in dilute acid. (a) Suggest a mechanism for the derivatization, which is pictured below. (b) Why are such derivatives relatively stable to base-catalyzed hydrolysis?

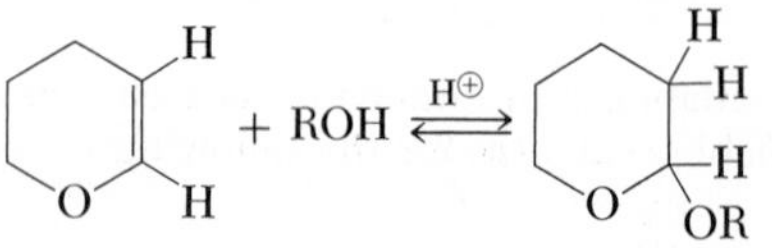

$$\text{dihydropyran (ring O; H, H on C=C)} + ROH \overset{H^{\oplus}}{\rightleftharpoons} \text{tetrahydropyranyl ether (ring O; H, H, H; OR)}$$

19. (a) Both guanidines and amidines are very strong organic bases, much stronger than simple amines. Account for this enhanced basicity.

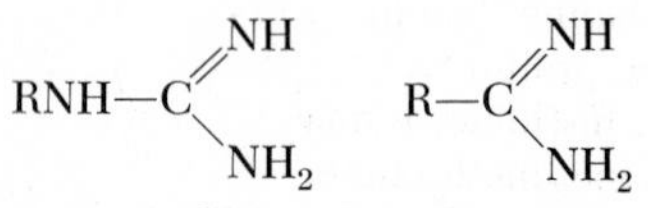

$$\underset{\text{a guanidine}}{RNH{-}C(=NH){-}NH_2} \qquad \underset{\text{an amidine}}{R{-}C(=NH){-}NH_2}$$

(b) While amidines usually act as bases, *extremely* potent bases could conceivably make them behave as acids. If an amidine were to act as an acid, what would be the structure of the resultant anion?

(c) Suggest the "proper" structure for the zwitterionic form of L-arginine; show 3-D characteristics where appropriate.

(d) Arginine must have its guanidino group, —NH—C(=NH)—NH_2, protected during peptide syntheses. This can be done by nitration (see below). The protecting group can be removed by catalytic hydrogenolysis. How do you account for the guanidino group's ability to undergo facile electrophilic substitution?

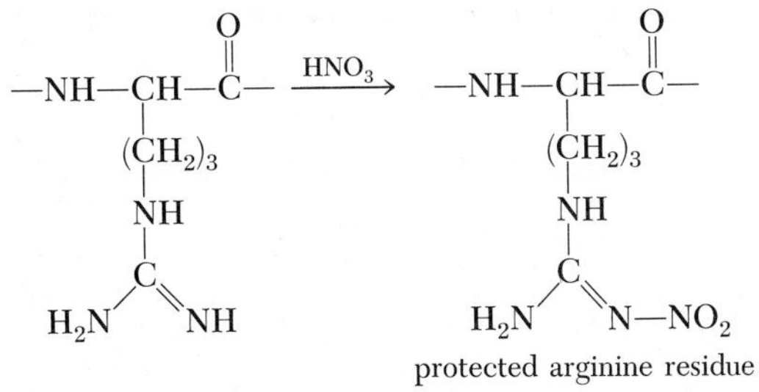

—NH—CH(—$(CH_2)_3$—NH—C(=NH)—NH_2)—C(=O)— $\xrightarrow{HNO_3}$ —NH—CH(—$(CH_2)_3$—NH—C(=N—NO_2)—NH_2)—C(=O)—

protected arginine residue

20. The reagent 1-cyclohexyl-3-(2-morpholinoethyl)carbodiimide, **A**, has one advantage over DCC as a condensing agent in peptide synthesis: the urea produced from **A** is soluble in dilute acid and can be washed away from the peptide. (a) What are the structures of the urea derivatives of 1-cyclohexyl-3-(2-morpholinoethyl)carbodiimide and of DCC? (b) What accounts for the difference in solubility in dilute hydrochloric acid? (c) Is **A** capable of exhibiting optical activity? Explain.

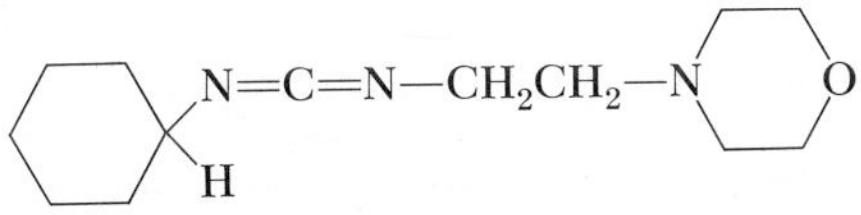

Cyclohexyl(H)—N=C=N—CH_2CH_2—N(morpholine ring, O)

A

21. A critical feature in the determination of the structure of pituitary human growth hormone, HGH, involved the selective cleavage of peptide bonds involving methionine. The sequence that was employed is represented below. Provide mechanisms for these transformations.

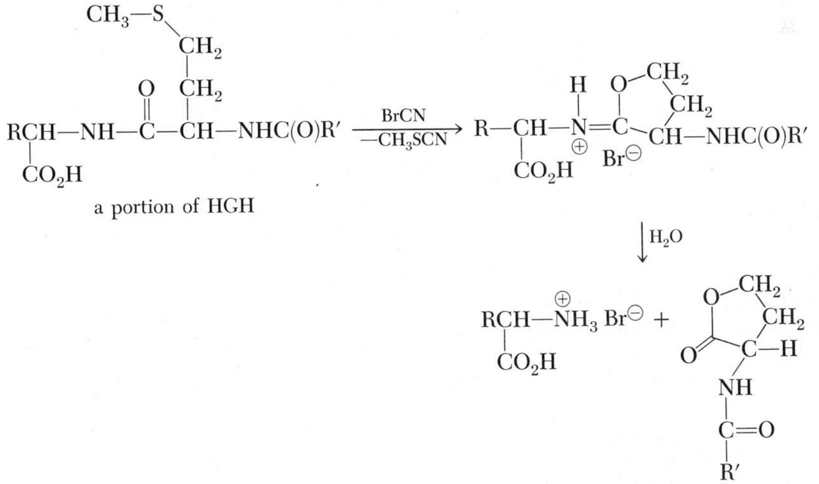

RCH(CO_2H)—NH—C(=O)—CH(CH_2CH_2—S—CH_3)—NHC(O)R′ (a portion of HGH) $\xrightarrow[-CH_3SCN]{BrCN}$ R—CH(CO_2H)—N(H)$^{\oplus}$=C(—O—CH_2—CH_2—)CH—NHC(O)R′ $Br^{\ominus}$ (five-membered ring)

$\downarrow H_2O$

RCH(CO_2H)—$\overset{\oplus}{N}H_3$ $Br^{\ominus}$ + lactone: O—CH_2—CH_2—C(H)(NH—C(=O)—R′)—C(=O)—O (ring)

22. DTNB, 5,5′-dithio*bis*-(2-nitrobenzoic acid), is a reagent that has been used to assay for thiols, RSH.
(a) Suggest the product of the reaction of this, the **Ellman reagent**, with cysteine.

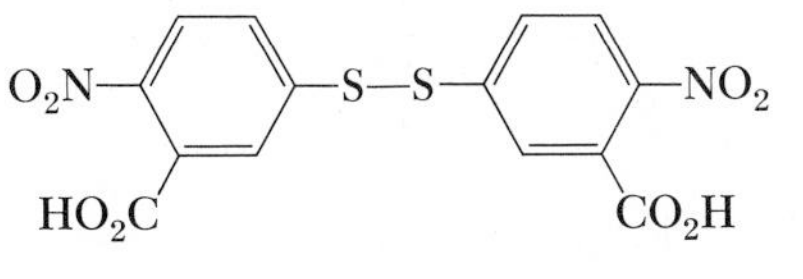

O_2N—C_6H_3(CO_2H)—S—S—C_6H_3(CO_2H)—NO_2

Part of the determination of the number of disulfide linkages in a protein requires (1) reduction of the disulfide linkages to thiols, (2) derivatization of the thiol groups present with DTNB, (3) hydrolysis of the peptide linkages, and (4) determination of the amount of DTNB derivative produced. (b) What other determination must be carried out on the protein in order to have this procedure accurately reflect the number of disulfide linkages in the protein? (c) How might this be accomplished?

23. "Even though no 'racemization' of any chiral atoms has occurred, the ord curve of a globular protein in the native state will differ from the ord curve of the corresponding denatured protein." Explain this in terms of the primary and secondary structure of proteins. (Note the quotation marks around *racemization*. When a given dissymmetric atom in a molecule containing many dissymmetric atoms undergoes configurational inversion, the product is *not* an enantiomer of the starting material; it is a diastereomer.)

24. On the basis of the following data, deduce the structure of ornithine and identify compounds **A** through **D**.

$$CH_3\overset{\overset{\displaystyle O}{\|}}{C}NHCH(CO_2C_2H_5)_2 \xrightarrow[(b)\ CH_2{=}CHCN]{(a)\ \overset{\oplus\ominus}{Na\,OC_2H_5}} \mathbf{A}(C_{12}H_{18}N_2O_5)$$

$$\mathbf{A} \xrightarrow[Ni]{H_2} \mathbf{B}(C_{12}H_{22}N_2O_5)$$

$$\mathbf{B} \xrightarrow[\text{warming}]{\text{gentle}} \mathbf{C}(C_{10}H_{16}N_2O_4)$$

$$\mathbf{C} \xrightarrow[\text{heat}]{HCl(aq.)} \mathbf{D}(C_5H_{14}N_2O_2Cl_2)$$

ornithine dihydrochloride

Ornithine reacts with $H_2N{-}CN$, in basic solution, to produce arginine. Provide a mechanism for this conversion.

25. Heating glycine, an α-amino acid, produces diketopiperazine, $C_4H_6N_2O_2$. However, heating 3-aminobutanoic acid, a β-amino acid, produces **A** ($C_4H_6O_2$). Heating 4-aminopentanoic acid leads to **B**, a lactam, C_5H_9NO. Suggest structures for diketopiperazine and for **A** and **B**. What is a lactam? (*Hint:* These amino acids behave like the corresponding hydroxy acids.)

26. Dansyl chloride (1-dimethylaminonaphthalene-5-sulfonyl chloride) is highly fluorescent, and it produces derivatives of amino acids that can be detected and quantitated in very small amounts. Suggest a structure for the dansyl derivative of glycine.

$N(CH_3)_2$

SO_2Cl

dansyl chloride

Dansyl chloride can be used to label *N*-terminal amino acids by treating a peptide with the reagent and then hydrolyzing the peptide bonds with dilute acid. Why does the dansyl derivative of the *N*-terminal amino acid survive this hydrolysis?

27. The α-helix described by Pauling and Corey is a "right-handed" helix. Why would the "left-handed" helical arrangement of a group of L-amino acids be expected to be different from the "right-handed" helical arrangement of the same sequence of those same amino acids?

28. How, if at all, could the "syndiotactic-isotactic-atactic" terminology be applied to polypeptides containing L-amino acids?

29. The dipeptide met-gly reacts with hydrogen peroxide to form two compounds in unequal amounts. Upon cleavage of the peptide bond, this mixture produces three substances, one of which is glycine. The other two compounds are both optically active but have different rotations. (a) Account for these observations.

(b) How would you expect cys-gly to behave in the presence of one equivalent of hydrogen peroxide?

CARBOHYDRATES—MONOSACCHARIDES

25A.1 INTRODUCTION

Carbohydrates are natural products that usually have the general formula $C_n(H_2O)_n$. This very large family includes monomers, called **simple sugars** or **monosaccharides,** as well as extremely complex polymers of these simple sugars. Included among such **polysaccharides** are *starch, cellulose, dextrin,* and *glycogen.*

The plant kingdom is Earth's richest source of carbohydrates, with nearly three-quarters of the dry weight of most plants ascribable to polysaccharides. Very large numbers invariably defy conceptualization, and "number dropping" loses its impact when astronomical figures are discussed. Nonetheless, it is worth knowing that the plants on this planet produce about 10^{11} (that's 100,000,000,000) tons of cellulose per year. Sunlight, via photosynthesis, converts carbon dioxide and water into carbohydrates, making carbohydrates the major storehouse of the sun's energy on this Earth of ours. In addition to the gross structural support provided to plants by cellulose, carbohydrates are a source of energy for all living systems, are important in cell wall formation, are involved in a number of cellular components responsible for growth and function, and provide man with several important chemotherapeutic materials.

Our treatment of carbohydrates is divided into two parts. In this first part (Chapter 25A) we examine the classification, nomenclature, structure, and reactivity of the monosaccharides. Because of its biological importance, and because many of its reactions can be extrapolated to other monosaccharides, we will use glucose ("dextrose") as a representative monosaccharide in many of these discussions. Chapter 25B begins with a discussion of disaccharides and weaves its way into a treatment of some very complex carbohydrates. During this sojourn we will encounter some subjects that, at first glance, might appear to be unlikely neighbors in one chapter. However, all of them (*e.g.*, billiard balls, crab shells, and guncotton) are related in some way to the chemistry of polysaccharides.

25A.2 CLASSIFICATION AND NOMENCLATURE OF MONOSACCHARIDES

Carbohydrates have been known for many years and, consequently, every carbohydrate has its own non-systematic name. Unfortunately, these names tell us little about the structures of the compounds. (This facet of carbohydrate nomenclature is similar to that of amino acid nomenclature.) In order to assist chemists in thinking about the properties of carbohydrates, these compounds may be placed into sometimes-overlapping categories that are based upon the functional groups that they possess (or lack) and upon their stereochemistry. It will help, in reading this chapter, to think of a name and a formula and their relationship to the various kinds of carbohydrates that are discussed.

Aldoses and Ketoses. Monosaccharides usually have trivial names that end in "ose." Since monosaccharides are polyhydroxy aldehydes or polyhydroxy ketones, or their derivatives, they are divided into **aldoses** (**ald**ehyde + **ose**) and **ketoses** (**ket**one + **ose**). For example, glucose (shown below) is an aldose, but fructose (also shown) is a ketose.

$$\begin{array}{ccc} {}_1CHO & \quad & {}_1CH_2OH \\ H-{}_2C-OH & & {}_2C{=}O \\ HO-{}_3C-H & & HO-{}_3C-H \\ H-{}_4C-OH & & H-{}_4C-OH \\ H-{}_5C-OH & & H-{}_5C-OH \\ {}_6CH_2OH & & {}_6CH_2OH \end{array}$$

glucose (an aldose) — fructose (a ketose)

Fischer projections of linear monosaccharides are drawn with the carbonyl group at or near the top of the structure. When written in this way, the uppermost carbon atom is numbered 1. Thus, the carboxaldehyde carbon (*i.e.*, —**CHO**) of an aldose is written at the top of the Fischer projection and is numbered C1. (This was followed in drawing glucose and fructose, above.)

Simple sugars also can be categorized according to the number of skeletal carbons that they contain. Thus, since they both contain six carbons, glucose and fructose are both **hexoses.** These two types of classification can be combined, making glucose an *aldohexose* and fructose a *ketohexose*. The simplest carbohydrates, glyceraldehyde and dihydroxyacetone, are an aldotriose and a ketotriose, respectively.

$$\begin{array}{ccc} CHO & \quad & CH_2OH \\ H-C-OH & & C{=}O \\ CH_2OH & & CH_2OH \end{array}$$

glyceraldehyde (an aldotriose) — dihydroxyacetone (a ketotriose)

D and L Monosaccharides. Emil Fischer, a great German carbohydrate chemist, arbitrarily assigned the configurations shown below to (+)- and (−)-glyceraldehyde, calling the former "D" and the latter "L". Fortunately, it was shown much later that D- and L-glyceraldehyde do, indeed, have the configurations that were assigned by Fischer. Thus, D- and L-glyceraldehyde may now be described as (**R**)- and (**S**)-glyceraldehyde, respectively.

$$\begin{array}{ccc} CHO & \quad & CHO \\ H-C-OH & & HO-C-H \\ CH_2OH & & CH_2OH \end{array}$$

(+)-D-glyceraldehyde — (−)-L-glyceraldehyde

One can begin to create a series of aldoses by starting with one enantiomer of glyceraldehyde and inserting a —CHOH— moiety between the carboxaldehyde group and the chiral carbon of glyceraldehyde. This creates two diastereomeric tetroses, both of which have the **R** configuration at C3 but which differ in configuration at C2. Since these were both derived from D-glyceraldehyde, they both are considered to be D tetroses. Beginning with L-glyceraldehyde would produce two L tetroses, both of which have the **S** configuration at C3 but which differ in configuration at C2.

CHO
H—C—OH
CH_2OH
D-glyceraldehyde (a D triose)

insert one —CHOH— →

$\overset{1}{C}HO$
H—$\overset{2}{C}$—OH
H—$\overset{3}{C}$—OH
$\overset{4}{C}H_2OH$
D-erythrose

+

$\overset{1}{C}HO$
HO—$\overset{2}{C}$—H
H—$\overset{3}{C}$—OH
$\overset{4}{C}H_2OH$
D-threose

D tetroses

CHO
HO—C—H
CH_2OH
L-glyceraldehyde (an L triose)

insert one —CHOH— →

CHO
H—C—OH
HO—C—H
CH_2OH
L-threose

+

CHO
HO—C—H
HO—C—H
CH_2OH
L-erythrose

L tetroses

The construction of the series of D and L sugars is continued by inserting additional —CHOH— units between the carboxaldehyde group and the adjacent —CHOH— group. This procedure has been used to construct the series of D sugars shown in Figure 25A-1 on page 804. Note that most of the sugars found in nature are members of the D series.

The simplest ketose, dihydroxyacetone, lacks a chiral center. However, insertion of a —CHOH— moiety between two carbons creates erythrulose, the D isomer of which serves as the basis for the D ketose sugar series. The construction of this series is shown in Figure 25A-2.

CH_2OH
C=O
HCOH
CH_2OH
D-erythrulose

CH_2OH
C=O
HCOH
HCOH
CH_2OH
D-ribulose

CH_2OH
C=O
HOCH
HCOH
CH_2OH
D-xylulose

CH_2OH
C=O
HCOH
HCOH
HCOH
CH_2OH
D-psicose

CH_2OH
C=O
HOCH
HCOH
HCOH
CH_2OH
D-fructose

CH_2OH
C=O
HCOH
HOCH
HCOH
CH_2OH
D-sorbose

CH_2OH
C=O
HOCH
HOCH
HCOH
CH_2OH
D-tagatose

Figure 25A-2 The D-ketose series.

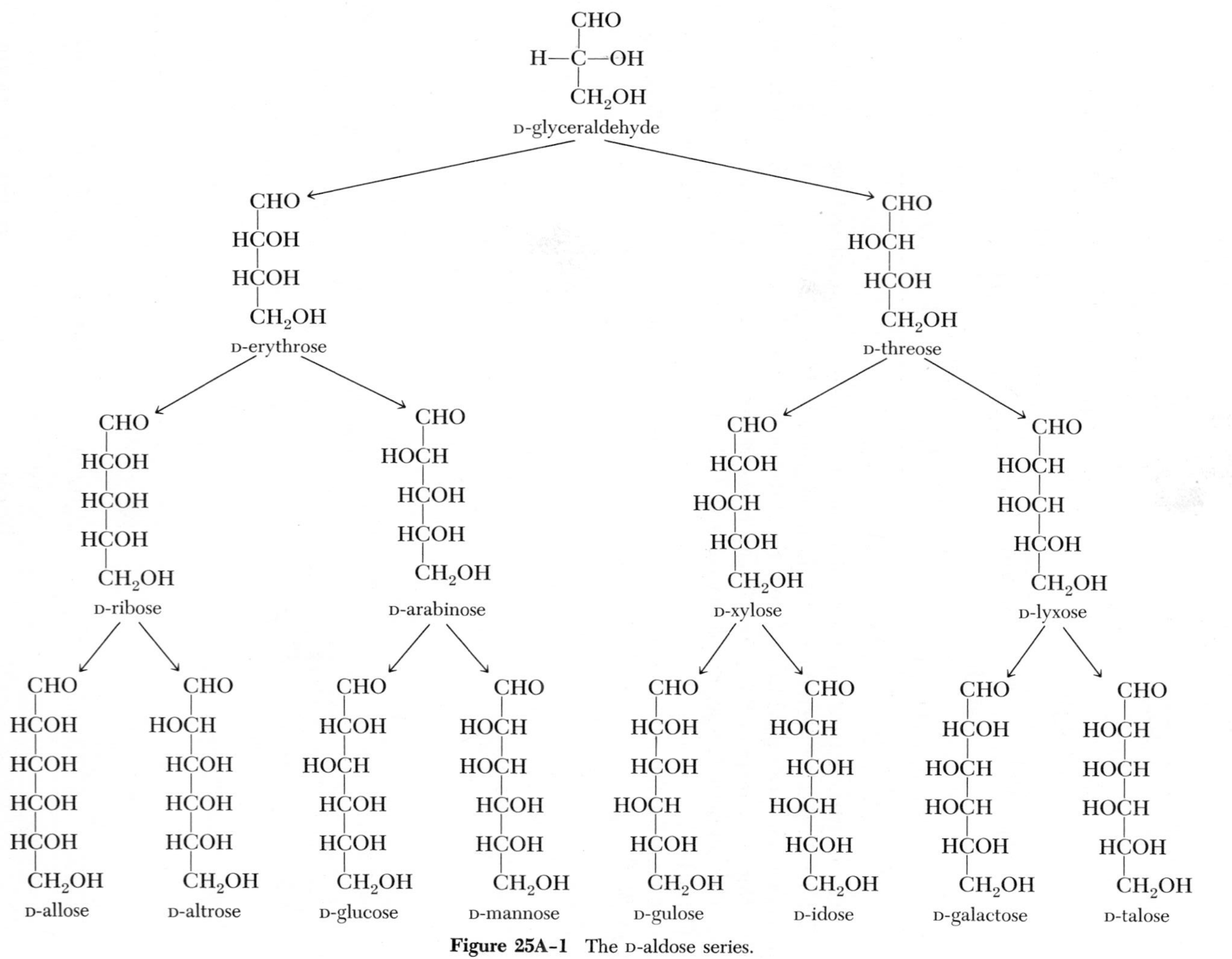

Figure 25A-1 The D-aldose series.

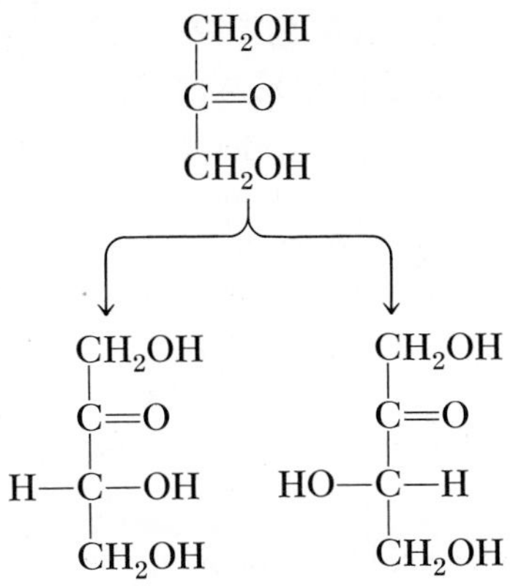

$$\begin{array}{c} CH_2OH \\ | \\ C{=}O \\ | \\ CH_2OH \end{array}$$

$$\begin{array}{cc} \begin{array}{c} CH_2OH \\ | \\ C{=}O \\ | \\ H{-}C{-}OH \\ | \\ CH_2OH \end{array} & \begin{array}{c} CH_2OH \\ | \\ C{=}O \\ | \\ HO{-}C{-}H \\ | \\ CH_2OH \end{array} \end{array}$$

D-erythrulose L-erythrulose

1. Discuss the appropriateness of considering formaldehyde to be the simplest carbohydrate.
2. Identify each of the following as an aldose or ketose and as a triose, tetrose, etc.
   (a) idose (e) galactose
   (b) talose (f) ribose
   (c) erythrulose (g) ribulose
   (d) glucose

Deoxy and Amino Sugars. While most sugars are either polyhydroxy aldehydes or polyhydroxy ketones, there are some that have at least one —OH group replaced by some other substituent. The two most common groups that replace —OH in these monosaccharides are —H and —NH_2. **Deoxy sugars** have a —CH_2— group in place of a —CHOH— group. The most important deoxy sugar, 2-deoxy-D-ribose, is an integral part of deoxyribonucleic acid, DNA (Chapter 26).

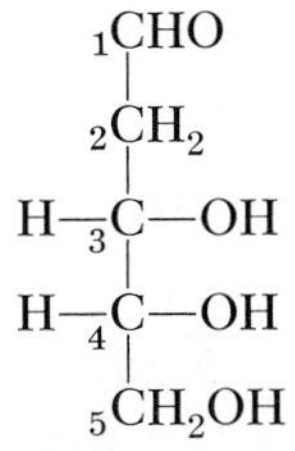

$$\begin{array}{c} {}_1CHO \\ | \\ {}_2CH_2 \\ | \\ H{-}{}_3C{-}OH \\ | \\ H{-}{}_4C{-}OH \\ | \\ {}_5CH_2OH \end{array}$$

2-deoxy-D-ribose

Amino sugars have a —$CHNH_2$— group in place of a —CHOH— group. D-Glucosamine and D-galactosamine, the 2-amino analogs of D-glucose and D-galactose, are among the more important sugars. Amino sugars may also occur in nature as some amino derivative, *e.g.*, as an acetamide.

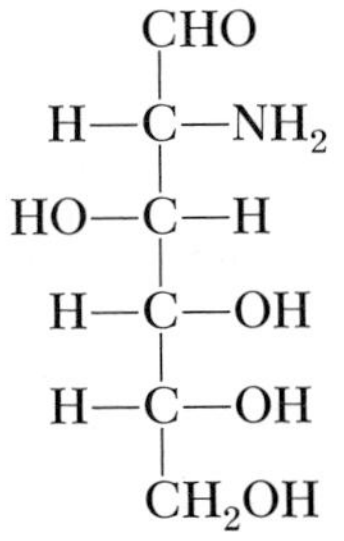

$$\begin{array}{ccc} \begin{array}{c} CHO \\ | \\ H{-}C{-}NH_2 \\ | \\ HO{-}C{-}H \\ | \\ H{-}C{-}OH \\ | \\ H{-}C{-}OH \\ | \\ CH_2OH \end{array} & \begin{array}{c} CHO \\ | \\ H{-}C{-}NH_2 \\ | \\ HO{-}C{-}H \\ | \\ HO{-}C{-}H \\ | \\ H{-}C{-}OH \\ | \\ CH_2OH \end{array} & \begin{array}{c} CHO \\ | \\ H{-}C{-}NHC(O)CH_3 \\ | \\ HO{-}C{-}H \\ | \\ H{-}C{-}OH \\ | \\ H{-}C{-}OH \\ | \\ CH_2OH \end{array} \end{array}$$

D-glucosamine (2-amino-2-deoxy-D-glucose) 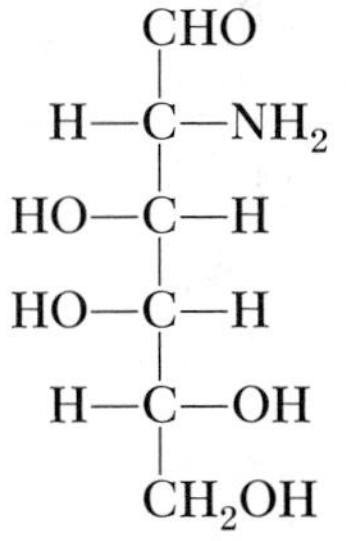 D-galactosamine (2-amino-2-deoxy-D-galactose) 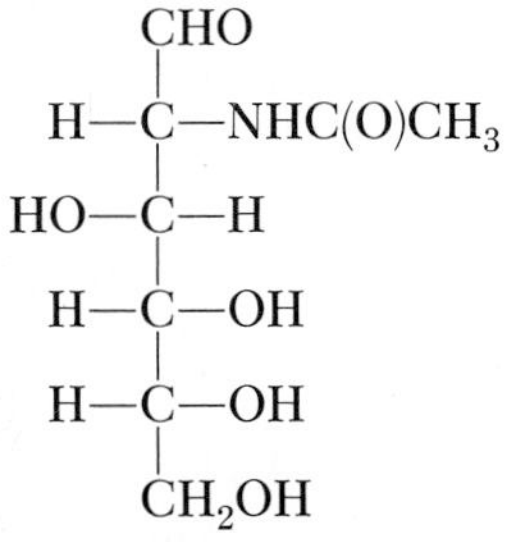 2-acetamido-2-deoxy-D-glucose

25A.3 RING STRUCTURE OF MONOSACCHARIDES

The spectral properties of monosaccharides such as glucose are inconsistent with the presence of a significant amount of a free carbonyl group. However, these same monosaccharides often react as if they were α-hydroxy carbonyl compounds. It is now clear that glucose and other sugars exist as cyclic hemiacetals or hemiketals formed by intramolecular reaction of a hydroxy group with a carbonyl group (see pp. 472 to 476).

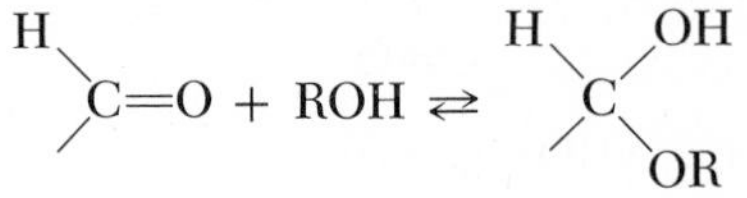

$$\mathrm{H(\!-\!)C{=}O + ROH \rightleftarrows H(\!-\!)C(OH)(OR)}$$

generalized hemiacetal formation

Glucose forms its most stable acetal by employing the hydroxy group at C5. The resultant six-membered ring is called the **pyranose** form of glucose because of its formal resemblance to tetrahydropyran. To emphasize its cyclic nature, this form of D-glucose is called D-*glucopyranose.*

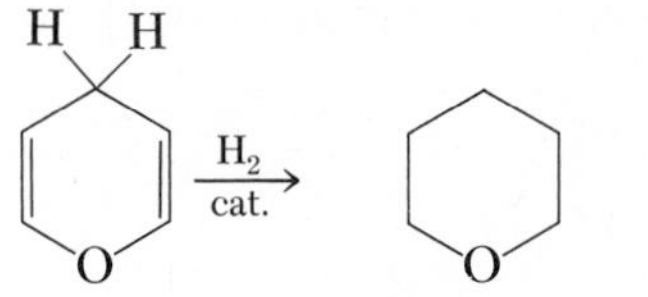

γ-pyran $\xrightarrow[\text{cat.}]{H_2}$ tetrahydropyran

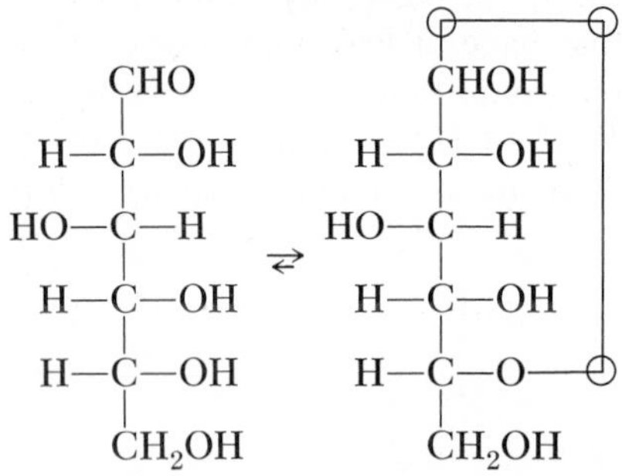

D-glucose		D-glucopyranose
CHO		CHOH
H—C—OH		H—C—OH
HO—C—H	⇄	HO—C—H
H—C—OH		H—C—OH
H—C—OH		H—C—O—
CH_2OH		CH_2OH

no atoms present where circles are located

Hemiacetal formation creates a new chiral center, usually called the **anomeric carbon** or **anomeric center,** which in turn results in two diastereomeric products differing in configuration only at C1. These anomers are designated α or β, depending upon the relative configuration at C1 and at the carbon used to establish the sugar as D or L. In the Fischer representation of α-D-glucopyranose, the C1—OH group (the hemiacetal hydroxy group) is on the same side of the carbon skeleton as is the oxygen bonded to the carbon that labels the sugar as D or L. (In glucose, this carbon is C5.) In β-D-glucopyranose, these two groups are on opposite sides of the skeleton.

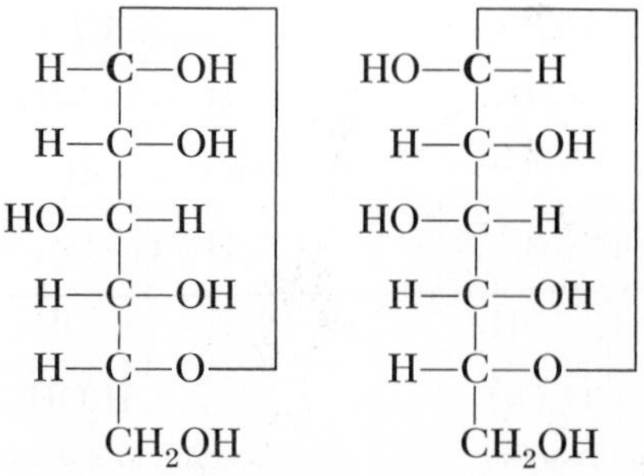

α-D-glucopyranose	β-D-glucopyranose
H—**C**—OH	HO—**C**—H
H—C—OH	H—C—OH
HO—C—H	HO—C—H
H—C—OH	H—C—OH
H—C—O—	H—C—O—
CH_2OH	CH_2OH

anomeric carbon in boldface

Fischer projection

3. Draw a Fischer projection of each of the following.
   (a) D-glucose (c) α-L-glucopyranose
   (b) L-glucose (d) β-L-glucopyranose

A representation of the pyranose form that does not create the unusual, odd-appearing bond found in the Fischer projection is the **Haworth structure.** Haworth structures of the pyranose sugars are constructed around an idealized, planar tetrahydropyran ring.

α-D-glucopyranose β-D-glucopyranose

Haworth structures of glucose anomers

In Haworth structures drawn with the heterocyclic oxygen in the upper right corner, the α form has the hydroxy group on C1 pointing "down," while the β form has the same group pointing "up."

One might ask why glucose does not preferentially form a pair of anomeric, cyclic hemiacetals using the hydroxy group on C4, rather than the one on C5. (This, of course, would produce five-membered ring systems, commonly termed **furanose** forms.) The reason is that a substituted five-membered ring, with its angle strain and eclipsing interactions, is less stable than is a substituted six-membered ring. Therefore, equilibrium favors the pyranose form at the expense of the furanose form.

furanose form of D-glucose ⇄ open chain form of D-glucose ⇄ pyranose form of D-glucose

Are there any monosaccharides that form substantial quantities of cyclic five-membered hemiacetals (or hemiketals)? The answer is "yes," but even for these compounds the pyranose form usually is more stable than is the furanose form. Sugars like threose and erythrose, because of the limited number of carbons, can form only a furanose ring.

L-threose ⇄ α- and β-furanose form ≡ α-L-threofuranose + β-L-threofuranose

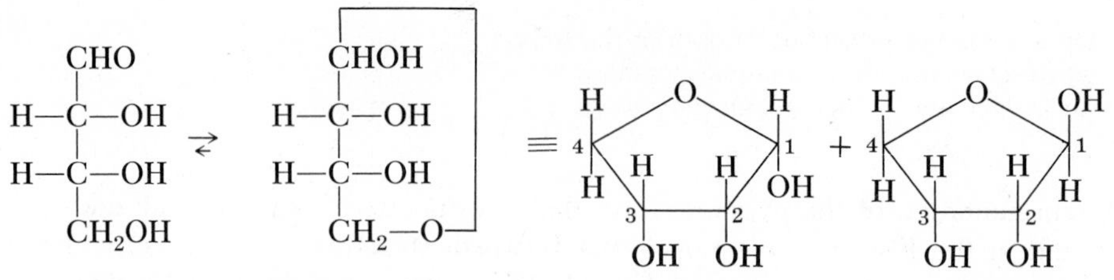

D-erythrose ⇄ α- and β-furanose form ≡ α-D-erythrofuranose + β-D-erythrofuranose

If the aldohexoses exist in pyranose forms but the aldotetroses (must) exist in furanose forms, what about aldopentoses? To begin with, aldopentoses, like tetroses and hexoses, exist as cyclic hemiacetals. Like hexoses, they form both furanose and pyranose forms but exist largely as the more stable pyranose. Ribose, for example, contains less than 25% of the furanose form in solution.

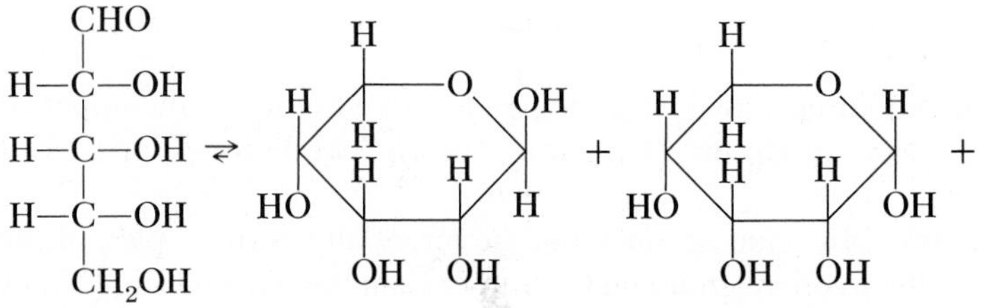

D-ribose ⇄ β-D-ribopyranose (56%) + α-D-ribopyranose (20%) +

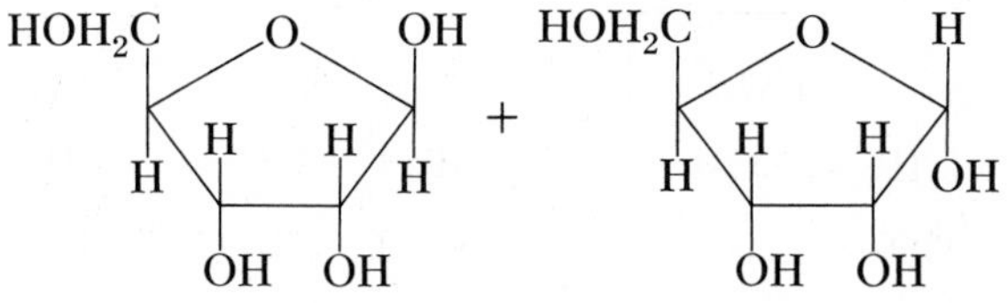

β-D-ribofuranose (18%) + α-D-ribofuranose (6%)

Actually, the amount of furanose form produced by pentoses can be much less than 25%. Arabinose, for example, contains less than 3% of the furanose form at equilibrium.

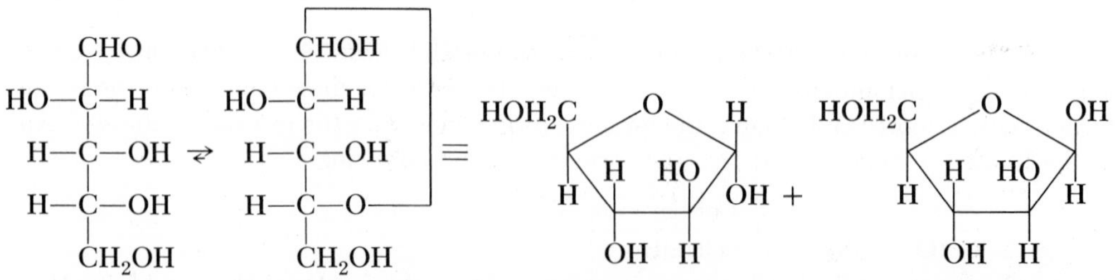

D-arabinose ⇄ α- and β-furanose form ≡ α-D-arabinofuranose + β-D-arabinofuranose

Like aldopentoses, 2-ketohexoses also may exist in furanose forms. The most famous example of this is fructose, whose β-fructofuranose form will be encountered in Chapter 25B in our discussion of the disaccharide sucrose.

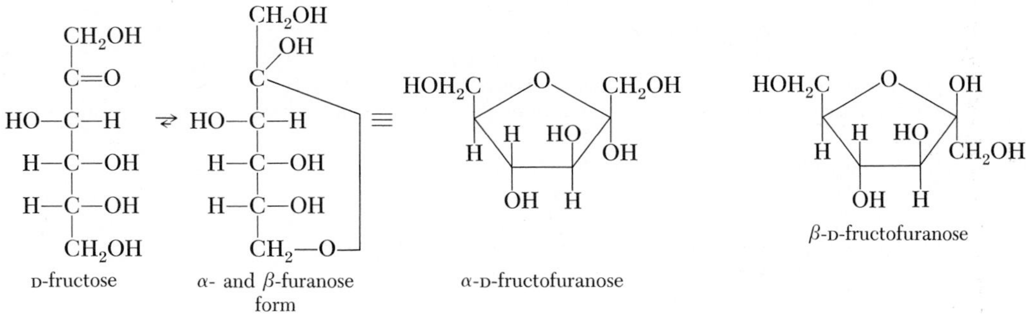

D-fructose ⇌ α- and β-furanose form ≡ α-D-fructofuranose; β-D-fructofuranose

In the examples above, we have drawn the cyclic forms of monosaccharides as both Fischer projections and Haworth structures. In Figure 25A–3 we see how one can convert a given Fischer projection of a linear monosaccharide directly into a Haworth structure.

A: D-arabinose (Fischer projection) — turn structure 90° — rotate C3–C4 bond 120° — use C4–OH to form hemiacetal — α-D-arabinofuranose + β-D-arabinofuranose

B: D-glucose — turn structure 90° — rotate C4–C5 bond 120° — use C5–OH to form hemiacetal — α-D-glucopyranose + β-D-glucopyranose

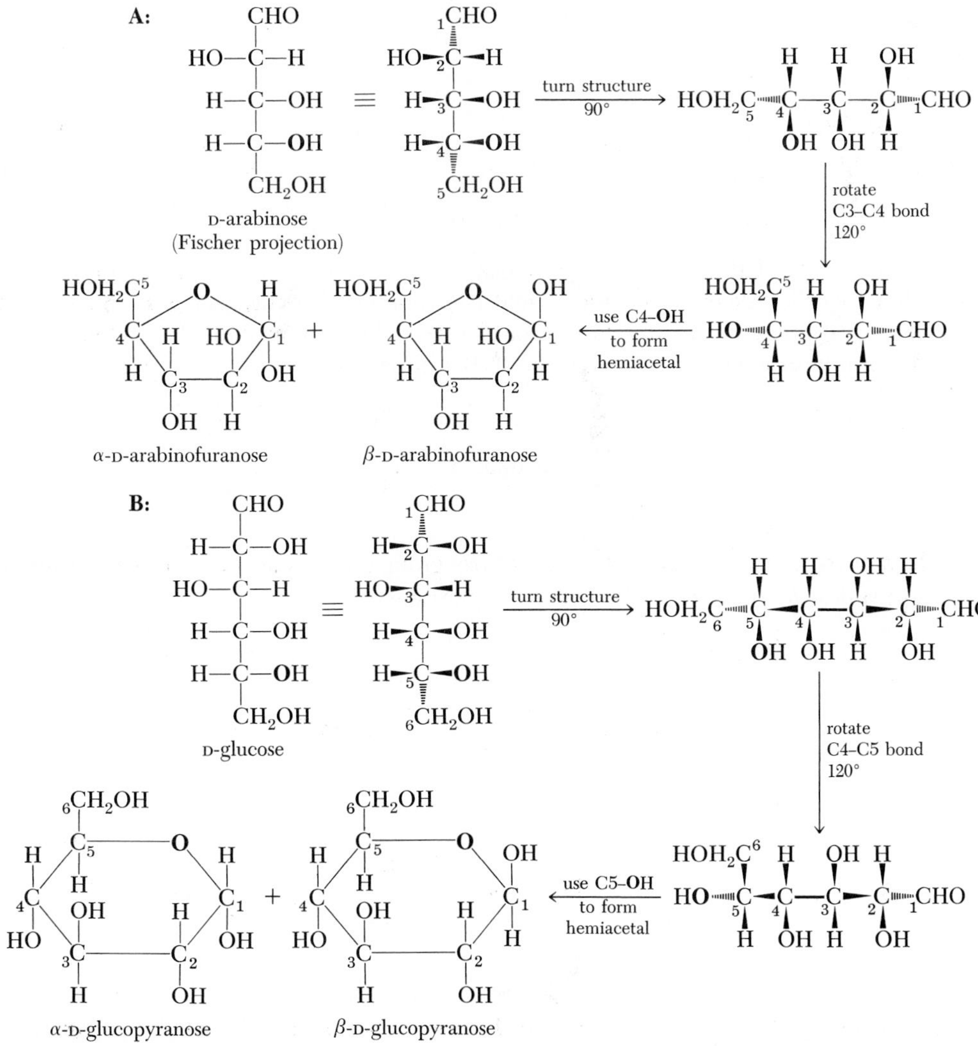

Figure 25A–3 Sequence **A** describes the conversion of an aldopentose from the Fischer projection to a Haworth structure; sequence **B** treats an aldohexose similarly.

4. Draw a Haworth representation of each of the following.
   (a) α-D-glucopyranose (c) α-D-arabinopyranose
   (b) α-L-glucopyranose (d) β-D-arabinopyranose
5. Which, if any, of the following are enantiomers? Use Haworth structures of each to support your answers.
   (a) α-D-glucopyranose and β-D-glucopyranose
   (b) α-L-glucopyranose and β-L-glucopyranose
   (c) α-L-glucopyranose and β-D-glucopyranose
   (d) α-L-glucopyranose and α-D-glucopyranose
   (e) α-L-fructofuranose and β-L-fructofuranose

Haworth structures were devised before the influence of conformation upon reactivity was fully appreciated. An even better representation of the pyranose form of a sugar is one that is based upon the chair form of the six-membered ring. D-Glucose, whose more stable chair form is shown below, is the most abundant of the monosaccharides and the basic unit in most polysaccharides. This certainly reflects the fact that only glucose can exist with all of its ring substituents in the equatorial position.

60° rotation around axis

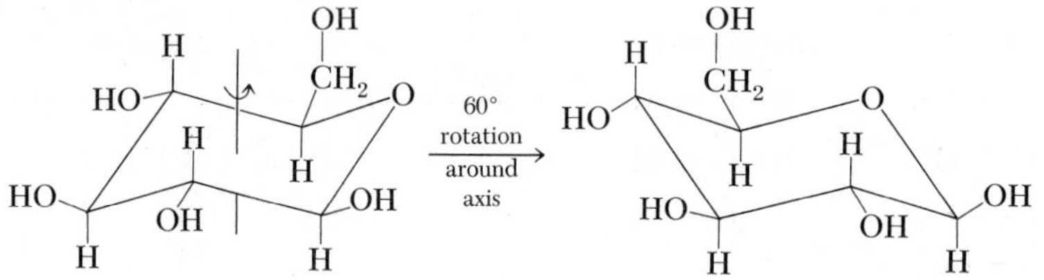

chair representations of β-D-glucopyranose

The conversion of the Haworth to the chair representation of a pyranose form begins with the heterocyclic oxygen in the upper right-hand corner of the hexagon that characterizes the Haworth projection; the hemiacetal carbon (C1) is on the extreme right. This is then related to a chair form with its heterocyclic oxygen in the upper right.

becomes

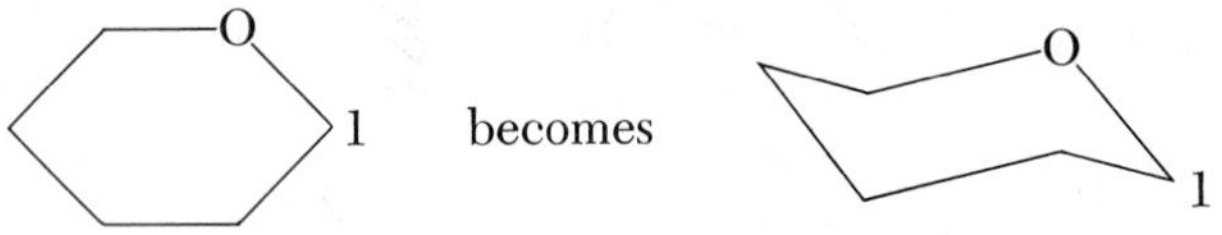

If the monosaccharide is an α form, the hydroxy group on C1 is drawn in the *axial* position; the β form is drawn with OH in the *equatorial* position.

becomes

α

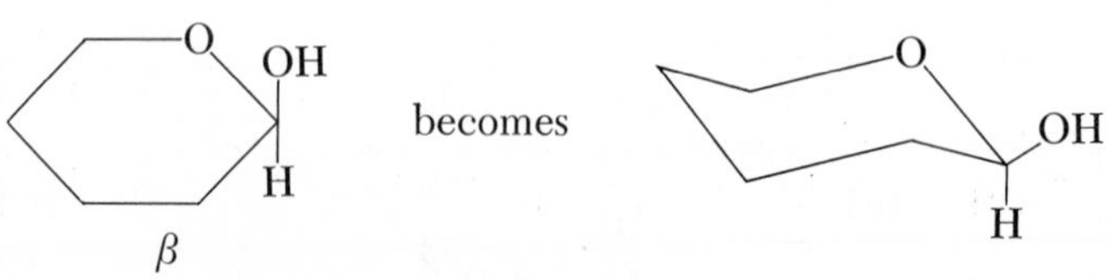

becomes

β

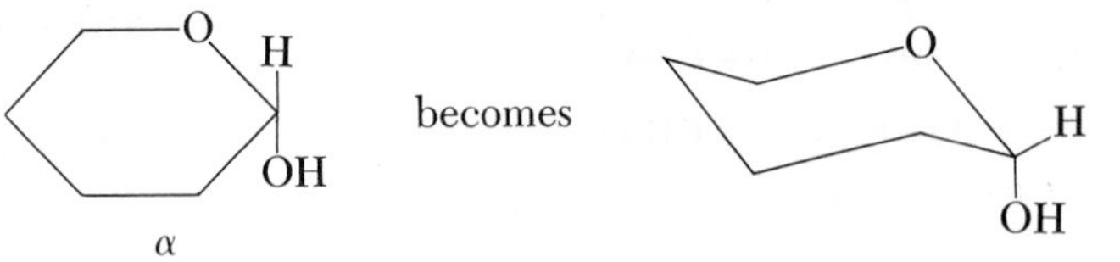

The remaining substituents in the Haworth structure are identified as being either *cis* or *trans* to the C1—OH group and are then drawn either *axial* or *equatorial* (as required) in the chair form.

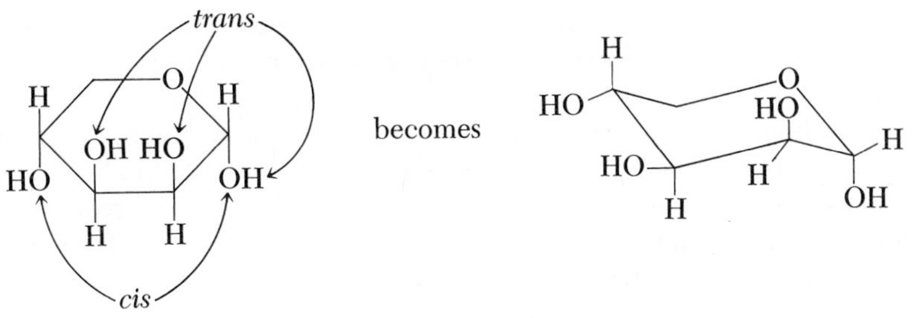

becomes

The near-planarity of cyclopentane permits adequate representation of furanose forms by Haworth structures.

6. Draw each pyranose form in problem 5 in a chair form.

Mutarotation. D-Glucose can be crystallized in a form with a specific rotation of +112°. A second form of D-glucose, $[\alpha] + 19°$, can be obtained under different crystallization conditions. When dissolved in water, either form slowly changes its optical rotation, a process called **mutarotation,** to +53°. The "112° glucose" is the α anomer, while the "19° glucose" is the β anomer. Mutarotation stems from the conversion of either anomer to an equilibrium mixture of both ("53° glucose"). The composition of this equilibrium mixture is approximately 64% β, 36% α, and a trace (0.02%) of the free carbonyl form. Equilibration of these anomers occurs *via* the open chain aldehyde.

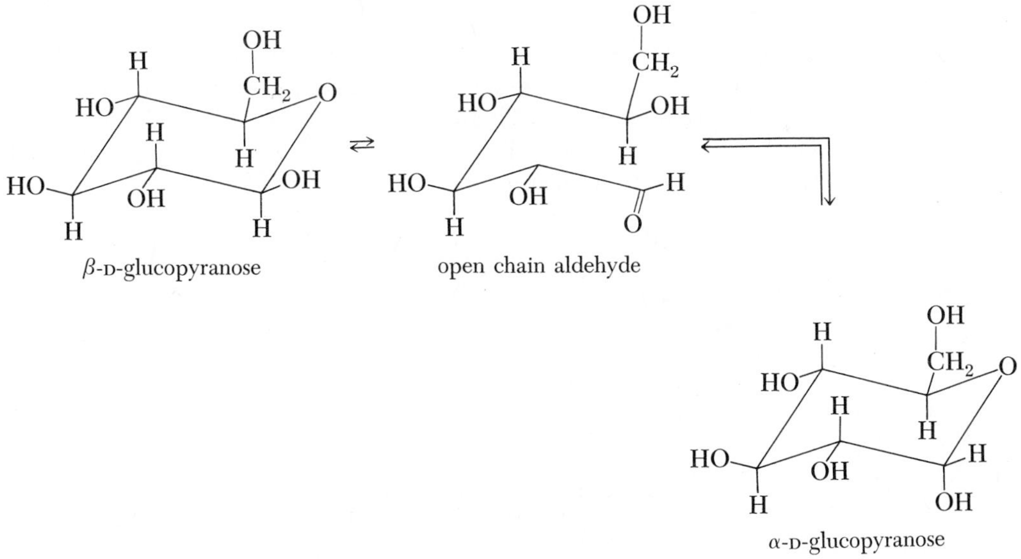

β-D-glucopyranose ⇄ open chain aldehyde

α-D-glucopyranose

All saccharides that contain hemiacetal or hemiketal linkages will exhibit anomer equilibration ("anomerization") in solution, each proceeding through its particular aldehyde or ketone. However, if a very large polysaccharide contains only very few hemiacetal linkages (see polysaccharide discussion in Chapter 25B), then even though equilibration occurs it will be beyond the limits of detection and "mutarotation" will not be observed.

7. Why should you not expect α- and β-D-glucopyranose to be present in equimolar amounts at equilibrium?

Anomer vs. Epimer. Before moving on in this discussion of carbohydrates, it will help us to be sure that we understand the difference between two related terms: **epimer** and **anomer.** If a compound contains n different chiral centers, it can exist as 2^n stereoisomers. Those stereoisomers which are not enantiomers must, of course, be diastereomers. The terms epimer and anomer both describe a pair of stereoisomers, possessing several chiral centers, that differ in configuration at only one of these centers. Thus, each member of a pair of epimers or a pair of anomers is a diastereomer of the other member of that pair. How, then, do they differ? To qualify as epimers, two diastereomers must differ in configuration at one, and only one, of the compounds' several chiral centers. Anomers must meet the same condition and, in addition, the carbon that differs in configuration must be the anomeric carbon. An anomeric pair, from a slightly different point of view, is just a special type of epimeric pair. By way of illustration, D-glucosamine and D-galactosamine are epimers because they differ in configuration at C4. α-D-Glucosamine and β-D-glucosamine are epimers also, *but are more accurately described as anomers* because they differ in configuration at C1, the anomeric carbon.

8. Although D-idose and D-talose differ in configuration at only one carbon, and even though both of them exhibit mutarotation, they are not interconverted on dissolution in water. Explain.
9. Although it is the configuration at C5 that makes D-glucose "D", inversion of the configuration at C5 does not convert D-glucose to L-glucose. Why? What does such an interconversion of configuration create?

25A.4 GLYCOSIDES

Simple hemiacetals or hemiketals react with alcohols to produce acetals or ketals, respectively (see p. 473). Aldoses and ketoses also form acetals and ketals, these both being called **glycosides.** The term *glucosides* refers specifically to acetals of glucose. Glycosides, like hemiacetals and hemiketals, exist in α and β forms. By analogy with the corresponding hemiacetals and hemiketals, cyclic glycosides are termed *pyranosides* or *furanosides.*

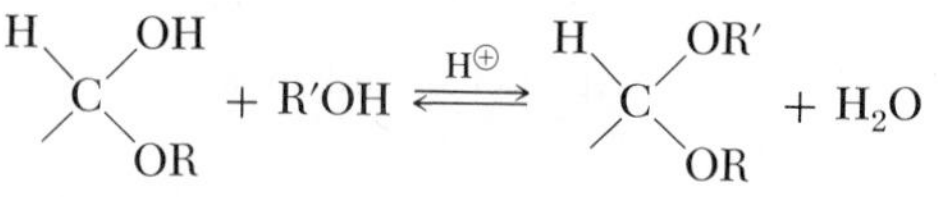
H
OH
C
OR
+ R′OH
H⊕
H
OR′
C
OR
+ H₂O

hemiacetal acetal

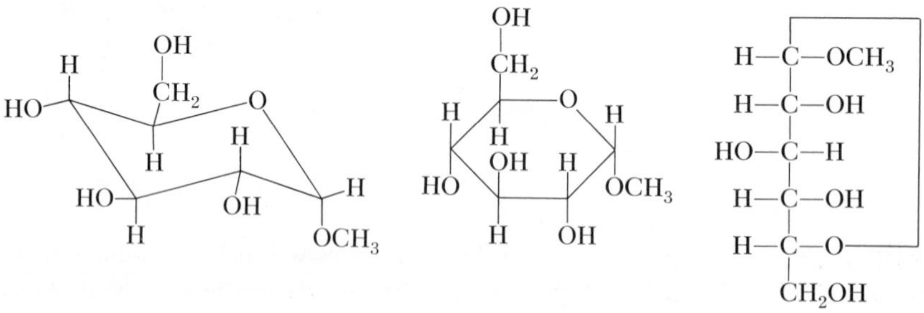
H
OH
CH₂
O
HO
H
H
HO
OH
H
H
OCH₃
OH
CH₂
O
H
H
H
OH
H
HO
OCH₃
H
OH
H—C—OCH₃
H—C—OH
HO—C—H
H—C—OH
H—C—O
CH₂OH

equivalent views of methyl α-D-glucopyranoside

The most common synthetic glycosides are mixed acetals or ketals made by the reaction of monosaccharides with alcohols in the presence of acid. The unit being added, the **aglycone** ("without sugar"), is bonded to C1 of the sugars *via* oxygen; hence, such glycosides are more properly termed *O*-glycosides. Since the reaction conditions used to prepare a glycoside usually are too mild to convert an alcohol to an ether, only the hydroxy group at C1 reacts under the conditions described.

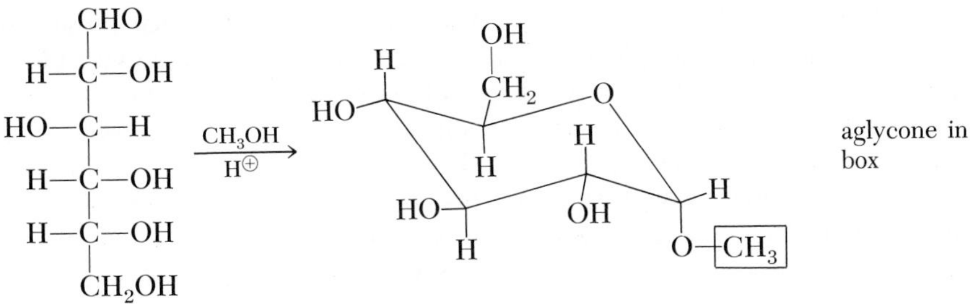

D-glucose

methyl α-D-glucopyranoside
(β anomer also formed)

aglycone in box

The glycosidic bond is ubiquitous in naturally occurring carbohydrates (see p. 475) and the methods used to achieve its cleavage, therefore, are important in carbohydrate analyses. Like any other acetal or ketal, glycosides are inert to mildly alkaline solutions but are easily hydrolyzed by acids. A reasonable mechanism for the acid-catalyzed hydrolysis of a glycoside is presented below.

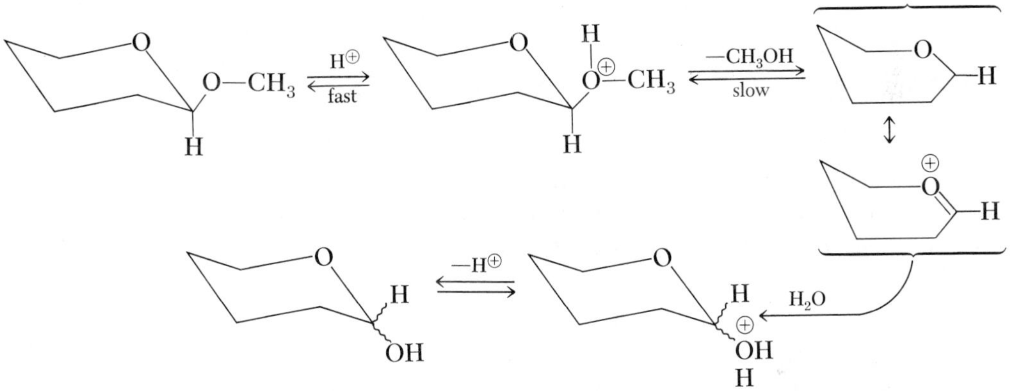

10. Suggest an alternate mechanism for the acid-catalyzed hydrolysis of a glycoside, one involving initial protonation of the heterocyclic oxygen. What important feature does this mechanism share with the one in the text, if any?

Although acid-catalyzed hydrolysis is fast, *enzymatic hydrolysis* is sometimes more desirable because enzymes are not only fast-acting but, in addition, can be selective about which glycosidic bonds they cleave. For example, *α-D-glucosidase* from yeast cleaves only glycosidic bonds originating at C1 of the α anomer of glucose. Almond emulsin *β-D-glucosidase,* on the other hand, is specific for β-glucosidic linkages.

In contrast to the behavior of hemiacetals and hemiketals, acetals and ketals do *not* undergo spontaneous cleavage. Therefore, anomeric glycosides do not equilibrate spontaneously, and this means that furanosides and pyranosides do not exhibit mutarotation.

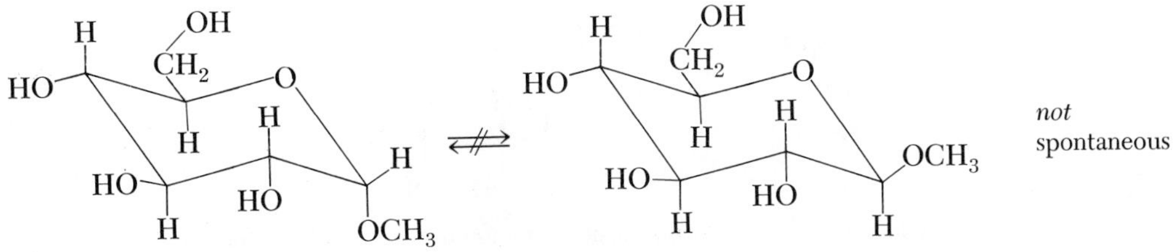

not spontaneous

Interesting Aglycones. Many natural products contain sugars that possess glycosidic linkages to interesting aglycones. Among the medically important simple glycosides are the *cardiac glycosides*, compounds that possess a steroid-like aglycone and that can be used to give the heart a strong, powerful beat. The poisonous foxglove plant (*Digitalis purpurea*) produces, upon extraction, a mixture known as *digitalis*. This complex material is rich in glycosides, a principal aglycone of which is digitoxigenin. This polycyclic alcohol is used in the treatment of certain heart disorders. Toad poisons are similar in both structure and behavior to cardiac glycosides. The skin glands of toads (*Bufo vulgaris*) produce a toxic substance that is called *bufotalin*.

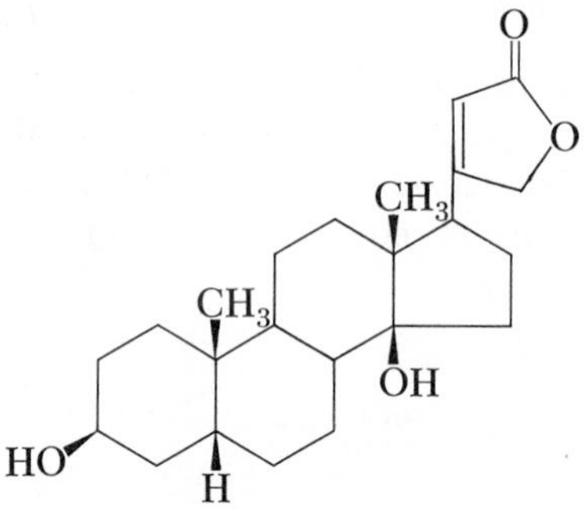

digitoxigenin

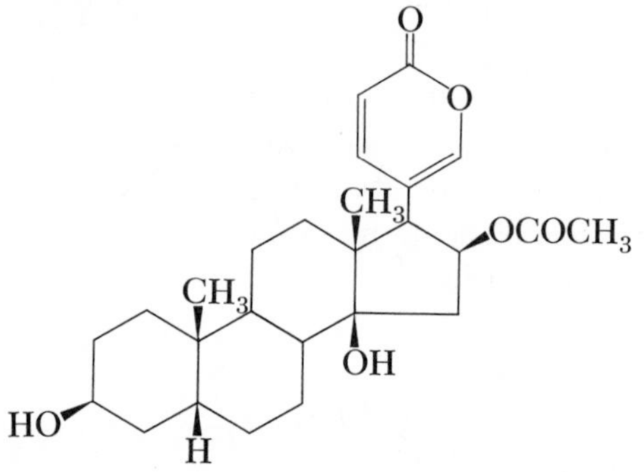

bufotalin

Most plant pigments are also glycosides, the aglycone being the fragment responsible for the color. For example, *cyanin*, a pigment found in roses, contains a substituted anthocyanidin cation as an aglycone. Such glycosides are called *anthocyanins*.

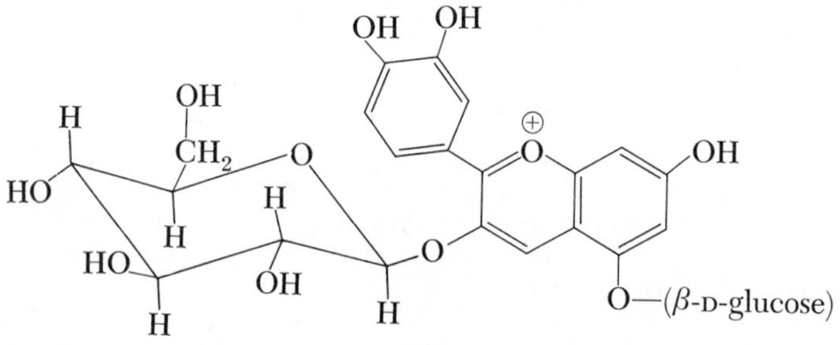

cyanin

While our discussion has been limited, thus far, to *O*-glycosides, there are other bridging atoms that occur in nature and, therefore, other types of glycosides. For example, *N*-glycosides are compounds in which the aglycone is bound to C1 *via* nitrogen instead of oxygen. Among the most important *N*-glycosides are the *nucleosides*, composed of ribose or deoxyribose and one of several heterocyclic amines. Adenine and ribose form the nucleoside shown below. Much more will be said of such compounds in the following chapter.

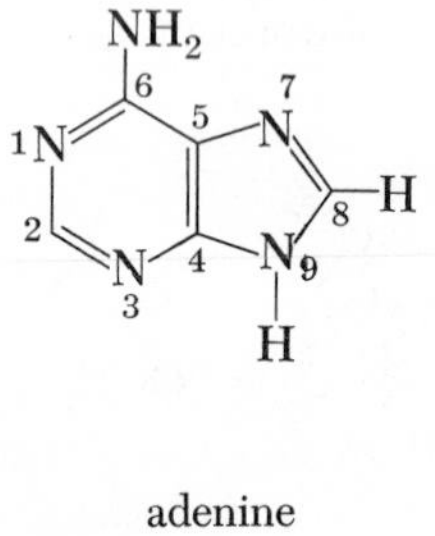

adenine

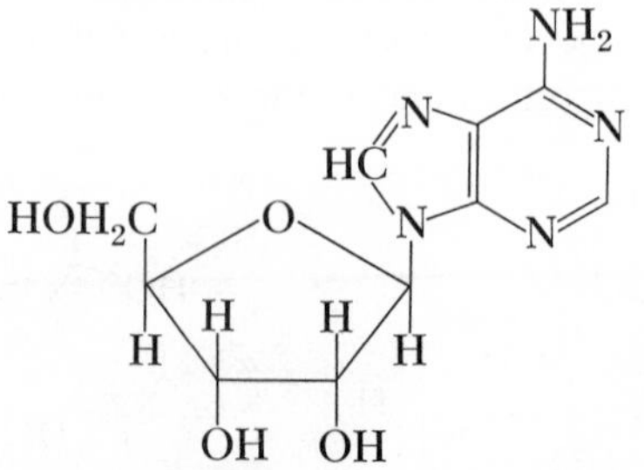

a nucleoside composed of ribose and adenine

25A.5 REACTIONS OF MONOSACCHARIDES

Even though monosaccharides exist largely as cyclic hemiacetals and hemiketals, rapid equilibration with even trace quantities of carbonyl-containing compounds means that a discussion of the reactions of most sugars must consider both the acyclic and the cyclic structures.

Reduction. Early studies on the simple sugars were aimed, in part, at establishing the number of carbons in a given monosaccharide. This was usually done by reducing the sugar to a hydrocarbon with a mixture of hydriodic acid and red phosphorus.

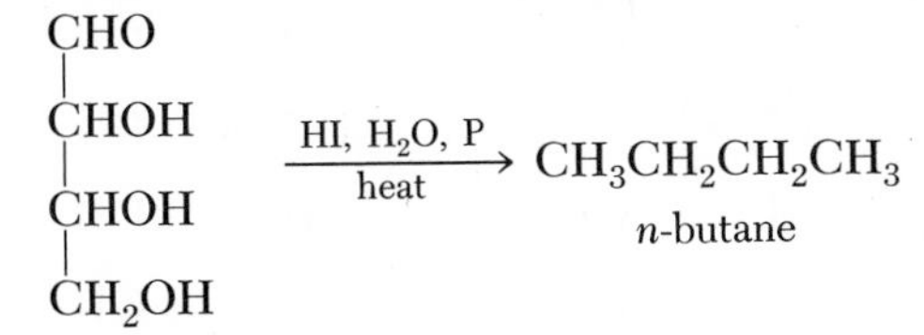

$$\begin{array}{c} CHO \\ | \\ CHOH \\ | \\ CHOH \\ | \\ CH_2OH \\ \text{an aldotetrose} \end{array} \xrightarrow[\text{heat}]{HI,\ H_2O,\ P} \begin{array}{c} CH_3CH_2CH_2CH_3 \\ n\text{-butane} \end{array}$$

Both sodium amalgam and sodium borohydride reduce the carbonyl group to a hydroxy group but do not reduce hydroxy groups further. Therefore, they convert a monosaccharide into a perhydroxyalkane, or *glycitol.* Sodium borohydride reductions are particularly valuable because both the substrate and the reductant are water-soluble, and the reduction may be carried out in an aqueous solution. The most important glycitol is *sorbitol,* prepared by reduction of glucose. Sorbitol occurs in fruits such as cherries and pears, and not just in "diet" candies.

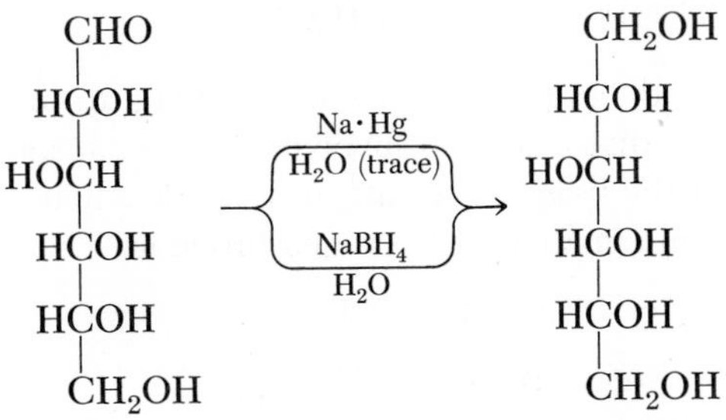

$$\begin{array}{c} CHO \\ | \\ HCOH \\ | \\ HOCH \\ | \\ HCOH \\ | \\ HCOH \\ | \\ CH_2OH \\ \text{D-glucose} \end{array} \xrightarrow{\left\{\begin{array}{c} Na\cdot Hg \\ H_2O\ (\text{trace}) \\ \\ NaBH_4 \\ H_2O \end{array}\right\}} \begin{array}{c} CH_2OH \\ | \\ HCOH \\ | \\ HOCH \\ | \\ HCOH \\ | \\ HCOH \\ | \\ CH_2OH \\ \text{sorbitol} \end{array}$$

Even as simple a reaction as a borohydride reduction can be used to distinguish between some monosaccharides. Thus, both D-(−)-erythrose and D-(−)-threose produce 1,2,3,4-tetrahydroxybutanes; but only the tetrol from D-threose will be optically active.

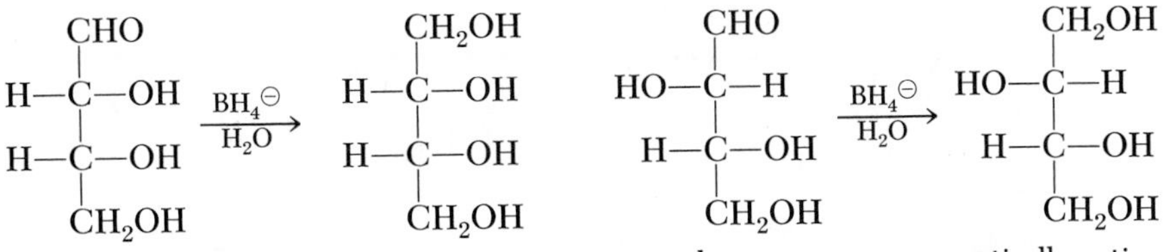

$$\begin{array}{c} CHO \\ | \\ H{-}C{-}OH \\ | \\ H{-}C{-}OH \\ | \\ CH_2OH \\ \text{D-erythrose} \end{array} \xrightarrow[H_2O]{BH_4^{\ominus}} \begin{array}{c} CH_2OH \\ | \\ H{-}C{-}OH \\ | \\ H{-}C{-}OH \\ | \\ CH_2OH \\ \text{optically inactive} \end{array} \qquad \begin{array}{c} CHO \\ | \\ HO{-}C{-}H \\ | \\ H{-}C{-}OH \\ | \\ CH_2OH \\ \text{D-threose} \end{array} \xrightarrow[H_2O]{BH_4^{\ominus}} \begin{array}{c} CH_2OH \\ | \\ HO{-}C{-}H \\ | \\ H{-}C{-}OH \\ | \\ CH_2OH \\ \text{optically active} \end{array}$$

Oxidation. Simple sugars are replete with oxidizable functional groups and are oxidized by a variety of reagents. Nitric acid oxidizes aldoses and ketoses to dicarboxylic acids, called

glycaric or **saccharic acids** (or **aldaric acids**). The glycaric acid derived from D-glucose is called D-glucaric acid.

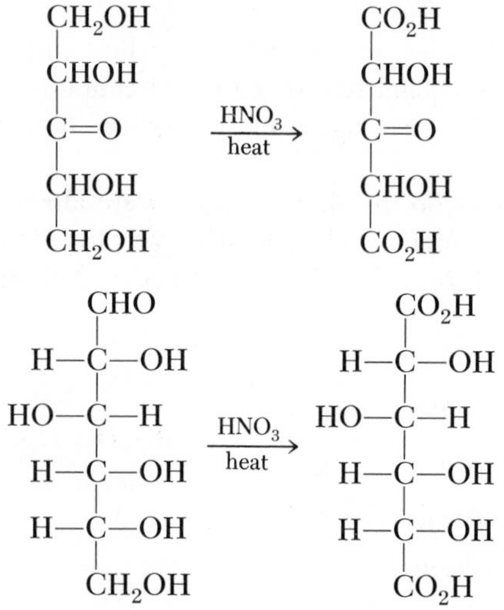

$$\begin{array}{c} CH_2OH \\ | \\ CHOH \\ | \\ C{=}O \\ | \\ CHOH \\ | \\ CH_2OH \end{array} \xrightarrow[\text{heat}]{HNO_3} \begin{array}{c} CO_2H \\ | \\ CHOH \\ | \\ C{=}O \\ | \\ CHOH \\ | \\ CO_2H \end{array}$$

$$\underset{\text{D-glucose}}{\begin{array}{c} CHO \\ | \\ H{-}C{-}OH \\ | \\ HO{-}C{-}H \\ | \\ H{-}C{-}OH \\ | \\ H{-}C{-}OH \\ | \\ CH_2OH \end{array}} \xrightarrow[\text{heat}]{HNO_3} \underset{\text{D-glucaric acid}}{\begin{array}{c} CO_2H \\ | \\ H{-}C{-}OH \\ | \\ HO{-}C{-}H \\ | \\ H{-}C{-}OH \\ | \\ H{-}C{-}OH \\ | \\ CO_2H \end{array}}$$

Bromine water, in the presence of strontium carbonate, oxidizes aldoses to monocarboxylic acids called **glyconic acids** (or **aldonic acids**). Ketoses do not react with bromine water, and the oxidation of aldoses with bromine water can be used to distinguish aldoses from ketoses.

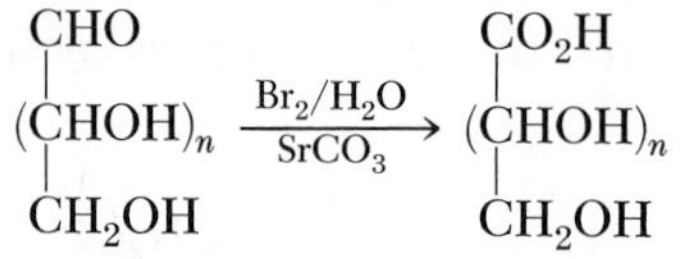

$$\begin{array}{c} CHO \\ | \\ (CHOH)_n \\ | \\ CH_2OH \end{array} \xrightarrow[SrCO_3]{Br_2/H_2O} \begin{array}{c} CO_2H \\ | \\ (CHOH)_n \\ | \\ CH_2OH \end{array} \quad \text{glyconic acid formation}$$

In reality, oxidation produces the lactone corresponding to the glyconic acid rather than the acid itself. This suggests that oxidation involves the hemiacetal rather than the carbonyl compound. The function of the base is to create an alkoxide anion, which attacks the halogen. This is followed by an E2 elimination of hydrogen bromide.

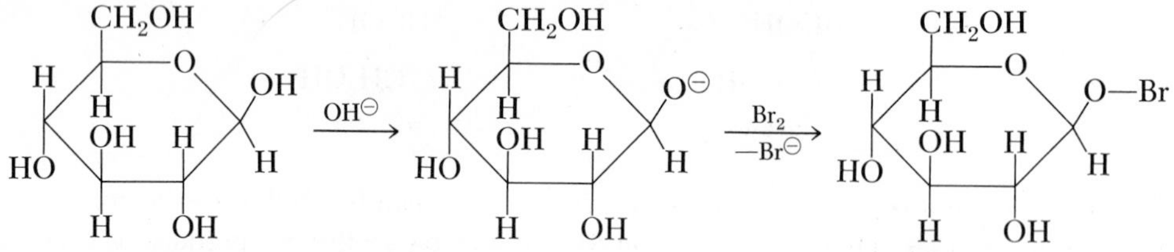

β-D-glucopyranose $\xrightarrow{OH^\ominus}$ (alkoxide, $O^\ominus$ at C-1) $\xrightarrow[-Br^\ominus]{Br_2}$ (O—Br at C-1) $\xrightarrow[-Br^\ominus]{-H_2O}$ D-gluconolactone

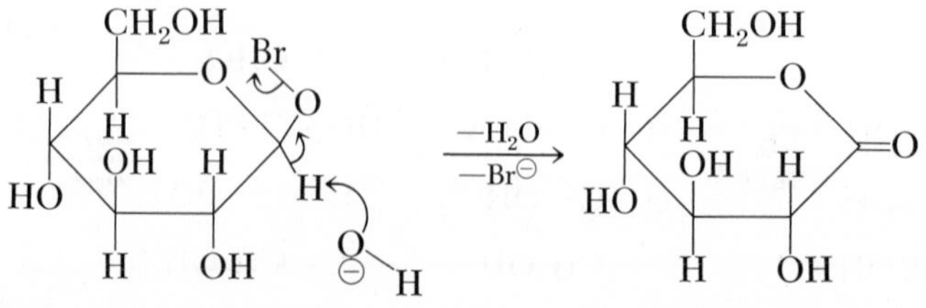

α-D-Glucopyranose is oxidized more slowly than is the β anomer, reflecting the presence of unfavorable axial interactions in that conformer of the α anomer which is necessary for the *trans* elimination to occur. Such interactions are absent in the conformer of the β anomer required for *trans* elimination. Therefore, the energy of activation leading to a *trans* elimination is lower in the β anomer than in the α anomer.

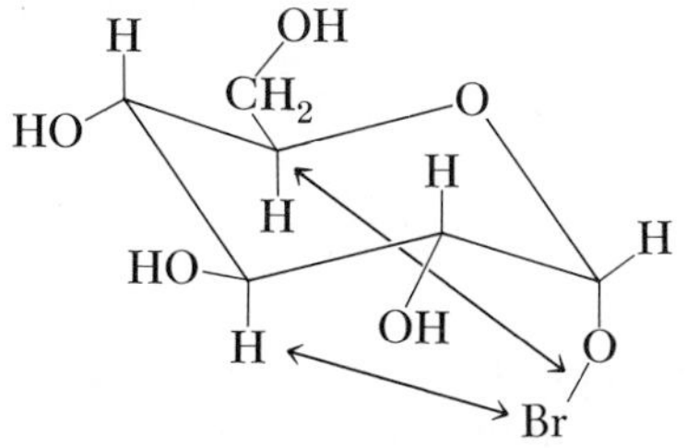

conformer necessary for E2 reaction in α anomer

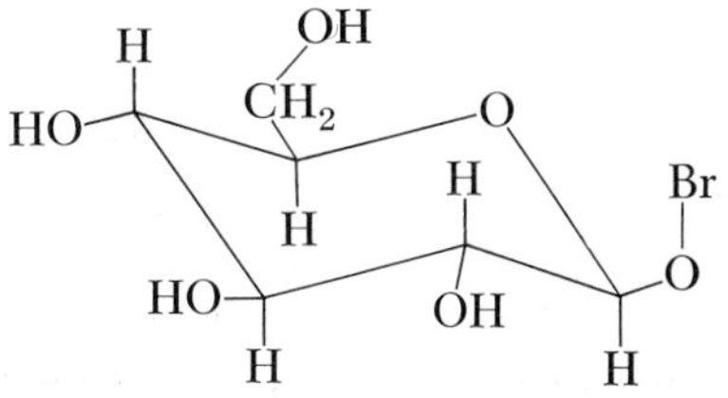

conformer necessary for E2 reaction in β anomer

Glycuronic acids (also called **uronic acids**) are compounds that have the terminal primary hydroxy group of a monosaccharide oxidized to a carboxy group while the carbonyl group is unaltered. They are not easily made in the laboratory but are present in nature. The most common of these, D-glucuronic acid and L-iduronic acid, are made *in vivo* by the enzymatic oxidation of a complex molecule: *uridine diphosphate-α-D-glucose* (UDP glucose). D-Glucuronic acid is particularly important because many toxic substances are excreted in the urine as derivatives of D-glucuronic acid, called *glucuronides*.

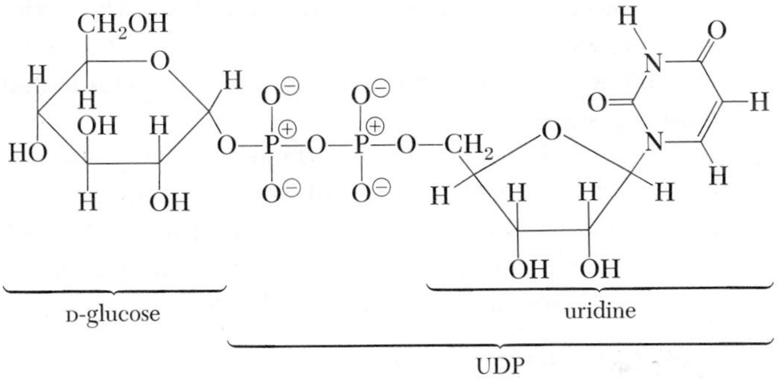

UDP-glucose

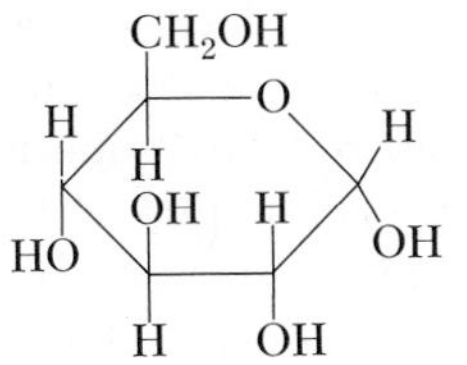

D-glucose — uridine — UDP

glycuronic acids written in their hemiacetal forms

α-D-glucuronic acid

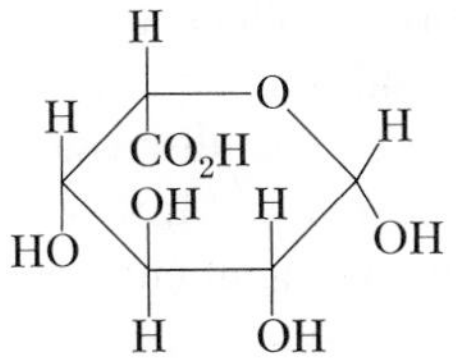

α-L-iduronic acid

Oxidation of diols with periodate ion, $IO_4^{\ominus}$, has already been discussed (p. 338). This reaction, which is quite useful in carbohydrate analysis, is reviewed below:

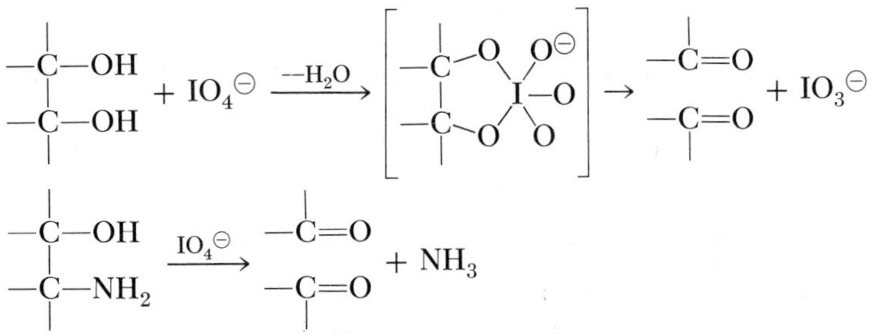

$$\begin{matrix}-\text{C}-\text{OH}\\-\text{C}-\text{OH}\end{matrix} + IO_4^{\ominus} \xrightarrow{-H_2O} \left[\text{cyclic periodate ester}\right] \rightarrow \begin{matrix}-\text{C}=\text{O}\\-\text{C}=\text{O}\end{matrix} + IO_3^{\ominus}$$

$$\begin{matrix}-\text{C}-\text{OH}\\-\text{C}-\text{NH}_2\end{matrix} \xrightarrow{IO_4^{\ominus}} \begin{matrix}-\text{C}=\text{O}\\-\text{C}=\text{O}\end{matrix} + NH_3$$

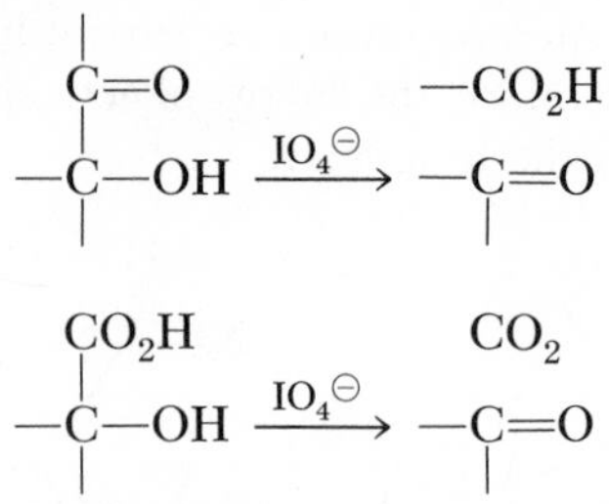

$$\begin{array}{c} | \\ C{=}O \\ | \\ -C{-}OH \\ | \end{array} \xrightarrow{IO_4^{\ominus}} \begin{array}{c} -CO_2H \\ \\ -C{=}O \\ | \end{array}$$

$$\begin{array}{c} CO_2H \\ | \\ -C{-}OH \\ | \end{array} \xrightarrow{IO_4^{\ominus}} \begin{array}{c} CO_2 \\ \\ -C{=}O \\ | \end{array}$$

Sodium metaperiodate oxidizes glucose, for example, to five moles of formic acid and one mole of formaldehyde. Which carbon in glucose is found in the formaldehyde?

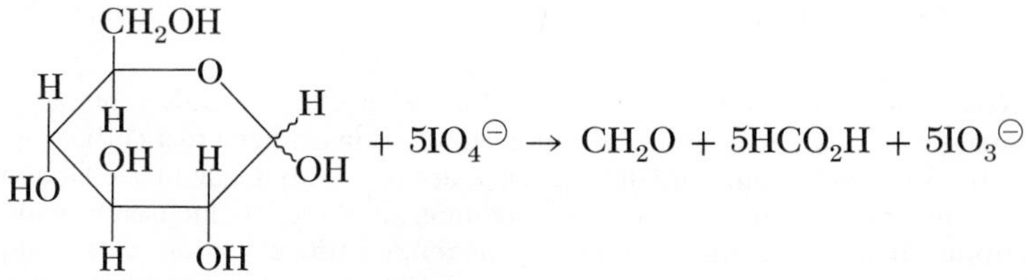

$$\text{D-glucose} + 5IO_4^{\ominus} \rightarrow CH_2O + 5HCO_2H + 5IO_3^{\ominus}$$

D-glucose

Reaction with Benedict's, Fehling's, and Tollens' Reagents. The action of these three reagents is based upon the oxidation of aldoses, and they usually are used to test for the presence of sugars with free aldehyde groups. Such sugars reduce certain metallic cations and are, therefore, called "reducing sugars."

Tollen's reagent is prepared by mixing solutions of sodium hydroxide and silver nitrate to form a precipitate of silver oxide. Careful addition of aqueous ammonia dissolves the precipitate and forms soluble $Ag(NH_3)_2^{\oplus}$. Addition of an aldose to Tollen's reagent produces a precipitate of metallic silver, usually in the form of a mirror-like coating on the walls of the test tube. For this reason the test sometimes is called the "silver mirror" test.

$$Ag(NH_3)_2^{\oplus} + \text{D-glucose} \rightarrow \text{D-gluconic acid} + Ag\downarrow$$

Interestingly, fructose gives a positive Tollens' test even though it lacks a carboxaldehyde group. The reason for this is a base-catalyzed equilibration between glucose, mannose, and fructose, similar to that discussed on p. 512. (That discussion considered the equilibrium between glucose 6-phosphate and fructose 6-phosphate. However, the equilibration of fructose with mannose and glucose is quite similar, involving an enediol as a critical intermediate. This latter equilibration is called the Lobry de Bruyn-van Ekenstein rearrangement.) Thus, the positive Tollens' test actually is due to the mannose and glucose formed from fructose by the action of the alkaline reagent.

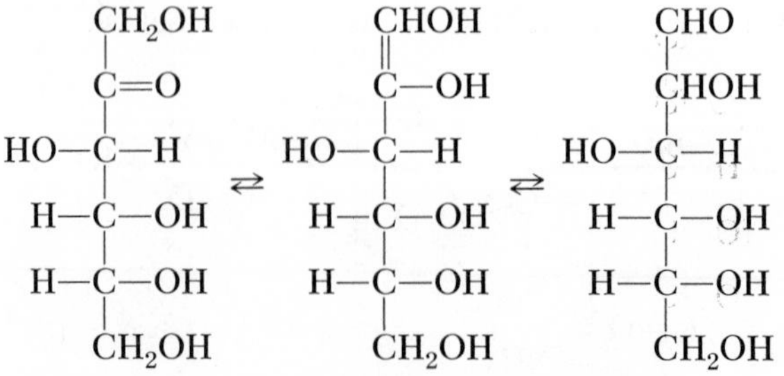

$$\begin{array}{c} CH_2OH \\ | \\ C{=}O \\ | \\ HO{-}C{-}H \\ | \\ H{-}C{-}OH \\ | \\ H{-}C{-}OH \\ | \\ CH_2OH \end{array} \rightleftarrows \begin{array}{c} CHOH \\ \| \\ C{-}OH \\ | \\ HO{-}C{-}H \\ | \\ H{-}C{-}OH \\ | \\ H{-}C{-}OH \\ | \\ CH_2OH \end{array} \rightleftarrows \begin{array}{c} CHO \\ | \\ CHOH \\ | \\ HO{-}C{-}H \\ | \\ H{-}C{-}OH \\ | \\ H{-}C{-}OH \\ | \\ CH_2OH \end{array}$$

D-fructose — enediol intermediate — D-glucose and D-mannose

Lobry de Bruyn-van Ekenstein rearrangement

11. How many enediols can be derived directly from D-fructose? Explain.

Fehling's solution is made by mixing a slightly acidic solution of copper sulfate with a basic solution of sodium potassium tartrate (Rochelle salt). When it is warmed in the presence of an aldehyde, this solution produces a red precipitate of cuprous oxide, Cu_2O.

Also based upon the reduction of Cu(II) to Cu(I) and the precipitation of the corresponding oxide, Benedict's solution gives the same observable result as does Fehling's solution. The unique difference between these is that Benedict's solution is a "single solution" reagent containing Cu(II) stabilized in alkaline solution by citrate ion.

$$RCHO + 2Cu(II) + 4OH^{\ominus} \rightarrow RCO_2H + \underset{\text{red}}{Cu_2O} + 2H_2O$$

As with Tollens' reagent, fructose gives a positive test with Fehling's and Benedict's solutions. Glycosides do not give a positive result with any of these reagents. (Why?)

Osazone Formation. α-Hydroxy carbonyl compounds react with excess phenylhydrazine to form *bis*phenylhydrazones, commonly called *osazones*. A reasonable mechanism for osazone formation is presented in Figure 25A-4.

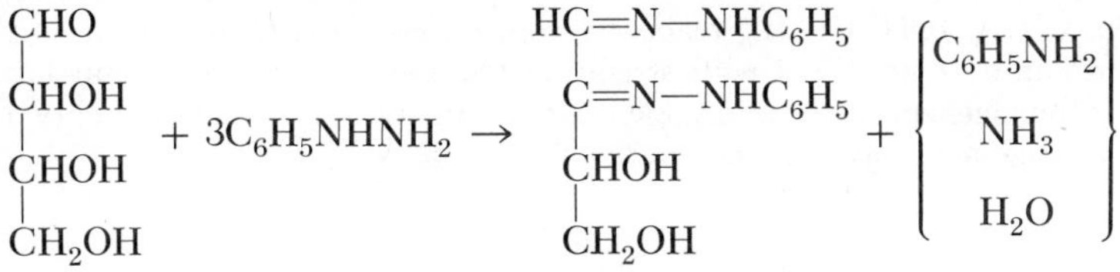

$$\begin{matrix} CHO \\ | \\ CHOH \\ | \\ CHOH \\ | \\ CH_2OH \end{matrix} + 3C_6H_5NHNH_2 \rightarrow \begin{matrix} HC{=}N{-}NHC_6H_5 \\ | \\ C{=}N{-}NHC_6H_5 \\ | \\ CHOH \\ | \\ CH_2OH \end{matrix} + \left\{ \begin{matrix} C_6H_5NH_2 \\ NH_3 \\ H_2O \end{matrix} \right\}$$

an osazone

$$\begin{matrix} H \\ \backslash \\ C{=}NNH{-}C_6H_5 \\ | \\ CHOH \\ | \\ R \end{matrix} \underset{}{\overset{A}{\rightleftharpoons}} \begin{matrix} H \\ \backslash \\ C{-}NHNHC_6H_5 \\ \| \\ C{-}OH \\ | \\ R \end{matrix} \overset{B}{\rightleftharpoons} \begin{matrix} CH_2NH{-}NHC_6H_5 \\ | \\ C{=}O \\ | \\ R \end{matrix}$$

a hydrazone

$$\begin{matrix} H \\ | \\ CH{-}NH{-}NHC_6H_5 \\ | \\ C{=}O \\ | \\ R \end{matrix} \xrightarrow[C]{H_2NNHC_6H_5} \begin{matrix} H \\ | \\ CH{-}NH{-}NHC_6H_5 \\ | \\ C{=}NNHC_6H_5 \\ | \\ R \end{matrix}$$

$$\begin{matrix} H \\ C_6H_5NHN{:}\ \ H \\ H \quad | \\ CH{-}NH{-}NHC_6H_5 \\ | \\ C{=}NNHC_6H_5 \\ | \\ R \end{matrix} \xrightarrow{D} \begin{matrix} H \\ \backslash \\ C{=}NH \\ | \\ C{=}NNHC_6H_5 \\ | \\ R \end{matrix} \xrightarrow[E]{C_6H_5NHNH_2} \begin{matrix} H \\ \backslash \\ C{=}NNHC_6H_5 \\ | \\ C{=}NNHC_6H_5 \\ | \\ R \end{matrix} + NH_3$$

osazone

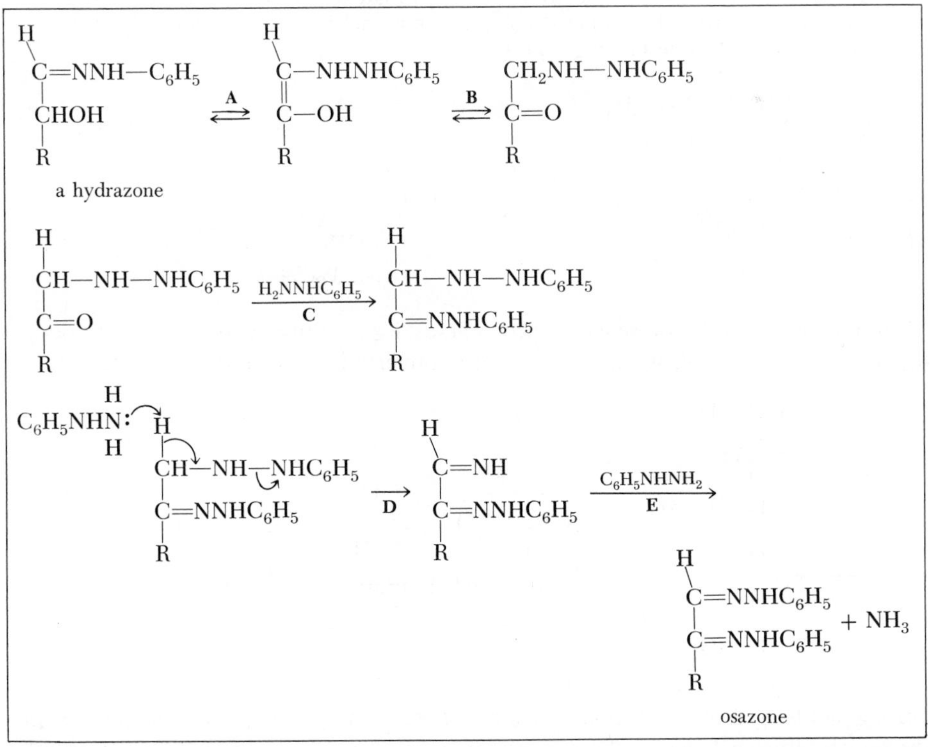

Figure 25A-4 The conversion of the hydrazone of a monosaccharide to its osazone. The mechanism for hydrazone formation is illustrated on p. 483. Step **D** is an elimination, while step **E** follows an addition-elimination path. Can you draw out the mechanism of step **E** in detail?

Since osazone formation destroys the chirality at C2, it is possible for several monosaccharides to produce the same osazone.

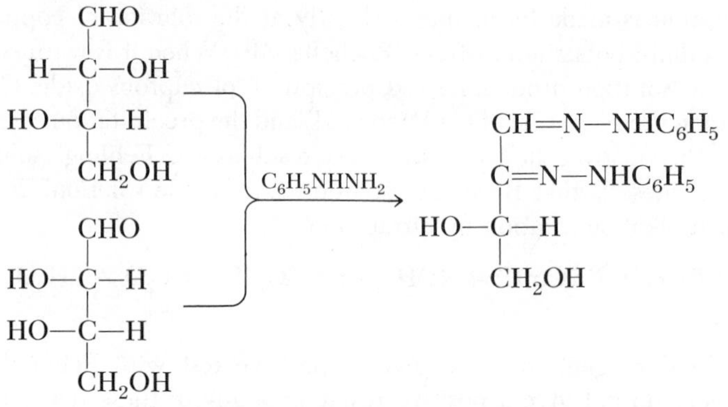

CHO
H—C—OH
HO—C—H
CH_2OH

CHO
HO—C—H
HO—C—H
CH_2OH

$\xrightarrow{C_6H_5NHNH_2}$

$CH{=}N{-}NHC_6H_5$
$C{=}N{-}NHC_6H_5$
HO—C—H
CH_2OH

12. Osazones react with two moles of benzaldehyde (in the presence of acid) to form two moles of benzaldehyde phenylhydrazone and one mole of a type of compound called an *osone*. D-Glucose, for example, forms an osone with the formula $C_6H_{10}O_6$. Suggest a structure for this.
13. Benzoin, $C_6H_5CHOHCOC_6H_5$, reacts with an excess of phenylhydrazine in the presence of acid to form an osazone. What is its structure? Outline a mechanism for its formation.
14. What do the mechanisms for osazone formation (Figure 25A–4) and the Lobry de Bruyn-van Ekenstein rearrangement have in common?

Osazones are converted to highly crystalline *osatriazoles* by heating with aqueous cupric sulfate. These are more useful than are osazones as sugar derivatives, since osatriazoles do not decompose upon melting but, instead, give well-defined, reproducible melting points.

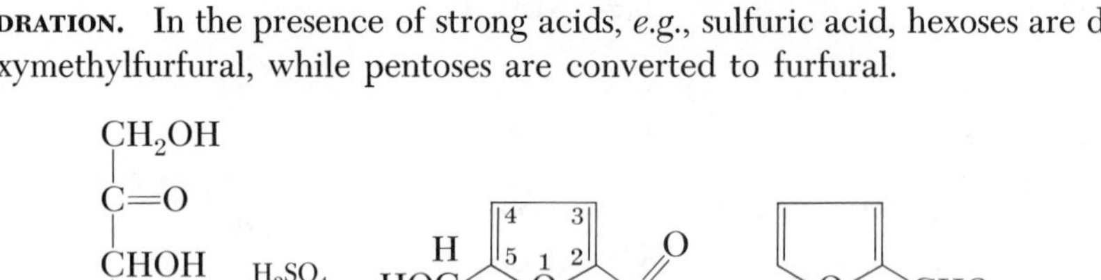

$CH{=}N{-}NHC_6H_5$
$C{=}N{-}NHC_6H_5$
HO—C—H
CH_2OH

$\xrightarrow[\text{heat}]{Cu(II),\ H_2O}$

N, N, N (triazole ring bearing phenyl) + $C_6H_5NH_2$
HO—C—H
CH_2OH

L-tetrulose phenylosatriazole

Dehydration. In the presence of strong acids, *e.g.*, sulfuric acid, hexoses are dehydrated to 5-hydroxymethylfurfural, while pentoses are converted to furfural.

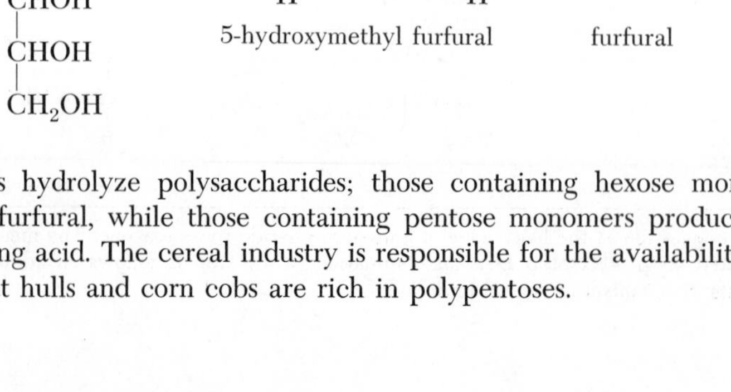

CH_2OH
C=O
CHOH
CHOH
CHOH
CH_2OH

$\xrightarrow{H_2SO_4}$

H
HOC (ring positions 4, 3, 5, 1, 2; O) C(=O)H
H

5-hydroxymethyl furfural

O CHO

furfural

Strong acids hydrolyze polysaccharides; those containing hexose monomers produce 5-hydroxymethylfurfural, while those containing pentose monomers produce furfural when treated with strong acid. The cereal industry is responsible for the availability of inexpensive furfural, since oat hulls and corn cobs are rich in polypentoses.

Ester and Ether Formation. The hydroxy groups of sugars are easily esterified. Acetylation, the most common esterification reaction, usually is achieved using acetic anhydride with acidic (*e.g.*, sulfuric acid or zinc chloride) or basic (*e.g.*, sodium acetate) catalysts. Consistent with their cyclic structures, hexoses yield *cyclic* pentaacetates. These acetates, which exist in α and β forms, are not spontaneously interconverted because they cannot open up to form a free carbonyl group.

The proportion of α- and β-glucose pentaacetates produced by acetylation can be controlled by varying experimental conditions. Above room temperature α and β acetates are interconverted by acids to produce a mixture of 90% α and 10% β. At temperatures below 0°, and in the presence of a basic catalyst, the rate of acetylation is much faster than is the rate of acetate anomer equilibration. Since the equatorial hydroxy group is more reactive than the axial hydroxy group, base-catalyzed acetylation produces mainly the β anomer. These interconversions are summarized in Figure 25A–5.

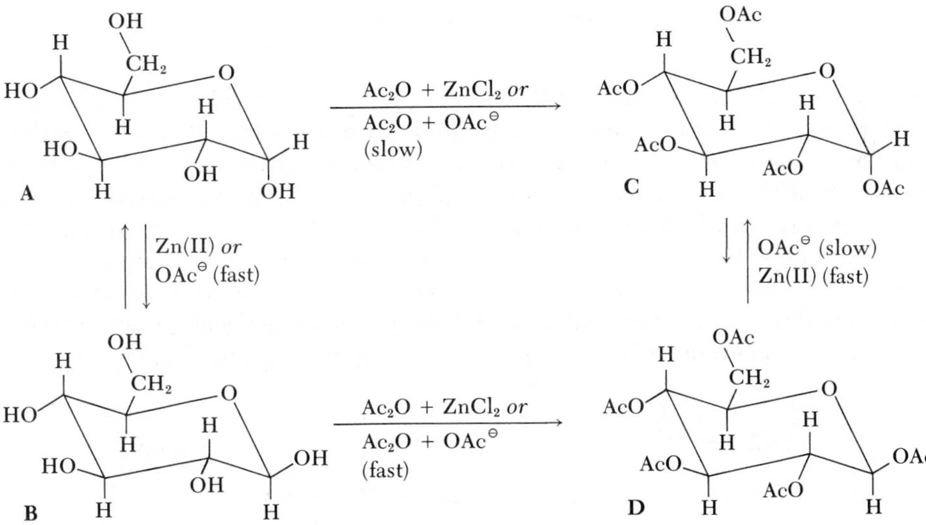

A $\xrightarrow{Ac_2O + ZnCl_2 \text{ or } Ac_2O + OAc^{\ominus} \text{ (slow)}}$ C

A $\rightleftarrows$ B: Zn(II) *or* $OAc^{\ominus}$ (fast)

C $\rightleftarrows$ D: $OAc^{\ominus}$ (slow), Zn(II) (fast)

B $\xrightarrow{Ac_2O + ZnCl_2 \text{ or } Ac_2O + OAc^{\ominus} \text{ (fast)}}$ D

Figure 25A-5 The reaction of glucose with acetic anhydride. Compounds are as follows: **A** = α-D-glucopyranose; **B** = β-D-glucopyranose; **C** = penta-*O*-acetyl-α-D-glucopyranose; **D** = penta-*O*-acetyl-β-D-glucopyranose.

As expected, sugar acetates are easily hydrolyzed by base. However, the acetyl groups are often removed by transesterification.

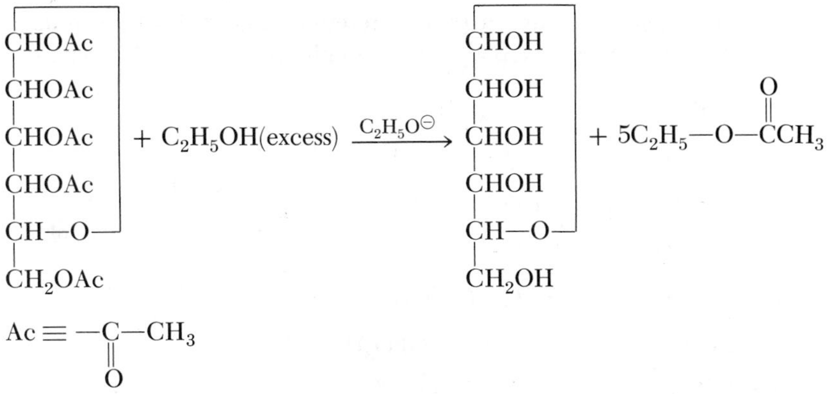

$$\begin{array}{l} \text{CHOAc} \\ | \\ \text{CHOAc} \\ | \\ \text{CHOAc} \\ | \\ \text{CHOAc} \\ | \\ \text{CH—O} \\ | \\ \text{CH}_2\text{OAc} \end{array} + C_2H_5OH(\text{excess}) \xrightarrow{C_2H_5O^{\ominus}} \begin{array}{l} \text{CHOH} \\ | \\ \text{CHOH} \\ | \\ \text{CHOH} \\ | \\ \text{CHOH} \\ | \\ \text{CH—O} \\ | \\ \text{CH}_2\text{OH} \end{array} + 5C_2H_5\text{—O—}\overset{O}{\overset{\|}{C}}CH_3$$

$$Ac \equiv \text{—}\underset{\underset{O}{\|}}{C}\text{—}CH_3$$

The glycosidic acetate group can be replaced with bromine, using dry hydrogen bromide dissolved in glacial acetic acid. The resultant acetobromoglycose is useful in synthesizing glycosides. The reaction proceeds with configurational inversion at the anomeric carbon. However, even penta-*O*-acetyl-α-D-glucose gives the α bromide, since the β bromide is rapidly converted to the α bromide under the experimental conditions.

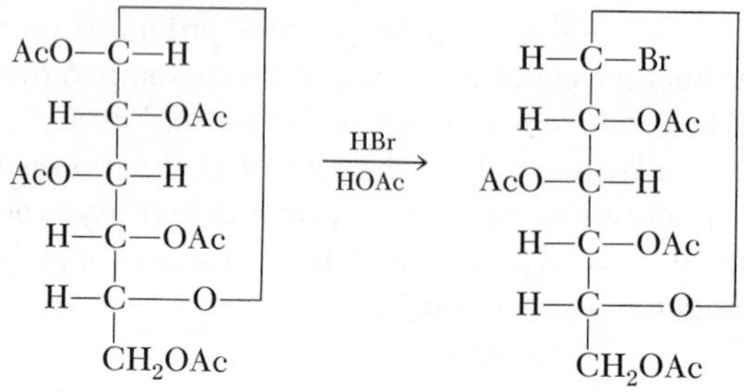

AcO—C—H / H—C—OAc / AcO—C—H / H—C—OAc / H—C——O / CH_2OAc —(HBr, HOAc)→ H—C—Br / H—C—OAc / AcO—C—H / H—C—OAc / H—C——O / CH_2OAc

penta-*O*-acetyl-β-D-glucose   tetra-*O*-acetyl-α-D-glucosyl bromide

Substituted cyclohexanes are usually more stable when their substituents are equatorial rather than axial. The same is true for the pyranose form of sugars and, as noted earlier, this explains why β-D-glucose is the most prevalent monosaccharide in nature. However, in discussing the formation of the pentaacetates of glucose (Figure 25A–5) and the reaction of penta-*O*-acetyl-α-D-glucose with hydrogen bromide, we introduced two interesting exceptions—both of these reactions preferentially afford the α (axial) geometry at C1, the anomeric carbon. This is an example of the *anomeric effect*, the preference for the axial geometry at C1 brought about because dipolar repulsions between the heterocyclic oxygen and the substituent at C1 destablize certain substituents (*e.g.*, methoxy, acetoxy) in the equatorial geometry.

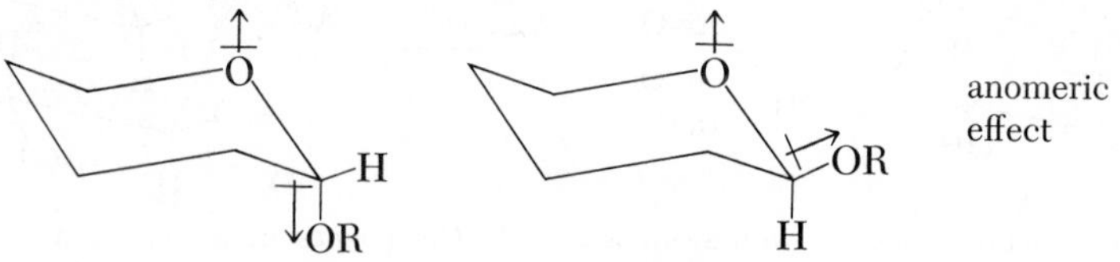

O, H, OR   O, OR, H   anomeric effect

more stable   less stable

Of great importance to living systems are various **phosphate esters** of carbohydrates; almost all biological interconversions, degradations, polymerizations, *etc.*, of monosaccharides employ these esters. For example, the initial step in the metabolism of glucose involves the use of ATP to phosphorylate (see p. 512, *et seq.*) glucose, thus converting it to glucose 6-phosphate. Indeed, the intermediates that are involved in the eventual conversion of glucose to pyruvic acid (a process called *glycolysis*) are all phosphate esters (Figure 17–1). The degradation of glucose to pyruvic acid provides the body with a significant portion of its energy.

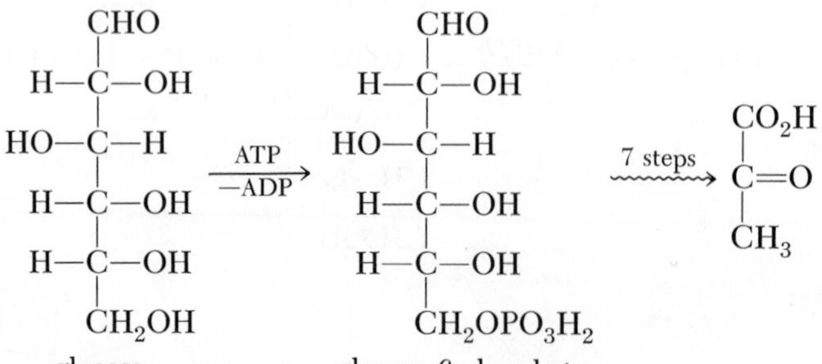

CHO / H—C—OH / HO—C—H / H—C—OH / H—C—OH / CH_2OH —(ATP, −ADP)→ CHO / H—C—OH / HO—C—H / H—C—OH / H—C—OH / $CH_2OPO_3H_2$ —(7 steps)→ CO_2H / C=O / CH_3 glycolysis

glucose   glucose 6-phosphate

Let us now turn to ether derivatives of sugars. Methanol, in the presence of acid, methylates only the hemiacetal or hemiketal hydroxy group. Etherification of the remaining hydroxyl groups requires reaction conditions such as are found in the Williamson ether synthesis.

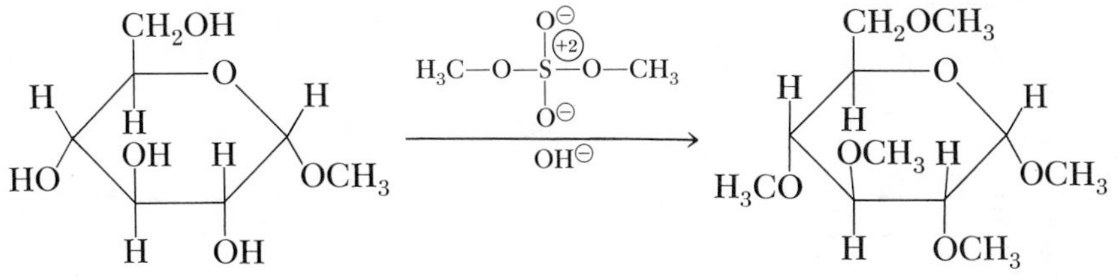

methyl 2,3,4,6-tetra-*O*-methyl-α-D-glucoside

Since acids hydrolyze acetals and ketals but not ethers, it becomes possible to locate the carbonyl group within a carbohydrate and to determine which hydroxy group was employed in hemiacetal formation by employing a series of simple reactions. The sequence necessary to do this is: (1) acid-catalyzed methylation, (2) Williamson etherification, and (3) glycoside hydrolysis. Application of these steps to glucose produces 2,3,4,6-tetra-*O*-methyl D-glucose. Vigorous oxidation with nitric acid converts this, *via* a keto acid, to a trimethoxyglutaric acid and a dimethoxysuccinic acid. Such a mixture of acids could arise only if the keto group of the keto acid was located at C5. In turn, the free hydroxy group must have been at C5 and the ring must have been pyranose rather than furanose.

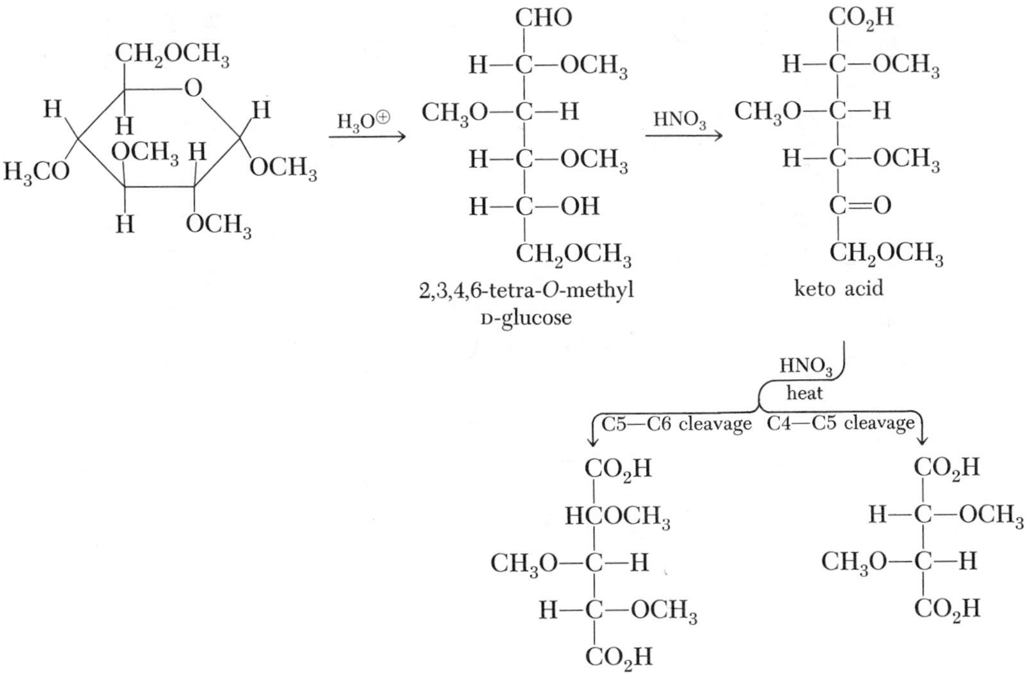

2,3,4,6-tetra-*O*-methyl D-glucose

keto acid

HNO_3 heat

C5—C6 cleavage C4—C5 cleavage

25A.6 SYNTHESIS AND DEGRADATION

The Kiliani-Fischer Synthesis. The Kiliani-Fischer synthesis is a method for lengthening the carbon chain of an aldose by one carbon, with the production of two diastereomeric aldoses.

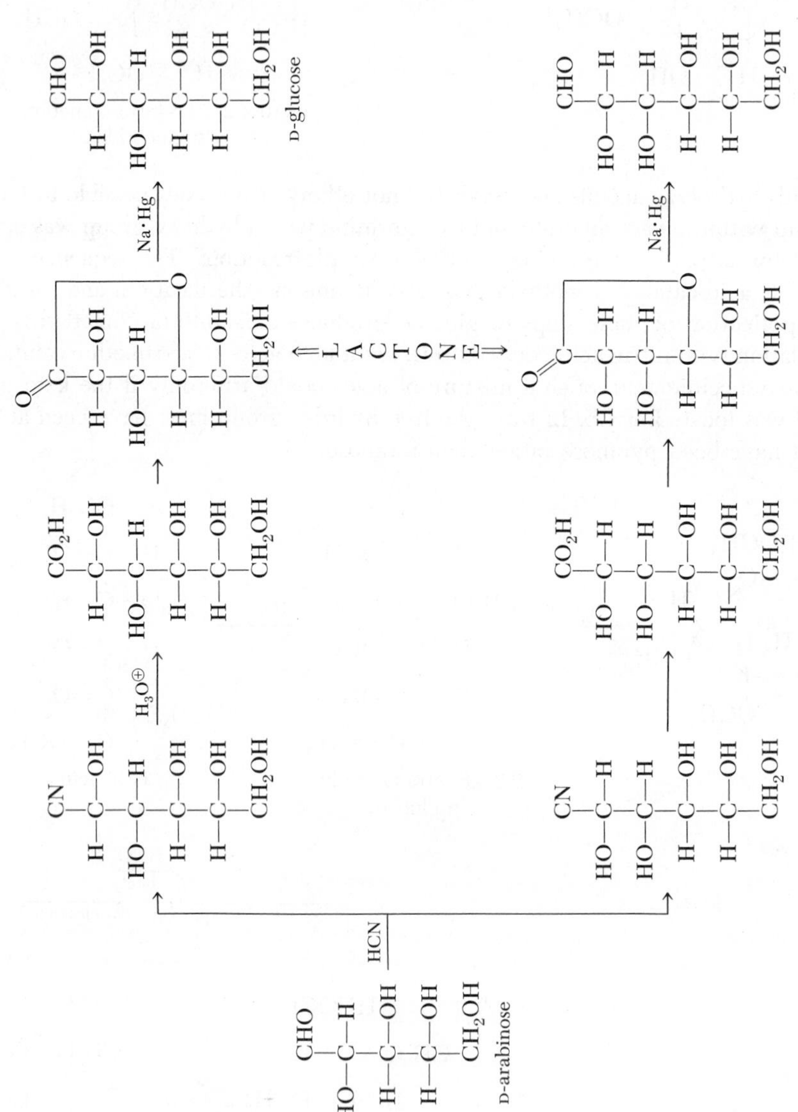

Figure 25A-6 The Kiliani-Fischer synthesis of D-glucose and D-mannose.

Diastereoisomerism is introduced in the first step of the sequence, with the production of isomeric cyanohydrins. Hydrolysis produces, after dehydration, a lactone that is reduced to an aldose containing one more carbon than the starting material. The preparation of D-glucose and D-mannose from an aldopentose is shown in Figure 25A–6. While the mixture of diastereomeric products could be separated at several different places in the sequence, the separation is usually made *before* the final reduction because the two resultant sugars may be difficult to separate.

This reaction can be run in reverse in order to remove a skeletal carbon atom from a monosaccharide. The *retro*-Kiliani-Fischer reaction begins with the conversion of an aldose to an oxime by reaction with hydroxylamine. The oxime is dehydrated to a nitrile with acetic anhydride, a reaction that is accompanied by the acetylation of the free hydroxyl groups in the sugar. Transesterification converts the acetylated product to the cyanohydrin of a monosaccharide. This can lose hydrogen cyanide to produce an aldose containing one less carbon than the original sugar. This sequence, outlined in Figure 25A–7, does not generate two products because no new chiral centers are created. While the Kiliani-Fischer synthesis employs excess hydrogen cyanide to insure a high yield of cyanohydrin, the reverse reaction succeeds because hydrogen cyanide is removed as it is produced.

```
   CHO                    CH=NOH                  CN
H—C—OH                 H—C—OH                 H—C—OAc
HO—C—H    --H2NOH-->   HO—C—H    --Ac2O-->  AcO—C—H
H—C—OH                 H—C—OH                 H—C—OAc
H—C—OH                 H—C—OH                 H—C—OAc
   CH2OH                  CH2OH                  CH2OAc
D-glucose
                                                   | NaOCH3, CH3OH
                                                   v
      CHO                              CN
   HO—C—H       <--(—HCN)--        H—C—OH
    H—C—OH                         HO—C—H
    H—C—OH                          H—C—OH
      CH2OH                         H—C—OH
  D-arabinose                         CH2OH
```

Figure 25A–7 The *retro*-Kiliani-Fischer conversion of D-glucose to D-arabinose.

Ruff Degradation. The Ruff degradation of aldoses employs the free-radical decarboxylation of a salt of a glyconic acid to decrease the length of an aldose by one carbon. The decarboxylation is accomplished using a mixture of hydrogen peroxide and Fe(III), a combination that sometimes is called Fenton's reagent. Unfortunately, this procedure affords only about a 30 to 40% yield. The Ruff degradation of glucose is shown in Figure 25A–8.

```
   CHO                    CO2H                  CO2⊖
H—C—OH                 H—C—OH                H—C—OH                               CHO
HO—C—H  --Br2/H2O-->   HO—C—H   --CaCO3-->   HO—C—H   --H2O2/Fe(III), —CO2-->  HO—C—H
H—C—OH                 H—C—OH                H—C—OH                            H—C—OH
H—C—OH                 H—C—OH                H—C—OH                            H—C—OH
   CH2OH                  CH2OH                 CH2OH                             CH2OH
D-glucose            D-gluconic acid                                          D-arabinose
```

Figure 25A–8 Ruff degradation of D-glucose to D-arabinose.

15. Outline the reactions needed to convert D-glucose into each of the following.
    (a) D-mannose
    (b) D-arabinose
    (c) methyl β-D-glucopyranoside
    (d) hexane
    (e) **(R,S)**-tartaric acid
    (f) α-D-glucose pentaacetate (penta-*O*-acetyl-α-D-glucose)

IMPORTANT TERMS

Aglycone: A non-sugar residue bonded to C1 of the acetal (or ketal) form of a sugar. Aglycones may be bonded to the sugar by elements other than oxygen (for example, nitrogen or carbon).

Aldonic acid: A monocarboxylic acid derived by oxidation of the carboxaldehyde group of an aldose. Since the usual oxidant (Br_2/H_2O) does not oxidize ketones, aldonic acid formation can be used to distinguish aldoses from ketoses. Also called *glyconic acids*.

Aldose: A monosaccharide containing an aldehyde group. Glucose is an aldose.

Amino sugar: A sugar in which a hydroxy group has been replaced by an amino group.

Anomeric carbon: The carbon in the cyclic form of a sugar that used to be the carbonyl carbon.

Anomers: Stereoisomers that differ in configuration only at the anomeric carbon. Anomers usually are classified as α or β, depending upon the relative configurations at the anomeric carbon and at the carbon used to classify the compounds as D or L. The anomeric carbon is present only in the cyclic form of a sugar, since only cyclic forms have hemiacetal (or hemiketal) linkages. The interconversion of anomers (α and β forms) is called *anomerization*.

Anomeric effect: Unusual conformational (and chemical) behavior exhibited by sugars with methoxy, acetoxy, or similar groups serving as aglycones. The most obvious illustration is that methoxy groups at C1 tend to be axial rather than equatorial in pyranose forms. The effect has its origin in the repulsive interaction between the heterocyclic oxygen and the oxygen of the aglycone.

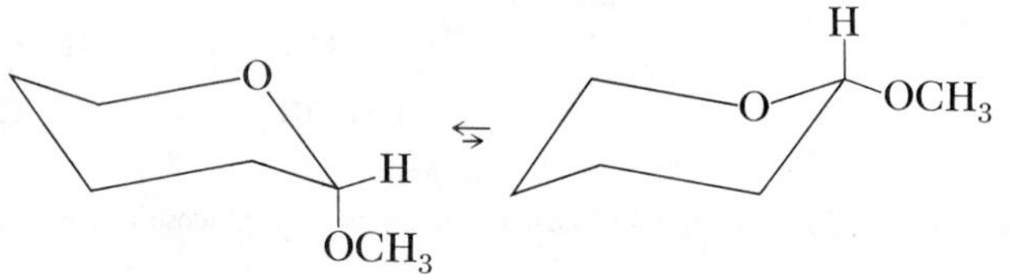

O, H, OCH_3 ⇆ O, H, OCH_3

Benedict's solution: A solution used to test for the presence of an aldose (a red precipitate of Cu_2O is formed). Unfortunately, ketoses also may give a positive test (see p. 819). The reagent, which is blue in color, consists of cupric ion complexed with citrate ion.

Deoxy sugar: A sugar in which a hydroxy group has been replaced by hydrogen. Deoxy sugars normally can be recognized by the presence of either a methylene or a methyl group.

D Sugar: A sugar in which the chiral center farthest away from the carbonyl group has the configuration shown below.

R
|
H—C—OH
|
CH_2OH

Dextrose: A synonymn for glucose. The most abundant organic compound on this planet.

Epimers: Two stereoisomers, of a structure containing several chiral centers, that differ in configuration at only one chiral center (the so-called *epimeric center*). While not usually thought of this way, *meso*- and (+)-2,3-dibromobutane are epimers. Anomers are special examples of epimers. The interconversion of epimers is called *epimerization*.

Fehling's solution: Similar in purpose and action to Benedict's solution, Fehling's solution contains cupric ion complexed to tartrate ion.

Furanose form: The cyclic form of a sugar in which the ring is tetrahydrofuran (oxacyclopentane), *i.e.*, a five-membered ring.

Furanoside: An acetal (or ketal) derived from the furanose form of a sugar.

Glucose: An alternate, and more commonly used, name for dextrose.

Glucoside: An acetal derived from glucose. Glucosides may exist as isomeric α and β forms.

Glycaric acid: A dicarboxylic acid derived from a sugar. The carboxy groups are found at the two ends of the carbon skeleton. These also are called *saccharic* or *aldaric acids*.

Glycitol: A perhydroxyalkane, $HOCH_2(CHOH)_nCH_2OH$.

Glycoside: The acetal (or ketal) of a sugar. Glycosides exist in isomeric α and β forms. Glucosides are a specific type of glycoside.

Haworth structure: A quasi-three-dimensional representation of a cyclic sugar, which portrays the ring as an idealized planar polygon. The ring oxygen of the pyranose form is commonly represented as being in the upper right-hand corner. The Haworth structure (or Haworth projection) does not depict axial and equatorial relationships in a pyranose, although it does present correct *cis,trans* relationships.

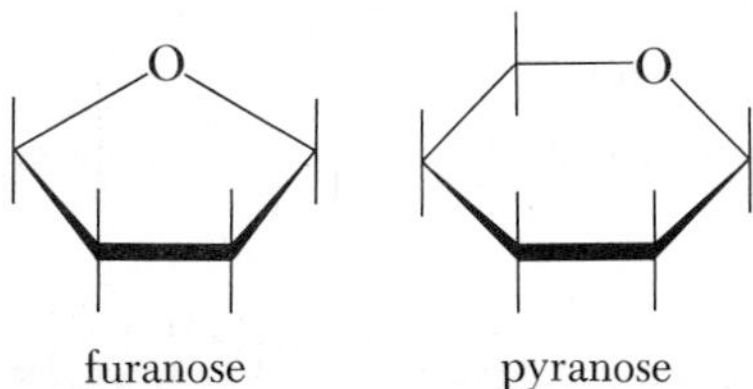

Ketose: A sugar containing a ketonic carbonyl group. Fructose is a ketose. Monosaccharides are either ketoses or aldoses.

Kiliani-Fischer synthesis: A means of converting a given aldose into the two diastereomeric next-higher homologs (*e.g.*, D-xylose into D-gulose and D-idose). The details of the method are found in Figure 25A–6.

L Sugar: A sugar in which the chiral center farthest from the carbonyl group has the configuration shown below.

$$\begin{array}{c} R \\ | \\ HO{-}C{-}H \\ | \\ CH_2OH \end{array}$$

Monosaccharide: A "simple" sugar. A polyhydroxy aldehyde or ketone *or* a derivative of one of these.

Mutarotation: The change in the optical rotation of a sample. Mutarotation ceases after equilibrium has been established at a given set of experimental conditions. Mutarotation is not expected to lead to an ultimate rotation of 0° unless the species in equilibrium are enantiomers (as compared to epimers or anomers).

Osazone: A compound containing the functional group shown below. Further information may be found in Figure 25A–4 as well as in problems 12 and 13.

$$\begin{array}{l} \diagdown \\ C{=}N{-}NHC_6H_5 \\ | \\ C{=}N{-}NHC_6H_5 \\ \diagup \end{array}$$

Polysaccharide: A polymer made of monosaccharide units connected by glycosidic bonds.

Pyranose form: The cyclic form of a sugar in which the ring is tetrahydropyran (oxacyclohexane), *i.e.*, a six-membered ring.

Pyranoside: An acetal (or ketal) derived from the pyranose form of a sugar.

Ruff degradation: A method for shortening the chain of an aldose by one carbon (see Figure 25A–8).

Sugar: Any carbohydrate, regardless of complexity. The term *saccharide* is a synonymn. The term "simple sugar" usually refers to a monosaccharide. Not every sugar is sweet!

Tollens' test: A positive test is indicated by the precipitation of metallic silver (often as a mirror) from a solution of silver ammonia ion ($Ag(NH_3)_2^{\oplus}$). This is usually used to test for aldehydes, but ketones may also give a positive test.

Since the carboxaldehyde group is oxidized to a carboxy group ($-CHO \rightarrow -CO_2H$) in the reaction, this "test" also can be used to synthesize carboxylic acids from aldehydes. Since double bonds are inert to the reagent, this is a particularly valuable method for synthesizing α,β-unsaturated acids from α,β-unsaturated aldehydes. (The latter can be made by aldol condensations.)

$$\underset{\alpha,\beta\text{-unsaturated aldehyde}}{>C{=}C{<}^{CHO}} \xrightarrow[H_2O]{Ag(NH_3)_2^{\oplus}} \underset{\alpha,\beta\text{-unsaturated acid}}{>C{=}C{<}^{CO_2H}}$$

PROBLEMS

16. Suggest a reason for the high water solubility of most simple sugars.
17. How many straight-chain aldoheptoses are possible (a) excluding α and β forms and (b) including α and β forms of pyranoses?
18. Provide a specific example, including name and structure, of each of the following. Do not use the same example more than once.
    (a) aldose
    (b) ketose
    (c) aldohexose
    (d) aldopentose
    (e) ketohexose
    (f) monosaccharide
    (g) triose
    (h) aminosugar
    (i) deoxysugar
19. Draw the chair forms of α- and β-D-glucopyranose. What is the intermediate in their interconversion? Why is β-D-glucopyranose more stable than the α isomer?
20. (a) Compare and contrast the terms *mutarotation, anomerization,* and *epimerization.* (b) Explain how a molecule could undergo anomerization, yet not display mutarotation. (c) Why is it that α-D-glucopyranose undergoes anomerization while methyl α-D-glucopyranoside does not?
21. A pentose of the D family gave an optically active glycaric acid on oxidation. When it was degraded to a tetrose and the tetrose was oxidized, *meso*-tartaric acid was formed. What is the configuration and name of the starting pentose?
22. A pentose gave an optically inactive glycaric acid on oxidation. It also was converted to a pair of diastereomeric hexoses, one of which was oxidized to a glycaric acid identical with that derived from D-glucose. What are the configurations and names of the hexoses and the pentose?
23. A D pentose gave an optically active dibasic acid on oxidation. It also was converted to a pair of diastereomeric hexoses, of which one gave an optically active glycaric acid when oxidized and the

other gave an optically inactive glycaric acid. What are the configurations and names of the hexoses and the pentose?

24. Write the Haworth representation of β-L-talopyranose and the conformational representations of both conformers. Which should be the more stable?

25. (a) Explain, with the aid of equations, why the Kiliani-Fischer synthesis of D-galactose from D-lyxose must be accompanied by the formation of D-talose. (b) Why would you not expect D-talose and D-galactose to be formed in equal amounts? (c) Why would you not observe the formation of L-galactose during this synthesis?

26. Using the structures in Figures 25A–1 and 25A–2, assign absolute configurations to the chiral centers of the following.

(a) D-ribose and D-arabinose
(b) D-ribose and L-ribose
(c) D-glucose and D-mannose

27. What is the relationship, if any, between the stereochemical designations *threo* and *erythro*, and the structures of the sugars threose and eythrose?

28. (a) What steps are involved in the Ruff degradation? (b) Show how the Ruff degradation could be used to distinguish D-glucose from D-galactose. (c) Could it distinguish between D-glucose and D-mannose? (d) D-glucose and L-glucose?

29. Sugar **A**, which is optically active, produces the same osazone as does D-allulose. Nitric acid oxidation of **A** produces a dicarboxylic acid, **B**, which is optically inactive. Suggest structures for **A** and **B.**

30. Indicate the product(s) of the reaction of D-glucose with each of the following.

(a) bromine water and strontium carbonate
(b) excess phenylhydrazine
(c) nitric acid
(d) acetic anhydride
(e) sodium borohydride
(f) periodic acid, HIO_4
(g) hydrogen iodide and phosphorus
(h) hydroxylamine, H_2NOH
(i) dimethylsulfate and aqueous base
(j) ethanol and hydrogen chloride
(k) benzoyl chloride and pyridine
(l) methanol and hydrogen chloride followed by periodic acid
(m) aqueous bromine; then calcium carbonate followed by hydrogen peroxide and Fe(III)

31. Describe any relationship that may exist between the Ruff degradation and the Hunsdiecker reaction.

32. L-Ascorbic acid is made in nature from D-glucose. Ascorbic acid is a stronger acid ($pK_a = 4.2$) than is acetic acid ($pK_a = 4.7$). Account for this enhanced acidity, indicating the acidic proton(s).

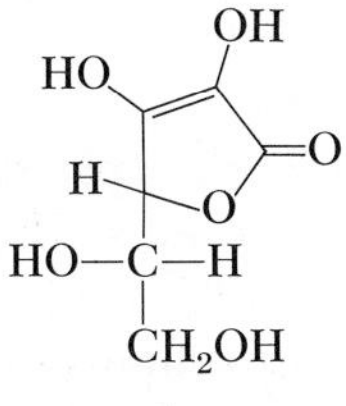

OH
HO
=O
H
O
HO—C—H
CH$_2$OH

L-ascorbic acid

33. Treatment of either D-glyceraldehyde or dihydroxyacetone with aqueous base produces D-fructose and D-sorbose. Suggest a mechanism for their formation. Also formed in this reaction is D,L-dendroketose; account for its formation. How would these products differ, if at all, if L-glyceraldehyde had replaced D-glyceraldehyde?

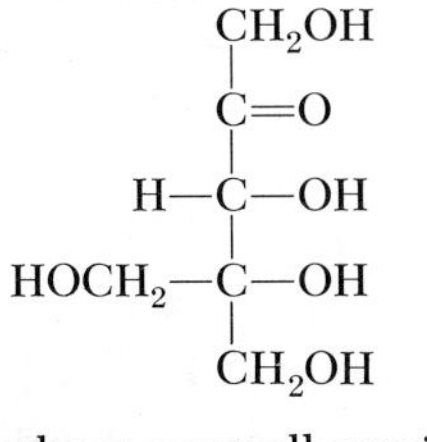

CH$_2$OH
C=O
H—C—OH
HOCH$_2$—C—OH
CH$_2$OH

D-dendroketose

34. D-Glucose 1-phosphate hydrolyzes unusually rapidly in dilute acid. Moreover, hydrolysis involves cleavage of the carbon-oxygen bond. Both observations are not typical of simple alkyl phosphates. Suggest an explanation for this unusual behavior.

OLIGOSACCHARIDES AND POLYSACCHARIDES

25B.1 INTRODUCTION

Carbohydrates are most often found in nature as oligosaccharides, polymers containing from two to ten monosaccharide units, or polysaccharides, polymers containing more than ten monomers. In this chapter we will focus upon some of the more important disaccharides and polysaccharides. As noted on p. 475, these polymers arise from a condensation between the hydroxy group on the hemiacetal carbon of one monosaccharide and a hydroxy group on a second monosaccharide unit. While these bonds usually involve C1 of one aldose and C4 of a second, other linkages found in these polymers are between C1 and C2, between C1 and C3, and between C1 and C6. The last of these is usually responsible for branching of a polymer chain (see p. 836.)

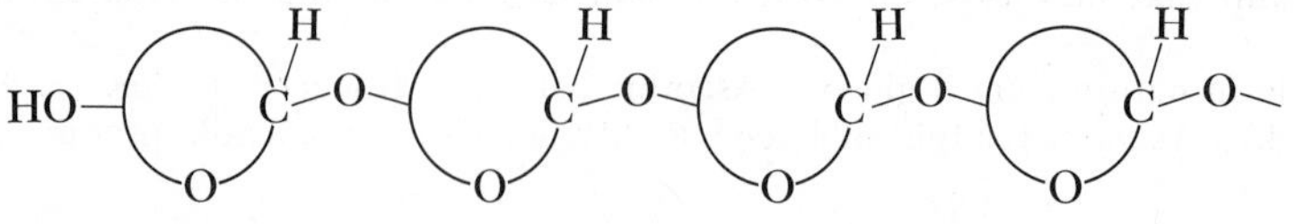

model of a linear polysaccharide

Animal tissues contain very few oligosaccharides compared to plant tissues. The only major one is lactose, a disaccharide found in mammalian milk. Human milk differs somewhat in composition from cow's milk, one of the important differences being that human milk contains oligosaccharides that include L-fucose (6-deoxy-L-galactose) or *N*-acetyl-D-glucosamine or both. The most common of these is the trisaccharide fucosyl-lactose.

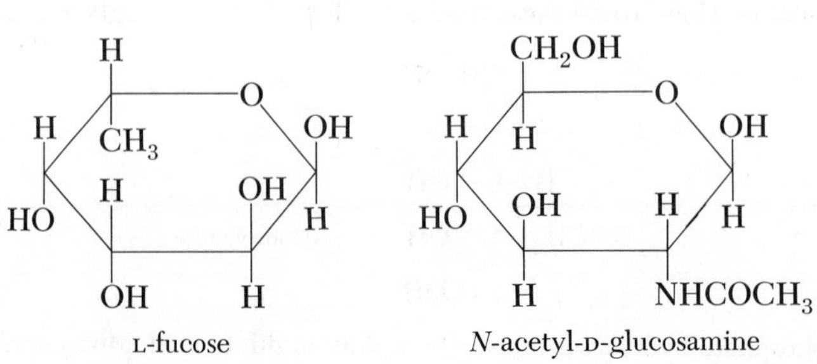

L-fucose

N-acetyl-D-glucosamine

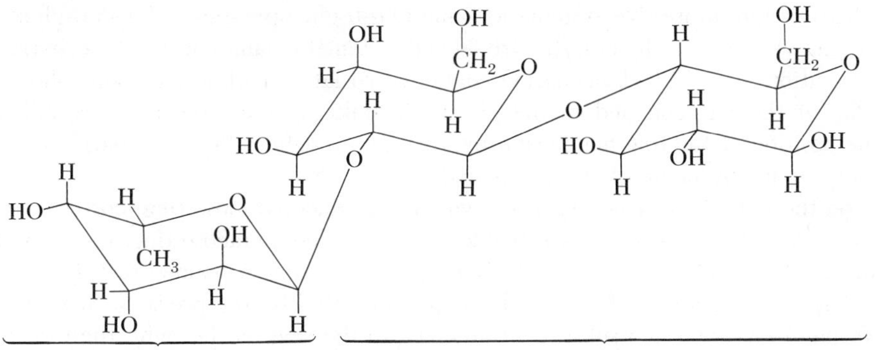

fucosyl-lactose

fucose lactose

There are two major types of polysaccharides: *homopolysaccharides* (or homoglycans), which are composed almost entirely of one monosaccharide, and *heteropolysaccharides* (or heteroglycans), which contain significant amounts of two or more different monosaccharides. The storage forms of polysaccharides in plants and animals are mainly homoglycans. Polysaccharides that are used for support purposes in the plant kingdom include both homo- and heteroglycans, while those in the animal kingdom are almost always heteroglycans.

1. Is the structure of L-fucose shown on p. 830 the α or the β form? Draw the Haworth structures for α- and β-D-fucose. Draw the Fischer projections for acyclic D- and L-fucose.

25B.2 DISACCHARIDES

Disaccharides consist of two simple sugars held together by a glycosidic bond. The most common of these involve the anomeric carbon of one sugar and a non-anomeric carbon of another sugar. Disaccharides usually are made by the partial hydrolysis of polysaccharides; before discussing the latter, we must become familiar with several common disaccharides.

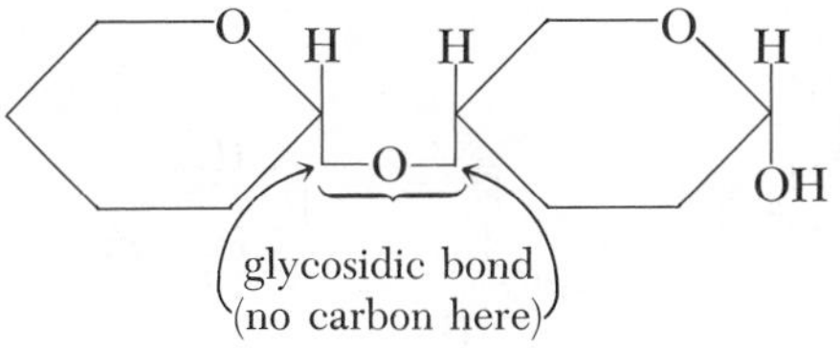

glycosidic bond (no carbon here)

idealized representative disaccharide

Maltose.

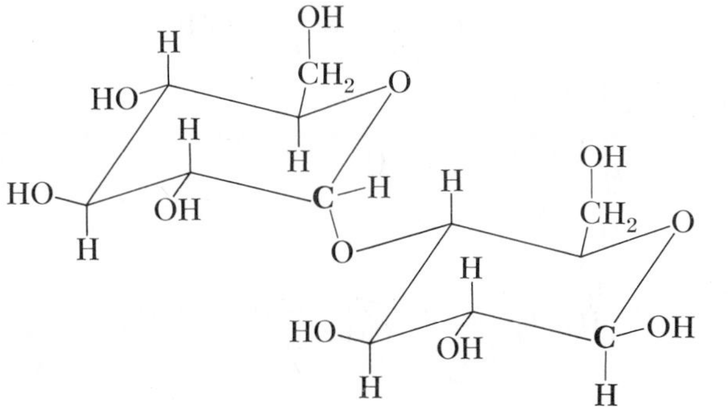

β-maltose

β-Maltose carries the impressive systematic name *O*-α-D-glucopyranosyl-(1,4)-β-D-glucopyranose.[a] How can we explain why both the trivial and systematic names of this disaccharide include the designation "β"? This disaccharide contains two latent carbonyl carbons, shown in boldface. One of these is confined in the glycosidic linkage, while the other is still a hemiacetal. The position of the free hemiacetal hydroxy group defines the disaccharide as β, and this must appear in any name of the compound.

The sugar on the left in the structure as drawn, an α-D-glucopyranose ring, corresponds to the sugar on the left in the name. This is bonded to a β-D-glucopyranose ring, the second sugar in the name. The "(1,4)" indicates that C1 of the "first" ring is bonded to C4 of the "second" ring. The bridging atom is oxygen, this being indicated by the prefix *O*. Since the ring on the left may be viewed as a substituent attached to the ring on the right, the former is named "osyl" while the latter is named "ose."

Maltose is the major product of the hydrolysis of starch by an enzyme in the saliva (*salivary amylase*). The name *maltose* or *malt sugar* stems from the formation of this disaccharide by enzymatic hydrolysis of the starch found in malt.

2. Draw the structure of α-maltose and provide its systematic name.
3. Oxidation of (+)-maltose with aqueous bromine yields D-maltobionic acid (or the corresponding lactone), **A.** Methylation of **A** with dimethyl sulfate produces octa-*O*-methyl-D-maltobionic acid, **B.** Acid-catalyzed hydrolysis of **B** produces one mole each of 2,3,4,6-tetra-*O*-methyl-α-D-glucopyranose, **C,** and 2,3,5,6-tetra-*O*-methyl-D-gluconic acid, **D.** Draw three-dimensional structures for compounds **A** through **D.** How do these data support the structure assigned to maltose?

Lactose. Milk sugar, or lactose, is synthesized in the mammary glands from D-glucose. The commercial source is milk whey, lactose being important in the preparation of baby food.

α-Lactose bears the systematic name *O*-β-D-galactopyranosyl-(1,4)-α-D-glucopyranose. As with maltose, the designation α reflects the configuration of the free hydroxy group at the anomeric center of the "second" sugar. The disaccharide consists of β-D-galactopyranose linked to a glucose residue through a glycosidic linkage. Lactose also forms a β anomer, which differs from the α anomer in the configuration at the carbon bearing the free hydroxy group at the anomeric center of the "second" sugar.

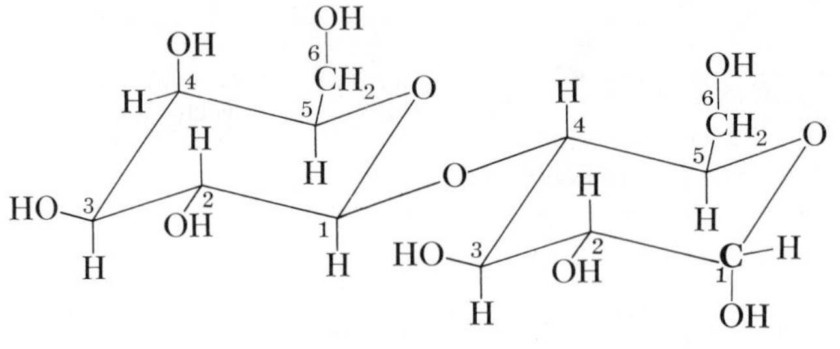

α-lactose

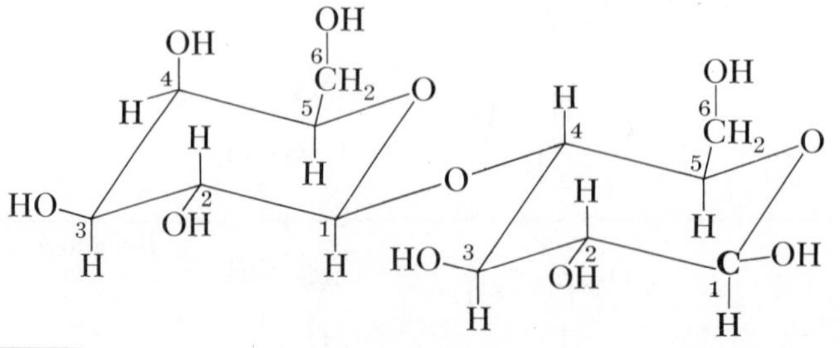

β-lactose

[a] The comma used to separate numbers (*e.g.*, (1,4)) sometimes is replaced by an arrow (1 → 4).

4. Lactobionic acid, $C_{12}H_{21}O_{12}$, is produced by the oxidation of lactose with bromine water. Suggest a structure for lactobionic acid. What will be produced by the acid-catalyzed hydrolysis of lactobionic acid?

Gentiobiose and Cellobiose. These two disaccharides contain β-D-glucose bonded to a second D-glucose residue *via* a glycosidic bond. Since they are not interconverted upon standing in aqueous solutions, they cannot be α and β forms of the same basic molecule. Actually, *cellobiose* possesses a glycosidic bond between C1 and C4 of two simple sugars, while *gentiobiose* has a glycosidic bond between C1 and C6 of two simple sugars.

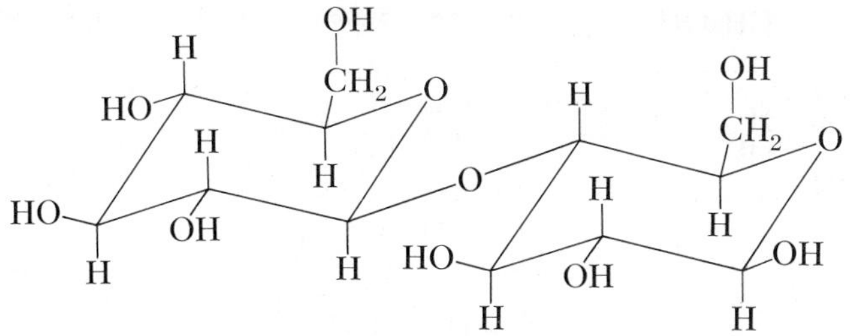

cellobiose
(*O*-β-D-glucopyranosyl-(1,4)-β-D-glucopyranose)

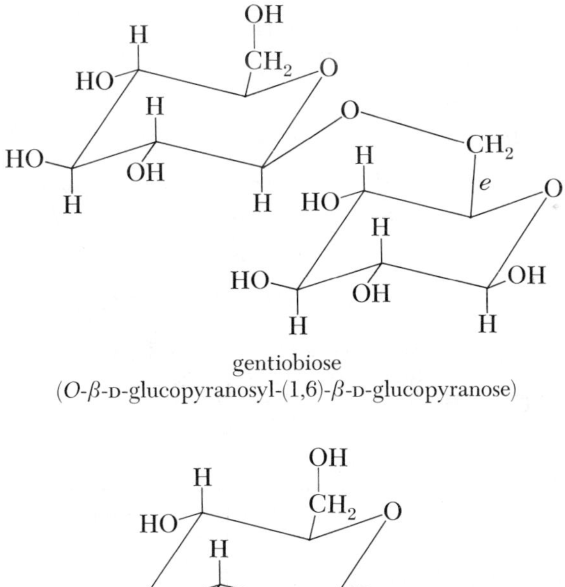

gentiobiose
(*O*-β-D-glucopyranosyl-(1,6)-β-D-glucopyranose)

Sucrose.

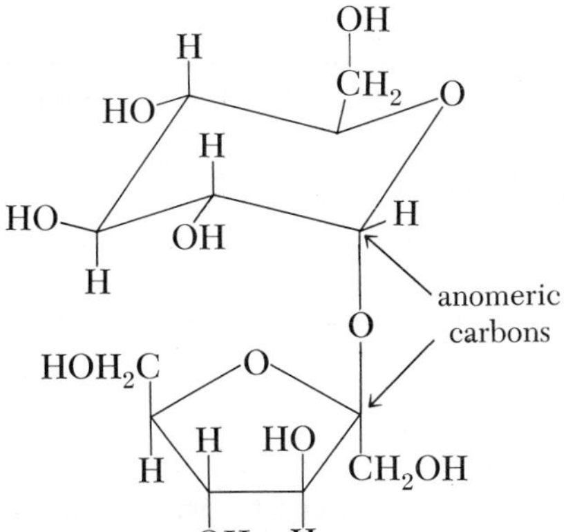

anomeric carbons

sucrose (common sugar)

Cane sugar and *beet sugar* are alternative names for **sucrose,** *O*-α-D-glucopyranosyl-(1,2)-β-D-fructofuranoside. In addition to the existence of a fructofuranose ring, sucrose is different from the other disaccharides covered thus far because the glycosidic linkage involves the anomeric centers of *both* sugars. Sucrose, unlike maltose, lactose, cellobiose, and gentiobiose, does not exist as readily interconverted α and β forms. The fructose portion of the name ends in "oside" in order to indicate that C2 of fructose is involved in the glycoside bond.

Relatively simple glycosides containing a disaccharide bonded to an aglycone are known. Two interesting ones are *amygdalin*, isolated from bitter almonds, and *streptomycin*. Amygdalin is a glycoside of gentiobiose whose aglycone is the cyanohydrin of benzaldehyde.[a] Streptomycin, an antibiotic that inhibits protein synthesis on ribosomes (discussed in the next chapter), is a glycoside whose aglycone is a derivative of *inositol* (*i.e.*, a hexahydroxycyclohexane) that has two hydroxy groups replaced by guanidine residues.

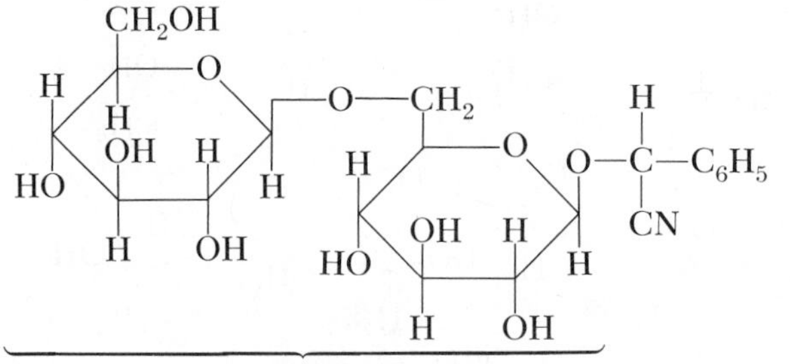

amygdalin

gentiobiose

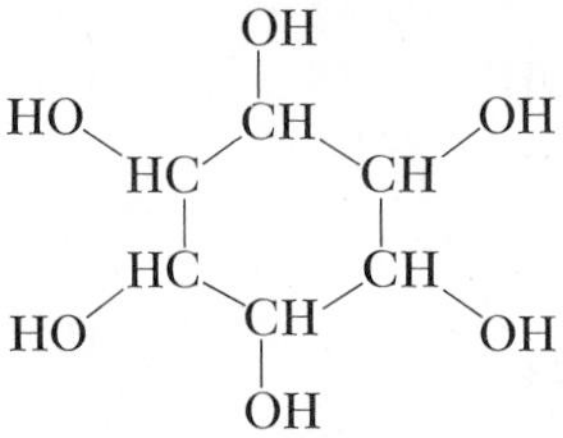

an inositol

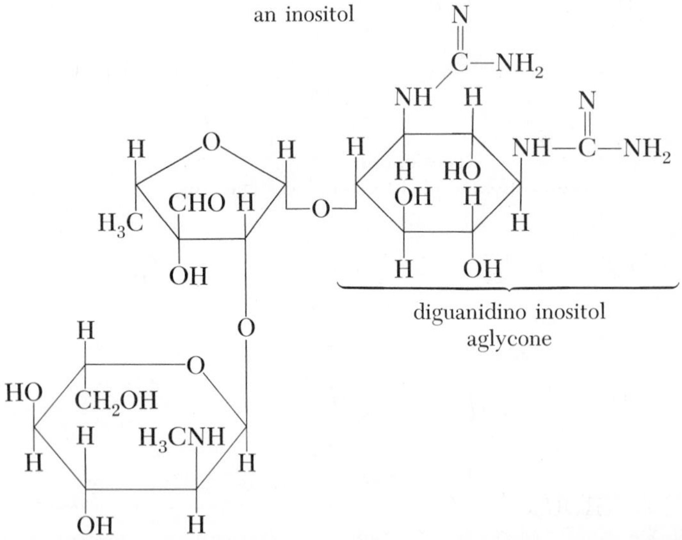

streptomycin

diguanidino inositol aglycone

5. (+)-Sucrose has a specific rotation of +66.5°, while D-(+)-glucose and D-(−)-fructose have specific rotations of +52.7° and −92.4°, respectively. Assuming complete hydrolysis of (+)-sucrose, calculate the "specific" rotation of the hydrolysate, a mixture commonly called "invert sugar."

[a] See discussion of bitter almond toxicity, p. 479.

6. Would you expect sucrose to be oxidized by aqueous bromine? Explain.

25B.3 POLYSACCHARIDES

Polysaccharides are polymers containing as many as several thousand monosaccharide units per molecule. As with other polymers, it is important to know which specific monomers are involved in polymer formation, the method of linkage between the monomers, and the gross structure of the polymer. If the polymer contains more than one type of monosaccharide, *i.e.*, if it is a *heteropolysaccharide*, the sequence of sugars is also important.

While there are many naturally occurring polysaccharides, the two most important are *starch* and *cellulose*. Both of these are products of photosynthesis. Cellulose plays a tremendous role in our society. As wood, cellulose provides shelter; as wood pulp, it is the basis of paper. It is hard to imagine our civilization without paper, books, and libraries, all of which depend upon trees. Cellulose is also the major constituent of cotton, the most important natural fiber. Starch is the mainstay of many diets, since it is the major compound in rice, potatoes, wheat, and corn.

Starch. Starch is the major source of energy in plant cells. It occurs as starch granules which, in turn, contain two major fractions: **amylose** (approximately 20%) and **amylopectin** (approximately 80%). Although both of these can be hydrolyzed in acid to give only D-glucose, amylose is water-soluble while amylopectin is not. This behavior suggests that, since both polymers contain the same monomer, the important differences between them must exist in the bonding within these polymers.

Amylose. (+)-Maltose is the only disaccharide produced upon hydrolysis of amylose. The absence of any (+)-cellobiose suggests that amylose is a linear polymer of D-glucose molecules, each bonded by an α-glycosidic linkage to C4 of the adjacent glucose unit. (Had cellobiose been produced, it would have indicated that some of the bonds between adjacent glucose units were β-glycosidic linkages.)

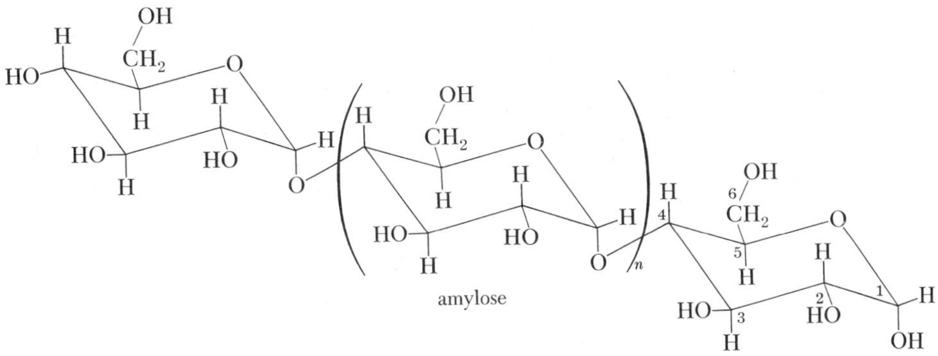

How long is the amylose chain? Physical measurements suggest a molecular weight of about 40,000, *i.e.*, a polymer containing upwards of 200 monosaccharide units. This is supported by chemical analysis, which we will now summarize.

A given amylose chain contains three types of glucose units. The most abundant are those within the body of the chain. These units have free hydroxy groups at C2, C3, and C6. In addition, each chain will have a glucose unit at each end, but these will differ in which carbons bear free hydroxy groups. At one end (the left end in the picture above) there will be a glucose bonded to the chain only through C1 and, therefore, having free hydroxy groups at C2, C3, C4, and C6. (Remember that *every* glucose unit has the hydroxy group of C5 tied up in hemiacetal formation.) At the other end (the right end in the picture on p. 835 there will be a glucose bonded to the chain only through C4 and, therefore, having free hydroxy groups at C1, C2, C3, and C6. If one could determine the ratio of, for example, glucose at the left end of the polymer to glucose within the body of the polymer, one could calculate the chain length of amylose. This can be done, simply enough, by methylating all of the free hydroxy groups in amylose.

The major product formed when amylose is methylated with dimethyl sulfate and then hydrolyzed is 2,3,6-tri-*O*-methyl-D-glucose. This results from alkylation of the internal glucose units. However, about 0.5% of the product is 2,3,4,6-tetra-*O*-methyl-D-glucose, formed by reaction of the glucose at the left end of the polymer. This ratio establishes a chain length of about 200.

The famous *starch-iodine test,* the development of a blue color when starch is treated with iodine, is better considered an "amylose-iodine test" since it is the amylose in starch that is responsible for most of the color. It is believed that amylose adopts a helical macro structure and that iodine is lodged within this helix to give a blue charge-transfer complex.

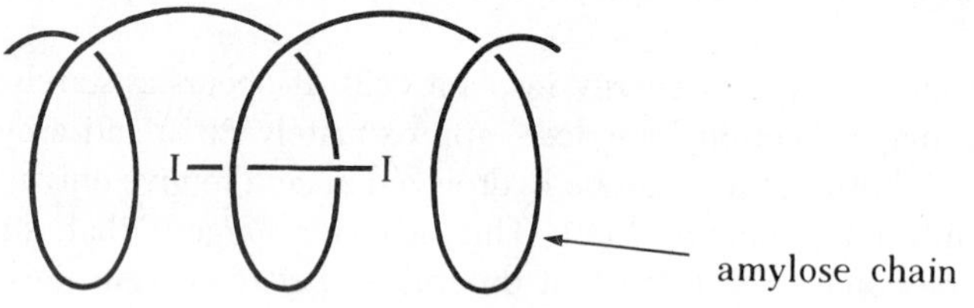

I—I

amylose chain

blue amylose-iodine complex

Amylopectin. Amylopectin is a branched polymer containing about 1000 D-glucose units. The main chain consists of an α-1,4-glycosidic linkage, while branching occurs with an α-1,6-glycosidic bond. Branching is moderate, with perhaps twenty or twenty-five glucose units occurring between branch points. There is a sufficient amount of helicity in this section of polymer to provide a blue-purple color upon addition of iodine.

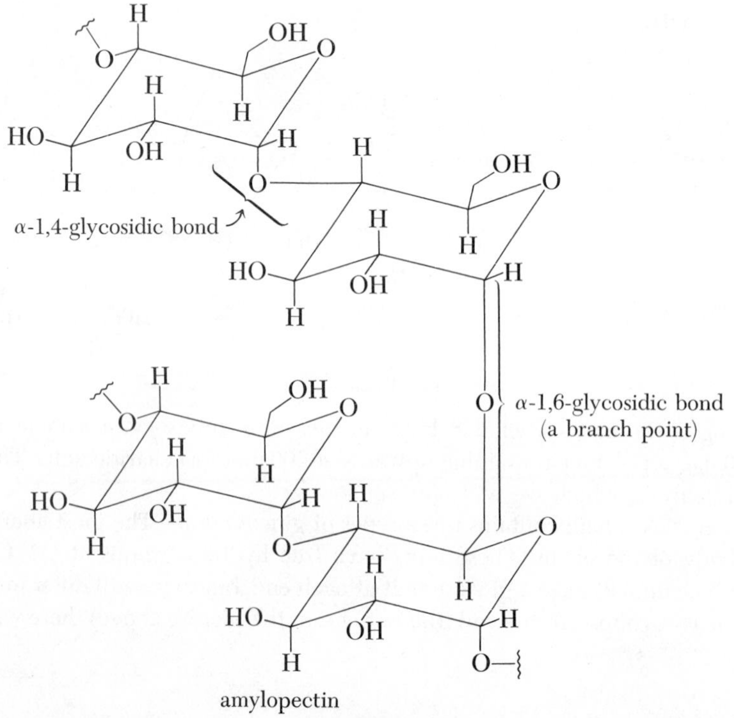

α-1,4-glycosidic bond

α-1,6-glycosidic bond (a branch point)

amylopectin

The partial hydrolysis of amylopectin produces rather large molecules called *dextrins.* Dextrins are used to prepare mucilage, pastes, and fabric sizing. (*Sizes* are materials used to fill the pores in cloth, paper, *etc.*) Printing inks are often thickened by the addition of dextrins.

While the major disaccharide produced by the hydrolysis of amylopectin is maltose, the glucose unit at each branch point has C4 and C6 hydroxy groups involved in glycosidic linkages. This leads to a small amount of *isomaltose* upon hydrolysis.

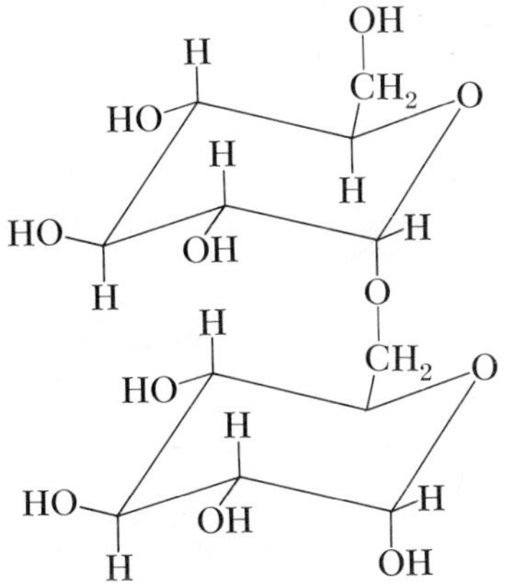

isomaltose

Glycogen. Glycogen is the polysaccharide that supplies the body with glucose during periods of exertion and between meals. It is stored mainly in the liver and in skeletal muscles. Chemically, it is quite similar to amylopectin except that branching is much more extensive in glycogen. *Glycogen is structurally and functionally the animal's counterpart of a plant's starch.*

The body synthesizes glycogen in a sequence that we will pick up at a convenient point, the enzymatic conversion of glucose to α-D-glucose 6-phosphate. This ester is isomerized to α-D-glucose 1-phosphate with the aid of magnesium ion and the enzyme *phosphoglucomutase.* Glucose 1-phosphate then is converted to UDP-glucose which, in turn, transfers a D-glucose residue to a straight chain identical to a small section of amylose (Figure 25B–1).

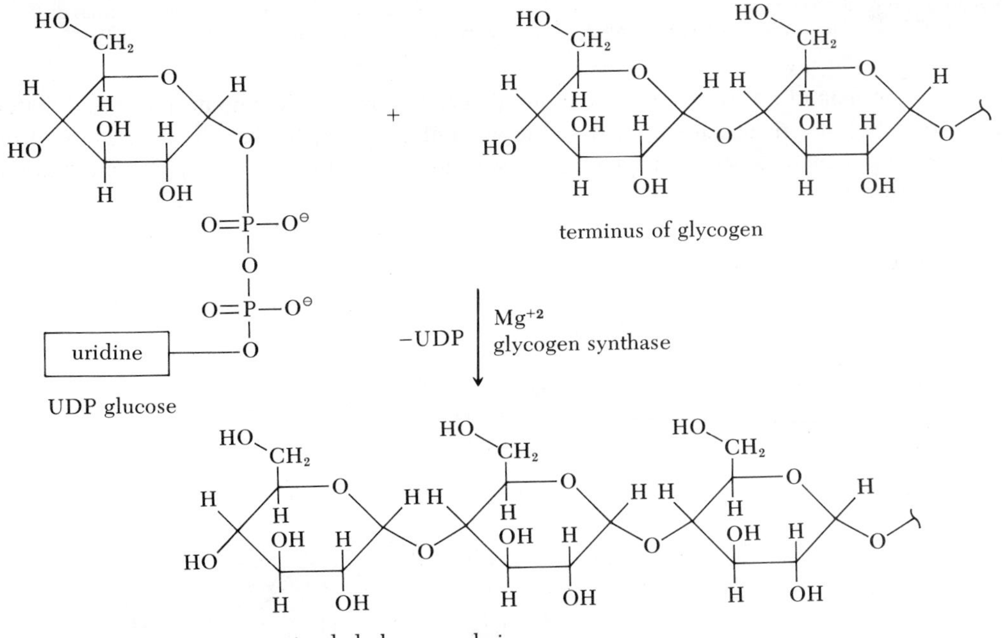

uridine

UDP glucose

terminus of glycogen

−UDP

Mg^{+2}
glycogen synthase

extended glycogen chain

Figure 25B–1 Growth of the glycogen chain. Terminal amylose chains on glycogen are extended by transfer of additional glucosyl residues from the phosphate of UDP glucose to the free hydroxyl group at C4 of the chain. The process may then be repeated to extend the chain still further. (From R.W. McGilvery, *Biochemistry—A Functional Approach,* W.B. Saunders Company, Philadelphia, 1970.)

Continued addition of the type shown in Figure 25B–1 would eventually lead to amylose. However, a branching enzyme, a *glycosyl 4:6-transferase,* catalyzes the transfer of segments of the growing chain onto C6—OH of an adjacent chain (Figure 25B–2).

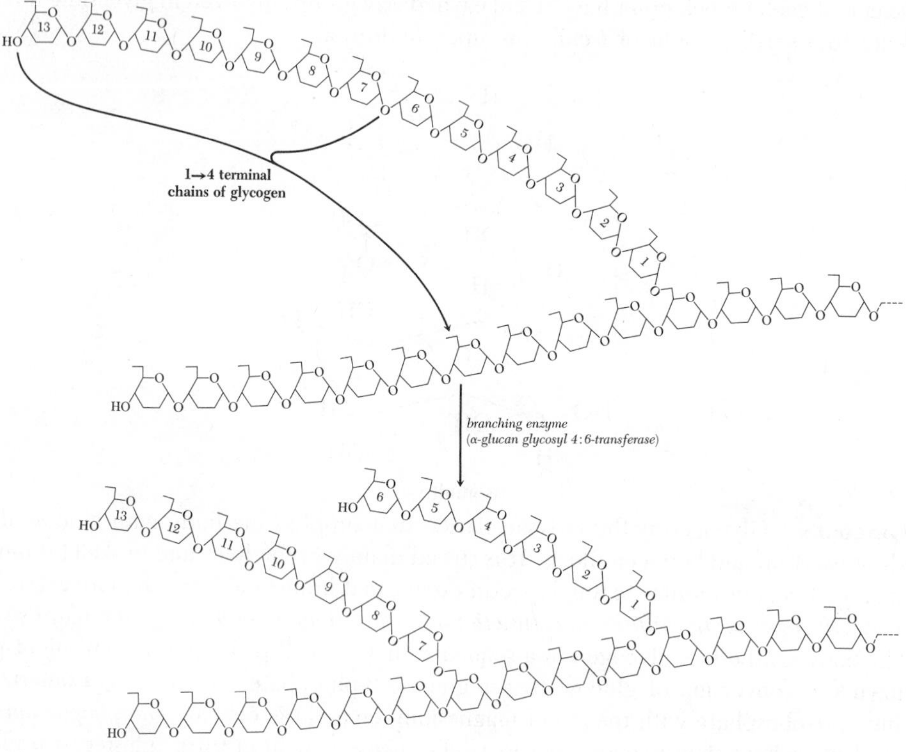

Figure 25B–2 Branching in the glycogen chain. A branch is created by transfer of a seven-residue segment of an amylose terminal chain to a hydroxy group on C6 of a glucose residue that is four residues removed from an existing branch. A terminal branch must be at least eleven residues long before a segment can be transferred from it. (From R.W. McGilvery, *Biochemistry—A Functional Approach,* W.B. Saunders Company, Philadelphia, 1970.)

The problem of fat (triglyceride) buildup ("obesity") is an epidemic in the United States. This arises because tissues are limited in the amount of glycogen that they can store; after producing up to 50 to 60 g of glycogen/kg of tissue, glucose is diverted from glycogen synthesis to fat synthesis.

Cellulose.

cellulose

Cellulose, an unbranched polymer of β-D-glucose, occurs in most plants. Most animals, including man and cattle, cannot hydrolyze the β-glycosidic link in cellulose although they can digest the α-glycosidic link found in amylose; such is the specificity of enzymes! However, many microorganisms can digest cellulose, and these are found in the soil and in the intestinal tracts of those animals which can "digest" wood and leaves. These microorganisms not only are important in our food cycle (they are critical to the beef industry), but they also are essential to maintaining the balance in our ecosystem. The soil bacteria, termites, carpenter ants, and similar organisms aid in the removal of forest debris (dead leaves, fallen trees, man's left-behind rubbish) and are a vital part of our planet's attempt to cleanse itself.

The most important natural fiber, cotton, is about 98% cellulose. Acetal linkages in cellulose are hydrolyzed by acids but not by bases. This explains why a cotton blouse is quickly destroyed if acid is spilled on it but survives contact with an equivalent amount of base.

That "thick, rich and creamy" shampoo or lotion which is "superior" because of its viscosity may have been thickened artificially by the addition of a cellulose derivative. *Methocel*, cellulose containing a limited number of methylated hydroxy groups, is an excellent example of such a thickener. Cellulose that has many of its hydroxy groups converted to methoxy groups is used in varnishes, enamels, and packaging films.

Billiards and elephants used to be closely associated because early billiard balls were made from elephant ivory. Even a century ago the obvious disadvantages of this source of billiard ball material were apparent, and chemists developed *Celluloid* as a potential ivory substitute. Celluloid is made by molding together a mixture of *pyroxylin*, ethanol, and camphor. The key ingredient, pyroxylin, is *cellulose dinitrate*, cellulose containing two nitrate ester groups per glucose unit.

Guncotton, the trinitrate of cellulose, looks like ordinary cotton but is extremely explosive. Mixed with nitroglycerin in varying proportions, it is used in high-powered explosives and as propellant for artillery shells.

Cellophane is made by converting cellulose to its xanthate by reaction with carbon disulfide and base. The resulting viscous solution is extruded through a thin slit into a dilute solution of acid to form cellophane film. Extrusion of the solution through small holes (spinnerets) produces filaments, which are processed into the fiber known as *viscous rayon*.

$$\underset{\text{cellulose}}{\text{R—OH}} + CS_2 + NaOH \rightarrow \underset{\text{viscous solution}}{\text{R—O—}\overset{\overset{\text{S}}{\|}}{\text{C}}\text{—S}^{\ominus}\ \text{Na}^{\oplus}} \xrightarrow{H^{\oplus}} \underset{\substack{\text{regenerated}\\ \text{cellulose}\\ \text{(cellophane } or \text{ rayon)}}}{\text{R—OH}} + CS_2$$

"I think that I shall never see,
Anything as valuable as a tree."
(with apologies to Joyce Kilmer)

25B.4 AMINO SUGARS

For a long period in carbohydrate chemistry, amino sugars were considered as no more than curiosa. However, it is now clear that amino sugars are important to many animals, including mammals and arthropods.

Chitin. Members of the class Crustacea of the phylum Arthropoda are characterized by a hard, crusty *exoskeleton*. This covering is composed mainly of chitin, a polymer of 2-acetamido-2-deoxy-D-glucose. The rigid, non-cellular nature of this polymer requires periodic molting as the crustacean grows. The function of chitin, aside from structural support, is to act as a cuticle, preventing the entry or loss of excess water. It is, therefore, functionally analogous to keratin. The chitin of crustaceans and that of insects differ in a remarkable way. The former is impregnated with calcium carbonate and other salts, while the latter is impregnated with a complex material known as "insect wax." In both, this impregnation is a type of biological sizing, serving to fill the pores in the polymer matrix.

Crab shells serve as an excellent source of 2-amino-2-deoxy-D-glucose, producing this amino sugar in about 70% yield upon hydrolysis in concentrated hydrochloric acid. From the structure below, it can be seen that chitin and cellulose are rather similar.

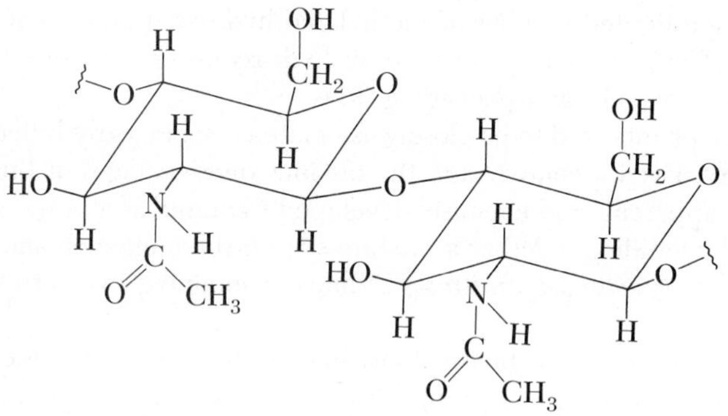

chitin

Chondroitin. Connective tissues consist of collagen fibers embedded in a *ground substance*. This ground substance consists of protein covalently bonded to carbohydrate, a complex often called a proteinpolysaccharide. The most abundant example of this type of polymer is given the general name *chondroitin*. The chondroitin carbohydrate chain is unbranched, unlike that of glycogen. Furthermore, the glycosidic bond in chondroitin has the β-configuration, again unlike glycogen. The deviation from the structure of the other polysaccharides is still greater in that chondroitin is a heteropolymer, consisting of alternating residues of D-glucuronic acid and *N*-acetyl-D-galactosamine bearing a sulfate unit. The particular chondroitin previously called "chondroitin A" has a sulfate at C4 of the amino sugar and is termed chondroitin 4-sulfate. Similarly, "chondroitin C" is now termed chondroitin 6-sulfate.

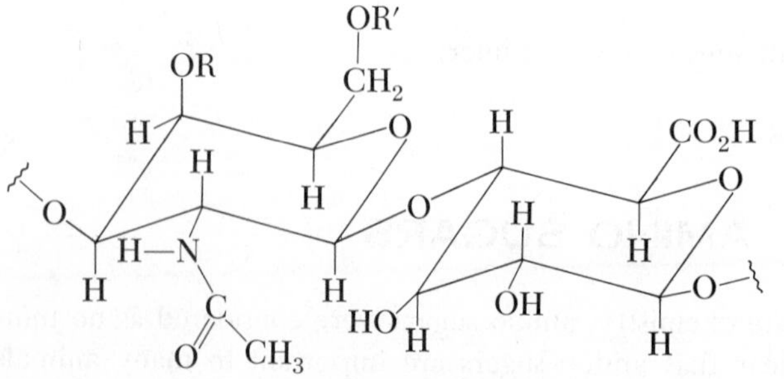

the chondroitin polymer

	R	R′
chondroitin A:	SO_3H	H
chondroitin C:	H	SO_3H

Dermatan sulfates (formerly "chondroitin B") are unusual analogs of chondroitin sulfate, having some D-glucuronate replaced by L-iduronate. This *mucopolysaccharide* (a polysaccharide consisting of alternating "ordinary" sugars and amino sugars) was originally found in the skin but was later discovered in the umbilical cord as a mixture of C4 and C6 sulfates. Its biological function is unclear, but it may be related to the spreading of skin.

Hyaluronic Acid. Hyaluronic acid consists of a long heteropolysaccharide containing alternating residues of D-glucuronic acid and *N*-acetyl-D-glucosamine; it lacks sulfate groups. The chain structure is abbreviated below (protein is associated with the polymer but has been deleted from the structure). This polysaccharide is found in connective tissue and is important in joint lubrication (sinovial fluid). It also is found in the fluids in the eye.

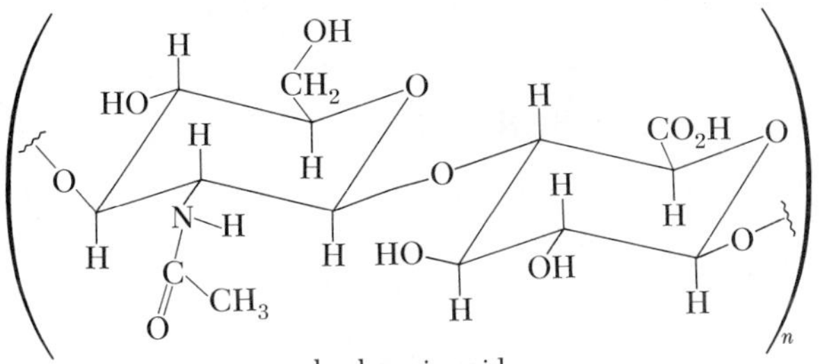

hyaluronic acid

both chondroitin and hyaluronic acid have alternating 1,3- and 1,4-glycosidic bonds

Heparin. Heparin is a small proteinpolysaccharide that is found in the liver and lungs and in mast cells, rather than in connective tissue. This small polymer (approximately 50 monosaccharide units) acts as an anticoagulant, preventing the conversion of prothrombin to thrombin. It may have other biological functions as well.

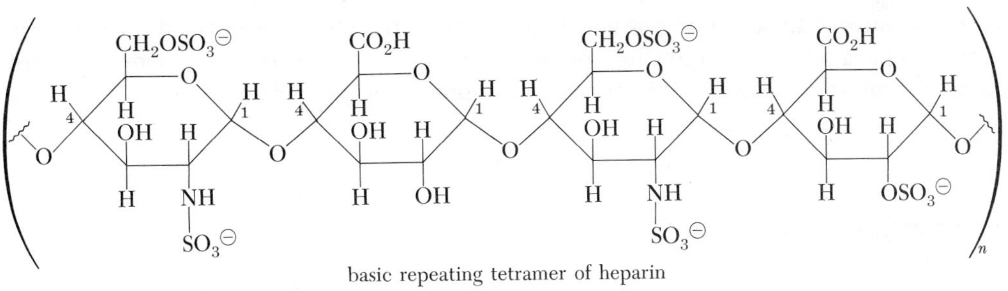

basic repeating tetramer of heparin

25B.5 BLOOD-GROUP SUBSTANCES

Blood-group substances are those compounds that are responsible for your having a certain letter-designated blood "type," *e.g.*, type A or type B. These complex molecules contain between 80 and 90% carbohydrate, the remainder being peptide. As an insoluble material on the surface of red blood cells, they constitute some of the substances (technically, agglutinogens) responsible for the clumping of red blood cells in incompatible serum of a different blood type. It is now known that blood-group substances are secreted, in water-soluble form, in semen, saliva, and gastric juices. It may be important that high concentrations of these substances occur in the fluid of ovarian cysts. We should also add that only 75% of persons of A, B, or O(H) blood types actually secrete A, B, or O(H) substances; the remainder secrete a different, specific substance termed Le[a].

The carbohydrates found in the blood-group substances are D-galactose, *N*-acetyl-D-glucosamine, *N*-acetyl-D-galactosamine, L-fucose, and *N*-acetyl-D-neuraminic acid (often abbreviated NANA). They form highly branched oligosaccharides that are bonded at frequent intervals to a peptide backbone, most likely through serine or threonine units. The exact structures of these substances are unknown, but their determination is a most important challenge because it is believed that the sugar sequence, stereochemistry, and nature of the glycosidic linkages control the high degree of immunological specificity of these molecules.

N-acetyl-α-D-neuraminic acid
NANA

While there are no known pathological conditions that have yet been directly linked with the blood-group substances, it has been established that type O individuals have a greater predilection toward duodenal ulcers, while type A individuals have a greater predilection toward pernicious anemia and gastric carcinoma. Research on the behavior of carbohydrates, far from being a subject of largely historic interest, is right at the forefront of modern chemistry and at the interface between chemistry and the practicing physician.

IMPORTANT TERMS

Arthropods: Invertebrate animals with jointed appendages and a horny external covering. This phylum includes spiders, crustaceans, and insects.

Carpenter ants: For the purist: phylum Arthropoda, subphylum Mandibulata, class Insecta, order Hymenoptera, family Formicidae, subfamily Formicinae. They are about one-half inch long and often nest in galleries excavated in wood (*e.g.*, house timbers). Unlike fire ants and harvester ants, the female carpenter ant does not sting.

Cellulose: A polysaccharide that surrounds the cells of all terrestrial plants. This coat is termed the *cell wall*. Cotton is almost pure cellulose. If treated with base, and then washed with water and dried, cotton increases in strength and ability to absorb dyes. Cotton treated in this way is said to be *mercerized*.

Dextrin: A gummy, sticky material produced by the partial hydrolysis of starch. Usually dextrorotatory.

Exoskeleton: The tough, external skeleton found on arthropods.

Glycogen: The polysaccharide used for carbohydrate storage in humans and other animals.

Guanidine: A strong base that may be viewed as the imine of urea.

$$\underset{\text{urea}}{H_2N-\overset{\overset{\displaystyle O}{\|}}{C}-NH_2} \qquad \underset{\text{guanidine}}{H_2N-\overset{\overset{\displaystyle NH}{\|}}{C}-NH_2}$$

Inositol: A general term representing one of the hexahydroxycyclohexanes. An even more general term, *cyclitol*, refers to any perhydroxycycloalkane.

$$\underset{\text{an inositol}}{(CHOH)_6} \qquad \underset{\text{a cyclitol}}{(CHOH)_n}$$

Mast cell: A connective tissue cell whose specific physiological function remains unknown.

Molting: Periodic shedding of an exoskeleton in order to permit continued growth of an animal.

Oligosaccharide: A sugar containing a modest (*e.g.*, 2 to 8) number of simple sugars in the polymer.

Prothrombin: A substance that is converted to thrombin in the body. This conversion involves the loss of carbohydrate, although the nature of the carbohydrate is in dispute.

Thrombin: A material produced in damaged cells that, by converting fibrinogen to fibrin (a protein), is essential for the clotting of blood. The blood clot consists largely of molecules of fibrin that have become cross-linked to one another.

PROBLEMS

7. Suggest a convenient means of distinguishing between members of the following pairs of compounds.
   (a) glucose and arabinose
   (b) ribose and arabinose
   (c) glucose and mannose
   (d) glucose and maltose
   (e) cellobiose and sucrose
   (f) maltose and lactose
8. Explain the significance of all of the terms in the following names.
   (a) *O*-α-D-glucopyranosyl-(1,2)-β-D-fructofuranoside
   (b) *O*-β-D-glucopyranosyl-(1,4)-α-D-glucopyranose
   (c) 2-acetamido-2-deoxy-D-glucopyranose
9. What, if anything, is the difference between members of the following pairs?
   (a) α-lactose and β-lactose
   (b) glycoside and glucoside
   (c) cellobiose and gentiobiose
   (d) table sugar and sucrose
   (e) glyconic acid and glycaric acid
   (f) osazone and osatriazole
   (g) amylose and amylopectin
   (h) anomer and epimer
   (i) furanoside and pyranoside
   (j) *N*-glycoside and *O*-glycoside

10. Paper napkins, tissues, and similar items are composed largely of cellulose. What is the structure of cellulose? At one time there was a television commercial that showed how "concentrated stomach acid" (hydrochloric acid) could rapidly destroy such paper articles. What chemical reaction was occurring during this dissolution?

11. D-Raffinose is a trisaccharide that does not react with Fehling's solution. Upon hydrolysis it produces D-glucose, D-galactose, and D-fructose. Completely methylated raffinose is hydrolyzed to 2,3,4-tri-*O*-methylglucose, 1,3,4,6-tetra-*O*-methylfructose, and 2,3,4,6-tetra-*O*-methylgalactose. What structures for raffinose are consistent with these data? What further information, if any, is required to identify completely the structure of raffinose?

NUCLEIC ACIDS

26.1 INTRODUCTION

Nucleic acids were discovered in 1869; they received this name because of their prevalence in cell nuclei. It is now recognized that nucleic acids are found in both eukaryotic (nucleated) and prokaryotic (non-nucleated) cells, associated with proteins as nucleoproteins. This wide distribution is related to two vital roles played by nucleic acids: protein biosynthesis and the transmission of genetic information from generation to generation.

Nucleic acids are heteropolymers of high molecular weight that can be hydrolyzed to an equimolar mixture of *heterocyclic amines*, a *pentose*, and *phosphoric acid*. These giant molecules, with molecular weights up to ten million, are easily broken; early structural investigations were hampered by the chemist's inability to isolate intact, homogeneous nucleic acids. Technology now has advanced to the point where one can separate an intact nucleic acid from a mixture of nucleic acids.

The two known types of nucleic acids are **RNA** (ribonucleic acid) and **DNA** (deoxyribonucleic acid).[a] These names reflect the fact that RNA liberates the pentose D-ribose upon complete hydrolysis, while DNA liberates 2-deoxy-D-ribose instead. Incomplete hydrolysis of nucleic acids produces **nucleotides,** which can be further hydrolyzed to phosphoric acid and **nucleosides.** Hydrolysis of the nucleoside produces a heterocyclic amine (often called simply a "base") and the appropriate pentose. Various stages in the degradation of nucleoproteins are shown below.

nucleoproteins
↓
nucleic acid *and* protein
↓
nucleotides
↓
nucleosides *and* phosphoric acid
↓
pentose *and* heterocyclic amines

While DNA and RNA differ most obviously in their sugar component, it is their bases that appear to be most intimately related to their biological function. The bases that occur in nucleic acids are derivatives of **pyrimidine** and **purine.**

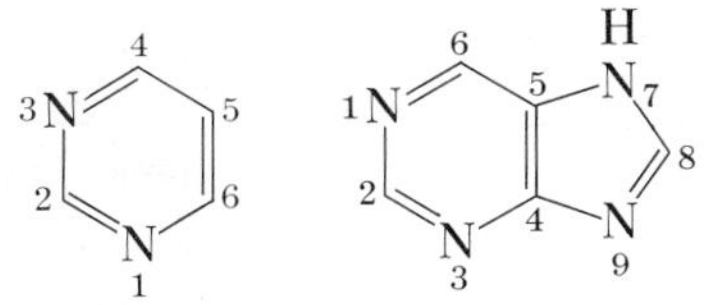

pyrimidine purine

---

[a] While these terms are usually used in the singular form, there are actually many different DNA and RNA molecules.

26.2 PYRIMIDINES AND PURINES

Pyrimidines and purines are important compounds in their own right as well as because of their role in nucleic acid chemistry. Therefore, we shall digress a bit and look at these compounds, with an emphasis on their reactions and their medical importance.

Pyrimidines. The synthesis of various pyrimidines is usually accomplished by a condensation reaction between urea, thiourea, or an amidine with a malonic acid derivative or a β-keto ester. Pyrimidine itself can be prepared *via* a sequence involving barbituric acid as an intermediate.

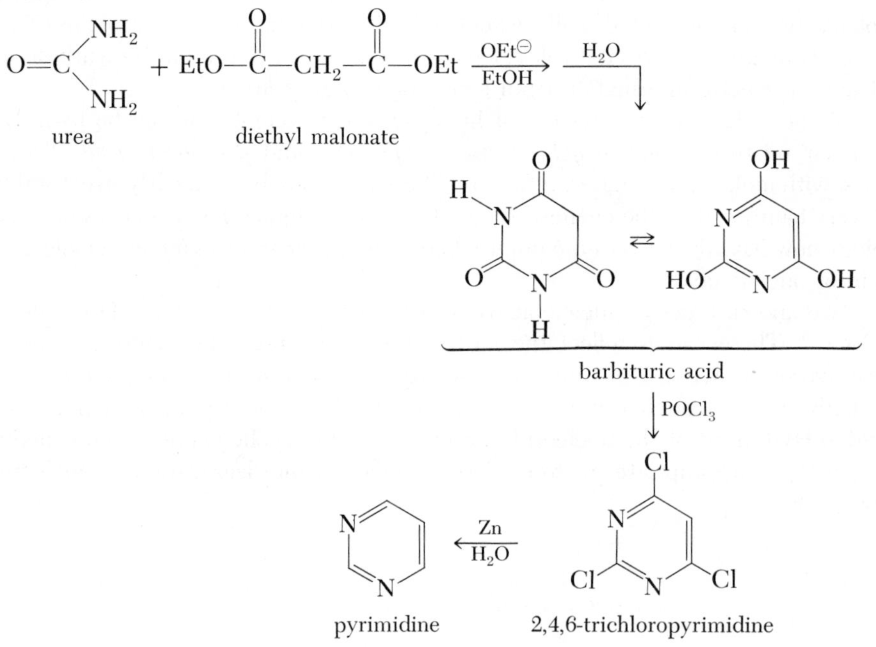

$$O{=}C(NH_2)_2 + EtO{-}\overset{O}{\overset{\|}{C}}{-}CH_2{-}\overset{O}{\overset{\|}{C}}{-}OEt \xrightarrow[EtOH]{OEt^{\ominus}} \xrightarrow{H_2O}$$

urea — diethyl malonate

barbituric acid (keto form $\rightleftarrows$ 4,6-dihydroxy-2-hydroxypyrimidine form: HO, OH, OH)

$\downarrow POCl_3$

2,4,6-trichloropyrimidine $\xrightarrow[H_2O]{Zn}$ pyrimidine

1. Suggest a mechanism for the conversion of urea and diethyl malonate to barbituric acid.

Condensation of urea with substituted malonic esters produces a class of compounds called **barbiturates.** As cyclic diimides, these compounds are acidic and form stable sodium salts. Both free barbiturates and the corresponding salts are used as sedatives, soporifics, and anesthetics. Barbituric acid itself has no hypnotic effect upon humans. The familiar drugs phenobarbital and pentobarbital sodium (Nembutal) are included in the class of barbiturates. Phenobarbital has been used to prevent epileptic seizures; unfortunately, the indiscriminate use of barbiturates, coupled with the synergistic effect of "booze," has provided a relatively simple route to unintentional suicide. Sodium pentothal, a salt of a *thiobarbiturate,* is used as a general anesthetic that is administered intravenously.

$$(C_2H_5O{-}C(=O))_2C(C_2H_5)C_6H_5 + H_2NC(O)NH_2 \xrightarrow[EtOH]{OEt^{\ominus}} \xrightarrow{H_2O} \text{phenobarbital}$$

diethyl ethylphenylmalonate

phenobarbital

$$(C_2H_5OC(=O))_2C(C_2H_5){-}C(CH_3)HCH_2CH_2CH_3 + H_2NC(O)NH_2 \xrightarrow[EtOH]{OEt^{\ominus}} \xrightarrow{H_2O} \text{pentobarbital}$$

diethyl ethyl-1-methylbutyl malonate

pentobarbital (*Nembutal*)

$$\text{diethyl ethyl(1-methylbutyl)malonate} + (H_2N)_2CS \xrightarrow{OEt^{\ominus}} \text{thiobarbiturate} \xrightarrow{NaOEt} \text{sodium pentothal}$$

thiourea

sodium pentothal

2. Starting with benzene and compounds containing no more than two carbons, suggest a synthesis of diethyl ethylphenylmalonate and diethyl ethyl-1-methylbutylmalonate.

The product of the condensation of guanidine with ethyl formylacetate can be converted to 2-aminopyrimidine. This pyrimidine derivative is used in the synthesis of the antibiotic *sulfadiazine.*

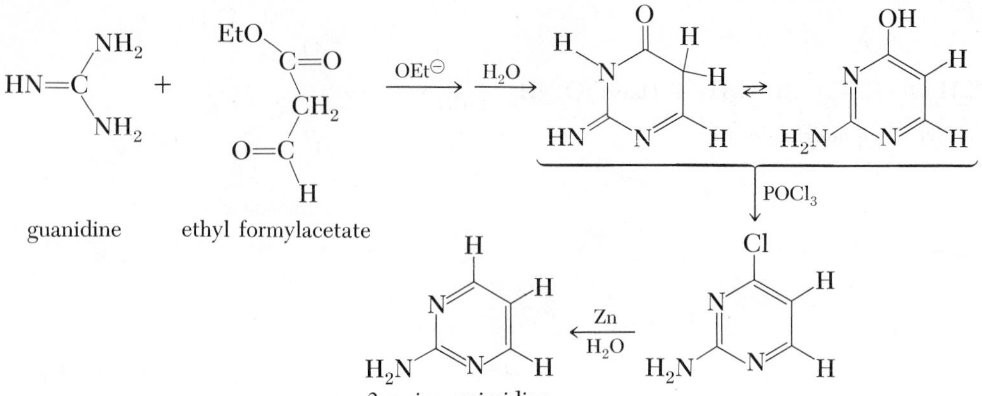

guanidine + ethyl formylacetate $\xrightarrow{OEt^{\ominus}}$ $\xrightarrow{H_2O}$ (tautomers) $\xrightarrow{POCl_3}$ $\xrightarrow{Zn, H_2O}$ 2-aminopyrimidine

The pyrimidine derivatives most commonly found in nucleic acids are uracil, thymine, and cytosine. These compounds are capable of tautomerism (see below); the most important tautomers are the non-aromatic ones, since these are involved in forming nucleosides and nucleotides. (Are all of the possible tautomers shown below?)

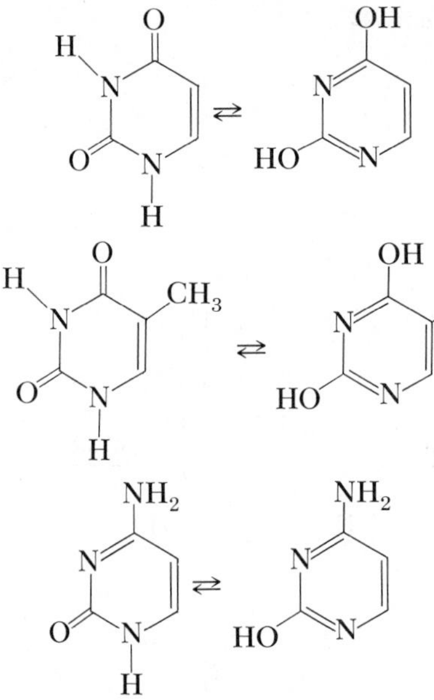

uracil

thymine

cytosine

Uracil can be prepared by the condensation of urea with ethyl formylacetate. Substitution of thiourea for urea produces *2-thiouracil,* used in treating hyperthyroidism and angina pectoris.

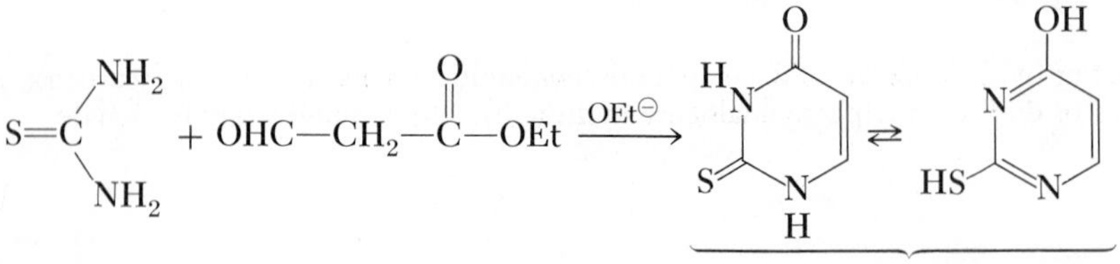

$$S{=}C(NH_2)_2 + OHC{-}CH_2{-}\overset{O}{\overset{\|}{C}}{-}OEt \xrightarrow{OEt^{\ominus}}$$

2-thiouracil

The condensation of S-methylthiourea and ethyl 2-formylpropionate can be used to prepare thymine.

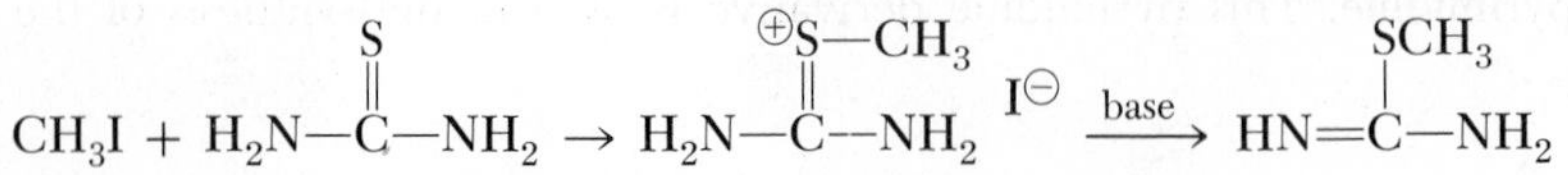

$$CH_3I + H_2N{-}\overset{S}{\overset{\|}{C}}{-}NH_2 \rightarrow H_2N{-}\overset{\oplus S-CH_3}{\overset{\|}{C}}{-}NH_2 \; I^{\ominus} \xrightarrow{base} HN{=}\overset{SCH_3}{\overset{|}{C}}{-}NH_2$$

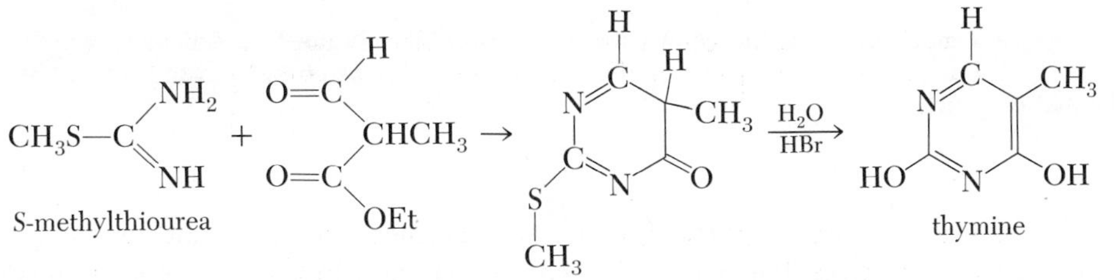

S-methylthiourea + … → … $\xrightarrow[HBr]{H_2O}$ thymine

Purines. The great and prolific Emil Fischer used barbituric acid to synthesize 2,6,8-trihydroxypurine, usually called *uric acid*. (Recall that the relationship between gout and and uric acid was discussed on p. 782.)

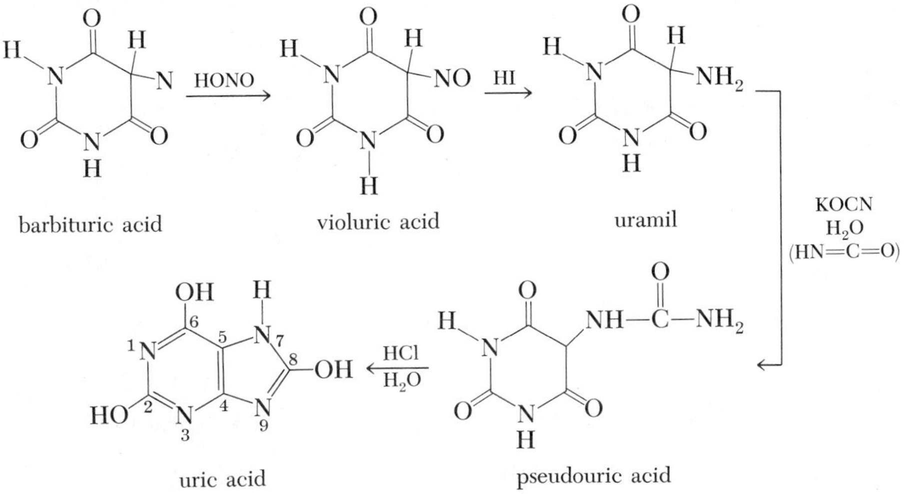

barbituric acid $\xrightarrow{HONO}$ violuric acid $\xrightarrow{HI}$ uramil $\xrightarrow[(HN{=}C{=}O)]{KOCN,\ H_2O}$ pseudouric acid $\xrightarrow[H_2O]{HCl}$ uric acid

3. What is the mechanism for the conversion of pseudouric acid to uric acid in the presence of hydrochloric acid?

Using uric acid as starting material, Fischer also synthesized *adenine, guanine,* and *xanthine*. Both adenine and guanine commonly occur in nucleic acids.

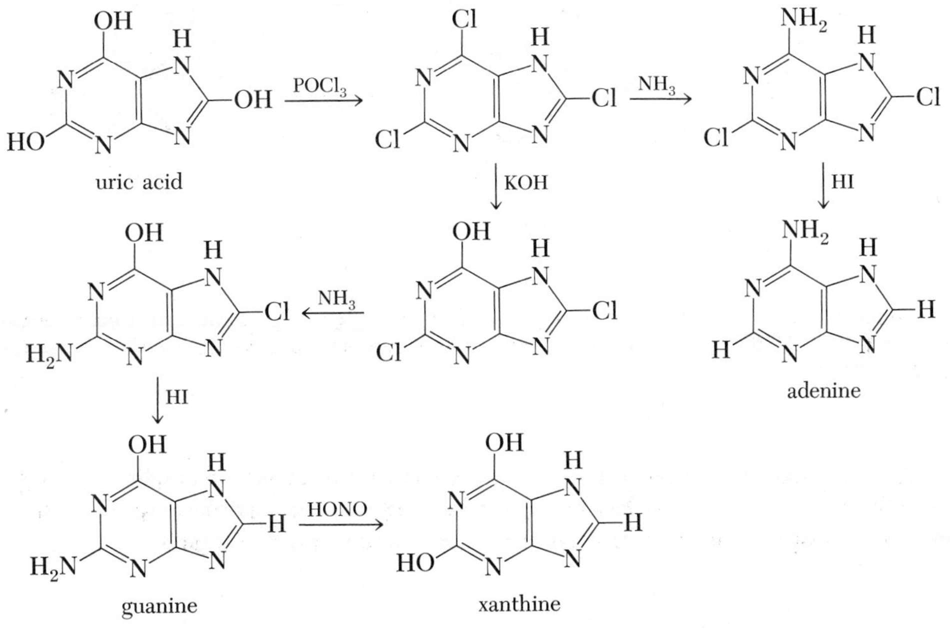

uric acid $\xrightarrow{POCl_3}$ … $\xrightarrow{NH_3}$ … $\xrightarrow{HI}$ adenine

… $\xrightarrow{KOH}$ … $\xrightarrow{NH_3}$ … $\xrightarrow{HI}$ guanine $\xrightarrow{HONO}$ xanthine

4. Suggest a mechanism for the conversion of 2,6,8-trichloropurine to 2,8-dichloro-6-hydroxypurine by the action of hydroxide ion. (*Hint:* The reaction may be considered to be an addition-elimination process.)

Although uric acid usually is thought of as a convenient starting point for the synthesis of many purine derivatives, the purine ring system may also be considered as an imidazole ring fused to a pyrimidine ring. In the synthesis of guanine shown below, a technique for synthesizing a diaminopyrimidine with adjacent amino groups is demonstrated, as is the use of this type of compound to create a substituted purine.

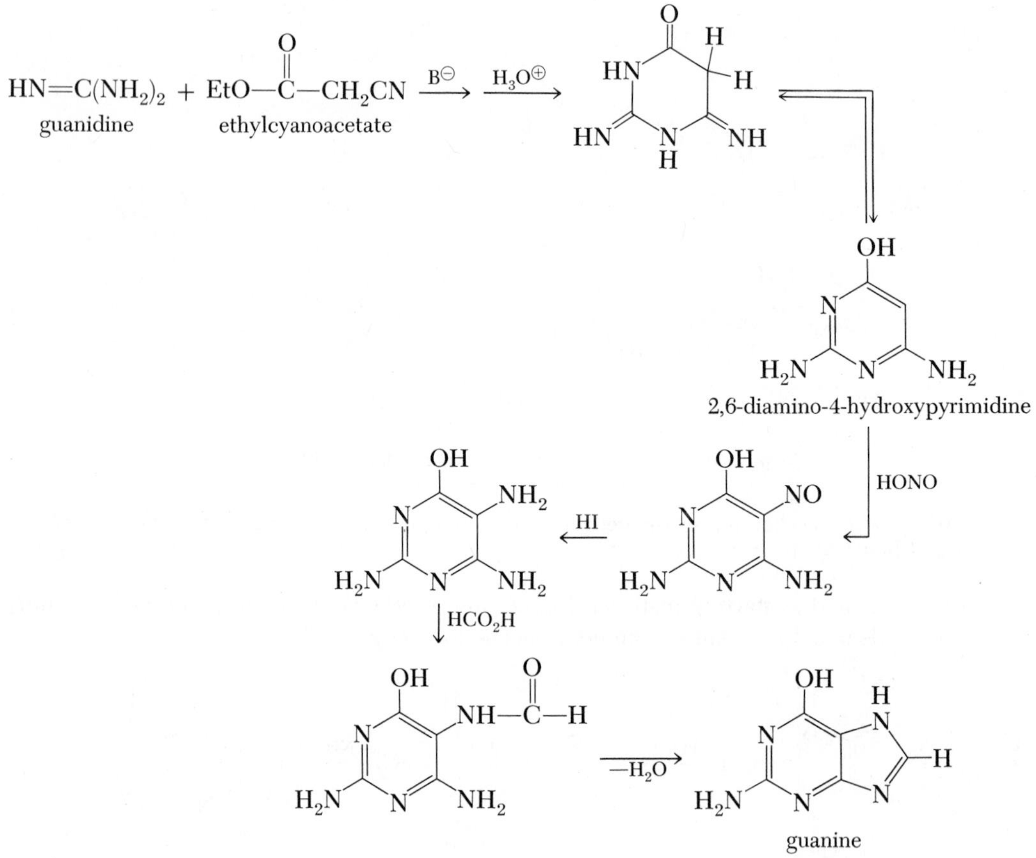

$$\underset{\text{guanidine}}{HN{=}C(NH_2)_2} + \underset{\text{ethylcyanoacetate}}{EtO{-}\overset{O}{\overset{\|}{C}}{-}CH_2CN} \xrightarrow{B^{\ominus}} \xrightarrow{H_3O^{\oplus}}$$

2,6-diamino-4-hydroxypyrimidine

HONO

HI

HCO_2H

$-H_2O$

guanine

5. Suggest a mechanism for the base-catalyzed condensation of guanidine with ethyl cyanoacetate to form 2,6-diamino-4-hydroxypyrimidine. Which protons of guanine should be exchanged in $D_2O/OD^{\ominus}$? In D_2O/DCl?

The sulfur analog of hypoxanthine, *6-mercaptopurine*, is an anti-metabolite that has been used in the treatment of acute leukemia. A related compound, azathioprine (*Imuran*), is an immunosuppressive agent used to prevent rejection of transplanted organs.

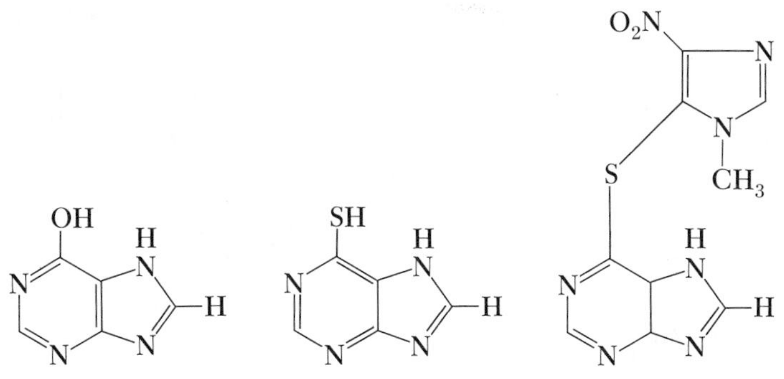

hypoxanthine 6-mercaptopurine *Imuran*

Certain plants are grown and valued for the purines they possess. Thus, the stimulant action of tea, cocoa, and coffee is due to three related methylated purines: *theophylline, theobromine,* and *caffeine.*

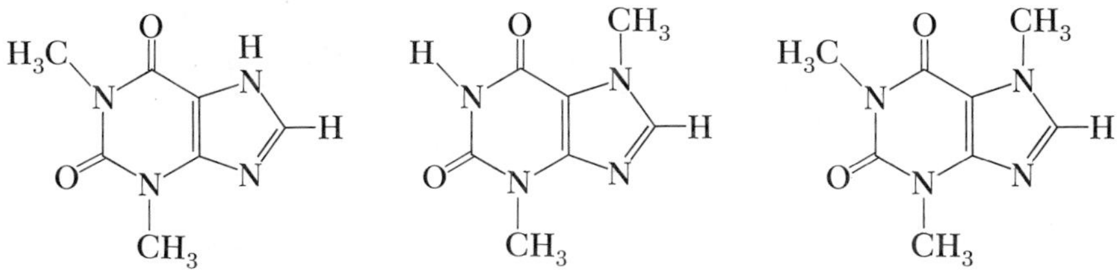

theophylline (tea) theobromine (cocoa) caffeine (coffee)

26.3 NUCLEOSIDES AND NUCLEOTIDES

A **nucleoside** is an *N*-glycoside whose aglycone is, with rare exception, a derivative of either pyrimidine or purine. Nucleosides are divided into *ribosides* and *deoxyribosides,* depending upon whether the sugar is ribose or deoxyribose. As seen from Table 26–1, nucleosides carry trivial names based upon their aglycones; the names of pyrimidine derivatives end in "idine," while those of purine derivatives end in "osine." The structures of representative

TABLE 26–1 Common Nucleosides

	Nucleoside	
Base	*Riboside*	*Deoxyriboside*
adenine	adenosine	deoxyadenosine
guanine	guanosine	deoxyguanosine
uracil	uridine	deoxyuridine
cytosine	cytidine	deoxycytidine
thymine	thymine riboside	thymidine
hypoxanthine	inosine	deoxyinosine

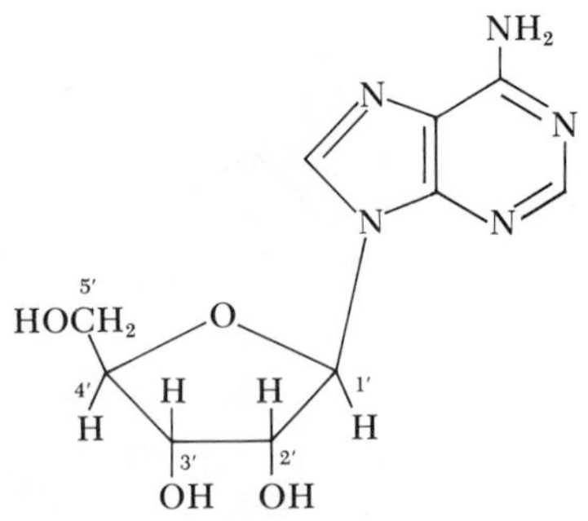

adenosine

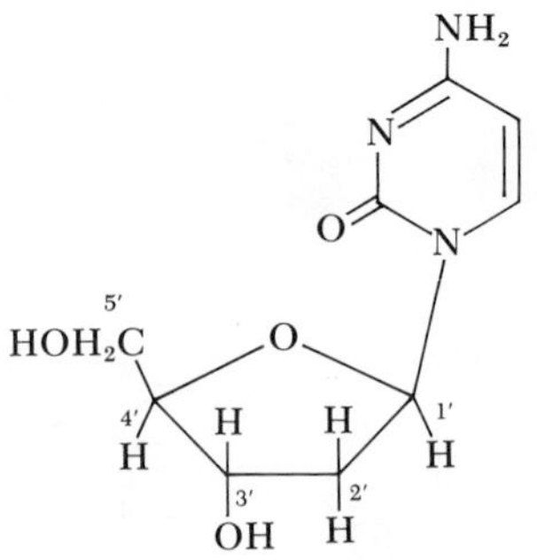

deoxycytidine

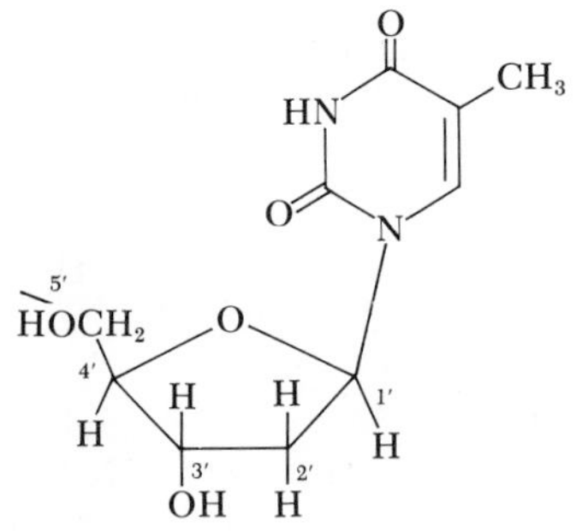

thymidine

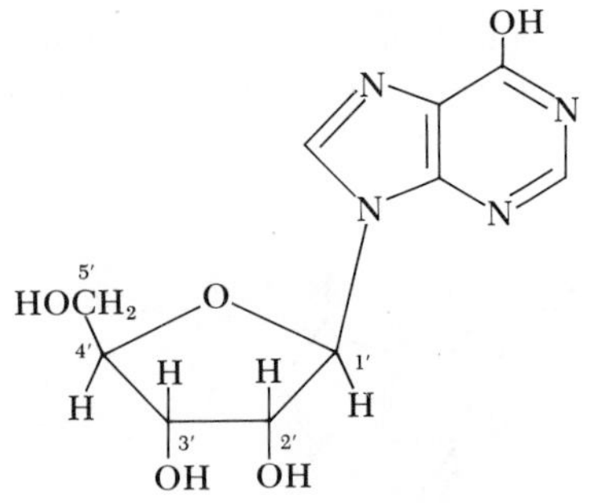

inosine

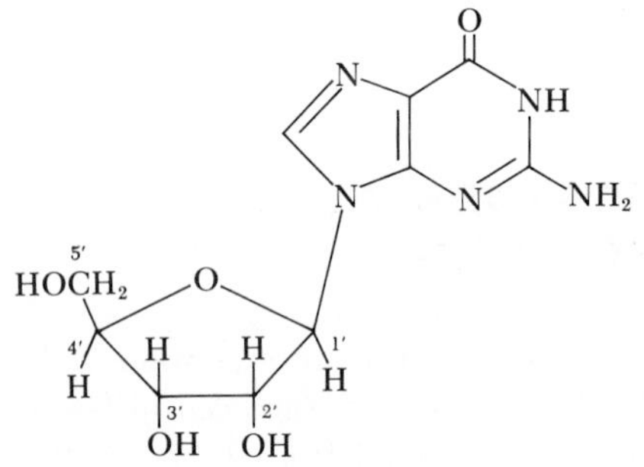

guanosine

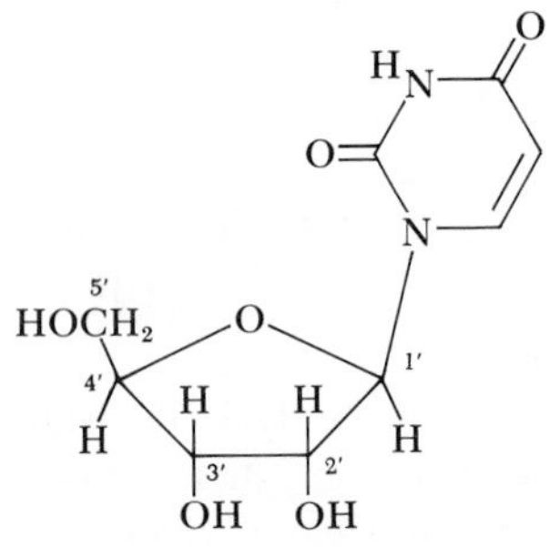

uridine

Figure 26–1 Nucleosides. Adenosine and inosine are ribosides, while deoxycytidine and thymidine are deoxyribosides. What are guanosine and uridine?

nucleosides are presented in Figure 26–1. Bases other than those included in Table 26–1 also occur in nature, often in very unusual or important systems. For example, nucleic acids from bacteriophages (*i.e.*, viruses that attack bacteria) contain 5-hydroxymethylcytosine and 5-hydroxymethyluracil. The unusual *C*-nucleoside, *pseudouridine*, is found in *t*RNA (see p. 867).

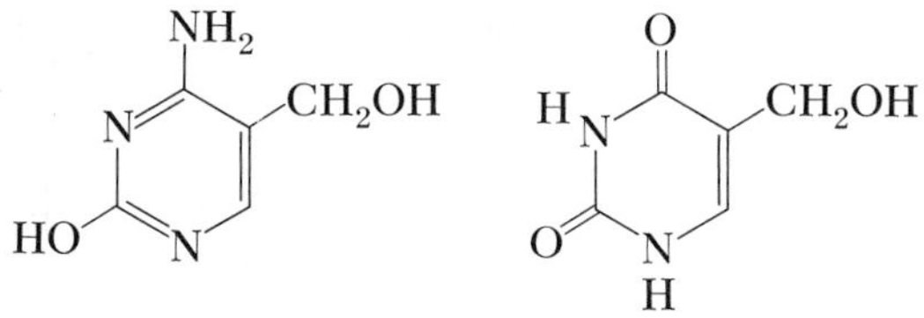

5-hydroxymethylcytosine 5-hydroxymethyluracil

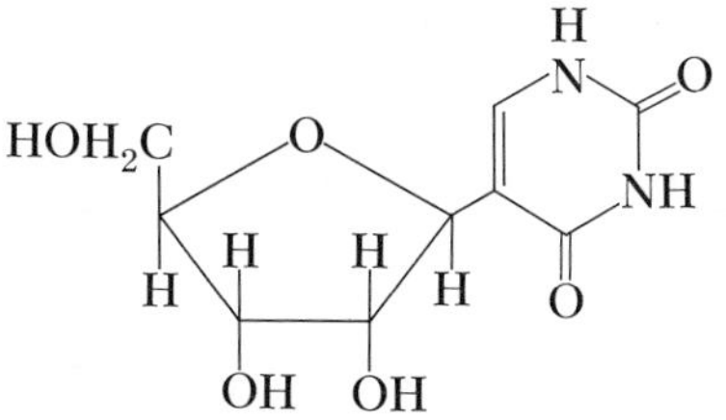

pseudouridine

A **nucleotide** is a nucleoside that bears a phosphate group, —OP(O)$(OH)_2$, bonded to the pentose. The phosphate is usually attached to C3′ or C5′ of the pentose.[a] Since these —OP(O)$(OH)_2$ groups are acidic, nucleotides can be named as phosphates or as acids (Table 26–2).

TABLE 26-2 Common Nucleotides

	NUCLEOTIDE[a]	
BASE	*Named as Acid*	*Named as Phosphate*
adenine	2′-adenylic acid	adenosine 2′-monophosphate[b]
adenine	3′-adenylic acid	adenosine 3′-monophosphate
adenine	5′-adenylic acid	adenosine 5′-monophosphate
uracil	3′-uridylic acid	uridine 3′-monophosphate
guanine	3′-guanylic acid	guanosine 3′-monophosphate
cytosine	3′-cytidylic acid	cytidine 3′-monophosphate
hypoxanthine	3′-inosinic acid	inosine 3′-monophosphate
adenine	deoxyadenylic acid	deoxyadenosine 5′-monophosphate
guanine	deoxyguanylic acid	deoxyguanosine 5′-monophosphate
thymine	thymidylic acid	thymidine 5′-monophosphate
cytosine	deoxycytidylic acid	deoxycytidine 5′-monophosphate

[a] The first seven entries are ribonucleotides, while the last four are deoxyribonucleotides. Where compounds form 3′-phosphates, they also form 5′-phosphates, diphosphates, and triphosphates (as shown, for example, for adenine in the structure on p. 854). Indeed, all 5′-phosphates may also form di- and triphosphates.

[b] The prefix "mono" is often deleted.

One of the most important nucleotides, and one involved in a variety of biological reactions, is adenosine 5′-phosphate, usually called *adenosine monophosphate* or AMP. AMP is a by-product of an energy-producing sequence that uses *adenosine 5′-triphosphate*, ATP, as

[a] In naming nucleosides and nucleotides, we assign "primed" numbers to the pentose atoms and "unprimed" numbers to the purine or pyrimidine atoms.

the energy source (see p. 756). The compound in an intermediate state of phosphorylation is ADP, adenosine 5′-diphosphate.

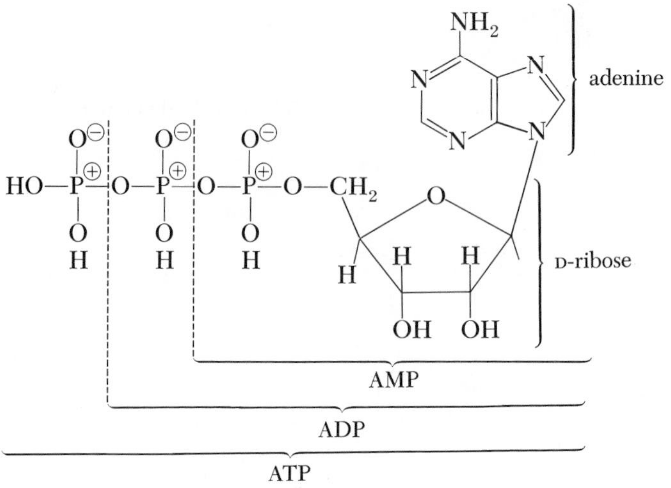

adenine

D-ribose

AMP

ADP

ATP

In addition to the simple monophosphates, *cyclic nucleotides,* involving two hydroxy groups of the pentose, are now recognized as being of great importance in biological processes. Of these, the most important is called *cyclic AMP* or adenosine 3′,5′-monophosphate.[a]

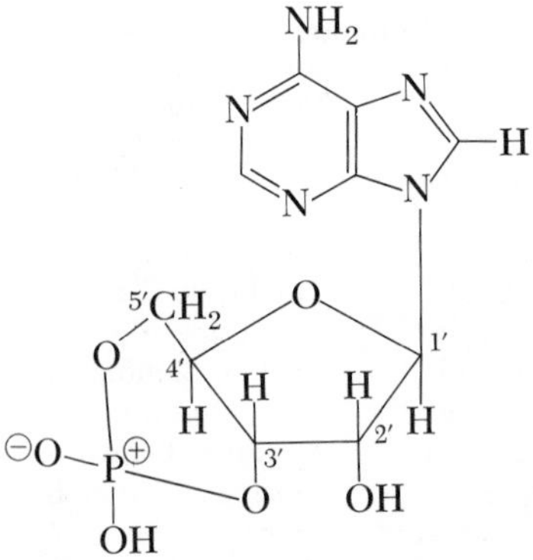

cyclic AMP

26.4 *IN VITRO* NUCLEOSIDE SYNTHESIS

There are three general routes to nucleosides: (a) a carbohydrate (or derivative) and a base (or derivative) can be combined directly; (b) a simple *N*-glycoside can be built up into a nucleoside; or (c) an existing nucleoside can be modified to produce a new nucleoside.

Direct Combination. A nucleoside can be prepared by the reaction of a glycosyl halide with an alkoxy derivative of a base. Use of an alkoxy derivative of the base, rather than the free base itself, serves to limit the nucleophilic centers on the base to nitrogen by tying up potential hydroxy groups. Moreover, the alkoxy group is easily converted into a hydroxy group. The synthesis of uridine illustrates this procedure.

[a] The biological role of cyclic AMP has been reviewed in an article by I. Pastan [*Sci. Amer.*, **227,** 97 (1972)].

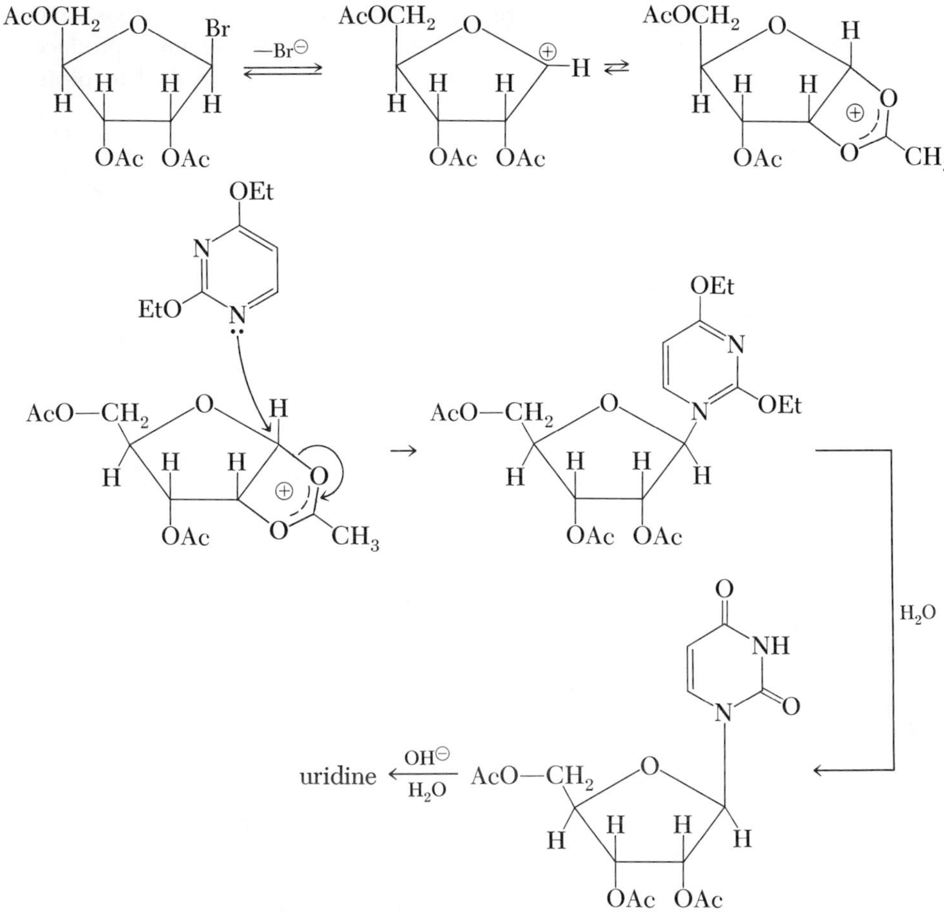

If the base has reactive $—NH_2$ groups, it is best added as a *mercuri* or *chloromercuri* derivative of the acetylated base. Cytosine preparation illustrates this point:

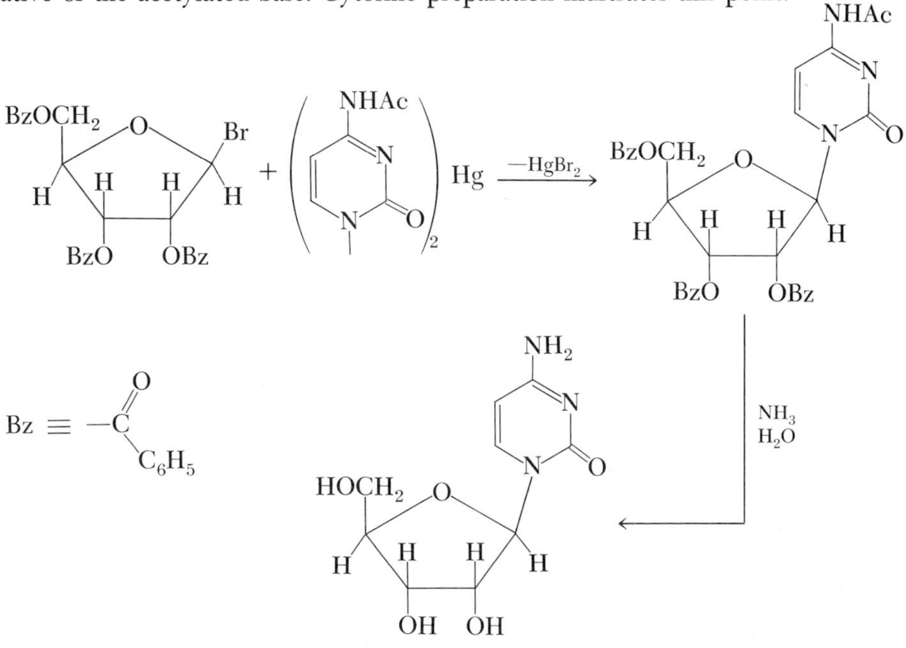

$Bz \equiv -C(=O)C_6H_5$

cytosine

Beginning with Amino Sugars. In the synthesis outlined below, β-ethoxy-*N*-ethoxycarbonylacrylamide is condensed with 2,3,5-tri-*O*-benzoyl-β-D-ribosylamine to produce, after hydrolysis, uridine. Here, as in the previous example, acyl groups are used to protect the hydroxy groups of the pentose.

net reaction:

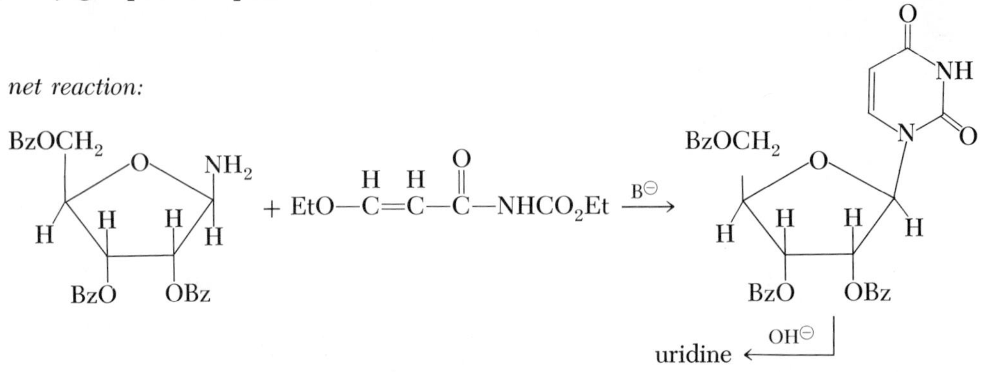

$BzOCH_2$ (ribosylamine: NH_2, BzO, OBz) $+ \ EtO—CH{=}CH—C(=O)—NHCO_2Et \xrightarrow{B^{\ominus}}$ $BzOCH_2$ (N-glycoside of the pyrimidine: O, NH, N, O; BzO, OBz)

uridine $\xleftarrow{OH^{\ominus}}$

mechanism:

$EtO—CH{=}CH—C(=O)—NH—C(=O)—OEt$ + $R—NH_2$ → $EtO—CH(NH_2^{\oplus}R)—CH^{\ominus}—C(=O)—NH—C(=O)—OEt$

⇅

$HC(NHR){=}CH—C(=O)—NH—C(=O)—OEt \xleftarrow{-EtOH} EtO—CH(NHR)—CH^{\ominus}—C(=O)—NH—C(=O)—OEt$

↓

(cyclic intermediate: $H_2N^{\oplus}$, C=C, C=O, C—N, $^{\ominus}O$, OEt, H) $\xrightarrow{-EtO^{\ominus}}$ (H, $N^{\oplus}$, C=C, C=O, C—N, O, H)

$\downarrow -H^{\oplus}$

(N-substituted uracil: H, H, C=C, N, C=O, C—N, O, H)

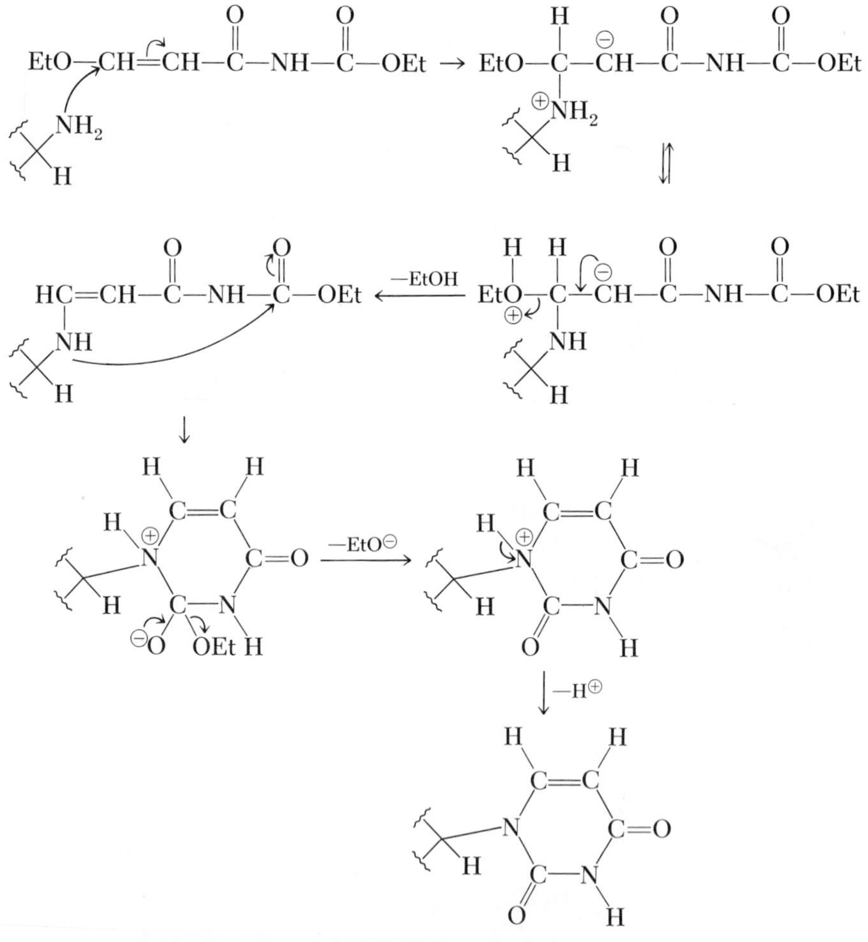

Nucleoside Modification. Interconversion of nucleosides may involve changes in the carbohydrate or the base or both. The examples that follow involve alterations of the base, the more common of the two possible modifications.

A:

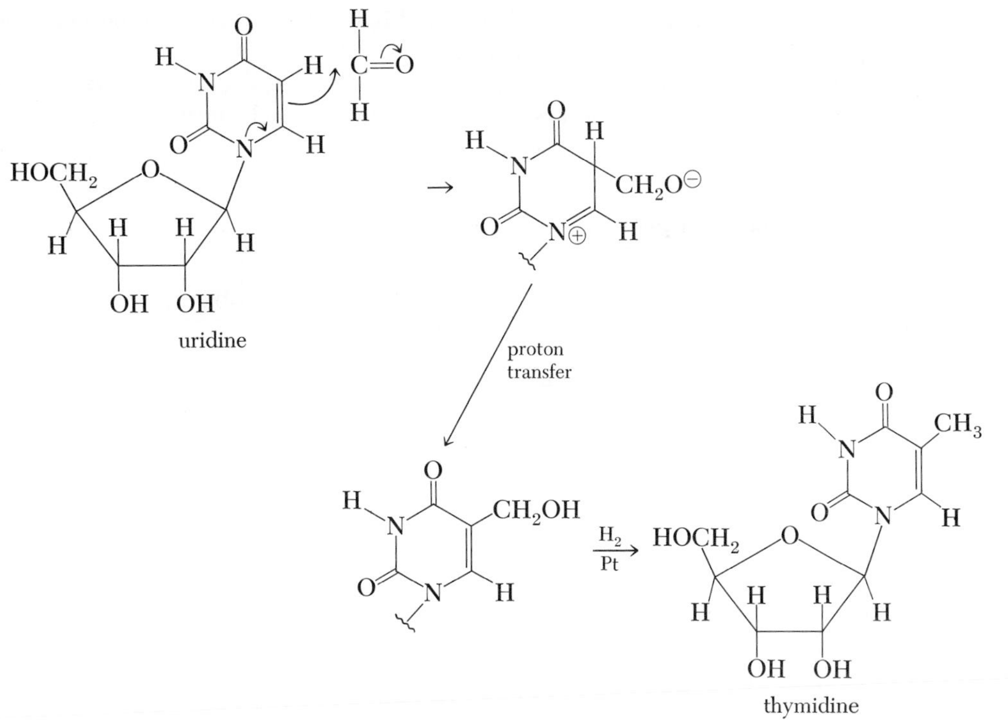
HOCH2
uridine
proton
transfer
H2
Pt
thymidine
CH2O⊖
CH2OH
CH3

B:

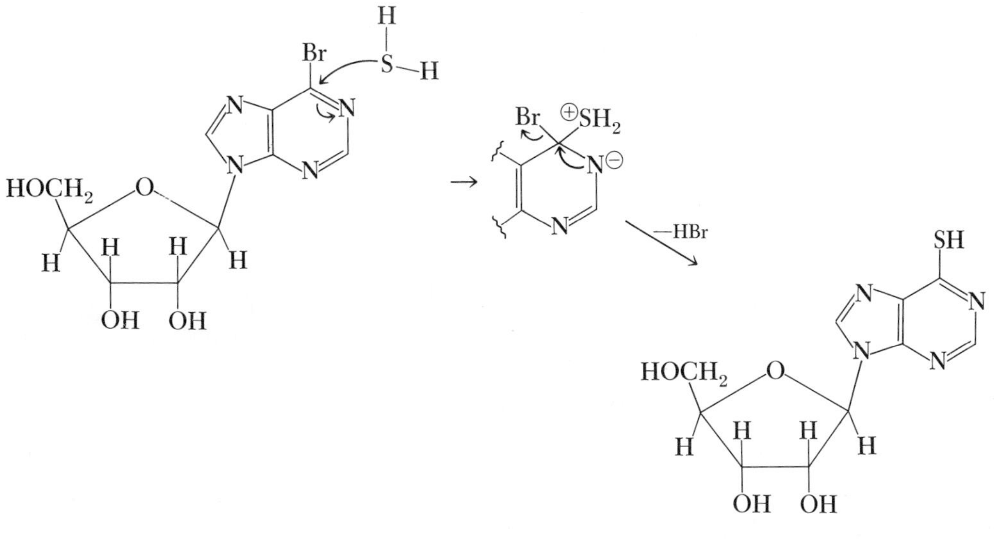
HOCH2
Br
—HBr
SH

26.5 THE STRUCTURE OF NUCLEIC ACIDS

As with other giant molecules, nucleic acids must be described in terms of a basic or primary structure, a set of ancillary bonds (*e.g.*, hydrogen bonding), and a gross morphology.

Since much more is known about DNA than about RNA (RNA is irregular and amorphous in structure), DNA will be the focus of our discussion.

Primary Structure. The backbone of both DNA and RNA molecules is a heteropolymer of carbohydrates and phosphate. The bases are pendant from each sugar unit. A section of DNA is shown schematically in Figure 26–2.

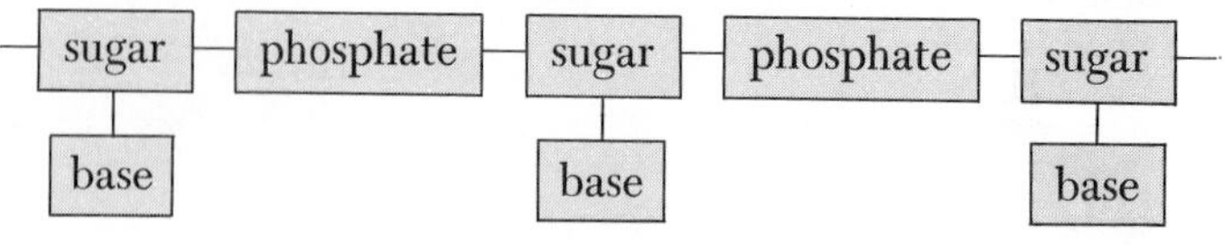

a polynucleotide or nucleic acid

Figure 26–2 A fragment of the DNA polymer. The phosphate group links two sugar residues at C5′ of one sugar and C3′ of another. **B** represents adenine, guanine, cytosine, or thymine.

The constitutional differences between DNA and RNA can be uncovered by identifying the products obtained by the complete hydrolysis of these nucleic acids. As noted from the data in Table 26–3, both DNA and RNA contain adenine, guanine, and cytosine. However, the thymine in DNA is replaced by uracil in RNA. A fact that is not evident from Table 26–3, but whose origin will become clear later, is that cytosine and guanine are present in a 1:1 ratio, as are adenine and thymine (or uracil).

6. Draw structures for adenosine, guanosine, uridine, cytidine, and thymidine.

A less cumbersome representation of polynucleotides than the one shown in Figure 26–2

TABLE 26-3 Products from Hydrolysis of Nucleic Acids[a]

Nucleic Acid	Base	Sugar	Other
RNA	adenine	D-ribose	phosphoric acid
	guanine	D-ribose	phosphoric acid
	cytosine	D-ribose	phosphoric acid
	uracil	D-ribose	phosphoric acid
DNA	adenine	2-deoxy-D-ribose	phosphoric acid
	guanine	2-deoxy-D-ribose	phosphoric acid
	cytosine	2-deoxy-D-ribose	phosphoric acid
	thymine	2-deoxy-D-ribose	phosphoric acid

[a] Only the most common bases are shown.

is created by abbreviating the nucleosides adenosine, guanosine, uridine, cytidine, and thymidine as A, G, U, C, and T, respectively. (These abbreviations also are used for the bases alone.) In this scheme, the phosphate linkage between sugars is represented by *p*. Since polynucleotides are made up by phosphate-bonding a C5′ of one nucleotide to a C3′ on an adjacent nucleotide, the polymer is written in the abbreviated fashion with the 5′ end of the chain on the left and the 3′ end on the right. It is clear, therefore, that the sequence -*p*C*p*A*p*- is not the same as the sequence -*p*A*p*C*p*-. These two are compared below:

—*p*C*p*A*p*— —*p*A*p*C*p*—

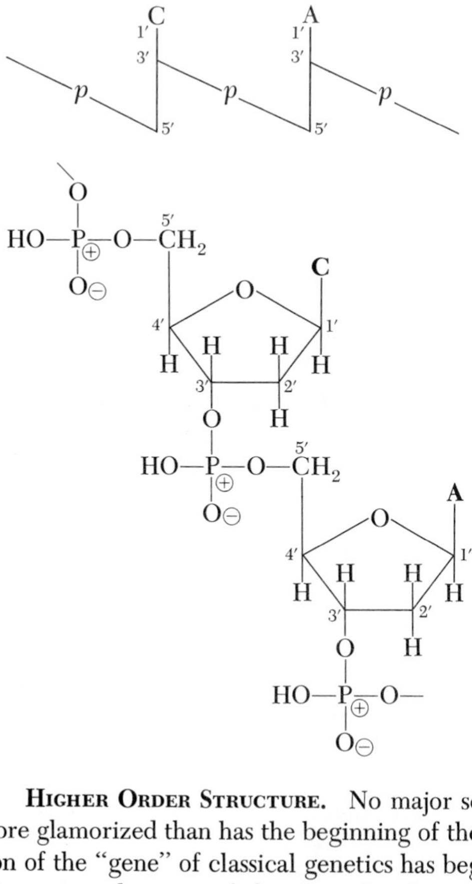

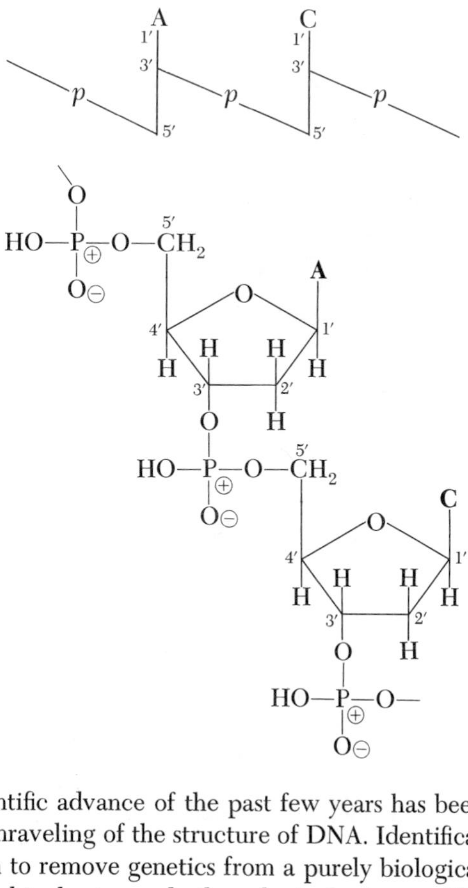

Higher Order Structure. No major scientific advance of the past few years has been more glamorized than has the beginning of the unraveling of the structure of DNA. Identification of the "gene" of classical genetics has begun to remove genetics from a purely biological science into the areas of chemistry, biochemistry, biophysics, and other physical sciences. The popularization of the "DNA story" is epitomized by the appearance of an interesting book by Nobel laureate James D. Watson.[a] The task of elucidating the *complete* structure of even a single DNA molecule is far from complete, perhaps because a single DNA molecule is so big. For example, the DNA of *Escherichia coli* (a bacterium inhabiting the human colon) has an extended length of almost one millimeter.

DNA consists of two helical coils, intertwined to form the now classical "double helix" of the Watson-Crick hypothesis. The backbone of each helix is the alternating carbohydrate-phosphate chain alluded to earlier. The two helices are oriented (Figure 26–3) so as to create two unequal helical grooves running parallel to the major helical axis. These grooves are occupied by proteins (histones) believed involved in biological activation of DNA.

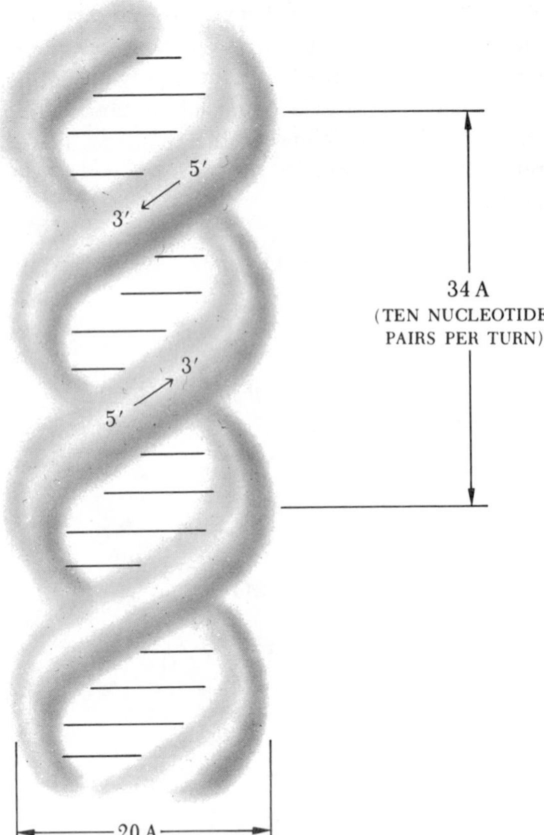

Figure 26–3 The Watson-Crick double helix structure of DNA. One turn of a given strand contains ten nucleotides and is about 34 Å in length (coiled). (From *Concepts in Biochemistry* by F.J. Reithel. Copyright 1967 by McGraw-Hill Book Co. Used with permission of McGraw-Hill Book Co.)

Ten nucleotides are found within one complete twist of one strand of the DNA chain. The integrity of the double helix is maintained largely by van der Waals' attractions between stacked bases sitting on top of one another. Hydrogen bonding between parallel bases also has a stabilizing effect upon the helical structure. Geometric requirements demand that adenine and thymine hydrogen-bond with one another, and that guanine and cytosine hydrogen-bond with one another. This base pairing (Figure 26–4) accounts for the appropriate bases being

[a] *The Double Helix*, J.D. Watson, New American Library (Signet Books), New York, 1969.

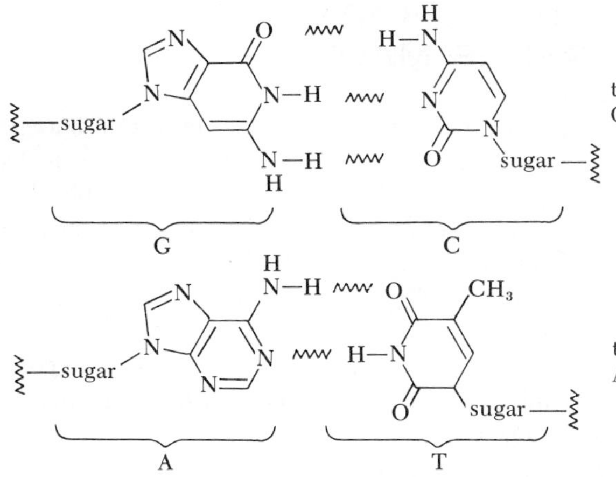

sugar
sugar
G
C
the guanine-cytosine, GC, base pair
sugar
sugar
A
T
the adenine-thymine, AT, base pair

Figure 26-4 Base pairing in nucleic acids.

present in a 1:1 ratio in DNA. Figure 26-5 suggests why two helices in DNA are said to be *complementary;* at a given point along the helix, one strand has a purine base while the other has the appropriate pyrimidine base. Using the abbreviations introduced earlier, base pairing is often indicated as being "AT" and "GC."

S—A T—S
P
P
S—G C—S
P
P
S—T A—S
P
P
S—C G—S

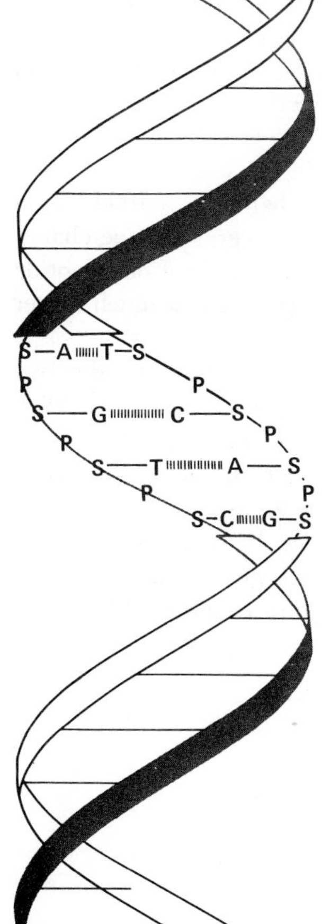

Figure 26-5 Base-pairing interactions in the double helix. The letters are used to symbolize the bases rather than the nucleosides since the sugars are included in the backbone structure. P represents phosphate, $-O-\underset{OH}{\overset{O^{\ominus}}{P^{\oplus}}}-O-$; S represents deoxyribose. (From M.M. Jones, J.T. Netterville, D.O. Johnston, and J.L. Wood, *Chemistry, Man and Society,* W.B. Saunders Company, 1972.)

26.6 OF SUCH STUFF ARE GENES MADE

Genes are the structures that permit the maintenance of species from one generation to the next, by passing along information from mother cell to daughter cell. A given DNA polymer contains several major units of genetic information. The only structural variable along the DNA chain, the one that is responsible for information storage, is the sequence of the four bases. The smallest unit of information on DNA, called the *anticodon*, consists of three contiguous nucleotide units; *it specifies the incorporation of a particular amino acid in a particular protein.*

Why amino acid? All of the chemical reactions carried out in the body are mediated by enzymes, all of which are proteins of one type or another. Clearly, in order to control growth, development, and death of an organism most directly, one must specify the nature of its proteins. The information encoded in DNA is eventually converted into protein synthesis with the aid of enzymes and the intervention of RNA. But we must not get ahead of ourselves. . . .

While a detailed treatment of the "genetic code" is beyond the scope of this text, we should note not only that three bases are required for an anticodon but, also, that a particular sequence and order of reading is important for specifying a given amino acid. For example, the triplet C-A-A (read left-to-right, *i.e.*, the 5′-to-3′ direction) on the active strand (see below) of DNA denotes leucine, while the sequence A-A-C specifies valine; A-C-A specifies cysteine. There are 64 (4^3) combinations of base triplets in DNA, of which sixty-one are codes for amino acids (some amino acids have more than one anticodon). The remainder (TTA, CTA, and TCA) are "nonsense" anticodons, their purpose being to stop protein synthesis.

Whatever mechanism reads the genetic code must also start at a specific point in order to get the correct information. For example, the sequence CAACAACAACAA codes leu-leu-leu-leu if read from the extreme left. However, if it is read starting one nucleotide in from the left it means val-val-val and, if read beginning two in from the left, it means cys-cys-cys.

DNA of mammalian cells contains at least 30,000 nucleotides.[a] Given a triplet code, this suggests that one DNA molecule can transmit information for coding at least 10,000 amino acids. Since most polypeptide chains are much shorter than this (perhaps up to 1000 amino acid residues), one DNA molecule contains the information for coding several peptide chains. *Cistron* is the term used to describe the information required for the synthesis of one peptide chain; one DNA molecule contains several cistrons. Therefore, the cistron is a much better approximation of the "gene" than is the DNA molecule as a whole.

The structure of DNA is a bit more complex than we have suggested, since the two polymeric strands that compose DNA are twisted in opposition, *i.e.*, they are anti-parallel. Moving along both strands *in the same direction,* one finds phosphate linking sugars -5′,3′-5′,3′-5′,3′- on one chain and -3′,5′-3′,5′-3′,5′- on the other. *During protein synthesis,* one strand is the active information source for the cell, providing a template for the synthesis of the nucleic acid called **messenger RNA** (*m*RNA). *When the cell divides,* both strands serve as templates for complementary DNA strands. Thus, after division each nucleus has a pair of DNA strands (or several pairs of strands) identical to the one(s) in the progenitor. The process is represented at two levels of detail in Figures 26–6 and 26–7.

---

[a] Some estimates place this as high as ten billion!

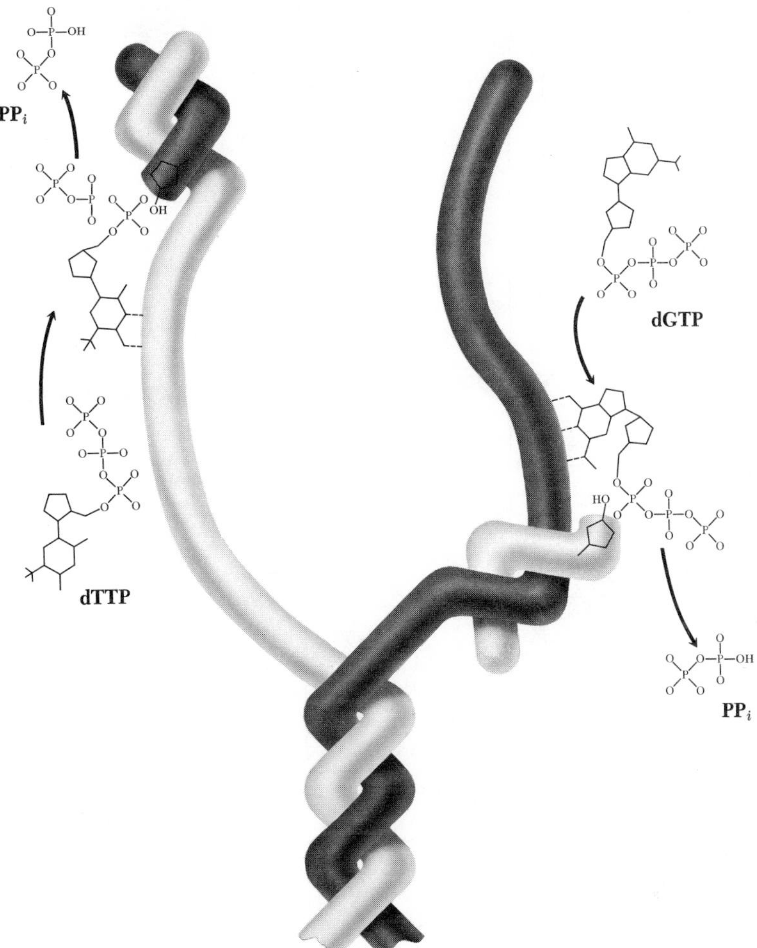

Figure 26-6 Highly schematic representation of the replication of DNA. As the strands of the parent molecule unwind, complementary nucleoside-5′-triphosphates are bound to the exposed bases, but at opposite ends of the separated strands. The triphosphates react with 3′-hydroxyl groups of the preceding nucleotide in the growing strands with the formation of a new 3′, 5′-phosphodiester linkage and the loss of inorganic pyrophosphate. As the new chains grow, new double helices are formed on each of the parent strands. The new polynucleotides are formed in segments, and are later joined in another reaction (not shown here) to make the very long finished strands. The unwinding was originally thought to occur only at the ends of the double helix, but is now believed to occur at several interior positions. (From R.W. McGilvery, *Biochemistry*, W.B. Saunders Company, 1970.)

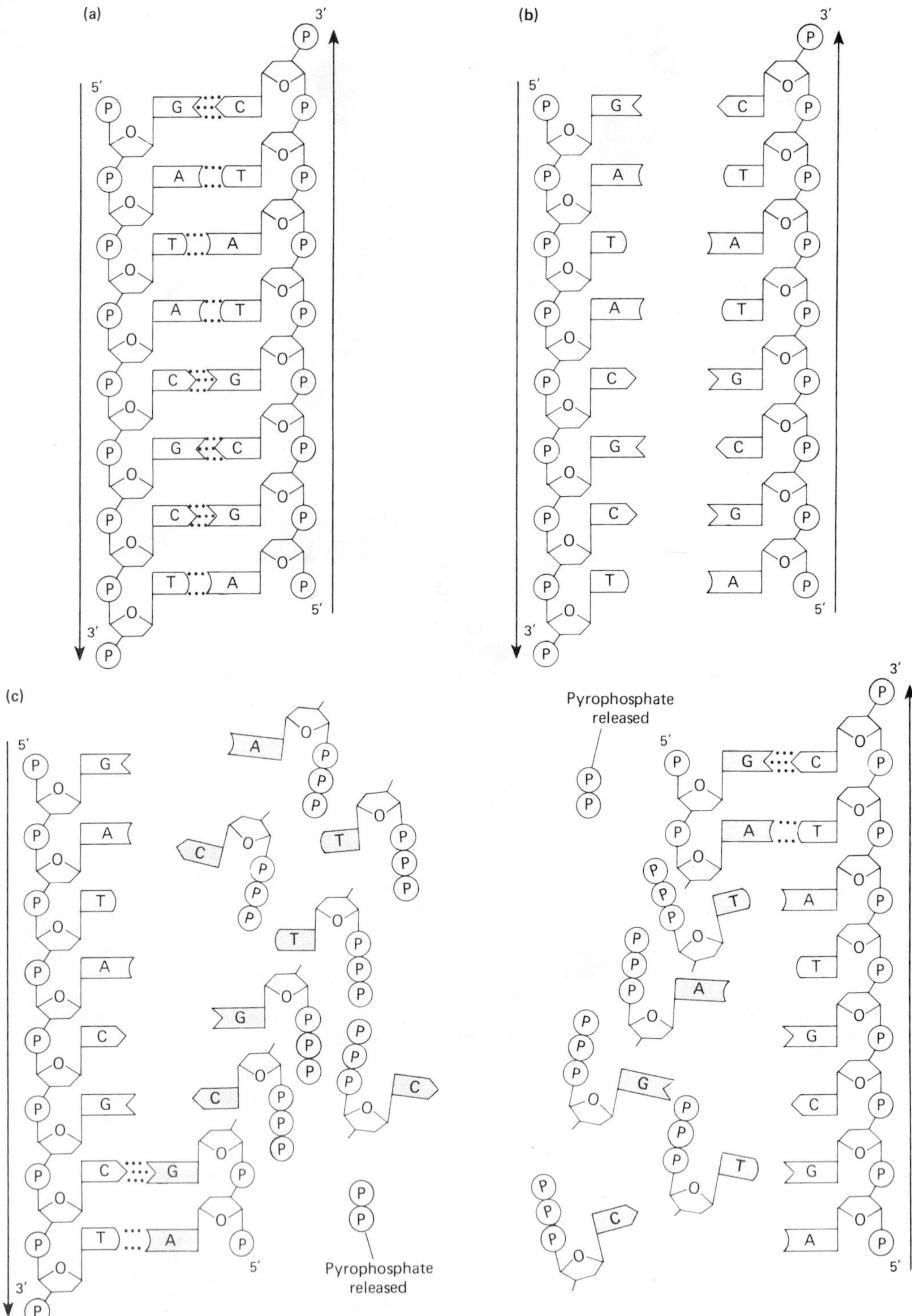

Figure 26–7 *Illustration continued on opposite page.*

Illustration continued on opposite page.

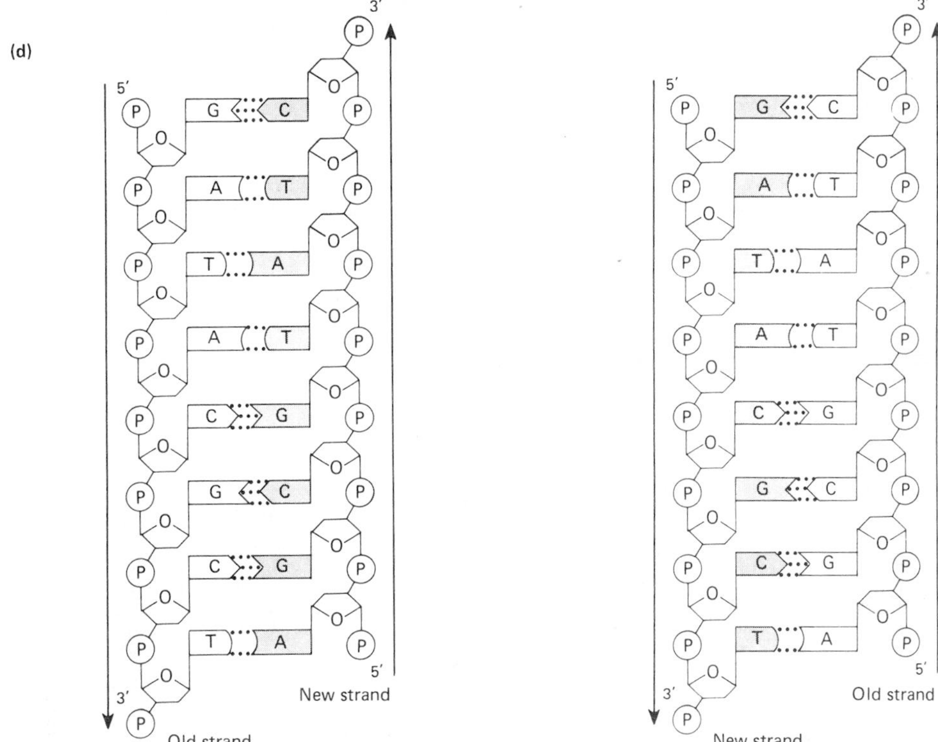

Figure 26-7 Replication of DNA. The fact that DNA consists of two antiparallel strands means that during replication the two new strands must be formed from opposite directions if the synthesis occurs by the successive additions of deoxyribonucleoside 5′-triphosphates to the 3′-hydroxyls of the growing polynucleotide chain. (a) Section of DNA before replication. (b) The two strands separate in the region to be duplicated. (c) New strands form from opposite directions. (d) When replication is complete there are two identical sections of DNA, each consisting of an old strand and a new strand. (A = adenine; C = cytosine; G = guanine; P = phosphate; T = thymine. (Reprinted with permission of Macmillan Publishing Co., Inc. from *Chemical Biology* by J. Ramsey Bronk. Copyright © 1973, J. Ramsey Bronk.)

Protein Biosynthesis. A detailed discussion of protein biosynthesis is beyond the scope of this text. However, in order to appreciate the interplay between DNA and RNA, we must examine briefly how DNA within the nucleus can control protein synthesis outside the nucleus.

Three types of RNA are known, and each is involved in some aspect of protein biosynthesis. Messenger RNA is synthesized on the active strand of DNA and then "migrates" from the nucleus to a ribosome located on the endoplasmic reticulum. The information in DNA is encoded on *m*RNA by the method of *m*RNA synthesis: ribonucleotides bearing bases complementary to those of the DNA are aligned, one by one, along the DNA strand and are attached, by condensation polymerization, to the end of the growing *m*RNA molecule (Figure 26–8).

The genetic code, the triplet codon, could be specified by either the nucleotide sequence in DNA or that in *m*RNA. Since the experimental work was done largely on *m*RNA, *codon* assignments for amino acids are given in the form present in *m*RNA (Table 26–4), while the corresponding DNA and *t*RNA sequences are called the "anticodons."

A ribosome (a collection of ribosomes is called a *polysome*) consists of two masses of dissimilar size. The major constituent of each is a specific RNA called **ribosomal RNA** (*r*RNA). In eukaryotic cells, *r*RNA is synthesized by a DNA located on the chromosome in the nucleoli.

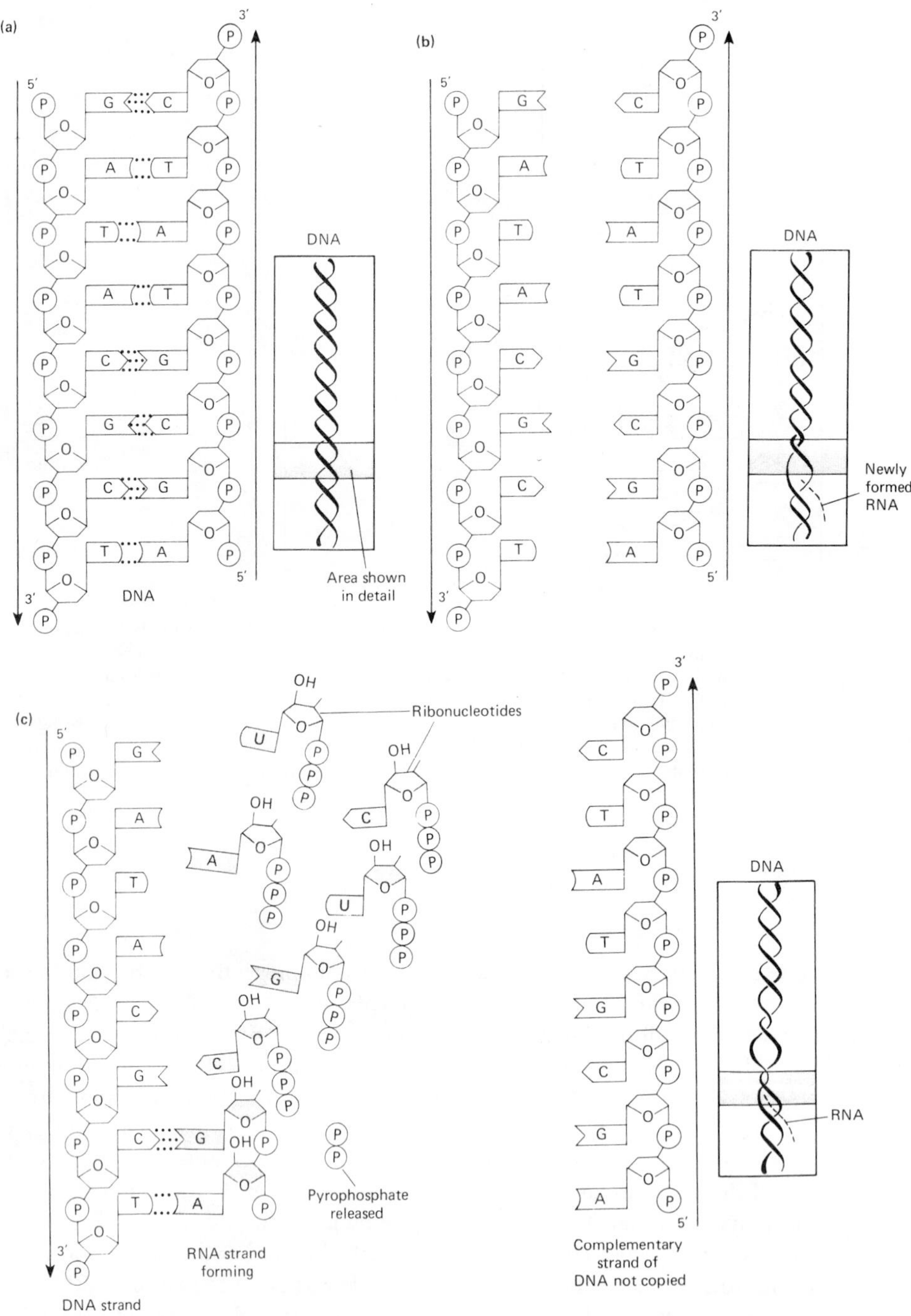

Figure 26-8 *Illustration continued on opposite page.*

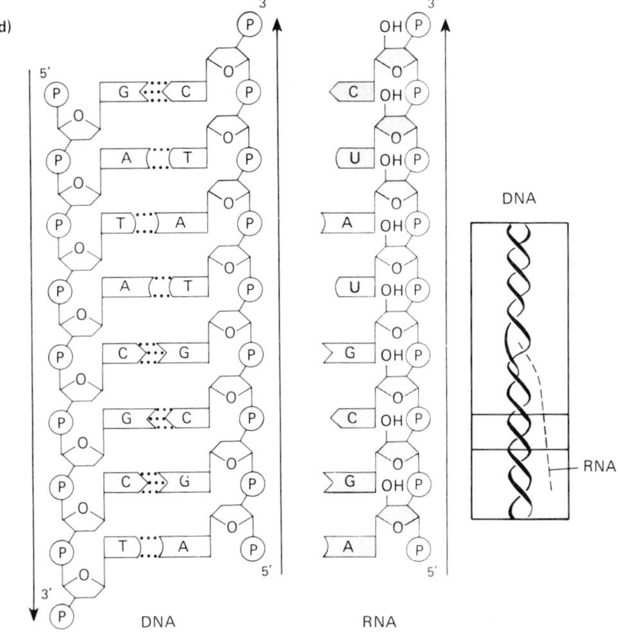

Figure 26–8 Schematic mechanism for the formation of RNA from a DNA template. According to this mechanism, a section of DNA would have to become single-stranded to serve as a template for the synthesis of RNA by the enzyme *RNA polymerase*. The small insets show how transcription could occur sequentially in a longer section of double-helical DNA. (a) Section of DNA before transcription. (b) Separation of the two strands in the area to be transcribed. (c) Formation of an RNA strand complementary to one strand of the DNA. (d) Following transcription, the usual double-stranded form of DNA is reestablished. (Reprinted with permission of Macmillan Publishing Co., Inc. from *Chemical Biology* by J. Ramsey Bronk. Copyright © 1973, J. Ramsey Bronk.)

TABLE 26–4 Messenger RNA Codons For Selected Amino Acids

AMINO ACID	CODONS[a]
phe	UUU, UUC
leu	UUA, UUG, CUU, CUC, CUA, CUG
ile	AUU, AUC, AUA
val	GUG, GUA, GUU, GUC
met	AUG
ser	AGU, AGC, UCU, UCC, UCA, UCG
pro	CCU, CCC, CCA, CCG
thr	ACU, ACC, ACA, ACG
ala	GCC, GCA, GCG, GCU
gly	GGU, GGC, GGA, GGG
trp	UGG
cys	UGU, UGC
lys	AAA, AAG
glu	GAA, GAG
gln	CAA, CAG

[a] A, adenine; C, cytosine; G, guanine; U, uracil.

Once synthesized, an *r*RNA strand migrates out of the nucleus, breaking in two to form the two strands of *r*RNA associated with each ribosome. The function of the *r*RNA appears to be to aid in bonding *m*RNA to those enzymes required for peptide bond formation.

The third type of RNA is **transfer RNA** (*t*RNA), which combines with specific amino acids (each amino acid has a specific *t*RNA) in the cytoplasm to form an **aminoacyl-*t*RNA complex** (Figure 26–9). At the proper time, a particular aminoacyl-*t*RNA complex combines with *m*RNA

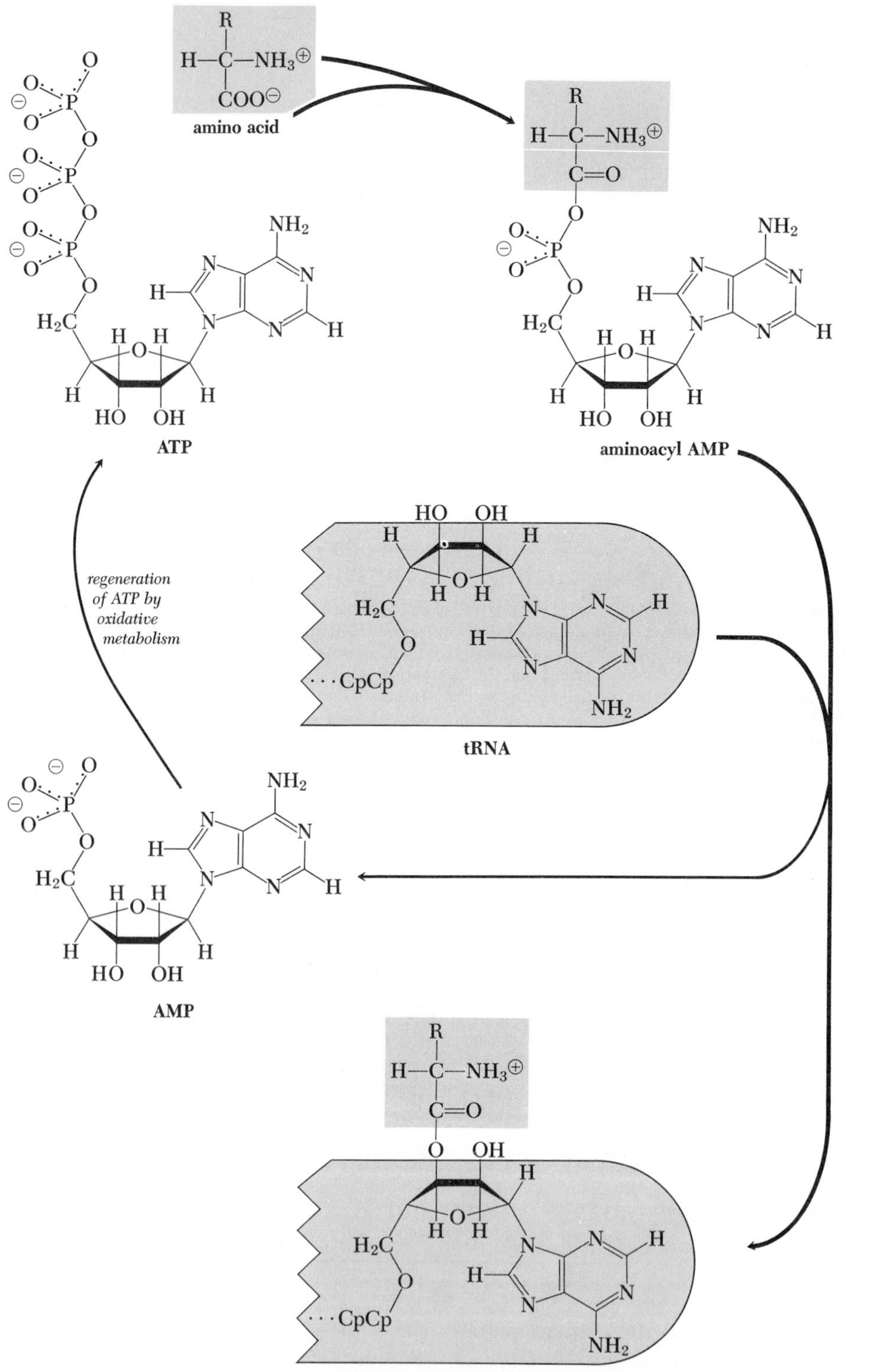

amino acid

ATP

aminoacyl AMP

regeneration of ATP by oxidative metabolism

tRNA

AMP

aminoacyl tRNA

Figure 26-9 Aminoacyl *t*RNA complex formation. The formation of an aminoacyl *t*RNA involves a reaction between the proper amino acid and adenosine triphosphate (top), releasing PP_i, and forming the aminoacyl adenosine monophosphate, which is a mixed anhydride of carboxylic and phosphoric acids. The aminoacyl group is then transferred to the terminal ribosyl moiety of the corresponding *t*RNA (lower right). The group is shown on the 3′ oxygen (bottom center), but there is in fact an equilibration between the 2′ and 3′ positions. The AMP that is released (center left) is phosphorylated by the processes of oxidative metabolism, regenerating the original ATP. (From R.W. McGilvery, *Biochemistry*, copyright © 1970, W.B. Saunders Company, Philadelphia.)

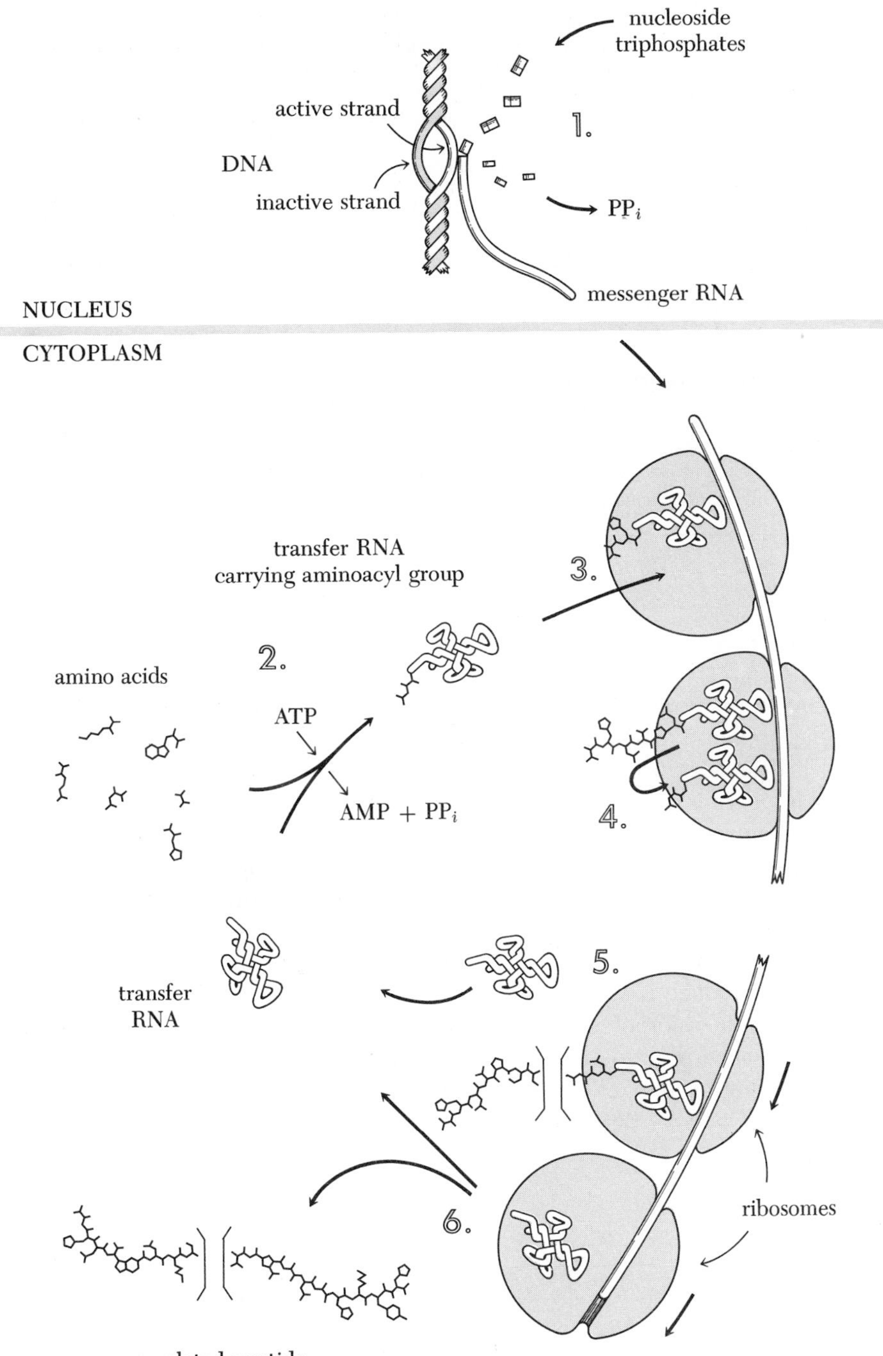

Figure 26-10 Schematic summary of protein synthesis. *Top, step 1.* A molecule of DNA in the nucleus unfolds, and one of its strands is used as a template to direct the formation of messenger RNA from nucleoside triphosphates, which lose inorganic pyrophosphate (PP_i) as they attach to the growing RNA chain. The completed *m*RNA moves to the cytoplasm (*bottom*), where it binds ribosomes into a polysome, and acts as a template for protein synthesis.

Legend continued at top of following page

Figure 26–10 *Continued.*

The following steps are shown on separate ribosomes for clarity, but in fact they are repeated in sequence on each ribosome. The successive ribosomes grow longer and longer peptide chains as they move down the molecule of *m*RNA.

Step 2. Meanwhile, amino acids are combined with specific molecules of transfer RNA (*t*RNA) in the cytoplasm by a reaction that also involves the cleavage of adenosine triphosphate (ATP) into adenosine monophosphate (AMP) and PP_i.

Step 3. The *t*RNA molecules, carrying the amino acids in the form of aminoacyl groups, diffuse to the polysome, where the growing peptide chain is on another molecule of *t*RNA already attached. The incoming *t*RNA, which bears the next group required for the growing peptide (in this case a leucyl residue), has the proper configuration to complex with *m*RNA on the ribosome.

Step 4. When the proper *t*RNA is in place, the peptide chain is transferred onto the amino group of the new residue brought in by *t*RNA, so that the chain is now one residue longer.

Step 5. When the transfer of the previous step is completed, the previously bound *t*RNA no longer carries a peptide chain and is free to dissociate from the ribosome, returning to the mixed pool of *t*RNA in the soluble cytoplasm, where it is available for transport of another molecule of its specific amino acid. The ribosome now moves along the *m*RNA molecule to the position where the placement of the next amino acid will be directed.

Step 6. Steps 3, 4 and 5 are repeated. As each amino acid residue adds to the peptide chain, the ribosome moves down the *m*RNA molecule. When a ribosome has reached the end of the molecule, the peptide is completed and is detached into the soluble cytoplasm. The ribosome itself can then move free of the *m*RNA and be available for attachment to the beginning of yet another molecule of *m*RNA (not shown).

(From R.W. McGilvery, *Biochemistry*, copyright © 1970, W.B. Saunders Company, Philadelphia.)

(already bound to a ribosome) at the spot on the *m*RNA that is coded for that particular amino acid. After the aminoacyl-*t*RNA complex is in place, a second aminoacyl-*t*RNA complex becomes attached and an enzyme creates the peptide bond between the two amino acids. Once an amino acid is incorporated, the *t*RNA that carried it to the ribosome is released and diffuses away in order to shuttle over another molecule of the same amino acid for use later in the synthesis. The process continues until the codon reads STOP (Figures 26–10 and 26–11).

Inhibition of Protein Synthesis. Puromycin is an antibiotic that functions by substituting for an incoming aminoacyl-*t*RNA in the ribosome. The result is a peptide bond from the growing protein chain to puromycin rather than to another amino acid residue. The peptide-puromycin adduct is released from the ribosome, thereby halting synthesis of the protein required by the microorganism. The structural similarity of puromycin and the adenosine end of aminoacyl-*t*RNA is responsible for this facile substitution. Not all antibiotics function by inhibiting protein synthesis, and the mechanism of action of some of those that do is still unknown. One very potent protein synthesis inhibitor in bacteria is chloramphenicol

Figure 26–11 Formation of protein from aminoacyl *t*RNA molecules on the ribosome. (a) Schematic indication of the way in which the sequence of amino acids is dictated by the sequence of bases on the *m*RNA template which is being translated. The *m*RNA is moving to the left and, as each new amino acid is added, the growing polypeptide chain is transferred from the *t*RNA on the donor site (D) to the newly arrived aminoacyl *t*RNA on the acceptor site (A). (b) Details of the peptide elongation cycle during translation of an artificial *m*RNA (poly U). The third base (G) in the anticodon of the phenylalanine *t*RNA is marked with an asterisk to emphasize that it is pairing with U instead of C. This is allowed for by the wobble hypothesis, since the anticodon can recognize the UUU triplet as well as the UUC triplet. (1) The cycle begins with the peptidyl *t*RNA on the donor site. (2) The next phenylalanine *t*RNA arrives at the acceptor site and the peptide is transferred to the free amino group of the aminoacyl *t*RNA. (3) The free *t*RNA is now displaced from site D. (4) The cycle is completed with the movement of the peptidyl *t*RNA from site A to site D carrying the *m*RNA along one codon to the left. (Reprinted with permission of Macmillan Publishing Co., Inc. from *Chemical Biology* by J. Ramsey Bronk. Copyright © 1973, J. Ramsey Bronk.)

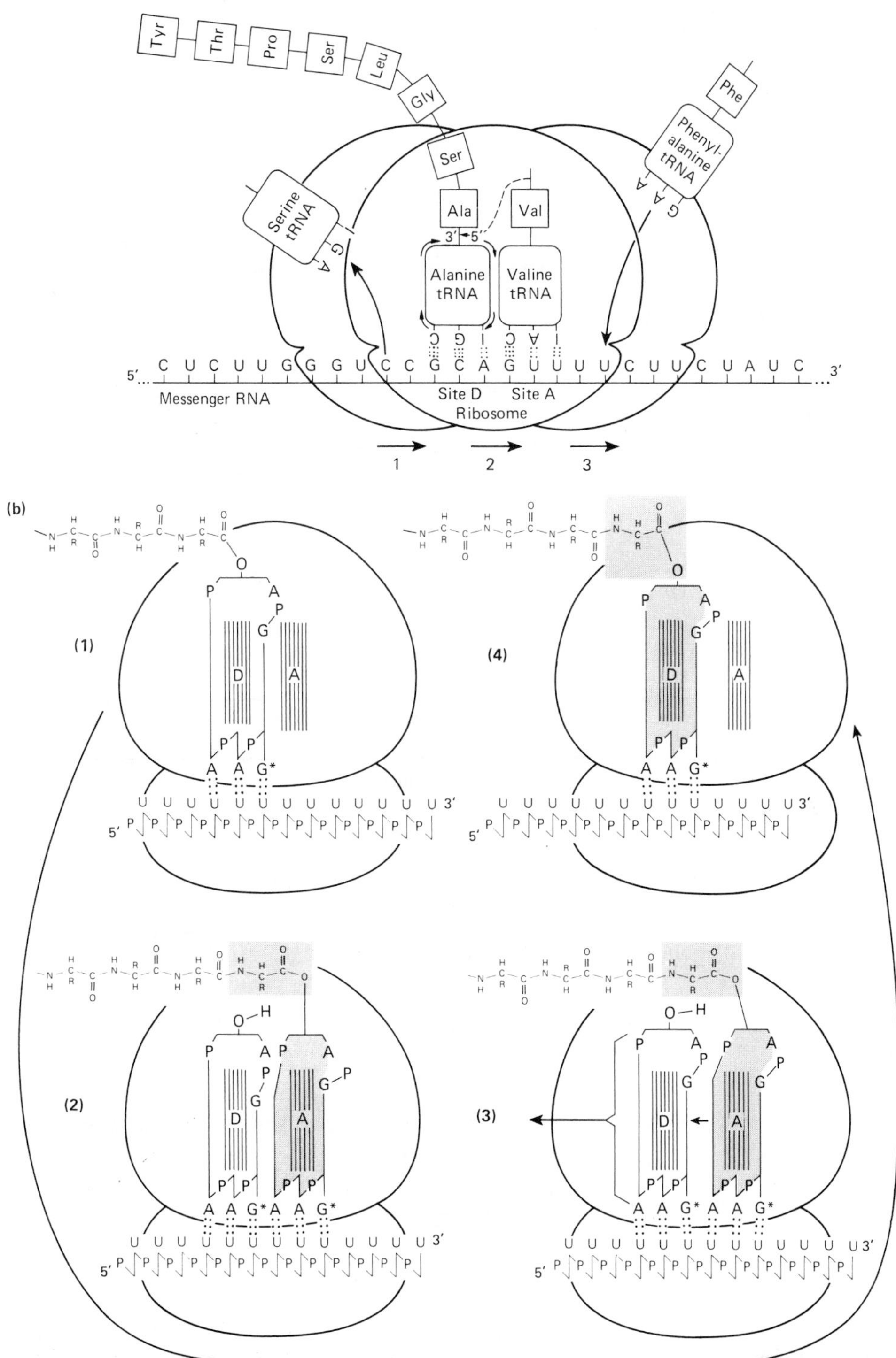

Figure 26-11 *See opposite page for legend.*

(Chloromycetin). Its mode of action is still unclear, although it may influence the synthesis of the ribosome itself.

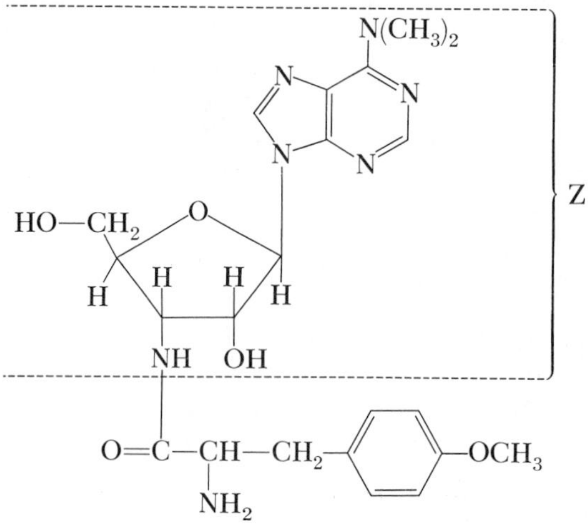

N(CH$_3$)$_2$ HO—CH$_2$ NH OH O=C—CH—CH$_2$—C$_6$H$_4$—OCH$_3$ NH$_2$ Z

puromycin

NH$_2$ R—O—CH$_2$ O OH O=C—CH—R NH$_2$

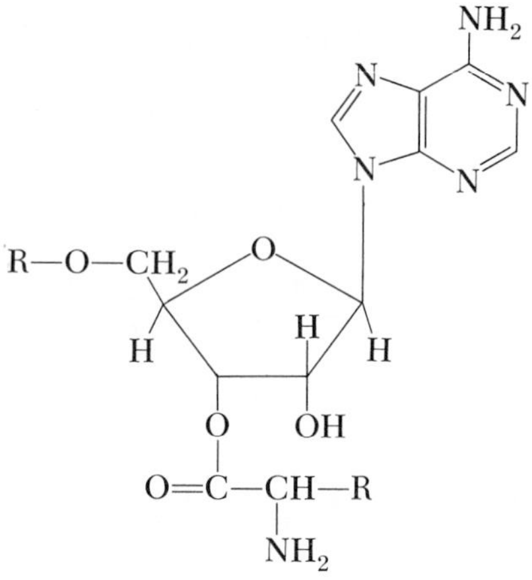

adenosine terminus of *t*RNA

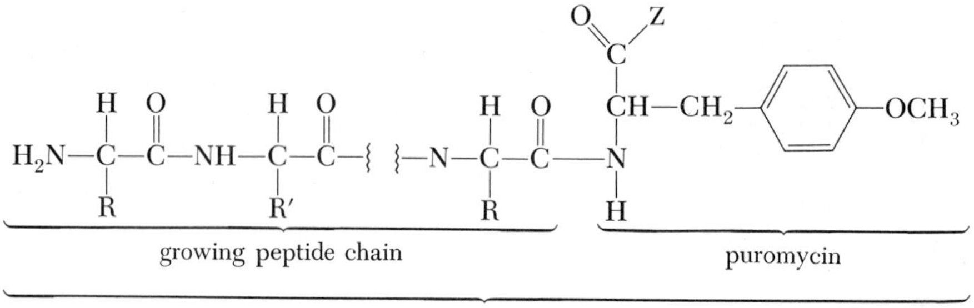

H$_2$N—C(H)(R)—C(=O)—NH—C(H)(R′)—C(=O)— … —N—C(H)(R)—C(=O)—N(H)—CH(C(=O)Z)—CH$_2$—C$_6$H$_4$—OCH$_3$

growing peptide chain puromycin

peptide-puromycin adduct

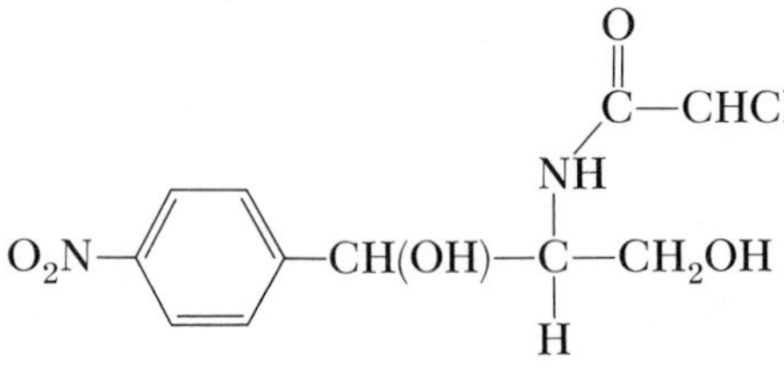

O$_2$N—C$_6$H$_4$—CH(OH)—C(H)(NH—C(=O)—CHCl$_2$)—CH$_2$OH

chloramphenicol

7. Suggest a synthesis for the compound shown below, using benzene and ethanol as your only organic starting materials.

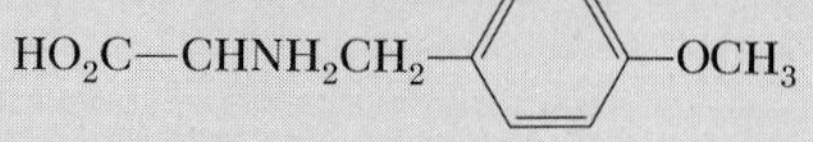

HO$_2$C—CHNH$_2$CH$_2$—C$_6$H$_4$—OCH$_3$

Sickle-Cell Anemia—The Effect of Codon Change. Obviously we can now translate the word **mutation** into molecular terms. Any change in a codon will result in a mutation, the vast majority of which are lethal. One of the most actively studied groups of mutations consists of those that have altered the structure of hemoglobin.

Hemoglobin carries oxygen from the lungs to the tissues and removes carbon dioxide from tissues to lungs. It is made up of four peptide units, or globins, of which two are designated α and the other two β; each globin is associated with a heme unit. (A picture of hemoglobin can be found in Figure 27–7.) All known genetic variations in hemoglobin biosynthesis involve the globin and not the heme. Over fifty abnormal varieties of hemoglobin have been reported. One of these abnormalities involves replacement of one glutamic acid residue by a valine residue in the β chains. This seemingly minor change decreases the ionic charge of the molecule and the extent of dissociation between the heme and the globin. The decreased polarity appears to facilitate crystallization of unsymmetric, deoxygenated hemoglobin molecules, thereby forcing the red cells to assume unusual shapes. These misshapen cells are rapidly destroyed by the spleen, causing a hemolytic anemia. The resultant *molecular disease* (a term introduced by Linus Pauling) is commonly called *sickle-cell anemia.*

normal hemoglobin: val-his-leu-threo-pro-*glu*-glu-lys
sickle-cell hemoglobin: val-his-leu-threo-pro-*val*-glu-lys
N-terminal amino acid of β chain

The disease is particularly prevalent among individuals of African origin. It appears that sickle-cell hemoglobin has a beneficial effect if one lives in malaria-infested regions: *Plasmodium falciparum,* a dangerous malarial parasite, spends a portion of its life within human erythrocytes, and the chances of a child surviving malaria are known to be enhanced if he possesses this abnormal hemoglobin. Thus, a greater percentage of those with sickle-cell trait than of those with normal hemoglobin survive to child-bearing age, and the mutation remains a significant part of the gene pool. People who leave regions where malaria is endemic no longer need fear *P. falciparum* but are still plagued by the deleterious effects of the anemia.

Mitomycin. As already implied in our discussion of puromycin, not all chemically-induced variations of the "DNA-RNA-protein synthesis combine" are deleterious. A further illustration is found in mitomycin, an anti-tumor, antibiotic complex produced by *Streptomyces caespitosus (griseovinaceseus).* From this complex can be identified three compounds: mitomycin A, B, and C.

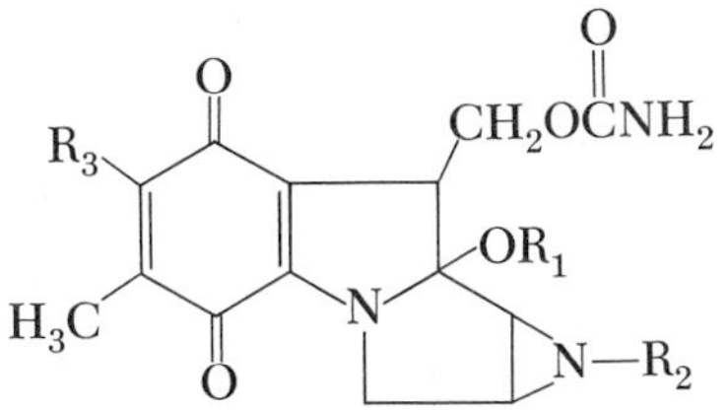

	R_1	R_2	R_3
A:	CH_3	H	OCH_3
B:	H	CH_3	OCH_3
C:	CH_3	H	NH_2

mitomycin

Mitomycin is believed to act as an anti-tumor agent by alkylating and cross-linking DNA strands, thus preventing their participation in cell division (mitosis). Mitomycin C is the most potent of the three forms.

26.7 EPILOGUE

It would be improper to finish this all-too-brief discussion of nucleic acids without one last observation. A hallmark of many early "monster movies" (and some of the recent ones as well) was a scene near the end of the film in which the hero, or sometimes the wizened scientist, comments: "There are some things in our nature which Man was never meant to explore!"

It seems unlikely, in spite of such comments, that Man has come this far in understanding his own heredity only to stop short of "experimenting" upon himself. In spite of the horrors conceived by Robert Louis Stevenson and Mary Shelley, there is little doubt that we have already begun to modify ourselves. An illustration of this is genetic counselling of potential parents about Tay-Sachs disease, sickle-cell anemia, and Rh problems. Genetic control of disease, a logical extension of genetic counselling, holds out for Man great promise and the accompanying massive responsibilities.

IMPORTANT TERMS

Angina pectoris: A disease caused by obstruction of coronary arteries, which results in oxygen deprivation of the heart muscle. Angina pectoris is not fatal.

Anticodon: A combination of three nucleotides that, when read in the proper direction, specifies a particular amino acid. The sequence is found on DNA (and *t*RNA).

Anti-metabolite: A molecule that resembles the usual substrate for an enzyme and competes with that substrate for the enzyme, but produces a compound that cannot be used by the body. Anti-metabolites often are used to treat malignancies.

Barbiturates: A class of drugs that act as general depressants of the central nervous system. Derivatives of barbituric acid.

Cistron: A region of the DNA molecule that is coded for the synthesis of a particular protein. Perhaps the closest approximation, in molecular terms, to the classical "gene."

Codon: A combination of three nucleotides that, when read in the proper direction, specifies a specific amino acid. Codon and anticodon for a specific amino acid are complementary.

Deoxyribonucleic acid (DNA): A polynucleotide containing genetic information. The sugar that is part of the polymer backbone is 2-deoxyribose.

Endemic: The property of being peculiar to a given region or group of individuals.

Endoplasmic reticulum: A system of tubules distributed throughout the cytoplasm of a cell.

Eukaryotic cells: Cells with relatively complex internal organization and a true nucleus. Such cells are found in all animals as well as in most plants (including protozoa, fungi, and most algae). The DNA of eukaryotic cells is confined almost entirely to the structure called the nucleus.

Histones: Proteins that are believed to reside in the shallow groove of the DNA double helix and control nucleic acid replication.

Hyperthyroidism: Overactivity of the thyroid gland.

Messenger RNA (*m*RNA): A polynucleotide whose function is to transmit information from within the nucleus (on DNA) to ribosomes outside the nucleus.

Nucleoli: Plural of nucleolus. A small body found within the cell nucleus. The nucleolus is composed largely of RNA.

Nucleoside: An *N*-glycoside in which the sugar is either ribose or deoxyribose, and the aglycone is one of several derivatives of either pyrimidine or purine.

Nucleotide: A phosphate ester of a nucleoside. The nucleotide is the fundamental repeating unit in nucleic acids (polynucleotides).

Polynucleotide: A synonym for nucleic acid. A polymer of nucleotides that do not necessarily contain the same base but that always contain the same sugar (ribose or deoxyribose).

Prokaryotic cell: The extremely unstructured cell characteristic of bacteria and blue-green algae. DNA in such cells is not confined to a sub-cellular particle but is distributed in the cytoplasm.

Ribonucleic acid (RNA): A group of polynucleotides that control the synthesis of proteins. These are grouped according to function: messenger RNA (*m*RNA), ribosomal RNA (*r*RNA), and transfer RNA (*t*RNA). All are characterized by the presence of ribose in the nucleotide units of the polymer.

In 1968, R.W. Holley shared a Nobel Prize for establishing the structure of alanine *t*RNA of yeast. Because of the relatively small size of these molecules, much effort is now being devoted to the complete structural elucidation of various *t*RNA's.

Ribosomal RNA (*r*RNA): The RNA found in ribosomes. Its function in peptide synthesis is less clearly understood than is that of either *m*RNA or *t*RNA.

Ribosome: The ribosomes of all cells generally consist of two parts of unequal size. Protein synthesis occurs *within* the "space" between these two portions (see Figure 26–11). Ribosomes of prokaryotic cells consist of about 60% *r*RNA and about 40% protein. In eukaryotic cells the ribosomes are about 50% *r*RNA and 50% protein. Ribosomes are found in many locations throughout a cell, often assembled in a bead-like string called a polysome. The polysome is, in fact, a group of ribosomes simultaneously bonded to a given RNA molecule.

Transfer RNA (*t*RNA): A ribonucleotide whose function is to transfer a particular amino acid to the ribosome at the proper time for peptide synthesis. It does this by forming an aminoacyl-*t*RNA ester. This ester formation may be considered as "activation" of the amino acid for protein biosynthesis. In the older literature *t*RNA sometimes is called *s*RNA (*soluble* RNA).

PROBLEMS

8. Illustrate each of the following.
   - (a) ribonucleic acid
   - (b) deoxyribonucleic acid
   - (c) riboside
   - (d) deoxyriboside
   - (e) nucleoside
   - (f) nucleotide
   - (g) barbiturate
   - (h) *N*-glycoside
   - (i) double helix
   - (j) cistron
   - (k) codon
   - (l) anticodon
   - (m) molecular disease
9. Draw the structure of each of the following.
   - (a) pyrimidine (include numbering scheme)
   - (b) purine (include numbering scheme)
   - (c) 2-deoxy-D-ribose
   - (d) D-ribose
   - (e) barbituric acid
   - (f) phenobarbital
   - (g) Nembutal
   - (h) uridine
   - (i) uracil
   - (j) adenine
   - (k) adenosine
   - (l) adenosine monophosphate
   - (m) 2′-adenylic acid
   - (n) caffeine
   - (o) ATP
10. Starting with acyclic compounds of your choice, devise a synthesis for each of the following.
    - (a) phenobarbital
    - (b) Nembutal
    - (c) barbituric acid
    - (d) hypoxanthine
11. The amino acid sequence in the terminal trinucleotide of all *t*RNA's studied thus far is -pCpCpA. Draw the structure of this terminus. Draw the structure of -pApCpC.

12. What are the mechanisms for the steps leading to the conversion of methyl iodide and thiourea to thymine by the scheme outlined in the text? Why is thiourea used instead of urea?
13. Suggest a mechanism for the following reaction.

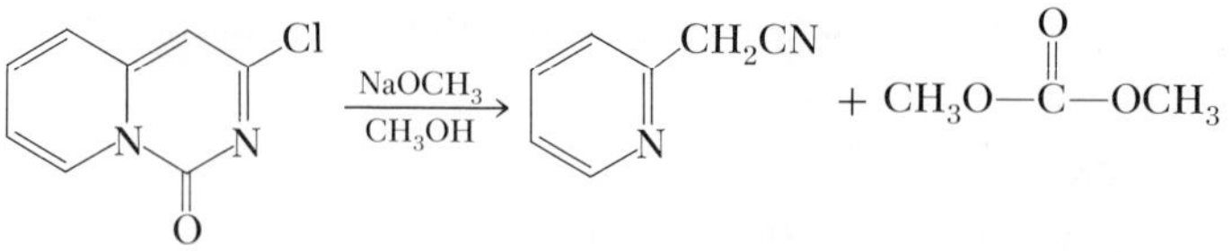

$$\xrightarrow[CH_3OH]{NaOCH_3} \quad (CH_2CN \text{ pyridine}) + CH_3O{-}\overset{O}{\overset{\|}{C}}{-}OCH_3$$

[°] 14. Given unlabeled samples of theophylline, theobromine, and caffeine, how could they be identified spectroscopically?
15. The bases in DNA and RNA exist in the keto form rather than in the enol form. Suggest a reason for this.
16. What is the major structural difference between DNA and RNA?
17. Compare and contrast the three types of RNA.
18. (a) What is the difference between a nucleoside and a nucleotide? (b) What would you anticipate to be the difference between a riboside and a ribotide? (c) How would they be similar?
19. Of the three possible monohydroxypyridines, only 3-hydroxypyridine behaves like a normal phenol. Explain.
20. Table 26–4 contains codons for a number of amino acids. What are the anticodons (located on DNA) for (a) valine, (b) leucine, and (c) glycine? (Remember that the direction of reading is important.)
21. What, if anything, is the significance of the dotted lines in the structures shown in Figure 26–7?
22. What is the minimum number of nucleotides that would have to be misread or altered in order to result in the synthesis of a globin characteristic of sickle-cell anemia (HbS) rather than that of normal hemoglobin (HbA)?
23. The anticodons for "STOP" are TTA, CTA and TCA. What are the corresponding codons on *m*RNA? (Remember that both the codon and the anticodon are read in the 5′-to-3′ direction.)
24. Of the three letters in the codon, two seem to be much more important in determining specificity than a third. With the information in Table 26–4, suggest whether it is the first, second, or third letter of the code that is least important in fixing specificity. (The theory that explains this is called the "wobble" hypothesis.)
25. What is the molecular basis for sickle-cell anemia? How does this disease get its name?
26. (a) Draw the structure of the fragment -CpApTp-. (b) If this were found in one strand of DNA, what would be the arrangement in the complementary strand? (c) Would the fragment be expected to occur in RNA? Why?
27. Account for the following observation:

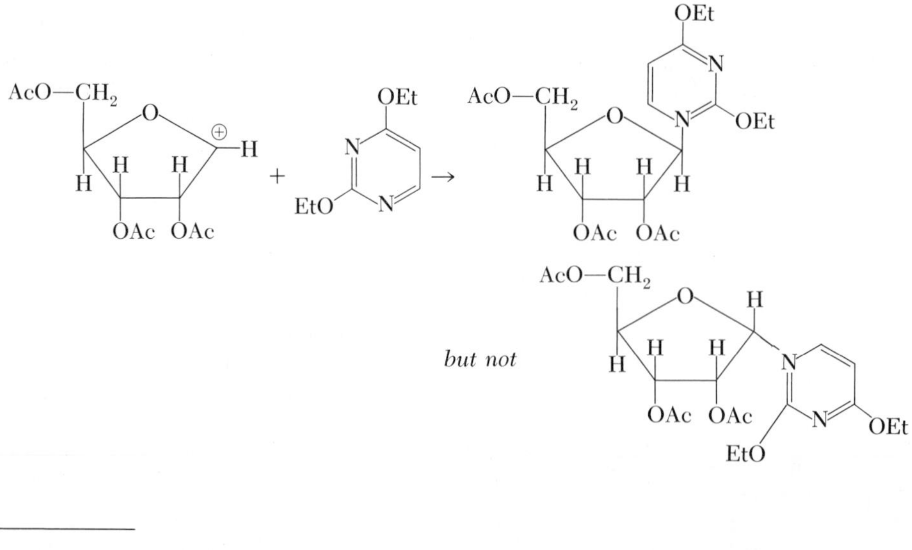

but not

[°] This problem requires a knowledge of spectroscopy.

THE CATALYST AND THE CONTAINER—A SAGA OF THE CELL

27.1 INTRODUCTION

A chemist synthesizing a compound has available a variety of starting materials, solvents, and reaction conditions. Once the desired reaction has been completed, the reaction product can be isolated and purified by many combinations of elegant procedures. If the product is highly reactive, it can be stored at temperatures as low as $-196°C$ (liquid nitrogen) and in an inert atmosphere (*e.g.*, argon). Is it any wonder, then, that the chemist has become fascinated by the reactions occurring in living systems? In humans, the "poor" cell is restricted to water as its reaction medium and to 37°C as the reaction temperature. Furthermore, cells lack the multitude of devices contrived by man and considered by him to be essential for the isolation and purification of reaction products. In spite of these restrictions, cells manage to perform their duties with remarkable ease—and this is most fortunate. If a laboratory reaction fails, the chemist can discard the material and begin again. If a reaction *within* the chemist fails, it could mark the death of the chemist.

What permits cells, tissues, organs or organisms to perform myriad organic and inorganic reactions at the proper relative rates, with minimum by-product formation, when every reaction is occurring at the same temperature? ENZYMES! These proteinaceous biological catalysts mediate all reactions *in vivo*, including those that are spontaneous *in vitro*. As with any catalyst, enzymes cannot induce reactions if they would not occur in the absence of the catalyst. However, the enzyme can take a comparatively slow reaction and greatly increase the speed at which it occurs. Put differently, the enzyme functions by increasing the rate at which a system reaches equilibrium. Since the equilibrium constant, K_{eq}, can be expressed in terms of the ratio of the rates of the forward and reverse reactions, k_1 and k_{-1}, enzymes must, in fact, accelerate both forward and reverse reactions.

$$\mathrm{A} \underset{k_{-1}}{\overset{k_1}{\rightleftarrows}} \mathrm{B} \qquad K_{eq} = \frac{k_1}{k_{-1}} = \frac{[\mathrm{B}]}{[\mathrm{A}]}$$

Products of most biological reactions are consumed in subsequent steps. Therefore, in discussing enzyme-catalyzed reactions we will generally consider only the forward component of these equilibria. The diagram in Figure 27–1 shows the effect of the enzyme hog pancreatic lipase (a catalyst for ester hydrolysis or ester formation) upon *n*-butyl butyrate (lower curve) and upon a mixture of *n*-butanol and butanoic acid (upper curve).

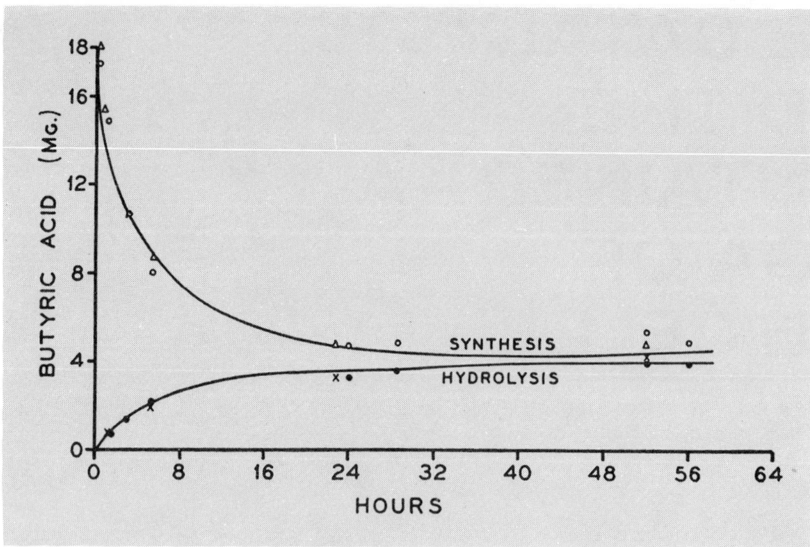

Figure 27-1 Synthesis and hydrolysis of *n*-butyl-*n*-butyrate by hog pancreatic lipase. (From A. Mazur and B. Harrow, *Textbook of Biochemistry,* 10th ed., W.B. Saunders Company, 1971.)

27.2 WHAT FACTORS INFLUENCE THE RATE OF ENZYME-CATALYZED REACTIONS?

In this section we will see how varying the enzyme concentration *or* the reactant concentration *or* the reaction temperature *or* the pH can influence the rate of an enzyme-catalyzed reaction and, by extension, every reaction within our bodies. Since enzymes can, under proper circumstances, be made to function outside of living systems, some of these data reflect experiments carried out *in vitro.*

Enzyme Concentration. The velocity of an enzyme-mediated reaction is proportional to the enzyme concentration; as the enzyme concentration increases, so does the rate of reaction.

$$\text{velocity} = -\frac{d[\text{S}]}{dt} = k[\text{E}][\text{S}]$$

where $-\frac{d[\text{S}]}{dt}$ = rate of disappearance of substrate (the substance upon which the enzyme acts)

[E] = enzyme concentration

[S] = substrate concentration

This dependence is illustrated in Figure 27–2. The reaction depicted, the conversion of

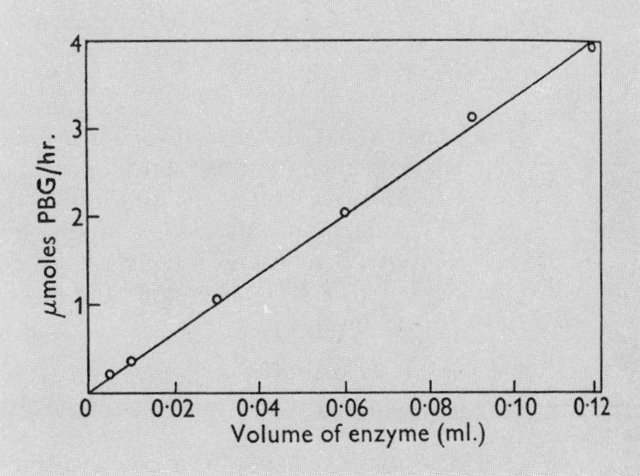

Figure 27-2 Proportionality between reaction velocity and enzyme concentration. The enzyme is aminolevulinic acid dehydrase, which converts aminolevulinic acid to porphobilinogen (PBG). (From A. Mazur and B. Harrow, *Textbook of Biochemistry*, 10th ed., W.B. Saunders Company, 1971.)

δ-aminolevulinic acid to porphobilinogen (PBG), is important because PBG is a precursor to *heme,* a key component of hemoglobin.

CO_2H–CH_2–CH_2–C(=O)–CH_2–NH_2 + CO_2H–CH_2–CH_2–C(=O)–CH_2–NH_2 $\xrightarrow{\delta\text{-aminolevulinic acid dehydrase}}$ porphobilinogen (pyrrole ring, N–H, bearing $CH_2CH_2CO_2H$, CH_2CO_2H and CH_2CO_2H groups)

δ-aminolevulinic acid

porphobilinogen

Substrate Concentration. If one begins with a low substrate concentration, increasing the substrate concentration will increase the reaction rate. This, of course, is consistent with the rate equation given in the previous section. Beyond a certain concentration of substrate, however, further increases no longer produce an increased rate of reaction. The latter situation is described by $-d[S]/dt = k[E]$, and it normally indicates saturation of the capability of the enzyme to handle the substrate. With some systems, a high substrate concentration actually produces a decrease in the rate of reaction. This is a phenomenon called "substrate inhibition." The overall behavior of reaction rate *vs.* substrate concentration is presented schematically in Figure 27-3.

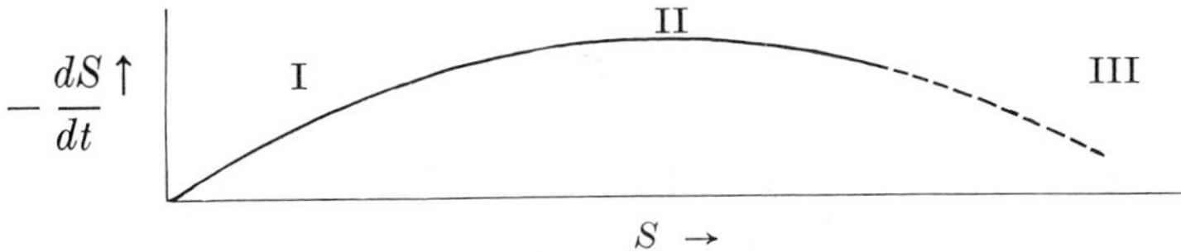

Figure 27-3 The effect of substrate concentration upon enzyme activity. Region III reflects decreased reactivity caused by substrate inhibition. (From A. Mazur and B. Harrow, *Textbook of Biochemistry*, 10th ed., W.B. Saunders Company, 1971.)

The Michaelis-Menten Equation. A fundamental assumption about the mode of action of enzymes is that an enzyme, E, reacts with its substrate, S, to form an enzyme-substrate complex, ES, prior to conversion to product, P.

$$E + S \underset{k_{-1}}{\overset{k_1}{\rightleftarrows}} ES \xrightarrow{k_2} P + E$$

In order to look a bit more quantitatively at the rate of an enzyme-catalyzed reaction, we shall define symbols for several quantities derived from the reaction scheme above: *total enzyme concentration*, [E]; *enzyme-substrate complex concentration*, [ES]; *free enzyme concentration*, [E − ES]; and *substrate concentration*, [S]. Now, the rate of *formation* of ES equals k_1[E − ES][S], while the rate of *reversion* of ES to E and S equals k_{-1}[ES]. The rate of *decomposition* of ES to product equals k_2[ES]. Therefore, the rate of change of the concentration of ES, $d[ES]/dt$, is given by

$$\frac{d[ES]}{dt} = k_1[E - ES][S] - k_{-1}[ES] - k_2[ES]$$

The rate of change in the concentration of ES becomes zero very quickly after the reaction has begun. Given $d[ES]/dt = 0$,

$$[ES] = \frac{k_1[E][S]}{k_1[S] + k_{-1} + k_2}$$

or

$$[ES] = \frac{[E][S]}{[S] + \left(\frac{k_{-1} + k_2}{k_1}\right)}$$

The ratio of rate constants, $(k_{-1} + k_2)/k_1$, is called the **Michaelis constant,** abbreviated K_m. Its value varies from 1 to about 10^{-8} and reflects the affinity of the enzyme for its substrate. The smaller the value of the Michaelis constant for a given substrate, the greater the affinity of the enzyme for that substrate.

Since the rate of product formation has already been defined as equal to k_2[ES], we can now make the following substitutions:

$$[ES] = \frac{[E][S]}{[S] + (k_{-1} + k_2)/k_1}$$

becomes

$$[ES] = \frac{[E][S]}{[S] + K_m}$$

and, substituting in "rate = k_2[ES],"

$$\text{rate} = \frac{k_2[E][S]}{[S] + K_m}$$

When the substrate concentration is very high, K_m is negligible compared to [S], and the rate of reaction becomes approximately equal to k_2[E]. The maximum velocity of this reaction, V_{max}, is defined as k_2[E] and is related to the Michaelis constant by the **Michaelis-Menten equation.**

$$\text{rate} = \frac{V_{max}[S]}{[S] + K_m} \qquad \text{Michaelis-Menten equation}$$

From this equation we can see that the Michaelis constant equals the particular substrate concentration that allows the reaction to proceed at one-half of its maximum possible velocity. When k_2 is much smaller than k_1 and k_{-1} (that is, under conditions where E and S have an opportunity to equilibrate with ES), K_m equals K_s, the dissociation constant of ES. A typical Michaelis-Menten plot of rate *vs.* substrate concentration is shown in Figure 27–4.

The Michaelis-Menten treatment of the kinetics of enzyme-catalyzed reactions may be something of an approximation to what actually is occurring. For example, some enzymic reactions involve several different intermediate enzyme-substrate complexes, a possibility not considered in this treatment.

$$E + S \rightleftarrows (ES)_1 \rightleftarrows (ES)_2 \rightleftarrows EP \rightarrow E + P$$

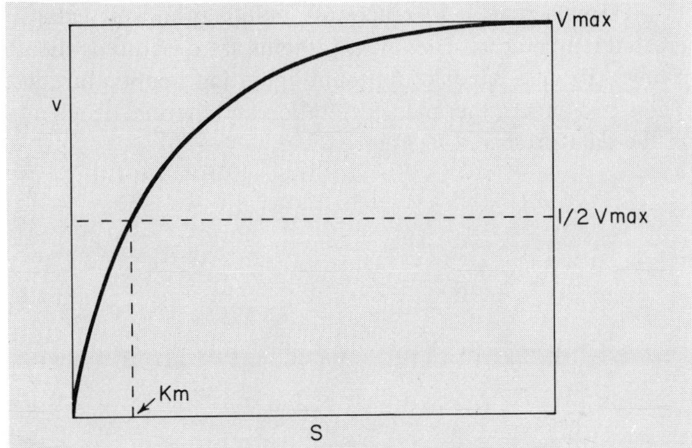

Figure 27–4 Plot of v against S according to the Michaelis equation. K_m is the Michaelis constant. (From A. Mazur and B. Harrow, *Textbook of Biochemistry,* 10th ed., W.B. Saunders Company, 1971.)

pH. Enzymes are proteins and, like all proteins, are denatured by extremes of pH. This denaturation causes them to become inactive as catalysts. The effect of pH upon enzyme activity is illustrated in Figure 27–5, in which the activity of tyramine oxidase (also called monoamine oxidase or MAO) is plotted against pH (curve B). The generally bell-shaped pattern of this pH profile is observed for most enzymes. This pH dependence suggests that an enzyme's activity is in some way related to acid-base catalysis by that enzyme.

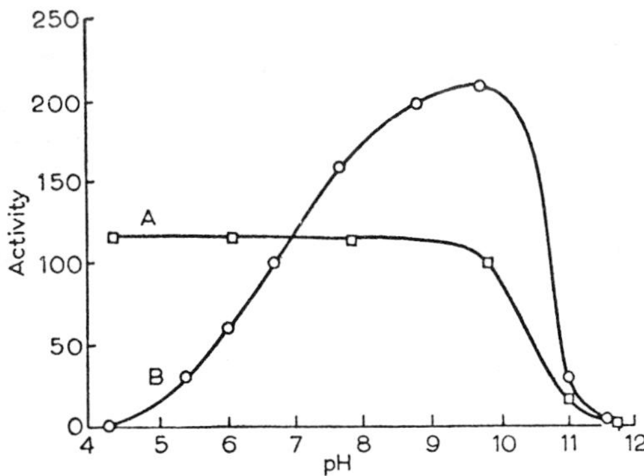

Figure 27–5 Influence of pH on the activity of monamine (tyramine) oxidase. Curve A shows the activity of the enzyme at pH 7.3 after being exposed for 5 minutes to the various pH values shown. Curve B is the activity when tested at the different pH values. The apparent optimum pH is clearly caused by irreversible inactivation of the enzyme. (From A. Mazur and B. Harrow, *Textbook of Biochemistry,* 10th ed., W.B. Saunders Company, 1971.)

Some enzymes possess atypical pH optima. Thus, pancreatic lipase has an optimum pH of 8, while lipase from the stomach has an optimum pH of approximately 4. The peptide bond-hydrolyzing enzyme pepsin has an optimum pH of 1.6!

Aside from its pedagogic value, MAO is important in maintaining sound mental health. MAO deaminates amines of the type RCH_2NH_2, including dopamine and serotonin, and aids in the control and regulation of endogenous supplies of such amines. Materials that interfere with MAO (*MAO inhibitors*), such as *trans*-2-phenylcyclopropylamine (tranylcypromine, Parnate), are termed "psychic energizers" and are used as antidepressants. Administration of MAO inhibitors to individuals who consume large quantities of cheese can lead to their death. The reason is that cheese is rich in tyramine, a sympathomimetic amine with potent cardiovascular activity. Normally, ingested tyramine is rapidly destroyed by MAO in the body; however, treatment with MAO inhibitors permits buildup of tyramine levels to the point at which a cardiovascular "crisis" (*e.g.*, brain hemorrhage) may occur. We should add that Chianti, canned figs, and chocolate are also rich in tyramine and must be avoided by patients receiving MAO inhibitors.

Temperature. All reactions, including enzyme-catalyzed ones, are accelerated by increased temperature. However, proteins are denatured when temperatures are elevated excessively. Enzymes have an optimum operating temperature just as they have an optimum pH. These two effects, thermal acceleration and thermal deceleration due to denaturation, are seen in the data presented in Figure 27–6.

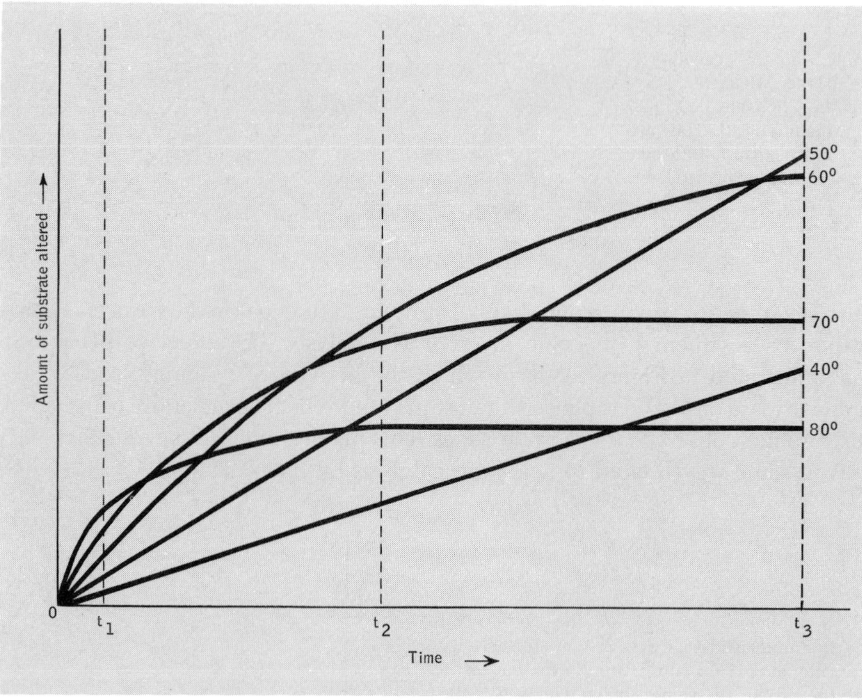

Figure 27–6 An enzymatic reaction carried out at various temperatures. (From A. Mazur and B. Harrow, *Textbook of Biochemistry*, 10th ed., W.B. Saunders Company, 1971.)

27.3 THE ACTIVE SITE AND ENZYME INHIBITION

The major portion of an enzyme does not participate directly in the reaction undergone by the substrate. A comparatively small locus (or loci) on the enzyme, the **active site** (or sites), is directly involved in the substrate transformation, while portions of the "unused" polypeptide chain form non-covalent bonds to the substrate and hold it in the geometry necessary for catalysis. Both of these features are essential for enzyme activity and are collectively termed the **active center.** The key to understanding enzyme activity rests in identifying the amino acids in the active site and the active center. Although a great deal of effort is being expended to determine the structures of enzymes and enzyme-substrate complexes by x-ray diffraction analysis, chemical analysis can still provide information important to the modern enzymologist.

Simple Inhibition. Substances that interfere with an enzyme's catalytic activity are **inhibitors** of that enzyme. Enzyme inhibitors, acting by blocking any portion of the active center or by so distorting other portions of the molecule that the enzyme-substrate complex cannot form, often are extremely poisonous if ingested by an organism. One example of a potent and toxic enzyme inhibitor is diisopropylfluorophosphate, DFP. Among other things, DFP blocks the protein-hydrolyzing action of the enzyme chymotrypsin. DFP acts as an *irreversible* enzyme inhibitor toward chymotrypsin; the activity of the enzyme is not restored when its normal substrate is added to an already inhibited enzyme.

Chymotrypsin reacts with DFP to form an adduct that can be isolated and degraded, *in vitro,* without the loss of DFP. When the individual amino acids formed from this degradation are analyzed, the DFP is found covalently bonded to only one of twenty-six serine residues within chymotrypsin. This and other data suggest that the derivatized serine, ser-195[a], is part of the active site of chymotrypsin. Other experiments, including the dependence of the activity of chymotrypsin upon pH, indicate that his-157 also is part of the active site of chymotrypsin.

$$((CH_3)_2CHO)_2\overset{\oplus}{P}(O^{\ominus})F + HOCH_2-CH(NH-)(C{=}O) \rightarrow ((CH_3)_2CHO)_2\overset{\oplus}{P}(O^{\ominus})-O-CH_2-CH(NH-)(C{=}O)$$

DFP ser-195 in chymotrypsin

$$\xrightarrow{H_2O\ (degradation)} ((CH_3)_2CHO)_2\overset{\oplus}{P}(O^{\ominus})-O-CH_2-CH(NH_2)CO_2H$$

DFP derivative of serine-195

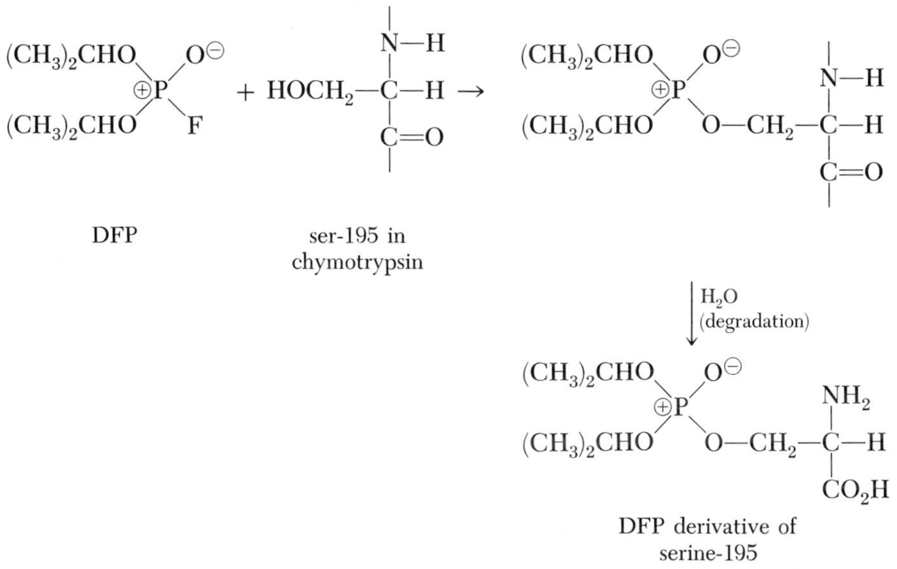

Later in this chapter we will examine the way in which chymotrypsin catalyzes peptide bond hydrolysis. However, it is interesting to note here that nature appears to have "favorite" units in the active sites of some enzymes. Thus, trypsin, chymotrypsin, elastase, and thrombin all have serine and histidine at the active site, and in each enzyme these amino acids are flanked by the same amino acids, *i.e.*, -asp-*ser*-gly- and -ala-*his*-cys-.

The inhibition of chymotrypsin by DFP represents enzyme inhibition by a material that bears no obvious resemblance to the normal enzyme substrate (the peptide bond). An enzyme may also be inhibited when it forms a complex with a molecule that resembles the normal substrate but that cannot undergo the reaction catalyzed by the enzyme. If subsequent addition of the normal substrate regenerates enzyme activity (by displacing the mock substrate), the material that caused the inhibition is termed a *reversible* inhibitor. By way of illustration, succinic dehydrogenase is an enzyme that oxidizes succinic acid. When pre-treated with malonic acid, the enzyme fails to oxidize small quantities of the normal substrate. However, if a large excess of succinic acid is added, the malonic acid–enzyme complex is replaced by the succinic acid–enzyme complex and the reaction proceeds as expected.

Carefully controlled enzyme inhibition has its place in modern medicine. For example, the disease *myasthenia gravis* is related to the malfunctioning of receptor sites on the post-

---

[a] Each successive amino acid in a polypeptide chain is numbered 1,2,3, . . . , beginning with the *N*-terminal amino acid (see p. 783).

synaptic muscle membrane. (Review the discussion beginning on p. 757 if necessary.) An exciting nerve impulse releases the normal amount of acetylcholine (a neurotransmitter), but this is not enough to produce the desired action potential. The disease can sometimes be treated by very careful administration of physostigmine or neostigmine, compounds that inhibit the enzyme cholinesterase in its attempt to destroy acetylcholine. This inhibition permits a buildup of acetylcholine in the synaptic gap and, thereby, partially compensates for the malfunctioning receptor site.

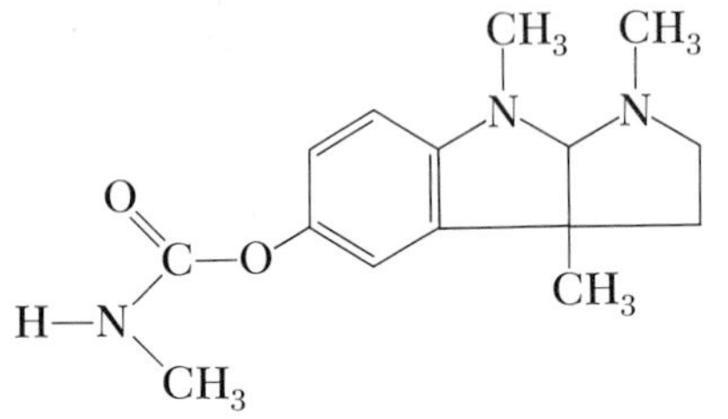

physostigmine

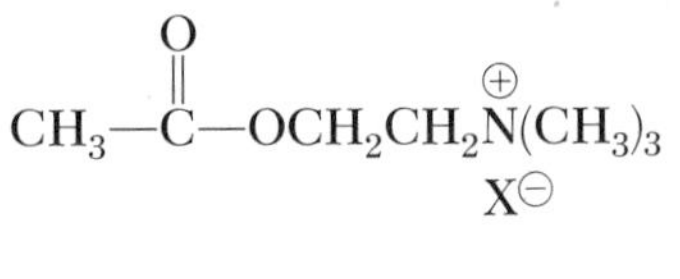

$$CH_3-\overset{O}{\overset{\|}{C}}-OCH_2CH_2\overset{\oplus}{N}(CH_3)_3\ X^{\ominus}$$

acetylcholine

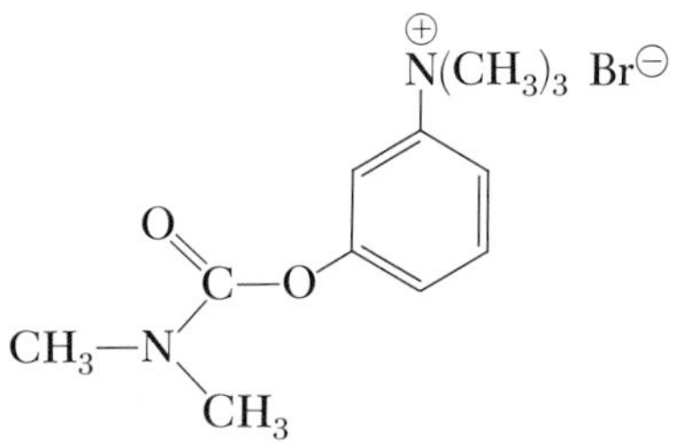

neostigmine bromide

The role of phosphorus-containing enzyme inhibitors as insecticides and war gases was discussed in Chapter 23.

Feedback Inhibition. Numerous biochemical reactions involve a series of consecutive steps, *e.g.*,

$$A \xrightarrow{E_A} B \xrightarrow{E_B} C \xrightarrow{E_C} D \xrightarrow{E_D} P$$

Such reactions usually may be controlled by a **feedback mechanism,** which is very important because it prevents the over-production of P. This control sometimes is accomplished by having the final product form a complex with one of the enzymes used early in the sequence. This changes that enzyme's conformation and prevents it from forming any more of the enzyme-substrate complex leading to the final product.

27.4 SPECIFICITY OF ENZYMES

A given enzyme is not free to catalyze *any* reaction; indeed, most enzymes can catalyze only one kind of process (*e.g.*, decarboxylation). In addition to this type of specificity, enzymes often are limited to certain specific molecules as substrates for these reactions. Finally, stereochemical aspects of enzymic reactions usually are subject to severe restrictions. In this section we will focus upon substrate specificity, as it applies to protein digestion, and upon some rather amazing stereospecificity.

Substrate Specificity. Enzymes vary in their ability to catalyze more than one reaction and/or to convert more than one precisely defined substrate into product. One extreme is exemplified by urease, an enzyme whose sole function is to convert urea into ammonia and carbon dioxide.

$$\underset{\text{urea}}{H_2N{-}\overset{\overset{\displaystyle O}{\|}}{C}{-}NH_2} + H_2O \xrightarrow{\text{urease}} 2NH_3 + CO_2$$

We can learn about some less stringently controlled enzymic reactions and, at the same time, about ourselves by looking briefly at the process of protein digestion. Most animals, including man, are not continuous feeders. Instead, they ingest food, process it, excrete the residue, and sooner or later repeat this sequence. The reasons for the existence of this feeding pattern are complex and undoubtedly related to environmental problems. One of these ecological considerations revolves around the need to be able to flee from an enemy at a moment's notice—and it is not easy to eat and run simultaneously.

An equally important reason for intermittent feeding in man is that food requires a certain time for internal processing, and little is gained if partially digested food is excreted. In order to derive the maximum nutritional value from a given amount of protein, it must be processed in a certain order and for particular time periods. The *proteolytic enzymes* control this remarkable process, the conversion of proteins into amino acids.

The stepwise digestion of proteins that occurs in mammals is facilitated to some extent by the division of the alimentary tract into different regions. The hydrolysis of proteins is initiated in the stomach by the "gastric juices." The stomach wall secretes hydrochloric acid and pepsinogen, which, below pH 6, is autocatalytically converted to the enzyme pepsin. Pepsin hydrolyzes peptide bonds within the protein chain as well as those at the ends of the chain and, for this reason, is called an *endopeptidase.*

Secretions from the pancreas and liver bring the partially digested food to a pH near 7; then, in the intestine, the enzyme trypsin continues the hydrolysis of peptide bonds. Secreted by the pancreas as the inactive precursor trypsinogen, trypsin is an endopeptidase with an optimum pH of between 7 and 9.

Endopeptidases produce short peptide chains, which are further degraded to dipeptides by a group of enzymes called *exopeptidases* because they act on only the terminal peptide bonds. Carboxypeptidase (which, you will recall, is an enzyme that removes the *C*-terminal amino acid) is secreted by the pancreas. By way of contrast, aminopeptidase, an enzyme that removes the *N*-terminal amino acid, is secreted by the intestine.

The last act of this scenario occurs when *dipeptidases* cleave dipeptides to individual amino acids. (While the majority of dipeptidases are secreted by the intestine, it is believed that the epithelial cells of the intestine also contain some dipeptidases that will cleave any dipeptides absorbed by these cells.) Finally, the individual amino acids enter the cells of the intestine by the poorly understood process called "active transport."

In summary, we see that the digestion of proteins is a complex sequential hydrolysis of peptide bonds. Perhaps of greatest interest to the potential physiologist is the fact that the various enzymes are made available by different organs and tissues along the digestive tract.

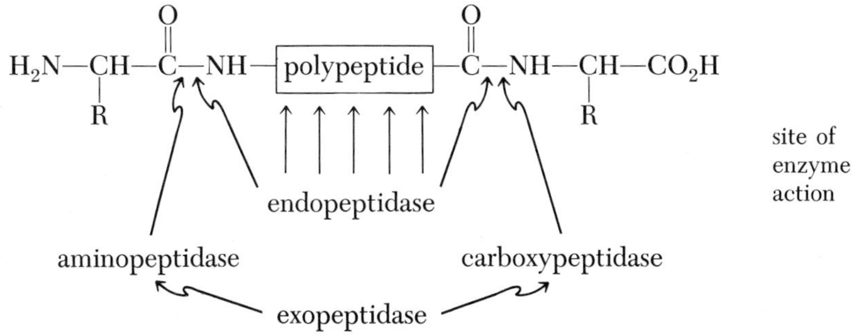

$$H_2N{-}\underset{\displaystyle R}{\underset{|}{C}H}{-}\overset{\overset{\displaystyle O}{\|}}{C}{-}NH{-}\boxed{\text{polypeptide}}{-}\overset{\overset{\displaystyle O}{\|}}{C}{-}NH{-}\underset{\displaystyle R}{\underset{|}{C}H}{-}CO_2H$$

site of enzyme action

endopeptidase

aminopeptidase

carboxypeptidase

exopeptidase

Stereospecificity. Enzymes exhibit several types of stereospecificity. For example, most enzymes can employ only one member of a *d,l* pair as a substrate. A more interesting type of stereospecificity involves the recognition of enantiotopic functionalities. The enzyme *glycerol kinase*[a] catalyzes the conversion of glycerol to L-phosphoglycerol. The fact that this enzyme does not produce any D-phosphoglycerol establishes its ability to distinguish among the enantiotopic $—CH_2OH$ groups of glycerol.

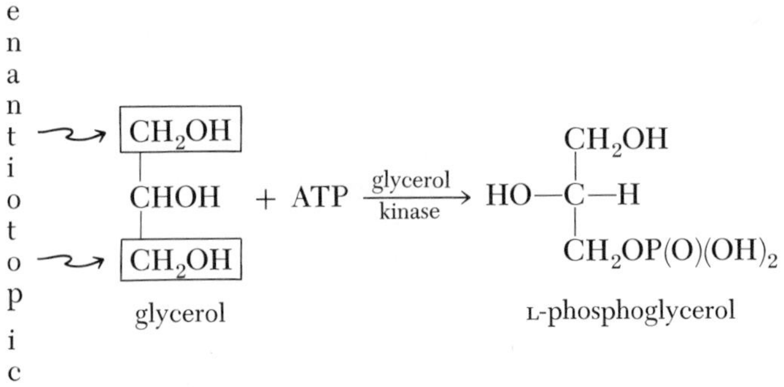

enantiotopic → CH_2OH (boxed), $CHOH$, CH_2OH (boxed)

$$\underset{\text{glycerol}}{\begin{array}{c}\boxed{CH_2OH}\\|\\CHOH\\|\\\boxed{CH_2OH}\end{array}} + ATP \xrightarrow[\text{kinase}]{\text{glycerol}} \underset{\text{L-phosphoglycerol}}{\begin{array}{c}CH_2OH\\|\\HO—C—H\\|\\CH_2OP(O)(OH)_2\end{array}}$$

A classic example of this type of enzyme stereospecificity occurs in citric acid metabolism. When oxaloacetic acid is metabolized, it is converted into citric acid. Labeling the starting material with ^{14}C at the carboxy group next to the methylene group produces citric acid that is labeled in one carboxy group. If this labeled citric acid is treated with enzymes from mitochondria (in which many oxidations occur), the citric acid is converted to α-ketoglutaric acid that is labeled exclusively at the carboxy group adjacent to the carbonyl group.

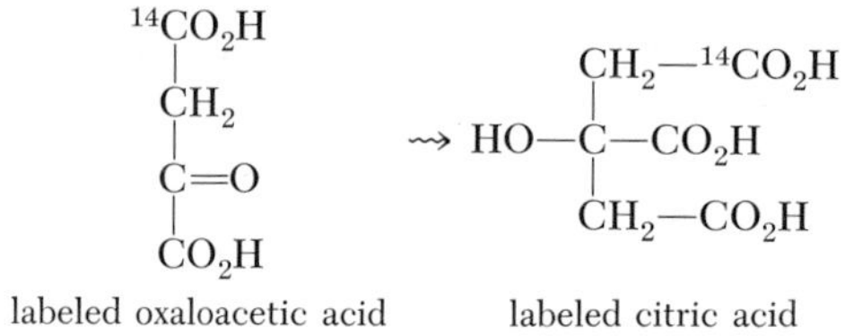

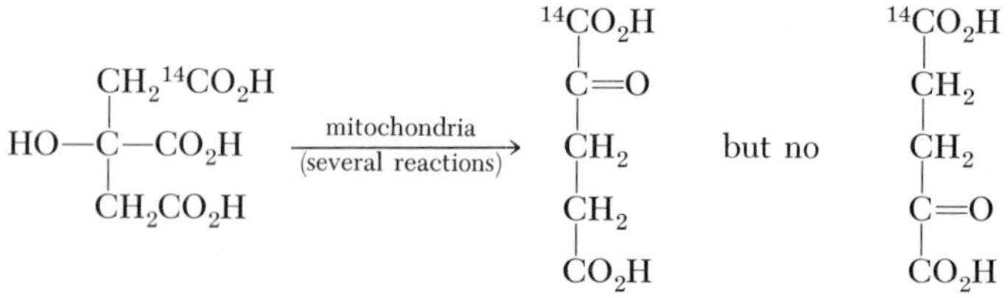

$$\underset{\text{labeled oxaloacetic acid}}{\begin{array}{c}^{14}CO_2H\\|\\CH_2\\|\\C{=}O\\|\\CO_2H\end{array}} \rightsquigarrow \underset{\text{labeled citric acid}}{\begin{array}{c}CH_2—^{14}CO_2H\\|\\HO—C—CO_2H\\|\\CH_2—CO_2H\end{array}}$$

$$\begin{array}{c}CH_2{}^{14}CO_2H\\|\\HO—C—CO_2H\\|\\CH_2CO_2H\end{array} \xrightarrow[\text{(several reactions)}]{\text{mitochondria}} \begin{array}{c}^{14}CO_2H\\|\\C{=}O\\|\\CH_2\\|\\CH_2\\|\\CO_2H\end{array} \quad \text{but no} \quad \begin{array}{c}^{14}CO_2H\\|\\CH_2\\|\\CH_2\\|\\C{=}O\\|\\CO_2H\end{array}$$

How can an enzyme distinguish between two "identical" methylene groups? As noted in Chapter 4, when two similar groups on a prochiral atom are associated with an optically active substance, the products and the transition states leading to them are diastereomeric and will be formed at different rates. If the rate difference is sufficiently large, we may say that one group reacts and one does not. Apparently, the particular complex leading to the observed labeled α-ketoglutaric acid is the only one involved in the reaction.

While the exact structure of the enzyme-substrate complex may be unknown (one exception to this is presented much later in this chapter), one principle is always the same: *once a prochiral center becomes associated with a dissymmetric substance, it is no longer prochiral but chiral; and those functional groups that were enantiotopic become diastereotopic.* These functional groups will, therefore, react at different rates.

[a] A *kinase* is any phosphorylating enzyme that uses ATP as a phosphate source.

27.5 ENZYME COFACTORS AND ENZYME ACTIVATION

Enzymes usually consist of two different components—the **apoenzyme** and the **cofactor.** The apoenzyme is the protein portion of the enzyme, while the cofactor is often some non-proteinaceous material. There are three types of cofactors: **prosthetic groups, metal ions,** and **coenzymes.** The biologically active adduct of apoenzyme and cofactor sometimes is called the **holoenzyme.**

Prosthetic Groups. Prosthetic groups are distinguished by being firmly bound to the apoenzyme. Hemoproteins such as hemoglobin contain an iron-porphyrin prosthetic group, heme, which is responsible for the introduction of oxygen into our bodies. This prosthetic group, pictured below, contains iron coordinated to four nitrogens, and it is nearly encased in protein in the holoenzyme (Figure 27–7).

Figure 27–7 *Left,* Complete hemoglobin molecule. The heme groups are indicated by gray disks. *Right,* Hemoglobin molecule viewed from another direction. (From M.F. Perutz, Science, *140:*863, 1963. Copyright 1963 by the American Association for the Advancement of Science.)

porphyrin

heme (a complex between Fe^{+2} and protoporphyrin)

Prosthetic groups may, through their biosynthesis or degradation, play important roles in maintaining (or destroying) our good health. One example of this is heme. The major catabolic pathway for hemoglobin converts heme into a group of linear tetrapyrroles called *bile pigments*. Among the more familiar are bilirubin and biliverdin. These are, as a group, dark brown in color and account for most of the color of feces.

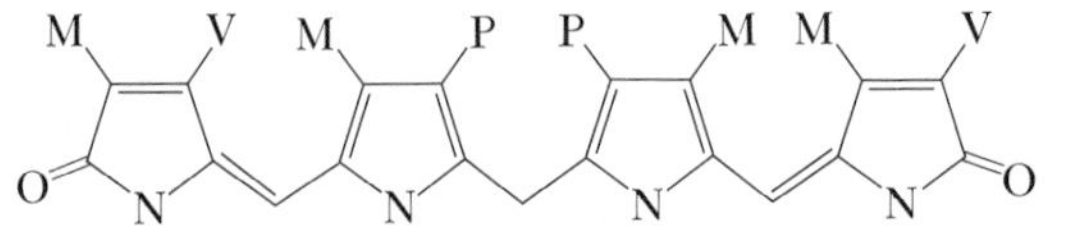

bilirubin (orange)

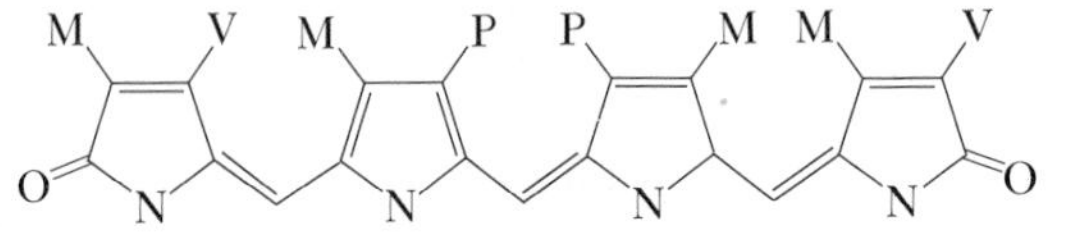

biliverdin (blue-green)

M: $-CH_3$
V: $-CH{=}CH_2$
P: $-CH_2CH_2CH_3$

Premature babies often are deficient in the enzyme that converts the comparatively insoluble bilirubin into its much more soluble glucuronide (see Chapter 25) and they cannot, therefore, excrete it as quickly as it is produced. (Newborn babies produce bilirubin at an accelerated rate.) However, it is important to avoid the accumulation of bilirubin in the brain, since this produces the often fatal condition known as *kernicterus*. Under normal conditions bilirubin is bound to plasma proteins, and this aids in preventing infiltration of bilirubin into the body tissues. Unfortunately, many sulfa drugs (review the discussion beginning on p. 737) compete effectively for the binding sites on the plasma proteins that are used by bilirubin. Therefore, administration of sulfa drugs to newborns has produced mortalities in which kernicterus could be identified at autopsy. The genetic disorder *bilirubin UDP-glucuronyl transferase deficiency* ("familial nonhemolytic jaundice") is characterized by the absence of the enzyme necessary to form the glucuronide of bilirubin.

Metal Ions. Although as many as twenty metals are essential for good health, we will mention only three of them—magnesium, manganese, and zinc.

Magnesium is required by living systems, its function often being to assist in the conversion of ATP to AMP and pyrophosphate anion, PP (see discussion beginning on p. 754). This rate enhancement is accomplished by coordination of the magnesium with two adjacent phosphate groups. Complexation with magnesium improves the leaving group ability of PP by dispersing some of its negative charge.

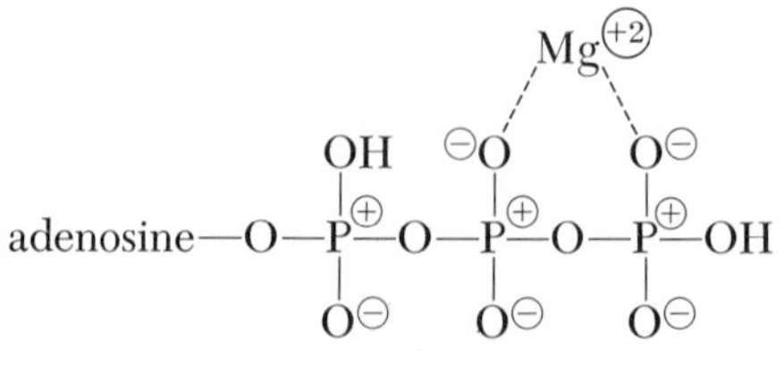

The formation of mixed anhydrides of phosphoric acid and carboxylic acids is catalyzed by an enzyme that requires magnesium. This type of process is an essential feature of the metabolism of straight chain carboxylic acids, since the mixed anhydride then reacts with coenzyme A to produce an acylated coenzyme A (a thiol ester).

$$\mathrm{R{-}CO_2H + E + ATP \xrightarrow{Mg^{+2}} E{\cdot}R{-}\overset{\overset{\displaystyle O}{\|}}{C}{-}O{-}\overset{\overset{\displaystyle O^{\ominus}}{|}}{\underset{\underset{\displaystyle O^{\ominus}}{|}}{P^{\oplus}}}{-}O{-}adenosine + P_2O_7^{-2}}$$

enzyme

$$\mathrm{R{-}\overset{\overset{\displaystyle O}{\|}}{C}{-}O{-}\overset{\overset{\displaystyle O^{\ominus}}{|}}{\underset{\underset{\displaystyle O^{\ominus}}{|}}{P^{\oplus}}}{-}O{-}adenosine + HS{-}CoA \rightarrow R{-}\overset{\overset{\displaystyle O}{\|}}{C}{-}S{-}CoA + AMP}$$

coenzyme A — acylated coenzyme A (a thiol ester)

The importance of manganese to living systems is due, in part, to its utilization in the enzymic *decarboxylation* of oxaloacetic acid to pyruvic acid.

$$\text{oxaloacetate-manganese complex } [{}^{\ominus}O(O{=})C{-}C({=}O)CH_2{-}CO_2^{\ominus}\cdots Mn^{+2}] \xrightarrow{\text{enzyme}} [{}^{\ominus}O(O{=})C{-}C(O^{\ominus}){=}CH_2\cdots Mn^{+2}] \xrightarrow[H^{\oplus}]{\text{enzyme}} {}^{\ominus}O(O{=})C{-}C({=}O)CH_3 + Mn^{+2}$$

pyruvate

Zinc, as Zn^{+2}, is essential for the action of the enzyme carboxypeptidase, and for this reason carboxypeptidase is sometimes called a *metalloenzyme*. The zinc actually binds together two histidine residues in carboxypeptidase, and the enzyme becomes inactive if the zinc is removed.

```
-leu-gly-ile-his-ser-
               ⋮
             Zn(II)      a portion of carboxypeptidase
               ⋮
-leu-ser-ile-his-ser
```

Coenzymes. Many enzymes, and particularly those involved in redox reactions, require the presence of small, non-proteinaceous organic molecules called **coenzymes** before they can function properly. The need for vitamins in our diet stems largely from the fact that many vitamins either act as coenzymes directly or, more often, are a part of a larger coenzyme molecule. Unfortunately, we will not have time to cover all of the vitamins or all of the coenzymes involved with the vitamins that are discussed.

Vitamin B6. The vitamin B6 group, as exemplified by pyridoxal phosphate, plays several critical roles in the metabolism of amino acids. The success of these reactions, which include transamination, deamination, and decarboxylation, hinges upon imine formation and upon tautomerism within these imines.

$$\mathrm{RCH_2{-}CH{=}NR' \rightleftarrows RCH{=}CH{-}NHR'} \quad \text{tautomerism in imines}$$

The mechanism by which transamination can be achieved using pyridoxal phosphate (associated with an enzyme) was presented in Chapter 16, while the utility of the process was discussed in Chapter 24. The process is summarized in Figure 27–8 for the sake of review.

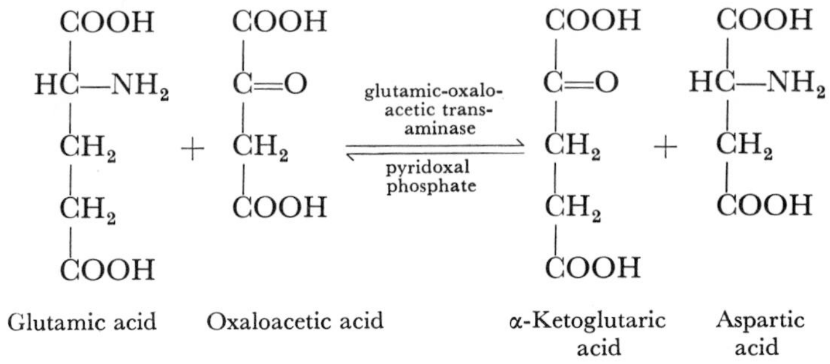

$$\underset{\text{Glutamic acid}}{HOOC-CH(NH_2)-CH_2-CH_2-COOH} + \underset{\text{Oxaloacetic acid}}{HOOC-C(=O)-CH_2-COOH} \xrightleftharpoons[\text{pyridoxal phosphate}]{\text{glutamic-oxalo-acetic trans-aminase}} \underset{\alpha\text{-Ketoglutaric acid}}{HOOC-C(=O)-CH_2-CH_2-COOH} + \underset{\text{Aspartic acid}}{HOOC-CH(NH_2)-CH_2-COOH}$$

Figure 27–8

The decarboxylation of amino acids to amines requires an enzyme (a *decarboxylase*) which, in turn, requires a pyridoxal derivative as a coenzyme. A general mechanism for the decarboxylation of α-amino acids is shown below.

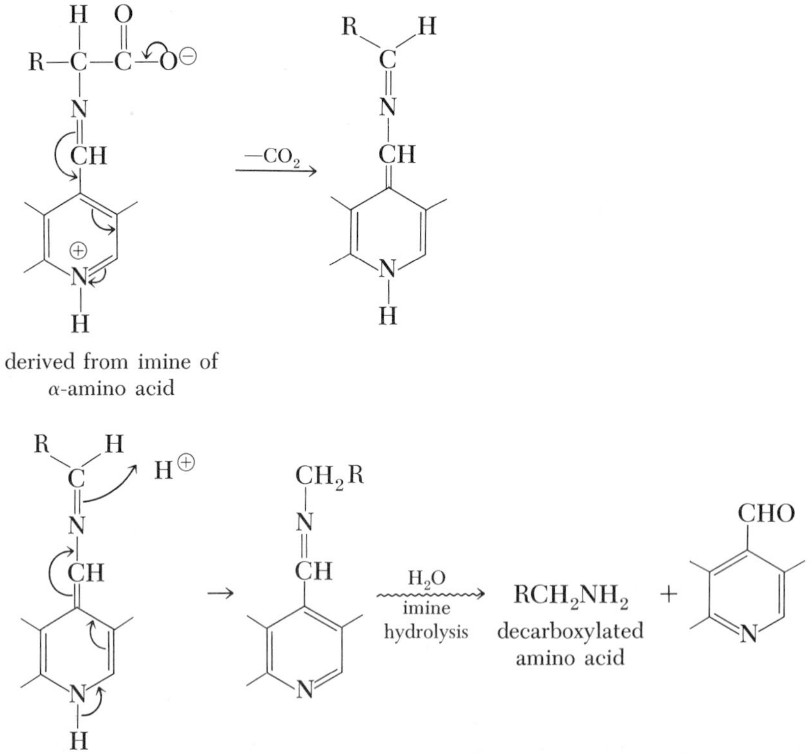

$$R-CH(-COO^{\ominus})-N{=}CH-(\text{pyridinium ring, } N^{\oplus}-H) \xrightarrow{-CO_2} RHC{=}N-CH{=}(\text{dihydropyridine ring, } N-H)$$

derived from imine of α-amino acid

$$RHC{=}N-CH{=}(\text{ring, } N-H) \xrightarrow{H^{\oplus}} RCH_2-N{=}CH-(\text{pyridine ring}) \xrightarrow[\text{imine hydrolysis}]{H_2O} \underset{\text{decarboxylated amino acid}}{RCH_2NH_2} + (\text{pyridine ring})-CHO$$

Niacin. Niacin, the official name for nicotinic acid or nicotinamide, occurs in nature in the pyridine nucleotide coenzymes, the most important of which is NAD, **nicotinamide adenine dinucleotide.** In the past this was called DPN, diphosphopyridine nucleotide.

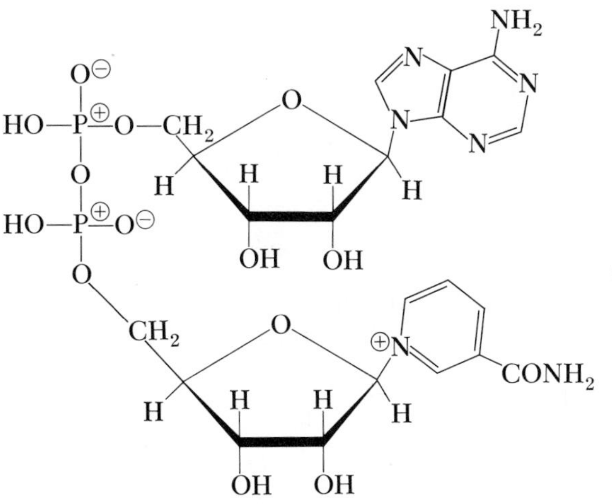

NAD
(DPN)

The biological role of NAD in serving as a coenzyme for the enzymes called *dehydrogenases* (a group of oxidizing enzymes) was discussed on p. 489. That discussion should be reviewed now.

Thiamin. Thiamin, vitamin B1, occurs in high concentration in cereal grain. Dietary insufficiency in humans produces the deficiency disease termed beri-beri. In the body, thiamin often is found as part of the more complex *thiamin pyrophosphate.* The ability of thiamin pyrophosphate to participate as a coenzyme in a variety of reactions reflects the acidity of C2—H in the thiazolium ring and the nucleophilicity of the resultant anion. A scheme whereby pyruvic acid is decarboxylated to acetaldehyde with the aid of thiamin pyrophosphate is shown in Figure 27-9.

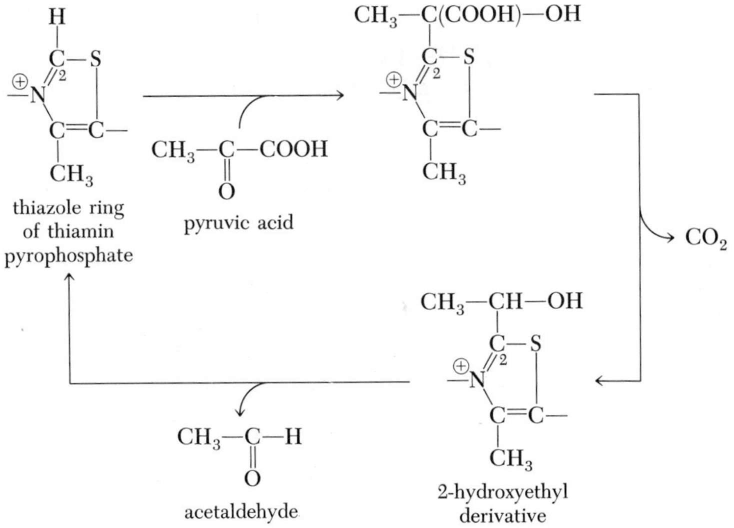

thiazole ring of thiamin pyrophosphate

$CH_3—C(=O)—COOH$
pyruvic acid

$CH_3—C(COOH)—OH$

CO_2

$CH_3—CH—OH$
2-hydroxyethyl derivative

$CH_3—C(=O)—H$
acetaldehyde

Figure 27-9

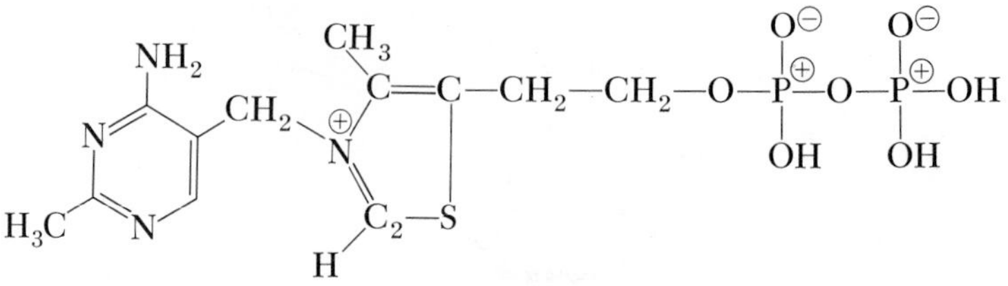

thiamin pyrophosphate

Pantothenic Acid. Other vitamins that appear to have coenzyme activity include **B12** and **pantothenic acid.** The latter occurs in nature primarily in another adenine-based nucleotide, **coenzyme A** (abbreviated CoA or CoA—SH).[a]

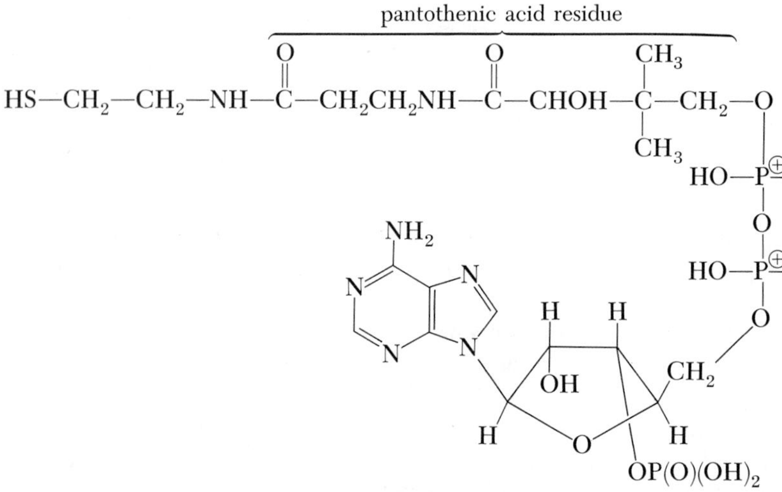

pantothenic acid residue

coenzyme A
(CoA—SH)

Perhaps the most important biological synthesis that requires the participation of CoA—SH is the preparation of long-chain carboxylic acids. In the following section we will discuss the biological synthesis of these acids.

27.6 THE BIOSYNTHESIS OF FATTY ACIDS

In our discussion of terpenes (Chapter 12), we noted that the term "lipid" is used by biochemists to refer to a variety of compounds that are insoluble in water but soluble in typical organic solvents (*e.g.*, carbon tetrachloride and hexane). Included within this group are steroids, fats, oils, and terpenes such as carotene. Lipids, because of their structural inhomogeneity, play a variety of biochemical roles. One of these, the maintenance of the cell membrane, will be discussed shortly. Lipids are also important to us because they are related to America's prime cause of death, cardiovascular disease. While the verbal battle over the significance of cholesterol in causing cardiovascular disease continues, there is mounting evidence that blood triacylglycerol levels may be even more important in indicating which individuals are prone to such disease.

---

[a] This was introduced on p. 362, in our discussion of the biosynthesis of terpenes.

Long-chain carboxylic acids occur in *fats*,[a] which are the naturally occurring esters of glycerol ($HOCH_2CHOHCH_2OH$, also called 1,2,3-trihydroxypropane) with such acids. For this reason, straight-chain carboxylic acids are sometimes called *fatty acids*. Among the most common fatty acids are palmitic and stearic acids (C_{16} and C_{18} acids, respectively). We shall demonstrate one biological role of coenzyme A by observing its function in the biosynthesis of stearic acid (Figure 27-10).

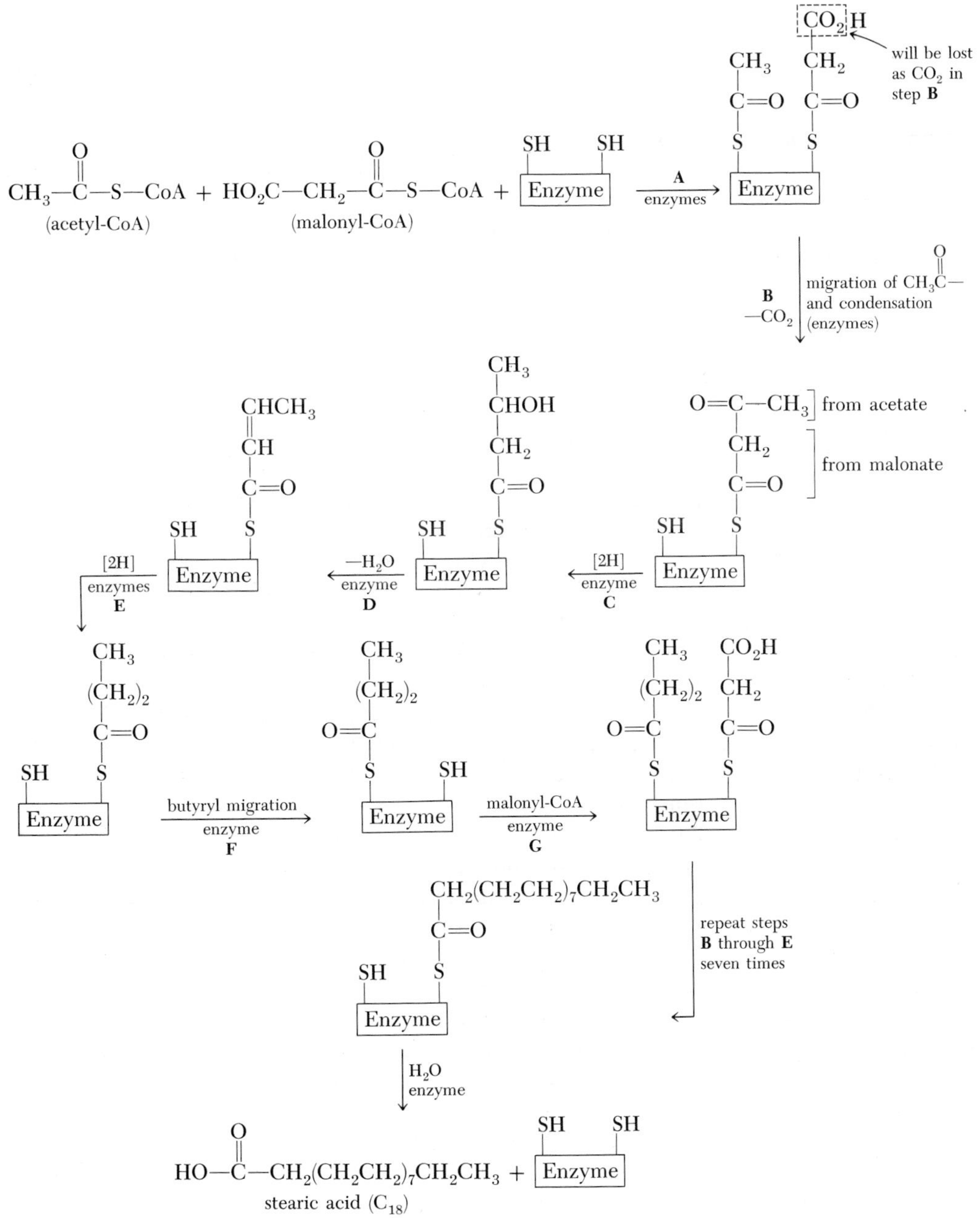

$$CH_3-\overset{O}{\overset{\|}{C}}-S-CoA + HO_2C-CH_2-\overset{O}{\overset{\|}{C}}-S-CoA + (HS)_2\text{Enzyme} \xrightarrow[\text{enzymes}]{\text{A}} \text{Enzyme}(S-\overset{O}{\overset{\|}{C}}-CH_3)(S-\overset{O}{\overset{\|}{C}}-CH_2-CO_2H)$$

(acetyl-CoA) (malonyl-CoA)

CO_2 (boxed, of the CO_2H group): will be lost as CO_2 in step **B**

B, $-CO_2$: migration of $CH_3\overset{O}{\overset{\|}{C}}-$ and condensation (enzymes)

$$\text{Enzyme}(SH)(S-\overset{O}{\overset{\|}{C}}-CH_2-\overset{O}{\overset{\|}{C}}-CH_3)$$

$O{=}C-CH_3$ from acetate; CH_2, $C{=}O$ from malonate

$$\xrightarrow[\text{enzyme, C}]{[2H]} \text{Enzyme}(SH)(S-\overset{O}{\overset{\|}{C}}-CH_2-CHOH-CH_3) \xrightarrow[\text{enzyme, D}]{-H_2O} \text{Enzyme}(SH)(S-\overset{O}{\overset{\|}{C}}-CH{=}CHCH_3)$$

$$\xrightarrow[\text{enzymes, E}]{[2H]} \text{Enzyme}(SH)(S-\overset{O}{\overset{\|}{C}}-(CH_2)_2-CH_3) \xrightarrow[\text{enzyme, F}]{\text{butyryl migration}} \text{Enzyme}(S-\overset{O}{\overset{\|}{C}}-(CH_2)_2-CH_3)(SH)$$

$$\xrightarrow[\text{enzyme, G}]{\text{malonyl-CoA}} \text{Enzyme}(S-\overset{O}{\overset{\|}{C}}-(CH_2)_2-CH_3)(S-\overset{O}{\overset{\|}{C}}-CH_2-CO_2H)$$

repeat steps **B** through **E** seven times

$$\text{Enzyme}(SH)(S-\overset{O}{\overset{\|}{C}}-CH_2(CH_2CH_2)_7CH_2CH_3)$$

$$\xrightarrow[\text{enzyme}]{H_2O} HO-\overset{O}{\overset{\|}{C}}-CH_2(CH_2CH_2)_7CH_2CH_3 + (HS)_2\text{Enzyme}$$

stearic acid (C_{18})

Figure 27-10 Synthesis of fatty acids.

---

[a] You may wish to review the relevant material on pp. 574–575.

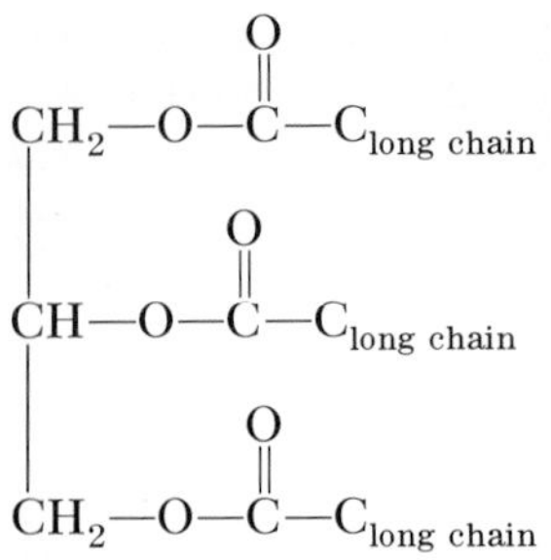

$$\begin{array}{l} CH_2-O-\overset{\displaystyle O}{\overset{\|}{C}}-C_{\text{long chain}} \\ | \\ CH-O-\overset{\displaystyle O}{\overset{\|}{C}}-C_{\text{long chain}} \\ | \\ CH_2-O-\overset{\displaystyle O}{\overset{\|}{C}}-C_{\text{long chain}} \end{array}$$

a "fat" or "triacylglycerol" or "triglyceride"

The first step in fatty acid synthesis is really one of the last steps in carbohydrate degradation, since it is carbohydrate metabolism that produces acetyl coenzyme A (acetyl-CoA), the starting compound in our biosynthesis.

Acetyl-CoA is converted in the body to the malonic acid thioester of coenzyme A, **malonyl-CoA,** by reaction with carbon dioxide. (The acetyl-CoA → malonyl-CoA conversion is known to require biotin, another B vitamin, and is not a simple, direct reaction of acetyl-CoA with carbon dioxide.) Armed with acetyl-CoA and malonyl-CoA, we begin!

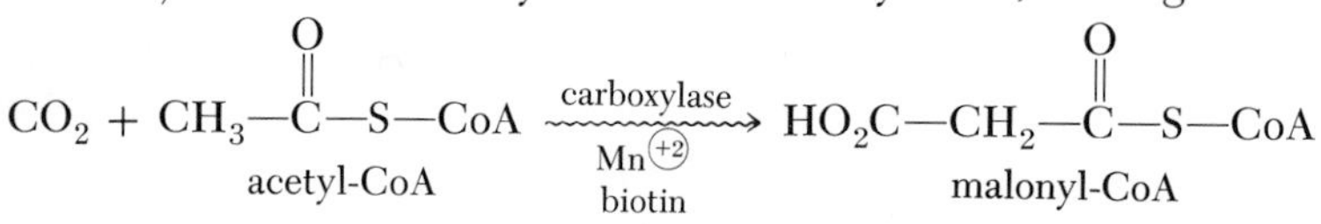

$$CO_2 + \underset{\text{acetyl-CoA}}{CH_3-\overset{\displaystyle O}{\overset{\|}{C}}-S-CoA} \xrightarrow[\text{Mn}^{(+2)}\ \text{biotin}]{\text{carboxylase}} \underset{\text{malonyl-CoA}}{HO_2C-CH_2-\overset{\displaystyle O}{\overset{\|}{C}}-S-CoA}$$

The enzyme called fatty acid synthetase contains two proximal thiol groups, which react with acetyl-CoA and malonyl-CoA, respectively. These adjacent groups undergo a condensation, with the loss of carbon dioxide, to liberate the thiol group that carried the acetyl moiety *and* to create a four-carbon chain bonded to the other thiol group. A series of three steps, each of which requires at least one enzyme, converts this $CH_3—C(O)—CH_2—C(O)—$ group into a butyryl group, $CH_3(CH_2)_2C(O)—$. This butyryl group then migrates back to the thiol group that initially bore the acetyl group. The newly liberated thiol group proceeds to acquire a new malonyl group, and the process then goes through another complete cycle. Each such cycle adds two carbons to the growing carbon chain and, after eight cycles have been completed, the product stearic acid is removed from the enzyme system.

It is common for chemists and biochemists to speak of the "two-carbon" unit that is used to build up fatty acids and of how this type of construction explains why the common fatty acids have an even number of carbons. However, this line of reasoning appears to be inconsistent with the scheme presented in Figure 27–10, which shows that the three-carbon malonyl unit is the one that is responsible for the buildup of the fatty acid chain. This conflict can be reconciled by noting (a) that the malonyl unit (C_3) was synthesized from an acetyl unit (C_2) and carbon dioxide, and (b) that the chain-lengthening condensation between the malonyl unit and the growing fatty acid chain involves the loss of carbon dioxide from the malonyl unit. Since the carbon lost as carbon dioxide is the same one that converted the acetyl group to the malonyl group, this conflict is more apparent than real. From a pragmatic point of view one could argue that fatty acids are made by condensing acetyl (C_2) units while, from a mechanistic point of view, one could argue that only the first C_2 unit is derived directly from acetyl and that the other C_2 units are derived from acetyl groups that must first be converted to malonyl groups.

There are, in fact, several different ways for living systems to synthesize fatty acids. The sequence described above is one of two extramitochondrial pathways and, moreover, is the only one that involves a non-microsomal multienzyme system. It also happens to be the best understood route to fatty acids and the one responsible for most *in vivo* fatty acid synthesis.

27.7 BIFUNCTIONAL CATALYSIS

Bifunctional catalysis involves the *cooperative and simultaneous* interaction of two functionalities, or two sites within one functional group, with a substrate. A typical example of such a process is the mutarotation of 2,3,4,6-tetramethyl-α-D-glucose in benzene. While mutarotation is catalyzed by a mixture of phenol and pyridine, the reaction rate is fifty times greater in the presence of 2-hydroxypyridine. Mutarotation involves the hydrolysis of the hemiacetal linkage; the role of pyridine and phenol in accelerating the process is presented below.

When 2-hydroxypyridine is substituted for phenol and pyridine, the reaction is further accelerated because both the requisite proton donor and the proton acceptor occur in the same molecule; thus, the entropic requirement present when a mixture is used (*i.e.*, the need to align three molecules properly) is diminished.

Simultaneous transfer of two protons explains why benzoic acid also catalyzes the mutarotation.

a hemiacetal a hydroxyaldehyde

It is believed that many enzyme-catalyzed reactions involve bifunctional catalysis. One example is the action of the enzyme ribonuclease (RNAase), which catalyzes the hydrolysis of RNA. Cleavage of the biopolymer occurs in one step, as presented below. Since the nucleophile that cleaves the P—O bond between sugar and phosphate is on the sugar, one intermediate contains a cyclic 2′,3′-phosphate. The cyclic phosphate ring is opened in the second step.

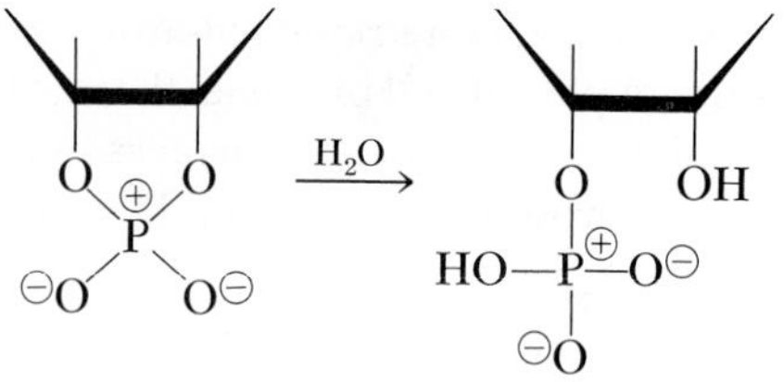

$\xrightarrow{H_2O}$

The enzyme RNAase provides both the base, which accepts a proton from one of the hydroxy groups of the sugar, and the acid, which donates a proton destined to become part of one of the sugar's hydroxy groups. It appears that both the acidic and the basic sites on the enzyme are histidine residues, the basic site being the imidazole ring of histidine-119 and the acid being the protonated imidazole ring of histidine-12. It is not surprising to find two histidine residues within one molecule that act separately as acid and base in a given reaction. After all, the residues are in different environments, and these environments will alter their pK_a's by different amounts from some common value.

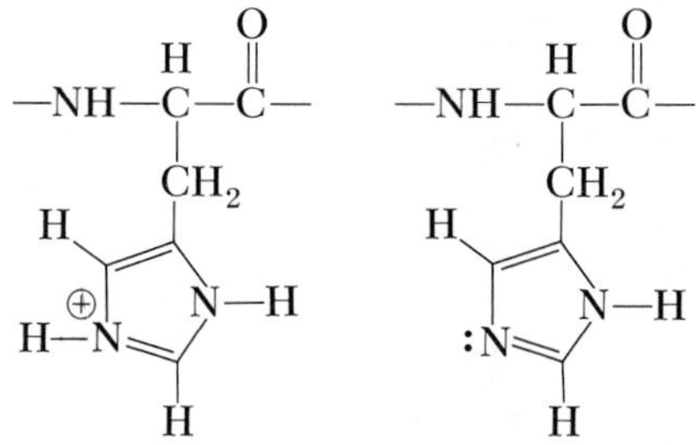

histidine-12
(proton donor)

histidine-119
(proton acceptor)

For the sake of comparison, a partial mechanism to explain the action of chymotrypsin is shown below. Remember that we have already indicated that ser-195 and his-157 are both parts of the active site (p. 883).

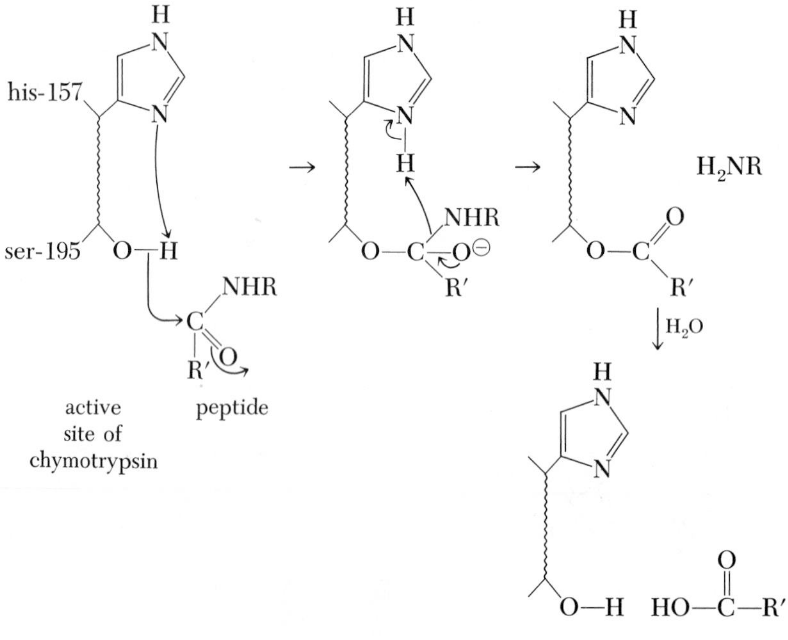

his-157

ser-195

active
site of
chymotrypsin

peptide

An important difference between the mechanisms of chymotrypsin and RNAase action is that RNAase employs a nucleophilic site (a pentose hydroxy group) *within the substrate* to achieve polymer cleavage, while chymotrypsin uses a nucleophilic site (a serine hydroxy group) within the *enzyme* to achieve cleavage. However, they share the common, and apparently universal, characteristic of having "hydrolysis" occur in two steps, that is, transesterification followed by hydrolysis. It can be shown that employing water in the second of two bimolecular steps produces a process that is lower in free energy than is the equivalent process involving the enzyme, the substrate, and water in a single *termolecular* step. Since the process that is lower in free energy will occur more rapidly, it is the one that occurs *in vivo.*

27.8 THE CONTAINER—CELL MEMBRANES

Prokaryotes, those non-nucleated organisms that include the bacteria and the blue-green algae, are composed of cells that contain both a cell membrane and a cell wall. However, while cells of prokaryotes show internal structure, they are not subdivided by internal membranes. Eukaryotes, characterized by cells with a nucleus that is distinct from the cytoplasm, show much more internal cellular organization than do prokaryotes and contain internal membranes. While cells may vary in the degree to which they are organized, they *all* possess an outer *plasma* or *cell membrane* through which must pass all substances entering or leaving the cell. The cell membrane is most remarkable in that it exhibits selective permeability, permitting some, but not all, substances to pass through it. Large molecules, such as proteins, usually cannot pass through the plasma membrane (although they can enter eukaryotic cells by a process called *pinocytosis*[a]). Small molecules and ions can pass through the cell membrane, and some are even forced through against a concentration gradient by the process of active transport.

Since membranes surround all cells and play vital roles in the functioning of such diverse structures as mitochondria, lysosomes, nerve fibers, Golgi apparatus, and the endoplasmic reticulum, we shall devote the remainder of this chapter to the cell membrane, the bacterial cell wall, and to the compounds that constitute these barriers.

The Unit Membrane Hypothesis. The earliest important hypothesis about the structure of the biological membrane was put forward by Davson and Danielli in 1935. Their idea of a membrane containing a continuous hydrocarbon-like phase composed of lipids was modified, particularly by Robertson, until it became what is now termed the **unit membrane hypothesis.** According to this hypothesis, the membrane consists of three layers—two outer protein layers with a central core of lipid. The protein layer is pictured as extending in the plane of the membrane, while the lipid molecules are arrayed perpendicularly to that plane.

The lipids that we have considered thus far—triacylglycerols, terpenes, and steroids—are not the most important constituents of the lipid core of the unit membrane. Rather, the unit membrane contains the highly polar **phospholipids,** compounds that can be viewed as esters of phosphoric acid. While several different types of phospholipids occur in living systems, we will limit the following discussion to *ethanolamine phosphoglyceride* (also called phosphatidyl ethanolamine or cephalin), *choline phosphoglyceride* (also called phosphatidyl choline or lecithin), and *sphingomyelin.*

Lecithin is abundant in egg yolks and also is the most abundant phospholipid in human plasma.[b] The alternate name of choline phosphoglyceride suggests that it is a phosphoric acid derivative of glycerol which bears a choline unit, $HOCH_2CH_2\overset{\oplus}{N}(CH_3)_3\ X^{\ominus}$. Indeed, lecithin

[a] This process is somewhat akin to the engulfing procedure of amoebas.

[b] It is used commercially in "non-stick" cooking sprays.

is an unsymmetric diester of phosphoric acid in which one alcohol is a *di*acylglycerol and the other is choline. Since the diacylglycerol is non-polar and the choline is highly polar, choline phosphoglyceride can be viewed as a long, non-polar substance with a polar terminus.

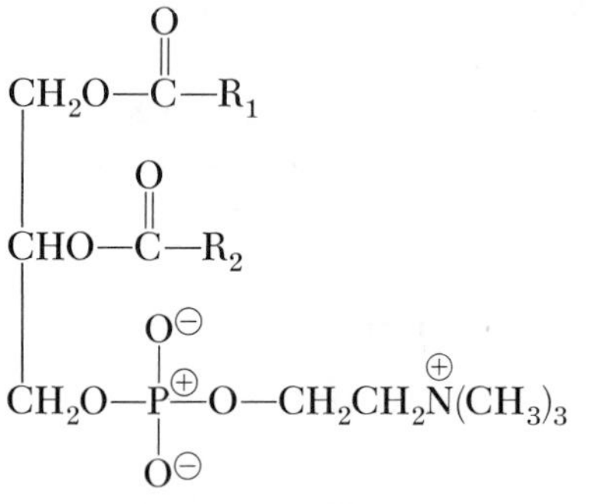

$$\begin{array}{l} CH_2O{-}\overset{\displaystyle O}{\overset{\|}{C}}{-}R_1 \\ | \\ CHO{-}\overset{\displaystyle O}{\overset{\|}{C}}{-}R_2 \\ | \\ CH_2O{-}\overset{O^{\ominus}}{\overset{|}{\underset{\underset{O^{\ominus}}{|}}{P^{\oplus}}}}{-}O{-}CH_2CH_2\overset{\oplus}{N}(CH_3)_3 \end{array} \qquad \text{usually} \begin{cases} R_1 \text{ saturated long chain acid} \\ R_2 \text{ unsaturated long chain acid} \end{cases}$$

lecithin
(choline phosphoglyceride)

As a class, cephalins differ from lecithins only in the nature of the nitrogenous unit that is bonded to the phosphoric acid residue. The most common of the cephalins is cephalin itself, ethanolamine phosphoglyceride.

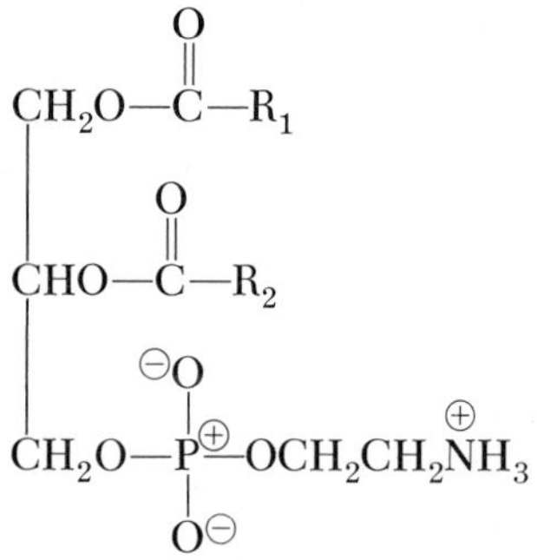

$$\begin{array}{l} CH_2O{-}\overset{\displaystyle O}{\overset{\|}{C}}{-}R_1 \\ | \\ CHO{-}\overset{\displaystyle O}{\overset{\|}{C}}{-}R_2 \\ | \\ CH_2O{-}\overset{^{\ominus}O}{\overset{|}{\underset{\underset{O^{\ominus}}{|}}{P^{\oplus}}}}{-}OCH_2CH_2\overset{\oplus}{N}H_3 \end{array} \qquad \text{usually} \begin{cases} R_1 \text{ saturated long chain acid} \\ R_2 \text{ unsaturated long chain acid} \end{cases}$$

cephalin
(ethanolamine phosphoglyceride)

These first two phospholipids can be viewed as derivatives of glycerol as well as derivatives of phosphoric acid. Sphingomyelin is a *sphingolipid,* that is, a phospholipid containing the complex alcohol sphingosine but lacking a backbone of glycerol. Like the other phospholipids, it contains a polar end and a non-polar end. Sphingomyelin is found in the membranes of plants and animals but occurs in highest concentrations in the brain and in nerve tissue.

$$CH_3(CH_2)_{12}CH{=}CHCH(OH)CHNH_2CH_2OH$$

sphingosine

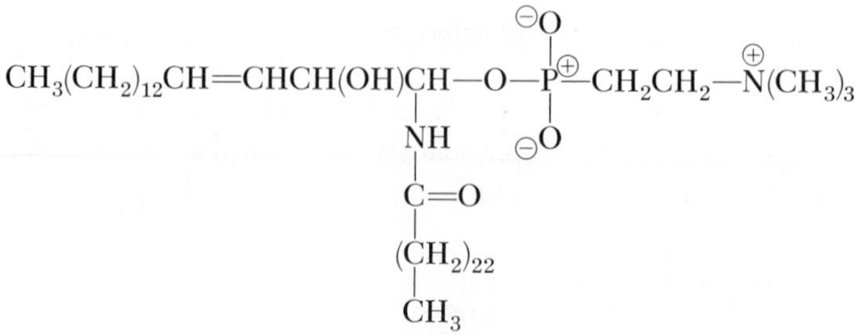

$$CH_3(CH_2)_{12}CH{=}CHCH(OH)\underset{\underset{\underset{\underset{\underset{\underset{\underset{CH_3}{|}}{(CH_2)_{22}}}{|}}{C{=}O}}{|}}{NH}}{|}}{CH}{-}O{-}\overset{^{\ominus}O}{\overset{|}{\underset{\underset{^{\ominus}O}{|}}{P^{\oplus}}}}{-}CH_2CH_2{-}\overset{\oplus}{N}(CH_3)_3$$

sphingomyelin

Why do we need to know that phospholipids have these structures? Because the picture of the unit membrane does not include a haphazard arrangement of lipids within the lipid core but, instead, envisions a bilayer of lipid in which the non-polar "tails" are parallel to one another and the polar "heads" are against the protein coat. An electron micrograph of a cell membrane and a schematic representation of the unit membrane are shown in Figure 27–11.

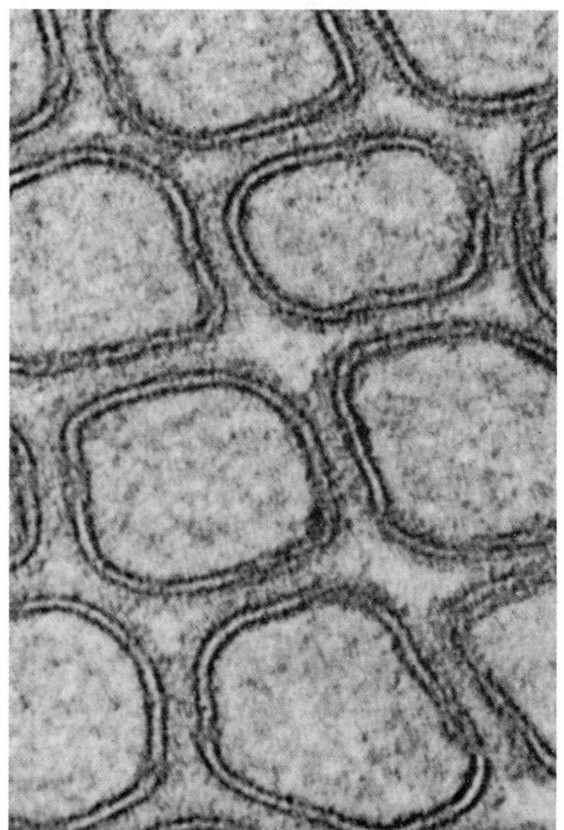

Figure 27–11 **A,** Intestinal microvilli of the cat in cross section, 230,000×. (From D.W. Fawcett, *The Cell,* W.B. Saunders Company, 1966.) **B,** Schematic diagram of the organization of unit membranes: bars, hydrophobic lipid chains; circles, hydrophilic groups of lipid molecules; zigzag lines, nonlipid component, primarily protein. (From E.J. Masoro, *Physiological Chemistry of Lipids in Mammals,* W.B. Saunders Company, 1968.)

The Modified Unit Membrane. A variety of studies have forced a change in the original unit membrane hypothesis. For example, it now is known that many membranes have enzymes bound to them and that these enzymes probably are associated with lipid material in a lipoprotein complex (a non-covalently bonded complex between lipids and proteins). Since these enzymes may act at the surface of the membrane, a reasonable picture of the membrane must have regions in which large amounts of lipid come close to the surface. Direct evidence for variations in the chemical composition of cell membranes also demands that all membranes cannot have identical structures (Table 27–1).

TABLE 27–1 Percentage of Protein and Various Lipids in Membranes from Different Sources

	Source of Membrane			
Component	*Myelin* (%)	*Chloroplasts* (%)	*Erythrocytes* (%)	*Mitochondria* (%)
Protein	25	50	60	70
Phospholipids	24	5	22	29
Cholesterol	19	0	10	1
Sphingolipids	23	0	7	0
Glycolipids[a]	0	20	0	0
Other lipids	9	25	1	0

[a] Glycolipids are compounds containing carbohydrate groups as the polar, hydrophilic end and long-chain fatty acids as the hydrophobic end. Like the triacylglycerols and the phospholipids, they are derivatives of glycerol. The simplest glycolipids, containing one carbohydrate moiety and two fatty acids bound to glycerol, are called glycosyldiacylglycerols. One representative is the monogalactosyldiacylglycerol shown below.

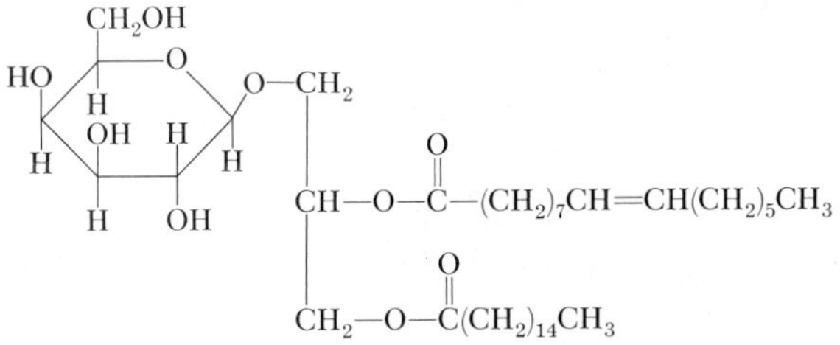

CH_2OH; HO; H; OH; H; H; H; OH (galactose ring) —O—CH_2

$CH—O—\overset{O}{\overset{\|}{C}}—(CH_2)_7CH{=}CH(CH_2)_5CH_3$

$CH_2—O—\overset{O}{\overset{\|}{C}}(CH_2)_{14}CH_3$

(Reprinted with permission of Macmillan Publishing Co., Inc. from *Chemical Biology* by J. Ramsey Bronk. Copyright © 1973, J. Ramsey Bronk.)

It is now believed that cell membranes are made up of discrete subunits, one of which is quite similar to the Davson-Danielli-Robertson unit membrane. Another subunit is believed to be a double thickness of lipoprotein complex, the protein portion being either structural or enzymic in function, depending upon the membrane. Membranes appear to be easily synthesized *in vivo*, and this rapid synthesis may be due to the pre-existence of the subunits and the easy merger of them into the final membrane. This more complicated picture of the membrane helps to explain why the plasma membrane presents a barrier to many water-soluble materials but not to lipid-soluble materials.

One possible picture of the plasma membrane is shown in Figure 27–12. While it seems likely that this will be changed in the future, one aspect of its structure seems firmly established—the membrane is held together largely by van der Waals attractions and hydrogen bonding, rather than by covalent bonds.

The Axon. The membrane around an axon (nerve fiber), commonly called the *myelin sheath*, consists of a set of concentric, deformed unit membranes generated from a structure called the Schwann cell (Figure 27–13). In addition to supplying structural strength, it provides the nerve cell with a type of electrical insulation from its surroundings. Once formed, the myelin sheath survives for the life of the individual. Diseases associated with faulty myelin sheaths include multiple sclerosis, Niemann-Pick disease, and infantile Gaucher's disease.

Sphingolipids, which are important in nerve tissue, contain fatty acid chains longer than

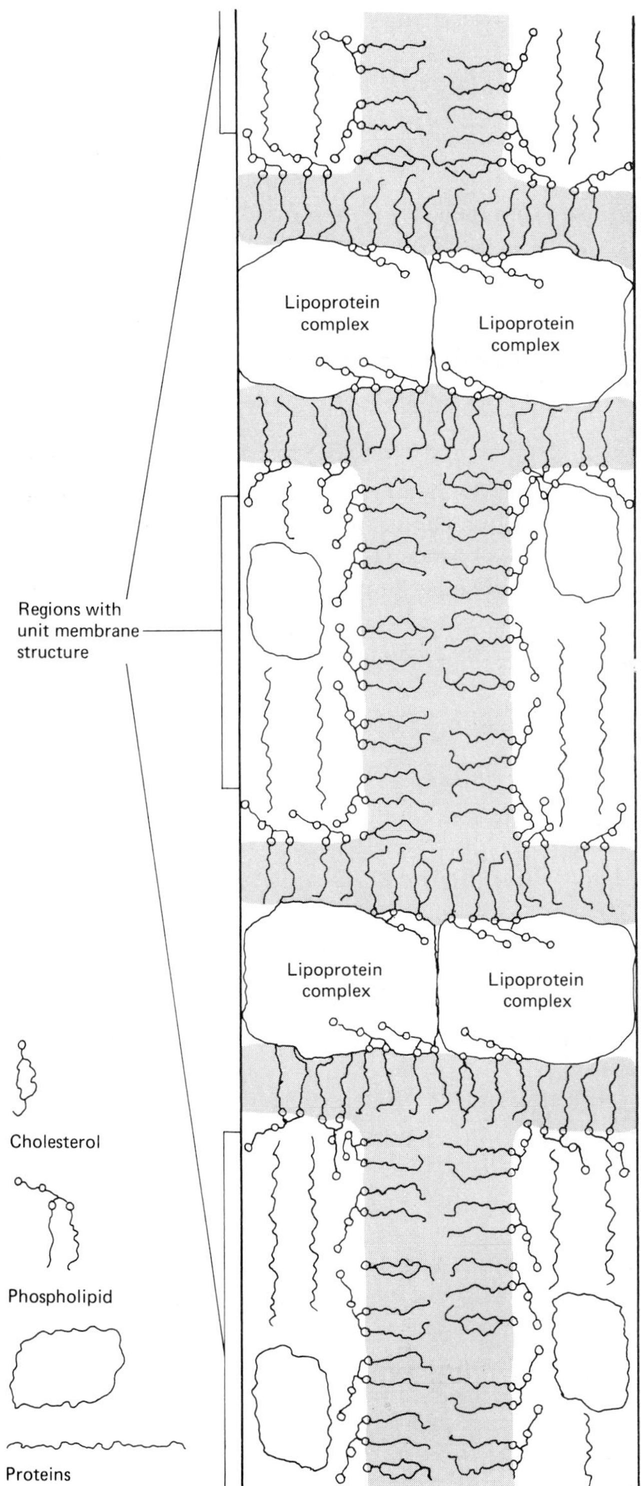

Figure 27-12 One possible arrangement of protein and lipid components in a membrane. According to this model, most of the membrane consists of a bimolecular layer of phospholipids coated on both sides with protein. In some regions, however, this "unit membrane" structure is broken by the presence of lipoprotein complexes, which may consist of a number of subunits. The proteins in these subunits could be enzymes of specific transporting agents, and their identity would vary from one type of membrane to another. The relative proportions of specific lipoprotein complexes and unit membrane regions would reflect the characteristics of the particular membrane. Carbohydrate is likely to adhere to the membrane surface in some regions, possibly in the form of glycoprotein. (Reprinted with permission of Macmillan Publishing Company, Inc., from *Chemical Biology* by J. Ramsey Bronk. Copyright © 1973, J. Ramsey Bronk.)

those associated with triglycerides. Some of the strength of the myelin sheath is believed due to the intertwining of these long chains. If shorter (less than nineteen carbons) chains replace these long ones, or if the chains contain unsaturation, the intertwining is diminished and the nerve is structurally inferior. The myelin of individuals afflicted with the diseases mentioned is deficient in very long chain fatty acids, suggesting that these are molecular disorders.

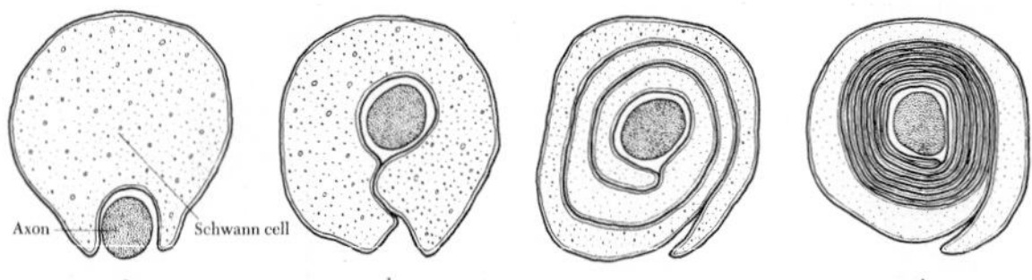

Figure 27–13 Schematic representation of the progressive envelopment of an axon by the membranes of the Schwann cell, as described by Dr. Betty B. Geren. Such an axon is said to be myelinated. (From C.P. Swanson, *The Cell,* 3rd ed., Prentice-Hall, 1969.)

27.9 THE CONTAINER—THE BACTERIAL CELL WALL

Bacterial cells are surrounded by a rigid polysaccharide-oligopeptide wall, which may be viewed as *one* gigantic molecule called a murein (Latin *murus* = wall). The basic unit of this sack-like polymer contains a disaccharide and a tetrapeptide and is called a *muropeptide.* The disaccharide portion of the muropeptide consists of *N*-acetyl-D-glucosamine bonded $\beta(1 \rightarrow 4)$ to *N*-acetylmuramic acid. (*Muramic acid* is the 3-*O* ether formed from glucosamine and lactic acid, $HOCH(CH_3)CO_2H$.)

N-acetylglucosamine — *N*-acetylmuramic acid

$^6CH_2OH$ (5), O, H (4), H, OH, H (1), H (3), H (2), $NHCOCH_3$ — β O — CH_2OH, O, H (4), H, H (1), H, H, $NHCOCH_3$, O

$H_3C—CH—C{=}O$

L-alanine { NH, $CH—CH_3$, C=O }

D-isoglutamic acid { NH, $\overset{\alpha}{C}H—COOH$, CH_2, CH_2, $\overset{\gamma}{C}{=}O$ }

L-lysine { NH, $CH—(CH_2)_4—NH_2$, C=O }

D-alanine { NH, $CH—CH_3$, COOH }

a muropeptide

As seen from Figure 27–14, the disaccharide units of murein are connected to one another by 1,4-glycosidic linkages to form a linear ("two-dimensional") polymer. Moreover, the peptide

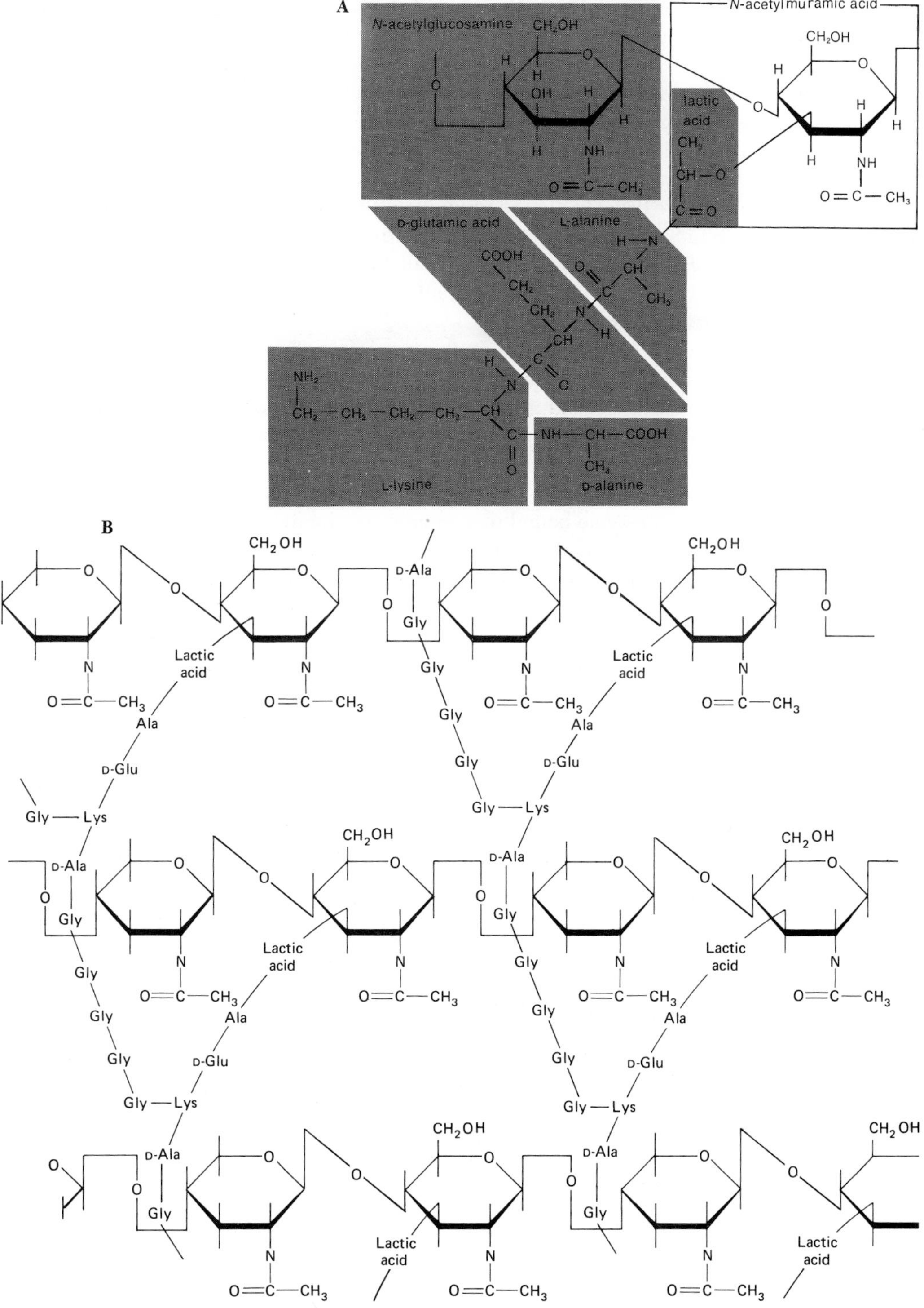

Figure 27-14 Bacterial cell wall polysaccharide. **A,** Muropeptide repeating unit consisting of a tetrapeptide and a disaccharide made up of *N*-acetyl-D-glucosamine and *N*-acetylmuramic acid joined by a $\beta(1 \rightarrow 4)$ linkage. **B,** Portions of several polysaccharide chains, showing the glycine pentapeptide cross linkages. (Reprinted with permission of Macmillan Publishing Company, Inc., from *Chemical Biology* by J. Ramsey Bronk. Copyright © 1973, J. Ramsey Bronk.)

chains are cross-linked to give a three-dimensional, net-like appearance to the final murein. It is this extensive cross-linking that gives the polymer, and the cell wall, its great mechanical strength.

Recently it has become clear that the bacterial cell wall is more than an inert shield against a hostile environment. For example, sexual conjugation of bacteria occurs through the contact of certain elaborations of the cell wall. Secondly, certain viruses that destroy bacteria have specific loci on the cell wall to which they become attached. Many of the immunities to bacterial infections that we develop are the result of antigens on the bacterial cell wall. Finally, the time-honored division of bacteria into gram-positive[a] and gram-negative classes is based upon the amount of *lipid* associated with the cell wall. Gram-positive organisms (*e.g.*, *Staphylococcus albus*) contain very little lipid, while gram-negative organisms (*e.g.*, *Escherichia coli*) have walls that are rich in lipids.

Lysozyme. Lysozyme is an enzyme containing 129 amino acids, which catalyzes the hydrolysis of the glycosidic linkage between *N*-acetylglucosamine and *N*-acetylmuramic acid. Since these are the saccharides found in mureins, lysozyme is capable of lysing bacterial cell walls. This may account for the occurrence of lysozyme in tears and in mucous from the nasal passages.

The complete amino acid sequence and structure of lysozyme have been determined. Very recently, the structure of lysozyme bound to a competitive inhibitor also has been determined by using x-ray diffraction analysis. This last study revealed that the enzyme changes its shape significantly when it passes from the free state to the enzyme-substrate complex. Such an observation is extremely important, since it had been believed that an enzyme and its substrate fit together like "hand and glove" or "lock and key." Now the view is changing, and it is believed that for some enzymes the proper substrate actually may *induce* the necessary fit. Certainly, this points out that, when possible, one should compare the structure of the free enzyme with that of the enzyme-substrate complex before drawing detailed conclusions about the active site and the active center.

Knowing the detailed structure of the enzyme-substrate complex (Figure 27–15) has made

Figure 27–15 The lysozyme-substrate complex. The substrate is the six-ringed molecule passing diagonally through the center of the complex. (From The Three-dimensional Structure of an Enzyme Molecule, by David C. Phillips. Copyright © 1966 by Scientific American, Inc. All rights reserved.)

---

[a] *I.e.*, giving a positive response to the staining test devised by H.C.J. Gram, a Danish physician (1853–1938).

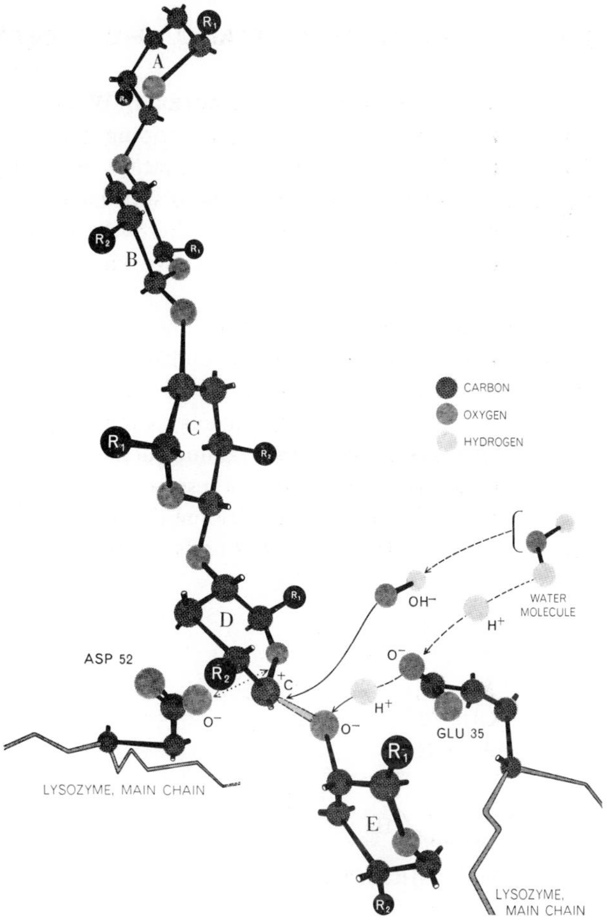

Figure 27-16 A magnified view of the active site of the lysozyme-substrate complex. It is proposed that a hydrogen ion leaves the carboxy group of residue 35 (glutamic acid) and combines with the oxygen joining rings *D* and *E* of the substrate, breaking the bond. The carbocation thus formed at C1 of the *D* ring is stabilized by the negative carboxyl group of residue 52 (aspartic acid). A water molecule supplies the $OH^{\ominus}$ to react with the carbocation and an $H^{\oplus}$ to replace that lost by residue 35. The two parts of the substrate then leave the enzyme, which is free to bind another substrate molecule. (From The Three-dimensional Structure of an Enzyme Molecule, by David C. Phillips. Copyright © 1966 by Scientific American, Inc. All rights reserved.)

it clear that glutamic acid-35 and aspartic acid-52 are a part of the active site of lysozyme. The mechanism of the action of this enzyme is suggested in Figure 27-16 and is shown in a more conventional way in Figure 27-17.

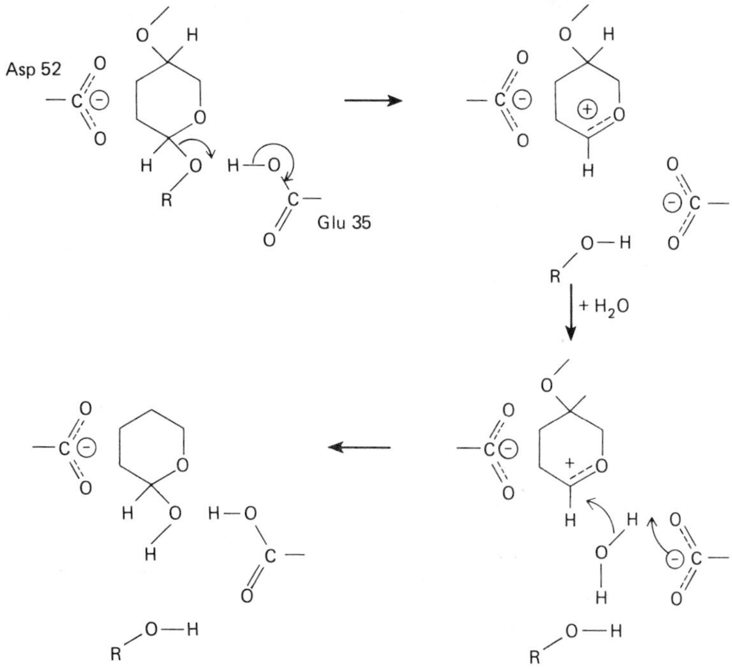

Figure 27-17 A possible mechanism for the hydrolysis of a polysaccharide by lysozyme. The ring shown here is ring *D* of Figures 27-15 and 27-16; R is the fragment containing rings *E* and *F*. While this format makes more apparent the electron movements involved in the hydrolysis, it gives no feeling for the manner in which the structure of the enzyme is responsible for its activity. (Reprinted with permission of Macmillan Publishing Company from *Chemical Biology* by J. Ramsey Bronk. Copyright © 1973, J. Ramsey Bronk.)

Penicillin, an Inhibitor of Bacterial Wall Synthesis. It is fitting that a course which many students will use as a critical stepping stone to a medical education or to careers in biochemistry should end with a presentation of what we know about how one of the most important antibiotics, penicillin, acts to stop bacterial infections. Volumes have been written about penicillin, and it is unfortunate that we cannot go into greater detail about this subject. It is, however, more unfortunate that we know so little about how many other antibiotics function.

The synthesis of the cell wall is inherently intriguing because it is mediated by enzymes on the plasma membrane and occurs outside of the cell. Several antibiotics, including penicillin, are known to prevent the synthesis of bacterial cell walls. We will begin a step-by-step examination of how the cell constructs its murein and discover that penicillin acts only after the bacterium has made a large commitment to cell wall construction.

The first major step in the synthesis of the cell wall occurs within the cell; it is a complex synthesis that leads to the UDP (uridine diphosphate; see Figure 26–1) derivative of *N*-acetylmuramyl pentapeptide, one of the units used to construct the murein (Figure 27–18). The last

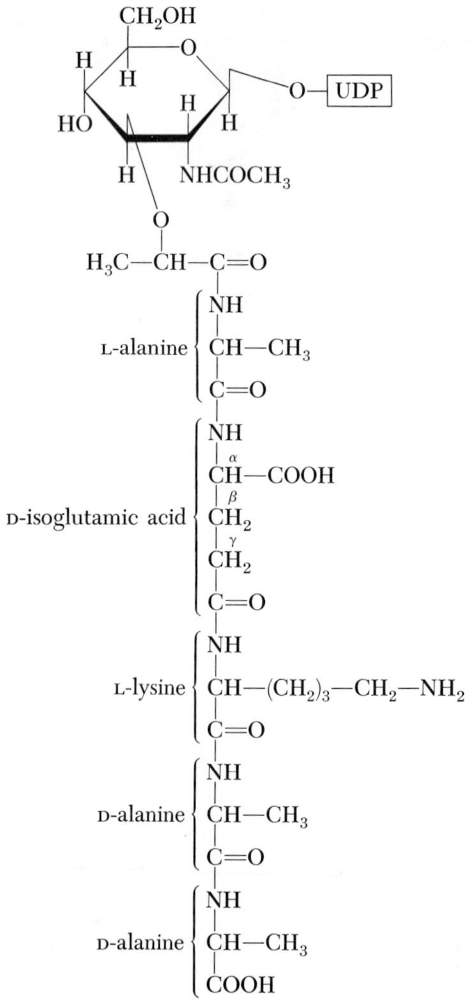

CH_2OH
O
H
H
O—UDP
H
HO
H
H
$NHCOCH_3$
O
$H_3C—CH—C{=}O$

L-alanine: NH, $CH—CH_3$, $C{=}O$

D-isoglutamic acid: NH, $CH—COOH$ (α), CH_2 (β), CH_2 (γ), $C{=}O$

L-lysine: NH, $CH—(CH_2)_3—CH_2—NH_2$, $C{=}O$

D-alanine: NH, $CH—CH_3$, $C{=}O$

D-alanine: NH, $CH—CH_3$, COOH

Figure 27–18 UDP-*N*-acetylmuramylpentapeptide.

step in the first phase of murein synthesis is the addition of the ala-ala (both unnatural in configuration!) to the terminal lysine. (The antibiotic D-cycloserine prevents this step.)

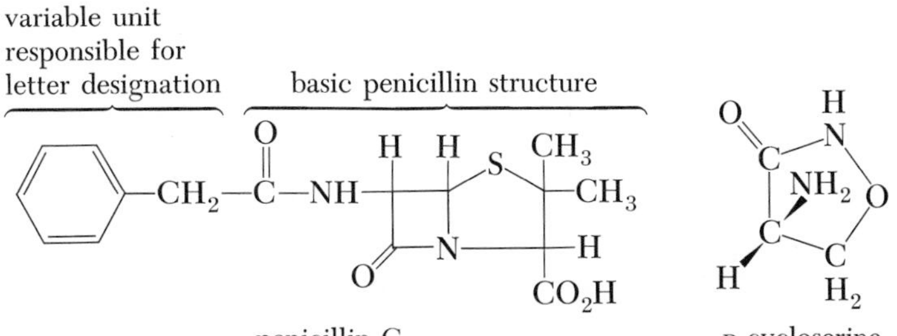

variable unit responsible for letter designation — basic penicillin structure

$C_6H_5-CH_2-\overset{O}{\overset{\|}{C}}-NH-$ (β-lactam/thiazolidine ring: H, H, S, CH_3, CH_3, H, N, O, CO_2H)

penicillin G

D-cycloserine (ring: O=C, N–H, NH_2, O, C, H_2, H)

D-cycloserine

The *N*-acetylmuramylpentapeptide is now transferred from UDP to a lipid, which delivers it to that point on the *outside* of the plasma membrane at which the wall is being synthesized. Here the *N*-acetylglucosamine becomes attached to the *N*-acetylmuramylpentapeptide to form a species quite similar to the murein's repeating unit (a "muropeptide"). After this, the new disaccharide is attached to the growing polysaccharide chain.

The next step in the process involves the construction of short peptide chains at the L-lysine residue of the muropentapeptide. These short chains will, in the last phase of murein synthesis, be used to cross-link the murein polysaccharide chains *via* their peptide units. The length of this potential cross-linking fragment depends upon the specific bacterium; in this illustration, it is the pentapeptide $(gly)_5$.

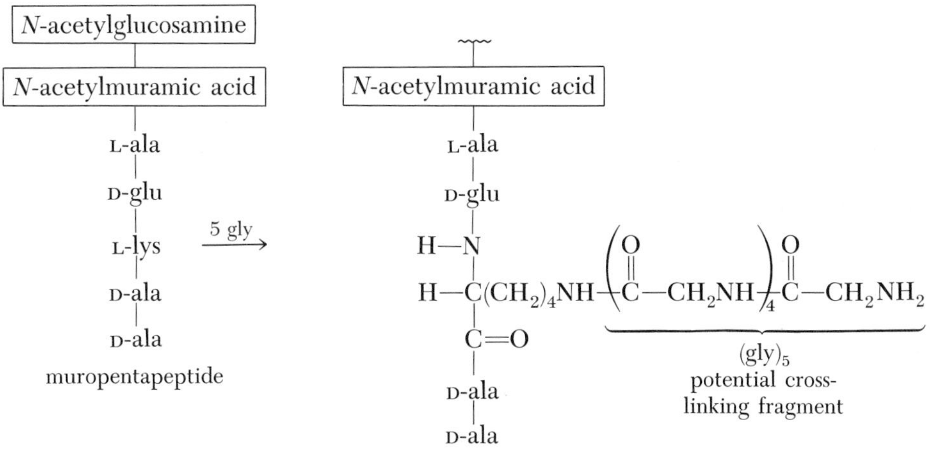

N-acetylglucosamine
|
N-acetylmuramic acid
|
L-ala
|
D-glu
|
L-lys
|
D-ala
|
D-ala

muropentapeptide

$\xrightarrow{\text{5 gly}}$

N-acetylmuramic acid
|
L-ala
|
D-glu
|
H—N
|
$H-C(CH_2)_4NH\left(\overset{O}{\overset{\|}{C}}-CH_2NH\right)_4\overset{O}{\overset{\|}{C}}-CH_2NH_2$
|
C=O
|
D-ala
|
D-ala

$(gly)_5$
potential cross-linking fragment

In the last phase of cell wall synthesis, the one that is inhibited by penicillin, the terminal glycine (the one with the free amino group) of the cross-linking fragment enzymically displaces the last D-ala unit from the pentapeptide of the muropentapeptide and cross-links the polysaccharide chains *via* the peptide chains (Figure 27–19).

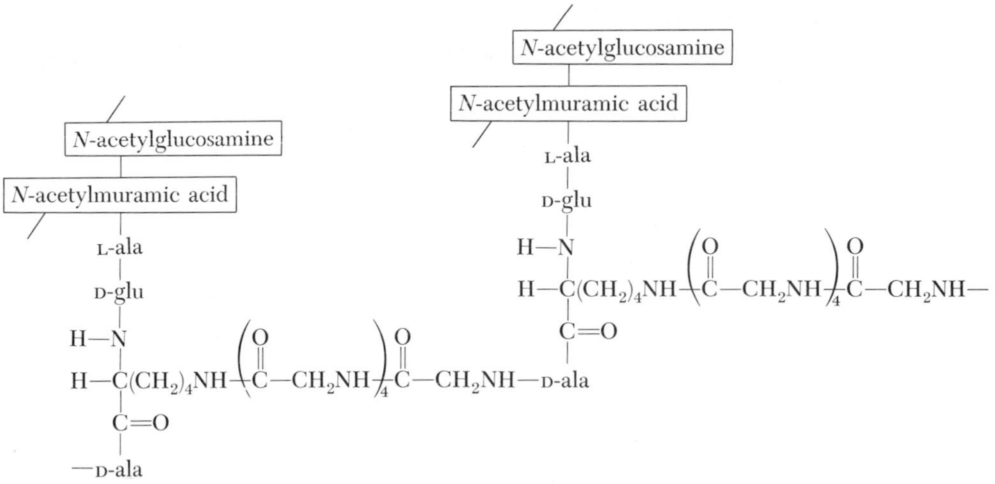

N-acetylglucosamine
N-acetylmuramic acid
L-ala
D-glu
H—N
H—C(CH$_2$)$_4$NH—(C(=O)—CH$_2$NH)$_4$—C(=O)—CH$_2$NH—
C=O

N-acetylglucosamine
N-acetylmuramic acid
L-ala
D-glu
H—N
H—C(CH$_2$)$_4$NH—(C(=O)—CH$_2$NH)$_4$—C(=O)—CH$_2$NH—D-ala
C=O
—D-ala

Figure 27-19 Cross-linking between adjacent chains of murein by glycine pentapeptides. The reaction that produces the cross-link, which occurs in the last stage of bacterial cell wall synthesis, is inhibited by penicillin.

One suggestion of how penicillin works hinges upon the structural similarity between penicillin and the D-alanyl-D-alanine fragment in the incipient murein. It is believed that penicillin competes effectively for the active site of the enzyme that removes the terminal D-ala from the initial pentapeptide, and thereby prevents cross-linking of the macromolecule. Penicillin is an enzyme inhibitor!

Why is it possible to give massive doses of penicillin to most individuals without doing them harm? The answer to this question is inherent in the mode of action of penicillin. Since mammalian cells are not protected by a murein-based cell wall as bacteria are, penicillin can have little direct effect upon the cells of humans but can prevent bacterial growth. The reason that *all* people cannot receive penicillin is related to the fact that some bacteria can destroy penicillin before the bacteria are, themselves, destroyed. This type of decomposition is mentioned in our discussion of infrared spectroscopy (Chapter 28); we note here that most drug reactions to "penicillin" are not really reactions to penicillin at all, but to its decomposition products. Therefore, one of the active areas of research concerning penicillins is aimed at the discovery of drugs that will avoid such allergic reactions by being more stable to decomposition by bacteria.

IMPORTANT TERMS

Active center: All of the portions of an enzyme that are necessary for the biological activity of the enzyme. This includes the portion(s) at which the actual chemical transformations occur, as well as the portions that maintain the substrate-enzyme complex.

Active site: Those portions of the active center of an enzyme at which the chemical transformations occur.

Apoenzyme: Many enzymes are complex molecules and consist of both a protein portion and a smaller molecule or ion (usually called a cofactor). The protein portion of this complex is called an apoenzyme.

Bile pigments: A group of colored compounds containing the pyrrole ring system, which are produced by the breakdown of the hemoglobin in red blood cells. (The average life-span of a red blood cell is about four months.) One of the first pigments produced, verdoglobin, is green and retains both the protein portion (globin) and the iron (Fe^{+3}) associated with the porphyrin ring. Verdoglobin readily loses both iron and the globin to form biliverdin, which, in turn, is readily reduced to bilirubin (red color). These are further processed in the body to form other pyrrole derivatives, which also are called "bile pigments" even though not all of them are colored.

Bile pigments are found in many places other than in mammals, including the wings of some insects (as pigments) and some sea algae.

Cofactor: A substance whose presence is required in order for an enzyme to be able to function as a catalyst.

Enzyme: A biological catalyst consisting largely, but not necessarily exclusively, of a protein. While many enzymes carry trivial names (*e.g.*, pepsin, emulsin), systematic names of enzymes end in *ase*. These names indicate the normal substrate of the enzyme and the type of reaction that is catalyzed (*e.g.*, cytochrome oxidase, aminolevulinic acid dehydrase; see Figure 27–2).

Extramitochondrial: A process (or substance) occurring outside of the mitochondria.

Feedback inhibition: Slowing of an enzyme-catalyzed process by the interaction of the end product of the reaction sequence with an enzyme that leads to the end product.

Feedback inhibition often operates through a (reversible) process called allosteric inhibition. In this mechanism the inhibitor is bound to the enzyme at a location that is different from the active site. This, in turn, alters the shape of the enzyme and prevents it from binding to its normal substrate.

Golgi apparatus: An organelle, located fairly close to the cell nucleus, which usually consists of a number of individual, closely spaced sacs. Its detailed functions are not well understood, but it is known that it participates in glycoprotein and mucopolysaccharide synthesis.

Heme: A complex between iron and protoporphyrin (a specific type of porphyrin). When ferrous ion (Fe^{+2}) is inserted into the center of the protoporphyrin ring, the two protons on the pyrrole nitrogens of this ring are lost and the result (heme or ferroprotoporphyrin) is an uncharged complex capable of coordinating with two molecules of a Lewis base (*e.g.*, O_2).

Holoenzyme: The completely active enzyme. It consists of the apoenzyme and, if necessary, the cofactor.

Inhibitor: Anything that decreases the activity of an enzyme.

Lysosomes: Microscopic particles that contain enzymes (called hydrolases) which hydrolyze various functional groups. These enzymes are separated from the cell's cytoplasm by the membrane of the lysosome.

Microsome: A small inclusion in the cell's protoplasm. Microsomes include ribosomes and fragments from the endoplasmic reticulum.

Murein: A complex macromolecule that makes up a bacterial cell wall. It consists, in very simple terms, of a polysaccharide chain cross-linked by short peptide chains. Mureins are hydrolyzed at the glycosidic bond by the enzyme lysozyme.

Prosthetic group: A non-protein portion of an enzyme. Heme is a prosthetic group.

PROBLEMS

1. Coenzyme A must be stored dry and in evacuated ampules. Contact with air destroys all catalytic activity and causes a near doubling of the molecular weight. Suggest a structure for the deactivated enzyme.

2. The Merck Index (8th Edition) suggests that D(+)-*N*-(2,4-dihydroxy-3,3-dimethylbutyryl)-β-alanine, also called chick antidermatitis factor, can be prepared ". . . by the direct condensation of β-alanine with the optically resolved form of the lactone of pantoic acid (2,4-dihydroxy-3,3-dimethylbutyric acid) which is [prepared] from isobutyraldehyde. . . ." Using the route suggested, outline the preparation of this compound. What is the common name for this substance? (*Hint:* It is given in this chapter.)

3. Pantolactone, an important intermediate in the synthesis of pantothenic acid, is prepared as outlined below. Suggest structures for intermediates and pantolactone, giving mechanisms where possible.

$$\text{isobutyraldehyde} + \text{formaldehyde} \xrightarrow{\text{base}} \mathbf{A}\ (C_5H_{10}O) \xrightarrow{\text{HCN}} \mathbf{B}\ (C_6H_{11}NO)$$

$$\text{pantolactone}\ (C_6H_{10}O_3) \xleftarrow{H_2O/CaCl_2} \mathbf{B}$$

4. Biotin is a prosthetic group involved in the activation of carbon dioxide in biological reactions. Sketch all of the stereoisomers (including enantiomers) of biotin. What isomers are possible for the corresponding sulfoxide?

O

HN NH

S $(CH_2)_4CO_2H$

biotin

5. Bethanechol chloride is a parasympathomimetic drug with side effects similar to those associated with neostigmine. Its synthesis is outlined below. Identify the intermediates and give the structure of the drug. Explain the difference between the reactivities of the two chlorines in **A**. (*Note:* Bethanechol chloride is highly water soluble and reacts instantly with aqueous silver nitrate.)

$$\text{propylene chlorohydrin} + \text{phosgene} \rightarrow \mathbf{A}\ (C_4H_6Cl_2O_2) \xrightarrow[\text{ether}]{NH_3} \mathbf{B}$$

$$\text{bethanechol chloride}\ (C_7H_{17}ClN_2O_2) \xleftarrow[\text{heat}]{N(CH_3)_3} \mathbf{B}\ (C_4H_8ClO_2)$$

6. Closely related to NAD$^{\oplus}$ is the coenzyme NADP, which bears an additional phosphate group on one of the sugars of NAD$^{\oplus}$. How many structures are possible for NADP? NADP bears this phosphate group on C2′ of the sugar bearing the purine ring. What is the correct structure of NADP?

INFRARED SPECTROSCOPY, ULTRAVIOLET SPECTROSCOPY, AND MASS SPECTROMETRY

28.1 INTRODUCTION

One of the tasks faced by many chemists is the determination of the structure of new compounds. A *structure proof,* as it is called, is usually accomplished by employing an assortment of procedures, some of which are chemical tests (so-called "wet tests") and some of which involve instrument-oriented methods. It is to a discussion of several of these instrumental methods that this and the following chapter are devoted.

There are four instrumental techniques that the chemist (and not just the organic chemist) commonly employs in attempts to elucidate the structure of a compound: infrared spectroscopy (ir), ultraviolet spectroscopy (uv), mass spectrometry (ms), and nuclear magnetic resonance spectroscopy (nmr). These may be used individually, although it is common practice to obtain data about a single compound by using several different methods. Many volumes have been written on each of these techniques, but the almost universal acceptance of nmr spectroscopy as the major structural tool has resulted in the devotion of a complete chapter (Chapter 29) to this single technique. The three remaining areas (ir, uv, and ms) are discussed in this chapter.

Another factor that has influenced the relative lengths of presentation of these four areas is the ability of students to get some direct exposure to these techniques in the undergraduate laboratory. Of the four methods, students are less likely to see a mass spectrometer than any of the other instruments required to get these data. Therefore, mass spectrometry is played down in a manner that is completely out of proportion with the importance of ms to the chemist. (Perhaps the recent advent of low-cost mass spectrometers will alter this situation in the near future.) Ultraviolet spectroscopy has received a somewhat larger space than it might otherwise get because it is the procedure that is most useful in determining whether two (or more) groups are conjugated to one another. Be aware that the length of coverage in this text does not reflect the relative or absolute value that a given chemist places on a given technique in determining the structure of a particular compound.

Beginning with Chapter 6, almost every chapter of this book ends with a discussion of spectroscopy. This has been done so that much of the material of analytical value, as well as representative spectra, can be found close to the other discussions of the various functional groups. Chapters 28 and 29 present the rudiments of these techniques. You are advised to read the relevant material in these two chapters before attempting to study the spectroscopic portions of earlier chapters.

28.2 THE ELECTROMAGNETIC SPECTRUM

The electromagnetic spectrum consists of a number of different "types" of radiation, including *ultraviolet* radiation, *infrared* radiation, and *radio waves*. These differ in their **wavelengths** (the distance from wavecrest to wavecrest) and in their **frequencies** (the number of waves passing a given point per unit time). Wavelength, λ, is inversely proportional to frequency, ν, the relation being $\lambda = c/\nu$ where c is the velocity of light. Since the **energy,** E, of radiation can be related to its frequency by $E = h\nu$ (where h is a constant), the different "kinds" of radiation must also have different energies. The major regions of the electromagnetic spectrum are presented in Figure 28–1, while some terms and units used to describe electromagnetic radiation are presented in Table 28–1.

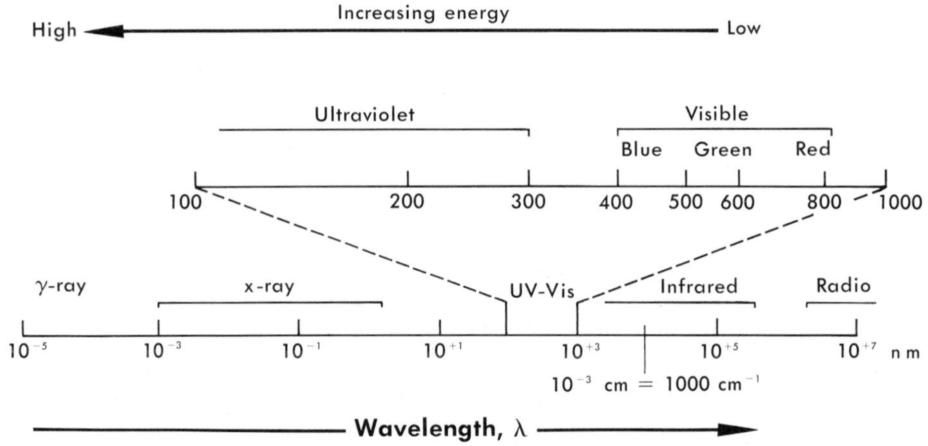

Figure 28–1 The electromagnetic spectrum. The *exact* wavelength separating one region from another is somewhat arbitrary. Wavelength increases from left to right, while both frequency and energy increase from right to left. (From Moore, J.A.: *Elementary Organic Chemistry*. Philadelphia, W.B. Saunders Company, 1974.)

TABLE 28–1 Definitions and Units[a]

SYMBOL	DEFINITION
wavelength (λ)[b]	
Å	*Ångstrom unit;* equivalent to 10^{-10} meter (m)
μm	*micrometer* (previously *micron,* μ); equivalent to 10^{-6} m or 10^4 Å
nm	*nanometer* (previously *millimicron,* mμ); equivalent to 10^{-9} m
frequency (ν)	
Hz	*Hertz* (previously *cycles per second,* cps); also called the "true frequency"
cm^{-1}	*wavenumber;* equivalent to the reciprocal of the wavelength in cm (e.g., 10^{-3} cm = 1000 cm^{-1}).

[a] The preferred units are those of Le Système International d'Unites, commonly called SI units. *J. Chem. Doc.*, **11,** 3 (1971).

[b] Along with *wavenumber, micron* is still the most commonly encountered unit used to describe infrared absorptions. Discussions in this text will be limited to using wavenumbers.

Molecules absorb energy from throughout the entire spectrum, with strikingly different results in various regions. Absorption of the energetic X-rays usually results in cleavage of some bonds, while absorption of the comparatively non-energetic radio waves results in small, temporary changes in subatomic particles. The increase in the energy of a molecule (ΔE) upon absorption of radiation is given by:

$$\Delta E = h\nu = hc/\lambda$$

where h = Planck's constant (6.62×10^{-27} erg/sec)
ν = frequency of absorbed radiation (Hz)
c = velocity of light (3×10^{10} cm/sec)
λ = wavelength of absorbed radiation (cm)

These equations show that ΔE is directly proportional to frequency but inversely proportional to wavelength.

Absorption. All of the various excitation processes that interest us are *quantized;* that is, the excited states are separated from the ground states by discrete, specific amounts of energy. Therefore, *a molecule is not free to absorb all of the frequencies within the spectrum of electromagnetic radiation; it can absorb only those frequencies corresponding to the energy required for excitation from one energy level to a higher one.* In turn, this allows us to assign specific absorption frequencies to specific excitations! The study of the effect of structural change upon the frequencies that correspond to various excitation processes is a major portion of that area of chemistry termed *absorption spectroscopy.*

Each of the major areas of the electromagnetic spectrum is associated with at least one specific type of excitation. For example, absorption of infrared radiation of the proper frequency causes bonded atoms to vibrate about their mean positions, while absorption of ultraviolet "light" causes electrons to be excited from the electronic ground state to an electronic excited state.

Our aims in this chapter will be modest. We shall try to glean a small bit of information about the structures of molecules and leave the bulk of information that is in these spectra behind. A great deal of expertise is required to completely interpret any type of molecular spectrum, and it is often unnecessary to interpret all of the spectra of a compound completely in order to identify it.

28.3 INFRARED SPECTROSCOPY—THE RUDIMENTS

The infrared region of the spectrum extends from 4000 cm^{-1} to 625 cm^{-1}. The regions to either side of the infrared are known as the *near infrared* (12,500 to 4000 cm^{-1}) and the *far infrared* (625 to 50 cm^{-1}) [the "near" and "far" refer to proximity to the visible region].

The absorption of infrared radiation produces either a *bond stretching* or a *bond angle* vibration. The specific bond being stretched or the specific angle being bent depends upon the frequency of radiation being absorbed.

Stretching vibrational modes are those corresponding to a change (positive or negative) in bond distance between bonded atoms with no displacement from the internuclear axis; *bending* vibrational modes are those in which atoms are displaced from the internuclear axis.

bond stretching

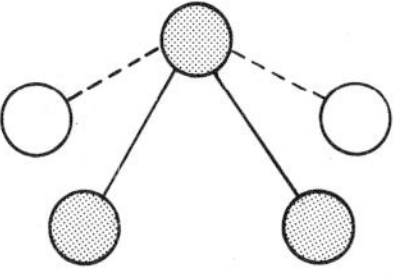

bond angle bending

Vibrations corresponding to excitation from the ground state to the lowest energy excited state are **fundamentals,** while those corresponding to higher excitations are **overtones.** Overtones occur at frequencies that are near-integral multiples of the frequency of the fundamental. **Combination bands** arise from the interaction of two vibrations (*e.g.*, $\nu_{\text{comb}} = \nu_1 + \nu_2$), as do **difference bands** (*e.g.*, $\nu_{\text{diff}} = \nu_1 - \nu_2$). Although overtones, combination bands, and difference bands complicate an infrared spectrum, they are usually much weaker in intensity than are the fundamentals.

The intensity (*i.e.*, the area under the "curve") of a particular absorption is dependent upon the difference between the dipole moment of the molecule in the ground state and that in the vibrational excited state; the greater the difference, the more intense the absorption. Conversely, if no change in the dipole moment accompanies a vibration, then that mode is "infrared inactive" and does not give rise to a band in the infrared spectrum.

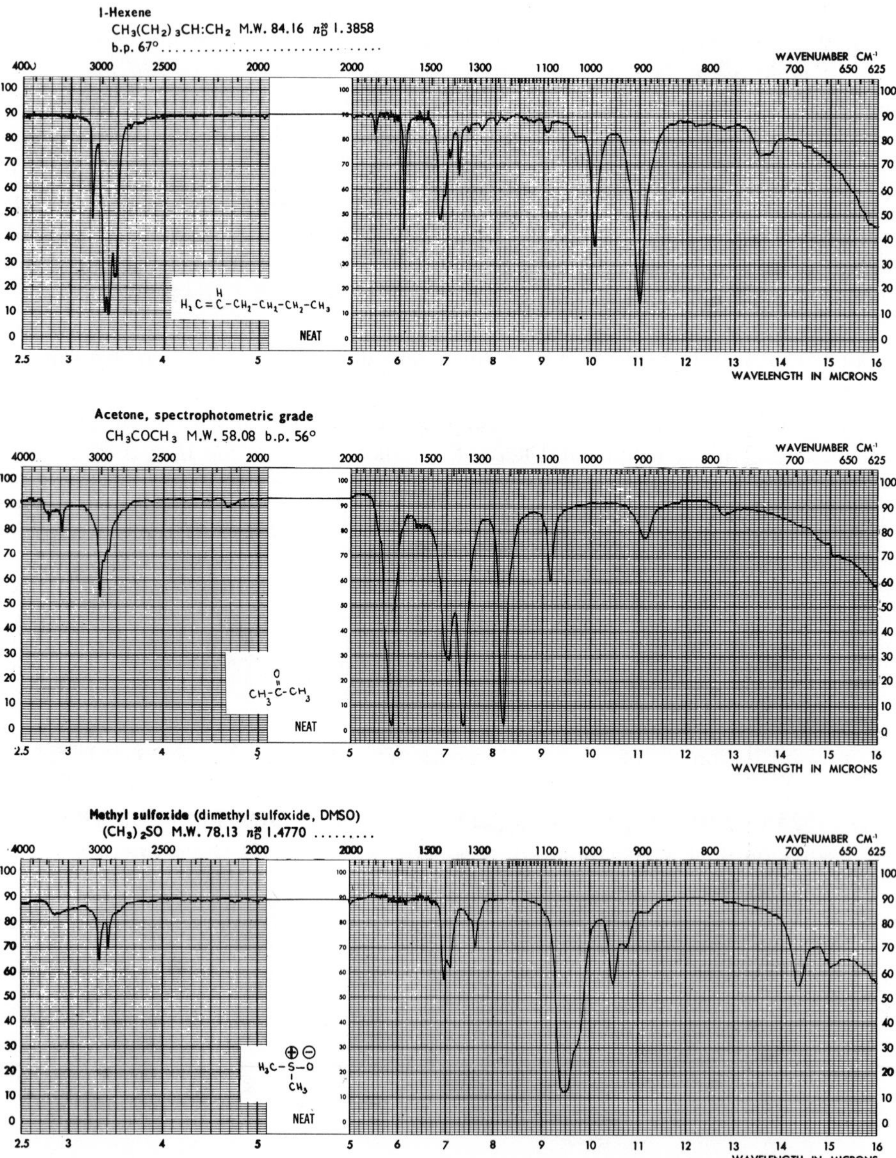

Figure 28–2 The infrared spectra of 1-hexene, acetone, and dimethyl sulfoxide.

The C=C bond stretching mode of ethene is an example of an infrared inactive vibration; the net change in the dipole moment is zero because the molecule has a point of symmetry within the bond under study. Bonds with highly polar ground states (*e.g.*, >C=O and >S⊕—O⊖) show particularly strong infrared absorptions. The effect of polarity is clarified with the aid of Figure 28–2, which shows the infrared spectra of 1-hexene (contains >C=C<), acetone (contains >C=O), and dimethyl sulfoxide (contains >S⊕—O⊖).

The region below about 1500 cm^{-1} usually is very complex, since angle bending modes appear only below 1500 cm^{-1}. (Stretching modes occur at <4000 cm^{-1}.) While specific vibrational assignments in this region are difficult, the overall appearance of the region below about 1400 cm^{-1} is unique for a given compound and is called the *fingerprint region.*

Fundamental Stretching Frequency Calculations. The bond between two atoms is, in the simplest models, analogous to (and approximated by) a vibrating spring. The vibrations of the bond are governed by the same laws of physics that dictate the motion of a spring.

This approximation, often called the "Hooke's Law" approximation, permits the calculation of the fundamental stretching frequency of a bond between two atoms, A and B, using the following equation:

$$\nu = \frac{1}{2\pi c}(f/\mu)^{1/2}$$

where ν = stretching frequency (cm^{-1})
c = velocity of light (cm/sec)
f = the bond stretching force constant (dynes/cm)
μ = reduced mass (grams) of the atoms involved in the bond

The reduced mass is defined as

$$\mu = \frac{M_A M_B}{M_A + M_B}$$

where M = mass (grams) of the appropriate atom.

Force constants for single, double, and triple bonds are approximately 5, 10, and 15 × 10^5 dynes/cm, respectively. Although the analogy is approximate, this "explains" why a triple bond is harder to stretch than is a double bond—the greater the multiplicity of the bond, the stiffer is the spring used to approximate the behavior of the bond. The exact force constant for a bond depends upon its specific molecular environment and is usually determined from infrared spectra.

Problem: Calculate the stretching force constant for the carbonyl group in cyclopentanone, which absorbs at 1742 cm^{-1}. On the basis of that result, which of the electronic arrangements, **A** or **B**, is a better description of the carbonyl group?

>C=Ö: (**A**)   >C⊕—Ö:⊖ (**B**)

Solution:

$$\nu = \frac{1}{2\pi c}(f/\mu)^{1/2}$$

$$1742 = \frac{1}{2 \times 3.14 \times 3 \times 10^{10}}(f/\mu)^{1/2}$$

$$3.28 \times 10^{14} = (f/\mu)^{1/2}$$

$$\mu = \frac{M_C M_O}{M_C + M_O}$$

$$M_C = 12.0/6.02 \times 10^{23} = 1.99 \times 10^{-23}$$

$$M_O = 16.0/6.02 \times 10^{23} = 2.66 \times 10^{-23}$$

$$\mu = 1.14 \times 10^{-23}$$

$$3.28 \times 10^{14} = (f/1.14 \times 10^{-23})^{1/2}$$

$$f = 12.3 \times 10^5 \text{ dynes/cm}$$

This value is consistent with a double bond and not with a single bond; therefore, **A** is the better description.

Problem: Quantitate the effect of isotopic substitution ($^1H \rightarrow {}^2H$) upon the C—H stretching vibration, assuming a force constant of 5×10^5 dynes/cm for both C—H and C—D bonds.

Solution: There is no need to calculate the actual stretching frequencies (although that might be a valuable exercise), as seen from the following calculation.

$$\frac{\nu_{C-H}}{\nu_{C-D}} = \frac{\dfrac{1}{2\pi c}\left(\dfrac{f}{\mu_{C-H}}\right)^{1/2}}{\dfrac{1}{2\pi c}\left(\dfrac{f}{\mu_{C-D}}\right)^{1/2}} = \frac{\left(\dfrac{f}{\nu_{C-H}}\right)^{1/2}}{\left(\dfrac{f}{\mu_{C-D}}\right)^{1/2}}$$

$$\left(\frac{\nu_{C-H}}{\nu_{C-D}}\right)^2 = \frac{\dfrac{f}{\mu_{C-H}}}{\dfrac{f}{\mu_{C-D}}} = \frac{\mu_{C-D}}{\mu_{C-H}}$$

$$\left(\frac{\nu_{C-H}}{\nu_{C-D}}\right)^2 = \frac{\dfrac{M_C M_D}{M_C + M_D}}{\dfrac{M_C M_H}{M_C + M_H}} = \frac{M_D M_C + M_D M_H}{M_D M_H + M_C M_H} \approx 2$$

$$\frac{\nu_{C-H}}{\nu_{C-D}} \approx \sqrt{2}$$

28.4 GROUP FREQUENCIES

Because bonds interact in a fairly constant fashion within a functional group, with only minor dependence upon the nature of the carbon skeleton bearing the functional group, it is possible to assign *group* frequencies to various functional groups. For this reason, the major use of infrared spectroscopy is in determining which functional groups are present in a molecule. Often it is difficult to envision simple stretching or bending modes that can account for the many characteristic vibrations of some complex functional groups. However, because of their applicability to structural analysis, the practicing organic chemist must be familiar with the frequencies associated with the common functional groups; Table 28–2 is a compilation of such data. Appropriate chapters should be consulted for more detail. Because of their ubiquitous nature, we will at this point examine the infrared spectra of simple hydrocarbons.

TABLE 28-2 Infrared Vibrations of Analytical Utility[a]

GROUP	FREQUENCY (cm^{-1})[b]
O—H	3650–3200 (v)
N—H	3500–2900 (m)
C—H	3300–2700 (s-m)
S—H	~2550 (m-w)
C≡C	~2200 (w)
C≡N	~2200 (m-w)
C=O	1850–1650 (s)
C=C	~1650 (m-w)
C—NO_2	~1550 (s) and ~1350 (s)
	~900–850 (m)
C—O—	1300–1000 (s-m)
C—F	1400–1000 (s)
C—Cl	800–600 (s)
C—Br	650–500 (s)
C—I	600–500 (s)
$-\overset{\oplus}{\underset{\vert}{S}}-O^{\ominus}$	1070–1030 (s)
$-\underset{\vert}{SO_2}$	~1150 (s) and ~1330 (s)

[a] A detailed chart of various frequencies is found inside the back cover of this book.

[b] Value in parentheses is the intensity of the absorption, *i.e.*, s: strong, m: medium, w: weak, v: variable. The ranges include most absorptions, but specific chapters should be consulted for more detail.

Infrared Spectra of Alkanes and Cycloalkanes. By examining the spectral characteristics of alkanes and cycloalkanes in some detail, we can begin to appreciate the complexities that can result from even simple interactions between vibrations.

Alkanes and cycloalkanes shown C—H stretching vibrations below 3000 cm^{-1}. In general, C_{sp^3}—H stretching vibrations occur below 3000 cm^{-1}, while C_{sp^2}—H and C_{sp}—H stretching vibrations occur above 3000 cm^{-1}. (An important exception is cyclopropane, which absorbs at approximately 3050 cm^{-1}.) The C—H stretching modes of —CH_2— and —CH_3 groups occur as doublets (two absorptions from the same group, whose peak frequencies are separated by an interval) because of the existence of *symmetric* and *anti-symmetric* vibrational contributors. Unfortunately, many of the less expensive ir spectrophotometers cannot resolve (separate) these doublets, and they appear as a single absorption. A methine hydrogen ($-\overset{|}{\underset{|}{C}}_{sp^3}-H$) exhibits a stretching frequency of approximately 2890 cm^{-1}.

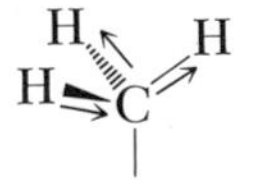

2962 cm^{-1} 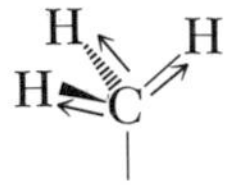 2872 cm^{-1}

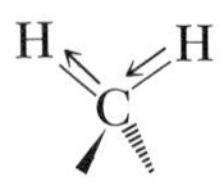

2926 cm^{-1} 2853 cm^{-1}

asymmetric stretching modes symmetric stretching modes

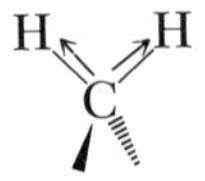

Methyl and methylene groups also absorb at approximately 1465 to 1460 cm^{-1}. These absorptions arise from the *methylene scissoring* vibration and the *methyl asymmetric bending* vibration. Cycloalkanes lacking a methyl group show a band at 1460 cm^{-1} that is rather sharp compared to those of alkanes and cycloalkanes bearing an alkyl group, because cycloalkanes lacking a methyl group exhibit a methylene scissoring vibration but not the methyl asymmetric bending vibration.

methyl asymmetric bend — methyl symmetric bend — methylene scissoring

An absorption at 1380 cm^{-1} arises from the *methyl symmetric bending* vibration and is a valuable diagnostic tool. It will be absent from the spectra of compounds lacking a methyl group. Furthermore, this absorption will be split (*i.e.*, appear as two bands) when either an isopropyl group or a *t*-butyl group is the absorbing group. One can distinguish between the latter two, since the presence of a *t*-butyl group results in two peaks of which the lower frequency absorption is the more intense, while the doublet produced by an isopropyl group is symmetric in intensity. (Absorptions at 1255 cm^{-1} and 1210 cm^{-1} serve to confirm the presence of the *t*-butyl group.) The infrared spectra of hexane, cyclohexane, and methylcyclohexane are presented in Figure 28–3.

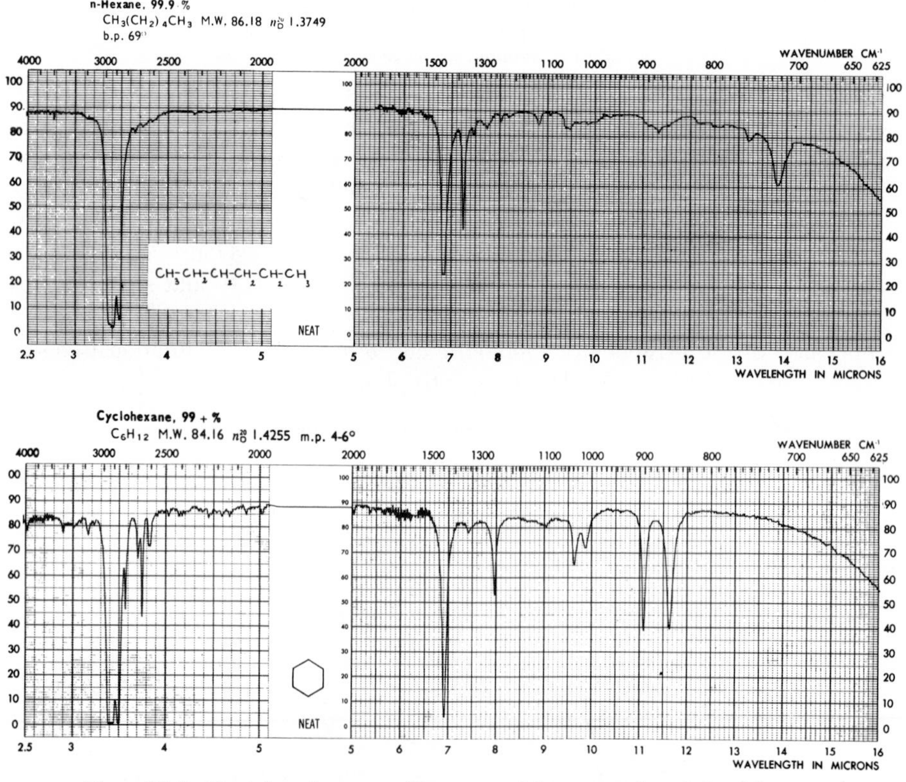

Figure 28–3 The infrared spectra of hexane, cyclohexane, and methylcyclohexane.

Illustration continued on opposite page.

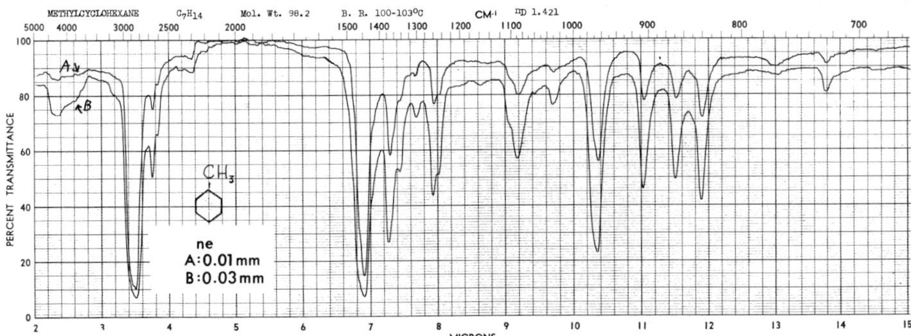

Figure 28-3 *Continued.* © Sadtler Research Laboratories, Inc., 1976.

A set of connected methylene groups ($—(CH_2)_n—$) can be identified by a weak band near 750 cm^{-1}. More precisely, $\nu \approx 720$ cm^{-1} when $n = 4$, $\nu \approx 740$ cm^{-1} when $n = 3$, and $\nu \approx 780$ cm^{-1} when $n = 2$.

Alkenes. The C=C bond of all of the common alkenes absorbs at 1650 ± 20 cm^{-1}. This absorption, while of moderate intensity, is sufficiently insensitive to substitution to be of little diagnostic value.

The vinyl hydrogens of an alkene absorb above 3000 cm^{-1} (=C—H stretch), and this can be used to substantiate the presence of a double bond and at least one vinyl hydrogen.

TABLE 28-3 Principal Alkene Absorptions in the Infrared[a]

Type[b]	C—H Stretch[c]	C=C Stretch[d]	C=C—H in-Plane Bending	C=C—H out-of-Plane Bending
$RCH=CH_2$ (R, H / H, H)	>3000	1645 (m)	CH_2 1420 (m-w) CH 1300 (m-w)	CH 990 (s) CH_2 910 (s)
$R_2C=CH_2$ (R, H / R, H)	>3000	1655 (m)	1415 (m-w)	890 (s)
cis-RCH=CHR (R, R / H, H)	>3000	1660 (m)	1415 (m-w)	735–670 (s-m)
trans-RCH=CHR (R, H / H, R)	>3000	1675 (w)	1300 (m-w)	965 (s)
$R_2C=CHR$ (R, R / R, H)	>3000	1670 (m-w)	1415 (m-w)	840–805 (s)
$R_2C=CR_2$ (R, R / R, R)	—	1670 (w)	—	—

[a] All frequencies in cm^{-1}.
[b] R groups need not be the same; however, the higher the symmetry, the weaker will be certain absorptions.
[c] All of moderate intensity.
[d] Conjugation lowers the C=C stretching frequency. The shift is greater for a system such as C=C—C=C than for Ar—C=C.

The most valuable vibration associated with alkenes is the out-of-plane C═C—H bending mode, occurring in the 1000 to 800 cm^{-1} region. The exact frequency of this absorption is dependent upon the substitution around the double bond, as seen from the data in Table 28–3, and it is valuable in determining the kind of alkene that is present.

28.5 INTERPRETED SPECTRA

The following spectra are interpreted in order to demonstrate how much information you might hope to get from an ir spectrum without any other spectral data. A great deal of time and reference to other spectra might lead to more information than is given here.

CYCLOHEXANONE (Figure 28–4)

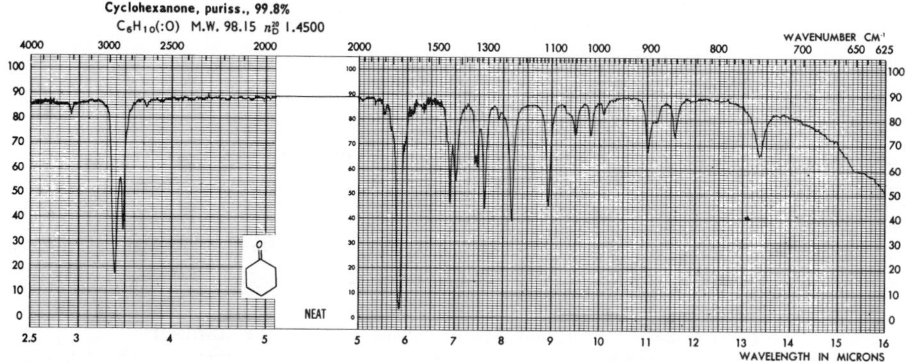

Figure 28–4 The infrared spectrum of cyclohexanone.

1. The lack of absorptions above 3000 cm^{-1} excludes the presence of a variety of groups, such as —OH, —NHR, —NH_2, —C_6H_5, —C≡C—H, and R_2C═CRH. The absence of the double bond is further supported by the lack of absorption of even moderate intensity near 900 cm^{-1}. All C—H stretching frequencies suggest only C_{sp^3}—H bonds.

2. The absence of an absorption at 1380 cm^{-1} excludes methyl, ethyl, isopropyl, and *t*-butyl groups.

3. The *sharp* absorption at 1460 cm^{-1} suggests a *cyclo*alkyl skeleton. (Why?)

4. The very strong absorption at 1715 cm^{-1} is assigned as a C═O stretch. Since only open-chain ketones and cyclohexanones absorb at 1715 cm^{-1}, the compound is a cyclohexanone. (The aldehyde group, —CHO, can be discounted by arguments presented in the following discussion.)

UNDECANAL (Figure 28–5)

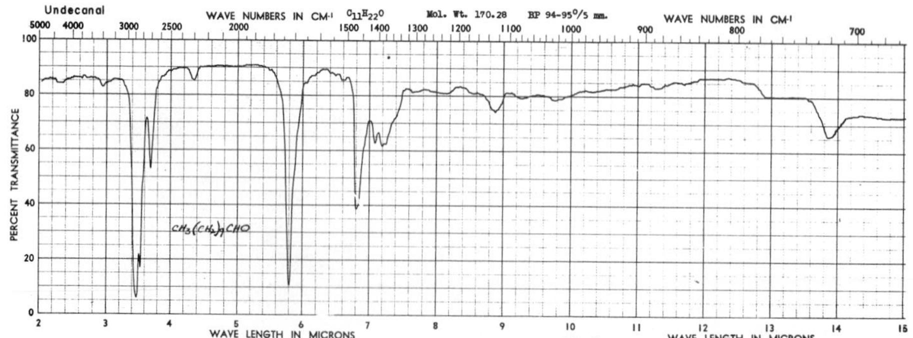

Figure 28-5 The infrared spectrum of undecanal. © Sadtler Research Laboratories, Inc., 1976.

1. The prominent absorption at 1710 cm^{-1} is assignable to a carbonyl group. Since the C═O frequency is rather sensitive to proximal electron-donating or electron-withdrawing

groups (*e.g.*, $—OCH_3$ or C═C), and since this value of 1710 cm^{-1} is rather "normal," such groups cannot be bonded to the carbonyl group.

2. The absence of such groups as —OH is evident from the lack of absorption above 3000 cm^{-1} (see item 1 of the cyclohexanone discussion).

3. The C—H stretching region shows a rather sharp absorption at ~2720 cm^{-1}. This is actually one half of a doublet (the other half occurs at ~2820 cm^{-1} and is buried in the other C—H vibrations) that is characteristic of aldehydes (—CHO).

4. The region near 1380 cm^{-1} is complex, and it is difficult to tell whether this region contains a singlet or a doublet.

5. The absorption near 1460 cm^{-1} is due to both the methylene scissoring and the methyl asymmetric bending modes (the signal is broad). Therefore, at least one methyl group is present.

6. Although the broad absorption near 715 cm^{-1} suggests a $—(CH_2)_n—$ group in which $n > 3$, it is impossible to deduce that this is $CH_3—(CH_2)_9—CHO$.

Toluene (Figure 28-6)

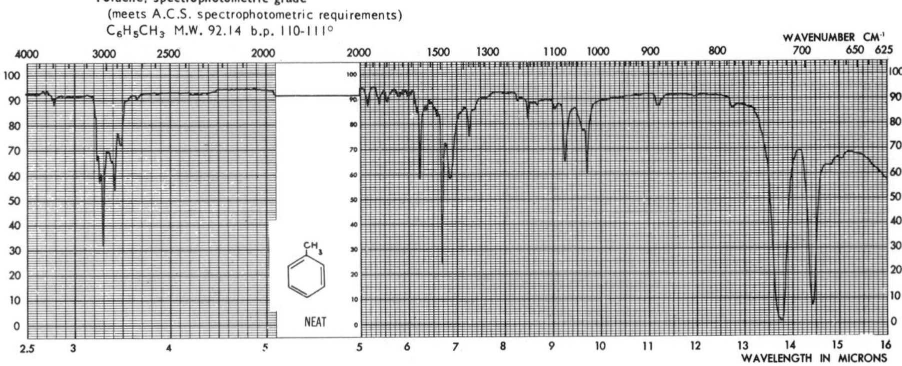

Figure 28-6 The infrared spectrum of toluene.

1. The suggestion of unsaturation (C═C, *etc.*) stems from the C—H absorption at 3090 cm^{-1}. (The weakness, sharpness, and position of this band argue against $—NH_2$, —OH, *etc.*)

2. Confirmation that the benzene ring is present comes from the weak, finger-like bands in the 2000 to 1700 cm^{-1} region. The observed pattern suggests a monosubstituted benzene.

3. The sharp absorptions at 1600 and 1500 cm^{-1} are ring vibrations of the benzene system and are ususlly present in all derivatives of benzene.

4. The region near 1380 cm^{-1} lacks a clearly defined doublet, suggesting the absence of $—CH(CH_3)_2$ and $—C(CH_3)_3$.

Limonene (Figure 28-7)

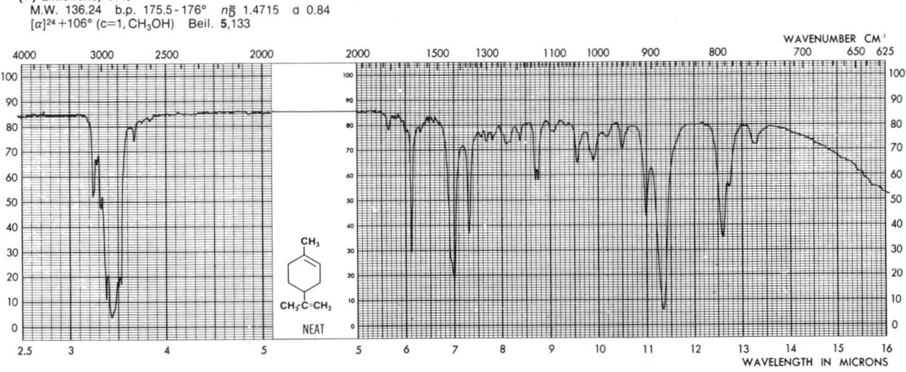

Figure 28-7 The infrared spectrum of (+)-limonene.

1. The presence of the sharp absorption at 1645 cm^{-1} is characteristic of a double bond. We could not tell from this region that there are, in fact, two types of double bonds present.

2. The weak absorption band above 3000 cm^{-1} suggests that the double bond is not tetrasubstituted. (If two C=C groups were known to be present, one of the double bonds could be tetrasubstituted and the compound could still produce the same type of absorption near 3000 cm^{-1}.)

3. The broad absorption near 1460 cm^{-1} implies the presence of both methyl and methylene groups.

4. The sharp symmetric methyl bending vibration (1380 cm^{-1}) indicates the absence of isopropyl and *t*-butyl groups.

5. The absorption at ~890 cm^{-1} is due to a terminal methylene group (=CH_2), while the absorption at ~800 cm^{-1} is the out-of-plane C—H bending mode for a trisubstituted alkene (R_2C=CRH).

6. One could not deduce from the ir spectrum that this is limonene.

2-Mercaptodiphenylmethane (Figure 28-8)

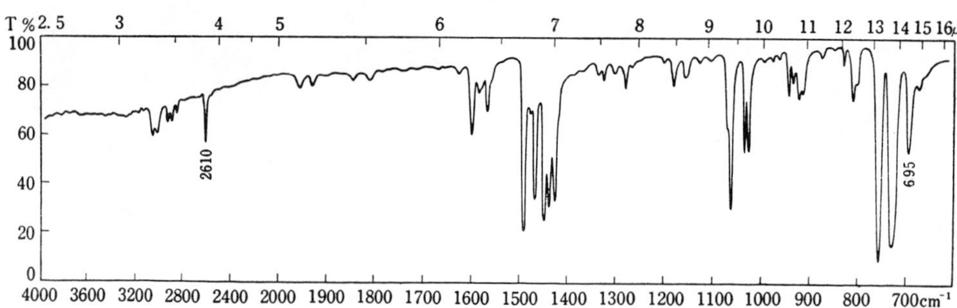

Figure 28-8 The infrared spectrum of 2-mercaptodiphenylmethane. (From K. Nakanishi, *Infrared Absorption Spectroscopy*, Holden-Day, Inc., 1962.)

1. The C_{sp^3}—H absorption (<3000 cm^{-1}) is very weak and suggests relatively few alkyl groups. The absence of a methyl group is established by the absence of an absorption at 1380 cm^{-1}.

2. The sharp absorption at 2610 cm^{-1} is not hydrogen-bonded and is consistent with an —S—H stretching mode.

3. The lack of C=C stretching (1650 cm^{-1}) suggests that the 3050 cm^{-1} absorption is due to aryl-H stretching.

4. The absorptions at 760, 734 and 695 cm^{-1} are assigned to aryl-H bending modes and are characteristic *substitution patterns* for a monosubstituted and an *ortho*-disubstituted benzene derivative.

H
H R
H R
H

an *ortho*-disubstituted benzene ring

5. The large number of absorptions between 1600 and 1400 cm^{-1} suggests the presence of more than one benzene ring.

2-Phenyl-3-butyn-2-ol (Figure 28–9)

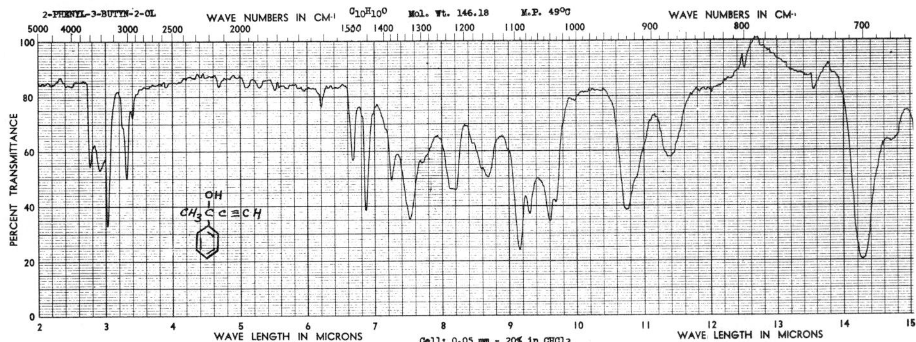

Figure 28–9 The infrared spectrum of 2-phenyl-3-butyn-2-ol. © Sadtler Research Laboratories, Inc., 1976.

1. The broad absorption (~3200 cm^{-1}) is due to an —OH stretching mode. The absorption near 1090 cm^{-1} is the C—O stretching mode of a 3° alcohol.

2. The absorption at 2165 cm^{-1} is due to the C≡C stretching vibration; it is weaker in intensity than usual. The ≡C—H stretch (~3300 cm^{-1}) is masked by the broad —OH signal.

3. The presence of a methyl group is indicated by the sharp 1380 cm^{-1} absorption.

28.6 APPLICATIONS OF INFRARED SPECTROSCOPY

As already noted, ir spectra are not used to deduce the *total* structure of a compound. Rather, ir is used to answer the question, "Is *that* particular functional group present?" In the examples that follow, we will examine two specific uses of ir: the structural analysis of penicillin and the conformational analysis of cyclohexanol.

The Structure of Penicillin. Many penicillins are known, some of the more important being penicillins F, G, K, and X.

penicillin nucleus (R—C(=O)—NH— on the β-lactam/thiazolidine ring system bearing two CH_3 groups and CO_2H)

penicillin designation	*R*
F	$CH_3—CH_2—CH{=}CH—CH_2—$
G	$C_6H_5—CH_2—$
K	$CH_3—(CH_2)_5—CH_2—$
X	$HO—C_6H_4—CH_2—$

The structures above are now known to be correct; however, near the end of World War II (when the need for antibiotics was especially great) there was a strong belief that such a highly strained fused ring system was unlikely and that the correct ring system was that shown below.

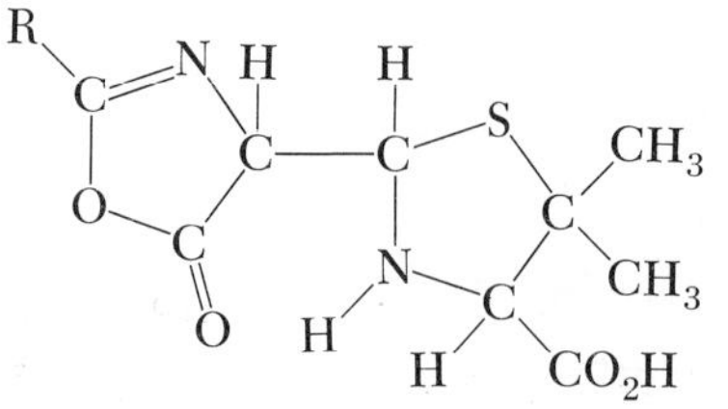

Since chemical reactions did not provide an unequivocal answer to this problem, infrared spectroscopy was used to determine the correct structure. (Remember that at this time it was the most powerful structural tool readily available to organic chemists.)

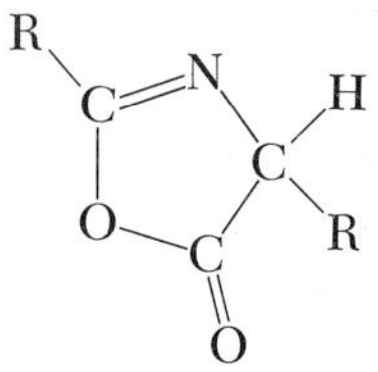

A

Derivatives of ring system **A** show two characteristic bands: 1825 cm^{-1} (the C═O stretching frequency) and 1675 cm^{-1} (the C═N stretching frequency). *Neither* of these absorptions appears in the ir spectra of penicillins, thus eliminating what was an otherwise reasonable structure. This example also illustrates the common procedure of examining the spectra of model compounds (*i.e.*, compounds of known structure) in order to elucidate the structure of unknown materials.

That there is a fair amount of strain present in the fused ring system of penicillin is supported by the mechanism by which staphylococci deactivate penicillins before the antibiotic can destroy the pathogen. These "resistant" bacteria secrete an enzyme, *penicillinase*, which opens the four-membered ring and converts active penicillins into inactive penicilloic acids. Some of the newer penicillins (*e.g.*, methicillin) are more resistant to this mode of decomposition.

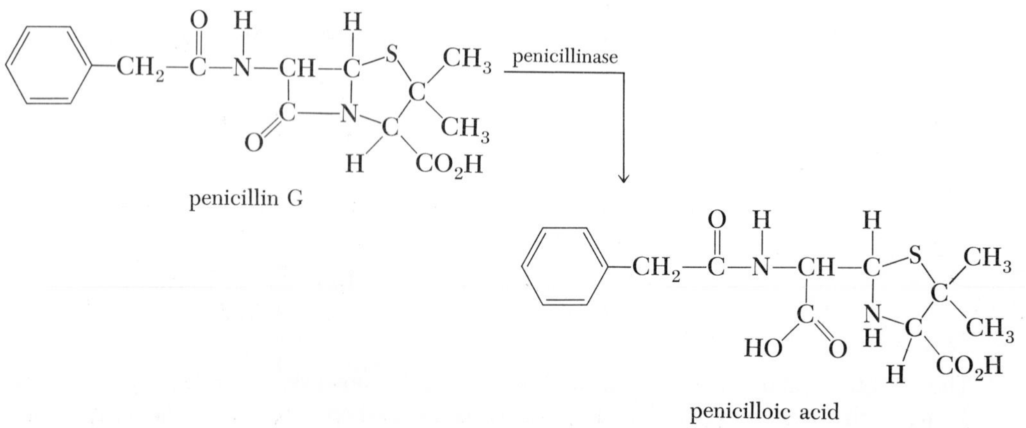

penicillinase

penicillin G

penicilloic acid

methicillin

Conformational Analysis of Cyclohexanol. Axial and equatorial substituents on a cyclohexyl ring show different *ring-to-substituent* stretching frequencies; an equatorial substituent usually absorbs at a higher frequency than does the same substituent in the axial position. For example, *trans*-4-*t*-butylcyclohexanol shows a C—O stretching mode at ~1060 cm^{-1}, while *cis*-4-*t*-butylcyclohexanol shows a C—O stretching mode at ~950 cm^{-1}. Because of the conformational rigidity of the ring, the former is considered to be the *equatorial* C—O vibration, while the latter is the *axial* C—O vibration.

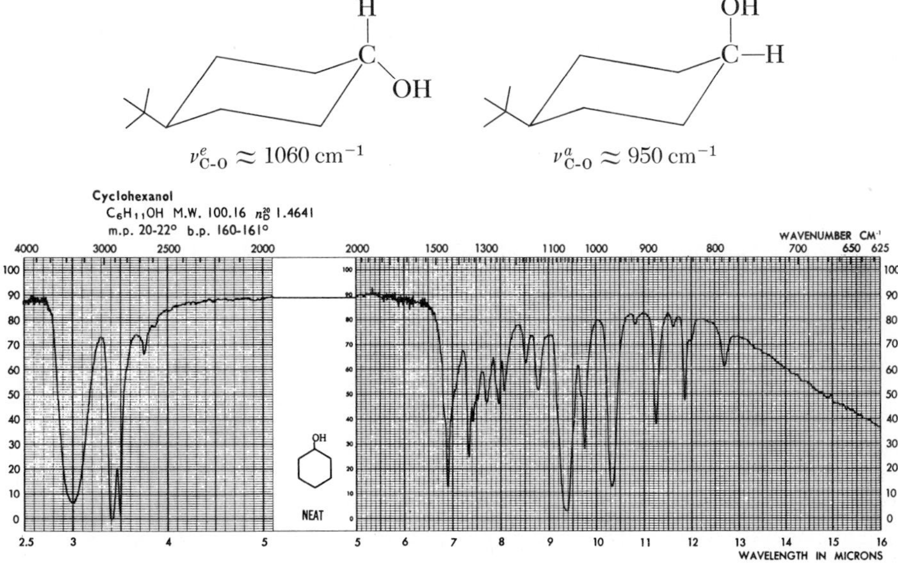

Figure 28–10 The infrared spectrum of cyclohexanol.

The infrared spectrum of cyclohexanol (Fig. 28–10) shows both of these absorptions because it exists as a mixture of conformers.

$$a \rightleftarrows e \qquad K_{eq} = \frac{[e]}{[a]}$$

If it is assumed that solutions of the same concentration of *cis*- and *trans*-4-*t*-butylcyclohexanol give rise to C—O absorptions of the same intensities, it is possible to compare the intensities of the C—O absorptions of cyclohexanol and thus determine K_{eq}.

Using this approximation, K_{eq} can be shown to be 2.0. Since the difference in the free energy between these conformers, ΔG, equals $-RT \ln K_{eq}$, the equatorial conformer is more stable than the axial by 0.4 kcal/mole.

$$\Delta G = -RT \ln K_{eq}$$
$$\Delta G = -2.3RT \log 2.0 = 0.4 \text{ kcal/mole}$$

Inherent in this approach is the fact that the ir method has a "fast shutter speed"; that is, it presents separately the spectral properties of all conformers that are present (intensities are proportional both to the "inherent" intensity of a group and to its concentration). Infrared spectroscopy does not give an average picture, as nmr so often does (see Chapter 29).

28.7 INSTRUMENTATION

A brief description of an idealized double-beam infrared spectrophotometer is included here. In a double-beam instrument, one beam of radiation passes through the sample compartment while the other passes through the reference compartment. This permits the determination of difference spectra (*i.e.*, the difference between the spectral characteristics of two materials). The sample, if run as a pure compound, is compared to air; pure solvent is the reference if the spectrum is obtained from a solution.

The principal components of a typical ir instrument are: radiation source, sample compartments, monochromator (either prism, grating, or both), reference beam attenuator, radiation detector, and recorder. A typical ir instrument is outlined in Figure 28–11.

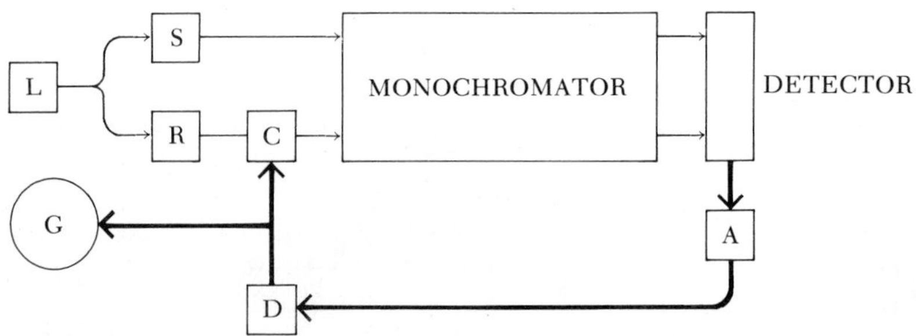

Figure 28–11 Schematic representation of a double-beam infrared spectrophotometer. L, lamp (infrared source); S, sample; R, reference blank; A, electronic amplifier for signal generated by detector; D, drive mechanism that moves reference beam attenuator (C); G, recorder. Light arrows represent optical paths; heavy arrows represent electrical or mechanical connections.

1. One source (most simply a resistor, electrically heated to incandescence) emits radiation that is split and reflected through the sample and reference compartments, ensuring that identical beams enter these compartments.

2. Radiation passes through the sample and the reference, some frequencies being absorbed more strongly by the sample than by the reference.

3. The monochromator is subjected to a rapidly alternating light beam, first from the sample cell and then from the reference, and so on; it slowly scans through the range of wavelengths.

4. At each wavelength at which the sample absorbs more than does the reference, the difference is sensed by the detector (*e.g.*, a thermistor), which responds by generating a small current. This current, after amplification, drives the attenuator into the reference beam in order to make the intensities of both beams identical (the *optical null* principle).

5. The pen of the chart recorder moves in exactly the same manner as the attenuator, either through mechanical linkage to the attenuator drive mechanism or through electrical connection to the same amplified signal from the detector. The chart paper is moved synchronously with the monochromator, at right angles to the pen travel, so that a standard distance along the length of the paper corresponds to the proper frequency interval.

28.8 ULTRAVIOLET SPECTRA

Electronic absorption spectra result from the absorption of ultraviolet (uv) and visible (vis) radiation, accompanied by the excitation of an outer electron from a given level to one of higher energy. Because of this, electronic absorption spectroscopy is often dubbed "uv-vis" or "uv" spectroscopy.

Of the vast panorama of electromagnetic radiation, man can see only the "visible" portion, with wavelengths of 400 to 800 nm. The ultraviolet region extends from 1 nm to 400 nm but, because atmospheric constituents absorb below 200 nm, the region from 200 to 400 nm is the portion of the spectrum normally covered by the term *ultraviolet* (more correctly called the *near ultraviolet*). The region from 1 nm to 200 nm can be studied in evacuated systems and is termed the "vacuum ultraviolet." This atmospheric absorption below 200 nm is a blessing to all, including spectroscopists, since it prevents very dangerous (high energy) ultraviolet radiation in sunlight from striking the earth's surface. The near ultraviolet will be the focus of this chapter.

28.9 EXCITATION AND RELAXATION

Electronic excitation may promote σ bonding, π bonding, or non-bonding electrons (abbreviated n) into various anti-bonding orbitals. These *transitions*, reading from ground state to excited state, include: $n \rightarrow \pi^*$, $\pi \rightarrow \pi^*$, and $n \rightarrow \sigma^*$. A $\sigma \rightarrow \sigma^*$ transition requires more energy than does any of the others, and it occurs in the short wavelength region of the vacuum ultraviolet. The *relative* energies of these transitions are presented in Figure 28–12.

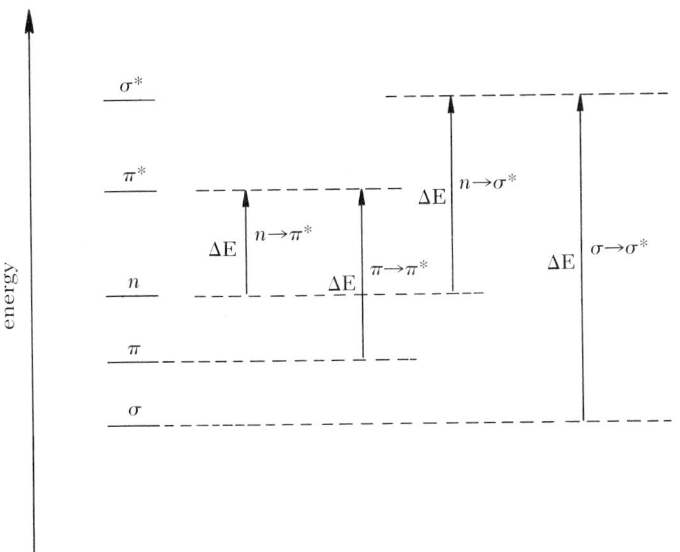

Figure 28–12 Relative energies of various electronic transitions.

Once a molecule becomes electronically excited, it may lose this excess energy by any of the several pathways listed below. The one to be followed is a function of (a) the specific molecule being studied and (b) the type of transition that accompanied the initial absorption.

1. *Bond Cleavage:* usually leads to an irreversible chemical reaction.

2. *Emission:* the same frequency that was absorbed is re-emitted.
3. *Fluorescence:* radiation of longer wavelength than the "exciting" radiation is emitted after excitation. Fluorescence occurs rapidly, often less than 10^{-6} sec after absorption.
4. *Phosphorescence:* radiation of longer wavelength than the exciting radiation is emitted after excitation. Emission may continue for several hours after excitation. (*Bio-phosphorescence,* as displayed by some algae, is not "phosphorescence" as defined here but is due to chemical reactions.)

Although fluorescence and phosphorescence emission spectroscopy are beyond the scope of this text, these phenomena have many important practical applications. Fluorescent compounds are used in "pop art" posters and as "optical brighteners," to mention but two uses. An optical brightener absorbs in the uv but re-emits this energy as a red-blue fluorescence, thus masking the less desirable yellow colors of some fabrics. Because fluorescence emission can be very intense, and entails absorption and emission of characteristic wavelengths, it is used in many analytical procedures, *e.g.*, the estimation of adrenaline levels in blood and urine.

28.10 THE BEER-BOUGER-LAMBERT LAW

Ultraviolet spectra normally are obtained by using a solution of the sample (in a non-absorbing solvent) held in cells possessing uv-transparent walls (*e.g.*, fused silica). Spectra normally are obtained on a double-beam instrument, using the pure solvent as reference.

The amount of absorbed radiation is directly proportional to the number of solute molecules in the "light" path and will, therefore, increase with increasing concentration and/or with increasing sample thickness (*i.e.*, *pathlength* of the cell).

At any wavelength, the intensity of light entering the solution (I_0) is related to the intensity of the light leaving the solution (I) by the **Beer-Bouger-Lambert Law** (also called the **Beer-Lambert Law**):

$$\log(I_0/I) = abc$$

where a = absorptivity, a "constant" dependent upon λ but independent of concentration
b = pathlength
c = concentration

With c expressed in moles/liter of solution and b in cm, a becomes the **molar absorptivity,** ϵ. Older literature termed ϵ the **molar extinction coefficient.** Although ϵ has units (What are they?), it is usually expressed as a dimensionless quantity.

The amount of radiation absorbed by a solution may be reported as absorbance, A (called "optical density" in older works), or as transmittance, T. These quantities are defined as:

$$A = \log(I_0/I)$$

or

$$T = I/I_0$$

$$A = -\log T$$

The absorptivity of 1 cm of a solution containing 1 g of solute per 100 ml of solution, $E^{1\%}_{1\text{ cm}}$, is used instead of ϵ if the molecular weight of a sample is unknown or if the sample is a mixture.

Normally, only absorption maxima and shoulders, and the corresponding values of ϵ, are reported; for example, $\lambda_{max}^{hexane} = 235$ nm (ϵ 5400). The solvent is indicated, since it may influence both λ and ϵ.

28.11 CHROMOPHORES

Saturated hydrocarbons, ethers (alkyl-O-alkyl), and alcohols (alkyl-O—H) do not absorb above approximately 200 nm and, therefore, are useful uv solvents. The most common solvents include 2,2,4-trimethylpentane, cyclohexane, methylcyclohexane, and 95% ethanol.

A simple functional group that gives rise to an absorption with characteristic values of λ and ϵ is called a **chromophore** (Table 28–4). Molecules containing the same chromophores can usually be expected to have similar uv spectra. If a molecule contains two chromophores that are separated by more than one single bond, the spectrum of the compound will be the sum of the spectral characteristics of the individual chromophores. When two chromophores are separated by only one single bond (that is, when they are *conjugated*), however, the spectrum of the molecule is not the sum of the spectra of the separate chromophores. Instead, the two simple groups create a new, larger chromophore with new spectral characteristics. (See, for example, the data in Table 12–2, p. 352).

TABLE 28–4 Ultraviolet Chromophores

FUCTIONAL GROUP		
Structure	*Name*	λ_{max}, nm (ϵ)
—C≡C—	acetylene	175 (5000)
>C=C=C<	allene	185 (10,000)
—C≡N	cyano	340 (100)
$—NO_2$	nitro	210 (15,000)
$—NO_2$	nitro	280 (20)
$—C_6H_5$	phenyl	208 (3000)
$—C_6H_5$	phenyl	265 (150)
>C=O	carbonyl	280 (20)
>C=N—	azomethine	190 (5000)
$—CO_2H$	carboxy	205 (50)
—S—S—	disulfide	215 (400)[a]
—I	iodo	260 (400)

[a] Shoulder

Functional groups that do not themselves absorb in the near uv may influence the behavior of a chromophore if the two are conjugated. Such functional groups, called **auxochromes,** usually cause the appearance of an absorption at longer wavelength and with greater ϵ than is normally ascribed to a given chromophore. Representative auxochromes include —SH, $—NH_2$, and —OH.

Perhaps the greatest utilization of these complex chromophores has been in the area of steroid structural analysis. The empiricisms presented below are largely the result of the work of L. Fieser and R.B. Woodward. The data in Tables 28–5 and 28–6 allow the calculation of λ_{max} for conjugated dienes (C=C—C=C) and conjugated enones (C=C—C=O), respectively.

TABLE 28-5 Rules for Conjugated Diene Absorption

Group	Contribution (nm)
parent heteroannular diene	214
parent homoannular diene	253
add for each substituent:	
double bond extending conjugation	30
alkyl substituent on double bond	5
exocyclic double bond	5
—N(alkyl)$_2$	60
—S-alkyl	30
—O-alkyl	6
—Cl, —Br	5

TABLE 28-6 Rules for Conjugated Enone Absorption

Group	Contribution (nm)
parent 6-membered ring or acycle	215
parent 5-membered ring	202
parent α,β-unsaturated aldehyde	207

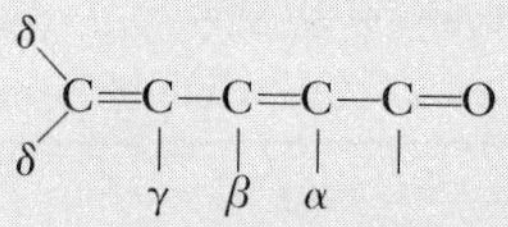

δ, δ C=C—C=C—C=O (γ, β, α)

SUBSTITUENT CONTRIBUTIONS

Substituent	*Location*	*Contribution* (nm)
alkyl	α	10
	β	12
	γ	18
	δ	18
hydroxy	α	35
	β	30
	γ	—
	δ	50
exocyclic double bond		5
heteroannular double bond extending conjugation		30
homoannular double bond extending conjugation		68

Problem: Predict the position of λ_{max} for **A**.

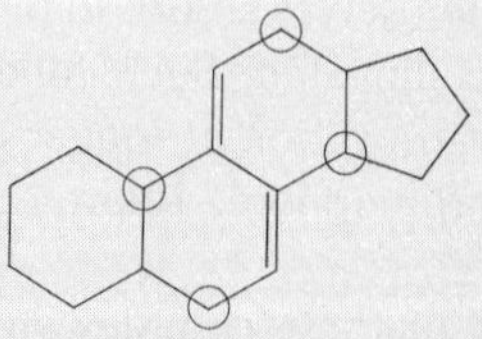

A

Solution: Since the double bonds occur in two different rings, the base value of 214 nm (heteroannular diene) is used. There are four alkyl groups (○) attached to the double bonds, increasing the λ_{max} by 4×5, or 20 nm. Finally, each double bond is exocyclic to the "other" ring, adding 2×5, or 10 nm. Thus, the predicted λ_{max} is $214 + 20 + 10 = 244$ nm; the observed value is 245 nm.

Problem: On the basis of their uv spectra, identify the samples as cholesta-4-ene-3-one (**A**) or cholesta-4,6-diene-3-one (**B**). Sample **1** exhibits an absorption maximum at 241 nm, and sample **2** has a maximum at 281 nm.

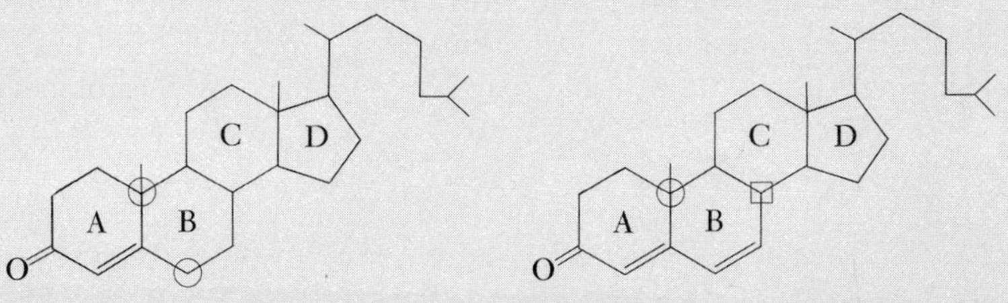

A **B**

Solution: **A** has a parent enone in a 6-membered ring (215 nm). There are two substituents at β positions (marked ○) "worth" 2×12 nm. The double bond is exocyclic to the steroid B ring, a 1×5 nm contribution. The calculated λ_{max} is $215 + 24 + 5$, or 244 nm.

B also has a parent enone in a 6-membered ring (215 nm). Furthermore, it has one β and one δ alkyl group (○ and □), contributing 12 and 18 nm, respectively. The double bond in ring A is exocyclic to ring B (worth 5 nm). The double bond in ring B is heteroannular to the parent ring (ring A) *and* extends conjugation (worth a 30 nm contribution). The calculated λ_{max} is $215 + 30 + 5 + 30$, or 280 nm.

Thus, **1** is assigned to **A** while **2** is assigned to **B.**

28.12 APPLICATIONS OF UV SPECTROSCOPY

Electronic spectroscopy has been applied to the study of organic structures for a long period of time, and a large number of applications of uv to organic chemistry are known. This discussion will arbitrarily delete mention of many of these and will concentrate upon proof-of-structure and degree-of-purity applications.

A. INCLUSION AND EXCLUSION OF FUNCTIONAL GROUPS

If a compound shows no λ_{max} above approximately 200 nm, it *cannot* have a conjugated chromophore (*e.g.*, C=C—C=C); an aldehyde (R—CHO) or keto (R—CO—R) group; a benzene ring (R—C_6H_5); or bromine or iodine. However, it may possess fluorine or chlorine; an isolated double bond (C=C); a carboxy group (R—CO_2H); a cyano group (—C≡N); a hydroxy group (R—OH); a mercapto group (R—SH); or an amino group (—NH_2).

B. NATURE AND EXTENT OF CONJUGATION

The calculations of the λ_{max} of steroids illustrate this point. Furthermore, the extent of conjugation in simple polyolefins of the type R—$(C{=}C)_n$—R can be estimated by noting that additional unsaturation shifts the longest wavelength transition to an even longer wavelength.

$CH_3(CH{=}CH)_8CH_3$	λ_{max} 420 nm	all geometries
$CH_3(CH{=}CH)_9CH_3$	λ_{max} 440 nm	around the double
$CH_3(CH{=}CH)_{10}CH_3$	λ_{max} 476 nm	bonds are *trans*

These three alkenes absorb in the visible region and are, therefore, colored. If they did not absorb in the visible region they would appear white (also called "colorless"). Conjugation as a means of shifting λ_{max} into the visible range is epitomized by β-carotene, λ_{max} 447 nm.

β-carotene

β-Carotene is important in human nutrition, since it is a precursor to the A group of vitamins. Also, β-carotene is converted in the body to a derivative of 11-*cis*-retinal, the substance responsible for light absorption during the process of vision.

H— O

11-*cis*-retinal

The existence of a specific, highly conjugated unsaturated system results in maximal absorption by 11-*cis*-retinal in the middle of the visible spectrum.

C. SAMPLE HOMOGENEITY

Compounds with large values of λ_{max} often can be detected, even at low concentrations, as impurities in samples of substances having weak absorptions in the region of λ_{max} of the impurity. A practical example is the detection of benzene as a low concentration impurity in ethanol.

Ethanol that contains water can be made anhydrous by distilling the aqueous ethanol from a flask containing benzene. Because of the toxicity of benzene, it is important that all alcohol for use in prescription compounding, and for internal use in general, be free of benzene. One method for detecting residual benzene is by examining the uv spectrum of the sample near 260 nm. At this wavelength, ethanol is transparent but benzene has a λ_{max} with ϵ of 230.

Problem: Calculate A at ~260 nm for a solution containing 1 g of benzene in 100 liters of ethanol. The pathlength of the sample cell is 1.0 cm.

Solution:

$$A = abc$$

$$c = \frac{1 \text{ gram}}{(78 \text{ grams/mole}) \times (100 \text{ liters})} = 1.28 \times 10^{-4} \text{ gram/liter}$$

$$A = 230 \times 1 \times 1.28 \times 10^{-4} = 0.0295$$

Even an absorbance that is this small can be detected by a research-grade spectrophotometer.

28.13 MASS SPECTROMETRY— AN INTRODUCTION

Unlike infrared or ultraviolet spectroscopy, which are nondestructive analytical procedures, mass spectrometry is a destructive procedure. It measures the extent of damage sustained by a molecule that is subjected to an electron beam. When a low energy (about 10 eV) electron beam strikes a molecule, which is in the vapor state in the mass spectrometer, that molecule normally loses one electron and produces a molecular ion, $M^{\cdot\oplus}$. When a high energy (about 70 eV) electron beam strikes a molecule, the initially produced molecular ion will fragment into smaller pieces. Some of these will be charged and some will be uncharged. Only the charged ones are studied. Because of the low pressure within the mass spectrometer (about 10^{-7} torr), only *intra*molecular reactions follow bombardment of a molecule by an electron beam of high energy. The equations below indicate, in a schematic way, some of the types of processes that may occur after bombardment. Figure 28–13 describes the mass spectrometer.

$$\text{A-B-C-D-E} \xrightarrow[\text{10 eV}]{\text{electron beam}} \underset{\text{molecular ion}}{[\text{A-B-C-D-E}]^{\cdot\oplus}} + e^{\ominus}$$

$$\text{A-B-C-D-E} \xrightarrow[\text{70 eV}]{\text{electron beam}} \underset{\text{molecular ion}}{[\text{A-B-C-D-E}]^{\cdot\oplus}} + e^{\ominus}$$

$$[\text{A-B-C-D-E}]^{\cdot\oplus} \xrightarrow{\text{fragmentation}} [\text{A-B}]^{\oplus} + [\text{C-D-E}]^{\cdot}$$

$$[\text{A-B-C-D-E}]^{\cdot\oplus} \xrightarrow{\text{rearrangement}} [\text{A-B-E-C-D}]^{\cdot\oplus} \xrightarrow{\text{fragmentation}} [\text{A-B}]^{\cdot} + [\text{E-C-D}]^{\oplus}$$

$$[\text{A-B-E-C-D}]^{\cdot\oplus} \xrightarrow{\text{fragmentation}} [\text{A-B-E}]^{\cdot\oplus} + \text{CD}$$

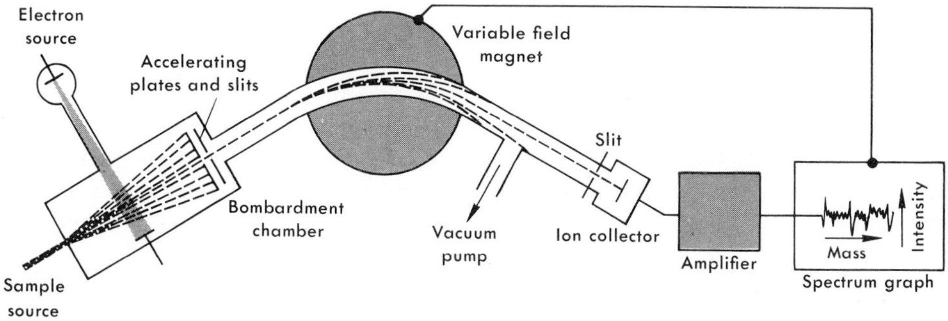

Figure 28–13 Schematic diagram of a mass spectrometer. (From Moore, J.A.: *Elementary Organic Chemistry.* Philadelphia, W.B. Saunders Company, 1974.)

The two major applications of mass spectrometry are to (a) exact molecular weight and molecular formula determination and (b) structural analysis. The former application depends upon the identification of the molecular ion and usually is carried out using a low energy electron beam. The latter requires the decomposition of the molecular ion and analysis of the ensuing fragments, and is carried out using a high energy beam.

28.14 MOLECULAR WEIGHT AND FORMULA DETERMINATION

The mass spectrum of a substance is plotted with the intensity of the signal on the ordinate and the ratio of mass-to-charge (m/e) for the various charged species produced by decomposition on the abscissa. Since the charge, e, on these ions usually is 1, the ratio m/e actually corresponds to the mass of the ion. Illustrative of this is the mass spectrum shown in Figure 28–14.

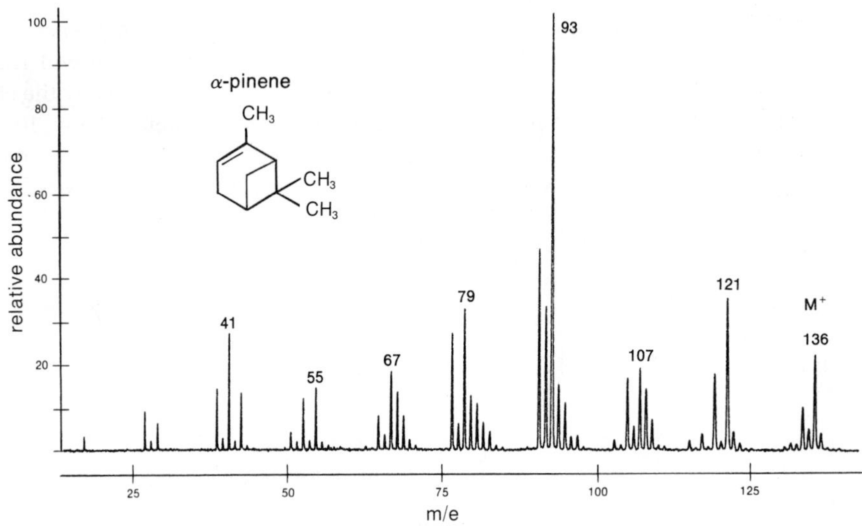

Figure 28–14 The mass spectrum of α-pinene. The base peak has an m/e of 93, while the molecular ion (parent peak) has an m/e of 136.

When the electron beam is of low energy, the most intense peak in the spectrum (the so-called **base peak**) usually corresponds to the molecular ion of the parent molecule. The molecular ion peak sometimes is called the **parent peak.** Since the charge on the parent peak is unity, the m/e ratio of the parent peak gives the molecular weight of the starting material. The parent peak in the spectrum presented in Figure 28–14 appears at an m/e of 136; in this case, however, the base peak is at m/e 93.

With the usual ionizing energy (70 eV), the $M^{\cdot\oplus}$ peak will be of diminished intensity because it will have undergone subsequent reactions. Therefore, the base peak is not always the $M^{\cdot\oplus}$ peak. Furthermore, the highest mass peak may not be due to $M^{\cdot\oplus}$ because of (a) the total fragmentation of $M^{\cdot\oplus}$ or (b) the existence of isotopes that will produce small peaks with m/e values of M + 1, M + 2, and so forth. These "M+" peaks will be the highest mass peaks in the spectrum. If you examine Figure 28–14 carefully, you will note the existence of such peaks above m/e 136.

Molecular formulas can be obtained from a mass spectrum in several ways. Using a high resolution instrument, one that measures m/e values accurately to the fourth decimal place, one can get an exact molecular weight and, by use of appropriate tables, obtain a molecular formula for the substance. For example, while both N_2 and C_2H_4 have nominal masses of 28.0, the former is exactly 28.0061 while the latter is exactly 28.0313. These are easily distinguished using high resolution mass spectrometry.

Molecular formulas also can be obtained from low-to-medium resolution mass spectral data by comparing the intensities of the various "M+" peaks to that of the parent peak. From a knowledge of the natural abundance of isotopes and their masses, one can calculate (therefore,

TABLE 28-7 "M+" Peaks for $M^{\cdot\oplus} = 66$

	Intensities		
Formula	$M^{\cdot\oplus}$	$M + 1$	$M + 2$
$C_3H_2N_2$	100	4.04	0.06
C_4H_4N	100	4.77	0.09
C_5H_6	100	5.50	0.12

predict!) the anticipated intensities of M + 1 and M + 2 peaks relative to the parent peak. (This is the proper time to note that the parent peak, $M^{\cdot\oplus}$, is that peak which is produced from the most abundant isotope of each element in the molecule.) Extensive tables of such data have been collected, and an abbreviated version is included in the text by Silverstein, Bassler, and Morrill.[a] A sample of the kind of data one would find in such compilations is included here as Table 28–7. You can see from this table that even though $C_3H_2N_2$, C_4H_4N, and C_5H_6 have nominal masses of 66, a comparison of $M^{\cdot\oplus}$, M + 1, and M + 2 would permit the distinction of each from the others.

28.15 SULFUR, NITROGEN, AND THE HALOGENS

The presence and amount of sulfur in a compound often can be inferred from the intensity of the M + 2 peak. Sulfur is about 95% ^{32}S and 4.2% ^{34}S; one would expect, therefore, that a compound with one sulfur atom should have an M + 2 peak with an intensity of about 4% of that of the parent peak. In the section of a mass spectrum shown in Figure 28–15, we see the molecular ion (m/e 135) and the so-called satellite peaks (the "M+" peaks) of benzothiazole. The M + 2 peak intensity is about 4% of the intensity of the $M^{\cdot\oplus}$ peak, suggesting that the molecule has only one sulfur atom. (This assumes that there are no other major contributors to the M + 2 peak, a valid assumption for this compound.)

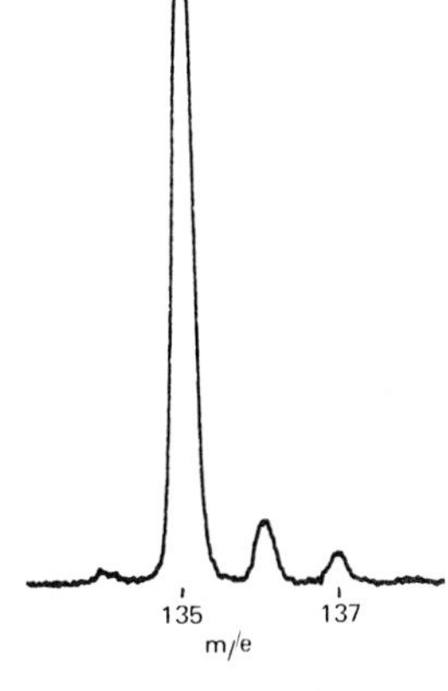

Figure 28-15 The molecular ion and two satellite peaks from the spectrum of benzothiazole. These are sufficient to determine the molecular formula C_7H_5NS.

---

[a] Silverstein, R.M., Bassler, G.C., and Morrill, T.C.: *Spectrometric Identification of Organic Compounds,* 3rd ed. New York, John Wiley and Sons, 1974.

The intensity of the M + 1 peak is about 8% of that of the $M^{\cdot\oplus}$ peak. Assuming carbon to be a major contributor to M + 1, and since ^{13}C has a natural abundance of 1.11%, we calculate that the benzothiazole contains seven carbons (8.0/1.11 = 7.2). The parent peak has an odd value of m/e, and this often signifies an odd number of nitrogens in the molecule. (See "**Nitrogen rule**" on p. 941.)

Putting all of these data together leads to the conclusion that benzothiazole, with a molecular weight of 135, has the formula C_7H_5NS. The complete mass spectrum (medium resolution) of benzothiazole, along with its structure, is presented in Figure 28–16.

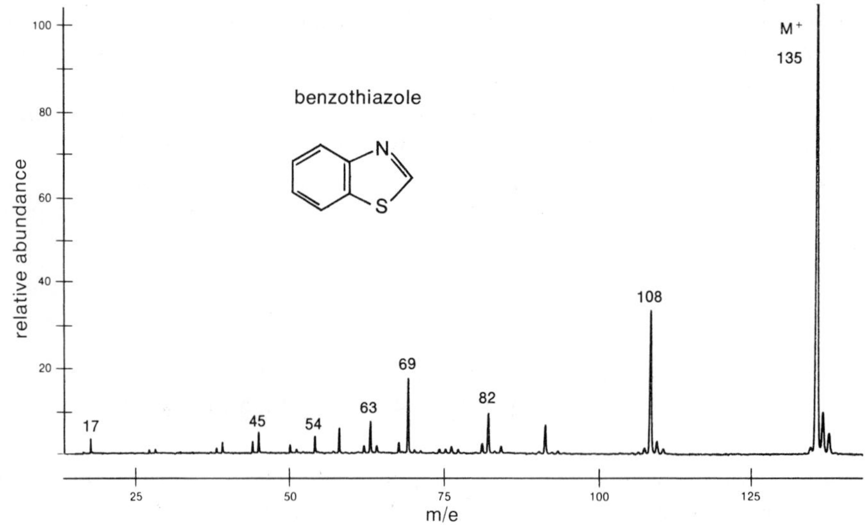

Figure 28–16 The mass spectrum of benzothiazole.

Bromine exists as two stable isotopes (^{79}Br and ^{81}Br) separated by two mass numbers; they are present in nearly equal abundance. Chlorine exists as two stable isotopes (^{35}Cl and ^{37}Cl) separated by two mass numbers; but they are present in about a 3:1 ratio. The presence of bromine and/or chlorine is suggested by the presence of comparatively intense M + 2 peaks. Figure 28–17, the mass spectrum of 1-chloro-2-nitrobenzene, illustrates this effect.

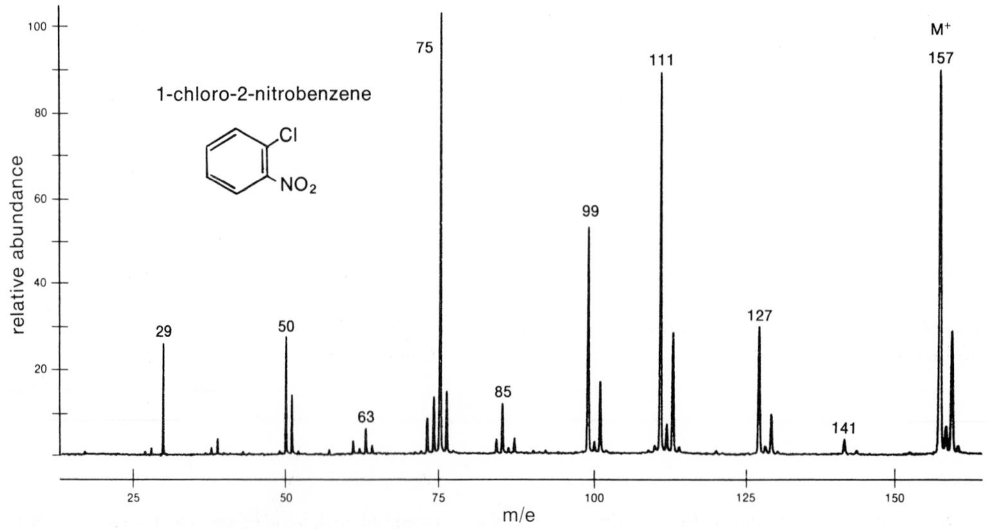

Figure 28–17 The mass spectrum of 1-chloro-2-nitrobenzene.

28.16 DECOMPOSITION

Most of the recent effort in mass spectrometry has been devoted to studying the reactions of molecular ions. Unfortunately, we shall be able to spend very little time on this subject. Indeed, we shall examine the mass spectrum of only three compounds: *n*-hexane, *n*-hexadecane, and methyl salicylate ("oil of wintergreen").

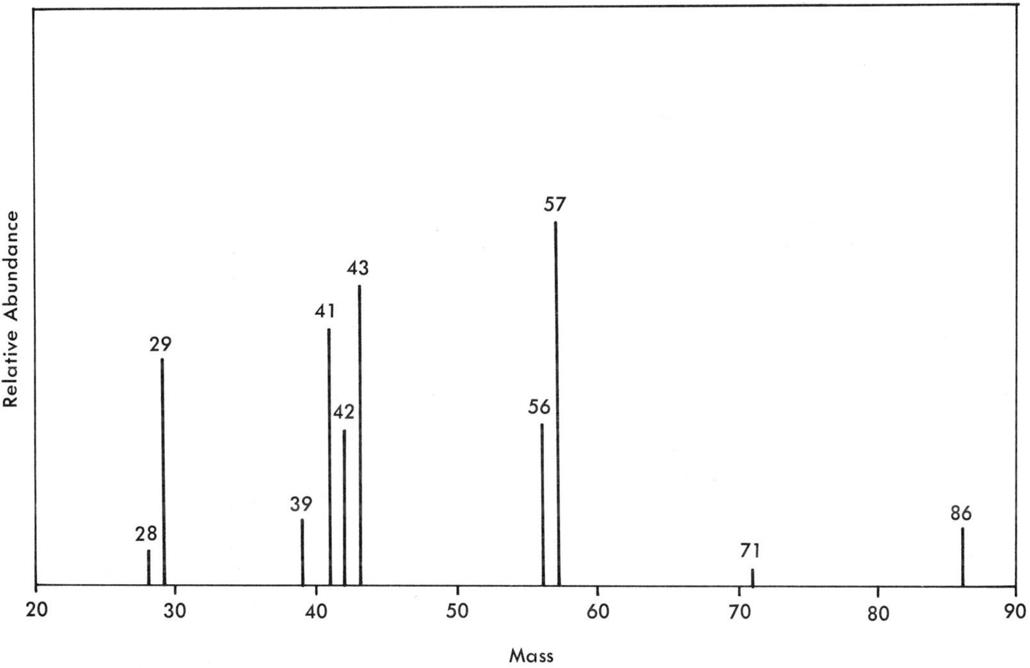

Figure 28-18 The mass spectrum of hexane. (From Moore, J.A.: *Elementary Organic Chemistry*. Philadelphia, W.B. Saunders Company, 1974.)

***n*-Hexane.** The mass spectrum of *n*-hexane is shown, schematically, in Figure 28–18. The pattern of fragment ions shows groups of peaks at intervals of 14 mass units, representing the loss of methylene (CH_2) units. This pattern is typical, as we shall see in the next example, of unbranched, saturated hydrocarbons. The major peak in each cluster is due to the cation formed by loss of an alkyl radical. The less intense peaks arise from further loss of hydrogen atoms. In general, the differences in abundance (intensity) reflect differences in the stability of the various ions.

$$\underset{\text{molecular ion}}{[CH_3CH_2CH_2CH_2CH_2CH_2]^{\cdot\oplus}} \rightarrow \begin{cases} \underset{m/e\ 57}{CH_3CH_2CH_2CH_2^{\oplus}} + CH_3CH_2^{\cdot} \\ \underset{m/e\ 43}{CH_3CH_2CH_2^{\oplus}} + CH_3CH_2CH_2^{\cdot} \\ \underset{m/e\ 29}{CH_3CH_2^{\oplus}} + CH_3CH_2CH_2CH_2^{\cdot} \end{cases}$$

***n*-Hexadecane.** The typical fragmentation of straight chain hydrocarbons, alluded to in the discussion above, is seen quite clearly in the mass spectrum of *n*-hexadecane (Figure 28–19).

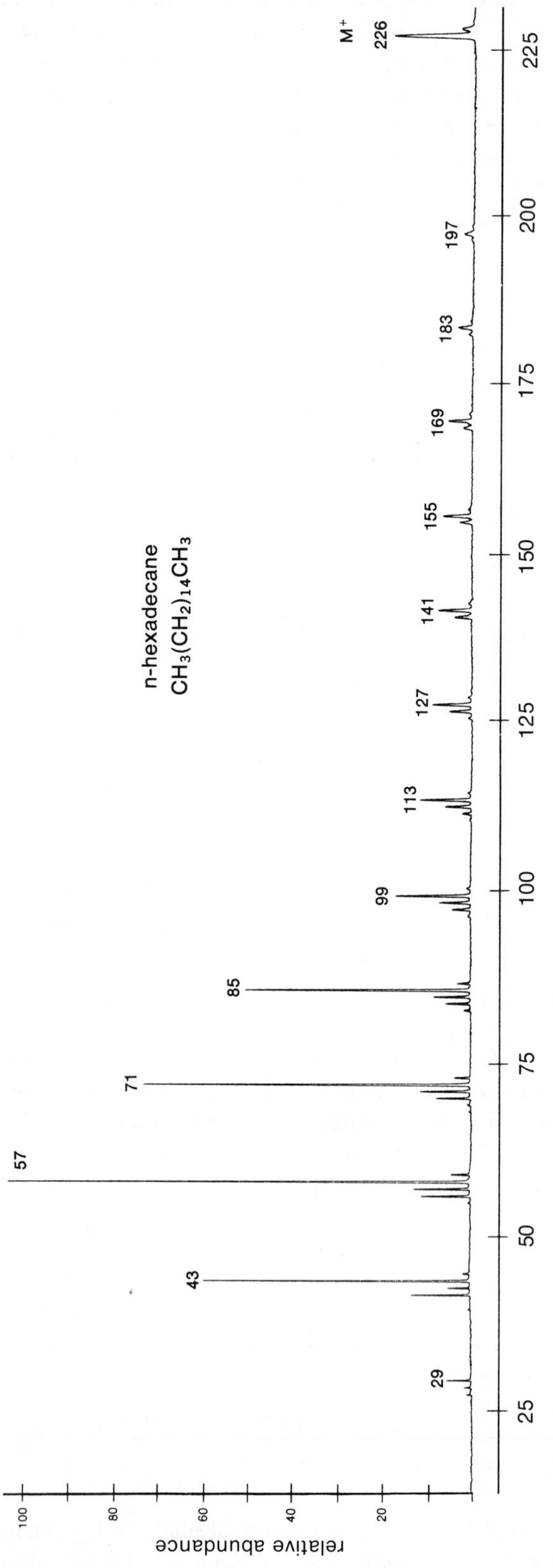

Figure 28-19 The mass spectrum of *n*-hexadecane. The spectrum shows a familiar pattern for saturated hydrocarbons or for a long-chain alkyl moiety of a complex compound. This homologous series of ions can be represented by the formula $(C_nH_{2n+1})^{\oplus}$. Note that each of the groups of ions is separated by 14 amu, indicating the continuous loss of methylene groups.

Methyl Salicylate. Methyl salicylate's molecular ion undergoes a rapid loss of methanol to form a species of m/e 120. This is believed to be formed *via* a six-membered transition state. This type of transition state is typical of any benzoic acid or benzoate ester that has an *ortho* substituent bearing an α hydrogen. For this reason it sometimes is called the "*ortho* effect." The mass spectrum of methyl salicylate is shown in Figure 28–20.

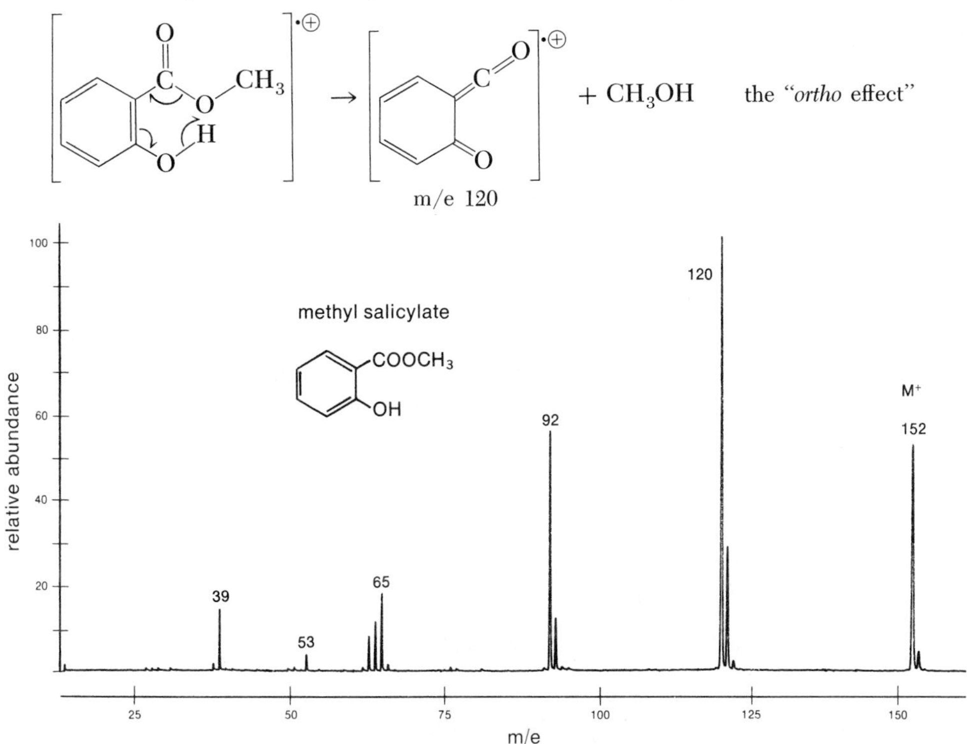

Figure 28–20 The mass spectrum of methyl salicylate.

28.27 WHAT WE HAVE NOT SAID

We have barely scratched the surface of the applications of mass spectrometry to organic chemistry. We have, for example, not pointed out that it is possible to connect ("interface") the mass spectrometer with the gas chromatograph and thereby produce the most powerful separation-analysis tool known to chemistry. We have not said that most mass spectra may now be interpreted with the aid of computers. Indeed, there exist systems in which you inject a sample into a gas chromatograph and, at the other "end," a computer prints out a list of what is in your sample.

We have not said that sugars, compounds that are quite nonvolatile, may be studied by mass spectrometry by first converting them to volatile derivatives such as trimethylsilyl ethers.

a sugar—difficult to study by ms

trimethylsilyl derivative of sugar—easy to study by ms

We have not said that amino acids and small peptides can be studied by mass spectrometry if they first are converted to volatile derivatives.

$$\overset{\oplus}{H_3N}-\underset{|}{\overset{R}{CH}}-CO_2^{\ominus} \xrightarrow{CF_3COCl} CF_3CONH-\overset{R}{CH}-CO_2H \xrightarrow{CH_2N_2} CF_3CONH-\overset{R}{CH}-CO_2CH_3$$

an amino acid (nonvolatile)

a volatile derivative

We have not said anything about how mass spectrometry can be used to follow isotope exchange reactions or to locate isotopes in molecules.

We have not said anything about how mass spectrometry can be used to determine the ionization potential of a substance.

The interested student is referred to the citations in the bibliography for information concerning these and other applications of mass spectrometry to organic chemistry.

IMPORTANT TERMS

Ångstrom unit: A very small unit of distance, equal to 10^{-8} cm. It is named after the Swedish physicist Anders Jonas Ångström (1814–1874).

Base peak: The most intense peak in a mass spectrum.

Beer-Bouger-Lambert law: A quantitative relationship between the intensity of light incident upon a sample (I_0) and the light transmitted by that sample (I).

$$\epsilon = \frac{\log (I_0/I)}{cb}$$

In this equation, c is the molar concentration of substrate in solution, b is the thickness of the solution in centimeters (this corresponds to the thickness of the sample cell), and ϵ is the molar absorptivity (molar extinction coefficient). The molar extinction coefficient (or simply "extinction coefficient") may be viewed as a proportionality constant between cb and $\log (I_0/I)$.

This law frequently is called the Beer law or the Beer-Lambert law, and it may be given as $A = \epsilon\, cb$, where A is the *absorbance* of the solution (*optical density* is an old synonymn for absorbance).

Chromophore: A functional group that shows a characteristic absorption in the ultraviolet or visible region of the spectrum.

Emission: The giving off of electromagnetic radiation.

Fingerprint region: The most complex, and therefore most characteristic, region of the infrared spectrum of a compound (1400–800 cm^{-1}).

Frequency: The number of waves passing by a particular point during a specific period of time.

Fundamental vibration: A vibration corresponding to a fundamental transition, which is the excitation of the molecule from one energy level to the next higher level. The most important fundamentals are those involving transitions from the ground state to the first excited state.

Gamma ray: Electromagnetic radiation of very high frequency (therefore, very high energy).

Microwave: Radiation with a wavelength in the range from approximately 1 mm to 50 cm. Microwave radiation is important in many radar devices and in certain cooking devices as well as in structural chemistry.

Molar absorptivity: A synonym for molar extinction coefficient. Its magnitude is proportional to the probability that a particular transition will occur. Transitions with a high probability will have a high extinction coefficient ($\approx 10^5$), while those with a low probability will have a low extinction coefficient (10^0 to 10^3).

Molar extinction coefficient: Experimentally, it may be viewed as a proportionality constant between the amount of radiation absorbed by a particular sample and the quantity of sample. Theoretically, it is a measure of the probability of occurrence of a particular transition. See **Molar absorptivity** and **Beer-Bouger-Lambert law.**

Molecular ion: The cation produced after a molecule has lost an electron due to electron bombardment. The molecular ion is a cation radical. The molecular ion, if present in a given mass spectrum, will have the highest m/e (neglecting peaks due to the presence of isotopes). The molecular ion of benzene is considered to have an m/e of 78 ($6 \times {}^{12}C + 6 \times {}^{1}H$).

Monochromator: A device for sorting out light of a single wavelength from light of many wavelengths.

Nitrogen rule: The molecular ion of a compound will have an even mass number if that compound contains an even number of nitrogens. (The presence of an odd number of nitrogens leads to a molecular ion with an odd mass number.) The examples below illustrate this rule.

name	structure	formula	m/e of $M^{\cdot\oplus}$
hydrazine	$H_2N—NH_2$	N_2H_4	32
4-aminopyridine	H_2N–(pyridine ring)–N	$C_5H_6N_2$	94
ammonia	NH_3	NH_3	17
dimethylamine	$(CH_3)_2NH$	C_2H_7N	45
pyridine	(pyridine ring) N	C_5H_5N	79

Parent peak: An alternate name for the peak corresponding to the molecular ion.

Thermistor: A resistor whose resistance changes with temperature.

Wavelength: The distance between two crests (peaks) of a wave.

PROBLEMS

1. The amount of energy associated with one mole of photons is called one Einstein and equals $Nh\nu$, where N is Avogadro's number. Calculate the value of the Einstein for (a) 4000 cm^{-1} and (b) 625 cm^{-1} radiation. Which end of the infrared spectrum is richer in energy, based upon this calculation?
2. The fundamental vibration of I_2 occurs at 214 cm^{-1}, while that of Cl_2 occurs at 565 cm^{-1}. Which vibration requires more energy? What are the force constants for these vibrations? Why are these vibrations infrared inactive?

3. Cyclopropane has an absorption at 3050 cm^{-1}. Calculate the expected frequencies of the corresponding absorptions in $^{13}C{-}^{1}H$, $^{13}C{-}^{2}H$, and $^{12}C{-}^{2}H$ cyclopropane. Which of these shows the greatest isotope effect?

4. Gamma rays vary in wavelength from 0.25 Å to 0.001 Å. What is the energy range of gamma rays? (*Note:* Energy may be expressed in ergs, joules, or calories; 1 joule = 10^7 ergs = 0.24 cal.)

5. Suggest a reason for the appearance of *two* S—O stretching frequencies (~1320 and ~1130 cm^{-1}) in the ir spectrum of dimethyl sulfone, $(CH_3)_2SO_2$. Dimethyl sulfoxide, $(CH_3)_2SO$, shows only one S—O stretching frequency (~1050 cm^{-1}).

6. Describe one way that ir spectroscopy could be used to distinguish between the various members of the following sets:

Set 1: (a) 2,4-dimethylpentane
(b) *n*-heptane
Set 2: (a) acetylene
(b) 1-hexyne
(c) 2-butyne
Set 3: (a) ethylene
(b) propylene
(c) tetramethylethylene
Set 4: (a) cyclohexane
(b) cyclohexene
(c) 1-hexene
Set 5: (a) *cis*-2-butene
(b) *trans*-2-butene
(c) 1-butene
Set 6: (a) chlorocyclohexane
(b) chlorobenzene
(c) 2-chlorobutane

7. The free-radical chlorination of *t*-butylcyclohexane produced a mixture of monochlorinated products. The infrared spectrum of this mixture showed four different C—Cl stretching frequencies. Explain.

8. The spectra reproduced below are those of (a) isovaleraldehyde, (b) monodeuterioethanol, (c) *p*-nitrophenylacetic acid, (d) vinyl sulfone, and (e) amphetamine. Given the structures shown below, account for as many absorptions in the ir as you can.

(a) 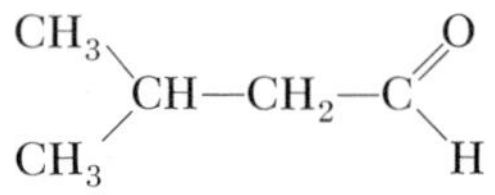 $(CH_3)_2CH{-}CH_2{-}C(=O)H$ (b) CH_3CH_2OD (c) 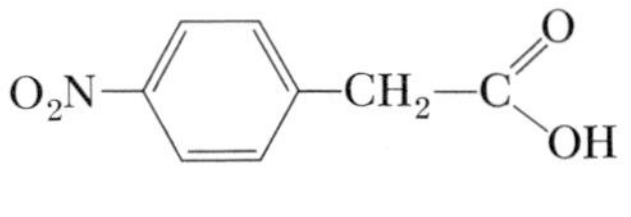 $O_2N{-}C_6H_4{-}CH_2{-}C(=O)OH$

(d) $(CH_2{=}CH)_2SO_2$ (e) 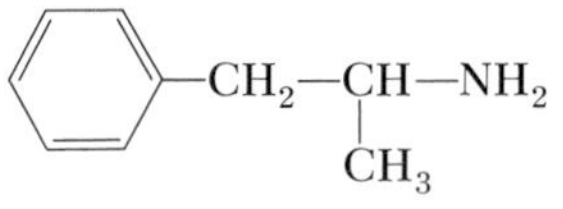 $C_6H_5{-}CH_2{-}CH(CH_3){-}NH_2$

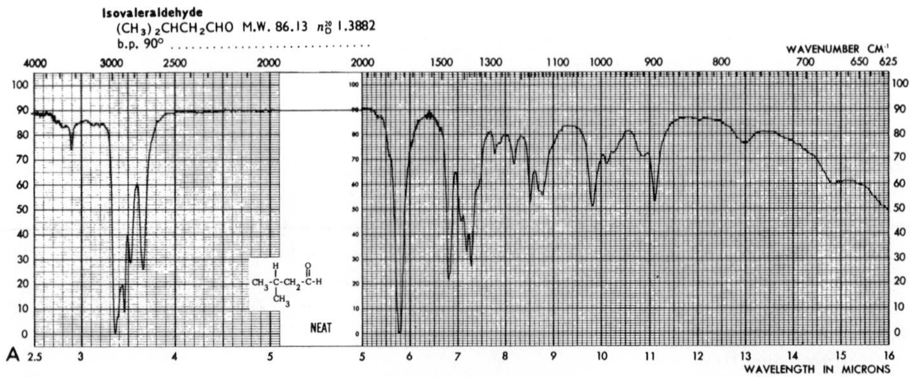

Ethyl alcohol-*d* (ethanol-*d*), **99%**
C_2H_5OD M.W. 47.08 n_D^{20} 1.3595

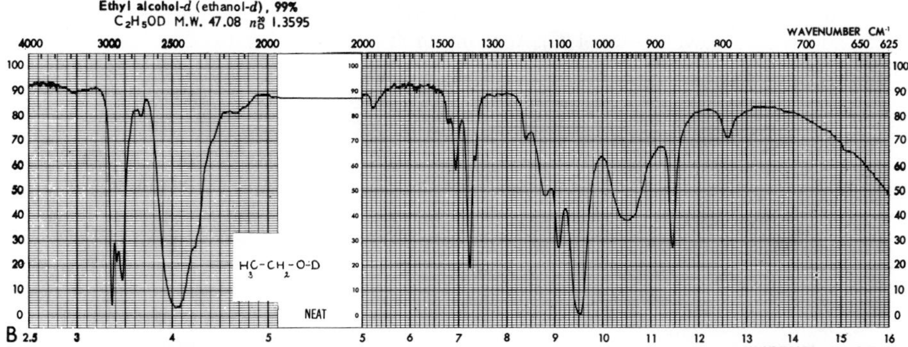

p-Nitrophenylacetic acid
$O_2NC_6H_4CH_2CO_2H$ M.W. 181.15
m.p. 154-155°

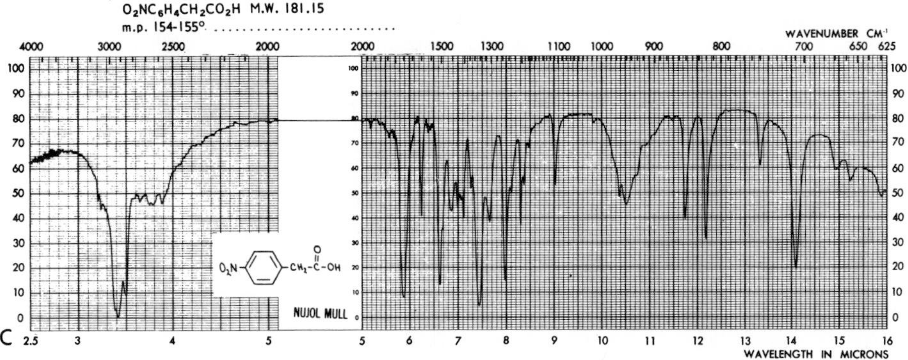

Vinyl sulfone (divinyl sulfone)
$(H_2C{:}CH)_2SO_2$ M.W. 118.15 n_D^{20} 1.4765
b.p. 102-105°/10 mm.

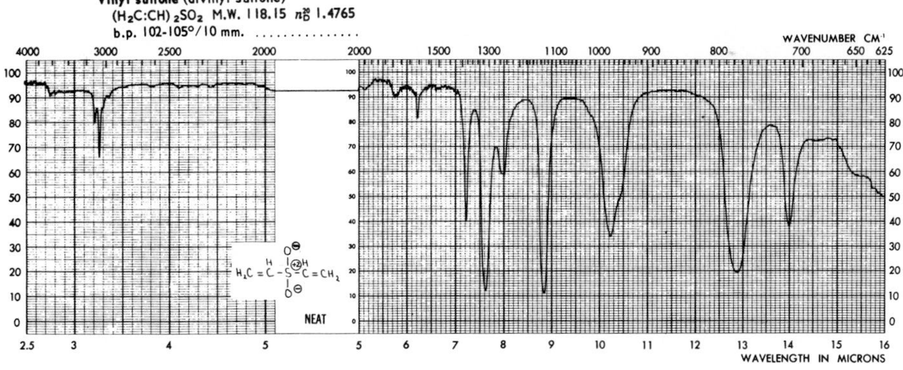

***d*-Amphetamine** (*d*-α-methylphenethylamine), **puriss.**
$C_6H_5CH_2CH(CH_3)NH_2$ M.W. 135.21
n_D^{20} 1.5163 $[\alpha]_D^{20}$ + 33.0° (neat)

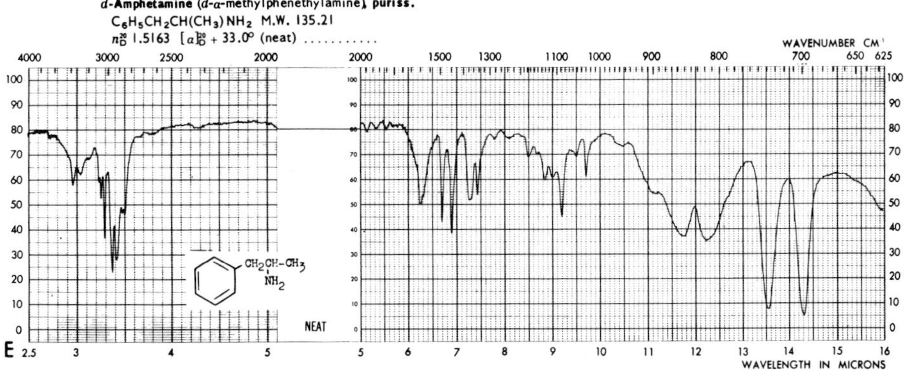

9. The infrared spectra below are those of *cis*-4-methylcyclohexanol (**A**), *trans*-4-methylcyclohexanol (**B**), and *cis*-3,3,5-trimethylcyclohexanol (**C**). Account for the differences in the C—O absorptions of these compounds.

HO, CH_3 — **B**

H_3C, H_3C, CH_3, OH — **C**

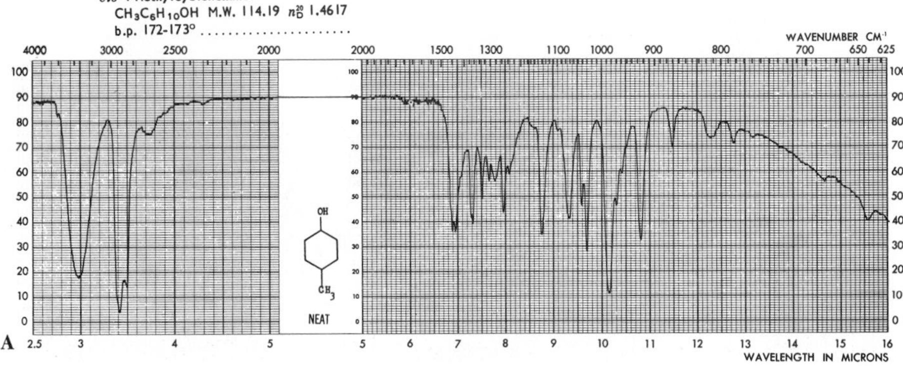

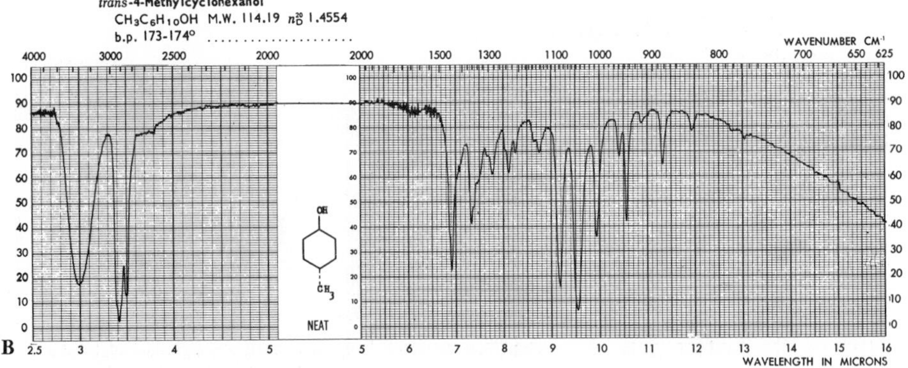

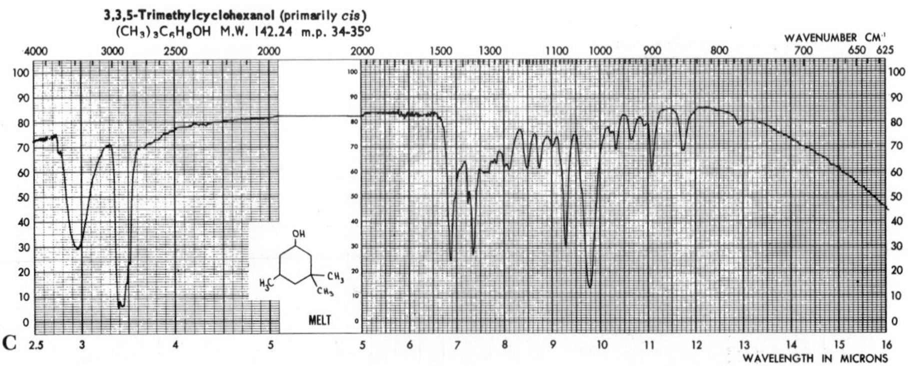

10. Define with the aid of an equation, each of the following:
    (a) transmittance
    (b) absorbance
    (c) molar absorptivity
    (d) molar extinction coefficient
    (e) absorptivity
    (f) optical density

11. Azulene has an absorption maximum at 700 nm (ϵ 300), the red end of the visible spectrum; the next shortest wavelength transition occurs at 357 nm (ϵ 4000). What color would you predict azulene to have?

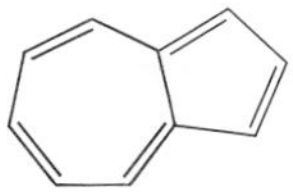

azulene

12. The ultraviolet spectra of isomeric conjugated polyenes show the longest wavelength maximum and the most intense absorption for that isomer with the largest number of *trans* linkages; the opposite is true for the "all-*cis*" isomer—it has the lowest λ and ϵ. Explain.

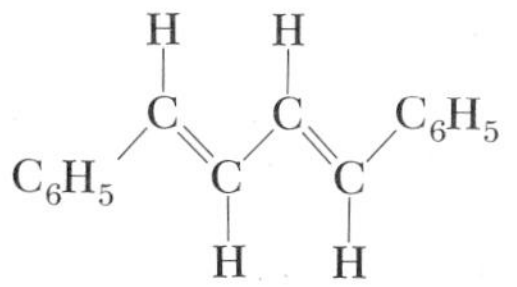

all-*trans* 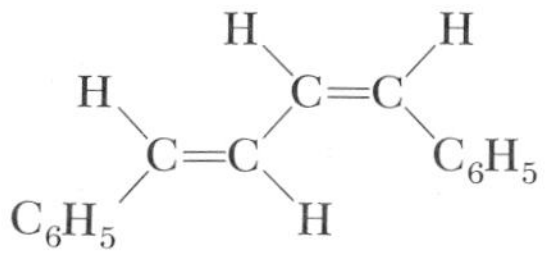 *trans-cis* 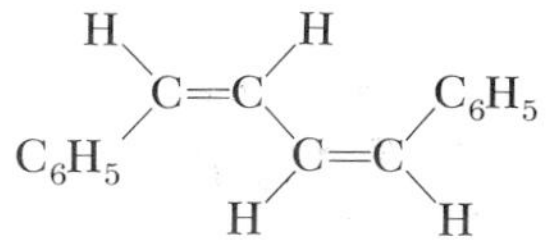 all-*cis*

13. Calculate λ_{max} for each of the following.

(a) 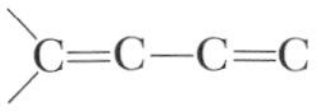 C=C—C=C (d) 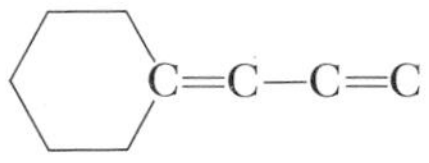 C=C—C=C

(b) 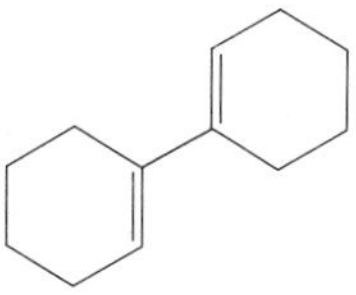 (e)

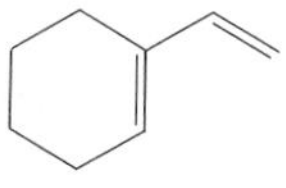

(c) 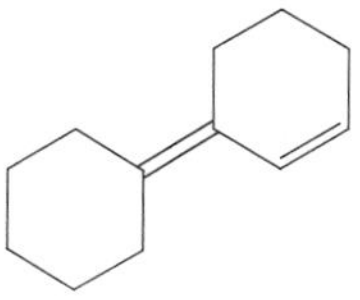 (f)

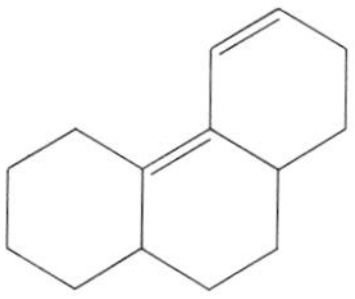

14. When a double bond and a carbonyl group are conjugated, the complex chromophore is described by a molecular orbital picture which is different from that of the isolated double bond or of the isolated carbonyl group. A representative m.o. picture of a conjugated enone is presented on p. 946. A conjugated enone shows an $n \rightarrow \pi^*$ transition that occurs at longer wavelengths than does that of an isolated carbonyl compound. It also exhibits a $\pi \rightarrow \pi^*$ transition at longer wavelengths than is found in either isolated chromophore. (These $\pi \rightarrow \pi^*$ transitions occur below 210 nm and are quite intense.) With the aid of the m.o. diagram, indicate which transitions correspond to the complex chromophore's $n \rightarrow \pi^*$ and $\pi \rightarrow \pi^*$ transitions.

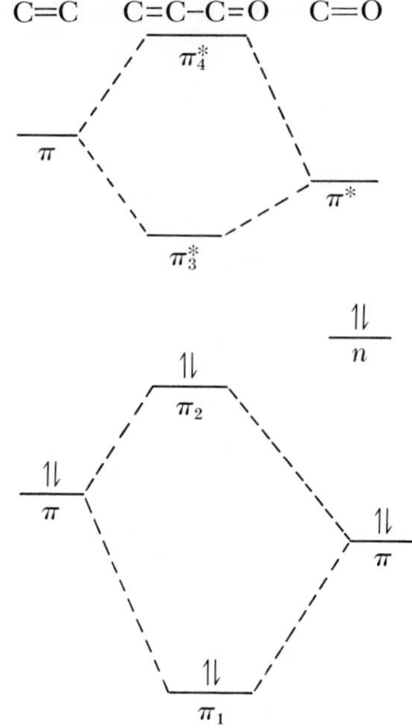

15. Since unconjugated chromophores can be treated additively, the uv spectra of complex molecules can sometimes be anticipated with the aid of model compounds—simpler structures containing the same chromophores. As an example, thalidomide (the hypnotic that may have been responsible for as many as 10,000 birth defects) can be viewed as being made up of two simpler chromophores:

thalidomide A + B

This hypothetical separation (i) does not separate any conjugated chromophores and/or auxochromes, and (ii) maintains the same degree of alkylation in the fragments as in the real compound. Portion **A** is actually a poorer model than is **B.** Explain.

16. In the spectrum below, identify (a) the base peak, (b) the parent peak, and (c) the peak due to the molecular ion.

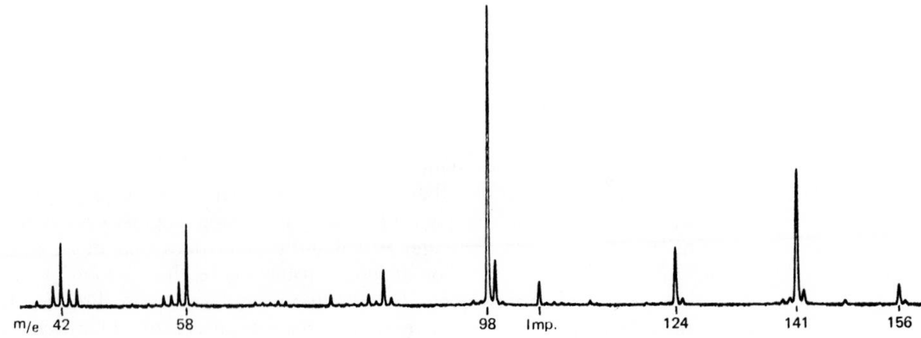

17. The compound whose spectrum is shown in problem 16 contains nitrogen. Does it contain an even number of nitrogen atoms?

18. The following molecules have been discovered in interstellar space, often in clouds as much as ten light years across: OH, NH_3, H_2O, H_2CO, CO, CH_3OH, HCN, HCO_2H, HNCO. (The concentration of molecules in these clouds is as low as one *molecule* per milliter.) Analysis was made by microwave spectroscopy. Assume that the analysis had been made by molecular ion identification on an instrument incapable of separating masses that differ by less than one mass unit. Under these conditions, which (if any) species would not have been resolved?

19. Why is it possible to find both NH_3 and HCO_2H in the same region in space (see problem 18) although they react instantly to form ammonium formate?

$$NH_3 + HCO_2H \rightarrow \underset{\text{ammonium formate}}{NH_4^{\oplus}\ HCO_2^{\ominus}}$$

20. The mass spectrum of 2,2,3,3-tetramethylbutane is virtually devoid of a peak for the molecular ion. However, it contains an intense peak at m/e 57. Suggest a structure for this fragment.

21. Rationalize the following:
   (a) The relative abundance of the $M^{\cdot\oplus}$ peak is greatest for a straight chain hydrocarbon and decreases with increased branching.
   (b) Cleavage of a molecular ion is favored at branched carbons; the greater the branching, the more likely the cleavage.
   (c) The M+ peaks of compounds containing fluorine and phosphorus are "surprisingly" weak.

22. When methyl alcohol, CH_3OH, is analyzed by mass spectrometry, it is found to give rise to about one dozen peaks, four of which have a significant relative abundance (RA): m/e 15, RA 13; m/e 29, RA 64; m/e 31, RA 100; m/e 32, RA 66. (a) Match the equations shown below to the production of the species responsible for these four peaks. (b) Which of these is the base peak? (c) Which of these is the parent peak? (d) Suggest structures for the charged products of **C** and **D.**

A $CH_3OH \rightarrow CH_3OH^{\cdot\oplus} + e^{\ominus}$
B $CH_3OH^{\cdot\oplus} \rightarrow CH_3^{\oplus} + OH^{\cdot}$
C $CH_2OH^{\oplus} \rightarrow CHO^{\oplus} + H_2$
D $CH_3OH^{\cdot\oplus} \rightarrow CH_2OH^{\oplus} + H^{\cdot}$

NUCLEAR MAGNETIC RESONANCE SPECTROSCOPY

29.1 INTRODUCTION

Nuclear magnetic resonance spectroscopy, or **nmr** as it is usually called, was first reported in 1946. Since that time it has become the most important structural tool available to the organic chemist. Indeed, this outgrowth of World War II technology may be considered one of the great scientific advances of this era. Nmr has not replaced infrared or ultraviolet spectroscopy; these continue to be significant procedures. However, for a given investment of time one usually gets more information from nmr than from either ir or uv.

What are some of the capabilities of nmr spectroscopy? Nmr does not "see" every bond in a molecule, as ir usually does, nor does it "see" loosely held electron pairs, as uv usually does; nmr "sees" only *nuclei*, and only one type of nucleus at a time (*e.g.*, all ^{1}H *or* ^{19}F *or* ^{13}C nuclei). Furthermore, there are advantages to the fact that some nuclei which are quite common in organic compounds, *e.g.*, ^{12}C and ^{16}O, cannot be "seen" by the procedure. Nmr signals can be easily *integrated* (the areas under the peaks can be measured), allowing convenient determination of the relative amounts of different "kinds" (see subsequent discussion) of protons and other "nmr active" nuclei. Samples being studied are easily maintained over a wide range of temperatures, permitting kinetic studies that are difficult with ir or uv. Finally, application of computer technology now permits the analysis and prediction of complex spectra and the detection of low concentrations of certain nmr active nuclei, especially ^{13}C.

The single major drawback to proton nmr, and one that is easily surmounted, is that the solvents used must be free of protons (that is, ^{1}H). This usually means the use of relatively expensive deuterated analogs of protonated solvents, *e.g.*, deuteriochloroform, $CDCl_3$. While not as good a solvent as $CDCl_3$, carbon tetrachloride often has been used as an nmr solvent because of its complete lack of hydrogen.

29.2 THE NMR PHENOMENON

The proton and the neutron, like the electron, have a spin quantum number of $\frac{1}{2}$. The *total* spin of a particular nucleus is equal to the vector sum of the spins of the constituent protons and neutrons. Since every atom has a particular number of protons and neutrons, the net nuclear spin varies from element to element and even among isotopes of one element.

While it is impossible to predict the nuclear spin of a specific isotope, the empirical rules presented below narrow the possible values of the nuclear spin quantum number, *I*. *Only nuclei with values of I other than zero may give rise to an nmr signal.* (The nuclear spin quantum numbers for selected nuclei were included in Table 1–2, p. 3.)

1. I equals zero for nuclei with even numbers of protons and neutrons.
2. I has integral values (1, 2, 3 . . .) for nuclei with odd numbers of protons and neutrons.
3. I has half-integral values ($\frac{1}{2}$, $\frac{3}{2}$, $\frac{5}{2}$ . . .) for nuclei with an even number of protons but an odd number of neutrons and *vice versa*.

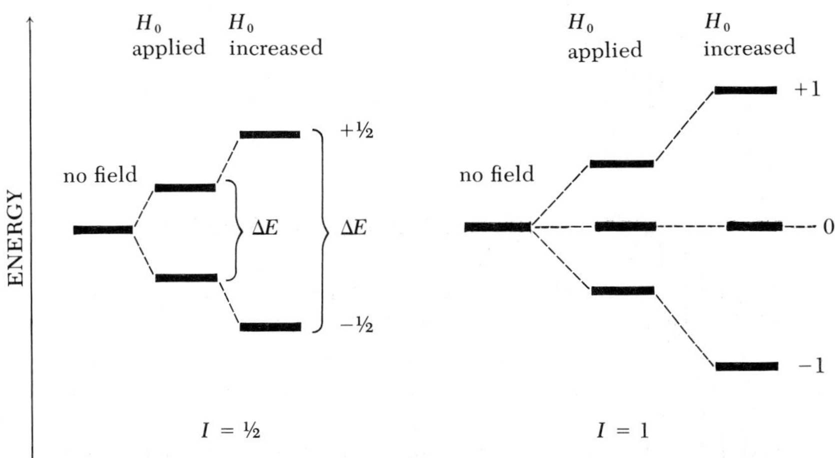

Figure 29–1 The generation of nuclear energy levels *via* application of an external magnetic field (H_0).

A nucleus with a spin quantum number I may adopt $2I + 1$ orientations, or energy levels, in an applied magnetic field of strength H_0. While the amount of energy separating these levels increases with increasing H_0, the amount of energy separating adjacent levels is constant for a given value of H_0 (Figure 29–1). The specific amount of energy separating adjacent levels, ΔE, is given by

$$\Delta E = \frac{H_0 \gamma h}{2\pi}$$

where

γ = magnetogyric ratio, a constant for a given isotope
H_0 = strength of the applied magnetic field
h = Planck's constant

Since $\Delta E = h\nu$, the frequency of electromagnetic radiation, ν, corresponding to this energy difference equals $\gamma H_0/2\pi$.

The remainder of this chapter will focus upon the nmr spectra of protons, ^{1}H, sometimes called *proton magnetic resonance spectroscopy* (pmr), although many of the principles to be presented apply to the study of other nuclei as well.

Nuclear Magnetic Resonance. When hydrogen nuclei are placed in a magnetic field, H_0, the immediate result is alignment of all of the nuclei in the direction of H_0, the nuclei behaving like small bar magnets. Thermally induced agitation quickly reduces the number of nuclei that remain aligned with the field to slightly more than 50% of the total number of nuclei present. This slight excess of protons will absorb radiation of the particular frequency that corresponds to the difference in energy between the ground state (aligned with H_0) and the excited state (aligned against H_0). This absorption produces the signal that eventually becomes the spectrum traced on a sheet of chart paper. The radiation that is absorbed occurs in the radio frequency range, and its frequency equals the *resonance frequency* of the nucleus.

It is possible to induce **nuclear magnetic resonance** (excitation from one energy level to the next higher one) by keeping H_0 constant and sweeping through a range of radio frequencies until the one corresponding to ΔE is reached. (At this point the radiation is absorbed and the excitation is accomplished.) However, since ν and H_0 are related ($\nu = \gamma H_0/2\pi$), it is also possible to achieve resonance by keeping ν constant and varying the applied magnetic field until H_0 is reached. Many nmr spectrometers employ a fixed radio frequency generator (usually 60 or 100 MHz) and vary the applied magnetic field. However, no matter which way the spectrum is obtained, the data are always plotted as *intensity of absorption of radio frequency radiation vs. frequency of radiation*. Why not plot *changes in the applied magnetic field* as the abscissa if the experiment is conducted so as to get that type of data? Largely, tradition! Chemists are used to plotting their spectral data with an abscissa scale in frequency or wavelength units.

29.3 THE CHEMICAL SHIFT

As described thus far, pmr spectroscopy would not be particularly valuable because all ^1H nuclei would absorb at the same value, *e.g.*, 60 MHz at 1.41×10^4 gauss or 100 MHz at 2.35×10^4 gauss. Fortunately, the magnetic field at a given proton bound into a molecule is rarely exactly H_0. Instead, each proton experiences an *effective field*, H_{eff}, which is slightly different from H_0 because the very application of H_0 to the molecule causes that molecule to generate small magnetic fields of its own. These small fields, in turn, may add to or subtract from H_0.

The applied field, H_0, causes the electrons around any nucleus to circulate around that nucleus and, thereby, induces a magnetic field which opposes H_0 (Figure 29–2). The result of this is that the nucleus is *shielded* from the full strength of the applied field, the magnitude of this **shielding effect** being directly proportional to the magnitude of H_0. While every proton in a molecule is subjected to this particular shielding effect, there are other factors (to be discussed shortly) which alter the value of H_{eff} by varying amounts for different nuclei. Suffice it to note at this point that protons absorb over a frequency range of approximately 700 Hz (near 60 MHz) at a field strength of 14,100 gauss.

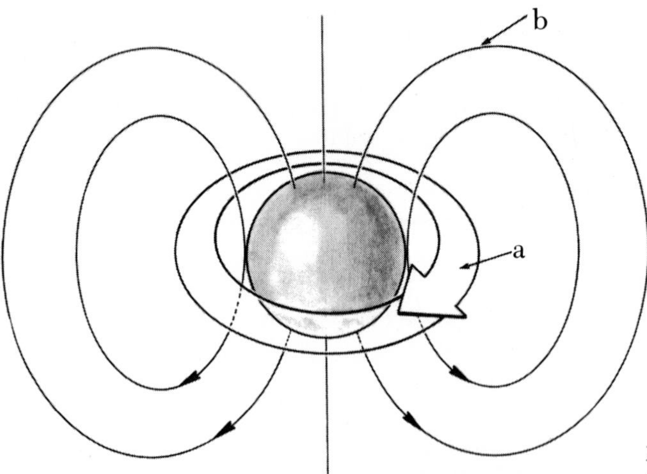

Figure 29–2 Electronic circulation and resultant induced magnetic field caused by exposing an atom to an applied magnetic field (H_0): a, direction of electron circulation; b, lines of force of the induced magnetic field.

Since it is extremely difficult to accurately measure frequencies in the radio frequency range, the nmr experiment does not involve the determination of the absolute value of any frequency. Instead, it is common practice to add an inert standard substance to the sample under study (the so-called "internal standard") and to refer the resonance frequency of a particular signal back to the signal from that standard. This permits the reporting of differences in resonance frequencies, $\Delta\nu$, something that can be done with the necessary accuracy.

The difference between the resonance frequency of a specific signal and that of the standard is called the *chemical shift* of that signal. For proton resonance studies of compounds soluble in organic solvents the usual reference compound is tetramethylsilane, TMS. (The standard used in D_2O is discussed on p. 974.)

$$
\begin{array}{c}
CH_3 \\
| \\
CH_3\text{—}Si\text{—}CH_3 \\
| \\
CH_3
\end{array}
$$

tetramethylsilane, TMS

The resonance frequency of a given nucleus, expressed in Hz, is a function of the magnitude of the applied field. In order to avoid having to use two numbers to characterize a given proton (that is, the difference of the resonance frequencies in Hz *and* the magnitude of the magnetic field strength), the chemical shift is commonly expressed as δ, in parts per million (ppm).

$$\delta = \frac{\Delta\nu \times 10^6}{\text{oscillator frequency (Hz)}}$$

Problem: A proton absorbs 120 Hz downfield from the internal TMS standard at 60 MHz. (a) What is the chemical shift in ppm? (b) Where would this signal appear at 100 MHz?

Solution:

$$\delta = \frac{\Delta\nu \times 10^6}{60 \times 10^6} = \frac{120}{60} = 2.0$$

Therefore, the chemical shift at 60 MHz is 2.0 δ.

At 100 MHz the chemical shift is still 2.0 δ; however, at 100 MHz this equals 200 Hz downfield from TMS.

In an alternate (but less popular) procedure, the chemical shift is reported as the *tau value*, τ, equal to $10 - \delta$.

Pmr spectra are plotted with the magnetic field strength increasing to the right. This places TMS, which is highly shielded, at the extreme right of the spectrum with a *defined* chemical shift of 0.00 δ. If the induced magnetic fields influencing a given proton add to the applied field (*i.e.*, if $H_{eff} > H_0$), that proton will come into resonance at a value of H_0 slightly lower than would otherwise have been expected. Such a proton is *deshielded* and occurs *downfield* of the standard. The greater the deshielding, the larger will be the value of δ. Most signals occur downfield (by convention, to the left) of TMS. These and related terms are summarized in Figure 29–3.

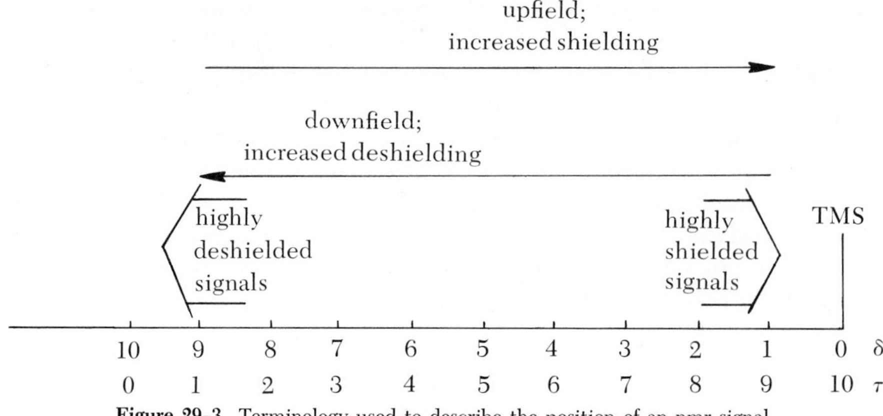

Figure 29-3 Terminology used to describe the position of an nmr signal.

Factors Controlling Chemical Shift. The effect of H_0 upon the electrons around a nucleus, and the concomitant shielding of the nucleus, has already been noted. It follows that the extent of shielding by the induced field should decrease as the electron density around the nucleus decreases. This prediction is supported by the effect of different halogens upon the resonance of the methyl protons of methyl halides. As we pass through the series of methyl iodide, methyl bromide, methyl chloride, and methyl fluoride, the inductive effect of the halogen causes the proton resonance to occur farther downfield with increasing electronegativity (Figure 29–4).

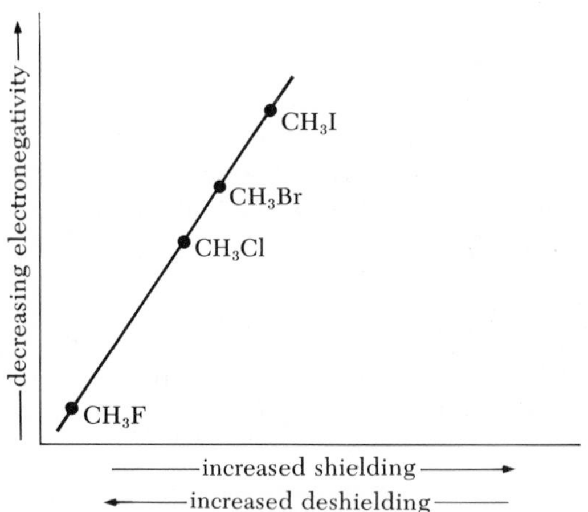

Figure 29-4 Influence of halogens on the chemical shift of methyl halides.

Silicon is less electronegative than is carbon and, consequently, methyl groups bonded to silicon are richer in electron density than are methyl groups bonded to carbon. The result is a *shielding* of Si—CH_3 protons which is sufficiently great that most organic compounds absorb downfield from this group. It is this shielding that helps to make TMS, $(CH_3)_4Si$, such a valuable nmr standard.

In addition to these through-bond *inductive effects*, H_0 induces unsymmetrical magnetic fields about bonds. The effect of these induced fields may be large or small; they act to shield or deshield a proton, operate through space, and are more important for π bonds than for σ bonds. Such effects are called *anisotropic effects*.

The anisotropic effects of multiple bonds are summarized in Figure 29–5. In this figure

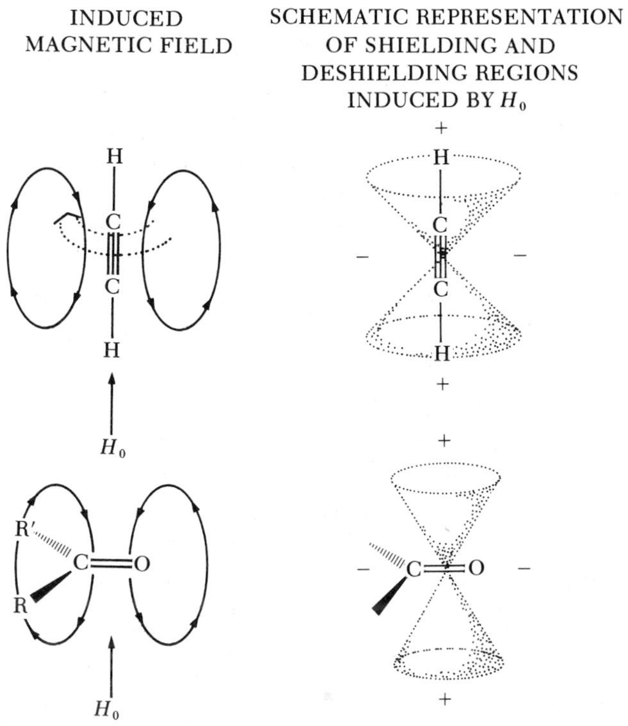

INDUCED MAGNETIC FIELD

SCHEMATIC REPRESENTATION OF SHIELDING AND DESHIELDING REGIONS INDUCED BY H_0

Figure 29-5 Long range shielding and deshielding effects ascribed to multiple bonds. The model of the carbon-carbon triple bond is valid for the carbon-nitrogen triple bond; that of the carbon-oxygen double bond is valid for the carbon-carbon double bond. The + regions are shielding, and the − regions are deshielding, to protons within those volumes.

each functional group is depicted relative to H_0 in a way that emphasizes the maximum sensitivity of the π system to H_0.

The benzene ring contains a large, closed loop of π electrons. When it is placed in a magnetic field, electronic circulation (here given the special name *ring current*) over the entire π system causes a powerful induced field which adds to H_0 in the region that includes the protons bonded to the ring (see Figure 14–5). Consequently, signals from protons on benzene rings occur quite far downfield, at about 7 ppm. This ring current effect is not limited to benzene but also is seen in the nmr spectra of other "benzenoid" molecules, such as naphthalene and pyridine (a nitrogen analog of benzene) (Figure 29–6).[a]

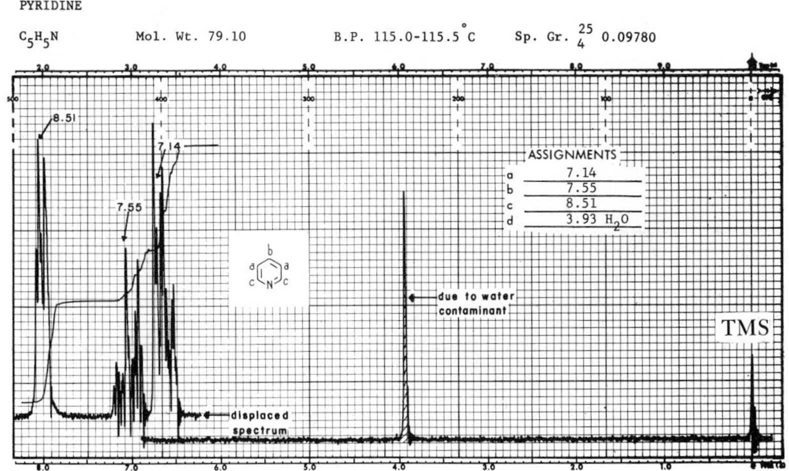

Figure 29-6 *Illustration continued on following page.*

[a] Review the discussion beginning on p. 404.

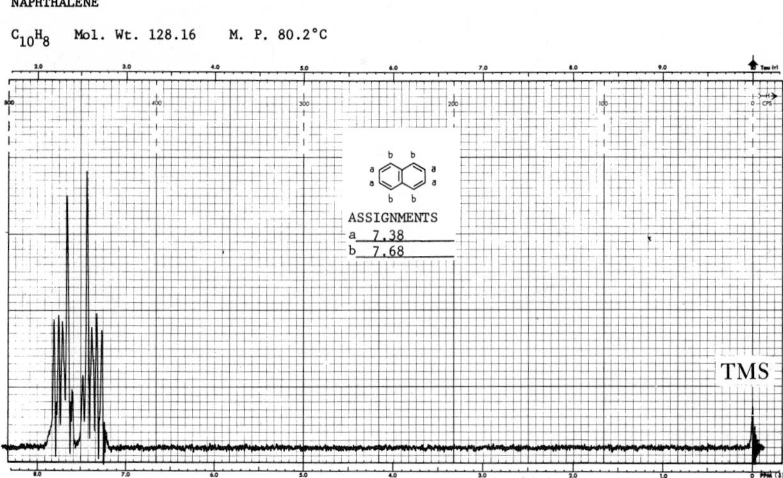

Figure 29-6 *Continued.* The nmr spectra of (a) pyridine and (b) napthalene. Note that a portion of the pyridine spectrum is "offset." This signal, which has a chemical shift of 8.51δ, would not be observed during a normal sweep (from 0.0δ to 8.3δ) of the spectrum. However, by adjusting the instrument so that 0.0δ on the chart really represents 0.4δ (*i.e.*, by *offsetting* the instrument), a signal that absorbs at 8.5δ will appear to absorb at 8.1δ. Even with rather simple instruments it is possible to offset a spectrum by 500 Hz (at 60 MHz), thus permitting the observation of virtually all types of protons. © Sadtler Research Laboratories, Inc., 1976.

Analyses of chemical shifts based upon anisotropic effects have helped to make distinctions between similar groups within one molecule. Figure 29–7 shows how three different methyl groups of α-pinene can be assigned in this way.

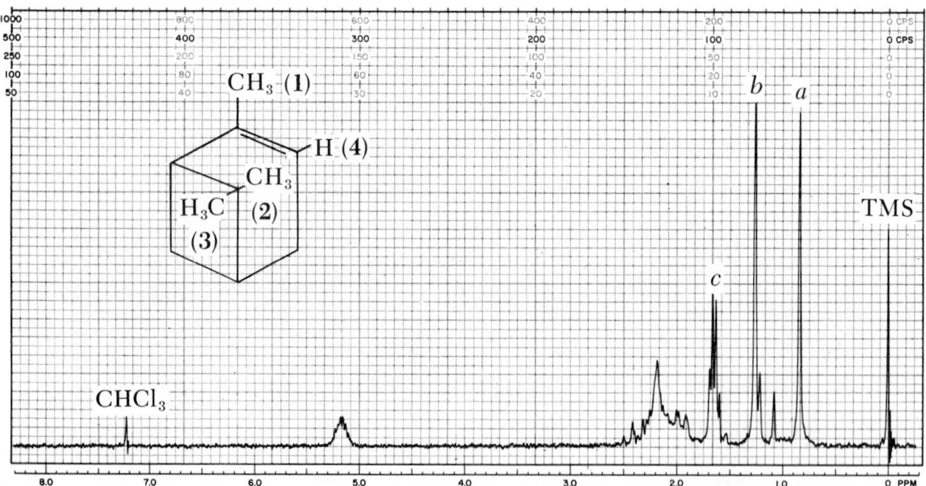

Figure 29-7 The nmr spectrum of α-pinene. Notice three methyl signals: *a*, 0.84δ; *b*, 1.27δ; and *c*, 1.65δ. The one furthest downfield, *c*, is assigned to the vinylic methyl group, **1**, because that group should be deshielded by the anisotropy of the double bond. Methyl group **2** should be shielded by the same double bond (build a model) and is assigned to *a*. Methyl group **3**, being relatively unaffected by the double bond, is shielded more than **1** and less than **2**; it is assigned to *b*. Methyl **1** is split (see subsequent text discussion) because of coupling to the vinylic proton (**4**). Can you deduce the operating frequency of the spectrometer used to obtain this spectrum?

The chemical shifts of representative groupings are presented in Table 29–1. The fact that chemical shifts can be tabulated without footnoting a myriad of exceptions indicates that they are fairly insensitive to changes in solvent or to the presence of impurities. However,

TABLE 29-1 Chemical Shifts of Proton Groupings*

Group	Shift, δ ppm
Methyl Groups CH_3—	
CH_3—R	0.8–1.2
CH_3—CR=C<	1.6–1.9
CH_3—Ar	2.2–2.5
CH_3—C(=O)—R	2.1–2.4
CH_3—C(=O)—Ar	2.4–2.6
CH_3—C(=O)—OR	1.9–2.2
CH_3—C(=O)—OAr	2.0–2.5
CH_3—N<	2.2–2.6
CH_3—OR	3.2–3.5
CH_3—OAr	3.7–4.0
CH_3—OC(=O)R	3.6–3.9
Unsaturated Groups	
RCH=C<	5.0–5.7
Ar—**H**	6.0–7.5
R—C(=O)—**H**	9.4–10.4

Group	Shift, δ ppm
Methylene Groups RCH_2—	
R—CH_2—R	1.1–1.5
R—CH_2—Ar	2.5–2.9
R—CH_2—C(=O)—R	2.5–2.9
R—CH_2OH	3.2–3.5
R—CH_2OAr	3.9–4.3
R—CH_2—OC(=O)R	3.7–4.1
R—CH_2—Cl	3.5–3.7
Methine Groups R_2CH—	
R_3CH	1.4–1.6
R_2CHOH	3.5–3.8
Ar_2CHOH	5.7–5.8
Other Groups	
ROH	3–6
ArOH	6–8
RCO_2H	10–12
RNH—	2–4

* In this table, R = saturated carbon (CH_3—, —CH_2—, —CH<, —C<); Ar = aromatic ring.

(From Moore, J.A.: *Elementary Organic Chemistry*. Philadelphia, W.B. Saunders Company, 1974.)

when one adds a small amount of a paramagnetic complex of a trivalent rare earth metal [*e.g.*, a complex between the 2,2,6,6-tetramethyl-3,5-heptanedionato ligand, abbreviated "dpm," and europium—Eu$(dpm)_3$], a great change occurs in the chemical shifts of protons near electronegative substituents (*e.g.*, —NH_2, —OH). These protons are most affected because association between the organic substrate and the complex occurs *via* these groups.

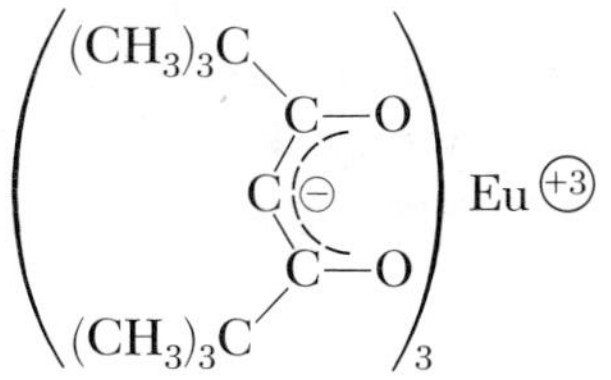

$((CH_3)_3C$—C(—O)=C$^{\ominus}$—C(—O)—$C(CH_3)_3)_3$ Eu^{+3}

Eu$(dpm)_3$
also called
tris-(dipivalomethanato)
europium

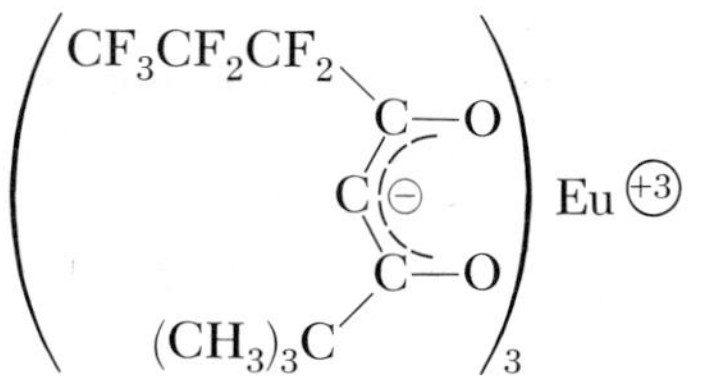

$(CF_3CF_2CF_2$—C(—O)=C$^{\ominus}$—C(—O)—$C(CH_3)_3)_3$ Eu^{+3}

Eu$(fod)_3$
also called *tris*-1,1,1,2,2,3,3-heptafluoro-7,7-dimethyl-3,5-octanedionato europium

common chemical shift reagents

Complexes of Eu, Yb, Er, and Tm move the signals downfield, while those of Pr, Nd, Sm, Tb, Dy, and Ho move the signals upfield. Of all of these, Eu and Pr complexes are the most widely used. These "inorganic" complexes, commonly called *chemical shift reagents,* sometimes can transform a cluttered, unresolvable spectrum into one that is easily analyzed by computer or, in simple cases, by inspection (Figure 29–8).

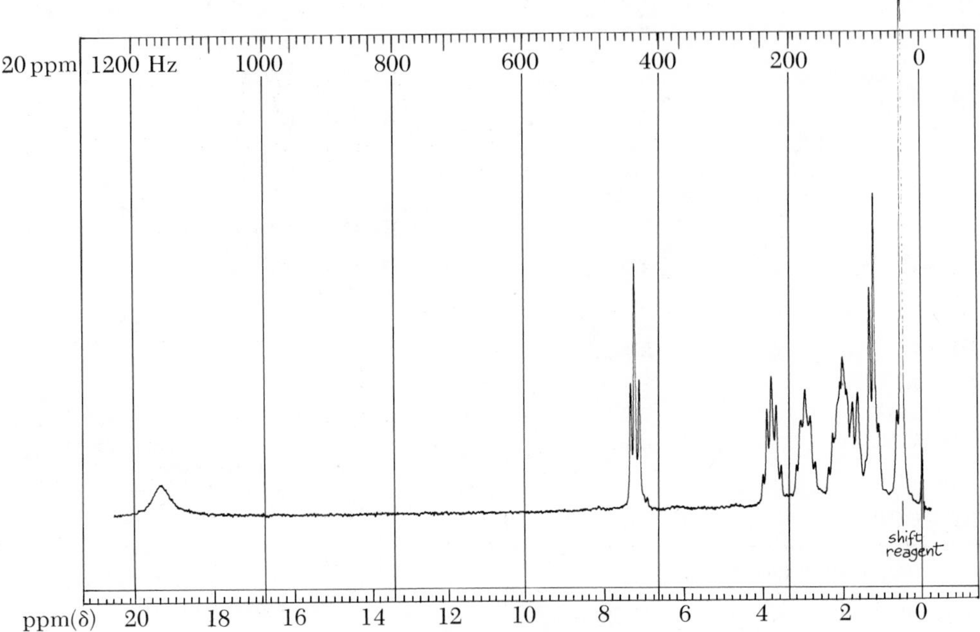

Figure 29–8 The nmr spectrum of 1-hexanol in the presence of a chemical shift reagent [Eu(fod)$_3$].

29.4 SPIN-SPIN COUPLING

At this point one might imagine the pmr spectrum of ethyl iodide to consist of two lines: an upfield signal (3H) and a downfield signal (2H),[a] representing the methyl and methylene groups, respectively. (The methylene hydrogens are expected to occur downfield of the methyl hydrogens because of the inductive effect of the iodine.) In fact, the spectrum consists of seven distinct lines—a cluster of three upfield (*total* intensity 3H) and a cluster of four downfield (total intensity 2H) (Figure 29–9). These are termed a "triplet" and a "quartet," respectively. The increase in the multiplicity of signals in the nmr is a very common observation, ascribed to *mutual* **spin-spin coupling** (also called *spin-spin splitting*) between nuclei.

While usage varies, in this text we shall limit the use of the word "line" when describing nmr spectra to a single absorption peak on the chart. When we speak of the *resonance* or *signal,* or the *absorption due to a certain group,* we may be speaking either of a singlet or of a cluster of lines making up a multiplet of some kind.

---

[a] The symbol in parentheses is used to indicate the relative intensity of the resonance under discussion. Intensity refers to the area under a peak, not to the height of the peak. In nmr, the intensity of a signal is proportional to the number of protons producing that signal.

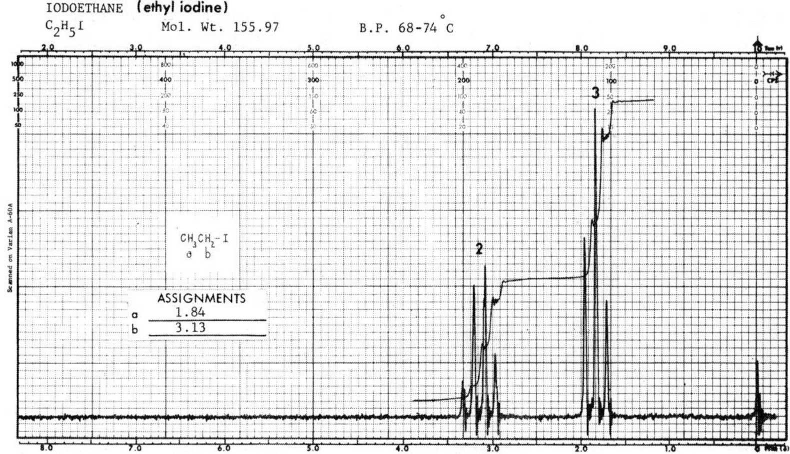

Figure 29-9 The nmr spectrum of ethyl iodide. Note that the multiplets are not perfectly symmetric but, instead, *lean* toward one another. This is indicative of signals that are splitting each other. © Sadtler Research Laboratories, Inc., 1976.

What Is the Origin of Spin-Spin Splitting? Recall that the field experienced by a proton is modified by induced magnetic fields. One of these local effects is the small field set up by the circulation of electrons around an adjacent proton. The magnetic field produced by proton H_a (refer to Figure 29–10) can be either aligned with H_0 or opposed to H_0; it will be one way in some atoms and the other way in the rest of the atoms, and this will result in two different effective fields at proton H_b, one slightly greater and one slightly less than H_0. (Actually, this is a major over-simplification, since coupling involves the electrons in the intervening bonds and is transmitted through bonds rather than through space.) As a result of this, the signal corresponding to H_b appears as two lines, *i.e.*, a doublet. Since splitting is a mutual undertaking, a similar argument leads to the conclusion that H_a will also appear as a doublet, being split by H_b.

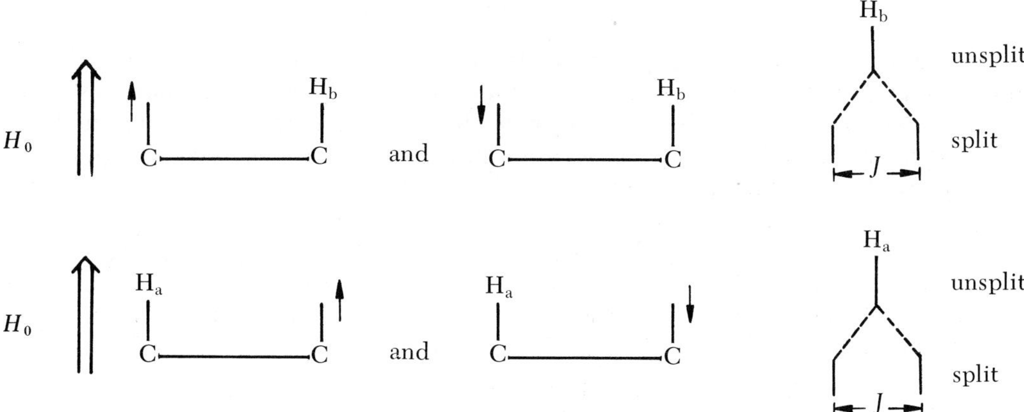

Figure 29-10 A simplified picture of the production of spin-spin splitting in vicinal protons. The actual mechanism of spin transmission involves the σ electrons. *J*, the "coupling constant," measures the extent of interaction between the two nuclei.

We noted earlier that we can measure the area under the peaks in an nmr spectrum. The ratio of these areas reflects the relative numbers of protons giving rise to each resonance signal. For example, if the areas under the peaks corresponding to H_a and H_b in Figure 29–10 could be measured before splitting (an impossibility), they would be in the ratio of 1:1. Splitting does not change the *total* area under each doublet. However, that area is now distributed among both halves of the doublet, so that each half has an intensity of one-half. You must remember the following: THE AREA UNDER A MULTIPLET WILL REFLECT THE NUMBER OF PROTONS GIVING RISE TO THAT SIGNAL, WHILE THE MULTIPLICITY OF THE SIGNAL WILL REFLECT THE NUMBER OF PROTONS SPLITTING THE PROTON(S) WHOSE RESONANCE IS BEING DISCUSSED.

The degree of multiplicity (the number of lines within a multiplet) and relative intensities of the lines within a multiplet are derived from the various probabilities of nuclear spin alignment with H_0. An analysis of the splitting pattern for the ethyl group is presented in Figure 29–11.

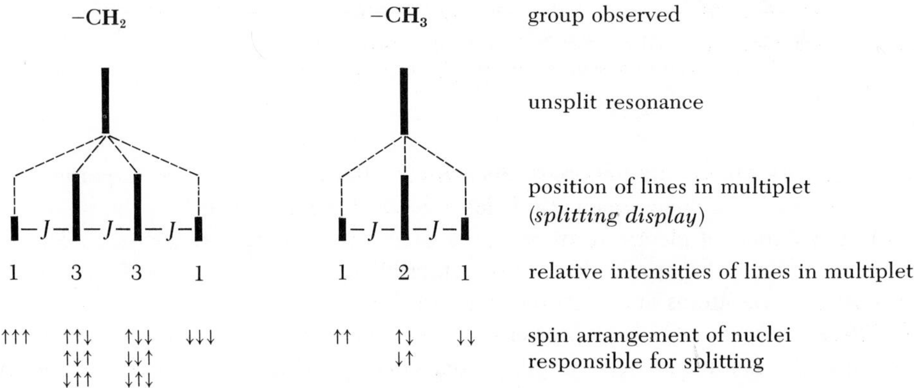

Figure 29–11 Spin-spin splitting diagram for the ethyl group. The arrows under the methylene group depict the spin arrangements (relative to H_0) that can be taken by the methyl hydrogens; arrows under the methyl group indicate the spin arrangements that can be taken by the methylene hydrogens. Arrangements of CH_3 such as ↑↑↓, ↑↓↑, and ↓↑↑ produce the same net field at CH_2 (they all have a net spin of ↑) and will cause an inner line of the quartet to be three times as intense as an outer line. This detailed analysis of spin combinations leads to the representations immediately above it; the latter (so-called "splitting diagrams") are commonly used to represent coupling interactions.

It can be shown that if one group of n protons (set A) splits n' protons of set B, the resonance of A will consist of $n' + 1$ lines and the resonance of set B will consist of $n + 1$ lines. Any line in either multiplet will be separated from an adjacent line in the same multiplet by the same number of Hz. The intensities of the lines within such a multiplet follow a simple pattern, one that is summarized in Figure 29–12.

NUMBER OF EQUIVALENT NUCLEI RESPONSIBLE FOR SPLITTING	MULTIPLICITY OF OBSERVED SIGNAL	RELATIVE INTENSITY AND LINE DISTRIBUTION WITHIN OBSERVED MULTIPLET
0	singlet	1
1	doublet	1 1
2	triplet	1 2 1
3	quartet	1 3 3 1
4	quintet (pentet)	1 4 6 4 1
5	sextet	1 5 10 10 5 1
		δ

Figure 29–12 Relative intensities and line distribution in multiplets derived from spin-spin splitting of one group of equivalent nuclei by a second group of equivalent nuclei. Note that only when the multiplet contains an odd number of lines does a line appear at the actual chemical shift of the proton being observed.

The Coupling Constant. The measure of the degree of interaction between nuclei, the *spin-spin splitting constant* (also called the *coupling constant*), *J*, is independent of H_0 and so is reported in Hz rather than in ppm. This is very important, since one can distinguish, for example, between two singlets and a doublet simply by recording the spectrum at two different radio frequencies. If the separation, in Hz, between the lines does not change, then the signal is a doublet. However, if the line separation increases with increasing frequency, then "the signal" is, in fact, two singlets.

The magnitude of *J* is dependent upon several factors, including the geometry between the coupled nuclei and the number of intervening bonds. Coupling is usually not observed between protons separated by more than three single bonds. Intervening multiple bonds may increase the number of bonds through which coupling may be observed.

That there is a geometric constraint upon the magnitude of *J* can be seen from the data

TABLE 29-2 Representative Coupling Constants

Group	J(Hz)	Group	J(Hz)
$H_2C<$ (H–C–H)	12–18	H–C–C(=O)H	1–3
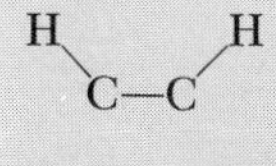 H–C–C–H	2–10[a]	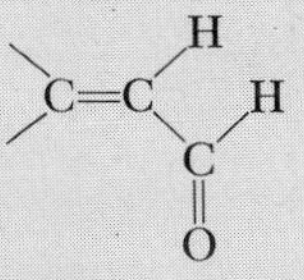 >C=C(H)–C(=O)H	6–8
H–C–(C)$_n$–C–H	0		9–10
	3		3–4
	5–10		0–2
	3		
>C=C(H)H	1–3		
H–C=C–H (cis)	7–12		
H–C=C–H (trans)	13–18		
HC–C≡C–H	3		

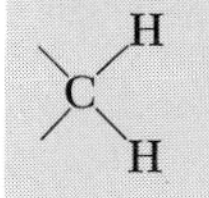

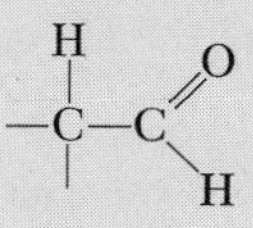

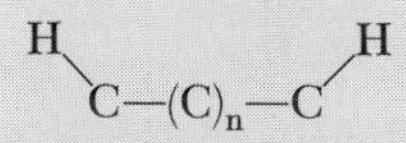

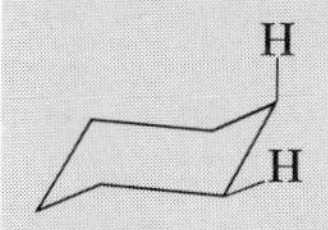

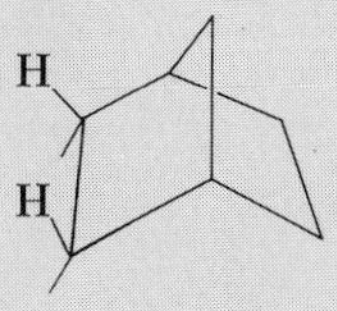

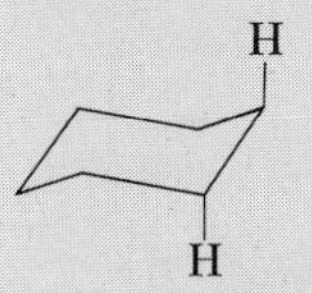

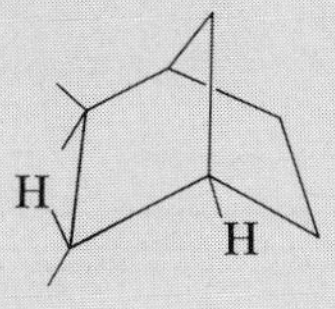

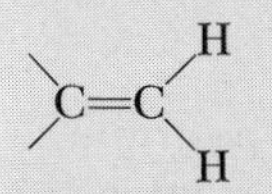

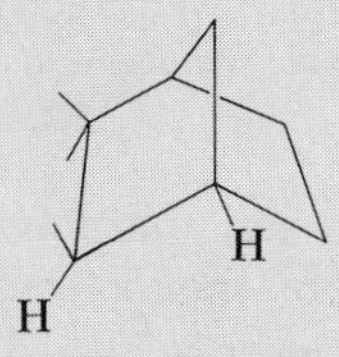

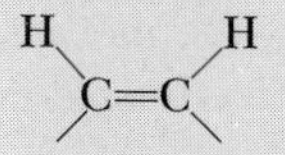

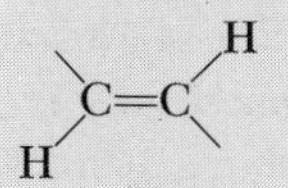

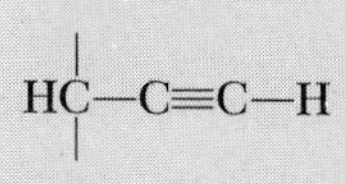

[a] $J \approx 7$ Hz with free rotation.

in Table 29–2. With relatively free rotation, *vicinal* (*vic*) protons (**H**—C—C—**H**) have a J of about 7 Hz. This represents an averaged J value, weighted to consider the populations of the various conformers and the J's associated with each.

$$J_{obs} = n_a J_a + n_b J_b + n_c J_c \cdots n_i J_i$$

where n = mole fraction of a given conformer
J = *vic*-coupling constant in that conformer

Splitting Patterns. Some skeletal fragments are easily recognized in the nmr—a *t*-butyl group appears as a sharp, intense, single line near 0.9 δ; the isopropyl group appears as an upfield doublet (CH_3, 1.2 δ, J = 7 Hz) and a downfield heptet ($(CH_3)_2CH$—, 4.0 δ, J = 7 Hz); the ethyl group has already been discussed. Isolated methyl groups (*e.g.*, —OCH_3) occur as single lines with δ dependent upon the specific molecular environment (Table 29–1).

Once we pass beyond these simple patterns, more complex multiplets sometimes can still be analyzed with the aid of simple splitting diagrams. Some multiplets, on the other hand, can be analyzed only with the assistance of computers. An illustration of a complex spectrum that can be analyzed without the aid of a computer is that of 1-nitropropane, $CH_3CH_2CH_2NO_2$, shown in Figure 29–13.

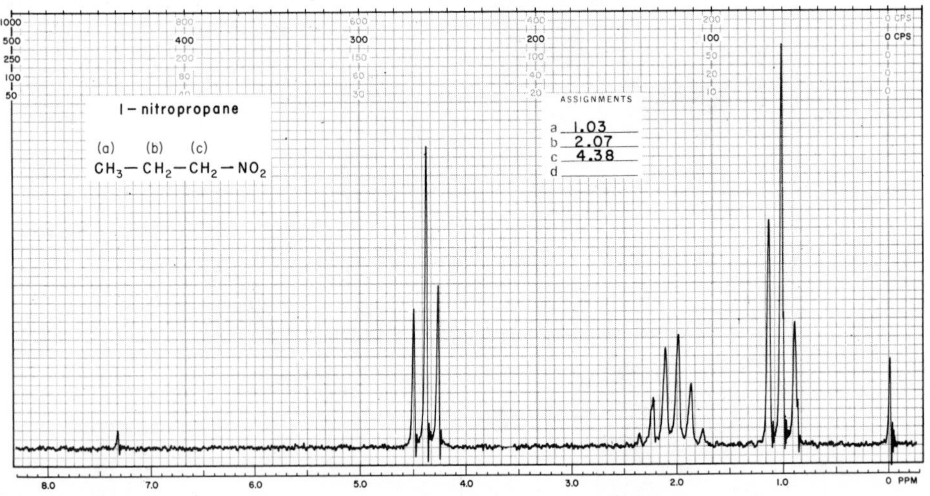

Figure 29–13 The nmr spectrum of 1-nitropropane. The solvent was $CDCl_3$ containing a trace of $CHCl_3$. TMS was added to the solution to provide the chemical shift reference.

The methyl group and the downfield methylene group appear as triplets, which are expected from splitting by the molecule's central methylene group. This latter group appears as a multiplet, which can be analyzed by considering splitting first by the methyl group and then by the methylene group bearing the nitro substituent. This analysis has been done in Figure 29–14.

Because of the similarities in the coupling constants (see Figure 29–14), a multiplet that might otherwise contain twelve lines appears as a somewhat broadened sextet. (This, then, resembles a signal split by five equivalent nuclei.) If coupling constants were significantly different, the pattern would be more complex but could still be analyzed by a cascade of splitting diagrams.

The three methyl protons of 1-nitropropane do not appear to split one another because quantum mechanical restrictions prevent the observation of splitting between protons with the same chemical shift. You might do well at this point to re-examine our analysis of the

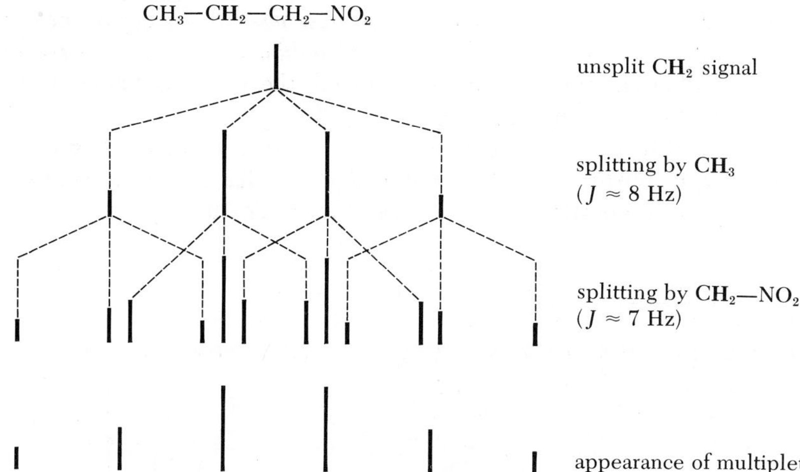

Figure 29-14 Analysis of the splitting pattern of the central methylene multiplet in 1-nitropropane.

spectrum of ethyl iodide and note that we never considered any extra lines appearing because the methyl hydrogens split one another. It should now be clear that since both methylene protons in ethyl iodide have the same environment, just as do the three methyl protons, they must not split one another. Furthermore, if you examine Figure 29–6 you will see that the methyl resonance of tetramethylsilane is a strong *singlet!* Nine protons, all with the same chemical shift, must give rise to only one line, and that is what makes a *t*-butyl group so easily recognized in the nmr.

Spin Decoupling. Two nuclei that are coupled may be *decoupled* by irradiating the

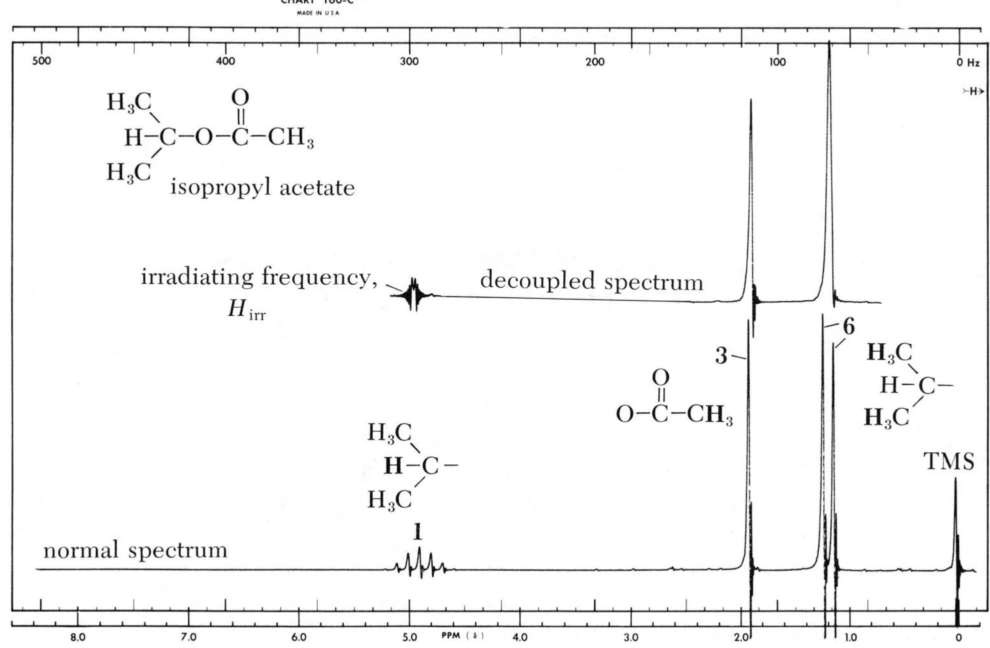

Figure 29-15 Decoupling of the isopropyl methyl hydrogens by irradiation of the methine signal. Note that one methyl group $\left(—O—\overset{O}{\overset{\|}{C}}—CH_3\right)$ of isopropyl acetate is not split in either spectrum, but that the isopropyl methyl groups become unsplit (upper spectrum) when the center of the methine signal is irradiated.

sample with the radio frequency corresponding to the resonance frequency of one of the nuclei. When a spectrum is scanned under these *double irradiation* conditions, the irradiated signal becomes obliterated but the other signal is simplified, the coupling between them having disappeared. This is illustrated in Figure 29–15.

Since protons and deuterons couple only weakly, one can achieve a similar effect by examining a compound in which one of the coupled protons has been replaced by deuterium. The result of a deuterium substitution is shown in Figure 29–16.

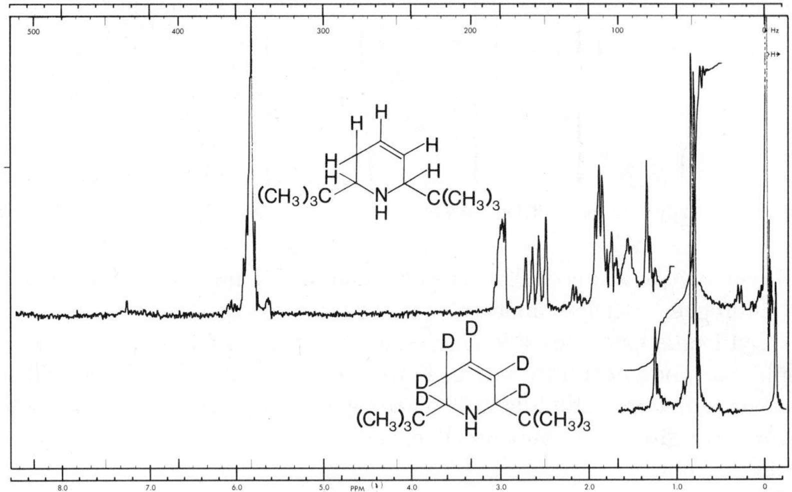

Figure 29–16 Simplification as a result of deuterium substitution. The upper trace displays a wide variety of alkyl and vinyl resonances. Use of deuterated pyridine, rather than "ordinary" pyridine, in the reaction produced a product containing only *t*-butyl and N—H resonances. (Courtesy of Dr. R. F. Francis.)

29.5 HOW SYMMETRY REARS ITS HEAD IN ORDER TO COMPLICATE NMR SPECTRA

Perhaps the thing about nmr spectra that most confuses and angers the student is the appearance of "all those lines." Because of the depth of mathematical training that is necessary, we cannot hope to explain fully why fairly simple molecules often have very complex spectra. For example, with only five protons, pyridine gives rise to a spectrum containing well over a dozen clearly defined lines (see Figure 29–6). However, if we pause long enough to examine the degree of "sameness" or "difference" possessed by protons within a molecule, we will be able to employ tables that help to predict the maximum number of lines that a multiplet may contain. Chapter 4 served to introduce some basic ideas of molecular symmetry, and the forthcoming discussion is an extension of Chapter 4 into the area of nmr spectroscopy. The object is to permit you to determine whether an nmr spectrometer will recognize two protons, or groups of protons, as being the same or as being different (in the nmr sense).

Nmr equivalence or non-equivalence of various protons hinges upon two nmr parameters, chemical shift and coupling constant. In order for two protons to be *identical* in the nmr experiment, they must (a) have the same chemical shift and (b) be spin coupled equally to all other nuclei in the molecule. Protons with the same chemical shift are said to be **chemical shift equivalent** or **isochronous.** Protons with the same coupling constants are said to be **coupling constant equivalent.** We will now examine these two aspects of *magnetic* equivalence, looking particularly for those instances in which equivalence is absent, since those are the circumstances that lead to increased multiplicity.

Chemical Shift Non-Equivalence. In our analysis of the nmr spectrum of 1-nitropropane we observed that the methyl protons do not split one another, and that this is one example of the precept that protons with the same chemical shift do not appear to split one another. What remained unsaid at that time is equally important; that is, protons that do not have the same chemical shift *do* appear to split one another. (Normally, such protons must be separated by three or fewer bonds in order for the splitting to be detectable.)

How do you recognize chemical shift *non*-equivalent protons? It can be done rather easily, if you will review pages 110 to 112. Chemical shift non-equivalent protons, sometimes called *anisochronous* protons, are diastereotopic, and *vice versa.* Protons that are enantiotopic or identical usually have the same chemical shift.

One of the classic types of chemical shift non-equivalence is that which is exhibited by two protons, or groups of protons, that are rendered diastereotopic by the proximity of an achiral center. The most common examples of this include methylene groups and isopropyl groups adjacent to a chiral center (**F** and **G,** respectively).

Y H_2 — CH_3 X

X—C—C—R *diastereotopic* H—C——C—Y

Z H_1 — CH_3 Z

F **G**

The chemical shift non-equivalence of the methylene protons in **F** or the isopropyl methyl groups in **G** is a phenomenon that is independent of the conformational distribution within **F** and **G.** Enantiomers have identical nmr spectra in achiral solvent (and the solvents commonly used in nmr spectroscopy are achiral); therefore, the following discussion is equally valid for a single enantiomer or for a racemic modification.

Using **F** as our example, we can see from the equations below that the chemical shifts that are observed for the two methylene protons, δH_1^{obs} and δH_2^{obs}, are weighted averages of the chemical shifts of these protons. It can be argued that, because of the different environments in which the protons are found, the chemical shift of H_1 in conformer *a*, δH_1^{a}, does not equal the chemical shift of H_2 in conformer *a*, δH_2^{a}. In conformer *a*, H_1 is gauche to X and Z, while H_2 is gauche to X and Y. Furthermore, when you study the one conformer (*c*) in which H_2 *is* gauche to X and Z, you find that R is gauche to X and Y, rather than to Z and Y as it is in conformer *a*. This means that conformers *a* and *c* do not really put H_1 and H_2 in *identical* environments between X and Z, because the orientation of the remaining groups has changed in going from *a* to *c*. A similar argument can be constructed for H_1 and H_2 skew to Z and Y *and* to Y and X. Even very rapid rotation around the C—C bond will not make H_1^{obs} equal H_2^{obs}.

X R

Y—C—C—H_2 an eclipsed conformation of **F**

Z H_1

the three conformers of **F**

$$\delta H_1^{obs} = n_a \delta H_1^{a} + n_b \delta H_1^{b} + n_c \delta H_1^{c}$$
$$\delta H_2^{obs} = n_a \delta H_2^{a} + n_b \delta H_2^{b} + n_c \delta H_2^{c}$$
$$n_a \neq n_b \neq n_c$$

where n = mole fraction of conformer indicated by subscript
δH_1 = chemical shift of H_1 in conformer indicated by superscript
δH_2 = chemical shift of H_2 in conformer indicated by superscript
δH^{obs} = observed chemical shift of proton indicated by subscript

We must conclude that only the most fortuitous of conditions can make δH_1^{obs} equal δH_2^{obs}. When this does happen, the situation is described as "accidental" equivalence (p. 975).

Coupling Constant Non-Equivalence. Diastereotopic protons always exhibit coupling constant non-equivalence. An important example can be developed by beginning with the protons of a methylene group adjacent to a chiral center. Having just established that these have different chemical shifts, and since they are separated by only two σ bonds, we can conclude that the methylene protons will split one another—and, indeed, they do. However, the resultant signal for the methylene group appears as a four-line multiplet, commonly called an *AB quartet* but more correctly called a *pair of AB doublets*. None of these four lines correspond to the chemical shift of either proton; these values must be obtained from calculations based upon the observed spectrum (see problem 16).

The interaction of methylene protons with one another is so great that the separation of the four lines and the relative intensities of the inner lines *vs.* the outer lines in a pair of AB doublets is a function of the coupling constant between the two protons, J_{AB}, and the difference in the chemical shifts of H_A and H_B. The variation that can be anticipated as a function of a change in these two parameters is shown in Figure 29–17.

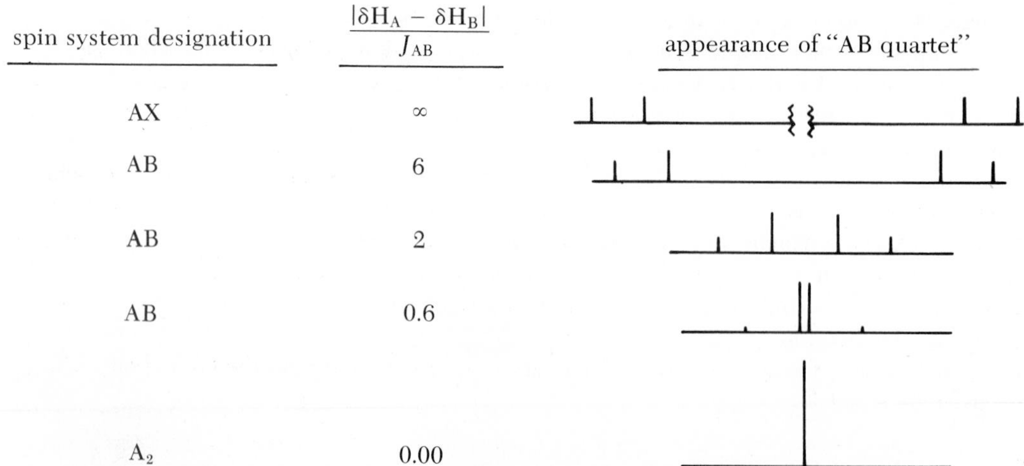

spin system designation
$\frac{|\delta H_A - \delta H_B|}{J_{AB}}$
appearance of "AB quartet"
AX
∞
AB
6
AB
2
AB
0.6
A_2
0.00

Figure 29–17 The dependence of an "AB quartet" upon the nmr parameters of H_A and H_B. The respective chemical shifts are δH_A and δH_B; the mutual coupling constant is J_{AB}. (See Section 29.6.) The systems AX and A_2 (*i.e.*, two identical nuclei) are considered "first order" systems.

Our discussion of coupling constant non-equivalence can now be entered by referring back to structure **F** and replacing one of the ill-defined substituents on the chiral carbon with a proton, H_3. Protons H_1 and H_2 are vicinal to H_3 and will couple to it. However, the coupling between H_1 and H_3 ($J_{1,3}$) will not be identical to the coupling between H_2 and H_3 ($J_{2,3}$) and, as a result, each of these will split H_3 differently. *Since H_1 and H_2 are coupled differently to a third proton (H_3), they are coupling constant non-equivalent.*

```
  X  H1              H3 H1
  |  |               |  |
Y—C—C—R            Y—C—C—R
  |  |               |  |
  Z  H2              Z  H2
```

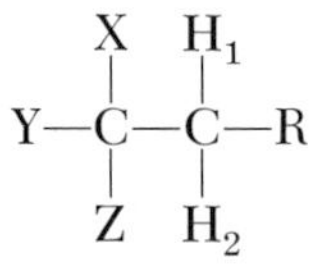

F

H_1 and H_2 anisochronous

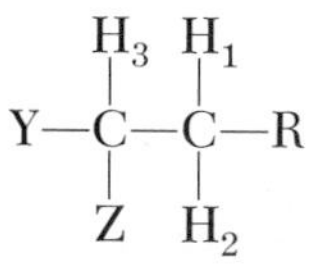

H_1 and H_2 anisochronous *and* coupling constant non-equivalent $J_{1,3} \neq J_{2,3}$

Another illustration of coupling constant non-equivalence is found in monosubstituted benzene derivatives, C_6H_5—R. In this structure the protons that are adjacent to the substituent, the *ortho* protons (H_o), are in the same environment and, therefore, have the same chemical shift. However, they are disposed differently relative to the reference proton shown, H_m, and are, therefore, coupling constant non-equivalent. Thus, it *is* possible to have protons that are chemical shift equivalent but coupling constant non-equivalent. This situation leads to more spin-spin splitting than would have been imagined had only chemical shift equivalence been considered.

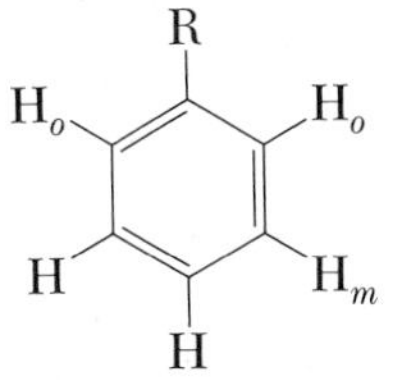

R; H_o, H_o; H, H_m; H

a monosubstituted benzene

both H_o's have the same chemical shift but are coupling constant non-equivalent

The extent to which this coupling constant non-equivalence will split open a spectrum into many lines depends, in part, upon the electronegativity of the substituent on benzene. This point will be illustrated later in this chapter.

29.6 A NOTE ON SYMBOLS

Protons with different chemical shifts can be represented by letter designations, the most downfield being "A," the next most downfield "B," and so on. Once all of a group of protons of similar chemical shifts has been named (A, B, C, . . .), the next higher field group is labeled beginning with "X" (*i.e.*, X, Y, Z). Since the methylene protons adjacent to a chiral center have only *slightly* different chemical shifts, they represent an AB system, not an AX system.

This type of nomenclature is not used to describe all of the protons in a molecule at one time, just those interacting protons that are under discussion at a given time. For that reason, one will often encounter phrases like ". . . the AB portion of the spectrum . . ."

Chemical shift equivalent protons exhibiting coupling constant non-equivalence receive identical letter designations, since the letter refers only to chemical shift. However, to distinguish coupling constant non-equivalence, the letter is "primed." This is illustrated below for the AA′BB′ and AA′XX′ systems.

```
     R   HA  HA'  R              HA      HA'
     |   |   |    |                \    /
S—C——C——C——C—S                  C
     |   |   |    |                  ||
     T   HB  HB'  T                  C
                                    /    \
                                  FX      FX'
         1                          2
  AA′BB′ spin system        AA′XX′ spin system
```

In **1**, A and A′ have identical chemical shifts but couple differently to B (or B′), while B and B′ have identical chemical shifts but couple differently to A (or A′). In **2**, A and A′ have identical chemical shifts but couple differently to either fluorine. (^{1}H and ^{19}F do split one another in the same way that ^{1}H and ^{1}H split one another, except that J is larger for the former couplings.)

The main value of this system is that one can predict the *maximum* number of lines that a given group of protons will exhibit in the nmr by properly assigning the spin system and by using the information in Table 29–3.

TABLE 29-3 Multiplicity as a Function of Spin System

Spin System	Number of Lines		
	Total	*A*	*B or X*
A_3	1	1	—
A_4	1	1	—
A_2X_2	6	3	3
A_2B_2	14	7	7
AA′XX′	20	10	10
AA′BB′	24	12	12

An Example. 2-Iodopropanoic acid represents an A_2B_2 spin system. While one might have naively expected its two methylene groups to appear as two triplets, Figure 29–18 reveals

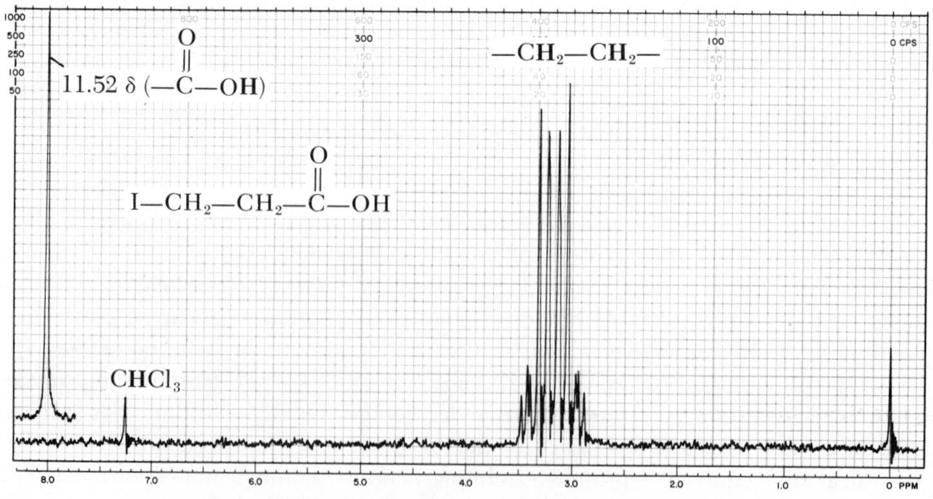

Figure 29-18 The nmr spectrum of 2-iodopropanoic acid.

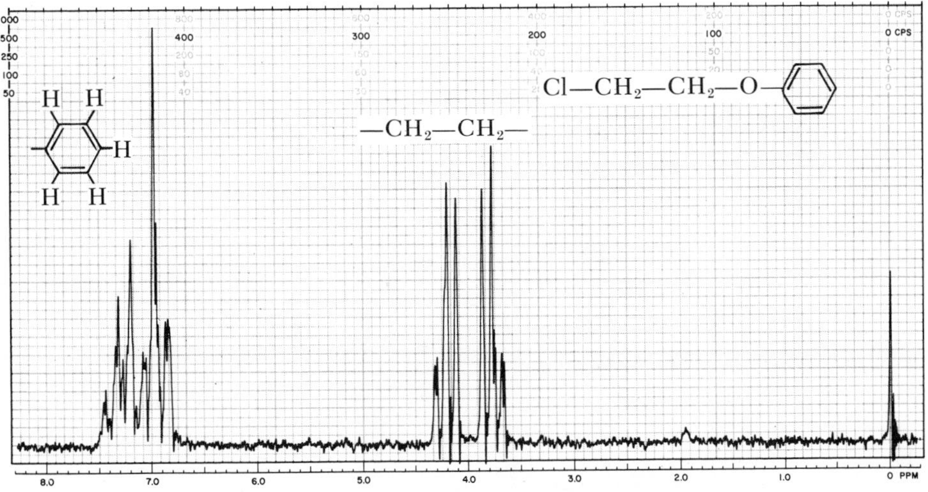

Figure 29-19 The nmr spectrum of β-chlorophenetole.

this not to be so. The A_2B_2 system is quite complex, and its appearance is not the same in all A_2B_2 systems. The ethano region (—CH_2CH_2—) of the nmr spectrum of β-chlorophenetole is also an A_2B_2 region. Its spectrum, displayed in Figure 29–19, is not the same as that of 2-iodopropanoic acid.

A_2B_2 over I—CH_2—CH_2—C(=O)—O—H

2-iodopropanoic acid

C_6H_5—O—CH_2—CH_2—Cl, with A_2B_2 over the —CH_2—CH_2— group

β-chlorophenetole

If the chemical shifts of the two methylene groups had been quite different; that is, if these two molecules had been A_2X_2 systems, then they both would have spectra containing two sets of triplets.

29.7 EXCHANGE PROCESSES

Conformational Exchange. The nmr spectrum of dimethylformamide, $HC(O)N(CH_3)_2$, shows two signals for the methyl groups. Their relative intensities compared to that of the remaining hydrogen, and the lack of splitting of that lone hydrogen, indicates that each upfield line represents one methyl group.

As the temperature of the sample in the spectrometer increases, these two sharp lines broaden, coalesce, and finally, at approximately 165°, produce one sharp line. This change, shown in Figure 29–20, occurs because at room temperature the nmr spectrometer "sees" two different methyl groups but at higher temperature only a single, average resonance spectrum is recorded. The reason for the observation of two lines at room temperature and one line at elevated temperatures is, on the molecular level, the restricted rotation around the C—N bond (see below). When rotation is slow, the nmr senses one methyl group *cis*, and one methyl group *trans*, to the lone hydrogen. At elevated temperature, rotation is more rapid and the instrument records only the average environment. It is important to note that bond rotation

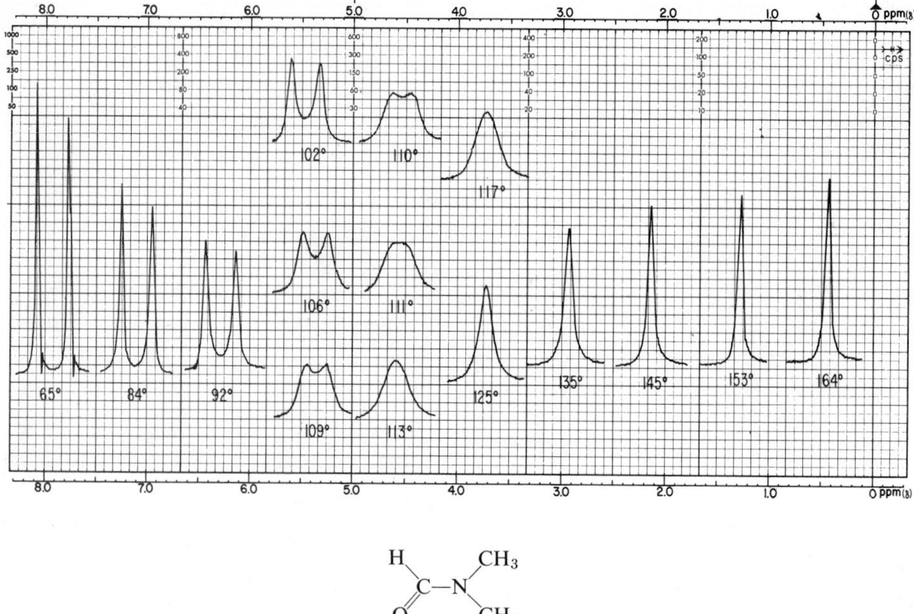

Figure 29-20 The temperature dependence of the methyl resonance(s) of dimethylformamide (DMF). Sweep width is 250 Hz. (From John R. Dyer, *Applications of Absorption Spectroscopy of Organic Compounds*, © 1965. Reprinted by permission of Prentice-Hall, Inc., Englewood Cliffs, N.J.)

does not have to cease in order for nmr to detect two different methyl groups; it simply has to decrease to any speed less than what might be called the "shutter speed" of the instrument.

$$H-\overset{O}{\overset{\|}{C}}-N(CH_3)_2 \quad \text{(DMF)} \qquad \underset{O}{\overset{H}{}}C\leftrightarrows N\underset{CH_3(2)}{\overset{CH_3(1)}{}} \rightleftarrows \underset{O}{\overset{H}{}}C\leftrightarrows N\underset{CH_3(1)}{\overset{CH_3(2)}{}}$$

DMF — N is sp^2

Cyclohexane and its Derivatives. At room temperature, the nmr spectrum of cyclohexane consists of only one line (2.44 δ), although both axial and equatorial hydrogens are present (and should give rise to different signals, since they are in different environments). The reason for the presence of only a single line is that at room temperature the rate of chair-chair interconversion is too rapid to permit detection of individual species. Only a time-averaged spectrum is recorded, and axial and equatorial protons both have the same average environment.

H_a, H_e $\rightleftarrows$ $H_{e'}$, $H_{a'}$

$H_e \rightarrow H_{a'}$
$H_a \rightarrow H_{e'}$

Similarly, even though chlorocyclohexane exists as a mixture of *a* and *e* conformers, the nmr records only a single C1—**H** resonance. (This resonance is split because of coupling to adjacent non-equivalent protons.) The observed chemical shift at room temperature is the weighted average of the chemical shifts of C1—H in the individual conformers.

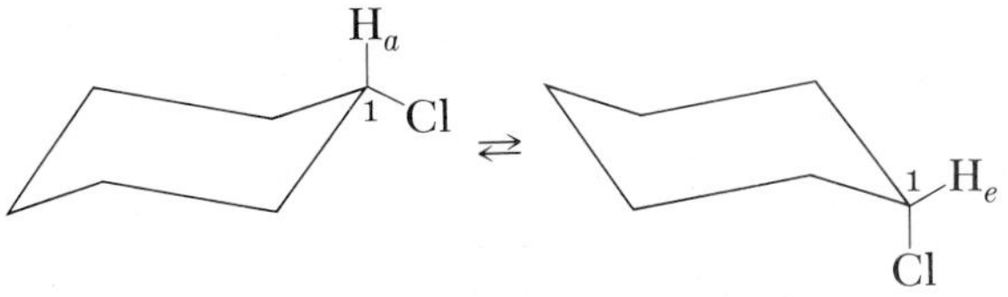

$$\delta^{obs}_{C1-H} = n_a\delta^a_{C1-H_a} + n_e\delta^e_{C1-H_e}$$

The mole fractions of the equatorial and axial forms could be calculated if $\delta^a{}_{C1-H_a}$ and $\delta^e{}_{C1-H_e}$ were known. They can be obtained by using model compounds in which the conformations must be fixed. Useful models for this type of work are the 1-substituted 4-*t*-butylcyclohexanes. The *t*-butyl group demands the *e* position, and the position (*a vs. e*) of C1—H is then controlled simply by the nature of the isomer under study (*cis* or *trans*). The *cis* compound provides the chemical shift of C1—H_e and the *trans* isomer provides the chemical shift of C1—H_a. (You must also remember that the sum of the mole fractions of all forms must equal unity, *i.e.*, $n_a + n_e = 1$.)

Chemical Exchange. We have already seen how, by rotation around a single bond, two protons may end up producing a signal in the nmr corresponding to the average environment for these two nuclei. Moreover, whether we observe one average signal or two independent signals is a function of the rate at which these two nuclei are exchanging environments. This explains the temperature dependence of the nmr spectrum of dimethylformamide. Protons bonded to carbon behave differently from protons bonded to heteroatoms (*e.g.*, oxygen). The greatest differences are that protons bonded to heteroatoms are (a) exchangeable and (b) subject to the effects of hydrogen bonding. The remainder of this section will deal with these two, paying special attention to exchange processes.

Alcohols show a hydroxy resonance that is quite sensitive to solvent, concentration, and temperature. For example, the **OH** resonance of ethanol becomes deshielded by 1.5 ppm as the temperature decreases from +75°C to −115°C. This change is due to the increasing importance of hydrogen bonding at lower temperatures; increased hydrogen bonding should decrease electron density around a proton and cause that proton to absorb at lower fields.

When a small amount of D_2O is added to a solution of an alcohol, the **OH** resonance of the alcohol disappears because of the incorporation of deuterons in place of protons. This, then, is a convenient way of identifying which signal in the original nmr belonged to R**OH.** In order for the R**OH** resonance to disappear completely one must, of course, use an excess of D_2O. (Depending upon the exact experimental conditions, you may expect a resonance due to HOD to appear.) The exchange process, depicted below, is acid- and base-catalyzed (see p. 315).

$$\text{ROH} + \text{DOD} \rightleftarrows \text{ROD} + \text{HOD}$$
$$2\text{HOD} \rightleftarrows \text{HOH} + \text{DOD}$$

What does the nmr spectrum of a mixture of two compounds, each of which contains a hydroxy group, look like? To answer this we will consider the nmr spectrum of a mixture of phenol, C_6H_5OH, and methanol. A solution of phenol in deuteriochloroform exhibits an **OH** resonance near 5.8 δ. The spectrum of methanol in the same solvent exhibits its **OH** resonance near 3.0 δ. A solution of approximately equimolar amounts of these two in deuteriochloroform contains a new resonance (**OH**) at 4.5 δ and lacks a signal at either 3.0 δ or 5.8 δ. Only one hydroxy resonance is observed in this mixture because the hydroxy groups in phenol and methanol are exchanging so rapidly that the nmr "sees" only one kind of proton, and that is in an average environment. When two nuclei exchange environments at a sufficiently rapid rate so that only one average signal is observed, this average signal will have a chemical shift which is (a) related to the relative concentrations of the two solutes and (b) between the corresponding resonances in the pure solutes. (You should review the discussion on pp. 286 and 314 if you have not already done so.)

As described on p. 314, a rapid proton exchange in alcohols prevents the observation of vicinal coupling in the **H—C—O—H** fragment of alcohols.

Carboxylic acids, RCO_2H, like alcohols, also undergo rapid exchange with D_2O. What is rather unique is that the CO_2**H** resonance of carboxylic acids dissolved in non-polar solvents (*e.g.*, carbon tetrachloride) is relatively insensitive to dilution. The reason for this is that, at moderate concentrations, carboxylic acids exist as dimers, held together by hydrogen bonds, and dilution does not diminish the total amount of these dimers present.

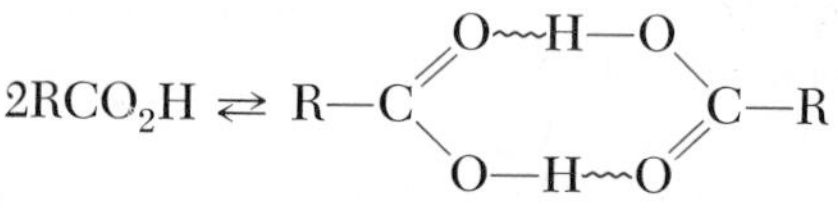

```
                 O~~H—O
               //      \
2RCO2H ⇄ R—C            C—R
               \      //
                 O—H~~O
```

Aliphatic amines, alkylNH_2 and $(alkyl)_2NH$, undergo rapid exchange of the **NH** proton and, so, will not show **H**—C—N—**H** spin-spin splitting except under unusual circumstances. Such compounds will, therefore, produce a sharp singlet for the **NH** proton. Some amines show an intermediate rate of exchange and are partially decoupled. This situation produces a broad **NH** resonance, such as is found in the nmr spectrum of *N*-methyl-*p*-nitroaniline, *p*-$CH_3C_6H_4NHCH_3$. Under such circumstances the adjacent **CH** signal (CH_3 in this specific case) is not split.

Perhaps the most interesting cases are those in which **NH** protons are undergoing slow chemical exchange. In such systems (which include amides and pyrroles), coupling between **NH** and **CH** in **H**—C—N—**H** fragments is observed *if you examine the* CH *resonance.* However, such systems still do not show the sharp multiplet structure that you might expect from the **NH** portion of an **H**—C—N—**H** fragment in which the **CH** *is* split. Instead, the **NH** resonance appears as a broad signal. The reason for this is complex, being due to electrical properties of the nitrogen atom, and will not be discussed further.

29.8 INTERPRETED NMR SPECTRA

The following explanations are offered in order (a) to present a bit more information about the interpretation of nmr spectra and (b) to bring together much of this chapter and focus it upon spectral interpretation.

ISOBUTANOL (Figure 29–21)

1. The isopropyl methyl group signal is split by the methine hydrogen into a doublet.
2. The methylene signal is also a doublet, but is easily distinguished from the former because of the relative intensities (**2H** *vs.* **6H**) and the chemical shifts.
3. The methine signal is more complex than that in Figure 29–11 because of the additional splitting by the methylene group.
4. The O—**H** signal is a singlet, indicating that it is not coupled to the methylene protons. Such behavior is typical of alcohols in $CDCl_3$ that is not *extremely* pure.
5. The relative intensities of the various signals are shown by the step-like integration curve. The relative intensities in this case are obtained by dividing each of the numbers in the integration by 10, the lowest number. *While this step function is the manner in which*

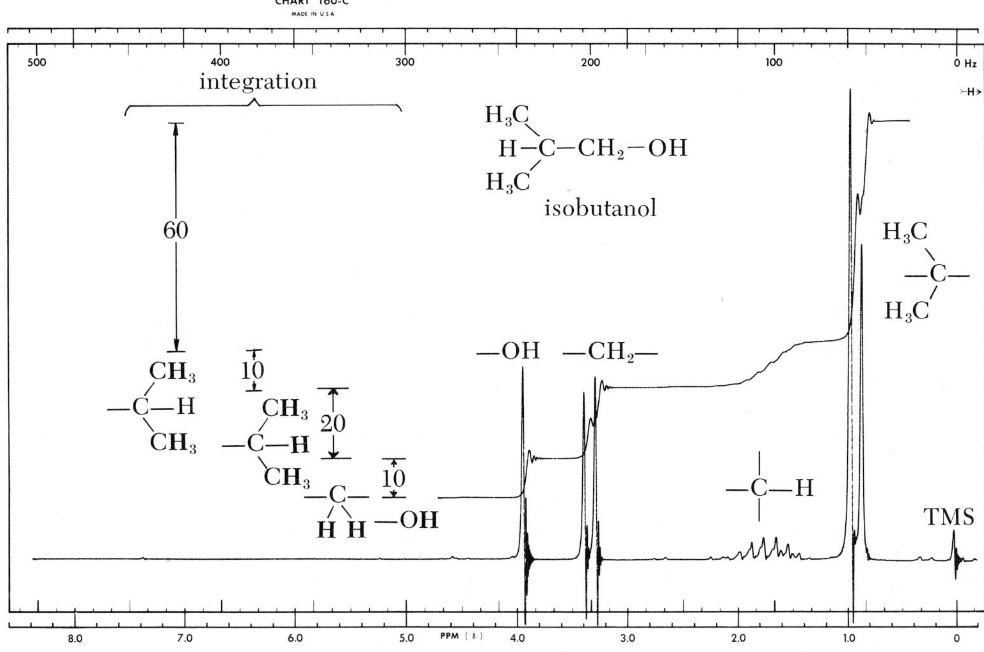

Figure 29–21 The nmr spectrum of isobutanol.

integrations are obtained from the nmr spectrometer, throughout this text integrated intensities may be presented as integers shown above the various signals.

BENZYL ALCOHOL, PHENYLACETYLENE, AND ANISOLE
(Figure 29–22)

1. The signals from the protons on the phenyl groups all occur at about the same place (about 7.2 δ), but the appearance varies considerably. In general, the more electronegative the

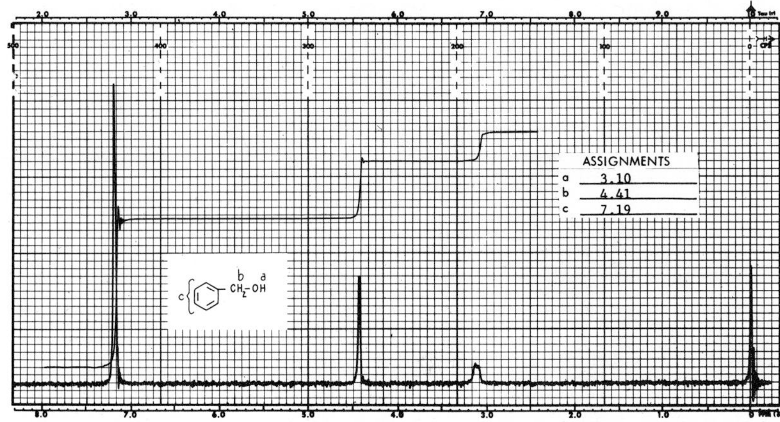

Figure 29–22 The nmr spectra of benzyl alcohol, phenylacetylene, and anisole. © Sadtler Research Laboratories, Inc., 1976.

Illustration continued on following page.

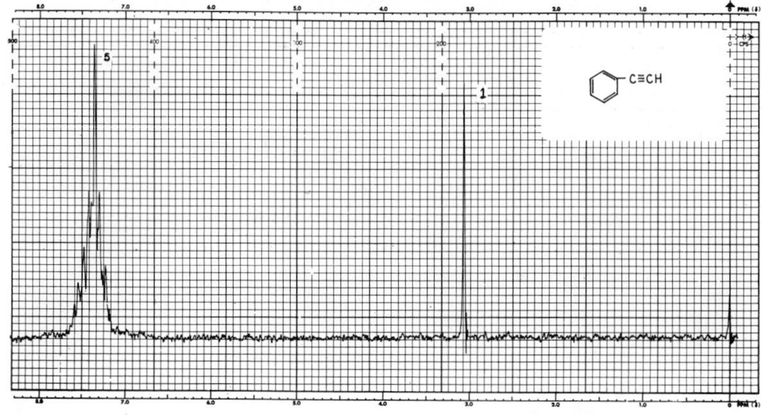

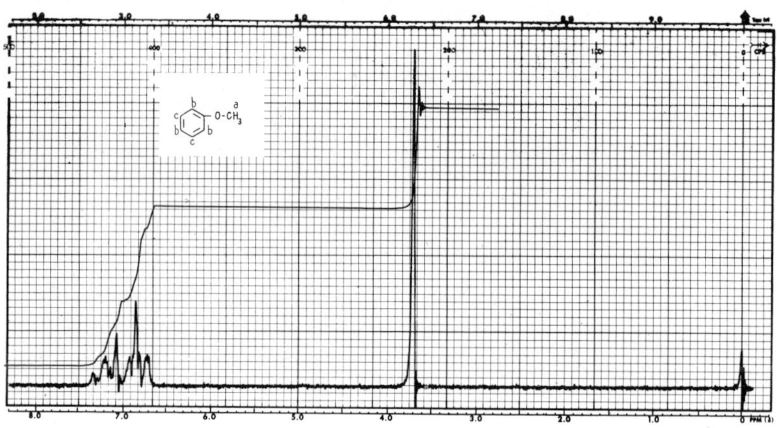

Figure 29-22 *Continued.*

atom bonded to the phenyl group, the more complex will be the spectrum of those protons. The order of electronegativity is O > C_{sp} > C_{sp^3}.

2. The O—H and CH_2 resonances are unsplit (typical of alcohols in most solvents). Note that, because of pH, temperature, and concentration differences, the O—H resonances of benzyl alcohol and isobutanol (Figure 29-21) differ by about 1.0 ppm.

ETHYL CHLOROACETATE AND ETHYL DICHLOROACETATE
(Figure 29-23)

1. Both spectra are similar in the appearance of an upfield triplet (methyl group). However, the combination of the C=O anisotropy and the Cl inductive effect cause one methylene resonance (singlet) of ethyl chloroacetate to overlap with the uppermost signal of the methylene quartet. In ethyl dichloroacetate, the additional inductive effect of the second Cl moves this resonance (now a methine group) sufficiently downfield so that a normal ethyl resonance is observed.

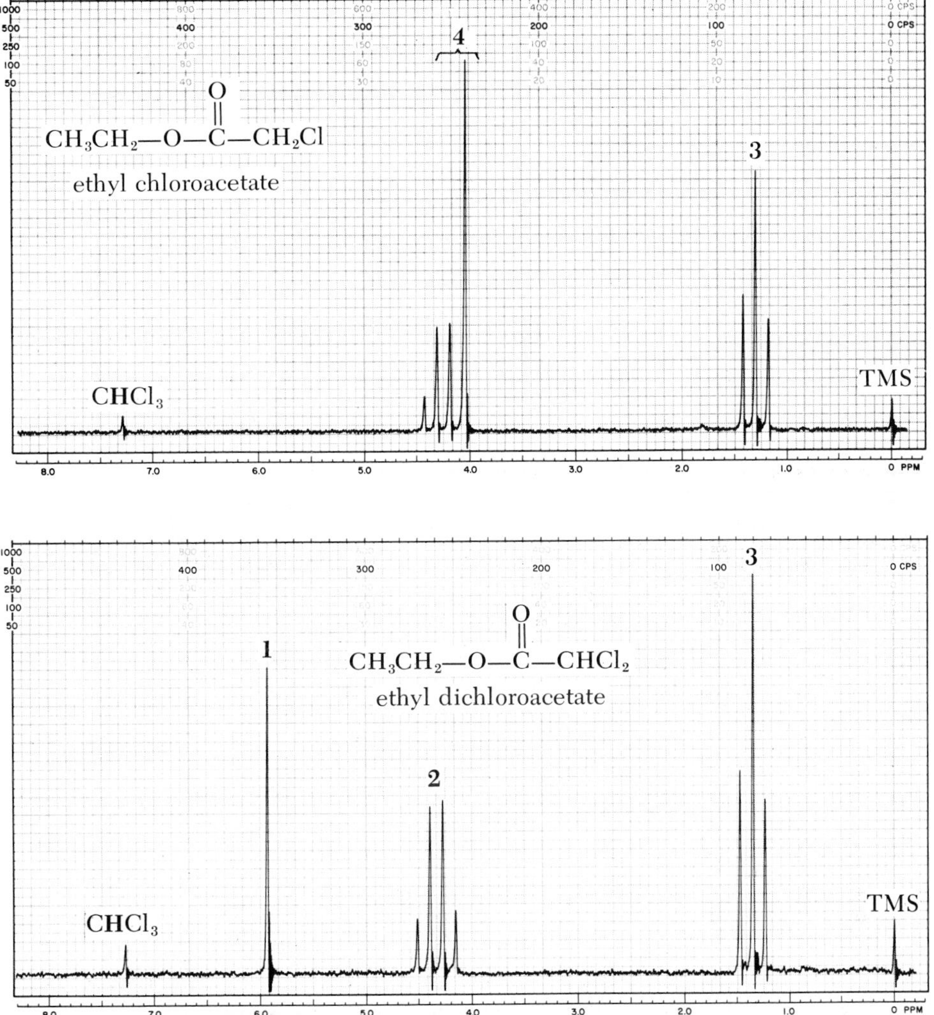

Figure 29-23 The nmr spectra of ethyl chloroacetate and ethyl dichloroacetate. The methylene singlet, arising from —C(O)—CH_2—Cl, overlaps the highest field line of the methylene quartet, arising from CH_3—CH_2—O—, in the spectrum of ethyl chloroacetate. Replacement of a methylene hydrogen by a second chlorine decreases the electron density around what is now the methine hydrogen, and shifts its signal downfield.

2. Each spectrum displays a weak $CHCl_3$ signal, indicating that the spectra were measured in $CDCl_3$ of less than 100% isotopic purity.

19-NORPROGESTERONE (Figure 29–24)

1. In this extremely complex spectrum one can still see three recognizable signals. Two methyl signals appear as singlets (uncoupled!). One is rather far downfield (2.13 δ), indicating that it is the one bonded to the carbonyl group.

2. The broad signal at 5.87 δ must be the vinylic proton further deshielded by the carbonyl group.

3. The remainder of the spectrum cannot be analyzed.

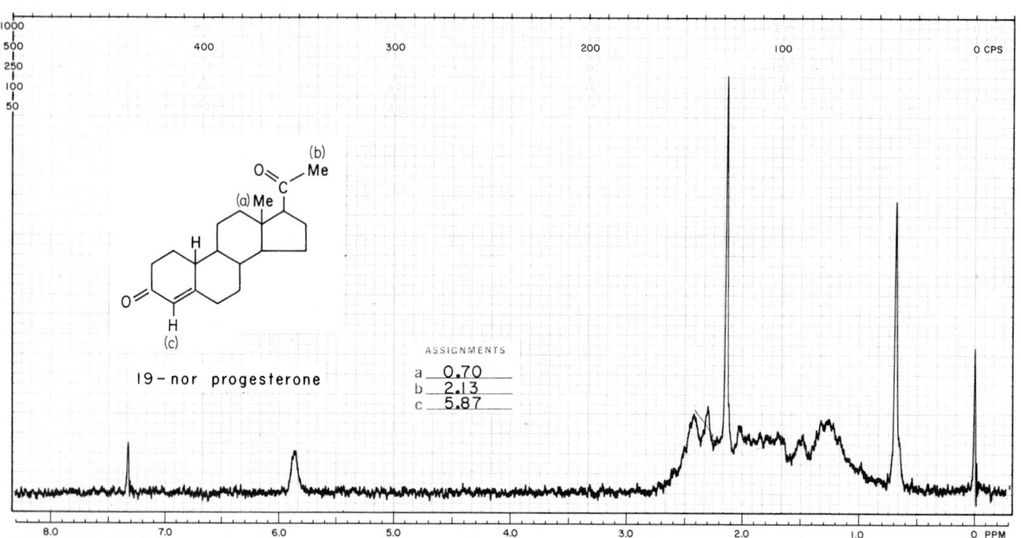

Figure 29-24 The nmr spectrum of 19-norprogesterone. The multitude of splittings and different types of protons permit the ready detection of only three signals. Two of these belong to the different methyl groups, and one belongs to a vinyl proton. The vinyl signal is broadened by couplings to protons more than three bonds away. This long-range coupling is facilitated by the presence of the double bond.

VALINE (Figure 29–25)

α-Amino acids (compounds containing the $-\overset{\text{NH}_2}{\overset{|}{\underset{|}{\text{C}}}}-\text{CO}_2\text{H}$ group) are usually much more soluble in water than in the common organic solvents. These two spectra were, therefore, taken in D_2O solutions immediately after the solutions were prepared. The first spectrum is in a slightly basic solution, and the second is in a slightly acidic solution. Since TMS is insoluble in water, the reference compound is sodium 2,2-dimethyl-2-silapentane-5-sulfonate, $(CH_3)_3SiCH_2CH_2CH_2SO_3Na$ (DSS). (The most upfield signal of DSS is quite close to that of TMS.)

Alkaline Solution. In basic solution, valine exists as a carboxylate salt, $(CH_3)_2CH{-}CH(NH_2)CO_2^{\ominus}\ M^{\oplus}$.

1. The methyl groups are anisochronous (they are adjacent to a chiral center). Each of these is split into a doublet by the methine hydrogen of the isopropyl group, producing the pair of doublets near 0.9 δ.

2. The isopropyl methine hydrogen is split by each of the methyl groups. Because of the similar couplings between the methine hydrogen and the two methyl groups, the methine signal appears as a heptet (the two outer signals of the heptet are lost in "noise"). This multiplet is further split by the remaining methine hydrogen (—**CH**—NH_2).

3. Cα—**H** is split into a doublet by coupling with the isopropyl methine hydrogen.

4. The amine (—NH_2) resonance is unsplit (4.8 δ). (Amines as well as alcohols often lack the vicinal coupling that H—C—C—H systems display.)

Acidic Solution. In acidic solution, valine exists as an ammonium salt, $(CH_3)_2CH{-}CH(CO_2H)\overset{\oplus}{N}H_3\ X^{\ominus}$.

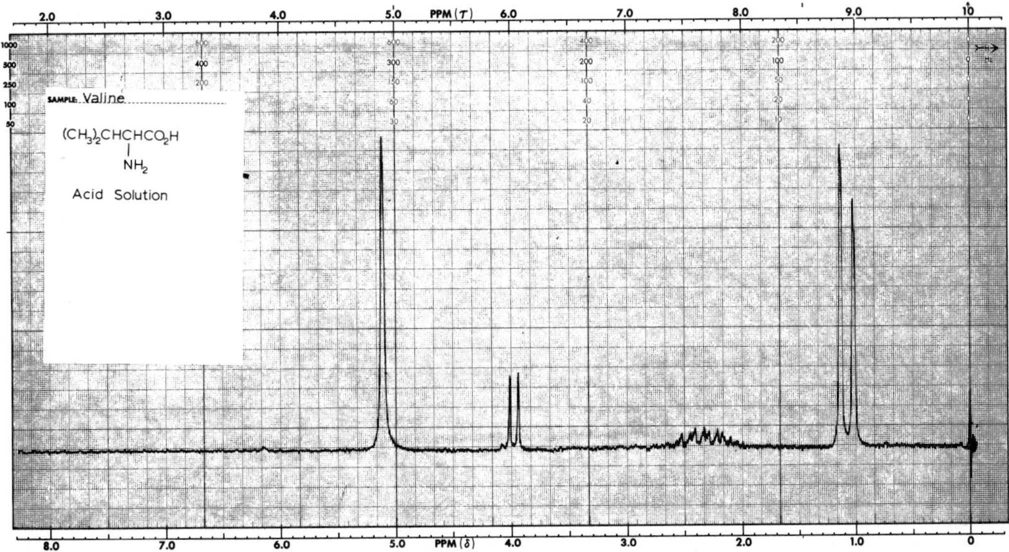

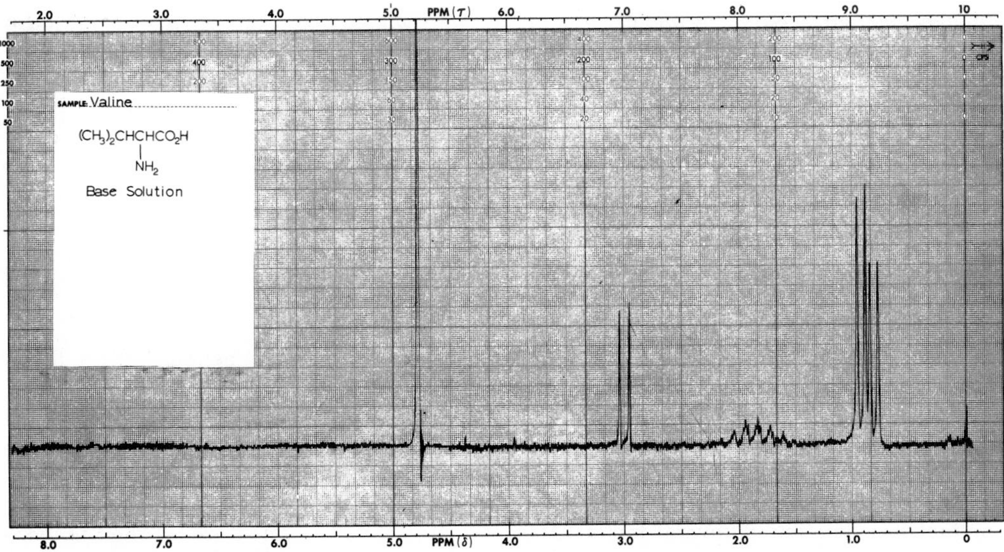

Figure 29-25 The nmr spectra of valine in acidic and basic media. Solvent, D_2O; reference, DSS (sodium 2,2-dimethyl-2-silapentane-5-sulfonate). (Courtesy of Texas Christian University, from W.B. Smith and A.M. Ihrig, *The NMR Spectra of Eighteen Essential Amino Acids.*)

1. The isopropyl portion of the spectrum reveals a barely detectable non-equivalence of the methyl groups (note the shoulder on the upfield side of what *appears to be* a doublet.)

2. The Cα—**H** resonance is downfield relative to the spectrum of the basic solution because of the electron withdrawing inductive effect of —$NH_3^{\oplus}$.

3. The $NH_3^{\oplus}$ resonance is less deshielded than might have been imagined from a simple consideration of the inductive effect of a positively charged N. There is no *simple* explanation for this comparatively minor deshielding.

IMPORTANT TERMS

Anisochronous: Chemical shift non-equivalent. Protons that do not have the same chemical shift. Diastereotopic protons are anisochronous.

Anisotropic effect: Electron circulation induces a magnetic field that may add to, or subtract from, the applied field, H_0, at a given point. When the effect (magnitude and direction) of the induced field upon a given proton is a function of the orientation of that proton relative to the induced field, this is described as an anisotropic effect. In general, the term "anisotropic" simply means "unsymmetric in space." One might expect the anisotropic effect to be conformation-dependent and, indeed, conformational analysis by nmr requires the presence of some anisotropic induced magnetic field within the molecule.

Applied magnetic field: The magnetic field, H_0, into which a molecule is placed during the nmr experiment. This is the field produced by the instrument. (Compare induced magnetic effect.)

Chemical shift: The resonance frequency of a given proton compared to that of the methyl protons of tetramethylsilane, TMS, under the same experimental conditions. When this number is divided by the operating (radio) frequency of the instrument, the chemical shift, δ, is expressed in parts per million (downfield from TMS).

Chemical shift equivalent: Isochronous protons. Protons which have the same chemical shift. Such protons are not observed to split one another.

Chemical shift reagent: Any of a group of compounds that increase the chemical shift difference between protons that are anisochronous.

Coupling constant: A measure of the extent of interaction (spin-spin coupling or splitting) between nuclei. In general, when nuclei are separated by more than three sigma bonds, the interaction decreases rapidly and the coupling constant, J, approaches zero. The coupling constant is a function of several factors, the most important of which are (a) the number of bonds separating the nuclei and (b) the torsional angle that these bonds make with one another. The nmr records an average J when free rotation around a bond is possible.

Coupling constant equivalent: Two, or more, nuclei that couple equally to all other nuclei within the molecule.

Deshielding region: That region around a bond, or group of bonds, which will cause a proton resonance to appear farther downfield than might otherwise have been expected. The outer perimeter of the benzene ring is such a deshielding region (review Figure 14–5).

Double irradiation: A technique for simplifying a spin-spin splitting pattern in an nmr spectrum. It is accomplished by irradiating a sample with a (radio) frequency corresponding to the resonance frequency of one of the nuclei involved in the spin-spin splitting *while the normal* *nmr experiment is being performed.* The result is two-pronged: (i) obliteration of the signal corresponding to the proton being irradiated by the "extra" radio frequency and (ii) simplification of the resonance signal of the proton that is coupled to this proton.

Downfield: Absorbing at a lower applied field. "To the left" in the usual presentation of an nmr spectrum. Almost all protons absorb downfield of the protons in TMS. The more deshielded a proton is, the farther downfield it will appear in a spectrum.

Effective magnetic field: The magnetic field at a particular proton. This field, H_{eff}, is the sum of H_0 and all of the induced fields.

Exchange process: Any process that interconverts the magnetic environment of nuclei. While the most common exchange process is bond rotation, ionization-recombination also results in exchange. This exchange (intermolecular or intramolecular) by ionization-recombination can return a nucleus of either spin ($+\frac{1}{2}$ or $-\frac{1}{2}$), and this *normally leads to spin decoupling.* (Now is the time to try problem 8b if you have not already done so: You should also review pp. 314–315.)

$$R^*\text{—O—}H^{(+1/2)} + R\text{—O—}H^{(-1/2)} \rightleftharpoons R^*\text{—O}^{\ominus} + R\text{—}\overset{\oplus}{O}(H^{(+1/2)})(H^{(-1/2)})$$

$$\rightleftharpoons R^*\text{—O—}H^{(-1/2)} + R\text{—O—}H^{(+1/2)}$$

spin decoupling by ionization-recombination[a]

Induced magnetic field: A magnetic field that is set up by motion of electrons within a molecule responding to H_0, the applied magnetic field.

Integrated intensity: The *area* under a particular nmr signal. The integrated intensity of a signal is directly related to the number of protons giving rise to that signal. The term is only indirectly related to the amplitude (height) of a given signal. *The height of a signal should never be taken as a measure of the number of protons giving rise to a signal.*

As shown in Figure 29–21, the nmr instrument plots out integrated intensity as a step function, and the chemist measures the integrated intensity of a given signal by measuring the height of the appropriate step. However, the height of a given step does not indicate the *exact* number of protons giving rise to a signal. Instead, by comparing the sizes of the steps corresponding to the various signals in the spectrum, one obtains a measure of the *relative* numbers of protons responsible for the respective signals. For example, if an integration showed two steps with intensities of 2 and 3, it simply means that the protons giving rise to those two signals are present in a *ratio* of 2:3, respectively. But there could be 6 of one type and 9 of the other!

Isochronous: Protons with the same chemical shift. Protons that are identical or enantiotopic are isochronous in achiral solvents.

Magnetic equivalence: A rather diffuse term indicating that nuclei are in the same magnetic environment. It is best thought of as having a chemical shift component and a coupling constant component.

Shielding region: That region around a bond, or group of bonds, which will cause a proton to appear farther upfield than might otherwise have been expected. The region above and below the center of the benzene ring is such a region (see Figure 14–5).

[a] R and R* are identical, but one is starred to facilitate discussion.

Spin decoupling: Nullifying the effects of spin-spin interactions and, thereby, reducing the multiplicity of signals within an nmr spectrum. Decoupling may be accomplished in several ways, including double irradiation and exchange. The effect of deuterium substitution is similar to that of spin decoupling.

Spin coupling (spin-spin coupling): The interaction of two or more protons, most often through bonds connecting them, which increases the multiplicity of nmr signals.

Upfield: Absorbing at a higher applied field. "To the right" in the usual presentation of an nmr spectrum. The more shielded a proton is, the farther upfield it will appear. Very few protons absorb upfield of TMS.

PROBLEMS

1. Define, in your own words, the following:
   (a) coupling constant
   (b) chemical shift
   (c) internal standard
   (d) upfield shift
   (e) deshielded
   (f) effective magnetic field
   (g) anisotropic effect
   (h) spin decoupling
   (i) double irradiation experiment
   (j) isochronous
   (k) anisochronous
   (l) coupling constant non-equivalence
   (m) chemical shift reagent

2. (a) Construct splitting diagrams for the boldface **H** in the system —CH_2—CH_2—C(<)—**H**, assuming each of the sets of coupling constants below. Graph paper will facilitate the construction of these diagrams.

—CH_2—CH_2—C(<)—H

	J (CH_2–CH_2)	J (CH_2–C–H)
(1)	7 Hz	7 Hz
(2)	5 Hz	7 Hz
(3)	3 Hz	7 Hz
(4)	7 Hz	3 Hz

(b) If the instrument being used cannot resolve (*i.e.*, separate) signals closer together than 2 Hz, how many signals would each of the above sets produce in the final spectrum? (c) How many signals would appear if the resolving power were 1 Hz?

3. Provide a specific illustration of a compound which contains protons that are:
   (a) chemical shift equivalent and coupling constant equivalent
   (b) chemical shift equivalent and coupling constant non-equivalent
   (c) chemical shift and coupling constant non-equivalent.

4. A professor treated **A** (below) with Br_2 and produced a product which, according to the professor, could not have been **B** (below) because "there are too many methyl resonances in the nmr of the product." A student disagreed (and correctly so). Explain to the professor the error of his ways.

$(CH_3)_2$CH—CH=CH_2 **A**

$(CH_3)_2$CH—CHBr—CH_2Br **B**

5. A student treated ethylene with ICl and produced a product whose nmr consisted of only a complex multiplet. The student concluded that the expected product (ICH_2CH_2Cl) was not produced, since ". . . the product should exhibit two triplets in the nmr . . ." Comment upon the student's argument.

6. Label the pairs of protons (or proton sets) shown in boldface in each of the following compounds as *diastereotopic, enantiotopic,* or *identical,* as required. Assume normal rotational barriers and observations at room temperature.

(a) H / Br—C—Br / H

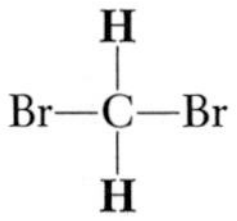

(b) H / Br—C—Cl / H

(c) H / $Br—CH_2—C—Br$ / H

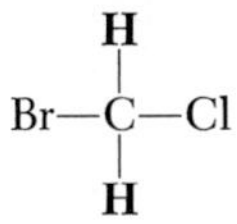

(d) H / $Br—CH_2—C—Cl$ / H

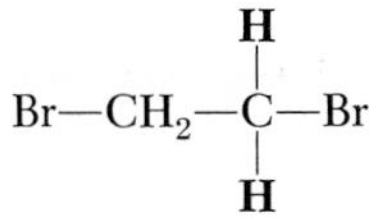

(e) H H / Br—C—C—Cl / H Br

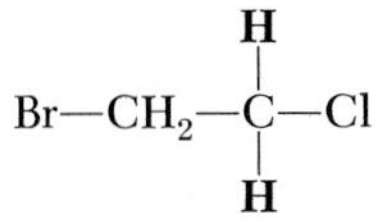

(f) H H / Br—C—C—Br / Br Br

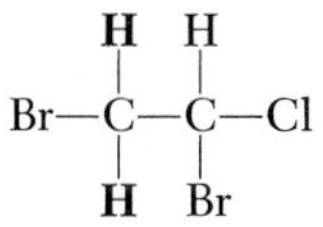

(g) H / $C(CH_3)_3$ / H / H

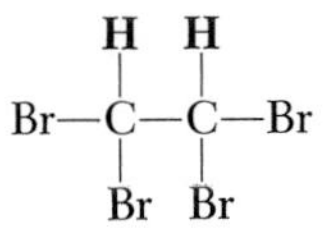

(h) H H H / Br—C—C—C—Cl / H H H

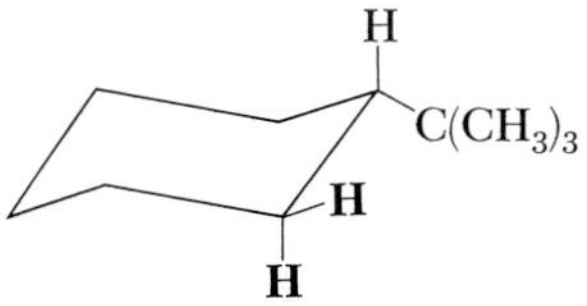

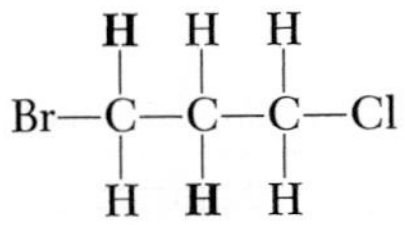

(i) H H H / Br—C—C—C—Br / H H H

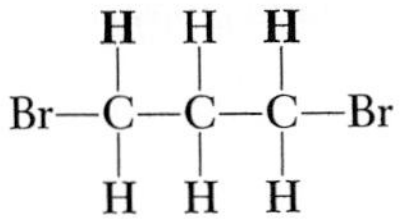

(j) H / C=S⊕ / H / O⊖

(k) H Br / C=C / Br H

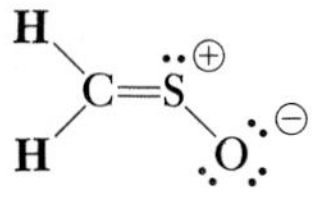

(l) H / C—H / Br Br

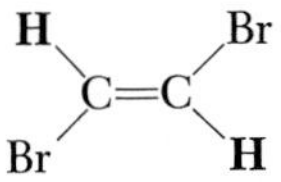

(m) H H / Cl H / H H / H Cl / H

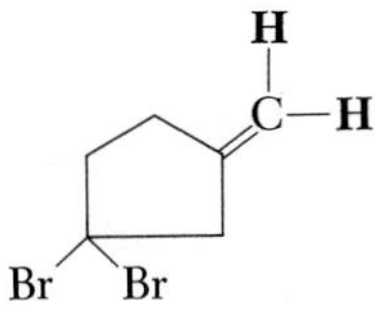

(n) Cl H / H H / H H / Cl / H

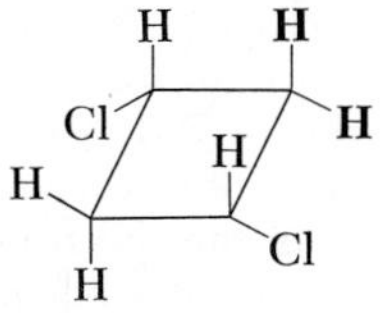

(o) Cl H / Cl Cl H / H / H / H

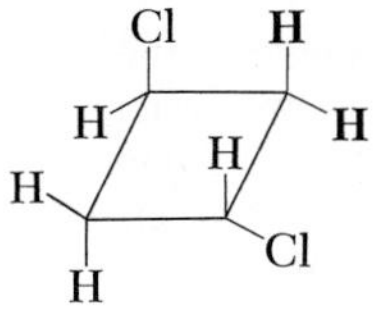

(p) Cl H / Cl Cl H / H / H / H

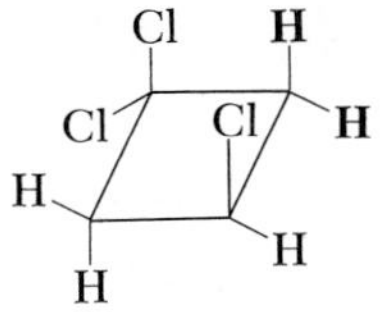

(q) Cl H / Cl Cl H / H / H / H

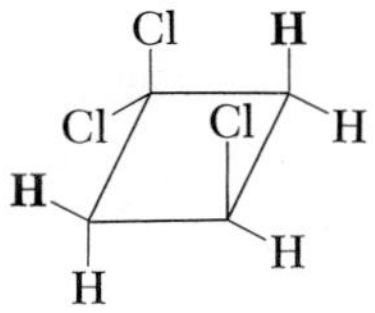

(r) ⊖:Ö: H H :Ö:⊖ / $CH_3—S—C—C—S—C_2H_5$ / ⊕ H H ⊕

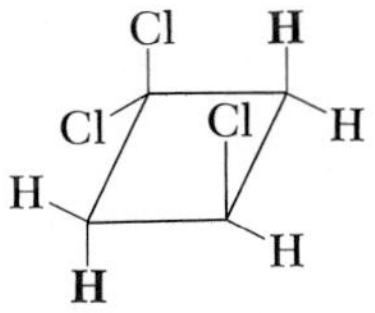

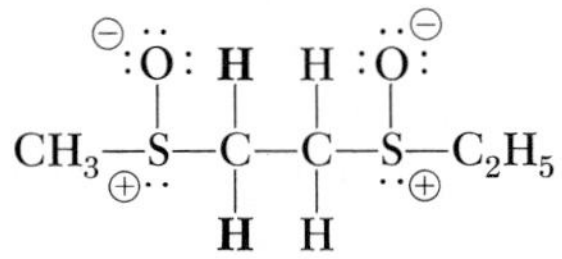

(s) H H / Cl / H H

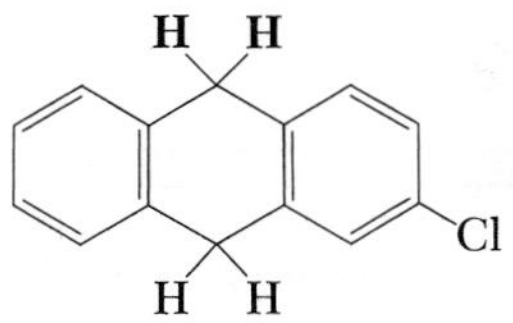

(t) H_3C CH_3 / H_3C H

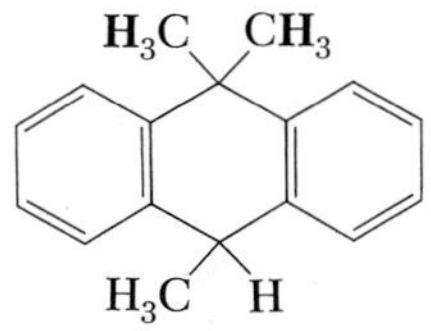

(u) H H / H_3C CH_3

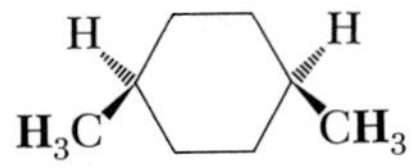

(v) H_3C H / H CH_3

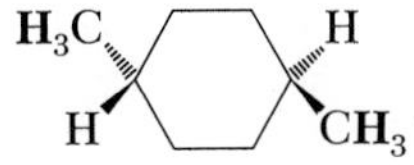

7. Could nmr spectroscopy have been used to distinguish between the two possible structures originally considered for the penicillin ring system (see Chapter 28)?

8. (a) A solution of methanol (CH_3OH) in "heavy water" (D_2O) shows only a single line in the nmr (near 2.2 δ), although solutions of methanol in pure CCl_4 show two absorptions (3H and 1H). Account for the solvent dependence of the spectrum.

(b) The nmr spectrum of pure ethanol in very pure $CDCl_3$ does contain a triplet for the OH resonance and a multiplet for the methylene signal. If a *trace* of HCl or H_2O is added to the solution, the triplet collapses to a single (rather broad) line while the methylene region collapses to a quartet. Suggest a reason for these observations.

9. The nmr spectrum of an unknown material, which contains only ^1H as an nmr-active nucleus, consists of two sharp lines that are separated by 10 Hz. Is this a doublet or just two singlets that happen to have similar chemical shifts? Explain.

10. The spectrum below is of a mixture of isopropyl formate and isopropanol. Assign the resonances in the spectrum of the mixture.

$$\underset{\text{isopropyl formate}}{H-\overset{\overset{O}{\|}}{C}-O-\overset{\overset{CH_3}{|}}{\underset{\underset{CH_3}{|}}{C}}-H} \qquad \underset{\text{isopropanol}}{H_3C-\overset{\overset{OH}{|}}{\underset{\underset{H}{|}}{C}}-CH_3}$$

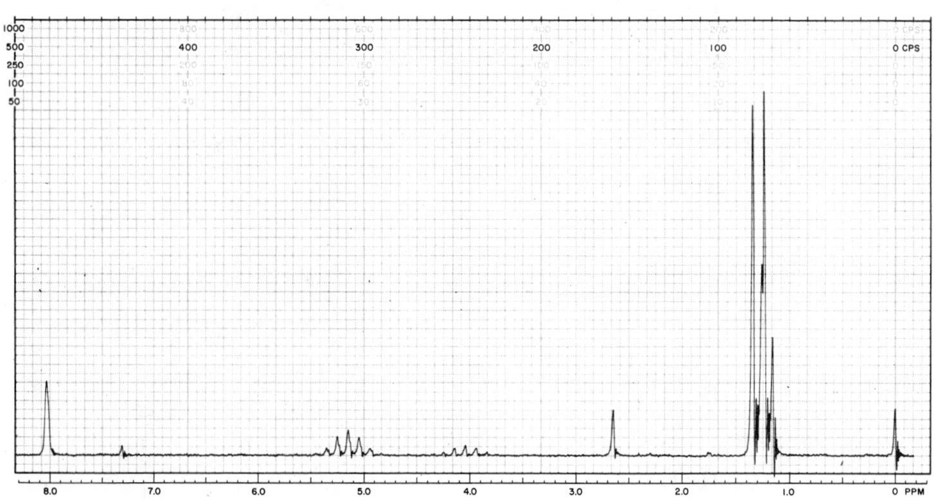

Problem 10: A mixture of isopropyl formate and isopropanol.

11. Assign structures consistent with the following data:

(a) $C_4H_6O_2$
multiplet 1.97 δ 3H
multiplet 5.72 δ 1H
multiplet 6.30 δ 1H
singlet 11.57 δ 1H

(b) $C_4H_7BrO_2$
triplet 1.97 δ 3H
quintet 2.07 δ 2H
triplet 4.23 δ 1H
singlet 10.97 δ 1H

(c) $C_4H_8O_2$
triplet 1.25 δ 3H
singlet 2.03 δ 3H
quartet 4.12 δ 2H

(d) $C_8H_{11}N$
doublet 1.38 δ 3H
singlet 1.58 δ 2H
quartet 4.10 δ 1H
singlet 7.30 δ 5H

(e) CH_4O
singlet 1.43 δ 1H
singlet 3.47 δ 3H

(f) $C_2H_3Cl_3$
doublet 3.95 δ 2H
triplet 5.77 δ 1H

(g) $C_2H_3F_3O$
singlet 3.38 δ 1H
quartet 3.93 δ 2H

(h) $C_3H_6Cl_2$
quintet 2.20 δ 2H
triplet 3.70 δ 4H

(i) C_3H_6O
quintet 2.72 δ 2H
triplet 4.73 δ 4H

12. The nmr spectrum below is that of acetaldoxime in $CDCl_3$. Account for all of the spectral features.

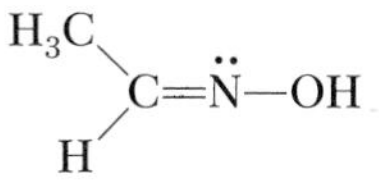

H_3C—C(H)=N̈—OH

acetaldoxime

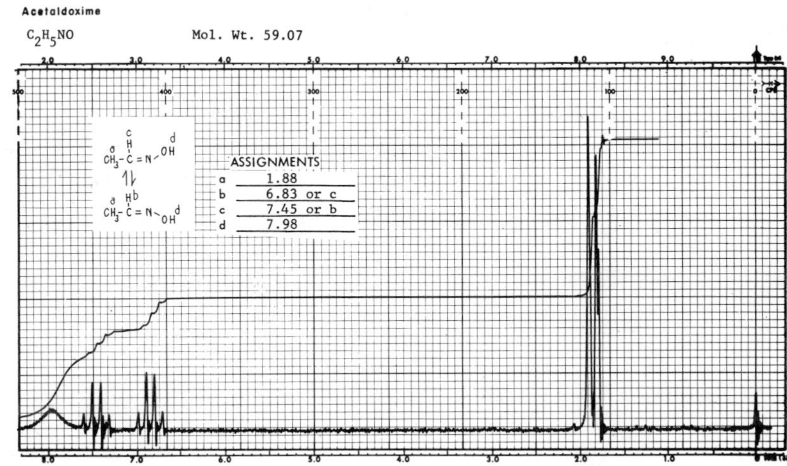

Problem 12: The nmr spectrum of acetaldoxime. © Sadtler Research Laboratories, Inc., 1976.

13. At −130° the *fluorine* nmr spectrum of 1,2-difluorotetrachloroethane consists of two singlets of *unequal* intensity, while at room temperature only one singlet is observed. Account for these results.

14. The following nmr spectra are those of (a) 1,1,2,3,3-pentachloropropane, (b) 1,2-dichloro-1,1-difluoroethane, (c) 15,16-dihydro-15,16-dimethylpyrene, and (d) 1,2-epoxy-2-phenylbutane. Account for as many of the spectral characteristics as you can.

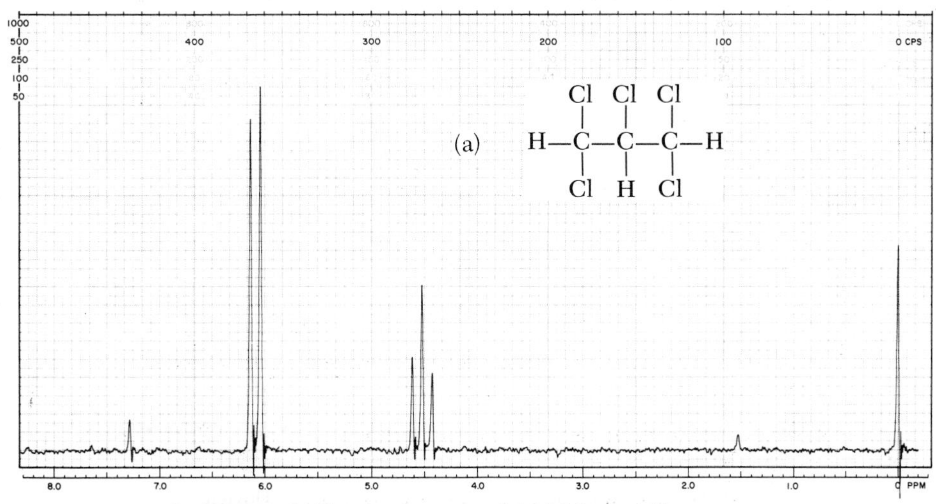

A

Problem 14: **A,** The nmr spectrum of 1,1,2,3,3-pentachloropropane.

Illustration continued on following page.

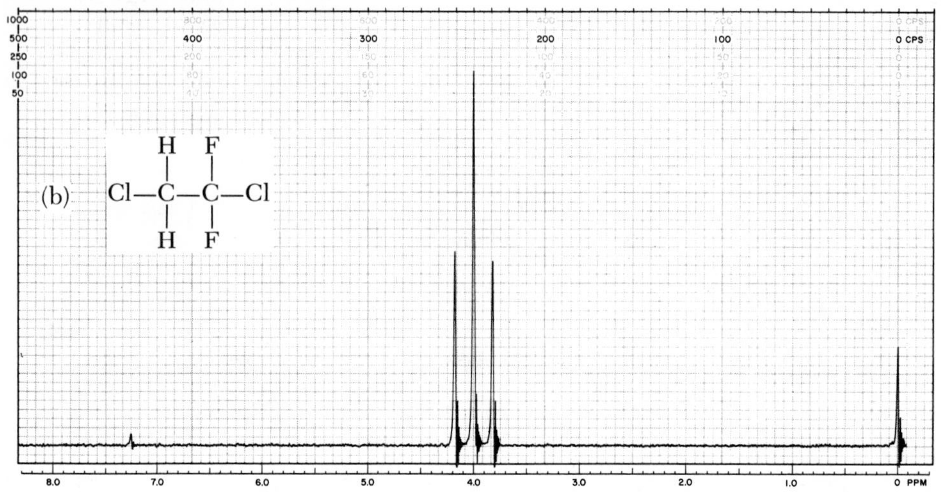

B

B, The nmr spectrum of 1,2-dichloro-1,1-difluoroethane.

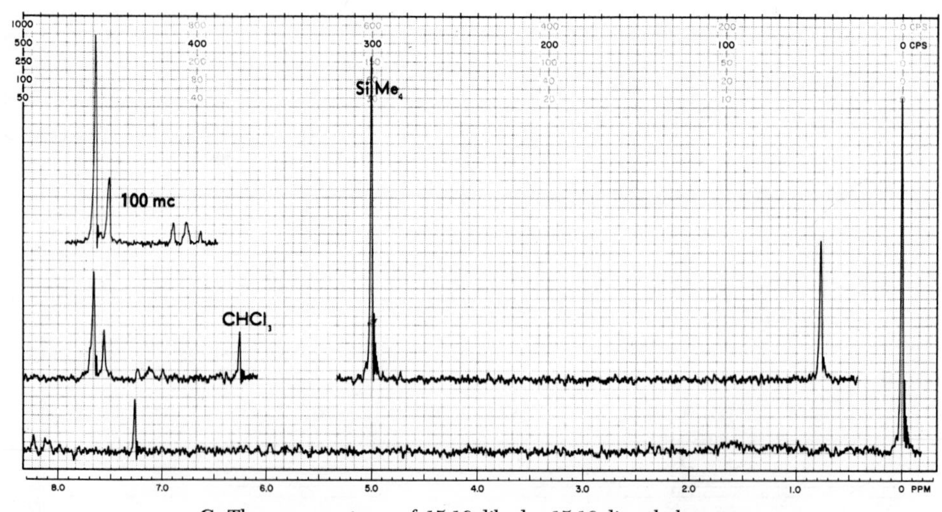

C

C, The nmr spectrum of 15,16-dihydro-15,16-dimethylpyrene.

Illustration continued on facing page.

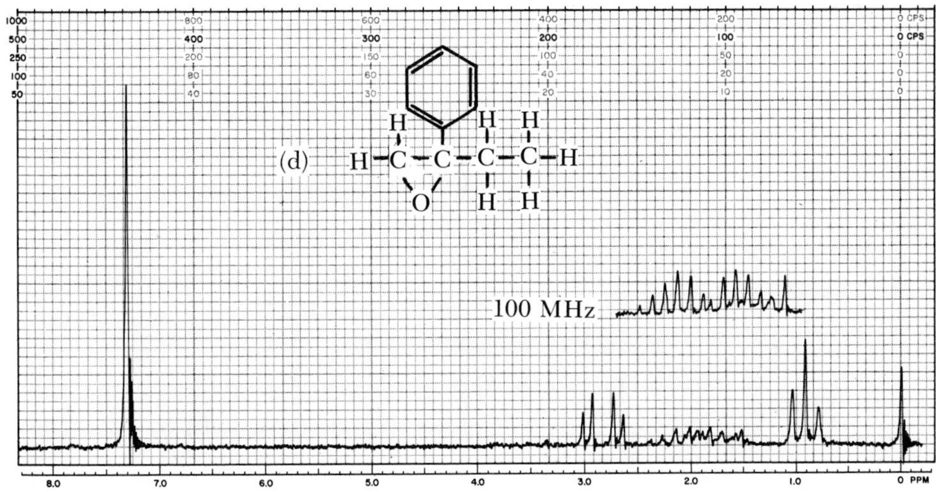

D

D, The nmr spectrum of 1,2-epoxy-2-phenylbutane. This spectrum contains a scan of the complex multiplet at 100 MHz.

15. Compound **A,** $C_2H_6N_2O$, shows two sharp resonances of equal intensity at 3.09 δ and 3.82 δ (room temperature, $CDCl_3$ solvent). At high temperatures this material shows only one resonance. Assign a structure to **A** that is consistent with these data.

16. If the four signals representing a pair of AB doublets are numbered 1 to 4 (going upfield), the value of J_{AB} equals $\delta H_1 - \delta H_2$ *or* $\delta H_3 - \delta H_4$ (in Hz). The value of $|\delta H_A - \delta H_B| = \sqrt{(\delta H_1 - \delta H_4)(\delta H_2 - \delta H_3)}$, while the actual values of δH_A and δH_B can be gotten by the addition and subtraction of $|\delta H_A - \delta H_B|/2$ from the mid-point of the AB quartet.

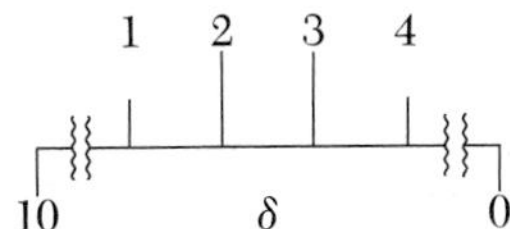

The AB system of cyclopropyl bromide has δH_A of 0.88 δ and δH_B of 1.00 δ. Assuming that J_{AB} equals 15 Hz, and that the operating frequency is 60 MHz, calculate the line positions of the four signals of this AB pattern.

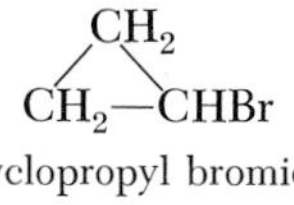

CH_2
CH_2—CHBr
cyclopropyl bromide

17. Suggest as many ways as you can to distinguish between 1-butene, *cis*-2-butene, *trans*-2-butene, and isobutylene by nmr.

18. Sketch the appearance of the nmr spectra of the following deuterated analogs of ethyl iodide.
    (a) CH_3CD_2I (d) CHD_2CHDI
    (b) CD_3CH_2I (e) CH_2DCHDI
    (c) CD_2HCH_2I

19. Convert the chemical shifts in Table 29–1 to Hz for operating frequencies of (a) 60 MHz, (b) 100 MHz, and (c) 300 MHz.

20. The spectrum on p. 984 is that of phenacetin (also called "acetophenetidin," an analgesic antipyretic). Assign the signals in the nmr.

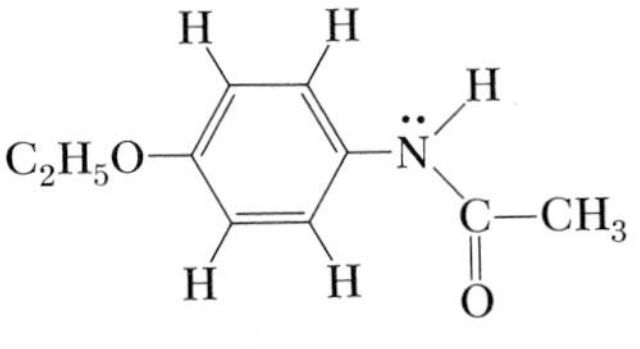

H H H
C_2H_5O—[ring]—$\ddot{N}$—C—CH_3
H H ‖ O

phenacetin

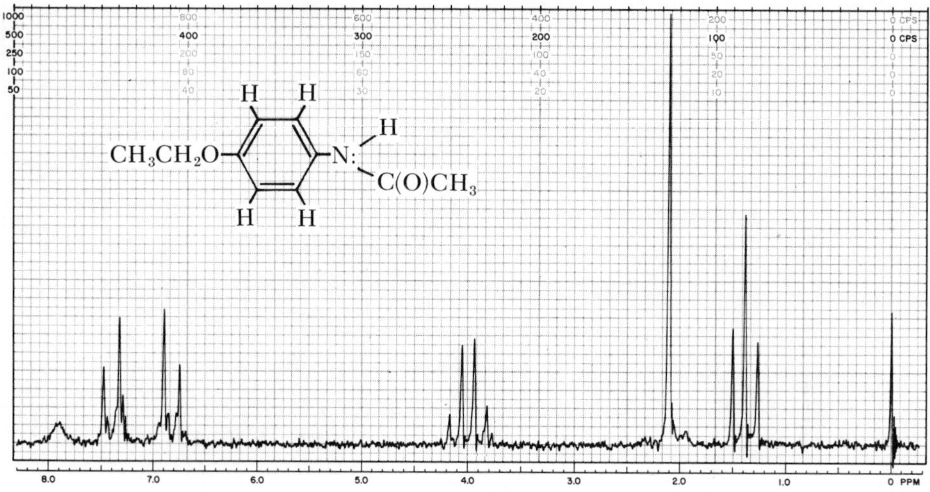

Problem 20: The nmr spectrum of phenacetin.

21. Although the coupling constant between "trans" protons can be as high as 12 Hz, the vinyl region of diethyl fumarate shows only a sharp single line (6.83 δ). Explain.

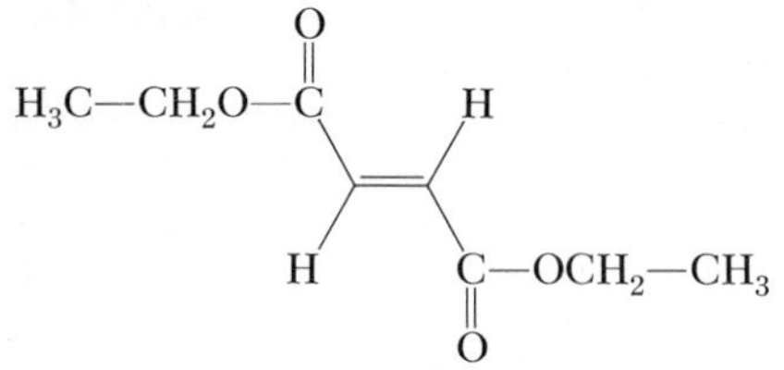

O
‖
$H_3C—CH_2O—C$ H

H $C—OCH_2—CH_3$
‖
O

diethyl fumarate

22. Do you find the following argument reasonable? "The equatorial proton of cyclohexane is deshielded relative to the axial proton because of the anisotropic effect of the C—C single bonds *once removed* (bonds *a* in diagram below)."

The anisotropy of the C—C single bond is as shown below, where + and − regions are shielding and deshielding, respectively.

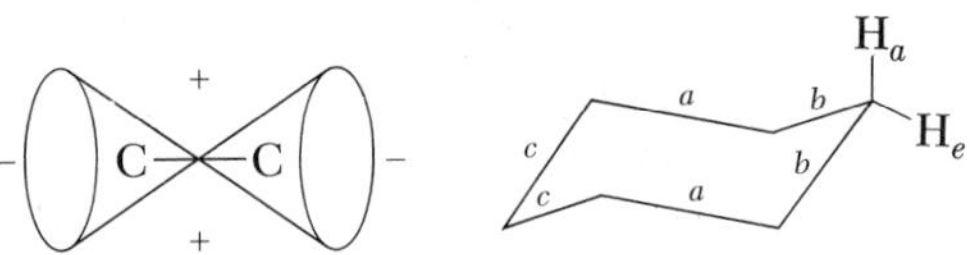

If you agree with the initial statement, you must explain why differences between H_e and H_a are not ascribed to bonds *b*.

BIBLIOGRAPHY

Chapter 1

1. Pierce, J. B., The Chemistry of Matter. Houghton-Mifflin, Boston, 1970 (esp. Chs. 1–4).
2. Dickerson, R. E., Gray, H. B., and Haight, G. P., Chemical Principles, 2nd ed. W. A. Benjamin, Reading, 1974 (esp. Chs. 8 and 9).
3. Ander, P., and Sonessa, A. J., Principles of Chemistry. Macmillan, New York, 1965 (esp. Ch. 2).
4. Masterton, W. L., and Slowinski, E. J., Chemical Principles, 3rd ed. W. B. Saunders, Philadelphia, 1973.
5. Pasachoff, J. M., and Fowler, W. A., "Deuterium in the Universe," *Scientific American,* **230**(5):108, 1974.

Chapter 2

1. (a) Ferguson, L. N., The Modern Structural Theory of Organic Chemistry. Prentice-Hall, Englewood Cliffs, 1963 (Chs. 1 and 2).
   (b) Ferguson, L. N., Organic Molecular Structure. Willard Grant, Boston, 1975 (Chs. 1 and 5).
2. Sienko, M. J., and Plane, R. A., Chemical Principles and Properties, 2nd ed. McGraw-Hill, New York, 1974 (Ch. 3).
3. Dickerson, R. E., Gray, H. B., and Haight, G. P., Chemical Principles, 2nd ed. W. A. Benjamin, Reading, 1974 (Ch. 10).
4. March, J., Advanced Organic Chemistry. McGraw-Hill, New York, 1968 (Ch. 1).
5. Pauling, L., The Nature of the Chemical Bond, 3rd ed. Cornell University Press, Ithaca, 1960.
6. Wheland, G. W., Advanced Organic Chemistry, 3rd ed. John Wiley, New York, 1960 (Ch. 1).
7. Syrkin, Y. K., and Dyatkina, M. E., Structure of Molecules and the Chemical Bond. Tr. by M. A. Partridge and D. O. Jordan. Dover, New York, 1950.
8. Allinger, N. L., and Allinger, J., Structures of Organic Molecules. Prentice-Hall, Englewood Cliffs, 1971.
9. Lagowski, J. J., The Chemical Bond. Houghton-Mifflin, Boston, 1966.
10. Wheland, G. W., Resonance in Organic Chemistry. John Wiley, New York, 1955.
11. Benfey, O. T., ed., Classics in the Theory of Chemical Combination. Dover, New York, 1963.

Chapter 3

1. Fletcher, J. H., Dermer, O. C., and Fox, R. B., eds., Advances in Chemistry, Ser. No. 126. American Chemical Society, Washington, 1974 (Chs. 1 and 2).
2. Banks, J., Naming Organic Compounds, 2nd ed. W. B. Saunders, Philadelphia, 1976.
3. Campbell, J. A., Why Do Chemical Reactions Occur? Prentice-Hall, Englewood Cliffs, 1965.
4. Pryor, W. A., Introduction to Free Radical Chemistry. Prentice-Hall, Englewood Cliffs, 1965.
5. Huyser, E. S., Free-Radical Chain Reaction. John Wiley, New York, 1970.
6. Hirsch, J. A., Concepts in Theoretical Organic Chemistry. Allyn and Bacon, Boston, 1974 (Ch. 11).
7. Wheland, G. W., Advanced Organic Chemistry, 3rd ed. John Wiley, New York, 1960 (Ch. 15).
8. Dawber, J. G., and Moore, A. T., Chemistry for the Life Sciences. McGraw-Hill, New York, 1973 (Ch. 3).
9. Bolin, B., "The Carbon Cycle." *In* Calvin, M., and Pryor, W. A., eds., Organic Chemistry of Life. W. H. Freeman, San Francisco, 1973.

Chapter 4

1. Mislow, K., Introduction to Stereochemistry. W. A. Benjamin, New York, 1965.
2. Natta, G., and Farina, M., Stereochemistry. Harper and Row, New York, 1973.
3. Eliel, E. L., Stereochemistry of Carbon Compounds. McGraw-Hill, New York, 1962.

4. Eliel, E. L., Elements of Stereochemistry. John Wiley, New York, 1969.
5. Flügge, J., The Principles of Polarimetry. Carl Zeiss Optical Co.
6. Heller, W., and Fitts, D. D., "Polarimetry." *In* Weissberger, A., ed., Technique of Organic Chemistry, Vol. 1, Part III. John Wiley, New York, 1960 (Ch. XXXIII).
7. Dawber, J. G., and Moore, A. T., Chemistry for the Life Sciences. McGraw-Hill, New York, 1973 (Ch. 2).
8. Mills, J. A., and Klyne, W., "The Correlation of Configuration." *In* Progress in Stereochemistry, Vol. 1. Butterworths, London, 1954 (Ch. 5).
9. Symposium: Three Dimensional Chemistry. *J. Chem. Ed.*, **41**:61, 1964.
10. Mislow, K., "Stereoisomerism." *In* Florkin, M., and Stotz, E. H., eds., Comprehensive Biochemistry, Vol. 1. Elsevier, New York, 1962.
11. Bentley, R., Molecular Asymmetry in Biology, Vol. 1. Academic Press, New York, 1969.
12. Korolkovas, A., Essentials of Molecular Pharmacology. John Wiley, New York, 1970 (Ch. 5, stereochemical aspects of drugs).

Chapter 5

1. Bunton, C. A., Nucleophilic Substitution at a Saturated Carbon Atom. Elsevier, New York, 1963.
2. Streitwieser, A., Solvolytic Displacement Reactions. McGraw-Hill, New York, 1962.
3. Gould, E. S., Mechanism and Structure in Organic Chemistry. Holt, New York, 1959 (Ch. 8).
4. Hammett, L. P., Physical Organic Chemistry, 2nd ed. McGraw-Hill, New York, 1970 (Ch. 6).
5. Ingold, C. K., Structure and Mechanism in Organic Chemistry, 2nd ed. Cornell University Press, Ithaca, 1969 (Ch. VII).
6. de la Mare, P. B., "Stereochemical Factors in Reaction Mechanisms and Kinetics." *In* Progress in Stereochemistry, Vol. 1. Butterworths, London, 1954.
7. Breslow, R., Organic Reaction Mechanisms, 2nd ed. W. A. Benjamin, New York, 1969 (Ch. 3 and Special Topic 3).
8. Olah, G. A., and Schleyer, P. V., Carbonium Ions, Vol. 1. John Wiley, New York, 1968 (part of a four-volume series).
9. Wheland, G. W., Advanced Organic Chemistry, 3rd ed. John Wiley, New York, 1960 (Ch. 8, esp. pp. 378–391; and Ch. 12, esp. pp. 561–569).
10. Saunders, W. H., Ionic Aliphatic Reactions. Prentice-Hall, Englewood Cliffs, 1964 (Ch. 3).
11. March, J., Advanced Organic Chemistry. McGraw-Hill, New York, 1968 (Ch. 10).
12. Wilen, S. H., "Resolving Agents and Resolutions in Organic Chemistry." *In* Allinger, N. L., and Eliel, E. L., eds., Topics in Stereochemistry, Vol. 6. John Wiley, New York, 1971.
13. Wheland, G. W., Resonance in Organic Chemistry. John Wiley, New York, 1955.
14. Hartshorn, S. R., Aliphatic Nucleophilic Substitution. Cambridge, New York, 1973.

Chapter 6

1. Banthorpe, D. V., Elimination Reaction. Elsevier, New York, 1963.
2. Saunders, W. H., Ionic Aliphatic Reactions. Prentice-Hall, Englewood Cliffs, 1965 (Chs. 4, 5, and 7).
3. March, J., Advanced Organic Chemistry. McGraw-Hill, New York, 1968 (Ch. 17).
4. Wagner, R. B., and Zook, H. D., Synthetic Organic Chemistry. John Wiley, New York, 1953 (Ch. 2).
5. Norman, R. O. C., Principles of Organic Synthesis. Barnes and Noble, New York, 1968 (Ch. 6, organometallic reagents).
6. Buehler, C. A., and Pearson, D. E., Survey of Organic Synthesis. Interscience, New York, 1970 (Ch. 2).
7. Breslow, R., Organic Reaction Mechanisms, 2nd ed. W. A. Benjamin, New York, 1969 (Ch. 4).
8. Buchanan, G. L., "Bredt's Rule," *Chem. Soc. Rev.*, **3**:41, 1974.
9. Fry, A., "Isotope Effect Studies of Elimination Reactions," *Chem. Soc. Rev.*, **1**:163, 1972.
10. Kharasch, M. S., and Reinmuth, O., Grignard Reagents of Non-Metallic Substances. Prentice-Hall, Englewood Cliffs, 1954.

Chapter 7

1. Hanack, M., Conformation Theory. Academic Press, New York, 1965.
2. Hirsch, J. A., Concepts in Theoretical Organic Chemistry. Allyn and Bacon, Boston, 1974 (Ch. 12).
3. Chiurdoglu, G., ed., Conformational Analysis. Academic Press, New York, 1971.
4. Ihde, A. J., "The Development of Strain Theory." *In* Kekulé Centennial, Advances in Chemistry, Ser. No. 66. American Chemical Society, Washington, 1966.
5. Wheland, G. W., Advanced Organic Chemistry, 3rd ed. John Wiley, New York, 1960 (Chs. 10 and 12).

6. Eliel, E. L., Allinger, N. L., Angyal, S. J., and Morrison, G. A., Conformational Analysis. John Wiley, New York, 1965.
7. Orloff, H. D., "The Stereoisomerism of Cyclohexane Derivatives," *Chem. Rev.*, **54**:347, 1954.
8. McKenna, J., Conformational Analysis of Organic Compounds. Lecture Series, No. 1, The Royal Institute of Chemistry, London, 1966.
9. Ferguson, L. N., Organic Molecular Structure. Willard Grant, Boston, 1975 (Chs. 17–19).
10. Lambert, J. B., "The Shapes of Organic Molecules." *In* Calvin, M., and Pryor, W. A., eds., Organic Chemistry of Life. W. H. Freeman, San Francisco, 1973.
11. L. F. Feiser, "Steroids." *In* Calvin, M., and Pryor, W. A., *ibid.*

Chapter 8

1. March, J., Advanced Organic Chemistry. McGraw-Hill, New York, 1968 (Ch. 15).
2. Hine, J., Physical Organic Chemistry, 2nd ed. McGraw-Hill, New York, 1962 (Ch. 9).
3. de la Mare, P. B., and Bolton, R., Electrophilic Addition to Unsaturated Systems. Elsevier, New York, 1966.
4. Breslow, R., Organic Reaction Mechanisms, 2nd ed. W. A. Benjamin, New York, 1969 (Ch. 4).
5. Bond, G. C., "Mechanism of Catalytic Hydrogenation and Related Reactions," *Quart. Rev. (London),* **8**:279, 1954.
6. Goodman, M., "Concepts of Polymer Stereochemistry." *In* Allinger, N. L., and Eliel, E. L., eds., Topics in Stereochemistry, Vol. 2. John Wiley, New York, 1967.
7. Natta, G., and Farina, M., Stereochemistry. Harper and Row, New York, 1972 (Ch. VI, polymer stereochemistry).
8. Smith, K., "Preparation of Organoboranes: Reagents for Organic Synthesis," *Chem. Soc. Rev.*, **3**:443, 1974.
9. Stewart, R., Oxidation Mechanisms. W. A. Benjamin, New York, 1964 (Ch. 5).
10. Brown, H. C., "Recent Developments in Hydroboration and Organoboranes." *In* Milligan, W. O., ed., Organic-Inorganic Reagents in Synthetic Chemistry, Vol. XVII of Proceedings of the Robert A. Welch Foundation Conferences on Chemical Research, Houston, 1974.
11. Bethell, D., "Structure and Mechanism in Carbene Chemistry." *In* Gold, V., ed., Advances in Physical Organic Chemistry, Vol. 7. Academic Press, New York, 1969.
12. Mark, H. F., "Giant Molecules." *In* Calvin, M., and Jorgensen, M. J., eds., Bio-Organic Chemistry. W. H. Freeman, San Francisco, 1968.

Chapter 9

1. Rinehart, K. L., Oxidation and Reduction of Organic Compounds. Prentice-Hall, Englewood Cliffs, 1973 (Chs. 2 and 3).
2. Buehler, C. A., and Pearson, D. E., Survey of Organic Synthesis. John Wiley, New York, 1970 (Ch. 3).
3. Hickenbottom, W. J., Reactions of Organic Compounds. Longmans, London, 1957 (pp. 61–69).
4. Sandler, S. R., and Karo, W., Organic Functional Group Preparation. Academic Press, New York, 1968 (Ch. 3).
5. Wagner, R. B., and Zook, H. D., Synthetic Organic Chemistry. John Wiley, New York, 1953 (Ch. 3).
6. Eglinton, G., and McCrae, W., "The Coupling of Acetylenic Compounds." *In* Raphael, R. A., Taylor, E. C., and Wynberg, H., eds., Advances in Organic Chemistry, Vol. 4. John Wiley, New York, 1963.

Chapter 10

1. Wagner, R. B., and Zook, H. D., Synthetic Organic Chemistry. John Wiley, New York, 1953 (Ch. 3).
2. Buehler, C. A., and Pearson, D. E., Survey of Organic Synthesis. John Wiley, New York, 1970 (Ch. 4).
3. Sandler, S. R., and Karo, W., Organic Functional Group Preparation. Academic Press, New York, 1968 (Ch. 4).
4. Zweifel, G., and Brown, H. C., "Hydration of Olefins, Dienes and Acetylenes via Hydroboration." *In* Adams, R., *et al.*, eds., Organic Reactions, Vol. 13. John Wiley, New York, 1963.
5. Rinehart, K. L., Oxidation and Reduction of Organic Compounds. Prentice-Hall, Englewood Cliffs, 1973 (Chs. 5 and 6, redox reactions of alcohols and carbonyl compounds).
6. Stewart, R., Oxidation Reactions. W. A. Benjamin, New York, 1964 (Chs. 4, 5, and 10).
7. Clapp, L. B., The Chemistry of the OH Group. Prentice-Hall, Englewood Cliffs, 1967 (a blend of inorganic and organic chemistry).
8. Mendelson, J. H., "Alcohol." *In* Clark, W. G., and Del Giudice, J., eds., Principles of Psychopharmacology. Academic Press, New York, 1970.

Chapter 11

1. Sandler, S. R., and Karo, W., Organic Functional Group Preparation. Academic Press, New York, 1968 (Ch. 5).
2. Buehler, C. A., and Pearson, D. E., Survey of Organic Synthesis. John Wiley, New York, 1970 (Ch. 6).
3. Wagner, R. B., and Zook, H. D., Synthetic Organic Chemistry. John Wiley, New York, 1953 (Ch. 6).
4. Hickenbottom, W. J., Reactions of Organic Compounds. Longmans, London, 1957 (Ch. IV).
5. Patai, S., ed., Chemistry of the Ether Linkage. John Wiley, New York, 1966.
6. Catchpool, J. F., "The Pauling Theory of General Anesthesia." *In* Rich, A., and Davidson, N., eds., Structural Chemistry and Molecular Biology. W. H. Freeman, San Francisco, 1969 (p. 343).
7. Ferguson, L. N., Organic Chemistry: A Science and an Art. Willard Grant, Boston, 1972 (Sec. 4.1, ether and anesthesia).

Chapter 12

1. Wheland, G. W., Resonance in Organic Chemistry. John Wiley, New York, 1955.
2. Hamer, J., ed., 1,4-Cycloaddition Reactions: The Diels-Alder Reaction in Heterocyclic Syntheses. Academic Press, New York, 1967.
3. Adams, R., *et al.*, eds., Organic Reactions, Vol. IV. John Wiley, New York, 1948 (Chs. 1 and 2, the Diels-Alder reaction).
4. Wilson, E. D., "Pheromones." *In* Calvin, M., and Jorgensen, M. J., eds., Bio-Organic Chemistry. W. H. Freeman, San Francisco, 1968.
5. Jacobson, M., and Beroza, M., "Insect Attractants." *In* Calvin, M., and Jorgensen, M. J., *ibid.*
6. Eglinton, G., and Calvin, M., "Chemical Fossils." *In* Calvin, M., and Jorgensen, M. J., *ibid.*
7. Hubbard, R., and Kropf, A., "Molecular Isomers in Vision." *In* Calvin, M., and Jorgensen, M. J., *ibid.*
8. Natta, G., "Precisely Constructed Polymers." *In* Calvin, M., and Jorgensen, M. J., *ibid.*
9. Khudari, A. K., "The Ripening of Tomatoes," *American Scientist,* **60**(6): 696, 1972.
10. Bentley, R., Molecular Asymmetry in Biology, Vol. 2. Academic Press, New York, 1970 (Chs. 4 and 6).
11. Evans, D. A., and Green, C. L., "Insect Attractants of Natural Origin," *Chem. Soc. Rev.*, **2**:75, 1973.

Chapter 13

1. Lehr, R. E., and Marchand, A. P., Orbital Symmetry: A Problem-Solving Approach. Academic Press, New York, 1972.
2. Hirsch, J. A., Concepts in Theoretical Organic Chemistry. Allyn and Bacon, Boston, 1974 (Ch. 3).
3. Woodward, R. B., and Hoffmann, R., The Conservation of Orbital Symmetry. Academic Press, New York, 1970.
4. Gilchrist, T. L., and Storr, R. C., Organic Reactions and Orbital Symmetry. Cambridge University Press, London, 1972.
5. Depuy, C. H., and Chapman, O. L., Molecular Reactions and Photochemistry. Prentice-Hall, Englewood Cliffs, 1972 (Ch. 6).
6. Volmer, J. J., and Service, K. L., "Woodward-Hoffmann Rules: Electrocyclic Reactions," *J. Chem. Ed.*, **45**:214, 1968; "Woodward-Hoffmann Rules: Cycloaddition Reactions," *J. Chem. Ed.*, **47**:491, 1970.
7. Breslow, R., Organic Reaction Mechanisms, 2nd ed. W. A. Benjamin, New York, 1969 (p. 256).
8. Coyle, J. D., "The Photochemistry of Olefinic Compounds," *Chem. Soc. Rev.*, **3**:329, 1974.
9. Coxon, J. M., and Halton, B., Organic Photochemistry. Cambridge, New York, 1974 (Chs. 2 and 4).

Chapter 14

1. Hirsch, J. A., Concepts in Theoretical Organic Chemistry. Allyn and Bacon, Boston, 1974 (Ch. 2).
2. Breslow, R., "The Nature of Aromatic Molecules," *Scientific American*, **227**(2): 32, 1972.
3. Kekulé Centennial, Advances in Chemistry, Ser. No. 66. American Chemical Society, Washington, 1966.
4. Aromaticity, Special Publication No. 21. The Chemical Society, London, 1967.
5. Ginsburg, D., Non-Benzenoid Aromatic Compounds. John Wiley, New York, 1959.
6. Ferguson, L. N., Organic Molecular Structure. Willard Grant, Boston, 1975 (Chs. 5 and 6).

Chapter 15

1. Stock, L. M., Aromatic Substitution Reactions. Prentice-Hall, Englewood Cliffs, 1968.
2. Tomlinson, M., An Introduction to the Chemistry of Benzenoid Compounds. Pergamon Press, New York, 1971 (Chs. 1–3).

3. Breslow, R., Organic Reaction Mechanisms, 2nd ed. W. A. Benjamin, New York, 1969 (Ch. 5).
4. Berliner, E., "Electrophilic Aromatic Substitution Reactions." *In* Cohen, S. G., Streitwieser, A., Jr., and Taft, R. W., eds., Progress in Physical Organic Chemistry, Vol. 2. John Wiley, New York, 1964.
5. de la Mare, P. B. D., and Ridd, J. H., Aromatic Substitution: Nitration and Halogenation. Butterworths, London, 1959.
6. Norman, R. O. C., Principles of Organic Synthesis. Barnes and Noble, New York, 1968 (Ch. 11).
7. Norman, R. O. C., and Taylor, R., Electrophilic Aromatic Substitution in Benzenoid Compounds. Elsevier, New York, 1965.
8. Olah, G., ed., Friedel-Crafts and Related Reactions. John Wiley, New York, 1963 (a four-volume set).

Chapter 16

1. Gutsche, C. D., The Chemistry of Carbonyl Compounds. Prentice-Hall, Englewood Cliffs, 1967.
2. Royals, E. E., Advanced Organic Chemistry. Prentice-Hall, Englewood Cliffs, 1956.
3. Patai, S., ed., Chemistry of the Carbonyl Group. John Wiley, New York, 1965.
4. Hickenbottom, W. J., Reactions of Organic Compounds. Longmans, London, 1957 (Ch. V).
5. Rinehart, K. L., Jr., Oxidation and Reduction of Organic Compounds. Prentice-Hall, Englewood Cliffs, 1973 (Ch. 6, reduction of carbonyl compounds to alcohols).
6. Jencks, W. P., "Mechanism and Catalysis of Simple Carbonyl Group Reactions." *In* Cohen, S. G., Streitwieser, A., Jr., and Taft, R. W., eds., Progress in Physical Organic Chemistry, Vol. 2. John Wiley, New York, 1964.
7. House, H. O., Modern Synthetic Reactions, 2nd ed. W. A. Benjamin, Menlo Park, 1972 (Ch. 2, metal hydride reductions of carbonyl compounds).
8. Kosower, E. M., Molecular Biochemistry. McGraw-Hill, New York, 1962 (Sec. 2.13, pyridine nucleotides; Sec. 1.6, transamination).
9. Bronk, J. R., Chemical Biology. Macmillan, New York, 1973 (Ch. 8, overall view of cellular metabolism).
10. Steiner, R. F., Life Chemistry. Von Nostrand, New York, 1968 (Ch. 9, amino acid metabolism).
11. Schepartz, B., Regulation of Amino Acid Metabolism In Mammals. W. B. Saunders, Philadelphia, 1973.

Chapter 17

1. Wheland, G. W., Advanced Organic Chemistry, 3rd ed. John Wiley, New York, 1960 (Ch. 14, tautomerism and enolization).
2. Norman, R. O. C., Principles of Organic Synthesis. Barnes and Noble, New York, 1968 (Ch. 7, base-catalyzed condensations).
3. House, H. O., Modern Synthetic Reactions, 2nd ed. W. A. Benjamin, Menlo Park, 1972 (pp. 459–478, α-halo ketones; Ch. 10, aldol condensation).
4. Kosower, E. M., Molecular Biochemistry. McGraw-Hill, New York, 1962 (Sec. 2.3, enolization and aldol condensation).
5. Reithel, F. J., Concepts in Biochemistry. McGraw-Hill, New York, 1967 (Ch. 5, glycolysis and fermentation).
6. Djerassi, C., Optical Rotatory Dispersion. McGraw-Hill, New York, 1960.
7. Crabbe, P., Optical Rotatory Dispersion and Circular Dichroism in Organic Chemistry. Holden-Day, San Francisco, 1965.

Chapter 18

1. Hickenbottom, W. J., Reactions of Organic Compounds. Longmans, London, 1957 (Chs. VI and VII).
2. Rinehart, K. L., Jr., Oxidation and Reduction of Organic Compounds. Prentice-Hall, Englewood Cliffs, 1973 (Ch. 7, interconversion of aldehydes and ketones with acid derivatives).
3. Sonntag, N. O. V., "Reactions of Aliphatic Acid Chlorides," *Chem. Rev.*, **52**:237, 1953.
4. Bruice, T. C., and Benkovic, S., Bioorganic Chemistry, Vol. 1, W. A. Benjamin, New York, 1966 (Ch. 3, thiolesters).
5. Westheimer, F. H., "Mechanism of Enzymic Decarboxylation." *In* Milligan, O., ed., Bio-Organic Chemistry and Mechanism, Vol. XV of Proceedings of the Robert A. Welch Foundation Conferences on Chemical Research, Houston, 1972.
6. Seymour, R. B., Introduction to Polymer Chemistry. McGraw-Hill, New York, 1971.
7. Bachmann, W. E., and Struve, S. W., "The Arndt-Eistert Synthesis." *In* Adams, R., *et al.*, eds., Organic Reactions, Vol. 1. John Wiley, New York, 1942.
8. Bender, M. L., "Mechanisms of Catalysis of Nucleophilic Reactions of Carboxylic Acid Derivatives," *Chem. Rev.*, **60**:53, 1960.

Chapter 19

1. Bergmann, E. D., Ginsburg, D., and Pappo, R., "The Michael Reaction." *In* Adams, R., *et al.*, eds., Organic Reactions, Vol. 10. John Wiley, New York, 1959.
2. Hauser, C. R., and Hudson, B. E., "The Acetoacetic Ester Condensation." *In* Adams, R., *et al.*, eds., Organic Reactions, Vol. 1. John Wiley, New York, 1942.
3. Johnson, J. R., "The Perkin Reaction." *In* Adams, R., *et al.*, eds., *ibid.*
4. Bruson, H. A., "Cyanoethylation." *In* Adams, R., *et al.*, eds., Organic Reactions, Vol. 5. John Wiley, New York, 1949.
5. House, H. O., Modern Synthetic Reactions, 2nd ed. W. A. Benjamin, Menlo Park, 1972 (Ch. 9, alkylation of active methylene groups; pp. 671–709, Reformatsky reaction; pp. 682–709, Wittig reaction).
6. Maercker, A., "The Wittig Reaction." *In* Adams, R., *et al.*, eds., Organic Reactions, Vol. 14. John Wiley, New York, 1965.
7. Bruice, T. C., and Benkovic, S., Bioorganic Mechanisms, Vol. 1. W. A. Benjamin, New York, 1966 (Ch. 4, hydration and specificity; discussion of steps in the Krebs cycle and related processes).
8. Bronk, J. R., Chemical Biology. Macmillan, New York, 1973 (pp. 347–361, Krebs cycle).
9. McGilvery, R. W., Biochemical Concepts. W. B. Saunders, Philadelphia, 1975 (Ch. 12).
10. Crabbe, P., "Prostaglandins: Production and Prospects," *Chemistry in Britain,* **11**:132, 1975.

Chapter 20

1. Sandler, S. R., and Karo, W., Organic Functional Group Preparation. Academic Press, New York, 1968 (Ch. 13).
2. Paquette, L. A., Principles of Modern Heterocyclic Chemistry. W. A. Benjamin, New York, 1968.
3. Cook, A. G., Enamines: Synthesis, Structure and Reactions. Marcel Dekker, New York, 1969.
4. Smith, P. A. S., The Organic Chemistry of Open-Chain Nitrogen Compounds. W. A. Benjamin, New York, 1966 (two volumes).
5. Elderfield, R. C., Heterocyclic Compounds, Vols. 1–7. John Wiley, New York, 1950–1961.
6. House, H. O., Modern Synthetic Reactions, 2nd ed. W. A. Benjamin, Menlo Park, 1972 (pp. 570–586, enamines).
7. Manske, R. H. F., and Holmes, H. L., The Alkaloids, Vols. 1–9. Academic Press, New York, 1950–1967.
8. Ferguson, L. N., Organic Chemistry: A Science and an Art. Willard Grant, Boston, 1972 (pp. 44–51, amines as pain killers).
9. Farnsworth, N. R., "Hallucinogenic Plants," *Science,* **162**:1086, 1968.
10. Hammond, A. L., "Narcotic Antagonists: New Methods to Treat Heroin Addiction," *Science,* **173**:503, 1971.

Chapter 21

1. Sandler, S. R., and Karo, W., Organic Functional Group Preparation. Academic Press, New York, 1968 (Chs. 14–16, aryl nitrogen compounds).
2. Zollinger, H., Azo and Diazo Chemistry. John Wiley, New York, 1961.
3. Crampton, M. R., "Meisenheimer Complexes." *In* Gold, V., ed., Advances in Physical Organic Chemistry, Vol. 7. Academic Press, New York, 1969.
4. Hoffmann, R. W., Dehydrobenzene and Cycloalkynes. Academic Press, New York, 1967.
5. Fields, E. K., and Meyerson, S., "Mechanism of Formation and Reactions of Arynes at High Temperatures." *In* Gold, V., ed., Advances in Physical Organic Chemistry, Vol. 6. Academic Press, New York, 1968.

Chapter 22

1. Sandler, S. R., and Karo, W., Organic Functional Group Preparation. Academic Press, New York, 1968 (Ch. 4).
2. Kosower, E. M., Molecular Biochemistry. McGraw-Hill, New York, 1962 (Sec. 2.14, flavins; Sec. 1.3, oxidation).
3. Reithel, F. J., Concepts in Biochemistry. McGraw-Hill, New York, 1967 (Ch. 3, the mitochondrion and biological oxidation; Ch. 28, metalloporphyrins).
4. McGilvery, R. W., Biochemical Concepts. W. B. Saunders, Philadelphia, 1975 (Ch. 11).

Chapter 23

1. Senning, A., ed., Sulfur in Organic and Inorganic Chemistry, Vols. 1–3. Marcel Dekker, New York, 1972.
2. Reid, E. E., Organic Chemistry of Bivalent Sulfur, Vols. 1–5. Chemical Publishing Co., New York, 1958–1963.
3. Young, L., and Maw, G. A., The Metabolism of Sulphur Compounds. John Wiley, New York, 1958.
4. House, H. O., Modern Synthetic Reactions, 2nd ed. W. A. Benjamin, Menlo Park, 1972 (pp. 709–733, sulfur ylids).
5. Challenger, F., Aspects of the Organic Chemistry of Sulphur. Academic Press, New York, 1959 (takes a more biological slant).
6. Durst, T., "Dimethylsulfoxide (DMSO) in Organic Synthesis." *In* Taylor, E. C., and Wynberg, H., eds., Advances in Organic Chemistry, Vol. 6. John Wiley, New York, 1969.
7. Trost, B. M., and Melvin, L. S., Jr., Sulfur Ylides. Academic Press, New York, 1975.
8. Goldstein, A., Aronow, L., and Kalman, S. M., Principles of Drug Action. Harper and Row, New York, 1969 (Ch. 12, chemical teratogenesis; Ch. 14, drug selection for clinical use).
9. Kirby, A. J., and Warren, S. G., The Organic Chemistry of Phosphorus. Elsevier, New York, 1967.
10. Hudson, R. F., Structure and Mechanism in Organo-Phosphorus Chemistry. Academic Press, New York, 1965.
11. Quin, L. D., "The Natural Occurrence of Compounds with the Carbon-Phosphorus Bond." *In* Grayson, M., and Griffith, E. J., eds., Topics in Phosphorus Chemistry, Vol. 4. John Wiley, New York, 1967.
12. Kosower, E. M., Molecular Biochemistry. McGraw-Hill, New York, 1962 (Sec. 2.15, phosphates).
13. Korolkovas, A., Essentials of Molecular Pharmacology. John Wiley, New York, 1970 (pp. 197–231, cholinergic receptors).
14. Bruice, T. C., and Benkovic, S., Bioorganic Chemistry, Vol. 2. W. A. Benjamin, New York, 1966 (Chs. 5–7, phosphate esters).

Chapter 24

1. Barker, R., Organic Chemistry of Biological Compounds. Prentice-Hall, Englewood Cliffs, 1971 (Ch. 4).
2. Kaldor, G., Physiological Chemistry of Proteins and Nucleic Acids in Mammals. W. B. Saunders, Philadelphia, 1969.
3. Rich, A., and Davidson, N., eds., Structural Chemistry and Molecular Biology. W. H. Freeman, San Francisco, 1968 (first two sections, "The Structure of Proteins" and "The Chemistry of Proteins").
4. Kopple, K. D., Peptides and Amino Acids. W. A. Benjamin, New York, 1966.
5. Bentley, R., Molecular Asymmetry in Biology, Vol. 2. Academic Press, New York, 1970 (Ch. 5, Sec. III, configurations of amino acids).
6. Reithel, F. J., Concepts in Biochemistry. McGraw-Hill, New York, 1967 (Chs. 11–15, 20–24).
7. Bronk, J. R., Chemical Biology. Macmillan, New York, 1973 (Ch. 3).
8. Scheraga, H. A., "Calculations of Conformations of Polypeptides." *In* Gold, V., ed., Advances in Physical Organic Chemistry, Vol. 6. Academic Press, New York, 1968.
9. Doty, P., "Proteins." *In* Calvin, M., and Jorgensen, M. J., eds., Bio-Organic Chemistry. W. H. Freeman, San Francisco, 1968.
10. Stein, W. H., and Moore, S., "The Chemical Structure of Proteins." *In* Calvin, M., and Jorgensen, M. J., eds., *ibid.*
11. Neurath, H., "Protein-Digesting Enzymes." *In* Calvin, M., and Jorgensen, M. J., eds., *ibid.*
12. Collier, O. J., "Kinins." *In* Calvin, M., and Jorgensen, M. J., eds., *ibid.*
13. Thompson, E. O. P., "The Insulin Molecule." *In* Calvin, M., and Jorgensen, M. J., eds., *ibid.*
14. Stroud, R. M., "A Family of Protein-Cutting Proteins," *Scientific American,* **231**(1):74, 1974.
15. Lehninger, A. L., Biochemistry, 2nd ed. Worth, New York, 1975.

Chapter 25 A and B

1. Barker, R., Organic Chemistry of Biological Compounds. Prentice-Hall, Englewood Cliffs, 1972 (Ch. 5).
2. Shreeve, W. W., Physiological Chemistry of Carbohydrates in Mammals. W. B. Saunders, Philadelphia, 1974 (Ch. 1).
3. Bronk, J. R., Chemical Biology. Macmillan, New York, 1973 (Ch. 2).
4. Bassham, J. A., "The Path of Carbon in Photosynthesis." *In* Calvin, M., and Jorgensen, M. J., eds., Bio-Organic Chemistry. W. H. Freeman, San Francisco, 1968.
5. Kretchmer, N., "Lactose and Lactase," *Scientific American,* **227**(4):70, 1972.
6. Nordsick, F. W., "The Sweet Tooth," *American Scientist,* **60**(1):41, 1972.
7. Hall, F. K., "Wood Pulp," *Scientific American,* **230**(4):52, 1974.

8. Sharon, N., "Glycoproteins," *Scientific American,* **230**(5): 78, 1974.
9. Kennedy, J. F., "Chemical Aspects of Glycoproteins, Proteoglycans and Carbohydrate Protein Complexes of Human Tissues," *Chem. Soc. Rev.,* **2**:355, 1973.
10. Pigman, W. W., and Horton, D., The Carbohydrates. Academic Press, New York, 1970. (The first in a several-volume treatise.)

Chapter 26

1. Barker, R., Organic Chemistry of Biological Compounds. Prentice-Hall, Englewood Cliffs, 1971 (Ch. 8).
2. Spencer, J. H., The Physics and Chemistry of DNA and RNA. W. B. Saunders, Philadelphia, 1972.
3. Reithel, F. J., Concepts in Biochemistry. McGraw-Hill, New York, 1967 (Chs. 17–20).
4. Kaldor, G., Physiological Chemistry of Proteins and Nucleic Acids in Mammals. W. B. Saunders, Philadelphia, 1969.
5. Kosower, E. M., Molecular Biochemistry. McGraw-Hill, New York, 1962 (Sec. 1.7, purines and pyrimidines).
6. Tatum, E. L., "Molecular Biology, Nucleic Acids and the Future of Medicine." *In* Lyght, C. E., ed., Reflections on Research and the Future of Medicine. McGraw-Hill, New York, 1967.
7. Calvin. M., "Chemical Evolution," *American Scientist,* **63**(2):169, 1975.
8. Asimov, I., The Genetic Code. Signet Books, New York, 1962.
9. Watson, J. D., The Double Helix. Signet Books, New York, 1969 (an interesting comparison to Asimov's book).

Chapter 27

1. Koshland, D. E., Jr., "Protein Shape and Biological Control," *Scientific American,* **229**(4):52, 1973.
2. Koshland, D. E., Jr., "The Catalytic Power of Enzymes." *In* Milligan, W. O., ed., Bio-Organic Chemistry and Mechanism, Vol. XV of Proceedings of the Robert A. Welch Foundation Conferences on Chemical Research, Houston, 1972.
3. Lipscomb, W. N., "Structures and Mechanisms of Enzymes." *In* Milligan, W. O., ed., *ibid.*
4. Cleland, W. W., "What Limits the Rate of an Enzyme-Catalyzed Reaction?" *Accounts of Chemical Research,* **8**:145, 1975.
5. Korolkovas, A., Essentials of Molecular Pharmacology. John Wiley, New York, 1970 (pp. 290–305, action of drugs on enzymes).
6. Suckling, C. J., and Suckling, K. E., "Enzymes in Organic Synthesis," *Chem. Soc. Rev.,* **3**:387, 1974.
7. Reithel, F. J., Concepts in Biochemistry. McGraw-Hill, New York, 1967 (Ch. 30, cell membranes and cell walls).
8. Bloomfield, V. A., and Harrington, R. E., eds., Biophysical Chemistry. W. H. Freeman, San Francisco, 1975 (twenty-one interesting articles from *Scientific American*).

Chapter 28

1. Silverstein, R. M., Bassler, G. C., and Morrill, T. C., Spectrometric Identification of Organic Compounds, 3rd ed. John Wiley, New York, 1974.
2. Nakanishi, K., Infrared Absorption Spectroscopy. Holden-Day, San Francisco, 1964.
3. Conley, R. T., Infrared Spectroscopy, 2nd ed. Allyn and Bacon, Boston, 1972.
4. Pasto, D. J., and Johnson, C. R., Organic Structure Determination. Prentice-Hall, Englewood Cliffs, 1969 (Ch. 3, ultraviolet spectroscopy; Ch. 4, infrared spectroscopy; Ch. 8, mass spectrometry).
5. Jaffé, H. H., and Orchin, M., Theory and Applications of Ultraviolet Spectroscopy. John Wiley, New York, 1962.
6. Schrader, S. R., Introductory Mass Spectrometry. Allyn and Bacon, Boston, 1971.
7. McLafferty, F. W., Interpretation of Mass Spectra, 2nd ed. W. A. Benjamin, Reading, 1973.
8. Bentley, T. W., and Johnstone, R. A. W., "Mechanism and Structure in Mass Spectrometry: A Comparison with Other Chemical Processes." *In* Gold, V., ed., Advances in Physical Organic Chemistry, Vol. 8. Academic Press, New York, 1970.

Chapter 29

1. Silverstein, R. M., Bassler, G. C., and Morrill, T. C., Spectrometric Identification of Organic Compounds, 3rd ed. John Wiley, New York, 1974.

2. Jackman, L. M., and Sternhell, S., Applications of Nuclear Magnetic Resonance Spectroscopy in Organic Chemistry, 2nd ed. Pergamon Press, New York, 1969.
3. Bible, R. H., Jr., Interpretation of NMR Spectra, an Empirical Approach. Plenum Press, New York, 1965.
4. Paudler, W. W., Nuclear Magnetic Resonance. Allyn and Bacon, Boston, 1971.
5. Pasto, D. J., and Johns, C. R., Organic Structure Determination. Prentice-Hall, Englewood Cliffs, 1969 (Ch. 5).
6. Mays, B. C., "Lanthanide Shift Reagents in Nuclear Magnetic Resonance Spectroscopy," *Chem. Soc. Rev.*, **2**:49, 1973.
7. Cudby, M. E. A., and Willis, H. A., "The Nuclear Magnetic Resonance Spectra of Polymers." *In* Mooney, E. F., ed., Annual Reports on NMR Spectroscopy, Vol. 4. Academic Press, New York, 1971.
8. James, T. L., Nuclear Magnetic Resonance in Biochemistry. Academic Press, New York, 1975.

ANSWERS TO SELECTED PROBLEMS

Chapter 1

1. (a) mass number: number of protons + neutrons in nucleus; atomic number: number of protons in nucleus; (b) atomic weight: average weight of an atom of an element; atomic number: number of protons in nucleus; (c) they are identical; (d) electronegativity: tendency of an atom to attract its own outer electrons; electron affinity: energy released upon addition of an electron to an atom; (e) cation: a positively charged species; anion: a negatively charged species; (f) deuterium: isotope of hydrogen with a mass number of 2; tritium: isotope of hydrogen with a mass number of 3; (g) 2*s* orbital: spherically symmetric orbital of the second main quantum level; 2*p* orbital: "dumbbell"-shaped orbital of the second main quantum level; (h) orbit: circular path around the nucleus which the electron was postulated as taking; orbital: volume around a nucleus in which there is a high probability of finding an electron.

2. (a) atomic number 82-91; (b) atomic number 82-91.

3. (a) suggested that no two electrons in the same atom can have the same four quantum numbers; (b) provided a quantitative relationship for the dualistic (wave-particle) theory of the electron; (c) presented the "planetary model" of the atom and suggested that electrons occupy discrete energy levels; (d) discovered the nucleus by studying the bombardment of gold foil with α particles; (e) demonstrated that electrons possess a magnetic moment (related to their spin about their axes); (f) discovered that normal spectral lines may be split into several closely spaced lines when the atoms providing the spectral lines are placed in a magnetic field (the Zeeman effect); (g) described atomic particles in terms of wave properties, their mass and energy, and conceived of the "wave equation"; (h) established the most commonly used scale of electronegativity.

4. ^{16}O: $n = 2, l = 1, m = -1, s = +\frac{1}{2}$; ^{14}N: $n = 2, l = 1, m = +1, s = -\frac{1}{2}$; ^{32}S: $n = 3$, $l = 1$, $m = -1$, $s = +\frac{1}{2}$; ^{31}P: $n = 3$, $l = 1$, $m = +1$, $s = -\frac{1}{2}$; U: $n = 6$; $l = 2$, $m = -1$, $s = -\frac{1}{2}$ (see Table 1–3 for filling order). Note that these answers do *not* depend upon any specific isotope.

5. Same quantum numbers as sulfur, after deletion of the last two electrons (see Table 1–4). Outer electrons in carbon and silicon are similar except for main quantum levels ($n = 2$ for C and $n = 3$ for Si).

6. O: 16.00; B: 10.80; N: 14.01.

7. $(n - l) - 1$ = number of nodes.

8. LiBr.

9. The electron that is lost most readily is the one most shielded (and farthest removed) from the nucleus. The rules that dictate aufbau filling do not necessarily place the "last" electron the farthest from the nucleus.

10. Within a given quantum level the greater the penetration, the greater the electronegativity; *s* orbitals penetrate more than do *p* orbitals (Figure 1–2).

11. The data suggest that one electron in an outer quantum level is more mobile and, therefore, provides good conductivity. Since $d^{10}s^1$ contains a completely filled and a half-filled level, and d^9s^2 contains a completely filled and a partly (but not half) filled level, one concludes that half-filled levels are more stable than partly filled (but not half-filled) levels.

12. Shielding of outer electrons from the nucleus by the inner electrons increases as atomic number increases.

13. The electronegative portion is written on the right both for simple molecules (*e.g.*, NaCl and $CaCO_3$) and for complex anions (*e.g.*, CrO_4^{-2}).

Chapter 2

1. All bonds lower the energy of the system, *i.e.*, all systems are stabilized by bonds. These bonds are also non-directional. However, they vary in strength: ionic > ion-dipole > dipole-dipole.

2. Order of stability He > $He_2^{\oplus}$ > He_2.

3. NO contains an unpaired electron (in π^*); CO does not contain an unpaired electron.

4. (a) H—$\ddot{\underset{\cdot\cdot}{\text{Cl}}}$: (b) H—N(H)(H): (c) :$\ddot{\underset{\cdot\cdot}{\text{Br}}}$—$\ddot{\underset{\cdot\cdot}{\text{Br}}}$: (d) H—$\ddot{\underset{\cdot\cdot}{\text{O}}}$—H

(e) :$\ddot{\underset{\cdot\cdot}{\text{Cl}}}$—C(H)(H)—H (f) (:$\ddot{\text{Cl}}$)₂C=C(:$\ddot{\text{Cl}}$:)₂ (g) H—C≡C—H

5. Individual bond moments cancel.

6. Great electronegativity of F renders H—F quite polar.

7. (a) trigonal (planar); (b) digonal (linear); (c) and (d) 4; (e) 4; (f) 1; (g) 2; (h) 4; (i) 2; (j) 109.5°; (k) 180°.

8. H's bonded to *p* orbitals; non-bonding electron pairs in *sp* orbitals.

9. All species contain non-bonding electron pairs and are, therefore, Lewis bases. The positive charge makes $H_3O^{\oplus}$ the poorest.

10. Lewis bases; they possess non-bonding electron pairs.

11. Lewis bases transfer two electrons to one atom to form a covalent bond.

12. (a) $H_2C{=}CH_2$ (b) H—C(H)(H)—H (c) $H_2C{=}CH$—$\ddot{\underset{\cdot\cdot}{\text{F}}}$: (d) CH_3—$\overset{\oplus}{\text{N}}(CH_3)_2$—$\ddot{\underset{\cdot\cdot}{\text{O}}}$:$^{\ominus}$

a-c: because of minimum charge separation; *d:* other has 10 electrons around nitrogen.

13. (a) hydroxy, carboxy; (b) vinyl, phenyl; (c) bromo, amino, (substituted) phenyl; (d) cyano, hydroxy; (e) methyl, ethyl, carbonyl (keto); (f) phenyl, azo.

14. Using the abbreviations in the text, for hydrogen: $A = 1$, $B = 0$, $C = 1$, $D = 0$,

$E = +1$. For sodium: $A = 1$, $B = 0$, $C = 1$, $D = 0$, $E = +1$. For chlorine: $A = 7$, $B = 0$, $C = 7$, $D = 8$, $E = -1$.

15. Items A–E (as used in text) are identical for both oxygens: $A = 6$, $B = 2$, $C = 8$, $D = 8$, $E = 0$.

Chapter 3

1. Since each point on the curve represents a conformation, an infinite number.

2. All maxima are transition states. They are of identical structure; hence, of equal energy.

3. **C** and **E** are mirror images of one another (same energy; similar geometry). **D** is transition state between them.

4. Increasing temperature leads toward identical concentrations of all conformations. However, temperature necessary for this is sometimes beyond decomposition temperature of compound. *Conformer distribution* of ethane unchanged; that of butane approaches equimolar.

5. (a) butane (a butane); (b) 2,3-dimethylbutane (a hexane); (c) 4-ethyl-2,6-dimethylheptane (an undecane); (d) 2,2,3,3-tetramethylpentane (a nonane); (e) 3,3,4,4-tetramethylhexane (a decane); (f) 2,3,4,5-tetramethylhexane (a decane).

6. $K = 10^{156}$ at $T = 25°C$.

7. Ions will attract one another and require energy to be kept apart.

8. N_2 and varying amount of $\left(CH_3{-}\overset{\overset{CN}{|}}{\underset{\underset{CH_3}{|}}{C}}{-}\right)_2$, $CH_3{-}\overset{\overset{CN}{|}}{C}{=}CH_2$ and $CH_3{-}\overset{\overset{CN}{|}}{\underset{\underset{H}{|}}{C}}{-}CH_3$ derived from $CH_3{-}\overset{\overset{CN}{|}}{\underset{\underset{CH_3}{|}}{C}}\cdot$

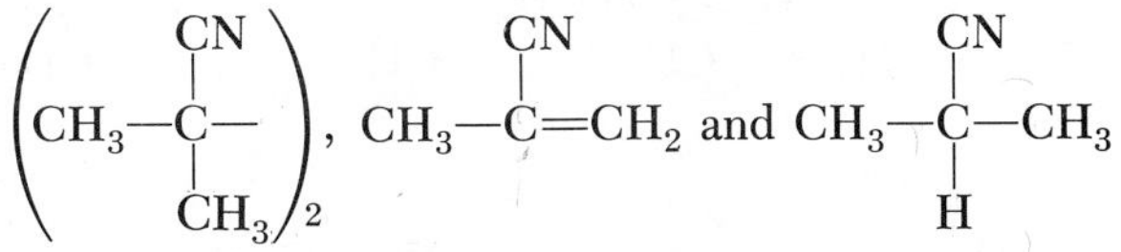

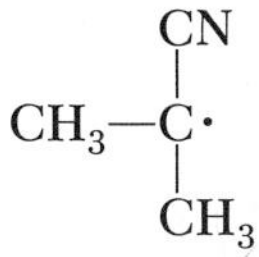

9. $(CH_3)_4C$. Derivatives: $(ClCH_2)_2C(CH_3)_2$ and $Cl_2CHC(CH_3)_3$.

Chapter 4

1. a, c, d, f, g, h, i, k.

2. (a) plane through center and valve; (b) vertical plane through stem; (c) point in center/any plane through center; (d) plane through center of tape parallel to edge of roll; (e) plane passing lengthwise through center of chalk; (f) plane passing through glass and stem through center; (g) plane of the rings. (Other answers are possible for some of these.)

3. No. It may have a plane passing through the spot. Spot precludes a point of symmetry. The Earth is chiral.

4. b, d, i.

5. 0.083 g for 1 dm; 0.17 g for 2 dm.

6. (a) H (top), Cl—(left), —D (right), CH_3 (bottom); (b) H (top), D—(left), —Cl (right), CH_3 (bottom); (c) H (top), D—(left), —Cl (right), CH_3 (bottom); (d) H (top), CH_3—(left), —Cl (right), Br (bottom);
(e) H (top), Br—(left), —CH_3 (right), D (bottom); (f) CH_3 (top), H—(left), —D (right), Br (bottom); (g) H (top), Br—(left), —CH_3 (right), D (bottom); (h) NH_2 (top), CH_3—(left), —CO_2H (right), H (bottom).

7. (a) Cl (top), CH_3 (left), Br (right), H (bottom); (c) Br (top), CH_3 (left), Cl (right), H (bottom); (e) Cl (top), Br (left), CH_3 (right), H (bottom);
(b) Br (top), Cl (left), CH_3 (right), H (bottom); (d) Br (top), H (left), Cl (right), CH_3 (bottom); (f) H (top), Cl (left), Br (right), CH_3 (bottom).

8. (a) Br > Cl > H; (b) I > S > N; (c) $^3H > ^2H > ^1H$;
(d) Sn > Br > H > lone pair.

9. (a) $—C(CH_3)_3 > —CH(CH_3)_2 > —C_2H_5 > —CH_3 > —H$;
(b) $—CH(CH_2CH_3)_2 > —CH_2CH(CH_3)_2 > —CH_2CH_2CH_3$;
(c) $—C(CH_3)_2CH_2CH_2CH_3 > —C(CH_3)_2CH_2CH_3 > —CHCH_3CH_2CH_3$.

10. (a) $—CCH_3{=}CH_2 > —CH{=}CHCH_3$;
(b) $—C{\equiv}C—H > —C(CH_3)_3 > —CH_2CH_2CH_2C(CH_3)_3$;
(c) $—OCH_3 > —CH_2SH > —CH_2OCH_3 > —CH_2OH$.

11. (−)-(**S**)-malic acid: CO_2H (top), HO—C—H, CH_2CO_2H (bottom)

12. *4.6:* (a) **S**; (b) **R**; (c) **R**; (d) **S**; (e) **R**; (f) **S**; (g) **R**; (h) **R**; *4.7:*
(a) **S**; (b) **S**; (c) **R**; (d) **S**; (e) **R**; (f) **S**.

13. (a) **R**; (b) **S**; (c) **S**; (d) **R**; (e) **S**; (f) **R**; (g) **S**; (h) **S**; (i) **S**.

14. (a) Cl—C(H)(H)—Br, both H circled; (b) Cl—C(H)(CH₃)(CH₃), both CH_3 circled; (c) Cl—C(H)(CH_2CH_3)(CH_2CH_3), both CH_2CH_3 circled;
(d) Cl—C(H)(H)—C(H)(H)—Br, the two H on the first carbon circled; (e) Cl—C(H)(H)—C(H)(H)—Cl, the two H on the first carbon circled.

15. (a) 2; (b) 4; (c) 7; (d) 3; (e) 2; (f) 7; (g) 11.

16. (a) 2; (b) 5; (c) 10; (d) 3; (e) 2; (f) 9; (g) 17.

Chapter 5

1. (a) No racemization; only one molecule present.

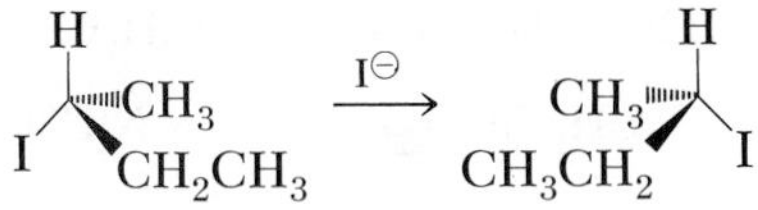

$$\text{H}(\text{I})\text{C}(\text{CH}_3)(\text{CH}_2\text{CH}_3) \xrightarrow{\text{I}^\ominus} \text{H}(\text{CH}_3)(\text{CH}_3\text{CH}_2)\text{C}\text{I}$$

(b) One (**S**)-2-iodobutane formed; racemization.
(c) Must assume an even number of halide molecules present.

2. $\begin{matrix}\text{R—Cl}\\ \text{R—Cl}\end{matrix} + \text{I}^\ominus \rightarrow \begin{matrix}\text{S—I}\\ \text{R—Cl}\end{matrix} + \text{Cl}^\ominus \rightarrow \begin{matrix}\text{S—I}\\ \text{S—Cl}\end{matrix} + \text{Cl}^\ominus \rightarrow \begin{matrix}\text{R—Cl}\\ \text{S—Cl}\end{matrix} + \text{I}^\ominus$

3. (2**R**, 3**R**)-2-chloro-3-methylpentane (*erythro*) $\xrightarrow{\text{Cl}^\ominus}$ (2**S**, 3**R**)-2-chloro-3-methylpentane (*threo*). S_N2 inverts configuration only at C2; racemization does not occur. Yes, this is optically active because of C3.

4. (a) 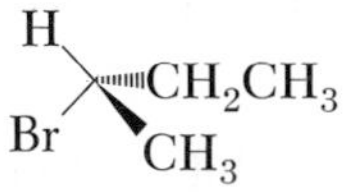 H, Br, CH_2CH_3, CH_3 — (**S**)-2-bromobutane
(b) 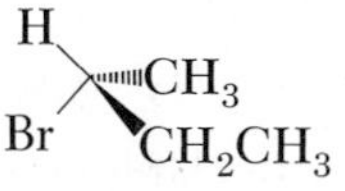 H, Br, CH_3, CH_2CH_3 — (**R**)-2-bromobutane
(c) 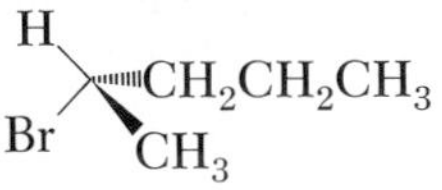 H, Br, $CH_2CH_2CH_3$, CH_3 — (**S**)-2-bromopentane
(d) 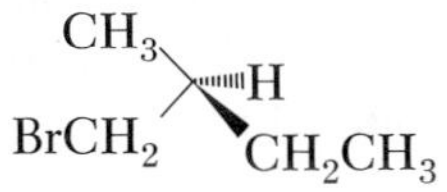 CH_3, $BrCH_2$, H, CH_2CH_3 — (**S**)-1-bromo-2-methylbutane

5. Allyl chloride is accelerated, others react normally (see discussion, above, in small type).

6. Steric hindrance increases as G's size increases.

7. Steric hindrance (problem 6) prevents any S_N2.

8. (a) (±)-$CH_3CHCl(CH_2)_5CH_3$; (b) (±)-$CH_3CHI(CH_2)_5CH_3$;
(c) (±)-$CH_3CHSH(CH_2)_5CH_3$; (d) (±)-$CH_3CH(\overset{\oplus}{N}H_3)(CH_2)_5CH_3$ $I^\ominus$;
(e) (±)-$CH_3CH(CH_2COCH_3)(CH_2)_5CH_3$;

(f) 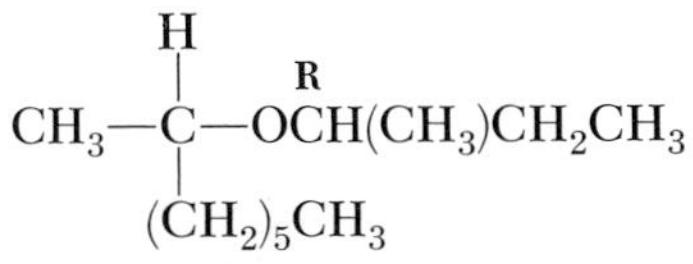 $CH_3—\underset{(CH_2)_5CH_3}{\overset{H}{C}}—\overset{R}{O}CH(CH_3)CH_2CH_3$ and $CH_3—\underset{\underset{R}{OCH(CH_3)CH_2CH_3}}{\overset{H}{C}}—(CH_2)_5CH_3$;

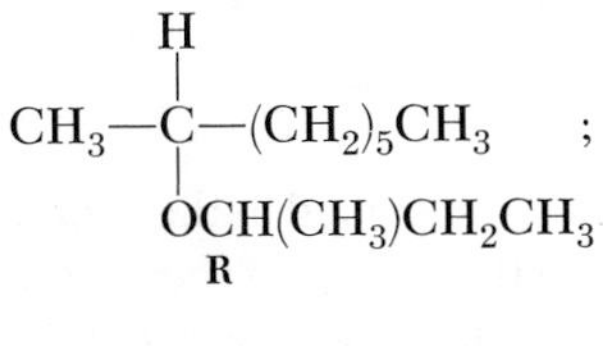

(g) 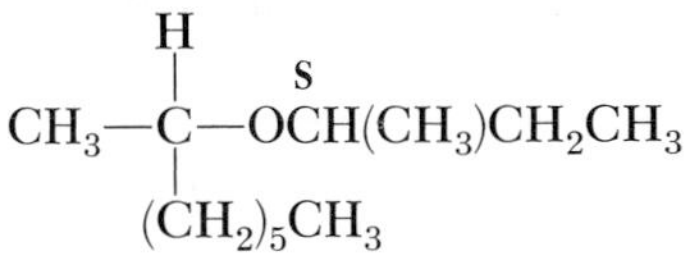 $CH_3—\underset{(CH_2)_5CH_3}{\overset{H}{C}}—\overset{S}{O}CH(CH_3)CH_2CH_3$ and $CH_3—\underset{\underset{S}{OCH(CH_3)CH_2CH_3}}{\overset{H}{C}}—(CH_2)_5CH_3$;

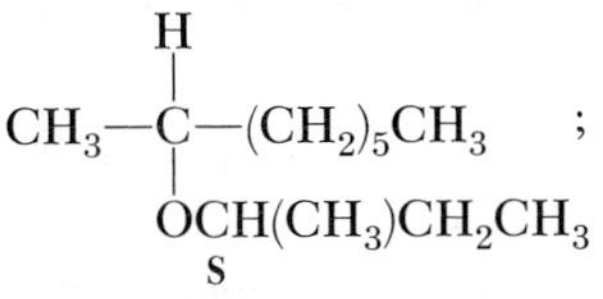

(h) $CH_3\overset{*}{C}H(O\overset{*}{C}H(CH_3)CH_2CH_3)(CH_2)_5CH_3$ with configurations (**R,R**), (**R,S**), (**S,R**) and (**S,S**).

9. (a) —$S(O)_2CH_3$ leaves as $^\ominus OS(O)_2CH_3$; (b) —$OS(O)_2OCH_3$ leaves as $^\ominus OS(O)_2OCH_3$ but can react again:

$$^\ominus OS(O)_2OCH_3 + Nu^\ominus \rightarrow CH_3Nu + SO_4^{\ominus 2};$$

(c) $-\overset{\oplus}{O}H_2$ leaves as H_2O; (d) $-\overset{\oplus}{O}(CH_3)_2$ leaves as CH_3OCH_3 *and* $-\overset{\oplus}{O}(CH_3)CH_2CH_2CH(CH_3)_2$ leaves as $CH_3OCH_2CH_2CH(CH_3)_2$; (e) $-\overset{\oplus}{O}(CH_3)C_2H_5$ leaves as $CH_3OC_2H_5$ *and* $-\overset{\oplus}{O}(CH_3)CH_2CH_2C(CH_3)_3$ leaves as $CH_3OCH_2CH_2C(CH_3)_3$ *and* $-\overset{\oplus}{O}(C_2H_5)CH_2CH_2C(CH_3)_3$ leaves as $C_2H_5OCH_2CH_2C(CH_3)_3$; (f) $-\overset{\oplus}{N}(CH_3)_3$ leaves as $N(CH_3)_3$.

10. (a) $CH_3CH_2CH_2I$; (b) CH_3I; (c) CH_3CH_2I; (d) CH_3I; (e) CH_3I; (f) CH_3I. Reactions *d* and *e* will give lesser amounts of $(CH_3)_2CHCH_2CH_2I$ *and* $(CH_3)_3CCH_2CH_2I + CH_3CH_2I$, respectively.

11. (a) $CH_3O^{\ominus} + CH_3Cl$; (b) $C_6H_5CO_2^{\ominus} + C_6H_5CH_2OS(O)_2CH_3$; (c) CH_3Cl + (**R**)-$CH_3CH(O^{\ominus})CH_2CH_3$ *or* (**S**)-$CH_3CHClCH_2CH_3 + CH_3O^{\ominus}$; (d) $(CH_3)_3N + CH_3Cl$; (e) $CH_2{=}CHCH_2^{\ominus}\ Na^{\oplus} + CH_2{=}CHCH_2Br$; (f) $Na^{\oplus}\ ClCH_2CH_2CH_2CH_2O^{\ominus}$; (g) $2Na^{\oplus}\ ClCH_2CH_2O^{\ominus}$.

12. (a) $(CH_3)_3COH$; (b) $(CH_3)_3COH$; (c) (**R,S**)-$CH_3CH_2C(OH)CH_3CH_2CH_2CH_3$; (d) same as *c*; (e) (3**R**,5**S**)- and (3**S**,5**S**)-$CH_3CH_2CCH_3(OH)CH_2C(CH_3)CH_2CH_3$.

13. "Quite stable" refers to one particular group of reactions, while "extremely reactive" refers to a different reaction. It is possible for a compound to undergo one type of reaction readily and another only with great difficulty. (Note: Some chemists prefer to limit the discussion of ease of reaction to the term "reactivity" and to leave "stability" to discussions of the ground state energy of the starting material.)

14. Allyl chloride contains a covalent bond (C—Cl) which the separated ions lack. It takes energy to keep oppositely charged ions apart.

15. (a) $CH_3\overset{\oplus}{C}CH_3CH_2CH_3$; (b) $(CH_3)_2\overset{\oplus}{C}CH(CH_3)_2$; (c) $CH_3CH_2CH_2CH_2\overset{\oplus}{C}(CH_3)C(CH_3)_3$; (d) $CH_2{=}CH\overset{\oplus}{C}HCH_2CH_2CH_3$; (e) $CH_3O\overset{\oplus}{C}HCH_2C(CH_3)_3$.

16. 82%.

Chapter 6

1. (a) β; (b) β; (c) γ; (d) δ; (f) β.

2. More bonds are formed in the β-elimination. Assuming that the energy of the activated complex is related to the energy of the products, the activated complex for β-elimination should have a lower energy than that for α-elimination. Thus, the E_{act} for β-elimination should be lower.

3. $R = k[CH_3CHBrCH_3][CH_3CH_2O^{\ominus}]$ and
$R = k'[CH_3CHBrCH_3][CH_3CH_2O^{\ominus}]$
These differ in k *vs.* k'. Doubling base concentration doubles both rates.

4. A *trans* (or *anti*) elimination. The word *trans* is used to mean two different things: (a) conformation about a single bond and (b) configuration about a double bond. Because of this, the term *anti* is preferred in describing the appropriate conformàtion.

5. (a) Hofmann; (b) Saytzeff; (c) Hofmann; (d) Hofmann (see problem 27, Chapter 20); (e) Saytzeff.

6. *meso* affords *trans; d,l* affords *cis*.

7. (a) $CH_3CH_2Br + OH^{\ominus}$; (b) $CH_3CHClCH_3 + OH^{\ominus}$;
(c) $(CH_3)_2CHCH_2Br + OH^{\ominus}$; (d) $CH_3CH_2CHClCH_2CH_3 + OH^{\ominus}$;
(e) $CH_3CH_2CHClCH_3 + OH^{\ominus}$; (f) $(CH_3)_2CHC(CH_3)_2\overset{\oplus}{S}(CH_3)_2 + OH^{\ominus}$;
(g) $(CH_3)_3\overset{\oplus}{N}CH_2CH_2CH_2CH_2CH_3 + OH^{\ominus}$.
In *a-e* the solvent is ethanol (Section 6.6, #3). Heat accelerates all eliminations (Section 6.6, #5), and *f* and *g* can be accomplished simply by heating the salt.

8. Increasing the size of the base (and substrate) increases Hofmann elimination at the expense of Saytzeff. (See p. 166.)

9. 1,2-Dimethoxyethane is higher boiling (85°) and can coordinate with both oxygens.

10. $C_6H_5CH_2Br + C_6H_5CH_2MgBr \xrightarrow{S_N2} C_6H_5CH_2CH_2C_6H_5$

11. One mole (46 grams).

12. Both (+)- and (−)- yield *cis*-2-butene; *meso* yields *trans*-2-butene.

Chapter 7

1. (a) monocyclic

```
   H  H
   |  |
H—C—C—H
   |  |
H—C—C—H
   |  |
   H  H
```

(b) monocyclic

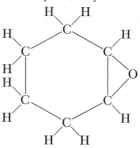

(c) acyclic

```
   H  H  H  H  H
   |  |  |  |  |
H—C—C—C—C—C—H
   |  |  |  |  |
   H  H  H  H  H
```

(d) heterocyclic, bicyclic

(e) monocyclic

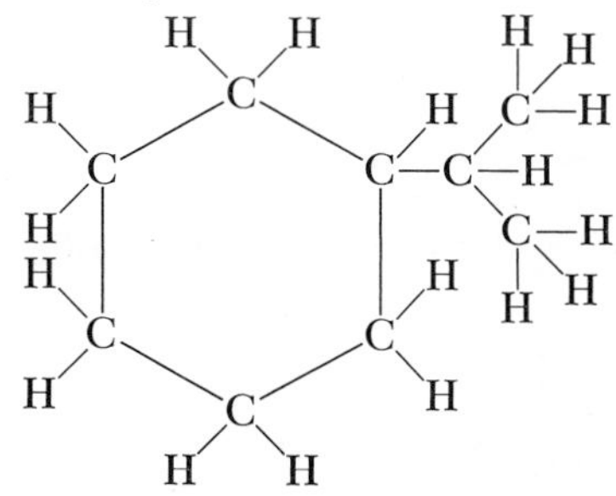

(f) spiro

(g) bicyclic

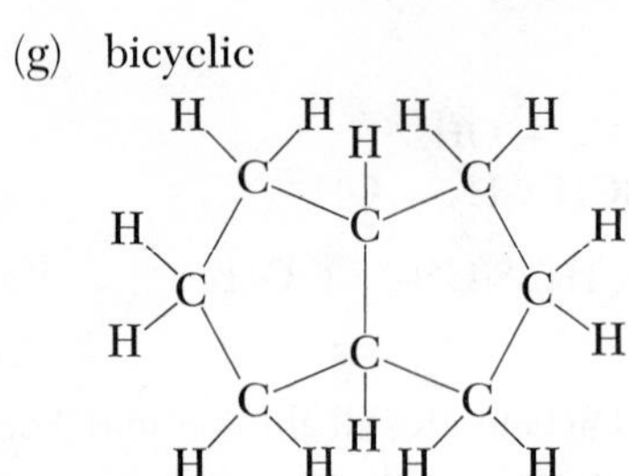

(h) monocyclic, alkene

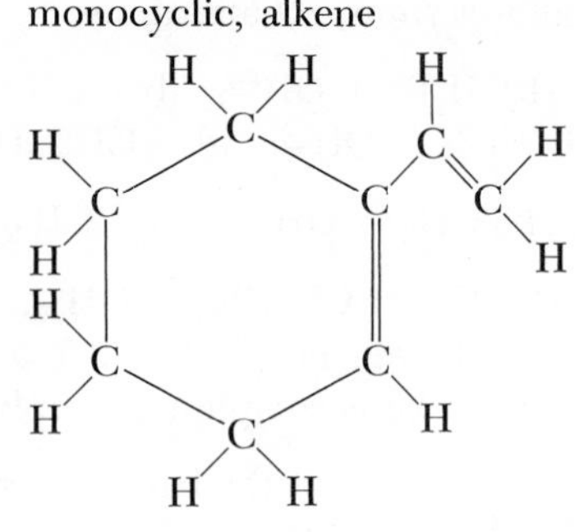

2. (a) 8; (b) 0.

3. (a)

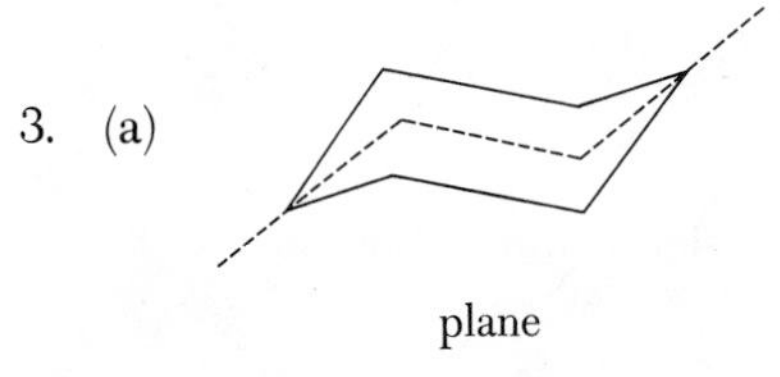

(b)

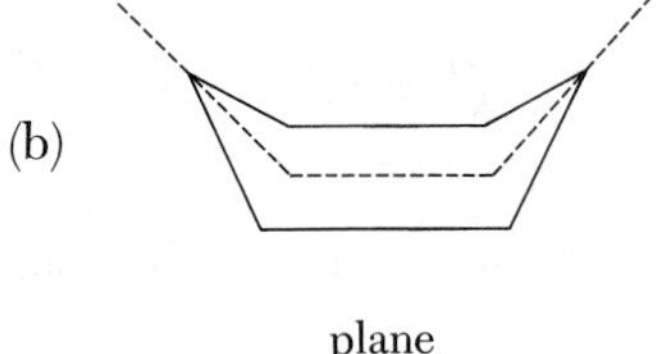

(c)

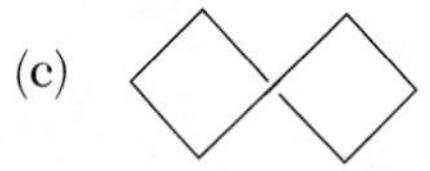

(d)

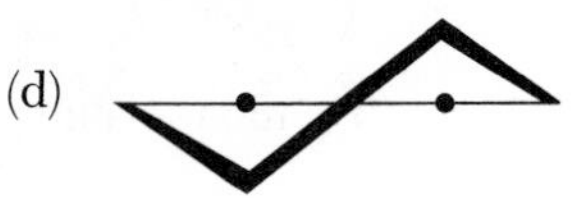

4. (a) 2-methyltetrahydrofuran; (b) 2,2-dimethyltetrahydrofuran; (c) 2,3,3-trimethyltetrahydrofuran; (d) 2-ethyl-2-methylmorpholine; (e) 4,4-dichloro-3-methyl-3-isopropylpiperidine.

5. Acidity varies with stability of resulting carbanion. The more *s* character, the greater the stability of the orbital bearing the negative charge. C_{sp} orbital, being more electronegative (greater *s* character), forms most stable carbanion.

6. **B, C, D** and **E** (see text) possess chiral centers; **C** and **E** are chiral; **B** and **D** are achiral

7. *trans*-1,3-di-*t*-butylcyclohexane would have an axial *t*-butyl group if ring had chair form.

8.

9.

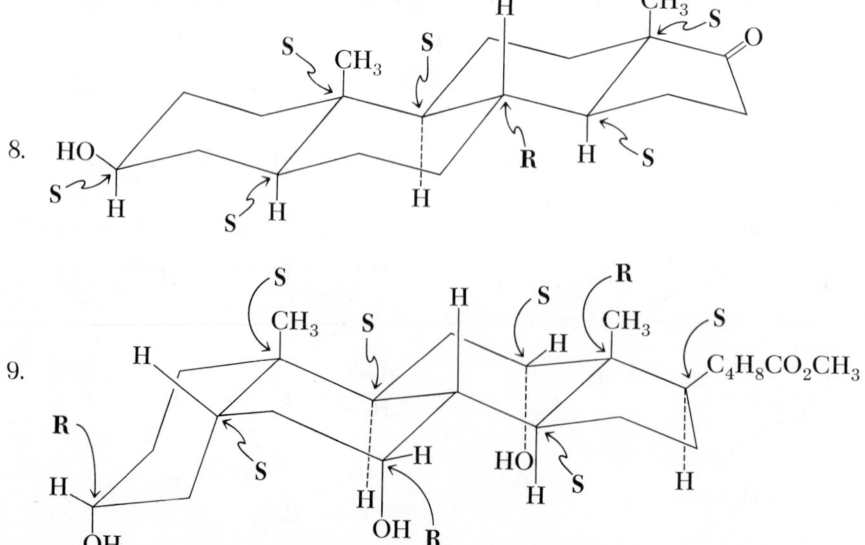

10.

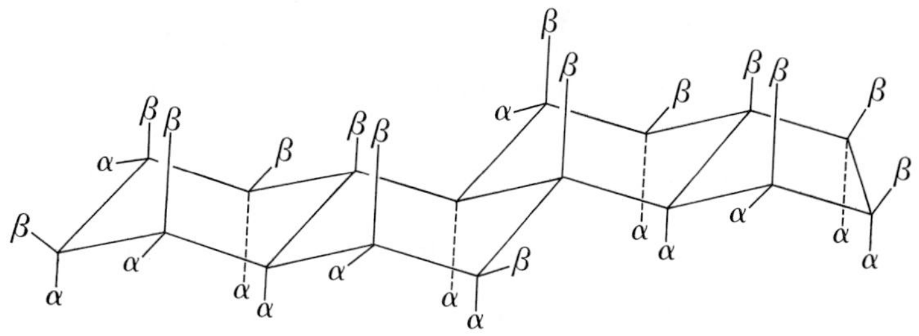

11. All bridge lengths identical. **A** is **S** at C bearing Cl; **B** is **R** at same carbon.

12. Neomenthyl chloride has higher percentage of molecules in conformation required for E2 (H and Cl *anti*).

13. E1, producing most stable alkene.

14. Increased strain in ground state of *cis* lowers E_{act}.

Chapter 8

1. (a) (**E**)-3-methyl-2-trifluoromethyl-2-pentene; (b) 1-chloro-2-methylcyclohexene (**E**, but unnecessary to specify since ring permits only **E**); (c) 2,3,3-trimethyl-1-isopropylcyclohexene (**E**, but unnecessary to specify since ring permits only **E**); (d) (**Z**)-3-methyl-1,3,5-hexatriene; (e) isopropylidenecyclohexane (no configuration).

2. They produce different cycloalkanes.

3. (a) $\overset{-3}{H_3C}—\overset{-1}{CH}{=}\overset{-2}{CH_2}$ (b) $\overset{-3}{H_3C}—\overset{-2}{CH_2}—\overset{-1}{CH}{=}\overset{-2}{CH_2}$

(c) $\overset{-3}{H_3C}—\overset{-1}{CH}{=}\overset{-1}{CH}—\overset{-3}{CH_3}$ (d) $\overset{-1}{Cl}—\overset{-1}{CH_2}—\overset{-1}{CH}{=}\overset{-1}{CH}—\overset{-3}{CH_3}$

(e) $\overset{-3}{H_3C}—\overset{+1}{C}(\overset{-1}{Cl}){=}\overset{-2}{CH_2}$ (f) $\overset{-3}{H_3C}—\overset{-1}{Cl}\overset{+2}{C}\overset{-1}{Cl}—\overset{-3}{CH_3}$

(g) $\overset{-3}{H_3C}—\underset{+3}{C}(=O^{-2})—O^{-2}—H$ (h) $H_3\overset{-2}{C}—\overset{-2}{S}—H$

All hydrogens are +1.

4. A series of hydrogenation-dehydrogenation-rehydrogenation reactions.

5. (a) butane; 1,2-dideuteriobutane; (b) butane; *d,l*-2,3-dideuteriobutane; (c) butane; *meso*-2,3-dideuteriobutane; (d) pentane; *threo*-2,3-dideuteriopentane; (e) pentane; *erythro*-2,3-dideuteriopentane; (f) cyclohexane; *cis*-1,2-dideuteriocyclohexane.

6. (a) $CH_3CH{=}CH_2$; (b) $CH_3CH_2CH{=}CH_2$ or $CH_3CH{=}CHCH_3$; (c) $(CH_3)_2CHCH{=}CH_2$; (d) $(CH_3)_3CCH_2CH{=}CH_2$ or $(CH_3)_3CCH{=}CHCH_3$; (e) (**Z**)-$CH_3CD{=}CDCH_3$; (f) (**E**)-$CH_3CD{=}CDCH_3$; (g) 3-methylcyclohexene or 4-methylcyclohexene; (h) vinylcyclohexane or 3-ethylcyclohexene or 4-ethylcyclohexene.

7. React 0.1 mole of alkene with slightly less and with slightly more than 0.1 mole of diimide. Unreacted alkene, present in first experiment, detected with Br_2/CCl_4.

8. 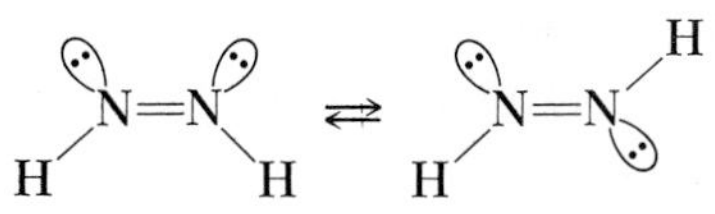 H—N=N—H (*cis*) ⇄ H—N=N—H (*trans*)   *N sp*^{2}, bond angles 120°

Cis form should be reductant; geometry permits cyclic activated complex.

9. Use B_2H_6 but RCO_2D. This gives HD addition, not HH or DD.

10. Br acts like nucleophile, donating electron pair to a carbocation. (This is like second step in S_N1.)

11. **A.** Greater ring strain. **A** forms 1,2-dibromoethane while **B** forms 1,3-dibromopropane.

12. No. It could have begun with $Cl^{\ominus}$ attacking $CH_2{=}CH_2$ to give $ClCH_2{-}CH_2^{\ominus}$.

13. *cis*-2-butene forms *d,l*-2,3-dibromobutane; *trans*-2-butene forms *meso*-2,3-dibromobutane. Identify products by gas chromatography.

14. (a) Open should have free rotation; (b) open should have most positive charge on carbon, symmetric the least; (c) greatest for symmetric, least for open; (d) open might produce some; (e) equal in symmetric, greatest difference in open.

15. Since H employs only a 1*s* orbital, a Lewis structure with two bonds to H is impossible. Bromine has eight outer electrons and can have two bonds to it (although this gives it a formal +1 charge). The bromonium ion should be more stable.

16. (a) $CH_3CHOHCH_2CH_3$; (b) $CH_3CH_2CHOHCH_3$; (c) $C_6H_5CHOHCH_3$;

(d) 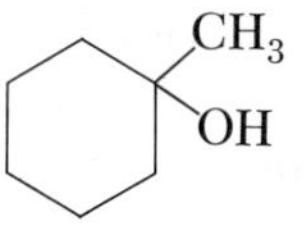 CH_3 OH (e) 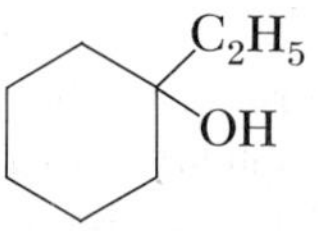 C_2H_5 OH

17. Formation of **D** requires energy-rich 1° carbocation. **B** formed after methyl group migration.

18. They all favor most stable intermediate.

19. Both involve ring opening of strained three-membered ring by nucleophile (S_N2). In cyclic bromonium ion, ring carbons bear some positive charge (which should accelerate attack by nucleophile).

20. Products for OsO_4 and $KMnO_4$ identical (after work-up). (a) $CH_3CHOHCH_2OH$; (b) $HOCH_2CHOHCH_2CH_3$; (c) (2**S**,3**R**)-$HOCH_2CHOHCHClCH_3$ + (2**R**,3**R**)-$HOCH_2CHOHCHClCH_3$ (unequal amounts); (d) (2**S**,3**S**)-$HOCH_2CHOHCHClCH_3$ + (2**R**,3**S**)-$HOCH_2CHOHCHClCH_3$;

(e) 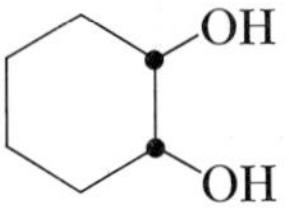 OH OH (f) 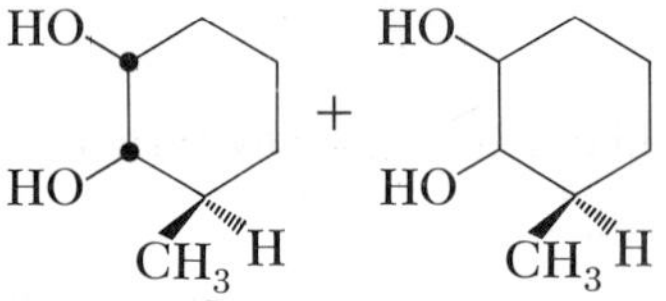 HO HO CH_3 H + HO HO CH_3 H (g) 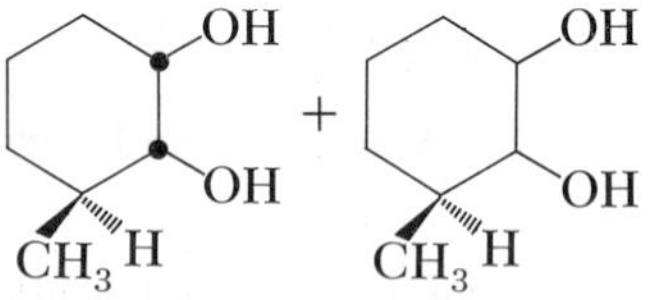 OH OH CH_3 H + OH OH CH_3 H

21. (a) $CH_3CO_2H + CO_2$; (b) $CH_3CH_2CO_2H + CO_2$; (c) (**R**)-$CH_3CHClCO_2H + CO_2$; (d) (**S**)-$CH_3CHClCO_2H + CO_2$; (e) $HO_2C(CH_2)_5CO_2H$; (f) (**R**)-$HO_2CCH_2CH_2CH_2CHCH_3CO_2H$; (g) (**S**)-$HO_2CCH_2CH_2CH_2CHCH_3CO_2H$.

Chapter 9

1. (a) propyne; (b) 1-deuteriopropyne; (c) 1,3,3,3-tetradeuteriopropyne (perdeuteriopropyne); (d) 3-chloropropyne (propargyl chloride); (e) phenylacetylene (ethynylbenzene); (f) 1-buten-3-yne (butenyne/vinylacetylene); (g) 2-methyl-1-buten-3-yne; (h) 2,2-dimethyl-3-hexyne (*t*-butylethylacetylene).

2. (a) acids: HCl, $H_3O^{\oplus}$; bases: $Cl^{\ominus}$, H_2O; (b) acids: H_2O, HCl; bases: $Cl^{\ominus}$, $OH^{\ominus}$; (c) acids: NH_3, C_2H_2; bases: $HC{\equiv}C^{\ominus}$, $NH_2^{\ominus}$; (d) acids: C_2H_2, $NH_4^{\oplus}$; bases NH_3, $HC{\equiv}C^{\ominus}$.

3. Such statements about basic or acidic behavior are relative. Since the nitrogen of ammonia has an available electron pair, it could function as a base toward a sufficiently strong acid (*e.g.*, $H_3O^{\oplus}$).

4. Bubble through a strongly basic solution—1-butyne is "trapped" as a salt while 2-butyne passes through. Recover 1-butyne by acidification.

5. Only 1-butyne will form a salt when reacted with $NaNH_2$ (and liberate ammonia). 1-Butene will decolorize Br_2/CCl_4. Octane will not react with either reagent.

6. Produce $CH_3CH_3 + NH_2^{\ominus}$ since $CH_3CH_2^{\ominus}$ is a stronger base than $NH_2^{\ominus}$.

7. (a) $Ag(NH_3)_2^{\oplus}$; (b) $Ag(NH_3)_2^{\oplus}$; (c) Br_2/CCl_4 or $MnO_4^{\ominus}$; (d) $Ag(NH_3)_2^{\oplus}$ or $MnO_4^{\ominus}$.

8. (a) $Br_2/h\nu$ followed by KOH/EtOH; (b) ethene (prepared in *a*) + Br_2/CCl_4 followed by $NH_2^{\ominus}$; (c) same as *b*; (d) $BrCH_2C{\equiv}CH$ (must use two moles of C_6H_5MgBr because of acidity of terminal alkyne hydrogen) followed by dilute acid.

9. Chlorine is a better leaving group.

10. 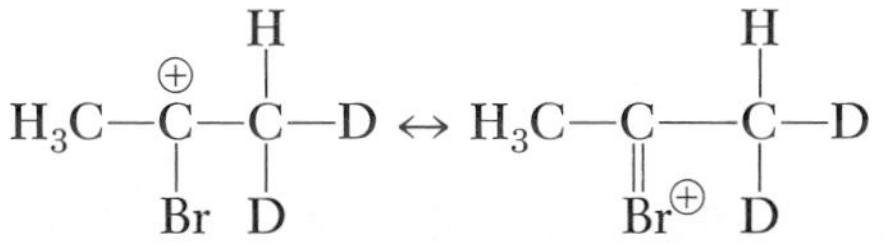 $H_3C-\overset{\oplus}{C}(Br)-CH(D)-D \leftrightarrow H_3C-C(={Br}^{\oplus})-CH(D)-D$

11. *cis, trans* isomers about the C=N bond.

12. (a) $CH_3CH_2CHBr_2$; (b) $CH_3CHBrCH_2Br$; (c) $CH_3CBr_2CH_3$; (d) $CH_3CH_2CHBr_2$.

Chapter 10

1. (a) 2-buten-1-ol, 1°; (b) 2-pentanol, 2°; (c) (**R**)-3-phenyl-1-butanol, 1°; (d) (2**S**,5**R**)-2,5-heptanediol, 2°; (e) (2**S**,5**R**)-5-methyl-5-phenyl-2-heptanol, 2°; (f) (**Z**)-3-penten-1-ol, 1°.

2. $CH_3(CH_2)_5CH_2OH$, 1-heptanol; $CH_3CHOH(CH_2)_4CH_3$, 2-heptanol; $CH_3CH_2C(OH)CH_3CH_2CH_2CH_3$, 3-methyl-3-hexanol.

3. 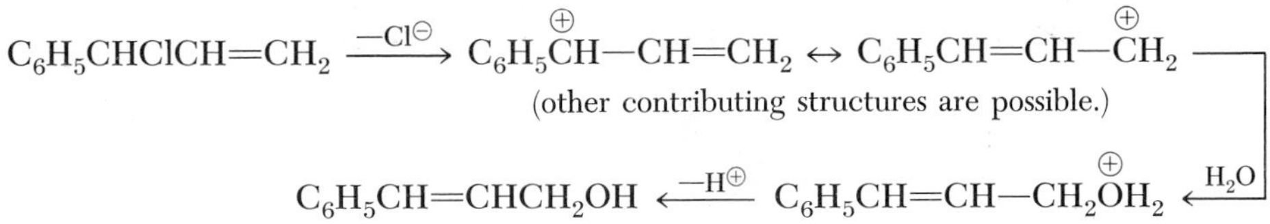 $C_6H_5CHClCH{=}CH_2 \xrightarrow{-Cl^{\ominus}} C_6H_5\overset{\oplus}{C}H-CH{=}CH_2 \leftrightarrow C_6H_5CH{=}CH-\overset{\oplus}{C}H_2$
(other contributing structures are possible.)

$\xrightarrow{H_2O} C_6H_5CH{=}CH-CH_2\overset{\oplus}{O}H_2 \xrightarrow{-H^{\oplus}} C_6H_5CH{=}CHCH_2OH$

4. (a) $HOCD_2CH_2CH_3$; (b) and (c) $CH_3CHOHCH_2CH_3$; (d) 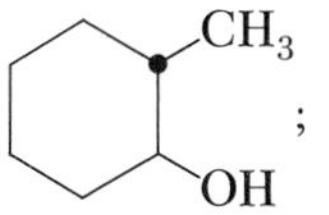 2-methylcyclohexanol (cyclohexane ring bearing CH_3 and OH on adjacent carbons); (e) 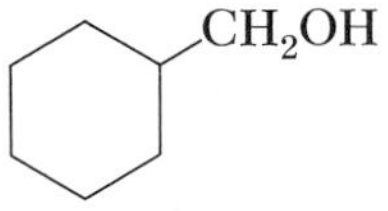 cyclohexylmethanol (cyclohexane ring bearing CH_2OH)

5. (a) $CH_3MgCl + CH_3COC_2H_5$; (b) $(CH_2)_3CHMgCl + CH_2O$; (c) $CH_3MgBr + (CH_2)_3CO$; (d) m-$CH_3C_6H_4MgBr + CH_3CDO$; (e) $BrMgC_6H_{10}MgBr + 2CH_2O$; (f) $(CH_3)_3CMgBr + OCHCH(CH_3)_2$. Each must be followed by the usual (dil. acid) work-up.

6. The two chlorosulfites are diastereomers. The sulfur is chiral and configurationally stable. (The stereochemistry of sulfur compounds is discussed in detail in Chapter 23.)

7. (a) $(CH_3)_2CHOH + CrO_3 \cdot (C_5H_5N)_2$; (b) $CH_3CH{=}CHCH_2OH + MnO_2$; (c) $C_6H_{11}OH + MnO_4^{\ominus}$; (d) $C_6H_{10}(OH)_2 + Cr_2O_7^{\,2-} + H^{\oplus}$;

(e) 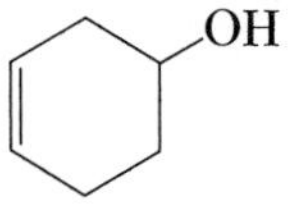 OH $+ CrO_3 \cdot (C_5H_5N)_2$.

8. (a) $CH_3CHOHCH_3$, doublet; (b) $CH_3CH{=}CHCH_2OH$, triplet; (c) $C_6H_{11}OH$, doublet; (d) *cis*- and *trans*-$C_6H_{10}(OH)_2$, each **OH** a doublet;

(e) 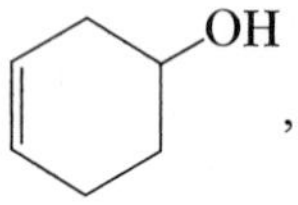 OH , doublet.

Chapter 11

1. (a) methyl propyl ether; (b) vinyl ether; (c) allyl ethynyl ether; (d) isobutylene oxide; (e) *cis*-2,6-dihydroxytetrahydropyran (*cis*-2,6-dihydroxyoxacyclohexane); (f) (1**R**,2S,3S)-3-methylcyclohexene oxide.

2. (a) $CH_2{=}CH_2 + H_2O \xrightarrow{H^{\oplus}} CH_3CH_2OH \xrightarrow{H_2SO_4} CH_3CH_2OCH_2CH_3$;

(b) $CH_3(CH_2)_3CH{=}CH_2 + H_2O \xrightarrow{H^{\oplus}} CH_3(CH_2)_3CHCH_3OH \xrightarrow{Na}$

$CH_3(CH_2)_3CHCH_3O^{\ominus}\ Na^{\oplus}$

$CH_3CH{=}CH_2 + HCl \rightarrow CH_3CHClCH_3$

$CH_3(CH_2)_3CHCH_3O^{\ominus}\ Na^{\oplus} + CH_3CHClCH_3 \rightarrow$

$CH_3(CH_2)_3CHCH_3OCH(CH_3)_2$;

(c) $CH_2{=}CH_2 \xrightarrow{O_3} \xrightarrow{Zn/H^{\oplus}} CH_2O \xrightarrow{BH_4^{\ominus}} CH_3OH \xrightarrow{Na} CH_3O^{\ominus}\ Na^{\oplus}$

$CH_3OH + SOCl_2 \rightarrow CH_3Cl \xrightarrow{NaOCH_3} CH_3OCH_3$

(d) 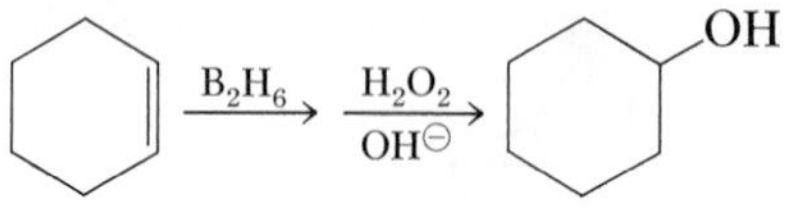 $\xrightarrow{B_2H_6} \xrightarrow[OH^{\ominus}]{H_2O_2}$ OH

OH $+$ $\xrightarrow{Hg(CF_3CO_2)_2} \xrightarrow{BH_4^{\ominus}}$ O

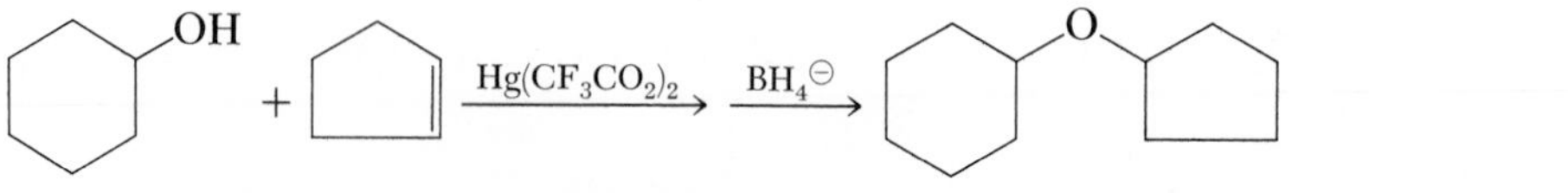

(e) $(CH_3)_2CH{=}CH_2 + CH_3CH_2OH$ (part *a*) $\xrightarrow{Hg(CF_3CO_2)_2} \xrightarrow{BH_4^{\ominus}} (CH_3)_3COC_2H_5$.

3. (a) $CH_3CH_2I + CH_3CH_2OH$;
(b) $(CH_3)_2CHI + (CH_3)_2CHOH + CH_3CHI(CH_2)_3CH_3 + CH_3CHOH(CH_2)_3CH_3$;

(c) $CH_3I + CH_3OH$; (d) $C_5H_9I + C_5H_9OH + C_6H_{11}I + C_6H_{11}OH$;
(e) $(CH_3)_3CI + CH_3CH_2OH$.

4. (a) $\overset{O}{CH_2—CH}(CH_2)_3CH_3$ (b) *cis* and *trans* $CH_3\overset{O}{CH—CH}CH_2CH_2CH_3$;

(c) *cis* and *trans* Et—$\overset{O}{CH—CH}$—Et (d)

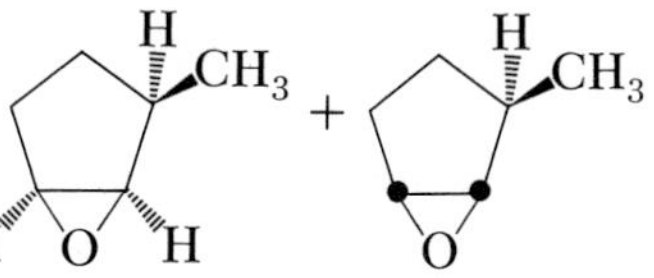

(e)

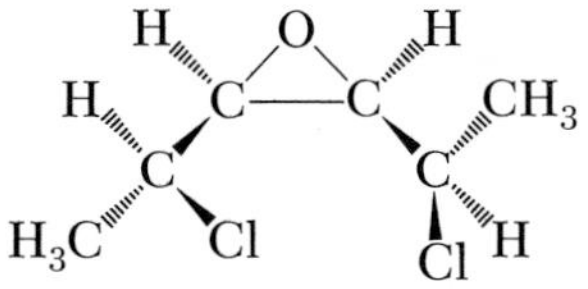

5. (a) $HOCH_2CH_2OH$; (b) $HOCH_2CH_2OH$; (c) $ClCH_2CH_2OH$;
(d) $CH_3SCH_2CH_2O^{\ominus}\ Na^{\oplus}$; (e) $CH_3CH_2CH_2CH_2CH_2OH$; (f) $CH_3CH_2CH_2OH$;
(g) $NCSCH_2CH_2O^{\ominus}\ K^{\oplus}$; (h) $HOCH_2CH_2OSO_3H$; (i) $CH_3C{\equiv}CCH_2CH_2O^{\ominus}\ Na^{\oplus}$.

6. Cannot form structure of type shown below with C-P and C-O eclipsed.

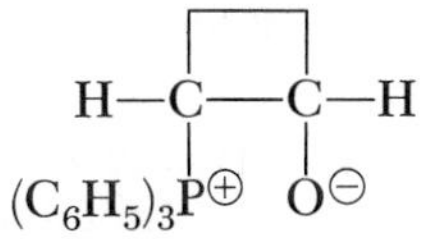

H—C—C—H
$(C_6H_5)_3P^{\oplus}$ $O^{\ominus}$

7. (a) $CH_3S^{\ominus}\ Na^{\oplus} + \overset{O}{CH_2—CH_2} \rightarrow \xrightarrow{\text{epoxide}} \xrightarrow{H_3O^{\oplus}}$

(b) $CH_3O^{\ominus}\ Na^{\oplus} \xrightarrow{\text{epoxide}} \xrightarrow{\text{epoxide}} \xrightarrow{\text{epoxide}} \xrightarrow{\text{epoxide}} \xrightarrow{H_3O^{\oplus}}$

(c) $CH_3(CH_2)_3O^{\ominus}\ Na^{\oplus}$ + oxetane $\xrightarrow{\text{heat}} \xrightarrow{\text{epoxide}} \xrightarrow{H_3O^{\oplus}}$

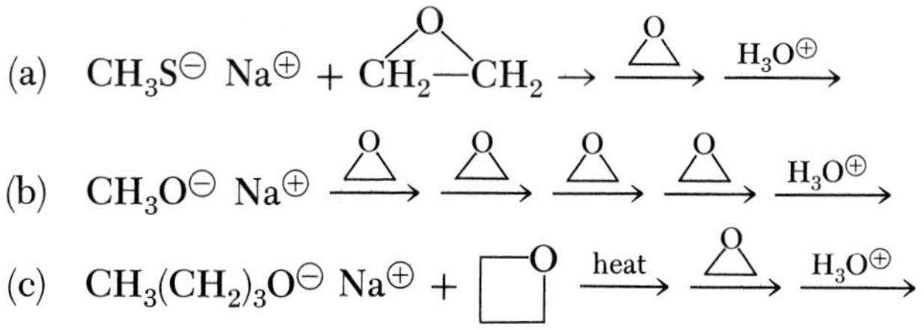

Chapter 12

1. Bond order: 0.12.

2. C9,10.

3. Steric repulsion prevents coplanarity of both double bonds.

4. 1,3-Cyclohexadecadiene is large enough to accomodate an *s-trans* diene fragment.

5. **A.** $\xrightarrow{H^{\oplus}}$ $\leftrightarrow$ $\xrightarrow{Cl^{\ominus}}$ +

identical

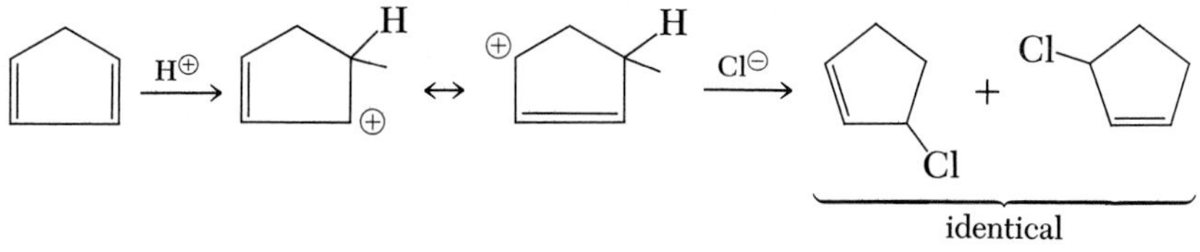

B. $\xrightarrow{H^{\oplus}}$ $H^{\oplus}$ $\xrightarrow{Cl^{\ominus}}$ +

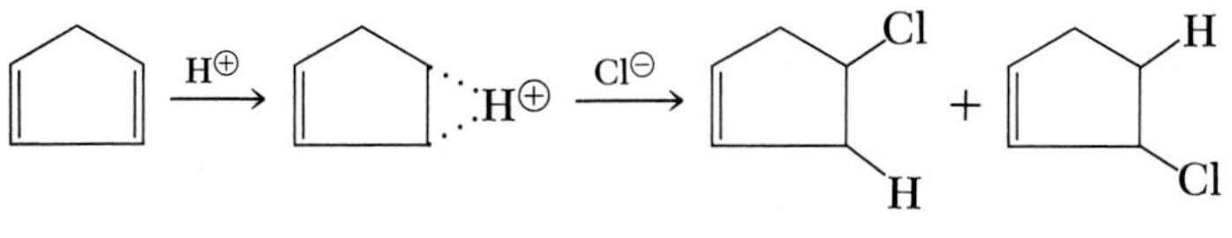

(i) Mechanism **B,** bridged protonium ion intermediate, produces two products (not necessarily in equal amounts). (ii) Add DCl instead of HCl. Mechanism **A** should afford both *cis* and *trans* adducts, while **B** should give predominately *trans*. (iii) Add DCl instead of HCl. Mechanism **B** puts D and Cl on adjacent carbons. (Other answers are possible.)

6. Levulinic acid: $CH_3COCH_2CH_2CO_2H$; with $BH_4^{\ominus}$ produces $CH_3CHOHCH_2CH_2CO_2H$; with CH_3MgCl produces one mole of CH_4 per mole of levulinic acid.

7. Ecdysone is a steroid. Catalytic hydrogenation creates three new chiral centers, and four diastereomeric products.

Chapter 13

1. Ground state: π_1:2; π_2:2; π_3^* and π_4^*:0; first excited state: π_1:2; π_2:1; π_3^*:1; π_4^*:0.

2. It was prepared by illumination and should be protected from a photochemical reverse-reaction. However, it could be decomposed, in a non-concerted fashion, by heat. Photochemical cyclization of 1,3-butadiene is a disrotatory process and this product's stereochemistry requires a disrotatory cyclization.

3. Upon illumination π_4^* becomes HOMO and conrotatory motion is required. In a thermally induced reaction HOMO is π_3 and a disrotatory motion is required. (**2E,4Z,6E**)-Octatriene will produce *cis*-5,6-dimethyl-1,3-cyclohexadiene by disrotatory closure, but *trans*-5,6-dimethyl-1,3-cyclohexadiene by conrotatory closure.

4. Not readily. The two ends of the molecule are not as close together as are the ends of 1,3-butadiene (due to *sp* carbon).

Chapter 14

1. Three: [ortho-disubstituted benzene, X X] (*or* [Kekulé form]); [meta-disubstituted benzene, X X] (*or* [Kekulé form]); [para-disubstituted benzene, X X] (*or* [Kekulé form])

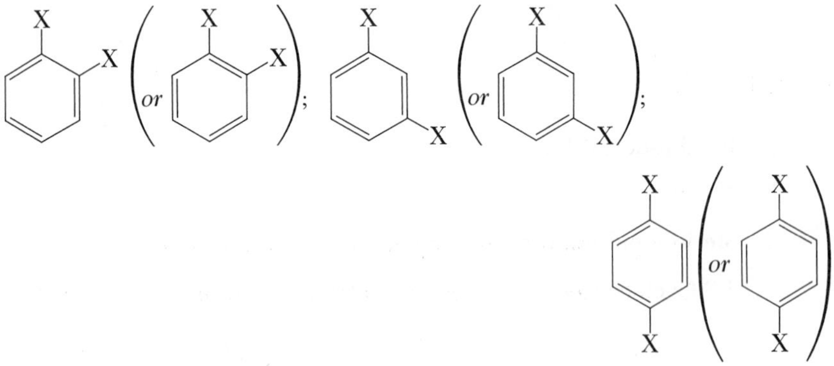

2. All carbons sp^2 hybridized; all vertical lines represent p orbitals.

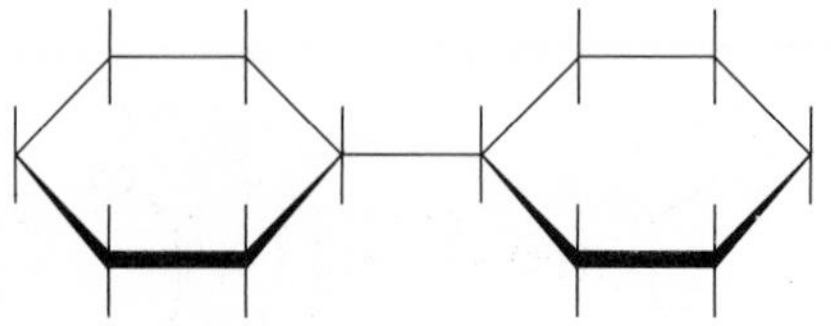

estimated heat of hydrogenation of biphenyl: 100–103 kcal/mole.

3. One pair of electrons, in an sp^2 orbital, is perpendicular to the π network. Thus, there are only 6 π electrons that are delocalized.

4. The "extra" bond is perpendicular to the π network. Thus, as in problem 3, there are only 6 π electrons which are delocalized (see Figure 21–5).

"extra" bond

It is reactive because of the great strain present in this extra bond. Aromaticity refers to ground state stability (due to π electron delocalization), while reactivity deals with the energy of activation for a specific process. These statements are not inconsistent.

5. No. It contains $4n$ π electrons.

6. No. It would then be anti-aromatic (planar, $4n$ electrons).

7. contains 6 π electrons

8. Both planar, but oxygen is perpendicular to plane in the ether.

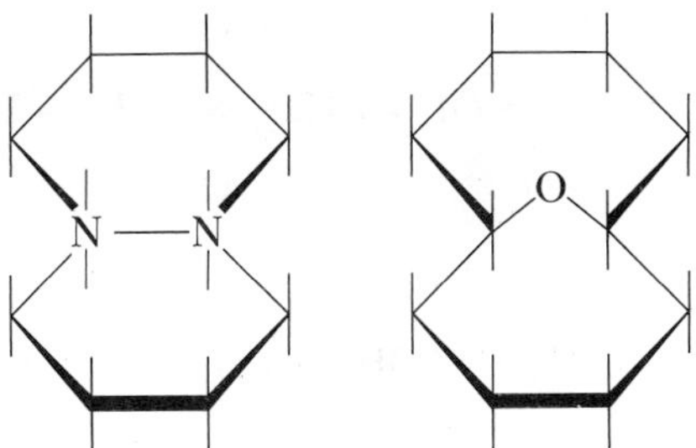

9. All are sp^2.

10. The resultant cation is aromatic:

11. The heterocyclic ring system is aromatic (an oxygen heterolog of naphthalene), containing 10 π electrons. Delphinidin chloride is a pigment of delphinium, violet pansies, and purple grapes.

Chapter 15

1. (a) 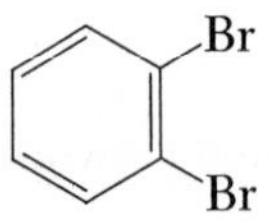 Br, Br (b) 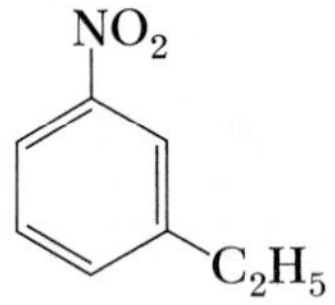 NO_2, C_2H_5 (c) Br, Br, C_6H_5 (d) 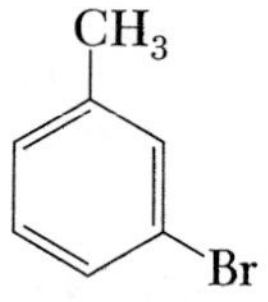 CH_3, Br

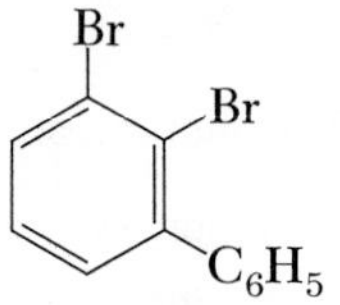

(e) 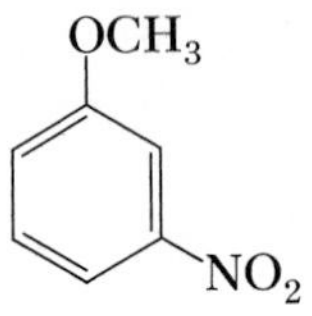 OCH_3, NO_2 (f) 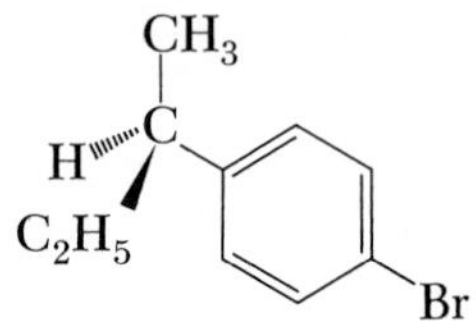 CH_3, H, C, C_2H_5, Br (g) 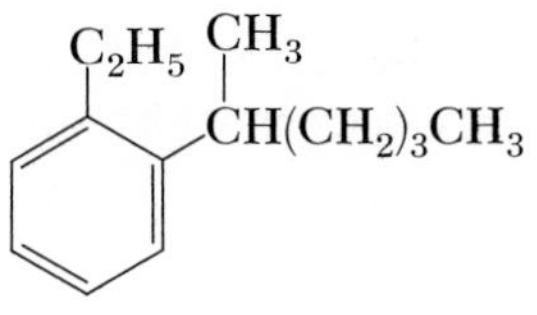 C_2H_5, CH_3, $CH(CH_2)_3CH_3$

2. $^{\ominus}\!:\!\ddot{O}\!-\!\overset{(+2)}{N}\!=\!\ddot{O} \leftrightarrow \ddot{O}\!=\!\overset{(+2)}{N}\!-\!\ddot{O}\!:^{\ominus}$

3. (a) $C_6H_6 + CH_3Cl + AlCl_3$; $C_6H_6 + CH_3OH + H_2SO_4$;
(b) $C_6H_6 + (CH_3)_3CCl + AlCl_3$; $(CH_3)_2C{=}CH_2 + C_6H_6 + H_3PO_4$;
(c) $C_6H_6 + CH_3CHClCH_2CH_3 + AlCl_3$; $C_6H_6 + CH_3CH{=}CHCH_3 + H_2SO_4$;
(d) $C_6H_6 + CH_2{=}CH_2 + H_2SO_4$; $C_6H_6 + CH_3CH_2Cl + AlCl_3$.

4. Fluoride is less nucleophilic than bromide and, consequently, there will be less addition of HX across the double bond.

5. $CH_3\!-\!\overset{\oplus}{C}\!=\!\ddot{O} \leftrightarrow CH_3\!-\!C\!\equiv\!O\!:^{\oplus}$

6. (a) 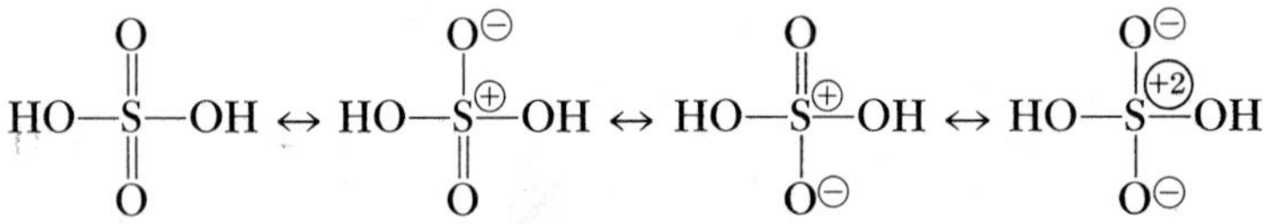 $HO\!-\!S(=O)_2\!-\!OH \leftrightarrow HO\!-\!\overset{\oplus}{S}(=O)(O^{\ominus})\!-\!OH \leftrightarrow HO\!-\!\overset{\oplus}{S}(=O)(O^{\ominus})\!-\!OH \leftrightarrow HO\!-\!\overset{+2}{S}(O^{\ominus})_2\!-\!OH$

(b) $C_6H_5\!-\!S(=O)_2\!-\!OH \leftrightarrow C_6H_5\!-\!\overset{\oplus}{S}(=O)(O^{\ominus})\!-\!OH \leftrightarrow C_6H_5\!-\!\overset{\oplus}{S}(=O)(O^{\ominus})\!-\!OH \leftrightarrow C_6H_5\!-\!\overset{+2}{S}(O^{\ominus})_2\!-\!OH$

(c) $^{\ominus}O\!-\!S(=O)_2\!-\!O^{\ominus} \leftrightarrow {}^{\ominus}O\!-\!S(=O)(O^{\ominus})\!=\!O \leftrightarrow {}^{\ominus}O\!-\!S(O^{\ominus})(=O)\!=\!O \leftrightarrow$
$O\!=\!S(O^{\ominus})(=O)\!-\!O^{\ominus} \leftrightarrow O\!=\!S(=O)(O^{\ominus})\!-\!O^{\ominus} \leftrightarrow O\!=\!S(O^{\ominus})_2\!=\!O$

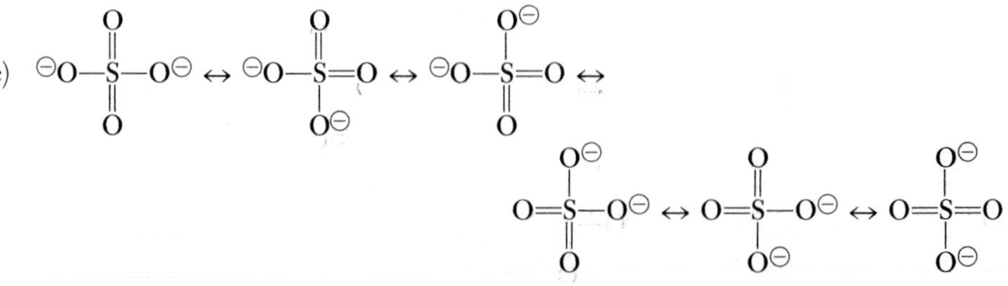

(d) 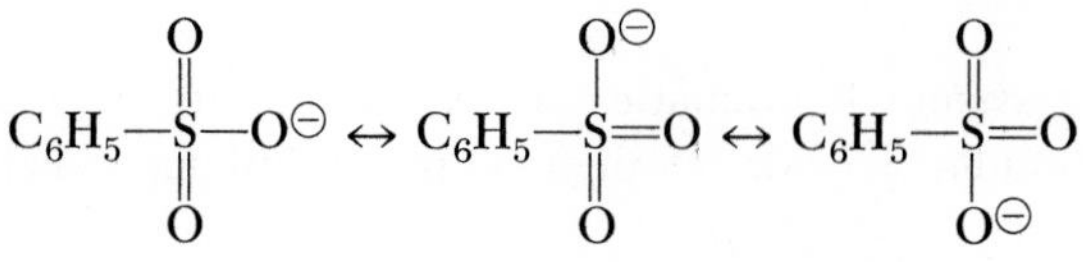 $C_6H_5\!-\!S(=O)_2\!-\!O^{\ominus} \leftrightarrow C_6H_5\!-\!S(O^{\ominus})(=O)\!=\!O \leftrightarrow C_6H_5\!-\!S(=O)(O^{\ominus})\!=\!O$

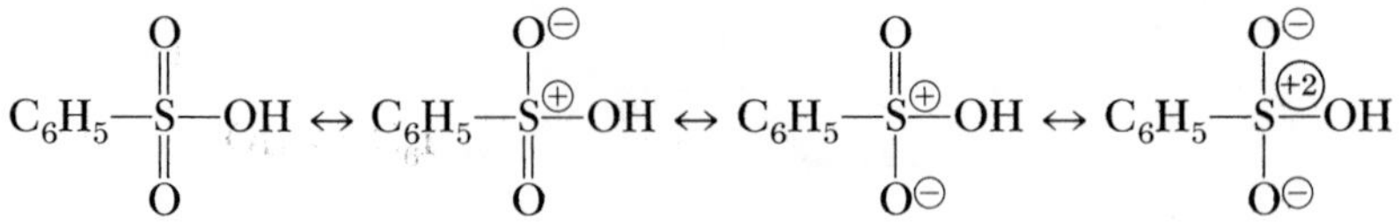

7. 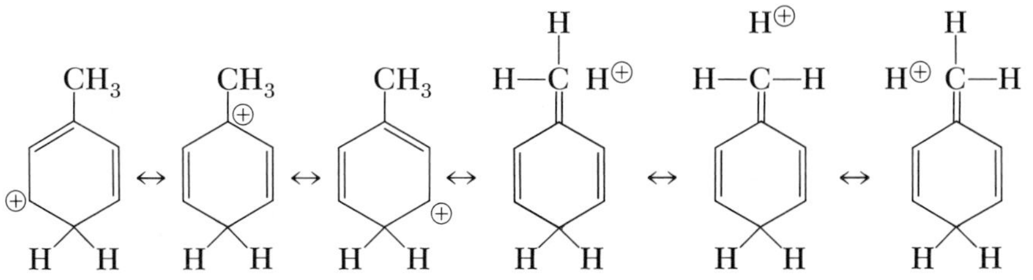 [resonance structures of protonated toluene]

8. 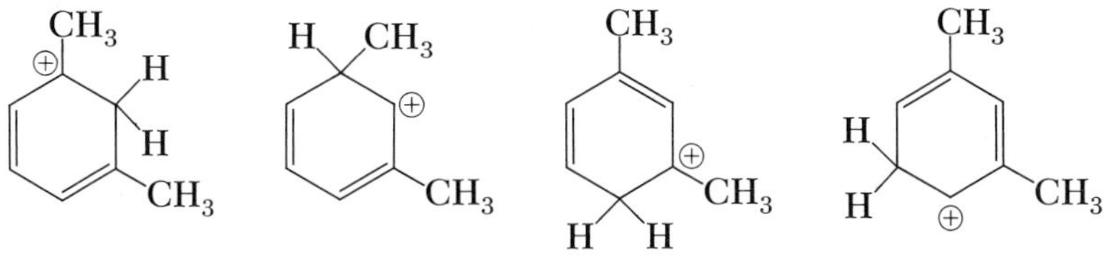 A B C D

A and **C** most stable, since positive charge may reside on both carbons bearing a methyl group. (See problem 7.)

9. 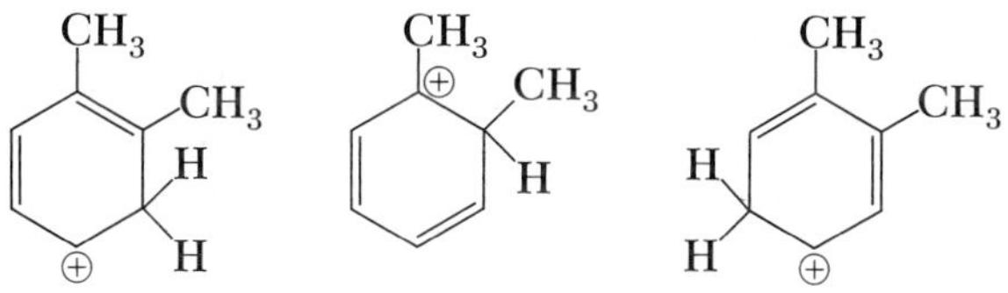 [structures]

No single protonated *o*-xylene can utilize both methyl groups to delocalize positive charge (compare to **A** and **C** in problem 8).

10. One. Diprotonation places two positive charges on the same π system.

11. Both OH and O$^\ominus$ are *o,p*-directors (O$^\ominus$ more potent), since both can donate a non-bonding electron pair to positively charged ring.

12. Although present in low concentration, the —NH_2 group is sufficiently activating that it is responsible for the bulk of the product.

13. (a) 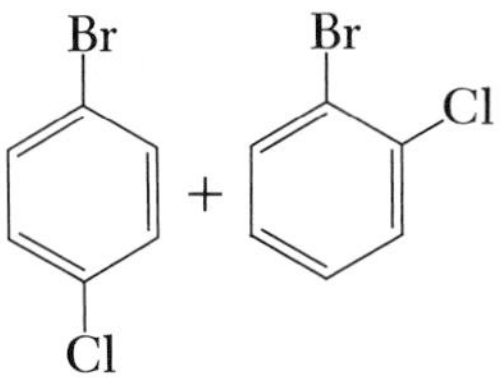 [p-bromochlorobenzene] + [o-bromochlorobenzene] (b) 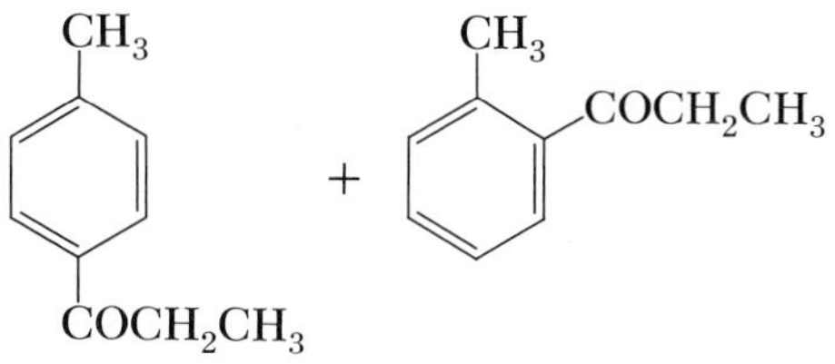 [p-$CH_3C_6H_4COCH_2CH_3$] + [o-$CH_3C_6H_4COCH_2CH_3$]

(c) 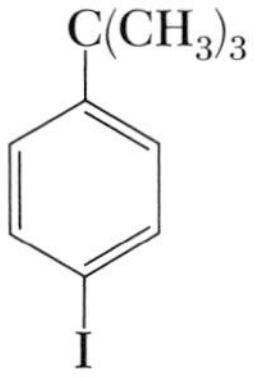 [p-iodo-$C(CH_3)_3$ benzene] (d) 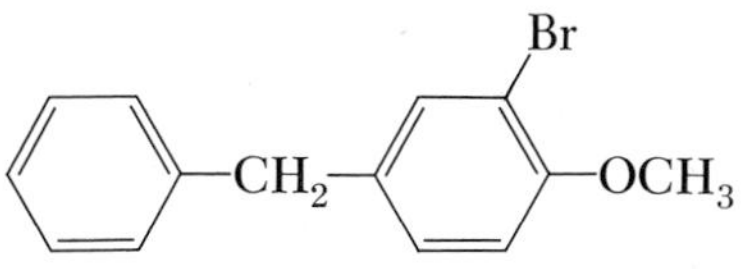 C_6H_5—CH_2—[3-bromo-4-methoxyphenyl] (e) 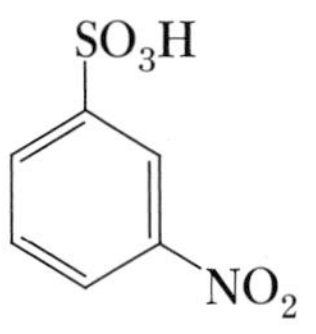 [m-nitrobenzenesulfonic acid, SO_3H, NO_2]

14. (a) 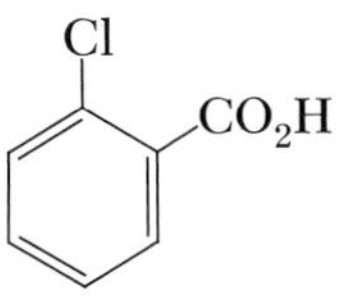 [o-chlorobenzoic acid, Cl, CO_2H] (b) 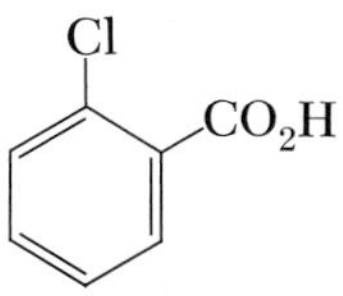 [Cl, CO_2H] (c) 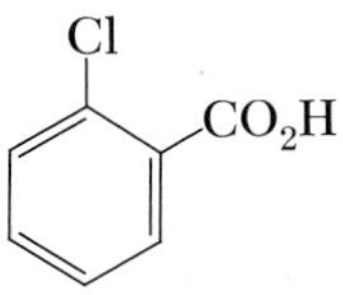 [Cl, CO_2H]

(d) [benzene ring with two adjacent CO_2H groups] (e) [benzene ring with CO_2H] (f) [benzene ring with CO_2H para to CO_2H] + [benzene ring with CO_2H para to NO_2]

(g) [benzene ring with CO_2H] + [benzene ring with CO_2H para to NO_2] (h) [benzene ring with CO_2H] + [benzene ring with two adjacent CO_2H groups and NO_2]

15. Thiophene is converted to a water-soluble sulfonic acid derivative and removed by washing. Fuming sulfuric acid would sulfonate benzene as well as thiophene, and benzenesulfonic acid is water-soluble. This would remove the desired benzene.

16. (a) The following compounds: 1,2-, 1,3-, 1,4-, 2,3-, 1,5-, 1,6-, 1,7-, 1,8-, 2,6-, 2,7-dichloronaphthalene; (b) the following compounds: 1,2-, 1,3-, 1,4-, 1,5-, 1,6-, 1,7-, 1,8-, 1,9-, 1,10-, 2,3-, 2,4-, 2,5-, 2,6-, 2,7-, 2,8-, 2,9-, 2,10-, 3,4-, 3,5-, 3,6-, 3,7-, 3,8-, 3,9-, 3,10-, 4,5-, 4,6-, 4,7-, 4,8-, 4,9-, 4,10-, 9,10-dichlorophenanthrene; (c) the following compounds: 1,2-, 1,3-, 1,4-, 1,5-, 1,6-, 1,7-, 1,8-, 1,9-, 1,10-, 2,3-, 2,4-, 2,5-, 2,6-, 2,7-, 2,8-, 2,9-, 2,10-, 9,10-dichloroanthracene.

Chapter 16

1. Hydrogen bonding between carbonyl oxygen and water.

2. (a) acetaldehyde; (b) propenal; (c) 3-phenylpropenal; (d) 1,1,1-trichloroethanal; (e) dimethyl ketone; (f) 2-butanone; (g) 4-methyl-3-penten-2-one.

3. (a) $C_6H_5CH_3 \xrightarrow[H_2SO_4]{CrO_3/Ac_2O} \xrightarrow{HOH}$

(b) $C_6H_5CH_3 + HCN + HCl \xrightarrow[100°]{ZnCl_2} \xrightarrow{H_2O}$

(c) [benzene ring with CH_3 para to CHO] $\xrightarrow[H_2O]{BH_4^{\ominus}}$ [benzene ring with CH_3 para to CH_2OH]

(part *b*)

(d) $C_6H_5CH_3 \xrightarrow[AlCl_3]{Br_2} \xrightarrow[THF]{Mg} \xrightarrow{C_6H_5CHO} \xrightarrow{H_2O}$

(e) [benzene ring with CH_3 para to CHO] $\xrightarrow[HCl]{Zn \cdot Hg}$

(f) $CH_3-C_6H_4-CH_3 \xrightarrow[H_2SO_4]{CrO_3/Ac_2O} \xrightarrow{H_2O}$

(g) $C_6H_5CH_2OH \xrightarrow{PBr_3}$

4. $CH_3-C(OH)(Et)-C(OH)(Et)-CH_3$ (*meso* and *d,l*)

5. (a) $C_6H_5COCH_3$ (b) $CH_3C(=O)-C_6H_4-C(=O)CH_3$ (c) $CH_3CH_2NH_2$

(d) CH_3CHO (e) $CH_2{=}CH-C_6H_4-C(=O)C_6H_5$ (f) $(CH_3)_2COHC_6H_5$

6. Ethyl magnesium iodide removes H adjacent to the cyano group in CH_3CH_2CN. This hydrogen is acidic because the resultant carbanion, $CH_3\overset{\ominus}{C}HCN$, is resonance stabilized.

7. $R-C(=NR)-R' \xrightarrow{H_3O^{\oplus}} R-C(=\overset{\oplus}{N}RH)-R' \xrightarrow{H_2O} R-C(NRH)(\overset{\oplus}{O}H_2)-R' \rightarrow R-C(\overset{\oplus}{N}RH_2)(OH)-R'$

$R-C(\overset{\oplus}{N}RH_2)(OH)-R' \;(+\; H_2O) \rightarrow R-C(=O)-R' + RNH_2 + H_3O^{\oplus}$

8. Formation of the hemiacetal creates a new chiral center (C1). The resultant diastereomers equilibrate via the free aldehyde (acyclic form).

9. Other suggestions are possible.

$R'-C(=O)-H \xrightarrow{H^{\oplus}} R'-\overset{\oplus}{C}(OH)-H \;(\leftarrow RO-C(OR)_2R) \rightarrow R'-C(OH)(H)-\overset{\oplus}{O}(R)-C(OR)_2R \rightarrow R'-C(OH)(OR)-H + R-\overset{\oplus}{C}(OR)_2$

$R'-C(OR)(OH)-H + R-\overset{\oplus}{C}(OR)(OR) \rightarrow R'-C(OR)(H)-\overset{\oplus}{O}(H)R + R-C(=O)-OR$

$\xrightarrow{-H^{\oplus}} R'-C(OR)_2-H$

10. Acid-catalyzed cleavage accelerated by a resonance-stabilized cationic intermediate.

$$CH_3-\underset{OCH_3}{\overset{OCH_3}{C}}-CH_3 \xrightarrow{H^\oplus} CH_3-\underset{OCH_3}{\overset{\oplus\,HOCH_3}{C}}-CH_3 \rightarrow CH_3OH + CH_3-\underset{OCH_3}{\overset{\oplus}{C}}-CH_3 \leftrightarrow CH_3-\underset{\oplus OCH_3}{C}-CH_3$$

11. H—C(OCH$_3$)— / H—C—OH / HO—C—H / H—C—OH / H—C— / CH$_2$OH (ring O) → H—$\overset{\oplus}{O}$(H)—CH$_3$ on C1 $\xrightarrow{-CH_3OH}$ $\overset{\oplus}{C}$ (C1 cation) $\xrightarrow[-H^\oplus]{H_2O}$ H—C(OH)— / H—C—OH / HO—C—H / H—C—OH / H—C— / CH$_2$OH (ring O)

12. **B.** It involves formation of a resonance-stabilized cation, analogous to that formed in acetal formation.

13. (a) 4-methylcyclohexanone + Al—O—CD(CH$_3$)$_2$ → (Al—O, D)-4-methylcyclohexane $\xrightarrow{H_3O^\oplus}$ (HO, D)-4-methylcyclohexane

(b) $R_2CHOH + Al(OCH(CH_3)_2)_3 \rightleftarrows (R_2CH{-}O)_3Al$

$R_2CH{-}O{-}Al + O{=}C(CH_3)_2 \rightarrow O{=}CR_2 + Al{-}O{-}CH(CH_3)_2$

In the MPV reduction, reaction is driven to completion by use of an excess of reductant (isopropanol). Use of an excess of acetone ensures that the Openauer oxidation goes to completion.

Chapter 17

1. $CH_3-\overset{O}{\overset{\|}{C}}-CH_3 + D_2{}^{18}O \rightleftarrows CH_3-\overset{\oplus OD}{\overset{\|}{C}}-CH_3 \leftrightarrow CH_3-\underset{\oplus}{\overset{OD}{C}}-CH_3 \rightleftarrows$ ($D_2{}^{18}O$)

$\rightleftarrows CH_3-\underset{^{18}\overset{\oplus}{O}-D\,(D)}{\overset{OD}{C}}-CH_3 \rightleftarrows CH_3-\underset{^{18}O\ (D\ OD_2)}{\overset{OD_2}{C}}-CH_3 \leftrightarrows CH_3-\underset{^{18}O}{\overset{\|}{C}}-CH_3$

and

$$CH_3-\overset{\overset{\large O-D}{|}}{\underset{\oplus}{C}}-CH_2-H \quad O(D)_2 \rightleftarrows CH_3-\overset{\overset{\large O-D}{|}}{C}=CH_2 \quad D-\overset{\oplus}{O}(H)(D) \rightleftarrows CH_3-\overset{\overset{\large O}{\|}}{C}-CDH_2 \rightarrow \textit{etc.}$$

2. $H_2C=C=O$ (+ OH_2) $\xrightarrow{\textbf{A}}$ $\left[H_2\overset{\ominus}{C}-C(=O)-\overset{\oplus}{O}H_2 \leftrightarrow H_2C=C(-O^{\ominus})-\overset{\oplus}{O}H_2\right]$

$\overset{\ominus}{C}H_2-C(=O)-\overset{\oplus}{O}H_2$ (+ H_2O) $\xrightarrow{\textbf{B}}$ $CH_3-C(=O)-O-H + H_2O$

(The second step, **B**, may be non-concerted.).

$$CH_3-\overset{\overset{\large O}{\|}}{C}-CH_3 \xrightleftharpoons{H^{\oplus}} CH_2=\overset{\overset{\large O-H}{|}}{C}-CH_3 \xrightarrow{CH_2=C=O} CH_3-\overset{\overset{\large O}{\|}}{C}-O-C(CH_3)=CH_2$$

3. The negative charge is delocalized over oxygen and one carbon:

$$-\overset{\overset{:\ddot{O}:^{\ominus}}{|}}{C}=\overset{|}{C}- \leftrightarrow -\overset{\overset{:\ddot{O}:}{\|}}{C}-\overset{\ddot{}}{\underset{|}{C}}^{\ominus}-$$

4. The rate-determining step in each is the same, *i.e.*, enol formation. Halogenation occurs after the rate-determining step.

5. (a) 2-methylcyclohex-1-en-1-ol (ring C(OH)=C(CH₃)) (b) 3-hydroxycyclohex-2-en-1-one (ring with C=O and C(OH)=CH) (c) $CH_3-\overset{\overset{\large OH}{|}}{C}=CH-CH_3$

(d) $CH_2=\overset{\overset{\large OH}{|}}{C}H-C(CH_3)_3$ (e) $CH_3-\overset{\overset{\large OH}{|}}{C}=CH-CH=CH_2$

6. The enol, commonly called phenol, is aromatic

cyclohexa-2,4-dienone (C=O, CH_2) $\xrightarrow{\text{base}}$ carbanion $\leftrightarrow$ phenoxide ($O^{\ominus}$) $\xrightarrow{H^{\oplus}}$ phenol (OH)

cyclohexa-2,5-dienone (C=O, CH_2) $\xrightarrow{\text{base}}$ carbanion $\leftrightarrow$ phenoxide ($O^{\ominus}$) $\xrightarrow{H^{\oplus}}$ phenol

7. (a) $C_6H_5CO_2H$; (b) $HO_2C(CH_2)_5CO_2H$; (c) $(CH_3)_2C{=}CHCO_2H$;

(d) [benzene ring with CO_2H and CH_3 ortho] + [benzene ring with CO_2H and CO_2H ortho] (e) [benzene ring with CO_2H and CH_3 meta]

8. $CH_3{-}\overset{O}{\overset{\|}{C}}{-}CH_2{-}\overset{CH_3}{\overset{|}{\underset{O-H \leftarrow \overset{\ominus}{O}H}{C}}}{-}CH_3 \rightleftarrows CH_3{-}\overset{O}{\overset{\|}{C}}{-}\overset{\ominus}{C}H_2 + CH_3{-}\overset{O}{\overset{\|}{C}}{-}CH_3 + H_2O \rightleftarrows 2CH_3{-}\overset{O}{\overset{\|}{C}}{-}CH_3 + H_2O$

9. *BASE:* $RCH_2{-}\overset{O}{\overset{\|}{C}}{-}R' + \overset{\ominus}{O}H \rightleftarrows R\overset{\ominus}{C}H{-}\overset{O}{\overset{\|}{C}}{-}R' + H_2O$

$RCH_2{-}\overset{O}{\overset{\|}{C}}{-}R'$ + $R\overset{\ominus}{C}H{-}\overset{O}{\overset{\|}{C}}{-}R'$ $\rightleftarrows$ $RCH_2{-}\overset{O^{\ominus}}{\overset{|}{C}}{-}R'$ (bonded to $RCH{-}\overset{O}{\overset{\|}{C}}{-}R'$) $\xrightleftharpoons{HOH}$ $RCH_2{-}\overset{OH}{\overset{|}{C}}{-}R'$ (bonded to $RCH{-}\overset{O}{\overset{\|}{C}}{-}R'$) $+ \overset{\ominus}{O}H$

ACID: $RCH_2{-}\overset{O}{\overset{\|}{C}}{-}R' \xrightleftharpoons{H^{\oplus}} RCH_2{-}\overset{OH}{\overset{|}{\underset{\oplus}{C}}}{-}R' \xrightleftharpoons{-H^{\oplus}} RCH{=}\overset{OH}{\overset{|}{C}}{-}R'$

$RCH_2{-}\overset{OH}{\overset{|}{\underset{\oplus}{C}}}{-}R'$ + $RCH{=}C(O{-}H)(R')$ $\xrightleftharpoons{-H^{\oplus}}$ $RCH_2{-}\overset{OH}{\overset{|}{C}}{-}R'$ (bonded to $RCH{-}\overset{O}{\overset{\|}{C}}{-}R'$)

for (a): $R{=}CH_3$, $R'{=}H$
(b): $R{=}H$; $R'{=}C_6H_5$
(c): $R{=}R'{=}CH_2CH_2CH_2$
(d): $RCH_2 = $ (H)(C_2H_5)(CH_3)C; $R' = H$ {*Note:* chiral centers in product will be both **R** and **S**. In starting material configuration is **R**, as shown.

Acids tend to induce dehydration. This is worst in *b* since the double bond is conjugated in styrene (product of dehydration).

10. Carbanion formation is slow. As quickly as it is formed, it adds to a carbonyl group of a second acetaldehyde molecule. As the concentration of acetaldehyde diminishes, rate of addition to carbonyl group decreases (statistical factor) but abstraction of deuteron from solvent is unchanged (perhaps a slight increase might be expected). Acetone shows more incorporation because the rate of addition of carbanion to the carbonyl group of acetone is slower. (Increased strain as carbonyl carbon goes from sp^2 to sp^3 during the addition step.)

11. (a) $C_6H_5CHO + CH_3COCH_3 \xrightarrow{OH^{\ominus}} \xrightarrow[heat]{-H_2O} \xrightarrow{H_2/Pt}$

(b) $C_6H_5CHO + CH_3CHO \xrightarrow{OH^{\ominus}} \xrightarrow[heat]{-H_2O} \xrightarrow{H_2/Pt}$

(c) $C_6H_5CHO + C_6H_5COCH_3 \xrightarrow{OH^\ominus} \xrightarrow[\text{heat}]{-H_2O} \xrightarrow{H_2/Pt}$

(d) $CH_3CHO \xrightarrow{OH^\ominus} \xrightarrow[\text{heat}]{-H_2O} \xrightarrow{H_2/Pt}$

(e) $CH_3CH_2CH_2CHO \xrightarrow{OH^\ominus} \xrightarrow[\text{heat}]{-H_2O} \xrightarrow{H_2/Pt}$

(f) $CH_3COCH_3 \xrightarrow[\text{heat}]{H^\oplus} \xrightarrow{H_2/Pt}$

(g) $(C_6H_5CH_2)_2CO + CH_2O \xrightarrow[\text{heat}]{H^\oplus} \xrightarrow{H_2/Pt}$

Chapter 18

1. (a) 2-methylbutanoic acid; α-methylbutyric acid; (b) 4-methylpentanoic acid; γ-methylvaleric acid; (c) same as *a*; (d) 4-bromo-2-methylbutanoic acid; α-methyl-γ-bromobutyric acid; (e) 3-(2-bromoethyl)hexanoic acid; (f) *p*-nitrobenzoic acid; (g) 3-(4-chlorophenyl)-4-methylpentanoic acid.

2. (a) calcium acetate; (b) sodium formate; (c) 1,4-dicyanobenzene; (d) phenyl acetate; (e) 3-pyridinecarboxamide (nicotinamide); (f) benzoic anhydride; (g) benzyl benzoate; (h) benzyl phenyl carbonate; (i) diphenylmethyl α-furoate; (j) α-aminopropionic acid (2-aminopropanoic acid *or* alanine).

3. Electron withdrawing.

4. 2.17 and 3.49. *p*-Nitrobenzoic acid: $pK_a = 3.98 \times 10^{-4}$. The acidities of *m*- and *p*-nitrobenzoic acid are almost identical, supporting a lack of resonance interaction with the nitro groups. The slightly enhanced acidity of *p*-nitrobenzoic acid may be due to stabilizing effect of structure below.

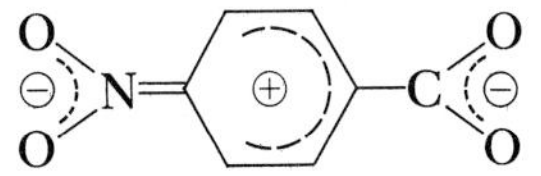

$^\ominus(O)_2N{=}C_6H_4^{\oplus}{-}C(O)_2{}^\ominus$

5. (a) $CH_2{=}CH_2 \xrightarrow[\text{(b) } H_2O_2]{\text{(a) } O_3}$ (b) $CH_2{=}CH_2 \xrightarrow{BH_3} \xrightarrow[OH^\ominus]{H_2O_2} \xrightarrow[\text{heat}]{CrO_3/H^\oplus}$

(c) $CH_3CH{=}C(CH_3)_2 \xrightarrow{HBr} \xrightarrow[\text{ether}]{Mg} \xrightarrow{CO_2} \xrightarrow{H^\oplus}$

(d) $C_6H_5CH{=}CHC_6H_5 \xrightarrow{Br_2} \xrightarrow[\text{excess}]{NH_2^\ominus} \xrightarrow{O_3} \xrightarrow{H_3O^\oplus}$ (e) 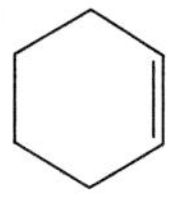 cyclohexene $\xrightarrow[H_3O^\ominus/\text{heat}]{MnO_4^\ominus}$

6. 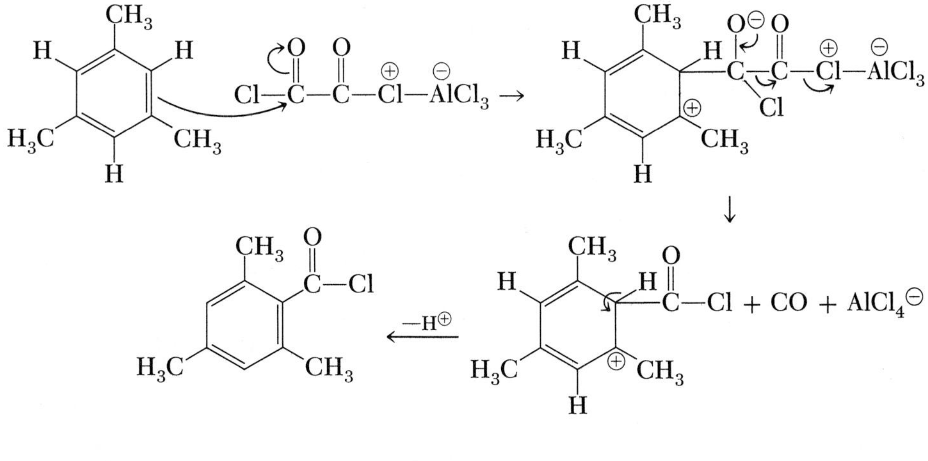 mesitylene (1,3,5-trimethylbenzene) + $Cl{-}CO{-}CO{-}Cl^\oplus{-}AlCl_3^\ominus \rightarrow$ σ-complex with $-C(O^\ominus)(Cl){-}CO{-}Cl^\oplus{-}AlCl_3^\ominus$ $\downarrow$ σ-complex with $-C(O){-}Cl$ + CO + $AlCl_4^\ominus$ $\xrightarrow{-H^\oplus}$ 2,4,6-trimethylbenzoyl chloride ($C{-}Cl$ with $C{=}O$)

7. $Br(CH_2)_4Br \xrightarrow[DMSO]{CN^{\ominus}} NC(CH_2)_4CN \xrightarrow[heat]{H_3PO_4/H_2O} HO_2C(CH_2)_4CO_2H$

$\downarrow BH_3$

$Br(CH_2)_6Br \xleftarrow{PBr_3} HO(CH_2)_6OH$

8. 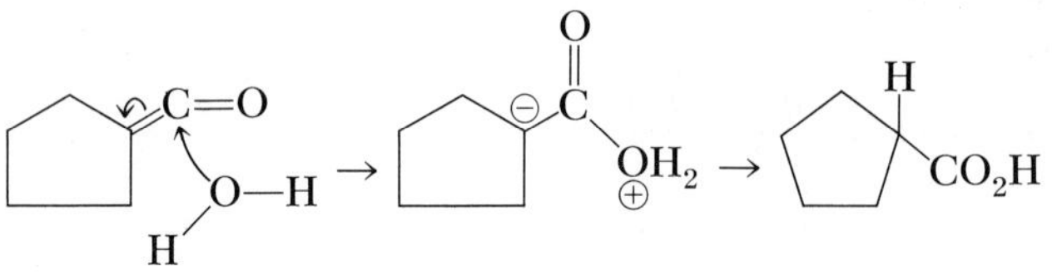 (cyclopentylidene ketene C=O attacked by H_2O) → (carbanion with $C(=O)OH_2^{\oplus}$) → (cyclopentane-H with CO_2H)

(see problem 2, Chapter 17)

9. $RCOCH_2OCH_3$

10. (a) 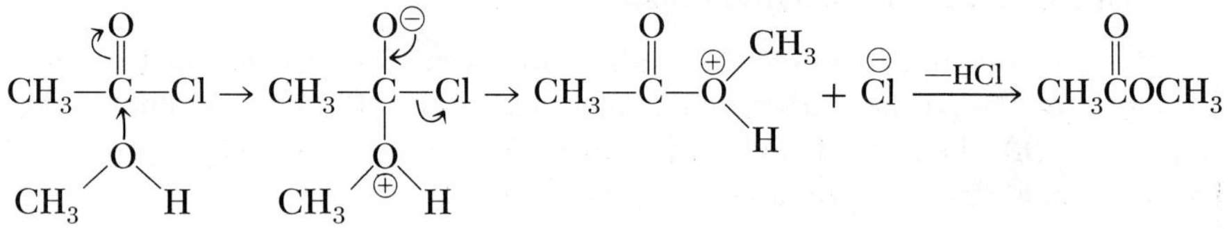 $CH_3-C(=O)-Cl$ + $HOCH_3$ → $CH_3-C(O^{\ominus})(Cl)-O^{\oplus}(H)CH_3$ → $CH_3-C(=O)-O^{\oplus}(H)CH_3$ + $Cl^{\ominus}$ $\xrightarrow{-HCl}$ $CH_3C(=O)OCH_3$

(b) as above with $—OC(O)CH_3$ replacing —Cl.

The acetyl chloride reaction, since it produces hydrogen chloride (which is much more volatile than acetic acid). The methanol is the nucleophile while the acid derivative is the electrophile.

11. (a) $CH_2O + BD_3 \rightarrow CH_2DOD \xrightarrow{CH_3COCl}$;

(b) $HCO_2H + LiAlD_4 \rightarrow \xrightarrow{HOH} HCD_2OH \xrightarrow{CH_3COCl}$;

(c) $CH_2{=}C{=}O + D_2O \rightarrow \xrightarrow{HOH} CH_2DCO_2H$.

12. 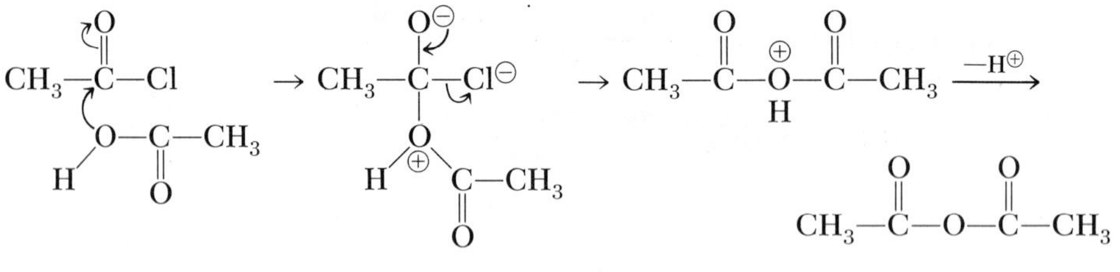 $CH_3-C(=O)-Cl$ + $HO-C(=O)-CH_3$ → $CH_3-C(O^{\ominus})(Cl)-O^{\oplus}(H)-C(=O)-CH_3$ → $CH_3-C(=O)-O^{\oplus}(H)-C(=O)-CH_3$ $\xrightarrow{-H^{\oplus}}$ $CH_3-C(=O)-O-C(=O)-CH_3$

13. (a) $HO_2C(CH_2)_3CO_2H$; (b) $HO_2C(CH_2)_3CO_2CH_3$;

(c) (**R**)-$CH_3CH_2CHCH_3OC(=O)(CH_2)_3CO_2H$; (d) $HO_2C(CH_2)_3CONH_2$;

(e) $HO_2C(CH_2)_3C(O)OC(O)CH_3$; (f) 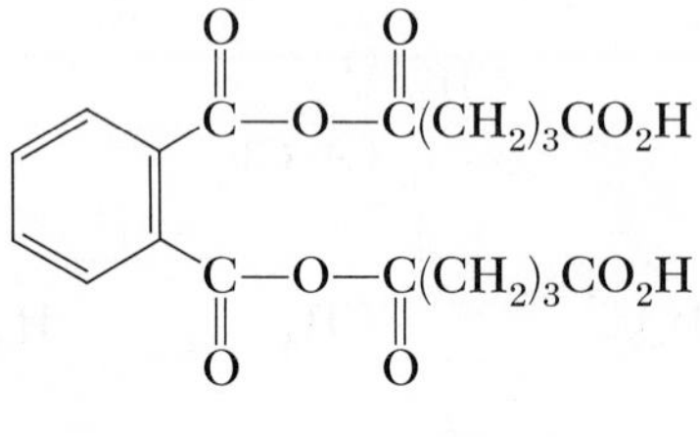 $C_6H_4[C(=O)-O-C(=O)(CH_2)_3CO_2H]_2$ (ortho)

14. (a) Rate increases with increasing ionic strength of solution; (b) rate independent of nucleophile concentration.

15. (**A**) $C_6H_5\overset{O}{\overset{\|}{C}}\text{—O—}C(CH_3)_3 \xrightarrow{H^{\oplus}} C_6H_5\overset{\oplus OH}{\overset{\|}{C}}\text{—O—}C(CH_3)_3 \rightarrow$

$C_6H_5CO_2H + \overset{\oplus}{C}(CH_3)_3 \xrightarrow{-H^{\oplus}}$ isobutylene

(**B**) $C_6H_5\text{—}\overset{\oplus OH}{\overset{\|}{C}}\text{—O—}C(CH_3)_2\text{—}CH_2\text{—H}\cdots O(CH_3)(H) \rightarrow$

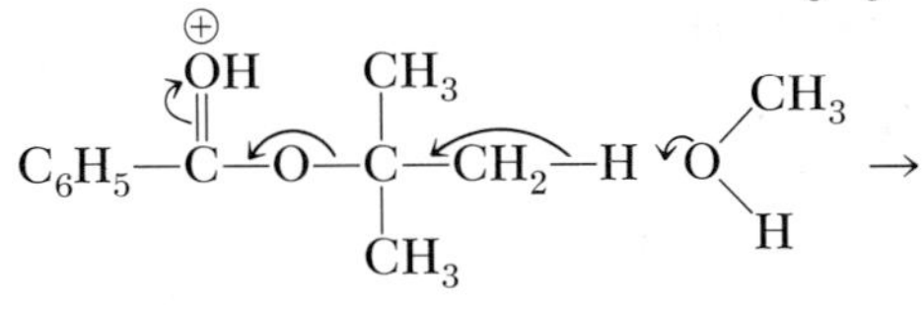

$C_6H_5CO_2H + CH_3\overset{\oplus}{O}H_2 +$ isobutylene

Rate of E1-type process (**A**) should be independent of methanol concentration.

16. Greater resonance delocalization in amide linkage because of lower electronegativity of nitrogen compared to oxygen. (*N* more likely to bear positive charge than *O*.)

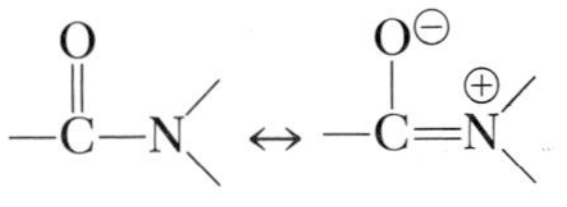

$\text{—}\overset{O}{\overset{\|}{C}}\text{—N}\langle \leftrightarrow \text{—}\overset{O^{\ominus}}{\overset{|}{C}}\text{=}\overset{\oplus}{N}\langle$ more important than

$\text{—}\overset{O}{\overset{\|}{C}}\text{—O—} \leftrightarrow \text{—}\overset{O^{\ominus}}{\overset{|}{C}}\text{=}\overset{\oplus}{O}\text{—}$

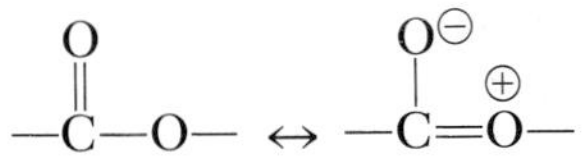

Chapter 19

1. Free radical chlorination.

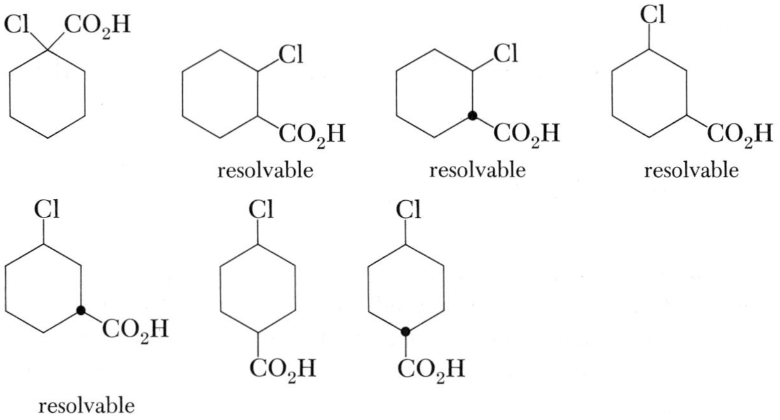

1-chloro-cyclohexanecarboxylic acid; 2-Cl (resolvable); 2-Cl (resolvable); 3-Cl (resolvable); 3-Cl (resolvable); 4-Cl; 4-Cl

resolvable resolvable resolvable

resolvable

2. Diminished delocalization in **A** compared to **B** permits greater enolization in **A** compared to **B.**

A $\text{—}\overset{O}{\overset{\|}{C}}\text{—Cl} \leftrightarrow \text{—}\overset{O^{\ominus}}{\overset{|}{C}}\text{=}Cl^{\oplus}$ **B** $\text{—}\overset{O}{\overset{\|}{C}}\text{—OH} \leftrightarrow \text{—}\overset{O^{\ominus}}{\overset{|}{C}}\text{=}\overset{\oplus}{O}H$

3. Review Figure 19–1. Parts *a* and *b* are S_N2 processes.

(a) $-\underset{\text{H}}{\overset{\text{Cl}}{\text{C}}}-CO_2H \xrightarrow{NH_2^{\ominus}} -\underset{\text{H}}{\overset{\text{NH}_2}{\text{C}}}-CO_2^{\ominus}$

(b) $-\underset{\text{H}}{\overset{\text{Cl}}{\text{C}}}-CO_2H \xrightarrow{CN^{\ominus}} -\underset{\text{H}}{\overset{\text{CN}}{\text{C}}}-CO_2H$

(c) $-\overset{\text{H}}{\text{C}}-\underset{\text{Cl}}{\text{C}}-CO_2H \xrightarrow{R_3N:} \;\rangle C{=}C\langle^{CO_2H}$

(d) $-\underset{\text{H}}{\overset{\text{Cl}}{\text{C}}}-CO_2H \xrightarrow{SOCl_2} -\underset{\text{H}}{\overset{\text{Cl}}{\text{C}}}-COCl$

(e) $-\underset{\text{H}}{\overset{\text{Cl}}{\text{C}}}-CO_2H \xrightarrow[\text{ether}]{CH_2N_2} -\underset{\text{H}}{\overset{\text{Cl}}{\text{C}}}-CO_2CH_3$

(f) $-\underset{\text{H}}{\overset{\text{Cl}}{\text{C}}}-CO_2H \xrightarrow{OH^{\ominus}}$ cyclic dimer (six-membered ring: –C–C(=O)–O–C–C(=O)–O–)

4. $CH_3I \xrightarrow[\text{ether}]{Mg} CH_3MgI$

$$H_2C\overset{O}{\frown}CH_2 + CH_3MgI \rightarrow \xrightarrow{HOH} CH_3CH_2CH_2OH \xrightarrow[H_3O^{\oplus}]{Cr_2O_7^{2-}} CH_3CH_2CO_2H$$

$$\xrightarrow{Cl_2/PCl_3} \xrightarrow{H_3O^{\oplus}} CH_3CHClCO_2H \xrightarrow[\text{(b) } H_3O^{\oplus}]{\text{(a) } NH_3} CH_3CHNH_2CO_2H$$

5. Condensation of esters by weak bases (*e.g.*, ethoxide ion) is reversible, and formation of enolate anion of starting ester is not favored. The reaction must be driven forward, under these circumstances, by formation of the enolate anion of the product. If *this* enolate anion is not very stable, then a strong base is required to convert the final β-keto ester to its enolate anion.

6. Thiol esters are not as stabilized by resonance ($O{=}\overset{|}{C}-S- \leftrightarrow \overset{\ominus}{O}-\overset{|}{C}{=}\overset{\oplus}{S}-$) as are ordinary esters ($O{=}\overset{|}{C}-O- \leftrightarrow \overset{\ominus}{O}-\overset{|}{C}{=}\overset{\oplus}{O}-$) because sulfur's outer electrons are in a different main quantum level than are those of carbon. (See question 2.)

7. (a) $(EtO_2CH_2)_2 + HCO_2Et \xrightarrow{OEt^{\ominus}}$

(b) $C_6H_5CO_2Et + CH_3CO_2Et \xrightarrow{OEt^{\ominus}}$

8. 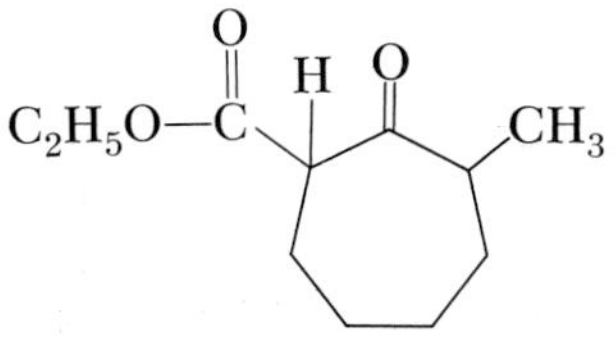 [structure: cycloheptanone ring bearing $C_2H_5O—C(=O)—$ and H on one α-carbon and CH_3 on the other α-carbon]

9. (a) $CH_3CH_2CHO + (EtO_2C)_2CH_2 \xrightarrow[heat]{piperidine}$

(b) $C_6H_5CHO + CH_3COCH_2CO_2C_2H_5 \xrightarrow[heat]{piperidine}$

10. $CH_3C(O)CH{=}C(CH_3)_2$ **A** $CH_3C(O)CH_2C(CH_3)_2CH(CO_2C_2H_5)_2$ **B**

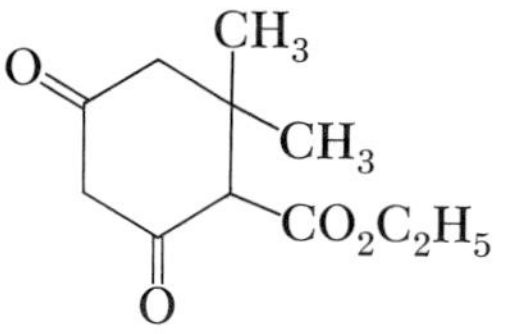

[structures: cyclohexane-1,3-dione ring with gem-dimethyl (CH_3, CH_3) carbon adjacent to a carbon bearing $CO_2C_2H_5$ — **C**; same ring with $CO_2^{\ominus}$ $K^{\oplus}$ — **D**]

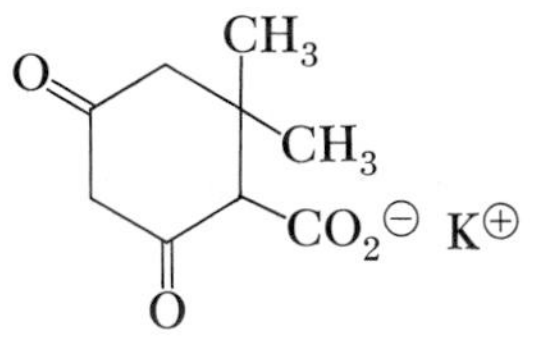

11. (a) $CH_3CH_2CHO + (C_6H_5)_3\overset{\oplus\ominus}{P}CH_2$ (b) $CH_3CHO + (C_6H_5)_3\overset{\oplus\ominus}{P}CHCH_3$
(c) $(CH_2)_5C{=}O + (C_6H_5)_3\overset{\oplus\ominus}{P}CHCH_3$ (d) $(C_6H_5)_2C{=}O + (C_6H_5)_3\overset{\oplus\ominus}{P}CH_2$
(e) $C_6H_5CHO + (C_6H_5)_3\overset{\oplus\ominus}{P}CHC_6H_5$ (f) $(CH_3)_2C{=}O + (CH_3CH_2O)_2\overset{\oplus}{P}(O)\overset{\ominus}{C}HCO_2CH_3$

12. Number of double bonds present.

Chapter 20

1. (a) methylamine (b) dimethylammonium chloride (c) *N*-ethylaniline (ethylphenylamine) (d) dimethyldiphenylammonium hydroxide (e) 2-aminopyridine (f) 1-chloro-2,2-dimethylaziridine (g) 2-azaanthracene

2. Resonance reduces electron density in ring of amide. The amine is aromatic, containing 10 π electrons.

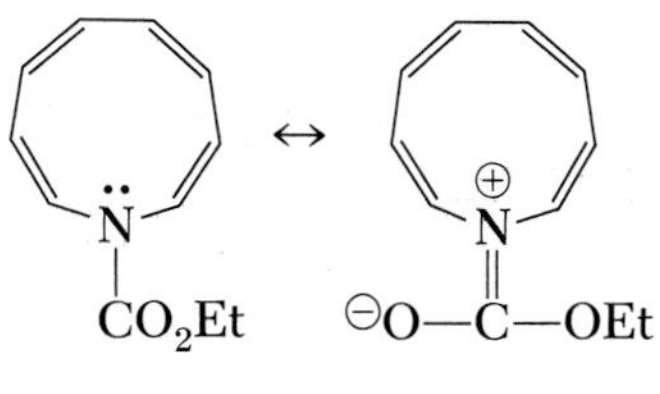

[structures: azonine ring with $\ddot{N}$—CO_2Et $\leftrightarrow$ azonine ring with $\overset{\oplus}{N}{=}C(O^{\ominus})—OEt$]

Decomposition involves electrocyclic closure:

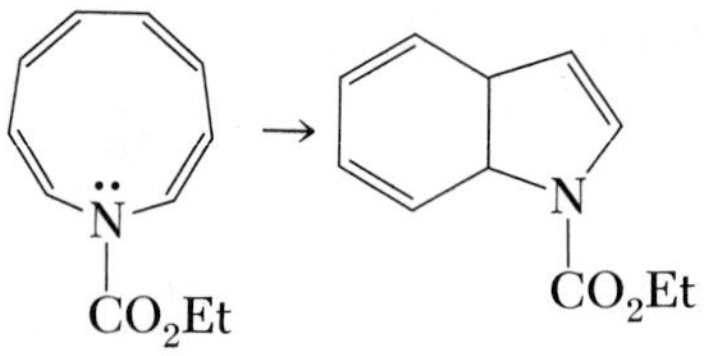
N
CO_2Et
→
N
CO_2Et

3. If configurationally stable, the $C_6H_5CH_2$— protons should be chemical shift non-equivalent and appear as an AB system. If configurationally unstable, they should be a single line.

4. (a) $CH_3CH(C_2H_5)CH_2NH_3^{\oplus}$ $CH_3CH(OH)CO_2^{\ominus}$

$CH_3CH(C_2H_5)CH_2NH_3^{\oplus}$	$CH_3CH(OH)CO_2^{\ominus}$
R	**R**
S	**R**

(b) $CH_3CH(C_2H_5)CH_2NH(CH_3)C_2H_5^{\oplus}$ $CH_3CH(OH)CO_2^{\ominus}$

$CH_3CH(C_2H_5)$	$CH_2NH(CH_3)C_2H_5^{\oplus}$	$CH_3CH(OH)CO_2^{\ominus}$
R	**R**	**R**
R	**S**	**R**
S	**R**	**R**
S	**S**	**R**

5. Not aromatic. Lacks a circuit of 6 pi electrons.

6. *d*.

7. The answer below is one of several possibilities.

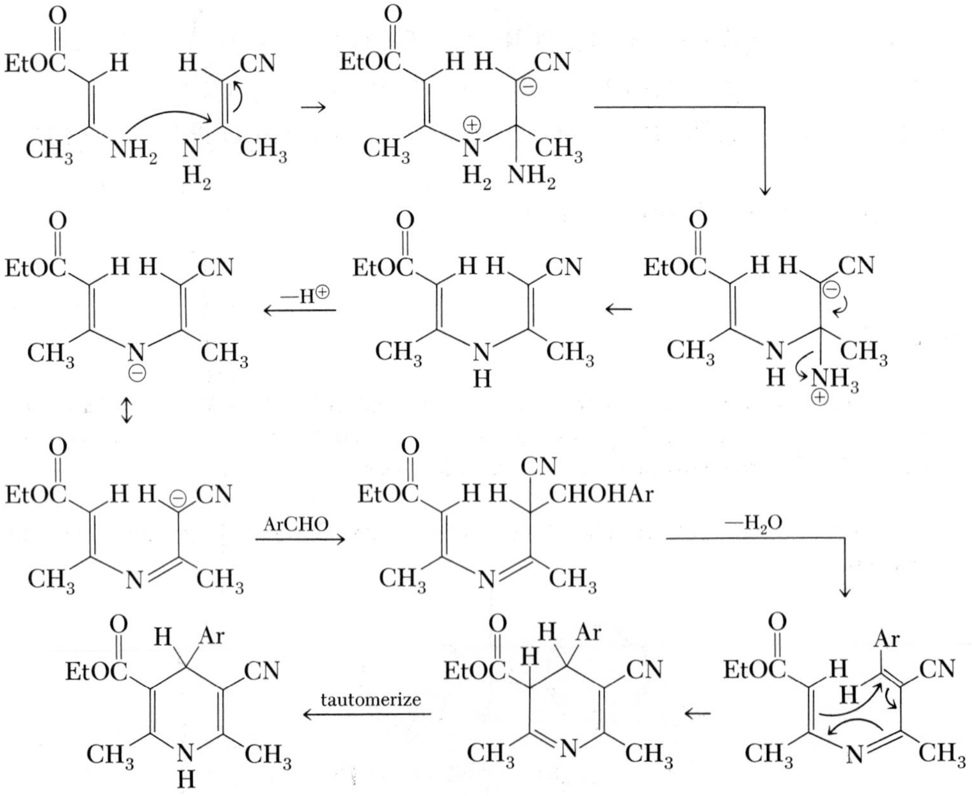
EtOC
CN
CH_3
NH_2
H_2
−H⊕
ArCHO
CHOHAr
$-H_2O$
tautomerize
Ar

8. The carbonyl group of the dimer undergoes very facile addition-elimination to regenerate the starting material.

$$\text{N}\langle\text{C}_5\text{H}_4\rangle\text{—C(=O)—N}\langle\text{C}_5\text{H}_4\rangle\text{—C(=O)—Cl} + Cl^{\ominus} \rightarrow \text{—C(O}^{\ominus}\text{)(Cl)—N}^{\oplus} \rightarrow 2\,\text{N}\langle\text{C}_5\text{H}_4\rangle\text{—C(=O)—Cl}$$

9. $$HO{-}CH_2{-}CH(OH){-}CH_2{-}OH \xrightarrow{H^{\oplus}} HO{-}CH_2{-}CH(\overset{\oplus}{O}H_2){-}CH_2{-}OH \rightarrow HO{-}CH_2{-}CH{=}CH{-}OH \rightleftarrows HO{-}CH_2{-}CH_2{-}CH{=}O$$

$$CH_2{=}CH{-}CHO \leftarrow H_2\overset{\oplus}{O}{-}CH_2{-}CH_2{-}CH{=}O \xleftarrow{H^{\oplus}} HO{-}CH_2{-}CH_2{-}CH{=}O$$

10. $C_6H_6 \xrightarrow{Br_2/AlCl_3} C_6H_5Br \xrightarrow[THF]{Mg} C_6H_5MgBr \xrightarrow[\text{(b) } H_2O]{\text{(a) aziridine (HN)}} C_6H_5CH_2CH_2NH_2$;

$CH_3COCl + H_2NCH_2CH_2C_6H_5 \rightarrow CH_3CONHCH_2CH_2C_6H_5$

11. (a) Conversion of carboxylic acids to methyl esters; (b) reaction with acyl halides in the Arndt-Eistert reaction.

$$H_2\overset{\ominus}{C}{-}\overset{\oplus}{N}{\equiv}N \leftrightarrow H_2C{=}\overset{\oplus}{N}{=}\overset{\ominus}{N}$$

12. $-\ddot{N}{-}C_{\alpha}{=}C_{\beta}{-}$; C_{β} is nucleophilic, nitrogen provides the electrons.

13. Reacting groups on vicinal carbons are eclipsed in the transition state.

14. The ester pyrolysis is concerted and stereospecific, and procedes without isomerization of the product alkene.

$$\text{H—C—C—O—C(=O)CH}_3 \text{ (cyclic transition state)} \rightarrow \rangle C{=}C\langle + CH_3CO_2H$$

Chapter 21

1. (a) $\xrightarrow{Sn/HCl} \xrightarrow{OH^{\ominus}}$

(b) $\xrightarrow[NaOH]{CH_3OH} \xrightarrow{Fe/heat} \xrightarrow[heat]{ZN/OH^{\ominus}}$

(c) $C_6H_5NH-NHC_6H_5$ (from *b*) $\xrightarrow{O_2}$

(d) $\xrightarrow[H_2O]{Zn/NH_4Cl}$

(e) $C_6H_5NH_2$ (from *a*) $\xrightarrow{D_3O^{\oplus}} \xrightarrow[D_2O]{OD^{\ominus}}$

2. $(CH_3)_2CHCN \qquad [(CH_3)_2C(CN)]_2 \qquad CH_3C(CN){=}CH_2$

3. (a) $C_6H_5^{\oplus} + BF_4^{\ominus} \rightarrow [C_6H_5\cdots F\cdots BF_3] \rightarrow C_6H_5F + BF_3$

$C_6H_5^{\oplus} + H_2O \rightarrow C_6H_5{-}\overset{\oplus}{O}H_2 \rightarrow C_6H_5OH + H^{\oplus}$

(b) Methane is not sufficiently acidic.

4. $Ar{-}N{=}N{-}NH{-}Alkyl \xrightarrow{H^{\oplus}} Ar{-}\underset{\oplus}{\overset{H}{N}}{=}N{-}NH{-}Alkyl \xrightarrow{-H^{\oplus}} Ar{-}NH{-}N{=}N{-}Alkyl$

$\downarrow H^{\oplus}$

$ArNH_2 + N_2 + AlkylBr \xleftarrow{SN_2} Ar{-}\overset{\oplus}{N}H_2{-}N{=}N{-}Alkyl \quad Br^{\ominus}$

The use of aniline would prevent the last step (S_N2).

5. (a) Acceleration by a radical initiator; decrease in rate due to presence of radical inhibitor; general insensitivity to changes in solvent polarity.

(b) $H{-}\overset{\oplus}{P}(H)(O^{\ominus}){-}O{-}H$ *or* $H{-}P(O{-}H){-}O{-}H$ (discussed in Chapter 23).

6. $C_6D_5N{=}N{-}NH{-}C_6H_5$ (*cis* and *trans*) *and* $C_6D_5NH{-}N{=}N{-}C_6H_5$ (*cis* and *trans*) (tautomers)

7. Anion of 1,1-dimethoxy-2,4,6-trinitrocyclohexadienyl: ring carbon bearing CH_3O and OCH_3 is sp^3; ring substituents O_2N, NO_2, NO_2; all others sp^2

8. One pair of electrons occupies an sp^2 orbital and is perpendicular to the pi network.

9. One is a ground state argument ("aromatic") while the other deals with the difference in energy between the ground state and the transition state.

10. 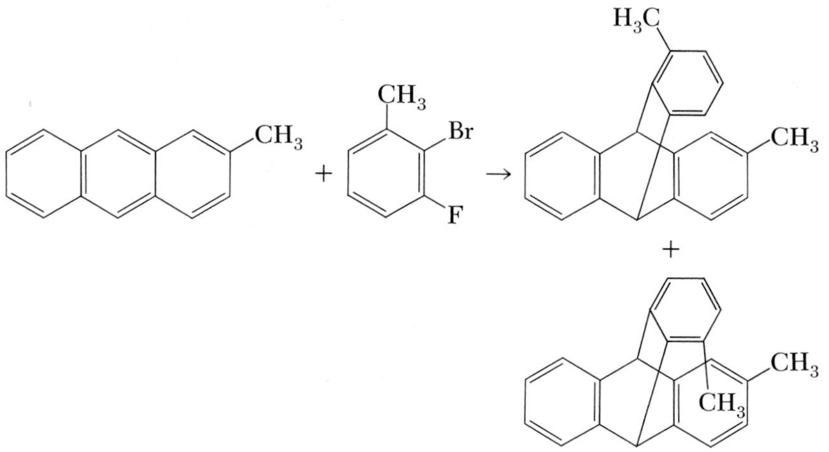 CH_3 + CH_3, Br, F → H_3C, CH_3 (chiral) + CH_3, CH_3 (chiral)

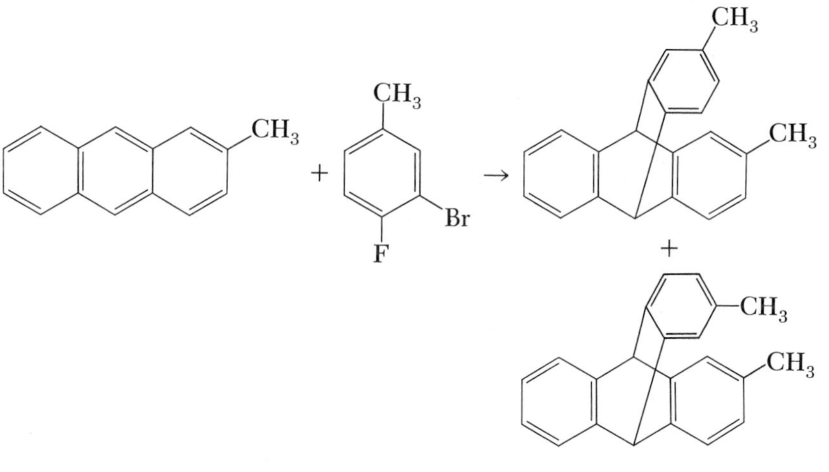

CH_3 + CH_3, Br, F → CH_3, CH_3 (achiral) + CH_3, CH_3 (chiral)

Chapter 22

1. (a) Extraction with NaOH/H_2O followed by acidification of aqueous phase produces acetic acid. (Actually, you do not even need the base because acetic acid is water soluble. However, it will remove any acetic acid that might dissolve in the cyclohexanol.) Distillation of organic phase would separate hexane (bp 68°) and cyclohexanol (bp 161°).

(b) Extraction with $NaHCO_3/H_2O$ followed by acidification of the aqueous phase produces *p*-chlorobenzoic acid. Extraction of remaining organic material with NaOH/H_2O, followed by acidification of the aqueous phase, produces phenol. The residue is hexane.

(c) Since acetic acid is water soluble while *p*-chlorobenzoic acid is not, extraction of the organic phase with water should separate acetic acid. Hexane and *p*-chlorobenzoic acid can be separated by extraction with base or by simple distillation.

2. 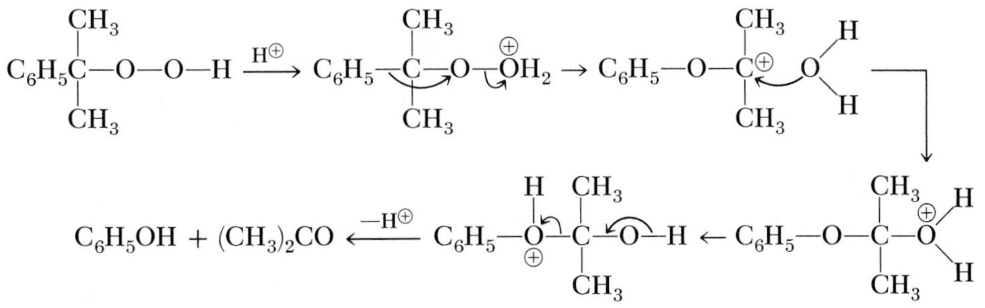 $C_6H_5C(CH_3)_2{-}O{-}O{-}H \xrightarrow{H^{\oplus}} C_6H_5{-}C(CH_3)_2{-}O{-}\overset{\oplus}{O}H_2 \rightarrow C_6H_5{-}O{-}\overset{\oplus}{C}(CH_3)_2 \;\; O(H)_2 \rightarrow C_6H_5{-}O{-}C(CH_3)_2{-}\overset{\oplus}{O}H_2$

$\rightarrow C_6H_5{-}\overset{\oplus}{O}(H){-}C(CH_3)_2{-}O{-}H \xrightarrow{-H^{\oplus}} C_6H_5OH + (CH_3)_2CO$

3. *c*, *d* and *e*. Grignard reagent reacts with other functional group on benzene ring.

4. One, OD.

5. $\xrightarrow{OH^{\ominus}}$ $\xrightarrow{ClCH_2CO_2Na}$ $\xrightarrow{H_3O^{\oplus}}$

6. (a) OD, D, $CH_2—CH{=}CH_2$ (b) OH, H, $CD_2—CH{=}CH_2$

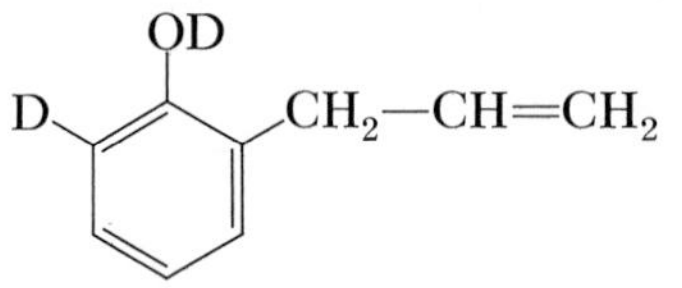

(c) OH, H, $CH_2—CD{=}CH_2$ (d) OH, H, $CH_2—CH{=}CD_2$

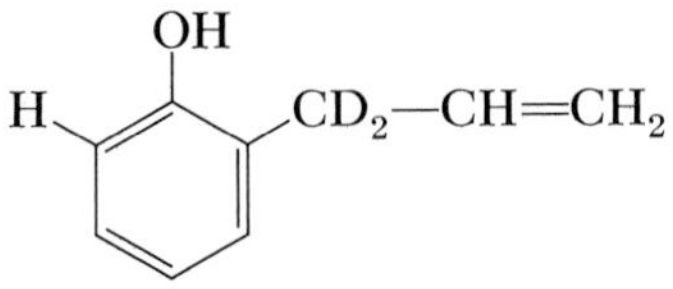

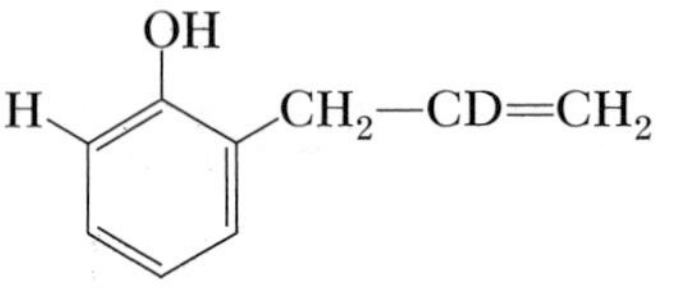

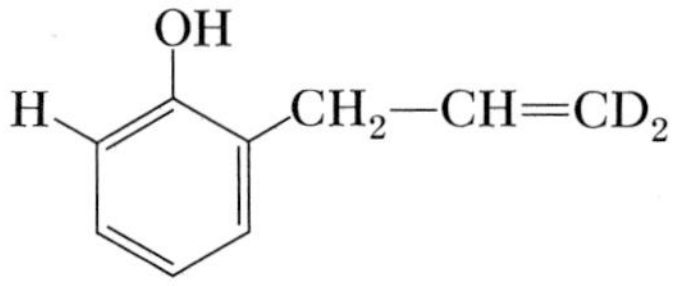

7. If an allyl cation were produced, an appropriate deuterium label would become scrambled and give rise to two products. The product mixture could be analyzed by mass spectrometry.

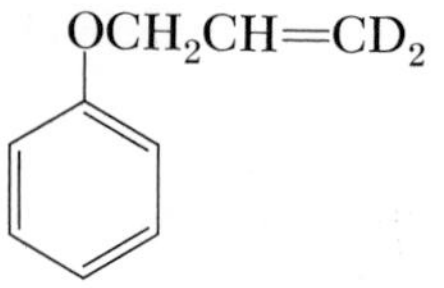

$OCH_2CH{=}CD_2$ does *not* form OH $CH_2CH{=}CD_2$ even though it might be expected to result from $\overset{\oplus}{C}H_2—CH{=}CD_2 \leftrightarrow CH_2{=}CH—CD_2^{\oplus}$

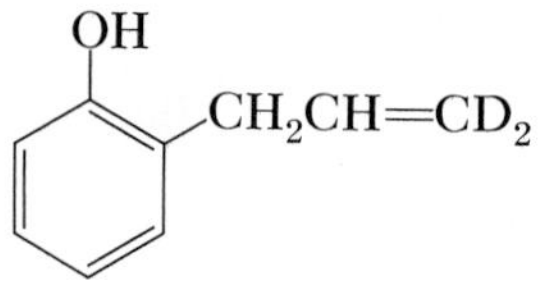

8. O O O $\xrightarrow{H^{\oplus}}$ O ⊕ OH $\leftrightarrow$ O O⊕ OH

O O ⊕ OH OH → O O HO ⊕ H OH $\xrightarrow{-H^{\oplus}}$

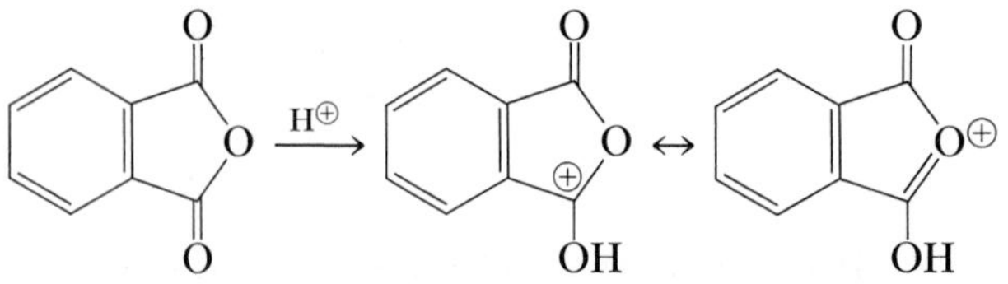

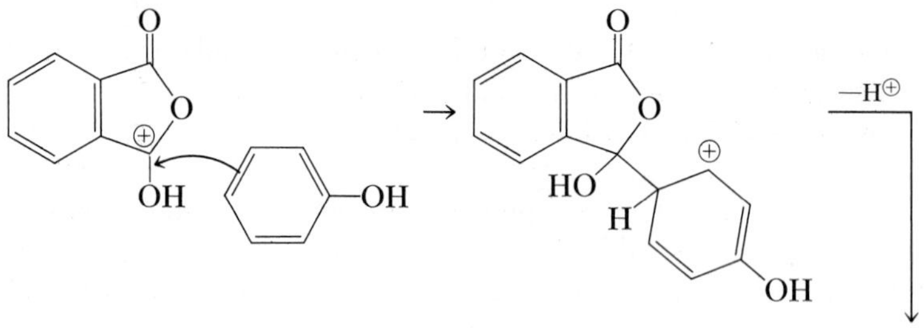

repeat, now protonating **OH** and forming new cation ($—H_2O$) which couples with a second C_6H_5OH ← O O **HO** OH

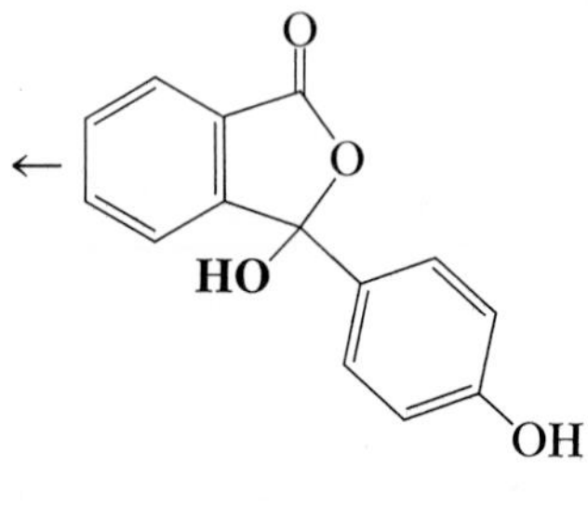

9. 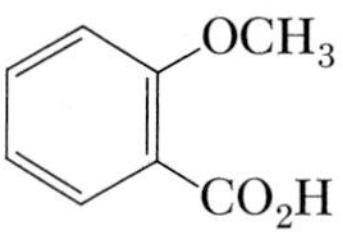 [structure: benzene ring with OCH_3 and CO_2H ortho]

If the reaction is S_N1, this would require formation of either [structure: phenyl cation with ortho CO_2H] or $CH_3^{\oplus}$, both of which should be difficult to form compared to [structure: benzene ring with OH and ortho $\overset{\oplus}{C}O_2H_2$], the intermediate leading to methyl salicylate.

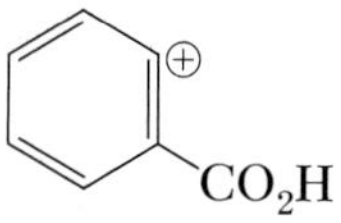

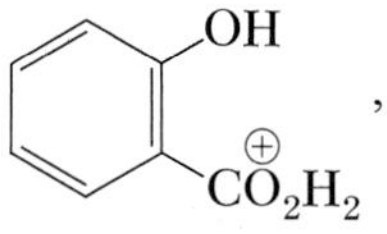

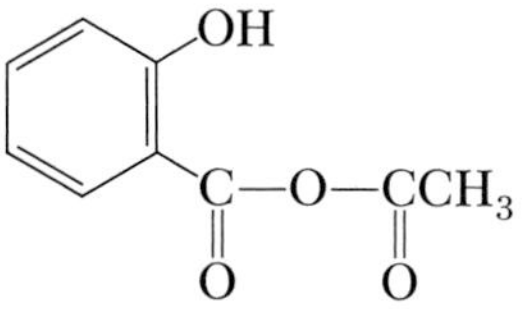

[structure: benzene ring with OH and ortho C(=O)—O—C(=O)CH_3]

The nucleophilicity of the hydroxy portion of the carboxy group is reduced by delocalization with the carbonyl portion.

10. Diels-Alder reactions are concerted and lead to *cis* addition to double bonds. The new chiral centers are formed with the methine hydrogens *trans* to the existing methine hydrogens because diene approaches dienophile from less-hindered face. (See p. 380.)

11. 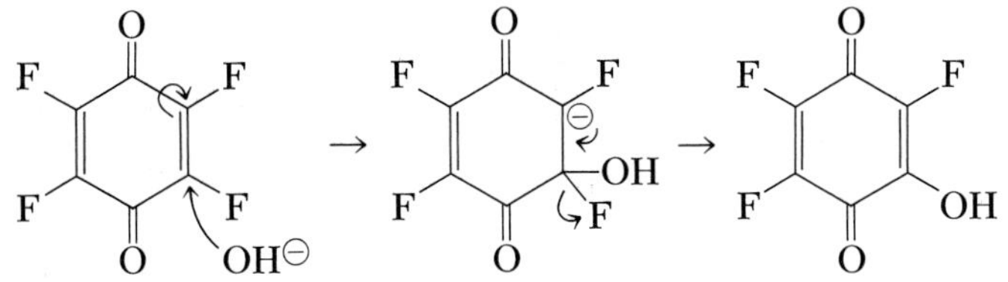 [scheme: tetrafluoro-p-benzoquinone + $OH^{\ominus}$ → anionic adduct (C bearing OH and F) → trifluorohydroxy-p-benzoquinone (loss of F)]

12. 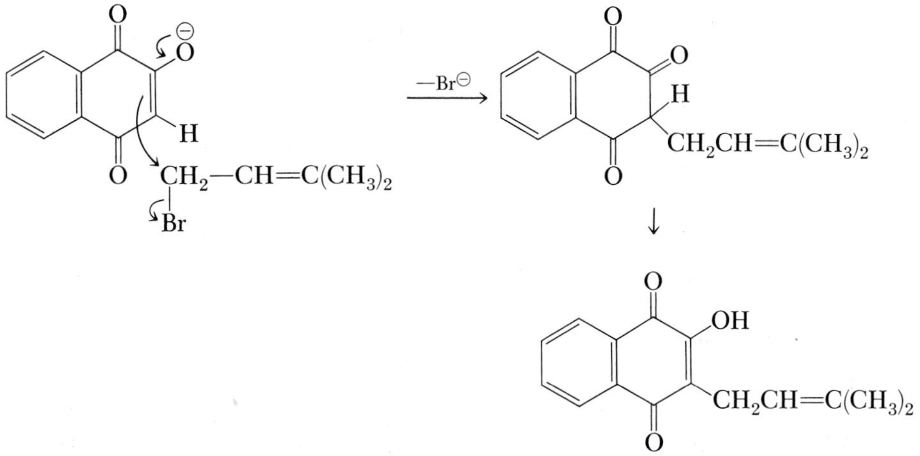 [scheme: 2-oxido-1,4-naphthoquinone anion attacks $BrCH_2—CH=C(CH_3)_2$; $\xrightarrow{-Br^{\ominus}}$ 2-oxo-3-H-3-$CH_2CH=C(CH_3)_2$ naphthalenetrione; ↓ 2-hydroxy-3-$CH_2CH=C(CH_3)_2$-1,4-naphthoquinone]

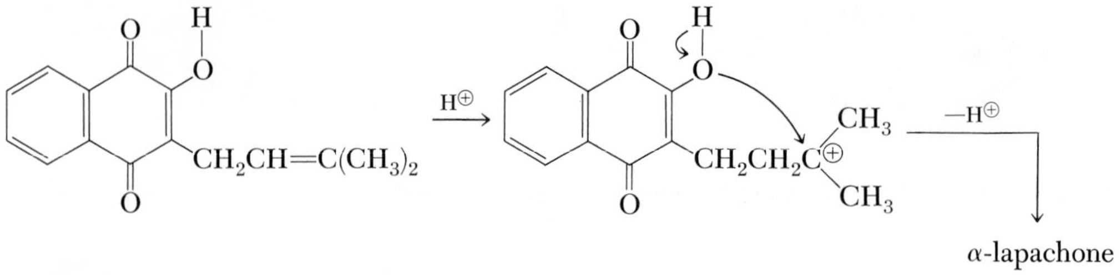

$\xrightarrow{H^{\oplus}}$ $CH_2CH=C(CH_3)_2$ ... $CH_2CH_2C^{\oplus}(CH_3)_2$ $\xrightarrow{-H^{\oplus}}$ α-lapachone

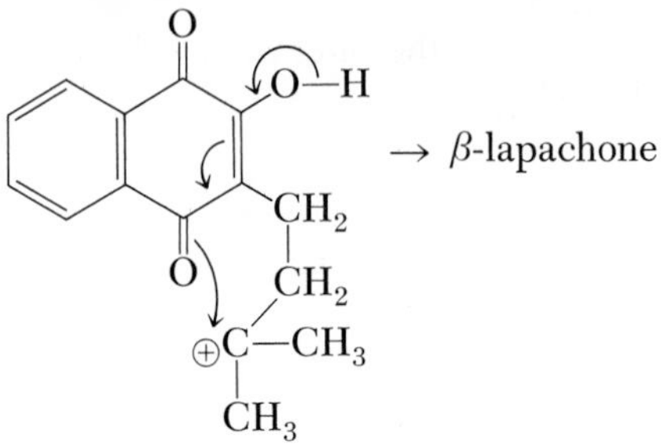

→ β-lapachone

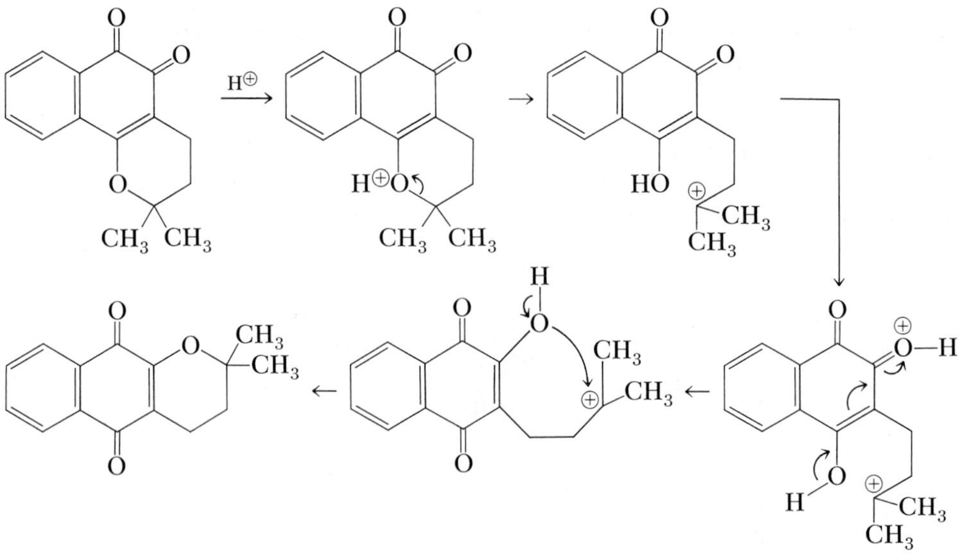

α-Lapachone is more stable, perhaps because of mutual carbonyl repulsion in β-lapachone (see Section 17.2).

Chapter 23

1. (a) sulfone; (b) sulfonate; (c) sulfide, sulfoxide; (d) sulfinate; (e) sulfide, sulfoxide; (f) sulfenate; (g) disulfide; (h) thiosulfinate; (i) sulfide; (j) thiosulfinate; (k) thiol; (l) sulfinyl chloride.

2. (a) methyl phenyl disulfide; (b) 1-chloro-2-naphthalenethiol; (c) *m*-toluenethiol; (d) ethyl methanesulfinate; (e) phenyl *p*-toluenesulfonate; (f) *o*-phenylthioanisole; (g) 5-methyl-4,6-dithianonane; (h) 2-phenyl-1,3-dithiolane (2-phenyl-1,3-dithiacyclopentane); (i) isopropylsulfonyl chloride; (j) 3-thia-1-butanol.

3. Trimethylsulfonium cation forms an ylid ($\overset{\oplus}{S}$-$\overset{\ominus}{C}$) in base. This ylid may react with D_2O to incorporate deuterium. Proton loss from tetramethylammonium cation does not produce a resonance-stabilized anion (nitrogen lacks available *d*-orbitals). You may find it valuable to re-examine part *d* of question 12 in Chapter 2 (p. 43).

4. (a) $\xrightarrow[\text{EtOH}]{\text{KOH}} \xrightarrow{CH_3I}$

(b) $\xrightarrow[\text{EtOH}]{\text{KOH}/I_2}$

(c) $C_6H_5SCH_3$ (part *a*) $\xrightarrow[0°]{N_2O_4}$

(d) $C_6H_5SCH_3$ (part *a*) $\xrightarrow[\text{heat}]{CH_3Br/\text{EtOH}}$

(e) $\xrightarrow[\text{EtOH}]{\text{KOH}} \xrightarrow[\text{excess}]{CH_3SSCH_3}$

(f) $\xrightarrow[\text{heat}]{MnO_4^{\ominus}/H^{\oplus}}$

(g) $C_6H_5SSC_6H_5$ (part *b*) $\xrightarrow[\text{(excess)}]{(CH_3)_3CS^{\ominus}}$

(h) $C_6H_5SSC_6H_5$ (part *b*) $\xrightarrow[\text{ether}]{(CH_3)_3CMgBr}$

5. Two alternative paths are shown. Each gives the same product, so that beginning with optically active starting material produces optically active product. (Note: path *b* is similar to path *a*.)

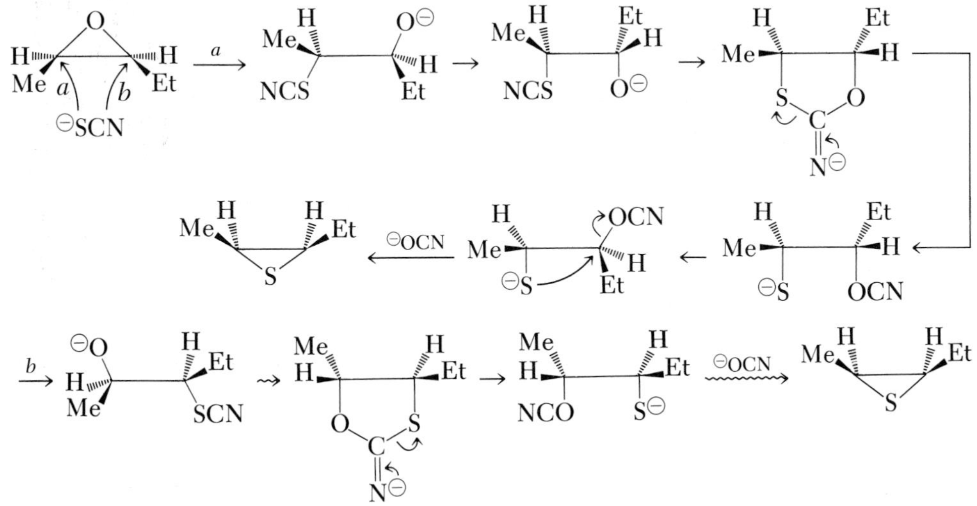

6. (a) 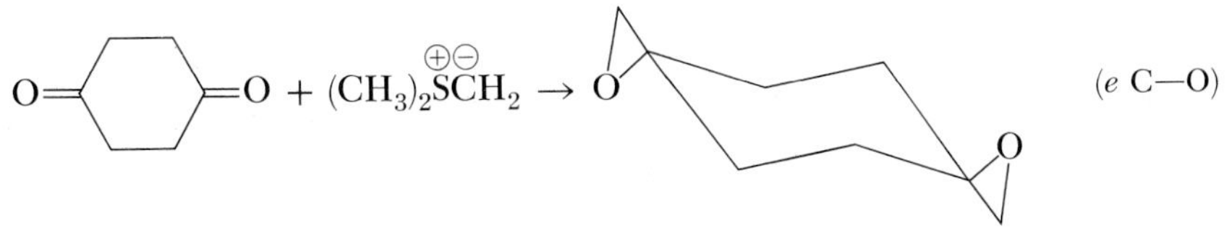 O=⟨cyclohexane⟩=O + $(CH_3)_2\overset{\oplus\ominus}{S}CH_2$ → (*e* C—O)

(b) $O{=}C_6H_8{=}O + (CH_3)_2\overset{\oplus}{S}(O)\overset{\ominus}{C}H_2 \rightarrow$ (diepoxide) (*a* C—O)

Products are diastereomers of one another (p. 734).

7. $CH_3S(O)_2CH_3$; dimethyl sulfone. An S_N2 attack by the carbanion on *t*-butyl bromide is impossible. However, this bromide will provide a proton and convert the dimsyl anion ($CH_3S(O)CH_2^{\ominus}$) to DMSO, which will be oxidized to sulfone by H_2O_2.

8. Rapid racemization by formation of an achiral sulfinate anion.

$$R\text{—}\overset{\oplus}{S}(O^{\ominus})(OH) \rightarrow R\text{—}\overset{\oplus}{S}(O^{\ominus})(O^{\ominus}) \quad H^{\oplus} \rightarrow R\text{—}\overset{\oplus}{S}(O^{\ominus})(OH) + R\text{—}\overset{\oplus}{S}(OH)(O^{\ominus})$$

chiral achiral racemic

9. (a) phosphine; (b) phosphorus trichloride; (c) phosphorous acid; (d) phosphine oxide; (e) phosphonous acid; (f) triphenyl phosphite; (h) triphenyl phosphate.

10. 73 kcal/mole.

11. Non-ylid multiple bonds containing phosphorus are not very stable. P,P double bonds are weaker than N,N double bonds because of the greater covalent radius of P (resulting in poorer overlap). P═O and C≡P analogs of N═O and C≡N bonds are unstable because the former multiple bonds require overlap of orbitals of different main quantum levels.

12. $$H_3P \rightarrow CH_2{=}O \rightarrow H_3\overset{\oplus}{P}\text{—}CH_2O^{\ominus} \xrightarrow[-Cl^{\ominus}]{HCl} H_3\overset{\oplus}{P}\text{—}CH_2OH \xrightarrow{-H^{\oplus}} H_2P\text{—}CH_2OH \xrightarrow[H^{\oplus}]{CH_2O} H_2\overset{\oplus}{P}(CH_2OH)_2 \xrightarrow{-H^{\oplus}} HP(CH_2OH)_2$$

$$HP(CH_2OH)_2 \xrightarrow[H^{\oplus}]{CH_2OH} H\overset{\oplus}{P}(CH_2OH)_3 \xrightarrow{-H^{\oplus}} P(CH_2OH)_3 \xrightarrow[HCl]{CH_2O} \overset{\oplus}{P}(CH_2OH)_4\ Cl^{\ominus}$$

13. **A.** $$R\text{—}C{\equiv}C\text{—}Br + {}^{\ominus}P(C_6H_5)_2 \rightarrow R\text{—}\overset{\ominus}{C}{=}C(Br)\text{—}P(C_6H_5)_2 \rightarrow R\text{—}C{\equiv}C\text{—}P(C_6H_5)_2$$

B. $$R\text{—}C{\equiv}C\text{—}Br + (C_6H_5)_2P^{\ominus} \rightarrow (C_6H_5)_2P(R)C{=}\overset{\ominus}{C}\text{—}Br \rightarrow (C_6H_5)_2P\text{—}C{\equiv}C\text{—}R$$

14. This reaction is favored by the strong P—O bond that is produced in the by-product.

$$R-\overset{O}{\overset{\|}{C}}-O \longrightarrow \underset{Cl}{\overset{H\ \ Cl}{P}}-Cl \rightarrow R-\overset{O}{\overset{\|}{C}}-\overset{\oplus}{O}-\underset{}{\overset{H\ \ Cl}{P}}-Cl \quad Cl^{\ominus} \rightarrow R-\underset{Cl}{\overset{O^{\ominus}}{C}}-\overset{\ominus}{O}-\overset{H\ \ Cl}{P}-Cl$$

$$\downarrow$$

$$R-\overset{O}{\overset{\|}{C}}-Cl + HO-\overset{Cl}{P}-Cl$$

$HOPCl_2$ continues to react further with RCO_2H to produce, ultimately, $HP(OH)_2O$ and three moles of RCOCl.

Chapter 24

1. (structure: H, R, NH_2, HO_2C on stereocenter)

leucine	R: $-C(CH_3)_3$
serine	R: $-CH_2OH$
valine	R: $-CH(CH_3)_2$
alanine	R: $-CH_3$

2. $-NH_2$ is more basic, in part because $-CO_2^{\ominus}$ is resonance stabilized. Monoprotonation of $H_2NCHRCO_2^{\ominus}$ produces $H_3\overset{\oplus}{N}CHRCO_2^{\ominus}$.

3. (hydantoin) $\xrightarrow{OH^{\ominus}}$ (tetrahedral intermediate) $\rightarrow R-\underset{\ominus NH}{\overset{H}{C}}-\overset{O}{\overset{\|}{C}}-NH-\overset{O}{\overset{\|}{C}}-O-H \rightarrow R-\underset{NH_2}{\overset{H}{C}}-\overset{O}{\overset{\|}{C}}-NH-\overset{O}{\overset{\|}{C}}-O^{\ominus} \xrightarrow{-CO_2} R-\underset{NH_2}{\overset{H}{C}}-\overset{O}{\overset{\|}{C}}-\overset{\ominus}{NH} \xrightarrow{H_2O} R-\underset{NH_2}{\overset{H}{C}}-\overset{O}{\overset{\|}{C}}-NH_2$

repeat amide hydrolysis

$$R-\underset{NH_2}{\overset{H}{C}}-CO_2^{\ominus} \xrightarrow{H_3O^{\oplus}} R-\underset{NH_2}{\overset{H}{C}}-CO_2H$$

4. The imido hydrogen (—C(O)NHC(O)—) is most acidic. The resultant anion is resonance stabilized by two carbonyl groups.

5. **A,** $Cl-\overset{O}{\overset{\|}{C}}-S-C_6H_5$; **B,** $(CH_3)_3C-O-\overset{O}{\overset{\|}{C}}-S-C_6H_5$;

C, $(CH_3)_3C-O-\overset{O}{\overset{\|}{C}}-NH-NH_2$

6. (a) $C_6H_5NO_2 \xrightarrow{Sn/H^{\oplus}} \xrightarrow{OH^{\ominus}} \xrightarrow{Ac_2O} C_6H_5NHAc \xrightarrow[H_2SO_4]{HNO_3}$ (NHAc, NO_2, NO_2 benzene) $\xrightarrow[heat]{OH^{\ominus}}$ (NH$_2$, NO_2, NO_2 benzene) $\xrightarrow[0°]{[HONO]} \xrightarrow{HBF_4} \xrightarrow{heat}$

(b) C_6H_5NHAc (part *a*) $\xrightarrow[AlCl_3]{CH_3Cl}$ (NHAc, CH_3 benzene) $\xrightarrow{NO_2{}^{\oplus}BF_4{}^{\ominus}}$ (O_2N, NHAc, NO_2, CH_3 benzene) $\xrightarrow[heat]{H_3O^{\ominus}} \xrightarrow[0°]{[HONO]} \xrightarrow{EtOH}$ (O_2N, NO_2, CH_3 benzene) $\xrightarrow[OH^{\ominus}\ heat]{MnO_4{}^{\ominus}} \xrightarrow{H^{\oplus}} \xrightarrow{SOCl_2} \xrightarrow{NH_3}$ (O_2N, NO_2, $CONH_2$ benzene) $\xrightarrow[OH^{\ominus}]{Br_2}$ (O_2N, NO_2, NH_2 benzene) $\xrightarrow[0°]{[HONO]} \xrightarrow{HBF_4} \xrightarrow{heat}$

7. Resonance stabilization of anion formed by proton loss.

8. A: Ala; S: Ser; G: Gly; V: Val in this answer set.

(a) A-S-G-V; A-S-V-G; A-G-S-V; A-G-V-S; A-V-S-G; A-V-G-S; S-G-V-A; S-V-G-A; G-S-V-A; G-V-S-A; V-S-G-A; V-G-S-A; S-A-G-V; S-A-V-G; S-G-A-V; S-V-A-G; G-S-A-V; G-V-A-S; G-A-S-V; G-A-V-S; V-S-A-G; V-G-A-S; V-A-G-S; V-A-S-G.

(b) In each, the *N*-terminal amino acid is on the extreme left and the *C*-terminal amino acid is on the right.

(c) A-V-S-G; A-V-G-S; S-V-G-A; S-V-A-G; G-V-A-S; G-V-S-A

(d) A-G-S-V (others are possible)

$$\underbrace{H_2N\text{—}\underset{CH_3}{CH}\text{—}\overset{O}{\overset{\|}{C}}}_{ala(1)}\text{—}\underbrace{NH\text{—}\underset{H}{CH}\text{—}\overset{O}{\overset{\|}{C}}}_{gly(2)}\text{—}\underbrace{NH\text{—}\underset{CH_2OH}{CH}\text{—}\overset{O}{\overset{\|}{C}}}_{ser(3)}\text{—}\underbrace{NH\text{—}\underset{CH(CH_3)_2}{CH}\text{—}CO_2H}_{val(4)}$$

9. (a) 5040; (b) 40320; (c) 362880.

10. 294.

11. Phe-Val-Asp-Glu-His.

12. (a) $C_6H_5CH_2OCOCl + H_2NCHCH_3CO_2H \xrightarrow[\text{(b) } H^{\oplus}]{\text{(a) } OH^{\ominus}/5°}$

$C_6H_5CH_2OCONHCHCH_3CO_2H$
N-protected alanine

$C_6H_5CH_2OCONHCHCH_3CO_2H + ClCO_2Et \rightarrow$

$C_6H_5CH_2OCONHCHCH_3CO_2CO_2Et$
N-protected carboxy-activated alanine

$C_6H_5CH_2OCONHCHCH_3CO_2CO_2Et + H_2NCH_2CO_2Et \xrightarrow[-H_2O]{-CO_2}$

$\xrightarrow[OH^{\ominus}]{H_2O} \xrightarrow{H_3O^{\ominus}} C_6H_5CH_2OCONHCHCH_3CONHCH_2CO_2Et$

$C_6H_5CH_2OCONHCHCH_3CONHCH_2CO_2H \xrightarrow{H_2/Pd}$

$H_2NCHCH_3CONHCH_2CO_2H + CO_2 + C_6H_5CH_3$

(b) $C_6H_5CH_2OCOCl + H_2NCH_2CONHCHCH_3CO_2H \rightarrow \xrightarrow{ClCO_2Et}$
(Figure 24–7)

$C_6H_5CH_2OCONHCH_2CONHCHCH_3CO_2CO_2Et$

$C_6H_5CH_2OCONHCH_2CONHCHCH_3CO_2CO_2Et \xrightarrow[-CO_2;\ -H_2O]{H_2NCH_2CO_2Et}$

$\xrightarrow[\text{(b) } H_3O^{\oplus}]{\text{(a) } OH^{\ominus}} C_6H_5CH_2OCONHCH_2CONHCHCH_3CONHCH_2CO_2H$

$C_6H_5CH_2OCONHCH_2CONHCHCH_3CONHCH_2CO_2H \xrightarrow{H_2/Pd}$

$\underbrace{H_2NCH_2CO}_{\text{gly}}\underbrace{NHCHCH_3CO}_{\text{ala}}\underbrace{NHCH_2CO_2H}_{\text{gly}} + CO_2 + C_6H_5CH_3$

Chapter 25A

1. Since it lacks a hydroxy group, formaldehyde should not be considered the simplest carbohydrate.

2. (a) aldose, hexose (aldohexose); (b) aldose, hexose (aldohexose); (c) ketose, tetrose (ketotetrose); (d) aldose, hexose (aldohexose); (e) aldose, hexose (aldohexose); (f) aldose; pentose (aldopentose); (g) ketose, pentose (ketopentose).

3. (a)

	CHO	
H		OH
HO		H
H		OH
H		OH
	CH_2OH	

(b)

	CHO	
HO		H
H		OH
HO		H
HO		H
	CH_2OH	

(NOT

	CHO	
H		OH
HO		H
H		OH
HO		H
	CH_2OH	

)

(c) 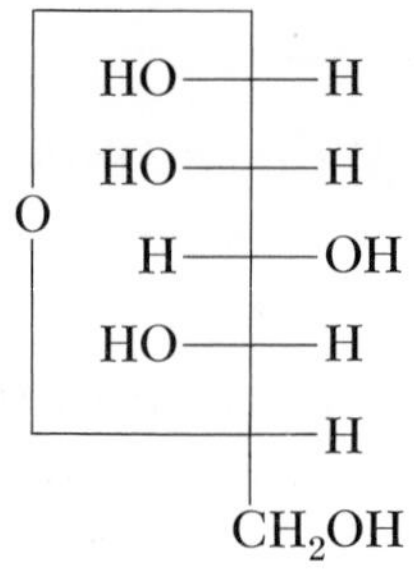 O
HO—H
HO—H
H—OH
HO—H
—H
CH_2OH

(d) 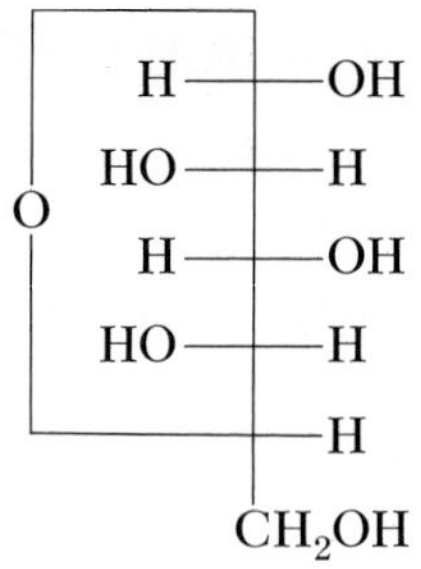 O
H—OH
HO—H
H—OH
HO—H
—H
CH_2OH

Examine Problem 9 for a comparison with part *b*.

4. (a) 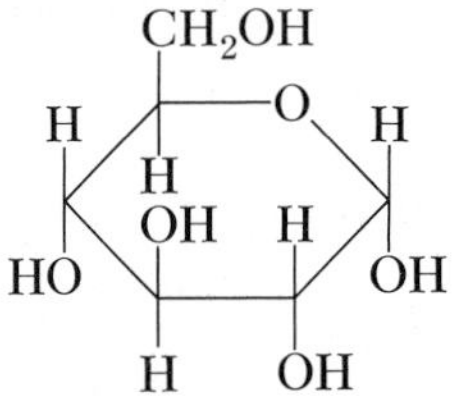
CH_2OH
H O H
H
OH H
HO OH
H OH

(b) 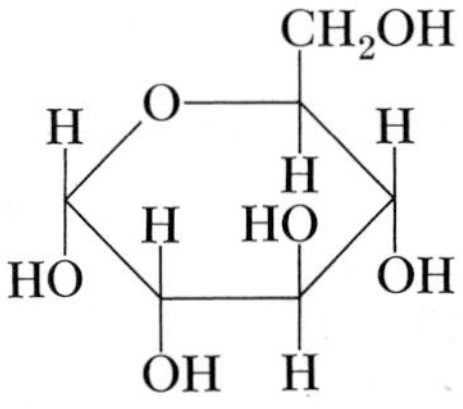
CH_2OH
H O H
H
H HO
HO OH
OH H

(c) 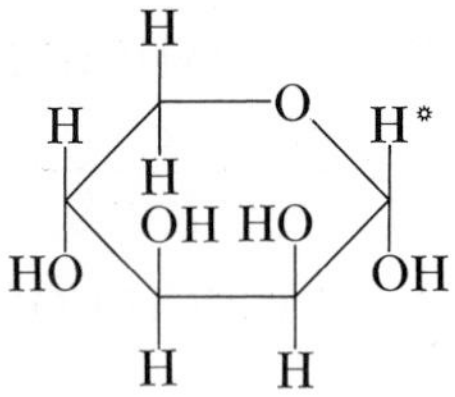
H
H O H*
H
OH HO
HO OH
H H

(d) 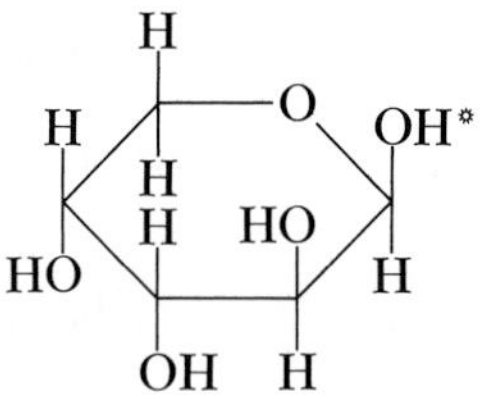
H
H O OH*
H
H HO
HO H
OH H

5. The α and β forms of any sugar are diastereomers, not enantiomers. The only pair of enantiomers in the set consists of α-L-glucopyranose and α-D-glucopyranose.

6. α-D-glucopyranose
H
HO CH_2OH O
H
H
HO H
OH
H OH

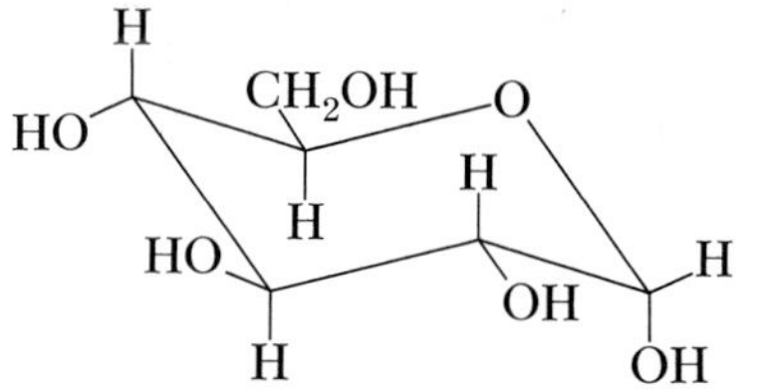

β-D-glucopyranose
H
HO CH_2OH O
H
H
HO OH
OH
H H

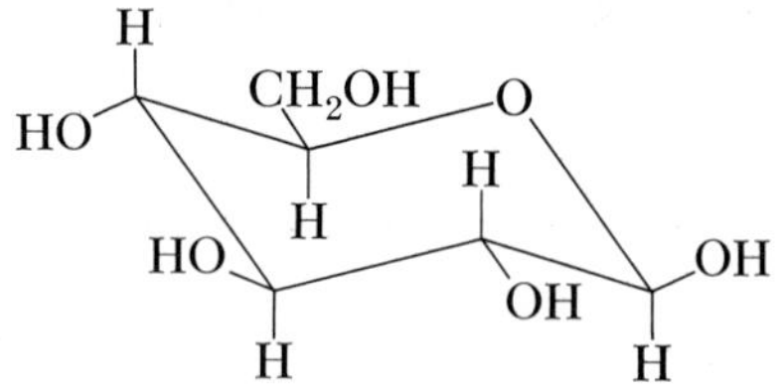

α-L-glucopyranose
OH
H
CH_2
O OH
H
H
H OH
OH
HO H

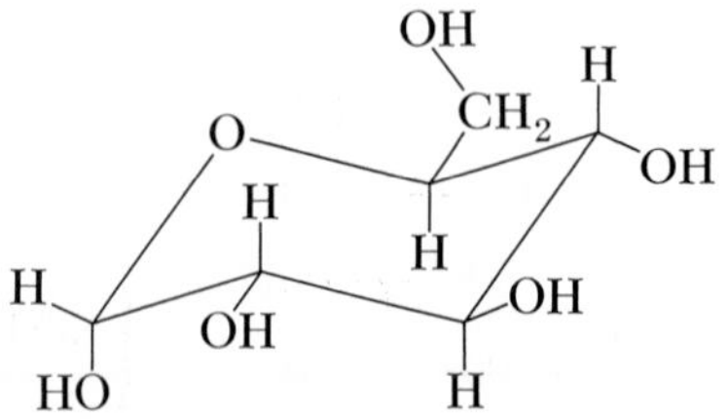

* These can be constructed easily by beginning with sequence **A** in Figure 25A–3 and forming the pyranose ring immediately after turning the structure 90°.

β-L-glucopyranose

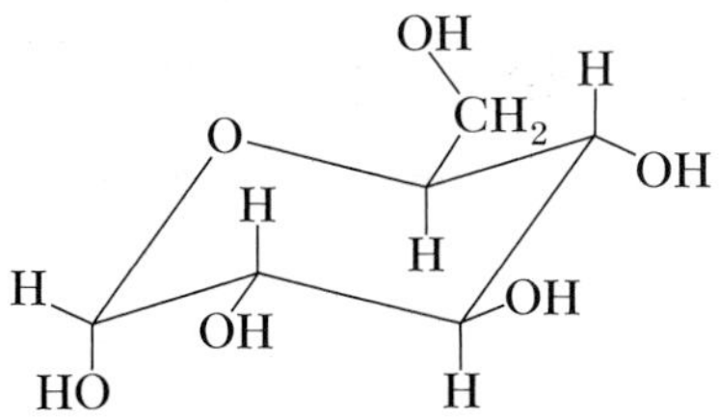

OH H CH$_2$ OH O H H H OH OH HO H

7. They are diastereomers, and of unequal energy. Since $\Delta G = -RT \ln K$, K can equal 1 only when $\Delta G = 0$.

8. The carbon at which D-idose and D-talose differ (C3) is not involved in hemiacetal formation. Mutarotation signals a change in configuration at the anomeric carbon, C1 in these compounds.

9. Inversion of configuration at C5 of D-glucose produces an L-aldohexose. However, since the configuration at C5 has been changed relative to all of the other chiral centers, we have produced L-idose and not L-glucose.

10. The mechanism presented on p. 813 is believed to be correct. However, both involve a resonance stabilized carbocation.

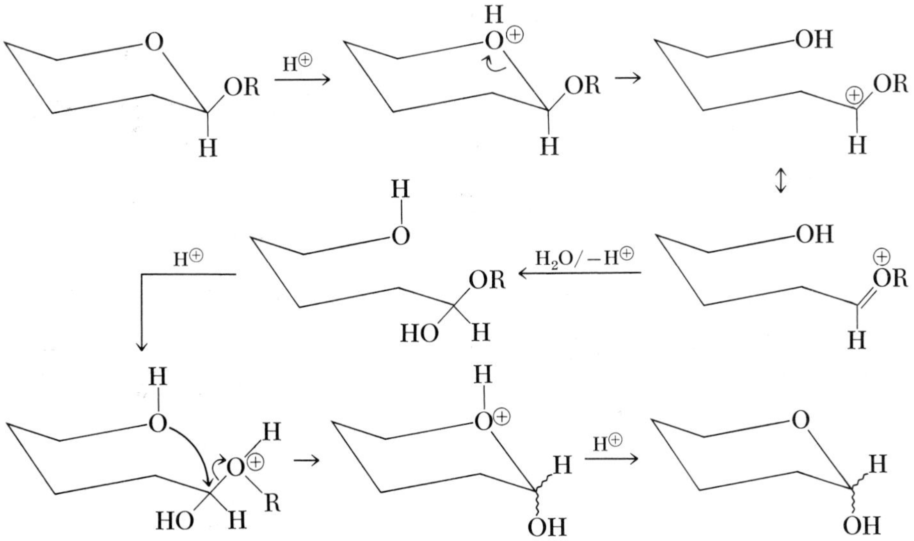

O OR H $\xrightarrow{H^{\oplus}}$ H O⊕ OR H → OH ⊕ OR H ↕ OH ⊕OR H

H O OR HO H $\xleftarrow{H_2O/-H^{\oplus}}$ H⊕

H O HO H H O⊕ R → H O⊕ H OH $\xrightarrow{H^{\oplus}}$ O H OH

11. Four. *Cis* and *trans* isomers at either carbon adjacent to C2.

12.

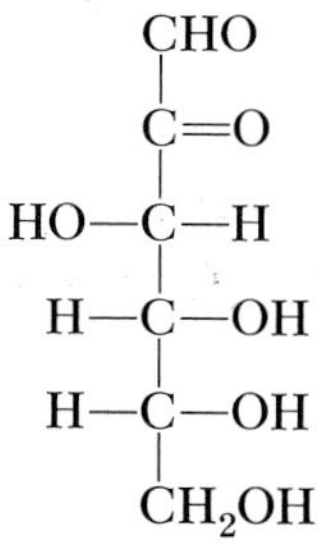

```
      CHO
       |
       C=O
       |
  HO—C—H
       |
   H—C—OH
       |
   H—C—OH
       |
      CH2OH
```

13. $C_6H_5-CH(OH)-C(=O)-C_6H_5$ + $NH_2NHC_6H_5$ $\xrightarrow{1}$ $C_6H_5-CH(OH)-C(O^{\ominus})(\overset{\oplus}{N}H_2NHC_6H_5)-C_6H_5$ $\xrightarrow[H^{\oplus}]{2}$ $C_6H_5-CH(OH)-C(OH)(\overset{\oplus}{N}H_2NHC_6H_5)-C_6H_5$

$\overset{3}{\rightleftharpoons}$ $C_6H_5-CH(OH)-C(\overset{\oplus}{O}H_2)(HN-NHC_6H_5)-C_6H_5$

$\longrightarrow$ $C_6H_5-CH(OH)-C(=N-NHC_6H_5)-C_6H_5$ $\xrightarrow{4}$ $C_6H_5-C(OH)=C(NH-NHC_6H_5)-C_6H_5$ $\overset{5}{\rightleftharpoons}$ $C_6H_5-C(=O)-CH(NH-NHC_6H_5)-C_6H_5$

$\xrightarrow{6}$ $C_6H_5-C(O^{\ominus})(C_6H_5NH\overset{\oplus}{N}H_2)-CH(NH-NHC_6H_5)-C_6H_5$ $\xrightarrow[\text{repeat 2}]{7}$ $\xrightarrow[\text{repeat 3}]{8}$ $C_6H_5-C(=N-NHC_6H_5)-CH(NH-NHC_6H_5)-C_6H_5$

$\xrightarrow[H^{\oplus}]{9}$ $C_6H_5-C(=N-NHC_6H_5)-CH(NH-\overset{\oplus}{N}H_2C_6H_5)-C_6H_5$ (+ $C_6H_5NHNH_2$)

$\xrightarrow{10}$ $C_6H_5-C(=N-NHC_6H_5)-C(=NH)-C_6H_5$ + $NH_2NHC_6H_5$

$\xrightarrow{11}$ $C_6H_5-C(=N-NHC_6H_5)-C(\overset{\ominus}{N}H)(\overset{\oplus}{N}H_2NHC_6H_5)-C_6H_5$ $\xrightarrow[12]{H^{\oplus}}$ $C_6H_5-C(=N-NHC_6H_5)-C(\overset{\oplus}{N}H_3)(HN-NHC_6H_5)-C_6H_5$ (+ $C_6H_5NHNH_2$)

$\xrightarrow[-H^{\oplus}]{-NH_3}$ $C_6H_5-C(=N-NHC_6H_5)-C(=N-NHC_6H_5)-C_6H_5$

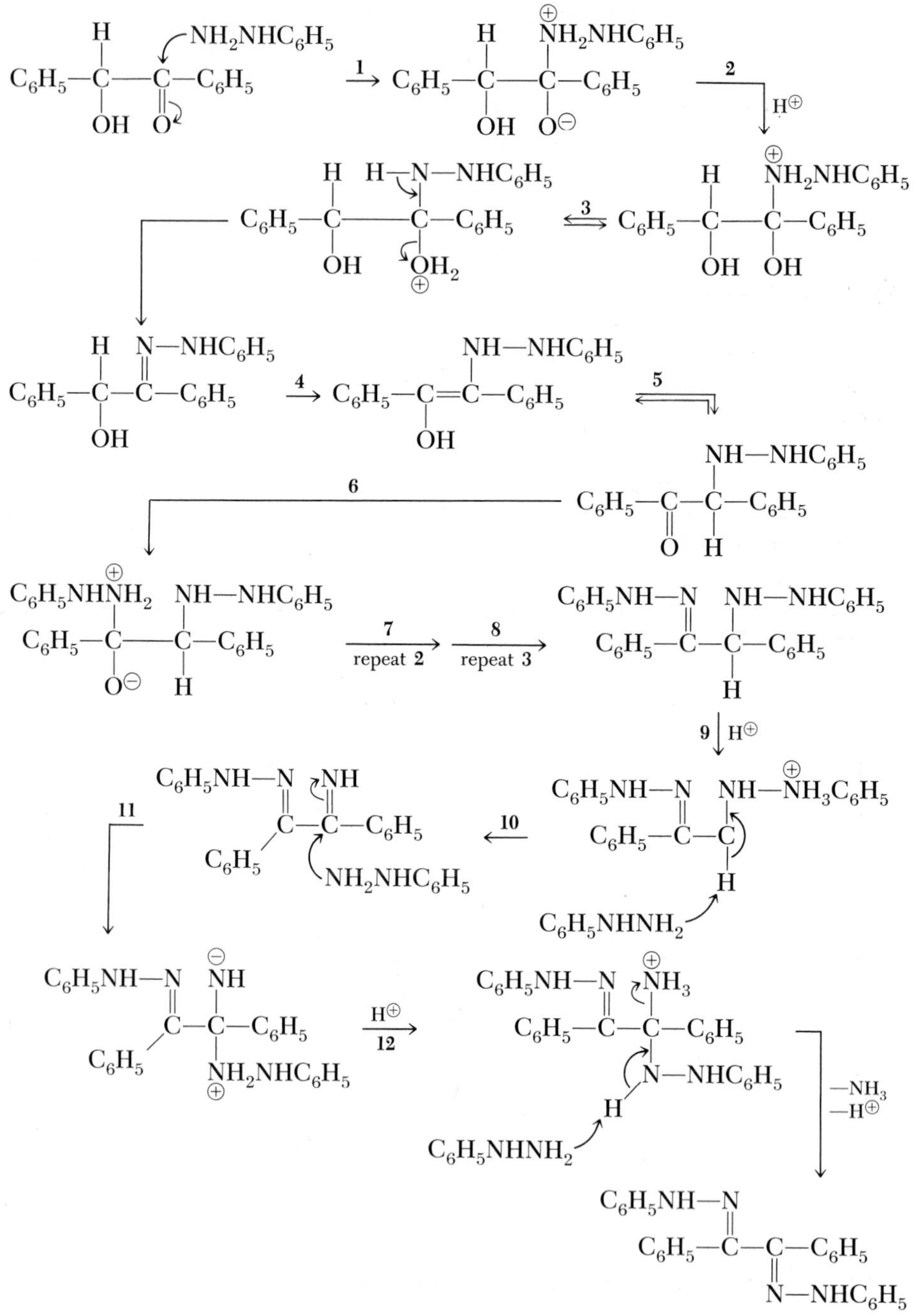

the osazone of benzoin

14. Tautomerism plays a critical role in both. In osazone formation the mechanism requires both an imine-enamine isomerization and a keto-enol isomerization. Only keto-enol isomerization is required in the Lobry de Bruyn-Alberda von Eckenstein rearrangement.

15. (a) $\xrightarrow[H_2O]{OH^{\ominus}}$ (D-glucose and D-fructose also formed)

(b) $\xrightarrow{H_2NOH}$ $\xrightarrow{Ac_2O}$ $\xrightarrow[CH_3OH]{CH_3O^{\ominus}}$ $\xrightarrow{-HCN}$ (see Figure 25A-7)

(c) $\xrightarrow{CH_3OH/H^{\oplus}}$ (α also formed)

(d) $\xrightarrow[H_2O/heat]{HI/P}$

(e) $\xrightarrow[\text{(Figure 25A–8)}]{\text{Ruff degradation}}$ D-arabinose $\xrightarrow{\text{repeat}}$ D-erythrose $\xrightarrow[\text{heat}]{HNO_3}$

(f) $\xrightarrow[\text{heat}]{Ac_2O/HOAc}$

Chapter 25B

1. α

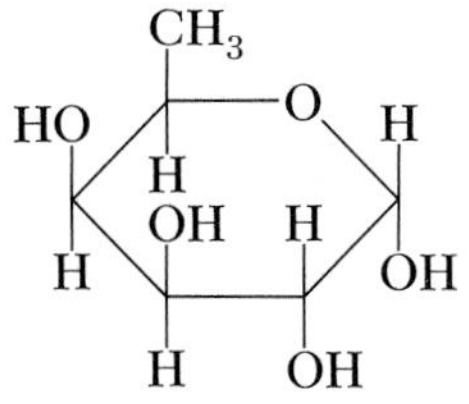

CH$_3$, HO, O, H, H, OH, H, H, OH, H, OH

α-D-fucose

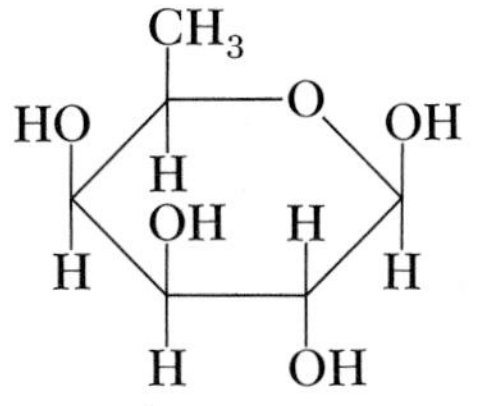

CH$_3$, HO, O, OH, H, OH, H, H, H, H, OH

β-D-fucose

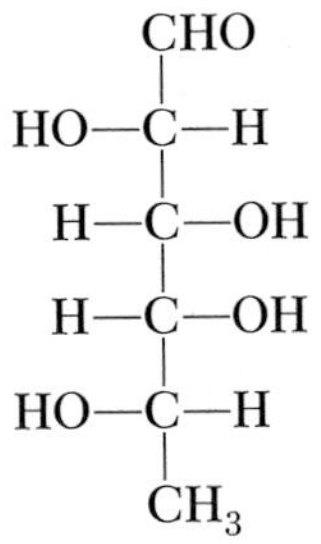

CHO
HO—C—H
H—C—OH
H—C—OH
HO—C—H
CH$_3$

L-fucose

CHO
H—C—OH
HO—C—H
HO—C—H
H—C—OH
CH$_3$

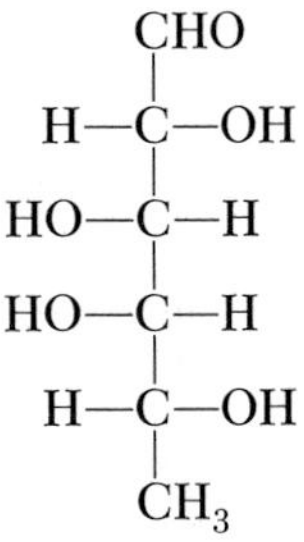

D-fucose

2. *O*-α-D-glucopyranosyl-(1,4)-α-D-glucopyranose

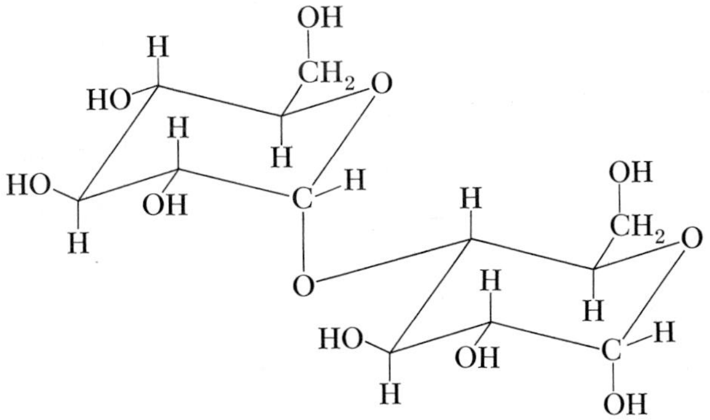

OH, H, CH$_2$, O, HO, H, H, HO, OH, C—H, H, O, OH, H, CH$_2$, O, H, H, HO, OH, C—H, H, OH

3.

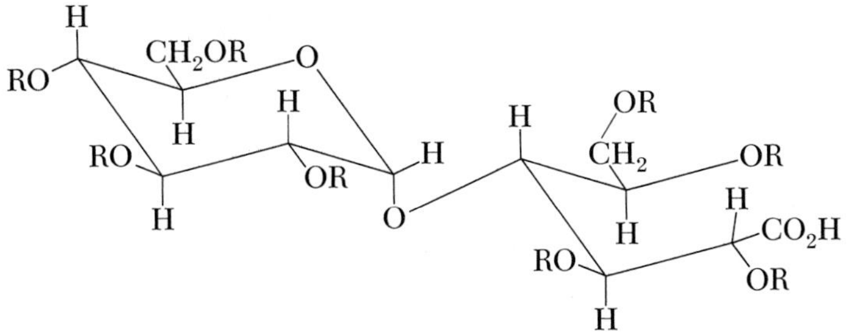

H, CH$_2$OR, O, RO, H, H, RO, OR, H, H, O, H, OR, CH$_2$, OR, H, H, CO$_2$H, RO, OR, H

A: R = H
B: R = CH$_3$

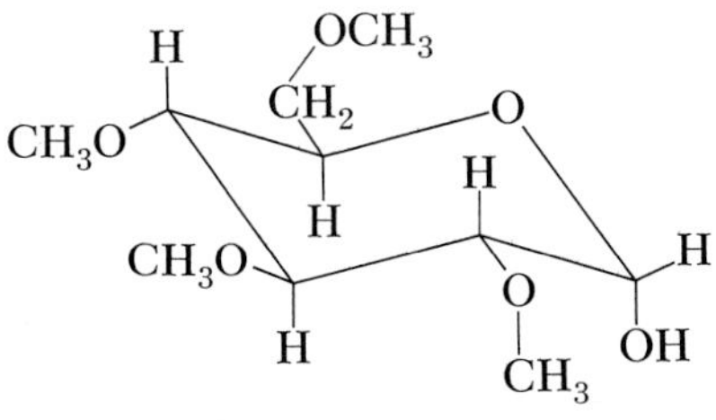

H, OCH$_3$, CH$_2$, O, CH$_3$O, H, H, CH$_3$O, H, O, H, OH, CH$_3$

C

CO$_2$H
H—C—OCH$_3$
CH$_3$O—C—H
H—C—OH
H—C—OCH$_3$
CH$_2$OCH$_3$

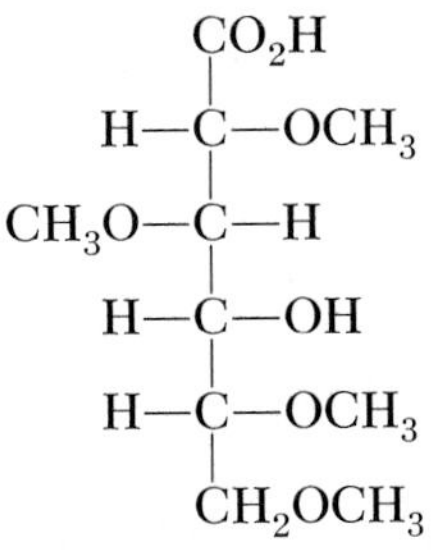

D

(i) α anomer of **C** establishes stereochemistry between both rings; (ii) free —OH at C4 of **D** establishes location of ring attachment.

5. $-19.9°$. Invert sugar, the major constituent of honey, is about as sweet as sucrose. Because of its greater tendency not to crystallize, invert sugar is preferred over sucrose in the preparation of some syrups and confections. "Inversion" in bees is accomplished by the enzyme invertase.

6. No. Both anomeric centers are involved in the glycosidic bond.

Chapter 26

1. 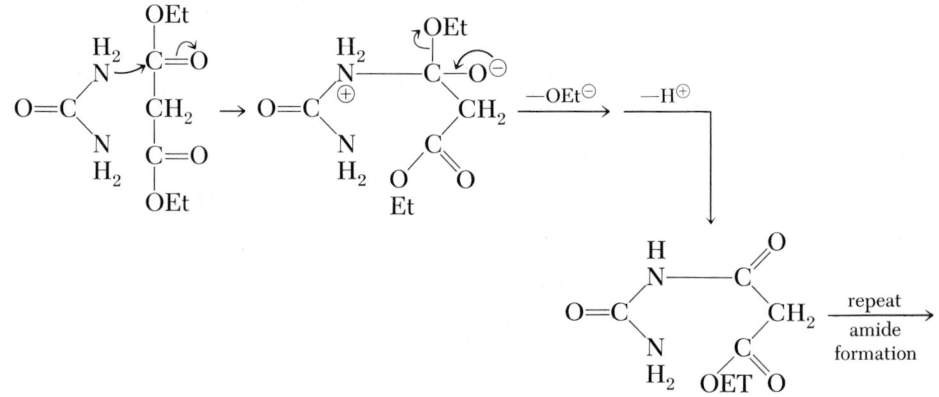 [Mechanism: urea $(O{=}C(NH_2)_2)$ attacks diethyl malonate $(EtO_2C\text{—}CH_2\text{—}CO_2Et)$ → tetrahedral intermediate $\xrightarrow{-OEt^{\ominus}}\ \xrightarrow{-H^{\oplus}}$ $O{=}C(NH_2)\text{—}NH\text{—}C(=O)\text{—}CH_2\text{—}C(=O)OET$ $\xrightarrow{\text{repeat amide formation}}$]

2. (a) $C_6H_5MgBr + \overline{CH_2CH_2O} \rightarrow \xrightarrow{H_3O^{\oplus}} \xrightarrow[\text{heat}]{MnO_4/H^{\oplus}} \xrightarrow{Cl_2/PCl_3}$

$C_6H_5CHClCO_2H \xrightarrow{OH^{\ominus}} \xrightarrow{CN^{\ominus}} \xrightarrow[\text{heat}]{EtOH/H^{\oplus}} C_6H_5CH(CO_2Et)_2 \xrightarrow[\text{(b) EtBr}]{\text{(a) EtO}^{\ominus}}$

(b) $CH_3CH_2MgBr + \overline{CH_2CH_2O} \rightarrow \xrightarrow{H_3O^{\oplus}} CH_3CH_2CH_2CH_2OH \xrightarrow[\text{pyridine}]{CrO_3}$

$\xrightarrow{CH_3MgCl} \xrightarrow{H_3O^{\oplus}} \xrightarrow{SOCl_2} \xrightarrow[\text{ether}]{Mg} CH_3CH_2CH_2CHCH_3MgCl$

Now repeat synthesis in part *a* using $CH_3CH_2CH_2CHCH_3MgCl$ instead of C_6H_5MgBr.

3. 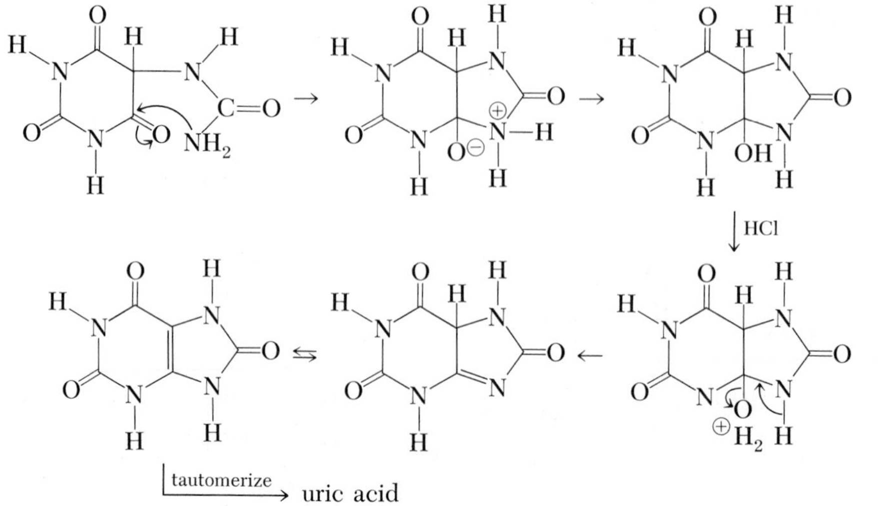 [Mechanism: ureido-substituted pyrimidine; NH_2 attacks ring C=O → zwitterionic tetrahedral intermediate → OH intermediate $\xrightarrow{HCl}$ protonated $OH_2^{\oplus}$ loses water → imine $\leftrightarrows$ purine structure] $\xrightarrow{\text{tautomerize}}$ uric acid

4. 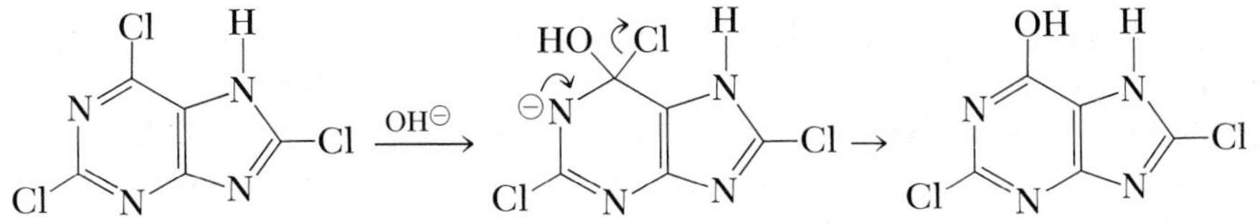 [Reaction scheme: 2,6,8-trichloropurine $\xrightarrow{OH^\ominus}$ tetrahedral intermediate (HO, Cl at C6; $N^\ominus$) → 2,8-dichloro-6-hydroxypurine]

5. 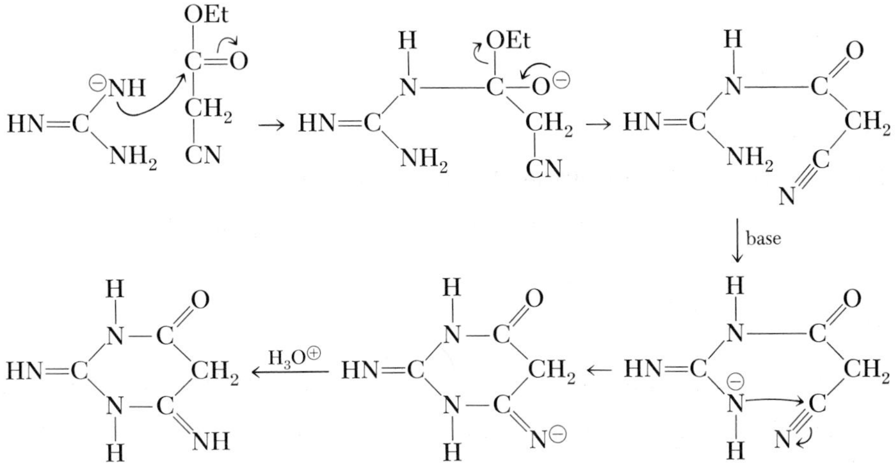 $HN{=}C(NH^\ominus)(NH_2)$ + $EtO{-}C({=}O){-}CH_2{-}CN$ → $HN{=}C(NH{-}C(OEt)(O^\ominus){-}CH_2CN)(NH_2)$ → $HN{=}C(NH{-}C({=}O){-}CH_2{-}C{\equiv}N)(NH_2)$ $\xrightarrow{\text{base}}$ cyclization ($N^\ominus$ attacks C≡N) → ring with $C{=}N^\ominus$ $\xrightarrow{H_3O^\oplus}$ $HN{=}C$ ring with $N{-}H$, $C{=}O$, CH_2, $C{=}NH$

All protons on heteroatoms exchangeable.

6. See Figure 26–1 for adenosine, guanosine, iridine, and thymidine. Cytidine is similar to deoxycytidine (Figure 26–1) except that C2′ is $-\overset{H}{\underset{OH}{C}}-$ rather than $-\overset{H}{\underset{H}{C}}-$.

7. 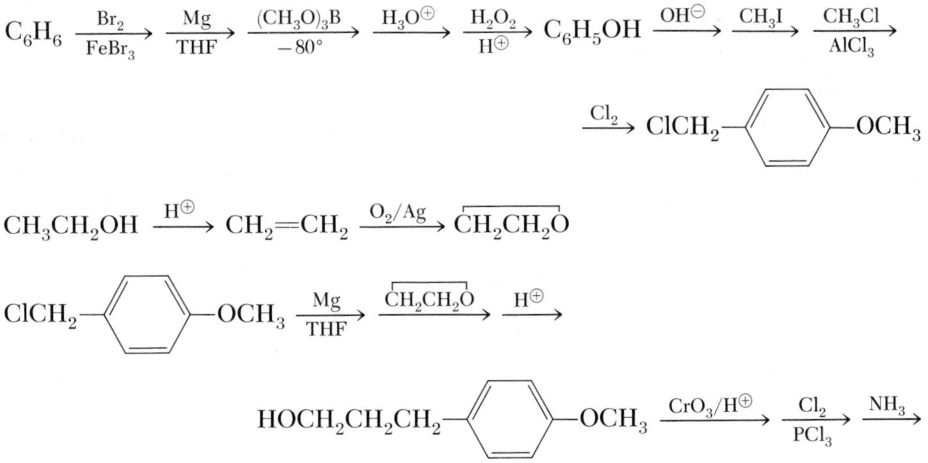 $C_6H_6 \xrightarrow[FeBr_3]{Br_2} \xrightarrow[THF]{Mg} \xrightarrow[-80^\circ]{(CH_3O)_3B} \xrightarrow{H_3O^\oplus} \xrightarrow[H^\oplus]{H_2O_2} C_6H_5OH \xrightarrow{OH^\ominus} \xrightarrow{CH_3I} \xrightarrow[AlCl_3]{CH_3Cl}$

$\xrightarrow{Cl_2} ClCH_2{-}C_6H_4{-}OCH_3$

$CH_3CH_2OH \xrightarrow{H^\oplus} CH_2{=}CH_2 \xrightarrow{O_2/Ag} \overline{CH_2CH_2O}$

$ClCH_2{-}C_6H_4{-}OCH_3 \xrightarrow[THF]{Mg} \xrightarrow{\overline{CH_2CH_2O}} \xrightarrow{H^\oplus}$

$HOCH_2CH_2CH_2{-}C_6H_4{-}OCH_3 \xrightarrow{CrO_3/H^\oplus} \xrightarrow[PCl_3]{Cl_2} \xrightarrow{NH_3}$

Chapter 28

5. Asymmetric and symmetric stretching modes.

6. Set 1: (a) presence of 1380 cm^{-1} doublet; (b) 1380 cm^{-1} absorption a singlet; Set 3: (a) absence of C=C stretch (approx. 1650 cm^{-1}); (b) presence of C=C stretch; (c) absence of C=C stretch, absorption at 1380 cm^{-1}, absence of absorption above 3000 cm^{-1}; Set 5: (a) absorption near 1380, 1660, and 700 cm^{-1}; (b) very weak absorption near 1670 cm^{-1}, absorption at 965 cm^{-1}; (c) absorption at 1645, 990, and 910 cm^{-1}.

8. (a) C═O stretch near 1720 cm^{-1}, aldehyde C—H doublet near 2720 and 2820 cm^{-1}; (b) O—D stretch near 2475 cm^{-1}; (c) C═O stretch near 1700 cm^{-1}, broad absorption 3000–2500 cm^{-1} carboxylic O—H stretch; (d) intense absorptions near 1350 and 1150 cm^{-1} due to SO_2, absence of C—H stretch below 3000 cm^{-1} indicates absence of C_{sp^3} —H; (e) broad N—H stretch near 3400 cm^{-1}.

11. Blue.

16. (a) 98; (b) 156; (c) 156.

17. Yes.

19. Low probability of collision of molecules in space.

22. (b) 31; (d) **C:** protonated carbon monoxide; **D:** protonated formaldehyde.

Chapter 29

3. (a) methane; 1,3,5-tribromobenzene; (b) 1,4-dichlorobenzene; (c) CH_2═CFH

5. Only if the system is A_2X_2 will it appear as two triplets. An A_2B_2 system will produce a more complex spectrum.

6. (a) equivalent; (b) enantiotopic; (c) enantiotopic; (d) enantiotopic; (e) diastereotopic; (f) equivalent; (g) diastereotopic.

8. (a) In CCl_4 the exchange rate is sufficiently slow that $\mathbf{CH_3}$ and **OH** are seen separately. Coupling is observed. In D_2O, —**OH** is rapidly converted to —OD and only $\mathbf{CH_3}$ (decoupled from **OH**) is observed.
(b) In $CDCl_3$ the exchange rate is sufficiently slow that the —**OH** resonance appears as a separate signal, split by the adjacent methylene group. The methylene group is split by both CH_3 and **OH.** Addition of a proton source catalyzes the exchange and results in decoupling.

11. (a) methacrylic acid, CH_2═$C(CH_3)CO_2H$; (b) α-bromobutyric acid; (c) ethyl acetate; (d) α-methyl benzylamine; (e) methanol; (f) 1,1,1-trichloroethane; (g) 2,2,2-trifluoroethane; (h) 1,3-dichloropropane; (i) trimethylene oxide (oxetane). Each of the spectra will be found in Volume 1 of the Varian Catalog of nmr spectra, published by Varian Associates.

18. (e) Since —CHDI is chiral, the protons of —CH_2D are non-equivalent.

21. The two nuclei are isochronous (chemical shift equivalent) and will not appear to split one another.

INDEX

Entries are alphabetized by disregarding prefixes and spaces between words. Page numbers in *italics* indicate problems; those in **bold type** indicate entries in "Important Terms" sections.